W9-BYT-094

IEEE Standard Dictionary of Electrical and Electronics Terms

Dictionary Subcommittee
of the
IEEE Standards Committee

H. P. Westman, *Chairman*

J. J. Anderson, *Secretary* (1965–1970)*
H. R. Mimno, *Technical Consultant*

M. W. Baldwin (1965–1967)†
D. C. Fleckenstein (1969–1971)
J. A. Goetz (1969–1971)
W. Y. Lang (1965–1969)
W. A. Lewis (1965–1969)
D. T. Michael (1969–1971)
C. H. Page (1965–1969)
F. von Roeschlaub (1965–1968)
W. T. Wintringham (1965–1970)

Approved October 13, 1971

IEEE Standards Committee

B. B. Barrow, *Chairman*

J. Forster, *Vice Chairman*
S. I. Sherr, *Secretary*

S. J. Angello
E. C. Barnes
F. K. Becker
W. H. Cook
W. H. Devenish
C. J. Essel
R. F. Estoppey
J. A. Goetz
A. D. Hasley
G. E. Hertig
A. R. Hileman
H. Lance
D. T. Michael
J. B. Owens
R. H. Rose, II
S. V. Soanes
L. Van Rooij
R. V. Wachter
B. O. Weinschel
C. E. White
W. T. Wintringham

* Retired.
† Deceased.

IEEE Standard Dictionary of Electrical and Electronics Terms

Approved by the Standards Committee of The Institute of Electrical and Electronics Engineers, Inc.

Wiley-Interscience a division of John Wiley & Sons, Inc.
New York • London • Sydney • Toronto

Editorial Staff

E. K. Gannett, *Director of Editorial Services,* E. C. Day†, *Manager of Dictionary Production,* H. J. Carter, *Managing Editor (Transactions),* Sandra Greer, *Senior Editor,* Sophia Martynec, *Associate Editor,* Lee Cole, Arline Jacob, Audrey Schneider, Marjorie Stephan, *Editorial Assistants*

† Deceased.

Consultants

Julius Green, Bertram Stanleigh, *Standards Engineers,* Robert E. Whitlock, *Senior Editor, Spectrum*

Introduction

From their earliest years, both the American Institute of Electrical Engineering (AIEE) (1884) and the Institute of Radio Engineers (IRE) (1912) published standards defining technical terms. They have maintained this practice since they were combined in 1963 to become the IEEE (Institute of Electrical and Electronics Engineers).

In 1928 the AIEE organized Sectional Committee C42 on Definitions of Electrical Terms under the procedures of the American Standards Association, now the American National Standards Institute. In 1941 AIEE published its first edition of *American Standard Definitions of Electrical Terms* in a single volume. However, by the time a second edition was ready, the highly accelerated development of new terms made it impracticable to publish in a single volume, and 17 separate documents, each limited to a specific field, were published from 1956 to 1959.

Over the years, IRE published a large number of standards that either included definitions or were devoted entirely to definitions. In 1961 it published all of its then-approved definitions in an alphabetically arranged single volume.

This Dictionary represents a different approach from that of the 1961 document. It is more than a compilation of what has been published in the past. It includes not only substantial amounts of new material but also extensive revisions of previously published definitions to eliminate duplications when two or more earlier definitions existed with equivalent technical meaning but different wording.

The directness with which a term can be found, the availability of all terms in a single volume, and the ease of comparing the meanings that certain terms have in two or more fields are recommendations for the single-volume alphabetically arranged dictionary. However, with an alphabetical arrangement, one forfeits the ability to browse conveniently among significantly related terms. To alleviate this loss, substantial cross-indexing has been provided to suggest other terms in a particular technical category.

While every attempt has been made to reflect latest usage and to include new terms as they have developed, no regular review has been made of standards published after January 31, 1968 to extract definitions. It is planned that this and other new material will be reflected in subsequent editions of the Dictionary.

How To Use This Dictionary

This Dictionary defines technical terms used by electrical and electronics engineers. A knowledge of the following editorial rules used in compiling the Dictionary will be helpful to the user.

1. Style. As in all general dictionaries, terms are arranged in alphabetical order and definitions follow the listed term in a series without repeating the term. If a term consists of two or more words, and its modifiers and the generic word form a single concept, the terms have been alphabetized with the modifier as the initial word.

2. Variations in Meaning. The meanings of some terms differ with the field in which they are used, thus requiring more than one definition. In addition most definitions have been taken from existing standards and not infrequently may be restrictively worded to emphasize the special purpose of a particular standard that may concern, for example, measurements or tests. An alphanumeric identification code follows each definition to identify its source and to refer the reader to the document in which the term was originally defined.

3. Preferred and Deprecated Terms. A preferred term appears first, followed by variations of the term in parentheses in descending order of preference. Deprecated terms follow within their own parentheses together with reference marks indicating that they are deprecated.

A definition appears only with a preferred term. Variations, both permitted and deprecated, are included in the alphabetical list with cross reference to the preferred term. The designation *See also* following a definition draws attention to related terms.

4. Abbreviations. Many abbreviations appear in the Dictionary with cross references to the full terms. Their inclusion is solely for the convenience of the reader, and they are not meant to establish standard abbreviations for terms or to reflect preferred practice.

5. Sources. Definitions are derived from three major sources:

IEEE Standards

American National Standards

IEC Recommendations

Many of the definitions have been altered in whole or in part by the Technical Committees of the Institute, particularly in those fields in which major technical developments have occurred during the past few years, to include both the older meanings of the terms as well as their present-day significance.

American National Standard Definitions of Electrical Terms, C42, a long-term project under IEEE Secretariat, consists of a series of documents devoted to definitions in specific technical fields. Most of the definitions from the C42 series are included in this Dictionary, and it is anticipated that this Dictionary will also receive designation as an American National Standard.

The International Electrotechnical Commission (IEC), whose membership includes most of the technically developed nations of the world, prepares and publishes Recommendations that reflect substantial international support and that hopefully serve as the basis for national standards in individual countries. In the interests of international standardization, IEEE Technical Committees have been urged to give preference to IEC Recommendations; thus many definitions of long standing in IEEE Standards have been replaced by definitions derived from IEC Recommendations. That all IEC Recommendations have not been adopted by IEEE indicates that there are still differences of opinion regarding the meanings of some terms or the exact wording of some definitions. A significant factor is the tendency of IEC to prefer simple, limited definitions that are readily translatable into many languages without distortion rather than more complicated definitions that encompass all possible usages of a term.

In addition to the three major sources noted above, a number of definitions have been credited to other documents. Sources are shown in Table 5, and permission for use of this material is greatly appreciated.

6. Identification Codes. A code group made up of letters, numbers, and signs follows each definition to identify its source, which in turn indicates the field for which the definition was developed.

6.1 Signs. A hyphen separates the two major parts of each code group. The first part, which is to the left of the hyphen, identifies an existing source document or, if this part of the code is zero, that it is a new or revised definition. The second part, to the right of the hyphen, identifies a sponsoring committee that worked specifically on the Dictionary. Not all existing definitions were formally reaffirmed by these working committees, and absence of such reaffirmation, indicated by a zero to the right of the hyphen, is not significant. A solidus (/) separates the codes for two or more source documents in the first part, or committees in the second part, of a code group. Where more than one code group is listed, the groups are separated by semicolons.

6.2 Numbers. Numbers identify both source documents and committees as listed in the appended tables.

6.3 Letters. A single letter indicates an organization. Only three such letters are used.

A = ANSI (American National Standards Institute)
E = IEEE (Institute of Electrical and Electronics Engineers)
I = IEC (International Electrotechnical Commission)

6.3.1 *Letter A*. The C series of American National Standards are designated by the single letter A in place of the decimal—for example, 42A65 for C42.65; the full titles identifying the fields are given in Table 1. Documents from other American National Standard series are identified by the initial letter in addition to the A—such as Z7A1 for American National Standard Z7.1.

6.3.2 *Letter E*. Table 2 lists standards that have been published by IEEE. They are identified by E and their number to the left of the hyphen.

Table 3 gives numbers for IEEE Groups and their Technical Committees. Sponsorship of a definition by a specific Technical Committee is shown in the right half of the code, the first number being that of the IEEE Group and the second number that assigned to its Technical Committee, these two numbers being separated by the letter E designating IEEE. If the second number is a 0, the Group, not one of its Technical Committees, is indicated.

The letter E alone to the right of a hyphen indicates that to fill a need the editing group prepared a new definition or selected one from the identified source.

6.3.3 *Letter I.* Table 4 lists source documents of the International Electrotechnical Commission. The number 50 preceding the letter I for the source indicates it is from the International Electrotechnical Vocabulary series. The final number identifies the specific document in the series, each on a different subject.

6.3.4 *Two or More Letters Together.* Combinations of letters indicate sources listed in Table 5.

6.4 Examples.

42A35-31E13

The letter A in the code to the left of the hyphen refers to ANSI, and Table 1 identifies it as C42.35, Generation, Transmission, and Distribution. The letter E in the second half refers to a sponsoring committee of IEEE; from Table 3 we find that Group 31 is Power and its Technical Committee 13 is Transmission and Distribution.

E165-0

The E indicates a source document from IEEE, and Table 2 identifies Standards Publication 165 which is on Analog Computers. The zero at the right means that this previously approved definition was not specifically reaffirmed by any committee working on this Dictionary.

0-10E6

The zero to the left of the hyphen indicates no source document, so this is a new definition or a revision of a previously published definition that is now withdrawn. The E to the right of the hyphen indicates a sponsoring committee of IEEE; from Table 3 we find that Group 10 is Aerospace and Electronics Systems and its Technical Committee 6 is Navigation Aids.

50I07-15E6

The I to the left of the hyphen indicates a source document issued by IEC; Table 4 shows it to be Publication 50(07) Electronics. The sponsoring committee indicated to the right of the hyphen is IEEE Group 15, Electron Devices, Technical Committee 6, Standards on Electron Tubes, as given in Table 3.

ITU-2E2

From Table 5 we find ITU identifies the International Telecommunications Union as the publisher of the source document, and the sponsoring committee of IEEE to be Group 2, Broadcasting, and Technical Committee 2, Video Techniques, from Table 3.

Table 1

American National Standards Identification Letter A

C1	National Electrical Code
C2.2	Installation and Maintenance of Electric Supply and Communication Lines
C5.1	Lightning Protection Code
C8.1	Wire and Cable
C8.16	Rubber-Insulated Tree Wire
C12	Electricity Metering
C29.1	Electric Power Insulators
C34.1	Pool-Cathode Mercury-Arc Power Converters
C37.1	Relays and Relay Systems Associated with Electric Power Apparatus
C37.100	Power Switchgear
C39.1	Electric Indicating Instruments
C39.2	Direct-Acting Electric Recording Instruments
C39.4	Automatic Null-Balancing Electric Measuring Instruments
C42.10	Rotating Machinery
C42.15	Transformers, Regulators, Reactors, and Rectifiers
C42.20	Lightning Arresters
C42.25	Control Equipment
C42.30	Instruments, Meters and Meter Testing
C42.35	Transmission and Distribution
C42.40	Transportation—General
C42.41	Transportation—Air
C42.42	Transportation—Land
C42.43	Transportation—Marine
C42.45	Electromechanical Devices
C42.55	Illuminating Engineering
C42.60	Electrochemistry and Electrometallurgy
C42.65	Communications
C42.70	Electron Devices
C42.80	Electrobiology, Including Electrotherapeutics
C42.85	Mining
C42.95	Miscellaneous
C57.12.75	Removable Air-Filled Junction Boxes for Cable Termination for Power Transformers
C57.12.76	Integral Air-Filled Junction Boxes for Cable Termination for Power Transformers

C57.12.80	Terminology
C57.14	Constant-Current Transformers of the Moving-Coil Type
C57.15	Step-Voltage and Induction-Voltage Regulators
C57.16	Current-Limiting Reactors
C57.18	Pool-Cathode Mercury-Arc Rectifier Transformers
C63.4	Radio-Noise Voltage and Radio-Noise Field Strength, 0.015 to 25 Megahertz Low-Voltage Electric Equipment, and Nonelectric Equipment
C64.1	Brushes for Electric Machines
C67.1	Preferred Nominal Voltages, 100 Volts and Under
C70.1	Household Automatic Electric Flatirons
C71.1	Household Electric Ranges
C76.1	Outdoor Apparatus Bushings
C78.180	Fluorescent Lamp Starters
C78.385	Glow Lamps
C79.1	Glass Bulbs Intended for Use with Electron Tubes and Electric Lamps
C79.2	Molded Glass Flares Intended for Use with Electron Tubes and Electric Lamps
C80.1	Rigid Steel Conduit, Zinc Coated
C80.4	Fittings for Rigid Metal Conduit and Electrical Metallic Tubing
C82.1	Fluorescent Lamp Ballasts
C82.3	Fluorescent Lamp Reference Ballasts
C82.4	Mercury Lamp Ballasts (Multiple-Supply Type)
C82.8	Incandescent Filament Lamp Transformers, Constant-Current (Series) Supply Type
C83.14	Rigid Coaxial Transmission Lines—50 Ohms
C83.16	Relays
C84.1	Preferred Voltage Ratings for A-C Systems and Equipment
C87.1	Electric Arc-Welding Apparatus
C89.1	Specialty Transformers
C92.1	Voltage Values for Preferred Basic Impulse Insulation Levels
C95.1	Electromagnetic Radiation with Respect to Personnel
C99.1	Highly Reliable Soldered Connections in Electronic and Electrical Applications
Z7.1	Illuminating Engineering

Table 2

IEEE Standards
Identification Letter E

1	Temperature Limits in the Ratings of Electric Equipment
16	Electric Control Apparatus for Land Transportation Vehicles
18	Shunt Power Capacitors
28	Lightning Arresters for A-C Power Circuits
31	Outdoor Coupling Capacitors and Capacitance Potential Devices
32	Neutral Grounding Devices
45	Electric Installations on Shipboard
48	Potheads
49	Roof, Floor, and Wall Bushings
54	Induction and Dielectric Heating Equipment
59	Semiconductor Rectifier Components
74	Industrial Control (600 Volts or Less)
81	Ground Resistance and Potential Gradients in the Earth
94	Automatic Generation Control on Electric Power Systems
95	Insulation Testing of Large A-C Rotating Machinery with High Direct Voltage
102	Transistors
106	Toroidal Magnetic Amplifier Cores
107	Magnetic Amplifiers
111	Low-Power Wide-Band Transformers
145	Antennas
146	Antennas and Waveguides
147	Waveguide Components
149	Antennas
151	Audio
153	Network Topology
154	Linear Varying Parameter and Nonlinear Circuits
155	Linear Signal Flow Graphs
156	Linear Passive Reciprocal Time Invariant Networks
157	Electroacoustics
160	Electron Tubes
161	Electron Tubes
162	Electronic Digital Computers
163	Static Magnetic Storage
165	Analog Computers

168	Facsimile
169	Industrial Electronics
170	Modulation Systems
171	Information Theory
175	Scintillation Counter Field
180	Ferroelectric Crystals
182	Radio Transmitters
182A	Supplement to 182
188	Radio Receivers
191	Noise
193	Flutter Content in Sound Recorders and Reproducers
194	Pulses
196	Transducers
201	Television: Color
203	Television: Signal Measurement
204	Video Techniques: Television
206	Television: Differential Gain and Differential Phase
207	Television: Time of Rise, Pulse Width, and Pulse Timing of Video Pulse
208	Video Techniques: Resolution of Camera Systems
209	Television: Electronically Regulated Power Supplies
210	Radio Wave Propagation: Guided Waves
216	Semiconductors
217	Superconductive Electronics
221	Thermoelectric Devices
222	Optoelectronic Devices
223	Thyristors
224	Electrostatographic Devices
226	Nonlinear Capacitors
253	Semiconductor Tunnel (Esaki) Diodes and Backward Diodes
254	Parametric Devices
257	Burst Measurements in the Time Domain
258	Close-Talking Pressure-Type Microphones
264	High-Power Wide-Band Transformers
265	Burst Measurements in the Frequency Domain
267	Symbols
269	Telephone Sets
270	General (Fundamental and Derived) Electrical and Electronics Terms
274	Integrated Electronics

Table 3

Institute of Electrical and Electronics Engineers
Codes Identifying Groups and Technical Committees

Group		Technical Committee	
1	Audio and Electroacoustics		
		1	Standards
2	Broadcasting		
		1	Television Systems
		2	Video Techniques
3	Antennas and Propagation		
		1	Antennas and Waveguides
		2	Wave Propagation
4	Circuit Theory		
		1	Circuits
		2	Solid-State Circuits
5	Nuclear Science		
		1	Instrumentation
		2	Plasma and High-Energy Physics
		3	Radiation Detectors
		4	Radiation Effects
		5	Reactor and Reactor Control
		6	Standards
6	Vehicular Technology		
		1	Mobile Communication System Standards
7	Reliability		
		1	Definitions and Standards
		2	Maintainability
		3	Reliability Physics
		4	Technical Education
8	Broadcast and Television Receivers		

Table 3 *(Continued)*

Group		Technical Committee	
9	Instrumentation and Measurement		
		1	Electromagnetic Measurement State-of-the-Art
		2	Frequency and Time
		3	Fundamental Electrical Standards
		4	High-Frequency Instrumentation and Measurements
		5	Low-Frequency Instrumentation and Measurements
		6	Materials Measurements
10	Aerospace and Electronics Systems		
		1	Energy Conversion
		2	Field Integration and Support
		3	Flight Vehicle Systems
		4	Instrumentation
		5	Instrumentation in Aerospace Simulation Facilities
		6	Navigation Aids
		7	Standards
		8	Support Systems
12	Information Theory		
		1	Information Theory and Modulation
13	Industrial Electronics and Control Instrumentation		
		1	Control Instrumentation
		2	Electronic Techniques
		3	Industrial Electronics
		4	Process Test Instrumentation
		5	Special Technologies
		6	Standards Coordinating
14	Engineering Management		
15	Electron Devices		
		1	Electron Tubes
		2	Energy Sources
		3	Integrated Electron Devices
		4	Quantum Devices
		5	Solid-State Devices
		6	Standards on Electron Tubes
		7	Standards on Solid-State Devices
16	Computer		
		1	Computer Elements
		2	Computer Programming
		3	Computer Systems
		4	Data Acquisition and Control
		5	Design Automation
		6	Management Data Processing Applications
		7	Pattern Recognition
		8	Switching and Automata Theory
		9	Standards

Table 3 *(Continued)*

Group		**Technical Committee**	
17	Microwave Theory and Techniques		
		1	Standards Coordinating
18	Engineering in Medicine and Biology		
		1	Standards
19	Communication Technology		
		1	Communication Switching
		2	Communication Systems Discipline
		3	Communication Theory
		4	Data Communication Systems
		5	Radio Communication
		6	Space Communication
		7	Telemetering
		8	Wire Communication
20	Sonics and Ultrasonics		
		1	Resonators and Transducers
21	Parts, Materials, and Packaging		
		1	Electronics Transformers
		2	Interconnections
		3	Packaging
		4	Standards
23	Automatic Control		
		1	Applications and Systems Evaluation
		2	Control Components
		3	Discrete Systems
		4	Feedback Control Systems
		5	Linear Systems
		6	Optimal Systems
		7	Simulation
		8	Stability Theory and Nonlinear Systems
		9	Stochastic Systems
25	Education		
26	Engineering Writing and Speech		
27	Electromagnetic Compatibility		
		1	Standards
28	Man–Machine Systems		
29	Geoscience Electronics		
31	Power		
		1	Insulated Conductors
		2	Power Generation
		3	Power System Communications
		4	Power System Engineering
		5	Power System Instrumentation and Measurement
		6	Power System Relaying
		7	Protective Devices
		8	Rotating Machinery

Table 3 *(Continued)*

Group		Technical Committee
31 Power *(Continued)*		
	9	Standards Coordinating
	10	Substations
	11	Switchgear
	12	Transformers
	13	Transmission and Distribution
32 Electrical Insulation		
	1	Materials
	2	Phenomena
33 Magnetics		
	1	Combined Semiconductor–Magnetic Devices and Circuits
	2	Education
	3	Environmental Effects and Reliability
	4	Magnetic Materials
	5	Magnetic Recording
	6	Magnetic Transducers
	7	Magnetics in Memory
	8	Magnetism
	9	Microwave Magnetics
	10	Power Conversion
	11	Standards
	12	Superconducting Materials and Devices
34 Industry and General Applications		
	1	Cement Industry
	2	Corrosion and Cathodic Protection
	3	Domestic Appliance
	4	Electric Process Heating
	5	Electric Space Heating and Air Conditioning
	6	Electric Welding
	7	Electrostatic Processes
	8	General Industry Applications
	9	Industrial and Commercial Power Systems
	10	Industrial Control
	11	Land Transportation
	12	Machine Tools Industry
	13	Marine Transportation
	14	Metal Industry
	15	Mining Industry
	16	Petroleum and Chemical Industry
	17	Power Semiconductor
	18	Production and Application of Light
	19	Pulp and Paper Industry
	20	Rubber and Plastics Industry
	21	Rural Electric
	22	Safety Liaison
	23	Standards Coordinating

Table 3 *(Continued)*

Group		Technical Committee	
34	Industry and General Applications *(Continued)*		
		24	Static Power Converter
		25	Textile Industry
35	Systems Science and Cybernetics		
		1	Cybernetics
		2	Definitions
		3	Systems Science

Table 4

International Electrotechnical Commission
Identification Letter I

50(05)	Fundamental
50(07)	Electronics
50(08)	Electro-Acoustics
50(10)	Machines and Transformers
50(11)	Static Converters
50(12)	Transductors
50(15)	Switchboards and Apparatus for Connection and Regulation
50(16)	Protective Relays
50(20)	Scientific and Industrial Measuring Instruments
50(25)	Generation, Transmission and Distribution of Electrical Energy
50(26)	Nuclear Power Plants for Electric Energy Generation
50(30)	Electric Traction
50(31)	Signalling and Security Apparatus for Railways
50(35)	Electromechanical Applications
50(37)	Automatic Controlling and Regulating Systems
50(40)	Electro-Heating Applications
50(45)	Lighting
50(50)	Electrochemistry and Electrometallurgy
50(62)	Waveguides
50(65)	Radiology and Radiological Physics
50(66)	Detection and Measurement of Ionizing Radiation by Electric Means
50(70)	Electro-Biology

Table 5

Codes to Identify Other Sources

Code	Source
AD8	American Society for Testing and Materials publication D8
AD16	American Society for Testing and Materials publication D16
AD123	American Society for Testing and Materials publications D123
AD883	American Society for Testing and Materials publication D883
AD1566	American Society for Testing and Materials publication D1566
AS1	National Electrical Manufacturers Association publication AS1
CISPR	International Special Committee on Radio Interference
CM	Corrosion Magazine
CTD	Chambers Technical Dictionary
CV1	National Electrical Manufacturers Association publication CV 1
EIA3B	Electronic Industries Association publication 3B
IC1	National Electrical Manufacturers Association publication IC 1
ISA	Instrument Society of America
ITU	International Telecommunications Union
KPSH	Kepco Power Supply Handbook
LA1	National Electrical Manufacturers Association publication LA 1
MA1	National Electrical Manufacturers Association publication MA 1
MDE	Modern Dictionary of Electronics
MG1	National Electrical Manufacturers Association publication MG 1
SCC	IEEE Standards Coordinating Committee

A

AA (transformer classification). *See:* **transformer, oil-immersed.**

***aa* auxiliary switch.** *See:* **auxiliary switch; *aa* contact.**

***aa* contact.** A contact that is open when the operating mechanism of the main device is in the standard reference position and that is closed when the operating mechanism is in the opposite position. *See:* **standard reference position.** 37A100-31E11

***A* and *R* display (electronic navigation).** An *A* display, any portion of which may be expanded. *See also:* **navigation.** E172-10E6

***a* auxiliary switch.** *See:* ***a* contact; auxiliary switch.**

abampere. The unit of current in the centimeter-gram-second (cgs) electromagnetic system. The abampere is 10 amperes. E270-0

***A* battery.** A battery designed or employed to furnish current to heat the filaments of the tubes in a vacuum-tube circuit. *See also:* **battery (primary or secondary).** 42A60-0

abbreviation. A shortened form of a word or expression. *See:* **functional designation; graphic symbol; letter combination; mathematical symbol; reference designation; symbol for a quantity; symbol for a unit.** E267-0

abnormal decay (charge-storage tubes). The dynamic decay of multiply-written, superimposed (integrated) signals whose total output amplitude changes at a rate distinctly different from that of an equivalent singly-written signal. *Note:* Abnormal decay is usually very much slower than normal decay and is observed in bombardment-induced conductivity type of tubes. *See also:* **charge-storage tube.** E158-15E6

abnormal glow discharge (gas). The glow discharge characterized by the fact that the working voltage increases as the current increases. *See also:* **discharge.** 50I07-15E6

absolute accuracy. Accuracy as measured from a reference that must be specified. EIA3B-34E12

absolute address (computing machines). (1) An address that is permanently assigned by the machine designer to a storage location. (2) A pattern of characters that identifies a unique storage location without further modification. *See:* **electronic digital computer.** *See also:* **machine address.** X3A12-16E9

absolute altimeter (electronic navigation). A device that measures altitude above local terrain. E172-10E6

absolute block. A block governed by the principle that no train shall be permitted to enter the block while it is occupied by another train. *See also:* **railway signal and interlocking.** 42A42-0

absolute capacitivity (absolute dielectric constant) (permittivity). Of a homogeneous, isotropic, insulating material or medium, in any system of units, the product of its relative capacitivity and the electric constant appropriate to that system of units. *See also:* **electric constant.** E270-0

absolute delay (loran). The interval of time between the transmission of a signal from the master station and transmission of the next signal from the slave station. *See also:* **navigation.** E172-10E6

absolute dielectric constant. *See:* **absolute capacitivity.**

absolute dimension. A dimension expressed with respect to the initial zero point of a coordinate axis. *See:* **coordinate dimension word.** EIA3B-34E12

absolute error. (1) The amount of error expressed in the same units as the quantity containing the error. (2) Loosely, the absolute value of the error, that is, the magnitude of the error without regard to its algebraic sign. X3A12-16E9

absolute measurement (system of units). Measurement in which the comparison is directly with quantities whose units are basic units of the system. *Notes:* (1) For example, the measurement of speed by measurements of distance and time is an absolute measurement, but the measurement of speed by a speedometer is not an absolute measurement. (2) The word absolute implies nothing about precision or accuracy. E270-0

absolute permissive block. A term used for an automatic block signal system on a track signaled in both directions. For opposing movements the block is from siding to siding and the signals governing entrance to this block indicate STOP. For following movements the section between sidings is divided into two or more blocks and train movements into these blocks, except the first one, are governed by intermediate signals usually displaying STOP; then proceed at restricted speed, as their most restrictive indication. *See also:* **railway signal and interlocking.** 42A42-0

absolute refractory state (medical electronics). The portion of the electrical recovery cycle during which a biological system will not respond to an electric stimulus. *See also:* **medical electronics.** 0-18E1

absolute steady-state deviation (control). The numerical difference between the ideal value and the final value of the directly controlled variable (or another variable if specified). *See:* **deviation (control); percent steady-state deviation.** AS1-34E10

absolute system deviation (control). At any given point on the time response, the numerical difference between the ideal value and the instantaneous value of the directly controlled variable (or another variable if specified). *See:* **deviation.** AS1-34E10

absolute threshold. The luminance threshold or minimum perceptible luminance (photometric brightness) when the eye is completely dark adapted. *See also:* **visual field.** Z7A1-0

absolute transient deviation (control). The numerical difference between the instantaneous value and the final value of the directly controlled variable (or another variable if specified). *See:* **deviation; percent transient deviation.** AS1-34E10

absolute value (number). The absolute value of a number u (real or complex) is that positive real number $|u|$ given by

$$|u| = +(u_1^2 + u_2^2)^{1/2}$$

where u_1 and u_2 are respectively the real and imaginary parts of u in the equation

$$u = u_1 + ju_2.$$

If u is a real number, $u_2 = 0$. E270-0

absolute-value device. A transducer that produces an output signal equal in magnitude to the input signal but always of one polarity. *See also:* **electronic analog computer.** E165-16E9

absorptance (illuminating engineering). The ratio of the absorbed flux to the incident flux. *Note:* The sum of the hemispherical reflectance, the hemispherical transmittance, and the absorptance is one. *See also:* **lamp.** Z7A1-0

absorption (1) (transmission of waves). The loss of energy over radio or wire paths due to conversion into heat or other forms of energy. In wire transmission, the term is usually applied only to loss of energy in extraneous media. *See also:* **transmission characteristics.** 42A65-0

(2)(radio wave propagation). The irreversible conversion of the energy of an electromagnetic wave into another form of energy as a result of its interaction with matter. *See also:* **radio wave propagation; transmission characteristics.** E211-3E2/27E1/31E3

(3)(illuminating engineering). A general term for the process by which incident flux is dissipated. *Note:* All of the incident flux is accounted for by the processes of reflection, transmission, and absorption. *See also:* **lamp; transmission characteristics.** Z7A1-0

absorption current (rotating machinery). A reversible component of the measured current, which changes with time of voltage application, resulting from the phenomenon of dielectric absorption within the insulation when stressed by direct voltage. 0-31E8

absorption frequency meter (reaction frequency meter)(waveguide). A one-port cavity frequency meter that, when tuned, absorbs electromagnetic energy from a waveguide. *See:* **waveguide.** 0-3E1

absorption modulation. A method for producing amplitude modulation of the output of a radio transmitter by means of a variable-impedance (principally resistive) device inserted in or coupled to the output circuit. E145/E182A/42A65-0

absorptive attenuator (waveguide). *See:* **resistive attenuator.**

abstract quantity. *See:* **mathematico-physical quantity.**

ac. *See:* **alternating current.**

ACA. *See:* **adjacent channel attenuation.**

accelerated life test. A test in which certain factors such as voltage, temperature, etcetera, to which a cable is subjected are increased in magnitude above normal operating values to obtain observable deterioration in a reasonable period of time, and thereby afford some measure of the probable cable life under operating voltage, temperature, etcetera. *See also:* **power distribution, underground construction.** 42A35-31E13

accelerated test (reliability). A test in which the applied-stress level is chosen to exceed that stated in the reference conditions, in order to shorten the time required to observe the stress response of the item, or magnify the response in a given time. To be valid, an accelerated test must not alter the basic modes and/or mechanisms of failure, or their relative prevalence. *See also:* **reliability.** 0-7E1

accelerating (rotating machinery). The process of running a motor up to speed after breakaway. *See also:* **asynchronous machine; direct-current commutating machine; synchronous machine.** 0-31E8

accelerating electrode. An electrode to which a potential is applied to increase the velocity of the electrons or ions in the beam. E175-0;E160-15E6

accelerating grid (electron tubes). *See:* **accelerating electrode.**

accelerating relay (rotating electric equipment). A programming relay whose function is to control the acceleration. 37A100-31E11/31E6

accelerating time (control)(industrial control). The time in seconds for a change of speed from one specified speed to a higher specified speed while accelerating under specified conditions. *See:* **electric drive.** AS1-34E10

accelerating torque (rotating machinery). Difference between the input torque to the rotor (electromagnetic for a motor or mechanical for a generator) and the sum of the load and loss torques; the net torque available for accelerating the rotating parts. *See:* **rotor.** 0-31E8

accelerating voltage (oscilloscopes). The cathode-to-viewing-area voltage applied to a cathode-ray tube for the purpose of accelerating the electron beam. *See:* **oscillograph.** 0-9E4

acceleration (electric drive). Operation of raising the motor speed from zero or a low level to a higher level. *See also:* **electric drive.** 42A25-34E10

acceleration factor (reliability). The ratio between the times necessary to obtain a stated proportion of failures for two different sets of stress conditions involving the same failure modes and/or mechanisms. *See also:* **reliability.** 0-7E1

acceleration, programmed. A controlled velocity increase to the programmed rate. EIA3B-34E12

acceleration space (velocity-modulated tube). The part of the tube following the electron gun in which the emitted electrons are accelerated to reach a determined velocity. *See also:* **velocity-modulated tube.** 50I07-15E6

acceleration, timed (industrial control). A control function that accelerates the drive by automatically controlling the speed change as a function of time. *See:* **electric drive.** AS1-34E10

accelerator, electron, linear. *See:* **linear electron accelerator.**

accelerator, particle. *See:* **particle accelerator.**

accelerometer (electronic navigation). A device that senses inertial reaction to measure linear or angular acceleration. *Note:* In its simplest form, an accelerometer consists of a case-mounted spring and mass arrangement where displacement of the mass from its rest position relative to the case is proportional to the total nongravitational acceleration experienced along the instrument's sensitive axes. *See also:* **navigation.** E172-10E6

accent lighting. Directional lighting to emphasize a particular object or draw attention to a part of the field of view. *See also:* **general lighting.** Z7A1-0

acceptance angle (phototube housing). The solid angle within which all received light reaches the phototube cathode. *See also:* **electronic controller.** 42A25-34E10

acceptance proof test. A test applied to new insulated winding before commercial use. *Note:* It may be performed at the factory and/or after installation. *See also:* **insulation testing.** E95-0

acceptance test. (1) A test to demonstrate the degree of compliance of a device with purchaser's requirements. (2) A **conformance test** demonstrates the quality of the units of a consignment, without implication of contractual relations between buyer and seller. *See:* **routine test; test (instrument or meter).** E270/12A0/42A15-31E12;37A100-31E8/31E11

acceptor (semiconductor). *See:* **impurity, acceptor.** *See also:* **semiconductor.**

access. *See:* **random access; serial access.** *See also:* **electronic digital computer.**

access fitting. A fitting permitting access to the conductors in a raceway at locations other than at a box. *See also:* **raceway.** 42A95-0

accessible (1) (equipment). Admitting close approach because not guarded by locked doors, elevation, or other effective means. 42A95-0

(2) (wiring methods). Not permanently closed in by the structure or finish of the building; capable of being removed without disturbing the building structure or finish. 42A95-0

accessible terminal (network). A network node that is available for external connections. *See also:* **network analysis.** E153/E270-0

accessories. Devices that perform a secondary or minor duty as an adjunct or refinement to the primary or major duty of a unit of equipment. 37A100-31E11

access time. (1) A time interval that is characteristic of a storage device, and is essentially a measure of the time required to communicate with that device. *Note:* Many definitions of the beginning and ending of this interval are in common use. *See also:* **electronic computation; electronic digital computer.** E162/E270-0

(2) (A) The time interval between the instant at which data are called for from a storage device and the instant delivery is completed, that is, the read time. (B) The time interval between the instant at which data are requested to be stored and the instant at which storage is completed, that is, the write time. X3A12-16E9

accommodation. The process by which the eye changes focus from one distance to another. *See also:* **visual field.** Z7A1-0

accommodation, electrical (biology)(electrobiology). A rise in the stimulation threshold of excitable tissue due to its electrical environment, often observed following a previous stimulation cycle. *See also:* **excitability.** 0-18E1

accumulating stimulus (electrotherapy). A current that increases so gradually in intensity as to be less effective than it would have been if the final intensity had been abruptly attained. *See also:* **electrotherapy.** 42A80-18E1

accumulator. (1) A device that retains a number (the augend) adds to it another number (the addend) and replaces the augend with the sum. (2) Sometimes only the part of (1) that retains the sum. *Note:* The term is also applied to devices that function as described but that also have other properties. *See also:* **electronic digital computer.** E162/E270-0

accuracy (1). The quality of freedom from mistake or error, that is, of conformity to truth or to a rule. *Notes:* (1) Accuracy is distinguished from precision as in the following example: A six-place table is more precise than a four-place table. However, if there are errors in the six-place table, it may be more or less accurate than the four-place table. (2) The accuracy of an indicated or recorded value is expressed by the ratio of the error of the indicated value to the true value. It is usually expressed in percent. Since the true value cannot be determined exactly, the measured or calculated value of highest available accuracy is taken to be the true value or reference value. Hence, when a meter is calibrated in a given echelon, the measurement made on a meter of a higher-accuracy echelon usually will be used as the reference value. Comparison of results obtained by different measurement procedures is often useful in establishing the true value. *See also:* **dynamic accuracy; electronic analog computer; measurement system; static accuracy.** E284-9E1

(2) (A) Conformity of a measured value to an accepted standard value. (B) A measure of the degree by which the actual output of a device approximates the output of an ideal device nominally performing the same function. *See also:* **electronic analog computer.** E165-0

(3) (power supplies). Used as a specification for the output voltage of power supplies, accuracy refers to the absolute voltage tolerance with respect to the stated nominal output. *See also:* **power supply.** KPSH-10E1

(4) (numerically controlled machines). Conformity of an indicated value to the true value, that is, an actual or an accepted standard value. *Note:* Quantitatively, it should be expressed as an error or an uncertainty. The property is the joint effect of method, observer, apparatus, and environment. Accuracy is impaired by mistakes, by systematic bias such as abnormal ambient temperature, or by random errors (imprecision). The accuracy of a control system is expressed as the system deviation (the difference between the ultimately controlled variable and its ideal value), usually in the steady state or at sampled instants. *See:* **precision and reproducibility.** *See also:* **numerically controlled machines.** 85A1-34E12

(5) (electronic navigation). Generally, the quality of freedom from mistake or error; that is, of conformity to truth or a rule. Specifically, the difference between the mean value of a number of observations and the true value. *Note:* Often refers to a composite character including both accuracy and precision. *See also:* **navigation; precision.** E172-10E6

(6) (signal-transmission system). Conformity of an indicated value to an accepted standard value, or true value. *Note:* Quantitatively, it should be expressed as an error or uncertainty. The accuracy of a determination is affected by the method, observer, environment, and apparatus, including the working standard used for the determination. *See also:* **signal.** 85A1-13E6

(7) (measurement) (control equipment). The degree of correctness with which a measurement device yields the true value of measured quantity. *See also:* **power systems, low-frequency and surge testing.** E94-0

(8) (indicated or recorded value). The accuracy of an indicated or recorded value is expressed by the ratio of the error of the indicated value to the true value. It is usually expressed in percent. *See also:* **accuracy rating of an instrument.** 42A30-0

(9) (instrument transformer). Means of expressing the degree of conformity of the actual values obtained from the secondaries of the instrument transformers to the values that would have been obtained with the marked ratio. Performance characteristics associated with accuracy of an instrument transformer are expressed either in terms of correction factors or in terms of percent errors. 57A13-31E12

accuracy burden rating. A burden that can be carried

at a specified accuracy for an unlimited period without causing the established limitations to be exceeded. *See:* **instrument transformer.** 42A15-31E12

accuracy class (instrument transformer). The highest of the standard accuracy classes, the requirements of which are fulfilled by the values of the instrument-transformer correction factor under specified standard conditions. *See:* **instrument transformer.** 12A0/42A15/42A30-0

accuracy classes for metering. Limits of a transformer correction factor, in terms of percent error, that have been established to cover specific performance ranges for line power-factor conditions between 1.0 and 0.6 lag. 57A13-31E12

accuracy classes for relaying. Limits, in terms of percent ratio error, that have been established. 57A13-31E12

accuracy rating (class) (electric instrument). The accuracy classification of the instrument. It is given as the limit, usually expressed as a percentage of full-scale value, that errors will not exceed when the instrument is used under reference conditions. *Notes:* (1) The accuracy rating is intended to represent the tolerance applicable to an instrument in an "as-received condition." Additional tolerances for the various influences are permitted when applicable. It is required that the accuracy, as received, be directly in terms of the indications on the scale and without the application of corrections from a curve, chart, or tabulation. Over that portion of the scale where the accuracy tolerance applies, all marked division points shall conform to the stated accuracy class. (2) Generally the accuracy of electrical indicating instruments is stated in terms of the electrical quantities to which the instrument responds. In instruments with the zero at a point other than one end of the scale, the arithmetic sum of the end-scale readings to the right and to the left of the zero point shall be used as the full-scale value. Exceptions: (A) The accuracy of frequency meters shall be expressed on the basis of the percentage of actual scale range. Thus, an instrument having a scale range of 55 to 65 hertz would have its error expressed as a percentage of 10 hertz. (B) The accuracy of a power-factor meter shall be expressed as a percentage of scale length. (C) The accuracy of instruments that indicate derived quantities, such as series type ohmmeters, shall be expressed as a percentage of scale length. (3) In the case of instruments having nonlinear scales, the stated accuracy only applies to those portions of the scale where the divisions are equal to or greater than two-thirds the width they would be if the scale were even divided. The limit of the range at which this accuracy applies may be marked with a small isosceles triangle whose base marks the limit and whose point is directed toward the portion of the scale having the specified accuracy. (4) Instruments having an accuracy rating of 0.1 percent are frequently referred to as laboratory standards. Portable instruments having an accuracy rating of 0.25 percent are frequently referred to as portable standards. (5) For an extensive list of cross references, see *Appendix A.* 39A1/42A30-0

accuracy ratings for metering. The accuracy class followed by a standard burden for which the accuracy class applies. Accuracy rating applies only over the specified current or voltage range and at the stated frequency. 57A13-31E12

accuracy ratings for relaying. The relay accuracy class is described by a letter denoting whether the accuracy can be obtained by calculation or must be obtained by test, followed by the maximum secondary terminal voltage that the transformer will produce at 20 times secondary current with one of the standard burdens, without exceeding the relay accuracy class limit. A13-31E12

accuracy ratings of instrument transformers. Means of classifying transformers in terms of percent error limits under specified conditions of operation. 57A13-31E12

accuracy test. A test to determine the degree to which the value of the quantity obtained from the secondary reflects the value of the quantity applied to the primary. 57A13-31E12

accuracy, synchronous-machine regulating system. The degree of correspondence (or ratio) between the actual and the ideal values of a controlled variable of the synchronous-machine regulating system under specified conditions, such as load changes, drift, ambient temperature, humidity, frequency, and supply voltage. *See also:* **synchronous machine.** 0-31E8

acetate disks. Mechanical recording disks, either solid or laminated, that are made of various acetate compounds. *See also:* **electroacoustics.** 0-1E1

achromatic locus (achromatic region). Chromaticities that may be acceptable reference standards under circumstances of common occurrence are represented in a chromaticity diagram by points in a region which may be called the achromatic locus. *Note:* The boundaries of the achromatic locus are indefinite, depending on the tolerances in any specific application. Acceptable reference standards of illumination (commonly referred to as white light) are usually represented by points close to the locus of Planckian radiators having temperatures higher than about 2000 kelvins*. While any point in the achromatic locus may be chosen as the reference point for the determination of dominant wavelength, complementary wavelength and purity for specification of object colors, it is usually advisable to adopt the point representing the chromaticity of the luminator. Mixed qualities of illumination, and luminators with chromaticities represented very far from the Planckian locus, require special consideration. Having selected a suitable reference point, dominant wavelength may be determined by noting the wavelength corresponding to the intersection of the spectrum locus with the straight line drawn from the reference point through the point representing the sample. When the reference point lies between the sample point and the intersection, the intersection indicates the complementary wavelength. Any point within the achromatic locus, chosen as a reference point, may be called an achromatic point. Such points have also been called white points. **See:* **kelvin.** *See also:* **color terms.** E201-2E2

acid-resistant (industrial control). So constructed that it will not be injured readily by exposure to acid fumes. 42A95/IC1-34E10

acknowledger (forestaller). A manually operated electric switch or pneumatic valve by means of which, on a locomotive equipped with an automatic train stop or train control device, an automatic brake application can be forestalled, or by means of which, on a locomotive equipped with an automatic cab signal device, the sounding of the cab indicator can be silenced. *See also:* **railway signal and interlocking.** 42A42-0

acknowledging device. *See:* **acknowledger (forestaller).**

acknowledging (forestalling). The operating by the engineman of the acknowledger associated with the vehicle-carried equipment of an automatic speed control or cab signal system to recognize the change of the aspect of the vehicle-carried signal to a more restrictive indication. The operation stops the sounding of the warning whistle, and in a locomotive equipped with speed control it also forestalls a brake application. *See also:* **railway signaling and interlocking.** 42A42-0

acknowledging switch. *See:* **acknowledger (forestaller).**

acknowledging whistle. An air-operated whistle that is sounded when the acknowledging switch is operated. Its purpose is to inform the fireman that the engineman has recognized a more restrictive signal indication. 42A42-0

***a* contact (front contact).** A contact that is open when the main device is in the standard reference position, and that is closed when the device is in the opposite position. *Notes:* (1) *a* contact has general application. However, this meaning for front contact is restricted to relay parlance. (2) For indication of the specific point of travel at which the contact changes position, an additional letter or percentage figure may be added to *a* as detailed in Sections 9.4.4.1. and 9.4.4.2 of American National Standard C37.2-1962. *See:* **standard reference position.** 37A100-31E11

acoustic, acoustical. Used as qualifying adjectives to mean containing, producing, arising from, actuated by, related to, or associated with sound. Acoustic is used when the term being qualified designates something that has the properties, dimensions, or physical characteristics associated with sound waves; acoustical is used when the term being qualified does not designate explicitly something that has such properties, dimensions, or physical characteristics. *Notes:* (1) The following examples qualify as having the properties or physical characteristics associated with sound waves and hence would take acoustic: impedance, inertance, load (radiation field), output (sound power), energy, wave, medium, signal, conduit, absorptivity, transducer. (2) The following examples do not have the requisite physical characteristics and therefore take acoustical: society, method, engineer, school, glossary, symbol, problem, measurement, point of view, end-use, device. (3) As illustrated in the preceding notes, usually the generic term is modified by acoustical, whereas the specific technical implication calls for acoustic. *See also:* **electroacoustics.** 0-1E1

acoustical. *See:* **acoustic.**

acoustical reciprocity theorem. In an acoustic system comprising a fluid medium having bounding surfaces $S_1, S_2, S_3, \ldots$, and subject to no impressed body forces, if two distributions of normal velocities v_n' and v_n'' of the bounding surfaces produce pressure fields p' and p'', respectively, throughout the region, then the surface integral of $(p''v_n' - p'v_n'')$ over all the bounding surfaces $S_1, S_2, S_3, \ldots$ vanishes. *Note:* If the region contains only one simple source, the theorem reduces to the form ascribed to Helmholtz, namely, in a region as described, a simple source at *A* produces the same sound pressure at another point *B* as would have been produced at *A* had the source been located at *B*. *See also:* **electroacoustics.** 0-1E1

acoustical units. In different sections of the field of acoustics at least three systems of units are in common use: the meter-kilogram-second (mks), the centimeter-gram-second (cgs), and the British. The following table facilitates conversion from one system of units to another. *See also:* electroacoustics. 0-1E1

acoustic center, effective (acoustic generator). The point from which the spherically divergent sound waves appear to diverge when observed at remote points. *See also:* **loudspeaker.** 0-1E1

acoustic delay line. A delay line whose operation is based on the time of propagation of sound waves. *See:* **sonic delay line.** *See also:* **electronic digital computer.** X3A12-16E9

acoustic horn (horn). A tube of varying cross section having different terminal areas that provides a change of acoustic impedance and control of the direction pattern. *See also:* **loudspeaker.** 0-1E1

acoustic impedance. The acoustic impedance of a sound medium on a given surface lying in a wave front is the complex quotient of the sound pressure (force per unit area) on that surface by the flux (volume velocity, or linear velocity multiplied by the area), through the surface. When concentrated, rather than distributed, impedances are considered, the impedance of a portion of the medium is defined by the complex quotient of the pressure difference effective in driving that portion, by the flux (volume velocity). The acoustic impedance may be expressed in terms of mechanical impedance, acoustic impedance being equal to the mechanical impedance divided by the square of the area of the surface considered. The commonly used unit is the acoustical ohm. *Notes:* (1) Velocities in the direction along which the impedance is to be specified are considered positive. (2) The terms and definitions to which this note is appended pertain to single-frequency quantities in the steady state, and to systems whose properties are independent of the magnitudes of these quantities. E157-0

acoustic interferometer. An instrument for the measurement of wavelength and attenuation of sound. Its operation depends upon the interference between reflected and direct sound at the transducer in a standing-wave column. *See also:* **electroacoustics; instrument.** 42A30-0

acoustic output (telephone set). The sound pressure level developed in an artificial ear, measured in decibels referred to 0.00002 newton per square meter. *See also:* **telephone station.** E269-19E8

acoustic pickup (sound box) (phonograph). A device that transforms groove modulations directly into acoustic vibrations. *See also:* **loudspeaker; phonograph pickup.** 0-1E1

acoustic properties of water, representative. Numerical values for some important acoustic properties of water at representative temperatures and salinities are listed in the following table. 0-1E1

acoustic radiating element. A vibrating surface in a transducer that can cause or be actuated by sound waves. *See also:* **electroacoustics.** E157-1E1

acoustic radiation pressure. A unidirectional steady-state pressure component exerted upon a surface by an acoustic wave. *See:* **electroacoustics.** 0-1E1

Relations among various acoustical units

Quantity	Dimension	centimeter-gram-second (cgs) Unit	meter-kilogram-second (mks) Unit	Conversion Factor*	British Unit	Conversion Factor†
Mass	M	gram	kilogram	10^{-3}	slug	6.854×10^{-5}
Velocity (linear)	LT^{-1}	centimeter per second	meter per second	10^{-2}	foot per second	3.281×10^{-2}
Force	MLT^{-2}	dyne	newton	10^{-5}	pound-force	2.248×10^{-5}
Sound pressure	$ML^{-1}T^{-2}$	dyne per square centimeter [microbar]	newton per square meter	10^{-1}	pound per square foot	2.089×10^{-3}
Volume velocity	L^3T^{-1}	cubic centimeter per second	cubic meter per second	10^{-6}	cubic foot per second	3.531×10^{-5}
Sound energy	ML^2T^{-2}	erg	joule	10^{-7}	foot-pound	7.376×10^{-1}
Sound-energy density	$ML^{-1}T^{-2}$	erg per cubic centimeter	joule per cubic meter	10^{-1}	foot-pound per cubic foot	2.089×10^{-1}
Sound-energy flux [sound power of source]	ML^2T^{-3}	erg per second	watt	10^{-7}	foot-pound per second	7.376×10
Sound-energy-flux density [sound intensity]	MT^{-3}	erg per second per square centimeter	watt per square meter	10^{-3}	(foot-pound per second) per square foot	6.847×10^{-5}
Mechanical impedance	MT^{-1}	mechanical ohm [dyne second per centimeter]	mks mechanical ohm [newton second per meter]	10^{-3}	pound-second per foot	6.854×10^{-5}
Acoustic impedance [resistance, reactance]	$ML^{-4}T^{-1}$	acoustical ohm	mks acoustical ohm	10^5	(pound per square foot) per (cubic foot per second)	59.16
Specific acoustic impedance	$ML^{-2}T^{-1}$	rayl [acoustical ohm × square centimeter]	mks rayl [mks acoustical ohm × square meter]	10	(pound per square foot) per (foot per second)	6.366×10^{-2}
Acoustic inertance	ML^{-4}	gram per centimeter to the fourth power	kilogram per meter to the fourth power	10^5	slug per (foot to the fourth power)	59.16

Acoustic stiffness	$ML^{-4}T^{-2}$	(gram per centimeter to the fourth power) per square second	(kilogram per meter to the fourth power) per square second	10^5	(slug per foot to the fourth power) per square second	59.16
Acoustic compliance	$M^{-1}L^4T^2$	(centimeter to the fifth power) per dyne	(meter to the fifth power) per newton	10^5	(foot to the fifth power) per pound	1.690×10^{-2}

* Multiply a magnitude expressed on centimeter-gram-second (cgs) units by the tabulated conversion factor to obtain magnitude in meter-kilogram-second (mks) units.

† Multiply a magnitude expressed in centimeter-gram-second (cgs) units by the tabulated conversion factor to obtain magnitude in British units. These conversion factors were calculated on the basis of standard acceleration due to gravity.

NOTE: *M*, *L*, *T* represent mass, length, and time, respectively, in the sense of the theory of dimensions. For meter-kilogram-second (mks) mechanical ohm and meter-kilogram-second (mks) acoustical ohm, rayl and meter-kilogram-second (mks) rayl are proposed terms. Alternative terms and units are in square brackets.

Speed of sound, density, and characteristic impedance of water at atmospheric pressure for various temperatures and salinities

	Fresh Water				Sea Water							
Salinity in per mil (0/00)	0				30				35			
Temperature in degrees Celsius	0	4	15	20	0	4	15	20	0	4	15	20
Speed of sound in meters per second	1403	1422	1466	1483	1443	1461	1501	1516	1449	1467	1507	1522
Density in kilograms per cubic meter	0999.8	1000.0	0999.1	0998.2	1024.1	1023.8	1022.2	1021.0	1028.1	1027.8	1026.0	1024.8
Characteristic impedance $\times 10^{-6}$ meter-kilogram-second units (rayls)	1.402	1.422	1.465	1.480	1.478	1.496	1.534	1.548	1.490	1.508	1.546	1.560

acoustic radiator. A means for radiating acoustic waves. *See also:* **loudspeaker.** 42A65-0

acoustic radiometer. An instrument for measuring acoustic radiation pressure. *See also:* **electroacoustics; instrument.** 42A30-1E1

acoustic refraction. The process by which the direction of sound propagation is changed due to spatial variation in the speed of sound in the medium. *See also:* **electroacoustics.** 0-1E1

acoustic scattering. The irregular reflection, refraction, or diffraction of a sound in many directions. *See also:* **electroacoustics.** 0-1E1

acoustic transmission system. An assembly of elements for the transmission of sound. *See also:* **electroacoustics.** 0-1E1

acoustic wave filter. A filter designed to separate acoustic waves of different frequencies. *Note:* Through electroacoustic transducers such a filter may be associated with electric circuits. *See also:* **filter.** 0-42A65

acoustics. The science of sound or the application thereof. *See also:* **electroacoustics.** 0-1E1

across-the-line starter. A device that connects the motor to the supply without the use of a resistance or autotransformer to reduce the voltage. *Note:* It may consist of a manually operated switch or a master switch that energizes an electromagnetically operated contactor. *See also:* **starter.** E45-0

ACSR. *See:* **aluminum cable steel reinforced.**

action potential (action current) (medical electronics). The instantaneous value of the potential observed between excited and resting portions of a cell or excitable living structure. *Note:* It may be measured direct or through a volume conductor. *See also:* **medical electronics.** 0-18E1

action spike (medical electronics). The greatest in magnitude and briefest in duration of the characteristic negative waves seen during the observation of action potentials. *See also:* **medical electronics.** 0-18E1

activation (cathode) (thermionics). The treatment applied to a cathode in order to create or increase its emission. *See also:* **electron emission.** 50I07-15E6

activation polarization. The difference between the total polarization and the concentration polarization. *See also:* **electrochemistry.** 42A60-0

active (corrosion). A state wherein passivity is not evident. *See also:* **corrosion terms.** CM-34E2

active area (1) (semiconductor rectifier cell). The portion of the rectifier junction that effectively carries forward current. *See also:* **semiconductor.** E59-34E17

(2) (solar cell). The illuminated area normal to light incidence, usually the face area less the contact area. *Note:* For the purpose of determining efficiency, the area covered by collector grids is considered a part of the active area. *See also:* **semiconductor.** 0-10E1

active current (rotating machinery). The component of the alternating current that is in phase with the voltage. *See also:* **asynchronous machine; synchronous machine.** 0-31E8

active-current compensator (rotating machinery). A compensator that acts to modify the functioning of a voltage regulator in accordance with active current. *See also:* **synchronous machine.** 0-31E8

active electric network. An electric network containing one or more sources of power. *See also:* **network analysis.** 42A65-0

active electrode (electrobiology). (1) A pickup electrode that, because of its relation to the flow pattern of bioelectric currents, shows a potential difference with respect to ground or to a defined zero, or to another (reference) electrode on related tissue. (2) Any electrode, in a system of stimulating electrodes, at which excitation is produced. (3) A stimulating electrode (different electrode) applied to tissue for stimulation and distinguished from another (inactive, dispersive, diffuse, or indifferent) electrode by having a smaller area of contact thus affording a higher current density. *See also:* **electrobiology.** 42A80-18E1

active materials (storage battery). The materials of the plates that react chemically to produce electric energy when the cell discharges and that are restored to their original composition, in the charged condition, by oxidation and reduction processes produced by the charging current. *See also:* **battery.** 42A60-0

active power (1) (rotating machinery). A term used for power when it is necessary to distinguish among apparent power, complex power and its components, active and reactive power. *See also:* **asynchronous machine; synchronous machine.** 0-31E8

(2) (general). *See:* **power, active.**

active-power relay. A power relay that responds to active power. *See also:* **relay.** 0-31E6

active redundancy (reliability). *See:* **redundancy, active.**

active transducer. *See:* **transducer, active.**

actual transient recovery voltage. The transient recovery voltage that actually occurs across the terminals of a pole of a circuit-interrupting device on a particular interruption. *Note:* This is the modified circuit transient recovery voltage with whatever distortion may be introduced by the circuit-interrupting device. 37A100-31E11

actuating current (automatic line sectionalizer). The root-mean-square current that actuates a counting operation or an automatic operation. 37A100-31E11

actuation time, relay. *See:* **relay actuation time.**

actuating device (protective signaling). A manually or automatically operated mechanical or electric device that operates electric contacts to effect signal transmission. *See also:* **protective signaling.** 42A65-0

actuating signal (industrial control). The reference input signal minus the feedback signal. *See also:* **control system, feedback.** 85A1/AS1-34E10/23E0

actuator (automatic train control). A mechanical or electric device used for automatic operation of a brake valve. *See also:* **railway signaling and interlocking.** 42A42-0

actuator, centrifugal (rotating machinery). Rotor-mounted element of a centrifugal starting switch. *See also:* **centrifugal starting switch.** 0-31E8

actuator, relay. *See:* **relay actuator.**

actuator valve. An electropneumatic valve used to control the operation of a brake valve actuator. 42A42-0

acyclic machine (homopolar machine*) (unipolar machine*) (rotating machinery). A direct-current machine in which the voltage generated in the active conductors maintains the same direction with respect to those conductors. *See also:* **direct-current commutating machine.**

*Deprecated 42A10-31E8

adaptation. The process by which the retina becomes

accustomed to more or less light than it was exposed to during an immediately preceding period. It results in a change in the sensitivity of the photoreceptors to light. *Note:* Adaptation is also used to refer to the final state of the process, as reaching a condition of dark adaptation or light adaptation. *See:* **scotopic vision; photopic vision.** *See also:* **visual field.** Z7A1-0

adapter. A device for connecting parts that will not mate. An accessory to convert a device to a new or modified use. 0-9E4

adapter, waveguide. A two-port device for joining two waveguides having nonmating connectors. *See also:* **waveguide.** 0-9E4

adapting. *See:* **self-adapting.**

adaptive antenna system. An antenna system having circuit elements associated with its radiating elements such that some of the antenna properties are controlled by the received signal. *See also:* **antenna.** 0-3E1

adaptive system. A system that has a means of monitoring its own performance and a means of varying its own parameters by closed-loop action to improve its performance. *See also:* **system science.** 0-35E2

Adcock antenna. A pair of vertical antennas separated by a distance of one-half wavelength or less and connected in phase opposition to produce a radiation pattern having the shape of a figure of eight. *See also:* **antenna.** E145-3E1

add and subtract relay. A stepping relay that can be pulsed to rotate the movable contact arm in either direction. *See also:* **relay.** 83A16-0

adder. A device whose output is a representation of the sum of the two or more quantities represented by the inputs. *See:* **half-adder.** *See also:* **electronic analog computer; electronic digital computer.** E162/E270/X3A12-16E9

addition agent (electroplating). A substance that, when added to an electrolyte, produces a desired change in the structure or properties of an electrodeposit, without producing any appreciable change in the conductivity of the electrolytes, or in the activity of the metal ions or hydrogen ions. *See also:* **electroplating.** 42A60-0

address (electronic computation and data processing). (1) An identification, as represented by a name, label, or number, for a register, location in storage, or any other data source or destination such as the location of a station in a communication network. (2) Loosely, any part of an instruction that specifies the location of an operand for the instruction. (3) (electronic machine-control system). A means of identifying information or a location in a control system. *Example:* The x in the command x 12345 is an address identifying the numbers 12345 as referring to a position on the x axis.
See:
absolute address;
base address;
content addressed storage;
direct address;
effective address;
four address;
four-plus-one address;
immediate address;
indirect address;
machine address;
multilevel address;
one-level address;
one-plus-one address;
relative address;
single-address;
symbolic address;
three-address;
three-plus-one address;
triple address;
two-address;
two-plus-one address;
zero-level address.
See also: **electronic digital computer.** EIA3B-34E12;X3A12-16E9

address, effective (computing systems). The address that is derived by applying any specified rules (such as rules relating to an index register or indirect address) to the specified address and that is actually used to identify the current operand. 0-16E9

address format (computing machines). The arrangement of the address parts of an instruction. *Note:* The expression plus-one is frequently used to indicate that one of the addresses specifies the location of the next instruction to be executed, such as one-plus-one, two-plus-one, three-plus-one, four-plus-one. *See also:* **electronic digital computer.** X3A12-16E9

address part. A part of an instruction that usually is an address, but that may be used in some instructions for another purpose. *See also:* **electronic computation; electronic digital computer; instruction code.** E162-0

address register (computing machines). A register in which an address is stored. *See also:* **electronic digital computer.** X3A12-16E9

ADF. *See:* **automatic direction finder.**

***A* display (radar).** A display in which targets appear as vertical deflections from a line representing a time base. Target distance is indicated by the horizontal position of the deflection from one end of the time base. The amplitude of the vertical deflection is a function of the signal intensity. See accompanying figure. *See also:* **navigation.** E172-10E6

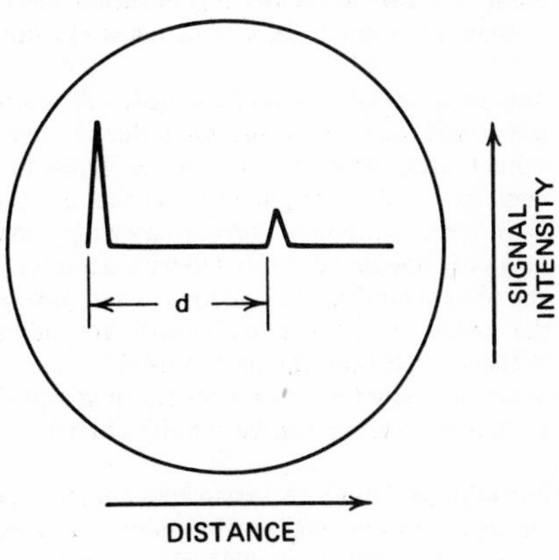

A display.

adjacent-channel attenuation (receivers). *See:* **selectance.**

adjacent-channel interference. Interference in which the extraneous power originates from a signal of assigned (authorized) type in an adjacent channel. *See also:* **interference; radio transmission.** E188-0

adjacent-channel selectivity and desensitization (receiver performance) (receiver). A measure of the ability to discriminate against a signal at the frequency of the adjacent channel. Desensitization occurs when the level of any off-frequency signal is great enough to alter the useable sensitivity. *See also:* **receiver performance.** 0-6E1

adjoint system. (1) A method of computation based on the reciprocal relation between a system of ordinary linear differential equation and its adjoint. *Note:* By solution of the adjoint system it is possible to obtain the weighting function (response to a unit impulse) $W(T, t)$ of the original system for fixed T (the time of observation) as a function of t (the time of application of the impulse). Thus, this method has particular application to the study of systems with time-varying coefficients. The weighting function then may be used in convolution to give the response of the original system to an arbitrary input. *See also:* **electronic analog computer.** E165-16E9

(2) For a system whose state equations are $dx(t)/dt = f(x(t), u(t), t)$, the adjoint system is defined as that system whose state equations are $dy(t)/dt = -A^*y(t)$, where A^* is the conjugate transpose of the matrix whose i,j element is $\partial f_i / \partial x_j$. *See also:* **control system.** 0-23E0

adjust (instrument). Change the value of some element of the mechanism, or the circuit of the instrument or of an auxiliary device, to bring the indication to a desired value, within a specified tolerance for a particular value of the quantity measured. *See also:* **instrument.** 42A30-0

adjustable capacitor. A capacitor, the capacitance of which can be readily changed. E270-0

adjustable constant-speed motor. A motor, the speed of which can be adjusted to any value in the specified range, but when once adjusted the variation of speed with load is a small percentage of that speed. For example, a direct-current shunt motor with field-resistance control designed for a specified range of speed adjustment. *See also:* **asynchronous machine; direct-current commutating machine; synchronous machine.** 0-31E8

adjustable impedance-type ballast. A reference ballast consisting of an adjustable inductive reactor and a suitable adjustable resistor in series. These two components are usually designed so that the resulting combination has sufficient current-carrying capacity and range of impedance to be used with a number of different sizes of lamps. The impedance and power factor of the reactor-resistor combination are adjusted and checked each time the unit is used. 82A9-0

adjustable inductor. An inductor in which the self or mutual inductance can be readily changed. E270-0

adjustable resistor. A resistor so constructed that its resistance can be changed by moving and setting one or more movable contacting elements. E270-0

adjustable-speed drive (industrial control). An electric drive designed to provide easily operable means for speed adjustment of the motor, within a specified speed range. *See also:* **electric drive.** 42A25-34E10

adjustable-speed motor. A motor the speed of which can be varied gradually over a considerable range, but when once adjusted remains practically unaffected by the load; such as a direct-current shunt-wound motor with field resistance control designed for a considerable range of speed adjustment. *See also:* **asynchronous machine; direct-current commutating machine.** 42A10-0

adjustable varying-speed motor. A motor the speed of which can be adjusted gradually, but when once adjusted for a given load will vary in considerable degree with change in load; such as a direct-current compound-wound motor adjusted by field control or a wound-rotor induction motor with rheostatic speed control. *See also:* **asynchronous machine; direct-current commutating machine.** 42A10-0

adjustable varying-voltage control (industrial control). A form of armature-voltage control obtained by impressing on the armature of the motor a voltage that may be changed by small increments, but when once adjusted for a given load will vary considerably with change in load with a consequent change in speed, such as may be obtained from a differentially compound-wound generator with adjustable field current or by means of an adjustable resistance in the armature circuit. *See also:* **control.** 42A25-34E10

adjustable voltage control. A form of armature-voltage control obtained by impressing on the armature of the motor a voltage that may be changed in small increments; but when once adjusted, it, and consequently the speed of the motor, are practically unaffected by a change in load. *Note:* Such a voltage may be obtained from an individual shunt-wound generator with adjustable field current, for each motor. *See also:* **control.** 42A25-34E10

adjusted speed (industrial control). The speed obtained intentionally through the operation of a control element in the apparatus or system governing the performance of the motor. *Note:* The adjusted speed is customarily expressed in percent (or per unit) of base speed (for direct-current shunt motors) or of rated full-load speed (for all other motors). *See also:* **electric drive.** 42A25-34E10

adjuster (rotating machinery). An element or group of elements associated with a feedback control system by which manual adjustment of the level of a controlled variable can be made. *See also:* **synchronous machine.** 0-31E8

adjuster, synchronous-machine voltage-regulator. An adjuster associated with a synchronous-machine voltage regulator by which manual adjustment of the synchronous-machine voltage can be made. *See also:* **synchronous machine.** 0-31E8

adjustment accuracy of instrument shunts (electric power systems). The limit of error, expressed as a percentage of the rated voltage drop, of the initial adjustment of the shunt by resistance or low-current methods. *See also:* **power system, low frequency and surge testing.** 0-31E5

adjustment, relay. *See:* **relay adjustment.**

Adler tube*. *See:* **beam parametric-amplifier.**

*Deprecated

admissible control input set (control system). A set of control inputs that satisfy the control constraints. *See also:* **control system.** 0-23E0

admittance (1) (linear constant-parameter system). (1) The corresponding admittance function with p replaced by $j\omega$ in which ω is real. (2) The ratio of the phasor equivalent of a steady-state sine-wave current or current-like quantity (response) to the phasor equivalent of the corresponding voltage or voltage-like quantity (driving force). The real part is the conductance and the imaginary part is the susceptence. *Note:* Definitions (1) and (2) are equivalent. *Editor's Note:* The ratio Y is commonly expressed in terms of its orthogonal components, thus:

$$Y = G + jB$$

where Y, G, and B are respectively termed the admittance, conductance, and susceptance, all being measured in mhos (reciprocal ohms). In a simple circuit consisting of R, L. and C all in parallel, Y becomes

$$Y = G + j\left(\omega C - \frac{1}{\omega L}\right),$$

where $\omega = 2\pi f$ and f is the frequency. In this special case, $G = 1/R$. Historically, some authors have preferred an opposite sign convention for susceptance, thus: $Y = G - jB$ (now deprecated). Thus, in the simple parallel circuit

$$Y = G - j\left(\frac{1}{\omega L} - \omega C\right).$$

The reader will note that according to either convention, the end result is that any predominantly capacitive susceptance will advance the phase of the response, relative to the driving force. However, the $Y = G - jB$ convention* requires that a predominantly capacitive susceptance B, like a predominantly capacitive reactance X, shall be denoted as negative. The reader is therefore advised to become aware of his author's preference. *See:* **conductance; susceptance.** *See also:* **transmission characteristics.** E270-E4

*Deprecated

(2) (electric machine). A linear operator expressing the relation between current (incrementals) and voltage (incrementals). Its inverse is called the impedance of an electric machine. *Notes:* (1) If a matrix has admittances as its elements, it is usually referred to as admittance matrix. Frequently the admittance matrix is called admittance for short. (2) Most admittances are usually defined with the mechanical angular velocity of the machine at steady state. *See also:* **asynchronous machine; direct-current commutating machine; synchronous machine.** 0-31E8

admittance, effective input (electron tube or valve). The quotient of the sinusoidal component of the control-grid current by the corresponding component of the control voltage, taking into account the action of the anode voltage on the grid current; it is a function of the admittance of the output circuit and the interelectrode capacitance. *Note:* It is the reciprocal of the effective input impedance. *See:* **electron-tube admittances.** 50I07-15E6

admittance, effective output (electron tube or valve). The quotient of the sinusoidal component of the anode current by the corresponding component of the anode voltage, taking into account the output admittance and the interelectrode capacitance. *Note:* It is the reciprocal of the effective output impedance. *See* **electron-tube admittances.** 50I07-15E6

admittance, electrode (*j*th electrode of the *n*-electrode electron tube). *See:* **electrode admittance.**

admittance function (defined for linear constant-parameter systems or parts of such systems). That mathematical function of p that is the ratio of a current or current-like quantity (response) to the corresponding voltage-like quantity (driving force) in the hypothetical case in which the latter is e^{pt} (where e is the base of the natural logarithms, p is arbitrary but independent of t, and t is an independent variable that physically is usually time), and the former is a steady-state response in the form of $Y(p)e^{pt}$. *Note:* In electric circuits voltage is always the driving force and current is the response even though as in nodal analysis the current may be the independent variable; in electromagnetic radiation electric field strength is always considered the driving force and magnetic field strength the response, and in mechanical systems mechanical force is always considered as a driving force and velocity as a response. In a general sense the dimension (and unit) of admittance in a given application may be whatever results from the ratio of the dimensions of the quantity chosen as the response to the dimensions of the quantity chosen as the driving force. However, in the types of systems cited above any deviation from the usual convention should be noted. *See also:* **network analysis.** E270-E

admittance, short-circuit driving-point (*j*th terminal of an *n*-terminal network). The driving-point admittance between that terminal and the reference terminal when all other terminals have zero alternating components of voltage with respect to the reference point. *See also:* **electron-tube admittances.** 42A70-15E6

admittance, short-circuit feedback (electron-device transducer). The short-circuit transfer admittance from the physically available output terminals to the physically available input terminals of a specified socket, associated filters, and electron device. *See also:* **electron-tube admittances.** 42A70-15E6

admittance, short-circuit forward (electron-device transducer). The short-circuit transfer admittance from the physically available input terminals to the physically available output terminals of a specified socket, associated filters, and electron device. *See also:* **electron-tube admittances.** 42A70-15E6

admittance, short-circuit input (electron-device transducer). The driving-point admittance at the physically available input terminals of a specified socket, associated filters, and tube. All other physically available terminals are short-circuited. *See also:* **electron-tube admittances.** E160-15E6

admittance, short-circuit output (electron-device transducer). The driving-point admittance at the physically available output terminals of a specified socket, associated filters, and tube. All other physically available terminals are short-circuited. *See also:* **electron-tube admittances.** E160-15E6

admittance, short-circuit transfer (from the *j*th terminal to the *l*th terminal of an *n*-terminal network). The transfer admittance from terminal j to terminal l when all terminals except j have zero complex alternating components of voltage with respect to the reference point. *See also:* **electron-tube admittances.** E160-15E6

advance ball (mechanical recording). A rounded support (often sapphire) attached to a cutter that rides on the surface of the recording medium so as to maintain a uniform depth of cut and correct for small irregularities of the disk surface. *See also:* **electroacoustics.** 0-1E1

adverse weather (electric power systems). Weather conditions that cause an abnormally high rate of forced outages for exposed components during the periods such conditions persist. *Note:* Adverse weather conditions can be defined for a particular system by selecting the proper values and combinations of conditions reported by the Weather Bureau: thunderstorms, tornadoes, wind velocities, precipitation, temperature, etcetera. *See also:* **outage.** 0-31E4

adverse-weather persistent-cause forced-outage rate (electric power systems) (particular type of component). The mean number of outages per unit of adverse-weather time per component. *See also:* **outage.** 0-31E4

aeolight (optical sound recording). A glow lamp employing a cold cathode and a mixture of permanent gases in which the intensity of illumination varies with the applied signal voltage. 0-1E1

aeration cell. *See:* **differential aeration cell.**

aerial lug. *See* **external connector (pothead).**

aerial cable. An assembly of insulated conductors installed on a pole line or similar overhead structures; it may be self-supporting or installed on a supporting messenger cable. *See also:* **cable.** 42A35-31E13

aerodrome beacon. An aeronautical beacon used to indicate the location of an aerodrome. *Note:* An aerodrome is any defined area on land or water, including any buildings, installations, and equipment intended to be used either wholly or in part for the arrival, departure, and movement of aircraft. *See also:* **signal lighting.** Z7A1-0

aeronautical beacon. An aeronautical ground light visible at all azimuths, either continuously or intermittently, to designate a particular location on the surface of the earth. *See also:* **signal lighting.** Z7A1-0

aeronautical ground light. Any light specially provided as an aid to air navigation, other than a light displayed on an aircraft. *See also:* **signal lighting.** Z7A1-0

aeronautical light. Any luminous sign or signal that is specially provided as an aid to air navigation. *See also:* **signal lighting.** Z7A1-0

aerophare (air operations). A name for radio beacon. *See also:* **navigation.** E172-10E6

aerosol development (electrostatography). Development in which the image-forming material is carried to the field of the electrostatic image by means of a suspending gas. *See also:* **electrostatography.** E224-15E7

AF. *See:* **analog-to-frequency converter.**

AFC. *See:* **automatic frequency control.**

afterimage. A visual response that occurs after the stimulus causing it has ceased.*See also:* **visual field.** Z7A1-0

afterpulse (photo multipliers). A spurious pulse induced in a photomultiplier by a previous pulse. *See also:* **phototube.** E175-0

AGC. *See:* **automatic gain control.**

aggressive carbon dioxide (corrosion). Free carbon dioxide in excess of the amount necessary to prevent precipitation of calcium as calcium carbonate. *See also:* **corrosion terms.** CM-34E2

aging (metallic rectifier) (semiconductor rectifier cell). Any persisting change (except failure) that takes place for any reason in either the forward or reverse resistance characteristic. *See also:* **rectification.** 42A15-0;E59-34E17

agitator (hydrometallurgy) (electrowinning). A receptacle in which ore is kept in suspension in a leaching solution. *See also:* **electrowinning.** 42A60-0

aided tracking. A system of tracking a target signal in bearing, elevation, or range, or any combination of these variables, in which manual correction of the tracking error automatically corrects the rate of motion of the tracking mechanism. *See also:* **radio navigation.** 42A65-0

air (1) (industrial control) (prefix). Applied to a device that interrupts an electric circuit, indicates that the interruption occurs in air. 42A65-0/IC1-34E10
(2) (rotating machinery). *Note:* When a definition mentions the term air, this term can be replaced, where appropriate, by the name of another gas (for example, hydrogen). Similarly the term water can be replaced by the name of another liquid. *See also:* **asynchronous machine; direct-current commutating machines; synchronous machine.** 0-31E8

AI radar (airborne intercept radar). *See:* **fire-control radar.** *See also:* **navigation.**

air baffle (rotating machinery). *See:* **air guide.**

air-blast circuit breaker. *See:* Note under **circuit breaker.**

air cell. A gas cell in which depolarization is accomplished by atmospheric oxygen. *See also:* **electrochemistry.** 42A60-0

air circuit breaker. *See:* Note under **circuit breaker.**

air conduction (hearing). The process by which sound is conducted to the inner ear through the air in the outer ear canal as part of the pathway. *See also:* **electroacoustics.** 0-1E1

air-cooled (rotating machinery). Cooled by air at atmospheric pressure. *See also:* **asynchronous machine; direct-current commutating machine; synchronous machine.** 0-31E8

air cooler (rotating machinery). A cooler using air as one of the fluids. *See also:* **fan (rotating machinery)** 0-31E8

air-core inductance (winding inductance). The effective self-inductance of a winding when no ferromagnetic materials are present. *Note:* The winding inductance is not changed when ferromagnetic materials are present. 0-21E1

aircraft aeronautical light. Any aeronautical light specially provided on an aircraft. *See also:* **signal lighting.** Z7A1-0

aircraft bonding. The process of electrically interconnecting all parts of the metal structure of the aircraft as a safety precaution against the buildup of isolated static charges and as a means of reducing radio interference. 42A41-0

aircraft induction motor (rotating machinery). A motor designed for operation in the environment seen by aircraft yet having minimum weight and utmost reliability for a limited life. *See also:* **asynchronous machine.** 0-31E8

air-derived navigation data. Data obtained from measurements made at an airborne vehicle. *See also:* **navigation.** 0-10E6

air duct (air guide) (ventilating duct) (rotating machinery). Any passage designed to guide ventilating air. *See also:* **cradle base (rotating machinery).** 0-31E8

air ducting (rotating machinery). Any separate structure mounted on a machine to guide ventilating air to or from a heat exchanger, filter, fan or other device mounted on the machine. *See also:* **cradle base (rotating machinery).** 0-31E8

air filter (rotating machinery). Any device used to remove suspended particles from air. *See also:* **fan (rotating machinery).** 0-31E8

air gap (gap) (rotating machinery). A separating space between two parts of magnetic material, the combination serving as a path for magnetic flux. *Note:* This space is normally filled with air or hydrogen and represents clearance between rotor and stator of an electric machine. *See also:* **direct-current commutating machine.** 0-31E8

air-gap factor (fringing coefficient) (rotating machinery). A factor used in machine design calculations to determine the effective length of the air gap from the actual separation of rotor and stator. *Note:* The use of the fringing coefficient is necessary to account for several geometric effects such as the presence of slots and cooling ducts in the rotor and stator. *See also:* **rotor (rotating machinery); stator.** 0-31E8

air-gap line. The extended straight line part of the no-load saturation curve. *See also:* **synchronous machine.** 42A10-31E8

air gap, relay. *See:* **relay air gap.**

air guide (air duct) (rotating machinery). Any structure designed to direct the flow of ventilating air. *See:* **cradle base (rotating machinery).** 0-31E8

air horn. A horn having a diaphragm that is vibrated by the passage of compressed air. *See also:* **protective signaling.** 42A65-0

air-insulated terminal box. A terminal box so designed that the protection of phase conductors against electrical failure within the terminal box is by adequately spacing bare conductors with appropriate insulation supports. *See also:* **cradle base (rotating machinery).** 0-31E8

air opening (rotating machinery). A port for the passage of ventilation air. *See also:* **cradle base (rotating machinery).** 0-31E8

air pipe (rotating machinery). Any separate structure designed for attaching to a machine to guide the inlet ventilating air to the machine or the exhaust air away from the machine. *See also:* **cradle base (rotating machinery).** 0-31E8

airport surface detection equipment (ASDE) (electronic navigation). A radar for observation of the positions of aircraft on the surface of an airport. *See also:* **navigation.** E172-10E6

airport surveillance radar (ASR) (electronic navigation). *See:* **surveillance radar.** *See also:* **navigation; radar.**

air-position indicator (API). A dead-reckoning computer that integrates headings and speeds to give a continuous indication of position with respect to the air mass in which the vehicle is moving. *See also:* **radio navigation.** 42A65-0

air-route surveillance radar (ARSR). *See:* **surveillance radar.**

air shield (rotating machinery). An air guide used to prevent ventilating air from returning to the blower inlet before passing through the ventilating circuit. *See also:* **cradle base (rotating machinery).** 0-31E8

air speed. The rate of motion of a vehicle relative to the air mass. *See also:* **navigation.** E172-10E6

air switch. *See:* Note under **mechanical switching device.**

air terminal (lightning protection). The combination of elevation rod and brace, or footing placed on upper portions of structures, together with tip or point if used. *See also:* **arrester; lightning protection and equipment.** 42A95-0

air transportation. *See:* **air-transportation electric equipment; air-transportation instruments; air-transportation wiring and associated equipment; electric equipment.**

air-transportation electric equipment.
See:
air transportation;
aircraft bonding;
altitude-treated current-carrying brush;
automatic pilot;
automatic pilot servo-motor;
auxiliary generator set;
booster coil;
electric capacitance altimeter;
electric gun heater;
electric parachute flare launching tube;
electrically heated airspeed tube;
electrically heated flying suit;
engine-driven generator for aircraft;
induction vibrator;
radio shielding;
reverse-current cutout;
shielded ignition harness;
tachometer generator;
vapor-safe electric equipment;
wind-driven generator for aircraft;
windshield wiper for aircraft.

air-transportation electronic equipment.
See:
air transportation;
automatic approach control;
fairlead for aircraft;
flush antenna for aircraft;
glide-path receiver;
interphone equipment;
localizer receiver;
marker-beacon receiver;
mast-type antenna for aircraft;
trailing-type antenna for aircraft.

air-transportation instruments. *Note:* For an extensive list of cross references, see *Appendix A.*

air-transportation wiring and associated equipment.
See:
air transportation;
bomb control switch;
conduit;
electric pin-and-socket coupler;
electric power distribution panel;
gun control switch.

airway beacon. An aeronautical beacon used to indicate a point on the airway. *See also:* **signal lighting.** Z7A1-0

alarm point (power-system communication). A supervisory control status point considered to be an alarm. *See also:* **supervisory control system.** 0-31E3

alarm signal. A signal for attracting attention to some

abnormal condition. *See also:* **telephone switching system.** 42A65-19E1

alarm (signal) relay. A monitoring relay whose function is to operate an audible or visual signal to announce the occurrence of an operation or a condition needing personal attention, and usually provided with a signaling cancellation device. *See also:* **relay.** 0-31E6

alarm switch (switching device). An auxiliary switch that actuates a signaling device upon the automatic opening of the switching device with which it is associated. 37A100-31E11

alarm system (protective signaling). An assembly of equipment and devices arranged to signal the presence of a hazard requiring urgent attention. *See also:* **protective signaling.** 42A65-0

albedo (photovoltaic power system). The reflecting power expressed as the ratio of light reflected from an object to the total amount falling on it. *See also:* **photovoltaic power system; solar cells (photovoltaic power system).** 0-10E1

ALC. *See:* **automatic load (level) control.**

Alford loop (electronic navigation). A multielement antenna, having approximately equal amplitude currents that are in phase and uniformly distributed along each of its peripheral elements, producing a substantially circular radiation pattern in the plane of polarization; originally developed as a four-element horizontally polarized very-high-frequency loop antenna. *See also:* **navigation.** E172-10E6

algorithm. A prescribed set of well-defined rules or processes for the solution of a problem in a finite number of steps, for example, a full statement of an arithmetic procedure for evaluating sin x to a stated precision. *See:* **heuristic.** X3A12-16E9

aligned-grid tube (or valve). A vacuum multigrid tube or valve in which at least two of the grids are aligned the one behind the other so as to obtain a particular effect (canalizing an electron beam, suppressing noise, etcetera). *See also:* **electron tube.** 50I07-15E6

alignment (1) (communication practice). The process of adjusting a plurality of components of a system for proper interrelationship. *Note:* The term is applied especially to (1) the adjustment of the tuned circuits of an amplifier for desired frequency response, and (2) the synchronization of components of a system. *See also:* **radio transmission.** 42A65-31E3

(2) (inertial navigation equipment). The orientation of the measuring axes of the inertial components with respect to the coordinate system in which the equipment is used. *Note:* Initial alignment refers to the result of the process of bringing the measuring axes into a desired orientation with respect to the coordinate system in which the equipment is used, prior to departure. The initial alignment can be refined by the use of noninertial sensors while in operational use. *See also:* **navigation.** E174-10E6

alive (electric system). Electrically connected to a source of potential difference, or electrically charged so as to have a potential different from that of the ground. *Note:* The term **alive** is sometimes used in place of the term **current-carrying**, where the intent is clear, to avoid repetitions of the longer term. *See also:* **insulated; live.** 42A95/2A2-0

alkaline cleaning (electroplating). Cleaning by means of alkaline solutions. *See also:* **electroplating.** 42A60-0

alkaline storage battery. A storage battery in which the electrolyte consists of an alkaline solution, usually potassium hydroxide. *See also:* **battery (primary or secondary).** 42A60-0

allocation (computing machines). *See:* **storage allocation.** *See also:* **electronic digital computer.**

alloy or fused junction (semiconductors). A junction formed by recrystallization on a base crystal from a liquid phase of one or more components and the semiconductor. *See also:* **semiconductor.** E59-34E17

alloy plate. An electrodeposit that contains two or more metals codeposited in combined form or in intimate mixtures. *See also:* **electroplating.** 42A60-0

all-pass function (linear passive networks). A transmittance that provides only phase shift, its magnitude characteristic being constant. *Notes:* (1) For lumped-parameter networks, this is equivalent to specifying that the zeros of the function are the negatives of the poles. (2) A realizable all-pass function exhibits nondecreasing phase lag with increasing frequency. (3) A trivial all-pass function has zero phase at all frequencies. *See also:* **linear passive networks.** E156-0

all-pass network (all-pass transducer). A network designed to introduce phase shift or delay without introducing appreciable attenuation at any frequency. *See also:* **network analysis.** 42A65-0

all-pass transducer. *See:* **all-pass network.**

all-relay system. An automatic telephone switching system in which all switching functions are accomplished by relays. *See also:* **telephone switching system.** 42A65-0

alphabet. A character set arranged in certain order. *Note:* Character sets are finite quantities of letters of the normal alphabet, digits, punctuation marks, control signals, such as carriage return and other ideographs. Characters are usually represented by letters (graphics) or technically realized in the form of combinations of punched holes, sequences of electric pulses, etcetera. 0-19E4

alphameric. *See:* **alphanumeric.**

alphanumeric. Pertaining to a character set that contains both letters and digits, and usually other characters. Synonymous with alphameric. 0-16E9

alteration (elevator, dumbwaiter, or escalator). Any change or addition to the equipment other than ordinary repairs or replacements. *See also:* **elevator.** 42A45-0

alternate-channel interference (second-channel interference). Interference caused in one communication channel by a transmitter operating in a channel next beyond an adjacent channel. *See also:* **radio transmission.** 42A65-0

alternate display (oscillography). A means of displaying output signals of two or more channels by switching the channels in sequence. *See:* **oscillograph.** 0-9E4

alternating charge characteristic (nonlinear capacitor). The function relating the instantaneous values of the alternating component of transferred charge, in a steady state, to the corresponding instantaneous values of a specified applied periodic capacitor-voltage. *Note:* The nature of this characteristic may depend upon the nature of the applied voltage. *See also:* **nonlinear capacitor.** E226-15E7

alternating current. A periodic current the average value of which over a period is zero. *Note:* Unless distinctly specified otherwise, the term **alternating**

current refers to a current that reverses at regularly recurring intervals of time and that has alternately positive and negative values. E45-0

alternating-current analog computer. *See:* **analog computer, alternating current.**

alternating-current circuit. A circuit that includes two or more interrelated conductors intended to be energized by alternating current. E270-0

alternating-current commutator motor. An alternating-current motor having an armature connected to a commutator and included in an alternating-current circuit. *See also:* **asynchronous machine.** 0-31E8

alternating-current component. The current remaining when the average value has been subtracted from an alternating current. *See:* **symmetrical component (total current).** E270-0

alternating-current–direct-current general-use snap-switch. A form of general-use snap-switch suitable for use on either direct- or alternating-current circuits for controlling the following: (1) Resistive loads not exceeding the ampere rating at the voltage involved. (2) Inductive loads not exceeding one-half the ampere rating at the voltage involved, except that switches having a marked horsepower rating are suitable for controlling motors not exceeding the horsepower rating of the switch at the voltage involved. (3) Tungsten filament lamp loads not exceeding the ampere rating at 125 volts, when marked with the letter "T". Alternating-current–direct-current general-use snap-switches are not generally marked alternating-current–direct-current, but are always marked with their electrical rating. *See also:* **switch.** 1A0-0

alternating-current–direct-current ringing. Ringing in which a combination of alternating and direct currents is utilized, the direct current being provided to facilitate the functioning of the relay that stops the ringing. *See also:* **telephone switching system.** 42A65-0

alternating-current distribution. The supply to points of utilization of electric energy by alternating current from its source or one or more main receiving stations. *Notes:* (1) Generally a voltage is employed that is not higher than that which could be delivered or utilized by rotating electric machinery. Step-down transformers of a capacity much smaller than that of the line are usually employed as links between the moderate voltage of distribution and the lower voltage of the consumer's apparatus. (2) For an extensive list of cross references, see *Appendix A.* 42A35-31E13

alternating-current electric locomotive. An electric locomotive that collects propulsion power from an alternating-current distribution system. *See also:* **electric locomotive.** 42A42-0

alternating-current erasing head (magnetic recording). A head that uses alternating current to produce the magnetic field necessary for erasing. *Note:* Alternating-current erasing is achieved by subjecting the medium to a number of cycles of a magnetic field of a decreasing magnitude. The medium is, therefore, essentially magnetically neutralized. *See also:* **electroacoustics.** 0-1E1

alternating-current floating storage-battery system. A combination of alternating-current power supply, storage battery, and rectifying devices connected so as to charge the storage battery continuously and at the same time to furnish power for the operation of signal devices. *See also:* **railway signal and interlocking.** 42A42-0

alternating-current general-use snap-switch. A form of general-use snap-switch suitable only for use on alternating-current circuits for controlling the following: (1)Resistive and inductive loads (including electric discharge lamps) not exceeding the ampere rating at the voltage involved. (2) Tungsten filament lamp loads not exceeding the ampere rating at 120 volts. (3)Motor loads not exceeding 80 percent of the ampere rating of the switches at the rated voltage. *Note:* All alternating-current general-use snap-switches are marked ac in addition to their electrical rating. *See also:* **switch.** 1A0-0

alternating-current generator. A generator for the production of alternating-current power. *See also:* **synchronous machine.** 0-31E8

alternating-current magnetic biasing (magnetic recording). Magnetic biasing accomplished by the use of an alternating current, usually well above the signal-frequency range. *Note:* The high-frequency linearizing (biasing) field usually has a magnitude approximately equal to the coercive force of the medium. *See also:* **electroacoustics.** 0-1E1

alternating-current motor. An electric motor for operation by alternating current. *See also:* **synchronous machine.** 0-31E8

alternating-current pulse. An alternating-current wave of brief duration. *See also:* **pulse.** 42A65-0

alternating-current relay. *See:* **relay, alternating-current.**

alternating-current root-mean-square voltage rating (semiconductor rectifiers). The maximum root-mean-square value of applied sinusoidal voltage permitted by the manufacturer under stated conditions. *See:* **semiconductor rectifier stack.** 0-34E24

alternating-current transmission. The transfer of electric energy by alternating current from its source to one or more main receiving stations for subsequent distribution. *Note:* Generally a voltage is employed that is higher than that which would be delivered or utilized by electric machinery. Transformers of a capacity comparable to that of the line are usually employed as links between the high voltage of transmission and the lower voltage used for distribution or utilization. *See also:* **alternating-current distribution.** 42A35-31E13

alternating-current transmission (television). That form of transmission in which a fixed setting of the controls makes any instantaneous value of signal correspond to the same value of brightness only for a short time. *Note:* Usually this time is not longer than one field period and may be as short as one line period. *See also:* **television.** 42A65-0

alternating-current winding (rectifier transformer). The primary winding that is connected to the alternating-current circuit and usually has no conductive connection with the main electrodes of the rectifier. *See also:* **rectifier transformer.** 42A15-31E12

alternating function. A periodic function whose average value over a period is zero. For instance, $f(t) = B \sin \omega t$ is an alternating function (ω, B assumed constants). E270-0

alternating sparkover voltage (arrester). *See:* **arrester, alternating sparkover voltage.**

alternating voltage. *See:* **alternating current.**

alternator transmitter. A radio transmitter that utilizes power generated by a radio-frequency alternator. *See also:* **radio transmitter.** E145/E182/42A65-0

altitude (astronomy). The angular distance of a heavenly body measured on that great circle that passes perpendicular to the plane of the horizon through the body and through the zenith. *Note:* It is measured positively from the horizon to the zenith, from 0 to 90 degrees. *See also:* **light; sunlight.** Z7A1-0

altitude-treated current-carrying brush. A brush specially fabricated or treated to improve its wearing characteristics at high altitudes (over 6000 meters). *See also:* **air-transportation electric equipment.** 42A41-0

aluminum cable steel reinforced (ACSR). A composite conductor made up of a combination of aluminum and steel wires. In the usual construction the aluminum wires surround the steel. *See also:* **conductor.** 42A35-31E13

aluminum conductor. A conductor made wholly of aluminum. *See also:* **conductor.** 42A35-31E13

aluminum-covered steel wire (power distribution underground cables). A wire having a steel core to which is bonded a continuous outer layer of aluminum. *See also:* **power distribution, underground construction.** 0-31E1

AMA. *See:* **automatic message accounting system.**

AM to FS converter. *See:* **transmitting converter, facsimile.** *See also:* **facsimile (electrical communication).**

amalgam (electrolytic cells). The product formed by mercury and another metal in an electrolytic cell. 42A60-0

ambient conditions. Characteristics of the environment, for example, temperature, humidity, pressure. *See also:* **measurement system.** 0-9E4

ambient level (electromagnetic compatibility). The values of radiated and conducted signal and noise existing at a specified test location and time when the test sample is not activated. *See also:* **electromagnetic compatibility.** 0-27E1

ambient noise (1) (room noise). Acoustic noise existing in a room or other location. Magnitudes of ambient noise are usually measured with a sound level meter. *Note:* The term room noise is commonly used to designate ambient noise at a telephone station. *See also:* **circuit noise; circuit noise level; line noise.** 42A65-31E3

(2) (mobile communication). The average noise power in a given location that is the integrated sum of atmospheric, galactic, and man-made noise. *See also:* **telephone station.** 0-6E1

ambient operating-temperature range (power supplies). The range of environmental temperatures in which a power supply can be safely operated. For units with forced-air cooling, the temperature is measured at the air intake. *See also:* **power supply.** KPSH-10E1

ambient radio noise. *See:* **ambient level.**

ambient temperature. The temperature of the medium such as air, water, or earth into which the heat of the equipment is dissipated. *Notes:* (1) For self-ventilated equipment, the ambient temperature is the average temperature of the air in the immediate neighborhood of the equipment. (2) For air- or gas-cooled equipment with forced ventilation or secondary water cooling, the ambient temperature is taken as that of the ingoing air or cooling gas. (3) For self-ventilated enclosed (including oil-immersed) equipment considered as a complete unit, the ambient temperature is the average temperature of the air outside of the enclosure in the immediate neighborhood of the equipment. IC1-34E10

ambient temperature rating at quarter-rated thermal burden. Maximum ambient temperature at which a potential transformer can be operated without exceeding the specified temperature limitations, when operated at rated voltage and frequency while supplying 25 percent of the rated thermal burden. 0-31E12

ambiguity (navigation) (electronic navigation). The condition when navigation coordinates define more than one point, direction, line of position, or surface of position. *See also:* **navigation.** E172-10E6

American Morse Code. *See:* **Morse Code.**

ammeter. An instrument for measuring the magnitude of an electric current. *Note:* It is provided with a scale, usually graduated in either amperes, milliamperes, microamperes, or kiloamperes. If the scale is graduated in milliamperes, microamperes, or kiloamperes, the instrument is usually designated as a milliammeter, a microammeter, or a kiloammeter. *See also:* **instrument.** 42A30-0

amortisseur. A permanently short-circuited winding consisting of conductors embedded in the pole shoes of a synchronous machine and connected together at the ends of the poles, but not necessarily connected between poles. *Note:* This winding when used in salient-pole machines sometimes includes bars that do not pass through the pole shoes, but are supported in the interpolar spaces between the pole tips. *See also:* **synchronous machine.** 42A10-0

amortisseur bar (damper bar) (rotating machinery). A single conductor that is a part of an amortisseur winding or starting winding. *See also:* **rotor (rotating machinery); stator.** 0-31E8

amortisseur winding. *See:* **damper winding; damping winding.**

ampacity. Current-carrying capacity, expressed in amperes, of a wire or cable under stated thermal conditions. 1A0-31E1

ampere. That constant current that, if maintained in two straight parallel conductors of infinite length, of negligible circular cross section, and placed 1 meter apart in vacuum, would produce between these conductors a force equal to 2×10^{-7} newton per meter of length. *See also:* **abampere.** CGPM SCC 14

ampere-conductors (distributed winding) (rotating machinery). The product of the number of conductors round the periphery of the winding and the current (in amperes, root-mean-square) circulating in these conductors. *See also:* **rotor (rotating machinery); stator.** 0-31E8

ampere-hour capacity (storage battery). The number of ampere-hours that can be delivered under specified conditions as to temperature, rate of discharge, and final voltage. *See also:* **battery (primary or secondary).** 42A60-0

ampere-hour efficiency (storage cell) (storage battery). The electrochemical efficiency expressed as the ratio of the ampere-hours output to the ampere-hours input required for the recharge. *See also:* **charge.** 42A60-0

ampere-hour meter. An electricity meter that measures and registers the integral, with respect to time, of the current of the circuit in which it is connected. *Note:* The unit in which this integral is measured is usually the ampere-hour. *See also:* **electricity meter (meter).** 42A30-0

ampere-turn per meter. The unit of magnetic field strength in SI units (International System of Units). The ampere-turn per meter is the magnetic field strength in the interior of an elongated uniformly wound solenoid that is excited with a linear current density in its winding of one ampere per meter of axial distance. E270-0

ampere-turns (rotating machinery). The product of the number of turns of a coil or a winding (distributed or concentrated), and the current in amperes circulating in these turns. *See also:* **asynchronous machine; direct-current commutating machine; synchronous machine.** 0-31E8

Ampere's law. *See:* **magnetic field strength produced by an electric current.**

amplification (signal-transmission system). (1) The ratio of output magnitude to input magnitude in a device which is intended to produce an output that is an enlarged reproduction of its input. *Note:* It may be expressed as a ratio or, by extension of the term, in decibels. (2) The process causing this increase. *See also:* **signal; transmission characteristics.** E151/42A65-31E3/13E6

amplification, current. *See:* **current amplification.**

amplification factor. The μ factor for a specified electrode and the control grid of an electron tube under the condition that the anode current is held constant. *Notes:* (1) In a triode this becomes the μ factor for the anode and control-grid electrodes. (2) In multielectrode tubes connected as triodes the term anode applies to the combination of electrodes used as the anode. *See also:* **electron-tube admittances.** E160-15E6

amplification factor, gas (gas phototube). *See:* **gas amplification factor (gas phototube).**

amplification, voltage. *See:* **voltage amplification.**

amplifier. A device that enables an input signal to control power from a source independent of the signal and thus be capable of delivering an output that bears some relationship to, and is generally greater than, the input signal. *Note:* For an extensive list of cross references, see *Appendix A.* E145/E151-0; E165-16E9; 0-42A65

amplifier, balanced (push-pull amplifier). An amplifier in which there are two identical signal branches connected so as to operate in phase opposition and with input and output connections each balanced to ground. E145/E151/E182/42A65-31E3

amplifier, bridging. An amplifier with an input impedance sufficiently high so that its input may be bridged across a circuit without substantially affecting the signal level of the circuit across which it is bridged. *See also:* **amplifier.** E151-0

amplifier, buffer (signal-transmission system). *See:* **amplifier, isolating.**

amplifier, carrier (signal-transmission system). An alternating-current amplifier capable of amplifying a prescribed carrier frequency and information sidebands relatively close to the carrier frequency. *See also:* **signal.** 0-13E6

amplifier class ratings (electron tubes).

(1) class-A amplifier. An amplifier in which the grid bias and alternating grid voltages are such that anode current in a specific tube flows at all times. *Note:* The suffix 1 is added to the letter or letters of the class identification to denote that grid current does not flow during any part of the input cycle. The suffix 2 is used to denote that current flows during some part of the cycle. *See also:* **amplifier.** E160-15E6; E145/42A65-0

(2) class-AB amplifier. An amplifier in which the grid bias and alternating grid voltages are such that anode current in a specific tube flows for appreciably more than half but less than the entire electrical cycle. *Note:* The suffix 1 is added to the letter or letters of the class identification to denote that grid current does not flow during any part of the input cycle. The suffix 2 is used to denote that current flows during some part of the cycle. *See also:* **amplifier.** E160-15E6;E145/42A65-0

(3) class-B amplifier. An amplifier in which the grid bias is approximately equal to the cutoff value so that the anode current is approximately zero when no exciting grid voltage is applied, and so that anode current in a specific tube flows for approximately one half of each cycle when an alternating grid voltage is applied. *Note:* The suffix 1 is added to the letter or letters of the class identification to denote that grid current does not flow during any part of the input cycle. The suffix 2 is used to denote that current flows during some part of the cycle. *See also:* **amplifier.** E160-15E6;E145/42A65-0

(4) class-C amplifier. An amplifier in which the grid bias is appreciably greater than the cutoff value so that the anode current in each tube is zero when no alternating grid voltage is applied, and so that anode current in a specific tube flows for appreciably less than one half of each cycle when an alternating grid voltage is applied. *Note:* The suffix 1 is added to the letter or letters of the class identification to denote that grid current does not flow during any part of the input cycle. The suffix 2 is used to denote that current flows during some part of the cycle. *See also:* **amplifier.** E160-15E6; E145/42A65-0

amplifier, chopper (signal-transmission system). A modulated amplifier in which the modulation is achieved by an electronic or electromechanical chopper, the resultant wave being substantially square. *See also:* **signal.** 0-13E6

amplifier, clipper. An amplifier designed to limit the instantaneous value of its output to a predetermined maximum. *See also:* **amplifier.** E151-0

amplifier, difference. *See:* **differential amplifier.**

amplifier, differential. *See:* **differential amplifier.**

amplifier, distribution. A power amplifier designed to energize a speech or music distribution system and having sufficiently low output impedance so that changes in load do not appreciably affect the output voltage. *See also:* **amplifier.** E151-0

amplifier ground (signal-transmission system). *See:* **receiver ground.**

amplifier, horizontal. *See:* **horizontal amplifier.**

amplifier, intensity. *See:* **intensity amplifier.**

amplifier, isolating (signal-transmission system). An amplifier employed to minimize the effects of a following circuit on the preceding circuit. *Example:* An amplifier having effective direct-current resistance and/or alternating-current impedance between any part of its input circuit and any other of its circuits that is high compared to some critical resistance or impedance value in the input circuit. *See also:* **signal.** E151-13E6

amplifier, isolation (buffer). An amplifier employed to minimize the effects of a following circuit on the preceding circuit. *See also:* **amplifier.** E151-2E2

amplifier, line. An amplifier that supplies a transmission line or system with a signal at a stipulated level. *See also:* **amplifier.** E151-2E2

amplifier, modulated (signal-transmission signal). *See:* **modulated amplifier.**

amplifier, monitoring (electroacoustics). An amplifier used primarily for evaluation and supervision of a program. *See also:* **amplifier.** E151-2E2

amplifier, peak limiting. *See:* **peak limiter.**

amplifier, power. An amplifier that drives a utilization device such as a loudspeaker. *See also:* **amplifier.** E151-0

amplifier, program. *See:* **amplifier, line.**

amplifier, vertical. *See:* **vertical amplifer.**

amplifier, *X*-axis. *See:* **horizontal amplifier.**

amplifier, *Y*-axis. *See:* **vertical amplifier.**

amplifier, *Z*-axis. *See:* ***Z*-axis amplifier; intensity amplifier.**

amplitude (sine wave). A in $A \sin (\omega t + \theta)$ where A, ω, θ are not necessarily constants, but are specified functions of t. In amplitude modulation, for example, the amplitude A is a function of time. In electrical engineering, the term **amplitude** is often used for the modulus of a complex quantity. Amplitude with a modifier, such as peak or maximum, minimum, root-mean-square, average, etcetera, denotes values of the quantity under discussion that are either specified by the meanings of the modifiers or otherwise understood. *See:* **amplitude (simple sine wave).** E270-0

amplitude (simple sine wave). The positive real A in $A \sin (\omega t + \theta)$, where A, ω, θ are constants. In this case, amplitude is synonymous with maximum or peak value. *See:* **amplitude (sine wave).** E270-0

amplitude balance control (electronic navigation). The portion of a system that may be varied to adjust the relative output levels of two related signals. Originally used in instrument landing systems and later in loran. *See also:* **navigation.** E172-10E6

amplitude characteristic. *See:* **amplitude-frequency characteristic.**

amplitude discriminator. A circuit whose output is a function of the relative magnitudes of two signals. *See also:* **navigation.** E172-10E6

amplitude distortion.
See:
amplitude-frequency distortion;
distortion;
distortion, intermodulation;
harmonic distortion;
nonlinear distortion;
waveform-amplitude distortion.

amplitude factor (restriking voltage) (transient recovery voltage) (lightning arrester). The ratio between the peak restriking voltage and the peak value ($\sqrt{2}$ times the root-mean-square value) of the recovery voltage. *See also:* **lightning arrester (surge diverter).** 99I2-31E7

amplitude factor (transient recovery voltage). The ratio of the highest peak of the transient recovery voltage to the peak value of the normal-frequency recovery voltage. *Note:* In tests made under one condition to simulate duty under another, as in single-phase tests made to simulate duty on three-phase ungrounded faults, the amplitude factor is expressed in terms of the duty being simulated. 37A100-31E11

amplitude-frequency characteristic (amplitude characteristic). The variation with frequency of the amplitude (that is, modulus or absolute magnitude) of a phasor quantity. *Note:* The magnitudes of transfer admittances, transfer ratios, amplification, etcetera, plotted against frequency are a few examples; measures in decibels or other units of these same quantities plotted against frequency are included. E270-0

amplitude-frequency distortion. Distortion due to an undesired amplitude-frequency characteristic. *Notes:* (1) The usual desired characteristic is flat over the frequency range of interest. (2) Also sometimes called amplitude distortion or frequency distortion. *See also:* **distortion.** E154-0

amplitude-frequency response. The variation of gain, loss, amplification, or attenuation as a function of frequency. *Note:* This response is usually measured in the region of operation in which the transfer characteristic of the system or transducer is essentially linear. *See also:* **transmission characteristics.** E151-2E2; 42A65-31E3

amplitude gate. *See:* **slicer.**

amplitude locus (control system, feedback) (for a nonlinear system or element whose gain is amplitude dependent). A plot of the describing function, in any convenient coordinate system. *See also:* **control system, feedback.** 85A1-23E0

amplitude-modulated transmitter. A transmitter that transmits an amplitude-modulated wave. *Note:* In most amplitude-modulated transmitters, the frequency is stabilized. *See also:* **radio transmitter.** E145/E182/42A65-0

amplitude modulation (signal-transmission system). (1) The process, or the result of the process, whereby the amplitude of one electrical quantity is varied in accordance with some selected characteristic of a second quantity, which need not be electrical in nature. *See also:* **modulating systems; signal.** 0-13E6
(2) Modulation in which the amplitude of a wave is the characteristic varied.
See:
carrier-to-noise ratio;
improvement threshold;
modulation factor;
sidebands;
side frequency;
suppressed-carrier operation;
transmitted-carrier operation;
vestigial sideband;
vestigial-sideband transmission.
See also: **modulating systems.**
E145/E170/E188/42A65-19E4/31E3

amplitude-modulation noise. The noise produced by undesired amplitude variations of a radio-frequency signal. *See also:* **radio transmission.** E182A/42A65-0

amplitude-modulation noise level. The noise level produced by undesired amplitude variations of a radio-frequency signal in the absence of any intended modulation. E145-0

amplitude noise (radar). The noise-like variation of the amplitude of the received echo from a target caused by change of target aspect; a component of scintillation. *See also:* **navigation.** 0-10E6

amplitude pulse. A general term indicating the magnitude of a pulse. *Note:* Pulse amplitude is measured with respect to the nominally constant baseline, unless otherwise stated. For specific designation, adjectives such as average, instantaneous, peak, root-mean-square, etcetera, should be used to indicate the particular meaning intended. *See also:* **pulse.** 0-9E4

amplitude range (electroacoustics). The ratio, usually expressed in decibels, of the upper and lower limits of program amplitudes that contain all significant energy contributions. *See also:* **electroacoustics.** E151-0

amplitude reference level (pulse techniques). The arbitrary reference level from which all amplitude measurements are made. *Note:* The arbitrary reference level normally is considered to be at an absolute amplitude of zero but may, in fact, have any magnitude of either polarity. If this arbitrary reference level is other than zero, its value and polarity must be stated. *See also:* **pulse.** 0-9E4

amplitude resonance. Resonance in which amplitude is stationary with respect to frequency. E270-0

amplitude response (camera tubes). The ratio of (1) the peak-to-peak output from the tube resulting from a spatially periodic test pattern, to (2) the difference in output corresponding to large-area blacks and large-area whites, having the same illuminations as the test pattern minima and maxima, respectively. *Note:* The amplitude response is referred to as modulation transfer (sine-wave response) when a sinusoidal test pattern is used and as square-wave response when the pattern consists of alternate black and white bars of equal width. *See also:* **camera tube.** 0-15E6

amplitude response characteristic (camera tubes). The relation between (1) amplitude response and (2) television line number (camera tubes) or (image tubes) test-pattern spatial frequency, usually in line pairs per millimeter. *See also:* **camera tube.** 0-15E6

amplitude selection. A summation of one or more variables and a constant resulting in a sudden change in rate or level at the output of a computing element as the sum changes sign. *See also:* **electronic analog computer.** E165-16E9

amplitude suppression ratio (frequency modulation). The ratio of the undesired output to the desired output of a frequency-modulation receiver when the applied signal has simultaneous amplitude modulation and frequency modulation. *Note:* This ratio is generally measured with an applied signal that is amplitude modulated 30 percent at a 400-hertz rate and is frequency modulated 30 percent of maximum system deviation at a 1000-hertz rate. *See also:* **frequency modulation.** 42A65-0

anaerobic. Free of uncombined oxygen. *See:* **corrosion terms.**

analog (1). Pertaining to data in the form of continuously variable physical quantities. *See:* **digital.** X3A12-16E9

(2) (adjective). Used to describe a physical quantity, such as voltage or shaft position, that normally varies in a continuous manner, or devices such as potentiometers and synchros that operate with such quantities. *See also:* **electronic analog computer.** E165-0

(3) (industrial control). Pertains to information content that is expressed by signals dependent upon magnitude. *See also:* **control system, feedback.** AS1-34E10

(4) (electronic computers). A physical system on which the performance of measurements yields information concerning a class of mathematical problems. *See also:* **electronic computation.** E270-0

analog and digital data. Analog data implies continuity as contrasted to digital data that is concerned with discrete states. *Note:* Many signals can be used in either the analog or digital sense, the means of carrying the information being the distinguishing feature. The information content of an analog signal is conveyed by the value or magnitude of some characteristics of the signal such as the amplitude, phase, or frequency of a voltage, the amplitude or duration of a pulse, the angular position of a shaft, or the pressure of the fluid. To extract the information, it is necessary to compare the value or magnitude of the signal to a standard. The information content of the digital signal is concerned with discrete states of the signal, such as the presence or absence of a voltage, a contact in the open or closed position, or a hole or no hole in certain positions on a card. The signal is given meaning by assigning numerical values or other information to the various possible combinations of the discrete states of the signal. *See also:* **analog data; digital data; numerically controlled machines.** EIA3B-34E12

analog computer (1)(general). A computer that operates on analog data by performing physical processes on these data. *See:* **digital computer.** X3A12-16E9

(2) (direct-current). An analog computer in which computer variables are represented by the instantaneous values of voltages. *See also:* **electronic analog computer.** E165-16E9

(3) (alternating-current). An analog computer in which electric signals are of the form of amplitude-modulated suppressed-carrier signals where the absolute value of a computer variable is represented by the amplitude of the carrier and the sign of a computer variable is represented by the phase (0 or 180 degrees) of the carrier relative to the reference alternating-current signal. *See also:* **electronic analog computer.** E165-16E9

analog device (control equipment). A device that operates with variables represented by continuously measured quantities such as voltages, resistances, rotations, pressures, etcetera. *See also:* **power systems, low-frequency and surge testing.** E94-0

analog output. One type of continuously variable quantity used to represent another; for example, in temperature measurement, an electric voltage or current output represents temperature input. *See also:* **signal.** 0-13E6

analog signal (control) (industrial control). A signal that is solely dependent upon magnitude to express information content. *See also:* **control system, feedback.** AS1-34E10

analog telemetering. Telemetering in which some characteristic of the transmitter signal is proportional to the quantity being measured. 37A100-31E11

analog-to-digital converter (1)(data processing). A device that converts a signal that is a function of a continuous variable into a representative number sequence. *See also:* **electronic digital computer.** 0-23E3

(2) (A–D). A circuit whose input is information in analog form and whose output is the same information in digital form. *See also:* **analog; digital.** 0-31E3

(3) (digitizer). A device or a group of devices that converts an analog quantity or analog position input signal into some type of numerical output signal or code. *Note:* The input signal is either the measurand or a signal derived from it. 37A100-31E11

analog-to-frequency (A–F) converter. A circuit whose input is information in an analog form other than frequency and whose output is the same information as a frequency proportional to the magnitude of the information. *See also:* **analog.** 0-31E3

analysis. *See:* **numerical analysis.**

analytic inertial-navigation equipment. The class of inertial-navigation equipment in which geographic navigational quantities are obtained by means of computations (generally automatic) based upon the outputs of accelerometers whose orientations are maintained fixed with respect to inertial space. *See also:* **navigation.** E174-10E6

analytic signal (pre-envelope). A complex function of a real variable (time, for example) whose imaginary part is the Hilbert transform of its real part. *Note:* The analytic signal associated with a real signal is that one whose real part is the given real signal. *See also:* **Hilbert transform; network analysis.** 0-12E1

analyzer. *See:* **differential analyzer; digital differential analyzer; network analyzer.**

anchor guy guard. A protective cover over the guy, usually a length of sheet metal shaped to a semicircular or tubular section and equipped with means of attachment to the guy. *See also:* **tower.** 42A35-31E13

anchor light. (1) A lantern hung in the rigging at a prescribed height to indicate to navigators of nearby vessels that a ship is at anchor. 42A43-0

(2) An aircraft light designed for use on a seaplane or amphibian to indicate its position when at anchor or moored. *See also:* **signal lighting.** Z7A1-0

anchor log (dead man). A piece of rigid material such as timber, metal, or concrete, usually several feet in length, buried in earth in a horizontal position and at right angles to anchor rod attachment. *See also:* **tower.** 42A35-31E13

anchor rod. A steel or other metal rod designed for convenient attachment to a buried anchor and also to provide for one or more guy attachments above ground. *See also:* **tower.** 42A35-31E13

AND. A logic operator having the property that if *P* is a statement, *Q* is a statement, *R* is a statement, . . . , then the AND of *P, Q, R*,. . . is true if all statements are true, false if any statement is false. *P* AND *Q* is often represented by $P \cdot Q$, PQ, $P \wedge Q$. X3A12-16E9

AND-circuit. *See:* AND**-gate.** *See also:* **electronic computation.**

AND gate (1) (general). A combinational logic element such that the output channel is in its ONE state if and only if each input channel is in its ONE state. *See also:* **electronic digital computer.** E162-0

(2) A gate that implements the logic AND operator. X3A12-16E9

(3) (AND circuit). A gate whose output is energized when and only when every input is in its prescribed state. An AND gate performs the function of the logical AND. *See also:* **electronic computation.** E270-0

Anderson bridge. A 6-branch network in which an outer loop of 4 arms is formed by three nonreactive resistors and the unknown inductor, and an inner loop of 3 arms is formed by a capacitor and a fourth resistor in series with each other and in parallel with the arm that is opposite the unknown inductor, the detector being connected between the junction of the capacitor and the fourth resistor and that end of the unknown inductor that is separated from a terminal of the capacitor by only one resistor, while the source is connected to the other end of the unknown inductor and to the junction of the capacitor with two resistors of the outer loop. *Note:* Normally used for the comparison of self-inductance with capacitance. The balance is independent of frequency. *See also:* **bridge.** 42A30-0

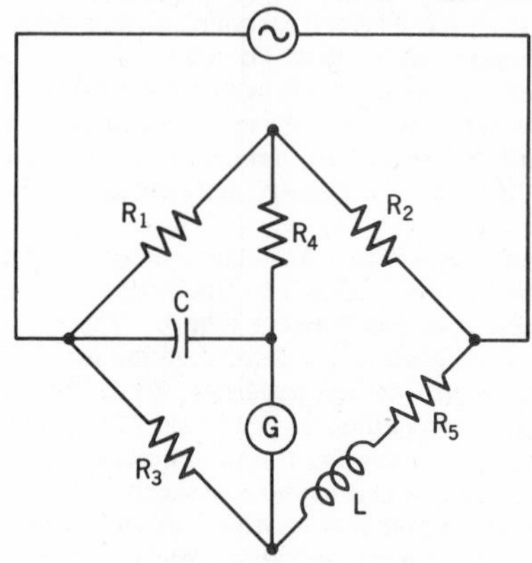

$$R_1R_5 = R_3R_2$$

$$L = CR_3\left[R_4\left(1 + \frac{R_2}{R_1}\right) + R_2\right]$$

Anderson bridge.

anechoic chamber. An enclosure especially designed with boundaries that absorb sufficiently well the sound incident thereon to create an essentially free-field condition in the frequency range of interest. *See also:* **electroacoustics.** 0-1E1

anechoic enclosure (radio frequency). An enclosure whose internal walls have low reflection characteristics. *See also:* **electromagnetic compatibility.** 0-27E1

anelectrotonus (electrobiology). Electrotonus produced in the region of the anode. *See also:* **excitability.** 42A80-18E1

angle, bunching (electron stream). *See:* **bunching angle.**

angle, effective bunching (reflex klystrons). *See:* **effective bunching angle.**

angle, flow (gas tubes). That portion, expressed as an angle, of the cycle of an alternating voltage during which current flows. *See also:* **gas tubes.** 0-15E6

angle, maximum-deflection. The maximum plane angle subtended at the deflection center by the usable screen area. *Note:* In this term the hyphen is frequently omitted. *See also:* **beam tubes; circuit characteristics of electrodes.** E160-2E2/15E6

angle modulation. Modulation in which the angle of a sine-wave carrier is the characteristic varied from its reference value. *Notes:* (1) Frequency modulation and phase modulation are particular forms of angle modulation; however, the term frequency modulation is often used to designate various forms of angle modulation. (2) The reference value is usually taken to be the angle of the unmodulated wave. *See also:* **modulating systems; modulation index.** E145/E170/42A65-0

angle noise (radar). The noise-like variation in the apparent angle of arrival of an echo received from a target, because of change in target aspect; a component of scintillation. *See also:* **navigation.** 0-10E6

angle of advance (1) (power inverter). The time interval in electrical degrees by which the beginning of anode conduction leads the moment at which the anode voltage would attain a negative value equal to that of the succeeding anode in the commutating group. *See also:* rectification. 34A1-0

(2) (semiconductor rectifiers) (semiconductor power converter). The angle by which foward conduction is advanced by the control means only, in the incoming circuit element, ahead of the instant in the cycle at which the incoming commutating voltage passes through zero in the direction to produce forward conduction in the outgoing circuit element. *See also:* **rectification; semiconductor rectifier stack.** 0-34E24

angle-of-approach lights. Aeronautical ground lights arranged so as to indicate a desired angle of descent during an approach to an aerodrome runway. Also called **optical glide-path lights.** *See also:* **signal lighting.** Z7A1-0

angle of cut (navigation). The angle at which two lines of position intersect. *See also:* **navigation.** 0-10E6

angle of extinction (industrial control). The phase angle of the stopping (extinction) instant of anode-current flow in a gas tube with respect to the starting instant of the corresponding positive half cycle of the anode voltage of the tube. *See also:* **electronic controller.** 42A25-34E10

angle of ignition (industrial control). The phase angle of the starting instant of anode-current flow in a gas tube with respect to the starting instant of the corresponding positive half cycle of the anode voltage of the tube. *See also:* **electronic controller.** 42A25-34E10

angle of lag. The angle of lag of one simple sine wave with respect to a second having the same period is the angle by which the second must be assumed to be shifted backward to make it coincide in position with the first. E270-0

angle of lead. The angle of lead of one simple sine wave with respect to a second having the same period is the angle by which the second function must be assumed to be shifted forward to make it coincide in position with the first. E270-0

angle of protection (lightning). The angle between the vertical plane and a plane through the ground wire, within which the line conductors must lie in order to ensure a predetermined degree of protection against direct lightning strokes. *See also:* **lightning arrester (surge diverter).** 50I25-31E7

angle of retard (1) (semiconductor rectifiers) (semiconductor rectifier operating with phase control). The angle by which forward conduction is delayed by the control means only, beyond the instant in the cycle at which the incoming commutating voltage passes through zero in the direction to produce forward conduction in the incoming circuit element. *See also:* **semiconductor rectifier stack.** 0-34E24

(2) (power rectifier). The angle by which the beginning of anode conduction leads the point at which the incoming and outgoing rectifying elements of a commutating group have equal negative fundamental components of alternating voltage. *See also:* **rectification.** 42A15-0

angle optimum bunching. *See:* **optimum bunching.**

angle or phase (sine wave). The measure of the progression of the wave in time or space from a chosen instant or position or both. *Notes:* (1) In the expression for a sine wave, the angle or phase is the value of the entire argument of the sine function. (2) In the representation of a sine wave by a phasor or rotating vector, the angle or phase is the angle through which the vector has progressed. *See also:* **modulating systems; wavefront.** E170/E145-0

angle, overlap. *See:* **overlap angle.**

angle tower. A tower located where the line changes horizontal direction sufficiently to require special design of the tower to withstand the resultant pull of the wires and to provide adequate clearance. *See also:* **tower.** 42A35-31E13

angle, transit. *See:* **transit angle.**

angular deviation loss (acoustic transducer). The ratio of the response in a specific direction to the response on the principal axis, usually expressed in decibels. *See also:* **loudspeaker.** 0-1E1

angular deviation sensitivity (electronic navigation). The ratio of change of course indication to the change of angular displacement from the course line. *See also:* **navigation.** E172-10E6

angular displacement (1) (polyphase regulator). The phase angle expressed in degrees between the line-to-neutral voltage of the reference identified source voltage terminal and the line-to-neutral voltage of the corresponding identified load voltage terminal. *Note:* The connection and arrangement of terminal markings for three-phase regulators is a wye-wye connection having an angular displacement of zero degrees. *See also:* **voltage regulator.** 57A15-0

(2) (polyphase transformer). The phase angle expressed in degrees between the line-to-neutral voltage of the reference identified high-voltage terminal and the line-to-neutral voltage of the corresponding identified low-voltage terminal. *Note:* The preferred connection and arrangement of terminal markings for polyphase transformers are those which have the smallest possible phase-angle displacements and are measured in a clockwise direction from the line-to-neutral voltage of the reference identified high-voltage terminal. Thus, standard three-phase transformers have angular displacements of either zero or 30 degrees. *See:* **routine test.** 42A15-31E12

(3) (polyphase rectifier transformer). The time angle expressed in degrees between the line-to-neutral voltage of the reference identified alternating-current winding terminal and the line-to-neutral voltage of the corresponding identified direct-current winding terminal. *See also:* **rectifier transformer.** 57A18-0

angular frequency (periodic function). 2π times the frequency. *Note:* This definition applies to any definition of frequency. *See also:* **network analysis.** E270-3E2/12E1

angular resolution (radar). The ability to distinguish between two targets solely by the measurement of angles; generally expressed in terms of the minimum angle by which two targets must be spaced to be separately distinguishable. *See also:* **navigation.** E172-10E6

angular variation (alternating–current circuits). The maximum angular displacement, expressed in electrical degrees, of corresponding ordinates of the voltage wave and of a wave of constant frequency, equal to the average frequency of the alternating-current circuit in question. 42A10-0

angular variation (synchronous generators). *See:* **angular variation (alternating-current circuits).** *See also:* **synchronous machine.**

angular variation (synchronous generator). The maximum angular displacement, expressed in electrical degrees, of corresponding ordinates of the voltage wave and of a wave of absolutely constant frequency, equal to the average frequency of the synchronous generator in question. *See also:* **synchronous machine.** 0-31E8

angular width (electronic navigation). *See:* **course.** *See also:* **width; navigation.**

anion. A negatively charged ion or radical that migrates toward the anode under the influence of a potential gradient. *See also:* **ion.** CM-34E2

annotation. An added descriptive comment or explanatory note. X3A12-16E9

announcement system. A general arrangement for supplying information by means of periodic announcements distributed to the various central offices over one-way distribution circuits. *Note:* It is used for time of day, weather, etcetera. *See also:* **telephone switching system.** 42A65-0

annunciator. A visual signal device consisting of a number of pilot lights or drops, each one indicating the condition that exists or has existed in an associated circuit, and being labeled accordingly. *See also:* **circuits and devices.** 42A65-0

annunciator relay. A relay that indicates visually whether a current in flowing or has flowed in one or more circuits. *See also:* **relay.** 83A16-0

anode (1). An electrode through which current enters any conductor of the nonmetallic class. Specifically, an electrolytic anode is an electrode at which negative ions are discharged, or positive ions are formed, or at which other oxidizing reactions occur. 42A60-0
(2). An electrode or portion of an electrode at which a net oxidation-reaction occurs. *See also:* **electrochemical cell.** CV1-10E1/34E2
(3) (thyristor). The electrode by which current enters the thyristor, when the thyristor is in the ON state with the gate open-circuited. *Note:* This term does not apply to bidirectional thyristors.
See:
anode terminal;
cathode;
collector junction;
gate;
gate terminal;
junction;
main terminal 1;
main terminal 2;
main terminals;
semiconductor device.
See also: **electrolytic cell.**
E223-34E17/34E24/15E7
(4) (electron tube or valve). An electrode through which a principal stream of electrons leaves the interelectrode space. *See also:* **electrode (electron tube).** 42A70-15E6;E160/E175-0
(5) (semiconductor rectifier cell). The electrode from which the forward current flows within the cell. *See also:* **semiconductor.** E59-34E17
(6) (X-ray tube). *See:* **target (X-ray tube).**

anode breakdown voltage (glow-discharge cold-cathode tube). The anode voltage required to cause conduction across the main gap with the starter gap not conducting and with all other tube elements held at cathode potential before breakdown. *See also:* **electrode voltage (electron tube).** 42A70-15E6

anode butt. A partially consumed anode. *See also:* **fused electrolyte.** 42A60-0

anode characteristic. *See:* **anode-to-cathode voltage-current characteristic.**

anode circuit (industrial control). A circuit that includes the anode-cathode path of an electron tube in series connection with other elements. *See also:* **electronic controller.** 42A25-34E10

anode cleaning (reverse-current cleaning) (electroplating). Electrolytic cleaning in which the metal to be cleaned is made the anode. *See also:* **battery (primary or secondary).** 42A60-0

anode corrosion efficiency. The ratio of the actual corrosion of an anode to the theoretical corrosion calculated from the quantity of electricity that has passed. *See also:* **corrosion terms.** CM-34E2

anode current (electron tubes). *See:* **electrode current; electronic controller.**

anode dark space (gas tubes) (gas). A narrow dark zone next to the surface of the anode. *See also:* **discharge (gas).** 50I07-15E6

anode differential resistance. *See:* **anode resistance.**

anode effect. A phenomenon occurring at the anode, characterized by failure of the electrolyte to wet the anode and resulting in the formation of a more or less continuous gas film separating the electrolyte and anode and increasing the potential difference between them. *See also:* **fused electrolyte.** 42A60-0

anode efficiency. The current efficiency of a specified anodic process. *See also:* **electrochemistry.** 42A60-0

anode, excitation (pool-cathode tube). *See:* **excitation anode.**

anode (potential) fall (gas). The fall of potential due to the space charge near the anode. *See also:* **discharge (gas).** 50I07-15E6

anode firing (industrial control). The method of initiating conduction of an ignitron by connecting the ignitor through a rectifying element to the anode of the ignitron to obtain power for the firing current pulse. *See also:* **electronic controller.** 42A25-34E10

anode glow (gas tubes) (gas). A very bright narrow zone situated at the near end of the positive column with respect to the anode. *See also:* **discharge (gas).** 50I07-15E6

anode layer. A molten metal or alloy, serving as the anode in an electrolytic cell, that floats on the fused electrolyte or upon which the fused electrolyte floats. *See also:* **fused electrolyte.** 42A60-0

anode, main (pool-cathode tube). *See:* **main anode.**

anode mud. *See:* **slime.**

anode paralleling reactor. A reactor with a set of mutually coupled windings connected to anodes operating in parallel from the same transformer terminal. 42A15-31E12

anode power supply (electron tube) (plate power sup-

ply). The means for supplying power to the plate at a voltage that is usually positive with respect to the cathode. *See also:* **power pack.** 42A65-0

anode region (gas tubes) (gas). The group of regions comprising the positive column, anode glow, and anode dark space. *See also:* **discharge (gas).** 50I07-15E6

anode relieving (pool-cathode tube) (gas tube). An anode that provides an alternative conducting path to reduce the current to another electrode. *See also:* **electrode (electron tube); gas filled rectifier.** 50I07-15E6

anode resistance (anode differential resistance) (electron tube). The quotient of a small change in anode voltage by a corresponding small change of the anode current, all the other electrode voltages being maintained constant. It is equal to the reciprocal of the anode conductance. *See also:* **ON period (electron tubes).** 50I07-15E6

anode scrap. That portion of the anode remaining after the schedule period for the electrolytic refining of the bulk of its metal content has been completed. *See also:* **electrorefining.** 42A60-0

anode slime. *See:* **slime.**

anode strap (magnetron). A metallic connector between selected anode segments of a multicavity magnetron, principally for the purpose of mode separation. *See also:* **magnetrons.** E160-15E6;42A70-0

anode supply voltage (industrial control). The voltage at the terminals of a source of electric power connected in series in the anode circuit. *See also:* **electronic controller.** 42A25-34E10

anode terminal (1)(semiconductor device). The terminal by which current enters the diode. *See also:* **semiconductor; semiconductor device.** E216-34E17

(2) (semiconductor diode). The terminal that is positive with respect to the other terminal when the diode is biased in the forward direction. *See also:* **semiconductor devices.** E270-0

(3) (semiconductor rectifier diode or rectifier stack). The terminal to which forward current flows from the external circuit. *Note:* In the semiconductor rectifier components field, the anode terminal is normally marked negative. *See also:* **semiconductor device; semiconductor rectifier cell.** E59-34E17/34E24

(4) (thyristor). The terminal that is connected to the anode. *Note:* This term does not apply to bidirectional thyristors. *See also:* **anode.** E223-34E17/15E7

anode-to-cathode voltage (anode voltage) (thyristor). The voltage between the anode terminal and the cathode terminal. *Note:* It is called positive when the anode potential is higher than the cathode potential and called negative when the anode potential is lower than the cathode potential. *See also:* **electronic controller; principal voltage-current characteristic (principal characteristic).** E223-34E17/34E24/15E7

anode-to-cathode voltage-current characteristic (anode characteristic) (thyristor). A function, usually represented graphically, relating the anode-to-cathode voltage to the principal current with gate current, where applicable, as a parameter. *Note:* This term does not apply to bidirectional thyristors. *See also:* **principal voltage-current characteristic (principal characteristic).** E223-34E17/15E7

anode voltage. *See:* **anode-to-cathode voltage.**

anode voltage (electron tubes). *See:* **electrode voltage; electronic controller.**

anode voltage drop (glow-discharge cold-cathode tube). The main gap voltage drop after conduction is established in the main gap. 42A70-15E6

anode voltage, forward, peak. *See:* **peak forward anode voltage.**

anode voltage, inverse, peak. *See:* **peak inverse anode voltage.**

anodic polarization. Polarization of an anode. *See also:* **electrochemistry.** 42A60-0

anolyte. The portion of an electrolyte in an electrolytic cell adjacent to an anode. If a diaphragm is present, it is the portion of electrolyte on the anode side of the diaphragm. *See also:* **electrolytic cell.** 42A60-34E2

A-N radio range. *See:* **aural radio range.**

answering plug and cord. A plug and cord used to answer a calling line. *See also:* **telephone switching system.** 42A65-0

antenna. A means for radiating or receiving radio waves. *Note:* For an extensive list of cross references, see *Appendix A.* E145/E149-3E1

antenna conflict (wiring system). The situation in which an antenna or its guy wire is at a higher level than a supply or communication conductor and approximately parallel thereto, provided the breaking of the antenna or its support will be likely to result in contact between the antenna or guy wire and the supply or communication conductor. 2A2-0

antenna effect (radio direction-finding). (1) The presence of output signals having no directional information and caused by the directional array acting as a simple nondirectional antenna; the effect is manifested by (1)angular displacement of the nulls, or (2)a broadening of the nulls. *See also:* **antenna; navigation.** 0-10E6

(2) (loop antenna) (old usage). Any spurious effect resulting from the capacitance of the loop to ground. *See also:* **antenna.** E145-3E1

antenna, effective area (in a given direction). The ratio of the power available at the terminals of an antenna to the incident power density of a plane wave from that direction polarized coincident with the polarization that the antenna would radiate. *See also:* **antenna.** 0-3E1

antenna, effective height. (1) The height of its center of radiation above the effective ground level. (2) In low-frequency applications the term effective height is applied to loaded or nonloaded vertical antennas and is equal to the moment of the current distribution in the vertical section, divided by the input current. *Note:* For an antenna with symmetrical current distribution the center of radiation is the center of distribution. For an antenna with asymmetrical current distribution the center of radiation is the center of current moments when viewed from directions near the direction of maximum radiation. *See also:* **antenna.** 42A65-3E1

(3) The height of the center of a vertical antenna (of at least ¼ wavelength) above the effective ground plane of the vehicle on which the antenna is mounted. *See also:* **antenna; mobile communication system.** 0-6E1

antenna, effective height base station (mobile communication). The height of the physical center of the antenna above the effective ground plane. *See also:* **mobile communication system.** 0-6E1

antenna, effective length (effective height, low-frequency usage). (1) For an antenna radiating linearly polarized waves, the length of a thin straight conductor oriented perpendicular to the direction of maximum radiation, having a uniform current equal to that at the antenna terminals and producing the same far field strength as the antenna. (2) Alternatively, for the same antenna receiving linearly polarized waves from the same direction, the ratio of the open-circuit voltage developed at the terminals of the antenna to the component of the electric field strength in the direction of antenna polarization. *Notes:* (1) The two definitions yield equal effective lengths. (2) In low-frequency usage the effective length of a ground-based antenna is taken in the vertical direction and is frequently referred to as effective height. Such usage should not be confused with effective height (of an antenna, high-frequency usage). *See also:* **antenna.** 0-3E1

antenna, effective length (electromagnetic compatibility). The ratio of the antenna open-circuit voltage to the strength of the field component being measured. *See also:* **electromagnetic compatibility.** 0-27E1

antenna factor (field-strength meter). That factor that, when properly applied to the meter reading of the measuring instrument, yields the electric field strength in volts/meter or the magnetic field strength in amperes/meter. *Notes:* (1) This factor includes the effects of antenna effective length and mismatch and transmission line losses. (2) The factor for electric field strength is not necessarily the same as the factor for the magnetic field strength. *See also:* **electromagnetic compatibility.** 0-27E1

antenna resistance. The ratio of the power accepted by the entire antenna circuit to the square of the root-mean-square antenna current referred to a specified point. *Note:* Antenna resistance is made up of such components as radiation resistance, ground resistance, radio-frequency resistance of conductors in the antenna circuit, and equivalent resistance due to corona, eddy currents, insulator leakage, and dielectric power loss. *See also:* **antenna.** E149-0; 0-3E1

antenna terminal conducted interference (electromagnetic compatibility). Any undesired voltage or current generated within a receiver, transmitter, or their associated equipment appearing at the antenna terminals. *See also:* **electromagnetic compatibility.** 0-27E1

anticathode (X-ray tube)*. *See:* **anode.**

*Deprecated.

anticlutter circuits (radar). Circuits that attenuate undesired reflections to permit detection of targets otherwise obscured by such reflections. *See also:* **navigation.** E172-10E6

anticlutter gain control (radar). A device that automatically and smoothly increases the gain of a radar receiver from a low level to the maximum, within a specified period after each transmitter pulse, so that short-range echoes producing clutter are amplified less than long-range echoes. *See also:* **radar.** 42A65-0

anticoincidence (radiation counters). The occurrence of a count in a specified detector unaccompanied simultaneously or within an assignable time interval by a count in one or more other specified detectors.
See:
anticoincidence circuit;
coincidence;
count (counting) rate;
counter tube;
dead-time correction;
detector, radiation;
end-window counter tube;
gas counter tube;
gas-filled radiation-counter tubes;
liquid counter tube;
liquid-flow counter tube;
quenching circuit;
resolution (resolving) time;
resolution time correction;
scaling circuit;
scaler;
starting voltage;
thin-wall counter;
window. 0-15E6

anticoincidence circuit (pulse techniques). A circuit that produces a specified output pulse when one (frequently predesignated) of two inputs receives a pulse and the other receives no pulse within an assigned time interval. *See also:* **pulse, anticoincidence (radiation counter).** E175-0

anticollision light. A flashing aircraft aeronautical light or system of lights designed to provide a red signal throughout 360 degrees of azimuth for the purpose of giving long-range indication of an aircraft's location to pilots of other aircraft. *See also:* **signal lighting.** Z7A1-0

antiferroelectric crystals. Crystals that exhibit structural phase changes and anomalies in the dielectric permittivity as do ferroelectrics, but that exhibit no net spontaneous polarization, and hence, no associated hysteresis phenomena. *Note:* In some cases it is possible to apply electric fields sufficiently high to induce the ferroelectric state and a corresponding double hysteresis loop. *See also:* **ferroelectric domain.** E180-0

antifouling. Pertaining to the prevention of marine organism attachment and growth on a submerged metal surface (through the effects of chemical action). *See also:* **corrosion terms.** CM-34E2

antifreeze pin, relay. *See:* **relay antifreeze pin.**

antifriction bearing (rotating machinery). A bearing incorporating a peripheral assembly of separate parts that will support the shaft and reduce sliding friction by rolling as the shaft rotates. *See also:* **bearing.** 0-31E8

antinode (standing wave). *See:* **loop (standing wave).**

antinoise microphone. A microphone with characteristics that discriminate against acoustic noise. *See also:* **microphone.** 42A65-0

anti-overshoot (industrial control). The effect of a control function or a device that causes a reduction in the transient overshoot. *Note:* Anti-overshoot may apply to armature current, armature voltage, field current, etcetera. *See also:* **control system, feedback.** IC1-34E10

antiplugging protection (industrial control). The effect of a control function or a device that operates to prevent application of counter torque by the motor until the motor speed has been reduced to an acceptable value. *See also:* **control system, feedback.** IC1-34E10

antipump device (pump-free device). A device that prevents reclosing after an opening operation as long as the device initiating closing is maintained in the position for closing. 37A100-31E11

antiresonant frequency (crystal unit). The frequency for a particular mode of vibration at which, neglecting dissipation, the effective impedance of the crystal unit is infinite. *See also:* **crystal.** 42A65-0

antisidetone induction coil. An induction coil designed for use in an antisidetone telephone set. *See also:* **telephone station.** 42A65-0

antisidetone telephone set. A telephone set that includes a balancing network for the purpose of reducing sidetone. *See also:* **telephone station; sidetone.** 42A65-0

anti-transmit-receive box (ATR box) (electronic navigation). *See:* **anti-transmit-receive switch.**

anti-transmit-receive switch (ATR switch) (radar). An electronic switch that automatically decouples the transmitter from the antenna during the receiving period; it is employed when a common transmitting and receiving antenna is used. *See also:* **radar.** E172-10E6

anti-transmit-receive tube (ATR tube). A gas-filled radio-frequency switching tube used to isolate the transmitter during the interval for pulse reception. *See also:* **gas tubes.** E160-15E6

***A* operator.** An operator assigned to an *A* switchboard. *See also:* **telephone system.** 42A65-0

APC. *See:* **automatic phase control.**

aperiodic circuit. A circuit in which it is not possible to produce free oscillations. *See also:* **oscillatory circuit.** 50I05-31E3

aperiodic component of short-circuit current (rotating machinery). The component of current in the primary winding immediately after it has been suddenly short-circuited when all components of fundamental and high frequencies have been subtracted. *See also:* **asynchronous machine; synchronous machine.** 0-31E8

aperiodic damping*. *See:* **overdamping.**

*Deprecated.

aperiodic function. Any function that is not periodic. E270-0

aperiodic time constant (rotating machinery). The time constant of the aperiodic component (when it is practically exponential) or of the exponential, which envelops it when it shows an appreciable periodicity. *See also:* **asynchronous machine; direct current commutating machine; synchronous machine.** 0-31E8

aperture (antenna). A surface, near or on an antenna, on which it is convenient to make assumptions regarding the field values for the purpose of computing fields at external points. *Notes:* (1) In some cases the aperture may be considered as a line. (2) In the case of a unidirectional antenna the aperture is often taken as that portion of a plane surface near the antenna, perpendicular to the direction of maximum radiation, through which the major part of the radiation passes. *See also:* **antenna.** E149-3E1

aperture compensation (television). *See:* **aperture equalization.**

aperture correction (television). *See:* **aperture equalization.**

aperture efficiency (antenna)(for an antenna aperture). The ratio of its directivity to the directivity obtained when the aperture illumination is uniform. *See also:* **antenna.** 0-3E1

aperture equalization (1) (television). Electrical compensation for the distortion introduced by the size of a scanning aperture. *See also:* **television.** 0-2E2 **(2) (aperture correction, aperture compensation).** Compensation for the distortion introduced by the finite size of the aperture of a scanning beam by means of circuits inserted in the signal channel. 0-42A65

aperture illumination. The amplitude, phase, and polarization of the field distribution over the aperture. *See also:* **antenna.** 0-3E1

API. *See:* **air-position indicator.**

APL. *See:* **average picture level.**

apparatus (electric). The terms appliance and device generally refer to small items of electric equipment, usually located at or near the point of utilization, but with no established criterion for the meaning of small. The term apparatus then designates the large items of electric equipment, such as generators, motors, transformers, or circuit breakers. *See also:* **generating station.** 84A1-0

apparatus insulator (cap and pin, post). An assembly of one or more apparatus-insulator units, having means for rigidly supporting electric equipment. *See also:* **insulator.** 29A1-0

apparatus insulator unit. The assembly of one or more elements with attached metal parts, the function of which is to support rigidly a conductor, bus, or other conducting elements on a structure or base member. *See also:* **tower.** 42A35-31E13

apparent bearing (direction finding). A bearing from a direction-finder site to a target transmitter determined by averaging the readings made on a calibrated direction-finder test standard; the apparent bearing is then used in the calibration and adjustment of other direction-finders at the same site. *See also:* **navigation.** 0-10E6

apparent candlepower (extended source). At a specified distance, the candlepower of a point source that would produce the same illumination at that distance. Z7A1-0

apparent inductance. The reactance between two terminals of a device or circuit divided by the angular frequency at which the reactance was determined. This quantity is defined only for frequencies at which the reactance is positive. *Note:* Apparent inductance includes the effects of the real and parasitic elements that comprise the device or circuit and is therefore a function of frequency and other operating conditions. 0-21E1

apparent power (1) (rotating machinery). The product of the root-mean-square current and the root-mean-square voltage. *Note:* It is a scalar quantity equal to the magnitude of the phasor power. *See also:* **asynchronous machine; synchronous machine.** 0-31E8 **(2) (general).** *See:* **power, apparent.**

apparent-power loss (volt-ampere loss) (electric instrument). Of the circuit for voltage-measuring instruments, the product of end-scale voltage and the resulting current; and for current-measuring instruments, the product of the end-scale current and the resulting voltage. *Notes:* (1) For other than current-measuring or voltage-measuring instruments, for example, wattmeters, the apparent-power loss of any circuit is expressed at a stated value of current or of voltage. (2) Computation of loss: for the purpose of computing the loss of alternating-current instruments having current circuits at some selected value other than that for which it is rated, the actual loss at the rated current is multiplied by the square of the ratio of the selected current to the rated current. *Example:* A current transformer with a ratio of 500:5 amperes is used with an instrument having a scale of 0-300 amperes and, therefore, a 3-ampere field coil, and the allowable loss at end scale is as stated on the Detailed Requirement Sheet. The allowable loss of the instrument referred to a 5-amperes basis is as follows: Allowable loss in volt-amperes equals (allowable loss end-scale volt-amperes) $(5/3)^2$. *See also:* **accuracy rating (instrument).** 39A1-42A30

apparent sag (wire in a span). (1) The maximum departure in the vertical plane of the wire in a given span from the straight line between the two points of support of the span, at 60 degrees Fahrenheit, with no wind loading. *Note:* Where the two supports are at the same level this will be the sag. *See also:* **tower.** 42A35-31E13;2A2-0

(2) The departure in the vertical plane of the wire at the particular point in the span from the straight line between the two points of support. *See also:* **power distribution overhead construction.** 42A35-31E13

apparent time constant (thermal converter) (63-percent response time) (characteristic time). The time required for 63 percent of the change in output electromotive force to occur after an abrupt change in the input quantity to a new constant value. *See:* Note 1 of **Response time of a thermal converter.** *See also:* **thermal converter.** 39A1-0

apparent vertical (electronic navigation). The direction of the vector sum of the gravitational and all other accelerations. *See also:* **navigation.** E174-10E6

appliance (electric). A utilization item of electric equipment, usually complete in itself, generally other than industrial, normally built in standardized sizes or types that transforms electric energy into another form, usually heat or mechanical motion, at the point of utilization. For example, a toaster, flatiron, washing machine, dryer, hand drill, food mixer, air conditioner. 1A0/2A2/84A1/E270-0

appliances (including portable) (interior wiring).
See:
appliance;
appliance, fixed;
appliance, portable;
appliance, stationary;
convection heater;
dielectric heating;
electric sign;
flasher;
heating unit;
hot plate;
induction heating;
radiant heater;
space heater.

appliance branch circuit. A circuit supplying energy to one or more outlets to which appliances are to be connected; such circuits to have no permanently connected lighting fixtures not a part of an appliance.
See:
appliance;
branch circuit;
cooking unit, counter-mounted;
low-energy power circuit;
oven, wall-mounted;
remote-control circuit;
sealed refrigeration compressor;
signal circuit. 42A95-0

appliance, fixed (electric system). An appliance that is fastened or otherwise secured at a specific location. *See also:* **appliances.** E270/1A0-0

appliance outlet (household electric ranges). An outlet mounted on the range and to which a portable appliance may be connected by means of an attachment plug cap.
See:
appliances;
design voltage;
grounding jumper;
heating cycle;
heating element;
heating unit;
indicator light;
maximum operating voltage;
rated watts input;
supply circuit;
voltage rating. 71A1-0

appliance, portable (electric system). An appliance that is actually moved or can easily be moved from one place to another in normal use. *See also:* **appliances.** E270/1A0-0

appliance, stationary (electric system). An appliance that is not easily moved from one place to another in normal use. *See also:* **appliances.** E270/1A0-0

application valve (brake application valve). An air valve through the medium of which brakes are automatically applied. 42A42-0

applicator (applicator electrodes) (dielectric heating). Appropriately shaped conducting surfaces between which is established an alternating electric field for the purpose of producing dielectric heating. *See also:* **dielectric heating.** E54/E169-0

applicator impedance, loaded (dielectric heating). *See:* **loaded applicator impedance.**

applicator impedance, unloaded (dielectric heating). *See:* **unloaded applicator impedance.**

applied fault protection. A protective method in which, as a result of relay action, a fault is intentionally applied at one point in an electric system in order to cause fuse blowing or further relay action at another point in the system. 37A100-31E11/31E6

applied-potential tests (electric power). Dielectric tests in which the test voltages are low-frequency alternating voltages from an external source applied between conducting parts, and between conducting parts and ground. *See also:* **power systems, low-frequency and surge testing.** E32-0

approach circuit. A circuit used to announce the approach of trains at block or interlocking stations. *See also:* **railway signal and interlocking.** 42A42-0

approach indicator. A device used to indicate the approach of a train. *See also:* **railway signal and interlocking.** 42A42-0

approach-light beacon. An aeronautical ground light placed on the extended centerline of the runway at a fixed distance from the runway threshold to provide an early indication of position during an approach to a runway. *Note:* The runway threshold is the beginning of the runway usable for landing. *See also:* **signal lighting.** Z7A1-0

approach lighting. An arrangement of circuits so that the signal lights are automatically energized by the approach of a train. *See also:* **railway signal and interlocking.** 42A42-0

approach-lighting relay. A relay used to close the lighting circuit for signals upon the approach of a train. *See also:* **railway signal and interlocking.** 42A42-0

approach lights. A configuration of aeronautical ground lights located in extension of a runway or channel before the threshold to provide visual approach and landing guidance to pilots. *See also:* **signal lighting.** Z7A1-0

approach locking (electric approach locking). Electric locking effective while a train is approaching, within a specified distance, a signal displaying an aspect to proceed, and that prevents, until after the expiration of a predetermined time interval after such signal has been caused to display its most restrictive aspect, the movement of any interlocked or electrically locked switch, movable-point frog, or derail in the route governed by the signal, and that prevents an aspect to proceed from being displayed for any conflicting route. *See also:* **interlocking.** 42A42-0

approach navigation. Navigation during the time that the approach to a dock, runway, or other terminal facility is of immediate importance. *See also:* **navigation.** E172-10E6

approach path (electronic navigation). The portion of the flight path between the point at which the descent for landing is normally started and the point at which the aircraft touches down on the runway. *See also:* **navigation; radio navigation.** 0-10E6

approach signal. A fixed signal used to govern the approach to one or more other signals. *See also:* **railway signal and interlocking.** 42A42-0

approval plate (mining). A label that the United States Bureau of Mines requires manufacturers to attach to every completely assembled machine or device sold as permissible mine equipment. *Note:* By this means, the manufacturer certifies to the permissible nature of the machine or device. *See also:* **mining.** 42A85-0

approval test (acceptance test). The testing of one or more meters or other items under various controlled conditions to ascertain the performance characteristics of the type of which they are a sample. *See also:* **service test (field test).** 12A0/42A30-0

approved. Approved by the enforcing authority. *See also:* **elevator.** 42A45-0

APT (numerically controlled machines). *See:* **automatic programmed tools.**

arc. (1) A discharge of electricity through a gas, normally characterized by a voltage drop in the immediate vicinity of the cathode approximately equal to the ionization potential of the gas. *See:* **gas tubes.** E160/42A70-0

(2) A continuous luminous discharge of electricity across an insulating medium, usually accompanied by the partial volatilization of the electrodes. E270-0

arc-back (gas tube). A failure of the rectifying action that results in the flow of a principal electron stream in the reverse direction, due to the formation of a cathode spot on an anode. *See also:* **gas tubes; rectification.** 42A15/42A70-0

arc cathode (gas tube). A cathode the electron emission of which is self-sustaining with a small voltage drop approximately equal to the ionization potential of the gas. *See also:* **gas-filled rectifier.** 50I07-15E6

arc chute (switching device). A structure affording a confined space or passageway, usually lined with arc-resisting material, into or through which an arc is directed to extinction. *See also:* **contactor.** 37A100-31E11; 42A25-34E10

arc, clockwise (numerically controlled machines). An arc generated by the coordinated motion of two axes in which curvature of the path of the tool with respect to the workpiece is clockwise, when viewing the plane of motion in the negative direction of the perpendicular axis. *See also:* **numerically controlled machines.** EIA3B-34E12

arc converter. A form of negative-resistance oscillator utilizing an electric arc as the negative resistance. *See also:* **radio transmission.** E182-0

arc, counterclockwise (numerically controlled machines). An arc generated by the coordinated motion of two axes in which curvature of the path of the tool with respect to the workpiece is counterclockwise, when viewing the plane of motion in the negative direction of the perpendicular axis. *See also:* **numerically controlled machines.** EIA3B-34E12

arc discharge. An electric discharge characterized by high cathode current densities and a low voltage drop at the cathode. *See also:* **lamp.** Z7A1-0

arc-discharge tube (valve). A gas-filled tube or valve in which the required current is that of an arc discharge. *See also:* **tube definitions.** 0-15E6

arc-drop loss (gas tube). The product of the instantaneous values of arc-drop voltage and current averaged over a complete cycle of operation. *See:* **gas tubes.** 42A70-0

arc-drop voltage (gas tube). The voltage drop between the anode and cathode of a rectifying device during conduction. *See also:* **electrode voltage (electron tube); tube voltage drop.** 42A70-15E6

arc-extinguishing medium (fuse filler) (fuse). Material included in the fuse to facilitate current interruption. 37A100-31E11

arc furnace. An electrothermic apparatus the heat energy for which is generated by the flow of electric current through one or more arcs internal to the furnace. *See also:* **electrothermics.** 42A60-0

arcing chamber (expulsion-type arrester). The part of an expulsion-type arrester that permits the flow of discharge current to the ground and interrupts the follow current. *See:* **lightning arrester (surge diverter).** 99I2-31E7

arcing contacts (switching device). The contacts on which the arc is drawn after the main (and intermediate, where used) contacts have parted. 37A100-31E11

arcing horn. A pair of diverging electrodes on which an arc is extended to the point of extinction after the main contacts of the switching device have parted. *Note:* Arcing horns are sometimes referred to as arcing runners. 37A100-31E11

arcing time (1) (fuse). The time elapsing from the severance of the current-responsive element to the final interruption of the circuit. 37A100-31E11

(2) (mechanical switching device). The interval of time between the instant of the first initiation of the arc and the instant of final arc extinction in all poles. *Note:* For switching devices that embody switching resistors, a distinction should be made between the arcing time up to the instant of the extinction of the main arc, and the arcing time up to the instant of the breaking of the resistance current. 37A100-31E11

arc loss (switching tubes). The decrease in radio-frequency power measured in a matched termination when a fired tube, mounted in a series or shunt junction with a waveguide, is inserted between a matched generator and the termination. *Note:* In the case of a pre-transmit-receive tube, a matched output termination is also required for the tube. *See also:* **gas tubes.** E160-15E6

arc-shunting-resistor-current arcing time. The interval between the parting of the secondary arcing contacts and the extinction of the arc-shunting-resistor current. 37A100-31E11

arc suppression (rectifier). The prevention of the recurrence of conduction, by means of grid or ignitor action, or both, during the idle period, following a current pulse. *See also:* **rectification.** 34A1-0

arc-through (gas tube). A loss of control resulting in the flow of a principal electron stream through the rectifying element in the normal direction during a scheduled nonconducting period. *See also:* **rectification.** 42A15-0; 42A70-15E6

arc-tube relaxation oscillator. *See:* **gas-tube relaxation oscillator.**

arc-welding engine-generator. A device consisting of an engine mechanically connected to and mounted with one or more arc-welding generators. *See also:* **electric arc-welding apparatus.** 87A1-0

arc-welding motor-generator. A device consisting of a motor mechanically connected to and mounted with one or more arc-welding generators. *See also:* **electric arc-welding apparatus.** 87A1-0

area assist action (electric power systems). The component of area supplementary control that involves the temporary assignment of generation changes to minimize the area control error prior to the assignment of generation changes on an economic dispatch basis. *See also:* **speed-governing system.** E94-0

area control error (isolated-power system consisting of a single control area). The frequency deviation (of a control area on an interconnected system) is the net interchange minus the biased scheduled-net interchange. *Note:* The above polarity is that which has been generally accepted by electric power systems and is in wide use. It is recognized that this is the reverse of the sign of control error generally used in servomechanism and control literature, which defines control error as the reference quantity minus the controlled quantity. *See also:* **power systems, low-frequency and surge testing.** 0-31E4

area frequency-response characteristic (control area). The sum of the change in total area generation caused by governor action and the change in total area load, both of which result from a sudden change in system frequency, in the absence of automatic control action. *See also:* **power systems, low-frequency and surge testing.** E94-0

area load-frequency characteristic (control area). The change in total area load that results from a change in system frequency. *See also:* **power systems, low-frequency and surge testing.** E94-0

area moving-target indication (electronic navigation). A moving target indicator (MTI) system in which the moving target is selected on the basis of its change in position between looks, rather than its Doppler frequency. *See also:* **navigation.** 0-10E6

area supplementary control (electric power systems). The control action applied, manually or automatically, to area generator speed governors in response to changes in system frequency, tie-line loading, or the relation of these to each other, so as to maintain the scheduled system frequency and/or the established net interchange with other control areas within predetermined limits. *See also:* **power systems, low-frequency and surge testing.** E94-0

area tie line (electric power systems). A transmission line connecting two control areas. *Note:* Similar to interconnection tie. *See also:* **transmission line.** E94-0

argument (information processing). An independent variable, for example, in looking up a quantity in a table, the number, or any of the numbers, that identifies the location of the desired value. X3A12-16E9

arithmetic apparent power (polyphase circuit). At the terminals of entry of a polyphase circuit, a scalar quantity equal to the arithmetic sum of the apparent powers for the individual terminals of entry. *Notes:* (1) The apparent power for each terminal of entry is determined by considering each phase conductor and the common reference terminal as a single-phase circuit, as described for distortion power. The common reference terminal shall be taken at the neutral terminal of entry, if one exists, otherwise as the true neutral point. (2) If the ratios of the components of the vector powers, for each of the terminals of entry, to the corresponding apparent power are the same for every terminal of entry, the arithmetic apparent power is equal to the apparent power for the polyphase circuit; otherwise the arithmetic apparent power is greater than the apparent power. (3) If the voltages have the same wave form as the corresponding currents, the arithmetic apparent power is equal to the amplitude of the phasor power. (4) Arithmetic apparent power is expressed in voltamperes when the voltages are in volts and the currents in amperes. E270-0

arithmetic element. That part of a computer that performs arithmetic operations. *See also:* **arithmetic unit; electronic computation.** E270-0

arithmetic reactive factor. The ratio of the reactive power to the arithmetic apparent power. E270-0

arithmetic shift. (1) A shift that does not affect the sign position. (2) A shift that is equivalent to the multiplication of a number by a positive or negative integral power of the radix. *See also:* **electronic digital computer.** X3A12-16E9

arithmetic unit. The unit of a computing system that contains the circuits that perform arithmetic operations. *See also:* **arithmetic element; electronic digital computer.** X3A12-16E9

arm. *See:* **branch.** *See also:* **network analysis.**

arm, thermoelectric. The part of a thermoelectric device in which the electric-current density and temperature gradient are approximately parallel or antiparallel and that is electrically connected only at its extremities to a part having the opposite relation between the direction of the temperature gradient and the electric-current density. *Note:* The term thermoelement is ambiguously used to refer to either a thermoelectric arm or to a thermoelectric couple, and its use is therefore not recommended. *See also:* **thermoelectric device.** E221-15E7

arm, thermoelectric, graded. A thermoelectric arm whose composition changes continuously along the direction of the current density. *See also:* **thermoelectric device.** E221-15E7

arm, thermoelectric, segmented. A thermoelectric arm composed of two or more materials having different compositions. *See also:* **thermoelectric device.** E221-15E7

armature (1) (rotating machinery). The member of an electric machine in which an alternating voltage is generated by virtue of relative motion with respect to a magnetic flux field. In direct-current universal, alternating-current series, and repulsion-type machines, the term is commonly applied to the entire rotor. *See:*
armature band;
armature-band insulation;
armature bar;
armature coil;
armature core;
armature spider;
armature terminal;
armature-winding equalizer;
coil pitch;
distributed winding;
frog-leg winding;
initial alternating short-circuit current;
journal;
pitch factor;
pole pitch;
primary winding;
shaft extension;
slotted armature;
spacer shaft;
turn insulation. 0-31E8
(2) (relay) (electromechanical relay). The moving element that contributes to the designed response of the relay and that usually has associated with it a part of the relay contact assembly. *See:* **relay armature.** *See also:* **relay.** 37A100-31E6

armature band (rotating machinery). A thin circumferential structural member applied to the winding of a rotating armature to restrain and hold the coils so as to counteract the effect of centrifugal force during rotation. *Note:* Armature bands may serve the further purpose of arch-binding the coils. They may be on the end windings only or may be over the coils within the core. 0-31E8

armature band insulation (rotating machinery). An insulation member placed between a rotating armature winding and an armature band. *See also:* **armature.** 0-31E8

armature bar (half coil) (rotating machinery). Either of two similar parts of an armature coil, comprising an embedded coil side and two end sections, that when connected together form a complete coil. *See also:* **armature.** 0-31E8

armature coil (rotating machinery). A unit of the armature winding composed of one or more insulated conductors. *See:* **armature; asynchronous machine; direct-current commutating machine; synchronous machine.** 42A10-0

armature core (rotating machinery). A core on or around which armature windings are placed. *See also:* **armature.** 0-31E8

armature I^2R loss (synchronous machine). The sum of the I^2R losses in all of the armature current paths. *Note:* The I^2R loss in each current path shall be the product of its resistance in ohms, as measured with direct current and corrected to a specified temperature, and the square of its current in amperes. *See also:* **synchronous machine.** 50A10-31E8

armature (magnet). A piece or an assembly of pieces of ferromagnetic material associated with the pole pieces of a magnet in such a manner that it and the pole pieces can move in relation to each other. E270-0

armature quill. *See:* **armature spider.**

armature-reaction excited machine. A machine having a rotatable armature, provided with windings and a commutator, whose load-circuit voltage is generated by flux that is produced primarily by the magnetomotive force of currents in the armature winding. *Notes:* (1) By providing the stationary member of the machine with various types of windings different characteristics may be obtained, such as a constant-current characteristic or a constant-voltage characteristic. (2) The machine is normally provided with two sets of brushes, displaced around the commutator from one another, so as to provide primary and secondary circuits through the armature. (3) The primary circuit carrying the excitation armature current may be completed externally by a short-circuit connection, or through some other external circuit, such as a field winding or a source of power supply; and the secondary circuit is adapted for connection to an external load. *See also:* **direct-current commutating machine.** 42A10-0

armature reaction (rotating machinery). The magnetomotive force due to armature-winding current. 0-31E8

armature sleeve. *See:* **armature spider.**

armature spider (armature sleeve) (armature quill). A support upon which the armature laminations are mounted and which in turn is mounted on the shaft. *See also:* **armature (rotating machinery).** 42A10-0

armature terminal (rotating machinery). A terminal connected to the armature winding. *See also:* **armature.** 0-31E8

armature-voltage control. A method of controlling the speed of a motor by means of a change in the magnitude of the voltage impressed on its armature winding. *See:* **control.** 42A25-34E10

armature winding (rotating machinery). The winding in which alternating voltage is generated by virtue of relative motion with respect to a magnetic flux field. *See:* **asynchronous machine; direct-current commutating machine; synchronous machine.** 0-31E8

armature winding cross connection. *See:* **armature winding equalizer.**

armature winding equalizer (armature winding cross connection) (rotating machinery). An electric connection to normally equal-potential points in an armature circuit having more than two parallel circuits. *See also:* **armature (rotating machinery).** 42A10-31E8

armed sweep. *See:* **single sweep.**

armor clamp. A fitting for gripping the armor of a cable at the point where the armor terminates or where the cable enters a junction box or other piece of apparatus. *See also:* **power distribution, underground construction.** 42A35-31E13

armored cable (interior wiring). A fabricated assembly of insulated conductors and a flexible metallic covering. *Note:* Armored cable for interior wiring has its flexible outer sheath or armor formed of metal strip, helically wound and with interlocking edges. Armored cable is usually circular in cross section but may be oval or flat and may have a thin lead sheath between the armor and the conductors to exclude moisture, oil, etcetera, where such protection is needed. *See:* **nonmetallic sheathed cable.** *See also:* **interior wiring; power distribution, underground construction; tree wire.** 42A95-0

array (photovoltaic converters). A combination of panels coordinated in structure and function. *See also:* **semiconductor.** 0-10E1

array antenna. An antenna comprising a number of radiating elements, generally similar, that are arranged and excited to obtain directional radiation patterns. *See also:* **antenna.** 0-3E1

array element (array antenna) (antenna). A single radiating element or a convenient grouping of radiating elements that have a fixed relative excitation. *See also:* **antenna.** 0-3E1

arrester. *Note:* For an extensive list of cross references, see *Appendix A.*

arrester alternating sparkover voltage. The root-mean-square value of the minimum 60-hertz sine-wave voltage that will cause sparkover when applied between its line and ground terminals. *See also:* **current rating, 60-hertz (arrester); lightning arrester (surge diverter).** 42A20-31E7

arrester discharge capacity. The crest value of the maximum discharge current of specified wave shape that the arrester can withstand without damage to any of its parts. *See:* **lightning arrester (surge diverter).** 42A20-31E7

arrester discharge voltage-current characteristic. The variation of the crest values of discharge voltage with respect to discharge current. *Note:* This characteristic is normally shown as a graph based on three or more current surge measurements of the same wave shape but of different crest values. *See also:* **lightning; lightning arrester (surge diverter); current rating, 60-hertz (arrester).** E28/42A20-0;62A1-31E7

arrester discharge voltage-time curve. A graph of the discharge voltage as a function of time while discharging a current surge of given wave shape and magnitude. *See:* **lightning arrester (surge diverter).** 42A20-31E7

arrester disconnector. A means to disconnect a circuit through an arrester in anticipation of or after a failure in order to prevent a permanent fault on the circuit and to give indication of a failed arrester. *Note:* Clearing of the power current through the arrester during disconnection generally is a function of the nearest source-side overcurrent protective device. *See:* **lightning arrester (surge diverter).** 0-31E7

arrester, expulsion-type. An arrester having an arcing chamber in which the follow-current arc is confined and brought into contact with gas-evolving or other arc-extinguishing material in a manner that results in the limitation of the voltage at the line terminal and the interruption of the follow current. *Note:* The term **expulsion arrester** includes any external series-gap or current-limiting resistor if either or both are used as a part of the complete device as installed for service. *See also:* **arrester.** 42A20/99I2-31E7

arrester ground. An intentional electric connection of the arrester ground terminal to the ground. *See:* **arrester; lightning arrester (surge diverter).** 42A20-31E7

arrester, valve-type. An arrester having a characteristic element consisting of a resistor with a nonlinear volt-ampere characteristic that limits the follow current to a value that the series gap can interrupt. *Note:* If the arrester has no series gap the characteristic element limits the follow current to a magnitude that does not interfere with the operation of the system. *See:* **nonlinear-resistor type arrester (valve type).** *See also:* **arrester; lightning arrester (surge diverter).** 42A20-31E7

ARSR (air route surveillance radar). *See:* **surveillance radar.**

articulation (percent articulation) and intelligibility (percent intelligibility). The percentage of the speech units spoken by a talker or talkers that is correctly repeated, written down, or checked by a listener or listeners. *Notes:* (1) The word articulation is used when the units of speech material are meaningless syllables or fragments; the word intelligibility is used when the units of speech material are complete, meaningful words, phrases, or sentences. (2) It is important to specify the type of speech material and the units into which it is analyzed for the purpose of computing the percentage. The units may be fundamental speech sounds, syllables, words, sentences, etcetera. (3) The percent articulation or percent intelligibility is a property of the entire communication system; talker, transmission equipment or medium, and listener. Even when attention is focused upon one component of the system (for example, a talker, a radio receiver), the other components of the system should be specified. (4) The kind of speech material used is identified by an appropriate adjective in phrases such as syllable articulation, individual sound articulation, vowel (or consonant) articulation, monosyllabic word intelligibility, discrete word intelligibility, discrete sentence intelligibility. *See also:* **electroacoustics; volume equivalent.** 42A65-1E1

articulation equivalent (complete telephone connection). A measure of the articulation of speech reproduced over it. The articulation equivalent of a complete telephone connection is expressed numerically in terms of the trunk loss of a working reference system when the latter is adjusted to give equal articulation. *Note:* For engineering purposes, the articulation equivalent is divided into articulation losses assignable to (1) the station set, subscriber line, and battery supply circuit that are on the transmitting end, (2) the station set, subscriber line, and battery supply circuit that are on the receiving end, (3) the trunk, and (4) interaction effects arising at the trunk terminals. *See also:* **volume equivalent.** 42A65-0

artificial antenna (dummy antenna). A device that has the necessary impedance characteristics of an antenna and the necessary power-handling capabilities, but which does not radiate or receive radio waves. *See also:* **antenna.** E145-3E1

artificial dielectric. A three-dimensional arrangement of metallic conductors, which are usually small compared with a wavelength, the resulting medium having the property of acting as though it were a more or less uniform dielectric to electromagnetic waves. *See:* **antenna; waveguide.** 50I62-3E1

artificial ear. A device for the measurement of the acoustic output of earphones in which the artificial ear presents to the earphone an acoustic impedance approximating the impedance presented by the average human ear and is equipped with a microphone for measurement of the sound pressures developed by the earphone. *See also:* **electroacoustics.** E269-19E8;0-1E1

artificial hand (electromagnetic compatibility). A device simulating the impedance between an electric appliance and the local earth when the appliance is grasped by the hand. *See also:* **electromagnetic compatibility.** CISPR-27E1

artificial language. A language based on a set of prescribed rules that are established prior to its usage. *See:* **natural language.** X3A12-16E9

artificial line. An electric network that simulates the electrical characteristics of a line over a desired frequency range. *Note:* Although the term basically is applied to the case of simulation of an actual line, by extension it is used to refer to all periodic lines that may be used for laboratory purposes in place of actual lines, but that may represent no physically realizable line. For example, an artificial line may be composed of pure resistances. *See also:* **network analysis.** E145/E270;42A65-31E3

artificial load. A dissipative but essentially nonradiating device having the impedance characteristics of an antenna, transmission line, or other practical utilization circuit. *See also:* **circuits and devices.** E145/E270-0;42A65-31E3

artificial mains network (electromagnetic compatibility). A network inserted in the supply mains lead of the apparatus to be tested that provides a specified measuring impedance for interference voltage measurements and isolates the apparatus from the supply mains at radio frequencies. *See also:* **electromagnetic compatibility.** CISPR-27E1

artificial mouth (telephony). An electroacoustic transducer that produces a sound field, with respect to a reference point, simulating that of a typical human talker. *Note:* The reference point corresponds to the center of the lips of the talker. *See also:* **telephone station.** E269-19E8

artificial pupil. A diaphragm or other limitation that confines the light entering the eye to a smaller aperture than does the normal pupil. *See also:* **visual field.** Z7A1-0

articulated unit substation. A unit substation in which the incoming, transforming, and outgoing sections are manufactured as one or more subassemblies intended for connection in the field. 37A100-31E11

artificial voice (close-talking pressure-type microphone). A sound source for microphone measurements consisting of a small loudspeaker mounted in a shaped baffle proportioned to simulate the acoustic constants of the human head. *See also:* **close-talking pressure-type microphone; loudspeaker.** E258-0

***A* scan (electronic navigation).** *See:* ***A* and *R* display.**

***A* scope.** A cathode-ray oscilloscope arranged to present an *A* display. *See:* ***A* and *R* display.** *See also:* **radar.** 42A65-0

ASDE. *See:* **airport surface detection equipment.**

askarel. A synthetic nonflammable insulating liquid that, when decomposed by the electric arc, evolves only nonflammable gaseous mixtures. *See also:* **insulation.** 42A95-0

aspect ratio (television). The ratio of the frame width to the frame height. *See also:* **television.** E204-0;E188-2E2

asphalt (rotating machinery). A dark brown to black cementitious material, solid or semisolid in consistency, in which the predominating constituents are bitumens that occur in nature as such or are obtained as residue in refining of petroleum. AD8-31E8

ASR (airport surveillance radar). *See:* **surveillance radar.**

assemble (computing machines). To prepare a machine language program from a symbolic language program by substituting absolute operation codes for symbolic operation codes and absolute or relocatable addresses for symbolic addresses. *See also:* **electronic digital computer.** X3A12-16E9

assembler (computing machines). A program that assembles. X3A12-16E9

assembly (industrial control). A combination of all or of a portion of component parts included in an electric apparatus, mounted on a supporting frame or panel, and properly interwired. 42A25-34E10

associative storage. A storage device in which storage locations may be identified by specifying part or all of their contents. *Note:* Also called parallel-search storage or content-addressed storage. *See also:* **electronic digital computer.** E162-0;0-16E9

aster rectifier circuit. A circuit that employs twelve or more rectifying elements with a conducting period of 30 electrical degrees plus the commutating angle. *See also:* **rectification.** 42A15-0

astigmatism (electron optical). In an electron-beam tube, a focus defect in which electrons in different axial planes come to focus at different points. *See also:* **beam tubes; oscillograph.** E160-0;42A70-15E6

Aston dark space (gas). The dark space in the immediate neighborhood of the cathode, in which the emitted electrons have a velocity insufficient to excite the gas. *See also:* **discharge (gas).** 50I07-15E6

astro-inertial navigation equipment. *See:* **celestial inertial navigation equipment.**

***A* switchboard.** A manual telephone switchboard in a local central office, primarily to receive subscriber calls and to complete connections either directly or through some other switching equipment. 42A65-19E1

asymmetric terminal voltage (electromagnetic compatibility). Terminal voltage measured with a delta network between the midpoint of the resistors across the mains lead and ground. *See also:* **electromagnetic compatibility.** CISPR-27E1

asymmetrical cell. A cell in which the impedance to the flow of current in one direction is greater than in the other direction. *See also:* **electrolytic capacitor.** 42A60-0

asymmetrical current. *See:* **total asymmetrical current.**

asynchronous computer. A computer in which each event or the performance of each operation starts as a result of a signal generated by the completion of the previous event or operation, or by the availability of the parts of the computer required for the next event or operation. *See also:* **electronic digital computer.** X3A12-16E9

asynchronous impedance (rotating machinery). The quotient of the voltage, assumed to be sinusoidal and balanced, supplied to a rotating machine out of synchronism, and the same frequency component of the current. *Note:* The value of this impedance depends on the slip. *See also:* **asynchronous machine.** 0-31E8

asynchronous machine. An alternating-current machine in which the rotor does not turn at synchronous speed. *Note:* For an extensive list of cross references, see *Appendix A.* 0-31E8

asynchronous operation (rotating machinery). Operation of a machine where the speed of the rotor is other than synchronous speed. *See also:* **asynchronous machine.** 0-31E8

atmospheric absorption. The loss of energy in transmission of radio waves, due to dissipation in the atmosphere. *See also:* **radiation.** 42A65-0

atmospheric duct (radio wave propagation). A stratum of the troposphere within which the variation of refractive index is such as to confine within the limits of the stratum the propagation of an abnormally large proportion of any radiation of sufficiently high frequency. *Note:* The duct extends from the level of a local minimum of the modified index of refraction as a function of height down to the level where the minimum value is again encountered, or down to the surface bounding the atmosphere if the minimum value is not again encountered. *See also:* **radiation; radio wave propagation.** 42A65-0;0-3E2

atmospheric radio noise (electromagnetic compatibility). Noise having its source in natural atmospheric phenomena. *See also:* **electromagnetic compatibility.** 0-27E1

atmospheric radio wave. A radio wave that is propagated by reflection in the atmosphere. *Note:* It may include either the ionospheric wave or the tropospheric wave, or both. *See also:* **radiation.** 42A65-0

atmospherics. *See:* **static (atmospherics).**

atmospheric transmissivity. The ratio of the directly transmitted flux incident on a surface after passing through unit thickness of the atmosphere to the flux which would be incident on the same surface if the flux had passed through a vacuum. *See also:* **signal lighting.** Z7A1-0

A-trace (loran). The first (upper) trace on the scope display. *See also:* **navigation.** 0-10E6

ATR box. *See:* **anti-transmit-receive box.**

ATR switch. *See:* **anti-transmit-receive switch.**

ATR tube. *See:* **anti-transmit-receive tube.**

attachments. Accessories to be attached to switchgear apparatus, as distinguished from auxiliaries. 37A100-31E11

attachment plug (plug cap*) (cap*). A device, that by insertion in a receptacle, establishes connection between the conductors of the attached flexible cord and the conductors connected permanently to the receptacle. *See also:* **interior wiring.**
*Deprecated 42A95-0

attack time (electroacoustics). The interval required, after a sudden increase in input signal amplitude to a system or transducer, to attain a stated percentage (usually 63 percent) of the ultimate change in amplification or attenuation due to this increase. *See:* **electroacoustics.** E151-42A65

attenuating pad. *See:* **pad.**

attenuation. **(1)** A general term used to denote a decrease in magnitude in transmission from one point to another. *Note:* It may be expressed as a ratio or, by extension of the term, in decibels. *See also:* **transmission characteristics.** 42A65-27E131E3

(2) (quantity associated with a traveling wave in a homogeneous medium). The decrease of its amplitude with increasing distance from the source, excluding the decrease due to spreading. *See also:* **radio wave propagation.** 0-3E2

(3) (signal-transmission system). A decrease in signal magnitude. *Note:* Attenuation may be expressed as the scalar ratio of the input magnitude to the output magnitude or in decibels as 20 times the logarithm of that ratio. 0-13E6

(4) (per unit length) (coaxial transmission line). The decrease with distance in the direction of propagation of a quantity associated with a traveling transverse electromagnetic wave. *See also:* **wave front.** 83A14-0

(5) (waveguide) (quantity associated with a traveling waveguide or transmission-line wave). The decrease with distance in the direction of propagation. *Note:* Attenuation of power is usually measured in terms of decibels or decibels per unit length.
See :
attenuation, current;
attenuation, voltage;
compandor;
compressor;
expander;
filter;
interaction loss.
See also: **waveguide.** E146-3E2

attenuation band (uniconductor waveguide). Rejection band. *See also:* **waveguide.** E146-3E1

attenuation constant. The real part of the propagation constant. *Note:* Unit: Neper per unit length. (1 neper equals 8.686 decibels). *See also:* **transmission characteristics; radio transmission.** E270-3E2/9E4;42A65-0

attenuation, current. Either (1) a decrease in signal current magnitude, in transmission from one point to another, or the process thereof, or (2) of a transducer, the scalar ratio of the signal input current to the signal output current. *Note:* By incorrect extension of the term decibel, this ratio is sometimes expressed in decibels by multiplying its common logarithm by 20. It may be correctly expressed in decilogs. *See:* **decibel.** *See also:* **attenuation.** 42A65/E151-0

attenuation distortion (frequency distortion*). Attenuation distortion is either (1) departure, in a circuit or system, from uniform amplification or attenuation over the frequency range required for transmission, or (2) the effect of such departure on a transmitted signal.
*Deprecated. 42A65-0

attenuation equalizer. A corrective network that is designed to make the absolute value of the transfer impedance, with respect to two chosen pairs of terminals, substantially constant for all frequencies within a desired range. *See also:* **network analysis.** 42A65-0

attenuation loss. The reduction in radiant power surface density, excluding the reduction due to spreading. E270-0

attenuation network. A network providing relatively little phase shift and substantially constant attenuation over a range of frequency. E270-0

attenuation ratio. The magnitude of the propagation ratio. *See also:* **radiation; radio wave propagation.** 42A65-0;E211-3E2

attenuation vector (field quantity) (radio wave propagation). The vector pointing in the direction of maximum decrease of amplitude, whose magnitude is the attenuation constant. *See also:* **radio wave propagation.** 0-3E2

attenuation, voltage. Either (1) a decrease in signal voltage magnitude in transmission from one point to another, or the process thereof, or (2) a transducer, the scalar ratio of the signal input voltage to the signal output voltage. *Notes:* (1) By incorrect extension of the term decibel, this ratio is sometimes expressed in decibels by multiplying its common logarithm by 20. It may be correctly expressed in decilogs. (2) If the input and/or output power consist of more than one component, such as multifrequency signal or noise, then the particular components used and their weighting must be specified *See:* **decibel.** *See also:* **attenuation; signal; transducer.** E151/E270/42A65-0E151-0

attenuator. (1) An adjustable transducer for reducing the amplitude of a wave without introducing appreciable distortion. 42A65-31E3

(2) An adjustable passive network that reduces the power level of a signal without introducing appreciable distortion. *See also:* **transducer.** E151-0

attenuator tube (electron tubes). A gas-filled radio-frequency switching tube in which a gas discharge, initiated and regulated independently of radio-frequency power, is used to control this power by reflection or absorption. *See also:* **gas tubes.** E160-15E6

attenuator, waveguide. A waveguide or transmission-line device for the purpose of producing attenuation by any means, including absorption and reflection. *See:* **waveguide.** E147-3E1

attitude-effect error (electronic navigation). A manifestation of polarization error; an error in indicated bearing that is dependent upon the attitude of the vehicle with respect to the direction of signal propagation. *See:* **heading-effect error.** *See also:* **navigation.** 0-10E6

attitude gyro-electric indicator. An electrically driven device that provides a visual indication of an aircraft's roll and pitch attitude with respect to the earth. *Note:* It is used in highly maneuverable aircraft and differs from the gyro-horizon electric indicator in that the gyro is not limited by stops and has complete freedom about the roll and pitch axes. *See also:* **air transportation instruments.** 42A41-0

audible busy signal (busy tone). An audible signal connected to the calling line to indicate that the called line is in use. *See also:* **telephone switching system.** 42A65-0

audible cab indicator. A device (usually an air whistle, bell, or buzzer) located in the cab of a vehicle equipped with cab signals or continuous train control designed to sound when the cab signal changes to a more restrictive indication. 42A42-0

audible signal device (protective signaling). A general term for bells, buzzers, horns, whistles, sirens, or other devices that produce audible signals. *See also:* **protective signaling.** 42A65-0

audio. Pertaining to frequencies corresponding to a normally audible sound wave. *Note:* These frequencies range roughly from 15 hertz to 20 000 hertz. *See also:* **signal wave.** 42A65-31E3

audio frequency. Any frequency corresponding to a normally audible sound wave. *Notes:* (1) Audio frequencies range roughly from 15 hertz to 20 000 hertz. (2) This term is frequently shortened to audio and used as a modifier to indicate a device or system intended to operate at audio frequencies, for example, audio amplifier. E145/E151/E157/E188-1E1/21E1/42A65

audio-frequency distortion. The form of wave distortion in which the relative magnitudes of the different frequency components of the wave are changed on either a phase or amplitude basis.
See:
audio input power;
audio input signal;
carrier power output;
hum and noise level;
modulation;
modulation and noise spectrum;
modulation bandwidth;
modulation limiting;
modulation stability;
power-supply voltage range;
radiated power output;
spurious emissions. 0-6E1

audio-frequency harmonic distortion. The generation in a system of integral multiples of a single audio-frequency input signal. *See also:* **modulation.** E145-0

audio-frequency (interference terminology). Components of noise having frequencies in the audio range. *See also:* **signal.** 0-13E6

audio-frequency noise. *See:* **noise, audio frequency.**

audio-frequency oscillator (audio oscillator). A nonrotating device for producing an audio-frequency sinusoidal electric wave, whose frequency is determined by the characteristics of the device. *See:* **oscillatory circuit.** E151-0

audio-frequency peak limiter. A circuit used in an audio-frequency system to cut off peaks that exceed a predetermined value. *See also:* **modulating systems.** E145-0

audio-frequency response (receiver performance). The variation of gain or loss as a function of frequency within a specified bandwidth at the output of a receiver. *See also:* **receiver performance; amplitude-frequency response.** 0-6E1

audio-frequency spectrum (audio spectrum). The continuous range of frequencies extending from the lowest to the highest audio frequency. *See:* **electroacoustics.** E151-0

audio-frequency transformer. A transformer for use with audio-frequency currents. *See also:* **circuits and devices.** 42A65-21E1

audio input power (transmitter performance). The input power level to the modulator, expressed in decibels referred to a 1 milliwatt power level. *See also:* **audio-frequency distortion.** 0-6E1

audio input signal (transmitter performance). That composite input to the transmitter modulator that consists of frequency components normally audible to the human ear. *See also:* **audio-frequency distortion.** 0-6E1

audio oscillator. See: **audio-frequency oscillator.**

audio output power (receiver performance) (receiver). The audio-frequency power dissipated in a load across the output terminals. *See also:* **receiver performance.** 0-6E1

audio-tone channel. *See:* **voice-frequency carrier-telegraph.**

audiogram (threshold audiogram). A graph showing hearing level as a function of frequency. *See also:* **electroacoustics.** 0-1E1

audiometer. An instrument for measuring hearing level. *See also:* **electroacoustic; instrument.** 0-1E1

auditory sensation area. (1) The region enclosed by the curves defining the threshold of feeling and the threshold of audibility each expressed as a function of frequency. (2) The part of the brain (temporal lobe of the cortex) that is responsive to auditory stimuli. *See also:* **electroacoustics.** 0-1E1

***A* unit.** A motive power unit so designed that it may be used as the leading unit of a locomotive, with adequate visibility in a forward direction, and which includes a cab and equipment for full control and observation of the propulsion power and brake applications for the locomotive and train. *See also:* **electric locomotive.** 42A42-0

aural harmonic. A harmonic generated in the auditory mechanism. *See also:* **electroacoustics.** E157-1E1

aural radio range. A radio range providing four radial lines of position identified aurally as a continuous tone resulting from the interleaving of equal amplitude A and N International Morse code letters. The sense of deviation from these lines is indicated by deterioration of the steady tone into audible A or N code signals. *See also:* **navigation; radio navigation.** E172-10E6

aural transmitter. The radio equipment used for the transmission of the aural (sound) signals from a television broadcast station. *See also:* **television.** E182A/E145-0

austenitic. The face-centered cubic crystal structure of ferrous metals. *See:* **corrosion terms.** CM-34E2

authorized bandwidth (mobile communication). The frequency band containing those frequencies upon which a total of 99 percent of the radiated power appears. *See also:* **mobile communication system.** 0-6E1

auto alarm. A radio receiver that automatically produces an audible alarm when a prescribed radio signal is received. 42A43-0

autocondensation* (electrotherapy). A method of applying alternating currents of frequencies exceeding 100 kilohertz to limited areas near the surface of the human body through the use of one very large and one small electrode, the patient becoming part of the capacitor. *See also:* **electrotherapy.** 42A80-18E1

*Deprecated

autoconduction* (electrotherapy). A method of applying alternating currents, of frequencies exceeding 100 kilohertz for therapeutic purposes, by electromagnetic induction, the patient being placed inside a large solenoid. *See also:* **electro-therapy.** 42A80-18E1

*Deprecated

autodyne reception. A system of heterodyne reception through the use of a device that is both an oscillator and a detector. *See also:* **modulating systems.** 42A65-0

automatic (industrial control). Self-acting, operating by its own mechanism when actuated by some impersonal influence; as, for example, a change in current strength, pressure, temperature, or mechanical configuration. 42A25-34E10

automatic acceleration (1) (vehicle). Acceleration under the control of devices that function automatically to maintain, within relatively close predetermined values or schedules, current passing to the traction motors, the tractive force developed by them, the rate of vehicle acceleration, or similar factors affecting acceleration. *See also:* **electric drive; multiple-unit control.** 42A42-0

(2) (industrial control). Acceleration under the control of devices that function automatically to raise the motor speed. *See also:* **electric drive; multiple unit control.** 42A25-34E10

automatically regulated (rotating machinery). Applied to a machine that can regulate its own characteristics when associated with other apparatus in a suitable closed-loop circuit. *See also:* **direct-current commutating machine; synchronous machine; asynchronous machine.** 0-31E8

automatically reset relay. *See:* **self-reset relay.**

automatic approach control. A system that integrates signals, received by localizer and glide path receivers, into the automatic pilot system, and guides the airplane down the localizer and glide path beam intersection. *See also:* **air transportation electronic equipment.** 42A41-0

automatic block signal system. A series of consecutive blocks governed by block signals, cab signals, or both, operated by electric, pneumatic, or other agency actuated by a train or by certain conditions affecting the use of a block. *See also:* **block signal system.** 42A42-0

automatic cab signal system. A system that provides for the automatic operation of cab signals. *See also:* **automatic train control.** 42A42-0

automatic carriage. A control mechanism for a typewriter or other listing device that can automatically control the feeding, spacing, skipping, and ejecting of paper or preprinted forms. *See also:* **electronic digital computer.** X3A12-16E9

automatic chart-line follower (electronic navigation). A device that automatically derives error signals proportional to the deviation of the position of a vehicle from a predetermined course line drawn on a chart. *See also:* **navigation.** E172-10E6

automatic check. *See:* **check, automatic.**

automatic circuit recloser. A self-controlled device for automatically interrupting and reclosing an alternating-current circuit, with a predetermined sequence of opening and reclosing followed by resetting, hold-closed, or lockout operation. 37A100-31E11

automatic combustion control. A method of combustion control that is effected automatically by mechanical or electric devices. 42A35-31E13

automatic computer. A computer that can perform a sequence of operations without intervention by a human operator. X3A12-16E9

automatic control (electrical controls). An arrangement that provides for switching or otherwise controlling, or both, in an automatic sequence and under predetermined conditions, the necessary devices comprising an equipment. *Note:* These devices thereupon maintain the required character of service and provide adequate protection against all usual operating emergencies. 37A100-31E11

automatic controller. *See:* **controller, automatic.**

automatic direction-finder (electronic navigation). A direction-finder that automatically and continuously provides a measure of the direction of arrival of the received signal. Data are usually displayed visually. *See also:* **navigation.** E172-10E6

automatic dispatching system (electric power sys-

tems). A controlling means for maintaining the area control error or station control error at zero by automatically loading generating sources, and it also may include facilities to load the sources in accordance with a predetermined loading criterion. *See also:* **power systems, low-frequency and surge testing.** E94-0

automatic equipment (for a specified type of power circuit or apparatus). Equipment that provides automatic control. 37A100-31E11

automatic fire-alarm system. A fire-alarm system for automatically detecting the presence of fire and initiating signal transmission without human intervention. *See also:* **protective signaling.** 42A65-0

automatic frequency control. An arrangement whereby the frequency of an oscillator or the tuning of a circuit is automatically maintained within specified limits with respect to a reference frequency. *See also:* **circuits and devices; radio transmitter; receiver performance.** E182-0

automatic–frequency–control synchronization. A process for locking the frequency (phase) of a local oscillator to that of an incoming synchronizing signal by the use of a comparison device whose output continuously corrects the local-oscillator frequency (phase). 0-42A65

automatic gain control (AGC). (1) A process or means by which gain is automatically adjusted in a specified manner as a function of input or other specified parameters. *See also:* **circuits and devices.** E151-2E2;E188-0;0-42A65

(2) A method of automatically obtaining a substantially constant output of some amplitude characteristic of the signal over a range of variation of that characteristic at the input. The term is also applied to a device for accomplishing this result. *See also:* **circuits and devices.** 42A65-31E3

automatic generation control. The regulation of the power output of electric generators within a prescribed area in response to change in system frequency, tie-line loading, or the relation of these to each other, so as to maintain the scheduled system frequency or the established interchange with other areas within predetermined limits or both. 0-31E4

automatic grid bias. Grid-bias voltage provided by the difference of potential across resistance(s) in the grid or cathode circuit due to grid or cathode current or both. *See also:* **radio receiver; circuits and devices.** E145/42A65-0

automatic hold (analog computer). Attainment of the hold condition automatically through amplitude comparison of a problem variable or through an overload condition. *See also:* **electronic analog computer.** E165-16E9

automatic holdup alarm system. An alarm system in which the signal transmission is initiated by the action of the robber. *See also:* **protective signaling.** 42A65-0

automatic interlocking. An arrangement of signals, with or without other signal appliances, that functions through the exercise of inherent powers as distinguished from those whose functions are controlled manually, and that are so interconnected by means of electric circuits that their movements must succeed one another in proper sequence. *See also:* **interlocking (interlocking plant).** 42A42-0

automatic keying device. A device that, after manual initiation, controls automatically the sending of a radio signal that actuates the auto alarm. *Note:* The prescribed signal is a series of twelve dashes, each of four seconds duration, with one-second interval between dashes, transmitted on the radiotelegraph distress frequency in the medium-frequency band. This signal is used only to precede distress calls or urgent warnings. 42A43-0

automatic line sectionalizer. A self-controlled circuit-opening device that opens the main electric circuit through it while disconnected from the source of power after the device has sensed and responded to a predetermined number of successive main-current impulses of predetermined or greater magnitude. *Note:* It may have provision to be manually operated to interrupt loads up to its rated continuous current. 37A100-31E11

automatic load (armature current division) (industrial control). The effect of a control function or a device to automatically divide armature currents in a prescribed manner between two or more motors or two or more generators connected to the same load. *See:* **control system, feedback.** IC1-34E10

automatic load (level) control (ALC) (power-system communication). A method of automatically maintaining the peak power of a single-sideband suppressed-carrier transmitter at a constant level. *See also:* **radio transmitter.** 0-31E3

automatic message accounting system (AMA). An arrangement of apparatus for recording and processing on continuous tapes the data required for computing telephone charges on certain classes of calls. *Note:* This system may include provision for compiling all charges and credits that affect the customer's bill and for automatic printing of the bill. *See also:* **telephone switching system.** 42A65-0

automatic opening (switching device) (automatic tripping). The opening under predetermined conditions without the intervention of an attendant. 37A100-31E11

automatic operation (1) (elevators). Operation wherein the starting of the elevator car is effected in response to the momentary actuation of operating devices at the landing, and/or of operating devices in the car identified with the landings, and/or in response to an automatic starting mechanism, and wherein the car is stopped automatically at the landings. *See:* **control.** 42A45-0

(2) (switching device). The ability to complete an assigned sequence of operations by automatic control without the assistance of an attendant. 37A100-31E11

automatic phase control (television). A process or means by which the phase of an oscillator signal is automatically maintained within specified limits by comparing its phase to the phase of an external reference signal and thereby supplying correcting information to the controlled source or a device for accomplishing this result. *Note:* Automatic phase control is sometimes used for accurate frequency control and under these conditions is often called automatic frequency control. *See:* **television.** 0-42A65

automatic pilot (electronic navigation). Equipment that automatically controls the attitude of a vehicle about one or more of its rotational axes (pitch, roll, and yaw), and may be used to respond to manual or electronic commands. *See also:* **air transportation electric equipment; navigation.** 0-10E6

automatic-pilot servo motor. A device that converts electric signals to mechanical rotation so as to move the control surfaces of an aircraft. *See also:* **air transportation electric equipment.** 42A41-0

automatic programmed tools (APT) (numerically controlled machines). A computer-based numerical control programming system that uses English-like symbolic descriptions of part and tool geometry and tool motion. *See also:* **numerically controlled machines.** EIA3B-34E12

automatic programming (analog computer). A method of computer operation using auxiliary automatic equipment to perform computer control state selections, switching operations, or component adjustments in accordance with previously selected criteria. *See also:* **electronic analog computer.** E165-16E9

automatic reclosing equipment. An automatic equipment that provides for reclosing a switching device as desired after it has opened automatically under abnormal conditions. *Note:* Automatic reclosing equipment may be actuated by conditions sensed on either or both sides of the switching device as designed. 37A100-31E11

automatic-reset manual release of brakes (control) (industrial control). A manual release that, when operated, will maintain the braking surfaces in disengagement but will automatically restore the braking surfaces to their normal relation as soon as electric power is again applied. *See:* **control system, feedback.** IC1-34E10

automatic-reset relay. *See:* **relay, automatic-reset.**

automatic-reset thermal protector (rotating machinery). A thermal protector designed to perform its function by opening the circuit to or within the protected machine and then automatically closing the circuit after the machine cools to a satisfactory operating temperature. *See also:* **starting switch assembly.** 0-31E8

automatic reversing (industrial control). Reversing of an electric drive, initiated by automatic means. *See also:* **electric drive.** 42A25-34E10

automatic signal. A signal controlled automatically by the occupancy or certain other conditions of the track area that it protects. *See also:* **railway signal and interlocking.** 42A42-0

automatic smoke alarm system. An alarm system designed to detect the presence of smoke and to transmit an alarm automatically. *See also:* **protective signaling.** 42A65-0

automatic speed adjustment (industrial control). Speed adjustment accomplished automatically. *See:* **automatic.** *See also:* **electric drive.** 42A25-34E10

automatic starter (industrial control). A starter in which the influence directing its performance is automatic. *See also:* **starter.** 42A25-34E10

automatic station (generating stations and substations). A station (usually unattended) that under predetermined conditions goes into operation by an automatic sequence; that thereupon by automatic means maintains the required character of service within its capability; that goes out of operation by automatic sequence under other predetermined conditions, and includes protection against the usual operating emergencies. *Note:* An automatic station may go in and out of operation in response to predetermined voltage, load, time, or other conditions, or in response to supervisory control or to a remote or local manually operated control device. 37A100-31E11

automatic switchboard. A switchboard in which the connections are made by apparatus controlled from remote calling devices. *See also:* **telephone switching system.** 0-19E1

automatic switching system (machine switching system) (dial system) (telephony). A system in which connections between customers station equipment are ordinarily established as a result of signals produced by a calling device. *See also:* **telephone switching system.** 0-19E1

automatic system. A system in which the operations are performed by electrically controlled devices without the intervention of operators. 0-19E4

automatic telegraphy. That form of telegraphy in which transmission or reception of signals, or both, are accomplished automatically. *See also:* **telegraphy.** 42A65-19E4

automatic throw-over equipment. *See:* **automatic transfer equipment.**

automatic throw-over equipment of the fixed preferential type. *See:* **automatic transfer equipment of the fixed preferential type.**

automatic throw-over equipment of the nonpreferential type. *See:* **automatic transfer equipment of the nonpreferential type.**

automatic throw-over equipment of the selective-preferential type. *See:* **automatic transfer equipment of the selective-preferential type.**

automatic track-follower (electronic navigation). *See:* **automatic chart-line follower.**

automatic tracking (1) (radar). The process whereby a mechanism actuated by the echo automatically keeps the radar beam set on a target and usually determines the range of the target simultaneously. *See also:* **radar.** 42A65-0

(2) (electronic navigation). Tracking in which a servomechanism automatically follows some characteristic of the signal. *See also:* **radar.** E172-10E6

automatic train control (automatic speed control) (train control). A system or an installation so arranged that its operation on failure to forestall or acknowledge will automatically result in either one or the other or both of the following conditions: (1) Automatic train stop: The application of the brakes until the train has been brought to a stop; (2) Automatic speed control: The application of the brakes when the speed of the train exceeds a prescribed rate and continued until the speed has been reduced to a predetermined and prescribed rate.
See:
acknowledger forestaller;
acknowledging forestalling;
actuator;
automatic cab signal system;
automatic train control application;
cab signal;
continuous train control;
delayed application;
delay time;
false proceed operation;
false restrictive operation;
intermittent inductive train control;
intermittent train control;
operation;

potential false proceed operation;
reset device;
roadway element;
train control territory. 42A42-0

automatic train control application. An application of the brake by the automatic train control device. *See also:* **railway signal and interlocking.** 42A42-0

automatic transfer equipment (automatic throw-over equipment). An equipment that automatically transfers a load to another source of power when the original source to which it has been connected fails, and that will automatically retransfer the load to the original source under desired conditions. *Notes:* (1) It may be of the nonpreferential, fixed-preferential, or selective-preferential type. (2) Compare with transfer switch where transfer is accomplished without current interruption. 37A100-31E11

automatic transfer equipment of the fixed-preferential type (automatic throw-over equipment of the fixed preferential type). Equipment in which the original source always serves as the preferred source and the other source as the emergency source. The automatic transfer equipment will retransfer the load to the preferred source upon its reenergization. *Note:* The restoration of the load to the preferred source from the emergency source upon the reenergization of the preferred source after an outage may be of the continuous-circuit restoration type or the interrupted-circuit restoration type. 37A100-31E11

automatic transfer equipment of the nonpreferential type (automatic throw-over equipment of the nonpreferential type). Equipment that automatically retransfers the load to the original source only when the other source, to which it has been connected, fails. 37A100-31E11

automatic transfer equipment of the selective-preferential type (automatic throw-over equipment of the selective-preferential type). Equipment in which either source may serve as the preferred or the emergency source of preselection as desired, and that will retransfer the load to the preferred source upon its reenergization. *Note:* The restoration of the load to the preferred source from the emergency source upon the reenergization of the preferred source after an outage may be of the continuous-circuit restoration type or the interrupted-circuit restoration type. 37A100-31E11

automatic triggering (oscilloscopes). A mode of triggering in which one or more of the triggering-circuit controls are preset to conditions suitable for automatically displaying repetitive waveforms. *Note:* The automatic mode may also provide a recurrent trigger or recurrent sweep in the absence of triggering signals. *See:* **oscillograph.** 0-9E4

automatic tripping. *See:* **automatic opening.**

automatic volume control (AVC). A method of automatically maintaining a substantially constant audio output volume over a range of variation of input volume. *Note:* The term is also applied to a device for accomplishing this result in a system or transducer. *See also:* **circuits and devices.** E151-0; 42A65-31E3;0-42A65

automation. (1) The implementation of processes by automatic means. (2) The theory, art, or technique of making a process more automatic. (3) The investigation, design, development, and application of methods of rendering processes automatic, self-moving, or self-controlling. X3A12-16E9

autonavigator. Navigation equipment that includes means for coupling the output navigational data derived from the navigation sensors to the control system of the vehicle. *See also:* **navigation.** 0-10E6

autopilot. *See:* **automatic pilot.**

autopilot coupler (electronic navigation). The means used to link the navigation system receiver output to the automatic pilot. *See also:* **navigation.** E172-10E6

autoradar plot (electronic navigation). A particular chart comparison unit using a radar presentation of position. *See also:* **navigation.** 0-10E6

autoregulation induction heater. An induction heater in which a desired control is effected by the change in characteristics of a magnetic charge as it is heated at or near its Curie point. *See also:* **dielectric heating; induction heating.** E54/E169-0

autotransformer. (1) A transformer consisting of one electrically continuous winding with one or more fixed or movable taps that is intended for use in such a manner that part of the winding is common to both primary and secondary circuits. (2) Any multiwinding transformer connected in such a manner that it has one electrically continuous winding and part of that winding is common to both primary and secondary circuits. *See:*

common or shunt winding of an autotransformer;
interphase transformer;
main transformer;
Scott-connected transformer assembly;
series winding of an autotransformer;
teaser transformer;
transformer. 42A15-31E12;57A15-21E1

autotransformer, individual-lamp. A series autotransformer that transforms the primary current to a higher or lower current as required for the operation of an individual street light. *See:* **transformer, speciality.** 42A15-31E12

autotransformer starter (industrial control). A starter that includes an autotransformer to furnish reduced voltage for starting of an alternating-current motor. It includes the necessary switching mechanism and it is frequently called a compensator or autostarter. *See also:* **starter.** 42A25-34E10

autotransformer starting (rotating machinery). The process of starting a motor at reduced voltage by connecting the primary winding to the supply initially through an autotransformer and reconnecting the winding directly to the supply at rated voltage for the running conditions. *See:* **asynchronous machine; synchronous machine.** 0-31E8

auxiliaries (switchgear). Accessories to be used with switchgear apparatus but not attached to it, as distinguished from attachments. 37A100-31E11

auxiliary anode (industrial control). An anode located adjacent to the pool cathode in an ignitron to facilitate the maintenance of a cathode spot under conditions adverse to its maintenance by the main anode circuit. *See also:* **electronic controller.** 42A25-34E10

auxiliary burden (capacitance potential device). A variable burden furnished, when required, for adjustment purposes. *See:* **outdoor coupling capacitor.** E31-0

auxiliary device to an instrument. A separate piece of equipment used with an instrument to extend its range, increase its accuracy, or otherwise assist in making a measurement or to perform a function additional to the primary function of measurement. *Note:* For an extensive list of cross references, see *Appendix A.* 42A30-0

auxiliary function (numerically controlled machines). A function of a machine other than the control of the coordinates of a workpiece or tool. Includes functions such as miscellaneous, feed, speed, tool selection, etcetera. *Note:* Not a preparatory function. *See also:* **numerically controlled machines.** EIA3B-34E12

auxiliary generator. A generator, commonly used on electric motive power units, for serving the auxiliary electric power requirements of the unit. *See also:* **traction motor.** 42A42-0

auxiliary generator set. A device usually consisting of a commonly mounted electric generator and a gasoline engine or gas turbine prime mover designed to convert liquid fuel into electric power. *Note:* It provides the aircraft with an electric power supply independent of the aircraft propulsion engines. *See also:* **air transportation electric equipment.** 42A41-0

auxiliary lead (rotating machinery). A conductor joining an auxiliary terminal to the auxiliary device. *See:* **cradle base of rotating machinery.** 0-31E8

auxiliary means. A system element or group of elements that changes the magnitude but not the nature of the quantity being measured to make it more suitable for the primary detector. In a sequence of measurement operations it is usually placed ahead of the primary detector. *See also:* **measurement system.** 42A30-0

auxiliary operation. An operation performed by equipment not under continuous control of the central processing unit. X3A12-16E9

auxiliary or shunt capacitance (capacitance potential device). The capacitance between the network connection and ground, if present. *See also:* **outdoor coupling capacitor.** E31-0

auxiliary power supply (industrial control). A power source supplying power other than load power as required for the proper functioning of a device. *See also:* **electronic controller.** 42A25-34E10

auxiliary power transformer. A transformer having a fixed phase position used for supplying excitation for the rectifier station and essential power for the operation of rectifier equipment auxiliaries. *See also:* **transformer.** 57A18-0

auxiliary relay. A relay whose function is to assist another relay or control device in performing a general function by supplying supplementary actions. *Notes:* (1) Some of the specific functions of an auxiliary relay are:

(A) Reinforcing contact current-carrying capacity of another relay or device.

(B) Providing circuit seal-in functions.

(C) Increasing available number of independent contacts.

(D) Providing circuit-opening instead of circuit-closing contacts or vice-versa.

(E) Providing time delay in the completion of a function.

(F) Providing simple functions for interlocking or programming.

(G) Controls signals, lights, or other secondary circuit functions.

(2) The operating coil or the contacts of an auxiliary relay may be used in the control circuit of another relay or other control device. *Example:* An auxiliary relay may be applied to the auxiliary contact circuits of a circuit breaker in order to coordinate closing and tripping control sequences. (3) A relay that is auxiliary in its functions even though it may derive its driving energy from the power system current or voltage is a form of auxiliary relay. *Example:* A timing relay operating from current or potential transformers. (4) Relays that, by direct response to power system input quantities, assist other relays to respond to such quantities with greater discrimination are *not* auxiliary relays. *Example:* Fault-detector relay. (5) Relays that are limited in function by a control circuit, but are actuated primarily by system input quantities, are *not* auxiliary relays. *Example:* Torque-controlled relays. *See also:* **relay.** 37A100-31E11/31E6; 83A16-0

auxiliary rope-fastening device (elevators). A device attached to the car or counterweight or to the overhead dead-end rope-hitch support that will function automatically to support the car or counterweight in case the regular wire-rope fastening fails at the point of connection to the car or counterweight or at the overhead dead-end hitch. *See also:* **elevators.** 42A45-0

auxiliary secondary terminals. The auxiliary secondary terminals provide the connections to the auxiliary secondary winding, when furnished. *See:* **auxiliary secondary winding.** E31-0

auxiliary secondary winding (capacitance potential device). The auxiliary secondary winding is an additional winding that may be provided in the capacitance potential device when practical considerations permit. *Note:* It is a separate winding that provides a potential that is substantially in phase with the potential of the main winding. The primary purpose of this winding is to provide zero sequence voltage by means of a broken delta connection of three single-phase devices. *See:* **auxiliary secondary terminals.** *See also:* **outdoor coupling capacitor.** E31-0

auxiliary storage (computing machines). A storage that supplements another storage. X3A12-16E9

auxiliary switch. A switch mechanically operated by the main device for signaling, interlocking, or other purposes. *Note:* Auxiliary switch contacts are classed as *a, b, aa, bb, LC,* etcetera, for the purpose of specifying definite contact positions with respect to the main device. 37A100-31E11

auxiliary terminal (rotating machinery). A termination for parts other than the armature or field windings. *See also:* **cradle base (rotating machinery).** 0-31E8

auxiliary winding (single-phase induction motor). A winding that produces poles of a magnetic flux field that are displaced from those of the main winding, that serves as a means for developing torque during starting operation, and that, in some types of design, also serves as a means for improvement of performance during running operation. An auxiliary winding may have a resistor or capacitor in series with it and may be connected to the supply line or across a portion of the main winding. *See also:* **asynchronous machine.** 0-31E8

availability. The fraction of time that the system is actually capable of performing its mission. *See also:* **system.** 0-35E2

availability factor. The ratio of the time a generating unit or piece of equipment is ready for or in service to the total time interval under consideration. *See also:* **generation station.** 42A35-31E13

available. *See also:* **generic term being modified.**

available conversion power gain (conversion transducer). The ratio of the available output-frequency power from the output terminals of the transducer to the available input-frequency power from the driving generator with terminating conditions specified for all frequencies that may affect the result. *Notes:* (1) This applies to outputs of such magnitude that the conversion transducer is operating in a substantially linear condition. (2) The maximum available conversion power gain of a conversion transducer is obtained when the input termination admittance, at input frequency, is the conjugate of the input-frequency driving-point admittance of the conversion transducer. *See also:* **transducer.** E196-0

available current (lightning arresters). *See:* **prospective current.**

available current (circuit) (industrial control) (prospective current). The current that would flow if each pole of the breaking device under consideration were replaced by a link of negligible impedance without any change of the circuit or the supply. *See:* **contactor.** 50I15-34E10

available line. The portion of the scanning line that can be used specifically for picture signals. E168-0

available power efficiency (electroacoustics). Of an electroacoustic transducer used for sound reception, the ratio of the electric power available at the electric terminals of the transducer to the acoustic power available to the transducer. *Notes:* (1) For an electroacoustic transducer that obeys the reciprocity principle, the available power efficiency in sound reception is equal to the transmitting efficiency. (2) In a given narrow frequency band the available power efficiency is numerically equal to the fraction of the open-circuit mean-square thermal noise voltage present at the electric terminals that contributed by thermal noise in the acoustic medium. *See also:* **microphone.** E157-1E1

available power gain (1) (linear transducer). The ratio of the available power from the output terminals of the transducer, under specified input termination conditions, to the available power from the driving generator. *Note:* The maximum available power gain of an electric transducer is obtained when the input termination admittance is the conjugate of the driving-point admittance at the input terminals of the transducer. It is sometimes called completely matched power gain. *See also:* **transmission characteristics; transducer.** E196/E270-0

(2) (two-port linear transducer). At a specified frequency, the ratio of (1) the available signal power from the output port of the transducer, to (2) the available signal power from the input source. *Note:* The available signal power at the output port is a function of the match between the source impedance and the impedance of the input port. *See also:* **network analysis.** E160-0

available power response (electroacoustics) (electroacoustic transducer used for sound emission). The ratio of the mean-square sound pressure apparent at a distance of 1 meter in a specified direction from the effective acoustic center of the transducer to the available electric power from the source. *Notes:* (1) The sound pressure apparent at a distance of 1 meter can be found by multiplying the sound pressure observed at a remote point where the sound field is spherically divergent by the number of meters from the effective acoustic center to that point.

(2) The available power response is a function not only of the transducer but also of some source impedances, either actual or nominal, the value of which must be specified. *See also:* **loudspeaker.** 0-1E1

available short-circuit current (at a given point in a circuit) (prospective short-circuit current). The maximum current that the power system can deliver through a given circuit point to any negligible-impedance short-circuit applied at the given point, or at any other point that will cause the highest current to flow through the given point. *Notes:* (1) This value can be in terms of either symmetrical or asymmetrical; peak or root-mean-square current, as specified. (2) In some resonant circuits the maximum available short-circuit current may occur when the short circuit is placed at some other point than the given one where the available current is measured. 37A100-31E11

available short-circuit test current (at the point of test) (prospective short-circuit test current). The maximum short-circuit current for any given setting of a testing circuit that the test power source can deliver at the point of test, with the test circuit short-circuited by a link of negligible impedance at the line terminals of the device to be tested. *Note:* This value can be in terms of either symmetrical or asymmetrical, peak or root-mean-square current, as specified. 37A100-31E11

available signal-to-noise ratio (at a point in a circuit). The ratio of the available signal power at that point to the available random noise power. *See also:* **signal-to-noise ratio.** 42A65-0

available time (electric drive) (industrial control). The period during which a system has the power turned on, is not under maintenance, and is known or believed to be operating correctly or capable of operating correctly. *See:* **electric drive.** AS1-34E10

avalanche (gas-filled radiation counter tube). The cumulative process in which charged particles accelerated by an electric field produce additional charged particles through collision with neutral gas molecules or atoms. It is therefore a cascade multiplication of ions. *See:* **amplifier.** E160-15E6;42A70-0

avalanche breakdown (semiconductor device). A breakdown that is caused by the cumulative multiplication of charge carriers through field-induced impact ionization. *See also:* **semiconductor; semiconductor device.** E216/E270-34E17

avalanche impedance* (semiconductor). *See:* **breakdown impedance.** *See also:* **semiconductor.**

*Deprecated

average absolute burst magnitude (audio and electroacoustics). The average of the instantaneous burst magnitude taken over the burst duration. See: figure under **burst duration.** *See also:* **burst (audio and electroacoustics).** E257-1E1

average absolute pulse amplitude. The average of the absolute value of the instantaneous amplitude taken over the pulse duration. *See also:* **pulse terms.** E194-0

average absolute value (periodic quantity). The average of the instantaneous absolute value $|x|$ of the quantity over one period:

$$\frac{1}{T}\int_{t}^{t+T}|x|\,\mathrm{d}t$$

where the lower limit of the integral may be any value of t, and the upper limit T greater. *Note:* Average value of a periodic quantity and average absolute value of a periodic quantity are not equivalent, and care must be taken to specify the definition being used. E270-0

average active power (single-phase, two-wire or polyphase circuit) (average power). The time average of the values of active power when the active power varies slowly over a specified period of time (long in comparison with the period of a single cycle). This situation is normally encountered because electric system voltages or currents or both are regularly quasi-periodic. The average active power is readily obtained by dividing the energy flow during the specified period of time, by the time. *See:* **power, active; power, instantaneous.** E270-0

average current (periodic current). The value of the current averaged over a full cycle unless otherwise specified. *See also:* **rectification.** E59-34E17

average detector. A detector, the output voltage of which approximates to the average value of the envelope of an applied signal. *See also:* **electromagnetic compatibility.** CISPR-27E1

average electrode current (electron tubes). The value obtained by integrating the instantaneous electrode current over an averaging time and dividing by the averaging time. *See also:* **electrode current (electron tube).** E160-15E6;42A70-0

average forward-current rating (rectifier circuit). The maximum average value of forward current averaged over a full cycle, permitted by the manufacturer under stated conditions.
See:
continuous rating;
direct-current voltage rating;
maximum surge current rating;
nonrepetitive peak reverse-voltage rating;
peak forward-current rating (repetitive);
rated output current;
rated output voltage;
repetitive peak reverse-voltage rating;
root-mean-square reverse-voltage rating;
short-time rating;
temperature derating;
working peak reverse-voltage rating.
E59-34E17/34E24

average information content (per symbol) (information rate from a source, per symbol). The average of the information content per symbol emitted from a source. *Note:* The terms entropy and negentropy are sometimes used to designate average information content. *See also:* **information theory.** E171-0

average inside air temperature (enclosed switchgear). The average temperature of the surrounding cooling air that comes in contact with the heated parts of the apparatus within the enclosure. 37A100-31E11

average magner. *See:* **average reactive power (single-phase, two-wire, or polyphase circuit).**

average noise factor. *See:* **average noise figure.**

average noise figure (transducer) (average noise factor). The ratio of total output noise power to the portion thereof attributable to thermal noise in the input termination, the total noise being summed over frequencies from zero to infinity, and the noise temperature of the input termination being standard (290 kelvins). *See:* **noise figure.** *See also:* **signal-to-noise ratio.** 42A65-0

average picture level (APL) (television). The average signal level, with respect to the blanking level, during the active picture scanning time (averaged over a frame period, excluding blanking intervals) expressed as a percentage of the difference between the blanking and reference white levels. *See:* **television.** 0-2E2/42A65

average power. *See:* **average active power (single-phase, two-wire or polyphase circuit).**

average power factor (single-phase, two-wire, or a polyphase circuit, when the voltages and currents are sinusoidal). The ratio of the average active power to the amplitude of the average phasor power, for a specified period of time. E270-0

average power output (amplitude-modulated transmitter). The radio-frequency power delivered to the transmitter output terminals averaged over a modulation cycle. *See also:* **radio transmitter.** E145/42A650-0

average pulse amplitude. The average of the instantaneous amplitude taken over the pulse duration. *See also:* **pulse terms.** E-194/E270-0

average reactive factor (single-phase, two-wire, or a polyphase circuit, when the voltages and currents are sinusoidal). The ratio of the average reactive power to the amplitude of the average phasor power, for a specified period of time. E270-0

average reactive power (single-phase, two-wire, or polyphase circuit) (average magner). The time average of the values of reactive power when the reactive power varies slowly over a specified period of time (long in comparison with the period of a single cycle). This situation is normally encountered because electric power system voltages or currents, or both, are regularly quasi-periodic. *Note:* The average reactive power is readily obtained by dividing the quadergy flow during the specified period, by the time. *See:* **power, instantaneous (two-wire circuit)** E270-0

average transinformation (output symbols and input symbols). Transinformation averaged over the ensemble of pairs of transmitted and received symbols. *See also:* **information theory.** E171-0

average value of a function. The average value of a function, over an interval $a \leq x \leq b$, is

$$\frac{1}{b-a}\int_{a}^{b} f(x)\mathrm{d}x$$

E270-0

average value of a periodic quantity. The average of the instantaneous value x of the quantity over one period:

$$\frac{1}{T}\int_{t}^{t+T} x \, dt$$

where the lower limit of the integral may be any value of t, and the upper limit T greater. *See also:* **average absolute value (periodic quantity).** E270-0

average voltage (1) (periodic voltage). The value of the voltage averaged over a full cycle unless otherwise specified. *See also:* **battery; rectification.** E59-34E17

(2) (storage battery) (storage cell). The average value of the voltage during the period of charge or discharge. It is conveniently obtained from the time integral of the voltage curve. *See also:* **battery.** 42A60-0

averaging time, electrode current. *See:* **electrode current averaging time.**

axial ratio. The ratio of the major axis to the minor axis of the polarization ellipse. *Note:* This is preferred to ellipticity because mathematically ellipticity is 1 minus the reciprocal of the axial ratio. *See also:* **radio transmission.** E146/3E1-0

axially extended interaction tube (microwave tubes). A klystron tube utilizing an output circuit having more than one gap. *See also:* **microwave tube or valve.** 0-15E6

axis. *See:* **magnetic axis.**

axle bearing. A bearing that supports a portion of the weight of a motor or a generator on the axle of a vehicle. *See:* **bearing; traction motor.** 42A42-0

axle-bearing cap. The member bolted to the motor frame supporting the bottom half of the axle bearing.- *See:* **bearing.** 42A10-0

axle-bearing-cap cover. A hinged or otherwise applied cover for the waste and oil chamber of the axle bearing. *See:* **bearing.** 42A10-0

axle circuit. The circuit through which current flows along one of the track rails to the train, through the wheels and axles of the train, and returns to the source along the other track rail. 42A42-0

axle current. The electric current in an axle circuit. 42A42-0

axle generator. An electric generator designed to be driven mechanically from an axle of a vehicle. *See also:* **axle-generator system.** 42A42-0

axle-generator pole changer. A mechanically or electrically actuated changeover switch for maintaining constant polarity at the terminals of an axle generator when the direction of the rotation of the armature is reversed due to a change in direction of movement of a vehicle on which the generator is mounted. *See also:* **axle-generator system.** 42A42-0

axle-generator regulator. A control device for automatically controlling the voltage and current of a variable-speed axle generator. *See also:* **axle-generator system.** 42A42-0

axle-generator system. A system in which electric power for the requirements of a vehicle is supplied from an axle generator carried on the vehicle, supplemented by a storage battery.
See:
axle generator;
axle-generator pole changer;
axle-generator regulator;
battery-current regulation;
body generator suspension;
combination current and voltage regulation;
constant-voltage regulation;
controlled-speed axle generator;
engine-generator system;
head-end system;
straight storage system;
total-current regulation;
truck generator system suspension;
variable-speed axle generator. 42A42-0

axle-hung motor (or generator). A traction motor (or generator), a portion of the weight of which is carried directly on the axle of a vehicle by means of axle bearings. *See also:* **traction motor.** 42A42-0

azimuth (electronic navigation). (1) The direction of a celestial point from a terrestrial point, expressed as the angle in the horizontal plane between a reference line and the horizontal projection of the line joining the two points. *Note:* True north is usually but not always implied when no reference direction is stated. (2) In directional antenna systems, the pointing direction of an antenna, expressed as the angle in the horizontal plane between a reference direction and the horizontal component of the pointing direction of the boresight of the antenna. (3) Bearing. *See also:* **navigation; sunlight.** E172-10E6

azimuth discrimination* (electronic navigation). *See:* **angular resolution.**

*Deprecated

azimuth marker (electronic navigation). *See:* **calibration mark.**

azimuth-stabilized plan-position indicator (electronic navigation). A plan-position indicator on which the reference bearing remains fixed with respect to the indicator, regardless of the vehicle orientation. *See also:* **radar.** E172-10E6;42A65-0

B

babble. The aggregate crosstalk from a large number of interfering channels. *See also:* **signal-to-noise ratio.** E151-0

back (motor or generator) (turbine or drive end). The end that carries the largest coupling or driving pulley. *See:* **armature.** 0-31E8

back contact (relay). *See:* ***b*** **contact.** *See also:* **relay** entries.

back course (instrument landing systems). The course that is located on the opposite side of the localizer from the runway. *See also:* **navigation.** 0-10E6

back flashover (lightning). A flashover of insulation resulting from a lightning stroke to part of a network or electric installation that is normally at ground potential. *See also:* **direct-stroke protection (lightning).** 0-31E13

back light (television). Illumination from behind the subject in a direction substantially parallel to the plane through the optical axis of the camera. *See also:* **television.** Z7A1-0

back plate (signal plate) (camera tubes). The electrode in an iconoscope or orthicon camera tube to which the stored charge image is capacitively coupled. *See:* **beam tubes; television.** 0-2E2

back porch (1) (monochrome composite picture signal). The portion that lies between the trailing edge of a horizontal synchronizing pulse and the trailing edge of the corresponding blanking pulse.

(2) (National Television System Committee composite color-picture signal). The portion that lies between the color burst and the trailing edge of the corresponding blanking pulse. *See also:* **television.** 0-2E2

back scatter (primary radar). Energy reflected with a direction component toward the transmitter. *See also:* **navigation.** 0-10E6

back-scattering cross section. The scattering cross-section in the direction towards the source. *See:* **back-scattering coefficient; monostatic cross section; radar cross section; target echoing area.** *See also:* **antenna.** 0-3E1

back-shunt keying. A method of keying a transmitter in which the radio-frequency energy is fed to the antenna when the telegraph key is closed and to an artificial load when the key is open. *See also:* **radio transmission.** E145/E182A-42A65

back wave. A signal emitted from a radio telegraph transmitter during spacing portions of the code characters and between the code characters. *See also:* **radio transmission.** E182A/42A65-0

backed stamper (phonograph techniques) (mechanical recording). A thin metal stamper that is attached to a backing material, generally a metal disk of desired thickness. *See also:* **phonograph pickup.** S1A1-1E1

background counts (radiation counters). Counts caused by ionizing radiation coming from sources other than that to be measured; and by any electric disturbance in the circuitry that is used to record the counts. *See also:* **ionizing radiation.** E175/42A70-0;0-15E6

background ionization voltage (lightning arresters). A high-frequency voltage appearing at the terminals of the apparatus to be tested that is generated by ionization extraneous to the apparatus. *Note:* While this voltage does not add arithmetically to the radio influence or internal ionization voltage, it affects the sensitivity of the test. *See:* **lightning arrester (surge diverter).** LA1-31E7

background noise (1) (general). Noise due to audible disturbances of periodic and/or random occurence. *See also:* **transmission characteristics; modulation.** E145-0

(2) (electroacoustics) (recording and reproducing). The total system noise in the absence of a signal. *See also:* **transmission characteristics; phonograph pickup.** 0-1E1

(3) (receivers). The noise in the absence of signal modulation on the carrier. E188-0

background response (radiation detectors). Response caused by ionizing radiation coming from sources other than that to be measured. *See also:* **ionizing radiation.** E175-0

background returns (radar). *See:* **clutter.**

backing (planar structure) (rotating machinery). A fabric, mat, film, or other material used in intimate conjunction with a prime material and forming a part of the composite for mechanical support or to sustain or improve its properties. *See:* **cradle base (rotating machinery).** 0-31E8

backing lamp (backing light) (backup light). *See:* **backup lamp.** 42A42-0

backlash (1) (general). A relative movement between interacting mechanical parts, resulting from looseness. *See also:* **control system, feedback; industrial control; numerically controlled machines.** 85A1-34E12; AS1-34E10

(2) (signal generator). The difference in actual value of a parameter when the parameter is set to an indicated value by a clockwise rotation of the indicator, and when it is set by a counterclockwise rotation. *See also:* **signal generator.** 0-9E4

(3) (tunable microwave tube). The amount of motion of the tuner control mechanism (in a mechanically tuned oscillator) that produces no frequency change upon reversal of the motion. *See also:* **tunable microwave oscillators.** E158-15E6

backstop, relay. *See:* **relay backstop.**

backup lamp. A lighting device mounted on the rear of a vehicle for illuminating the region near the back of the vehicle while moving in reverse. *Note:* It normally can be used only while backing up. *See also:* **headlamp.** Z7A1-0

backup protection (relay system). A form of protection that operates independently of specified components in the primary protective system and that is intended to operate if the primary protection fails or is temporarily out of service. 37A100-31E11/31E6

backward-acting regulator. A transmission regulator in which the adjustment made by the regulator affects the quantity that caused the adjustment. *See also:* **transmission regulator.** 42A65-0

backward wave (traveling-wave tubes). A wave whose group velocity is opposite to the direction of electron-stream motion. *See:* **amplifier.** *See also:* **electron devices, miscellaneous .** E160-15E6

backward-wave oscillator. *See:* **carcinotron.**

backward-wave (BW) structure (microwave tubes). A slow-wave structure whose propagation is characterized on an ω/β diagram (sometimes called a Brillouin diagram) by a negative slope in the region $0<\beta<\pi$ (in which the phase velocity is therefore of opposite sign to the group velocity). *See also:* **microwave tube (or valve).** O-15E6

bactericidal (germicidal) effectiveness (illuminating engineering). The capacity of various portions of the ultraviolet spectrum to destroy bacteria, fungi, and viruses. *See also:* **ultraviolet radiation.** Z7A1-0

bactericidal (germicidal) efficiency (radiant flux). For a particular wavelength, the ratio of the bactericidal effectiveness of that wavelength to that of wavelength 265.0 nanometers, which is rated as unity. *Note:*

Tentative bactericidal efficiency of various wavelengths of radiant flux is given in the following table.

Erythemal and bactericidal efficiency of ultraviolet radiation

Wavelength (nanometers)	Erythemal Efficiency	Tentative Bactericidal Efficiency
*235.3	------	0.35
240.0	0.56	--------
*244.6	0.57	0.58
*248.2	0.57	0.70
250.0	0.57	--------
*253.7	0.55	0.85
*257.6	0.49	0.94
260.0	0.42	--------
265.0	------	1.00
*265.4	0.25	0.99
*267.5	0.20	0.98
*270.0	0.14	0.95
*275.3	0.07	0.81
*280.4	0.06	0.68
285.0	0.09	--------
*285.7	0.10	0.55
*289.4	0.25	0.46
290.0	0.31	--------
*292.5	0.70	0.38
295.0	0.98	--------
*296.7	1.00	0.27
300.0	0.83	--------
*302.2	0.55	0.13
305.0	0.33	--------
310.0	0.11	--------
*313.0	0.03	0.01
315.0	0.01	--------
320.0	0.005	--------
325.0	0.003	--------
330.0	0.000	--------

*Emission lines in the mercury spectrum; other values interpolated.

See also: **ultraviolet radiation.** Z7A1-0

bactericidal (germicidal) exposure. The product of bactericidal flux density on a surface and time. It usually is measured in bactericidal microwatt-minutes per square centimeter or bactericidal watt-minutes per square foot. *See also:* **ultraviolet radiation.** Z7A1-0

bactericidal (germicidal) flux. Radiant flux evaluated according to its capacity to produce bactericidal effects. It usually is measured in microwatts of ultraviolet radiation weighted in accordance with its bactericidal efficiency. Such quantities of bactericidal flux would be in bactericidal microwatts. *Note:* Ultraviolet radiation of wavelength 253.7 nanometers usually is referred to as "ultraviolet microwatts" or "UV watts." These terms should not be confused with "bactericidal microwatts." *See also:* **ultraviolet radiation.** Z7A1-0

bactericidal (germicidal) flux density. The bactericidal flux per unit area of the surface being irradiated. *Note:* It is equal to the quotient of the incident bactericidal flux divided by the area of the surface when the flux is uniformly distributed. It usually is measured in microwatts per square centimeter or watts per square foot of bactericidally weighted ultraviolet radiation (bactericidal microwatts per square centimeter or bactericidal watts per square foot). *See also:* **ultraviolet radiation.** Z7A1-0

baffle (1) (audio and electroacoustics). A shielding structure or partition used to increase the effective length of the transmission path between two points in an acoustic system as for example, between the front and back of an electroacoustic transducer. *Note:* In the case of a loudspeaker, a baffle is often used to increase the acoustic loading of the diaphragm. *See also:* **loudspeaker.** 0-1E1

(2) (light sources). A single opaque or translucent element to shield a source from direct view at certain angles or to absorb unwanted light. *See also:* **bare (exposed) lamp.** Z7A1-0

(3) (gas tube). An auxiliary member, placed in the arc path and having no separate external connection. *Note:* A baffle may be used for: (A) controlling the flow of mercury vapor or mercury particles; (B) controlling the flow of radiant energy; (C) forcing a distribution of current in the arc path; or (D) deionizing the mercury vapor following conduction. It may be of either conducting or insulating material. *See also:* **electrode (electron tube).** 42A70-15E6

bag-type construction (dry cell) (primary cell). A type of construction in which a layer of paste forms the principal medium between the depolarizing mix, contained within a cloth wrapper, and the negative electrode. *See also:* **electrolytic cell.** 42A60-0

balance beam (relay). A lever form of relay armature, one end of which is acted upon by one input and the other end restrained by a second input. 37A100-31E11/31E6

balance check (operational amplifiers). The computer control state in which all amplifier summing functions are connected to the computer zero reference level (usually signal ground) to permit zero balance of the operational amplifier. *Note:* Integrator capacitors may be shunted by a resistor to permit the zero balance of an integrator. *See also:* **electronic analog computer.** E165-16E9

balanced (1) (general). Used to signify proper relationship between two or more things, such as stereophonic channels.

(2) (communication practice). Usually signifies (A) electrically alike and symmetrical with respect to a common reference point, usually ground, or (B) arranged to provide conjugacy between certain sets of terminals. E151/42A65-31E3

balanced amplifier (push-pull amplifier). *See:* **amplifier, balanced.**

balanced capacitance (between two conductors) (mutual capacitance between two conductors*). The capacitance between two conductors when the changes in the charges on the two are equal in magnitude but opposite in sign and the potentials of the other $n-2$ conductors are held constant. *See:* **direct capacitances (system of conductors).**

*Deprecated. E270-9E4

balanced circuit (1) (signal-transmission system). A circuit, in which two branches are electrically alike and symmetrical with respect to a common reference point, usually ground. *Note:* For an applied signal difference at the input, the signal relative to the reference at equivalent points in the two branches must be opposite in polarity and equal in amplitude. 0-9E4

(2) (electric power system). A circuit in which there are substantially equal currents, either alternating or direct, in all main wires and substantially equal voltages between main wires and between each main wire and neutral (if one exists). *See also:* **center of distribution.** 42A35-31E13

balanced conditions (1) (time domain) (rotating machinery). A set of polyphase quantities (phase currents, phase voltages, etcetera) that are sinusoidal in time, that have identical amplitudes, and that are shifted in time with respect to each other by identical phase angles.
(2) (space domain). In space, a set of coils (for example, of a rotating machine) each having the same number of effective turns, with their magnetic axes shifted by identical angular displacements with respect to each other. *Notes:* (1) The impedance (matrix) of a balanced machine is balanced. A balanced set of currents will produce a balanced set of voltage drops across a balanced set of impedances. (2) If all sets of windings of a machine are balanced and if the magnetic structure is balanced, the machine is balanced. *See also:* **asynchronous machine; direct-current commutating machine; synchronous machine.** 0-31E8

balanced currents (on a balanced line). Currents flowing in the two conductors of a balanced line that, at every point along the line, are equal in magnitude and opposite in direction. *See also:* **transmission line.** E146-3E1

balanced line. A transmission line consisting of two single or two interconnected groups of conductors capable of being operated in such a way that when the voltages of the two groups of conductors at any transverse plane are equal in magnitude and opposite in polarity with respect to ground, the total currents along the two groups of conductors are equal in magnitude and opposite in direction; for example, a line in which the two conductors are identical in cross section and symmetrically positioned with respect to the electric ground. *Note:* A balanced line may be operated under unbalanced conditions and the aggregate then does not form a balanced line system. *See also:* **transmission line.** E146-3E1/9E4

balanced line system. A system consisting of generator, balanced line, and load adjusted so that the voltages of the two conductors at all transverse planes are equal in magnitude and opposite in polarity with respect to ground. *Note:* Balanced line system is frequently shortened to balanced line. Care should be taken not to confuse this abbreviated terminology with the standard definition of balanced line. *See also:* **transmission line.** E146-3E1

balanced mixer. A hybrid junction with crystal receivers in one pair of uncoupled arms, the arms of the remaining pair being fed from a signal source and a local oscillator. *Note:* The resulting intermediate-frequency signals from the crystals are added in such a manner that the effect of local-oscillator noise is minimized. *See:* **converter; hybrid junction; radio receiver; waveguide.** 50I62-3E1

balanced modulator (signal-transmission system). A modulator, specifically a push-pull circuit, in which the carrier and modulating signal are so introduced that after modulation takes place the output contains the two sidebands without the carrier. *See also:* **modulation; modulator, symmetrical.** E145-2E2

balanced oscillator. An oscillator in which at the oscillator frequency the impedance centers of the tank circuit are at ground potential and the voltages between either end and their centers are equal in magnitude and opposite in phase. *See also:* **oscillatory circuit.** E145/E182A/42A65-0

balanced polyphase load. A load to which symmetrical currents are supplied when it is connected to a system having symmetrical voltages. *Note:* The term balanced polyphase load is applied also to a load to which are supplied two currents having the same wave form and root-mean-square value and differing in phase by 90 electrical degrees when it is connected to a quarter-phase (or two-phase) system having voltages of the same wave form and root-mean-square value. *See also:* **generating station.** 42A35-31E13

balanced polyphase system. A polyphase system in which both the currents and voltages are symmetrical. *See also:* **alternating-current distribution.** 42A35-31E13

balanced telephone-influence factor (three-phase synchronous machine). The ratio of the square root of the sum of the squares of the weighted root-mean-square values of the fundamental and the non-triple series of harmonics to the root-mean-square value of the normal no-load voltage wave. *See:* **synchronous machine.** 42A10-31E8

balanced termination (system or network having two output terminals). A load presenting the same impedance to ground for each of the output terminals. *See also:* **network analysis.** E146-3E1; 0-9E4

balanced three-wire system. A three-wire system in which no current flows in the conductor connected to the neutral point of the supply. *See:* **three-wire system.** *See also:* **alternating-current distribution.** 42A35-31E13

balanced voltages (1) (on a balanced line). Voltages (relative to ground) on the two conductors of a balanced line that, at every point along the line, are equal in magnitude and opposite in polarity. *See also:* **transmission line.** E/146-3E1
(2) (signal-transmission system). The voltages between corresponding points of a balanced circuit (voltages at a transverse plane) and the reference plane relative to which the circuit is balanced. *See also:* **signal.** 0-13E6

balanced wire circuit. A circuit whose two sides are electrically alike and symmetrical with respect to ground and other conductors. *Note:* The term is commonly used to indicate a circuit whose two sides differ only by chance. *See also:* **transmission line.** 42A65-31E3

balancer. That portion of a direction-finder that is used for the purpose of improving the sharpness of the direction indication. *See also:* **radio receiver.** 42A65-0

balance relay. A relay that operates by comparing the magnitudes of two similar input quantities. *Note:* The balance may be effected by counteracting electromagnetic forces on a common armature, or by counteracting magnetomotive forces in a common magnetic circuit, or by similar means, such as springs, levers, etcetera. 37A100-31E11/31E6

balance test (rotating machinery). A test taken to enable a rotor to be balanced within specified limits. *See also:* **rotor (rotating machinery).** 0-31E8

balancing network. An electric network designed for use in a circuit in such a way that two branches of the circuit are made substantially conjugate, that is, such that an electromotive force inserted in one branch produces no current in the other branch. *See also:* **network analysis.** 42A65-31E3

balancing of an operational amplifier. The act of adjusting the output level of an operational amplifier to coincide with its input reference level, usually ground or zero voltage. *See also:* **electronic analog computer.** E165-16E9

ballast (1) (general). An impedor connected in series with a circuit or device, such as an arc, that normally is unstable when connected across a constant-voltage supply, for the purpose of stabilizing the current. E270-0

(2) (fluorescent lamps or mercury lamps). Devices that by means of inductance, capacitance, or resistance, singly or in combination, limit the lamp current of fluorescent or mercury lamps, to the required value for proper operation, and also, where necessary, provide the required starting voltage and current and, in the case of ballasts for rapid-start lamps, provide for low-voltage cathode heating. *Note:* Capacitors for power-factor correction and capacitor-discharge resistors may form part of such a ballast. 82A1/82A4-0

(3) (fixed-impedance type) (reference ballast). Designed for use with one specific type of lamp that, after adjustment during the original calibration, is expected to hold its established impedance throughout normal use. 82A3/82A9-0

(4) (variable-impedance type). An adjustable inductive reactor and a suitable adjustable resistor in series. *Note:* These two components are usually designed so that the resulting combination has sufficient current-carrying capacity and range of impedance to be used with a number of different sizes of lamps. The impedance and power factor of the reactor-resistor combination are adjusted, or rechecked, each time the unit is used. 82A3/82A9-0

ballast leakage. The leakage of current from one rail of a track circuit to another through the ballast, ties, earth, etcetera. *See also:* **railway signal and interlocking.** 42A42-0

ballast resistance. The resistance offered by the ballast, ties, earth, etcetera, to the flow of leakage current from one rail of a track circuit to another. *See also:* **railway signal and interlocking.** 42A42-0

ballast tube (ballast lamp). A current-controlling resistance device designed to maintain substantially constant current over a specified range of variation in the applied voltage or the resistance of a series circuit. *See also:* **circuits and devices.** 42A65-0

ball bearing (rotating machinery). A bearing incorporating a peripheral assembly of balls. *See also:* **bearing.** 0-31E8

ball burnishing. Burnishing by means of metal balls. *See also:* **electroplating.** 42A60-0

ballistic focusing (microwave tubes). A focusing system in which static electric fields cause an initial convergence of the beam and the electron trajectories are thereafter determined by momentum and space charge forces only. *See also:* **microwave tube or valve.** 0-15E6

ball lightning. A type of lightning discharge reported from visual observations to consist of luminous, ball-shaped regions of ionized gases. *Note:* In reality ball lightning may or may not exist. *See also:* **direct-stroke protection (lightning).** 0-31E13

balun. (1) A passive device having distributed electrical constants used to couple a balanced system or device to an unbalanced system or device. *Note:* The term is derived from **balance to unbalance transformer.** 0-21E1

(2) A network for the transformation from an unbalanced transmission line or system to a balanced line or system, or vice versa. 0-9E4

banana plug. A single-conductor plug with a spring metal tip that somewhat resembles a banana in shape. *See also:* **circuits and devices.** 42A65-21E0

band (electronic computers). A group of circular recording tracks, on a moving storage device such as a drum or disc. X3A12-16E9

band, effective (facsimile). The frequency band of a facsimile signal wave equal in width to that between zero frequency and maximum keying frequency. *Note:* The frequency band occupied in the transmission medium will in general be greater than the effective band. *See also:* **facsimile signal (picture signal).** E168-0

band-elimination filter. *See:* **filter, band-elimination.**

banding insulation (rotating machinery). Insulation between the end winding-overhang and the binding bands. *See also:* **rotor (rotating machinery); stator.** 0-31E8

band of regulated voltage. The band or zone, expressed in percent of the rated value of the regulated voltage, within which the excitation system will hold the regulated voltage of an electric machine during steady or gradually changing conditions over a specified range of load. *See:* **direct-current commutating machine; synchronous machine.** 42A10-0

band-pass filter. *See:* **filter, band-pass.**

bandpass tube (microwave gas tubes). *See:* **broadband tube.**

band spreading. (1) The spreading of tuning indications over a wide scale range to facilitate tuning in a crowded band of frequencies. (2) The method of double-sideband transmission in which the frequency band of the modulating wave is shifted upward in frequency so that the sidebands produced by modulation are separated in frequency from the carrier by an amount at least equal to the bandwidth of the original modulating wave, and second-order distortion products may be filtered from the demodulator output. *See also:* **radio receiver.** 42A65-0

band switch. A switch used to select any one of the frequency bands in which an electric transmission apparatus may operate. *See also:* **circuits and devices.** 42A65-0

bandwidth (1) (continuous frequency band). The difference between the limiting frequencies. E270-0

(2) (device). The range of frequencies within which performance, with respect to some characteristic, falls within specific limits. *See also:* **radio receiver.**

(3) (wave). The least frequency interval outside of which the power spectrum of a time-varying quantity is everywhere less than some specified fraction of its value at a reference frequency. *Warning:* This definition permits the spectrum to be less than the specified fraction within the interval. *Note:* Unless otherwise stated, the reference frequency is that at which the spectrum has its maximum value. E188-0

(4) **(burst) (burst measurements)** The smallest frequency interval outside of which the integral of the energy spectrum is less than some designated fraction of the total energy of the burst. *See also:* **burst.** E265-0

(5) **(antenna).** The range of frequencies within which its performance, in respect to some characteristic, conforms to a specified standard. *See also:* **antenna.** E145-3E1

(6) **(facsimile system).** The difference in hertz between the highest and the lowest frequency components required for adequate transmission of the facsimile signals. *See also:* **facsimile (electrical communication).** E168-0

(7) **(industrial control).** The interval separating two frequencies between which both the gain and the phase difference (of sinusodial output referred to sinusoidal input) remain within specified limits. *Note:* For control systems and many of their components, the lower frequency often approaches zero. *See also:* **control system, feedback.** 85A1-23E0;AS1-34E10

(8) **(signal-transmission system).** The range of frequencies within which performance, with respect to some characteristic, falls within specific limits. *Note:* For systems capable of transmitting at zero frequency: the frequency at which the system response is less than that at zero frequency by a specified ratio. For carrier-frequency systems: the difference in the frequencies at which the system response is less than that at the frequency of reference response by a specified ratio. For both types of systems, bandwidth is commonly defined at the points where the response is 3 decibels less than the reference value (0.707 root-mean-square voltage ratio). *See:* **equivalent noise bandwidth.** E188-13E6

(9) **(oscilloscope).** The difference between the upper and lower frequency at which the response is 0.707 (−3 decibels) of the response at the reference frequency. Usually both upper and lower limit frequencies are specified rather than the difference between them. When only one number appears, it is taken as the upper limit. *Notes:* (1) The reference frequency shall be at least 20 times greater for the lower bandwidth limit and at least 20 times less for the upper bandwidth limit than the limit frequency. The upper and lower reference frequencies are not required to be the same. In cases where exceptions must be made, they shall be noted. (2) This definition assumes the amplitude response to be essentially free of departures from a smooth roll-off characteristic. (3) If the lower bandwidth limit extends to zero frequency, the response at zero frequency shall be equal to the response at the reference frequency, not −3 decibels from it. 0-9E4

bandwidth, effective (bandpass filter in a signal transmission system). The width of an assumed rectangular bandpass filter having the same transfer ratio at a reference frequency and passing the same mean square of a hypothetical current and voltage having even distribution of energy over all frequencies. *Note:* For a nonlinear system, the bandwidth at a specified input level. *See also:* **network analysis; signal.** E145-13E6

bank. An aggregation of similar devices (for example, transformers, lamps, etcetera) connected together and used in cooperation. *Note:* In automatic switching, a bank is an assemblage of fixed contacts over which one or more wipers or brushes move in order to establish electric connections. *See:* **relay level.** 42A65-0

bank-and-wiper switch. A switch in which an electromagnetic ratchet or other mechanisms are used, first, to move the wipers to a desired group of terminals, and second, to move the wipers over the terminals of this group to the desired bank contacts. *See also:* **telephone switching system.** 42A65-0

banked winding. *See:* **bank winding.**

bank winding (banked winding). A compact multilayer form of coil winding, for the purpose of reducing distributed capacitance, in which single turns are wound successively in each of two or more layers, the entire winding proceeding from one end of the coil to the other, without return. *See also:* **circuits and devices.** 42A65-21E0

bar (lights). A group of three or more aeronautical ground lights placed in a line transverse to the axis, or extended axis, of the runway. *See also:* **signal lighting.** Z7A1-0

bare conductor. A conductor not covered with insulating material. 42A35-31E13

bare (exposed) lamp. A light source with no shielding.
See:
baffle;
bowl;
cutoff angle;
diffusing panel;
globe;
lamp shield angle;
louver;
louver shielding angle;
luminaire;
matte surface;
reflector;
refractor;
shade;
shielding angle;
specular surface.
See also: **light.** 27A1-0

barette. A short bar in which the lights are closely spaced so that from a distance they appear to be a linear light. *Note:* Barettes are usually less than 15 feet in length. *See:* **bar (lights).** *See also:* **signal lighting.** Z7A1-0

Barkhausen-Kurz oscillator. An oscillator of the retarding-field type in which the frequency of oscillation depends solely upon the electron transit-time within the tube. *See also:* **oscillatory circuit.** E145-0

Barkhausen tube. *See:* **positive-grid oscillator tube.**

bar generator (television). A generator of pulses that are uniformly spaced in time and are synchronized to produce a stationary bar pattern on a television screen. *See also:* **television.** E188-0

bar pattern (in television). A pattern of repeating lines or bars on a television screen. When such a pattern is produced by pulses that are equally separated in time, the spacing between the bars on the television screen can be used to measure the linearity of the horizontal or vertical scanning systems. *See also:* **television.** 42A65-0

barrel plating. Mechanical plating in which the cathodes are kept loosely in a container that rotates. *See also:* **electroplating.** 42A60-0

barretter. *See:* **bolometric detector.**

barrier. (1) A partition for the insulation or isolation of electric circuits or electric arcs. 37A100-31E11

(2) (in a semiconductor) (obsolete). *See:* **depletion layer.**

barrier grid (charge-storage tubes). A grid, close to or in contact with a storage surface, which grid establishes an equilibrium voltage for secondary-emission charging and serves to minimize redistribution. *See also:* **charge-storage tube.** E158-15E6

barring hole (rotating machinery). A hole in the rotor to permit insertion of a pry bar for the purpose of turning the rotor slowly or through a limited angle. *See also:* **rotor (rotating machine).** 0-31E8

bar, rotor (rotating machinery). *See:* **rotor bar.**

base (1) (number system). An integer whose successive powers are multiplied by coefficients in a positional notation system. *See:* **electronic digital computer; positional notation; radix.** E162-0

(2) (rotating machinery). A structure, normally mounted on the foundation, that supports a machine or a set of machines. In single-phase machines rated up through several horsepower, the base is normally a part of the machine and supports it through a resilient or rigid mounting to the end shields. *See also:* **cradle base (rotating machinery).** 0-31E8

(3) (electron tube or valve). The part attached to the envelope, carrying the pins or contacts used to connect the electrodes to the external circuit and that plugs in to the holder. *See also:* **electron tube.** 50I07-15E6

(4) (basis or base metals) (electroplating). The object upon which the metal is electroplated. *See also:* **electroplating.** 42A60-0

(5) (transistor). A region that lies between an emitter and a collector of a transistor and into which minority carriers are injected. *See also:* **transistor.** E270-0; E216-34E17

base address. A given address from which an absolute address is derived by combination with a relative address. X3A12-16E9

base ambient temperature (power distribution underground cables) (cable or duct). The no-load temperature in a group with no load on any cable or duct in the group. *See also:* **power distribution, underground construction.** 0-31E1

base electrode (transistor). An ohmic or majority-carrier contact to the base region. *See also:* **transistor base region; base.** E102/42A70-0

base light (television). Uniform, diffuse illumination approaching a shadowless condition, that is sufficient for a television picture of technical acceptibility, and that may be supplemented by other lighting. *See also:* **television.** Z7A1-0

base load (electric power utilization). The minimum load over a given period of time. *See also:* **generating station.** 42A35-31E13/31E4

base load control (electric generating unit or station). A mode of operation in which the unit or station generation is held constant. *See also:* **speed-governing system.** E94-0

base-minus-ones complement. A number representation that can be derived from another by subtracting each digit from one less than the base. Nines complements and ones complements are base-minus-ones complements. *See also:* **electronic digital computer.** E162-0

base-mounted electric hoist. A hoist similar to an overhead electric hoist except that it has a base or feet and may be mounted overhead, on a vertical plane, or in any position for which it is designed. *See also:* **hoist.** 42A45-0

base region (transistor). The interelectrode region of a transistor into which minority carriers are injected. *See also:* **transistor.** E102/42A70-0

base speed of an adjustable-speed motor. The lowest rated speed obtained at rated load and rated voltage at the temperature rise specified in the rating. *See:* **asynchronous machine; direct-current commutating machine; synchronous machine.** 42A10-0

base station (mobile communication). A land station in the land-mobile service carrying on a radio communication service with mobile and fixed radio stations. *See also:* **mobile communication system.** 0-6E1

base value (rotating machinery). A normal or nominal or reference value in terms of which a quantity is expressed in per unit or percent. *See also:* **asynchronous machine; direct-current commutating machines; synchronous machine.** 0-31E8

baseband (carrier or subcarrier wire or radio transmission system). The band of frequencies occupied by the signal before it modulates the carrier (or subcarrier) frequency to form the transmitted line or radio signal. *Note:* The signal in the baseband is usually distinguished from the line or radio signal by ranging over distinctly lower frequencies, which at the lower end relatively approach or may include direct current (zero frequency). In the case of a facsimile signal before modulation on a subcarrier, the baseband includes direct current. *See also:* **facsimile transmission.** E168-2E2

baseline (1) (electronic navigation). The line joining the two points between which electrical phase or time is compared in determining navigation coordinates; for two ground stations, this is the great circle joining the two stations, and, in the case of a rotating collector system, it is the line joining the two sides of the collector. *See also:* **navigation.** E172-10E6

(2) (pulse techniques). That amplitude level from which the pulse waveform appears to originate. *See also:* **pulse.** 0-9E4

baseline delay (loran). The time interval needed for a signal from a loran master station to travel to the slave station. *See also:* **navigation.** 0-10E6

baseline offset (pulse techniques). The algebraic difference between the amplitude of the baseline and the amplitude reference level. *See also:* **pulse.** 0-9E4

baseline overshoot (pulse techniques). *See:* **distortion, pulse.**

basic alternating voltage (power rectifier). The sustained sinusoidal voltage that must be impressed on the terminal of the alternating-current winding of the rectifier transformer, when set on the rated voltage tap, to give rated output voltage at rated load with no phase control. *See also:* **rectification.** 42A15-0

basic device (supervisory system). *See:* **common device.**

basic element (measurement system). A measurement component or group of components that performs one necessary and distinct function in a sequence of measurement operations. *Note:* Basic elements are single-purpose units and provide the smallest steps into which the measurement sequence can be classified conveniently. Typical examples of basic elements are: a permanent magnet, a control spring, a coil, and a pointer and scale. *See:* **measurement system.** 42A30-0

basic frequency. Of an oscillatory quantity having sinusoidal components with different frequencies, the frequency of the component considered to be the most important. *Note:* In a driven system, the basic frequency would, in general, be the driving frequency, and in a periodic oscillatory system, it would be the fundamental frequency. *See also:* **electroacoustics.** E157-1E1

basic functions (industrial control) (controller). The functions of those of its elements that govern the application of electric power to the connected apparatus. *See also:* **electric controller.** 42A25-34E10

basic impulse insulation levels (1) (general). Reference insulation levels expressed as the impulse crest value of withstand voltage of a specified full impulse voltage wave. Nominal 1.2 × 50-microsecond wave. *Notes:* (1) Impulse waves are defined by a combination of two numbers. The first number is the time from the start of the wave to the instant of crest value, and the second number is the time from the start to the instant of half-crest value on the tail of the wave. (2) In practice it is necessary to determine the starting point of the wave in a prescribed manner, set forth in the test code. The starting point so determined is called the virtual time zero. (3) Various specifications for test-voltage-waves include: 1.2 × 50 microseconds and 1.5 × 40 microseconds. 42A15-31E12
(2) (electric power). Basic impulse insulation levels are reference levels expressed in impulse crest voltage with a standard wave not longer than 1.5 × 40-microsecond wave. *See also:* **insulation.** E32-0
(3) (BIL) (lightning arrester). A reference impulse insulation strength expressed in terms of the crest value of withstand voltage of a standard full impulse voltage wave.
See:
coordination of insulation;
impulse insulation level;
impulse protection level;
insulation level;
lightning arrester;
preferred basic impulse insulation level;
rated withstand voltage;
standard full impulse voltage wave;
withstand voltage.
See also: **insulation.**
E28-0; 37A100-31E11; 62A1-31E7; 82A7/82A8/92A1-0

basic metallic rectifier. That in which each rectifying element consists of a single metallic rectifying cell. *See also:* **rectification.** 42A15-0

basic network. An electric network designed to simulate the iterative impedance, neglecting dissipation, of a line at a particular termination. *See also:* **network analysis.** 42A65-0

basic numbering plan USA (telephony). The plan whereby every telephone station is identified for nationwide dialing by a code for routing and a number of digits. *See also:* **telephone switching system.** 0-19E1

basic reference standards (laboratory). Those standards with which the values of the electrical units are maintained in the laboratory, and that serve as the starting point of the chain of sequential measurements carried out in the laboratory. *See also:* **measurement system.** 12A0-0

basic repetition frequency (loran). The lowest pulse repetition frequency of each of the several sets of closely spaced repetition frequencies employed. *See also:* **navigation.** 0-10E6

basic repetition rate. *See:* **basic repetition frequency.**

basic units (system of units). Those particular units arbitrarily selected to serve as a basis, in terms of which the other units of the system may be conveniently derived. *Note:* A basic unit is usually determined by a specified relationship to a single prototype standard. E270-0

bass boost. An adjustment of the amplitude-frequency response of a system or transducer to accentuate the lower audio frequencies. 0-42A65

bath voltage. The total voltage between the anode and cathode of an electrolytic cell during electrolysis. It is equal to the sum of (1) equilibrium reaction potential, (2) *IR* drop, (3) anode polarization, and (4) cathode polarization. *See:* **tank voltage.** *See also:* **electrolytic cell.** 42A60-0

battery (primary or secondary). Two or more cells electrically connected for producing electric energy. (Common usage permits this designation to be applied also to a single cell used independently. In this Dictionary, unless otherwise specified, the term battery will be used in this dual sense.) *Note:* For an extensive list of cross references, see *Appendix A.* 42A60-0

battery chute. A small cylindrical receptacle for housing track batteries and so set in the ground that the batteries will be below the frost line. *See also:* **railway signal and interlocking.** 42A42-0

battery-current regulation (generator). That type of automatic regulation in which the generator regulator controls only the current used for battery charging purposes. *See also:* **axle generator system.** 42A42-0

battery, electric. A device that transforms chemical energy into electric energy. *See also:* **battery.** 0-31E3

battery eliminator. A device that provides direct-current energy from an alternating-current source in place of a battery. *See also:* **battery.** 0-31E3

battery, power station (1) (communications). A battery that is a separate source of energy for communication equipment in power stations.
(2) (control). A battery that is a separate source of energy for the control of power apparatus in a power station. *See also:* **battery.** 0-31E3

baud (1) (general). A unit of signalling speed equal to the number of discrete conditions or signal events per second. For example, one baud equals one half dot cycle per second in Morse code, one bit per second in a train of binary signals, and one 3-bit value per second in a train of signals each of which can assume one of 8 different states. *See also:* **electronic digital computer; telegraphy.** X3A12-16E9
(2) (telegraphy). The unit of telegraph signaling speed, derived from the duration of the shortest signaling pulse. A telegraphic speed of one baud is one pulse per second. *Note:* The term unit pulse is often used for the same meaning as the baud. A related term, the dot

cycle, refers to an ON-OFF or MARK-SPACE cycle in which both mark and space intervals have the same length as the unit pulse. E145-0

***b* auxiliary switch.** *See:* **auxiliary switch; *b* contact.**

bay (computing system). *See:* **patch bay.** *See also:* **electronic analog computer.**

***B* battery.** A battery designed or employed to furnish the plate current in a vacuum-tube circuit. *See also:* **battery (primary or secondary).** 42A60-0

***bb* auxiliary switch.** *See:* **auxiliary switch; *bb* contact.**

***bb* contact.** A contact that is closed when the operating mechanism of the main device is in the standard reference position and that is open when the operating mechanism is in the opposite position. *See:* **standard reference position.** 37A100-31E11

b* contact (back contact) (relay).** A contact that is closed when the main device is in the standard reference position and that is open when the device is in the opposite position. *Notes:* *b* contact has general application. However, this meaning for **back contact** is restricted to relay parlance. (2) For indication of the specific point of travel at which the contact changes position, an additional letter or percentage figure may be added to ***b as detailed in ANSI C37.2-1962, or revisions thereof in which see **standard reference position.** *See also:* **railway signal and interlocking; relay.** 37A100-31E11

***B* display (radar).** A rectangular display in which each target appears as a blip, with bearing indicated by the horizontal coordinate and distance by the vertical coordinate. *See also:* **navigation.** E172-10E6

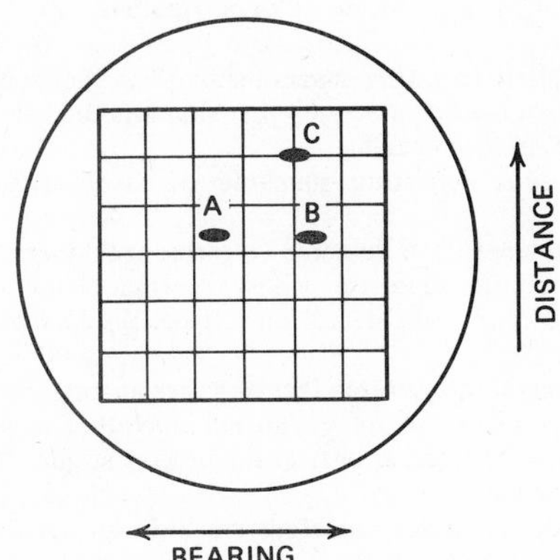

B display.

beacon. A light (or mark) used to indicate a geographic location. *See also:* **signal lighting.** Z7A1-0

beacon receiver. A radio receiver for converting waves, emanating from a radio beacon, into perceptible signals. *See also:* **radio receiver; radio beacon.** 42A65-0

beam (antenna). The major lobe of the radiation pattern. *See also:* **radiation.** 0-3E1

beam alignment (camera tubes). An adjustment of the electron beam, performed on tubes employing low-velocity scanning, to cause the beam to be perpendicular to the target at the target surface. *See:* **beam tubes.** E160-15E6

beam bending (camera tubes). Deflection of the scanning beam by the electrostatic field of the charges stored on the target. *See:* **beam tubes.** E160-15E6; 0-2E2

beam current (storage tubes). The current emerging from the final aperture of the electron gun. *See also:* **storage tube.** E158-15E6

beam-deflection tube. An electron-beam tube in which current to an output electrode is controlled by the transverse movement of an electron beam. *See also:* **beam tubes; tube definitions.** E160-15E6

beam error (navigational systems using directionally propagated signals). The lateral or angular difference between the mean position of the actual course and the desired course position. *Note:* Sometimes called course error. *See also:* **navigation.** 0-10E6

beam-indexing color tube. A color-picture tube in which a signal, generated by an electron beam after deflection, is fed back to a control device or element in such a way as to provide an image in color. *See also:* **beam tubes.** E160-2E2/15E6

beam landing error (camera tube). A signal non-uniformity resulting from beam electrons arriving at the target with a spatially varying component of velocity parallel to the target. *See also:* **camera tube.** 0-15E6

beam modulation, percentage (image orthicons). One hundred times the ratio of (1) the signal output current for highlight illumination on the tube to (2) the dark current. *See also:* **beam tubes.** E160-2E2

beam noise (navigational systems using directionally propagated signals). Extraneous disturbances tending to interfere with ideal system performance. *Note:* Beam noise is the aggregate effect of bends, scalloping, roughness, etcetera. *See also:* **navigation.** 0-10E6

beam parametric amplifier. A parametric amplifier that uses a modulated electron beam to provide a variable reactance. *See also:* **parametric device.** E254-15E7

beam pattern. *See:* **directional response pattern.**

beam power tube. An electron-beam tube in which use is made of directed electron beams to contribute substantially to its power-handling capability, and in which the control grid and the screen grid are essentially aligned. *See also:* **beam tube; tube definitions.** E160-15E6;42A70-0

beam rider guidance. That form of missile guidance wherein a missile, through a self-contained mechanism, automatically guides itself along a beam. *See also:* **guided missile.** 42A65-0

beam spread (any plane) (illuminating engineering). The angle between the two directions in the plane in which the candlepower is equal to a stated percent (usually 10 percent) of the maximum candlepower in the beam. *See also:* **lamp.** Z7A1-0

beam steering (antenna). Changing the direction of the major lobe of a radiation pattern. *See also:* **radiation.** 0-3E1

beam tube. *Note:* For an extensive list of cross references, see *Appendix A.*

beam waveguide. A structure consisting of a sequence of lenses or mirrors that can guide an electromagnetic wave. *See:* **waveguide.** 0-3E1

bearing (1) (electronic navigation). (A) The horizontal direction of one terrestrial point from another, expressed as the angle in the horizontal plane between a reference line and the horizontal projection of the line

joining the two points. (B) Azimuth. *See also:* **navigation.** 0-10E6
(2) (rotating machinery). (A) A stationary member or assembly of stationary members in which a shaft is supported and may rotate. (B) In a ball or roller bearing, a combination (frequently preassembled) of stationary and rotating members containing a peripheral assembly of balls or rollers, in which a shaft is supported and may rotate. *Note:* For an extensive list of cross references, see *Appendix A.* 0-31E8

bearing accuracy, instrumental (direction-finding). (1) The difference between the indicated and the apparent bearings in a measurement of the same signal source. (2) As a statement of overall system performance, a difference between indicated and correct bearings whose probability of being exceeded in any measurement made on the system is less than some stated value. *See also:* **navigation.** 0-10E6

bearing bracket (rotating machinery). A part secured to the stator to support the bearing, but including no part thereof. A bearing bracket is not specifically constructed to provide protection for the windings or rotating parts. *See also:* **end shield.** 0-31E8

bearing cap (bearing bracket cap) (rotating machinery). A cover for the bearing enclosure of a bearing bracket type machine or the removable upper half of the enclosure for a bearing. *See:* **bearing.** 42A10-31E8

bearing cartridge (rotating machinery). A complete enclosure for a ball or roller bearing, separate from the bearing bracket or end shield. *See:* **bearing.** 42A10-31E8

bearing clearance (rotating machinery). (1) The difference between the bearing inner diameter and the journal diameter. (2) The total distance for axial movement permitted by a double-acting thrust bearing. *See:* **bearing.** 0-31E8

bearing dust-cap (rotating machinery). A removable cover to prevent the entry of foreign material into the bearing. *See also:* **bearing.** 42A10-31E8

bearing error curve (direction-finding). (1) A plot of the combined instrumental bearing error (of the equipment) and site error versus indicated bearings. (2) A plot of the instrumental bearing errors versus either indicated or correct bearings. *See also:* **navigation.** E173-10E6

bearing housing (rotating machinery). A structure supporting the actual bearing liner or ball or roller bearing in a bearing assembly. *See also:* **bearing.** 0-31E8

bearing insulation (rotating machinery). Insulation that prevents the circulation of stray currents by electrically insulating the bearing from its support. *See also:* **bearing.** 0-31E8

bearing liner (rotating machinery). The assembly of a bearing shell together with its lining. *See also:* **bearing.** 0-31E8

bearing lining (rotating machinery). The element of the journal bearing assembly in which the journal rotates. *See also:* **bearing.** 0-31E8

bearing locknut (rotating machinery). A nut that holds a ball or roller bearing in place on the shaft. *See also:* **bearing.** 42A10-31E8

bearing lock washer (rotating machinery). A washer between the bearing locknut and the bearing that prevents the locknut from turning. *See also:* **bearing.** 42A10-31E8

bearing offset, indicated (direction-finding) (electronic navigation). The mean difference between the indicated and apparent bearings of a number of signal sources, the sources being substantially uniformly distributed in azimuth. *See also:* **navigation.** E173-10E6

bearing oil seal (bearing seal) (rotating machinery). *See:* **oil seal.**

bearing oil system (oil-circulating system) (rotating machinery). All parts that are provided for the flow, treatment, and storage of the bearing oil. *See also:* **oil cup (rotating machinery).** 0-31E8

bearing pedestal (rotating machinery). A structure mounted from the bedplate or foundation of the machine to support a bearing, but not including the bearing. *See:* **bearing.** 42A10-31E8

bearing-pedestal cap (rotating machinery). The top part of a bearing pedestal. *See also:* **bearing.** 42A10-31E8

bearing, reciprocal (electronic navigation). *See:* **reciprocal bearing.**

bearing reservoir (oil tank) (oil well) (rotating machinery). A container for the oil supply for the bearing. It may be a sump within the bearing housing. *See also:* **oil cup (rotating machinery).** 0-31E8

bearing seal (bearing oil seal) (rotating machinery). *See:* **oil seal.**

bearing seat (rotating machinery). The surface of the supporting structure for the bearing shell. *See also:* **bearing.** 0-31E8

bearing sensitivity (electronic navigation). The minimum field strength input to a direction-finder system to obtain repeatable bearings within the bearing accuracy of the system. *See also:* **navigation.** 0-10E6

bearing shell (rotating machinery). The element of the journal bearing assembly that supports the bearing lining. *See also:* **bearing.** 0-31E8

bearing shoe (rotating machinery). *See:* **segment shoe.**

bearing-temperature detector (rotating machinery). A temperature detector whose sensing element is mounted at or near the bearing surface. *See also:* **bearing.** 0-31E8

bearing-temperature relay (bearing thermostat) (rotating machinery). A relay whose temperature sensing element is mounted at or near the bearing surface. *See also:* **bearing.** 0-31E8

bearing thermometer (rotating machinery). A thermometer whose temperature sensing element is mounted at or near the bearing surface. *See also:* **bearing.** 0-31E8

bearing thermostat (rotating machinery). *See:* **bearing-temperature relay.**

beating. A phenomenon in which two or more periodic quantities of different frequencies produce a resultant having pulsations of amplitude. *See also:* **beats; signal wave.** E188-0

beat note. The wave of difference frequency created when two sinusoidal waves of different frequencies are supplied to a nonlinear device. *See also:* **radio receiver.** E188-0

beat reception. *See:* **heterodyne reception.**

beats. Periodic variations that result from the superposition of waves having different frequencies. *Note:* The term is applied both to the linear addition of two waves, resulting in a periodic variation of amplitude, and to the nonlinear addition of two waves, resulting in new frequencies, of which the most important usually are the sum and difference of the original frequencies. *See also:* **signal wave.** 42A65/50I05-31E3

bel. The fundamental division of a logarithmic scale for expressing the ratio of two amounts of power, the number of bels denoting such a ratio being the logarithm to the base 10 of this ratio. *Note:* With P_1 and P_2 designating two amounts of power and N the number of bels denoting their ratio, $N=\log_{10}(P_1/P_2)$ bels. E145-0

bell box (ringer box). An assemblage of apparatus, associated with a desk stand or hand telephone set, comprising a housing (usually arranged for wall mounting) within which are those components of the telephone set not contained in the desk stand or hand telephone set. These components are usually one or more of the following: induction coil, capacitor assembly, signaling equipment, and necessary terminal blocks. In a magneto set a magneto and local battery may also be included. *See also:* **telephone station.** 42A65-0

belt (rotating machinery). A continuous flexible band of material used to transmit power between pulleys by motion. *See also:* **cradle base (rotating machinery).** 0-31E8

belt-drive machine (elevators). An indirect-drive machine having a single belt or multiple belts as the connecting means. *See also:* **driving machine (elevators).** 42A45-0

belt insulation (rotating machinery). A form of overhang packing inserted circumferentially between adjacent layers in the winding overhang. *See also:* **rotor (rotating machinery); stator.** 0-31E8

belt leakage flux (rotating machinery). The low-order harmonic airgap flux attributable to the phase belts of a winding. The magnitude of this leakage flux varies with winding pitch. *See also:* **rotor (rotating machinery); stator.** 0-31E8

belted-type cable. A multiple-conductor cable having a layer of insulation over the assembled insulated conductors. *See also:* **power distribution, underground construction.** 42A35-31E13

benchboard. A combination of a control desk and a vertical or enclosed switchboard in a common assembly. 37A100-31E11

benchmark problem (computers). A problem used to evaluate the performance of computers relative to each other. X3A12-16E9

bend (electronic navigation). A departure of the course line from the desired direction at such a rate that it can be followed by the vehicle. *Note:* Sometimes spuriously produced by defects in the design of the receiver. *See also:* **navigation.** 0-10E6

bend amplitude (navigation) (electronic navigation). The measured maximum amount of course deviation due to bend; measurement is made from the nominal or bend-free position of the course. *See also:* **bend; bend frequency; navigation.** 0-10E6

bend frequency (electronic navigation). The frequency at which the course indicator oscillates when the vehicle track is straight and the course contains bends; bend frequency is a function of vehicle velocity. *See also:* **navigation.** 0-10E6

bend reduction factor (electronic navigation). The ratio of bend amplitude existing before the introduction of bend-reducing features to that existing afterward. *See also:* **navigation** 0-10E6

bend, waveguide. A section of waveguide or transmission line in which the direction of the longitudinal axis is changed. *See:* **waveguide.** 0-3E1

beta (β) circuit (feedback amplifier). That circuit that transmits a portion of the amplifier output back to the input. *See also:* **feedback.** 42A65-0

betatron. An electric device in which electrons revolve in a vacuum enclosure in a circular or a spiral orbit normal to a magnetic field and have their energies continuously increased by the electric force resulting from the variation with time of the magnetic flux enclosed by their orbits. *See also:* **electron devices, miscellaneous.** 42A70-15E6

bevatron. A synchrotron designed to produce ions of a billion (10^9) electron-volts energy or more. *See also:* **electron devices, miscellaneous.** 42A70-15E6

beveled brush corners (electric machines). Where material has been removed from a corner, leaving a triangular surface. *See also:* **brush.** 64A1-0

beveled brush edges (electric machines). The removal of an edge to provide a slanting surface from which a shunt connection can be made or for clearance of pressure fingers or for any other purpose. *See also:* **brush.** 64A1-0

beveled brush ends and toes (electric machines). The angle included between the beveled surface and a plane at right angles to the length. The toe is the uncut or flat portion on the beveled end. When a brush has one or both ends beveled,the front of the brush is the short side or the side exposing the face level. *See also:* **brush.** 64A1-0

Beverage antenna (wave antenna). A directional antenna composed of a system of parallel, horizontal conductors from one-half to several wavelengths long, and terminated to ground at the far end in its characteristic impedance. *See also:* **antenna.** E145-3E1

BF *See:* **ballistic focusing.**

bias (1) (general). To influence or dispose to one direction, as, for example, with a direct voltage or with a spring.

(2) (telegraph transmission) A uniform displacement of like signal transitions resulting in a uniform lengthening or shortening of all marking signal intervals. *See also:* **telegraphy.** 42A65-19E4

bias distortion. The distortion of nonreturn-to-zero (binary) code by a transmission path that requires a time for a 0-1 (space-to-mark) transition that is different than the time for a (mark-to-space) 1-0 transition. *See also:* **digital; bias telegraph distortion.** 0-31E3

bias, grid, direct. *See:* **direct grid bias.**

bias telegraph distortion. Distortion in which all mark pulses are lengthened (positive bias) or shortened (negative bias). It may be measured with a steady stream of unbiased reversals, square waves having equal-length mark and space pulses. The average lengthening or shortening gives true bias distortion only if other types of distortion are negligible. *See also:* **modulation.** E145-0

bias winding (1) (saturable reactor). A control winding through which a biasing magnetomotive force is applied. 42A65-0

(2) (relay). *See:* **relay bias winding.**

biased ferroelectric materials. Ferroelectric materials that exhibit hysteresis loops that are unsymmetrical with respect to one or both of the normal coordinate axes ($E = 0$, $D = 0$) but that are symmetrical with respect to a second set of axes parallel to the first set. *Note:* The coordinates of the origin of this second set of axes, with respect to the first, define the bias. *See also:* **ferroelectric domain.** E180-0

biased induction. At a point in a magnetic material that is subjected simultaneously to a periodically varying magnetizing force and a biasing magnetizing force, the mean value of the magnetic induction at the point. E270-0

biased scheduled net interchange of a control area (electric power systems). The scheduled net interchange plus the frequency and/or other bias. E94-0;0-31E4

biased telephone ringer. A telephone ringer whose clapper-driving element is normally held toward one side by mechanical forces or by magnetic means, so that the ringer will operate on alternating current or on electric pulses in one direction, but not on pulses in the other direction. *See also:* **telephone station.** 42A65-0

biasing magnetizing force. At a point in a magnetic material that is subjected simultaneously to a periodically varying magnetizing force and a constant magnetizing force, the mean value of the combined magnetizing forces. E270-0

biconical antenna. An antenna formed by two conical conductors having a common axis and vertex and excited at the vertex. When the vertex angle of one of the cones is 180 degrees, the antenna is called a discone. *See also:* **antenna.** 42A65-3E1

bidirectional antenna. An antenna having two directions of maximum response. *See also:* **antenna.** 42A65-3E1

bidirectional diode-thyristor. A two-terminal thyristor having substantially the same switching behavior in the first and third quadrants of the principal voltage-current characteristic. *See also:* **thyristor.** E223-34E17/15E7

bidirectional pulses. Pulses, some of which rise in one direction and the remainder in the other direction. *See also:* **modulating systems; pulse.** E145/42A65-0

bidirectional relay or add-and-subtract relay. A stepping relay in which the rotating wiper contacts may move in either direction. 0-21E0

bidirectional triode-thyristor. A three-terminal thyristor having substantially the same switching behavior in the first and third quadrants of the principal voltage-current characteristic. *See also:* **thyristor.** E223-34E17/15E7

bidirectional transducer (bilateral transducer). A transducer that is not a unidirectional transducer. *See also:* **transducer.** E270/42A65-0

bifilar suspension. A suspension employing two parallel ligaments, usually of conducting material, at each end of the moving element. 39A1-0

BIL (1). *See:* **basic impulse insulation levels.**
(2) (insulation strength). *See:* **preferred basic impulse insulation level.**

bilateral-area track (electroacoustics). A photographic sound track having the two edges of the central area modulated according to the signal. *See also:* **phonograph pickup.** E157-1E1

bilateral network. A network that is not a unilateral network. *See also:* **network analysis.** E270-0

bilateral transducer. A transducer capable of transmission simultaneously in both directions between at least two terminations. E-196-0

bilevel operation. Operation of a storage tube in such a way that the output is restricted to one or the other of two permissible levels. *See also:* **storage tube.** E158-15E6

billing demand (electric power utilization). The demand that is used to determine the demand charges in accordance with the provisions of a rate schedule or contract. *See also:* **alternating-current distribution.** 0-31E4

bimetallic element. An actuating element consisting of two strips of metal with different coefficients of thermal expansion bound together in such a way that the internal strains caused by temperature changes bend the compound strip. *See also:* **relay.** 83A16-0

binary (computers). (1) Pertaining to a characteristic or property involving a selection, choice, or condition in which there are two possibilities. (2) Pertaining to the numeration system with a radix of two. *See also:* **column binary; electronic digital computer; positional notation; row binary.** X3A12-16E9

binary cell (1) (computers). An elementary unit of storage that can be placed in either of two stable states. *Note:* It is therefore a storage cell of one binary digit capacity, for example, a single-bit register. *See also:* **electronic computation; electronic digital computer.** E270/X3A12-16E9

binary code. (1) A code in which each code element may be either of two distinct kinds or values, for example, the presence or absence of a pulse. 42A65-31E3/19E4
(2) A code that makes use of members of an alphabet containing exactly two characters, usually 0 and 1. The binary number system is one of many binary codes. *See also:* **electronic digital computer; information theory; pulse; reflected binary code.** 0-16E9;EIA3B-34E12

binary-coded-decimal. Pertaining to a number-representation system in which each decimal digit is represented by a unique arrangement of binary digits (usually four), for example, in the 8-4-2-1 binary-coded-decimal notation, the number 23 is represented as 0010 0011 whereas in binary notation, 23 is represented as 10111. *See also:* **electronic digital computer.** E162-0;X3A12-16E9

binary-coded-decimal number (BCD) The representation of the cardinal numbers 0 through 9 by 10 binary codes of any length. Note that the minimum length is four and that there are over 29×10^9 possible four-bit binary-coded-decimal codes.
Note: An example of 8-4-2-1 binary-coded decimal code follows for numbers 0 through 9.

Number	*r*4	*r*3	*r*2	*r*1
0	0	0	0	0
1	0	0	0	1
2	0	0	1	0
3	0	0	1	1
4	0	1	0	0
5	0	1	0	1
6	0	1	1	0
7	0	1	1	1
8	1	0	0	0
9	1	0	0	1

Where *r*1 is termed the **least significant binary digit (bit).** *See also:* **digital.** 0-31E3

binary digit. A character used to represent one of the two digits in the numeration system with a radix of two. Abbreviated **bit.** *See:* **equivalent binary digits.** *See also:* **electronic digital computer.** X3A12-16E9

binary number. Loosely, a binary numeral. *See also:* **electronic digital computer.** X3A12-16E9

binary number system. *See:* **positional notation.**

binary numeral. The binary representation of a number, for example, 101 is the binary numeral and V is the Roman numeral of the number of fingers on one hand. *See also:* **electronic digital computer.** X3A12-16E9

binary point. *See:* **point.**

binary search. A search in which a set of items is divided into two parts, one part is rejected, and the process is repeated on the accepted part until those items with the desired property are found. *See:* **dichotomizing search.** *See also:* **electronic digital computer.** X3A12-16E9

binary word (power-system communication). A binary code of stated length given a specific meaning. *See also:* **code character; digital.** 0-31E3

binder (bond) (1) (rotating machinery). A solid, liquid, or semiliquid composition that exhibits marked ability to act as an adhesive, and that, when applied to wires, insulation components, or other parts, will solidify, hold them in position, and strengthen the structure. *See also:* **cradle base (rotating machinery).** 0-31E8

(2) (electroacoustics) A resinous material that causes the various materials of a record compound to adhere to one another. *See also:* **phonograph pickup.** S1A1-1E1

binding band (rotating machinery). A band of material, encircling stator or rotor windings to restrain them against radial movement. *See:* **rotor (rotating machinery).** 0-31E8

binding post. *See:* **binding screw.**

binding screw (binding post) (terminal screw) (clamping screw). A screw for holding a conductor to the terminal of a device or equipment. *See also:* **interior wiring.** 42A95-0

binocular visual field. That portion of the visual field where the fields of the two eyes overlap. *Note:* It has a half angle of roughly 60 degrees. *See also:* **visual field.** Z7A1-0

binomial array. A linear array in which the currents in successive elements are made proportional to the binomial coefficients of $(x + y)^{n-1}$ for the purpose of reducing minor lobes. *See also:* **antenna.** 42A65-3E1

bioelectric null (zero lead) (medical electronics). A region of tissue or other area in the system, which has such electric symmetry that its potential referred to infinity does not significantly change. *Note:* This may or may not be ground potential. *See also:* **medical electronics.** 0-18E1

Biot-Savart law. *See:* **magnetic field strength produced by an electric current.**

biparting door (elevators). A vertically sliding or a horizontally-sliding door, consisting of two or more sections so arranged that the sections or groups of sections open away from each other and so interconnected that all sections operate simultaneously. *See also:* **hoistway (elevator or dumbwaiter).** 42A45-0

bipolar (power supplies). Having two poles, polarities, or directions. *Note:* Applied to amplifiers or power supplies, it means that the output may vary in either polarity from zero; as a symmetrical program it need not contain a direct-current component. *See:* **unipolar.** *See also:* **power supply.** KPSH-10E1

bipolar electrode. An electrode, without metallic connection with the current supply, one face of which acts as an anode surface and the opposite face as a cathode surface when an electric current is passed through the cell. *See also:* **electrolytic cell.** 42A60-0

bipolar electrode system (electrobiology). Either a pickup or stimulating system consisting of two electrodes whose relation to the tissue currents is roughly symmetrical. *See also:* **electrobiology.** 42A80-18E1

biquinary. Pertaining to the number representation system in which each decimal digit N is represented by the digit pair AB, where $N = 5A + B$, and where $A = 0$ or 1 and $B = 0,1,2,3,$ or 4; *for example,* decimal 7 is represented by biquinary 12. This system is sometimes called a mixed-radix system having the radices 2 and 5. X3A12-16E9

bistable. The ability of a device to assume either of two stable states. AS1-34E10

bistable (industrial control). Pertaining to a device capable of assuming either one or two stable states. *See:* **control system, feedback.** X3A12-16E9

bistable amplifier (industrial control). An amplifier with an output that can exist in either of two stable states without a sustained input signal and can be switched abruptly from one state to the other by specified inputs. *See:* **control system, feedback; rating and testing magnetic amplifiers.** AS1-34E10

bistable operation. Operation of a charge-storage tube in such a way that each storage element is inherently held at either of two discrete equilibrium potentials. *Note:* Ordinarily this is accomplished by electron bombardment. *See also:* **charge-storage tube.** E158-15E6

bit. A binary digit. *See:* **check bit; parity bit.**

bit (1) (electronic computers). (A) An abbreviation of binary digit. (B) A single occurrence of a character in a language employing exactly two distinct kinds of characters. (C) A unit of storage capacity. The capacity, in bits, of a storage device is the logarithm to the base two of the number of possible states of the device. *See also:* **electronic computation; storage capacity.** E162/E270-0

(2) (information theory). A unit of information content equal to the information content of a message the *a priori* probability of which is one-half. *Note:* If, in the definition of information content, the logarithm is taken to the base two, the result will be expressed in bits. One bit equals $\log_{10} 2 \times$ hartley.

See:

gross information content;
hartley;
net information content;
sequential;
serial operation;
serial transmission;
service bits.

See also: **electronic computation; information theory.** E171-0;0-19E4

black and white. *See:* **monochrome.** *See also:* **color terms.**

blackbody. A temperature radiator of uniform temperature whose radiant exitance in all parts of the spectrum is the maximum obtainable from any temperature radiator at the same temperature. *Notes:* (1) Such a radiator is called a blackbody because it will absorb all the radiant energy that falls upon it. All other temperature radiators may be classed as nonblackbodies. They radiate less in some or all wavelength intervals than a blackbody of the same size and the same temperature. (2) The blackbody is practically realized in the form of a cavity with opaque walls at a uniform temperature and with a small opening for observation purposes. It is variously called a standard radiator or an ideal radiator. *See also:* **radiant energy.** Z7A1-0

blackbody (Planckian) locus. The locus of points on a chromaticity diagram representing the chromaticities of blackbodies having various color temperatures. *See also:* **color.** Z7A1-0.

black compression (black saturation) (television). The reduction in gain applied to a picture signal at those levels corresponding to dark areas in a picture with respect to the gain at that level corresponding to the mid-range light value in the picture. *Notes:* (1) The gain referred to in the definition is for a signal amplitude small in comparison with the total peak-to-peak picture signal involved. A quantitative evaluation of this effect can be obtained by a measurement of differential gain. (2) The over-all effect of black compression is to reduce contrast in the low lights of the picture as seen on a monitor. *See also:* **television.** E203-2E2

black level (television). The level of the picture signal corresponding to the maximum limit of black peaks. *See also:* **television.** E203-2E2

black light. The popular term for ultraviolet energy near the visible spectrum. *Note:* For engineering purposes the wavelength range 320 to 400 nanometers has been found useful for rating lamps and their effectiveness upon fluorescent materials (excluding phosphors used in fluorescent lamps). By confining black-light applications to this region, germicidal and erythemal effects are, for practical purposes, eliminated. *See also:* **ultraviolet radiation.** Z7A1-0

black-light flux. Radiant flux within the wavelength range 320 to 400 nanometers. It is usually measured in milliwatts. *Note:* The fluoren is used as a unit of black-light flux and is equal to one milliwatt of radiant flux in the wavelength range 320 to 400 nanometers. Because of the variability of the spectral sensitivity of materials irradiated by black-light in practice, no attempt is made to evaluate black-light flux according to its capacity to produce effects. *See also:* **ultraviolet radiation.** Z7A1-0

black-light flux density. Black-light flux per unit area of the surface being irradiated. It is equal to the incident black-light flux divided by the area of the surface when the flux is uniformly distributed. It usually is measured in milliwatts per square foot of black-light flux. *See also:* **ultraviolet radiation.** Z7A1-0

black peak (television). A peak excursion of the picture signal in the black direction. *See also:* **television.** E203/42A65-2E2

black recording (1) (amplitude-modulation facsimile system). The form of recording in which the maximum received power corresponds to the maximum density of the record medium.
(2) (frequency-modulation facsimile system). The form of recording in which the lowest received frequency corresponds to the maximum density of the record medium. *See also:* **recording (facsimile).** E168-0

black saturation (television). *See:* **black compression.**

black signal (at any point in a facsimile system). The signal produced by the scanning of a maximum-density area of the subject copy. *See also:* **facsimile signal (picture signal).** E168-0

black transmission (1) (amplitude-modulation facsimile system). The form of transmission in which the maximum transmitted power corresponds to the maximum density of the subject copy.
(2) (frequency-modulation facsimile system). The form of transmission in which the lowest transmitted frequency corresponds to the maximum density of the subject copy. *See also:* **facsimile transmission.** E168-0

blade (disconnecting blade) (switching device). The moving contact member that enters or embraces the contact clips. *Note:* In cutouts the blade may be a fuse carrier or fuseholder on which a nonfusible member has been mounted in place of a fuse link. When so used the nonfusible member alone is also called a blade in fuse parlance. *See also:* **relay blades.** 37A100-31E11

blade latch (stick-operated switch). A latch used to hold the switch blade in the closed position. 37A100-31E11

blank character. A character used to produce a character space on an output medium. X3A12-16E9

blanked picture signal (television). The signal resulting from blanking a picture signal. *Note:* Adding synchronizing signal to the blanked picture signal forms the composite picture signal. *See also:* **television.** E203-2E2

blanketing. The action of a powerful radio signal or interference in rendering a receiving set unable to receive desired signals. *See also:* **radiation.** 42A65-0

blanking (1) (general). The process of making a channel or device noneffective for a desired interval.
(2) (television). Blanking is the substitution for the picture signal, during prescribed intervals, of a signal whose instantaneous amplitude is such as to make the return trace invisible. *See also:* **television.** 42A65-0
(3) (oscilloscopes). Extinguishing of the spot. Retrace blanking is the extinction of the spot during the retrace portion of the sweep waveform. The term does not necessarily imply blanking during the holdoff interval or while waiting for a trigger in a triggered sweep system. 0-9E4

blanking level (television). The level of a composite picture signal that separates the range containing picture information from the range containing synchronizing information. *Note:* The setup region is regarded as picture information. *See also:* **television.** E203-2E2

blanking signal (television). A wave constituted of recurrent pulses, related in time to the scanning process, used to effect blanking. *Note:* In television, this signal is composed of pulses at line and field frequencies, which usually originate in a central synchronizing generator and are combined with the picture signal at the pickup equipment in order to form the blanked picture signal. The addition of synchronizing signal completes the composite picture signal. The blanking portion of the composite picture signal is intended primarily to make the return trace on a picture tube invisible. The same blanking pulses or others of somewhat shorter duration are usually used to blank the pickup device also. *See also:* **television.** E203-0

blaster. *See:* **blasting unit.**

blasting circuit. A shot-firing cord together with connecting wires and electric blasting caps used in preparation for the firing of a blast in mines, quarries, and tunnels. *See also:* **blasting unit.** 42A85-0

blasting switch. A switch used to connect a power source to a blasting circuit. *Note:* A blasting switch is sometimes used to short-circuit the leading wires as a safeguard against premature blasts. *See also:* **blasting unit.** 42A85-0

blasting unit (blaster) (exploder) (shot-firing unit). A portable device including a battery or a hand-operated generator designed to supply electric energy for firing explosive charges in mines, quarries, and tunnels. *See:*
blasting circuit;
blasting switch;
delay electric blasting cap;
electric blasting cap;
electric squib;
leading wire;
leg wire;
misfire;
multiple-shot blasting unit;
shot-firing (blasting) cord;
single-shot blasting unit;
waterproof electric blasting-cap.
See also: **mining.** 42A85-0

bleeder. A resistor connected across a power source to improve voltage regulation, to drain off the charge remaining in capacitors when the power is turned off, or to protect equipment from excessive voltages if the load is removed or substantially reduced. *See also:* **circuits and devices.** 42A65-0

blemish charge (storage tubes). A localized imperfection of the storage assembly that produces a spurious output. *See also:* **storage tube.** E158-15E6

blind speed (radar systems using moving target indicators). The radial velocity of a target with respect to the radar for which the response is approximately zero. *Note:* In a coherent moving target indicator system using a uniform repetition rate, the blind speed is the radial velocity at which the moving target changes distance by one-half wavelength (or multiples thereof) during each pulse period. *See also:* **navigation.** 0-10E6

blinding glare. Glare that is so intense that for an appreciable length of time no object can be seen. *See also:* **visual field.** Z7A1-0

blinker signal. *See:* **Morse signal light.**

blinking (pulse navigation systems). A method of providing information by modifying the signal at its source so that the signal presentation on the display at the receiver alternately appears and disappears; for example, in loran, blinking is used to indicate that a station is malfunctioning. *See also:* **navigation.** E172-10E6

blip (radar display). A deflection, or a spot of contrasting luminescence, caused by the presence of a target. *See also:* **navigation.** E172-10E6

block (1) (data processing and computation). (A) A set of things, such as words, characters, or digits, handled as a unit. (B) A collection of contiguous records recorded as a unit. *Notes:* (1) Blocks are separated by interblock gaps and each block may contain one or more records. (C) In data communication, a group of contiguous characters formed for transmission purposes. (2) The groups are separated by interblock characters. 0-16E9

(2) (railway practice). A length of track of defined limits on which the movement of trains is governed by block signals, cab signals, or both. *See also:* **absolute block; railway signal and interlocking.** 42A42-0

block cable (communication practice). A distribution cable installed on poles or outside building walls, in the interior of a block, including cable run within buildings from the point of entrance to a cross-connecting box, terminal frame, or point of connection to house cable. *See also:* **cable.** 42A65-0

block count readout. Display of the number of blocks that have been read from the tape derived by counting each block as it is read. *See:* **sequence number readout.** EIA3B-34E12

block diagram. A diagram of a system, instrument, computer, or program in which selected portions are represented by annotated boxes and interconnecting lines. X3A12-16E9

blocked impedance (transducer). The input impedance of the transducer when its output is connected to a load of infinite impedance. *Note:* For example, in the case of an electromechanical transducer, the blocked electric impedance is the impedance measured at the electric terminals when the mechanical system is blocked or clamped; the blocked mechanical impedance is measured at the mechanical side when the electric circuit is open-circuited. *See also:* **self-impedance.** E157-1E1;42A65-0

block indicator. A device used to indicate the presence of a train in a block. *See also:* **railway signal and interlocking.** 42A42-0

blocking (1) (tube rectifier). The prevention of conduction by means of grid or ignitor action, or both, when forward voltage is applied across a tube. 34A1-0

(2) (semiconductor rectifier). The action of a semiconductor rectifier cell that essentially prevents the flow of current. *See also:* **rectification .** E59-34E17

(3) (relay system). A relaying function which prevents action that would otherwise be initiated by the relay system. *See also:* **relay.** 37A1-31E6

(4) (rotating machinery). A structure or combination of parts, usually of insulating material, formed to hold coils in relative position for mechanical support. *Note:* Usually inserted in the end turns to resist forces during running and abnormal conditions. *See also:* **stator.** 0-31E8

blocking capacitor (1) (blocking condenser*). A capacitor that introduces a comparatively high series impedance for limiting the current flow of low-frequency alternating current or direct current without materially affecting the flow of high-frequency alternating current. *See also:* **circuits and devices; electrolytic capacitor.**

*Deprecated 42A65-21E0

(2) (check valve) An asymmetrical cell used to prevent flow of current in a specified direction. *See also:* **circuits and devices; electrolytic capacitor.** 42A60-0

blocking condenser. *See:* **blocking capacitor.**

blocking oscillator (1). A relaxation oscillator consisting of an amplifier (usually single-stage) with its output coupled back to its input by means that include capacitance, resistance, and mutual inductance. *See also:* **oscillatory circuit.** 42A65-0

(2) (squegging oscillator). An electron-tube oscillator operating intermittently with grid bias increasing during oscillation to a point where oscillations stop, then decreasing until oscillation is resumed. *Note:* Squegge rhymes with wedge. *See also:* **oscillatory circuit.** E145-0

blocking period (rectifier-circuit element) (1) (rectifier circuit). The part of an alternating-voltage cycle during which reverse voltage appears across the rectifier-circuit element. *Note:* The blocking period is not necessarily the same as the reverse period because of the effect of circuit parameters and semiconductor rectifier cell characteristics. *See also:* **rectifier circuit element.** E59-34E17

(2) (gas tube). The part of the idle period corresponding to the commutation delay due to the action of the control grid. *See also:* **gas-filled rectifier.** 50I07-15E6

blocking relay. An auxiliary relay whose function is to render another relay or device ineffective under specified conditions. 37A100-31E6/31E11

block-interval demand meter. *See:* **integrated-demand meter.** *See also:* **demand meter.**

block signal. A fixed signal installed at the entrance of a block to govern trains entering and using that block. *See also:* **railway signal and interlocking.** 42A42-0

block-signal system. A method of governing the movement of trains into or within one or more blocks by block signals or cab signals.
See:
automatic block-signal system;
centralized traffic-control system;
code system;
controlled manual block-signal system;
manual block-signal system;
normal-clear system;
normal-stop system;
traffic-control system. 42A42-0

block station. A place at which manual block signals are displayed. *See also:* **railway signal and interlocking.** 42A42-0

Blondel diagram (rotating machinery). A phasor diagram intended to illustrate the currents and flux linkages of the primary and secondary windings of a transformer, and the components of flux due to primary and secondary winding currents acting alone. *Note:* This diagram is also useful as an aid in visualizing the fluxes in an induction motor. *See also:* **asynchronous machine.** 0-31E8

blooming (1) (radar). An increase in the blip size on the display as a result of an increase in signal intensity or its duration; may be employed in navigational systems with intensity modulation displays for the purpose of conveying information. *See also:* **navigation.** 0-10E6

(2) (television picture tube). Excessive luminosity of the spot due to an excessive beam current. *See also:* **cathode-ray tubes.** 50I07-15E6

blower blade (rotating machinery). An active element of a fan or blower. *See also:* **fan (rotating machinery).** 0-31E8

blower housing. *See:* **fan housing.**

blowoff valve (gas turbines). A device by means of which a part of the air flow bypasses the turbine(s) and/or the regenerator to reduce the rate of energy input to the turbine(s). *Note:* It may be used in the speed governing system to control the speed of the turbine(s) at rated speed when fuel flow permitted by the minimum fuel limiter would otherwise cause the turbine to operate at a higher speed. *See also:* **asynchronous machine; direct-current commutating machine; synchronous machine.** E282-31E2

blowout coil. An electromagnetic device that establishes a magnetic field in the space where an electric circuit is broken and helps to extinguish the arc by displacing it, for example, into an arc chute. *See also:* **contactor; relay.** 83A16-0;50I15-34E10

blowout magnet. A permanent-magnet device that establishes a magnetic field in the space where an electric circuit is broken and helps to extinguish the arc by displacing it. *See also:* **relay.** 83A16-0

blue dip (electroplating). A solution containing a mercury compound, and used to deposit mercury upon an immersed metal, usually prior to silver plating. *See also:* **electroplating.** 42A60-0

blur (null-type direction-finding systems). The output (including noise) at the bearing of minimum response expressed as a percentage of the output at the bearing of maximum response. *See also:* **navigation.** E173-10E6

board (computing system). *See:* **problem board.** *See also:* **electronic analog computer.**

bobbin (1) (primary cell). A body in a dry cell consisting of depolarizing mix molded around a central rod of carbon and constituting the positive electrode in the assembled cell. *See also:* **electrolytic cell.** 42A60-0

(2) (rotating machinery). Spool-shaped ground insulation fitting tightly on a pole piece, into which field coil is wound or placed. *See also:* **rotor (rotating machinery); stator.** 0-31E8

bobbin core. A tape-wound core in which the ferromagnetic tape has been wrapped on a form or bobbin that supplies mechanical support to the tape. *Note:* The dimensions of a bobbin are illustrated in the accompanying figure. Bobbin I.D. is the center-hole diameter (D) of the bobbin. Bobbin O.D. is the over-all diameter (E) of the bobbin. The bobbin height is the over-all axial dimension (F) of the bobbin. Groove diameter is the diameter (G) of the center portion of the bobbin on

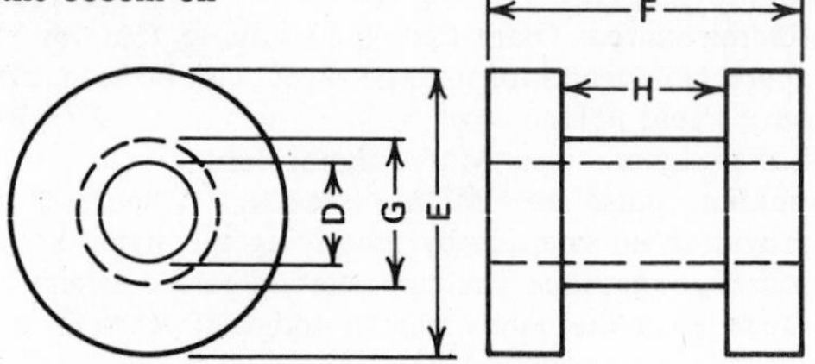

Dimensions of a bobbin.

which the first tape wrap is placed. The groove width is the axial dimension (H) of the bobbin measured inside the groove at the groove diameter. *See also:* **static magnetic storage.** E163-0

bobbin height. *See:* **bobbin core; tape-wound core.** *See also:* **static magnetic storage.**

bobbin I.D. *See:* **bobbin core; tape-wound core.** *See also:* **static magnetic storage.**

bobbin O.D. *See:* **bobbin core; tape-wound core.** *See also:* **static magnetic storage.**

Bode diagram. A plot of log-gain and phase-angle values on a log-frequency base, for an element transfer function $G(j\omega)$, a loop transfer function $GH(j\omega)$, or an output transfer function $G(j\omega)/[1 + GH(j\omega)]$. The generalized Bode diagram comprises similar plots of functions of the complex variable $s = \sigma + j\omega$. *Note:* Except for functions containing lightly damped quadratic factors, the gain characteristic may be approximated by asymptotic straight-line segments that terminate at corner frequencies. The ordinate may be expressed as a gain, a log-gain, or in decibels as 20 times log-gain; the abscissa as cycles per unit time, radians per unit time, or as the ratio of frequency to an arbitrary reference frequency. *See also:* **control system, feedback.** 85A1-23E0

body capacitance. Capacitance introduced into an electric circuit by the proximity of the human body. *See also:* **transmission characteristics.** 42A65-0

body-capacitance alarm system. A burglar alarm system for detecting the presence of an intruder through his body capacitance. *See also:* **protective signaling.** 42A65-0

body generator suspension. A design of support for an axle generator in which the generator is supported by the vehicle body. *See also:* **axle generator system.** 42A42-0

bolometer. *See:* **bolometric instrument; bolometric detector.**

bolometer bridge. A bridge circuit with provisions for connecting a balometer in one arm and for converting bolometer-resistance changes to indications of power. *See also:* **bolometric power meter.** 0-9E4

bolometer bridge, balanced. A bridge in which the bolometer is maintained at a prescribed value of resistance before and after radio-frequency power is applied, or after a change in radio-frequency power, by keeping the bridge in a state of balance. *Note:* The state of balance can be achieved automatically or manually by decreasing the bias power when the radio-frequency power is applied or increased and by increasing the bias power when the radio-frequency power is turned off, or decreased. The change in the bias power is a measure of the applied radio-frequency power. *See also:* **bolometric power meter.** 0-9E4

bolometer bridge, unbalanced. A bridge in which the resistance of the bolometer changes after the radio-frequency power is applied and unbalances the bridge. The degree of bridge unbalance is a measure of the radio-frequency power dissipated in the bolometer. *See also:* **bolometric power meter.** 0-9E4

bolometer-coupler unit. A directional coupler with a bolometer unit attached to either the side arm or the main arm, normally used as a feed-through power-measuring system. *Note:* Typically, a bolometer unit is attached to the side arm of the coupler so that the radio-frequency power at the output port can be determined from a measurement of the substitution power in the side arm. This system can be used as a terminating power meter by terminating the output port of the directional coupler. *See also:* **bolometric power meter.** 0-9E4

bolometer element. *See:* **bolometric detector.**

bolometer mount. A waveguide or transmission-line termination that houses a bolometer element(s). *Note:* It normally contains internal matching devices or other reactive elements to obtain specified impedance conditions when a bolometer element is inserted and appropriate bias power is applied. Bolometer mounts may be subdivided into tunable, fixed-tuned, and broad-band untuned types. *See also:* **bolometric power meter.** 0-9E4

bolometer unit. An assembly consisting of a bolometer element or elements and bolometer mount in which they are supported. *See also:* **bolometric power meter.** 0-9E4

bolometer unit, dual element. An assembly consisting of two bolometer elements and a bolometer mount in which they are supported. *Note:* The bolometer elements are effectively in series to the bias power and in parallel to the radio frequency power. 0-9E4

bolometric detector (bolometer). The primary detector in a bolometric instrument for measuring power or current and consists of a small resistor, the resistance of which is strongly dependent on its temperature. *Notes:* (1) Two forms of bolometric detector are commonly used for power or current measurement: (A) The barretter that consists of a fine wire or metal film; and (B) the thermistor that consists of a very small bead of semiconducting material having a negative temperature-coefficient of resistance; either is usually mounted in a waveguide or coaxial structure and connected so that its temperature can be adjusted and its resistance measured. (2) Bolometers for measuring radiant energy usually consist of blackened metal-strip temperature-sensitive elements arranged in a bridge circuit including a compensating arm for ambient temperature compensation. *See also:* **instrument; bolometric instrument.** 42A30-9E4

bolometric instrument (bolometer). An electrothermic instrument in which the primary detector is a resistor, the resistance of which is temperature sensitive, and that depends for its operation on the temperature difference maintained between the primary detector and its surroundings. Bolometric instruments may be used to measure nonelectrical quantities, such as gas pressure or concentration, as well as current and radiant power. *See also:* **instrument.** 42A30-0

bolometric technique (power measurement). A technique wherein the heating effect of an unknown amount of radio-frequency power is compared with that of a measured amount of direct-current or audio-frequency power dissipated within a temperature sensitive resistance element (bolometer). *Note:* The bolometer is generally incorporated into a bridge network, so that a small change in its resistance can be sensed. This technique is applicable to the measurement of low levels of radio-frequency power, that is, below 100 milliwatts. 0-9E4

bolometric power meter. A device consisting of a bolometer unit and associated bolometer-bridge circuit(s).

See:

bolometer bridge;

bolometer bridge, balanced;

bolometer bridge, unbalanced;

bolometer coupler unit;

bolometer mount;
bolometer unit;
efficiency;
efficiency, effective;
instrument;
substitution error;
substitution power;
thermoelectric-effect error. 0-9E4

bolt leakage (rotating machinery). The low-order harmonic air-gap flux attributable to the phase belts of a winding. *Note:* The magnitude of this leakage flux varies with winding pitch. *See also:* **rotor (rotating machinery); stator.** 0-31E8

bomb-control switch. A switch that closes an electric circuit, thereby tripping the bomb-release mechanism of an aircraft, usually by means of a solenoid. *See also:* **air transportation wiring and associated equipment.** 42A41-0

bombardment-induced conductivity (storage tubes). An increase in the number of charge carriers in semiconductors or insulators caused by bombardment with ionizing particles. *See also:* **storage tube.** E158-15E6

bond (rotating machinery). *See:* **binder.**

bonded motor (rotating machinery). A complete motor in which the stator and end shields are held together by a cement, or by welding or brazing. *See also:* **cradle base (rotating machinery).** 0-31E8

bonding (electric cables). The electric interconnecting of cable sheaths or armor to sheaths or armor of adjacent conductors. *See:* **cable bond cross-cable bond; continuity-cable bond.** *See also:* **power distribution, underground construction.** 42A35-31E13

bone conduction (hearing). The process by which sound is conducted to the inner ear through the cranial bones. *See also:* **electroacoustics.** S1A1-1E1

Boolean. (1) Pertaining to the processes used in the algebra formulated by George Boole. (2) Pertaining to the operations of formal logic. X3A12-16E9

boost. The act of increasing the power output capability of an operational amplifier by circuit modification in the output stage. *See also:* **electronic analog computer.** E165-0

boost charge (quick charge) (storage battery). A partial charge, usually at a high rate for a short period. *See also:* **charge.** 42A60-0

booster. An electric generator inserted in series in a circuit so that it either adds to or subtracts from the voltage furnished by another source. *See:* **direct-current commutating machine; synchronous machine.** 42A10-0

booster coil. An induction coil utilizing the aircraft direct-current supply to provide energy to the spark plugs of an aircraft engine during its starting period. *See also:* **air transportation electric equipment.** 42A41-0

booster dynamotor. A dynamotor having a generator mounted on the same shaft and connected in series for the purpose of adjusting the output voltage. *See:* **converter.** 42A10-0

bootleg (railway techniques). A protection for track wires when the wires leave the conduit or ground near the rail. *See also:* **railway signal and interlocking.** 42A42-0

bootstrap (computers). A technique or device designed to bring itself into a desired state by means of its own action, for example, a machine routine whose first few instructions are sufficient to bring the rest of itself into the computer from an input device. *See also:* **electronic digital computer.** X3A12-16E9

bootstrap circuit. A single-stage electron-tube amplifier circuit in which the output load is connected between cathode and ground or other common return, the signal voltage being applied between the grid and the cathode. *Note:* The name bootstrap arises from the fact that a change in grid voltage changes the potential of the input source with respect to ground by an amount equal to the output signal. *See also:* **circuits and devices.** E182A-0

***B* operator.** An operator assigned to a *B* switchboard. *See also:* **telephone system.** 42A65-0

bore (rotating machinery). The surface of a cylindrical hole (for example, stator bore). *See also:* **stator.** 0-31E8

bore-hole lead insulation (rotating machinery). Special insulation surrounding connections that pass through a hollow shaft. *See:* **rotor (rotating machinery).** 0-31E8

borehole cable (mining). A cable designed for vertical suspension in a borehole or shaft and used for power circuits in mines. *See also:* **mine feeder circuit.** 42A85-0

boresight error (antenna). The angular deviation of the electrical boresight of an antenna from its reference boresight. *See also:* **antenna.** E149-3E1

boresighting (directional radio). The process of aligning the axis of a directional antenna system, usually by an optical procedure. *See also:* **navigation.** 0-10E6

borrow. In direct subtraction, a carry that arises when the result of the subtraction in a given digit place is less than zero. *See also:* **electronic digital computer.** E162-0

bottom-car clearance (elevators). The clear vertical distance from the pit floor to the lowest structural or mechanical part, equipment, or device installed beneath the car platform, except guide shoes or rollers, safety jaw assemblies, and platform aprons or guards, when the car rests on its fully compressed buffers. *See also:* **hoistway (elevator or dumbwaiter).** 42A45-0

bottom-coil slot (radially outer-coil side) (rotating machinery). The coil side of a stator slot farthest from the bore of the stator or from the slot wedge. *See also:* **stator.** 0-31E8

bottom-half bearing (rotating machinery). The bottom half of a split-sleeve bearing. *See also:* **bearing.** 0-31E8

bottom-terminal landing (elevators). The lowest landing served by the elevator that is equipped with a hoistway door and hoistway-door locking device that permits egress from the hoistway side. *See also:* **elevator landing.** 42A45-0

bounce (television). A transient disturbance affecting one or more parameters of the display and having duration much greater than the period of one frame. *Note:* The term is usually applied to changes in vertical position or in brightness. *See:* **television.** 42A65-2E2

boundary (region). Comprises all the boundary points of that region. E270-0

boundary lights. Aeronautical ground-lights delimiting the boundary of a land aerodrome without runways. *See also:* **signal lighting.** Z7A1-0

boundary marker (instrument landing systems). A

radio transmitting station, near the approach end of the landing runway, that provides a fix on the localizer course. *See also:* **radio navigation.** 0-10E6

boundary, *p-n* (semiconductor). A surface in the transition region between *p*-type and *n*-type material at which the donor and acceptor concentrations are equal. *See also:* **semiconductor; transistor.** E102/E270-0;E216-34E17

boundary point (region). A point any neighborhood of which contains points that belong and points that do not belong to the region. E270-0

boundary potential. The potential difference, of whatever origin, across any chemical or physical discontinuity or gradient. *See also:* **electrobiology.** 42A80-18E1

bounded function. A function whose absolute value never exceeds some finite number *M.* E270-0

bowl (illumination techniques). An open-top diffusing glass or plastic enclosure used to shield a light source from direct view and to redirect or scatter the light. *See also:* **bare (exposed) lamp.** Z7A1-0

box frame (rotating machinery). A stator frame in the form of a box with ends and sides and that encloses the stator core. *See also:* **rotor (rotating machinery).** 0-31E8

bracket function [*x*]. Equal to x if x is an integer and is the lesser of the two integers between which x falls if x is not an integer. *Note:* Thus, if $x = 2$ then $[x] = 2$, if $x = 1.5$ then $[x] = 1$, if $x = -3.7$ then $[x] = -4$ (noting that lesser means less positive). E270-0

bracket (mast arm) (illumination techniques). An attachment to a lamp post or pole from which a luminaire is suspended. *See also:* **street-lighting luminaire.** Z7A1-0

bracket-type handset telephone (suspended-type handset telephone). *See:* **hang-up hand telephone set.**

brake assembly (rotating machinery). All parts that are provided to apply braking to the rotor. *See also:* **rotor (rotating machinery).** 0-31E8

brake control (industrial control). The provision for controlling the operation of an electrically actuated brake. *Note:* Electrical energizing of the brake may either release or set the brake, depending upon its design. *See:* **electric controller.** IC1-34E10

brake drum. *See:* **brake ring.**

brake ring (brake drum) (rotating machinery). A rotating ring mounted on the rotor that provides a bearing surface for the brake shoes. *See also:* **rotor (rotating machinery).** 0-31E8

braking (industrial control). The control function of retardation by dissipating the kinetic energy of the drive motor and the driven machinery. *See also:* **electric drive.** 42A25-34E10

braking magnet (drag magnet). *See:* **retarding magnet.**

braking resistor. A resistor commonly used in some types of dynamic braking systems, the prime purpose of which is to convert the electric energy developed during dynamic braking into heat and to dissipate this energy to the atmosphere. *See also:* **dynamic braking.** 42A42-0

braking test (rotating machinery). (1) A test in which the mechanical power output of a machine acting as a motor is determined by the measurement of the shaft torque, by means of a brake, dynamometer, or similar device, together with the rotational speed. (2) A test performed on a machine acting as a generator, by means of a dynamometer or similar device, to determine the mechanical power input. *See also:* **asynchronous machine; direct-current commutating machine; synchronous machine.** 0-31E8

braking torque (synchronous motor). Any torque exerted by the motor in the same direction as the load torque so as to reduce its speed. 0-31E8

branch (1) (arm) (network). A portion of a network consisting of one or more two-terminal elements in series, comprising a section between two adjacent branch-points. E153/E270/42A65-0

(2) (network analysis). A line segment joining two nodes, or joining one node to itself. *See also:* **directed branch; network analysis; linear signal flow graphs.** E155-0

(3) (electronic computers). (A) A set of instructions that are executed between two successive decision instructions. (B) To select a branch as in (A). (C) Loosely, a conditional jump. *See:* **conditional jump.** *See also:* **electronic digital computer.** X3A12-16E9

branch circuit. That portion of a wiring system extending beyond the final overcurrent device protecting the circuit. *See:* **appliance branch circuit; combination lighting and appliance; lighting branch circuit; motor branch circuit; receptacle circuit.** 42A95-0

branch-circuit distribution center. A distribution center at which branch circuits are supplied. *See also:* **distribution center.** 42A95-0

branch circuit, general purpose. A branch circuit that supplies a number of outlets for lighting and appliances. 1A0-0

branch circuit, individual. A branch circuit that supplies only one utilization equipment. 1A0-0

branch circuit, multiwire. A circuit consisting of two or more ungrounded conductors having a potential difference between them, and an identified grounded conductor having equal potential difference between it and each ungrounded conductor of the circuit and that is connected to the neutral conductor of the system. 1A0-0

branch conductor (lightning protection). A conductor that branches off at an angle from a continuous run of conductor. *See also:* **lightning protection and equipment.** 42A95-0

branch current. The current in a branch of a network. E270-0

branch impedance. The impedance of a passive branch when no other branch is electrically connected to the branch under consideration. E270-0

branch input signal (network analysis). The signalx_j at the input end of branch *jk*. *See also:* **linear signal flow graphs.** E155-0

branch joint (T joint) (tee joint) (Y joint) (multiple joint). A joint used for connecting a branch conductor or cable to a main conductor or cable, where the latter continues beyond the branch. *Note:* A branch joint may be further designated by naming the cables between which it is made, for example, single-conductor cables; three-conductor main cable to single-conductor branch cable, etcetera. With the term **multiple joint** it is customary to designate the various kinds as 1-way, 2-way, 3-way, 4-way, etcetera, multiple joint. *See:* **cable joint; straight joint; reducing joint.** *See also:* **power distribution, underground construction.** 42A35-31E13

branch output signal (branch *jk*) (network analysis). The component of signal x_k contributed to node k via branch *jk*. *See also:* **linear signal flow graphs.** E155-0

branch point (1) (electric network). A junction where more than two conductors meet. *See also:* **network analysis; node.** 42A65-0

(2) (computers). A place in a routine where a branch is selected. *See also:* **electronic digital computer; network analysis.** X3A12-12-16E9

branch, thermoelectric. Alternative term for thermoelectric arm. *See also:* **thermoelectric device.** E221-15E7

branch transmittance (network analysis). The ratio of branch output signal to branch input signal. *See also:* **linear signal flow graphs.** E155-0

branch voltage. The voltage between the two nodes terminating the branch, the path of integration coinciding with the branch. E270-0

breadboard construction (communication practice). An arrangement in which components are fastened temporarily to a board for experimental work. 42A65-0

break (1) (circuit-opening device). The minimum distance between the stationary and movable contacts when these contacts are in the open position. (A) The length of a single break is as defined above. (B) The length of a multiple break (breaks in series) is the sum to two or more breaks. *See:* **contactor.** IC1-34E10

(2) (communication circuit). For the receiving operator or listening subscriber to interrupt the sending operator or talking subscriber and take control of the circuit. *Note:* The term is used especially in connection with half-duplex telegraph circuits and two-way telephone circuits equipped with voice-operated devices. *See also:* **telegraphy.** 42A65-0

breakaway. The condition of a motor at the instant of change from rest to rotation. *See also:* **asynchronous machine; direct-current commutating machine; synchronous machine.** 0-31E8

breakaway torque (rotating machinery). The torque that a motor is required to develop to break away its load from rest to rotation. *See:* **asynchronous machine; direct-current commutating machine; synchronous machine.** 0-31E8

breakdown (1) (gas) (electron device). The abrupt transition of the gap resistance from a practically infinite value to a relatively low value. *See:* **gas tubes.** 0-15E6

(2) (semiconductor diode). A phenomenon occurring in a reverse-biased semiconductor diode, the initiation of which is observed as a transition from a region of high dynamic resistance to a region of substantially lower dynamic resistance for increasing magnitude of reverse current. *See also:* **rectification; semiconductor.** E216-34E17;E270-0

(3) (rotating machinery). The condition of operation when a motor is developing breakdown torque. *See also:* **asynchronous machine; direct-current commutating machine; synchronous machine.** 0-31E8

(4) (puncture) (lightning). A disruptive discharge through insulation. *See also:* **lightning arrester (surge diverter); power systems, low-frequency and surge testing.** 42A35-31E13/31E7

breakdown impedance (semiconductor diode). The small-signal impedance at a specified direct current in the breakdown region. *See also:* **semiconductor.** E216-34E17;E270-0

breakdown region (semiconductor diode characteristic). The entire region of the volt-ampere characteristic beyond the initiation of breakdown for increasing magnitude of reverse current. *See also:* **rectification; semiconductor.** E59-34E17;E216/E270-0

breakdown strength. *See:* **dielectric strength.**

breakdown torque (rotating machinery). The maximum shaft-output torque that an induction motor (or a synchronous motor operating as an induction motor) develops when the primary winding is connected for running operation, at normal operating temperature, with rated voltage applied at rated frequency. *Note:* A motor with a continually increasing torque as the speed decreases to standstill, is not considered to have a breakdown torque. *See:* **synchronous machine.** 0-31E8

breakdown-torque speed (rotating machinery). The speed of rotation at which a motor develops breakdown torque. *See also:* **asynchronous machine; direct-current commutating machine; synchronous machine.** 0-31E8

breakdown transfer characteristic (gas tube). A relation between the breakdown voltage of an electrode and the current to another electrode. *See:* **gas tubes.** E160-15E6

breakdown voltage (1) (insulation). The voltage at which a disruptive discharge takes place through or over the surface of the insulation. *See also:* **insulation testing (large alternating-current rotating machinery).** E95-0

(2) (gas) (electron device). The voltage necessary to produce a breakdown. *See:* **gas tubes.** 0-15E6

(3) (electrode of a gas tube). The voltage at which breakdown occurs to that electrode. *Notes:* (1) The breakdown voltage is a function of the other electrode voltages or currents and of the environment. (2) In special cases where the breakdown voltage of an electrode is referred to an electrode other than the cathode, this reference electrode shall be indicated. (3) This term should be used in preference to pickup voltage, firing voltage, starting voltage, etcetera, which are frequently used for specific types of gas tubes under specific conditions. *See:* **critical grid voltage (multielectrode gas tubes).** *See also:* **gas tubes.** E160-15E6

(4) (glow lamp) (starting or igniting voltage). The voltage across the lamp at which an abrupt increase in current between operating electrodes occurs and is accompanied by a negative glow. 78A385-0

(5) (electrochemical valve). The lowest voltage at which failure of an electrochemical valve occurs. *See also:* **electrochemical valve.** 42A60-0

(6) (semiconductor diode). The voltage measured at a specified current in the breakdown region. *See also:* **semiconductor; semiconductor device.** E216-34E17;E270-0

breakdown voltage, anode (glow-discharge cold-cathode tube). *See:* **anode breakdown voltage.**

breakdown voltage starter (glow-discharge cold-cathode tube). The starter voltage required to cause conduction across the starter gap with all other tube elements held at cathode potential before breakdown. *See also:* **electrode voltage (electron tube).** 42A70-15E6

breaking capacity (interrupting capacity) (industrial control). The current that the device is capable of breaking at a stated recovery voltage under prescribed conditions of use and behavior. *See:* **control.** 50I15-34E10

breaking current (pole of a breaking device) (industrial control). The current in that pole at the instant of contact separation, expressed as a root-mean-square value. *See:* **contactor.** 50I15-34E10

breaking point (transmission system or element thereof). A level at which there occurs an abrupt change in distortion or noise that renders operation unsatisfactory. *See also:* **level.** 42A65-0

break-in keying. A method of operating a continuous-wave radio telegraph communication system in which the receiver is capable of receiving signals during transmission of spacing intervals. *See also:* **modulation; radio transmission.** 42A65-0

breakover current (thyristor). The principal current at the breakover point. *See also:* **principal current.** E223-15E7/34E17/34E24

breakover point (thyristor). Any point on the principal voltage-current characteristic for which the differential resistance is zero and where the principal voltage reaches a maximum value. *See also:* **principal voltage-current characteristic (principal characteristic).** E223-15E7/34E17

breakover voltage (thyristor). The principal voltage at the breakover point. *See also:* **principal voltage-current characteristic (principal characteristic).** E223-15E7/34E17/34E24

breakpoint (1) (plotted curve). The junction of two confluent straight-line segments of a plotted curve. *Note:* In the asymptotic approximation of a log-gain versus log-frequency relation in a Bode diagram, the value of the abscissa is called the corner frequency. *See also:* **control system, feedback.** 85A1-23E0
(2) (computer routine). (A) Pertaining to a type of instruction, instruction digit, or other condition used to interrupt or stop a computer at a particular place in a routine when manually requested. (B) A place in a routine where such an interruption occurs or can be made to occur. *See also:* **electronic computation; electronic digital computer.** E162-0;X3A12-16E9

breather. A device fitted in the wall of an explosion-proof compartment, or connected by piping thereto, that permits relatively free passage of air through it, but that will not permit the passage of incendiary sparks or flames in the event of gas ignition inside the compartment. *See also:* **mining.** 42A85-0

breathing (carbon microphones). The phenomenon manifested by a slow cyclic fluctuation of the electric output due to changes in resistance resulting from thermal expansion and contraction of the carbon chamber. *See also:* **close-talking pressure-type microphone.** E258-0

breezeway (television synchronizing waveform for color transmission). The time interval between the trailing edge of the horizontal synchronizing pulse and the start of the color burst. 42A65-2E2

bridge (bridge network). (1) A network with a minimum of two ports or terminal pairs capable of being operated in such a manner that when power is fed into one port, by suitable adjustment of the elements in the network or of elements connected to one or more other ports, zero output can be obtained at another port. Under these conditions, the bridge is balanced. *Note:* The described networks include hybrids and two-channel nulling circuits. 0-9E4
(2) An instrument or an intermediate means in a measurement system that embodies part or all of a bridge circuit, and by means of which one or more of the electrical constants of a bridge circuit may be measured. *Note:* The operation of a bridge consists of the insertion of a suitable electromotive force and a suitable detecting device in branches that can be made conjugate and that do not include the branch whose constants are to be measured, followed by the adjustment of one or more of the branches until the response of the detecting device becomes zero or an amount measurable by the detector for the purpose of interpolation.
See also:
Anderson bridge;
conjugate bridge;
Hay bridge;
Heaviside-Campbell mutual inductance bridge;
Heaviside mutual inductance bridge;
instrument;
Kelvin bridge;
Maxwell bridge;
Maxwell direct-current commutator bridge;
Maxwell inductance bridge;
Maxwell mutual inductance bridge;
Owen bridge;
resonance bridge;
Schering bridge;
Wheatstone bridge;
Wien capacitance bridge;
Wien inductance bridge. 42A30-0

bridge control. Apparatus and arrangement providing for direct control from the bridge or wheelhouse of the speed and direction of a vessel. 42A43-0

bridge current (power supplies). The circulating control current in the comparison bridge. *Note:* Bridge current equals the reference voltage divided by the reference resistor. Typical values are 1 milliampere and 10 milliamperes, corresponding to control ratios of 1000 ohms per volt and 100 ohms per volt, respectively. *See also:* **power supply.** KPSH-10E1

bridged-T network. A T network with a fourth branch connected across the two series arms of the T, between an input terminal and an output terminal. *See also:* **network analysis.** E153/E270-0

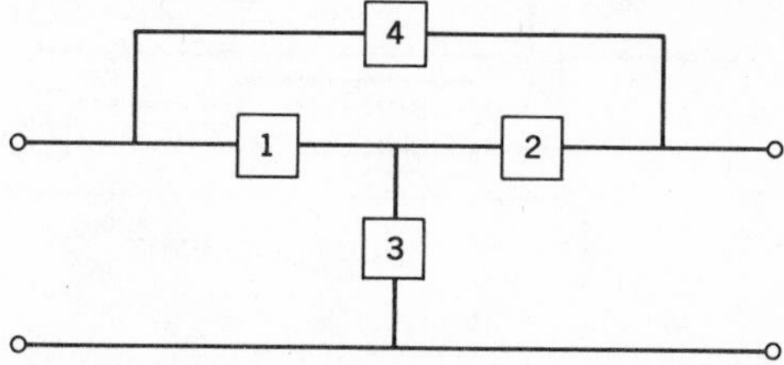

Bridged-*T* network. Branches 1, 2, and 3 comprise the T network and branch 4 is the fourth branch.

bridge duplex system. A duplex system based on the Wheatstone bridge principle in which a substantial neutrality of the receiving apparatus to the sent currents is obtained by an impedance balance. *Note:* Received currents pass through the receiving relay that is bridged between the points that are equipotential for the sent currents. *See also:* **telegraphy.** 42A65-0

bridge limiter. *See:* **limiter circuit.** *See also:* **electronic analog computer.**

bridge network. Any network (of relatively few branches), containing only two-terminal elements, that is not a series-parallel network. *Note:* Elementary bridges such as the Wheatstone bridge are special cases of a bridge network. *See also:* **network analysis.** E270-0

bridge rectifier. A full-wave rectifier with four rectifying elements connected as the arms of a bridge circuit. *See also:* **rectifier.** E145-42A65

bridge transition. A method of changing the connection of motors from series to parallel in which all of the motors carry like currents throughout the transfer due to the Wheatstone bridge connection of motors and resistors. *See also:* **multiple-unit control.** 42A42-0

bridging (1) (signal circuits). The shunting of one signal circuit by one or more circuits usually for the purpose of deriving one or more circuit branches. *Note:* A bridging circuit often has an input impedance of such a high value that it does not substantially affect the circuit bridged. E151-0

(2) (soldered connections). Solder that forms an unwanted conductive path. *See also:* **soldered connections (electronic and electrical applications).** 99A1-0

(3) (relays). *See:* **relay bridging.**

bridging amplifier. *See:* **amplifier, bridging.**

bridging connection. A parallel connection by means of which some of the signal energy in a circuit may be withdrawn, frequently with imperceptible effect on the normal operation of the circuit. *See also:* **circuits and devices.** 42A65-0

bridging gain. The ratio of the signal power a transducer delivers to its load (Z_B) to the signal power dissipated in the main circuit load (Z_M) across which the input of the transducer is bridged. *Note:* Bridging gain is usually expressed in decibels. See accompanying figure. *See also:* **transmission characteristics.** E151-0

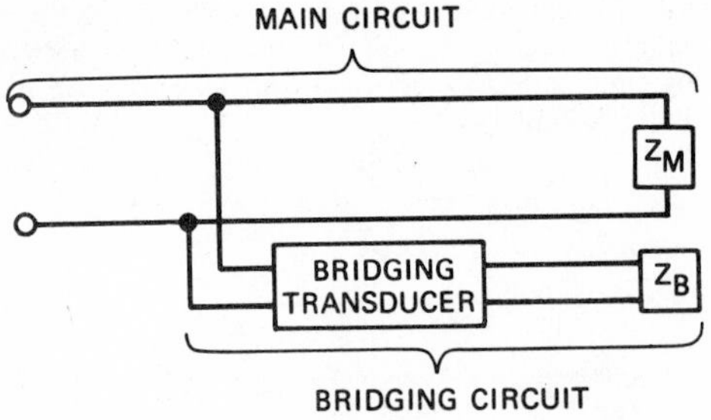

Bridging gain.

bridging loss. (1) The ratio of the signal power delivered to that part of the system following the bridging point before the insertion of the bridging element to the signal power delivered to the same part after the bridging. (2) The ratio of the power dissipated in a load *B* across which the input of a transducer is bridged, to the power the transducer delivers to its load *A*. *Notes:* (1) The inverse of bridging gain. Bridging loss is usually expressed in decibels. (2) It may be considered a special case of insertion loss. *See also:* **transmission loss.** E151-0;42A65-31E3

bridle wire. Insulated wire for connecting conductors of an open wire line to associated pole-mounted apparatus. *See also:* **open wire.** 42A65-0

bright dip (electroplating). A dip used to produce a bright surface on a metal. *See also:* **electroplating.** 42A60-0

brightener (electroplating). An addition agent used for the purpose of producing bright deposits. *See also:* **electroplating.** 42A60-0

brightness. The attribute of visual perception in accordance with which an area appears to emit more or less light. *Note:* Luminance is recommended for the photometric quantity that has been called brightness. Luminance is a purely photometric quantity. Use of this name permits brightness to be used entirely with reference to the sensory response. The photometric quantity has been often confused with the sensation merely because of the use of one name for two distinct ideas. Brightness will continue to be used, properly, in nonquantitative statements, especially with reference to sensations and perceptions of light. Thus, it is correct to refer to a brightness match, even in the field of a photometer, because the sensations are matched and only by inference are the photometric quantities (luminances) equal. Likewise, a photometer in which such matches are made will continue to be called an equality-of-brightness photometer. A photoelectric instrument, calibrated in foot-lamberts, should not be called a brightness meter. If correctly calibrated, it is a luminance meter. A troublesome paradox is eliminated by the proposed distinction of nomenclature. The luminance of a surface may be doubled, yet it will be permissible to say that the brightness is not doubled, since the sensation that is called brightness is generally judged to be not doubled. *See also:* **color terms.** E201-2E2/9E4

brightness (perceived light source color). The attribute in accordance with which the source seems to emit more or less luminous flux per unit area. *See also:* **color.** Z7A1-0

brightness channel* (television). *See:* **luminance channel.** *See also:* **television.**

*Deprecated

brightness control (television). A control, associated with a picture display device, for adjusting the average luminance of the reproduced picture. *Note:* In a cathode-ray tube the adjustment is accomplished by shifting bias. This affects both the average luminance and the contrast ratio of the picture. In a color-television system, saturation and hue are also affected. 42A65-0

brightness signal.* *See:* **luminance signal.**

*Deprecated

brine. A salt solution. 42A60-0

broadband interference (measurement) (electromagnetic compatibility). A disturbance that has a spectral energy distribution sufficiently broad, so that the response of the measuring receiver in use does not vary significantly when tuned over a specified number of receiver bandwidths. *See also:* **electromagnetic compatibility.** 0-27E1

broadband radio noise. Radio noise having a spectrum broad in width as compared to the nominal bandwidth of the radio-noise meter, and whose spectral components are sufficiently close together and uniform so that the receiver cannot resolve them. *See also:* **radio-noise field strength.** 63A4-0

broadband tube (microwave gas tubes). A gas-filled fixed-tuned tube incorporating a bandpass filter of geometry suitable for radio-frequency switching. *See:* **gas tubes; pretransmit-receive tube; transmit-receive tube.** E160-15E6

broadside array. A linear or planar array antenna whose direction of maximum radiation is perpendicular to the line or plane of the array. *See also:* **antenna.** 0-3E1

bronze conductor. A conductor made wholly of an alloy of copper with other than pure zinc. *Note:* The copper may be alloyed with tin, silicon, cadmium, manganese, or phosphorus, for instance, or several of these in combination. *See also:* **conductor.** 42A35-31E13

bronze leaf brush (rotating machinery). A brush made up of thin bronze laminations. *See also:* **brush.** 0-31E8

brush (1) (electric machines). A conductor, usually composed in part of some form of the element carbon, serving to maintain an electric connection between stationary and moving parts of a machine or apparatus. *Note:* Brushes are classified according to the types of material used, as follows: carbon, carbon-graphite, electrographitic, graphite, and metal-graphite. *Note:* For an extensive list of cross references, see *Appendix A.* 64A1-0
(2) (relay). *See:* **relay wiper.**

brush box (rotating machinery). The part of a brush holder that contains a brush. *See also:* **brush.** 0-31E8

brush chamfers (electric machines). The slight removal of a sharp edge. *See also:* **brush.** 64A1-0

brush contact loss (rotating machinery). The I^2R loss in brushes and contacts of the field collector ring or the direct-current armature commutator. *See also:* **brush.** 0-31E8

brush convex and concave ends (electric machines). Partially cylindrical surfaces of a given radius. *Note:* When concave bottoms are applied to bevels, both bevel angle and radius shall be given. *See also:* **brush.** 64A1-0

brush corners (electric machines). The point of intersection of any three surfaces. *Note:* They are designated as top or face corners. *See also:* **brush.** 64A1-0

brush diameter (electric machines). The dimension of the round portion that is at right angles to the length. *See also:* **brush.** 64A1-0

brush discharge (gas). An intermittent discharge of electricity having the form of a mobile brush, that starts from a conductor when its potential exceeds a certain value, but remains too low for the formation of an actual spark. *Note:* It is generally accompanied by a whistling or crackling noise. *See also:* **discharge (gas).** E270-0

brush edges (electric machines). The intersection of any two brush surfaces. *See also:* **brush.** 64A1-0

brush ends (electric machines). The surfaces defined by the width and thickness of the brush. *Note:* They are designated as top or holder end and bottom or commutator end. The end that is in contact with the commutator or ring is also known as the brush face. *See also:* **brush.** 64A1-0

brush friction loss (rotating machinery). The mechanical loss due to friction of the brushes normally included as part of the friction and windage loss. *See also:* **brush.** 50A10-31E8

brush hammer, lifting, or guide clips (electric machines). Metal parts attached to the brush that serve to accommodate the spring finger or hammer or to act as guides. *Note:* Where these serve to prevent the wear of the carbon due to the pressure finger, they are called hammer or finger clips. Rotary converter brushes may have clips that serve the dual purpose of lifting the brushes and of preventing wear from the spring finger. These are generally called lifting clips. *See also:* **brush.** 64A1-0

brush holder (rotating machinery). A structure that supports a brush and that enables it to be maintained in contact with the sliding surface. *See also:* **brush (rotating machinery).** 0-31E8

brush-holder bolt insulation (rotating machinery). A combination of members of insulating materials that insulate the brush yoke mounting bolts, brush yoke, and brush holders. *See also:* **brush.** 0-31E8

brush-holder insulating barriers (rotating machinery). Pieces of sheet insulation installed in the brush yoke assembly to provide longer leakage paths between live parts and ground, or between live parts of different polarities. *See also:* **brush.** 0-31E8

brush-holder spindle insulation (rotating machinery). Insulation members that (when required by design) insulate the spindle on which the brush spring is mounted from the brush yoke and the brush holder. *See also:* **brush.** 0-31E8

brush-holder spring. That part of the brush holder that provides pressure to hold the brush against the collector ring or commutator. *See also:* **brush (rotating machinery).** 42A10-0

brush-holder stud (rotating machinery). An intermediate member between the brush holder and the supporting structure. *See also:* **brush (rotating machinery).** 42A10-31E8

brush-holder-stud insulation (rotating machinery). An assembly of insulating material that insulates the brush holder or stud from the supporting structure. *See also:* **brush (rotating machinery).** 42A10-31E8

brush-holder support (rotating machinery). The intermediate member between the brush holder or holders and the supporting structure. *Note:* This may be in the form of plates, spindles, studs, or arms. *See also:* **brush.** 0-31E8

brush-holder yoke. A rocker arm, ring, quadrant, or other support for maintaining the brush holders or brush-holder studs in their relative positions. *See also:* **brush (rotating machinery).** 42A10-0

brush length (electric machines). The maximum overall dimension of the carbon only, measured in the direction in which the brush feeds to the commutator or collector ring. *See also:* **brush.** 64A1-0

brushless (rotating machinery). Applied to machines with primary and secondary or field windings that are constructed such that all windings are stationary, or in which the conventional brush gear is eliminated by the use of transformers having both moving and stationary windings, or by the use of rotating rectifiers. *See also:* **brush.** 0-31E8

brushless exciter. An alternating-current rotating-armature-type exciter whose output is rectified by a semiconductor device to provide excitation to an electric machine. *See:* **synchronous machine; asynchronous machine.** E45-0

brushless synchronous machine. A synchronous machine having a brushless exciter with its rotating armature and semiconductor devices on a common shaft with the field of the main machine. *Note:* This type of machine with its exciter has no collector, commutator, or brushes. *See:* **synchronous machine.** E45-0

brush or sponge plating (electroplating). A method of plating in which the anode is surrounded by a brush or sponge or other absorbent to hold electrolyte while it is moved over the surface of the cathode during the plating operation. *See also:* **electroplating.** 42A60-0

brush rigging (rotating machinery). The complete assembly of parts whose main function is to position and support all of the brushes for a commutator or collector. *See also:* **brush.** 0-31E8

brush rocker (brush yoke) (rotating machinery). A structure from which the brush holders are supported and fixed relative to each other and so arranged that the whole assembly may be moved circumferentially. *See also:* **brush.** 0-31E8

brush-rocker gear (brush-yoke gear) (rotating machinery). The worm wheel or other gear by means of which the position of the brush rocker may be adjusted. *See also:* **brush.** 0-31E8

brush shoulders (electric machines). When the top of the brush has a portion cut away by two planes at right angles to each other, this is designated as a shoulder. *See also:* **brush.** 64A1-0

brush shunt (rotating machinery). A stranded cable or other flexible conductor attached to a brush to connect it electrically to the machine or apparatus. *Note:* Its purpose is to conduct the current that would otherwise flow from the brush to the brush holder or brush-holder finger. *See:* **brush (rotating machinery).** 42A10-31E8

brush shunt length (electric machines). The distance from the extreme top of the brush to the center of the hole or slot in the terminal, or the center of the inserted portion of a plug terminal or, if there is no terminal, to the end of the shunt. *See also:* **brush.** 64A1-0

brush sides (electric machines). (1) Front and back (bounded by width and length). *Note:* (1) If the brush has one or both ends beveled, the short side of the brush is the front. (2) If there are no top or bottom bevels and width is greater than thickness and there is a top clip, the side to which the clip is attached is the back, except in the case of angular clips where the front or short side is determined by the slope of the clip and not by the side to which it is attached. (3) Left side and right side (bounded by thickness and length). *See also:* **brush.** 64A1-0

brush slots, grooves, and notches (electric machines). Hollows in the brush. *See also:* brush. 64A1-0

brush spring (rotating machinery). The portion of a brush holder that exerts pressure on the brush to hold it in contact with the sliding surface. 0-31E8

brush thickness (electric machines). The dimension at right angles to the length in the direction of rotation. *See also:* **brush.** 64A1-0

brush width (electric machines). The dimension at right angles to the length and to the direction of rotation. *See also:* **brush.** 64A1-0

brush yoke. *See:* **brush rocker.**

***B* scan (electronic navigation).** *See:* ***B* display.**

***B* scope (class-*B* oscilloscope) (electronic navigation).** A cathode-ray oscilloscope arranged to present a type *B* display. *See also:* ***B* display; radar.** 42A65-0

***B* stage.** An intermediate stage in the reaction of certain thermosetting resins in which the material swells when in contact with certain liquids and softens when heated, but may not entirely dissolve or fuse. *Note:* The resin in an uncured thermosetting moulding compound is usually in this stage. D883-31E8

***B*-station* (electronic navigation).** *See:* ***B*-trace.**

*Deprecated

***B* switchboard.** A manual telephone switchboard in a local central office, primarily to receive and complete connections from other central offices. *See also:* **telephone switching system.** 42A65-19E1

B-trace (loran). The second (lower) trace on the scope display. *See also:* **navigation.** 0-10E6

Buchmann-Meyer pattern (mechanical recording). *See:* **light pattern.**

buck arm. A crossarm placed approximately at right angles to the line crossarm and used for supporting branch or lateral conductors or turning large angles in line conductors. *See also:* **tower.** 42A35-31E13

buffer (1) (circuit design). An isolating circuit used to avoid reaction of a driven circuit upon the corresponding driving circuit.

(2) (data processing and computation). A storage device used to compensate for a difference in rate of flow of information or time of occurrence of events when transmitting information from one device to another. *See also:* **electronic computation.** E162/E270-0;X3A12-16E9

(3) (elevator design). A device designed to stop a descending car or counterweight beyond its normal limit of travel by storing or by absorbing and dissipating the kinetic energy of the car or counterweight. *See also:* **elevator.** 42A45-0

(4) (relay). *See:* **relay spring stud.**

buffer amplifier. An amplifier in which the reaction of output-load-impedance variation on the input circuit is reduced to a minimum for isolation purposes. *See also:* **amplifier; electronic analog computer; unloading amplifier.** E145-0;E165-16E9

buffer salts. *See:* **buffers.**

buffer storage. An intermediate storage medium between data input and active storage. EIA3B-34E12

buffers (buffer salts). Salts or other compounds that reduce the changes in the pH of a solution upon the addition of an acid or alkali. *See also:* **ion.** 42A60-0

buffing (electroplating). The smoothing of a metal surface by means of flexible wheels, to the surface of which fine abrasive particles are applied, usually in the form of a plastic composition or paste. *See also:* **electroplating.** 42A60-0

bug. A semiautomatic telegraph key in which movement of a lever to one side produces a series of correctly spaced dots and movement to the other side produces a single dash. *See also:* **telegraphy.** 42A65-0

bugduster. An attachment used on shortwall mining machines to remove cuttings (bugdust) from back of the cutter and to pile them at a point that will not interfere with operation. *See also:* **mining.** 42A85-0

building. A structure that stands alone or that is cut off from adjoining structures by fire walls with all openings therein protected by approved fire doors. 1A0-0

building bolt (rotating machinery). A bolt used to insure alignment and clamping of parts. *See also:* **cradle base (rotating machinery).** 0-31E8

building out (communication practice). The addition to an electric structure of an element or elements electrically similar to an element or elements of the structure, in order to bring a certain property or characteristic to a desired value. *Note:* Examples are building-out capacitors, building-out sections of line, etcetera. *See also:* **communication.** 42A65-0

building-out capacitor (building-out condenser*). A capacitor employed to increase the capacitance of an electric structure to a desired value. *See also:* **circuits and devices.**

*Deprecated 42A65-21E0

building-out network. An electric network designed to be connected to a basic network so that the combination will simulate the sending-end impedance, neglecting dissipation, of a line having a termination other than that for which the basic network was designed. *See also:* **network analysis.** 42A65-0

building pin (rotating machinery). A dowel used to insure alignment of parts. *See also:* **cradle base (rotating machinery).** 0-31E8

building up (electroplating). Electroplating for the purpose of increasing the dimensions of an article. *See also:* **electroplating.** 42A60-0

built-in ballast (mercury lamp). A ballast specifically designed to be built into a lighting fixture. 82A4-0

built-in check (computers). *See:* **automatic check.** *See also:* **electronic digital computer.**

built-in transformer. A transformer specifically designed to be built into a luminaire. 82A7-0

built-up connection. A toll call that has been relayed through one or more switching points between the originating operator and the receiving exchange. *See also:* **telephone system.** 42A65-0

bulb (electron tubes and electric lamp). (1) The glass envelope used in the assembly of an electron tube or an electric lamp. (2) The glass component part used in a bulb assembly. 79A1-0

bull ring. A metal ring used in overhead construction at the junction point of three or more guy wires. *See also:* **tower.** 42A35-31E13

bump. *See:* **distortion pulse.**

bumper (elevators). A device other than an oil or spring buffer designed to stop a descending car or counterweight beyond its normal limit of travel by absorbing the impact. *See also:* **elevator.** 42A45-0

buncher space (velocity-modulated tube). The part of the tube following the acceleration space where there is a high-frequency field, due to the input signal, in which the velocity modulation of the electron beam occurs. *Note:* It is the space between the input resonator grids. *See also:* **velocity-modulated tube.** 50I07-15E6

bunching. The action in a velocity-modulated electron stream that produces an alternating convection-current component as a direct result of the differences of electron transit time produced by the velocity modulation. *See:* **optimum bunching; overbunching; reflex bunching; space-charge debunching; underbunching.** *See also:* **electron devices, miscellaneous.** E160-15E6

bunching angle (electron stream) (given drift space). The average transit angle between the processes of velocity modulation and energy extraction at the same or different gaps. *See also:* **bunching angle, effective (reflex klystrons); electron devices, miscellaneous.** E160-15E6

bunching angle, effective (reflex klystrons) (given drift space). The transit angle that would be required in a hypothetical drift space in which the potentials vary linearly over the same range as in the given space and in which the bunching action is the same as in the given space. *See:* **magnetrons.** *See also:* **electron devices, miscellaneous.** E160-15E6

bunching, optimum. *See:* **optimum bunching.**

bunching parameter. One-half the product of: (1) the bunching angle in the absence of velocity modulation and (2) the depth of velocity modulation. *Note:* In a reflex klystron the effective bunching angle must be used. *See also:* **electron devices, miscellaneous.** E160-15E6

bunching time, relay. *See:* **relay bunching time.**

***B* unit.** A motive power unit designed primarily for use in multiple with an *A* unit for the purpose of increasing locomotive power, but not equipped for use as the leading unit of a locomotive or for full observation of the propulsion power and brake applications for a train. *Note: B* units are normally equipped with a single control station for the purpose of independent movement of the unit only, but are not usually provided with adequate instruments for full observation of power and brake applications. *See also:* **electric locomotive.** 42A42-0

burden (1) (instrument transformer). That property of the circuit connected to the secondary winding that determines the active and reactive power at the secondary terminals. *Note:* It is expressed either as total ohms impedance together with the effective resistance and reactance components of the impedance or as the total volt-amperes and power factor of the secondary devices and leads at the specified values of frequency and current or voltage. *See:* **instrument transformer.** 12A042A30-0;42A15-31E12

(2) (relay). Load imposed by a relay on an input circuit, expressed in ohms or volt-amperes. *See also:* **relay.** 0-31E6

burden regulation (capacitance potential device). Refers to the variation in voltage ratio and phase angle of the secondary voltage of the capacitance potential device as a function of burden variation over a specified range, when energized with constant, applied primary line-to-ground voltage. *See also:* **outdoor coupling capacitor.** E31-0

burglar-alarm system. An alarm system signaling an entry or attempted entry into the area protected by the system. *See also:* **protective signaling.** 42A65-0

buried cable. A cable installed under the surface of the ground in such a manner that it cannot be removed without disturbing the soil. *See also:* **cable; power distribution, underground construction.** 42A65-0

burn-in (reliability). The operation of items prior to their ultimate application intended to stabilize their characteristics and to identify early failures. *See also:* **reliability.** 0-7E1

burnishing. The smoothing of metal surfaces by means of a hard tool or other article. *See also:* **electroplating.** 42A60-0

burnishing surface (mechanical recording). The portion of the cutting stylus directly behind the cutting edge, that smooths the groove. *See also:* **phonograph pickup.** S1A1/1E1

burnt deposit. A rough or noncoherent electrodeposit produced by the application of an excessive current density. *See also:* **electroplating.** 42A60-0

burst (1) (pulse techniques). A wave or waveform composed of a pulse train or repetitive waveform that starts at a prescribed time and/or amplitude, continues for a relatively short duration and/or number of cycles, and upon completion returns to the starting amplitude. *See also:* **pulse.** 0-9E4

(2) (audio and electroacoustics). An excursion of a quantity (voltage, current, or power) in an electric system that exceeds a selected multiple of the long-time average magnitude of this quantity taken over a period of time sufficiently long that increasing the length of this period will not change the result appreciably. This multiple is called the upper burst reference. *Notes:* (1) If measurements are made at different points in a system, or at different times, the same quantity must be measured consistently. (2) The excursion may be an electrical representation of a change of some other physical variable such as pressure, velocity, displacement, etcetera. (3) For an extensive list of cross references, see *Appendix A.* E257-1E1

burst build-up interval. The time interval between the burst leading-edge time and the instant at which the upper burst reference is first equaled. *See:* The figure attached to the definition of **burst duration**. *See also:* **burst.** E257-1E1

burst decay interval (audio and electroacoustics). The time interval between the instant at which the peak burst magnitude occurs and the burst trailing-edge time. *See:* The figure attached to the definition of **burst duration**. *See also:* **burst.** E257-1E1

burst duration (audio and electroacoustics). The time interval during which the instantaneous magnitude of the quantity exceeds the lower burst reference, disregarding brief excursions below the reference, provided the duration of any individual excursion is less than a burst safeguard interval of selected length. *Notes:* (1) If the duration of an excursion is equal to or greater than the burst safeguard interval, the burst has ended. (2) These terms, as well as those defined below, are illustrated in the accompanying figure.
(A) A burst is found with the aid of a "window" that is slid horizontally to the right with its base resting on the lower burst reference. The width of the window equals the burst safeguard interval and the height of the window equals the difference between the upper and lower burst references. The window is slid to the right until the trace crosses the top of the window. The upper burst reference has then been reached and a burst has occurred. (B) The burst leading-edge time is found by sliding the window to the left until the trace disappears from the window. The right-hand side of the window marks the burst leading-edge time. (C) The burst trailing edge time is found by a similar procedure. The window is slid to the right past its position in (A) until the trace disappears from the window. The left-hand side of the window marks the burst trailing-edge time. (D) Terms used in defining a burst: **burst leading-edge time, t_1; burst build-up interval, $t_2 - t_1$; burst rise interval, $t_3 - t_1$; burst trailing-edge time, t_5; burst decay interval, $t_5 - t_3$; burst fall-off interval,**

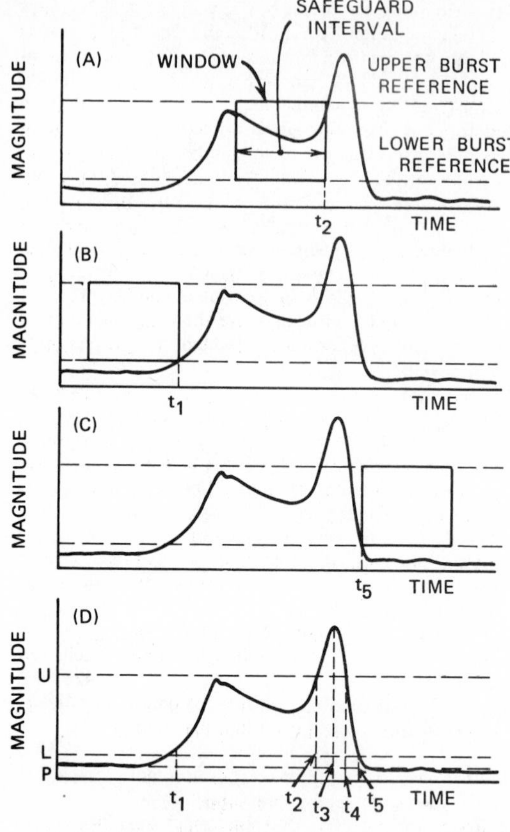

Plot of instantaneous magnitude versus time to illustrate terms used in defining a burst.

$t_5 - t_4$; burst duration, $t_5 - t_1$; upper burst reference, U; lower burst reference, L; long-time average power, P. *See also:* **burst.** E257-1E1

burst duty factor (audio and electroacoustics). The ratio of the average burst duration to the average spacing. *Note:* This is equivalent to the product of the average burst duration and the burst repetition rate. *See:* The figure attached to the definition of **burst duration**. *See also:* **burst.** E257-1E1

burst fall-off interval (audio and electroacoustics). The time interval between the instant at which the upper burst reference is last equaled and the burst trailing edge time. *See:* The figure attached to the definition of **burst duration**. *See also:* **burst.** E257-1E1

burst flag (television). A keying or gating signal used in forming the color burst from a chrominance subcarrier source. *See:* **television.** 0-2E2

burst gate (television). A keying or gating device or signal used to extract the color burst from a color picture signal. *See:* **television.** 0-2E2

burst keying signal (television). *See:* **burst flag; television.**

burst leading-edge time (audio and electroacoustics). The instant at which the instantaneous burst magnitude first equals the lower burst reference. *See:* The figure attached to the definition of **burst duration**. *See also:* **burst (audio and electroacoustics).** E257-1E1

burst magnitude, instantaneous. *See:* **instantaneous burst magnitude.**

burst measurements. *See:* **energy density spectrum.**

burst quiet interval (audio and electroacoustics). The time interval between successive burst during which the instantaneous magnitude does not equal the upper burst reference. *See:* The figure attached to the definition of **burst duration.** *See also:* **burst (audio and electroacoustics).** E257-1E1

burst repetition rate (audio and electroacoustics). The average number of bursts per unit of time. *See:* The figure attached to the definition of **burst duration.** *See also:* **burst (audio and electroacoustics).** E257-1E1

burst rise interval (audio and electroacoustics). The time interval between the burst leading-edge time and the instant at which the peak burst magnitude occurs. *See:* The figure attached to the definition of **burst duration**. *See also:* **burst (audio and electroacoustics).** E257-1E1

burst safeguard interval (audio and electroacoustics). A time interval of selected length during which excursions below the lower burst reference are neglected; it is used in determining those instants at which the lower burst reference is first and last equaled during a burst. *See:* The figure attached to the definition of **burst duration**. *See also:* **burst (audio and electroacoustics).** E257-1E1

burst spacing (audio and electroacoustics). The time interval between the burst leading-edge times of two consecutive bursts. *See:* The figure attached to the definition of **burst duration**. *See also:* **burst (audio and electroacoustics).** E257-1E1

burst trailing-edge time (audio and electroacoustics). The instant at which the instantaneous burst magnitude last equals the lower burst reference. *See:* The figure attached to the definition of **burst duration**. *See also:* **burst (audio and electroacoustics).** E257-1E1

burst train (audio and electroacoustics). A succession of similar bursts having comparable adjacent burst quiet intervals. *See:* The figure attached to the definition of **burst duration**. *See also:* **burst (audio and electroacoustics.** E257-1E1

bus (1) (electric power). A conductor, or group of conductors, that serve as a common connection for two or more circuits. 37A100-31E11
(2) (electronic computers). One or more conductors used for transmitting signals or power from one or more sources to one or more destinations. *See also:* **electronic digital computer.** E162-0

bushing (1) (outdoor electric apparatus). An insulating structure including a central conductor, or providing a central passage for a conductor, with provision for mounting on a barrier, conducting or otherwise, for the purpose of insulating the conductor from the barrier and conducting current from one side of the barrier to the other. *See:* **floor bushing; indoor wall bushing; outdoor wall bushing; power distribution, overhead construction; roof bushing; wall bushing.** E49/76A1-0;37A100-31E11
(2) (rotating machinery). (A) (electrical). Preformed insulating casing, with mechanical seal, permitting passage of a high-potential conductor through the generator housing, usually acting as armature terminal. (B) (mechanical). A sleeve or quill on which a spider is mounted and which, in turn, is mounted on a shaft. *See also:* **cradle base (rotating machinery).** 0-31E8
(3) (relay). *See:* **relay spring stud.**

bushing potential tap (outdoor electric apparatus). An insulated connection to one of the conducting layers of a bushing providing a capacitance voltage divider to indicate the voltage on the bushing. The tap is accessible from outside the bushing. *See also:* **power distribution, overhead construction.** 76A1-0

bushing, rotor. *See:* **rotor bushing.**

bushing test tap (outdoor electric apparatus). An insulated connection to one of the conducting layers of a bushing for the purpose of making ungrounded specimen power-factor tests. The tap is accessible from outside the bushing. *See also:* **power distribution, overhead construction.** 76A1-0

bus line (railway terminology). A continuous electric circuit other than the electric train line, extending through two or more vehicles of a train, for the distribution of electric energy. *See also:* **multiple-unit control.** 42A42-0

bus structure. An assembly of bus conductors, with associated connection joints and insulating supports. 37A100-31E11

bus support. An insulating support for a bus. *Note:* It includes one or more insulator units with fittings for fastening to the mounting structure and for receiving the bus. 37A100-31E11

bus-type shunts (electric power systems). Instrument shunts intended for switchboard use that are designed to be installed in the bus or connection bar structure of the circuit whose current is to be measured. *See also:* **power system, low-frequency and surge testing.** 0-31E5

busy test. A test made to find out whether certain facilities that may be desired, such as a subscriber line or trunk, are available for use. *See also:* **telephone system.** 42A65-19E1

busy tone. *See:* **audible busy signal.**

butt contacts (industrial control). An arrangement in which relative movement of the cooperating members is substantially in a direction perpendicular to the surface of contact. *See:* **contactor.** 50I15-34E10

butt joint (waveguides). A connection between two waveguides or transmission lines that provides physical contact between the ends of the waveguides in order to maintain electric continuity. *See:* **waveguides.** E147-0;0-3E1

buzz (electromagnetic compatibility). A disturbance of relatively short duration, but longer than a specified value as measured under specified conditions. *Note:* For the specified values and conditions, guidance should be found in documents of the International Special Committee on Radio Interference. *See also:* **electromagnetic compatibility.** CISPR-27E1

buzz stick. A device for testing suspension insulator units for fault when the units are in position on an energized line. *Note:* It consists of an insulating stick, on one end of which are metal prongs of the proper dimensions for spanning and short-circuiting the porcelain of one insulator unit at a time, and thereby checking conformity to normal voltage gradient. *See also:* **tower.** 42A35-31E13

buzzer. A signaling device for producing a buzzing sound by the vibration of an armature. *See also:* **circuits and devices.** 42A65-0

BW. *See:* **backward wave structure.**

bypass capacitor (bypass condenser*). A capacitor for providing an alternating-current path of comparatively low impedance around some circuit element. *See also:* **circuits and devices.**

*Deprecated 42A65-0

byte. A sequence of adjacent binary digits operated upon as a unit and usually shorter than a word. X3A12-16E9

C

cab signal. A signal located in the engineman's compartment or cab indicating a condition affecting the movement of a train or engine and used in conjunction with interlocking signals and in conjunction with or in lieu of block signals. *See also:* **automatic train control.** 42A42-0

cabinet. An enclosure designed either for surface or flush mounting, and provided with a frame, mat, or trim in which swinging doors are hung.
See:
conductor-support box;
cutout box;
fixture stud;
fuse holder;
interior wiring;
junction box;
knockout;
lampholder;
outlet box;
pull box;
receptacle;
rosette;
sealable equipment. 42A95-0

cabinet for safe (burglar-alarm system). Usually a wood enclosure, having protective linings on all inside surfaces and traps on the doors, built to surround a safe and designed to produce an alarm condition in a protection circuit if any attempt is made to attack the safe. *See also:* **protective signaling.** 42A65-0

cable (1) (electric power). Either a stranded conductor (single-conductor cable), or a combination of conductors insulated from one another (multiple-conductor cable). *Note:* The first kind of cable is a single conductor, while the second kind is a group of several conductors. The component conductors of the second kind of cable may be either solid or stranded, and this kind of cable may or may not have a common insulating covering. The term cable is applied by some manufacturers to a solid wire heavily insulated and lead covered; this usage arises from the manner of the insulation, but such a conductor is not included under this definition of cable. The term cable is a general one, and in practice, it is usually applied only to the larger sizes. A small cable is called a stranded wire or a cord. Cables may be bare or insulated, and the latter may be sheathed with lead, or armored with wires or bands. E30-0

(2) (signal-transmission system). A transmission line or group of transmission lines mechanically assembled into a complex flexible form. *Note:* The conductors are insulated and are closely spaced and usually have a common outer cover which may be a conductor or an insulator. If a conductor, the outer cover may be an electric portion of the cable. This definition also includes a twisted pair. *Note:* For an extensive list of cross references, see *Appendix A.* E146-13E6

cable bedding (power distribution, underground cables). A relatively thick layer of material, such as a jute serving, between two elements of a cable to provide a cushion effect, or gripping action, as between the lead sheath and wire armor of a submarine cable. *See also:* **power distribution, underground construction.** 0-31E1

cable bond. An electric connection across a joint in the armor or lead sheath of a cable, or between the armor or lead sheath and the earth, or between the armor or sheath of adjacent cables. *See:* **continuity cable bond; and cross cable bond.** *See also:* **power distribution, underground construction.** 42A35-0

cable complement (communication practice). A group of pairs in a cable having some common distinguishing characteristic. *See also:* **cable.** 42A65-0

cable core (cable). The portion lying under other elements of a cable. *See also:* **power distribution, underground construction.** 42A35-31E13

cable core binder. A wrapping of tapes or cords around the several conductors of a multiple-conductor cable used to hold them together. *Note:* Cable core binder is usually supplemented by an outer covering of braid, jacket, or sheath. *See also:* **power distribution, underground construction.** 42A35-31E13

cable coupler (rotating machinery). A form of termination in which the ends of the machine winding are connected to the supply leads by means of a plug-and-socket device. *See also:* **cradle base (rotating machinery).** 0-31E8

cable entrance fitting (pothead). A fitting used to seal or attach the cable sheath or armor to the pothead. *Note:* A cable entrance fitting is also used to attach and support the cable sheath or armor where a cable passes into a transformer removable cable terminating box without the use of potheads. *See also:* **transformer.** 57A12.75/57A12.76-0

cable fill. The ratio of the number of pairs in use to the total number of pairs in a cable. *Note:* The maximum cable fill is the percentage of pairs in a cable that may be used safely and economically without serious interference with the availability and continuity of service. *See also:* **cable.** 42A65-0

cable filler. The material used in multiple-conductor cables to occupy the interstices formed by the assembly of the insulated conductors, thus forming a cable core of the desired shape (usually circular). *See also:* **power distribution, underground construction.** 42A35-31E13

cableheads. *See:* **submersible entrance terminals (distribution oil cutouts).**

cable joint (splice). A connection between two or more separate lengths of cable with the conductors in one length connected individually to conductors in other lengths and with the protecting sheaths so connected as to extend protection over the joint. *Note:*

Cable joints are designated by naming the conductors between which the joint is made, for example, 1 single-conductor to 2 single-conductor cables; 1 single-conductor to 3 single-conductor cables; 1 concentric to 2 concentric cables; 1 concentric to 1 single-conductor cable; 1 concentric to 2 single-conductor cables; 1 concentric to 4 single-conductor cables; 1 three-conductor to 3 single-conductor cables. *See:* **branch joint; regular straight joint; reducing joint.** *See also:* **power distribution, underground construction.** 42A35-31E13

cable Morse code. A three-element code, used mainly in submarine cable telegraphy, in which dots and dashes are represented by positive and negative current impulses of equal length, and a space by absence of current. *See also:* **telegraphy.** 42A65-0

cable rack (hanger) (shelf*). A device usually secured to the wall of a manhole, cable raceway, or building to provide support for cables. *See also:* **power distribution, underground construction.**

*Deprecated 42A35-31E13

cable reel (mining). A drum on which conductor cable is wound, including one or more collector rings and associated brushes, by means of which the electric circuit is made between the stationary winding on the locomotive or other mining device and the trailing cable that is wound on the drum. *Note:* The drum may be driven by an electric motor, a hydraulic motor, or mechanically from an axle on the machine. *See also:* **mine feeder circuit.** 42A85-0

cable separator (power distribution, underground cables). A serving of threads, tapes, or films to separate two elements of the cable, usually to prevent contamination or adhesion. *See also:* **power distribution, underground construction.** 0-31E1

cable sheath. A tubular impervious metallic protective covering applied directly over the cable core. *See also:* **power distribution, underground construction.** 42A35-31E13;0-31E1

cable sheath insulator (pothead). An insulator used to insulate an electrically conductive cable sheath or armor from the metallic parts of the pothead or transformer removable cable terminating box in contact with the supporting structure for the purpose of controlling cable sheath currents. *See also:* **transformer; transformer removable cable terminating box.** 57A12.75/57A12.76-0

cable shield. *See:* **duct edge shield.**

cable splice. A connection between two or more separate lengths of cable with the conductors in one length individually connected to conductors in the other length and with the protecting sheaths so connected as to extend protection over the joint. *See also:* **cable.** 42A65-0

cable splicer (mining). A short piece of tubing or a specially formed band of metal generally used without solder in joining ends of portable cables for mining equipment. *See also:* **mine feeder circuit.** 42A85-0

cable terminal. A structure adapted to be associated with a cable, by means of which electric connection is made available for any predetermined group of cable conductors in such a manner as to permit of their individual selection and extension by conductors outside of the cable. *See also:* **cable.** 42A65-0

cable terminal (pothead) (end bell) (power work). A device that seals the end of a cable and provides insulated egress for the conductors. *See also:* **power distribution, underground construction.** 42A35-31E13

cable vault. *See also:* **manhole; power distribution, underground construction; splicing chamber.**

CADF (electronic navigation). *See:* **commutated-antenna direction-finder.**

cage (Faraday cage). A system of conductors forming an essentially continuous conducting mesh or network over the object protected and including any conductors necessary for interconnection to the object protected and an adequate ground. *See also:* **lightning protection and equipment.** 42A95-0

cage winding. *See:* **squirrel-cage winding.**

calculator. (1) A device capable of performing arithmetic. (2) A calculator as in (1) that requires frequent manual intervention. (3) Generally and historically, a device for carrying out logic and arithmetic digital operations of any kind. X3A12-16E9

calibrate (instrument). To ascertain, usually by comparison with a standard, the locations at which scale graduations should be placed to correspond to a series of values of the quantity which the instrument is to measure. *Note:* This definition of **calibrate** is to be considered as reflecting usage only in the electric instrument and meter industry and in electric metering practice. 42A30-0

calibrated (industrial control). Checked for proper operation at selected points on the operating characteristic. 42A25-34E10

calibrated-driving-machine test (rotating machinery). A test in which the mechanical input or output of an electric machine is calculated from the electric input or output of a calibrated machine mechanically coupled to the machine on test. *See also:* **asynchronous machine; direct-current commutating machine; synchronous machine.** 0-31E8

calibration (device). The adjustment to have the designed operating characteristics and the subsequent marking of the positions of the adjusting means, or the making of adjustments necessary to bring operating characteristics into substantial agreement with standardized scales or marking. 37A100-31E11/31E6

calibration error (relay). In operation the departure under specified conditions, of actual performance from performance indicated by scales, dials, or other markings on the relay. *Note:* The indicated performance may be by calibration markings in terms of input or performance quantities (amperes, ohms, seconds, etcetera) or by reference to specific performance data recorded elsewhere. *See also:* **relay; setting error.** 37A1-31E6

calibration factor (bolometer-coupler unit). The ratio of the substitution power in the bolometer attached to the side arm of the directional coupler to the microwave power incident on a nonreflecting load connected to the output port of the main arm of the directional coupler. *Note:* If the bolometer unit is attached to the main arm of the directional coupler, the calibration factor is the ratio of the substitution power in the bolometer unit attached to the main arm of the directional coupler to the microwave power incident upon a nonreflecting load connected to the output port of the side arm of the directional coupler. 0-9E4

calibration factor (bolometer unit). The ratio of the substitution power to the radio-frequency power incident upon the bolometer unit. The ratio of the bolometer-unit calibration factor to the effective efficiency is determined by the reflection coefficient of the bolometer unit. The two terms are related as follows

$$K_b/\eta_e = 1 - |\Gamma|^2$$

where K_b, η_e, and Γ are the calibration factor, effective efficiency, and reflection coefficient of the bolometer unit, respectively. 0-9E4

calibration level (signal generator). The level at which the signal generator output is calibrated against a standard. *See also:* **signal generator.** 0-9E4

calibration marks (electronic navigation). Indications superimposed on a display to provide a numerical scale of the parameters displayed. *See also:* **navigation.** 0-10E6

calibration or conversion factor (calibration) (loosely called antenna factor). The factor or set of factors that, at given frequency, expresses the relationship between the field strength of an electromagnetic wave impinging upon the antenna of a field-strength meter and the indication of the field-strength meter. *Note:* The composite of antenna characteristic, balun and transmission line effects, receiver sensitivity and linearity, etcetera. *See also:* **measurement system.** E284-9E1

calibration programming (power supplies). Calibration with reference to power-supply programming describes the adjustment of the control-bridges current to calibrate the programming ratio in ohms per volt. *Note:* Many programmable supplies incorporate a calibrate control as part of the reference resistor that performs this adjustment. *See also:* **power supply.** KPSH-10E1

calibration scale. A set of graduations marked to indicate values of quantities, such as current, voltage, or time at which an automatic device can be set to operate. 37A100-31E11/31E6

calibration voltage. The voltage applied during the adjustment of a meter. *See also:* **test an instrument or meter.** 12A0-0

call (1) (computers). To transfer control to a specified closed subroutine.
(2) (communications). The action performed by the calling party, or the operations necessary in making a call, or the effective use made of a connection between two stations. *See also:* **electronic digital computer.** X3A12-16E9

call announcer (automatic telephone office). A device for receiving pulses and audibly reproducing the corresponding number in words, so that it may be heard by a manual operator. *See also:* **telephone switching system.** 42A65-0

call circuit (manual switching). A communication circuit between switching points used by the traffic forces for the transmission of switching instructions. *See also:* **telephone system.** 42A65-0

call indicator. A device for receiving pulses from an automatic switching system and displaying the corresponding called number before an operator at a manual switchboard. *See also:* **telephone switching system.** 42A65-0

calling device. An apparatus that generates the signals required for establishing connections in an automatic telephone switching system. *See also:* **telephone switching system.** 42A65-19E1

calling plug and cord. A plug and cord that are used to connect to a called line. *See also:* **telephone switching system.** 42A65-0

calling sequence (computers). A specified arrangment of instructions and data necessary to set up and call a given subroutine. *See also:* **electronic digital computer.** X3A12-16E9

calomel electrode. *See:* **calomel half-cell.**

calomel half-cell (calomel electrode). A half-cell containing a mercury electrode in contact with a solution of potassium chloride of specified concentration that is saturated with mercurous chloride of which an excess is present. *See also:* **electrochemistry.** 42A60-0

calorimetric test (rotating machinery). A test in which the losses in a machine are deduced from the heat produced by them. The losses are calculated from the temperature rises produced by this heat in the coolant or in the surrounding media. *See also:* **asynchronous machine; direct-current commutating machine; synchronous machine.** 0-31E8

cam contactor or cam switch. A contactor or switch actuated by a cam. *See also:* **control switch.** E16-0

camera storage tube. A storage tube into which the information is introduced by means of electromagnetic radiation, usually light, and read at a later time as an electric signal. *See also:* **storage tube.** E158-15E6

camera tube (television). An electron-beam tube in which an electron-current or charge-density image is formed from an optical image and scanned in a predetermined sequence to provide an electric signal.
See:
amplitude response;
amplitude response characteristic;
beam landing error;
channel multiplier;
current amplification;
electrode dark current;
fiber optic plate;
image-converter tube;
image-intensifier tube;
image tube;
isocon mode;
lag;
microchannel plate;
monoscope;
numerical aperture of a fiber optic plate;
resolution;
retained image;
return-beam mode;
secondary-electron conduction;
signal-to-noise ratio.
See also: **tube definitions.** 42A70-15E6

cam-operated switch (industrial control). A switch consisting of fixed contact elements and movable contact elements operated in sequence by a camshaft. *See:* **switch.** 50I15-34E10

cam-shaft position (electric power systems). The angular position of the main shaft directly operating the governor-controlled valves. *See also:* **speed-governing system.** E94-0

can (dry cell). A metal container, usually zinc, in which the cell is assembled and that serves as its negative electrode. *See also:* **electrolytic cell.** 42A60-0

cancel (numerically controlled machines). A command that will discontinue any fixed cycles or sequence commands. *See also:* **numerically controlled machines.** EIA3B-34E12

cancellation ratio (radar moving-target indicator). The ratio of (1) canceller-voltage amplification for fixed-target echoes received with a fixed antenna to (2) the gain for a single pulse passing through the undelayed channel of the canceller. *See also:* **navigation.** 0-10E6

cancelled video (radar moving-target indicator). The video output remaining after the cancellation process. *See also:* **navigation.** E172-10E6

canceller (radar moving-target indicator). The portion of the system in which cancellation of the fixed-target signals is accomplished. *See:* **moving-target indicator.** *See also:* **navigation.** 0-10E6

candela. The luminous intensity, in the perpendicular direction, of a surface of 1/600 000 square meter of a black body at the temperature of freezing platinum under a pressure of 101 325 newtons per square meter. *Notes:* (1) Values for standards having other spectral distributions are derived by the use of accepted spectral luminous efficiency data for photopic vision. (2) From 1909 until the introduction of the present photometric system on January 1, 1948, the unit of luminous intensity in the United States, as well as in France and Great Britain, was the **international candle** which was maintained by a group of carbon-filament vacuum lamps. For the present unit as defined above, the internationally accepted term is **candela.** The difference between the candela and the old **international candle** is so small that only measurements of high precision are affected. *See:* **spectral luminous efficiency of radiant flux; spectral luminous efficiency for photopic vision.** *See also:* **light.** CGPM-SCC14;Z7A1-0

C and I (C & I). *See:* **supervisory control point, control and indication.**

candle. *See:* **candela.**

candlepower. Luminous intensity expressed in candelas. *See also:* **color terms; light.** E201/Z7A1-0

candlepower distribution curve. A curve, generally polar, that represents the variation of luminous intensity of a lamp or luminaire in a plane through the light center. *Note:* A vertical candlepower distribution curve is obtained by taking measurements at various angles of elevation in a vertical plane through the light center; unless the plane is specified, the vertical curve is assumed to represent an average such as would be obtained by rotating the lamp or luminaire about its vertical axis. A horizontal candlepower distribution curve represents measurements made at various angles of azimuth in a horizontal plane through the light center. *See also:* **colorimetry.** Z7A1-0

can loss (rotating machinery). Electric losses in a can used to protect electric components from the environment. *See also:* **asynchronous machine; direct-current commutating machine; synchronous machine.** 0-31E8

canned (rotating machinery). Completely enclosed and sealed by a metal sheath. *See also:* **cradle base (rotating machinery).** 0-31E8

canned cycle (numerically controlled machines). *See:* **fixed cycle.** *See also:* **numerically controlled machines.**

capability (electric power supply). The maximum generation that a machine, station or power source can be expected to supply to a system under specified conditions for a given time interval. *See also:* **generating station.** 0-31E4

capability margin (electric power supply). The difference between (1) total capability for load and (2) system load responsibility. It is the margin of capability available to provide for scheduled maintenance, emergency outages, adverse system operating requirements, and unforeseen loads. *See also:* **generating station.** 0-31E4

capacitance (capacity*) (1) (general). The property of a system of conductors and dielectrics that permits the storage of electrically separated charges when potential differences exist between the conductors. *Notes:* (A) Its value is expressed as the ratio of a quantity of electricity to a potential difference. (B) As defined elsewhere, self-, direct, balanced, and plenary are used to indicate the capacitances corresponding to different groupings of the n conductors of a multiple-conductor system; and the names are used even if, in addition to the n conductors constituting the working system, additional conductors are present that, either intentionally or because they are inaccessible, are not explicitly considered either experimentally or theoretically. Two or more conductors when connected conductively together are regarded as one conductor. Ground is regarded as one of the n conductors of the system. *See also:* **circuit characteristics of electrodes; electron-tube admittances.**

*Deprecated E45/E270-0

(2) (semiconductor diode). The small-signal capacitance measured between the terminals of the diode under specified conditions of bias and frequency. *See also:* **rectification; semiconductor; semiconductor device.** E59-34E17;E216-34E17;E270-0

capacitance between two conductors. The capacitance between two conductors, although it is a function only of the total geometry and the absolute capacitivity of the medium, is defined most conveniently as the ratio of the change in charge on one conductor to the corresponding change in potential difference between it and the second conductor with the charges or the potentials of the other conductors of the system specified. *Notes:* (1) In the ideal case where one, and only one, conductor is completely surrounded by another, the capacitance between them is not affected by the presence of the other $n-2$ conductors. (2) The capacitance between two conductors may be made independent of all except a third if the third conductor completely surrounds the two under discussion. E270-0

capacitance between two conductors, balanced. *See:* **balanced capacitance.**

capacitance coupling (signal-transmission system). *See:* **coupling, capacitance.**

capacitance current (or component). A reversible component of the measured current on charge or discharge of the winding and is due to the geometrical capacitance, that is, the capacitance as measured with alternating current of power or higher frequencies. With high direct voltage this current has a very short time constant and so does not affect the usual measurements. *See also:* **insulation testing (large alternating-current rotating machinery).** E95-0

capacitance, discontinuity (waveguide or transmission line). The shunt capacitance that, when inserted in a uniform waveguide or transmission line, would cause reflected waves of the dominant mode equal to those resulting from the given discontinuity. *See also:* **waveguide.** 0-9E4

capacitance, effective. The imaginary part of a capacitive admittance divided by the angular frequency. *See also:* **transmission characteristics.** 0-9E4

capacitance, input (*n*-terminal electron tubes). The short-circuit transfer capacitance between the input terminal and all other terminals, except the output terminal, connected together. *Note:* This quantity is equivalent to the sum of the interelectrode capacitances between the input electrode and all other electrodes except the output electrode. *See also:* **circuit characteristics of electrodes; electron-tube admittances.** E160-15E6;42A70-0

capacitance meter. An instrument for measuring capacitance. *Note:* If the scale is graduated in microfarads the instrument is usually designated as a microfaradmeter. *See also:* **instrument.** 42A30-0

capacitance of a bushing (outdoor electric apparatus). (1) The main capacitance C_1 of a bushing is the value in picofarads between the high-voltage conductor and the potential tap or the test tap. (2) The tap capacitance C_2 of a bushing is the value in picofarads between the potential tap and ground. *Note:* The ratio $C_1 / (C_1 + C_2)$ is the ratio of the potential tap voltage to ground to the voltage of the high-voltage terminal to ground. (3) The capacitance C of a bushing without a potential or test tap is the value in picofarads between the high-voltage conductor and the ground flange. *See also:* **power distribution, overhead construction.** 76A1-0

capacitance, output (*n*-terminal electron tube). The short-circuit transfer capacitance between the output terminal and all other terminals, except the input terminal, connected together. *See also:* **circuit characteristics of electrodes; electron-tube admittances.** 42A70/E160-15E6

capacitance potential device. A voltage-transforming equipment or network connected to one conductor of a circuit through a capacitance, such as a coupling capacitor or suitable high-voltage bushing, to provide a low voltage such as required for the operation of instruments and relays. *Notes:* (1) The term **potential device** applies only to the network and is exclusive of the coupling capacitor or high-voltage bushing. (2) The term **coupling-capacitor potential device** indicates use with coupling capacitators. (3) The term **bushing potential device** indicates use with bushings. (4) The term **capacitance potential device** indicates use with any type of capacitance coupling. (5) Capacitance potential devices and their associated coupling capacitors or bushings are designed for line-to-ground connection, and not line-to-line connection. The potential device is a single-phase device, and, in combination with its coupling capacitor or bushing, is connected line-to-ground. The low voltage thus provided is a function of the line-to-ground voltage and the constants of the capacitance potential device. Two or more capacitance potential devices, in combination with their coupling capacitors or bushings, may be connected line-to-ground on different high-voltage phases to provide low voltages of other desired phase relationships. (6) Zero-sequence voltage may be obtained from the broken-delta connection of the auxiliary windings or by the use of one device with three coupling capacitors or bushings. In the latter case, the three operating-tap connection-points are joined together and one device connected between this common point and ground. Although used in combination with three coupling capacitors or bushings, the device output and accuracy rating standards are based on the single-phase conditions. *See also:* **outdoor coupling capacitor.** E31-0

capacitance ratio (nonlinear capacitor). The ratio of maximum to minimum capacitance over a specified voltage range, as determined from a capacitance characteristic, such as a differential capacitance characteristic, or a reversible capacitance characteristic. *See also:* **nonlinear capacitor.** E226-15E7

capacitance, short-circuit input (*n*-terminal electron device). The effective capacitance determined from the short-circuit input admittance. *See also:* **electron-tube admittances.** E160-15E6

capacitance, short-circuit output (*n*-terminal electron device). The effective capacitance determined from the short-circuit output admittance. *See also:* **electron-tube admittances.** E160-15E6

capacitance, short-circuit transfer (electron tubes). The effective capacitance determined from the short-circuit transfer admittance. *See also:* **electron-tube admittances.** E160/42A70-15E6

capacitance, signal electrode. *See:* **electrode capacitance.**

capacitance, stray (electric circuits). Capitance arising from proximity of component parts, wires, and ground. *Note:* It is undesirable in most circuits, although in some high-frequency applications it is used as the tuning capacitance. In bridges and other measuring equipment, its effect must be eliminated by preliminary balancing out, or known and included in the results of any measurement performed. *See also:* **measurement system.** 0-9E4

capacitance, target (camera tubes). *See:* **target capacitance (camera tubes).**

capacitivity (free space). *See:* **electric constant.** *See also:* **absolute capacitivity.**

capacitor (condenser*). A device consisting of two electrodes separated by a dielectric, which may be air, for introducing capacitance into an electric circuit. *See:* **adjustable capacitor.**
*Deprecated 0-21E0

capacitor antenna (condenser antenna*). An antenna consisting of two conductors or systems of conductors, the essential characteristic of which is its capacitance. *See also:* **antenna.**
*Deprecated 42A65-3E1

capacitator braking (rotating machinery). A form of dynamic braking for induction motors in which a capacitator is used to magnetize the motor. *See:* **asynchronous machine.** 0-31E8

capacitor bushing (condenser bushing*). A bushing in which cylindrical conducting layers are arranged coaxially with the conductor within the insulating material. *Note:* The length and diameter of the cylinders are designed with the object of controlling the distribution of the electric field in the bushing. Capacitor bushings may be one of several types: (1) resin-bonded paper-insulated (2) oil-impregnated paper insulated, or (3) other. *See also:* **power distribution, overhead construction.**
*Deprecated 76A1-0

capacitor enclosure. The case in which the capacitor is mounted. 42A10-0

capacitor, ideal. *See:* **ideal capacitor.**

capacitor loudspeaker (condenser loudspeaker). *See:* **electrostatic loudspeaker.**

capacitor microphone (condenser microphone). *See:* **electrostatic microphone.**

capacitor motor. A single-phase induction motor with a main winding arranged for direct connection to a source of power and an auxiliary winding connected in series with a capacitor. The capacitor may be directly in the auxiliary circuit or connected into it through a transformer. *See also:* **asynchronous machine; capacitor-start motor; permanent-split capacitor; two-value capacitor motors.** 42A10-0;0-31E8

capacitor mounting strap. A device by means of which the capacitor is affixed to the motor. 42A10-0

capacitor pickup. A phonograph pickup that depends for its operation upon the variation of its electric capacitance. *See also:* **phonograph pickup.** S1A1/42A65-0

capacitor start-and-run motor. A capacitor motor in which the auxiliary primary winding and series-connected capacitors remain in circuit for both starting and running. *See:* **permanent-split capacitor motor.** *See also:* **asynchronous machine.** 0-31E8

capacitor-start motor. A capacitor motor in which the auxiliary winding is energized only during the starting operation. *Note:* The auxiliary-winding circuit is open-circuited during running operation. *See also:* **asynchronous machine.** 0-31E8

capacitor unit. A single assembly of dielectric and electrodes in a container with terminals brought out. *See:* **alternating-current distribution; indoor capacitor unit; outdoor capacitor unit.** E18/E270-0

capacitor voltage. The voltage across two terminals of a capacitor. E226-15E7

capacity* (1) (capacitor*). *See:* **capacitance.**

(2) (electric-power devices). (A) The capacity of a machine, apparatus, or appliance is the rated load (rated capacity). (B) The capacity of a machine, apparatus, or appliance is the maximum load of which it is capable under existing service conditions (maximum capacity). E270-0

(3) (data processing and computation). *See:* **storage capacity.** *See also:* **electronic digital computer.**

*Deprecated for (1) E162-0-X3U12-16E9

capacity factor. The ratio of the average load on a machine or equipment for the period of time considered, to the rating of the machine or equipment. *See also:* **generating station.** 42A35-31E13

cap-and-pin insulator. An assembly of one or more shells with metallic cap and pin, having means for direct and rigid mounting. *See also:* **insulator.** 29A1-0

capture effect (modulation systems). The effect occuring in a transducer (usually a demodulator) whereby the input wave having the largest magnitude controls the output. *See also:* **modulating systems.** E170-0

car annunciator. An electric device in the car that indicates visually the landings at which an elevator-landing signal-registering device has been actuated. *See also:* **elevator.** 42A45-0

carbon-arc lamp. An electric-discharge lamp employing an arc discharge between carbon electrodes. *Note:* One or more of these electrodes may have cores of special chemicals which contribute importantly to the radiation. *See also:* **lamp.** Z7A1-0

carbon brush (motors and generators). (1) A specific type of brush composed principally of amorphous carbon. *Note:* This type of brush is usually hard and is adapted to low speeds and moderate currents. (2) A broader classification of brush, containing carbon in appreciable amount. *See:* **brush (rotating machinery).** 42A10-31E8

carbon-consuming cell (carbon-combustion cell). A cell for the production of electric energy by galvanic oxidation of carbon. *See also:* **electrochemistry.** 42A60-0

carbon-contact pickup. A phonograph pickup that depends for its operation upon the variation in resistance of carbon contacts. *See also:* **phonograph pickup.** 42A65-0

carbon-dioxide system (rotating machinery). A fire-protection system using carbon-dioxide gas as the extinguisher. *See also:* **cradle base (rotating machinery).** 0-31E8

carbon-graphite brush (electric machines). A carbon brush to which graphite is added. This type of brush can vary from medium hardness to very hard. It can carry only moderate currents and is adapted to moderate speeds. *See:* **brush (rotating machinery).** 64A1-0

carbon microphone. A microphone that depends for its operation upon the variation in resistance of carbon contacts. *See also:* **microphone.** 42A65-0

carbon noise (carbon microphones). The inherent noise voltage of the carbon element. *See also:* **close-talking pressure-type microphone.** E258-0

carbon-pressure recording facsimile. That type of electromechanical recording in which a pressure device acts upon carbon paper to register upon the record sheet. *See also:* **recording (facsimile).** E168-0

carbon telephone transmitter. A telephone transmitter that depends for its operation upon the variation in resistance of carbon contacts. *See also:* **telephone station.** 42A65-0

carcinotron (*M*-type backward-wave oscillator) (microwave tubes). A crossed-field oscillator tube in which an electron stream interacts with a backward wave on the nonreentrant circuit. The oscillation frequency is a function of anode-to-sole voltage. *See also:* **microwave tube.** 0-15E6

card (computers). *See:* **magnetic card, punched card, tape to card.** *See also:* **electronic digital computer.**

card hopper (computers). A device that holds cards and makes them available to a card-feed mechanism. *See:* **input magazine; card stacker.** *See also:* **electronic digital computer.** X3A12-16E9

card image (computers). A one-to-one representation of the contents of a punched card, for example, a matrix in which a 1 represents a punch and a 0 represents the absence of a punch. *See also:* **electronic digital computer.** X3A12-16E9

cardiogram. *See:* **electrocardiogram.**

car-door or gate electric contact. An electric device, the function of which is to prevent operation of the driving machine by the normal operating device unless the car door or gate is in the closed position. *See also:* **elevator.** 42A45-0

car-door or gate power closer. A device or assembly of devices that closes a manually opened car door or gate by power other than by hand, gravity, springs, or the movement of the car. *See also:* **elevator.** 42A45-0

card, relay. *See:* **relay armature card.**

card stacker (computers). An output device that accumulates punched cards in a deck. *See:* **card hopper.** *See also:* **electronic digital computer.** X3A12-16E9

car enclosure (elevator). Consists of the top and the walls resting on and attached to the car platform. *See also:* **elevator.** 42A45-0

car-frame sling. The supporting frame to which the car platform, upper and lower sets of guide shoes, car safety, and the hoisting ropes or hoisting-rope sheaves, or the plunger of a direct-plunger elevator are attached. *See also:* **elevator.** 42A45-0

car or counterweight safety. A mechanical device attached to the car frame or to an auxiliary frame, or to the counterweight frame, to stop and hold the car or counterweight in case of predetermined overspeed or free fall, or if the hoisting ropes slacken. *See also:* **elevator.** 42A45-0

car or hoistway door or gate (elevators). The sliding portion of the car or the hinged or sliding portion in the hoistway enclosure that closes the opening giving access to the car or to the landing. *See also:* **hoistway (elevator or dumbwaiter).** 42A45-0

car platform (elevators). The structure that forms the floor of the car and that directly supports the load. *See also:* **hoistway (elevator or dumbwaiter).** 42A45-0

car retarder. A braking device, usually power operated, built into a railway track and used to reduce the speed of cars by means of brake shoes that when set in braking position press against the sides of the lower portions of the wheels. *See:* **control machine; master controller (pressure regulator); switch machine; trimmer signal.** 42A42-0

carriage (typewriter). *See:* **automatic carriage.**

carriage return (typewriter). The operation that causes the next character to be printed at the left margin. *See also:* **electronic digital computer.** X3A12-16E9

carrier (1) (signal transmission system). (A) A wave having at least one characteristic that may be varied from a known reference value by modulation. (B) That part of the modulated wave that corresponds in a specified manner to the unmodulated wave, having, for example, the carrier-frequency spectral components. *Note:* Examples of carriers are a sine wave and a recurring series of pulses.
See:
carrier current;
carrier frequency;
carrier repeater;
carrier transmission;
crest factor;
intermediate subcarrier;
pulse carrier;
subcarrier.
See also: **modulating systems.** E145/E170-0
(2) (semiconductor). A mobile conduction electron or hole. *See also:* **semiconductor device.** E102-0
(3) (electrostatography). The substance in a developer that conveys a toner, but does not itself become a part of the viewable record. *See also:* **electrostatography.** E224-15E7

carrier-amplitude regulation. The change in amplitude of the carrier wave in an amplitude-modulated transmitter when modulation is applied under conditions of symmetrical modulation. *Note:* The term **carrier shift,** often applied to this effect, is deprecated. E145/42A65-0

carrier beat (facsimile). The undesirable heterodyne of signals each synchronous with a different stable reference oscillator causing a pattern in received copy. *Note:* Where one or more of the oscillators is fork controlled, this is called fork beat. *See also:* **facsimile transmission.** E168-0

carrier bypass (power-system communication). A path for carrier current of comparatively low impedance around some circuit element. *See also:* **power-line carrier.** 0-31E3

carrier chrominance signal. *See:* **chrominance signal.**

carrier-controlled approach system (CCA) (electronic navigation). An aircraft-carrier radar system providing information by which aircraft approaches may be directed via radio communication. *See also:* **navigation.** E172-10E6

carrier current. The current associated with a carrier wave. *See also:* **carrier.** 42A65-31E3

carrier-current choke coil (capacitance potential device). A reactor or choke coil connected in series between the potential tap of the coupling capacitor and the potential device transformer unit, to present a low impedance to the flow of power current and a high impedance to the flow of carrier-frequency current. Its purpose is to limit the loss of carrier-frequency current through the potential-device circuit. *See also:* **outdoor coupling capacitor.** E31-0

carrier-current coupling capacitor (power-system communication). An assembly of capacitor units for coupling carrier-current receiving and transmitting equipment to a power line. *See also:* **power-line carrier.** 0-31E3

carrier-current drain coil (capacitance potential device). A reactor or choke coil connected between the carrier-current lead and ground, to present a low impedance to the flow of power current and a high impedance to the flow of carrier-frequency current. *Notes:* Its purpose is to prevent high voltage at power frequency from being impressed on the carrier-current lead and to limit the loss of carrier-frequency energy to ground. *See also:* **outdoor coupling capacitor.** E31-0;0-31E3

carrier-current grounding-switch and gap. Consists of a protective gap for limiting the voltage impressed on the carrier-current equipment and the line tuning unit (if used); and a switch that, when closed, solidly grounds the carrier equipment for maintenance or adjustment without interrupting either high-voltage line or potential-device operation. *See also:* **outdoor coupling capacitor.** E31-0

carrier-current lead (power-system communication). A cable for interconnecting carrier-current coupling capacitor, tuning units, carrier-current transmitter, and carrier-current receiver equipment. *See also:* **outdoor coupling capacitor; power-line carrier.** 0-31E3

carrier-current line trap (power-system communication). A network of inductance and capacitance inserted into a power line that offers a high impedance to one or more carrier-current frequencies and a low impedance to power-frequency current. *See also:* **power-line carrier.** 0-31E3

carrier-current line trap, single-frequency (power-system communication). A carrier-current line trap that offers high impedance to only one carrier-current frequency. *See also:* **power-line carrier.** 0-31E3

carrier-current line trap, two-frequency (power-system communication). A carrier-current line trap that offers high impedance to two separate carrier-current frequencies. *See also:* **power-line carrier.** 0-31E3

carrier frequency (1) (in a periodic carrier). The reciprocal of its period. *Note:* The frequency of a periodic pulse carrier often is called the pulse-repetition frequency in a signal-transmission system.
(2) (modulated amplifier). The frequency that is used to modulate the input signal for amplification. *See also:* **carrier.** E145-0;E170-13E6;42A65-31E3

carrier-frequency choke coil, power line. A reactor that is connected between a carrier-current coupling capacitor and a grounded shunt capacitor that may be used for power-frequency voltage measurement. *Note:* The coil is a low impedance to the flow of power-frequency current and a high impedance to the flow of carrier current. *See also:* **power-line carrier.** 0-31E3

carrier-frequency pulse. A carrier that is amplitude modulated by a pulse. *Notes:* (1) The amplitude of the modulated carrier is zero before and after the pulse. (2) Coherence of the carrier (with itself) is not implied. *See also:* **pulse terms.** E194-0

carrier frequency range (transmitter). The continuous range of frequencies within which the transmitter may be adjusted for normal operation. A transmitter may have more than one carrier-frequency range. *See also:* **radio transmitter.** E145/E182/42A65-31E3

carrier frequency stability (radio transmitter) (transmitter performance). The measure of the ability to remain on its assigned channel as determined on both a short term (1-second) and a long term (24-hour) basis. 0-6E1

carrier, fuse. *See:* **fuse carrier.**

carrier isolating choke coil. An inductor inserted, in series with a line on which carrier energy is applied, to impede the flow of carrier energy beyond that point. *See also:* **circuits and devices.** 42A65-31E3/21E0

carrier noise level (residual modulation). The noise produced by undesired variations of radio-frequency signal in the absence of any intended modulation. *See also:* **radio transmission; modulation.** 42A65/E145-0

carrier power output (transmitter performance). The radio-frequency power available at the antenna terminal when no modulating signal is present. *See also:* **audio-frequency distortion.** 0-6E1

carrier reinsertion (power-system communication). The process of mixing a locally generated carrier frequency with a received single-sideband suppressed-carrier signal to reestablish a desired amplitude-modulated signal that can be demodulated. *See also:* **power-line carrier.** 0-31E3

carrier relaying protection. A form of pilot protection in which high-frequency current is used over a metallic circuit (usually the line protected) for the communicating means between the relays at the circuit terminals. 37A100-31E11

carrier repeater. A repeater for use in carrier transmission. 42A65-31E3

carrier start time (power-system communication). The interval between the instant that a keyer causes a carrier transmitter to increase its output and the instant the output reaches a specified upper amplitude, namely 90 percent of the peak amplitude. *See also:* **power-line carrier.** 0-31E3

carrier stop time (power-system communication). The interval between the instant that a keyer causes a carrier transmitter to decrease its output and the instant the output reaches a specified lower amplitude, namely 10 percent of the peak amplitude. *See also:* **power-line carrier.** 0-31E3

carrier suppression (radio communication). The method of operation in which the carrier wave is not transmitted. *See also:* **modulation.** E145-0

carrier tap choke coil. A carrier-isolating choke coil inserted in series with a line tap. 42A65-21E0

carrier telephone channel. A telephone channel employing carrier transmission. *See also:* **channel.** 42A65-0

carrier telephony. The form of telephony in which carrier transmission is used, the modulating wave being a voice-frequency wave. *Note:* This term is ordinarily applied only to wire telephony. *See also:* **communication.** 42A65-31E3

carrier telegraphy. The form of telegraphy in which, in order to form the transmitted signals, alternating-current is supplied to the line after being modulated under the control of the transmitting apparatus. *See also:* **telegraphy.** 42A65-31E3-19E4

carrier terminal. The assemblage of apparatus at one end of a carrier transmission system, whereby the processes of modulation, demodulation, filtering, amplification, and associated functions are effected. *See also:* **modulating systems.** 42A65-0-31E3

carrier terminal grounding switch (power-system communication). A switch provided with power-line carrier coupling capacitors to protect personnel during equipment adjustment. *Note:* The switch closes a circuit across the power-line carrier terminal between the power-line coupling capacitor and ground. *See also:* **power-line carrier.** 0-31E3

carrier terminal protective gap (power-system communication). A gap used to limit voltage impressed on power-line carrier terminal equipment by electric disturbances on the power line. *See also:* **power-line carrier.** 0-31E3

carrier test switch (CTS) (power-system communication). A switch on power-line carrier transmitters for applying carrier energy to the power line for test purposes. *See also:* **power-line carrier.** 0-31E3

carrier-to-noise ratio. The ratio of specified measures of the carrier and the noise after specified band limiting and before any nonlinear process such as amplitude limiting and detection. *Note:* This ratio is expressed in many different ways, for example, in terms of peak values in the case of impulse noise and in terms of mean-square or root-mean-square values for other types of noise. *See also:* **amplitude modulation.** E170-0

carrier transmission. That form of electric transmission in which the transmitted electric wave is a wave resulting from the modulation of a single-frequency wave by a modulating wave. *See also:* **carrier.** 42A65-0

carrier wave. *See:* **carrier.**

carry. (1) A character or characters, produced in connection with an arithmetic operation on one digit place of two or more number representations in positional notation, and forwarded to another digit place for processing there. (2) The number represented by the character or characters in (1). (3) Usually, a signal or expression as defined in (1) which arises in adding, when the sum of two digits in the same digit place equals or exceeds the base of the number system in use. *Note:* If a carry into a digit place will result in a carry out of the same digit place, and if the normal adding circuit is bypassed when generating this new carry, it is called a high-speed carry, or standing-on-nines carry. If the normal adding circuit is used in such a case, the carry is called a cascaded carry. If a carry resulting from the addition of carries is not allowed to propagate (for example, when forming the partial product in one step of a multiplication process), the process is called a partial carry. If it is allowed to propagate, the process is called a complete carry. If a carry generated in the most-significant-digit place is sent directly to the least-significant place (for example, when adding two negative numbers using nines complements) that carry is called an end-around carry. (4) A carry, in direct subtraction, is a signal or expression as defined in (1) that arises when the difference between the digits is less than zero. Such a carry is frequently called a borrow. (5) To carry is the action of forwarding a carry. (6) A carry is the command directing a carry to be forwarded. *See:* **cascaded carry; complete carry; end-around carry; high-speed carry; partial carry; standing-on-nines carry.** *See also:* **electronic computation; electronic digital computer.** E162/E270-0

car-switch automatic floor-stop operation (elevators). Operation in which the stop is initiated by the operator from within the car with a definite reference to the landing at which it is desired to stop, after which the slowing down and stopping of the elevator is effected automatically. *See:* **control.** 42A45-0

car-switch operation (elevators). Operation wherein the movement and direction of travel of the car are directly and solely under the control of the operator by means of a manually operated car switch or of continuous-pressure buttons in the car. *See:* **control.** 42A45-0

cartridge fuse. A low-voltage fuse consisting of a current-responsive element inside a fuse tube with terminals on both ends. 37A100-31E11

cartridge size (cartridge fuse). The range of voltage and ampere ratings assigned to a fuse cartridge with specific dimensions and shape. 37A100-31E11

cartridge-type bearing (rotating machinery). A complete ball or roller bearing assembly consisting of a ball or roller bearing and bearing housing that is intended to be inserted into a machine endshield. *See also:* **bearing.** 0-31E8

car-wiring apparatus. *See:* **electric train-line; train-line coupler; wire.** *See also:* **multiple-unit control.**

cascade. *See:* **tandem.**

cascade (electrolyte cells). A series of two or more electrolytic cells or tanks so placed that electrolyte from one flows into the next lower in the series, the flow being favored by differences in elevation of the cells, producing a cascade at each point where electrolyte drops from one cell to the next. *See also:* **electrowinning.** 42A60-0

cascade connection (cascade). A tandem arrangement of two or more similar component devices in which the output of one is connected to the input of the next. *See:* **tandem.** 42A65-0

cascade control (street lighting system). A method of turning street lights on and off in sections, each section being controlled by the energizing and de-energizing of the preceding section. *See also:* **alternating-current distribution; direct-current distribution.** 42A35-31E13

cascaded carry (parallel addition). A carry process in which the addition of two numerals results in a partial-sum numeral and a carry numeral that are in turn added together, this process being repeated until no new carries are generated. *See:* **carry; electronic digital computer; high-speed carry.** X3A12-16E9

cascade development (electrostatography). Development in which the image-forming material is carried to the field of the electrostatic image by means of gravitational forces, usually in combination with a granular carrier. *See also:* **electrostatography.** E224-15E7

cascaded thermoelectric device. A thermoelectric device having two or more stages arranged thermally in series. *See also:* **thermoelectric device.** E221-15E7

cascade node (branch) (network analysis). A node (branch) not contained in a loop. *See also:* **linear signal flow graphs.** E155-0

cascade rectifier (cascade rectifier circuit). A rectifier in which two or more similar rectifiers are connected in such a way that their direct voltages add, but their commutations do not coincide. *Note:* When two or more rectifiers operate so that their commutations coincide, they are said to be in parallel if the direct currents add, and in series if the direct voltages add. *See also:* **power rectifier; rectification; rectifier circuit element.** E59/34A1-34E24/34E17

cascading (switching devices). The application in which the devices nearest the source of power have interrupting ratings equal to, or in excess of, the available short-circuit current, while devices in succeeding steps farther from the source, have successively lower interrupting ratings. 37A100-31E11

case (1) (storage battery) (storage cell). A multiple compartment container for the elements and electrolyte of two or more storage cells. Specifically wood cases are containers for cells in individual jars. *See also:* **battery (primary or secondary).** 42A60-0
(2) (electrotyping). A metal plate to which is attached a layer of wax to serve as a matrix. *See also:* **electroforming).** 42A60-0

case shift (telegraphy). The change-over of the translating mechanism of a telegraph receiving machine from letters-case to figures-case or vice versa. *See also:* **telegraphy.** 0-19E4

casting (electrotyping). The pouring of molten electrotype metal upon tinned shells. *See also:* **electroforming.** 42A60-0

catastrophic failure (reliability). *See:* **failure, catastrophic.**

catcher (electron tubes). *See:* **output resonator.**

catcher space (velocity-modulated tube). The part of the tube following the drift space, and where the density-modulated electron beam excites oscillations in the output resonator. It is the space between the output-resonator grids. *See also:* **velocity-modulated tube.** 50I07-15E6

catelectrotonus (electrobiology). Electrotonus produced in the region of the cathode. *See also:* **excitability, electrotonus (electrobiology).** 42A80-18E1

cathode (1) (electron tube or valve). (1) An electrode

through which a primary stream of electrons enters the interelectrode space. *See:* **electrode (electron tube) anode.**
See also:
cold cathode;
filament;
hot cathode;
indirectly heated cathode;
ionic-heated cathode;
photocathode;
semitransparent photocathode.
E160-15E6;42A70-0
(2) (semiconductor rectifier cell). The electrode to which the forward current flows within the cell. *See also:* **semiconductor.** E59-34E17
(3) (electrolytic). An electrode through which current leaves any conductor of the nonmetallic class. Specifically, an electrolytic cathode is an electrode at which positive ions are discharged, or negative ions are formed, or at which other reducing reactions occur. *See also:* **electrolytic cell.** 42A60-0
(4) (thyristor). The electrode by which current leaves the thyristor when the thyristor is in the ON state with the gate open-circuited. *Note:* This term does not apply to bidirectional thyristors.
E223-15E7/34E17/34E24

cathode border (gas) (gas tube). The distinct surface of separation between the cathode dark space and the negative glow. *See also:* **discharge (gas).**
50I07-15E6

cathode cleaning (electroplating). Electrolytic cleaning in which the metal to be cleaned is the cathode. *See also:* **battery (primary or secondary).** 42A60-0

cathode coating impedance (electron tube). The impedance, excluding the cathode interface (layer) impedance, between the base metal and the emitting surface of a coated cathode. *See:* **circuit characteristics of electrodes.** E160-15E6

cathode, cold. *See:* **cold cathode.**

cathode current. *See:* **electrode current (electron tube); electronic controller.**

cathode current, peak (1) (fault). The highest instantaneous value of a nonrecurrent pulse of cathode current occurring under fault conditions. *See also:* **electrode current (electron tube).** 42A70-15E6
(2) (steady state). The maximum instantaneous value of a periodically recurring cathode current. *See also:* **electrode current (electron tube).** 42A70-15E6
(3) (surge). The highest instantaneous value of a randomly recurring pulse of cathode current. *See also:* **electrode current (electron tube).** 42A70-15E6

cathode dark space (Crookes dark space) (gas tube). The relatively nonluminous region in a glow-discharge cold-cathode tube between the cathode glow and the negative glow. *See:* **gas tube.** 42A70-15E6

cathode efficiency. The current efficiency of a specified cathodic process. *See also:* **electrochemistry.**
42A60-0

cathode (potential) fall (gas). The difference of potential due to the space charge near the cathode. *See also:* **discharge (gas).** 50I07-15E6

cathode follower. A circuit in which the output load is connected in the cathode circuit of an electron tube and the input is applied between the control grid and the remote end of the cathode load, which may be at ground potential. *Note:* The circuit is characterized by low output impedance, high input impedance, gain less than unity, and negative feedback. *See also:* **circuits and devices.** E145/E182A-42A65

cathode glow (gas tube). The luminous glow that covers all, or part, of the surface of the cathode in a glow-discharge cold-cathode tube, between the cathode and the cathode dark space. *See:* **gas tube.**
E160-15E6;42A70-0

cathode heating time (vacuum tube). The time required for the cathode to attain a specified condition, for example: (1) a specified value of emission, or (2) a specified rate of change of emission. *Note:* All electrode voltages are to remain constant during measurement. The tube elements must all be at room temperature at the start of the test. *See:* **operation time.**
E160-0;42A70-15E6

cathode interface (layer) capacitance (electron tube). A capacitance that, in parallel with a suitable resistance, forms an impedance approximating the cathode interface impedance. *Note:* Because the cathode interface impedance cannot be represented accurately by the two-element resistance-capacitance circuit, this value of capacitance is not unique. *See:* **circuit characteristics of electrodes.** E160-15E6

cathode interface (layer) impedance (electron tube). An impedance between the cathode base and coating. *Note:* This impedance may be the result of a layer of high resistivity or a poor mechanical bond between the cathode base and coating. *See:* **circuit characteristics of electrodes.** E160-15E6

cathode interface (layer) resistance (electron tube). The low-frequency limit of cathode interface impedance. *See:* **circuit characteristics of electrodes.**
E160-15E6

cathode, ionic-heated. A hot cathode that is heated primarily by ionic bombardment of the emitting surface. 42A70-15E6

cathode layer. A molten metal or alloy forming the cathode of an electrolytic cell and which floats on the fused electrolyte, or upon which fused electrolyte floats. *See also:* **fused electrolyte.** 42A60-0

cathode luminous sensitivity (multiplier phototube). *See:* **sensitivity, cathode luminous.**

cathode modulation. Modulation produced by application of the modulating voltage to the cathode of any electron tube in which the carrier is present. *Note:* Modulation in which the cathode voltage contains externally generated pulses is called cathode pulse modulation. *See also:* **modulating systems.** 42A65-0

cathode (or anode) sputtering (gas). The emission of fine particles from the cathode (or anode) produced by positive ion (or electron) bombardment. *See also:* **discharge (gas).** 50I07-15E6

cathode, pool. A cathode at which the principal source of electron emission is a cathode spot on a metallic pool electrode. 42A70-15E6

cathode preheating time (electron tube). The minimum period of time during which the heater voltage should be applied before the application of other electrode voltages. *See:* **circuit characteristics of electrodes.** *See also:* **heater current (electron device).**
E160-15E6

cathode pulse modulation. Modulation produced in an amplifier or oscillator by application of externally generated pulses to the cathode circuit. *See also:* **modulating systems.** E145-0

cathode-ray charge-storage tube. A charge-storage tube in which the information is written by means of a cathode-ray beam. *Note:* Dark-trace tubes and cathode-ray tubes with a long persistence are examples of cathode-ray storage tubes that are not charge-storage tubes. Most television camera tubes are examples of charge-storage tubes that are not cathode-ray storage tubes. *See also:* **charge-storage tube.** E158-15E6

cathode-ray instrument. *See:* **electron beam instrument.**

cathode-ray oscillograph. An oscillograph in which a photographic or other record is produced by means of the electron beam of a cathode-ray tube. *Note:* The term cathode-ray oscillograph has frequently been applied to a cathode-ray oscilloscope but this usage is deprecated. *See also:* **oscillograph.** 42A30-0

cathode-ray oscilloscope. An oscilloscope that employs a cathode-ray tube as the indicating device. *See also:* **oscillograph.** 42A30-0

cathode-ray storage tube. A storage tube in which the information is written by means of a cathode-ray beam. *See also:* **storage tube.** E158-15E6

cathode-ray tube. An electron-beam tube in which the beam can be focused to a small cross section on a luminescent screen and varied in position and intensity to produce a visible pattern.
See:
beam tubes;
blooming;
cone;
crossover;
dark-trace screen;
dark-trace tube;
deflecting voltage;
deflection;
deflection defocusing;
electron-ray indicator tube;
electrostatic deflection;
faceplate;
halation;
ion trap;
luminescent-screen tube;
magnetic deflection;
metalized screen;
neck;
skiatron;
spot;
sticking voltage;
trapezium distortion.
See also: **tube definitions.** 42A70-0

cathode-ray-tube display area. *See:* **graticule area.**

cathode region (gas) (gas tube). The group of regions that extends from the cathode to the Faraday dark space inclusively. *See also:* **discharge (gas).** 50I07-15E6

cathode spot (arc). An area on the cathode of an arc from which electron emission takes place at a current density of thousands of amperes per square centimeter and where the temperature of the electrode is too low to account for such currents by thermionic emission. *See:* **gas tube.** 42A70-15E6

cathode sputtering. *See:* **sputtering (electroacoustics).**

cathode terminal (1) (semiconductor rectifier diode or rectifier stack). The terminal from which forward current flows to the external circuit. *Note:* In the semiconductor recifier components field, the cathode terminal is normally marked positive. *See also:* **semiconductor; semiconductor rectifier cell.** E59-34E17/34E24

(2) (thyristor). The terminal that is connected to the cathode. *Note:* This term does not apply to bidirectional thyristors. *See also:* **anode.** E223-34E17/15E7

cathodic corrosion. An increase in corrosion of a metal by making it cathodic. *See also:* **stray current corrosion.** CM-34E2

cathodic polarization. Polarization of a cathode. *See also:* **electrochemistry.** 42A60/CM-34E2

cathodic protection. Reduction or prevention of corrosion by making a metal the cathode in a conducting medium by means of a direct electric current (which is either impressed or galvanic). *See:* **corrosion terms.** CM-34E2

catholyte. The portion of an electrolyte in an electrolytic cell adjacent to a cathode. If a diaphragm is present, it is the portion of electrolyte on the cathode side of the diaphragm. *See also:* **electrolytic cell.** 42A60-0

cation. A positively charged ion or radical that migrates toward the cathode under the influence of a potential gradient. *See also:* **ion.** CM-34E2

catwhisker. A small, sharp-pointed wire used to make contact with a sensitive point on the surface of a semiconductor. *See also:* **circuits and devices.** 42A65-0

caustic embrittlement. Stress-corrosion cracking in alkaline solutions. *See:* **corrosion terms.** CM-34E2

caustic soda cell. A cell in which the electrolyte consists primarily of a solution of sodium hydroxide. *See also:* **electrochemistry.** 42A60-0

cavitation (liquid). Formation, growth, and collapse of gaseous and vapor bubbles due to the reduction of pressure of the cavitation point below the vapor pressure of the fluid at the working temperature. *See also:* **electroacoustics.** 0-1E1

cavitation damage. Deterioration caused by formation and collapse of cavities in a liquid. *See:* **corrosion terms.** CM-34E2

cavity magnetron*. *See:* **magnetron, cavity resonator.**
*Deprecated

cavity resonator. A space normally bounded by an electrically conducting surface in which oscillating electromagnetic energy is stored, and whose resonant frequency is determined by the geometry of the enclosure. *See also:* **waveguide.** E145-0

cavity-resonator frequency meter (electromagnetic wave). A cavity resonator used to determine frequency. *See also:* **waveguide.** E147-3E1

CAX. *See:* **unattended automatic exchange.**

***C* battery.** A battery designed or employed to furnish voltage used as a grid bias in a vacuum-tube circuit. *See also:* **battery (primary or secondary).** 42A60-0

CCA (electronic navigation). *See:* **carrier-controlled approach system.**

***C* display (radar).** A rectangular display in which each target appears as a blip with bearing indicated by the horizontal coordinate and angle of elevation by the vertical coordinate. *See also:* **navigation.**

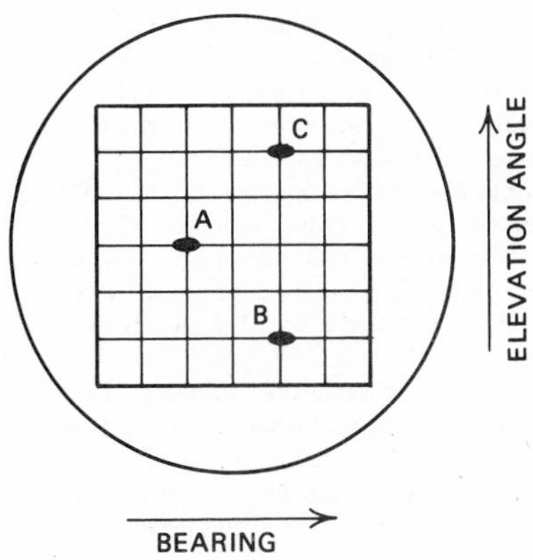

C display.

E172-10E6

CDO. *See:* **unattended automatic exchange.**

ceiling area lighting. A general lighting system in which the entire ceiling is, in effect, one large luminaire. *Note:* Ceiling area lighting includes luminous ceilings and louvered ceilings. *See also:* **general lighting.** Z7A1-0

ceiling direct voltage (direct potential rectifier unit). The average direct voltage at rated direct current with rated sinusoidal voltage applied to the alternating-current line terminals, with the rectifier transformer set on rated voltage tap and with voltage regulating means set for maximum output. *See also:* **power rectifier; rectification.** 34A1/42A15-34E24

ceiling ratio (illuminating engineering). The ratio of the luminous flux that reaches the ceiling directly to the upward component from the luminaire. *See also:* **inverse-square law.** Z7A1-0

ceiling voltage (synchronous-machine excitation system). The maximum output voltage that may be attained by an excitation system under specified conditions. 0-31E8

celestial-inertial navigation equipment. An equipment employing both celestial and inertial sensors. *See also:* **navigation.** 0-10E6

cell (information storage). An elementary unit of storage, for example, binary cell, decimal cell. *See also:* **electronic computation; electronic digital computer.** E162/E270-0

cell cavity (electrolysis). The container formed by the cell lining for holding the fused electrolyte. *See also:* **fused electrolyte.** 42A60-0

cell connector (storage cell). An electric conductor used for carrying current between adjacent storage cells. *See also:* **battery (primary or secondary).** 42A60-0

cell constant (electrolytic cell). The resistance in ohms of that cell when filled with a liquid of unit resistivity. 42A60-0

cell-type tube (microwave gas). A gas-filled radio-frequency switching tube that operates in an external resonant circuit. *Note:* A tuning mechanism may be incorporated in either the external resonant circuit or the tube. *See :* **gas tubes.** E160-15E6

cellular metal floor raceway. A hollow space of a cellular metal floor suitable for use as a raceway. *See also:* **raceway.** 42A95-0

cent (acoustics). The interval between two sounds whose basic frequency ratio is the twelve-hundredth root of 2. *Note:* The interval, in cents, between any two frequencies is 1200 times the logarithm to the base 2 of the frequency ratio. Thus 1200 cents equal 12 equally tempered semitones equal 1 octave. *See also:* **electroacoustics.** E157-1E1

center-break switching device. A mechanical switching device in which both contacts are movable and engage at a point substantially midway between their supports. 37A100-31E11

center frequency (1) (frequency modulation). The average frequency of the emitted wave when modulated by a symmetrical signal. *See also:* **frequency modulation.** E145-0;42A65-31E3

(2) (burst measurements). The arithmetic mean of the two frequencies that define the bandwidth of a filter. *See also:* **burst.** E265-0

center of distribution (primary distribution). The point from which the electric energy must be supplied if the minimum weight of conducting material is to be used. *Notes:* (1) The center of distribution is commonly considered to be that fixed point that, in practice, most nearly meets the ideal conditions stated above. (2) For an extensive list of cross references, see *Appendix A.* 42A35-31E13

centimeter-gram-second (system of units). *See:* **cgs.**

central processing unit (computing system). The unit of a computing system that includes the circuits controlling the interpretation and execution of instructions. *See also:* **electronic digital computer.** X3A12-16E9

central station (protective signaling). An office to which remote alarm and supervisory signaling devices are connected, where operators supervise the circuits, and where guards are maintained continuously to investigate signals. *Note:* Facilities may be provided for transmission of alarms to police and fire departments or other outside agencies. *See also:* **protective signaling.** 42A65-0

central station equipment (protective signaling). The signal receiving, recording, or retransmission equipment installed in the central station. *See also:* **protective signaling.** 42A65-0

central station switchboard (protective signaling). That portion of the central station equipment on or in which are mounted the essential control elements of the system. *See also:* **protective signaling.** 42A65-0

central station system (protective signaling) (central office system). A system in which the operations of electric protection circuits and devices are signaled automatically to, recorded in, maintained, and supervised from a central station having trained operators and guards in attendance at all times. *See also:* **protective signaling.** 42A65-0

central (foveal) vision. The seeing of objects in the central or foveal part of the visual field, approximately two degrees in diameter. *Note:* It permits seeing much finer detail than does peripheral vision. *See also:* **visual field.** Z7A1-0

central visual field. That region of the visual field that corresponds to the foveal portion of the retina. *See also:* **visual field.** Z7A1-0

centralized traffic-control machine (railway practice).

A control machine for operation of a specific type of traffic control system of signals and switches. *See also:* **centralized traffic-control system.** 42A42-0

centralized traffic-control system (railway practice). A specific type of traffic control system in which the signals and switches for a designated section of track are controlled from a remotely located centralized traffic control machine. *See also:* **block signal system; centralized traffic-control machine; control machine; electropneumatic interlocking machine.** 42A42-0

centrifugal actuator. *See:* **actuator, centrifugal.**

centrifugal-mechanism pin (governor pin). A component of the linkage between the centrifugal mechanism weights and the short-circuiting device. *See:* **centrifugal starting switch.** 42A10-0

centrifugal-mechanism spring (governor spring). A spring that opposes the centrifugal action of the centrifugal-mechanism weights in determining the motor speed at which the switch or short-circuiting device is actuated. *See:* **centrifugal starting switch.** 42A10-0

centrifugal-mechanism weights (governor weights). Moving parts of the centrifugal-mechanism assembly that are acted upon by centrifugal force. *See:* **centrifugal starting switch.** 42A10-0

centrifugal relay. An alternating-current frequency-selective relay in which the contacts are operated by a fly-ball governor or centrifuge driven by an induction motor. *See also:* **railway signal and interlocking.** 42A42-0

centrifugal starting-switch (rotating machinery). A centrifugally operated automatic mechanism used to perform a circuit-changing function in the primary winding of a single-phase induction motor after the rotor has attained a predetermined speed, and to perform the reverse circuit-changing operation prior to the time the rotor comes to rest. *Notes:* (1) One of the circuit changes that is usually performed is to open or disconnect the auxiliary winding circuit. (2) In the usual form of this device, the part that is mounted to the stator frame or end shield is the starting switch, and the part that is mounted on the rotor is the centrifugal actuator.
See:
actuator, centrifugal (rotating machinery);
centrifugal mechanism;
centrifugal-mechanism pin;
centrifugal-mechanism spring;
centrifugal-mechanism weights;
moving contact assembly. 0-31E8

CEP (electronic navigation). *See:* **circular probable error.**

cgs electromagnetic system of units. A system in which the basic units are the centimeter, gram, second, and abampere. *Notes:* (1) The abampere is a derived unit defined by assigning the magnitude 1 to the unrationalized magnetic constant (sometimes called the permeability of space). (2) Most electrical units of this system are designated by prefixing the syllable "ab-" to the name of the corresponding unit in the mksa system. Exceptions are the maxwell, gauss, oersted, and gilbert. E270-0

cgs electrostatic system of units. The system in which the basic units are the centimeter, gram, second, and statcoulomb. *Notes:* (1) The statcoulomb is a derived unit defined by assigning the magnitude 1 to the unrationalized electric constant (sometimes called the permittivity of space). (2) Each electrical unit of this system is commonly designated by prefixing the syllable "stat-" to the name of the corresponding unit in the International System of Units. E270-0

cgs system of units. A system in which the basic units are the centimeter, gram, and second. E270-0

chad. The piece of material removed when forming a hole or notch in a storage medium such as punched tape or punched cards. *See also:* **electronic digital computer.** X3A12-16E9

chadded. Pertaining to the punching of tape in which chad results. *See also:* **electronic digital computer.** X3A12-16E9

chadless. Pertaining to the punching of tape in which chad does not result. *See also:* **electronic digital computer.** X3A12-16E9

chadless tape. A punched tape wherein only partial perforation is completed and the chad remains attached to the tape. EIA3B-34E12

chafing strip. *See:* **drive strip.**

chain (electronic navigation). A network of similar stations operating as a group for determination of position or for furnishing navigational information. E172-10E6

chain code (computing system). An arrangement in a cyclic sequence of some or all of the possible different *n*-bit words, in which adjacent words are related such that each word is derivable from its neighbor by displacing the bits one digit position to the left, or right, dropping the leading bit, and inserting a bit at the end. The value of the inserted bit needs only to meet the requirement that a word must not recur before the cycle is complete, for example, 000 001 010 101 011 111 110 100 000... *See also:* **electronic digital computer.** X3A12-16E9

chain-drive machine (elevators). An indirect-drive machine having a chain as the connecting means. *See also:* **driving machine (elevators).** 42A45-0

chalking (corrosion). The development of loose removable powder at or just beneath a coating surface. *See:* **corrosion terms.** CM-34E2

challenge (radar). *See:* **interrogation.**

challenger (radar). *See:* **interrogator.**

changeover switch. A switching device for changing electric circuits from one combination to another. *Note:* It is usual to qualify the term changeover switch by stating the purpose for which it is used, such as a series-parallel changeover switch, trolley-shoe changeover switch, etcetera. *See also:* **multiple-unit control.** 42A42-0

channel (1) (electric communication). (A) A single path for transmitting electric signals, usually in distinction from other parallel paths. (B) A band of frequencies. *Note:* The word path is to be interpreted in a broad sense to include separation by frequency division or time division. The term channel may signify either a one-way path, providing transmission in one direction only, or a two-way path, providing transmission in two directions. 42A65-0
See:
bandwidth;
carrier telephone channel;
channel group;
channel supergroup;
frequency band;

interference guard bands;
link;
telegraph channel;
telephone channel;
television channel.
(2) (electronic computers). (A) A path along which signals can be sent, for example, data channel, output channel. (B) The portion of a storage medium that is accessible to a given reading station. *See also:* **electronic digital computer; track.** E162-0
(3) (information theory). A combination of transmission media and equipment capable of receiving signals at one point and delivering related signals at another point. *See also:* **information theory.** E171-0
(4) (illuminating engineering). An enclosure containing the ballast, starter, lamp holders, and wiring for a fluorescent lamp, or a similar enclosure on which filament lamps (usually tubular) are mounted. *See also:* **luminaire.** Z7A1-0

channel capacity (information theory). The maximum possible information rate through a channel subject to the constraints of that channel. *Note:* channel capacity may be either per second or per symbol. *See also:* **information theory.** E171-0

channel failure alarm (power-system communication). A circuit to give an alarm if a communication channel should fail. *See also:* **power-line carrier.** 0-31E3

channel group (group). A number of channels regarded as a unit. *Note:* The term is especially used to designate part of a larger number of channels. *See also:* **channel.** 42A65-0

channel lights. Aeronautical ground lights arranged along the sides of a channel of a water aerodrome. *See also:* **signal lighting.** Z7A1-0

channel, melting. *See:* **melting channel.**

channel multiplier. A tubular electron-multiplier with a continuous interior surface of secondary-electron emissive material. *See also:* **amplifier; camera tube.** 0-15E6

channel, radio. *See:* **radio channel.**

channel spacing (radio communication). The frequency increment between the assigned frequency of two adjacent radio-frequency channels. *See also:* **dispatch operation; radio transmission; single-frequency simplex operation; two-frequency simplex operation.** 0-6E1

channel supergroup (supergroup). A number of channel groups regarded as a unit. *Note:* The term is especially used to designate part of a larger number of channels. *See also:* **channel.** 42A65-0

channel utilization index (information theory). The ratio of the information rate (per second) through a channel to the channel capacity (per second). *See also:* **information theory.** E171-0

character (electronic computers). (1) An elementary mark or event that may be combined with others, usually in the form of a linear string, to form data or represent information. If necessary to distinguish from (2) below, such a mark may be called a character event. (2) A class of equivalent elementary marks or events as in (1) having properties in common, such as shape or amplitude. If necessary to distinguish from (1) above, such a class may be called a character design. There are usually only a finite set of character designs in a given language. *Notes:* (1) In "bookkeeper" there are six character designs and ten character events, while in "1010010" there are two character designs and seven character events. (2) A group of characters, in one context, may be considered as a single character in another, as in the binary-coded-decimal system. *See:* **blank character; check character; control character; escape character; numerical control; special character.** *See also:* **electronic digital computer; electronic computation** E162-0

character-indicator tube (electron device). A glow-discharge tube in which the cathode glow displays the shape of a character, for example, letter, number, or symbol. *See also:* **tube definitions.** 0-15E6

characteristic angular phase difference. The characteristic angular phase difference of a set of polyphase voltages or currents of m phases is the minimum phase angle difference by which each member of a polyphase symmetrical set of polyphase voltages or currents may lag the preceding member of the set, when the members are arranged in the normal sequence. The characteristic angular phase difference is $2\pi/m$ radians, where m is the number of phases. E270-0

characteristic curve (illuminating engineering). A curve that expresses the relationship between two variable properties of a light source, such as candle-power and voltage, flux and voltage, etcetera. *See also:* **colorimetry.** Z7A1-0

characteristic curves (rotating machinery). The graphical representation of the relationships between certain quantities used in the study of electric machines. *See also:* **asynchronous machine; direct-current commutating machine; synchronous machine.** 0-31E8

characteristic distortion (telegraphy). A displacement of signal transitions resulting from the persistence of transients caused by preceding transitions. *See also:* **telegraphy.** 42A65-19E4

characteristic element (lightning arresters). The element that in a valve-type arrester determines the discharge voltage and the follow current, and in an expulsion-type arrester determines the discharge voltage and interrupts the follow current. *See:* **arrester; lightning arrester (surge diverter).** 42A20-31E7

characteristic equation (feedback control system) (control system, feedback). The relation formed by equating to zero the denominator of a transfer function of a closed loop. *See also:* **control system, feedback.** 0-23E0

characteristic impedance (1) (circular waveguide.) For a traveling wave in the dominant ($TE_{1,1}$) mode of a lossless circular waveguide at a specified frequency above the cutoff frequency, (A) the ratio of the square of the root-mean-square voltage along the diameter where the electric vector is a maximum to the total power flowing when the guide is match terminated, (B) the ratio of the total power flowing to the square of the total root-mean-square longitudinal current flowing in one direction when the guide is match terminated, (C) the ratio of the root-mean-square voltage along the diameter where the electric vector is a maximum to the total root-mean-square longitudinal current flowing along the half surface bisected by this diameter when the guide is match terminated. *Note:* Under definition (A) the power $W = V^2/Z_{(W,V)}$ where V is the voltage and $Z_{(W,V)}$ is the characteristic impedance defined in (A). Under definition (B) the power $W = I^2 Z_{(W,I)}$

where I is the current and $Z_{(W, I)}$ is the characteristic impedance defined in (B). The characteristic impedance $Z_{(V, I)}$ as defined in (C) is the geometric mean of the values given by (A) and (B). Definition (C) can be used also below the cutoff frequency. *See also:* **self-impedance; waveguide.**

(2) (rectangular waveguide). For a traveling wave in the dominant ($TE_{1,0}$) mode of a lossless rectangular waveguide at a specified frequency above the cutoff frequency, (A) the ratio of the square of the root-mean-square voltage between midpoints of the two conductor faces normal to the electric vector, to the total power flowing when the guide is match terminated, (B) the ratio of the total power flowing to the square of the root-mean-square longitudinal current, flowing on one face normal to the electric vector when the guide is match terminated, (C) the ratio of the root-mean-square voltage, between midpoints of the two conductor faces normal to the electric vector, to the total root-mean-square longitudinal current, flowing on one face when the guide is match terminated. *Note:* Under definition (A) the power $W = V^2/Z_{(W, V)}$ where V is the voltage, and $Z_{(W, V)}$ the characteristic impedance defined in (A). Under definition (B) the power $W = I^2 Z_{(W, I)}$ where I is the current and $Z_{(W, I)}$ the characteristic impedance defined in (B). The characteristic impedance $Z_{(V, I)}$ as defined in (C) is the geometric mean of the values given by (A) and (B). Definition (C) can be used also below the cutoff frequency. *See:* **waveguide.** *See also:* **self-impedance.**

(3) (two-conductor transmission line) (for a traveling transverse electromagnetic wave). The ratio of the complex voltage between the conductors to the complex current on the conductors in the same transverse plane with the sign so chosen that the real part is positive. *See also:* **self-impedance; transmission line; waveguide.**

(4) (coaxial transmission line). The driving impedance of the forward-traveling transverse electromagnetic wave. *See also:* **self-impedance; transmission line.** E146/83A14-3E1

(5) (surge impedance) (lightning arrester). The driving-point impedance that the line would have if it were of infinite length. *Note:* It is recommended that this term be applied only to lines having approximate electric uniformity. For other lines or structures the corresponding term is iterative impedance. *See also:* **self-impedance.** E270-31E7/3E1/9E4;42A65-31E3

characteristic insertion loss (waveguide and transmission line). The insertion loss in a transmission system that is reflectionless looking toward both the source and the load from the inserted transducer. *Notes:* (1) This loss is a unique property of the inserted transducer. (2) The frequency, internal impedance, and available power of the source and the impedance of the load have the same value before and after the transducer is inserted. *See also:* **waveguide.** 0-9E4

characteristic insertion loss, incremental. The change in the characteristic insertion loss of an adjustable device between two settings. *See also:* **waveguide.** 0-9E4

characteristic insertion loss, residual. The characteristic insertion loss of an adjustable device at an indicated minimum position. *See also:* **waveguide.** 0-9E4

characteristic phase shift. For a 2-port device inserted into a stable, nonreflecting system between the generator and its load, the magnitude of the phase change of the voltage wave incident upon the load before and after insertion of the device, or change of the device from initial to final condition. *Note:* The following conditions apply: (1) The frequency, the load impedance, and the generator characteristics, internal impedance and available power, initially have the same values as after the device is inserted; (2) the joining devices, connectors or adapters belonging to the system conform to some set of standard specifications, the same specifications to be used by different laboratories, if measurements are to agree precisely; (3) the nonreflecting conditions are to be obtained in uniform, standard sections of waveguide on the system sides of the connectors at the place of insertion. *See also:* **measurement system.** E285-9E1

characteristic telegraph distortion. Distortion that does not affect all signal pulses alike, the effect on each transition depending upon the signal previously sent, due to remnants of previous transitions or transients that persist for one or more pulse lengths. *Note:* Lengthening of the mark pulse is positive, and shortening, negative. Characteristic distortion is measured by transmitting biased reversals, square waves having unequal mark and space pulses. The average lengthening or shortening of mark pulses, expressed in percent of unit pulse length, gives a true measure of characteristic distortion only if other types of distortion are negligible. See also: **modulation.** E145-0

characteristic time. *See:* **apparent time constant (of a thermal converter).**

characteristic wave impedance. *See:* **wave impedance, characteristic.**

character recognition. The identification of graphic, phonic, or other characters by automatic means. *See:* **magnetic-ink character recognition; optical character recognition.** *See also:* **electronic digital computer.** X3A12-16E9

charge (1) (storage battery) (storage cell). The conversion of electric energy into chemical energy within the cell or battery. *Note:* This restoration of the active materials is accomplished by maintaining a unidirectional current in the cell or battery in the opposite direction to that during discharge; a cell or battery that is said to be charged is understood to be fully charged. *See:*

ampere-hour efficiency;
battery;
boost charge;
charging rate;
constant-current charge;
constant-voltage charge;
cycle of operation;
discharge;
electric charge;
equalizing charge;
finishing rate;
floating;
modified constant-voltage charge;
rating of storage batteries;
reversal;
service life;
trickle charge;
volt efficiency;
watthour efficiency. 42A60-0

(2) (induction and dielectric-heating usage). *See:* **load.**

charge, space. *See:* **space charge.**

charge carrier (semiconductor). A mobile conduction

electron or mobile hole. *See also:* **semiconductor.** E59/E216-34E17;E270-0

charge-resistance furnace. A resistance furnace in which the heat is developed within the charge acting as the resistor. *See also:* **electrothermics.** 42A60-0

charge-storage tube (electrostatic memory tube). A storage tube in which the information is retained on the storage surface in the form of a pattern of electric charges. *Note:* For an extensive list of cross references, see *Appendix A.* E158-15E6

charge transit time. *See:* **transit time.**

charging (electrostatography). *See:* **sensitizing.** *See also:* **electrostatography.**

charging circuit (surge generator) (lightning arresters). The portion of the surge generator connections through which electric energy is stored up prior to the production of a surge. *See:* **lightning arrester (surge diverter); power systems, low-frequency and surge testing.** 42A35-31E7/31E13

charging current (transmission line). The current that flows in the capacitance of a transmission line when voltage is applied at its terminals. *See also:* **transmission line.** 42A35-31E13

charging rack (mining). A device used for holding batteries for mining lamps and for connecting them to a power supply while the batteries are being recharged. *See also:* **mine feeder circuit.** 42A85-0

charging rate (storage battery) (storage cell). The current expressed in amperes at which a battery is charged. *See also:* **charge.** 42A60-0

charles or kino gun. *See:* **end injection.**

chart (recording instrument). The paper or other material upon which the graphic record is made. *See also:* **moving element (instrument).** 42A30-0

chart comparison unit (electronic navigation). A device for the simultaneous viewing of a navigational position presentation and a navigational chart in such a manner that one appears superimposed upon the other. *See also:* **navigation.** E172-10E6

chart mechanism (recording instrument). The parts necessary to carry the chart. *See also:* **moving element (instrument).** 39A2-0

chart scale (recording instrument). The scale of the quantity being recorded, as marked on the chart. *Note:* Independent of and generally in quadrature with the chart scale is the time scale which is graduated and marked to correspond to the principal rate at which the chart is advanced in making the recording. This quadrature scale may also be used for quantities other than time. *See also:* **moving element (instrument).** 42A30-0

chart scale length (recording instrument). The shortest distance between the two ends of the chart scale. *See also:* **instrument.** 42A30-0

chatter, relay. *See:* **relay chatter time; relay contact chatter.**

check (1) (standardize) (instrument or meter). Ascertain the error of its indication, recorded value, or registration. *Note:* The use of the word **standardize** in place of **adjust** to designate the operation of adjusting the current in the potentiometer circuit to balance the standard cell is deprecated. *See also:* **test (instrument or meter).** 42A30-0

(2) (computer-controlled machines). A process of partial or complete testing of (1) the correctness of machine operations, or (2) the existence of certain prescribed conditions within the computer. A check of any of these conditions may be made automatically by the equipment or may be programmed.
See:
automatic check;
balance check;
duplication check;
dynamic check;
dynamic computer check;
echo check;
machine check;
marginal check;
mathematical check;
modulo N check;
odd-even check;
parity check;
problem check;
programmed check;
residue check;
selection check;
self-checking code;
summation check;
transfer check.
See also: **electronic analog computer; electronic digital computer; error-detecting code; forbidden combination; self-checking code; test; verify.** E162/E270-0

check, automatic (electronic computation). A check performed by equipment built into the computer specifically for that purpose and automatically accomplished each time the pertinent operation is performed. *Note:* Sometimes referred to as a **built-in check**. Machine check can refer to an automatic check or to a programmed check of machine functions. *See also:* **electronic computation.** E270-0

check bit. A binary check digit, for example, a parity bit. *See also:* **electronic digital computer.** 0-16E9

check bits (data transmission). Associated with a code character or block for the purpose of checking the absence of error within the code character or block. *See also:* **data processing.** 0-19E4

check character. A character used for the purpose of performing a check, but often otherwise redundant. 0-16E9

check digit. A digit used for the purpose of performing a check, but often otherwise redundant. *See:* **check, forbidden-combination.** *See also:* **electronic digital computer.** 0-16E9

check, forbidden-combination (electronic computation). A check (usually an automatic check) that tests for the occurrence of a nonpermissible code expression. *Notes:* (1) A self-checking code (or error-detecting code) uses a code expression such that one (or more) error(s) in a code expression produces a forbidden combination. (2) A parity check makes use of a self-checking code employing binary digits in which the total number of 1's (or 0's) in each permissible code expression is always even or always odd. A check may be made for either even parity or odd parity. (3) A redundancy check employs a self-checking code that makes use of redundant digits called check digits. *See also:* **electronic computation.** E270-0

checkpoint (1) (electronic computation). A place in a routine where a check, or a recording of data for restart purposes, is performed. See also: **electronic digital computer.**

(2) (electronic navigation). *See:* **way point.** X3A12-16E9

check problem (electronic computation). A routine or

problem that is designed primarily to indicate whether a fault exists in the computer, without giving detailed information on the location of the fault. Also called check routine. *See:* **diagnostic; test.** *See also:* **electronic digital computer; programmed check.** E162-0

check, programmed (electronic computation). *See:* **programmed check.**

check routine. Same as check problem. *See also:* **electronic digital computer.**

check, selection (electronic computation). *See:* **selection check.**

check solution. A solution to a problem obtained by independent means to verify a computer solution. *See also:* **electronic analog computer.** E165-16E9

check, transfer (electronic computation). *See:* **transfer check.**

check valve. *See:* **blocking capacitor.**

check, field-coil flange (washer). *See:* **collar.**

cheese antenna. A cylindrical reflector enclosed by two parallel conducting plates perpendicular to the cylinder, so spaced and excited as to permit the propagation of more than one mode in the narrow dimension of the parallel-plate region. *See also:* **antenna.** 0-3E1

chemical conversion coating. A protective or decorative coating produced in situ by chemical reaction of a metal with a chosen environment. See: **corrosion terms.** CM-34E2

Child-Langmuir equation (thermionics). An equation representing the cathode current of a thermionic diode in a space-charge-limited-current state.

$$I = GV^{3/2}$$

where I is the cathode current, V is the anode voltage of a diode or the equivalent diode of a triode or of a multielectrode valve or tube, and G is a constant (perveance) depending on the geometry of the diode or equivalent diode. *See also:* **electron emission.** 50I07-15E6

chip (mechanical recording). The material removed from the recording medium by the recording stylus while cutting the groove. *See also:* **phonograph pickup.** E157-1E1

choke (waveguide). A device for preventing energy within a waveguide in a given frequency range from taking an undesired path. *See:* **waveguide.** 0-3E1

choke coil. An inductor used in a special application to impede the current in a circuit over a specified frequency range while allowing relatively free passage of the current at lower frequencies. 0-21E1

choke flange (microwave technique). A flange in whose surface is cut a groove so dimensioned that the flange may form part of a choke joint. *See also:* **waveguide.** 42A65-3E1

choke joint (1) (waveguides). A connection between two waveguides that provides effective electric continuity without metallic continuity at the inner walls of the waveguide. *See also:* **waveguide.** E147-0;0-3E1

(2) (microwave transmission lines) (microwave technique). A connector between two sections of transmission line in which the gap between sections to be connected is built out to form a series-branching transmission line carrying a standing wave in which actual contact falls at or near a current minimum. *Note:* The series-branching line is typically one-half wave in length. The connection then occurs at a quarter-wave point, and the closed end of the line is contained wholly within one of the sections. Such joints are used to prevent radio-frequency leakage and high ohmic losses, and for other purposes. 42A65-0

choke piston (choke plunger) (noncontact plunger) (waveguide). A piston in which there is no metallic contact with the walls of the waveguide at the edges of the reflecting surface; the short-circuit to high-frequency currents is achieved by a choke system. *Note:* This definition covers a number of configurations: dumbbell; Z-slot; inverted bucket; etcetera. *See also:* **waveguide.** 0-3E1

choke plunger. *See:* **choke piston.**

chopped display (oscilloscopes). A time-sharing method of displaying output signals of two or more channels with a single cathode-ray-tube gun, at a rate that is higher than and not referenced to the sweep rate. *See:* **oscillograph.** 0-9E4

chopped impulse voltage. A transient voltage derived from a full impulse voltage that is interrupted by the disruptive discharge of an external gap or the external portion of the test specimen causing a sudden collapse in the voltage, practically to zero value. *Note:* The collapse can occur on the front, at the peak, or on the tail. *See also:* **test voltage and current.**

chopped impulse wave (lightning arresters). An impulse wave that has been caused to collapse suddenly by a flashover. *See:* **lightning arrester (surge diverter).** 50I25-31E7

chopped wave. A voltage impulse that is terminated intentionally by sparkover of a gap. *See also:* **power systems, low-frequency and surge testing.** 42A350-31E13

chopper. (1) A device for interrupting a current or a light beam at regular intervals. Choppers are frequently used to facilitate amplification. *See also:* **circuits and devices.** 42A65-0

(2) A special form of pulsing relay having contacts arranged to rapidly interrupt, or alternately reverse, the direct-current polarity input to an associated circuit. *See also:* **circuits and devices.** 0-21E0

(3) (signal-transmission system). Any mechanical, electric, electronic, or electromechanical circuit components designed to convert a direct-current or low-frequency input signal to an alternating-current signal of higher frequency and of substantially square waveform. *Note:* As applied to a direct-coupled operational amplifier, it is a modulator used to convert the direct current at the summing junction to alternating current for amplification and reinsertion as a correcting voltage to reduce drift. *See also:* **circuits and devices; electronic analog computer.** 0-13E6;E165-16E9

chopping frequency. *See:* **chopping rate.**

chopping rate (oscilloscopes). The rate at which channel switching occurs in chopped-mode operation. *See:* **oscillograph.** 0-9E4

chroma. *See:* **Munsell chroma.**

chromaticity. The color quality of light definable by its chromaticity coordinates, or by its dominant (or complementary) wavelength and its purity, taken together. *See also:* **color terms; color.** E201-0;50I45-2E2

chromaticity coordinate (light). The ratio of any one of the tristimulus values of a sample to the sum of the three tristimulus values. *Note:* In the standard colorimetric system CIE (1931) the symbols *x, y, z* are recommended for the chromaticity coordinates. *See also:* **CIE; color terms.** E201-0;50I45-2E2

chromaticity diagram. A plane diagram formed by plotting one of the three chromaticity coordinates against another. *Note:* The most common chromaticity diagram at present is the CIE *(x,y)* diagram plotted in rectangular coordinates. *See also:* **CIE; color terms; color.** E201-2E2;Z7A1-0

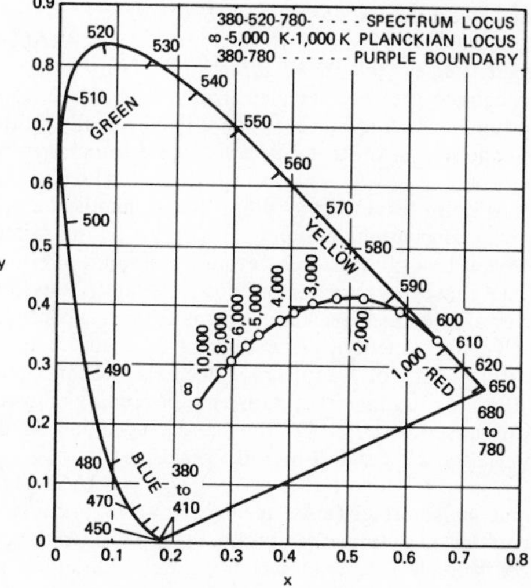

Chromaticity diagram.

chromaticity flicker. The flicker that results from fluctuation of chromaticity only. *See also:* **color terms.** E201-2E2

chrominance. The colorimetric difference between any color and a reference color of equal luminance, the reference color having a specified chromaticity. *Notes:* (1) In three-dimensional color space, chrominance is a vector that lies in a plane of constant luminance. In that plane it may be resolved into components, called chrominance components. (2) In color television transmission, for example, the chromaticity of the reference color may be that of a specified white. *See also:* **color terms.** E201-2E2

chrominance channel (color television system) (television). Any path that is intended to carry the chrominance signal. *See also:* **television.** E204-2E2;42A65

chrominance channel bandwidth (television). The bandwidth of the path intended to carry the chrominance signal. *See also:* **television.** E204-2E2

chrominance components. *See:* **chrominance.**

chrominance demodulator (color television reception). A demodulator used for deriving video-frequency chrominance components from the chrominance signal and a sine wave of chrominance subcarrier frequency. *See also:* **color terms.** E201-2E2

chrominance modulator (color television transmission). A modulator used for generating the chrominance signal from the video-frequency chrominance components and the chrominance subcarrier. *See also:* **color terms.** E201-2E2

chrominance primary (color television). A transmission primary that is one of two whose amounts determine the chrominance of a color. *Note:* Chrominance primaries have zero luminance and are nonphysical. *See also:* **color terms.** E201-2E2

chrominance signal (carrier chrominance signal) (in color television). The sidebands of the modulated chrominance subcarrier that are added to the monochrome signal to convey color information. *See also:* **color terms.** E201-0

chrominance signal component (television). A signal resulting from suppressed-carrier modulation of a chrominance subcarrier voltage at a specified phase, by a chrominance primary signal. *See also:* **television.** E204-2E2;42A65

chrominance signal (television). *See:* **television.**

chrominance subcarrier (color television). The carrier whose modulation sidebands are added to the monochrome signal to convey color information. *See also:* **color terms.** E201-0

chronaxie (medical electronics). The minimum duration of time required to stimulate with a current of twice the rheobase. *See also:* **medical electronics.** 0-18E1

CIE. Abbreviation for Commission Internationale de l'Eclairage. *Note:* These are the initials of the official French name of the International Commission on Illumination. This translated name is approved for usage in English-speaking countries, but at its 1951 meeting the Commission recommended that only the initials of the French name be used. The initials ICI, which have been used commonly in this country, are deprecated because they conflict with an important trademark registered in England and because the initials of the name translated into other languages are different. *See also:* **color terms.** E201-2E2

CIE standard chromaticity diagram. One in which the *x* and *y* chromaticity coordinates are plotted in rectangular coordinates. *See also:* **color.** Z7A1-0

CIE (1931) standard colorimetric observer. Receptor of radiation whose colorimetric characteristics correspond to the distribution coefficients $\bar{x}_\lambda$, $\bar{y}_\lambda$, $\bar{z}_\lambda$ adopted by the International Commission on Illumination in 1931. *See also:* **color.** 50I45-2E2

circle diagram (rotating machinery). (1) Circular locus describing performance characteristics (current, impedance, etcetera) of a machine or system. In case of rotating machinery, the term **circle diagram** has, in addition, some specific usages: The locus of the armature current phasor of an induction machine, or of some other type of asynchronous machine, displayed on the complex plane, with the shaft speed as the variable (parameter), when the machine operates at a constant voltage and at a constant frequency. (2) The locus of the current vector(s) of a nonsalient-pole synchronous machine, displayed in a synchronously rotating reference frame (Park transform, *d-q* coordinates), with the active component of the load, hence with the rotor displacement angle, as the variable (parameter), when the machine operates at a constant voltage, at a constant frequency, and at a constant field current. (3) The locus of the current phasor(s) of (2). *See:* **asynchronous machine; synchronous machine.** 0-31E8

circling guidance lights. Aeronautical ground lights provided to supply additional guidance during a circling approach when the circling guidance furnished by the approach and runway lights is inadequate. *See also:* **atmospheric transmissivity.** Z7A1-0

circuit (1). A conductor or system of conductors through which an electric current is intended to flow. *See also:* **center of distribution; service.** 2A2-0;42A35-31E13

(2). A network providing one or more closed paths. *See also:* **network analysis.** E153/E270-0

(3) (machine winding). The element of a winding that comprises a group of series-connected coils. A single-phase winding or one phase of a polyphase winding may comprise one circuit or several circuits connected in parallel. *See:* **synchronous machine.** 0-31E8

circuit analyzer (multimeter). The combination in a single enclosure of a plurality of instruments or instrument circuits for use in measuring two or more electrical quantities in a circuit. *See also:* **instrument.** 42A30-0

circuitation. *See:* **circulation.**

circuit, balanced. *See:* **balanced circuit.**

circuit breaker. (1) A device designed to open and close a circuit by nonautomatic means, and to open the circuit automatically on a predetermined overload of current, without injury to itself when properly applied within its rating. 1A0-0

(2) A mechanical switching device capable of making, carrying, and breaking currents under normal circuit conditions and also, making, carrying for a specified time, and breaking currents under specified abnormal circuit conditions such as those of short-circuit.

Notes: (1) A circuit breaker is usually intended to operate infrequently, although some types are suitable for frequent operation. (2) The medium in which circuit interruption is performed may be designated by suitable prefix, such as, air-blast circuit breaker, air circuit breaker, compressed-air circuit breaker, gas circuit breaker, oil circuit breaker, vacuum circuit breaker, etcetera. (3) Circuit breakers are classified according to their application or characteristics and these classifications are designated by the following modifying words or clauses delineating the several fields of application, or pertinent characteristics:

High-voltage power—Rated 1000 volts alternating current or above.

Molded-case—See separate definition.

Low-voltage power—Rated below 1000 volts alternating current or 3000 volts direct current and below, but not including molded-case circuit breakers.

Direct-current low-voltage power circuit breakers are subdivided according to their specified ability to limit fault-current magnitude by being called general purpose, high-speed, semi-high-speed, or anode. For specifications of these restrictions see the latest revision of the applicable American National Standard. *See also:* **alternating-current distribution; switch.** 37A100-31E11

circuit characteristics of electrodes. *Note:* For an extensive list of cross references, see *Appendix A.*

circuit-commutated turn-off time (thyristor). The time interval between the instant when the principal current has decreased to zero after external switching of the principal voltage circuit and the instant when the thyristor is capable of supporting a specified principal voltage without turning on. *See also:* **principal voltage-current characteristic (principal characteristic).** E223-34E17/34E24/15E7

circuit controller. A device for closing and opening electric circuits. *See also:* **railway signal and interlocking.** 42A42-0

circuit efficiency (output circuit of electron tubes). The ratio of (1) the power at the desired frequency delivered to a load at the output terminals of the output circuit of an oscillator or amplifier to (2) the power at the desired frequency delivered by the electron stream to the output circuit. *See also:* **circuit characteristics of electrodes; network analysis.** E160-15E6

circuit element. A basic constituent part of a circuit, exclusive of interconnections. *See also:* **circuits and devices.** 42A65-0

circuit noise (telephone practice). Noise that is brought to the receiver electrically from a telephone system, excluding noise picked up acoustically by the telephone transmitters. *See also:* **signal-to-noise ratio.** 42A65-31E3

circuit noise level (at any point in a transmission system) (telecommunication). The ratio of the circuit noise at that point to some arbitrary amount of circuit noise chosen as a reference. *Note:* This ratio is usually expressed in decibels above reference noise, abbreviated "dBrn," signifying the reading of a circuit noise meter, or in adjusted decibels, abbreviated "dBa," signifying circuit noise meter reading adjusted to represent interfering effect under specified conditions. *See also:* **signal-to-noise ratio.** 42A65-31E3

circuit noise meter (noise measuring set). An instrument for measuring circuit noise level. Through the use of a suitable frequency-weighting network and other characteristics, the instrument gives equal readings for noises that are approximately equally interfering. The readings are expressed as circuit noise levels in decibels above reference noise. *See:* **circuit noise level.** *See also:* **instrument.** 42A30-0

circuit parameters. The values of physical quantities associated with circuit elements. *Note:* For example, the resistance (parameter) of a resistor (element), the amplification factor and plate resistance (parameters) of a tube (element), the inductance per unit length (parameter) of a transmission line (element), etcetera. *See also:* **network analysis.** E270-0

circuit switching element (inverters). A group of one or more simultaneously conducting thyristors, connected in series or parallel or any combination of both, bounded by no more than two main terminals and conducting principal current between these main terminals. *See also:* **self-commutated inverters.** 0-34E24

circuit transient recovery voltage. The transient recovery voltage characterizing the circuit and obtained with 100-percent normal-frequency recovery voltage, symmetrical current, and no modifying effect of the interrupting device. *Note:* This voltage indicates the inherent severity of the circuit with respect to recovery voltage phenomena. 37A100-31E11

circuit voltage class (electric power system). A phase-to-phase reference voltage that is used in the selection of insultation class designations for neutral grounding devices. *See also:* **power systems, low-frequency and surge testing.** E32-0

circuits and devices. *Note:* For an extensive list of cross references, see *Appendix A.*

circular electric wave. A transverse electric wave for which the lines of electric force form concentric circles. *See also:* **waveguide.** E146-3E1

circular interpolation (numerically controlled machines). A mode of contouring control that uses the information contained in a single block to produce an arc of a circle. *Note:* The velocities of the axes used to generate this arc are varied by the control. *See also:* **numerically controlled machines.** EIA3B-34E12

circular magnetic wave. A transverse magnetic wave for which the lines of magnetic force form concentric circles. *See also:* **waveguide.** E146-3E1

circular mil. A unit of area equal to $\pi/4$ of a square mil (= 0.7854 square mil). The cross-sectional area of a circle in circular mils is therefore equal to the square of its diameter in mils. A circular inch is equal to one million circular mils. *Note:* A mil is one-thousandth part of an inch. There are 1974 circular mils in a square millimeter. E30-0

circular probable error (CPE or CEP) (two-dimensional error distribution). The radius of a circle encompassing half of all errors. *See also:* **navigation.** 0-10E6

circular scanning (radio). Scanning in which the direction of maximum response generates a plane or a right circular cone whose vertex angle is close to 180 degrees.
See:
antenna;
monopulse;
phase-corrected horn;
radar;
radar cross-section;
radome;
sector scanning. 42A65-3E1

circularly polarized wave (1) (general). An elliptically polarized wave in which the ellipse is a circle in a plane perpendicular to the direction of propagation. *See also:* **radiation.** 42A65-0
(2) (radio wave propagation). An electromagnetic wave for which either the electric or the magnetic field vector at a fixed point describes a circle. *Note:* This term is usually applied to transverse waves. *See also:* **radiation; radio wave propagation.** 0-3E2

circulating memory. *See:* **circulating register.**

circulating register (1) (data processing and computation). A register that retains data by inserting it into a delaying means, and regenerating and reinserting the data into the register. *See also:* **electronic digital computer.** E162/E270-0
(2) Shift register in which data moved out of one end of the register are reentered into the other end as in a closed loop. *See:* **cyclic shift.** *See also:* **electronic digital computer.** X3A12-16E9

circulation (circuitation). The circulation of a vector field is its line integral over a closed curve *C*. E270-0

circulation of electrolyte. A constant flow of electrolyte through a cell to facilitate the maintenance of uniform conditions of electrolysis. *See also:* **electrorefining.** 42A60-0

circulator (waveguide system). A passive waveguide junction of three or more arms in which the arms can be listed in such an order that when power is fed into any arm it is transferred to the next arm on the list, the first arm being counted as following the last in order. *See also:* **transducer; waveguide.** 50I62-3E1

CISPR. International Special Committee on Radio Interference.

clamp. *See:* **clamping circuit.**

clamper. *See:* **clamping circuit.**

clamping (control) (industrial control). A function by which the extreme amplitude of a waveform is maintained at a given level. *See:* **control system, feedback.** AS1-34E10

clamping circuit (clamper) (clamp) (1) (electronic circuits). A circuit that adds a fixed bias to a wave at each occurrence of some predetermined feature of the wave so that the voltage or current of the feature is held at or "clamped" to some specified level. The level may be fixed or adjustable. *See also:* **circuits and devices.** 42A65-0
(2) (analog computers). A circuit used to provide automatic hold and reset action electronically for the purposes of switching or supplying repetitive operation in an analog computer. *See also:* **electronic analog computer.** E165-0

clamping screw. *See:* **binding screw.**

clapper. An armature that is hinged or pivoted. 83A16-0

class (electric instrument). *See also:* **accuracy rating (electric instrument).**

class-A amplifier (electron tubes). *See:* **amplifier, class ratings.**

class-AB amplifier (electron tubes). *See:* **amplifier, class ratings.**

class-AB operation. *See:* **amplifier, class ratings.**

class-A insulation. *See:* **insulation, class ratings.**

class-A modulator. A class-A amplifier that is used specifically for the purpose of supplying the necessary signal power to modulate a carrier. See also: **modulation.** E145/E182/42A65-0

class-A operation. *See:* **amplifier, class ratings.**

class-A push-pull sound track. A class-A push-pull photographic sound track consists of two single tracks side by side, the transmission of one being 180 degrees out of phase with the transmission of the other. Both positive and negative halves of the sound wave are linearly recorded on each of the two tracks. *See also:* **phonograph pickup.** 0-1E1

class-B amplifier (electron tubes). *See:* **amplifier, class ratings.**

class-B insulation. *See:* **insulation, class ratings.**

class-B modulator. A class-B amplifier that is used specifically for the purpose of supplying the necessary signal power to modulate a carrier. *Note:* In such a modulator the class-B amplifier is normally connected in push-pull. *See also:* **modulation.** E145/E182/42A65-0

class-B operation. *See:* **amplifier, class ratings.**

class-B push-pull sound track. A class-B push-pull photographic sound track consists of two tracks side by side, one of which carries the positive half of the signal only, and the other the negative half. *Note:* During the inoperative half cycle, each track transmits little or no light. *See also:* **phonograph pickup.** 0-1E1

class-C amplifier (electron tubes). *See:* **amplifier, class ratings.**

class-C insulation. *See:* **insulation, class ratings.**

class-C operation. *See:* **amplifier, class ratings.**

class designation of a watthour meter. Denotes the maximum of the load range in amperes. *See also:* **watthour meter.** 12A0-0

class-H insulation. *See:* **insulation, class ratings.**

class-O insulation. *See:* **insulation, class ratings.**

class-90 insulation. *See:* **insulation, class ratings.**

class-105 insulation. *See:* **insulation, class ratings.**
class-130 insulation. *See:* **insulation, class ratings.**
class-155 insulation. *See:* **insulation, class ratings.**
class-180 insulation. *See:* **insulation, class ratings.**
class-220 insulation. *See:* **insulation, class ratings.**
class-over-220 insulation. *See:* **insulation, class ratings.**

classification lamp (classification light). A signal lamp placed at the side of the front end of a train or vehicle, displaying light of a particular color to identify the class of service in which the train or vehicle is operating. 42A42-0

classification light. *See:* **classification lamp.**

cleaner (electroplating). A compound or mixture used in degreasing, which is usually alkaline. 42A60-0

cleaning (electroplating). The removal of grease or other foreign material from a metal surface, chiefly by physical means. *See also:* **electroplating.** 42A60-0

clear (electronic computers). (1) To preset a storage or memory device to a prescribed state, usually that denoting zero. (2) To place a binary cell in the zero state. *See:* **reset; set.** *See also:* **electronic digital computer.** E162/E270-0;X3A12-16E9

clearance (1) (electronic navigation). (A) In instrument landing systems (ILS), a difference in depth of modulation (DDM) in excess of that required to produce full-scale deflection of the course deviation indicator in flight areas outside the on-course sector; when the difference in depth of modulation is too low the indicator falls below full-scale deflection and the condition of low clearance exists. (B) In air-traffic control, permission by a control facility to the pilot to proceed in a mutually understood manner. *See also:* **navigation.** 0-10E6

(2) (transmission and distribution). The minimum separation between two conductors, between conductors and supports or other objects, or between conductors and ground. *See also:* **tower.** 42A35-31E13

clearance antenna array (directional localizer). The antenna array that radiates a localizer signal on a separate frequency within the pass band of the receiver and provides the required signals in the clearance sectors as well as a back course. *See also:* **navigation.** 0-10E6

clearance lamp. A lighting device for the purpose of indicating the width of a vehicle. *See also:* **headlamp.** Z7A1-0

clearance point. The location on a turnout at which the carrier's specified clearance is provided between tracks. *See also:* **railway signal and interlocking.** 42A42-0

clearance sector (instrument landing systems). The sector extending around either side of the localizer from the course sector to the back course sector, and within which the deviation indicator provides the required off-course indication. *See also:* **navigation.** 0-10E6

clearing circuit. A circuit used for the operation of a signal in advance of an approaching train. *See also:* **railway signal and interlocking.** 42A42-0

clearing time (fuse) (total clearing time). The time elapsing from the beginning of an overcurrent to the final circuit interruption. *Note:* The clearing time is equal to the sum of melting time and arcing time. 37A100-31E11

clearing-out drop (cord circuit or trunk circuit). A drop signal that is operated by ringing current to attract the attention of the operator. *See also:* **telephone switching system.** 42A65-0

clear sky. A sky that has less than 30-percent cloud cover. *See also:* **sunlight.** Z7A1-0

cleat. An assembly of two pieces of insulating material provided with grooves for holding one or more conductors at a definite spacing from the surface wired over and from each other, and with screw holes for fastening in position. *See also:* **raceway.** 42A95-0

clerestory. That part of a building that rises clear of the roofs or other parts and whose walls contain windows for lighting the interior. *See also:* **sunlight.** Z7A1-0

click. A disturbance of a duration less than a specified value as measured under specified conditions. *Note:* For the specified values and conditions, guidance should be found in International Special Committee on Radio Interference (CISPR) publications. *See also:* **electromagnetic compatibility.** CISPR-27E1

climbing space (wiring system). The vertical space reserved along the side of a pole or tower to permit ready access for linemen to equipment and conductors located thereon. *See also:* **power distribution, overhead construction; tower.** 42A35-31E13; 2A2-0

clipper (peak chopper). A device that automatically limits the instantaneous value of the output to a predetermined maximum value. *Note:* The term is usually applied to devices that transmit only portions of an input wave lying on one side of an amplitude boundary. *See also:* **circuits and devices.** 42A65-0

clipper amplifier. *See:* **amplifier, clipper.**

clipper limiter. A transducer that gives output only when the input lies above a critical value and a constant output for all inputs above a second higher critical value. *Note:* This is sometimes called an amplitude gate, or slicer. *See also:* **transducer.** E145-0

clipping (voice-operated telephone circuit). The loss of initial or final parts of words or syllables due to nonideal operation of the voice-operated devices. *See also:* **telephone switching system.** 42A65-0

clips. *See:* **contact clip; fuse clips.**

clock. (1) A device that generates periodic signals used for synchronization. (2) A device that measures and indicates time. (3) A register whose content changes at regular intervals in such a way as to measure time. *See:* **electronic digital computer.** 0-16E9

clocked logic (power-system communication). The technique whereby all the memory cells (flip-flops) of a logic network are caused to change in accordance with logic input levels but at a discrete time. *See also:* **digital.** 0-31E3

clocking (data transmission). The generation of periodic signals used for synchronization. *See also:* **data processing.** 0-19E4

close coupling (tight coupling). Any degree of coupling greater than the critical coupling. *See also:* **coupling; critical coupling.** 42A65-0

closed air circuit (rotating machinery). A term referring to duct-ventilated apparatus used in conjunction with external components so constructed that while it is not necessarily airtight, the enclosed air has no deliberate connection with the external air. *Note:* The term must be qualified to describe the means used to circulate the cooling air and to remove the heat produced in the apparatus. *See:* **cradle base (rotating machinery).** 0-31E8

closed amortisseur. An amortisseur that has the end connections connected together between poles by bolted or otherwise separable connections. *See:* **synchronous machine.** 42A10-0

closed-circuit signaling (telecommunication). That type of signaling in which current flows in the idle condition, and a signal is initiated by increasing or decreasing the current. *See also:* **circuits and devices.** 42A65-0

closed-circuit principle. The principle of circuit design in which a normally energized electric circuit, on being interrupted or de-energized, will cause the controlled function to assume its most restrictive condition. *See also:* **railway signal and interlocking.** 42A42-0

closed-circuit transition (industrial control). As applied to reduced-voltage controllers, including star-delta controllers, a method of starting in which the power to the motor is not interrupted during the starting sequence. *See:* **electric controller.** IC1- 34E10

closed-circuit voltage (working voltage) (battery). The voltage at its terminals when a specified current is flowing. *See also:* **battery (primary or secondary); working voltage.** 42A60-0

closed loop (feedback loop) (industrial control). A signal path that includes a forward path, a feedback path, and a summing point and that forms a closed circuit. *See:* **control system, feedback.** AS1-34E10

closed-loop control system (control system, feedback). A control system in which the controlled quantity is measured and compared with a standard representing the desired performance. *Note:* Any deviation from the standard is fed back into the control system in such a sense that it will reduce the deviation of the controlled quantity from the standard. *See:* **control; network analysis; numerically controlled machines.** E111/E254-0; 42A25-34E10

closed-loop gain (operational gain) (power supplies). The gain, measured with feedback, is the ratio of the voltage appearing across the output terminal pair to the causative voltage required at the input resistor. If the open-loop gain is sufficiently large, the closed-loop gain can be satisfactorily approximated by the ratio of the feedback resistor to the input resistor. *See:* **open-loop gain.** *See also:* **power supply.** KPSH-10E1

closed-loop series street lighting system. Street lighting system that employs two-wire series circuits in which the return wire is always adjacent. *See also:* **alternating-current distribution; direct-current distribution.** 42A35-0

closed region. An open region and its boundary. E270-0

closed subroutine (computing system). A subroutine that can be stored at one place and can be connected to a routine by linkages at one or more locations. *See:* **open subroutine.** *See also:* **electronic digital computer.** X3A12-16E9

close-open operation (switching device). A close operation followed immediately by an open operation without purposely delayed action. *Note:* The letters CO signify this operation: close-open. 37A100-31E11

close operation (switching device). The movement of the contacts from the normally open to the normally closed position. *Note:* The letter C signifies this operation: close. 37A100-31E11

close-talking microphone. A microphone designed particularly for use close to the mouth of the speaker. *See also:* **microphone.** 42A65-0

close-talking pressure-type microphones. An acoustic transducer that is intended for use in close proximity to the lips of the talker and is either hand-held or boom-mounted. *Notes:* (1) Various types of microphones are currently used for close-talking applications. These include carbon, dynamic, magnetic, piezoelectric, electrostrictive, and capacitor types. Each of these microphones has only one side of its diaphragm exposed to sound waves, and its electric output substantially corresponds to the instantaneous sound pressure of the impressed sound wave. (2) Since a close-talking microphone is used in the near sound field produced by a person's mouth, it is necessary when measuring the performance of such microphones to utilize a sound source that approximates the characteristics of the human sound generator.
See:
artificial voice;
breathing;
carbon noise;
microphone;
nonlinear distortion;
positional response;
power response;
principal axis;
sensitivity;
voltage response. E258-0

close-time delay-open operation (switching device). A close operation followed by an open operation after a purposely delayed action. *Note:* The letters CTO signify this operation: close-time delay-open. 37A100-31E11

closing coil (switching device). A coil used in the electromagnet that supplies power for closing the device. *Note:* In an air-operated, or other stored-energy-operated device, the closing coil may be the coil used to release the air or other stored energy that in turn closes the device. 37A100-31E11

closing relay (electrically operated device). A form of auxiliary relay used to control the closing and opening of the closing circuit of the device so that the main closing current does not pass through the control switch or other initiating device. 37A100-31E11/31E6

closing time (mechanical switching device). The interval of time between the initiation of the closing operation and the instant when metallic continuity is established in all poles. *Notes:* (1) It includes the operating time of any auxiliary equipment necessary to close the switching device, and that forms an integral part of the switching device. (2) For switching devices that embody switching resistors, a distinction should be made between the closing time up to the instant of establishing a circuit at the secondary arcing contacts, and the closing time up to the establishment of a circuit at the main or primary arcing contacts, or both. 37A100-31E11

cloud pulse (charge-storage tubes). The output resulting from space-charge effects produced by the turning on or off of the electron beam. *See also:* **beam tubes; charge-storage tube.** E158/E160-15E6

cloudy sky. A sky that has more than 70-percent cloud cover. *See also:* **sunlight.** Z7A1-0

CLR. *See:* **recording-completing trunk (combined line and recording trunk).**

clutter (radar). Confused unwanted echoes on a radar display. *See also:* **radar.** 42A65-0

clutter attenuation (radar moving-target indicator). The ratio of clutter power at the canceller input to clutter residue at the output, normalized to the attenuation for a single pulse passing through the undelayed channel of the canceller. *See also:* **navigation.** 0-10E6

clutter reflectivity (radar). The backscattering coefficient of a unit volume, or of a unit area, containing clutter sources. *See also:* **navigation.** 0-10E6

clutter residue (radar moving-target indicator). The clutter power remaining at the output of a moving-target indicator system. *Note:* It is the sum of several (generally uncorrelated) components resulting from radar instabilities, antenna scanning, relative motion of the radar with respect to the sources of clutter and fluctuations of the clutter reflectivity. *See also:* **navigation.** 0-10E6

clutter visibility factor (radar). The signal-to-clutter ratio at the receiver output that provides stated probabilities of detection and false alarm on a display; in moving-target indicator systems, it is the ratio after cancellation or Doppler filtering. When the clutter residue has the spectrum of thermal noise, this factor is the same as the visibility factor. *See also:* **navigation.** 0-10E6

CMRR. *See:* **common-mode rejection ratio.**

***C* network.** A network composed of three impedance branches in series, the free ends being connected to one pair of terminals, and the junction points being connected to another pair of terminals. *See also:* **network analysis.** 42A65-0

CO. *See:* **close-open operation (switching device).**

coagulating current. *See:* **Tesla current (electrotherapy).**

coal cleaning equipment. Equipment generally electrically driven, to remove impurities from the coal as mined, such as slate, sulphur, pyrite, shale, fire clay, gravel, and bone. *See also:* **mining.** 42A85-0

coarse chrominance primary (color television system at present standardized for broadcasting in the United States). One of the two chrominance primaries that is associated with the lesser transmission bandwidth. *See also:* **color terms.** E201-2E2

coated fabric (coated mat) (rotating machinery). A fabric or mat in which the elements and interstices may or may not in themselves be coated or filled but that has a relatively uniform compound or varnish finish on either one or both surfaces. *See also:* **rotor (rotating machinery); stator.** 0-31E8

coated magnetic tape (magnetic powder-coated tape). A tape consisting of a coating of uniformly dispersed, powdered ferromagnetic material (usually ferromagnetic oxides) on a nonmagnetic base. *See also:* **magnetic tape; phonograph pickup.** 0-1E1

coated mat. *See:* **coated fabric.**

coating (electroplating). The layer deposited by electroplating. *See also:* **electroplating.** 42A60-0

coaxial (coaxial line) (concentric line) (coaxial pair*). A transmission line formed by two coaxial conductors. *See also:* **cable; transmission line.** E270/42A65

*Deprecated

coaxial antenna. An antenna comprised of a quarter-wavelength extension to the inner conductor of a coaxial line and a radiating sleeve which in effect is formed by folding back the outer conductor of the coaxial line for approximately one-quarter wavelength. *See also:* **antenna.** E145-0;42A65-3E1

coaxial cable. A cable containing one or more coaxial lines. *See also:* **cable.** 42A65-31E3

coaxial conductor. An electric conductor comprising outgoing and return current paths having a common axis, one of the paths completely surrounding the other throughout its length. E54-0

coaxial line. *See:* **coaxial.**

coaxial pair*. *See:* **coaxial.**

*Deprecated

coaxial relay. A relay that opens and closes an electric contact switching high-frequency current as required to maintain minimum losses. *See also:* **relay.** 0-21E0

coaxial stop filter (electromagnetic compatibility). A tuned movable filter set round a conductor in order to limit the radiating length of the conductor for a given frequency. *See also:* **electromagnetic compatibility.** CISPR-27E1

coaxial stub. A short length of coaxial that is joined as a branch to another coaxial. *Note:* Frequently a coaxial stub is short-circuited at the outer end and its length is so chosen that a high or low impedance is presented to the main coaxial in a certain frequency range. *See also:* **waveguide.** 42A65-0

coaxial switch. A switch used with and designed to simulate the critical electric properties of coaxial conductors. 0-E

coaxial transmission line. A transmission line consisting of two coaxial cylindrical conductors. *See also:* **radio receiver; transmission line.** E146/E188-3E1

co-channel interference. Interference caused in one communication channel by a transmitter operating in the same channel. *See also:* **radio transmission.** 42A65-2E2

code (1) (general). A plan for representing each of a finite number of values or symbols as a particular arrangement or sequence of discrete conditions or events. 42A65-0

(2) (electronic computers). (A) The characters or expressions of an originating or source language, each correlated with its equivalent expression in an intermediate or target language, for example, alphanumeric characters correlated with their equivalent 6-bit expressions in a binary machine language. *Note:* For punched or magnetic tape; a predetermined arrangement of possible locations of holes or magnetized areas and rules for interpreting the various possible patterns. (B) Frequently, an expression in the target language. (C) Frequently, the set of expressions in the target language that represent the set of characters of the source language. (D) To encode is to express given information by means of a code. (E) To translate the program for the solution of a problem on a given computer into a sequence of machine-language or pseudo instructions acceptable to that computer. E162-0;EIA3B-34E12

See:

American Morse code;

binary code;

cable Morse code;

chain code;
computer code;
error-correcting code;
error-detecting code;
excess-three code;
Gray code;
instruction code;
International Morse code;
machine code;
minimum distance code;
operation code;
reflected binary code;
self-checking code;
two-out-of-five code.
See also: **electronic digital computer; information theory; telegraphy.**

code character. A particular arrangement of code elements representing a specific symbol or value. 42A65-19E4/31E3

coded. *See:* **binary-coded decimal.** *See also:* **electronic digital computer.**

coded-decimal code. The decimal number system with each decimal digit expressed by a code. EIA3B-34E12

coded fire-alarm system. A local fire-alarm system in which the alarm signal is sounded in a predetermined coded sequence. *See also:* **protective signaling.** 42A65-0

coded track circuit. A track circuit in which the energy is varied or interrupted periodically. *See also:* **railway signal and interlocking.** 42A42-0

code element. One of the discrete conditions or events in a code, for example, the presence or absence of a pulse. *See also:* **data processing; information theory; pulse.** 42A65-19E4

code letter (locked-rotor kilovolt-amperes). A letter designation under the caption "code" on the nameplate of alternating-current motors (except wound-rotor motors) rated 1/20 horsepower and larger to designate the locked-rotor kilovolt-amperes per horsepower as measured at rated voltage and frequency. 0-31E8

coder (1) (general). A device that sets up a series of signals in code form. *See also:* **circuits and devices.** 42A65-0

(2) (code transmitter). A device used to interrupt or modulate the track or line current periodically in various ways in order to establish corresponding controls in other apparatus. *See also:* **circuits and devices: railway signal and interlocking.** 42A42-0

code ringing. Party-line ringing wherein the number of rings, or the duration, or both, indicate which station is being called. *See also:* **telephone switching system.** 42A65-19E1/31E3

code system. A system of control of wayside signals, cab signals, train stop or continuous train control in which electric currents of suitable character are supplied to control apparatus, each function being controlled by its own distinctive code. *See also:* **block signal system.** 42A42-0

code translator. *See:* **digital converter.**

coding (1) (communication). The process of transforming messages or signals in accordance with a definite set of rules. *See also:* **modulating systems.** E170-0

(2) (computing systems). Loosely, a routine. *See:* **relative coding; straight-line coding; symbolic coding.** *See also:* **electronic digital computer.** 0-16E9

coding delay (loran). An arbitrary time delay in the transmission of pulse signals from the slave station to permit the resolution of ambiguities. *Note:* The term **suppressed time delay** more accurately represents what is being accomplished and should be used instead of **coding delay.** *See also:* **navigation.** E172-10E6

coding fan. *See:* **electrode radiator.**

coding siren. A siren having an auxiliary mechanism to interrupt the flow of air through the device, thereby enabling it to produce a series of sharp blasts as required in code signaling. *See also:* **protective signaling.** 42A65-0

coefficient. *See:* **switching coefficient.** *See also:* **static magnetic storage.**

coefficient of attenuation (illuminating engineering). The decrement in flux per unit distance in a given direction within a medium and is defined by the relation: $\Phi_x = \Phi_0 e^{-\mu x}$ where Φ_x is the flux at any distance x from a reference point having flux Φ_0. *See also:* **lamp.** Z7A1-0

coefficient of beam utilization. The ratio of (1) the luminous flux (lumens) reaching a specified area directly from a floodlight or projector to (2) the total beam luminous flux. *See also:* **inverse-square law.** Z7A1-0

coefficient of capacitance (system of conductors) (coefficients of capacitance and induction—Maxwell). The coefficients in the array of linear equations that express the charges on the conductors in terms of their potentials. For a set of n conductors, $n-1$ of which are insulated from each other and mounted on insulating supports within a closed conducting shell or on one side of a conducting plane of infinite extent, the shell or plane as a common conductor (or ground) is taken to be at zero potential and the charge on each insulated conductor is regarded as having been transferred from the common conductor. Hence the algebraic sum of the charges on the $n-1$ insulated conductors is equal and opposite in sign to the charge on the common conductor. The charges on the insulated conductors are given by the equations

$$Q_1 = c_{11}V_1 + c_{12}V_2 + c_{13}V_3 \cdots + c_{1(n-1)}V_{n-1}$$
$$Q_2 = c_{21}V_1 + c_{22}V_2 + c_{23}V_3 \cdots + c_{2(n-1)}V_{n-1}$$
$$\cdot$$
$$\cdot$$
$$\cdot$$
$$Q_{n-1} = c_{(n-1)1}V_1 + c_{(n-1)2}V_2 + c_{(n-1)3}V_3 + \cdots + c_{n-1)(n-1)}V_{n-1}$$

The coefficients c_{rr} are the self-capacitances and the coefficients c_{rp} are the negatives of the direct (or mutual) capacitances of the system. *Note:* Under the conventions stated both self and direct capacitances have positive signs; also $c_{rp} = c_{pr}$. If the mth conductor completely surrounds the kth conductor and no other, then $c_{kp} = 0$ for $p \neq k$ or m; and $c_{kk} = c_{km}$. *See also:* **network analysis.** E270-0

coefficient of coupling. *See:* **coupling coefficient.**

coefficient of grounding (lightning arrester). The ratio (E_{L-G}/E_{L-L}), expressed as a percentage, of the highest root-mean-square line-to-ground power-frequency voltage (E_{L-G}) on a sound phase, at a selected location, during a fault to earth affecting one or more phases to the line-to-line power-frequency voltage ($E_{L\text{-}L}$) that would be obtained, at the selected location, with the fault removed. *Notes:* (1) Coefficients of grounding for three-phase systems are calculated from the phase-sequence impedance components as viewed from the selected location. For machines use the subtransient reactance. (2) The coefficient of grounding is useful in the determination of an arrester rating for a selected location. (3) A value not exceeding 80 percent is obtained approximately when for all system conditions the ratio of zero-sequence reactance to positive-sequence reactance is positive and less than three and the ratio of zero-sequence resistance to positive-sequence reactance is positive and less than one. *See also:* **lightning protection and equipment.** E28/99I2-31E7

coefficient of performance (1) (thermoelectric cooling couple). The quotient of (1) the net rate of heat removal from the cold junction by the thermoelectric couple by (2) the electric power input to the thermoelectric couple. *Note:* This is an idealized coefficient of performance assuming perfect thermal insulation of the thermoelectric arms. *See also:* **thermoelectric device.** E221-15E7

(2) (thermoelectric cooling device). The quotient of (1) the rate of heat removal from the cooled body by (2) the electric power input to the device. *See also:* **thermoelectric device.** E221-15E7

(3) (thermoelectric heating device). The quotient of (1) the rate of heat addition to the heated body by (2) the electric power input to the device. *See also:* **thermoelectric device.** E221-15E7

(4) (thermoelectric heating couple). The quotient of (1) the rate of heat addition to the hot junction by the thermoelectric couple by (2) the electric power input to the thermoelectric couple. *Note:* This is an idealized coefficient of performance assuming perfect thermal insulation of the thermoelectric arms. *See also:* **thermoelectric device.** E221-15E7

coefficient of performance, reduced (thermoelectric device). The ratio of (1) a specified coefficient of performance to (2) the corresponding coefficient of performance of a Carnot cycle. *See also:* **thermoelectric device.** E221-15E7

coefficient of trip point repeatability. *See:* **trip-point repeatability coefficient.**

coefficient of utilization (illuminating engineering). The ratio of (1) luminous flux (lumens) calculated as received on the work plane to (2) the rated lumens emitted by the lamps alone. *See also:* **inverse-square law.** Z7A1-0

coefficient potentiometer. *See also:* **electronic analog computer.**

coefficients of capacitance and induction—Maxwell. *See:* **coefficients of capacitance (system of conductors).** E270-0

coefficients of elastance (system of conductors) (coefficients of potential—Maxwell). The coefficients in the array of linear equations that express the potentials of the conductors in terms of their charges. For the system of n conductors (including ground) described in **coefficients of capacitance (system of conductors)** the equations for the potentials of the insulated conductors are

$$V_1 = s_{11}Q_1 + s_{12}Q_2 + s_{13}Q_3 + \cdots + s_{1(n-1)}Q_{n-1}$$
$$V_2 = s_{21}Q_1 + s_{22}Q_2 + s_{23}Q_3 + \cdots + s_{2(n-1)}Q_{n-1}$$
$$\cdot$$
$$\cdot$$
$$\cdot$$
$$V_{n-1} = s_{(n-1)1}Q_1 + s_{(n-1)2}Q_2 + s_{(n-1)3}Q_3 + \cdots + s_{(n-1)(n-1)}Q_{n-1}$$

The coefficients s_{rr} are the self-elastances and the coefficients s_{rp} are the negatives of the mutual elastances of the system. *Note:* The matrix of the coefficients of elastance is the reciprocal of the matrix of the coefficients of capacitance. *See also:* **network analysis.** E270-0

coefficients of potential—Maxwell. *See:* **coefficients of elastance (system of conductors); elastances (system of conductors).** E270-0

coercive electric field (ferroelectric material). The value of the electric field at which the displacement on a continuously cycled major hysteresis loop equals zero. *Note:* See the accompanying figure. The value of the coercive field E_c is dependent on the frequency, amplitude, and wave shape of the applied alternating electric field. To facilitate comparison of data, the coercive field, unless otherwise specified, should be determined with a 60-hertz sinusoidal electric field of sufficient amplitude to insure saturation. Saturation is insured when the peak voltage is substantially greater than the value at which the rising and falling portions of the D-versus-E plot coalesce. The peak amplitude of the applied field should be given with the numerical data. The coercive field in the meter-kilogram-second-ampere (mksa) system is expressed in volts per meter. E180-0

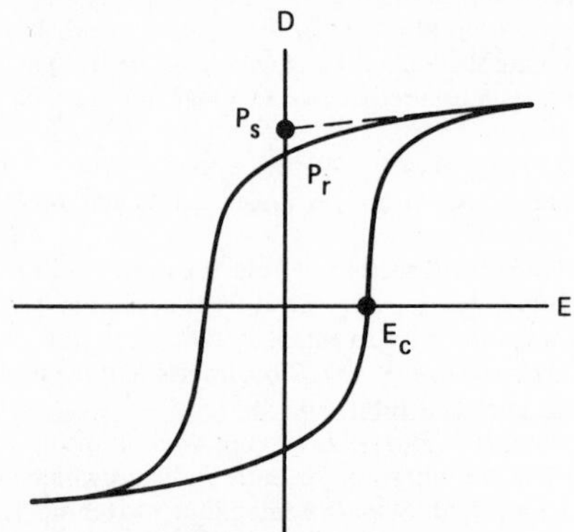

Ferroelectric hysteresis loop.

coercive force. The magnetizing force at which the magnetic flux density is zero when the material is in a symmetrically cyclically magnetized condition. *Note:* Coercive force is not a unique property of a magnetic material but is dependent upon the conditions of measurement. *See also:* **static magnetic storage; coercivity.** E163/E270-0

coercive voltage (ferroelectric device). The voltage at which the charge on a continuously cycled hysteresis loop is zero. *Note:* Coercive voltage (V_c) is dependent on the frequency, amplitude, and wave shape of the alternating signal. In order to have a basis of comparison, this quantity should be determined from a 60-hertz sinusoidal hysteresis loop with the applied signal of sufficient amplitude to insure saturation. The peak applied voltage and the sample thickness should accompany the data. *See:* figure under **total charge.** E180-0

coercivity. The property of a magnetic material measured by the coercive force corresponding to the saturation induction for the material. *Note:* This is a quasi-static property only. *See also:* **static magnetic storage.** E163/E270-0

cofactor (or path cofactor) (network analysis). *See:* **path (loop) factor.**

coffer (illuminating engineering). A recessed panel or dome in the ceiling. *See also:* **luminaire.** Z7A1-0

cogging (rotating machinery). Variations in motor torque at very low speeds caused by variations in magnetic flux due to the alignment of the rotor and stator teeth at various positions of the rotor. *See also:* **rotor (rotating machinery); stator.** 0-31E8

cohered video (in radar moving-target indicator). Video-frequency signal output employed in a coherent system. *See also:* **navigation.** E172-10E6

coherent interrupted waves. Interrupted continuous waves occurring in wave trains in which the phase of the waves is maintained through successive wave trains. *See also:* **wave front.** 42A65-0

coherent moving-target indicator (radar moving-target indicator). A system in which the target echo is selected on the basis of its Doppler frequency when compared to a local reference frequency maintained by a coherent oscillator. *See also:* **navigation.** 0-10E6

coherent oscillator (coho) (radar moving-target indicator). An oscillator that provides a reference phase by which changes in the radio-frequency phase of successively received pulses may be recognized. *See also:* **navigation.** E172-10E6

coherent pulse operation. The method of pulse operation in which a fixed phase relationship is maintained from one pulse to the next. *See also:* **pulse.** 42A65-0

coho. Abbreviation for coherent oscillator. E172-10E6

coil (1) (general). An assemblage of successive convolutions of a conductor.

(2) (rotating machinery). A unit of a winding consisting of one or more insulated conductors connected in series and surrounded by common insulation, and arranged to link or produce magnetic flux. *See also:* **rotor (rotating machinery); stator.** E270-31E8

(3) (relay). *See:* **relay coil.**

coil brace (1) (coil support). A structure for the support or restraint for one or more coils.

(2) (V wedge, salient-pole construction). A trapezoidal insulated insert clamped between field poles, to provide radial restraint for the field coil turns against centrifugal force and to brace the coils tangentially. *See also:* **stator.** 0-31E8

coil end-bracing (rotating machinery). *See:* **end-winding support.** *See also:* **rotor (rotating machinery); stator.**

coil insulation (rotating machinery). The main insulation to ground or between phases surrounding a coil, additional to any conductor or turn insulation. *See also:* **rotor (rotating machinery); stator.** 0-31E8

coil lashing (rotating machinery). The binding used to attach a coil end to the supporting structure. *See also:* **rotor (rotating machinery); stator.** 0-31E8

coil loading. Loading in which inductors, commonly called loading coils, are inserted in a line at intervals. *Note:* The loading coils may be inserted either in series or in shunt. As commonly understood, coil loading is a series loading in which the loading coils are inserted at uniformly spaced recurring intervals. *See also:* **loading.** 42A65-0

coil pitch (rotating machinery). The distance between the two active conductors (coil sides) of a coil, usually expressed as a percentage of the pole pitch. *See:* **armature.** 0-31E8

coil *Q* (dielectric heating usage). Ratio of reactance to resistance measured at the operating frequency. *Note:* The loaded-coil Q is that of a heater coil with the charge in position to be heated. Correspondingly, the unloaded-coil Q is that of a heater coil with the charge removed from the coil. *See also:* **dielectric heating.** E54-0

coil shape factor (dielectric heating usage). A correction factor for the calculation of the inductance of a coil based on its diameter and length. *See also:* **dielectric heating.** E54-0

coil side (rotating machinery). Either of the two normally straight parts of a coil that lie in the direction of the axial length of the machine. *See also:* **rotor (rotating machinery); stator.** 0-31E8

coil-side separator (rotating machinery). Additional insulation used to separate embedded coil sides. *See also:* **rotor (rotating machinery); stator.** 0-31E8

coil space factor. The ratio of the cross-sectional area of the conductor metal in a coil to the total cross-sectional area of the coil. *Note:* If the overall insulation, such as spool bodies or stop linings, is omitted from consideration when the space factor is calculated, this omission should be specifically stated. *See:* **asynchronous machine; direct-current commutating machine; synchronous machine.** 42A10-0

coil span (rotating machinery). *See:* **coil pitch.** *See also:* **rotor (rotating machinery); stator.**

coil support bracket (rotating machinery). A bracket used to mount a coil support ring or binding band. *See also:* **rotor (rotating machinery); stator.** 0-31E8

coin box. A telephone set equipped with a device for collecting coins in payment for telephone messages. *See also:* **telephone station.** 42A65-0

coincidence (radiation counters). The practically simultaneous production of signals from two or more counter tubes. *Note:* A genuine or true coincidence is due to signals from related events (passage of one particle or of two or more related particles through the counter tubes); an accidental, spurious, or chance coincidence is due to unrelated signals that coincide accidentally. *See:* **anticoincidence (radiation counters).** 0-15E6

coincidence circuit. A circuit that produces a specified output pulse when and only when a specified number (two or more) or a specified combination of input terminals receives pulses within an assigned time interval. *See also:* **anticoincidence (radiation counters); pulse.** E175-0

coincidence factor. The ratio of the maximum coincident total demand of a group of consumers to the sum of the maximum power demands of individual consumers comprising the group both taken at the same point of supply for the same time. *See also:* **generating station.** 42A35-31E13

coincident-current selection. The selection of a magnetic cell for reading or writing, by the simultaneous application of two or more currents. *See also:* **static magnetic storage.** E163-16E9

coincident demand (electric power utilization). Any demand that occurs simultaneously with any other demand, also the sum of any set of coincident demands. *See also:* **alternating-current distribution.** 0-31E4

cold cathode. A cathode that functions without the application of heat. *See also:* **electrode (electron tube).** 42A70-0

cold-cathode glow-discharge tube (glow tube). A gas tube that depends for its operation on the properties of a glow discharge. *See also:* **tube definitions.** 42A70-15E6

cold-cathode lamp. (1) An electric-discharge lamp whose mode of operation is that of a glow discharge, and that has electrodes so spaced that most of the light comes from the positive column between them. (2) An electric-discharge lamp in which the electrodes, operating at less than incandescent temperatures, furnish an electron current by field emission, and in which the cathode drop is relatively high (75—150 volts). *Note:* The current density at the cathodes is relatively low, and cathodes become impracticably large for currents greater than a few hundred milliamperes. *See also:* **lamp.** Z7A1-0/82A1-0

cold-cathode stepping tube (electron device). A glow discharge tube having several main gaps with or without associated auxiliary gaps, and in which the main discharge has two or more stable positions and can be made to step in sequence, when a suitable shaped signal is applied to an input electrode, or a group of input electrodes. *See also:* **tube definitions.** 0-15E6

cold-cathode tube. An electron tube containing a cold cathode. *See also:* **tube definitions.** E160-15E6;42A70-0

cold reserve. Thermal generating capacity available for service but not maintained at operating temperature. 42A35-31E13

collar (cheek, field-coil flange) (washer) (rotating machinery). Insulation between the field coil and the pole shoe (top collar) and between the field coil and the member carrying the pole body (bottom collar). *See also:* **rotor (rotating machinery).** 0-31E8

collate. To compare and merge two or more similarly ordered sets of items into one ordered set. *See also:* **electronic digital computer.** X3A12-16E9

collating sequence. An ordering assigned to a set of items, such that any two sets in that assigned order can be collated. *See also:* **electronic digital computer.** X3A12-16E9

collator. A device to collate sets of punched cards or other documents into a sequence. *See also:* **electronic digital computer.** X3A12-16E9

collector (1) (rotating machinery). An assembly of collector rings, individually insulated, on a supporting structure. *See:* **asynchronous machine; synchronous machine.** 42A10-31E8

(2) (electron tube). An electrode that collects electrons or ions that have completed their functions within the tube. *See also:* **electrode (electron tube).** 42A70-15E6

(3) (transistor). A region through which primary flow of charge carriers leaves the base. E175-0;E216-34E17;E270-0

collector grid (photoelectric converter). A specially designed metallic conductor electrically bonded to the *p* or *n* layer of a photoelectric converter to reduce the spreading resistance of the device by reducing the mean path of the current carriers within the semiconductor. *See also:* **semiconductor.** 0-10E1

collector junction. *See:* **junction, collector.**

collector plates. Metal inserts embedded in the cell lining to minimize the electric resistance between the cell lining and the current leads. *See also:* **fused electrolyte.** 42A60-0

collector ring (slip ring). A metal ring suitably mounted on an electric machine that (through stationary brushes bearing thereon) conducts current into or out of the rotating member. *See:* **asynchronous machine; synchronous machine.** 42A10-0

collector-ring (slip-ring) lead insulation (rotating machinery). Additional insulation, applied to the leads that connect the collector rings to the windings of the rotating member, to prevent grounding to the metallic parts of the rotating members, and to provide electrical separation between leads. *See also:* **rotor (rotating machinery).** 0-31E8

collector-ring (slip-ring) shaft insulation (rotating machinery). The combination of insulating members that insulate the collector rings from the parts of the structure that are mounted on the shaft. *See also:* **rotor (rotating machinery).** 0-31E8

collimate (storage tubes). To modify the paths of electrons in a flooding beam or of various rays of a scanning beam in order to cause them to become more nearly parallel as they approach the storage assembly. *See also:* **storage tube.** E158-15E6

collimating lens (storage tubes). An electron lens that collimates an electron beam. *See also:* **storage tube.** E158-15E6

collinear array. A linear array of radiating elements, usually dipoles, with their axes lying in a straight line. *See also:* **antenna.** 0-3E1

collision frequency (plasma) (radio wave propagation). The average number of collisions per second of a charged particle of a given species with particles of another or the same species. *See also:* **radio wave propagation.** 0-3E2

colloidal ions. Ions suspended in a medium, that are larger than atomic or molecular dimensions but sufficiently small to exhibit Brownian movement. 42A60-0

color. The characteristics of light other than spatial and temporal inhomogeneities. *Notes:* (1) The measure of color is three dimensional. One of the many ways of measuring color is in terms of luminance, dominant wavelength, and purity. (2) Inhomogeneities, for example, particular distributions and variations of light, and characteristics of objects that are revealed by variations such as gloss, lustre, sheen, texture, sparkle, opalescence, and transparency, are not included among the color characteristics of objects. *See also:* **color terms.** E201-0

color breakup (color television). Any fleeting and partial separation of a color picture into its display primary components caused by a rapid change in the condition of viewing. *Note:* Illustrations of rapid changes in the condition of viewing are (1) fast movement of the head, (2) fast interruption of the line of sight, and (3) blinking of the eyes. *See also:* **color terms.** E201-2E2

color burst (color television). The portion of the composite color signal, comprising a few cycles of a sine wave of chrominance-subcarrier frequency, that is used to establish a reference for demodulating the chrominance signal. *See also:* **color terms.** E201-2E2

color-burst flag (color-burst keying signal) (television). A keying signal used to form the color burst from a color-subcarrier signal source. *See:* **burst flag.** 0-42A65

color-burst gate (television). A keying or gating signal used to extract the color burst from a color-television signal. *See:* **burst gate; television.** 0-42A65

color-burst keying signal. *See:* **color-burst flag.**

color carrier (color television). *See:* **chrominance subcarrier.**

color cell (repeating pattern of phosphors on the screen of a color-picture tube). The smallest area containing a complete set of all the primary colors contained in the pattern. *Note:* If the cells are described by only one dimension as in the line type of screen, the other dimension is determined by the resolution capabilities of the tube. *See:* **beam tubes.** E160-15E6

color center (color-picture tubes). A point or region (defined by a particular color-selecting electrode and screen configuration) through which an electron beam must pass in order to strike the phosphor array of one primary color. *Note:* This term is not to be used to define the color-triad center of a color-picture tube screen. *See:* **beam tubes.** E160-15E6

color code (electrical). A system of standard colors adopted for identification of conductors for polarity, etcetera, and for identification of external terminals of motors and starters to facilitate making power connections between them. *See also:* **mine feeder circuit.** 42A85-0

color coder (color-television transmission). An apparatus for generating the color-picture signal (and possibly the color burst) from camera signals and the chrominance subcarrier. *See also:* **color terms.** E201-2E2

color contamination (color television). An error of color rendition due to incomplete separation of paths carrying different color components of the picture. *Note:* Such errors can arise in the optical, electronic, or mechanical portions of a color-television system as well as in the electric portions. *See also:* **color terms.** E201-2E2

color coordinate transformation (color television). Computation of the tristimulus values of colors in terms of one set of primaries from the tristimulus values of the same colors in another set of primaries. *Note:* This computation may be performed electrically in a color-television system. *See also:* **color terms.** E201-2E2

color decoder (color television). An apparatus for deriving the signals for the color-display device from the color-picture signal and the color-burst. *See also:* **color terms.** E201-2E2

color-difference signal (color television). An electric signal that when added to the monochrome signal produces a signal representative of one of the tristimulus values (with respect to a stated set of primaries) of the transmitted color. *See also:* **color terms.** E201-2E2

color discrimination. The perception of differences between two or more colors. *See also:* **visual field.** Z7A1-0

color encoder (color television). *See:* **color coder.**

color-field corrector (electron tubes). A device located external to the tube producing an electric or magnetic field that affects the beam after deflection as an aid in the production of uniform color fields. *See:* **beam tubes.** E160-15E6

color flicker. The flicker that results from fluctuation of both chromaticity and luminance. *See also:* **color terms.** E201-2E2

color fringing (color television). Spurious chromaticity at boundaries of objects in the picture. *Note:* Color fringing can be caused by the change in relative position of the televised object from field to field or by misregistration and, in the case of small objects, may even cause them to appear separated into different colors. *See also:* **color terms.** E201-2E2

colorimetric purity (light). The ratio L_1/L_2 where L_1 is the luminance (photometric brightness) of the single frequency component that must be mixed with a reference standard to match the color of the light and L_2 is the luminance (photometric brightness) of the light. *See also:* **color.** Z7A1-0

colorimetry. The techniques for the measurement of color and for the interpretation of the results of such measurements. *Note:* The measurement of color is made possible by the properties of the eye and is based on a set of conventions. *See also:* **color terms.** E201-0;50I45-2E2

coloring (electroplating) (1) (chemical). The production of desired colors on metal surfaces by appropriate chemical action. *See also:* **electroplating.** 42A60-0

(2) (buffing). Light buffing of metal surfaces, for the purpose of producing a high luster. *See also:* **electroplating.** 42A60-0

color light signal. A fixed signal in which the indications are given by the color of a light only. *See also:* **railway signal and interlocking.** 42A42-0

color match. The condition in which the two halves of a structureless photometric field are judged by the observer to have exactly the same appearance. *Note:* A color match for the standard observer may be calculated. *See also:* **color terms.** E201-2E2

color mixture. Color produced by the combination of light of different colors. *Notes:* (1) The combination may be accomplished by successive presentation of the components, provided the rate of alternation is sufficiently high, or the combination may be accomplished by simultaneous presentation, either in the same area or on adjacent areas, provided they are small enough and close enough together to eliminate pattern effects. (2) A color mixture as here defined is sometimes denoted as an additive color mixture, to distinguish it from combinations of dyes, pigments, and other absorbing substances. Such mixtures of substances are sometimes called subtractive color mixtures, but might more appropriately be called colorant mixtures. *See also:* **color terms.** E201-2E2

color-mixture data. *See:* **tristimulus values.**

color-picture tube. An electron tube used to provide an image in color by the scanning of a raster and by varying the intensity of excitation of phosphors to produce light of the chosen primary colors. *See:* **beam tubes.** E160-15E6

color plane (multibeam color-picture tubes). A surface approximating a plane containing the color centers. *See:* **beam tubes.** E160-15E6

color-position light signal. A fixed signal in which the indications are given by the color and the position of two or more lights. *See also:* **railway signal and interlocking.** 42A42-0

color-purity magnet. A magnet in the neck region of a color-picture tube to alter the electron beam path for the purpose of improving color purity. *See:* **beam tubes.** E160-15E6

color-selecting-electrode system. A structure containing a plurality of openings mounted in the vicinity of the screen of a color-picture tube (electron tubes), the function of this structure being to cause electron impingement on the proper screen area by using either masking, focusing, deflection, reflection, or a combination of these effects. *Note:* For examples see **shadow mask.** *See also:* **beam tubes.** E160-15E6

color-selecting-electrode system transmission (electron tubes). The fraction of incident primary electron current that passes through the color-selecting-electrode system. *See:* **beam tubes.** E160-15E6

color signal (color-television system). Any signal at any point for wholly or partially controlling the chromaticity values of a color-television picture. *Note:* This is a general term that encompasses many specific connotations, such as are conveyed by the words, color-picture signal, chrominance signal, carrier color signal, monochrome signal (in color television), etcetera. *See also:* **color terms.** E201-2E2

color-synchronizing signal (television). A signal used to establish and to maintain the same color relationships that are transmitted. *Note:* In Rules covering Radio Broadcast Services, Part 3, of the Federal Communications Commission, the color synchronizing signal consists of a sequence of color bursts that are continuous except for a specified time interval during the vertical interval, each burst occurring on the back porch. *See also:* **television.** E204-2E2;42A65

color temperature (light source). The absolute temperature at which a blackbody radiator must be operated to have a chromaticity equal to that of the light source. *See also:* **color.** Z7A-1-0

color terms. *Note:* For an extensive list of cross references, see *Appendix A.*

color tracking (television). (1) The degree to which color balance is maintained over the complete range of the achromatic (neutral gray) scale. 0-42A65
(2) A qualitative term indicating the degree to which constant chromaticity within the achromatic region in the chromaticity diagram is achieved on a color-display device over the range of luminances produced from a monochrome signal. *See:* **television.** 0-2E2

color transmission (color television). The transmission of a signal wave for controlling both the luminance values and the chromaticity values in a picture. *See also:* **color terms; television.** E201-2E2

color triad (phosphor-dot screen). A color cell of a three-color phosphor-dot screen. *See:* **beam tubes.** E160-15E6

color triangle. A triangle drawn on a chromaticity diagram, representing the entire range of chromaticities obtainable as additive mixtures of three prescribed primaries, represented by the corners of the triangle. *See also:* **color terms.** E201-2E2

Colpitts oscillator. An electron-tube circuit in which the parallel-tuned tank circuit is connected between grid and plate, the capacitive portion of the tank circuit being comprised of two series elements, the connection between the two being at cathode potential with the feedback voltage obtained across the grid-cathode portion of the capacitor. *Note:* In an electron-tube oscillator, when the two voltage-dividing capacitances are the plate-to-cathode and the grid-to-cathode capacitances of the tube, the circuit has been known as the ultraaudion oscillator. *See also:* **oscillatory circuit.** E54/E145/E182A-0

column (positional notation). (1) A vertical arrangement of characters or other expressions. (2) Loosely, a digit place. *See:* **electronic digital computer; place.** E162-0;X3A12-16E9

column binary. Pertaining to the binary representation of data on punched cards in which adjacent positions in a column correspond to adjacent bits of data, for example, each column in a 12-row card may be used to represent 12 consecutive bits of a 36-bit word. *See also:* **electronic digital computer.** X3A12-16E9

column, positive. *See:* **positive column.**

combination. *See:* **forbidden combination.** *See also:* **electronic digital computer.**

combinational logic element. (1) A device having zero or more input channels and one output channel, each of which is always in one of exactly two possible physical states, except during switching transients. *Note:* On each of the input channels and the output channel, a single state is designated arbitrarily as the "one" state, for that input channel or output channel, as the case may be. For each input channel and output channel, the other state may be referred to as the "zero" state. The device has the property that the output channel state is determined completely by the contemporaneous input-channel-state combination, to within switching transients. (2) By extension, a device similar to (1) except that one or more of the input channels or the output channel, or both, have a finite number, but more than two, possible physical states each of which is designated as a distinct logic state. The output channel state is determined completely by the contemporaneous input-channel-state combination, to within switching transients. (3) A device similar to (1) or (2) except that it has more than one output channel. *See:* AND **gate;** OR **gate; electronic digital computer.** E162-0

combination controller (industrial control). A full magnetic or semimagnetic controller with additional externally operable disconnecting means contained in a common enclosure. The disconnecting means may be a circuit breaker or a disconnect switch. *See:* **electric controller.** IC1-34E10

combination current and voltage regulation. That type of automatic regulation in which the generator regulator controls both the voltage and current output of the generator. *Note:* This type of control is designed primarily for the purpose of insuring proper charging of storage batteries on cars or locomotives. *See also:* **axle generator system.** 42A42-0

combination electric locomotive. An electric locomotive, the propulsion power for which may be drawn from two or more sources, either located on the locomotive or elsewhere. *Note:* The prefix combination may be applied to cars, buses, etcetera, of this type. *See also:* **electric locomotive.** 42A42-0

combination lighting and appliance branch circuit. A circuit supplying energy to one or more lighting outlets and to one or more appliance outlets. *See also:* **branch circuit.** 42A95-0

combination microphone. A microphone consisting of a combination of two or more similar or dissimilar microphones. *Examples:* Two oppositely phased pressure microphones acting as a gradient microphone; a pressure microphone and velocity microphone acting as a unidirectional microphone. *See also:* **microphone.** 42A65-0

combination rubber tape. The assembly of both rubber and friction tape into one tape that provides both insulation and mechanical protection for joints. *See also:* **interior wiring.** 42A95-0

combination thermoplastic tape. An adhesive tape composed of a thermoplastic compound that provides both insulation and mechanical protection for joints. *See also:* **interior wiring.** 42A95-0

combination watch-report and fire-alarm system. A coded manual fire-alarm system, the stations of which are equipped to transmit a single watch-report signal or repeated fire-alarm signals. *See also:* **protective signaling.** 42A65-0

combined mechanical and electrical strength (insulator). The loading in pounds at which the insulator fails to perform its function either electrically or mechanically, voltage and mechanical stress being applied simultaneously. *Note:* The value will depend upon the conditions under which the test is made. *See also:* **insulator; tower.** 42A35-31E13

combined telephone set. A telephone set including in a single housing all the components required for a complete telephone set except the handset which it is arranged to support. *Note:* Wall hand telephone sets are of this type, but the term is usually reserved for a self-contained desk telephone set to distinguish it from desk telephone sets requiring an associated bell box. A desk local-battery telephone set may be referred to as a combined set if it includes in its mounting all components except its associated local batteries. *See also:* **telephone station.** 42A65-0

combined voltage and current influence (wattmeter). The percentage change (of full-scale value) in the indication of an instrument that is caused solely by a voltage and current departure from specified references while constant power at the selected scale point is maintained. *See also:* **accuracy rating (instrument).** 39A1-0

combustion control. The regulation of the rate of combination of fuel with air in a furnace. 42A35-31E13

command (1) (electronic computation). (A) One of a set of several signals (or groups of signals) that occurs as a result of interpreting an instruction; the commands initiate the individual steps that form the process of executing the instruction's operation. (B) Loosely: an instruction in machine language. (C) Loosely: a mathematical or logic operator. (D) Loosely: an operation. *See also:* **electronic computation; electronic digital computer.** E162/E270-0;X3A12-16E9

(2) (industrial control). An input variable established by means external to, and independent of, the feedback control system. It sets, is equivalent to, and is expressed in the same units as the ideal value of the ultimately controlled variable. *See:* **control system, feedback.** 85A1-23E0;AS1-34E10

command control (electric power systems). A control mode in which each generating unit is controlled to reduce unit control error. *See also:* **speed-governing system.** E94-0

command guidance. That form of missile guidance wherein control signals transmitted to the missile from an outside agency cause it to traverse a directed path through space. *See also:* **guided missile.** 42A65-0

command reference (servo or control system) (power supplies). The voltage or current to which the feedback signal is compared. As an independent variable, the command reference exercises complete control over the system output. *See:* **operational programming.** *See also:* **power supply.** KPSH-10E1

commercial tank (electrorefining). An electrolytic cell in which the cathode deposit is the ultimate electrolytically refined product. *See also:* **electrorefining.** 42A60-0

commissioning tests (rotating machinery). Tests applied to a machine at site under normal service conditions to show that the machine has been erected and connected in a correct manner and is able to work satisfactorily. *See also:* **asynchronous machine; direct-current commutating machine; synchronous machine.** 0-31E8

common device (supervisory system) (basic device). A device in either the master or remote station that is required for the basic operation and is not part of the equipment for the individual points. 37A100-31E11

common-battery central office. A central office that supplies transmitter and signaling currents for its associated stations and current for the central office equipment from batteries located in the central office. *See also:* **telephone switching system.** 42A65-19E1

common-battery switchboard. A telephone switchboard for serving common-battery telephone sets. *See also:* **telephone switching system.** 42A65-19E1

common-mode conversion (interference terminology). The process by which differential-mode interference is produced in a signal circuit by common-mode interference applied to the circuit. *Note:* See the accompanying figure.

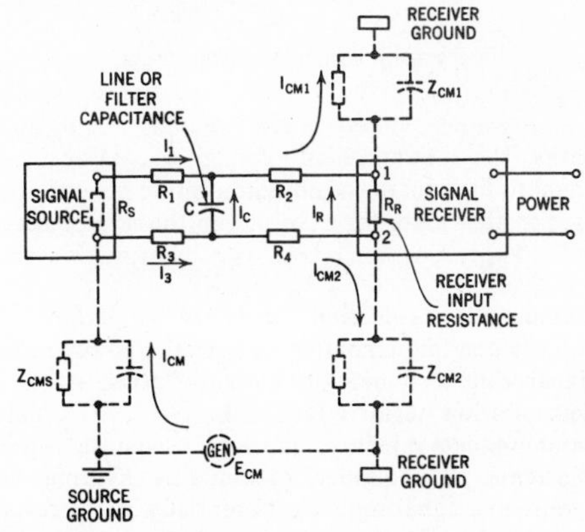

Common-mode conversion.

Common-mode currents are converted to differential-mode voltages by impedances R_1, R_2, R_3, R_4, R_S, R_R, and c. The differential-mode voltage at the receiver resulting from the conversion is the algebraic summation of the voltage drops produced by the various currents in these impedances. Various of the impedances may be neglected at particular frequencies. At direct current,

$$V_{CM} = I_r R_r \approx I_{CM1}(R_s + R_1 + R_2) - I_{CM2}(R_3 + R_4)$$

At

$$f > \frac{I}{c(R_1 + R_3 + R_s)},$$

$$V_{CM} \approx I_c X_c \frac{R_R}{R_2 + R_4 + R_R}$$

See also: **interference.** 0-13E6

common-mode interference (automatic null-balancing electrical instruments). (1) Interference that appears between both signal leads and a common reference plane (ground) and causes the potential of both sides of the transmission path to be changed simultaneously and by the same amount relative to the common reference plane (ground). *See:* **interference.** 0-13E6

(2). A form of interference that appears between any measuring circuit terminal and ground. *See also:* **accuracy rating (instrument).** 39A4-0

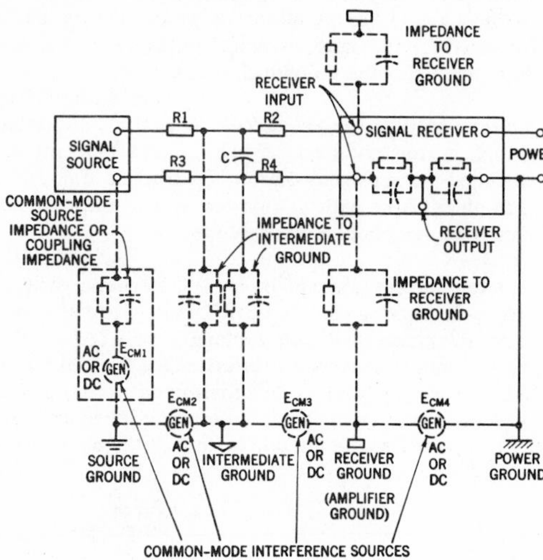

Common-mode interference—sources and current paths. The common-mode voltage V_{CM} in any path is equal to the sum of the common-mode generator voltages in that path; for example, in the source-receiver path, $V_{CM} = E_{CM1} + E_{CM2} + E_{CM3}$.

common-mode rejection (in-phase rejection). The ability of certain amplifiers to cancel a common-mode signal while responding to an out-of-phase signal. *See:* **degeneration negative feedback.** 0-18E1

common-mode rejection quotient (in-phase rejection quotient). The quotient obtained by dividing the response to a signal applied differentially by the response to the same signal applied in common mode, or the relative magnitude of a common-mode signal that produces the same differential response as a standard differential input signal. 0-18E1

common-mode rejection ratio (CMRR) (1) (signal transmission system). The ratio of the common-mode interference voltage at the input terminals of the system to the effect produced by the common-mode interference, referred to the input terminals for an amplifier. For example,

$$\text{CMRR} = \frac{V_{CM}(\text{root-mean-square) at input}}{\text{effect at output/amplifier gain}}$$

See also: **interference.** 0-13E6

(2) (oscilloscopes). The ratio of the deflection factor for a common-mode signal to the deflection factor for a differential signal applied to a balanced-circuit input. *See:* **oscillograph.**

common-mode signal (1) (general). The instantaneous algebraic average of two signals applied to a balanced circuit, both signals referred to a common reference. *See:* **oscillograph.** 0-9E4

(2) (in-phase signal) (medical electronics). A signal applied equally and in phase to the inputs of a balanced amplifier or other differential device. *See also:* **medical electronics.** 0-18E1

common, or shunt, winding of an autotransformer. That part of the autotransformer winding that is common to both the primary and the secondary circuits. *See:* **autotransformer.** 42A15-31E12;57A15-0

common return. A return conductor common to several circuits. *See also:* **center of distribution.** 42A35-31E13

common use (wiring system). Simultaneous use by two or more utilities of the same kind. 2A2-0

communication (telecommunication) (electric system). The transmission of information from one point to another by means of electromagnetic waves. *Notes:* (1) While there is no sharp line of distinction between the terms "communication" and "telecommunication", the latter is usually applied where substantial distances are involved. (2) For an extensive list of cross references, see *Appendix A.* 42A65-0

communication band. *See:* **frequency band of emission.**

communication circuits. Circuits used for audible and visual signals and communication of information from one place to another, within or on the vessel. E45-0

communication lines (electric system). Conductors and their supporting or containing structures that are located outside of buildings and are used for public or private signal or communication service, and that operate at not exceeding 400 volts to ground or 750 volts between any two points of the circuit, and the transmitted power of which does not exceed 150 watts. *Notes:* (1) When operating at less than 150 volts no limit is placed on the capacity of the system. (2) Telephone, telegraph, railroad-signal, messenger-call, clock, fire, police-alarm, community television antennas, and other systems conforming with the above are included. Lines used for signaling purposes, but not included under the above definition, are considered as supply lines of the same voltage and are to be so run. Exception is made under certain conditions for communication circuits used in the operation of supply lines. *See also:* **communication.** 2A2-0

communication reliability (mobile communication). A specific criterion of system performance related to the percentage of times a specified signal can be received in a defined area during a given interval of time. *See also:* **mobile communication system.** 0-6E1

communication theory (information theory). The mathematical theory underlying the communication of messages from one point to another. *See also:* **communication.** 42A65-0

commutated-antenna direction-finder (CADF). A system using a multiplicity of antennas in a circular array and a receiver that is connected to the antennas in sequence through a commutating device for finding the direction of arrival of radio waves; the directional sensing is related to phase shift that occurs as a result of the commutation. *See also:* **navigation.** 0-10E6

commutating angle (rectifier circuits). The time, expressed in degrees, during which the current is commutated between two rectifying elements. *See also:* **rectification; rectifier circuit element.** E59-34E17/34E24;42A15-0

commutating-field winding. An assembly of field coils located on the commutating poles, that produces a field strength approximately proportional to the load current. The commutating field is connected in direction and adjusted in strength to assist the reversal of current in the armature coils for successful commutation. This field winding is used alone, or supplemented by a compensating winding. *See:* **asynchronous machine; direct-current commutating machine.** 42A10-0

commutating group (rectifier circuit). A group of rectifier-circuit elements and the alternating-voltage supply elements conductively connected to them in which the direct current of the group is commutated between individual elements that conduct in succession. *See also:* **circuit element; rectification; rectifier.** E59-34E17/34E24;42A15-0

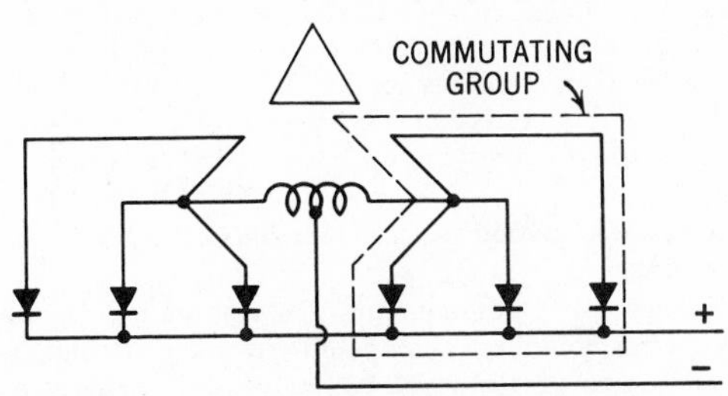

Commutating group in a single-way rectifier circuit.

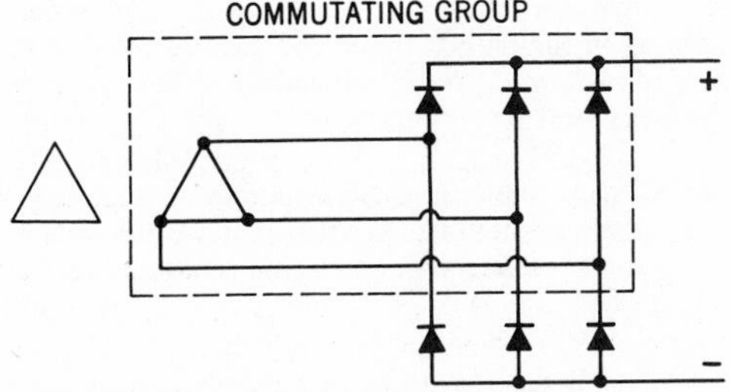

Commutating group in a double-way rectifier circuit.

commutating impedance (rectifier transformer). The impedance that opposes the transfer of current between two direct-current winding terminals of a commutating group, or a set of commutating groups. *See also:* **rectifier transformer.** 57A18-0

commutating period (inverters). The time during which the current is commutated. *See also:* **self-commutated inverters.** 0-34E24

commutating pole (interpole). An auxiliary pole placed between the main poles of a commutating machine. Its exciting winding carries a current proportional to the load current and produces a flux in such a direction and phase as to assist the reversal of the current in the short-circuited coil. 42A10-0

commutating reactance (rectifier transformer). Reactive component of the commutating impedance. *Note:* This value is defined as one-half the total reactance in the commutating circuit expressed in ohms referred to the total direct-current winding. For wye, star, and multiple-wye circuits, this is the same value as derived in ohms on a phase-to-neutral voltage base; while with diametric and zig-zag circuits it must be expressed as one-half the total due to both halves being mutually coupled on the same core leg or phase. *See also:* **commutating impedance; rectification; rectifier transformer.** 57A18-0

commutating reactance factor (rectifier circuit). The line-to-neutral commutating reactance in ohms multiplied by the direct current commutated and divided by the effective (root-mean-square) value of the line-to-neutral voltage of the transformer direct-current winding. *See also:* **circuit element; rectification; rectifier circuit.** E59/42A15-34E24/34E17

commutating reactance transformation constant. A constant used in transforming line-to-neutral commutating reactance in ohms on the direct-current winding to equivalent line-to-neutral reactance in ohms referred to the alternating-current winding. *See also:* **rectification.** 42A15-34E24

commutating reactor. A reactor used primarily to modify the rate of current transfer between rectifying elements. *See:* **reactor.** 42A15-31E12

commutating resistance (rectifier transformer). The resistance component of the commutating impedance. *See also:* **rectifier transformer.** 57A18-0

commutating voltage (rectifier circuit). The phase-to-phase alternating-current voltage of a commutating group. *See also:* **rectifier circuit element.** E59-34E17

commutation (rectifier) (rectifier circuit). The transfer of unidirectional current between rectifier circuit elements that conduct in succession. *See also:* **rectification; rectifier circuit element.** E59-34E17/34E24;34A1/42A15-0

commutation elements (semiconductor rectifiers). The circuit elements used to provide circuit-commutated turnoff time. *See also:* **semiconductor rectifier stack.** 0-34E24

commutation factor (1) (rectifier circuits). The product of the rate of current decay at the end of conduction, in amperes per microsecond, and the initial reverse voltage, in kilovolts. *See also:* **element; rectification; rectifier circuit.** E50-34E17/34E24;34A1-0

(2) (gas tubes). The product of the rate of current decay and the rate of the inverse voltage rise immediately following such current decay. *Note:* The rates are commonly stated in amperes per microsecond and volts per microsecond. *See also:* **heterodyne conversion transducer (converter); gas tubes; rectification.** E160-15E6

commutator (rotating machinery). A cylindrical ring

or disc assembly of conducting members, individually insulated in a supporting structure, with an exposed surface for contact with current-collecting brushes. *Note:* It is mounted on the moving element of the machine.
See:
asynchronous machine;
commutating pole;
commutator bore;
commutator brush track diameter;
commutator core;
commutator-core extension;
commutator-inspection cover;
commutator-insulating tube;
commutator nut;
commutator riser;
commutator segments;
commutator-segment assembly;
commutator shell;
commutator-shell insulation;
commutator shrink ring;
commutator vee ring;
commutator-vee-ring insulation;
commutator-vee-ring insulation extension;
direct-current commutating machine.
42A10-31E8

commutator bars. *See:* **commutator segments.**

commutator bore. Diameter of the finished hole in the core that accommodates the armature shaft. *See:* **commutator.** 42A10-0

commutator brush track diameter. That diameter of the commutator segment assembly that after finishing on the armature is in contact with the brushes. *See:* **commutator.** 42A10-0

commutator core. The complete assembly of all of the retaining members of a commutator. *See:* **commutator.** 42A10-0

commutator-core extension. That portion of the core that extends beyond the commutator segment assembly. *See:* **commutator.** 42A10-0

commutator inspection cover. A hinged or otherwise attached part that can be moved to provide access to commutator and brush rigging for inspection and adjustment. *See:* **commutator.** 42A10-0

commutator insulating segments (rotating machinery). The insulating members between adjacent commutator segments. *See:* **commutator.** 42A10-31E8

commutator insulating tube (rotating machinery). The insulation between the underside of the commutator segment assembly and the core. *See also:* **commutator.** 42A10-31E8

commutator motor meter. A motor type of meter in which the rotor moves as a result of the magnetic reaction between two windings, one of which is stationary and the other assembled on the rotor and energized through a commutator and brushes. *See also:* **electricity meter (meter).** 42A30-0

commutator nut. The retaining member that is used in combination with a vee ring and threaded shell to clamp the segment assembly. *See:* **commutator.** 42A10-0

commutator riser. That portion of the commutator segment assembly that is larger than the brush track diameter. *See:* **commutator.** 42A10-0

commutator-segment assembly. A cylindrical ring or disc assembly of commutator segments and insulating segments that are bound and ready for installation. *Note:* The binding used may consist of wire, temporary assembly rings, shrink rings, or other means. *See:* **commutator.** 42A10-0

commutator segments (commutator bars). Metal current-carrying members that are insulated from one another by insulating segments and that make contact with the brushes. *See:* **commutator.** 42A10-0

commutator shell. The support on which the component parts of the commutator are mounted. *Note:* The commutator may be mounted on the shaft, on a commutator spider, or it may be integral with a commutator spider. *See:* **commutator.** 42A10-0

commutator-shell insulation (rotating machinery). The insulation between the under (or in the case of a disc commutator, the back) side of the commutator assembled segments and the commutator shell. *See also:* **commutator.** 42A10-31E8

commutator shrink ring. A member that holds the commutator-segment assembly together and in place by being shrunk on an outer diameter of and insulated from the commutator-segment assembly. *See:* **commutator.** 42A10-0

commutator vee ring. The retaining member that, in combination with a commutator shell, clamps or binds the commutator segments together. *See:* **commutator.** 42A10-0

commutator vee-ring insulation (rotating machinery). The insulation between the end of the commutator-segment assembly and the core. *See also:* **commutator.** 42A10-31E8

commutator vee-ring insulation extension (rotating machinery). The portion of the vee-ring insulation that extends beyond the commutator segment assembly. *See:* **commutator.** 42A10-31E8

companding (modulation systems). A process in which compression is followed by expansion. *Note:* Companding is often used for noise reduction, in which case the compression is applied before the noise exposure and the expansion after the exposure. *See also:* **modulating systems; transmission characteristics.** E170/42A65-0

companding, instantaneous. *See:* **instantaneous companding.**

compandor. A combination of a compressor at one point in a communication path for reducing the amplitude range of signals, followed by an expander at another point for a complementary increase in the amplitude range. *Note:* The purpose of the compandor is to improve the ratio of the signal to the interference entering in the path between the compressor and expander. *Editorial Note:* **Compandor** is the preferred spelling rather than **compander.** E151-31E3;42A65-0

comparator. A circuit for performing amplitude selection between either two variables or between a variable and a constant. *See also:* **electronic analog computer.** E165-16E9

comparison amplifier (power supplies). A high-gain noninverting direct-current amplifier that, in a bridge-regulated power supply, has as its input the voltage between the null junction and the common terminal. The output of the comparison amplifier drives the series pass elements. *See also:* **power supply.** KPSH-10E1

comparison bridge (power supplies). A type of voltage-comparison-circuit whose configuration and principle of operation resemble a four-arm electric bridge. See the accompanying figure.

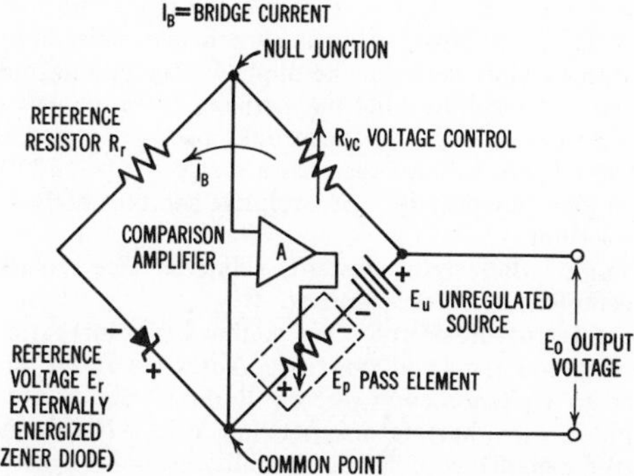

Comparison bridge connected as a voltage regulator. U.S. Patent No. 3,028,538 - Forbro Design Corp.

Note: The elements are so arranged that, assuming a balance exists in the circuit, a virtual zero error signal is derived. Any tendency for the output voltage to change, in relation to the reference voltage, creates a corresponding error signal, that by means of negative feedback, is used to correct the output in the direction toward restoring bridge balance. *See:* **error signal.** *See also:* **power supply.** KPSH-10E1

comparison lamp. A light source having a constant, but not necessarily known, luminous intensity with which standard and test lamps are successively compared. *See also:* **primary standards.** Z7A1-0

compartment-cover mounting plate (pothead). A metal plate used for covering the flange opening. It may be altered in the field to accommodate various cable entrance fittings. It may also be used to attach single pothead mounting plates to the compartment flange. 57A12.75-0

compass bearing. Bearing as indicated by a magnetic compass. *See also:* **navigation.** 0-10E6

compass-controlled directional gyro. A device that uses the earth's magnetic field as a reference to correct a directional gyro. *Note:* The direction of the earth's field is sensed by a remotely located compass that is connected electrically to the gyro. *See also:* **air transportation instruments.** 42A41-0

compass course. Course as indicated by a magnetic compass. *See also:* **navigation.** 0-10E6

compass declinometer. *See:* **declinometer.**

compass deviation. *See:* **magnetic deviation.**

compass heading. Heading as indicated by a magnetic compass. *See also:* **navigation.** 0-10E6

compass locator (electronic navigation). *See:* **non-directional beacon.**

compass repeater. An indicator, located remotely from a master compass, so controlled that it continuously indicates (repeats) the bearing shown by the master compass. 42A43-0

compatibility (color television). The property of a color television system that permits substantially normal monochrome reception of the transmitted signal by typical unaltered monochrome receivers. *See also:* **color terms.** E201-2E2

compensated-loop direction-finder. A direction-finder employing a loop antenna and a second antenna system to compensate polarization error. *See also:* **radio receiver.** 42A65-0

compensated repulsion motor. A repulsion motor in which the primary winding on the stator is connected in series with the rotor winding via a second set of brushes on the commutator in order to improve the power factor and commutation. *See:* **synchronous machine.** 0-31E8

compensated series-wound motor. A series-wound motor with a compensating-field winding. The compensating-field winding and the series-field winding may be combined into one field winding. *See also:* **asynchronous machine; direct-current commutating machine.** MA1-31E8

compensating-field winding (rotating machinery). Conductors embedded in the pole shoes and their end connections. It is connected in series with the commutating-field winding and the armature circuit. *Note:* A compensating-field winding supplements the commutating-field winding, and together they function to assist the reversal of current in the armature coils for successful commutation. *See also:* **asynchronous machine; direct-current commutating machine.** 42A10-0

compensating-rope sheave switch. A device that automatically causes the electric power to be removed from the elevator driving-machine motor and brake when the compensating sheave approaches its upper or lower limit of travel. *See also:* **hoistway (elevator or dumbwaiter).** 42A45-0

compensation (control system, feedback). A modifying or supplementary action (also, the effect of such action) intended to improve performance with respect to some specified characteristic. *Note:* In control usage, this characteristic is usually the system deviation. Compensation is frequently qualified as series, parallel, feedback, etcetera, to indicate the relative position of the compensating element. *See also:* **control system, feedback; equalization.** 85A1-23E0

compensation theorem. States that if an impedance ΔZ is inserted in a branch of a network, the resulting current increment produced in any branch in the network is equal to the current that would be produced at that point by a compensating voltage, acting in series with the modified branch, whose value is $-I\Delta Z$, where I is the original current that flowed where the impedance was inserted before the insertion was made. *See also:* **communication.** 42A65-0

compensator (1) (rotating machinery). An element or group of elements that acts to modify the functioning of a device in accordance with one or more variables. *See also:* **asynchronous machine; synchronous machine.** 0-31E8

(2) (radio direction-finders). That portion of a direction-finder that automatically applies to the direction indication all or a part of the correction for the deviation. *See also:* **radio receiver.** 42A65-0

compensatory leads. Connections between an instrument and the point of observation so contrived that variations in the properties of leads, such as variations of resistance with temperature, are so compensated that they do not affect the accuracy of the instrument readings. *See also:* **auxiliary device to an instrument.** 42A30-0

compile (computing system). To prepare a machine language program from a computer program, written in another programming language, by making use of the overall logic structure of the program, or generating more than one machine instruction for each symbolic statement, or both, as well as performing the function of an assembler. *See also:* **electronic digital computer.** X3A12-16E9

compiler. A program that compiles. X3A12-16E9

complement. (1) A number whose representation is derived from the finite positional notation of another by one of the following rules: (A) true complement —subtract each digit from one less than the base; then add one to the least-significant digit, executing all carries required, (B) base minus ones complement—subtract each digit from one less than the base (for example, "nines complement" in the base 10, "ones complement" in the base 2, etcetera). (2) To complement is to form the complement of a number. *Note:* In many machines, a negative number is represented as a complement of the corresponding positive number. *See:* **nines complement; ones complement; radix complement; radix-minus-ones complement; tens complement; true complement; twos complement.** *See also:* **electronic computation.** E270-0

complementary functions. Two driving-point functions whose sum is a positive constant. *See also:* **linear passive networks.** E156-0

complementary tracking (power supplies). A system of interconnection of two regulated supplies in which one (the master) is operated to control the other (the slave). The slave supply voltage is made equal (or proportional) to the master supply voltage and of opposite polarity with respect to a common point. See the accompanying figure. *See also:* **power supply.** KPSH-10E1

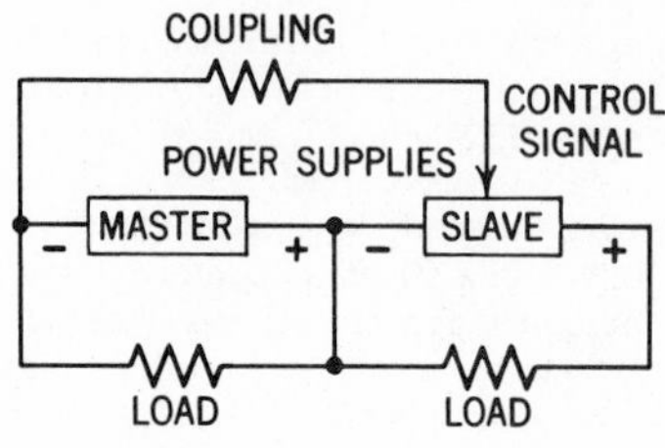

Complementary tracking.

complementary wavelength (color). The wavelength of light of a single frequency that matches the reference standard light when combined with a sample color in suitable proportions. *Notes:* (1) The wide variety of purples that have no dominant wavelengths, including nonspectral violet, purple, magenta, and nonspectral red colors, are specified by use of their complementary wavelengths. (2) Refer to **dominant wavelength**. *See also:* **color terms.** E201-0

complete carry. A carry process in which a carry resulting from addition of carries is allowed to propagate. Contrasted with partial carry. *See also:* **carry; electronic digital computer.** E162-0

complete diffusion (illuminating engineering). Diffusion in which the diffusing medium redirects the flux incident upon it so that none is in an image-forming state. *See also:* **lamp.** Z7A1-0

completely mesh-connected circuit. A mesh-connected circuit in which a current path extends directly from the terminal of entry of each phase conductor to the terminal of entry of every other phase conductor. *See also:* **network analysis.** E270-0

complex apparent permeability. The complex (phasor) ratio of induction to applied magnetizing force. *Notes:* (1) In ordinary ferromagnetic materials, there is a phase shift between the applied magnetizing force and the resulting total magnetizing force, caused by eddy currents. (2) In anisotropic media, complex apparent permeability becomes a matrix. E270-0

complex capacitivity. *See:* **relative complex dielectric constant.**

complex dielectric constant, relative. *See:* **relative complex dielectric constant.**

complex permeability. The complex (phasor) ratio of induction to magnetizing force. *Notes:* (1) This is related to a phenomenon wherein the induction is not in phase with the total magnetizing force. (2) In anisotropic media, complex permeability becomes a matrix. E270-0

complex permittivity. *See:* **relative complex dielectric constant.**

complex power (rotating machinery). An expression in which the active power is the real part and the reactive power, with sign reversed, is the imaginary part. *See also:* **asynchronous machine; synchronous machine.** 50I05-31E8

complex target (radar). A target composed of a number of reflecting surfaces, such as an aircraft at short distance, which surfaces in aggregate are smaller in any dimension than the resolution capabilities of the radar. *See also:* **navigation.** E172-10E6

complex tone. (1) A sound containing simple sinusoidal components of different frequencies. (2) A sound sensation characterized by more than one pitch. *See also:* **electroacoustics.** 0-1E1

compliance (industrial control). A property reciprocal to stiffness. *See:* **control system, feedback.** AS1-34E10

compliance extension (power supplies). A form of master/slave interconnection of two or more current-regulated power supplies to increase their compliance voltage range through series connection. *See also:* **power supply; compliance voltage.** KPSH-10E1

compliance voltage (power supplies). The output voltage of a direct-current power supply operating in constant-current mode. *Note:* The compliance range is the range of voltages needed to sustain a given value of constant current throughout a range of load resistances. *See also:* **power supply.** KPSH-10E1

component (1). A piece of equipment, a line, a section of line, or a group of items that is viewed as an entity for purposes of reporting, analyzing, and predicting outages. *See also:* **computer component; solid-state component.** 0-31E4

(2) (power-system communication). An elementary device. *See also:* **circuits and devices.** 0-31E3

components (vector). The components of a vector are any vectors the sum of which gives the original vector. The components of a vector **V** in a specified coordinate system are the directed projections of **V** on that system. Thus if V_x, V_y, and V_z are the scalar values of the projections of **V** on the x, y, and z axes, respectively and **i**, **j**, **k**, are unit vectors along these axes, then $\mathbf{i}V_x$, $\mathbf{j}V_y$, and $\mathbf{k}V_z$ are the components of V. Moreover

$$\mathbf{V} = \mathbf{i}V_x + \mathbf{j}V_y + \mathbf{k}V_z.$$

Again if V is the magnitude of the vector **V**, and θ its polar angle with the x axis and ϕ its azimuth with the x-y plane, then

$$V_x = V \cos \theta$$
$$V_y = V \sin \theta \cos \phi$$
$$V_z = V \sin \theta \sin \phi$$
$$\text{and } \mathbf{V} = \mathbf{i}V_x + \mathbf{j}V_y + \mathbf{k}V_z$$

Note: A right-handed system of coordinates for a rectangular system is such that if **i, j, k** are unit vectors measured along the *x, y,* and *z* axes, respectively, then the 90-degree rotation from **i** to **j** of a right-handed screw causes that screw to progress along the direction of **k**. E270-0

composite bushing (electric apparatus). A bushing in which the insulation consists of several coaxial layers of different insulating materials. *See also:* **power-distribution, overhead construction.** 76A1-0

composite cable (communication practice). A cable in which conductors of different gauges or types are combined under one sheath. *Note:* Differences in length of twist are not considered here as constituting different types. *See also:* **cable.** 42A65-0

composite color-picture signal (television). The electric signal that represents complete color-picture information, including setup, the color burst (in the NTSC system), and all other synchronizing signals. *See:* **television.** 0-2E2

composite color signal (color television). The color-picture signal plus blanking and all synchronizing signals. *See also:* **color terms.** E201-0

composite color synchronizing (color television). The signal comprising all the synchronizing signals necessary for proper operation of a color receiver. *Note:* This includes the deflection synchronizing signals to which the color synchronizing signal is added in the proper time relationship. *See also:* **color terms.** E201-2E2

composite conductor. A composite conductor consists of two or more strands. *See also:* **conductor.** 42A35-31E13

composite controlling voltage (electron tube). The voltage of the anode of an equivalent diode combining the effects of all individual electrode voltages in establishing the space-charge-limited current. *See:* **excitation (drive).** *See also:* **circuit characteristics of electrodes.** E160-15E6;42A70-0

composited circuit. A circuit that can be used simultaneously for telephony and direct-current telegraphy or signaling, separation between the two being accomplished by frequency discrimination. *See also:* **transmission line.** 42A65-0

composite picture signal (television). The signal that results from combining a blanked picture signal with the synchronizing signal. *See also:* **television.** E203-2E2

composite plate (electroplating). An electrodeposit consisting of two or more layers of metals deposited separately. *See also:* **electroplating.** 42A60-0

composite pulse (pulse navigation systems). A pulse composed of a series of overlapping pulses received from the same signal source but via different paths. *See also:* **navigation.** E172-10E6

composite set. An assembly of apparatus designed to provide one end of a composited circuit. *See also:* **circuits and devices.** 42A65-0

composite supervision. The use of a composite signaling channel for transmitting supervisory signals between two points in a connection. *See also:* **telephone switching system.** 42A65-0

compound (rotating machinery). (1) A definite substance resulting from the combination of specific elements or radicals in fixed proportions: distinguished from mixture. (2) The intimate admixture of resin with ingredients such as fillers, softeners, plasticizers, catalysts, pigments, or dyes. *See also:* **rotor (rotating machinery); stator.** AD883-31E8

compound-filled (reactor, transformer). Having the coils encased in an insulating fluid that becomes solid or remains slightly plastic at normal operating temperatures. *See also:* **instrument transformer; reactor.** 42A15/57A15/57A18-31E12

compound-filled bushing (outdoor electric apparatus). A bushing in which the space between the inside surface of the porcelain and the major insulation (or conductor where no major insulation is used) is filled with compound. *See also:* **power-distribution-overhead construction.** 76A1-0

compounding curve (direct-current generator). A regulation curve of a compound-wound direct-current generator. *Note:* The shunt field may be either self or separately excited. *See:* **direct-current commutating machine.** 42A10-0

compound target (radar). *See:* **complex target.**

compound-wound. A qualifying term applied to a direct-current machine to denote that the excitation is supplied by two types of windings, shunt and series. *Note:* When the electromagnetic effects of the two windings are in the same direction, it is termed cumulative compound wound; when opposed, differential compound wound. *See:* **direct-current commutating machine.** 42A10-0

compound-wound generator. A direct-current generator that has two separate field windings—one, supplying the predominating excitation, is connected in parallel with the armature circuit and the other, supplying only partial excitation, is connected in series with the armature circuit and of such proportion as to require an equalizer connection for satisfactory parallel operation. *See also:* **direct-current commutating machine.** E45-0

compound-wound motor. A direct-current motor that has two separate field windings—one, usually the predominating field, connected in parallel with the armature circuit, and the other connected in series with the armature circuit. *See:* **direct-current commutating machine.** 42A10-0

compressed-air circuit breaker. *See:* Note under **circuit breaker.**

compression (1) (modulation systems). A process in which the effective gain applied to a signal is varied as a function of the signal magnitude, the effective gain being greater for small than for large signals. *See also:* **modulating systems; transmission characteristic.** E170-0

(2) (television). The reduction in gain at one level of a picture signal with respect to the gain at another level of the same signal. *Note:* The gain referred to in the definition is for a signal amplitude small in comparison with the total peak-to-peak picture signal involved. A quantitative evaluation of this effect can be obtained by a measurement of differential gain. *See:* **black com-**

pression; white compression. *See also:* **television.** E203-2E2
(3) **(oscillography).** An increase in the deflection factor, usually as the limits of the quality area are exceeded. *See:* **oscillograph.** 0-9E4

compression ratio (gain or amplification). The ratio of (1) the magnitude of the gain (or amplification) at a reference signal level to (2) its magnitude at a higher stated signal level. *See:* **amplifier.** E160-15E6

compressional wave. A wave in an elastic medium that is propagated by fluctuation in elemental volume, accompanied by velocity components along the direction of propagation only. *Note:* A compressional plane wave is a longitudinal wave. *See also:* **electroacoustics.** 0-1E1

compressor. A transducer that for a given input amplitude range produces a smaller output range. *See also:* **attenuation.** E151-31E3;42A65-0

compressor stator-blade-control system (gas turbines). A means by which the turbine compressor stator blades are adjusted to vary the operating charcteristics of the compressor. *See also:* **speed-governing system (gas turbines).** E282-31E2

computation. *See:* **implicit computation.**

computer. (1) A machine for carrying out calculations. (2) By extension, a machine for carrying out specified transformations on information. (3) A stored-program data-processing system.
See:
alternating-current analog computer;
analog computer;
asynchronous computer;
automatic computer;
digital computer;
direct-current analog computer;
dynamic computer check;
general-purpose computer;
incremental computer;
special-purpose computer;
stored-program computer;
synchronous computer;
See also:
electronic analog computer;
electronic digital computer;
static magnetic storage. E162/E165/E270-0

computer code. A machine code for a specific computer. *See also:* **electronic digital computer.** X3A12-16E9

computer component. Any part, assembly, or subdivision of computer, such as resistor, amplifier, power supply, or rack. *See also:* **electronic analog computer.** E165-16E9

computer control (physical process) (electric power systems). A mode of control wherein a computer, using as input the process variables, produces outputs that control the process. *See also:* **power system.** 0-31E4

computer control state. One of several distinct and selectable conditions of the computer control circuits. *See:* **balance check; hold; operate; reset.** *See also:* **electronic analog computer.** E165-16E9

computer equation (machine equation). An equation derived from a mathematical model for more convenient use on a computer. *See also:* **electronic analog computer.** E165-16E9

computer diagram. A functional drawing showing interconnections between computing elements. *See also:* **electronic analog computer.** E165-0

computer instruction. A machine instruction for a specific computer. *See also:* **electronic digital computer.** X3A12-16E9

computer network. A complex consisting of two or more interconnected computing units. *See also:* **electronic digital computer.** X3A12-16E9

computer program. A plan or routine for solving a problem on a computer, as contrasted with such terms as fiscal program, military program, and development program. *See also:* **electronic digital computer.** X3A12-16E9

computer variable (1). A dependent variable as represented on the computer. *See also:* **time.** E165-0
(2) (machine variable). *See:* **scale factor.** *See also:* **electronic analog computer.**

computer word. A sequence of bits or characters treated as a unit and capable of being stored in one computer location. *See:* **machine word.** X3A12-16E9

computing element. A computer component that performs the mathematical operations required for problem solution. *See also:* **electronic analog computer; electronic digital computer.** E165-16E9

concealed. Rendered inaccessible by the structure or finish of the building. Wires in concealed raceways are considered concealed, even though they may become accessible by withdrawing them. 1A0-42A95-0

concealed knob-and-tube wiring. A wiring method using knobs, tubes, and flexible tubing for the protection and support of insulated conductors in the hollow spaces of walls and ceilings of buildings. *See also:* **interior wiring.** 42A95-0

concentrate (metallurgy). The product obtained by concentrating disseminated or lean ores by mechanical or other processes thereby eliminating undesired minerals or constituents. *See also:* **electrowinning.** 42A60-0

concentrated winding (rotating machinery). A winding, the coils of which occupy one slot per pole; or a field winding mounted on salient poles. *See also:* **asynchronous machine; direct-current commutating machine; synchronous machine.** 0-31E8

concentration cell. (1) An electrolyte cell, the electromotive force of which is due to differences in composition of the electrolyte at anode and cathode areas. CM-34E2
(2) A cell of the two-fluid type in which the same dissolved substance is present in differing concentrations at the two electrodes. *See also:* **electrochemistry.** 42A60-0

concentration polarization. (1) That part of the total polarization that is caused by changes in the activity of the potential-determining components of the electrolyte. *See also:* **electrochemistry.** 42A60-0
(2) That portion of the polarization of an electrode produced by concentration changes at the metal-environment interface. CM-34E2

concentric electrode system (electrobiology) (coaxial electrode system). An electrode system that is geometrically coaxial but electrically unsymmetrical. *Example:* One electrode may have the form of a cylindrical shell about the other so as to afford electrical shielding. *See also:* **electrobiology.** 42A80-18E1

concentric groove (disk recording). *See:* **locked groove.**

concentric-lay cable. (1) A concentric-lay conductor as defined below or (2) a multiple-conductor cable composed of a central core surrounded by one or more layers of helically laid insulated conductors. *See also:* **power distribution, underground construction.** E30-0;42A35-31E13

concentric-lay conductor. A conductor composed of a central core surrounded by one or more layers of helically laid wires. *Note:* In the most common type of concentric-lay conductor, all wires are of the same size and the central core is a single wire. *See also:* **conductor.** E30/42A35-31E13

concentric line. *See:* **coaxial.**

concentric winding (rotating machinery). A winding in which the two coil sides of each coil of a phase belt, or of a pole of a field winding, are symmetrically located so as to be equidistant from a common axis. *See also:* **asynchronous machine; synchronous machine.** 0-31E8

concrete quantity. *See:* **physical quantity.**

concrete-tight fitting (for conduit). A fitting so constructed that embedment in freshly mixed concrete will not result in the entrance of cement into the fitting. 80A4-0

condensed-mercury temperature (mercury-vapor tube). The temperature measured on the outside of the tube envelope in the region where the mercury is condensing in a glass tube or at a designated point on a metal tube. *See:* **gas tube.** E160-15E6;42A70-0

condenser*(1). *See:* **capacitor.**

*Deprecated

(2) (fuse). *See:* **fuse condenser.**

condenser antenna*. *See:* **capacitor antenna.**

*Deprecated

condenser box*. *See:* **subdivided capacitor.**

*Deprecated

condenser bushing*. *See:* **capacitor bushing.**

*Deprecated

condenser loudspeaker* (capacitor loudspeaker). *See:* **electrostatic loudspeaker.**

*Deprecated

condenser microphone* (capacitor microphone). *See:* **electrostatic microphone.**

*Deprecated

condition (computing system). *See:* **initial condition.**

conditional information content (first symbol given a second symbol). The negative of the logarithm of the conditional probability of the first symbol, given the second symbol. *Notes:* (1) The choice of logarithmic base determines the unit of information content. *See:* **bit** and **hartley**. (2) The conditional information content of an input symbol given an output symbol, averaged over all input-output pairs, is the equivocation. (3) The conditional information content of output symbols relative to input symbols, averaged over all input-output pairs, has been called spread, prevarication, irrelevance, etcetera. *See also:* **information theory.** E171-0

conditional jump. To cause, or an instruction that causes, the proper one of two (or more) addresses to be used in obtaining the next instruction, depending upon some property of one or more numerical expressions or other conditions. Sometimes called a branch. *See also:* **electronic digital computer; jump.** E162/E270-0

conditional transfer of control. Same as **conditional jump.** *See also:* **electronic digital computer.** E162/E270-0

conditioning stimulus (medical electronics). A stimulus of given configuration applied to a tissue before a test stimulus. *See also:* **medical electronics.** 42A80-18E1

conductance. (1) That physical property of an element, device, branch, network or system, that is the factor by which the mean square voltage must be multiplied to give the corresponding power lost by dissipation as heat or as other permanent radiation or loss of electromagnetic energy from the circuit. (2) The real part of admittance. *Note:* Definitions (1) and (2) are not equivalent but are supplementary. In any case where confusion may arise, specify the definition being used. E270-9E4

conductance, electrode. *See:* **electrode conductance.**

conductance for rectification (electron tube). The quotient of (1) the electrode alternating current of low frequency by (2) the in-phase component of the electrode alternating voltage of low frequency, a high frequency sinusoidal voltage being applied to the same or another electrode and all other electrode voltages being maintained constant. *See also:* **circuit characteristics of electrodes; rectification factor.** E160-15E6;42A70-0

conductance relay. A relay for which the center of the operating characteristic on the R-X diagram is on the R axis. *Note:* The equation that describes such a characteristic is $Z = K \cos \theta$ where K is a constant and θ is the phase angle by which the input voltage leads the input current. 37A100-31E11/31E6

conducted heat. The thermal energy transported by thermal conduction. *See also:* **thermoelectric device.** E221-15E7

conducted interference (electromagnetic compatibility). Interference resulting from conducted radio noise or unwanted signals entering a transducer (receiver) by direct coupling. *See also:* **electromagnetic compatibility.** 0-27E1

conducted radio noise (electromagnetic compatibility). Radio noise propagated along circuit conductors. *Note:* It may enter a transducer (receiver) by direct coupling or by an antenna as by subsequent radiation from some circuit element. *See also:* **electromagnetic compatibility.** 0-27E1

conducting element (fuse) (fuse link). The conducting means, including the current-responsive element, for completing the electric circuit between the terminals of a fuseholder or fuse unit. 37A100-31E11

conducting material. A conducting medium in which the conduction is by electrons, and whose temperature coefficient of resistivity is, except for certain alloys, nonnegative at all temperatures below the melting point. E270-0

conducting mechanical joint. The juncture of two or more conducting surfaces held together by mechanical means. *Note:* Parts jointed by fusion processes, such as welding, brazing, or soldering, are excluded from this definition. 37A100-31E11

conducting paint (rotating machinery). A paint in which the pigment or a portion of pigment is a conductor of electricity and the composition is such that when it is converted to a solid film, the electric conductivity of the film approaches that of metallic substances. *See also:* **cradle base (rotating machinery).** 0-31E8

conducting parts (industrial control). The parts that are designed to carry current or that are conductively connected therewith. IC1-34E10

conducting period (1) (rectifier circuit). The part of an alternating-voltage cycle during which the current flows in the forward direction. *See:* Note under **forward period.** *See also:* **rectifier circuit element.** E59-34E17/34E24

(2) (gas tube). That part of an alternating-voltage cycle during which a certain arc path is carrying current. *See also:* **gas-filled rectifier.** 50I07-15E6

(3) (rectifying element). That part of an alternating-voltage cycle during which current flows from the anode to the cathode. *See also:* **rectification.** 42A15-0

conducting salts. Salts that, when added to a plating solution, materially increase its conductivity. *See also:* **electroplating.** 42A60-0

conduction band (semiconductor). A range of states in the energy spectrum of a solid in which electrons can move freely. *See also:* **semiconductor.** E102/E216-34E17;E270-0;0-10E1

conduction current. Through any surface, the integral of the normal component of the conduction current density over that surface. *Notes:* (1) Conduction current is a scalar and hence has no direction. (2) Current does not flow. In speaking of conduction current through a given surface it is appropriate to refer to the charge flowing or moving through that surface, but not to the current "flow". (3) For the use on circuit diagrams of arrows or other graphic symbols in connection with currents, *see* **reference direction (current).** E270-0

conduction electron. *See:* **electrons, conduction.**

conductive coating (rotating machinery). Conducting paint applied to the slot portion of a coil-side, to carry capacitive and leakage currents harmlessly between insulation and grounded iron. *See also:* **cradle base (rotating machinery).** 0-31E8

conductive coupling (interference terminology). *See:* **coupling, conductance.**

conductivity (material). A factor such that the conduction-current density is equal to the electric-field intensity in the material multiplied by the conductivity. *Note:* In the general case it is a complex tensor quantity. *See also:* **transmission line.** E270-9E4

conductivity modulation (semiconductor). The variation of the conductivity of a semiconductor by variation of the charge-carrier density. *See also:* **semiconductor; semiconductor device.** E102-0;E216-34E17;E270-0;42A70-0

conductivity, *n*-type (semiconductor). The conductivity associated with conduction electrons in a semiconductor. *See also:* **semiconductor.** E102/E216-34E17;E270-10E1

conductivity, *p*-type (semiconductor). The conductivity associated with holes in a semiconductor. *See also:* **semiconductor.** E102/E216-34E17;E270-10E1

conductor. (1) A substance or body that allows a current of electricity to pass continuously along it. *See:* **conducting material.** E270-0;50I05-31E8

(2) A wire or combination of wires not insulated from one another, suitable for carrying an electric current. It may be, however, bare or insulated. E30-0

(3) The portion of a lightning-protection system designed to carry the lightning discharge between air terminal and ground.

See:

aluminium conductor;
aluminium-steel conductor;
arresters;
bare conductor;
bronze conductor;
composite conductor;
concentric-lay conductor;
copper conductor;
copper-covered steel wire;
covered;
cross-sectional area;
final unloaded conductor tension;
hollow-core annular conductor;
initial conductor tension;
insulated resistance;
insulation resistance;
line conductor;
plain conductor;
resistive conductor;
rope-lay conductor or cable;
round conductor;
segmental conductor;
self-supporting aerial cable;
service-entrance conductors;
solid conductor;
strand;
stranded conductor;
tie wire;
tree wire;
twin wire;
twisted pair;
vertical conductor;
wire.

See also: **lightning protection and equipment; tower.** 5A/42A95-0

conductor, bare. One having no covering or insulation whatsoever. 1A0-0

conductor conflict. The situation in which a conductor is so situated with respect to a conductor of another line at a lower level that the horizontal distance between them is less than the sum of the following values: (1) five feet, (2) one-half the difference of level between the conductors concerned, and (3) the value required in the accompanying tables for horizontal separation between conductors on the same support for the highest voltage carried by either conductor concerned. See accompanying figure. *See also:* **conflict (wiring system).** 2A2-0

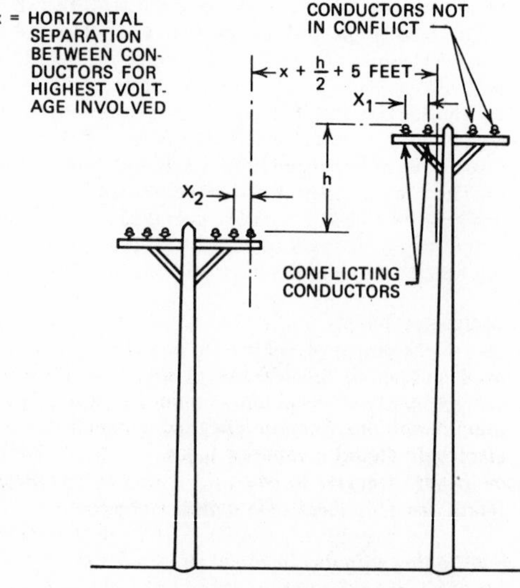

Conductor conflict.

Minimum horizontal separation at supports between line conductors of the same or different circuits

(All voltages are between conductors except for railway feeders, which are to ground)

Class of circuit	Separation	Notes
	Inches	
Communication conductors	6	Preferable minimum. Does not apply at conductor transposition points.
	3	Permitted where pin spacings less than 6 inches have been in regular use. Does not apply at conductor transposition points.
Railway feeders:		
0 to 750 volts, Number 4/0 or larger	6	
0 to 750 volts, smaller than Number 4/0	12	Where 10- to 12-inch separation has already been established by practice, it may be continued for conductors having apparent sags not over 3 feet and for voltages not exceeding 8700.
750 volts to 8700 volts	12	
Other supply conductors:		
0 to 8700 volts	12	
For all conductors of more than 8700 volts add for each 1000 volts in excess of 8700 volts	0.4	

Separation in inches required for line conductors smaller than Number 2 American Wire Gauge

Voltages between conductors	Sag (in inches)						
	36	48	72	96	120	180	240
2 400	14.5	20.5	28.5	35.0	40.5	51.5	60.0
7 200	16.0	22.0	30.0	36.5	42.0	52.5	61.5
13 200	18.0	24.0	32.0	38.5	43.5	54.5	63.5
23 000	21.0	27.0	35.0	41.5	46.5	57.5	66.5
34 500	24.5	30.5	38.5	44.5	50.5	61.0	70.0
46 000	28.0	34.0	42.0	48.0	53.5	64.5	73.0
69 000	–	40.5	48.5	55.0	60.5	71.0	80.0

Separation in inches required for line conductors Number 2 American Wire Gauge or larger

Voltages between conductors	Sag (in inches)						
	36	48	72	96	120	180	240
2 400	14.5	16.5	20.5	23.5	26.0	31.5	36.5
7 200	16.0	18.0	22.0	25.0	27.5	33.0	38.0
13 200	18.0	20.0	23.5	26.5	29.5	35.0	39.5
23 000	21.0	23.0	26.5	29.5	32.0	38.0	42.5
34 500	24.0	26.5	30.0	33.0	35.5	41.5	46.0
46 000	27.5	30.0	33.5	36.5	39.0	45.0	49.5
69 000	–	36.5	40.5	43.5	46.0	51.5	56.5

conductor-cooled (rotating machinery). A term referring to windings in which coolant flows in close contact with the conductors so that the heat generated within the principal portion of the windings reaches the cooling medium without flowing through the major ground insulation. *See also:* **cradle base (rotating machinery).** 50A10-31E8

conductor, covered. A conductor having one or more layers of nonconducting materials that are not recognized as insulation under the electric code. 1A0-0

conductor insulation (rotating machinery). The insulation surrounding a conductor which may consist of one or more uninsulated laminations or strands. *See also:* **rotor (rotating machinery); stator; strand insulation.** 0-31E8

conductor loading (mechanical). The combined load per unit length of a conductor due to the weight of the wire plus the wind and ice loads. *See also:* **tower.** 42A35-31E13

conductor shielding (power distribution underground cables). A conducting or semiconducting element in direct contact with the conductor and in intimate contact with the inner surface of the insulation so that the potential of this element is the same as the conductor. Its function is to eliminate ionizable voids at the conductor and provide uniform voltage stress at the inner surface of the insulating wall. *See also:* **power distribution, underground construction.** 0-31E1

conductor support box. A box that is inserted in a vertical run of raceway to give access to the conductors for the purpose of providing supports for them. *See also:* **cabinet.** 42A95-0

conduit (1) (electric power). A structure containing one or more ducts. *Note:* Conduit may be designated as iron pipe conduit, tile conduit, etcetera. If it contains one duct only it is called "single-duct conduit," if it contains more than one duct it is called "multiple-duct conduit," usually with the number of ducts as a prefix, namely, two-duct multiple conduit. *See also:* **cable.** 42A35/42A65-31E13

(2) (aircraft). An enclosure used for the radio shielding or the mechanical protection of electric wiring in an aircraft. *Note:* It may consist of either rigid or flexible, metallic or nonmetallic tubing. Conduit differs from pipe and metallic tubing in that it is not normally used to conduct liquids or gases. *See also:* **air transportation wiring and associated equipment; flexible metal conduit; rigid metal conduit.** 42A41-0

conduit fitting. An accessory that serves to complete a conduit system, such as bushings and access fittings. *See also:* **raceways.** 42A95/80A4-0

conduit run. *See:* **duct bank.**

cone (1) (cathode-ray tube). The divergent part of the envelope of the tube. *See also:* **cathode-ray tubes.** 50I07-15E6

(2) (vision). Retinal elements that are primarily concerned with the perception of detail and color by the light-adapted eye. *See also:* **retina.** Z7A1-0

cone of ambiguity (electronic navigation). A generally conical volume of airspace above a navigation aid within which navigational information from that facility is unreliable. *See also:* **navigation.** 0-10E6

cone of nulls. A conical surface formed by directions of negligible radiation. *See also:* **antenna.** 42A65-3E1

cone of protection (lightning). The space enclosed by a cone formed with its apex at the highest point of a lightning rod or protecting tower, the diameter of the base of the cone having a definite relation to the height of the rod or tower. *Note:* This relation depends on the height of the rod and the height of the cloud above the earth. The higher the cloud, the larger the radius of the base of the protecting cone. The ratio of radius of base to height varies approximately from one to two. When overhead ground wires are used, the space protected is called a zone of protection or protected zone. *See also:* **lightning protection and equipment.** 5A1-31E13;42A95-0

cone of silence (electronic navigation). A conically shaped region above an antenna where the field strength is relatively weak because of the configuration of the antenna system. *See also:* **navigation.** E172-10E6

conference connection. A special connection for a telephone conversation among more than two stations. *See also:* **telephone switching system.** 42A65-0

configuration (group of electric conductors). Their geometrical arrangement, including the size of the wires and their relative positions with respect to other conductors and the earth. *See also:* **inductive coordination.** 42A65-0

conflict, antenna. *See:* **antenna conflict.**

conformance tests (acceptance tests*) (lightning arrester). (1) Tests made, when required in addition to design and routine tests, to demonstrate certain performance characteristics of a product or representative samples thereof. *See:* **acceptance test; arresters.**
*Deprecated E28-0;62A1-31E7

(2) Tests that are specifically made to demonstrate conformity with applicable standards. 37A100-31E11

conformity (potentiometer) The accuracy of its output; used especially in reference to a function potentiometer. *See also:* **electronic analog computer.** E165-16E9

conical horn. A horn whose cross-sectional area increases as the square of the axial length. *See also:* **loudspeaker.** E157-1E1;42A65-0

conical scanning (antenna) (electronic navigation). A form of sequential lobing in which the direction of maximum radiation generates a cone where the vertex angle is of the order of the antenna half-power beamwidth. *Note:* Such scanning may be either rotating or nutating according to whether the direction of polarization rotates or remains unchanged. *See also:* **antenna.** E149-10E16;0-3E1

conical wave (radio wave propagation). A wave whose equiphase surfaces asymptotically form a family of coaxial circular cones. *See also:* **radio wave propagation.** 0-3E2

conjugate branches. Any two branches of a network such that a driving force impressed in either branch produces no response in the other. *See also:* **network analysis.** E270-0

conjugate bridge. The detector circuit and the supply circuit are interchanged as compared with a normal bridge of the given type. *See also:* **bridge.** 42A30-0

connected (network). A network is connected if there exists at least one path, composed of branches of the network, between every pair of nodes of the network.

See also: **network analysis.** E153-0

connected load. The sum of the continuous ratings of the load-consuming apparatus connected to the system or any part thereof. *See also:* **generating station.** 42A35-31E13

connected position (switchgear-assembly removable element). That position of the removable element in which both primary and secondary disconnecting devices are in full contact. 37A100-31E11

connecting wire (mining). A wire generally of smaller gauge than the shot-firing cord and used for connecting the electric blasting-cap wires from one drill hole to those of an adjoining one in mines, quarries, and tunnels. *See also:* **mine feeder circuit.** 42A85-0

connection (rotating machinery). Any low-impedance tie between electrically conducting components. 0-31E8

connection box (mine type). A piece of apparatus with enclosure within which electric connections between sections of cable can be made. *See also:* **mine feeder circuit.** 42A85-0

connection diagram. A diagram that shows the connection of an installation or its component devices, controllers, and equipment. *Notes:* (1) It may cover internal or external connections, or both, and shall contain such detail as is needed to make or trace connections that are involved. It usually shows the general physical arrangement of devices and device elements and also accessory items such as terminal blocks, resistors, etcetera. (2) A connection diagram excludes mechanical drawings, commonly referred to as wiring templates, wiring assemblies, cable assemblies, etcetera. E270-0;IC1-34E10;89A1-0

connection insulation (joint insulation) (rotating machinery). The insulation at an electric connection such as between turns or coils or at a bushing connection. *See also:* **stator.** 0-31E8

connections of polyphase circuits (rotating machinery). *See:* **mesh-connected circuit; star-connected circuit; zig-zag connection of polyphase circuits.**

connector (1). A coupling device employed to connect conductors of one circuit or transmission element with those of another circuit or transmission element. *See also:* **auxiliary device to an instrument.** 0-9E4

(2) (wires). A device attached to two or more wires or cables for the purpose of connecting electric circuits without the use of permanent splices. E16-0

(3) (splicing sleeve). A metal sleeve, usually copper, that is slipped over and secured to the butted ends of the conductors in making up a joint. 42A35-31E13

(4) (waveguides). A mechanical device, excluding an adapter, for electrically joining separable parts of a waveguide or transmission-line system. 0-3E1/9E4

connector base (motor plug*) (motor attachment plug cap*). A device, intended for flush or surface mounting on an appliance, that serves to connect the appliance to a cord connector. *See also:* **interior wiring.**

*Deprecated 42A95-0

connector, precision (waveguide or transmission line). A connector that has the property of making connections with a high degree of repeatability without introducing significant reflections, loss, or leakage. *See also:* **auxiliary device to an instrument.** 0-9E4

connector switch (connector). A remotely controlled switch for connecting a trunk to the called line. *See also:* **telephone switching system.** 42A65-0

conservator (expansion tank) system (transformer or regulator). A method of oil preservation in which the oil in the main tank is sealed from the atmosphere, over the temperature range specified, by means of an auxiliary tank partly filled with oil and connected to the completely filled main tank. *See:* **transformer, oil-immersed.** 42A15-31E12;57A15-0

consol (electronic navigation). A keyed continuous-wave short-baseline radio navigation system operating in the low- and medium-frequency bands, generally useful to about 1500 nautical miles (2800 kilometers), and using three radiators to provide a multiplicity of overlapping lobes of dot-and-dash patterns that form equisignal hyperbolic lines of position. *Note:* These lines of position are moved slowly in azimuth by changing radio-frequency phase, thus allowing a simple listening and counting or timing operation to be used to determine a line of position within the sector bounded by any pair of equisignal lines. *See also:* **navigation.** 0-10E6

consolan (electronic navigation). A form of consol using two radiators instead of three. *See also:* **navigation.** 0-10E6

console (telephony) (1) (switchgear). A control cabinet located apart from the associated switching equipment arranged to control those functions for which an attendant or an operator is required. *See also:* **telephone switching system.** 0-19E1

(2) (computing system). The part of a computer used for communication between the operator or maintenance engineer and the computer. X3A12-16E9

consonant articulation (percent consonant articulation). The percent articulation obtained when the speech units considered are consonants (usually combined with vowels into meaningless syllables). *See also:* **volume equivalent.** 42A65-0

conspicuity. The capacity of a signal to stand out in relation to its background so as to be readily discovered by the eye. *See also:* **atmospheric transmissivity.** Z7A1-0

constancy (probe coupling). *See:* **residual probe pickup.**

constant. *See:* **time constant of integrator.**

constant-amplitude recording (mechanical recording). A characteristic wherein, for a fixed amplitude of a sinusoidal signal, the resulting recorded amplitude is independent of frequency. *See also:* **phonograph pickup.** 0-1E1

constant-available-power source (telephony). A signal source with a purely resistive internal impedance and a constant open-circuit terminal voltage, independent of frequency. *See also:* **telephone station.** E269-19E8

constant-current arc-welding power supply. A power supply that has characteristically drooping volt-ampere curves producing relatively constant current with a limited change in load voltage. *Note:* This type of supply is conventionally used in connection with manual-stick-electrode or tungsten-inert-gas arc welding. *See also:* **electric arc-welding apparatus.** 87A1-0

constant-current characteristic (electron tubes). The relation, usually represented by a graph, between the voltages of two electrodes, with the current to one of them as well as all other voltages maintained constant. *See also:* **circuit characteristics of electrodes.** 42A70-0;E160-15E6

constant-current charge (storage battery) (storage cell). A charge in which the current is maintained at a constant value. *Note:* For some types of lead-acid batteries this may involve two rates called the starting and finishing rates. *See also:* **charge.** 42A60-0

constant-current (Heising) modulation. A system of amplitude modulation wherein the output circuits of the signal amplifier and the carrier-wave generator or amplifier are directly and conductively coupled by means of a common inductor that has ideally infinite impedance to the signal frequencies and that therefore maintains the common plate-supply current of the two devices constant. *Note:* The signal-frequency voltage thus appearing across the common inductor appears also as modulation of the plate supply to the carrier generator or amplifier with corresponding modulation of the carrier output. *See also:* **modulating systems.** E145-0

constant-current power supply. A power supply that is capable of maintaining a preset current through a variable load resistance. *Note:* This is achieved by automatically varying the load voltage in order to maintain the ratio V_{load}/R_{load} constant. *See also:* **power supply.** KPSH-10E1

constant-current regulation (generator). That type of automatic regulation in which the regulator maintains a constant-current output from the generator. *See also:* **axle generator system.** 42A42-0

constant-current (series) incandescent filament lamp transformer. *See:* **incandescent filament lamp transformer (series type).**

constant-current street-lighting system (series street-lighting system). A street-lighting system employing a series circuit in which the current is maintained substantially constant. *Note:* Special generators or rectifiers are used for direct current while suitable regulators or transformers are used for alternating current. *See also:* **alternating-current distribution; direct-current distribution.** 42A35-31E13

constant-current (series) mercury-lamp transformer. A transformer that receives power from a current-regulated series circuit and transforms the power to another circuit at the same or different current from that in the primary circuit. *Note:* It also provides the required starting and operating voltage and current for the specified lamp. Further, it provides protection to the secondary circuit, casing, lamp, and associated luminaire from the high voltage of the primary circuit. 82A7-0

constant-current transformer. A transformer that automatically maintains an approximately constant current in its secondary circuit under varying conditions of load impedance when supplied from an approximately constant-potential source.
See:
current regulation;
dry-type;
impedance voltage;
indoor;
insulation, class ratings;
oil-immersed;
pole-type;
rated kilowatts;
rated primary voltage;
rated secondary current;
relative lead polarity;
station-type;
submersible;
subway-type transformer;
transformer;
vault-type. 42A15-0

constant cutting speed (numerically controlled machine). The condition achieved by varying the speed of rotation of the workpiece relative to the tool inversely proportional to the distance of the tool from the center of rotation. *See also:* **numerically controlled machines.** EIA3B-34E12

constant-delay discriminator (electronic navigation). *See:* **pulse decoder.**

constant-failure period (reliability). The period during which failures of some items occur at an approximately uniform rate. *Note:* The curve below shows the failure pattern when this definition applies to an item. *See also:* **reliability.** 0-7E1

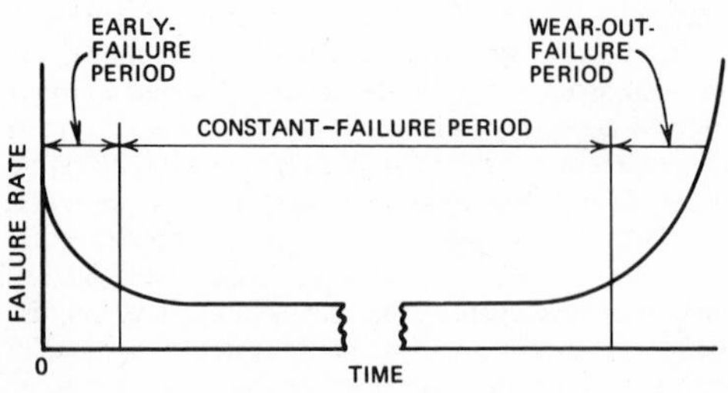

Constant-failure period.

constant-frequency control (power system). A mode of operation under load-frequency control in which the area control error is the frequency deviation. *See also:* **speed-governing system.** E94-0

constant-horsepower motor. *See:* **constant-power motor.**

constant-horsepower range (electric drive). The portion of its speed range within which the drive is capable of maintaining essentially constant horsepower. *See also:* **electric drive.** 42A25-34E10

constant-*K* network. A ladder network for which the product of series and shunt impedances is independent of frequency within the range of interest. *See also:* **network analysis.** E270/42A65-0

constant-luminance transmission (color television). The type of transmission in which the transmission primaries are a luminance primary and two chrominance primaries. *See also:* **color terms.** E201-2E2

constant-net-interchange control (power system). A mode of operation under load-frequency control in which the area control error is determined by the net interchange deviation. *See also:* **speed-governing system.** E94-0

constant-power motor (constant-horsepower motor). A multispeed motor that develops the same rated power output at all operating speeds. The torque then is inversely proportional to the speed. *See also:* **asynchronous machine.** 0-31E8

constant-resistance (conductance) network. A network having at least one driving-point impedance (admittance) that is a positive constant. *See also:* **linear passive networks.** E156-0

constant-speed motor. A motor in which the speed is constant or substantially constant over the normal range of loads. *Note:* For example, a synchronous motor, an induction motor with small slip, or a direct-current shunt-wound motor with constant excitation. *See also:* **asynchronous machine; direct-current commutating machine; synchronous machine.** 42A10-31E8

constant-torque motor. Multispeed motor that is capable of developing the same torque for all design speeds. The rated power output varies directly with the speed. *See also:* **asynchronous machine.** 0-31E8

constant-torque range (electric drive). The portion of its speed range within which the drive is capable of maintaining essentially constant torque. *See also:* **electric drive.** 42A25-34E10

constant-torque resistor (motors). A resistor for use in the armature or rotor circuit of a motor in which the current remains practically constant throughout the entire speed range. *See also:* **asynchronous machine; direct-current commutating machine; electric drive; synchronous machine.** E45-0;IC1-34E10

constant-torque speed range (industrial control). The portion of the speed range of a drive within which the drive is capable of maintaining essentially constant torque. AS1-34E10

constant-velocity recording (mechanical recording). A characteristic wherein for a fixed amplitude of a sinusoidal signal, the resulting recorded amplitude is inversely proportional to the frequency. *See also:* **phonograph pickup.** 0-1E1

constant-voltage arc-welding power supply. Power supply (arc welder) that has characteristically flat volt-ampere curves producing relatively constant voltage with a change in load current. This type of power supply is conventionally used in connection with welding processes involving consumable electrodes fed at a constant rate. 87A1-0

constant-voltage charge (storage battery) (storage cell) A charge in which the voltage at the terminals of the battery is held at a constant value. *See also:* **charge.** 42A60-0

constant-voltage power supply (voltage regulator). A power supply that is capable of maintaining a preset voltage across a variable load-resistance. This is achieved by automatically varying the output current in order to maintain the product of load current times load resistance constant. *See also:* **power supply.** KPSH-10E1

constant-voltage regulation (generator). That type of automatic regulation in which the regulator maintains constant voltage of the generator. *See also:* **axle generator system.** 42A42-0

constant-voltage transformer. A transformer that maintains an approximately constant voltage ratio over the range from zero to rated output. *See:* **transformer.** 42A15-0

constraints (1). Limits on the ranges of variables or system parameters because of physical or system requirements. *See also:* **system.** 0-35E2
(2) (control system). A restriction placed on the control signal, control law, or state variables. *See also:* **control system.** 0-23E0

construction diagram (industrial control). A diagram that shows the physical arrangement of parts, such as wiring, buses, resistor units, etc. *Example:* A diagram showing the arrangement of grids and terminals in a grid-type resistor. E270-34E10

contact. A conducting part that co-acts with another conducting part to make or break a circuit. 37A100-31E11

contact area (photoelectric converter). The area of ohmic contact provided on either the *p* or *n* faces of a photoelectric converter for electrical circuit connections. *See also:* **semiconductor.** 0-10E1

contact chatter, relay. *See:* **relay contact chatter.**

contact clip (mechanical switching device). The clip that the switchblade enters or embraces. 37A100-31E11

contact conductor (1) (electric traction). The part of the distribution system other than the track rails, that is in immediate electric contact with current collectors of the cars or locomotives. *See:* **contact wire (trolley wire); trolley; underground collector or plow.** *See also:* **multiple-unit control.** E16-0
(2) (contact electrode) (electrochemistry). A device to lead electric current into or out of a molten or solid metal or alloy that itself serves as the active electrode in the cell. *See also:* **fused electrolyte.** 42A60-0

contact corrosion. *See:* **crevice corrosion.**

contact current-carrying rating (relay). The current that can be carried continuously, or for stated periodic intervals, without impairment of the contact structure or interrupting capability. *See also:* **relay.** 42A20-31E6

contact current-closing rating (relay). The current that the device can close successfully with prescribed operating duty and circuit conditions without significant impairment of the contact structure. *See also:* **relay.** 42A20-31E6

contact follow-up (relays, switchgear, and industrial control). The distance between the position one contact face would assume, were it not blocked by the second (mating) contact, and the position the second contact face would assume were the first contact removed, when the actuating member is fixed in its final contact-closed position. *See:* **electric controller.** *See also:* **initial contact pressure.** E74-0;IC1-34E10

contact gap (break) (industrial control). The final length of the isolating distance of a contact in the open position. *See:* **contactor.** 50I16-34E10

contact, high recombination rate (semiconductor). A semiconductor-semiconductor or metal-semiconductor contact at which thermal equilibrium charge-carrier concentrations are maintained substantially independent of current density. *See also:* **semiconductor; semiconductor device.** E270/E102;E216-34E17

contact interrupting rating (relay). The current that the device can interrupt successfully, with prescribed operating duty and circuit conditions without significant impairment of the contact structure. *See also:* **relay.** 42A20-31E6

contactless vibrating bell. A vibrating bell whose continuous operation depends upon application of alternating-current power without circuit-interrupting contacts. *See also:* **protective signaling.** 42A65-0

contact, majority carrier (semiconductor). An electric contact across which the ratio of majority-carrier current to applied voltage is substantially independent of the polarity of the voltage while the ratio of minority-carrier current to applied voltage is not independent of the polarity of the voltage. *See also:* **semiconductor.** E102-0

contact-making clock (demand meter). A device designed to close momentarily an electric circuit to a demand meter at definite and consecutive intervals. *See also:* **demand meter.** 12A0/42A30-0

contact mechanism (demand meter). A device for attachment to an electricity meter or to a demand-totalizing relay for the purpose of providing electric impulses for transmission to a demand meter or relay. *See also:* **demand meter.** 42A30-0

contact nomenclature. *See:* **relay.**

contact opening time (relay). The time a contact remains closed, while in process of opening, following a specified change of input. *Note:* This term is properly restricted to the inherent delay while the contact is actually being operated and is not to include delay that may occur before contact operation begins. *For example:* contact-opening time while opening results from the flexing and wiping together of contacts when they are closed. *See also:* **drop-out time.** 37A100-31E11-31E6

contactor (industrial control). A device for repeatedly establishing and interrupting an electric power circuit.
See:
arc chute;
auxiliary contactor;
available current of a circuit;
blowout coil;
break;
breaking current;
butt contacts;
contact gap;
contact pressure, final;
contact-wear allowance;
current-carrying capacity;
drop-out voltage;
electronic contactor;
electropneumatic contactor;
initial contact pressure;
magnetic blowout;
magnetic contactor;
making capacity;
normally open and normally closed;
pick-up voltage;
rolling contacts;
sealing voltage;
setting;
sliding contacts;
thermal protection;
thermal protector. 42A25-34E10

contactor, load. *See:* **load switch (load contactor).**

contactor or unit switch. A device operated other than by hand for repeatedly establishing and interrupting an electric power circuit under normal conditions. *See also:* **control switch.** E16-0

contact parting time (mechanical switching device). The interval between the time when the actuating quantity in the main circuit reaches the value causing actuation of the release and the instant when the primary arcing contacts have parted in all poles. *Note:* Contact parting time is the numerical sum of release delay and opening time. 37A100-31E11

contact piston (contact plunger) (waveguide). A piston with sliding metallic contact with the walls of a waveguide. *See:* **waveguide.** 0-3E1

contact plating. The deposition, without the application of an external electromotive force, of a metal coating upon a base metal, by immersing the latter in contact with another metal in a solution containing a compound of the metal to be deposited. *See also:* **electroplating.** 42A60-0

contact plunger. *See:* **contact piston.**

contact potential. The difference in potential existing at the contact of two media or phases. *See also:* **biological contact potential; electrolytic cell.** *See:* **depolarization (biological); depolarization front; injury potential (electrobiology); negative after-potential (electrobiology); positive after-potential (electrobiology).** 42A60-0

contact-potential difference. The difference between the work functions of two materials divided by the electronic charge. E160-15E6

contact pressure, final (industrial control). The force exerted by one contact against the mating contact when the actuating member is in the final contact-closed position. *Note:* Final contact pressure is usually measured and expressed in terms of the force that must be exerted on the yielding contact while the actuating member is held in the final contact-closed position, and with the mating contact fixed in position, in order to separate the mating contact surfaces. *See:* **contactor.** IC1-34E10

contact pressure, initial (industrial control). The force exerted by one contact against the mating contact when the actuating member is in the initial contact-touch position. *Note:* The initial contact pressure is usually measured and expressed in terms of the force that must be exerted on the yielding contact while the actuating member is held in the initial contact-touch position in order to separate the mating contact surface against the action of the spring or other contact pressure device. *See:* **electric controller.** IC1-34E10

contact rectifier. A rectifier consisting of two different solids in contact, in which rectification is due to greater conductivity across the contact in one direction than in the other. *See also:* **rectifier.** E145/42A65-0

contacts (1). Conducting parts which co-act to complete or to interrupt a circuit. 42A25-0
(2) (nonoverlapping) (industrial control). Combinations of two sets of contacts, actuated by a common means, each set closing in one of two positions, and so arranged that the contacts of one set open before the contacts of the other set close. *See:* **electric controller.** IC1-34E10
(3) (auxiliary) (industrial control) (switching device). Contacts in addition to the main circuit contacts that function with the movement of the latter. *See also:* **contactor.** 42A25-34E10
(4) (overlapping, industrial control). Combinations of two sets of contacts, actuated by a common means, each set closing in one of two positions, and so arranged that the contacts of one set open after the contacts of the other set have been closed. *See:* **electric controller.** IC1-34E10

contact surface. That surface of a contact through which current is transferred to the co-acting contact. 37A100-31E11

contact voltage (human safety). A voltage accidentally appearing between two points with which a person can simultaneously make contact. *See:* **lightning arrester (surge diverter).** 50I25-31E7

contact-wear allowance (industrial control). The total thickness of material that may be worn away before the co-acting contacts cease to perform adequately. *See also:* **contactor.** 37A100-31E11;42A25-34E10

contact wire (trolley wire). A flexible conductor, customarily supported above or to one side of the vehicle. *See also:* **contact conductor.** E16-0

content addressed storage (computing system). *See:* **associative storage.** *See also:* **electronic digital computer.** X3A12-16E9

content, average information. *See:* **average information content.** *See also:* **information theory.**

content, conditional information. *See:* **conditional information content.** *See also:* **information theory.**

continuing current (lightning). The low-magnitude current that may continue to flow between components of a multiple stroke. *See also:* **direct-stroke protection (lightning).** 0-31E13

continuity cable bond. A cable bond used for bonding of cable sheaths and armor across joints between continuous lengths of cable. *See* **cable bond; cross cable bond.** *See also:* **power distribution, underground construction.** 42A35-31E13

continuous-current tests. Tests made at a rated current, until temperature rise ceases, to determine that the device or equipment can carry its rated continuous current without exceeding its allowable temperature rise. 37A100-31E11

continuous data. Data of which the information content can be ascertained continuously in time. EIA3B-34E12

continuous duty (rating of electric equipment). A duty that demands operation at a substantially constant load for an indefinitely long time. *See also:* **asynchronous machine; direct-current commutating machine; synchronous machine; voltage regulator.** E45/E270/57A15-0;E9 6-SSC4;42A15-31E12;IC1-34E10

continuous-duty rating. The rating applying to operation for an indefinitely long time. E145-0

continuous electrode. A furnace electrode that receives successive additions in length at the end remote from the active zone of the furnace to compensate for the length consumed therein. *See also:* **electrothermics.** 42A60-0

continuous inductive train control. *See:* **continuous train control.**

continuous lighting (railway practice). An arrangement of circuits so that the signal lights are continuously energized. *See also:* **railway signal and interlocking.** 42A42-0

continuous load. A load where the maximum current is expected to continue for three hours or more. 1A0-0

continuous load rating (power inverter unit). Defines the maximum load that can be carried continuously without exceeding established limitations under prescribed conditions of test, and within the limitations of established standards. *See also:* **self-commutated inverters.** 0-34E24

continuously adjustable capacitor (continuously variable capacitor*). An adjustable capacitor in which the capacitance can have every possible value within its range.

*Deprecated E270-0

continuously adjustable inductor (continuously variable inductor*) (variable inductor*). An adjustable inductor in which the inductance can have every possible value within its range.

*Deprecated E270-0

continuously adjustable resistor (continuously variable resistor*). An adjustable resistor in which the resistance can have every possible value within its range.

*Deprecated E270-0

continuous noise (electromagnetic compatibility). Noise, the effect of which is not resolvable into a succession of discrete impulses. *See also:* **electromagnetic compatibility.** 0-27E1

continuous periodic rating (industrial control). The load that can be carried for the alternate periods of load and rest specified in the rating and repeated continuously without exceeding the specified limitation. 42A25-34E10

continuous-pressure operation (elevators). Operation by means of buttons or switches in the car and at the landings, any one of which may be used to control the movement of the car as long as the button or switch is manually maintained in the actuating position. *See:* **control.** 42A45-0

continuous rating (electric equipment). The maximum constant load that can be carried continuously without exceeding established temperature-rise limitations under prescribed conditions of test and within the limitations of established standards. *See also:* **duty; rectification.** 42A15-31E12/34E24;57A14/57A15/57A18-0

continuous-scan system, supervisory control. *See:* **supervisory control system, continuous-scan.**

continuous-speed adjustment (industrial control). Refers to an adjustable-speed drive capable of being adjusted with small increments, or continuously, between minimum and maximum speed. *See also:* **electric drive.** 42A25-34E10

continuous test (battery). A service test in which the battery is subjected to an uninterrupted discharge until the cutoff voltage is reached. *See also:* **battery (primary or secondary); cutoff voltage.** 42A60-0

continuous-thermal-current rating factor (current transformer). The factor by which the rated primary current is multiplied to obtain the maximum allowable primary current based on the limiting temperature rise on a continuous basis. *Note:* The rated primary current is established in many instances as a basis for accuracy classification and to place the ratio on the basis of a 5-ampere secondary. In many designs the rated primary current will produce the limiting temperature rise; in such cases, the factor is unity. Some transformers have an additional continuous thermal-current capacity in excess of the rated primary current; in such cases the factor will be greater than unity. *See:* **instrument transformer.** 12A0/42A15/42A30-0

continuous train control (continuous inductive train control). A type of train control in which the locomotive apparatus is constantly in operative relation with the track circuit and is immediately responsive to a change in the character of the current flowing in the track circuit of the track on which the locomotive is traveling. *See also:* **automatic train control.** 42A42-0

continuous-type control (electric power systems). A control mode that provides a continuous relation between the deviation of the controlled variable and the position of the final controlling element. *See also:* **speed-governing system.** E94-0

continuous-voltage-rise test (rotating machinery). A controlled overvoltage test in which voltage is increased in continuous function of time, linear or otherwise. *See also:* **asynchronous machine; direct-current commutating machines; synchronous machine.** 0-31E8

continuous waves (CW). Waves, the successive oscillations of which are identical under steady-state conditions. *See also:* **radio transmission; wave front.** E145-0/42A65-31E3

contouring control system (numerically controlled machines). A system in which the controlled path can result from the coordinated, simultaneous motion of two or more axes. *See also:* **numerically controlled machines.** EIA3B-34E12

contract demand (electric-power utilization). The demand that the supplier of electric service agrees to have available for delivery. *See also:* **alternating-current distribution.** 0-31E4

contrast (display presentation). The subjective assessment of the difference in appearance of two parts of a field of view seen simultaneously or successively. (Hence: luminosity contrast, lightness contrast, color contrast, simultaneous contrast, successive contrast). *See also:* **photometry; television.** 50I45-2E2

contrast control. A control, associated with a picture-display device, for adjusting the contrast ratio of the reproduced picture. *Note:* The contrast control is normally an amplitude control for the picture signal. In a monochrome-television system, both average luminance and the contrast ratio are affected. In a color-television system, saturation and hue also may be affected. *See also:* **television.** 0-42A65

contrast ratio (television). The ratio of the maximum to the minimum luminance values in a television picture or a portion thereof. *Note:* Generally the entire area of the picture is implied but smaller areas may be specified as in detail contrast. *See also:* **television.** E204-2E2;42A65-0

contrast sensitivity. The ability to detect the presence of luminance (photometric brightness) differences. *Note:* Quantitatively, it is equal to the reciprocal of the contrast threshold. *See also:* **visual field.** Z7A1-0

contrast threshold. (1) The minimal perceptible contrast for a given state of adaption of the eye. (2) The luminance contrast that can be detected during some specific fraction of the times it is presented to an observer, usually 50 percent. *See also:* **visual field.** Z7A1-0

control (1) (industrial). (A) Broadly, the methods and means of governing the performance of any electric apparatus, machine, or system. 42A25-34E10
(B) The system governing the starting, stopping, direction of motion, acceleration, speed, and retardation of the moving member. 42A45-0
(C) A designation of how the equipment is governed, that is, by an attendant, by automatic means, or partially by automatic means and partially by an attendant. *Note:* The word **control** is often used in a broad sense to include **indication** also. 37A100-31E11
(D) Frequently, one or more of the components in any mechanism responsible for interpreting and carrying out manually-initiated directions. *Note:* Sometimes called manual control.
(E) (verb). To execute the function of control as defined above. 42A25-34E10
(2) (electronic computation). (A) Usually, those parts of a digital computer that effect the carrying out of instructions in proper sequence, the interpretation of each instruction, and the application of the proper signals to the arithmetic unit and other parts in accordance with this interpretation. (B) In some business applications of mathematics, a mathematical check. *See:* **computer control state; electronic digital computer.** *Note:* For an extensive list of cross references, see *Appendix A.* E162/E270-0
(3) (cryotron). An input element of a cryotron. *See also:* **super conductivity.** E217-15E7

control accuracy (industrial control). The degree of correspondence between the final value and the ideal value of the directly controlled variable. *See:* **control system, feedback.** AS1-34E10

control action, derivative (industrial control). The component of control action for which the output is proportional to the rate of change of the input. *See:* **control system, feedback.** AS1-34E10

control action, integral (industrial control). Control action in which the output is proportional to the time integral of the input. *See also:* **control system, feedback.** AS1-34E10

control action, proportional. Control action in which there is a continuous linear relation between the output and the input. *Note:* This condition applies when both the output and input are within their normal operating ranges. AS1-34E10

control and indication point, supervisory control. *See:* **supervisory control point, control and indication.**

control apparatus. A set of control devices used to accomplish the intended control functions. *See:* **control.** 42A25-34E10

control area (electric power). A part of a power system or a combination of systems to which a common generation control scheme is applied. *See also:* **power system, low-frequency and surge testing.** E94-0

control battery (industrial control). A battery used as a source of energy for the control of an electrically operated device. 42A25-34E10

control bus. A bus used to distribute power for operating electrically controlled devices. 37A100-31E11

control character. A character whose occurrence in a particular context initiates, modifies, or stops a control operation, for example, a character to control carriage return. X3A12-16E9

control characteristic (gas tube). A relation, usually shown by a graph, between critical grid voltage and anode voltage. *See:* **gas tube.** E160-15E6;42A70-0

control circuit (industrial control). The circuit that carries the electric signals directing the performance of the controller but does not carry the main power circuit. *See:* **control.** E45/42A25-34E10

control-circuit limit switch. A limit switch the contacts of which are connected only into the control circuit. *See:* **control; switch.** 42A25-34E10

control-circuit transformer. A voltage transformer utilized to supply a voltage suitable for the operation of control devices. *See:* **control.** 42A25-34E10

control cutout switch (motors). An isolating switch that isolates the control circuit of a motor controller from the source of energy. *See:* **control.** *See also:* **control switch.** E16-0

control desk. An assembly of master switches, indicators, instruments, and the like on a structure in the form of a desk for convenience in manipulation and control supervision by the operator. *See:* **control.** 42A25-34E10

control device. An individual device used to execute a control function. *See:* **control.** 42A25-34E10

control electrode (electron tubes). An electrode used to initiate or vary the current between two or more electrodes. *See also:* **circuit characteristics of electrodes; electrode (electron tube).** E160-15E6

control-electrode discharge recovery time (attenuator tubes). The time required for the control-electrode discharge to deionize to a level such that a specified fraction of the critical high-power level is required to ionize the tube. *See:* **gas tube.** *See also:* **circuit characteristics of electrodes.** E160-15E6

control exciter (rotating machinery). An exciter that acts as a rotary amplifier in a closed-loop circuit. *See also:* **asynchronous machines; synchronous machines.** 0-31E8

control generator. A generator, commonly used on electric motive power units for the generation of electric energy in proportion to vehicle speed, prime mover speed, or some similar function, thereby serving as a guide for initiating appropriate control functions. *See also:* **traction motor.** 42A42-0

control grid (electron tube). A grid, ordinarily placed between the cathode and an anode, for use as a control electrode. *See also:* **electrode (electron tube); grid.** E160-15E6;42A70-0

control initiation. The function introduced into a measurement sequence for the purpose of regulating any subsequent control operations in relation to the quantity measured. *Note:* The system element comprising the control initiator is usually included in the end device but may be associated with the primary detector or the intermediate means. *See also:* **measurement system.** 42A30-0

controllable. A property of a component of a state whereby, given an initial value of the component at a given time, there exists a control input that can change this value to any other value at a later time. *See also:* **control system.** 0-23E0

controllable, completely. The property of a plant whereby all components of the state are controllable within a given time interval. *See also:* **control system.** 0-23E0

control law. A function of the state of a plant and possibly of time, generated by a controller to be applied as the control input to a plant. *See also:* **control system.** 0-23E0

control law, closed-loop. A control law specified in terms of some function of the observed state. *See also:* **control system.** 0-23E0

control law, open-loop. A control law specified in terms of the initial state only and possibly of time. *See also:* **control system.** 0-23E0

controlled carrier (floating carrier) (variable carrier). A system of compound modulation wherein the carrier is amplitude modulated by the signal frequencies in any conventional manner, and, in addition, the carrier is simultaneously amplitude modulated in accordance with the envelope of the signal so that the percentage of modulation, or modulation factor, remains approximately constant regardless of the amplitude of the signal. *See also:* **modulating systems.** E145/E182A/42A65-0

controlled manual block signal system. A series of consecutive blocks governed by block signals, controlled by continuous track circuits, operated manually upon information by telegraph, telephone, or other means of communication, and so constructed as to require the cooperation of the signalmen at both ends of the block to display a clear or permissive block signal. *See also:* **block signal system.** 42A42-0

controlled overvoltage test (1) (rotating machinery). A test of an insulation system, in which a voltage exceeding the normal operating value is applied and increased by manual or automatic control, according to a prearranged time schedule or function, under designated conditions of temperature, humidity, and frequency. *Note:* Usually a direct-current test. *See also:* **asynchronous machine; direct-current commutating machine; synchronous machine.** 0-31E8

(2) (direct-current leakage, voltage test). A test in which the increase of applied direct voltage is controlled, and measured currents continuously observed for abnormalities, with the intention of stopping the test before breakdown occurs. *See also:* **insulation testing (large alternating-current rotating machinery).** E95-0

controlled rectifier. A rectifier in which means for controlling the current flow through the rectifying devices is provided. *See also:* **electronic controller; rectification.** 42A15-0

controlled-speed axle generator. An axle generator in which the speed of the generator is maintained approximately constant at all vehicle speeds above a predetermined minimum. *See also:* **axle generator system.** 42A42-0

controller. The part of a control system that implements the control law. *See also:* **control system; electric controller.** 0-23E0

controller, automatic (process control). A device that operates automatically to regulate a controlled variable in response to a command and a feedback signal. *Note:* The term originated in process control usage. Feedback elements and final control elements may also be part of the device. *See also:* **control system, feedback.** 85A1-23E0

controller diagram (electric-power devices). A diagram that shows the electric connections between the parts comprising the controller and that shows the external connections. E270-34E10

controllers for steel-mill accessory machines. Controllers for machines that are not used directly in the processing of steel, such as pumps, machine tools, etcetera. *See also:* **electric controller.** IC1-34E10

controllers for steel-mill auxiliaries. Controllers for machines that are used directly in the processing of steel, such as screwdowns and manipulators but not cranes and main rolling drives. *See also:* **electric controller.** IC1-34E10

controlling elements (control system, feedback). The functional components of a controlling system. *See* *also:* **control system, feedback.** 85A1-23E0

controlling elements, forward (control system, feedback). The elements in the controlling system that change a variable in response to the actuating signal. *See also:* **control system, feedback.** 85A1-23E0

controlling means (of an automatic control system). Consists of those elements that are involved in producing a corrective action. E94-0

controlling section. A length of track consisting of one or more track circuit sections, by means of which the roadway elements or the device that governs approach to or movement within a block are controlled. 42A42-0

controlling system (1) (automatic control system without feedback). That portion of the control system that manipulates the controlled system. 85A1-23E0

(2) (control system, feedback). The portion that compares functions of a directly controlled variable and a command and adjusts a manipulated variable as a function of the difference. *Note:* It includes the reference input elements; summing point; forward and final controlling elements; and feedback elements. *See also:* **control system, feedback.** 85A1-23E0

controlling voltage, composite. *See:* **composite controlling voltage.**

control machine (railroad practice). (1) An assemblage of manually operated levers or other devices for the control of signals, switches, or other units, without mechanical interlocking, usually including a track diagram with indication lights. (2) A group of levers or equivalent devices used to operate the various mechanisms and signals that constitute the car retarder installation. *See also:* **car retarder; centralized traffic control system.** 42A42-0

control metering point (tie line) (electric power systems). The location of the metering equipment that is used to measure power on the tie line for the purpose of control. *See also:* **center of distribution; power system.** E94-31E4

control module (rotating machinery). Control-circuit subassembly. *See also:* **starting switch assembly.** 0-31E8

control panel. A part of a computer console that contains manual controls. *See:* **plugboard.** *See also:* **electronic analog computer; electronic digital computer.** X3A12-16E9

control position electric indicator. A device that provides an indication of the movement and position of the various control surfaces or structural parts of an aircraft. It may be used for wing flaps, cowl flaps, trim tabs, oil-cooler shutters, landing gears, etcetera. *See also:* **air transportation instruments.** 42A41-0

control positioning accuracy, precision, or reproducibility (numerically controlled machines). Accuracy, precision, or reproducibility of position sensor or transducer and interpreting system and including the machine positioning servo. *Note:* May be the same as machine positioning accuracy, precision, or reproducibility in some systems. *See also:* **numerically controlled machines.** EIA3B-34E12

control problem. Given the plant dynamics, problem constraints, and specifications (which may be represented by the performance index), determine the control law that satisfies the given specifications. *See also:* **control system.** 0-23E0

control ratio (1) (gas tube). The ratio of the change in anode voltage to the corresponding change in critical grid voltage, with all other operating conditions maintained constant. *See:* **gas tubes.** E160-15E6;42A70-0

(2) (power supplies). The required change in control resistance to produce a one-volt change in the output voltage. The control ratio is expressed in ohms per volt and is reciprocal of the bridge current. *See also:* **power supply.** KPSH-10E1

control relay. An auxiliary relay whose function is to initiate or permit the next desired operation in a control sequence. 37A100-31E11-31E6

control ring. *See:* **grading ring.**

control sequence table (electric-power devices). A tabulation of the connections that are made for each successive position of the controller. E270-34E10

control station (mobile communication). A base station, the transmission of which is used to control automatically the emission or operation of another radio station. *See also:* **mobile communication system.** 0-6E1

control switch. A manually operated switching device for controlling power-operated devices. *Notes:* (1) It may include signaling, interlocking, etcetera, as dependent functions. (2) Control switches derive their names from the function of the apparatus that they control, such as master control switches, reset switches, etcetera.
See:
cam contactor or cam switch;
contactor or unit switch;
control cutout switch;
drop-out voltage;
drum switch;
motor cutout switch;
multiple-unit control;
pneumatic switch;
sealing voltage. E16-0;37A100-31E11

control switchboard. A type of switchboard including control, instrumentation, metering, protective (relays), or regulating equipment for remotely controlling other equipment. *Note:* Control switchboards do not include the primary power circuit-switching devices or their connections. 37A100-31E11

control switch point (telephone networks). A switching point (arranged for routing and control) in the direct distance network at which intertoll trunks are connected to other intertoll trunks. *See:* **telephone system.** 0-19E1

control system (1) (broadly). An assemblage of control apparatus coordinated to execute a planned set of controls. *See:* **control.** 42A25-34E10

(2). A system in which a desired effect is achieved by operating on the various inputs to the system until the output, which is a measure of the desired effect, falls within an acceptable range of values. *Note:* For an extensive list of cross references, see *Appendix A*. *See:* **closed-loop control system (system, feedback control); control; network analysis; open-loop control system; transfer function.** E111/E264-0

control system, adaptive (industrial control). A control system within which automatic means are used to change the system parameters in a way intended to improve the performance of the control system. *See:* **control system, feedback.** AS1-34E10;85A1-23E0

control system, automatic. A control system that operates without human intervention. *See also:* **control system, feedback.** 85A1-23E0

control system, automatic feedback. A feedback control system that operates without human intervention. *See also:* **control system, feedback.** 85A1-23E0

control system, cascade. A control system in which the output of one subsystem is the input for another subsystem. *See also:* **control system, feedback.** 85A1-23E0

control system, coarse-fine (industrial control). A control system that uses some elements to reduce the difference between the directly controlled variable and its ideal value to a small value and that uses other elements to reduce the remaining difference to a smaller value. AS1-34E10;0-23E0

control system, dual-mode. A control system in which control alternates between two predetermined modes. *Note:* The condition for change from one mode to the other is often a function of the actuating signal. One use of dual-mode action is to provide rapid recovery from large deviations without incurring large overshoot. *See also:* **control system, feedback.** 85A1-23E0

control system, feedback. A control system that operates to achieve prescribed relationships between selected system variables by comparing functions of these variables and using the comparison to effect control. See the accompanying figures.

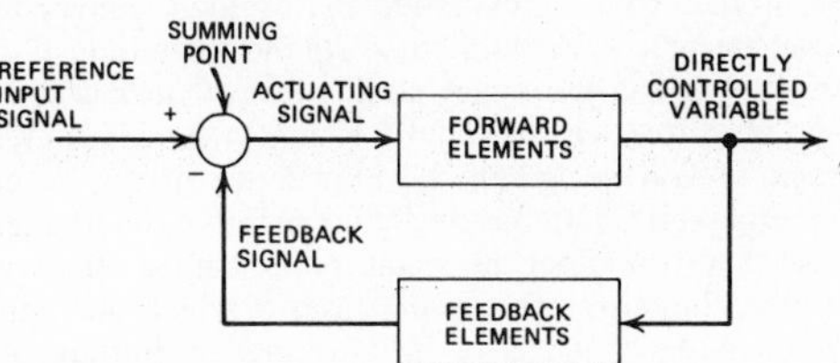

Simplified block diagram indicating essential elements of an automatic control system.

Note: For an extensive list of cross references, see *Appendix A.* 0-23E0

control system, on-off. A two-step control system in which a supply of energy to the controlled system is either on or off. *See also:* **control system, feedback.** 85A1-23E0

control system, sampling. Control using intermittently observed values of signals such as the feedback signal or the actuating signal. *Note:* The sampling is often done periodically. *See also:* **control system, feedback.** 85A1-23E0

control system, two-step. A control system in which the manipulated variable alternates between two predetermined values. *Note:* A control system in which the manipulated variable changes to the other predetermined value whenever the actuating signal passes through zero is called a two-step single-point control system. A two-step neutral-zone control system is one in which the manipulated variable changes to the other predetermined value when the actuating signal passes through a range of values known as the neutral zone. The neutral zone may be produced by a mechanical differential gap. The neutral zone is also called overlap, and two-step neutral-zone control overlap control. *See also:* **control system, feedback.** 85A1-23E0

control track (electroacoustics). A supplementary track usually placed on the same medium with the record carrying the program material. *Note:* Its purpose is to control, in some respect, the reproduction of the program, or some related phenomenon. Ordinarily, the control track contains one or more tones, each of which may be modulated either as to amplitude, frequency, or both. *See also:* **phonograph pickup.** 0-1E1

control terminal (base station) (mobile communication). Equipment for manually or automatically supervising a multiplicity of mobile and/or radio stations including means for calling or receiving calls from said stations. *See also:* **mobile communication system.** 0-6E1

control unit (1) (digital computer). The parts that effect the retrieval of instructions in proper sequence, the

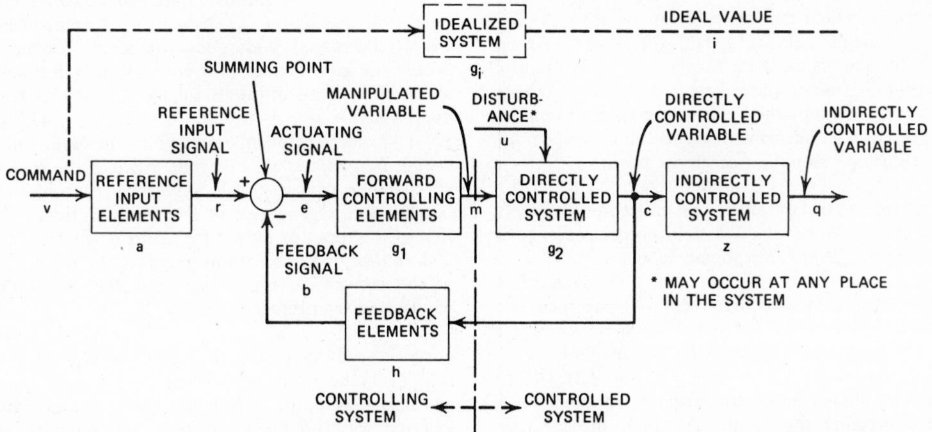

Block diagram of an automatic control system illustrating expansion of the simplified block diagram to a more complex system.

interpretation of each instruction, and the application of the proper signals to the arithmetic unit and other parts in accordance with this interpretation. *See also:* **electronic digital computer.** X3A12-16E9
(2) (mobile station) (mobile communication). Equipment including a microphone and/or handset and loudspeaker together with such other devices as may be necessary for controlling a mobile station. *See also:* **mobile communication system.** 0-6E1

control voltage (device). The voltage applied to the operating mechanism to actuate it. 37A100-31E11

control winding (1) (rotating machinery). An excitation winding that carries a current controlling the performance of a machine. *See also:* **asynchronous machines; synchronous machines.** 0-31E8
(2) (saturable reactor). A winding by means of which a controlling magnetomotive force is applied to the core. *See also:* **magnetic amplifier.** 42A65-0

convection current. In an electron stream, the time rate at which charge is transported through a given surface. *See:* **electron emission.** *See also:* **circuit characteristics of electrodes.** E160-15E6

convection-current modulation. The time variation in the magnitude of the convection current passing through a surface, or the process of directly producing such a variation. *See:* **electron emission.** *See also:* **circuit characteristics of electrodes.** E160-15E6

convection heater. A heater that dissipates its heat mainly by convection and conduction. *See also:* **appliances (including portable).** 42A95-0

convective discharge (effluve*) (electric wind) (static breeze) (medical electronics). The movement of a visible or invisible stream of particles carrying away charges from a body that has been charged to a sufficiently high voltage. *See also:* **medical electronics.**
*Deprecated 42A80-18E1

convenience outlet. *See:* **receptacle (electric distribution).**

conventionally cooled (rotating machinery). A term referring to windings in which the heat generated within the principal portion of the windings must flow through the major ground insulation before reaching the cooling medium. *See also:* **cradle base (rotating machinery).** 50A10-31E8

convergence (multibeam cathode-ray tubes). A condition in which the electron beams intersect at a specified point. *See:* **beam tubes.** E160-15E6

convergence, dynamic (multibeam cathode-ray tubes). The process whereby the locus of the point of convergence of electron beams is made to fall on a specified surface during scanning. *See:* **beam tubes.** E160-15E6

convergence electrode (multibeam cathode-ray tubes). An electrode whose electric field converges two or more electron beams. *See:* **beam tubes.** E160-15E6

convergence magnet (multibeam cathode-ray tubes). A magnet assembly whose magnetic field converges two or more electron beams. *See:* **beam tubes.** E160-15E6

convergence plane (multibeam cathode-ray tubes). A plane containing the points at which the electron beams appear to experience a deflection applied for the purpose of obtaining convergence. *See:* **beam tubes.** E160-15E6

convergence surface (multibeam cathode-ray tubes). The surface generated by the point of intersection of two or more electron beams during the scanning process. *See:* **beam tubes.** E160-15E6

conversion efficiency (1) (electrical conversion). In alternating-current to direct-current conversion equipment, the ratio of the product of output direct-current and voltage to input watts expressed in percent. *Note:* It reflects alternating-current power capacity required for a given voltage and current output and does not necessarily reflect watts lost.

$$\text{Conversion Efficiency} = \frac{(E_{\text{dc}})(I_{\text{dc}})}{P}\,(100\ \text{percent})$$

See also: **electric conversion.** 0-10E1
(2) (overall) (photoelectric converter). The ratio of available power output to total ncident radiant power in the active area for photovoltaic operation. *Note:* This depends on the spectral distribution of the source and junction temperature. *See also:* **semiconductor.** 0-10E1
(3) (klystron oscillator). The ratio of the high-frequency output power to the direct-current power supplied to the beam. *See also:* **velocity-modulated tube.** 50I07-15E6
(4) (metallic rectifier). The ratio of the product of (1) the average values of the unidirectional voltage and current (2) the total power input on the alternating-voltage side. *See also:* **rectification.** 42A15-0

conversion transconductance (heterodyne conversion transducer). The quotient of (1) the magnitude of the desired output-frequency component of current by (2) the magnitude of the input-frequency (signal) component of voltage, when the impedance of the output external termination is negligible for all of the frequencies that may affect the result. *Note:* Unless otherwise stated, the term refers to the cases in which the input-frequency voltage is of infinitesimal magnitude. All direct electrode voltages, and the magnitude of the local-oscillator voltage, must remain constant. *See:* **modulation; transducer.** E160-15E6;E196/E270/42A70-0

conversion transducer (1) (broadly). A transducer in which the signal undergoes frequency conversion. *Note:* The gain or loss of a conversion transducer is specified in terms of the useful signal. *See also:* **transducer.** 42A65-0
(2) An electric transducer in which the input and the output frequencies are different. *Note:* If the frequency-changing property of a conversion transducer depends upon a generator of frequency different from that of the input or output frequencies, the frequency and voltage or power of this generator are parameters of the conversion transducer. *See also:* **heterodyne conversion transducer (converter).** E160-15E6;E196-0

conversion voltage gain (conversion transducer). The ratio of (1) the magnitude of the output-frequency voltage across the output termination, with the transducer inserted between the input-frequency generator and the output termination, to (2) the magnitude of the input-frequency voltage across the input termination of the transducer. 47A70-15E6

convert (data processing). To change the representation of data from one form to another, for example, to change numerical data from binary to decimal or from cards to tape. *See also:* **electronic digital computer.** X3A12-16E9

converter (1) (broadly). A machine that changes alternating-current power to direct-current power or vice versa, or from one frequency to another. *Note:* The term converter is sometimes used in a narrow sense to designate a machine that changes alternating current to direct current in distinction to an inverter. 42A10-0

(2) (frequency converter) (heterodyne reception). The portion of the receiver that converts the incoming signal to the intermediate frequency. 42A65-31E3

(3) (facsimile). A device that changes the type of modulation. *See also:* **facsimile (electrical communication).** E168-0

(4) (industrial control). A network or device for changing the form of information or energy.
See:
control system, feedback;
circuits and devices;
direct-current balancer;
dynamotor;
frequency changer;
frequency converter;
heterodyne conversion transducer;
induction frequency converter;
inductor dynamotor;
inductor frequency converter;
inverter;
mercury hydrogen spark-gap converter;
motor-converter;
motor-generator set;
phase advancer;
phase converter;
phase modifier;
pool-cathode mercury-arc converter;
quenched spark-gap converter;
rectifier assembly;
synchronous-booster converter;
synchronous-booster inverter;
synchronous capacitor;
synchronous converter;
synchronous inverter;
torque margin. AS1-34E10

converter tube. An electron tube that combines the mixer and local-oscillator functions of a heterodyne conversion transducer. *See also:* **heterodyne conversion transducer (converter); tube definitions.** E160-15E6;42A70-0

conveyor. A mechanical contrivance, generally electrically driven, that extends from a receiving point to a discharge point and conveys, transports, or transfers material between those points. *See:* **conveyor, belt-type; conveyor, chain-type; conveyor, shaker-type; conveyor, vibrating-type.** *See also:* **mining.** 42A85-0

conveyor, belt-type. A conveyor consisting of an endless belt used to transport material from one place to another. *See also:* **conveyor.** 42A85-0

conveyor, chain-type. A conveyor using a driven endless chain or chains, equipped with flights that operate in a trough and move material along the trough. *See also:* **conveyor.** 42A85-0

conveyor, shaker-type. A conveyor designed to transport material along a line of troughs by means of a reciprocating or shaking motion. *See also:* **conveyor.** 42A85-0

conveyor, vibrating-type. A conveyor consisting of a movable bed mounted at an angle to the horizontal, that vibrates in such a way that the material advances. *See also:* **conveyor.** 42A85-0

convolution function (burst measurements). The integral of the function $x(\tau)$ multiplied by another function $y(-\tau)$ shifted in time by t

$$\int_{-\infty}^{\infty} x(\tau)y(t-\tau)\mathrm{d}\tau$$

See also: **burst.** E265-0

cooking unit, counter-mounted. An assembly of one or more domestic surface heating elements for cooking purposes, designed for flush mounting in, or supported by, a counter, and which assembly is complete with inherent or separately mountable controls and internal wiring. *See:* **oven, wall-mounted.** 1A0-0

coolant. *See:* **cooling medium.**

cooler (heat exchanger) (rotating machinery). A device used to transfer heat between two fluids without direct contact between them. *See also:* **cradle base (rotating machinery).** 0-31E8

Coolidge tube. An X-ray tube in which the needed electrons are produced by a hot cathode. *See also:* **electron devices, miscellaneous.** 42A70-15E6

cooling (power supplies). The cooling of regulator elements refers to the method used for removing heat generated in the regulating process. *Note:* Methods include radiation, convection, and conduction or combinations thereof. *See also:* **power supply.** KPSH-10E1

cooling coil (rotating machinery). A tube through whose wall, heat is transferred between two fluids without direct contact between them. *See also:* **cradle base (rotating machinery).** 0-31E8

cooling, convection (power supplies). A method of heat transfer that uses the natural upward motion of air warmed by the heat dissipators. *See also:* **power supply.** KPSH-10E1

cooling duct. *See:* **ventilating duct (rotating machinery).**

cooling fin (electron device). A metallic part or fin extending the cooling area to facilitate the dissipation of the heat generated in the device. *See:* **electron device.** 0-15E6

cooling, lateral forced-air (power supplies). An efficient method of heat transfer by means of side-to-side circulation that employs blower movement of air through or across the heat dissipators. *See also:* **power supply.** KPSH-10E1

cooling medium (rotating machinery) (coolant). A fluid, usually air, hydrogen, or water, used to remove heat from a machine or from certain of its components. *See also:* **cradle base (rotating machinery).** 0-31E8

cooling system (rectifier). Equipment, that is, parts and their interconnections, used for cooling a rectifier. *Note:* It includes all or some of the following: rectifier water jacket, cooling coils or fins, heat exchanger, blower, water pump, expansion tank, insulating pipes, etcetera. *See also:* **rectification.** 42A15-34E24

cooling-water system (rotating machinery). All parts that are provided for the flow, treatment, or storage of cooling water. *See also:* **cradle base (rotating machinery).** 0-31E8

coordinate dimension word (numerically controlled machines). A word defining an absolute dimension. *See also:* **numerically controlled machines.** EIA3B-34E12

coordinated transpositions (electric supply or communication circuits). Transpositions that are installed for the purpose of reducing inductive coupling, and that are located effectively with respect to the discontinuities in both the electric supply and communication circuits. *See also:* **inductive coordination.** 42A65-0

coordination of insulation (lightning insulation strength). The steps taken to prevent damage to electric equipment due to overvoltages and to localize flashovers to points where they will not cause damage. *Note:* In practice, coordination consists of the process of correlating the insulating strengths of electric equipment with expected overvoltages and with the characteristics of protective devices. *See also:* **basic impulse insulation level (insulation strength); lightning arrester (surge diverter).** 50I25-31E7

copper brush (rotating machinery). A brush composed principally of copper. *See also:* **brush.** 0-31E8

copper-clad steel. Steel with a coating of copper welded to it, as distinguished from copper-plated or copper-sheathed material. *See also:* **arresters; lightning protection and equipment.** 42A95-0

copper conductor. A conductor made wholly of copper. *See also:* **conductor.** 42A35-31E13

copper-covered steel wire. A wire having a steel core to which is bonded a continuous outer layer of copper. *See also:* **conductor.** 42A35-31E13/31E1

copper losses. *See also:* **load losses.**

copy (electronic data processing). (1) To reproduce data leaving the original data unchanged. X3A12-16E9

(2) To produce a sequence of character events equivalent, character by character, to another sequence of character events.

(3) The sequence of character events produced in (2). *See also:* **electronic digital computer; transfer.** E162-0

cord. One or a group of flexible insulated conductors, enclosed in a flexible insulating covering and equipped with terminals. *See also:* **circuits and devices.** 42A65-0

cord adjuster. A device for altering the pendant length of the flexible cord of a pendant. *Note:* This device may be a ratchet reel, a pulley and counterweight, a tent-rope stick, etcetera. *See also:* **interior wiring.** 42A95-0

cord circuit. A connecting circuit terminating in a plug at one or both ends and used at switchboard positions in establishing telephone connections. *See also:* **telephone switching system.** 42A65-19E1

cord-circuit repeater. A repeater associated with a cord circuit so that it may be inserted in a circuit by an operator. *See also:* **repeater.** 42A65-0

cord grip (strain relief). A device by means of which the flexible cord entering a device or equipment is gripped in order to relieve the terminals from tension in the cord. *See also:* **interior wiring.** 42A95-0

cord connector (cord connector body*) (table tap*). A plug receptacle provided with means for attachment to flexible cord. *See also:* **interior wiring.**

*Deprecated 42A95-0

cordless switchboard. A manual telephone switchboard that uses manually operated keys to make connections. *See also:* **telephone switching system.** 42A65-19E1

core (1) (magnetic core) (rotating machinery). An element made of magnetic material, serving as part of a path for magnetic flux. *Note:* In a rotating machine, this is frequently part of the stator, a hollow cylinder of laminated magnetic steel, slotted on the inner surface for the purpose of containing the stator windings. Such a stator core frequently contains ducts through which a cooling medium may flow to remove heat generated in the core and windings. *See:* **magnetic core.** 0-31E8

(2) (electronic information storage). *See:* **electronic digital computer.**

(3) (mechanical recording). The central layer or basic support of certain types of laminated media. E157-1E1

(4) (electromagnet). The part of the magnetic structure around which the magnetizing winding is placed. E270-0

core ducts (rotating machinery). The space between or through core laminations provided to permit the radial or axial flow of ventilating air. *See also:* **rotor (rotating machinery).** 0-31E8

core end plate (end plate) (flange) (rotating machinery). A plate or structure at the end of a laminated core to maintain axial pressure on the laminations. 0-31E8

core length (rotating machinery). The dimension of the stator, or rotor, core measured in the axial direction. *See also:* **rotor (rotating machinery); stator.** 0-31E8

coreless-type induction heater or furnace. A device in which a charge is heated by induction and no magnetic core material links the charge. *Note:* Magnetic material may be used elsewhere in the assembly for flux guiding purposes. *See also:* **dielectric heating; induction heating.** E54/E169-0

core loss (1). The power dissipated in a magnetic core subjected to a time-varying magnetizing force. *Note:* core loss includes hysteresis and eddy-current losses in the core. 0-21E1

(2) (synchronous machine). The difference in power required to drive the machine at normal speed, when excited to produce a voltage at the terminals on open circuit corresponding to the calculated internal voltage, and the power required to drive the unexcited machine at the same speed. *Note:* The internal voltage shall be determined by correcting the rated terminal voltage for the resistance drop only. *See also:* **synchronous machine.** 50A10-31E8

core-loss current. The in-phase component (with respect to the induced voltage) of the exciting current supplied to a coil. *Note:* It may be regarded as a hypothetical current, assumed to flow through the equivalent core-loss resistance. E270-21E1

core loss, open-circuit (rotating machinery). The difference in power required to drive a machine at normal speed, when excited to produce a specified voltage at the open-circuited armature terminals, and the power required to drive the unexcited machine at the same speed. *See:* **synchronous machine.** 50A10-31E8

core-loss test (rotating machinery). A test taken on a built-up (usually unwound) core of a machine to determine its loss characteristic. *See also:* **stator.** 0-31E8

core, relay. *See:* **relay core.**

core test (rotating machinery). A test taken on a built-up (usually unwound) core of a machine to determine its loss characteristics or its magnetomotive force characteristics, or to locate short-circuited laminations. *See also:* **rotor (rotating machinery); stator.** 0-31E8

core-type induction heater or furnace. A device in which a charge is heated by induction and a magnetic core links the inducing winding with the charge. *See also:* **dielectric heating; induction heating.** E54/E169-0

Coriolis correction (navigation). A correction, that must be applied to measurements made with respect to a coordinate system in translation, to compensate for the effect of any angular motion of the coordinate system with respect to inertial space. *See also:* **navigation.** E174-10E6

corner (waveguide technique). An abrupt change in the direction of the axis of a waveguide. *Note:* Also termed **elbow.** *See also:* **waveguide.** 42A65-0;50I62-3E1

corner reflector (1). A reflecting object consisting of two or three mutually intersecting conducting surfaces. *Note:* Dihedral forms of corner reflectors are frequently used in antennas; trihedral forms are more often used as radar targets. *See also:* **antenna; radar.** 0-3E1

(2) (radar). Three conducting surfaces mutually intersecting at right angles, designed to return electromagnetic radiations toward their sources and used to render a target more conspicuous to radar observations. *See also:* **radar.** E172-10E6

corner-reflector antenna. An antenna consisting of a primary radiating element and a corner reflector. *See also:* **antenna.** 0-3E1

cornice lighting. Light sources shielded by a panel parallel to the wall and attached to the ceiling and distributing light over the wall. *See also:* **general lighting.** Z7A1-0

corona (1) (air). A luminous discharge due to ionization of the air surrounding a conductor caused by a voltage gradient exceeding a certain critical value. *See also:* **tower.** 42A35-31E13

(2) (gas). A discharge with slight luminosity produced in the neighborhood of a conductor, without greatly heating it, and limited to the region surrounding the conductor in which the electric field exceeds a certain value. *See also:* **discharge (gas).** 50I07-15E6

corona charging (electrostatography). Sensitizing by means of gaseous ions of a corona. *See also:* **electrostatography.** E224-15E7

corona-discharge tube. A low-current gas-filled tube utilizing the corona-discharge properties. *See also:* **tube definitions.** 0-15E6

corona effect. The particular form of the glow discharge that occurs in the neighborhood of electric conductors where the insulation is subject to hight electric stress. E270-0

corona inception test. *See:* **discharge inception test.**

corona level (power distribution, underground cables). *See:* **ionization extinction voltage.**

corona shielding (corona grading) (rotating machinery). A means adapted to reduce potential gradients along the surface of coils. *See also:* **asynchronous machine; direct-current commutating machine; synchronous machine.** 0-31E8

corona voltmeter. A voltmeter in which the crest value of voltage is indicated by the inception of corona. *See also:* **instrument.** 42A30-0

corrected compass course. *See:* **magnetic course.**

corrected compass heading. *See:* **magnetic heading.**

correcting signal. *See:* **synchronizing signal.**

correction. A quantity (equal in absolute value to the error) added to a calculated or observed value to obtain the true value. *See also:* **accuracy rating (instrument); electronic digital computer.** E162-0

correction rate (control). The velocity at which the control system functions to correct error in register. 42A25-34E10

corrective network. An electric network designed to be inserted in a circuit to improve its transmission properties, its impedance properties, or both. *See also:* **network analysis.** 42A65-0

correct relaying-system performance. The satisfactory operation of all equipment associated with the protective-relaying function in a protective-relaying system. *Note:* It includes the satisfactory presentation of system input quantities to the relaying equipment, the correct operation of the relays in response to these input quantities, and the successful operation of the assigned switching device or devices. 37A100-31E11/31E6

correct relay operation. An output response by the relay that agrees with the operating characteristic for the input quantities applied to the relay. *See also:* **correct relaying-system performance.** 37A100-31E11/31E6

correlated color temperature (light source). The absolute temperature of a blackbody whose chromaticity most nearly resembles that of the light source. *See also:* **color.** Z7A1-0

correlation detection (modulation systems). Detection based on the averaged product of the received signal and a locally generated function possessing some known characteristic of the transmitted wave. *Notes:* (1) The averaged product can be formed, for example, by multiplying and integrating, or by the use of a matched filter whose impulse response, when reversed in time, is the locally generated function. (2) Strictly, the above definition applies to detection based on cross correlation. The term correlation detection may also apply to detection involving autocorrelation, in which case the locally generated function is merely a delayed form of the received signal. *See also:* **modulating systems.** E170-0

corrosion. The deterioration of a substance (usually a metal) because of a reaction with its environment. *See:* **corrosion terms.** CM-34E2

corrosion fatigue. Reduction in fatigue life in a corrosive environment. *See:* **corrosion terms.** CM-34E2

corrosion fatigue limit. The maximum repeated stress endured by a metal without failure in a stated number of stress applications under defined conditions of corrosion and stressing. *See:* **corrosion terms.** CM-34E2

corrosion rate. The rate at which corrosion proceeds. *See:* **corrosion terms.** CM-34E2

corrosion-resistant parts. *Notes:* (1) General: Where essential to minimize deterioration due to marine atmospheric corrosion, corrosion-resisting materials or other materials treated in a satisfactory manner to render them adequately resistant to corrosion should be used. (2) Corrosion-resisting materials: Silver, corrosion-resisting steel, copper, brass, bronze, copper-nickel, certain nickel-copper alloys, and certain aluminum alloys are considered satisfactory corrosion-resisting materials within the intent of the foregoing. (3) Corrosion-resistant treatments: The following treatments when properly done and of a sufficiently heavy coating are considered satisfactory corrosion-resistant treatments within the intent of the foregoing. Electroplating of: cadmium, chromium, and copper; sherardizing and galvanizing of: nickel, silver, and zinc. Dipping and painting with phosphate or suitable cleaning, followed by the application of zinc chromate primer or equivalent. (4) These provisions should apply to the following: Parts—Interior small parts that are normally expected to be removed in service, such as bolts, nuts, pins, screws, cap screws, terminals, brush-holder studs, springs, etcetera. Assemblies, subassemblies and other units—Where necessary due to the unit function, or for interior protection, such as shafts within a motor or generator enclosure, and surface of stator and rotor. Enclosures and their fastenings and fittings—Such as enclosing cases for control apparatus and outer cases for signal and communication systems (both outside and inside), and all their fastenings and fittings that would be seriously damaged or rendered ineffective by corrosion. E45-0

corrosion terms. *Note:* For an extensive list of cross references, see *Appendix A.*

cosecant-squared antenna. A shaped-beam antenna in which the radiation intensity over a part of its pattern in some specified plane (usually the vertical) is proportional to the square of the cosecant of the angle measured from a specified direction in that plane (usually the horizontal). *Note:* Its purpose is to lay down a uniform field along a line that is parallel to the specified direction but that does not pass through the antenna. *See also:* **antenna.** 42A65-3E1

cosecant-squared pattern (electronic navigation). An antenna field in which the signal-power pattern in the vertical plane varies as the square of the cosecant of the elevation angle. *Note:* Similar objects at the same altitude give equal response regardless of distance. *See also:* **navigation.** E172-10E6

cosine law (illuminating engineering). A law stating that the illumination on any surface varies as the cosine of the angle of incidence. The angle of incidence θ is the angle between the normal to the surface and the direction of the incident light. The inverse-square law and the cosine law can be combined as $E = (I \cos \theta)/d^2$. *See also:* **inverse-square law (illuminating engineering).** Z7A1-0

cosine-cubed law (illuminating engineering). An extension of the cosine law in which the distance d between the source and surface is replaced by $h/\cos \theta$, where h is the perpendicular distance of the source from the plane in which the point is located. It is expressed by $E = (I \cos^3 \theta)/h^2$. *See also:* **inverse-square law (illuminating engineering).** Z7A1-0

cosine function. A periodic function of the form $A \cos x$, or $A \cos \omega t$. The amplitude of the cosine function is A and its argument is x or ωt. E270-0

costate (control system). The state of the adjoint system. *See also:* **control system.** 0-23E0

cost of incremental fuel (electric power systems) (usually expressed in cents per million British thermal units). The ultimate replacement cost of the fuel that would be consumed to supply an additional increment of generation. *See also:* **power system, low-frequency and surge testing.** E94-0

coulomb. The unit of electric charge in SI units (International System of Units). The coulomb is the quantity of electric charge that passes any cross section of a conductor in one second when the current is maintained constant at one ampere. E270-0

Coulomb friction. A constant velocity-independent force that opposes the relative motion of two surfaces in contact. *See also:* **control system, feedback.** 0-23E0

Coulomb's law (electrostatic attraction). The force of repulsion between two like charges of electricity concentrated at two points in an isotropic medium is proportional to the product of their magnitudes and inversely proportional to the square of the distance between them and to the dielectric constant of the medium. *Note:* The force between unlike charges is an attraction. E270-0

coulometer (voltameter). An electrolytic cell arranged for the measurement of a quantity of electricity by the chemical action produced. *See also:* **electricity meter (meter).** 42A30-0

count (radiation counters). A single response of the counting system. *See also:* **gas-filled radiation-counter tubes; scintillation counter; tube count.** E160/E175/42A70-15E6

count-down (transponder). The ratio of the number of interrogation pulses not answered to the total number of interrogation pulses received. *See also:* **navigation.** E172-10E6

counter cells. *See:* **counter-electromotive-force cells.**

counterclockwise polarized wave (radio wave propagation). *See:* **left-handed polarized wave.**

counter electromotive force (any system). The effective electromotive force within the system that opposes the passage of current in a specified direction. *See also:* **electrolytic cell.** 42A60-0

counter-electromotive-force cells (counter cells). Cells of practically no ampere-hour capacity used to oppose the battery voltage. *See also:* **battery (primary or secondary).** 42A60-0

counter (electronic computation). (1) A device, capable of changing from one to the next of a sequence of distinguishable states upon each receipt of an input signal. *Note:* One specific type is a circuit that produces one output pulse each time it receives some predetermined number of input pulses. The same term may also be applied to several such circuits connected in cascade to provide digital counting. (2) Less frequently, a counter is an accumulator. (3) (ring). A loop of interconnected bistable elements such that one and only one is in a specified state at any given time and such that, as input signals are counted, the position of the one specified state moves in an ordered sequence around the loop. *See also:* **electronic computation; electronic digital computer; trigger circuit.** E270/42A65-0

counterpoise (1) (antenna). A system of wires or other conductors, elevated above and insulated from the ground, forming the lower system of conductors of an

antenna. *See also:* **antenna.** E149-0;42A65-3E1
(2) (overhead line) (lightning protection). A conductor or system of conductors, arranged beneath the line, located on, above, or most frequently below the surface of the earth, and connected to the footings of the towers or poles supporting the line. *See also:* **antenna; ground.** 42A35-31E13

counter, radiation. *See:* **radiation counter.**

counter tube (1) (radiation counters). A device that reacts to individual ionizing events, thus enabling them to be counted. *See:* **anticoincidence (radiation counters).** *See also:* **gas-filled radiation-counter tube.** 0-15E6

counting efficiency (1) (radiation-counter tubes). The average fraction of the number of ionizing particles or quanta incident on the sensitive area that produce tube counts. *Note:* The operating conditions of the counter and the condition of irradiation must be specified. *See:* **gas-filled radiation-counter tubes.** E160-15E6
(2) (scintillation counters). The ratio of (A) the average number of photons or particles of ionizing radiation that produce counts to (B) the average number incident on the sensitive area. *Note:* The operating conditions of the counter and the conditions of irradiation must be specified. *See also:* **scintillation counter.** E175-0

counting mechanism (automatic line sectionalizer). A device that counts the number of electric impulses and, following a predetermined number of successive electric impulses, actuates a releasing mechanism. *Note:* It resets if the total predetermined number of successive impulses do not occur in a predetermined time. 37A100-31E11

counting operation (automatic line sectionalizer). Each advance of the counting mechanism towards an opening operation. 37A100-31E11

counting-operation time (automatic line sectionalizer). The time between the cessation of a current above the minimum actuating current value and the completion of a counting operation. 37A100-31E11

counting rate. Number of counts per unit time. *See:* **anticoincidence (radiation counters).** 0-15E6

counting-rate meter (pulse techniques). A device that indicates the time rate of occurrence of input pulses averaged over a time interval. *See also:* **scintillation counter.** E175-0

counting rate versus voltage characteristic (gas-filled radiation-counter tube). The counting rate as a function of applied voltage for a given constant average intensity of radiation. *See:* **gas-filled radiation-counter tubes.** 42A70-0;E160-15E6

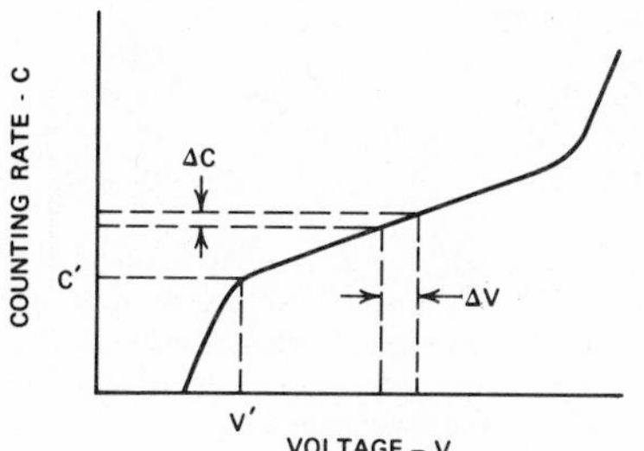

Counting rate–voltage characteristic in which

$$\text{Relative plateau slope} = 100\,\frac{\Delta C/C}{\Delta V}$$

$$\text{Normalized plateau slope} = \frac{\Delta C/\Delta V}{C'/V'} = \frac{\Delta C/C'}{\Delta V/V'}$$

counts background. *See:* **background counts.**

counts, tube, multiple (radiation-counter tubes). *See:* **multiple tube counts.**

counts, tube, spurious (radiation-counter tubes). *See:* **spurious tube counts.**

couple (1) (storage cell). An element of a storage cell consisting of two plates, one positive and one negative. *Note:* The term couple is also applied to a positive and a negative plate connected together as one unit for installation in adjacent cells. *See also:* **battery (primary or secondary); galvanic cell.** 42A60-0
(2) (thermoelectric). A thermoelectric device having two arms of dissimilar composition. *Note:* The term thermoelement is ambiguously used to refer to either a thermoelectric arm or to a thermoelectric couple, and its use is therefore not recommended. *See also:* **thermoelectric device.** E221-15E7

coupled circuit. A network containing at least one element common to two meshes. *See also:* **network analysis.** E270-0

coupler (navigation). The portion of a navigational system that receives signals of one type from a sensor and transmits signals of a different type to an actuator. *See:* **autopilot coupler.** *See also:* **navigation.** E172-10E6

coupler, 3-decibel. *See:* **hybrid control.**

coupling (1) (electric circuits). The circuit element or elements, or the network, that may be considered common to the input mesh and the output mesh and through which energy may be transferred from one to the other.
(2) The association of two or more circuits or systems in such a way that power or signal information may be transferred from one to another. *Note:* Coupling is described as close or loose. A close-coupled process has elements with small phase shift between specified variables; close-coupled systems have large mutual effect shown mathematically by cross-products in the system matrix. *See also:* **network analysis.** E270-34E10;42A65-31E3;0-21E1
(3) (interference terminology) (electric circuits). The effect of one system or subsystem upon another. (A) For interference, the effect of an interfering source on a signal transmission system. (B) The mechanism by which an interference source produces interference in a signal circuit. *See also:* **interference.** 85A1-13E6
(4) (induction heating). The percentage of the total magnetic flux, produced by an inductor, that is effective in heating a load or charge. *See also:* **dielectric heating; induction heating.** E54/E169-0
(5) (metal raceway). A fitting intended to connect two lengths of metal raceway or perform a similar function. 42A95-0
(6) (rotating machinery). A part or combination of parts that connects two shafts for the purpose of transmitting torque or maintaining alignment of the two shafts.
See:
capacitance coupling;
close coupling;
coupling coefficient;

coupling, electric;
coupling, magnetic particle;
critical coupling;
cross coupling;
crosstalk;
crosstalk coupling;
crosstalk unit;
decoupling;
direct coupling;
far-end crosstalk;
inductive coupling;
interaction crosstalk coupling;
loose coupling;
near-end coupling;
resistance-capacitance coupling;
resistance coupling;
transverse crosstalk coupling;
unbalance.
See also: **rotor (rotating machinery).** 0-31E8

coupling aperture (coupling hole, coupling slot) (waveguides). An aperture in the wall of a transmission line or waveguide or cavity resonator designed to transfer energy to or from an external circuit. *See:* **waveguide.** E147-3E1

coupling, broadband, aperiodic (power-system communication). Coupling that is equally responsive to all frequencies (over a stated range). As applied to power-line-carrier coupling devices, a broadband or an aperiodic coupler does not require tuning to the particular carrier frequencies being used. *See also:* **power-line carrier.** 0-31E3

coupling capacitance (1) (general). The association of two or more circuits with one another by means of capacitance mutual to the circuits. *See also:* **coupling.** 42A65-31E3

(2) (interference terminology). The type of coupling in which the mechanism is capacitance between the interference source and the signal system, that is, the interference is induced in the signal system by an electric field produced by the interference source. *See also:* **interference.** 0-13E6

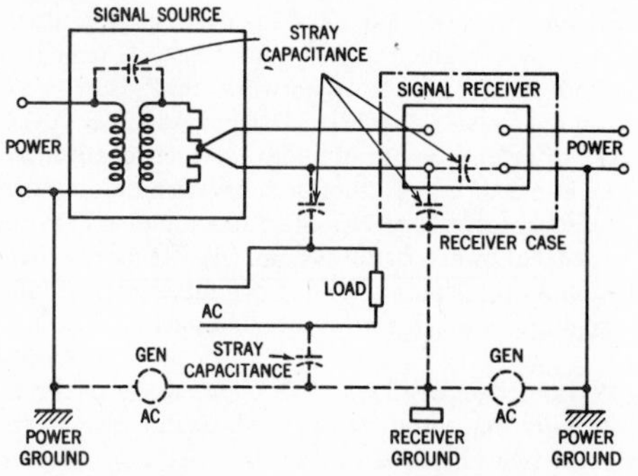

Capacitance coupling (interference).

coupling capacitor (power-system communication). A capacitor employed to connect the carrier lead-in conductor to the high-voltage power-transmission line. *See also:* **power-line carrier.** 0-31E3

coupling coefficient (coefficient of coupling). The ratio of impedance of the coupling to the square root of the product of the total impedances of similar elements in the two meshes. *Notes:* (1) Used only in the case of resistance, capacitance, self-inductance, and inductance coupling. (2) Unless otherwise specified, coefficient of coupling refers to inductance coupling, in which case it is equal to $M/(L_1L_2)^{1/2}$, where M is the mutual inductance, L_1 the total inductance of one mesh, and L_2 the total inductance of the other. *See also:* **network analysis.** E270/42A65-9E4

coupling coefficient, small-signal (electron stream). The ratio of (1) the maximum change in energy of an electron traversing the interaction space to (2) the product of the peak alternating gap voltage by the electronic charge. *See also:* **circuit characteristics of electrodes; coupling; coupling coefficient; electron emission.** E160-15E6;0-9E4

coupling, common (joint) (power-system communication). Line coupling employing tuning circuits that accommodate more than one carrier channel through a common set of coupling capacitors. *See also:* **power-line carrier.** 0-31E3

coupling, conductance (interference terminology). The type of coupling in which the mechanism is conductance between the interference source and the signal system. *See also:* **interference; raceway.** 0-13E6

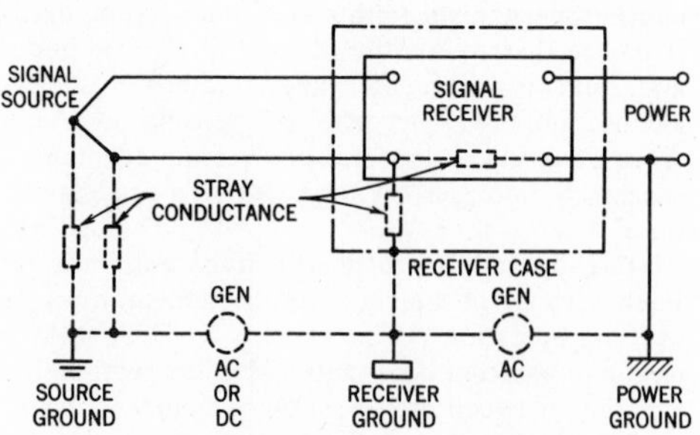

Conductance coupling (interference).

coupling, electric (rotating machinery). (1) A device for transmitting torque by means of electromagnetic force in which there is no mechanical torque contact between the driving and driven members. *Note:* The slip-type electric coupling has poles excited by direct current on one rotating member, and an armature winding, usually of the double-squirrel-cage type, on the other rotating member. (2) A rotating machine that transmits torque by electric or magnetic means or in which the torque is controlled by electric or magnetic means. E45-0;42A10-31E8

coupling factor (1) (lightning). The ratio of the induced voltage to the inducing voltage on parallel conductors. *See also:* **direct-stroke protection (lightning).** 0-31E13

(2) (directional coupler). The ratio of the incident power fed into the main port, and propagating in the preferred direction, to the power output at an auxiliary port, all ports being terminated by reflectionless terminations. *See also:* **waveguide.** 0-9E4

coupling flange (flange) (rotating machinery). The disc-shaped element of a half coupling that permits attachment to a mating half coupling. *See also:* **rotor (rotating machinery).** 0-31E8

coupling hole. *See:* **coupling aperture.**

coupling, hysteresis. An electric coupling in which torque is transmitted from the driving to the driven member by magnetic forces arising from the resistance to reorientation of established magnetic flux fields within ferromagnetic material usually of high coercivity. *Note:* The magnetic flux field is normally produced by current in the excitation winding, provided by an external source. 0-31E8

coupling, inductance (interference terminology). The type of coupling in which the mechanism is mutual inductance between the interference source and the signal system, that is, the interference is induced in the signal system by a magnetic field produced by the interference source. See the following figure. *See also:* **interference.** 0-13E6

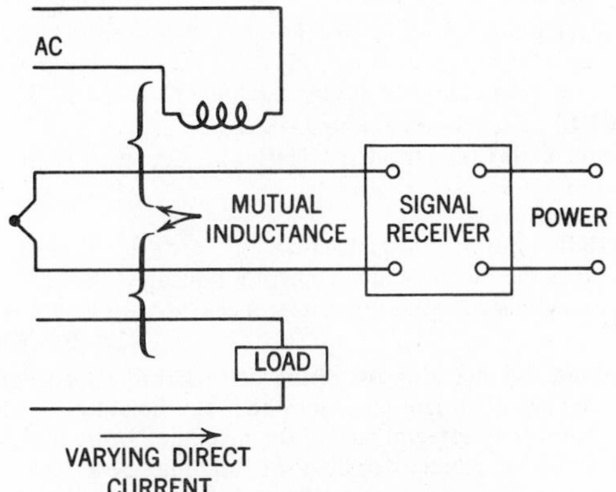

Inductance coupling (interference).

coupling, induction. An electric coupling in which torque is transmitted by the interaction of the magnetic field produced by magnetic poles on one rotating member and due to an induced voltage in the other rotating member. *Note:* The magnetic poles may be produced by direct-current excitation, permanent-magnet excitation, or alternating-current excitation. Currents due to the induced voltages may be carried in a wound armature, cylindrical cage, or may be present as eddy currents in an electrically conductive disc or cylinder. Couplings utilizing a wound armature or a cylindrical cage are known as slip or magnetic couplings. Couplings utilizing eddy-current effects are known as eddy-current couplings. 0-31E8

coupling loop (waveguides). A conducting loop projecting into a waveguide transmission line or cavity resonator, designed to transfer energy to or from an external circuit. *See:* **waveguide.** E147-3E1

coupling, magnetic friction. An electric coupling in which torque is transmitted by means of mechanical friction. Pressure normal to the rubbing surfaces is controlled by means of an electromagnet and a return spring. *Note:* Couplings may be either magnetically engaged or magnetically released depending upon application. 0-31E8

coupling, magnetic-particle. A type of electric coupling in which torque is transmitted by means of a fluid whose viscosity is adjustable by virtue of suspended magnetic particles. *Note:* The coupling fluid is incorporated in a magnetic circuit in which the flux path includes the two rotating members, the fluid, and a magnetic yoke. Flux density, and hence the fluid viscosity, are controlled through adjustment of current in a magnet coil linking the flux path. 0-31E8

coupling (power-system communication) (1) (phase to ground). A coupling circuit employing a coupling capacitor connected to one phase of a power line (or power lines) with the return path completed through ground or earth. *See also:* **power-line carrier.** 0-31E3

(2) (phase to phase). A coupling circuit employing coupling capacitors connected to two phases of a power line (or power lines). *Note:* It is a metallic circuit, that is, one in which ground or earth forms no intentional part. *See also:* **power-line carrier.** 0-31E3

coupling probe (waveguides). A probe projecting into a waveguide transmission line or cavity resonator designed to transfer energy to or from an external circuit. *See:* **waveguide.** E147-3E1

coupling, radiation (interference terminology). The type of coupling in which the interference is induced in the signal system by electromagnetic radiation produced by the interference source. *See also:* **interference.** 0-13E6

couplings (pothead). Entrance fittings which may be provided with a rubber gland to provide a hermetic seal at the point where the cable enters the box and may have, in addition, a threaded portion to accommodate the conduit used with the cable or have an armor clamp to clamp and ground the armored sheath on armor-covered cable. 57A12.76-0

coupling slat. *See:* **coupling aperture.**

coupling, synchronous (rotating machinery). A type of electric coupling in which torque is transmitted at zero slip, either between two electromagnetic members of like number of poles, or between one electromagnetic member and a reluctance member containing a number of saliencies equal to the number of poles. *Note:* Synchronous couplings may have induction members or other means for providing torque during nonsynchronous operation such as starting. *See also:* **electric coupling.** 0-31E8

course-deviation indicator. *See:* **course-line deviation indicator.**

course (navigation). (1) The intended direction of travel, expressed as an angle in the horizontal plane between a reference line and the course line, usually measured clockwise from the reference line. (2) The intended direction of travel as defined by a navigational facility. (3) Common usage for course line. *See also:* **navigation.** E172-10E6

course line (electronic navigation). The projection in the horizontal plane of a flight path (proposed path of travel). *See also:* **navigation.** E172-10E6

course-line computer (electronic navigation). A device, usually carried aboard a vehicle, to convert navigational signals such as those from very-high-frequency omnidirectional range and distance-measuring equipment into courses extending between any desired points regardless of their orientation with respect to the source of the signals. *See also:* **navigation.** 0-10E6

course-line deviation (electronic navigation). The amount by which the track of a vehicle differs from its course line, expressed in terms of either an angular or linear measurement. *See also:* **navigation.** 0-10E6

course-line deviation indicator (course-deviation indicator) (electronic navigation). A device providing a visual display of the direction and amount of deviation from the course. Also called flight-path deviation indicator. *See also:* **navigation.** 0-10E6

course linearity (instrument landing systems). A term used to describe the change in the difference in depth of modulation of the two modulation signals with respect to displacement of the measuring position from the course line but within the course sector. *See also:* **navigation.** 0-10E6

course made good (navigation). The direction from the point of departure to the position of the vehicle. *See also:* **navigation.** 0-10E6

course push (or pull) (electronic navigation). An erroneous deflection of the indicator of a navigational aid, produced by altering the attitude of the receiving antenna. *Note:* This effect is a manifestation of polarization error and results in an apparent displacement of the course line. *See also:* **navigation.** E172-10E6

course roughness (electronic navigation). A term used to describe the imperfections in a visually indicated course when such imperfections cause the course indicator to make rapid erratic movements. *See:* **scalloping.** *See also:* **navigation.** 0-10E6

course scalloping (electronic navigation). *See:* **scalloping.**

course sector (instrument landing systems). A wedge-shaped section of airspace containing the course line and spreading with distance from the ground station; it is bounded on both sides by the loci of points at which the difference in depth of modulation is a specified amount, usually of the difference in depth of modulation giving full-scale deflection of the course-deviation indicator. *See also:* **course linearity; difference in depth of modulation; navigation.** 0-10E6

course sector width (instrument landing systems). The transverse dimension at a specified distance, or the angle in degrees between the sides of the course sector. *See also:* **navigation.** 0-10E6

course sensitivity (electronic navigation systems). The relative response of a course-line deviation indicator to the actual or simulated departure of the vehicle from the course line. *Note:* In very-high-frequency omnidirectional ranges, tacan, or similar omnirange systems, course sensitivity is often taken as the number of degrees through which omnibearing selector must be moved to change the deflection of the course-line deviation indicator from full scale on one side to full scale on the other, while the receiver omnibearing input signal is held constant. *See also:* **navigation.** 0-10E6

course softening (electronic navigation). The intentional decrease in course sensitivity upon approaching a navigational aid such that the ratio of indicator deflection to linear displacement from the course line tends to remain constant. *See also:* **navigation.** E172-10E6

course width (electronic navigation). Twice the displacement (of the vehicle), in degrees, to either side of a course line, that produces a specified amount of indication on the course-deviation indicator. *Note:* Usually the specified amount is full scale. *See also:* **navigation.** 0-10E6

cove lighting. Light sources shielded by a ledge or horizontal recess, and distributing light over the ceiling and upper wall. *See also:* **general lighting.** Z7A1-0

cover (electric machine). A protective covering used to enclose, or to enclose partially, parts that are external to the stator frame and end shields. *Note:* In general, the word **cover** will be preceded by the name of the part that is covered. *See also:* **cradle base (rotating machinery).** 0-31E8

coverage area (mobile communication). The area surrounding the base station that is within the signal-strength contour that provides a reliable communication service 90 percent of the time. *See also:* **mobile communication system.** 0-6E1

covered plate (storage cell). A plate bearing a layer of oxide between perforated sheets. *See also:* **battery (primary or secondary).** 42A60-0

cover plate (rotating machinery). An essentially flat removable part used as a closure for an opening. *See also:* **cradle base (rotating machinery).** 0-31E8

CPE. *See:* **circular probable error.**

crab angle* (electronic navigation). *See:* **drift correction angle.**

*Deprecated

cradle base (rotating machinery). A device that supports the machine at the bearing housings. *Note:* For an extensive list of cross references, see *Appendix A.* 42A10-31E8

crane. A machine for lifting or lowering a load and moving it horizontally, in which the hoisting mechanism is an integral part of the machine. *Note:* It may be driven manually or by power and may be a fixed or a mobile machine. *See also:* **elevators.** 42A45-0

crawling (rotating machinery). The stable but abnormal running of a synchronous or asynchronous machine at a speed near to a submultiple of the synchronous speed. *See also:* **asynchronous machine; synchronous machine.** 0-31E8

creep (watthour meter). A meter creeps if, with the load wires removed, and with test voltage applied to the voltage circuits of the meter, the rotor moves continuously. *Note:* For the practical recognition of creep, a meter in service is considered to creep when, with all load wires disconnected, the rotor makes one complete revolution in ten minutes or less. *See also:* **accuracy rating (instrument).** 12A0-0

creepage. The travel of electrolyte up the surface of electrodes or other parts of the cell above the level of the main body of electrolyte. *See also:* **electrolytic cell.** 42A60-0

creepage distance. The shortest distance between two conducting parts measured along the surface or joints of the insulating material between them. 37A100-31E11

creepage surface (rotating machinery). An insulating-material surface extending across the separating space between components at different electric potential, where the physical separation provides the electrical insulation. *See also:* **asynchronous machine; direct-current commutating machine; synchronous machine.** 0-31E8

creeping stimulus. *See:* **accumulating stimulus.**

crest (peak) (wave, surge, or impulse) (lightning arresters). The maximum value that it attains. *See:* **lightning; lightning arrester (surge diverter).** E28/62A1-31E7

crest factor (1) (periodic function). The ratio of its crest (peak, maximum) value to its root-mean-square value. E270-0
(2) (pulse carrier). The ratio of the peak pulse amplitude to the root-mean-square amplitude. *See also:* **carrier.** E145-0
(3) (pulse). The ratio of the peak-pulse amplitude to the root-mean-square pulse amplitude. *See also:* **pulse terms.** E194-0

crest value (peak value). The maximum absolute value of a function when such a maximum exists. E270-0

crest voltage. The maximum instantaneous value that an alternating voltage attains during a cycle. 89A1-0

crest voltmeter. A voltmeter depending for its indications upon the crest or maximum value of the voltage applied to its terminals. *Note:* Crest voltmeters should have clearly marked on the instrument whether readings are in equivalent root-mean-square values or in true crest volts. It is preferred that the marking should be root-mean-square values of the sinusoidal wave having the same crest value as that of the wave measured. *See also:* **instrument.** 42A30-0

crest working voltage (between two points) (semiconductor rectifiers). The maximum instantaneous difference of voltage, excluding oscillatory and transient overvoltages, that exists during normal operation. *See also:* **crest; rectification; semiconductor rectifier stack.** 34A1-34E24

crevice corrosion. Localized corrosion as a result of the formation of a crevice between a metal and a nonmetal, or between two metal surfaces. *See:* **corrosion terms.** CM-34E2

critical anode voltage (multielectrode gas tubes). Synonymous with anode breakdown voltage (gas tubes). *See:* **gas tube.** E160-15E6

critical build-up speed (rotating machinery). The limiting speed below which the machine voltage will not build up under specified condition of field-circuit resistance. *See:* **direct-current commutating machine.** 0-31E8

critical controlling current (cryotron). The current in the control that just causes direct-current resistance to appear in the gate, in the absence of gate current and at a specified temperature. *See also:* **superconductivity.** E217-15E7

critical coupling. That degree of coupling between two circuits, independently resonant to the same frequency, that results in maximum transfer of energy at the resonance frequency. *See also:* **coupling.** 42A65-0

critical current (superconductor). The current in a superconductive material above which the material is normal and below which the material is superconducting, at a specified temperature and in the absence of external magnetic fields. *See also:* **superconductivity.** E217-15E7

critical damping. *See:* **damped harmonic system (2) (critical damping).**

critical dimension (waveguide). The dimension of the cross-section that determines the cutoff frequency. *See also:* **waveguide.** 42A65-0

critical field (magnetrons). The smallest theoretical value of steady magnetic flux density, at a steady anode voltage, that would prevent an electron emitted from the cathode at zero velocity from reaching the anode. *See also:* **magnetrons.** E160-15E6;42A70-0

critical frequency (radio propagation by way of the ionosphere). The limiting frequency below which a wave component is reflected by, and above which it penetrates through, an ionospheric layer at vertical incidence. *Note:* The existence of the critical frequency is the result of electron limitation, that is, the inadequacy of the existing number of free electrons to support reflection at higher frequencies. *See also:* **radiation; radio wave propagation.** 42A65-3E2

critical grid current (gas tube). The instantaneous value of grid current at which the anode current starts to flow. *See also:* **electrode current (electron tube).** 42A70-15E6

critical high-power level (attenuator tubes). The radio-frequency power level at which ionization is produced in the absence of a control-electrode discharge. E160-15E6

critical humidity (corrosion). The relative humidity above which the atmospheric corrosion rate of a given metal increases sharply. *See:* **corrosion terms.** CM-34E2

critical impulse (relay). The maximum impulse in terms of duration and input magnitude that can be applied suddenly to a relay without causing pickup. 37A100-31E11/31E6

critical impulse flashover voltage (insulator). The crest value of the impulse wave that, under specified conditions, causes flashover through the surrounding medium on 50 percent of the applications. *See also:* **impulse flashover voltage; lightning arrester (surge diverter); power systems, low-frequency and surge testing.** 29A1-0

critical impulse time (relay). The duration of a critical impulse under specified conditions. 37A100-31E11/31E6

critical magnetic field (superconductor). The field below which a superconductive material is superconducting and above which the material is normal, at a specified temperature and in the absence of current. *See also:* **superconductivity.** E217-15E7

critical overtravel time (relay). The time following a critical impulse until movement of the responsive element ceases just short of pickup. 37A100-31E11/31E6

critical point (feedback control system) (1) (Nichols chart). The bound of stability for the $GH(j\omega)$ plot; the intersection of $|GH| = 1$ with ang $GH = -180$ degrees.
(2) (Nyquist diagram). The bound of stability for the locus of the loop transfer function $GH(j\omega)$; the $(-1, j0)$ point. 85A1-23E0

critical rate-of-rise of OFF-state voltage (thyristor). The minimum value of the rate-of-rise of principal voltage that will cause switching from the OFF-state to the ON-state. *See also:* **principal voltage-current characteristic (principal characteristic).** E223-34E17/34E24/15E7

critical rate-of-rise of ON-state current (thyristor). The maximum value of the rate-of-rise of ON-state current that a thyristor can withstand without deleterious effect. *See also:* **principal current.** E223-34E17/34E24/15E7

critical speed (rotating machinery). A speed at which the amplitude of the vibration of a rotor due to shaft transverse vibration reaches a maximum value. *See* *also:* **rotor (rotating machinery).** 0-31E8

critical temperature (1) (superconductor). The temperature below which a superconductive material is superconducting and above which the material is normal, in the absence of current and external magnetic fields. *See also:* **superconductivity.** E217-15E7
(2) (storage cell or battery). The temperature of the electrolyte at which an abrupt change in capacity occurs. *See also:* **initial test temperature.** 42A60-0

critical torsional speed (rotating machinery). A speed at which the amplitude of the vibration of a rotor due to shaft torsional vibration reaches a maximum value. *See also:* **rotor (rotating machinery).** 0-31E8

critical travel (relay). The amount of movement of the responsive element of a relay during a critical impulse, but not subsequent to the impulse. 37A100-31E11/31E6

critical voltage (1) (magnetron). The highest theoretical value of steady anode voltage, at a given steady magnetic flux density, at which electrons emitted from the cathode at zero velocity would fail to reach the anode. *See also:* **magnetrons.** E160-15E6
(2) (relay). *See:* **relay critical voltage.**

critical-voltage parabola (cutoff parabola) (magnetrons). The curve representing in Cartesian coordinates the variation of the critical voltage as a function of the magnetic induction. *See:* **magnetron.** 50I07-15E6

critical wavelength. *See:* **cutoff wavelength.**

critical withstand current (surge). The highest crest value of a surge of given waveshape and polarity that can be applied without causing disruptive discharge on the test specimen. *See also:* **lightning arrester (surge diverter); power systems, low-frequency and surge testing.** 42A35-31E7

critical withstand voltage (impulse). The highest crest value of an impulse, of given waveshape and polarity, that can be applied without causing disruptive discharge on the test specimen. *See:* **lightning arrester (surge diverter).** *See also:* **power systems, low-frequency and surge testing.** 42A35-31E7

critically damped (industrial control). Damping that is sufficient to prevent any overshoot of the output following an abrupt stimulus. *See:* **control; damping (note).** AS1-34E10

Crookes dark space. *See:* **cathode dark space.**

crossarm. A horizontal member (usually wood or steel) attached to a pole, post, tower or other structure and equipped with means for supporting the conductors. *Note:* The crossarm is placed at right angles to conductors on straight line poles, but splits the angle on light corners. *See also:* **tower.** 42A35-31E13

crossarm guy. A tensional support for a crossarm used to offset unbalanced conductor stress. 42A35-31E13

crossbar switch. A switch having a plurality of vertical paths, a plurality of horizontal paths, and electromagnetically-operated mechanical means for interconnecting any one of the vertical paths with any one of the horizontal paths. *See also:* **telephone switching system.** 42A65-0

crossbar system. An automatic telephone switching system that is generally characterized by the following features: (1) The selecting mechanisms are crossbar switches. (2) Common circuits select and test the switching paths and control the operation of the selecting mechanisms. (3) The method of operation is one in which the switching information is received and stored by controlling mechanisms that determine the operations necessary in establishing a telephone connection. *See also:* **telephone switching system.** 42A65-0

cross cable bond. A cable bond used for bonding between the armor or lead sheath of adjacent cables. *See:* **cable bond; continuity cable bond.** *See also:* **power distribution, underground construction.** 42A35-31E13

cross coupling (transmission medium). A measure of the undesired power transferred from one channel to another. *See also:* **coupling; transmission line.** E146-27E1

crossed-field amplifier (microwave tubes). A crossed-field tube or valve, with a nonreentrant slow-wave structure, used as an amplifier. *See also:* **microwave tube or valve.** 0-15E6

crossed-field tube (microwave). A high-vacuum electron tube in which a direct, alternating, or pulsed voltage is applied to produce an electric field perpendicular both to a static magnetic field and to the direction of propagation of a radio-frequency delay line. *Note:* The electron beam interacts synchronously with a slow wave on the delay line. *See also:* **microwave tube (or valve).** 0-15E6

crossfire. Interfering current in one telegraph or signaling channel resulting from telegraph or signaling currents in another channel. *See also:* **telegraphy.** 42A65-0

crossing angle (electronic navigation). *See:* **angle of cut.**

cross light (television). Equal illumination in front of the subject from two directions at substantially equal and opposite angles with the optical axis of the camera and a horizontal plane. *See also:* **television lighting.** Z7A1-0

cross modulation. A type of intermodulation due to modulation of the carrier of the desired signal by an undesired signal wave. *See also:* **modulating systems; transmission characteristics.** E145-0;42A65-31E3

cross neutralization. A method of neutralization used in push-pull amplifiers whereby a portion of the plate-cathode alternating voltage of each tube is applied to the grid-cathode circuit of the other tube. *See also:* **amplifier; feedback.** E145/E182A/42A65-0

crossover (cathode-ray tube). The first focusing of the beam that takes place in the electron gun. *See also:* **cathode-ray tubes.** 50I07-15E6

crossover, automatic voltage-current (power supplies). The characteristic of a power supply that automatically changes the method of regulation from constant voltage to constant current (or vice versa) as dictated by varying load conditions (see the following figure). *Note:* The constant-voltage and constant-current levels can be independently adjusted within the specified voltage and current limits of the power supply. The intersection of constant-voltage and constant-current lines is called the crossover point *E, I* and may be located anywhere within the volt-ampere range of the power supply.

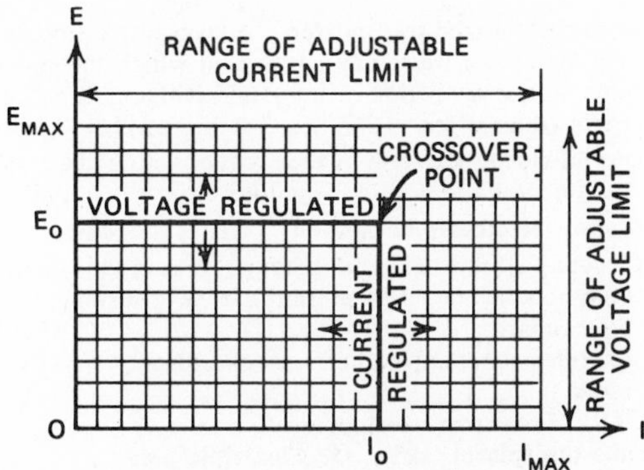

Automatic voltage-current crossover.

See also: **power supply.** KPSH-10E1

crossover characteristic curve (navigation systems such as very-high-frequency omnidirectional ranges and instrument landing systems). The graphical representation of the indicator current variation with change of position in the crossover region. *See also:* **navigation.** E172-10E6

crossover frequency (frequency-dividing networks). The frequency at which equal power is delivered to each of two adjacent channels when all channels are properly terminated. *See also:* **loudspeaker; transition frequency.** 0-1E1

crossover network. *See:* **dividing network.**

crossover region (navigation systems). A loosely defined region in space containing the course line and within which a transverse flight yields information useful in determining course sensitivity and flyability. *See also:* **navigation.** 0-10E6

crossover spiral. *See:* **lead-over groove.**

crossover voltage, secondary-emission (charge-storage tubes). The voltage of a secondary-emitting surface, with respect to cathode voltage, at which the secondary-emission ratio is unity. The crossovers are numbered in progression with increasing voltage. See the following figure. *Note:* The qualifying phrase **secondary-emission** is frequently dropped in general usage.

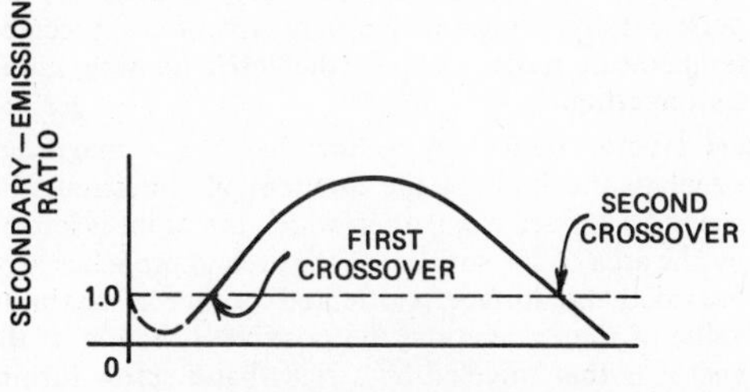

Typical secondary-emission curve.

See also: **charge-storage tube.** E158-15E6

cross polarization. The polarization orthogonal to a reference polarization component. *See also:* **antenna.** 0-3E1

cross protection. An arrangement to prevent the improper operation of devices from the effect of a cross in electric circuits. *See also:* **railway signal and interlocking.** 42A42-0

cross product. *See:* **vector product.**

cross rectifier circuit. A circuit that employs four or more rectifying elements with a conducting period of 90 electrical degrees plus the commutating angle. *See also:* **rectification.** 42A15-0

cross section (radar). *See:* **effective echo area.**

cross-sectional area (conductor) (cross section of a conductor). The sum of the cross-sectional areas of its component wires, that of each wire being measured perpendicular to its individual axis. 42A35-31E3

crosstalk (1). Undesired energy appearing in one signal path as a result of coupling from other signal paths. *Note:* Path implies wires, waveguides, or other localized or constrained transmission systems. *See also:* **coupling.** E151/42A65-0

(2) (electroacoustics). The unwanted sound reproduced by an electroacoustic receiver associated with a given transmission channel resulting from cross coupling to another transmission channel carrying sound-controlled electric waves or, by extension, the electric waves in the disturbed channel that result in such sound. *Note:* In practice, crosstalk may be measured either by the volume of the overheard sounds or by the magnitude of the coupling between the disturbed and disturbing channels. In the latter case, to specify the volume of the overheard sounds, the volume in the disturbing channel must also be given. *See also:* **coupling.** 42A65-31E3

crosstalk coupling (crosstalk loss). Cross coupling between speech communication channels or their component parts. *Note:* Crosstalk coupling is measured between specified points of the disturbing and disturbed circuits and is preferably expressed in decibels. *See also:* **coupling.** 42A65-0

crosstalk, electron beam (charge-storage tubes). Any spurious output signal that arises from scanning or from the input of information. *See also:* **charge-storage tube.** E158-15E6

crosstalk loss. *See:* **crosstalk coupling.**

crosstalk unit*. Crosstalk coupling is sometimes expressed in crosstalk units through the relation

$$\text{Crosstalk units} = 10^{[6 - (L/20)]}$$

where L = crosstalk coupling in decibels. *Note:* For two circuits of equal impedance, the number of crosstalk units expresses the current in the disturbed circuit as millionths of the current in the disturbing circuit. *See also:* **coupling.**

*Obsolescent 42A65-0

crude metal. Metal that contains impurities in sufficient quantities to make it unsuitable for specified purposes or that contains more valuable metals in sufficient quantities to justify their recovery. *See also:* **electrorefining.** 42A60-0

crust. A layer of solidified electrolyte. *See also:* **fused electrolyte.** 42A60-0

cryogenics. The study and use of devices utilizing properties of materials near absolute-zero temperature. X3A12-16E9

cryotron. A superconductive device in which current in one or more input circuits magnetically controls the superconducting-to-normal transition in one or more output circuits, provided the current in each output circuit is less than its critical value. *See also:* **superconductivity.** E217-15E7

crystal (communication practice). (1) A piezoelectric crystal. (2) A piezoelectric crystal plate. (3) A crystal rectifier.
See:
antiresonant frequency of a crystal unit;
circuits and devices;
equivalent circuit;
fundamental-type piezoelectric crystal unit;
mode of vibration;
overtone-type piezoelectric crystal unit;
piezoelectric crystal element;
piezoelectric crystal plate;
piezoelectric crystal unit;
resonance frequency;
type of piezoelectric crystal cut. 42A65-21E0

crystal-controlled oscillator. *See:* **crystal oscillator.**

crystal-controlled transmitter. A transmitter whose carrier frequency is directly controlled by a crystal oscillator. *See also:* **radio transmitter.** 42A65-31E3

crystal diode. A rectifying element comprising a semiconducting crystal having two terminals designed for use in circuits in a manner analogous to that of electron-tube diodes. *See also:* **rectifier.** 42A65-0

crystal filter. An electric wave filter employing piezoelectric crystals for its reactive elements. *See also:* **circuits and devices.** 0-31E3

crystal loudspeaker (piezoelectric loudspeaker). A loudspeaker in which the mechanical displacements are produced by piezoelectric action. *See also:* **loudspeaker.** 42A65-0

crystal microphone (piezoelectric microphone). A microphone that depends for its operation on the generation of an electric charge by the deformation of a body (usually crystalline) having piezoelectric properties. *See also:* **microphone.** 42A65-0

crystal mixer (mixer). A crystal receiver that can be fed simultaneously from a local oscillator and signal source, for the purpose of frequency changing. *See:* **waveguide.** 50I62-3E1

crystal oscillator (crystal-controlled oscillator). An oscillator in which the principal frequency-determining factor is the mechanical resonance of a piezoelectric crystal. *See also:* **oscillatory circuit.** E182A-0

crystal pickup (piezoelectric pickup). A phonograph pickup that depends for its operation on the generation of an electric charge by the deformation of a body (usually crystalline) having piezoelectric properties. *See also:* **phonograph pickup.** 42A65-0

crystal pulling. A method of crystal growing in which the developing crystal is gradually withdrawn from a melt. *See also:* **electron devices, miscellaneous.** E102-0

crystal receiver. A waveguide incorporating a crystal detector for the purpose of rectifying received electromagnetic signals. *See:* **waveguide.** 50I62-3E1

crystal spots. Spots produced by the growth of metal sulfide crystals upon metal surfaces with a sulfide finish and lacquer coating. The appearance of crystal spots is called spotting in. *See also:* **electroplating.** 42A60-0

crystal-stabilized transmitter. A transmitter employing automatic frequency control, in which the reference frequency is that of a crystal oscillator. *See also:* **radio transmitter.** E145/42A65-0

crystal-video receiver. A receiver consisting of a crystal detector and a video amplifier. 42A65-0

***C* scan (electronic navigation).** *See:* ***C* display.**

***C* scope (1)(class *C* oscilloscope).** A cathode-ray oscilloscope arranged to present a type *C*-display. *See also:* **radar.** 42A65-0
(2) (electronic navigation). *See:* ***C* display.** *See also:* **radar.**

CTS. *See:* **carrier test switch.**

cube tap (plural tap*). *See:* **multiple plug.**
*Deprecated

cumulative compounded (rotating machinery). Applied to a compound machine to denote that the magnetomotive forces of the series and the shunt field windings are in the same direction. *See:* **direct-current commutating machine; magnetomotive force.** 0-31E8

cumulative demand meter (or register). An indicating demand meter in which the accumulated total of maximum demands during the preceding periods is indicated during the period after the meter has been reset and before it is reset again. *Note:* The maximum demand for any one period is equal or proportional to the difference between the accumulated readings before and after reset. *See also:* **electricity meter (meter).** 42A30/12A0-0

cuprous chloride cell. A primary cell in which depolarization is accomplished by cuprous chloride. *See also:* **electrochemistry.** 42A60-0

Curie point. A temperature marking a transition between ferromagnetic (ferroelectric, antiferromagnetic) properties and nonferromagnetic (nonferroelectric, nonantiferromagnetic) properties of a material. *Notes:* (1) Ferromagnetic materials have a single Curie point, above which they are paramagnetic. (2) The susceptibility of antiferromagnetic material increases with temperature, up to the Curie point, decreasing thereafter. (3) Some ferroelectric materials have both upper and lower Curie points, and some have different Curie points associated with different crystal axes. (4) Experimentally, the transition may cover an appreciable temperature range. *See also:* **dielectric heating; induction heating.** E270-0

curl (vector field). A vector that has a magnitude equal to the limit of the quotient of the circulation around a surface element on which the point is located by the area of the surface, as the area approaches zero, provided the surface is oriented to give a maximum value of the circulation; the positive direction of this vector is that traveled by a right-hand screw turning about an axis normal to the surface element when an integration around the element in the direction of the turning of the screw gives a positive value to the circulation. If the vector **A** of a vector field is expressed in terms of its three rectangular components A_x, A_y, and A_z, so that the values of A_x, A_y, and A_z are each given as a function of x, y, and z, the curl of the vector field (abbreviated curl **A** or $\nabla \times \mathbf{A}$) is the vector sum of the partial derivatives of each component with respect to the axes that are perpendicular to it, or

$$\text{curl } \mathbf{A} = \nabla \times \mathbf{A} = \begin{vmatrix} \mathbf{i} & \mathbf{j} & \mathbf{k} \\ \frac{\delta}{\delta x} & \frac{\delta}{\delta y} & \frac{\delta}{\delta z} \\ A_x & A_y & A_z \end{vmatrix}$$

$$= \mathbf{i}\left(\frac{\delta A_x}{\delta y} - \frac{\delta A_y}{\delta z}\right) + \mathbf{j}\left(\frac{\delta A_x}{\delta z} - \frac{\delta A_z}{\delta x}\right)$$

$$+ \mathbf{k}\left(\frac{\delta A_y}{\delta x} - \frac{\delta A_x}{\delta y}\right)$$

$$= \mathbf{i}(D_y A_z - D_z A_y)$$

$$+ \mathbf{j}(D_z A_x - D_x A_z) + \mathbf{k}(D_x A_y - D_y A_z)$$

where **i, j,** and **k** are unit vectors along the *x, y,* and *z* axes, respectively. Example: The curl of the linear velocity of points in a rotating body is equal to twice the angular velocity. The curl of the magnetic field strength at a point within an electric conductor is equal to *k* times the current density at the point where *k* is a constant depending on the system of units. E270-0

current (electric) (1) (general). A generic term used when there is no danger of ambiguity to refer to any one or more of the currents specifically described. *Notes:* (1) For example, in the expression "the current in a simple series circuit," the word current refers to the conduction current in the wire of the inductor and the displacement current between the plates of the capacitor. (2) A direct current is a unidirectional current in which the changes in value are either zero or so small that they may be neglected. A given current would be considered a direct current in some applications, but would not necessarily be so considered in other applications. E270-0

(2) (modified by an adjective). The use of certain adjectives before "current" is often convenient, as in convection current, anode current, electrode current, emission current, etcetera. The definition of conducting current usually applies in such cases and the meaning of adjectives should be defined in connection with the specific applications. E270-0

current amplification (1) (general). An increase in signal current magnitude in transmission from one point to another or the process thereof.

(2) (transducer). The scalar ratio of the signal output current to the signal input current. *Warning:* By incorrect extension of the term decibels, this ratio is sometimes expressed in decibels by multiplying its common logarithm by 20. It may be correctly expressed in decilogs. *See also:* **transmission characteristics.** 0-42A65

(3) (multiplier phototube). The ratio of the output current to the cathode current due to photoelectric emission at constant electrode voltages. *Notes:* (A) The term output current and photocathode current as here used does not include the dark current. (B) This characteristic is to be measured at levels of operation that will not cause saturation. *See also:* **phototube.** 42A70-15E6

(4) (magnetic amplifier). The ratio of differential output current to differential control current. *See also:* **rating and testing magnetic amplifiers.** E107-0

(5) (electron multipliers). The ratio of the signal output current to the current applied to the input. *See also:* **amplifier; transmission characteristics.** 0-15E6

current, anode. *See:* **electrode current.**

current attenuation. *See:* **attenuation, current.**

current balance ratio. The ratio of the metallic-circuit current or noise-metallic (arising as a result of the action of the longitudinal-circuit induction from an exposure on unbalances outside the exposure) to the longitudinal circuit current or noise-longitudinal in sigma at the exposure terminals. It is expressed in microamperes per milliampere or the equivalent. *See also:* **inductive coordination.** 42A65-0

current-balance relay. A relay that operates by comparing the magnitudes of two current inputs. *See also:* **relay.** 0-31E6

current-balancing reactor. *See:* **reactor, current-balancing.**

current-carrying capacity (contacts). The maximum current that a contact is able to carry continuously or for a specified period of time. *See:* **contactor.** 50I16-34E10

current-carrying part. A conducting part intended to be connected in an electric circuit to a source of voltage. *Note:* Non-current-carrying parts are those not intended to be so connected. *See also:* **center of distribution; power distribution, overhead construction.** 2A2/42A35-31E13;37A100-31E11

current circuit (electric instrument). The combination of conductors and windings of the instrument that carries the current of the circuit in which a given electrical quantity is to be measured, or a definite fraction of that current, or a current dependent upon it. *See also:* **moving element of an instrument; watthour meter.** 39A1/42A30-0

current crest factor (mercury-lamp–ballast combination). The ratio of the peak value of lamp current to the root mean-square value of lamp current. 82A9-0

current cutoff (power supplies). An overload protective mechanism designed into certain regulated power supplies to reduce the load current automatically as the load resistance is reduced. This negative resistance characteristic reduces overload dissipation to negligible proportions and protects sensitive loads. The accompanying figure shows the *E-I* characteristic of a power supply equipped with a current cutoff overload protector. *See also:* **power supply.** KPSH-10E1

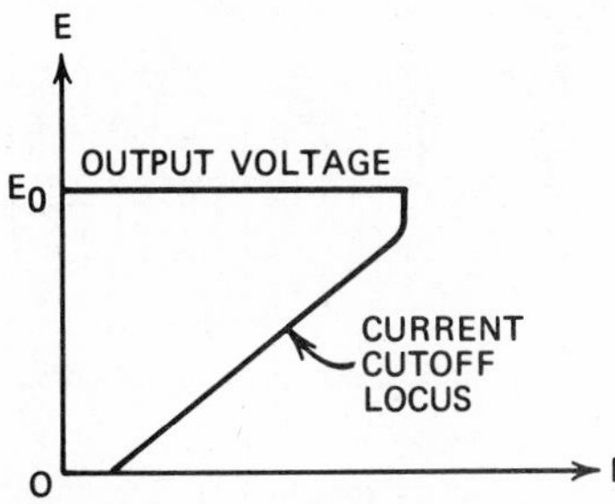

Output characteristic of a power supply equipped with a current cutoff overload protector.

current density (1). A generic term used when there is no danger of ambiguity to refer either to conduction-

current density or to displacement-current density, or to both. E270-0
(2) (electrolytic cells). On a specified electrode in an electrolytic cell, the current per unit area of that electrode. *See also:* **electrolytic cell.** 42A60-0
(3) (semiconductor rectifier). The current per unit active area. *See also:* **rectification.** E59-34E17

current efficiency (specified electrochemical process). The proportion of the current that is effective in carrying out that process in accordance with Faraday's law. *See also:* **electrochemistry.** 42A60-0

current generator (signal-transmission system). A two-terminal circuit element with a terminal current substantially independent of the voltage between its terminals. *Note:* An ideal current generator has zero internal admittance. *See also:* **network analysis; signal.** E160-13E6

current-limit (control) (industrial control). A control function that prevents a current from exceeding its prescribed limits. *Note:* Current-limit values are usually expressed as percent of rated-load value. If the current-limit circuit permits the limit value to increase somewhat instead of being a single value, it is desirable to provide either a curve of the limit value of current as a function of some variable such as speed or to give limit values at two or more conditions of operation. *See:* **control system, feedback.** AS1-34E10

current-limit acceleration (electric drive) (industrial control). A system of control in which acceleration is so governed that the motor armature current does not exceed an adjustable maximum value. *See also:* **electric drive.** 42A25-34E10

current-limit control (electric drive). A system of control in which acceleration, or retardation, or both, are so governed that the armature current during speed changes does not exceed a predetermined value. *See also:* **electric drive.** 42A25-0

current limiting, automatic (power supplies). An overload protection mechanism that limits the maximum output current to a preset value, and automatically restores the output when the overload is removed. See the accompanying figure. *See:* **short-circuit protection.** *See also:* **power supply.** KPSH-10E1

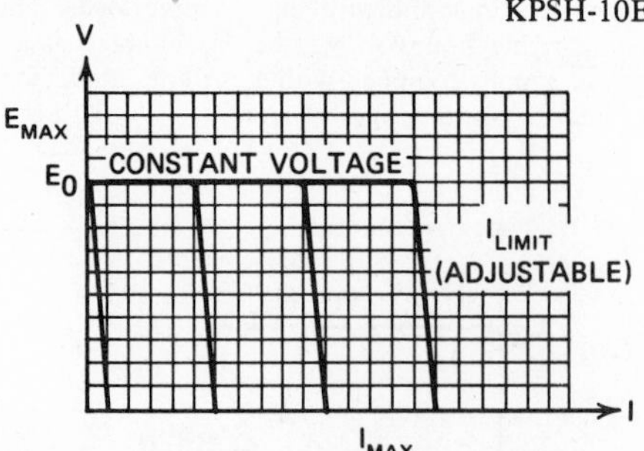

Plot of typical current-limiting curves.

current-limiting fuse. A fuse that, when it is melted by a current within its specified current-limiting range, abruptly introduces a high arc voltage to reduce the current magnitude and duration. *Note:* The values specified in standards for the threshold ratio, peak let-through current, and $I^2 t$ characteristic are used as the measures of current-limiting ability. 37A100-31E11

current-limiting characteristic curve (current-limiting fuse) (peak let-through characteristic curve) (cutoff characteristic curve). A curve showing the relationship between the maximum peak current passed by a fuse and the correlated root-mean-square available current magnitudes under specified voltage and circuit impedance conditions. *Note:* The root-mean-square available current may be symmetrical or asymmetrical. 37A100-31E11

current-limiting range (current-limiting fuse). That specified range of currents between the threshold current and the rated interrupting current within which current limitation occurs. 37A100-31E11

current-limiting resistor (industrial control). A resistor inserted in an electric circuit to limit the flow of current to some predetermined value. *See:* **control system, feedback.** 42A20-0

current loss (electric instrument) (voltage circuit current drain) (parallel loss of an electric instrument). In a voltage-measuring instrument, the value of the current when the applied voltage corresponds to nominal end-scale indication. *Note:* In other instruments it is the current in the voltage circuit at rated voltage. *See also:* **accuracy rating (instrument).** 39A1/42A30-0

current margin (neutral direct-current telegraph system). The difference between the steady-state currents flowing through a receiving instrument, corresponding, respectively, to the two positions of the telegraph transmitter. *See also:* **telegraphy.** 42A65-0

current of traffic. The movement of trains on a main track in one direction specified by the rules. *See also:* **railway signal and equipment.** 42A42-0

current phase-balance protection. A method of protection in which an abnormal condition within the protected equipment is detected by the current unbalance between the phases of a normally balanced polyphase system. 37A100-31E11/31E6

current pulsation (rotating machinery). The difference between maximum and minimum amplitudes of the motor current during a single cycle corresponding to one revolution of the driven load expressed as a percentage of the average value of the current during this cycle. *See also:* **asynchronous machine; synchronous machine.** 0-31E8

current, rated. *See:* **rated current.**

current rating (rectifier transformer). The root-mean-square equivalent of a rectangular current waveshape based on direct-current rated load commutated with zero commutating angle. *See also:* **rectifier transformer.** 57A18-0

current rating, 60-hertz (arrester). A designation of the range of the symmetrical root-mean-square fault currents of the system for which the arrester is designed to operate. *Notes:* (1) An expulsion arrester is given a maximum current rating and may also have a minimum current rating. (2) The designation of the maximum and minimum current ratings of an expulsion arrester not only specifies the useful operating range of the arrester between those extreme values for symmetrical root-mean-square short-circuit current, but indicates that at the point of application of the arrester the root-mean-square short-circuit current for the system should neither be greater than the maximum nor less than the minimum current rating.

See:
alternating sparkover voltage;
arrester discharge voltage-current characteristic;
discharge current;
discharge withstand current rating;
impulse sparkover voltage;
power-frequency current interrupting rating;
power-frequency sparkover voltage;
rating;
unit operation;
voltage rating. 42A20-0

current ratio (series transformer) (mercury lamp). The ratio of the (root-mean-square) primary current to the root-mean-square secondary current under specified conditions of load. 42A15-31E12;82A7-0

current-recovery ratio (arc-welding apparatus). With a welding power supply delivering current through a short-circuited resistor whose resistance is equivalent to the load setting on the power supply, and with the short-circuit suddenly removed, the ratio of (1) the minimum transient value of current upon the removal of the short-circuit to (2) the final steady-state value is the current-recovery ratio. *See also:* **electric arc-welding apparatus.** 87A1-0

current regulation (constant-current transformer). The maximum departure of the secondary current from its rated value, with rated primary voltage at rated frequency applied, and at rated secondary power factor, and with the current variation taken between the limits of a short-circuit and rated load. *Note:* This regulation may be expressed in per unit, or percent, on the basis of the rated secondary current. *See:* **constant-current transformer.** 42A15-31E12

current relay. A relay that functions at a predetermined value of current. *Note:* It may be an overcurrent, undercurrent, or reverse-current relay. *See also:* **relay.** E16-0

current-responsive element (fuse). That part with predetermined characteristics, the melting and severance or severances of which initiate the interrupting function of the fuse. *Note:* The current-responsive element may consist of one or more fusible elements combined with a strain element and/or other component(s) that affect(s) the current-responsive characteristic. 37A100-31E11

current-sensing resistor (power supplies). A resistor placed in series with the load to develop a voltage proportional to load current. A current-regulated direct-current power supply regulates the current in the load by regulating the voltage across the sensing resistor. *See also:* **power supply.** KPSH-10E1

current tap.* *See:* **multiple lampholder; plug adapter lampholder.**

*Deprecated

current terminals of instrument shunts (electric power systems). Those terminals that are connected into the line whose current is to be measured and that will carry the current of the shunt. *See also:* **power system, low-frequency and surge testing.** 0-31E5

current transformer (1) (general). An instrument transformer intended to have its primary winding connected in series with a circuit carrying the current to be measured or controlled. *Note:* The term **primary winding** when applied to certain types of current transformer denotes the cable or bus bar that links with the magnetic circuit to form a single effective primary turn. *See also:* **instrument transformer.** 12A0/42A30-0;42A15-31E12

(2) (wound-type). A transformer that has a fixed primary winding mechanically encircling the core; it may have one or more primary turns. *Note:* The primary and secondary windings are completely insulated and permanently assembled on the core as an integral structure. *See also:* **instrument transformer.** 0-31E12

(3) (window-type). A transformer that has a secondary winding insulated from and permanently assembled on the core, but has no primary winding as an integral part of the structure. *Note:* Complete or partial insulation is provided for a primary winding in the window through which one or more turns of the line conductor can be threaded to provide the primary winding. 0-31E12

(4) (bushing-type). Has a secondary winding completely insulated and permanently assembled on the core, but has no primary winding or insulation for a primary winding. *Note:* It is intended for use with a primary winding consisting of a completely insulated conductor. This conductor is usually a component part of other apparatus with which the design of the bushing-type current transformer must be coordinated. 42A15-31E12

(5) (double-secondary). One that has two secondary coils each on a separate magnetic circuit with both magnetic circuits excited by the same primary winding. 57A13-31E12

(6) (multiple-secondary). Three or more secondary coils each on a separate magnetic circuit with all magnetic circuits excited by the same primary winding. 57A13-31E12

(7) (multiratio). More than one ratio can be obtained by the use of taps on the secondary winding. 57A13-31E12

(8) (three-wire-type). A transformer that has two primary windings each completely insulated for the rated insulation class of the transformer. This type of current transformer is for use in a three-wire, single phase installation. *Note:* The primary windings and secondary winding are permanently assembled on the core as an integral structure. The secondary current is proportional to the phasor sum of the primary currents. 0-31E12

current-type telemeter. A telemeter that employs the magnitude of a single current as the translating means. *See also:* **telemetering.** 42A30/37A100-31E11

curvature (coaxial transmission line). The radial departure from a straight line between any two points on the external surface of a conductor. *See also:* **waveguide.** 83A14-0

curve follower. *See also:* **electronic analog computer.**

cushion clamp (rotating machinery). A device for securing the cushion to the supporting member. *See:* **cradle base (rotating machinery).** 42A10-0

cut-in loop. A circuit on the roadway energized to automatically cut in the train control or cab signal apparatus on a passing vehicle. 42A42-0

cutoff. *See:* **cutoff frequency.**

cutoff angle (luminaire). The angle, measured up from nadir, between the vertical axis and the first line of sight at which the bare source is not visible. *See also:* **bare (exposed) lamp.** Z7A1-0

cutoff attenuator. An adjustable length of waveguide used below its cutoff frequency to introduce variable nondissipative attenuation. *See also:* **waveguide.** 42A65-0

cutoff characteristic (current-limiting fuse). *See:* **current-limiting characteristic curve.**

cutoff frequency (1) (general). The frequency that is identified with the transition between a pass band and an adjacent attenuation band of a system or transducer. *Note:* It may be either a theoretical cutoff frequency or an effective cutoff frequency. *See also:* **transmission characteristics.** E151-0

(2) (waveguide). For a given transmission mode in a nondissipative uniconductor waveguide, the frequency below which the propagation constant is real. *See also:* **transmission characteristics; waveguide.** E146-3E1

cutoff frequency, effective (electric circuit). A frequency at which its insertion loss between specified terminating impedances exceeds by some specified amount the loss at some reference point in the transmission band. *Note:* The specified insertion loss is usually 3 decibels. *See also:* **cutoff frequency; network analysis.** 42A65-0

cutoff relay (telephony). A relay associated with a subscriber line, that disconnects the line relay from the line when the line is called or answered. *See also:* **telephone switching system.** 42A65-0

cutoff voltage (1) (final voltage) (battery). The prescribed voltage at which the discharge is considered complete. *Note:* The cutoff or final voltage is usually chosen so that the useful capacity of the battery is realized. The cutoff voltage varies with the type of battery, the rate of discharge, the temperature, and the kind of service. The term **cutoff voltage** is applied more particularly to primary batteries, and **final voltage** to storage batteries. *See also:* **battery (primary or secondary).** 42A60-0

(2) (electron tube). The electrode voltage that reduces the value of the dependent variable of an electron-tube characteristic to a specified low value. *Note:* A specific cutoff characteristic should be identified as follows: current versus grid cutoff voltage, spot brightness versus grid cutoff voltage, etcetera. *See also:* **electrode voltage (electron tube).** 42A70-15E6

(3) (magnetrons). *See:* **critical voltage (magnetrons).**

cutoff waveguide (evanescent waveguide). A waveguide used at a frequency below its cutoff frequency. *See:* **waveguide.** 50I62-3E1

cutoff wavelength (1) (mode in a waveguide) (critical wavelength). That wavelength, in free space or in the unbounded guide medium, as specified, above which a traveling wave in that mode cannot be maintained in the guide. *Note:* For $TE_{m,n}$ or $TM_{m,n}$ waves in hollow rectangular cylinders,

$$\lambda_c = 2/[(m/a)^2 + (n/b)^2]^{1/2}$$

where a is the width of the waveguide along the x coordinate and b is the height of the waveguide along the y coordinate.

The following table gives the ratio of cutoff wavelengths to diameters for $TM_{n,m}$ waves in hollow circular metal cylinders:

	n =	0	1	2	3	4	5
m =	1	1.307	0.820	0.613	0.483	0.414	0.358
	2	0.569	0.448	0.373	0.322	0.284	0.2547
	3	0.363	0.309	0.270	0.241	0.219	0.200
	4	0.267	0.2375	0.2127	0.1938		
	5	0.2106	0.1910	0.1750			

The following table gives the ratio of cutoff wavelengths to diameters for $TE_{m,n}$ waves in hollow circular metal cylinders:

	n =	0	1	2	3	4
m =	1	0.820	1.708	1.030	0.748	0.590
	2	0.448	0.59	0.468	0.382	0.338
	3	0.309	0.368	0.315	0.277	0.247
	4	0.2375	0.27	0.24	0.22	0.198
	5	0.1910	0.21	0.194	0.18	

See: **guided wave.** *See also:* **waveguide.** E210/50I62-3E1

(2)(uniconductor waveguide). The ratio of the velocity of electromagnetic waves in free space to the cutoff frequency. *See also:* **waveguide.** E146-0

cutout (1) (general). An electric device used manually or automatically to interrupt the flow of current through any particular apparatus or instrument. *See also:* **railway signal and interlocking.** 42A42-0

(2) (electric distribution systems). An assembly of a fuse support with either a fuseholder, fuse carrier, or disconnecting blade. *Notes:* (1) The fuseholder or fuse carrier may include a conducting element (fuse link), or may act as a disconnecting blade by the inclusion of a nonfusible member. (2) The term **cutout** as defined here is restricted in practice to equipment used on distribution systems. *See:* **distribution** and **distribution cutout.** For fuses having similar components used on power systems. *See* **power and power fuse.** 37A100-31E11

cutout base.* *See:* **fuse holder.**

*Deprecated

cutout box (interior wiring). An enclosure designed for surface mounting and having swinging doors or covers secured directly to and telescoping with the walls of the box proper. *See also:* **cabinet; interior wiring.** 42A95-0

cutout loop (railway practice). A circuit in the roadway that cooperates with vehicle-carried apparatus to cut out the vehicle train control or cab signal apparatus. 42A42-0

cut paraboloidal reflector. A reflector that is not symmetrical with respect to its axis. *See also:* **antenna.** E145-0

cut-section. A location within a block other than a signal location where two adjacent track circuits end. 42A42-0

cut-set (network). A set of branches of a network such that the cutting of all the branches of the set increases the number of separate parts of the network, but the cutting of all the branches except one does not. *See also:* **network analysis.** E153/E270-0

cutter (audio and electroacoustics) (mechanical recording head). An electromechanical transducer that transforms an electric input into a mechanical output, that is typified by mechanical motions that may be inscribed into a recording medium by a cutting stylus. *See also:* **phonograph pickup.** 0-1E1

cutter compensation (numerically controlled machines). Displacement, normal to the cutter path, to adjust for the difference between actual and programmed cutter radii or diameters. *See also:* **numerically controlled machines.** EIA3B-34E12

cutting down (metal or electrodeposit) (electroplating). Polishing for the purpose of removing roughness or irregularities. *See also:* **electroplating.** 42A60-0

cutting stylus (electroacoustics). A recording stylus with a sharpened tip that, by removing material, cuts a groove into the recording medium. *See also:* **phonograph pickup.** 0-1E1

cybernetics. *See:* **system science.**

cycle. (1) (A) An interval of space or time in which one set of events or phenomena is completed. (B) Any set of operations that is repeated regularly in the same sequence. The operations may be subject to variations on each repetition. *See also:* **electronic digital computer.** X3A12-16E9

(2) The complete series of values of a periodic quantity that occurs during a period. *Note:* It is one complete set of positive and negative values of an alternating current. *See also:* **electronic digital computer.** E45/E270-0

cycle. *See:* **major cycle; minor cycle.** *See also:* **electronic digital computer.**

cycle of operation (storage cell or battery). The discharge and subsequent recharge of the cell or battery to restore the initial conditions. *See also:* **charge.** 42A60-0

cyclically magnetized condition. A condition of a magnetic material when, under the influence of a magnetizing force that is a cyclic (but not necessarily periodic) function of time having one maximum and one minimum per cycle, it follows identical hysteresis loops on successive cycles. *See also:* **static magnetic storage.** E163/E270-0

cyclic code error detection (power-system communication). The process of cyclically computing bits to be added at the end of a word such that an identical computation will reveal a large portion of errors that may have been introduced in transmission. *See also:* **digital.** 0-31E3

cyclic duration factor (rotating machinery). The ratio between the period of loading including starting and electric braking, and the duration of the duty cycle, expressed as a percentage. *See also:* **asynchronous machine; direct-current commutating machine; synchronous machine.** 0-31E8

cyclic function. A function that repetitively assumes a given sequence of values at an arbitrarily varying rate. *Note:* That is, if y is a periodic function of x and x in turn is a monotonic nondecreasing function of t, then y is said to be a cyclic function of t. E270-0

cyclic irregularity (rotating machinery). The periodic fluctuation of speed by irregularity of the prime-mover torque. *See also:* **asynchronous machine; direct-current commutating machine; synchronous machine.** 0-31E8

cyclic shift (digital computers). (1) An operation that produces a word whose characters are obtained by a cyclic permutation of the characters of a given word. *See also:* **electronic digital computer.** E162/E270-0

(2) A shift in which the data moved out of one end of the storing register are reentered into the other end, as in a closed loop. *See:* **circulating register.** X3A12-16E9

cyclometer register. A set of 4 or 5 wheels numbered from zero to 9 inclusive on their edges, and so enclosed and connected by gearing that the register reading appears as a series of adjacent digits. *See also:* **watthour meter.** 42A30-0

cyclotron. A device for accelerating positively charged particles (for example, protons, deuterons, etcetera.) to high energies. The particles in an evacuated tank are guided in spiral paths by a static magnetic field while they are accelerated many times by an electric field of mixed frequency. *See also:* **electron devices, miscellaneous.** 42A70-15E6

cyclotron frequency (1). The frequency at which an electron traverses an orbit in a steady uniform magnetic field and zero electric field. *Note:* It is given by the product of the electron charge and the magnetic flux density, divided by 2π times the electron mass.

(2) (radio wave propagation). Gyrofrequency. *See:* **gyrofrequency; magnetrons.** 42A70/E160-15E6

cyclotron, frequency-modulated. *See:* **frequency-modulated cyclotron.**

cyclotron-frequency magnetron oscillations. Those oscillations whose frequency is substantially the cyclotron frequency. 42A70-0

cylindrical reflector. A reflector that is a portion of a cylinder. *Note:* This cylinder is usually parabolic, although other shapes may be used. *See also:* **antenna.** E145-3E1

cylindrical-rotor generator (solid-iron cylindrical-rotor generator) (turbine generator*). An alternating-current generator driven by a high-speed turbine (usually steam) and having an exciting winding embedded in a cylindrical steel rotor. *See:* **generating station.**

*Deprecated. Favor steam-turbine or gas-turbine generator as distinguished from hydraulic-turbine generator. 0-31E8

cylindrical-rotor machine. A machine having a cylindrically-shaped rotor the periphery of which is slotted to accommodate the coil sides of a field winding. *See:* **rotor (rotating machinery).** 0-31E8

cylindrical wave (radio wave propagation). A wave whose equiphase surfaces form a family of coaxial or confocal cylinders. *See also:* **radiation; radio wave propagation.** E211-3E2;42A65-0

D

damped harmonic system (1) (linear system with one degree of freedom). A physical system in which the internal forces, when the system is in motion, can be represented by the terms of a linear differential equation with constant coefficients, the order of the equation being higher than the first. Example: The differential equation of a damped system is often of the form

$$M\frac{d^2x}{dt^2}+F\frac{dx}{dt}+Sx=f(t)$$

where M, F, and S are positive constants of the system; x is the dependent variable of the system (displacement in mechanics, quantity of electricity, etcetera); and $f(t)$ is the applied force. Examples: A tuning fork is a damped harmonic system in which M represents a mass; F a coefficient of damping; S a coefficient of restitution; and x a displacement. Also an electric circuit containing constant inductance, resistance, and capacitance is a damped harmonic system, in which case M represents the self-inductance of the circuit; F the resistance of the circuit; S the reciprocal of the capacitance; and x the charge that has passed through a cross section of the circuit.
(2) (critical damping). The name given to that special case of damping that is the boundary between underdamping and overdamping. A damped harmonic system is critically damped if $F^2 = 4MS$. *See also:* **network analysis.**
(3) (overdamping) (aperiodic damping*). The special case of a damping in which the free oscillation does not change sign. A damped harmonic system is overdamped if $F^2 > 4MS$. *See also:* **network analysis.**
*Deprecated
(4) (underdamping) (periodic damping*). The special case of damping in which the free oscillation changes sign at least once. A damped harmonic system is underdamped if $F^2 < 4MS$.
**Deprecated* E270-0
(5) (underdamped). Damped insufficiently to prevent oscillation of the output following an abrupt input stimulus. *Note:* In an underdamped linear second-order system, the roots of the characteristic equation have complex values. *See:* **control; critical damping; damping.** *See also:* **control system, feedback.** ASI-34E10;0-23E0

damped oscillation (damped vibration). An oscillation in which the amplitude of the oscillating quantity decreases with time. *Note:* If the rate of decrease can be expressed by means of a mathematical function, the name of the function may be used to describe the damping. Thus if the rate of decrease is exponential, the system is said to be an exponentially damped system. E270-0

damped sinusoidal function. A function having the form $f_1(t)\sin(\omega t+\theta)$ where $f_1(t)$ is a monotonic decreasing function, and ω and θ are constants. E270-0

damped vibration. *See:* **damped oscillation.**

damped wave. A wave in which, at every point, the amplitude of each sinusoidal component is a decreasing function of the time. *See also:* **signal; wave front.** E270-0

damper bar. *See:* **amortisseur bar.**

damper segment (rotating machinery). One portion of a short-circuiting end ring (of an amortisseur winding) that can be separated into parts for mounting or removal without access to a shaft end. *See also:* **rotor (rotating machinery).** 0-31E8

damper winding (amortisseur winding) (rotating machinery). A winding consisting of a number of conducting bars short-circuited at the ends by conducting rings or plates and distributed on the field poles of a synchronous machine to suppress pulsating changes in magnitude or position of the magnetic field linking the poles. 0-31E8

damping (1) (noun). The temporal decay of the amplitude of a free oscillation of a system, associated with energy loss from the system.
(2) (adjective). Pertaining to or productive of damping. *Note:* The damping of many physical systems is conveniently approximated by a viscous damping coefficient in a second-order linear differential equation (or a quadratic factor in a transfer function). In this case the system is said to be critically damped when the time response to an abrupt stimulus is as fast as possible without overshoot; underdamped (oscillatory) when overshoot occurs; overdamped (aperiodic) when response is slower than critical. The roots of the quadratic are, respectively, real and equal; complex, and real and unequal. *See also:* **control system, feedback.**
(3) (relative underdamped system). A number expressing the quotient of the actual damping of a second-order linear system or element by its critical damping. *Note:* For any system whose transfer function includes a quadratic factor $s^2+2z\omega_n s+\omega_n^2$, relative damping is the value of z, since $z = 1$ for critical damping. Such a factor has a root $-\sigma+j\omega$ in the complex s plane, from which $z = \sigma/\omega_n = \sigma/(\sigma^2+\omega^2)^{1/2}$. *See also:* **control system, feedback.**
(4) (Coulomb). That due to Coulomb friction. *See also:* **control system, feedback.**
(5) (instrument). Term applied to the performance of an instrument to denote the manner in which the pointer settles to its steady indication after a change in the value of the measured quantity. Two general classes of damped motion are distinguished as follows: (A) periodic (underdamped) in which the pointer oscillates about the final position before coming to rest, and (B) aperiodic (overdamped) in which the pointer comes to rest without overshooting the rest position. The point of change between periodic and aperiodic damping is called critical damping. *Note:* An instrument is considered for practical purposes to be critically damped when overshoot is present but does not exceed an amount equal to one-half the rated accuracy of the instrument. *See also:* **accuracy rating (instrument); moving element (instrument).** E270-34E10;85A1-23E0;39A1/42A30-0

damping amortisseur. An amortisseur the primary function of which is to oppose rotation or pulsation of the magnetic field with respect to the pole shoes. *See:* **synchronous machine.** 42A10-0

damping coefficient, viscous. In the characteristic equation of a second-order linear system, say $Mx'' + Bx' + Kx = 0$, the coefficient B of the first derivative of the damped variable. *See also:* **control system, feedback.** 85A1-23E0

damping factor (1) (free oscillation of a second-order linear system). A measure of damping, expressed (without sign) as the quotient of the greater by the lesser of a pair of consecutive swings of the variable (in opposite directions) about an ultimate steady-state value. *Note:* The ratio is a constant whose Napierian logarithm (used with a negative sign) is called the logarithmic decrement, whose square relates adjoining swings in the same direction, and whose value following a step input from zero to a value x_u is conveniently measured as the ratio of x_u to the initial overshoot. *See also:* **control system, feedback; moving element (instrument).**

(2) (electric instruments). (A) Since, in some instruments, the damping factor depends upon the magnitude of the deflection, it is measured as the ratio in angular degrees of the steady deflection to the difference between angular maximum momentary deflection and the angular steady deflection produced by a sudden application of a constant electric power. The damping factor shall be determined with sufficient constant electric power applied to carry the pointer to end-scale deflection on the first swing. The damping shall be due to the instrument and its normal accessories only. For purposes of testing damping, in instances where the damping is influenced by the circuit impedance, the impedance of the driving source shall be at least 100 times the impedance of the instrument and its normal accessories. (B) In the determination of the damping factor of instruments not having torque-spring control, such as a power-factor meter or a frequency meter, the instrument shall be energized to give center-scale deflection. The pointer shall then be moved mechanically to end scale and suddenly released. *See also:* **moving element (instrument).**

(3) (underdamped harmonic system). The quotient of the logarithmic decrement by the natural period. *Note:* The damping factor of an underdamped harmonic system is $F/2M$. *See:* **damped harmonic system.** 39A1/85A1/E270-23E0

damping magnet. A permanent magnet so arranged in conjunction with a movable conductor such as a sector or disk as to produce a torque (or force) tending to oppose any relative motion between them. *See also:* **moving element (instrument).** 42A30-0

damping torque (synchronous machine). The torque produced, such as by action of the amortisseur winding, that opposes the relative rotation, or changes in magnitude, of the magnetic field with respect to the rotor poles. *See also:* **synchronous machine.** 0-31E8

damping torque coefficient (synchronous machine). A proportionality constant that, when multiplied by the angular velocity of the rotor poles with respect to the magnetic field, for specified operating conditions, results in the damping torque. *See also:* **synchronous machine.** 0-31E8

damp location (electric system). A location subject to a moderate degree of moisture, such as some basements, some barns, some cold-storage warehouses, and the like. *See also:* **distribution center.** 1A0-0

dark adaptation. The process by which the retina becomes adapted to a luminance less than about 0.01 footlambert (0.03 nit). *See also:* **visual field.** Z7A1-0

dark current (photoelectric device) (electron device). The current flowing in the absence of irradiation. *See:* **photoelectric effect; dark current; electrode; dark-current pulses (phototubes).** 50I07-15E6

dark-current pulses (phototubes). Dark-current excursions that can be resolved by the system employing the phototube. *See also:* **phototube.** E175-0

darkening (electroplating). The production by chemical action, usually oxidation, of a dark colored film (usually a sulfide) on a metal surface. *See also:* **electroplating.** 42A60-0

dark space, cathode. *See:* **cathode dark space.**

dark space, Crookes. *See:* **cathode dark space.**

dark-trace screen (cathode-ray tube). A screen giving a spot darker than the remainder of the surface. *See also:* **cathode-ray tubes.** 50I07-15E6

dark-trace tube (skiatron) (electronic navigation). A cathode-ray tube having a special screen that changes color but does not necessarily luminesce under electron impact, showing, for example, a dark trace on a bright background. *See also:* **cathode-ray tubes; navigation.** 50I07-15E6

D'Arsonval current* (solenoid current*) (medical electronics). The current of intermittent and isolated trains of heavily damped oscillations of high frequency, high voltage, and relatively low amperage. *See also:* **medical electronics.** *See:* Note under **D'Arsonvalization.**

*Deprecated 42A80-18E1

D'Arsonvalization* (medical electronics). The therapeutic use of intermittent and isolated trains of heavily damped oscillations of high frequency, high voltage, and relatively low amperage. *Note:* This term is deprecated because it was initially ill-defined and because the technique is not of contemporary interest. *See also:* **medical electronics.**

*Deprecated 42A80-18E1

data. Any representations such as characters or analog quantities to which meaning might be assigned. X3A12-16E9;0-19E4

data-hold (data processing). A device that converts a sampled function into a function of a continuous variable. The output between sampling instants is determined by an extrapolation rule or formula from a set of past inputs. *See also:* **electronic digital computer.** 0-23E3

data link (data transmission). The communications lines, modems, and communication controls of all stations connected to the line, used in the transmission of information between two or more stations. *See also:* **data processing.** 0-19E4

data logger (power-system communication). A system to measure a number of variables and make a written tabulation and/or record in a form suitable for computer input. *See also:* **digital.** 0-31E3

data-logging equipment. Equipment for numerical recording of selected quantities on log sheets or paper or magnetic tape or the like, by means of an electric typewriter or other suitable device. 37A100-31E11

data processing. Pertaining to any operation or combination of operations on data.

See:
alphanumeric;
analog data;
block;
break;
buffer;
check bits;
clocking;
code;
code element;
data;
data link;
hold;
reading speed. X3A12-16E9

data processor. Any device capable of performing operations on data, for example, a desk calculator, a tape recorder, an analog computer, a digital computer. X3A12-16E9

data reconstruction (data processing). The conversion of a signal defined on a discrete-time argument to one defined on a continuous-time argument. *See also:* **electronic digital computer.** 0-23E3

data reduction. The transformation of raw data into a more useful form, for example, smoothing to reduce noise. X3A12-16E9

data stabilization (vehicle-borne navigation systems). The stabilization of the output signals with respect to a selected reference invariant with vehicle orientation. *See also:* **navigation.** 0-10E6

data transmission.
Note: For an extensive list of cross references, see *Appendix A.*

daylight factor. A measure of daylight illumination at a point on a given plane expressed as a ratio of the illumination on the given plane at that point to the simultaneous exterior illumination on a horizontal plane from the whole of an unobstructed sky of assumed or known luminance (photometric brightness) distribution. *Note:* Direct sunlight is excluded from both interior and exterior values of illumination. *See also:* **sunlight.** Z7A1-0

dB. *See:* **decibel.**

dBm. A unit for expression of power level in decibels with reference to a power of one milliwatt (0.001 watt). *See:* **power level (dBm).** E151/42A65

dBV. *See:* **voltage level.**

dc. *See:* **direct current.**

D cable. A two-conductor cable, each conductor having the shape of the capital letter D with insulation between the conductors themselves and between conductors and sheath. *See also:* **power distribution, underground construction.** 42A35-31E13

DDA (computing systems). *See:* **digital differential analyzer.** *See also:* **electronic digital computer.**

***D* dimension (motor).** The standard designation of the distance from the centerline of the shaft to the plane through the mounting surface bottom of the feet, in National Electrical Manufacturers Association approved designations. *See also:* **cradle base (rotating machinery).** 0-31E8

***D* display (radar).** A *C* display in which the blips extend vertically to give a rough estimate of distance. *See also:* **navigation.** E172-10E6

DDM (electronic navigation). See: **difference in depth of modulation.**

deactivation (corrosion). The process of removing active constituents from a corroding liquid (as removal of oxygen from water). *See:* **corrosion terms.** CM-34E2

dead (electric system). Free from any electric connection to a source of potential difference and from electric charge; not having a potential different from that of the ground. The term is used only with reference to current-carrying parts that are sometimes alive. *See also:* **insulated; power distribution, overhead construction.** 2A2/42A95-0

dead band. The range through which the measured signal can be varied without initiating response. *Notes:* (1) This is usually expressed in percent of span. (2) Threshold and ultimate sensitivity are defined as one-half the dead band. When the instrument is balanced at the center of the dead band, they denote the minimum change in measured signal required to initiate response. Since this condition of balance is not normally achieved and is not readily recognizable, these terms are deprecated and the various requirements should preferably be stated in terms of dead band. Sensitivity is frequently used to denote the same quantity. However, its usage in this sense is also deprecated since it is not in accord with accepted standard definitions of the term. (3) The presence of dead band may be intentional, as in the case of a differential gap or neutral zone, or undesirable, as in the case of play in linkages or backlash in gears. (4) With reference to gas turbines, speed dead band is expressed in percent of rated speed. *See also:* **control system, feedback; instrument; numerically controlled machines.** E282-31E2;39A4-0;AS1-34E10

dead-band rating. The limit that the dead band will not exceed when the instrument is used under rated operating conditions. *See also:* **accuracy rating (instrument).** 39A4-0

dead-end guy. An installation of line or anchor guys to hold the pole at the end of a line. *See:* **pole guy.** *See also:* **tower.** 42A35-31E13

dead-end tower. A tower designed to withstand unbalanced pull from all of the conductors in one direction together with wind and vertical loads. *See also:* **tower.** 42A35-31E13

dead-front (industrial control). So constructed that there are no exposed live parts on the front of the assembly. *See:* **dead-front switchboard.** 42A25-34E10

dead-front mounting (switching device). A method of mounting in which a protective barrier is interposed between all live parts and the operator, and all exposed operating parts are insulated or grounded. *Note:* The barrier is usually grounded metal. 37A100-31E11

dead-front switchboard. A switchboard that has no exposed live parts on the front. *Note:* The switchboard panel is usually grounded metal and provides a barrier between the operator and all live parts. 37A100-31E11

dead man (overhead construction). *See:* **anchor log.**

deadman's handle (industrial control). A handle of a controller or master switch that is designed to cause the controller to assume a preassigned operating condition if the force of the operator's hand on the handle is released. *See:* **electric controller.** IC1-34E10

deadman's release (industrial control). The effect of that feature of a semiautomatic or nonautomatic control system that acts to cause the controlled apparatus to assume a preassigned operating condition if the operator becomes incapacitated. *See:* **control.** IC1-34E10

dead reckoning (navigation). The determining of the position of a vehicle at one time with respect to its position at a different time by the application of vector(s) representing course(s) and distance(s). *See also:* **navigation.** 0-10E6

dead room (audio and electroacoustics) (acoustics). A room that has an unusually large amount of sound absorption. *See also:* **electroacoustics.** 0-1E1

dead spots (mobile communication). Those locations completely within the coverage area where the signal strength is below the level needed for reliable communication. *See also:* **mobile communicating system.** 0-6E1

dead time (1) (automatic control). The interval of time between initiation of an input change or stimulus and the start of the resulting response. *Notes:* (1) It may be qualified as effective if extended to the start of build-up time; theoretical if the dead band is negligible; apparent if it includes the time spent within an appreciable dead band. (2) The interval of inaction is a result that may arise from any of several causes; for example, the quotient distance/velocity when the speed of signal propagation is finite; the presence of a hysteretic zone, a resetting action, Coulomb friction, play in linkages, or backlash in gears; the purposeful introduction of a periodic sampling interval, or a neutral zone. *See also:* **control system, feedback.** 85A1-23E0

(2) (electronic navigation). The time interval in an equipment's cycle of operation during which the equipment is prevented from providing normal response, for example, in a radar display, the portion of the interpulse interval that is not displayed, or, in secondary radar, the interval immediately following the transmission of a pulse reply, during which the transponder is insensitive to interrogations. *See also:* **navigation.** 0-10E6

(3) (recovery time) (radiation counter). The time from the start of a counted pulse until an observable succeeding pulse can occur. *See also:* **gas-filled radiation-counter tubes.** 42A70-15E6

(4) (circuit breaker on a reclosing operation). The interval between interruption in all poles on the opening stroke and reestablishment of the circuit on the reclosing stroke. *Notes:* In breakers using arc-shunting resistors, the following intervals are recognized and the one referred to should be stated. (1) Dead time from interruption on the primary arcing contacts to reestablishment through the primary arcing contacts. (2) Dead time from interruption on the primary arcing contacts to reestablishment through the secondary arcing contacts. (3) Dead time from interruption on the secondary arcing contacts to reestablishment on the primary arcing contacts. (4) Dead time from interruption on the secondary arcing contacts to reestablishment on the secondary arcing contacts. (5) The dead time of an arcing fault on a reclosing operation is not necessarily the same as the dead time of the circuit breakers involved, since the dead time of the fault is the interval during which the faulted conductor is deenergized from all terminals. 37A100-31E11

dead-time correction (radiation counters). A correction to the observed counting rate to allow for the probability of the occurrence of events within the dead time of the system. *See:* **anticoincidence (radiation counters).** 0-15E6

dead zone (industrial control). The period(s) in the operating cycle of a machine during which corrective functions cannot be initiated. 42A25-34E10

debug (1). To examine or test a procedure, routine, or equipment for the purpose of detecting and correcting errors. 0-19E4

(2) (computing systems). To detect, locate, and remove mistakes from a routine or malfunctions from a computer. *See:* **troubleshoot.** X3A12-16E9

debugging (reliability). The operation of an equipment or complex item prior to use to detect and replace parts that are defective or expected to fail, and to correct errors in fabrication or assembly. *See also:* **reliability.** 0-7E1

Debye length (radio wave propagation). The distance in a plasma over which a free electron may move under its own kinetic energy before it is pulled back by the electrostatic restoring forces of the polarization cloud surrounding it. *Note:* Over this distance a net charge density can exist in an ionized gas. The Debye length is given by

$$l_{\mathrm{D}} = \left(\frac{\epsilon_0 k T_{\mathrm{e}}}{e^2 N_{\mathrm{e}}}\right)^{1/2}$$

where ϵ_0 is the permittivity of vacuum, k is Boltzmann's constant, e is the charge of the electron, T_{e} is the electron temperature, and N_{e} is the electron number density. *See also:* **radio wave propagation.** 0-3E2

decade (electric circuit theory). The frequency interval between two frequencies having a ratio of 10 to 1. *Note:* This definition does not conform with the usual dictionary definition and does not agree with other uses of decade in electrical engineering: for example, as in decade resistance box. E270-0

decalescent point (induction heating usage). The temperature at which there is a sudden absorption of heat when metals are raised in temperature. *See also:* **dielectric heating; induction heating.** E54/E169-0

decay (storage tubes). A decrease in stored information by any cause other than erasing or writing. *Note:* Decay may be caused by an increase, a decrease, or a spreading of stored charge. *See also:* **circuit characteristic of electrodes; storage tube.** E158-15E6

decay characteristic (luminescent screen). *See:* **persistence characteristic.**

decay, dynamic (storage tubes). Decay caused by an action such as that of the reading beam, ion currents, field emission, or holding beam. *See also:* **storage tube.** E158-15E6

decay, static (charge-storage tubes). Decay that is a function only of the target properties, such as lateral and transverse leakage. *See also:* **charge-storage tube.** E158-15E6

decay time (storage tubes). The time interval during which the stored information decays to a stated fraction of its initial value. *Note:* Information may not decay exponentially. *See also:* **storage tube.** E158-15E6

Decca. A radio navigation system transmitting on several related frequencies near 100 kilohertz useful to about 200 nautical miles (350 kilometers), in which sets of hyperbolic lines of position are determined by comparison of the phase of (1) one reference continuous wave signal from a centrally located master with (2) each of several continuous wave signals from slave transmitters located in a star pattern, each about 70 nautical miles (130 kilometers) from the master. *See also:* **navigation.** 0-10E6

decelerating electrode (electron-beam tubes). An electrode the potential of which provides an electric field to decrease the velocity of the beam electrons. E160-15E6

decelerating relay (industrial control). A relay that functions automatically to maintain the armature current or voltage within limits, when decelerating from speeds above base speed, by controlling the excitation of the motor field. *See also:* **relay.** IC1-34E10

decelerating time (industrial control). The time in seconds for a change of speed from one specified speed to a lower specified speed while decelerating under specified conditions. *See:* **control system, feedback.** AS1-34E10

deceleration (industrial control). *See:* **retardation.**

deceleration, programmed (numerically controlled machines). A controlled velocity decrease to a fixed percent of the programmed rate. *See also:* **numerically controlled machines.** EIA3B-34E12

deceleration, timed (industrial control). A control function that decelerates the drive by automatically controlling the speed change as a function of time. *See:* **control system, feedback.** AS1-34E10

decibel. One-tenth of a bel, the number of decibels denoting the ratio of the two amounts of power being ten times the logarithm to the base 10 of this ratio. *Note:* The abbreviation dB is commonly used for the term decibel. With P_1 and P_2 designating two amounts of power and n the number of decibels denoting their ratio,

$$n = 10 \log_{10}(P_1/P_2) \text{ decibel.}$$

When the conditions are such that ratios of currents or ratios of voltages (or analogous quantities in other fields) are the square roots of the corresponding power ratios, the number of decibels by which the corresponding powers differ is expressed by the following equations:

$$n = 20 \log_{10}(I_1/I_2) \text{ decibel}$$
$$n = 20 \log_{10}(V_1/V_2) \text{ decibel}$$

where I_1/I_2 and V_1/V_2 are the given current and voltage ratios, respectively. By extension, these relations between numbers of decibels and ratios of currents or voltages are sometimes applied where these ratios are not the square roots of the corresponding power ratios; to avoid confusion, such usage should be accompanied by a specific statement of this application. Such extensions of the term described should preferably be avoided. *See also:* **transmission characteristics.** E145/E270-21E1/31E3

decibel meter. An instrument for measuring electric power level in decibels above or below an arbitrary reference level. *See also:* **instrument.** 42A30-0

decilog (dg). A division of the logarithmic scale used for measuring the logarithm of the ratio of two values of any quantity. *Note:* Its value is such that the number of decilogs is equal to 10 times the logarithm to the base 10 of the ratio. One decilog therefore corresponds to a ratio of $10^{0.1}$ (that is 1.25892+). *See also:* **transmission characteristics.** 42A65-0

decimal. (1) Pertaining to a characteristic or property involving a selection, choice, or condition in which there are ten possibilities. (2) Pertaining to the numeration system with a radix of ten. *See:* **binary-coded decimal; electronic digital computer.** X3A12-16E9

decimal code. A code in which each allowable position has one of 10 possible states. The conventional decimal number system is a decimal code. EIA3B-34E12

decimal number system. *See:* **positional notation.**

decimal point. *See:* **point.**

decineper. One-tenth of a neper. E145-0

decision gate (electronic navigation). A specified point near the lower end of an instrument-landing-system approach at which a pilot must make a decision either to complete the landing or to execute a missed-approach procedure. *See also:* **navigation.** 0-10E6

decision instruction (computing systems). An instruction that effects the selection of a branch of a program, for example, a conditional jump instruction. *See also:* **electronic digital computer.** X3A12-16E9

decision table (computing systems). A table of all contingencies that are to be considered in the description of a problem, together with the actions to be taken. Decision tables are sometimes used in place of flow charts for problem description and documentation. X3A12-16E9

deck (computing systems). A collection of punched cards. X3A12-16E9

declinometer (compass declinometer). A device for measuring the direction of a magnetic field relative to astronomical or survey coordinates. *See also:* **magnetometer.** 42A30-0

decode. (1) To recover the original message from a coded form of the message.
(2) To apply a code so as to reverse some previous encoding. X3A12-16E9
(3) To produce a single output signal from each combination of a group of input signals. *See also:* **translate; matrix; encode; electronic digital computer.** E162-0

decoder (1) (telephone practice). A device for decoding a series of coded signals. *Note:* In automatic telephone switching, a decoder is a relay-type translator that determines from the office code of each call the information required for properly recording the call through the switching train. Each decoder has means, such as a cross-connecting field, for establishing the controls desired and for readily changing them. *See also:* **telephone switching system.** 42A65-0
(2) (electronic computation and control). A matrix of logic elements that selects one or more output channels according to the combination of input signals present. *See also:* **telephone switching system; electronic digital computer.** E270-0;X3A12-16E9
(3) (railway practice). A device adopted to select controls for wayside signal apparatus, train control apparatus, or cab signal apparatus in accordance with the code received from the track or the line circuit. *See*

also: **telephone switching system; railway signal and interlocking.** 42A42-0

decomposition potential (decomposition voltage). The minimum potential (excluding *IR* drop) at which an electrochemical process can take place continuously at an appreciable rate. *See also:* **electrochemistry.** 42A60-0

decoupling. The reduction of coupling. *See also:* **coupling.** 42A65-0

Dectra. An experimental adaptation of the Decca low-frequency radio navigation system in which two pairs of continuous-wave transmitters are oriented so that the center lines of both pairs are along and at opposite ends of the same great-circle path, to provide course guidance along and adjacent to the great-circle path. *Note:* Distance along track may be indicated by synchronized signals from one transmitter of each pair. *See also:* **navigation.** 0-10E6

de-emphasis (post emphasis) (post equalization). The use of an amplitude-frequency characteristic complementary to that used for pre-emphasis earlier in the system. *See also:* **circuits and devices; pre-emphasis.** 0-42A65

de-emphasis network. A network inserted in a system in order to restore the pre-emphasized frequency spectrum to its original form. *See also:* **modulating system.** E145-0

de-energize (relay). To disconnect the relay from its power source. 83A16-0

deep-bar rotor. A squirrel-cage induction-motor rotor having a winding that is narrow and deep giving the effect of varying secondary resistance, large at standstill and decreasing as the speed rises. *See also:* **rotor (rotating machinery).** 0-31E8

deepwell pump. An electrically-driven pump located at the low point in the mine to discharge the water accumulation to the surface. *See also:* **mining.** 42A85-0

defeater (industrial control). *See:* **interlocking deactivating means.**

definite minimum-time relay. An inverse-time relay in which the operating time becomes substantially constant at high values of input. 0-31E6

definite-purpose controller (industrial control). Any controller having ratings, operating characteristics, or mechanical construction for use under service conditions other than usual or for use on a definite type of application. *See:* **electric controller.** IC1-34E10

definite-purpose motor. Any motor designed, listed, and offered in standard ratings with standard operating characteristics or mechanical construction for use under service conditions other than usual or for use on a particular type of application. *Note:* Examples: crane, elevator, and oil-burner motors. *See:* **asynchronous machine; direct-current commutating machine; synchronous machine.** 42A10-0

definite time (relays). A qualifying term indicating that there is purposely introduced a delay in action, which delay remains substantially constant regardless of the magnitude of the quantity that causes the action. *See:* **relay.** 37A100-31E11

definite-time acceleration (electric drive) (industrial control). A system of control in which acceleration proceeds on a definite-time schedule. *See also:* **electric drive.** 42A25-34E10

definite-time relay. A relay in which the operating time is substantially constant regardless of the magnitude of the input quantity. *See also:* **relay.** 37A1-31E6

definition. The fidelity with which the detail of an image is reproduced. *See also:* **recording (facsimile).** 42A65-0

deflecting electrode (electron-beam tubes). An electrode the potential of which provides an electric field to produce deflection of an electron beam. *See also:* **electrode (electron tube).** E160-15E6

deflecting force (direct-acting recording instrument). At any part of the scale (particularly full scale), the force for that position, measured at the marking device, and produced by the electrical quantity to be measured, acting through the mechanism. *See also:* **accuracy rating (instrument).** 42A30-0

deflecting voltage (cathode-ray tube). Voltage applied between the deflector plates to create the deflecting electric field. *See also:* **cathode-ray tubes.** 50I07-15E6

deflecting yoke. An assembly of one or more coils that provide a magnetic field to produce deflection of an electron beam. *See also:* **beam tubes.** 42A70-15E6

deflection (cathode-ray tube) (electron tubes). The displacement of the beam or spot on the screen under the action of the deflecting field. *See also:* **cathode-ray tubes.** 50I07-15E6

deflection blanking (oscilloscopes). Blanking by means of a deflection structure in the cathode-ray tube electron gun that traps the electron beam inside the gun to extinguish the spot, permitting blanking during retrace and between sweeps regardless of intensity setting. *See:* **oscillograph.** 0-9E4

deflection center (electron-beam tube). The intersection of the forward projection of the electron path prior to deflection and backward projection of the electron path in the field-free space after deflection. *See:* **beam tubes.** E160-15E6

deflection coefficient. *See:* **deflection factor.**

deflection defocusing (cathode-ray tube). A fault of a cathode-ray tube characterized by the enlargement, usually nonuniform, of the deflected spot which becomes progressively greater as the deflection is increased. *See also:* **cathode-ray tubes.** 50I07-15E6

deflection factor (1) (inverse sensitivity) (general). The reciprocal of sensitivity. *Note:* It is, for example, often used to describe the performance of a galvanometer by expressing this in microvolts per millimeter (or per division) and for a mirror galvanometer at a specified scale distance, usually 1 meter. *See also:* **accuracy rating (instrument).** 42A30-0

(2) (oscilloscopes). The ratio of the input signal amplitude to the resultant displacement of the indicating spot, for example, volts/division. *See also:* **beam tubes; oscillograph.** 0-9E4

deflection plane (cathode-ray tubes). A plane perpendicular to the tube axis containing the deflection center. *See:* **beam tubes.** E160-15E6

deflection polarity (oscilloscopes). The relation between the polarity of the applied signal and the direction of the resultant displacement of the indicating spot. *Note:* Conventionally a positive-going voltage causes upward deflection or deflection from left to right. *See:* **oscillograph.** 42A30-9E4

deflection sensibility (oscilloscopes). The number of trace widths per volt that can be simultaneously resolved anywhere within the quality area. *See:* **oscillograph.** 0-9E4

deflection sensitivity (magnetic-deflection cathode-ray tube and yoke assembly). The quotient of the spot displacement by the change in deflecting-coil current. *See also:* **beam tubes.** 42A70-15E6

deflection yoke (television). An assembly of one or more coils, whose magnetic field deflects an electron beam. *See:* **deflecting yoke.** *See also:* **television.** E204-2E2

deflection-yoke pull-back (cathode-ray tubes). (1) **color**. The distance between the maximum possible forward position of the yoke and the position of the yoke to obtain optimum color purity. (2) **monochrome**. The maximum distance the yoke can be moved along the tube axis without producing neck shadow. *See:* **beam tubes.** E160-15E6

deflector (lightning arrester). Means for directing the flow of the gas discharge from the vent of the arrester. *See also:* **arresters; lightning arrester (surge diverter).** E28/42A20/62A1-31E7

deflector plates. *See:* **deflecting electrode.**

degassing (electron tube) (rectifier). The process of driving out and exhausting occluded and remanent gases within the vacuum tank or tube, anodes, cathode, etcetera, that are not removed by evacuation alone. *See also:* **rectification.** 42A15-0

degauss. To neutralize or bias a ship's magnetic polarity by electric means so as to approximate the effect of a nonmagnetic ship. *See:* **degaussing coil; degaussing generator; deperm; magnetic mine; nonmagnetic ship.** 42A43-0

degaussing coil. A single conductor or a multiple-conductor cable so disposed that passage of current through it will neutralize or bias the magnetic polarity of a ship or portion of a ship. *Note:* Continuous application of current, with adjustment to suit changes of position or heading of the ship, is required to maintain a degaussed condition. *See also:* **degauss.** 42A43-0

degaussing generator. An electric generator provided for the purpose of supplying current to a degaussing coil or coils. *See also:* **degauss.** 42A43-0

degeneracy (resonant device). The condition where two or more modes have the same resonance frequency. *See also:* **waveguide.** 42A65-3E1

degenerate gas. A gas formed by a system of particles whose concentration is very great, with the result that the Maxwell-Boltzmann law does not apply. *Example:* An electronic gas made up of free electrons in the interior of the crystal lattice of a conductor. *See also:* **discharge (gas).** 50I07-15E6

degenerate parametric amplifier. An inverting parametric device for which the two signal frequencies are identical and equal to one-half the frequency of the pump. *Note:* This exact but restrictive definition is often relaxed to include cases where the signals occupy frequency bands that overlap. *See also:* **parametric device.** E254-15E7

degeneration. Same as negative feedback. *See also:* **feedback.** E145-0

degradation failure (reliability). *See:* **failure, degradation.**

degreasing machine. An electrically driven machine including high-pressure pump and special cleaning solution for removing grease and oil from underground mine machines as a prevention of mine fires. *See also:* **mining.** 42A85-0

degree of asymmetry (current at any time). The ratio of the direct-current component to the peak value of the symmetrical component determined from the envelope of the current wave at that time. *Note:* This value is 100 percent when the direct-current component equals the peak value of the symmetrical component. 37A100-31E11

degrees of freedom (1) (mesh basis). *See:* **nullity. (2) (node basis).** *See:* **rank.**

deionization time (gas tube). The time required for the grid to regain control after anode-current interruption. *Note:* To be exact the deionization time of a gas tube should be presented as a family of curves relating such factors as condensed-mercury temperature, anode and grid currents, anode and grid voltages, and regulation of the grid current. *See:* **gas tubes.** 42A70-15E6

deionizing grid (gas tubes). A grid accelerating deionization in its vicinity in a gas-filled valve or tube, and forming a screen between two regions within the envelope. 50I07-15E6

delay. (1) The amount of time by which an event is retarded. X3A12-16E9
(2) The amount of time by which a signal is delayed. *Note:* It may be expressed in time (milliseconds, microseconds, etcetera) or in number of characters (pulse times, word times, major cycles, minor cycles, etcetera). *See also:* **electronic digital computer.** E162-0

delay, absolute (loran). *See:* **absolute delay.**

delay circuit (pulse techniques). A circuit that produces an output signal that is delayed intentionally with respect to the input signal. *See also:* **pulse.** E175-0

delay coincidence circuit (pulse techniques). A coincidence circuit that is actuated by two pulses, one of which is delayed by a specified time interval with respect to the other. *See also:* **pulse.** E175-0

delay distortion. Either (1) phase-delay distortion (also called phase distortion) that is either departure from the flatness in the phase delay of a circuit or system over the frequency range required for transmission or the effect of such departure on a transmitted signal, or (2) envelope-delay distortion, that is either departure from flatness in the envelope delay of a circuit or system over the frequency range required for transmission or the effect of such departure on a transmitted signal. *See:* **distortion; distortion, envelope delay.** 42A65-31E3

delayed application (railway practice). The application of the brakes by the automatic train control equipment after the lapse of a predetermined interval of time following its initiation by the roadway apparatus. *See also:* **automatic train control.** 42A42-0

delayed overcurrent trip. *See:* **delayed release (delayed trip); overcurrent release (overcurrent trip).**

delayed plan-position indicator (radar). A plan-position indicator in which the initiation of the time base is delayed. *See also:* **navigation.** E172-10E6

delayed release (delayed trip). A release with intentional delay introduced between the instant when the activating quantity reaches the release setting and the instant when the release operates. 37A100-31E11

delayed sweep (oscilloscopes). (1) A sweep that has been delayed either by a predetermined period or by a period determined by an additional independent variable. (2) A mode of operation of a sweep, as defined above. *See:* **radar.** 0-9E4

delayed test. A service test of a battery made after a specified period of time, which is usually made for comparison with an initial test to determine shelf depreciation. *See also:* **battery (primary or secondary).** 42A60-0

delay electric blasting cap. An electric blasting cap with a delay element between the priming and detonating composition to permit firing of explosive charges in sequence with but one application of the electric current. *See also:* **blasting unit.** 42A85-0

delay, envelope. *See:* **envelope delay.**

delay equalizer facsimile. A corrective network that is designed to make the phase delay or envelope delay of a circuit or system substantially constant over a desired frequency range. *See also:* **facsimile transmission; network analysis.** E270-0;42A65-2E2

delaying sweep (oscilloscopes). A sweep used to delay another sweep. *See:* **delayed sweep; oscillograph.** 0-9E4

delay line (electronic computers). (1) Originally, a device utilizing wave propagation for producing a time delay of a signal. E270-0
(2) Commonly, any real or artificial transmission line or equivalent device designed to introduce delay. *See also:* **circuits and devices.** 42A65-0
(3) A sequential logic element or device with one input channel in which the output-channel state at a given instant t is the same as the input-channel state at the instant $t-n$, that is, the input sequence undergoes a delay of n units. There may be additional taps yielding output channels with smaller values of n. *See:* **acoustic delay line; electromagnetic delay line; magnetic delay line; sonic delay line.** *See also:* **circuits and devices; electronic digital computer; pulse.** E162-0;X3A12-16E9

delay-line memory. *See:* **delay-line storage.**

delay-line storage (delay-line memory) (electronic computation). A storage or memory device consisting of a delay line and means for regenerating and reinserting information into the delay line. *See also:* **electronic computation.** E270-0

delay, phase. The ratio of the total phase shift (radians) experienced by a sinusoidal signal in transmission through a system or transducer, to the frequency (radians/second) of the signal. *Note:* The unit of phase delay is the second. E151-0

delay pickoff (oscilloscopes). A means of providing an output signal when a ramp has reached an amplitude corresponding to a certain length of time (delay interval) since the start of the ramp. The output signal may be in the form of a pulse, a gate, or simply amplification of that part of the ramp following the pickoff time. *See:* **circuits and devices.** 0-9E4

delay, pulse (1) (general). The time interval by which a pulse is time retarded with respect to a reference time. *Note:* Pulse delay may be of either polarity, but **negative delay** is sometimes known as **pulse advance.** *See also:* **pulse.** 0-9E4
(2) (transducer). The interval of time between a specified point on the input pulse and a specified point on the related output pulse. *Notes:* (A) This is a general term which applies to the pulse delay in any transducer, such as receiver, transmitter, amplifier, oscillator, etcetera. (B) Specifications may require illustrations. *See also:* **transducer.** E194-31E3;E270-0

delay relay. A relay having an assured time interval between energization and pickup or between de-energization and dropout. *See also:* **relay.** 83A16-21E0

delay, signal. *See:* **signal delay.**

delay time (railway practice). The period or interval after the initiation of an automatic train-control application by the roadway apparatus and before the application of the brakes becomes effective. *See also:* **automatic train control.** 42A42-0

delimited region (electric-circuit purposes). A space from which, or into which, energy is transmitted by an electric circuit, so chosen that the boundary surface satisfies certain conditions. *Note:* In order that conventional measurements of electric voltage, current, and power may be correctly interpreted, it is necessary: (1) that the surface bounding the region cut each current-carrying conductor that enters the region in a cross-section that is perpendicular to the electric-field intensity within the conductor; (2) that the component of the magnetizing force normal to the boundary surface at every point be constant or negligibly small; (3) that the component of the displacement current normal to the boundary surface at every point (not included within a conductor) be negligibly small; (4) that the paths along which voltages are determined must lie in the boundary surface or be chosen so that equal voltages are obtained. Examples of regions that may be delimited for practical purposes include: a generator, a generating station, a transformer, a substation, a manufacturing establishment, a motor, a lamp, a geographical area, etcetera. *See also:* **network analysis.** E270-0

delimiter (computing systems). A flag that separates and organizes items of data. *See:* **separator.** *See also:* **electronic digital computer.** X3A12-16E9

Dellinger effect. *See:* **radio fadeout.**

delta. The difference between a partial-select output of a magnetic cell in a ONE state and a partial-select output of the same cell in a ZERO state. *See:* **coincident-current selection.** *See also:* **static magnetic storage.** E163-0

delta *B* (ΔB). *See:* **delta induction.**

delta-connected (Δ-connected) circuit. A three-phase circuit that is (simply or completely) mesh connected. *See also:* **network analysis.** E270-0

delta induction (delta flux density) (toroidal magnetic amplifier cores). The change in induction (flux density) when a core is in a cyclically magnetized condition. E106-0

delta, minimum (power supplies). A qualifier, often appended to a percentage specification to describe that specification when the parameter in question is a variable, and particularly when that variable may approach zero. The qualifier is often known as the minimum delta V, or minimum delta I, as the case may be. *See also:* **power supply.** KPSH-10E1

delta network (1). A set of three branches connected in series to form a mesh. *See also:* **network analysis.** E153/E270-0
(2) (electromagnetic compatibility). An artificial mains network of specified symmetric and asymmetric impedance used for two-wire mains operation and comprising resistors connected in delta formation be-

tween the two conductors, and each conductor and earth. *See also:* **electromagnetic compatibility.** CISPR-27E1

demand (1) (electric power utilization). The load integrated over a specified interval of time. *See also:* **alternating-current distribution.** 0-31E4

(2) (installation or system). The load at the receiving terminals averaged over a specified interval of time. *Note:* Demand is expressed in kilowatts, kilovolt-amperes, kilovars, amperes, or other suitable units. *See also:* **alternating-current distribution.** 12A0/42A35-31E13

demand charge (electric power utilization). That portion of the charge for electric service based upon a customer's demand. *See also:* **alternating-current distribution.** 0-31E4

demand deviation (demand meter or register). The difference between the indicated or recorded demand and the true demand, expressed as a percentage of the full-scale value of the demand meter or demand register. *See also:* **demand meter.** 12A0-0

demand factor (system demand factor). The ratio of the maximum demand of a system to the total connected load of the system. *Note:* The demand factor of a part of the system may be similarly defined as the ratio of the maximum demand of the part of the system to the total connected load of the part of the system under consideration. *See also:* **alternating-current distribution; direct-current distribution; power system, low-frequency and surge testing.** 42A35-31E13

demand, instantaneous. *See:* **instantaneous demand.**

demand interval (demand meter or register). The length of the interval of time upon which the demand measurement is based. *Note:* The demand interval of a block-interval demand meter is a specific period of time such as 15, 30, or 60 minutes during which the electric energy flow is averaged. The demand interval of a lagged-demand meter is the time required to indicate 90 percent of the full value of a constant load suddenly applied. Some meters record the highest instantaneous load. *See also:* **demand meter; electricity meter (meter).** 12A0-31E4

demand-interval deviation (demand meter or register). The difference between the actual (measured) demand interval and the rated demand interval, expressed as a percentage of the rated demand interval. *See also:* **demand meter.** 12A0-0

demand meter. A device that indicates or records the demand or maximum demand. *Note:* Since demand involves both an electrical factor and a time factor, mechanisms responsive to each of these factors are required as well as in indicating or recording mechanism. These mechansims may be either separate from or structurally combined with one another. Demand meters may be classified as follows: Class 1, curve-drawing meters; Class 2, integrated-demand meters; Class 3, lagged-demand meters.

See:

block-interval demand meter;
contact-making clock;
contact mechanism;
demand deviation;
demand interval;
demand-interval deviation;
demand register;
demand-totalizing relay;
electric mechanism;
electricity meter;
indicating or recording mechanism;
integrated demand meter;
maximum demand;
maximum demand pointer;
pointer pusher;
timing deviation;
timing mechanism.

*Deprecated 12A/42A30-0

demand register. A mechanism, intended for mounting in an integrating electricity meter, that indicates maximum demand and that also registers electric energy (or other integrated quantity). *See also:* **demand meter.** 12A0-0

demand-totalizing relay. A device designed to receive and totalize electric pulses from two or more sources for transmission to a demand meter or to another relay. *See also:* **demand meter.** 12A0/42A30-0

demarcation potential (demarcation current*) (current of injury*) (electrobiology). *See:* **injury potential.**

*Deprecated

demineralization. The process of removing dissolved minerals (usually by chemical means). *See:* **corrosion terms.** CM-34E2

demodulation. A modulation process wherein a wave resulting from previous modulation is employed to derive a wave having substantially the characteristics of the original modulating wave. *Note:* The term is sometimes used to describe the action of a frequency converter or mixer, but this practice is deprecated. *See:* **demodulator.** *See also:* **modulating systems.** 42A65-0

demodulator. A device to effect the process of demodulation. *See also:* **demodulation.** E170/42A65-0

densitometer. A photometer for measuring the optical density (common logarithm of the reciprocal of the transmittance) of materials. *See also:* **photometry.** Z7A1-0

density (1) (facsimile). A measure of the light-transmitting or -reflecting properties of an area. *Notes:* (1) It is expressed by the common logarithm of the ratio of incident to transmitted or reflected light flux. (2) There are many types of density that will usually have different numerical values for a given material; for example, diffuse density, double diffuse density, specular density. The relevant type of density depends upon the geometry of the optical system in which the material is used. *See also:* **scanning.** E168-0

(2) (electron or ion beam). The density of the electron or ion current of the beam at any given point. *See:* **beam tubes.**

(3) (computing systems). *See:* **packing density.** *See also:* **static magnetic storage.** 50I07-15E6

density-modulated tube (space-charge-control tube) (microwave tubes). Microwave tubes or valves characterized by the density modulation of the electron stream by a gating electrode. *Note:* The electron stream is collected on those electrodes that form a part of the microwave circuit, principally the anode. These electrodes are often small compared to operating wavelength so that for this reason space-charge-control tubes or valves are often not considered to be microwave tubes even though they are used at microwave frequencies. *See:* **tube definitions.** 0-15E6

density modulation (electron beam). The process whereby a desired time variation in density is impressed on the electrons of a beam. *See also:* **velocity-modulated tube.** 50I07-15E6

denuder. That portion of a mercury cell in which the metal is separated from the mercury. 42A60-0

dependability (relay or relay system). A facet of reliability pertaining to the certainty of correct operation of the relay or relay system. *Note:* The distinction between the two aspects of reliability, dependability and security, is usually made when a communication channel is involved in the relay system and noise or extraneous signals are a potential hazard to the correct performance of the system. This distinction may be extended to other systems. 37A100-31E11/31E6

dependent contact. A contacting member designed to complete any one of two or three circuits, depending on whether a two- or a three-position device is considered. *See also:* **railway signal and interlocking.** 42A42-0

dependent manual operation. An operation solely by means of directly applied manual energy, such that the speed and force of the operation are dependent upon the action of the attendant. 37A100-31E11

dependent node (network analysis). A node having one or more incoming branches. *See also:* **linear signal flow graphs.** E155-0

dependent power operation. An operation by means of energy other than manual, where the fulfillment of the operation is dependent upon the continuity of the power supply (to solenoids, electric or pneumatic motors, etcetera). 37A100-31E11

deperm. To remove, as far as practicable, the permanent magnetic characteristic of a ship's hull by powerful external demagnetizing coils. *See also:* **degauss.** 42A43-0

depletion layer (semiconductor). A region in which the charge-carrier charge density is insufficient to neutralize the net fixed charge density of donors and acceptors. *See also:* **semiconductor.** E102/E270-0;E216-34E17

depolarization (1) (electrochemistry). A decrease in the polarization of an electrode at a specified current density. *See also:* **electrochemistry.** 42A60-0
(2) (medical electronic biology). A reduction of the voltage between two sides of a membrane or interface below an initial value. *See also:* **contact potential; electrochemistry; medical electronics.** 42A80-18E1
(3) (corrosion). Reduction or elimination of polarization by physical means or by change in environment. CM-34E2

depolarization front (medical electronics). The border of a wave of electric depolarization, traversing an excitable tissue that has appreciable width and thickness as well as length. *See also:* **contact potential; medical electronics.** 42A80-18E1

depolarizer (1). A substance or a means that produces depolarization. 42A60-0
(2) (primary cell). A cathodic depolarizer that is adjacent to or a part of the positive electrode. *See also:* **electrochemistry; electrolytic cell.** 42A60-0

depolarizing mix (primary cell). A mixture containing a depolarizer and a material to improve conductivity. *See also:* **electrolytic cell.** 42A60-0

deposit attack (deposition corrosion). Pitting corrosion resulting from deposits on a metallic surface. *See also:* **corrosion terms.** CM-34E2

deposited-carbon resistor. A resistor containing a thin coating of carbon deposited on a supporting material. 0-21E0

deposition corrosion. *See:* **deposit attack.**

depth of current penetration (induction-heating usage). The thickness of a layer extending inward from the surface of a conductor that has the same resistance to direct current as the conductor as a whole has to alternating current of a given frequency. *Notes:* (1) About 87 percent of the heating energy of an alternating current is dissipated in the so-called **depth of penetration.** (2) This term is useful only in cases where the surface is substantially flat. *See also:* **dielectric heating; induction heating.** E54/E169-0

depth of heating (dielectric-heating usage). The depth below the surface of a material in which effective dielectric heating can be confined when the applicator electrodes are applied adjacent to one surface only. *See also:* **dielectric heating** E54/E169-0

depth of penetration (induction-heating usage). *See:* **depth of current penetration.**

depth of velocity modulation (electron beams). The ratio of (1) the amplitude of a stated frequency component of the varying velocity of an electron beam, to (2) the average beam velocity. *See:* **beam tubes.** 0-15E6

derating (reliability). The intentional reduction of stress/strength ratio in the application of an item, usually for the purpose of reducing the occurrence of stress-related failures. *See also:* **reliability.** 0-7E1

derivative control action (electric power systems). *See:* **control action, derivative.**

derived envelope (loran-C). The waveform equivalent to the summation of the video envelope of the radio-frequency received pulse and the negative of its derivative, in proper proportion. *Note:* The resulting envelope has a zero crossing at a standard point (for example, 25 microseconds) from the pulse beginning, serving as an accurate reference point for envelope time-difference measurements and as a gating point in rejecting the latter part of the received pulse, which may be contaminated by sky-wave transmissions. *See also:* **navigation.** 0-10E62

derived pulse (loran-C). A pulse derived by summing the received radio-frequency pulse and an oppositely phased radio-frequency pulse so that it has an envelope that is the derivative of the received radio-frequency pulse envelope. *Note:* The resultant envelope has a zero point and a radio-frequency phase reversal at a stan-dard interval (for example, 25 microseconds) from the pulse beginning and it serves as an accurate reference for cycle time-difference measurements and as a gating point in rejecting the latter part of the received pulse, which may be contaminated by sky-wave transmissions. *See also:* **navigation.** 0-10E62

derived units. The derived units of a system of units are those that are not selected as basic but are derived from the basic units. *Note:* A derived unit is defined by specifying an operational procedure for its realization. E270-0

derrick. An apparatus consisting of a mast or equivalent members held at the top by guys or braces, with or without a boom, for use with a hoisting mechanism and operating ropes. *See also:* **elevators.** 42A45-0

design. *See:* **logic design.** *See also:* **electronic digital computer.**

design *A* motor. An integral-horsepower polyphase squirrel-cage induction motor designed for full-voltage starting with normal values of locked-rotor torque and breakdown torque, and with locked-rotor current higher than that specified for design *B, C,* and *D* motors. 0-31E8

design *B* motor. An integral-horsepower polyphase squirrel-cage induction motor, designed for full-voltage starting, with normal locked-rotor and breakdown torque and with locked-rotor current not exceeding specified values. 0-31E8

design *C* motor. An integral-horsepower polyphase squirrel-cage induction motor, designed for full-voltage starting with high locked-rotor torque and with locked-rotor current not exceeding specified values. 0-31E8

design current (glow lamp). The value of current flow through the lamp upon which rated-life values are based. 78A385-0

design *D* motor. An integral-horsepower polyphase squirrel-cage induction motor with rated-load slip of at least 5 percent and designed for full-voltage starting with locked-rotor torque at least 275 percent of rated-load torque and with locked-rotor currents not exceeding specified values. 0-31E8

design letters. Terminology established by the National Electrical Manufacturers Association to describe a standard range of characteristics. The characteristics covered are slip at rated load, locked-rotor and breakdown torque, and locked-rotor current. 0-31E8

design limits. Design aspects of the instrument in terms of certain limiting conditions to which the instrument may be subjected without permanent physical damage or impairment of operating characteristics. *See also:* **instrument.** 39A4-0

design *L* motor. An integral-horsepower single-phase motor, designed for full-voltage starting with locked-rotor current not exceeding specified values, which are higher than those for design *M* motors. 0-31E8

design *M* motor. An integral-horsepower single-phase motor, designed for full-voltage starting with locked-rotor current not exceeding specified values, which are lower than those for design *L* motors. 0-31E8

design *N* motor. A fractional-horsepower single-phase motor, designed for full-voltage starting with locked-rotor current not exceeding specified values, which are lower than those for design *O* motors. 0-31E8

design *O* motor. A fractional-horsepower single-phase motor, designed for full-voltage starting with locked-rotor current not exceeding specified values, which are higher than those for design *N* motors. The limiting locked-rotor current values are given in amperes in American National Standard C50.2. 0-31E8

design tests (1) (lightning arresters). Those tests made to determine the adequacy of the design of a particular type, style, or model of switchgear or its component parts to meet its assigned ratings and to operate satisfactorily under normal service conditions or under special conditions if specified; and to demonstrate compliance with appropriate standards of the industry. *Note:* Design tests are made only on representative apparatus to substantiate the ratings assigned to all other apparatus of basically the same design. These tests are not intended to be used as a part of normal production. The applicable portion of these design tests may also be used to evaluate modifications of a previous design and to assure that performance has not been adversely affected. Test data from previous similar designs may be used for current designs, where appropriate. Once made, the tests need not be repeated unless the design is changed so as to modify performance. *See also:* **lightning arrester (surge diverter); power distribution, overhead construction.** E28/37A100/62A1-31E7/31E11

(2) (less specifically). Tests made by the manufacturer to obtain data for design or application, or to obtain information on the performance of each type of device. *Note:* Design tests may be made on selected specimens during production, at the request of the purchaser, when specified and agreed upon. *See also:* **pothead; power distribution, overhead construction.** E48/E49/E270-0

design voltage. The voltage at which the device is designed to draw rated watts input. *See also:* **appliance outlet.** 71A1-0

desired track (electronic navigation). *See:* **course line.**

deskstand. A movable pedestal or stand (adapted to rest on a desk or table) that serves as a mounting for the transmitter of a telephone set and that ordinarily includes a hook for supporting the associated receiver when not in use. *See also:* **telephone station.** 42A65-0

deskstand telephone set. A telephone set having a deskstand. *See also:* **telephone station.** 42A65-0

destructive read (computing systems). A read process that also erases the data in the source. *See also:* **electronic digital computer.** X3A12-16E9

destructive reading (charge-storage tubes). Reading that partially or completely erases the information as it is being read. *See also:* **charge-storage tube.** E158-15E6

detail contrast (television). *See:* **resolution response.**

detectability factor (1) (radar). In pulsed radar, the ratio of single-pulse signal energy to noise power per unit bandwidth that provides stated probabilities of detection and false alarm, measured in the intermediate-frequency amplifier and using an intermediate-frequency filter matched to the single pulse, followed by optimum video integration. *See:* **navigation.**

(2) (continuous-wave radar). The ratio of single-look signal energy to noise power per unit bandwidth, using a filter matched to the time on target. 0-10E62

detecting means. The first system element or group of elements that responds quantitatively to the measured variable and performs the initial measurement operation. The detecting means performs the initial conversion or control of measurement energy. *See also:* **instrument.** 39A4-0

detection. (1) Determination of the presence of a signal. (2) Demodulation. The process by which a wave corresponding to the modulating wave is obtained in response to a modulated wave. *See also:* **linear detection; modulating systems; power detection; square-law detection.** E145-0;E170-31E3

detector (receivers) (1). A device to effect the process of detection.

(2) A mixer in a superheterodyne receiver. *Note:* In definition (2), the device is often referred to as a first detector and the device is not used for detection as defined above. E188-0

(3) (electromagnetic energy). A device for the indication of the presence of electromagnetic fields. *Note:* In combination with an instrument, a detector may be employed for the determination of the complex field amplitudes. *See also:* **auxiliary device to an instrument.** 0-9E4

detector, average. *See:* **average detector.**

determinant (network). (1) (branch basis). The determinant formed by the coefficients of the branch currents after the application of Kirchhoff's two laws. This determinant is not necessarily symmetrical in the case of a passive network.

(2) (mesh or loop basis). The determinant formed by the impedance coefficients of the mesh or loop currents in a complete set of mesh or loop equations.

(3) (node basis) The determinant formed by the (admittance) coefficients of the node voltages in a complete set of node equations. *See also:* **network function.** E270-0

developer (electrostatography). A material or materials that may be used in development. *See also:* **electrostatography.** E224-15E7

development (electrostatography). The act of rendering an electrostatic image viewable. *See also:* **electrostatography.** E224-15E7

deviation (automatic control). Any departure from a desired or expected value or pattern.
See:
absolute steady-state deviation;
absolute system deviation;
absolute transient deviation;
deviation, steady-state;
deviation, system;
deviation, transient;
percent steady-state deviation;
percent system deviation;
percent transient deviation.
See also: **control system, feedback.**
85A1-23E0;AS1-34E10

deviation, frequency. *See:* **frequency deviation.**

deviation distortion. Distortion in a frequency-modulation receiver caused by inadequate bandwidth, inadequate amplitude-modulation rejection, or inadequate discriminator linearity. *See also:* **distortion.** E188/42A65-31E3

deviation factor (wave) (rotating machinery). The ratio of the maximum difference between corresponding ordinates of the wave and of the equivalent sine wave to the maximum ordinate of the equivalent sine wave when the waves are superposed in such a way as to make this maximum difference as small as possible. *Note:* The equivalent sine wave is defined as having the same frequency and the same root-mean-square value as the wave being tested. *See:* **direct-axis synchronous impedance (rotating machinery); synchronous machine.** 42A10-31E8

deviation ratio (frequency-modulation system). The ratio of the maximum frequency deviation to the maximum modulating frequency of the system. *See also:* **frequency modulation.** E145/E170/42A65-0

deviation sensitivity (1) (frequency-modulation receivers). The least frequency deviation that produces a specified output power. E188-0

(2) (electronic navigation). The rate of change of course indication with respect to the change of displacement from the course line. *See also:* **navigation.** E172-10E6

deviation, steady-state (control). The system deviation after transients have expired. *Note:* For the purpose of this definition, drift is not considered to be a transient. *See also:* **deviation.**
85A1-23E0;AS1-34E10

deviation, system (control). The instantaneous value of the ultimately controlled variable minus the command. *Note:* The use of system error to mean a system deviation with its sign changed is deprecated. *See also:* **deviation.** 85A1-23E0;AS1-34E10

deviation, transient (control). The instantaneous value of the ultimately controlled variable minus its steady-state value. *See also:* **deviation.**
85A1-23E0;AS1-34E10

device (electric system) (1). A mechanical or an electric contrivance to serve a useful purpose. *See also:* **absolute-value device; control; electronic digital computer; power distribution, overhead construction; storage device.** E270-0

(2) A smallest subdivision of a system that still has a recognizable function of its own. *See also:* **circuits and devices.** 0-31E3

(3) A unit of an electric system that is intended to carry but not utilize electric energy.
E270/1A0/2A2-0

(4) (electric). An item of electric equipment that is used in connection with, or as an auxiliary to, other items of electric equipment. For example, thermostat, relay, push button or switch, or instrument transformers. *See also:* **control.** 84A1-0

dezincification. Parting of zinc from an alloy (parting is the preferred term). *Note:* Other terms in this category, such as denickelification, dealuminification, demolybdenization, etcetera, should be replaced by the term parting. *See:* **electrometallurgy.** CM-34E2

DF. *See:* **direction-finder; radio direction-finder.**

dg. *See:* **decilog.**

diagnostic. Pertaining to the detection and isolation of either a malfunction or mistake. *See also:* **check problem; electronic digital computer.**
E162-0;X3A12-16E9

diagnostic routine. A routine designed to locate either a malfunction in the computer or a mistake in coding. *See also:* **programmed check.** E270-0

diagram. *See:* **block diagram; logic diagram; Venn diagram.** *See also:* **electronic digital computer.**

dial (1) (industrial control). A plate or disc, suitably marked, that serves to indicate angular position, as for example the position of a handwheel.
42A25-34E10

(2) (automatic switching). A type of calling device used in automatic switching that, when wound up and released, generates pulses required for establishing connections. *See also:* **telephone switching system.**
42A65-19E1

dial-mobile telephone system (mobile communication). A mobile communication system that can be interconnected with a telephone network by dialing, or a mobile communication system connected on a dial basis with a telephone network. *See also:* **mobile communication system.** 0-6E1

dial pulsing (telephony). This consists of regular, momentary interruptions of the direct-current path at the sending end. *See also:* **telephone switching system.** 0-19E1

dial system. *See:* **telephone switching system.**

dial telephone set. A telephone set equipped with a dial. *See also:* **telephone station.** 42A65-0

dial train (register). All the gear wheels and pinions used to interconnect the dial pointers. *See also:* **watthour meter.** 42A30-0

diamagnetic material. A material whose relative permeability is less than unity. E270-0

diametric rectifier circuit. A circuit that employs two or more rectifying elements with a conducting period of 180 electrical degrees plus the commutating angle. *See also:* **rectification.** 42A15-0

diaphragm (electrolytic cells). A porous or permeable membrane separating anode and cathode compartments of an electrolytic cell from each other or from an intermediate compartment for the purpose of preventing admixture of anolyte and catholyte. *See also:* **electrolytic cell.** 42A60-0

diathermy (medical electronics). The therapeutic use of alternating currents to generate heat within some part of the body, the frequency being greater than the maximum frequency for neuromuscular response. *See also:* **medical electronics.** 42A80-18E1

dichotomizing search. *See:* **binary search.** *See also:* **electronic digital computer.**

dichromate cell. A cell having an electrolyte consisting of a solution of sulphuric acid and a dichromate. *See also:* **electrochemistry.** 42A60-0

dielectric (lightning arresters). A medium in which it is possible to maintain an electric field with little or no supply of energy from outside sources. E270-31E7

dielectric absorption. A phenomenon that occurs in imperfect dielectrics whereby positive and negative charges are separated and then accumulated at certain regions within the volume of the dielectric. This phenomenon manifests itself usually as a gradually decreasing current with time after the application of a fixed direct voltage. E270-0

dielectric attenuation constant. Per unit of length, the reciprocal of the distance that a plane electromagnetic wave travels through a dielectric before its amplitude is reduced to 1/e of the original amplitude, where e is the base of the natural logarithms. *Note:* The dielectric attenuation constant is related to the dielectric dissipation factor by the equation

$$\alpha = \frac{2\pi}{\lambda}\left[\frac{\epsilon'\mu}{2}\left((1+\tan^2\delta)^{1/2}-1\right)\right]^{1/2}$$

(neper/meter)

where ϵ' is the relative dielectric constant, μ is the relative magnetic permeability, λ is the wavelength in a vacuum, and $\tan\delta$ is the dissipation factor. This expression is valid for dielectric in free space or a coaxial transmission line, but must be modified if the wave is in a dielectric-filled waveguide. E270-0

dielectric constant (dielectric). The property that determines the electrostatic energy stored per unit volume for unit potential gradient. *Note:* This numerical value usually is given relative to a vacuum. *See also:* **dielectric heating.** E54-0

dielectric dispersion. The phenomenon in which the magnitude of the dielectric constant of a material decreases or increases with increasing frequency. E270-0

dielectric dissipation factor. (1) The cotangent of the dielectric phase angle of a dielectric material or the tangent of the dielectric loss angle. *See also:* **dielectric heating.** (2) The ratio of the loss index ϵ'' to the relative dielectric constant ϵ'. *See:* **relative complex dielectric constant.** E54/E169/E270-0

dielectric guide. A waveguide in which the waves travel through solid dielectric material. *See also:* **waveguide.** 42A65-0

dielectric heater.) A device for heating normally insulating material by applying an alternating-current field to cause internal losses in the material. *Note:* The normal frequency range is above 10 megahertz. *See also:* **interference.** 0-13E6

dielectric heating. The heating of a nominally insulating material in an alternating electric field due to its internal losses. *Note:* For an extensive list of cross references, see *Appendix A.* E54/E269-0

dielectric, imperfect. A dielectric in which a part of the energy required to establish an electric field in the dielectric is not returned to the electric system when the field is removed. *Note:* The energy that is not returned is converted into heat in the dielectric. E270-0

dielectric lens. A lens made of dielectric material and used for refraction of radio-frequency energy. *See:* **antenna; waveguide.** 50I62-3E1

dielectric loss. The time rate at which electric energy is transformed into heat in a dielectric when it is subjected to a changing electric field. *Note:* The heat generated per unit volume of material upon the application of a sinusoidal electric field is given by the expression

$$W = \frac{5}{9}\epsilon'' f E^2 \times 10^{-12} \text{ watt/centimeter}^3$$

where E is the root-mean-square field strength in volts per centimeter, f is the frequency, and ϵ'' is the loss index. E270-0

dielectric loss angle. The angle whose tangent is the dissipation factor, or arc tan ϵ''/ϵ'. *See:* **dielectric dissipation factor; relative complex dielectric constant.** E270-0

dielectric loss factor*. *See:* **dielectric loss index.**
*Deprecated

dielectric loss index ϵ'' (dielectric loss factor*) (homogeneous isotropic material). The negative of the imaginary part of the relative complex dielectric constant. *See:* **relative complex dielectric constant.**
*Deprecated E270-0

dielectric, perfect (ideal dielectric). A dielectric in which all of the energy required to establish an electric field in the dielectric is recoverable when the field or impressed voltage is removed. Therefore, a perfect dielectric has zero conductivity and all absorption phenomena are absent. A complete vacuum is the only known perfect dielectric. E270-0

dielectric phase angle. (1) The angular difference in phase between the sinusoidal alternating voltage applied to a dielectric and the component of the resulting alternating current having the same period as the voltage. *See also:* **dielectric heating. (2) The angle whose cotangent is the dissipation factor, or arc cot ϵ''/ϵ'.** *See:* **dielectric dissipation factor; relative complex dielectric constant; dielectric heating.** E54/E169/E270-0

dielectric phase constant. Per unit of length in a dielectric, 2π divided by the wavelength of the electromagnetic wave in the dielectric. *Note:* The phase constant is related to the dielectric constant by the following equation

$$\beta = \frac{2\pi}{\lambda}\left[\frac{\epsilon'\mu}{2}\ [(1+\tan^2\delta)^{1/2}+1]\right]^{1/2}$$

where λ is the wavelength of the wave in a vacuum, ϵ' is the relative dielectric constant, μ is the relative magnetic permeability, and tan δ is the dissipation factor. This expression is valid for dielectric in free space or a coaxial transmission line but must be modified if the wave is in a dielectric-filled waveguide. E270-0

dielectric power factor. The cosine of the dielectric phase angle (or the sine of the dielectric loss angle). *See also:* **dielectric heating.** E54/E169/E270-0

dielectric rod antenna. An antenna that employs a shaped dielectric rod as the significant part of a radiating element. *See:* **dielectric lens.** *See also:* **antenna.** 0-3E1

dielectric strength (material) (electric strength) (breakdown strength). The potential gradient at which electric failure or breakdown occurs. To obtain the true dielectric strength the actual maximum gradient must be considered, or the test piece and electrodes must be designed so that uniform gradient is obtained. The value obtained for the dielectric strength in practical tests will usually depend on the thickness of the material and on the method and conditions of test. E270-31E7

dielectric waveguide. A waveguide consisting of a dielectric structure. *See also:* **waveguide.** E146-3E1

dielectric withstand-voltage tests. Tests made to determine the ability of insulating materials and spacings to withstand specified overvoltages for a specified time without flashover or puncture. *Note:* The purpose of the tests is to determine the adequacy against breakdown of insulating materials and spacings under normal conditions. *See also:* **power systems, low-frequency and surge testing; routine test; service test.** E32/12A0/57A15-0;37A100-31E11;42A15-31E12;I-C1-34E10

diesel-electric drive (oil-electric drive). A self-contained system of power generation and application in which the power generated by a diesel engine is transmitted electrically by means of a generator and a motor (or multiples of these) for propulsion purposes. *Note:* The prefix diesel-electric is applied to ships, locomotives, cars, buses, etcetera, that are equipped with this drive. *See also:* **electric locomotive.** 42A40-0

difference amplifier. *See:* **differential amplifier.**

difference detector. A detector circuit in which the output is a function of the difference of the peak amplitudes or root-mean-square amplitudes of the input waveforms. *See also:* **navigation.** E172-10E6

difference frequency (parametric device). The absolute magnitude of the difference between a harmonic nf_p of the pump frequency f_p and the signal frequency f_s, where n is a positive integer. *Note:* Usually n is equal to one. *See also:* **parametric device.** E254-15E7

difference-frequency parametric amplifier*. *See:* **inverting parametric device.**

*Deprecated

difference signal. *See:* **differential signal.**

difference in depth of modulation (DDM) (electronic navigation) (directional systems employing overlapping lobes with modulated signals (such as instrument landing systems)). A fraction obtained by subtracting from the percentage of modulation of the larger signal the percentage of modulation of the smaller signal and dividing by 100. *See also:* **navigation.** E172-10E6

difference limen (differential threshold) (just noticeable difference). The increment in a stimulus that is just noticeable in a specified fraction of trials. *Note:* The relative difference limen is the ratio of the difference limen to the absolute magnitude of the stimulus to which it is related. *See also:* **phonograph pickup.** 0-1E1

differential (photoelectric lighting control). The difference in foot-candles between the light levels for turn-on and turn-off operation. *See also:* **photoelectric control.** 42A25-34E10

differential aeration cell. An oxygen concentration cell. *See:* **electrolytic cell.** CM-34E2

differential amplifier (1). An amplifier whose output signal is proportional to the algebraic difference between two input signals. *See also:* **amplifier.** 0-9E4

(2) (signal-transmission system). An amplifier that produces an output only in response to a potential difference between its input terminals (differential-mode signal) and in which outputs from common-mode interference voltages on its input terminals are suppressed. *Note:* An ideal differential amplifier produces neither a differential-mode nor a common-mode output in response to a common-mode interference input. *See also:* **amplifier; signal.** 0-13E6

differential analyzer. A mechanical or electric analog device primarily designed and used to solve differential equations. *See also:* **electronic analog computer; digital differential analyzer; instrument.** X3A12-16E9;42A30-0

differential capacitance (nonlinear capacitor). The derivative with respect to voltage of a charge characteristic, such as an alternating charge characteristic or a mean charge characteristic, at a given point on the characteristic. *See also:* **nonlinear capacitor.** E226-15E7

differential-capacitance characteristic (nonlinear capacitor). The function relating differential capacitance to voltage. *See also:* **nonlinear capacitor.** E226-15E7

differential compounded (rotating machinery). Applied to a compound machine to denote that the magnetomotive forces of the series field winding is opposed to that of the shunt field winding. *See:* **direct-current commutating machine.** 0-31E8

differential control. A system of load control for self-propelled electrically driven vehicles wherein the action of a differential field wound on the field poles of a main generator (or of an exciter) and connected in circuit between the main generator and the traction motors, serves to limit the power demand from the prime mover. *See also:* **multiple-unit control.** 42A42-0

differential control current (magnetic amplifier). The total absolute change in current in a specified control winding necessary to obtain differential output voltage when the control current is varied very slowly (a quasi-static characteristic). *See also:* **rating and testing magnetic amplifiers.** E107-0

differential control voltage (magnetic amplifier). The total absolute change in voltage across the specified control terminals necessary to obtain differential output voltage when the control voltage is varied very slowly (a quasistatic characteristic). *See also:* **rating and testing magnetic amplifiers.** E107-0

differential duplex system. A duplex system in which the sent currents divide through two mutually inductive sections of the receiving apparatus, connected respectively to the line and to a balancing artificial line, in opposite directions so that there is substantially no net effect on the receiving apparatus; whereas the received currents pass mainly through one section, or through the two sections in the same direction, and operate the apparatus. *See also:* **telegraphy.** 42A65-0

differential gain (video transmission system). The difference between (1) the ratio of the output amplitudes of a small high-frequency sine-wave signal at two stated levels of a low-frequency signal on which it is superimposed, and (2) unity. *Notes:* (A) Differential gain may be expressed in percent by multiplying the above difference by 100. (B) Differential gain may be expressed in decibels by multiplying the common logarithm of the ratio described in (1) above by 20. (C) In this definition, level means a specified position on an amplitude scale applied to a signal waveform. (D) The low- and high-frequency signals must be specified. *See also:* **television.** E206-2E2

differential-gain control (gain sensitivity control). A device for altering the gain of a radio receiver in accordance with an expected change of signal level, to reduce the amplitude differential between the signals at the output of the receiver. *See also:* **radio receiver.** 42A65-0

differential-gain-control circuit (electronic navigation). The circuit of a receiving system that adjusts the gain of a single radio receiver to obtain desired relative output levels from two alternately applied or sequentially unequal input signals. *Note:* This may be accomplished automatically or manually; if automatic, it is referred to as automatic differential-gain control. Example: Loran circuits that adjust gain between successive pulses from different ground stations. 0-10E6

differential-gain-control range. The maximum ratio of signal amplitudes (usually expressed in decibels), at the input of a single receiver, over which the differential-gain-control circuit can exercise proper control and maintain the desired output levels. *See also:* **navigation.** 0-10E6

differential-mode interference (interference terminology). *See:* **interference, differential mode; interference, normal-mode.** *See also:* **interference; accuracy rating (instrument).**

differential nonreversible output voltage. *See:* **differential output voltage.**

differential output current (magnetic amplifier). The ratio of differential output voltage to rated load impedance. *See also:* **rating and testing magnetic amplifiers.** E107-0

differential output voltage (magnetic amplifier) (1) (nonreversible output). The voltage equivalent to the algebraic difference between maximum test output voltage and minimum test output voltage. *See also:* **rating and testing magnetic amplifiers.** E107-0

(2) (reversible output). The voltage equivalent to the algebraic difference between positive maximum test output voltage and negative maximum test output voltage. *See also:* **rating and testing magnetic amplifiers.** E107-0

differential permeability. The derivative of the scalar magnitude of the magnetic flux density with respect to the scalar magnitude of the magnetizing force. *Notes:* (1) This term has also been used for the slope of the normal induction curve. (2) In anisotropic media, differential permeability becomes a matrix. E270-0

differential permittivity (ferroelectric material). The slope of the hysteresis loop (electric displacement versus electric field) at any point. *See also:* **ferroelectric materials.** E180-0

differential phase (video transmission system). The difference in output phase of a small high-frequency sine-wave signal at the two stated levels of a low-frequency signal on which it is superimposed. *Note:* Notes 3 and 4 appended to **differential gain** apply also to **differential phase.** *See also:* **television.** E206-2E2

differential phase shift. A change in phase of a field quantity at the output port of a network that is produced by an adjustment of the electrical properties, or characteristics, of the network. *Note:* Differential phase shift may also be the difference between the insertion phase shifts of two 2-port networks. *See also:* **measurement system.** E285-9E1

differential-phase-shift keying (modulation systems). A form of phase-shift keying in which the reference phase for a given keying interval is the phase of the signal during the preceding keying interval. *See also:* **modulating systems.** E170-0

differential protection. A method of apparatus protection in which an internal fault is identified by comparing electrical conditions at all terminals of the apparatus. 37A100-31E11/31E6

differential relay (1). A relay that by its design or application is intended to respond to the difference between incoming and outgoing electrical quantities associated with the protected apparatus. *See also:* **relay.** 37A100-31E11/31E6

(2). A relay with multiple windings that functions when the voltage, current, or power difference between the windings reaches a predetermined value. *See also:* **relay.** 83A16-0

(3). A relay with multiple windings that functions when the power developed by the individual windings is such that pickup or dropout results from the algebraic summation of the fluxes produced by the effective windings. 0-21E0

(4) (industrial control). A relay that operates when the vector difference of two or more similar electrical quantities exceeds a predetermined amount. *Note:* This term includes relays heretofore known as **ratio balance relays, biased relays,** and **percentage differential relays.** IC1-34E10

differential resistance (semiconductor rectifiers). The differential change of forward voltage divided by a stated increment of forward current producing this change. *See also:* **semiconductor rectifier stack.** 0-34E24

differential reversible output voltage. *See:* **differential output voltage.**

differential signal. The instantaneous, algebraic difference between two signals. *See:* **oscillograph.** 0-9E4

differential threshold. *See:* **difference limen.**

differential trip signal (magnetic amplifier). The absolute magnitude of the difference between trip OFF and trip ON control signal. *See also:* **rating and testing magnetic amplifiers.** E107-0

differentiating network. *See:* **differentiator.**

differentiator (1). A device producing an output proportional to the derivative of one variable with respect to another, usually time. *See also:* **circuits and devices; electronic analog computer.** E165-0

(2) (electronic computers). A device, usually of the analog type, whose output is proportional to the derivative of an input signal. *See also:* **network analysis.** E270-0

(3) (electronic circuits). A device whose output function is reasonably proportional to the derivative of the input function with respect to one or more variables, for example, a resistance-capacitance network used to select the leading and trailing edges of a pulse signal. *See also:* **circuits and devices.** X3A12-16E9

(4) (modulation circuits and industrial control) (differentiating circuit) (differentiating network). A transducer whose output waveform is substantially the time derivative of its input waveform. *Note:* Such a transducer preceding a frequency modulator makes the combination a phase modulator; or following a phase detector makes the combination a frequency detector. Its ratio of output amplitude to input amplitude is proportional to frequency and its output phase leads its input phase by 90 degrees. *See also:* **circuits and devices; modulating systems.** E270-34E10;E145/42A65-0

diffracted wave (1). When a wave in a medium of certain propagation characteristics is incident upon a discontinuity or a second medium, the diffracted wave is the wave component that results in the first medium in addition to the incident wave and the waves corresponding to the reflected rays of geometrical optics. *See also:* **radiation.** 42A65-0

(2) (audio and electroacoustics). A wave whose front has been changed in direction by an obstacle or other nonhomogeneity in a medium, rather than by reflection or refraction. *See also:* **electroacoustics; radiation.** 0-1E1

diffraction. A process that produces a diffracted wave. *See also:* **electroacoustics; radiation.** 42A65-1E1

diffused junction. *See:* **junction, diffused.**

diffused lighting. Lighting that provides on the work plane or on an object, light that is not incident predominantly from any particular direction. *See also:* **general lighting.** Z7A1-0

diffuser. A device to redirect or scatter the light from a source, primarily by the process of diffuse transmission. Z7A1-0

diffuse reflectance. The ratio of (1) the flux leaving a surface or medium by diffuse reflection to (2) the incident flux. *See also:* **lamp.** Z7A1-0

diffuse reflection. Diffuse reflection is that process by which incident flux is redirected over a range of angles. *See also:* **lamp.** Z7A1-0

diffuse sound field. A sound field in which the time average of the mean-square sound pressure is everywhere the same and the flow of energy in all directions is equally probable. *See also:* **loudspeaker.** 0-1E1

diffuse transmission (illuminating engineering). That process by which the incident flux passing through a surface or medium is scattered. *See also:* **transmission.** Z7A1-0

diffuse transmission density. The value of the photographic transmission density obtained when the light flux impinges normally on the sample and all the transmitted flux is collected and measured. 0-1E1

diffuse transmittance (illuminating engineering). The ratio of the diffusely transmitted flux leaving a surface or medium to the incident flux. *See also:* **transmission (illuminating engineering).** Z7A1-0

diffusing panel. A translucent material covering the lamps in a luminaire in order to reduce the brightness by distributing the flux over an extended area. *See also:* **bare (exposed) lamp.** Z7A1-0

diffusing surfaces and media. Those that redistribute some of the incident flux by scattering in all directions. *See also:* **lamp.** Z7A1-0

diffusion capacitance (semiconductor junction). The rate of change of stored minority-carrier charge with the voltage across the junction. *See also:* **semiconductor.** E216-34E17;E270-0

diffusion constant (charge carrier) (homogeneous semiconductor). The quotient of diffusion current density by the charge-carrier concentration gradient. It is equal to the product of the drift mobility and the average thermal energy per unit charge of carriers. *See also:* **semiconductor.** E102/E270-0;E216-34E17

diffusion length, charge-carrier (homogeneous semiconductor). The average distance to which minority carriers diffuse between generation and recombination. *Note:* The diffusion length is equal to the square root of the product of the charge-carrier diffusion constant and the volume lifetime. *See also:* **semiconductor.** E102-10E1;E216-34E17;E270-0

digit (positional notation) (notation). (1) (A) A character that stands for an integer. (B) Loosely, the integer that the digit stands for. (C) Loosely, any character. *See:*
binary digit;
check digit;
electronic digital computer;
equivalent binary digits;
positional notation;
sign digit;
significant digit.

(2) A character used to represent one of the nonnegative integers smaller than the radix, for example, in decimal notation, one of the characters 0 to 9. X3A12-16E9

digital. (1) Pertaining to data in the form of digits. *See:* **analog.** *See also:* **electronic digital computer.** E162-0;X3A12-16E9

(2) Information in the form of one of a discrete number of codes.
See:
analog-to-digital converter;
binary word;
binary-coded-decimal number;
bias distortion;
clock;
clocked logic;
cyclic-code error detection;
data logger;
digital-to-analog converter;
enabling signal;
fan-in network;
fan-out network;
logic board;
logic, 0-1;

nonreturn-to-zero (NRZ) code;
priority string;
return-to-zero (RZ) code;
shift register. 0-31E3

digital computer. A computer that operates on discrete data by performing arithmetic and logic processes on these data. Contrast with analog computer. *See also:* **electronic digital computer.** X3A12-16E9

digital controller (data processing). A controller that accepts an input sequence of numbers and processes them to produce an output sequence of numbers. *See also:* **electronic digital computer.** 0-23E3

digital converter (code translator). A device, or group of devices, that converts an input numerical signal or code of one type into an output numerical signal or code of another type. 37A100-31E11

digital data. Pertaining to data in the form of digits, or integral quantities. Contrast with analog data. *See:* **analog and digital data.** *See also:* **electronic computation.** 0-19E4

digital device (control equipment). A device that operates on the basis of discrete numerical techniques in which the variables are represented by coded pulses or states. *See also:* **power system, low-frequency and surge testing.** E94-0

digital differential analyzer (DDA). A special purpose digital computer that performs integration by means of a suitable integration code on incremental quantities and that can be programmed for the solution of differential equations in a manner similar to an analog computer. *See also:* **electronic analog computer.** E165-0

digital readout clock. A clock that gives (usually with visual indication) a voltage or contact-closure pattern of electric circuitry for a readout of time. *Note:* A digital readout calendar clock also includes a readout of day, month, and year, usually also with indication. 37A100-31E11

digital telemeter indicating receiver. A device that receives the numerical signal transmitted from a digital telemeter transmitter and gives a visual numerical display of the quantity measured. 37A100-31E11

digital telemetering. Telemetering in which a numerical representation, as for example some form of pulse code, is generated and transmitted; the number being representative of the quantity being measured. 37A100-31E11

digital telemeter receiver. A device that receives the numerical signal transmitted by a digital telemeter transmitter and stores it and/or converts it to a usable form, or both, for such purposes as recording, indication, or control. 37A100-31E11

digital telemeter transmitter. A device that converts its input signal to a numerical form for transmission to a digital telemeter receiver over an interconnecting channel. 37A100-31E11

digital-to-analog converter (1) (power-system communication). A circuit or device whose input is information in digital form and whose output is the same information in an analog form. *See also:* **analog; digital.** 0-31E3
(2) (data processing). A device that converts an input number sequence into a function of a continuous variable. *See also:* **electronic digital computer.** 0-23E3

digitize. (1) To express data in a digital form. X3A12-16E9
(2) The conversion of analog to digital data. (A/D). EIA3B-34E12

dimension (physical quantity). A convenience label that is determined by the kind of quantity (for example, mass, length, voltage) and that is an element of multiplicative group. *Notes:* (1) The product of any pair of dimensions is a dimension. (2) The unit element of the group is "numeric" or "pure number" which is therefore a dimension. *See:* **incremental dimension; normal dimension; short dimension.** E270-0

dimensional system. A multiplicative group of dimensions characterized by the independent dimensions generating the group. *Notes:* (1) Derived dimensions are often given individual names (for example, energy and speed, when mass, length, and time are the generators). (2) The basic dimensions and basic units may be chosen independently of each other. In the special case in which the basic dimensions are the dimensions of the quantities to which the basic units are assigned, dimensional analysis indicates the manner in which the measure of a quantity would change with an assumed virtual variation in the basic units. E270-0

dimming resistor. A resistor that may be inserted in a lamp circuit at will for reducing the luminous intensity of the lamp. *Note:* Dimming resistors are normally used to dim headlamps, but may be applied to other circuits, such as gauge lamp circuits. 42A42-0

diode (1) (electron tube). A two-electrode electron tube containing an anode and a cathode. *See also:* **equivalent diode; tube definitions.** 42A70-15E6
(2) (semiconductor). A semiconductor device having two terminals and exhibiting a nonlinear voltage-current characteristic; in more-restricted usage, a semiconductor device that has the asymmetrical voltage-current characteristic exemplified by a single *p-n* junction. *See also:* **semiconductor.** E216-34E17;E270-0

diode characteristic (multielectrode tube). The composite electrode characteristic taken with all electrodes except the cathode connected together. *See also:* **circuit characteristics of electrodes.** E160-15E6;42A70-0

diode equivalent. The imaginary diode consisting of the cathode of a triode or multigrid tube and a virtual anode to which is applied a composite controlling voltage such that the cathode current is the same as in the triode or multigrid tube. *See also:* **circuit characteristics of electrodes.** E160-15E6;42A70-0

diode fuses (semiconductor rectifiers). Fuses of special characteristics connected in series with one or more semiconductor rectifier diodes to disconnect the semiconductor rectifier diode in case of failure and protect the other components of the rectifier. *Note:* Diode fuses may also be employed to provide coordinated protection in case of overload or short-circuit. *See also:* **semiconductor rectifier stack.** 0-34E24

dip (electroplating). A solution used for the purpose of producing a chemical reaction upon the surface of a metal. *See also:* **electroplating.** 42A60-0

diplex operation. The simultaneous transmission or reception of two signals using a specified common feature, such as a single antenna or a single carrier. *See also:* **telegraphy.** 42A65-0

diplex radio transmission. The simultaneous transmission of two signals using a common carrier wave. *See also:* **radio transmission.** E145-0

dip needle. A device for indicating the angle between the magnetic field and the horizontal. *See also:* **magnetometer.** 42A30-0

dipole antenna. Any one of a class of antennas producing the radiation pattern approximating that of an elementary electric dipole. *Note:* Common usage considers a dipole to be a metal radiating structure that supports a line current distribution similar to that of a thin straight wire, a half wavelength long, so energized that the current has two nodes, one at each of the far ends. *See also:* **antenna.** E149-3E1

dipole molecule. A molecule that possesses a dipole moment as a result of the permanent separation of the centroid of positive charge from the centroid of negative charge for the molecule as a whole. E270-0

dip plating. *See:* **immersion plating.**

dip soldering (soldered connections). The process whereby assemblies are brought in contact with the surface of molten solder for the purpose of making soldered connections. *See also:* **soldered connections (electronic and electrical applications).** 99A1-0

direct-acting machine voltage regulator. A machine voltage regulator having a voltage-sensitive element that acts directly without interposing power-operated means to control the excitation of an electric machine. 37A100-31E11

direct-acting recording instrument. A recording instrument in which the marking device is mechanically connected to, or directly operated by, the primary detector. *See also:* **instrument.** 42A30-0

direct address (computing systems). An address that specifies the location of an operand. *See:* **one-level address.** X3A12-16E9

direct-arc furnace. An arc furnace in which the arc is formed between the electrodes and the charge. *See also:* **electrothermics.** 42A60-0

direct axis (synchronous machine). The axis that represents the direction of the plane of symmetry of the no-load magnetic-flux density, produced by the main field winding current, normally coinciding with the radial plane of symmetry of a field pole. *See:* **direct-axis synchronous reactance; synchronous machine.** 0-31E8

direct-axis component of armature current. That component of the armature current that produces a magnetomotive force distribution that is symmetrical about the direct axis. *See:* **synchronous machine.** 42A10-0

direct-axis component of armature voltage. That component of the armature voltage of any phase that is in time phase with the direct-axis component of current in the same phase. *Note:* A direct-axis component of voltage may be produced by: (1) Rotation of the quadrature-axis component of magnetic flux, (2) Variation (if any) of the direct-axis component of magnetic flux, (3) Resistance drop caused by flow of the direct-axis component of armature current. As shown in the phasor diagram, the direct-axis component of terminal voltage, assuming no field magnetization in the quadrature-axis, is given by

$$\mathbf{E}_{ad} = -R\ \mathbf{I}_{ad} - jX_q\mathbf{I}_{aq}$$

See: **synchronous machine.** 42A10-0

direct-axis component of magnetomotive force (rotating machinery). The component of a magnetomotive force that is directed along the axis of the magnet poles. *See also:* **asynchronous machine; direct-axis synchronous impedance (rotating machinery); direct-current commutating machine; synchronous machine.** 0-31E8

direct-axis current (rotating machinery). The current that produces direct-axis magnetomotive force. *See:* **direct-axis synchronous reactance.** 0-31E8

direct-axis magnetic-flux component (rotating machinery). The magnetic-flux component direct along the direct axis. *See:* **direct-axis synchronous reactance; synchronous machine.** 0-31E8

direct-axis subtransient open-circuit time constant. The time in seconds required for the rapidly decreasing component (negative) present during the first few cycles in the direct-axis component of symmetrical armature voltage under suddenly removed symmetrical short-circuit condition, with the machine running at rated speed to decrease to $1/e \approx 0.368$ of its initial value. *Note:* If the rotor is made of solid steel no single subtransient time constant exists but a spectrum of time constants will appear in the subtransient region. *See:* **synchronous machine; direct-axis synchronous impedance (rotating machinery).** 42A10-31E8

direct-axis subtransient reactance (rotating machinery). The quotient of the initial value of a sudden change in that fundamental alternating-current component of armature voltage, which is produced by the total direct-axis primary flux, and the value of this simultaneous change in fundamental alternating-current component of direct-axis armature current, the machine running at rated speed. *Note:* The rated current value is obtained from the tests for the rated current value of direct-axis transient reactance. The rated voltage value is that obtained from a sudden short-circuit test at the terminals of the machine at rated armature voltage, no load. *See:* **synchronous machine; direct-axis synchronous reactance (rotating machinery).** 0-31E8

direct-axis subtransient short-circuit time constant. The time required for the rapidly changing component, present during the first few cycles in the direct-axis alternating component of short-circuit armature current, following a sudden change in operating conditions, to decrease to $1/e \approx 0.368$ of its initial value, the machine running at rated speed. *Note:* The rated current value is obtained from the test for the rated current value of the direct-axis transient reactance. The rated voltage value is obtained from the test for the rated voltage value of direct-axis transient reactance. *See:* **direct-axis synchronous reactance; synchronous machine (rotating machinery).** 0-31E8

direct-axis subtransient voltage (rotating machinery). The direct-axis component of the armature voltage that appears immediately after the sudden opening of the external circuit when running at a specified load. *See also:* **asynchronous machine; synchronous machine (rotating machinery).** 0-31E8

direct-axis synchronous impedance (synchronous machine) (rotating machinery). The impedance of the armature winding under steady-state conditions where the axis of the armature current and magnetomotive force coincides with the direct axis. In large machines where the armature resistance is negligibly small, the direct-axis synchronous impedance is equal to the direct-axis synchronous reactance. *Note:* For an extensive list of cross references, see *Appendix A.* 0-31E8

direct-axis synchronous reactance. The quotient of a sustained value of that fundamental alternating-current component of armature voltage that is produced by the total direct-axis flux due to direct-axis armature current and the value of the fundamental alternating-current component of this current, the machine running at rated speed. Unless otherwise specified, the value of synchronous reactance will be that corresponding to rated armature current. For most machines, the armature resistance is negligibly small compared to the synchronous reactance. Hence the synchronous reactance may be taken also as the synchronous impedance.
See:
direct axis;
direct-axis current;
direct-axis magnetic flux;
direct-axis subtransient reactance;
direct-axis subtransient short-circuit time constant;
direct-axis synchronous impedance;
direct-axis transient reactance;
direct-axis transient voltage;
direct-axis voltage;
initial alternating short-circuit current;
quadrature-axis;
quadrature-axis current;
quadrature-axis magnetic flux;
quadrature-axis subtransient voltage;
synchronous machine;
transient internal voltage;
zero-sequence impedance;
zero-sequence reactance. 0-31E8

direct-axis transient open-circuit time constant. The time in seconds required for the root-mean-square alternating-current value of the slowly decreasing component present in the direct-axis component of symmetrical armature voltage on open-circuit to decrease to $1/e \approx 0.368$ of its initial value when the field winding is suddenly short-circuited with the machine running at rated speed. *See:* **synchronous machine; direct-axis synchronous impedance (rotating machinery).** 42A10-0

direct-axis transient reactance (rotating machinery). The quotient of the initial value of a sudden change in that fundamental alternating-current component of armature voltage, which is produced by the total direct-axis flux, and the value of the simultaneous change in fundamental alternating-current component of direct-axis armature current, the machine running at rated speed and the high-decrement components during the first cycles being excluded. *Note:* The rated current value is that obtained from a three-phase sudden short-circuit test at the terminals of the machine at no load, operating at a voltage such as to give an initial value of the alternating component of current, neglecting the rapidly decaying component of the first few cycles, equal to the rated current. This requirement means that the per-unit test voltage is equal to the rated current value of transient reactance (per unit). In actual practice, the test voltage will seldom result in initial transient current of exactly rated value, and it will usually be necessary to determine the reactance from a curve of reactance plotted against voltage. The rated voltage value is that obtained from a three-phase sudden short-circuit test at the terminals of the machine at rated voltage, no load. *See:* **direct-axis synchronous reactance; synchronous machine.** *See also:* **direct-axis synchronous impedance (rotating machinery).** 0-31E8

direct-axis transient short-circuit time constant. The time in seconds required for the root-mean-square value of the slowly decreasing component present in the direct-axis component of the alternating-current component of the armature current under suddenly applied symmetrical short-circuit conditions with the machine running at rated speed, to decrease to $1/e \approx 0.368$ of its initial value. 42A10-0

direct-axis transient voltage (rotating machinery). The direct-axis component of the armature voltage that appears immediately after the sudden opening of the external circuit when running at a specified load, the components that decay very fast during the first few cycles, if any, being neglected. *See:* **direct-axis synchronous reactance.** 0-31E8

direct-axis voltage (rotating machinery). The component of voltage that would produce direct-axis current when resistance-limited. *See:* **direct-axis synchronous reactance; synchronous machines.** 0-31E8

direct capacitance (between two conductors). The ratio of the change in charge on one conductor to the corresponding change in potential of the second conductor when the second conductor is the only one whose potential is permitted to change. *See:* **direct capacitances (system of conductors).** E270-0

direct capacitances (system of conductors). The direct capacitances of a system of n conductors such as that considered in **coefficients of capacitance (system of conductors)** are the coefficients in the array of linear equations that express the charges on the conductors in terms of their differences in potential, instead of potentials relative to ground.

$$Q_1 = 0 + C_{12}(V_1 - V_2) + C_{13}(V_1 - V_3) \cdots C_{1(n-1)}(V_1 - V_{n-1}) + C_{10} V_1$$

$$Q_2 = C_{21}(V_2 - V_1) + 0 + C_{23}(V_2 - V_3) \cdots C_{2(n-1)}(V_2 - V_{n-1}) + C_{20} V_2$$

$$Q_{n-1} = C_{(n-1)1}(V_{n-1} - V_1 + C_{(n-1)2}(V_{n-1} - V_2) \cdots 0 + C_{(n-1)0} V_{n-1}$$

with $C_{rp} = C_{pr}$ and C_{re} not involved but defined as zero. *Note:* The coefficients of capacitance c are related to the direct capacitances C as follows

$$c_{rp} = -C_{rp}, \text{ for } r \neq p$$

and

$$c_{rr} = \sum_{p=1}^{p=n} C_{rp}$$

Note: The relationships of the direct capacitances in a four-conductor system to the other defined capacitances are given in the following table. As in the diagram, the direct capacitances between the conductors of the system, including ground, are indicated as C_{10}, C_{12}, C_{13}, etcetera. Here the subscript *0* is used to denote the ground (*n*th) conductor.

Conductor	Self Capacitance
1	$C_{10} + C_{12} + C_{13}$
2	$C_{20} + C_{21} + C_{23}$
3	$C_{30} + C_{31} + C_{32}$

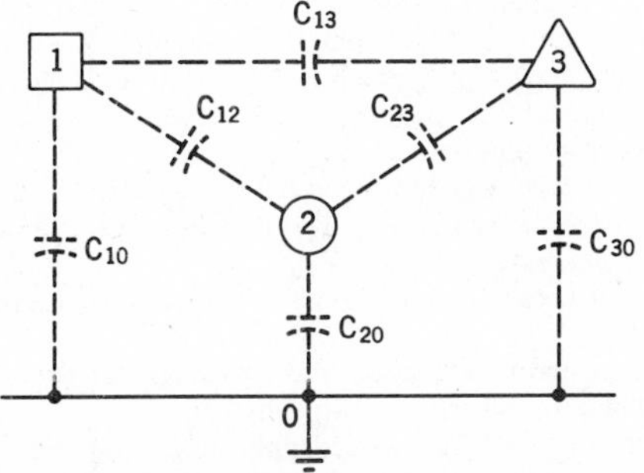

Capacitance diagram showing the equivalent direct capacitance network of a 4-conductor system.

Pair of Conductors: 1 and 2

$$\text{Plenary Capacitance} = C_{12} + \frac{C_{30}(C_{10}+C_{13})(C_{20}+C_{23})+C_{10}C_{20}(C_{31}C_{32})+C_{31}C_{32}(C_{10}+C_{20})}{C_{30}(C_{10}+C_{13}+C_{20}+C_{23})+(C_{10}+C_{20})(C_{31}+C_{32})}.$$

$$\text{Balanced Capacitance} = C_{12} + \frac{(C_{13}+C_{10})(C_{23}+C_{20})}{C_{13}+C_{10}+C_{23}+C_{20}}.$$

Pair of Conductors: 2 and 3

$$\text{Plenary Capacitance} = C_{23} + \frac{C_{10}(C_{20}+C_{21})(C_{30}+C_{31})+C_{20}C_{30}(C_{31}+C_{21})+C_{31}C_{21}(C_{20}+C_{30})}{C_{10}(C_{20}+C_{21}+C_{30}+C_{31})+(C_{20}+C_{30})(C_{31}+C_{21})}.$$

$$\text{Balanced Capacitance} = C_{23} + \frac{(C_{21}+C_{20})(C_{31}+C_{30})}{C_{21}+C_{20}+C_{31}+C_{30}}.$$

Pair of Conductors: 3 and 1

$$\text{Plenary Capacitance} = C_{31} + \frac{C_{20}(C_{10}+C_{12})(C_{30}+C_{32})+C_{10}C_{30}(C_{32}+C_{12})+C_{12}C_{32}(C_{10}+C_{30})}{C_{20}(C_{10}+C_{12}+C_{30}+C_{32})+(C_{10}+C_{30})(C_{12}+C_{32})}.$$

$$\text{Balanced Capacitance} = C_{31} + \frac{(C_{32}+C_{30})(C_{12}+C_{10})}{C_{32}+C_{30}+C_{12}+C_{10}}$$

E270-0

direct component (illuminating engineering). That portion of the light from a luminaire that arrives at the work plane without being reflected by room surfaces. *See also:* **inverse-square law (illuminating engineering).** Z7A1-0

direct-connected exciter (rotating machinery). An exciter mounted on or coupled to the main machine shaft so that both machines operate at the same speed. *See also:* **asynchronous machine; synchronous machine.** 0-31E8

direct-connected system. *See:* **headquarters system.**

direct-coupled amplifier (signal-transmission system). A direct-current amplifier in which all signal connections between active channels are conductive. *See also:* **signal.** 0-13E6

direct-coupled attenuation (transmit-receive, pretransmit-receive, and attenuator tubes). The insertion loss measured with the resonant gaps, or their functional equivalent, short-circuited. E160-0

direct coupling. The association of two or more circuits by means of self-inductance, capacitance, resistance, or a combination of these, that is common to the circuits. *Note:* **Resistance coupling** is the case in which the common branch contains only resistance; **capacitance coupling** that in which the branch contains only capacitance; **direct-inductance** or **self-inductance coupling** that in which the branch contains only self-inductance. *See also:* **coupling.** E270/42A65-0

direct current. A unidirectional current in which the changes in value are either zero or so small that they may be neglected. *Note:* As ordinarily used, the term designates a practically nonpulsating current. A given current would be considered a direct current in some applications, but would not necessarily be so considered in other applications. E45/E270-0

direct-current amplifier (1). An amplifier capable of amplifying waves of infinitesimal frequency. *See also:* **amplifier.** E145-0

(2) (signal-transmission system). An amplifier capable of producing a sustained single-valued, unidirectional output in response to a similar but smaller input. *Note:* It generally employs between stages either resistance coupling alone or resistance coupling combined with other forms of coupling. *See also:* **amplifier.** 42A65-31E3/13E6

direct-current analog computer. *See:* **analog computer.**

direct-current balance (amplifiers). An adjustment to avoid a change in direct-current level when changing gain. *See:* **amplifier.** 0-9E4

direct-current balancer (direct-current compensator*). A machine that comprises two or more similar direct-current machines (usually with shunt or compound

excitation) directly coupled to each other and connected in series across the outer conductors of a multiple-wire system of distribution, for the purpose of maintaining the potentials of the intermediate conductors of the system, which are connected to the junction points between the machines. *See:* **converter.**
*Deprecated 42A10-0

direct-current blocking voltage rating (rectifier circuit). The maximum continuous direct-current reverse voltage permitted by the manufacturer under stated conditions. *See also:* **rectifier circuit element.** E59-34E17/34E24

direct-current circuit. A circuit that includes two or more interrelated conductors intended to be energized by direct current. E270-0

direct-current commutating machine. Comprises a magnetic field excited from a direct-current source or formed of permanent magnets, an armature, and a commutator connected therewith. *Notes:* (1) Specific types of direct-current commutating machines are: **direct-current generators, motors, synchronous converters, boosters, balancers,** and **dynamotors.** (2) For an extensive list of cross references, see *Appendix A.* 42A10-0

direct-current compensator. *See:* **direct-current balancer.**

direct-current component (total current). That portion of the total current that constitutes the asymmetry. 37A100-31E11

direct-current component of a composite picture signal, blanked picture signal, or picture signal (television). The difference in level between the average value, taken over a specified time interval, and the peak value in the black direction, which is taken as zero. *Note:* The averaging period is usually one line interval or greater. *See:* **television.** 0-2E2

direct-current distribution. The supply, to points of utilization, of electric energy by direct current from its point of generation or conversion.
See:
cascade control of a street-lighting system;
center of distribution;
closed-loop series of a street-lighting system;
constant-current street-lighting system;
demand factor;
direct-current transmission;
diversity factor;
Edison distribution system;
electric power substation;
low-voltage system;
maximum demand;
mixed-loop series street-lighting system;
multiple street-lighting system;
neutral point;
open-loop series lighting system;
radical system;
series distribution system;
system interconnection;
three-wire system;
Thury transmission system;
transmission line;
two-wire system;
ungrounded system;
utilization factor. 42A35-31E13

direct-current drift. *See:* **stability.**

direct-current dynamic short-circuit ratio. The ratio of the maximum transient value of a current, after a suddenly applied short circuit, to the final steady-state value. *See also:* **electric arc welding apparatus.** 87A1-0

direct-current electric locomotive. An electric locomotive that collects propulsion power from a direct-current distribution system. *See also:* **electric locomotive.** 42A42-0

direct-current electron-stream resistance (electron tubes). The quotient of electron-stream potential and the direct-current component of stream current. *See also:* **circuit characteristics of electrodes; electrode current (electron tube).** E160-15E6

direct-current erasing head (magnetic recording). One that uses direct current to produce the magnetic field necessary for erasing. *Note:* Direct-current erasing is achieved by subjecting the medium to a unidirectional field. Such a medium is, therefore, in a different magnetic state than one erased by alternating current. *See also:* **phonograph pickup.** 0-1E1

direct-current generator. A generator for production of direct-current power. *See:* **direct-current commutating machine.** 0-31E8

direct-current leakage. *See:* **controlled overvoltage test.**

direct-current magnetic biasing (magnetic recording). Magnetic biasing accomplished by the use of direct current. *See also:* **phonograph pickup.** E157-1E1

direct-current neutral grid. A network of neutral conductors, usually grounded, formed by connecting together within a given area of all the neutral conductors of a low-voltage direct-current supply system. *See also:* **center of distribution.** 42A35-31E13

direct-current offset (amplifiers). A direct-current level that may be added to the input signal, referred to the input terminals. *See:* **amplifier.** 0-9E4

direct-current quadruplex system. A direct-current telegraph system that affords simultaneous transmission of two messages in each direction over the same line, operation being obtained by superposing neutral telegraph upon polar telegraph. *See also:* **telegraphy.** 42A65-0

direct-current relay. *See:* **relay, direct-current.**

direct-current restoration (television). The re-establishment by a sampling process of the direct-current and the low-frequency components of a video signal that have been suppressed by alternating-current transmission. *See also:* **television.** E204-0

direct-current restorer (1) (general). A means, used in a circuit incapable of transmitting slow variations but capable of transmitting components of higher frequency, by which a direct-current component is reinserted after transmission, and in some cases other low-frequency components are also reinserted. *See also:* **circuits and devices.** 0-42A65
(2) (television). A device for re-establishing by a sampling process the direct-current and low-frequency components of a video signal that have been suppressed by alternating-current transmission. *Note:* The sampling process can be accomplished either by the video signal itself or by external pulses. *See also:* **television.** E204-0

direct-current self-synchronous system. A system for transmitting angular position or motion, comprising a transmitter and one or more receivers. The transmitter is an arrangement of resistors that furnishes the receiver with two or more voltages that are functions of transmitter shaft position. The receiver has two or more stationary coils that set up a magnetic field causing a rotor to take up an angular position corresponding to the angular position of the transmitter shaft. *See:* **synchro system.** 42A10-0

direct-current telegraphy. That form of telegraphy in which, in order to form the transmitted signals, direct current is supplied to the line under the control of the transmitting apparatus. *See also:* **telegraphy.** 42A65-0

direct-current transmission (1) (electric energy). The transfer of electric energy by direct current from its source to one or more main receiving stations. *Note:* For transmitting large blocks of power, high voltage may be used such as obtained with generators in series, rectifiers, etcetera. *See also:* **direct-current distribution.** 42A35-31E13
(2) (television). A form of transmission in which the direct-current component of the video signal is transmitted. *Note:* In an amplitude-modulated signal with direct-current transmission, the black level is represented always by the same value of envelope. In a frequency-modulated signal with direct-current transmission, the black level is represented always by the same value of the instantaneous frequency. *See also:* **television.** E204-2E2/42A65

direct-current winding (rectifier transformer). The secondary winding that is conductively connected to the main electrodes of the rectifier, and that conducts the direct current of the rectifier. *See:* **rectifier transformer.** 42A15-31E12

direct digital control (electric power systems). A mode of control wherein digital computer outputs are used to directly control a process. *See also:* **power system.** 0-31E4

direct-drive machine. An electric driving machine the motor of which is directly connected mechanically to the driving sheave, drum, or shaft without the use of belts or chains, either with or without intermediate gears. 42A45-0

directed branch (network analysis). A branch having an assigned direction. *Note:* In identifying the branch direction, the branch *jk* may be thought of as outgoing from node *j* and incoming at node *k*. Alternatively, branch *jk* may be thought of as originating or having its input at node *j* and terminating or having its output at node *k*. The assigned direction is conveniently indicated by an arrow pointing from node *j* toward node*k*. *See also:* **linear signal flow graphs.** E155-0

directed reference flight (electronic navigation). The type of stabilized flight that obtains control information from external signals that may be varied as necessary to direct the flight. For example, flight of a guided missile or a target aircraft. *See also:* **navigation.** E172-10E6

direct feeder. A feeder that connects a generating station, substation, or other supply point to one point of utilization. *See:* **radial feeder.** *See also:* **center of distribution.** 42A35-31E13

direct glare. Glare resulting from high brightnesses or insufficiently shielded light sources in the field of view or from reflecting areas of high brightness. *Note:* It usually is associated with bright areas, such as luminaires, ceilings, and windows that are outside the visual task or region being viewed. *See also:* **visual field.** Z7A1-0

direct grid bias. The direct component of grid voltage. *Note:* This is commonly called grid bias. *See also:* **electrode voltage (electron tube).** E160/42A70-15E6

direct-indirect lighting. A variant of general diffuse lighting in which the luminaires emit little or no light at angles near the horizontal. *See also:* **general lighting.** Z7A1-0

direct inductance coupling. *See:* **inductance coupling (communication circuits).**

direct interelectrode capacitance (electron tubes). The direct capacitance between any two electrodes excluding all capacitance between either electrode and any other electrode or adjacent body. 50I07-15E6

direction (navigation). The position of one point in space relative to another without reference to the distance between them. *Notes:* (1) Direction may be either three dimensional or two dimensional, and it is not an angle but is often indicated in terms of its angular difference from a reference direction. (2) Five terms used in navigation: azimuth, bearing, course, heading, and track, involve measurement of angles from reference directions. To specify the reference directions, certain modifiers are used. These are: true, magnetic, compass, relative grid, and gyro. *See also:* **navigation.** E172-10E6

directional antenna. An antenna having the property of radiating or receiving radio waves more effectively in some directions than others. *See also:* **antenna.** E145-3E1

directional-comparison protection (relays). A form of pilot protection in which the relative positions of the directional units at the line terminals are compared to determine whether a fault is in the protected line section. *Note:* Commonly, the directional units are relays, and the comparison is made by noting the relative positions of the contacts of the relay directional units. *See also:* **relay.** 37A100-31E11;0-31E6

directional control (protective relay or relay scheme). A qualifying term that indicates a means of controlling the operating force in a nondirectional relay so that it will not operate until the two or more phasor quantities used to actuate the controlling means (directional relay) are in a predetermined band of related phase positions. 37A100-31E11/31E6

directional coupler (1) (transmission lines). A transmission coupling device for separately (ideally) sampling (through a known coupling loss for measuring purposes) either the forward (incident) or the backward (reflected) wave in a transmission line. *Notes:* (1) Similarly, it may be used to excite in the transmission line either a forward or backward wave. (2) A unidirectional coupler has available terminals or connections for sampling only one direction of transmission; a bidirectional coupler has available terminals for sampling both directions. *See also:* **auxiliary device to an instrument.** 42A30-0
(2) (waveguides). A four-branch junction consisting of two waveguides coupled together in a manner such that a single traveling wave in either guide will induce a single traveling wave in the other, direction of latter wave being determined by direction of the former. *See also:* **auxiliary device to an instrument; waveguide.** E147-3E1

directional-current tripping. *See:* **directional-overcurrent protection; directional-overcurrent relay.**

directional gain directivity index (transducer) (audio

and electroacoustics). In decibels, 10 times the logarithm to the base 10 of the directivity factor. *See also:* **electroacoustics.** 42A65-0;1E1-0

directional gyro electric indicator. An electrically driven device for use in aircraft for measuring deviation from a fixed heading. *See also:* **air transportation instruments.** 42A41-0

directional homing (navigation). The process of homing wherein the navigational quantity maintained constant is the bearing. *See also:* **radio navigation.** E172-10E6

directional lighting. Lighting provides on the work plane or on an object light that is predominantly from a preferred direction. *See also:* **general lighting.** Z7A1-0

directional localizer (instrument landing systems). A localizer in which maximum energy is directed close to the runway centerline, thus minimizing extraneous reflections. *See also:* **navigation.** 0-10E6

directional microphone. A microphone the response of which varies significantly with the direction of sound incidence. *See also:* **microphone.** 42A65-0

directional-overcurrent protection. A method of protection in which an abnormal condition within the protected equipment is detected by the current being in excess of a predetermined amount and in a predetermined phase relation with a reference input. 37A100-31E11

directional-overcurrent relay. A relay consisting of an overcurrent unit and a directional unit combined to operate jointly. 37A100-31E11/31E6

directional pattern (radiation pattern). The directional pattern of an antenna is a graphical representation of the radiation or reception of the antenna as a function of direction. *Note:* Cross sections in which directional patterns are frequently given are vertical planes and the horizontal planes or the principal electric and magnetic polarization planes. *See also:* **antenna.** 0-42A65

directional phase shifter (directional phase changer) (nonreciprocal phase shifter). A passive phase changer in which the phase change for transmission in one direction differs from that for transmission in the opposite direction. *See:* **transmission line.** 0-3E1

directional-power relay. A relay that operates in conformance with the direction of power flow. 37A100-31E11/31E6

directional-power tripping. *See:* **directional-power relay.**

directional relay. A relay that responds to the relative phase position of a current with respect to another current or voltage reference. *Note:* The above definition, which applies basically to a single-phase directional relay, may be extended to cover a polyphase directional relay. 37A100-31E11/31E6

directional response pattern (beam pattern) (electroacoustics) (transducer used for sound emission or reception). A description, often presented graphically, of the response of the transducer as a function of the direction of the transmitted or incident sound waves in a specified plane and at a specified frequency. *Notes:* (1) A complete description of the directional response pattern of a transducer would require a three-dimensional presentation. (2) The directional response pattern is often shown as the response relative to the maximum response. *See also:* **loudspeaker.** 0-1E1

direction-finder (DF). *See:* **navigation; radio direction-finder.**

direction-finder antenna. Any antenna used for radio direction finding. *See also:* **navigation.** 0-10E6

direction-finder antenna system. One or more direction-finder antennas, their combining circuits and feeder systems, together with the shielding and all electrical and mechanical items up to the termination at the receiver input terminals. *See also:* **navigation.** E173-10E6

direction-finder deviation. The amount by which an observed radio bearing differs from the corrected bearing. *See also:* **navigation.** 0-10E6

direction-finder noise level (in the absence of the desired signals). The average power or root-mean-square voltage at any specified point in a direction-finder system circuit. *Note:* In radio-frequency and audio channels, the direction-finder noise level is usually measured in terms of the power dissipated in suitable termination. In a video channel, it is customarily measured in terms of voltage across a given impedance or of the cathode-ray deflection. *See also:* **navigation.** E173-10E6

direction-finder sensitivity. The field strength at the direction-finder antenna, in microvolts per meter, that produces a ratio of signal-plus-noise to noise equal to 20 decibels in the receiver output, the direction of arrival of the signal being such as to produce maximum pickup in the direction-finder antenna system. *See also:* **navigation.** E173-10E6

direction finding. *See:* **radio direction-finder.**

direction of energy flow (specified circuit). With reference to the boundary (of a delimited region), the direction in which electric energy is being transmitted past the boundary, into or out of the region. E270-0

direction of lay (cables). The lateral direction, designated as left-hand or right-hand, in which the elements of a cable run over the top of the cable as they recede from an observer looking along the axis of the cable. *See also:* **power distribution, underground construction.** E30-31E1;42A35-31E13

direction of polarization (radio propagation) (1) (linearly polarized wave). The direction of the electric intensity. 42A65-0

(2) (elliptically polarized wave). The direction of the major axis of the electric vector ellipse. *See also:* **radiation; radio wave propagation.** 0-3E2

direction of propagation (point in a homogeneous isotropic medium) (1). The normal to an equiphase surface taken in the direction of increasing phase lag. *See also:* **radiation.** 42A65-0

(2). The direction of time-average energy flow. *Notes:* (1) In a uniform waveguide the direction of propagation is often taken along the axis. (2) In the case of a uniform lossless waveguide the direction of propagation at every point is parallel to the axis and in the direction of time-average energy flow. *See also:* **radiation; waveguide.** E146-3E1

directive gain (antenna) (given direction). 4π times the ratio of the radiation intensity in that direction to the total power radiated by the antenna. *Note:* The directive gain is fully realized on reception only when the incident polarization is the same as the polarization of the antenna on transmission. *See also:* **antenna.** 10-3E1

directivity (1) (antenna). The value of the directive gain in the direction of its maximum value. *See also:* **antenna.** E145-0;E149-3E1
(2) (directional coupler). The ratio of the power output at an auxiliary port, when power is fed into the main waveguide or transmission line in the preferred direction, to the power output at the same auxiliary port when power is fed into the main guide or line in the opposite direction, the incident power fed into the main guide or line being the same in each case, and reflectionless terminations being connected to all ports. *Note:* The ratio is usually expressed in decibels. 0-9E4

directivity factor (audio and electroacoustics) (1) (transducer used for sound emission). The ratio of the sound pressure squared, at some fixed distance and specified direction, to the mean-square sound pressure at the same distance averaged over all directions from the transducer. *Note:* The distance must be great enough so that the sound pressure appears to diverge spherically from the effective acoustic center of the transducer. Unless otherwise specified, the reference direction is understood to be that of a maximum response. The frequency must be stated. *See also:* **loudspeaker.** 0-1E1
(2) (transducer used for sound reception). The ratio of the square of the open-circuit voltage produced in response to sound waves arriving in a specified direction to the mean-square voltage that would be produced in a perfectly diffused sound field of the same frequency and mean-square sound pressure. *Notes:* (1) This definition may be extended to cover the case of finite frequency bands whose spectrum may be specified. (2) The average free-field response may be obtained in various ways, such as (A) by the use of a spherical integrator, (B) by numerical integration of a sufficient number of directivity patterns corresponding to different planes, or (C) by integration of one or two directional patterns whenever the pattern of the transducer is known to possess adequate symmetry. *See also:* **microphone.** 0-1E1

direct lighting. Lighting involving luminaires that distribute 90 to 100 percent of the emitted light in the general direction of the surface to be illuminated. *Note:* The term usually refers to light emitted in a downward direction. *See also:* **general lighting.** Z7A1-0

direct liquid cooling system (semiconductor rectifiers). A cooling system in which a liquid, received from a constantly available supply, is passed directly over the cooling surfaces of the semiconductor power converter and discharged. *See also:* **semiconductor rectifier stack.** 0-34E24

direct liquid cooling system with recirculation (semiconductor rectifiers). A direct liquid cooling system in which part of the liquid passing over the cooling surfaces of the semiconductor power converter is recirculated and additional liquid is added as needed to maintain the required temperature, the excess being discharged. *See also:* **semiconductor rectifier stack.** 0-34E24

directly controlled variable (industrial control). The variable in a feedback control system whose value is sensed to originate the primary feedback signal. *See:* **control system, feedback.** AS1-34E10

directly grounded. *See:* **grounded solidly.**

direct on-line starting (rotating machinery). The process of starting a motor by connecting it directly to the supply at a rated voltage. *See:* **asynchronous machine; direct-current commutating machine; synchronous machine.** 0-31E8

direct operation (mechanism). Operation by means connected directly to the main operating shaft or an extension of the same. 37A100-31E11

director element. A parasitic element located forward of the driven element of an antenna, intended to increase the directive gain of the antenna in the forward direction. *See also:* **antenna.** 0-3E1

directory number (telephony). The full complement of digits required to designate a subscriber in the directory. *See also:* **telephone switching system.** 0-19E1

direct-plunger driving machine (elevators). A machine in which the energy is applied by a plunger or piston directly attached to the car frame or platform and that operates in a cylinder under hydraulic pressure. *Note:* It includes the cylinder and plunger or piston. *See also:* **driving machine (elevators).** 42A45-0

direct-plunger elevator. A hydraulic elevator having a plunger or piston directly attached to the car frame or platform. *See also:* **elevators.** 42A45-0

direct-point repeater. A telegraph repeater in which the receiving relay, controlled by the signals received over a line, repeats corresponding signals directly into another line or lines without the interposition of any other repeating or transmitting apparatus. *See also:* **telegraphy.** 42A65-0

direct ratio (illuminating engineering). The ratio of the luminous flux that reaches the work plane directly to the downward component from the luminaire. *See also:* **inverse-square law (illuminating engineering).** Z7A1-0

direct raw-water cooling system (rectifier). A cooling system in which water, received from a constantly available supply, such as a well or water system, is passed directly over the cooling surfaces of the rectifier and discharged. *See:* **direct liquid cooling system.** *See also:* **rectification.** 42A15-34E24

direct raw-water cooling system with recirculation. A direct raw-water cooling system in which part of the water passing over the cooling surfaces of the rectifier is recirculated and raw water is added as needed to maintain the required temperature, the excess being discharged. *See:* **direct liquid cooling system with recirculation.** *See also:* **rectification.** 42A15-34E24

direct-recording facsimile. The type of recording in which a visible record is produced, without subsequent processing, in response to the received signals. *See also:* **recording (facsimile).** E168-0

direct release (series trip). A release directly energized by the current in the main circuit of a switching device. 37A100-31E11

direct stroke. A lightning stroke direct to any part of a network or electric installation. *See:* **lightning arrester (surge diverter).** 50I25-31E7

direct-stroke protection (lightning). Lightning protection designed to protect a network or electric installation against direct strokes.
See:
back flashover;
ball lightning;

continuing current;
coupling factor;
down lead;
ground plane;
indirect-stroke protection;
induced current;
induced voltage;
isokeraunic level;
lightning outage;
lightning protection and equipment;
lightning stroke;
lightning-stroke component;
lightning-stroke current;
lightning-stroke voltage;
line lightning performance;
multiple lightning stroke;
negative-polarity lightning stroke;
overhead ground wire;
pilot streamer;
positive-polarity lightning stroke;
prestrike current;
return stroke;
shielding angle;
shielding failure;
side flash;
side flashover;
stepped leader;
tower footing resistance;
traveling wave. 0-31E13

direct vacuum-tube current (medical electronics). A current obtained by applying to the part to be treated an evacuated glass electrode connected to one terminal of a generator of high-frequency current (100 to 10 000 kilohertz), the other terminal being grounded. *Note:* Deprecated as confusing and as representing an ill-defined and obsolescent procedure. *See also:* **medical electronics.** 42A80-18E1

direct-voltage high-potential test (rotating machinery). A test that consists of the application of a specified unidirectional voltage higher than the rated root-mean-square value for a specified time for the purpose of determining (1) the adequacy against breakdown of the insulation system under normal conditions, or (2) the resistance characteristic of the insulation system. *See also:* **asynchronous machines; direct-current commutating machines; synchronous machines.** 0-31E8

direct wave (radio wave propagation). A wave propagated directly from a source to a point. *See also:* **radiation; radio wave propagation.** 0-3E2

direct-wire circuit (one-wire circuit). A supervised circuit, usually consisting of one metallic conductor and a ground return, and having signal receiving equipment responsive to either an increase or a decrease in current. *See also:* **protective signaling.** 42A65-0

disability glare. Glare that reduces visual performance and visibility and often is accompanied by discomfort. *See:* **veiling brightness.** *See also:* **visual field.** Z7A1-0

disc (disk). *See:* **magnetic disc; disc recorder.** *See also:* **electronic digital computer.**

disc-and-wiper-lubricated bearing (rotating machinery). A bearing in which a disc mounted on and concentric with the shaft dips into a reservoir of oil. *Note:* As the shaft rotates the oil is diverted from the surface of the disc by a scraper action into the bearing. *See also:* **bearing.** 0-31E8

discharge (storage battery) (1) (storage cell). The conversion of the chemical energy of the battery into electric energy. *See also:* **charge.** 42A60-0
(2) (gas). The passage of electricity through a gas. *Note:* For an extensive list of cross references, see *Appendix A.* E160/50I07-15E6

discharge capacity (arrester). *See:* **arrester discharge capacity.**

discharge circuit (surge generator). That portion of the surge-generator connections in which exist the current and voltage variations constituting the surge generated. *See also:* **power systems, low-frequency and surge testing; lightning arrester (surge diverter).** 42A35-31E13/31E7

discharge counter (lightning arrester). A means or device for recording the number of arrester discharge operations. *See also:* **arrester.** E28/42A20-0

discharge current (lightning arrester). The surge current that flows through the arrester after a sparkover. *See:* **lightning; lightning arrester (surge diverter); current rating, 60-hertz (arrester).** E28/42A20-0;99I2-31E7

discharge detector (ionization or corona detector) (rotating machinery). An instrument that can be connected in or across an energized insulation circuit to detect current or voltage pulses produced by electric discharges within the circuit. *See also:* **instrument.** 0-31E8

discharge-energy test (rotating machinery). A test for determining the magnitude of the energy dissipated by a discharge or discharges within the insulation. *See also:* **asynchronous machine; direct-current commutating machine; synchronous machine.** 0-31E8

discharge extinction voltage (ionization or corona extinction voltage) (rotating machinery). The voltage at which discharge pulses that have been observed in an insulation system, using a discharge detector of specified sensitivity, cease to be detectable as the voltage applied to the system is decreased. *See also:* **asynchronous machines; direct-current commutating machine; synchronous machines.** 0-31E8

discharge inception test (corona inception test) (rotating machinery). A test for measuring the lowest voltage at which discharges of a specified magnitude recur in successive cycles when an increasing alternating voltage is applied to insulation. *See also:* **asynchronous machine; direct-current commutating machine; synchronous machine.** 0-31E8

discharge inception voltage (ionization or corona inception voltage) (1) (rotating machinery). The voltage at which discharge pulses in an insulation system become observable with a discharge detector of specified sensitivity, as the voltage applied to the system is raised. *See also:* **asynchronous machine; direct-current commutating machine; synchronous machine.** 0-31E8
(2) (lightning arrester). The root-mean-square value of the power-frequency voltage at which discharges start, the measurement of their intensity being made under specified conditions. *See:* **lightning arrester (surge diverter).** 50A25-31E7

discharge indicator (arrester). An attachable accessory for the purpose of indicating that the arrester has discharged. See: **arrester; lightning arrester (surge diverter).** 42A20-0

discharge opening (rotating machinery). A port for the exit of ventilation air. *See also:* **cradle base (rotating machinery).** 0-31E8

discharge probe (ionization or corona probe) (rotating machinery). A portable antenna, safely insulated, and designed to be used with a discharge detector for locating sites of discharges in an energized insulation system. *See also:* **instrument.** 0-31E8

discharge resistor. A resistor connected across the field windings of a generator, motor, synchronous capacitor, or an exciter to limit the transient voltage in the field circuit and to hasten the decay of field current of these machines upon interruption of excitation. 37A100-31E11

discharge tube. An evacuated enclosure containing a gas at low pressure that permits the passage of electricity through the gas upon application of sufficient voltage. *Note:* The tube is usually provided with metal electrodes, but one form permits an electrodeless discharge with induced voltage. *See also:* **tube definitions.** 42A70-0

discharge voltage (arrester). The voltage that appears across its terminals during passage of discharge current. See: **lightning; lightning arrester (surge diverter).** 42A20/62A1-31E7

discharge voltage-current characteristic (lightning arrester). *See:* **arrester discharge voltage-current characteristic.**

discharge voltage-time curve (arrester). *See:* **arrester discharge voltage-time curve.**

discharge withstand current rating (lightning arrester). The specified magnitude and wave shape of a discharge current that can be applied to the arrester a specified number of times without causing damage to it. *See also:* **lightning arrester (surge diverter); current rating, 60-hertz (arrester).** E28/62A1-31E7

discomfort glare. Glare that produces discomfort. It does not necessarily interfere with visual performance or visibility. *See also:* **visual field.** Z7A1-0

discomfort-glare factor. The numerical assessment of the capacity of a single source of brightness, such as a luminaire, in a given visual environment for producing discomfort. *See:* **glare; discomfort glare.** *See also:* **inverse-square law (illuminating engineering).** Z7A1-0

discomfort-glare rating. A numerical assessment of the capacity of a number of sources of brightness, such as luminaires, in a given visual environment for producing discomfort. *Note:* It usually is derived from the discomfort-glare factors of the individual sources. *See also:* **inverse-square law (illuminating engineering).** Z7A1-0

disconnect (release) (telephony). To disengage the apparatus used in a telephone connection and to restore it to its condition when not in use. *See also:* **telephone switching system.** 42A65-19E1

disconnected position (switchgear-assembly removable element). That position in which the primary and secondary disconnecting devices of the removable element are separated by a safe distance from the stationary element contacts. *Note:* Safe distance, as used here, is a distance at which the equipment will meet its withstand-voltage ratings, both low-frequency and impulse, between line and load terminals with the switching device in the closed position. 37A100-31E11

disconnecting blade. *See:* **blade (disconnecting blade) (switching device).**

disconnecting cutout. A cutout having a disconnecting blade for use as a disconnecting or isolating switch. 37A100-31E11

disconnecting fuse. *See:* **fuse-disconnecting switch.**

disconnecting means. A device, group of devices, or other means whereby the conductors of a circuit can be disconnected from their source of supply. *See also:* **switch.** 1A0-0

disconnecting or isolating switch (disconnector, isolator). A mechanical switching device used for changing the connections in a circuit or for isolating a circuit or equipment from the source of power. *Note:* It is required to carry normal load current continuously and also abnormal or short-circuit currents for short intervals as specified. It is also required to open or close circuits either when negligible current is broken or made or when no significant change in the voltage across the terminals of each of the switch poles occurs. 37A100-31E11

disconnection (control) (industrial control). Connotes the opening of a sufficient number of conductors to prevent current flow. IC1-34E10

disconnector. A switch that is intended to open a circuit only after the load has been thrown off by some other means. *Note:* Manual switches designed for opening loaded circuits are usually installed in circuit with disconnectors, to provide a safe means for opening the circuit under load. 2A2-0

disconnect signal (telephony). A signal transmitted from one end of a subscriber line or trunk to indicate that the relevant party has released. *See also:* **telephone switching system.** 0-19E1

disconnect-type pothead. A pothead in which the electric continuity of the circuit may be broken by physical separation of the pothead parts, part of the pothead being on each conductor end after the separation. *See also:* **pothead.** E48-0

discontinuity (1). An abrupt nonuniformity in a uniform waveguide or transmission line that causes reflected waves. *See:* **capacitance, discontinuity.** *See also:* **waveguide.** 0-9E4

(2) (inductive coordination). An abrupt change at a point, in the physical relations of electric supply and communication circuits or in electrical parameters of either circuit, that would materially affect the coupling. *Note:* Although technically included in the definition, transpositions are not rated as discontinuities because of their application to coordination. *See also:* **inductive coordination.** 42A65-0

disc recorder (phonograph techniques). A mechanical recorder in which the recording medium has the geometry of a disc. *See also:* **phonograph pickup.** E157-1E1

discrete sentence intelligibility. The percent intelligibility obtained when the speech units considered are sentences (usually of simple form and content). *See also:* **volume equivalent.** 42A65-0

discrete word intelligibility. The percent intelligibility obtained when the speech units considered are words (usually presented so as to minimize the contextual relation between them). *See also:* **volume equivalent.** 42A65-0

discrimination (any system or transducer). The difference between the losses at specified frequencies, with the system or transducer terminated in specified impedances. *See also:* **transmission loss.** 42A65-0

discrimination ratio (ferroelectric device). The ratio of signal charge to induced charge. *Note:* Discrimination ratio is dependent on the magnitude of the applied voltage, which therefore should be specified. *See also:* **ferroelectric domain.** E180-0

discriminator (1). A device in which amplitude variations are derived in response to frequency or phase variations. *Note:* The device is termed a frequency discriminator or phase discriminator according to whether it responds to variations of frequency or phase. *See also:* **circuits and devices; modulating systems.** 42A65-31E3

(2) (electronic navigation). A circuit in which the output is dependent upon how an input signal differs in some aspect from a standard or from another signal. *See also:* **circuits and devices.** E172-10E6

discriminator, amplitude (pulse techniques). *See:* **discriminator, pulse-height.**

discriminator, pulse-height (pulse techniques). A circuit that produces a specified output pulse if and only if it receives an input pulse whose amplitude exceeds an assigned value. *See also:* **pulse.** E175-0

dish (radio practice). A reflector the surface of which is concave as, for example, a part of a sphere or of a paraboloid of revolution. *See also:* **antenna.** 42A65-0

disk. *See:* **disc.**

dispatching system (mining practice). A system employing radio, telephones, and/or signals (audible or light) for orderly and efficient control of the movements of trains of cars in mines. *See:* **mine fan signal system; mine radio telephone system.** *See also:* **mining.** 42A85-0

dispatch operation (radio-communication circuit). A method for permitting a maximum number of terminal devices to have access to the same two-way radio communication circuit. *See also:* **channel spacing.** 0-6E1

dispenser cathode (electron tubes). A cathode that is not coated but is continuously supplied with suitable emission material from a separate element associated with it. *See also:* **electron tube.** 50I07-15E6

dispersed magnetic powder tape (impregnated tape). *See:* **magnetic powder-impregnated tape.**

displacement current (any surface). The integral of the normal component of the displacement current density over that surface. *Note:* Displacement current is a scalar and hence has no direction. E270-0

displacement current density (any point in an electric field). The time rate of change in SI units (International System of Electrical Units) of the electric flux density vector at that point. E270-0

displacement power factor (rectifier) (phasor power factor) (rectifier unit). The displacement component of power factor; the ratio of the active power of the fundamental wave, in watts, to the apparent power of the fundamental wave, in root-mean-square voltamperes (including the exciting current of the rectifier transformer). *See also:* **power rectifier; rectification.** 34A1-34E24

display (navigation systems). The visual representation of output data. *Note:* See display by type designation, for example, ***B*** display, ***C*** display, etcetera. 0-10E6

display primaries (receiver primaries). The colors of constant chromaticity and variable luminance produced by the receiver or any other display device that, when mixed in proper proportions, are used to produce other colors. *Note:* Usually three primaries are used: red, green and blue. *See also:* **color terms.** E201-2E2

display storage tube. A storage tube into which the information is introduced as an electric signal and read at a later time as a visible output. *See also:* **storage tube.** E158-15E6

display tube. A tube, usually a cathode-ray tube, used to display data. X3A12-16E9

disruptive discharge. The phenomena associated with the failure of insulation, under electric stress, that include a collapse of voltage and the passage of current; the term applies to electrical breakdown in solid, liquid, and gaseous dielectrics and combinations of these. *Note:* A disruptive discharge in a solid dielectric produces permanent loss of electric strength; in a liquid or gaseous dielectric the loss may be only temporary. *See:* **lightning arrester (surge diverter); low-frequency and surge testing; power systems; test voltage and current.** 68AI-31E5;0-31E7

disruptive discharge voltage (1). The value of the test voltage for which disruptive discharge takes place. *Note:* The disruptive discharge voltage is subject to statistical variation which can be expressed in different ways as, for example, by the mean, the maximum, and the minimum values of series of observations, or by the mean and standard deviation from the mean, or by a relation between voltage and probability of a disruptive discharge. *See also:* **test voltage and current.** 68A1-31E1

(2) (50 percent) (lightning arresters). The voltage that has a 50-percent probability of producing a disruptive discharge. *Note:* The term applies mostly to impulse tests and has significance only in cases when the loss of electric strength resulting from a disruptive discharge is temporary. *See:* **lightning arrester (surge diverter).** 68A1-31E5;60I0-31E7

(3) (100 percent) (lightning arresters). The specified voltage that is to be applied to a test object in a 100-percent disruptive discharge test under specified conditions. *Note:* The term applies mostly to impulse tests and has significance only in cases when the loss of electric strength resulting from a disruptive discharge is temporary. During the test, in general, all voltage applications should cause disruptive discharge. 60I0-31E7;68A1-31E5

dissector. *See:* **image dissector.**

dissector tube. A camera tube having a continuous photochathode on which is formed a photoelectric-emission pattern that is scanned by moving its electron optical image over an aperture. *See also:* **tube definitions; image dissector tube.** 42A70-0

dissipation (electrical energy). Loss of electric energy as heat. MDE-21E1

dissipation, electrode. *See:* **electrode dissipation.**

dissipation-factor test (rotating machinery). *See:* **loss-tangent test.**

dissymmetrical transducer. Dissymmetrical with respect to a specified pair of terminations when the interchange of that pair of terminations will affect the transmission. *See also:* **transducer.** 42A65-0

distance. *See:* **Hamming distance; signal distance.**

distance mark (range mark) (on a radar display). A calibration marker used on a cathode-ray screen in determining target distance. *See also:* **radar.** 42A65-10E6

distance-measuring equipment. A radio aid to navigation that provides distance information by measuring total round-trip time of transmission from an interrogator to a transponder and return. *Note:* Commonly abbreviated DME. 42A65-0;E172-10E6

distance protection. A method of line protection in which an abnormal condition within a predetermined electrical distance of a line terminal on the protected circuit is detected by measurement of system conditions at that terminal. 37A100-31E11

distance relay. A generic term covering those forms of protective relays in which the response to the input quantities is primarily a function of the electrical circuit distance between the relay location and the point of fault. *Note:* Distance relays may be single-phase devices or they may be polyphase devices. 37A100-31E11/31E6

distance resolution (radar). The ability to distinguish between two targets solely by the measurement of distances; generally expressed in terms of the minimum distance by which two targets must be spaced to be separately distinguishable. *See also:* **navigation.** 0-10E6

distortion. (1) An undesired change in waveform. *Note:* The principal sources of distortion of a waveform are: (A) nonlinear relationship between input and output, (B) nonuniform transmission at different frequencies, and (C) phase shift not proportional to frequency. Nonlinear effects due to heating of components may not appreciably change the output waveform. *See:* **electrical noise; fortuitous telegraph distortion; frequency distortion; nonlinear distortion; total telegraph distortion.** (2) Any departure from a specified input-output relationship over a range of frequencies, amplitudes, or phase shifts, or during a time interval.
See:
amplitude distortion;
attenuation distortion;
delay distortion;
deviation distortion;
harmonic distortion;
intermodulation distortion;
nonlinear distortion;
phase distortion. *See also:* **amplitude-frequency distortion; frequency distortion; transmission characteristics; close-talking pressure-type microphone.**
(3) Any undesired change in a transmitted pattern or picture.
(4) **(electrical conversion).** The portion of an alternating-current waveform caused by frequencies other than the fundamental. It provides an indication of the harmonic content of the alternating-current wave and is expressed as a percent of the fundamental.
Distortion Factor =

$$\left[\frac{\text{(sum of squares of amplitudes of all harmonics)}}{\text{(square of amplitude of fundamental)}}\right]^{1/2} (100\%)$$

Note: For distortion factors less than 10 percent a distortion-factor meter will provide a method of accurate measurement.

distortion, amplitude-frequency (electroacoustics). *See:* **amplitude-frequency distortion; distortion.**

distortion, barrel (camera tubes or image tubes). A distortion that results in a progressive decrease in radial magnification in the reproduced image away from the axis of symmetry of the electron optical system. *Note:* For a camera tube, the reproducer is assumed to have no geometric distortion. *See:* **beam tubes.**

distortion, envelope delay (1) (general). Of a system or transducer, the difference between the envelope delay at one frequency and the envelope delay at a reference frequency. E151-0
(2) (facsimile). The form of distortion that occurs when the rate of change of phase shift with frequency of a circuit or system is not constant over the frequency range required for transmission. *Note:* In facsimile, envelope delay distortion is usually expressed as one-half the difference in microseconds between the maximum and the minimum envelope delays existing between the two extremes of frequency defining the channel used. *See also:* **facsimile transmission.** E168-0

distortion factor (1) (power-system communication). The ratio of the root-mean-square value of the harmonic content to the root-mean-square value of the non-sinusoidal quantity. *See also:* **distortion.** 50I05-31E3
(2) (wave) (rotating machinery). The ratio of the root-mean-square value of the residue of a voltage wave after the elimination of the fundamental to the root-mean-square value of the original wave. *See:* **synchronous machine.** 0-31E8

distortion, frequency. *See:* **amplitude-frequency distortion.**

distortion, harmonic. Nonlinear distortion of a system or transducer characterized by the appearance in the output of harmonics other than the fundamental component when the input wave is sinusoidal. *Note:* Subharmonic distortion may also occur. *See:* **distortion.** E151/E154/E188/E270 -0

distortion, intermodulation. Nonlinear distortion of a system or transducer characterized by the appearance in the output of frequencies equal to the sums and differences of integral multiples of the two or more component frequencies present in the input wave. *Note:* Harmonic components also present in the output are usually not included as part of the intermodulation distortion. When harmonics are included, a statement to that effect should be made. *See also:* **distortion.** E151-42A65

distortion, percent harmonic (electroacoustics). A measure of the harmonic distortion in a system or transducer, numerically equal to 100 times the ratio of the square root of the sum of the squares of the root-mean-square voltages (or currents) of each of the individual harmonic frequencies, to the root-mean-square voltage (or current) of the fundamental. *Note:* It is practical to measure the ratio of the root-mean-square amplitude of the residual harmonic voltages (or currents), after the elimination of the fundamental, to the root-mean-square amplitude of the fundamental and harmonic voltages (or currents) combined. This measurement will indicate percent harmonic distortion with an error of less than 5 percent if the magnitude of the distortion does not exceed 30 percent. *See:* **distortion.** E151-0

distortion, phase delay (system or transducer). The difference between the phase delay at one frequency and the phase delay at a reference frequency. E151-0

distortion, phase-frequency. *See:* **distortion, phase delay.**

distortion, pincushion (camera tubes or image tubes). A distortion that results in a progressive increase in radial magnification in the reproduced image away from the axis of symmetry of the electron optical system. *Note:* For a camera tube, the reproducer is assumed to have no geometric distortion. *See:* **beam tubes; distortion, amplitude-frequency (electroacoustics); distortion factor; distortion, percent harmonic (electroacoustics); hiss (in an electron device).**
42A65/E160-2E2/10E1/15E6

distortion power (single-phase two-wire circuit). At the two terminals of entry of a single-phase two-wire circuit into a delimited region, a scalar quantity having an amplitude equal to the square root of the difference of the squares of the apparent power and the amplitude of the phasor power. *Note:* Mathematically the amplitude of the distortion power D is given by the equation

$$D = (U^2 - S^2)^{1/2}$$
$$= (U^2 - P^2 - Q^2)^{1/2}$$
$$= \left(\sum_{r=1}^{r=\infty} \sum_{q=1}^{q=\infty} \{E_r^2 I_q^2 - E_r E_q I_r I_q \cos[(\alpha_r - \beta_r) - (\alpha_q - \beta_q)]\} \right)^{1/2}$$

where the symbols are as in **power, apparent (single-phase two-wire circuit).** If the voltage and current are quasi-periodic and the amplitudes are slowly varying, the distortion power at any instant may be taken as the value derived from the amplitude of the apparent power and phasor power at that instant. By this definition the sign of distortion power is not definitely determined, and it may be given either sign. In the absence of other definite information, it is to be taken the same as for the active power. Distortion power is expressed in volt-amperes when the voltage is in volts and the current in amperes. The distortion power is zero if the voltage and the current have the same waveform. This condition is fulfilled when the voltage and current are sinusoidal and have the same period, or when the circuit consists entirely of noninductive resistors.
E270-0

distortion power (polyphase circuit). At the terminals of entry of a polyphase circuit, equal to the sum of the distortion powers for the individual terminals of entry. *Notes:* (1) The distortion power for each terminal of entry is determined by considering each phase conductor, in turn, with the common reference point as a single-phase, two-wire circuit and finding the distortion power for each in accordance with the definition of **distortion power (single-phase two-wire circuit).** The common reference terminal shall be taken as the neutral terminal of entry, if one exists, otherwise as the true neutral point. The sign given to the distortion power for each single-phase current, and therefore to the total for the polyphase circuit, shall be the same as that of the total active power. (2) Distortion power is expressed in volt-amperes when the voltages are in volts and the current in amperes. (3) The distortion power is zero if each voltage has the same wave form as the corresponding current. This condition is fulfilled, of course, when all the currents and voltages are sinusoidal.
E270-0

distortion, pulse (pulse techniques). The unwanted deviation of a pulse waveform from a reference waveform. *Note:* Some specific forms of pulse distortion have specific names. They include, but are not exclusive to, the following: **overshoot, ringing, preshoot, tilt (droop*), rounding (undershoot and dribble-up*), glitch, bump, spike,** and **backswing.** For further explanation of the forms of pulse distortion, see the following illustrations and IEEE Standard 194 (1968). *See also:* **pulse.**
0-9E4

*Deprecated

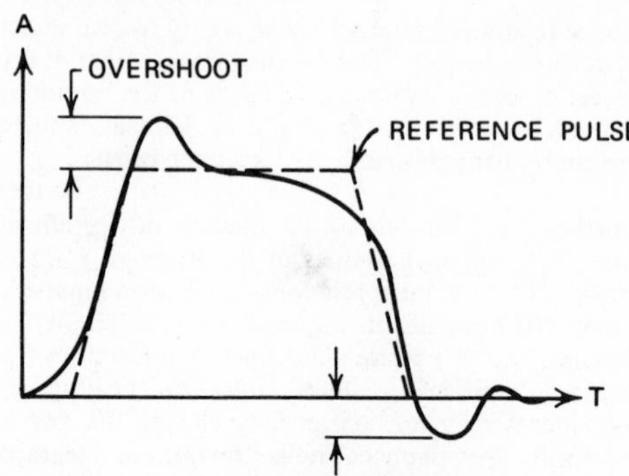

Overshoot.

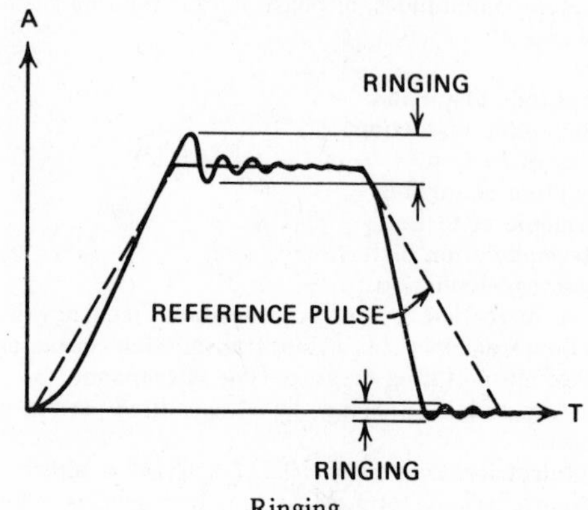

Ringing.

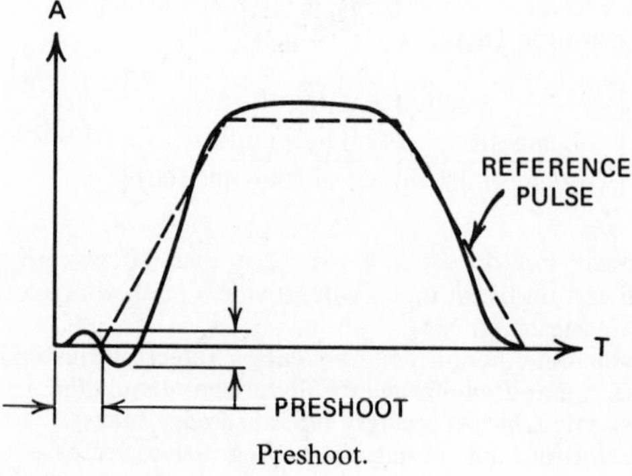

Preshoot.

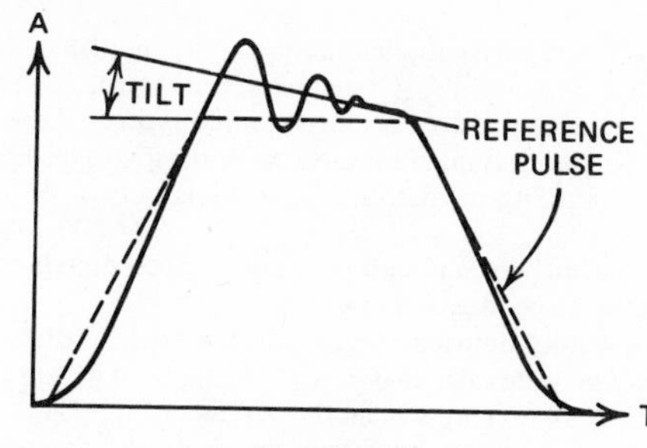

Negative tilt.

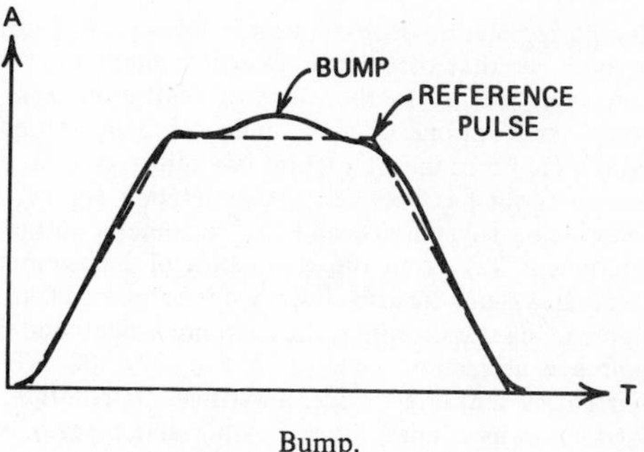

Bump.

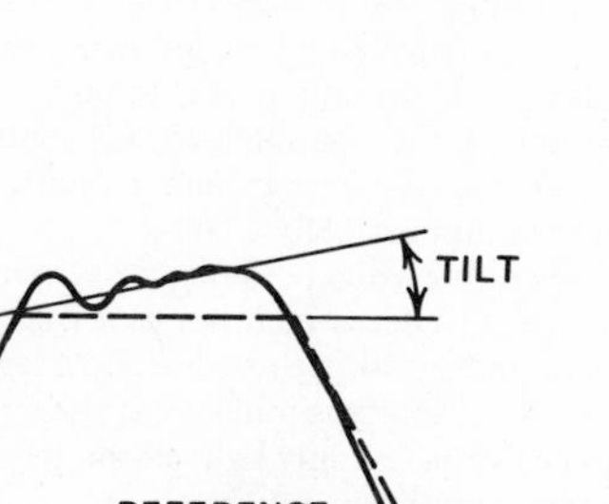

Positive tilt.

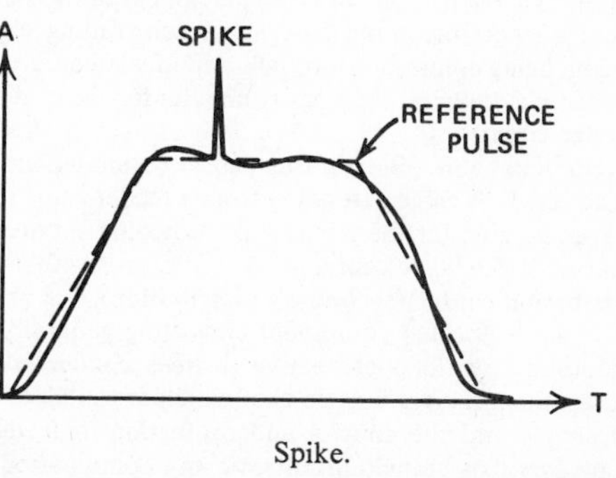

Spike.

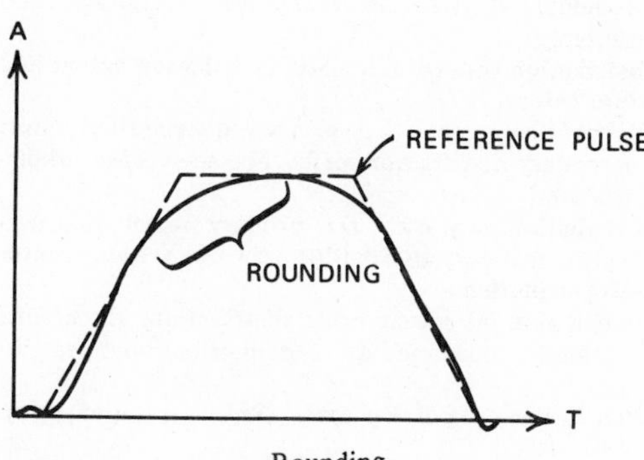

Rounding.

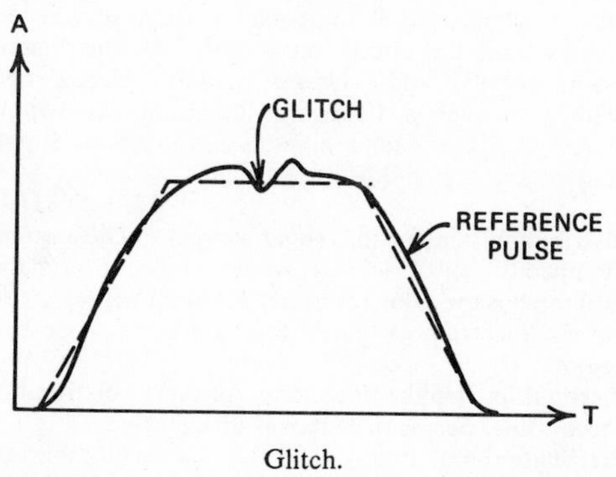

Glitch.

distortion, spiral (camera tubes or image tubes using magnetic focusing). A distortion in which image rotation varies with distance from the axis of symmetry of the electron optical system. *See also:* **beam tubes.** E160-2E2/15E6

distortion tolerance (telegraph receiver). The maximum signal distortion that can be tolerated without error in reception. *See also:* **telegraphy.** 42A65-19E4

distributed. Spread out over an electrically significant length or area. 42A65-0

distributed constant (waveguide). A circuit parameter that exists along the length of a waveguide or transmission line. *Note:* For a transverse electromagnetic wave on a two-conductor transmission line, the distributed constants are series resistance, series inductance, shunt conductance, and shunt capacitance per unit length of line. *See also:* **waveguide.** E146-3E1

distributed element circuit (microwave tubes). A circuit whose inductance and capacitance are distributed over a physical distance that is comparable to a wavelength. *See also:* **microwave tube or valve.** 0-15E6

distributed winding (rotating machinery). A winding, the coils of which occupy several slots per pole. *See also:* **armature; rotor (rotating machinery); stator.** 0-31E8

distributing cable. *See:* **distribution cable.**

distributing frame. A structure for terminating permanent wires of a central office, private branch exchange, or private exchange and for permitting the easy change of connections between them by means of cross-connecting wires. *See also:* **telephone switching system.** 42A65-19E1

distribution (adjective). A general term used, by reason of specific physical or electrical characteristics, to denote application or restriction of the modified term, or both, to that part of an electric system used for conveying energy to the point of utilization from a source or from one or more main receiving stations. *Notes:* (1) From the standpoint of a utility system, the area described is between the generating source, or intervening substations, and the customer's entrance equipment. (2) From the standpoint of a customer's internal system, the area described is between a source or receiving station within the customer's plant and the points of utilization. 37A100-31E11

distribution amplifier. *See:* **amplifier, distribution.**

distribution box (mine type). A portable piece of apparatus with enclosure by means of which an electric circuit is carried to one or more machine trailing cables from a single incoming feed line, each trailing cable circuit being connected through individual overcurrent protective devices. *See also:* **distributor box; mine feeder circuit.** 42A85-0

distribution cable (distributing cable) (communication practice). A cable extending from a feeder cable into a specific area for the purpose of providing service to that area. *See also:* **cable.** 42A65-0

distribution center (secondary distribution). A point at which is located equipment consisting generally of automatic overload protective devices connected to buses, the principal functions of which are subdivision of supply and the control and protection of feeders, subfeeders, or branch circuits, or any combination of feeders, subfeeders, or branch circuits.
See:
branch-circuit distribution center;
damp location;
dry location;
electric-supply system;
equipment;
explosionproof;
exposed;
externally operable;
feeder distribution center;
inclosed;
main distribution center;
manual;
panelboard;
service equipment;
switchboard;
utilization equipment;
watertight;
wet location.
See also: **center of distribution; interior wiring.** E45-0

distribution coefficients (color). The tristimulus values of monochromatic radiations of equal power. *Note:* Generally represented by overscored, lower-case letters, such as $\bar{x}$, $\bar{y}$, $\bar{z}$ in the CIE system. *See also:* **CIE color terms.** E201-2E2

distribution cutout. A fuse or disconnecting device consisting of any one of the following assemblies: (1) A fuse support and fuseholder which may or may not include the conducting element (or fuse link). (2) A fuse support and disconnecting blade. (3) A fuse support and fuse carrier which may or may not include the conducting element (fuse link) or disconnecting blade. *Note:* In addition the distribution cutout is identified by the following characteristics: (A) dielectric withstand (basic impulse insulation level strengths at distribution levels; (B) application primarily on distribution feeders and circuits; (C) mechanical construction basically adapted to pole or crossarm mounting except for the distribution oil cutout; (D) operating voltage limits corresponding to distribution systems voltage. 37A100-31E11

distribution disconnecting cutout. *See:* **distribution cutout; disconnecting cutout.**

distribution enclosed single-pole air switch (distribution enclosed air switches). A single-pole disconnecting switch in which the contacts and blade are mounted completely within an insulated enclosure. (Cannot be converted into a distribution cutout or disconnecting fuse). *Note:* The distribution enclosed air switch is identified by the following characteristics: (1) dielectric withstand (basic impulse insulation level) strengths at distribution level; (2) application primarily on distribution feeders and circuits; (3) mechanical construction basically adapted to cross-arm mounting; (4) operating voltage limits correspond to distribution voltage.(5) Unless incorporating load-break means, it has no interrupting (load-break current) rating. (Some load-break ability is inherent in the device. This ability can be evaluated only by the user, based on experience under operating conditions). 37A100-31E11

distribution factor (rotating machinery). The ratio of the resultant voltage induced in the group of series-connected coils forming a complete circuit to the arithmetic sum of the magnitudes of the voltages induced in the coils. *See also:* **rotor (rotating machinery); stator.** 0-31E8

distribution feeder. *See:* **primary distribution feeder; secondary distribution feeder.** *See also:* **distribution center.**

distribution fuse cutout. *See:* **distribution cutout and fuse cutout.**

distribution main. *See:* **primary distribution mains; secondary distribution mains.** *See also:* **center of distribution.**

distribution network. *See:* **primary distribution network; secondary distribution network.** *See also:* **center of distribution.**

distribution oil cutout. *See:* **distribution; oil cutout.**

distribution open cutout. *See:* **distribution; open cutout.**

distribution open-link cutout. *See:* **distribution; open-link cutout.**

distribution switchboard. A power switchboard used for the distribution of electric energy at the voltages common for such distribution within a building. *Note:* Knife switches, air circuit breakers, and fuses are generally used for circuit interruption on distribution switchboards, and voltages seldom exceed 600. However, such switchboards often include switchboard equipment for a high-tension incoming supply circuit and a step-down transformer. 37A100-31E11

distribution temperature (light source). The absolute temperature of a blackbody whose relative spectral distribution is the same (or nearly so) in the visible region of the spectrum as that of the light source. *See also:* **color.** Z7A1-0

distribution trunk line. *See:* **primary distribution trunk line.** *See also:* **center of distribution.**

distributor box. A box or pit through which cables are inserted or removed in a draw-in system of mains. It contains no links, fuses, or switches and its usual function is to facilitate tapping into a consumer's premises. *See also:* **distribution box; mining; tower.** 42A35-31E13

distributor duct. A duct installed for occupancy of distribution mains. *See:* **service pipe.** 42A35-31E13

distributor suppressor (internal-combustion engine terminology). A suppressor designed for direct connection to the high-voltage terminals of a distributor cap. *See also:* **electromagnetic compatibility.** CISPR-27E1

disturbance (1) (general). An undesired variable applied to a system that tends to affect adversely the value of a controlled variable. *See also:* **transmission characteristics.** 85A1-23E0

(2) (communication practice). Any irregular phenomenon associated with transmission that tends to limit or interfere with the interchange of intelligence. *See also:* **transmission characteristics.** 42A65-0

(3) (interference terminology). *See:* **interference.** *See also:* **transmission characteristics.**

(4) (industrial control). An undesired input variable that may occur at any point within a feedback control system. *See also:* **control system, feedback; transmission characteristics.** AS1-34E10

(5) (storage tubes). That type of spurious signal generated within a tube that appears as abrupt variations in the amplitude of the output signal. *Notes:* (1) These variations are spatially fixed with reference to the target area. (2) The distinction between this and shading. (3) A blemish, a mesh pattern, and moire present in the output are forms of disturbance. Random noise is not a form of disturbance. *See also:* **storage tube; transmission characteristics.** E158-15E6

disturbed-ONE output (magnetic cell). A ONE output to which partial-read pulses have been applied since that cell was last selected for writing. *See:* **coincident-current selection.** *See also:* **static magnetic storage; one output.** E163-0

disturbed-ZERO output (magnetic cell). A ZERO output to which partial-write pulses have been applied since that cell was last selected for reading. *See:* **coincident-current selection.** *See also:* **static magnetic storage.** E163-0

dither (control circuits). A useful oscillation of small amplitude, introduced to overcome the effects of friction, hysteresis, or clogging. *See:* **control system, feedback.** 85A1-23E0;AS1-34E10

divergence (vector field at a point). A scalar equal to the limit of the quotient of the outward flux through a closed surface that surrounds the point by the volume within the surface, as the volume approaches zero. *Note:* If the vector **A** of a vector field is expressed in terms of its three rectangular components A_x, A_y, A_z, so that the values of A_x, A_y, A_z are each given as a function of x, y, and z, the divergence of the vector field **A** (abbreviated $\nabla \cdot \mathbf{A}$) is the sum of the three scalars obtained by taking the derivatives of each component in the direction of its axis, or

$$\mathrm{div}\mathbf{A} \equiv \nabla \cdot \mathbf{A} = \frac{\partial A_x}{\partial x} + \frac{\partial A_y}{\partial y} + \frac{\partial A_z}{\partial z}$$

Examples: If a vector field **A** represents velocity of flow such as the material flow of water or a gas or the imagined flow of heat or electricity, the divergence of **A** at any point is the net outward rate of flow per unit volume and per unit time. It is the time rate of decrease in density of the fluid at that point. Because the density of an incompressible fluid cannot change, the divergence of an incompressible fluid is always zero. The divergence of the flow of heat at a point in a body is equal to the rate of generation of heat per unit volume at the point. The divergence of the electric field strength at a point is proportional to the volume density of charge at the point. E270-0

divergence loss (acoustic wave). The part of the transmission loss that is due to the divergence or spreading of the sound rays in accordance with the geometry of the system (for example, spherical waves emitted by a point source). *See also:* **loudspeaker.** E157-1E1

diversity factor (system diversity factor). The ratio of the sum of the individual maximum demands of the various subdivisions of a system to the maximum demand of the whole system. *Note:* The diversity factor of a part of the system may be similarly defined as the ratio of the sum of the individual maximum demands of the various subdivisions of the part of the system to the maximum demand of the part of the system under consideration. *See also:* **alternating-current distribution; direct-current distribution.** 42A35-31E13

diversity reception. That method of radio reception whereby, in order to minimize the effects of fading, a resultant signal is obtained by combination or selection, or both, of two or more sources of received-signal energy that carry the same modulation or intelligence, but that may differ in strength or signal-to-noise ratio at any given instant. *See also:* **radio receiver.** 42A65-0

divided code ringing (divided ringing). A method of code ringing that provides partial ringing selectivity by connecting one-half of the ringers from one side of the line to ground and the other half from the other side of the line to the ground. This term is not ordinarily applied to selective and semiselective ringing systems. *See also:* **telephone switching system.** 42A65-0

divided ringing. *See:* **divided code ringing.**

divider. (1) A device capable of dividing one variable by another. (2) A device capable of attenuating a variable by a constant or adjustable amount, as an attenuator. *See also:* **electronic analog computer.** E165-16E9

dividing network (crossover network) (loudspeaker dividing network). A frequency selective network that divides the spectrum into two or more frequency bands for distribution to different loads. *See also:* **loudspeaker.** E151/42A65-1E1

***D* layer (radio wave propagation).** An ionized layer in the *D* region. *See also:* **radio wave propagation.** 0-3E2

DME. *See:* **distance-measuring equipment.**

document. (1) A medium and the data recorded on it for human use, for example, a report sheet, a book. (2) By extension, any record that has permanence and that can be read by man or machine. X3A12-16E9

Doherty amplifier. A particular arrangement of a radio-frequency linear power amplifier wherein the amplifier is divided into two sections whose inputs and outputs are connected by quarter-wave (90-degree) networks and whose operating parameters are so adjusted that, for all values of the input signal voltage up to one-half maximum amplitude, Section No. 2 is inoperative and Section No. 1 delivers all the power to the load, which presents an impedance at the output of Section No. 1 that is twice the optimum for maximum output. At one-half maximum input level, Section No. 1 is operating at peak efficiency, but is beginning to saturate. Above this level, Section No. 2 comes into operation, thereby decreasing the impedance presented to Section No. 1, which causes it to deliver additional power into the load until, at maximum signal input, both sections are operating at peak efficiency and each section is delivering one-half the total output power to the load. *See also:* **amplifier.** E145-0

domestic induction heater. A cooking device in which the utensil is heated by current, usually of commercial line frequency, induced in it by a primary inductor associated with it. *See also:* **induction heating.** E54/E169-0

dominant mode (fundamental mode) (waveguide transmission). The mode of propagation with the lowest cutoff frequency. *Note:* Designations for this mode are $TE_{1,0}$ and $TE_{1,1}$ for rectangular and circular waveguides, respectively. *See also:* **waveguide.** 42A65-0

dominant wave (uniconductor waveguide). The guided wave having the lowest cutoff frequency. *Note:* It is the only wave that will carry energy when the excitation frequency is between the lowest cutoff frequency and the next higher cutoff frequency. *See also:* **guided waves; waveguide.** E146-3E1;E210-0

dominant wavelength (colored light, not purple). The wavelength of the spectrum light that, when combined in suitable proportions with the specified achromatic light, yields a match with the light considered. *Notes:* (1) When the dominant wavelength cannot be given (this applies to purples) its place is taken by the complementary wavelength. (2) Light of a single frequency is approximated in practice by the use of a range of wavelengths within which there is no noticeable difference of color. Although this practice is ambiguous in principle, the dominant wavelength is usually taken as the average wavelength of the band used in the mixture with the reference standard matching the sample. Many different qualities of light are used as reference standards under various circumstances. Usually the quality of the prevailing illumination is acceptable as the reference standard in the determination of the dominant wavelength of the color of objects. *See also:* **color terms; color.** E201-0;50I45-2E2

Donnan potential (electrobiology). The potential difference across an inert semipermeable membrane separating mixtures of ions, attributed to differential diffusion. *See also:* **electrobiology.** 42A80-18E1

donor (semiconductor). *See:* **impurity, donor.** *See also:* **semiconductor.**

door contact (burglar-alarm system). An electric contacting device attached to a door frame and operated by opening or closing the door. *See also:* **protective signaling.** 42A65-0

door or gate closer. A device that closes a manually opened hoistway door, a car door, or gate by means of a spring or by gravity. *See also:* **hoistway (elevator or dumbwaiter).** 42A45-0

door or gate power operator. A device, or assembly of devices, that opens a hoistway door and/or a car door or gate by power other than by hand, gravity, springs, or the movement of the car; and that closes them by power other than by hand, gravity, or the movement of the car. *See also:* **hoistway (elevator or dumbwaiter).** 42A45-0

doping (semiconductor). Addition of impurities to a semiconductor or production of a deviation from stoichiometric composition, to achieve a desired characteristic. *See also:* **semiconductor.** E102/E216/E270-34E17

doping compensation (semiconductor). Addition of donor impurities to a *p*-type semiconductor or of acceptor impurities to an *n*-type semiconductor. *See also:* **semiconductor.** E102-0

Doppler effect. The phenomenon evidenced by the change in the observed frequency of a wave in a transmission system caused by a time rate of change in the effective length of the path of travel between the source and the point of observation. *See also:* **electroacoustics.** E157-1E1

Doppler-inertial navigation equipment. Hybrid navigation equipment that employs both Doppler navigation radar and inertial sensors. *See also:* **navigation.** 0-10E6

Doppler navigator. A self-contained dead-reckoning navigation aid transmitting two or more beams of electromagnetic or acoustic energy outward and downward from the vehicle and utilizing the Doppler effect of the reflected energy, a reference direction, and the relationship of the beams to the vehicle to determine speed and direction of motion over the reflecting surface, *See also:* **navigation.** 0-10E6

Doppler radar. A radar that utilizes the Doppler effect to determine the relative radar-target velocity. *See also:* **navigation; radar.** 0-10E6

Doppler shift. The magnitude of the change in the observed frequency of a wave due to the Doppler effect. The unit is the hertz. *See also:* **electroacoustics.** 0-1E1

Doppler system, pulsed (electronic navigation). *See:* **pulsed Doppler system.**

Doppler very-high-frequency omnidirectional range. A very-high-frequency radio range, operationally compatible with conventional very-high-frequency omnidirectional ranges, less susceptible to siting difficulties because of its increased aperture; in it the variable signal (the signal producing azimuthal information) is developed by sequentially feeding a radio-frequency signal to a multiplicity of antennas disposed in a ring-shaped array. *Note:* The array usually surrounds the central source of reference signal. *See also:* **navigation.** 0-10E6

dosage. *See:* **dose.**

dose (dosage) (photovoltaic power system). The radiation delivered to a specified area or the whole body. *Note:* Units of dose are rads or roentgens for X or gamma rays and rads for beta rays and protons. *See also:* **photovoltaic power system; solar cells (in a photovoltaic power system).** 0-10E1

dose rate (photovoltaic power system). Radiation dose delivered per unit time. *See also:* **photovoltaic power system; solar cells (in a photovoltaic system).** 0-10E1

dot cycle. One cycle of a periodic alternation between two signaling conditions, each condition having unit duration. *Note:* Thus, in two-condition signaling, it consists of a dot, or marking element, followed by a spacing element. *See also:* **telegraphy.** 42A65-31E3/19E4

dot product. *See:* **scalar product.**

dot-product line integral. *See:* **line integral.**

dot-sequential (color television). Pertaining to the association of the several primary colors in sequence with successive picture elements. *Examples:* Dot-sequential pickup, dot-sequential display, dot-sequential system, dot-sequential transmission. *See also:* **color terms.** E201-2E2

double-break switch (circuit). A switch that opens a conductor at two points. 37A100-31E11

double bridge (Thomson bridge). *See:* **Kelvin bridge.**

double-circuit system (protective signaling). A system of protective wiring in which both the positive and the negative sides of the battery circuit are employed, and that utilizes either an open or a short circuit in the wiring to initiate an alarm. *See also:* **protective signaling.** 42A65-0

double-current generator. A machine that supplies both direct and alternating currents from the same armature winding. *See:* **direct-current commutating machine.** 42A10-31E8

double diode (electron device). An electron tube or valve containing two diode systems. *See:* **multiple tube (valve).** *See also:* **circuit characteristics of electrodes.** 0-15E6

double-end control. A control system in which provision is made for operating a vehicle from either end. *See also:* **multiple-unit control.** 42A42-0

double-faced tape. Fabric tape finished on both sides with a rubber or synthetic compound. *See also:* **power distribution, underground construction.** 42A35-31E13

double-fed asynchronous machine. An asynchronous machine of which the stator winding is fed by a certain supply frequency and of which the rotor winding is fed by a variable frequency. *See also:* **asynchronous machine.** 0-31E8

double-gun cathode-ray tube. A cathode-ray tube containing two separate electron-gun systems. *See also:* **tube definitions.** 0-15E6

double length. Pertaining to twice the normal length of a unit of data or a storage device in a given computing system. *Note:* For example, a double-length register would have the capacity to store twice as much data as a single-length or normal register; a double-length word would have twice the number of characters or digits as a normal or single-length word. *See also:* **double precision; electronic digital computer.** E162-0

double-length number (double-precision number). A number having twice as many digits as are ordinarily used in a given computer. *See also:* **electronic computation.** E270-0

double modulation. The process of modulation in which a carrier wave of one frequency is first modulated by a signal wave and a resultant wave is then made to modulate a second carrier wave of another frequency. *See also:* **modulating systems.** 42A65-0

double pole-piece magnetic head (electroacoustics). A magnetic head having two separate pole pieces in which pole faces of opposite polarity are on opposite sides of the medium. *Note:* One or both of these pole pieces may be provided with an energizing winding. *See also:* **phonograph pickup.** E157-1E1

double-pole relay. *See:* **relay, double-pole.**

double precision (electronic computation). Pertaining to the use of two computer words to represent a number in order to preserve or gain precision. 0-16E9

double-precision number. *See:* **double-length number.**

double-sideband transmitter. A transmitter that transmits the carrier frequency and both sidebands resulting from the modulation of the carrier by the modulating signal. *See also:* **radio transmitter.** E145-0

double squirrel cage (rotating machinery). A combination of two squirrel-cage windings mounted on the same induction-motor rotor, one at a smaller diameter than the other. *Note:* It is common but not essential for the two windings to have the same number of slots. In any case, each bar of the lower (smaller-diameter) winding is located at the bottom of a slot containing a bar of the upper winding. A narrow portion of the slot (called the **leakage slot**) is provided in the radial separation between the two bars. *See also:* **asynchronous machine.** 0-31E8

double-superheterodyne reception (triple detection). The method of reception in which two frequency converters are employed before final detection. *See also:* **radio receiver.** 42A65-0

doublet antenna. An antenna consisting of two elevated conductors substantially in the same straight line, of substantially equal length, with the power delivered at the center. *See also:* **antenna.** 42A65-3E1

double-throw (mechanical switching device). A qualifying term indicating that the device can change the circuit connections by utilizing one or the other of its two operating positions. *Note:* A double-throw air switch changes circuit connections by moving the switchblade from one of two sets of contact clips into the other. 37A100-31E11

double-tuned amplifier. An amplifier of one or more stages in which each stage utilizes coupled circuits having two frequencies of resonance, for the purpose of obtaining wider bands than those obtainable with single tuning. *See also:* **amplifier.** 42A65-0

double-tuned circuit. A circuit whose response is the same as that of two single-tuned circuits coupled together. *See also:* **circuits and devices.** 42A65-0

double-way rectifier. A rectifier in which the current between each terminal of the alternating-voltage circuit and the rectifier circuit elements conductively connected to it flows in both directions. *Note:* The terms single-way and double-way provide a means for describing the effect of the rectifier circuit on current flow in transformer windings connected to rectifiers. Most rectifier circuits may be classified into these two general types. Double-way rectifiers are also referred to as bridge rectifiers. *See also:* **rectification; rectifier circuit element; power rectifier.** 34A1;34E24

double-winding synchronous generator. A generator that has two similar windings, in phase with one another, mounted on the same magnetic structure but not connected electrically, designed to supply power to two independent external circuits. *See:* **synchronous machine.** 42A10-31E8

doughnut (electronic device). *See:* **toroid.**

dovetail projection. A tenon, commonly flared; used for example, to fasten a pole to the spider. *See also:* **stator.** 0-31E8

dovetail slot. (1) A recess along the side of a coil slot into which a coil-slot wedge is inserted. (2) A flaring slot into which a dovetail projection is engaged; used for example, to fasten a pole to the spider. *See also:* **stator.** 0-31E8

dowel (dowel pin). A pin fitting with close tolerance into a hole in abutting pieces to establish and maintain accurate alignment of parts. Frequently designed to resist a shear load at the interface of the abutting pieces. *See also:* **cradle base (rotating machinery).** 0-31E8

down conductor. The vertical portion of a run of conductor that ends at the ground. *See also:* **lightning protection and equipment.** 42A95-0

down lead (lightning protection). The conductor connecting an overhead ground wire or lightning conductor with the grounding system. *See also:* **direct-stroke protection (lightning).** 0-31E13

downlight. A small direct lighting unit which can be recessed, surface mounted, or suspended. *See also:* **luminaire.** Z7A1-0

down time. The period during which a system or device is not operating due to internal failures, scheduled shut down, or servicing. *See:* **electric drive.** AS1-34E10

downward component (illuminating engineering). That portion of the luminous flux from a luminaire that is emitted at angles below the horizontal. *See also:* **inverse-square law (illuminating engineering).** Z7A1-0

downward modulation. Modulation in which the instantaneous amplitude of the modulated wave is never greater than the amplitude of the unmodulated carrier. *See also:* **modulating systems.** E188-0

drag-in (electroplating). The quantity of solution that adheres to cathodes when they are introduced into a bath. *See also:* **electroplating.** 42A60-0

drag magnet (braking magnet). *See:* **retarding magnet.**

drag-out (electroplating). The quantity of solution that adheres to cathodes when they are removed from a bath. *See also:* **electroplating.** 42A60-0

drain. The current supplied by a cell or battery when in service. *See also:* **battery (primary or secondary).** 42A60-0

drainage (corrosion). Conduction of current (positive electricity) from an underground metallic structure by means of a metallic conductor. *See:* **corrosion terms.** CM-34E2

drain coil (power-system communication). A reactor or choke connected between the carrier-current lead-in terminal of a coupling capacitor and ground, to present a low impedance to the flow of power current and a high impedance to the flow of carrier-frequency current. *Note:* Its purpose is to prevent high voltage at power frequency from being impressed on the carrier-current lead and to limit the loss of carrier-frequency energy to ground. *See also:* **power-line carrier.** 0-31E3

drain line (rotating machinery) (bearing oil system). A return pipe line using gravity flow. *See also:* **oil cup (rotating machinery).** 0-31E8

drawbridge coupler. *See:* **movable-bridge coupler.**

drawout-mounted device. A device having disconnecting devices and in which the removable portion may be removed from the stationary portion without the necessity of unbolting connections or mounting supports. *Note:* Compare with **stationary-mounted device.** 37A100-31E11

***D* region (radio wave propagation).** The region of the terrestrial ionosphere between about 40 and 90 kilometers altitude responsible for most of the attenuation of radio waves in the range 1 to 100 megahertz. *See also:* **radiation; radio wave propagation.** 0-3E2

dribble-up*. *See:* **distortion, pulse.**

*Deprecated

drift (1) (rotating machinery). A long-time change in synchronous-machine regulating system error resulting from causes such as aging of components, self-induced temperature changes, and random phenomena. *Note:* Maximum acceptable drift is normally a specified change for a specified period of time, for specified conditions. *See:* **synchronous machine.** 0-31E8

(2) (industrial control). An undesired but relatively slow change in output over a period of time, with a fixed reference input. *Note:* Drift is usually expressed in percent of the maximum rated value of the variable being measured. *See:* **control system, feedback.** AS1-34E10

(3) (sound recording and reproducing). Designates random-frequency variations that are continuously in one direction or the other for periods of the order of a second or more. *See also:* **sound recording and reproducing.** E193-0

(4) (analog computer). A slowly varying error in a computer element. *See:* **zero drift.** *See also:* **electronic analog computer.**

(5) (electronic navigation). *See:* ***G* drift; G^2 drift.**

(6) (power supplies). *See:* **stability, long term.** E165-16E9

drift angle (navigation). The angular difference between the heading and the track. *See also:* **navigation.** 0-10E6

drift band of amplification (magnetic amplifier). The maximum change in amplification due to uncontrollable causes for a specified period of time during which all controllable quantities have been held constant. *Note:* The units of this drift band are the amplification units per the time period over which the drift band was determined. *See also:* **rating and testing magnetic amplifiers.** E107-0

drift compensation (industrial control). The effect of a control function, device, or means to decrease overall systems drift by minimizing the drift in one or more of the control elements. *Note:* Drift compensation may apply to feedback elements, reference input, or other portions of a system. *See:* **control system, feedback.** IC1-34E10

drift correction angle (navigation). The angular difference between the course and the heading. Sometimes called the crab angle. *See also:* **navigation.** E172-10E6

drift, kinematic (electronic navigation). *See:* **misalignment drift.**

drift mobility (homogeneous semiconductor). The ensemble average of the drift velocities of the charge-carriers per unit electric field. *Note:* In general, the mobilities of electrons and holes are different. *See also:* **semiconductor device.** E102/E270-10E1;E216-34E17

drift offset (magnetic amplifier). The change in quiescent operating point due to uncontrollable causes over a specified period of time when all controllable quantities are held constant. *See also:* **rating and testing magnetic amplifiers.** E107-0

drift rate (voltage regulators or reference tubes). The slope at a stated time of the smoothed curve of tube voltage drop with time at constant operating conditions. *See also:* **circuit characteristics of electrodes.** E160-15E6

drift space (electron tube). A region substantially free of externally applied alternating fields, in which a relative repositioning of the electrons takes place. *See:* **beam tubes; circuit characteristics of electrodes.** E160-15E6

drift, stability (electric conversion). Gradual shift or change in the output over a period of time due to change or aging of circuit components. (All other variables held constant.) *See also:* **electric conversion equipment.** 0-10E1

drift stabilization. Any automatic method used to minimize the drift of a direct-current amplifier. *See also:* **electronic analog computer.** E165-16E9

drift tunnel (velocity-modulated tube). A piece of metal tubing, held at a fixed potential, that forms the drift space. *Note:* The drift tunnel may be divided into several parts, which constitute the drift electrodes. *See also:* **velocity-modulated tube.** 50I07-15E6

drift, zero. Drift with zero input. *See also:* **electronic analog computer.** E165-16E9

dripproof (enclosure). So constructed or protected that successful operation is not interfered with when falling drops of liquid or solid particles strike or enter the enclosure at any angle from 0 to 15 degrees from the downward vertical unless otherwise specified. *See also:* **asynchronous machine; direct-current commutating machine; synchronous machine.** E45/42A95-0;MG1-31E8;37A100-31E11

dripproof machine. An open machine in which the ventilating openings are so constructed that drops of liquid or solid particles falling on the machine at any angle not greater than 15 degrees from the vertical cannot enter the machine either directly or by striking and running along a horizontal or inwardly inclined surface. *See:* **asynchronous machine; direct-current commutating machine; synchronous machine.** 42A10-0

driptight enclosure. An enclosure so constructed that falling drops of liquid or solid particles striking the enclosure at any angle within a specified variation from the vertical cannot enter the enclosure either directly or by striking and running along a horizontal or inwardly inclined surface. 37A100-31E11;42A95-0

drive (1) (industrial control). The equipment used for converting available power into mechanical power suitable for the operation of a machine. *See:* **electric drive.** IC1-34E10

(2) (electronic computation and recording). *See:* **tape drive.** *See also:* **static magnetic storage.**

driven element (antenna). A radiating element coupled directly to the feed line.*See also:* **antenna.** 0-3E1

drive pattern (facsimile). Density variation caused by periodic errors in the position of the recording spot. *Note:* When caused by gears this is called gear pattern. *See also:* **recording (facsimile).** E168-0

drive pin (disc recording). A pin similar to the center pin, but located to one side thereof, that is used to prevent a disc record from slipping on the turntable. *See also:* **phonograph pickup.** E157-1E1

drive-pin hole (disc recording). A hole in a disc record that accommodates the turntable drive pin. *See also:* **electroacoustics.** E157-1E1

drive pulse (static magnetic storage). A pulsed magnetomotive force applied to a magnetic cell from one or more sources. *See also:* **static magnetic storage.** E163-0

driver (communication practice). An electronic circuit that supplies input to another electronic circuit. *See also:* **circuits and devices.** 42A65-0

drive strip (chafing strip) (rotating machinery). An insulating strip located in the coil slots between the wedge and the top of the slot armor or the top coil side, to provide protection during assembly of the wedges. *See also:* **cradle base (rotating machinery); stator.** 0-31E8

driving machine. The power unit that applies the energy necessary to raise and lower an elevator or dumbwaiter car or to drive an escalator or a private-residence inclined lift.
See:
belt-drive machine;
chain-drive machine;
direct-drive machine;
direct-plunger driving machine;
electric driving machine;
geared-drive machine;
geared traction machine;
gearless traction machine;
hydraulic driving machine;
indirect-driving machine;
roped-hydraulic driving machine;
screw machine;
traction machine;
winding-drum machine;
worm-geared machine. 42A45-0

driving-point admittance (between the *j*th terminal and the reference terminal of an *n*-terminal network). The quotient of (1) the complex alternating component I_j of the current flowing to the *j*th terminal from its external termination by (2) the complex alternating component V_j of the voltage applied to the *j*th terminal with respect to the reference point when all other terminals have arbitrary external terminations. *Note:* In specifying the driving-point admittance of a given pair of terminals of a network or transducer having two or more pairs of terminals, no two pairs of which contain a common terminal, all other pairs of terminals are connected to arbitrary admittances. *See also:* **electron-tube admittances; linear passive networks; network analysis.** 42A70/E160-15E6

driving-point function (linear passive network). A response function for which the variables are measured at the same port (terminal pair). *See also:* **linear passive network.** E156-0

driving-point impedance (network). At any pair of terminals, the ratio of an applied potential difference to the resultant current at these terminals, all terminals being terminated in any specified manner. *See also:* **linear passive networks; self-impedance.** 42A65-0

driving power, grid. *See:* **grid driving power.**

driving signals (television). Signals that time the scanning at the pickup point. *Note:* Two kinds of driving signals are usually available from a central synchronizing generator. One is composed of pulses at line frequency and the other is composed of pulses at field frequency. *See also:* **television.** E203-2E2;42A65-0

droop* (pulse techniques). *See:* **distortion, pulse.**
*Deprecated

drop (drop signal). A visual shutter device consisting of an electromagnet and a visual target either moved or tripped magnetically to a position indicating the condition supervised. *See also:* **circuits and devices.** 42A65-0

drop-away. The electrical value at which the movable member of an electromagnetic device will move to its de-energized position. *See also:* **railway signal and interlocking.** 42A42-0

dropout (relay). A term for contact operation (opening or closing) as a relay just departs from pickup. Also indentifies the maximum value of an input quantity that will allow the relay to dropout. *Note:* Dropout has been used to denote reset for electromechanical relays that have no intermediate steady-state position between pickup and reset and that reset instantaneously. *See also:* **relay.** 37A100-31E11

dropout fuse. A fuse in which the fuseholder or fuse unit automatically drops into an open position after the fuse has interrupted the circuit. 37A100-31E11

dropout ratio. The ratio of the dropout value to the pickup value of an input quantity. *Note:* This term has been used mostly with relays for which the reset value is not differentiated from the dropout value. Hence a similar term, reset ratio, the ratio of reset value to pickup value is not generally used, though technically correct. 37A100-31E11/31E6

dropout time (relay). The time interval to dropout following a specified change of input conditions. *Note:* When the change of input conditions is not specified it is intended to be a sudden change from pickup value of input to zero input. 37A100-31E11/31E6

dropout voltage (or current) (magnetically operated device). The maximum voltage (or current) at which the device will release to its de-energized position. *See also:* **contactor; control switch; extinguishing voltage.** 42A25-34E10

drop, voltage, anode (glow-discharge cold-cathode tube). *See:* **anode voltage drop.**

drop, voltage, starter (glow-discharge cold-cathode tube). *See:* **starter voltage drop.**

drop, voltage, tube (glow-discharge cold-cathode tube). *See:* **tube voltage drop.**

drop wire (drop). Wire suitable for extending an open wire or cable pair from a pole or cable terminal to a building. *See also:* **cable; open wire.** 42A65-0

drum. *See:* **magnetic drum.** *See also:* **electronic digital computer.**

drum controller (industrial control). An electric controller that utilizes a drum switch as the main switching element. *Note:* A drum controller usually consists of a drum switch and a resistor. *See also:* **electric controller.** 42A25-34E10

drum factor (facsimile). The ratio of usable drum length to drum diameter. *Note:* Before a picture is transmitted, it is necessary to verify that the ratio of used transmitter drum length to transmitter drum diameter is not greater than the receiver drum factor if the receiver is of the drum type. *See also:* **facsimile (electrical communication).** 0-19E4

drum speed (facsimile). The angular speed of the transmitter or recorder drum. *Note:* This speed is measured in revolutions per minute. *See also:* **recording (facsimile); scanning.** E168-0

drum switch (industrial control). A switch in which the electric contacts are made of segments or surfaces on the periphery of a rotating cynlinder or sector, or by the operation of a rotating cam. *See also:* **control switch; switch.** E160/42A25-34E10

dry-arcing distance (insulator). The shortest distance through the surrounding medium between terminal electrodes, or the sum of the distances between intermediate electrodes, whichever is the shorter, with the insulator mounted for dry flashover test. *See also:* **insulator.** 29A1-0

dry cell. A cell in which the electrolyte is immobilized. *See also:* **electrochemistry.** 42A60-0

dry contact. One through which no direct current flows. *See also:* **telephone switching system.** 42A65-19E1

dry location (electric system). A location not normally subject to dampness or wetness. *Note:* A location classified as dry may be temporarily subject to dampness or wetness, as in the case of a building under construction. *See also:* **distribution center.** 1A0-0

dry reed relay. A reed relay with dry (nonmercury-wetted) contacts. 0-21E0

dry-type (1) (current-limiting reactor). Having the coils immersed in an insulating gas. *See also:* **reactor.** 57A16-0

(2) (regulator). Having the core and coils not immersed in an insulating liquid. *See also:* **voltage regulator.** 57A15-0

(3) (transformer). Having the core and coils neither impregnated with an insulating fluid nor immersed in an insulating oil. *See:* **dry-type self-cooled forced-air-cooled (Class AA/FA); transformer, oil-immersed.** *See also:* **transformer.** 42A15-31E12

dry-type forced-air-cooled (Class AFA). Cooled by the forced circulation of the cooling air through the core and coils. *See also:* **reactor; transformer; voltage regulator.** 57A15-0

dry-type self-cooled (Class AA). Cooled by the natural circulation of the cooling air. *See also:* **reactor; transformer; voltage regulator.** 57A15-0

dry-type self-cooled/forced-air-cooled (Class AA/FA). Cooled at the self-cooled rating by natural circulation of the cooling air through the core and coils, and cooled at the forced-air-cooled rating by the forced circulation of the cooling air through the core and coils. *See:* **dry-type.** *See also:* **reactor; transformer; voltage regulator.** 57A15-0

***D*-scan (electronic navigation).** *See:* ***D*-display.**

***D*-scope (electronic navigation).** *See:* ***D*-display.**

dual-beam oscilloscope. A multibeam oscilloscope in which the cathode-ray tube produces two separate electron beams that may be individually or jointly controlled. *See:* **multibeam oscilloscope.** *See also:* **oscillograph.** 0-9E4

dual benchboard. A combination assembly of a benchboard and a vertical-hinged panel switchboard placed back to back (no aisle) and enclosed with a top and ends. *Note:* No primary switching devices are located between the benchboard and panels. 37A100-31E11

dual control. A term applied to signal appliances provided with two authorized methods of operation. *See also:* **railway signal and interlocking.** 42A42-0

dual-element fuse (single fuse). A fuse having current-responsive elements of two different fusing characteristics in series. 37A100-31E11

dual-frequency induction heater or furnace. A heater in which the charge receives energy by induction, simultaneously or successively, from a work coil or coils operating at two different frequencies. *See also:* **induction heating.** E54/E169-0

dual-headlighting system. A system consists of two double headlighting units, one mounted on each side of the front end of a vehicle. *Note:* Each unit consists of two sealed-beam lamps mounted in a single housing. The upper or outer lamps have two filaments that supply the lower beam and part of the upper beam, respectively. The lower or inner lamps have one filament that provides the primary source of light for the upper beam. *See also:* **headlamp.** Z7A1-0

dual modulation (facsimile). The process of modulating a common carrier wave or subcarrier by two different types of modulation. *Note:* For example, amplitude and frequency modulation, each conveying separate information. *See also:* **facsimile transmission.** E168-0

dual networks. *See:* **structurally dual networks.**

dual overcurrent trip. *See:* **dual release (dual trip); overcurrent release (overcurrent trip).**

dual release (dual trip). A release that combines the function of a delayed and an instantaneous release. 37A100-31E11

dual service (plural service). Two separate services, usually of different characteristics, supplying one consumer. *Note:* A dual service might consist of an alternating-current and direct-current service, or of 120-208-volt 3-phase, 4-wire service for light and some power and a 13.2-kilovolt service for power, etcetera. *See:* **service; duplicate service; emergency service; loop service.** 42A35-31E13

dual switchboard. A control switchboard with front and rear panels separated a comparatively short distance and enclosed at both ends and top. *Note:* The panels on at least one side are hinged for access to the panel wiring. *Note:* No primary switching devices are located between the panels. 37A100-31E11

dual trace (displays). A multitrace operation in which a single beam in a cathode-ray tube is shared by two signal channels. *See:* **alternate display; chopped display; multitrace; oscillograph.** 0-9E4

dubbing (electroacoustics). A term used to describe the combining of two or more sources of sound into a complete recording, at least one of the sources being a recording. *See also:* **phonograph pickup; re-recording.** E157-1E1

duck tape. Tape of heavy cotton fabric, such as duck or drill, which may be impregnated with an asphalt, rubber, or synthetic compound. *See also:* **power distribution, underground construction.** 42A35-31E13

duct (underground electric systems). A single enclosed runway for conductors or cables. *See also:* **power distribution, underground construction; cable.** 42A35-31E13;42A65-0

duct bank (conduit run). An arrangement of conduit providing one or more continuous ducts between two points. *Note:* An underground runway for conductors or cables, large enough for workmen to pass through, is termed a gallery or tunnel. *See also:* **power distribution, underground construction.** 42A35-31E13

duct edge shield (cable shield). A collar or thimble, usually flared, inserted at the duct entrance in a manhole for the purpose of protecting the cable sheath or insulation from being worn away by the duct edge. *See also:* **power distribution, underground construction.** 42A35-31E13

duct entrance. The opening of a duct at a manhole, distributor box, or other accessible space. 42A35-31E13

duct rodding (rodding a duct). The threading of a duct by means of a jointed rod of suitable design for the purpose of pulling in the cable-pulling rope, mandrel, or the cable itself. *See also:* **power distribution, underground construction.** 42A35-31E13

duct sealing. The closing of the duct entrance for the purpose of excluding water, gas, or other undesirable substances. 42A35-31E13

duct spacer (vent finger) (rotating machinery). A spacer between adjacent packets of laminations to provide a radial ventilating duct. *See also:* **cradle base (rotating machinery).** 0-31E8

duct ventilated (pipe ventilated) (rotating machinery). A term applied to apparatus that is so constructed that a cooling gas can be conveyed to or from it through ducts. *See:* **cradle base (rotating machinery).** 0-31E8

dumbwaiter. A hoisting and lowering mechanism equipped with a car that moves in guides in a substantially vertical direction, the floor area of which does not exceed 9 square feet, whose total inside height whether or not provided with fixed or removable shelves does not exceed 4 feet, the capacity of which does not exceed 500 pounds, and which is used exclusively for carrying materials. 42A45-0

dummy antenna. A device that has the necessary impedance characteristics of an antenna and the necessary power-handling capabilities, but that does not radiate or receive radio waves. *Note:* In receiver practice, that portion of the impedance not included in the signal generator is often called **dummy antenna.** *See also:* **radio receiver.** E188-0

dummy-antenna system. An electric network that simulates the impedance characteristics of an antenna system. *See also:* **navigation.** E173-10E6

dummy coil (rotating machinery). A coil that is not required electrically in a winding, but that is installed for mechanical reasons and left unconnected. *See also:* **rotor (rotating machinery); stator.** 0-31E8

dummy load (radio transmission). A dissipative but essentially nonradiating substitute device having impedance characteristics simulating those of the substituted device. *See also:* **artificial load; radio transmission.** 0-9E4;E188-0

dump (computing systems). (1) To copy the contents of all or part of a storage, usually from an internal storage into an external storage. (2) A process as in (1). (3) The data resulting from the process as in (1). *See:*

dynamic dump; postmortem dump; selective dump; snapshot dump; static dump. *See also:* **electronic digital computer.** X3A12-16E9

dump power. Power generated from water, gas, wind, or other source that cannot be stored or conserved and that is beyond the immediate needs of the electric system producing the power. *See also:* **generating station.** 42A35-31E13

duodecimal. (1) Pertaining to a characteristic or property involving a selection, choice, or condition in which there are twelve possibilities. (2) Pertaining to the numeration system with a radix of twelve. *See also:* **electronic digital computer.** X3A12-16E9

duolateral coil. *See:* **honeycomb coil.**

duplex (communications). Pertaining to a simultaneous two-way independent transmission in both directions. *See:* **half duplex; full duplex.** X3A12-16E9

duplex artificial line (balancing network). A balancing network, simulating the impedance of the real line and distant terminal apparatus, that is employed in a duplex circuit for the purpose of making the receiving device unresponsive to outgoing signal currents. *See also:* **telegraphy.** 42A65-0

duplex benchboard. A combination assembly of a benchboard and a vertical control switchboard placed back to back and enclosed with a top and ends (not grille). *Notes:* (1) Access space with entry doors is provided between the benchboard and vertical control switchboard. (2) No primary switching devices are located between the benchboard and panels. 37A100-31E11

duplex cable. (1) A cable composed of two insulated stranded conductors twisted together. *Note:* They may or may not have a common insulating covering. *See also:* **power distribution, underground construction.** E30-0

(2) A cable composed of two insulated single-conductor cables twisted together. *Note:* The assembled conductors may or may not have a common covering of binding or protecting material. *See also:* **power distribution, underground construction.** 42A35-31E13

duplex cavity (radar). *See:* **transmit-receive cavity (radar).**

duplexer (radar practice). A device that utilizes the finite delay between the transmission of a pulse and the echo thereof so as to permit the connection of the transmitter and receiver to a common antenna. *Note:* A duplexer commonly employs a transmit-receive switch and an antitransmit-receive switch, though the latter is sometimes omitted. *See also:* **radar.** 42A65-0

duplexing assembly, radar. *See:* **transmit-receive switch.**

duplex operation (1) (general). The operation of transmitting and receiving apparatus at one location in conjunction with associated transmitting and receiving equipment at another location, the processes of transmission and reception being concurrent. *See also:* **telegraphy.** E145-0;42A65-31E3

(2) (radio communication) (two-way radio communication circuit). The operation utilizing two radio-frequency channels, one for each direction of transmission, in such manner that intelligence may be transmitted concurrently in both directions. 0-6E1

duplex switchboard. A control switchboard consisting of panels placed back to back and enclosed with a top and ends (not grille). *Notes:* (1) Access space with entry doors is provided between the rows of panels. (2) No primary switching devices are located between the panels. 37A100-31E11

duplex system. A telegraph system that affords simultaneous independent operation in opposite directions over the same channel. *See also:* **telegraphy.** 42A65-0

duplicate. *See:* **copy.**

duplicate lines (power transmission). Lines of substantially the same capacity and characteristics, normally operated in parallel, connecting the same supply point with the same distribution point. *See also:* **center of distribution.** 42A35-31E13

duplicate service (power transmission). Two services, usually supplied from separate sources, of substantially the same capacity and characteristics. *Note:* The two services may be operated in parallel on the consumer's premises, but either one alone is of sufficient capacity to carry the entire load. *See:* **service; dual service; emergency service; loop service.** 42A35-31E13

duplication check. A check based on the consistency of two independent performances of the same task. X3A12-16E9

dust-ignitionproof machine. A totally enclosed machine whose enclosure is designed and constructed in a manner that will exclude ignitable amounts of dusts or amounts that might affect performance or rating, and that, when installation and protection are in conformance with the National Electrical Code, will not permit arcs, sparks, or heat otherwise generated or liberated inside of the enclosure to cause ignition of exterior accumulations or atmospheric suspensions of a specific dust on or in the vicinity of the enclosure. *See:* **asynchronous machine; direct-current commutated machine; synchronous machine.** 42A10-0;MG1-31E8

dustproof (1) (general). So constructed or protected that the accumulation of dust will not interfere with successful operation. IC1-34E10;42A95-0;37A100-31E11

(2) (enclosure). An enclosure so constructed or protected that any accumulation of dust that may occur within the enclosure will not prevent the successful operation of, or cause damage to, the enclosed equipment. E45-0

(3) (luminaire). Luminaire so constructed or protected that dust will not interfere with its successful operation. *See also:* **luminaire.** Z7A1-0

dust seal (rotating machinery). A sealing arrangement intended to prevent the entry of a specified dust into a bearing. *See also:* **asynchronous machine; direct-current commutating machine; synchronous machine.** 0-31E8

dusttight (1) (enclosure). An enclosure so constructed that dust will not enter the enclosing case. E45-0;IC1-34E10

(2) (luminaire). So constructed that dust will not enter the enclosing case under specified conditions. *See also:* **luminaire.** 37A100-31E11;Z7A1-0

duty (1) (general). A statement of loads including no-load and rest and deenergized periods, to which the machine or apparatus is subjected including their duration and sequence in time. 0-31E8

(2) **(rating of electric equipment).** A statement of the operating conditions to which the machine or apparatus is subjected, their respective durations, and their sequence in time. E96-SCC4

(3) (industrial control) (transformers). A requirement of service that defines the degree of regularity of the load.
See:
asynchronous machine;
continuous duty;
continuous rating;
direct-current commutating machine;
intermittent duty;
load;
nominal rating;
periodic duty;
primary voltage rating;
rated current of equipment;
rated kilowatts of a constant-current transformer;
rated kVA of a transformer;
rated primary voltage;
rated secondary current;
rated secondary voltage;
rated voltage of equipment;
rating of interphase transformer;
rating of transformer equipment;
secondary voltage rating;
short-time duty;
short-time rating;
synchronous machine;
varying duty;
voltage regulator;
winding voltage rating. 42A15-31E12

duty cycle (1) (general). The time interval occupied by a device on intermittent duty in starting, running, stopping, and idling. E145-0
(2) (rotating machinery). A variation of load with time which may or may not be repeated, and in which the cycle time is too short for thermal equilibrium to be attained. *See also:* **asynchronous machine; direct-current commutating machine; synchronous machine.** 0-31E8
(3) (pulse systems). The ratio of the sum of all pulse durations to the total period, during a specified period of continuous operation. *See also:* **navigation.** 0-10E6
(4) (radio transmitter performance). A criterion defining the ratio of average to peak power output from a transmitter, as a function of carrier-on time versus time available. 0-6E1
(5) (arc-welding apparatus). The ratio of arc time to total time. *See also:* **electric arc-welding apparatus.** 87A1-0
(6) (relay). *See:* **relay duty cycle.**

duty-cycle rating (rotating machinery). The statement of the loads and conditions assigned to the machine by the manufacturer, at which the machine may be operated on duty cycles. *See also:* **asynchronous machine; direct-current commutating machine; synchronous machine.** 0-31E8

duty factor (1) (pulse techniques). The ratio of the pulse duration to the pulse period of a periodic pulse train. *See also:* **pulse; pulse carrier.** 0-9E4
(2) (electron tubes). The ratio of the ON period to the total period during which an electronic valve or tube is operating. *See also:* ON **period (electron tubes); pulse.** 50I07-15E6

duty ratio (pulse system). The ratio of average to peak pulse power. *See also:* **navigation.** E172-10E6

dwell (numerically controlled machines). A timed delay of programmed or established duration, not cyclic or sequential, that is, not an interlock or hold. *See also:* **numerically controlled machines.** EIA3B-34E12

dyadic operation. An operation on two operands. X3A12-16E9

dynamic (industrial control). A state in which one or more quantities exhibit appreciable change within an arbitrarily short time interval. *See:* **control system, feedback.** AS1-34E10

dynamic accuracy. Accuracy determined with a time-varying output. Contrast with **static accuracy.** *See also:* **electronic analog computer.** E165-16E9

dynamic braking. A system of electric braking in which the traction motors, when used as generators, convert the kinetic energy of the vehicle into electric energy, and in so doing, exert a retarding force on the vehicle.
See also:
asynchronous machine;
braking resistor;
direct-current commutating machine;
dynamic braking envelope;
dynamic holding brake;
electric drive;
regenerative braking;
resistance braking;
rheostatic braking;
synchronous machine. 42A42-0

dynamic braking envelope. A curve that defines the dynamic braking limits in terms of speed and tractive force as restricted by such factors as maximum current flow, maximum permissible voltage, minimum field strength, etcetera. *See also:* **dynamic braking.** 42A42-0

dynamic characteristic (electron tube) (operating characteristic). *See:* **load (dynamic) characteristic (electron tube).** *See also:* **circuit characteristics of electrodes.**

dynamic check. *See:* **problem check.** *See also:* **electronic analog computer.**

dynamic computer check. *See:* **problem check.** *See also:* **electronic analog computer.**

dynamic deviation (control) (industrial control). The difference between the ideal value and the actual value of a specified variable when the reference input is changing at a specified constant rate and all other transients have expired. *Note:* Dynamic deviation is expressed either as a percentage of the maximum value of the directly controlled variable or, in absolute terms, as the numerical difference between the ideal and the actual values of the directly controlled variable. *See:* **control system, feedback.** AS1-34E10

dynamic dump (computing systems). A dump that is performed during the execution of a program. *See also:* **electronic digital computer.** X3A12-16E9

dynamic electrode potential. An electrode potential when current is passing between the electrode and the electrolyte. *See also:* **electrochemistry.** 42A60-0

dynamic energy sensitivity (photoelectric devices). *See:* **sensitivity dynamic.**

dynamic equilibrium (electromagnetic system). (1) Any two circuits carrying current tend to dispose themselves so that the magnetic flux linking the two will be a maximum. (2) Every electromagnetic system tends to change its configuration so that the magnetic flux will be a maximum. *See also:* **network analysis.** E270-0

dynamic error. An error in a time-varying signal resulting from inadequate dynamic response of a transducer. Contrast with **static error.** *See also:* **electronic analog computer.** E165-16E9

dynamic holding brake. A braking system designed for the purpose of exerting maximum braking force at a fixed speed only and used primarily to assist in maintaining this fixed speed when a train is descending a grade, but not to effect a deceleration. *See also:* **dynamic braking.** 42A42-0

dynamic load line (electron device). The locus of all simultaneous values of total instantaneous output electrode current and voltage for a fixed value of load impedance. *See:* **circuit characteristics of electrodes.** 0-15E6

dynamic loudspeaker. *See:* **moving-coil loudspeaker.**

dynamic microphone. *See:* **moving-coil microphone.**

dynamic problem check. *See:* **problem check.** *See also:* **electronic analog computer.**

dynamic range. The difference, in decibels, between the overload level and the minimum acceptable signal level in a system or transducer. *Note:* The minimum acceptable signal level of a system or transducer is ordinarily fixed by one or more of the following: noise level, low-level distortion, interference, or resolution level. *See also:* **electronic analog computer; signal; transmission characteristics.** E151-13E6;E165-16E9

dynamic range, reading (storage tubes). The range of output levels that can be read, from saturation level to the level of the minimum discernible output signal. *See also:* **storage tube.** E158-15E6

dynamic range, writing (storage tubes). The range of input levels that can be written under any stated condition of scanning, from the input that will write the minimum usable signal. *See also:* **storage tube.** E158-15E6

dynamic regulation. Expresses the maximum or minimum output variations occurring during transient conditions, as a percentage of the final value. *Note:* Typical transient conditions are instanteous or permanent input or load changes.

$$\text{Dynamic Regulation} = \frac{E_{\text{max}} - E_{\text{final}}}{E_{\text{final}}} (100)\%$$

$$= \frac{E_{\text{final}} - E_{\text{min}}}{E_{\text{final}}} (100\%)$$

0-1E1

dynamic regulator. A transmission regulator in which the adjusting mechanism is in self-equilibrium at only one or a few settings and requires control power to maintain it at any other setting. *See also:* **transmission regulator.** 42A65-0

dynamic slowdown (industrial control). Dynamic braking applied for slowing down, rather than stopping, a drive. *See:* **electric drive.** AS1-34E10

dynamic variable brake. A dynamic braking system designed to allow the operator to select (within the limits of the electric equipment) the braking force best suited to the operation of a train descending a grade and to increase or decrease this braking force for the purpose of reducing or increasing train speed. *See also:* **dynamic braking.** 42A42-0

dynamic vertical. *See:* **apparent vertical.**

dynamometer, electric (rotating machinery). An electric generator or motor equipped with means for indicating torque. *Note:* When used for determining power input or output of a coupled machine, means for indicating speed are also provided. 0-31E8

dynamometer test (rotating machinery). A braking test in which a dynamometer is used. *See also:* **asynchronous machine; direct-current commutating machine; synchronous machine.** 0-31E8

dynamotor (rotating machinery). A converter that combines both motor and generator action, with one magnetic field and with two armatures, or with one armature having separate windings. *See:* **converter.** 42A10-31E8

dynatron effect (electron tubes) (dynatron characteristic). An effect equivalent to a negative resistance, which results when the electrode characteristic (or transfer characteristic) has a negative slope. *Example:* Anode characteristic of a tetrode, or tetrode-connected valve or tube. *See also:* **electronic tube.** 50I07-15E6

dynatron oscillation. Oscillation produced by negative resistance due to secondary emission. *See also:* **oscillatory circuit.** E145-0

dynatron oscillator. A negative-resistance oscillator in which negative resistance is derived between plate and cathode of a screen-grid tube operating so that secondary electrons produced at the plate are attracted to the higher potential screen grid. *See:* **oscillatory circuit.** E145/42A65-0

dyne. The unit of force in the cgs (centimeter-gram-second) systems. The dyne is 10^{-5} newton. E270-0

dynode (electron tubes). An electrode that performs a useful function, such as current amplification, by means of secondary emission. *See also:* **electrode (electron tube); electron tube; tube definitions.** E175-15E6

dynode spots (image orthicons). A spurious signal caused by variations in the secondary-emission ratio across the surface of a dynode that is scanned by the electron beam. *See:* **beam tubes.** E160-15E6

E

early-failure period (reliability). The early period, beginning at some stated time and during which the failure rate of some items is decreasing rapidly. *Note:* The curve below shows the failure pattern when this definition applies to an item. *See also:* **reliability.** 0-7E1

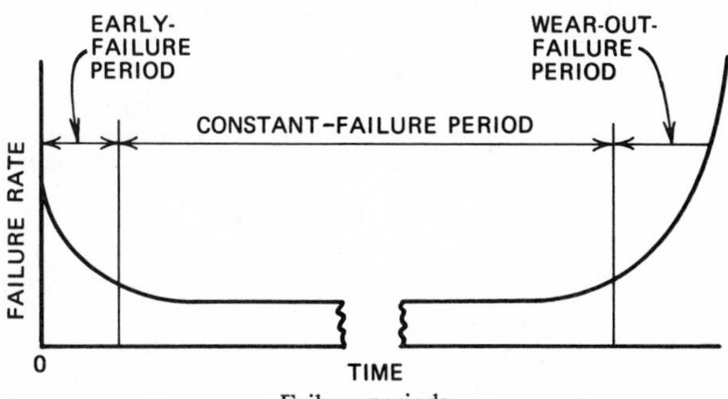

Failure periods.

early warning radar. Radar employed to search for distant enemy aircraft. *See also:* **radar.** 42A65-0

earphone (receiver). An electroacoustic transducer intended to be closely coupled acoustically to the ear. *Note:* The term receiver should be avoided when there is risk of ambiguity. *See also:* **loudspeaker.** E157-1E1;42A65-0

earphone coupler. A cavity of predetermined size and shape that is used for the testing of earphones. The coupler is provided with a microphone for the measurement of pressures developed in the cavity. *Note:* Couplers generally have a volume of 6 cubic centimeters for testing regular earphones and a volume of 2 cubic centimeters for testing insert earphones. *See also:* **loudspeaker.** 0-1E1

earth, effective radius (radio wave propagation). A value for the radius of the earth that is used in place of the geometrical radius to correct approximately for atmospheric refraction when the index of refraction in the atmosphere changes linearly with height. *Note:* Under conditions of standard refraction the effective radius of the earth is 8.5×10^6 meters, or 4/3 the geometrical radius. *See also:* **radiation; radio wave propagation.** 42A65-0;0-3E2

earth-fault protection. *See:* **ground protection.**

earth inductor. *See:* **generating magnetometer.**

earth terminal. *See:* **ground terminal.**

***E* bend (*E*-plane bend) (waveguide technique).** A smooth change in the direction of the axis of a waveguide, throughout which the axis remains in a plane parallel to the direction of polarization. *See also:* **waveguide.** 42A65-0

eccentric groove (eccentric circle) (disc recording). A locked groove whose center is other than that of the disc record (generally used in connection with mechanical control of phonographs). *See also:* **phonograph pickup.** E157-1E1

eccentricity (power distribution, underground cables) (1) (general). The ratio of the difference between the minimum and average thickness to the average thickness of an annular element, expressed in percent. *See also:* **power distribution, underground construction.** 0-31E1

(2) (disc recording). The displacement of the center of the recording groove spiral, with respect to the record center hole. *See also:* **phonograph pickup.** E157-1E1

Eccles-Jordan circuit. A flip-flop circuit consisting of a two-stage resistance-coupled electron-tube amplifier with its output similarly coupled back to its input, the two conditions of permanent stability being provided by the alternate biasing of the two stages beyond cut-off. *See also:* **trigger circuit.** 42A65-0

echelon (calibration). A specific level of accuracy of calibration in a series of levels, the highest of which is represented by an accepted national standard. *Note:* There may be one or more auxiliary levels between two successive echelons. *See also:* **measurement system.** E285-9E1

echo (1) (general). A wave that has been reflected or otherwise returned with sufficient magnitude and delay to be perceived in some manner as a wave distinct from that directly transmitted. *Note:* Echoes are frequently measured in decibels relative to the directly transmitted wave. *See also:* **electroacoustics; recording (facsimile); transmission characteristics.** E151/E168-1E1;42A65-31E3

(2) (radar). The portion of energy of the transmitted pulse that is reflected to a receiver. *See also:* **navigation.** E172-10E6

echo area, effective (radar). The area of a fictitious perfect electromagnetic reflector that would reflect the same amount of energy back to the radar as the target. *See also:* **navigation.** 0-10E6

echo box (radar). A device for checking the overall performance of a radar system. *Note:* It comprises a resonant cavity of high quality factor that receives a portion of the pulse energy from the transmitting system, and then retransmits this energy as a slowly decaying transient to the radar receiving system. The time required for this transient response to decay below the minimum detectable level observable on the radar indicator is known as the ring time, and is indicative of the overall performance of the radar set. *See also:* **instrument; navigation.** 42A30-0

echo check. A method of checking the accuracy of transmission of data in which the received data are returned to the sending end for comparison with the original data. *See also:* **electronic digital computer.** X3A12-16E9

echo, second-time-around (electronic navigation). *See:* **second-time-around echo.**

echo sounding system (depth finder). A system for determination of the depth of water under a ship's keel, based on the measurement of elapsed time between the propagation and projection through the water of a sonic or supersonic signal, and reception of the echo reflected from the bottom. 42A43-0

echo suppressor (1) (telephony). A voice-operated device for connection to a two-way telephone circuit to attenuate echo currents in one direction caused by telephone currents in the other direction. *See also:* **voice-frequency telephony.** 42A65-31E3

(2) (electronic navigation). A circuit component

that desensitizes the receiving equipment for a period after the reception of one pulse, for the purpose of rejecting pulses arriving later over indirect reflection paths. *See also:* **navigation.** 0-10E6

economic dispatch (electric power systems). The distribution of total generation requirements among alternative sources for optimum system economy with due consideration of both incremental generating costs and incremental transmission losses. *See also:* **power system, low-frequency and surge testing.** E94-0

economy power. Power produced from a more economical source in one system and substituted for less economical power in another system. *See also:* **generating station.** 42A35-31E13

eddy-current braking (rotating machinery). A form of electric braking in which the energy to be dissipated is converted into heat by eddy currents produced in a metallic mass. *See also:* **asynchronous machine; direct-current commutating machine; synchronous machine.** 0-31E8

eddy-current loss. Power dissipated due to eddy currents. *Note:* The eddy-current loss of a magnetic device includes the eddy-current losses in the core, windings, case, and associated hardware. 0-21E1

eddy currents (Foucault currents) (1) (general). Those currents that exist as a result of voltages induced in the body of a conducting mass by a variation of magnetic flux. *Note:* The variation of magnetic flux is the result of a varying magnetic field or of a relative motion of the mass with respect to the magnetic field. E270-0

(2) (transformers). The currents that are induced in the body of a conducting mass by the time variation of magnetic flux penetrating the mass of the transformer. *Note:* In transformers, eddy currents in windings, core, or other conducting parts contribute to the exciting current but not to the load current. 0-21E1

Edison distribution system. A three-wire direct-current system, usually about 120-240 volts, for combined light and power service from a single set of mains. *See also:* **direct-current distribution.** 42A35-31E13

Edison effect. *See:* **thermionic emission.**

Edison storage battery. An alkaline storage battery in which the positive active material is nickel oxide and the negative an iron alloy. *See also:* **battery (primary or secondary).** 42A60-0

***E* display (radar).** A rectangular display in which targets appear as blips with distance indicated by the horizontal coordinate and elevation by the vertical coordinate. *See also:* **navigation.** E172-10E6

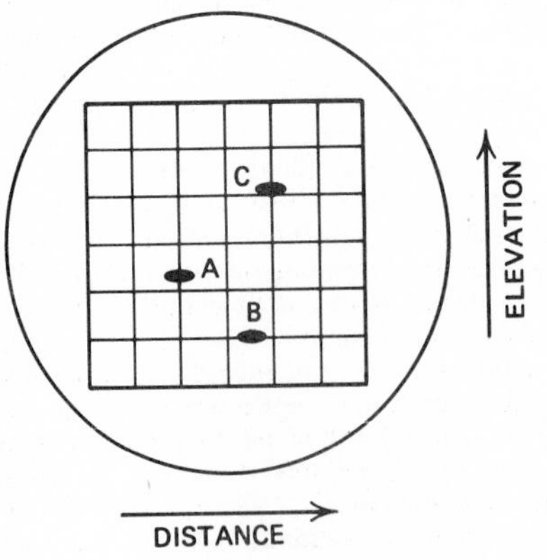

E display.

edit (computing systems). To modify the form or format of data, for example, to insert or delete characters such as page numbers or decimal points. *See also:* **electronic digital computer.** X3A12-16E9

EDR. *See:* **electrodermal reaction.**

effective value*. *See:* **root-mean-square value of a periodic function.** E270-0

*Deprecated

efficiency (1) (engineering). The ratio of the useful output to the input (energy, power, quantity of electricity, etcetera). *Note:* Unless specifically stated otherwise, the term **efficiency** means efficiency with respect to power. E270-0

(2) (electric conversion). The ratio of output power to input power expressed in percent; Eff $= P_o/P_{in}$ (100%). *Note:* It is an evaluation of power losses within the conversion equipment and may be also expressed as a ratio of the output power to the sum of the output power and the power losses, and is expressed in percent.

$$Eff = \frac{P_o}{P_o + P_L}\ (100\%).$$

0-10E1

(3) (by direct calculation) (rotating machinery). The method by which the efficiency is calculated from the input and output, these having been measured directly. *See also:* **asynchronous machine; direct-current commutating machine; synchronous machine.** 0-31E8

(4) (conventional) (rotating machinery). The efficiency calculated from the summation of the component losses measured separately. *See:* **asynchronous machine; direct-current commutating machine; synchronous machine.** 0-31E8

(5) (from total loss) (rotating machinery). The method of indirect calculation of efficiency from the measurement of total loss. *See also:* **asynchronous machine; direct-current commutating machine; synchronous machine.** 0-31E8

(6) (transformer). The ratio of the useful power output to the total power input.

See:
excitation current;
excitation losses;
impedance drop;
impedance kilovolt-amperes;
impedance voltage;
load losses;
per-unit resistance;
reactance drop;
resistance drop;
Scott-connected transformer interlacing impedance voltage;
Scott-connected transformer per-unit interlacing impedance voltage;
total loss.

See also: **mechanical back-to-back test (rotating machinery).** 42A15-31E12

(7) (rectifier). The ratio of the power output to the total power input. *Note:* It may also be expressed as the ratio of the power output to the sum of the output and the losses. 34A1-34E24

(8) (rectification). Ratio of the direct-current component of the rectified voltage at the input terminals of the apparatus to the maximum amplitude of the applied sinusoidal voltage in the specified conditions. 50I07-15E6

(9) (arc-welding apparatus) (electric arc-welding power supply). The ratio of the power output at the welding terminals to the total power input. *See also:* **electric arc-welding apparatus; losses, excitation current, and impedance voltage.** 87A1-0

(10) (bolometer units). The ratio of the radio-frequency power absorbed by the bolometer element to the total radio-frequency power dissipated within the bolometer unit. *Note:* The bolometer unit efficiency is a measure of the radio-frequency losses in the bolometer unit and is a function of the amount of energy dissipated in the dielectric supports, metal surfaces, etcetera, of the bolometer mount. Bolometer unit efficiency is generally independent of power level. *See also:* **bolometric power meter.** 0-9E4

(11) (storage battery). The ratio of the output of the cell or battery to the input required to restore the initial state of charge under specified conditions of temperature, current rate, and final voltage. 42A60-0

(12) (station or system). The ratio of the energy delivered from the station or system to the energy received by it under specified conditions. *See also:* **generating station.** 42A35-31E13

(13) (radiation-counter tube). The probability that a tube count will take place with a specified particle or quantum incident in a specified manner. *See also:* **gas-filled radiation-counter tubes.** 42A70-15E6

efficiency, effective (bolometer units). The ratio of the substitution power to the total radio-frequency power dissipated within the bolometer unit. *Note:* Effective efficiency includes the combined effect of the direct-current–radio-frequency substitution error and bolometer unit efficiency. *See also:* **bolometric power meter.** 0-9E4

efficiency, generator (thermoelectric device). *See:* **generator efficiency (thermoelectric couple).**

efficiency, generator, overall (thermoelectric device). *See:* **overall generator efficiency (thermoelectric couple).**

efficiency, generator, reduced (thermoelectric device). *See:* **reduced generator efficiency.**

efficiency, load circuit. *See:* **load circuit efficiency.**

efficiency, overall electrical. *See:* **overall electrical efficiency.**

efficiency, quantum (phototubes). *See:* **quantum efficiency.**

effluve. *See:* **convective discharge.**

E-H **T (waveguides).** A junction composed of a combination of *E* and *H*-plane T junctions having a common point of intersection with the main guide. *See also:* **waveguide.** E147-3E1

E-H **tuner (waveguides).** An *E-H* T used for impedance transformation having two arms terminated in adjustable plungers. *See also:* **waveguide.** E147-3E1

EI. *See:* **end injection.**

eight-hour rating (magnetic contactor). The rating based on its current-carrying capacity for eight hours, starting with new clean contact surfaces, under conditions of free ventilation, with full-rated voltage on the operating coil, and without causing any of the established limitations to be exceeded. 42A25-34E10

einschleichender stimulus. *See:* **accumulating stimulus.**

Einstein's law (photoelectric device). The law according to which the absorption of a photon frees a photo-electron with a kinetic energy equal to that of the photon less the work function

$$\frac{1}{2}mv^2 = hv - p \text{ (if } hv > p).$$

See also: **photoelectric effect.** 50I07-15E6

elastance. The reciprocal of capacitance. E270-0

elastances (system of conductors) (coefficients of potential—Maxwell). A set of n conductors of any shape that are insulated from each other and that are mounted on insulating supports within a conducting shell, or on one side of a conducting sheet of infinite extent or above the surface of the earth constitutes a system of n capacitors having mutual elastances and capacitances. *Note:* If the shell (or the earth) is regarded as the electrode common to all n capacitors and the transfers of charge as taking place between shell and the individual electrodes, the sum of the charges on the conductors will be equal and opposite in sign to the charge on the common electrode. The shell (or the earth) is taken to be at zero potential. Let Q_r represent the value of the charge that has been transferred from the shell to the other electrode of the r th capacitor, and let V_r represent the algebraic value of the potential of this electrode resulting from the charges in all n capacitors. If the charges are known the values of the potentials can be computed from the equations:

$$V_1 = S_{11}Q_1 + S_{12}Q_2 + S_{13}Q_3 + \cdots$$
$$V_2 = S_{21}Q_1 + S_{22}Q_2 + S_{23}Q_3 + \cdots$$
$$V_3 = S_{31}Q_1 + S_{32}Q_3 + S_{33}Q_3 + \cdots$$
$$\vdots$$
$$V_r = \sum_{c=1}^{c=n} S_{r,c}Q_c.$$

The multiplying operators $S_{r,r}$ are the self-elastances and the multipliers $S_{r,c}$ are the mutual elastances of the system. Maxwell termed them the coefficients of potential of the system. Their values can be measured by noting that the defining equation for the mutual elastance $S_{r,c}$ is

$$S_{r,c}(\text{daraf}) = \frac{V_r \text{ (volt)}}{Q_c \text{ (coulomb)}}$$
$$\text{(every } Q \text{ except } Q_c \text{ being zero).}$$

It can be shown that $S_{r,c} = S_{c,r}$ and that under the conventions stated all the elastances have positive values. E270-0

elastomer (rotating machinery). Macromolecular material that returns rapidly to approximately the initial dimensions and shape after substantial deformation by a weak stress and release of the stress. *See also:* **asynchronous machine; direct-current commutating machine; synchronous machine.** D1566-31E8

E **layer (radio wave propagation).** An ionized layer in the *E* region. *Note:* The principal layer corresponds roughly to what was formerly called the Kennelly-Heaviside layer. In addition, areas of abnormally intense ionization frequently occur, which are called **sporadic *E*.** *See also:* **radiation; radio wave propagation.** E211-3E2;42A65-0

elbow (sharp bend) (interior wiring). (1) A short curved piece of metal raceway of comparatively short radius. *See also:* **raceway.** 42A95-0
(2) A curved section of rigid steel conduit threaded on each end. 80A1-0
(3) (waveguide techniques). *See:* **corner.**

electret. A dielectric body possessing separated electric poles of opposite sign of a permanent or semipermanent nature. *Note:* It is the electrostatic analog of a permanent magnet. E270-0

electric. Containing, producing, arising from, actuated by, or carrying electricity, or designed to carry electricity and capable of so doing. Examples: Electric eel, energy, motor, vehicle, wave. *Note:* Some dictionaries indicate electric and electrical as synonymous but usage in the electrical engineering field has in general been restricted to the meaning given in the definitions above. It is recognized that there are borderline cases wherein the usage determines the selection. *See also:* **electrical.** 37A100-31E11;42A95-0

electric air-compressor governor. A device responsive to variations in air pressure that automatically starts or stops the operation of a compressor for the purpose of maintaining air pressure in a reservoir between predetermined limits. 42A42-0

electrical. Related to, pertaining to, or associated with electricity, but not having its properties or characteristics. *Examples:* Electrical engineer, handbook, insulator, rating, school, unit. *Note:* Some dictionaries indicate electric and electrical as synonymous but usage in the electrical engineering field has in general been restricted to the meaning given in the definitions above. It is recognized that there are borderline cases wherein the usage determines the selection. *See also:* **electric.** 37A100-31E11;42A95-0

electrical anesthesia (medical electronics). More or less complete suspension of general or local sensibility produced by electric means. *See also:* **medical electronics.** 42A80-18E1

electrical boresight (antenna). The tracking axis as determined by an electrical indication, such as the null direction of a conical-scanning or monopulse antenna system or the beam-maximum direction of a highly directive antenna. *See also:* **antenna.** E149-3E1

electrical codes (1) (general). A compilation of rules and regulations covering electric installations.
(2) official electrical code. One issued by a municipality, state, or other political division, and which may be enforced by legal means.
(3) unofficial electrical code. One issued by other than political entities such as engineering societies, and the enforcement of which depends on other than legal means.
(4) National Electrical Code (N.E.C.). The code of rules and regulations as recommended by the National Fire Protection Association and approved by the American National Standards Institute. *Note:* This code is the accepted minimum standard for electric installations and has been accepted by many political entities as their official code, or has been incorporated in whole or in part in their official codes.
(5) National Electrical Safety Code. A set of rules, sponsored by the National Bureau of Standards and approved by the American National Standards Institute governing: (A) Protective grounding. (B) Installation and maintenance of electric supply and communication lines. (C) Installation and maintenance of electrically operated equipment. (D) Operation of electric equipment and lines. (E) Radio installations. (F) Electric fences. E270-0

electrical conversion equipment.
See:
amplitude modulation;
conversion efficiency;
distortion;
drift, stability;
efficiency;
frequency modulation;
inherent harmonics;
input-dependent overshoot and undershoot;
line or input regulation;
load regulation;
maximum excursion;
output-dependent overshoot and undershoot;
output impedance;
percent ripple;
percent unbalance of phase voltages;
phase shift;
power factor;
recovery time;
reflected harmonics;
regulation;
response time;
ripple;
semiconductor;
temperature stability;
transmitted or induced harmonics;
zener-diode regulator.

electrical degree (rotating machinery). The 360th part of the angle subtended, at the axis of a machine, by two consecutive field poles of like polarity. One mechanical degree is thus equal to as many electrical degrees as there are pairs of poles in the machine. *See;* **direct-current commutating machine; synchronous machine.** 42A10-31E8

electrical distance (electronic navigation). The distance between two points expressed in terms of the duration of travel of an electromagnetic wave in free space between the two points. *Note:* A convenient unit of electrical distance is the light-microsecond, approximately 300 meters (983 feet). This unit is widely used in radar technology. *See also:* **electrical range; navigation; signal wave.** E172-10E6

electrical interchangeability (fuse links or fuse units). The characteristic that permits the designs of various manufacturers to be used interchangeably so as to provide a uniform degree of overcurrent protection and fuse coordination. 37A100-31E11

electrical length (1) (general). The physical length expressed in wavelengths, radians, or degrees. *Note:* When expressed in angular units, it is distance in wavelengths multiplied by 2π to give radians or by 360 to give degrees. *See also:* **radio wave propagation; signal wave; wave guide.**
(2) (two-port network at a specified frequency). The length of an equivalent lossless reference waveguide or reference air line (which in the ideal case would be evacuated) introducing the same total phase shift as the two-port when each is terminated in a reflectionless termination. *Note:* It is usually expressed in fractions or multiples of waveguide wavelength. When expressed in radians or degrees it is equal to the phase

angle of the transmission coefficient $+2n\pi$. *See also:* **waveguide.** 42A65/E146-3E1/9E4/31E3

electrically connected. Connected by means of a conducting path or through a capacitor, as distinguished from connection merely through electromagnetic induction. *See also:* **inductive coordination.** 42A65-0

electrically heated airspeed tube. A Pitot-static or Pitot-Venturi tube utilizing a heating element for deicing purposes. *See also:* **air transportation electric equipment.** 42A41-0

electrically heated flying suit. A garment that utilizes sewn-in heating elements energized by electric means designed to cover the torso and all or part of the limbs. *Note:* It may be a one-piece garment or consist of a coat, trousers, and the like. The lower portion of the one-piece suit is in trouser form. *See also:* **air transportation electric equipment.** 42A41-0

electrically interlocked manual release of brakes (control) (industrial control). A manual release provided with a limit switch that is operated when the braking surfaces are disengaged manually. *Note:* The limit switch may operate a signal, open the control circuit, or perform other safety functions. *See:* **switch.** IC1-34E10

electrically release-free (electrically operated switching device) (electrically trip-free). A term indicating that the release can open the device even though the closing control circuit is energized. *Note:* Electrically release-free switching devices are usually arranged so that they are also anti-pump. With such an arrangement the closing mechanism will not reclose the switching device after opening until the closing control circuit is opened and again closed. 37A100-31E11

electrically trip-free. *See:* **electrically release-free.**

electrical metallic tubing. A thin-walled metal raceway of circular cross section constructed for the purpose of the pulling in or the withdrawing of wires or cables after it is installed in place. *See also:* **raceways.** 42A95-0

electrical range. The range expressed in equivalent electrical units. *See also:* **electrical distance; instrument.** 39A4-0

electric arc (gas). A discharge characterized by a cathode drop that is small compared with that in a glow discharge. *Note:* The electron emission of a cathode is due to various causes (thermionic emission, high-field emission, etcetera) acting simultaneously or separately, but secondary emission plays only a small part. *See also:* **discharge (gas).** 50I07-15E6

electric arc-welding apparatus. *Note:* For an extensive list of cross references, see *Appendix A.*

electric back-to-back test. *See:* **pump-back test.**

electric bell. An audible signal device consisting of one or more gongs and an electromagnetically actuated striking mechanism. *Note:* The gong is the resonant metallic member that produces an audible sound when struck. However, the term gong is frequently applied to the complete electric bell. *See also:* **circuits and devices.** 42A65-0

electric bias, relay. *See:* **relay electric bias.**

electric blasting cap. A device for detonating charges of explosives electrically. *See also:* **blasting unit.** 42A85-0

electric braking. A system of braking wherein electric energy, either converted from the kinetic energy of vehicle movement or obtained from a separate source, is one of the principal agents for the braking of the vehicle or train. *See:* **electromagnetic braking (magnetic braking); electropneumatic brake; magnetic track braking; regenerative braking.** 42A42-0

electric bus. A passenger vehicle operating without track rails, the propulsion of which is effected by electric motors mounted on the vehicle. *Note:* A prefix diesel-electric, gas-electric, etcetera, may replace the word electric. *See:* **trolley coach (trolley bus) (trackless trolley coach).** 42A42-0

electric-cable-reel mine locomotive. An electric mine locomotive equipped with a reel for carrying an electric conductor cable that is used to conduct power to the locomotive when operating beyond the trolley wire. *See also:* **electric mine locomotive.** 42A85-0

electric capacitance altimeter. An altimeter, the indications of which depend on the variation of an electric capacitance with distance from the earth's surface. *See also:* **air transportation, electric equipment.** 42A41-0

electric center (power system out of synchronism). A point at which the voltage is zero when a machine is 180 degrees out of phase with the rest of the system. *Note:* There may be one or more electric centers depending on the number of machines and the interconnections among them. 37A100-31E11/31E6

electric charge (charge) (quantity of electricity). Electric charge, like mass, length, and time, is accepted as a fundamentally assumed concept, required by the existence of forces measurable experimentally; other definitions of electromagnetic quantities are developed on the basis of these four concepts. *Note:* The electric charge on (or in) a body or within a closed surface is the excess of one kind of electricity over the other kind. A plus sign indicates that the positive electricity is in excess, a minus sign indicates that the negative is in excess. E270-0

electric charge time constant (detector). The time required, after the instantaneous application of a sinusoidal input voltage of constant amplitude, for the output voltage across the load capacitor of a detector circuit to reach 63 percent of its steady-state value. *See also:* **electromagnetic compatibility.** CISPR-27E1

electric coal drill. An electric motor-driven drill designed for drilling holes in coal for placing blasting charges. *See also:* **mining.** 42A85-0

electric conduction and convection current density. At any point at which there is a motion of electric charge, a vector quantity whose direction is that of the flow of positive charge at this point, and whose magnitude is the limit of the time rate of flow of net (positive) charge across a small plane area perpendicular to the motion, divided by this area, as the area taken approaches zero in a macroscopic sense, so as to always include this point. *Note:* The flow of charge may result from the movement of free electrons or ions but is not, in general, except in microscopic studies, taken to include motions of charges resulting from the polarization of the dielectric. E270-0

electric console lift. An electrically driven mechanism for raising and lowering an organ console and the organist. *See also:* **elevators.** 42A45-0

electric constant (permittivity or capacitivity of free space) (pertinent to any system of units). The scalar ϵ_0 that in that system relates the electric flux density D, in empty space, to the electric field strength E (D

$= \epsilon_0 E$). *Note:* It also relates the mechanical force between two charges in empty space to their magnitudes and separation. Thus in the equation

$$F = Q_1 Q_2 / (n \epsilon_0 r^2)$$

for the force F between charges Q_1 and Q_2 separated by a distance r, ϵ_0 is the electric constant, and n is a dimensionless factor that is unity in unrationalized systems and 4π in a rationalized system. *Note:* In the centimeter-gram-second (cgs) electrostatic system ϵ_0 is assigned the magnitude unity and the dimension "numeric." In the centimeter-gram-second (cgs) electromagnetic system the magnitude of ϵ_0 is that of $1/c^2$ and the dimension is $[L^{-2}T^2]$. In the International system of Units (SI) the magnitude of ϵ_0 is that of $10^7/(4\pi c^2)$ and the dimension is $[L^{-3}M^{-1}T^4I^2]$. Here c is the speed of light expressed in the appropriate system of units. E270-0

electric contact. The junction of conducting parts permitting current to flow. 42A25-0

electric controller. A device or a group of devices, that serves to govern, in some predetermined manner, the electric power delivered to the apparatus to which it is connected. *Note:* For an extensive list of cross references, see *Appendix A.* 42A25-34E10

electric-control trail car. A trail car used in a multiple-unit train, provided at one or both ends with a master controller and other apparatus necessary for controlling the train. *See also:* **electric motor car; electric trail car.** 42A42-0

electric coupler. A group of devices (plugs, receptacles, cable, etcetera) that provides for readily connecting or disconnecting electric circuits. 42A42-0

electric coupler plug. The removable portion of an electric coupler. 42A42-0

electric coupler receptacle (electric coupler socket). The fixed portion of an electric coupler. 42A42-0

electric coupler socket. *See:* **electric coupler receptacle.**

electric coupling. *See:* **coupling, electric.**

electric course recorder. A device that operates, under control of signals from a master compass, to make a continuous record of a ship's heading with respect to time. 42A43-0

electric crab-reel mine locomotive. An electric mine locomotive equipped with an electrically driven winch, or crab reel, for the purpose of hauling cars by means of a wire rope from places beyond the trolley wire. *See also:* **electric mine locomotive.** 42A85-0

electric depth recorder. A device for continuously recording, with respect to time, the depth of water determined by an echo sounding system. 42A43-0

electric dipole. (1) An elementary radiator consisting of a pair of equal and opposite oscillating electric charges an infinitesimal distance apart. It is equivalent to a linear current element. *See also:* **antenna.** E149-3E1

(2) The limit of an electric doublet as the separation approaches zero while the moment remains constant. E270-0

electric dipole moment (two point charges, *q* and *-q*, a distance *a* apart). A vector at the midpoint between them, whose magnitude is the product qa and whose direction is along the line between the charges from the negative toward the positive charge. E270-0

electric-discharge lamp. A lamp in which light (or radiant energy near the visible spectrum) is produced by the passage of an electric current through a vapor or a gas. *Note:* Electric-discharge lamps may be named after the filling gas or vapor that is responsible for the major portion of the radiation; that is, mercury lamps, sodium lamps, neon lamps, argon lamps, etcetera. A second method of designating electric-discharge lamps is by physical dimensions or operating parameters; that is short-arc lamps, high-pressure lamps, low-pressure lamps, etcetera. A third method of designating electric-discharge lamps is by their application; in addition to lamps for illumination there are photochemical lamps, bactericidal lamps, black-light lamps, sun lamps, etcetera. *See also:* **lamp.** Z7A1-0

electric-discharge time constant (detector). The time required, after the instantaneous removal of a sinusoidal input voltage of constant amplitude, for the output voltage across the load capacitor of the detector circuit to fall to 37 percent of its initial value. *See also:* **electromagnetic compatibility.** CISPR-27E1

electric displacement. *See:* **electric flux.** E270-0

electric displacement density. *See:* **electric flux density.** E270-0

electric doublet. A separated pair of equal and opposite charges. E270-0

electric drive (industrial control). A system consisting of one or several electric motors and of the entire electric control equipment designed to govern the performance of these motors. The control equipment may or may not include various rotating electric machines. *Note:* For an extensive list of cross references, see *Appendix A.* 42A25-34E10

electric driving machine. A machine where the energy is applied by an electric motor. *Note:* It includes the motor and brake and the driving sheave or drum together with its connecting gearing, belt, or chain, if any. *See also:* **driving machine (elevators).** 42A45-0

electric elevator. A power elevator where the energy is applied by means of an electric motor. *See also:* **elevators.** 42A45-0

electric energy (energy). The electric energy delivered by an electric circuit during a time interval is the integral with respect to time of the instantaneous power at the terminals of entry of the circuit to a delimited region. *Note:* If the reference direction for energy flow is selected as into the region, the net delivery of energy will be into the region when the sign of the energy is positive and out of the region when the sign is negative. If the reference direction is selected as out of the region, the reverse will apply. Mathematically where

$$W = \int_{t_0}^{t + t_0} p \, dt$$

W = electric energy
p = instantaneous power
t = time during which energy is determined.

When the voltages and currents are periodic, the electric energy is the product of the active power and the time interval, provided the time interval is one or more complete periods or is quite long in comparison with the time of one period. The energy is expressed by

$$W = Pt$$

where
P = active power
t = time interval.
If the instantaneous power is constant, as is true when the voltages and currents form polyphase symmetrical sets, there is no restriction regarding the relation of the time interval to the period. If the voltages and currents are quasi-periodic and the amplitudes of the voltages and currents are slowly varying, the electric energy is the integral with respect to time of the active power, provided the integration is for a time that is one or more complete periods or that is quite long in comparison with the time of one period. Mathematically

$$W = \int_{t_0}^{t_0 + t} P \, dt$$

where P= active power determined for the condition of voltages and currents having slowly varying amplitudes. Electric energy is expressed in joules (watt-seconds) or watthours when the voltages are in volts and the currents in amperes, and the time interval is in seconds or hours, respectively. E270-0

electric explosion-tested mine locomotive. An electric mine locomotive equipped with explosion-tested equipment. *See also:* **electric mine locomotive.** 42A85-0

electric field (1) (general). A vector field of electric field strength or of electric flux density. *Note:* The term is also used to denote a region in which such vector fields have a significant magnitude. *See:* **vector field.** E270-0
(2) (static). A state of the region in which stationary charged bodies are subject to forces by virtue of their charges. 0-3E2
(3) (signal-transmission system). A state of a medium characterized by spatial potential gradients (electric field vectors) caused by conductors at different potentials, that is, the field between conductors at different potentials that have capacitance between them. *See also:* **signal.** 0-13E6

electric field intensity.* *See:* **electric field strength.**
*Deprecated E270-0

electric field strength (electric field) (electric field intensity*) (electric force*) (1) (general). At a given point, the vector limit **E** of the quotient of the force that a small stationary charge at that point will experience, by virtue of its charge, to the charge as the charge approaches zero in a macroscopic sense. *Note:* The concept of a tunnel-like (Kelvin) cavity, properly oriented along **E,** is frequently employed to visualize and compute the **E** vector in material media. E270-0
*Deprecated
(2) (signal-transmission system). The magnitude of the potential gradient in an electric field expressed in units of potential difference per unit length in the direction of the gradient. *See also:* **signal.** 0-13E6
(3) (radio wave propagation). The magnitude of the electric field vector. *Note:* This term has sometimes been called the **electric field intensity,** but such use of the word intensity is deprecated in favor of field strength, since intensity connotes power in optics and radiation. *See also:* **radio wave propagation.** E211-27E1/3E2

electric field vector (at a point in an electric field). The force on a stationary positive charge per unit charge. *Note:* This may be measured either in newtons per coulomb or in volts per meter. This term is sometimes called the **electric field intensity,** but such use of the word intensity is deprecated since intensity connotes power in optics and radiation. *See also:* **radio wave propagation; waveguide.** E146-3E2

electric filter (electric wave filter). A filter designed to separate electric waves of different frequencies. *Note:* An electric wave filter may be classified in terms of the reactive elements that it includes, for example, inductors and capacitors, piezoelectric crystal units, coaxial lines, resonant cavities, etcetera. *See also:* **filter.** 42A65-31E3

electrix flux (through a surface) (electric displacement). The surface integral of the normal component of the electric flux density over the surface. E270-0

electric flux density (electric displacement density) (electric induction*). A quantity related to the charge displaced within the dielectric by application of an electric field. *Notes:* (1) Electric flux density at any point in an isotropic dielectric is a vector that has the same direction as the electric field strength and a magnitude equal to the product of the electric field strength and the absolute capacitivity. The electric flux density is that vector point function whose divergence is the charge density, and that is proportional to the electric field in regions free of polarized matter. The electric flux density is given by

$$\mathbf{D} = \epsilon_0 \epsilon \mathbf{E}$$

where **D** is the electric flux density, $\epsilon_0\epsilon$ is the absolute capacitivity, and **E** is the electric field strength. (2) In a nonisotropic medium, ϵ becomes a tensor represented by a matrix and **D** is not necessarily parallel to **E.** (3) The concept of a disk-like (Kelvin) cavity, properly oriented normal to **D,** is frequently employed to visualize and compute the **D** vector in material media. (4) The electric flux density at a point is equal to the charge per unit area that would appear on one face of a small thin metal plate introduced in the electric field at the point and so oriented that this charge is a maximum. (5) The symbol Γ_ϵ is often used in modern practice in place of ϵ_0; the symbol ϵ_v has occasionally been used. E270-3E2
*Deprecated

electric focusing (microwave tubes). The combination of electric fields that acts upon the electron beam in addition to the forces derived from momentum and space charge. *See also:* **microwave tube or valve.** 0-15E6

electric force* *See:* **electric field strength.** E270-0
*Deprecated

electric freight locomotive. An electric locomotive, commonly used for hauling freight trains and generally designed to operate at higher tractive force values and lower speeds than a passenger locomotive of equal horsepower capacity. *Note:* A prefix diesel-electric, gas-electric, turbine-electric, etcetera, may replace the word electric. *See also:* **electric locomotive.** 42A42-0

electric gathering mine locomotive. An electric mine locomotive, the chief function of which is to move empty cars into, and remove loaded cars from, the working places. *See also:* **electric mine locomotive.** 42A85-0

electric generator. A machine that transforms mechanical power into electric power. *See also:* **direct-current commutating machine; synchronous machine.** 42A10-0

electric gun heater. An electrically heated element attached to the gun breech to prevent the oil from congealing or the gun mechanism from freezing. *See also:* **air transportation, electric equipment.** 42A41-0

electric haulage mine locomotive. An electric mine locomotive used for hauling trains of cars, that have been gathered from the working faces of the mine, to the point of delivery of the cars. *See also:* **electric mine locomotive.** 42A85-0

electric horn. A horn having a diaphragm that is vibrated electrically. *See also:* **protective signaling.** 42A65-0

electric hygrometer. An instrument for indicating by electric means the humidity of the ambient atmosphere. *Note:* Electric hygrometers usually depend for their operation on the relation between the electric conductance of a film of hygroscopic material and its moisture content. *See also:* **instrument.** 42A30-0

electric incline railway. A railway consisting of an electric hoist operating a single car with or without counterweights, or two cars in balance, which car or cars travel on inclined tracks. *See also:* **elevators.** 42A45-0

electric indication lock. An electric lock connected to a lever of an interlocking machine to prevent the release of the lever or latch until the signals, switches, or other units operated, or directly affected by such lever, are in the proper position. *See also:* **interlocking (interlocking plant).** 42A42-0

electric indication locking. Electric locking adapted to prevent manipulation of levers that would bring about an unsafe condition for a train movement in case a signal, switch, or other operated unit fails to make a movement corresponding with that of its controlling lever; or adapted directly to prevent the operation of one unit in case another unit to be operated first, fails to make the required movement. *See also:* **interlocking (interlocking plant)** 42A42-0

electric induction*. *See:* **electric flux density.**

*Deprecated

electric interlocking machine. An interlocking machine designed for the control of electrically operated functions. *See also:* **interlocking (interlocking plant).** 42A42-0

electricity meter (meter). A device that measures and registers the integral of an electrical quantity with respect to time. *Notes:* (1) The term **meter** is also used in a general sense to designate any type of measuring device, including all types of electric measuring instruments. Such use as a suffix or as part of a compound word (for example, voltmeter, frequency meter) is universally accepted. Meter may be used alone with this wider meaning when the context is such as to prevent confusion with the narrower meaning here defined. (2) For an extensive list of cross references, see *Appendix A.* 42A30-0

electric larry car. A burden-bearing car for operation on track rails used for short movements of materials, the propulsion of which is effected by electric motors mounted on the vehicle. *Note:* A prefix diesel-electric, gas-electric, etcetera, may replace the word electric. *See also:* **electric motor car.** 42A42-0

electric loading (rotating machinery). The average ampere-conductors of the primary winding per unit length of the air-gap periphery. *See also:* **rotor (rotating machinery); stator.** 0-31E8

electric lock. A device to prevent or restrict the movement of a lever, a switch, or a movable bridge unless the locking member is withdrawn by an electric device such as an electromagnet, solenoid, or motor. *See also:* **interlocking (interlocking plant).** 42A42-0

electric locking. The combination of one or more electric locks and controlling circuits by means of which levers of an interlocking machine, or switches, or other units operated in connection with signaling and interlocking, are secured against operation under certain conditions, as follows: (1) approach locking, (2) indication locking, (3) switch-lever locking, (4) time locking, (5) traffic locking. *See also:* **interlocking (interlocking plant).** 42A42-0

electric locomotive. A vehicle on wheels, designed to operate on a railway for haulage purposes only, the propulsion of which is effected by electric motors mounted on the vehicle. *Note:* While this is a generic term covering any type of locomotive driven by electric motors, it is usually applied to locomotives receiving electric power from a source external to the locomotive . The prefix electric may also be applied to cars, buses, etcetera, driven by electric motors. A prefix diesel-electric, gas-electric, turbine-electric, etcetera, may replace the word electric.
See:
alternating-current electric locomotive;
***A* unit;**
***B* unit;**
combination electric locomotive;
diesel-electric drive (oil-electric drive);
direct-current electric locomotive;
electric freight locomotive;
electric motive power unit;
electric passenger locomotive;
electric road locomotive;
electric road-transfer locomotive;
electric switching locomotive;
electric transfer locomotive;
electrified track;
gas-electric drive;
gas-turbine–electric drive;
industrial electric locomotive;
motor-generator electric locomotive;
multiple-unit electric locomotive;
multiple-unit electric motive power unit;
rectifier electric locomotive;
self-propelled electric locomotive;
single-phase electric locomotive;
split-phase electric locomotive;
steam-turbine–electric locomotive;
third-rail clearance line;
three-phase electric locomotive;
tractive force;
trolley locomotive. 42A42-0

electric-machine regulator (rotating machinery). A specified element or a group of elements that is used within an electric-machine regulating system to perform a regulating function by acting to maintain a designated variable (or variables) at a predetermined value, or to vary it according to a predetermined plan. *See also:* **synchronous machine.** 0-31E8

electric-machine regulating system (rotating machinery). A feedback control system that includes one or more electric machines and the associated control. *See also:* **synchronous machine.** 0-31E8

electric mechanism (demand meter). That portion, the action of which, in response to the electric quantity to be measured, gives a measurement of that quantity. *Note:* For example, the electric mechanism of certain demand meters is similar to the ordinary ammeter or wattmeter of the deflection type; in others it is a watthour meter or other integrating meter; and in still others it comprises an electric circuit that heats temperature-responsive elements, such as bimetallic spirals, that deflect to move the indicating means. The electrical quantity may be measured in kilowatts, kilowatt-hours, kilovolt-amperes, kilovolt-ampere-hours, amperes, ampere-hours, kilovars, kilovar-hours, or other suitable units. *See also:* **demand meter.** 42A30-0

electric mine locomotive (1) (general). An electric locomotive designed for use underground; for example, in such places as coal, metal, gypsum, and salt mines, tunnels, and in subway construction.
(2) (storage-battery type). An electric locomotive that receives its power supply from a storage battery mounted on the chassis of the locomotive.
(3) (trolley type). An electric locomotive that receives its power supply from a trolley-wire distribution system.
(4) (combination type). An electric locomotive that receives power either from a trolley-wire distribution system or from a storage battery carried on the locomotive.
(5) (separate tandem). An electric mine locomotive consisting of two locomotive units that can be coupled together or operated from one controller as a single unit, or else separated and operated as two independent units.
(6) (permanent tandem). A locomotive consisting of two locomotive units permanently connected together and provided with one set of controls so that both units can be operated by a single operator.
See:
electric-cable-reel mine locomotive;
electric crab-reel mine locomotive;
electric explosion-tested mine locomotive;
electric gathering mine locomotive;
electric haulage mine locomotive.
See also: **mining.** 42A85-0

electric moment (electric doublet or dipole). The product of the magnitude of either charge by the distance between them. E270-0

electric motive power unit. A self-contained electric traction unit, comprising wheels and a superstructure capable of independent propulsion from a power supply system, but not necessarily equipped with an independent control system. *Note:* While this is a generic term covering any type of motive power driven by electric motors, it is usually applied to locomotives receiving electric power from an external source. A prefix diesel-electric, gas-electric, turbine-electric, etcetera, may replace the word electric. *See also:* **electric locomotive.** 42A42-0

electric motor. A machine that transforms electric power into mechanical power. *See:* **asynchronous machine; direct-current commutating machine; synchronous machine.** 42A10-0

electric motor car. A vehicle for operating on track rails, used for the transport of passengers or materials, the propulsion of which is effected by electric motors mounted on the vehicle. *Note:* A prefix diesel-electric, gas-electric, etcetera, may replace the word electric.
See:
electric-control trail car;
electric larry car;
electric tower car;
electric trail car;
multiple-unit electric car;
multiple-unit electric train;
rectifier electric motor car;
self-propelled electric car;
third-rail electric car;
trolley car. 42A42-0

electric motor controller. A device or group of devices that serve to govern, in some predetermined manner, the electric power delivered to the motor. *Note:* An electric motor controller is distinct functionally from a simple disconnecting means whose principal purpose in a motor circuit is to disconnect the circuit, together with the motor and its controller, from the source of power. *See:* **electric controller.** IC1-34E10

electric movable-bridge (drawbridge) lock. A device used to prevent the operation of a movable bridge until the device is released. *See also:* **interlocking (interlocking plant).** 42A42-0

electric network. *See:* **network.**

electric noise (1) (general). Unwanted electrical energy other than crosstalk present in a transmission system.
(2) (interference terminology). A form of interference introduced into a signal system by natural sources that constitutes for that system an irreducible limit on its signal-resolving capability. *Note:* Noise is characterized by randomness of amplitude and frequency distribution and therefore cannot be eliminated by band-rejection filters tuned to preselected frequencies. *See also:* **distortion; interference.** E145-13E6

electric operation. Power operation by electric energy. 37A100-31E11

electric orchestra lift. An electrically driven mechanism for raising and lowering the musicians' platform and the musicians. *See also:* **elevators.** 42A45-0

electric parachute-flare-launching tube. A tube mounted on an aircraft through which a metal container carrying a parachute flare is launched, the tube being so designed that, as the parachute-flare container passes through the tube, an electric circuit is completed that ignites a slow-burning fuse in the container, the fuse being so designed as to permit the container to clear the aircraft before it ignites the parachute flare. *See also:* **air transportation, electric equipment.** 42A41-0

electric passenger locomotive. An electric locomotive, commonly used for hauling passenger trains and generally designed to operate at higher speeds and lower tractive-force values than a freight locomotive of equal horsepower capacity. *Note:* A prefix diesel-electric, gas-electric, turbine-electric, etcetera, may replace the word electric. *See also:* **electric locomotive.** 42A42-0

electric permissible mine locomotive. An electric locomotive carrying the official approval plate of the United States Bureau of Mines. *See also:* **electric mine locomotive.** 42A85-0

electric pin-and-socket coupler (connector). A readily disconnective assembly used to connect electric circuits between components of an aircraft electric system by means of mating pins and sockets. *See also:* **air transportation wiring and associated equipment.** 42A41-0

electric polarizability. Of an isotropic medium for which the direction of electric polarization and electric field strength are the same at any point in the medium, the magnitude P of the electric polarization at that point divided by the electric field strength there, **E**. *Note:* In a rationalized system, the electric polarizability $P_e = P/\mathbf{E} = \epsilon_0 (\epsilon - 1)$. E270-0

electric polarization (electric field). At any point, the vector difference between the electric flux density at that point and the electric flux density that would exist at that point for the same electric field strength there, if the medium were a vacuum there. *Note:* Electric polarization is the vector limit of the quotient of the vector sum of electric dipole moments in a small volume surrounding a given point, and this volume, as the volume approaches zero in a microscopic sense. E270-0

electric port (optoelectronic device). A port where the energy is electric. *Note:* A designated pair of terminals may serve as one or more electric ports. *See also:* **optoelectronic device.** E222-15E7

electric power distribution panel. A metallic or nonmetallic, open or enclosed, unit of an electric system. The operable and the indicating components of an electric system, such as switches, circuit breakers, fuses, indicators, etcetera, usually are mounted on the face of the panel. Other components, such as terminal strips, relays, capacitors, etcetera, usually are mounted behind the panel. *See also:* **air transportation wiring and associated equipment.** 42A41-0

electric power substation. An assemblage of equipment for purposes other than generation or utilization, through which electric energy in bulk is passed for the purpose of switching or modifying its characteristics. Service equipment, distribution transformer installations, or other minor distribution or transmission equipment are not classified as substations. *Note:* A substation is of such size or complexity that it incorporates one or more buses, a multiplicity of circuit breakers, and usually is either the sole receiving point of commonly more than one supply circuit, or it sectionalizes the transmission circuits passing through it by means of circuit breakers. *See also:* **alternating-current distribution; direct-current distribution.** 42A35-31E13

electric power systems. *See:* **energy metering point.**

electric propulsion apparatus. Electric apparatus (generators, motors, control apparatus, etcetera) provided primarily for ship's propulsion. *Note:* For certain applications, and under certain conditions, auxiliary power may be supplied by propulsion apparatus. *See also:* **electric propulsion system.** 42A43-0

electric propulsion system. A system providing transmission of power by electric means from a prime mover to a propeller shaft with provision for control, partly or wholly by electric means, of speed and direction. *Note:* An electric coupling (which see) does not provide electric propulsion.

See:

electric coupling;

electric propulsion apparatus;

propulsion set-up switch;

torque margin;

torsionmeter;

shaft-revolution indicator;

steam-turbine–electric drive. 42A43-0

electric reset relay. A relay that is so constructed that it remains in the picked-up condition even after the input quantity is removed; an independent electric input is required to reset the relay. 37A100-31E11/31E6

electric resistance-type temperature indicator. A device that indicates temperature by means of a resistance bridge circuit. *See also:* **air transportation instruments.** 42A41-0

electric road locomotive. An electric locomotive designed primarily for hauling dispatched trains over the main or secondary lines of a railroad. *Note:* A prefix diesel-electric, gas-electric, turbine-electric, etcetera, may replace the word electric. *See also:* **electric locomotive.** 42A42-0

electric road-transfer locomotive. An electric locomotive designed primarily so that it may be used either for hauling dispatched trains over the main or secondary lines of a railroad or for transferring relatively heavy cuts of cars for short distances within a switching area. *Note:* A prefix diesel-electric, gas-electric, turbine-electric, etcetera, may replace the word electric. *See also:* **electric locomotive.** 42A42-0

electric sign. A fixed or portable self-contained electrically illuminated appliance with words or symbols designed to convey information or attract attention. *See also:* **appliance.** 42A95-0

electric-signal storage tube. A storage tube into which the information is introduced as an electric signal and read at a later time as an electric signal. *See also:* **storage tube.** E158-15E6

electric sounding machine. A motor-driven reel with wire line and weight for determination of depth of water by mechanical sounding. 42A43-0

electric squib. A device similar to an electric blasting cap but containing a gunpowder composition that simply ignites but does not detonate an explosive charge. *See also:* **blasting unit.** 42A85-0

electric stage lift. An electrically driven mechanism for raising and lowering various sections of a stage. 42A45-0

electric strength (dielectric strength). The maximum potential gradient that the material can withstand without rupture. *See:* **dielectric strength.** *See also:* **insulation testing (large alternating-current rotating machinery).** E95-0

electric stroboscope. An instrument for observing rotating or vibrating objects or for measuring rotational speed or vibration frequency, or similar periodic quantities, by electrically produced periodic changes in illumination. *See also:* **instrument.** 42A30-0

electric-supply equipment. Equipment that produces, modifies, regulates, controls, or safeguards a supply of electric energy. *Note:* Similar equipment, however, is not included where used in connection with signaling systems under the following conditions: (1) where the voltage does not exceed 150, and (2) where the voltage is between 150 and 400 and the power transmitted does not exceed 3 kilowatts. *See also:* **distribution center.** 2A2-0

electric-supply lines. The conductors and their necessary supporting or containing structures that are located entirely outside of buildings and are used for transmitting a supply of electric energy. *Note:* This does not include open wiring on buildings, in yards, or similar locations where spans are less than 20 feet, and all the precautions required for stations or utilization equipment, as the case may be, are observed. Railway signal lines of more than 400 volts to ground are always supply lines within the meaning of these rules, and those of less than 400 volts may be considered as supply lines, if so run and operated throughout. *See also:* **center of distribution; power distribution, overhead construction.** 2A2-0,42A35-31E13

electric-supply station. Any building, room, or separate space within which electric-supply equipment is located and the interior of which is accessible, as a rule, only to properly qualified persons. *Note:* This includes generating stations and substations and generator, storage-battery, and transformer rooms, but excludes manholes and isolated transformer vaults on private premises. *See:* **transformer vault.** *See also:* **generating station.** 2A2-0

electric susceptibility. Of an isotropic medium, for which the direction of electric polarization and electric field strength are the same, at any point in the medium, the magnitude of the electric polarization at that point of the medium, divided by the electric flux density that would exist at that point for the same electric field strength, if the medium there were a vacuum. *Note:* In a rationalized system the electric susceptibility $\chi_\epsilon = P/D(\epsilon - 1)$. E270-0

electric switching locomotive. An electric locomotive designed for yard movements of freight or passenger cars, its speed and continuous electrical capacity usually being relatively low. *Note:* A prefix diesel-electric, gas-electric, turbine-electric, etcetera, may replace the word electric. *See also:* **electric locomotive.** 42A42-0

electric switch-lever lock. An electric lock used to prevent the movement of a switch lever or latch in an interlocking machine until the lock is released. *See also:* **interlocking (interlocking plant).** 42A42-0

electric switch-lever locking. A general term for route or section locking. *See also:* **interlocking (interlocking plant).** 42A42-0

electric switch lock. An electric lock used to prevent the operation of a switch or a switch movement until the lock is released. *See also:* **interlocking (interlocking plant).** 42A42-0

electric tachometer (marine usage). An instrument for measuring rotational speed by electric means. *See also:* **instrument.** 42A30-0

electric telegraph. A telegraph having the relationship of the moving parts of the transmitter and receiver maintained by the use of self-synchronous motors or equivalent devices. 42A43-0

electric telemeter. The measuring, transmitting, and receiving apparatus, including the primary detector, intermediate means (excluding the channel), and end devices for electric telemetering. *Note:* A telemeter that measures current is called a teleammeter; voltage, a televoltmeter; power, a telewattmeter; one that measures angular or linear position, a position telemeter. The names of the various component parts making up the telemeter are, in general, self-defining; for example, the transmitter, receiver, indicator, etcetera. *See also:* **telemetering.** 37A100-31E11;42A30-0

electric telemetering (electric telemetry*). Telemetering performed by an electric translating means separate from the measurand. *See also:* **telemetering.**

*Deprecated 37A100-31E11;42A30-0

electric thermometer (temperature meter). An instrument that utilizes electric means to measure temperature. *Note:* An electric thermometer may employ any electrical or magnetic property that is dependent on temperature. The most commonly used properties are the thermoelectric effects in a circuit of two or more metals and the change of electric resistance of a metal.
See:
instrument;
microradiometer;
optical pyrometer;
pyrometer;
radiation pyrometer;
resistance thermometer;
resistance temperature detector;
thermal converter;
thermocouple;
thermoelectric thermometer;
thermojunction;
thermopile;
temperature detector. 42A30-0

electric tower car. A rail vehicle, the propulsion of which is effected by electric means and that is provided with an elevated platform, generally arranged to be raised and lowered, for the installation, inspection, and repair of a contact wire system. *Note:* A prefix diesel-electric, gas-electric, etcetera, may replace the word electric. *See also:* **electric motor car.** 42A42-0

electric trail car (electric trailer). A car not provided with motive power that is used in a train with one or more electric motor cars. *Note:* A prefix diesel-electric, gas-electric, etcetera, may replace the word electric to identify the motor cars. *See also:* **electric-control trail car, electric motor car.** 42A42-0

electric train-line (multiple-unit control system). A continuous multiple electric circuit, extending between vehicles, provided with control stations to permit the control of traction motors and other equipment from any of several points on the train (or in special cases one point only). *Note:* An electric train-line may include circuits for electric brakes, telephones, engine controls, and other devices sometimes carried between units in a multiple-unit system. *See also:* **car wiring apparatus; multiple-unit control.** E16-42A42-0

electric transducer. A transducer in which all of the waves concerned are electric. *See also:* **transducer.** E-196/E270/42A65-0

electric transfer locomotive. An electric locomotive designed primarily for transferring relatively heavy cuts of cars for short distances within a switching area. *Note:* A prefix diesel-electric, gas-electric, turbine-electric, etcetera, may replace the word electric. *See also:* **electric locomotive.** 42A42-0

electric-tuned oscillator. An oscillator whose frequency is determined by the value of a voltage, current, or power. Electric tuning includes electronic tuning, electrically activated thermal tuning, electromechanical tuning, and tuning methods in which the properties of the medium in a resonant cavity are changed by an external electric means. An example is the tuning of a ferrite-filled cavity by changing an external magnetic field. *See also:* **tunable microwave oscillators.** E158-15E6

electric turn-and-bank indicator. A device that utilizes an electrically driven gyro for turn determination and a gravity-actuated inclinometer for bank determination. *See also:* **air transportation instruments.** 42A41-0

electric vector (radio wave propagation). Electric field vector. *See also:* **radio wave propagation.** 0-3E2

electric wave filter. *See:* **electric filter.**

electric wind. *See:* **convective discharge.**

electrification by friction. *See:* **triboelectrification.**

electrified track. A railroad track suitably equipped in association with a contact conductor or conductors for the operation of electrically propelled vehicles that receive electric power from a source external to the vehicle. *See also:* **electric locomotive.** 42A42-0

electroacoustical reciprocity theorem. For an electroacoustic transducer satisfying the reciprocity principle, the quotient of the magnitude of the ratio of the open-circuit voltage at output terminals (or the short-circuit current) of the transducer, when used as a sound receiver, to the free-field sound pressure referred to an arbitrarily selected reference point on or near the transducer, divided by the magnitude of the ratio of the sound pressure apparent at a distance δ from the reference point to the current flowing at the transducer input terminals (or the voltage applied at the input terminals), when used as a sound emitter, is a constant, called the reciprocity constant, independent of the type or constructional details of the transducer. *Note:* The reciprocity constant is given by

$$\left|\frac{M_o}{S_s}\right| = \left|\frac{M_s}{S_s}\right| = \frac{2\delta}{\rho f}$$

where

M_o = open free-field voltage response, as a sound receiver, in open-circuit volts per newton per square meter, referred to the arbitrary reference point on or near the transducer.

M_s = free-field current response in short-circuit amperes per newton per square meter, referred to the arbitrary reference point on or near the transducer

S_o = sound pressure in newtons per square meter per ampere of input current produced at a distance δ meters from the arbitrary reference point

S_s = sound pressure in newtons per square meter per volt applied at the input terminals produced at a distance δ meters from the arbitrary reference point

f = frequency in hertz

ρ = density of the medium in kilograms per cubic meter

δ = distance in meters from the arbitrary reference point on or near the transducer to the point in which the sound pressure established by the transducer when emitting is evaluated.

See also: **loudspeaker.** 0-1E1

electroacoustics. *Note:* For an extensive list of cross references, see *Appendix A.*

electroacoustic transducer (electric system). A transducer for receiving waves and delivering waves to an acoustic system, or vice versa. *See also:* **electroacoustics; loudspeaker; transducer.** E157-1E1

electroanalysis. The electrodeposition of an element or compound for the purpose of determining its quantity in the solution electrolyzed. *See also:* **electrodeposition.** 42A60-0

electrobiology. The study of electrical phenomena in relation to biological systems.
See:
boundary potential;
Donnan potential;
electroosmotic potential;
electrophoretic potential;
electrotaxis;
ephapse;
membrane potential;
neuroelectricity;
partial-body irradiation;
polarization potential;
radiation protection guide;
sedimentation potential;
streaming potential;
synapse electrobiology;
whole-body irradiation. 42A80-18E1

electrocapillary phenomena. The phenomena depending on the variation in surface tension, at the boundary of two liquids, with the potential difference established between these two liquids. E270-0

electrocardiogram. The graphic record of the variation with time of the voltage associated with cardiac activity. *See:* **electrocorticogram (electrobiology); electrodermogram (electrobiology); Galvani's experiment (electrobiology); spindle wave (electrobiology); vector electrocardiogram (electrobiology).** 0-18E1

electrocautery (electrotherapy). An instrument for cauterizing the tissues by means of a conductor brought to a high temperature by an electric current. *See also:* **electrotherapy.** 42A80-18E1

electrochemical cell. A system consisting of an anode, cathode, and an electrolyte plus such connections (electric and mechanical) as may be needed to allow the cell to deliver or receive electric energy.
See:
anode;
cathode;
electrode;
electrolyte;
fuel;
fuel cell;
oxidant;
oxidation;
reduction. CV1-10E1

electrochemical equivalent (element, compound, radical, or ion) (1) (general). The weight of that substance involved in a specified electrochemical reaction during the passage of a specified quantity of electricity, such as a faraday, ampere-hour, or coulomb. 42A60-0

(2) (oxidation). The weight of an element or group of

elements oxidized or reduced at 100-percent efficiency by a unit quantity of electricity. *See also:* **electrochemistry.** CM-34E2

electrochemical recording (facsimile). Recording by means of a chemical reaction brought about by the passage of signal-controlled current through the sensitized portion of the record sheet. *See also:* **recording (facsimile).** E168-0

electrochemical series. *See:* **electromotive series.**

electrochemical valve. An electric valve consisting of a metal in contact with a solution or compound across the boundary of which current flows more readily in one direction than in the other direction and in which the valve action is accompanied by chemical changes.
See:
breakdown voltage;
electrochemical valve metal;
electrolytic rectifier;
film;
formation voltage;
forming;
valve action;
valve ratio. 42A60-0

electrochemical valve metal. A metal or alloy having properties suitable for use in an electrochemical valve. *See also:* **electrochemical valve.** 42A60-0

electrochemistry. That branch of science and technology that deals with interrelated transformations of chemical and electric energy. *Note:* For an extensive list of cross references, see *Appendix A.* 42A60-0

electrocoagulation (medical electronics). The clotting of tissue by heat generated within the tissue by impressed electric currents. *See also:* **medical electronics.** 42A80-18E1

electrocorticogram (medical electronics). A graphic record of the variation with time of voltage taken from exposed cortex cerebra. *See also:* **medical electronics.** 42A80-18E1

electroculture (medical electronics). The stimulation of growth, flowering, or seeding by electric means. *See also:* **medical electronics.** 42A80-18E1

electrocution. The destruction of life by means of electric current. 42A80-18E1

electrode (1) (general). A conductor through which an electric current enters or leaves an electrolyte, gas, or vacuum. 42A60-0

(2) (electrochemistry). An electric conductor for the transfer of charge between the external circuit and the electroactive species in the electrolyte. *Note:* Specifically, in an electrolytic cell, an electrode is a conductor at the surface of which a change occurs from conduction by electrons to conduction by ions or colloidal ions. *See also:* **electrolytic cell; electrochemical cell.** 42A60-0;CV1-10E1

(3) (electron tube). A conducting element that performs one or more of the functions of emitting, collecting, or controlling by an electric field the movements of electrons or ions. *Note:* For an extensive list of cross references, see *Appendix A.* E160-15E6;42A70-0

(4) (semiconductor device). An element that performs one or more of the functions of emitting or collecting electrons or holes, or of controlling their movements by an electric field. *See also:* **semiconductor.** E102/42A70-0

(5) (biological electronics) (reference, inactive, diffuse, dispersive, indifferent electrode). (A) A pickup electrode that, because of averaging, shunting, or other aspects of the tissue-current pattern to which it connects, shows potentials not characteristic of the region near the active electrode. (B) Any electrode, in a system of stimulating electrodes, at which due to its dispersive action, excitation is not produced. (C) An electrode of relatively large area applied to some inexcitable or distant tissue in order to complete the circuit with the active electrode that is used for stimulation. *See also:* **medical electronics.** 0-18E1

electrode, accelerating (electron-beam tube). *See:* **accelerating electrode.**

electrode admittance (*j*th electrode of an *n*-electrode electron tube). The short-circuit driving-point admittance between the *j*th electrode and the reference point measured directly at the *j*th electrode. *Note:* To be able to determine the intrinsic electronic merit of an electron tube, the driving-point and transfer admittances must be defined as if measured directly at the electrodes inside the tube. The definitions of electrode admittance and electrode impedance are included for this reason. *See also:* **electron-tube admittances.** E160/42A70-15E6

electrode alternating-current resistance (electron device). The real component of the electrode impedance. *See:* **self-impedance.** 0-15E6

electrode bias (electron tubes). The voltage at which an electrode is stabilized under operating conditions with no incoming signal, but taking into account the voltage drops in the connected circuits. *See:* **electrode voltage.** 50I07-15E6

electrode capacitance (*n*-terminal electron tube). The capacitance determined from the short-circuit driving-point admittance at that electrode. *See also:* **circuit characteristics of electrodes; electron-tube admittances.** E160-15E6;42A70-0

electrode characteristic. A relation, usually shown by a graph, between the electrode voltage and the current of an electrode, all other electrode voltages being maintained constant. *See also:* **circuit characteristics of electrodes.** 42A70-15E6

electrode conductance. The real part of the electrode admittance. E160/42A70-15E6

electrode, control. *See:* **control electrode.**

electrode current (electron tube). The current passing to or from an electrode through the interelectrode space. *Note:* The terms cathode current, grid current, anode current, plate current, etcetera, are used to designate electrode currents for these specific electrodes. Unless otherwise stated, an electrode current is measured at the available terminal.
See:
average electrode current;
critical grid current;
direct-current electron-stream resistance;
electrode-current averaging time;
fault electrode current;
gas ratio;
inverse electrode current;
leakage current;
peak cathode current;
peak electrode current;
space current;
space-charge-limited current. E160-15E6

electrode current, average. *See:* **average electrode current.**

electrode-current averaging time (electron tubes). The time interval over which the current is averaged in defining the operating capabilities of the electrode. *See also:* **electrode current (electron tube); circuit characteristics of electrodes.**
42A70-0;E160-15E6

electrode dark current (1) (phototubes). The component of electrode current remaining when ionizing radiation and optical photons are absent. *Notes:* (1) Optical photons are photons with energies corresponding to wavelengths between 2000 and 15 00 angstroms. (2) Since the dark current may change considerably with temperature, the temperature should be specified. *See also:* **phototube.**
(2) (camera tubes). The current from an electrode in a photoelectric tube under stated conditions of radiation shielding. *See also:* **beam tube; camera tube.**
42A70/E160/E175-2E2/15E6

electrode dissipation. The power dissipated in the form of heat by an electrode as a result of electron or ion bombardment, or both, and radiation from other electrodes. *See:* **grid driving power; modes, degenerate.** 42A70-15E6

electrode drop (arc-welding apparatus). The voltage drop in the electrode due to its resistance (or impedance). *See also:* **electric arc-welding apparatus.**
87A1-0

electrode economizer. A collar that makes a seal between an electrode and the roof of a covered electric furnace with substantial exclusion of air, thereby preventing serious oxidation of the part of the electrode within the furnace. *See also:* **electrothermics.**
42A60-0

electrode impedance. The reciprocal of the electrode admittance. *See also:* **electron-tube admittances.**
E160/42A70-15E6

electrode impedance, biological. The ratio between two vectors, the numerator being the vector that represents the potential difference between the electrode and biological material, and the denominator being the vector that represents the current between the electrode and the biological material. *See:* **loss angle (biological); polarization capacitance (biological); polarization reactance (biological); polarization resistance (biological).** 42A80-18E1

electrodeposition. The process of depositing a substance upon an electrode by electrolysis or electrophoresis. Electrodeposition includes electroplating, electroforming, electrorefining, and electrowinning.
See:
electroanalysis;
electrodissolution;
pits;
reguline;
re-solution;
slime;
sponge;
trees and nodules. *See also:* **electrochemistry; electroforming; electrorefining; electrowinning.**
42A60-0

electrode potential. The difference in potential between an electrode and the immediately adjacent electrolyte, referred to an arbitrary zero of potential. *See also:* **electrolytic cell.** 42A60-0

electrode potential, biological. The potential between an electrode and biological material. *See also:* **medical electronics.** 0-18E1

electrode radiator (cooling fin) (electron tubes). A metallic piece, often of large area, extending the electrode to facilitate the dissipation of the heat generated in the electrode. *See also:* **electron tube.**
50I07-15E6

electrode reactance (electron device). The imaginary component of the electrode impedance. *See:* **self-impedance.** 0-15E6

electrode resistance (1) (general). The reciprocal of the electrode conductance. *Note:* This is the effective parallel resistance and is not the real component of the electrode impedance. E160/42A70-15E6
(2) (at a stated operating point) (electron device). The quotient of the direct electrode voltage by the direct electrode current. *See:* **self-impedance.**
0-15E6

electrodermal reaction (EDR) (medical electronics). The change in electric resistance of the skin during emotional stress. *See also:* **medical electronics.**
0-18E1

electrodermogram (electromyogram) (electroretinogram) (electrobiology). A graphic record of the variation with time of voltage taken from the given anatomical structure (skin, muscle, and retina, respectively). *See also:* **electrocardiogram.** 42A80-18E1

electrodesiccation (fulguration). The superficial destruction of tissue by electric sparks from a movable electrode. *See also:* **electrotherapy.** 0-18E1

electrode, signal (camera tube). *See:* **signal electrode.**

electrode susceptance (electron device). The imaginary component of the electrode admittance. *See:* **self-impedance.** 0-15E6

electrode voltage. The voltage between an electrode and the cathode or a specified point of a filamentary cathode. *Note:* The terms grid voltage, anode voltage, plate voltage, etcetera, are used to designate the voltage between these specific electrodes and the cathode. Unless otherwise stated, electrode voltages are understood to be measured at the available terminals.
See:
anode breakdown voltage;
anode voltage drop;
arc-drop voltage;
cutoff voltage;
direct grid bias;
electrode bias;
initial inverse voltage;
peak forward anode voltage;
peak inverse anode voltage;
starter voltage drop;
supply voltage;
tube voltage drop. 42A70-15E6

electrodiagnosis. The study of functional states of parts of the body either by studying their responses to electric stimulation or by studying the electric potentials (or currents) that they spontaneously produce.
42A80-18E1

electrodissolution. The process of dissolving a substance from an electrode by electrolysis. *See also:* **electrodeposition.** 42A60-0

electrodynamic instrument. An instrument that depends for its operation on the reaction between the current in one or more movable coils and the current in one or more fixed coils. *Note:* Electrodynamic instruments may or may not have iron cores or shields. They may or may not have control or restoring torque springs. An example of the latter is the crossed coil design used in ratio meters, power-factor meters, etcetera, when two sets of moving coils are set at an angle to each other to produce the necessary operating torque. *See also:* **instrument.** 42A30-0

electroencephalogram (medical electronics). A graphic record of the changes with time of the voltage obtained by means of electrodes applied to the skin over the cerebrum. *See also:* **medical electronics.** 0-18E1

electroextraction. The extraction by electrochemical processes of metals or compounds from ores and intermediate compounds. *See also:* **electrowinning.** 42A60-0

electroforming. The production or reproduction of articles by electrodeposition.
See:
case;
casting;
electrotyping;
finishing;
form;
master form;
matrix;
mold;
negative matrix;
oxidizing;
positive matrix;
shell;
silvering;
tinning. 42A60-0

electrographic recording (electrostatography). The branch of electrostatic electrography that employs a charge transfer between two or more electrodes to form directly electrostatic-charge patterns on an insulating medium for producing a viewable record. *See also:* **electrostatography.** E224-15E7

electrographitic brush (rotating machinery). A brush composed of selected amorphous carbon that, in the process of manufacture, is carried to a temperature high enough to convert the carbon to the graphitized form. *Note:* This type of brush is exceedingly versatile in that it can be made soft or very hard, also nonabrasive or slightly abrasive. Grades of brushes of this type have a high current-carrying capacity, but differ greatly in operating speed from low to high. *See:* **brush (rotating machinery).** 42A10-31E8;64A1-0

electrohydraulic elevator. A direct-plunger elevator where liquid is pumped under pressure directly into the cylinder by a pump driven by an electric motor. *See also:* **elevators.** 42A45-0

electrokinetic potential (zeta potential) (medical electronics). A set of four electric or velocity potentials that accompany relative motion between solids and liquids. *See also:* **medical electronics.** 0-18E1

electroluminescence. The emission of light from a phosphor excited by an electromagnetic field. *See also:* **lamp.** Z7A1-0

electroluminescent display panel. A thin, usually flat, electroluminescent display device. *See also:* **optoelectronic device.** E222-15E7

electroluminescent display device. An optoelectronic device with a multiplicity of electric ports, each capable of independently producing an optic output from an associated electroluminator element. *See also:* **optoelectronic device.** E222-15E7

electrolysis (1) (general). The production of chemical changes by the passage of current from an electrode to an electrolyte, or vice versa. *See also:* **electrolytic cell.** 42A60-0

(2) (underground structures). The destructive chemical action caused by stray or local electric currents to pipes, cables, and other metalwork. *See also:* **corrosion; power distribution, underground construction.** 42A35-31E13

electrolyte. A conducting medium in which the flow of electric current takes place by migration of ions. *Note:* Many physical chemists define electrolyte as a substance that when dissolved in a specified solvent, usually water, produces an ionically conducting solution. *See also:* **electrolytic cell.** E270/42A60-0

electrolyte cells. *See:* **cascade.**

electrolytic. *See:* **cathode.**

electrolytic capacitor. A combination of two capacitors at least one of which is a valve metal, separated by an electrolyte, and between which a dielectric film is formed adjacent to the surface of one or both of the capacitors. *See:* **asymmetrical cell; blocking capacitor; electrochemical valve; nonpolarized electrolytic capacitor; polarized electrolytic capacitor; unit-area capacitance.** 42A60-0

electrolytic cell. A cell in which electrochemical reactions are produced by applying electric energy, or conversely, that supplies electric energy as a result of electrochemical action. The latter cell may be called a galvanic cell. Each electrolytic cell comprises two or more electrodes and one or more electrolytes contained in a suitable vessel. The cell is a unit which may be used singly or electrically connected to other like units. *Note:* For an extensive list of cross references, see *Appendix A.* 42A60-0

electrolytic cleaning. The process of degreasing or descaling a metal by making it an electrode in a suitable bath. *See:* **corrosion terms.** CM-34E2

electrolytic dissociation. In a solution, a process whereby some fraction of the molecules of a solute is ionized to form ions. *See also:* **electrochemistry.** 42A60-0

electrolytic oxidation. An anodic process by which electrons are removed from or positive charges added to atoms or ions, or by which when applied to organic molecules, the oxygen or acidic proportion is increased. *See also:* **electrochemistry; electrolytic reduction.** 42A60-0

electrolytic parting (electrorefining). The electrolysis of a silver-gold alloy anode in which the silver is deposited in a pure form at the cathode, and the gold remains in the anode slime. *See also:* **electrorefining.** 42A60-0

electrolytic recording (facsimile). The type of electrochemical recording in which the chemical change is made possible by the presence of an electrolyte. *See also:* **recording (facsimile).** E168-0

electrolytic rectifier. A rectifier in which rectification of an alternating current is accompanied by electrolytic action. *See also:* **electrochemical valve; rectification.** 42A60-0

electrolytic reduction. A cathode process by which electrons are added to or positive charges removed from atoms or ions, or by which when applied to organic molecules, the hydrogen or basic proportion is increased. *See also:* **electrochemistry; electrolytic oxidation.** 42A60-0

electrolytic tank. A vessel containing a poorly conducting liquid, in which are inserted conductors that are scale models of an electrode system. *Note:* It is used to obtain potential diagrams. *See also:* **electron optics.** 50I07-15E6

electrolyzer. An electrolytic cell for the production of chemical products. 42A60-0

electromagnet. A device consisting of a ferromagnetic core and a coil, that produces appreciable magnetic effects only when an electric current exists in the coil. E270-0

electromagnetic braking (magnetic braking). A system of electric braking in which a mechanical braking system is actuated by electromagnetic means. *See also:* **electric braking.** 42A42-0

electromagnetic compatibility. The capability of electronic equipments or systems to be operated in the intended operational electromagnetic environment at designed levels of efficiency. *Note:* For an extensive list of cross references, see *Appendix A.* 0-27E1

electromagnetic delay line. A delay line whose operation is based on the time of propagation of electromagnetic waves through distributed or lumped capacitance and inductance. *See also:* **electronic digital computer.** X3A12-16E9

electromagnetic disturbance (electromagnetic compatibility). An electromagnetic phenomenon that may be superimposed on a wanted signal. *See also:* **electromagnetic compatibility.** 0-27E1

electromagnetic environment. The electromagnetic field or fields existing in a transmission medium. *See also:* **electromagnetic compatibility.** 0-27E1

electromagnetic induction. The production of an electromotive force in a circuit by a change in the magnetic flux linking with that circuit. CTD-21E1

electromagnetic interference (electromagnetic compatibility). Impairment of the reception of a wanted electromagnetic signal caused by an electromagnetic disturbance. *See also:* **electromagnetic compatibility.** 0-27E1

electromagnetic lens. An electronic lens in which the result is obtained by a magnetic field. *See also:* **electron optics.** 50I07-15E6

electromagnetic noise (electromagnetic compatibility). An electromagnetic disturbance that is not of a sinusoidal character. *See also:* **electromagnetic compatibility.** CISPR-27E1

electromagnetic oscillograph. An oscillograph in which the record is produced by means of a mechanical motion derived from electromagnetic forces. *See also:* **oscillograph.** 42A30-0

electromagnetic relay. An electromagnetically operated switch, ordinarily composed of one or more coils that control one or more armatures, each of which actuates electric contacts. *See:* **heat coil; neutral relay; polarized relay (polar relay); slow-operating relay; slow-release relay.** *See also:* **circuits and devices.** 42A65-0

electromagnetic wave. A wave characterized by variations of electric and magnetic fields. *Note:* Electromagnetic waves are known as radio waves, heat rays, light rays, etcetera, depending on the frequency. *See also:* **radio wave propagation; waveguide.** E146/E211-3E1/3E2

electromechanical bell. A bell having a prewound spring-driven mechanism, operation of which is initiated by actuation of an electric tripping mechanism. *See also:* **protective signaling.** 42A65-0

electromechanical device (control equipment). A device that is electrically operated and has mechanical motion such as relays, servos, etcetera. *See also:* **power system, low-frequency and surge testing.** E94-0

electromechanical interlocking machine. An interlocking machine designed for the control of both mechanically and power-operated functions. *See also:* **interlocking (interlocking plant).** 42A42-0

electromechanical recorder. An equipment for transforming electric signals into mechanical motion of approximately like form and inscribing such motion in an appropriate medium by cutting or embossing. *See also:* **electroacoustics.** 42A65-0

electromechanical recording (facsimile). Recording by means of a signal-actuated mechanical device. *See also:* **recording (facsimile).** E168-0

electromechanical relay. A relay that operates by physical movement of parts resulting from electromagnetic, electrostatic, or electrothermic forces created by the input quantities. 37A100-31E11/31E6

electromechanical transducer. A transducer for receiving waves from an electric system and delivering waves to a mechanical system, or vice versa. *See also:* **transducer.** E196/E270/42A65-0;E157-1E1

electrometallurgy. The branch of science and technology that deals with the application of electric energy to the extraction or treatment of metals.
See:
anode scrap;
circulation of electrolyte:
commercial tank;
crude metal;
dezincification;
electrolytic parting;
electrorefining;
foul electrolyte;
liberator tank;
multiple system;
nest or section;
purification of electrolyte;
regeneration of electrolyte;
series system;
starting sheet;
starting sheet blank;
stripper tank. 42A60-0

electrometer. An electrostatic instrument for measuring a potential difference or an electric charge by means of the mechanical forces exerted between electrically charged surfaces. *See also:* **instrument.** 42A30-0

electrometer tube. A vacuum tube having a very low control-electrode conductance to facilitate the measurement of extremely small direct current or voltage. *See also:* **tube definitions.** 42A70-15E6

electromotive force. *See:* **voltage.**

electromotive force series. A list of elements arranged according to their standard electrode potentials. CM-34E2

electromotive series (electrochemical series). A table that lists in order the standard potentials of specified electrochemical reactions. *See also:* **electrochemistry.** 42A60-0

electromyograph (medical electronics). An instrument for recording action potentials or physical movements of muscles. *See also:* **medical electronics.** 0-18E1

electron (1) (noun). An elementary particle containing the smallest negative electric charge. *Note:* The mass of the electron is approximately equal to 1/1837 of the mass of the hydrogen atom. 50I05-E
(2) (adjective). Operated by, containing, or producing electrons. *Examples:* Electron tube, electron emission, and electron gun. *See:* **electronic; electronics.** 42A70-0

electron accelerator, linear. *See:* **linear electron accelerator.**

electronarcosis. The production of transient insensibility by means of electric current applied to the cranium at intensities insufficient to cause generalized convulsions. *See also:* **electrotherapy.** 42A80-18E1

electron (ion) beam. A beam of electrons (ions) emitted from a single source and moving in neighboring paths that are confined to a desired region. *See:* **beam tube.** 50I07-15E6

electron-beam instrument (cathode-ray instrument). An instrument that depends for its operation on the deflection of a beam of electrons by an electric or magnetic field or by both. *Note:* The deflection is usually observed by causing the beam to strike a fluorescent screen. *See also:* **instrument.** 42A30-0

electron-beam magnetometer. A magnetometer that depends for its operation upon the change of the intensity or direction of an electron beam immersed in the field to be measured. *See also:* **magnetometer.** 42A30-0

electron-beam tube. An electron tube, the performance of which depends upon the formation and control of one or more electron beams. *See also:* **tube definitions.** E160/42A70-0

electron collector (microwave tube). The electrode that receives the electron beam at the end of its path. *Note:* The power of the beam is used to produce some desired effect before it reaches the collector. *See also:* **velocity-modulated tube.** 50I07-15E6

electron-coupled oscillator. An oscillator employing a multigrid tube with the cathode and two grids operating as an oscillator in any conventional manner, and in which the plate circuit load is coupled to the oscillator through the electron stream. *See also:* **oscillatory circuit.** E145/E182/42A65-0

electron device. A device in which conduction is principally by electrons moving through a vacuum, gas, or semiconductor.
See:
electron;
electron optics;
electron tube;
electronic;
electronics;
expansion orbit;
linear electron accelerator;
magnetic spectrograph;
mass spectrograph;
toroid;
transmit-receive switch;
trigatron;
tube. E160-15E6;E94/42A70-0

electron-device transducer. *See:* **admittance, short-circuit forward.**

electron emission. The liberation of electrons from an electrode into the surrounding space. *Notes:* (1) Quantitatively, it is the rate at which electrons are emitted from an electrode. (2) For an extensive list of cross references, see *Appendix A.* 42A70-15E6

electron gun (electron tubes). An electrode structure that produces and may control, focus, deflect, and converge one or more electron beams. *See also:* **beam tubes; electrode (electron tube.)** E160-15E6;42A70-0

electron-gun density multiplication (electron tubes). The ratio of the average current density at any specified aperture through which the stream passes to the average current density at the cathode surface. *See:* **beam tubes.** E160-15E6

electronic. Electronic signifies of, or pertaining to, devices, circuits, or systems utilizing electron devices. *Examples:* Electronic control, electronic equipment, electronic instrument, and electronic circuit. *See also:* **electron device; electronics.** 42A70-15E6

electronic analog computer. An automatic computing device that operates in terms of continuous variation of some physical quantities, such as electric voltages and currents, mechanical shaft rotations, or displacements, and that is used primarily to solve differential equations. *Notes:* (1) The equations governing the variation of the physical quantities have the same or very nearly the same form as the mathematical equations under investigation and therefore yield a solution analogous to the desired solution of the problem. Results are measured on meters, dials, oscillograph recorders, or oscilloscopes. (2) For an extensive list of cross references, see *Appendix A.* E165-0

electronic computation. *Note:* For an extensive list of cross references, see *Appendix A.*

electronic contactor (industrial control). A contactor whose function is performed by electron tubes. *See also:* **contactor.** 42A25-34E10

electronic controller (industrial control). An electric controller in which the major portion or all of the basic functions are performed by electron tubes. *Note:* For an extensive list of cross references, see *Appendix A.* 42A25-34E10

electronic digital computer. *Note:* For an extensive list of cross references, see *Appendix A.*

electronic direct-current motor controller (industrial control). A phase-controlled rectifying system using tubes of the vapor- or gas-filled variety for power conversion to supply the armature circuit or the armature and shunt-field circuits of a direct-current motor, to provide adjustable-speed, adjustable- and compensated-speed, or adjustable- and regulated-speed characteristics. *See* **electronic controller.** IC1-34E10

electronic direct-current motor drive (industrial control). The combination of an electronic direct-current motor controller with its associated motor or motors. *See:* **electronic controller.** IC1-34E10

electronic efficiency (electron tubes). The ratio of (1) the power at the desired frequency delivered by the electron stream to the circuit in an oscillator or amplifier to (2) the average power supplied to the stream. *See also:* **circuit characteristics of electrodes.** E160-15E6

electronic frequency changer. An electric power converter for changing the frequency of electric power. *See also:* **rectification.** 42A15-0

electronic instrument. An instrument that utilizes for its operation the action of one or more electron tubes. 42A30-0

electronic keying. A method of keying whereby the control is accomplished solely by electronic means. *See also:* **modulating systems; telegraphy.** E145-0

electronic line scanning (facsimile). The method of scanning that provides motion of the scanning spot along the scanning line by electronic means. *See also:* **scanning.** E168-2E2

electronic microphone. A microphone that depends for its operation on a change in the terminal electrical characteristic of an active device when a force is applied to some part of the device. *See also:* **microphone.** 0-1E1

electronic navigation. *See:* **navigation.**

electronic power converter. Electronic devices for transforming electric power. *See also:* **rectification.** 42A15-0

electronic raster scanning (facsimile). The method of scanning in which motion of the scanning spot in both dimensions is accomplished by electronic means. *See also:* **scanning.** E168-2E2

electronic rectifier (industrial control). A rectifier in which electron tubes are used as rectifying elements. *See also:* **electronic controller; rectification of an alternating current.** 42A25-34E10

electronics (1) (adjective). Of, or pertaining to, the field of electronics. *Examples:* Electronics engineer, electronics course, electronics laboratory, and electronics committee. *See:* **electron; electronic.**
(2) (noun). That field of science and engineering that deals with electron devices and their utilization. *See also:* **electron device.** 42A70-15E6

electronic scanning (inertialess scanning) (antenna). Scanning an antenna beam by electronic or electric means without moving parts. *See also:* **radiation.** 0-3E1

electronic transformer. Any transformer intended for use in a circuit or system utilizing electron or solid-state devices. *Note:* Mercury-arc rectifier transformers and luminous-tube transformers are normally excluded from this classification. 0-21E1

electronic trigger circuit (industrial control). A network containing electron tubes in which the output changes abruptly with an infinitesimal change in input at one or more points in the operating range. 42A25-34E10

electronic voltmeter (vacuum-tube voltmeter). A voltmeter that utilizes the rectifying and amplifying properties of electron tubes and their associated circuits to secure desired characteristics, such as high input impedance, wide frequency range, crest indications, etcetera. *See also:* **instrument.** 42A30-0

electron injector. The electron gun of a betatron. 0-15E6

electron lens. A device for the purpose of focusing an electron (ion) beam. *See also:* **electron optics.** 50I07-15E6

electron microscope. An electron-optical device that produces a magnified image of an object. *Note:* Detail may be revealed by virtue of selective transmission, reflection, or emission of electrons by the object. *See also:* **electron devices, miscellaneous.** 42A70-15E6

electron mirror. An electronic device causing the total reflection of an electron beam. *See also:* **electron optics.** 50I07-15E6

electron multiplier. A structure, within an electron tube, that employs secondary electron emission from solids to produce current amplification. *See:* **amplifier; electron emission.** E160-15E6

electron multiplier section (electron tube). A section of an electron tube in which an electron current is amplified by one or more successive dynode stages. *See also:* **electrode (electron tube).** 42A70-0

electron optics. The branch of electronics that deals with the operation of certain electronic devices, based on the analogy between the path of electron (ion) beams in magnetic or electric fields and that of light rays in refractive media.
See:
electrolytic tank;
electromagnetic lens;
electron device;
electron lens;
electron microscope;
electron mirror;
electron telescope;
electrostatic electron microscope;
electrostatic lens;
hodoscope;
ion gun;
magnetic electron microscope;
potential diagram;
proton microscope. 50I07-15E6

electron-ray indicator tube. An elementary form of cathode-ray tube used to indicate a change of voltage. *Note:* Such a tube used to indicate the tuning of a circuit is sometimes called a **magic eye.** *See also:* **cathode-ray tubes.** 50I07-15E6

electrons, conduction (semiconductor). The electrons in the conduction band of a solid that are free to move under the influence of an electric field. *See also:* **semiconductor.** E102/E270/10E1

electron sheath (gas) (ion sheath). A film of electrons (or of ions) that has formed on or near a surface that is held at a potential different from that of the discharge. *See also:* **discharge (in a gas).** 50I07-15E6

electron-stream potential (electron tubes) (any point in an electron stream). The time average of the potential difference between that point and the electron-emitting surface. *See:* **electron emission.** *See also:* **circuit characteristics of electrodes.** E160-15E6

electron-stream transmission efficiency (electron tubes) (electrode through which the electron stream passes). The ratio of (1) the average stream current through the electrode to (2) the average stream current approaching the electrode. *Note:* In connection with multitransit tubes, the term electron stream should be taken to include only electrons approaching the electrode for the first time. *See:* **electron emission.** *See also:* **circuit characteristics of electrodes.** E160-15E6

electron telescope. An optical instrument for astronomy including an electronic image transformer associated with an optical telescope. *See also:* **electron optics.** 50I07-15E6

electron tube. An electron device in which conduction by electrons takes place through a vacuum or gaseous medium within a gastight envelope. *Note:* The envelope may be either pumped during operation or sealed off. *Note:* For an extensive list of cross references, see *Appendix A.* 42A70-0

electron-tube admittances. The cross-referenced terms generalize the familiar electron-tube coefficients so that they apply to all types of electron devices operated at any frequency as linear transducers. *Note:* The generalizations include the familiar low-frequency tube concepts. In the case of a triode, for example, at relatively low frequencies the short-circuit input admittance reduces to substantially the grid admittance, the short-circuit output admittance reduces to substantially the plate admittance, the short-circuit forward admittance reduces to substantially the grid-plate transconductance, and the short-circuit feedback admittance reduces to substantially the admittance of the grid-plate capacitance. When reference is made to alternating-voltage or -current components, the components are understood to be small enough so that linear relations hold between the various alternating voltages and currents. Consider a generalized network or transducer having n available terminals to each of which is flowing a complex alternating component I_j of the current and between each of which and a reference point (which may or may not be one of the n network terminals) is applied a complex alternating voltage V_j. This network represents an n-terminal electron device in which each one of the terminals is connected to an electrode. The cross-referenced terms found in *Appendix A* refer to this network. 42A70-0

electron-tube amplifier. An amplifier that obtains its amplifying properties by means of electron tubes. 42A25-34E10

electron-wave tube. An electron tube in which mutually interacting streams of electrons having different velocities cause a signal modulation to change progressively along their length. *See:* **beam tubes.** *See also:* **tube definition.** E160-15E6

electrooptical effect (dielectrics) (Kerr electrostatic effect). Certain transparent dielectrics when placed in an electric field become doubly refracting. *Note:* The strength of the electrooptical effect for unit thickness of the dielectric varies directly as the square of the electric field strength. E270-0

electroosmosis. The movement of fluids through diaphragms that is as a result of the application of an electric current. 42A60-0

electroosmotic potential (electrobiology). The electrokinetic potential gradient producing unit velocity of liquid flow through a porous structure. *See also:* **electrobiology.** 42A80-18E1

electrophonic effect. The sensation of hearing produced when an alternating current of suitable frequency and magnitude from an external source is passed through an animal. *See also:* **electroacoustics.** E157-1E1

electrophoresis. A movement of colloidal ions as a result of the application of an electric potential. *See also:* **ion.** 42A60-0

electrophoretic potential (electrobiology). The electrokinetic potential gradient required to produce unit velocity of a colloidal or suspended material through a liquid electrolyte. *See also:* **electrobiology.** 42A80-18E1

electroplating. The electrodeposition of an adherent coating upon an object for such purposes as surface protection or decoration. *Note:* For an extensive list of cross references, see *Appendix A.* 42A60-0

electropneumatic brake. An air brake that is provided with electrically controlled valves for control of the application and release of the brakes. *Note:* The electric control is usually in addition to a complete air brake equipment to provide a more prompt and synchronized operation of the brakes on two or more vehicles. *See also:* **electric braking.** 42A42-0

electropneumatic contactor (1) (industrial control). A contactor actuated by air pressure. *See also:* **contactor.** 42A25-34E10
(2) (electropneumatic unit switch). A contactor or switch controlled electrically and actuated by air pressure. *See also:* **contactor; control switch.** E16-0

electropneumatic controller. An electrically supervised controller having some or all of its basic functions performed by air pressure. *See:* **electric controller.** *See also:* **multiple-unit control.** 42A25-34E10;E160-0

electropneumatic interlocking machine. An interlocking machine designed for electric control of electropneumatically operated functions. *See also:* **centralized traffic control system.** 42A42-0

electropneumatic valve. An electrically operated valve that controls the passage of air. *See also:* **railway signal and interlocking.** 42A42-0

electropolishing (electroplating). The smoothing or brightening of a metal surface by making it anodic in an appropriate solution. *See also:* **electroplating.** 42A60-0

electrorefining. The process of electrodissolving a metal from an impure anode and depositing it in a more pure state.
See:
anode scrap;
anode slime;
circulation of electrolyte;
commercial tank;
crude metal;
electrolytic parting;
foul electrolyte;
liberator tank;
multiple system;
nest or section;
purification of electrolyte;
regeneration of electrolyte;
series system;
starting sheet;
starting sheet blank;
stripper tank. 42A60-0

electroretinogram. *See:* **electrodermogram.**

electroscope. An electrostatic device for indicating a potential difference or an electric charge. *See also:* **instrument.** 42A30-0

electroshock therapy. The production of a reaction in the central nervous system by means of electric current applied to the cranium. *See also:* **electrotherapy.** 42A80-18E1

electrostatic actuator. An apparatus constituting an auxiliary external electrode that permits the application of known electrostatic forces to the diaphragm of a microphone for the purpose of obtaining a primary calibration. *See also:* **microphone.** E157-1E1

electrostatic coupling (interference terminology). *See:* **coupling, capacitive.** *See also:* **signal.**

electrostatic deflection (cathode-ray tube). Deflecting an electron beam by the action of an electric field. *See also:* **cathode-ray tubes.** 50I07-15E6

electrostatic electrography. The branch of electrostatography that employs an insulating medium to form, without the aid of electromagnetic radiation, latent electrostatic-charge patterns for producing a viewable record. *See also:* **electrostatography.** E224-15E7

electrostatic electron microscope. An electron microscope with electrostatic lenses. *See also:* **electron optics.** 50I07-15E6

electrostatic electrophotography. The branch of electrostatography that employs a photoresponsive medium to form, with the aid of electromagnetic radiation, latent electrostatic-charge patterns for producing a viewable record. *See also:* **electrostatography.** E224-15E7

electrostatic focusing (electron beam). *See:* **focusing, electrostatic.**

electrostatic instrument. An instrument that depends for its operation on the forces of attraction and repulsion between bodies charged with electricity. *See also:* **instrument.** 42A30-0

electrostatic lens. An electron lens in which the result is obtained by an electrostatic field. *See also:* **electron optics.** 50I07-15E6

electrostatic loudspeaker (capacitor loudspeaker) (condenser loudspeaker*). A loudspeaker in which the mechanical forces are produced by the action of electrostatic fields. *See also:* **loudspeaker.**
*Deprecated 42A65/E157-1E1

electrostatic microphone (capacitor microphone) (condenser microphone*). A microphone that depends for its operation upon variations of its electrostatic capacitance. *See also:* **microphone.**
*Deprecated 42A65/E157-1E1

electrostatic potential (any point). The potential difference between that point and an agreed-upon reference point, usually the point at infinity. E270-0

electrostatic potential difference (between two points). The scalar-product line integral of the electric field strength along any path from one point to the other in an electric field resulting from a static distribution of electric charge. E270-0

electrostatic recording (facsimile). Recording by means of a signal-controlled electric field. *See also:* **recording (facsimile).** E168-0

electrostatic relay. A relay in which operation depends upon the application or removal of electrostatic charge. 0-21E0

electrostatics. The branch of science that treats of the electric phenomena associated with electric charges at rest in the frame of reference. E270-0

electrostatic storage. A storage device that stores data as electrostatically charged areas on a dielectric surface. X3A12-16E9

electrostatic voltmeter. A voltmeter depending for its action upon electric forces. An electrostatic voltmeter is provided with a scale, usually graduated in volts or kilovolts. *See also:* **instrument.** 42A30-0

electrostatography. The formation and utilization of latent electrostatic-charge patterns for the purpose of recording and reproducing patterns in viewable form.
See:
aerosol development;
carrier;
cascade development;
charging;
corona charging;
developer;
development;
electrographic recording;
electrostatic electrography;
electrostatic electrophotography;
fixing;
liquid development;
magnetic-brush development;
sensitizing;
toner;
transfer;
xerography;
xeroprinting;
xeroradiography. *See also:* **electron devices, miscellaneous.** E224-15E7

electrostenolysis. The discharge of ions or colloidal ions in capillaries through the application of an electric potential. *See also:* **ion.** 42A60-0

electrostriction (electric field). The variation of the dimensions of a dielectric under the influence of an electric field. E270-0

electrotaxis (electrobiology) (galvanotaxis). The act of a living organism in arranging itself in a medium in such a way that its axis bears a certain relation to the direction of the electric current in the medium. *See also:* **electrobiology.** 42A80-18E1

electrotherapy. The use of electric energy in the treatment of disease.
See:
accumulating stimulus;
autocondensation*;
autoconduction*;
coagulating current*;
electrocautery;
electronarcosis;
electroshock therapy;
Faradic current;
Faradization;
Galvanism;
ion transfer;
Leduc current*;
microwave therapy;
Morton wave current*;
phosphene;
spike train;
static induced current*;
static wave current*;
Tesla current*;
tetanizing current.
*Deprecated 42A80-18E1

electrothermal efficiency. The ratio of energy usefully employed in a furnace to the total energy supplied. *See also:* **electrothermics.** 42A60-0

electrothermal recording (facsimile). The type of recording that is produced principally by signal-controlled thermal action. *See also:* **recording (facsimile)** E168-0

electrothermic instrument. An instrument that depends for its operation on the heating effect of a current or currents. *Note:* Among the several possible types are (1) the expansion type, including the hot-wire and hot-strip instruments; (2) the thermocouple type; and (3) the bolometric type. *See also:* **instrument.** 42A30-0

electrothermics. The branch of science and technology that deals with the direct transformations of electric energy and heat.

See:
arc furnaces;
charge-resistance furnace;
continuous electrode;
direct-arc furnace;
electrode economizer;
electrothermal efficiency;
high-frequency furnace;
indirect-arc furnace;
induction furnace;
low-frequency furnace;
noncontinuous electrode;
pinch effect;
pyroconductivity;
refractory;
resistance furnace;
resistor furnace;
smothered-arc furnace. 42A60-0

electrotonic wave (electrobiology). A brief nonpropagated change of potential on an excitable membrane in the vicinity of an applied stimulus; it is often accompanied by a propagated response and always by electrotonus. *See also:* **excitability (electrobiology).** 42A80-18E1

electrotonus (1) (physical). The change in distribution of membrane potentials in nerve and muscle during or after the passage of an electric current. *See also:* **excitability (electrobiology).** 42A80-18E1
(2) (physiological). The change in the excitability of a nerve or muscle during the passage of an electric current. *See also:* **excitability (electrobiology).** 42A80-18E1

electrotyping. The production or reproduction of printing plates by electroforming. *See also:* **electroforming.** 42A60-0

electrowinning. The electrodeposition of metals or compounds from solutions derived from ores or other materials using insoluble anodes.
See:
agitator;
cascade;
concentrate;
electroextraction;
extraction liquor;
gas grooves;
hydrometallurgy;
regenerated leach liquor;
stripping compound;
tailing;
thickener. 42A60-0

element (1) (electrical engineering). Any electric device (such as inductor, resistor, capacitor, generator, line, electron tube) with terminals at which it may be directly connected to other electric devices. *See also:* **network analysis.** E153/E270-0
(2) (electron tubes). A constituent part of the tube that contributes directly to its electrical operation.
See:
baffle;
electrode;
electron multiplier section;
getter;
heater;
phosphor;
screen;
storage element;
target;
vacuum envelope;
(3) (semiconductor device). Any integral part that contributes to its operation. *See also:* **semiconductor device.** E102/E216/E270-0;42A70-34E17
(4) (integrated circuit). A constituent part of the integrated circuit that contributes directly to its operation. *See also:* **integrated circuit.** E274-15E7/21E0
(5) (computing systems). *See:* **combinational logic element; logic element; sequential logic element; threshold element.** *See also:* **electronic digital computer.**
(6) (storage cell). Consists of the positive and negative groups with separators, or separators and retainers, assembled for one cell. *See also:* **battery (primary or secondary).** 42A60-0

elemental area (facsimile). Any segment of a scanning line of the subject copy the dimension of which along the line is exactly equal to the nominal line width. *Note:* Elemental area is not necessarily the same as the scanning spot. *See also:* **scanning.** E168-0

elementary diagram. *See:* **schematic diagram.** E270-0

elements, feedback (control system). The elements in the controlling system that change the feedback signal in response to the directly controlled variable. See Figure B. *See:* **control system, feedback.** 85A1-23E0;AS1-34E10

elements, forward (automatic control). Those elements situated between the actuating signal and the controlled variable in the closed loop being considered. See Figure B. *See also:* **control system, feedback.** 85A1-23E0

elements, loop (control system, feedback) (a closed loop). All elements in the signal path that begins with the loop error signal and ends with the loop return signal. See Figures A and B. *See also:* **control system, feedback.** 85A1-23E0

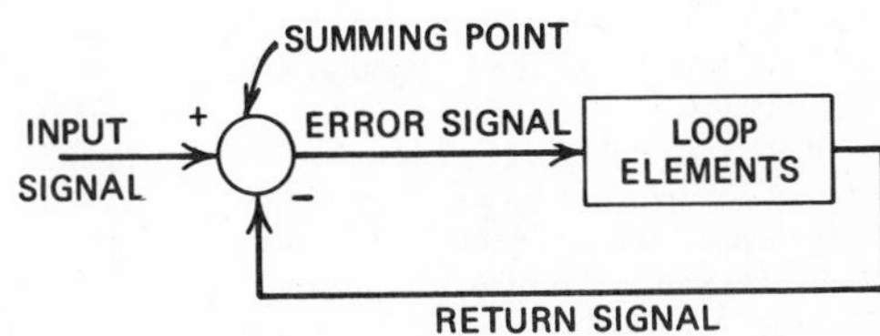

Figure A. Block diagram of a closed loop.

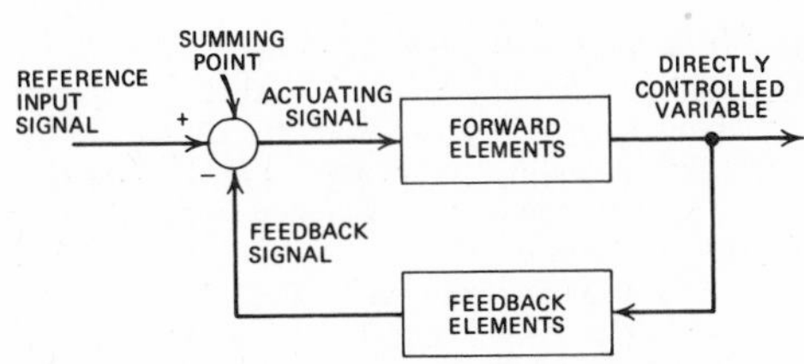

Figure B. Simplified block diagram indicating essential elements of an automatic control system.

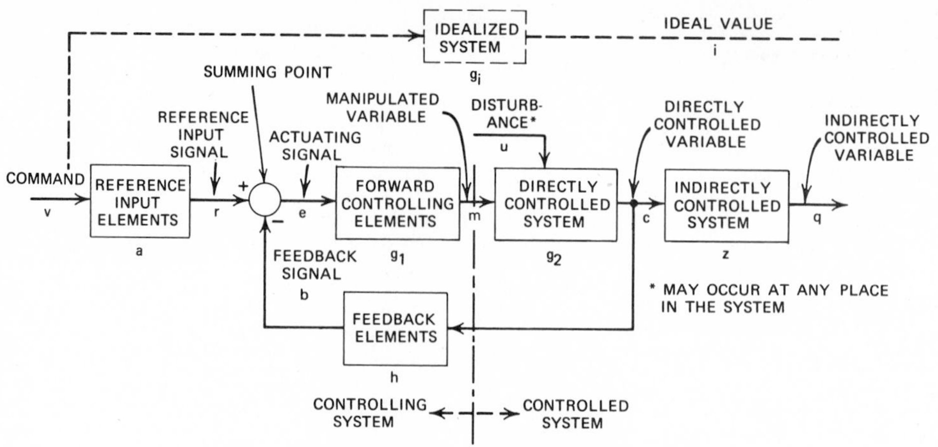

Figure C. Block diagram of an automatic control system illustrating expansion of the simplified block diagram to a more complex system.

elements of a fix (electronic navigation). The specific values of the navigation coordinates necessary to define a position. *See also:* **navigation.** E172-10E6

elements, reference-input. The portion of the controlling system that changes the reference input signal in response to the command. See Figure C. *See:* **control system, feedback.** AS1-34E10;85A1-23E0

elevated-zero range. A range where the zero value of the measured variable, measured signal, etcetera, is greater than the lower range value. *Note:* The zero may be between the lower and upper range values, at the upper range value, or above the upper range value. For example: (a) −20 to 100, (b) −40 to 0, and (c) −50 to −10. *See also:* **instrument.** 39A4-0

elevation (illuminating engineering). The angle between the axis of a searchlight drum and the horizontal. *Note:* For angles above the horizontal, elevation is positive, and below the horizontal negative. *See also:* **searchlight.** Z7A1-0

elevation rod (lightning protection). The vertical portion of conductor in an air terminal by means of which it is elevated above the object to be protected. *See also:* **lightning protection and equipment.** 42A95-0

elevator. A hoisting and lowering mechanism equipped with a car or platform that moves in guides in a substantially vertical direction, and that serves two or more floors of a building or structure. *Note:* For an extensive list of cross references, see *Appendix A.* 42A45-0

elevator automatic dispatching device. A device, the principal function of which is to either: (1) operate a signal in the car to indicate when the car should leave a designated landing; or (2) actuate its starting mechanism when the car is at a designated landing. *See also:* **control (elevators).** 42A45-0

elevator automatic signal transfer device. A device by means of which a signal registered in a car is automatically transferred to the next car following, in case the first car passes a floor, for which a signal has been registered, without making a stop. *See also:* **control.** 42A45-0

elevator car. The load-carrying unit including its platform, car frame, enclosure, and car door or gate. *See also:* **elevator.** 42A45-0

elevator car bottom runby (elevator car). The distance between the car buffer striker plate and the striking surface of the car buffer when the car floor is level with the bottom terminal landing. *See also:* **elevator.** 42A45-0

elevator-car flash signal device. One providing a signal light, in the car, that is illuminated when the car approaches the landings at which a landing-signal-registering device has been actuated. *See:* **control.** 42A45-0

elevator car-leveling device. Any mechanism that will, either automatically or under the control of the operator, move the car within the leveling zone toward the landing only, and automatically stop it at the landing. *Notes:* (1) Where controlled by the operator by means of up-and-down continuous-pressure switches in the car, this device is known as an **inching device.** (2) Where used with a hydraulic elevator to correct automatically a change in car level caused by leakage in the hydraulic system, this device is known as an **anticreep device.** *See:* **elevator; leveling zone (elevators); one-way automatic leveling device; (elevators); two-way automatic maintaining leveling device (elevators); two-way automatic nonmaintaining leveling device (elevators).** 42A45-0

elevator counterweight bottom runby. The distance between the counterweight buffer striker plate and the striking surface of the counterweight buffer when the car floor is level with the top terminal landing. *See also:* **elevator.** 42A45-0

elevator landing. That portion of a floor, balcony, or platform used to receive and discharge passengers or freight. *See also:* **bottom terminal landing (elevators); elevators; landing zone; top terminal landing (elevators).** 42A45-0

elevator-landing signal registering device. A button or other device, located at the elevator landing that, when actuated by a waiting passenger, causes a stop signal to be registered in the car. *See:* **control.** 42A45-0

elevator-landing stopping device. A button or other device, located at an elevator landing that, when actuated, causes the elevator to stop at that floor. *See:* **control.** 42A45-0

elevator parking device. An electric or mechanical device, the function of which is to permit the opening, from the landing side, of the hoistway door at any landing when the car is within the landing zone of that landing. The device may also be used to close the door. *See:* **control.** 42A45-0

elevator pit. That portion of a hoistway extending from the threshold level of the lowest landing door to the floor at the bottom of the hoistway. *See also:* **elevators.** 42A45-0

elevator separate-signal system. A system consisting of buttons or other devices located at the landings that, when actuated by a waiting passenger, illuminate a flash signal or operate an annunciator in the car, indicating floors at which stops are to be made. *See also:* **control (elevators).** 42A45-0

elevator signal-transfer switch. A manually operated switch, located in the car, by means of which the operator can transfer a signal to the next car approaching in the same direction, when he desires to pass a floor at which a signal has been registered in the car. *See:* **control.** 42A45-0

elevator starter's control panel. An assembly of devices by means of which the starter may control the manner in which an elevator, or group of elevators, functions. *See:* **control.** 42A45-0

elevator truck zone. The limited distance above an elevator landing within which the truck-zoning device permits movement of the elevator car. *See:* **control.** 42A45-0

elevator truck-zoning device. A device that will permit the operator in the car to move a freight elevator, within the truck zone, with the car door or gate and a hoistway door open. *See:* **control.** 42A45-0

elliptically polarized wave (1) (general). A transverse wave in which the displacement vector at any point describes an ellipse. E270-0
(2) (radio wave propagation) (given frequency). An electromagnetic wave for which the component of the electric vector in a plane normal to the direction of propagation describes an ellipse. *See also:* **electromagnetic wave; radiation; radio transmitter; waveguide.** E146-3E1

ellipticity. *See note under* **axial ratio.** *See also:* **waveguide.**

embedded temperature detector (rotating machinery). An element, usually a resistance thermometer or thermocouple, built into apparatus for the purpose of measuring temperature. *Note:* This is ordinarily installed in a stator slot between coil sides at a location at which the highest temperature is anticipated. 0-31E8

embossing stylus. A recording stylus with a rounded tip that displaces the material in the recording medium to form a groove. *See also:* **phonograph pickup.** 0-1E1

embrittlement. Severe loss of ductility of a metal or alloy. *See:* **corrosion terms.** CM-34E2

emergency announcing system. A system of microphones, amplifier, and loud speakers (similar to a public address system) to permit instructions and orders from a ship's officers to passengers and crew in an emergency and particularly during abandon-ship operations. 42A43-0

emergency cells (storage cell). End cells that are held available for use exclusively during emergency discharges. *See also:* **battery (primary or secondary).** 42A60-0

emergency electric system (marine). All electric apparatus and circuits the operation of which, independent of ship's service supply, may be required under casualty conditions for preservation of a ship or personel.
See also:
emergency generator;
emergency lighting storage battery;
emergency power feedback;
emergency switchboard;
final emergency circuits;
final emergency lighting;
temporary emergency circuits;
temporary emergency lighting. 42A43-0

emergency generator (marine). An internal-combustion-engine-driven generator so located in the upper part of a vessel as to permit operation as long as the ship can remain afloat, and capable of operation, independent of any other apparatus on the ship, for supply of power to the emergency electric system upon failure of a ship's service power.*See also:* **emergency electric system.** 42A43-0

emergency lighting. Lighting designed to supply illumination essential to the safety of life and property in the event of failure of the normal supply.*See also:* **general lighting.** Z7A1-0

emergency lighting storage battery (marine transportation). A storage battery for instant supply of emergency power, upon failure of a ship's service supply, to certain circuits of special urgency (principally temporary emergency lighting). *See also:* **emergency electric system.** 42A43-0

emergency power feedback. An arrangement permitting feedback of emergency-generator power to a ship's service system for supply of any apparatus on the ship within the limit of the emergency-generator rating. *See also:* **emergency electric system.** 42A43-0

emergency service. An additional service intended only for use under emergency conditions. *See:* **dual service; duplicate service; loop service.** *See also:* **service.** 42A35-31E13

emergency stop switch (elevators). A device located in the car that, when manually operated, causes the electric power to be removed from the driving-machine motor and brake of an electric elevator or from the electrically operated valves and/or pump motor of a hydraulic elevator. *See:* **control.** 42A45-0

emergency switchboard. A switchboard for control of sources of emergency power and for distribution to all emergency circuits. *See also:* **emergency electric system.** 42A43-0

emergency-terminal stopping device (elevators). A device that automatically causes the power to be removed from an electric elevator driving-machine motor and brake, or from a hydraulic elevator machine, at a predetermined distance from the terminal landing, and independently of the functioning of the operating device and the normal-terminal stopping device, if the normal-termimal stopping device does not slow down the car as intended. *See:* **control.** 42A45-0

emergency (maximum) transfer capability (transmission) (electric power supply). The maximum amount of power that can be interchanged without causing additional transmission outages following a loss of transmission or generation capacity. *See also:* **generating station.** 0-31E4

EMF. *See:* **voltage (electromotive force).**

emission characteristic. A relation, usually shown by a graph, between the emission and a factor controlling the emission (such as temperature, voltage, or current of the filament or heater). *See also:* **electron emission.** 42A70-15E6

emission current. The current resulting from electron emission. 50I07-15E6

emission current, field-free (cathode). *See:* **field-free emission current (cathode).**

emission efficiency (thermionics). The quotient of the saturation current by the heating power absorbed by the cathode. *See also:* **electron emission.** 50I07-15E6

emissivity (photovoltaic power system). The emittance of a specimen of material with an optically smooth, clean surface and sufficient thickness to be opaque. *See also:* **photovoltaic power system; solar cells (photovoltaic power system).** 0-10E1

emittance (photovoltaic power system). The ratio of the radiant flux-intensity from a given body to that of a black body at the same temperature. *See also:* **photovoltaic power system; solar cells (photovoltaic power system).** 0-10E1

emitter (transistor). A region from which charge carriers that are minority carriers in the base are injected into the base. E270/E216-34E17

emitter, majority (transistor). An electrode from which a flow of majority carriers enters the interelectrode region. *See also:* **transistor.** E102/42A70-0

emitter, minority (transistor). An electrode from which a flow of minority carriers enters the interelectrode region. *See also:* **semiconductor; transistor.** E102/42A70-0

emitting sole (microwave tubes). An electron source in crossed-field amplifiers that is extensive and parallel to the slow-wave circuit and that may be a hot or cold electron-emitter. *See also:* **microwave tube or valve.** 0-15E6

empirical propagation model (electromagnetic compatibility). A propagation model that is based solely on measured path-loss data. *See also:* **electromagnetic compatibility.** 0-27E1

enabling pulse (1) (general). A pulse that prepares a circuit for some subsequent action. E172-10E6

(2). A pulse that opens an electric gate normally closed, or otherwise permits an operation for which it is a necessary but not a sufficient condition. *See also:* **pulse.** 42A65-0

enabling signal. A logic (binary) signal, one of whose states permits a number of other logic (binary) events to occur. *See also:* **digital.** 0-31E3

enamel (1) (general). A paint that is characterized by an ability to form an especially smooth film. AD16-31E8

(2) (wire) (rotating machinery). A smooth film applied to wire usually by a coating process. *See also:* **rotor (rotating machinery); stator.** 0-31E8

encapsulated (rotating machinery). A machine in which one or more of the windings is completely encased by molded insulation. *See also:* **asynchronous machine; direct-current commutating machine; synchronous machine.** 0-31E8

enclosed (inclosed). Surrounded by a case that will prevent a person from accidentally contacting live parts. 42A95-0

enclosed brake (industrial control). A brake that is provided with an enclosure that covers the entire brake, including the brake actuator, the brake shoes, and the brake wheel. *See:* **electric drive.** IC1-34E10

enclosed cutout. A cutout in which the fuse clips and fuseholder or disconnecting blade are mounted completely within an insulating enclosure. 37A100-31E11

enclosed relay. A relay that has both coil and contacts protected from the surrounding medium. *See also:* **relay.** 83A16-0

enclosed self-ventilated machine. A machine having openings for the admission and discharge of the ventilating air that is circulated by means integral with the machine, the machine being otherwise totally enclosed. *Note:* These openings are so arranged that inlet and outlet ducts or pipes may be connected to them. *Note:* Such ducts or pipes, if used, must have ample section and be so arranged as to furnish the specified volume of air to the machine, otherwise the ventilation will not be sufficient. E45-0

enclosed separately ventilated machine. A machine having openings for the admission and discharge of the ventilating air that is circulated by means external to and not a part of the machine, the machine being otherwise totally enclosed. *Note:* These openings are so arranged that inlet and outlet duct pipes may be connected to them. E45-0

enclosed switch (industrial control) (safety switch). A switch either with or without fuse holders, meter-testing equipment, or accommodation for meters, having all current-carrying parts completely enclosed in metal, and operable without opening the enclosure. *See also:* **switch.** 42A25-34E10

enclosed switchboard. A dead-front switchboard that has an overall sheet metal enclosure (not grille) covering back and ends of the entire assembly. *Note:* Access to the enclosure is usually provided by doors or removable covers. The top may or may not be covered. 37A100-31E11

enclosed switchgear assembly. A switchgear assembly that is enclosed on all sides and top. 37A100-31E11

enclosed ventilated (rotating machinery). A term applied to an apparatus with a substantially complete enclosure in which openings are provided for ventilation only. *See also:* **asynchronous machine; direct-current commutating machine; synchronous machine.** 0-31E8

enclosed ventilated apparatus. Apparatus totally enclosed except that openings are provided for the admission and discharge of the cooling air. *Note:* These openings may be so arranged that inlet and outlet ducts or pipes may be connected to them. An enclosed ventilated apparatus or machine may be separately ventilated or self-ventilated. 42A95-0

enclosure. A surrounding case or housing used to protect the contained equipment and prevent personnel from accidentally contacting live parts. *Note:* Material and finish shall conform to the standards for the switchgear enclosed. 37A100-31E11

encode (1) (general). To express a single character or a message in terms of a code. E162-0

(2) (electronic control). To produce a unique combination of a group of output signals in response to each of a group of input signals. E162-0

(3) (computing systems). To apply the rules of a code. *See:* **code.** *See also:* **decode; electronic digital computer; matrix; translate.** X3A12-16E9

encoder (electronic computation). A network or system in which only one input is excited at a time and each input produces a combination of outputs. *Note:* Sometimes called matrix. E270-0

end-around carry (computing systems). A carry generated in the most significant place and forwarded directly to the least significant place, for example, when adding two negative numbers, using nines complement. *See:* **carry.** *See also:* **electronic digital computer.** E162-0;X3A12-16E9

end bell. *See:* **cable terminal.**

end bracket (rotating machinery). A beam or bracket attached to the frame of a machine and intended for supporting a bearing. *See also:* **cradle base (rotating machinery).** 0-31E8

end cells (storage battery) (storage cell). Cells that may be cut in or cut out of the circuit for the purpose of adjusting the battery voltage. *See also:* **battery (primary or secondary).** 42A60-0

end device (telemeter). The final system element that responds quantitatively to the measurand through the translating means and performs the final measurement operation. *Note:* An end device performs the final conversion of measurement energy to an indication, record, or the initiation of control. *See:* **instrument; measurement system.** 37A100-31E11;39A4/42A30-0

end distortion (start-stop teletypewriter signals). The shifting of the end of all marking pulses from their proper positions in relation to the beginning of the start pulse. *See also:* **telegraphy.** 42A65-0

end finger (outside space-block) (rotating machinery). A radially extending finger piece at the end of a laminated core to transfer pressure from an end clamping plate or flange to a tooth. *See also:* **rotor (rotating machinery); stator.** 0-31E8

end-fire array. A linear or planar array antenna whose direction of maximum radiation lies along the line or the plane of the array. *See also:* **antenna.** 0-3E1

end injection (Charles or Kino gun (EI)) (microwave tubes). A gun used in the presence of crossed electric and magnetic fields to inject an electron beam into the end of a slow-wave structure. *See also:* **microwave tube.** 0-15E6

end-of-block signal (numerically controlled machines). A symbol or indicator that defines the end of one block of data. *See also:* **numerically controlled machines.** EIA3B-34E12

end-of-copy signal (facsimile). A signal indicating termination of the transmission of a complete subject copy. *See also:* **facsimile signal (picture signal).** E168-0

end office (telephone networks). The central office where customers' telephones are terminated for purposes of interconnection to each other. End offices are classified as Class 5 offices. *See:* **telephone system.** 0-19E1

end of program (numerically controlled machines). A miscellaneous function indicating completion of workpiece. *Note:* Stops spindle, coolant, and feed after completion of all commands in the block. Used to reset control and/or machine. Resetting control may include rewinding of tape or progressing a loop tape through the splicing leader. The choice for a particular case must be defined in the format classification sheet. *See also:* **numerically controlled machines.** EIA3B-34E12

end of tape (numerically controlled machines). A miscellaneous function that stops spindle, coolant, and feed after completion of all commands in the block. *Note:* Used to reset control and/or machine. Resetting control will include rewinding of tape, progressing a loop tape through the splicing leader, or transferring to a second tape reader. The choice for a particular case must be defined in the format classification sheet. *See also:* **numerically controlled machines.** EIA3B-34E12

end-on armature relay. *See:* **armature, end-on.** *See also:* **relay.**

end-play washers (rotating machinery). Washers of various thicknesses and materials used to control axial position of the shaft. 42A10-31E8

end rail (rotating machinery). A rail on which a bearing pedestal can be mounted. *See also:* **bearing.** 0-31E8

end ring, rotor (rotating machinery). The conducting structure of a squirrel-cage or amortisseur (damper) winding that short-circuits all of the rotor bars at one end. *See also:* **rotor (rotating machinery).** 0-31E8

end-scale value (electric instrument). The value of the actuating electrical quantity that corresponds to end-scale indication. *Notes:* (1) When zero is not at the end or at the electrical center of the scale, the higher value is taken. (2) Certain instruments such as power-factor meters, ohmmeters, etcetera, are necessarily excepted from this definition. (3) In the specification of the range of multiple-range instruments, it is preferable to list the ranges in descending order, as 750/300/150. *See also:* **accuracy rating (instrument); instrument.** 39A1/42A30-0

end shield (1) (rotating machinery). A solid or skeletal structure, mounted at one end of a machine, for the purpose of providing a specified degree of protection for the winding and rotating parts or to direct the flow of ventilating air. *Note:* Ordinarily a machine has an end shield at each end. For certain types of machine, one of the end shields may be constructed as an integral part of the stator frame. The end shields may be used to align and support the bearings, oil deflectors, and, for a hydrogen-cooled machine, the hydrogen seals. 0-31E8

(2) (magnetrons). A shield for the purpose of confining the space charge to the interaction space. *See also:* **magnetrons.** 42A70-15E6

end-shift frame (rotating machinery). A stator frame so constructed that it can be moved along the axis of the machine shaft for purposes of inspection. *See also:* **stator.** 0-31E8

endurance limit. The maximum stress a metal can withstand without failure during a specified large number (usually 10 million) cycles of stress. *See:* **corrosion terms.** CM-34E2

end winding (rotating machinery). That portion of a winding extending beyond the slots. *Note:* It is outside the major flux path and its purpose is to provide connections between parts of the winding within the slots of the magnetic circuit. *See:* **asynchronous machine; direct-current commutating machine; rotor (rotating machinery); stator; synchronous machine.** 42A10-31E8

end-winding cover (winding shield) (rotating machinery). A cover to protect an end winding against mechanical damage and/or to prevent inadvertent contact with the end winding. *See also:* **cradle base (rotating machinery).** 0-31E8

end-winding support (rotating machinery). The structure by which coil ends are braced against gravity and electromagnetic forces during start-up (for motors), running, and abnormal conditions such as sudden short-circuit, for example, by blocking and lashings between coils and to brackets or rings. *See also:* **cradle base (rotating machinery); stator.** 0-31E8

end-window counter tube (radiation). A counter tube designed for the radiation to enter at one end. *See:* **anticoincidence (radiation counters).** 0-15E6

end-wire insulation (rotating machinery). Insulation members placed between the end wires of individual coils, such as between main and auxiliary windings. *See also:* **rotor (rotating machinery); stator.** 0-31E8

end wire, winding (rotating machinery). The portion of a random-wound winding that is not inside the core. *See also:* **rotor (rotating machinery); stator.** 0-31E8

energy (system). The available energy is the amount of work that the system is capable of doing. *See:* **electric energy.** E270-0

energy density (point in a field) (audio and electroacoustics). The energy contained in a given infinitesimal part of the medium divided by the volume of that part of the medium. *Notes:* (1) The term energy density may be used with prefatory modifiers such as instantaneous, maximum, and peak. (2) In speaking of average energy density in general, it is necessary to distinguish between the space average (at a given instant) and the time average (at a given point). *See also:* **electroacoustics.** 0-1E1

energy density spectrum (burst measurements) (finite energy signal). The square of the magnitude of the Fourier transform of a burst. *See also:* **burst; network analysis.** E265-12E1

energy efficiency (specified electrochemical process). The product of the current efficiency and the voltage efficiency. *See also:* **electrochemistry.** 42A60-0

energy flux (audio and electroacoustics). The average rate of flow of energy per unit time through any specified area. *Note:* For a sound wave in a medium of density ρ and for a plane or spherical free wave having a velocity of propagation c, the sound-energy flux through the area S corresponding to an effective sound pressure p is

$$\mathbf{J} = \frac{p^2 S}{\rho c} \cos\theta$$

where θ is the angle between the direction of propagation of the sound and the normal to the area S. *See also:* **electroacoustics.** 0-1E1

energy gap (semiconductor). The energy range between the bottom of the conduction band and the top of the valence band. *See also:* **semiconductor device.** E102/E216-34E17;E270/10E1

energy metering point (tie line) (electric power systems). The location of the integrating metering equipment used to measure energy transfer on the tie line. *See also:* **center of distribution.** E94-0

engine-driven generator for aircraft. A generator mechanically, hydraulically, or pneumatically coupled to an aircraft propulsion engine to provide power for the electric and electronic systems of an aircraft. It may be classified as follows: (1) Engine-mounted; (2) Remote-driven: (A) Flexible-shaft-driven; (B) Variable-ratio-driven; (C) Air-turbine-driven. *See also:* **air transportation electric equipment.** 42A41-0

engine-generator system (electric power supply). A system in which electic power for the requirements of a railway vehicle (other than propulsion) is supplied by an engine-driven generator carried on the vehicle, either as an independent source of electric power or supplemented by a storage battery. *See also:* **axle generator system.** 42A42-0

engine-room control. Apparatus and arrangement providing for control in the engine room, on order from the bridge, of the speed and direction of a vessel. 42A43-0

engine synchronism indicator. A device that provides a remote indication of the relative speeds between two or more engines. *See also:* **air transportation instruments.** 42A41-0

engine-temperature thermocouple-type indicator. A device that indicates temperature of an aircraft engine cylinder by measuring the electromotive force of a thermocouple. *See also:* **air transportation instruments.** 42A41-0

engine-torque indicator. A device that indicates engine torque in pound-feet. *Note:* It is usually converted to horsepower with reference to engine revolutions per minute. *See also:* **air transportation instruments.** 42A41-0

entropy (information theory). *See:* **average information content.** *See also:* **information theory.**

entry point (routine) (computing systems). Any place to which control can be passed. *See also:* **electronic digital computer.** X3A12-16E9

envelope (wave). The boundary of the family of curves obtained by varying a parameter of the wave. For the special case

$$y = E(t)\sin(\omega t + \theta),$$

variation of the parameter θ yields $E(t)$ as the envelope. E270-0

envelope delay (television radio wave propagation, and facsimile). The time of propagation, between two points, of the envelope of a wave. *Note:* It is equal to the rate of change with angular frequency of the difference in phase between these two points. It has significance over the band of frequencies occupied by the wave only if this rate is approximately constant over that band. If the system distorts the envelope, the envelope delay at a specified frequency is defined with reference to a modulated wave that occupies a frequency bandwidth approaching zero. *See:* **television.** *See also:* **facsimile transmission; radio wave propagation; transmission characteristic.** E168-0;E211-2E2

envelope delay distortion. *See:* **distortion, envelope delay.**

envelope, vacuum (electron tubes). *See:* **bulb.**

environment. The universe within which the system must operate. All the elements over which the designer has no control and that affect the system or its inputs and outputs. *See also:* **system.** 0-35E2

environmental change of amplification (magnetic amplifier). The change in amplification due to a specified change in one environmental quantity while all other environmental quantities are held constant. *Note:* Use of a coefficient implies a reasonable degree of linearity of the considered quantity with respect to the specified environmental quantity. If significant deviations from linearity exist within the environmental range over which the amplifier is expected to operate, particularly if the amplification, for example, is not a monotonic function of the environmental quantity, the existence of such deviations should be noted. *See also:* **rating and testing magnetic amplifiers.** E107-0

environmental coefficient of amplification (magnetic amplifier). The ratio of the change in amplification to the change in the specified environmental quantity when all other environmental quantities are held constant. *Note:* The units of this coefficient are the amplification units per unit of environmental quantity. *See also:* **rating and testing magnetic amplifiers.** E107-0

environmental coefficient of offset (magnetic amplifier). The ratio of the change in quiescent operating point to the change in the specified environmental quantity when all other environmental quantities are held constant. *Note:* The units of this coefficient are the output units per unit of environmental quantity. *See also:* **rating and testing magnetic amplifiers.** E107-0

environmental coefficient of trip-point stability (magnetic amplifier). The ratio of the change in trip point to the change in the specified environmental quantity when all other environmental quantities are held constant. *Notes:* (1) The units of this coefficient are the control signal units per unit of environmental quantity. (2) Use of a coefficient implies a reasonable degree of linearity of the considered quantity with respect to the specified environmental quantity. If significant deviations from linearity exist within the environmental range over which the amplifier is expected to operate, particularly if the amplification, for example, is not a monotonic function of the environmental quantity, the existence of such deviations should be noted. *See also:* **rating and testing magnetic amplifiers.** E107-0

environmental offset (magnetic amplifier). The change in quiescent operating point due to a specified change in one environmental quantity (such as line voltage) while all other environmental quantities are held constant. *See also:* **rating and testing magnetic amplifiers.** E107-0

environmental trip-point stability (magnetic amplifier). The change in the magnitude of the trip point (either trip OFF or trip ON, as specified) control signal due to a specified change in one environmental quantity (such as line voltage) while all other environmental quantities are held constant. *See also:* **rating and testing magnetic amplifiers.** E107-0

ephapse. The electric junction of two parallel or crossing nerve fibers at which there may occur phenomena similar to those occurring at a synapse. 42A80-18E1

***E*-plane bend (waveguides) (rectangular uniconductor waveguide operating in the dominant mode).** A bend in which the longitudinal axis of the guide remains in a plane parallel to the electric field vector throughout the bend. E147-3E1

***E*-plane, principal (linearly polarized antenna).** The plane containing the electric field vector and the direction of maximum radiation. *See also:* **antenna; radiation.** 0-3E1

***E*-plane T junction (series T junction).** A waveguide T junction in which the change in structure occurs in the plane of the electric field. *See also:* **waveguide.** 42A65-0

E-plane T junction (waveguides) (rectangular uniconductor waveguide). A T junction of which the electric field vector of the dominant wave of each arm is parallel to the plane of the longitudinal axes of the guides. E147-3E1

equal-energy source (color). A light source for which the time rate of emission of energy per unit of wavelength is constant throughout the visible spectrum. *See also:* **color terms.** E201-2E2

equal interval (isophase) (light). A rhythmic light in which the light and dark periods are equal. *See also:* **signal lighting.** Z7A1-0

equalization. *See:* **frequency-response equalization.**

equalizer (1) (general). A device designed to compensate for an undesired characteristic of a system or component. *See also:* **circuits and devices.** 0-2E2

(2) (signal circuits). A device designed to compensate for an undesired amplitude-frequency or phase-frequency characteristic, or both, of a system or transducer. 0-42A65

(3) (rotating machinery). A connection made between points on a winding to minimize any undesirable voltage between these points. *See also:* **asynchronous machine; direct-current commutating machine; synchronous machine.** 0-31E8

equalizing charge (storage battery) (storage cell). An extended charge to a measured end point that is given to a storage battery to insure the complete restoration of the active materials in all the plates of all the cells. *See also:* **charge.** 42A60-0

equalizing pulses (television). Pulse trains in which the pulse-repetition frequency is twice the line frequency and that occur just before and just after a vertical synchronizing pulse. *Note:* The equalizing pulses minimize the effect of line-frequency pulses on the interlace. *See also:* **pulse terms; television.** E194/E203-0;42A65-2E2

equally tempered scale. A series of notes selected from a division of the octave (usually) into 12 equal intervals, with a frequency ratio between any two adjacent notes equal to the twelfth root of two.

Equally tempered intervals

Name of Interval	Frequency Ratio *	Cents
Unison	1:1	0
Minor second or semitone	1.059463:1	100
Major second or whole tone	1.122462:1	200
Minor third	1.189207:1	300
Major third	1.259921:1	400
Perfect fourth	1.334840:1	500
Augmented fourth / Diminished fifth	1.414214:1	600
Perfect fifth	1.498307:1	700
Minor sixth	1.587401:1	800
Major sixth	1.681793:1	900
Minor seventh	1.781797:1	1000
Major seventh	1.887749:1	1100
Octave	2:1	1200

* The frequency ratio is $[(2)^{1/12}]^n$ where n equals the number of the interval. (The number of the interval is its value in cents divided by 100.)

0-1E1

equal vectors. Two vectors are equal when they have the same magnitude and the same direction. E270-0

equation. *See:* **computer equation.**

equilibrium electrode potential. A static electrode potential when the electrode and electrolyte are in equilibrium with respect to a specified electrochemical reaction. *See also:* **electrochemistry.** 42A60-0

equilibrium point (control system). A point in state space of a system where the time derivative of the state vector is identically zero. *See also:* **control system.** 0-23E0

equilibrium potential. The electrode potential at equilibrium. CM-34E2

equilibrium reaction potential. The minimum voltage at which an electrochemical reaction can take place. *Note:* It is equal to the algebraic difference of the equilibrium potentials of the anode and cathode with respect to the specified reaction. It can be computed from the free energy of the reaction. Thus

$$\Delta F = -n\,F\,E$$

where ΔF is the free energy of the reaction, n is the number of chemical equivalents involved in the reaction, F is the value of the Faraday expressed in calories per volt gram-equivalent (23 060.5) and E is the equilibrium reaction potential (in volts). *See also:* **electrochemistry.** 42A60-0

equilibrium voltage. *See:* **storage-element equilibrium voltage.** *See also:* **storage tube.**

equiphase surface (radio wave propagation). Any surface in a wave over which the field vectors at the same instant are in the same phase or 180 degrees out of phase. *See also:* **radio wave propagation.** E211-3E2

equiphase zone (electronic navigation). The region in space within which difference in phase of two radio signals is indistinguishable. *See also:* **radio navigation.** E172-10E6;42A65-0

equipment (electrical engineering). A general term including materials, fittings, devices, appliances, fixtures, apparatus, machines, etcetera, used as a part of, or in connection with, an electrical installation. *See also:* **interior wiring.** E270/1A0/2A2/84A1-0

equipment ground (1) (general). A ground connection to noncurrent-carrying metal parts of a wiring installation or of electric equipment, or both. *See:* **ground.** 42A35-31E13
(2) (lightning arresters). A grounding system connected to parts of an installation, normally not alive, with which persons may come into contact. *See:* **lightning arrester (surge diverter).** 50I25-31E7

equipment noise (sound recording and reproducing system). The noise output that is contributed by the elements of the equipment during recording and reproducing, excluding the recording medium, when the equipment is in normal operation. *Note:* Equipment noise usually comprises hum, rumble, tube noise, and component noise. *See also:* **noise (sound recording and reproducing system).** E191-0

equipotential cathode. *See:* **cathode, indirectly heated.**

equipotential line or contour. The locus of points having the same potential at a given time. *See also:* **ground.** E81-0

equisignal localizer (electronic navigation) (tone localizer). A type of localizer in which two modulation frequencies are compared in amplitude to obtain lateral guidance. *See also:* **radio navigation.** 42A65-0

equisignal zone (radio navigation). The region in space within which the difference in amplitude of two radio signals (usually emitted by a single station) is indistinguishable. *See also:* **radio navigation.** 42A65-0

equivalence (computing systems). A logic operator having the property that if P is a statement, Q is a statement, R is a statement, ..., then the equivalence of $P, Q, R, \ldots$ is true if and only if all statements are true or all statements are false. X3A12-16E9

equivalent binary digits (computing systems). The number of binary places required to count the elements of a given set. *See also:* **electronic digital computer.** X3A12-16E9

equivalent circuit (1) (general). An arrangement of circuit elements that has characteristics, over a range of interest, electrically equivalent to those of a different circuit or device. *Note:* In many useful applications, the equivalent circuit replaces (for convenience of analysis) a more-complicated circuit or device. *See also:* **network analysis.** E270/42A65-0
(2) (piezoelectric crystal unit). An electric circuit that has the same impedance as the unit in the frequency region of resonance. *Note:* It is usually represented by an inductance, capacitance, and resistance in series, shunted by the direct capacitance between the terminals of the crystal unit. *See also:* **crystal.** 42A65-0

equivalent concentration (ion type). The concentration equal to the ion concentration divided by the valency of the ion considered. *See also:* **ion.** 42A60-0

equivalent conductance (1) (acid, base, or salt). The conductance of the amount of solution that contains one gram equivalent of the solute when measured between parallel electrodes that are one centimeter apart and large enough in area to include the necessary volume of solution. *Note:* Equivalent conductance is numerically equal to the conductivity multiplied by the volume in cubic centimeters containing one gram equivalent of the acid, base, or salt. *See also:* **electrochemistry.** 42A60-0

(2) (microwave gas). The normalized conductance of the tube in its mount measured at its resonance frequency. *Note:* Normalization is with respect to the characteristic impedance of the transmission line at its junction with the tube mount. *See:* **electron-tube admittances; element (electron tube).** E160-15E6

equivalent core-loss resistance. A hypothetical resistance, assumed to be in parallel with the magnetizing inductance, that would dissipate the same power as that dissipated in the core of the transformer winding for a specified value of excitation. 0-21E1

equivalent dark-current input (phototubes). The incident luminous (or radiant) flux required to give a signal output current equal to the output electrode dark current. *Note:* Since the dark current may change considerably with temperature, the temperature should be specified. *See:* **phototubes.** E160-15E6;42A70-0

equivalent diode (triode or a multielectrode tube). *See:* **diode, equivalent.**

equivalent diode voltage. *See:* **composite controlling voltage.**

equivalent flat-plate area (scattering object). For an object large compared to the wavelength of plane waves scattered by it, the area of a flat, perfectly reflecting plate, parallel to the incident wave front, that has the same backscattering cross section as the object. *See also:* **antenna.** 0-3E1

equivalent network. A network that, under certain conditions of use, may replace another network without substantial effect on electrical performance. *Note:* If one network can replace another network in any system whatsoever without altering in any way the electrical operation of that portion of the system external to the networks, the networks are said to be **networks of general equivalence.** If one network can replace another network only in some particular system without altering in any way the electrical operation of that portion of the system external to the networks, the networks are said to be **networks of limited equivalence.** Examples of the latter are networks that are equivalent only at a single frequency, over a single band, in one direction only, or only with certain terminal conditions (such as *H* and *T* networks). *See also:* **network analysis.** E270/42A65-0

equivalent noise bandwidth (interference terminology) (signal system). The frequency interval, determined by the response-frequency characteristics of the system, that defines the noise power transmitted from a noise source of specified characteristics. *Note:* For Gaussian noise

$$\Delta f = \int_0^\infty y(f)^2\, df$$

where $y(f) = Y(0)/Y(f)$ is the relative frequency dependent response characteristic. *See also:* **interference.** 0-13E6

equivalent noise conductance (interference terminology). A quantitative representation in conductance units of the spectral density of a noise-current generator at a specified frequency. *Notes:* (1) The relation between the equivalent noise conductance G_n and the spectral density W_i of the noise-current generator is

$$G_n = \pi W_i/(kT_0)$$

where k is Boltzmann's constant and T_0 is the standard noise temperature (290 kelvins) and $kT_0 = 4.00 \times 10^{-21}$ watt-seconds. (2) The equivalent noise conductance in terms of the mean-square noise-generator current $\overline{i^2}$ within a frequency increment Δf is

$$G_n = \overline{i^2}/(4kT_0\Delta f).$$

See: **electron-tube admittances; signal-to-noise signal.** E160-13E6/15E6

equivalent noise current (electron tubes) (interference terminology). A quantitative representation in current units of the spectral density of a noise current generator at a specified frequency. *Notes:* (1) The relation between the equivalent noise current I_n and the spectral density W_i of the noise-current generator is

$$I_n = (2\pi W_i)/e$$

where e is the magnitude of the electron charge. (2) The equivalent noise current in terms of the mean-square noise-generator current $\overline{i^2}$ within a frequency increment Δf is

$$I_n = \overline{i^2}/(2e\Delta f).$$

See also: **circuit characteristics of electrodes; interference; signal-to-noise ratio.** E160-13E6/15E6

equivalent noise input (phototube). The value of incident luminous (or radiant) flux that, when modulated in a stated manner, produces a root-mean-square signal output current equal to the root-mean-square dark-current noise both in the same specified bandwidth (usually 1 hertz). *See also:* **phototube.** E158-15E6

equivalent noise resistance. A quantitative representation in resistance units of the spectral density of a noise voltage generator at a specified frequency. *Notes:* (1) The relation between the equivalent noise resistance R_n and the spectral density W_e of the noise-voltage generator is

$$R_n = (\pi W_e)/(kT_0)$$

where k is Boltzmann's constant and T_0 is the standard noise temperature (290 kelvins) and $kT_0 = 4.00 \times 10^{-21}$ watt-seconds. (2) The equivalent noise resistance in terms of the mean-square noise-generator voltage $\overline{e^2}$ within a frequency increment Δf is

$$R_n = \overline{e^2}/(4kT_0\Delta f).$$

See also: **circuit characteristics of electrodes; interference; signal-to-noise ratio.** E160-13E6/15E6

equivalent periodic line (uniform line). A periodic line having the same electrical behavior, at a given frequency, as the uniform line when measured at its terminals or at corresponding section junctions. *See also:* **transmission line.** E270/42A65-0

equivocation (information theory). The conditional information content of an input symbol given an output symbol, averaged over all input-output pairs. *See also:* **information theory.** E171-0

erase (charge-storage tubes). To reduce by a controlled operation the amount of stored information. *See also:* **storage tube.** E158-15E6

erasing head (electroacoustics). A device for obliterating any previous magnetic recordings. *See:* **alternating-current erasing head; direct-current erasing**

head; permanent-magnet erasing head; phonograph pickup. 0-1E1

erasing rate (charge-storage tubes). The time rate of erasing a storage element line or area, from one specified level to another. Note the distinction between this and erasing speed. *See also:* **storage tube.** E158-15E6

erasing, selective (storage tubes). Erasing of selected storage elements without disturbing the information stored on other storage elements. *See also:* **storage tube.** E158-15E6

erasing speed (charge-storage tubes). The lineal scanning rate of the beam across the storage surface in erasing. Note the distinction between this and erasing rate. *See also:* **electron devices, miscellaneous; storage tube.** E158-15E6

erasing time, minimum usable (storage tubes). The time required to erase stored information from one specified level to another under stated conditions of operation and without rewriting. *Note:* The qualifying adjectives **minimum usable** are frequently omitted in general usage when it is clear that the minimum usable erasing time is implied. *See also:* **storage tube.** E158-15E6

***E* region.** The region of the terrestrial ionosphere between about 90 and 160 kilometers above the earth's surface. *See also:* **radiation; radio wave propagation.** 42A65-0

erg. The unit of work and of energy in the centimeter-gram-second systems. The erg is 10^{-7} joule. E270-0

erosion (corrosion). Deterioration by the abrasive action of fluids, usually accelerated by the presence of solid particles of matter in suspension. When deterioration is further increased by corrosion, the term erosion-corrosion is often used. *See also:* **corrosion terms.** CM-34E2

error (1) (mathematics). Any discrepancy between a computed, observed, or measured quantity and the true, specified, or theoretically correct value or condition. *See:* **absolute error; inherited error.** *Note:* A positive error denotes that the indication of the instrument is greater than the true value. Error = Indication −True. *See:* **absolute error; correction; inherited error.** *See also:* **accuracy rating of an instrument; low-frequency testing; power-system testing; surge testing.** X3A12-16E9

(2) (computer or data-processing system). Any incorrect step, process, or result. *Note:* In the computer field the term commonly is used to refer to a machine malfunction as a machine error (or computer error) and to a human mistake as a human error (or operator error). Frequently it is helpful to distinguish between these errors as follows: an **error** results from approximations used in numerical methods; a **mistake** results from incorrect programming, coding, data transcription, manual operation, etcetera; a **malfunction** results from a failure in the operation of a machine component such as a gate, a flip-flop, or an amplifier. *See:* **dynamic error; linearity error; loading error; resolution error; static error.** *See also:* **electronic analog computer; electronic computation.** E165/E270-16E9

error and correction. The difference between the indicated value and the true value of the quantity being measured. *Note:* It is the quantity that algebraically subtracted from the indicated value gives the true value. A positive error denotes that the indicated value of the instrument is greater than the true value. The correction has the same numerical value as the error of the indicated value, but the opposite sign. It is the quantity that algebraically added to the indicated value gives the true value. If T, I, E and C represent, respectively, the true value, the indicated value, the error, and the correction, the following equations hold:

$$E = I - T; \; C = T - I.$$

Example: A voltmeter reads 112 volts when the voltage applied to its terminals is actually 110 volts.

$$\text{Error} = 112 - 110 = +2 \text{ volts};$$

$$\text{correction} = 110 - 112 = -2 \text{ volts}.$$

See also: **accuracy rating (instrument).** 42A30-0

error burst. A group of bits in which two successive erroneous bits are always separated by less than a given number of correct bits. 0-19E4

error coefficient (control system, feedback). The real number C_n by which the nth derivative of the reference input signal is multiplied to give the resulting nth component of the actuating signal. *Note:* The error coefficients may be obtained by expanding in a Maclaurin series the error transfer function as follows:

$$\frac{1}{1 + GH(s)} = C_0 + C_1 s + C_2 s^2 + \cdots + C_n s^n.$$

See also: **control system, feedback** 0-23E0

error constant (control system, feedback). The real number K_n by which the nth derivative of the reference input signal is divided to give the resulting nth component of the actuating signal.

Note: $K_n = 1/C_n$
$K_0 = 1 + K_p$, where K_p is position constant
$K_1 = K_v$ velocity constant
$K_2 = K_a$ acceleration constant
$K_3 = K_j$ jerk constant

In some systems these constants may equal infinity. *See also:* **control system, feedback.** 85A1-23E0

error-correcting code. A code in which each telegraph or data signal conforms to specific rules of construction so that departures from this construction in the received signals can be automatically detected, and permits the automatic correction, at the received terminal, or some or all of the errors. *Note:* Such codes require more signal elements than are necessary to convey the basic information. *See also:* **error-detecting code; error-detecting and feedback system; error-detecting system.** 0-19E4

error-detecting code. A code in which each expression conforms to specific rules of construction, so that if certain errors occur in an expression the resulting expression will not conform to the rules of construction and thus the presence of the errors is detected. *Note:* Such codes require more signal elements than are necessary to convey the fundamental information. *See also:* **check, forbidden-combination; electronic computation; electronic digital computer; error correcting code; data transmission.** X3A12-16E9;0-19E4

error-detecting and feedback system. A system employing an error-detecting code and so arranged that a character or block detected as being in error automatically initiates a request for retransmission of the signal detected as being in error. *See also:* **data transmission.** 0-19E4

error-detecting system (data transmission). A system employing an error-detecting code and so arranged that any signal detected as being in error is either (1) deleted from the data delivered to the receiver, in some cases with an indication that such deletion has taken place, or (2) delivered to the receiver together with an indication that it has been detected as being in error. *See also:* **data transmission.** 0-19E4

error, matching (analog computer). An error resulting from inaccuracy in matching (two resistors) or mating (a resistor and a capacitor) passive elements. *See also:* **electronic analog computer.** Z3E165-0

error range. The difference between the highest and lowest error values. X3A12-16E9

error rate (data transmission). Ratio of the number of characters of a message incorrectly received to the number of characters of the message received. *See also:* **data transmission.** 0-19E4

error signal. *See:* **signal, error.**

erythema. The temporary reddening of the skin produced by exposure to ultraviolet energy. *Note:* The degree of erythema is used as a guide to dosages applied in ultraviolet therapy. *See also:* **ultraviolet radiation.** Z7A1-0

erythemal effectiveness. The capacity of various portions of the ultraviolet spectrum to produce erythema. *See also:* **ultraviolet radiation.** Z7A1-0

erythemal efficiency (radiant flux for a particular wavelength). The ratio of the erythemal effectiveness of that wavelength to that of wavelength 296.7 nanometers which is rated as unity. *Note:* This term formerly was called **relative erythemal factor.** The erythemal efficiency of various wavelengths of radiant flux for producing a minimum perceptible erythema is given in the table. These values have been accepted for evaluating the erythemal effectiveness of sun lamps. *See also:* **ultraviolet radiation.**

erythemal exposure. The product of erythemal flux density on a surface, and time. It usually is measured in erythemal microwatt-minutes per square centimeter. *Note:* For average untanned skin a minimum perceptible erythema requires about 300 microwatt-minutes per square centimeter of radiation at 296.7 nanometers. *See also:* **ultraviolet radiation.** Z7A1-0

erythemal flux. Radiant flux evaluated according to its capacity to produce erythema of the untanned human skin. *Notes:* (1) It usually is measured in microwatts of ultraviolet radiation weighted in accordance with its erythemal efficiency. Such quantities of erythemal flux would be in erythemal microwatts. (2) A commonly used practical unit of erythemal flux is the erythemal unit (EU) or E-viton (erythema) that is equal to the amount of radiant flux that will produce the same erythemal effect as 10 microwatts of radiant flux at wavelength 296.7 nanometers. *See also:* **ultraviolet radiation.** Z7A1-0

erythemal flux density. The erythemal flux per unit area of the surface being irradiated. *Notes:* (1) It is equal to the quotient of the incident erythemal flux divided by the area of the surface when the flux is uniformly distributed. It usually is measured in microwatts per square centimeter erythemally weighted ultraviolet radiation (erythemal microwatts per square centimeter). (2) A suggested practical unit of erythemal flux density is the finsen, which is equal to one E-viton per square centimeter. *See also:* **ultraviolet radiation.** Z7A1-0

Erythemal and bactericidal efficiency of ultraviolet radiation

Wavelength (nanometers)	Erythemal Efficiency	Tentative Bactericidal Efficiency
*235.3	------	0.35
240.0	0.56	-------
*244.6	0.57	0.58
*248.2	0.57	0.70
250.0	0.57	-------
*253.7	0.55	0.85
*257.6	0.49	0.94
260.0	0.42	-------
265.0	------	1.00
*265.4	0.25	0.99
*267.5	0.20	0.98
*270.0	0.14	0.95
*275.3	0.07	0.81
*280.4	0.06	0.68
285.0	0.09	-------
*285.7	0.10	0.55
*289.4	0.25	0.46
290.0	0.31	-------
*292.5	0.70	0.38
295.0	0.98	-------
*296.7	1.00	0.27
300.0	0.83	-------
*302.2	0.55	0.13
305.0	0.33	-------
310.0	0.11	-------
*313.0	0.03	0.01
315.0	0.01	-------
320.0	0.005	-------
325.0	0.003	-------
330.0	0.000	-------

* Emission lines in the mercury spectrum; other values interpolated. Z7A1-0

escalator. A power-driven, inclined, continuous stairway used for raising or lowering passengers. *See also:* **elevators.** 42A45-0

***E* scan (electronic navigation).** *See:* ***E* display.**

escape character (computing systems). A character used to indicate that the succeeding one or more characters are expressed in a code different from the code currently in use. *See also:* **electronic digital computer.** X3A12-16E9

escape ratio (charge-storage tubes). The average number of secondary and reflected primary electrons leaving the vicinity of a storage element per primary electron entering that vicinity. *Note:* The escape ratio is less than the secondary-emission ratio when, for example, some secondary electrons are returned to the secondary-emitting surface by a retarding field. *See also:* **charge-storage tube.** E158-15E6

***E* scope (electronic navigation).** *See:* ***E* display.**

EU. *See:* **erythemal flux.**

evacuating equipment. The assembly of vacuum pumps, instruments, and other parts for maintaining and indicating the vacuum. *See also:* **rectification.** 42A15-0

evanescent mode (cutoff mode) (waveguide). A field configuration in a waveguide such that the amplitude of the field diminishes along the waveguide, but the phase is unchanged. The frequency of this mode is less than the critical frequency. *See:* **waveguide.** 0-3E1

evanescent waveguide. *See:* **cutoff waveguide.**

***E*-viton.** *See:* **erythemal flux.**

exalted carrier reception. *See:* **reconditioned carrier reception.**

excess meter. An electricity meter that measures and registers the integral, with respect to time, of those portions of the active power in excess of the predetermined value. *See also:* **electricity meter (meter).** 42A30-0

excess-three code (electronic computation). Number code in which the decimal digit n is represented by the four-bit binary equivalent of $n+3$. Specifically

decimal digit	excess-three code
0	0011
1	0100
2	0101
3	0110
4	0111
5	1000
6	1001
7	1010
8	1011
9	1100

See also: **binary-coded-decimal system; electronic computation; electronic digital computer.** E162/E270/X3A12-16E9

exchangeable power (per unit bandwidth, at a port). The extremum value of the power flow per unit bandwidth from or to a port under arbitrary variations of its terminating impedance. *Notes:* (1) The exchangeable power p_e at a port with a mean-square open-circuit voltage spectral density $\overline{e^2}$ and an internal impedance with a real part R is given by the relation

$$p_e = \frac{\overline{e^2}}{4R}.$$

(2) The exchangeable power is equal to the available power when the internal impedance of the port has a positive real part. *See:* **signal-to-noise ratio; waveguide.** 0-15E6

exchangeable power gain (two-port linear transducer). At a pair of selected input and output frequencies, the ratio of (1) the exchangeable signal power of the output port of the transducer to (2) the exchangeable signal power of the source connected to the input port. *Note:* The exchangeable power gain is equal to the available power gain when the internal impedances of the source and the output port of the transducer have positive real parts. *See:* **signal-to-noise ratio; waveguide.** 0-15E6

excitability (electrobiology) (irritability). The inherent ability of a tissue to start its specific reaction in response to an electric current.
See:
anelectrotonus;
catelectrotonus;
electrotonic wave;
electrotonus;
latent period;
nerve block;
recovery cycle;
relatively refractory state;
subnormality;
threshold;
unpropagated potential. 42A80-18E1

excitability curve (medical electronics). A graph of the excitability of a given tissue as a function of time, where excitability is expressed either as the reciprocal of the intensity of an electric current just sufficient at a given instant to start the specific reaction of the tissue, or as the quotient of the initial (or conditioning) threshold intensity for the tissue by subsequent threshold intensities. *See also:* **medical electronics.** 0-18E1

excitation (drive). A signal voltage applied to the control electrode of an electron tube. *See also:* **circuits and devices; composite controlling voltage (electron tube).** E145/42A65-0

excitation anode (pool-cathode rectifier tube). An electrode that is used to maintain an auxiliary arc in the vacuum tank. *See also:* **electrode (electron tube); rectification.** 42A15-0

excitation coefficients (array antenna) (feeding coefficients) (antenna). The relative values of the excitation currents of the radiating elements. *See also:* **antenna.** 0-3E1

excitation current (1) (transformer). The current that maintains the excitation of the transformer. *Note:* It is usually expressed in per unit, or in percent, of the rated current of the winding in which it is measured. *See also:* **efficiency.** 42A15-31E12;57A14-0

(2) **(voltage regulator).** The current that maintains the excitation of the regulator. *Note:* It is usually expressed in per unit or in percent of the rated series-winding current of the regulator. *See also:* **efficiency; voltage regulator.** 57A15-0

excitation equipment (rectifier). The equipment for starting, maintaining, and controlling the arc. *See also:* **rectification.** 42A15-0

excitation losses (1) (transformer or regulator). Losses that are incident to the excitation of the transformer at its alternating-current winding terminals. *Note:* Excitation losses include core loss, dielectric loss, conductor loss in the windings due to excitation current, and conductor loss due to circulating current in parallel windings. Interphase transformer losses are not included in excitation losses. *See also:* **rectifier transformer; voltage regulator.** 42A15-31E12;57A14/57A18-0

(2) **(series transformer).** The losses in the transformer with the secondary winding open-circuited when the primary winding is excited at rated frequency and at a voltage that corresponds to the primary voltage obtained when the transformer is operating at nominal rated load. *Note:* The measurement should be made

with a constant voltage source of supply with not more than 3-percent harmonic deviation from sine wave. 82A7-0

excitation purity (purity) (color). The ratio of: (1) the distance from the reference point to the point representing the sample, to (2) the distance along the same straight line from the reference point to the spectrum locus or to the purple boundary, both distances being measured (in the same direction from the reference point) on the CIE chromaticity diagram. *Note:* The reference point is the point in the chromaticity diagram that represents the reference standard light mentioned in the definition of dominant wavelength. *See also:* **color; color terms.** E201/Z7A1-0

excitation response. *See:* **voltage response, exciter.**

excitation system (rotating machinery). The source of field current for the excitation of a principal electric machine, including means for its control. *See:* **direct-current commutating machine; synchronous machine.** 42A10/31E8

excitation-system stability (rotating machinery). The ability of the excitation system to control the field voltage of the principal electric machine so that transient changes in the regulated voltage are effectively suppressed and sustained oscillations in the regulated voltage are not produced by the excitation system during steady-load conditions following a change to a new steady-load condition. *See:* **direct-current commutating machine; synchronous machine.** 42A10/31E8

excitation voltage. The nominal voltage of the excitation circuit. 42A35-31E13

excitation winding. *See:* **field winding.**

excite (rotating machinery). To initiate or develop a magnetic field in (such as in an electric machine). *See also:* **asynchronous machine; direct-current commutating machine; synchronous machine.** 0-31E8

excited-field loudspeaker. A loudspeaker in which the steady magnetic field is produced by an electromagnet. *See also:* **loudspeaker.** 42A65-0

exciter (rotating machinery). The source of all or part of the field current for the excitation of an electric machine. *Note:* Familiar sources include direct-current commutator machines; alternating-current generators whose output is rectified; and batteries. *See also:* **asynchronous machine; direct-current commutating machine; synchronous machine.** 42A10/31E8

exciter ceiling voltage. The maximum voltage that may be attained by an exciter under specified conditions. *See:* **direct-current commutating machine; synchronous machine.** 42A10/31E8

exciter ceiling voltage, nominal (rotating machinery). The ceiling voltage of an exciter loaded with a resistor having an ohmic value equal to the resistance of the field winding to be excited and with this field winding at a temperature of: (1) 75 degrees Celsius for field windings designed to operate at rating with a temperature rise of 60 degrees Celsius or less. (2) 100 degrees Celsius for field windings designed to operate at rating with a temperature rise greater than 60 degrees Celsius. *See:* **direct-current commutating machine; synchronous machine.** 42A10-0

exciter dome (rotating machinery). Exciter housing for a vertical machine. *See also:* **cradle base (rotating machinery).** 0-31E8

exciter losses (synchronous machine). The total of the electric and mechanical losses in the equipment supply excitation. *See also:* **synchronous machine.** 50A10-31E8

exciter platform (rotating machinery). A deck on which to stand while inspecting the exciter. *See also:* **cradle base (rotating machinery).** 0-31E8

exciter response. *See:* **voltage response, exciter.**

exciter voltage response ratio (rotating machinery). *See:* **voltage response ratio.**

exciter voltage-time response (rotating machinery). *See:* **voltage-time response, synchronous-machine excitation system.**

exciting current. (1) The total current applied to a coil that links a ferromagnetic core. E270-0
(2) The component of the primary current of a transformer that is sufficient by itself to cause the counter electromotive force to be induced in the primary winding. 0-21E1

excitron. A single-anode pool tube provided with means for maintaining a continuous cathode spot. *See also:* **tube definition.** 42A70-15E6

exclusive OR. A logic operator having the property that if *P* is a statement and *Q* is a statement, then *P* exclusive OR *Q* is true if either but not both statements are true, false if both are true or both are false. *Note:* *P* exclusive OR *Q* is often represented by $P \oplus Q$, $P \not\vee Q$. *See:* **OR.** X3A12-16E9

excursion (computing system). *See:* **reference excursion.** E165-16E9

executive routine (computing systems). A routine that controls the execution of other routines. *See:* **supervisory routine.** X3A12-16E9

exfoliation (corrosion). A thick layer-like growth of corrosion product. *See:* **corrosion terms.** CM-34E2

existing installation (elevators). An installation, prior to the effective date of a code: (1) all work of installation was completed, or (2) the plans and specifications were filed with the enforcing authority and work begun not later than three months after the approval of such plans and specifications. *See also:* **elevators.** 42A45-0

expanded sweep. A sweep of the electron beam of a cathode-ray tube in which the movement of the beam is speeded up during a part of the sweep. *See also:* **magnified sweep; radar.** 42A65-0

expander. A transducer that for a given amplitude range of input voltages produces a larger range of output voltages. *Note:* One important type of expander employs the envelope of speech signals to expand their volume range. *See also:* **attenuation.** E151-0;42A65-31E3

expansion (1) (modulation systems). A process in which the effective gain applied to a signal is varied as a function of the signal magnitude, the effective gain being greater for large than for small signals.
(2) (oscillography). A decrease in the deflection factor, usually as the limits of the quality area are exceeded. *See also:* **modulating systems; transmission characteristics.** E170-0;42A65-0/0-9E4

expansion chamber (oil cutout). A sealed chamber separately attachable to the vent opening to provide additional air space into which the gases developed during circuit interruption can expand and cool. 37A100-31E11

expansion orbit (electronic device). The last part of the electron path that terminates at the target. It is outside the equilibrium orbit. *See also:* **electron device.** 0-15E6

expendable cap (expendable-cap cutout). A replacement part or assembly for clamping the button head of a fuse link and closing one end of the fuseholder. *Note:* It includes a pressure-responsive section that opens to relieve the pressure within the fuseholder, when a predetermined value is exceeded during circuit interruption. 37A100-31E11

expendable-cap cutout. An open cutout having a fuse support designed for and equipped with a fuseholder having an expendable cap. 37A100-31E11

exploder. *See:* **blasting unit.**

explosionproof (1) (enclosure). One that is designed and constructed to withstand an explosion of a specified gas or vapor that may occur within it and to prevent the ignition of the specified gas or vapor surrounding the enclosure by sparks, flashes, or explosions of the specified gas or vapor that may occur within the enclosure. *Notes:* (A) It must operate at such an external temperature that a surrounding flammable atmosphere will not be ignited thereby. (B) Without any significant change in the text of this definition, the word enclosure is frequently replaced by more specific terms, such as luminaire, machine, apparatus, etcetera. In some contexts, flameproof appears as a synonym of explosionproof. (C) Explosionproof apparatus should bear Underwriters' Laboratories approval ratings of the proper class and group consonant with the spaces in which flammable volatile liquids, highly flammable gases, mixtures, or highly flammable substances may be present.
(2) (mine apparatus). Apparatus capable of withstanding explosion tests as established by the United States Bureau of Mines, namely, internal explosions of methane-air mixtures, with or without coal dust present, without ignition of surrounding explosive methane-air mixtures and without damage to the enclosure or discharge of flame. *See:* **hazardous area groups; hazardous area classes.** *See also:* **distribution center; luminaire; mining.**
E45/1A0/2A2/42A10/42A85/42A95/Z7A1-0; 37A100-31E11;MG1-31AG;1A0-34E16

explosion-tested equipment. Equipment in which the housings for the electric parts are designed to withstand internal explosions of methane-air mixtures without causing ignition of such mixtures surrounding the housings. *See also:* **mining.** 42A85-0

exponential function. One of the form $y = ae^{bx}$, where a and b are constants and may be real or complex. An exponential function has the property that its rate of change with respect to the independent variable is proportional to the function, or $dy/dx = by$. E270-0

exponential horn. A horn the cross-sectional area of which increases exponentially with axial distance. *Note:* If S is the area of a plane section normal to the axis of the horn at a distance x from the throat of the horn, S_0 is the area of the plane section normal to the axis of the horn at the throat, and m is a constant that determines the rate of taper or flare of the horn, then

$$S = S_0 e^{mx}$$

See also: **loudspeaker.** E157-1E1;42A65-0

exponentially damped sine function. A generalized sine function of the form $Ae^{-bx} \sin (x+a)$ where $b>0$. E270-0

exponential transmission line. A tapered transmission line whose characteristic impedance varies exponentially with electrical length along the line. *See:* **transmission line.** *See also:* **waveguide.** E146/3E1

exposed (1) (live parts). A live part that can be inadvertently touched or approached nearer than a safe distance by a person. *Note:* It is applied to parts not suitably guarded, isolated, or insulated. *See:* **accessible; concealed.** *See also:* **distribution center.** 1A0/42A95-0
(2) (wiring methods). Not concealed. 42A95-0
(3) (circuits or lines). In such a position that in case of failure of supports or insulation contact with another circuit or line may result. 2A2-0
(4) (equipment). An object or device that can be inadvertently touched or approached nearer than a safe distance by any person. *Note:* It is applied to objects not suitably guarded or isolated. 2A2-0

exposed installation (lightning). An installation in which the apparatus is subject to overvoltages of atmospheric origin. *Note:* Such installations are usually connected to overhead transmission lines either directly or through a short length of cable. *See:* **lightning arrester (surge diverter).** 50I25-31E7

expression. An ordered set of one or more characters. E162-16E9

expulsion element (arrester). A chamber in which an arc is confined and brought into contact with gas-evolving material. *See:* **arrester; lightning arrester (surge diverter).** E28/62A1-31E7

expulsion fuse. *See:* **expulsion-fuse unit.**

expulsion-fuse unit (expulsion fuse). A vented fuse unit in which the expulsion effect of gases produced by the arc and lining of the fuseholder, either alone or aided by a spring, extinguishes the arc. 37A100-31E11

expulsion-type lightning arrester (expulsion-type arrester). *See:* **arrester, expulsion-type.**

extended capability (electric power supply). The generating capability increment in excess of net dependable capability that can be obtained under emergency operating procedures. *See also:* **generating station.** 0-31E4

extended-time rating (neutral grounding device) (electric power). A rating in which the time is sufficiently long to permit the temperature rise to become constant, but is expected to be limited in duration. *See also:* **grounding device.** E32-0

extension cord. An assembly of a flexible cord with an attachment plug on one end and a cord connector on the other. *See also:* **interior wiring.** 42A95-0

extension station. A telephone station associated with a main station through connection to the same subscriber line and having the same call number designation as the associated main station. *See also:* **telephone station.** 42A65-0

external connector (aerial lug) (pothead). A connector that joins the external conductor to the other current-carrying parts of a pothead. *See also:* **pothead; transformer.** E48/57A12.75/57A12.76-0

external field influence (electric instrument). The percentage change (of full-scale value) in indication caused solely by a specified external field. Such a field is produced by a standard method with a current of the same kind and frequency as that which actuates the mechanism. This influence is determined with the most unfavorable phase and position of the field in relation to the instrument. *Note:* The coil used in the standard method shall be approximately 40 inches in diameter not over 5 inches long, and carrying sufficient current to produce the required field. The current to produce a field to an accuracy of ± 1 percent in air shall be calculated without the instrument in terms of the specific dimensions and turns of the coil. In this coil, 400 ampere-turns will produce a field of approximately 5 oersteds. The instrument under test shall be placed in the center of the coil. *See also:* **accuracy rating (instrument).** 39A1/39A2-0

external insulation (apparatus) (lightning arresters). The external insulating surfaces and the surrounding air. *Note:* The dielectric strength of external insulation is dependent on atmospheric conditions. *See:* **lightning arrester (surge diverter).** 60I0-31E7

externally commutated inverters. An inverter in which the means of commutation is not included within the power inverter. *See:* **self-commutated inverters.** 0-34E24

externally heated arc (gas). An electric arc characterized by the fact that the thermionic cathode is heated by an external source. *See also:* **discharge (gas).** 50I07-15E6

externally operable. Capable of being operated without exposing the operator to contact with live parts. *Note:* This term is applied to equipment, such as a switch, that is enclosed in a case or cabinet. *See also:* **distribution center; insulated.** 2A2/42A95-0

externally quenched counter tube. A radiation counter tube that requires the use of an external quenching circuit to inhibit reignition. *See also:* **gas-filled radiation-counter tube.** E160-15E6;42A70-15E6

externally ventilated machine (rotating machinery). A machine that is ventilated by means of a separate motor-driven blower, usually mounted on the machine enclosure. *See also:* **synchronous machine.** 42A10-31E8

external series gap (expulsion-type arrester). An intentional gap between spaced electrodes, in series with the gap or gaps in the arcing chamber. *See:* **lightning arrester (surge diverter).** 99I2-31E7

external temperature influence (1) (electric instrument). The percentage change (of full-scale value) in the indication of an instrument that is caused solely by a difference in ambient temperature from the reference temperature after equilibrium is established. *See also:* **accuracy rating (instrument).** 39A1-0

(2) (electric recording instrument). The change in the recorded value that is caused solely by a difference in ambient temperature from the reference temperature. *Note:* It is to be expressed as a percentage of the full-scale value. *See also:* **accuracy rating (instrument).** 39A2-0

external termination (*j*th terminal of an *n*-terminal network). The passive or active two-terminal network that is attached externally between the *j*th terminal and the reference point. *See also:* **electron-tube admittances.** E160/42A70-15E6

extinction voltage (gas tube). The anode voltage at which the discharge ceases when the supply voltage is decreasing. *See also:* **gas-filled rectifier.** 50I07-15E6

extinguishing voltage (drop-out voltage) (glow lamp). Dependent upon the impedance in series with the lamp, the voltage across the lamp at which an abrupt decrease in current between operating electrodes occurs and is accompanied by the disappearance of the negative glow *Note:* In recording or specifying extinguishing voltage, the impedance must be specified. 78A385-0

extract (electronic computation). To form a new word by juxtaposing selected segments of given words. *See also:* **electronic computation.** E270-0

extract instruction (electronic digital computation). An instruction that requests the formation of a new expression from selected parts of given expressions. E162-0;X3A12-16E9

extraction liquor. The solvent used in hydrometallurgical processes for extraction of the desired constituents from ores or other products. *See also:* **electrowinning.** 42A60-0

extraordinary wave (X wave) (1) (radio wave propagation). The magneto-ionic wave component that, when viewed below the ionosphere in the direction of propagation, has clockwise or counterclockwise elliptical polarization; respectively, according as the earth's magnetic field has a positive or negative component in the same direction. *See also:* **radiation.**

(2) (radio wave propagation). The magneto-ionic wave component in which the electric vector rotates in the opposite sense to that for the ordinary-wave component. *See:* **ordinary-wave component.** *See also:* **radiation.** 42A65-0/E211-3E2

extreme operating conditions (automatic null-balancing electrical instrument). The range of operating conditions within which a device is designed to operate and under which operating influences are usually stated. *See also:* **measurement system.** 39A14-0

extrinsic properties (semiconductor). The properties of a semiconductor as modified by impurities or imperfections within the crystal. *See also:* **semiconductor; semiconductor device.** E102/E216/E270-34E17;0-10E1

extrinsic semiconductor. A semiconductor whose charge-carrier concentration is dependent upon impurities. *See also:* **semiconductor.** E59-34E17

eye bolt (rotating machinery). A bolt with a looped head used to engage a lifting hook. *See also:* **cradle base (rotating machinery).** 0-31E8

eyelet (soldered connections). A hollow tube inserted in a printed circuit or terminal board to provide electric connection or mechanical support for component leads. *See also:* **soldered connections (electronic and electric applications).** 99A1-0

eye light (television). Illumination on a person to provide a specular reflection from the eyes (and teeth) without adding a significant increase in light on the subject. *See also:* **television lighting.** Z7A1- 0

F

FA. *See:* **transformer, oil-immersed.**

fabric (rotating machinery). A planar structure comprising two or more sets of fiber yarns interlaced in such a way that the elements pass each other essentially at right angles and one set of elements is parallel to the fabric axis. AD123-31E8

faceplate (cathode-ray tube). The large transparent end of the envelope through which the image is viewed or projected. *See also:* **cathode-ray tubes.** 50I07-15E6

faceplate controller (industrial control). An electric controller consisting of a resistor and a faceplate switch in which the electric contacts are made between flat segments, arranged on a plane surface, and a contact arm. *See also:* **electric controller.** 42A25-34E10

faceplate rheostat (industrial control). A rheostat consisting of a tapped resistor and a panel with fixed contact members connected to the taps, and a lever carrying a contact rider over the fixed members for adjustment of the resistance. 42A25-34E10

facilitation. The brief rise of excitability above normal either after a response or after a series of subthreshold stimuli. *See also:* **biological.** 42A80-18E1

facility (communication). Anything used or available for use in the furnishing of communication service. *See also:* **communication.** 42A65-0

facing (planar structure) (rotating machinery). A fabric, mat, film, or other material used in intimate conjunction with a prime material and forming a relatively minor part of the composite for the purpose of protection, handling, or processing. *See also:* **asynchronous machine: direct-current commutating machine; synchronous machine.** 0-31E8

facsimile (electrical communication). The process, or the result of the process, by which fixed graphic material including pictures or images is scanned and the information converted into signal waves that are used either locally or remotely to produce, in record form, a likeness (facsimile) of the subject copy.
See:
bandwidth;
converter;
drum factor;
facsimile signal;
facsimile system;
facsimile transmission;
frame;
receiver;
recorder;
scanning;
solid-state scanning;
subject copy;
synchronizing;
transmitter, facsimile;
transmitting converter, facsimile. E168-0

facsimile signal (picture signal). A signal resulting from the scanning process.
See:
black signal;
effective band;
end-of-copy signal;
facsimile-signal level;
framing signal;
phasing signal;
ready-to-receive signal;
signal contrast;
signal frequency shift;
start record signal;
start signal;
stop record signal;
stop signal;
synchronizing signal;
tailing;
white signal.
See also: **facsimile.** E168-0

facsimile-signal level. The maximum facsimile signal power or voltage (root-mean-square or direct-current) measured at any point in a facsimile system. *Note:* It may be expressed in decibels with respect to some standard value as 1 milliwatt. *See also:* **facsimile signal (picture signal).** E168-0

facsimile system. An integrated assembly of the elements used for facsimile. *See also:* **facsimile (electrical communication).** E168-0

facsimile transient. A damped oscillatory transient occurring in the output of the system as a result of a sudden change in input. *See also:* **facsimile transmission.** E168-0

facsimile transmission. The transmission of signal waves produced by the scanning of fixed graphic material, including pictures, for reproduction in record form.
See:
baseband;
black transmission;
carrier beat;
delay equalizer;
dual modulation;
envelope delay;
envelope delay distortion;
facsimile transient;
maximum modulating frequency;
multipath transmission;
noise;
phase delay;
phase-frequency distortion;
picture inversion;
receiving converter;
subcarrier;
transmitting converter;
vestigial sideband;
vestigial-sideband transmission;
white transmission.
See also: **communication; facsimile.** E145/E168/42A65-0

factor of assurance (wire or cable insulation). The ratio of the voltage at which completed lengths are tested to that at which they are used. *See also:* **power distribution, underground construction.** E30-0;42A35-31E13

factory-renewable fuse. A fuse that, after circuit interruption, must be returned to the manufacturer to be restored for service. 37A100-31E11

fade in. To increase signal strength gradually in a sound or television channel. 42A65-0

fade out. To decrease signal strength gradually in a sound or television channel. 42A65-0

fading (radio wave propagation). The variation of radio field intensity caused by changes in the transmission medium, and transmission path, with time. *See also:* **radiation; radio wave propagation.** 42A65-3E2

failure (1) (reliability). The termination of the ability of an item to perform its required function.
(2) (catastrophic). Failures that are both sudden and complete.
(3) (complete). Failures resulting from deviations in characteristic(s) beyond specified limits such as to cause complete lack of the required function. *Note:* The limits referred to in this category are special limits specified for this purpose.
(4) (degradation). Failures that are both gradual and partial.
(5) (gradual). Failures that could be anticipated by prior examination.
(6) (inherent-weakness). Failures attributable to weakness inherent in the item itself when subjected to stresses within the stated capabilities of that item.
(7) (misuse). Failures attributable to the application of stresses beyond the stated capabilities of the item.
(8) (partial). Failures resulting from deviations in characteristic(s) beyond specified limits but not such as to cause complete lack of the required function.
(9) (random). Any failure whose cause and/or mechanism make its time of occurrence unpredictable.
(10) (secondary). Failure of an item caused either directly or indirectly by the failure of another item.
(11) (sudden). Failures that could not be anticipated by prior examination.
(12) (wear-out). A failure that occurs as a result of deterioration processes or mechanical wear and whose probability of occurrence increases with time. *See also:* **reliability.** 0-7E1

failure cause (reliability). The circumstance that induces or activates a failure mechanism. *See also:* **reliability.** 0-7E1

failure criteria (reliability). Rules for failure relevancy such as specified limits for the acceptability of an item. *See also:* **reliability.** 0-7E1

failure mechanism (reliability). The physical, chemical, or other process that results in a failure. *See also:* **reliability.** 0-7E1

failure mode (reliability). The effect by which a failure is observed; for example, an open- or short-circuit condition, or a gain change. *See also:* **reliability.** 0-7E1

failure rate (any point in the life of an item) (reliability). The incremental change in the number of failures per associated incremental change in time.

failure rate, assessed. The failure rate of an item determined within stated confidence limits from the observed failure rates of nominally identical items. *Note:* Alternatively, point estimates may be used, the basis of which must be defined. 0-7E1

failure rate, extrapolated. Extension by a defined extrapolation or interpolation of the assessed failure rate for durations or stress conditions different from those applying to the conditions of that assessed rate.

failure rate, instantaneous. *See:* **instantaneous failure rate.**

failure rate, observed. The ratio of the total number of failures in a single population to the total cumulative observed time on that population. *Notes:* (1) The observed failure rate is to be associated with particular, and stated, time intervals (or summation of intervals) and stress conditions. (2) The criteria for what constitutes a failure should be stated. (3) Cumulative observed time is a product of items and time or the sum of their products. 0-7E1

failure rate, predicted. The failure rate of an equipment or complex item computed from its design considerations and from the reliability of its parts in the intended conditions of use. *See also:* **reliability.** 0-7E1

failure-rate acceleration factor (reliability). The ratio of the accelerated testing failure rate to the failure rate under stated reference test conditions and time period. *See also:* **reliability.** 0-7E1

failure to trip (relay or relay system). In the performance the lack of tripping that should have occurred, considering the objectives of the relay system design. *See also:* **relay.** 37A1-31E6

fairlead (aircraft). A tube through which a trailing wire antenna is fed from an aircraft, with particular care in the design as to voltage breakdown and corona characteristics. *Note:* An antenna reel and counter are frequently a part of the assembly. *See also:* **air transportation electronic equipment.** 42A41-0

fall time. The time, in seconds, for the pointer to reach 0.1 (plus or minus a specified tolerance) of the end scale from a steady end-scale deflection when the instrument is short-circuited. *See also:* **moving element (instrument).** 39A1-0

fall time of a pulse. The time interval of the trailing edge of a pulse between stated limits. *See also:* **pulse.** 0-9E4

false course (navigation systems normally providing one or more course lines). A spurious additional course line indication due to undesired reflections or to a maladjustment of equipment. E172-10E6

false-proceed operation. The creation or continuance of a condition of the vehicle apparatus in an automatic train control or cab signal installation that is less restrictive than is required by the condition of the track of the controlling section, when the vehicle is at a point where the apparatus, is or should be, in operative relation with the controlling track elements. *See also:* **automatic train control.** 42A42-0

false-restrictive operation. The creation or continuance of a condition of the automatic train control or cab signal vehicle apparatus that is more restrictive than is required by the condition of the track of the controlling section when the vehicle apparatus is in operative relation with the controlling track elements, or which is caused by failure or derangement of some part of the apparatus. *See also:* **automatic train control.** 42A42-0

false tripping (relay or relay system). In the performance the tripping that should not have occurred, considering the objectives of the relay system design. *See also:* **relay.** 37A1-31E6

fan (blower) (rotating machinery). The part that provides an air stream for ventilating the machine.
See:
air cooler;
air filter;
blower blade;
fan cover;
inclined-blade;

propeller-type blower; radial-blade blower. 42A10-31E8

fan-beam antenna. A unidirectional antenna so designed that transverse cross sections of the major lobe are approximately elliptical. *See also:* **antenna.** 42A65-3E1

fan cover (rotating machinery). An enclosure for the fan that directs the flow of air. *See:* **fan (rotating machinery).** 0-31E8

fan-duty resistor (motors). A resistor that is intended for use in the armature or rotor circuit of a motor in which the current is approximately proportional to the speed of the motor. *See:* **asynchronous machine; direct-current commutating machine; electric drive; synchronous machine.** E45-0;IC1-34E10

fan housing (blower housing) (rotating machinery). The structure surrounding a fan and which forms the outer boundary for the air passing through the fan. *See also:* **cradle base (rotating machinery).** 0-31E8

fan-in network (power-system communication). A logic network whose output is a binary code in parallel form of n bits and having up to 2^n inputs with each input producing one of the output codes. *See also:* **digital.** 0-31E3

fan marker (electronic navigation). A marker having a vertically directed fan beam intersecting an airway to provide a position fix. *See also:* **radio navigation.** E172-10E6;42A65-0

fan-out network (power-system communication). A logic network taking n input bits in parallel and producing a unique logic output on the one and only one of up to 2^n outputs that corresponds to the input code. *See also:* **digital.** 0-31E3

fan shroud (rotating machinery). A structure, either stationary or rotating, that restricts leakage of gas past the blades of a fan. *See also:* **cradle base (rotating machinery).** 0-31E8

farad. The unit of capacitance in the International System of Electrical Units (SI units). The farad is the capacitance of a capacitor in which a charge of 1 coulomb produces 1 volt potential difference between its terminals. E270-0

faraday. The number of coulombs (96 485) required for an electrochemical reaction involving one chemical equivalent. *See also:* **electrochemistry.** 42A60-0

Faraday dark space (gas tube). The relatively nonluminous region in a glow-discharge cold-cathode tube between the negative glow and the positive column. *See:* **gas tubes.** 42A70-15E6

Faraday effect. *See:* **magnetic rotation (polarized light).**

Faraday rotation (radio wave propagation). The process of rotation of the polarization ellipse of an electromagnetic wave in a magnetoionic medium. *See:* **elliptically polarized wave.** *See also:* **radio wave propagation.** 0-3E2

Faraday's law (electromagnetic induction, circuit). The electromotive force induced is proportional to the time rate of change of magnetic flux linked with the circuit. E270-0

faradic current (electrotherapy). An asymmetrical alternating current obtained from or similar to that obtained from the secondary winding of an induction coil operated by repeatedly interrupting a direct current in the primary. *See also:* **electrotherapy.** 42A80-18E1

faradism. *See:* **faradization.**

faradization (faradism) (electrotherapy). The use of a faradic current to stimulate muscles and nerves. *See:* **faradic current.** *See also:* **electrotherapy.** 42A80-18E1

far-end crosstalk. Crosstalk that is propagated in a disturbed channel in the same direction as the direction of propagation of the current in the disturbing channel. The terminal of the disturbed channel at which the far-end crosstalk is present and the energized terminal of the disturbing channel are ordinarily remote from each other. *See also:* **coupling.** 42A65-0

far-field region. The region of the field of an antenna where the angular field distribution is essentially independent of the distance from the antenna. *Notes:* (1) If the antenna has a maximum over-all dimension D that is large compared to the wavelength, the far-field region is commonly taken to exist at distances greater than $2D^2/\lambda$ from the antenna, λ being the wavelength. (2) For an antenna focused at infinity, the far-field region is sometimes referred to as the Fraunhofer region on the basis of analogy to optical terminology. *See also:* **antenna.** 0-3E1

fastener (lightning protection). A device used to secure the conductor to the structure that supports it. *See also:* **lighting protection and equipment.** 42A95/5A1-0

fast groove (disk recording) (fast spiral). An unmodulated spiral groove having a pitch that is much greater than that of the recorded grooves. *See also:* **phonograph pickup.** E157-1E1

fast-operate, fast-release relay. A high-speed relay specifically designed for both short operate and short release time. 0-21E0

fast-operate relay. A high-speed relay specifically designed for short operate time but not necessarily short release time. 0-21E0

fast-operate, slow-release relay. A relay specifically designed for short operate time and long release time. 0-21E0

fast spiral. *See:* **fast groove.**

fast-time-constant circuit (radar). A circuit with short time-constant used to emphasize signals of short duration to produce discrimination against low frequency components of clutter. *See also:* **navigation.** E172-10E6

fatigue. The tendency for a metal to fracture in a brittle manner under conditions of repeated cyclic stressing at stress levels below its tensile strength. *See:* **corrosion terms.** CM-34E2

fault (1) (wire or cable). A partial or total local failure in the insulation or continuity of a conductor. *See also:* **center of distribution.** 42A35-31E13

(2) (components). A physical condition that causes a device, a component, or an element to fail to perform in a required manner, for example, a short-circuit, a broken wire, an intermittent connection. *See:* **pattern-sensitive fault; program-sensitive fault.** X3A12-16E9

(3) (lightning arresters). A disturbance that impairs normal operation, for example, insulation failure or conductor breakage. *See:* **lightning arrester (surge diverter).** 50I25-31E7

fault bus (fault ground bus). A bus connected to normally grounded parts of electric equipment, so insulated that all of the ground current passes to ground through fault-detecting means. 37A100-31E11/31E6

fault-bus protection. A method of ground-fault protection that makes use of a fault bus. 0-31E6

fault current. A current that flows from one conductor to ground or to another conductor owing to an abnormal connection (including an arc) between the two. *Note:* A fault current flowing to ground may be called a ground fault current. *See also:* **center of distribution; lightning arrester (surge diverter).** 42A35-31E13;50I25-31E7

fault-detector relay. A monitoring relay whose function is to limit the operation of associated protective relays to specific system conditions. 37A100-31E11/31E6

fault electrode current (1) (electron tubes) (surge electrode current*). The peak current that flows through an electrode under fault conditions, such as arc-backs and load short-circuits. *See also:* **electrode current (electron tube; gas-filled rectifier.**
*Deprecated 50I07-15E6;42A70-0

fault ground bus. *See:* **fault bus.**

fault-initiating switch. A mechanical switching device used in applied-fault protection to place a short-circuit on an energized circuit and to carry the resulting current until the circuit has been deenergized by protective operation. *Notes:* (1) This switch is operated by a stored-energy mechanism capable of closing the switch within a specified rated closing time at its rated making current. The switch may be opened either manually or by a power-operated mechanism. (2) The applied short-circuit may be intentionally limited to avoid excessive system disturbance. 37A100-31E11

fault resistance (lightning arresters). The resistance of that part of the fault path associated with the fault itself. *See:* **lightning arrester (surge diverter).** 50I25-31E7

Faure plate (storage cell) (pasted plate). A plate consisting of electroconductive material, which usually consists of lead-antimony alloy covered with oxides or salts of lead, that is subsequently transformed into active material. *See also:* **battery (primary or secondary).** 42A60-0

***F* display (radar).** A rectangular display in which a target appears as a centralized blip when the radar antenna is aimed at it. *Note:* Horizontal and vertical aiming errors are respectively indicated by the horizontal and vertical displacement of the blip. *See also:* **navigation.** E172-10E6

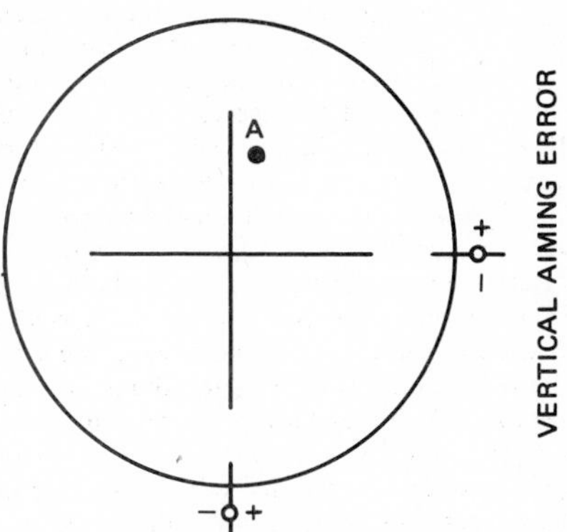

F display.

feed (1) (machines). (A) To supply the material to be operated upon to a machine. (B) A device capable of feeding as in (A). *See also:* **electronic digital computer.** E162-0

(2) (antenna). The portion of an antenna coupled to the terminals that functions to produce the aperture illumination. *See also:* **antenna.** 0-3E1

feedback (transmission system or section thereof). The returning of a fraction of the output to the input. *See:*
beta (β) circuit;
cross neutralization;
degeneration;
grid neutralization;
inductive neutralization;
mu (μ) circuit;
negative feedback;
neutralization;
Nyquist diagram;
plate neutralization;
positive feedback;
stabilized feedback.
See also: **forward path.** E145/E182/42A65-0

feedback admittance, short-circuit (electron-device transducer). *See:* **admittance, short-circuit feedback.**

feedback control system. *See:* **control system, feedback.**

feedback limiter. *See:* **limiter circuit.** *See also:* **electronic analog computer.** E165-16E9

feedback loop (numerically controlled machines). The part of a closed-loop system that provides controlled response information allowing comparison with a referenced command. *See also:* **numerically controlled machines.** EIA3B-34E12

feedback node (branch) (network analysis). A node (branch) contained in a loop. *See also:* **linear signal flow graphs.** E155-0

feedback oscillator. An oscillating circuit, including an amplifier, in which the output is coupled in phase with the input, the oscillation being maintained at a frequency determined by the parameters of the amplifier and the feedback circuits such as inductance-capacitance, resistance-capacitance, and other frequency-selective elements. *See also:* **oscillatory circuit.** E145-0

feedback signal (control) (industrial control). *See:* **signal, feedback.**

feedback winding (saturable reactor). A control winding to which a feedback connection is made. 42A65-0

feeder (power distribution). A set of conductors originating at a main distribution center and supplying one or more secondary distribution centers, one or more branch-circuit distribution centers, or any combination of these two types of equipment. *Notes:* (1) Bus tie circuits between generator and distribution switchboards, including those between main and emergency switchboards, are not considered as feeders. (2) Feeders terminate at the overcurrent device that protects the distribution center of lesser order.
See:
center of distribution;
direct feeder;
distribution;
interconnection;
lightning feeder;
loop feeder;
multiple feeder;

network feeder;
power feeder;
radial feeder;
subfeeder;
teed feeder;
tie feeder;
transmission feeder;
trunk feeder. E45-0

feeder cable (communication practice). A cable extending from the central office along a primary route (main feeder cable) or from a main feeder cable along a secondary route (branch feeder cable) and providing connections to one or more distribution cables. *See also:* **cable.** 42A65-0

feeder distribution center. A distribution center at which feeders or subfeeders are supplied. *See also:* **distribution center.** 42A95-0

feed function (numerically controlled machines). The relative velocity between the tool or instrument and the work due to motion of the programmed axis (axes). *See also:* **numerically controlled machines.** EIA3B-34E12

feed groove (rotating machinery). A groove provided to direct the flow of oil in a bearing. *See also:* **bearing.** 0-31E8

feeding point. The point of junction of a distribution feeder with a distribution main or service connection. *See also:* **center of distribution.** 42A35-31E13

feed line (rotating machinery). A supply pipe line. *See also:* **oil cup (rotating machinery).** 0-31E8

feed rate bypass (numerically controlled machines). A function directing the control system to ignore programmed feed rate and substitute a selected operational rate. *See also:* **numerically controlled machines.** EIA3B-34E12

feed rate override (numerically controlled machines). A manual function directing the control system to modify the programmed feed rate by a selected multiplier. *See also:* **numerically controlled machines.** EIA3B-34E12

fenestra method (lighting calculation). The procedure for predicting the interior illumination received from daylight through windows. *See also:* **inverse-square law (illuminating engineering).** Z7A1-0

fenestration. Any opening or arrangement of openings (normally filled with media for control) for the admission of daylight. *See also:* **sunlight.** Z7A1-0

Fermi level. The value of the electron energy at which the Fermi distribution function has the value one-half. *See also:* **electron emission.** E102-0

ferreed relay. Coined name (Bell Telephone Laboratories) for a special form of dry reed switch having a return magnetic path of high remanence material that provides a bistable, or latching, transfer contact. 0-21E0

ferritic. The body-centered cubic crystal structure of ferrous metals. *See:* **corrosion terms.** CM-34E2

ferrodynamic instrument. An electrodynamic instrument in which the forces are materially argmented by the presence of ferromagnetic material. *See also:* **instrument.** 42A30-0

ferroelectric axis. The crystallographic direction is parallel to the spontaneous polarization vector. *Note:* In some materials the ferroelectric axis may have several possible orientations with respect to the macroscopic crystal. *See also:* **ferroelectric domain.** E180-0

ferroelectric Curie temperature. The temperature above which ferroelectric materials do not exhibit reversible spontaneous polarization. *Note:* As the temperature is lowered from above the ferroelectric Curie temperature spontaneous polarization is detected by the onset of a hysteresis loop. The ferroelectric Curie temperature should be determined only with unstrained crystals, at atmospheric pressure, and with no externally applied direct-current fields. (In some ferroelectrics multiple hysteresis-loop patterns may be observed at temperatures slightly higher than the ferroelectric Curie temperature under alternating-current fields.) *See also:* **ferroelectric domain.** E180-0

ferroelectric domain. A region in a crystal throughout which the spontaneous polarization is uniformly directed.
See:
antiferroelectric crystal;
biased ferroelectric materials;
coercive electric field;
coercive voltage;
differential permittivity;
discrimination ratio;
ferroelectric axis;
ferroelectric Curie temperature;
induced charge;
measured switching time;
paraelectric Curie temperature;
polarization;
remanent charge;
reversible permittivity;
signal charge;
spontaneous polarization;
total charge;
total switching time. E180-0

ferromagnetic material. Material whose relative permeability is greater than unity and depends upon the magnetizing force. A ferromagnetic material usually has relatively high values of relative permeability and exhibits hysteresis. E270-0

ferrule (cartridge fuse). A fuse terminal of cylindrical shape at the end of a cartridge fuse. 37A100-31E11

fiber-optic plate (camera tubes). An array of fibers, individually clad with a lower index-of-refraction material, that transfers an optical image from one surface of the plate to the other. *See also:* **camera tube; circuit characteristics of electrodes.** 0-15E6

fibrillation (medical electronics). A continued, uncoordinated activity in the fibres of the heart, diaphragm, or other muscles consisting of rhythmical but asynchronous contraction and relaxation of individual fibres. *See also:* **medical electronics.** 0-18E1

fictitious power (polyphase circuit). At the terminals of entry, a vector equal to the (vector) sum of the fictitious powers for the individual terminals of entry. *Note:* The fictitious power for each terminal of entry is determined by considering each phase conductor and the common reference point as a single-phase circuit, as described for distortion power. The sign given to the distortion power in determining the fictitious power for each single-phase circuit shall be the same as that of the total active power. Fictitious power for a polyphase circuit has as its two rectangular components the reactive power and the distortion power. If the voltages have the same waveform as the corresponding currents, the magnitude of the fictitious power becomes the same as the reactive power. Fictitious power is expressed in volt-amperes when the voltages are in volts and the currents in amperes. E270-0

fictitious power (single-phase two-wire circuit). At the two terminals of entry into a delimited region, a vector quantity having as its rectangular components the reactive power and the distortion power. *Note:* Its magnitude is equal to the square root of the difference of the squares of the apparent power and the amplitude of the active power. Its magnitude is also equal to the square root of the sum of the squares of the amplitudes of reactive power and distortion power. If voltage and current have the same waveform, the magnitude of the fictitious power is equal to the reactive power. The magnitude of the fictitious power is given by the equation

$$F = (U^2 - p^2)^{1/2}$$
$$= (Q^2 + D^2)^{1/2}$$

$$= \left\{ \sum_{r=1}^{r=\infty} \sum_{q=1}^{q=\infty} \left[E_r^{\,2} I_q^{\,2} - E_r E_q I_r I_q \cos(\alpha_r - \beta_r) \cos(\alpha_q - \beta_q) \right] \right\}^{1/2}$$

where the symbols are those of **power, apparent (single-phase two-wire circuit).** In determining the vector position of the fictitious power, the sign of the distortion power component must be assigned arbitrarily. Fictitious power is expressed in volt-amperes when the voltage is in volts and the current in amperes. *See:* **distortion power (single-phase two-wire circuit).** E270-0

fidelity. The degree with which a system, or a portion of a system, accurately reproduces at its output the essential characteristics of the signal that is impressed upon its input. *See also:* **transmission characteristic.** E145/E188-0;42A65-31E3

field (1) (television). One of the two (or more) equal parts into which a frame is divided in interlaced scanning. *See also:* **television.** E203-2E2;42A65-0
(2) (record) (computing systems). A specified area used for a particular category of data, for example, a group of card columns used to represent a wage rate or a set of bit locations in a computer word used to express the address of the operand. *See:* **threshold field.** *See also:* **electronic digital computer.** X3A12-16E9

field accelerating relay (industrial control). A relay that functions automatically to maintain the armature current within limits, when accelerating to speeds above base speed, by controlling the excitation of the motor field. 42A25-34E10

field coil (rotating machinery) (1) (direct-current and salient-pole alternating-current machines). A suitably insulated winding to be mounted on a field pole to magnetize it.
(2) (cylindrical-rotor synchronous machines). A group of turns in the field winding, occupying one pair of slots. *See:* **asynchronous machine; direct-current commutating machine; synchronous machine.** 0-31E8

field-coil flange (rotating machinery). Insulation between the field coil and the pole shoe, and between the field coil and the member carrying the pole body, in a salient-pole machine. *See also:* **rotor (rotating machinery); stator.** 0-31E8

field control (motor) (industrial control). A method of controlling a motor by means of a change in the magnitude of the field current. *See:* **control.** AS1-34E10

field, critical (magnetrons). *See:* **critical field.**

field, cutoff (magnetrons). *See:* **critical field.**

field decelerating relay (industrial control). A relay that functions automatically to maintain the armature current or voltage within limits, when decelerating from speeds above base speed, by controlling the excitation of the motor field. *See also:* **relay.** 42A25-34E10

field discharge (switching device). A qualifying term indicating that the switching device has main contacts for energizing and deenergizing the field of a generator, motor, synchronous capacitor, or exciter; and has auxiliary contacts for short-circuiting the field through a discharge resistor at the instant preceding the opening of the main contacts. The auxiliary contacts also disconnect the field from the discharge resistor at the instant following the closing of the main contacts. *Note:* For direct-current generator operation the auxiliary contacts may open before the main contacts close. 37A100-31E11

field discharge protection (industrial control). A control function or device to limit the induced voltage in the field when the field circuit is disrupted or when an attempt is made to change the field current suddenly. *See:* **control.** AS1-34E10

field emission. Electron emission from a surface due directly to high-voltage gradients at the emitting surface. *See also:* **electron emission.** 42A70-15E6

field-enhanced photoelectric emission. The increased photoelectric emission resulting from the action of a strong electric field on the emitter. *See:* **phototubes.** E160-15E6

field-enhanced secondary emission. The increased secondary emission resulting from the action of a strong electric field on the emitter. *See:* **electron emission.** E160-15E6

field-failure protection (industrial control). The effect of a device, operative on the loss of field excitation, to cause and maintain the interruption of power in the motor armature circuit. 42A25-34E10

field-failure relay (industrial control). A relay that functions to disconnect the motor armature from the line in the event of loss of field excitation. *See also:* **relay.** 42A25-34E10

field forcing (industrial control). A control function that temporarily overexcites or underexcites the field of a rotating machine to increase the rate of change of flux. *See:* **control.** IC1-34E10

field forcing relay (industrial control). A relay that functions to increase the rate of change of field flux by underexciting or overexciting the field of a rotating machine. *See also:* **relay.** 42A25-34E10

field frame. *See:* **frame yoke.**

field-free emission current (1) (general). The emission current from an emitter when the electric gradient at the surface is zero. 50I07-15E6
(2) (cathode). The electron current drawn from the cathode when the electric gradient at the surface of the

cathode is zero. *See also:* **electron emission.** 42A70-15E6

field frequency (television). The product of frame frequency multiplied by the number of fields contained in one frame. *See also:* **television.** E203/42A65-2E2

field I^2R loss. The product of the measured resistance, in ohms, of the field winding, corrected to a specified temperature, and the square of the field current in amperes. *See also:* **synchronous machine.** 50A10-31E8

field-intensity meter*. *See:* **field-strength meter.**

*Deprecated

field-lead insulation (rotating machinery). The dielectric material applied to insulate the enclosed conductor connecting the collector rings to the coil end windings. *Note:* Field leads also include the pole jumpers forming the series connection between the concentric windings on each pole. Where rectangular strap leads are employed, the insulation may consist of either taped mica and glass or moulded mica and glass composites. Where circular rods are used, moulded laminate tubing is frequently employed as the primary insulation. *See also:* **asynchronous machine; collector-ring lead insulator; direct-current commutating machine; synchronous machine.** 0-31E8

field-limiting adjusting means (industrial control). The effect of a control function or device (such as a resistor) that limits the maximum or minimum field excitation of a motor or generator. *See:* **control.** IC1-34E10

field loss relay (industrial control). *See:* **motor-field failure relay.**

field pole (rotating machinery). A structure of magnetic material on which a field coil may be mounted. *Note:* There are two types of field poles: main and commutating. *See also:* **asynchronous machine; direct-current commutating machine; synchronous machine.** 42A10-0

field protection (industrial control). The effect of a control function or device to prevent overheating of the field excitation winding by reducing or interrupting the excitation of the shunt field while the machine is at rest. *See:* **control.** IC1-34E10

field protective relay (industrial control). A relay that functions to prevent overheating of the field excitation winding by reducing or interrupting the excitation of the shunt field. *See also:* **relay.** 42A25-34E10

field-renewable fuse. *See:* **renewable fuse.**

field-reversal permanent-magnet focusing (microwave tubes). Magnetic focusing by a limited series of field reversals, not periodic, whose location is usually related to breaks in the slow-wave circuit. *See:* **magnetrons.** 0-15E6

field rheostat. A rheostat designed to control the exciting current of an electric machine. 42A25-34E10

field-sequential (color television). Pertaining to the association of individual primary colors with successive fields. *Examples:* Field-sequential pickup, field-sequential display, field-sequential system, field-sequential transmission. *See also:* **color terms.** E201-2E2

field shunting control (shunted-field control) A system of regulating the tractive force of an electrically driven vehicle by shunting, and thus weakening, the traction motor series fields by means of a resistor. *See also:* **multiple-unit control.** 42A42-0

field spool (rotating machinery). A structure for the support of a field coil in a salient-pole machine, either constructed of insulating material or carrying field-spool insulation. *See also:* **asynchronous machine; direct-current commutating machine; synchronous machine.** 0-31E8

field-spool insulation (rotating machinery). Insulation between the field spool and the field coil in a salient-pole machine. *See also:* **asynchronous machine; direct-current commutating machine; synchronous machine.** 0-31E8

field strength (electromagnetic wave). A general term that usually means the magnitude of the electric field vector, commonly expressed in volts per meter, but that may also mean the magnitude of the magnetic field vector, commonly expressed in amperes (or ampere-turns) per meter. *Note:* At frequencies above about 100 megahertz, and particularly above 1000 megahertz, field strength in the far zone is sometimes identified with power flux density P. For a linearly polarized wave in free space $P = E^2/(\mu_v/\epsilon_v)$, where E is the electric field strength, and μ_v and ϵ_v are the magnetic and electric constants of free space, respectively. When P is expressed in watts per square meter and E in volts per meter, the denominator is often rounded off to 120π. *See:* **electric field strength; magnetic field strength.** *See also:* **measurement system.** E284-9E1

field-strength meter. A calibrated radio receiver for measuring field strength. *See also:* **induction heater; intensity meter.** E54/E169-0

field system (rotating machinery). The portion of a direct-current or synchronous machine that produces the excitation flux. *See also:* **asynchronous machine; direct-current commutating machine; synchronous machine.** 0-31E8

field terminal (rotating machinery). A termination for the field winding. *See also:* **rotor (rotating machinery); stator.** 0-31E8

field tests (1) (cable systems). Dielectric tests which may be made on a cable system (including the pothead) by the user after installation, as an acceptance or proof test. *See also:* **pothead.** E48-0

(2) (switchgear). Tests made on operating systems usually for the purpose of investigating the performance of switchgear or its component parts under conditions that cannot be duplicated in the factory. *Note:* Field tests are usually supplementary to factory tests and therefore may not provide a complete investigation of capabilities. 37A100-31E11

field-turn insulation (rotating machinery). Insulation in the form of strip or tape separating the individual turns of a field winding. *See also:* **asynchronous machine; direct-current commutating machine; synchronous machine.** 0-31E8

field winding (excitation winding) (rotating machinery). A winding on either the stationary or the rotating part of a synchronous machine whose sole purpose is the production of the main electromagnetic field of the machine. *See also:* **synchronous machine.** 0-31E8

fifth voltage range. *See:* **voltage range.**

figure of merit (1) (magnetic amplifier). The ratio of

power amplification to time constant in seconds. *See also:* **rating and testing magnetic amplifiers.** E107-0

(2) (thermoelectric couple).

$$\alpha^2[(\rho_1\kappa_1)^{1/2} + (\rho_2\kappa_2)^{1/2}]^{-2}$$

where α is the Seebeck coefficient of the couple and ρ_1, ρ_2, κ_1, and κ_2 are the respective electric resistivities and thermal conductivities of materials 1 and 2. *Note:* This figure of merit applies to materials for thermoelectric devices whose operation is based on the Seebeck effect or the Peltier effect. *See also:* **thermoelectric device.**

(3) (thermoelectric couple, ideal).

$$\bar{\alpha}^{-2}[(\overline{\rho_1\kappa_1})^{1/2} + (\overline{\rho_2\kappa_2})^{1/2}]^{-2}$$

where $\bar{\alpha}$ is the average value of the Seebeck coefficient of the couple and $\overline{\rho_1\kappa_1}$ and $\overline{\rho_2\kappa_2}$ are the average values of the products of the respective electric resistivities and thermal conductivities of materials 1 and 2, where the averages are found by integrating the parameters over the specified temperature range of the couple. *See also:* **thermoelectric device.**

(4) (thermoelectric material). The quotient of (A) the square of the absolute Seebeck coefficient α by (B) the product of the electric resistivity ρ and the thermal conductivity κ.

$$\alpha^2/\rho\kappa.$$

Note: This figure of merit applies to materials for thermoelectric devices whose operation is based on the Seebeck effect or the Peltier effect. *See also:* **thermoelectric device.** E221-15E7

filament (electron tube). A hot cathode, usually in the form of a wire or ribbon, to which heat may be supplied by passing current through it. *Note:* This is also known as a filamentary cathode. *See also:* **electrode (electron tube).** 42A70-15E6

filament current. The current supplied to a filament to heat it. *See also:* **electronic controller; heater current.** 42A70-15E6

filament power supply (electron tube). The means for supplying power to the filament. *See also:* **power pack.** 42A65-0-15E6

filament voltage. The voltage between the terminals of a filament. *See also:* **electrode voltage (electron tube); electronic controller.** 42A70

file (computing systems). A collection of related records treated as a unit. *Note:* Thus in inventory control, one line of an invoice forms an item, a complete invoice forms a record, and the complete set of such records forms a file. *See also:* **electronic digital computer.** X3A12-16E9

file gap (computing systems). An area on a storage medium, such as tape, used to indicate the end of a file. X3A12-16E9

file maintenance (computing systems). The activity of keeping a file up to date by adding, changing, or deleting data. *See also:* **electronic digital computer.** X3A12-16E9

filiform corrosion. *See:* **underfilm corrosion.**

filled-core annular conductor. A conductor composed of a plurality of conducting elements disposed around a nonconducting supporting material that substantially fills the space enclosed by the conducting elements. *See also:* **conductor.** 42A35-31E13

filled tape. Fabric tape that has been thoroughly filled with a rubber or synthetic compound, but not necessarily finished on either side with this compound. *See also:* **conductor.** 42A35-31E13

filler (filler strip) (1) (rotating machinery). Additional insulating material used to insure a tight depthwise fit in the slot. *See also:* **rotor (rotating machinery); stator.** 0-31E8

(2) (mechanical recording). The inert material of a record compound as distinguished from the binder. *See also:* **phonograph pickup.** E157-1E1

filler strip. *See:* **filler.**

fill light (television). Supplementary illumination to reduce shadow or contrast range. *See also:* **television lighting.** Z7A1-0

film (1) (rotating machinery). Sheeting having a nominal thickness not greater than 0.030 centimeters and being substantially homogeneous in nature. *See also:* **asynchronous machine; direct-current commutating machine; electrochemical valve; synchronous machine.** 0-31E8

(2) (electrochemical valve). The layer adjacent to the valve metal and in which is located the high-potential drop when current flows in the direction of high impedance. *See also:* **electrochemical valve.** 42A60-0

film integrated circuit. An integrated circuit whose elements are films formed in situ upon an insulating substrate. *Note:* To further define the nature of a film integrated circuit, additional modifiers may be prefixed. Examples are (1) thin-film integrated circuit, (2) thick-film integrated circuit. *See:* **magnetic thin film; thin film.** *See also:* **electrochemical valve; electronic digital computer; integrated circuit.** E274-15E7/21E0

filter (1) (wave filter). A transducer for separating waves on the basis of their frequency. *Note:* A **filter** introduces relatively small insertion loss to waves in one or more frequency bands and relatively large insertion loss to waves of other frequencies.

See:

acoustic wave filter;
circuits and devices;
electric filter;
filter, band-elimination;
filter, band-pass;
filter, high-pass;
filter impedance compensator;
filter, low-pass;
filter, sound effects;
mechanical wave filter;
rejection filter;
ripple filter. E151-42A65

(2) (computing systems). (A) A device or program that separates data, signals, or material in accordance with specified criteria. (B) A mask. X3A12-16E9

filter attenuation band (filter stop band). A frequency band of attenuation; that is a frequency band in which, if dissipation is neglected, the attenuation constant is not zero. *See also:* **attenuation.** 42A65-0

filter, band-elimination (signal-transmission system). A filter that has a single attenuation band, neither of the cutoff frequencies being zero or infinite. *See also:* **rejection filter; filter.** 42A65-31E3;E151-0

filter, band-pass. A filter that has a single transmission band, neither of the cutoff frequencies being zero or infinite. *See also:* **filter.** E151-0

filter capacitor. A capacitor used as an element of an electric wave filter. *See also:* **electronic controller.** 42A25-34E10

filter factor (illuminating engineering). The transmittance of *black light* by a filter. *Note:* The relationship among these terms is illustrated by the following formula for determining the luminance (photometric brightness) of fluorescent materials exposed to black light:

$$\text{footlamberts} = \frac{\text{fluorens}}{\text{square feet}} \times \text{glow factor} \times \text{filter factor.}$$

When integral-filter black-light lamps are used, the filter factor is dropped from the formula because it already has been applied in assigning fluoren ratings to these lamps. *See also:* **ultraviolet radiation.** Z7A1-0

filter, high-pass. A filter having a single transmission band extending from some cutoff frequency, not zero, up to infinite frequency. *See also:* **filter.** E151-0

filter impedance compensator. An impedance compensator that is connected across the common terminals of electric wave filters when the latter are used in parallel in order to compensate for the effects of the filters on each other. *See also:* **network analysis; filter.** E270/42A65-0

filter inductor An inductor used as an element of an electric wave filter. *See also:* **electronic controller.** 42A25-34E10

filter, low-pass. A filter having a single transmission band extending from zero to some cutoff frequency, not infinite. *See also:* **filter.** E151-0

filter pass band. *See:* **filter transmission band.**

filter, rejection (signal-transmission system). *See:* **rejection filter.** *See also:* **filter.**

filters (power supplies). Resistance-capacitance or inductance-capacitance networks arranged as low-pass devices to attenuate the varying component that remains when alternating-current voltage is rectified. *Note:* In power supplies without subsequent active series regulators, the filters determine the amount of ripple that will remain in the direct-current output. In supplies with active-feedback series regulators, the regulator mainly controls the ripple, with output filtering serving chiefly for phase-gain control as a lag element. *See also:* **power supply.** KPSH-10E1

filter, sound effects (electroacoustics). A filter used to adjust the frequency response of a system for the purpose of achieving special aural effects. *See:* **filter.** E151-0

filter stop band. *See:* **filter attenuation band.**

filter transmission band (filter pass band). A frequency band of free transmission; that is, a frequency band in which, if dissipation is neglected, the attenuation constant is zero. *See also:* **transmission characteristics.** 42A65-0

final approach path (electronic navigation). *See:* **approach path.**

final contact pressure. *See:* **contact pressure, final.**

final controlling element (electric power systems). The controlling element that directly changes the value of the manipulated variable. E94-0

final emergency circuits. All circuits (including temporary emergency circuits) that, after failure of a ship's service supply, may be supplied by the emergency generator. *See also:* **emergency electric system.** 42A43-0

final emergency lighting. Temporary emergency lighting plus manually controlled lighting of the boat deck and overside to facilitate lifeboat loading and launching. *See also:* **emergency electric system.** 42A43-0

final-terminal stopping device (elevators). A device that automatically causes the power to be removed from an electric elevator or dumbwaiter driving-machine motor and brake or from a hydraulic elevator or dumbwaiter machine independent of the functioning of the normal-terminal stopping device, the operating device, or any emergency terminal stopping device, after the car has passed a terminal landing. *See:* **control.** 42A45-0

final unloaded conductor tension (electric systems). The longitudinal tension in a conductor after the conductor has been stretched by the application for an appreciable period, and subsequent release, of the loadings of ice and wind, and temperature decrease, assumed for the loading district in which the conductor is strung (or equivalent loading). *See also:* **conductor; initial conductor tension.** 2A2-0

final unloaded sag. The sag of a conductor after it has been subjected for an appreciable period to the loading prescribed for the loading district in which it is situated, or equivalent loading, and the loading removed. *See:* **sag; power distribution; overhead construction.** 2A2-0

final value (industrial control). The steady-state value of a specified variable. *See:* **control.** AS1-34E10

final voltage. *See:* **cutoff voltage.**

finder switch. An automatic switch for finding a calling subscriber line or trunk and connecting it to the switching apparatus. *See also:* **telephone switching system.** 42A65-0

fine chrominance primary (color-television system at present standardized for broadcasting in the United States). One of the two chrominance primaries that is associated with the greater transmission bandwidth. *See also:* **color terms.** E201-2E2

finger (rotating machinery). *See:* **end finger.**

finishing (electrotype). The operation of bringing all parts of the printing surface into the same plane, or, more strictly speaking, into positions having equal printing values. *See also:* **electroforming.** 42A60-0

finishing rate (storage battery) (storage cell). The rate of charge expressed in amperes to which the charging current for some types of lead batteries is reduced near the end of charge to prevent excessive gassing and temperature rise. *See also:* **charge.** 42A60-0

finite energy signal. *See:* **energy density signal.**

finite-time stability. *See:* **stability, finite-time.**

finsen. The recommended practical unit of erythemal flux or intensity of radiation. It is equal to one unit of erythemal flux per square centimeter. 42A55-0

fire-alarm system. An alarm system signaling the presence of fire. *See also:* **protective signaling.** 42A65-0

fire-alarm thermostat. An electric switch designed to operate in response to the application of heat. *See also:* **protective signaling.** 42A65-0

fire-control radar. A radar whose prime function is to provide information for the manual or automatic control of artillery or other weapons. *See also:* **radar.** 0-10E6

fire-door magnet. An electromagnet for holding open a self-closing fire door. 42A43-0

fire-door release system. A system providing remotely controlled release of self-closing doors in fire-resisting bulkheads to check the spread of fire. *See also:* **marine electric apparatus.** 42A43-0

fired tube (microwave gas tubes). The condition of the tube during which a radio-frequency glow discharge exists at either the resonant gap, resonant window, or both. *See:* **gas tubes.** E160-15E6

fireproofing (cables) . The application of a fire-resisting covering to protect them from arcs in an adjacent cable or from fires from any cause. *See also:* **power distribution, underground construction.** 42A35-31E13

fire-resistance rating. The measured time, in hours or fractions thereof, that the material or construction will withstand fire exposure as determined by fire tests conducted in conformity to recognize standards. 42A45-0

fire-resistant (fire-resistive). So constructed or treated that it will not be injured readily by exposure to fire. 42A95-0

fire-resistive construction. A method of construction that prevents or retards the passage of hot gases or flames as defined by the fire-resistance rating. 42A45-0

firm capacity purchases or sales (electric power supply). That firm capacity that is purchased, or sold, in transactions with other systems and that is not from designated units, but is from the over-all system of the seller. *Note:* It is understood that the seller provides reserve capacity for this type of transaction. *See also:* **generating station.** 0-31E4

firm power. Power intended to be always available even under emergency conditions. *See also:* **generating station.** 42A35-31E13

firm transfer capability (transmission) (electric power supply). The maximum amount of power that can be interchanged continuously, over an extended period of time. *See also:* **generating station.** 0-31E4

first dial (register). That graduated circle or cyclometer wheel, the reading on which changes most rapidly. The test dial or dials, if any, are not considered. *See also:* **watthour meter.** 42A30-0

first Fresnel zone (optics and radio communication). The circular portion of a wave front, transverse to the line between an emitter and a more-distant point where the resultant disturbance is being observed, whose center is the intersection of the front with the direct ray and whose radius is such that the shortest path from the emitter through the periphery to the receiving point is one half-wave longer than the ray. *Note:* A second zone, a third, etcetera, are defined by successive increases of the path by half-wave increments. *See also:* **radiation.** 42A65-0

first Townsend discharge (gas). A semi-self-maintained discharge in which the additional ions that appear are due solely to the ionization of the gas by electron collisions. *See also:* **discharge (gas).** 50I07-15E6

first voltage range. *See:* **voltage range.**

fishing wire. *See:* **fish tape.**

fish tape (fishing wire) (snake). A tempered steel wire, usually of rectangular cross section, that is pushed through a run of conduit or through an inaccessible space, such as a partition, and that is used for drawing in the wires. *See also:* **interior wiring.** 42A95-0

fitting (electric system). An accessory such as a locknut, bushing, or other part of a wiring system that is intended primarily to perform a mechanical rather than an electrical function. *See also:* **raceways.** 1A0-0;42A95-0

fix (1) (navigation). A position determined without reference to any former position. *See also:* **radio navigation.** E172-10E6;42A65-0

(2) (interference) (electromagnetic compatibility). A device or equipment modification to prevent interference or to reduce an equipment's susceptibility to interference. *See also:* **electromagnetic compatibility.** 0-27E1

fixed block format (numerically controlled machines). A format in which the number and sequence of words and characters appearing in successive blocks is constant. *See also:* **numerically controlled machines.** EIA3B-34E12

fixed cycle (numerically controlled machines). A preset series of operations that direct machine axis movement and/or cause spindle operation to complete such actions as boring, drilling, tapping, or combinations thereof. *See also:* **numerically controlled machines.** EIA3B-34E12

fixed-cycle operation. An operation that is completed in a specified number of regularly timed execution cycles. X3A12-16E9

fixed-frequency transmitter. A transmitter designed for operation on a single carrier frequency. *See also:* **radio transmitter.** E145/E182/42A65-31E3

fixed impedance-type ballast. A reference ballast designed for use with one specific type of lamp and, after adjustment during the original calibration, is expected to hold its established impedance throughout normal use. 82A9-0

fixed light. A light having a constant luminous intensity when observed from a fixed point. *See also:* **signal lighting.** Z7A1-0

fixed motor connections. A method of connecting electric traction motors wherein there is no change in the motor interconnections throughout the operating range. *Note:* This term is used to indicate that a transition from series to parallel relation is not provided. *See also:* **traction motor.** 42A42-0

fixed point (electronic computation). Pertaining to a numeration system in which the position of the point is fixed with respect to one end of the numerals, according to some convention. *See:* **floating point; variable point.** *See also:* **electronic digital computer.** E162-0;X3A12-16E9

fixed-point fire-alarm thermostat. A fire-alarm thermostat designed to operate at a predetermined temperature. *See also:* **protective signaling.** 42A65-0

fixed-point system (electronic computation). *See:* **point.**

fixed sequential format. A means of identifying a word by its location in the block. *Note:* Words must be presented in a specific order and all possible words preceding the last desired word must be present in the block. EIA3B-34E12

fixed signal. A signal of fixed location indicating a condition affecting the movement of a train or engine. *See also:* **railway signal and interlocking.** 42A42-0

fixed storage (computing systems). A storage device that stores data not alterable by computer instructions, for example, magnetic core storage with a lockout feature or punched paper tape. *See:* **nonerasable storage; permanent storage; read-only storage.** *See also:* **static magnetic storage.** X3A12-16E9

fixed transmitter. A transmitter that is operated in a fixed or permanent location. *See also:* **radio transmitter.** 42A65/E145/E182-0

fixing (electrostatography). The act of making a developed image permanent. *See also:* **electrostatography.** E224-15E7

fixture stud (stud). A threaded fitting used to mount a lighting fixture in an outlet box. *See also:* **cabinet.** 42A95-0

flag (computing systems). (1) Any of various types of indicators used for indentification, for example, a wordmark. (2) A character that signals the occurrence of some condition, such as the end of a word. *See:* **mark; sentinal; tag.** *See also:* **electronic digital computer.** X3A12-16E9

flag alarm. An indicator in certain types of navigation instruments used to warn when the readings are unreliable. *See also:* **navigation.** E172-10E6

flameproof. *See:* **explosionproof.**

flameproof apparatus. Apparatus so treated that it will not maintain a flame or will not be injured readily when subjected to flame. 42A95-0

flameproof terminal box. A terminal box so designed that it may form part of a flameproof enclosure. *See also:* **cradle base (rotating machinery).** 0-31E8

flame protection of vapor openings. Self-closing gauge hatches, vapor seals, pressure-vacuum breather valves, flame arresters, or other reasonably effective means to minimize the possibility of flame entering the vapor space of a tank. *Note:* Where such a device is used, the tank is said to be flameproofed. 5A1-0

flame-resistant cable. A portable cable that will meet the flame test requirements of the United States Bureau of Mines. *See also:* **mine feeder circuit.** 42A85-0

flame resisting. *See:* **flame retarding.**

flame-retardant. So constructed or treated that it will not support or convey flame. 37A100-31E11;42A95-0

flame-retarding (flame resisting). Flame-retarding materials and structures should have such fire-resisting properties that they will not convey flame nor continue to burn for longer times than specified in the appropriate flame test. *Note:* Compliance with the requirements of the preceding paragraph should be determined with apparatus and according to the methods described in the Underwriters' Laboratory Standards for the materials and structures. E45-0

flammable air-vapor mixtures. When flammable vapors are mixed with air in certain proportions, the mixture will burn rapidly when ignited. *Note:* The combustion range for ordinary petroleum products, such as gasoline, is from 1½ to 6 percent of vapor by volume, the remainder being air. 5A1-0

flammable vapors. The vapors given off from a flammable liquid at and above its flash point. 5A1-0

flange. *See:* **coupling flange.**

flare-out (navigation). The portion of the approach path of an aircraft in which the slope is modified to provide the appropriate rate of descent at touchdown. *See also:* **navigation.** 0-10E6

flarescan (electronic navigation). A ground-based navigation system used in conjunction with an instrument approach system to provide flare-out vertical guidance to an aircraft by the use of a pulse-space-coded vertically scanning fan beam that provides elevation angle data. *See also:* **navigation.** 0-10E6

flash barrier (rotating machinery). A screen of fire-resistant material to prevent the formation of an arc or to minimize the damage caused thereby. *See also:* **cradle base (rotating machinery).** 0-31E8

flash current (primary cell) The maximum electric current indicated by an ammeter of the dead-beat type when connected directly to the terminals of the cell or battery by wires that together with the meter have a resistance of 0.01 ohm. *See also:* **electrolytic cell.** 42A60-0

flasher. A device for alternately and automatically lighting and extinguishing electric lamps. *See also:* **appliances (including portable).** 42A95-0

flasher relay. A relay that is so designed that when energized its contacts open and close at predetermined intervals. *See also:* **appliances (including portable); railway signal and interlocking; relay.** 42A42-0

flashing light. A rhythmic light in which the periods of light are of equal duration and are clearly shorter than the periods of darkness. *See also:* **signal lighting.** Z7A1-0

flashing-light signal. A railroad-highway crossing signal the indication of which is given by two red lights spaced horizontally and flashed alternately at predetermined intervals to give warning of the approach of trains, or a fixed signal in which the indications are given by color and flashing of one or more of the signal lights. *See also:* **railway signal and equipment.** 42A42-0

flashlight battery. A battery designed or employed to light a lamp of an electric hand lantern or flashlight. *See also:* **battery (primary or secondary).** 42A60-0

flashover (1) (lightning arresters). A distruptive discharge through air around or over the surface of solid or liquid insulation, between parts of different potential or polarity, produced by the application of voltage wherein the breakdown path becomes sufficiently ionized to maintain an electric arc. *See also:* **lightning arrester (surge diverter); power system, low-frequency and surge testing; test voltage and current.** E49/E270-0;42A35-31E13;62A1-31E7;68A1-31E5

(2) (pothead). A disruptive discharge around or over the surface of the pothead insulator, between parts of different potential or polarity, produced by the application of voltage, wherein the breakdown path becomes sufficiently ionized to maintain an electric arc. *See also:* **pothead.** E48-0

flash plate. A thin electrodeposited coating produced in a short time. 42A60-0

flash point. The minimum temperature at which a liquid will give off vapor in sufficient amount to form a flammable air-vapor mixture that can be ignited under specified conditions. 5A1-0

flat-compounded. A qualifying term applied to a compound-wound generator to denote that the series winding is so proportioned that the terminal voltage at rated load is the same as at no load. *See:* **direct-current commutating machine.** 42A10-0

flat leakage power (microwave gas tubes). The peak radio-frequency power transmitted through the tube after the establishment of the steady-state radio-frequency discharge. *See:* **gas tubes.** E160-15E6

flat-strip conductor. *See:* **strip (-type) transmission line.**

***F* layer.** An ionized layer in the *F* region. *See also:* **radiation.** 42A65-0

F^1 layer (radio wave propagation). The lower of the two ionized layers normally existing in the *F* region in the day hemisphere. *See also:* **radiation; radio wave propagation.** E211-3E2;42A65-0

F^2 layer (radio wave propagation). The single ionized layer normally existing in the *F* region in the night hemisphere and the higher of the two layers normally existing in the *F* region in the day hemisphere. *See also:* **radiation; radio wave propagation.** E211-3E2

flection-point emission current. That value of current on the diode characteristic for which the second derivative of the current with respect to the voltage has its maximum negative value. *Note:* This current corresponds to the upper flection point of the diode characteristic. *See also:* **circuit characteristics of electrodes; electron emission.** E160-15E6

flexible connector (rotating machinery). An electric connection that permits expansion, contraction, or relative motion of the connected parts. *See also:* **cradle base (rotating machinery).** 0-31E8

flexible coupling (rotating machinery). A coupling having relatively high transverse or torsional compliance. *Notes:* (1) May be used to reduce or eliminate transverse loads or deflections of one shaft from being carried, or felt by the other coupled shaft. (2) May be used to reduce the torsional stiffness between two rotating masses in order to change torsional natural frequencies of the shaft system or to limit transient or pulsating torques carried by the shafts. *See also:* **rotor (rotating machinery).** 0-31E8

flexible metal conduit. A flexible raceway of circular cross section specially constructed for the purpose of the pulling in or the withdrawing of wires or cables after the conduit and its fittings are in place. *See also:* **raceway.** 42A95-0

flexible mounting (rotating machinery). A flexible structure between the core and foundation used to reduce the transmission of vibration. *See also:* **cradle base (rotating machinery).** 0-31E8

flexible nonmetallic tubing (loom). A mechanical protection for electric conductors that consists of a flexible cylindrical tube having a smooth interior and a single or double wall of nonconducting fibrous material. *See also:* **raceways.** 42A95-0

flexible tower (frame). A tower that is dependent on the line conductors for longitudinal stability but is designed to resist transverse and vertical loads. *See also:* **tower.** 42A35/31E13

flexible waveguide. A waveguide constructed to permit bending and/or twisting without appreciable change in its electrical properties. *See:* **waveguide.** 50I62-3E1

flicker (1) (general). Impression of fluctuating brightness or color, occurring when the frequency of the observed variation lies between a few hertz and the fusion frequencies of the images. *See also:* **color; television.** 50I45-2E2

(2) (television). The sensation produced when the field frequency is insufficient to produce complete fusion of the visual images. *See also:* **television.** 42A65-0

flicker effect (electron tubes). The random variations of the output-current in a valve or tube with an oxide-coated cathode. *Note:* Its value varies inversely with the frequency. *See also:* **electron tube.** 50I07-15E6

flicker fusion frequency. The frequency of intermittent stimulation of the eye at which flicker disappears. *Note:* It also is called **critical fusion frequency** or **critical flicker frequency.** *See also:* **visual field.** Z7A1-0

flight path (navigation). A proposed route in three dimensions. *See also:* **course line; navigation.** E172-10E6

flight-path computer (electronic navigation). Equipment providing outputs for the control of the motion of a vehicle along a flight path. *See also:* **navigation.** 0-10E6

flight-path deviation (electronic navigation). The amount by which the flight track of a vehicle differs from its flight path expressed in terms of either angular or linear measurement. *See also:* **navigation.** 0-10E6

flight-path-deviation indicator (electronic navigation). A device providing a visual display of flight-path deviation. *See also:* **navigation.** 0-10E6

flight track (electronic navigation). The path in space actually traced by a vehicle. *See also:* **track.** E172-10E6

flip-flop (electronic computation or control). (1) A circuit or device, containing active elements, capable of assuming either one of two stable states at a given time, the particular state being dependent upon (A) the nature of an input signal, for example, its polarity, amplitude, and duration, and (B) which of two input terminals last received the signal. *Note:* The input and output coupling networks, and indicators, may be considered as an integral part of the flip-flop. (2) A device, as in (1) above, that is capable of counting modulo 2, in which case it might have only one input terminal. (3) A sequential logic element having properties similar to (1) or (2) above. *See also:* **control system, feedback; electronic digital computer; toggle.** E162-0

flip-flop circuit. A trigger circuit having two conditions of permanent stability, with means for passing from one to the other by an external stimulus. *See also:* **trigger circuit.** 42A65-31E3

floating. A method of operation for storage batteries in which a constant voltage is applied to the battery terminals sufficient to maintain an approximately constant state of charge. *See:* **trickle charge.** *See also:* **charge.** 42A60-0

floating battery. A storage battery that is kept in operating condition by a continuous charge at a low rate. *See also:* **railway signal and equipment.** 42A42-0

floating carrier. *See:* **controlled carrier.**

floating grid (electron tubes). An insulated grid, the potential of which is not fixed. *See also:* **electronic tube.** 50I07-15E6

floating neutral. One whose voltage to ground is free to vary when circuit conditions change. *See also:* **center of distribution.** 42A35-31E3

floating point. Pertaining to a system in which the location of the point does not remain fixed with respect to one end of the numerical expressions, but is regularly recalculated. The location of the point is usually given by expressing a power of the base. *See also:* **electronic digital computer; fixed point; variable point.** E162-0

floating-point system (electronic computation). *See:* **point.**

floating zero (numerically controlled machines). A characteristic of a numerical machine control permitting the zero reference point on an axis to be established readily at any point in the travel. *Note:* The control retains no information on the location of any previously established zeros. *See:* **zero offset.** *See also:* **numerically controlled machines.** EIA3B-34E12

float switch (liquid-level switch) (industrial control). A switch in which actuation of the contacts is effected when a float reaches a predetermined level. *See:* **switch.** 50I16-34E10

flood (charge-storage tubes) (verb). To direct a large-area flow of electrons, containing no spatially distributed information, toward a storage assembly. *Note:* A large-area flow of electrons with spatially distributed information is used in image-converter tubes. *See also:* **charge-storage tube.** E158-15E6

floodlight. A projector designed for lighting a scene or object to a brightness considerably greater than its surroundings. *Notes:* (1) It usually is capable of being pointed in any direction and is of weatherproof construction. (2) The beam spread of floodlights may range from relatively narrow (10 degrees) to wide (more than 100 degrees). *See also:* **floodlighting.** Z7A1-0

floodlighting. A system designed for lighting a scene or object to a brightness greater than its surroundings. *Note:* It may be for utility, advertising, or decorative purposes.
See:
floodlight;
general-purpose floodlight;
ground-area open floodlight;
ground-area open floodlight with reflector insert;
heavy-duty floodlight;
light;
protective lighting. Z7A1-0

flood-lubricated bearing (rotating machinery). A bearing in which a continuous flow of lubricant is poured over the top of the bearing or journal at about normal atmospheric pressure. *See also:* **bearing.** 0-31E8

flood projection (facsimile). The optical method of scanning in which the subject copy is floodlighted and the scanning spot is defined in the path of the reflected or transmitted light. *See also:* **scanning (facsimile).** E168-0

floor bushing. A bushing intended primarily to be operated entirely indoors in a substantially vertical position to carry a circuit through a floor or horizontal grounded barrier. Both ends must be suitable for operating in air. *See also:* **bushing.** E49-0

floor lamp. A portable luminaire on a high stand suitable for standing on the floor. *See also:* **luminaire.** Z7A1-0

floor trap (burglar-alarm system). A device designed to indicate an alarm condition in an electric protective circuit whenever an intruder breaks or moves a thread or conductor extending across a floor space. *See also:* **protective signaling.** 42A65-0

flowchart (computing systems). A graphical representation for the definition, analysis, or solution of a problem, in which symbols are used to represent operations, data, flow, and equipment. X3A12-16E9

flow diagram (electronic computers). Graphic representation of a program or a routine. *See also:* **electronic computation.** E270-0

flow relay. A relay that responds to a rate of fluid flow. 37A100-31E11/31E6

flow soldering. *See:* **dip soldering.**

fluctuating power (rotating machinery). A phasor quantity of which the vector represents the alternating part of the power, and that rotates at a speed equal to double the angular velocity of the current. *See also:* **asynchronous machine; synchronous machine.** 0-31E8

fluctuation noise. *See:* **random noise.**

fluid loss (rotating machinery). That part of the mechanical losses in a machine having liquid in its air gap that is caused by fluid friction. *See also:* **asynchronous machine: direct-current commutating machine; synchronous machine.** 0-31E8

fluidly delayed overcurrent trip. *See:* **fluidly delayed release (fluidly delayed trip) and overcurrent release (overcurrent trip).**

fluidly delayed release (fluidly delayed trip). A release delayed by fluid displacement or adhesion. 37A100-31E11

fluorescence. Emission of light from a substance (a phosphor) as the result of, and only during, excitation by radiant energy. *See* **lamp; oscillograph.** Z7A1-9E4

fluorescent lamp. A low-pressure mercury electric-discharge lamp in which a fluorescing coating (phosphor) transforms some of the ultraviolet energy generated by the discharge into light.
See:
ballast;
instant-start fluorescent lamp;
lamp;
phosphor;
rapid-start fluorescent lamp;
starter. Z7A1-0

fluorescent-mercury lamp. An electric-discharge lamp having a high-pressure mercury arc in an arc tube, and an outer envelope coated with a fluorescing substance (phosphor) that transforms some of the ultraviolet energy generated by the arc into light. *See also:* **lamp.** Z7A1-0

flush antenna (aircraft). An antenna having no projections outside the streamlined surface of the aircraft. In general, flush antennas may be considered as slot antennas. *See also:* **air transportation electronic equipment.** 42A41-0

flush-mounted device. A device in which the body projects only a small specified distance in front of the mounting surface. 37A100-31E11

flush mounted or recessed (luminaire). A luminaire that is mounted above the ceiling (or behind a wall or other surface) with the opening of the luminaire level with the surface. *See also:* **suspended (pendant).** Z7A1-0

flutter (communication practice). (1) Distortion due to variations in loss resulting from the simultaneous transmission of a signal at another frequency. (2) A similar effect due to phase distortion. (3) Distortion that occurs in sound reproduction as a result of undesired speed variations during the recording, duplicating, or reproducing. *Notes:* (1) One important usage is to denote the effect of variation in the transmission characteristics of a loaded telephone circuit caused by the action of telegraph direct currects on the loading coils. (2) The colloquial term wow is defined in the same way, but is commonly applied to relatively slow variations (for example, one to five or six repetitions per second) that are recognized aurally as pitch fluctuations, in contradistinction to the roughening of tones that is the most noticeable effect of rapid fluctuations. (3) A constant difference in pitch such as results from a difference in the average speeds during recording and reproduction is not included in the meanings of the terms wow, flutter, and drift. (4) By an extension of their meanings, the terms flutter and wow are used to designate variations in speed itself or variations in recorded wavelengths. (5) Although most recorded sound comprises multitudes of tones, it is convenient to refer to flutter as variations in frequency, assuming the recorded sound to have been a single steady tone. *See also:* **transmission characteristics.** E188/E193/42A65-0

flutter echo. A rapid succession of reflected pulses resulting from a single initial pulse. *See also:* **electroacoustics.** E157-1E1

flutter rate (sound recording and reproducing). The number of frequency excursions in hertz, in a tone that is frequency-modulated by flutter. *Notes:* (1) Each cyclical variation is a complete cycle of deviation, for example, from maximum-frequency to minimum-frequency and back to maximum-frequency at the rate indicated. (2) If the over-all flutter is the resultant of several components having different repetition rates, the rates and magnitudes of the individual components are of primary importance. *See also:* **sound recording and reproducing.** E193-0

flux (1) (photovoltaic power system). The rate of flow of energy through a surface. *See also:* **photovoltaic power system; solar cells (photovoltaic power system).** 0-10E1

(2) (soldering) (connections). A liquid or solid which when heated exercises a cleaning and protective action upon the surfaces to which it is applied. *See also:* **soldered connections (electronic and electric applications).** 99A1-0

flux guide (induction heating usage). Magnetic material to guide electromagnetic flux in desired paths. *Note:* The guides may be used either to direct flux to preferred locations or to prevent the flux from spreading beyond definite regions. *See also:* **induction heater.** E54/E169-0

flux linkages. The sum of the fluxes linking the turns forming the coil, that is, in a coil having N turns the flux linkage is

$$\lambda = \phi_1 + \phi_2 + \phi_3 \cdots \phi_N$$

where ϕ_1 = flux linking turn 1, ϕ_2 = flux linking turn 2, etcetera, and ϕ_N = flux linking the Nth turn. 0-21E1

fluxmeter. An instrument for use with a test coil to measure magnetic flux. It usually consists of a moving-coil galvanometer in which the torsional control is either negligible or compensated. *See also:* **magnetometer.** 42A30-0

flyback (television). The rapid return of the beam in the direction opposite to that used for scanning. *See also:* **television.** E204-2E2;42A65-0

flying spot scanner (optical character recognition). A device employing a moving spot of light to scan a sample space, the intensity of the transmitted or reflected light being sensed by a photoelectric transducer. *See also:* **electronic digital computer.** X3A12-16E9

flywheel ring (rotating machinery). A heavy ring mounted on the spider for the purpose of increasing the rotor moment of inertia. *See also:* **rotor (rotating machinery).** 0-31E8

FM. *See:* **frequency modulation.**

FOA. *See:* **transformer, oil-immersed.**

focus (oscilloscopes). Maximum convergence of the electron beam manifested by minimum spot size on the phosphor screen. *See:* **astigmatism; oscillograph.** 0-9E4

focusing (electron tubes). The process of controlling the convergence of the electron beam. *See also:* **beam tubes.** 50I07-2E2

focusing coil. *See:* **focusing magnet.**

focusing device. An instrument used to locate the filament of an electric lamp at the proper focal point of lens or reflector optical systems. *See also:* **railway signal and interlocking.** 42A42-0

focusing, dynamic (picture tubes). The process of focusing in accordance with a specified signal in synchronism with scanning. *See:* **beam tubes.** E160-2E2/15E6

focusing electrode (beam tube). An electrode the potential of which is adjusted to focus an electron beam. *See also:* **electrode (electron tube).** 50I07-2E2

focusing, electrostatic (electron beam). A method of focusing an electron beam by the action of an electric field. *See also:* **beam tubes.** E160/42A70-0

focusing grid (pulse techniques). *See:* **focusing electrode.**

focusing magnet. An assembly producing a magnetic field for focusing an electron beam. *See also:* **beam tubes.** 42A70-15E6

focusing, magnetic (electron beam). A method of focusing an electron beam by the action of a magnetic field. *See also:* **beam tubes.** 42A70-15E6;E160-2E2

fog (adverse-weather) lamps. Lamps that may be used in lieu of headlamps to provide road illumination under conditions of rain, snow, dust, or fog. *See also:* **headlamp.** Z7A1-0

fog-bell operator. A device to provide automatically the periodic bell signals required when a ship is anchored in fog. 42A43-0

foil (foil tape) (burglar-alarm system). A fragile strip of conducting material suitable for fastening with an adhesive to glass, wood, or other insulating material in order to carry the alarm circuit and to initiate an alarm when severed. *See also:* **protective signaling.** 42A65-0

folded-dipole antenna. An antenna composed of two or more parallel, closely spaced dipole antennas connected together at their ends with one of the dipole antennas fed at its center. *See also:* **antenna.** 0-3E1

follow current (lightning arresters). The current from the connected power source that flows through an arrester following the passage of discharge current. *See:* **lightning; lightning arrester (surge diverter).** 99I2-31E7;42A20-0

follower drive (slave drive) (industrial control). A drive in which the reference input and operation are direct functions of another drive, called the master drive. *See:* **control system, feedback.** AS1-34E10

font (computing systems). A family or assortment of characters of a given size and style. *See:* **type font.** *See also:* **electronic digital computer.** X3A12-16E9

foot (rotating machinery). The part of the stator structure, end shield, or base, that provides means for mounting and fastening a machine to its foundation. *See also:* **stator.** 0-31E8

footcandle. A unit of illuminance when the foot is taken as the unit of length. It is the illuminance on a surface one square foot in area on which there is a uniformly distributed flux of one lumen, or the illuminance at a surface all points of which are at a distance of one foot from a uniform source of one candle. *See also:* **color terms; light.** E201-2E2;Z7A1-0

footings (foundations). Structures set in the ground to support the bases of towers, poles, or other overhead structures. *Note:* Footings are usually skeleton steel pyramids, grilles, or piers of concrete. *See also:* **tower.** 42A35-31E13

footlambert. A unit of luminance (photometric brightness) equal to $1/\pi$ candela per square foot, or to the uniform luminance of a perfectly diffusing surface emitting or reflecting light at the rate of one lumen per square foot, or to the average luminance of any surface emitting or reflecting light at that rate. *Note:* The average luminance of any reflecting surface in footlamberts is, therefore, the product of the illumination in footcandles by the luminous reflectance of the surface. *See also:* **light.** Z7A1-0

foot switch (industrial control). A switch that is suitable for operation by an operator's foot. *See:* **switch.** AS1-34E10

forbidden combination. A code expression that is defined to be nonpermissible and whose occurrence indicates a mistake or malfunction. *See also:* **electronic digital computer.** E162-0

forbidden-combination check (electronic computation). *See:* **check, forbidden-combination.**

force. Any physical cause that is capable of modifying the motion of a body. The vector sum of the forces acting on a body at rest or in uniform rectilinear motion is zero. E270-0

forced-air cooling system (rectifier). An air cooling system in which heat is removed from the cooling surfaces of the rectifier by means of a flow of air produced by a fan or blower. *See also:* **rectification.** 34A1-34E24

forced drainage (underground metallic structures). A method of controlling electrolytic corrosion whereby an external source of direct-current potential is employed to force current to flow to the structure through the earth, thereby maintaining it in a cathodic condition. *See also:* **inductive coordination.** 42A65-0

forced interruption (electric power systems). An interruption caused by a forced outage. *See also:* **outage.** 0-31E4

forced-lubricated bearing (rotating machinery). A bearing in which a continuous flow of lubricant is forced between the bearing and journal. *See:* **cradle base (rotating machinery).** 0-31E8

forced oscillation (linear constant-parameter system). The response to an applied driving force. *See also:* **network analysis.** E270-0

forced outage (electric power systems). An outage that results from emergency conditions directly associated with a component, requiring that component to be taken out of service immediately, either automatically or as soon as switching operations can be performed; or an outage caused by improper operation of equipment or human error. *See also:* **outage.** 0-31E4

forced-ventilated machine. *See:* **open pipe-ventilated machine.**

force factor (1) (electroacoustic transducer). (A) The complex quotient of the pressure required to block the acoustic system divided by the corresponding current in the electric system; (B) the complex quotient of the resulting open-circuit voltage in the electric system divided by the volume velocity in the acoustic system. *Note:* Force factors (A) and (B) have the same magnitude when consistent units are used and the transducer satisfies the principle of reciprocity. *See also:* **electroacoustics: transmission characteristic.** E157-1E1;42A65-0

(2) (electromechanical transducer). (A) The complex quotient of the force required to block the mechanical system divided by the corresponding current in the electric system; (B) the complex quotient of the resulting open-circuit voltage in the electric system divided by the velocity in the mechanical system. *Notes:* (1) Force factors (A) and (B) have the same magnitude when consistent units are used and the transducer satisfies the principle of reciprocity. (2) It is sometimes convenient in an electrostatic or piezoelectric transducer to use the ratios between force and charge or electric displacement, or between voltage and mechanical displacement. 42A65-0

forcing (industrial control). The application of control impulses to initiate a speed adjustment, the magnitude of which is greater than warranted by the desired controlled speed, in order to bring about a greater rate of speed change. *Note:* Forcing may be obtained by directing the control impulse so as to effect a change in the field or armature circuit of the motor, or both. *See also:* **electric drive.** 42A25-34E10

foreign exchange line. A subscriber line by means of which service is furnished to a subscriber at his request from an exchange other than the one from which service would normally be furnished. *See also:* **telephone system.** 42A65-19E1

forestalling switch. *See:* **acknowledger; forestaller.**

fork beat (facsimile). *See:* **carrier beat.**

form. Any article such as a printing plate, that is used as a pattern to be reproduced. *See also:* **electroforming.** 42A60-0

formal logic. The study of the structure and form of valid argument without regard to the meaning of the terms in the argument. X3A12-16E9

format (1) (computing systems). The general order in which information appears on the input medium.

(2) (data transmission). Arrangement of code characters within a group, such as a block or message. *See also:* **data transmission.** 0-19E4

(3) Physical arrangement of possible locations of holes or magnetized areas. *See:* **address format; electronic digital computer.** EIA3B-34E12

format classification (numerically controlled machines). A means, usually in an abbreviated notation, by which the motions, dimensional data, type of control system, number of digits, auxiliary functions, etcetera, for a particular system can be denoted. *See also:* **numerically controlled machines.** EIA3B-34E12

format detail (numerically controlled machines). Describes specifically which words and of what length are used by a specific system in the format classification. *See also:* **numerically controlled machines.** EIA3B-34E12

formation light. A navigation light specially provided to facilitate formation flying. *See also:* **signal lighting.** Z7A1-0

formation voltage. The final impressed voltage at which the film is formed on the valve metal in an electrochemical valve. *See also:* **electrochemical valve.** 42A60-0

formette (rotating machinery). *See:* **form-wound motorette.**

form factor (periodic function). The ratio of the root-mean-square value to the average absolute value, averaged over a full period of the function. *See also:* **power rectifier; rectification.** E59-34E17;E270-34E24

forming (1) (electrical) (semiconductor devices). The process of applying electric energy to a semiconductor device in order to modify permanently the electrical characteristics. *See also:* **semiconductor.** E102/42A70-0

(2) (semiconductor rectifier). The electrical or thermal treatment, or both, of a semiconductor rectifier cell for the purpose of increasing the effectiveness of the rectifier junction. *See also:* **rectification.** E59-34E17

(3) (electrochemical). The process that results in a change in impedance at the surface of a valve metal to the passage of current from metal to electrolyte, when the voltage is first applied. *See also:* **electrochemical valve.** 42A60-0

form-wound (rotating machinery). Applied to a winding whose coils are formed essentially to their final shape prior to assembly into the machine. *See also:* **rotor (rotating machinery); stator.** 0-31E8

form-wound motorette (formette) (rotating machinery). A motorette for form-wound coils. *See also:* **asynchronous machine; direct-current commutating machine; synchronous machine.** 0-31E8

fortuitous distortion. A random distortion of telegraph signals such as that commonly produced by interference. *See also:* **telegraphy.** 42A65-19E4

fortuitous telegraph distortion. Distortion that includes those effects that cannot be classified as bias or characteristic distortion and is defined as the departure, for one occurrence of a particular signal pulse, from the average combined effects of bias and characteristic distortion. *Note:* Fortuitous distortion varies from one signal to another and is measured by a process of elimination over a long period. It is expressed in percent of unit pulse. *See also:* **distortion.** E145-0

forward-acting regulator. A transmission regulator in which the adjustment made by the regulator does not affect the quantity that caused the adjustment. *See also:* **transmission regulator.** 42A65-0

forward admittance, short-circuit (electron-device transducer). *See:* **admittance, short-circuit forward.**

forward current (1) (metallic rectifier). The current that flows through a metallic rectifier cell in the forward direction. *See also:* **rectification.** 42A15-0

(2) (semiconductor rectifier). The current that flows through a semiconductor rectifier cell in the forward direction. *See also:* **circuits and devices; rectification.** E59-34E17

(3) (reverse-blocking or reverse-conducting thyristor). The principal current for a positive anode-to-cathode voltage. *See also:* **circuits and devices; principal current.** E223-34E17/34E24/15E7

forward current, average, rating. *See:* **average forward current rating.**

forward direction (1) (metallic rectifier). The direction of lesser resistance to current flow through the cell; that is, from the negative electrode to the positive electrode. *See also:* **rectification.** 42A15-0

(2) (semiconductor rectifier cell). The direction of lesser resistance to steady direct-current flow through the cell; for example, from the anode to the cathode. *See also:* **semiconductor; semiconductor rectifier stack.** E59-34E17/34E24

forward error-correcting system. A system employing an error-correcting code and so arranged that some or all signals detected as being in error are automatically corrected at the receiving terminal before delivery to the data sink or to the telegraph receiver. 0-19E4

forward gate current (thyristor). The gate current when the junction between the gate region and the adjacent anode or cathode region is forward biased. *See also:* **principal current.** E223-34E17/34E24/15E7

forward gate voltage (thyristor). The voltage between the gate terminal and the terminal of an adjacent region resulting from forward gate current. *See also:* **principal voltage-current characteristic (principal characteristic).** E223-34E17/34E24/15E7

forward path (signal-transmission system) (feedback-control loop). The transmission path from the loop-error signal to the loop-output signal. *See:* **feedback.** 0-13E6

forward period (rectifier circuit element) (rectifier circuit). The part of an alternating-voltage cycle during which forward voltage appears across the rectifier circuit element. *Note:* The forward period is not necessarily the same as the conducting period because of the effect of circuit parameters and semiconductor rectifier cell characteristics. *See also:* **rectifier circuit element.** E59-34E17

forward power loss (semiconductor rectifiers). The power loss within a semiconductor rectifier diode resulting from the flow of forward current. *See also:* **rectification; semiconductor rectifier stack.** E59-34E17/34E24

forward recovery time (semiconductor diode). The time required for the current or voltage to recover to a specified value after instantaneous switching from a stated reverse voltage condition to a stated forward current or voltage condition in a given circuit. *See also:* **rectification.** E59-34E17

forward resistance (metallic rectifier). The resistance measured at a specified forward voltage drop or a specified forward current. *See also:* **rectification.** 42A15-0

forward voltage (1) (rectifiers). Voltage of the polarity that produces the larger current.

(2) (reverse blocking or reverse conducting thyristor).

A positive anode-to-cathode voltage. *See also:* **circuits and devices; principal characteristic (principal voltage-current characteristic).** E223-34E17/34E24/15E7;42A65-0

forward voltage drop (1) (metallic rectifier). The voltage drop in the metallic rectifying cell resulting from the flow of current through a metallic rectifier cell in the forward direction.
(2) (semiconductor rectifier). The voltage drop in a semiconductor rectifier cell, rectifier diode, or rectifier stack resulting from the flow of forward current. *See also:* **rectification; semiconductor rectifier stack.** E59-34E17/34E24;42A15-0

forward wave (traveling-wave tubes). A wave whose group velocity is in the same direction as the electron stream motion. *See also:* **beam tubes; electron devices, miscellaneous.** E160-15E6

forward-wave structure (microwave tubes). A slow-wave structure whose propagation is characterized on a ω/β diagram (ω versus phase shift/section) by a positive slope in the region $O<\beta<\pi$ (in which the group and phase velocity therefore have the same sign). *See also:* **microwave tube or valve.** 0-15E6

Foster's reactance theorem. States that the driving-point impedance of a finite two-terminal network composed of pure reactances is a reactance that is an odd rational function of frequency and that is completely determined, except for a constant factor, by assigning the resonant and antiresonant frequencies. *Note:* In other words, the driving-point impedance consists of segments going from minus infinity to plus infinity (except that at zero or infinite frequency, a segment may start or stop at zero impedance). The frequencies at which the impedance is infinite are termed poles and those at which the impedance is zero are termed zeros. *See also:* **communication.** 42A65-0

Foucault currents. *See:* **eddy currents.**

foul electrolyte. An electrolyte in which the amount of impurities is sufficient to cause an undesirable effect on the operation of the electrolytic cells in which it is employed. *See also:* **electrorefining.** 42A60-0

fouling. The accumulation and growth of marine organisms on a submerged metal surface. *See:* **corrosion terms.** CM-34E2

fouling point (railway practice). The location in a turnout back of a frog at or beyond the clearance point at which insulated joints or derails are placed. *See also:* **railway signal and interlocking.** 42A42-0

foundation (rotating machinery). The structure on which the feet or base of a machine rest and are fastened. *See also:* **cradle base (rotating machinery).** 0-31E8

foundation bolt (rotating machinery). A bolt used to fasten a machine to a foundation. *See also:* **cradle base (rotating machinery).** 0-31E8

foundation-bolt cone (rotating machinery). A cone placed around a foundation bolt when imbedded in a concrete foundation to provide clearance for adjustment during erection. *See also:* **cradle base (rotating machinery).** 0-31E8

four-address. Pertaining to an instruction code in which each instruction has four address parts. *Note:* In a typical four-address instruction the addresses specify the location of two operands, the destination of the result, and the location of the next instruction to be interpreted. *See also:* **electronic digital computer; three-plus-one address.** E162-0

four-address code (electronic computation). *See:* **instruction code.**

Fourier series. A single-valued periodic function (that fulfills certain mathematical conditions) may be represented by a Fourier series as follows

$$f(x) = 0.5A_0 + \sum_{n=1}^{n=\infty} [A_n \cos nx + B_n \sin nx$$

$$= 0.5A_0 + \sum_{n=1}^{n=\infty} C_n \sin(nx+\theta_n)$$

$$\text{where } A_n = \frac{1}{\pi}\int_0^{2\pi} f(x) \cos nx\, dx$$

$$n = 0,1,2,3,\cdots$$

$$B_n = \frac{1}{\pi}\int_0^{2} f(x) \sin nx\, dx$$

$$C_n = +(A_n^2+B_n^2)^{1/2}$$

$$\text{and } \theta_n = \text{arc tan } A_n/B_n.$$

Note: $0.5A_0$ is the average of a periodic function *f(x)* over one primitive period. E270-0

four-plus-one address (computing systems). Pertaining to an instruction that contains four operand addresses and a control address. *See also:* **electronic digital computer.** X3A12-16E9

four-pole. *See:* **two-terminal pair network.**

four-terminal network. A network with four accessible terminals. *Note:* See **two-terminal-pair network** for an important special case. *See:* **two-terminal-pair network; quadripole.** E270-0

fourth voltage range (railway signal and interlocking). *See:* **voltage range.**

four-wire circuit. A two-way circuit using two paths so arranged that the electric waves are transmitted in one direction only by one path and in the other direction only by the other path. *Note:* The transmission paths may or may not employ four wires. *See also:* **transmission line.** 42A65-0

four-wire repeater. A telephone repeater for use in a four-wire circuit and in which there are two amplifiers, one serving to amplify the telephone currents in one side of the four-wire circuit and the other serving to amplify the telephone currents in the other side of the four-wire circuit. *See also:* **repeater.** 42A65-31E3

four-wire terminating set. A hybrid set for interconnecting a four-wire and two-wire circuit. *See also:* **circuits and devices.** 42A65-31E3

fovea. A small region at the center of the retina, subtending about two degrees, that contains only cones and forms the site of most distinct vision. *See also:* **retina.** Z7A1-0

foveal vision. *See:* **central vision.**

FOW. *See:* **transformer, oil-immersed.**

fractional-horsepower brush (electric machines). A brush with a cross-sectional area of ¼ square inch (thickness × width) or less and not exceeding 1½ inches in length but larger than a miniature brush. *See:* **brush; brush (rotating machinery).** 42A10-31E8;64A1-0

fractional-horsepower motor. A motor built in a frame smaller than that of a motor of open construction having a continuous rating of 1 horsepower at 1700-1800 revolutions per minute. *See:* **asynchronous machine; direct-current commutating machine; synchronous machine.** 42A10-31E8

fractional-slot winding (rotating machinery). A distributed winding in which the average number of slots per pole per phase is not integral, for example 3 2/7 slots per pole per phase. *See also:* **asynchronous machine; direct-current commutating machine; synchronous machine.** 0-31E8

frame (1) (television). The total area, occupied by the picture, that is scanned while the picture signal is not blanked.
(2) (facsimile). A rectangular area, the width of which is the available line and the length of which is determined by the service requirements. E168/E203-0;42A65-2E2

framed plate (storage cell). A plate consisting of a frame supporting active material. *See also:* **battery (primary or secondary).** 42A60-0

frame frequency (television). The number of times per second that the frame is scanned. *See also:* **television.** E203/42A65-0

frame, intermediate distributing (telephony). A frame in a central-office building, the primary purpose of which is to serve as a place where the subscriber line multiple is cross connected to the subscriber line circuit. *See also:* **telephone switching system.** 0-19E1

frame, main distributing (telephony). A frame in a central-office building, on one part of which terminate the permanent outside lines entering the building and on another part of which terminate the customer line multiple cabling, and cabling from trunk circuits, etcetera, used for associating any outside line with any desired terminal in such cabling or with any other outside line or other trunk termination. *Note:* It usually carries the central office protective devices and functions as a test point between lines, trunks, and the office. *See also:* **telephone switching system.** 0-19E1

frame ring (rotating machinery). A plate or assembly of flat plates forming an annulus in a radial plane and serving as a part of the frame to stiffen it. *See also:* **cradle base (rotating machinery).** 0-31E8

frame split (rotating machinery). A joint at which a frame may be separated into parts. *See also:* **cradle base (rotating machinery).** 0-31E8

framework (rotating machinery). A stationary supporting structure. *See also:* **cradle base (rotating machinery).** 0-31E8

frame yoke (field frame) (rotating machinery). The annular support for the poles of a direct-current machine. *Note:* It may be laminated or of solid metal and forms part of the magnetic circuit. *See also:* **cradle base (rotating machinery).** 0-31E8

framing (facsimile). The adjustment of the picture to a desired position in the direction of line progression. *See also:* **recording (facsimile).** E168-0

framing signal (facsimile). A signal used for adjustment of the picture to a desired position in the direction of line progression. *See also:* **facsimile signal (picture signal).** E168-0

Fraunhofer pattern (antenna). A radiation pattern obtained in the Fraunhofer region. *See also:* **antenna.** 0-3E1

Fraunhofer region. The region in which the field of an antenna is focused. *See:* Note 2 of **far-field region** for a more restricted usage. *See also:* **antenna.** 0-3E1

free capacitance (1) (conductor). The limiting value of its self-capacitance when all other conductors, including isolated ones, are infinitely removed.
(2) (between two conductors). The limiting value of the plenary capacitance as all other, including isolated, conductors are infinitely removed. E270-0

free cyanide (electrodepositing solution) (electroplating). The excess of alkali cyanide above the minimum required to give a clear solution, or above that required to form specified soluble double cyanides. *See also:* **electroplating.** 42A60-0

free field. A field (wave or potential) in a homogeneous, isotropic medium free from boundaries. In practice, a field in which the effects of the boundaries are negligible over the region of interest. *Note:* The actual pressure impinging on an object (for example, electroacoustic transducer) placed in an otherwise free sound field will differ from the pressure that would exist at that point with the object removed, unless the acoustic impedance of the object matches the acoustic impedance of the medium. *See also:* **electroacoustics.** E157-1E1

free-field current response (receiving current sensitivity) (electroacoustic transducer used for sound reception). The ratio of the current in the output circuit of the transducer when the output terminals are short-circuited to the free-field sound pressure existing at the transducer location prior to the introduction of the transducer in the sound field. *Notes:* (1) The free-field response is defined for a plane progressive sound wave whose direction of propagation has a specified orientation with respect to the principal axis of the transducer. (2) The free-field current response is usually expressed in decibels, namely, 20 times the logarithm to the base 10 of the quotient of the observed ratio divided by the reference ratio, usually 1 ampere per newton per square meter. *See also:* **loudspeaker.** 0-1E1

free-field voltage response (receiving voltage sensitivity) (electroacoustic transducer used for sound reception). The ratio of the voltage appearing at the output terminals of the transducer when the output terminals are open-circuited to the free-field sound pressure existing at the transducer location prior to the introduction of the transducer in the sound field. *Notes:* (1) The free-field response is determined for a plane progressive sound wave whose direction of propagation has a specified orientation with respect to the principal axis of the transducer. (2) The free-field voltage response is usually expressed in decibels, namely, 20 times the logarithm to the base 10 of the quotient of the observed ratio divided by the reference ratio, usually 1 volt per newton per square meter. *See also:* **loudspeaker.** 0-1E1

free impedance (transducer). The impedance at the input of the transducer when the impedance of its load is made zero. *Note:* The approximation is often made that the free electric impedance of an electroacoustic transducer designed for use in water is that measured with the transducer in air. *See also:* **loudspeaker; self-impedance.** 42A65-1E1

free motional impedance (transducer) (electroacoustics). The complex remainder after the blocked impedance has been subtracted from the free impedance. *See also:* **electroacoustics; self-impedance.** E157-1E1;42A65-0

free oscillation. The response of a system when no external driving force is applied and energy previously stored in the system produces the response. *Note:* The frequency of such oscillations is determined by the parameters in the system or circuit. The term shock-excited oscillation is commonly used. *See:* **oscillatory circuit.** E145/E270-0

free progressive wave (free wave) (acoustics). A wave in a medium free from boundary effects. A free wave in a steady state can only be approximated in practice. *See also:* **electroacoustics.** E157-1E1

free-radiation frequencies for industrial, scientific, or medical (ISM) apparatus (electromagnetic compatibility). Center of a band of frequencies assigned to industrial, scientific, or medical equipment either nationally or internationally for which no power limit is specified. *See:* **ISM apparatus.** *See also:* **electromagnetic compatibility.** CISPR-Z7A1

free-running frequency. The frequency at which a normally synchronized oscillator operates in the absence of a synchronizing signal. *See also:* **circuits and devices.** E188-2E2;42A65-0

free-running sweep (oscilloscopes). A sweep that recycles without being triggered and is not synchronized by any applied signal. *See:* **oscillograph.** 0-9E4

free-space field intensity. The radio field intensity that would exist at a point in a uniform medium in the absence of waves reflected from the earth or other objects. *See also:* **radiation.** 42A65-0

free-space transmission (mobile communication). Electromagnetic radiation that propagates unhindered by the presence of obstructions, and whose power or field intensity decreases as a function of distance squared. *See also:* **mobile communication system.** 0-6E1

free wave (acoustics). *See:* **free progressive wave.**

freeze-out (telephone circuit). A short-time denial to a subscriber by a speech-interpolation system. *See also:* **telephone switching system.** 42A65-0

***F* region (radio wave propagation).** The region of the terrestrial ionosphere above about 160 kilometers altitude. *See also:* **radiation; radio wave propagation.** 42A65-3E2

freight elevator. An elevator primarily used for carrying freight on which only the operator and the persons necessary for loading and unloading the freight are permitted to ride. *See also:* **elevators.** 42A45-0

frequency (periodic function) (wherein time is the independent variable). The number of periods per unit time. E45/E270-0

frequency allocation (table) (electromagnetic compatibility). (1) The process of designating radio-frequency bands for use by specific radio services; or (2) The resulting table (of frequency allocations). *See also:* **electromagnetic compatibility.** 0-27E1

frequency allotment (plan) (electromagnetic compatibility). (1) The process of designating radio frequencies within an allocated band for use within specific geographic areas. (2) The resulting plan (of frequency allotment). *See also:* **electromagnetic compatibility.** 0-27E1

frequency assignment (list) (electromagnetic compatibility). (1) The process of designating radio frequency for use by a specific station under specified conditions of operations. (2) The resulting list of frequency assignments. *See also:* **electromagnetic compatibility.** 0-27E1

frequency band. A continuous range of frequencies extending between two limiting frequencies. *Note:* The term frequency band or band is also used in the sense of the term bandwidth. *See also:* **channel; signal; signal wave.** E145-13E6;E270/E182/42A65-0

frequency-band number. The number N in the expression 0.3×10^N that defines the range of band N. Frequency band N extends from 0.3×10^N hertz to 3×10^N hertz, the lower limit exclusive, the upper limit inclusive. E270-0

frequency band of emission (communication band). The band of frequencies effectively occupied by that emission, or the type of transmission and the speed of signaling used. *See also:* **radio transmission.** E145/42A65-0

frequency bands (mobile communication). The frequency allocations that have been made available for land mobile communications by the Federal Communications Commission, including the spectral bands: 25.0 to 50.0 megahertz, 150.8 to 173.4 megahertz, and 450.0 to 470.0 megahertz. *See also:* **mobile communication system.** 0-6E1

frequency bias (electric power systems). An offset in the scheduled net interchange power of a control area that varies in proportion to the frequency deviation. *Note:* This offset is in a direction to assist in restoring the frequency to schedule. *See also:* **power system; power systems, low-frequency and surge testing.** E94-31E4

frequency bias setting (control area). A factor with negative sign that is multiplied by the frequency deviation to yield the frequency bias. *See also:* **power system.** 0-31E4

frequency changer. A motor-generator set that changes power of an alternating-current system from one frequency to one or more different frequencies, with or without a change in the number of phases, or in voltage. *See:* **converter.** 42A10-0

frequency-changer set (rotating machinery). A motor-generator set that changes the power of an alternating-current system from one frequency to another. 0-31E8

frequency-change signaling (telecommunication). A method in which one or more particular frequencies correspond to each desired signaling condition. *Note:* The transition from one set of frequencies to the other may be either a continuous or a discontinuous change in frequency or in phase. *See also:* **frequency modulation.** 0-19E4

frequency control. The regulation of frequency within a narrow range. *See also:* **generating station.** 42A35-31E13

frequency-conversion transducer. *See:* **conversion transducer.**

frequency converter. *See:* **frequency changer.**

frequency converter, commutator type (rotating machinery). A polyphase machine the rotor of which has one or two windings connected to slip rings and to a commutator. *Note:* By feeding one set of terminals with a voltage of given frequency, a voltage of another frequency may be obtained from the other set of terminals. *See also:* **asynchronous machine; synchronous machine.** 0-31E8

frequency, corner (asymptotic form of Bode diagram) (control system, feedback). The frequency indicated by a breakpoint, that is, the junction of two confluent straight lines asymptotic to the log gain curve. *Note:* One breakpoint is associated with each distinct real root of the characteristic equation, one with each set of repeated roots, and one with each pair of complex roots. For a single real root, corner frequency (in radians per second) is the reciprocal of the corresponding time constant (in seconds), and the corresponding phase angle is halfway between the phase angles belonging to the asymptotes extended to infinity. *See also:* **control system, feedback.** 85A1-23E0

frequency, cyclotron. *See:* **cyclotron frequency.**

frequency, damped (automatic control). The apparent frequency of a damped oscillatory time response of a system resulting from a nonoscillatory stimulus. *Note:* The value of the frequency in a particular system depends somewhat on the subsidence ratio. *See also:* **control system, feedback.** 85A1-23E0

frequency departure (telecommunication). The amount of variation of a carrier frequency or center frequency from its assigned value. *Note:* The term frequency deviation, which has been used for this meaning, is in conflict with this essential term as applied to phase and frequency modulation and is therefore deprecated for future use in the above sense. *See also:* **radio transmission.** E145/E188/42A65-0

frequency deviation (1) (power system). System frequency minus the scheduled frequency. *See also:* **frequency modulation; power systems, low-frequency and surge testing; frequency departure.** E94-0
(2) (telecommunication; frequency modulation). The peak difference between the instantaneous frequency of the modulated wave and the carrier frequency. *See also:* **frequency modulation.** E145/E170/E188/42A65-9E4

frequency distortion. A term commonly used for that form of distortion in which the relative magnitude of the different frequency components of a complex wave are changed in transmission. *Note:* When referring to the distortion of the phase-versus-frequency characteristic, it is recommended that a more specific term such as phase-frequency distortion or delay distortion be used. *See also:* **amplitude distortion; distortion; distortion, amplitude-frequency.** E145-0

frequency diversity (telecommunication). *See:* **frequency diversity reception.**

frequency-diversity reception (frequency diversity telecommunication). That form of diversity reception that utilizes transmission at different frequencies. *See also:* **radio receiver.** 42A65-0

frequency divider. A device fo, delivering an output wave whose frequency is a proper fraction, usually a submultiple, of the input frequency. *Note:* Usually the output frequency is an integral submultiple or an integral proper fraction of the input frequency. *See also:* **circuits and devices; harmonic conversion transducer.** E145-0;42A65-31E3

frequency-division multiplex (telecommunication). The process or device in which each modulating wave modulates a separate subcarrier and the subcarriers are spaced in frequency. *Note:* Frequency division permits the transmission of two or more signals over a common path by using different frequency bands for the transmission of the intelligence of each message signal. *See also:* **modulating systems.** E145/E170/42A65-0

frequency doubler. A device delivering output voltage at a frequency that is twice the input frequency. *See also:* **circuits and devices.** E145-0

frequency, image (heterodyne frequency converters in which one of the two sidebands produced by beating is selected). An undesired input frequency capable of producing the selected frequency by the same process. *Note:* The word **image** implies the mirrorlike symmetry of signal and image frequencies about the beating-oscillator frequency or the intermediate frequency, whichever is the higher. *See also:* **radio receiver.** E188/42A65-0

frequency influence (electric instrument) (instruments other than frequency meters). The percentage change (of full-scale value) in the indication of an instrument that is caused solely by a frequency departure from a specified reference frequency. *Note:* Because of the dominance of 60 hertz as the common frequency standard in the United States, alternating-current (power-frequency) instruments are always supplied for that frequency unless otherwise specified. *See also:* **accuracy rating (instrument).** 39A1/39A2-0

frequency, instantaneous. *See:* **instantaneous frequency.**

frequency interlace (television). The relationship of intermeshing between the frequency spectrum of an essentially periodic interfering signal and the spectrum of harmonics of the scanning frequencies, which relationship minimizes the visibility of the interfering pattern by altering its appearance on successive scans. *See also:* **color terms.** E201-0

frequency lock (power-system communication). A means of recovering in a single-sideband suppressed-carrier receiver the exact modulating frequency that is applied to a single-sideband transmitter. *See also:* **power-line carrier.** 0-31E3

frequency meter. An instrument for measuring the frequency of an alternating current. *See also:* **instrument.** 42A30-0

frequency-modulated cyclotron. A cyclotron in which the frequency of the accelerating electric field is modulated in order to hold the positively charged particles in synchronism with the accelerating field despite their increase in mass at very high energies. *See also:* **electron devices, miscellaneous.** 42A70-15E6

frequency-modulated radar (FM radar). A form of radar in which the radiated wave is frequency modulated and the returning echo beats with the wave being radiated, thus enabling the range to be measured. *See also:* **radar.** 42A65-0

frequency-modulated transmitter. A transmitter that transmits a frequency-modulated wave. *See also:* **radio transmitter.** E145/42A65/E182-0

frequency modulation (1) (electrical conversion). The cyclic or random dynamic variation, or both, of instantaneous frequency about a mean frequency during steady-state electric system operation. *See also:*

electric conversion equipment; modulating systems. 0-10E1

(2) (FM) (telecommunication). Angle modulation in which the instantaneous frequency of a sine-wave carrier is caused to depart from the carrier frequency by an amount proportional to the instantaneous value of the modulating wave. *Note:* Combinations of phase and frequency modulation are commonly referred to as **frequency modulation.**

See:

amplitude-suppression ratio;
center frequency;
deviation ratio;
frequency-change signaling;
frequency deviation;
frequency-shift keying or FSK;
frequency stabilization;
frequency swing;
maximum-deviation sensitivity;
maximum system deviation;
reactance modulator.

See also: **modulating systems.** E145/42A65-31E3/19E4;E188-0;E170-0

frequency monitor. An instrument for indicating the amount of deviation of a frequency from its assigned value. *See also:* **circuits and devices; instrument.** 42A30/42A65-0

frequency multiplier. A device for delivering an output wave whose frequency is an exact integral multiple of the input frequency. *Note:* Frequency doublers and triplers are common special cases of frequency multipliers. *See also:* **circuits and devices; harmonic conversion transducer.** E145-0;42A65-31E3

frequency pulling (oscillator). A change of the generated frequency of an oscillator caused by a change in load impedance. *See also:* **circuit characteristics of electrodes; oscillatory circuit; waveguide.** E160-15E6;42A65-0

frequency, pulse repetition. The number of pulses per unit time of a periodic pulse train or the reciprocal of the pulse period. *Note:* This term also includes the average number of pulses per unit time of aperiodic pulse trains where the periods are of random duration. *See also:* **pulse.** 0-9E4

frequency range (1) (general). A specifically designated part of the frequency spectrum.

(2) (transmission system). The frequency band in which the system is able to transmit power without attenuating or distorting it more than a specified amount.

(3) (device). The range of frequencies over which the device may be considered useful with various circuit and operating conditions. *Note:* Frequency range should be distinguished from bandwidth, which is a measure of useful range with fixed circuits and operating conditions. *See also:* **signal wave.** 42A65/E270-0:E160-15E6

frequency record (electroacoustics). A recording of various known frequencies at known amplitudes, usually for the purpose of testing or measuring. *See also:* **phonograph pickup.** E157-1E1

frequency relay. A relay that functions at a predetermined value of frequency. *Note:* It may be an overfrequency relay, an underfrequency relay, or a combination of both. *See also:* **relay.** 37A100-31E11/31E6

frequency response (power supplies). The measure of an amplifier or power supply's ability to respond to a sinusoidal program. *Notes:* (1) The frequency response measures the maximum frequency for full-output voltage excursion. (2) Frequency response connotes amplitude-frequency response, which should be used in full, particularly if phase-frequency response is significant. This frequency is a function of the slewing rate and unity-gain bandwidth. *See also:* **amplitude-frequency response; power supply.** KPSH-10E1

frequency-response characteristic (signal-transmission system, industrial control) (linear system). The frequency-dependent relation, in both gain and phase difference, between steady-state sinusoidal inputs and the resulting steady-state sinusoidal outputs. *Notes:* (1) With nonlinearity, as evidenced by distortion of a sinusoidal input of specified amplitudes, the relation is based on that sinusoidal component of the output having the frequency of the input. (2) Mathematically, the frequency-response characteristic is the complex function of $S = j\omega$:

$$A_o(j\omega)/A_i(j\omega)\exp\{j[\theta_o(j\omega)-\theta_i(j\omega)]\}$$

where: A_i = input amplitude
A_o = output amplitude
θ_i = input phase angle (relative to fixed reference)
θ_o = output phase angle (relative to same reference)

See also: **control system, feedback; signal.** 85A1-13E6;85A1-23E0;AS1-34E10

frequency-response equalization (equalization). The effect of all frequency discriminative means employed in a transmission system to obtain a desired over-all frequency response. *See also:* **electroacoustics.** E157-1E1

frequency-selective ringing (telephony). Selective ringing that employs currents of several frequencies to activate ringers that are tuned mechanically or electrically to the frequency of one of the ringing currents so that only the desired ringer responds. *See also:* **telephone switching system.** 0-19E1

frequency-selective voltmeter. A selective radio receiver, with provisions for output indication. E263-27E1

frequency selectivity (selectivity). (1)A characteristic of an electric circuit or apparatus in virtue of which electric currents or voltages of different frequencies are transmitted with different attenuation. (2) The degree to which a transducer is capable of differentiating between the desired signal and signals or interference at other frequencies. *See also:* **transducer; transmission characteristics.** 42A65-0

frequency-sensitive relay. A relay that operates when energized with voltage, current, or power within specific frequency bands. *See also:* **relay.** 83A16-0

frequency-shift keying (FSK) (telecommunication). The form of frequency modulation in which the modulating wave shifts the output frequency between predetermined values, and the output wave has no phase discontinuity. *Note:* Commonly, the instantaneous frequency is shifted between two discrete values termed the mark and space frequencies. *See also:* **modulating systems; telegraphy.** E170/E145;42A65-31E3

frequency stability. *See:* **carrier frequency stability of a transmitter.**

frequency stabilization. The process of controlling the center or carrier frequency so that it differs from that of a reference source by not more than a prescribed amount. *See also:* **frequency modulation.** E145/42A65-0

frequency standard (electric power systems). A device that produces a standard frequency. *See:* **standard frequency.** *See also:* **speed-governing system.** E94-0

frequency swing (frequency modulation). The peak difference between the maximum and the minimum values of the instantaneous frequency. *Note:* The term frequency swing is sometimes used to describe the maximum swing permissible under specified conditions. Such usage should preferably include a specific statement of the conditions. *See also:* **frequency modulation.** E145/42A65-0

frequency tolerance (radio transmitter). The extent to which a characteristic frequency of the emission, for example, the carrier frequency itself or a particular frequency in the sideband, may be permitted to depart from a specified reference frequency within the assigned band. *Note:* The frequency tolerance may be expressed in hertz or as a percentage of the reference frequency. *See also:* **radio transmitter.** 0-42A65

frequency tripler. A device delivering output voltage at a frequency that is three times the input frequency. *See also:* **circuits and devices.** E145-0

frequency-type telemeter. A telemeter that employs the frequency of a periodically recurring electric signal as the translating means. *See also:* **telemetering.** 37A100-31E11;42A30-0

frequency, undamped (frequency, natural). (1) Of a second-order linear system without damping, the frequency of free oscillation in radians per unit time or in hertz. (2) Of any system whose transfer function contains the quadratic factor $s^2 + 2\zeta\omega_n s + \omega_n^2$ in the denominator, the value $\omega_n (0 \leq \zeta < 1)$. 0-23E0

frequently-repeated overload rating (power converter). The maximum direct current that can be supplied by the converter on a repetitive basis under normal operating conditions. *See:* **power rectifier.** 0-34E24

Fresnel pattern (antenna). A radiation pattern obtained in the Fresnel region. *See also:* **antenna.** 0-3E1

Fresnel region. The region (or regions) adjacent to the region in which the field of an antenna is focused (that is, just outside the Fraunhofer region). *See:* Note 2 of **radiating near-field region** for a more restricted usage. *See also:* **antenna.** 0-3E1

fretting (corrosion). Deterioration resulting from repetitive slip at the interface between two surfaces. *Note:* When deterioration is further increased by corrosion, the term fretting-corrosion is used. *See:* **corrosion terms.** CM-34E2

friction and windage loss (rotating machinery). The power required to drive the unexcited machine at rated speed with the brushes in contact, deducting that portion of the loss that results from: (1) Forcing the gas through any part of the ventilating system that is external to the machine and cooler (if used). (2) The driving of direct-connected flywheels or other direct-connected apparatus. *See also:* **asynchronous machine; direct-current commutating machine; synchronous machine.** 50A10-31E8

friction electrification. *See:* **triboelectrification.**

friction tape. A fibrous tape impregnated with a sticky moisture-resistant compound that provides a protective covering for insulation. *See also:* **interior wiring.** 42A95-0

fritting, relay. *See:* **relay fritting.**

frog-leg winding (rotating machinery). A composite winding consisting of one lap winding and one wave winding placed on the same armature and connected to the same commutator. *See:* **direct-current commutating machine.** 0-31E8

front (motor or generator). The front of a normal motor or generator is the end opposite the largest coupling or driving pulley. *See also:* **asynchronous machine; direct-current commutating machine; synchronous machine.** 0-31E8

front contact. A part of a relay against which, when the relay is energized, the current-carrying portion of the movable neutral member is held so as to form a continuous path for current. *See also:* ***a* contact; railway signal and interlocking.** 42A42-0

front-of-wave impulse sparkover voltage (arrester). The impulse sparkover voltage with a wavefront that rises at a uniform rate and causes sparkover on the wavefront. *See also:* **lightning arrester (surge diverter).** 0-31E7

front porch (television). The portion of a composite picture signal that lies between the leading edge of the horizontal blanking pulse and the leading edge of the corresponding synchronizing pulse. *See also:* **television.** E203/42A65-2E2

fruit. *See:* **fruit pulse.**

fruit pulse (fruit*). A pulse reply received as the result of interrogation of a transponder by interrogators not associated with the responsor in question. *See also:* **pulse terms.** E194-0

*Deprecated

***F* scan (electronic navigation).** *See:* ***F* display.**

***F* scope (electronic navigation).** *See:* ***F* display.**

FSK (telecommunication). *See:* **frequency-shift keying.**

FS to AM converter (facsimile). *See:* **receiving converter, facsimile.**

FTC (radar). *See:* **fast-time-constant circuit.**

front-to-back ratio. The ratio of the directivity of an antenna to the directive gain in a specified direction toward the back. *See also:* **antenna.** 0-3E1

fuel (fuel cells). A chemical element or compound that is capable of being oxidized. *See also:* **electrochemical cell.** CV1-10E1

fuel-and-oil quantity electric gauge. A device that measures, by means of bridge circuits and an indicator with separate pointers and scales, the quantity of fuel and oil in the aircraft tanks. *See also:* **air transportation instruments.** 42A41-0

fuel battery. An energy-conversion device consisting of more than one fuel cell connected in series, parallel, or both. *See also:* **fuel cell.** CV1-10E1

fuel-battery power-to-volume ratio. The kilowatt output per envelope volume of the fuel battery (exclusive of the fuel, oxidant, storage, and auxiliaries). *See also:* **fuel cell.** CV10-10E1

fuel-battery power-to-weight ratio. The kilowatt output per unit weight of the fuel battery (exclusive of the fuel, oxidant, storage, and auxiliaries). *See also:* **fuel cell.** CV1-10E1

fuel cell. An electrochemical cell that can continuously change the chemical energy of a fuel and oxidant to electric energy by an isothermal process involving an essentially invariant electrode-electrolyte system.
See:
electrochemical cell;
fuel battery;
fuel-battery power-to-volume ratio;
fuel-battery power-to-weight ratio;
fuel-cell Coulomb efficiency;
fuel-cell standard voltage;
fuel-cell system;
fuel-cell-system energy-to-volume ratio;
fuel-cell-system energy-to-weight ratio;
fuel-cell-system power-to-volume ratio;
fuel-cell-system power-to-weight ratio;
fuel-cell-system standard thermal efficiency.
fuel-cell-system working voltage;
fuel-cell working voltage;
regenerative fuel-cell system. CV1-10E1

fuel-cell Coulomb efficiency. The ratio of the number of electrons obtained from the consumption of a mole of the fuel to the electrons theoretically available from the stated reaction.

$$\text{Coulomb Efficiency} = \frac{\int_0^{t_m} i\,dt}{nF} \times 100.$$

t_m = time required to consume a mole of fuel
i = instantaneous current
n = number of electrons furnished in the stated reaction by the fuel molecule
F = Faraday's constant = 96485.3 ± 10.0 absolute joules per absolute volt gram equivalent.

See also: **fuel cell.** CV1-10E1

fuel-cell standard voltage (at 25 degrees Celsius). The voltage associated with the stated reaction and determined from the equation

$$E^0 = \frac{-J\,\Delta G^0}{nF}$$

E^o = fuel-cell standard voltage
J = Joule's equivalent = 4.1840 absolute joules per calorie
ΔG^o = standard free energy change in kilocalories/mole of fuel
n = number of electrons furnished in the stated reaction by the fuel molecule
F = Faraday's constant = 96485.3 ± 10.0 absolute joules per absolute volt gram equivalent.

See also: **fuel cell.** CV1-10E1

fuel-cell system. An energy conversion device consisting of one or more fuel cells and necessary auxiliaries. *See also:* **fuel cell.** CV1-10E1

fuel-cell-system energy-to-volume ratio. The kilowatt-hour output per displaced volume of the fuel-cell system (including the fuel, oxidant, and storage). *See also:* **fuel cell.** CV1-10E1

fuel-cell-system energy-to-weight ratio. The kilowatt-hour output per unit weight of the fuel-cell system (including the fuel, oxidant, and storage). *See also:* **fuel cell.** CV1-10E1

fuel-cell-system power-to-volume ratio. The kilowatt output per displaced volume of the fuel-cell system (exclusive of the fuel, oxidant, and storage). *See also:* **fuel cell.** CV1-10E1

fuel-cell-system power-to-weight ratio. The kilowatt output per unit weight of the fuel-cell system (exclusive of the fuel, oxidant, and storage). *See also:* **fuel cell.** CV1-10E1

fuel-cell-system standard thermal efficiency. The efficiency of a system made up of a fuel cell and auxiliary equipment. *Note:* This efficiency is expressed as the ratio of (1) the electric energy delivered to the load circuit to (2) the enthalpy change for the stated cell reaction.

$$\text{Thermal Efficiency} = \frac{\int_0^{t_m} (E_{IL} \times i_L)dt}{\Delta H^0}$$

t_m = time required to consume a mole of fuel
E_{IL} = fuel-cell-system working voltage
i_L = instantaneous current into the load
ΔH^o = enthalpy change for the stated cell reaction at standard conditions.

See also: **fuel cell.** CV1-10E1

fuel-cell-system working voltage. The voltage at the load terminals of a fuel-cell system delivering current into the load. *See also:* **fuel cell.** CV1-10E1

fuel-cell working voltage. The voltage at the terminals of a single fuel-cell delivering current into system auxiliaries and load. *See also:* **fuel cell.** CV1-10E1

fuel-control mechanism (gas turbines). All devices, such as power-amplifying relays, servomotors, and interconnections required between the speed governor and the fuel-control valve. *See also:* **speed-governing system.** E282-31E2

fuel-control system (gas turbines). Devices that include the fuel-control valve and all supplementary fuel-control devices and interconnections necessary for adequate control of the fuel entering the combustion system of the gas turbine. *Note:* The supplementary fuel-control devices may or may not be directly actuated by the fuel-control mechanism. *See also:* **speed-governing system.** E282-31E2

fuel-control valve (gas turbines). A valve or any other device operating as a final fuel-metering element controlling fuel input to the gas turbine. *Notes:* (1) This valve or device may be directly or indirectly controlled by the fuel-control mechanism. (2) Variable-displacement pumps, or other devices that operate as the final fuel-control element in the fuel-control system, and that control fuel entering the combustion system are fuel-control valves. *See also:* **speed-governing system.** E282-31E2

fuel economy. The ratio of the chemical energy input to a generating station to its net electric output. *Note:* Fuel economy is usually expressed in British thermal units per kilowatthour. *See also:* **generating station.** 42A35-31E13

fuel-pressure electric gauge. A device that measures the fuel pressure (usually in pounds per square inch) at the carburetor of an aircraft engine. *Note:* It provides remote indication by means of a self-synchronous generator and motor. *See also:* **air transportation instruments.** 42A41-0

fuel stop valve (gas turbines). A device that, when actuated, shuts off all fuel flow to the combustion system, including that provided by the minimum fuel limiter. *See also:* **speed-governing system.** E282-31E2

fulguration. *See:* **electrodesiccation.**

full automatic plating. Mechanical plating in which the cathodes are automatically conveyed through successive cleaning and plating tanks. 42A60-0

full duplex (communication circuit) (telecommunication). Method of operation where each end can simultaneously transmit and receive. *See:* **duplex.** *See also:* **communication.** 0-19E4

full-field relay (industrial control). A relay that functions to maintain full field excitation of a motor while accelerating on reduced armature voltage. *See also:* **relay.** 42A25-34E10

full impulse voltage. An aperiodic transient voltage that rises rapidly to a maximum value and falls, usually less rapidly, to zero. See the following figure. *See also:* **test voltage and current; full-wave voltage impulse.**

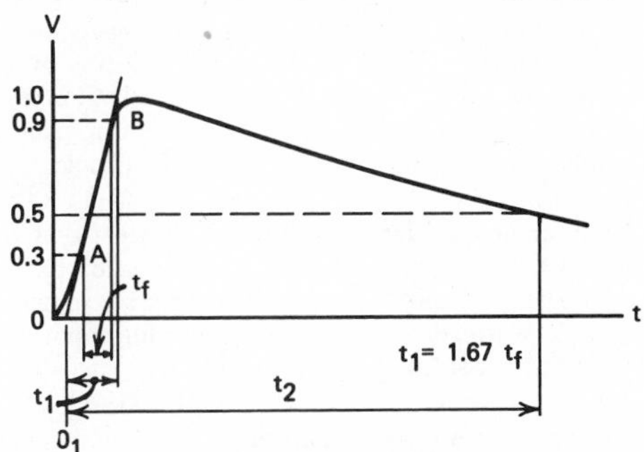

Full impulse voltage.

68A1-31E5

full-impulse wave (lightning arresters). An impulse wave in which there is no sudden collapse. *See:* **lightning arrester (surge diverter).** 50I25-31E7

full-load speed (electric drive) (industrial control). The speed that the output shaft of the drive attains with rated load connected and with the drive adjusted to deliver rated output at rated speed. *Note:* In referring to the speed with full load connected and with the drive adjusted for a specified condition other than for rated output at rated speed, it is customary to speak of the full-load speed under the (stated) conditions. *See also:* **electric drive.** 42A25-34E10

full-magnetic controller (industrial control). An electric controller having all of its basic functions performed by devices that are operated by electromagnets. *See also:* **electric controller.** 42A25-34E10

full-pitch winding (rotating machinery). A winding in which the coil pitch is 100 percent, that is, equal to the pole pitch. *See also:* **asynchronous machine; direct-current commutating machine; synchronous machine.** 0-31E8

full scale (analog computer). The nominal maximum value of a computer variable or the nominal maximum value at the output of a computing element. *See also:* **electronic analog computer.** E165-16E9

full-scale value. The largest value of the actuating electrical quantity that can be indicated on the scale or, in the case of instruments having their zero between the ends of the scale, the full-scale value is the arithmetic sum of the values of the actuating electrical quantity corresponding to the two ends of the scale. *Note:* Certain instruments, such as power-factor meters, are necessarily excepted from this definition. *See also:* **accuracy rating (instrument); instrument.** 39A1/39A2/42A30-0

full-voltage starter (industrial control). A starter that connects the motor to the power supply without reducing the voltage applied to the motor. *Note:* Full-voltage starters are also designated as across-the-line starters. *See also:* **starter.** 42A25-34E10

full-wave rectification (rectifying process) (power supplies). Full-wave rectification inverts the negative half-cycle of the input sinusoid so that the output contains two half-sine pulses for each input cycle. A pair of rectifiers arranged as shown with a center-tapped transformer or a bridge arrangement of four rectifiers and no center tap are both methods of obtaining full-wave rectification. *See also:* **rectification; rectifier.** 42A15-34E24;KPSH-10E1

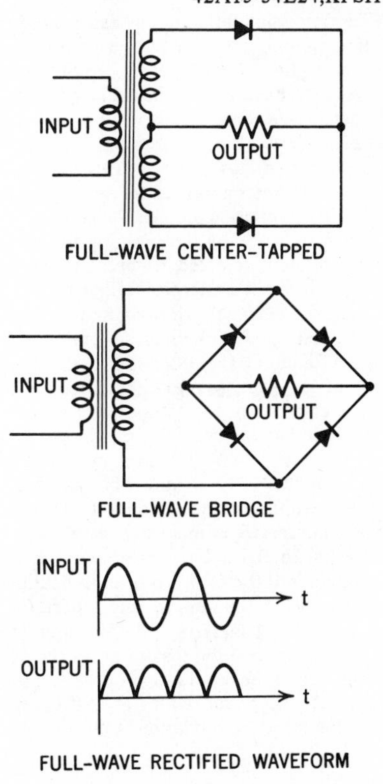

Full-wave rectification.

full-wave rectifier. *See:* **full-wave rectification.**

full-wave rectifier circuit. A circuit that changes single-phase alternating current into pulsating unidirectional current, utilizing both halves of each cycle. *See also:* **rectification.** E59-34E17

full-wave voltage impulse (lightning arresters). A voltage impulse that is not interrupted by sparkover, flashover, or puncture. *See:* **lightning arrester (surge diverter); full-impulse voltage.** 99I1-31E7

full width at half maximum (FWHM) (scintillation counters). The full width of a distribution measured at half the maximum ordinate. For a normal distribution, it is equal to $2(2 \ln 2)^{1/2}$ times the standard deviation (σ).

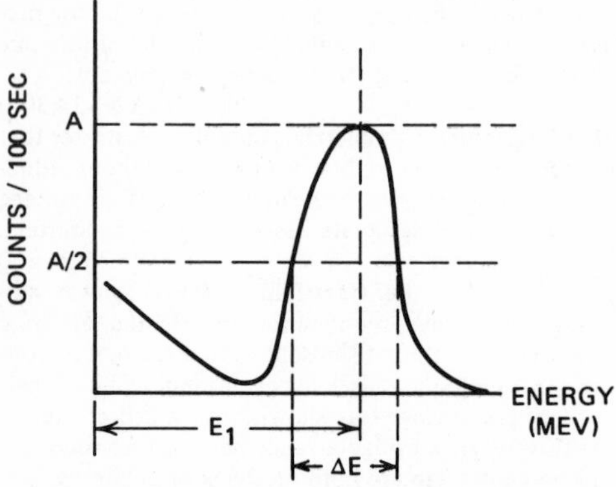

Full width at half maximum (in this case, ΔE).

Note: The expression **full width at half maximum**, given either as an absolute value or as a percentage of the value of the argument at the maximum of the distribution curve, is frequently used in nuclear physics as an approximate description of a distribution curve. Its significance can best be made clear by reference to a typical distribution curve, shown in the figure, of the measurement of the energy of the gamma rays from Cs^{137} with a scintillation counter spectrometer. The measurement is made by determining the number of gamma-ray photons detected in a prescribed interval of time, having measured energies falling within a fixed energy interval (channel width) about the values of energy (channel position) taken as argument of the distribution function. The abscissa of the curve shown is energy in megaelectronvolts (MeV) units and the ordinate is counts per given time interval per megaelectronvolt energy interval. The maximum of the distribution curve shown has an energy E_1 megaelectronvolts. The height of the peak is A_1 counts/100 seconds/megaelectronvolts. The full width at half maximum ΔE is measured at a value of the ordinate equal to $A_1/2$. The percentage full width at half maximum is $100\ \Delta E/E_1$. It is an indication of the width of the distribution curve, and where (as in the example cited) the gamma-ray photons are monoenergetic, it is a measure of the resolution of the detecting instrument. When the distribution curve is a Gaussian curve, the percentage full width at half maximum is related to the standard deviation σ by

$$100 \frac{\Delta E}{E_1} = 100 \times 2(2 \ln 2)^{1/2} \times \sigma.$$

See also: **scintillation counter.** E175-0

fume-resistant (industrial control). So constructed that it will not be injured readily by exposure to the specified fume. 37A100-31E11;42A25-34E10;42A95-0

function (1) (general). When a mathematical quantity u depends on a variable quantity x so that to each value of x (within the interval of definition) there correspond one or more values of u, then u is a function of x written $u = f(x)$. The variable x is known as the independent variable or the argument of the function. When a quantity u depends on two or more variables $x_1, x_2, \cdots, x_n$ so that for every set of values of $x_1, x_2, \cdots, x_n$ (within given intervals for each of the variables) there correspond one or more values of u, then u is a function of $x_1, x_2, \cdots, x_n$ and is written $u = f(x_1, x_2, \cdots, x_n)$. The variables $x_1, x_2, \cdots, x_n$ are the independent variables or arguments of the function. E270-0
(2) (vector). When a scalar or vector quantity u depends upon a variable vector **V** so that if for each value of **V** (within the region of definition) there correspond one or more values of u, then u is a function of the vector **V**. E270-0

functional designation (abbreviation). Letters, numbers, words, or combinations thereof, used to indicate the function of an item or a circuit, or of the position or state of a control or adjustment. Compare with: letter combination, reference designation, symbol for a quantity. *See also:* **abbreviation.** E267-0

functional unit. A system element that performs a task required for the successful operation of the system. *See also:* **system.** 0-35E2

function, describing (nonlinear element under periodic input) (control system, feedback). A transfer function based solely on the fundamental, ignoring other frequencies. *Note:* This equivalent linearization implies amplitude dependence with or without frequency dependence. *See also:* **control system, feedback.** 85A1-23E0

function, error transfer (closed loop) (control system, feedback). The transfer function obtained by taking the ratio of the Laplace transform of the error signal to the Laplace transform of its corresponding input signal. *See also:* **control system, feedback.** 85A1-23E0

function generator (1) (computers). A computing element with an output a specified nonlinear function of its input or inputs. *Note:* Normal usage excludes multipliers and resolvers. *See also:* **electronic analog computer.** E165-16E9
(2) (electric power systems). A device in which a mathematical function such as $y = f(x)$ can be stored so that for any input equal to x, an output equal to $f(x)$ will be obtained. *See also:* **speed-governing system.** E94-0

function generator, bivariant. A function generator having two input variables. *See also:* **electronic analog computer.** E165-0

function generator, curve-follower. A function generator that operates by automatically following a curve drawn or constructed on a surface. *See also:* **electronic analog computer.** E165-0

function generator, diode. A function generator that uses the transfer characteristics of resistive networks containing biased diodes. *Note:* The desired function is approximated by linear segments. *See also:* **electronic analog computer.** E165-0

function generator, map-reader. A bivariant function generator using a probe to detect the voltage at a point on a conducting surface and having coordinates proportional to the inputs. *See also:* **electronic analog computer.** E165-0

function generator, servo. A function generator consisting of a position servo driving a function potentiometer. *See also:* **electronic analog computer.** E165-0

function generator, switch-type. A function generator using a multitap switch rotated in accordance with the input and having its taps connected to suitable voltage sources. *See also:* **electronic analog computer.** E165-0

function, loop-transfer (closed loop) (control system, feedback). The transfer function obtained by taking the ratio of the Laplace transform of the return signal to the Laplace transform of its corresponding error signal. *See also:* **control system, feedback.** 85A1-23E0

function, output-transfer (closed loop) (control system, feedback). The transfer function obtained by taking the ratio of the Laplace transform of the output signal to the Laplace transform of the input signal. *See also:* **control system, feedback.** 85A1-23E0

function relay (analog computer). A relay used as a computing element. *See also:* **electronic analog computer.** E165-16E9

function, return-transfer (closed loop) (control system, feedback). The transfer function obtained by taking the ratio of the Laplace transform of the return signal to the Laplace transform of its corresponding input signal. *See also:* **control system, feedback.** 85A1-23E0

function switch (analog computer). A switch used as a computing element. *See also:* **electronic analog computer.** E165-16E9

function, system-transfer (automatic control). The transfer function obtained by taking the ratio of the Laplace transform of the signal corresponding to the ultimately controlled variable to the Laplace transform of the signal corresponding to the command. *See also:* **control system, feedback.** 85A1-23E0

function, transfer (control system, feedback). A mathematical, graphic, or tabular statement of the influence that a system or element has on a signal or action compared at input and at output terminals. *Note:* For a linear system, general usage limits the transfer function to mean the ratio of the Laplace transform of the output to the Laplace transform of the input in the absence of all other signals, and with all initial conditions zero. *See also:* **control system, feedback.** 85A1-23E0

function, weighting (control system, feedback). A function representing the time response of a linear system, or element to a unit-impulse forcing function; the derivative of the time response to a unit-step forcing function. *Notes:* (1) The Laplace transform of the weighting function is the transfer function of the system or element. (2) The time response of a linear system or element to an arbitrary input is described in terms of the weighting function by means of the convolution integral. *See also:* **control system, feedback.** 85A1-23E0

function, work. *See:* **work function.**

fundamental component. The fundamental frequency component in the harmonic analysis of a wave. *See also:* **signal wave.** E154-0

fundamental frequency (1) (signal-transmission system). The reciprocal of the period of a wave.
(2) (mathematically). The lowest frequency component in the Fourier representation of a periodic quantity. *See also:* **signal wave.** E154-13E6;E157-1E1

fundamental mode. *See:* **dominant mode.**

fundamental-type piezoelectric crystal unit. A unit designed to utilize the lowest frequency of resonance for a particular mode of vibration. *See also:* **crystal.** 42A65-0

fuse. An overcurrent protective device with a circuit-opening fusible part that is heated and severed by the passage of overcurrent through it. *Note:* A fuse comprises all the parts that form a unit capable of performing the prescribed functions. It may or may not be the complete device necessary to connect it into an electric circuit. *See:* **control.** 37A100-31E11;42A20-34E10

fuse blade (cartridge fuse). A cartridge-fuse terminal having a substantially rectangular cross section. 37A100-31E11

fuse carrier (oil cutout). An assembly of a cap that closes the top opening of an oil-cutout housing, an insulating member, and fuse contacts with means for making contact with the conducting element and for insertion into the fixed contacts of the fuse support. *Note:* The fuse carrier does not include the conducting element (fuse link). 37A100-31E11

fuse clips. The contacts on a base, fuse support, or fuse mounting for connecting into the circuit the current-responsive element with its holding means, if such means are used for making a complete device. 37A100-31E11

fuse condenser. A device that, added to a vented fuse, converts it to a nonvented fuse by providing a sealed chamber for condensation of the gases developed during circuit interruption. 37A100-31E11

fuse contact. *See:* **fuse terminal.**

fuse cutout. A cutout having a fuse link or fuse unit. *Note:* A fuse cutout is a fuse disconnecting switch. 37A100-31E11

fused electrolyte (bath) (fused salt) (electrolyte). A molten anhydrous electrolyte.
See:
anode butt;
anode effect;
anode layer;
cathode layer;
cell cavity;
collector plates;
contact conductor;
crust;
electrolytic cell;
internal heating;
metal fog;
shell;
working. 42A60-0

fused-electrolyte cell. A cell for the production of electric energy when the electrolyte is in a molten state. *See also:* **electrochemistry.** 42A60-0

fuse-disconnecting switch (disconnecting fuse). A disconnecting switch in which a fuse unit or fuse holder and fuse link forms all or a part of the blade. 37A100-31E11

fused salt (bath). *See:* **fused electrolyte.**

fused trolley tap (mining). A specially designed holder with enclosed fuse for connecting a conductor of a portable cable to the trolley system or other circuit supplying electric power to equipment in mines. *See also:* **mine feeder circuit.** 42A85-0

fuse filler. *See:* **arc-extinguishing medium.**

fuseholder (cutout base*). A device intended to support a fuse mechanically and connect it electrically in a circuit. *See also:* **cabinet.** 42A95-0

*Deprecated

fuseholder (1) (high-voltage fuse). An assembly of a fuse tube or tubes together with parts necessary to enclose the conducting element and provide means of making contact with the conducting element and the fuse clips. *Note:* The fuseholder does not include the conducting element (fuse link or refill unit). 37A100-31E11

(2) (low-voltage fuse). An assembly of base, fuse clips, and necessary insulation for mounting and connecting into the circuit the current-responsive element, with its holding means if used for making a complete device. *Notes:* (1) For low-voltage fuses the current-responsive element and holding means are called a fuse. (2) For high-voltage fuses the general type of assembly defined above is called a fuse support or fuse mounting. The holding means (fuseholder) and the current-responsive or conducting element are called a fuse unit. 37A100-31E11

fuse hook. A hook provided with an insulating handle for opening and closing fuses or switches and for inserting the fuseholder, fuse unit, or disconnecting blade into and removing it from the fuse support. 37A100-31E11

fuselage lights. Aircraft aeronautical lights, mounted on the top and bottom of the fuselage, used to supplement the navigation lights. *See also:* **signal lighting.** Z7A1-0

fuse link. A replaceable part or assembly, comprised entirely or principally of the conducting element, required to be replaced after each circuit interruption to restore the fuse to operating condition. 37A100-31E11

fuse support (high-voltage fuse) (fuse mounting). An assembly of base, mounting support or oil-cutout housing, fuse clips, and necessary insulation for mounting and connecting into the circuit the current-responsive element with its holding means if such means are used for making a complete device. *Notes:* (1) For high-voltage fuses the holding means is called a fuse carrier or fuseholder and in combination with the current-responsive or conducting element is called a fuse unit. (2) For low-voltage fuses the general type of assembly defined above is called a fuseholder. 34A100-31E11

fuse terminal (fuse contact). The means for connecting the current-responsive element or its holding means, if such means are used for making a complete device, to the fuse clips. 37A100-31E11

fuse time-current characteristic. The correlated values of time and current that designate the performance of all or a stated portion of the functions of the fuse. *Note:* The time-current characteristics of a fuse are usually shown as a curve. 37A100-31E11

fuse time-current tests. Tests that consist of the application of current to determine the relation between the root-mean-square alternating current or direct current and the time for the fuse to perform the whole or some specified part of its interrupting function. 37A100-31E11

fuse tongs. Tongs provided with an insulating handle and jaws. Fuse tongs are used to insert the fuseholder or fuse unit into the fuse support or to remove it from the support. 37A100-31E11

fuse tube. A tube of insulating material that surrounds the current-responsive element the conducting element, or the fuse link. 37A100-31E11

fuse unit. An assembly comprising a conducting element mounted in a fuseholder with parts and materials in the fuseholder essential to the operation of the fuse. 37A100-31E11

fusible element (fuse). That part, having predetermined current-responsive melting characteristics, which may be all or part of the current-responsive element. 37A100-31E11

fusible enclosed (safety) switch (industrial control). A switch complete with fuse holders and either with or without meter-testing equipment or accommodation for meters, having all current-carrying parts completely enclosed in metal, and operable without opening the enclosure. *See also:* **switch.** 42A25-34E10

future point (for supervisory control or indication or telemeter selection). Provision for the future installation of equipment required for a point. *Note:* A future point may be provided with (1) space only (2) drilling or other mounting provisions only, or (3) drilling or other mounting provisions and wiring only. 37A100-31E11

FW. *See:* **forward wave.**

FWHM. *See:* **full-width at half maximum.**

G

gain (1) (transmission gain). The increase in signal power in transmission from one point to another under stated conditions. *Note:* Power gain is usually expressed in decibels. *See also:* **related transmission terms; transmission characteristic.**

(2) (magnitude ratio) (linear system or element). The ratio of the magnitude (amplitude) of a steady-state sinusoidal output relative to the causal input; the length of a phasor from the origin to a point of the transfer locus in a complex plane. *Note:* The quantity may be separated into two factors: (A) a proportional amplification often denoted as K, which is frequency independent and associated with a dimensioned scale factor relating the units of input and output; (B) a dimensionless factor often denoted as $G(j\omega)$, which is frequency dependent. Frequency, conditions of operation, and conditions of measurement must be specified. A loop gain characteristic is a plot of log gain versus log frequency. In nonlinear systems, gains are often amplitude dependent. *See also:* **signal; transmission characteristic.**

(3) (closed-loop) (industrial control). The gain of a

closed-loop system, expressed as the ratio of the directly controlled output to the reference input. *See:* **control system, feedback.**
E151-2E2;0-42A65;42A65-31E3;85A1-23E0;85A1-23E0;AS1-34E10

gain (photomultipliers). *See:* **current amplification.** *See also:* **transmission characteristics.**

gain, available conversion power. *See:* **available conversion power gain.**

gain, available-power (two-port linear transducer). *See:* **available-power gain.**

gain, available-power, maximum (two-port linear transducer). *See:* **available-power gain, maximum.**

gain characteristic, loop (closed loop) (control system, feedback). The magnitude of the loop transfer function for real frequencies. *See also:* **control system, feedback.** 85A1-23E0

gain control. A device for adjusting the gain of a system or transducer. E151-2E2

gain control, instantaneous automatic. *See:* **instantaneous automatic gain control.**

gain crossover frequency (transfer function of an element or system). The frequency at which the gain becomes unity (and its decibel value zero). *See also:* **control system, feedback.** 0-23E0

gain, insertion. *See:* **insertion gain.**

gain, insertion voltage (electric transducer). The complex ratio of (1) the alternating component of voltage across the external termination of the output with the transducer inserted between the generator and the output termination, to (2) the voltage across the external termination of the output when the generator is connected directly to the output termination. *See also:* **network analysis; transducer.** E160-15E6

gain, loop (gain, open loop). *See:* **loop gain.**

gain margin (control system, feedback) (loop transfer function for a stable feedback system). The reciprocal of the gain at the frequency at which the phase angle reaches minus 180 degrees. *Note:* Gain margin, sometimes expressed in decibels, is a convenient way of estimating relative stability by Nyquist, Bode, or Nichols diagrams. For systems with similar gain and phase characteristics in a conditionally stable feedback system, gain margin is understood to refer to the highest frequency at which the phase angle is minus 180 degrees. *See also:* **control system, feedback.**
85A1-23E0

gain of an antenna. *See:* **directive gain; power gain; relative gain; directivity.** *See also:* **antenna.**

gain, power. *See:* **power gain.**

gain sensitivity control. *See:* **differential gain control.**

gain, static (control system, feedback) (gain of an element or loop). Gain of a system at zero frequency. *See also:* **control system, feedback.** 0-23E0

gain time control (electronic navigation). *See:* **sensitivity time control.** *See also:* **navigation.**

gain, transducer. *See:* **transducer gain.**

gain turn-down (transponder). The automatic receiver gain control incorporated for the purpose of protecting the transmitter from overload when the number of interrogations exceeds the design limitations. *See also:* **navigation.** 0-10E6

gain, zero frequency. *See:* **gain, static.**

Galvanic cell. A cell in which chemical change is the source of electric energy. *Note:* It usually consists of two dissimilar conductors in contact with each other and with an electrolyte, or two similar conductors in contact with each other and with dissimilar electrolytes. *See also:* **electrochemistry.** CM-34E2

Galvanic current (medical electronics). An essentially steady unidirectional current (direct current). *See also:* **medical electronics.** 0-18E1

Galvanic series. A list of metals and alloys arranged according to their relative potentials in a given environment. CM-34E2

Galvani's experiment (electrobiology). Consists of bringing a muscle, or a motor nerve connected with a muscle, into simultaneous contact with two dissimilar metals, such as copper and iron, and producing a muscular contraction by closing the circuit. *See also:* **electrocardiogram.** 42A80-18E1

galvanism (galvanization) (voltaisation) (electrotherapy). The use of a Galvanic current for its biological or medical effects. *See:* **Galvanic current.** *See also:* **electrotherapy.** 42A80-18E1

galvanization (voltaisation) (electrotherapy). *See:* **galvanism.**

galvanometer. An instrument for indicating or measuring a small electric current, or a function of the current, by means of a mechanical motion derived from electromagnetic or electrodynamic forces that are set up as a result of the current. *See also:* **instrument.** 42A30-0

galvanometer recorder (photographic recording). A combination of mirror and coil suspended in a magnetic field. The application of a signal current to the coil causes a reflected light beam from the mirror to pass across a slit in front of a moving photographic film, thus providing a photographic record of the signal.
0-1E1

gamma (1) (photographic material). The gamma is the slope of the straight-line portion of the Hurter and Driffield curve. It represents the rate of change of photographic density with the logarithm of exposure. *Notes:* (A) Gamma is a measure of the contrast properties of the film. (B) Both gamma and density specifications are commonly used as controls in the processing of photographic film. E157-1E1
(2) (television). The exponent of that power law that is used to approximate the curve of output magnitude versus input magnitude over the region of interest. *See also:* **color terms; television.**
E160-15E6; E201-2E2

gamma correction (television). The introduction of a nonlinear output-input characteristic for the purpose of changing the effective value of gamma. *See also:* **color terms.** E201-2E2

gamma ray (photovoltaic power system) (Γ ray). Penetrating short-wavelength electromagnetic radiation of nuclear origin. *See also:* **photovoltaic power system; solar cells (photovoltaic power system).**
0-10E1

gangway cable (mining). A cable designed to be installed horizontally (or nearly so) for power circuits in mine gangways and entries. *See also:* **mine feeder circuit.** 42A85-0

gap (computing systems).
See:
file gap;
input gap;

interaction gap;
main gap;
output gap;
record gap;
resonant gap;
starter gap.

gap (rotating machinery). *See:* **air gap.**

gap admittance (electronic) (electron tubes). The difference between (1) the gap admittance with the electron stream traversing the gap and (2) the gap admittance with the stream absent. *See:* **admittance gap (electronic); electron-tube admittances.** E160-15E6

gap admittance circuit (electron tubes). The admittance of the circuit at a gap in the absence of an electron stream. *See:* **admittance, gap (circuit); electron-tube admittances; gas tubes.** E160-15E6

gap capacitance, effective (electron tubes). One half the rate of change with angular frequency of the resonator susceptance, measured at the gap, for frequencies near resonance. *See:* **electron-tube admittances; gas tubes.** E160-15E6

gap coding (navigation) (electronic). A process of communicating in which a normally continuous signal (such as a radio-frequency carrier) is interrupted so as to form a telegraph-type message. *See also:* **navigation.** 0-10E6

gap length (longitudinal magnetic recording). The gap length is the physical distance between the adjacent surfaces of the poles of a magnetic head. *Note:* The value of gap length necessary in certain studies exceeds that defined above and usually is referred to as the effective gap length. *See:* **magnetic head.** *See also:* **phonograph pickup.** 0-1E1

gap loading multipactor (electron tubes). The electronic gap admittance, resulting from a sustained secondary-emission discharge existing within a gap as a result of the motion of the secondary electrons in synchronism with the electric field in the gap. *See:* **electron-tube admittances; gas tubes.** E160-15E6

gap loading, primary transit-angle (electron tubes). The electronic gap admittance that results from the traversal of the gap by an initially unmodulated electron stream. *Note:* This is exclusive of secondary emission in the gap. *See:* **electron-tube admittances; gas tubes.** E160-15E6

gap loading, secondary electron (electron tubes). The electronic gap admittance that results from the traversal of a gap by secondary electrons originating in the gap. *See:* **electron-tube admittances; gas tubes.** E160-15E6

gas (rotating machinery). Any gas or gas mixture used as a cooling medium. *Note:* The term is usually not applied to air. *See also:* **cradle base (rotating machinery).** 0-31E8

gas-accumulator relay. A relay that is so constructed that it accumulates all or a fixed proportion of gas released by the protected equipment and operates by measuring the volume of gas so accumulated. 37A100-31E11/31E6

gas amplification (gas-filled radiation-counter tube). The ratio of the charge collected to the charge liberated by the initial ionizing event. *See also:* **counter tubes; gas-filled radiation; gas tubes.** 42A70-15E6

gas amplification factor (gas phototube). The ratio of radiant or luminous sensitivities with and without ionization of the contained gas. *See also:* **gas tubes; phototubes.** 42A70-15E6

gas cell. A cell in which the action of the cell depends on the absorption of gases by the electrodes. *See also:* **electrochemistry.** 42A60-0

gas counter tube (radiation). A counter tube into which the sample to be measured is introduced in the form of a gas. *See:* **anticoincidence (radiation counters).** 0-15E6

gas (ionization) current (vacuum tube). The ionization current flowing to a negatively biased electrode and composed of positive ions that are produced by an electron current flowing between other electrodes. *Note:* These positive ions are a result of collisions between electrons and molecules of the residual gas. *See also:* **electron device.** 42A70-15E6

gas dryer (rotating machinery). Any device used to remove moisture from gas. *See also:* **cradle base (rotating machinery).** 0-31E8

gas-electric drive (gasoline-electric drive). A self-contained system of power generation and application in which the power generated by a gasoline engine is transmitted electrically by means of a generator and a motor (or multiples of these) for propulsion purposes. *Note:* The prefix gas-electric is applied to ships, locomotives, cars, buses, etcetera, that are equipped with this drive. *See also:* **electric locomotive.** 42A40-0

gaseous discharge. The emission of light from gas atoms excited by an electric current. *See also:* **lamp.** Z7A1-0

gaseous tube generator. A power source comprising a gas-filled electron-tube oscillator, a power supply, and associated control equipment. *See also:* **induction heating; oscillatory circuit.** E54/E169-0

gas-filled cable. A self-contained pressure cable in which the pressure medium is an inert gas having access to the insulation. *See:* **pressure cable; self-contained pressure cable.** *See also:* **power distribution, underground construction.** 42A35-31E13

gas-filled phototube. A phototube into which a quantity of gas has been introduced, usually for the purpose of increasing its sensitivity. *See also:* **tube definitions.** 42A70-15E6

gas-filled pipe cable. A pipe cable in which the pressure medium is an inert gas having access to the insulation. *See:* **pressure cable; pipe cable.** *See also:* **power distribution, underground construction.** 42A35-31E13

gas-filled radiation-counter tube. A gas tube, in a radiation counter, used for the detection of radiation by means of gas ionization. *See also:* **tube definitions.** *Note:* For an extensive list of cross references, see *Appendix A.* 42A70-0

gas-filled rectifier. A gas-filled tube whose function is to rectify an alternating current.
See:
arc cathode;
blocking period;
conducting period;
extinction voltage;
fault electrode current;
gas tubes;
idle period;
mercury-arc rectifier;
mercury-pool cathode;

overlap angle;
pool rectifier;
preheating time;
relieving anode;
restriking voltage;
starter gap;
starting electrode;
striking current. 50I07-15E6

gas-flow counter tube. A radiation-counter tube in which an appropriate atmosphere is maintained by a flow of gas through the tube. *See also:* **gas-filled radiation-counter tube.** E160-15E6

gas focusing (electron-beam tubes). A method of concentrating an electron beam by gas ionization within the beam. *See:* **beam tubes; focusing, gas; gas tubes.** E160-15E6

gas grooves (electrometallurgy). The hills and valleys in metallic deposits caused by streams of hydrogen or other gas rising continuously along the surface of the deposit while it is forming. *See also:* **electrowinning.** 42A60-0

gasket-sealed relay. A relay in an enclosure sealed with a gasket. *See also:* **relay.** 83A16-0

gas-oil-sealed system (regulator). A system in which the interior of the tank is sealed from the atmosphere, over the temperature range specified, by means of an auxiliary tank or tanks to form a gas-oil seal operating on the manometer principle. *See also:* **transformer, oil-immersed.** 42A15/57A15-31E12

gasoline-electric drive. *See:* **gas-electric drive.**

gas-pressure relay. A relay so constructed that it operates by the gas pressure in the protected equipment. 37A100-31E11/31E6

gasproof. So constructed or protected that the specified gas will not interfere with successful operation. 37A100-31E11;42A95-0;ICI-34E10

gasproof or vaporproof (rotating machinery). So constructed that the entry of a specified gas or vapor under prescribed conditions cannot interfere with satisfactory operating of the machine. *See also:* **asynchronous machine; direct-current commutating machine; synchronous machine.** 0-31E8

gas ratio. The ratio of the ion current in a tube to the electron current that produces it. *See also:* **electrode current.** 42A70-15E6

gas seal (rotating machinery). A sealing arrangement intended to minimize the leakage of gas to or from a machine along a shaft. *Note:* It may be incorporated into a ball or roller bearing assembly. *See also:* **cradle base (rotating machinery).** 0-31E8

gassing. The evolution of gases from one or more of the electrodes during electrolysis. *See also:* **electrolytic cell.** 42A60-0

gas system (rotating machinery). The combination of parts used to ventilate a machine with any gas other than air, including facilities for charging and purging the gas in the machine. *See also:* **cradle base (rotating machinery).** 0-31E8

gastight. So constructed that gas or air can neither enter nor leave the structure except through vents or piping provided for the purpose. 5A1-0

gas tube. An electron tube in which the pressure of the contained gas or vapor is such as to affect substantially the electrical characteristics of the tube. *See also:* **tube definitions.** *Note:* For an extensive list of cross references, see *Appendix A.* 42A70-15E6

gas-tube relaxation oscillator (arc-tube relaxation oscillator). A relaxation oscillator in which the abrupt discharge is provided by the breakdown of a gas tube. *See also:* **oscillatory circuit.** 42A65-0

gas-turbine-electric drive. A self-contained system of power generation and application in which the power generated by a gas turbine is transmitted electrically by means of a generator and a motor (or multiples of these) for propulsion purposes. *Note:* The prefix gas-turbine-electric is applied to ships, locomotives, cars, buses, etcetera, that are equipped with this drive. *See also:* **electric locomotive.** 42A40-0

gate (1) (industrial control). A device or element that, depending upon one or more specified inputs, has the ability to permit or inhibit the passage of a signal. *See also:* **circuits and devices; control.** AS1-34E10

(2) (electronic computers). (A) A device having one output channel and one or more input channels, such that the output channel state is completely determined by the contemporaneous input channel states, except during switching transients. (B) A combinational logic element having at least one input channel. (C) An AND gate. (D) An OR gate. *See also:* **circuits and devices; electronic digital computer; electronic computation.** E162/E270-0;X3A12-16E9

(3) (cryotron). An output element of a cryotron. *See also:* **circuits and devices; superconductivity.** E217-15E7

(4) (thyristor). An electrode connected to one of the semiconductor regions for introducing control current. *See also:* **anode.** E223-34E17/34E24/15E7

(5) (navigation systems). (A) An interval of time during which some portion of the circuit or display is allowed to be operative, or (B) the circuit that provides gating. *See also:* **navigation.** 0-10E6

gate-controlled delay time (thyristor). The time interval, between a specified point at the beginning of the gate pulse and the instant when the principal voltage (current) has dropped (risen) to a specified value near its initial value during switching of a thyristor from the OFF state to the ON state by a gate pulse. *See also:* **principal voltage-current characteristic.** E223-34E1/34E24/15E7

gate-controlled rise time (thyristor). The time interval between the instants at which the principal voltage (current) has dropped (risen) from a specified value near its initial value to a specified low (high) value, during switching of a thyristor from the OFF state to the ON state by a gate pulse. *Note:* This time interval will be equal to the rise time of the ON-state current only for pure resistive loads. *See also:* **principal voltage-current characteristic.** E223-34E1/34E24/15E7

gate-controlled turn-off time (turn-off thyristor). The time interval, between a specified point at the beginning of the gate pulse and the instant when the principal current has decreased to a specified value, during switching from the ON state to the OFF state by a gate pulse. *See also:* **principal voltage-current characteristic.** E223-34E1/34E24/15E7

gate-controlled turn-on time (thyristor). The time interval, between a specified point at the beginning of the gate pulse and the instant when the principal voltage (current) has dropped (risen) to a specified low (high) value during switching of a thyristor from the OFF state to the ON state by a gate pulse. *See also:* **principal voltage-current characteristic.** E223-34E17/34E24/15E7

gated sweep (oscilloscopes). A sweep controlled by a gate waveform. Also, a sweep that will operate recurrently (free-running, synchronized, or triggered) during the application of a gating signal. *See:* **oscillograph.** 0-9E4

gate nontrigger current (thyristor). The maximum gate current that will not cause the thyristor to switch from the OFF state to the ON state. *See also:* **principal current.** E223-34E17/34E24/15E7

gate nontrigger voltage (thyristor). The maximum gate voltage that will not cause the thyristor to switch from the OFF state to the ON state. *See also:* **principal voltage-current characteristic (principal characteristic).** E223-34E17/34E24/15E7

gate terminal (thyristor). A terminal that is connected to a gate. *See also:* **anode.** E223-34E17/15E7

gate trigger current (thyristor). The minimum gate current required to switch a thyristor from the OFF state to the ON state. *See also:* **principal current.** E223-34E17/34E24/15E7

gate trigger voltage (thyristor). The gate voltage required to produce the gate-trigger current. *See also:* **principal voltage-current characteristic.** E223-34E17/34E24/15E7

gate turn-off current (turn-off thyristor). The minimum gate current required to switch a thyristor from the ON state to the OFF state. *See also:* **principal current.** E223-34E17/34E24/15E7

gate turn-off voltage (turn-off thyristor). The gate voltage required to produce the gate turn-off current. *See also:* **principal voltage-current characteristic.** E223-34E17/34E24/15E7

gate voltage (thyristor). The voltage between a gate terminal and a specified main terminal. *See also:* **principal voltage-current characteristic.** E223-34E17/34E24/15E7

gating. (1) The process of selecting those portions of a wave that exist during one or more selected time intervals or (2) that have magnitudes between selected limits. *See also:* **modulation; modulating systems; wave front.** E145-0;42A65-0

gating signal (keying signal). A signal that activates or deactivates a circuit during selected time intervals. 42A65-0

gauss (centimeter-gram-second electromagnetic system). The gauss is 10^{-4} webers per square meter or 1 maxwell per square centimeter. E270-0

Gaussian response (amplifiers). A particular frequency-response characteristic following the curve $y(f) = e^{-af^2}$. *Note:* Typically, the frequency response approached by an amplifier having good transient response characteristics. *See:* **amplifier.** 0-9E4

Gaussian system (units). A system in which centimeter-gram-second electrostatic units are used for electric quantities and centimeter-gram-second electromagnetic units are used for magnetic quantities. *Note:* When this system is used, the factor c (the speed of light) must be inserted at appropriate places in the electromagnetic equations. E270-0

Gauss law (electrostatics). States that the integral over any closed surface of the normal componant of the electric flux density is equal in a rationalized system to the electric charge Q_0 within the surface. Thus

$$\underset{\text{closed surface}}{\int (\mathbf{D}\cdot\mathbf{n})\,dA} = \underset{\text{volume enclosed}}{\int \rho_0\,dV} = Q_0.$$

Here, **D** is the electric flux density, **n** is a unit normal to the surface, dA the element of area, ρ_0 is the space charge density in the volume V enclosed by the surface. E270-0

gaussmeter. A magnetometer provided with a scale graduated in gauss or kilogauss. *See also:* **magnetometer.** 42A30-0

GCA. *See:* **ground-controlled approach.**

GCI. *See:* **ground-controlled interception.**

***G* display (radar).** A rectangular display in which a target appears as a laterally centralized blip when the radar antenna is aimed at it in azimuth, and wings appear to grow on the blip as the distance to the target is diminished; horizontal and vertical aiming errors are respectively indicated by horizontal and vertical displacement of the blip. *See also:* **navigation.** E172-10E6

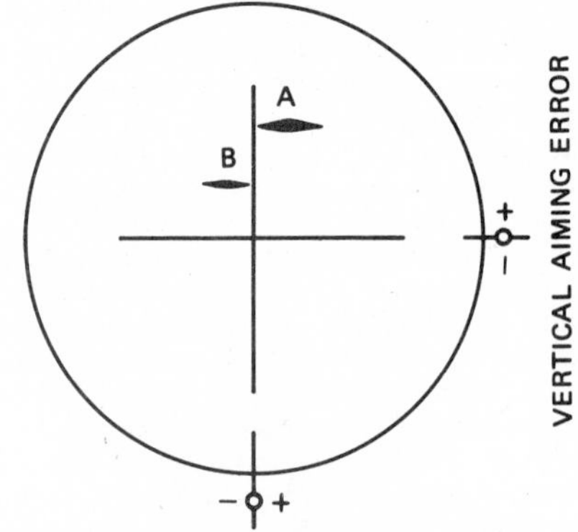

G display.

***G* drift (electronic navigation).** A drift component in gyros (sometimes in accelerometers) proportional to the nongravitational acceleration and caused by torques resulting from mass unbalance. Jargon. *See also:* **navigation.** E172-10E6

G^2 drift (electronic navigation). A drift component in gyros (sometimes in accelerometers) proportional to the square of the nongravitational acceleration and caused by anisoelasticity of the rotor supports. Jargon. *See also:* **navigation.** E174-10E6

geared-drive machine. A direct-drive machine in which the energy is transmitted from the motor to the driving sheave, drum, or shaft through gearing. *See also:* **driving machine (elevators).** 42A45-0

geared traction machine (elevators). A geared-drive traction machine. *See also:* **driving machine (elevators).** 42A45-0

gearless motor. A traction motor in which the armature is mounted concentrically on the driving axle, or is carried by a sleeve or quill that surrounds the axle, and drives the axle directly without gearing. *See also:* **traction motor.** 42A42-0

gearless traction machine (elevators). A traction machine, without intermediate gearing, that has the traction sheave and the brake drum mounted directly on the motor shaft. *See also:* **driving machine (elevators).** 42A45-0

gear pattern (facsimile). *See:* **drive pattern.**

gear ratio (watthour meter). The number of revolutions of the rotor for one revolution of the first dial pointer. *Note:* This is commonly denoted by the symbol R_g. See also: **electricity meter (meter); moving element of an instrument.** 12A0/42A30-0

Geiger-Mueller counter tube. A radiation-counter tube designed to operate in the Geiger-Mueller region. *See also:* **tube definitions.** 42A70-0

Geiger-Mueller region (radiation counter tube). The range of applied voltage in which the charge collected per isolated count is independent of the charge liberated by the initial ionizing event. *See also:* **gas-filled radiation-counter tubes.** 42A70-15E6

Geiger-Mueller threshold (radiation-counter tube). The lowest applied voltage at which the charge collected per isolated tube count is substantially independent of the nature of the initial ionizing event. *See also:* **gas-filled radiation-counter tubes.** 42A70-15E6

Geissler tube. A special form of gas-filled tube for showing the luminous effects of discharges through rarefied gases. *Note:* The density of the gas is roughly one-thousandth of that of the atmosphere. *See:* **gas tubes.** 50I07-15E6

general alarm system. A system of bells energized from a storage battery (the emergency lighting battery, if provided) controlled from the wheelhouse and so located as to be audible to all persons on a ship, to give an alarm in case of a major casualty. 42A43-0

general coordinated methods (general application to electric supply or communication systems). Those methods reasonably available that contribute to inductive coordination without specific consideration of the requirements for individual inductive exposures. *See also:* **inductive coordination.** 42A65-0

general diffuse lighting. Lighting involving luminaires that distribute 40 to 60 percent of the emitted light downward and the balance upward sometimes with a strong component at 90 degrees (horizontal). *See also:* **general lighting.** Z7A1-0

generalized entity. *See:* **generalized property.**

generalized property (generalized quantity) (generalized entity). Any of the physical concepts in terms of examples of which observable physical systems and phenomena are described qualitatively. *Notes:* (1) Examples are the abstract concepts of length, electric current, energy, etcetera. (2) A generalized property is characterized by the qualitative attribute of physical nature, or dimensionality, but not by a quantitative magnitude. E270-0

generalized quantity. *See:* **generalized property.**

general lighting. Lighting designed to provide a substantially uniform level of illumination throughout an area, exclusive of any provision for special local requirements.
See:
accent lighting;
ceiling-area lighting;
cornice lighting;
cove lighting;
diffused lighting;
direct-indirect lighting;
direct lighting;
directional lighting;
emergency lighting;
general diffuse lighting;
indirect lighting;
light;
local lighting;
localized general lighting;
luminaire;
portable lighting;
semi-direct lighting;
semi-indirect lighting;
supplementary lighting;
valance lighting. Z7A1-0

general-purpose computer. A computer that is designed to solve a wide class of problems. X3A12-16E9

general-purpose controller (industrial control). Any controller having ratings, characteristics, and mechanical construction for use under usual service conditions. *See:* **electric controller.** IC1-34E10

general-purpose enclosure (1) (general). An enclosure used for usual service applications where special types of enclosures are not required. 37A100-31E11
(2) (rotating machinery). An enclosure that primarily protects against accidental contact and slight indirect splashing but is neither dripproof nor splash-proof. *See also:* **asynchronous machine; direct-current commutating machine; synchronous machine.** E45-0

general-purpose floodlight. A weatherproof unit so constructed that the housing forms the reflecting surface. *Note:* The assembly is enclosed in a cover glass. *See also:* **floodlighting.** Z7A1-0

general-purpose induction motor (rotating machinery). Any open motor having a continuous rating of 50-degrees Celsius rise by resistance for Class A insulation, or of 80-degrees Celsius rise for Class B, a service factor as listed in the following tabulation, and designed, listed, and offered in standard ratings with standard operating characteristics and mechanical construction, for use under usual service conditions without restrictions to a particular application or type of application.

Service Factor

	Synchronous Speed, revolutions per minute			
Horsepower	3600	1800	1200	900
1/20	1.4	1.4	1.4	1.4
1/12	1.4	1.4	1.4	1.4
1/8	1.4	1.4	1.4	1.4
1/6	1.35	1.35	1.35	1.35
1/4	1.35	1.35	1.35	1.35
1/3	1.35	1.35	1.35	1.35
1/2	1.25	1.25	1.25	
3/8	1.25	1.25		
1	1.25			

See also: **asynchronous machine.** 0-31E8

general-purpose motor. Any open motor having a continuous 40-degree Celsius rating and designed, listed, and offered in standard ratings with standard operating characteristics and mechanical construction for use under usual service conditions without restrictions to a particular application or type of application. *See:* **asynchronous machine; direct-current commutating machine; synchronous machine.** 42A10-0

general-purpose relay. A relay that is adaptable to a variety of applications. *See also:* **relay.** 83A16-0

general-use snap switch. A form of general-use switch so constructed that it can be installed in flush device boxes, or on outlet box covers, or otherwise used in conjunction with wiring systems (recognized by the National Electric Code). 1A0-0

general-use switch (industrial control). A switch intended for use in general distribution and branch circuits. *Note:* It is rated in amperes, and it is capable of interrupting the rated current at the rated voltage. *See also:* **switch.** 42A25-34E10

generate (computing systems). To produce a program by selection of subsets from a set of skeletal coding under the control of parameters. *See also:* **electronic digital computers.** X3A12-16E9

generated voltage (rotating machinery). A voltage produced in a closed path or circuit by the relative motion of the circuit or its parts with respect to magnetic flux. *See:* **asynchronous machine; direct-current commutating machine; synchronous machine.** 0-31E8

generating electric field meter (gradient meter). A device in which a flat conductor is alternately exposed to the electric field to be measured and then shielded from it. *Note:* The resulting current to the conductor is rectified and used as a measure of the potential gradient at the conductor surface. *See also:* **instrument.** 42A30-0

generating magnetometer (earth inductor). A magnetometer that depends for its operation on the electromotive force generated in a coil that is rotated in the field to be measured. *See also:* **magnetometer.** 42A30-0

generating station (electric power supply). A plant wherein electric energy is produced from some other form of energy (for example, chemical, mechanical, or hydraulic) by means of suitable apparatus. *Note:* For an extensive list of cross references, see *Appendix A.* 42A35-31E13

generating-station auxiliaries. Accessory units of equipment necessary for the operation of the plant. *Example:* Pumps, stokers, fans, etcetera. *Note:* Auxiliaries may be classified as essential auxiliaries or those that must not sustain service interruptions of more than 15 seconds to 1 minute, such as boiler feed pumps, forced draft fans, pulverized fuel feeders, etcetera; and nonessential auxiliaries that may, without serious effect, sustain service interruptions of 1 to 3 minutes or more, such as air pumps, clinker grinders, coal crushers, etcetera. *See also:* **generating station.** 42A35-31E13

generating-station auxiliary power. The power required for operation of the generating station auxiliaries. *See also:* **generating station.** 42A35-31E13

generating-station efficiency. *See:* **efficiency.**

generating-station reserve. *See:* **reserve equipment.**

generating voltmeter (rotary voltmeter). A device in which a capacitor is connected across the voltage to be measured and its capacitance is varied cyclically. The resulting current in the capacitor is rectified and used as a measure of the voltage. *See also:* **instrument.** 42A30-0

generation rate (semiconductor). The time rate of creation of electron-hole pairs. *See also:* **semiconductor device.** E102/E270-0;E216-34E17

generator (1) (rotating machinery). A machine that converts mechanical power into electric power. *See also:* **asynchronous machine; direct-current commutating machine; synchronous machine.** 0-31E8
(2) (computing systems). A controlling routine that performs a generate function, for example, report generator, input-output generator. *See also:* **electronic digital computer; function generator; noise generator.** X3A12-16E9

generator, alternating-current. *See:* **alternating-current generator.**

generator, arc welder (1) (generator, alternating-current arc welder). An alternating-current generator with associated reactors, regulators, control, and indicating devices required to produce alternating current suitable for arc welding.
(2) (generator-rectifier, alternating-current–direct current arc welder). A combination of static rectifiers and the associated alternating-current generator, reactors, regulators, control, and indicating devices required to produce either direct or alternating current suitable for arc welding.
(3) (generator, direct-current arc welder). A direct-current generator with associated reactors, regulators, control, and indicating devices required to produce direct current suitable for arc welding. *See also:* **electric arc-welding apparatus.** 87A1-0
(4) (generator-rectifier, direct-current arc welder). A combination of static rectifiers and the associated alternating-current generator, reactors, regulators, controls, and indicating devices required to produce direct current suitable for arc welding. *See also:* **electric arc-welding apparatus.** 87A1-0

generator efficiency (thermoelectric couple). The ratio of (1) the electric power output of a thermoelectric couple to (2) its thermal power input. *Note:* This is an idealized efficiency assuming perfect thermal insulation of the thermoelectric arms. *See also:* **thermoelectric device.** E221-15E7

generator-field accelerating relay (industrial control). A relay that functions automatically to maintain the armature current within prescribed limits when a motor supplied by a generator is accelerated to any speed, up to base speed, by controlling the generator field current. *Note:* This definition applies to adjustable-voltage direct-current drives. *See:* **relay.** IC1-34E10

generator-field control. A system of control that is accomplished by the use of an individual generator for each elevator or dumbwaiter wherein the voltage applied to the driving-machine motor is adjusted by varying the strength and direction of the generator field. *See also:* **control (elevators).** 42A45-0

generator field decelerating relay (industrial control). A relay that functions automatically to maintain the armature current within prescribed limits when a motor, supplied by a generator, is decelerated from base speed, or less, by controlling the generator field-current. *Note:* This definition applies to adjustable-voltage direct-current drives. *See:* **relay.** IC1-34E10

generator/motor. A machine that may be used as either a generator or a motor usually by changing rotational direction. *Note:* (1) This type of machine has particular application in a pumped-storage operation, in which water is pumped into a reservoir during off-peak periods and released to provide generation for peaking loads. (2) This definition eliminates the confusion of terminology for this type of machine. A slant is used between the terms to indicate their equality, and also the machine serves one function or the other and not both at the same time. The word **generator** is placed first to provide a distinction in speech between this term and the commonly used term **motor-generator**, which has an entirely different meaning. *See also:* **asynchronous machine; synchronous machine.** 0-31E8

generator set. A unit consisting of one or more generators driven by a prime mover. *See also:* **asynchronous machine; direct-current commutating machine; synchronous machine.** 0-31E8

generette (rotating machinery). A test jig designed on the principle of a motorette, for endurance tests on sample lengths of coils or bars for large generators. *See also:* **asynchronous machine; direct-current commutating machines; synchronous machine.** 0-31E8

geocentric latitude (navigation). The acute angle between (1) a line joining a point with the earth's geometric center and (2) the earth's equatorial plane. E174-10E6

geocentric vertical. *See:* **geometric vertical.**

geodesic. The shortest line between two points measured on any mathematically derived surface that includes the points. *See also:* **navigation.** 0-10E6

geodetic latitude (navigation). The angle between the normal to the spheroid and the earth's equatorial plane; the latitude generally used in maps and charts. Also called **geographic latitude**. *See also:* **navigation.** 0-10E6

geographic latitude. *See:* **geodetic latitude.**

geographic (map) vertical. The direction of a line normal to the surface of the geoid. *See also:* **navigation.** E174-10E6

geoid. The shape of the earth as defined by the hypothetical extension of mean sea level continuously through all land masses. *See also:* **navigation.** E174-10E6

geometrical factor (navigation). The ratio of the change in a navigational coordinate to the change in distance, taken in the direction of maximum navigational coordinate change; the magnitude of the gradient of the navigational coordinate. *See also:* **navigation.** E172-10E6

geometric distortion (television). The displacement of elements in the reproduced picture from the correct relative positions in the perspective plane projection of the original scene. *See also:* **television.** E204-0;0-42A65

geometric factor (cable calculations) (power distribution, underground cables). A parameter used and determined solely by the relative dimensions and geometric configuration of the conductors and insulation of a cable. *See also:* **power distribution, underground construction.** 0-31E1

geometric inertial navigation equipment. The class of inertial navigation equipment in which the geographic navigational quantities are obtained by computations (generally automatic) based upon the outputs of accelerometers whose vertical axes are maintained parallel to the local vertical, and whose azimuthal orientations are maintained in alignment with a predetermined geographic direction (for example, north). *See also:* **navigation.** E174-10E6

geometric (geocentric) vertical (navigation). The direction of the radius vector drawn from the center of the earth through the location of the observer. *See also:* **navigation.** E174-10E6

geometry (oscilloscopes). The degree to which a cathode-ray tube can accurately display a rectilinear pattern. *Note:* Generally associated with properties of a cathode-ray tube; the name may be given to a cathode-ray-tube electrode or its associated control. *See:* **oscillograph.** 0-9E4

getter (electron tubes). A substance introduced into an electron tube to increase the degree of vacuum by chemical or physical action on the residual gases. *See also:* **electrode (of an electron tube); electronic tube.** 50I07-15E6;42A70-0

ghost (television). A spurious image resulting from an echo. *See also:* **television.** 42A65-0

ghost pulse. *See:* **ghost signals.**

ghost signals (loran). (1) Identification pulses that appear on the display at less than the desired loran station full pulse repetition frequency. (2) Signals appearing on the display that have a basic repetition frequency other than that desired. *See also:* **navigation.** 0-10E6

gilbert (centimeter-gram-second electromagnetic system). The unit of magnetomotive force. The gilbert is one oersted-centimeter. E270-0

Gill-Morrell oscillator. An oscillator of the retarding-field type in which the frequency of oscillation is dependent not only on electron transit time within the tube, but also on associated circuit parameters. *See also:* **oscillatory circuits.** E145-0

gland seal (rotating machinery). A seal used to prevent leakage between a moveable and a fixed part. *See also:* **cradle base (rotating machinery).** 0-31E8

glare. The sensation produced by brightnesses within the visual field that are sufficiently greater than the luminance to which the eyes are adapted to cause annoyance, discomfort, or loss in visual performance and visibility. *Note:* The magnitude of the sensation of glare depends upon such factors as the size, position, and luminance of a source, the number of sources, and the luminance to which the eyes are adapted. *See also:* **visual field.** Z7A1-0

glass half cell (glass electrode). A half cell in which the potential measurements are made through a glass membrane. *See also:* **electrolytic cell.** 42A60-0

GLC circuit. *See:* **simple parallel circuit.**

glide path (electronic navigation). The path used by an aircraft in approach procedures as defined by an instrument landing facility. *See also:* **navigation.** E172-10E6

glide-path receiver. An airborne radio receiver used to detect the transmissions of a ground-installed glide-path transmitter. *Note:* It furnishes a visual, audible, or electric signal for the purpose of vertically guiding an aircraft using an instrument landing system. *See also:* **air transportation; electronic equipment.** 42A41-0

glide slope (electronic navigation). An inclined surface generated by the radiation of electromagnetic waves and used with a localizer in an instrument landing system to create a glide path. *See also:* **navigation.** 0-10E6

glide-slope angle (electronic navigation). The angle in the vertical plane between the glide slope and the horizontal. *See also:* **navigation.** 0-10E6

glide-slope deviation (electronic navigation). The vertical location of an aircraft relative to a glide slope, expressed in terms of the angle measured at the intersection of the glide slope with the runway; or the linear distance above or below the glide slope. *See also:* **navigation.** 0-10E6

glide-slope facility. A radio transmitting facility that provides signals for vertical guidance of aircraft along an inclined surface extending upward at an angle to the horizontal from the point of desired ground contact. *See also:* **radio navigation.** 42A65-0

glide-slope sector (instrument landing system). A vertical sector containing the glide slope and within which the pilot's indicator gives a quantitative measure of the deviation above and below the glide slope; the sector is bounded above and below by a specified difference in depth of modulation, usually that which gives full-scale deflection of the glide-slope deviation indicator. *See also:* **navigation.** 0-10E6

glitch. A perturbation of the pulse waveform of relatively short duration and of uncertain origin. *See:* **distortion, pulse.** 0-9E4

global stability. *See:* **stability, global.**

globe. A transparent or diffusing enclosure intended to protect a lamp, to diffuse and redirect its light, or to change the color of the light. *See also:* **bare (exposed) lamp.** Z7A1-0

glossmeter. An instrument for measuring gloss as a function of the regular and diffuse reflection characteristics of a material. *See also:* **photometry.** Z7A1-0

glow discharge (1) (general). The phenomenon of electric conduction in gases shown by a slight luminosity, without great hissing or noise, and without appreciable heating or volatilization of the electrode, when the voltage gradient exceeds a certain value. *See also:* **abnormal glow discharge.** E270-0
(2) (electron tubes). A discharge of electricity through gas characterized by (A) a change of space potential, in the immediate vicinity of the cathode, that is much higher than the ionization potential of the gas. (B) a low, approximately constant, current density at the cathode, and a low cathode temperature. (C) the presence of a cathode glow. *See:* **gas tubes.** *See also:* **lamp.** E160-15E6;42A70/Z7A1-0

glow factor (illuminating engineering). A measure of the visible light response of a fluorescent material to black light. It is equal to the luminance (photometric brightness) in footlamberts produced on the material divided by the incident black-light flux density in milliwatts per square foot. *Notes:* (1) It may be measured in lumens per milliwatt. (2) See note after **filter factor.** *See also:* **ultraviolet radiation.** Z7A1-0

glow lamp. An electric-discharge lamp whose mode of operation is that of a glow discharge and in which light is generated in the space close to the electrodes. *See also:* **lamp.** Z7A1-0

glow, negative. *See:* **negative glow.**

glow-switch. An electron tube containing contacts operated thermally by means of a glow discharge. *See also:* **tube definitions.** 42A70-15E6

glow-tube. *See:* **glow discharge tube.**

glue-line heating (dielectric heating usage). An arrangement of electrodes designed to give preferential heating to a thin film of material of relatively high loss factor between alternate layers of relatively low loss factor. *See also:* **dielectric heating.** E54/E169-0

goniometer (electronic navigation). A combining device used with a plurality of antennas so that the direction of maximum radiation or of greatest response may be rotated in azimuth without physically moving the antenna array. 0-10E6

goniophotometer. A photometer for measuring the directional light distribution characteristics of sources, luminaires, media, and surfaces. *See also:* **photometry.** Z7A1-0

governor-controlled gates (hydro-turbine). Include the gates that control the energy input to the turbine and that are normally actuated by the speed governor directly or through the medium of the speed-control mechanism. *See also:* **speed-governing system.** E94-0

governor-controlled valves (steam turbine). Include the valves that control the energy input to the turbine and that are normally actuated by the speed governor directly or through the medium of the speed-control mechanism. *See also:* **speed-governing system.** E94-0

governor dead band. The magnitude of the total change in steady-rate speed within which there is no resulting measurable change in the position of the governor-controlled valves. *Note:* Dead band is the measure of the insensitivity of the speed-governing system and is expressed in percent of rated speed. *See also:* **speed-governing system.** E94-0

governor pin. *See:* **centrifugal-mechanism pin.**

governor speed-changer. A device by means of which the speed-governing system may be adjusted to change the speed or power output of the turbine while the turbine is in operation. *See also:* **speed-governing system.** E94-0

governor speed-changer position. The position of the speed changer indicated by the fraction of its travel from the position corresponding to minimum turbine speed to the position corresponding to maximum speed and energy input. It is usually expressed in percent. *See also:* **speed-governing system.** E94-0

governor spring. *See:* **centrifugal-mechanism spring.**

governor weights. *See:* **centrifugal-mechanism weights.**

GPI. *See:* **ground-position indicator.**

graded-time step-voltage test (rotating machinery). A controlled overvoltage test in which calculated voltage increments are applied at calculated time intervals. *Note:* Usually, a direct-voltage test with the increments and intervals so calculated that dielectric absorption appears as a constant shunt-conductance; to simplify interpretation. *See also:* **asynchronous machine; direct-current commutating machine; synchronous machine.** 0-31E8

gradient (scalar field). At a point, a vector (denoted by ∇u) equal to, and in the direction of, the maximum space rate of change of the field. It is obtained as a vector field by applying the operator ∇ to a scalar function. Thus if $u = f(x,y,z)$

$$\nabla u = \text{grad } u = \mathbf{i}\frac{\partial u}{\partial x} + \mathbf{j}\frac{\partial u}{\partial y} + \mathbf{k}\frac{\partial u}{\partial z}.$$

E270-0

gradient meter. *See:* **generating electric field meter.**

gradient microphone. A microphone the output of which corresponds to a gradient of the sound pressure. *Note:* Gradient microphones may be of any order as, for example, zero, first, second, etcetera. A pressure microphone is a gradient microphone of zero order. A velocity microphone is a gradient microphone of order one. Mathematically, from a directivity standpoint for plane waves the root-mean-square response is proportional to $\cos^n \theta$, where θ is the angle of incidence and n is the order of the microphone. *See also:* **microphone.** 42A65-0

grading ring (arrester) (control ring*). A metal part, usually circular or oval in shape, mounted to modify electrostatically the voltage gradient or distribution. *See:* **arresters; lightning arrester (surge diverter).**
* Deprecated E28/62A1-31E7

graduated (control) (industrial control). Marked to indicate a number of operating positions. 42A25-34E10

grain (photographic material). A small particle of metallic silver remaining in a photographic emulsion after development and fixing. *Note:* In the agglomerate, these grains form the dark area of a photographic image. E157-1E1

graininess (photographic material). The visible coarseness under specified conditions due to silver grains in a developed photographic film. E157-1E1

granular-filled fuse unit. A fuse unit in which the arc is drawn through powdered, granular, or fibrous material. 37A100-31E11

graph determinant (network analysis). One plus the sum of the loop-set transmittances of all nontouching loop sets contained in the graph. *Notes:* (1) The graph determinant is conveniently expressed in the form

$$\Delta = (1 - \sum L_i + \sum L_iL_j - \sum L_iL_jL_k + \cdots)$$

where L_i is the loop transmittance of the ith loop of the graph, and the first summation is over all of the different loops of the graph, the second is over all of the different pairs of nontouching loops, and the third is over all the different triplets of nontouching loops, etcetera. (2) The graph determinant may be written alternatively as

$$\Delta = [(1 - L_1)(1 - L_2) \cdots (1 - L_n)]\dagger$$

where $L_1, L_2, \cdots, L_n$, are the loop transmittances of the n different loops in the graph, and where the dagger indicates that, after carrying out the multiplications within the brackets, a term will be dropped if it contains the transmittance product of two touching loops. (3) The graph determinant reduces to the return difference for a graph having only one loop. (4) The graph determinant is equal to the determinant of the coefficient equations. *See also:* **linear signal flow graphs.** E155-0

graphic symbol (abbreviation). A geometric representation used to depict graphically the generic function of an item as it normally is used in a circuit. *See also:* **abbreviation.** E267-0

graphite brush. A brush composed principally of graphite. *Note:* This type of brush is soft. Grades of brushes of this type differ greatly in current-carrying capacity and in operating speed from low to high. *See also:* **brush (rotating machinery).** 42A10-31E8;64A1-0

graphitic corrosion. Corrosion of gray cast iron in which the metallic constituents are converted to corrosion products leaving the graphite intact. *See:* **corrosion terms.** CM-34E2

graphitization (corrosion). Decomposition of a metal carbide to matrix-metal and graphite. *See:* **corrosion terms; graphitic corrosion.** CM-34E2

graph transmittance (network analysis). The ratio of signal at some specified dependent node, to the signal applied at some specified source node. *Note:* The graph transmittance is the weighted sum of the path transmittances of the different open paths from the designated source node to the designated dependent node, where the weight for each path is the path factor divided by the graph determinant. *See also:* **linear signal flow graphs.** E155-0

grass. The pattern produced by random noise on an *A* scope. *See also:* **radar.** 42A65-10E6

graticule (oscilloscopes). A scale for measurement of quantities displayed on the cathode-ray tube of an oscilloscope. *See:* **oscillograph.** 0-9E4

graticule area. The area enclosed by the continuous outer graticule lines. *Note:* Unless otherwise stated the graticule area shall be equal to or less than the viewing area. *See:* **oscillograph.** *See also:* **quality-area; viewing area.** 0-9E4

grating. *See:* **ultrasonic space grating.**

gravity elevator. An elevator utilizing gravity to move the car. *See also:* **elevators.** 42A45-0

gravity vertical (navigation). *See:* **mass-attraction vertical.**

graybody. A temperature radiator whose spectral emissivity is less than unity and the same at all wavelengths. *See also:* **radiant energy (illuminating engineering).** Z7A1-0

Gray code (computing systems). A binary code in which sequential numbers are represented by binary expressions, each of which differs from the preceding expression in one place only. *See:* **reflected binary code.** *See also:* **electronic digital computer.** X3A12-16E9

gray scale (television). An optical pattern in discrete steps between light and dark. *Note:* A gray scale with ten steps is usually included in resolution test charts. *See also:* **television.** E204-42A65

greasing truck. An electrically driven service vehicle to transport greases and oil for servicing the underground mine machinery. *Note:* It may include a compressor, air storage tank, and fittings to place lubricant at the proper points in the mining machinery. *See also:* **mining.** 42A85-0

grid (1) (electron tube). An electrode having one or more openings for the passage of electrons or ions. *See:*
control grid;
electrode;
electron tube;
screen grid;
shield grid;
space-charge grid;
suppressor grid. 50I07-15E6

(2) (storage cell). A metallic framework employed in a storage cell for supporting the active material and conducting the electric current. For an alkaline battery the grid may be called a frame. *See also:* **primary or secondary cells.** 42A60-0

grid bearing (navigation). Bearing relative to grid north. *See also:* **navigation.**

grid circuit (industrial control). A circuit that includes the grid-cathode path of an electron tube in series connection with other elements. *See also:* **electronic controller.** 42A25-34E10

grid control (industrial control). Control of anode current of an electron tube by means of proper variation (control) of the control-grid potential with respect to the cathode of the tube. *See also:* **electronic controller.** 42A25-34E10

grid-controlled mercury-arc rectifier. A mercury-arc rectifier in which one or more electrodes are employed exclusively to control the starting of the discharge. *See also:* **rectifier.** E145-0

grid course (navigation). Course relative to grid north. *See also:* **navigation.** 0-10E6

grid current (analog computer). The current flowing between the summing junction and the grid of the first amplifying stage of an operational amplifier. *Note:* Grid current results in an error voltage at the amplifier output. *See also:* **electronic analog computer; electronic controller.** E165-0

grid-drive characteristic (electron tubes). A relation, usually shown by a graph, between electric or light output and control-electrode voltage measured from cutoff. *See also:* **circuit characteristics of electrodes.** E160-15E6;42A70-0

grid driving power (electron tubes). The average of the product of the instantaneous values of the alternating components of the grid current and the grid voltage over a complete cycle. *Note:* This power comprises the power supplied to the biasing device and to the grid. *See also:* **circuit characteristics of electrodes; electrode dissipation.** E160-15E6;42A70-0

grid emission. Electron or ion emission from a grid. *See also:* **electron emission.** 42A70-15E6

grid emission, primary (thermionic). Current produced by electrons or ions thermionically emitted from a grid. *See also:* **electron emission.** 42A70-15E6

grid emission, secondary. Electron emission from a grid due directly to bombardment of its surface by electrons or other charged particles. *See also:* **electron emission.** 42A70-15E6

grid-glow tube. A glow-discharge cold-cathode tube in which one or more control electrodes initiate but do not limit the anode current, except under certain operating conditions. *Note:* This term is used chiefly in the industrial field. *See also:* **tube definitions.** 42A70-15E6

grid heading (navigation). Heading relative to grid north. *See also:* **navigation.** 0-10E6

grid-leak detector. A triode or multielectrode tube in which rectification occurs because of electron current to the grid. *Note:* The voltage associated with this flow through a high resistance in the grid circuit appears in amplified form in the plate circuit. *See also:* **modulating systems.** 42A65-0

grid modulation (electron tubes). Modulation produced by the application of the modulating voltage to the control grid of any tube in which the carrier is present. *Note:* Modulation in which the grid voltage contains externally generated pulses is called **grid pulse modulation.** *See also:* **modulating systems.** E145/42A65-0

grid neutralization (electron tubes). The method of neutralizing an amplifier in which a portion of the grid-cathode alternating-current voltage is shifted 180 degrees and applied to the plate-cathode circuit through a neutralizing capacitor. *See also:* **amplifier; feedback.** E145-0

grid north (navigation). An arbitrary reference direction used in connection with a system of rectangular coordinates superimposed over a chart. *See also:* **navigation.** 0-10E6

grid number *n* (electron tubes). A grid occupying the *n*th position counting from the cathode. *See also:* **electron tube.** 50I07-15E6

grid pitch (electron tubes). The pitch of the helix of a helical grid. *See also:* **electron tube.** 50I07-15E6

grid pulse modulation. Modulation produced in an amplifier or oscillator by application of one or more pulses to a grid circuit. *See also:* **modulating system.** E145-0

grid (circuit) resistor (industrial control). A resistor used to limit grid current. *See also:* **electronic controller.** 42A25-34E10

grids (high-power rectifier). Electrodes that are placed in the arc stream and to which a control voltage may be applied. *See also:* **rectification.** 42A15-0

grid transformer (industrial control). Supplies an alternating voltage to a grid circuit or circuits. 42A25-34E10

grid voltage. *See:* **electrode voltage; electronic controller.**

grid voltage, critical. *See:* **control-grid voltage.**

grid voltage supply (electron tube). The means for supplying to the grid of the tube a potential that is usually negative with respect to the cathode. *See also:* **power pack.** 42A65-0

groove (mechanical recording). The track inscribed in the record by the cutting or embossing stylus. *See also:* **phonograph pickup.** E157-1E1

groove angle (disk recording). The angle between the two walls of an unmodulated groove in a radial plane perpendicular to the surface of the recording medium. *See also:* **phonograph pickup.** E157-1E1

groove diameter. *See:* **tape-wound core.** *See also:* **static magnetic storage.**

groove shape (disk recording). The contour of the groove in a radial plane perpendicular to the surface of the recording medium. *See also:* **phonograph pickup.** E157-1E1

groove speed (disk recording). The linear speed of the groove with respect to the stylus. *See also:* **phonograph pickup.** E157-1E1

groove width. *See:* **tape-wound core.** *See also:* **static magnetic storage.**

gross demonstrated capacity. The gross steady output that a generating unit or station has produced while demonstrating its maximum performance under stipulated conditions. *See also:* **generating station.** 42A35-31E13

gross generation (electric power systems). The generated output power at the terminals of the generator. *See also:* **power systems, low-frequency and surge testing.** E94-0

gross information content. A measure of the total information, redundant or otherwise, contained in a message. *Note:* It is expressed as the number of bits or hartleys required to transmit the message with specified accuracy over a noiseless medium without coding. *See also:* **bit.** 42A65-0

gross rated capacity. The gross steady output that a generating unit or station can produce for at least two hours under specified operating conditions. *See also:* **generating station.** 42A35-31E13

ground (earth) (electric system). A conducting connection, whether intentional or accidental, by which an electric circuit or equipment is connected to the earth, or to some conducting body of relatively large extent that serves in place of the earth. *Note:* It is used for establishing and maintaining the potential of the earth (or of the conducting body) or approximately that potential, on conductors connected to it, and for conducting ground current to and from the earth (or the conducting body). *Note:* For an extensive list of cross references, see *Appendix A.* 1A0-0;42A35-31E13

groundable parts. Those parts that may be connected to ground without affecting operation of the device. 37A100-31E11

ground absorption. The loss of energy in transmission of radio waves due to dissipation in the ground. *See also:* **radiation.** 42A65-0

ground-area open floodlight (illumination). A unit providing a weatherproof enclosure for the lamp socket and housing. *Note:* No cover glass is required. *See also:* **floodlighting.** Z7A1-0

ground-area open floodlight with reflector insert. A weatherproof unit so constructed that the housing forms only part of the reflecting surface. *Note:* An auxiliary reflector is used to modify the distribution of light. No cover glass is required. *See also:* **floodlighting.** Z7A1-0

ground bar (lightning). A conductor forming a common junction for a number of ground conductors. *See:* **lightning arrester (surge diverter).** 50I25-31E7

ground-based navigation aid. An aid that requires facilities located upon land or sea. *See also:* **navigation.** 0-10E6

ground bus (electric system). A bus to which the grounds from individual pieces of equipment are connected, and that, in turn, is connected to ground at one or more points. *See:* **ground.** 37A100-31E11;42A35-31E13

ground cable bond. A cable bond used for grounding the armor or sheaths of cables or both. *See:* **ground.** 42A35-31E13

ground clamp (grounding clamp). A clamp used in connecting a grounding conductor to a grounding electrode or to a thing grounded. *See:* **ground.** 42A35-31E13

ground clutter (radar). The pattern produced on the screen of a radar indicator by undesired ground return. *See also:* **radar.** 42A65-0

ground conductor (lightning). A conductor providing an electric connection between part of a system, or the frame of a machine or piece of apparatus, and a ground electrode or a ground bar. *See:* **lightning arrester (surge diverter); grounded conductor.** 50I25-31E7

ground conduit. A conduit used solely to contain one or more grounding conductors. *See:* **ground.** 42A35-31E13

ground connection. *See:* **grounding connection.**

ground contact (switchgear assembly). A self-coupling separable contact provided to connect and disconnect the ground connection between the removable element and the ground bus of the housing and so constructed that it remains in contact at all times except when the primary disconnecting devices are separated by a safe distance. *Note:* Safe distance, as used here, is a distance at which the equipment will meet its withstand-voltage ratings, both low-frequency and impulse, between line and load terminals with the switching device in the closed position. 37A100;31E11

ground-controlled approach (GCA) (electronic navigation). A ground radar system providing information by which aircraft approaches to landing may be directed via radio communications; the system consists of a precision approach radar (PAR) and an airport surveillance radar (ASR). *See also:* **navigation.** 42A65-10E6

ground-controlled interception (GCI). A ground radar system by means of which a controller at the radar may direct an aircraft to intercept another aircraft. *See also:* **radar.** 42A65-10E6

ground current. Current flowing in the earth or in a grounding connection. *See:* **ground; lightning arrester (surge diverter).** 42A35-31E13

ground-derived navigation data (air navigation). Data obtained from measurements made on land or sea at locations external to the vehicle. *See also:* **navigation.** 0-10E6

ground detection rings (rotating machinery). Collector rings connected to a winding and its core to facilitate the measurement of insulation resistance on a rotor winding. *See also:* **rotor (rotating machinery).** 0-31E8

ground detector. An instrument or an equipment used for indicating the presence of a ground on an ungrounded system. *See:* **ground.** 42A35-31E13

grounded (electric systems). Connected to earth or to some extended conducting body that serves instead of the earth, whether the connection is intentional or accidental.
See:
ground;
ground-fault neutralizer grounded;
grounded circuit;
grounded, effectively;
grounded, solidly;
grounded system;
grounding transformer;
guarded;
impedance grounded;
neutral ground;
reactance grounded;
resistance grounded;
ungrounded;
voltage to ground. 1A0/2A2-0

grounded capacitance. *See:* **self-capacitance of a conductor; ground.**

grounded-cathode amplifier. An electron-tube amplifier with the cathode at ground potential at the operating frequency, with input applied between the control grid and ground, and the output load connected between plate and ground. *Note:* This is the conventional amplifier circuit. *See also:* **amplifier; circuits and devices.** E145/42A65-0

grounded circuit. A circuit in which one conductor or point (usually the neutral conductor or neutral point of transformer or generator windings) is intentionally grounded, either solidly or through a grounding device. *See:* **ground; grounded conductor; grounded system.** 42A15/42A35-31E13

grounded concentric wiring system. A grounded system in which the external (outer) conductor is solidly grounded and completely surrounds the internal (inner) conductor through its length. The external conductor is usually uninsulated. *See:* **ground.** 42A35-31E13

grounded conductor (electric system). A conductor that is intentionally grounded, either solidly or through a current limiting device. *See also:* **ground.** 1A0-0

grounded, effectively. An expression that means grounded through a grounding connection of sufficiently low impedance (inherent or intentionally added or both) that fault grounds that may occur cannot build up voltages in excess of limits established for apparatus, circuits, or systems so grounded. *Note:* An alternating-current system or portion thereof may be said to be effectively grounded when, for all points on the system or specified portion thereof, the ratio of zero-sequence reactance to positive-sequence reactance is not greater than three and the ratio of zero-sequence resistance to positive-sequence reactance is not greater than one for any condition of operation and for any amount of connected generator capacity. *See:* **ground; grounded.** 42A15-31E12;42A35-31E13

grounded-grid amplifier. An electron-tube amplifier circuit in which the control grid is at ground potential at the operating frequency, with input applied between cathode and ground, and output load connected between plate and ground. *Note:* The grid-to-plate impedance of the tube is in parallel with the load instead of acting as a feedback path. *See also:* **amplifier.** E145/42A65-0

grounded neutral system (lightning arresters). A system in which the neutral is connected to ground, either solidly or through a resistance or reactance of low value. *See:* **lightning arrester (surge diverter).** 50I25-31E7

grounded parts (industrial control). Parts that are intentionally connected to ground. 37A100-31E11;42A35-31E13

grounded-plate amplifier (cathode-follower). An electron-tube amplifier circuit in which the plate is at ground potential at the operating frequency, with input applied between control grid and ground, and the output load connected between cathode and ground. *See also:* **amplifier.** E145-0

grounded, solidly (directly grounded). Grounded through an adequate ground connection in which no impedance has been inserted intentionally. *Note:* Adequate as used herein means suitable for the purpose intended. *See:* **grounded.** 42A15/42A35-31E12

grounded system (electric power). A system of conductors in which at least one conductor or point (usually the middle wire or neutral point of transformer or generator windings) is intentionally grounded, either solidly or through a current-limiting device. *See also:* **ground; grounded; grounded circuit.** 2A2-0;42A35-31E13

ground electrode (lightning arresters). A conductor or group of conductors in intimate contact with the ground for the purpose of providing a connection with the ground. *See:* **lightning arrester (surge diverter).** 50I25-31E7

ground end (grounding device). The end or terminal of the device that is grounded directly or through another device. *See:* **grounding device.** 42A35-31E13

ground equalizer inductors (antenna). Coils of relatively low inductance, placed in the circuit connected to one or more of the grounding points of an antenna to distribute the current to the various points in any desired manner. *Note:* Broadcast usage only and now in disuse. *See also:* **antenna.** E145-3E1

ground fault (lightning arresters). An insulation fault between a conductor and ground or frame. *See:* **lightning arresters (surge diverter).** 50I25-31E7

ground-fault neutralizer grounded (resonant grounded). Reactance grounded through such values of reactance that, during a fault between one of the conductors and earth, the rated-frequency current flowing in the grounding reactances and the rated-frequency capacitance current flowing between the unfaulted conductors and earth shall be substantially equal. *Notes:* (1) In the fault these two components of fault current will be substantially 180 degrees out of phase. (2) When a system is ground-fault neutralizer grounded, it is expected that the quadrature component of the rated-frequency single-phase-to-ground fault current will be so small that an arc fault in air will be self-extinguishing. *See:* **ground; grounding device.** E32-0;42A15-31E12;42A35-31E13

ground grid. A system of grounding electrodes consisting of interconnected bare cables buried in the earth to provide a common ground for electric devices and metallic structures. *Note:* It may be connected to auxiliary grounding electrodes to lower its resistance. *See also:* **grounding device.** E81-0

ground indication. An indication of the presence of a ground on one or more of the normally ungrounded conductors of a system. *See:* **ground.** 42A35-31E13

grounding cable. A cable used to make a connection to ground. *See:* **cradle base (rotating machinery); grounding conductor.** 0-31E8

grounding-cable connector. The terminal mounted on the end of a grounding cable. *See:* **cradle base (rotating machinery.** 0-31E8

grounding clamp. *See:* **ground clamp.**

grounding conductor (ground conductor) (1) (general). The conductor that is used to establish a ground and that connects an equipment, device, wiring system, or another conductor (usually the neutral conductor) with the grounding electrode or electrodes. *See:* **conductor; ground; grounding device.** 1A0/2A2-0;42A35-31E13

(2) (mining) (safety ground conductor) (safety ground) (frame ground). A metallic conductor used to connect the metal frame or enclosure of an equipment, device, or wiring system with a mine track or other effective grounding medium. *See also:* **mine feeder circuit.** 42A85-0

grounding connection (ground connection). A connection used in establishing a ground and consists of a grounding conductor, a grounding electrode, and the earth (soil) that surrounds the electrode. *See:* **ground.** 42A35-31E13

grounding device (electric power). An impedance device used to connect conductors of an electric system to ground for the purpose of controlling the ground current or voltages to ground. *Notes:* (1) The grounding device may consist of a grounding transformer or a neutral grounding device, or a combination of these. Protective devices, such as lightning arresters, may also be included as an integral part of the device. (2) For an extensive list of cross references, see *Appendix A.* E32/42A35-31E13

grounding electrode (ground electrode). A conductor used to establish a ground. *See:* **ground.** 42A35-31E13

grounding jumper (electric appliances). A strap or wire to connect the frame of the range to the neutral conductor of the supply circuit. *See also:* **appliance outlet.** 71A1-0

grounding outlet (safety outlet*). An outlet equipped with a receptacle of the polarity type having, in addition to the current-carrying contacts, one grounded contact that can be used for the connection of an equipment grounding conductor. *Note:* This type of outlet is used for connection of portable appliances. *See:* **ground.**
*Deprecated 42A35-31E13

grounding pad (rotating machinery). A contact area, usually on the stator frame, provided to permit the connection of a grounding terminal. *See also:* **stator.** 0-31E8

grounding switch (general). A mechanical switching device by means of which a circuit or piece of apparatus may be electrically connected to ground. *Note:* A grounding switch is used to ground a piece of equipment to permit maintenance personnel to work with safety. 37A100-31E11

grounding system (1) (general). Consists of all interconnected grounding connections in a specific area. *See also:* **grounding device.** E81-0
(2) (lightning arresters). A complete installation comprising one or more ground electrodes, ground conductors, and ground bars as required. *See:* **lightning arrester (surge diverter).** 50I25-31E7

grounding terminal (rotating machinery). A terminal used to make a connection to a ground. *See also:* **stator.** 0-31E8

grounding transformer (electric power). A transformer intended primarily to provide a neutral point for grounding purposes. *Note:* It may be provided with a delta winding in which resistors or reactors are connected. *See also:* **grounded; grounding device; transformer.** E32/42A15-31E12

ground insulation (rotating machinery). Insulation used to insure the electric isolation of a winding from the core and mechanical parts of a machine. *See also:* **asynchronous machine; coil insulation; direct-current commutating machine; synchronous machine.** 0-31E8

ground level (mobile communication). The elevation of the ground above mean sea level at the antenna site or other point of interest. *See also:* **mobile communication system.** 0-6E1

ground light. Visible radiation from the sun and sky reflected by surfaces below the plane of the horizon. *See also:* **sunlight.** Z7A1-0

ground loop (interference terminology). A potentially detrimental loop formed when two or more points in an electric system that are nominally at ground potential are connected by a conducting path. *Note:* The term usually is employed when, by improper design or by accident, unwanted noisy signals are generated in the common return of relatively low-level signal circuits by the return currents or by magnetic fields from relatively high-powered circuits or components. *See also:* **electronic analog computer; interference.** E165-16E9

ground lug. A lug used in connecting a grounding conductor to a grounding electrode or to a thing grounded. *See:* **ground.** 42A35-31E13

ground mat. A system of bare conductors, on or below the surface of the earth, connected to a ground or a ground grid to provide protection from dangerous touch voltages. *Note:* Plates and gratings of suitable area are common forms of **ground mats.** *See also:* **grounding device.** E81-0

ground plane. (1) An assumed plane of true ground or zero potential. *See also:* **direct-stroke protection (lightning).** 0-31E13
(2) A conducting surface or plate used as a common reference point for circuit returns and electric or signal potentials. *See also:* **electromagnetic compatibility.** 0-27E1

ground plane, effective (mobile communication). The height of the average terrain above mean sea level as measured for a distance of 100 meters out from the base of the antenna in the desired direction of communication. It may be considered the same as ground level only in open flat country. *See also:* **mobile communication system.** 0-6E1

ground plate (grounding plate). A plate of conducting material buried in the earth to serve as a grounding electrode. *See:* **ground.** 42A35-31E13

ground-position indicator (GPI) (electronic navigation). A dead-reckoning tracer or computer similar to an air position indicator (API) with provision for taking account of drift. *See also:* **navigation; radio navigation.** 42A65-0;0-10E6

ground protection (ground-fault protection). A method of protection in which faults to ground within the protected equipment are detected irrespective of system phase conditions. 37A100-31E11/31E6

ground-referenced navigation data. Data in terms of a coordinate system referenced to the earth or to some specified portion thereof. *See also:* **navigation.** 0-10E6

ground-reflected wave. The component of the ground wave that is reflected from the ground.*See also:* **radiation.** 42A65-0

ground relay. A relay that by its design or application is intended to respond primarily to system ground faults. 37A100-31E11/31E6

ground resistance (grounding electrode). The ohmic resistance between the grounding electrode and a remote grounding electrode of zero resistance. *Note:* By remote is meant at a distance such that the mutual resistance of the two electrodes is essentially zero. *See also:* **ground.** E81-0

ground return (radar practice). The aggregate of received echoes due to reflection from the surface of the earth and fixed objects thereon. *See also:* **radar; ground clutter.** 42A65-0

ground-return circuit (earth-return circuit). A circuit in which the earth is utilized to complete the circuit. *See also:* **ground; telegraphy; transmission line.** 42A35-31E13

ground-return current (line residual current) (electric supply line). The vector sum of the currents in all conductors on the electric supply line. *Note:* Actually the ground-return current in this sense may include components returning to the source in wires on other pole lines, but from the inductive coordination standpoint these components are substantially equivalent to components in the ground. *See also:* **inductive coordination.** 42A65-0

ground-return system. A system in which one of the conductors is replaced by ground. 50I25-31E7

ground rod (ground electrode). A rod that is driven into the ground to serve as a ground terminal, such as a copper-clad rod, solid copper rod, galvanized iron rod, or galvanized iron pipe. *See also:* **arresters.** 5A1-0

ground speed (navigation). The speed of a vehicle along its track. *See also:* **navigation.** E172-10E6

ground surveillance radar. A radar set operated at a fixed point for observation and control of the position of aircraft or other vehicles in the vicinity. *See also:* **navigation.** 0-10E6

ground system of an antenna. That portion of an antenna closely associated with and including an extensive conducting surface which may be the earth itself. *See also:* **antenna.** 42A65-3E1

ground terminal (earth terminal) (1) (industrial control). A terminal intended to ensure, by means of a special connection, the grounding (earthing) of part of an apparatus. 50I16-34E10
(2) (lightning protection system). The portion extending into the ground, such as a ground rod, ground plate, or the conductor itself, serving to bring the lightning protection system into electric contact with the ground. *See:* **arrester; lightning arrester (surge diverter).** E28/5A1/42A20-0;50I25-31E7

ground transformer. *See:* **grounding transformer.**

ground wave. A radio wave that is propagated over the earth and is ordinarily affected by the presence of the ground and troposphere. *Notes:* (1) The ground wave includes all components of a radio wave over the earth except ionospheric and tropospheric waves. (2) The ground wave is refracted because of variations in the dielectric constant of the troposphere including the condition known as a surface duct. *See also:* **radiation; radio wave propagation.** E149/42A65-0

ground wire (1) (telecommunication). A conductor leading to an electric connection with the ground. *See also:* **antenna; grounded.** 42A65-3E1
(2) (overhead power line). A conductor having grounding connections at intervals, that is suspended usually above but not necessarily over the line conductor to provide a degree of protection against lightning discharges. *See:* **ground; overhead ground wire.** 42A35-31E13

group (1) (communications). *See:* **channel group.**
(2) (storage cell). An assembly of plates of the same polarity burned to a connecting strap. *See also:* **battery (primary or secondary).** 42A60-0

group ambient temperature (cable or duct) (power distribution, underground cables). The no-load temperature in a group with all other cables or ducts in the group loaded. *See also:* **power distribution, underground construction.** 0-31E1

group delay time. The rate of change, with angular frequency, of the total phase shift through a network. *Notes:* (1) Group delay time is the time interval required for the crest of a group of interfering waves to travel through a 2-port network, where the component wave trains have slightly different individual frequencies. (2) Group delay time is usually very close in value to envelope delay and transmission time delay, and in the case of vanishing spectrum bandwidth of the signal these quantities become identical. *See also:* **measurement system.** E285-9EI

group flashing light (navigation). A flashing light in which the flashes are combined in groups, each including the same number of flashes, and in which the groups are repeated at regular intervals. *Note:* The duration of each flash is clearly less than the duration of the dark periods between flashes, and the duration of the dark periods between flashes is clearly less than the duration of the dark periods between groups. *See also:* **signal lighting.** Z7A1-0

grouping (1) (facsimile). Periodic error in the spacing of recorded lines. *See also:* **facsimile signal (picture signal).** E168-0
(2) (electroacoustics). Nonuniform spacing between the grooves of a disk recording. *See also:* **electroacoustics.** E157-1E1

group operation (switchgear). The operation of all poles of a multipole switching device by one operating mechanism. 37A100-31E11

group velocity (traveling wave). The velocity of propagation of the envelope, provided that this moves without significant change of shape. *Notes:* (1) The magnitude of the group velocity is equal to the reciprocal of the change of phase constant with angular frequency. (2) Group velocity differs in magnitude from phase velocity if the phase velocity varies with frequency and differs in direction from phase velocity if the phase velocity varies with direction. *See also:* **radio wave propagation; transmission characteristics.** 0-3E2

grout (rotating machinery). A very rich concrete used to bond the feet, sole plates, bedplate, or rail of a machine to its foundation. *See:* **cradle base (rotating machinery).** 0-31E8

grown junction. *See:* **junction, grown.**

***G* scan (electronic navigation).** *See:* ***G* display.**

***G* scope (electronic navigation).** *See:* ***G* display.**

guard (interference terminology). A conductor situated between a source of interference and a signal path in such a way that interference currents are conducted to the return terminal of the interference source without entering the signal path. *See also:* **interference.** 0-13E6

guard band (interference guard band). A frequency band left vacant between two channels to give a margin of safety against mutual interference. *See also:* **radio transmission.** 42A65-0

guard circle (disk recording). An inner concentric groove inscribed, on disk records, to prevent the pick-up from being damaged by being thrown to the center of the record. E157-1E1

guarded (electric system). Covered, shielded, fenced, inclosed, or otherwise protected, by means of suitable covers or casings, barrier rails or screens, mats or platforms, to remove the likelihood of dangerous contact or approach by persons or objects to a point of danger. *Note:* Wires that are insulated, but not otherwise protected, are not considered as guarded. *See also:* **grounded; power distribution, overhead construction.** 1A0/2A2/42A95-0

guarded input (amplifiers). Means of connecting an input signal so as to prevent any common-mode signal from causing current to flow in the input, thus differences of source impedance do not cause conversion of the common-mode signal into a differential signal. *See:* **amplifier.** 0-9E4

guarded machine (rotating machinery). An open machine in which all openings giving direct access to live or rotating parts (except smooth shafts) are limited in size by the design of the structural parts, or by screens, grilles, expanded metal, etcetera, to prevent accidental contact with such parts. Such openings are of such size as not to permit the passage of a cylindrical rod ½ inch in diameter, except where the distance from the guard to the live or rotating parts is more than 4 inches; they are of such size as not to permit the passage of a cylindrical rod ¾ inch in diameter. *See:* **asynchronous machine; direct-current commutating machine; synchronous machine.** 42A10-31E8

guard electrode (testing of electric power system components). One or more electrically conducting elements, arranged and connected in an electric instrument or measuring circuit so as to divert unwanted conduction or displacement currents from, or confine wanted currents to, the measurement device. *See also:* **power system, low frequency and surge testing.** 0-31E5

guard-ground system (interference terminology). A combination of guard shields and ground connections that protects all or part of a signal transmission system from common-mode interference by eliminating ground loops in the protected part. *Note:* Ideally the guard shield is connected to the source ground. The source is usually grounded also to the source ground by bonding of the transducer to the test body. The filter, signal receiver, etcetera, are floating with respect to their own grounded cases. This necessitates physically isolating the signal receiver and filter chassis from the cases and using isolation transformers in power supplies, or isolating input circuits from cases and using isolating input transformers. This arrangement in effect places the signal receiver and filter electrically at the source. By means of a similar guard, the load can be placed effectively at the source. See the accompanying figures. 0-13E6

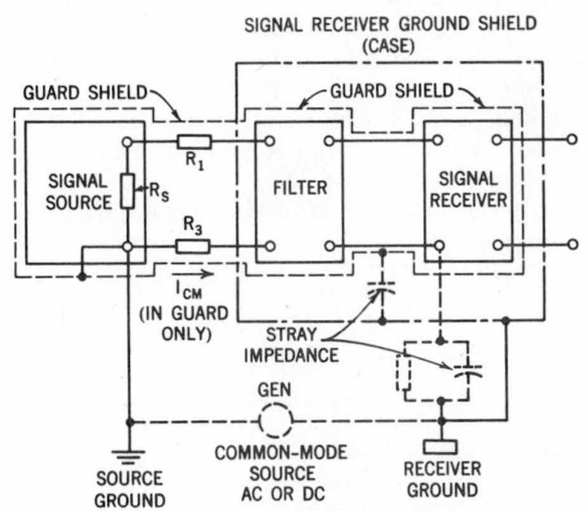

Ideal guard shield. The guard shield consists of the signal source shield (if present), the signal shield, the filter shield, and the signal receiver shield.

See also: **interference.**

guard shield (interference terminology). A guard that is in the form of a shielding enclosure surrounding all or part of a signal path. *Note:* A guard shield is effective against both capacitively coupled and conductively coupled interference whereas a simple guard conductor is usually effective only against conductively coupled interference. *See also:* **guard-ground system; interference .** 0-13E6

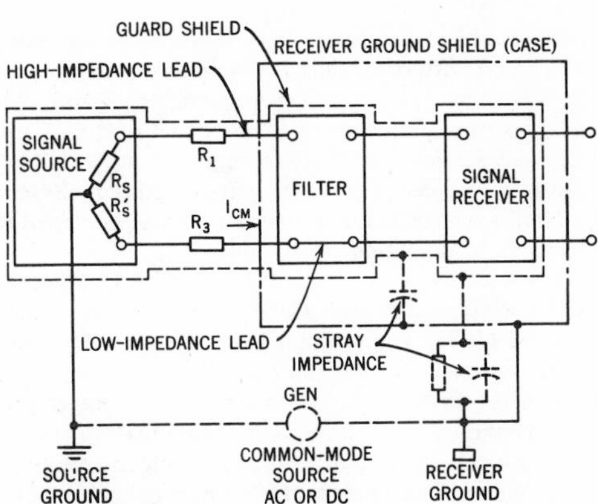

Guard shield connections when connection at source is not convenient. When $(R_3+R_S') \ll (R_1+R_S)$, this arrangement causes common-mode current in the low-impedance lead, but protects the more critical high-impedance lead from current flow.

guard signal. A signal sent over a communication channel to make the system secure against false information by preventing or guarding against the relay operation of a circuit breaker or other relay action until the signal is removed and replaced by a releasing (trip) or permissive signal. 37A100-31E11/31E6

guard wire. A grounded wire erected near a lower-voltage circuit or public crossing in such a position that a high (or higher) voltage overhead conductor cannot come into accidental contact with the lower-voltage circuit, or with persons or objects on the crossing without first becoming grounded by contact with the guard wire. *See:* **ground.** 42A35-31E13

guidance (missile). The process of controlling the flight path through space through the agency of a mechanism within the missile. *See also:* **guided missile.** 42A65-0

guide bearing (rotating machinery). A bearing arranged to limit the transverse movement of a vertical shaft. *See also:* **bearing.** 0-31E8

guided missile. An unmanned device whose flight path through space may be controlled by a self-contained mechanism. *See:* **beam rider guidance; command guidance; guidance; homing guidance; preset guidance.** 42A65-0

guided wave. A wave whose energy is concentrated within or near boundaries between materials of different properties and that is propagated along the path so defined. *Notes:* (1) The waves that can be transmitted by cylindrical waveguides may be classified into two types: transverse electric (TE) for which the axial electric field is zero, and transverse magnetic (TM) for which the axial magnetic field is zero. If the line or lines which bound the cross section of the waveguide are lines for which one coordinate remains constant in a curvilinear coordinate system in which it is possible to separate the variables of the wave equation, the type of wave is designated by TE or TM followed by two subscripts indicating the appropriate solution in each of the two coordinates. The interpretation of the subscripts for waveguides of the more usual cross sections is given in their respective definitions. (2) If the cross section of the waveguide does not fulfill the aforementioned requirement, the type of wave is designated by TE or TM with one numerical subscript starting with the wave that has the lowest frequency as 1 and numbering the remaining waves in the order of their cutoff frequencies. *See:* **cutoff wavelength; dominant wave; resonant modes; resonant wavelength; transverse electric wave; transverse magnetic wave.** *See also:* **radio wave propagation; waveguide.** E210/42A65-3E1/3E2

guide wavelength. The wavelength in a waveguide, measured in the longitudinal direction. *See also:* **waveguide.** 0-3E1

gun-control switch. A switch that closes an electric circuit, thereby actuating the gun-trigger-operating mechanism of an aircraft, usually by means of a solenoid. *See also:* **air transportation wiring and associated equipment.** 42A41-0

guy. A tension member having one end secured to a fixed object and the other end attached to a pole, crossarm, or other structural part that it supports. *See:* **guy wire.** *See also:* **tower.** 42A35-31E13

guy anchor. The buried element of a guy assembly that provides holding strength or resistance to guy wire pull. *Note:* The anchor may consist of a plate, a screw or expanding device, a log of timber, or a mass of concrete installed at sufficient depth and of such size as to develop strength proportionate to weight of earth or rock it tends to move. The anchor is designed to provide attachment for the anchor rod which extends above surface of ground for convenient guy connection. *See:* **guy.** *See also:* **tower.** 42A35-31E13

guy insulator. An insulating element, generally of elongated form with transverse holes or slots for the purpose of insulating two sections of a guy or provide insulation between structure and anchor and also to provide protection in case of broken wires. Porcelain guy insulators are generally designed to stress the porcelain in compression, but wood insulators equipped with suitable hardware are generally used in tension. *See also:* **tower.** 42A35-31E13

guy wire. A stranded cable used for a semiflexible tension support between a pole or structure and the anchor rod, or between structures. *See:* **guy.** *See also:* **tower.** 42A35-31E13

gyrator. (1) A directional phase changer in which the phase changes in opposite directions differ by π radians or 180 degrees. (2)* Any nonreciprocal passive element employing gyromagnetic properties. *See:* **waveguide.**

*Deprecated 50I62-3E1

gyro (gyroscope) (inertial navigation). A device using angular momentum (usually a spinning rotor) to establish a frame of reference in inertial space. *See also:* **navigation.** E174-10E6

gyrocompass. A compass consisting of a continuously driven Foucault gyroscope whose supporting ring normally confines the spinning axis to a horizontal plane, so that the earth's rotation causes the spinning axis to assume a position in a plane passing through the earth's axis, and thus to point to true north. 42A43-0

gyrocompass alignment (inertial systems). A process of self-alignment in azimuth based upon measurements of misalignment drift about the nominal east-west axis of the system. *See also:* **navigation.** E174-10E6

gyrocompassing. *See:* **gyrocompass alignment.**

gyro flux-gate compass (gyro flux-valve compass). A device that uses saturable reactors in conjunction with a vertical gyroscope, to sense the direction of the magnetic north with respect to the aircraft heading. *See also:* **air transportation instruments.** 42A41-0

gyro flux-valve compass. *See:* **gyro flux-gate compass.**

gyro frequency (radio wave propagation). The lowest natural frequency at which charged particles spiral in a fixed magnetic field. *Note:* It is a vector quantity expressed by

$$f_k = \frac{1}{2\pi}\ \frac{q\mathbf{B}}{m}$$

where q is the charge of the particles, **B** is the magnitude magnetic induction vector, and m is the mass of the particles. *See also:* **radio wave propagation.** 0-3E2

gyro horizon electric indicator. An electrically driven device for use in aircraft to provide the pilot with a fixed artificial horizon. *Note:* It indicates deviation from level flight. *See also:* **air transportation instruments.** 42A41-0

H

***H* (*H* beacon) (electronic navigation).** A designation applied to two types of facilities: (1) A nondirectional radio beacon for homing by means of an airborne direction finder. (2) A radar air navigation system using an airborne interrogator to measure the distances from two ground transponders. *See also:* **navigation.** 0-10E6

halation (cathode-ray tube). An annular area surrounding a spot, that is due to the light emanating from the spot being reflected from the front and rear sides of the face plate. *See also:* **cathode-ray tubes.** 50I07-15E6

half-adder. A combinational logic element having two outputs, S and C, and two inputs, A and B, such that the outputs are related to the inputs according to the following equations:

$S = A$ OR B (exclusive OR).

$C = A + B$.

S denotes sum without carry, C denotes carry. Two half-adders may be used for performing binary addition. *See also:* **electronic computation; electronic digital computer.** 0-16E9

half cell. An electrode immersed in a suitable electrolyte. *See also:* **electrolytic cell.** 42A60-0

half duplex (communications). Pertaining to an alternate, one way at a time, independent transmission. *See:* **duplex.** *See also:* **communication.** X3A12-16E9/19E4

half-duplex operation (telegraph system). Operation of a duplex system arranged to permit operation in either direction but not in both directions simultaneously. *See also:* **telegraphy.** 42A65-0

half-duplex repeater. A duplex telegraph repeater provided with interlocking arrangements that restrict the transmission of signals to one direction at a time. *See also:* **telegraphy.** 42A65-0

half-period average-value (symmetrical periodic-quantity). The absolute value of the algebraic average of the values of the quantity taken throughout a half-period, beginning with a zero value. If the quantity has more than two zeros during a cycle, that zero shall be taken that gives the largest half-period average value. E270-0

half-power beamwidth (plane containing the direction of the maximum value of a beam). The angle between the two directions in which the radiation intensity is one-half the maximum value of the beam. *See also:* **antenna.** 0-3E1

halftone characteristic (facsimile). A relation between the density of the recorded copy and the density of the subject copy. *Note:* The term may also be used to relate the amplitude of the facsimile signal to the density of the subject copy or the record copy when only a portion of the system is under consideration. In a frequency-modulation system an appropriate parameter is to be used instead of the amplitude. *See also:* **recording (facsimile).** E168-0

halftones*(storage tubes). *See:* **level.** *See also:* **storage tube.**

*Deprecated

half-wave rectification (power supplies). In the rectifying process, half-wave rectification passes only one-half of each incoming sinusoid, and does not pass the opposite half-cycle. The output contains a single half-sine pulse for each input cycle. A single rectifier provides half-wave rectification. Because of its poorer efficiency and larger alternating-current component, half-wave rectification is usually employed in noncritical low-current circumstances. See the accompanying figure. *See also:* **power supply; rectifier circuit element; rectification; rectifier.** KPSH-10E1

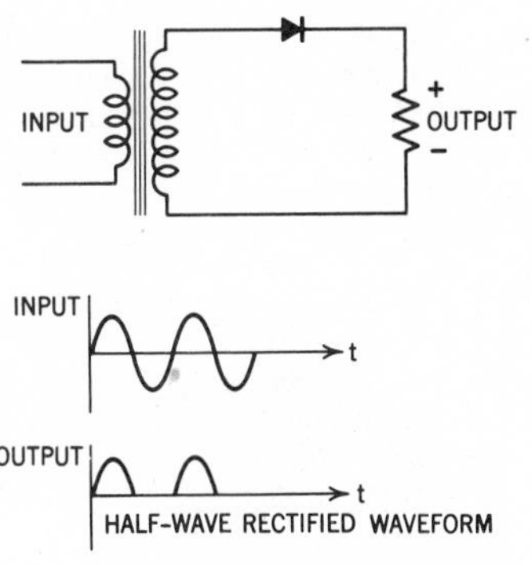

Half-wave rectification.

Hall coefficient (Hall constant). *See:* **Hall effect.**

Hall effect. When a conductor or semiconductor carrying a current is placed in a magnetic field that has a component normal to the direction of the current density, the distribution of the current density over the cross section of the conductor or semiconductor is changed and a potential gradient is set up in a direction normal to the directions of both the current density and the field. *Notes:* (1) The potential gradient is, at each point, a function, of the vector product of the current density and the magnetic field strength. Thus: $\mathbf{g} = f[\mathbf{J} \times \mathbf{H}]$ where **g** is the component of the potential gradient due to the Hall effect, **J** is the vector current density, and **H** is the vector magnetic field strength. In the simplest case the Hall effect component of the potential gradient is proportional to the vector product, so that $\mathbf{g} = R\,[\mathbf{J} \times \mathbf{H}]$. In this case, R is a parameter called the Hall coefficient. (2) The sign of the majority carrier can be inferred from the sign of the **Hall coefficient.** *See also:* **semiconductor.** E270-0

Hall mobility (electric conductor). The quantity μ_H in the relation $\mu_H = R\sigma$, where R = Hall coefficient and σ = conductivity. *See also:* **semiconductor.** E102/E216/E270-34E17

halogen-quenched counter tube. A self-quenched counter tube in which the quenching agent is a halogen, usually bromine or chlorine. *See also:* **tube definitions.** 0-15E6

Hamming distance (computing systems). *See:* **signal distance.** *See also:* **electronic digital computer.**

hand burnishing (electroplating). Burnishing done by a hand tool, usually of steel or agate. *See also:* **electroplating.** 42A60-0

hand (head or butt) cable (mining). A flexible cable used principally in making electric connections between a mining machine and a truck carrying a reel of portable cable. *See also:* **mine feeder circuit.** 42A85-0

***H* and *D* curve.** *See:* **Hurter and Driffield curve.**

hand elevator. An elevator utilizing manual energy to move the car. *See also:* **elevators.** 42A45-0

handhole. An opening in an underground run or system into which workmen reach, but do not enter. *Note:* This affords facilities for placing and maintaining in the runs, conductors, cables, and any associated apparatus. *See also:* **power distribution, underground construction.** 2A2/42A35-31E13

handling device (metal-clad switchgear). The accessory that is used for the removal, replacement, or transportation of the removable element. 37A100-31E11

hand operation (industrial control). Actuation of an apparatus by hand without auxiliary power. *See:* **switch.** 50I16-34E10

hand receiver. An earphone designed to be held to the ear by the hand. *See also:* **loudspeaker.** 42A65-0

hand-reset relay (mechanically reset relay). A relay that is so constructed that it remains in the picked-up condition even after the input quantity is removed; specific manual action is required to reset the relay. 37A100-31E11/31E6

handset. A combination of a telephone transmitter and a telephone receiver mounted on a handle. *See also:* **telephone station.** 42A65-0

handset telephone. *See:* **hand telephone set.**

hand telephone set (handset telephone). A telephone set having a handset and a mounting that serves to support the handset when the latter is not in use. *Note:*

The prefix desk, wall, drawer, etcetera, may be applied to the term **hand telephone set** to indicate the type of mounting. *See also:* **telephone station.** 42A65-0

handwheel (industrial control). A wheel the rim of which serves as a handle for manual operation of a rotary device. 42A25-34E10

hand winding (rotating machinery). A winding placed in slots or around poles by a human operator. *See also:* **rotor (rotating machinery); stator.** 0-31E8

hangover (facsimile). *See:* **tailing.**

hang-up hand telephone set (suspended-type handset telephone) (bracket-type handset telephone). A hand telephone set in which the mounting is arranged for attachment to a vertical surface and is provided with a switch bracket from which the handset is suspended. *See also:* **telephone station.** 42A65-0

hard limiting. *See:* **limiter circuit.** *See also:* **electronic analog computer.**

hardware. (1) Physical entities such as computers, circuits, tape readers, etcetera. Contrasted with software. (2) Parts made of metal such as fasteners, hinges, etcetera. *See also:* **electronic digital computer; software.** E162-0

harmful interference (electromagnetic compatibility). Any emission, radiation, or induction that endangers the functioning of a radio-navigation service or of other safety services, or seriously degrades, obstructs, or repeatedly interrupts a radiocommunication service operating in accordance with regulations. *See also:* **electromagnetic compatibility.** ITA-27E1

harmonic. A sinusoidal component of a periodic wave or quantity having a frequency that is an integral multiple of the fundamental frequency. *Note:* For example, a component the frequency of which is twice the fundamental frequency is called the second harmonic. *See also:* **signal wave.** E145-0;42/A65-31E3;0-1E1

harmonic analyzer. A mechanical device for measuring the amplitude and phase of the various harmonic components of a periodic function from its graph. *See also:* **instrument; signal wave; wave analyzer.** 42A30-0

harmonic components (harmonics). The harmonic components of a Fourier Series are the terms $C_n \sin (nx + \theta_n)$. *Note:* For example, the component that has a frequency twice that of the fundamental ($n = 2$) is called the second harmonic. E270-0

harmonic conjugate. *See:* **Hilbert transform.**

harmonic content (nonsinusoidal periodic wave). The deviation from the sinusoidal form, expressed in terms of the order and magnitude of the Fourier series terms describing the wave. *See also:* **power rectifier; rectification.** 0-34E24

harmonic conversion transducer (frequency multiplier, frequency divider). A conversion transducer in which the output-signal frequency is a multiple or submultiple of the input frequency. *Notes:* (1) In general, the output-signal amplitude is a nonlinear function of the input-signal amplitude. (2) Either a frequency multiplier or a frequency divider is a special case of harmonic conversion transducer. *See also:* **heterodyne conversion transducer (converter); transducer.** E160/E270/42A65-0;E196-15E6

harmonic leakage power (transmit-receive and pretransmit-receive tubes). Obsolete term.

harmonic-restraint relay. A relay so constructed that its operation is restrained by selecting harmonic components of one or more separate input quantities. *See also:* **relay.** 37A1-31E6;37A100-31E11

harmonics. *See:* **harmonic components.**

harmonic series. A series in which each component has a frequency that is an integral multiple of a fundamental frequency. *See also:* **electroacoustics.** 0-1E1

harmonic telephone ringer. A telephone ringer that responds only to alternating current within a very narrow frequency band. *Note:* A number of such ringers, each responding to a different frequency, are used in one type of selective ringing. *See also:* **telephone station.** 42A65-0

harmonic test (rotating machinery). A test to determine directly the value of one or more harmonics of the waveform of a quantity associated with a machine, relative to the fundamental of that quantity. *See also:* **asynchronous machine; synchronous machine.** 0-31E8

hartley (information theory). A unit of information content, equal to one decadal decision, or the designation of one of ten possible and equally likely values or states of anything used to store or convey information. *Notes:* (1) A hartley may be conveyed by one decadal code element. One hartley equals (log of 10 to base 2) times one bit. (2) If, in the definition of information content, the logarithm is taken to the base ten, the result will be expressed in hartleys. *See also:* **bit.** 42A65-0

Hartley oscillator. An electron-tube circuit in which the parallel-tuned tank circuit is connected between grid and plate, the inductive element of the tank having an intermediate tap at cathode potential, and the necessary feedback voltage obtained across the grid-cathode portion of the inductor. *See also:* **oscillatory circuit.** E54/E145-0

hauptnutzzeit. *See:* **utilization time.**

Hay bridge. A 4-arm alternating-current bridge in which the arms adjacent to the unknown impedance are nonreactive resistors and the opposite arm comprises a capacitor in series with a resistor. *Note:* Normally used for the measurement of inductance in terms of capacitance, resistance, and frequency. Usually, the bridge is balanced by adjustment of the resistor that is in series with the capacitor, and of one of the nonreactive arms. The balance depends upon the frequency. It differs from the Maxwell bridge in that in the arm opposite the inductor, the capacitor is in series with the resistor. *See also:* **bridge.** 42A30-1

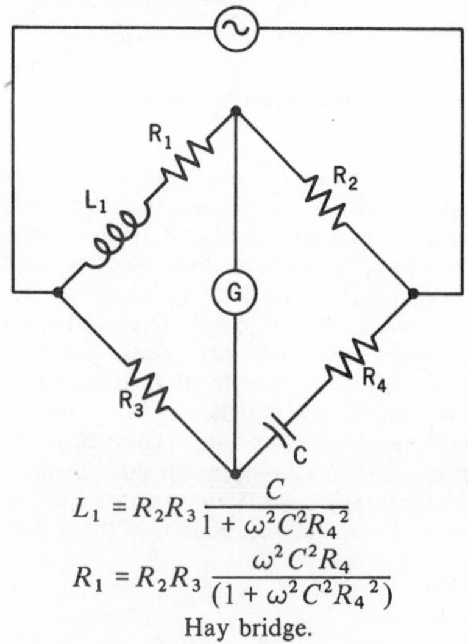

$$L_1 = R_2 R_3 \frac{C}{1 + \omega^2 C^2 R_4{}^2}$$

$$R_1 = R_2 R_3 \frac{\omega^2 C^2 R_4}{(1 + \omega^2 C^2 R_4{}^2)}$$

Hay bridge.

hazard or obstruction beacon. An aeronautical beacon used to designate a danger to air navigation. *See also:* **signal lighting.** Z7A1-0

hazardous area class I. The locations in which flammable gases or vapors are or may be present in the air in quantities sufficient to produce explosive or ignitible mixtures. *See also:* **explosionproof apparatus.** 1A0-34E16

hazardous area class I, division I. Locations (A) in which hazardous concentrations of flammable gases or vapors exist continuously, intermittently, or periodically under normal operating conditions, (B) in which hazardous concentrations of such gases or vapors may exist frequently because of repair or maintenance operations or because of leakage, or (C) in which breakdown or faulty operation of equipment or processes which might release hazardous concentrations of flammable gases or vapors, might also cause simultaneous failure of electric equipment. *See also:* **explosionproof apparatus.** 1A0-34E16

hazardous area class I, division II. Locations (A) in which volatile flammable liquids or flammable gases are handled, processed, or used, but in which the hazardous liquids, vapors, or gases will normally be confined within closed containers or closed systems from which they can escape only in case of accidental rupture or breakdown of such containers or systems, or in case of abnormal operation of equipment, (B) in which hazardous concentrations of gases or vapors are normally prevented by positive mechanical ventilation, but which might become hazardous through failure or abnormal operation of the ventilating equipment, or (C) which are adjacent to class I, division I locations, and to which hazardous concentrations of gases or vapors might occasionally be communicated unless such communication is prevented by adequate positive-pressure ventilation from a source of clean air, and effective safeguards against ventilation failure are provided. *See also:* **explosionproof apparatus.** 1A0-34E16

hazardous area class II. The locations that are hazardous because of the presence of combustible dust. *See also:* **explosionproof apparatus.** 1A0-34E16

hazardous area class II, division I. Locations (A) in which combustible dust is or may be in suspension in the air continuously, intermittently, or periodically under normal operating conditions, in quantities sufficient to produce explosive or ignitible mixtures, (B) where mechanical failure or abnormal operation of machinery or equipment might cause such mixtures to be produced, and might also provide a source of ignition through simultaneous failure of electric equipment, operation of protection devices, or from other causes, or (C) in which dusts of an electrically conducting nature may be present. *See also:* **explosionproof apparatus.** 1A0-34E16

hazardous area class II, division II. Locations in which combustible dust will not normally be in suspension in the air, or will not be likely to be thrown into suspension by the normal operation of equipment or apparatus, in quantities sufficient to produce explosive or ignitible mixtures, but (A) where deposits or accumulations of such dust may be sufficient to interfere with the safe dissipation of heat from electric equipment or apparatus, or (B) where such deposits or accumulations of dust on, in, or in the vicinity of electric equipment might be ignited by arcs, sparks, or burning material from such equipment. *See also:* **explosionproof apparatus.** 1A0-34E16

hazardous area group A. Atmospheres containing acetylene. *See also:* **explosionproof apparatus.** 1A0-34E16

hazardous area group B. Atmospheres containing hydrogen, or gases or vapors of equivalent hazard such as manufactured gas. *See also:* **explosionproof apparatus.** 1A0-34E16

hazardous area group C. Atmospheres containing ethyl ether vapors, ethylene, or cyclopropane. *See also:* **explosionproof apparatus.** 1A0-34E16

hazardous area group D. Atmospheres containing gasoline, hexane, naphtha, benzine, butane, propane, alcohol, acetone, benzol, lacquer solvent vapors, or natural gas. *See also:* **explosionproof apparatus.** 1A0-34E16

hazardous area group E. Atmospheres containing metal dust, including aluminum, magnesium, and their commercial alloys, and other metals of similarly hazardous characteristics. *See also:* **explosionproof apparatus.** 1A0-34E16

hazardous area group F. Atmospheres containing carbon black, coal or coke dust. *See also:* **explosionproof apparatus.** 1A0-34E16

hazardous area group G. Atmospheres containing flour, starch, or grain dusts. *See also:* **explosionproof apparatus.** 1A0-34E16

hazardous location. An area where ignitible vapors or dust may cause a fire or explosion created by energy emitted from lighting or other electric equipment or by electrostatic generation. *See also:* **luminaire.** Z7A1-0

***H* bend (waveguide technique).** *See:* ***H*-plane bend.**

HCL. *See:* **relay, high, common, low.**

***H* display (radar).** A *B* display modified to include indication of angle of elevation. *Note:* The target appears as two closely spaced blips that approximate a short bright line, the slope of which is in proportion to the sine of the angle of target elevation. *See also:* **navigation.** E172-10E6

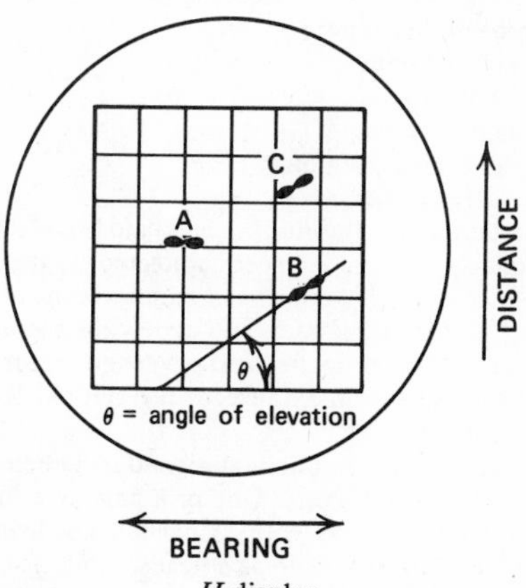

H display.

head (computing system). A device that reads, records, or erases data on a storage medium. *Note:* For example, a small electromagnet used to read, write, or erase data on a magnetic drum or tape, or the set of perforating, reading, or marking devices used for punching, reading, or printing on paper tape. *See also:* **electronic digital computer.** X3A12-16E9

headed brush (rotating machinery). A brush having a top (cylindrical, conical, or rectangular) with a smaller cross section than the cross section of the body of the brush. *Note:* The length of the head shall not exceed 25 percent of the overall length. *See also:* **brush.** 64A1-0

head-end system (railways). A system in which the electrical requirements of a train are supplied from a generator or generators, located on the locomotive or in one of the cars, customarily at the forward part of the train. *Note:* The generators may be driven by steam turbine, internal-combustion engine, or, if located in one of the cars, by a mechanical drive from a car axle. *See also:* **axle generator system.** 42A42-0

heading (navigation). The horizontal direction in which a vehicle is pointed, expressed as an angle between a reference line and the line extending in the direction the vehicle is pointed, usually measured clockwise from the reference line. *See also:* **navigation.** 0-10E6

heading-effect error (navigation). A manifestation of polarization error causing an error in indicated bearing that is dependent upon the heading of a vehicle with respect to the direction of signal propagation. *Note:* Heading-effect error is a special case of attitude-effect error where the vehicle is in straight level flight; it is sometimes referred to as course push (or pull). *See also:* **navigation.** 0-10E6

headlamp. A major lighting device mounted on a vehicle and used to provide illumination ahead of it. Also called **headlight.**
See:
back-up lamp;
clearance lamp;
dual headlighting system;
foglamp;
light;
lower (passing) beams;
multiple-beam headlamp;
parking lamp;
retro-reflector (reflex reflector);
sealed-beam headlamp;
side marker lamp;
stop lamp;
tail lamp;
turn-signal operating unit;
upper (driving) beams. Z7A1-0

head or butt cable (mining). *See:* **hand cable.**

headquarters system (direct-connected system). A local system to which has been added means of transmitting system signals to and receiving them at an agency maintained by the local government, for example, in a police precinct house, or fire station. *See also:* **protective signaling.** 42A65-0

head receiver. An earphone designed to be held to the ear by a headband. *Note:* One or a pair (one for each ear) of head receivers with associated headband and connecting cord is known as a headset. *See also:* **loudspeaker.** 42A65-0

hearing level. *See:* **hearing loss.**

hearing loss (hearing level) (1) (for speech). The difference in decibels between the speech levels at which the average normal ear and the defective ear, respectively, reach the same intelligibility, often arbitrarily set at 50 percent.

(2) (hearing-threshold level) (ear at a specified frequency). The amount, in decibels, by which the threshold of audibility for that ear exceeds a standard audiometric threshold. *Notes:* (1) See: Current issue of American Standard Specification for Audiometers for General Diagnostic Purposes. (2) This concept was at one time called deafness; such usage is now deprecated. (3) Hearing loss and deafness are both legitimate qualitative terms for the medical condition of a moderate or severe impairment of hearing, respectively. Hearing level, however, should only be used to designate a quantitative measure of the deviation of the hearing threshold from a prescribed standard. *See also:* **electroacoustics; loudspeaker.** 0-1E1

heat coil. A protective device that grounds or opens a circuit, or does both, by means of a mechanical element that is allowed to move when the fusible substance that holds it in place is heated above a predetermined temperature by current in the circuit. *See also:* **electromagnetic relay.** 42A65-31E3/21E0

heat detector (burglar-alarm system). A temperature-sensitive device mounted on the inside surface of a vault to initiate an alarm in the event of an attack by heat or burning. *See also:* **protective signaling.** 42A65-0

heater (electron tube). An electric heating element for supplying heat to an indirectly heated cathode. *See also:* **electrode (electron tube).** 42A70-15E6

heater coil. *See:* **load coil (induction heating usage).**

heater connector (heater plug). A cord connector designed to engage the male terminal pins of a heating or cooking appliance. *See also:* **interior wiring.** 42A95-0

heater current. The current flowing through a heater. *See:* **cathode preheating time; filament current.** *See also:* **electronic controller.** 42A70-15E6

heater transformer (industrial control). Supplies power for electron-tube filaments or heaters of indirectly heated cathodes. *See also:* **electronic controller.** 42A25-34E10

heater voltage. The voltage between the terminals of a heater. *See also:* **electronic controller; electrode voltage (electron tube).** 42A70-15E6

heater warm-up time. *See:* **cathode heating time.**

heat exchanger. *See:* **cooler.**

heat-exchanger cooling system (rectifier). A cooling system in which the coolant, after passing over the cooling surfaces of the rectifier, is cooled in a heat exchanger and recirculated. *Note:* The coolant is generally water of a suitable purity, or water that has been treated by a corrosion-inhibitive chemical. Antifreeze solutions may also be used where there is exposure to low temperatures. The heat exchanger is usually either: (1) water-to-water where the heat is removed by raw water, (2) water-to-air where the heat is removed by air supplied by a blower, (3) air-to-water, (4) air-to-air, (5) refrigeration cycle. The liquid in the closed system may be other than water, and the gas in the closed system may be other than air. *See also:* **rectification; rectifier.** 42A15-34E24

heating cycle. One complete operation of the thermostat from ON to ON or from OFF to OFF. 71A1-0

heating element. A length of resistance material connected between terminals and used to generate heat electrically. *See also:* **appliance outlet.** 71A1-0

heating pattern. The distribution of temperature in a load or charge. *See also:* **induction heating; industrial electronics.** E54/E169-0

heating station (dielectric heating usage). The location that includes the work coil or applicator and its associated production equipment. *See also:* **induction heating; industrial electronics.** E54/E169-0

heating time, tube (mercury-vapor tube). *See:* **preheating time.**

heating unit (electrical appliances). An assembly containing one or more heating elements, electric terminals or leads, electrical insulation, and a frame, casing, or other suitable supporting means. *See also:* **appliance outlet; appliances (including portable).** 71A1-0

heat loss. The part of the transmission loss due to the conversion of electric energy into heat. *See also:* **waveguide.** E146-0

heat-shield (cathode) (electron tubes). A metallic surface surrounding a hot cathode, in order to reduce the radiation losses. *See also:* **electron tube.** 50I07-15E6

heat sink (1) (general). A part used to absorb heat. *See:* **cradle base (rotating machinery).** 0-31E8
(2) (semiconductor rectifier diode). A mass of metal generally having much greater thermal capacity than the diode itself and intimately associated with it. It encompasses that part of the cooling system to which heat flows from the diode by thermal conduction only and from which heat may be removed by the cooling medium. *See:* **semiconductor rectifier stack.** 0-34E24
(3) (photovoltaic power system). A material capable of absorbing heat; a device utilizing such material for the thermal protection of components or systems. *See also:* **photovoltaic power system; solar cells (photovoltaic power system).** 0-10E1

Heaviside-Campbell mutual-inductance bridge. A mutual-inductance bridge of the Heaviside type in which one of the inductive arms contains a separate inductor that is included in the bridge arm during the first of a pair of measurements and is short-circuited during the second. *Note:* The balance is independent of the frequency. *See:* **Heaviside mutual-inductance bridge.** *See also:* **bridge.** 42A30-0

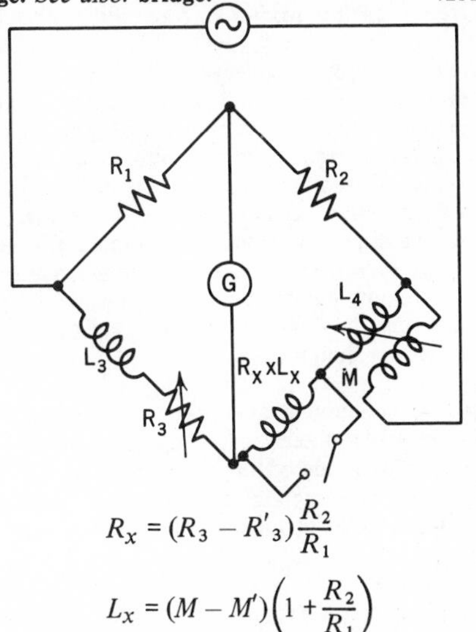

$$R_X = (R_3 - R'_3)\frac{R_2}{R_1}$$

$$L_X = (M - M')\left(1 + \frac{R_2}{R_1}\right)$$

Heaviside-Campbell mutual-inductance bridge.

Heaviside-Lorentz system of units. A rationalized system based on the centimeter, gram, and second and is similar to the Gaussian system but differs in that a factor 4π is explicitly inserted to multiply r^2 in each of the formulations of the Coulomb Laws. E270-0

Heaviside mutual-inductance bridge. An alternating-current bridge in which two adjacent arms contain self-inductance, and one or both of these have mutual inductance to the supply circuit, the other two arms being normally nonreactive resistors. *Note:* Normally used for the comparison of self- and mutual inductances. The balance is independent of the frequency. *See also:* **bridge.** 42A30-0

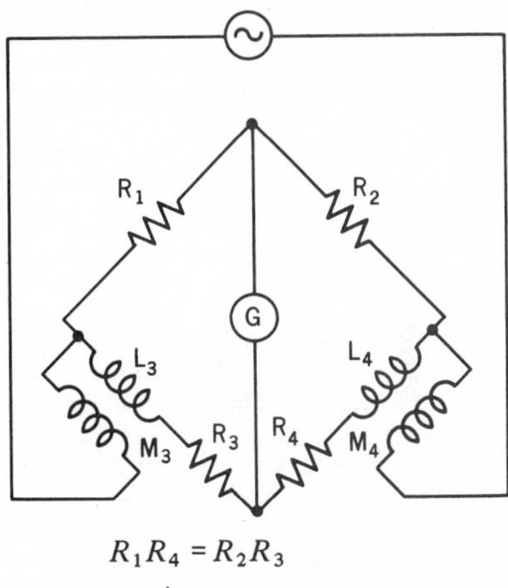

$$R_1R_4 = R_2R_3$$

$$L_3 - L_4\left(\frac{R_1}{R_2}\right) = -(M_3 - M_4)\left(1 + \frac{R_1}{R_2}\right)$$

Heaviside mutual-inductance bridge.

heavy-duty floodlight. A weatherproof unit having a substantially constructed metal housing into which is placed a separate and removable reflector. *Note:* A weatherproof hinged door with cover glass encloses the assembly but provides an unobstructed light opening at least equal to the effective diameter of the reflector. *See also:* **floodlighting.** Z7A1-0

height marker (radar). *See:* **calibration marks.**

height, pulse*. *See:* **amplitude, pulse.**

*Deprecated

helical antenna. An antenna whose configuration is that of a helix. *Note:* The diameter, pitch, and number of turns in relation to the wavelength provide control of the polarization state and directivity of helical antennas. *See also:* **antenna.** 0-3E1

helical plate (storage cell). A plate of large area formed by helically wound ribbed strips of soft lead inserted in supporting pockets or cells of hard lead. *See also:* **battery (primary or secondary).** 42A60-0

HEM wave. *See:* **hybrid electromagnetic wave.**

hemispherical reflectance. The ratio of all the flux leaving a surface or medium by reflection to the incident flux. *Note:* If reflectance is not preceded by an adjective descriptive of the angles of view, hemispherical reflectance is implied. *See also:* **lamp.** Z7A1-0

hemispherical transmittance (illuminating engineering). The ratio of the transmitted flux leaving a surface or medium to the incident flux. *Note:* If transmittance is not preceded by an adjective descriptive of the angles of view, hemispherical transmittance is implied. *See also:* **transmission (illuminating engineering).** Z7A1-0

henry (circuit). The unit of inductance in the International System of Units (SI units). The inductance for which the induced voltage in volts is numerically equal to the rate of change of current in amperes per second. E270-0

heptode. A seven-electrode electron tube containing an anode, a cathode, a control electrode, and four additional electrodes that are ordinarily grids. *See also:* **tube definitions.** 42A70-15E6

hermetically sealed relay. A relay in a gastight enclosure that has been completely sealed by fusion or other comparable means to insure a low rate of gas leakage over a long period of time. *See also:* **relay.** 83A16-0

hermetic motor. A stator and rotor without shaft, end shields, or bearings for installation in refrigeration compressors of the hermetically sealed type. 0-31E8

hertz. The unit of frequency, one cycle per second. E45-0

heterodyne conversion transducer (converter). A conversion transducer in which the useful output frequency is the sum or difference of (1) the input frequency and (2) an integral multiple of the frequency of another wave usually derived from a local oscillator. *Note:* The frequency and voltage or power of the local oscillator are parameters of the conversion transducer. Ordinarily, the output-signal amplitude is a linear function of the input-signal amplitude over its useful operating range.
See:
available conversion gain;
conversion transducer;
converter tube;
gain available conversion;
gain, conversion voltage;
harmonic conversion transducer.
See also: **transducer.**
E160-15E6;E196/E270-0;42A65-0

heterodyne frequency. *See:* **beats.**

heterodyne reception (beat reception). The process of reception in which a received high-frequency wave is combined in a nonlinear device with a locally generated wave, with the result that in the output there are frequencies equal to the sum and difference of the combining frequencies. *Note:* If the received waves are continuous waves of constant amplitude, as in telegraphy, it is customary to adjust the locally generated frequency so that the difference frequency is audible. If the received waves are modulated the locally generated frequency is generally such that the difference frequency is superaudible and an additional operation is necessary to reproduce the original signal wave. *See:* **superheterodyne reception.** *See also:* **modulating systems.** 42A65-0

heteropolar machine (rotating machinery). A machine having an even number of magnetic poles with successive (effective) poles of opposite polarity. *See also:* **asynchronous machine; direct-current commutating machine; synchronous machine.** 0-31E8

heuristic. Pertaining to exploratory methods of problem solving in which solutions are discovered by evaluation of the progress made toward the final result. *See:* **algorithm.** X3A12-16E9

Hevea rubber. Rubber from the *Hevea brasiliensis* tree. *See also:* **insulation.** 42A95-0

hexadecimal. (1) Pertaining to a characteristic or property involving a selection, choice, or condition in which there are sixteen possibilities. (2) Pertaining to the numeration system with a radix of sixteen. (3) More accurately called sexadecimal. *See also:* **positional notation.** 0-16E9

hexode. A six-electrode electron tube containing an anode, a cathode, a control electrode, and three additional electrodes that are ordinarily grids. *See also:* **tube definitions.** 42A70-15E6

hickey. (1) A fitting used to mount a lighting fixture in an outlet box or on a pipe or stud. *Note:* It has openings through which fixture wires may be brought out of the fixture stem. (2) A pipe-bending tool. *See also:* **interior wiring.** 42A95-0

high direct voltage (insulation test). A direct voltage above 5000 volts supplied by portable test equipment of limited capacity. *See also:* **insulation testing (large alternating-current rotating machinery).** E95-0

higher-order mode (waveguide or transmission line). Any mode of propagation characterized by a field configuration other than that of the fundamental or first-order mode with lowest cutoff frequency. *See also:* **waveguide.** 0-9E4

high-field-emission arc (gas). An electric arc in which the electron emission is due to the effect of a high electric field in the immediate neighborhood of the cathode, the thermionic emission being negligible. *Sec also:* **discharge (gas).** 50I07-15E6

high-frequency carrier telegraphy. The form of carrier telegraphy in which the carrier currents have their frequencies above the range transmitted over a voice-frequency telephone channel. *See also:* **telegraphy.** 42A65-31E3

high-frequency furnace (coreless-type induction furnace). An induction furnace in which the heat is generated within the charge, or within the walls of the containing crucible, or in both, by currents induced by high-frequency flux from a surrounding solenoid. 42A60-0

high-frequency induction heater or furnace. A device for causing electric current flow in a charge to be heated, the frequency of the current being higher than that customarily distributed over commercial networks. *See also:* **induction heating.** E54/E169-0

high-frequency stabilized arc welder. A constant-current arc-welding power supply including a high-frequency arc stabilizer and suitable controls required to produce welding current primarily intended for tungsten-inert-gas arc welding. *See:* **constant-current arc-welding power supply.** *See also:* **electric arc-welding apparatus.** 87A1-0

high-gain direct-current amplifier. An amplifier that is capable of amplification substantially greater than required for a specified operation throughout a frequency band extending from zero to some maximum. *See also:* **electronic analog computer; operational amplifier.** E165-16E9

high-impedance rotor. An induction-motor rotor having a high-impedance squirrel cage, used to limit starting current. *See also:* **rotor (rotating machinery).** 0-31E8

high-key lighting (television). A type of lighting that, applied to a scene, results in a picture having graduations falling primarily between gray and white; dark grays or blacks are present, but in very limited areas. *See also:* **television lighting.** Z7A1-0

high-level firing time (microwave) (switching tubes). The time required to establish a radio-frequency discharge in the tube after the application of radio-frequency power. *See:* **gas tubes.** E160-15E6

high-level modulation. Modulation produced at a point in a system where the power level approximates that at the output of the system. *See also:* **modulating systems.** 42A65/E145/E182-0

high-level radio-frequency signal (microwave gas tubes). A radio-frequency signal of sufficient power to cause the tube to become fired. *See:* **gas tubes.** E160-15E6

high-level voltage standing-wave ratio (microwave switching tubes). The voltage standing-wave ratio due to a fired tube in its mount located between a generator and matched termination in the waveguide. *See:* **gas tubes.** E160-15E6

high lights (any metal article). Those portions that are most exposed to buffing or polishing operations, and hence have the highest luster. 42A60-0

high-pass filter. *See:* **filter, high-pass.**

high peaking. The introduction of an amplitude-frequency characteristic having a higher relative response at the higher frequencies. *See also:* **television.** 2E2/42A65

high pot. *See:* **high-potential test.**

high-potential test (overvoltage test) (rotating machinery). A test that consists of the application of a voltage higher than the rated voltage for a specified time for the purpose of determining the adequacy against breakdown of insulating materials and spacings under normal conditions. *Note:* The test is used as a proof test of new apparatus, a maintenance test on older equipment, or as one method of evaluating developmental insulation systems. *See also:* **asynchronous machine; direct-current commutating machine; synchronous machine.** MG1-31E8

high-power-factor mercury-lamp ballast. A multiple-supply type power-factor-corrected ballast, so designed that the input current is at a power factor of not less than 90 percent when the ballast is operated with center rated voltage impressed upon its input terminals and with a connected load, consisting of the appropriate reference lamp(s), operated in the position for which the ballast is designed. 82A9-0

high-power-factor transformer. *See:* **transformer, high-power-factor.**

high-pressure vacuum pump. A vacuum pump that discharges at atmospheric pressure. *See also:* **rectification.** 42A15-0

high-reactance rotor. An induction-motor rotor having a high-reactance squirrel cage, used where low starting current is required and where low locked-rotor and breakdown torques are acceptable. *See also:* **rotor (rotating machinery).** 0-31E8

high-speed carry (electronic computation). A carry process such that if the current sum in a digit place is exactly one less than the base, the carry input is bypassed to the next place. *Note:* The processing necessary to allow the bypass occurs before the carry input arrives. Further processing required in the place as a result of the carry input, occurs after the carry has passed by. Contrasted with **cascaded carry.** *See also:* **electronic digital computer; standing-on-nines carry.** E162-0

high-speed excitation system. An excitation system capable of changing its voltage rapidly in response to a change in the excited generator field circuit. *See also:* **generating station.** 42A35-31E13

high-speed grounding switch. *See:* **fault-initiating switch.**

high-speed regulator (power supplies). A power supply regulator circuit that, by the elimination of its output capacitor, has been made capable of much higher slewing rates than are normally possible. *Note:* High-speed regulators are used where rapid step-programming is needed; or as current regulators, for which they are ideally suited. *See:* **slewing rate.** *See also:* **power supply.** KPSH-10E1

high-speed relay (electric-power circuits). A relay that operates in less than a specified time. *Note:* The specified time in present practice is fifty milliseconds (three cycles on a 60-hertz basis). 37A100-31E11;0-31E6

high-speed short-circuiting switch. *See:* **fault-initiating switch.**

high-velocity camera tube (anode-voltage stabilized camera tube) (electron device). A camera tube operating with a beam of electrons having velocities such that the average target voltage stabilizes at a value approximately equal to that of the anode. *See also:* **tube definitions.** 0-15E6

high-voltage relay. (1) A relay adjusted to sense and function in a circuit or system at a specific maximum voltage. (2) A relay designed to handle elevated voltages on its contacts, coil, or both. 0-21E0

high-voltage time test. An accelerated life test on a cable sample in which voltage is the factor increased. *See also:* **power distribution, underground construction.** 42A35-31E13

highway crossing back light (railway practice). An auxiliary signal light used for indication in a direction opposite to that provided by the main unit of a highway crossing signal. *See also:* **railway signal and interlocking.** 42A42-0

highway crossing bell (railway practice). A bell located at a railroad-highway grade crossing and operated to give a characteristic and arrestive signal to give warning of the approach of trains. *See also:* **railway signal and interlocking.** 42A42-0

highway crossing signal. An electrically operated signal used for the protection of highway traffic at railroad-highway grade crossings. *See also:* **railway signal and interlocking.** 42A42-0

Hilbert transform (harmonic conjugate) (real function $x(t)$ of the real variable t). The real function $x(t)$ that is the Cauchy principal value of

$$\frac{1}{\pi}\int_{-\infty}^{\infty}\frac{x(\tau)d\tau}{t-\tau}.$$

See: **analytic signal; network analysis.** 0-12E1

hinge clip (switching device). The clip to which the blade is movably attached. 37A100-31E11

hinged-iron ammeter. A special form of moving-iron ammeter in which the fixed portion of the magnetic circuit is arranged so that it can be caused to encircle the conductor, the current in which is to be measured.

This conductor then constitutes the fixed coil of the instrument. *Note:* The combination of a current transformer of the split-core type with an ammeter is often used similarly to measure alternating current, but should be distinguished from the hinged-iron ammeter. *See also:* **instrument.** 42A30-0

hiss (electron device). Noise in the audio-frequency range, having subjective characteristics analogous to prolonged sibilant sounds. *See also:* **circuit characteristics of electrodes.** E151/15E6/42A65

***H* network.** A network composed of five branches, two connected in series between an input terminal and an output terminal, two connected in series between another input terminal and output terminal, and the fifth connected from the junction point of the first two branches to the junction point of the second two branches. *See also:* **network analysis.** E270-0

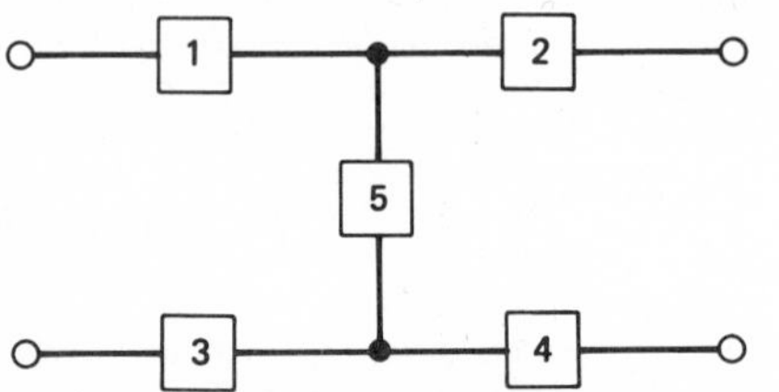

H network. Branches 1 and 2 are the first two branches between an input and an output terminal; branches 3 and 4 are the second two branches; and branch 5 is the branch between the junction points.

hodoscope. An apparatus for tracing the path of a charged particle in a magnetic field. *See also:* **electron optics.** 50I07-15E6

hoist. An apparatus for moving a load by the application of a pulling force, and not including a car or platform running in guides. *See:* **base-mounted electric hoist; elevator; overhead electric hoist.** 42A45-0

hoist back-out switch (mining). A switch that permits operation of the hoist only in the reverse direction in case of overwind. *See also:* **mine hoist.** 42A85-0

hoisting-rope equalizer. A device installed on an elevator car or counterweight to equalize automatically the tensions in the hoisting wire ropes. *See also:* **elevator.** 42A45-0

hoist overspeed device (mining). A device that can be set to prevent the operation of a mine hoist at speeds greater than predetermined values and usually causes an emergency brake application when the predetermined speed is exceeded. *See also:* **mine hoist.** 42A85-0

hoist overwind device (mining). A device that can be set to cause an emergency brake application when a cage or skip travels beyond a predetermined point into a danger zone. *See also:* **mine hoist.** 42A85-0

hoist signal code (mining). Consists of prescribed signals for indicating to the hoist operator the desired direction of travel and whether men or materials are to be hoisted or lowered in mines. *See also:* **mine hoist.** 42A85-0

hoist signal system (mining). A system whereby signals can be transmitted to the hoist operator (and in some instances by him to the cager) for control of mine hoisting operations. *See also:* **mine hoist.** 42A85-0

hoist slack-brake switch (mining). A device for automatically cutting off the power from the hoist motor and causing the brake to be set in case the links in the brake rigging require tightening or the brakes require relining. *See also:* **mine hoist.** 42A85-0

hoist trip recorder (mining). A device that graphically records information such as the time and number of hoists made as well as the delays or idle periods between hoists. *See also:* **mine hoist.** 42A85-0

hoistway (elevator or dumbwaiter). A shaftway for the travel of one or more elevators or dumbwaiters. *Note:* It includes the pit and terminates at the underside of the overhead machinery space floor or grating, or at the underside of the roof where the hoistway does not penetrate the roof.
See:
biparting door;
bottom car clearance;
car or hoistway door or gate;
car platform;
compensating-rope sheave switch;
door or gate closer;
door or gate power operator;
elevators;
hoistway-door combination mechanical lock and electric contact;
hoistway-door electric contact;
hoistway-door interlock;
hoistway-door or gate locking device;
hoistway enclosure;
hoistway-gate separate mechanical lock;
hoistway-unit system;
manually operated door or gate;
multiple hoistway;
overslung car frame;
power-operated door or gate;
self-closing door or gate;
semiautomatic gate;
single hoistway;
subpost car frame;
top car clearance;
top counterweight clearance. 42A45-0

hoistway access switch (elevators). A switch, located at a landing, the function of which is to permit operation of the car with the hoistway door at this landing and the car door or gate open, in order to permit access to the top of the car or to the pit. *See also:* **control (elevators).** 42A45-0

hoistway-door combination mechanical lock and electric contact (elevators). A combination mechanical and electric device, the two related, but entirely independent, functions of which are: (1) to prevent operation of the driving machine by the normal operating device unless the hoistway door is in the closed position, and (2) to lock the hoistway door in the closed position and prevent it from being opened from the landing side unless the car is within the landing zone. *Note:* As there is no positive mechanical connection between the electric contact and the door-locking mechanism, this device insures only that the door will be closed, but not necessarily locked, when the car leaves the landing. Should the lock mechanism fail to operate as intended when released by a stationary or retiring car-cam device, the door can be opened from the landing side even though the car is not at the landing. If operated by a stationary car-cam device, it does not prevent opening the door from the landing side as the car passes the floor. *See also:* **hoistway (elevator or dumbwaiter).** 42A45-0

hoistway-door electric contact (elevators). An electric device, the function of which is to prevent operation of the driving machine by the normal operating device unless the hoistway door is in the closed position. *See also:* **hoistway (elevator or dumbwaiter).** 42A45-0

hoistway-door interlock (elevators). A device having two related and interdependent functions that are (1) to prevent the operation of the driving machine by the normal operating device unless the hoistway door is locked in the closed position; and (2) to prevent the opening of the hoistway door from the landing side unless the car is within the landing zone and is either stopped or being stopped. *See also:* **hoistway (elevator or dumbwaiter).** 42A45-0

hoistway-door or gate locking device (elevators). A device that secures a hoistway or gate in the closed position and prevents it from being opened from the landing side except under specified conditions. *See also:* **hoistway (elevator or dumbwaiter).** 42A45-0

hoistway enclosure. The fixed structure, consisting of vertical walls or partitions, that isolates the hoistway from all other parts of the building or from an adjacent hoistway and in which the hoistway doors and door assemblies are installed. *See also:* **hoistway (elevator or dumbwaiter).** 42A45-0

hoistway-gate separate mechanical lock (elevators). A mechanical device, the function of which is to lock a hoistway gate in the closed position after the car leaves a landing and prevent the gate from being opened from the landing side unless the car is within the landing zone. *See also:* **hoistway (elevator or dumbwaiter).** 42A45-0

hoistway-unit system (elevators). A series of hoistway-door interlocks, hoistway-door electric contacts, or hoistway-door combination mechanical locks and electric contacts, or a combination thereof, the function of which is to prevent operation of the driving machine by the normal operating device unless all hoistway doors are in the closed position and, where so required, are locked in the closed position. *See also:* **hoistway (elevator or dumbwaiter).** 42A45-0

hold (1) (electronic digital computer). An untimed delay in the program, terminated by an operator or interlock action. *See also:* **electronic digital computer.** EIA3B-34E12

(2) (electronic analog computer). The computer control state in which the problem solution is stopped, usually by disconnecting integrator input signals. *See:* **automatic hold.** *See also:* **electronic analog computer.** E165-16E9

(3) (industrial control). A control function that arrests the further speed change of a drive during the acceleration or deceleration portion of the operating cycle. *See also:* **control system, feedback.** AS1-34E10

(4) (charge-storage tubes) (verb). To maintain storage elements at an equilibrium voltage by electron bombardment. *See also:* **charge-storage tube; data processing.** E158-15E6

hold-closed mechanism (automatic circuit recloser). A device that holds the contacts in the closed position following the completion of a predetermined sequence of operations as long as current flows in excess of a predetermined value. 37A100-31E11

hold-closed operation (automatic circuit recloser). An opening followed by the number of closing and opening operations that the hold-closed mechanism will permit before holding the contacts in the closed position. 37A100-31E11

holding current (thyristor). The minimum principal current required to maintain the thyristor in the ON-state. *See also:* **principal current.** E223-34E17/34E24/15E7

holding-down bolt. A bolt that fastens a machine to its bedplate, rails, or foundation. *See:* **cradle base (rotating machinery).** 0-31E8

holding frequency (take the swings). A condition of operating a generator or station to maintain substantially constant frequency irrespective of variations in load. *Note:* A plant so operated is said to be regulating frequency. *See also:* **generating station.** 42A35-31E13

holding load. A condition of operating a generator or station at substantially constant load irrespective of variations in frequency. *Note:* A plant so operated is said to be operating on base load. *See:* **base load.** *See also:* **generating station.** 42A35-31E13

holdoff. *See:* **sweep holdoff.**

holdup-alarm attachment. A general term for the various alarm-initiating devices used with holdup-alarm systems, including holdup buttons, footrails, and others of a secret or unpublished nature. *See also:* **protective signaling.** 42A65-0

holdup-alarm system. An alarm system signaling a robbery or attempted robbery. *See also:* **protective signaling.** 42A65-0

hole (semiconductor). A mobile vacancy in the electronic valence structure of a semiconductor that acts like a positive electron charge with a positive mass. *See also:* **semiconductor.** E59-34E17;E102/E216-34E17;E270-10E1

hollow-core annular conductor (hollow-core conductor). A conductor composed of a plurality of conducting elements disposed around a supporting member that does not fill the space enclosed by the elements; alternatively, a plurality of such conducting elements disposed around a central channel and interlocked one with the other or so shaped that they are self-supporting. *See also:* **conductor.** 42A35-31E13

hollow-core conductor. *See:* **hollow-core annular conductor.**

home signal (railway practice). A fixed signal at the entrance of a route or block to govern trains or engines entering or using that route or block. *See also:* **railway signal and interlocking.** 42A42-0

homing (navigation). Following a course directed toward a point by maintaining constant some navigational coordinate (other than altitude). *See also:* **radio navigation.** 42A65-0;E172-10E6

homing guidance. That form of missile guidance wherein the missile steers itself toward a target by means of a mechanism actuated by some distinguishing characteristic of the target. *See also:* **guided missile.** 42A65-0

homing relay. A stepping relay that returns to a specified starting position prior to each operating cycle. *See also:* **relay.** 83A16-0

homochromatic gain (optoelectronic device). The radiant gain or luminous gain for specified identical spectral characteristics of both incident and emitted flux. *See also:* **optoelectronic device.** E222-15E7

homodyne reception (zero-beat reception). A system of reception by the aid of a locally generated voltage of carrier frequency. *See also:* **modulating systems.** 42A65-0

homopolar machine* (rotating machinery). A machine in which the magnetic flux passes in the same direction from one member to the other over the whole of a single air-gap area. Preferred term is **acyclic machine**, which see.

*Deprecated 0-31E8

honeycomb coil (duolateral coil). A coil in which the turns are wound in crisscross fashion to form a self-supporting structure or to reduce distributed capacitance. *See also:* **circuits and devices.** 42A65-21E0

hook operation. *See:* **stick hook operation.**

hook stick. *See:* **switch stick (switch hook).**

hopper (computing systems). *See:* **card hopper.** *See also:* **electronic digital computer.**

horizontal amplifier (oscilloscopes). An amplifier for signals intended to produce horizontal deflection. *See:* **oscillograph.** 0-9E4

horizontal hold control (television). A synchronization control that varies the free-running period of the horizontal-deflection oscillator. E204-0

horizontally polarized wave (1) (general). A linearly polarized wave whose direction of polarization is horizontal. *See also:* **radiation.** 42A65-0

(2) (radio wave propagation). A linearly polarized wave whose electric field vector is horizontal. *See also:* **radio wave propagation.** E211-3E2

horizontal machine. A machine whose axis of rotation is approximately horizontal. *See:* **cradle base (rotating machinery).** 0-31E8

horizontal plane (searchlight). The plane that is perpendicular to the elevation axis and in which the train lies. *See also:* **searchlight; train.** Z7A1-0

horizontal-ring induction furnace. A device for melting metal comprising an annular horizontally-placed open trough or melting channel, a primary inductor winding, and a magnetic core that links the melting channel with the primary winding. *See also:* **induction heating.** E54/E169-0

horn (acoustic practice). A tube of varying cross-sectional area for radiating or receiving acoustic waves. *Note:* Normally it has different terminal areas that provide a change of acoustic impedance and control of the directional response pattern. *See also:* **electroacoustics; loudspeaker.** 42A65-1E1

horn antenna. A radiating element having the shape of a horn. *See also:* **antenna.** 0-3E1

horn-gap switch. A switch provided with arcing horns. 37A100-31E11

horn mouth. Normally the opening, at the end of a horn, with larger cross-sectional area. *See also:* **loudspeaker.** 42A65-1E1

horn reflector antenna. An antenna consisting of a section of a paraboloidal reflector fed with an offset horn that intersects the reflector surface. *Note:* The horn is usually pyramidal or conical. *See:* **antenna.** 0-3E1

horn throat (audio and electroacoustics). Normally the opening, at the end of a horn, with the smaller cross-sectional area. *See also:* **loudspeaker.** 42A65-1E1

horsepower rating, basis for single-phase motor. A system of rating for single-phase motors, whereby horsepower values are determined, for various synchronous speeds, from the minimum value of breakdown torque that the motor design will provide. 0-31E8

hose (liquid cooling) (rotating machinery) The flexible insulated or insulating hydraulic connections applied between the conductors and either a central manifold or coolant passage. *See:* **cradle base (rotating machinery).** 0-31E8

hoseproof (rotating machinery). *See:* **waterproof machine.**

hot cathode (thermionic cathode). A cathode that functions primarily by the process of thermionic emission. 42A70-15E6

hot-cathode lamp (fluorescent lamps). An electric discharge lamp in which the electrodes operate at incandescent temperatures and in which the cath ode drop is relatively low (10-20 volts). *Note:* The current density at the cathodes is relatively high, and lamps may be designed to carry any desired current up to several amperes. The energy to maintain the cathodes at incandescence may come either from the arc (arc heating), from circuit elements, or from both. 82A1-0

hot-cathode tube (thermionic tube). An electron tube containing a hot cathode. *See also:* **tube definitions.** E160/42A70-15E6

hot plate. An appliance fitted with heating elements and arranged to support a flat-bottomed utensil containing the material to be heated. *See also:* **appliances (including portable).** 42A95-0

hot reserve. The thermal reserve generating capacity maintained at a temperature and in a condition to permit it to be placed into service promptly. *See also:* **generating station.** 42A35-31E13

hottest-spot temperature allowance (equipment rating). A conventional value selected to approximate the degrees of temperature by which the limiting insulation temperature rise exceeds the limiting observable temperature rise. *See also:* **limiting insulation temperature.** E1-SCC4

hot-wire instrument. An electrothermic instrument that depends for its operation on the expansion by heat of a wire carrying a current. *See also:* **instrument.** 42A30-0

hot-wire microphone. A microphone that depends for its operation on the change in resistance of a hot wire produced by the cooling or heating effects of a sound wave. *See also:* **microphone.** 42A65-0

hot-wire relay. A relay in which the operating current flows directly through a tension member whose thermal expansion actuates the relay. *See also:* **relay.** 83A16-0

house cable (communication practice). A distribution cable within the confines of a single building or a series of related buildings but excluding cable run from the point of entrance to a cross-connecting box, terminal frame, or point of connection to a block cable. *See also:* **cable.** 42A65-0

house turbine. A turbine installed to provide a source of auxiliary power. *See also:* **generating station.** 42A35-31E13

housing (1) (rotating machinery). Enclosing structure, used to confine the internal flow of air or to protect a machine from dirt and other harmful material. *See:* **cradle base (rotating machinery).** 0-31E8

(2) (body; oil cutout). A part of the fuse support that contains the oil and provides means for mounting the fuse carrier, entrance terminals, and fixed contacts. *Note:* The housing includes the means for mounting the cutout on a supporting structure and openings for attaching accessories such as a vent or an expansion chamber. 37A100-31E11

***H*-plane bend (waveguides) (rectangular uniconductor waveguide operating in the dominant mode).** A bend in which the longitudinal axis of the guide remains in a plane parallel to the plane of the magnetic field vector throughout the bend. E147-3E1

***H* plane, principal (linearly polarized antenna).** The plane containing the magnetic field vector and the direction of maximum radiation. *See also:* **antenna; radiation.** 0-3E1

***H*-plane T junction (shunt T junction) (waveguides) (rectangular uniconductor waveguide).** A T junction of which the magnetic field vector of the dominant wave of each arm is parallel to the plane of the longitudinal axes of the guides. E147-3E1

***H* scan (electronic navigation).** *See:* ***H* display.**

***H* scope (electronic navigation).** *See:* ***H* display.**

hub (rotating machinery). (1) The central part of a rotor used for mounting it on a shaft. (2) A small axial extension of an end shield. *Note:* The hub is commonly, but not always, used in conjunction with a resilient or rigid mounting to attach and support the stator on the base. *See also:* **rotor (rotating machinery).** 0-31E8

hue. The attribute of color perception that determines whether it is red, yellow, green, blue, purple, or the like. *Notes:* (1) This is a subjective term corresponding to the psychophysical term dominant (or complementary) wavelength. (2) White, black, and gray are not considered as being hues. *See also:* **color; light; color terms.** E201-0

hum (interference terminology). A low-pitched droning noise, consisting of several harmonically related frequencies, resulting from an alternating-current power supply, or from ripple from a direct-current power supply, or from induction due to exposure to a power system. *Note:* By extension, the term is applied in visual systems to interference from similar sources. *See also:* **phonograph pickup.** 42A65-31E3

hum and noise (receiver performance) (receiver). The low-pitched composite tone at the receiver output caused by fluctuations in the power supply. *Note:* Noise is any audible undesired signal present at the output. *See also:* **receiver performance.** 0-6E1

hum modulation. Modulation of a radio-frequency or detected signal by hum. E188-0

hum noise level (transmitter performance). The ratio of the total average noise-power attributable to electrical disturbance at the power-supply frequency or harmonics thereof, audio-frequency noise, Johnson noise, shot noise, and thermal noise generated within the transmitter, to the average carrier-power output delivered by the system into an output termination. *See also:* **audio-frequency distortion.** 0-6E1

hump. *See:* **hump yard.**

hump cab signal system (railway practice). A system of signals displayed in the engine-cab and indicating the desired direction and speed of movement of the hump engine. 42A42-0

hump repeater signal. A signal that repeats the hump-signal indication. 42A42-0

hump signal. A signal located at the hump and used to indicate to the hump engineman the desired direction and speed of movement of his train. 42A42-0

hump-signal controller. A unit located at the hump that includes the hump signal lever, the trimmer signal lever, and the signal-repeater lights. 42A42-0

hump-signal emergency lever (hump-signal lever). A lever usually located in one of the control towers and used to operate electric contacts that control the hump-signal circuit to set up a stop indication on the hump signal, the trimmer signal, and the repeater signals when in an emergency the operator desires to stop operation. 42A42-0

hump-signal lever. *See:* **hump-signal emergency lever.**

hump yard. A railroad classification yard in which the classification of cars is accomplished by pushing them over a summit, known as a hump, beyond which they run by gravity into tracks selected by the operator. 42A42-0

hunting (1) (control systems). An undesired oscillation generally of low frequency that persists after external stimuli disappear. *Note:* In linear systems, hunting is evidence of operation at or near the stability limit; nonlinearities may cause hunting of well-defined amplitude and frequency. *See also:* **control system, feedback.** AS1-34E10/23E0

(2) (rotating machinery). A fluctuation of angular velocity about a state of uniform rotation due to electric circuit constants, governor instability, or the like. *See also:* **synchronous machine.** 0-31E8

Hurter and Driffield curve (*H* and *D* curve) (photographic techniques). A characteristic curve of a photographic emulsion that is a plot of density against the logarithm of exposure. *Note:* It is used for the control of photographic processing and for defining the response characteristics to light of photographic emulsions. *See also:* **electroacoustics.** 0-1E1

Huygens source. An elementary radiator consisting of an infinitesimal area of a given electromagnetic wavefront. *Note:* Its radiation properties are equivalent to those of a suitably oriented and excited combination of electric and magnetic dipoles. *See also:* **antenna.** 0-3E1

hybrid balance. A measure of the degree of balance between two impedances connected to two conjugate sides of a hybrid set; it is given by the formula for return loss. *See also:* **transmission characteristics.** 42A65-31E3

hybrid coil (1) (bridge transformer). A single transformer, having effectively three windings, that is designed to be connected to four branches of a circuit so as to render these branches conjugate in pairs. *See also:* **circuits and devices.** E151/42A65-31E3/21E0

(2) (powerline carrier). A hybrid coil used with power-line-carrier equipment to provide electric isolation between two carrier transmitters or transmitters and receivers using common line-tuning equipment. *See also:* **circuits and devices; power-line carrier.** 0-31E3

hybrid control (electric power systems). Utilizing both analog and digital computers in combination to control a process. *See also:* **power system.** 0-31E4

hybrid coupler (3-decibel coupler). A hybrid junction in the form of a directional coupler. The coupling factor is normally 3 decibels. *See:* **hybrid junction; waveguide.** 50I62-3E1

hybrid electromagnetic wave (HEM wave). An electromagnetic wave having components of both the electric and magnetic field vectors in the direction of propagation. *See also:* **waveguide.** E146-3E1

hybrid integrated circuit. An integrated circuit consisting of a combination of two or more integrated-circuit types or one integrated-circuit type and discrete elements. *See also:* **integrated circuit.** E274-15E7/21E0

hybrid junction. Waveguide or transmission line arrangement with four branches that when branches are properly terminated, has the property that energy can be transferred from any one branch into only two of the remaining three. *Note:* In common usage, this energy is equally divided between the two branches. *See:* **waveguide.** E147-3E1

hybrid navigation equipment. Navigation equipment that employs two or more types of sensors. *See also:* **navigation.** 0-10E6

hybrid network. A network with four pairs of terminals or four ports with such characteristics that when two are properly terminated, energy can be transferred from any one port into only two of the remaining three. Usually this energy is equally divided between the two ports. *Notes:* (1) 3-decibel couplers have hybrid properties when two ports are terminated by reflectionless terminations. (2) An advance form contains matching elements such that it is reflectionless for waves propagating into any port of the hybrid when the other three ports are terminated by reflectionless terminations. *See also:* **waveguide.** 0-9E4

hybrid power supply. A combination of disparate elements to form a common circuit. In power supplies, the combination of vacuum tubes and transistors in the regulating circuit. *See also:* **power supply.** KPSH-10E1

hybrid ring (rat race). A hybrid junction consisting of a waveguide or transmission line forming a closed ring into which lead four guides or lines appropriately spaced around the circle. *See also:* **waveguide.** 50I62-3E1

hybrid set. Two or more transformers interconnected to form a network having four pairs of accessible terminals to which may be connected four impedances so that the branches containing them may be made conjugate in pairs when the impedances have the proper values but not otherwise. *See also:* **circuits and devices.** E151-0;42A65-31E3

hybrid system (data processing). A system with the property that at least one of its dependent variables is a function of a continuous variable and at least one of its dependent variables is a function of a discrete variable. *See also:* **electronic digital computer.** 0-23E3

hybrid T (magic T). A waveguide hybrid junction, having four waveguide terminals, that is designed to be connected to four branches of a circuit so as to render these branches conjugate in pairs. *Note:* A common form of hybrid T is a combination of an *E*-plane and an *H*-plane *T* junction in rectangular waveguide. *See also:* **waveguide.** 42A65-0

hybrid traveling-wave multiple-beam klystron (microwave tubes). An *O*-type tube having three or more electron beams, in which the input and output circuits are lateral traveling-wave structures, and the intermediate circuits are cavities uncoupled either laterally or axially. *See also:* **microwave tube or valve.** 0-15E6

hybrid wave (radio wave propagation). An electromagnetic wave in which either the electric or magnetic field vector is linearly polarized normal to the plane of propagation and the other vector is elliptically polarized in this plane. *See:* **transverse-electric hybrid wave; transverse-magnetic hybrid wave.** *See also:* **radio wave propagation.** 0-3E2

hydraulically release-free (hydraulically operated switching device) (trip-free). A term indicating that by hydraulic control the switching device is free to open at any position in the closing stroke if the release is energized. *Note:* This release-free feature is operative even though the closing control switch is held closed. 37A100-31E11

hydraulic driving machine. A machine in which the energy is applied by means of a liquid under pressure in a cylinder equipped with a plunger or piston. *See also:* **driving machine (elevators).** 42A45-0

hydraulic elevator. A power elevator where the energy is applied by means of a liquid under pressure in a cylinder equipped with a plunger or piston. *See also:* **elevators.** 42A45-0

hydraulic operation. Power operation by movement of a liquid under pressure. 37A100-31E11

hydro capability (electric power supply). The capability supplied by hydroelectric sources under specified water conditions. *See also:* **generating station.** 0-31E4

hydrogen-cooled machine (rotating machinery). A machine designed to use hydrogen for cooling. It has a strong sealed gastight enclosure, and heat exchangers to remove heat from hydrogen circulated within the enclosure. *See also:* **cradle base (rotating machinery).** 0-31E8

hydrogen embrittlement. Embrittlement of a metal caused by absorption of hydrogen during a cleaning, pickling, or plating process. 42A60-0

hydro-generator. A generator driven by a hydraulic turbine or water wheel. *See:* **asynchronous machine; direct-current commutating machine; synchronous machine.** 0-31E8

hydrogen overvoltage (electrolytic cell). Overvoltage associated with the liberation of hydrogen gas. CM-34E2

hydrometallurgy. The extraction of metals from their ores by the use of aqueous solutions. *See also:* **electrowinning.** 42A60-0

hydrophone. An electroacoustic transducer that responds to water-borne sound waves and delivers essentially equivalent electric waves. *Note:* There are many types of hydrophones whose definitions are analogous to those of corresponding microphones, for example, crystal hydrophone, magnetic hydrophone, pressure hydrophone, etcetera. *See also:* **microphone.** 42A65-0

hypersonic* (sonic frequency). The preferred term is pretersonic to avoid confusion with hypersonic speed.
*Deprecated 0-20E0

hysteresigraph. A device for experimentally presenting or recording the hysteris loop for a magnetic specimen. *See also:* **magnetometer.** 42A30-0

hysteresis (1) (control system). The property of an element evidenced by the dependence of the value of the output, for a given excursion of the input, upon the history of prior excursions and the direction of the current traverse. *Note:* Some reversal of output may be expected for any small reversal of input; this distinguishes the hysteretic effect from a dead band. The property may be either useful, as in magnetic amplifiers and computer storage elements, or undesirable, as when it contributes to a dead band in elements expected to show a unique transfer function. Originally used to describe magnetic phenomena, the term has been extended to inelastic behavior. *See also:* **control system, feedback.** 85A1-23E0;AS1-34E10

(2) (oscillator). A behavior that may appear in an oscillator wherein multiple values of the output power and/or frequency correspond to given (usually cyclic) values of an operating parameter. *See:* **oscillatory circuit; tuning hysteresis.** E160-15E6

(3) (radiation-counter tube). The temporary change in the counting rate versus voltage characteristic caused by previous operation. *See also:* **gas-filled radiation-counter tubes; gas tubes.** 42A70-15E6

(4) (measurement or control). The difference between the increasing input value and the decreasing input value that effect the same output value. *Note:* This term applies only where the output value is a continuous function of the input value. *See also:* **power systems, low-frequency and surge testing.** E94-0

hysteresis coupling. An electric coupling in which torque is transmitted by forces arising from the resistance to reorientation of established magnetic fields within a ferromagnetic material, usually of high coercitivity. *See:* **electric coupling.** 42A10-31E8

hysteresis error (industrial control). *See:* **hysteretic error.**

hysteresis heater. An induction device in which a charge or a muffle about the charge is heated principally by hysteresis losses due to a magnetic flux that is produced in it. *Note:* A distinction should be made between hystersis heating and the enhanced induction heating in a magnetic charge. *See also:* **induction heating.** E54/E169-0

hysteresis loop. For a magnetic material in a cyclically magnetized condition, a curve (usually with rectangular coordinates) showing, for each value of the magnetizing force, two values of the magnetic flux density—one when the magnetizing force is increasing, the other when it is decreasing. *Note:* The cyclic magnetic force is not necessarily periodic. *See also:* **static magnetic storage.** E163/E270-0

hysteresis motor. A synchronous motor without salient poles and without direct-current excitation, that starts by virtue of the hysteresis losses induced in its hardened-steel secondary member of the revolving field of the primary and operates normally at synchronous speed due to the retentivity of the secondary core. *See:* **synchronous machine.** 42A10-31E8

hysteretic error (hysteresis error) (industrial control). The maximum separation due to hysteresis between up-scale-going and down-scale-going indications of the measured variable (during a full-range traverse, unless otherwise specified) after transients have decayed. *Note:* Use of the term to define half the separation is deprecated. *See:* **control system, feedback.** AS1-34E10

I

IAGC. *See:* **instantaneous automatic gain control.**

ice-detection light. An inspection light designed to illuminate the leading edge of the wing to check for ice formation. *See also:* **signal lighting.** Z7A1-0

ICI* (color television). *See:* **CIE.**

*Deprecated

iconoscope. A camera tube in which a beam of high-velocity electrons scans a photoemissive mosaic that is capable of storing an electric charge pattern. *See also:* **television; tube definitions.** E160-2E2/15E6;42A70-0

ICW *See:* **interrupted continuous wave.**

ideal capacitor (nonlinear capacitor). A capacitor whose transferred charge characteristic is single-valued. *See also:* **nonlinear capacitor.** E226-15E7

ideal dielectric. *See:* **dielectric, perfect.**

ideal noise diode. *See:* **noise diode, ideal.**

ideal paralleling (rotating machinery). Paralleling by adjusting the voltage, and frequency and phase angle for alternating-current machines, such that the conditions of the incoming machine are identical with those of the system with which it is being paralleled. *See also:* **asynchronous machine; synchronous machine.** 0-31E8

ideal transducer. *See:* **transducer, ideal.**

ideal transformer. *See:* **transformer, ideal.**

ideal value (1) (control systems; general) (control) (industrial control). The value of a selected variable that would result from a perfect system operating from the same command as the actual system under consideration. *See:* **control system, feedback.** AS1-34E10

(2) (synchronous-machine regulating system). The value of a controlled variable (for example, generator terminal voltage) that results from a desired or agreed-upon relationship between it and the commands (commands such as voltage regulator setting, limits, and reactive compensators). *See also:* **synchronous machine.** 0-31E8

identification (radar). The knowledge that a particular blip on the display is a specific vehicle; blips are identified by size, shape, timing, position, maneuvers, and by means of coded responses through secondary radar. *See also:* **navigation.** 0-10E6

identification beacon. An aeronautical beacon emitting a coded signal by means of which a particular point of reference can be identified. *See also:* **signal lighting.** Z7A1-0

identifier. A symbol whose purpose is to identify, indicate, or name a body of data. *See also:* **electronic digital computer.** X3A12-16E9

identity friend or foe (IFF). Equipment used for transmitting radio signals between two stations located on ships, aircraft, or ground, for automatic identification. *Notes:* (1) The usual basic parts of equipment are interrogators, transpondors, and responsors. (2) Usually the initial letters of the name (IFF) are used instead of the full name. *See also:* **radio transmission.** 42A65-0

***I* display (radar).** A display in which a target appears as a complete circle when the radar antenna is pointed at it and in which the radius of the circle is proportional to target distance; incorrect aiming of the antenna changes the circle to a segment whose length is inversely proportional to the magnitude of the pointing error and the position of the segment indicates the reciprocal of the direction of pointing error. *See also:* **navigation.** 0-10E6

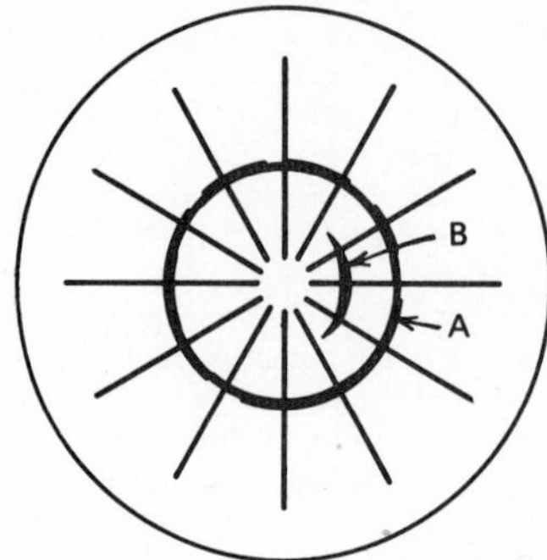

I display. Two targets A and B at different distances. Radar aimed at target A.

idle bar (rotating machinery). An open circuited conductor bar in the rotor of a squirrel-cage motor, used to give low starting current in a moderate torque motor. *See also:* **rotor (rotating machinery).** 0-31E8

idle period (gas tube). That part of an alternating-voltage cycle during which a certain arc path is not carrying current. *See also:* **gas-filled rectifier.** 50I07-15E6

idler circuit (parametric device). A portion of a parametric device that chiefly determines the behavior of the device at an idler frequency. *See also:* **parametric device.** E254-15E7

idler frequency (parametric device). A sum frequency (or difference frequency) generated within the parametric device other than the input, output, or pump frequencies that requires specific circuit consideration to achieve the desired device performance. *See also:* **parametric device.** E254-15E7

idle time (electric drive). The portion of the available time during which a system is believed to be in good operating condition but is not in productive use. *See:* **electric drive.** AS1-34E10

IF. *See:* **intermediate frequency.**

IFF. *See:* **identity friend or foe.**

ignition control (industrial control). Control of the starting instant of current flow in the anode circuit of a gas tube. *See also:* **electronic controller.** 42A25-34E10

ignition switch (industrial control). A manual or automatic switch for closing or interrupting the electric ignition circuit of an internal-combustion engine at the option of the machine operator, or by an automatic function calling for unattended operation of the engine. *Note:* Provisions for checking individual circuits of the ignition system for relative performance may be incorporated in such switches. *See also:* **switch.** 42A25-34E10

ignitor. A stationary electrode that is partly immersed in the cathode pool and has the function of initiating a cathode spot. *See also:* **electrode (electron tube); electronic controller.** 42A70-15E6

ignitor-current temperature drift (microwave gas tubes). The variation in ignitor electrode current caused by a change in ambient temperature of the tube. *See also:* **gas tube.** E160-15E6

ignitor discharge (microwave switching tubes). A direct-current glow discharge, between the ignitor electrode and a suitably located electrode, used to facilitate radio-frequency ionization. *See also:* **gas tube.** E160-15E6

ignitor electrode (microwave switching tubes). An electrode used to initiate and sustain the ignitor discharge. *See also:* **gas tube.** E160-15E6

ignitor firing time (microwave switching tubes). The time interval between the application of a direct voltage to the ignitor electrode and the establishment of the ignitor discharge. *See also:* **gas tube.** E160-15E6

ignitor interaction (microwave gas tubes). The difference between the insertion loss measured at a specified ignitor current and that measured at zero ignitor current. *See also:* **gas tube.** E160-15E6

ignitor leakage resistance (microwave switching tubes). The insulation resistance, measured in the absence of an ignitor discharge, between the ignitor electrode terminal and the adjacent radio-frequency electrode. *See also:* **gas tube.** E160-15E6

ignitor oscillations (microwave gas tubes). Relaxation oscillations in the ignitor circuit. *Note:* If present, these oscillations may limit the characteristics of the tube. *See also:* **gas tube.** E160-15E6

ignitor voltage drop (microwave switching tubes). The direct voltage between the cathode and the anode of the ignitor discharge at a specified ignitor current. *See also:* **gas tube.** E160-15E6

ignitron. A single-anode pool tube in which an ignitor is employed to initiate the cathode spot before each conducting period. *See also:* **electronic controller; tube definitions.** 42A70-15E6

ignored conductor. *See:* **isolated conductor.**

illuminance. *See:* **illumination.**

illumination (1) (general). The density of the luminous flux incident on a surface, it is the quotient of the luminous flux by the area of the surface when the latter is uniformly illuminated. *Note:* The term **illumination** also is commonly used in a qualitative or general sense to designate the act of illuminating or the state of being illuminated. Usually the context will indicate which meaning is intended, but occasionally it is desirable to use the expression level of illumination to indicate that the quantitative meaning is intended. The term **illuminance**, which sometimes is used in place of **illumination**, is subject to confusion with luminance and illuminants, especially when not clearly pronounced. *See also:* **color terms; light.** Z7A1-0;E201-0

(2) (at a point of a surface). The quotient of the luminous flux incident on an infinitesimal element of surface containing the point under consideration by the area of that element. *See also:* **light.** 50I45-2E2

illumination (footcandle) meter. An instrument for measuring the illumination on a surface. *Note:* Most such instruments consist of barrier-layer cells connected to a meter calibrated in footcandles. *See also:* **photometry.** Z7A1-0

illustrative diagram (industrial control). A diagram whose principal purpose is to show the operating principle of a device or group of devices without necessarily showing actual connections or circuits. Illustrative diagrams may use pictures or symbols to illustrate or represent devices or their elements. Illustrative diagrams may be made of electric, hydraulic, pneumatic, and combination systems. They are applicable chiefly to instruction books, descriptive folders, or other media whose purpose is to explain or instruct. *See also:* **control.** E270-34E10

ILS. *See:* **instrument landing system.**

image (1) (optoelectronic device). A spatial distribution of a physical property, such as radiation, electric charge, conductivity, or reflectivity, mapped from another distribution of either the same or another physical property. *Note:* The mapping process may be carried out by a flux of photons, electric charges, or other means. *See also:* **optoelectronic device.**
(2) (computing systems). *See:* **card image.** *See also:* **electronic digital computer.** E222-15E7

image antenna. The imaginary counterpart of an actual antenna, assumed for mathematical purposes to be located below the surface of the ground and symmetrical with the actual antenna above ground. *See also:* **antenna.** 42A65-3E1

image burn. *See:* **retained image.**

image camera tube. *See:* **image tube.**

image converter (solid state). An optoelectronic device capable of changing the spectral characteristics of a radiant image. *Note:* Examples of such changes are infrared to visible and X-ray to visible. *See also:* **optoelectronic device.** E222-15E7

image-converter panel. A thin, usually flat, multicell image converter. *See also:* **optoelectronic device.** E222-15E7

image-converter tube (camera tubes). An image tube in which an infrared or ultraviolet image input is converted to a visible image output. *See also:* **camera tube.** 0-15E6

image dissector (optical character recognition) (computing systems). A mechanical or electronic transducer that sequentially detects the level of light in different areas of a completely illuminated sample space. *See also:* **electronic digital computer.** X3A12-16E9

image dissector tube (dissector tube). A camera tube in which an electron image produced by a photoemitting surface is focused in the plane of a defining aperture and is scanned past that aperture. *See also:* **television; tube definitions.** E160-2E2/15E6

image element (optoelectronic device). The smallest portion of an image having a specified correlation with the corresponding portion of the original. *Note:* In some imaging systems the size of the image elements is determined by the structure of the image space, in others by the carrier employed for the mapping process. *See also:* **optoelectronic device.** E222-15E7

image frequency. *See:* **frequency, image.**

image iconoscope. A camera tube in which an electron image is produced by a photoemitting surface and focused on one side of a separate storage target that is scanned on the same side by an electron beam, usually of high-velocity electrons. *See also:* **television; tube definition.** E160-2E2/15E6

image impedances (transducer). The impedances that will simultaneously terminate all of its inputs and outputs in such a way that at each of its inputs and outputs the impedances in both directions are equal. *Note:* The image impedances of a four-terminal transducer are in general not equal to each other, but for any symmetrical transducer the image impedances are equal and are the same as the iterative impedances. *See also:* **self-impedance; transducer.** E151-9E4-E270-42A65

image intensifier (solid state). An optoelectronic amplifier capable of increasing the intensity of a radiant image. *See also:* **optoelectronic device.** E222-15E7

image-intensifier panel. A thin, usually flat, multicell image intensifier. *See also:* **optoelectronic device.** E222-15E7

image-intensifier tube. An image tube in which the output radiance is (1) in approximately the same spectral region as, and (2) substantially greater than, the photocathode irradiance. *See also:* **camera tube; image tube.** 0-15E6

image orthicon. A camera tube in which an electron image is produced by a photoemitting surface and focused on a separate storage target, which is scanned on its opposite side by a low-velocity electron beam. *See also:* **tube definitions.** 42A70-15E6

image phase constant. The imaginary part of the image transfer constant. *See also:* **transducer; transfer constant.** E270-0

image ratio (heterodyne receiver). The ratio of (1) the field strength at the image frequency to (2) the field strength at the desired frequency, each field being applied in turn, under specified conditions, to produce equal outputs. *See also:* **radio receiver.** E188-0

image response. Response of a superheterodyne receiver to the image frequency, as compared to the response to the desired frequency. *See also:* **radio receiver.** 42A65-0

image-storage device. An optoelectronic device capable of retaining an image for a selected length of time. *See also:* **optoelectronic device.** E222-15E7

image-storage panel (optoelectronic device). A thin, usually flat, multicell image-storage device. *See also:* **optoelectronic device.** E222-15E7

image-storage tube. A storage tube into which the information is introduced by means of radiation, usually light, and read at a later time as a visible output. *See also:* **storage tube.** E158-15E6

image transfer constant (electric transducer) (transfer constant). The arithmetic mean of the natural logarithm of the ratio of input to output phasor voltages and the natural logarithm of the ratio of the input to output phasor currents when the transducer is terminated in its image impedances. *Note:* For a symmetrical transducer the transfer constant is the same as the propagation constant. *See also:* **transducer.** E270-0

image tube. An electron tube that reproduces on its fluorescent screen an image of an irradiation pattern incident on its photosensitive surface. *See also:* **camera tube.** 0-15E6

imbedded temperature-detector insulation (rotating machinery). The insulation surrounding a temperature detector, taking the place of a coil separator in its area. *See:* **cradle base (rotating machinery).** 0-31E8

immediate address (computing systems). Pertaining to an instruction in which an address part contains the value of an operand rather than its address. *See:* **zero-level address.** *See also:* **electronic digital computer.** X3A12-16E9

immersed gun (microwave tubes). A gun in which essentially all the flux of the confining magnetic field passes perpendicularly through the emitting surface of the cathode. *See also:* **microwave tube.** 0-15E6

immersion plating (dip plating). The deposition, without application of an external electromotive force, of a thin metal coating upon a less noble metal by immersing the latter in a solution containing a compound of the metal to be deposited. *See also:* **electroplating.** 42A60-0

immittance (linear passive networks). A response function for which one variable is a voltage and the other a current. *Note:* Immittance is a general term for both impedance and admittance, used where the distinction is irrelevant. *See also:* **linear passive networks; transmission characteristic.** E156-0

immittance comparator. An instrument for comparing the impedance or admittance of the two circuits, components, etcetera. *See also:* **auxiliary device to an instrument.** 0-9E4

immunity to interference (electromagnetic compatibility). The property of a receiver enabling it to reject a radio disturbance. *See also:* **electromagnetic compatibility.** 0-27E1

impedance (1) (linear constant-parameter system). (A) The corresponding impedance function with p replaced by $j\omega$ in which ω is real. (B) The ratio of the phasor equivalent of a steady-state sine-wave voltage or voltagelike quantity (driving force) to the phasor equivalent of a steady-state sine-wave current or currentlike quantity (response). *Note:* Definitions (A) and (B) are equivalent. The *real* part of impedance is the resistance. The *imaginary* part is the susceptance. (C) A physical device or combination of devices whose impedance as defined in (A) or (B) can be determined. *Note:* This sentence illustrates the double use of the word impedance, namely for a physical characteristic of a device or system (definitions (A) and (B) and for a device (definition (C)). In the latter case the word impedor may be used to reduce confusion. Definition (C) is a second use of **impedance** and is independent of definitions (A) and (B). *Editor's Note:* The ratio Z is commonly expressed in terms of its orthogonal components, thus:

$$Z = R+jX$$

where Z, R, and X are respectively termed the impedance, resistance, and reactance, all being measured in ohms. In a simple circuit consisting of R, L, and C all in series, Z becomes

$$Z = R+j(\omega L - 1/\omega C),$$

where $\omega = 2\pi f$ and f is the frequency. *See:* **reactance; resistance.** *See also:* **network analysis; input impedance; feedback impedance; impedance function.** E270-0

(2) (electric machine). Linear operator expressing the relation between voltage (increments) and current (increments). Its inverse is called the admittance of an electric machine. *Notes:* (1) If a matrix has as its elements impedances, it is usually referred to as impedance matrix. Frequently the impedance matrix is called impedance for short. (2) Usually such impedances are defined with the mechanical angular velocity of the machine at steady state. *See:* **asynchronous machine; synchronous machine.** 0-31E8

(3) (two-conductor transmission line). The ratio of the complex voltage between the conductors to the complex current on one conductor in the same transverse plane. 0-3E1

(4) (circular or rectangular waveguide). A nonuniquely defined complex ratio of the voltage and current at a given transverse plane in the waveguide, which depends on the choice of representation of the characteristic impedance. *See:* **characteristic impedance; waveguide.** 0-3E1

impedance bond (railway practice). An iron-core coil of low resistance and relatively high reactance used on electrified railroads to provide a continuous direct-current path for the return propulsion current around insulated joints and to confine the alternating-current signaling energy to its own track circuit. *See also:* **railway signal and interlocking.** 42A42-0

impedance, characteristic wave. *See:* **wave impedance, characteristic.**

impedance compensator. An electric network designed to be associated with another network or a line with the purpose of giving the impedance of the combination a desired characteristic with frequency over a desired frequency range. *See also:* **network analysis.** E270/42A65-0

impedance, conjugate. An impedance the value of which is the complex conjugate of a given impedance. *Note:* For an impedance associated with an electric network, the complex conjugate is an impedance with the same resistance component and a reactance component the negative of the original. E151-0;E270-9E4

impedance drop (transformer or regulator). The phasor sum of the resistance drop and the reactance drop. *Note:* For transformers and regulators, the resistance drop, the reactance drop, and the impedance drop are, respectively, the sum of the primary and secondary drops reduced to the same terms. They are usually expressed in per-unit, or in percent, of the secondary terminal voltage of the transformer or the rated voltage of the regulator. *See:* **efficiency; impedance voltage of a regulator.** 42A15-31E12;57A15-0

impedance, effective input (output) (electron valve or tube). The quotient of the sinusoidal component of the control-electrode voltage (output-electrode voltage) by the corresponding component of the current for the given electrical conditions of all the other electrodes. *See also:* **ON period.** 50I07-15E6

impedance feedback (analog computer). A passive network connected between the output terminal of an operational amplifier and its summing junction. *See also:* **electronic analog computer.** E165-16E9

impedance function (defined for linear constant-parameter systems or parts of such systems). That mathematical function of p that is the ratio of a voltage or voltage-like quantity (driving force) to the corresponding current or current-like quantity (response) in the hypothetical case in which the former is e^{pt} (e is the natural log base, p is arbitrary but independent of t, t is an independent variable that physically is usually time) and the latter is a steady-state response of the form $e^{pt}/Z(p)$. *Note:* In electric circuits **voltage** is always the driving force and **current** is the response even though as in nodal analysis the current may be the independent variable; in electromagnetic radiation **electric field strength** is always considered the driving force and **magnetic field strength** the response, and in mechanical systems **mechanical force** is always considered as a driving force and **velocity** as a response. In a general sense the dimension (and unit) of **impedance** in a given application may be whatever results from the ratio of the dimensions of the quantity chosen as the driving force to the dimensions of the quantity chosen as the response. However, in the types of systems cited above any deviation from the usual convention should be noted. *See also:* **network analysis.** E270-0

impedance grounded (electric power). Means grounded through impedance. *Note:* The component of the impedance need not be at the same location. *See also:* **ground; grounded; grounded systems.** E32-0;42A15-31E12;42A35-31E13

impedance, image. *See:* **image impedances.**

impedance, input. *See:* **input impedance.**

impedance, iterative (transducer or a 2-port network). The impedance that, when connected to one pair of terminals, produces a like impedance at the other pair of terminals. *Notes:* (1) It follows that the iterative impedance of a transducer or a network is the same as the impedance measured at the input terminals when an infinite number of identically similar units are formed into an iterative or recurrent structure by connecting the output terminals of the first unit to the input terminals of the second, the output terminals of the second to the input terminals of the third, etcetera. (2) The iterative impedances of a four-terminal transducer or network are, in general, not equal to each other but for any symmetrical unit the iterative impedances are equal and are the same as the image impedances. The iterative impedance of a uniform line is the same as its characteristic impedance. *See also:* **transmission characteristics.** E151-9E4

impedance kilovolt-amperes (regulator). The kilovolt-amperes (kVA) measured in the shunt winding with the series winding short-circuited and with sufficient voltage applied to the shunt winding to cause rated current to flow in the windings. *See also:* **voltage regulator.** 57A15-0

impedance, load. The impedance presented by the load to a source or network. *See also:* **transmission characteristics.** 0-9E4

impedance matching. The connection across a source impedance of a matching impedance that allows optimum undistorted energy transfer. Types of matching are: (1) An impedance having the same magnitude and phase angle as the source impedance. (2) An impedance having the same magnitude and having a phase angle with the same magnitude but opposite signs as the source impedance. *See also:* **self-impedance.** 0-31E3

impedance matrix (multiport network). A matrix operator that interrelates the voltages at the various ports to the currents at the same and other ports. *See also:* **transmission characteristics.** 0-9E4

impedance, normalized. The ratio of an impedance to a specified reference impedance. *Note:* For a transmission line or a waveguide, the reference impedance is usually a characteristic impedance. *See also:* **transmission characteristics.** 0-9E4

impedance, output (device, transducer, or network). The impedance presented by the output terminals to a load. *Notes:* (1) Output impedance is sometimes incorrectly used to designate load impedance. (2) This is a frequency-dependent function, and is used to help describe the performance of the power supply and the degree of coupling between loads. *See also:* **electrical conversion; self-impedance.** 0-10E1

(2) (electron device). The output electrode impedance at the output electrodes. *See also:* **self-impedance.** 0-15E6

(3) (power supplies). The effective dynamic output impedance of a power supply is derived from the ratio of the measured peak-to-peak change in output voltage to a measured peak-to-peak change in load alternating current. Output impedance is usually specified throughout the frequency range from direct current to 100 kilohertz. *See also:* **power supply; self-impedance.** KPSH-10E1

impedance ratio (divider). The ratio of the impedance of the two arms connected in series to the impedance of the low-voltage arm. *Note:* In determining the ratio, account should be taken of the impedance of the measuring cable and the instrument. The impedance ratio is usually given for the frequency range within which it is approximately independent of frequency. For resistive dividers the impedance ratio is generally derived from a direct-current measurement such as by means of a Wheatstone bridge. 68A1-31E5

impedance relay. A distance relay in which the threshold value of operation depends only on the magnitude of the impedance and is substantially independent of the phase angle of the impedance. 37A100-31E11/31E6

impedance, source. The impedance presented by a source of energy to the input terminals of a device, or network. *See also:* **impedance, input; network analysis; self impedance.** E151/42A65-9E4

impedance voltage (1) (transformer). The voltage required to circulate rated current through one of two specified windings of a transformer when the other winding is short-circuited, with the windings connected as for rated voltage operation. *Note:* It is usually expressed in per unit, or percent, of the rated voltage of the winding in which the voltage is measured. 42A15-31E12;0-21E1

(2) (constant-current transformer). The measured primary voltage required to circulate rated secondary current through the short-circuited secondary coil for a particular coil separation. *Note:* It is usually expressed in per unit, or percent, of the rated primary voltage. *See:* **constant-current transformer.** 57A14-0;42A15-31E12

(3) (current-limiting reactor). The product of its rated ohms impedance and rated current. *See also:* **reactor.** 57A16-0

(4) (regulator). The voltage required to circulate rated current through one winding of the regulator when another winding is short circuited, with the respective windings connected as for rated voltage operation. *Note:* It is usually referred to the series winding, and then expressed in per unit, or percent, of the rated voltage of the regulator. *See also:* **voltage regulator.** 57A15-0

impedance, wave. *See:* **wave impedance.**

impedor. A device, the purpose of which is to introduce impedance into an electric circuit. *See also:* **network analysis.** E270-0

imperfect dielectric. *See:* **dielectric, imperfect.**

imperfection (crystalline solid). Any deviation in structure from that of an ideal crystal. *Note:* An ideal crystal is perfectly periodic in structure and contains no foreign atoms. *See also:* **miscellaneous electron devices; semiconductor.** E102/E270-0;E216-34E17

impingement attack (corrosion). Localized erosion-corrosion resulting from turbulent or impinging flow of liquids. *See also:* **corrosion terms.** CM-34E2

implicit computation. Computation using a self-nulling principle in which, for example, the variable sought first is assumed to exist, after which a synthetic variable is produced according to an equation and compared with a corresponding known variable and the difference between the synthetic and the known variable driven to zero by correcting the assumed variable. *Note:* Although the term applies to most analog circuits, even to a single operational amplifier, it is restricted usually to computation performed by (1) circuits in which a function is generated at the output of a single high-gain direct-current amplifier by inserting an element generating the inverse function in the feedback path, (2) circuits in which combinations of computing elements are interconnected in closed loops to satisfy implicit equations, or (3) circuits in which linear or nonlinear differential equations yield the solutions to a system of algebraic or transcendental equations in the steady state. *See also:* **electronic analog computer.** E165-16E9

impregnant (rotating machinery). A solid, liquid, or semiliquid material that, under conditions of application, is sufficiently fluid to penetrate and completely or partially fill or coat interstices and elements of porous or semiporous substances or composites. *See also:* **cradle base (rotating machinery).** 0-31E8

impregnate (rotating machinery). The act of adding impregnant (bond or binder material) to insulation or a winding. *Note:* The impregnant, if thermosetting, is usually cured in the process. *See also:* **cradle base (rotating machinery) vacuum-pressure impregnation.** 0-31E8

impregnated (fibrous insulation). A suitable substance replaces the air between the fibers, even though this substance does not fill completely the spaces between the insulated conductors. *Note:* To be considered suitable, the impregnating substance must have good insulating properties and must cover the fibers and render them adherent to each other and to the conductor. 42A95-0

impregnated tape (dispersed magnetic powder tape). *See:* **magnetic powder-impregnated tape.**

impregnation, winding (rotating machinery). The process of applying an insulating varnish to a winding and, when required, baking to cure the varnish. *See also:* **cradle base (rotating machinery).** 0-31E8

improvement threshold (angle-modulation systems). The condition of unity for the ratio of peak carrier voltage to peak noise voltage after selection and before any nonlinear process such as amplitude limiting and detection. *See also:* **amplitude modulation.** E145-0

impulse (1) (general). A waveform that, from the observer's frame of reference, approximates a Dirac delta function. 0-9E4

(2) (power systems). A surge of unidirectional polarity. *See also:* **power systems, low-frequency and surge testing.** 42A35-31E13

(3) (mathematics). A pulse that begins and ends within a time so short that it may be regarded mathematically as infinitesimal. *See also:* **pulse.** E270-0

(4) (lightning arresters). A unidirectional wave of voltage or current that, without appreciable oscillations rises rapidly to a maximum value and falls rapidly to zero. *Note:* The parameters that define an impulse voltage or current wave are polarity, peak value, wavefront, and wavetail. *See also:* **lightning arrester (surge diverter).** 99I1-31E7

impulse bandwidth (narrow-band filter). The peak value of the response envelope divided by the spectrum amplitude of an applied impulse. *See also:* **electromagnetic compatibility.** 0-27E1

impulse current (current testing). Ideally, an aperiodic transient current that rises rapidly to a maximum value and falls usually less rapidly to zero. A rectangular impulse current rises rapidly to a maximum value, remains substantially constant for a specified time and then falls rapidly to zero. *See also:* **test voltage and current.** 68A1-31E5

impulse (shock) excitation. A method of producing oscillator current in a circuit in which the duration of the impressed voltage is relatively short compared with the duration of the current produced. *See also:* **oscillatory circuit.** E145/E182-0

impulse flashover voltage (1) (insulator). The crest value of the impulse wave that, under specified conditions, causes flashover through the surrounding medium. 29A1-0

(2) (lightning arresters). The crest voltage of an impulse causing a complete disruptive discharge through the air between electrodes of a test specimen. *See also:* **critical impulse flashover voltage; insulator; lightning arrester (surge diverter); power system, low-frequency and surge testing.** 42A35-31E13/31E7

impulse flashover volt-time characteristic. A curve plotted between flashover voltage for an impulse and time to impulse flashover, or time lag of impulse flashover. *See also:* **insulator; power system, low-frequency and surge testing.** 42A35-31E13

impulse generator (electromagnetic compatibility). A standard reference source of broadband impulse energy. E263-27E1

impulse inertia (lightning arresters). The property of insulation whereby more voltage must be applied to produce disruptive discharge, the shorter the time of voltage application. *See also:* **lightning arrester (surge diverter); power system, low-frequency and surge testing.** 42A35-31E13/31E7

impulse insulation level. An insulation strength expressed in terms of the crest value of an impulse withstand voltage.*See also:* **basic impulse insulation level (BIL) (insulation strength).** 92A1-0

impulse margin (time margin) (operation of a relay). The difference in time between the duration of a given input and critical impulse time, with the same input values. See the following figure.*See also:* **relay.** 37A1-31E6

impulse noise. Noise characterized by transient disturbances separated in time by quiescent intervals. *Notes:* (1) The frequency spectrum of these disturbances must be substantially uniform over the useful pass band of the transmission system. (2) The same source may produce impulse noise in one system and random noise in a different system. *See also:* **modulating system; signal-to-noise ratio.** E145/42A65-0

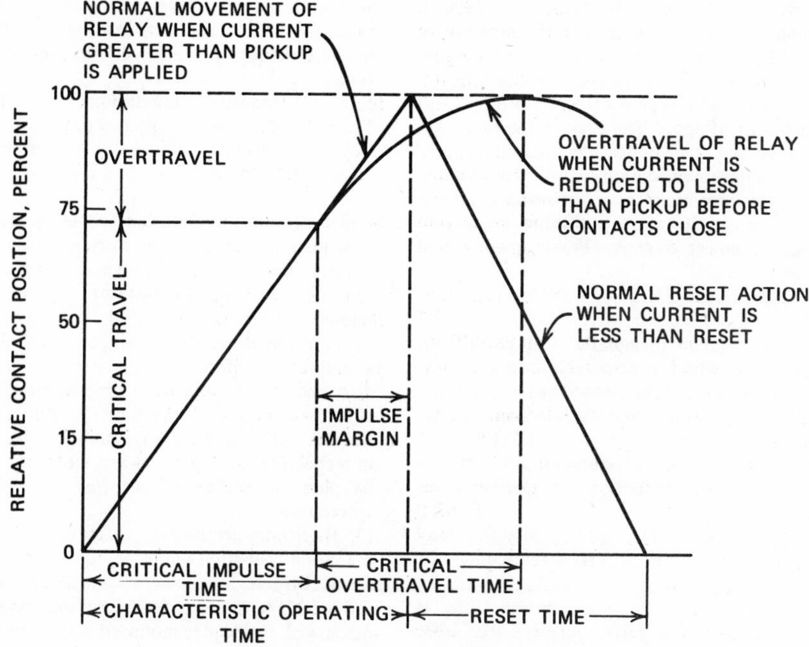

Relationship of relay operating times.

impulse-noise selectivity (receiver) (receiver performance). A measure of the ability to discriminate against impulse noise. *See also:* **receiver performance.** 0-6E1

impulse protection level (lightning arresters). The highest voltage (crest value) that appears at the terminals of overvoltage protective apparatus under specified impulse-test conditions. *See also:* **lightning arrester (surge diverter); basic impulse insulation level (insulation strength).** 50I25-31E7;92A1-0

impulse ratio (lightning arresters). The ratio of the flashover, sparkover, or breakdown voltage of an impulse to the crest value of the power-frequency, sparkover, or breakdown voltage. *See also:* **lightning arrester (surge diverter); power system, low-frequency and surge testing.** 42A35-31E13/31E7

impulse relay. (1) A relay that follows and repeats current pulses, as from a telephone dial. (2) A relay that operates on stored energy of a short pulse after the pulse ends. (3) A relay that discriminates between length and strength of pulses, operating on long or strong pulses and not operating on short or weak ones. (4) A relay that alternately assumes one of two positions as pulsed. (5) Erroneously used to describe an integrating relay. *See also:* **relay.** 0-21E0

impulse sparkover voltage (arrester). The highest value of voltage attained during an impulse of given wave shape and polarity applied between the line and earth terminals of an arrester prior to the flow of discharge current. *See:* **current rating, 60-hertz (arrester); lightning arrester (surge diverter).** E28-42A20-0;99I2-31E7

impulse sparkover voltage-time curve (arrester). A curve that relates the impulse sparkover voltage to the time to sparkover. *See also:* **lightning arrester (surge diverter).** 0-31E7

impulse test (1) (rotating machinery). A test for applying to a component an aperiodic transient voltage having predetermined polarity, amplitude, and waveform. *See also:* **asynchronous machine; direct-current commutating machine; power system, low-frequency and surge testing; synchronous machine.** E32-31E8 **(2) (insulation strength).** Dielectric tests consisting of the application of a high-frequency steep-wave-front voltage between windings and between windings and ground. *See also:* **routine test.** 89A1-0 **(3) (lightning arresters).** An insulation test in which the voltage applied is an impulse voltage of specified wave shape. *See also:* **lightning arrester (surge diverter).** 42A35-31E13/31E7

impulse transmission. That form of signaling, used principally to reduce the effects of low-frequency interference, that employs impulses of either or both polarities for transmission to indicate the occurrence of transitions in the signals. *Note:* The impulses are generally formed by suppressing the low-frequency components, including direct current, of the signals. *See also:* **telegraphy.** 42A65-0

impulse transmitting relay. A relay that closes a set of contacts briefly while going from the energized to the de-energized position or vice versa. *See also:* **relay.** 83A16-0

impulse-type telemeter. A telemeter that employs characteristics of intermittent electric signals, other than their frequency, as the translating means. *See also:* **telemetering.** 42A30-0

impulse voltage (current) (lightning arresters). Synonymous with voltage of an impulse wave (current of an impulse wave). *See:* **lightning arrester (surge diverter).** 50I25-31E7

impulse wave (1) (general). A unidirectional surge generated by the release of electric energy into an

impedance network. *See also:* **insulator.** 29A1-0
(2) (lightning arresters). A unidirectional wave of current or voltage of very short duration containing no appreciable oscillatory components. *See also:* **insulator; lightning arrester (surge diverter).** 50I25-31E7

impulse withstand voltage. The crest value of an applied impulse voltage that, under specified conditions, does not cause a flashover, puncture, or disruptive discharge on the test specimen. *See also:* **insulator; lightning arrester (surge diverter); lightning protection and equipment; power system, low-frequency and surge testing.**
E28/29A1-0;37A100-31E11;42A35-31E13;50I25-31E7

impulsive noise (electromagnetic compatibility). Noise, the effect of which is resolvable into a succession of discrete impulses in the normal operation of the particular system concerned. *See also:* **electromagnetic compatibility.** CISPR-27E1

impurity (1) (acceptor) (semiconductor). An impurity that may induce hole conduction. *See also:* **semiconductor device.** E102-0
(2) (donor) (semiconductor). An impurity that may induce electron conduction. *See also:* **semiconductor device.** E102-0
(3) (crystalline solid). An imperfection that is chemically foreign to the perfect crystal. *See also:* **semiconductor.** E216-34E17;E270-0
(4) (chemical) (semiconductor). An atom within a crystal, that is foreign to the crystal. *See also:* **electron devices, miscellaneous.** E102-0

impurity, stoichiometric. A crystalline imperfection arising from a deviation from stoichiometric composition. E102-0;E216-34E17;E270-0

inadvertent interchange (electric power systems) (control area). The time integral of the net interchange minus the time integral of the scheduled net interchange. *Note:* This includes the intentional interchange energy resulting from the use of frequency and/or other bias as well as the unscheduled interchange energy resulting from human or equipment errors. *See also:* **power system, low-frequency and surge testing.** E94-31E4

incandescence. The self-emission of radiant energy in the visible spectrum due to the thermal excitation of atoms or molecules. *See also:* **lamp.** Z7A1-0

incandescent-filament lamp. A lamp in which light is produced by a filament heated to incandescence by an electric current. *See also:* **lamp.** Z7A1-0

incandescent-filament-lamp transformer (series type). A transformer that receives power from a current-regulated series circuit and that transforms the power to another circuit at the same or different current from that in the primary circuit. *Note:* If of the insulating type, it also provides protection to the secondary circuit, casing, lamp, and associated luminaire from the high voltage of the primary circuit. 82A9-0

inching (rotating machinery). Electrically actuated angular movement or slow rotation of a machine, usually for maintenance or inspection. *See also:* **asynchronous machine; direct-current commutating machine; synchronous machine.** 0-31E8

incidental amplitude-modulation factor (signal generator). That modulation factor resulting unintentionally from the process of frequency modulation and/or phase modulation. *See also:* **signal generator.** 0-9E4

incidental and restricted radiation. Radiation in the radio-frequency spectrum from all devices excluding licensed devices. *See also:* **mobile communication system.** 0-6E1

incidental frequency modulation (signal generator). The ratio of the peak frequency deviation to the carrier frequency, resulting unintentionally from the process of amplitude modulation. *See also:* **signal generator.** 0-9E4

incidental phase modulation (signal generator). The peak phase deviation of the carrier, in radians, resulting unintentionally from the process of amplitude modulation. *See also:* **signal generator.** 0-9E4

incident wave (1) (general). A wave, traveling through a medium, that impinges on a discontinuity or a medium of different propagation characteristics. *See also:* **radiation; radio wave propagation.** 42A65-0
(2) (forward wave) (uniform guiding systems). A wave traveling along a waveguide or transmission line in a specified direction toward a discontinuity, terminal plane, or reference plane. *See also:* **reflected wave; waveguide.** 0-9E4
(3) (lightning arresters). A traveling wave before it reaches a transition point. *See also:* **lightning arrester (surge diverter).** 50I25-31E7

inclined-blade blower (rotating machinery). A fan made with flat blades mounted so that the plane of the blades is parallel to and displaced from the axis of rotation of the rotor. *See:* **fan (rotating machinery).** 0-31E8

inclosed. *See:* **enclosed.**

inclusive OR (computing systems). *See:* OR.

incoherent scattering (radio wave propagation). When waves encounter matter, a disordered change in the phase of the waves. *See also:* **radio wave propagation.** 0-3E2

incomplete diffusion (illuminating engineering). That in which the diffusing medium redirects some of the flux incident upon it in an image-forming state and another portion in a nonimage-forming state. *Note:* Also termed partial diffusion. *See also:* **lamp.** Z7A1-0

incorrect relay operation. Any output response or lack of output response by the relay that, for the applied input quantities, is not correct. 37A100-31E11/31E6

incorrect relaying-system performance (relays or associated equipment). Any operation or lack of operation that, under existing conditions, does not conform to correct relaying-system performance. 37A100-31E11/31E6

incremental computer. A special-purpose computer that is specifically designed to process changes in the variables as well as the absolute value of the variables themselves, for example, digital differential analyzer. *See also:* **electronic digital computer.** X3A12-16E9

incremental cost of delivered power (source). The additional per unit cost that would be incurred in supplying another increment of power from that source to the composite system load. *See also:* **power systems, low-frequency and surge testing.** E94-0

incremental cost of reference power (source). The additional per-unit cost that would be incurred in supplying another increment of power from that source to a designated reference point on a transmission system.

See also: **power systems, low-frequency and surge testing.** E94-0

incremental delivered power (electric power systems). The fraction of an increment in power from a particular source that is delivered to any specified point such as the composite system load, usually expressed in percent. *See also:* **power systems, low-frequency and surge testing.** E94-0

incremental dimension (numerically controlled machines). A dimension expressed with respect to the preceding point in a sequence of points. *See:* **dimension; long dimension; normal dimension; short dimension.** EIA3B-34E12

incremental feed (numerically controlled machines). A manual or automatic input of preset motion command for a machine axis. *See also:* **numerically controlled machines.** EIA3B-34E12

incremental fuel cost of generation (any particular source). The cost, usually expressed in mill/kilowatt-hour, that would be expended for fuel in order to produce an additional increment of generation at any particular source. *See also:* **power systems, low-frequency and surge testing.** E94-0

incremental generating cost (source at any particular value of generation) (electric power systems). The ratio of the additional cost incurred in producing an increment of generation to the magnitude of that increment of generation. *Note:* All variable costs should be taken into account including maintenance. *See also:* **power systems, low-frequency and surge testing.** E94-0

incremental heat rate (steam turbo-generator unit at any particular output). The ratio of a small change in heat input per unit time to the corresponding change in power output. *Note:* Usually, it is expressed in British thermal unit/kilowatt-hour. *See also:* **power systems, low-frequency and surge testing.** E94-0

incremental hysteresis loss (magnetic material). The hysteresis loss in a magnetic material when it is subjected simultaneously to a biasing and an incremental magnetizing force. E270-0

incremental induction. At a point in a material that is subjected simultaneously to a polarizing magnetizing force and a symmetrical cyclically varying magnetizing force, one-half the algebraic difference of the maximum and minimum values of the magnetic induction at that point. E270-0

incremental loading (electric power systems). The assignment of loads to generators so that the additional cost of producing a small increment of additional generation is identical for all generators in the variable range. *See also:* **power systems, low-frequency and surge testing.** E94-0

incremental magnetizing force (magnetic material). At a point that is subjected simultaneously to a biasing magnetizing force and a symmetrical cyclic magnetizing force, one-half of the algebraic difference of the maximum and minimum values of the magnetizing force at the point. E270-0

incremental maintenance cost (any particular source) (electric power systems). The additional cost for maintenance that will ultimately be incurred as a result of increasing generation by an additional increment. *See also:* **power systems, low-frequency and surge testing.** E94-0

incremental permeability (magnetic induction). The ratio of the cyclic change in the magnetic induction to the corresponding cyclic change in magnetizing force when the mean induction differs from zero. *Note:* In anisotropic media, incremental permeability becomes a matrix. E270-0

incremental sweep (oscilloscopes). A sweep that is not a continuous function, but that represents the independent variable in discrete steps. *See also:* **oscillograph; stairstep sweep.** 0-9E4

incremental transmission loss (electric power systems). The fraction of power loss incurred by transmitting a small increment of power from a point to another designated point. *Note:* One of these points may be mathematical (rather than physical), such as the composite system load. *See also:* **power systems, low-frequency and surge testing.** E94-0

incremental worth of power (designated point on a transmission system). The additional per-unit cost that would be incurred in supplying another increment of power from any variable source of a system in economic balance to such designated point. *Note:* When the designated point is the composite system load, the incremental worth of power is commonly called **lambda** or **Lagrangian multiplier.** *See also:* **power systems, low-frequency and surge testing.** E94-0

increment (network) starter (industrial control). A starter that applies starting current to a motor in a series of increments of predetermined value and at predetermined time intervals in closed-circuit transition for the purpose of minimizing line disturbance. One or more increments may be applied before the motor starts. *See:* **starter.** IC1-34E10

independent ballast (mercury lamp). A ballast that can be mounted separately outside a lighting fitting or fixture. 82A4-0

independent contact. A contacting member designed to close one circuit only. *See also:* **railway signal and interlocking.** 42A42-0

independent firing (industrial control). The method of initiating conduction of an ignitron by obtaining power for the firing pulse in the ignitor from a circuit independent of the anode circuit of the ignitron.*See also:* **electronic controller.** 42A25-34E10

independent ground electrode (lightning arresters). A ground electrode or system such that its voltage to ground is not appreciably affected by currents flowing to ground in other electrodes or systems. *See:* **lightning arrester (surge diverter).** 50I25-31E7

independent manual operation (switching device). A stored-energy operation where the energy originates from manual power, stored and released in one continuous operation, such that the speed and force of the operation are independent of the action of the operator. 37A100-31E11

independent transformer. A transformer that can be mounted separately outside a luminaire. 82A7-0

index (electronic computation). (1) An ordered reference list of the contents of a file or document, together with keys or reference notations for identification or location of those contents. (2) A symbol or a number used to identify a particular quantity in an array of similar quantities. For example, the terms of an array represented by X1, X2, ..., X100 have the indexes 1, 2,..., 100, respectively. (3) Pertaining to an index register. *See also:* **electronic digital computer.** X3A12-16E9

index of cooperation (facsimile in rectilinear scanning or recording). The product of the total length of a scanning or recording line by the number of scanning or recording lines per unit length. *Notes:* (1) The International Index of Cooperation (diametral index of cooperation) is based on drum diameter and is defined by the International Radio Consultative Committee (CCIR). It is $1/\pi$ times the scanning line index of cooperation. (2) For a scanner and recorder to be compatible the indexes of cooperation must be the same. *See also:* **recording (facsimile); scanning (facsimile).** E168-0

index register (computing systems). A register whose content is added to or subtracted from the operand address prior to or during the execution of an instruction. X3A12-16E9

indicated bearing (direction finding). A bearing from a direction-finder site to a target transmitter obtained by averaging several readings; the indicated bearing is compared to the apparent bearing to determine accuracy of the equipment. *See also:* **navigation.** 0-10E6

indicating circuit (control apparatus). The portion of its control circuit that carries intelligence to visual or audible devices that indicate the state of the apparatus controlled. *See also:* **control.** E45-0

indicating control switch. A switch that indicates its last control operation. 37A100-31E11

indicating demand meter (register). A demand meter equipped with a scale over which a friction pointer is advanced so as to indicate maximum demand. *See also:* **electricity meter (meter).** 12A0/42A30-0

indicating fuse. A fuse that automatically indicates that the fuse has interrupted the circuit. 37A100-31E11

indicating instrument. An instrument in which only the present value of the quantity measured is visually indicated. *See also:* **instrument.** 39A1/42A30-0

indicating or recording mechanism (demand meter). That mechanism that indicates or records the measurement of the electrical quantity as related to the demand interval. *Note:* This mechanism may be operated directly by and be a component part of the electric mechanism, or may be structurally separate from it. The demand may be indicated or recorded in kilowatts, kilovolt-amperes, amperes, kilovars, or other suitable units. This mechanism may be of an indicating type, indicating by means of a pointer related to its position on a scale or by means of the cumulative reading of a number of dial or cyclometer indicators; or a graphic type, recording on a circular or strip chart; or of a printing type, recording on a tape. It may record the demand for each demand interval or may indicate only the maximum demand. *See also:* **demand meter.** 42A30-0

indicating scale (recording instrument). A scale attached to the recording instrument for the purpose of affording an easily readable value of the recorded quantity at the time of observation. *Note:* For recording instruments in which the production of the graphic record is the primary function, the chart scale should be considered the primary basis for accuracy ratings. For instruments in which the graphic record is secondary to a control function the indicating scale may be more accurate and more closely related to the control than is the chart scale. *See also:* **moving element (instrument).** 42A30-0

indication point (railway practice). The point at which the train control or cab signal impulse is transmitted to the locomotive or vehicle apparatus from the roadway element. 42A42-0

indicator light. A light that indicates whether or not a circuit is energized. *See also:* **appliances outlet.** 71A1-0

indicator travel. The length of the path described by the indicating means or the tip of the pointer in moving from one end of the scale to the other. *Notes:* (1) The path may be an arc or a straight line. (2) In the case of knife-edge pointers and others extending beyond the scale division marks, the pointer shall be considered as ending at the outer end of the shortest scale division marks. *See also:* **moving element (instrument).** 39A4-0

indicator tube. An electron-beam tube in which useful information is conveyed by the variation in cross section of the beam at a luminescent target. *See also:* **tube definitions.** 42A70-15E6

indicial admittance. The instantaneous response to unit step driving force. *Note:* This is a time function that is not an admittance of the type defined under **admittance.** *See also:* **network analysis.** E270-0

indirect-acting machine voltage regulator. A machine voltage regulator having a voltage-sensitive element that acts indirectly, through the medium of an interposing device such as contactors or a motor, to control the excitation of an electric machine. *Note:* A regulator is called a generator voltage regulator when it acts in the field circuit of a generator and is called an exciter voltage regulator when it acts in the field circuit of the main exciter. 37A100-31E11

indirect-acting recording instrument. A recording instrument in which the level of measurement energy of the primary detector is raised through intermediate means to actuate the marking device. *Note:* The intermediate means are commonly either mechanical, electric, electronic, or photoelectric. *See also:* **instrument.** 42A30-0

indirect address (computing systems). An address that specifies a storage location that contains either a direct address or another indirect address. *See:* **multilevel address.** ***See also:*** **electronic digital computer.** X3A12-16E9

indirect-arc furnace. An arc furnace in which the arc is formed between two or more electrodes. 42A60-0

indirect component (illuminating engineering). That portion of the luminous flux from a luminaire that arrives at the work plane after being reflected by room surfaces. *See also:* **inverse-square law (illuminating engineering).** Z7A1-0

indirect-drive machine (elevators). An electric driving machine, the motor of which is connected indirectly to the driving sheave, drum, or shaft by means of a belt or chain through intermediate gears. *See also:* **driving machine (elevators).** 42A45-0

indirect lighting. Lighting involving luminaires that distribute 90 to 100 percent of the emitted light upward. *See also:* **general lighting.** Z7A1-0

indirectly controlled variable (control) (industrial control). A variable that is not directly measured for control but that is related to, and influenced by, the directly controlled variable. *See also:* **control system, feedback.** AS1-34E10

indirectly heated cathode (equipotential cathode) (unipotential cathode). A hot cathode to which heat is supplied by an independent heater. *See also:* **electrode (electron tube).** 42A70-0

indirect manual operation (switching device). Operation by hand through an operating handle mounted at a distance from, and connected to the switching device by, mechanical linkage. 37A100-31E11

indirect operation (switching device). Operating by means of an operating mechanism connected to the main operating shaft or an extension of it, through offset linkages and bearings. 37A100-31E11

indirect release (switching device) (indirect trip). A release energized by the current in the main circuit through a current transformer, shunt, or other transducing device. 37A100-31E11

indirect stroke (lightning arresters). A lightning stroke that does not strike directly any part of a network but that induces an overvoltage in it. *See:* **lightning arrester (surge diverters).** 50I25-31E7

indirect-stroke protection (lightning). Lightning protection designed to protect a network or electric installation against indirect strokes. *See also:* **direct-stroke protection (lightning).** 0-31E13

individual line. A subscriber line arranged to serve only one main station although additional stations may be connected to the line as extensions. *Note:* An individual line is not arranged for discriminatory ringing with respect to the stations on that line. *See also:* **telephone system.** 42A65-19E1

indoor (prefix). Not suitable for exposure to the weather. *Note:* For example, an indoor capacitor unit is designed for indoor service or for use in a weatherproof housing. *See also:* **outdoor.** 42A95/E270-0

indoor pothead. A pothead intended for use where the pothead insulator is protected from the weather. *Note:* Potheads designed for use in sealed enclosures where the external dielectric strength is dependent upon liquid or special gaseous dielectrics are included in this category. *See also:* **pothead.** E48-0

indoor wall bushing. A wall bushing of which both ends are suitable for operating only where protection from the weather is provided. *See also:* **bushing.** E49-0

induced charge (ferroelectric device). The charge that flows when the condition of the device is changed from that of zero applied voltage (after having previously been saturated with either a positive or negative voltage) to at least that voltage necessary to saturate in the same sense. *Note:* The induced charge is dependent on the magnitude of the applied voltage, which should be specified in describing this characteristic of ferroelectric devices. *See also:* **ferroelectric domain.** E180-0

induced current (1) (general). Current in a conductor due to the application of a time-varying electromagnetic field. *See also:* **induction heating.** E54/E169-0

(2) (interference terminology). The interference current flowing in a signal path as a result of coupling of the signal path with an interference field, that is, a field produced by an interference source. *See:* **interference.** 0-13E6

(3) (lightning strokes). The current induced in a network or electric installation by an indirect stroke. *See also:* **direct-stroke protection (lightning).** 0-31E13

induced electrification. The separation of charges of opposite sign onto parts of a conductor as a result of the proximity of charges on other objects. *Note:* The charge on a portion of such a conductor is often called an induced charge or a bound charge. E270-0

induced field current (synchronous machine). The current that will circulate in the field winding (assuming the circuit is closed) due to transformer action when an alternating voltage is applied to the armature winding, for example, during starting of a synchronous motor. *See also:* **synchronous machine.** 0-31E8

induced-potential tests (electric power). Dielectric tests in which the test voltages are suitable-frequency alternating voltages, applied or induced between the terminals. *See also:* **power systems, low-frequency and surge testing.** E32-0

induced voltage (1) (general). A voltage produced in a closed path or circuit by a change in magnetic flux linking that path. *See also:* **rotor (rotating machinery); stator.** 0-31E8

(2) (lightning strokes). The voltage induced on a network or electric installation by an indirect stroke. *See also:* **direct-stroke protection (lightning).** 0-31E13

inductance. The property of an electric circuit by virtue of which a varying current induces an electromotive force in that circuit or in a neighboring circuit. 0-21E1

inductance coil. *See:* **inductor.**

inductance, effective (1) (general). The imaginary part of an inductive impedance divided by the angular frequency. *See also:* **transmission characteristics.** 0-9E4

(2) (winding). The self-inductance at a specified frequency and voltage level, determined in such a manner as to exclude the effects of distributed capacitance and other parasitic elements of the winding but not the parasitic elements of the core. 0-21E1

induction coil. (1) A transformer used in a telephone set for interconnecting the transmitter, receiver, and line terminals. (2) A transformer for converting interrupted direct current into high-voltage alternating current. *See also:* **telephone station.** 42A65-21E0

induction compass. A device that indicates an aircraft's heading, in azimuth. Its indications depend on the current generated in a coil revolving in the earth's magnetic field. *See also:* **air transportation instruments.** 42A41-0

induction-conduction heater. A heating device in which electric current is conducted through but is restricted by induction to a preferred path in a charge of a material to be heated. *See also:* **induction heating.** E54/E169-0

induction coupling. An electric coupling in which torque is transmitted by the interaction of the magnetic field produced by magnetic poles on one rotating member and induced currents in the other rotating member. *Note:* The magnetic poles may be produced by direct-current excitation, permanent-magnet excitation, or alternating-current excitation. The induced currents may be carried in a wound armature, squirrel cage, or may appear as eddy currents. *See:* **electric coupling.** 42A10-0

induction cup (relay). A form of relay armature in the shape of a cylinder with a closed end that develops operating torque by its location within the fields of electromagnets that are excited by the input quantities. 37A100-31E11/31E6

induction cylinder (relay). A form of relay armature in the shape of an open-ended cylinder that develops operating torque by its location within the fields of electromagnets that are excited by the input quantities. 37A100-31E11/31E6

induction disc (relay). A form of relay armature in the shape of a disc that usually serves the combined function of providing an operating torque by its location within the fields of an electromagnet excited by the input quantities, and a restraining force by motion within the field of a permanent magnet. 37A100-31E11/31E6

induction frequency converter. A wound-rotor induction machine in which the frequency conversion is obtained by induction between a primary winding and a secondary winding rotating with respect to each other. *Notes:* (1) The secondary winding delivers power at a frequency proportional to the relative speed of the primary magnetic field and the secondary member. (2) In case the machine is separately driven, this relative speed is maintained by an external source of mechanical power. (3) In case the machine is self-driven, this relative speed is maintained by motor action within the machine obtained by means of additional primary and secondary windings with number of poles differing from the number of poles of the frequency-conversion windings. In special cases one secondary winding performs the function of two windings, being short-circuited with respect to the poles of the driving primary winding and open-circuited with respect to the poles of the primary-conversion winding. *See:* **converter.** 42A10-31E8

induction furnace. A transformer of electric energy to heat by electromagnetic induction. 42A60-0

induction generator. An induction machine, when driven above synchronous speed by an external source of mechanical power, used to convert mechanical power to electric power. *See also:* **asynchronous machine.** 0-31E8

induction heater (interference terminology). A device for causing electric current to flow in a charge of material to be heated. Types of induction heaters can be classified on the basis of the frequency of the induced current, for example, a low-frequency induction heater usually induces power-frequency current in the charge; a medium-frequency induction heater induces currents of frequencies between 180 and 540 hertz; a high-frequency induction heater induces currents having frequencies from 1000 hertz upward. *Note:* For an extensive list of cross references, see Appendix A. 0-13E6.

induction heating. The heating of a nominally conducting material in a varying electromagnetic field due to its internal losses. E54/E169-0

induction instrument. An instrument that depends for its operation on the reaction between a magnetic flux (or fluxes) set up by one or more currents in fixed windings and electric currents set up by electromagetic induction in movable conducting parts. *See also:* **instrument.** 42A30-0

induction loop (relay). A form of relay armature consisting of a single turn or loop that develops operating torque by its location within the fields of electromagnets that are excited by the input quantities. 37A100-31E11/31E6

induction loudspeaker. A loudspeaker in which the current that reacts with the steady magnetic field is induced in the moving member. *See also:* **loudspeaker.** 42A65-1E1

induction machine. An asynchronous alternating-current machine that comprises a magnetic circuit interlinked with two electric circuits, or sets of circuits, rotating with respect to each other and in which power is transferred from one circuit to another by electromagnetic induction. *Note:* Examples of induction machines are induction motors, induction generators, and certain types of frequency converters and phase converters. 42A10-0

induction motor. An alternating-current motor in which a primary winding on one member (usually the stator) is connected to the power source and a polyphase secondary winding or a squirrel-cage secondary winding on the other member (usually the rotor) carries induced current. *See also:* **asynchronous machine.** 42A10-0

induction-motor meter. A motor-type meter in which the rotor moves under the reaction between the currents induced in it and a magnetic field. *See also:* **electricity meter (meter).** 12A0/42A30-0

induction ring heater. A form of core-type induction heater adapted principally for heating electrically conducting charges of ring or loop form, the core being open or separable to facilitate linking the charge. *See also:* **induction heater.** E54/E169-0

induction vibrator. A device momentarily connected between the airplane direct-current supply and the primary winding of the magneto, thus converting the magneto to an induction coil. *Note:* It provides energy to the sparkplugs of an aircraft engine during its starting period. *See also:* **air transportation electric equipment.** 42A41-0

induction voltage regulator. A regulator having a primary winding in shunt and a secondary winding in series with a circuit, for gradually adjusting the voltage or the phase relation, or both, of the circuit by changing the relative position of the exciting and series windings of the regulator. *See also:* **voltage regulator.** 57A15-0

inductive coordination (electric supply and communication systems). The location, design, construction, operation, and maintenance in conformity with harmoniously adjusted methods that will prevent inductive interference.

See:
configuration;
coordinated transpositions;
current balance ratio;
discontinuity;
electrically connected;
forced drainage;
general coordinated methods;
ground-return current;
$I \cdot T$ product;
inductive coupling;
inductive exposer;
inductive influence;
inductive interference;
inductive susceptiveness;
kV·T product;
longitudinal circuit;
$m-l$ ratio;
noise-longitudinal;
noise-metallic;
noise-to-ground;

residual current;
residual voltage;
shield factor;
shield wire;
sigma;
specific coordinated methods;
telephone influence factor. 42A65-0

inductive coupling (communication circuits). The association of two or more circuits with one another by means of inductance mutual to the circuits or the mutual inductance that associates the circuits. *Note:* This term, when used without modifying words, is commonly used for coupling by means of mutual inductance, whereas coupling by means of self-inductance common to the circuits is called direct inductive coupling. *See also:* **coupling; inductive coordination; network analysis.** E270/42A65-0

inductive exposure. A situation of proximity between electric supply and communication circuits under such conditions that inductive interference must be considered. *See also:* **inductive coordination.** 42A65-0

inductive influence (electric supply circuit with its associated apparatus). Those characteristics that determine the character and intensity of the inductive field that it produces. *See also:* **inductive coordination.** 42A65-0

inductive interference (electric supply and communication systems). An effect, arising from the characteristics and inductive relations of electric supply and communication systems, of such character and magnitude as would prevent the communication circuits from rendering service satisfactorily and economically if methods of inductive coordination were not applied. *See also:* **inductive coordination.** 42A65-0

inductively coupled circuit. A coupled circuit in which the common element is mutual inductance. *See also:* **network analysis.** E270-0

inductive microphone. *See:* **inductor microphone.**

inductive neutralization (shunt neutralization) (coil neutralization). A method of neutralizing an amplifier whereby the feedback susceptance due to an interelement capacitance is canceled by the equal and opposite susceptance of an inductor. *See also:* **amplifier; feedback.** 42A65/E145-0

inductive susceptiveness (communication circuit with its associated apparatus). Those characteristics that determine, so far as such characteristics are able to determine, the extent to which the service rendered by the circuit can be adversely affected by a given inductive field. *See also:* **inductive coordination.** 42A65-0

inductor (1) (general). A device consisting of one or more associated windings, with or without a magnetic core, for introducing inductance into an electric circuit.

(2) (railway practice). A roadway element consisting of a mass of iron with or without a winding, that acts inductively on the vehicle apparatus of the train control, train stop, or cab signal system. *See also:* **circuits and devices.** 42A42-21E0

inductor alternator. An inductor machine for use as a generator, the voltage being produced by a variation of flux linking the armature winding without relative displacement of field magnet or winding and armature winding. *See:* **synchronous machine.** 0-31E8

inductor circuit (railway practice). A circuit including the inductor coil and the two lead wires leading therefrom taken through roadway signal apparatus as required. 42A42-0

inductor dynamotor (rotating machinery). A dynamotor inverter having toothed field poles and an associated stationary secondary winding for conversion of direct current to high-frequency alternating current by inductor-generator action. *See:* **convertor.** 42A10-31E8

inductor frequency-converter (rotating machinery). An inductor machine having a stationary input alternating-current winding, which supplies the excitation, and a stationary output winding of a different number of poles in which the output frequency is induced through change in field reluctance by means of a toothed rotor. *Note:* If the machine is separately driven, the rotor speed is maintained by an external source of mechanical power. If the machine is self-driven, the primary winding and rotor function as in a squirrel-cage induction motor or a reluctance motor. *See also:* **convertor.** 0-31E8

inductor machine (rotating machine). A synchronous machine in which one member, usually stationary, carries main and exciting windings effectively disposed relative to each other, and in which the other member, usually rotating, is without windings but carries a number of regular projections. (Permanent magnets may be used instead of the exciting winding). *See also:* **synchronous machine.** 0-31E8

inductor microphone (inductive microphone). A moving-conductor microphone in which the moving element is in the form of a straight-line conductor. *See also:* **microphone.** 42A65-0

inductor-type synchronous generator. A generator in which the field coils are fixed in magnetic position relative to the armature conductors, the electromotive forces being produced by the movement of masses of magnetic material. *See also:* **synchronous machine.** 42A10-0

inductor-type synchronous motor. An inductor machine for use as a motor, the torques being produced by forces between armature magnetomotive force and salient rotor teeth. *Note:* Such motors usually have permanent-magnet field excitation, are built in fractional-horsepower ratings, frames and operate at low speeds, 300 revolutions per minute or less. *See also:* **synchronous machine.** 0-31E8

industrial brush (rotating machinery). A brush having a cross-sectional area (width times thickness) of more than 1/4 square inch or a length of more than 1½ inches. *See also:* **brush (rotating machinery).** 42A10-31E8;64A1-0

industrial control. Broadly, the methods and means of governing the performance of an electric device, apparatus, equipment, or system used in industry. 42A25-34E10

industrial electric locomotive. An electric locomotive, used for industrial purposes, that does not necessarily conform to government safety regulations as applied to railroads. *Note:* This term is generally applied to locomotives operating in surface transportation and does not include mining locomotives. A prefix diesel-electric, gas-electric, etcetera, may replace the word electric.*See also:* **electric locomotive; industrial electronics.** 42A42-0

See:
dielectric heating;
heating pattern;

heating station;
induction heater;
ISM apparatus;
oscillator;
pool-cathode mercury-arc converter;
quenched-spark-gap converter;
radio-frequency converter;
radio-frequency generator;
rotary generator;
shield.

industrial process supervisory system. A supervisory system that initiates signal transmission automatically upon the occurrence of an abnormal or hazardous condition in the elements supervised, which include heating, air-conditioning, and ventilating systems, and machinery associated with industrial processes. *See also:* **protective signaling.** 42A65-0

inertance (automatic control). A property expressible by the quotient of a potential difference (temperature, sound pressure, liquid level) divided by the related rate of change of flow; the thermal or fluid equivalent of electrical inductance or mechanical moment of inertia. *See also:* **control system, feedback.** 85A1-23E0

inert-gas pressure system (regulator). A system in which the interior of the tank is sealed from the atmosphere, over the temperature range specified, by means of a positive pressure of inert gas maintained from a separate inert-gas source and reducing-valve system. *See also:* **transformer, oil-immersed.** 57A15-0

inertia compensation (industrial control). The effect of a control function during acceleration or deceleration to cause a change in motor torque to compensate for the driven-load inertia. *See also:* **control system, feedback.** IC1-34E10

inertia constant (machine). The energy stored in the rotor when operating at rated speed expressed as kilowatt-seconds per kilovolt-ampere rating of the machine. *Note:* The inertia constant is

$$h = \frac{0.231 \times Wk^2 \times n^2 \times 10^{-6}}{\text{kVA}}$$

where

h = inertia constant in kilowatt-seconds per kilovolt-ampere
Wk^2 = moment of inertia in pound-feet^2
n = speed in revolutions per minute
kVA = rating of machine in kilovolt-amperes

See: **asynchronous machine; direct-current commutating machine; synchronous machine.** 42A10-31E8

inertialess scanning. *See:* **electronic scanning.**

inertial navigation equipment. A type of dead-reckoning navigation equipment whose operation is based upon the measurement of accelerations; accelerations are sensed dynamically by devices stabilized with respect to inertial space, and the navigational quantities (such as vehicle velocity, angular orientation, or positional information) are determined by computers and/or other instrumentation. *See also:* **navigation.** E174-10E6

inertial navigator. A self-contained, dead-reckoning navigation aid using inertial sensors, a reference direction, and initial or subsequent fixes to determine direction, distance, and speed; single integration of acceleration provides speed information and a double integration provides distance information. *See also:* **navigation.** 0-10E6

inertial space (navigation). A frame of reference defined with respect to the fixed stars. *See also:* **navigation.** E174-10E6

inertia relay. A relay with added weights or other modifications that increase its moment of inertia in order either to slow it or to cause it to continue in motion after the energizing force ends. *See also:* **relay.** 83A16-0

inflection point (tunnel-diode characteristic). The point on the forward current-voltage characteristic at which the slope of the characteristic reaches its most negative value. *See also:* **peak point (tunnel-diode characteristic).** E253-15E7

inflection-point current (tunnel-diode characteristic). The current at the inflection point. *See also:* **peak point (tunnel-diode characteristic).** E253-15E7

inflection-point emission current (electron tubes). That value of current on the diode characteristic for which the second derivative of the current with respect to the voltage is zero. *Note:* This current corresponds to the inflection point of the diode characteristic and is, under suitable conditions, an approximate measure of the maximum space-charge-limited emission current. *See also:* **circuit characteristics of electrodes.** E160-15E6

inflection-point voltage (tunnel-diode characteristic). The voltage at which the inflection point occurs. *See also:* **peak point (tunnel-diode characteristic).** E253-15E7

influence (1) (upon an instrument) (specified variable or condition). The change in the indication of the instrument caused solely by a departure of the specified variable or condition from its reference value, all other variables being held constant. *See also:* **accuracy rating of an instrument.** 42A30-0;39A1-0
(2) (upon a recording instrument). The change in the recorded value caused solely by a departure of the specified variable or condition from its reference value, all other variables being held constant. *Note:* If the influences in any direction from reference conditions are not equal, the greater value applies. 39A2-0

information. The meaning assigned to data by known conventions. X3A12-16E9

information content (message or a symbol from a source). The negative of the logarithm of the probability that this particular message or symbol will be emitted from the source. *Notes:* (1) The choice of logarithmic base determines the unit of information content. (2) The probability of a given message or symbol being emitted may depend on one or more preceding messages or symbols. (3) The quantity has been called self-information. *See:* **bit; hartley.** *See also:* **information theory.** E171-0

information content, average. *See:* **average information content.**

information processing. (1) The processing of data that represents information. (2) Loosely, automatic data processing. *See also:* **electronic digital computer.** X3A12-16E9

information rate (1) (from a source, per second). The product of the average information content per symbol and the average number of symbols per second. *See also:* **information theory.** E171-0
(2) (from a source, per symbol). *See:* **average information content.**
(3) (through a channel, per second). The product of

the average transinformation per symbol and the average number of symbols per second. *See also:* **information theory.** E171-0

information retrieval. The methods and procedures for recovering specific information from stored data. *See also:* **electronic digital computer.** X3A12-16E9

information theory. (1) In the narrowest sense, is used to describe a body of work, largely about communication problems but not entirely about electrical communication, in which the information measures are central. (2)In a broader sense it is taken to include all statistical aspects of communication problems, including the theory of noise, statistical decision theory as applied to detection problems, and so forth. *Note:* This broader field is sometimes called "statistical communication theory." (3) In a still broader sense its use includes theories of measurement and observation that use other measures of information, or none at all, and indeed work on any problem in which information, in one of its colloquial senses, is important.
See:
average information content;
average transinformation;
binary code;
bit;
channel;
channel capacity;
channel utilization index;
code;
code character;
code element;
communication;
conditional information content;
entropy;
equivocation;
hartley;
information content;
information rate;
message;
message source;
***N*-ary code;**
redundancy;
signal;
ternary code;
transinformation. E171-0

information writing speed (oscilloscopes). The oscilloscope-recorder characteristic that is a measure of the maximum number of spots of information per second that can be recorded and identified on a single trace. Test conditions must be specified. *See also:* **oscillograph; writing speed (storage tubes).** 0-9E4

infrared radiation. For practical purposes any radiant energy within the wavelength range 780 to 10^5 nanometers is considered infrared energy. *Note:* In general, unlike ultraviolet energy, infrared energy is not evaluated on a wavelength basis but rather in terms of all of such energy incident upon a surface. Examples of these applications are industrial heating, drying, baking, and photoreproduction. However, some applications, such as infrared viewing devices, involve detectors that are sensitive to a restricted range of wavelengths; in such cases the spectral characteristics of the source and receiver are of importance. *See:* **light; photochemical radiation; units of wavelength.** Z7A1-0

infrasonic frequency (subsonic frequency*). A frequency lying below the audio-frequency range. *Notes:* (1) The word infrasonic may be used as a modifier to indicate a device or system intended to operate at infrasonic frequencies.(2) The term subsonic was once used in acoustics synonymously with infrasonic, such usage is now deprecated.
*Deprecated 0-1E1

inherent acceleration, by generator field (industrial control). The effect of accelerating the drive by the natural build-up of generator voltage without field forcing or intentional delay. *See:* **electric drive.** IC1-34E10

inherent harmonics (electrical conversion). Harmonics that are a by-product of the output of an inverter or frequency changer. *Note:* In most cases, these harmonics are the direct result of the methods of inversion using square-wave, quasisquare-wave, or semisinusoidal outputs. Harmonics may also be a by-product of various methods of regulation such as clipping and phase control. *See also:* **electrical conversion.** 0-10E1

inherent regulation (rotating machinery). The amount of change of voltage or speed resulting from a load change and due solely to the fundamental characteristics of the machine itself. *See also:* **asynchronous machine; direct-current commutating machine; synchronous machine.** 0-31E8

inherent restriking voltage (circuit transient-recovery voltage). The restriking voltage that is associated with a particular circuit and determined by the circuit parameters alone, its form being unmodified by the characteristics of the arrester. *Notes:* (1) It is expressed in terms of amplitude factor and rate-of-rise (or natural frequency). (2) It would be obtained if, in the particular circuit, a sinusoidal symmetrical current were broken by an ideal expulsion-type arrester (no arc voltage and no post-arc conductivity) at a stated recovery voltage. *See:* **lightning arrester (surge diverter).** 99I2-31E7

inherent voltage regulation (power rectifier). The voltage regulation without the use of regulating equipment, or other compensating means, and with no phase control. *See also:* **power rectifier; rectification; voltage regulation.** 42A15-0

inherited error. The error in the value of quantities that serve as the initial conditions at the beginning of a step in a step-by-step calculation. X3A12-16E9

inhibit. (1) To prevent an event from taking place. (2) To prevent a device or logic element from producing a specified output. *See also:* **electronic digital computer.** E162-0

inhibited oil. Mineral transformer oil to which a synthetic oxidation inhibitor has been added. *See:* **oil-immersed transformer.** 42A15-31E12

inhibiting input (electronic computation). A gate that, if in its prescribed state, prevents any output that might otherwise occur. *See also:* **electronic computation.** E270-0

inhibitor (electroplating). A substance added to a pickle for the purpose of reducing the rate of solution of the metal during the removal from it of oxides or other compounds. *See also:* **electroplating.** 42A60-0

inhibit pulse. A drive pulse that tends to prevent flux reversal of a magnetic cell by certain specified drive pulses. *See also:* **static magnetic storage.** E163-0

initial alternating short-circuit current (rotating machinery). The root-mean-square value of the current in the armature winding immediately after the winding has been suddenly short-circuited, the aperiodic component of current, if any, being neglected. *See:* **armature; direct-axis synchronous reactance.** 0-31E8

initial condition. The value of a variable at the start of computation. *See also:* **electronic analog computer; reset.** E165-16E9

initial conditions (inertial navigation). The values of position, velocity, level, azimuth, gyro bias, and accelerometer bias imposed on the system before departure. *See also:* **navigation.** E174-10E6

initial conductor tension (power lines). The longitudinal tension in a conductor prior to the application of any external load. *See also:* **conductor.** 2A2-0

initial contact pressure. *See:* **contact pressure, initial.**

initial differential capacitance (nonlinear capacitor). Differential capacitance at zero capacitor voltage. *See also:* **nonlinear capacitor.** E226-15E7

initial element (sensing element). *See:* **primary detector.**

initial excitation response (rotating machinery). The initial rate of increase in the excitation voltage when a sudden transition is made from the voltage at the rated conditions of the main machine to the conditions that enable the excitation ceiling voltage to be attained in the shortest possible time. *Note:* This rate can also be expressed by its relative value in relation to the excitation voltage for the rated conditions of the main machine. *See also:* **asynchronous machine; direct-current commutating machine; synchronous machine.** 0-31E8

initial inverse voltage (rectifier tube). The peak inverse anode voltage immediately following the conducting period. *See also:* **electrode voltage (electron tube); rectification.** 42A70-15E6

initial ionizing event (gas-filled radiation-counter tube). An ionizing event that initiates a tube count. *See also:* **counter tubes; gas-filled radiation.** 42A70-15E6

initialize (computing systems). To set counters, switches, and addresses to zero or other starting values at the beginning of or at prescribed points in a computer routine. X3A12-16E9

initial luminous exitance. The density of luminous flux leaving a surface within an enclosure before interreflections occur. *Note:* For light sources this is the luminous exitance as defined in luminous flux density (at a surface). For non-selfluminous surfaces it is reflected luminous exitance of the flux received directly from sources within the enclosure or from daylight. *See also:* **inverse-square law (illuminating engineering).** Z7A1-0

initial output test (storage battery). A service test begun shortly after the manufacture of the battery. (Usually less than one month.) *See also:* **battery (primary or secondary).** 42A60-0

initial reverse voltage (semiconductor rectifier). The instantaneous value of the reverse voltage that occurs across a rectifier circuit element immediately following the conducting period and including the first peak of oscillation. *See also:* **rectification.** E59-34E17/34E24

initial reversible capacitance (nonlinear capacitor). Reversible capacitance at a constant bias voltage of zero. *See also:* **nonlinear capacitor.** E226-15E7

initial test temperature (storage battery). The average temperature of the electrolyte in all cells at the beginning of discharge. *Note:* The standard reference temperature is 25 degree Celsius (77 degrees Fahrenheit) initially, without limit on the final temperature, but the ambient temperature on discharge shall be 5 degrees Celsius to 8 degrees Celsius lower than the temperature of the electrolyte at the beginning of the discharge and shall be kept constant during discharge. *See:* **critical temperature (storage cell); temperature coefficient of electromotive force (storage cell); temperature coefficient of capacity (storage cell).** 42A60-0

initial unloaded sag (power lines). The sag of a conductor prior to the application of any external load. *See:* **power distribution, overhead construction; sag.** 2A2-0

initial value (industrial control). The value of the time response of a system or element at the time a stimulus is applied. *See:* **control system, feedback.** AS1-34E10

initial voltage (battery). The closed-circuit voltage at the beginning of a discharge. *Note:* It is usually taken after current has been flowing for a sufficient period of time for the rate of change of voltage to become practically constant. *See also:* **battery (primary or secondary).** 42A60-0

initiating relay. A programming relay whose function is to constrain the action of dependent relays until after it has operated. 37A100-31E11/31E6

injury potential (electrobiology) (demarcation potential) (demarcation current*) (current of injury*). The difference in potential observed between injured and uninjured parts of a living structure such as a muscle or nerve. *See also:* **contact potential.**
*Deprecated 42A80-18E1

ink. *See:* **magnetic ink.**

ink-vapor recording (facsimile). The type of recording in which vaporized ink particles are directly deposited upon the record sheet. *See also:* **recording (facsimile).** E168-0

inkwell influence. The difference between the recorded values, at full-chart deflection, when the inkwell is full and when it is empty. *Note:* It is to be expressed as a percentage of the full-scale value. *See also:* **accuracy rating of an instrument.** 39A2-0

inner frame (rotating machinery). The portion of a frame in which the core and stator windings are mounted, which can be inserted and removed from an outer frame as a unit without disturbing the mounting on the foundation. *See:* **cradle base (rotating machinery).** 0-31E8

inner marker (electronic navigation). *See:* **boundary marker.**

in-phase rejection. *See:* **common-mode rejection.**

in-phase signal. *See:* **common-mode signal.**

input (1) (general). (A) The current, voltage, power, or driving force applied to a circuit or device. (B) The terminals or other places where current, voltage, power, or driving force may be applied to a circuit or device. 42A65-31E3/21E1
(2) (electronic digital computer). (A) The data to be processed. (B) The state or sequence of states occurring on a specified input channel. (C) The device or collective set of devices used for bringing data into another device. (D) A channel for impressing a state on a device or logic element. (E) The process of transfer-

ring data from an external storage to an internal storage. *See:* **manual input.** X3A12-16E9
(3) (rotating machinery). (A) **(for a generator)** The mechanical power transmitted to its shaft. (B) **(for a motor)** The power (active, reactive, or apparent) supplied to its terminals. *See:* **asynchronous machine; synchronous machine.** 0-31E8
(4) (relay). A physical quantity or quantities to which the relay is designed to respond. *Note:* A physical quantity that is not directly related to the prescribed response of a relay, though necessary to or in some way affecting the relay operation, is not considered part of input. Time is not considered a relay input, but is a factor in performance. 37A100-31E11/31E6

input admittance. *See:* **admittance, short-circuit input.**

input capacitance (*n*-terminal electron tubes). *See:* **capacitance, input.** *See also:* **electron-tube admittances.**

input circuit (electronic valve or tube). The external circuit connected to the input electrode and in which the control voltage appears. *See also:* **ON period (electron tubes).** 50I07-15E6

input, dark-current, equivalent (phototubes). *See:* **equivalent dark-current input.**

input-dependent overshoot and undershoot (electric conversion). Dynamic regulation for input changes. *See also:* **electric conversion equipment.** 0-10E1

input, driving-point or sending-end impedance. The impedance obtained when the response is determined at the point at which the driving force is applied. *Notes:* (1) **Point** is used for generality. In the case of any electric circuit it would be translated **pair of terminals.** (2) The definition does not imply that a driving force has to exist at the point. A hypothetical driving force may be assumed. *See also:* **network analysis.** E270-0

input electrode (electron tubes). The electrode to which is applied the voltage to be amplified, modulated, detected, etcetera. *See also:* **electron tube.** 50I07-15E6

input gap. An interaction gap used to initiate a variation in an electron stream. *See:* **gas tubes.** E160-15E6

input impedance (1) (antenna). The impedance presented by an antenna at its terminals. *See also:* **antenna.** E149-3E1
(2) (transducer device or network). The impedance presented by the transducer device or network to a source. *See also:* **output; self-impedance.** E151-0;E165-16E9;0-42A65;0-9E4;42A65-31E3
(3) (electron device). The input electrode impedance at the input electrodes. 0-15E6
(4) (transmission-line port). The impedance at the transverse plane represented by the port. *Note:* This impedance is independent of the generator impedance. 0-3E1
(5) (analog computer). A passive network connected between the input terminal of an operational amplifier and its summing junction. *See also:* **electronic analog computer.** E165-16E9
(6) (transmission line). The impedance between the input terminals with the generator disconnected. *See also:* **self-impedance; waveguide.** E146-0

input limiter. *See:* **limiter circuit.** *See also:* **electronic analog computer.**

input magazine. *See:* **card hopper.**

input noise temperature, effective (signal transmission system) (1) (two-port network or amplifier). The input termination (source) noise temperature that, when the input termination is connected to a noise-free equivalent of the network or amplifier, would result in the same output noise power as that of the actual network or amplifier connected to a noise-free termination (source). *Note:* In terms of the noise factor of the network or amplifier, the effective input noise temperature T_e is

$$T_e = 290\,(F - 1)$$

where F is the noise factor and 290 kelvins is the standard noise temperature. *See also:* **signal.** 0-13E6
(2) (linear multiport transducer). An effective temperature (in kelvins) defined as the noise temperature at a given port, designated as the output port in a linear multiport transducer, divided by the sum of the exchangeable power gains from each of the remaining ports to the designated output port, when these remaining ports, designated as input ports, are connected to specified noise-free impedances having real parts of the same sign. *Notes:* (A) If G_j is the exchangeable power gain from the *j*th input port to the designated output port, and T_o is the noise temperature at the output port, then the effective input noise temperature T_e is therefore given by the relation

$$T_e = T_o \Big/ \sum_j G_j.$$

(2) When the real parts of the input termination impedances do not have the same sign, the magnitude of T_e is given by the relation

$$|T_e| = |T_o| \Big/ \Big|\sum_j G_j\Big|.$$

(3) If the output place of access has other ports that are coupled either by mode conversion or by frequency conversion to the designated output port, the noise contributed at the selected output frequency by the terminations of these ports is to be included in computing the noise temperature T_o of the output port. The corresponding conversion gains are not to be included in the summation of gains from the input ports to the output port. (4) The effective input noise temperature is a function of the terminating impedances of all ports except the designated output port. *See also:* **transducer; waveguide.** 0-15E6

input-output equipment (data transmission). Any subscriber (user) equipment that introduces data into or extracts data from a data communications system. *See also:* **data transmission.** 0-19E4

input-output table. A plotting device used to generate or to record one variable as a function of another variable. *See also:* **electronic analog computer.** E165-0

input resonator buncher (electron tubes). A resonant cavity, excited by an external source, that produces velocity modulation of the electron beam. *See also:* **velocity-modulated tube.** 50I07-15E6

input transient energy (inverters). The product of the average power in the transient and the time the transient exists. *See:* **self-commutated inverters.** 0-34E24

input winding(s) (primary winding(s)). The winding(s) to which the input is applied. 0-21E1

insert earphones. Small earphones that fit partially inside the ear. *See also:* **loudspeaker.** 0-1E1

insertion gain. Resulting from the insertion of a transducer in a transmission system, the ratio of (1) the power delivered to that part of the system following the transducer to (2) the power delivered to that same part before insertion. *Notes:* (1) If the input and/or output power consist of more than one component, such as multifrequency signal or noise, then the particular components used and their weighting must be specified. (2) This gain is usually expressed in decibels. (3) The insertion of a transducer includes bridging of an impedance across the transmission system. *See also:* **transmission characteristics.** E196/E270-0;E151-42A65;0-21E1

insertion loss (1). Resulting from the insertion of a transducer in a transmission system, the ratio of (A) the power delivered to that part of the system following the transducer, before insertion of the transducer, to (B) the power delivered to that same part of the system after insertion of the transducer. *Notes:* (1) If the input and/or output power consist of more than one component, such as multifrequency signal or noise, then the particular components used and their weighting must be specified. (2) This loss is usually expressed in decibels. (3) The insertion of a transducer includes bridging of an impedance across the transmission system. *See also:* **transducer; transmission loss.** E146/E151/E196/E270/42A65-3E1/21E1/27E1

(2) (waveguide component). The change in load power, due to the insertion of a waveguide or transmission-line component at some point in a transmission system, where the specified input and output waveguides connected to the component are reflectionless looking in both directions from the component (match-terminated). This change in load power is expressed as a ratio, usually in decibels, of (A) the power received at the load before insertion of the waveguide or transmission-line component to (B) the power received at the load after insertion. *Notes:* (1) A more general definition of insertion loss does not specify match-terminated connecting waveguides, in which case the insertion loss would vary with the load and generator impedances. (2) When the input and output waveguides connected to the component are not alike or do not operate in the same mode, the change in load power is determined relative to an ideal reflectionless and lossless transition between the input and output waveguides. *See also:* **transmission loss; waveguide.** 0-3E1

(3) (microwave gas tubes). The decrease in power measured in a matched termination when the unfired tube, at a specified ignitor current, is inserted in the waveguide between a matched generator and the termination. *See also:* **beam tubes, gas tubes.** E160-15E6

insertion phase shift (electric structure). The change in phase caused by the insertion of the structure in a transmission system. *Note:* The usual convention is that shunt capacitance or series inductance produces a positive phase shift. *See also:* **transmission characteristics.** 42A65-3E1

(2) (waveguides). The change in phase of a field quantity or voltage or current, at a specified load port or terminal surface, caused by the insertion of a network at some point in a transmission system. *Note:* The most general definition of insertion phase shift does not specify match-terminated connecting waveguides, in which case the insertion phase shift would vary with the load and generator impedances. *See also:* **measurement system; transmission characteristics.** E285-9E1

inside air temperature. *See:* **average inside air temperature.**

inside plant (communication practice). That part of the plant within a central office, intermediate station, or subscriber's premises that is on the office or station side of the point of connection with the outside plant. *Note:* The plant in a central office is commonly referred to as central-office plant and that on station premises as station plant. *See also:* **communication.** 42A65-0

inspection (watt-hour meters). An observation made to obtain an approximate idea of the condition of the meter. *Note:* In such cases an examination is made of the meter and the conditions surrounding it, for the purpose of discovering defects or conditions that are likely to be detrimental to its accuracy. Such an examination may or may not include an approximate determination of the percentage registration of the meter. *See also:* **service test (field test).** 42A30-0

inspection hole (manhole). *See:* **inspection opening.**

inspection opening (inspection hole) (manhole). A port to permit observation by virtue of a transparent or removable cover. *See:* **cradle base (rotating machinery).** 0-31E8

installation (1) (industrial control). An assemblage of electric equipment in a given location, designed for coordinated operation, and properly erected and wired. 42A25-34E10

(2) (elevators). A complete elevator, dumbwaiter, or escalator, including its hoistway, hoistway enclosures and related construction, and all machinery and equipment necessary for its operation. *See also:* **elevators.** 42A45-0

installation test (electric power devices). A test made upon the consumer's premises within a limited period of time after installation. *See also:* **service test (field test).** 42A30-0

instantaneous (relay). A qualifying term applied to a relay or other device indicating that no delay is purposely introduced in its action. *See:* **relay.** 42A20/37A100-31E11/31E6

instantaneous automatic gain control (IAGC) (1) (electronic navigation). The portion of a system that automatically adjusts the gain of an amplifier for each pulse so as to obtain a substantially constant output pulse peak amplitude with varying input pulse peak amplitudes, the adjustment being sufficiently fast to operate during the time a pulse is passing through the amplifier. *See also:* **radar.** E172-10E6

(2) (radar). A quick-acting automatic gain control that responds to variations of mean clutter level, so that clutter is reduced. *See also:* **radar.** 42A65-0

instantaneous burst magnitude (audio and electroacoustics). The absolute value of the instantaneous voltage, current or power for a burst-like excursion measured at a specific time. *Note:* By **absolute value** is meant the numerical value regardless of sign. See the figure under **burst duration.** *See also:* **burst (audio and**

electroacoustics). E257-1E1

instantaneous companding (modulation systems). Companding in which the effective gain is determined by the instantaneous value of the signal wave. *See also:* **modulating systems; transmission characteristics.** E170-0

instantaneous demand (electric power utilization). The demand of any instant, usually determined from the readings of indicating or recording instruments. *See also:* **alternating-current distribution.** 0-31E4

instantaneous failure rate (hazard) (particular time) (reliability). The rate of change of the number of items that have failed divided by the number of items surviving. *See also:* **reliability.** 0-7E1

instantaneous frequency (modulation systems). The time rate of change of the angle of an angle-modulated wave. *Note:* If the angle is measured in radians, the frequency in hertz is the time rate of change of the angle divided by 2π. *See also:* **modulating systems; signal wave.** E170/E145/42A65-0

instantaneous value. The value of a variable quantity at a given instant. 50I05-31E3

instant-starting system (fluorescent lamps). The term applied to a system in which an electric discharge lamp is started by the application to the lamp of a voltage sufficiently high to eject electrons from the electrodes by field emission, initiate electron flow through the lamp, ionize the gases, and start a discharge through the lamp without previous heating of the electrodes. 82A1-0

instruction (electronic computation). A statement that specifies an operation and the values or locations of its operands. *Notes* (1) The instruction may specify some operands by the definition of the operation and may use one or more addresses to specify the location in storage of other operands, where the result is to be stored, the next instruction, etcetera. (2) In this context, the term instruction is preferable to the terms command or order, which are sometimes used synonymously; command should be reserved for electronic signals and order should be reserved for sequence, interpolation, and related usage.
See also:
computer instruction;
decision instruction;
electronic digital computer;
extract instruction;
instruction code;
logic instruction;
machine instruction;
macro instruction;
repetition instruction. E162-0;X3A12-16E9

instruction code (electronic computation). An artificial language for describing or expressing the instructions that can be carried out by a digital computer. *Note:* In automatically sequenced computers, the instruction code is used when describing or expressing sequences of instructions, and each instruction word usually contains a part specifying the operation to be performed and one or more addresses that identify a particular location in storage. Sometimes an address part of an instruction is not intended to specify a location in storage but is used for some other purpose. If more than one address is used, the code is called a multiple-address code. In a typical instruction of a four-address code the addresses specify the location of two operands, the destination of the result, and the location of the next instruction in the sequence. In a typical three-address code, the fourth address specifying the location of the next instruction is dispensed with and the instructions are taken from storage in a preassigned order. In a typical one-address or single-address code, the address may specify either the location of an operand to be taken from storage, the destination of a previously prepared result, or the location of the next instruction. The arithmetic element usually contains at least two storage locations, one of which is an accumulator. For example, operations requiring two operands may obtain one operand from the main storage and the other from a storage location in the arithmetic element that is specified by the operation part. *See also:* **electronic computation; electronic digital computer; operation code.** E270-0

instruction counter (computing systems). A counter that indicates the location of the next computer instruction to be interpreted. *See also:* **electronic digital computer.** X3A12-16E9

instruction register (computing systems). A register that stores an instruction for execution. *See also:* **electronic digital computer.** X3A12-16E9

instruction repertory. The set of operations that can be represented in a given operation code. *See also:* **electronic digital computer.** E162/X3A12-16E9

instrument. A device for measuring the value of the quantity under observation. *Notes:* (1)An instrument may be an indicating instrument or a recording instrument. The term instrument is used in two different senses: (A) instrument proper consisting of the mechanism and the parts built into the case or made a corporate part thereof; and (B) the instrument proper together with any necessary auxiliary devices, such as shunts, shunt leads, resistors, reactors, capacitors, or instrument transformers. The term meter is also used in a general sense to designate any type of measuring device, including all types of electric measuring instruments. Such use as a suffix or as part of a compound word (for example, voltmeter, frequency meter) is universally accepted. Meter may be used alone with this wider meaning when the context is such as to prevent confusion with the narrower meaning of **electricity meter.** (2) For an extensive list of cross references, see *Appendix A. See:* **meter.** 42A30-0

instrumental error (navigation). The error due to the inaccuracies introduced in any portion of the system by the mechanism that translates path-length differences into navigation coordinate information, including calibration errors and errors resulting from limited sensitivity of the indicating instruments. *See also:* **navigation.** 0-10E6

instrument approach (electronic navigation). The process of making an approach to a landing by the use of navigation instruments without dependence upon direct visual reference to the terrain. *See also:* **navigation.** 0-10E6

instrument approach system (electronic navigation). A system furnishing guidance in the vertical and horizontal planes to aircraft during descent from an initial approach altitude to a point near the landing area. *See also:* **navigation.** 0-10E6

instrument bulb. The sensing portion of certain types of instrument. *See:* **measurement system.** 0-31E8

instrument landing system (electronic navigation) (1) (general). A generic term for a system that provides in the aircraft the necessary lateral, longitudinal, and vertical guidance for a low approach or landing of the aircraft. *See also:* **radio navigation.** 42A65-10E6
(2) (ILS). An internationally adopted instrument landing system for aircraft, consisting of a very-high frequency localizer, an ultra-high-frequency glide slope, and 75-megahertz markers. *See also:* **instrument landing system reference point.** 0-10E6

instrument landing system marker beacon (electronic navigation). *See* **boundary marker.** *See also:* **navigation.**

instrument landing system reference point (electronic navigation). A point on the centerline of the instrument landing system runway designated as the optimum point of contact for landing; in standards of the International Civil Aviation Organization this point is from 500 to 1000 feet from the approach end of the runway. *See also:* **navigation.** 0-10E6

instrument multiplier. A particular type of series resistor that is used to extend the voltage range beyond some particular value for which the instrument is already complete. *See also:* **auxiliary device to an instrument; voltage-range multiplier (recording instrument).** 39A1-0

instrument relay. A relay whose operation depends upon principles employed in measuring instruments such as the electrodynamometer, iron vane, D'Arsonval galvanometer, and moving magnet. *See also:* **relay.** 83A16-0

instrument shunt. A particular type of resistor designed to be connected in parallel with a circuit of an instrument to extend its current range. *Note:* The shunt may be internal or external to the instrument proper. *See also:* **auxiliary device to an instrument; instrument.** 39A1/42A30-31E5

instrument switch. A switch used to connect or disconnect an instrument or to transfer it from one circuit or phase to another. 37A100-31E11

instrument terminals of shunts (electric power systems). Those terminals that provide a voltage drop proportional to the current in the shunt and to which the instrument or other measuring device is connected. *See also:* **power system, low-frequency and surge testing.** 0-31E5

instrument transformer. A transformer that is intended to reproduce in its secondary circuit, in a definite and known proportion suitable for utilization in measurements, control, or protective devices, the voltage or current of its primary circuit, with its phase relations substantially preserved. *Note:* For an extensive list of cross references, see *Appendix A.* 12A0/42A15-31E12;42A30-0

instrument-transformer correction factor (watt meter or watthour meter). The ratio of true watts, or watthours, to the measured watts, or watthours, divided by the marked ratio. *Note:* The transformer correction factor for a current or potential transformer is the ratio correction factor multiplied by the phase-angle correction factor for a specified primary-circuit power factor. The true primary watts or watthours are equal to the watts or watthours measured, multiplied by the transformer correction factor and the marked ratio. The true primary watts or watthours, when measured by both current and potential transformer, are equal to the current transformer correction factor times the potential transformer correction factor multiplied by the product of the marked ratios of the current and potential transformers multiplied by the observed watts or watthours. 42A15-31E12

instrument transformer, dry-type. *See:* **dry-type.**

instrument transformer, liquid-immersed. *See:* **liquid-immersed.**

instrument transformer, low-voltage winding. Winding that is intended to be connected to the measuring or control devices. 42A15-31E12

insulated (electric system). Separated from other conducting surfaces by a dielectric substance or air space permanently offering a high resistance to the passage of current and to disruptive discharge through the substance or space. *Note:* When any object is said to be insulated, it is understood to be insulated in suitable manner for the conditions to which it is subjected. Otherwise, it is, within the purpose of this definition, uninsulated. Insulating covering of conductors is one means for making the conductors insulated. *See:* **externally operable (equipment); alive (live) (equipment or wiring); dead (equipment or wiring); insulating; insulation.** 2A2-0

insulated bearing (rotating machinery). A bearing that is insulated to prevent the circulation of stray currents. *See:* **bearing.** 0-31E8

insulated bearing housing (rotating machinery). A bearing housing that is electrically insulated from its supporting structure to prevent the circulation of stray currents. *See also:* **bearing.** 0-31E8

insulated bearing pedestal (rotating machinery). A bearing pedestal that is electrically insulated from its supporting structure to prevent the circulation of stray currents. *See also:* **bearing.** 0-31E8

insulated bolt. A bolt provided with insulation. 42A95-0

insulated coupling (rotating machinery). A coupling whose halves are insulated from each other to prevent the circulation of stray current between shafts. *See:* **rotor (rotating machinery).** 0-31E8

insulated flange (piping). Element of a flange-type coupling, insulated to interrupt the electrically conducting path normally provided by metallic piping. *See:* **rotor (rotating machinery).** 0-31E8

insulated joint(1) (conduit). A coupling or joint used to insulate adjacent pieces of conduits, pipes, rods, or bars. E16-0
(2) (cable). A device that mechanically couples and electrically insulates the sheath and armor of contiguous lengths of cable. *See also:* **tower.** 42A35-31E13

insulated rail joint. A joint used to insulate abutting rail ends electrically from one another. *See also:* **railway signal and interlocking.** 42A42-0

insulated static wire. An insulated conductor on a power transmission line whose primary function is protection of the transmission line from lightning and one of whose secondary function is communications. *See also:* **power distribution, overhead construction.** 0-31E3

insulated supply system. *See:* **ungrounded system.**

insulated turnbuckle. An insulated turnbuckle is one so constructed as to constitute an insulator as well as a turnbuckle. *See also:* **tower.** 42A35-31E13

insulating (covering of a conductor, or clothing, guards,

rods, and other safety devices). A device that, when interposed between a person and current-carrying parts, protects the person making use of it against electric shock from the current-carrying parts with which the device is intended to be used; the opposite of conducting. *See also:* **insulated; insulation.** 2A2/42A95

insulating cell (rotating machinery). An insulating liner, usually to separate a coil-side from the grounded surface at a slot. *See also;* **rotor (rotating machinery); stator.** 0-31E8

insulating material (insulant) (rotating machinery). A substance or body, the conductivity of which is zero or, in practice, very small. *See also:* **asynchronous machine; direct-current commutating machine; synchronous machine.** 50I05-31E8

insulating materials. *See:* **insulation, class ratings.**

insulating spacer. Insulating material used to separate parts. *See:* **rotor (rotating machinery); stator.** 0-31E8

insulation (1) (general). *See:* **insulation testing; basic impulse insulation level.**

(2) (interior wiring).
See:
askarel;
hevea rubber;
insulated;
insulating;
interior wiring;
saturated sleeving;
varnished tubing. 42A95

(3) (rotating machinery) (electric system). Material or a combination of suitable nonconducting materials that provide electric isolation of two parts at different voltages.
See also:
askarel;
insulated;
insulating;
insulation, class ratings;
interior wiring;
rotor;
stator. 0-31E8

(4) (cable). That part that is relied upon to insulate the conductor from other conductors or conducting parts or from ground. *See also:* **cable; insulation, class ratings.** 42A35-31E13

insulation class (bushing). The voltage by which the bushing is identified and that designates the level on which the electrical performance requirements are based. *See also:* **power distribution, overhead construction.** 57A12.80-31E12;76A1-0

insulation, class ratings (electric-machine-windings and electric cables) (1) (temperature endurance). These temperatures are, and have been in most cases over a long period of time, benchmarks descriptive of the various classes of insulating materials, and various accepted test procedures have been or are being developed for use in their identification. They should not be confused with the actual temperatures at which these same classes of insulating materials may be used in the various specific types of equipment, nor with the temperatures on which specified temperature rise in equipment standards are based. (1) In the following definitions the words **accepted tests** are intended to refer to recognized test procedures established for the thermal evaluation of materials by themselves or in simple combinations. Experience or test data, used in classifying insulating materials, are distinct from the experience or test data derived for the use of materials in complete insulation systems. The thermal endurance of complete systems may be determined by test procedures specified by the responsible technical committees. A material that is classified as suitable for a given temperature may be found suitable for a different temperature, either higher or lower, by an insulation system test procedure. For example, it has been found that some materials suitable for operation at one temperature in air may be suitable for a higher temperature when used in a system operated in an inert gas atmosphere. Likewise some insulating materials when operated in dielectric liquids will have lower or higher thermal endurance than in air. (2) It is important to recognize that other characteristics, in addition to thermal endurance, such as mechanical strength, moisture resistance, and corona endurance, are required in varying degrees in different applications for the successful use of insulating materials. *See also:* **insulation.**

(A) class 90 insulation. Materials or combinations of materials such as cotton, silk, and paper without impregnation. *Note:* Other materials or combinations of materials may be included in this class if by experience or accepted tests they can be shown to have comparable thermal life at 90 degrees Celsius.

(B) class 105 insulation. Materials or combinations of materials such as cotton, silk, and paper when suitably impregnated or coated or when immersed in a dielectric liquid. *Note:* Other materials or combinations may be included in this class if by experience or accepted tests they can be shown to have comparable thermal life at 105 degrees Celsius.

(C) class 130 insulation. Materials or combinations of materials such as mica, glass fiber, asbestos, etcetera, with suitable bonding substances. *Note:* Other materials or combinations of materials may be included in this class if by experience or accepted tests they can be shown to have comparable thermal life at 130 degrees Celsius.

(D) class 155 insulation. Materials or combinations of materials such as mica, glass fiber, asbestos, etcetera, with suitable bonding substances. *Note:* Other materials or combinations of materials may be included in this class if by experience or accepted tests they can be shown to have comparable thermal life at 155 degrees Celsius.

(E) class 180 insulation. Materials or combinations of materials such as silicone elastomer, mica, glass fiber, asbestos, etcetera, with suitable bonding substances such as appropriate silicone resins. *Note:* Other materials or combinations of materials may be included in this class if by experience or accepted tests they can be shown to have comparable thermal life at 180 degrees Celsius.

(F) class 220 insulation. Materials or combinations of materials which by experience or accepted tests can be shown to have the required thermal life at 220 degrees Celsius.

(G) class over-220 insulation. Materials consisting entirely of mica, porcelain, glass, quartz, and similar inorganic materials. *Note:* Other materials or combinations of materials may be included in this class if by experience or accepted tests they can be shown to have the required thermal life at temperatures over 220 degrees Celsius. E1-0

(2) (letter symbols).

(A) class O insulation. *See:* **class 90 insulation.**

(B) class A insulation. (1) Cotton, silk, paper, and similar organic materials when either impregnated or immersed in a liquid dielectric. (2) Molded and laminated materials with cellulose filler, phenolic resins, and other resins of similar properties. (3) Films and sheets of cellulose acetate and other cellulose derivatives of similar properties. (4) Varnishes (enamel) as applied to conductors. *Note:* An insulation is considered to be impregnated when a suitable substance replaces the air between its fibers, even if this substance does not completely fill the spaces between the insulated conductors. The impregnating substances, in order to be considered suitable, must have good insulating properties; must entirely cover the fibers and render them adherent to each other and to the conductor; must not produce interstices within itself as a consequence of evaporation of the solvent or through any other cause; must not flow during the operation of the machine at full working load or at the temperature limit specified; and must not unduly deteriorate under prolonged action of heat.

(C) class B insulation. Mica, asbestos, glass fiber, and similar inorganic materials in built-up form with organic binding substances. *Note:* A small proportion of class A materials may be used for structural purposes only. Glass fiber or asbestos magnet-wire insulations are included in this temperature class. These may include supplementary organic materials, such as polyvinylacetal or polyamide films. The electrical and mechanical properties of the insulated winding must not be impaired by application of the temperature permitted for class B material. (The word **impaired** is here used in the sense of causing any change that could disqualify the insulating material for continuous service.) The temperature endurance of different class B insulation assemblies varies over a considerable range in accordance with the percentage of class A materials employed, and the degree of dependence placed on the organic binder for maintaining the structural integrity of the insulation.

(D) class H insulation. Insulation consisting of: (1) mica, asbestos, glass fiber, and similar inorganic materials in built-up form with binding substances composed of silicone compounds or materials with equivalent properties; (2) silicone compounds in rubbery or resinous forms or materials with equivalent properties. *Note:* A minute proportion of class A materials may be used only where essential for structural purposes during manufacture. The electrical and mechanical properties of the insulated winding must not be impaired by the application of the hottest-spot temperature permitted for the specific insulation class. The word **impaired** is here used in the sense of causing any change that could disqualify the insulating material for continuously performing its intended function, whether creepage spacing, mechanical support, or dielectric barrier action.

(E) class C insulation. Insulation consisting entirely of mica, porcelain, glass, quartz, and similar inorganic materials. 57A14/57A15/57A16/57A18/89A1-0

Editor's note: The detailed definitions above are arranged in two approximately equivalent groups: The first group defines seven classes of insulation in order of increasing ability to withstand stated limiting temperatures, while maintaining a satisfactory expectation of life. Illustrative examples of suitable materials are quoted, but the definitions are primarily temperature based, and are therefore general. The second (partially complete) group is arranged in similar sequence, but the texts include more specific reference to materials and forming processes that have already demonstrated probable ability to satisfy the several limiting-temperature specifications. In this (alphabetical) group, the editor has placed class O ahead of class A to maintain the logical physical sequence. Similarly, class C has been placed at the end of the sequence. The reader will note the pair-relationship between class 90 and class O; class 105 and class A; class 130 and class B; etcetera. Class over-220 pairs with class C. *Note:* These two overlapping sequences of terms, each sequence descriptive of a graded series of **insulation levels**, remain currently in general use. The sequence based upon letter symbols is particularly favored in reference to the insulation of electric cables. The system based upon numerical values of maximum temperature rating is particularly favored in reference to the insulation of the windings of electric machines. 0-E

insulation coordination (insulation strength). *See:* **coordination of insulation (insulation strength).**

insulation fault (lightning arresters). Accidental reduction or disappearance of the insulation resistance between conductor and ground or between conductors. *See:* **lightning arrester (surge diverter).** 50I25-31E7

insulation level (1) (insulation strength). An insulation strength expressed in terms of a withstand voltage. **(2) (lightning arresters).** A combination of voltage values (both power-frequency and impulse) that characterize the insulation of an equipment with regard to its capability of withstanding dielectric stresses. *See also:* **basic impulse insulation level (BIL) (insulation strength); lightning arrester (surge diverter).** 92A1-0:50I25-31E7

insulation power factor (1) (general). The ratio of the power dissipated in the insulation, in watts, to the product of the effective voltage and current in volt-amperes, when tested under a sinusoidal voltage and prescribed conditions. *Note:* If the current also is sinusoidal, the insulation power factor is equal to the cosine of the phase angle between the voltage and the resulting current.

(2) (rotating machinery). The ratio of dielectric loss in an insulation system to the applied apparent power, when measured at power frequency under designated conditions of voltage, temperature, and humidity. *Note:* Being the sine of an angle normally small, it is practically equal to loss tangent or dissipation factor, the tangent of the same angle. The angle is the complement of the angle whose cosine is the power factor. *See:* **asynchronous machine; direct-current commutating machine.** *See also:* **loss tangent; power distribution overhead construction; synchronous machine.** 76A1-31E8

insulation resistance, direct current (1) (insulated conductor). The resistance offered by its insulation to the flow of current resulting from an impressed direct voltage. *See also:* **conductor.** E30/42A35-31E3

(2) (between two electrodes in contact with or embedded in a specimen). The ratio of the direct voltage applied to the electrodes to the total current between them. *Note:* It is dependent upon both the volume and surface resistances of the specimen. E270-0
(3) (rotating machinery). The quotient of a specified direct voltage maintained on an insulation system divided by the resulting current at a specified time after the application of the voltage under designated conditions of temperature, humidity, and previous charge. *Note:* If steady state has not been reached the apparent resistance will be affected by the rate of absorption by the insulation of electric charge. *See also:* **asynchronous machine; direct-current commutating machine; synchronous machine.** 0-31E8

insulation-resistance test. A test for measuring the ohmic resistance of insulation at predetermined values of temperature and applied direct voltage. *See also:* **asynchronous machine; direct-current commutating machine; synchronous machine.** 0-31E8

insulation-resistance versus voltage test (rotating machinery). A series of insulation-resistance measurements, made at increasing direct voltages applied at successive intervals and maintained for designated periods of time, with the object of detecting insulation system defects by departures of the measured characteristic from a typical form. Usually this is a controlled overvoltage test. *See also:* **asynchronous machine; direct-current commutating machine; synchronous machine.** 0-31E8

insulation shielding (power distribution). Conducting and/or semiconducting elements applied directly over and in intimate contact with the outer surface of the insulation. Its function is to eliminate ionizable voids at the surface of the insulation and confine the dielectric stress to the underlying insulation. *See also;* **power distribution.** 0-31E1

insulation sleeving (tubing). A varnish-treated or resin-coated flexible braided tube providing insulation when placed over conductors, usually at connections or crossovers. *See:* **cradle base (rotating machinery).** 0-31E8

insulation testing.
See:
absorption current;
acceptance proof test;
breakdown voltage;
capacitive current;
controlled overvoltage test;
electric strength;
high direct voltage;
insulation;
leakage current;
maintenance proof test;
measured current;
overvoltage;
polarization index;
proof test.

insulator. A nonconducting support for an electric conductor.
See:
apparatus insulator;
cap-and-pin insulator;
combined mechanical strength;
critical impulse flashover voltage;
dry-arcing distance;
impulse flashover voltage;
impulse flashover volt-time characteristic;
impulse wave;
impulse withstand voltage;
leakage distance;
line insulator;
low-frequency flashover voltage;
low-frequency puncture voltage;
low-frequency withstand voltage;
mechanical-impact strength;
pin insulator;
post insulator;
radio-influence voltage;
shell;
spool insulator;
strain insulator;
suspension insulator;
time-load withstand strength;
wire holder. 29A1-0

insulator arcing horn. A metal part, usually shaped like a horn, placed at one or both ends of an insulator or of a string of insulators to establish an arcover path, thereby reducing or eliminating damage by arcover to the insulator or conductor or both. *See also:* **tower.** 42A35-31E13

insulator arcing ring. A metal part, usually circular or oval in shape, placed at one or both ends of an insulator or of a string of insulators to establish an arcover path, thereby reducing or eliminating damage by arcover to the insulator or conductor or both. *See also:* **tower.** 42A35-31E13

insulator arcing shield (insulator grading shield). An arcing ring so shaped and located as to improve the voltage distribution across or along the insulator or insulator string. *See also:* **tower.** 42A35-31E13

insulator arcover. A discharge of power current in the form of an arc, following a surface discharge over an insulator. *See also:* **tower.** 42A35-31E13

insulator grading shield. *See:* **insulator arcing shield.**

insulator string. Two or more suspension insulators connected in series. *See also:* **tower.** 42A35-31E13

insulator unit. An insulator assembled with such metal parts as may be necessary for attaching it to other insulating units or device parts. 37A100-31E11

intake opening (rotating machinery). A port for the entrance of ventilation air. *See:* **cradle base (rotating machinery).** 0-31E8

intake port (rotating machinery). An opening provided for the entrance of a fluid. *See:* **cradle base (rotating machinery).** 0-31E8

integral control action (electric power systems). *See:* **control action, integral.**

integral coupling (rotating machinery). A coupling flange that is a part of a shaft and not a separate piece. *See also:* **rotor (rotating machinery).** 0-31E8

integral-horsepower motor. A motor built in a frame as large as or larger than that of a motor of open construction having a continuous rating of 1 horsepower at 1700-1800 revolutions per minute. *See also:* **asynchronous machine; direct-current commutating machine; synchronous machine.** 42A10-0

integral-slot winding (rotating machinery). A distributed winding in which the number of slots per pole per phase is an integer and is the same for all poles. *See* *also:* **rotor (rotating machinery); stator.** 0-31E8

integral unit substation. A unit substation in which the incoming, transforming, and outgoing sections are manufactured as a single compact unit. 37A100-31E11

integrated circuit (solid state). A combination of interconnected circuit elements inseparably associated on or within a continuous substrate. *Note:* To further define the nature of an integrated circuit, additional modifiers may be prefixed. Examples are: (1) dielectric-isolated monolithic integrated circuit, (2) beam-lead monolithic integrated circuit, (3) silicon-chip tantalum thin-film hybrid integrated circuit. *See:* **circuits and devices; element (integrated circuit); integrated electronics.** *See also:* **monolithic integrated circuit; multichip integrated circuit; film integrated circuit; hybrid integrated circuit.** E274-15E7/21E0

integrated demand (electric power utilization). The demand averaged over a specified period usually determined by an integrating demand meter or by the integration of a load curve. *See also:* **alternating-current distribution; demand meter.** 0-31E4

integrated-demand meter (block-interval demand meter). A meter that indicates or records the demand obtained through integration. *See also:* **demand meter; electricity meter.** 12A0/42A300

integrated electronics. The portion of electronic art and technology in which the interdependence of material, device, circuit, and system-design consideration is especially significant; more specifically, that portion of the art dealing with integrated circuits. *See also:* **integrated circuit.** E274-15E7/21E0

integrated energy curve (electric power utilization). A curve of demand versus energy showing the amount of energy represented under a load curve, or a load-duration curve, above any point of demand. Also referred to as a **peak percent curve.** *See also:* **generating station.** 0-31E4

integrated mica (reconstituted mica). *See:* **mica paper.**

integrating amplifier. An operational amplifier that produces an output signal equal to the time integral of a weighted sum of the input signals. *Note:* In an analog computer, the term integrator is synonymous with integrating amplifier. *See also:* **electronic analog computer.** E165-16E9

integrating circuit (integrator) (integrating network). *See:* **integrator.**

integrating network. *See:* **integrator.**

integrating photometer. A photometer that enables total luminous flux to be determined by a single measurement. *Note:* The usual type is the Ulbricht sphere with associated photometric equipment for measuring the luminance (photometric brightness) of the inner surface of the sphere. Z7A1-0

integrating relay. A relay that operates on the energy stored from a long pulse or a series of pulses of the same or varying magnitude, for example, a thermal relay. *See:* **relay.** 0-21E0

integrator (1) (general). Any device producing an output proportional to the integral of one variable with respect to another, usually time. *See also:* **circuits and devices; electronic analog computer; integrating amplifier, watthour meter.** E165-0

(2) (integrating circuit) (integrating network). A transducer whose output waveform is substantially the time integral of its input waveform. *Notes:* (1) Such a transducer preceding a phase modulator makes the combination a frequency modulator; or, following a frequency detector, makes the combination a phase detector. Its ratio of output amplitude to input amplitude is inversely proportional to frequency, and its output phase lags its input phase by 90 degrees. (2) In a few instances, the approximation is crude. In pulse-techniques jargon, an elementary low-pass filter section (with series resistance and shunt capacitance) is referred to, loosely, as an integrator. *See also:* **integrating amplifier; time constant of integrator.** 42A65-0

(3) (digital differential analyzer). A device using an accumulator for numerically accomplishing an approximation to the mathematical process of integration. *See also:* **circuits and devices; electronic computation; electronic digital computer.** E162-0

intelligence bandwidth. The sum of the audio- (or video-) frequency bandwidths of the one or more channels. *See also:* **modulating systems.** E145-0

intelligibility. *See:* **articulation.**

intensifier electrode. An electrode causing post acceleration. *See:* **post-accelerating electrode.** *See also:* **electrode (electron tube).** E160-0

intensity (oscilloscopes). A term used to designate brightness or luminance of the spot. *See:* **oscillograph.** 0-9E4

intensity amplifier (oscilloscopes). An amplifier for signals controlling the intensity of the spot. *See:* **oscillograph.** 0-9E4

intensity level (specific sound-energy flux level) (sound-energy flux density level) (acoustics). In decibels, of a sound is 10 times the logarithm to the base 10 of the ratio of the intensity of this sound to the reference intensity. The reference intensity shall be stated explicitly. *Note:* In discussing sound measurements made with pressure or velocity microphones, especially in enclosures involving normal modes of vibration or in sound fields containing standing waves, caution must be observed in using the terms intensity and intensity level. Under such conditions it is more desirable to use the terms pressure level or velocity level since the relationship between the intensity and the pressure or velocity is generally unkown. *See also:* **electroacoustics.** E157-1E1

intensity modulation (1) (general). The process, or effect, of varying the electron-beam current in a cathode-ray tube resulting in varying brightness or luminance of the trace. *See also:* **oscillograph; television.** 0-9E4

(2) (radar). A process used in certain displays whereby the luminance of the signal indication is a function of the received signal strength. *See also:* **navigation.** E172-10E6

intensity of magnetization. *See:* **magnetization.**

interaction-circuit phase velocity (traveling-wave tubes). The phase velocity of a wave traveling on the circuit in the absence of electron flow. *See also:* **magnetrons; electron devices, miscellaneous.** E160-15E6

interaction crosstalk coupling (between a disturbing and a disturbed circuit in any given section). The vector summation of all possible combinations of crosstalk coupling, within one arbitrary short length, between the disturbing circuit and all circuits other than the disturbed circuit (including phantom and ground-return circuits) with crosstalk coupling, within another arbitrary short length, between the disturbed circuit and all circuits other than the disturbing circuit. *See also:* **coupling.** 42A65-0

interaction factor (transducer). The factor in the equation for the received current that takes into consideration the effect of multiple reflections at its terminals. For a transducer having a transfer constant θ, image impedances Z_{I_1} and Z_{I_2}, and terminating impedances Z_S and Z_R, this factor is

$$\frac{1}{1 - \dfrac{Z_{I_2} - Z_R}{Z_{I_2} + Z_R} \times \dfrac{Z_{I_1} - Z_S}{Z_{I_1} + Z_S} \times e^{-2\theta}}.$$

See also: **transmission characteristics.** E270/42A65-0

interaction gap. An interaction space between electrodes. *See also:* **beam tubes; electron devices, miscellaneous.** E160-15E6

interaction impedance (traveling-wave tubes). A measure of the radio-frequency field strength at the electron stream for a given power in the interaction circuit. It may be expressed by the following equation

$$K = \frac{E^2}{2\,(\omega/v)^2 P}$$

where E is the peak value of the electric field at the position of electron flow, ω is the angular frequency, v is the interaction-circuit phase velocity, and P is the propagating power. If the field strength is not uniform over the beam, an effective interaction impedance may be defined. *See also:* **beam tubes; electron devices, miscellaneous.** E160-15E6

interaction loss (transducer). The interaction loss expressed in decibels is 20 times the logarithm to base 10 of the scalar value of the reciprocal of the interaction factor. *See also:* **attenuation.** 42A65-0

interaction space (traveling-wave tubes). A region of an electron tube in which electrons interact with an alternating electromagnetic field. *See also:* **miscellaneous electron devices.** E160-15E6

intercalated tapes (insulation). Two or more tapes, generally of different composition, applied simultaneously in such a manner that a portion of each tape overlies a portion of the other tape. *See also:* **power distribution, underground construction.** 42A35-31E13

intercarrier sound system. A television receiving system in which use of the picture carrier and the associated sound-channel carrier produces an intermediate frequency equal to the difference between the two carrier frequencies. *Note:* This intermediate frequency is frequency modulated in accordance with the sound signal. *See also:* **television.** 42A65-0

intercept trunk (telephony). The termination of a telephone central-office connection to a vacant number, changed number, or line out of order. *See also:* **telephone switching system.** 0-19E1

interchangeable bushing (electric power system). A bushing designed for use in both power transformers and circuit breakers. 76A1-0

interchannel interference (modulation systems). In a given channel, the interference resulting from signals in one or more other channels. *See also:* **modulating systems.** E170-0

intercom (interphone). *See:* **intercommunicating system.**

intercommunicating system (intercom) (interphone). A privately owned two-way communication system without a central switchboard, usually limited to a single vehicle, building, or plant area. Stations may or may not be equipped to originate a call but can answer any call. 42A65-0

interconnected system (electric power systems). A system consisting of two or more individual power systems normally operating with connecting tie lines. *See also:* **power system; power system, low-frequency and surge testing.** E94-31E4

interconnected star connection of polyphase circuits. *See:* **zig-zag connection of polyphase circuits.**

interconnecting channel (supervisory system). The transmission link, such as the direct wire, carrier, or microwave channel (including the direct current, tones, etcetera) by which supervisory control or indication signals or selected telemeter readings are transmitted between the master station and the remote station, or stations, in a single supervisory system. 37A100-31E11

interconnection diagram (industrial control). A special form of connection diagram that shows only the external connections between controllers and associated machinery, equipment, and extraneous components. E270-34E10

interconnection tie. A feeder interconnecting two electric supply systems. *Note:* The normal flow of energy in such a feeder may be in either direction. *See also:* **center of distribution.** 42A35-31E13

interdendritic corrosion. Corrosive attack that progresses preferentially along interdendritic paths. *Note:* This type of attack results from local differences in composition, that is, coring, commonly encountered in alloy castings. *See:* **corrosion terms.** CM-34E2

interdigital magnetron. A magnetron having axial anode segements around the cathode, alternate segments being connected together at one end, remaining segments connected together at the opposite end. 42A70-15E6

interelectrode capacitance (*j-l* interelectrode capacitance C_{jl} of an *n*-terminal electron tube). The capacitance determined from the short-circuit transfer admittance between the *j* th and the *l* th terminals. *Note:* This quantity is often referred to as direct interelectrode capacitance. *See also:* **circuit characteristic of electrodes; electron-tube admittance.** E160/42A70-15E6

interelectrode transadmittance (*j-l* interelectrode transadmittance of an *n*-electrode electron tube). The short-circuit transfer admittance from the *j* th electrode to the *l* th electrode. *See also:* **electron-tube admittances.** 42A70-15E6

interelectrode transconductance (*j-l* interelectrode transconductance). The real part of the *j-l* interelectrode transadmittance. *See also:* **electron-tube admittances.** 42A70-15E6

interelement influence (polyphase wattmeters). The percentage change in the recorded value that is caused solely by the action of the stray field of one element upon the other element. *Note:* This influence is determined at the specified frequency of calibration with rated current and rated voltage in phase on both elements or such lesser value of equal currents in both elements as gives end-scale deflection. Both current and voltage in one element shall then be reversed, and, for rating purposes, one-half the difference in the readings in percent is the interelement influence. *See also:* **accuracy rating (instrument).** 39A1/39A2-0

interface (1) (general). A shared boundary. X3A12-16E9

(2)(equipment). Interconnection between two equipments having different functions. 0-19E4

interfacial connection (soldered connections). A conductor that connects conductive patterns on opposite sides of the base material. *See also:* **soldered connections (electronic and electrical applications).** 99A1-0

interference (1) (signal transmission system). Either extraneous power, that tends to interfere with the reception of the desired signals, or the disturbance of signals that results. *Note:* Interference can be produced by both natural and man-made sources either external or internal to the signal transmission system. E188-2E2/13E6

(2)(electric-power-system measurements). Any spurious voltage or current appearing in the circuits of the instrument. *Note:* The source of each type of interference may be within the instrument case or external. The instrument design should be such that the effects of interference arising internally are negligible. 39A4-0

(3) (induction or dielectric-heating usage). The disturbance of any electric circuit carrying intelligence caused by the transfer of energy from an induction- or dielectric-heating equipment. *Note:* For an extensive list of cross references, see *Appendix A*. E54/E169

interference, common-mode (signal-transmission system). *See:* **common-mode interference.**

interference coupling ratio (signal-transmission system). The ratio of the interference produced in a signal circuit to the actual strength of the interfering source (in the same units). *See:* **interference.** 0-13E6

interference, differential-mode (signal-transmission system). Interference that causes the potential of one side of the signal transmission path to be changed relative to the other side. *Note:* That type of interference in which the interference current path is wholly in the signal transmission path. *See:* **interference.** 0-13E6

interference field strength (electromagnetic compatibility). Field strength produced by a radio disturbance. *Note:* Such a field strength has only a precise value when measured under specified conditions. Normally, it should be measured according to publications of the International Special Committee on Radio Interference. *See also:* **electromagnetic compatibility.** CISPR-27E1

interference guard bands. The two bands of frequencies additional to, and on either side of, the communication band and frequency tolerance, which may be provided in order to minimize the possibility of interference. *See also:* **channel.** E145-0

interference, longitudinal (signal-transmission system). *See:* **interference, common-mode.**

interference measurement (induction or dielectric-heating). A measurement usually of field intensity to evaluate the probability of interference with sensitive receiving apparatus. *See also:* **dielectric heating; induction heating.** E54-0

interference, normal-mode (signal-transmission system). A form of interference that appears between measuring circuit terminals. *See:* **interferrence, differential-mode.** *See also:* **accuracy rating (instrument).** 39A4-0

interference pattern. The resulting space distribution when progressive waves of the same frequency and kind are superposed. *See also:* **wave front.** 42A65-0

interference power (electromagnetic compatibility). Power produced by a radio disturbance. *Note:* Such a power has only a precise value when measured under specified conditions. *See also:* **electromagnetic compatibility.** CISPR-27E1

interference, series-mode (signal-transmission system). *See:* **interference, differential-mode.**

interference susceptibility (mobile communication). A measure of the capability of a system to withstand the effects of spurious signals and noise that tend to interfere with reception of the desired intelligence. *See also:* **mobile communication system.** 0-6E1

interference, transverse. *See:* **interference, differential-mode.**

interference voltage (electromagnetic compatibility). Voltage produced by a radio disturbance. *Note:* Such a voltage has a precise value only when measured under specified conditions. Normally, it should be measured according to recommendations of the International Special Committee on Radio Interference. *See also:* **electromagnetic compatibility.** CISPR-27E1

interflectance method (lighting calculation). A lighting design procedure for predetermining the luminances (photometric brightnesses) of walls, ceiling, and floor and the average illumination on the work plane. It takes into account both direct and reflected flux. *See also:* **inverse-square law (in illuminating engineering).** Z7A1-0

interflected component. That portion of the luminous flux from a luminaire that arrives at the work plane after being reflected one or more times from room surfaces. *See also:* **inverse-square law (illuminating engineering).** Z7A1-0

interflection. The multiple reflection of light by the various room surfaces before it reaches the work plane or other specified surface of a room. *See also:* **inverse-square law (illuminating engineering).** Z7A1-0

intergranular corrosion. Corrosion that occurs preferentially at grain boundaries. *See:* **corrosion terms.** CM-34E2

interior communication systems (marine). Those systems providing audible or visual signals or transmission of information within or on a vessel. 42A43-0

interior wiring. *Note:* For an extensive list of cross references, see *Appendix A*.

interior wiring system ground. A ground connection to one of the current-carrying conductors of an interior wiring system. *See:* **ground.** 42A35-31E13

interlaboratory standards. Those standards that are used for comparing reference standards of one laboratory with those of another, when the reference standards are of such nature that they should not be shipped. *See also:* **measurement system.** E285-9E1

interlaced scanning (television). A scanning process in which the distance from center to center of successively scanned lines is two or more times the nominal line width, and in which the adjacent lines belong to different fields. *See also:* **television.** E203-2E2

interlace factor (television). A measure of the degree of interlace of nominally interlaced fields. *Note:* In a two-to-one interlaced raster, the interlace factor is the ratio of the smaller of two distances between the centers of adjacent scanned lines to one-half the distance between centers of sequentially scanned lines at a specified point. *See also:* **television.** E204-2E2

interleave. To arrange parts of one sequence of things or events so that they alternate with parts of one or more other sequences of things or events and so that each sequence retains its identity. X3A12-16E9

interlock (industrial control). A device actuated by the operation of some other device with which it is directly associated, to govern succeeding operations of the same or allied devices. *Note:* Interlocks may be either electric or mechanical. 42A25-34E10

interlock bypass. A command to temporarily circumvent a normally provided interlock. EIA3B-34E12

interlocking (interlocking plant) (railway). An arrangement of apparatus in which various devices for controlling track switches, signals, and related appliances are so interconnected that their movements must succeed one another in a predetermined order, and for which interlocking rules are in effect. *Note:* It may be operated manually or automatically.
See:
approach locking;
automatic locking;
electric approach locking;
electric indication lock;
electric interlocking machine;
electric lock;
electric locking;
electric movable bridge lock;
electric-switch lever lock;
electric-switch lock;
electromechanical interlocking machine;
interlocking limits;
interlocking machine;
mechanical interlocking machine;
movable bridge rail lock;
route locking;
section locking;
switch and lock movement;
time locking;
traffic locking. 42A42-0

interlocking deactivating means (defeater) (industrial control). A manually actuated provision for temporarily rendering an interlocking device ineffective, thus permitting an operation that would otherwise be prevented. For example, when applied to apparatus such as combination controllers or control centers, it refers to voiding of the mechanical interlocking mechanism between the externally operable disconnect device and the enclosure doors to permit entry into the enclosure while the disconnect device is closed. *See:* **electric controller.** IC1-34E10

interlocking limits (interlocking territory) (railway). An expression used to designate the trackage between the opposing home signals of an interlocking. *See also:* **interlocking.** 42A42-0

interlocking machine (railway). An assemblage of manually operated levers or equivalent devices, for the control of signals, switches, or other units, and including mechanical or circuit locking, or both, to establish proper sequence of movements. *See also:* **interlocking (interlocking plant).** 42A42-0

interlocking plant. *See:* **interlocking.**

interlocking relay (railway). A relay that has two independent magnetic circuits with their respective armatures so arranged that the dropping away of either armature prevents the other armature from dropping away to its full stroke. *See also:* **railway signal and interlocking.** 42A42-0

interlocking signals (railway). The fixed signals of an interlocking. *See also:* **railway signal and interlocking.** 42A42-0

interlocking station (railway). A place from which an interlocking is operated. *See also:* **railway signal and interlocking.** 42A42-0

interlocking territory. *See:* **interlocking limits.**

interlock relay. A relay with two or more armatures having a mechanical linkage, or an electric interconnection, or both, whereby the position of one armature permits, prevents, or causes motion of another armature. *See also:* **relay.** 83A16-0

intermediate contacts (switching device). Contacts in the main circuit that part after the main contacts and before the arcing contacts have parted. 37A100-31E11

intermediate frequency (IF) (1) (general). A frequency to which a signal wave is shifted locally as an intermediate step in transmission or reception. 42A65-31E3

(2) (superheterodyne reception). The frequency resulting from a frequency conversion before demodulation. *See also:* **radio transmission.** E188-0

intermediate-frequency-harmonic interference (superheterodyne receivers). Interference due to radio-frequency-circuit acceptance of harmonics of an intermediate-frequency signal. E188-0

intermediate-frequency interference ratio. *See:* **intermediate-frequency response ratio.** *See also:* **radio receiver.**

intermediate-frequency response ratio (superheterodyne receivers). The ratio of (1) the field strength at a specified frequency in the intermediate frequency band to (2) the field strength at the desired frequency, each field being applied in turn, under specified conditions, to produce equal outputs. *See also:* **radio receiver.** E188-0

intermediate-frequency transformer. A transformer used in the intermediate-frequency portion of a heterodyne system. *Note:* Intermediate-frequency transformers are frequently narrow-band devices. 0-21E1

intermediate means (measurement sequence). All system elements that are used to perform necessary and distinct operations in the measurement sequence between the primary detector and the end device. *Note:* The intermediate means, where necessary, adapts the operational results of the primary detector to the input requirements of the end device. *See also:* **measurement system.** 39A4/42A30-0

intermediate repeater. A repeater for use in a trunk or line at a point other than an end. *See also:* **repeater.** 42A65-0

intermediate subcarrier. A carrier that may be modulated by one or more subcarriers and that is used as a modulating wave to modulate a carrier or another intermediate subcarrier. *See also:* **carrier; subcarrier.** E145-0

intermittent duty (1) (general). A requirement of service that demands operation for alternate periods of (A) load and no load; or (B) load and rest; or (C) load, no load, and rest, such alternate intervals being definitely specified.
E96-SCC4;E270/57A15-0;42A15-31E12;IC1-34E10
(2) (rotating machinery). A duty in which the load changes but does not change regularly with time. *See also:* **asynchronous machine; direct-current commutating machine; synchronous machine; voltage regulator.** 0-31E8

intermittent-duty rating. The specified output rating of a device when operated for specified intervals of time other than continuous duty. E145-0

intermittent fault (lightning arresters). A fault that recurs in the same place and due to the same cause within a short period of time. *See:* **lightning arrester (surge diverter).** 50I25-31E7

intermittent inductive train control. Intermittent train control in which the impulses are communicated to the vehicle-carried apparatus inductively. *See also:* **automatic train control.** 42A42-0

intermittent rating. *See:* **periodic rating.**

intermittent test (battery). A service test in which the battery is subjected to alternate discharges and periods of recuperation according to a specified program until the cutoff voltage is reached. *See also:* **battery (primary or secondary).** 42A60-0

intermittent train control. A system of automatic train control in which impulses are communicated to the locomotive or vehicle at fixed points only. *See also:* **automatic train control; intermittent inductive train control.** 42A42-0

intermodulation (nonlinear transducer element). The modulation of the components of a complex wave by each other, as a result of which waves are produced that have frequencies equal to the sums and differences of integral multiples of those of the components of the original complex wave. *See also:* **modulation; transmission characteristics.** E145/42A65-27E1/31E3

intermodulation spurious response (receiver performance). The receiver audio output resulting from the mixing of *n*th-order frequencies, in the nonlinear elements of the receiver, in which the resultant carrier frequency is equivalent to the assigned frequency. *See:* **spurious response.** *See also:* **receiver performance.** 0-6E1

intermodulation interference (mobile communication). The modulation products attributable to the components of a complex wave that on injection into a nonlinear circuit produces interference on the desired signal. *See also:* **mobile communication system.** 0-6E1

internal bias (teletypewriter). Bias, either marking or spacing, that may occur within a start-stop printer receiving mechanism and that will have an effect on the margins of operation. *See also:* **data transmission.** 0-19E4

internal connector (pothead). A connector that joins the end of the cable to the other current-carrying parts of a pothead. *See also:* **pothead; transformer.** E48/57A12.75/57A12.76-0

internal correction voltage (electron tubes). The voltage that is added to the composite controlling voltage and is the voltage equivalent of such effects as those produced by initial electron velocity and contact potential. *See also:* **composite controlling voltage (electron tube).** 42A70-15E6

internal heating (electrolysis). The electrolysis of fused electrolytes is the method of maintaining the electrolyte in a molten condition by the heat generated by the passage of current through the electrolyte. *See also:* **fused electrolyte.** 42A60-0

internal impedance (rotating machinery). The total self-impedance of the primary winding under steady conditions. *Note:* For a three-phase machine, the primary current is considered to have only a positive-sequence component when evaluating this quantity. *See also:* **asynchronous machine; synchronous machine.** 0-31E8

internal impedance drop (rotating machinery). The product of the current and the internal impedance. *Note:* This is the phasor difference between the generated internal voltage and the terminal voltage of a machine. *See also:* **asynchronous machine; synchronous machine.** 0-31E8

internal insulation (apparatus) (lightning arresters). The insulation that is not directly exposed to atmospheric conditions. *See:* **lightning arrester (surge diverter).** 60I0-31E7

internal oxidation. *See:* **subsurface corrosion.**

internal resistance (battery). The resistance to the flow of an electric current within a cell or battery. *See also:* **battery (primary or secondary).** 42A60-0

internal triggering (oscilloscopes). The use of a portion of a deflection signal (usually the vertical deflection signal) as a triggering-signal source. *See:* **oscillograph.** 0-9E4

International Commission on Illumination. *See:* **CIE.**

International Morse code (Continental code). A system of dot and dash signals, differing from the American Morse code only in certain code combinations, used chiefly in international radio and wire telegraphy. *See also:* **telegraphy.** 42A65-0

International System of Electrical Units. A system that uses the **international ampere** and the **international ohm.** *Notes:* (1) The international ampere was defined as the current that will deposit silver at the rate of 0.00111800 gram per second; and the international ohm was defined as the resistance at 0 degrees Celsius of a column of mercury having a length of 106.300 centimeters and a mass of 14.4521 grams. (2) The International System of Electrical Units was in use between 1893 and 1947 inclusive. By international agreement it was discarded, effective January 1, 1948 in favor of the MKSA system. (3) Experiments have shown that as these units were maintained in the United States of America, 1 international ohm equalled 1.000495 ohm and that 1 international ampere equalled 0.999835 ampere. *See:* **International System of Units.** E270-0

International System of Units. A universal coherent system of units in which the following six units are considered basic: meter, kilogram, second, ampere, Kelvin degree, and candela. *Notes:* (1) The MKSA system of electrical units (MKSA System of Units) is a constituent part of this system adequate for mechanics and electromagnetism. (2) The electrical units of this system should not be confused with the units of the earlier International System of Electrical Units which was discarded January 1, 1948. (3) The International System of Units (abbreviated SI) was promulgated in 1960 by the Eleventh General Conference on Weights and Measures. *See also:* **kelvin.** E270-0

interoffice trunk (telephony). A direct trunk between local central offices in the same exchange. *See also:* **telephone system.** 42A65-0

interphase transformer. An autotransformer, or a set of mutually coupled reactors, used to obtain parallel operation between two or more simple rectifiers that have ripple voltages that are out of phase. *See:* **autotransformer; rectifier transformer.** 42A15-31E12

interphase-transformer loss (rectifier transformer). The losses in the interphase transformer that are incident to the carrying of rectifier load. *Note:* They include both magnetic core loss and conductor loss. *See also:* **rectifier transformer.** 57A18-0

interphase-transformer rating. Consists of the root-mean-square current, root-mean-square voltage, and frequency, at the terminals of each winding, for the rated load of the rectifier unit, and a designated amount of phase control, as assigned to it by the manufacturer. *See:* **duty; rectifier transformer.** 42A15-31E12

interphone (intercom). *See:* **intercommunicating system.**

interphone equipment (aircraft) . Equipment used to provide telephone communications between personnel at various locations in an aircraft. *See also:* **air transportation electronic equipment.** 42A41-0

interpolation (signal interpolation) (submarine cable telegraphy). A method of reception characterized by synchronous restoration of unit-length signal elements which are weak or missing in the received signals as a result of one or more of such factors as suppression at the transmitter, attenuation in transmission, or discrimination in the receiving networks. *Note:* This is sometimes referred to as local correction. *See also:* **telegraphy.** 42A65-0

interpolation function (burst measurements). A function that may be used to obtain additional values between sampled values. *See also:* **burst.** E265-0

interpole. *See:* **commutating pole.**

interposing relay (supervisory system). An auxiliary relay at the master or remote station, the contacts of which serve: (1) to energize a circuit (for closing, opening, or other purpose) of an element of remote station equipment when the selection of a desired point has been completed and when suitable operating signals are received through the supervisory equipment from the master station; or (2) to connect in the circuit the telemeter transmitting and receiving equipments, respectively, at the remote and master stations. *Note:* The interposing relays are considered part of a supervisory system. 37A100-31E11

interpreter (computing systems). (1) A program that translates and executes each source language expression before translating and executing the next one. (2) A device that prints on a punched card the data already punched in the card. X3A12-16E9

interrogation (transponder system). The signal or combination of signals intended to trigger a response. *See also:* **navigation.** 0-10E6

interrogator (electronic navigation). (1) A radio transmitter and receiver combined to interrogate a transponder and display the resulting replies. (2) The transmitting part of an interrogator-responsor. *See also:* **radio transmission.** 0-10E6

interrogator-responsor (IR). A combined radio transmitter and receiver for interrogating a transpondor and displaying the resulting replies. *See also:* **radio transmission.** 42A65-0

interrupt. To stop a process in such a way that it can be resumed. X3A12-16E9

interrupted continuous wave (ICW). A continuous wave that is interrupted at a constant audio-frequency rate. *See also:* **radio transmission.** E145/E182A-0;42A65-31E3;0-42A65

interrupted quick-flashing light. A quick-flashing light in which the rapid alternations are interrupted by periods of darkness at regular intervals. *See also:* **signal lighting.** Z7A1-0

interrupter. An element designed to interrupt specified currents under specified conditions. 37A100-31E11

interrupter blade (of an interrupter switch). A blade used in the interrupter for breaking the circuit. 37A100-31E11

interrupter switch. An air switch, equipped with an interrupter, for making or breaking specified currents, or both. *Note:* The nature of the current made or broken, or both, may be indicated by suitable prefix; that is, load interrupter switch, fault interrupter switch, capacitor current interrupter switch, etcetera. 37A100-31E11

interruptible loads (electric power utilization). Those loads that by contract can be interrupted in the event of a capacity deficiency on the supplying system. *See also:* **generating station.** 0-31E4

interruptible power. Power made available under agreed conditions that permit curtailment or cessation of delivery by the supplier. *See also:* **generating station.** 42A35-31E13

interrupting current (switching device) (breaking current). The current in a pole at the instant of the initiation of the arc. 37A100-31E11

interrupting tests. Tests that are made to determine or check the interrupting performance of a switching device. 37A100-31E11

interrupting time (mechanical switching device) (total break time). The interval between the time when the actuating quantity of the release circuit reaches the operating value, the switching device being in a closed position, and the instant of arc extinction on the primary arcing contacts. *Notes:* (1) Interrupting time is numerically equal to the sum of opening time and arcing time. (2) In multipole devices interrupting time may be measured for each pole or for the device as a whole, in which latter case the interval is measured to the instant of arc extinction in the last pole to clear. 37A100-31E11

interruption (electric power systems). The loss of service to one or more consumers or other facilities. *Note:* It is the result of one or more component outages, depending on system configuration. *See also:* **outage.** 0-31E4

interruption duration (electric power systems). The period from the initiation of an interruption to a consumer or other facility until service has been restored to that consumer or facility. *See also:* **outage.** 0-31E4

interruption duration index (electric power systems). The average interruption duration for consumers interrupted during a specified time period. It is estimated from operating history by dividing the sum of all consumer interruption durations during the specified period by the number of consumer interruptions during that period. *See also:* **outage.** 0-31E4

interruption frequency index (electric power systems). The average number of interruptions per consumer served per time unit. *Note:* It is estimated from operating history by dividing the number of consumer interruptions observed in a time unit by the number of consumers served. A consumer interruption is considered to be one interruption of one consumer. *See also:* **outage.** 0-31E4

interspersing (rotating machinery). Interchanging the coils at the edges of adjacent phase belts. *Note:* The purpose of interspersing depends on the type of machine in which it is done. In asynchronous motors it is used to reduce harmonics that can cause crawling. *See also:* **asynchronous machine; synchronous machine.** 0-31E8

intersymbol interference (transmission system) (modulation systems). Extraneous energy from the signal in one or more keying intervals that tends to interfere with the reception of the signal in another keying interval, or the disturbance that results. *See also:* **modulating systems.** E170-0

intertoll dialing (telephony). Dialing over intertoll trunks. *See also:* **telephone switching system.** 42A65-0

intertoll trunk (intertoll office trunk). A trunk between toll offices in different telephone exchanges. *See also:* **telephone system.** 42A65-19E1

interturn insulation (rotating machinery). The insulation between adjacent turns. *See also:* **asynchronous machine; direct-current commutating machine; synchronous machine.** 0-31E8

interturn test. *See:* **turn-to-turn test (rotating machin-ery).**

interval (acoustics). The spacing between two sounds in pitch or frequency, whichever is indicated by the context. *Note:* The frequency interval is expressed by the ratio of the frequencies or by a logarithm of this ratio. *See also:* **electroacoustics.** E157-1E1

intrinsically safe equipment. Equipment that is incapable of releasing sufficient electric or thermal energy under normal or abnormal conditions to cause ignition of a specific hazardous atmospheric mixture in its most-ignitible concentration. ISA-34E16

intrinsic coercive force. The magnetizing force at which the intrinsic induction is zero when the material is in a symmetrically cyclically magnetized condition. E270-0

intrinsic impedance (antenna). The theoretical input impedance for the basic radiating structure when idealized. *Note:* The idealized basic radiating structure usually consists of a uniform-cross-section radiating element, perfectly conducting ground or imaging planes, zero base capacitance (in the case of vertical radiators), and no internal losses. 0-3E1

intrinsic induction (magnetic polarization). At a point in a magnetized body, the vector difference between the magnetic induction at that point and the magnetic induction that would exist in a vacuum under the influence of the same magnetizing force. This is expressed by the equation $\mathbf{B}_i = \mathbf{B} - \mu_0\mathbf{H}$. *Note:* In the centimeter-gram-second electromagnetic-unit system, $\mathbf{B}_i/4\pi$ is often called magnetic polarization. E270-0

intrinsic permeability. The ratio of intrinsic normal induction to the corresponding magnetizing force. *Note:* In anisotropic media, intrinsic permeability becomes a matrix. E270-0

intrinsic properties (semiconductor). The properties of a semiconductor that are characteristic of the pure, ideal crystal. *See also:* **semiconductor.** E216-34E17;E102/E270-10E1

intrinsic semiconductor. *See:* **semiconductor, intrinsic.**

intrinsic temperature range (semiconductor). The temperature range in which the charge-carrier concentration of a semiconductor is substantially the same as that of an ideal crystal. *See also:* **semiconductor.** E216-34E17;E270-0

inverse electrode current. The current flowing through an electrode in the direction opposite to that for which the tube is designed. *See also:* **electrode current (electron tube).** 42A70-15E6

inverse magnitude contours. *See:* **magnitude contours.**

inverse networks. Two two-terminal networks are said to be inverse when the product of their impedances is independent of frequency within the range of interest. *See also:* **network analysis.** E270/42A65-0

inverse neutral telegraph transmission. That form of transmission employing zero current during marking intervals and current during spacing intervals. *See also:* **telegraphy.** 42A65-19E4

inverse Nyquist diagram. *See:* **Nyquist diagram.**

inverse-parallel connection (industrial control). An electric connection of two rectifying elements such that the cathode of the first is connected to the anode of the second, and the anode of the first is connected to the cathode of the second. 42A25-34E10

inverse period (rectifier element). The nonconducting part of an alternating-voltage cycle during which the anode has a negative potential with respect to the cathode. *See also:* **rectification.** 42A15-0

inverse-square law. A statement that the strength of the field due to a point source or the irradiance from a point source decreases as the square of the distance from the source. *Notes:* (1) For sources of finite size this gives results that are accurate within one-half percent when distance is at least five times the maximum dimension of the source (or luminaire) as viewed by the observer. (2) For an extensive list of cross references, see *Appendix A.* E270/Z7A1-0

inverse time (industrial control). *See:* **inverse-time relay.**

inverse-time relay. A relay in which the input quantity and operating time are inversely related throughout at least a substantial portion of the performance range. *Note:* Types of inverse-time relays are frequently identified by such modifying adjectives as definite minimum time, moderately, very, and extremely to identify relative degree of inverseness of the operating characteristics of a given manufacturer's line of such relays. 37A100-31E11/31E6

inverse voltage (rectifier). The voltage applied between the two terminals in the direction opposite to the forward direction. This direction is called the backward direction. *See also:* **rectification of an alternating current.** 50I07-15E6

inversion efficiency. The ratio of output fundamental power to input direct power expressed in percent. *See:* **self-commutated inverters.** 0-34E24

inverted (rotating machinery). Applied to a machine in which the usual functions of the stationary and revolving members are interchanged. *Example:* An induction motor in which the primary winding is on the rotor and is connected to the supply through sliprings, and the secondary is on the stator. *See:* **asynchronous machine.** 0-31E8

inverted-turn transposition (rotating machinery). A form of transposition used on multiturn coils in which one or more turns are given a 180-degree twist in the end winding or at the coil nose or series loop. *See also:* **rotor (rotating machinery); stator.** 0-31E8

inverter (electric power). A machine, device, or system that changes direct-current power to alternating-current power. *See:* **inverting amplifier.** *See also:* **electronic analog computer.** 42A10-0

inverting amplifier. An operational amplifier that produces an output signal of nominally equal magnitude and opposite algebraic sign to the input signal. *Notes:* (1) In an analog computer, the term inverter is synonymous with inverting amplifier. (2) An operational direct-current power supply can also be described as a high-gain inverting amplifier. *See also:* **electronic analog computer; power supply.** E165/KPSH-16E9/10E1

inverting parametric device. A parametric device whose operation depends essentially upon three frequencies, a harmonic of the pump frequency and two signal frequencies, of which the higher signal-frequency is the difference between the pump harmonic and the lower signal frequency. *Note:* Such a device can exhibit gain at either of the signal frequencies provided power is suitably dissipated at the other signal frequency. It is said to be inverting because if one of the two signals is moved upward in frequency, the other will move downward in frequency. *See also:* **parametric device.** E254-15E7

I/O (input/output). Input or output or both. X3A12-16E9

ion. An electrically charged atom or radical.
See:
anion;
buffers;
cation;
corrosion terms;
electrophoresis;
electrostenolysis;
equivalent concentration;
ion activity;
ion concentration;
ionizing radiation;
ion migration;
isoelectric point;
pH. CM-34E2

ion activity (ion species). The thermodynamic concentration, that is, the ion concentration corrected for the deviation from the law of ideal solutions. *Note:* The activity of a single ion species cannot, however, be measured thermodynamically. *See also:* **ion.** 42A60-0

ion burn. *See:* **ion spot.**

ion charging (charge-storage tubes). Dynamic decay caused by ions striking the storage surface. *See also:* **charge-storage tube.** E158-15E6

ion concentration (species of ion). The concentration equal to the number of those ions, or of moles or equivalent of those ions, contained in a unit volume of an electrolyte. 42A60-0

ion gun. A device similar to an electron gun but in which the charged particles are ions. *Example:* proton gun. *See also:* **electron optics.** 50I07-15E6

ionic-heated-cathode tube. An electron tube containing an ionic-heated cathode. *See also:* **tube definitions.** 42A70-0

ionization (1) (general). The process or the result of any process by which a neutral atom or molecule acquires either a positive or a negative charge. *See also:* **photovoltaic power system; solar cells (in a photovoltaic power system).** 0-10E1

(2) (power lines). A breakdown that occurs in parts of a dielectric when the electric stress in those parts exceeds a critical value without initiating a complete breakdown of the insulation system. *Note:* Ionization can occur both on internal and external parts of a device. It is a source of radio noise and can damage insulation. *See also:* **power distribution, overhead construction; lightning arrester (surge diverter).** LA1-31E7

ionization current. *See:* **gas ionization current.**

ionization extinction voltage (corona level) (cables). The minimum value of falling root-mean-square voltage that sustains electric discharge within the vacuous or gas-filled spaces in the cable construction or insulation. *See also:* **power distribution underground construction.** 0-31E1

ionization factor (power distribution underground cables) (dielectric). The difference between percent power factors at two specified values of electric stress. The lower of the two stresses is usually so selected that the effect of the ionization on power factor at this stress is negligible. *See also:* **power distribution, underground construction.** 0-31E1

ionization-gauge tube. An electron tube designed for the measurement of low gas pressure and utilizing the relationship between gas pressure and ionization current. *See also:* **tube definitions.** 42A70-15E6

ionization measurement. The measurement of the electric current resulting from the movement of electric charges in an ionized medium under the influence of the prescribed electric field. *See also:* **power distribution, overhead construction.** 76A1-0

ionization or corona detector. *See:* **discharge detector.**

ionization or corona inception voltage. *See:* **discharge inception voltage.**

ionization or corona probe. *See:* **discharge probe.**

ionization time (gas tube). The time interval between the initiation of conditions for and the establishment of conduction at some stated value of tube voltage drop. *Note:* To be exact the ionization time of a gas tube should be presented as a family of curves relating such factors as condensed-mercury temperature, anode and grid currents, anode and grid voltages, and regulation of the grid current. *See:* **gas tubes.** 42A70-15E6

ionization vacuum gauge. A vacuum gauge that depends for its operation on the current of positive ions produced in the gas by electrons that are accelerated between a hot cathode and another electrode in the evacuated space. *Note:* It is ordinarily used to cover a pressure range of 10^{-4} to 10^{-10} conventional millimeters of mercury. *See also:* **instrument.** 42A30-0

ionization voltage (lightning arresters). A high-frequency voltage appearing at the terminals of the arrester, generated by all sources, but particularly by ionization within the arrester, when a power-frequency voltage is applied across the terminals. *Note:* This voltage provides an indication of internal ionization that might cause deterioration in service. The internal ionization voltage of parts of an arrester may differ from that of a complete arrester. *See:* **lightning arrester (surge diverter).** LA1-31E7

ionizing event (gas-filled radiation-counter tube). Any interaction by which one or more ions are produced. *See also:* **gas-filled radiation-counter tubes.** 42A70-15E6

ionizing radiation (1) (general). Any electromagnetic or particulate radiation capable of producing ions, directly or indirectly, in its passage through matter. *See also:* **photovoltaic power system; solar cells (photovoltaic power system).** 0-10E1
(2) (air). (A) Particles or photons of sufficient energy to produce ionization in their passage through air. (B) Particles that are capable of nuclear interactions with the release of sufficient energy to produce ionization in air.
See:
background counts;
background response;
ion;
resolving time;
scintillation;
scintillation counter;
sensitivity;
sensitivity, incremental;
sensitivity, threshold;
time distribution analyzer. E175-0

ion migration. A movement of ions in an electrolyte as a result of the application of an electric potential. *See also:* **ion.** 42A60-0

ionosphere. That part of the earth's outer atmosphere where ions and free electrons are normally present in quantities sufficient to affect the propagation of radio waves. *Notes:* (1) According to present opinion, the lowest level of the ionosphere is approximately 50 kilometers above the earth's surface. (2) Modern usage extends the term to the atmospheres of other planets. *See also:* **radiation; radio wave propagation.** 42A65-3E2

ionospheric disturbance. A variation in the state of ionization of the ionosphere beyond the normally observed random day-to-day variation from average values for the location, date, and time of day under consideration. *Note:* Since it is difficult to draw the line between normal and abnormal variations, this definition must be understood in a qualitative sense. *See also:* **radiation.** 42A65-0

ionospheric error (electronic navigation). The total systematic and random error resulting from the reception of the navigational signal via ionospheric reflections; this error may be due to (1) variations in transmission paths, (2) nonuniform height of the ionosphere, and (3) nonuniform propagation within the ionosphere. *See also:* **navigation.** E172-10E6

ionospheric-height error (electronic navigation). The systematic component of the total ionospheric error due to the difference in geometrical configuration between ground paths and ionospheric paths. *See also:* **navigation.** E172-10E6

ionospheric storm. An ionospheric disturbance characterized by wide variations from normal in the state of the ionosphere, including effects such as turbulence in the *F* region, increases in absorption, and often decreases in ionization density and increases in virtual height. *Note:* The effects are most marked in high magnetic latitudes and are associated with abnormal solar activity. *See also:* **radiation.** 42A65-0

ionospheric tilt error (electronic navigation). The component of the ionospheric error due to nonuniform height of the ionosphere. *See also:* **navigation.** 0-10E6

ionospheric wave (radio wave propagation). A radio wave propagated by reflection or refraction from an ionosphere. *Note:* This is sometimes called a sky wave. *See also:* **radiation; radio wave propagation.** 42A65-3E2

ion repeller (charge-storage tubes). An electrode that produces a potential barrier against ions. *See also:* **charge-storage tube.** E158-15E6

ion sheath. *See:* **electron sheath.**

ion spot (1) (camera tubes or image tubes). The spurious signal resulting from the bombardment or alteration of the target or photocathode by ions. *See also:* **television.** E160-2E2/15E6
(2) (cathode-ray-tube screen). An area of localized deterioration of luminescence caused by bombardment with negative ions. *See also:* **beam tubes.** E160/42A70-2E2/15E6

ion transfer (electrotherapy) (ionic medication*) (iontophoresis*) (medical ionization*) (ion therapy*) (ionotherapy*). The forcing of ions through biological interfaces by means of an electric field. *See also:* **electrotherapy.**
*Deprecated 42A80-18E1

ion trap (cathode-ray tube). A device to prevent ion burn by removing the ions from the beam. *See also:* **cathode-ray tubes.** 50I07-15E6

IR. *See:* **interrogator-respondor.**

IR drop (electrolytic cell). The drop equal to the product of the current passing through the cell and the resistance of the cell. 42A60-0

IR-drop compensation (electric drive) (industrial control). A provision in the system of control by which the voltage drop (and corresponding speed drop) due to armature current and armature-circuit resistance is partially or completely neutralized. 42A25-34E10

iris (waveguide technique). A metallic plate, usually of small thickness compared with the wavelength, perpendicular to the axis of a waveguide and partially blocking it. *Notes:* (1) An iris acts like a shunt element in a transmission line; it may be inductive, capacitive, or resonant. (2) When only a single mode can be supported an iris acts substantially as a shunt admittance. *See also:* **waveguide.** 42A65/E147-3E1

ironclad plate (storage cell). A plate consisting of an assembly of perforated tubes of insulating material and of a centrally placed conductor. *See also:* **battery (primary or secondary).** 42A60-0

irradiance. *See:* **radiant flux density at a surface.**

irreversible process. An electrochemical reaction in which polarization occurs. *See also:* **electrochemistry.** 42A60-0

***I* scan.** *See:* ***I* display.**

***I* scope.** *See:* ***I* display.**

island effect (electron tubes). The restriction of the emission from the cathode to certain small areas of it (islands) when the grid voltage is lower than a certain value. *See also:* **electronic tube.** 50I07-15E6

ISM apparatus (industrial, scientific, and medical apparatus; electromagnetic compatibility. Apparatus intended for generating radio-frequency energy for industrial, scientific or medical purposes. *See also:* **electromagnetic compatibility; industrial electronics.** CISPR-27E1

isochronous speed governing (gas turbines). Governing with steady-state speed regulation of essentially zero magnitude. E282-31E2

isocandela line. A line plotted on any appropriate coordinates to show directions in space about a source of light, in which the candlepower is the same. For a complete exploration the line always is a closed curve. A series of such curves, usually for equal increments of candlepower, is called an isocandela diagram. *See also:* **lamp.** Z7A1-0

isocon mode (camera tube). A low-noise return-beam mode of operation utilizing only back-scattered electrons from the target to derive the signal, with the beam electrons specularly reflected by the electrostatic field near the target being separated and rejected. *See also:* **camera tube.** 0-15E6

isoelectric point. A condition of net electric neutrality of a colloid, with respect to its surrounding medium. *See also:* **ion.** 42A60-0

isokeraunic level (lightning). The average annual number of thunderstorm days. *See also:* **direct-stroke protection (lightning).** 0-31E13

isolated (electric system). An object not readily accessible to persons unless special means for access are used. *See also:* **power distribution, overhead construction.** 2A2/42A95-0

isolated by elevation. Elevated sufficiently so that persons may walk safely underneath. *See also:* **power distribution, overhead construction.** 42A95-0

isolated conductor (ignored conductor). In a multiple-conductor system, a conductor either accessible or inaccessible, the charge of which is not changed and to which no connection is made in the course of the determination of any one of the capacitances of the remaining conductors of the system. E270-0

isolated-neutral system. A system that has no intentional connection to ground except through indicating, measuring, or protective devices of very-high impedance. *See:* **grounded system.** 50I25-31E7

isolated-phase bus. A bus in which each phase conductor is enclosed by an individual metal housing separated from adjacent conductor housings by an air space. *Note:* The bus may be self-cooled or may be forced cooled by means of circulating a gas or liquid. 37A100-31E11

isolated plant (electric power). An electric installation deriving energy from its own generator driven by a prime mover and not serving the purpose of a public utility. 42A95-0

isolating amplifier (signal-transmission system). *See:* **amplifier, isolating.**

isolating switch. A switch intended for isolating an electric circuit from the source of power. *Note:* It has no interrupting rating, and it is intended to be operated only after the circuit has been opened by some other means. *See also:* **switch.** 42A25-34E10

isolating time (sectionalizer). The time between the cessation of a current above the minimum actuating current value that caused the final counting and opening operation and the maximum separation of the contacts. 37A100-31E11

isolating transformer (signal-transmission system). *See:* **transformer, isolating.**

isolation (1) (general). Separation of one section of a system from undesired influences of other sections. 0-21E1

(2) (signal transmission). Physical and electrical arrangement of the parts of a signal-transmission system to prevent interference currents within or between the parts. *See:* **signal.** 0-13E6

(3) (antennas). A measure of undesired power transfer from one antenna to another. *Note:* The isolation between antennas is the ratio of power input to one antenna to the power received by the other, usually expressed in decibels. *See also:* **radiation.** 0-3E1

(4) (multiport device). Between two ports of a nonreciprocal or directional multiport device, the characteristic insertion loss in the nonpreferred direction of propagation, all ports being terminated by reflectionless terminations. *Note:* It is usually expressed in decibels. *See also:* **transmission characteristics.** 0-9E4

(5) (industrial control). Connotes the opening of all conductors connected to the source of power. IC1-34E10

isolation amplifier (buffer). *See:* **amplifier, isolation.**

isolation by elevation. *See:* **isolated by elevation.**

isolation transformer*. *See:* **transformer, isolating.**

*Deprecated

isolation voltage (power supplies). A rating for a power supply that specifies the amount of external voltage that can be connected between any output terminal and ground (the chassis). This rating is important when power supplies are connected in series. *See also:* **power supply.** KPSH-10E1

isolator (1) (switchgear). *See:* **disconnecting or isolating switch.**

(2) (waveguide). A passive attenuator in which the loss in one direction is much greater than that in the opposite direction. *See:* **waveguide.** 50I62-3E1

isolux (isocandela) (isofootcandle) line. A line plotted on any appropriate coordinates to show all the points on a surface where the illumination is the same. *Note:* For a complete exploration the line is a closed curve. A series of such lines for various illumination values is called an isolux (isofootcandle) diagram. *See also:* **lamp.** Z7A1-0

isophase. *See:* **equal interval.**

isotropic radiator. A hypothetical antenna radiating or receiving equally in all directions. *Notes:* (1) For sound waves, a pulsating sphere is an isotropic radiator. (2) For coherent electromagnetic waves, an isotropically radiating antenna does not exist physically but represents a convenient reference for expressing the directive properties of actual antennas. *See:* **antenna.** E149-3E1

$I \cdot T$ product (electric supply circuit current workbook). The inductive influence usually expressed in terms of the product of its root-mean-square magnitude in amperes I times its telephone influence factor (TIF), abbreviated as $I \cdot T$ product. *See also:* **inductive coordination.** 42A65-0

I^2t characteristic (fuse). The amount of ampere-squared seconds passed by the fuse during a specified period and under specified conditions. *Notes:* (1) The specified period may be the melting, arcing, or total clearing time. The sum of melting and arcing I^2t is the clearing I^2t. (2) The melting I^2t characteristic is related to a specified current wave shape, and the arcing I^2t to specified voltage and circuit impedance conditions. 37A100-31E11

item (1) (computing systems). A collection of related characters, treated as a unit. *See also:* **file.** X3A12-16E9

(2) (reliability). An all-inclusive term to denote any level of hardware assembly; that is, system, segment of a system, subsystem, equipment, component, part, etcetera. *Note:* Item includes items, population of items, sample, etcetera, where the context of its use so justifies. *See also:* **reliability.** 0-7E1

item, nonrepaired (reliability). An item that is not repaired after a failure. *See also:* **reliability.** 0-7E1

item, repaired (reliability). An item that is repaired after a failure. *See also:* **reliability.** 0-7E1

iterative impedance. *See:* **impedance, iterative.** *See also:* **self-impedance.**

J

jack (electric circuits). A connecting device, ordinarily designed for use in a fixed location, to which a wire or wires of a circuit may be attached and that is arranged for the insertion of a plug. *See also:* **circuits and devices.** 42A65-21E0

jack bolt (rotating machinery). A bolt used to position or load an object. *See:* **cradle base (rotating machinery).** 0-31E8

jacket (1) (primary dry cell). An external covering of insulating material, closed at the bottom. *See also:* **electrolytic cell.** 42A60-0

(2) (cable). A thermoplastic or thermosetting covering, sometimes fabric reinforced, applied over the insulation, core, metallic sheath, or armor of a cable. *See also:* **power distribution, underground construction.** 0-31E1

jack shaft (rotating machinery). A separate shaft carried on its own bearings and connected to the shaft of a machine. *See also:* **rotor (rotating machinery).** 0-31E8

jack system (rotating machinery). A system design to raise the rotor of a machine. *See:* **rotor (rotating machinery).** 0-31E8

jamming. The intentional transmission of interfering radio signals in order to disturb the reception of other signals. *See also:* **radio transmitter.** 42A65-0

jar (storage cell). The container for the element and electrolyte of a lead-acid storage cell and unattacked by the electrolyte. *See also:* **battery (primary or secondary).** 42A60-0

***J* display (radar).** A modified *A* display in which the time base is a circle and targets appear as radial deflections from the time base. *See also:* **navigation.** E172-10E6

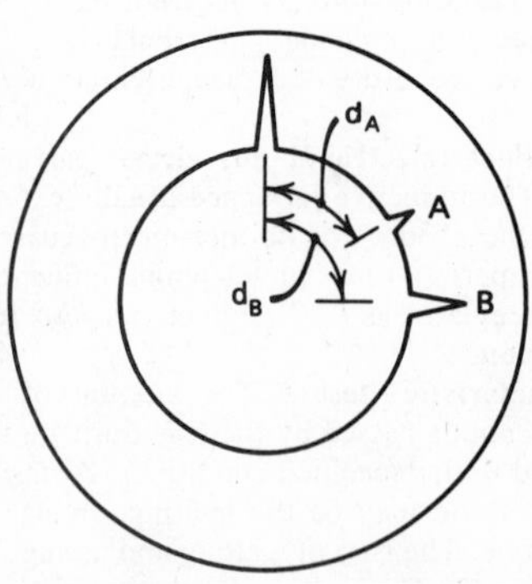

J display. Two targets A and B at different distances.

jitter (1) (repetitive wave). (A) Time-related, abrupt, spurious variations in the duration of any specified, related interval. (B) Amplitude-related, abrupt, spurious variations in the magnitude of successive cycles. (C) Frequency-related, abrupt, spurious variations in the frequency of successive pulses. (D) Phase related, abrupt, spurious variations in the phase of the frequency modulation of successive pulses as referenced to the phase of a continuous oscillator. *Note:* Qualitative use of jitter requires the use of a generic derivation of one of the above categories to identify whether the jitter is time, amplitude, frequency, or phase related and to specify which form within the category, for example, pulse-delay-time jitter, pulse-duration jitter, pulse-separation jitter, etcetera. Quantitative use of jitter requires that a specified measure of the time- or amplitude-related variation, for example, average, root-mean-square, peak-to-peak, etcetera, be included in addition to the generic term that specifies whether the jitter is time, amplitude, frequency, or phase related. *See also:* **pulse; pulse techniques.** 0-9E4

(2) (oscilloscopes, electronic navigation, and television). Small, rapid aberrations in the size or position of a repetitive display; indicating spurious deviations of the signal or instability of the display circuit. *Note:* Frequently caused by mechanical and electronic switching systems or faulty components. Generally continuous, but may be random or periodic. *See:* **fortuitous distortion; navigation; oscillograph; television.** E204-9E4/10E6

(3) (facsimile). Raggedness in the received copy caused by erroneous displacement of recorded spots in the direction of scanning. *See also:* **recording (facsimile).** E168-0

jog (inch) (control) (industrial control). A control function that provides for the momentary operation of a drive for the purpose of accomplishing a small movement of the driven machine. *See:* **electric drive.** IC1-34E10

jog control point, supervisory control. *See:* **supervisory control point, jog control.**

jogging (inching). The quickly repeated closure of the circuit to start a motor from rest for the purpose of accomplishing small movements of the driven machine. *See also:* **electric drive.** 42A25-0

jogging speed (industrial control). The steady-state speed that would be attained if the jogging pilot device contacts were maintained closed. *Note:* It may be expressed either as an absolute magnitude of speed or a percentage of maximum rated speed. *See:* **control system, feedback.** AS1-34E10

Johnson noise (interference terminology). The noise caused by thermal agitation (of electron charge) in a dissipative body. *Note:* The available thermal (Johnson) noise power N from a resistor at temperature T is $N = kT\,\Delta f$, where k is Boltzmann's constant and Δf is the frequency increment. *Note:* The noise power distribution is equal throughout the frequency spectrum, that is, the noise power is equal in all equal frequency increments. *See:* **signal.** E160-13E6

joint (interior wiring). A connection between two or more conductors. *See also:* **interior wiring.** 42A95-0

joint insulation. *See:* **connection insulation.**

joint use (electrical engineering). The simultaneous use of a pole or line by two or more kinds of utilities. *See also:* **transmission line.** 42A65/2A2-0

Jordan bearing. A sleeve bearing and thrust bearing combined in a single unit. *See:* **bearing.** 0-31E8

joule. The unit of work and energy in the International System of Units (SI). The joule is the work done by a force of 1 newton acting through a distance of 1 meter. E270-0

Joule effect. The evolution of thermal energy produced by an electric current in a conductor as a consequence of the electric resistance of the conductor. *See also:* **thermoelectric device; Joule's law.** E221-15E7

Joule heat. The thermal energy resulting from the Joule effect. *See also:* **thermoelectric device.** E221-15E7

Joule's law (heating effect of a current). The rate at which heat is produced in an electric circuit of constant resistance is equal to the product of the resistance and the square of the current. E270-0

journal (shaft). A cylindrical section of a shaft that is intended to rotate inside a bearing. *See also:* **bearing; armature.** 0-31E8

journal bearing (rotating machinery). A bearing that supports the cylindrical journal of a shaft. *See also:* **bearing.** 0-31E8

***J* scan (electronic navigation).** *See:* ***J* display.**

***J* scope (class-*J* oscilloscope).** A cathode-ray oscilloscope arranged to present a type-*J* display. *See:* ***J* display.** *See also:* **radar.** 42A65-0

jump (electronic computation). (1) To (conditionally or unconditionally) cause the next instruction to be obtained from a storage location specified by an address part of the current instruction when otherwise it would be specified by some convention. (2) An instruction that specifies a jump. *Note:* If every instruction in the instruction code specifies the location of the next instruction (for example, in a three-plus-one-address code), then each one is not called a jump instruction unless it has two or more address parts that are conditionally selected for the jump. *See also:* **conditional jump; electronic computation; electronic digital computer; transfer; unconditional jump.** E162-0

jumper (1) (general). A short length of cable used to make electric connections within, between, among, and around circuits and their associated equipment. *Note:* It is usually a temporary connection. *See also:* **conductor.** 42A35-31E13/31E1

(2) (electronic and electrical applications). A direct electric connection between two or more points on a printed-wiring or terminal board. *See also:* **soldered connections (electronic and electrical applications).** 99A1-0

(3) (electric vehicles) (bus line jumper) (train line jumper). A flexible conductor or group of conductors arranged for connecting electric circuits between adjacent vehicles. *Note:* As applied to vehicles, it usually consists of two electric coupler plugs with connecting cable, although sometimes one end of the connecting cable is permanently attached to the vehicle. *See also:* **multiple-control unit.** 42A42-0

junction (1). *See:* **summing junction.**

(2) (thermoelectric device). The transition region between two dissimilar conducting materials. *See also:* **thermoelectric device.** E221-15E7

(3). (A) A connection between two or more conductors or two or more sections of transmission line; (B) a contact between two dissimilar metals or materials, as in a rectifier or thermocouple. 42A65-0

(4) (semiconductor device). A region of transition between semiconductor regions of different electrical properties (for example, *n-n, p-n, p-p* semiconductors), or between a metal and a semiconductor. *See also:* **semiconductor; semiconductor device.** E59/E216/E223-34E17; E102/E270/42A70-10E1

junction, alloy (semiconductor). *See:* **alloy or fused junction.**

junction box (1) (interior wiring). A box with a blank cover that serves the purpose of joining different runs of raceway or cable and provides space for the connection and branching of the enclosed conductors. 42A95-0

(2). An enclosed distribution panel for connecting or branching one or more corresponding electric circuits without the use of permanent splices. *See also:* **cabinet.** 42A35-31E13

(3) (mine type). A stationary piece of apparatus with enclosure by means of which one or more electric circuits for supplying mining equipment are connected through overcurrent protective devices to an incoming feeder circuit. *See also:* **mine feeder circuit.** 42A85-0

junction, collector (semiconductor device). A junction normally biased in the high-resistance direction, the current through which can be controlled by the introduction of minority carriers. *Note:* The polarity of the voltage across the junction reverses when switching occurs. *See also:* **semiconductor; semiconductor device; transistor.** E102/E270/42A70-0;E216-34E17

junction depth (photoelectric converter). The distance from the illuminated surface to the junction in a photoelectric converter. *See also:* **semiconductor.** 0-10E1

junction, diffused (semiconductor). A junction that has been formed by the diffusion of an impurity within a semiconductor crystal. *See also:* **semiconductor.** E102/E216/E270/42A70-34E17

junction, doped (semiconductor). A junction produced by the addition of an impurity to the melt during crystal growth. *See also:* **semiconductor.** E102/E216/E270/42A70-34E17

junction, emitter (semiconductor device). A junction normally biased in the low-resistance direction to inject minority carriers into an interelectrode region. *See also:* **semiconductor; transistor.** E102/E270/42A70-0

junction, fused. *See:* **alloy or fused junction.**

junction, grown (semiconductor). A junction produced during growth of a crystal from a melt. *See also:* **semiconductor device.** E102/E216/E270/42A70-34E17

junction loss (wire communication). That part of the repetition equivalent assignable to interaction effects arising at trunk terminals. *See also:* **transmission loss.** 42A65-0

junction, *n-n* (semiconductor). A region of transition between two regions having different properties in *n*-type semiconducting material. *See also:* **semiconductor device.** E102/42A70-0

junction, *p-n* (semiconductor). A region of transition between *p*- and *n*-type semiconducting material. *See also:* **semiconductor device.** E102/42A70-0

junction point. *See:* **node.**

junction pole (wire communication). A pole at the end of a transposition section of an open wire line or the pole common to two adjacent transposition sections. *See also:* **open wire.** 42A65-0

junction, *p-p* (semiconductor). A region of transition between two regions having different properties in *p*-type semiconducting material. *See also:* **semiconductor device.** E102/42A70-0

junction, rate-grown (semiconductor). A grown junction produced by varying the rate of crystal growth. *See also:* **semiconductor device.** E102/E216/E270/42A70-34E17

junction resistance (thermoelectric device). The difference between the resistance of two joined materials and the sum of the resistances of the unjoined materials. *See also:* **thermoelectric device.** E221-15E7

junction transposition (*s*-pole transposition) (wire communication). A transposition located at the junction pole (*s* pole) between two transposition sections of an open wire line. *See also:* **open wire.** 42A65-0

junctor (crossbar systems) (wire communication). A circuit extending between frames of a switching unit and terminating in a switching device on each frame. *See also:* **switching system.** 42A65-0

just operate value, relay. *See:* **relay just operate value.**

just scale (acoustics). A musical scale formed by taking three consecutive triads each having the ratio 4:5:6, or 10:12:15. *Note:* Consecutive triads are triads such that the highest note of one is the lowest note of the next. *See also:* **electroacoustics.** E157-1E1

K

***K* display (radar).** A modified *A* display in which a target appears as a pair of vertical deflections; when the radar antenna is correctly pointed at the target the deflections (blips) are of equal height, and when not so pointed, the difference in blip height is an indication of the direction and magnitude of pointing error. *See also:* **navigation.** 0-10E6

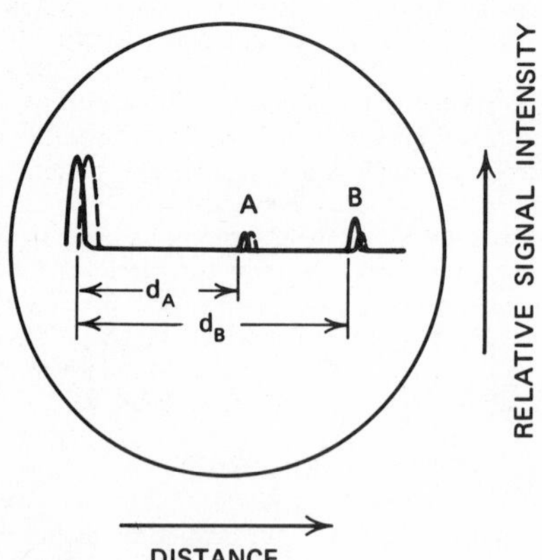

K display. Two targets A and B at different distances; radar aimed at target A.

keep-alive circuit (transmit-receive or antitransmit-receive switch). A circuit for producing residual ionization for the purpose of reducing the initiation time of the main discharge. *See also:* **navigation.** E172-10E6

kelvin. Unit of thermodynamic temperature, is the fraction 1/273.16 of the thermodynamic temperature of the triple point of water. *Note:* In 1967 the General Conference on Weights and Measures gave the name kelvin to the International Standard Unit (SI) of temperature (which had previously been called degree Kelvin and assigned the symbol K without the symbol °. CGPM SCC 14

Kelvin bridge (double bridge) (Thomson bridge). A 7-arm bridge intended for comparing the 4-terminal resistances of two 4-terminal resistors or networks, and characterized by the use of a pair of auxiliary resistance arms of known ratio that span the adjacent potential terminals of the two 4-terminal resistors that are connected in series by a conductor joining their adjacent current terminals. *See also:* **bridge.** 42A30-0

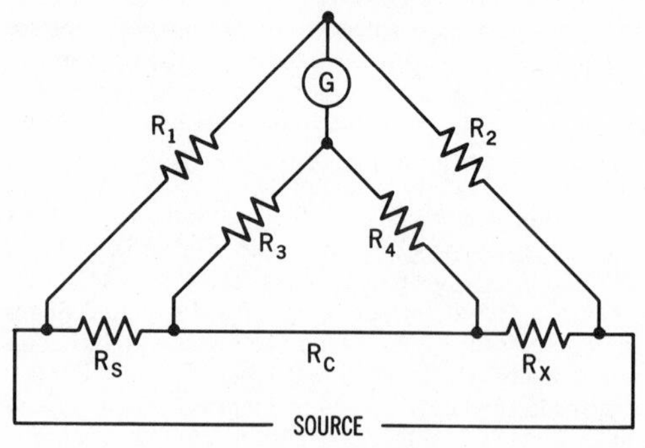

$$R_X = R_S\frac{R_2}{R_1} - \frac{R_C R_3 \left(\frac{R_4}{R_3} - \frac{R_2}{R_1}\right)}{R_3 + R_4 + R_C}$$

Kelvin bridge.

Kendall effect (facsimile). A spurious pattern or other distortion in a facsimile record, caused by unwanted modulation products arising from the transmission of a carrier signal, appearing in the form of a rectified baseband that interferes with the lower sideband of the carrier. *Note:* This occurs principally when the single sideband width is greater than half the facsimile carrier frequency. *See also:* **recording (facsimile).** E168-0

kenotron (tube or valve). A hot-cathode vacuum diode. *Note:* This term is used primarily in the industrial and X-ray fields. *See also:* **tube definitions.** 42A70-15E6

Kerr electrostatic effect. *See:* **electrooptical effect in dielectrics.**

key (1) (telephone switching system). A hand-operated switching device ordinarily comprising concealed spring contacts with an exposed handle or pushbutton, capable of switching one or more parts of a circuit. 42A65-19E1

(2) (rotating machinery). A bar that by being recessed partly in each of two adjacent members serves to transmit a force from one to the other. *See also:* **rotor (rotating machinery).** 0-31E8

(3) (computing systems). One or more characters used to identify an item of data. X3A12-16E9

keyer. A device that changes the output of a transmitter from one value of amplitude or frequency to another in accordance with the intelligence to be transmitted. *Note:* This applies generally to telegraph keying. *See also:* **radio transmission.** E145/42A65-0

keying (1) (modulating systems). Modulation involving a series of selections from a finite set of discrete states. *See also:* **telegraphy; modulating systems.** E170-0

(2) (telegraph). The forming of signals, such as those employed in telegraph transmission, by an abrupt modulation of the output of a direct-current or an alternating-current source as, for example, by interrupting it or by suddenly changing its amplitude or frequency or some other characteristic. *See also:* **telegraphy; modulating systems.** E145-0

(3) (television). A signal that enables or disables a network during selected time intervals. *See:* **television.** 0-2E2

keying interval (periodically keyed transmission system) (modulation systems). One of the set of intervals starting from a change in state and equal in length to the shortest time between changes of state. *See also:* **modulating systems.** E170-0

keying rate (modulation systems). The reciprocal of the duration of the keying interval. *See also:* **modulating systems.** E170-0

keying wave (telegraphic communication). *See:* **marking wave.**

keyless ringing (telephony). A form of machine ringing on manual switchboards that is started automatically by the insertion of the calling plug into the jack of the called line. *See also:* **telephone switching system.** 42A65-0

key light (television). The apparent principal source of directional illumination falling upon a subject or area. *See also:* **television lighting.** Z7A1-0

key pulsing. A switchboard arrangement using a nonlocking keyset and providing for the transmission of signals corresponding to the key depressions. *See also:* **telephone switching system.** 42A65-0

keypunch (electronic computation). A keyboard-actuated device that punches holes in a card to represent data. *See also:* **electronic digital computer.** X3A12-16E9

keyshelf (telephony). The horizontal shelf of a manual telephone switchboard on which are mounted the keys by means of which the operator switches one or more parts of the switchboard circuits. *See also:* **telephone switching system.** 42A65-19E1

keystone distortion (television). A form of geometric distortion that results in a trapezoidal display of a nominally rectangular raster or picture. *See also:* **television.** 0-42A65/2E2

keyway (rotating machinery). A recess provided for a key. *See:* **key (rotating machinery).** 0-31E8

kick-sorter (British) (pulse techniques). *See:* **pulse-height analyzer.** *See also:* **pulse.**

kilogram. The unit of mass; it is equal to the mass of the international prototype of the kilogram. CGPM-SCC14

kilovolt-ampere rating (voltage regulator). The product of the rated load amperes and the rated range of regulation in kilovolts. *Note:* The kilovolt-ampere rating of a three-phase voltage regulator is the product of the rated load amperes and the rated range of regulation in kilovolts multiplied by 1.732. *See also:* **voltage regulator.** 57A15-0

kinematic drift (electronic navigation). *See:* **misalignment drift.**

kinescope. *See:* **picture tube.**

kinetic energy. The energy that a mechanical system possesses by virtue of its motion. *Note:* The kinetic energy of a particle at any instant is $(1/2)mv^2$, where m is the mass of the particle and v is its velocity at that instant. The kinetic energy of a body at any instant is the sum of the kinetic energies of its several particles. E270-0

Kingsbury bearing. *See:* **tilting-pad bearing.**

kino gun. *See:* **end injection.**

Kirchhoff's laws (electric networks). (1) The algebraic sum of the currents toward any point in a network is zero. (2) The algebraic sum of the products of the current and resistance in each of the conductors in any closed path in a network is equal to the algebraic sum of the electromotive forces in that path. *Note:* These laws apply to the instantaneous values of currents and electromotive forces, but may be extended to the phasor equivalents of sinusoidal currents and electromotive forces by replacing algebraic sum by phasor sum and by replacing resistance by impedance. *See also:* **network analysis.** E270-0

klydonograph (surge voltage recorder). *See:* **Lichtenberg figure camera.**

klystron. A velocity-modulated tube comprising, in principle, an input resonator, a drift space, and an output resonator. *See also:* **reflex klystron; power klystron.** 50I07-0

knee of transfer characteristic (image orthicons). The region of maximum curvature in the transfer characteristic. *See also:* **beam tubes; television.** E160-15E6

knife switch. A form of switch in which the moving element, usually a hinged blade, enters or embraces the stationary contact clips. *Note:* In some cases, however, the blade is not hinged and is removable. 37A100-31E11

knob (1) (electronic devices) A mechanical element mounted on the end of a shaft which may be held by the fingers or within the hand for the purpose of rotating the shaft and which may include an engraved pointer or marking. 42A25-34E10

(2) (electric power distribution). An insulator, in one or two pieces, having a central hole for a nail or screw and one or more peripheral or chordal grooves for wire, used for supporting conductors at a definite spacing from the surface wired over. 42A95-0

knockout. A portion of the wall of a box or cabinet so fashioned that it may be removed readily by a hammer, screwdriver, and pliers, at the time of installation in order to provide a hole for the attachment of raceway, cable or fitting. *See also:* **cabinet.** 42A95-0

Kramer system (rotating machinery). A system of speed control below synchronous speed, for wound-rotor induction motors, giving constant power output over the speed range. *Note:* Slip power is recovered through the medium of an independently mounted rotary converter electrically connected between the secondary windings of the induction motor and a direct-current auxiliary motor that is directly coupled to the induction-motor shaft. *See:* **asynchronous machine.** 0-31E8

***K* scan (electronic navigation).** *See:* ***K* display.**

***K* scope (electronic navigation).** *See:* ***k* display.**

kV·T product (inductive coordination). Inductive influence usually expressed in terms of the product of its root-mean-square magnitude in kilovolts (kV) times its telephone influence factor (TIF) abbreviated as kV·T product. *See also:* **inductive coordination.** 42A65-0

L

label. A key attached to the item of data that it identifies. Synonymous with tag. 0-16E9

laboratory reference standards. The highest ranking order of standards at each laboratory. *Note:* Laboratory reference standards set the values of the standards for that laboratory, and they in turn have their values determined by the standards of another higher-ranking laboratory, or, sometimes, by means of the reference or working standards of another category or echelon. *See also:* **measurement system.** E285-9E1

laboratory test. *See:* **shop test.**

laboratory working standards. Those standards that are used for the ordinary calibration work of the standardizing laboratory. *Note:* Laboratory working standards are calibrated by comparison with the reference standards of that laboratory. *See also:* **measurement system.** E285-9E1;12A0-0

labyrinth seal ring (rotating machinery). Multiple oil catcher ring surrounding a shaft with small clearance. *See also:* **bearing.** 0-31E8

lacing, stator-winding end-wire (rotating machinery). Cord or other lacing material used to bind the stator-winding end wire, to hold in place labels and devices placed on or in the end wire, and to position lead cables at their take-off points on the end wire. *See also:* **stator.** 0-31E8

lacquer disks (cellulose nitrate disks) (electroacoustics). Mechanical recording disks usually made of metal, glass, or paper and coated with a lacquer compound (often containing cellulose nitrate). *See also:* **phonograph pickup.** E157-1E1

lacquer-film capacitor. A capacitor in which the dielectric is primarily a solid lacquer film and the electrodes are thin metallic coatings deposited thereon. 0-21E0

lacquer master*. *See:* **lacquer original.**

*Deprecated

lacquer original (electroacoustics) (lacquer master*). An original recording on a lacquer surface for the purpose of making a master. *See also:* **phonograph pickup.**

*Deprecated E157-1E1

lacquer recording (electroacoustics). Any recording made on a lacquer recording medium. *See also:* **phonograph pickup.** E157-1E1

ladder network. A network composed of a sequence of H, L, T, or pi networks connected in tandem. *See also:* **network analysis.** E153/E270-0

ladder-winding insulation (rotating machinery). An element of winding insulation in the form of a single sheet precut to fit into one or more slots and with a broad area at each end to provide end-wire insulation. *See also:* **rotor (rotating machinery); stator.** 0-31E8

lag (1) (general). The delay between two events. X3A12-16E9

(2) (control circuits). Any retardation of an output with respect to the casual input. *See also:* **telegraphy; control system; feedback.** 85A1-23E0;AS1-34E10

(3) (telegraph system). The time elapsing between the operation of the transmitting device and the response of the receiving device. *See also:* **telegraphy.** 42A65-0

(4) (distance/velocity) (automatic control). A delay attributable to the transport of material or the finite rate of propagation of a signal or condition. *Note:* Also termed transport lag and transportation lag. *See also:* **control system; feedback.** 85A1-23E0

(5) (first-order). The change in phase due to a linear element of transfer function. $1/(1+Ts)$. 0-23E0

(6) (second-order). The change in phase due to a linear element of transfer function $1/s^2+2\zeta\omega_n s+\omega_n^2$. 0-23E0

(7) (camera tubes). The retardation or relative slowness of the signal output current of a camera tube to follow temporal changes in input irradiance. *See also:* **camera tube.** 0-15E6

lagged-demand meter. A meter in which the indication of the maximun demand is subject to a characteristic time lag by either mechanical or thermal means. *See also:* **electricity meter.** 12A0/42A30-0

lag networks (power supplies). Resistance-reactance components, arranged to control phase-gain rolloff versus frequency. *Note:* Used to assure the dynamic stability of a power-supply's comparison amplifier. The main effect of a lag network is a reduction of gain at relatively low frequencies so that the slope of the remaining rolloff can be relatively more gentle. *See also:* **power supply.** KPSH-10E1

lambert. A unit of luminance (photometric brightness) equal to $1/\pi$ candela per square centimeter, and, therefore, is equal to the uniform luminance of a perfectly diffusing surface emitting or reflecting light at the rate of one lumen per square centimeter. *Note:* The lambert also is the average luminance of any surface emitting or reflecting light at the rate of one lumen per square centimeter. For the general case, the average must take account of variation of luminance with angle of observation; also of its variation from point to point on the surface considered. *See also:* **light; color terms.** E201-2E2;Z7A1-0

Lambert's cosine law. A law stating that the flux per solid angle in any direction from a plane surface varies as the cosine of the angle between that direction and the perpendicular to the surface. *See also:* **lamp.** Z7A1-0

laminated brush (industrial control). A contact part consisting of thin sheets of conducting material fastened together so as to secure individual contact by the edges of the separate sheets. 42A20-34E1

laminated core (rotating machinery). An assembly of laminations of magnetic material, for use as an element of a magnetic circuit. *See also:* **rotor (rotating machinery); stator.** 0-31E8

laminated frame (rotating machinery). A stator frame formed from laminations clamped, bolted, or riveted together with or without additional strengthening plates. *See also:* **stator.** 0-31E8

laminated record (electroacoustics). A mechanical recording medium composed of several layers of material. *Note:* Normally, it is made with a thin face of surface material on each side of a core. *See also:* **phonograph pickup.** E157-1E1

lamination. A relatively thin member, usually made of sheet material. A complete structure is made by assembling the laminations in the required number of layers. In a core that carries alternating magnetic flux, the core material is usually laminated to reduce eddy-current losses. 0-31E8

lamp. A generic term for a man-made source of light. By extension, the term is also used to denote sources that radiate in regions of the spectrum adjacent to the visible. *Notes:* (1) A lighting unit consisting of a lamp with shade, reflector, enclosing globe, housing, or other accessories is also called a lamp. In such cases, in order to distinguish between the assembled unit and the light source within it, the latter is often called a bulb or tube, if it is electrically powered. (2) For an extensive list of cross references, see *Appendix A*. *See also:* **luminaire.** Z7A1-0

lampholder (socket*). A device intended to support an electric lamp mechanically and connect it electrically to a circuit.
*Deprecated 42A95-0

lampholder adapter. A device that by insertion in a receptacle serves as a lampholder. 42A95-0

lampman (mining). A person having responsibility for cleaning, maintaining, and servicing of miners' lamps. *See also:* **mining.** 42A85-0

lamp post. A standard support provided with the necessary internal attachments for wiring and the external attachments for the bracket and luminaire. *See also:* **street-lighting luminaire.** Z7A1-0

lamp regulator. A device for automatically maintaining constant voltage on a lamp circuit from some higher variable voltage power source. 42A42-0

lamp shielding angle. The angle between the plane of the baffles or louver grid and the plane most nearly horizontal that is tangent to both the lamps and the louver blades. *Note:* The lamp shielding angle frequently is larger than the lower shielding angle, but never smaller. Z7A1-0

land (electroacoustics). The record surface between two adjacent grooves of a mechanical recording. *See also:* **phonograph pickup.** E157-1E1

landing direction indicator. A device to indicate visually the direction currently designated for landing and take-off. *See also:* **signal lighting.** Z7A1-0

landing light. An aircraft aeronautical light designed to illuminate a ground area from the aircraft. *See also:* **signal lighting.** Z7A1-0

landing zone (elevator). A zone extending from a point eighteen inches below a landing to a point eighteen inches above the landing. *See also:* **elevator landing.** 42A45-0

landmark beacon. An aeronautical beacon used to indicate the location of a landmark used by pilots as an aid to enroute navigation. *See also:* **signal lighting.** Z7A1-0

lane (navigation system) (electronic navigation). The projection of a corridor of airspace on a navigation chart, the right and left sides of the corridor being defined by the same (ambiguous) values of the navigation coordinate (phase or amplitude), but within which lateral position information is provided (for example, a Decca lane in which there is a 360-degree change of phase). *See also:* **navigation.** 0-10E6

language (1) (general). A set of representations, conventions, and rules used to convey information.
See:
artificial language;
machine language;
natural language;
object language;
problem oriented language;
procedure oriented language;
programming language;
source language;
target language. X3A12-16E9

(2) (electronic computers). (A) A system consisting of: (a) a well defined, usually finite, set of characters; (b) rules for combining characters with one another to form words or other expressions; (c) a specific assignment of meaning to some of the words or expressions, usually for communicating information or data among a group of people, machines, etcetera. (B) A system similar to the above but without any specific assignment of meanings. Such systems may be distinguished from (A) above, when necessary, by referring to them as formal or uninterpreted languages. Although it is sometimes convenient to study a language independently of any meanings, in all practical cases at least one set of meanings is eventually assigned. *See also:* **code; machine language; electronic computation; electronic digital computer.** E162/E270-0

lapel microphone. A microphone adapted to positioning on the clothing of the user. *See also:* **microphone.** E157/42A65-0

Laplace transform (unilateral) (function $f(t)$). The quantity obtained by performing the operation

$$F(s) = \int_0^\infty f(t)e^{-st}\,dt$$

where $s=\sigma+j\omega$. *See also:* **control system, feedback.** 85A1-23E0

Laplace's equation. The special form taken by Poisson's equation when the volume density of charge is zero throughout the isotropic medium. It is $\nabla^2 v=0$. E270-0

lap winding. A winding that completes all its turns under a given pair of main poles before proceeding to the next pair of poles. *Note:* In commutator machines the ends of the individual coils of a simplex lap winding are connected to adjacent commutator bars; those of a duplex lap winding are connected to alternate commutator bars etcetera. *See also:* **asynchronous machine; direct-current commutating machine; synchronous machine.** 42A10-0

larry (mine). A motor-driven burden-bearing track-mounted car designed for side or end dumping and used for hauling material such as coal, coke, or mine refuse. 42A85-0

latching current (switching device). The making current during a closing operation in which the device latches or the equivalent. 37A100-31E11

latching relay. A relay that is so constructed that it maintains a given position by means of a mechanical latch until released mechanically or electrically. *See also:* **latch-in relay; relay.** 37A100-31E11/31E6;83A16-0

latch-in relay. A relay that maintains its contacts in the last position assumed without the need of maintaining coil energization. 0-21E0

latency (1) (biological electronics). The condition in an excitable tissue during the interval between the application of a stimulus and the first indication of a response. 0-18E1

(2) (electronic computation). The time between the completion of the interpretation of an address and the start of the actual transfer from the addressed location. X3A12-16E9

latent period (electrobiology). The time elapsing between the application of a stimulus and the first indication of a response. *See also:* **excitability (electrobiology).** 42A80-18E1

lateral conductor (pole line work). A wire or cable extending sideways or at an angle to the general direction of the line. *Note:* Service wires either overhead or underground are considered laterals from the street mains. Short branches extended on poles at approximate right angles to lines are also known as laterals. *See also:* **tower; power distribution, overhead construction.** 2A2/42A35-31E13

lateral critical speeds (rotating machinery). The speeds at which the amplitudes of the lateral vibrations of a machine rotor due to shaft rotation reach their maximum values. *See also:* **rotor (rotating machinery).** 0-31E8

lateral-cut recording. *See:* **lateral recording.**

lateral insulator (storage cell). An insulator placed between the plates and the side wall of the container in which the element is housed. *See also:* **battery (primary or secondary).** 42A60-0

lateral recording (lateral-cut recording). A mechanical recording in which the groove modulation is perpendicular to the motion of the recording medium and parallel to the surface of the recording medium. *See also:* **electroacoustics.** 42A65-1E1

lateral width (light distribution). The lateral angle between the reference line and the width line, measured in the cone of maximum candlepower. *Note:* This angular width includes the line of maximum candlepower. *See also:* **street-lighting luminaire.** Z7A1-0

lateral working space (electric power distribution). The space reserved for working between conductor levels outside the climbing space, and to its right and left. *See also:* **conflict (wiring system); tower.** 2A2/42A35-31E13

lattice (navigation). A pattern of identifiable intersecting lines of position, which lines are laid down by a navigation aid. *See also:* **navigation.** 0-10E6

lattice network. A network composed of four branches connected in series to form a mesh, two nonadjacent junction points serving as input terminals, while the remaining two junction points serve as output terminals. *See also:* **network analysis.**

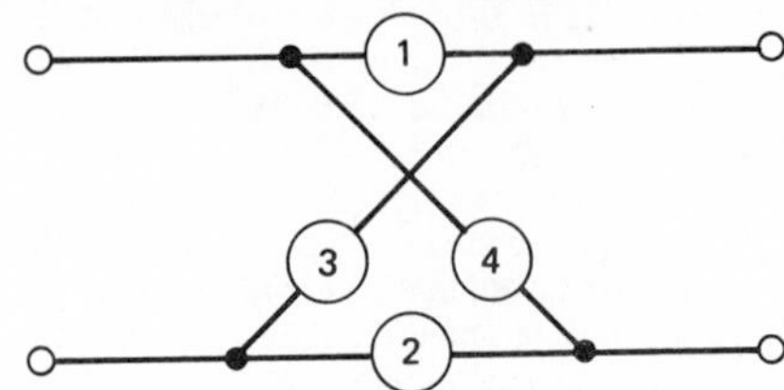

Lattice network. The junction points between branches 4 and 1 and between 3 and 2 are the input terminals; the junction points between branches 1 and 3 and between branches 2 and 4 are the output terminals. E153/E270/42A65-0

lay (1) (cables). The helical arrangement formed by twisting together the individual elements of a cable. 42A35-31E13/31E1

(2) (helical element of a cable). The axial length of a turn of the helix of that element. *Notes:* (1) Among the helical elements of a cable may be each strand in a concentric-lay cable, or each insulated conductor in a multiple-conductor cable. (2) Also termed pitch. *See also:* **power distribution, underground construction.** E30-0

LC auxiliary switch (railway practice). *See:* **auxiliary switch; contact.** 37A100-31E11

LC contact (railway practice). A latch-checking contact that is closed when the operating mechanism linkage is relatched after an opening operation of the switching device. 37A100-31E11

***L* display (radar).** A display in which a target appears as two horizontal blips, one extending to the right from a central vertical time base and the other to the left; both blips are of equal amplitude when the radar antenna is pointed directly at the target, any inequality representing relative pointing error, and distance upward along the baseline representing target distance. *See also:* **navigation.** 0-10E6

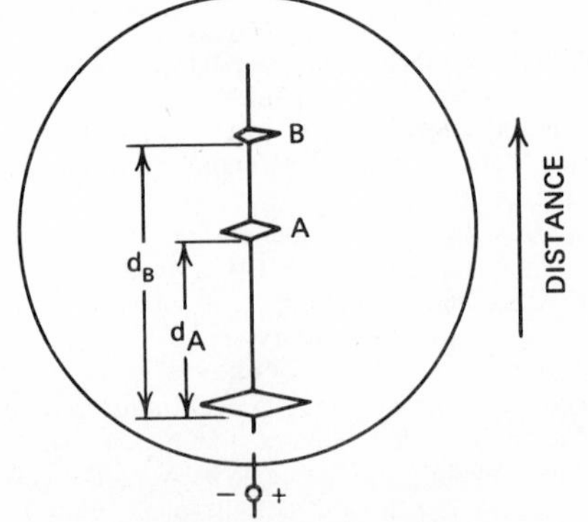

L display. Two targets A and B at different distances; radar aimed at target A.

lead (terminal connections). A conductor that connects a winding to its termination (that is, terminal, bushing, collector, or connection to another winding). *See:* **rotor (rotating machinery); stator.** 0-31E8

lead box (terminal housing) (rotating machinery). A box through which the leads are passed in emerging from the frame. *See:* **cradle base (rotating machinery).** 0-31E8

lead cable (rotating machinery). A cable type of conductor connected to the stator winding, used for making connections to the supply line or among circuits of the stator winding. *See:* **stator.** 0-31E8

lead clamp (salient-pole construction) (rotating machinery). A device used to retain and support the field leads between the hub and rotor rim along the rotor spider arms. *See:* **rotor (rotating machinery).** 0-31E8

lead collar (salient-pole construction) (rotating machinery). A bushing used to insulate field leads at a point of support between a collector ring and the rotor rim. *See:* **rotor (rotating machinery).** 0-31E8

lead-covered cable (lead-sheathed cable). A cable provided with a sheath of lead for the purpose of excluding moisture and affording mechanical protection. *See:* **armored cable.** *See also:* **power distribution, underground construction.** 42A35-31E13

leader (computing systems). The blank section of tape at the beginning and end of a reel of tape. *Note:* The otherwise blank section may, however, include a parity check. *See also:* **static magnetic storage.** X3A12-16E9;EIA3B-34E12

leader cable. A navigational aid consisting of a cable around which a magnetic field is established, marking the path to be followed. *See also:* **radio navigation.** 42A65-0

leader-cable system. A navigational aid in which a path to be followed is defined by the detection and comparison of magnetic fields emanating from a cable system that is installed on the ground or under water. *See also:* **navigation.** 0-10E6

lead, first-order (control system, feedback). The change in phase due to a factor $(1 + Ts)$ in the numerator of a transfer function. *See also:* **control system, feedback.** 0-23E0

lead-in. That portion of an antenna system that connects the elevated conductor portion to the radio equipment. *See also:* **antenna.** 42A65-3E1

leading edge (television). The major portion of the rise of a pulse. *See also:* **television.** E203-0

leading-edge pulse. The first major transition away from the pulse baseline occurring after a reference time. *See also:* **pulse; television.** 0-9E4

leading-edge pulse time. The time at which the instantaneous amplitude first reaches a stated fraction of the peak pulse amplitude. *See also:* **pulse terms; television.** E194/E203/E270-0

lead-in groove (lead-in spiral) (disk recording) (electroacoustics). A blank spiral groove at the beginning of a recording generally having a pitch that is much greater than that of the recorded grooves. *See also:* **phonograph pickup.** E157-0

leading wire (mining). An insulated wire strung separately or as a twisted pair, used for connecting the two free ends of the circuit of the blasting caps to the blasting unit. *See also:* **blasting unit.** 42A85-0

lead-in spiral (disk recording). *See:* **lead-in groove.**

lead-in wire. A conductor connecting an electrode to an external circuit. *See also:* **electron tube.** 50I07-15E6

lead networks (power supplies). Resistance-reactance components arranged to control phase-gain rolloff versus frequency. *Note:* Used to assure the dynamic stability of a power-supply's comparison amplifier. The main effect of a lead network is to introduce a phase lead at the higher frequencies, near the unity-gain frequency. *See also:* **power supply.** KPSH-10E1

lead-over groove (crossover spiral). In disk recording, a groove cut between successive short-duration recordings on the same disk, to enable the pickup stylus to travel from one cut to the next. *See also:* **phonograph pickup.** 0-1E1

lead-out groove (throw-out spiral) (disk recording). A blank spiral groove at the end of a recording generally of a pitch that is much greater than that of the recorded grooves and that terminates in either a locked or an eccentric groove. *See also:* **phonograph pickup.** 0-1E1

lead storage battery. A storage battery the electrodes of which are made of lead and the electrolyte consists of a solution of sulfuric acid. *See also:* **battery (primary or secondary).** 42A60-0

leakage (1) (general). (A) Undesirable conductive paths in certain components, specifically, in capacitors, a path through which a slow discharge may take effect; in problem boards, interaction effects between electric signals through insufficient insulation between patch bay terminals. (B) Current flowing through such paths. E165-16E9

(2) (signal-transmission system.) Undesired current flow between parts of a signal-transmission system or between the signal-transmission system and point(s) outside the system. *See:* **signal.** 0-13E6

(3) (transmission lines and waveguides). Radiation or conduction of signal power out of or into an imperfectly closed and shielded system. The leakage is usually expressed in decibels below a specified reference power. *See also:* **transmission characteristics.** 0-9E4

leakage (conduction) current (1) (insulation). The nonreversible constant current component of the measured current that remains after the capacitive current and absorption current have disappeared. *Note:* Leakage current passes through the insulation volume, through any defects in the insulation, and across the insulation surface of solid or liquid insulation. E95-0;42A35-31E13

(2) (electron tubes). A conduction current that flows between two or more electrodes by any path other than across the vacuous space between the electrodes. *See also:* **electrode current (of an electron tube); electron tube.** 50I07-15E6;42A70-0

leakage distance (insulator). The sum of the shortest distances measured along the insulating surfaces between the conductive parts, as arranged for dry flashover test. *Note:* Surfaces coated with semiconducting glaze shall be considered as effective leakage surfaces, and leakage distance over such surfaces shall be included in the leakage distance. *See also:* **insulator.** 29A1-0

leakage flux, relay. *See:* **relay leakage flux.**

leakage inductance (one winding with respect to a second winding). A portion of the inductance of a winding that is related to a difference in flux linkages in the two windings; quantitatively, the leakage inductance of winding 1 with respect to winding 2

$$L_{\text{lf}} = \frac{\partial\left(\lambda_{11} - \frac{N_1}{N_2}\lambda_{21}\right)}{\partial i_1}$$

where λ_{11} and λ_{21} are the flux linkages of windings 1 and 2, respectively, resulting from current i_1 in winding 1; and N_1 and N_2 are the number of turns of windings 1 and 2, respectively. 0-21E1

leakage power (microwave gas tubes). *See:* **flat leakage power; spike leakage energy.**

leakage radiation (radio transmitting system). Radiation from anything other than the intended radiating system. *See also:* **radio transmission.** 42A65/E145-0

learning system. An adaptive system with memory. *See:* **system science.** 0-35E2

Lecher wires. Two parallel wires on which standing waves are set up, frequently for the measurement of wavelength. *See also:* **transmission line.** 42A65-0

Leduc current* (electrotherapy). A pulsed direct current commonly having a duty cycle of 1:10. *See also:* **electrotherapy.**

*Deprecated 42A80-18E1

left-handed (counterclockwise) polarized wave (radio wave propagation). An elliptically polarized transverse electromagnetic wave in which the rotation of the electric field vector with time is counterclockwise for a stationary observer looking in the direction of the wave normal. *Note:* For an observer looking from a receiver toward the apparent source of the wave, the direction of rotation is reversed. *See also:* **radiation; radio wave propagation.** 42A65-3E2

leg (circuit). Any one of the conductors of an electric supply circuit between which is maintained the maximum supply voltage. *See also:* **center of distribution.** 42A35-31E13

leg, thermoelectric. Alternative term for thermoelectric arm. *See also:* **thermoelectric device.** E221-15E7

leg wire (mining). One of the two wires attached to and forming a part of an electric blasting cap or squib. *See also:* **blasting unit.** 42A85-0

Lenard tube. An electron beam tube in which the beam can be taken through a section of the wall of the evacuated enclosure. *See also:* **miscellaneous electron devices.** 42A70-15E6

length (electronic computers). (1) A measure of the magnitude of a unit of data, usually expressed as a number of subunits, for example, the length of a record is 32 blocks, the length of a word is 40 binary digits, etcetera. (2) The number of subunits of data, usually digits or characters, that can be simultaneously stored linearly in a given device, for example, the length of the register is 12 decimal digits or the length of the counter is 40 binary digits. (3) A measure of the amount of time that data are delayed when being transmitted from point to point, for example, the length of the delay line is 384 microseconds. *See:* **double length; word length.** *See also:* **electronic digital computer; storage capacity.** E162-0

length of lay (cable) (power distribution underground cables). The axial length of one turn of the helix of any helical element. *See also:* **power distribution, underground construction.** 0-31E1

lens and aperture (phototube housing) (industrial control). The cooperating arrangement of a light-refracting member and an opening in an opaque diaphragm through which all light reaching the phototube cathode must pass. *See also:* **electronic controller.** 42A25-34E10

lens antenna. An antenna consisting of a feed and an electromagnetic lens. *See also:* **antenna.** 0-3E1

lens multiplication factor (phototube housing) (industrial control). The maximum ratio of the light flux reaching the phototube cathode with the lens and aperture in place to the light flux with the lens and aperture removed. *See also:* **photoelectric control.** 42A25-34E10

Lenz's law (induced current). The current in a conductor as a result of an induced voltage is such that the change in magnetic flux due to it is opposite to the change in flux that caused the induced voltage. E270-0

letter. An alphabetic character used for the representation of sounds in a spoken language. X3A12-16E9

letter combination (abbreviation). One or more letters that form a part of the graphic symbol for an item and denote its distinguishing characteristic. *Compare with:* **functional designation; reference designation; symbol for a quantity.** *See also:* **abbreviation.** E267-0

level (quantity). Magnitude, especially when considered in relation to an arbitrary reference value. Level may be stated in the units in which the quantity itself is measured (for example, volts, ohms, etcetera) or in units (for example, decibels) expressing the ratio to a reference value. *Notes:* (1) Examples of kinds of levels in common use are electric power level, sound-pressure level, voltage level. (2) The level as here defined is measured in two common units; in decibels when the logarithmic base is 10, or in nepers when the logarithmic base is e. The decibel requires that k be 10 for ratios of power, or 20 for quantities proportional to the square root of power. The neper is used to represent ratios of voltage, current, sound pressure, and particle velocity. The neper requires that k be 1. (3) In symbols, $L = k \log_r(q/q_0)$ where L = level of kind determined by the kind of quantity under consideration. r = base of the logarithm of the reference ratio. q = the quantity under consideration. q_0 = reference quantity of the same kind. k = a multiplier that depends upon the base of the logarithm and the nature of the reference quantity. *See:* **blanking level; transmission level.** *See also:* **reference black level; reference white level; signal level (electroacoustics); transmission characteristic.** 42A65-31E3;0-1E1

level above threshold (sensation level) (sound) (acoustics). The pressure level of the sound in decibels above its threshold of audibility for the individual observer. *See also:* **electroacoustics.** E157-1E1

level band pressure (for a specified frequency band) (electroacoustics). The sound-pressure level for the sound contained within the restricted band. *Notes:* (1) The reference pressure must be specified. (2) The band may be specified by its lower and upper cutoff frequencies or by its geometric or arithmetic center frequency and bandwidth. The width of the band may be indicated by a prefatory modifier, for example, octave band (sound pressure) level, half-octave band level, third-octave band level, 50-hertz band level. *See also:* **elec-**

troacoustics; loudspeaker. 0-1E1

level compensator (signal transmission). An automatic transmission-regulating feature or device to minimize the effects of variations in amplitude of the received signal. *See also:* **telegraphy.** 42A65-0

leveling (electronic navigation). *See:* **platform erection.**

leveling block (rotating machinery). *See:* **leveling plate.**

leveling plate (leveling block) (rotating machinery). A heavy pad built into the foundation and used to support and align the bed plate or rails using shims for adjustment before grouting. *See:* **cradle base (rotating machinery).** 0-31E8

leveling zone (elevator). The limited distance above or below an elevator landing within which the leveling device may cause movement of the car toward the landing. *See also:* **elevator-car leveling device.** 42A45-0

level, relay. *See:* **relay level.**

levels, usable (storage tubes). The output levels, each related to a different input, that can be distinguished from one another regardless of location on the storage surface. *Note:* The number of usable levels is normally limited by shading and disturbance. *See also:* **storage tube.** E158-15E6

lever blocking device (railway signaling). A device for blocking a lever so that it cannot be operated. *See also:* **railway signal and interlocking.** 42A42-0

lever indication (railway signaling). The information conveyed by means of an indication lock that the movement of an operated unit has been completed. *See:* **railway signal and interlocking.** 42A42-0

liberator tank (electrorefining). Sometimes known as a depositing-out tank, an electrolytic cell equipped with insoluble anodes for the purpose of either decreasing, or totally removing the metal content of the electrolyte by plating it out on cathodes. *See also:* **electrorefining.** 42A60-0

library. A collection of organized information used for study and reference. *See:* **program library.** X3A12-16E9

library routine (computing systems). A proven routine that is maintained in a program library. *See also:* **electronic digital computer.** X3A12-16E9

Lichtenberg figure camera (klydonograph) (surge-voltage recorder). A device for indicating the polarity and approximate crest value of a voltage surge by the appearance and dimensions of the Lichtenberg figure produced on a photographic plate or film, the emulsion coating of which is in contact with a small electrode coupled to the circuit in which the surge occurs. *Note:* The film is backed by an extended plane electrode. *See also:* **instrument.** 42A30-0

life performance curve (illuminating engineering). A curve that represents the variation of a particular characteristic of a light source (luminous flux, candlepower, etcetera) throughout the life of the source. *Note:* Life performance curves sometimes are called maintenance curves as, for example, lumen maintenance curves. *See also:* **lamp.** Z7A1-0

life test. *See:* **accelerated test (reliability).**

life test of lamps. A test in which lamps are operated under specified conditions for a specified length of time, for the purpose of obtaining information on lamp life. *Note:* Measurements of photometric and electrical characteristics may be made at specified intervals of time during this test. *See also:* **lamp.** Z7A1-0

lifetime, volume (semiconductor). The average time interval between the generation and recombination of minority carriers in a homogeneous semiconductor. *See also:* **semiconductor; semiconductor device.** E102/E270-0;E216-34E17

lifting bar. A bar provided as part of an apparatus to permit attachment of a lifting device. 0-31E8

lifting hole. A hole provided to permit attachment of a lifting device. 0-31E8

lifting-insulator switch. A switch in which one or more insulators remain attached to the blade, move with it, and lift it to the open position. 37A100-31E11

light. For the purposes of illuminating engineering, light is visually evaluated radiant energy. *Note:* Light is psychophysical, neither purely physical nor purely psychological. Light is not synonymous with radiant energy, however restricted, nor is it merely sensation. In a general nonspecialized sense, light is the aspect of radiant energy of which a human observer is aware through the stimulation of the retina of the eye. The present basis for the engineering evaluation of light consists of the color-mixture data $\bar{x}, \bar{y}, \bar{z}$ (*see:* **spectral tristimulus values**) adopted in 1931 by the International Commisssion on Illumination (CIE). These data include the spectral luminous efficiency data for photopic vision. (*See:* **spectral luminous efficiency for photoptic vision**) adopted in 1924 by the CIE.
See:
altitude;
bare lamp;
candela;
candlepower;
color terms;
floodlighting;
footcandle;
footlambert;
general lighting;
headlamp;
illumination;
infrared radiation;
lambert;
lamp;
lumen;
lumen-hour;
luminance;
luminous density;
luminous efficacy;
luminous flux;
luminous flux density of a surface;
luminous intensity;
lux;
nit;
phot;
photochemical radiation;
photometry;
quantity of light;
radiant energy;
searchlight;
signal lighting;
spectral luminous efficiency;
stilb;
ultraviolet radiation;
units of luminance;
visual field. E201-0;Z7A1-0

light adaptation. The process by which the retina becomes adapted to a luminance greater than about one footlambert (3 nits). *See also:* **visual field.** Z7A1-0

light amplifier (optoelectronic device). An amplifier whose signal input and output consist of light. *See also:* **optoelectronic device.** E222-15E7

light-beam instrument. An instrument that utilizes the position of a beam of light on a scale as its indicating means. *See also:* **instrument.** 42A30-0

light carrier injection (facsimile). The method of introducing the carrier by periodic variation of the scanner light beam, the average amplitude of which is varied by the density changes of the subject copy. *See also:* **scanning (facsimile).** E168-0

light center (lamp). The center of the smallest sphere that would completely contain the light-emitting element of the lamp. *See also:* **lamp.** Z7A1-0

light center length (lamp). The distance from the light center to a specified reference point on the lamp. *See also:* **lamp.** Z7A1-0

lighting branch circuit. A circuit supplying energy to lighting outlets only. *See also:* **branch circuit.** 42A95-0

lighting feeder. A feeder supplying principally a lighting load. *See:* **feeder.** 42A95-0

lighting outlet. An outlet intended for the direct connection of a lampholder, a lighting fixture, or a pendent cord terminating in a lampholder. *See also:* **outlet.** 42A95-0

light-load test. A test on a machine connected to its normal driven or driving member in which the shaft power is restricted to the no-load loss of the member. 0-31E8

light microsecond. The distance over which light travels in free space in one microsecond, that is, about 983 feet (300 meters). This distance is employed as a unit for expressing electrical distance. *See:* **electrical distance.** *See also:* **signal wave.** 42A65-0

light modulator (audio and electroacoustics). A means for varying a light beam as a function of a controlling signal. *Note:* The resulting variation in the light beam may be used to produce a motion-picture sound track. *See also:* **electroacoustics.** 0-1E1

lightness (perceived object color). The attribute by which an object seems to transmit or reflect a greater or lesser fraction of the incident light. *See also:* **color.** Z7A1-0

lightning. An electric discharge that occurs in the atmosphere between clouds or between clouds and ground.
See:
arrester;
arrester discharge voltage-current characteristic;
arrester discharge voltage-time curve;
crest value;
discharge current;
discharge voltage;
fault current;
follow current;
indirect stroke;
lightning arrester;
lightning surge;
surge;
virtual duration of wave front;
wave;
wave front;
wave shape;
wave-shape designation;
wave tail. 42A20/62A1-31E7

lightning arrester (surge diverter). A device designed to protect electric apparatus from high transient voltage, to be connected usually between the electric conductors of a network and ground, and to limit the duration and frequently the amplitude of follow current to ground, and capable of repeating these functions as specified. *Notes:* (1) The term lightning arrester includes any external series gap and characteristic elements that are essential for the proper functioning of the devices as installed for service, regardless of whether or not they are supplied as an integral part of the device. (2) Sometimes lightning arresters are connected across the windings of apparatus or between electric conductors. (3) For an extensive list of cross references, see *Appendix A.* 2A2/42A20-31E7

lightning generator (impulse generator). *See:* **surge generator.**

lightning outage. A power outage following a lightning flashover that results in power follow current necessitating the operation of a switching device to clear the fault. *See also:* **direct-stroke protection (lightning).** 0-31E13

lightning protection and equipment.
See:
air terminal;
branch conductor;
cable;
cage;
coefficient of grounding;
conductor;
cone of protection;
copper-clad steel;
direct-stroke protection;
down conductor;
elevation rod;
faster;
impulse sparkover volt-time characteristic;
impulse withstand voltage;
maximum system voltage;
point;
power-frequency withstand voltage;
roof conductor;
sparkover;
system voltage;
time to impulse sparkover;
vapor openings.

lightning stroke. A single lightning discharge or series of discharges following the same path between cloud regions or between cloud regions and the earth. *See also:* **direct-stroke protection (lightning).** 0-31E13

lightning-stroke component. The high-current discharge of a lightning stroke. *See also:* **direct-stroke protection (lightning).** 0-31E13

lightning-stroke current. The crest magnitude of the current in any of the discharges of a lightning stroke. *See also:* **direct-stroke protection (lightning).** 0-31E13

lightning-stroke voltage. The crest magnitude of the leader voltage required to produce the stroke current. *See also:* **direct-stroke protection (lightning).** 0-31E13

lightning surge. A transient electric disturbance in an electric circuit caused by lightning. *See:* **lightning; lightning arrester (surge diverter).** 42A20/62A1-31E7

light pattern (optical pattern) (Buchmann-Meyer pattern) (mechanical recording). A pattern that is observed when the surface of the record is illuminated by a light beam of essentially parallel rays. *Note:* The width of the observed pattern is approximately proportional to the signal velocity of the recorded groove. 0-1E1

light pipe. An optical transmission element that utilizes unfocused transmission and reflection to reduce photon losses. *Note:* Light pipes have been used to distribute the light more uniformly over a photocathode. *See also:* **phototube.** E175-0

light source (industrial control). A device to supply radiant energy capable of exciting a phototube or photocell. *See also:* **photoelectric control.** 42A25-34E10

light-source color. The color of the light emitted by the source. *Note:* The color of a point source may be defined by its luminous intensity and chromaticity coordinates; the color of an extended source may be defined by its luminance and chromaticity coordinates. *See also:* **color.** Z7A1-0

light transition load (rectifier circuit). The transition load that occurs at light load, usually at less than 5 percent of rated load. *Note:* Light transition load is important in multiple rectifier circuits. A similar effect occurs in rectifier units using saturable-reactor control. *See also:* **rectification; rectifier circuit element.** E59-34E17;42A15-0

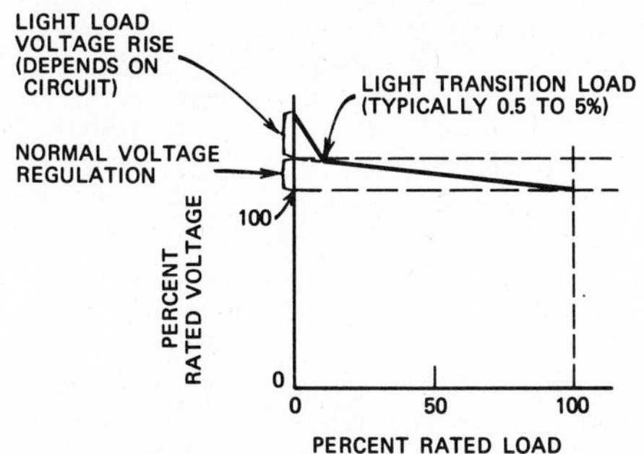

Voltage regulation characteristic showing light transition load.

light valve (electroacoustics). A device whose light transmission can be made to vary in accordance with an externally applied electrical quantity, such as voltage, current, electric field, magnetic field, or an electron beam. *See also:* **circuits and devices; phonograph pickup.** E157/42A65-1E1/2E2

lightwatt. *See:* **spectral luminous efficiency.**

limit (1) (mathematical). A boundary of a controlled variable.

(2) (synchronous machine regulating systems). The boundary at or beyond which a limiter functions. *See also:* **synchronous machine.** 0-31E8

(3) (industrial control). The designated quantity is controlled so as not to exceed a prescribed boundary condition. *See also:* **control system, feedback.** IC1-34E10

limit cycle (control system, feedback). A particular closed trajectory in state space such that all other trajectories in its neighborhood asymptotically approach it as time goes to $+\infty$ or $-\infty$. *See also:* **control system, feedback.** 0-23E0

limited proportionality, region of. *See:* **region of limited proportionality.**

limited signal (radar). A signal that is intentionally limited in amplitude by the dynamic range of the system. *See also:* **navigation.** E172-10E6

limited stability. A property of a system characterized by stability when the input signal falls within a particular range and by instability when the signal falls outside this range. E154-0

limiter (1) (general). A device in which some characteristic of the output is automatically prevented from exceeding a predetermined value. More specifically, a circuit in which the output amplitude is substantially linear with regard to the input up to a predetermined value and substantially constant thereafter. *Note:* For waves having both positive and negative values, the predetermined value is usually independent of sign. *See also:* **circuits and devices.** 0-31E3

(2) (rotating machinery). An element or group of elements that acts to limit by modifying or replacing the functioning of a regulator when predetermined conditions have been reached. *Note:* Examples are minimum excitation limiter, maximum excitation limiter, maximum armature-current limiter. *See also:* **circuits and devices; synchronous machine.** 0-31E8

(3) (radio receivers). A transducer whose output is constant for all inputs above a critical value. *Note:* A limiter may be used to remove amplitude modulation while transmitting angle modulation. *See also:* **circuits and devices; radio receiver; transducer.** E145/E188-0

limiter circuit. A circuit of nonlinear elements that restrict the electric excursion of a variable in accordance with some specified criteria. *Note:* Hard limiting is a limiting action with negligible variation in output in the range where the output is limited. Soft limiting is a limiting action with appreciable variation in output in the range where the output is limited. A bridge limiter is a bridge circuit used as a limiter circuit. In an analog computer, a feedback limiter is a limiter circuit usually employing biased diodes shunting the feedback component of an operational amplifier; an input limiter is a limiter circuit usually employing biased diodes in the amplifier input channel that operates by limiting the current entering the summing junction. *See also:* **electronic analog computer.** E165-16E9

limiting. The action performed upon a signal by a limiter. *See also:* **limiter circuit; radio receiver.** E188-0

limiting ambient temperature (equipment rating). An upper or lower limit of a range of ambient temperatures within which equipment is suitable for operation at its rating. Where the term is used without an adjective the upper limit is meant. *See also:* **limiting insulation temperature.** E1-SCC4

limiting hottest-spot temperature. *See:* **limiting insulation temperature.**

limiting insulation temperature (limiting hottest-spot temperature). The temperature selected for correlation with a specified test condition of the equipment

with the object of attaining a desired service life of the insulation system.
See:
ambient temperature;
hottest-spot temperature allowance;
limiting ambient temperature;
limiting insulation temperature;
limiting observable temperature;
observable temperature;
observable temperature rise. E1-SCC4

limiting insulation temperature rise (equipment rating) (limiting hottest-spot temperature rise). The difference between the limiting insulation temperature and the limiting ambient temperature. *See also:* **limiting insulation temperature.** E1-SCC4

limiting observable temperature rise (equipment rating). The limit of observable temperature rise specified in equipment standards. *See also:* **limiting insulation temperature.** E1-SCC4

limiting polarization (radio wave propagation). The resultant polarization of a wave after it has emerged from a magneto-ionic medium. *See also:* **radio wave propagation.** 0-3E2

limiting resolution (television). A measure of resolution usually expressed in terms of the maximum number of lines per picture height discriminated on a test chart. *Note:* For a number of lines N (alternate black and white lines) the width of each line is $1/N$ times the picture height. *See also:* **resolution; television.** E204-2E2;E208-0

limit of temperature rise (1) (contacts). The temperature rise of contacts, above the temperature of the cooling air, when tested in accordance with the rating shall not exceed the following values. All temperatures shall be measured by the thermometer method. Laminated contacts: 50 degrees Celsius; solid contacts: 75 degrees Celsius.
(2) (resistors). The temperature rise of resistors above the temperature of the cooling air, when test is made in accordance with the rating, shall not exceed the following temperatures for the several classes of resistors: Class A, cast resistors, 450 degrees Celsius; Class B, imbedded resistors, outside of imbedding material, 250 degrees Celsius; Class C, strap or ribbon wound on Class C insulation, 600 degrees Celsius continuous and 800 degrees Celsius intermittent; class D, enameled wire or strap wound resistance, 350 degrees Celsius. Temperatures to be measured by thermocouple method. E16-0

limits of interference (electromagnetic compatibility). Maximum permissible values of radio interference as specified in International Special Committee on Radio Interference recommendations or by other competent authorities or organizations. *See also:* **electromagnetic compatibility.** CISPR-27E1

limit switch. A switch that is operated by some part or motion of a power-driven machine or equipment to alter the electric circuit associated with the machine or equipment. *See also:* **switch.** 42A25-34E10

line (1) (electric power). A component part of a system extending between adjacent stations or from a station to an adjacent interconnection point. A line may consist of one or more circuits. *See:* **system.** *See also:* **power systems, low-frequency and surge testing.** E32-0
(2) (trace) (cathode-ray tube). The path of a moving spot. *See also:* **beam tubes.** 42A70-15E6
(3) (electromagnetic theory). *See:* **maxwell.**
(4) (acoustics). *See:* **acoustic delay line; delay line; electromagnetic delay line; magnetic delay line; sonic delay line.**

line amplifier. *See:* **amplifier, line.**

linear amplifier (magnetic). An amplifier in which the output quantity is essentially proportional to the input quantity. *Note:* This may be interpreted as an amplifier that has no intentional discontinuities in the output-input characteristic over the useful input range of the amplifier. *See also:* **rating and testing magnetic amplifiers.** E107-0

linear array. An array antenna having the centers of the radiating elements lying along a straight line. *See also:* **antenna.** 0-3E1

linear crossed-field amplifier (microwave tubes). A crossed-field amplifier in which a nonre-entrant beam interacts with a forward wave. *See also:* **microwave tube or valve.** 0-15E6

linear detection. That form of detection in which the output voltage is substantially proportional, over the useful range of the detecting device, to the voltage of the input wave. *See also:* **detection.** 42A65-0

linear electrical parameters (uniform line). The series resistance, series inductance, shunt conductance, and shunt capacitance per unit length of line. *Note:* The term **constant** is frequently used instead of **parameter.** *See also:* **transmission line.** 42A65-3E1

linear electron accelerator. An evacuated metal tube in which electrons are accelerated through a series of small gaps (usually in the form of cavity resonators in the high-frequency range) so arranged and spaced that at a specific excitation frequency, the stream of electrons on passing through successive gaps gains additional energy from the electric field in each gap. *See also:* **miscellaneous electron devices.** 42A70/50I07-15E6

linear-impedance relay. A distance relay for which the operating characteristic on an R-X diagram is a straight line. *Note:* It may be described by the equation $Z = K/\cos(\theta - \sigma)$ where K and σ are constants and θ is the phase angle by which the input voltage leads the input current. 37A100-31E11/31E6

linear interpolation (numerically controlled machines). A mode of contouring control that uses the information contained in a block to produce velocities proportioned to the distance moved in two or more axes simultaneously. *See also:* **numerically controlled machines.** EIA3B-34E12

linearity. (1)A property describing a constant ratio of incremental cause and effect. *Note:* Proportionality is a special case of linearity in which the straight line passes through the origin. Zero-error reference of a linear transducer is a selected straight-line function of the input from which output errors are measured. Zero-based linearity is transducer linearity defined in terms of a zero-error reference where zero input coincides with zero output. *See:* **normal linearity.** *See also:* **electronic analog computer.** E165-16E9
(2) The closeness with which a curve of a function approximates a straight line. *See also:* **control system, feedback.** AS1-34E10

linearity control (television). A control to adjust the variation of scanning speed during the trace interval to correct geometric distortion. *See also:* **television.** E204-2E2

linearity error (computing element). The deviation of the output quantity from a specified zero-error reference. *See also:* **electronic analog computer.** E165-16E9

linearity of a multiplier. *See:* **normal linearity.**

linearity of a potentiometer. *See:* **normal linearity.**

linearity, programming (power supplies). The linearity of a programming function refers to the correspondence between incremental changes in the input signal (resistance, voltage, or current) and the consequent incremental changes in power-supply output. *Note:* Direct programming functions are inherently linear for the bridge regulator and are accurate to within a percentage equal to the supply's regulating ability. *See also:* **power supply.** KPSH-10E1

linearity region (instrument approach system and similar guidance systems). The region in which the deviation sensitivity remains constant within specified values. *See also:* **navigation.** 0-10E6

linear light. A luminous signal having a perceptible physical length. *See also:* **signal lighting.** Z7A1-0

linearly polarized wave (plane-polarized wave)(1) (general). A transverse wave in which the displacements at all points along a line in the direction of propagation lie in a plane passing through this line. *See also:* **radiation.** E270-0
(2) (radio wave propagation). An electromagnetic wave whose electric and magnetic field vectors always lie along fixed lines at a given point. *See also:* **radio wave propagation.** 0-3E2

linear modulator. A modulator in which, for a given magnitude of carrier, the modulated characteristic of the output wave bears a substantially linear relation to the modulating wave. 42A65-0

linear passive networks.
See:
all-pass function;
complementary functions;
constant-resistance network;
driving-point admittance;
driving-point function;
driving-point impedance;
immittance;
magnitude characteristic;
minimum conductance function;
minimum-driving-point function;
minimum-phase function;
minimum-reactance function;
minimum-resistance function;
minimum-susceptance function;
phase characteristic;
reactance function;
realizable function;
response function;
susceptance function;
transfer admittance;
transfer current ratio;
transfer function;
transfer impedance;
transfer voltage ratio;
transmittance.
See also: **network analysis.**

linear power amplifier. A power amplifier in which the signal output voltage is directly proportional to the signal input voltage. *See also:* **amplifier.** E145-0

linear programming (computing systems). *See:* **programming, linear.**

linear pulse amplifier (pulse techniques). A pulse amplifier in which the peak amplitude of the output pulses is directly proportional to the peak amplitude of the corresponding input pulses, if the input pulses are alike in shape. *See also:* **pulse.** E175-0

linear rectifier. A rectifier, the output current or voltage of which contains a wave having a form identical with that of the envelope of an impressed signal wave. *See also:* **rectifier.** E145/E182/42A65-0

linear signal flow graphs.
See:
branch;
branch input signal;
branch output signal;
branch transmittance;
cascade node;
dependent node;
directed branch;
feedback node;
graph determinant;
graph transmittance;
loop;
loop graph;
loop-set transmittance;
loop transmittance;
network analysis;
node;
node absorption;
node signal;
nontouching loop set;
open path;
path;
path factor;
path transmittance;
return difference;
signal flow graph;
sink node;
source node;
split node. E155-0

linear system or element. A system or element with the properties that if y_1 is the response to x_1, and y_2 is the response to x_2, then (1) $(y_1 + y_2)$ is the response to $(x_1 + x_2)$, and (2) ky_1 is the response to kx_1. *See also:* **control system, feedback.** 0-23E0

linear system with one degree of freedom. *See:* **damped harmonic system.**

linear transducer. A transducer for which the pertinent measures of all the waves concerned are linearly related. *Notes:* (1) By linearly related is meant any relation of linear character whether by linear algebraic equation, by linear differential equation, by linear integral equation, or by other linear connection. (2) The term **waves concerned** connotes actuating waves and related output waves, the relation of which is of primary interest in the problem at hand. *See also:* **transducer.** E196/E270/42A65-0

linear-varying parameter (varying parameter). A parameter that varies with time or position or both, but not with any dependent variable. *Note:* Unless otherwise specified, **varying parameter** refers to a linear-varying parameter, not to a nonlinear parameter. E270-0

linear-varying-parameter network. A linear network in which one or more parameters vary with time. E154-0

line breaker (line switch) (electrically driven vehicles). A device that combines the functions of a contactor and of a circuit breaker. *Note:* This term is also used for circuit breakers that function to interrupt circuit faults and do not combine the function of a contactor. *See also:* **multiple-control unit.** 42A42-0

line-charging capacity (synchronous machine). The reactive power when the machine is operating synchronously at zero power factor, rated voltage, and with the field current reduced to zero. *Note:* This quantity has no inherent relationship to the thermal capability of the machine. *See also:* **synchronous machine.** E115-31E8

line charging current. The current supplied to an unloaded line or cable. 37A100-31E11

line circuit (railway signaling). A signal circuit on an overhead or underground line. *See also:* **railway signal and interlocking.** 42A42-0

line conductor (electric power). One of the wires or cables carrying electric current, supported by poles, towers, or other structures, but not including vertical or lateral connecting wires. *See also:* **conductor; tower.** 2A2/42A35-31E13

line-drop compensator (voltage regulator). A device that causes the voltage-regulating relay to increase the output voltage an amount that compensates for the impedance drop in the circuit between the regulator and a predetermined location on the circuit (sometimes referred to as the load center). *See:* **voltage regulator.** 42A15/57A15-31E12

line-drop signal (manual switchboard). A drop signal associated with a subscriber line. *See also:* **telephone switching system.** 42A65-0

line-drop voltmeter compensator. A device used in connection with a voltmeter that causes the latter to indicate the voltage at some distant point of the circuit. *See also:* **auxiliary device to an instrument.** 42A30-0

line end and ground end (electric power). (1) Line end is that end of a neutral grounding device that is connected to the line circuit directly or through another device. (2) Ground end is that end that is connected to ground directly or through another device. *See also:* **grounding device.** E32/42A35-31E13

line fill. The ratio of the number of connected main telephone stations on a line to the nominal main-station capacity of that line. *See also:* **cable.** 42A65-0

line-focus tube (X-ray tubes). An X-ray tube in which the focal spot is roughly a line. *See also:* **electron devices, miscellaneous.** 42A70-15E6

line frequency (television). The number of times per second that a fixed vertical line in the picture is crossed in one direction by the scanning spot. *Note:* Scanning during vertical return intervals is counted. *See also:* **television.** E203/42A65-2E2

line guy. Tensional support for poles or structures by attachment to adjacent poles or structures. *See also:* **tower.** 42A35-31E13

line impedance stabilization network. *See:* **artificial mains network.**

line insulator (pin, post). An assembly of one or more shells, having means for semirigidly supporting line conductors. *See also:* **insulator.** 29A1-0

line integral (dot-product line integral). The line integral between two points on a given path in the region occupied by a vector field is the definite integral of the dot product of a path element and the vector. Thus

$$I = \int_a^b \mathbf{V}\cos\theta \, \mathbf{ds}$$

$$= \int_a^b \mathbf{V} \cdot \mathbf{ds}$$

$$= \int_a^b (V_x \mathrm{d}x + V_y \mathrm{d}y + V_z \mathrm{d}z)$$

where **V** is the vector having a magnitude V, **ds** the vector element of the path, θ the angle between **V** and **ds**. *Example:* The magnetomotive force between two points on a line connecting two points in a magnetic field is the line integral of the magnetic field strength, that is, the definite integral between the two points of the dot product of a vector element of the length of the line and the magnetic strength at the element. E270-0

line lamp (wire communication). A switchboard lamp for indicating an incoming line signal. *See also:* **telephone switching system.** 42A65-0

line lightning performance. The performance of a line expressed as the annual number of lightning flashovers on a circuit mile or tower-line mile basis. *See also:* **direct-stroke protection (lightning).** 0-31E13

line microphone. A directional microphone consisting of a single straight line element, or an array of contiguous or spaced electroacoustic transducing elements disposed on a straight line, or the acoustical equivalent of such an array. *See also:* **microphone; line transducer.** 42A65-0

line noise. Noise originating in a transmission line. *See also:* **transmission characteristics.** 42A65-31E3

line number, television (measuring resolution). *See:* **television line number.**

line of position (LOP) (navigation). The intersection of two surfaces of position, normally plotted as lines on the earth's surface, each line representing the locus of constant indication of the navigational information. *See also:* **navigation.** 0-10E6

line or input regulation (electrical conversion). Static regulation caused by a change in input. *See also:* **electrical conversion.** 0-10E1

line or trace (cathode-ray tubes). The path of the moving spot on the screen or target. E160-2E2

line parameters. A sufficient set of parameters to specify the transmission characteristics of the line. E270-0

line printing. The printing of an entire line of characters as a unit. X3A12-16E9

liner (1) (rotating machinery). (A) A separate insulating member that is placed against a grounded surface. (B) A layer of insulating material that is deposited on a grounded surface. *See also:* **cradle base (rotating machinery); slot liner; pole cell insulation.** 0-31E8

(2) (dry cell) (primary cell). Usually a paper or pulpboard sheet covering the inner surface of the negative electrode and serving to separate it from the depolarizing mix. *See also:* **electrolytic cell.** 42A60-0

line regulation (power supplies). The maximum steady-state amount that the output voltage or current will change as the result of a specified change in line voltage (usually for a step change between 105-125 or 210-250 volts, unless otherwise specified). Regulation is given either as a percentage of the output voltage or current, or as an absolute change ΔE or ΔI. *See also:* **power supply.** KPSH-10E1

line relay (railway practice). A relay that receives its operating energy over a circuit that does not include the track rails. *See also:* **telephone switching system; railway signal and interlocking.** 42A42-0

line residual current. *See:* **ground-return current.**

lines. *See:* **communication lines; electric-supply lines.**

line side. Data terminal connections to a communications circuit. 0-19E4

line source (antenna). A radiator comprising a continuous distribution of current lying along a line segment. *See also:* **antenna.** 0-3E1

line stretcher (guided waves). A device for varying the electrical length of a waveguide or transmission line by varying its mechanical length. *See also:* **waveguide.** 0-9E4

line switch (wire communication). A switch attached to a subscriber line, that connects an originating call to an idle part of the switching apparatus. *See also:* **telephone switching system.** 42A65-0

line tap (1) (general). A radial branch connection to a main line. *See also:* **tower.** 42A35-31E13

(2) (system protection). A connection to a line with equipment that does not feed energy into a fault on the line in sufficient magnitude to require consideration in the relay plan. 37A100-31E11/31E6

line terminal (1) (lightning arrester). The conducting part provided for connecting the arrester to the circuit conductor. *Note:* When a line terminal is not supplied as an integral part of the arrester and the series gap is obtained by providing a specified air clearance between the line end of the arrester and a conductor or arcing electrode etcetera, the words **line terminal** used in the definition refer to the conducting part that is at line potential and that is used as the line electrode of the series gap. *See also:* **arresters.** E28/42A20-0;62A1-31E7

(2) (rotating machinery). A termination of the primary winding for connection to a line (not neutral or ground) of the power supply or load. *See:* **asynchronous machine; rotor (rotating machinery); stator; synchronous machine.** 0-31E8

(3) (power system protection). A connection to a line with equipment that can feed energy into a fault on the line in sufficient magnitude to a require consideration in the relay plan and that has means for automatic disconnection. 37A100-31E11/31E6

line-to-line voltage. *See:* **voltage sets.**

line-to-neutral voltage. *See:* **voltage sets.**

line transducer. A directional transducer consisting of a single straight-line element, or an array of contiguous or spaced electroacoustic transducing elements disposed on a straight line, or the acoustical equivalent of such an array. *See also:* **line microphone.** 0-1E1

line transformer. A transformer for interconnecting a transmission line and terminal equipment for such purposes as isolation, line balance, impedance matching, or for additional circuit connections. *See also:* **circuits and devices.** E151-42A65

line trap. *See:* **carrier-current line trap.**

line triggering. Triggering from the power-line frequency. *See:* **oscillograph.** 0-9E4

link. A channel or circuit designed to be connected in tandem with other channels or circuits. *Note:* In automatic switching, a link is a path between two units of switching apparatus within a central office. *See also:* **channel.** 42A65-0

linkage (programming) (computing systems). Coding that connects two separately coded routines. *See also:* **electronic digital computer.** X3A12-16E9

linkage voltage test, direct-current test (rotating machinery). A series of current measurements, made at increasing direct voltages, applied at successive intervals, and maintained for designated periods of time. *Note:* This may be a controlled overvoltage test. *See:* **asynchronous machine; direct-current commutating machines; synchronous machine.** 0-31E8

link-break cutout. A load-break fuse cutout that is operated by breaking the fuse link to interrupt the load current. 37A100-31E11

lin-log receiver (radar). A receiver having a linear amplitude response for small-amplitude signals and a logarithmic response for large-amplitude signals. *See also:* **navigation.** E172-10E6

lip microphone. A microphone adapted for use in contact with the lip. *See also:* **microphone.** 42A65-0

liquid controller (industrial control). An electric controller in which the resistor is a liquid. *See:* **electric controller.** 42A25-34E10

liquid cooling (rotating machinery). *See:* **manifold insulation.**

liquid counter tube (radiation counters). A counter tube suitable for the assay of liquid samples. It often consists of a thin glass-walled Geiger-Mueller tube sealed into a test tube providing an annular space for the sample. *See:* **anticoincidence (radiation counters).** 0-15E6

liquid development (electrostatography). Development in which the image-forming material is carried to the field of the electrostatic image by means of a liquid. *See also:* **electrostatography.** E224-15E7

liquid-filled fuse unit. A fuse unit in which the arc is drawn through a liquid. 37A100-31E11

liquid-flow counter tube (radiation counters). A counter tube specially constructed for measuring the radioactivity of a flowing liquid. *See:* **anticoincidence (radiation counters).** 0-15E6

liquid-function potential. The contact potential between two electrolytes. It is not susceptible of direct measurement. 42A60-0

liquid-immersed (regulator or transformer). Having the core and coils immersed in an insulating liquid. *See also:* **transformer; voltage regulator.** 42A15/57A15/89A1-31E12

liquid-level switch. *See:* **float switch.**

liquid resistor (industrial control). A resistor comprising electrodes immersed in a liquid. 50I16-34E10

Lissajous figure. A special case of an x-y plot in which the signals applied to both axes are sinusoidal functions, useful for determining phase and harmonic relationships. *See:* **oscillograph.** 0-9E4

list (computing systems). *See:* **push-down list; pushup list.** *See also:* **electronic digital computer.**

listener echo. Echo that reaches the ear of the listener. *See also:* **transmission characteristics.** 42A65-0

litz wire (litz). A conductor composed of a number of fine, separately insulated strands, usually fabricated in such a way that each strand assumes, to substantially the same extent, the different possible positions in the cross section of the conductor. Litz is an abbreviation for litzendraht. *See also:* **circuits and devices.** 42A65-0

live (electric system). *See:* **alive.**

live cable test cap. A protective structure at the end of a cable that insulates the conductors and seals the cable sheath. *See also:* **power distribution, underground construction.** 42A35-31E13

live-front (industrial control). So constructed that there are exposed live parts on the front of the assembly. *See:* **live-front switchboard.** 42A25-34E10

live-front switchboard. A switchboard having exposed live parts on the front. 37A100-31E11

live parts. Those parts that are designed to operate at voltage different from that of the earth. 37A100-31E11

live room (audio and electroacoustics). A room that has an unusually small amount of sound absorption. *See also:* **loudspeaker.** 0-1E1

live zone (industrial control). The period(s) in the operating cycle of a machine during which corrective functions can be initiated. 42A25-34E10

***L* network.** A network composed of two branches in series, the free ends being connected to one pair of terminals and the junction point and one free end being connected to another pair of terminals. *See also:* **network analysis.** E153/E270/42A65-0

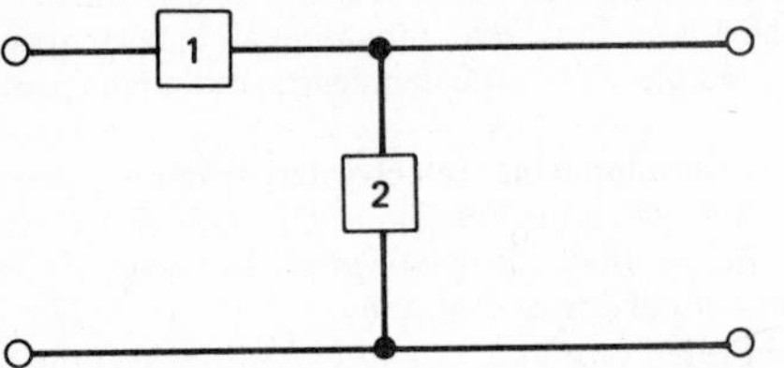

L network. The free ends are the left-hand terminal pair; the junction point and one free end are the right-hand terminal pair.

load (1) (general). (A) A device that receives power, or (B) the power or apparent power delivered to such a device. 0-42A65

(2) (signal-transmission system). The device that receives signal power from a signal-transmission system. *See:* **signal.** 0-13E6

(3) (induction and dielectric heating usage) (charge). The material to be heated. *See also:* **dielectric heating; induction heating.** E169-0

(4) (transformer). The power in kilovolt-amperes, or volt-amperes, supplied by the transformer. *See also:* **duty.** 42A15-31E12;89A1-0

(5) (rotating machinery). All the numerical values of the electrical and mechanical quantities that signify the demand to be made on a rotating machine by an electric circuit or a mechanism at a given instant. *See also:* **direct-current commutating machine; asynchronous machine; synchronous machine.** 0-31E8

(6) (programming). To place data into internal storage. X3A12-16E9

(7) (electric) (electric utilization). The electric power consumed by devices connected to an electrical generating system. *See also:* **generating station.** 0-31E4

load-and-go (computing systems). An operating technique in which there are no stops between the loading and execution phases of a program, and which may include assembling or compiling. *See also:* **electronic digital computer.** X3A12-16E9

load angle (synchronous machine). The angular displacement, at a specified load, of the center line of a field pole from the axis of the armature magnetomotive force pattern. *See also:* **synchronous machine.** 0-31E8

load-angle curve (load-angle characteristic) (synchronous machine). A characteristic curve giving the relationship between the rotor displacement angle and the load, for constant values of armature voltage, field current, and power factor. *See also:* **synchronous machine.** 42A10-31E8

load-band of regulated voltage (rotating machinery). The band or zone, expressed in percent of the rated value of the regulated voltage, within which the synchronous-machine regulating system will hold the regulated voltage of a synchronous machine during steady or gradually changing conditions over a specified range of load. *See also:* **synchronous machine.** 0-31E8

load-break cutout. A cutout with means for interrupting load currents. 37A100-31E11

load center (electric power utilization). A point at which the load of a given area is assumed to be concentrated. *See also:* **generating station.** 0-31E4

load (dynamic) characteristic (electron tube). For an electron tube connected in a specified operating circuit at a specified frequency, a relation, usually represented by a graph, between the instantaneous values of a pair of variables such as an electrode voltage and an electrode current, when all direct electrode supply voltages are maintained constant. *See also:* **circuit characteristics of electrodes.** 42A70-15E6

load circuit (industrial control). The complete circuit required to transfer power from a source to a load. *See:* **control.** *See also:* **circuits and devices; dielectric heating; induction heating.** E145-42A65;E270-0

load-circuit efficiency (induction and dielectric heating usage). The ratio of the power absorbed by the load to the power delivered at the generator output terminals. *See also:* **dielectric heating; induction heating; network analysis.** E54/E169-0

load-circuit power input. The power delivered to the load circuit. *Note:* It is the product of the alternating component of the voltage across the load circuit, the alternating component of the current passing through it (both root-mean-square values), and the power factor associated with these two quantities. *See also:* **network analysis.** E145-0

load coil (induction heating usage). An electric conductor that, when energized with alternating current, is adapted to deliver energy by induction to a charge to be heated. *See also:* **induction heating.** E169-0

load current (tube) (electron tubes). The current output utilized in an external load circuit. *See also:* **electron tube.** 50I07-15E6

load current output (arc-welding apparatus). The current in the welding circuit under load conditions. *See also:* **electric arc-welding apparatus.** 87A1-0

load curve. A curve of power versus time showing the value of a specific load for each unit of the period covered. *See also:* **generating station.** 42A35-31E13

load curves, daily (electric power supply). Curves of net 60-minute integrated demand for each clock hour of a 24-hour day. *See also:* **generating station.** 0-31E4

load diversity (electric power utilization). The difference between the sum of the maxima of two or more individual loads and the coincident or combined maximum load, usually measured in kilowatts over a specified period of time. *See also:* **generating station.** 42A35-31E13;0-31E4

load diversity power. The rate of transfer of energy necessary for the realization of a saving of system capacity brought about by load diversity. *See also:* **generating station.** 42A35-31E13

load division (load balance) (industrial control). A control function that divides the load in a prescribed manner between two or more power devices connected to the same load. *See also:* **control system, feedback.** AS1-34E10

load, dummy (artificial load). *See:* **dummy load.**

load duration curve. A curve showing the total time, within a specified period, during which the load equaled or exceeded the power values shown. *See also:* **generating station.** 42A35-31E13

loaded applicator impedance (dielectric heating). The complex impedance measured at the point of application with the load material at the proper position for heating at a specified frequency. *See also:* **dielectric heating.** E54/E169-0

loaded impedance (transducer). The impedance at the input of the transducer when the output is connected to its normal load. *See also:* **electroacoustics; self-impedance.** 42A65/E157-1E1

loaded Q (1) (working Q) (electric impedance). The value of Q of such impedance when coupled or connected under working conditions. *See also:* **transmission characteristics.** 42A65-0

(2) (switching tubes). The unloaded Q of the tube modified by the coupled impedances. *Note:* As here used, Q is equal to 2π times the energy stored at the resonance frequency divided by the energy dissipated per cycle in the tube, or for cell-type tubes, in the tube and its external resonant circuit. *Seeialso:* **transmission characteristics; gas tubes.** E160-15E6

load factor. The ratio of the average load over a designated period of time to the peak load occuring in that period. *See also:* **generating station.** 42A35-31E13/31E4

load-frequency control. The regulation of the power output of electric generators within a prescribed area in response to changes in system frequency, tie line loading, or the relation of these to each other, so as to maintain the scheduled system frequency or the established interchange with other areas within predetermined limits or both. *See also:* **generating station.** 42A35-31E13

load ground (signal-transmission system). The potential reference plane at the physical location of the load. *See:* **signal.** 0-13E6

load impedance. *See:* **impedance, load.**

load-impedance diagram (oscillators). A chart showing performance of the oscillator with respect to variations in the load impedance. *Note:* Ordinarily, contours of constant power and of constant frequency are drawn on a chart whose coordinates are the components of either the complex load impedance or of the reflection coefficient. *See:* **Rieke diagram.** *See also:* **oscillatory circuit.** E160-15E6

load-indicating automatic reclosing equipment (power distribution). An automatic reclosing equipment that provides for reclosing the circuit interrupter automatically in response to sensing of predetermined conditions of the load circuit. *Note:* This type of automatic reclosing equipment is generally used for direct-current load circuits. 37A100-31E11

load-indicating resistor. A resistor used, in conjunction with suitable relays or instruments, in an electric circuit for the purpose of determining or indicating the magnitude of the connected load. 37A100-31E11

loading (1) (communication practice). Insertion of reactance in a circuit for the purpose of improving its transmission characteristics in a given frequency band. *Note:* The term is commonly applied, in wire communication practice, to the insertion of loading coils in series in a transmission line at uniform intervals, and in radio practice, to the insertion of one or more loading coils anywhere in a transmission circuit. *See also:* **storm loading; transmission line.**
See:
coil loading;
loading coil;
loading-coil spacing;
phantom-circuit loading coil;
series loading;
shunt loading;
side-circuit loading coil. 42A65-0

(2) (antenna). The modification of a basic **antenna,** such as a **dipole** or **monopole,** by adding conductors or circuit elements that change the current distribution or input impedance. *See also:* **antenna.** 0-3E1

loading coil. An inductor inserted in a circuit to increase its inductance for the purpose of improving its transmission characteristics in a given frequency band. *See also:* **loading.** 42A65-21E0

loading-coil spacing. The line distance between the successive loading coils of a coil-loaded line. *See also:* **loading.** 42A65-0

loading error. An error due to the effect of a load upon the transducer or signal source driving it. *See also:* **electronic analog computer.** E165-16E9

loading machine (mining). A machine for loading materials such as coal, ore, or rock into cars or other means of conveyance for transportation to the surface of the mine. *See also:* **mining.** 42A85-0

load-interrupter switch (power distribution). An interrupter switch designed to interrupt currents not in excess of the continuous-current rating of the switch. *Notes:* (1) It may be designed to close and carry abnormal or short-circuit currents as specified. (2) In international practice (International Electrotechnical Commission), a device with such performance characteristics is called a switch. 37A100-31E11

load leads (induction and dielectric heating usage). The connections or transmission line between the power source or generator and load, load coil or applicator. *See also:* **dielectric heating; induction heating.** E54/E169-0

load-limit changer (speed-governing system). A device that acts on the speed-governing system to prevent the governor-controlled fuel valves from opening beyond the position for which the device is set. *See also:* **speed-governing system.** E94-0;E282-31E2

load losses (1) (constant-current transformer). The losses that are incident to the carrying of load. *Note:* Load losses include I^2R loss in the windings due to load current, stray loss due to stray fluxes in the windings, core clamps, etcetera; and also the loss due to circulating currents, if any, in parallel windings. *See also:* **routine test.** 57A14/57A15-0;42A15-31E12 **(2) (copper losses) (series transformer).** The load losses of a series transformer are the I^2R losses, computed from the rated currents for the windings and the measured direct-current resistances of the windings corrected to 75 degrees Celsius. 82A7-0

load matching. The process of adjustment of the load-circuit impedance to produce the desired energy transfer from the power source to the load. *See also:* **dielectric heating; induction heating.** E54/E169-0

load-matching network. An electric network for accomplishing load matching. *See also:* **induction heating; dielectric heating.** E54/E169-0

load-matching switch (induction and dielectric heating). A switch in the load-matching network to alter its characteristics to compensate for some sudden change in the load characteristics, such as passing through the Curie point. *See also:* **dielectric heating; induction heating.** E54/E169-0

load range (watthour meter). Denotes the range in amperes over which the meter is designed to operate continuously. *See also:* **watthour meter.** 12A0-0

load regulation (control) (industrial control). A steady-state decrease of the value of the specified variable resulting from a specified increase in load, generally from no-load to full-load unless otherwise specified. *Notes:* (1) In cases where the specified variable increases with increasing load, load regulation would be expressed as a negative quantity. (2) Regulation is given as a percentage of the output voltage or current or as an absolute change ΔV or ΔI. *See:* **control system, feedback; power supply.** AS1-34E10/10E1

load rheostat. A rheostat whose sole purpose is to dissipate electric energy. *Note:* Frequently used for load tests of generators. *See:* **control.** 42A25/50I16-34E10

load saturation curve (load characteristic) (synchronous machine). The saturation curve of a machine on a specified constant load current. *See also:* **synchronous machine.** 0-31E8

load-shifting resistor (electric circuit). A resistor used to shift load from one circuit to another. 37A100-31E11

load sliding. A load sliding inside or along a fixed length of waveguide or transmission line. *See also:* **waveguide.** 0-9E4

load switch or contactor (induction heating). The switch or contactor in an induction heating circuit that connects the high-frequency generator or power source to the heater coil or load circuit. *See also:* **induction heating.** E54/E169-0

load transfer switch. A switch used to connect a generator or power source optionally to one or another load circuit. *See also:* **dielectric heating; induction heating.** E54/E169-0

load variation within the hour (electric power utilization). The short-time (three minutes) net peak demand minus the net 60-minute clock-hour integrated peak demand of a supplying system. *See also:* **generating station.** 0-31E4

load voltage (arc-welding apparatus). The voltage between the output terminals of the welding power supply when current is flowing in the welding circuit. *See also:* **electric arc-welding apparatus.** 87A1-0

load, work, or heater coil (induction heating). An electric conductor that when energized with alternating current is adapted to deliver energy by induction to a charge to be heated. *See also:* **induction heating.** E54-0

lobe (antenna lobe) (directional lobe) (radiation lobe). A portion of the directional pattern bounded by one or two cones of nulls. *See also:* **antenna.** 42A65-3E1

lobe switching (electronic navigation). A means of direction finding in which a directive radiation pattern is periodically shifted in position so as to produce a variation of the signal at the target; the amount of signal variation is related to the amount of displacement of the target from the pattern mean position. *See also:* **antenna.** 0-10E6

local action (self discharge) (battery). The loss of otherwise usable chemical energy by spontaneous currents within the cell or battery regardless of its connections to an external circuit. *Note:* Corrosion may result from the action of local cells, that is, galvanic cells resulting from inhomogeneities between adjacent areas on a metal surface exposed to an electrolyte. *See also:* **battery (primary or secondary).** 42A60-0;CM-34E2

local backup (power systems). A form of backup protection in which the relays and circuit breakers are at the same station as the primary protective relays and circuit breakers. *See also:* **relay.** 37A100-31E11/31E6

local-battery-talking – common-battery-signaling telephone set. A local-battery telephone set in which current for signaling by the telephone station is supplied from a centralized direct-current power source. *See also:* **telephone station.** 42A65-0

local call (telephony). Any call (attempted or completed) for a destination within the local or extended service area of the calling customer. Local calls may be free or bulk billed. *See also:* **telephone system.** 0-19E1

local cell. A galvanic cell resulting from inhomogeneities between areas on a metal surface in an electrolyte. *Note:* The inhomogeneities may be of physical or chemical nature in either the metal or its environment. *See:* **electrolytic cell.** CM-34E2

local central office. A central office arranged for terminating subscriber lines and provided with trunks for establishing connections to and from other central offices. *See also:* **telephone system.** 42A65-19E1

local control (radio transmitters). A system or method of radio-transmitter control whereby the control functions are performed directly at the transmitter. *See also:* **radio transmitter.** E145-0

local correction (submarine cable telegraphy). A method of signal shaping used to restore deficient direct-current and low-frequency signal components that have been suppressed at the receiver principally for the purpose of minimizing susceptibility to low-frequency disturbances. *Note:* Local correction is also used to compensate for low-frequency signal distortion, commonly called **zero wander,** inherent in some types of receivers. The local correction network, comprising low-pass and delay elements, is energized under control of the received signals, and its output is applied to the receiver in feedback relationship. *See:* **interpolation.** *See also:* **telegraphy.** 42A65-0

localized general lighting. Lighting utilizing luminaires above the visual task and contributing also to the illumination of the surround. *See also:* **general lighting.** Z7A1-0

localizer (electronic navigation). A radio transmitting facility that provides signals for lateral guidance of aircraft with respect to a runway center line. *See also:* **navigation; radio navigation.** 42A65/E172-10E6

localizer on-course line. A line in a vertical plane passing through a localizer and on either side of which indications of opposite sense are received. *See also:* **radio navigation.** 42A65-0

localizer receiver. An airborne radio receiver used to detect the transmissions of a ground-installed localizer transmitter. *Note:* It furnishes a visual, audible, or electric signal for the purpose of laterally guiding an aircraft using an instrument landing system. *See also:* **air-transportation electronic equipment.** 42A41-0

localizer sector (equisignal localizer). The sector included between two radial lines from the localizer, the lines being defined by specified equal differences in depth of modulation (usually full-scale right and left, respectively, of the flight-path deviation indicator). *See also:* **navigation.** 0-10E6

local level (navigation). The plane normal to the local vertical. *See also:* **navigation.** E174-10E6

local lighting. Lighting that provides illumination over a relatively small area or confined space without providing any significant general surrounding lighting. *See also:* **general lighting.** Z7A1-0

local manual fire-alarm system (1) (general alarm type). A local fire-alarm system in which the alarm signal is sounded on all sounding devices installed.
(2) (presignal type). A local fire-alarm system in which the initial alarm signal is sounded by selected sounding devices, with provision for the subsequent sounding of a general alarm at the option of those responsible for system operation. *See also:* **protective signaling.** 42A65-0

local oscillator (1) (general). An oscillator whose output is mixed with a wave for frequency conversion. E188-0
(2) (superheterodyne circuit). An oscillator whose output is mixed with the received signal to produce a sum or difference frequency equal to the intermediate frequency of the receiver. *See also:* **oscillatory circuit.** 42A65-31E3

local-oscillator tube. An electron tube in a heterodyne conversion transducer to provide the local heterodyning frequency for a mixer tube. *See:* **tube definitions.** E160-15E6

local service area. That area within which are located the stations that a customer may call at rates in accordance with the local tariff. *See also:* **telephone system.** 42A65-0

local side (data transmission). Data terminal connections to input-output devices. *See also:* **data transmission.** 0-19E4

local system (protective signaling). A system in which the alarm or supervisory signal is sounded, as by a bell, horn, or whistle, locally at the protected premises. *See also:* **protective signaling.** 42A65-0

local vertical (navigation). The vertical at the location of the observer. *See also:* **navigation.** 0-10E6

locating bearing (rotating machinery). A bearing arranged to limit the axial movement of a horizontal shaft but that is not intended to carry any continuous thrust load. It may be combined with the load-carrying bearing. *See also:* **bearing.** 0-31E8

location (computing systems). Loosely, any place in which data may be stored. *See:* **protected location.** *See also:* **electronic digital computer.** X3A12-16E9

locator (electronic navigation). *See:* **nondirectional beacon.**

locator joint (rotating machinery). Any joint used to position a part when assembled. *See:* **cradle base (rotating machinery).** 0-31E8

locked groove (concentric groove) (disk recording). A blank and continuous groove at the end of modulated grooves whose function is to prevent further travel of the pickup. *See also:* **electroacoustics.** E157-1E1

locked rotor (rotating machinery). The condition existing when the circuits of a motor are energized, but the rotor is not turning. *See:* **rotor (rotating machinery).** 0-31E8

locked-rotor current (1) (motor general). The steady-state current taken from the line with the rotor locked and with rated voltage (and rated frequency in the case of alternating-current motors) applied to the motor. *See:* **asynchronous machine; direct-current commutating machine.** 42A10-0
(2) (motor and starter). The current taken from the line with the rotor locked, with the starting device in the starting position, and with rated voltage and frequency applied. 42A10-0
(3) (synchronous motor). The maximum measured steady-state current taken from the line with the motor at rest, for all angular positions of its rotor, with rated voltage and frequency applied. *See:* **synchronous machine.** 0-31E8

locked-rotor temperature-rise rate (rotor end ring or of a winding). The average rate of temperature rise of the rotor end rings, or a winding, at locked rotor with rated voltage applied at rated frequency to the primary winding, in increasing from an initial (usually ambient) temperature to a specified ultimate temperature, expressed in degrees Celsius per second or degrees Fahrenheit per second. *See also:* **rotor (rotating machinery).** 0-31E8

locked-rotor test (rotating machinery). A test on an energized machine with its rotor held stationary. *See also:* **asynchronous machine; direct-current commutating machine; synchronous machine.** 0-31E8

locked-rotor torque. The minimum torque that a motor will provide with locked rotor, at any angular position of the rotor, at a winding temperature of 25 degrees Celsius plus or minus 5 degrees Celsius, with rated voltage applied at rated frequency. *See:* **asynchronous machine; direct-current commutating machine; synchronous machine.** 42A10-31E8

locking-in. The shifting and automatic holding of one or both of the frequencies of two oscillating systems that are coupled together, so that the two frequencies have the ratio of two integral numbers. *See also:* **circuits and devices.** 42A65-0

locking plate (rotating machinery). *See:* **lockplate.**

locking ring (rotating machinery). A ring used to prevent motion of a second part. *See:* **cradle base (rotating machinery).** 0-31E8

lockout (1) (general). *See:* **sweep lockout.** *See also:* **telephone switching system.** 0-9E4
(2) (telephone circuit controlled by two voice-operated devices). The inability of one or both subscribers to get through, either because of excessive local circuit noise or continuous speech from either or both subscribers. *See also:* **telephone switching system.** 42A65-0

lockout-free (recloser or sectionalizer). A general term denoting that the lockout mechanism can operate even though the manual operating lever is held in the closed position. 37A100-31E11

lockout mechanism (automatic circuit recloser). A device that locks the contacts in the open position following the completion of a predetermined sequence of operations. 37A100-31E11

lockout operation (recloser). An opening operation followed by the number of closing and opening operations that the mechanism will permit before locking the contacts in the open position. 37A100-31E11

lockout relay. An electrically reset or hand-reset auxiliary relay whose function is to hold associated devices inoperative until it is reset. 37A100-31E11/31E6

lockplate (locking plate) (rotating machinery). A plate used to prevent motion of a second part (for example, to prevent a bolt or nut from turning). *See:* **cradle base (rotating machinery).** 0-31E8

lock-up relay. (1) A relay that locks in the energized position by means of permanent magnetic bias (requiring a reverse pulse for releasing) or by means of a set of auxiliary contacts that keep its coil energized until the circuit is interrupted. *Note:* Differs from a latching relay in that locking is accomplished magnetically or electrically rather than mechanically. (2) Sometimes used for latching relay. *See also:* **relay.** 83A16-0

logarithmic decrement (underdamped harmonic system). The natural logarithm of the ratio of a maximum of the free oscillation to the next following maximum. The logarithmic decrement of an underdamped harmonic system is

$$\ln\left(\frac{X_1}{X_2}\right) = \frac{2\pi F}{(4MS - F^2)^{1/2}}$$

where X_1 and X_2 are the two maxima. E270-0

logic. (1) The result of planning a data-processing system or of synthesizing a network of logic elements to perform a specified function. (2) Pertaining to the type or physical realization of logic elements used, for example, diode logic, AND logic. *See also:* **electronic digital computer; formal logic; logic design; symbolic logic.** E162-0

logic board (power-system communication). An assembly of decision-making circuits on a printed-circuit mounting board. *See also:* **digital.** 0-31E3

logic design (electronic computation). (1) The planning of a computer or data-processing system prior to its detailed engineering design. (2) The synthesizing of a network of logic elements to perform a specified function. (3) The result of (1) and (2) above, frequently called the logic of the system, machine, or network. *See also:* **electronic computation; electronic digital computer.** E162/E270-0

logic diagram. A diagram representing the logic elements and their interconnections without necessarily expressing construction or engineering details. *See also:* **electronic computation; electronic digital computer.** E162/E270-0

logic element (1) (general). A combinational logic element or sequential logic element. E162-0
(2) (computer or data-processing system). The smallest building blocks that can be represented by operators in an appropriate system of symbolic logic. Typical logic elements are the AND gate and the flip-flop, which can be represented as operators in a suitable symbolic logic. *See also:* **electronic computation; electronic digital computer.** E270-0

logic instruction (computing systems). An instruction that executes an operation that is defined in symbolic logic, such as AND, OR, NOR. X3A12-16E9

logic, 0-1. The representation of information by two states termed 0 and 1. *See also:* **bit; dot cycle; mark or space; digital.** 0-31E3

logic operation (electronic computation) (1) (general). Any nonarithmetical operation. *Note:* Examples are: extract, logical (bit-wise) multiplication, jump, data transfer, shift, compare, etcetera.
(2) (sometimes). Only those nonarithmetical operations that are expressible bit-wise in terms of the propositional calculus or two-valued Boolean algebra. *See also:* **electronic computation; electronic digital computer.** E162/E270-0

logic operator. *See:* **AND; exclusive OR; NAND; NOT; OR.**

logic shift (computing systems). A shift that affects all positions. X3A12-16E9

logic symbol (electronics computation). (1) A symbol used to represent a logic element graphically. (2) A symbol used to represent a logic connective. *See also:* **electronic computation; electronic digital computer.** E162/X3A12-16E9

log-periodic antenna. Any one of a class of **antennas** having a structural geometry such that its electrical characteristics repeat periodically as the logarithm of frequency. *See also:* **antenna.** 0-3E1

long dimension (numerically controlled machines). Incremental dimensions whose number of digits is one more to the left of the decimal point than for a normal dimension, and the last digit shall be zero, that is, XX.XXX0 for the example under normal dimension. *See also:* **numerically controlled machines.** EIA3B-34E12

long-distance navigation. Navigation utilizing self-contained or external reference aids or methods usable at comparatively great distances. *Note:* Examples of long-distance aids are loran, Doppler, inertial, and celestial navigation. *See:* **short-distance navigation; approach navigation.** *See also:* **navigation.** 0-10E6

longitudinal circuit (telephony). A circuit formed by one telephone wire (or by two or more telephone wires in parallel) with return through the earth or through any other conductors except those that are taken with the original wire or wires to form a metallic telephone circuit. *See also:* **inductive coordination.** 42A65-0

longitudinal interference. *See:* **common-mode interference.** *See also:* **accuracy rating (instrument); signal.**

longitudinal magnetization (magnetic recording). Magnetization of the recording medium in a direction essentially parallel to the line of travel. E157-1E1

longitudinal mode (interference terminology). *See:* **interference, common-mode.**

longitudinal redundancy check (LRC) (data transmission). A system of error control based on the formation of a block check following preset rules. *Note:* The check formation rule is applied in the same manner to each character. *See also:* **data transmission.** 0-19E4

longitudinal wave. A wave in which the direction of displacement at each point of the medium is the same as the direction of the propagation. E270-0

long-line current (corrosion). Current (positive electricity) flowing through the earth from an anodic to a cathodic area that returns along an underground metallic structure. *Note:* Usually used only where the areas are separated by considerable distance and where the current results from concentration cell action. *See also:* **stray-current corrosion.** CM-34E2

long-pitch winding (rotating machinery). A winding in which the coil pitch is greater than the pole pitch. *See:* **direct-current commutating machine.** 0-31E8

longwall machine (mining). A power-driven machine used for undercutting coal on relatively long faces. *See also:* **mining.** 42A85-0

long-wire antenna. A linear antenna that, by virtue of its considerable length in comparison with the operating wavelength, provides a directional pattern. *See also:* **antenna.** 42A65-3E1

look-up (computing systems). *See:* **table look-up.** *See also:* **electronic digital computer.**

loom. *See:* **flexible nonmetallic tubing.**

loop (1) (signal-transmission system and network analysis). A set of branches forming a closed current path, provided that the omission of any branch eliminates the closed path. *See:* **signal; ground loop; mesh.** *See also:* **linear signal-flow graphs.** E155-13E6

(2) (standing wave). A point at which the amplitude is a maximum. E270-0

(3) (computing systems). A sequence of instructions that is executed repeatedly until a terminal condition prevails. *See also:* **electronic digital computer.** X3A12-16E9

loop antenna. An antenna consisting of one or more complete turns of conductor, excited so as to provide essentially uniform circulatory current, and producing a radiation pattern approximating that of an elementary magnetic dipole. *See also:* **antenna.** E149-3E1

loop circuit (railway signaling). A circuit that includes a source of electric energy, a line wire that conducts current in one direction, and connections to the track rails at both ends of the line to complete the circuit through the two rails in parallel in the other direction. 42A42-0

loop control (industrial control). The effect of a control function or a device to maintain a specified loop of material between two machine sections by automatic speed adjustment of at least one of the driven sections. *See:* **control system, feedback.** IC1-34E10

loop current. The electric current in a loop circuit. *See:* **mesh current.** 42A42-0

loop (leakage) current (power supplies). A direct current flowing in the feedback loop (voltage control) independent of the control current generated by the reference Zener diode source and reference resistor. *Note:* The loop (leakage) current remains when the reference current is made zero. It may be compensated for, or nulled, in special applications to achieve a very-high impedance (zero current) at the feedback (voltage control) terminals. *See also:* **power supply.** KPSH-10E1

loop equations. *See:* **mesh or loop equations.**

loop factor (network analysis). *See:* **path factor.**

loop feeder (power distribution). A number of tie feeders in series, forming a closed loop. *Note:* There are two routes by which any point on a loop feeder can receive electric energy, so that the flow can be in either direction. *See also:* **center of distribution.** 42A35-31E13

loop gain (1) (open-loop gain) (industrial control). The ratio, under specified conditions, of the steady-state sinusiodal magnitude of the feedback signal to that of the actuating signal when the loop is open at the summing point. *See:* **control system, feedback.** AS1-34E10

(2) (power supplies). A measure of the feedback in a closed-loop system, being equal to the ratio of the open-loop to the closed-loop gains, in decibels. *Note:* The magnitude of the loop gain determines the error attenuation and, therefore, the performance of an amplifier used as a voltage regulator. *See:* **open-loop and closed-loop gain.** *See also:* **power supply.** KPSH-10E1

loop graph (network analysis). A signal flow graph each of whose branches is contained in at least one loop. *Note:* Any loop graph embedded in a general graph can be found by removing the cascade branches. *See also:* **linear signal-flow graphs.** E155-0

looping-in (interior wiring). A method of avoiding splices by carrying the conductor or cable to and from the outlet to be supplied. *See also:* **interior wiring.** 42A95-0

loop service (power distribution). Two services of substantially the same capacity and characteristics supplied from adjacent sections of a loop feeder. *Note:* The two sections of the loop feeder are normally tied together on the consumer's bus through switching devices. *See also:* **service.** 42A35-31E13

loop-service (ring) feeder (power distribution). A feeder that supplies a number of separate loads distributed along its length and that terminates at the same bus from which it originated. 37A100-31E11

loop-set transmittance (network analysis). The product of the negatives of the loop transmittances of the loops in a set. *See also:* **linear signal-flow graphs.** E155-0

loop test. A method of testing employed to locate a fault in the insulation of a conductor when the conductor can be arranged to form part of a closed circuit or loop. 42A95-0

loop transmittance (1) (network analysis). The product of the branch transmittances in a loop.

(2) (branch). The loop transmittance of an interior node inserted in that branch. *Note:* A branch may always be replaced by an equivalent sequence of branches, thereby creating interior nodes.

(3) (node). The graph transmittance from the source node to the sink node created by splitting the designated node. *See also:* **linear signal-flow graphs.** E155-0

loose coupling. Any degree of coupling less than the critical coupling. *See also:* **coupling.** 42A65-0

LOP (electronic navigation). *See:* **line of position.**

loran. A long-distance radio-navigation system in which hyperbolic lines of position are determined by measuring arrival-time differences of pulses transmitted in fixed time relationship from two fixed-base transmitters. *Note:* **Loran A,** generally useful to distances of 500 to 1500 nautical miles (900 to 2800 kilometers) over water, depending upon the availability of sky wave, uses a baseline of about 300 nautical miles (550 kilometers), operated at approximately 2 megahertz and gives time-difference measurement by the matching of the leading edges of the pulses, usually with the aid of an oscilloscope. **Loran C,** generally useful to distances of 1000 to 1500 nautical miles (1850 to 2800 kilometers) over water, uses a baseline of about 500 nautical miles, operates at approximately 100 kilohertz; it provides a coarse measurement of time-difference through the matching of pulse envelopes, and a fine measurement by the comparing of phase between the carrier waves. **Loran D** is a shorter-baseline and lower-power adaptation of loran C for tactical applications. *See also:* **radio direction-finder (radio compass); radio navigation.** E172-10E6

loran repetition rate (electronic navigation). *See:* **pulse repetition frequency.**

Lorenz number. The quotient of (1) the electronic thermal conductivity by (2) the product of the absolute temperature and the component of the electric conductivity due to electrons and holes. *See also:* **thermoelectric device.** E221-15E7

lorhumb line (navigation system chart, such as a loran chart with its overlapping families of hyperbolic lines). A line drawn so that it represents a path along which the change in values of one of the families of lines retains a constant relation to the change in values of another of the families of lines. *See also:* **navigation.** 0-10E6

loss (1) (general). (A) Power expended without accomplishing useful work. Such loss is usually expressed in watts. (B) The ratio of the signal power that could be delivered to the load under specified reference conditions to the signal power delivered to the load under actual operating conditions. Such loss is usually expressed in decibels. *Note:* Loss is generally due to dissipation or reflection due to an impedance mismatch or both. *See:* **transmission loss.** 0-21E1

(2) (network analysis). *See:* **related transmission terms.**

loss angle (biological). The complement Φ of the phase angle θ (between the electrode potential vector and the current vector).

$$\Phi = \frac{\pi}{2} - \theta$$

See also: **electrode impedance (biological).** 42A80-18E1

loss compensator. *See:* **transformer-loss compensator.**

losses (grounding device) (electric power). I^2R loss in the windings, core loss, dielectric loss (for capacitors), losses due to stray magnetic fluxes in the windings and other metallic parts of the device, and in cases involving parallel windings, losses due to circulating currents. *Note:* The losses as here defined do not include any losses produced by the grounding device in adjacent apparatus or materials not part of the device. Losses will normally be considered at the maximum rated neutral current but may in some cases be required at other current ratings, if more than one rating is specified, or at no load, as for grounding transformers. *Note:* The losses may be given at 25 degrees Celsius or at 75 degrees Celsius. *See also:* **grounding device.** E32-0

losses of a current-limiting reactor. Losses that are incident to the carrying of current. They include: (1) The resistance and eddy-current in the winding due to load current. (2) Losses caused by circulating current in parallel windings. (3) Stray losses caused by magnetic flux in other metallic parts of the reactor and in the reactor enclosure when the enclosure is supplied as an integral part of the reactor installation. (4) The losses produced by magnetic flux in adjacent apparatus or material not an integral part of the reactor or its enclosure are not included. *See also:* **reactor.** 57A16-0

loss factor (1) (electric power generation). The ratio of the average power loss to the peak-load power loss during a specified period of time. *See also:* **generating station.** 42A35-31E13

(2) (dielectric heater material). The product of its dielectric constant and the tangent of its dielectric loss angle. *See also:* **dielectric heating.** E54-0

loss function (control system). An instantaneous measure of the cost of being in state x and of using control u at time t. *See also:* **performance index.** 0-23E0

loss insertion. *See:* **insertion loss.**

loss of forming (semiconductor rectifier). A partial loss in the effectiveness of the rectifier junction. *See also:* **rectification.** E59-34E17

loss return. *See:* **return loss.**

loss tangent (1) (general). The ratio of the imaginary part of the complex dielectric constant of a material to its real part. *See also:* **transmission characteristics.** 0-9E4

(2) (tan δ) (rotating machinery). The ratio of dielectric loss in an insulation system, to the apparent power required to establish an alternating voltage across it of a specified amplitude and frequency, the insulation being at a specified temperature. *Note:* It is the cotangent of the power-factor angle. *See also:* **asynchronous machine; direct-current commutating machine; synchronous machine.** 0-31E8

loss-tangent test (dissipation-factor test) (rotating machinery). A test for measuring the dielectric loss of insulation at predetermined values of temperature, frequency, and voltage or dielectric stress, in which the dielectric loss is expressed in terms of the tangent of the complement of the insulation power-factor angle. *See also:* **asynchronous machine; direct-current commutating machines; synchronous machine.** 0-31E8

loss, total (1) (rotating machinery). The difference between the power input and the power output. *See:* **asynchronous machine; direct-current commutating machine; synchronous machine.** 0-31E8

(2) (transformer or voltage regulator). The sum of the excitation losses and the load losses. *See also:* **efficiency; voltage regulator.** 42A15/57A14-31E12

loss, transmission. *See:* **transmission loss.**

loudness. The intensive attribute of an auditory sensation in terms of which sound may be ordered on a scale extending from soft to loud. *Note:* Loudness depends primarily upon the sound pressure of the stimulus, but it also depends upon the frequency and waveform of the stimulus. *See also:* **loudspeaker.** E157-1E1

loudness contour. A curve that shows the related values of sound pressure level and frequency required to produce a given loudness sensation for the typical listener. *See also:* **loudspeaker.** 0-1E1

loudness level (sound). A value, expressed in phons, that is numerically equal to the median sound pressure level, in decibels, relative to 2×10^{-5} newton per square meter, of a free progressive wave of frequency 1000 hertz presented to listeners facing the source, which in a number of trials is judged by the listeners to be equally loud. *Note:* The manner of listening to the unknown sound, which must be stated, may be considered to be one of the characteristics of that sound. *See also:* **loudspeaker.** 0-1E1

loudspeaker (speaker). An electroacoustic transducer intended to radiate acoustic power into the air to be audible at a distance. *Note:* For an extensive list of cross references, see *Appendix A.* E188-42A65

loudspeaker system. A combination of one or more loudspeakers and all associated baffles, horns, and dividing networks arranged to work together as a coupling means between the driving electric circuit and the acoustic medium. *See also:* **loudspeaker.** E157/42A65-1E1

loudspeaker voice-coil. The moving coil of a dynamic loudspeaker. *See also:* **loudspeaker.** 42A65/1E1

louver (louver grid) (illuminating engineering). A series of baffles used to shield a source from view at certain angles or to absorb unwanted light. *Note:* The baffles usually are arranged in a geometric pattern. *See also:* **bare (exposed) lamp.** Z7A1-0

louvered ceiling (illuminating engineering). A ceiling area-lighting system comprising a wall-to-wall installation of multicell louvers shielding the light sources mounted above it. *See also:* **luminaire.** Z7A1-0

louver shielding angle (illuminating engineering). The angle between the horizontal plane of the baffles or louver grid and the plane at which the louver conceals all objects above. *Note:* The planes usually are so chosen that their intersection is parallel with the louvered blade. *See also:* **bare (exposed) lamp.** Z7A1-0

low-clearance area (instrument approach system). Any area containing only low-clearance points. *See also:* **navigation.** E172-10E6

low-clearance points (instrument landing systems). Locations in space outside the course sector at which course deviation indicator current is below some arbitrary value, usually the full-scale deflection value. *See also:* **navigation.** 0-10E6

low-energy power circuit (interior wiring). A circuit that is not a remote-control or signal circuit but that has the power supply limited in accordance with the requirements of Class 2 remote-control circuits. *Note:* Such circuits include electric door openers and circuits used in the operation of coin-operated phonographs. *See also:* **appliances.** 1A0-0

lower (passing) beams (motor vehicle). One or more beams directed low enough on the left to avoid glare in the eyes of oncoming drivers, and intended for use in congested areas and on highways when meeting other vehicles within a distance of 1000 feet. Formally **traffic beam.** *See also:* **headlamp.** Z7A1-0

lower bracket (rotating machinery). A bearing bracket mounted below the level of the core of a vertical machine. *See:* **cradle base (rotating machinery).** 0-31E8

lower burst reference (audio and electroacoustics). A selected multiple of the long-time average magnitude of a quantity, smaller than the upper burst reference. See the figure under **burst duration.** *See also:* **burst (audio and electroacoustics).** E257-1E1

lower coil support (rotating machinery). A support to restrain field-coil motion in the direction away from the air gap. *See:* **rotor (rotating machinery); stator.** 0-31E8

lower guide bearing (rotating machinery). A guide bearing mounted below the level of the core of a vertical machine. *See:* **cradle base (rotating machinery).** 0-31E8

lower-half bearing bracket (rotating machinery). The bottom half of a bracket that can be separated into halves for mounting or removal without access to a shaft end. *See:* **cradle base (rotating machinery).** 0-31E8

lower-range value. The lowest quantity that a device is adjusted to measure. *Note:* The following compound terms are used with suitable modifications in the units: **measured-variable lower-range value, measured signal lower-range value,** etcetera. *See also:* **instrument.** 39A4-0

lower-sideband parametric down-converter*. An inverting parametric device used as a parametric down-converter. *See also:* **parametric device.**
*Deprecated E254-15E7

lower-sideband parametric up-converter*. An inverting parametric device used as a parametric up-converter. *See also:* **parametric device.**
*Deprecated E254-15E7

lowest useful high frequency(radio wave propagation). The lowest high frequency effective for ionospheric propagation of radio waves between two specified points, under specified ionospheric conditions, and under specified factors such as absorption, transmitter power, antenna gain, receiver characteristics, type of service, and noise conditions. *See also:* **radiation; radio wave propagation.** 0-3E2

low-frequency dry-flashover voltage. The root-mean-square voltage causing a sustained disruptive discharge through the air between electrodes of a clean dry test specimen under specified conditions. *See also:* **power systems, low-frequency and surge testing.** 42A35-31E13

low-frequency flashover voltage (insulator). The root-mean-square value of the low-frequency voltage that, under specified conditions, causes a sustained disruptive discharge through the surrounding medium. *See also:* **insulator.** 29A1-0

low-frequency furnace (core-type induction furnace). An induction furnace that includes a primary winding, a core of magnetic material, and a secondary winding of one short-circuited turn of the material to be heated. 42A60-0

low-frequency impedance corrector. An electric network designed to be connected to a basic network, or to a basic network and a building-out network, so that the combination will simulate at low frequencies the sending-end impedance, including dissipation, of a line. *See also:* **network analysis.** 42A65-0

low-frequency induction heater or furnace. A device for inducing current flow of commercial power-line frequency in a charge to be heated. *See also:* **dielectric heating, induction heating.** E54/E169-0

low-frequency puncture voltage (insulator). The root-mean-square value of the low-frequency voltage that, under specified conditions, causes disruptive discharge through any part of the insulator. *See also:* **insulator.** 29A1-0

low-frequency wet-flashover voltage. The root-mean-square voltage causing a sustained disruptive discharge through the air between electrodes of a clean test specimen on which water of specified resistivity is being sprayed at a specified rate. *See also:* **low-frequency and surge testing; power systems.** 42A35-31E13

low-frequency withstand voltage (insulator). The root-mean-square value of the low-frequency voltage that, under specified conditions, can be applied without causing flashover or puncture. *See also:* **insulator.** 29A1-0

low-key lighting (television). A type of lighting that, applied to a scene, results in a picture having graduations falling primarily between middle gray and black, with comparatively limited areas of light grays and whites. *See also:* **television lighting.** Z7A1-0

low-level modulation (communication). Modulation produced at a point in a system where the power level is low compared with the power level at the output of the system. *See also:* **modulating systems.** E145/E182/42A65-0

low-pass filter. *See:* **filter, low-pass.**

low (normal) power-factor mercury lamp ballast. A ballast of the multiple-supply type that does not have a means for power-factor correction. 82A9-0

low-pressure vacuum pump. A vacuum pump that compresses the gases received directly from the evacuated system. *See also:* **rectification.** 42A15-0

low-velocity camera tube (electron device) (cathode-voltage-stabilized camera tube). A camera tube operating with a beam of electrons having velocities such that the average target voltage stabilizes at a value approximately equal to that of the electron-gun cathode. *See also:* **tube definitions.** 0-15E6

low-voltage protection (electric systems). The effect of a device, operative on the reduction or failure of voltage, to cause and maintain the interruption of power supply to the equipment protected. *See also:* **generating station; undervoltage protection.** 2A2-0

low-voltage release. The effect of a device, operative on the reduction or failure of voltage, to cause the interruption of power supply to the equipment, but not preventing the reestablishment of the power supply on return of voltage. *See also:* **generating system.** 2A2-0

low-voltage system (electric power). An electric system having an operating voltage less than 750 volts. *See also:* **alternating-current distribution; direct-current distribution.** 42A35-31E13

LRC (data transmission). *See:* **longitudinal redundancy check.**

***L* scan (electronic navigation).** *See:* ***L* display.**

***L* scope (electronic navigation).** *See:* ***L* display.**

LS dividing network: *See:* **dividing network.**

LTS (long-term stability). *See:* **stability, long-term.**

lug, stator mounting (rotating machinery). A part attached to the outer surface of a stator core or a stator shell to provide a means for the bolting or equivalent attachment to the appliance, machine, or other foundation. *See:* **stator.** 0-31E8

lumen. The unit of luminous flux. It is equal to the flux through a unit solid angle (steradian), from a uniform point source of one candela (candle), or to the flux on a unit surface all points of which are at unit distance from a uniform point source of one candela. *Note:* For some purposes, the kilolumen, equal to 1000 lumens, is a convenient unit. *See also:* **color terms; light.** E201/Z7A1-0;50I45-2E2

lumen hour. A unit of quantity of light (luminous energy). It is the quantity of light delivered in one hour by a flux of one lumen. *See also:* **light.** Z7A1-0

lumen (or flux) method (lighting calculation). A lighting design procedure used for predetermining the number and types of lamps or luminaires that will provide a desired average level of illumination on the work plane. *Note:* It takes into account both direct and reflected flux. *See also:* **inverse-square law (illuminating engineering).** Z7A1-0

luminaire. A complete lighting unit consisting of a lamp or lamps together with the parts designed to distribute the light, to position and protect the lamps, and to connect the lamps to the power supply.
See:
bare lamp;
channel;
coffer;
downlight;
dusttight;
explosionproof;
floor lamp;
general lighting;
hazardous location;
lowered ceiling;
luminous ceiling;
narrow-angle luminaire;
portable luminaire;
projector;
street-lighting luminaire;
suspended;
table lamp;
torchere;
troffer;
valance;
vaportight;
wide-angle luminaire. Z7A1-0

luminaire efficiency. The ratio of luminous flux (lumens) emitted by a luminaire to that emitted by the lamp or lamps used therein. *See also:* **inverse-square law (in illuminating engineering).** Z7A1-0

luminance (1) (at a point of a surface and in a given direction). The quotient of the luminous intensity in the given direction of an infinitesimal element of the surface containing the point under consideration, by the orthogonally projected area of the element on a plane perpendicular to the given direction. *Notes:* (A) Typical units in this system are the candela per square meter and the candela per square inch. (B) The term **luminance** is recommended for the photometry quality, which has been called **brightness.** Use of this term permits brightness to be used entirely with reference to sensory response. The photometric quantity has been confused often with the sensation merely because of the use of one name for two distinct ideas. Brightness will continue to be used properly in nonquantitative statements, especially with reference to sensations and perceptions of light. *See also:* **color terms; light.** E201-9E4;50I45-2E2

(2) (photometric brightness). In a direction, at a point of the surface, of a receiver, or of any other real or virtual surface, the quotient of the luminous flux leaving, passing through, or arriving at an element of the surface surrounding the point, and propagated in directions defined by an elementary cone containing the given direction, by the product of the solid angle of the cone and the area of the orthogonal projection of the element of the surface on a plane perpendicular to the given direction; or the luminous intensity of any surface in a given direction per unit of projected area of the surface as viewed from that direction.

$$L=\frac{d^2\Phi}{d\omega}(dA \cos \theta)$$

$$=\frac{dI}{dA \cos \theta}$$

where I = luminous intensity, Φ = luminous flux, and ω = solid angle. *Note:* In the defining equation θ is the angle between the direction of observation and the normal to the surface. In common usage the term **brightness** usually refers to the intensity of **sensation** that results from viewing surfaces or spaces from which light comes to the eye. This sensation is determined in part by the definitely measurable luminance (photometric brightness) defined above and in part by conditions of observation such as the state of adaptation of the eye. In much of the literature the term **brightness,** used alone, refers to both luminance and sensation. The context usually indicates which meaning is intended. *See:* **subjective brightness.** *See also:* **light.** Z7A1-0

(3) (average luminance) (average photometric brightness). The total lumens actually leaving the surface per unit area. *Notes:* (A) Average luminance specified in this way is identical in magnitude with **luminous exitance,** which is the preferred term. (B) In general, the concept of average luminance is useful only when the luminance is reasonably uniform throughout a very wide angle of observation and over a large area of the surface considered. It has the advantage that it can be readily computed for reflecting surfaces by multiplying the incident luminous flux density (illumination) by the luminous reflectance of the surface. For a transmitting body it can be computed by multiplying the incident luminous flux density by the luminous transmittance of the body. Z7A1-0

luminance channel (color-television system). Any path that is intended to carry the luminance signal. *See also:* **television.** E204/42A65-2E2

luminance channel bandwidth (television). The bandwidth of the path intended to carry the luminance signal. *See also:* **television.** E204/42A65-2E2

luminance (photometric brightness) contrast. The relationship between the luminances (photometric brightnesses) of an object and its immediate background. It is equal to $(L_1-L_2)/L_1$ or $(L_2-L_1)/L_1$, where L_1 and L_2 are the luminances of the background and object, respectively. *Note:* The form of the equation must be specified. Because of the relationship among luminance, illumination, and reflectance, contrast often is expressed in terms of reflectance when only reflecting surfaces are involved. Thus contrast is equal to $(\rho_1-\rho_2)/\rho_1$ or $(\rho_2-\rho_1)/\rho_1$ where ρ_1 and ρ_2 are the reflectances of the background and object, respectively. This method of computing contrast holds only for perfectly diffusing surfaces; for other surfaces it is only an approximation unless the angles of incidence and view are taken into consideration. *See also:* **reflectance; visual field.** Z7A1-0

luminance (photometric brightness) difference. The difference in luminance (photometric brightness) between two areas. *Note:* It usually is applied to contiguous areas, such as the detail of a visual task and its immediate background, in which case it is quantitatively equal to the numerator in the formula for luminance contrast. See the note under **luminance.** ***See also:*** **visual field.** Z7A1-0

luminance factor. The ratio of the luminance (photometric brightness) of a surface or medium under specified conditions of incidence, observation, and light source, to the luminance (photometric brightness) of a perfectly reflecting or transmitting, perfectly diffusing surface or medium under the same conditions. *Note:* Reflectance or transmittance cannot exceed unity, but luminance factor may have any value from zero to values approaching infinity. *See also:* **lamp.** Z7A1-0

luminance (photometric brightness) factor of room surfaces. Factor by which the average work-plane illumination is multiplied to obtain the average luminances of walls, ceilings, and floors. *See also:* **inverse-square law (illuminating engineering).** Z7A1-0

luminance flicker. The flicker that results from fluctuation of luminance only. *See also:* **color terms.** E201-2E2

luminance primary (color television). One of a set of three transmission primaries whose amount determines the luminance of a color. *See also:* **color terms.** E201-2E2

luminance (photometric brightness) ratio. The ratio between the luminances (photometric brightnesses) of any two areas in the visual field. See the note under **luminance.** ***See also:*** **visual field.** Z7A1-0

luminance signal (color television). A signal wave that is intended to have exclusive control of the luminance of the picture. *See also:* **color terms.** E201-2E2

luminance (photometric brightness) threshold. The minimum perceptible difference in luminance for a given state of adaptation of the eye. See the note under **luminance.** *See also:* **visual field.** Z7A1-0

luminescence. Any emission of light not ascribable directly to incandescence. *See also:* **lamp.** Z7A1-0

luminescent-screen tube. A cathode-ray tube in which the image on the screen is more luminous than the background. *See also:* **cathode-ray tubes.** 50I07-15E6

luminosity. Ratio of luminous flux to the corresponding radiant flux at a particular wavelength. It is expressed in lumens per watt. *See also:* **color terms.** E201-2E2

luminosity coefficients. The constant multipliers for the respective tristimulus values of any color, such that the sum of the three products is the luminance of the color. *See also:* **color terms.** E201-2E2

luminous ceiling. A ceiling area-lighting system comprising a continuous surface of transmitting material of a diffusing or light-controlling character with light sources mounted above it. *See also:* **luminaire.** Z7A1-0

luminous density. Quantity of light (luminous energy) per unit volume. *See also:* **light.** Z7A1-0

luminous efficacy (1) (radiant flux). The quotient of the total luminous flux by the total radiant flux. It is expressed in lumens per watt. *See also:* **light.** Z7A1-0

(2) (source of light). The quotient of the total luminous flux emitted by the total lamp power input. *Note:* The term **luminous efficiency** has in the past been extensively used for this concept. *See also:* **light.** Z7A1-0

luminous efficiency. The ratio of the luminous flux to the radiant flux. *Note:* Luminous efficiency is usually expressed in lumens per watt of radiant flux. It should not be confused with the term efficiency as applied to a practical source of light, since the latter is based upon the power supplied to the source instead of the radiant flux from the source. For energy radiated at a single wavelength, luminous efficiency is synonymous with luminosity. *See also:* **color terms.** E201-2E2

luminous flux. The time rate of flow of light. *See also:* **color terms; light.** E201/Z7A1-2E2

luminous flux density (surface). Luminous flux per unit area of the surface. *Note:* When referring to luminous flux emitted from a surface, this has been called **luminous emittance** (symbol *M*). The preferred term for luminous flux leaving a surface is **luminous exitance** (symbol *M*). When referring to flux incident on a surface, it is identical with **illumination** (symbol *E*). Z7A1-0

luminous gain (optoelectronic device). The ratio of the emitted luminous flux to the incident luminous flux. *Note:* The emitted and incident luminous flux are both determined at specified ports. *See also:* **optoelectronic device.** E222-15E7

luminous intensity (source of light in a given direction). The luminous flux per unit solid angle in the direction in question. Hence, it is the luminous flux on a small surface normal to that direction, divided by the solid angle (in steradians) that the surface subtends at the source. *Note:* Mathematically a solid angle must have a point as its apex; the definition of luminous intensity, therefore, applies strictly only to a point source. In practice, however, light emanating from a source whose dimensions are negligible in comparison with the distance from which it is observed may be considered as coming from a point. For extended sources *see also:* **color terms; light.** E201/Z7A1-0;50I45-2E2

luminous sensitivity (phototube). *See:* **sensitivity (camera tube or phototube).**

lumped. Effectively concentrated at a single point. 42A65-0

lumped element circuit (microwave tubes). A circuit consisting of discrete inductors and capacitors. *See also:* **microwave tube or valve.** 0-15E6

Luneburg lens antenna. An antenna with a lens of circular cross section having an index of refraction varying only in the radial direction such that a feed located on or near a surface or edge of the lens produces a major lobe diametrically opposite the feed. *See also:* **antenna.** 0-3E1

lux. In the International System of Units (SI) the unit of illumination on a surface one square meter in area on which there is a uniformly distributed flux of one lumen, or the illumination produced at a surface all points of which are at a distance of one meter from a uniform point source of one candela. *See also:* **light.** Z7A1-0

Luxemburg effect. A nonlinear effect in the ionosphere as a result of which the modulation on a strong carrier wave is transferred to another carrier passing through the same region. *See also:* **radiation.** 42A65-0

Lyapunov function (control system) (equilibrium point $\mathbf{x}_e$ of a system). A scalar differentiable function $V(\mathbf{x})$ defined in some open region including $\mathbf{x}_e$ such that in that region

(1) $V(\mathbf{x}) > 0$ for $\mathbf{x} \neq \mathbf{x}_e$

(2) $V(\mathbf{x}_e) = 0$

(3) $V'(\mathbf{x}) \leq 0$.

Notes: (1) The open region may be defined by (norm of $\mathbf{x} - \mathbf{x}_e$) < constant. (2) For the system $\mathbf{x}' = \mathbf{f}(\mathbf{x})$, $V'(\mathbf{x}) \equiv$ [grad $V(\mathbf{x} \cdot \mathbf{f}(\mathbf{x})$]. *See also:* **control system, feedback.** 0-23E0

M

machine (1) (general). An article of equipment consisting of two or more resistant, relatively constrained parts that, by a certain predetermined intermotion, may serve to transmit and modify force, motion, or electricity so as to produce some given effect or transformation or to do some desired kind of work. E270-0

(2) (computing systems). *See:* **Turing machine; universal Turing machine.** *See also:* **electronic digital computer.**

machine address (computing systems). *See:* **absolute address.** *See also:* **electronic digital computer.**

machine, aircraft electric. An electric machine designed for operation aboard aircraft. *Note:* Minimum weight and utmost reliability for a specified (usually short) life are required while operating under specified conditions of coolant temperature and, for air-cooled machines, pressure and humidity. 0-31E8

machine check. (1) An automatic check. (2) A programmed check of machine functions. *See also:* **check; automatic; electronic digital computer.** E162-0

machine code (computing systems). An operation code that a machine is designed to recognize. *See also:* **electronic digital computer.** X3A12-16E9

machine, electric. An electric apparatus depending on electromagnetic induction for its operation and having one or more component members capable of rotary and/or linear movement. *See also:* **asynchronous machine; synchronous machine.** 0-31E8

machine equation. *See:* **computer equation.**

machine final-terminal stopping device (stop-motion switch) (elevators). A final-terminal stopping device operated directly by the driving machine. *See:* **control.** 42A45-0

machine instruction (computing systems). An instruction that a machine can recognize and execute. *See also:* **electronic digital computer.** X3A12-16E9

machine language. (1) A language, occurring within a machine, ordinarily not perceptible or intelligible to persons without special equipment or training. (2) A translation or transliteration of (1) above into more conventional characters but frequently still not intelligible to persons without special training. *See also:* **electronic digital computer.** E162/E270-0

machine positioning accuracy, precision, or reproducibility (numerically controlled machines). Accuracy, precision, or reproducibility of position sensor or transducer and interpreting system, the machine elements, and the machine positioning servo. *Note:* Cutter, spindle, and work deflection, and cutter wear are not included. (May be the same as control positioning accuracy, precision, or reproducibility in some systems.) *See also:* **numerically controlled machines.** EIA3B-34E12

machine ringing (telephony). Ringing that once started continues automatically until the call is answered or abandoned. *See also:* **telephone switching system.** 0-19E1

machine switching system. *See:* **telephone switching system.**

machine winding (rotating machinery). A winding placed in slots or around poles directly by a machine. *See:* **rotor (rotating machinery); stator.** 0-31E8

machine word (computing systems). *See:* **computer word.** *See also:* **electronic digital computer.**

machining accuracy, precision, or reproducibility. Accuracy, precision, or reproducibility obtainable on completed parts under normal operating conditions. EIA3B-34E12

macro instruction (computing systems). An instruction in a source language that is equivalent to a specified sequence of machine instructions. *See also:* **electronic digital computer.** X3A12-16E9

magazine (computing systems). *See:* **input magazine.** *See also:* **electronic digital computer.**

magic tee (waveguides). *See:* **hybrid tee.**

magner. *See:* **reactive power.**

magnesium cell. A primary cell with the negative electrode made of magnesium or its alloy. *See also:* **electrochemistry.** 42A60-0

magnet. A body that produces a magnetic field external to itself. E270-0

magnet, focusing. *See:* **focusing magnet.**

magnetic (switching device). A term indicating that interruption of the circuit takes place between contacts separable in an intense magnetic field. *Note:* With respect to contactors, this term indicates the means of operation. 37A100-31E11

magnetically shielded type instrument. An instrument in which the effect of external magnetic fields is limited to a stated value. The protection against this influence may be obtained either through the use of a physical magnetic shield or through the instrument's inherent construction. *See also:* **instrument.** 39A1-0

magnetic amplifier. A device using one or more saturable reactors, either alone or in combination with other circuit elements, to secure power gain. Frequency conversion may or may not be included.
See:
circuits and devices;
control winding;
delta induction;
maximum sine-current differential permeability;
output winding;
peak induction;
peak magnetizing force;
power winding;
residual induction;
saturable reactor;
saturating reactor;
sine-current coercive force;
sine-current differential permeability;
sine-current magnetizing force. 42A65/21E0

magnetic area moment. *See:* **magnetic moment.**

magnetic-armature loudspeaker. A magnetic loudspeaker whose operation involves the vibration of a ferromagnetic armature. *See also:* **loudspeaker.** 42A65-0

magnetic axis (coil or winding) (rotating machinery). The line of symmetry of the magnetic-flux density produced by current in a coil or winding, this being the location of approximately maximum flux density, with the air gap assumed to be uniform. *See also:* **rotor (rotating machinery); stator.** 0-31E8

magnetic bearing (navigation). Bearing relative to magnetic north. *See also:* **navigation.** 0-10E6

magnetic biasing (magnetic recording). The simultaneous conditioning of the magnetic recording medium during recording by the superposing of an additional magnetic field upon the signal magnetic field. *Note:* In general, magnetic biasing is used to obtain a substantially linear relationship between the amplitude of the signal and the remanent flux density in the recording medium. *See:* **alternating-current magnetic biasing; direct-current magnetic biasing.** *See also:* **phonograph pickup.** 42A65-1E1

magnetic bias, relay. *See:* **relay magnetic bias.**

magnetic blowout (industrial control). A magnet, often electrically excited, whose field is used to aid the interruption of an arc drawn between contacts. *See also:* **contactor.** 42A25-34E10

magnetic brake (industrial control). A friction brake controlled by electromagnetic means. 42A25-34E10

magnetic-brush development (electrostatography). Development in which the image-forming material is carried to the field of the electrostatic image by means of ferromagnetic particles acting as carriers under the influence of a magnetic field. *See also:* **electrostatography.** E224-15E7

magnetic card (computing systems). A card with a magnetic surface on which data can be stored by selective magnetization of portions of the flat surface. *See also:* **static magnetic storage.** X3A12-16E9

magnetic circuit. A region at whose surface the magnetic induction is tangential. *Note:* The term is also applied to the minimal region containing essentially all the flux, such as the core of a transformer. E270-0

magnetic compass. A device for indicating the direction of the horizontal component of a magnetic field. *See also:* **magnetometer.** 42A30-0

magnetic-compass repeater indicator. A device that repeats the reading of a master direction indicator, through a self-synchronous coupling means. *See also:* **air-transportation instruments.** 42A41-0

magnetic constant (pertinent to any system of units) (permeability of free space). The magnetic constant is the scalar dimensional factor that in that system relates the mechanical force between two currents to their magnitudes and geometrical configurations. More specifically, μ_0 is the magnetic constant when the element of force dF of a current element $I_1\mathbf{dI}_1$ on another current element $I_2\mathbf{dI}_2$ at a distance r is given by

$$\mathbf{dF} = \mu_0 I_1 I_2 \mathbf{dI}_1 \times (\mathbf{dI}_2 \times \mathbf{r}_1)/nr^2$$

where $\mathbf{r}_1$ is a unit vector in the direction from $\mathbf{dI}_1$ to $\mathbf{dI}_2$, and n is a dimensionless factor which is unity in unrationalized systems and 4π in a rationalized system. *Note:* In the centimeter-gram-second (cgs) electromagnetic system μ_0 is assigned the magnitude unity and the dimension numeric. In the centimeter-gram-second (cgs) electrostatic system the magnitude of μ_0 is that of $1/c^2$ and the dimension is $[L^{-2}T^2]$. In the International System of Units (SI) μ_0 is assigned the magnitude $4\pi \cdot 10^{-7}$ and has the dimension $[LMT^{-2}I^{-2}]$. E270-0

magnetic contactor (industrial control). A contactor actuated by electromagnetic means. *See also:* **contactor.** 42A25-34E10

magnetic control relay. A relay that is actuated by electromagnetic means. *Note:* When not otherwise qualified, the term refers to a relay intended to be operated by the opening and closing of its coil circuit and having contacts designed for energizing and/or de-energizing the coils of magnetic contactors or other magnetically operated device. *See:* **relay.** IC1-34E10

magnetic core. A configuration of magnetic material that is, or is intended to be, placed in a rigid spatial relationship to current-carrying conductors and whose magnetic properties are essential to its use. *Note:* For example, it may be used: (1) to concentrate an induced magnetic field as in a transformer, induction coil, or armature; (2) to retain a magnetic polarization for the purpose of storing data; or (3) for its nonlinear properties as in a logic element. It may be made of iron wires, iron oxide, coils of magnetic tape, ferrite, thin film, etcetera. *See also:* **electronic digital computer.** E162/X3A12-16E9

magnetic course (navigation). Course relative to magnetic north. *See also:* **navigation.** 0-10E6

magnetic deflection (cathode-ray tube). Deflecting an electron beam by the action of a magnetic field. *See also:* **cathode-ray tubes.** 50I07-15E6

magnetic delay line (computing systems). A delay line whose operation is based on the time of propagation of magnetic waves. *See also:* **electronic digital computer.** X3A12-16E9

magnetic deviation. Angular difference between compass north and magnetic north caused by magnetic effects in the vehicle. *See also:* **navigation.** 0-10E6

magnetic dipole. An elementary radiator consisting of an infinitesimally small current loop. *See also:* **antenna.** E149-3E1

magnetic dipole moment* (centimeter-gram-second electromagnetic-unit system). The volume integral of **magnetic polarization** is often called magnetic dipole moment.

*Deprecated E270-0

magnetic direction indicator (MDI). An instrument providing compass indication obtained electrically from a remote gyro-stabilized magnetic compass or equivalent. *See also:* **radio navigation.** 42A65-0

magnetic disc. A flat circular plate with a magnetic surface on which data can be stored by selective polarization of portions of the flat surface. *See also:* **electronic digital computer.** E162/X3A12-16E9

magnetic dissipation factor (magnetic material). The cotangent of its loss angle or the tangent of its hysteretic angle. E270-0

magnetic drum. A right circular cylinder with a magnetic surface on which data can be stored by selective polarization of portions of the curved surface. *See also:* **electronic digital computer.** E162/X3A12-16E9

magnetic field (1) (generator action). A state produced in a medium, either by current flow in a conductor or by a permanent magnet, that can induce voltage in a second conductor in the medium when the state changes or when the second conductor moves in prescribed ways relative to the medium. *See:* **signal.** 0-13E6

(2) (motor action). A state of a region such that a moving charged body in the region would be accelerated in proportion to its charge and to its velocity.

(3) (general). The vector function **(B)** of position describing the magnetic state of a region. E270-0

magnetic field intensity*. *See:* **magnetic field strength.**

*Deprecated

magnetic field interference. A form of interference induced in the circuits of a device due to the presence of a magnetic field. *Note:* It may appear as common-mode or normal-mode interference in the measuring circuit. *See also:* **accuracy rating (instrument).** 39A4-0

magnetic field strength (1) (general). The magnitude of the magnetic field vector. E211-27E1/3E2

(2) (magnetizing force). That vector point function whose curl is the current density, and that is proportional to magnetic flux density in regions free of magnetized matter. *Note:* A consequence of this definition is that the familiar formula

$$\mathbf{H} = \frac{1}{4\pi}\int \mathbf{J} \times \nabla (1/r)\, dv - \frac{1}{4\pi}\nabla \int \mathbf{M} \cdot \nabla (1/r)\, dv$$

(where **H** is the magnetizing force, **J** is current density, and **M** is magnetization) is a mathematical identity. E270-0

magnetic field strength produced by an electric current (Biot-Savart law) (Ampere's law). The magnetic field strength, at any point in the neighborhood of a circuit in which there is an electric current *i*, can be computed on the assumption that every infinitesimal length of circuit produces at the point an infinitesimal magnetizing force and the resulting magnetizing force at the point is the vector sum of the contributions of all the elements of the circuit. The contribution, d**H**, to the magnetizing force at a point *P* caused by the current i in an element ds of a circuit that is at a distance r from P, has a direction that is perpendicular to both ds and r and a magnitude equal to

$$\frac{i\,\mathrm{d}s\sin\theta}{r^2}$$

where θ is the angle between the element ds and the line r. In vector notation

$$\mathrm{d}\mathbf{H} = \frac{i[\mathbf{r}\times\mathrm{d}\mathbf{s}]}{r^2}.$$

This law is sometimes attributed to Biot and Savart, sometimes to Ampere, and sometimes to Laplace, but no one of them gave it in its differential form. E270-0

magnetic field vector (any point in a magnetic field) (radio wave propagation). The magnetic induction divided by the permeability of the medium. *See also:* **radio wave propagation.** 0-3E2

magnetic figure of merit. The ratio of the real part of complex apparent permeability to magnetic dissipation factor. *Note:* The magnetic figure of merit is a useful index of the magnetic efficiency of a material in various electromagnetic devices. E270-0

magnetic flux (through an area). The surface integral of the normal component of the magnetic induction over the area. Thus

$$\phi_A = \int_A (\mathbf{B}\cdot\mathrm{d}A)$$

where ϕ_A is the flux through the area A, and $\mathbf{B}$ is the magnetic induction at the element dA of this area. *Note:* The net magnetic flux through any closed surface is zero. E270-0

magnetic flux density (magnetic induction). That vector quantity $\mathbf{B}$ producing a torque on a plane current loop in accordance with the relation $\mathbf{T} = IA\mathbf{n}\times\mathbf{B}$ where $\mathbf{n}$ is the positive normal to the loop and A is its area. *Note:* The concept of flux density is extended to a point inside a solid body by defining the flux density at such a point as that which would be measured in a thin disk-shaped cavity in the body centered at that point, the axis of the cavity being in the direction of the flux density. E270-0

magnetic focusing. *See:* **focusing, magnetic.**

magnetic freezing. *See:* **relay magnetic freezing.**

magnetic friction clutch. A friction clutch in which the pressure between the friction surfaces is produced by magnetic attraction. *See:* **electric coupling.** 42A10-0

magnetic head (magnetic recording). A transducer for converting electric variations into magnetic variations for storage on magnetic media, or for reconverting energy so stored into electric energy, or for erasing such stored energy. 42A65/E157-1E1

magnetic heading (navigation). Heading relative to magnetic north. *See also:* **navigation.** 0-10E6

magnetic hysteresis. The property of a ferromagnetic material exhibited by the lack of correspondence between the changes in induction resulting from increasing magnetizing force and from decreasing magnetizing force. E270-0

magnetic hysteresis loss (magnetic material). (1) The power expanded as a result of magnetic hysteresis when the magnetic induction is periodic. (2) The energy loss per cycle in a magnetic material as a result of magnetic hysteresis when the induction is cyclic (not necessarily periodic). *Note:* Definitions (1) and (2) are not equivalent; both are in common use. E270-0

magnetic hysteretic angle. The mean angle by which the exciting current leads the magnetizing current. *Note:* Because of hysteresis, the instantaneous value of the hysteretic angle will vary during the cycle; the hysteretic angle is taken to be the mean value. E270-0

magnetic induction (signal-transmission system). The process of generating currents or voltages in a conductor by means of a magnetic field. *See:* **magnetic flux density; signal.** 0-13E6

magnetic ink. An ink that contains particles of a magnetic substance whose presence can be detected by magnetic sensors. *See also:* **static magnetic storage.** X3A12-16E9

magnetic-ink character recognition. The machine recognition of characters printed with magnetic ink. *See:* **optical character recognition.** *See also:* **static magnetic storage.** X3A12-16E9

magnetic latching relay. (1) A relay that remains operated from remanent magnetism until reset electrically. (2) A bistable polarized (magnetically latched) relay. 0-21E0

magnetic loading (rotating machinery). The average flux per unit area of the air-gap surface. *See also:* **asynchronous machine; direct-current commutating machine; synchronous machine.** 0-31E8

magnetic loss angle (core). The angle by which the fundamental component of the core-loss current leads the fundamental component of the exciting current in an inductor having a ferromagnetic core. *Note:* The loss angle is the complement of the hysteretic angle. E270-0

magnetic loss factor, initial (material). The product of the real component of its complex permeability and the tangent of its magnetic loss angle, both measured when the magnetizing force and the induction are vanishingly small. *Note:* In anisotropic media, magnetic loss factor becomes a matrix. E270-0

magnetic loudspeaker. A loudspeaker in which acoustic waves are produced by mechanical forces resulting from magnetic reactions. *See also:* **loudspeaker.** 42A65-0

magnetic microphone. *See:* **variable-reluctance microphone.**

magnetic (electron) microscope. An electron microscope with magnetic lenses. *See also:* **electron optics.** 50I07-15E6

magnetic mine. A submersible explosive device with a detonator actuated by the distortion of the earth's magnetic field caused by the approach of a mass of magnetic material such as the hull of a ship. *See also:* **degauss.** 42A43-0

magnetic moment (1) (magnetized body). The volume integral of the magnetization

$$\mathbf{m} = \int \mathbf{M}\,\mathrm{d}v$$

E270-0

(2) (current loop).

$$\mathbf{m} = I\int \mathbf{n}\,\mathrm{d}a$$
$$= (I/2)\int \mathbf{r}\times\mathrm{d}\mathbf{r}$$

where **n** is the positive normal to a surface spanning the loop, and **r** is the radius vector from an arbitrary origin to a point on the loop. *Notes:* (1) The numerical value of the moment of a plane current loop is IA, where A is the area of the loop. (2) The reference direction for the current in the loop indicates a clockwise rotation, when the observer is looking through the loop in the direction of the positive normal. E270-0

magnetic north. The direction of the horizontal component of the earth's magnetic field toward the north magnetic pole. *See also:* **navigation.** 0-10E6

magnetic overload relay. An overcurrent relay the electric contacts of which are actuated by the electromagnetic force produced by the load current or a measure of it. *See:* **relay.** IC1-34E10

magnetic-particle coupling. An electric coupling that transmits torque through the medium of magnetic particles in a magnetic field between coupling members. *See:* **electric coupling.** 42A10-0

magnetic pickup. *See:* **variable-reluctance pickup.**

magnetic-plated wire. A magnetic wire having a core of nonmagnetic material and a plated surface of ferromagnetic material. E157-1E1

magnetic-platform influence (electric instrument). The change in indication caused solely by the presence of a magnetic platform on which the instrument is placed. *See also:* **accuracy rating (instrument).** 39A1-0

magnetic polarization.* In the centimeter-gram-second electromagnetic-unit system, the intrinsic induction divided by 4π is sometimes called **magnetic polarization** or **magnetic dipole moment per unit volume.** *See:* **intrinsic induction.**

*Deprecated E270-0

magnetic poles (magnet). Those portions of the magnet toward which or from which the external magnetic induction appears to converge or diverge, respectively. *Notes:* (1) By convention, the north-seeking pole is marked with *N*, or plus, or is colored red. (2) The term is also sometimes applied to a fictitious magnetic charge. E270-0

magnetic pole strength (magnet). The magnetic moment divided by the distance between its poles. *Note:* Many authors use the above quantity multiplied by the magnetic constant; the two choices are numerically equal in the centimeter-gram-second electromagnetic-unit system. E270-0

magnetic-powder-impregnated tape (impregnated tape) (dispersed-magnetic-powder tape). A magnetic tape that consists of magnetic particles uniformly dispersed in a nonmagnetic material. 0-1E1

magnetic power factor. The cosine of the magnetic hysteretic angle (the sine of the magnetic loss angle). E270-0

magnetic recorder. Equipment incorporating an electromagnetic transducer and means for moving a magnetic recording medium relative to the transducer for recording electric signals as magnetic variations in the medium. *Note:* The generic term **magnetic recorder** can also be applied to an instrument that has not only facilities for recording electric signals as magnetic variations, but also for converting such magnetic variations back into electric variations. *See also:* **phonograph pickup.** 42A65-1E1

magnetic recording (facsimile). Recording by means of a signal-controlled magnetic field. *See also:* **recording (facsimile).** E168-0

magnetic recording head. In magnetic recording, a transducer for converting electric currents into magnetic fields, in order to store the electric signal as a magnetic polarization of the magnetic medium. 0-1E1

magnetic recording medium. A material usually in the form of a wire, tape, cylinder, disk, etcetera, on which a magnetic signal may be recorded in the form of a pattern of magnetic polarization. 0-1E1

magnetic reproducer. Equipment incorporating an electromagnetic transducer and means for moving a magnetic recording medium relative to the transducer, for reproducing magnetic signals as electric signals. 0-1E1

magnetic reproducing head. In magnetic recording, a transducer for collecting the flux due to stored magnetic polarization (the recorded signal) and converting it into an electric voltage. 0-1E1

magnetic rotation (polarized light) (Faraday effect). When a plane polarized beam of light passes through certain transparent substances along the lines of a strong magnetic field, the plane of polarization of the emergent light is different from that of the incident light. E270-0

magnetic spectrograph. An electronic device based on the action of a constant magnetic field on the paths of electrons, and used to separate electrons with different velocities. *See also:* **electron device.** 50I07-15E6

magnetic storage. A method of storage that uses the magnetic properties of matter to store data by magnetization of materials such as cores, films, or plates, or of material located on the surfaces of tapes, discs, or drums, etcetera. *See also:* **electronic digital computer; magnetic-core; magnetic drum; magnetic tape.** E162-0

magnetic storm. A disturbance in the earth's magnetic field, associated with abnormal solar activity, and capable of seriously affecting both radio and wire transmission. *See also:* **radio transmitter.** 42A65-0

magnetic susceptibility (isotropic medium). In rationalized systems, the relative permeability minus unity.

$$k = \mu_r - 1 = \mathbf{B}_i / \mu_0 \mathbf{H}$$

Notes: (1) In unrationalized systems, $k = (\mu_r - 1)4\pi$. (2) The susceptibility divided by the density of a body is called the susceptibility per unit mass, or simply the mass susceptibility. The symbol is χ. Thus

$$\chi = k/\rho$$

where ρ is the density. χ multiplied by the atomic weight is called the atomic susceptibility. The symbol is χ_A. (3) In anisotropic media, susceptibility becomes a matrix. E270-0

magnetic tape (homogeneous or coated). (1) A tape with a magnetic surface on which data can be stored by selective polarization of portions of the surface. (2) A tape of magnetic material used as the constituent in some forms of magnetic cores. *See also:* **coated magnetic tape; electronic digital computer.** E162/X3A12-16E9

magnetic test coil (search coil) (exploring coil). A coil that, when connected to a suitable device, can be used to measure a change in the value of magnetic flux linked with it. *Note:* The change in the flux linkage may be produced by a movement of the coil or by a variation in the magnitude of the flux. Test coils used to measure magnetic induction **B** are often called **B** coils; those used to determine magnetizing force **H** may be called **H** coils. A coil arranged to rotate through an angle of 180 degrees about an axis of symmetry perpendicular to its magnetic axis is sometimes called a flip coil. *See also:* **magnetometer.** 42A30-0

magnetic thin film. A layer of magnetic material, usually less than 10 000 angstroms thick. *Note:* In electronic computers, magnetic thin films may be used for logic or storage elements. *See also:* **coated magnetic tape; electronic digital computer; magnetic core; magnetic tape.** E162-16E9

magnetic track braking. A system of braking in which a shoe or slipper is applied to the running rails by magnetic means. *See also:* **electric braking.** 42A42-0

magnetic variometer. An instrument for measuring differences in a magnetic field with respect to space or time. *Note:* The use of variometer to designate a continuously adjustable inductor is deprecated. *See also:* **magnetometer.** 42A30-0

magnetic vector (radio wave propagation). Magnetic field vector. *See also:* **radio wave propagation.** 0-3E2

magnetic vector potential. An auxiliary solenoidal vector point function characterized by the relation that its curl is equal to the magnetic induction and its divergence vanishes.

$$\text{Curl } \mathbf{A} = \mathbf{B} \qquad \text{Divergence } \mathbf{A} = \mathbf{0}$$

Note: These relations are satisfied identically by

$$A = (\mu_0/4\pi)\,[\int \mathbf{M}\times\nabla(1/r)dv + \int(\mathbf{J}/r)\,dv]$$

where v is the volume. E270-0

magnetization (intensity of magnetization) (at a point of a body). The intrinsic induction at that point divided by the magnetic constant of the system of units employed:

$$M = \mathbf{B}_i/\mu_0 = (\mathbf{B} - \mu_0\mathbf{H})/\mu_0.$$

Note: The magnetization can be interpreted as the volume density of magnetic moment. E270-0

magnetizing current (1) (transformers). A hypothetical current assumed to flow through the magnetizing inductance of a transformer. 0-21E1

(2) (rotating machinery). The quadrature (leading) component (with respect to the induced voltage) of the exciting current supplied to a coil. E270-31E8

magnetizing force. *See:* **magnetic field strength.**

magnetizing inductance. A hypothetical inductance, assumed to be in parallel with the core-loss resistance, that would store the same amount of energy as that stored in the core for a specified value of excitation. 0-21E1

magnet meter (magnet tester). An instrument for measuring the magnetic flux produced by a permanent magnet under specified conditions of use. It usually comprises a torque-coil or a moving-magnet magnetometer with a particular arrangement of pole-pieces. *See also:* **magnetometer.** 42A30-0

magneto. *See:* **magnetoelectric generator.**

magneto central office. A telephone central office for serving magneto telephone sets. *See also:* **telephone system.** 42A65-19E1

magnetoelectric generator (magneto). An electric generator in which the magnetic flux is provided by one or more permanent magnets. *See:* **direct-current commutating machine.** 42A10-0

magneto-ionic medium (radio wave propagation). An ionized gas that is permeated by a fixed magnetic field. *See also:* **radio wave propagation.** 0-3E2

magneto-ionic mode (radio wave propagation). Magneto-ionic wave component. *See also:* **radio wave propagation.** 0-3E2

magneto-ionic wave component (radio wave propagation) (1) (incidence of a linearly polarized wave upon a magneto-ionic medium). Either of the two elliptically polarized wave components into which a linearly polarized wave incident on the ionosphere is separated because of the earth's magnetic field. 42A65-0

(2) (at a given frequency) (wave-propagation within a magneto-ionic medium). Either of the two plane electromagnetic waves that can travel in a homogeneous magneto-ionic medium without change of polarization. *See also:* **radiation; radio wave propagation.** 0-3E2

magnetometer. An instrument for measuring the intensity or direction (or both) of a magnetic field or of a component of a magnetic field in a particular direction. *Note:* The term is more usually applied to instruments that measure the intensity of a component of a magnetic field, such as horizontal-intensity magnetometers, vertical-intensity magnetometers, and total-intensity magnetometers.
See:
declinometer;
dip needle;
electron-beam magnetometer;
fluxmeter;
gaussmeter;
generating magnetometer;
hysteresigraph;
instrument;
magnetic compass;
magnetic test coil;
magnetic variometer;
magnet meter;
moving-magnet magnetometer;
permeameter;
saturable-core magnetometer;
susceptibility meter;
torque-coil magnetometer. 42A30-0

magnetomotive force (acting in any closed path in a magnetic field). The line integral of the magnetizing force around the path. E270-0

magnetostriction. The phenomenon of elastic deformation that accompanies magnetization. E270-0

magnetostriction loudspeaker. A loudspeaker in which the mechanical displacement is derived from the deformation of a material having magnetostrictive properties. *See also:* **loudspeaker.** 42A65-0

magnetostriction microphone. A microphone that depends for its operation on the generation of an electromotive force by the deformation of a material having magnetostrictive properties. *See also:* **microphone.** 42A65-0

magnetostriction oscillator. An oscillator with the plate circuit inductively coupled to the grid circuit through a magnetostrictive element, the frequency of oscillation being determined by the magnetomechanical characteristics of the coupling element. *See also:* **oscillatory circuit.** E145-0

magnetostrictive relay. A relay in which operation depends upon dimensional changes of a magnetic material in a magnetic field. *See also:* **relay.** 83A16-21E0

magneto switchboard. A telephone switchboard for serving magneto telephone sets. *See also:* **telephone switching system.** 42A65-19E1

magneto telephone set. A local-battery telephone set in which current for signaling by the telephone station is supplied from a local hand generator, usually called a magneto. *See also:* **telephone station.** 42A65-0

magnetron. A vacuum tube, in which electrons, controlled by crossed steady electric and magnetic fields, interact with the field of a circuit element to produce alternating-current power output.
See:
anode strap;
critical field;
critical voltage;
critical-voltage parabola;
cyclotron frequency;
cyclotron-frequency magnetron oscillations;
effective bunching angle;
end shield;
field-reversal permanent-magnet focusing;
interaction-circuit phase velocity;
interdigital magnetron;
magnetron injection gun;
magnetron oscillator;
microwave tube (or valve);
mode;
multicavity magnetron;
multisegment magnetron;
oscillatory circuit;
packaged magnetron;
performance chart;
periodic permanent-magnet focusing;
permanent-magnet focusing;
rising-sun magnetron;
split-anode magnetron;
starting current;
synchronous voltage;
voltage-tunable magnetron. *See also:* **tube definitions.** E54/E169/42A70-0

magnetron injection gun (microwave tubes). A gun that produces a hollow beam of high total permeance that flows parallel to the axis of a magnetic field. *See:* **magnetrons; microwave tube (or valve).** *See also:* **tube definitions.** 0-15E6

magnetron oscillator. An electron tube in which electrons are accelerated by a radial electric field between the cathode and one or more anodes and by an axial magnetic field that provides a high-energy electron stream to excite the tank circuits. *See also:* **magnetron.** E145-0

magnet valve (electric controller). A valve controlling a fluid, usually air, operated by an electromagnet. *See also:* **multiple-unit control.** E16-0

magnet wire (rotating machinery). Single-strand wire with a thin flexible insulation, suitable for winding coils. *See:* **rotor (rotating machinery); stator.** 0-31E8

magnified sweep (oscilloscopes). A sweep whose time per division has been decreased by amplification of the sweep waveform rather than by changing the time constants used to generate it. *See:* **oscillograph.** 0-9E4

magnitude. The quantitative attribute of size, intensity, extent, etcetera, that allows a particular entity to be placed in order with other entities having the same attribute. *Notes:* (1) The magnitude of the length of a given bar is the same whether the length is measured in feet or in centimeters. (2) The word magnitude is used in other senses. The definition given here is the basic one needed for the logical buildup of later definitions. E270-0

magnitude characteristic (linear passive networks). The absolute value of a response function evaluated on the imaginary axis of the complex-frequency plane. *See also:* **linear passive networks.** E156-0

magnitude contours (control system, feedback). Loci of selected constant values of the magnitude of the return transfer function drawn on a plot of the loop transfer function for real frequencies. *Note:* Such loci may be drawn on the Nyquist or inverse Nyquist diagrams, or Nichols chart. *See also:* **control system, feedback.** 85A1-23E0

magnitude ratio. *See:* **gain.**

main (interior wiring). A feeder extending from the service switch, generator bus, or converter bus to the main distribution center. *See also:* **interior wiring.** 42A95-0

main anode (pool-cathode tube). An anode that conducts load current. *Note:* The word main is used only when it is desired to distinguish the anode to which it is applied from an auxiliary electrode such as an excitation anode. It is used only in connection with pool-tube terms. *See also:* **electrode (electron tube).** 42A70-15E6

main bang (radar). A transmitted pulse. *See also:* **pulse terms.** E194-0

main capacitance (capacitance potential device). The capacitance between the network connection and line. *See also:* **outdoor coupling capacitor.** E31-0

main contacts (switching device). Contacts that carry all or most of the current of the main circuit. 37A100-31E11

main distribution center. A distribution center supplied directly by mains. *See also:* **distribution center.** 42A95-0

main exciter (rotating machinery). An exciter that supplies all or part of the power required for the excitation of the principal electric machine or machines. *See:* **direct-current commutating machine; asynchronous machine; synchronous machine.** 42A10-31E8

main exciter response ratio (nominal exciter response). The numerical value obtained when the response, in volts per second, is divided by the rated-load field voltage; which response, if maintained constant, would develop, in one-half second, the same excitation voltage-time area as attained by the actual exciter. *Note:* The response is determined with no load on the exciter, with the exciter voltage initially equal to the rated-load field voltage, and then suddenly establishing circuit conditions that would be used to obtain nominal exciter ceiling voltage. For a rotating exciter, the response should be determined at the rated speed. This definition does not apply to main exciters having one or more series fields, except a light differential series field, or to electronic exciters. *See:* **direct-current commutating machine; synchronous machine.** 42A10-0

main gap (glow-discharge tubes). The conduction path between a principal cathode and a principal anode. 42A70-15E6

main lead (rotating machinery). A conductor joining a main terminal to the primary winding. *See:* **asynchronous machine; synchronous machine.** 0-31E8

main protection. *See:* **primary protection.**

mains. *See:* **primary distribution mains; secondary distribution mains.** *See also:* **center of distribution.**

mains coupling coefficient (electromagnetic compatibility). *See:* **mains decoupling factor.**

mains decoupling factor (mains coupling coefficient) (electromagnetic compatibility). The ratio of the radio-frequency voltage at the mains terminal to the interfering apparatus to the radio-frequency voltage at the aerial terminals of the receiver. *Note:* Generally expressed in logarithmic units. *See also:* **electromagnetic compatibility.** CISPR-27E1

main secondary terminals. The main secondary terminals provide the connections to the main secondary winding. *See:* **main secondary winding.** E31-0

main secondary winding (capacitance potential device). Provides the secondary voltage or voltages on which the potential device ratings are based. *See:* **main secondary terminals.** E31-0

mains-interference immunity (mains-interference ratio). The degree of protection of a radio receiver against interference conducted by its supply mains as measured under specified conditions. *Note:* See International Special Committee on Radio Interference recommendation 25/1 and International Electrotechnical Commission publication 69 or subsequent publications where the term mains-interference ratio is used. *See also:* **electromagnetic compatibility.** CISPR-27E1

main station. A telephone station with a distinct call number designation, directly connected to a central office. *See also:* **telephone station.** 42A65-0

main switchgear connections (primary switchgear connection). Those that electrically connect together devices in the main circuit, or connect them to the bus, or both. 37A100-31E11

maintaining voltage (glow lamp) (operating voltage). The voltage measured across the lamp electrodes when the lamp is operating. 78A385-0

maintenance (computing systems). Any activity intended to keep equipment or programs in satisfactory working condition, including tests, measurements, replacements, adjustments, and repairs. *See:* **file maintenance.** X3A12-16E9

maintenance factor (illuminating engineering). The ratio of the illumination on a given area after a period of time to the initial illumination on the same area. *Note:* The maintenance factor is used in lighting calculations as an allowance for the depreciation of lamps, light control elements, and room surfaces to values below the initial or design conditions, so that a minimum desired level of illumination may be maintained in service. The maintenance factor had formerly been widely interpreted as the ratio of average to initial illumination. *See also:* **inverse-square law (illuminating engineering).** Z7A1-0

maintenance proof test (insulated winding). A test applied to an insulated winding after being in service to determine that it is suitable for continued service. It is usually made at a lower voltage than the acceptance proof test. *See also:* **insulation testing (large alternating-current rotating machinery).** E95-0

main terminal (rotating machinery). A termination for the primary winding. *See:* **asynchronous machine; synchronous machine.** 0-31E8

main-terminal 1 (bidirectional thyristor). The main terminal that is named 1 by the device manufacturer. *See also:* **anode.** E223-34E17/15E7

main-terminal 2 (bidirectional thyristor). The main terminal that is named 2 by the device manufacturer.- *See also:* **anode.** E223-34E17/15E7

main terminals (thyristor). The terminals through which the principal current flows. *See also:* **anode.** E223-34E17/15E7

main transformer (polyphase power). The term, as applied to two single-phase Scott-connected units for three-phase to two-phase, or two-phase to three-phase operation, designates the transformer that is connected directly between two of the phase wires of the three-phase lines. *Note:* A tap is provided at the midpoint for connection to the teaser transformer. *See:* **autotransformer.** 42A15-31E12

main unit (two-core voltage-regulating transformer). The core and coil unit that furnishes excitation to the series unit. *See:* **voltage regulator.** 42A15-31E12

main winding, single-phase induction motor. A system of coils acting together, connected to the supply line, that determines the poles of the primary winding, and that serves as the principal winding for transfer of energy from the primary to the secondary of the motor. *Note:* In some multispeed motors, the same main winding will not be used for both starting operation and running operation. *See:* **asynchronous machine.** 0-31E8

major cycle (electronic computation). In a storage device that provides serial access to storage positions, the time interval between successive appearances of a given storage position. *See also:* **electronic digital computer.** E162/E270-0

majority (computing systems). A logic operator having the property that if P is a statement, Q is a statement, R is a statement, ..., then the majority of P, Q, R, ..., is true if more than half the statements are true, false if half or less are true. X3A12-16E9

majority carrier (semiconductor). The type of charge carrier constituting more than one half the total charge-carrier concentration. *See also:* **semiconductor; semiconductor device.** E102/E216/E270-10E1/34E17

major lobe. The radiation lobe containing the direction of maximum radiation. *Note:* In certain antennas, such as multilobed or split-beam antennas, there may exist more than one major lobe. *See also:* **antenna.** 0-3E1

major loop (control). A continuous network consisting of all of the forward elements and the primary feedback elements of the feedback control system. *See also:* **control system, feedback.** AS1-34E10

making capacity (industrial control). The maximum current or power that a contact is able to make under specified conditions. *See:* **contactor.** 50I16-34E10

making current (switching device). The value of the available current at the time the device closes. *Note:* Its root-mean-square value is measured from the envelope of the current wave at the time of the first major current peak. 37A100-31E11

malfunction. An error that results from failure in the hardware. *See also:* **electronic digital computer; error; mistake.** E162-0

manhole (electric systems) (1) (More accurately termed **splicing chamber** or **cable vault**). A subsurface chamber, large enough for a man to enter, in the route of one or more conduit runs, and affording facilities for placing and maintaining in the runs, conductors, cables, and any associated apparatus. *See also:* **cable vault; power distribution, underground construction; splicing chamber.** 42A35-31E13
(2) An opening in an underground system that workmen or others may enter for the purpose of installing cables, transformers, junction boxes, and other devices, and for making connections and tests. *See:* **cable vault; distribution center; power distribution, underground construction; splicing chamber.** 2A2-0

manhole chimney. A vertical passageway for workmen and equipment between the roof of the manhole and the street level. *See also:* **power distribution, underground construction.** 42A35-31E13

manhole cover frame. The structure that caps the manhole chimney at ground level and supports the cover. *See also:* **power distribution, underground construction.** 42A35-31E13

manifold insulation (liquid cooling) (rotating machinery). The insulation applied between ground and a manifold connecting several parallel liquid-cooling paths in a winding. *See:* **stator.** 0-31E8

manifold-pressure electric gauge. A device that measures the pressure of fuel vapors entering the cylinders of an aircraft engine. *Note:* The gauge is provided with a scale, usually graduated in inches of mercury, absolute. It provides remote indication by means of a self-synchronous generator and motor. *See also:* **air-transportation instruments.** 42A41-0

manipulated variable (control) (industrial control). A quantity or condition that is varied as a function of the actuating signal so as to change the value of the directly controlled variable. *Note:* In any practical control system, there may be more than one manipulated variable. Accordingly, when using the term it is necessary to state which manipulated variable is being discussed. In process control work, the one immediately preceding the directly controlled system is usually intended. *See:* **control system, feedback.** 85A1-23E0;AS1-34E10

man-made noise (electromagnetic compatibility). Noise generated in machines or other technical devices. *See also:* **electromagnetic compatibility.** 0-27E1

manual (electric systems). Operated by mechanical force, applied directly by personal intervention. *See also:* **distribution center.** 2A2/42A25/42A95-34E10

manual block-signal system. A block or a series of consecutive blocks governed by block signals operated manually upon information by telegraph, telephone, or other means of communication. *See also:* **block signal system.** 42A42-0

manual central office. A central office of a manual telephone system. *See also:* **telephone system.** 42A65-19E1

manual control. Control in which the main devices, whether manually or power operated, are controlled by an attendant. *See:* **control.** 37A100-31E11

manual controller (industrial control). An electric controller having all of its basic functions performed by devices that are operated by hand. *See:* **electric controller.** 42A25-34E10

manual data input (numerically controlled machines). A means for the manual insertion of numerical control commands. *See also:* **numerically controlled machines.** EIA3B-34E12

manual fire-alarm system. A fire-alarm system in which the signal transmission is initiated by manipulation of a device provided for the purpose. *See also:* **protective signaling.** 42A65-0

manual holdup-alarm system. An alarm system in which the signal transmission is initiated by the direct action of the person attacked or of an observer of the attack. *See also:* **protective signaling.** 42A65-0

manual input (computing systems). (1) The entry of data by hand into a device at the time of processing. (2) The data entered as in (1). *See also:* **electronic digital computer.** X3A12-16E9

manual load (armature current) division (industrial control). The effect of a manually operated device to adjust the division of armature currents between two or more motors or two or more generators connected to the same load. *See also:* **control system, feedback.** IC1-34E10

manual lockout device. A device that holds the associated device inoperative unless a predetermined manual function is performed to release the locking feature. 37A100-31E11

manually operated door or gate. A door or gate that is opened and closed by hand. *See also:* **hoistway (elevator or dumbwaiter).** 42A45-0

manually release-free (manually trip-free). *See:* **mechanically release-free.**

manually trip-free. *See:* **mechanically release-free.**

manual mobile telephone system. A mobile communication system manually interconnected with any telephone network, or a mobile communication system manually interconnected with a telephone network. 0-6E1

manual operation. Operation by hand without using any other source of power. 37A100-31E11

manual release (electromagnetic brake) (industrial control). A device by which the braking surfaces may be manually disengaged without disturbing the torque adjustment. *See:* **electric drive.** IC1-34E10

manual-reset manual release (control-brakes). A manual release that requires an additional manual action to re-engage the braking surfaces. IC1-34E10

manual-reset relay. *See:* **relay, manual-reset.**

manual-reset thermal protector (rotating machinery). A thermal protector designed to perform the function by opening the circuit to or within the protected machine, but requiring manual resetting to close the circuit. *See also:* **starting-switch assembly.** 0-31E8

manual ringing. Ringing that is started by the manual operation of a key and continues only while the key is held operated. *See also:* **telephone switching system.** 42A65-19E1

manual speed adjustment (industrial control). A speed adjustment accomplished manually. *See also:* **electric drive.** 42A25-34E10

manual switchboard. A telephone switchboard in which the connections are made manually, by plugs and jacks, or by keys. *See also:* **telephone switching system.** 42A65-19E1

manual telephone system. A telephone system in which telephone connections between customers are ordinarily established manually by telephone operators in accordance with orders given orally by the calling parties. *See also:* **telephone switching system.** 42A65-19E1

manuscript (numerically controlled machines). An ordered list of numerical control instructions. *See:* **programming.** *See also:* **numerically controlled machines.** EIA3B-34E12

map. To establish a correspondence between the elements of one set and the elements of another set. X3A12-16E9

map vertical (navigation). *See:* **geographic map vertical.**

margin (orientation margin) (printing telegraphy) (teletypewriter). That fraction of a perfect signal element through which the time of selection may be varied in one direction from the normal time of selection, without causing errors while signals are being received. *Note:* There are two distinct margins, determined by varying the time of selection in either direction from normal. *See also:* **telegraphy.** 42A65-19E4

marginal check (electronic computation). A preventive maintenance procedure in which certain operating conditions (for example, supply voltage or frequency) are varied about their nominal values in order to detect and locate incipient defective parts. *See also:* **check; electronic computation; electronic digital computer.** E162/270/X3A12-16E9

marginal checking (marginal testing). *See:* **marginal check.**

marginal relay. A relay that functions in response to predetermined changes in the value of the coil current or voltage. *See also:* **relay.** 83A16-0

marginal testing (electronic computation). *See:* **marginal check.**

marine distribution panel. A panel receiving energy from a distribution or subdistribution switchboard and distributing energy to energy-consuming devices or other distribution panels or panelboards of a ship. *See also:* **marine electric apparatus.** 42A43-0

marine electric apparatus. Electric apparatus designed especially for use on shipboard to withstand the conditions peculiar to such application.
See also:
fire-door release system;
marine distribution panel;
marine generator and distribution switchboard;
marine panelboard;
marine subdistribution switchboard;
ratproof electric installation;
ship's service electric system;
shockproof electric apparatus;
shore feeder;
watertight-door control system. 42A43-0

marine generator and distribution switchboard. Receives energy from the generating plant and distributes directly or indirectly to all equipment of a ship supplied by the generating plant. *See also:* **marine electric apparatus.** 42A43-0

marine panelboard. A single panel or a group of panel units assembled as a single panel, usually with automatic overcurrent circuit breakers or fused switches, in a cabinet for flush or surface mounting in or on a bulkhead and accessible only from the front, serving lighting branch circuits or small power branch circuits of a ship. *See also:* **marine electric apparatus.** 42A43-0

mariner's compass. A magnetic compass used in navigation consisting of two or more parallel polarized needles secured to a circular compass card that is delicately pivoted and enclosed in a glass-covered bowl filled with alcohol to support by flotation the weight of the moving parts. *Note:* The compass bowl is supported in gimbals mounted in the binnacle. The compass card is graduated to show the 32 points of the compass in addition to degrees. 42A43-0

marine subdistribution switchboard. Essentially a section of the marine generator and distribution switchboard (connected thereto by a bus feeder and remotely located) that distributes energy in a certain section of a vessel. *See also:* **marine electric apparatus.** 42A43-0

mark (computing systems). *See:* **flag.**

marked ratio (instrument transformer). The ratio of the primary current or voltage, as the case may be, to the secondary current or voltage, as stated on the nameplate. *See:* **instrument transformer.** 12A0-0;42A15/42A30-31E12

marker (air navigation). A radio transmitting facility whose signals are geographically confined so as to serve as a position fix. *See:* **boundary marker; fan marker; middle marker; outer marker; Z marker.** *See also:* **radio navigation.** 0-10E6

marker-beacon receiver. A receiver used in aircraft to receive marker-beacon signals that identify the position of the aircraft when over the marker-beacon station. *See also:* **air-transportation electronic equipment.** 42A41-0

marker lamp (marker light) (railway practice). A signal lamp placed at the side of the rear end of a train or vehicle, displaying light of a particular color to indicate the rear end and to serve for identification purposes. 42A42-0

marker light (railway practice). A light that by its color or position, or both, is used to qualify the signal aspect. *See also:* **railway signal and interlocking.** 42A42-0

marker signal (oscilloscopes). A signal introduced into the presentation for the purpose of identification, calibration, or comparison. 0-9E4

marking and spacing intervals (telegraph communication). Intervals that correspond, according to convention, to one condition or position of the originating transmitting contacts, usually a closed condition; spacing intervals are the intervals that correspond to another condition of the originating transmitting contacts, usually an open condition. *Note:* The terms **mark** and **space** are frequently used for the corresponding conditions. The waves corresponding to the marking and spacing intervals are frequently designated as marking and spacing waves, respectively. *See also:* **telegraphy.** 42A65-0

marking pulse (teletypewriter). The signal pulse that, in direct current, neutral, operation, corresponds to a circuit-closed or current-on condition. 0-19E4

marking wave (keying wave) (telegraph communication). The emission that takes place while the active portions of the code characters are being transmitted. *See also:* **radio transmitter.** E145-0

***M*-array glide slope (instrument landing systems).** A modified null-reference glide-slope antenna system in which the modification is primarily an additional antenna used to obtain a high degree of energy cancellation at the low elevation angles. *Note:* Called *M* because it was 13th in a series of designs. This system is used at locations where higher terrain exists in front of the approach end of the runway, in order to reduce unwanted reflections of energy into the glide-slope sector. *See also:* **navigation.** 0-10E6

mask (computing systems). (1) A pattern of characters that is used to control the retention or elimination of portions of another pattern of characters. (2) A filter. *See also:* **electronic digital computer.** X3A12-16E9

masking (1) (acoustics). (A) The process by which the threshold of audibility for one sound is raised by the presence of another (masking) sound. (B) The amount by which the threshold of audibility of a sound is raised by the presence of another (masking) sound. The unit customarily used is the decibel. *See also:* **electroacoustics; loudspeaker.**
(2) (color television). A process to alter color rendition in which the appropriate color signals are used to modify each other. *Note:* The modification is usually accomplished by suitable cross coupling between primary color-signal channels. *See:* **television.** 0-1E1/2E2

masking audiogram. A graphic presentation of the masking due to a stated noise. *Note:* This is plotted in decibels as a function of the frequency of the masked tone. *See also:* **loudspeaker.** E157-1E1

mass (body). The property that determines the acceleration the body will have when acted upon by a given force. E270-0

mass-attraction vertical. The normal to any surface of constant geopotential; it is the direction that would be indicated by a plumb bob if the earth were not rotating. *See also:* **navigation.** E172-10E6

mass spectrograph. An electronic device based on the action of a constant magnetic field on the paths of ions, used to separate ions of different masses. *See also:* **electron device.** 50I07-15E6

mast (power transmission and distribution). A column or narrow-base structure of wood, steel, or other material, supporting overhead conductors, usually by means of arms or brackets, span wires, or bridges. *Note:* Broad-base lattice steel supports are often known as towers; narrow-base steel supports are often known as masts. *See:* **pole.** *See also:* **tower.** 42A35-31E13

mast arm. *See:* **bracket.**

master compass. A magnetic or gyro compass arranged to actuate repeaters, course recorders, automatic pilots, or other devices. 42A43-0

master controller (load-frequency control system) (1) (electric power generators). The central device that develops corrective action, in response to the area control error, for execution at one or more generating units. E94-0
(2) (car retarders). A controller that governs the operation of one or more magnetic or electropneumatic controllers. *Note:* It is designed to coordinate the movement or the pressure of the retarder with the movement of the retarder level. *See also:* **car retarder; multiple-unit control.** E16/42A42-0

master direction indicator. A device that provides a remote reading of magnetic heading. It receives a signal from a magnetic sensing element. *See also:* **air-transportation instruments.** 42A41-0

master drive (industrial control). A drive that sets the reference input for one or more follower drives. *See also:* **control system, feedback.** AS1-34E10

master form. An original form from which, directly or indirectly, other forms may be prepared. *See also:* **electroforming.** 42A60-0

master oscillator. An oscillator so arranged as to establish the carrier frequency of the output of an amplifier. *See also:* **oscillatory circuit.** E182/42A65-31E3

master reference system for telephone transmission. Adopted by the International Advisory Committee for Long Distance Telephony (CCIF), a primary reference telephone system for determining, by comparison, the performance of other telephone systems and components with respect to the loudness, articulation, or other transmission qualities of received speech. *Note:* The determination is made by adjusting the loss of a distortionless trunk in the master reference system for equal performance with respect to the quality under consideration. *See also:* **transmission characteristics.** 42A65-0

master routine (electronic computation). *See:* **subroutine.**

master/slave operation (power supplies). A system of interconnection of two regulated power supplies in which one (the master) operates to control the other (the slave). *Note:* Specialized forms of the master/slave configuration are used in (1) complementary tracking (plus and minus tracking around a common point), (2) parallel operation to obtain increased current output for voltage regulation, (3) compliance extension to obtain increased voltage output for current regulation. *See also:* **power supply.** KPSH-10E1

master station (1) (supervisory system). The station from which remotely located units of switchgear or other equipment are controlled by supervisory control or that receives supervisory indications or selected telemeter readings. 37A100-31E11
(2) (electronic navigation). One station of a group of stations, as in loran, that is used to control or synchronize the emission of the other stations. *See also:* **radio navigation.** 42A65-10E6

master-station supervisory equipment. That part of a (single) supervisory system that includes all necessary supervisory control relays, keys, lamps, and associated devices located at the master station for selection, control, indication, and other functions to be performed. 37A100-31E11

master switch (industrial control). A switch that dominates the operation of contactors, relays, or other remotely operated devices. *See also:* **switch.** 42A25-34E10

mast-type antenna for aircraft. A rigid antenna of streamlined cross section consisting essentially of a formed conductor or conductor and supporting body. *See also:* **air-transportation electronic equipment.** 42A41-0

mat (rotating machinery). A randomly distributed unwoven felt of fibers in a sheetlike configuration having relatively uniform density and thickness. *See:* **rotor (rotating machinery); stator.** 0-31E8

matched condition. *See:* **termination, matched.** *See also:* **transmission characteristics.**

matched impedances. Two impedances are matched when they are equal. *Note:* Two impedances associated with an electric network are matched when their resistance components are equal and when their reactance components are equal. *See also:* **network analysis.** E270-0

matched termination (waveguide). A termination producting no reflected wave at any transverse section of a waveguide or transmission line. *See:* **transmission line; waveguide.** E146-3E1

matched transmission line. A line is matched at any transverse section if there is no wave reflection at that section. *See also:* **matched waveguide; transmission line; waveguide.** 42A65-0

matched waveguide. A waveguide having no reflected wave at any transverse section. *See also:* **waveguide.** E146/E148-3E1

matching section (transforming section) (waveguide transformer) (waveguide). A length of waveguide of modified cross section, or with a metal or dielectric insert, used for impedance transformation. *See:* **waveguide.** 0-3E1

matching transformer (induction heater). A transformer for matching the impedance of the load to the optimum output characteristic of the power source. E54-0

mathematical check. A programmed check of a sequence of operations that makes use of the mathematical properties of the sequence. Sometimes called a **control.** *See also:* **electronic digital computer; programmed check.** E162-0

mathematical model. A set of equations used to represent a physical system, process, device, or concept. *See also:* **electronic analog computer; model.** E165/X3A12-16E9

mathematical quantity. *See:* **mathematico-physical quantity.**

mathematical symbol (abbreviation). A graphic sign, a letter or letters (which may have letters or numbers, or both, as subscripts or superscripts, or both), used to denote the performance of a specific mathematical operation, or the result of such operation, or to indicate a mathematical relationship. *Compare with:* **symbol for a quantity, symbol for a unit.** *See also:* **abbreviation.** E267-0

mathematico-physical quantity (symbolic quantity) (mathematical quantity) (abstract quantity). A concept, amenable to the operations of mathematics, that is directly related on one (or more) physical quantity and is represented by a letter symbol in equations that are statements about that quantity. *Note:* Each mathematical quantity used in physics is related to a corresponding physical quantity in a way that depends on its defining equation. It is characterized by both a qualitative and a quantitative attribute (that is, dimensionality and magnitude). E270-0

matrix (1) (mathematics). (A) A two-dimensional rectangular array of quantities. Matrices are manipulated in accordance with the rules of matrix algebra. (B) By extension, an array of any number of dimensions. X3A12-16E9

(2) (electronic computers). A logic network whose configuration is an array of intersections of its input-output leads, with elements connected at some of these intersections. The network usually functions as an encoder or decoder. *Note:* A translating matrix develops several output signals in response to several input signals; a decoder develops a single output signal in response to several input signals (therefore sometimes called an AND matrix); an encoder develops several output signals in response to a single input signal and a given output signal may be generated by a number of different input signals (therefore sometimes called an OR matrix). *See also:* **decode; electronic digital computer; encode; translate.** E162/E270-0

(3)Loosely, any encoder, decoder, or translator. *See also:* **electronic computation.** E270-0

(4) (noun) (color television). An array of coefficients symbolic of a color coordinate transformation. *Note:* This definition is consistent with mathematical usage. *See also:* **color terms.** E201-2E2

(5) (verb) (color television). To perform a color coordinate transformation by computation or by electrical, optical, or other means. E201-2E2

(6) (electrochemistry). A form used as a cathode in electroforming . *See also:* **electroforming.** 42A60-0

matrix circuit (color television). *See:* **matrix unit.**

matrix, transition. A matrix that maps the state of a linear system at one instant of time into another state at a later instant of time provided that the system inputs are zero over the closed time interval between the two instants of time. *Note:* This is also the fundamental matrix of solutions of the homogeneous equations. *See also:* **control system.** 0-23E0

matrix unit (matrix circuit) (color television). A device that performs a color coordinate transformation by electrical, optical, or other means. *See also:* **color terms.** E201-0

matte dip (electroplating). A dip used to produce a matte surface on a metal. *See also:* **electroplating.** 42A60-0

matte surface. A surface from which the reflection is predominantly diffuse, with or without a negligible specular component. *See:* **diffuse reflection.** *See also:* **bare (exposed) lamp.** Z7A1-0

maximum asymmetric short-circuit current (rotating machinery). The instantaneous peak value reached by the current in the armature winding within a half of a cycle after the winding has been suddenly short-circuited, when conditions are such that the initial value of any aperiodic component of current is the maximum possible. 0-31E8

maximum average power output (television). The maximum radio-frequency output power that can occur under any combination of signals transmitted, averaged over the longest repetitive modulation cycle. *See also:* **television.** 42A65-0

maximum continuous rating (rotating machinery). The maximum values of electric and mechanical loads at which a machine will operate successfully and continuously. *Note:* An overload may be implied, along with temperature rises higher than normal standards for the machine. *See also:* **asynchronous machine; direct-current commutating machine; synchronous machine.** 0-31E8

maximum control current (magnetic amplifier). The maximum current permissible in each control winding either continuously or for designated operating intervals as specified by the manufacturer and shall be specified as either root-mean-square or average. *See also:* **rating and testing magnetic amplifiers.** E107-0

maximum current (instrument) (wattmeter or power-factor meter). A stated current that, if applied continuously at maximum stated operating temperature and with any other circuits in the instrument energized at rated values, will not cause electric breakdown or any observable physical degradation. *See also:* **instrument.** 39A1/42A30-0

maximum demand (installation or system). The greatest of all the demands that have occurred during the specified period of time. *Note:* The maximum demand is determined by measurement, according to specification, over a definitely prescribed time interval. *See also:* **alternating-current distribution; demand meter; direct-current distribution.** 12A0/42A35-31E13

maximum-demand pointer (demand meter) (friction pointer of a demand meter*). A means used to indicate the maximum demand that has occurred since its previous resetting. The maximum-demand pointer is advanced up the scale of an indicating demand meter by the pointer pusher. When not being advanced, it is held stationary, usually by friction, and it is reset manually when the meter is read for billing purposes. *See also:* **demand meter.**
*Deprecated 42A30-0

maximum design voltage (1) (device). The highest voltage at which the device is designed to operate. *Note:* When expressed as a rating this voltage is termed rated maximum voltage. 37A100-31E11
(2) (to ground) (outdoor electric apparatus). The maximum voltage at which the bushing is designed to operate continuously. *See also:* **power distribution, overhead construction.** 76A1-0

maximum-deviation sensitivity (in frequency-modulation receivers). Under maximum system deviation, the least signal input for which the output distortion does not exceed a specified limit. *See also:* **frequency modulation.** E188-0

maximum excursion (electric conversion). The maximum positive or negative deviation from the initial or steady value caused by a transient condition. *See also:* **electric conversion equipment.** 0-10E1

maximum instantaneous fuel change (gas turbines). The fuel change allowable for an instantaneous or sudden increased or decreased load or speed demand. *Note:* It is expressed in terms of equivalent load change in percent of rated load. E282-31E2

maximum keying frequency (fundamental scanning frequency) (facsimile). The frequency in cycles per second numerically equal to the spot speed divided by twice the scanning spot *X* dimension. *See also:* **scanning (facsimile).** E168-0

maximum modulating frequency (facsimile). The highest picture frequency required for the facsimile transmission system. *Note:* The maximum modulating frequency and the maximum keying frequency are not necessarily equal. *See also:* **facsimile transmission.** E168-0

maximum OFF voltage (magnetic amplifier). The maximum output voltage existing before trip ON control signal is reached as the control signal is varied from trip OFF to trip ON. *See also:* **rating and testing magnetic amplifiers.** E107-0

maximum operating voltage (household electric ranges). The maximum voltage to which the electric parts of the range may be subjected in normal operation. *See also:* **appliances outlet.** 71A1-0

maximum output (receivers). The greatest average output power into the rated load regardless of distortion. *See also:* **radio receiver.** E188-0

maximum output voltage (magnetic amplifier). The voltage across the rated load impedance with maximum control current flowing through each winding simultaneously in a direction that increases the output voltage. *Note:* Maximum output voltage shall be specified either as root-mean-square or average. *Note:* While specification may be either root-mean-square or average, it remains fixed for a given amplifier. *See also:* **rating and testing magnetic amplifiers.** E107-0

maximum rate of fuel change (gas turbines). The rate of fuel change that is allowable after the maximum instantaneous fuel change, when an instantaneous speed or load demand upon the turbine is greater than that corresponding to the maximum instantaneous fuel change. *Note:* It is expressed in percent of equivalent load change per second. E282-31E2

maximum sensitivity (frequency-modulation systems). The least signal input that produces a specified output power. E188-0

maximum sine-current differential permeability (toroidal magnetic amplifier cores). The maximum value of sine-current differential permeability obtained with a specified sine-current magnetizing force. E106-0

maximum sound pressure (for any given cycle of a periodic wave). The maximum absolute value of the instantaneous sound pressure occurring during that cycle. *Note:* In the case of a sinusoidal sound wave this maximum sound pressure is also called the pressure amplitude. *See also:* **electroacoustics.** 0-1E1

maximum speed (industrial control). The highest speed within the operating speed range of the drive. *See also:* **electric drive.** 42A25-34E10

maximum surge current rating (rectifier circuit) (nonrepetitive). The maximum forward current having a specified waveform and short specified time interval permitted by the manufacturer under stated conditions. *See also:* **average forward current rating (rectifier circuit).** E59-34E17/34E24

maximum system deviation (frequency-modulation systems). The greatest frequency deviation specified in the operation of the system. *Note:* In the case of frequency-modulation broadcast systems in the range from 88 to 108 megahertz, the maximum system deviation is 75 kilohertz. *See also:* **frequency modulation; radio transmission.** E188/42A65-0

maximum (highest) system voltage (lightning arrester). The highest voltage at which a system is operated. *Note:* This is generally considered to be the maximum tolerable system voltage as prescribed in Preferred Voltage Ratings for Alternating Current Systems and Equipment, American National Standard C84.1-1954, or latest revision. *See:* **lightning arrester (surge diverter).** *See also:* **lightning protection and equipment.** E28-0;82A1-31E7

maximum test output voltage (magnetic amplifier) (1) (nonreversible output). The output voltage equivalent to the summation of the minimum output voltage plus 66 2/3 percent of the difference between the rated and minimum output voltages. E107-0
(2) (reversible output). (A) Positive maximum test output voltage is the output voltage equivalent to 66 2/3 percent of the rated output voltage in the positive direction. (B) Negative maximum test output voltage is the output voltage equivalent to 66 2/3 percent of the rated output voltage in the negative direction. *See also:* **rating and testing magnetic amplifiers.** E107-0

maximum total sag (electric systems). The sag at the midpoint of the straight line joining the two points of support of the conductor. *See:* **power distribution, overhead construction; sag.** 2A2-0

maximum undistorted output (maximum useful output) (1) (radio receivers). For sinusoidal input, the greatest average output power into the rated load with distortion not exceeding a specified limit. *See:* **radio receiver.** E188-0

(2) (electroacoustics). The maximum power delivered under specified conditions with a total harmonic not exceeding a specified percentage. *See also:* **level.** 42A65-0

maximum usable frequency (radio transmission by ionospheric reflection) (MUF). The upper limit of frequencies that can be used at a particular time for transmission between two specified points by reflection from regular ionized layers. *Note:* Higher frequencies may be transmitted by sporadic and scattered reflections. The maximum usable frequency is usually but not always controlled by electron limitation. *See also:* **radiation; radio wave propagation.** 42A65-0

maximum useful output. *See:* **maximum undistorted output.**

maximum voltage (instrument) (wattmeter, power-factor meter, or frequency meter). A stated voltage that, if applied continuously at the maximum stated operating temperature and with any other circuits in the instrument energized at rated values, will not cause electric breakdown or any observable physical degradation. *See also:* **accuracy rating (instrument); instrument.** 39A1/42A30-0

maxwell (line). The unit of magnetic flux in the centimeter-gram-second electromagnetic system. *Note:* The maxwell is 10^{-8} weber. E270-0

Maxwell bridge (general). A 4-arm alternating-current bridge characterized by having in one arm an inductor in series with a resistor and in the opposite arm a capacitor in parallel with a resistor, the other two arms being normally nonreactive resistors. *Note:* Normally used for the measurement of inductance (or capacitance) in terms of resistance and capacitance (or inductance). The balance is independent of the frequency, and at balance the ratio of the inductance to the capacitance is equal to the product of the resistances of either pair of opposite arms. It differs from the Hay bridge in that in the arm opposite the inductor, the capacitor is shunted by the resistor. *See also:* **bridge.** 42A30-0

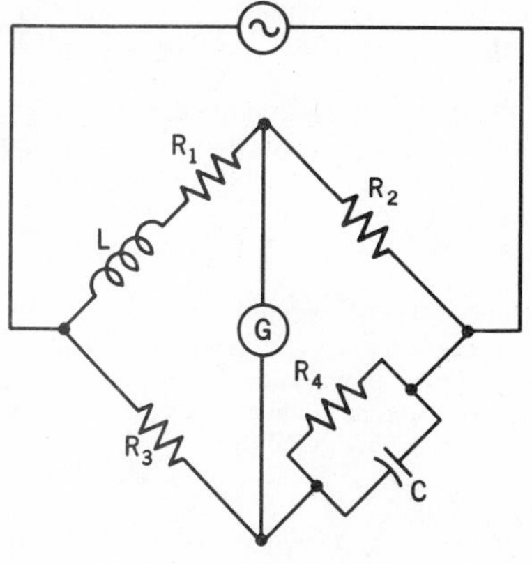

$$R_1R_4=R_2R_3=L/C$$

Maxwell bridge.

Maxwell direct-current commutator bridge. A 4-arm bridge characterized by the presence in one arm of a commutator, or 2-way contactor, that, with a known periodicity, alternately connects the unknown capacitor in series with the bridge arm and then opens the bridge arm while short-circuiting the capacitor, the other three arms being nonreactive resistors. *Note:* Normally used for the measurement of capacitance in terms of resistance and time. The bridge is normally supplied from a battery and the detector is a direct-current galvanometer. *See also:* **bridge.** 42A30-0

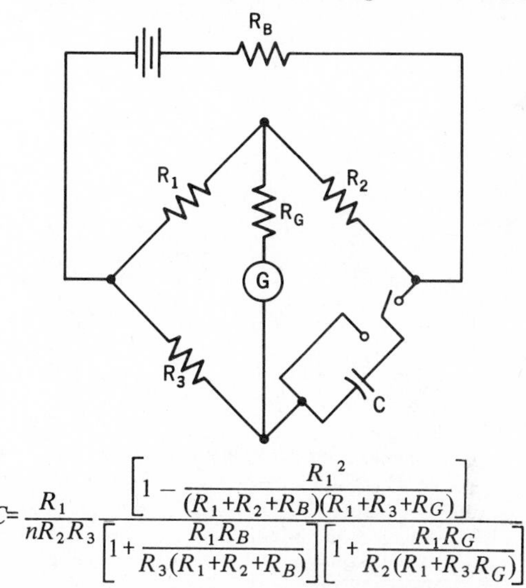

$$C=\frac{R_1}{nR_2R_3}\frac{\left[1-\dfrac{R_1^{\,2}}{(R_1+R_2+R_B)(R_1+R_3+R_G)}\right]}{\left[1+\dfrac{R_1R_B}{R_3(R_1+R_2+R_B)}\right]\left[1+\dfrac{R_1R_G}{R_2(R_1+R_3R_G)}\right]}$$

Maxwell direct-current commutator bridge.

Maxwell inductance bridge. A 4-arm alternating-current bridge characterized by having inductors in two adjacent arms and usually, nonreactive resistors in the other two arms. *Note:* Normally used for the comparison of inductances. The balance is independent of the frequency. *See also:* **bridge.** 42A30

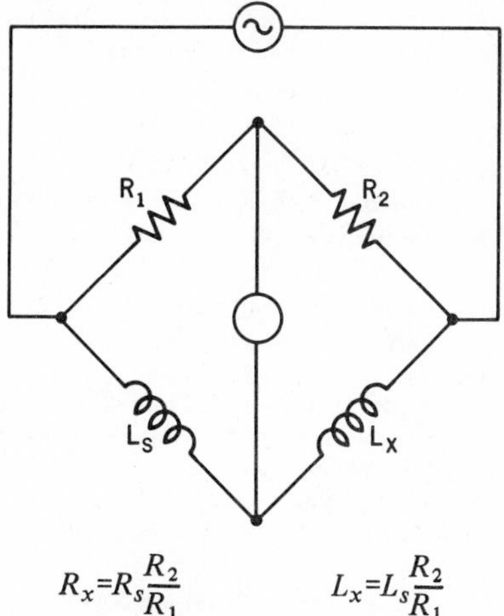

$$R_x=R_s\frac{R_2}{R_1} \qquad L_x=L_s\frac{R_2}{R_1}$$

Maxwell inductance bridge.

Maxwell mutual-inductance bridge. An alternating-current bridge characterized by the presence of mutual inductance between the supply circuit and that arm of the network that includes one coil of the mutual inductor, the other three arms being normally non-reactive resistors. *Note:* Normally used for the measurement of mutual inductance in terms of self-inductance. The balance is independent of the frequency. *See also:* **bridge.** 42A30-0

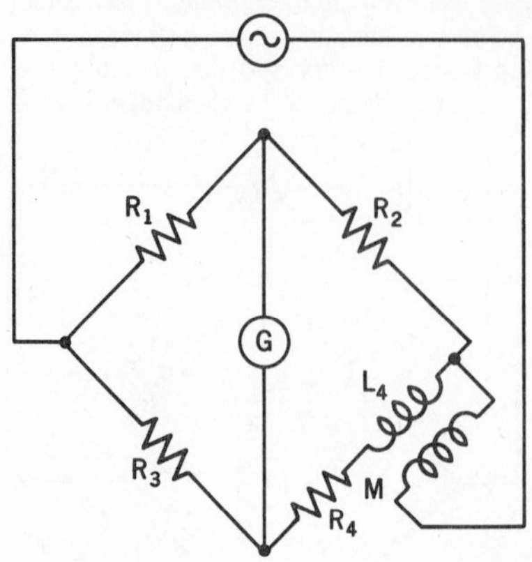

$R_1R_4=R_2R_3$ $L_4=-M\left(1+\frac{R_2}{R_1}\right)$

Maxwell mutual-inductance bridge.

Maxwell's equations (Maxwell's laws). The fundamental equations of macroscopic electromagnetic field theory. All real (physical) electric and magnetic fields satisfy Maxwell's equations, namely

$$\nabla \times \mathbf{E} = \frac{\partial \mathbf{B}}{\partial t} \qquad \nabla \cdot \mathbf{B} = 0$$

$$\nabla \times \mathbf{H} = \frac{\partial \mathbf{D}}{\partial t} + \mathbf{J} \qquad \nabla \cdot \mathbf{D} = q_r$$

where **E** is electric field strength, **D** is electric flux density, **H** is magnetic field strength, **B** is magnetic flux density, **J** is current density, and q_r is volume charge density. E270-0

Maxwell's laws. *See:* **Maxwell's equations.**

May Day. *See:* **radio distress signal.**

MBWO (reentrant field type microwave backward-wave oscillator) (microwave tubes). *See:* **carcinotron.**

McCulloh circuit. A supervised, metallic loop circuit having manually or automatically operated switching equipment at the receiving end, that, in the event of a break, a ground, or a combination of a break and a ground at any point in the metallic circuit, conditions the circuit, by utilizing a ground return, for the receipt of signals from suitable signal transmitters on both sides of the point of trouble. *See also:* **protective signaling.** 42A65-0

MCW. *See:* **modulated continuous wave.**

MDI. *See:* **magnetic direction indicator.**

***M* display (radar).** A type-*A* display in which the target distance is determined by moving an adjustable pedestal signal along the baseline until it coincides with the horizontal position of the target-signal deflections; the control that moves the pedestal is calibrated in distance. *See also:* **navigation.** E172-10E6

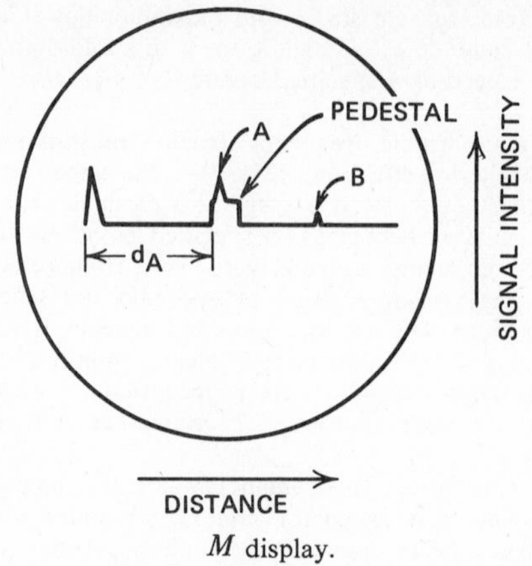

M display.

MEA (electronic navigation). *See:* **minimum en-route altitude.**

mean charge (nonlinear capacitor). The arithmetic mean of the transferred charges corresponding to a particular capacitor voltage, as determined from a specified alternating charge characteristic. *See also:* **nonlinear capacitor.** E226-15E7

mean-charge characteristic (nonlinear capacitor). The function relating mean charge to capacitor voltage. *Note:* Mean-charge characteristic is always single-valued. *See also:* **nonlinear capacitor.** E226-15E7

mean free path (acoustics). For sound waves in an enclosure, the average distance sound travels between successive reflections in the enclosure. *See also:* **electroacoustics.** E157-1E1

mean life (reliability). The arithmetic mean of the times to failure of a group of nominally identical items. *See also:* **reliability.** 0-7E1

mean life, assessed (reliability). The mean life of a nonrepaired item determined within stated confidence limits from the observed mean life of nominally identical items. *Note:* Alternatively, point estimates may be used, the basis of which must be defined. *See also:* **reliability.** 0-7E1

mean life, extrapolated (nonrepaired items) (reliability). Extension by a defined extrapolation or interpolation of the assessed mean life for durations or stress conditions different from those applying to the conditions of that assessed mean life. *See also:* **reliability.** 0-7E1

mean life, observed (reliability). The mean value of the lengths of observed times to failure of all specimens in a sample of items under stated stress conditions. *Note:* The criteria for what constitutes a failure should be stated. *See also:* **reliability.** 0-7E1

mean life, predicted (nonrepaired items) (reliability). The mean life of an item computed from its design considerations and, where appropriate, from the reliability of its part in the intended conditions of use. *See also:* **reliability.** 0-7E1

mean of reversed direct-current values (alternating-current instruments). The simple average of the indications when direct current is applied in one direction and then reversed and applied in the other direction. *See also:* **accuracy rating (instrument).** 39A1-0

mean pulse time. The arithmetic mean of the leading-edge pulse time and the trailing-edge pulse time. *Note:* For some purposes, the importance of a pulse is that it exists (or is large enough) at a particular instant of time. For such applications the important quantity is the mean pulse time. The leading-edge pulse time and trailing-edge pulse time are significant primarily in that they may allow a certain tolerance in timing. *See also:* **pulse terms.** E194/E270-0

mean time between errors (MTE). The average time between the generation of a single false code of a given set in use in a code transmission system, assuming a continuous source of random inputs and the bandwidth limited in a manner identical to the normally expected signal. 0-31E3

mean time between failures (MTF). The average time (preferably expressed in hours) between failures of a continuously operating device, circuit, or system. 0-31E3

mean time between (or before) failures (MTBF) (power supplies). A measure of reliability giving either the time before first failure or, for repairable equipment, the average time between repairs. Mean time between (before) failures may be approximated or predicted by summing the reciprocal failure rates of individual components in an assembly. *See also:* **power supply.** KPSH-10E1

mean time between failures (repairable items) (reliability). The product of the number of items and their operating time divided by the total number of failures. *See also:* **reliability.** 0-7E1

mean time between failures, assessed (reliability). The mean time between failures of a repairable item determined within stated confidence limits from the observed mean time between failures of nominally identical items. *Note:* Alternatively, point estimates may be used, the basis of which must be defined. *See also:* **reliability.** 0-7E1

mean time between failures, extrapolated (reliability). Extension by a defined extrapolation or interpolation of the assessed mean-time-between-failures for durations or stress conditions different from those applying to the conditions of that assessed mean time between failures. *See also:* **reliability.** 0-7E1

mean time between failures, observed (repairable items) (reliability) (stated period in the life of an item). The mean value of the lengths of observed times between consecutive failures under stated stress conditions. *Notes:* (1) The criteria for what constitutes a failure should be stated. (2) This is the reciprocal of the failure rate observed during the period. *See also:* **reliability.** 0-7E1

mean time between failures, predicted (reliability). The mean time between failures of a repairable complex item computed from its design considerations and from the failure rates of its parts for the intended conditions of use. *Note:* The method used for the calculation, the basis of the extrapolation, and the time and stress conditions should be stated. *See also:* **reliability.** 0-7E1

mean time to failure (nonrepaired items) (reliability). The total operating time of a number of items divided by the total number of failures. *See also:* **reliability.** 0-7E1

mean time to failure, assessed (reliability). The mean time to failure of a nonrepaired item determined within stated confidence limits from the observed mean time to failure of nominally identical items . *Note:* Alternatively, point estimates may be used, the basis of which must be defined. *See also:* **reliability.** 0-7E1

mean time to failure, extrapolated (reliability). Extension by a defined extrapolation or interpolation of the assessed mean time to failure for durations or stress conditions different from those applying to the conditions of that assessed mean time to failure. *See also:* **reliability.** 0-7E1

mean time to failure, observed (nonrepaired items) (truncated tests for a particular period) (reliability). The total cumulative observed time on a population divided by the total number of failures during the period and under stated stress conditions. *Notes:* (1) Cumulative observed time is a product of units and time, or the sum of these products. (2) The criteria for what constitutes a failure shall be stated. (3) This is the reciprocal of the failure rate observed during the period. *See also:* **reliability.** 0-7E1

mean time to failure, predicted (reliability). The mean time to failure of a nonrepaired complex item computed from its design considerations and from the failure rates of its parts for the intended conditions of use. *Note:* The method used for the calculation, the basis of the extrapolation, and the time and stress conditions should be stated. *See also:* **reliability.** 0-7E1

measurand. A physical or electrical quantity, property, or condition that is to be measured. *See:* **measurement system.** 37A100-31E11;42A30-0

measure. The number (real, complex, vector, etcetera) that expresses the ratio of the quantity to the unit used in measuring it. E270-0

measured current (insulation test). The total direct current resulting from the application of direct voltage to insulation and includes the leakage current, the absorption current, and theoretically the capacitive current. *Note:* The measured current is the value read on the microammeter during a direct high-voltage test of insulation. *See also:* **insulation testing (large alternating-current rotating machinery).** E95-0

measured service (telephony). Service in connection with which message use is measured in terms of messages or message units for purposes of charging for the service. *See also:* **telephone system.** 0-19E1

measured signal (automatic null-balancing electric instrument). The electrical quantity applied to the measuring-circuit terminals of the instrument. *Note:* It is the electrical analog of the measured variable. *See also:* **measurement system.** 39A4-0

measured switching time (ferroelectric device or material). The time required to reverse the sense of a specified major fraction of the signal charge or polarization. *Note:* Measured switching time is measured from the time of application of the applied voltage pulse or electric field, which must have a rise time much less than and a duration greater than measured switching time. The magnitude of the applied voltage or field, the sample thickness, and the fraction of the total change of charge used should be stated when measurements of measured switching time are reported. *See also:* **ferroelectric domain.** E180-0

measured variable (measurand) (automatic null-balancing electric instrument). The physical quantity, property, or condition that is to be measured. *Note:* Common measured variables are temperature, pressure, thickness, speed, etcetera. *See also:* **measurement system.** 39A4-0

measure equations. Equations in which the quantity symbols represent pure numbers, the measures of the physical quantities corresponding to the symbols. E270-0

measurement. The determination of the magnitude or amount of a quantity by comparison (direct or indirect) with the prototype standards of the system of units employed. *See:* **absolute measurement.** E270-0

measurement component. A general term applied to parts or subassemblies that are primarily used for the construction of measurement apparatus. *Note:* It is used to denote those parts made or selected specifically for measurement purposes and does not include standard screws, nuts, insulated wire, or other standard materials. *See:* **measurement system.** 42A30-0

measurement device. An assembly of one or more basic elements with other components and necessary parts to form a separate self-contained unit for performing one or more measurement operations. *Note:* It includes the protecting, supporting, and connecting, as well as the functioning, parts, all of which are necessary to fulfill the application requirements of the device. It should be noted that end devices (which see) are frequently but not always complete measurement devices in themselves, since they often are built-in with all or part of the intermediate means or primary detectors to form separate self-contained units. *See:* **measurement system.** 42A30-0

measurement energy. The energy required to operate a measurement device or system. *Note:* Measurement energy is normally obtained from the measurand or from the primary detector. *See:* **measurement system.** 42A30-0

measurement equipment. A general term applied to any assemblage of measurement components, devices, apparatus, or systems. *See:* **measurement system.** 42A30-0

measurement inverter (chopper). *See:* **measuring modulator.**

measurement mechanism. An assembly of basic elements and intermediate supporting parts for performing a mechanical operation in the sequence of measurement. *Note:* For example, it may be a group of components required to effect the proper motion of an indicating or recording means and does not include such parts as bases, covers, scales, and accessories. It may also be applied to a specific group of elements by substituting a suitable qualifying term; such as time-switch mechanism or chart-drive mechanism. *See:* **measurement system.** 42A30-0

measurement range (instrument). That part of the total range within which the requirements for accuracy are to be met. *See also:* **instrument.** 42A30-0

measurement system. One or more measurement devices and any other necessary system elements interconnected to perform a complete measurement from the first operation to the end result. *Notes:* (1) A measurement system can be divided into general functional groupings, each of which consists of one or more specific functional steps or basic elements. (2) For an extensive list of cross references, see *Appendix A.* 42A30-0

measurement voltage divider (voltage ratio box) (volt box*). A combination of two or more resistors, capacitors, or other circuit elements so arranged in series that the voltage across one of them is a definite and known fraction of the voltage applied to the combination, provided the current drain at the tap point is negligible or taken into account. *Note:* The term volt box is usually limited to resistance voltage dividers intended to extend the range of direct-current potentiometers. *See also:* **auxiliary device to an instrument.**
*Deprecated 42A30-0

measuring accuracy, precision, or reproducibility (numerically controlled machines). Accuracy, precision, or reproducibility of position sensor or transducer and interpreting system. *See also:* **numerically controlled machines.** EIA3B-34E12

measuring mechanism (recording instrument). The parts that produce and control the motion of the marking device. *See also:* **moving element (instrument).** 39A2-0

measuring modulator (measurement inverter) (chopper). An intermediate means in a measurement system by which a direct-current or low-frequency alternating-current input is modulated to give a quantitatively related alternating-current output usually as a preliminary to amplification. *Note:* The modulator may be of any suitable type such as mechanical, magnetic, or varistor. The mechanical types, which are actuated by vibrating or rotating members, may be classified as (1) contacting, (2) microphonic, or (3) generating (capacitive or inductive). *See also:* **auxiliary device to an instrument.** 42A30-0

mechanical back-to-back test (rotating machinery). A test in which two identical machines are mechanically coupled and the total losses of both machines are calculated from the electric input of one machine and the electric output of the other. *See:* **efficiency.** 0-31E8

mechanical bias. *See:* **relay mechanical bias.**

mechanical braking (industrial control). The kinetic energy of the drive motor and the driven machinery is dissipated by the friction of a mechanical brake. *See also:* **electric drive.** 42A25-34E10

mechanical current rating (neutral grounding device) (electric power). The symmetrical root-mean-square alternating-current component of the completely offset current wave that the device can withstand without mechanical failure. *Note:* The mechanical forces depend upon the maximum crest value of the current wave. However, for convenience, the mechanical current rating is expressed in terms of the root-mean-square value of the alternating-current component only of a completely displaced current wave. Specifically, the crest value of the completely offset wave will then be 2.82 times the mechanical current rating. *See also:* **grounding device.** E32/42A35-31E13

mechanical cutter (mechanical recorder). An equipment for transforming electric or acoustic signals into mechanical motion of approximately like form and inscribing such motion in an appropriate medium by cutting or embossing. 0-1E1

mechanical fatigue test (rotating machinery). A test designed to determine the effect of a specific repeated mechanical load on the life of a component. *See also:*

asynchronous machine; direct-current commutating machine; synchronous machine. 0-31E8

mechanical filter. *See:* **mechanical wave filter.**

mechanical-impact strength (insulator). The impact that, under specified conditions, the insulator can withstand without damage. *See also:* **insulator.** 29A1-0

mechanical impedance. The complex quotient of the effective alternating force applied to the system, by the resulting effective velocity. 0-1E1

mechanical interchangeability (fuse links). The characteristic that permits the designs of various manufacturers to be interchanged physically so they fit into and withstand the tensile stresses imposed by various types of prescribed cutouts made by different manufacturers. 37A100-31E11

mechanical interlocking machine. An interlocking machine designed to operate the units mechanically. *See also:* **interlocking (interlocking plant).** 42A42-0

mechanical latching relay. A relay in which the armature or contacts may be latched mechanically in the operated or unoperated position until reset manually or electrically. 0-21E0

mechanically delayed overcurrent trip. *See:* **mechanically delayed release (mechanically delayed trip); overcurrent release (overcurrent trip).**

mechanically delayed release (mechanically delayed trip). A release delayed by a mechanical device. 37A100-31E11

mechanically release-free (mechanically trip-free) (switching device). A term indicating that the release can open the device even though (1) in a manually operated switching device the operating lever is being moved toward the closed position; or (2) in a power-operated switching device, such as solenoid- or spring-actuated types, the operating mechanism is being moved toward the closed position either by continued application of closing power or by means of a maintenance closing lever. 37A100-31E11

mechanically reset relay. *See:* **hand-reset relay.**

mechanically timed relays. Relays that are timed mechanically by such features as clockwork, escapement, bellows, or dashpot. *See also:* **relay.** 83A16-0

mechanical modulator (electronic navigation). (1) A device that varies some characteristic of a carrier wave so as to transmit information, the variation being accomplished by physically moving or changing a circuit element. (2) In instrument landing systems, a particular arrangement of radio-frequency transmission lines and bridges with resonant sections coupled to the lines and motor-driven capacitor plates that alter the resonance so as to produce 90- and 150-hertz modulations. *See also:* **navigation.** 0-10E6

mechanical part (electric machine). Any part having no electric or magnetic function. *See:* **cradle base (rotating machinery).** 0-31E8

mechanical plating. Any plating operation in which the cathodes are moved mechanically during the deposition. *See also:* **electroplating.** 42A60-0

mechanical recorder. *See:* **mechanical cutter.**

mechanical rectifier. A rectifier in which rectification is accomplished by mechanical action. *See also:* **rectification.** 42A15-0

mechanical reproducer. *See:* **phonograph pickup.**

mechanical shock. A significant change in the position of a system in a nonperiodic manner in a relatively short time. *Note:* It is characterized by suddenness and large displacements and develops significant internal forces in the system. *See also:* **shock motion.** 0-1E1

mechanical short-time current rating (current transformer). The root-mean-square value of the alternating-current component of a completely displaced primary current wave that the transformer is capable of withstanding, with secondary short-circuited. *Note:* Capable of withstanding means that after a test the current transformer shows no visible sign of distortion and is capable of meeting the other specified applicable requirements. *See:* **instrument transformer.** 42A15/42A30-31E12

mechanical switching device. A switching device designed to close and open one or more electric circuits by means of guided separable contacts. *Note:* The medium in which the contacts separate may be designated by suitable prefix, that is, air, gas, oil, etcetera. 37A100-31E11

mechanical time constant (critically damped indicating instrument). The period of free oscillation divided by 2π. *See also:* **electromagnetic compatibility.** CISPR-27E1

mechanical transmission system. An assembly of elements adapted for the transmission of mechanical power. *See also:* **phonograph pickup.** E157-1E1

mechanical trip (trip arm) (railway practice). A roadway element consisting in part of a movable arm that in operative position engages apparatus on the vehicle to effect an application of the brakes by the train-control system. 42A42-0

mechanical wave filter (mechanical filter). A filter designed to separate mechanical waves of different frequencies. *Note:* Through electromechanical transducers, such a filter may be associated with electric circuits. *See also:* **filter.** 0-42A65

mechanical wrap or connection (soldered connections). The securing of a wire or lead prior to soldering. *See also:* **soldered connections (electronic and electrical applications).** 99A1-0

mechanism (1) (indicating instrument). The arrangement of parts for producing and controlling the motion of the indicating means. *Note:* It includes all the essential parts necessary to produce these results but does not include the base, cover, dial, or any parts, such as series resistors or shunts, whose function is to adapt the instrument to the quantity to be measured. *See also:* **instrument; moving element (instrument).** 39A1/42A30-0

(2) (recording instrument). Includes (A) the arrangement for producing and controlling the motion of the marking device; (B) the marking device; (C) the device (clockwork, constant-speed motor, or equivalent) for driving the chart; (D) the parts necessary to carry the chart. *Note:* It includes all the essential parts necessary to produce these results but does not include the base, cover, indicating scale, chart, or any parts, such as series resistors or shunts, whose function is to make the recorded value of the measured quantity correspond to the actual value. *See also:* **moving element (instrument).** 42A30-0

(3) (switching device). The complete assembly of levers and other parts that actuates the moving contacts of a switching device. 37A100-31E11

medical electronics. *Note:* For an extensive list of cross references, see *Appendix A.*

medium (computing systems). The material, or configuration thereof, on which data are recorded, for example, paper tape, cards, magnetic tape. X3A12-16E9

medium noise (sound recording and reproducing system). The noise that can be specifically ascribed to the medium. *See also:* **noise (sound recording and reproducing system).** E191-0

Meissner oscillator. An oscillator that includes an isolated tank circuit inductively coupled to the input and output circuits of an amplifying device to obtain the proper feedback and frequency. *See also:* **oscillatory circuit.** E145/42A65

mel. A unit of pitch. By definition, a simple tone of frequency 1000 hertz, 40 decibels above a listener's threshold, produces a pitch of 1000 mels. *Note:* The pitch of any sound that is judged by the listener to be *n* times that of the 1-mel tone is *n* mels. *See also:* **electroacoustics.** 0-1E1

melting channel. The restricted portion of the charge in a submerged resistor or horizontal-ring induction furnace in which the induced currents are concentrated to effect high energy absorption and melting of the charge. *See also:* **induction heating.** E54/E169-0

melting-speed ratio (fuse). A ratio of the current magnitudes required to melt the current-responsive element at two specified melting times. *Note:* Specification of the current wave shape is required for time less than one-tenth of a second. 37A100-31E11

melting time (fuse). The time required for overcurrent to sever the current-responsive element. 37A100-31E11

membrane potential. The potential difference, of whatever origin, between the two sides of a membrane. *See also:* **electrobiology.** 42A80-18E1

memory (electronic computation). *See:* **storage; storage medium.**

memory action (relay). A method of retaining an effect of an input after the input ceases or is greatly reduced, so that this input can still be used in producing the typical response of the relay. *Note:* For example, memory action in a high-speed directional relay permits correct response for a brief period after the source of voltage input necessary to such response is short-circuited. 37A100-31E11/31E6

memory capacity (electronic computation). *See:* **storage capacity.**

memory relay. (1) A relay having two or more coils, each of which may operate independent sets of contacts, and another set of contacts that remain in a position determined by the coil last energized. (2) Sometimes erroneously used for polarized relay. *See also:* **relay.** 83A16-0

mercury-arc converter, pool cathode. *See:* **pool cathode mercury-arc converter.** *See also:* **oscillatory circuit.**

mercury-arc rectifier. A gas-filled rectifier tube in which the gas is mercury vapor. *See also:* **gas-filled rectifier; rectification.** 50I07-15E6

mercury cells. Electrolytic cells having mercury cathodes with which deposited metals form amalgams. 42A60-0

mercury-contact relays. **(1) mercury plunger relay:** A relay in which the magnetic attraction of a floating plunger by a field surrounding a sealed capsule displaces mercury in a pool to effect contacting between fixed electrodes. **(2) mercury-wetted-contact relay:** A form of reed relay in which the reeds and contacts are glass enclosed and are wetted by a film of mercury obtained by capillary action from a mercury pool in the base of a glass capsule vertically mounted. **(3) mercury-contact relay:** A relay mechanism in which mercury establishes contact between electrodes in a sealed capsule as a result of the capsule's being tilted by an electromagnetically actuated armature, either on pick-up or dropout or both. *See also:* **mercury relay.** 0-21E0

mercury-hydrogen spark-gap converter (dielectric heater usage). A spark-gap generator or power source that utilizes the oscillatory discharge of a capacitor through an inductor and a spark gap as a source of radio-frequency power. The spark gap comprises a solid electrode and a pool of mercury in a hydrogen atmosphere. *See also:* **induction heating.** E54/E169-0

mercury-lamp ballast. *See:* **ballast.**

mercury-lamp transformer (series type). *See:* **constant-current (series) mercury-lamp transformer.**

mercury motor meter. A motor-type meter in which a portion of the rotor is immersed in mercury, which serves to direct the current through conducting portions of the rotor. *See also:* **electricity meter.** 42A30-0

mercury oxide cell. A primary cell in which depolarization is accomplished by oxide of mercury. *See also:* **electrochemistry.** 42A60-0

mercury-pool cathode (gas tube). A pool cathode consisting of mercury. *See also:* **gas-filled rectifier.** 50I07-15E6

mercury relay. A relay in which the movement of mercury opens and closes contacts. *See also:* **mercury-contact relay.** 82A16-21E0

mercury-vapor tube. A gas tube in which the active gas is mercury vapor. 42A70-15E6

merge (computing systems). To combine two or more sets of items into one, usually in a specified sequence. *See also:* **electronic digital computer.** X3A12-16E9

mesh. A set of branches forming a closed path in a network, provided that if any one branch is omitted from the set, the remaining branches of the set do not form a closed path. *Note:* The term **loop** is sometimes used in the sense of mesh. *See also:* **network analysis.** E153/E270-0

mesh-connected circuit. A polyphase circuit in which all the current paths of the circuit extend directly from the terminal of entry of one phase conductor to the terminal of entry of another phase conductor, without any intermediate interconnections among such paths and without any connection to the neutral conductor, if one exists. *Note:* In a three-phase system this is called the delta (or Δ) connection. *See also:* **network analysis.** E270-0

mesh current. A current assumed to exist over all cross sections of a given closed path in a network. *Note:* A mesh current may be the total current in a branch included in the path, or it may be a partial current such that when combined with others the total current is obtained. *See also:* **network analysis.** E270-0

mesh or loop equations. Any set of equations (of minimum number) such that the independent mesh or loop currents of a specified network may be determined from the impressed voltages. *Notes:* (1) For a given network, different sets of equations, equivalent to one another, may be obtained by different choices of mesh or loop currents. (2) The equations may be differential equations, or algebraic equations when impedances and phasor equivalents of steady-state single-frequency sine-wave quantities are used. *See also:* **network analysis.** E270-0

mesopic vision. Vision with luminance conditions between those of photopic and scotopic vision, that is, between about 1.0 footlambert (3.0 nits) and 0.01 footlambert (0.03 nit). *See also:* **visual field.** Z7A1-0

message. An arbitrary amount of information whose beginning and end are defined or implied. X3A12-16E9/19E4

message source. That part of a communication system where messages are assumed to originate. *See also:* **information theory.** E171-0

metal clad. The conducting parts are entirely enclosed in a metal casing. 42A95-0

metal-clad switchgear. Metal-enclosed power switchgear characterized by the following necessary features. (1) The main switching and interrupting device is of the removable type arranged with a mechanism for moving it physically between connected and disconnected positions and equipped with self-aligning and self-coupling primary and secondary disconnecting devices. (2) Major parts of the primary circuit, such as the circuit switching or interrupting devices, buses, potential transformers, and control power transformers are enclosed by grounded metal barriers. *Note:* Specifically included is an inner barrier in front of or a part of the circuit-interrupting device to insure that no primary-circuit components are exposed when the unit door is opened. (3) All live parts are completely enclosed within grounded metal compartments. *Note:* Automatic shutters prevent exposures of primary-circuit elements when the removable element is in the disconnected, test, or removed position. (4) Primary bus conductors and connections are covered with insulating material throughout. (5) Mechanical interlocks are provided to insure a proper and safe operating sequence. (6) Instruments, meters, relays, secondary control devices, and their wiring are isolated by grounded metal barriers from all primary-circuit elements with the exception of short lengths of wire such as at instrument-transformer terminals. (7) The door through which the circuit-interrupting device is inserted into the housing may serve as an instrument or relay panel and may also provide access to a secondary or control compartment within the housing. *Note:* Auxiliary units may be required for mounting devices or for use as bus transition. 37A100-31E11

metal distribution ratio (electroplating). The ratio of the thicknesses (weights per unit areas) of metal upon two specified parts of a cathode. *See also:* **electroplating.** 42A60-0

metal enclosed (switchgear assembly or components). Surrounded by a metal case or housing, usually grounded. 37A100-31E11

metal-enclosed bus. An assembly of rigid conductors with associated connections, joints, and insulating supports within a grounded metal enclosure. *Note:* In general, three basic types of construction are used: nonsegregated-phase, segregated-phase, and isolated-phase. 37A100-31E11

metal-enclosed interrupter switchgear. Metal-enclosed power switchgear including the following equipment as required: (1) interrupter switches; (2) power fuses; (3) bare bus and connections; (4) instrument transformers; and (5) control wiring and accessory devices. *Note:* The interrupter switches and power fuses may be stationary or removable type. When removable type, mechanical interlocks are provided to insure a proper and safe operating sequence. 37A100-31E11

metal-enclosed low-voltage power circuit-breaker switchgear. Metal-enclosed power switchgear including the following equipment as required: (1) low-voltage power circuit breaker (fused or unfused); (2) bare bus and connections; (3) instrument and control power transformers; (4) instruments, meters, and relays; and (5) control wiring and accessory devices. *Note:* The low-voltage power circuit breakers are contained in individual grounded metal compartments and controlled either remotely or from the front of the panels. The circuit breakers may be stationary or removable type . When removable type, mechanical interlocks are provided to insure a proper and safe operating sequence. 37A100-31E11

metal-enclosed power switchgear. A switchgear assembly completely enclosed on all sides and top with sheet metal (except for ventilating openings and inspection windows) containing primary power circuit switching or interrupting devices, or both, with buses and connections and may include control and auxiliary devices. *Note:* Access to the enclosure is provided by doors or removable covers. *Note:* Metal-clad switchgear, station-type cubicle switchgear, metal-enclosed interrupter switchgear, and low-voltage power circuit-breaker switchgear are specific types of metal-enclosed power switchgear. 37A100-31E11

metal fog (metal mist) (electrolysis). A fine dispersion of metal in a fused electrolyte. *See also:* **fused electrolyte.** 42A60-0

metal-graphite brush (rotating machinery). A brush composed of varying percentages of metal and graphite, copper being the metal generally used. *Note:* This type of brush is soft. Grades of brushes of this type have extremely high current-carrying capacities, but differ greatly in operating speed from low to high. *See also:* **brush (rotating machinery).** 42A10-31E8;64A1-0;0-31E8

metallic circuit. A circuit of which the ground or earth forms no part. *See also:* **transmission line.** 42A65-31E3

metallic longitudinal induction ratio. (M-L ratio). The ratio of the metallic-circuit current or noise-metallic arising in an exposed section of open wire telephone line, to the longitudinal-circuit current or noise-longitudinal in sigma. It is expressed in microamperes per milliampere or the equivalent. *See:* **noise-metallic.** *See also:* **inductive coordination.** 42A65-0

metallic rectifier. An integral assembly of one or more metallic rectifying cells. *See also:* **rectification.** 42A15-0

metallic rectifier cell. A device consisting of a conductor and a semiconductor forming a junction. *Notes:* (1) Synonymous with **metallic rectifying cells.**

(2) Such cells conduct current in each direction but provide a rectifying action because of the large difference in resistance to current flow in the two directions. (3) A metallic rectifier stack is a single columnar structure of one or more metallic rectifier cells. *See also:* **rectification.** E45/42A15-0

metallic rectifier stack assembly. The combination of one or more stacks consisting of all the rectifying elements used in one rectifying circuit. *See also:* **rectification.** 42A15-0

metallic rectifier unit. An operative assembly of a metallic rectifier, or rectifiers, together with the rectifier auxiliaries, the rectifier transformers, and the essential switchgear. *See also:* **rectification.** 42A15-0

metallized brush. *See:* **metal-graphite brush.**

metallized paper capacitor. A capacitor in which the dielectric is primarily paper and the electrodes are thin metallic coatings deposited thereon. 0-21E0

metallized screen (cathode-ray tube). A screen covered on its rear side (with respect to the electron gun) with metallic film, usually aluminized, transparent to electrons and with a high optical reflection factor, which passes on to the viewer a large part of the light emitted by the screen on the electron-gun side. *See also:* **cathode-ray tubes.** 50I07-15E6

metal master (metal negative) (no. 1 master) (disk recording) (electroacoustics). *See:* **original master.**

metal mist (electrolysis). *See:* **metal fog.**

metal negative (metal master) (no. 1 master) (disk recording) (electroacoustics). *See:* **original master.**

metamers. Lights of the same color but of different spectral energy distribution. *See also:* **color.** Z7A1-0

meter. The length equal to 1,650,763.73 wavelengths in vacuum of the radiation corresponding to the transition between the levels $2p_{10}$ and $5d_5$ of the krypton-86 atom. CGPM-SCC14

meter. *See:* **electricity meter.**

meter laboratory. A place where working standards are maintained for use in routine testing of electricity meters and auxiliary devices. *Note:* A meter laboratory may, but need not, assume some or all the functions of a standards laboratory. *See also:* **measurement system.** 12A0-0

meter relay. Sometimes used for instrument relay. *See also:* **relay.** 83A16-0

meter shop. A place where meters are inspected, tested, and repaired, and may be at a fixed location or may be mobile. *See also:* **measurement system.** 12A0-0

MEW. *See:* **microwave early warning.**

mho (siemens). The unit of conductance (and of admittance) in the International System of Units (SI). The mho is the conductance of a conductor such that a constant voltage of 1 volt between its ends produces a current of 1 ampere in it. E270-0

mho relay. A distance relay for which the inherent operating characteristic on an R-X diagram is a circle that passes through the origin. *Note:* The operating characteristic may be described by the equation $Z = K\cos(\theta - \alpha)$ where K and α are constants and θ is the phase angle by which the input voltage leads the input current. 37A100-31E11/31E6

MIC (electromagnetic compatibility). *See:* **mutual interference chart.**

mica flake (rotating machinery). Mica lamina in thickness not over approximately 0.0028 centimeter having a surface area parallel to the cleavage plane under 1.0 centimeter square. *See:* **rotor (rotating machinery); stator.** 0-31E8

mica folium (rotating machinery). A relatively thin flexible bonded sheet material composed of overlapping mica splittings with or without backing or facing. *See:* **rotor (rotating machinery); stator.** 0-31E8

mica paper (integrated mica) (reconstituted mica) (rotating machinery). Mica flakes having an area under approximately 0.200 centimeter square combined laminarly into a substantial sheet-like configuration with or without binder, backing, or facing. *See:* **rotor (rotating machinery); stator.** 0-31E8

mica sheet (rotating machinery). A composite of overlapping mica splittings bonded into a planar structure with or without backing or facing. *See:* **rotor (rotating machinery); stator.** 0-31E8

mica splitting (rotating machinery). Mica lamina in thickness approximately 0.0015 centimeter to 0.0028 centimeter having a surface area parallel to the basal cleavage plane of at least 1.0 centimeter square. *See:* **rotor (rotating machinery); stator.** 0-31E8

mica tape (rotating machinery). A composite tape composed of overlapping mica splittings bonded together with or without backing or facing. *See:* **rotor (rotating machinery); stator.** 0-31E8

MICR. *See:* **magnetic ink character recognition.**

microbar. A unit of pressure formerly in common usage in acoustics. One microbar is equal to 1 dyne per square centimeter and equals 0.1 newton per square meter. The newton per square meter is now the preferred unit. *Note:* The term bar properly denotes a pressure of 10^6 dynes per square centimeter. Unfortunately the bar was once used in acoustics to mean 1 dyne per square centimeter, but this is no longer correct. *See also:* **electroacoustics.** 0-1E1

microchannel plate (electron image tube). An array of small aligned channel multipliers usually used for intensification. *See also:* **amplifier; camera tube; circuit characteristics of electrodes.** 0-15E6

micron (metric system). The millionth part of a meter. *Note:* According to the set of submultiple prefixes now established in the International System of Units, the preferred term would be micrometer. However, use of the same word to denote a small length, and also to denote an instrument for measuring a small length, could occasionally invite confusion. Therefore it seems unwise to deprecate, at this time, the continued use of the word micron. 0-E

microphone. An electroacoustic transducer that responds to sound waves and delivers essentially equivalent electric waves.
See:
antinoise microphone;
available power efficiency;
carbon microphone;
close-talking microphone;
combination microphone;
crystal microphone;
directional microphone;
directivity factor;
electronic microphone;
electrostatic actuator;
electrostatic microphone;
free-field voltage response;
free impedance;
gradient microphone;

hot-wire microphone;
hydrophone;
inductor microphone;
lapel microphone;
line microphone;
lip microphone;
magnetostriction microphone;
moving-coil microphone;
moving-conductor microphone;
omnidirectional microphone;
pickup;
pressure microphone;
push-pull microphone;
ribbon microphone;
split hydrophone;
standard microphone;
thermophone;
throat microphone;
unidirectional microphone;
variable-reluctance microphone;
velocity microphone.
See also: **electroacoustics; telephone transmitter.** E157-1E1;E188/42A65-0

microphonics (1) (general). The noise caused by mechanical shock or vibration of elements in a system. E151-42A65
(2) **(interference terminology).** Electrical interference caused by mechanical vibration of elements in a signal transmission system. *See:* **signal.** 0-13E6
(3) **(microphonic effect) (microphonism) (electron device) (electron tubes).** The undesired modulation of one or more of the electrode currents resulting from the mechanical vibration of one or more of the valve or tube elements. *See also:* **electron tube.** 50I07-15E6

microphonism. *See:* **microphonics (electron tubes).**

microradiometer (radio-micrometer*). A thermosensitive detector of radiant power in which a thermopile is supported on and connected directly to the moving coil of a galvanometer. *Note:* This construction minimizes lead losses and stray electric pickup. *See also:* **electric thermometer (temperature meter).**
*Deprecated 42A30-0

microstrip. *See:* **strip (-type) transmission time.**

microwave early warning (MEW). A particular (World War 2) United States high-power, long-distance radar with a number of separate displays giving high resolution and large traffic-handling capacity in detecting and tracking targets. *See also:* **navigation.** 0-10E6

microwave pilot protection. A form of pilot protection in which the communication means between relays is a beamed microwave-radio channel. 37A100-31E11/31E6

microwaves. A term used rather loosely to signify radio waves in the frequency range from about 1000 megahertz upwards. *See also:* **radio transmission.** 42A65-0

microwave therapy. The therapeutic use of electromagnetic energy to generate heat within the body, the frequency being greater than 100 megahertz. *See also:* **electrotherapy.** 42A80-18E1

microwave tube (or valve). An electron tube or valve whose size is comparable to the wavelength of operation and that has been designed in conjunction with its circuits to provide effective electron interaction. *Notes:* (1) In many tubes the circuit is self-contained within the vacuum envelope so that the electrode terminals do not need to conduct radio-frequency signals. (2) For an extensive list of cross references, see *Appendix A.*

middle marker (electronic navigation). A marker facility in an instrument landing system that is installed approximately 3500 feet from the approach end of the runway on the localizer course line to provide a fix. *See also:* **navigation.** E172-10E6

MIG. *See:* **magnetic injection gun.**

mile of standard cable (MSC).** Two units, both loosely designated as a mile of standard cable, were formerly used as measures of transmission efficiency. One, correctly known as an 800-hertz mile of standard cable, signified an attenuation constant, independent of frequency, of 0.109. The other signified the effect upon speech volume of an actual mile of standard cable, equivalent to an attenuation constant of approximately 0.122. Both units are now obsolete, having been replaced by the decibel. One 800-hertz mile of standard cable is equal to approximately 0.95 decibel. One standard cable mile is equivalent in effect on speech volume to approximately 1.06 decibels. *See also:* **transmission characteristics.**
**Obsolete 42A65-0

mill scale (corrosion). The heavy oxide layer formed during hot fabrication or heat treatment of metals. Especially applied to iron and steel. *See:* **corrosion terms.** CM-34E2

mimic bus. A single-line diagram of the main connections of a system constructed on the face of a switchgear or control panel, or assembly. 37A100-31E11

mine-fan signal system. A system that indicates by electric light or electric audible signal, or both, the slowing down or stopping of a mine ventilating fan. *See also:* **dispatching system.** 42A85-0

mine feeder circuit. A conductor or group of conductors, including feeder and sectionalizing switches or circuit breakers, installed in mine entries or gangways and extending to the limits set for permanent mine wiring beyond which limits portable cables are used. *See:*
borehole cable;
cable reel;
cable splicer;
charging rack;
color code;
connecting wire;
connection box;
distribution box;
flame-resistant cable;
fused trolley tap;
gangway cable;
grounding conductor;
hand cable;
junction box;
packing gland;
portable concentric mine cable;
portable mine cable;
portable parallel duplex mine cable;
rail clamp;
splice box.
See also: **mining.** 42A85-0

mine hoist. A device for raising or lowering ore, rock, or coal from a mine and for lowering and raising men and supplies.
See:
hoist back-out switch;
hoist overspeed device;
hoist overwind device;
hoist signal code;
hoist signal system;
hoist slack-brake switch;
hoist trip recorder.
See also: **mining.** 42A85-0

mine jeep. A special electrically driven car for underground transportation of officials, inspectors, repair, maintenance, surveying crews, and rescue workers. *See also:* **mining.** 42A85-0

mine radio telephone system. A means to provide communication between the dispatcher and the operators on the locomotives where the radio impulses pass along the trolley wire and down the trolley pole to the radio telephone set. *See also:* **dispatching system.** 42A85-0

miner's electric cap lamp. A lamp for mounting on the miner's cap and receiving electric energy through a cord that connects the lamp with a small battery. 42A85-0

miner's hand lamp. A self-contained mine lamp with handle for convenience in carrying. 42A85-0

mine tractor. A trackless, self-propelled vehicle used to transport equipment and supplies and for general service work. *See also:* **mining.** 42A85-0

mine ventilating fan. A motor-driven disk, propeller, or wheel for blowing (or exhausting) air to provide ventilation of a mine. *See also:* **mining.** 42A85-0

miniature brush (electric machines). A brush having a cross-sectional area of less than 1/64 square inch with the thickness and width thereof less than 1/8 inch or, in the case of a cylindrical brush, a diameter less than 1/8 inch. *See also:* **brush.** 64A1-0

minimal perceptible erythema. The erythemal threshold. *See also:* **ultraviolet radiation.** Z7A1-0

minimum clearance between poles (phases) (switchgear). The shortest distance between any live parts of adjacent poles (phases). *Note:* Cautionary differentiation should be made between clearance and spacing or center-to-center distance. 37A100-31E11

minimum clearance to ground (switchgear). The shortest distance between any live part and adjacent grounded parts. 37A100-31E11

minimum conductance function (linear passive networks). *See:* **minimum resistance (conductance) function.** *See also:* **linear passive networks.**

minimum-distance code (computing systems). A binary code in which the signal distance does not fall below a specified minimum value. *See also:* **electronic digital computer.** X3A12-16E9

minimum-driving-point function (linear passive networks). A driving-point function that is a minimum-resistance, minimum-conductance, minimum-reactance, and minimum-susceptance function. *See also:* **linear passive networks.** E156-0

minimum en-route altitude (MEA) (electronic navigation). The lowest altitude between radio fixes that assures acceptable navigational signal coverage and meets obstruction clearance requirements for instrument flight. *See also:* **navigation.** 0-10E6

minimum firing power (microwave switching tubes). The minimum radio-frequency power required to initiate a radio-frequency discharge in the tube at a specified ignitor current. *See:* **gas tube.** E160-15E6

minimum flashover voltage (impulse). The crest value of the lowest voltage impulse, of a given wave shape and polarity that causes flashover. *See also:* **lightning arrester (surge diverter); power system, low-frequency and surge testing.** 42A35-31E13/31E7

minimum fuel limiter (gas turbines). A device by means of which the speed-governing system can be prevented from reducing the fuel flow below the minimum for which the device is set as required to prevent unstable combustion or blowout of the flame. E282-31E2

minimum illumination (sensitivity) (industrial control). The minimum level, in footcandles of a photoelectric lighting control, at which it will operate. *See also:* **photoelectric control.** 42A25-34E10

minimum ON-state voltage (thyristor). The minimum positive principal voltage for which the differential resistance is zero with the gate open-circuited. *See also:* **principal voltage-current characteristic (principal characteristic).** E223-34E17/34E24/15E7

minimum ON voltage (magnetic amplifier). The minimum output voltage existing before the trip OFF control signal is reached as the control signal is varied from trip ON to trip OFF. *See also:* **rating and testing magnetic amplifiers.** E107-0

minimum output voltage (magnetic amplifier). The minimum voltage attained across the rated load impedance as the control ampere-turns are varied between the limits established by (1) positive maximum control currents flowing through all the corresponding control windings simultaneously and (2) negative maximum control currents flowing through all the corresponding control windings simultaneously. *See also:* **rating and testing magnetic amplifiers.** E107-0

minimum-phase function (linear passive networks). A transmittance from which a nontrivial realizable all-pass function cannot be factored without leaving a non-realizable remainder. *Note;* For lumped-parameter networks, this is equivalent to specifying that the function has no zeros in the interior of the right half of the complex-frequency plane. *See also:* **linear passive networks.** E156-0

minimum-phase network (1). A network for which the phase shift at each frequency equals the minimum value that is determined uniquely by the attenuation-frequency characteristic in accordance with the following equation

$$B_{\mathrm{c}} = \frac{1}{\pi} \int_{-\infty}^{+\infty} \frac{\mathrm{d}A}{\mathrm{d}u} \log \coth \frac{|u|}{2} \,\mathrm{d}u$$

where B_c is phase shift in radians at a particular frequency f_c, A is attenuation in nepers as a function of frequency f, and u is log (f/f_c).

(2) (two specified terminals or two branches). A network for which the transfer admittance expressed as a function of p (*See:* **impedance function**) has neither poles nor zeros in the right-hand p plane. *Note:* A simple T section of real lumped constant parameters without coupling between branches is a minimum-phase network, whereas a bridged-T or lattice section of all-pass type may not be. *See also:* **network analysis; impedance function.** 42A65/E270-0

minimum-reactance function (linear passive networks). A driving-point impedance from which a reactance function cannot be subtracted without leaving a nonrealizable remainder. *Notes:* (1) For lumped-parameter networks, this is equivalent to specifying that the impedance function has no poles on the imaginary axis of the complex-frequency plane, including the point at infinity. (2) A driving-point impedance (admittance) having neither poles nor zeros on the imaginary axis is both a minimum-reactance and a minimum-susceptance function. *See also:* **linear passive networks.** E156-0

minimum reception altitude (MRA) (electronic navigation). The lowest en-route altitude at which adequate signals can be received to determine specific radio-navigation fixes. *See also:* **navigation.** 0-10E6

minimum-resistance (conductance) function (linear passive networks). A driving-point impedance (admittance) from which a positive constant cannot be subtracted without leaving a nonrealizable remainder. *See also:* **linear passive networks.** E156-0

minimum speed (adjustable-speed drive) (industrial control). The lowest speed within the operating speed range of the drive. *See also:* **electric drive.** 42A25-34E10

minimum-susceptance function (linear passive networks). A driving-point admittance from which a susceptance function cannot be subtracted without leaving a nonrealizable remainder. *Notes:* (1) For lumped-parameter networks, this is equivalent to specifying that the admittance function has no poles on the imaginary axis of the complex-frequency plane, including the point at infinity. (2) A driving-point immitance having neither poles nor zeros on the imaginary axis is both a minimum-susceptance and a minimum-reactance function. *See also:* **linear passive networks.** E156-0

minimum test output voltage (nonreversible output) (magnetic amplifier). The output voltage equivalent to the summation of the minimum output voltage plus 33 1/3 percent of the difference between the rated and minimum output voltages. *See also:* **rating and testing magnetic amplifiers.** E107-0

mining. *Note:* For an extensive list of cross references, see *Appendix A.*

minor cycle (electronic computation). In a storage device that provides serial access to storage positions, the time interval between the appearance of corresponding parts of successive words. *See also:* **electronic computation; electronic digital computer.** E162/E270-0

minority carrier (semiconductor). The type of charge carrier constituting less than one half the total charge-carrier concentration. *See also:* **semiconductor.** E216-34E17;E270-0

minor lobe (antenna pattern). Any lobe except a major lobe. *See also:* **antenna.** E149-3E1

minor loop (industrial control). A continuous network consisting of both forward elements and feedback elements and is only a portion of the feedback control system. *See:* **control system, feedback.** AS1-34E10

minor railway tracks. Railway tracks included in the following list. (1) Spurs less than 2000 feet long and not exceeding two tracks in the same span. (2) Branches on which no regular service is maintained or which are not operated during the winter season. (3) Narrow-gauge tracks or other tracks on which standard rolling stock cannot, for physical reasons, be operated. (4) Tracks used only temporarily for a period not exceeding 1 year. (5) Tracks not operated as a common carrier, such as industrial railways used in logging, mining, etcetera. 2A2-0

misalignment drift (gyros). The part of the total apparent drift component due to uncertainty of orientation of the gyro input axis with respect to the coordinate system in which the gyro is being used. *See also:* **navigation.** 0-10E6

miscellaneous function (numerically controlled machines). An on-off function of a machine such as spindle stop, coolant on, clamp. *See also:* **numerically controlled machines.** EIA3B-34E12

misfire (1) (gas tube). A failure to establish an arc between the main anode and cathode during a scheduled conducting period. *See also:* **gas tubes; rectification.** 42A70-15E6;42A15-0

(2) (mining). The failure of a blasting charge to explode when expected. *Note:* In electric firing, this usually is the result of a broken blasting circuit or insufficient current through the electric blasting cap. *See also:* **blasting unit.** 42A85-0

mismatch. The condition in which the impedance of a load does not match the impedance of the source to which it is connected. *See also:* **self-impedance; transmission characteristics.** 42A65-3E1

mismatch factor. *See:* **reflection factor.**

mission. The operating objective for which the system was intended. *See also:* **system.** 0-35E2

mistake (electronic computation). A human action that produces an unintended result. *Note:* Common mistakes include incorrect programming, coding, manual operation, etcetera. *See also:* **electronic analog computer; electronic digital computer; error; malfunction.** X3A12-16E9;E162-0

mixed highs (color television). Those high-frequency components of the picture signal that are intended to be reproduced achromatically in a color picture. *See also:* **color terms.** E201-2E2

mixed-loop series street-lighting system. A street-lighting system that comprises both open loops and closed loops. *See also:* **alternate-current distribution; direct-current distribution.** 42A35-31E13

mixed radix. Pertaining to a numeration system that uses more than one radix, such as the biquinary system. X3A12-16E9

mixed sweep (oscilloscopes). In a system having both a delaying sweep and a delayed sweep, a means of displaying the delaying sweep to the point of delay pickoff and displaying the delayed sweep beyond that point. *See:* **oscillograph.** 0-9E4

mixer (1) (general). A circuit having two or more input signals and an output signal that is some desired function of the input. *See also:* **circuits and devices.** 0-31E3

(2) (sound transmission, recording, or reproducing system.) A device having two or more inputs, usually adjustable, and a common output, that operates to combine linearly in a desired proportion the separate input signals to produce an output signal. *Note:* The term is sometimes applied to the operator of the above device. *See also:* **electroacoustics.** E151-0;E157-1E1

mixer tube. An electron tube that performs only the frequency-conversion function of a heterodyne conversion transducer when it is supplied with voltage or power from an external oscillator. *See also:* **tube definitions.** 42A70-15E6

MKSA system of units. A system in which the basic units are the meter, kilogram, and second, and the ampere is a derived unit defined by assigning the magnitude $4\pi \times 10^{-7}$ to the rationalized magnetic constant (sometimes called the permeability of space). *Notes:* (1) At its meeting in 1950 the International Electrotechnical Commission recommended that the MKSA system be used only in the rationalized form. (2) The electrical units of this system were formerly called the *practical* electrical units. (3) If the MKSA system is used in the unrationalized form the magnetic constant is 10^{-7} henry/meter and the electric constant is $10^{7}/c^{2}$ farads/meter. Here *c,* the speed of light, is approximately 3×10^{8} meters/second. (4) In this system, dimensional analysis is customarily used with the four independent (basic) dimensions: mass, length, time, current. E270-0

M-L ratio. *See:* **metallic-longitudinal induction ratio.**

mobile communication system. Combinations of interrelated devices capable of transmitting intelligence between two or more spatially separated radio stations, one or more of which shall be mobile. *Note:* For an extensive list of cross references, see *Appendix A.* **0-6E1**

mobile radio service. Radio service between a radio station at a fixed location and one or more mobile stations, or between mobile stations. *See also:* **radio transmission.** 42A65-0

mobile station (communication). A radio station designed for installation in a vehicle and normally operated when in motion. *See also:* **mobile communication system.** 0-6E1

mobile telemetering. Electric telemetering between points that may have relative motion, where the use of interconnecting wires is precluded. *Note:* Space radio is usually employed as an intermediate means for mobile telemetering, but radio may also be used for telemetering between fixed points. *See also:* **telemetering.** 42A30-0

mobile telephone system (automatic channel access). A mobile telephone system capable of operation on a plurality of frequency channels with automatic selection at either the base station or any mobile station of an idle channel when communication is desired. *See also:* **mobile communication system.** 0-6E1

mobile transmitter. A radio transmitter designed for installation in a vessel, vehicle, or aircraft, and normally operated while in motion. *See also:* **radio transmitter.** E145-0

mobile unit substation. A unit substation mounted and readily movable as a unit on a transportable device. 37A100-31E11

mobility (semiconductor). *See:* **drift mobility.**

mobility, Hall (electric conductor). *See:* **Hall mobility.**

modal analysis (power-system communication). A method of computing the propagation of a wave on a multiconductor power line. *See also:* **power distribution, overhead construction.** 0-31E3

mode. A state of a vibrating system to which corresponds one of the possible resonance frequencies (or propagation constants). *Note:* Not all dissipative systems have modes. *See:* **modes, degenerate; oscillatory circuit.** E160-15E6

mode coupler (waveguides). A coupler that provides preferential coupling to a sepcific wave mode. *See also:* **waveguide.** 0-9E4

mode filter (waveguides). A selective device designed to pass energy along a waveguide in one or more modes of propagation and substantially reduce energy carried by other modes. *See also:* **waveguide.** E147-3E1

mode, higher-order (waveguide or transmission line). Any mode of propagation characterized by a field configuration other than that of the fundamental or first-order mode with lowest cutoff frequency. 0-9E4

model. A mathematical or physical representation of the system relationships. *See also:* **mathematical model; system.** 0-35E2

modeling. Technique of system analysis and design using mathematical or physical idealizations of all or a portion of the system. Completeness and reality of the model are dependent on the questions to be answered, the state of knowledge of the system, and its environment. *See also:* **system.** 0-35E2

modem. A contraction of modulator-demodulator, an equipment that connects data terminal equipment to a communication line. 0-19E4

mode of operation (rectifier circuit). The characteristic pattern of operation determined by the sequence and duration of commutation and conduction. *Note:* Most rectifier circuits have several modes of operation, which may be identified by the shape of the current waves. The particular mode obtained at a given load depends upon the circuit constants. *See also:* **rectification; rectifier circuit element.** E59-34E17;42A15-0

mode of propagation (waveguides). A form of propagation of guided waves that is characterized by a particular field pattern in a plane transverse to the direction of propagation, which field pattern is independent of position along the axis of the waveguide. *Note:* In the case of uniconductor waveguides the field pattern of a particular mode of propagation is also independent of frequency. *See also:* **waveguide.** E146/E148-3E1

mode of resonance. A form of natural electromagnetic oscillation in a resonator characterized by a particular field pattern that is invariant with time. E146-3E1

mode of vibration (vibratory body, such as a piezoelectric crystal unit). A pattern of motion of the individual particles due to (1) stresses applied to the body; (2) its properties; and (3) the boundary conditions. Three common modes of vibration are (A) flexural, (B) extensional, and (C) shear. *See also:* **crystal.** 42A65-0

mode, π. *See:* **pi (π) mode.**

mode purity (waveguides). The ratio of power present in the forward traveling wave of a desired mode to the total power present in the forward traveling waves of all modes. E148-3E1

modes, degenerate. A set of modes having the same resonance frequency (or propagation constant). *Note:* The members of a set of degenerate modes are not unique. *See:* **electrode dissipation; oscillatory circuits.** E160-15E6

mode separation (oscillators). The frequency difference between resonator modes of oscillation. *See:* **oscillatory circuits.** E160-15E6

mode transducer (waveguides). A device for changing from one mode of guided propagation to another.

Note: Also termed **mode changer** or **mode transformer.** ***See also:*** **transducer; waveguide.**
50I62-3E1;E146/E147/E196/E270/42A65-0

mode transformer (transmission lines and waveguides). *See:* **mode transducer (waveguides).**

mode voltage. That voltage that occurs most frequently. It is of primary significance in evaluating certain critical factors affecting the design of equipment. *See also:* **power system, low-frequency and surge testing.** 84A1-0

modified autocorrelation function (burst measurements). The time integral of a function $x(t)$ multiplied by itself shifted in time by τ

$$\Gamma_s(\tau) = \int_{-\infty}^{\infty} x(t) \cdot x(t+\tau)dt$$

See also: **burst.** E265-0

modified circuit transient recovery voltage. The circuit transient recovery voltage modified in accordance with the normal-frequency recovery voltage and the asymmetry of the current wave obtained on a particular interruption. *Note:* This voltage indicates the severity of the particular interruption with respect to recovery-voltage phenomena. 37A100-31E11

modified constant-voltage charge (storage battery) (storage cell). A charge in which the bus voltage of the charging circuit is held substantially constant, but a fixed resistance is present in the battery circuit that results in an increase in voltage at the battery terminals as the charge progresses. *See also:* **charge.** 42A60-0

modified impedance relay. An impedance form of distance relay for which the operating characteristic of the distance unit on an *R-X* diagram is a circle having its center displaced from the origin. *Note:* It may be described by the equation

$$Z^2 - 2K_1 Z\cos(\theta - \alpha) = K_2^2 - K_1^2$$

where K_1, K_2, and α are constants and θ is the phase angle by which the input voltage leads the input current. 37A100-31E11/31E6

modified index of refraction (modified refractive index)(troposphere). The index of refraction at any height increased by h/a, where h is the height above sea level and a is the mean geometrical radius of the earth. *Note:* When the index of refraction in the troposphere is horizontally stratified, propagation over a hypothetical flat earth through an atmosphere with the modified index of refraction is substantially equivalent to propagation over a curved earth through the real atmosphere. *See also:* **radiation.** 42A65-3E2

modified Kramer system. A system of speed control below synchronous speed for wound-rotor induction motors. Slip power is recovered through the medium of a converting device electrically connected between the secondary of the induction motor and a direct-current auxiliary motor. The auxiliary motor may be directly coupled to the induction-motor shaft or may form part of a motor-generator set that can return slip power to the supply system. *Note:* This system is not necessarily associated with constant power output over the speed range. 0-31E8

modified refractive index (troposphere). *See:* **modified index of refraction.**

modular (power supplies). The term modular is used to describe a type of power supply designed to be built into other equipment either chassis or rack mount. *Note:* It is usually distinguished from laboratory bench equipment by a large choice of mounting configurations and by a lack of meters and controls. *See also:* **power supply.** KPSH-10E1

modulated amplifier. An amplifier stage in which the carrier is modulated by introduction of a modulating signal. *See also:* **amplifier; modulating systems; radio transmitter; signal.** E182A-13E6

modulated continuous wave (MCW). A wave in which the carrier is modulated by a constant audio-frequency tone. *Note:* In telegraphic service, it is understood that the carrier is keyed. *See also:* **modulation; wave front.** E145/42A65-31E3

modulated photoelectric system (protective signaling). A photoelectric system in which the light beam is interrupted or modulated in a predetermined manner and in which the receiving equipment is designed to accept only the modulated light. *See also:* **protective signaling.** 42A65-0

modulated wave. A wave, some characteristic of which varies in accordance with the value of a modulating function. *See also:* **modulation.** E170-0

modulating function. A function, the value of which is intended to determine a variation of some characteristic of a carrier. *See also:* **modulating systems.** E170-0

modulating signal (modulating wave). A wave which causes a variation of some characteristic of a carrier. E188-0

modulating systems. *Note:* For an extensive list of cross references, see *Appendix A.*

modulating wave. A wave that causes a variation of some characteristic of a carrier. *See also:* **modulation.** E145/42A65-0

modulation (1) (carrier). (1) The process by which some characteristic of a carrier is varied in accordance with a modulating wave. (2) The variation of some characteristic of a carrier. *See also:* **modulating systems.** E145-0

(2) (signal transmission system). (1) A process whereby certain characteristics of a wave, often called the carrier, are varied or selected in accordance with a modulating function; (2) the result of such a process. *See:*

audio-frequency harmonic distortion;
babble;
background noise;
balanced modulator;
baseband;
baud;
bias telegraph distortion;
break-in keying;
carrier noise level;
carrier suppression;
characteristic telegraph distortion;
class-A modulator;
class-B modulator;
conversion transconductance;
gating;
intermodulation;
linear modulator;
modulated continuous wave;
modulated wave;

modulating wave;
modulation rate;
modulator;
plate modulation;
product modulator;
pulse modulator;
reactance modulator;
single-sideband transmission.
See also: **modulating systems.**
E170/E188-0;42A65-31E3; 85A1-13E6;AS1-34E10/19E4

modulation acceptance bandwidth (receiver performance). A measure of the maximum permissible modulation deviation of a receiver radio-frequency input signal having twice the amplitude of a reference input level, which produces a 12-decibel sinad ratio. *See also:* **receiver performance.** 0-6E1

modulation and noise spectrum (transmitter performance). The composite signal output from a transmitter as measured on either a power or spectrum distribution basis. *See also:* **audio-frequency distortion.** 0-6E1

modulation bandwidth. *See:* **bandwidth, modulation.***See also:* **audio-frequency distortion.**

modulation capability (of a transmitter). The maximum percentage modulation that can be obtained without exceeding a given distortion figure. *See also:* **modulating systems.** E145/42A65-0

modulation, effective percentage. *See:* **percentage modulation, effective.**

modulation eliminator (electronic navigation). (1) A limiting device designed to remove the modulation components from a carrier. (2) In very-high-frequency omnidirectional radio range, a specific unit used to proportion the transmitter output between the reference and variable signal functions and to remove the modulation from that portion of carrier delivered to the goniometer for use as the variable signal. *See also:* **navigation.** 0-10E6

modulation factor (amplitude-modulated wave). The ratio (usually expressed in percent) of the peak variation of the envelope from its reference value, to the reference value. *Notes:* (1) For modulating functions having unequal positive and negative peaks, both positive and negative modulation factors are defined, respectively, in terms of the positive or negative peak variations. (2) The reference value is usually taken to be the amplitude of the unmodulated wave. (3) In conventional amplitude modulation the maximum design variation is considered that for which the instantaneous amplitude of the modulated wave reaches zero. *See also:* **amplitude modulation.**
E145/E170/E182A/E188-42A65

modulation index (angle modulation with a sinusoidal modulating function). The ratio of the frequency deviation of the modulated wave to the frequency of the modulating function. *Note:* The modulation index is numerically equal to the phase deviation expressed in radians. *See also:* **angle modulation.**
E145/E170/E182/42A65-0

modulation, intensity. *See:* **intensity modulation.**

modulation limiting (transmitter performance). That action, performed within the modulator, that intentionally contains the signal within the spectral and power limitations of the system. *See also:* **audio-frequency distortion.** 0-6E1

modulation meter (modulation monitor). An instrument for measuring the degree of modulation (modulation factor) of a modulated wave train, usually expressed in percent. *See also:* **instrument.** 42A30-0

modulation monitor. *See:* **modulation meter.**

modulation noise (1) (sound recording and reproducing system). Noise that exists only in the presence of a signal and is a function of the instantaneous amplitude of the recorded signal. (The signal is not to be included as part of the noise.) *Note:* This type of noise results from such characteristics of a recording medium as light transmission in a film, slope of a groove, magnetism in a wire or tape, etcetera. *See also:* **electroacoustics; noise (sound recording and reproducing system).** E191-0

(2) (noise behind the signal). The noise arising when the recorded signal randomly differs from the desired recording due to imperfections of the recording process. *Note:* Imperfections causing modulation noise include those of the recording medium, the recording and/or reproducing transducer, and the interaction of these elements. 0-1E1

modulation pulse (pulse train). The modulation of one or more of the parameters with the intent to transmit information. *See also:* **pulse.** 0-9E4

modulation rate. Reciprocal of the unit interval measured in seconds. This rate is expressed in bauds. *See also:* **modulation.** 0-19E4

modulation stability (transmitter performance). The measure of the ability of a radio transmitter to maintain a constant modulation factor as determined on both a short-term and long-term basis. *See also:* **audio-frequency distortion.** 0-6E1

modulation transfer (camera tubes). *See:* **amplitude response.**

modulator (1) general. A device to effect the process of modulation.

(2) (radar). A device for generating a succession of short pulses of energy that cause a transmitter tube to oscillate during each pulse. *See also:* **modulation; radar.** 42A65-31E3

modulator, asymmetrical (signal-transmission system). A modulator in which the signal attentuation is different for the alternate half-cycles of the carrier. *See:* **signal.** 0-13E6

modulator, balanced (signal-transmission system). *See:* **modulator, symmetrical.**

modulator electrode. A control electrode in the case of an electron gun. 50I07-15E6

modulator, full-wave (signal-transmission system). *See:* **modulator, symmetrical.**

modulator, half-wave (signal-transmission system). Extreme case of asymmetrical modulator in which the signal transmission is zero during one half-cycle. *See:* **signal.** 0-13E6

modulator, symmetrical (signal-transmission system). A modulator in which the signal attenuation is the same for alternate half-cycles of the carrier. (This implies also equal leakage impedances to the reference plane for the two half-cycles). *See:* **signal.** 0-13E6

modulator, unbalanced (signal-transmission system). *See:* **modulator, asymmetrical; signal.**

modulo N check (1) (data transmission). A form of check digits, such that the number of ones in each number A operated upon is compared with a check number B, carried along with A and equal to the re-

mainder of A when divided by N; for example, in a modulo 4 check, the check number will be 0, 1, 2, or 3 and the remainder of A when divided by 4 must equal the reported check number B, or else an error or malfunction has occurred; a method of verification by congruences; for example, casting out nines. *See also:* **data transmission; residue check.** 0-19E4
(2) (computing systems). *See:* **residue check.** *See also:* **electronic digital computer.**

modulus (phasor). Its absolute value. The modulus of a phasor is sometimes called its amplitude. E270-0

moiré (1) (television). The spurious pattern in the reproduced picture resulting from interference beats between two sets of periodic structures in the image. *Note:* Moirés may be produced, for example, by interference between regular patterns in the original subject and the target grid in an image orthicon, between patterns in the subject and the line pattern and the pattern of phosphor dots of a three-color kinescope, and between any of these patterns and the pattern produced by the chrominance signal. *See also:* **beam tubes; color terms; television.** 42A65/E160/E201-2E2/15E6
(2) (storage tubes). A spurious pattern resulting from interference beats between two sets of periodic structures or scan patterns, or between a periodic structure and a scan pattern. *Note:* In a storage tube, moiré may be produced by writing a resolved parallel-line scan pattern and reading using another parallel-line scan pattern. Since mesh elements usually have a periodic structure, moiré may also be produced between the mesh pattern and a scan pattern, or between two mesh patterns. *See also:* **storage tube.** E158-15E6

moisture-repellent. So constructed or treated that moisture will not penetrate. 42A95-0

moisture-resistant (industrial control). So constructed or treated that it will not be injured readily by exposure to a moist atmosphere. 37A100/42A95/IC1-31E11/34E10

molar conductance (of an acid, base, or salt). The conductance of a solution containing one gram mole of the solute when measured in a like manner to equivalent conductance. *See also:* **electrochemistry.** 42A60-0

mold (1) (disk recording). A mold is a metal part derived from a master by electroforming that results in a positive of the recording, that is, it has grooves similar to a recording and thus can be played in a manner similar to a record. *See also:* **phonograph pickup.** 0-1E1
(2) (electrotyping). A matrix made by taking an impression, for example, of a printing plate in wax, lead, or plastic. *See also:* **electroforming.** 42A60-0

molded-case circuit breaker. A breaker assembled as an integral unit in a supporting and enclosing housing of insulating material; the overcurrent and tripping means being of the thermal type, the magnetic type, or a combination of both. *See also:* **switch.** E45/37A100-31E11

molded glass flare (electron tubes and electric lamps). A pressed glass article of the general shape shown in the figure below. *Note:* It is intended for use in the assembly of an electron tube or an electric lamp for the purpose of sealing electric terminals, or an exhaust tube, or both.

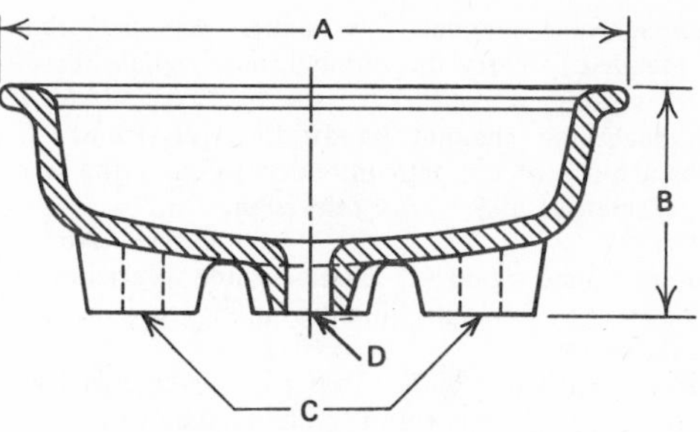

A–Major diameter
B–Over-all height
C–Holes for electrical terminals
D–Hole for exhaust tube

Molded glass flare.

momentary current. The current flowing in a device, an assembly, or a bus at the major peak of the maximum cycle as determined from the envelope of the current wave. *Note:* The current is expressed as the root-mean-square value, including the direct-current component, and may be determined by the method shown in American National Standard Methods for Determining the Values of a Sinusoidal Current Wave and Normal-Frequency Recovery Voltage for Alternating-Current High-Voltage Circuit Breakers, C37.05-1964 or revision thereof. 37A100-31E11

momentary interruption (electric power systems). An interruption that has a duration limited to the period required to restore service by automatic or supervisory-controlled switching operations or by manual switching at locations where an operator is immediately available. *Note:* Such switching operations are typically completed in a few minutes. *See also:* **outage.** 0-31E4

monadic operation (computing systems). An operation on one operand, for example, negation. *See:* **unary operation.** *See also:* **electronic digital computer.** X3A12-16E9

monitor (computing systems). Software or hardware that observes, supervises, controls, or verifies the operations of a system. *See also:* **electronic digital computer.** 0-16E9

monitoring (communication). Observation of the characteristics of transmitted signals. 42A65-19E4

monitoring key. A key that when operated makes it possible for an attendant or operator to listen on a telephone circuit without appreciably impairing transmission on the circuit. *See also:* **telephone switching system.** 42A65-0

monitoring radio receiver. A radio receiver arranged to permit a check to be made on the operation of a transmitting station. *See also:* **radio receiver.** 42A65-0

monitoring relay. A relay that has as its function to verify that system or control-circuit conditions conform to prescribed limits. 37A100-31E11/31E6

monochromatic (color). Referring to a negligibly small region of the spectrum. *See also:* **color terms.** E201-2E2

monochrome. Having only one chromaticity, usually achromatic. *See also:* **color terms.** E201-2E2

monochrome channel (television). Any path that is intended to carry the monochrome signal. *See also:* **television.** E204/42A65-2E2

monochrome channel bandwidth (television). The bandwidth of the path intended to carry the monochrome signal. *See also:* **television.** E204/42A65-2E2

monochrome signal (1) (monochrome television). A signal wave for controlling the luminance values in the picture.
(2) (color television). That part of the signal wave that has major control of the luminance values of the picture, whether displayed in color or in monochrome. *See also:* **color terms.** E201-2E2

monochrome transmission (television). The transmission of a signal wave for controlling the luminance values in the picture, but not the chromaticity values. *Note:* Also termed black-and-white transmission. *See also:* **television.** 42A65/E201-2E2

monocular visual field. The visual field of a single eye. Z7A1-0

monolithic conduit. A monolithic concrete structure built to the desired duct formation by an automatic conduit-forming machine or flexible tubular rubber forms. 42A35-31E13

monolithic integrated circuit. An integrated circuit whose elements are formed in situ upon or within a semiconductor substrate with at least one of the elements formed within the substrate. *Note:* To further define the nature of a monolithic integrated circuit, additional modifiers may be prefixed. Examples are: (1) *p-n*-junction-isolated monolithic integrated circuit, (2) dielectric-isolated monolithic integrated circuit, (3) beam-lead monolithic integrated circuit. *See also:* **integrated circuit.** E274-15E7/21E0

monopolar electrode system. *See:* **unipolar electrode system.**

monopole. Any one of a class of antennas that use an imaging plane to produce a radiation pattern approximating that of an electric dipole. *See also:* **antenna.** 0-3E1

monopulse (radar). Simultaneous lobing whereby complete direction-finding information is obtainable from a single pulse. *See also:* **circular scanning; radar.** E149-3E1

monoscope. A signal-generating electron-beam tube in which a picture signal is produced by scanning an electrode, parts of which have different secondary-emission characteristics. *See also:* **television; tube definitions.** E160/E188/42A70-2E2/15E6

monostable. Pertaining to a device that has one stable state. X3A12-16E9

monotonic decreasing function. A single-valued function that satisfies the relation $f(x_1)>f(x_2)$ if $x_1<x_2$. E270-0

monotonic increasing function. A single-valued function that satisfies the relation $f(x_1)<f(x_2)$ if $x_1<x_2$. E270-0

monotonic nondecreasing function. A single-valued function that satisfies the relation $f(x_1)\leq f(x_2)$ if $x_1<x_2$. E270-0

monotonic nonincreasing function. A single-valued function that satisfies the relation $f(x_1)\geq f(x_2)$ if $x_1<x_2$. E270-0

monthly peak duration curves (electric power utilization). Curves showing the total number of normal weekdays within the month during which a given net 60-minute clock-hour integrated peak demand equals or exceeds the percent of monthly peak values shown. Day one is 100 percent. *See also:* **generating station.** 0-31E4

Morse code (American Morse code). A system of dot and dash signals, invented by Samuel F. B. Morse, now used to a limited extent for wire telegraphy in North America. *See also:* **international Morse code; telegraphy.** 42A65-0

Morse signal light (blinker signal). A fixture, installed usually on a standard above the wheelhouse for best visibility all around, controlled by a telegraph key for visual communication by Morse or other code. 42A43-0

Morse telegraphy. That form of telegraphy in which the signals are formed in accordance with one of the so-called Morse codes, particularly the International (also called Continental) or American. *See also:* **telegraphy.** 42A65-0

Morton wave current* (electrotherapy). An interrupted current obtained from a static machine by applying to the part to be treated a flexible metal electrode connected to the positive terminal of the machine, the negative terminal being grounded and a suitable spark gap being employed between the terminals. (Deprecated as representing one of several types of current, ill-defined as to waveform, frequency, voltage, and tissue-intensity, formerly used in electrotherapy). *See also:* **electrotherapy.**
*Deprecated 42A80-18E1

most-probable position (navigation). A position obtained by using all available position information, weighted statistically in accordance with individual estimated errors. *See also:* **navigation.** 0-10E6

motional impedance (loaded motional impedance) (transducer). The complex remainder after the blocked impedance has been subtracted from the loaded impedance. *See:* Note (2) under **acoustic impedance.** *See also:* **electroacoustics; self-impedance.** E157/42A65-1E1

motor. A machine for the purpose of producing mechanical power or mechanical torque or force by means of a rotating shaft or through linear motion. *See also:* **asynchronous machine; direct-current commutating machine; synchronous machine.** 0-31E8

motorboating. An undesired oscillation in an amplifying system or transducer, usually of a pulse type, occurring at a subaudio or low audio frequency. *See also:* **transmission characteristics.** E151/42A65-2E2

motor branch circuit. A branch circuit supplying energy only to one or more motors and associated motor controllers. *See also:* **branch circuit.** 42A95-0

motor-circuit switch (industrial control). A switch intended for use in a motor branch circuit. *Note:* It is rated in horsepower, and it is capable of interrupting the maximum operating overload current of a motor of the same rating at the rated voltage. *See also:* **operating overload; switch.** 1A0/42A25-34E10

motor controller. A device or group of devices that serves to govern, in some predetermined manner, the electric power delivered to the motor or group of motors to which it is connected. *See also:* **multiple-unit control.** E16-0

motor-converter. A combination of an induction motor with a synchronous converter on a common shaft, the current produced in the rotor of the motor flowing through the armature of the synchronous converter. *See:* **converter.** 42A10-31E8

motor cutout switch. An isolating switch that isolates one or more of the traction motors from the supply circuits. *See also:* **control switch.** E16-0

motor-driven relay. A relay in which contact actuation is controlled through an electric motor, cams, and systems of gears. *Note:* An electromagnetic clutch may be used to engage and disengage the gear system. *See also:* **relay.** 83A16-0;0-21E0

motor effect (1) (general). The repulsion force exerted between adjacent conductors carrying currents in opposite directions. E169-0
(2) (induction heating). The force of repulsion or attraction between adjacent current-carrying conductors one of which may be the charge. *See also:* **induction heating.** E54-0

motor, electric. *See:* **electric motor.**

motor element (electroacoustic receiver). That portion that receives energy from the electric system and converts it into mechanical energy. *See also:* **loudspeaker.** 42A65-0

motorette (rotating machinery). A model for endurance tests on samples of rotating machine insulation. *Note:* It consists of a mounting plate, carrying two or more metal channels in which one or more test coils are wound as in the slots of a machine, and provided with suitable insulated terminals. It is designed for exposure to high temperature, moisture, vibration, or other chosen or specified insulation-aging influences, and for electric tests at designated stages of the aging processes. *See:* **asynchronous machine; direct-current commutating machines; synchronous machine.** 0-31E8

motor-field accelerating relay (industrial control). A relay that functions automatically to maintain the armature current within limits, when accelerating to speeds above base speed, by controlling the excitation of the motor field. *See:* **relay.** IC1-34E10

motor-field control. A method of controlling the speed of a motor by means of a change in the magnitude of the field current. *See:* **control.** 42A25-34E10

motor-field failure relay (field loss relay) (industrial control). A relay that functions to disconnect the motor armature from the line in the event of loss of field excitation. *See:* **relay.** IC1-34E10

motor-field induction heater. An induction heater in which the inducing winding typifies that of an induction motor of rotary or linear design. *See also:* **induction heating.** E54/E169

motor-field protective relay (industrial control). A relay that functions to prevent overheating of the field excitation winding by reducing the excitation of the shunt field. *See:* **relay.** IC1-34E10

motor-generator electric locomotive. An electric locomotive in which the main power supply for the traction motors is changed from one electrical characteristic to another by means of a motor-generator set carried on the locomotive. *See also:* **electric locomotive.** 42A42-0

motor-generator set. A machine that consists of one or more motors mechanically coupled to one or more generators. *See:* **converter; generator-motor.** 42A10-31E8

motor meter. A meter comprising a rotor, one or more stators, and a retarding element by which the resultant speed of the rotor is made proportional to the quantity being integrated (for example, power or current) and a register connected to the rotor by suitable gearing so as to count the revolutions of the rotor in terms of the accumulated integral (for example, energy or charge). *See also:* **electricity meter (meter).** 12A0/42A30-0

motor parts (electric). A term applied to a set of parts of an electric motor. Rotor shaft, conventional stator-frame (or shell), end shields, or bearings may not be included, depending on the requirements of the end product into which the motor parts are to be assembled. *See also:* **cradle base (rotating machinery).** 0-31E8

motor reduction unit. A motor with an integral mechanical means of obtaining a speed differing from the speed of the motor. *See also:* **asynchronous machine; direct-current commutating machine; synchronous machine.** 0-31E8

motor synchronizing. Synchronizing by means of applying excitation to a machine running at slightly below synchronous speed. *See also:* **asynchronous machine; synchronous machine.** 0-31E8

mount (switching tubes). The flange or other means by which the tube, or tube and cavity, are connected to a waveguide. *See:* **gas tubes.** E160-15E6

mounting lug, stator. *See:* **lug, stator mounting.**

mounting position (switch or fuse support). The position determined by and corresponding to the position of the base of the device. *Note:* The usual positions are (1) horizontal upright; (2) horizontal underhung; (3) vertical, and (4) angle. 37A100-31E11

mounting ring (rotating machinery). A ring of resilient or nonresilient material used for mounting an electric machine into a base at the end shield hub. *See:* **cradle base (rotating machinery).** 0-31E8

movable bridge coupler (drawbridge coupler). A device for engaging and disengaging signal or interlocking connections between the shore and a movable bridge span. *See also:* **railway signal and interlocking.** 42A42-0

movable bridge (drawbridge) rail lock. A mechanical device used to insure that the movable bridge rails are in proper position for the movement of trains. *See also:* **interlocking.** 42A42-0

moving-base-derived navigation data. Data obtained from measurements made at moving cooperative facilities located external to the navigated vehicle. *See also:* **navigation.** 0-10E6

moving-base navigation aid. An aid that requires ooperative facilities located upon a moving vehicle other than the one being navigated. *Notes:* (1) The cooperative facilities may move along a predictable path that is referenced to a specified coordinate system such as in the case of a nongeostationary navigation satellite. (2) Such an aid may also be designed solely to permit one moving vehicle to home upon another. *See also:* **navigation.** 0-10E6

moving-base-referenced navigation data. Data in terms of a coordinate system referenced to a moving vehicle other than the one being navigated. *See also:* **navigation.** 0-10E6

moving-coil loudspeaker (dynamic loudspeaker). A moving-conductor loudspeaker in which the moving conductor is in the form of a coil conductively connected to the source of electric energy. *See also:* **loudspeaker.** 42A65-0

moving-coil microphone (dynamic microphone). A moving-conductor microphone in which the movable conductor is in the form of a coil. *See also:* **microphone.** 42A65-0

moving-conductor loudspeaker (moving conductor). A loudspeaker in which the mechanical forces result from magnetic reactions between the field of the current and a steady magnetic field. *See also:* **loudspeaker.** 42A65-0

moving-conductor microphone. A microphone the electric output of which results from the motion of a conductor in a magnetic field. *See also:* **microphone.** 42A65-0

moving-contact assembly (rotating machinery). That part of the starting switch assembly that is actuated by the centrifugal mechanism. *See:* **centrifugal starting switch.** 42A10-0

moving element (instrument). Those parts that move as a direct result of a variation in the quantity that the instrument is measuring. *Notes:* (1) The weight of the moving element includes one-half the weight of the springs, if any. (2) The use of the term movement is deprecated.
See:
chart mechanism;
chart;
chart scale;
current circuit;
damping;
damping factor;
damping magnet;
fall time;
indicating scale;
indicator travel;
measuring mechanism;
mechanism;
null balance;
overshoot;
pen travel;
period;
pivot friction error;
pointer shift due to tapping;
repeatability;
response time;
rise time;
roll-in-jewel error;
stickiness;
taut band suspension;
timing mechanism;
voltage circuit;
zero adjuster;
zero-shift error.
See also:
instrument;
register;
register constant;
register ratio. 39A1/39A2/42A30-0

moving-iron instrument. An instrument that depends for its operation on the reactions resulting from the current in one or more fixed coils acting upon one or more pieces of soft iron or magnetically similar material at least one of which is movable. *Note:* Various forms of this instrument (plunger, vane, repulsion, attraction, repulsion-attraction) are distinguished chiefly by mechanical features of construction. *See also:* **instrument.** 42A30-0

moving-magnet instrument. An instrument that depends for its operation on the action of a movable permanent magnet in aligning itself in the resultant field produced either by another permanent magnet and by an adjacent coil or coils carrying current, or by two or more current-carrying coils, the axes of which are displaced by a fixed angle. *See also:* **instrument.** 42A30-0

moving-magnet magnetometer. A magnetometer that depends for its operation on the torques acting on a system of one or more permanent magnets that can turn in the field to be measured. *Note:* Some types involve the use of auxiliary magnets (Gaussian magnetometer), others electric coils (sine or tangent galvanometer). *See also:* **magnetometer.** 42A30-0

moving-target indicator (MTI) (electronic navigation). A device that limits the display of radar information primarily to moving targets. *See also:* **navigation: radar.** E172;42A65-10E6

moving-target indicator improvement factor (radar MTI). The average increase in signal-to-clutter ratio attributable to the moving-target-indicator system on targets distributed uniformly over the radial velocity spectrum; clutter attenuation normalized for uniformly distributed radial target velocities. *See also:* **navigation.** 0-10E6

***M* peak (closed loop) (control system, feedback).** The maximum value of the magnitude of the return transfer function for real frequencies, the value at zero frequency being normalized to unity. *See also:* **control system, feedback.** 85A1-23E0

***m*-phase circuit.** A polyphase circuit consisting of m distinct phase conductors, with or without the addition of a neutral conductor. *Note:* In this definition it is understood that m may be assigned the integral value of three or more. For a two-phase circuit see: **two-phase circuit; two-phase, three-wire circuit; two-phase, four-wire circuit; two-phase, five-wire circuit.** *See also:* **network analysis.** E270-0

MRA. *See:* **minimum reception altitude.**

MSC. *See:* **mile of standard cable.**

***M* scan (electronic navigation).** *See:* ***M* display.**

***M* scope (electronic navigation).** *See:* ***M* display.**

MTBF (reliability). *See:* **mean time between failures.**

MTE. *See:* **mean time between errors.**

MTF. *See:* **mean time between failures.**

MTI (electronic navigation). *See:* **moving-target indicator.**

mu (μ) circuit (feedback amplifier). That part that amplifies the vector sum of the input signal and the fed-back portion of the output signal in order to generate the output signal. *See also:* **feedback.** 42A65-0

MUF. *See:* **maximum usable frequency.**

mu factor (μ factor) (*n*-terminal electron tubes). The ratio of the magnitude of infinitesimal change in the voltage at the jth electrode to the magnitude of an infinitesimal change in the voltage at the lth electrode under the conditions that the current to the mth electrode remain unchanged and the voltages of all other electrodes be maintained constant. *See also:* **electron-tube admittances.** 42A7-15E6

muffler (fuse). An attachment for the vent of a fuse, or a vented fuse, that confines the arc and substantially reduces the venting from the fuse. 37A100-31E11

multiaddress (computers). *See:* **multiple-address.**

multianode tank (multianode tube). An electron tube having two or more main anodes and a single cathode. *Note:* This term is used chiefly for pool-cathode tubes. *See also:* **tube definitions.** 42A70-15E6

multibeam oscilloscope. An oscilloscope in which the cathode-ray tube produces two or more separate electron beams that may be individually, or jointly, controlled. *See:* **dual-beam oscilloscope; oscillograph.** 0-9E4

multicavity magnetron. A magnetron in which the circuit includes a plurality of cavities. *See also:* **magnetron.** 42A70-15E6

multicellular horn (electroacoustics). A cluster of horns with juxtaposed mouths that lie in a common surface. *Note:* The purpose of the cluster is to control the directional pattern of the radiated energy. *See also:* **loudspeaker.** 42A65-1E1

multichannel radio transmitter. A radio transmitter having two or more complete radio-frequency portions capable of operating on different frequencies, either individually or simultaneously. *See also:* **radio transmitter.** 42A65/E145-0

multichip integrated circuit. An integrated circuit whose elements are formed on or within two or more semiconductor chips that are separately attached to a substrate. *See also:* **integrated circuit.** E274-15E7/21E0

multidimensional system (control system). A system in which the state vector has more than one element. *See also:* **control system.** 0-23E0

multielectrode tube. An electron tube containing more than three electrodes associated with a single electron stream. 42A70-15E6

multifrequency transmitter. A radio transmitter capable of operating on two or more selectable frequencies, one at a time, using preset adjustments of a single radio-frequency portion. *See also:* **radio transmitter.** E145/E182/42A65-0

multilevel address (computing systems). *See:* **indirect address.**

multimeter. *See:* **circuit analyzer.**

multimode waveguide. A waveguide used to propagate more than one mode at the frequency of interest. *See:* **waveguide.** 0-3E1

multioffice exchange. A telephone exchange served by more than one local central office. *See also:* **telephone system.** 42A65-19E1

multioutlet assembly (interior wiring). A type of surface or flush raceway designed to hold conductors and attachment plug receptacles, assembled in the field or at the factory. *See also:* **raceway.** 1A0-0

multipath (facsimile). *See:* **multipath transmission.**

multipath error (electronic navigation). The error caused by the reception of a composite radio signal that arrives via two or more different paths. *See also:* **navigation.** 0-10E6

multipath transmission (radio propagation). The propagation phenomenon that results in signals reaching the receiving antenna by two or more paths. *Note:* In facsimile, multipath causes jitter. E168-0

multiple (1) (noun). A group of terminals arranged to make a circuit or group of circuits accessible at a number of points at any one of which connection can be made.

(2) (verb). To connect in parallel, or to render a circuit accessible at a number of points at any one of which connection can be made. 42A65-0

(3) (electronic analog computer). A junction into which patch cords may be plugged to form a common connection. *See also:* **electronic analog computer.** E165-16E9

multiple-address (multiaddress) (computers). Pertaining to an instruction that has more than one address part. *See also:* **electronic digital computer.** E162/X3A12-16E9

multiple-address code (electronic computation). *See:* **instruction code:** *See also:* **electronic computation.**

multiple-beam headlamp. A headlamp so designed as to permit the driver of a vehicle to use any one of two or more distributions of light on the road. *See also:* **headlamp.** Z7A1-0

multiple-beam klystron (microwave tubes). An *O*-type tube having more than one electron beam, and resonators coupled laterally but not axially. *See also:* **microwave tube or valve.** 0-15E6

multiple circuit. Two or more circuits connected in parallel. *See also:* **center of distribution.** 42A35-31E13

multiple-conductor cable. A combination of two or more conductors cabled together and insulated from one another and from sheath or armor where used. *Note:* Specific cables are referred to as 3-conductor cable, 7-conductor cable, 50-conductor cable, etcetera. *See also:* **power distribution, underground construction.** 42A35-31E13

multiple-conductor concentric cable. A cable composed of an insulated central conductor with one or more tubular stranded conductors laid over it concentrically and insulated from one another. *Note:* This cable usually has only two or three conductors. Specific cables are referred to as 2-conductor concentric cable, 3-conductor concentric cable, etcetera. *See also:* **power distribution, underground construction.** 42A35-31E13

multiple-current generator. A generator capable of producing simultaneously currents or voltages of different values, either alternating-current or direct-current. *See:* **direct-current commutating machine; synchronous machine.** 42A10-31E8

multiple feeder. (1) Two or more feeders connected in parallel. *See also:* **center of distribution.** 42A35-31E13

(2) One that is connected to a common load in multiple with one or more feeders from independent sources. 37A100-31E11

multiple-gun cathode-ray tube. A cathode-ray tube containing two or more separate electron-gun systems. *See also:* **tube definitions.** 0-15E6

multiple hoistway (elevators). A hoistway for more than one elevator or dumbwaiter. *See:* **hoistway (elevator or dumbwaiter).** 42A45-0

multiple lampholder (current tap*). A device that by insertion in a lampholder, serves as more than one lampholder. *See also:* **interior wiring.**

*Deprecated 42A95-0

multiple lightning stroke. A lightning stroke having two or more components. *See also:* **direct-stroke protection (lightning).** 0-31E13

multiple metallic rectifying cell. An elementary metallic rectifier having one common electrode and two or more separate electrodes of the opposite polarity. *See also:* **rectification.** 42A15-0

multiple modulation. A succession of processes of modulation in which the modulated wave from one process becomes the modulating wave for the next. *Note:* In designating multiple-modulation systems by their letter symbols, the processes are listed in the order in which the modulating function encounters them. For example, PPM-AM means a system in which one or more signals are used to position-modulate their respective pulse subcarriers which are spaced in time and are used to amplitude-modulate a carrier. *See also:* **modulating systems.** E145/E170/42A65-0

multiple plug (cube tap) (plural tap*). A device that, by insertion in a receptacle, serves as more than one receptacle. *See also:* **interior wiring.**
*Deprecated 42A95-0

multiple rectifier circuit. A rectifier circuit in which two or more simple rectifier circuits are connected in such a way that their direct currents add, but their commutations do not coincide. *See also:* **rectification; rectifier circuit element.** 42A15/E59-34E17/34E24

multiple rho (electronic navigation). A generic term referring to navigation systems based on two or more distance measurements for determination of position. *See also:* **navigation.** 0-10E6

multiple-shot blasting unit. A unit designed for firing a number of explosive charges simultaneously in mines, quarries, and tunnels. *See also:* **blasting unit.** 42A85-0

multiple sound track. Consists of a group of sound tracks, printed adjacently on a common base, independent in character but in a common time relationship, for example, two or more have been used for stereophonic sound recording. *See also:* **multitrack recording system; phonograph pickup.** 0-1E1

multiple spot scanning (facsimile). The method in which scanning is carried on simultaneously by two or more scanning spots, each one analyzing its fraction of the total scanned area of the subject copy. *See also:* **scanning (facsimile).** E168-0

multiple street-lighting system. A street-lighting system in which street lights, connected in multiple, are supplied from a low-voltage distribution system. *See also:* **alternating-current distribution; direct-current distribution.** 42A35-31E13

multiple-supply-type ballast. A ballast designed specifically to receive its power from an approximately constant-voltage supply circuit and that may be operated in multiple (parallel) with other loads supplied from the same source. *Note:* The deviation in source voltage ordinarily does not exceed plus or minus 5 percent, but in the case of ballasts designed for a stated input voltage range, the deviation may by greater as long as it is within the stated range. 82A9-0

multiple switchboard (telephony). A manual telephone switchboard in which each subscriber line is attached to two or more jacks, so as to be within reach of several operators. *See also:* **telephone switching system.** 42A65-19E1

multiple system (electrochemistry). The arrangement in a multielectrode electrolytic cell whereby in each cell all of the anodes are connected to the positive bus bar and all of the cathodes to the negative bus bar. *See also:* **electrorefining.** 42A60-0

multiple tube (or valve). A space-charge-controlled tube or valve containing within one envelope two or more units or groups of electrodes associated with independent electron streams, though sometimes with one or more common electrodes. Examples: Double diode, double triode, triode-heptode, etcetera. *See also:* **multiple-unit tube; tube definitions.** E160/42A70-15E6

multiple tube counts (radiation counter tubes). Spurious counts induced by previous tube counts. *See also:* **gas-filled radiation-counter tubes.** E160/42A70-15E6

multiple-tuned antenna. A low-frequency antenna having a horizontal section with a multiplicity of tuned vertical sections. *See also:* **antenna.** E145/42A65-3E1

multiple twin quad (telephony). A quad in which the four conductors are arranged in two twisted pairs, and the two pairs twisted together. *See also:* **cable.** 42A65-0

multiple-unit control (electric traction). A control system in which each motive-power unit is provided with its own controlling apparatus and arranged so that all such units operating together may be controlled from any one of a number of points on the units by means of a master controller.
See:
automatic acceleration;
bridge transition;
bus line;
car wiring apparatus;
changeover switch;
contact conductor;
control switch;
differential control;
double-end control;
electric train line;
electropneumatic controller;
field shunting control;
jumper;
line breaker;
magnet valve;
master controller;
motor controller;
open-circuit control;
open-circuit transition;
pneumatic controller;
reverser;
safety control feature;
safety control handle;
sequence switch;
sequence table;
series-parallel control;
shunt control;
shunt transition;
single-end control;
slip relay;
tapped field control;
transition;
weight-transfer compensation. 42A42-0

multiple-unit electric car. An electric car arranged either for independent operation or for simultaneous operation with other similar cars (when connected to form a train of such cars) from a single control station. *Note:* A prefix diesel-electric, gas-electric, etcetera, may replace the word electric. *See also:* **electric motor car.** 42A42-0

multiple-unit electric locomotive. A locomotive composed of two or more multiple-unit electric motive-power units connected for simultaneous operation of all such units from a single control station. *Note:* A prefix diesel-electric, turbine-electric, etcetera, may replace the word electric. *See also:* **electric locomotive.** 42A42-0

multiple-unit electric motive-power unit. An electric motive-power unit arranged either for independent operation or for simultaneous operation with other similar units (when connected to form a single locomotive) from a single control station. *Note:* A prefix diesel-electric, gas-electric, turbine-electric, etcetera, may replace the word electric. *See also:* **electric locomotive.** 42A42-0

multiple-unit electric train. A train composed of multiple-unit electric cars. *See also:* **electric motor car.** 42A42-0

multiple-unit tube. *See:* **multiple tube (or valve).**

multiplex (communication). To interleave or simultaneously transmit two or more messages on a single channel. X3A12-16E9

multiplexing (modulation systems). The combining of two or more signals into a single wave (the multiplex wave) from which the signals can be individually recovered. *See also:* **modulating systems.** E170-0

multiplex operation (communication). Simultaneous transmission of two or more messages in either or both directions over the same transmission path. *See also:* **telegraphy.** 42A65-31E3

multiplex printing telegraphy. That form of printing telegraphy in which a line circuit is employed to transmit in turn one character (or one or more pulses of a character) for each of two or more independent channels. *See:* **frequency-division multiplexing; time-division multiplexing.** *See also:* **telegraphy.** 42A65-0

multiplex radio transmission. The simultaneous transmission of two or more signals using a common carrier wave. *See also:* **radio transmission.** E145-0

multiplication factor (multiplier type of valve or tube) (thermionics). The ratio of the output current to the primary emission current. *See also:* **electron emission.** 50I07-15E6

multiplier (1) (general). A device that has two or more inputs and whose output is a representation of the product of the quantities represented by the input signals. *See also:* **electronic computation.** E270-0
(2) (analog computer). A device capable of multiplying one variable by another. *See also:* **electronic analog computer.** E165-0
(3) (linearity). *See:* **constant multiplier; normal linearity; servo multiplier.**

multiplier, constant (computing systems). A computing element that multiplies a variable by a constant factor. *See also:* **electronic analog computer; multiplier (linearity).** E165-16E9

multiplier, electronic. An all-electronic device capable of forming the product of two variables. *Note:* Examples are a time-division multiplier, a square-law multiplier, an amplitude-modulation–frequency-modulation (AM-FM) multiplier, and a triangular-wave multiplier. *See also:* **electronic analog computer.** E165-0

multiplier, four-quadrant. A multiplier in which operation is unrestricted as to the sign of the input variables. *See also:* **electronic analog computer.** E165-0

multiplier, one-quadrant. A multiplier in which operation is restricted to a single sign of both input variables. *See also:* **electronic analog computer.** E165-0

multiplier phototube. A phototube with one or more dynodes between its photocathode and output electrode. *See also:* **amplifier; photocathode.** E160/E175/42A70-15E6

multiplier section, electron (electron tubes). *See:* **electron multiplier.**

multiplier servo. An electromechanical multiplier in which one variable is used to position one or more ganged potentiometers across which the other variable voltages are applied. *See also:* **electronic analog computer; multiplier (linearity).** E165-16E9

multiplier, two-quadrant. A multiplier in which operation is restricted to a single sign of one input variable only. *See also:* **electronic analog computer.** E165-0

multipole fuse. *See second note under:* **pole (pole unit) (switching device or fuse).**

multiposition relay. A relay that has more than one operate or nonoperate position, for example, a stepping relay. *See also:* **relay.** 83A16-0

multiposition switches (industrial control). (1) **self-returning switch.** A switch that returns to a stated position when it is released from any one of a stated set of other positions. (2) **spring return switch.** A switch in which the self-returning function is effected by the action of a spring. (3) **gravity-return switch.** A switch in which the self-returning function is effected by the action of weight. (4) **self-positioning switch.** A switch that assumes a certain operating position when it is placed in the neighborhood of the position. *See:* **switch.** IC1-34E10

multipressure zone pothead (electric power distribution). A pressure-type pothead intended to be operated with two or more separate pressure zones that may be at different pressures. *See:* **pressure-type pothead; single-pressure zone potheads.** E48-0

multiprocessing (computing systems). Pertaining to the simultaneous or interleaved execution of two or more programs or sequences of instructions by a computer or computer network. *Note:* Multiprocessing may be accomplished by multiprogramming, parallel processing, or both. *See also:* **electronic digital computer.** X3A12-16E9

multiprocessor (computing systems). A computer capable of multiprocessing. *See also:* **electronic digital computer.** X3A12-16E9

multiprogramming (computing systems). Pertaining to the interleaved execution of two or more programs by a computer. *See:* **parallel processing.** *See also:* **electronic digital computer.** X3A12-16E9

multi-radio-frequency-channel transmitter. *See:* **multichannel radio transmitter.**

multirate meter. A meter that registers at different rates or on different dials at different hours of the day. *See also:* **electricity meter.** 42A30-0

multirestraint relay. A restraint relay that is so constructed that its operation is restrained by more than one input quantity. 37A100-31E11/31E6

multisegment magnetron. A magnetron with an anode divided into more than two segments, usually by slots parallel to its axis. *See also:* **magnetrons.** E160/42A70-15E6

multispeed motor. One that can be operated at any one of two or more definite speeds, each being practically independent of the load. *Note:* For example, a direct-current motor with two armature windings or an induction motor with windings capable of various pole groupings. *See:* **asynchronous machine; direct-current**

commutating machine. 42A10-0;0-31E8

multistage tube (X-ray tubes). An X-ray tube in which the cathode rays are accelerated by multiple ring-shaped anodes at progressively higher potential. *See also:* **electron devices, miscellaneous.** 42A70-15E6

multitrace (oscilloscopes). A mode of operation in which a single beam in a cathode-ray tube is shared by two or more signal channels: *See:* **alternate display; chopped display; dual trace; oscillograph.** 0-9E4

multitrack recording system. A system that provides two or more recording tracks on a medium, resulting in either related or unrelated recordings in common time relationship. *See also:* **multiple sound track; phonograph pickup.** 0-1E1

multivalent function. If to any value of u there corresponds more than one value of x (or more than one set of values of $x_1, x_2, \cdots, x_n$) then u is a multivalent function. Thus $u = \sin x$, $u = x^2$ are multivalent. E270-0

multivalued function. If to any value of x (or any set of values of $x_1, x_2, \cdots, x_n$) there corresponds more than one value of u, then u is a multivalued function. Thus $u = \cos^{-1} x$ is multivalued. E270-0

multivariable system (control system). A system in which the control input or output vector has more than one element. *See also:* **control system.** 0-23E0

multivibrator. A relaxation oscillator employing two electron tubes to obtain the in-phase feedback voltage by coupling the output of each to the input of the other through, typically, resistance-capacitance elements. *Notes:* (1) The fundamental frequency is determined by the time constants of the coupling elements and may be further controlled by an external voltage. (2) A multivibrator is termed free-running or driven, according to whether its frequency is determined by its own circuit constants or by an external sychronizing voltage. The name multivibrator was originally given to the free-running multivibrator, having been suggested by the large number of harmonics produced. (3) When such circuits are normally in a nonoscillating state and a trigger signal is required to start a single cycle of operation, the circuit is commonly called a one-shot, a flip-flop, or a start-stop multivibrator. *See also:* **oscillatory circuit.** E145/42A65-31E3

multivoltage control (elevators). A system of control that is accomplished by impressing successively on the armature of the driving-machine motor a number of substantially fixed voltages such as may be obtained from multicommutator generators common to a group of elevators. *See also:* **control (elevators).** 42A45-0

municipal fire-alarm system. A manual fire-alarm system in which the stations are accessibly located for operation by the public, and the signals of which register at a central station maintained and operated by the municipality. *See also:* **protective signaling.** 42A65-0

municipal police report system. A system of strategically located stations from any one of which a patrolling policeman may report his presence to a supervisor in a central office maintained and operated by the municipality. *See also:* **protective signaling.** 42A65-0

Munsell chroma. (1) The dimension of the Munsell system of color that corresponds most closely to saturation. *Note:* Chroma is frequently used, particulary in English works, as the equivalent of saturation. *See also:* **color terms; saturation.** E201-2E2
(2) The index of saturation of the perceived object color defined in terms of the Y value and chromaticity coordinates (x,y) of the color of light reflected or transmitted by the object. *See also:* **color terms; Munsell color system.** Z7A1-0

Munsell color system. A system of surface-color specification based on perceptually uniform color scales for the three variables: Munsell hue, Munsell value, and Munsell chroma. *Notes:* (1) For an observer of normal color vision, adapted to daylight and viewing the specimen when illuminated by daylight and surrounded with a middle gray to white background, the Munsell hue, value, and chroma of the color correlate well with the hue, lightness, and saturation of the perceived color. (2) A number of other color specification systems have been developed, usually for specific commercial purposes. *See also:* **color terms.** Z7A1-0

Munsell hue. The index of the hue of the perceived object color defined in terms of the Y value and chromaticity coordinates (x,y) of the color of the light reflected or transmitted by the object. *See also:* **color terms.** Z7A1-0

Munsell value. The index of the lightness of the perceived object color defined in terms of the Y value. *Note:* Munsell value is approximately equal to the square root of the reflectance expressed in percent. *See also:* **color terms.** Z7A1-0

musa antenna* (multiple-unit steerable antenna). *See:* **electronically scanned antenna.** *See also:* **antenna.**

*Deprecated

must operate value. *See:* **relay must operate value.**

mutual capacitance between two conductors*. *See:* **balanced capacitance between two conductors.**

*Deprecated

mutual conductance. The control-grid-to-anode transconductance. *See also:* **ON period (electron tubes).** 50I07-15E6

mutual impedance (1) (between any two pairs of terminals of a network). The ratio of the open-circuit potential difference between either pair of terminals, to the current applied at the other pair of terminals, all other terminals being open. *Note:* Mutual impedance may have either of two signs depending upon the assumed directions of input current and output voltage; the negative of the above ratio is usually employed. Mutual impedance is ordinarily additive if two coils of a transformer are connected in series or in parallel adding, and is subtractive if two coils of a transformer are in series or in parallel opposing. *See also:* **self-impedance.** 42A65-0
(2) (between two meshes). The factor by which the phasor equivalent of the steady-state sine-wave current in one mesh must be multiplied to give the phasor equivalent of the steady-state sine-wave voltage in the other mesh caused by the current in the first mesh. *See also:* **network analysis.** E270-0
(3) (antenna). The mutual impedance between any two terminal pairs in a multielement array antenna is equal to the open-circuit voltage produced at the first terminal pair divided by the current supplied to the second when all other terminal pairs are open circuited. *See also:* **antenna.** E149-3E1

mutual inductance. The common property of two electric circuits whereby an electromotive force is induced in one circuit by a change of current in the other circuit. *Notes:* (1) The coefficient of mutual inductance M between two windings is given by the following equation

$$M = \frac{\partial i}{\partial \lambda}$$

where λ is the total flux linkage of one winding and i is the current in the other winding. (2) The voltage e induced in one winding by a current i in the other winding is given by the following equation

$$e = -\left[M\frac{di}{dt} + i\,\frac{dM}{dt}\right]$$

If M is constant $e = -M\dfrac{di}{dt}$

0-21E1

mutual inductor. An inductor for changing the mutual inductance between two circuits. E270-0

mutual information. *See:* **transinformation.**

mutual interference chart (MIC) (electromagnetic compatibility). A plot or matrix, with ordinate and abscissa representing the tuned frequencies of a single transmitter-receiver combination, that indicates potential interference to normal receiver operation by reason of interaction of the two equipments under consideration at any combination of tuned transmit/receive frequencies. *Note:* This interaction includes transmitter harmonics and other spurious emissions, and receiver spurious responses and images. *See also:* **electromagnetic compatibility.** 0-27E1

mutual resistance of grounding electrodes. Equal to the voltage change in one of them produced by one ampere of direct current in the other, and is expressed in ohms. *See also:* **grounding device.** E81-0

mutual surge impedance (lightning arresters). The apparent mutual impedance between two lines, both of infinite length. *Note:* It determines the relationship between the surge voltage induced into one line by a surge current of short duration in the other. *See:* **lightning arrester (surge diverter).** 50I25-31E7

N

nameplate (rating plate) (rotating machinery). A plaque giving the manufacturer's name and the rating of the machine. *See:* **cradle base (rotating machinery).** 0-31E8

NAND. A logic operator having the property that if P is a statement, Q is a statement, R is a statement, . . . then the NAND of $P, Q, R, \ldots$ is true if at least one statement is false, false if all statements are true. X3A12-16E9

narrow-angle diffusion (illuminating engineering). Diffusion in which flux is scattered at angles near the direction that the flux would take by regular reflection or transmission. *See also:* **lamp.** Z7A1-0

narrow-angle luminaire. Luminaire that concentrates the light within a cone of comparatively small solid angle. *See also:* **luminaire.** Z7A1-0

narrow-band axis (color television) (phasor representation of the chrominance signal). The direction of the phasor representing the coarse chrominance primary. *See also:* **color terms.** E201-0

narrow-band interference (electromagnetic compatibility). For purposes of measurement, a disturbance of spectral energy lying within the bandpass of the measuring receiver in use. *See also:* **electromagnetic compatibility.** 0-27E1

narrow-band radio noise. Radio noise having a spectrum exhibiting one or more sharp peaks, narrow in width compared to the nominal bandwidth of the radio noise meter, and far enough apart to be resolvable by the receiver. *See also:* **radio-noise field strength.** 63A4-0

***N*-ary code (information theory).** A code employing N distinguishable types of code elements. *See also:* **information theory.** E171/42A65-0

nationwide toll dialing (telephony). A system of automatic switching whereby an outward toll operator can complete calls to any basic-numbering-plan area in the country covered by the system. *See also:* **telephone switching system.** 42A65-0

native system demand (electric power utilization). The monthly net 60-minute clock-hour integrated peak demands of the system. *See also:* **alternating-current distribution.** 0-31E4

natural air cooling system (rectifier). An air cooling system in which heat is removed from the cooling surfaces of the rectifier only by the natural action of the ambient air. *See also:* **rectification.** 34A1-34E24

natural frequency (1) (antenna). Its lowest resonance frequency without added inductance or capacitance. *See also:* **antenna.** 42A65-3E1

(2) (lightning arresters). The frequency or frequencies at which the circuit will oscillate if it is free to do so. *See:* **frequency, undamped.** 99I2-31E7

natural language. A language whose rules reflect and describe current usage rather than prescribe usage. *See:* **artificial language.** X3A12-16E9

natural noise (electromagnetic compatibility). Noise having its source in natural phenomena and not generated in machines or other technical devices. *See also:* **electromagnetic compatibility.** CISPR-27E1

natural period. The period of the periodic part of a free oscillation of the body or system. *Notes:* (1) When the period varies with amplitude, the natural period is the period as the amplitude approaches zero. (2) A body or system may have several modes of free oscillation, and the period may be different for each. E270-0

navigation. The process of directing a vehicle so as to reach the intended destination. *Note:* For an extensive list of cross references, see *Appendix A.* E172-10E6

navigational radar (surface search radar). A high-frequency radio transmitter-receiver for the detection, by means of transmitted and reflected signals, of any object (within range) projecting above the surface of the water and for visual indication of its bearing and distance. *See also:* **radio direction-finder (radio compass).** 42A43-0

navigation coordinate. Any one of a set of quantities, the set serving to define a position. *See also:* **navigation.** 0-10E6

navigation light system. A set of aircraft aeronautical lights provided to indicate the position and direction of motion of an aircraft to pilots of other aircraft or to ground observers. *See also:* **signal lighting.** Z7A1-0

navigation parameter. A measurable characteristic of motion or position used in the process of navigation. *See also:* **navigation.** 0-10E6

navigation quantity. A measured value of a navigation parameter. *See also:* **navigation.** 0-10E6

***n*-conductor cable (electric power distribution).** *See:* **multiple-conductor cable.**

***n*-conductor concentric cable (electric power distribution).** *See:* **multiple-conductor concentric cable.**

NDB (electronic navigation). *See:* **nondirectional beacon.**

***N*-display (radar).** A *K* display having an adjustable pedestal signal, as in the *M* display, for the measurement of distance. *See also:* **navigation.** 0-10E6

near-end crosstalk. Crosstalk that is propagated in a disturbed channel in the direction opposite to the direction of propagation of the current in the disturbing channel. *Note:* The terminal of the disturbed channel at which the near-end crosstalk is present is ordinarily near to or coincides with the energized terminal of the disturbing channel. *See also:* **coupling.** 42A65-0

near field. An area within the Fresnel region of an antenna system. *See also:* **navigation.** 0-10E6

near-field region, radiating. The region of the field of an antenna between the reactive near-field region and the far-field region wherein radiation fields predominate and wherein the angular field distribution is dependent upon distance from the antenna. *Notes:* (1) If the antenna has a maximum over-all dimension which is not large compared to the wavelength, this field region may not exist. (2) For an antenna focused at infinity, the radiating near-field region is sometimes referred to as the Fresnel region on the basis of analogy to optical terminology. *See also:* **radiation.** 0-3E1

near-field region, reactive. The region of the field immediately surrounding the antenna wherein the reactive field predominates. *Note:* For most antennas the outer boundary of the region is commonly taken to exist at a distance $\lambda/2\pi$ from the antenna surface, where λ is the wavelength. *See also:* **radiation.** 0-3E1

neck (cathode-ray tube). The small tubular part of the envelope near the base. *See also:* **cathode-ray tube.** 50I07-15E6

negate. To perform the logic operation NOT. *See also:* **electronic digital computer.** X3A12-16E9

negative after-potential (electrobiology). Relatively prolonged negativity that follows the action spike in a homogeneous fiber group. *See also:* **contact potential.** 42A80-18E1

negative conductor. A conductor connected to the negative terminal of a source of supply. *Note:* A negative conductor is frequently used as an auxiliary return circuit in a system of electric traction. *See also:* **center of distribution.** 42A35-31E13

negative-differential-resistance region (thyristor). Any portion of the principal voltage-current characteristic in the switching quadrant(s) within which the differential resistance is negative. *See also:* **principal voltage-current characteristic (principal characteristic).** E223-34E17/15E7

negative electrode (1) (primary cell). The anode when the cell is discharging. *Note:* The negative terminal is connected to the negative electrode. *See also:* **electrolytic cell.** 42A60-0

(2) (metallic rectifier). The electrode from which the forward current flows within the cell. *See also:* **rectification.** 42A15-0

negative feedback (1) (general). Feedback that results in decreasing the amplification. *See also:* **feedback.** E145-0

(2) (degeneration) (stabilized feedback). The process by which a part of the power in the output circuit of an amplifying device reacts upon the input circuit in such a manner as to reduce the initial power, thereby decreasing the amplification. *See also:* **feedback.** 42A65-0

(3) (control) (industrial control). A feedback signal in a direction to reduce the variable that the feedback represents. *See also:* **control system, feedback; feedback.** AS1-34E10

negative glow (gas tube). The luminous glow in a glow-discharge cold-cathode tube between the cathode dark space and the Faraday dark space. *See:* **gas tubes.** 42A70-15E6

negative matrix (negative). A matrix the surface of which is the reverse of the surface to be ultimately produced by electroforming. *See also:* **electroforming.** 42A60-0

negative modulation (amplitude-modulation television system). That form of modulation in which an increase in brightness corresponds to a decrease in transmitted power. *See also:* **television.** 42A65-0

negative-phase-sequence impedance (rotating machinery). The quotient of the negative-sequence rated-frequency component of the voltage, assumed to be sinusoidal, at the terminals of a machine rotating at synchronous speed, and the negative-sequence component of the current at the same frequency. *Note:* It is equal to the asynchronous impedance for a slip equal to 2. *See also:* **asynchronous machine.** 0-31E8

negative-phase-sequence reactance (rotating machinery). The quotient of the reactive fundamental component of negative-sequence primary voltage due to sinusoidal negative-sequence primary current of rated frequency, and the value of this current, the machine running at rated speed. *See also:* **asynchronous machine.** 0-31E2

negative-phase-sequence (phase-reversal) relay. A relay that responds to the negative-phase-sequence component of a polyphase input quantity. *Note:* Frequently employed in three-phase systems. *See also:* **relay.** 37A100-31E11/31E6

negative phase-sequence resistance (rotating machinery). The quotient of the in-phase fundamental component of negative-sequence primary voltage, due to sinusoidal negative-sequence primary current of rated frequency, and the value of this current, the machine running at rated speed. *See also:* **asynchronous machine.** 0-31E8

negative-phase-sequence symmetrical components. Of an unsymmetrical set of polyphase voltages or currents of *M* phases, that set of symmetrical components that have the $(m-1)$st phase sequence. That is, the angular phase lag from the first member of the set to the second, from every other member of the set to the succeeding one, and from the last member to the first, is equal to $(m-1)$ times the characteristic angular phase difference, or $(m-1)2\pi/m$ radians. The members of this set will reach their positive maxima uniformly but in the reverse order of their designations. *Note:* The negative-phase-sequence symmetrical components for a three-phase set of unbalanced sinusoidal voltages ($m=3$), having the primitive period, are represented by the equations

$$e_{a2} = (2)^{1/2}E_{a2}\cos(\omega t+\alpha_{a2})$$

$$e_{b2} = (2)^{1/2}E_{a2}\cos(\omega t+\alpha_{a2}-\frac{4\pi}{3})$$

$$e_{c2} = (2)^{1/2}E_{a2}\cos(\omega t+\alpha_{a2}-\frac{2\pi}{3})$$

derived from the equation of symmetrical components of a set of polyphase (alternating) voltages. Since in this case $r=1$ for every component (of first harmonic), the third subscript is omitted. Then k is 2 for $(m-1)$st sequence, and s takes on the algebraic values 1, 2, and 3 corresponding to phases *a, b,* and *c.* The sequence of maxima occurs in the order, *a, c, b,* which is the reverse or negative of the order for $k=1$. *See:* **polyphase alternating currents; symmetrical components.**
E270-0

negative plate (storage cell). The grid and active material to which current flows from the external circuit when the battery is discharging. *See also:* **battery (primary or secondary).** 42A60-0

negative-polarity lightning stroke. A stroke resulting from a negatively charged cloud that lowers negative charge to the earth. *See also:* **direct-stroke protection (lightning).** 0-31E13

negative-resistance device. A resistance in which an increase in current is accompanied by a decrease in voltage over the working range. E270-0

negative-resistance oscillator. An oscillator produced by connecting a parallel-tuned resonant circuit to a two-terminal negative-resistance device. (One in which an increase in voltage results in a decrease in current.) *Note:* Dynatron and transitron oscillators, arc converters, and oscillators of the semiconductor type are examples. *See also:* **oscillatory circuit.**
E145/E182A-42A65

negative-resistance repeater. A repeater in which gain is provided by a series or a shunt negative resistance or both. *See also:* **repeater.** 42A65-0

negative-sequence reactance. The ratio of the fundamental reactive component of negative-sequence armature voltage, resulting from the presence of fundamental negative-sequence armature current of rated frequency, to this current, the machine being operated at rated speed. *Notes:* (1) The rated current value of negative-sequence reactance is the value obtained from a test with a fundamental negative-sequence current equal to rated armature current. The rated voltage value of negative-sequence reactance is the value obtained from a line-to-line short-circuit test at two terminals of the machine at rated speed, applied from no load at rated voltage, the resulting value being corrected when necessary for the effect of harmonic components in the current. (2) For any unbalanced short-circuits, certain harmonic components of current, if present, may produce fundamental reactive components of negative-sequence voltage that modify the ratio of the total fundamental reactive component of negative-sequence voltage to the fundamental component of negative-sequence current. This effect can be included by multiplying the negative-sequence reactance, before it is used for short-circuit calculations, by a wave distortion factor, equal to or less than unity, that depends primarily upon the type of short-circuit, and upon the characteristics of the machine and the external circuit, if any, between the machine and the point of short-circuit. *See:* **synchronous machine.**
42A10-0

negative-sequence resistance. The ratio of the fundamental component of in-phase armature voltage, due to the fundamental negative-sequence component of armature current, to this component of current at rated frequency. *Note:* This resistance, which forms a part of the negative-sequence impedance for use in circuit calculations to establish relationships between voltages and currents, is not directly applicable in the calculations of the total loss in the machine caused by negative-sequence currents. This loss is the product of the square of the fundamental component of the negative-sequence current and the difference between twice the negative-sequence resistance and the positive-sequence resistance, that is, $I_2^2\ (2R_2 - R_1)$. *See:* **synchronous machine.** 42A10-0

negative terminal (of a battery). The terminal toward which positive electric charge flows in the external circuit from the positive terminal. *Note:* The flow of electrons in the external circuit is to the positive terminal and from the negative terminal. *See also:* **battery (primary or secondary).** 42A60-0

negative-transconductance oscillator. Oscillator in which the output of the device is coupled back to the input without phase shift, the condition for oscillation being satisfied by the negative conductance of the device. *See also:* **oscillatory circuit.** E145-42A65

negative vectors. Two vectors are mutually negative if their magnitudes are the same and their directions opposite. E270-0

negentropy (information theory). *See:* **average information content.**

neighborhood. Of any point u_0, in a three-dimensional space, the volume enclosed by a sphere drawn with u_0 as center. Of any point u_0, in a two-dimensional space, the area enclosed by a circle drawn with u_0 as center. E270-0

neon indicator (tube). A cold-cathode gas-filled tube containing neon, used as a visual indicator of a potential difference or a field. *See:* **gas tubes.** 50I07-15E6

neper. A division of the logarithmic scale such that the number of nepers is equal to the natural logarithm of the scalar ratio of two currents or two voltages. *Notes:* (1) With I_1 and I_2 designating the scalar value of two currents, and n the number of nepers denoting their scalar ratio: $n=\log_e(I_1/I_2)$. (2) 1 neper equals 0.8686 bel. (3) The neper is a dimensionless unit. *See also:* **transmission characteristics.**
E145/E270;42A65-31E3

nerve-block (electrobiology). The application of a current to a nerve so as to prevent the passage of a propagated potential. *See also:* **excitability (electrobiology).** 42A80-18E1

nest or section (multiple system). A group of electrolytic cells placed close together and electrically connected in series for convenience and economy of operation. *See also:* **electrorefining.** 42A60-0

net assured capability (electric power supply). The net dependable capability of all power sources of a system, including firm power contracts and emergency interchange agreements, less that assigned to provide for scheduled maintenance outages and forced outages of power sources. *See also:* **generating station.** 42A35-31E13/31E4

net dependable capability (electric power supply). The maximum system load, expressed in kilowatthours per hour that a generating unit, station, or power source can be depended upon to supply on the basis of average operating conditions. *Note:* This capability takes into account average conditions of weather, quality of fuel, degree of maintenance and other operating factors. It does not include provision for maintenance outages. *See also:* **generating station.** 42A35-31E13/31E4

net generation (electric power systems). Gross generation less station or unit power requirements. *See also:* **power system, low-frequency and surge testing.** E94-0

net information content. A measure of the essential information contained in a message. *Note:* It is expressed as the minimum number of bits or hartleys required to transmit the message with specified accuracy over a noiseless medium. *See also:* **bit.** 42A65-0

net interchange (power and/or energy) (control area). The algebraic sum of the power and/or energy on the area tie lines. *Note:* Positive net interchange is due to excess generation out of the area. *See also:* **power system, low-frequency and surge testing.** E94-0

net interchange deviation (control area) (electric power systems). The net interchange minus the scheduled net interchange. *See also:* **power system, low-frequency and surge testing.** E94-0

net interchange schedule programmer (speed-governing system). A means of automatically changing the net-interchange schedule from one level to another at a predetermined time and during a predetermined period or at a predetermined rate. *See also:* **speed-governing system.** E94-0

net load capability. The maximum system load expressed in kilowatthours per hour that a generating unit, station, or power source can be expected to supply under good operating conditions. *Notes:* (1) This capability provides for variations of load within the hour that it is assumed are to be spread among all of the power sources. (2) This is sometimes called net rated capability. *See also:* **generating station.** 42A35-31E13/31E4

net loss (circuit equivalent*) (circuit). The net loss is the sum of all the transmission losses occurring between the two ends of the circuit, minus the sum of all the transmission gains. *See also:* **transmission loss.**
*Obsolescent 42A65-0

network (1) (distribution of electric energy). An aggregation of interconnected conductors consisting of feeders, mains, and services. *See also:* **alternating-current distribution; network analysis.** 42A35-31E13
(2) **(computing systems).** *See:* **computer network.** *See also:* **electronic digital computer; stabilization network.**

network analysis (network). The derivation of the electrical properties, given its configuration and element values. *Note:* For an extensive list of cross references, see *Appendix A.* E270/42A65-0

network analyzer. An aggregation of electric circuit elements that can readily be connected so as to form models of electric networks. *Note:* Each model thus formed can be used to infer the electrical quantities at various points on the prototype system from corresponding measurements on the model. *See also:* **electronic analog computer; oscillograph.** 42A30-0

network feeder. A feeder that supplies energy to a network. *See also:* **center of distribution.** 42A35-31E13

network function. Any impedance function, admittance function, or other function of p that can be expressed in terms of or derived from the determinant of a network and its cofactors. *Notes:* (1) This includes not only impedance and admittance functions as previously defined, but also voltage ratios, current ratios, and numerous other quantities. (2) Certain network functions are sometimes classified together for a given purpose (for example, those with common zeros or common poles). These represent subgroups that should be specifically defined in each special case. (3) In the case of distributed-parameter networks, the determinant may be an infinite one; the definition still holds. *See also:* **network analysis.** E270-0

network limiter. An enclosed fuse for disconnecting a faulted cable from a low-voltage network distribution system and for protecting the unfaulted portions of that cable against serious thermal damage. 37A100-31E11

network master relay. A relay that functions as a protective relay by opening a network protector when power is back-fed into the supply system and as a programming relay by closing the protector in conjunction with the network phasing relay when polyphase voltage phasors are within prescribed limits. 37A100-31E11/31E6

network phasing relay. A monitoring relay that has as its function to limit the operation of a network master relay so that the network protector may close only when the voltages on the two sides of the protector are in a predetermined phasor relationship. 37A100-31E11/31E6

network primary distribution system. A system of alternating-current distribution in which the primaries of the distribution transformers are connected to a common network supplied from the generating station or substation distribution buses. *See also:* **alternating-current distribution.** 42A35-31E13

network protector. An assembly comprising a circuit breaker and its complete control equipment for automatically disconnecting a transformer from a secondary network in response to predetermined electric conditions on the primary feeder or transformer, and for connecting a transformer to a secondary network either through manual control or automatic control responsive to predetermined electrical conditions on the feeder and the secondary network. *Note:* The network protector is usually arranged to connect automatically its associated transformer to the network when conditions are such that the transformer, when connected, will supply power to the network and to automatically disconnect the transformer from the network when power flows from the network to the transformer. *See also:* **network tripping and reclosing equipment.** 37A100-31E11

network restraint mechanism. A device that prevents opening of a network protector on transient power reversals that do not either exceed a predetermined value or persist for a predetermined time. 37A100-31E11

network secondary distribution system. A system of alternating-current distribution in which the secondaries of the distribution transformers are connected to a common network for supplying light and power directly to consumers' services. *See also:* **alternating-current distribution.** 42A35-31E13

network synthesis (network). The derivation of the configuration and element values, with given electrical properties. *See also:* **network analysis.** E270/42A65-0

network tripping and reclosing equipment (networking equipment). An equipment that automatically connects its associated power transformer to an alternating-current network when conditions are such that the transformer, when connected, will supply power to the network and that automatically disconnects the transformer from the network when power flows from the network to the transformer. *See also:* **network protector.** 37A100-31E11

neuroelectricity. Any electric potential maintained or current produced in the nervous system. *See also:* **electrobiology.** 42A80-18E1

neutral (rotating machinery). The point along an insulated winding where the voltage is the instantaneous average of the line terminal voltages during normal operation. *See:* **asynchronous machine; synchronous machine.** 0-31E8

neutral conductor (when one exists) (circuit consisting of three or more conductors). The conductor that is intended to be so energized, that, in the normal steady state, the voltages from every other conductor to the neutral conductor, at the terminals of entry of the circuit into a delimited region, are definitely related and usually equal in amplitude. *Note:* If the circuit is an alternating-current circuit, it is intended also that the voltages have the same period and the phase difference between any two successive voltages, from each of the other conductors to the neutral conductor, selected in a prescribed order, have a predetermined value usually equal to 2π radians divided by the number of phase conductors m. *See also:* **center of distribution; network analysis.** E270-0

neutral direct-current telegraph system (single-current system) (single Morse system). A telegraph system employing current during marking intervals and zero current during spacing intervals for transmission of signals over the line. *See also:* **telegraphy.** 42A65-0

neutral ground. An intentional ground applied to the neutral conductor or neutral point of a circuit, transformer, machine, apparatus, or system. *See:* **ground.** 42A15/42A35-31E12/31E13

neutral grounding capacitor (electric power). A neutral grounding device the principal element of which is capacitance. *Note:* A neutral grounding capacitor is normally used in combination with other elements, such as reactors or resistors. *See also:* **grounding device.** E32/42A35-31E13

neutral grounding device (electric power). A grounding device used to connect the neutral point of a system of electric conductors to earth. *Note:* The device may consist of a resistance, inductance, or capacitance element, or a combination of them. *See also:* **grounding device; neutral grounding impedor.** E32/42A35-31E13

neutral grounding impedor (electric power). A neutral grounding device comprising an assembly of at least two of the elements resistance, inductance, or capacitance. *See also:* **grounding device; neutral grounding device.** E32/42A35-31E13

neutral grounding reactor. *See:* **reactor, neutral grounding.**

neutral grounding resistor (electric power). A neutral grounding device, the principal element of which is resistance. *See also:* **grounding device.** E32/42A35-31E13

neutralization. A method of nullifying the voltage feedback from the output to the input circuits of an amplifier through the tube interelectrode impedances. *Note:* Its principal use is in preventing oscillation in an amplifier by introducing a voltage into the input equal in magnitude but opposite in phase to the feedback through the interelectrode capacitance. *See also:* **amplifier; feedback.** E145/E182A/42A65-0

neutralizing indicator. An auxiliary device for indicating the degree of neutralization of an amplifier. (For example, a lamp or detector coupled to the plate tank circuit of an amplifier.) *See also:* **amplifier.** E145-0

neutralizing voltage. The alternating-current voltage specifically fed from the grid circuit to the plate circuit (or vice versa), deliberately made 180 degrees out of phase with, and equal in amplitude to, the alternating-current voltage similarly transferred through undesired paths, usually the grid-to-plate tube capacitance. *See also:* **amplifier.** E145-0

neutral lead (rotating machinery). A main lead connected to the common point of a star-connected winding. *See:* **asynchronous machine; synchronous machine.** 0-31E8

neutral point (of a system). The point that has the same potential as the point of junction of a group of equal nonreactive resistances if connected at their free ends to the appropriate main terminals or lines of the system. *Note:* The number of such resistances is 2 for direct-current or single-phase alternating-current; 4 for two-phase (applicable to 4-wire systems only) and 3 for three-, six- or twelve-phase systems. *See also:* **alternating-current distribution; center of distribution; direct-current distribution.** 42A35-31E13

neutral relay (1) (sometimes called nonpolarized relay). A relay in which the movement of the armature does not depend upon the direction of the current in the circuit controlling the armature. 42A42/42A65/83A16-21E0

(2) (power circuits). A relay that responds to quantities in the neutral of a power circuit. *See also:* **electromagnetic relay; relay.** 37A100-31E11/31E6

neutral terminal. The terminal connected to the neutral of a machine or apparatus. *See:* **asynchronous machine; synchronous machine.** 0-31E8

neutral wave trap (electric power). A neutral grounding device comprising a combination of inductance and capacitance designed to offer a very high impedance to a specified frequency or frequencies. *See also:* **grounding device.** E32/42A35-31E13

new installation (elevators). Any installation not classified as an existing installation by definition, or an existing elevator, dumbwaiter, or escalator moved to a new location subsequent to the effective date of a code. *See also:* **elevators.** 42A45-0

newton. The unit of force in the International System of Units (SI). The newton is the force that will impart an acceleration of 1 meter per second per second to a mass of 1 kilogram. One newton equals 10^5 dynes. E270-0

***n*-gate thyristor.** A thyristor in which the gate terminal is connected to the *n* region adjacent to the region to which the anode terminal is connected and which is normally switched to the ON state by applying a negative signal between gate and anode terminals. *See also:* **thyristor.** E223-34E17/34E24/15E7

Nichols chart (Nichols diagram) (control system, feedback). A plot showing magnitude contours and phase contours of the return transfer function referred to ordinates of logarithmic loop gain and to abscissae of loop phase angle. *See also:* **control system, feedback.** 85A1-23E0

nickel-cadmium storage battery. An alkaline storage battery in which the positive active material is nickel oxide and the negative contains cadmium. *See also:* **battery (primary or secondary).** 42A60-0

night. The hours between the end of evening civil twilight and the beginning of morning civil twilight. *Note:* Civil twilight ends in the evening when the center of the sun's disk is six degrees below the horizon and begins in the morning when the center of the sun's disk is six degrees below the horizon. *See also:* **sunlight.** Z7A1-0

night alarm. An electric bell or buzzer for attracting the attention of an operator to a signal when the switchboard is partially attended. *See also:* **telephone switching system.** 42A65-0

night effect (radio navigation systems). A special case of error occurring predominantly at night when sky-wave propagation is at the maximum. *See also:* **navigation.** 0-10E6

nines complement. The radix-minus-one complement of a numeral whose radix is ten. *See also:* **electronic digital computer.** X3A12-16E9

ninety-percent response time of a thermal converter (electric instrument). The time required for 90 percent of the change in output electromotive force to occur after an abrupt change in the input quantity to a new constant value. See note 1 of **response time of a thermal converter.** *See also:* **thermal converter.** 39A1-0

nipple (rigid metal conduit). A straight piece of rigid metal conduit not more than two feet in length and threaded on each end. *See also:* **raceways.** 42A95/80A1-0

nit. The unit of luminance (photometric brightness) equal to one candela per square meter. *Note:* Candela per square meter is the unit of luminance in the International System of Units (SI). Nit is the name recommended by the International Commission on Illumination. *See also:* **light.** Z7A1-0

***n*-level address (computing systems).** A multilevel address that specifies *n* levels of addressing. *See also:* **electronic digital computer.** X3A12-16E9

noble potential. A potential substantially cathodic to the standard hydrogen potential. *See also:* **stray current corrosion.** CM-34E2

node (network analysis). One of the set of discrete points in a flow graph. *See also:* **linear signal flow graphs.** E155-0

node (1) (network) (junction point) (branch point) (vertex). A terminal of any branch of a network or a terminal common to two or more branches of a network. *See also:* **network analysis.** E153/E270-0
(2) (standing wave). A point at which the amplitude is zero. *See also:* **radio transmission; transmission line.** E270-0

node absorption (network analysis). A flow-graph transformation whereby one or more dependent nodes disappear and the resulting graph is equivalent with respect to the remaining node signals. *Note:* For example, a circuit analog of node absorption is the star-delta transformation. *See also:* **linear signal flow graphs.** E155-0

node equations (network). Any set of equations (of minimum number) such that the independent node voltages of a specified network may be determined from the impressed currents. *Notes:* (1) The number of node equations for a given network is not necessarily the same as the number of mesh or loop equations for that network. (2) Notes for mesh or loop equations, with appropriate changes, apply here. *See also:* **network analysis.** E270-0

node signal (network analysis). A variable x_k associated with node *k*. *See also:* **linear signal flow graphs.** E155-0

node voltage (network). The voltage from a reference point to any junction point (node) in a network. *Note:* The assumptions of lumped-network theory are such that the path of integration is immaterial. E270-0

noise (1) (general). Unwanted disturbances superposed upon a useful signal that tend to obscure its information content. *See also:* **electric noise; electromagnetic noise; electronic analog computer; modulation systems; signal-to-noise ratio.** E165/85A1-23E0
(2) An undesired disturbance within the useful frequency band. *Note:* Undesired disturbances within the useful frequency band produced by other services may be called interference. E145-0
(3) (sound recording and reproducing system). Any output power that tends to interfere with the utilization of the applied signals except for output signals that consist of harmonics and subharmonics of the input signals, intermodulation products, and flutter or wow. *See:*
electroacoustics;
equipment noise;
medium noise;
modulation noise;
signal-frequency signal-to-noise ratio;
spectral-noise density;
system noise;
zero-modulation medium noise;
zero-modulation state. E191-0
(4) (phototubes). The random output that limits the

minimum observable signal from the phototube. *See also:* **phototube.** E175-0

(5) (facsimile). Any extraneous electric disturbance tending to interfere with the normal reception of a transmitted signal. *See also:* **facsimile transmission.** E168-0

noise, audio-frequency. Any unwanted disturbance in the audio-frequency range. E151-0

noise-current generator. A current generator, the output of which is described by a random function of time. *Note:* At a specified frequency, a noise-current generator can often be adequately characterized by its mean-square current within the frequency increment Δf or by its spectral density. If the circuit contains more than one noise-voltage generator or noise-current generator, the correlation coefficients among the generators must also be specified. *See also:* **network analysis; signal; signal-to-noise ratio.** E160-13E6/15E6

noise diode, ideal. A diode that has an infinite internal impedance and in which the current exhibits full shot noise fluctuations. *See also:* **signal-to-noise ratio; tube definitions.** E160-15E6

noise factor (interference terminology) (linear system) (1) (at a selected input frequency). The ratio of (A) the total noise power per unit bandwidth (at a corresponding output frequency) delivered by the system into an output termination to (B) the portion thereof engendered at the input frequency by the input termination, whose noise temperature is standard (290 kelvins at all frequencies). *Notes:* (a) Numerically, the noise factor F at frequency f can be expressed as

$$F(f)=\frac{P_{\text{noise out}}}{GP_{\text{noise in}}}$$

where $P_{\text{noise out}}$ and $P_{\text{noise in}}$ are taken at frequency f and 290 kelvins and G is the gain of the system at frequency f. (b) For heterodyne systems there will be, in principle, more than one output frequency corresponding to a single input frequency, and vice versa; for each pair of corresponding frequencies a noise factor is defined. (c) The phrase **available at the output terminals** may replace **delivered by the system into an output termination** without changing the sense of the definition. (d) The term **noise factor** is used where it is desired to emphasize that the noise factor is a point function to input frequency. *See:* **signal; two-port transducer.** E161-15E6;E188-13E6

(2) (average). The ratio of (A) the total noise power delivered by the system into its output termination when the noise temperature of its input termination is standard (290 kelvins) at all frequencies to (B) the portion thereof engendered by the input termination. *Notes:* (a) For heterodyne systems, portion (A) includes only that noise from the input termination that appears in the output via the principal frequency transformation of the system, and does not include spurious contributions such as those from image-frequency transformations. (b) A quantitative relation between **average noise factor** $\bar{F}$ and **spot noise factor** $F(f)$ is

$$\bar{F}=\frac{\int_0^\infty F(f)G(f)\mathrm{d}f}{\int_0^\infty G(f)\mathrm{d}f}$$

where f is the input frequency and $G(f)$ is the ratio of *(a)* the signal power delivered by the system into its output termination to *(b)* the corresponding signal power available from the input termination at the input frequency. For heterodyne systems, *(a)* comprises only power appearing in the output via the principal frequency transformation of the system. (c) To characterize a system by an average noise factor is meaningful only when the admittance (or impedance) of the input termination is specified. *See also:* **radio receiver; transducer; two-port transducer.** E161-15E6;E188-0

noise figure. *See:* **noise factor.**

noise-free equivalent amplifier (signal-transmission system). An ideal amplifier having no internally generated noise that has the same gain and input/output characteristics as the actual amplifier. *See:* **signal.** 0-13E6

noise generator (analog computer). A computing element used purposely to introduce noise of specified amplitude distribution, spectral density, and root-mean-square value into other computing elements. *See also:* **electronic analog computer.** E165-16E9

noise generator diode (electron device). A diode in which the noise is generated by shot effect and the noise power of which is a definite function of the direct current. *See also:* **tube definitions.** 0-15E6

noise killer (telegraph circuit). An electric network inserted usually at the sending end, for the purpose of reducing interference with other communication circuits. *See also:* **telegraphy.** 42A65-0

noise level. (1) The noise power density spectrum in the frequency range of interest, (2) the average noise power in the frequency range of interest, or (3) the indication on a specified instrument. *Notes:* (A) In (3), the characteristics of the instrument are determined by the type of noise to be measured and the application of the results thereof. (B) Noise level is usually expressed in decibels relative to a reference value. *See also:* **signal-to-noise ratio.** E151/42A65-0

noise-level test (rotating machinery). A test taken to determine the noise level produced by a machine under specified conditions of operation and measurement. *See also:* **asynchronous machine; direct-current commutating machine; synchronous machine.** 0-31E8

noise-longitudinal (telephone practice). The 1/1000th part of the total longitudinal-circuit noise current at any given point in one or more telephone wires. *See also:* **inductive coordination.** 42A65-0

noise measuring set. *See:* **circuit noise meter.**

noise-metallic (metallic circuit). The weighted noise current at a given point when the circuit is terminated at that point in the nominal characteristic impedance of the circuit. *See also:* **inductive coordination.** 42A65-0

noise pressure equivalent (electroacoustic transducer or system used for sound reception). The root-mean-square sound pressure of a sinusoidal plane progressive wave that, if propagated parallel to the principal axis of the transducer, would produce an open-circuit signal voltage equal to the root-mean-square of the inherent open-circuit noise voltage of the transducer in a transmission band having a bandwidth of 1 hertz and centered on the frequency of the plane sound wave. *Note:* If the equivalent noise pressure of the transducer is a function of secondary variables, such as ambient temperature or pressure, the applicable values of these quantities should be stated explicitly. *See also:* **phonograph pickup.** 0-1E1

noise quieting (receiver) (receiver performance). A measure of the quantity of radio-frequency energy, at a specified deviation from the receiver center frequency, required to reduce the noise output by a specified amount. *See also:* **receiver performance.** 0-6E1

noise reduction (photographic recording and reproducing). A process whereby the average transmission of the sound track of the print (averaged across the track) is decreased for signals of low level and increased for signals of high level. *Note:* Since the noise introduced by the sound track is less at low transmission, this process reduces film noise during soft passages. The effect is normally accomplished automatically. *See also:* **phonograph pickup.** 0-1E1

noise temperature (1) (general) (at a pair of terminals and at a specific frequency). The temperature of a passive system having an available noise power per unit bandwidth equal to that of the actual terminals. *Note:* Thus, the noise temperature of a simple resistor is the actual temperature of the resistor, while the noise temperature of a diode may be many times the observed absolute temperature. *See also:* **signal; signal-to-noise ratio.** E188-13E6;42A65-0

(2) (standard). The standard reference temperature T_0 for noise measurements is 290 kelvins. *Note:* $kT_0/e = 0.0250$ volt, where e is the magnitude of the electron charge and k is Boltzmann's constant. *See also:* **circuit characteristics of electrodes; transducer.** E188-0;E161-15E6

(3) (antenna). The temperature of a resistor having an available thermal noise power per unit bandwidth equal to that at the antenna output at a specified frequency. *Note:* Noise temperature of an antenna depends on its coupling to all noise sources in its environment as well as noise generated within the antenna. *See also:* **antenna.** 0-3E1

(4) (at a port and at a selected frequency). A temperature given by the exchangeable noise-power density divided by Boltzmann's constant, at a given port and at a stated frequency. *Notes:* (A) When expressed in units of kelvins, the noise temperature T is given by the relation

$$T = N/k$$

where N is the exchangeable noise-power density in watts per hertz at the port at the stated frequency and k is Boltzmann's constant expressed as joules per kelvin ($k \simeq 1.38 \times 10^{-23}$ joules per kelvin). (B) Both N and T are negative for a port with an internal impedance having a negative real part. *See also:* **signal-to-noise ratio; waveguide.** 0-15E6

noise-to-ground (telephone practice). The weighted noise current through the 100 000-ohm circuit of a circuit noise meter, connected between one or more telephone wires and ground. *See also:* **inductive coordination.** 42A65-0

noise transmission impairment (NTI). The noise transmission impairment that corresponds to a given amount of noise is the increase in distortionless transmission loss that would impair the telephone transmission over a substantially noisefree circuit by an amount equal to the impairment caused by the noise. Equal impairments are usually determined by judgment tests or intelligibility tests. *See also:* **transmission characteristics.** 42A65-0

noise unit*. An amount of noise judged to be equal in interfering effect to the one-millionth part of the current output of a particular type of standard generator of artificial noise, used under specified conditions. *Note:* This term was formerly used in connection with ear balance measurements, but has been largely superseded by dBa employed with indicating noise meter. Approximately seven noise units of noise on a telephone line are frequently taken as equivalent to reference noise. *See also:* **signal-to-noise ratio.**

*Obsolescent 42A65-0

noise-voltage generator (interference terminology). A voltage generator the output of which is described by a random function of time. *Note:* At a specified frequency, a noise-voltage generator can often be adequately characterized by its mean-square voltage with the frequency increment Δf or by its spectral density. If the circuit contains more than one noise-current generator or noise-voltage generator, the correlation coefficients among the generators must be specified. *See:* **network analysis; signal; signal-to-noise ratio.** E160-13E6/15E6

no-load (adjective) (rotating machinery). The state of a machine rotating at normal speed under rated conditions, but when no output is required of it. *See also:* **asynchronous machine; direct-current commutating machine; synchronous machine.** 0-31E8

no-load field voltage. The voltage required across the terminals of the field winding of an electric machine under conditions of no load, rated speed and terminal voltage, and with the field winding at 25 degrees Celsius. *See:* **direct-current commutating machine; synchronous machine.** 42A10-31E8

no-load saturation curve (no-load characteristic) (of a synchronous machine). The saturation curve of a machine on no-load. *See also:* **synchronous machine.** 0-31E8

no-load speed (industrial control) (of an electric drive). The speed that the output shaft of the drive attains with no external load connected and with the drive adjusted to deliver rated output at rated speed. *Note:* In referring to the speed with no external load connected and with the drive adjusted for a specified condition other than for rated output at rated speed, it is customary to speak of the no-load speed under the (stated) conditions. *See also:* **electric drive.** 42A25-34E10

no-load test (synchronous machine). A test in which the machine is run as a motor providing no useful mechanical output from the shaft. *See also:* **synchronous machine.** 0-31E8

nominal band of regulated voltage. The band of regulated voltage for a load range between any load requiring no-load field voltage and any load requiring rated-load field voltage with any compensating means used to produce a deliberate change in regulated voltage inoperative. *See:* **direct-current commutating machine; synchronous machine.** 42A10-0

nominal collector ring voltage (rotating machinery). *See:* **rated-load field voltage.**

nominal discharge current (arrester). The discharge current having a designated peak value and waveshape, that is used to classify an arrester with respect to protective characteristics. *Note:* It is also the discharge current that is used to initiate follow current in the operating duty test. *See:* **lightning arrester (surge diverter).** 0-31E7

nominal line pitch (television). The average separation between centers of adjacent lines forming a raster.

See also: **television.** E204/42A65-2E2

nominal line width (1) (television). The reciprocal of the number of lines per unit length in the direction of line progression. *See also:* **television.** 42A65-0

(2) (facsimile). The average separation between centers of adjacent scanning or recording lines. *See also:* **recording (facsimile); scanning (facsimile).** E168-0

nominal pull-in torque (synchronous motor). The torque it develops as an induction motor when operating at 95 percent of synchronous speed with rated voltage applied at rated frequency. *Note:* This quantity is useful for comparative purposes when the inertia of the load is not known. *See:* **synchronous machine.** 42A10-31E8

nominal rate of rise (impulse) (electric power). The slope of the line that determines the virtual zero. *Note:* It is usually expressed in volts or amperes per microsecond. *See also:* **lightning; lightning arrester (surge diverter).** E28/62A1-31E7

nominal rating. The maximum constant load that having been carried without causing further measurable increase in the temperature rise under prescribed conditions of test, may be increased by a specified amount for two hours without causing temperature-rise limitations established for nominal rated apparatus to be exceeded, and within the limitations of established standards for such equipment. *Note:* Usually 25 or 50 percent increase is specified. *See:* **duty.** 42A15-31E12

nominal synchronous-machine excitation-system ceiling voltage. The ceiling voltage of the excitation system with: (1) The exciter and all rotating elements at rated speed. (2) The auxiliary supply voltages and frequencies at rated values. (3) The excitation system loaded with a resistor having a value equal to the resistance of the field winding to be excited at a temperature of: (A) 75 degrees Celsius for field windings designed to operate at rating with a temperature rise of 60 degrees Celsius or less, and (B) 100 degrees Celsius for field windings designed to operate at rating with a temperature rise greater than 60 degrees Celsius. (4) The manual control means adjusted as it would be to produce the rated voltage of the excitation system, if this manual control means is not under the control of the voltage regulator when the regulator is in service, unless otherwise specified. (The means used for controlling the exciter voltage with the voltage regulator out of service is normally called the manual control means.) (5) The voltage sensed by the synchronous machine voltage regulator reduced to give the maximum output from the regulator. Note that the regulator action may be simulated in test by applying to the field of the exciter under regulator control the maximum output developed by the regulator. *See also:* **synchronous machine.** 0-31E8

nominal system voltage. A nominal value assigned to designate a system of a given voltage class. *Notes:* (1) See American National Standard Guide for Preferred Voltage Ratings for Alternating Current Systems and Equipment, C84.1-1954. (2) The actual voltage may be at variance with the nominal voltage. In fact, depending on the location on the circuit and the conditions of loading under which it is measured, the voltage usually will be other than the nominal value. (3) The term nominal voltage designates the line-to-line voltage, as distinguished from the line-to-neutral voltage. It applies to all parts of the system or circuit. *See also:* **lightning arrester (surge diverter); power systems, low frequency and surge testing; voltage regulator.** 37A100/62A1-31E7/31E11

nominal thickness (cable element) (power distribution, underground cables). The specified, indicated, or named thickness. *Note:* In general, measured thicknesses will approximate but will not necessarily be identical with nominal thicknesses. *See also:* **power distribution, underground construction.** 0-31E1

nonautomatic. The implied action that requires personal intervention for its control. *Note:* As applied to an electric controller, nonautomatic control does not necessarily imply a manual controller, but only that personal intervention is required. *See also:* **electric controller.** 42A25-34E10

nonautomatic opening (nonautomatic tripping). The opening of a switching device only in response to an act of an attendant. 37A100-31E11

nonautomatic operation. Operation controlled by an attendant. 37A100-31E11

nonautomatic tripping. *See:* **nonautomatic opening.**

noncode fire-alarm system. A local fire-alarm system in which the alarm signal is continuous and is usually sounded by vibrating bells. *See also:* **protective signaling.** 42A65-0

noncoincident demand (electric power utilization). The sum of the individual maximum demands regardless of time of occurrence within a specified period. *See also:* **alternating-current distribution.** 0-31E4

noncomposite color-picture signal (television). The electric signal that represents complete color-picture information, including setup and (in the National Television System Committee system) the color burst, but excluding all other synchronizing signals. (The term itself is considered a replacement for the earlier term color-picture signal. The new definition includes the setup and color burst that must be inherent parts of a color-picture signal whether composite or noncomposite.) *See:* **television.** 0-2E2

nonconforming load (electric power systems). A customer load, the characteristics of which are such as to require special treatment in deriving incremental transmission losses. *See also:* **power system, low-frequency and surge testing.** E94-0

noncontact plunger. *See:* **choke piston.**

noncontinuous electrode. A furnace electrode the residual length of which is discarded when too short for further effective use. *See also:* **electrothermics.** 42A60-0

nondestructive read (computing systems) (1) (general). A read process that does not erase the data in the source. *See also:* **electronic digital computer.** X3A12-16E9

(2) (magnetic cores). A method of reading the magnetic state of a core without changing its state. *See also:* **static magnetic storage.** E163-0

nondestructive reading (charge-storage tubes). Reading that does not erase the stored information. *See also:* **charge-storage tube.** E158-15E6

nondirectional beacon (NDB) (air navigation). A radio facility that can be used with an airborne direction-finder to provide a line of position; also known as a compass locator, *H*, *H* beacon. *See also:* **navigation.** 0-10E6

nondirectional microphone. *See:* **omnidirectional microphone.**

nondisconnecting fuse. An assembly consisting of a fuse unit or fuseholder and a fuse support having clips for directly receiving the associated fuse unit or fuseholder, and that has no provision for guided operation as a disconnecting switch. 37A100-31E11

nonerasable storage (computing systems). *See:* **fixed storage.** *See also:* **electronic digital computer.**

nonexposed installation (lightning). An installation in which the apparatus is not subject to overvoltages of atmospheric origin. *Note:* Such installations are usually connected to cable networks. *See:* **lightning arrester (surge diverter).** 50I25-31E7

noninverting parametric device. A parametric device whose operation depends essentially upon three frequencies, a harmonic of the pump frequency and two signal frequencies, of which one is the sum of the other plus the pump harmonic. *Note:* Such a device can never provide gain at either of the signal frequencies. It is said to be noninverting because if either of the two signals is moved upward in frequency, the other will move upward in frequency. *See also:* **parametric device.** E254-15E7

nonlinear capacitor. A capacitor having a mean-charge characteristic or a peak-charge characteristic that is not linear, or a reversible capacitance that varies with bias voltage.
See:
alternating charge characteristic;
capacitance ratio;
differential capacitance;
differential-capacitance characteristic;
ideal capacitor;
initial differential capacitance;
initial reversible capacitance;
mean charge;
mean-charge characteristic;
nonlinear ideal capacitor;
peak-charge characteristic;
reversible capacitance;
reversible-capacitance characteristic;
transferred charge;
transferred charge characteristic;
voltage coefficient of capacitance;
voltage sensitivity. *See also:* **electron devices, miscellaneous.** E226-15E7

nonlinear circuit. *See:* **nonlinear network.**

nonlinear distortion. Distortion caused by a deviation from a desired linear relationship between specified measures of the output and input of a system. *Note:* The related measures need not to be output and input values of the same quantity; for example, in a linear detector the desired relation is between the output signal voltage and the input modulation envelope; or the modulation of the input carrier and the resultant detected signal. *See also:* **close-talking pressure-type microphone; distortion.** E151/E154/E188-42A65

nonlinear ideal capacitor. An ideal capacitor whose transferred-charge characteristic is not linear. *See also:* **nonlinear capacitor.** E226-15E7

nonlinear network (signal-transmission system). (1) A network (circuit) not specifiable by linear differential equations with time and/or position coordinates as the independent variable. *Note:* It will not operate in accordance with the superposition theorem. (2) A network (circuit) in which the signal transmission characteristics depend on the input signal magnitude. *See also:* **network analysis; signal.** E270/E154-13E6

nonlinear parameter. A parameter dependent on the magnitude of one or more of the dependent variables or driving forces of the system. *Note:* Examples of dependent variables are current, voltage, and analogous quantities. E270-0

nonlinear resistor-type arrester (valve type). An arrester having a single or a multiple spark gap connected in series with nonlinear resistance. *Note:* If the arrester has no series gap, the characteristic element limits the follow current to a magnitude that does not interfere with the operation of the system. *See:* **arrester, valve type; lightning arrester (surge diverter).** 0-31E7

nonlinear series resistor (arrester). The part of the lightning arrester that, by its nonlinear voltage-current characteristics, acts as a low resistance to the flow of high discharge currents thus limiting the voltage across the arrester terminals, and as a high resistance at normal power-frequency voltage thus limiting the magnitude of follow current. *See:* **lightning arrester (surge diverter).** 0-31E7

nonlinear system or element. *See:* **linear system or element.**

nonlined construction (primary cell) (dry cell). A type of construction in which a layer of paste forms the only medium between the depolarizing mix and the negative electrode. *See also:* **electrolytic cell.** 42A60-0

nonloaded Q (basic Q) (of an electric impedance). The value of Q of such impedance without external coupling or connection. *See also:* **transmission characteristics.** 42A65-0

nonmagnetic relay armature stop. *See:* **relay armature stop, nonmagnetic.**

nonmagnetic ship. A ship constructed with an amount of magnetic material so small that it causes negligible distortion of the earth's magnetic field. *See also:* **degauss.** 42A43-0

nonmechanical switching device. A switching device designed to close or open, or both, one or more electric circuits by means other than by separable mechanical contacts. 37A100-31E11

nonmetallic sheathed cable. An assembly of two or more insulated conductors having an outer sheath of moisture-resistant, flame-retardant, nonmetallic material. *See also:* **armored cable (in interior wiring).** 42A95-0

nonmultiple switchboard. A manual telephone switchboard in which each subscriber line is attached to only one jack. *See also:* **telephone switching system.** 42A65-19E1

nonnumerical action (switch). That action that does not depend on the called number (such as hunting an idle trunk). *See also:* **telephone switching system.** 42A65-0

nonoperate value. *See:* **relay nonoperate value.**

nonphantomed circuit. A two-wire or four-wire circuit that is not arranged to form part of a phantom circuit. *See also:* **transmission line.** 42A65-0

nonphysical primary (color). A primary represented by a point outside the area of the chromaticity diagram enclosed by the spectrum locus and the purple boundary. *Note:* Nonphysical primaries cannot be produced because they require negative power at some wavelengths. However, they have properties that facilitate colorimetric calculation. Tristimulus values based upon them are derived from tristimulus values based upon physical primaries. *See also:* **color terms.** E201-0

nonplanar network. A network that cannot be drawn on a plane without crossing of branches. *See also:* **network analysis.** E153/E270

nonpolarized electrolytic capacitor. An electrolytic capacitor in which the dielectric film is formed adjacent to both metal electrodes and in which the impedance to the flow of current is substantially the same in both directions. *See also:* **electrolytic capacitor.** 42A60-0

***n*-on-*p* solar cells (photovoltaic power system).** Photovoltaic energy-conversion cells in which a base of *p*-type silicon (having fixed electrons in a silicon lattice and positive holes that are free to move) is overlaid with a surface layer of *n*-type silicon (having fixed positive holes in a silicon lattice with electrons that are free to move). *See also:* **photovoltaic power system; solar cells (photovoltaic power system).** 0-10E1

nonquadded cable. *See:* **paired cable.**

nonrenewable fuse (one-time fuse). A fuse not intended to be restored for service after circuit interruption. 37A100-31E11

nonrepaired item (reliability). *See:* **item, nonrepaired.**

nonrepetitive peak OFF-state voltage (thyristor). The maximum instantaneous value of any nonrepetitive transient OFF-state voltage that occurs across the thyristor. *See also:* **principal voltage-current characteristic (principal characteristic).** E223-34E17/34E24/15E7

nonrepetitive peak reverse voltage (1) (reverse-blocking thyristor). The maximum instantaneous value of any unrepetitive transient reverse voltage that occurs across a thyristor. *See also:* **principal voltage-current characteristic (principal characteristic).** E223-34E17/34E24/15E7

(2) (semiconductor rectifier). The maximum instantaneous value of the reverse voltage, including all nonrepetitive transient voltages but excluding all repetitive transient voltages, that occurs across a semiconductor rectifier cell, rectifier diode, or rectifier stack. *See also:* **rectification; semiconductor rectifier stack.** E59-34E24

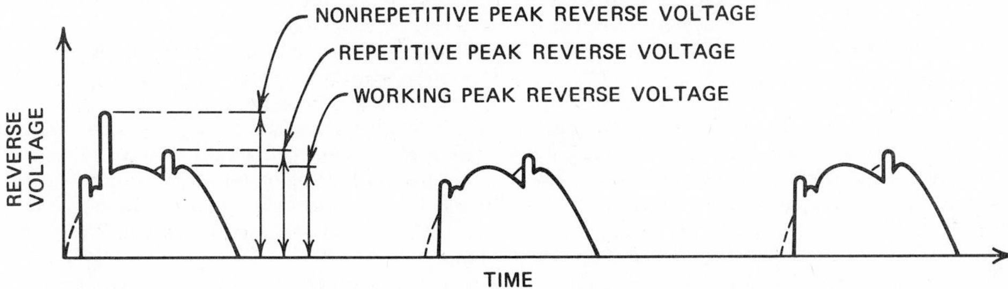

Peak reverse voltage.

nonrepetitive peak reverse voltage rating (rectifier circuit). The maximum value of nonrepetitive peak reverse voltage permitted by the manufacturer under stated conditions. *See also:* **average forward-current rating (of a rectifier circuit).** E59-34E17/34E24

non-return to zero (NRZ) code (power-system communication). A code form having two states termed zero and one, and no neutral or rest condition. *See also:* **digital.** 0-31E3

nonreversible output (magnetic amplifier). *See:* **maximum test output voltage.**

nonreversible power converter (semiconverter). An equipment containing assemblies of mixed power thyristor and diode devices that is capable of transferring energy in only one direction (that is, from the alternating-current side to the direct-current side). *See:* **power rectifier.** 0-34E24

nonreversing (industrial control). A control function that provides for operation in one direction only. *See also:* **control system, feedback.** IC1-34E10

nonsalient pole (rotating machinery). The part of a core, usually circular, that by virtue of direct-current excitation of a winding embedded in slots and distributed over the interpolar (and possibly over some or all of the polar) space, acts as a pole. *See also:* **asynchronous machine; synchronous machine.** 0-31E8

nonsalient-pole machine. *See:* **cylindrical-rotor machine.**

nonsegregated-phase bus. A bus in which all phase conductors are in a common metal enclosure without barriers between the phases. 37A100-31E11

nonselective collective automatic operation (elevators). Automatic operation by means of one button in the car for each landing level served and one button at each landing, wherein all stops registered by the momentary actuation of landing or car buttons are made irrespective of the number of buttons actuated or of the sequence in which the buttons are actuated. *Note:* With this type of operation the car stops at all landings for which buttons have been actuated, making the stops in the order in which the landings are reached after the buttons have been actuated, but irrespective of its direction of travel. *See also:* **control (elevators).** 42A45-0

nonself-maintained discharge (gas). A discharge characterized by the fact that the charged particles are produced solely by the action of an external ionizing agent. *See also:* **discharge (gas).** 50I07-15E6

nonstop switch (elevators). A switch that, when operated, will prevent the elevator from making registered landing stops. *See:* **control.** 42A45-0

nonstorage display (display storage tubes). Display of nonstored information in the storage tube without appreciably affecting the stored information. *See also:* **storage tube.** E158-15E6

nonsynchronous transmission (data transmission). A transmission process such that between any two significant instants in the same group, there is always an integral number of unit intervals. Between two significant instants located in different groups, there is not always an integral number of unit intervals. *Note:* In data transmission this group is a block or a character. In telegraphy this group is a character. *See also:* **data transmission.** 0-19E4

nontouching loop set (network analysis). A set of loops no two of which have a common node. *See also:* **linear signal flow graphs.** E155-0

nonuniformity (transmission lines and waveguides). The degree with which a characteristic quantity, for example, impedance, deviates from a constant value along a given path. *Note:* It may be defined as the maximum amount of deviation from a selected nominal value. For example, the nonuniformity of the characteristic impedance of a slotted coaxial line may be 0.05 ohm due to dimensional variations. *See also:* **transmission characteristics.** 0-9E4

nonvented fuse. A fuse without intentional provision for the escape of arc gases, liquids, or solid particles to the atmosphere during circuit interruption. 37A100-31E11

nonventilated enclosure. An enclosure designed to provide limited protection against the entrance of dust, dirt, or other foreign objects. *Note:* Doors or removable covers are usually gasketed and humidity control may be provided by filtered breathers. 37A100-31E11

no op (computing systems). An instruction that specifically instructs the computer to do nothing, except to proceed to the next instruction in sequence. *See also:* **electronic digital computer.** X3A12-16E9

NOR. A logic operator having the property that if *P* is a statement, *Q* is a statement, *R* is a statement, ... then the NOR of *P, Q, R,* ... is true if all statements are false, false if at least one statement is true. X3A12-16E9

normal (state of a superconductor). The state of a superconductor in which it does not exhibit superconductivity. *Example:* Lead is normal at temperatures above a critical temperature. *See:* **superconducting.** *See also:* **superconductivity.** E217-15E7

normal clear. A term used to express the normal indication of the signals in an automatic block system in which an indication to proceed is displayed except when the block is occupied. *See also:* **railway signal and interlocking.** 42A42-0

normal clear system. A term describing the normal indication of the signals in an automatic block signal system in which an indication to proceed is displayed except when the block is occupied. *See also:* **centralized traffic control system.** 42A42-0

normal contact. A contact that is closed when the operating unit is in the normal position. *See also:* **railway signal and interlocking.** 42A42-0

normal (through) dielectric heating applications. The metallic electrodes are arranged on opposite sides of the material so that the electric field is established through it. *Note:* The electrodes may be classified as plate electrodes, roller electrodes, or concentric electrodes. (1) Plate electrodes may have plane surfaces or surfaces of any desired shape to meet a particular condition and the spacing between them may be uniform or varied. (2) Roller electrodes are rollers separated by the material which moves between them. (3) Concentric electrodes consist of an enclosed and a surrounding electrode, with the material placed between the two. *See also:* **dielectric heating.** E54-0

normal dimension. Incremental dimensions whose number of digits is specified in the format classification. For example, the format classification would be plus 14 for a normal dimension; X.XXXX. *See:* **dimension; incremental dimension; long dimension; short dimension.** EIA3B-34E12

normal frequency. The frequency at which a device or system is designed to operate. 37A100-31E11

normal-frequency (low-frequency) dew withstand voltage. The normal-frequency withstand voltage applied to insulation completely covered with condensed moisture. *Note: See:* **normal-frequency withstand voltage.** 37A100-31E11

normal-frequency (low-frequency) dry withstand voltage. The normal-frequency withstand voltage applied to dry insulation. *See:* **normal-frequency (low-frequency) withstand voltage.** 37A100-31E11

normal-frequency line-to-line recovery voltage. The normal-frequency recovery voltage, stated on a line-to-line basis, that occurs on the source side of a three-phase circuit-interrupting device after interruption is complete in all three poles. 37A100-31E11

normal-frequency pole-unit recovery voltage. The normal-frequency recovery voltage that occurs across a pole unit of a circuit-interrupting device upon circuit interruption. 37A100-31E11

normal-frequency recovery voltage. The normal-frequency root-mean-square voltage that occurs across the terminals of an alternating-current circuit-interrupting device after the interruption of the current and after the high-frequency transients have subsided. *Note:* For determination of the normal-frequency recovery voltage, see American National Standard C37.05-1964. 37A100-31E11

normal-frequency (low-frequency) wet withstand voltage. The normal-frequency withstand voltage applied to wetted insulation. *See:* **normal-frequency withstand voltage.** 37A100-31E11

normal-frequency (low-frequency) withstand voltage. The normal-frequency voltage that can be applied to insulation under specified conditions for a specified time without causing flashover or puncture. *Note:* This value is usually expressed as a root-mean-square value. *See also:* **normal-frequency dew withstand voltage; normal-frequency dry withstand voltage; normal-frequency wet withstand voltage.** 37A100-31E11

normal-glow discharge (gas). The glow discharge characterized by the fact that the working voltage decreases or remains constant as the current increases. *See also:* **discharge (gas).** 50I07-15E6

normal induction (magnetic material). The maximum induction in a magnetic material that is in a symmetrically cyclically magnetized condition. E270-0

normalize (computing systems). To adjust the representation of a quantity so that the representation lies in a prescribed range. *See also:* **electronic digital computer.** X3A12-16E9

normalized admittance. The reciprocal of the normalized impedance. *See also:* **waveguide.** E146-3E1

normalized impedance (waveguide). The ratio of an impedance and the corresponding characteristic impedance. *Note:* The normalized impedance is independent of the convention used to define the characteristic impedance, provided that the same convention is also taken for the impedance to be normalized. *See:* **waveguide.** E146-3E1

normalized transimpedance (magnetic amplifier). The ratio of differential output voltage to the product of differential control current and control winding turns. *See also:* **rating and testing magnetic amplifiers.** E107-0

normal linearity (computing systems). Transducer linearity defined in terms of a zero-error reference, that is chosen so as to minimize the linearity error. *Note:* In this case, the zero input does not have to yield zero output and full-scale input does not have to yield full-scale output. The specification of normal linearity, therefore, is less stringent than zero-based linearity. Linearity of a multiplier is (1) The ability of an electromechanical or electronic multiplier to generate an output voltage that varies linearly with either one of its two inputs, provided the other input is held constant. (2) The accuracy with which the above requirement is met. Linearity of a potentiometer is the accuracy with which a potentiometer yields a linear but not necessarily a proportional relationship between the angle of rotation of its shaft and the voltage appearing at the output arm in the absence of loading errors. *See also:* **electronic analog computer; multiplier (linearity).** E165-16E9

normally closed. *See:* **normally open and normally closed.**

normally closed contact. A contact, the current-carrying members of which are in engagement when the operating unit is in its normal position. *See also:* **railway signal and interlocking.** 42A42-0

normally open contact (open contact). A contact, the current-carrying members of which are not in engagement when the operating unit is in the normal position. *See also:* **railway signal and interlocking.** 42A42-0

normally open and normally closed (industrial control). When applied to a magnetically operated switching device, such as a contactor or relay, or to the contacts thereof, these terms signify the position taken when the operating magnet is deenergized. Applicable only to nonlatching types of devices. *See also:* **contacter.** 42A25-34E10

normal-mode interference. *See:* **interference, normal-mode.**

normal permeability. The ratio of normal induction to the corresponding maximum magnetizing force. *Note:* In anisotropic media, normal permeability becomes a matrix. E270-0

normal position (of a device). A predetermined position that serves as a starting point for all operations. 42A42-0

normal stop system. A term used to describe the normal indication of the signals in an automatic block signal system in which the indication to proceed is given only upon the approach of a train to an unoccupied block. *See also:* **centralized traffic-control system.** 42A42-0

normal-terminal stopping device (elevators). A device, or devices, to slow down and stop an elevator or dumbwaiter car automatically at or near a terminal landing independently of the functioning of the operating device. *See:* **control.** 42A45-0

normal weather (electric power systems). All weather not designated as adverse. *See also:* **outage.** 0-31E4

normal weather persistent-cause forced-outage rate (for a particular type of component) (electric power systems). The mean number of outages per unit of normal weather time per component. *See also:* **outage.** 0-31E4

north-stabilized plan-position indicator. A special case of azimuth-stabilized plan-position indicator in which the reference direction is magnetic north. *See also:* **navigation.** 0-10E6

Norton's theorem. States that the voltage that will exist across an admittance Y', when connected to any two terminals of a linear network between which the short-circuit current previously was I and the admittance Y, is equal to the current I divided by the sum of Y and Y'. *See also:* **communication.** 42A65-0

nose suspension. A method of mounting an axle-hung motor or a generator to give three points of support, consisting of two axle bearings (or equivalent) and a lug or nose projecting from the opposite side of the motor frame, the latter supported by a truck or vehicle frame. *See also:* **traction motor.** 42A42-0

NOT. A logic operator having the property that if P is a statement, then the NOT of P is true if P is false, false if P is true. The NOT of P is often represented by $\overline{P}$, $\sim P$, $\neg P$, P'. X3A12-16E9

notation. *See:* **positional notation.** *See also:* **electronic digital computer.**

notching (relays). A qualifying term applied to a relay indicating that a predetermined number of separate impulses is required to complete operation. *See:* **relay.** 42A20

notching relay. A programming relay in which the response is dependent upon successive impulses of the input quantity. 37A100-31E11/31E6

note. A conventional sign used to indicate the pitch, or the duration, or both, of a tone. It is also the tone sensation itself or the oscillation causing the sensation. *Note:* The word serves when no distinction is desired among the symbol, the sensation, and the physical stimulus. *See also:* **electroacoustics.** E157-1E1

***N* scan (electronic navigation).** *See:* ***N* display.**

***N* scope (electronic navigation).** *See:* ***N* display.**

***n*-terminal electron tubes.** *See:* **capacitance, input.**

***N*-terminal network.** A network with N accessible terminals. *See also:* **network analysis.** E153/E270-0

***N*-terminal pair network.** A network with $2N$ accessible terminals grouped in pairs. *Note:* In such a network one terminal of each pair may coincide with a network node. *See also:* **network analysis.** E153/E270-0

***N*th harmonic.** The harmonic of frequency N times that of the fundamental component. E154-0

NTI. *See:* **noise transmission impairment.**

***n*-type crystal rectifier.** A crystal rectifier in which forward current flows when the semiconductor is negative with respect to the metal. *See also:* **rectifier.** 42A65-0

***n*-type semiconductor.** *See:* **semiconductor, *n*-type.**

null (1) (signal-transmission system). The condition of zero error-signal achieved by equality at a summing junction between an input signal and an automatically or manually adjusted balancing signal of phase or polarity opposite to the input signal. *See also:* **signal.** 0-13E6

(2) (direction-finding systems). The condition of minimum output as a function of the direction of arrival of the signal, or of the rotation of the response pattern of the direction-finder antenna system. *See also:* **navigation.** 0-10E6

null balance (automatic null-balancing electric instrument) (instruments). The condition that exists in the circuits of an instrument when the difference between an opposing electrical quantity within the instrument and the measured signal does not exceed the dead band. *Note:* The value of the opposing electrical quantity produced within the instrument is related to the position of the end device. *See also:* **control system, feedback; measurement system.** 39A4-0

null-balance system. A system in which the input is measured by producing a null with a calibrated balancing voltage or current. *See also:* **signal.** 0-13E6

nullity (degrees of freedom on mesh basis) (network). The number of independent meshes that can be selected in a network. The nullity N is equal to the number of branches B minus the number of nodes V plus the number of separate parts P. $N = B-V+P$. *See also:* **network analysis.** E153/E270-0

null junction (power supplies). The point on the Kepco bridge at which the reference resistor, the voltage-control resistance, and one side of the comparison amplifier coincide. *Note:* The null junction is maintained at almost zero potential and is a virtual ground. *See:* **summing point.** *See also:* **power supply.** KPSH-10E1

null-reference glide slope. A glide-slope system using a two-element array and in which the slope angle is defined by the first null above the horizontal in the field pattern of the upper antenna. *See also:* **navigation.** 0-10E6

number (electronic computation). (1) Formally, an abstract mathematical entity that is a generalization of a concept used to indicate quantity, direction, etcetera. In this sense a number is independent of the manner of its representation. (2) Commonly: A representation of a number as defined above (for example, the binary number 10110, the decimal number 3695, or a sequence of pulses). (3) An expression, composed wholly or partly of digits, that does not necessarily represent the abstract entity mentioned in the first meaning. *Note:* Whenever there is a possibility of confusion between meaning (1) and meaning (2) or (3), it is usually possible to make an unambiguous statement by using **number** for meaning (1) and **numerical expression** for meaning (2) or (3). *See also:* **electronic digital computer.** E162/E270-0

number of rectifier phases (rectifier circuit). The total number of successive, nonsimultaneous commutations occuring within that rectifier circuit during each cycle when operating without phase control. *Note:* It is also equal to the order of the principal harmonic in the direct-current potential wave shape. The number of rectifier phases influences both alternating-current and direct-current waveforms. In a simple single-way rectifier the number of rectifier phases is equal to the number of rectifying elements. *See also:* **rectification; rectifier circuit element.** 42A15/E59-34E17

number of scanning lines (television) (numerically). The ratio of line frequency to frame frequency. *Note:* In most television systems, the scanning lines that occur during the return intervals are blanked. 42A65/E204-2E2

number 1 master (metal master) (metal negative) (disk recording) (electroacoustics). *See:* **original master.**

number 1 mold (mother) (metal positive). A mold derived by electroforming from the original master. *See also:* **phonograph pickup.** E157-1E1

number system (electronic computation). Loosely, a numeration system. *See also:* **positional notation.** X3A12-16E9

number 2, number 3, etcetera master. A master produced by electroforming from a number 1, number 2, etcetera mold. *See also:* **phonograph pickup.** E157-1E1

number 2, number 3, etcetera mold. A mold derived by electroforming from a number 2, number 3, etcetera master. *See also:* **phonograph pickup.** E157-1E1

numeral. A representation of a number. *See:* **binary numeral.** X3A12-16E9

numeral system. *See:* **numeration system.**

numeration system. A system for the representation of numbers, for example, the decimal system, the Roman numeral system, the binary system. X3A12-16E9

numerical action (switch). That action that depends on at least part of the called number. *See also:* **telephone switching system.** 42A65-0

numerical analysis. The study of methods of obtaining useful quantitative solutions to problems that have been expressed mathematically, including the study of the errors and bounds on errors in obtaining such solutions. X3A12-16E9

numerical aperture (fiber-optic plate) (nominal) (camera tubes). The square root of the difference of (1) the square of the index of refraction of the fiber core, and (2) the square of the index of refraction of the cladding material. *Note:* Numerical aperture of a fiber-optic plate is a measure of the range of angles of incidence of entering radiation that can be conducted through the plate. *See:* **camera tube.** 0-15E6

numerical control. Pertaining to the automatic control of processes by the proper interpretation of numerical data. *See also:* **electronic digital computer.** X3A12-16E9

numerical control system (numerically controlled machines). A system in which actions are controlled by the direct insertion of numerical data at some point. *Note:* The system must automatically interpret at least some portion of these data. *See also:* **numerically controlled machines.** EIA3B-34E12

numerical data. Data in which information is expressed by a set of numbers or symbols that can only assume discrete values or configurations. EIA3B-34E12

numerically controlled machine. *Note:* For an extensive list of cross references, see *Appendix A.*

nutating feed (electronic navigation). *See:* **conical scan.**

nutation field (electronic navigation). *See:* **conical scan.**

Nyquist diagram (1) (general). A polar plot of the loop transfer function. *Note:* The inverse Nyquist diagram is a polar plot of the reciprocal function. The generalized Nyquist diagram comprises plots of the loop transfer function of the complex variable s, where $s=\sigma+j\omega$ and σ and ω are arbitrary constants including zero. 85A1-23E0

(2) (feedback amplifier). A plot, in rectangular coordinates, of the real and imaginary parts of the factor μ/β for frequencies from zero to infinity, where μ is the amplification in the absence of feedback, and β is the fraction of the output voltage that is superimposed on the amplifier input. *Note:* The criterion for stability of a feedback amplifier is that the curve of the Nyquist diagram shall not inclose the point $X=1$, $Y=0$, where μ/β equals $X+jY$. *See also:* **feedback.** 42A65-0

Nyquist interval (channel) (1) (general). The reciprocal of the Nyquist rate. E170-0
(2) The maximum separation in time that can be given to regularly spaced instantaneous samples of a wave of bandwidth W for complete determination of the waveform of the signal. Numerically, it is equal to ½ W seconds. *See also:* **transmission characteristics.** 42A65-0

Nyquist rate (channel). The maximum rate at which independent signal values can be transmitted over the specified channel without exceeding a specified amount of mutual interference. *Note:* The Nyquist rate of a channel is directly related to its bandwidth. For example, an ideal (physically unrealizable) low-pass filter of bandwidth W (constant amplitude and linear phase response within band W and zero transmission elsewhere) would permit a Nyquist rate of 2 W without interference. For a physical channel, the Nyquist rate, as defined, depends not only on the specified amount of mutual interference, but also on the amplitude and phase response of the channel. E170-0

O

OA. *See:* **transformer, oil-immersed.**

OBI. *See:* **omnibearing indicator.**

object color. The color of the light reflected or transmitted by the object when illuminated by a standard light source, such as source *A, B,* or *C* of the Commission Internationale de l'Eclairage (CIE). *See:* **standard source.** *See also:* **color.** Z7A1-0

object language. *See:* **target language.**

object program. *See:* **target program.**

observable (control system). A property of a component of a state whereby its value at a given time can be computed from measurements on the output over a finite past interval. *See also:* **control system.** 0-23E0

observable, completely (control system). The property of a plant whereby all components of the state are observable. *See also:* **control system; observable; plant.** 0-23E0

observable temperature (equipment rating). The temperature of equipment obtained on test or in operation. *See also:* **limiting insulation temperature.** E1-SCC4

observable temperature rise (equipment rating). The difference between the observable temperature and the ambient temperature at the time of the test. *See also:* **limiting insulation temperature.** E1-SCC4

obstruction lights. Aeronautical ground lights provided to indicate obstructions. *See also:* **signal lighting.** Z7A1-0

occulting light. A rhythmic light in which the periods of light are clearly longer than the periods of darkness. *See also:* **signal lighting.** Z7A1-0

occupied bandwidth (mobile communication). The frequency bandwidth such that below its lower and above its upper frequency limits, the mean powers radiated are each equal to 0.5 percent of the total power radiated. *See also:* **mobile communication system.** 0-6E1

OCR. *See:* **optical character recognition.**

octal. (1) Pertaining to a characteristic or property involving a selection, choice, or condition in which there are eight possibilities. (2) Pertaining to the numeration system with a radix of eight. *See also:* **electronic digital computer; positional notation.** X3A12-16E9

octantal error (electronic navigation). An error in measured bearing caused by the finite spacing of the antenna elements in systems using spaced antennas to provide bearing information (such as very-high-frequency omnidirectional radio ranges); this error varies in a sinusoidal manner throughout the 360 degrees and has four positive and four negative maximums. *See also:* **navigation.** 0-10E6

octave (communication) (1) (general). The interval between two frequencies having a ratio of 2:1. *See also:* **signal wave.** E151-0
(2) The interval between two sounds having a basic frequency ratio of two.
(3) The pitch interval between two tones such that one tone may be regarded as duplicating the musical import of the other tone at the nearest possible higher pitch. *Notes:* (A) The interval, in octaves between any two frequencies, is the logarithm to the base 2 (or 3.322 times the logarithm to the base 10) of the frequency ratio. (B) The frequency ratio corresponding to an octave pitch interval is approximately, but not always exactly, 2:1. *See also:* **electroacoustics; signal wave.** 0-1E1

octave-band pressure level (acoustics) (octave pressure level) (sound). The band pressure level for a frequency band corresponding to a specified octave. *Note:* The location of an octave-band pressure level on a frequency scale is usually specified as the geometric mean of the upper and lower frequencies of the octave. *See also:* **electroacoustics.** E157-1E1

octode. An eight-electrode electron tube containing an anode, a cathode, a control electrode, and five additional electrodes that are ordinarily grids. *See also:* **tube definitions.** 42A70-15E6

octonary (electronic computation). *See:* **octal.**

odd-even check (computing systems). *See:* **parity check.** *See also:* **electronic digital computer.**

***O* display (radar).** An *A* display modified by the inclusion of an adjustable notch for measuring distance. *See also:* **navigation.** 0-10E6

oersted. The unit of magnetic field strength in the unrationalized centimeter-gram-second (cgs) electromagnetic system. The oersted is the magnetic field strength in the interior of an elongated uniformly wound solenoid that is excited with a linear current density in its winding of one abampere per 4π centimeters of axial length. E270-0

off-center display. A plan-position-indicator display, the center of which does not correspond to the position of the radar antenna. *See also:* **radar.** 42A65-0

off-center plan-position indicator (radar). A plan-position indicator that has the zero position of the time base at a position other than at the center of the display, thus providing the equivalent of a larger display for a selected portion of the service area. E172-10E6

office class (telephone networks). A class designation (class 1, 2, 3, 4, 5) given to each office involved in the completion of toll calls. *Note:* The class is determined according to the office's switching function, its interrelation with other switching offices, and its transmission requirements. The class designation given to the switching points in the network determines the routing pattern for all calls. Class 1 is higher in rank than Class 2; Class 2 is higher than Class 3; etcetera. *See also:* **telephone system.** 0-19E1

office test (meter). A test made at the request or suggestion of some department of the company to determine the cause of seemingly abnormal registration. *See also:* **service test (field test).** 42A30-0

off-impedance (thyristor). The differential impedance between the terminals through which the principal current flows, when the thyristor is in the OFF state at a stated operating point. *See also:* **principal voltage-current characteristic (principal characteristic).** E223-34E17/15E7

off line (computing systems). Pertaining to equipment or devices not under direct control of the central processing unit. *See also:* **electronic digital computer.** X3A12-16E9

off-peak power. Power supplied during designated periods of relatively low system demands. *See also:* **generating station.** 42A35-31E13

OFF period: (1) (electron tubes). The time during an operating cycle in which the electronic tube is nonconducting. *See also:* **ON period (electron tubes).** 50I07-15E6

(2) (circuit switching element) (inverters). The part of an operating cycle during which essentially no current flows in the circuit switching element. *See also:* **self-commutated inverters.** 0-34E24

offset (1) (transducer). The component of error that is constant and independent of the inputs, often used to denote bias. *See also:* **electronic analog computer.** E165-16E9

(2) (control system, feedback). The steady-state deviation when the command is fixed. *Note:* The offset resulting from a no-load to full-load change (or other specified limits) is often called (speed) droop or load regulation. *See also:* **control system, feedback.** 85A1-23E0

(3) (distance relay). The displacement of the operating characteristic on an *R-X* diagram from the position inherent to the basic performance class of the relay. *Note:* A relay with this characteristic is called an offset relay. 37A100-31E11/31E6

(4) (course computer) (electronic navigation). An automatic computer that translates reference navigational coordinates into those required for a predetermined course. *See also:* **navigation.** 0-10E6

offset angle (electroacoustics) (lateral disk reproduction). The offset angle is the smaller of the two angles between the projections into the plane of the disk of the vibration axis of the pickup stylus and the line connecting the vertical pivot (assuming a horizontal disk) of the pickup arm with the stylus point. *See also:* **phonograph pickup.** E157-1E1

offset voltage (power supplies). A direct-current potential remaining across the comparison amplifier's input terminals (from the null junction to the common terminal) when the output voltage is zero. The polarity of the offset voltage is such as to allow the output to pass through zero and the polarity to be reversed. It is often deliberately introduced into the design of power supplies to reach and even pass zero-output volts. *See also:* **power supply.** KPSH-10E1

OFF state (thyristor). The condition of the thyristor corresponding to the high-resistance low-current portion of the principal voltage-current characteristic between the origin and the breakover point(s) in the switching quadrant(s). *See also:* **principal voltage-current characteristic (principal characteristic).** E223-34E17/34E24/15E7

OFF-state current (thyristor). The principal current when the thyristor is in the OFF state. *See also:* **principal current.** E223-34E17/34E24/15E7

OFF-state voltage (thyristor). The principal voltage when the thyristor is in the OFF state. *See also:* **principal voltage-current characteristic (principal characteristic).** E223-34E17/34E24/15E7

ohm. The unit of resistance (and of impedance) in the International System of Units (SI). The ohm is the resistance of a conductor such that a constant current of one ampere in it produces a voltage of one volt between its ends. E270-0

ohmic contact (1) (general). A purely resistive contact; that is, one that has a linear voltage-current characteristic throughout its entire operating range. *See also:* **semiconductor.** E216-34E17;E270-0

(2) (semiconductor). A contact between two materials, possessing the property that the potential difference across it is proportional to the current passing through. *See also:* **semiconductor.** E102-10E1

ohmic resistance test (rotating machinery). A test to measure the ohmic resistance of a winding, using direct current. *See also:* **asynchronous machine; synchronous machine.** 0-31E8

ohmmeter. A direct-reading instrument for measuring electric resistance. It is provided with a scale, usually graduated in either ohms, megohms, or both. If the scale is graduated in megohms, the instrument is usually called a megohmmeter. *See also:* **instrument.** 42A30-0

Ohm's law. The current in an electric circuit is inversely proportional to the resistance of the circuit and is directly proportional to the electromotive force in the circuit. *Note:* Ohm's law applies strictly only to linear constant-current circuits. E270-0

oil (1) (prefix). The prefix oil applied to a device that interrupts an electric circuit indicates that the interruption occurs in oil. 42A95-0;IC1-34E10

(2) (transformers). Includes synthetic insulating liquids as well as mineral transformer oil. *See:* **oil-immersed transformer.** 42A15-31E12;57A15-0

oil buffer (elevators). A buffer using oil as a medium that absorbs and dissipates the kinetic energy of the descending car or counterweight. *See also:* **elevators.** 42A45-0

oil-buffer stroke (oil buffer) (elevators). The oil-displacing movement of the buffer plunger or piston, excluding the travel of the buffer-plunger accelerating device. *See also:* **elevators.** 42A45-0

oil catcher (rotating machinery). A recess to carry off oil. *See also:* **oil cup (rotating machinery).** 0-31E8

oil cup (rotating machinery). An attachment to the oil reservoir for adding oil and controlling its upper level. *See:*
bearing oil system;
bearing reservoir;
drain line;
feed line;
oil catcher;
oil grooves;
oil-level gauge;
oil-lift system;
oil-overflow plug;
oil pot;
oil ring;
oil-ring guide;
oil-ring retainer;
oil seal;
oil slinger;
oil-well cover;
oil wick;
sediment separator. 42A10-31E8

oil cutout (oil-filled cutout). A cutout in which all or part of the fuse support and its fuse link or disconnecting blade are mounted in oil with complete immersion of the contacts and the fusible portion of the conducting element (fuse link), so that arc interruption by severing of the fuse link or by opening of the contacts will occur under oil. 37A100-31E11

oil feeding reservoirs. Oil storage tanks situated at intervals along the route of an oil-filled cable or at oil-filled joints of solid cable for the purpose of keeping the cable constantly filled with oil under pressure. *See also:* **power distribution, underground construction.** 42A35-31E13

oil-filled (designated liquid-filled) (prefix). The prefix oil-filled or designated liquid-filled as applied to equipment indicates that oil or the designated liquid is the surrounding medium. 42A95-0

oil-filled bushing (outdoor electric apparatus). A bushing in which the space between the inside surface of the porcelain and the major insulation (or conductor where no major insulation is used) is filled with oil. *See also:* **power distribution, overhead construction.** 76A1-0

oil-filled cable. A self-contained pressure cable in which the pressure medium is low-viscosity oil having access to the insulation. *See:* **pressure cable; self-contained pressure cable.** *See also:* **power distribution, underground construction.** 42A35-31E13

oil-filled pipe cable. A pipe cable in which the pressure medium is oil having access to the insulation. *See:* **pressure cable; pipe cable.** *See also:* **power distribution, underground construction.** 42A35-31E13

oil-fill stand pipe (rotating machinery). *See:* **oil overflow plug.**

oil groove (rotating machinery). A groove cut in the surface of the bearing lining or sometimes in the journal to help to distribute the oil over the bearing surface. *See:* **oil cup (rotating machinery).** 0-31E8

oil-immersed (transformers, reactors, regulators, and similar components). Having the coils immersed in an insulating liquid. *Note:* The insulating liquid is usually (though not necessarily) oil. *See:* **transformer, oil-immersed.** 57A15-0

oil-impregnated paper-insulated bushing. A bushing in which the major insulation is provided by paper impregnated with oil. *See also:* **power distribution, overhead construction.** 76A1-0

oil-level gauge (rotating machinery). An instrument showing oil level. *See also:* **oil cup (rotating machinery).** 0-31E8

oil-lift bearing (rotating machinery). A journal bearing in which high-pressure oil is forced under the shaft journal or thrust runner to establish a lubricating film. *See also:* **bearing.** 0-31E8

oil-lift system (rotating machinery). A system that lubricates a bearing before starting by forcing oil between the journal or thrust runner and bearing surfaces. *See also:* **oil cup (rotating machinery).** 0-31E8

oil-overflow plug (oil-fill stand-pipe) (rotating machinery). An attachment to the oil reservoir that can be opened to allow excess oil to escape, to inspect the oil level, or to add oil. 42A10-31E8

oil pot (oil reservoir) (rotating machinery). A bearing reservoir for a vertical-shaft bearing. *See also:* **oil cup (rotating machinery).** 0-31E8

oil-pressure electric gauge. A device that measures the pressure of oil in the line between the oil pump and the bearings of an aircraft engine. The gauge is provided with a scale, usually graduated in pounds per square inch. It provides remote indication by means of a self-synchronous generator and motor. *See also:* **air-transportation instruments.** 42A41-0

oilproof enclosure. An enclosure so constructed that oil vapors, or free oil not under pressure, which may accumulate within the enclosure, will not prevent successful operation of, or cause damage to, the enclosed equipment. *See also:* **asynchronous machine; direct-current commutating machine; synchronous machine.** E45-0

oil reservoir (rotating machinery). *See:* **oil pot.**

oil ring (rotating machinery). A ring encircling the shaft in such a manner as to bring oil from the oil reservoir to the sleeve bearing and shaft. *See also:* **bearing.** 42A10-31E8

oil-ring guide (rotating machinery). A part whose main purpose is the restriction of the motion of the oil ring. *See also:* **oil cup (rotating machinery).** 0-31E8

oil-ring lubricated bearing. A bearing in which a ring, encircling the journal, and rotated by it, raises oil to lubricate the bearing from a reservoir into which the ring dips. 0-31E8

oil-ring retainer (rotating machinery). A guard to keep the oil ring in position on the shaft. 42A10-31E8

oil seal (bearing seal) (bearing oil seal) (rotating machinery). A part or combination of parts in a bearing assembly intended to prevent leakage of oil from the bearing. 42A10-31E8

oil slinger. A rotating member mounted on the shaft, or integral with it, that expels oil by centrifugal force and tends to prevent it from creeping along adjacent surfaces. *See also:* **oil cup (rotating machinery).** 42A10-0

oil thrower (oil slinger) (rotating machinery). A peripheral ring or ridge on a shaft adjacent to the journal and which is intended to prevent any flow of oil along the shaft. *See also:* **oil cup (rotating machinery).** 0-31E8

oiltight enclosure. An enclosure so constructed that oil vapors, or free oil, not under pressure, which may be present in the surrounding atmosphere, cannot enter the enclosure. E45-0

oiltight pilot devices (industrial control). Devices such as push-button switches, pilot lights, and selector switches that are so designed that, when properly installed, they will prevent oil and coolant from entering around the operating or mounting means. *See also:* **switch.** IC1-34E10

oil, uninhibited. Mineral transformer oil to which no synthetic oxidation inhibitor has been added. *See:* **oil-immersed transformer.** 42A15-31E12

oil-well cover (rotating machinery). A cover for an oil reservoir. *See:* **oil cup (rotating machinery).** 42A10-31E8

oil wick (rotating machinery). Wool, cotton, or similar material used to bring oil to the journal surface by capillary action. *See:* **oil cup (rotating machinery).** 42A10-31E8

omega. A very-long-distance navigation system operating at approximately 10 kilohertz in which hyperbolic lines of position are determined by measurement of the differences in travel time of continuous-wave signals from two transmitters separated by 5000 to 6000 nautical miles (9000 to 11 000 kilometers), or in which changes in distances from the transmitters are measured by counting radio-frequency wavelengths in space or lanes as the vehicle moves from a known position, the lanes being counted by phase comparison with a stable oscillator aboard the vehicle. *See also:* **navigation.** 0-10E6

omnibearing (electronic navigation). A magnetic bearing indicated by a navigational receiver on transmissions from an omnidirectional radio range. *See also:* **navigation; radio navigation.** 42A65-10E6

omnibearing converter (electronic navigation). An electromechanical device that combines the omnibearings signal with vehicle heading information to furnish electric signals for the operation of the pointer of a radio-magnetic indicator. *See also:* **navigation.** 0-10E6

omnibearing-distance facility (electronic navigation). A combination of an omnidirectional radio range and a distance-measuring facility, so that both bearing and distance information may be obtained; tactical air navigation (tacan) and combined very-high-frequency omnidirectional radio range and distance-measuring equipment (VOR/DME) are omnibearing distance facilities. *See also:* **navigation.** 0-10E6

omnibearing-distance navigation. Radio navigation utilizing a polar-coordinate system as a reference, making use of omnibearing-distance facilities; often called rho-theta navigation. *See also:* **navigation.** E172-10E6

omnibearing indicator (OBI) (electronic navigation). An instrument that presents an automatic and continuous indication of an omnibearing. *See also:* **navigation.** E172-10E6

omnibearing selector (electronic navigation). A control used with an omnidirectional-radio-range receiver so that any desired omnibearing may be selected; deviations from on-course for any selected bearing is displayed on the course-line-deviations indicator. *See also:* **navigation.** 0-10E6

omnidirectional antenna. An antenna having an essentially nondirectional pattern in azimuth and a directional pattern in elevation. *See:* **isotropic radiator.** *See also:* **antenna.** 0-3E1

omnidirectional microphone (nondirectional microphone). A microphone the response of which is essentially independent of the direction of sound incidence. *See also:* **microphone.** 42A65-0

omnidirectional range (omnirange) (electronic navigation). A radio range that provides bearing information to vehicles at all azimuths within its service area. *See also:* **navigation; radio navigation.** E172/42A65-10E6

omnirange (electronic navigation). *See:* **omnidirectional range.**

on-course curvature (navigation). The rate of change of the indicated course with respect to distance along the course line or path. *See also:* **navigation.** E172-10E6

one-address. Pertaining to an instruction code in which each instruction has one address part. Also called single address. In a typical one-address instruction the address may specify either the location of an operand to be taken from storage, the destination of a previously prepared result, or the location of the next instruction to be interpreted. In a one-address machine, the arithmetic unit usually contains at least two storage locations, one of which is an accumulator. For example, operations requiring two operands may obtain one operand from the main storage and the other from the storage location in the arithmetic unit that is specified by the operation part. *See also:* **electronic digital computer; single-address.** E162-0

one-address code (electronic computation). *See:* **instruction code.**

one-fluid cell. A cell having the same electrolyte in contact with both electrodes. *See also:* **electrochemistry.** 42A60-0

one-level address (computing systems). *See:* **direct address.** *See also:* **electronic digital computer.** X3A12-16E9

one-line diagram (single-line) (industrial control). A diagram that shows, by means of single lines and graphic symbols, the course of an electric circuit or system of circuits and the component devices or parts used therein. E270-34E10

ONE output (magnetic cell). (1) The voltage response obtained from a magnetic cell in a ONE state by a reading or resetting process. (2) The integrated voltage response obtained from a magnetic cell in a ONE state by a reading or resetting process. *See also:* **distributed-ONE output; ONE state; static magnetic storage.** E163-0

one-plus-one address. Pertaining to an instruction that contains one operand address and a control address. *See also:* **electronic digital computer.** X3A12-16E9

ones complement. The radix-minus-one complement of a numeral whose radix is two. *Note:* In this context the terms **base** and **radix** are interchangeable. *See also:* **electronic digital computer.** X3A12-16E9

ONE state. A state of a magnetic cell wherein the magnetic flux through a specified cross-sectional area has a positive value, when determined from an arbitrarily specified direction of positive normal to that area. A state wherein the magnetic flux has a negative value, when similarly determined, is a ZERO state. A ONE output is (1) the voltage response obtained from a magnetic cell in a ONE state by a reading or resetting process, or (2) the integrated voltage response obtained from a magnetic cell in a ONE state by a reading or resetting process. A ZERO output is (1) the voltage response obtained from a magnetic cell in a ZERO state by a reading or resetting process, or (2) the integrated voltage response obtained from a magnetic cell in a ZERO state by a reading or resetting process. A ratio of a ONE output to a ZERO output is a ONE-to-ZERO ratio. A pulse, for example a drive pulse, is a **write pulse** if it causes information to be introduced into a magnetic cell or cells, or is a **read pulse** if it causes information to be acquired from a magnetic cell or cells. *See also:* **static magnetic storage.** E163-0

one-time fuse. *See:* **nonrenewable fuse.**

ONE-to-partial-select ratio. The ratio of a ONE output to a partial-select output. *See also:* **coincident-current selection; static magnetic storage.** E163-0

ONE-to-ZERO ratio. A ratio of a ONE output to a ZERO output. *See:* **ONE state.** *See also:* **static magnetic storage.** E163-0

***O* network.** A network composed of four impedance branches connected in series to form a closed circuit, two adjacent junction points serving as input terminals while the remaining two junction points serve as output terminals. *See also:* **network analysis.** 42A65-0

one-way automatic leveling device. A device that corrects the car level only in case of under-run of the car, but will not maintain the level during loading and unloading. *See also:* **elevator-car leveling device.** 42A45-0

one-way correction (industrial control). A method of register control that effects a correction in register in one direction only. 42A25-34E10

one-way trunk. A trunk between two central offices, used for calls that originate at one of those offices, but not for calls that originate at the other. At the originating office the one-way trunk is known as an outgoing trunk; at the other office, it is known as an incoming trunk. *See also:* **telephone system.** 42A65-19E1

one-wire circuit. *See:* **direct-wire circuit.**

one-wire line (open-wire lead). *See:* **open-wire pole line.**

ON impedance (thyristor). The differential impedance between the terminals through which the principal current flows, when the thyristor is in the ON state at a stated operating point. *See also:* **principal voltage-current characteristic (principal characteristic).** E223-34E17/15E7

online. (1) Pertaining to equipment or devices under direct control of the central processing unit. (2) Pertaining to a user's ability to interact with a computer. 0-16E9

ON-OFF keying (modulation systems). A binary form of amplitude modulation in which one of the states of the modulated wave is the absence of energy in the keying interval. *Note:* The terms mark and space are often used to designate, respectively, the presence and absence of energy in the keying interval. *See also:* **telegraphy; modulating systems.** E170-0

on-peak power. Power supplied during designated periods of relatively high system demands. *See also:* **generating station.** 42A35-31E13

ON period (electron tube or valve). The time during an operating cycle in which the electron tube or valve is conducting.
See:
admittance, effective input;
admittance, effective output;
anode differential resistance;
anode resistance;
capacitance, effective input;
circuit characteristic of electrodes;
direct interelectrode capacitance;
duty line;
impedance, effective input;
input circuit;
mutual conductance;
OFF period;
output circuit;
output power;
total electrode capacitance;
voltage factor. 50107-15E6

ON state (thyristor). The condition of the thyristor corresponding to the low-resistance low-voltage portion of the principal voltage-current characteristic in the switching quadrant(s). *Note:* In the case of reverse-conducting thyristors, this definition is applicable only for a positive anode-to-cathode voltage. *See also:* **principal voltage-current characteristic (principal characteristic).** E223-34E17/34E24/15E7

ON-state current (thyristor). The principal current when the thyristor is in the ON state. E223-34E17/34E24/15E7

ON-state voltage (thyristor). The principal voltage when the thyristor is in the ON state. *See also:* **principal voltage-current characteristic (principal characteristic).** E223-34E17/34E24/15E7

opacity (electroacoustics) (optical path). The reciprocal of transmission. *See also:* **transmission (transmittance).** *See also:* **electroacoustics.** E157-1E1

open amortisseur. An amortisseur that has no connections between poles. *See also:* **synchronous machine.** 42A10-0

open area (electromagnetic compatibility). *See:* **test site.**

open-center display. A plan-position-indicator display on which zero range corresponds to a ring around the center of the display. *See also:* **radar.** 42A65-0

open-center plan-position indicator (radar). A plan-position indicator in which the display of the initiation of the time base precedes that of the transmitted pulse. *See also:* **radar.** E172-10E6

open-circuit characteristic (synchronous machine). *See:* **open-circuit saturation curve.**

open-circuit control. A method of controlling motors employing the open-circuit method of transition from series to parallel connections of the motors. *See also:* **multiple-unit control.** 42A42-0

open-circuit impedance (1) (general). A qualifying adjective indicating that the impedance under consideration is for the network with a specified pair or group of terminals open-circuited. *See also:* **network analysis.** E270-0
(2) (line or four-terminal network). The driving-point impedance when the far end is open. *See also:* **self-impedance.** 42A65-31E3

open-circuit inductance (transformer). The apparent inductance of a winding with all other windings open-circuited. 0-21E1

open-circuit potential. The measured potential of a cell from which no current flows in the external circuit. CM-34E2

open-circuit saturation curve (open-circuit characteristic) (synchronous machine). The saturation curve of a machine with an open-circuited armature winding. *See also:* **synchronous machine.** 0-31E8

open-circuit signaling. That type of signaling in which no current flows while the circuit is in the idle condition. *See also:* **circuits and devices.** 42A65-0

open-circuit test (synchronous machine). A test in which the machine is run as a generator with its terminals open-circuited. *See also:* **synchronous machine.** 0-31E8

open-circuit transition (1) (multiple-unit control). A method of changing the connection of motors from series to parallel in which the circuits of all motors are open during the transfer. *See also:* **multiple-unit control.** 42A42-0
(2) (industrial control) (reduced-voltage controllers, including star-delta controllers). A method of starting in which the power to the motor is interrupted during normal starting sequence. *See also:* **electric controller.** IC1-34E10

open-circuit voltage (1) (battery). The voltage at its terminals when no appreciable current is flowing. *See also:* **battery (primary or secondary).** 42A60-0
(2) (arc-welding apparatus). The voltage between the output terminals of the welding power supply when no current is flowing in the welding circuit. *See also:* **electric arc-welding apparatus.** 87A1-0

open cutout. A cutout in which the fuse clips and fuseholder, fuse unit, or disconnecting blade are exposed. 37A100-31E11

open ended. Pertaining to a process or system that can be augmented. *See also:* **electronic digital computer.** X3A12-16E9

open-ended coil (rotating machinery). A partly preformed coil the turns of which are left open at one end to facilitate their winding into the machine. *See also:* **asynchronous machine; direct-current commutating machine; synchronous machine.** 0-31E8

opening operation (switching device). *See:* **open operation (switching device).**

opening time (mechanical switching device). The time interval between the time when the actuating quantity of the release circuit reaches the operating value and the instant when the primary arcing contacts have parted. Any time-delay device forming an integral part of the switching device is adjusted to its minimum setting or, if possible, is cut out entirely for the determination of opening time. *Note:* The opening time includes the operating time of an auxiliary relay in the release circuit when such a relay is required and supplied as part of the switching device. 37A100-31E11

opening time (sectionalizer). *See:* **isolating time (sectionalizer).**

open-link cutout. A cutout that does not employ a fuseholder and in which the fuse support directly receives an open-link fuse link or a disconnecting blade. 37A100-31E11

open-link fuse link. A replaceable part or assembly comprised of the conducting element and fuse tube, together with the parts necessary to confine and aid in extinguishing the arc and to connect it directly into the fuse clip of the open-link fuse support. 37A100-31E11

open-link fuse support. An assembly of base or mounting support, insulators or insulator unit, and fuse clips for directly mounting an open-link fuse link and for connecting it into the circuit. 37A100-31E11

open loop (automatic control). A signal path without feedback. *See also:* **control system, feedback.** 85A1-23E0

open-loop control system. A system in which the controlled quantity is permitted to vary in accordance with the inherent characteristics of the control system and the controlled power apparatus for any given adjustment of the controller. *Note:* No function of the controlled variable is used for automatic control of the system. It is not a feedback control system. *See also:* **control; control system; network analysis.** E-111/E264/42A25/IC1-34E10

open-loop gain (power supplies). The gain, measured without feedback, is the ratio of the voltage appearing across the output terminal pair to the causative voltage required at the (input) null junction. The open-loop gain is denoted by the symbol A in diagrams and equations. *See:* **closed loop; loop gain.** *See also:* **power supply.** KPSH-10E1

open-loop phase angle*. *See:* **loop phase angle.**
*Deprecated

open-loop series street-lighting system. A street-lighting system in which the circuits each consist of a single line wire that is connected from lamp to lamp and returned by a separate route to the source of supply. *See also:* **alternating-current distribution; direct-current distribution.** 42A35-31E13

open-loop system. A control system that has no means for comparing the output with input for control purposes. EIA3B-34E12

open machine (rotating machinery). A machine in which no mechanical protection as such is embodied and there is no restriction to ventilation other than that necessitated by good mechanical construction. *See:* **asynchronous machine; direct-current commutating machine; synchronous machine.** 0-31E8

open operation (switching device). The movement of the contacts from the normally closed to the normally open position. *Note:* The letter *O* signifies this operation: OPEN. 37A100-31E11

open path (network analysis). A path along which no node appears more than once. *See also:* **linear signal flow graphs.** E155-0

open-phase protection. A form of protection that operates to disconnect the protected equipment on the loss of current in one phase conductor of a polyphase circuit, or to prevent the application of power to the protected equipment on the absence of one or more phase voltages of a polyphase system. 37A100-31E11/31E6

open-phase relay. A polyphase relay designed to operate when one or more input phases of a polyphase circuit are open. *See also:* **relay.** 37A100-31E11/31E6

open pipe-ventilated machine. An open machine except that openings for the admission of the ventilating air are so arranged that inlet ducts or pipes can be connected to them. This air may be circulated by means integral with the machine or by means external to and not a part of the machine. In the latter case, this machine is sometimes known as a **separately ventilated machine** or a **forced-ventilated machine.** Mechanical protection may be defined as under **dripproof machine, splashproof machine, guarded machine,** or **semiguarded machine.** *See also:* **asynchronous machine; direct-current commutating machine; synchronous machine.** 42A10-0

open region (1) (three-dimensional space). A volume that satisfies the following conditions: (A) any point of the region has a neighborhood that lies within the region; (B) any two points of the region may be connect-

ed by a continuous space curve that lies entirely in the region.
(2) (two-dimensional space). An area that satisfies the following conditions: (A) any point of the region has a neighborhood that lies within the region; (B) any two points of the region may be connected by a continuous curve that lies entirely in the region.
E270-0

open relay. An unenclosed relay. *See also:* **relay.**
83A16-0

open-shut control system.* *See:* **control system, ON-OFF.**
*Deprecated

open subroutine (computing systems). A subroutine that must be relocated and inserted into a routine at each place it is used. *See:* **closed subroutine.** *See also:* **electronic digital computer.** X3A12-16E9

open terminal box (rotating machinery). A terminal box that is, normally, open only to the interior of the machine. *See also:* **cradle base (rotating machinery).**
0-31E8

open wire. A conductor separately supported above the surface of the ground. *Note:* An open wire is usually a conductor of a pole line.
See:
bridle wire;
drop wire;
junction pole;
junction transposition;
open-wire circuit;
open-wire pole line;
phantom group;
point transposition;
pole fixture;
pole line;
poling;
rolling transposition;
span;
storm loading;
suspension strand;
transposition;
transposition section.
See also: **power distribution, overhead construction.**
2A2/42A65-0

open-wire circuit. A circuit made up of conductors separately supported on insulators. *Note:* The conductors are usually bare wire, but they may be insulated by some form of continuous insulation. The insulators are usually supported by crossarms or brackets on poles. *See also:* **open wire.** 42A65-0

open-wire lead (open-wire line). *See:* **open-wire pole line.**

open-wire pole line (open-wire line) (open-wire lead). A pole line whose conductors are principally in the form of open wire. *See also:* **open wire.** 42A65-0

open wiring. A wiring method using cleats, knobs, tubes, and flexible tubing for the protection and support of insulated conductors run in or on buildings, and is not concealed by the building structure. *See also:* **interior wiring.** 42A95-0

operand. That which is, or is to be, operated upon. *Note:* An operand is usually identified by an address part of an instruction. *See also:* **electronic digital computer.** E162/X3A12-16E9

operate (analog computer). The computer control state in which input signals are connected to all appropriate computing elements, including integrators, for the generation of the solution. *See also:* **electronic analog computer.** E165-16E9

operated unit. A switch, signal, lock, or other device that is operated by a lever or other operating means. *See also:* **railway signal and interlocking.** 42A42-0

operating characteristic (relay). The response of the relay to the input quantities that result in relay operation. 37A100-31E11/31E6

operating conditions (1) (general). The whole of the electrical and mechanical quantities that characterize the work of a machine, apparatus, or supply network, at a given time. E270/E96-SCC4
(2) (measurement systems). Conditions (such as ambient temperature, ambient pressure, vibration, etcetera) to which a device is subjected, but not including the variable measured by the device. *See also:* **measurement system.** 39A4-0

operating device (elevators). The car switch, pushbutton, lever, or other manual device used to actuate the control. *See:* **control.** 42A45-0

operating duty (switching device). A specified number and kind of operations at stated intervals.
37A100-31E11

operating duty cycle (arrester). The operating duty cycle consists of one or more unit operations as specified. *See also:* **arrester.** 42A20-0

operating-duty test (lightning arresters). A test in which working conditions are simulated by the application to the arrester of a specified number of impulses while it is connected to a power supply of rated frequency and specified voltage. *See:* **lightning arresters (surge diverter).** 99I2-31E7

operating influence. The change in a designated performance characteristic caused solely by a prescribed change in a specified operating variable from its reference operating condition to its extreme operating condition, all other operating variables being held within the limits of reference operating conditions. *Notes:* (1) It is usually expressed as a percentage of span. (2) If the magnitude of the influence is affected by direction, polarity, or phase, the greater value shall be taken.
39A4-0

operating line (operating curve). *See:* **load characteristic.**

operating mechanism (switching device). That part of the mechanism that actuates all the main-circuit contacts of the switching device. 37A100-31E11

operating noise temperature. The temperature in kelvins given by

$$T_{\mathrm{op}} = \frac{N_0}{kG_s}$$

where N_0 is the output noise power per unit bandwidth at a specified output frequency flowing into the output circuit (under operating conditions), k is Boltzmann's constant, and G_s is the ratio of (1) the signal power delivered at the specified output frequency into the output circuit (under operating conditions) to (2) the signal power available at the corresponding input frequency or frequencies to the system (under operating conditions) at its accessible input terminations. *Notes:* (A) In a nonlinear system T_{op} may be a function of the signal level. (B) In a linear two-port transducer with a single input and a single output frequency, if the noise power originating in the output termination and reflected at the output port can be neglected, T_{op} is related to the noise temperature of the input termination T_i and the effective input noise temperature T_e by the equation

$$T_{op} = T_i + T_e.$$

See also: **transducer.** E161-15E6

operating line; operating curve (electron device). The locus of all simultaneous values of total instantaneous electrode voltage and current for given external circuit conditions. *See:* **circuit characteristics of electrodes.** 0-15E6

operating overload (industrial control). The overcurrent to which electric apparatus is subjected in the course of the normal operating conditions that it may encounter. *Notes:* (1) The maximum operating overload is to be considered six times normal full-load current for alternating-current industrial motors and control apparatus; four times normal full-load current for direct-current industrial motors and control apparatus used for reduced-voltage starting; and ten times normal full-load current for direct-current industrial motors and control apparatus used for full-voltage starting. (2) It should be understood that these overloads are currents that may persist for a very short time only, usually a matter of seconds. 42A25-34E10

operating point (working point) (electron device). The point on the family of characteristic curves corresponding to the average voltages or currents of the electrodes in the absence of a signal. *See also:* **circuit characteristics of electrodes; quiescent point.** 0-15E6

operating speed range. The range between the lowest and highest rated speeds at which the drive may perform at full load. *See also:* **electric drive.** 42A25-34E10

operating system (computing systems). An organized collection of techniques and procedures for operating a computer. *See also:* **electronic digital computer.** X3A12-16E9

operating tap voltage (capacitance potential devices). Indicates the root-mean-square voltage to ground at the point of connection (potential tap) of the device network to the coupling capacitor or bushing. This is the voltage on which certain insulation tests are based. *See also:* **outdoor coupling capacitor .** E31-0

operating temperature (power supplies). The range of environmental temperatures in which a power supply can be safely operated (typically, 20 to 50 degrees Celsius). *See:* **ambient operating temperature (range).** *See also:* **power supply.** KPSH-10E1

operating time (relay). The time interval from occurrence of specified input conditions to a specified operation. See the figure attached to **impulse margin.** 37A100-31E11/31E6

operating voltage (glow lamp). The voltage of the system on which a device is operated. *Note:* This voltage, if alternating, is usually expressed as a root-mean-square value. *See also:* **maintaining voltage.** 37A100-31E11

operation (1) (switching device). Action of the parts of the device to perform its normal function. E162/37A100-31E11

(2) (train control). The functioning of the automatic train-control or cab-signaling system that results from the movement of an equipped vehicle over a track element or elements for a block with the automatic train-control apparatus in service, or which results from the failure of some part of the apparatus. *See also:* **automatic train control.** 42A42-0

(3) (elevators). The method of actuating the control. *See:* **control.** 42A45-0

(4) (electronic digital computers). (1) A defined action, namely, the act of obtaining a result from one or more operands in accordance with a rule that completely specifies the result for any permissible combination of operands. (2) The set of such acts specified by such a rule, or the rule itself. (3) The act specified by a single computer instruction. (4) A program step undertaken or executed by a computer, for example, addition, multiplication, extraction, comparison, shift, transfer. The operation is usually specified by the operator part of an instruction. (5) The event or specific action performed by a logic element. (6) Loosely: command.
See:
auxiliary operation;
dyadic operation;
fixed-cycle operation;
monadic operation;
repetitive operation;
sequential operation;
serial operation;
unary operation. E162/X3A12-16E9

operational amplifier (computing systems). An amplifier, usually a high-gain direct-current amplifier, designed to be used with external circuit elements to perform (1) a specified computing operation or (2) to perform a specified computing operation or to provide a specified transfer function. *Notes:* (A) The gain and phase characteristics are generally designed to permit large variations of loss in the feedback circuit without instability. (B) The input terminal of an operational amplifier (2) is the summing junction of an operational amplifier (1) and is generally designed to draw current that is negligibly small relative to signal currents in the feedback impedance. (See the accompanying figure.) *See:* **balancing of an operational amplifier.** *See also:* **electronic analog computer.** E165-16E9

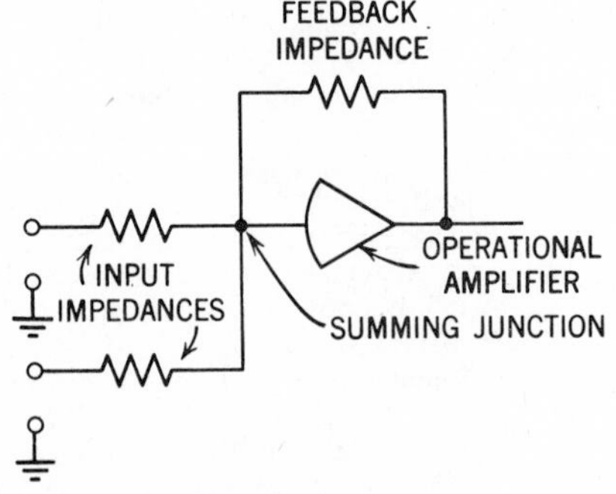

Operational amplifier, typical arrangement.

operational gain. *See:* **closed-loop gain.**

operational maintenance influence (instruments). The effect of routine operations that involve opening the case, such as to inspect or mark records, change charts, add ink, alter control settings, etcetera. *See also:* **accuracy rating (instrument).** 39A4-0

operational power supply. A power supply whose control amplifier has been optimized for signal-processing applications rather than the supply of steady-state power to a load. A self-contained combination of operational amplifier, power amplifier, and power supplies for higher-level operation applications. *See also:* **power supply.** KPSH-10E1

operational programming. The process of controlling the output voltage of a regulated power supply by means of signals (which may be voltage, current, resistance, or conductance) that are operated on by the power supply in a predetermined fashion. Operations may include algebraic manipulations, multiplication, summing, integration, scaling, and differentiation. *See also:* **power supply.** KPSH-10E1

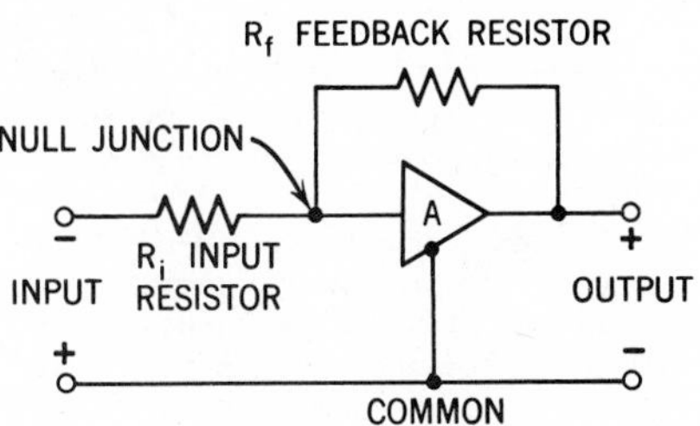

Operational programming.

operational relay. A relay that may be driven from one position or state to another by an operational amplifier or a relay amplifier. *See also:* **electronic analog computer.** E165-16E9

operational reliability. *See:* **reliability, operational.**

operation code. (1) The operations that a computing system is capable of executing, each correlated with its equivalent in another language; for example, the binary or alphanumeric codes in machine language along with their English equivalents; the English description of operations along with statements in a programming language such as **Cobol, Algol,** or **Fortran.** (2) The code that represents or describes a specific operation. The operation code is usually the operation part of the instruction. *See also:* **electronic digital computer.** E162/E270-0

operation factor. The ratio of the duration of actual service of a machine or equipment to the total duration of the period of time considered. *See also:* **generating station.** 42A35-31E13

operation influence (electrical influence). The maximum variation in the reading of an instrument from the initial reading, when continuously energized at a prescribed point on the scale under reference conditions over a stated interval of time, expressed as a percentage of full-scale value. *See also:* **accuracy rating (instrument).** 39A1-0

operation indicator (relay). *See:* **target (relay).**

operation part (instruction) (electronic computation). The part that usually specifies the kind of operation to be performed, but not the location of the operands. *See also:* **electronic computation; electronic digital computer; instruction code.** E162/E270-0

operation time (electron tubes). The time after simultaneous application of all electrode voltages for a current to reach a stated fraction of its final value. Conventionally the final value is taken as that reached after a specified length of time. *Note:* All electrode voltages are to remain constant during measurement. The tube elements must all be at room temperature at the start of the test. *See also:* **circuit characteristics of electrodes.** E160-15E6

operator (in the description of a process). (1) That which indicates the action to be performed on operands. (2) A person who operates a machine. *See also:* **electronic digital computer.** X3A12-16E9

operator's set. *See:* **operator's telephone set.**

operator's telephone set (operator's set). A set consisting of a telephone transmitter, a head receiver, and associated cord and plug, arranged to be worn so as to leave the operator's hands free. *See also:* **telephone station.** 42A65-0

opposition (electrical engineering). The relation between two periodic functions when the phase difference between them is one-half of a period. E270-0

optical ammeter. An electrothermic instrument in which the current in the filament of a suitable incandescent lamp is measured by comparing the resulting illumination with that produced when a current of known magnitude is used in the same filament. The comparison is commonly made by using a photoelectric cell and indicating instrument. *See also:* **instrument.** 42A30-0

optical character recognition (OCR). Machine identification of printed characters through use of light-sensitive devices. *See:* **magnetic-ink character recognition.** *See also:* **electronic digital computer.** X3A12-16E9

optical (optoelectronic device) coupling coefficient (between two designated ports). The fraction of the radiant or luminous flux leaving one port that enters the other port. *See also:* **optoelectronic device.** E222-15E7

optical density. *See :* **photographic transmission density.**

optical pattern (mechanical recording). *See:* **light pattern.**

optical pyrometer. A temperature-measuring device comprising a standardized comparison source of illumination and some convenient arrangement for matching this source, either in brightness or in color, against the source whose temperature is to be measured. The comparison is usually made by the eye. *See also:* **electric thermometer (temperature meter).** 42A30-0

optical scanner (character recognition). (1) A device that scans optically and usually generates an analog or digital signal. (2) A device that optically scans printed or written data and generates their digital representations. *See:* **visual scanner.** *See also:* **electronic analog computer; electronic digital computer.** X3A12-16E9

optical sound recorder. *See:* **photographic sound recorder.**

optical sound reproducer. *See:* **photographic sound reproducer.**

optic amplifier. An optoelectronic amplifier whose signal input and output ports are electric. *Note:* This is in accord with the accepted terminologies of other electric-signal input and output amplifiers such as dielectric, magnetic, and thermionic amplifiers. *See also:* **optoelectronic device.** E222-15E7

optic port. A port where the energy is electromagnetic radiation, that is, photons. *See also:* **optoelectronic device.** E222-15E7

optimal control. An admissible control law that gives a performance index an extremal value. *See also:* **control system.** 0-23E0

optimization. The procedure used in the design of a system to maximize or minimize some performance index. May entail the selection of a component, a principle of operation, or a technique. *See also:* **system.** 0-35E2

optimum bunching (electron tubes) (traveling-wave tube). The bunching condition that produces maximum power at the desired frequency in an output gap. *See:* **beam tubes.** *See also:* **electron devices, miscellaneous.** E160-15E6

optimum working frequency (radio transmission by ionospheric reflection). The frequency at which transmission between two specified points can be expected to be most effectively maintained regularly at a certain time of day. *Note:* A frequency 15 percent below the monthly median value of the $F2$ maximum usable frequency for a specified time of day and path is often taken as the optimum working frequency for propagation by way of the $F2$ layer. *See also:* **radiation.** 42A65-0

optional stop (numerically controlled machines). A miscellaneous function command similar to a program stop except that the control ignores the command unless the operator has previously pushed a button to validate the command. *See also:* **numerically controlled machines.** EIA3B-34E12

optoelectronic amplifier. An optoelectronic device capable of power gain, in which the signal ports are either all electric ports or all optic ports. *See also:* **optoelectronic device.** E222-15E7

optoelectronic cell. The smallest portion of an optoelectronic device capable of independently performing all the specified input and output functions. *Note:* An optoelectronic cell may consist of one or more optoelectronic elements. *See also:* **optoelectronic device.** E222-15E7

optoelectronic device. An electronic device combining optic and electric ports.
See:
electric port;
electroluminescent display device;
electroluminescent display panel;
electron devices, miscellaneous;
homochromatic gain;
image;
image converter;
image-converter panel;
image element;
image intensifier;
image-intensifier panel;
image-storage device;
image-storage panel;
light amplifier;
luminous gain;
optical coupling coefficient;
optic amplifier;
optic port;
optoelectronic amplifier;
optoelectronic cell;
optoelectronic element;
persistent-image device;
persistent-image panel;
port;
radiant gain;
spectral-conversion luminous gain;
spectral-conversion radiant gain;
spectral luminous gain;
spectral radiant gain. E222-15E7

optoelectronic element. A distinct constituent of an optoelectronic cell, such as an electroluminor, photoconductor, diode, optical filter, etcetera. *See also:* **optoelectronic device.** E222-15E7

OR. A logic operator having the property that if P is a statement, Q is a statement, R is a statement, ..., then the OR of $P, Q, R, \ldots$ is true if at least one statement is true, false if all statements are false. P OR Q is often represented by $P + Q$, $P \vee Q$. *See:* **exclusive OR.** X3A12-16E9

orbital stability. *See:* **stability of a limit cycle.**

OR circuit (electronic computation). *See:* **OR gate.**

order (1) (general). To put items in a given sequence. X3A12-16E9

(2) (in electronic computation). (A) Synonym for instruction. (B) Synonym for command. (C) Loosely, synonym for operation part. *Note:* The use of **order** in the computer field as a synonym for terms similar to the above is losing favor owing to the ambiguity between these meanings and the more-common meanings in mathematics and business. *See also:* **electronic computation; electronic digital computer.** E162/E270-0

order wire (communication practice). An auxiliary circuit for use in the line-up and maintenance of communication facilities. *See also:* **telephone system.** 42A65-19E1

ordinary wave (O wave). The magnetoionic wave component that, when viewed below the ionosphere in the direction of propagation, has counterclockwise or clockwise elliptical polarization, respectively, according as the earth's magnetic field has a positive or negative component in the same direction. *See also:* **radiation.** 42A65-0

ordinary-wave component (radio wave propagation). The magnetoionic wave component deviating the least, in most of its propagation characteristics, relative to those expected for a wave in the absence of a fixed magnetic field. More exactly, if at fixed electron density, the direction of the fixed magnetic field were rotated until its direction were transverse to the direction of phase propagation, the wave component whose propagation is then independent of the magnitude of the fixed magnetic field. *See also:* **radio wave propagation.** 0- 3E2

organizing. *See:* **self-organizing.**

OR gate (OR circuit) (electronic computation). A gate whose output is energized when any one or more of the inputs is in its prescribed state. An OR gate performs the function of the logical **inclusive OR.** *See also:* **electronic computation; electronic digital computer.** E270-0

orifice. An opening or window in a side or end wall of a waveguide or cavity resonator, through which energy is transmitted. *See also:* **waveguide.** 42A65-0

orifice plate (rotating machinery). A restrictive opening in a passage to limit flow. *See:* **cradle base (rotating machinery).** 0-31E8

original master (electroacoustics) (metal master) (metal negative) (number 1 master) (disk recording). The master produced by electroforming from the face of a wax or lacquer recording. *See also:* **electroacoustics.** E157-1E1

orthicon. A camera tube in which a beam of low-velocity electrons scans a photoemissive mosaic capable of storing an electric-charge pattern. *See also:* **television; tube definitions.** 42A70/E160-15E6

orthogonality (oscilloscopes). The extent to which traces parallel to the vertical axis of a cathode-ray-tube display are at right angles to the horizontal axis. *See:* **oscillograph.** 0-9E4

***O* scan (electronic navigation).** *See:* ***O* display.**

oscillating current. A current that alternately increases and decreases but is not necessarily periodic. E270-0

oscillation (1) (general). The variation, usually with time, of the magnitude of a quantity with respect to a specified reference when the magnitude is alternately greater and smaller than the reference. *See:* **electroacoustics.** *See also:* **vibration.** 0-1E1

(2) (vibration). A generic term referring to any type of a response that may appear in a system or in part of a system. *Note:* Vibration is sometimes used synonymously with oscillation, but it is more properly applied to the motion of a mechanical system in which the motion is in part determined by the elastic properties of the body. E270-0

(3) (gas turbines). The periodic variation of a function between limits above or below a mean value, for example, the periodic increase and decrease of position, speed, power output, temperature, rate of fuel input, etcetera within finite limits. E282-31E2

oscillator (1) (general). Apparatus intended to produce or capable of maintaining electric or mechanical oscillations. 50I05/E270-0

(2) (electronics). A nonrotating device for producing alternating current, the output frequency of which is determined by the characteristics of the device. *See also:* **industrial electronics.** E54/E145/E169/E188-0

oscillator starting time, pulsed. *See:* **pulsed oscillator starting time.** E194-0

oscillator tube, positive grid. *See:* **positive-grid oscillator tube.**

oscillatory circuit. A circuit containing inductance and capacitance so arranged that when shock excited it will produce a current or a voltage that reverses at least once. If the losses exceed a critical value, the oscillating properties will be lost. *Note:* For an extensive list of cross references, see *Appendix A.* 0-42A65

oscillatory surge (lightning arresters). A surge that includes both positive and negative polarity values. *See:* **lightning arrester (surge diverter).** *See also:* **power systems, low-frequency and surge testing.** 42A35/62A1-31E7/31E13

oscillogram. A record of the display presented by an oscillograph or an oscilloscope. *See also:* **oscillograph.** 0-9E4

oscillograph. An instrument primarily for producing a record of the instantaneous values of one or more rapidly varying electrical quantities as a function of time or of another electrical or mechanical quantity. *Note:* (1) Incidental to the recording of instantaneous values of electrical quantities, these values may become visible, in which case the oscillograph performs the function of an oscilloscope. (2) An oscilloscope does not have inherently associated means for producing records. (3) The term includes mechanical recorders. *Note:* For an extended list of cross references, see *Appendix A.* 42A30-0

oscillograph tube (oscilloscope tube). A cathode-ray tube used to produce a visible pattern that is the graphic representation of electric signals, by variations of the position of the focused spot or spots in accordance with these signals. *See also:* **beam tubes; tube definitions.** 42A70-0

oscillography. The art and practice of utilizing the oscillograph. *See also:* **oscillograph.** 0-9E4

oscilloscope. An instrument primarily for making visible the instantaneous values of one or more rapidly varying electrical quantities as a function of time or of another electrical or mechanical quantity. *See also:* **cathode-ray oscilloscope; oscillograph.** 42A30-0

***O* scope (electronic navigation).** *See:* ***O* display.**

***O*-type tube or valve (microwave tubes).** A microwave tube in which the beam, the circuit, and the focusing field have symmetry about a common axis. The interaction between the beam and the circuit is dependent upon velocity modulation, the suitably focused beam being launched by a gun structure outside one end of the microwave circuit and principally collected outside the other end of the microwave structure. *See also:* **microwave tube.** 0-15E6

Oudin current (desiccating current) (resonator current) (medical electronics). A brush discharge produced by a high-frequency generator that has an output range of 2 to 10 kilovolts and a current sufficient to evaporate tissue water without charring. It is usually applied through a small needlelike electrode with the reference or ground electrode being relatively large and diffuse. *See also:* **medical electronics.** 0-18E1

Oudin resonator. A coil of wire often with an adjustable number of turns, designed to be connected to a source of high-frequency current, such as a spark gap and induction coil, for the purpose of applying an effluve (convective discharge) to a patient. 0-18E1

outage (electric power systems). The state of a component when it is not available to perform its intended function due to some event directly associated with that component. *Note:* An outage may or may not cause an interruption of service to consumers; depending on system configuration.
See:
adverse weather;
adverse-weather persistent-cause forced-outage rate;
forced interruption;
forced outage;
interruption;
interruption duration;
interruption-duration index;
interruption frequency;
momentary interruption;
normal weather;
normal-weather persistent-cause forced-outage rate;
outage duration;
outage rate;
persistent-cause forced outage;
persistent-cause forced-outage duration;
scheduled interruption;
scheduled outage;
scheduled-outage duration;
sustained interruption;
switching time;

transient-cause forced outage;
transient-cause forced-outage duration. 0-31E4

outage duration (electric power systems). The period from the initiation of a component outage until the affected component once again becomes available to perform its intended function. *See also:* **outage.** 0-31E4

outage rate (electric power systems) (for a particular classification of outage and type of component). The mean number of outages per unit time per component. *See also:* **outage.** 0-31E4

outdoor (prefix). Designed for use outside buildings. 42A95/37A100-31E11

outdoor coupling capacitor. A capacitor designed for outdoor service that provides, as its primary function, capacitance coupling to a high-voltage line. *Note:* It is used in this manner to provide a circuit for carrier-frequency energy to and from a high-voltage line and to provide a circuit for power-frequency energy from a high-voltage line to a capacitance potential device or other voltage-responsive device.
See:
auxiliary burden;
auxiliary or shunt capacitance;
auxiliary secondary winding;
burden regulation;
capacitance potential device;
carrier-current choke coil;
carrier-current drain coil;
carrier-current grounding switch and gap;
carrier-current lead;
main capacitance;
main secondary winding;
operating tap voltage;
primary line-to-ground voltage;
rated burden;
rated circuit voltage;
rated insulation class;
rated secondary voltage;
secondary voltage;
transformer grounding switch and gap;
transmission line;
voltage ratio;
voltage regulation. E31-0

outdoor wall bushing. A wall bushing on which one or both ends (as specified) are suitable for operating continuously outdoors. *See also:* **bushing.** E49-0

outer frame (rotating machinery). The portion of a frame into which the inner frame with its assembled core and winding is installed. *See:* **cradle base (rotating machinery).** 0-31E8

outer marker (electronic navigation). A marker facility in an instrument landing system that is installed at approximately 5 nautical miles (9 kilometers) from the approach end of the runway on the localizer course line to provide height, distance, and equipment functioning checks to aircraft on intermediate and final approach. *See also:* **navigation.** E172-10E6

outlet. A point on the wiring system at which current is taken to supply utilization equipment. *See:* **interior wiring; lighting outlet; receptacle outlet.** 42A95-0

outlet box. A box used on a wiring system at an outlet. *See also:* **cabinet.** 42A95-0

out-of-roundness (conductor). The difference between the major and minor diameters at any one cross section. *See also:* **waveguide.** 83A14-0

out of step. A system condition in which two or more synchronous machines have lost synchronism with respect to one another and are operating at different average frequencies. 37A100-31E11/31E6

out-of-step protection (power system). A form of protection that separates the appropriate parts, or prevents separation that might otherwise occur, in the event of loss of synchronism. 37A100-31E6/31E11

output (1) (general). (A) The current, voltage, power, or driving force delivered by a circuit or device. (B) The terminals or other places where current, voltage, power or driving force may be delivered by a circuit or device. *See:* **signal, output; variable, output.** 42A65-31E3/21E1

(2) (rotating machinery). (A) (generator). The power (active, reactive, or apparent) supplied from its terminals. **(B) (motor).** The power supplied by its shaft. *See:* **asynchronous machine; direct-current commutating machine; synchronous machine.** 0-31E8

(3) (electronic digital computer). (A) Data that have been processed. (B) The state or sequence of states occurring on a specified output channel. (C) The device or collective set of devices used for taking data out of a device. (D) A channel for expressing a state of a device or logic element. (E) The process of transferring data from an internal storage to an external storage. E162/X3A12-16E9

output attenuation (signal generator). The ratio, expressed in decibels (dB), of any selected output, relative to the output obtained when the generator is set to its calibration level. *Note:* It may be necessary to eliminate the effect of carrier distortion and/or modulation feedthrough by the use of suitable filters. *See also:* **signal generator.** 0-9E4

output capacitance (*n*-terminal electron tube). The short-circuit transfer capacitance between the output terminal and all other terminals, except the input terminal, connected together. *See also:* **circuit characteristics of electrodes; electron-tube admittances.** 42A70/E160-15E6

output-capacitor discharge time (power supply). The interval between the time at which the input power is disconnected and the time when the output voltage of the unloaded regulated power supply has decreased to a specified safe value. *See also:* **regulated power supply.** E209-0

output circuit (electron tube or valve). The external circuit connected to the output electrode to provide the load impedance. *See also:* ON **period (electron tube or valve).** 50I07-15E6

output-dependent overshoot and undershoot. Dynamic regulation for load changes. *See also:* **electric conversion equipment.** 0-10E1

output electrode (electron tubes). The electrode from which is received the amplified, modulated, detected, etcetera, voltage. *See also:* **electron tube.** 50I07-15E6

output factor. The ratio of the actual energy output, in the period of time considered, to the energy output that would have occurred if the machine or equipment had been operating at its full rating throughout its actual hours of service during the period. *See also:* **generating station.** 42A35-31E13

output frequency stability (inverters). The deviation of the output frequency from a given set value. *See:* **self-commutated inverters.** 0-34E24

output gap (electron tubes) (traveling-wave tubes). An interaction gap by means of which usable power can be abstracted from an electron stream. *See:* **beam tubes; electron devices, miscellaneous.** E160-15E6

output impedance. *See:* **impedance, output.**

output phase displacement (power inverters that have polyphase output) (inverters). The angular displacement between fundamental phasors. *See:* **self-commutated inverters.** 0-34E24

output power (1) (general). The power delivered by a system or transducer to its load. E151-0
(2) (electron tube or valve). The power supplied to the load by the electron tube or valve at the output electrode. *See also:* **ON period (electron tube or valve).** 50I07-15E6

output resonator (catcher) (electron tubes). A resonant cavity, excited by density modulation of the electron beam, that supplies useful energy to an external circuit. *See also:* **velocity-modulated tube.** 50I07-15E6

output ripple voltage (regulated power supply). The portion of the output voltage harmonically related in frequency to the input voltage and arising solely from the input voltage. *Note:* Unless otherwise specified, percent ripple is the ratio of root-mean-square value of the ripple voltage to the average value of the total voltage expressed in percent. In television, ripple voltage is usually expressed explicitly in peak-to-peak volts to avoid ambiguity. *See also:* **regulated power supply.** E209-0

output voltage regulation (power supply). The change in output voltage, at a specified constant input voltage, resulting from a change of load current between two specified values. *See also:* **regulated power supply.** E209-0

output voltage stabilization (power supply). The change in output voltage, at a specified constant load current, resulting from a change of input voltage between two specified values. *See also:* **regulated power supply.** E209-0

output winding (1) (saturable reactor). A winding, other than a feedback winding, associated with the load and through which power is delivered to the load. *See also:* **magnetic amplifier.** 42A65-0
(2) (secondary winding(s)). The winding(s) from which the output is obtained. *See also:* **magnetic amplifier.** 0-21E1

outrigger (switching-device terminal). An attachment that is fastened to or adjacent to the terminal pad of a switching device to maintain electrical clearance between the conductor and other parts or, when fastened adjacent, to relieve mechanical strain on the terminal, or both. 37A100-31E11

outside plant (communication practice). That part of the plant extending from the line side of the main distributing frame to the line side of the station or private-branch-exchange protector or connecting block, or to the line side of the main distributing frame in another central office building. *See also:* **communication.** 42A65-0

outside space block. *See:* **end finger.**

oven. An enclosure and associated sensors and heaters for maintaining components at a controlled and usually constant temperature. E165-0

oven, wall-mounted. A domestic oven for cooking purposes designed for mounting in or on a wall or other surface. *See also:* **appliances.** 1A0-0

over-all electrical efficiency (dielectric and induction heater). The ratio of the power absorbed by the load material to the total power drawn from the supply lines. *See also:* **dielectric heating; induction heating.** E54/E169-0

over-all generator efficiency (thermoelectric device). The ratio of (1) electric power output to (2) thermal power input. *See also:* **thermoelectric device.** E221-15E7

overbunching (electron tubes). The bunching condition produced by the continuation of the bunching process beyond the optimum condition. *See also:* **beam tubes.** *See also:* **electron devices, miscellaneous.** E160-15E6

overcompounded. A qualifying term applied to a compound-wound generator to denote that the series winding is so proportioned that the terminal voltage at rated load is greater than at no load. *See:* **direct-current commutating machine.** 42A10-0

overcurrent (rotating machinery). An abnormal current greater than the full load value. *See also:* **asynchronous machine; direct-current commutating machine; synchronous machine.** 50I05-31E8

overcurrent protection. A form of protection that operates when current exceeds a predetermined value. *See also:* **circuit breaker; fuse.** 0-31E6

overcurrent relay. A relay that operates when the current through the relay during its operating period is equal to or greater than its setting. 37A100-31E11

overcurrent release (overcurrent trip). A release that operates when the current in the main circuit is equal to or exceeds the release setting. 37A100-31E11

overcutting (disk recording). The effect of excessive level characterized by one groove cutting through into an adjacent one. *See also:* **phonograph pickup.** E157-1E1

overdamped (industrial control). A degree of damping that is more than sufficient to prevent the oscillation of the output following an abrupt stimulus. *Note:* For a linear second-order system the facts of the characteristic equation are real and unequal. *See also:* **control system, feedback.** AS1-34E10/23E0

overdamping (aperiodic damping*). The special case of damping in which the free oscillation does not change sign. A damped harmonic system is overdamped if $F^2 > 4MS$. *See:* **damped harmonic system** for equation, definitions of letter symbols, and referenced terms.

*Deprecated E/270-0

overflow (electronic computation). (1) The condition that arises when the result of an arithmetic operation exceeds the capacity of the number representation in a digital computer. (2) The carry digit arising from this condition. *See also:* **electronic digital computer.** E162-0

overhang packing (rotating machinery). Insulation inserted in the end region of the winding to provide spacing and bracing. *See also:* **rotor (rotating machinery); stator.** 0-31E8

overhead electric hoist. A motor-driven hoist having one or more drums or sheaves for rope or chain, and supported overhead. It may be fixed or traveling. *See also:* **hoist.** 42A45-0

overhead ground wire (lightning protection). Multiply grounded wire or wires placed above phase conductors for the purpose of intercepting direct strokes in order to protect the phase conductors from the direct strokes. *See also:* **direct-stroke protection (lightning).** 0-31E13

overhead structure (elevators). All of the structural members, platforms, etcetera, supporting the elevator machinery, sheaves, and equipment at the top of the hoistway. *See also:* **elevators.** 42A45-0

overinterrogation control (electronic navigation). *See:* **gain turn-down.**

overlap. The distance the control of one signal extends into the territory that is governed by another signal or other signals. *See also:* **railway signal and interlocking.** 42A42-0

overlap angle (gas tube). The time interval, in angular measure, during which two consecutive arc paths carry current simultaneously. *See also:* **gas-filled rectifier.** 50I07-15E6

overlap *X* (facsimile). The amount by which the recorded spot *X* dimension exceeds that necessary to form a most nearly constant density line. *Note:* This effect arises in that type of equipment that responds to a constant density in the subject copy by a succession of discrete recorded spots. *See also:* **recording (facsimile).** E168-0

overlap *Y* (facsimile). The amount by which the recorded spot *Y* dimension exceeds the nominal line width. *See also:* **recording (facsimile).** E168-0

overlay (computing systems). The technique of repeatedly using the same blocks of internal storage during different stages of a problem. When one routine is no longer needed in storage, another routine can replace all or part of it. *See also:* **electronic digital computer.** X3A12-16E9

overload (analog computer). A condition existing within or at the output of a computing element that causes a substantial computing error because of the saturation of one or more of the parts of the computing element. *See also:* **electronic analog computer.** E165-16E9

overload capacity. The current, voltage, or power level beyond which permanent damage occurs to the device considered. This is usually higher than the rated load capacity. *Note:* To carry load greater than the continuous rating, may be acceptable for limited use. E145/42A25-34E10

overload factor (electromagnetic compatibility). The ratio of the maximum value of a signal for which the operation of the predetector circuits of the receiver does not depart from linearity by more than one decibel, to the value corresponding to full-scale deflection of the indicating instrument. *See also:* **electromagnetic compatibility.** CISPR-27E1

overload level (system or component). That level above which operation ceases to be satisfactory as a result of signal distortion, overheating, or damage. *See also:* **level.** E151/42A65-31E3

overload ON-state current (thyristor). An ON-state current of substantially the same wave shape as the normal ON-state current and having a greater value than the normal ON-state current. *See also:* **principal current.** E233-34E17/34E24/15E7

overload point, signal (electronic navigation). The maximum input signal amplitude at which the ratio of output to input is observed to remain within a prescribed linear operating range. *See also:* **navigation.** 0-10E6

overload protection (industrial control). The effect of a device operative on excessive current, but not necessarily on short-circuit, to cause and maintain the interruption of current flow to the device governed. *See:* **overcurrent protection.** 42A25-34E10

overload relay (1) (general). A relay that responds to electric load and operates at a preset value of overload. *Note:* Overload relays are usually current relays but they may be power, temperature, or other relays. 37A100-31E11

(2) (overcurrent). An overcurrent relay that functions at a predetermined value of overcurrent to cause disconnection of the load from the power supply. *Note:* An overload relay is intended to protect the load (for example, motor armature) or its controller, and does not necessarily protect itself. *See:* **overcurrent relay; undercurrent relay.** 42A25-34E10

overmoded waveguide. A waveguide used to propagate a single mode, but capable of propagating more than one mode at the frequency of interest. *See:* **waveguide.** 0-3E1

overpotential. *See:* **overvoltage.**

overreach (relay). The extension of the zone of protection beyond that indicated by the relay setting. 37A100-31E11/31E6

overreaching protection. A form of protection in which the relays at one terminal operate for faults beyond the next terminal. They may be constrained from tripping until an incoming signal from a remote terminal has indicated whether the fault is beyond the protected line section. 37A100-31E11/31E6

overshoot (instrument). The amount of the overtravel of the indicator beyond its final steady deflection when a new constant value of the measured quantity is suddenly applied to the instrument. The overtravel and deflection are determined in angular measure and the overshoot is expressed as a percentage of the change in steady deflection. *Notes:* (1) Since in some instruments the percentage depends on the magnitude of the deflection, a value corresponding to an initial swing from zero to end scale is used in determining the overshoot for rating purposes. (2) Overshoot and damping factor have a reciprocal relationship. The percentage overshoot may be obtained by dividing 100 by the damping factor. *See also:* **accuracy rating (instrument); moving element (instrument).** 42A30/39A1-0

(2) (power supplies). A transient rise beyond regulated output limits, occurring when the alternating-current power input is turned on or off, and for line or load step changes. See the accompanying figure. *See also:* **recovery time (voltage regulation); power supply.** KPSH-10E1

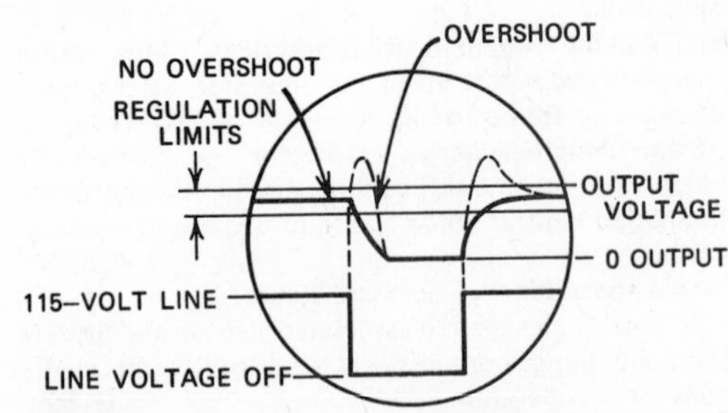

Scope view of turn-off/turn-on effects on a power supply, showing overshoot.

(3) **(television).** The initial transient response to a unidirectional change in input that exceeds the steady-state response. *See also:* **television.** 42A65-2E2

(4) (pulse techniques). *See:* **distortion, pulse.**

overshoot, system*. *See:* **deviation, system.**

*Deprecated

overshoot, transient*. *See:* **deviation, transient.**

*Deprecated

oversized waveguide. A waveguide operated in its dominant mode, but far above cutoff; sometimes termed **quasi-optical waveguide.** *See:* **waveguide.** 0-3E1

overslung car frame. A car frame to which the hoisting-rope fastenings or hoisting-rope sheaves are attached to the crosshead or top member of the car frame. *See also:* **hoistway (elevator or dumbwaiter).** 42A45-0

overspeed governor (gas turbines). A control element that is directly responsive to speed and that actuates the overspeed and overtemperature protection system when the turbine reaches the speed for which the device is set. E282-31E2

overspeed and overtemperature protection system (gas turbines). The overspeed governor, overtemperature detector, fuel stop valve(s), blow-off valve, other protective devices and their interconnections to the fuel stop valve, and to the blow-off valve, if used, required to shut off all fuel flow and shut down the gas turbine. E282-31E2

overspeed protection (industrial control). The effect of a device operative whenever the speed rises above a preset value to cause and maintain an interruption of power to the protected equipment or a reduction of its speed. *See:* **relay.** IC1-34E10

overspeed test (rotating machinery). A test on a machine rotor to demonstrate that it complies with specified overspeed requirements. *See also:* **rotor (rotating machinery).** 0-31E8

overtemperature detector (gas turbines). The primary sensing element that is directly responsive to temperature and that actuates the overspeed and overtemperature protection system when the turbine temperature reaches the value for which the device is set. E282-31E2

overtemperature protection (power supplies). A thermal relay circuit that turns off the power automatically should an overtemperature condition occur. *See also:* **power supply.** KPSH-10E1

overtone. *See:* **harmonic.**

overtone-type piezoelectric-crystal unit. A unit designed to utilize an overtone of the lowest frequency of resonance for a particular mode of vibration. *See also:* **crystal.** 42A65-0

overtravel (relay). The amount of continued movement of the responsive element after input is changed to a value that will cause this movement to ultimately cease. 37A100-31E11

overvoltage (overpotential) (1) (general). A voltage above the normal rated voltage or the maximum operating voltage of a device or circuit. A direct test overvoltage is a voltage above the peak of the line alternating voltage. *See also:* **insulation testing (large alternating-current rotating machinery).** E95-0

(2) (rotating machinery). An abnormal voltage higher than the service voltage. *See also:* **asynchronous machine; direct-current commutating machine; synchronous machine.** 50I05-31E8

(3) (lightning arresters). Abnormal voltage between two points of a system that is greater than the highest value appearing between the same two points under normal service conditions. *See:* **lightning arrester (surge diverter).** 50I25-31E7

(4) (radiation-counter tubes). The amount by which the applied voltage exceeds the Geiger-Mueller threshold. *See also:* **gas-filled radiation-counter tube; gas tube.** 42A70-15E6

(5) (electrochemistry). The displacement of an electrode potential from its equilibrium (reversible) value because of flow of current. *Note:* This is the irreversible excess of potential required for an electrochemical reaction to proceed actively at a specified electrode, over and above the reversible potential characteristic of that reaction. *See also:* **electrochemistry.** 42A60/CM-34E2

overvoltage due to resonance (lightning arresters). Overvoltage at the fundamental frequency of the installation, or of a harmonic frequency, resulting from oscillation of circuits. *See:* **lightning arrester (surge diverter).** 50I25-31E7

overvoltage protection. The effect of a device operative on excessive voltage to cause and maintain the interruption of power in the circuit or reduction of voltage to the equipment governed. 42A25-34E10

overvoltage relay. A relay that operates when the voltage applied to the relay is equal to or greater than its setting. 37A100-31E11

overvoltage release (overvoltage trip). A release that operates when the voltage of the main circuit is equal to or exceeds the release setting. 37A100-31E11

overvoltage test (rotating machinery). A high-voltage test at voltages above the rated operating voltage. *See:* **asynchronous machine; direct-current commutating machine; synchronous machine.** 0-31E8

overwriting (charge-storage tubes). Writing in excess of that which produces write saturation. *See also:* **charge-storage tube.** E158-15E6

OW. *See:* **transformer, oil-immersed.**

***O* wave (radio wave propagation).** *See:* **ordinary-wave component.** *See also:* **radio wave propagation.**

Owen bridge. A 4-arm alternating-current bridge in which one arm, adjacent to the unknown inductor, comprises a capacitor and resistor in series; the arm opposite the unknown consists of a second capacitor,

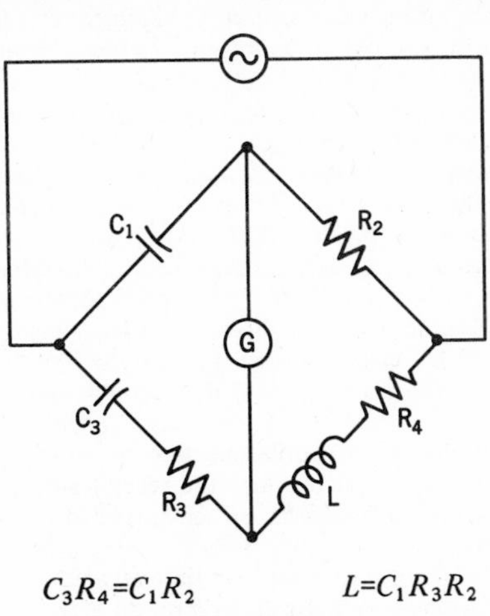

$C_3R_4=C_1R_2$ $L=C_1R_3R_2$

Owen bridge.

and the fourth arm of a resistor. *Note:* Normally used for the measurement of self-inductance in terms of capacitance and resistance. Usually, the bridge is balanced by adjustment of the resistor that is in series with the first capacitor and of another resistor that is inserted in series with the unknown inductor. The balance is independent of frequency. *See also:* **bridge.** 42A30-0

oxidant. A chemical element or compound that is capable of being reduced. *See also:* **electrochemical cell.** CV1-10E1

oxidation (electrochemical cells and corrosion). Loss of electrons by a constituent of a chemical reaction. *See:* **corrosion terms; electrochemical cell.** CV1/CM-34E2/10E1

oxide-cathode (thermionics). *See:* **oxide-coated cathode.**

oxide-coated cathode (oxide-cathode) (thermionics). A cathode whose active surface is a coating of oxides of alkaline earths on a metal. *See also:* **electron emission.** 50I07-15E6

oxidizing (electrotyping). The treatment of a graphited wax surface with copper sulfate and iron filings to produce a conducting copper coating. *See also:* **electroforming.** 42A60-0

oxygen-concentration cell. A galvanic cell resulting primarily from differences in oxygen concentration. *See:* **electrolytic cell.** CM-34E2

ozone-producing radiation. Ultraviolet energy shorter than about 220 nanometers that decomposes oxygen O_2 thereby producing ozone O_3. Some ultraviolet sources generate energy at 184.9 nanometers, which is particularly effective in producing ozone. *See also:* **ultraviolet radiation.** Z7A1-0

P

PABX. *See:* **private automatic branch exchange.**

pace voltage (lightning arresters). A voltage generated by ground current between two points on the surface of the ground at a distance apart corresponding to the conventional length of an ordinary pace. *See:* **lightning arrester (surge diverter).** 50I25-31E7

pack. To compress several items of data in a storage medium in such a way that the individual items can later be recovered. *See also:* **electronic digital computer.** X3A12-16E9

package, core (rotating machinery). The portion of core lying between two adjacent vent ducts or between an end plate and the nearest vent duct. *See:* **cradle base (rotating machinery).** 0-31E8

packaged magnetron. An integral structure comprising a magnetron, its magnetic circuit, and its output matching device. *See also:* **magnetron.** 42A70-15E6

packing density (computing systems). The number of useful storage cells per unit of dimension, for example, the number of bits per inch stored on a magnetic tape or drum track. *See also:* **electronic digital computer.** X3A12-16E9

packing gland. An explosionproof entrance for conductors through the wall of an explosionproof enclosure, to provide compressed packing completely surrounding the wire or cable, for not less than ½ inch measured along the length of the cable. *See also:* **mine feeder circuit.** 42A85-0

pad (attenuating pad). A nonadjustable passive network that reduces the power level of a signal without introducing appreciable distortion. *Note:* A pad may also provide impedance matching. *See also:* **transducer.** E151-2E2;42A65-31E3

pad electrode (dielectric heater usage). One of a pair of electrode plates between which a load is placed for dielectric heating. *See also:* **dielectric heating.** E54/E169-0

pad-type bearing (rotating machinery). A journal or thrust-type bearing in which the bearing surface is not continuous but consists of separate pads. *See also:* **bearing.** 0-31E8

pair. A term applied in electric transmission to two like conductors employed to form an electric circuit. *See also:* **cable.** 42A65-0

pair of brushes (rotating machinery). *See:* **paired brushes.**

paired brushes (pair of brushes) (rotating machinery). Two individual brushes that are joined together by a common shunt or terminal. *Note:* They are not to be confused with a split brush. *See also:* **brush (rotating machinery).** 64A1/42A10-31E8

paired cable (nonquadded cable). A cable in which all of the conductors are arranged in the form of twisted pairs, none of which is arranged with others to form quads. *See also:* **cable.** 42A65-0

pairing (scanning) (television). The condition in which lines appear in groups of two instead of equally spaced. *See also:* **television.** E204/42A65-2E2

PAM. *See:* **pulse-amplitude modulation.**

pancake coil. A coil having the shape of a pancake, usually with the turns arranged in the form of a flat spiral. *See also:* **circuits and devices.** 42A65-21E0

pancake motor. A motor that is specially designed to have an axial length that is shorter than normal. *See also:* **cradle base (rotating machinery).** 0-31E8

panel (1) (general). A unit of one or more sections of flat material suitable for mounting electric devices. 37A100-31E11

(2) (industrial control). An element of an electric controller consisting of a slab or plate on which various component parts of the controller are mounted and wired. 42A25-34E10

(3) (photovoltaic converter). Combination of shingles or subpanels as a mechanical and electric unit required to meet performance specifications. *See also:* **semiconductor.**

(4) (computing systems). *See:* **control panel; problem board.** *See also:* **electronic digital computer.** 0-10E1

panelboard (electric system). A single panel or a group of panel units designed for assembly in the form of a single panel, including buses and with or without switches and/or automatic overcurrent-protective devices for the control of light, heat, or power circuits of small individual as well as aggregate capacity; designed to be placed in a cabinet or cutout box placed in or against a wall or partition, and accessible only from the front. *See:* **switchboard.** *See also:* **distribution center.** 1A0/2A2-0

panel efficiency (photoelectric converter). The ratio of available power output to incident radiant power intercepted by a panel composed of photoelectric converters. *Note:* This is less than the efficiency of the individual photoelectric converters because of area not covered by photoelectric converters, Joule heating, and photoelectric-converter mismatches. *See also:* **semiconductor.** 0-10E1

panel-frame mounting (switching device). Mounting on a panel frame in the rear of a panel with the operating mechanism on the front of the panel. 37A100-31E11

panel system. An automatic telephone switching system that is generally characterized by the following features: (1) The contacts of the multiple banks over which selection occurs are mounted vertically in flat rectangular panels. (2) The brushes of the selecting mechanism are raised and lowered by a motor that is common to a number of these selecting mechanisms. (3) The switching pulses are received and stored by controlling mechanisms that govern the subsequent operations necessary in establishing a telephone connection. *See also:* **telephone switching system.** 42A65-0

paper-lined construction (dry cell) (primary cell). A type of construction in which a paper liner, wet with electrolyte, forms the principal medium between the negative electrode, usually zinc, and the depolarizing mix. (A layer of paste may lie between the paper liner and the negative electrode.) *See also:* **electrolytic cell.** 42A60-0

PAR. *See:* **precision approach radar.**

paraboloidal reflector. A reflector that is a portion of a paraboloid of revolution. *See also:* **antenna.** 42A65-3E1

paraelectric Curie temperature (of a ferroelectric material). The intercept of the linear portion of the plot of $1/\epsilon$ versus T, where ϵ is the small signal dielectric permittivity measured at zero bias field and T is the absolute temperature in the region above the ferroelectric Curie temperature where ϵ generally follows the Curie-Weiss relation. *See also:* **ferroelectric domain.** E180-0

parallel (parallel elements) (1) (network). (A) Two-terminal elements are connected in parallel when they are connected between the same pair of nodes. (B) Two-terminal elements are connected in parallel when any cut-set including one must include the others. *See also:* **network analysis.** E153/E270-0

(2) (electronic computers). (A) Pertaining to the simultaneity of two or more processes. (B) Pertaining to the simultaneity of two or more similar or identical processes. (C) Pertaining to simultaneous processing of the individual parts of a whole, such as the bits of a character and the characters of a word, using separate facilities for the various parts. *See:* **serial-parallel.** *See also:* **electronic computation; electronic digital computer.** E162/E270/X3A12-16E9

parallel connection. The arrangement of cells in a battery made by connecting all positive terminals together and all negative terminals together, the voltage of the group being only that of one cell and the current drain through the battery being divided among the several cells. *See also:* **battery (primary or secondary).** 42A60-0

parallel digital computer. One in which the digits are handled in parallel. Mixed serial and parallel machines are frequently called serial or parallel according to the way arithmetic processes are performed. An example of a parallel digital computer is one that handles decimal digits in parallel although it might handle the bits that comprise a digit either serially or in parallel. *See also:* **electronic computation; serial digital computer.** E270-0

parallel feeder. One that operates in parallel with one or more feeders of the same type from the same source. *Note:* These feeders may be of the stub-, multiple-, or tie-feeder type. *See:* **multiple feeder.** *See also:* **center of distribution.** 37A100/42A35-31E11/31E13

paralleling (rotating machinery). The process by which a generator is adjusted and connected to run in parallel with another generator or system. *See also:* **asynchronous machine; direct-current commutating machine; synchronous machine.** 0-31E8

parallel-mode interference (signal-transmission system). *See:* **interference, common-mode.**

parallel operation (power supplies). Voltage regulators, connected together so that their individual output currents are added and flow in a common load. Several methods for parallel connection are used: spoiler resistors, master/slave connection, parallel programming, and parallel padding. Current regulators can be paralleled without special precaution. *See also:* **power supply.** KPSH-10E1

parallel padding (power supplies). A method of parallel operation for two or more power supplies in which their current limiting or automatic crossover output characteristic is employed so that each supply regulates a portion of the total current, each parallel supply adding to the total and padding the output only when the load current demand exceeds the capability—or limit setting—of the first supply. *See also:* **power supply.** KPSH-10E1

parallel processing (computing systems). Pertaining to the simultaneous execution of two or more sequences of instructions by a computer having multiple arithmetic or logic units. *See:* **multiprogramming.** *See also:* **electronic digital computer.** X3A12-16E9

parallel programming (power supplies). A method of parallel operation for two or more power supplies in which their feedback terminals (voltage-control terminals) are also paralleled. These terminals are often connected to a separate programming source. *See also:* **power supply.** KPSH-10E1

parallel rectifier. A rectifier in which two or more similar rectifiers are connected in such a way that their direct currents add and their commutations coincide. *See:* **power rectifier.** 0-34E24

parallel rectifier circuit. A rectifier circuit in which two or more simple rectifier circuits are connected in such a way that their direct currents add and their commutations coincide. *See also:* **rectification; rectifier circuit element.** 42A15/E59-34E17

parallel resonance. The steady-state condition that exists in a circuit comprising inductance and capacitance connected in parallel, when the current entering the circuit from the supply line is in phase with the voltage across the circuit. 42A65-31E3

parallel search storage (computing systems). *See:* **associative storage.** *See also:* **electronic digital computer.** X3A12-16E9

parallel storage (computing systems). A storage device in which characters, words, or digits are dealt with simultaneously . *See also:* **electronic digital computer.** X3A12-16E9

parallel-T network (twin-T network). A network composed of separate T networks with their terminals connected in parallel. *See also:* **network analysis.** 42A65-0

parallel transmission (data transmission). Simultaneous transmission of the bits making up a character, either over separate channels or on different carrier frequencies on one channel. *See also:* **data transmission.** 0-19E4

parallel two-terminal pair networks. Two-terminal pair networks are connected in parallel at the input or at the output terminals when their respective input or output terminals are in parallel. *See also:* **network analysis.** E153-0

paramagnetic material. Material whose relative permeability is slightly greater than unity and practically independent of the magnetizing force. E270-0

parameter (1) (mathematical). A variable that is given a constant value for a specific purpose or process. X3A12-16E9

(2) (physical). One of the constants entering into a functional equation and corresponding to some characteristic property, dimension, or degree of freedom.

(3) (electrical). One of the resistance, inductance, mutual inductance, capacitance, or other element values included in a circuit or network. Also called **network constant.** E270-0

parametric amplifier. An inverting parametric device used to amplify a signal without frequency translation from input to output. *Note:* In common usage, this term is a synonym for reactance amplifier. *See also:* **parametric device.** E254-15E7

parametric converter. An inverting parametric device or noninverting parametric device used to convert an input signal at one frequency into an output signal at a different frequency. *See also:* **parametric device.** E254-15E7

parametric device. A device whose operation depends essentially upon the time variation of a characteristic parameter usually understood to be a reactance.
See:
beam parametric amplifier;
degenerate parametric amplifier;
difference frequency;
difference-frequency parametric amplifier;
electron devices, miscellaneous;
idler circuit;
idler frequency;
inverting parametric device;
lower-sideband parametric down-converter;
lower-sideband parametric up-converter;
noninverting parametric device;
parametric amplifier;
parametric converter;
parametric down-converter;
parametric up-converter;
pump;
quadrupole parametric amplifier;
reactance amplifier;
reactance frequency divider;
reactance frequency multiplier;
semiconductor-diode parametric amplifier;
sum frequency;
sum frequency parametric amplifier;
traveling-wave parametric amplifier;
upper-sideband parametric down-converter;
upper-sideband parametric up-converter.
E254-15E7

parametric down-converter. A parametric converter in which the output signal is at a lower frequency than the input signal. *See also:* **parametric device.** E254-15E7

parametric up-converter. A parametric converter in which the output signal is at a higher frequency than the input signal. *See also:* **parametric device.** E254-15E7

parasitic element. A radiating element that is not coupled directly to the feed lines of an antenna and that materially affects the radiation pattern and/or impedance of an antenna. *Note:* Compare with driven element. *See also:* **antenna.** 0-3E1

parasitic oscillations. Unintended self-sustaining oscillations, or transient impulses. *See also:* **oscillatory circuit; transmission characteristics.** E145/E182A/42A65-31E3

parcel plating. Electroplating upon only a part of the surface of a cathode. *See also:* **electroplating.** 42A60-0

parity. Pertaining to the use of a self-checking code employing binary digits in which the total number of ONES (or ZEROS) in each permissible code expression is always even or always odd. A check may be made for either even parity or odd parity. *See also:* **electronic digital computer.** E162-0

parity bit (computing systems). A binary digit appended to an array of bits to make the sum of all the bits always odd or always even. *See also:* **electronic digital computer.** X3A12-16E9

parity check (electronic computation). A summation check in which the bits in a character or block are added (modulo 2) and the sum checked against a single, previously computed parity digit; that is, a check that tests whether the number of ones is odd or even. *See also:* **electronic digital computer; odd-even check.** X3A12-16E9/19E4

parity code (power-system communication). A binary code so chosen that the count of ONEs is even (even parity) or odd (odd parity) and used for error detection. *See also:* **analog.** 0-31E3

parking lamp. A lamp placed on a vehicle to indicate its presence when parked. *See also:* **headlamp.** Z7A1-0

partial (audio and electroacoustics). (1) A physical component of a complex tone. (2) A component of a sound sensation that may be distinguished as a simple tone that cannot be further analyzed by the ear and that contributes to the timbre of the complex sound. *Notes:* (1) The frequency of a partial may be either higher or lower than the basic frequency and may or may not be an integral multiple or submultiple of the basic frequency. If the frequency is not a multiple or submultiple, the partial is inharmonic. (2) When a system is maintained in steady forced vibration at a basic frequency equal to one of the frequencies of the normal modes of vibration of the system, the partials in the resulting complex tone are not necessarily identical in frequency with those of the other normal modes of vibration. *See also:* **electroacoustics.** 0-1E1

partial-automatic station. A station that includes protection against the usual operating emergencies, but in which some or all of the steps in the normal starting or stopping sequence, or in the maintenance of the required character of service, must be performed by a station attendant or by supervisory control. 37A100-31E11

partial-automatic transfer equipment (partial-automatic throwover equipment). An equipment that automatically transfers load to another (emergency) source of power when the original (preferred) source to which it has been connected fails, but that will not automatically retransfer the load to the original source under any conditions. *Note:* The restoration of the load to the preferred source from the emergency source upon the reenergization of the preferred source after an outage may be of the continuous-circuit restoration type or the interrupted-circuit restoration type. 37A100-31E11

partial body irradiation (electrobiology). Pertains to the case in which part of the body is exposed to the incident electromagnetic energy. *See also:* **electrobiology.** 95A1-0

partial carry (parallel addition). A technique in which some or all of the carries are stored temporarily instead of being allowed to propagate immediately. *See:* **carry; electronic digital computer.** X3A12-16E9

partial-read pulse. Any one of the currents applied that cause selection of a cell for reading. *See:* **coincident-current selection.** *See also:* **static magnetic storage.** E163-0

partial-select output. (1) The voltage response of an unselected magnetic cell produced by the application of partial-read pulses or partial-write pulses. (2) The integrated voltage response of an unselected magnetic cell produced by the application of partial-read pulses or partial-write pulses. *See:* **coincident-current selection.** *See also:* **static magnetic storage.** E163-0

partial system test. *See:* **stimulation, physical.** *See also:* **electronic analog computer.** E165-0

partial-write pulse. Any one of the currents applied that cause selection of a cell for writing. *See:* **coincident-current selection.** *See also:* **static magnetic storage.** E163-0

particle accelerator. Any device for accelerating charged particles to high energies, for example, cyclotron, betatron, Van der Graaff generator, linear accelerator, etcetera. *See also:* **electron devices, miscellaneous.** 42A70-15E6

particle velocity (sound field). The velocity of a given infinitesimal part of the medium, with reference to the medium as a whole, due to the sound wave. *Note:* The terms **instantaneous particle velocity, effective particle velocity, maximum particle velocity,** and **peak particle velocity** have meanings that correspond with those of the related terms used for sound pressure. *See also:* **electroacoustics.** 42A65/E157-1E1;42A65-0

parting (corrosion). The selective corrosion of one or more components of a solid solution alloy. *See:* **corrosion terms.** CM-34E2

parting limit (corrosion). The maximum concentration of a more-noble component in an alloy, above which parting does not occur within a specific environment. *See:* **corrosion terms.** CM-34E2

partition noise (electron device). Noise caused by random fluctuations in the distribution of current between the various electrodes. *See:* **circuit characteristics of electrodes.** 0-15E6

part programming, computer (numerically controlled machines). The preparation of a manuscript in computer language and format required to accomplish a given task. The necessary calculations are to be performed by the computer. *See also:* **numerically controlled machines.** EIA3B-34E12

part programming, manual (numerically controlled machines). The preparation of a manuscript in machine control language and format required to accomplish a given task. The necessary calculations are to be performed manually. *See also:* **numerically controlled machines.** EIA3B-34E12

part-winding starter (industrial control). A starter that applies voltage successively to the partial sections of the primary winding of an alternating-current motor. *See:* **starter.** IC1-34E10

part-winding starting (rotating machinery). A method of starting a polyphase induction or synchronous motor, by which certain specially designed circuits of each phase of the primary winding are initially connected to the supply line. The remaining circuit or circuits of each phase are connected to the supply in parallel with initially connected circuits, at a predetermined point in the starting operation. *See also:* **synchronous machine.** 0-31E8

party line. A subscriber line arranged to serve more than one main station, with discriminatory ringing for each station. *See also:* **telephone switching system.** 42A65-19E1

Paschen's law (gas). The law stating that, at a constant temperature, the breakdown voltage is a function only of the product of the gas pressure by the distance between parallel plane electrodes. *See also:* **discharge (gas).** 50I07-15E6

pass element (power supplies). A controlled variable-resistance device, either a vacuum tube or power transistor, in series with the source of direct-current power. The pass element is driven by the amplified error signal to increase its resistance when the output needs to be lowered or to decrease its resistance when the output must be raised. *See:* **series regulator.** *See also:* **power supply.** KPSH-10E1

passenger elevator. An elevator used primarily to carry persons other than the operator and persons necessary for loading and unloading. *See also:* **elevators.** 42A45-0

passive-active cell (corrosion). A cell composed of passive and active areas. *See:* **electrolytic cell.** CM-34E2

passivation (corrosion). The process or processes (physical or chemical) by means of which a metal becomes passive. *See also:* **corrosion terms.** CM-34E2

passivator (corrosion). An inhibitor that changes the potential of a metal appreciably to a more cathodic or noble value. *See:* **corrosion terms.** CM-34E2

passive electric network. An electric network containing no source of energy. *See also:* **network analysis.** 42A65-0

passive transducer. *See:* **transducer, passive.**

passivity (1) (chemical). The condition of a surface that retards a specified chemical reaction at that sur-

face. *See also:* **electrochemistry.** 42A60-0
(2) (electrolytic or anodic). Such a condition of an anode that the normal anodic reaction is retarded. *See also:* **electrochemistry.** 42A60-0

paste (dry cell) (primary cell). A gelatinized layer containing electrolyte that lies adjacent to the negative electrode. *See also:* **electrolytic cell.** 42A60-0

pasted sintered plate (alkaline storage battery). A plate consisting of fritted metal powder in which the active material is impregnated. *See also:* **battery (primary or secondary).** 42A60-0

patch (1) (in general). To connect circuits together temporarily by means of a cord, known as a patch cord. 42A65-0
(2) (computing systems). To modify a routine in a rough or expedient way. *See also:* **electronic digital computer.** X3A12-16E9

patch bay (analog computer). A concentrated assembly of electric tie points that offers a means of electric connection between the inputs and outputs of computing elements, multiples, reference voltages, and ground. *See also:* **electronic analog computer.** E165-16E9

patch board. A board or panel where circuits are terminated in jacks for patching. *See:* **problem board.** *See also:* **electronic analog computer.** 42A65-0

patch panel. *See:* **problem board.** *See also:* **electronic analog computer.**

path (1) (navigation). A line connecting a series of points in space and constituting a proposed or traveled route. *See:* **flight path; flight track.** *See also:* **navigation.** E172-10E6
(2) (network analysis). Any continuous succession of branches, traversed in the indicated branch directions. *See also:* **linear signal flow graphs.** E155-0

path (loop) factor (network analysis). The graph determinant of that part of the graph not touching the specified path (loop). *Notes:* (1) A path (loop) factor is obtainable from the graph determinant by striking out all terms containing transmittance products of loops that touch that path (loop). (2) For loop L_k, the loop factor is

$$-\partial\Delta/\partial L_k.$$

See also: **linear signal flow graphs.** E155-0

path length. The length of a magnetic flux line in a core. *Note:* In a toroidal core with nearly equal inside and outside diameters, the value

$$l_m = \frac{\pi}{2}(\text{O.D.} + \text{I.D.})$$

where O.D. and I.D. are the outside and inside diameters of the core, is commonly used. *See also:* **static magnetic storage.** E163-0

path transmittance (network analysis). The product of the branch transmittances in that path. *See also:* **linear signal flow graphs.** E155-0

patina (corrosion). A green coating consisting principally of basic sulfate and occasionally containing small amounts of carbonate or chloride, that forms on the surface of copper or copper alloys exposed to the atmosphere a long time. *See:* **corrosion terms.** CM-34E2

pattern recognition. The identification of shapes, forms, or configurations by automatic means. *See also:* **electronic digital computer.** X3A12-16E9

pattern-sensitive fault. A fault that appears in response to some particular pattern of data. *See also:* **electronic digital computer.** X3A12-16E9

PAX (telephony). *See:* **private automatic exchange.**

pay station. *See:* **public telephone station.**

PBX. *See:* **private branch exchange.**

PBX trunk. *See:* **private-branch-exchange trunk.**

***P* display (electronic navigation).** *See:* **plan-position indicator.**

PDM. *See:* **pulse-duration modulation.**

peak. *See:* **crest.**

peak alternating gap voltage (electron tube) (traveling-wave tubes). The negative of the line integral of the peak alternating electric field taken along a specified path across the gap. *Note:* The path of integration must be stated. *See:* **beam tubes; electron devices, miscellaneous.** E160-15E6

peak anode current. The maximum instantaneous value of the anode current. *See also:* **electronic controller.** 42A25-34E10

peak burst magnitude (audio and electroacoustics). The maximum absolute peak value of voltage, current, or power for a burstlike excursion. *See:* The figure attached to the definition of **burst duration.** *See also:* **burst (audio and electroacoustics).** E257-1E1

peak-charge characteristic (nonlinear capacitor). The function relating one-half the peak-to-peak value of transferred charge in the steady state to one-half the peak-to-peak value of a specified applied symmetrical alternating capacitor voltage. *Note:* Peak-charge characteristic is always single-valued. *See also:* **nonlinear capacitor.** E226-15E7

peak detector. A detector, the output voltage of which approximates to the true peak value of an applied signal. *See also:* **electromagnetic compatibility.** CISPR-27E1

peak distortion. The largest total distortion of telegraph signals noted during a period of observation. *See also:* **telegraphy.** 42A65-19E4

peak electrode current (electron tube). The maximum instantaneous current that flows through an electrode. *See also:* **electrode current (electron tube).** 42A70/E160-15E6

peak flux density. The maximum flux density in a magnetic material in a specified cyclically magnetized condition. *See also:* **static magnetic storage.** E163-0

peak forward anode voltage (electron tube). The maximum instantaneous anode voltage in the direction in which the tube is designed to pass current. *See also:* **electrode voltage (electron tube); electronic controller.** E160/42A70-15E6

peak forward current rating (repetitive) (rectifier circuit). The maximum repetitive instantaneous forward current permitted by the manufacturer under stated conditions. *See also:* **average forward current rating (rectifier circuit).** E59-34E17/34E24

peak forward voltage (of a rectifying element). The maximum instantaneous voltage between the anode and cathode during the positive nonconducting period. *See also:* **rectification.** 42A15-0

peak induction (peak flux density) (of toroidal magnetic amplifier cores). The magnetic induction corresponding to the peak applied magnetizing force specified in this test. *Note:* It will usually be slightly less than the true saturation induction. E106-0

peaking circuit. A circuit capable of converting an input wave into a peaked waveform. *See also:* **circuits and devices.** 42A65-0

peaking network. A type of interstage coupling network in which an inductance is effectively in series (series peaking network) or in shunt (shunt peaking network) with the parasitic capacitance to increase the amplification at the upper end of the frequency range. *See also:* **network analysis.** 42A65-0

peak inverse anode voltage (electron tube). The maximum instantaneous anode voltage in the direction opposite to that in which the tube is designed to pass current. *See also:* **electrode voltage (electron tube); electronic controller.** E160/42A70-15E6

peak inverse voltage (PIV) (1) (semiconductor diode). The maximum instantaneous anode-to-cathode voltage in the reverse direction that is actually applied to the diode in an operating circuit. *Note:* This is an applications term not to be confused with **breakdown voltage,** which is a property of the device. *See also:* **semiconductor.** E216/E270-34E17

(2) (rectifying element). The maximum instantaneous voltage between the anode and cathode during the inverse period. *See also:* **rectification.** 42A15-0

peak inverse voltage, maximum rated (semiconductor diode). The recommended maximum instantaneous anode-to-cathode voltage that may be applied in the reverse direction. *See also:* **semiconductor.** E216/E270-34E17

peak let-through characteristic curve (current-limiting fuse). *See:* **current-limiting characteristic curve (current-limiting fuse).**

peak let-through current (current-limiting fuse) (peak cutoff current). The highest instantaneous current passed by the fuse during the interruption of the circuit. 37A100-31E11

peak limiter. A device that automatically limits the magnitude of a signal to a predetermined maximum value by changing its amplification. *Notes:* (1) The term is frequently applied to a device whose gain is quickly reduced and slowly restored when the instantaneous magnitude of the input exceeds a predetermined value. (2) In this context, the terms **instantaneous magnitude** and **instantaneous peak power** are used interchangeably. *See also:* **circuits and devices.** 42A65-2E2/31E3

peak load (1) (general). The maximum load consumed or produced by a unit or group of units in a stated period of time. It may be the maximum instantaneous load or the maximum average load over a designated interval of time. *Note:* Maximum average load is ordinarily used. In commercial transactions involving peak load (peak power) it is taken as the average load (power) during a time interval of specified duration occurring within a given period of time, that time interval being selected during which the average power is greatest. *See also:* **generating station.** 42A35-31E13

(2) (motor) (rotating machinery). The largest momentary or short-time load expected to be delivered by a motor. It is expressed in percent of normal power or normal torque. *See also:* **asynchronous machine; direct-current commutating machine; synchronous machine.** 0-31E8

peak load station (electric power supply). A generating station that is normally operated to provide power during maximum load periods. *See also:* **generating station.** 0-31E4

peak magnetizing force (1) (toroidal magnetic amplifier cores). The maximum value of applied magnetomotive force per mean length of path of the core. E106-0

(2) (peak field strength). The upper or lower limiting value of magnetizing force associated with a cyclically magnetized condition. *See also:* **static magnetic storage.** E163-0

peak point (tunnel-diode characteristic). The point on the forward current-voltage characteristic corresponding to the lowest positive (forward) voltage at which $dI/dV = 0$.
See:
electron devices, miscellaneous;
inflection point;
inflection-point current;
inflection-point voltage;
peak-point current;
peak-point voltage;
projected peak point;
projected peak-point voltage;
valley point;
valley-point current;
valley-point voltage. E253-15E7

peak-point current (tunnel-diode characteristic). The current at the peak point. *See also:* **peak point (tunnel-diode characteristic).** E253-15E7

peak-point voltage (tunnel-diode characteristic). The voltage at which the peak point occurs. *See also:* **peak point (tunnel-diode characteristic).** E253-15E7

peak power output (modulated carrier system). The output power, averaged over a carrier cycle, at the maximum amplitude that can occur with any combination of signals to be transmitted. *See also:* **radio transmitter; television.** 42A65/E145-0

peak pulse amplitude (television). The maximum absolute peak value of the pulse excluding those portions considered to be unwanted, such as spikes. *Note:* Where such exclusions are made, it is desirable that the amplitude chosen be illustrated pictorially. *See also:* **pulse; pulse terms.** E203/42A65-31E3

peak pulse power, carrier-frequency. The power averaged over that carrier-frequency cycle that occurs at the maximum of the pulse of power (usually one half the maximum instantaneous power). *See also:* **pulse terms.** E194-0

peak repetitive ON-state current (thyristor). The peak value of the ON-state current including all repetitive transient currents. *See also:* **principal current.** E223-34E17/34E24/15E7

peak responsibility. The load of a customer, a group of customers, or a part of the system at the time of occurrence of the system peak load. *See also:* **generating station.** 42A35-31E13

peak (crest) restriking voltage (lightning arresters). The maximum instantaneous voltage that is attained by the restriking voltage. *See:* **lightning arrester (surge diverter).** 99I2-31E7

peak reverse voltage (semiconductor rectifier). The maximum instantaneous value of the reverse voltage that occurs across a semiconductor rectifier cell, rectifier diode, or rectifier stack. *See also:* **rectification.** E59-34E17

peak sound pressure (for any specified time interval). The maximum absolute value of the instantaneous sound pressure in that interval. *Note:* In the case of a periodic wave, if the time interval considered is a complete period, the peak sound pressure becomes identical with the maximum sound pressure. *See also:* **electroacoustics.** 0-1E1

peak switching current (rotating machinery). The maximum peak transient current attained following a switching operation on a machine. *See also:* **asynchronous machine; direct-current commutating machine; synchronous machine.** 0-31E8

peak value (1) (general). The largest instantaneous value of a time function during a specified time period. *See:* **crest value.** 0-31E3

(2) (alternating voltage). The maximum value excluding small high-frequency oscillations arising, for instance, from partial discharges in the circuit. *See also:* **test voltage and current.** 68A1-31E5

(3) (crest value) (voltage or current) (lightning arresters). The maximum value of an impulse. If there are small oscillations superimposed at the peak, the peak value is defined by the maximum value and not the mean curve drawn through the oscillations. *See:* **lightning arrester (surge diverter).** 99I2-31E7

(4) (impulse current) (virtual peak value). Normally the maximum value. With some test circuits, overshoot or oscillations may be present on the current. The maximum value of the smooth curve drawn through the oscillations is defined as the virtual peak value. It will depend on the type of test whether the value of the test current shall be defined by the actual peak or a virtual peak value. *Note:* The term **peak value** is to be understood as including the term **virtual peak value** unless otherwise stated. *See also:* **test voltage and current.** 68A1-31E5

(5) (impulse voltage) (virtual peak value). Normally the maximum value. With some test circuits, oscillations or overshoot may be present on the voltage. If the amplitude of the oscillations is not greater than 5 percent of the peak value and the frequency is at least 0.5 megahertz or, alternatively, if the amplitude of the overshoot is not greater than 5 percent of the peak value and the duration not longer than 1 microsecond, then for the purpose of measurement a mean curve may be drawn, the maximum amplitude of which is defined as the virtual peak value. (See the accompanying figure.) If the frequency is less than or if the duration is greater than described above, the peak of the oscillation may be used as the peak value. *Note:* The term **peak value** is to be understood as including the term **virtual peak value** unless otherwise stated. *See also:* **test voltage and current.** 68A1-31E5

pedestal. A substantially flat-topped pulse that elevates the base level for another wave. *See also:* **pulse.** 42A65-0

pedestal bearing (rotating machinery). A complete assembly of a bearing with its supporting pedestal. 0-31E8

pedestal bearing insulation (rotating machinery). The insulation applied either below the bearing liner shell and the adjacent pedestal support or between the base of the pedestal and the machine bed plate, to break the current path that may be formed through the shaft to the outboard bearing to the frame to the drive-end bearing and thence back to the shaft. *Note:* The voltage is usually very low. However, very destructive bearing currents can flow in this path if some insulating break is not provided. High-pressure moulded laminates are usually employed for this type of insulation. *See:* **cradle base (rotating machinery).** 0-31E8

peeling. The unwanted detachment of a plated metal coating from the base metal. *See also:* **electroplating.** 42A60-0

Peltier coefficient, absolute. The product of the absolute temperature and the absolute Seebeck coefficient of the material; the sign of the Peltier coefficient is the same as that of the Seebeck coefficient. *Note:* The opposite sign convention has also been used in the technical literature. *See also:* **thermoelectric device.** E221-15E7

Peltier coefficient of a couple. The quotient of (1) the rate of Peltier heat absorption by the junction of the two conductors by (2) the electric current through the

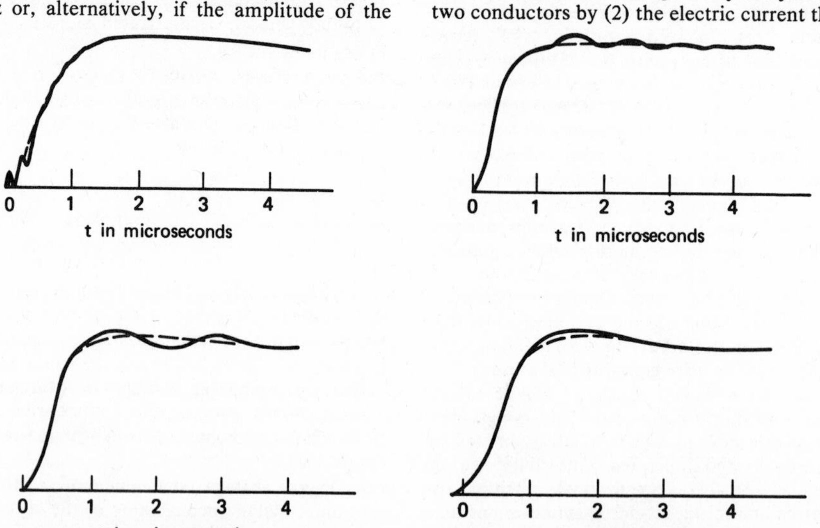

Construction for derivation of virtual peak values.

junction; the Peltier coefficient is positive if Peltier heat is absorbed by the junction when the electric current flows from the second-named conductor to the first conductor. *Notes:* (1) The opposite sign convention has also been used in the technical literature. (2) The Peltier coefficient of a couple is the algebraic difference of either the relative or absolute Peltier coefficients of the two conductors. *See also:* **thermoelectric device.** E221-15E7

Peltier coefficient, relative. The Peltier coefficient of a couple composed of the given material as the first-named conductor and a specified standard conductor. *Note:* Common standard conductors are platinum, lead, and copper. *See also:* **thermoelectric device.** E221-15E7

Peltier effect. The absorption or evolution of thermal energy, in addition to the Joule heat, at a junction through which an electric current flows; and in a nonhomogeneous, isothermal conductor, the absorption or evolution of thermal energy, in addition to the Joule heat, produced by an electric current. *Notes:* (1) For the case of a nonhomogeneous, nonisothermal conductor, the Peltier effect cannot be separated from the Thomson effect. (2) A current through the junction of two dissimilar materials causes either an absorption or liberation of heat, depending on the sense of the current, and at a rate directly proportional to it to a first approximation. *See also:* **thermoelectric device.** E270/E221-15E7

Peltier heat. The thermal energy absorbed or evolved as a result of the Peltier effect. *See also:* **thermoelectric device.** E221-15E7

penalty factor (electric power system). A factor that, when multiplied by the incremental cost of power at a particular source, produces the incremental cost of delivered power from that source. Mathematically, it is

$$\frac{1}{(1-\text{Incremental Transmission Loss})^*}$$

See also: **power system, low-frequency and surge testing.**

*Expressed as a decimal. E94-0

pencil beam. An antenna-system pattern whose energy is contained in a small solid angle. *See also:* **navigation.** 0-10E6

pencil-beam antenna. A unidirectional antenna having a narrow major lobe with approximately circular contours of equal radiation intensity in the region of the major lobe. *See also:* **antenna.** 0-3E1

pendant. A device or equipment that is suspended from overhead either by means of the flexible cord carrying the current or otherwise. *See also:* **interior wiring.** 42A95-0

penetration frequency (radio wave propagation). Critical frequency. *See also:* **radio wave propagation.** 0-3E2

pentode. A five-electrode electron tube containing an anode, a cathode, a control electrode, and two additional electrodes that are ordinarily grids. *See also:* **tube definitions.** E160/42A70-15E6

pen travel. The length of the path described by the pen in moving from one end of the chart scale to the other. The path may be an arc or a straight line. *See also:* **moving element (instrument).** 39A4-0

perceived light-source color. The color perceived to belong to a light source. *See also:* **color.** Z7A1-0

perceived object color. The color perceived to belong to an object, resulting from characteristics of the object, of the incident light, and of the surround, the viewing direction, and observer adaptation. *See also:* **color.** Z7A1-0

percentage differential relay. A differential relay in which the designed response to the phasor difference between incoming and outgoing electrical quantities is modified by a restraining action of one or more of the input quantities. *Note:* The relay operates when the magnitude of the phasor difference exceeds the specified percentage of one or more of the input quantities. 37A100-31E11/31E6

percentage error (watthour meter). The difference between its percentage registration and 100 percent. A meter whose percentage registration is 95 percent is said to be 5 percent slow, or its error is −5 percent. A meter whose percentage registration is 105 percent is 5 percent fast or its error is +5 percent. *See also:* **electricity meter (meter).** 12A0/42A30-0

percentage immediate appreciation (telephone transmission system). The percentage of the total number of spoken sentences that are immediately understood without conscious deductive effort when each sentence conveys a simple and easily understandable idea. *See also:* **volume equivalent.** 42A65-0

percentage modulation. (1) In angle modulation, the fraction of a specified reference modulation, expressed in percent. (2) In amplitude modulation, the modulation factor expressed in percent. *Note:* It is sometimes convenient to express percentage modulation in decibels below 100-percent modulation. *See also:* **modulating systems; radio transmission.** E182A-0

percentage modulation, effective (single, sinusoidal input component). The ratio of the peak value of the fundamental component of the envelope to the direct-current component in the modulated conditions, expressed in percent. *Note:* It is sometimes convenient to express percentage modulation in decibels below 100-percent modulation. *See also:* **modulating systems.** E145/42A65-0

percentage registration (watthour meter) (accuracy*) (percentage accuracy*). The ratio of the actual registration of the meter to the true value of the quantity measured in a given time, expressed as a percentage. *See also:* **electricity meter (meter).**

*Deprecated 12A0/42A30-0

percent articulation. *See:* **articulation.**

percent flutter (reproduced tone) (sound recording and reproducing). The root-mean-square deviation from the average frequency, expressed as a percentage of average frequency. *See also:* **sound recording and reproducing.** E193-0

percent harmonic distortion. *See:* **distortion, percent harmonic.**

percent hearing loss. *See:* **percent impairment of hearing.**

percent impairment of hearing (percent hearing loss). An estimate of a person's ability to hear correctly. It is usually based, by means of an arbitrary rule, on the pure-tone audiogram. The specific rule for calculating this quantity from the audiogram now varies from state to state according to a rule or law. *Note:* The term disability of hearing is sometimes used for impairment of hearing. Impairment refers specifically to a person's illness or injury that affects his personal efficiency in the activities of daily living. Disability has the additional medicolegal connotation that an impairment reduces a person's ability to engage in gainful activity. Impairment is only a contributing factor to the disability. 0-1E1

percent impedance (rectifier transformer). The percent of rated alternating-current winding voltage required to circulate current equivalent to rated line kilovolt-amperes in the alternating-current winding with all direct-current winding terminals short-circuited. *See also:* **rectifier transformer.** 57A18-0

percent intelligibility. *See:* **articulation.**

percent ratio. True ratio expressed in percent of the marked ratio. 57A13-31E12

percent ratio error (instrument transformer). The difference between the ratio correction factor and unity expressed in percent. *Note:* The percent ratio error is positive if the ratio correction factor is greater than unity. If the percent ratio error is positive, the measured secondary current or voltage will be less than the primary value divided by the marked ratio. 57A13-31E12

percent ripple (1) (general). The ratio of the effective (root-mean-square) value of the ripple voltage (current) to the average value of the total voltage (current), expressed in percent. *See:* **interference.** E145-13E6 **(2) (electrical conversion).** The percent ripple voltage is defined as the ratio of the root-mean-square (RMS) value of the voltage pulsations (E_{max} to E_{min}) to the average value of the total voltage.

$$\text{Percent Ripple} = \frac{\text{RMS Ripple}}{E_{av}} (100\%)$$

Note: In most applications the definition has been revised to simplify the calculations by defining percent ripple as the ratio of the root-mean-square (RMS) value of the voltage pulsations to the nominal no-load output voltage of the converter $E_{nominal}$

$$\text{Percent Ripple} = \frac{\text{RMS Ripple}}{E_{nominal}} (100\%)$$

See also: **electrical conversion.** 0-10E1

percent ripple voltage (or current). *See:* **percent ripple.**

percent steady-state deviation (control). The difference between the ideal value and the final value, expressed as a percentage of the maximum rated value of the directly controlled variable (or another variable if specified). *See:* **control system, feedback.** AS1-34E10

percent syllable articulation. *See:* **syllable articulation.**

percent system deviation (control). At any given point on the time response, the difference between the ideal value and the instantaneous value, expressed as a percentage of the maximum rated value of the directly controlled variable (or another variable if specified). *See:* **deviation (control).** *See also:* **control system, feedback.** AS1-34E10

percent total flutter (sound recording and reproducing). The value of flutter indicated by an instrument that responds uniformly to flutter of all rates from 0.5 up to 200 hertz. *Note:* Except for the most critical tests, instruments that respond uniformly to flutter of all rates up to 120 hertz are adequate, and their indications may be accepted as showing percent total flutter. *See also:* **sound recording and reproducing.** E193-0

percent transformer correction-factor error. Difference between the transformer correction factor and unity expressed in percent. *Note:* The percent transformer correction-factor error is positive if the transformer correction factor is greater than unity. If the percent transformer correction-factor error is positive, the measured watts or watthours will be less than the true value. 57A13-13E12

percent transient deviation (control). The difference between the instantaneous value and the final value, expressed as a percentage of the maximum rated value of the directly controlled variable (or another variable if specified). *See:* **control system, feedback.** AS1-34E10

percent unbalance of phase voltages (electrical conversion). The ratio of the maximum deviation of a phase voltage from the average of the total phases to the average of the phase voltages, expressed in percent.

$$\%\ \text{Unbalance} = \frac{\text{RMS Phase Voltage} - \text{RMS Average Phase Voltages}}{\text{RMS Average Phase Voltage}} \times 100\%$$

See also: **electrical conversion.** 0-10E1

perfect dielectric (ideal dielectric). A dielectric in which all of the energy required to establish an electric field in the dielectric is recoverable when the field or impressed voltage is removed. Therefore, a perfect dielectric has zero conductivity and all absorption phenomena are absent. A complete vacuum is the only known perfect dielectric. E270-0

perfect diffusion. That diffusion in which flux is scattered in accord with Lambert's cosine law. *See also:* **lamp.** Z7A1-0

perfect transformer. *See:* **ideal transformer.**

perforated tape. Tape in which a code hole(s) and a tape-feed hole have been punched in a row. EIA3B-34E12

perforator (telegraph practice). A device for punching code signals in paper tape for application to a tape transmitter. *Note:* A perforating device that is automatically controlled by incoming signals is called a reperforator. *See also:* **telegraphy; electronic digital computer.** 42A65/X3A12-16E9/19E4

performance characteristic (device). An operating characteristic, the limit or limits of which are given in the design test specifications. 37A100-31E11

performance chart (magnetron oscillators). A plot on coordinates of applied anode voltage and current showing contours of constant magnetic field, power output, and over-all efficiency. *See:* **magnetron.** E160-15E6

performance index (1) (system). The system function or objective to be optimized. *See also:* **system.** 0-35E2

(2) (control system). A scalar measure of the quality of system behavior. The performance index is usually a function of the plant output and control input over some time interval. *See also:* **control system.** 0-23E0

performance tests (rotating machinery). The tests required to determine the characteristics of a machine and to determine whether the machine complies with its specified performance. *See also:* **asynchronous machine; direct-current commutating machine; synchronous machine.** 0-31E8

perimeter lights. Aeronautical ground lights provided to indicate the perimeter of a landing pad for helicopters. *See also:* **signal lighting.** Z7A1-0

period (1) (primitive period) (function). The smallest number $k_1 \neq 0$ for which the following identity holds: $f(x) = f(x+k_1)$. *Note:* For example, the primitive period of sin x is 2π. Not all periodic functions have primitive periods. For example, $f(x) = A$ (a constant) has no primitive period. It is common engineering practice to exclude from the class of periodic functions, any function that has no primitive period. E270-0

(2) (periodic time) (electric instrument). The time between two consecutive transits of the pointer or indicating means in the same direction through the rest position. *See also:* **moving element (instrument).** 39A1-0

(3) (electric power systems). The minimum interval of the independent variable after which the same characteristics of a periodic phenomenon recur. 50I05-31E3

periodic-automatic-reclosing equipment. An equipment that provides for automatically reclosing a circuit-switching device a specified number of times at specified intervals between reclosures. *Note:* This type of automatic reclosing equipment is generally used for alternating-current circuits. 37A100-31E11

periodic damping*. *See:* **underdamping.**

*Deprecated

periodic duty. A type of intermittent duty in which the load conditions are regularly recurrent. *See:* **duty.** *See also:* **voltage regulator.** E270/42A15/57A15-31E12/34E10

periodic electromagnetic wave (radio wave propagation). A wave in which the electric field vector is repeated in detail in either of two ways: (1) At a fixed point, after the lapse of a time known as the period. (2) At a fixed time, after the addition of a distance known as the wavelength. *See also:* **radio wave propagation.** E211-3E2

periodic frequency modulation (inverters). The periodic variation of the output frequency from the fundamental. *See:* **self-commutated inverters.** 0-34E24

periodic function. A function that satisfies $f(x) = f(x+nk)$ for all x and for all integers n, k being a constant. For example,

$$\sin(x+a) = \sin(x+a+2n\pi).$$

E270-0

periodic line (transmission lines). A line consisting of successive identically similar sections, similarly oriented, the electrical properties of each section not being uniform throughout. *Note:* The periodicity is in space and not in time. An example of a periodic line is the loaded line with loading coils uniformly spaced. *See also:* **transmission line.** E270/42A65-0

periodic output voltage modulation (inverters). The periodic variation of output voltage amplitude at frequencies less than the fundamental output frequency. *See:* **self-commutated inverters.** 0-34E24

periodic permanent-magnet focusing (PPM) (microwave tubes). Magnetic focusing derived from a periodic array of permanent magnets. *See:* **magnetron.** 0-15E6

periodic rating (1) (electric power sources). The load that can be carried for the alternate periods of load and rest specified in the rating, the apparatus starting at approximately room temperature, and for the total time specified in the rating, without causing any of the specified limitations to be exceeded. *See also:* **asynchronous machine; direct-current commutating machine; relay; synchronous machine.** 42A25-34E10

(2) (relay). A rating that defines the current or voltage that may be sustained by the relay during intermittent periods of energization as specified, starting cold and operating for the total time specified without causing any of the prescribed limitations to be exceeded. 37A100-31E11/31E6

periodic slow-wave circuit (microwave tubes). A circuit whose structure is periodically recurring in the direction of propagation. *See also:* **microwave (tube or valve).** 0-15E6

periodic test. A test made at regular intervals to insure continued accuracy of registration. *See also:* **service test (field test).** 42A30-0

periodic wave. A wave in which the displacement at each point of the medium is a periodic function of the time. Periodic waves are classified in the same manner as periodic quantities. E270-0

periodic waveguide. A waveguide in which propagation is obtained by periodically arranged discontinuities or periodic modulations of the material boundaries. *See:* **waveguide.** 0-3E1

peripheral air-gap leakage flux (rotating machinery). That component of air-gap magnetic flux emanating from the rotor or stator, that flows from pole to pole without entering the radially opposite surface of the air gap. *See also:* **rotor (rotating machinery); stator.** 0-31E8

peripheral vision. The seeing of objects displaced from the primary line of sight and outside the central visual field. *See also:* **visual field.** Z7A1-0

peripheral visual field. That portion of the visual field that falls outside the region corresponding to the foveal portion of the retina. *See also:* **visual field.** Z7A1-0

permanent echo (primary radar system). A signal reflected from an object fixed with respect to the radar site. *See also:* **radar.** E172-10E6

permanent fault (lightning arresters). A fault that can be cleared only by action taken at the point of fault. *See:* **lightning arrester (surge diverter).** 50I25-31E7

permanent-field synchronous motor. A synchronous motor similar in construction to an induction motor in which the member carrying the secondary laminations and windings carries also permanent-magnet field poles that are shielded from the alternating flux by the laminations. It starts as an induction motor but operates normally at synchronous speed. *See:* **permanent-magnet synchronous motor; synchronous machine.** 42A10-31E8

permanent magnet (PM). A ferromagnetic body that maintains a magnetic field without the aid of external electric current. 50I05/E270-0

permanent-magnet erasing head (electroacoustics). A head that uses the fields of one or more permanent magnets for erasing. *See also:* **phonograph pickup.** 0-1E1

permanent-magnet focusing (microwave tubes). Magnetic focusing derived from the use of a permanent magnet. *See:* **magnetrons.** 0-15E6

permanent-magnet generator (magneto). A generator in which the open-circuit magnetic flux field is provided by one or more permanent magnets. 0-31E8

permanent-magnet loudspeaker. A moving-conductor loudspeaker in which the steady field is produced by means of a permanent magnet. *See also:* **loudspeaker.** 42A65-0

permanent-magnet moving-coil instrument (d'Arsonval instrument). An instrument that depends for its operation on the reaction between the current in a movable coil or coils and the field of a fixed permanent magnet. *See also:* **instrument.** 42A30-0

permanent-magnet moving-iron instrument (polarized-vane instrument). An instrument that depends for its operation on the action of an iron vane in aligning itself in the resultant magnetic field produced by a permanent magnet and the current in an adjacent coil of the instrument. *See also:* **instrument.** 42A30-0

permanent-magnet, second-harmonic, self-synchronous system. A remote-indicating arrangement consisting of a transmitter unit and one or more receiver units. All units have permanent-magnet rotors and toroidal stators using saturable ferromagnetic cores and excited with alternating current from a common external source. The coils are tapped at three or more equally spaced intervals, and the corresponding taps are connected together to transmit voltages that consist principally of the second harmonic of the excitation voltage. The rotors of the receiver units will assume the same angular position as that of the transmitter rotor. *See:* **synchro system.** 42A10-0

permanent-magnet synchronous motor. A synchronous motor with permanent-magnet field poles normally operating at synchronous speed. Such motors are often equipped with secondary cage windings embedded in laminations and effectively shielding the permanent magnets from alternating flux during start up as an induction motor. *See:* **permanent-field synchronous motor; synchronous machine.** 0-31E8

permanent-split capacitor motor. A capacitor motor with the same value of effective capacitance for both starting and running operations. *See:* **asynchronous machine.** 42A10/MG1-31E8

permanent storage (computing systems). *See:* **fixed storage.** *See also:* **electronic digital computer.** X3A12-16E9

permeability. A general term used to express various relationships between magnetic induction and magnetizing force. These relationships are either (1) absolute permeability, that in general is the quotient of a change in magnetic induction divided by the corresponding change in magnetizing force; or (2) specific (relative) permeability, which is the ratio of the absolute permeability to the magnetic constant. *Notes:* (1) Relative permeability is a pure number that is the same in all unit systems; the value and dimension of absolute permeability depend upon the system of units employed. (2) In anisotropic media, permeability becomes a matrix. E270-0

permeability of free space. *See:* **magnetic constant.**

permeameter. An apparatus for determining corresponding values of magnetizing force and flux density in a test specimen. From such values of magnetizing force and flux density, normal induction curves or hysteresis loops can be plotted and magnetic permeability can be computed. *See also:* **magnetometer.** 42A30-0

permeance. The reciprocal of reluctance. E270-0

permissible mine equipment. Equipment that complies with the requirements of and is formally approved by the United States Bureau of Mines after having passed the inspections and the explosion and/or other tests specified by that Bureau. *Note:* All equipment so approved must carry the official approval plate required as identification for permissible equipment. *See also:* **mining.** 42A85-0

permissible response rate (steam generating unit). The maximum assigned rate of change in generation for load-control purposes based on estimated and known limitations in the turbine, boiler, combustion control, or auxiliary equipment. The permissible response rate for a hydro-generating unit is the maximum assigned rate of change in generation for load-control purposes based on estimated and known limitations of the water column, associated piping, turbine, or auxiliary equipment. *See also:* **speed-governing system.** E94-0

permissive (relay system). A general term indicating that functional cooperation of two or more relays is required before control action can become effective. 37A100-31E11/31E6

permissive block. A block in manual or controlled manual territory, governed by the principle that a train other than a passenger train may be permitted to follow a train other than a passenger train in the block. *See:* **block signal system; controlled manual block signal system; railway signal and interlocking.** 42A42-0

permissive control (electric power systems). A control mode in which generating units are allowed to be controlled only when the change will be in the direction to reduce area-control error. *See also:* **speed-governing system.** E94-0

permittivity. *See:* **absolute capacitivity.**

permittivity of free space. *See:* **electric constant.**

perpendicular magnetization (magnetic recording). Magnetization of the recording medium in a direction perpendicular to the line of travel and parallel to the smallest cross-sectional dimension of the medium. *Note:* In this type of magnetization, either single pole-piece or double pole-piece magnetic heads may be used. *See also:* **phonograph pickup.** E157-1E1

persistence (oscilloscopes). *See:* **phosphor decay.**

persistence characteristic (1) (camera tubes). The temporal step response of a camera tube to illumination. *See:* **methods of measurement.** *See also:* **television.** E160-15E6

(2) (decay characteristic) (luminescent screen). A

relation, usually shown by a graph, between luminance (or emitted radiant power) and time after excitation is removed. *See:* **beam tubes.** 42A70/E160-15E6/2E2

persistent-cause forced outage (electric power systems). A component outage whose cause is not immediately self-clearing but must be corrected by eliminating the hazard or by repairing or replacing the affected component before it can be returned to service. *Note:* An example of a persistent-cause forced outage is a lightning flashover that shatters an insulator thereby disabling the component until repair or replacement can be made. *See also:* **outage.** 0-31E4

persistent-cause forced-outage duration (electric power systems). The period from the initiation of a persistent-cause forced outage until the affected component is replaced or repaired and made available to perform its intended function. *See also:* **outage.** 0-31E4

persistent current (superconducting material). A magnetically induced current that flows undiminished in a superconducting material or circuit. *See also:* **superconductivity.** E217-15E7

persistent-image device. An optoelectronic amplifier capable of retaining a radiation image for a length of time determined by the characteristics of the device. *See also:* **optoelectronic device.** E222-15E7

persistent-image panel (optoelectronic device). A thin, usually flat, multicell persistent-image device. *See also:* **optoelectronic device.** E222-15E7

per-unit quantity (rotating machinery). The ratio of the actual value of a quantity to the base value of the same quantity. The base value is always a magnitude, or in mathematical terms, a positive, real number. The actual value of the quantity in question (current, voltage, power, torque, frequency, etcetera) can be of any kind: root-mean-square, instantaneous, phasor, complex, vector, envelope, etcetera. *Note:* The base values, though arbitrary, are usually related to characteristic values, for example, in case of a machine, the base power is usually chosen to be the rated power (active or apparent), the base voltage to be the rated root-mean-square voltage, the base frequency, the rated frequency. Despite the fact that the choice of base values is rather arbitrary, it is of advantage to choose base values in a consistent manner. The use of a consistent per-unit system becomes a practical necessity when a complicated system is analyzed. *See:* **asynchronous machine; direct-current commutating machine; synchronous machine.** 0-31E8

per-unit resistance. The measured watts expressed in per-unit on the base of the rated kilovolt-amperes of the teaser winding. *See:* **efficiency.** 42A15-0

per-unit system (rotating machinery). The system of base values chosen in a consistent manner to facilitate analysis of a device or system, when per-unit quantities are used. Its importance becomes paramount when analog facilities (network analyzer, analog and hybrid computers) are utilized. *Note:* In electric network analysis and electromechanical system studies, usually four independent fundamental base values are chosen. The rest of the base values are derived from the fundamental ones. In most cases power, voltage, frequency, and time are chosen as fundamental base values. The base power must be the same for all types: apparent, active, reactive, instantaneous. The base time is usually 1 second. From the above, all other base values can be found, for example, base power times base time equals base energy, etcetera. The per-unit system can cover extensive networks because the base voltages of network sections connected by transformers can differ, in which case an **ideal per-unit transformer** is usually introduced having a turns ratio equal to the quotient of the effective turns ratio of the actual transformer and the ratio of base voltage values. By keeping the power, frequency, and time bases the same, only those base quantities will differ for different network sections that are directly or indirectly related to voltage (for example, current, impedance, reactance, inductance, capacitance, etcetera) but those related to power, frequency, and time only (for example, energy, torque, etcetera) will remain unchanged. *See:* **asynchronous machine; direct-current commutating machine; synchronous machine.** 0-31E8

perveance. The quotient of the space-charge-limited cathode current by the three-halves power of the anode voltage in a diode. *Note:* Perveance is the constant G appearing in the Child-Langmuir-Schottky equation

$$i_k = Ge_b^{3/2}.$$

When the term perveance is applied to a triode or multigrid tube, the anode voltage e_b is replaced by the composite controlling voltage e' of the equivalent diode. *See also:* **circuit characteristics of electrodes.** 42A70-15E6

PFM. *See:* **pulse-frequency modulation.**

***p* gate thyristor.** A thyristor in which the gate terminal is connected to the p region adjacent to the region to which the cathode terminal is connected and that is normally switched to the ON state by applying a positive signal between gate and cathode terminals. *See also:* **thyristor.** E223-34E17/34E24/15E7

pH (of a solution A). The pH is obtained from the measurements of the potentials E of a galvanic cell of the form H_2; solution A; saturated potassium chloride (KCl); reference electrode with the aid of the equation

$$\mathrm{pH} = \frac{E - E_0}{(RT/F)\ln 10} = \frac{E - E_0}{2.303\, RT/F}$$

in which E_0 is a constant depending upon the nature of the reference electrode, R is the gas constant in joules per mole per degree, T is the absolute temperature in kelvins, and F is the Faraday constant in coulombs per gram equivalent. Historically pH was defined by

$$\mathrm{pH} = \log \frac{1}{[\mathrm{H}^+]}$$

in which $[\mathrm{H}^+]$ is the hydrogen ion concentration. According to present knowledge there is no simple relation between hydrogen ion concentration or activity and pH. *See also:* **ion activity.** Values of pH may be regarded as a convenient scale of acidities. *See also:* **ion.** 42A60-0

phanotron. A hot-cathode gas diode. *Note:* This term is used primarily in the industrial field. *See also:* **tube definitions.** 42A70-15E6

phantom circuit. A superposed circuit derived from two suitably arranged pairs of wires, called side circuits, the two wires of each pair being effectively in parallel. *See also:* **transmission line.** 42A65-0

phantom-circuit loading coil. A loading coil for introducing a desired amount of inductance in a phantom circuit and a minimum amount of inductance in the constituent side circuits. *See also:* **loading.** 42A65-0

phantom-circuit repeat coil. *See:* **phantom-circuit repeating coil.**

phantom-circuit repeating coil (phantom-circuit repeat coil). A repeating coil used at a terminal of a phantom circuit, in the terminal circuit extending from the midpoints of the associated side-circuit repeating coils. *See also:* **telephone system.** 42A65-0

phantom group. A group of four open wire conductors suitable for the derivation of a phantom circuit. *See also:* **open wire.** 42A65-0

phantom target. (1) An echo box, or other reflection device, that produces a particular blip on the radar indicator. (2) A condition, maladjustment, or phenomenon (such as a temperature inversion) that produces a blip on the radar indicator resembling blips of targets for which the system is being operated. *See:* **echo box.** *See also:* **navigation.** 0-10E6

phase (of a periodic phenomenon $f(t)$, for a particular value of t). The fractional part t/P of the period P through which t has advanced relative to an arbitrary origin. *Note:* The origin is usually taken at the last previous passage through zero from the negative to the positive direction. *See:* **simple sine-wave quantity.** *See also:* **control system, feedback.** 0-23E0

phase advancer. A phase modifier that supplies leading reactive volt-amperes to the system to which it is connected. Phase advancers may be either synchronous or asynchronous. *See:* **converter.** 42A10-0

phase angle (1) (general). The measure of the progression of a periodic wave in time or space from a chosen instant or position. *Notes:* (A) The phase angle of a field quantity, or of voltage or current, at a given instant of time at any given plane in a waveguide is $(\omega t-\beta z+\theta)$, when the wave has a sinusoidal time variation. The term **waveguide** is used here in its most general sense and includes all transmission lines; for example, **rectangular waveguide, coaxial line, strip line,** etcetera. The symbol β is the imaginary part of the propagation constant for that waveguide, propagation is in the $+z$ direction, and θ is the phase angle when $z = t = 0$. At a reference time $t = 0$ and at the plane z, the phase angle $(-\beta z + \theta)$ will be represented by Φ.(B) Phase angle is obtained by multiplying the phase by 360 degrees or by 2π radians. *See also:* **sine-wave quantity.** E285/85A1-9E1/23E0

(2) (current transformer). The angle between the current leaving the identified secondary terminal and the current entering the identified primary terminal. *Note:* This angle is conveniently designated by the Greek letter beta (β) and is considered positive when the secondary current leads the primary current. *See also:* **instrument transformer.** 12A0/42A15/42A30-0

(3) (potential (voltage) transformer). The angle between the secondary voltage from the identified to the unidentified terminal and the corresponding primary voltage. *Note:* This angle is conveniently designated by the Greek letter gamma (γ) and is considered positive when the secondary voltage leads the primary voltage. *See:* **instrument transformer.** 12A0/42A15/42A30-0

(4) (instrument transformer). Phase displacement, in minutes, between the primary and secondary values. 57A13-31E12

phase-angle correction factor. That factor by which the reading of a wattmeter or the registration of a watthour meter, operated from the secondary of a current or a potential transformer or both, must be multiplied to correct for the effect of phase displacement of secondary current and voltage with respect to primary values due to instrument-transformer phase angles. *Note:* This factor equals the ratio of the true power factor to the apparent power factor and is a function of both the phase angle of the instrument transformer and the power factor of the primary circuit being measured. *See:* **instrument transformer; watthour meter.** 12A0/42A30-31E12

phase angle, loop (automatic control) (closed loop). The value of the loop phase characteristic at a specified frequency. *See:* **phase characteristic.** *See also:* **control system, feedback.** 85A1-23E0

phase-balance relay. A relay that responds to differences between quantities of the same nature associated with different phases of a normally balanced polyphase circuit. 37A100-31E11/31E6

phase belt (coil group). A group of adjacent coils in a distributed polyphase winding of an alternating-current machine that are ordinarily connected in series to form one section of a phase winding of the machine. Usually, there are as many such phase belts per phase as there are poles in the machine. *Note:* The adjacent coils of a phase belt do not necessarily occupy adjacent slots; the intervening slots may be occupied by coils of another winding on the same core. Such may be the case in a two-speed machine. *See also:* **rotor (rotating machinery); stator.** 0-31E8

phase center (in a given direction and in a given plane) (antenna). The center of curvature of the wave front of the radiation from an antenna. *Note:* The phase centers of an antenna are usually determined from the phase pattern of the major lobe. *See also:* **radiation.** 0-3E1

phase characteristic (1). The variation with frequency of the phase angle of a phasor quantity. E270-0

(2) (linear passive networks). The angle of a response function evaluated on the imaginary axis of the complex-frequency plane. *See also:* **linear passive networks.** E156-0

(3) (synchronous machine) (V curve). A characteristic curve between armature current and field current, with constant mechanical load (or no load) and with constant terminal voltage maintained. 42A10-0

phase characteristic, loop (automatic control) (closed loop). The phase angle of the loop transfer function for real frequencies. *See also:* **control system, feedback.** 85A1-23E0

phase-coil insulation (rotating machinery). Additional insulation between adjacent coils that are in different phases. *See also:* **asynchronous machine; synchronous machine.** 0-31E8

phase-comparison protection. A form of pilot protection that compares the relative phase-angle position of specified currents at the terminals of a circuit. 37A100-31E11/31E6

phase conductor (alternating-current circuit). The conductors other than the neutral conductor. *Note:* If an alternating-current circuit does not have a neutral conductor, all the conductors are phase conductors. E270-0

phase connections (rotating machinery). The insulated conductors (usually arranged in peripheral rings) that make the necessary connections between appropriate phase belts in an alternating-current winding. *See:* **rotor (rotating machinery); stator.** 0-31E8

phase constant (traveling plane wave at a given frequency) (wavelength constant). The imaginary component of the propagation constant. *Note:* This is the space rate of decrease of phase of a field component (or of the voltage or current), in the direction of propagation, in radians per unit length. *See also:* **transmission characteristics; waveguide.** 42A65/E146/E270-3E1/3E2/9E4

phase contours. Loci of the return transfer function at constant values of the phase angle. *Note:* Such loci may be drawn on the Nyquist, inverse Nyquist, or Nichols diagrams for estimating performance of the closed loop with unity feedback. In the complex plane plot of $KG(j\omega)$, these loci are circles with centers at $-1/2, j/2N$ and radiuses such that each circle passes through the origin and the point $-1, j0$. In the inverse Nyquist diagram they are straight lines $\gamma=-N(x+1)$ radiating from the point $-1, j0$. *See:* **Nichols chart; Nyquist diagram; inverse Nyquist diagram.** *See also:* **control system, feedback.** 0-23E0

phase control (rectifier circuits). The process of varying the point within the cycle at which forward conduction is permitted to begin. *Note:* The amount of phase control may be expressed in two ways: (1) the reduction in direct-current voltage obtained by phase control or (2) the angle of retard or advance. *See also:* **rectifier circuit element.** E59-34E17/34E24

phase converter (rotating machinery). A converter that changes alternating-current power of one or more phases to alternating-current power of a different number of phases but of the same frequency. *See:* **converter.** 42A10-31E8

phase-corrected horn. A horn designed to make the emergent electromagnetic wave front substantially plane at the mouth. *Note:* Usually this is achieved by means of a lens at the mouth. *See:* **circular scanning; waveguide.** 50I62-3E1

phase correction (telegraph transmission). The process of keeping synchronous telegraph mechanisms in substantially correct phase relationship. *See also:* **telegraphy.** 42A65-0

phase corrector. A network that is designed to correct for phase distortion. *See also:* **network analysis.** E270/42A65-0

phase crossover frequency (loop transfer function) (automatic control). The frequency at which the phase angle reaches plus or minus 180 degrees. *See also:* **control system, feedback.** 85A1-23E0

phased-array antenna. An array antenna whose beam direction or radiation pattern is controlled primarily by the relative phases of the excitation coefficients of the radiating elements. *See also:* **antenna.** 0-3E1

phase delay (facsimile) (in the transfer of a single-frequency wave from one point to another in a system). The time delay of a part of the wave identifying its phase. *Note:* The phase delay is measured by the ratio of the total phase shift in cycles to the frequency in hertz. *See also:* **facsimile transmission; transmission characteristics.** E168/42A65-0

phase delay time. In the transfer of a single-frequency wave from one point to another in a system, the time delay of a part of the wave identifying its phase. *Note:* The phase delay time is measured by the ratio of the total phase delay through the network, in cycles, to the frequency, in hertz. *See also:* **measurement system.** E285-9E1

phase deviation (angle modulation) (phase modulation). The peak difference between the instantaneous angle of the modulated wave and the angle of the carrier. *Note:* In the case of a sinusoidal modulating function, the value of the phase deviation, expressed in radians, is equal to the modulation index. *See also:* **phase modulation.** E145/E170/42A65-0

phase difference. The difference in phase between two sinusoidal functions having the same periods. *See:* **phase lag; phase lead.** E270-0

phase distortion. (1) Lack of direct proportionality of phase shift to frequency over the frequency range required for transmission, or (2) the effect of such departure on a transmitted signal. *See also:* **distortion; distortion, phase delay.** 42A65-31E3

phase-failure protection. *See:* **open-phase protection and phase-undervoltage protection.**

phase-frequency distortion (facsimile). Distortion due to lack of direct proportionality of phase shift to frequency over the frequency range required for transmission. *Notes:* (1) **delay distortion** is a special case. (2) This definition includes the case of a linear phase-frequency relation with the zero frequency intercept differing from an integral multiple of π. *See:* **distortion; distortion, phase delay; facsimile transmission; phase distortion.** 42A65/E154/E168-0

phase-insulated terminal box (rotating machinery). A terminal box so designed that the protection of phase conductors against electric failure within the terminal box is by insulation only. *See also:* **cradle base (rotating machinery).** 0-31E8

phase inverter. A stage whose chief function is to change the phase of a signal by 180 degrees, usually for feeding one side of a following push-pull amplifier. *See also:* **amplifier.** 42A65-31E3

phase lag (phase delay) (2-port network). The phase angle of the input wave relative to the output wave $(\Phi_{in}-\Phi_{out})$, or the initial phase angle of the output wave relative to the final phase angle of the output wave $(\Phi_i-\Phi_f)$. *Note:* Under matched conditions, **phase lag** is the negative of the angle of the transmission coefficient of the scattering matrix for a 2-port network. *See:* **phase difference.** *See also:* **measurement system.** E285-9E1

phase lead (phase advance) (2-port network). The phase angle of the output wave relative to the input wave $(\Phi_{out}-\Phi_{in})$, or the final phase angle of the output wave relative to the initial phase angle of the output wave $(\Phi_f-\Phi_i)$. *Note:* Under matched conditions, **phase lead** is the angle of the transmission coefficient of the scattering matrix for a 2-port network. *See:* **phase difference.** *See also:* **measurement system.** E285-9E1

phase localizer (electronic navigation). A localizer in which the on-course line is defined by the phase reversal of energy radiated by the sideband antenna system, a reference carrier signal being radiated and used for the detection of phase. *See also:* **radio navigation.** 0-10E6

phase locking. The control of an oscillator or periodic generator so as to operate at a constant phase angle relative to a reference signal source. *See also:* **oscillatory circuit.** 0-31E3

phase locus (for a loop transfer function, say *G(s) H(s)*. A plot in the *s* plane of those points for which the phase angle, ang *GH,* has some specified constant value. *Note:* The phase loci for 180 degrees plus or minus *n* 360 degrees are also root loci. *See also:* **control system, feedback.** 85A1-23E0

phase margin (of the loop transfer function for a stable feedback control system). 180 degrees minus the absolute value of the loop phase angle at a frequency where the loop gain is unity. *Note:* Phase margin is a convenient way of expressing relative stability of a linear system under parameter changes, in Nyquist, Bode, or Nichols diagrams. In a conditionally stable feedback control system where the loop gain becomes unity at several frequencies, the term is understood to apply to the value of phase margin at the highest of these frequencies. *See also:* **control system, feedback.** 85A1-23E0

phase meter (phase-angle meter). An instrument for measuring the difference in phase between two alternating quantities of the same frequency. *See also:* **instrument.** 42A30-0

phase modifier (rotating machinery). An electric machine, the chief purpose of which is to supply leading or lagging reactive power to the system to which it is connected. Phase modifiers may be either synchronous or asynchronous. *See:* **converter.** 42A10-31E8

phase-modulated transmitter. A transmitter that transmits a phase-modulated wave. E145/E182/42A65-0

phase modulation (PM). Angle modulation in which the angle of a carrier is caused to depart from its reference value by an amount proportional to the instantaneous value of the modulating function. *Notes:* (1) A wave phase modulated by a given function can be regarded as a wave frequency modulated by the time derivative of that function. (2) Combinations of phase and frequency modulation are commonly referred to as frequency modulation. *See also:* **modulating systems; phase deviation; pulse duration; reactance modulator.** E170/E145/E188/42A65-19E4

phase-modulation telemetering (electric power systems). A type of telemetering in which the phase difference between the transmitted voltage and a reference voltage varies as a function of the magnitude of the measured quantity. *See also:* **telemetering.** E94-0

phase recovery time (microwave gas tubes). The time required for a fired tube to deionize to such a level that a specified phase shift is produced in the low-level radio-frequency signal transmitted through the tube. *See:* **gas tubes.** E160-15E6

phase relay. A relay that by its design or application is intended to respond primarily to phase conditions of the power system. 37A100-31E11/31E6

phase resolution. The minimum change of phase that can be distinguished by a system. *See also:* **measurement system.** E285-9E1

phase-reversal protection. *See:* **phase-sequence reversal protection.**

phase-reversal relay. *See:* **negative-phase-sequence relay.**

phase-segregated terminal box. A terminal box so designed that the protection of phase conductors against electric failure within the terminal box is by insulation, and additionally by grounded metallic barriers forming completely isolated individual phase compartments so as to restrict any electric breakdown to a ground fault. *See also:* **cradle base (rotating machinery).** 0-31E8

phase-selector relay. A programming relay whose function is to select the faulted phase or phases thereby controlling the operation of other relays or control devices. 37A100-31E11/31E6

phase-separated terminal box. *See:* **phase-segregated terminal box.**

phase separator (rotating machinery). Additional insulation between adjacent coils that are in different phases. *See:* **rotor (rotating machinery); stator.** 0-31E8

phase sequence (of a set of polyphase voltages or currents). The order in which the successive members of the set reach their positive maximum values. *Note:* The phase sequence may be designated in several ways. If the set of polyphase voltages or currents is a symmetrical set, one method is to designate the phase sequence by specifying the integer that denotes the number of times that the angular phase lag between successive members of the set contains the characteristic angular phase difference for the number of phases *m.* If the integer is zero, the set is of zero phase sequence; if the integer is one, the set is of first phase sequence; and so on. Since angles of lag greater than 2π produce the same phase position for alternating quantities as the same angle decreased by the largest integral multiple of 2π contained in the angle of lag, it may be shown that there are only *m* distinct symmetrical sets, normally designated from 0 to $m-1$ phase sequence. It can be shown that only for the first phase sequence do all the members of the set reach their positive maximum in the order of identification at uniform intervals of time. E270-31E8

phase-sequence indicator. A device designed to indicate the sequence in which the fundamental components of a polyphase set of potential differences, or currents, successively reach some particular value, such as their maximum positive value. *See also:* **instrument.** 42A30-0

phase-sequence relay. A relay that responds to the order in which the phase voltages or currents successively reach their maximum positive values. *See also:* **relay.** 37A100/42A20-31E11/31E6

phase-sequence reversal (industrial control). A reversal of the normal phase sequence of the power supply. For example, the interchange of two lines on a three-phase system will give a phase reversal. 42A25-34E10

phase-sequence reversal protection. A form of protection that prevents energization of the protected equipment on the reversal of the phase sequence in a polyphase circuit. IC1/37A100-31E11/31E6/34E10

phase-sequence test (rotating machinery). A test to determine the phase sequence of the generated voltage of a three-phase generator when rotating in its normal direction. *See:* **asynchronous machine; synchronous machine.** 0-31E8

phase shift (1) (general). The absolute magnitude of the difference between two phase angles. *Notes:* (A) The phase shift between two planes of a 2-port network is the absolute magnitude of the difference between the phase angles at those planes. The total phase shift, or absolute phase shift, is expressed as the total number of cycles, including any fractional number, between the two planes, where one complete cycle is 2π radians or 360 degrees. Relative phase shift is the total or absolute phase shift less the largest integral number of 2π radians or 360 degrees. The unit of phase shift is, therefore, the radian or the electrical degree. The term **2-port network** is used in its most general sense to include structures of passive or active elements. This includes the case of a given length of waveguide but may also refer to any two ports of a multiport device, where it is understood that a signal is incident only at one port. (B) A phase shift can be either a phase lead (advance) or a phase lag (delay). *See also:* measurement system. E285-9E1

(2) (electrical conversion). The displacement between corresponding points in similar wave shapes and is expressed in degrees lead or lag. *See also:* **electrical conversion.** 0-10E1

(3) (transfer function). A change of phase angle with frequency, as between points on a loop phase characteristic. *See also:* **control system, feedback.**

(4) (signal). A change of phase angle with transmission. *See:* **transmission characteristics.** 0-23E0

phase-shift circuit (industrial control). A network that provides a voltage component shifted in phase with respect to a reference voltage. *See also:* **electronic controller.** 42A25-34E10

phase shifter. A device in which the output voltage (or current) may be adjusted, in use or in its design, to have some desired phase relation with the input voltage (or current). *See also:* **auxiliary device to an instrument; circuits and devices; phase-shifting transformer; waveguide.** 42A30-0

phase-shifting transformer (1) (general). A voltage regulator that advances or retards the phase-angle relationship of one circuit with respect to another. *Notes:* (A) The terms advance and retard describe the electrical angular position of the load voltage (or low voltage) with respect to the source voltage (or high voltage). (B) If the load voltage (or low voltage) reaches its positive maximum sooner than the source voltage (or high voltage), this is an advance position. (C) Conversely, if the load voltage (or low voltage) reaches its positive maximum later than the source voltage (or high voltage), this is a retard position. *See also:* **auxiliary device to an instrument; transformer; voltage regulator.** 42A15-0

(2) (measurement equipment). An assembly of one or more transformers intended to be connected across the phases of a polyphase circuit so as to provide voltages in the proper phase relations for energizing varmeters, varhour meters, or other measurement equipment. *See also:* **auxiliary device to an instrument; transformer.** 42A30/12A0-0

(3) (rectifier circuits). An autotransformer used to shift the phase position of the voltage applied to the alternating-current winding of a rectifier transformer for the purpose of decreasing the magnitude of the harmonic content in both the alternating-current and direct-current systems of a multiple-unit rectifier station. 57A18-0

phase-shift keying (PSK) (modulation systems). The form of phase modulation in which the modulating function shifts the instantaneous phase of the modulated wave between predetermined discrete values. *See also:* **modulating systems.** E170-0

phase-shift oscillator. An oscillator produced by connecting any network having a phase shift of an odd multiple of 180 degrees (per stage) at the frequency of oscillation, between the output and the input of an amplifier. When the phase shift is obtained by resistance-capacitance elements, the circuit is an *R-C* phase-shift oscillator. *See also:* **oscillatory circuit.** E145/42A65-0

phase space (control system). (1) The state space augmented by the independent time variable. (2) One used synonymously with the state space, usually with the state variables being successive time derivatives of each other. *See also:* **control system.** 0-23E0

phase splitter (phase-splitting circuit). A device that produces, from a single input wave, two or more output waves that differ in phase from one another. *See also:* **circuits and devices.** 42A65-0

phase-splitting circuit. *See:* **phase splitter.**

phase swinging (rotating machinery). Periodic variations in the speed of a synchronous machine above or below the normal speed due to power pulsations in the prime mover or driven load, possibly recurring every revolution. *See:* **synchronous machines.** 0-31E8

phase-tuned tube (microwave gas tubes). A fixed-tuned broad-band transmit-receive tube, wherein the phase angle through and the reflection introduced by the tube are controlled within limits. *See:* **gas tubes.** E160-15E5

phase-undervoltage protection. A form of protection that disconnects or inhibits connection of the protected equipment on deficient voltage in one or more phases of a polyphase circuit. 37A100-31E11/31E6

phase-undervoltage relay. A relay in which the contact operates when one or more phase voltages in a normally balanced polyphase circuit is equal to or less than its setting. 37A100-31E11/31E6

phase vector (of a wave). The vector in the direction of the wave normal, whose magnitude is the phase constant. 0-3E2

phase velocity (of a traveling plane wave at a single frequency). The velocity of an equiphase surface along the wave normal. *See also:* **radio wave propagation; transmission characteristics; waveguide.** 42A65/E146-3E1/3E2

phase-versus-frequency response characteristic. A graph or tabulation of the phase shifts occurring in an electric transducer at several frequencies within a band. *See also:* **transducer.** E145-0

phase voltage of a winding (machine or apparatus). The potential difference across one phase of the machine or apparatus. *See also:* **asynchronous machine; synchronous machine.** 50I05-31E8

phasing. The adjustment of picture position along the scanning line. *See also:* **scanning (facsimile).** E168-0

phasing signal. A signal used for adjustment of the picture position along the scanning line. *See also:* **facsimile signal (picture signal).** E168-0

phasing voltage (network protector). The voltage across the open contacts of a selected phase. *Note:*

This voltage is equal to the phasor difference between the transformer voltage and the corresponding network voltage. 37A100-31E11

phasor (vector*). A phasor is a complex number. Unless otherwise specified the term **phasor** is to be assumed to be used only in connection with quantities related to the steady alternating state in a linear network or system. *Notes:* (1) The term **phasor** is used instead of **vector** to avoid confusion with space vectors. (2) In polar form any phasor can be written $Ae^{j\theta A}$ or $A\angle\theta_A$, in which A, real, is the modulus, absolute value, or amplitude of the phasor and θ_A its phase angle (which may be abbreviated phase when no ambiguity will arise).

*Deprecated E270-0

phasor diagram (synchronous machine). The accompanying diagram shows the phasor relations among the voltages and currents, and their components, for a synchronous machine operating as a generator carrying a constant balanced polyphase load. The equations appearing in the cross-referenced definitions are phasor equations and are valid when the phasor quantities are expressed by complex numbers. In the equations j is the operator $(-1)^{1/2}$ and the product is 90 degrees ahead of the phasor quantity. Care must be used to assign the proper sign to the numerical quantities, as some of the components in the diagram are normally negative. *See:* **direct-axis component of armature voltage; quadrature-axis component of armature voltage; subtransient internal voltage; synchronous internal voltage; transient internal voltage.** *See also:* **synchronous machine.** 42A10-0

phasor difference. *See:* **phasor sum (difference).**

phasor function. A functional relationship that results in a phasor. E270-0

phasor power (rotating machinery). The phasor representing the complex power. *See:* **asynchronous machine; synchronous machine.** 0-31E8

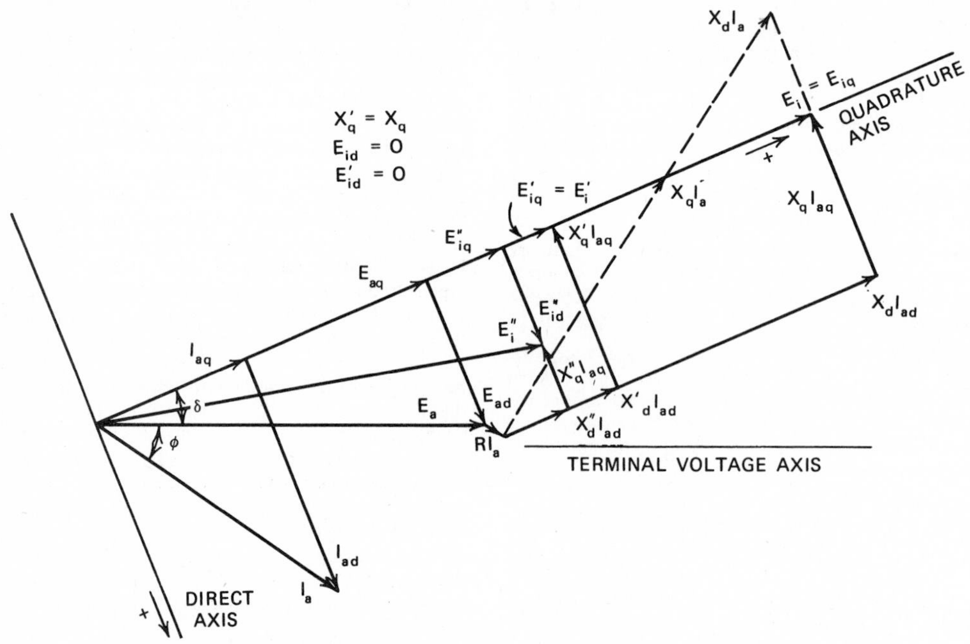

Phasor diagram of a synchronous machine.

Nomenclature

$\mathbf{I}_a$ = armature current

$\mathbf{I}_{ad}$ = direct-axis component of armature current

$\mathbf{I}_{aq}$ = quadrature-axis component of armature current

$\mathbf{E}_a$ = armature voltage

$\mathbf{E}_{ad}$ = direct-axis component of armature voltage

$\mathbf{E}_{aq}$ = quadrature-axis component of armature voltage

$\mathbf{E}_i$ = synchronous internal voltage

$\mathbf{E}_i'$ = transient internal voltage

$\mathbf{E}_i''$ = subtransient internal voltage

$\mathbf{X}_d$ = direct-axis synchronous reactance

$\mathbf{X}_d'$ = direct-axis transient reactance

$\mathbf{X}_d''$ = direct-axis subtransient reactance

$\mathbf{X}_q$ = quadrature-axis synchronous reactance

$\mathbf{X}_q'$ = quadrature-axis transient reactance

$\mathbf{X}_q''$ = quadrature-axis subtransient reactance

phasor power factor. The ratio of the active power to the amplitude of the phasor power. The phasor power factor is expressed by the equation

$$F_{pp} = \frac{P}{S}$$

where F_{pp} is the phasor power factor, P is the active power, S is the amplitude of phasor power. If the voltages and currents are sinusoidal and, for polyphase circuits, form symmetrical sets,

$$F_{pp} = \cos(\alpha - \beta).$$

See also: **displacement power factor (rectifier).** E270-0

phasor product (quotient). A phasor whose amplitude is the product (quotient) of the amplitudes of the two phasors and whose phase angle is the sum (difference) of the phase angles of the two phasors. If two phasors are

$$\mathbf{A} = |A|\, e^{j\theta_A}$$
$$\mathbf{B} = |B|\, e^{j\theta_B}$$

the phasor product is

$$AB = |AB|\, e^{j(\theta_A + \theta_B)}$$

and the quotient is

$$\frac{A}{B} = \left|\frac{A}{B}\right| e^{j(\theta_A - \theta_B)}$$

E270-0

phasor quantity. (1) A complex equivalent of a simple sine-wave quantity such that the modulus of the former is the amplitude A of the latter, and the phase angle (in polar form) of the former is the phase angle of the latter. (2) Any quantity (such as impedance) that is expressed in complex form. *Note:* In case (1), sinusoidal variation with t enters; in case (2), no time variation (in constant-parameter circuits) enters. The term **phasor quantity** covers both cases. E270-0

phasor quotient. *See:* **phasor product (quotient).**

phasor reactive factor. The ratio of the reactive power to the amplitude of the phasor power. The phasor reactive factor is expressed by the equation

$$F_{qp} = \frac{Q}{S}$$

where F_{qp} is the phasor reactive factor, Q is the reactive power, S is the amplitude of the phasor power. If the voltages and currents are sinusoidal and, for polyphase circuits, form symmetrical sets,

$$F_{pp} = \sin(\alpha - \beta).$$

E270-0

phasor sum (difference). A phasor of which the real component is the sum (difference) of the real components of two phasors and the imaginary component is the sum (difference) of the imaginary components of two phasors. It two phasors are

$$\mathbf{A} = a_1 + ja_2$$
$$\mathbf{B} = b_1 + jb_2$$

the phasor sum (difference) is

$$\mathbf{A} \pm \mathbf{B} = (a_1 \pm b_1) + j(a_2 \pm b_2).$$

E270-0

Philips gauge. A vacuum gauge in which the gas pressure is determined by measuring the current in a glow discharge. *See also:* **instrument.** 42A30-0

phi (Φ) polarization. The state of the wave in which the E vector is tangential to the lines of latitude of a given spherical frame of reference. *Note:* The usual frame of reference has the polar axis vertical and the origin at or near the antenna. Under these conditions, a vertical dipole will radiate only theta (θ) polarization, and a horizontal loop will radiate only phi (Φ) polarization. *See also:* **antenna.** 42A65/E145/E149-0

phon. The unit of loudness level as specified in the definition of loudness level. *See also:* **loudspeaker.** E157-1E1

phonograph pickup (mechanical reproducer). A mechanoelectrical transducer that is actuated by modulations present in the groove of the recording medium and that transforms this mechanical input into an electric output. *Note:* (1) Where no confusion is likely the term **phonograph pickup** may be shortened to **pickup.** (2) A phonograph pickup generally includes a pivoted mounting arm and the transducer itself (the pickup cartridge). *Note:* For an extensive list of cross references, see *Appendix A.* 42A65-1E1

phosphene (electrical) (electrotherapy). A visual sensation experienced by a human subject during the passage of current through the eye. *See also:* **electrotherapy.** 42A80-18E1

phosphor. A substance capable of luminescence. *See also:* **cathode-ray tube; fluorescent lamp; radio navigation; television.** 42A70-0

phosphor decay. A phosphorescence curve describing energy emitted versus time. *See also:* **oscillograph.** 0-9E4

phosphorescence. The emission of light as the result of the absorption of radiation, and continuing for a noticeable length of time after excitation, has been removed. *See also:* **lamp; oscillograph.** Z7A1-9E4

phosphor screen. All the visible area of the phosphor on the cathode-ray tube faceplate. *See also:* **oscillograph.** 0-9E4

phot. The unit of illumination when the centimeter is taken as the unit of length; it is equal to one lumen per square centimeter. *See also:* **light.** Z7A1-0

photocathode. An electrode used for obtaining a photoelectric emission when irradiated. *See also:* **electrode (electron tube); phototube.** 47A20/E160/E175-2E2/15E6

photocathode blue response. The photoemission current produced by a specified luminous flux from a tungsten filament lamp at 2870 kelvins color temperature when the flux is filtered by a specified blue filter. *See also:* **phototube.** E175-0

photocathode luminous sensitivity. *See:* **sensitivity, cathode luminous.**

photocathode radiant sensitivity. *See:* **sensitivity, cathode radiant.**

photocathode, semitransparent. *See:* **semitransparent photocathode.**

photocathode spectral-sensitivity characteristic. *See:* **spectral-sensitivity characteristic (photocathode).**

photocell (photoelectric cell). (1) A solid-state photosensitive electron device in which use is made of the variation of the current-voltage characteristic as a function of incident radiation. *See also:* **phototube.** E175-0

(2) A device exhibiting photovoltaic or photoconductive effects. *See:* **phototube.** 50I07-15E6

photochemical radiation. Energy in the ultraviolet, visible, and infrared regions to produce chemical changes in materials. *Note:* Examples of photochemical processes are accelerated fading tests, photography, photoreproduction, and chemical manufacturing. In many such applications a specific spectral region is of importance. *See also:* **infrared radiation; light; ultraviolet radiation.** Z7A1-0

photoconductive cell. A photocell in which the photoconductive effect is utilized. *See:* **phototube.** 50I07-15E6

photoconductive effect (photoconductivity). A photoelectric effect manifested as a change in the electric conductivity of a solid or a liquid and in which the charge carriers are not in thermal equilibrium with the lattice. *Note:* Many semiconducting metals and their compounds (notably selenium, selenides, and tellurides) show a marked increase in electric conductance when electromagnetic radiation is incident on them. *See:* **photoelectric effect; phototube; photovoltaic effect; photoemissive effect.** E270-15E6

photoelectric cathode. *See:* **photocathode.**

photoelectric color-register controller. A photoelectric control system used as a longitudinal position regulator for a moving material or web to maintain a preset register relationship between repetitive register marks in the first color and reference positions of the printing cylinders of successive colors. *See also:* **photoelectric control.** 42A25-34E10

photoelectric control (industrial control). Control by means of which a change in incident light effects a control function.
See:
differential;
lens multiplication factor;
light source;
minimum illumination;
photoelectric counter;
photoelectric cutoff register controller;
photoelectric directional counter;
photoelectric door opener;
photoelectric lighting controller;
photoelectric loop control;
photoelectric pinhole detector;
photoelectric pyrometer;
photoelectric relay;
photoelectric scanner;
photoelectric side-register controller;
photoelectric smoke-density control;
photoelectric smoke detector;
phototube housing;
reflected light scanning;
register marks or lines;
reset switch;
time of response;
transmitted light scanning. 42A25-34E10

photoelectric counter (industrial control). A photoelectrically actuated device used to record the number of times a given light path is intercepted by an object. *See also:* **photoelectric control.** 42A25-34E10

photoelectric current. The current due to a photoelectric effect. *See also:* **photoelectric effect.** 50I07-15E6

photoelectric cutoff register controller (industrial control). A photoelectric control system used as a longitudinal position regulator that maintains the position of the point of cutoff with respect to a repetitively referenced pattern on a moving material. *See also:* **photoelectric control.** 42A25-34E10

photoelectric directional counter. A photoelectrically actuated device used to record the number of times a given light path is intercepted by an object moving in a given direction. *See also:* **photoelectric control.** 42A25-34E1

photoelectric door opener. A photoelectric control system used to effect the opening and closing of a power-operated door. *See also:* **photoelectric control.** 42A25-34E1

photoelectric effect. Interaction between radiation and matter resulting in the absorption of photons and the consequent liberation of electrons. *Note:* This is a general term covering three separate phenomena in which an electric effect is influenced by incident electromagnetic radiation: the photoconductive effect, the photoemissive effect, and the photovoltaic effect.
See:
dark current;
Einstein's law;
photoconductive effect;
photoconductivity;
photoelectric current;
photoelectron;
photoemissive effect;
photon;
phototubes;
photovoltaic effect;
spectral selectivity;
threshold frequency;
threshold wavelength. E270/50I07-15E6

photoelectric emission. *See:* **electron emission; field-enhanced photoelectric emission; photoemissive effect; phototubes.**

photoelectric lighting controller. A photoelectric relay actuated by a change in illumination to control the illumination in a given area or at a given point. *See also:* **photoelectric control.** 42A25-34E10

photoelectric loop control (industrial control). A photoelectric control system used as a position regulator for a strip processing line that matches the average linear speed in one section to the speed in an adjacent section to maintain the position of the loop located between the two sections. *See also:* **photoelectric control.** 42A25-34E10

photoelectric pinhole detector. A photoelectric control system that detects the presence of minute holes in an opaque material. *See also:* **photoelectric control.** 42A25-34E1

photoelectric power system. *See:* **photovoltaic power system.**

photoelectric pyrometer (industrial control). An instrument that measures the temperature of a hot object by means of the intensity of radiant energy exciting a phototube. *See also:* **electronic control.** 42A25-34E10

photoelectric relay. A relay that functions at predetermined values of incident light. *See also:* **photoelectric control.** 42A25-34E10

photoelectric scanner (industrial control). A single-unit combination of a light source and one or more phototubes with a suitable optical system. *See also:* **photoelectric control.** 42A25-34E10

photoelectric side-register controller (industrial control). A photoelectric control system used as a lateral position regulator that maintains the edge of, or a line on, a moving material or web at a fixed position. *See also:* **photoelectric control.** 42A25-34E10

photoelectric smoke-density control. A photoelectric control system used to measure, indicate, and control the density of smoke in a flue or stack. *See also:* **photoelectric control.** 42A25-34E10

photoelectric smoke detector (industrial control). A photoelectric relay and light source arranged to detect the presence of more than a predetermined amount of smoke in air. *See also:* **photoelectric control.** 42A25-34E10

photoelectric system (protective signaling). An assemblage of apparatus designed to project a beam of invisible light onto a photoelectric cell and to produce an alarm condition in the protection circuit when the beam is interrupted. *See also:* **protective signaling.** 42A65-0

photoelectric tube. An electron tube, the functioning of which is determined by the photoelectric effect. *See also:* **phototube.** 0-15E6

photo-electron. An electron liberated by the photoemissive effect. *See also:* **photoelectric effect.** 50I07-15E6

photoemission spectrum (scintillator material). The relative numbers of optical photons emitted per unit wavelength as a function of wavelength interval. The emission spectrum may also be given in alternative units such as wave number, photon energies, frequency, etcetera. *Note:* Optical photons are photons with energies corresponding to wavelengths between 2000 and 15 000 angstroms. E175-0

photoemissive effect. Electromagnetic radiation (light, ultraviolet, X-rays, etcetera may cause the emission of electrons from matter, if the wavelength of the radiation is less than a critical maximum value that depends on the material. For many metals, the critical wavelength is in the visible spectrum. *See:* **photoelectric effect; photovoltaic effect; photoconductive effect (photoconductivity).** E270-0

photoflash lamp. A lamp in which combustible metal or other solid material is burned in an oxidizing atmosphere to produce light of high intensity and short duration for photographic purposes. *See also:* **lamp.** Z7A1-0

photoformer. A function generator that operates by means of a cathode-ray beam optically tracking the edge of a mask placed on a screen. *See also:* **electronic analog computer.** E165-0

photographic emulsion. The light-sensitive coating on photographic film consisting usually of a gelatin containing silver halide. E157-1E1

photographic sound recorder (optical sound recorder). Equipment incorporating means for producing a modulated light beam and means for moving a light-sensitive medium relative to the beam for recording signals derived from sound signals. *See also:* **electroacoustics.** E157-1E1

photographic sound reproducer (optical sound reproducer). A combination of light source, optical system, photoelectric cell, or other light-sensitive device such as a photoconductive cell, and a mechanism for moving a medium carrying an optical sound record (usually film), by means of which the recorded variations may be converted into electric signals of approximately like form. *See also:* **electroacoustics.** E157-1E1

photographic transmission density (optical density). The common logarithm of opacity. Hence, film transmitting 100 percent of the light has a density of zero, transmitting 10 percent a density of 1, and so forth. Density may be diffuse, specular, or intermediate. Conditions must be specified. 0-1E1

photometer. An instrument for measuring photometric quantities such as luminance (photometric brightness), luminous intensity, luminous flux, and illumination. *See also:* **photometry.** Z7A1-0

photometric brightness. *See:* **luminance.**

photometry (1) (general). The measurement of quantities associated with light. *Note:* Photometry may be visual in which the eye is used to make a comparison, or physical in which measurements are made by means of physical receptors. Z7A1-0
(2) (visual). The measurement of quantities referring to radiation evaluated according to the visual effect that it produces, as based on certain conventions. *See also:* **color, contrast.** 50I45-2E2
(3) (specifically). The techniques for the measurement of luminous flux and related quantities. *Note:* Such related quantities are luminous intensity, illuminance, luminance, luminosity, etcetera.
See:
densitometer;
glossmeter;
goniophotometer;
illumination meter;
integrating photometer;
light;
photometer;
physical photometer;
primary standards;
reflectometer;
spectophotometer;
transmissometer;
visual photometer.
See also: **color terms.** E201-0

photomultiplier. *See:* **multiplier phototube.**

photomultiplier tube. *See:* **multiplier phototube.**

photon. An elementary quantity of radiant energy (quantum) whose value is equal to the product of Planck's constant and the frequency of the electromagnetic radiation: $h\nu$. *See also:* **photoelectric effect; photovoltaic power system; solar cells (in a photovoltaic power system).** 50I07-15E6

photopic vision. Vision mediated essentially or exclusively by the cones. It is generally associated with adaptation to a luminance of at least 1.0 footlambert (3.0 nits). *See also:* **visual field.** Z7A1-0

photosensitive recording (facsimile). Recording by the exposure of a photosensitive surface to a signal-controlled light beam or spot. *See also:* **recording (facsimile).** E168-0

photosensitive tube. *See:* **photoelectric tube.**

phototube (photoelectric tube). An electron tube that contains a photocathode and has an output depending at every instant on the total photoelectric emission from the irradiated area of the photocathode.
See:
after pulse;
current amplification;
dark-current pulses;
electrode dark current;

equivalent dark current;
equivalent dark-current input;
equivalent noise input;
gas amplification;
light pipe;
luminous sensitivity;
multiplier phototubes;
noise;
photocathode;
photocathode blue response;
photocell;
photoconductive cell;
photoconductive effect;
photoelectric effect;
photoelectric emission;
photoelectric tube;
pulse-height resolution constant, electron;
quantum efficiency;
radiant sensitivity;
scintillation counter;
semitransparent photocathode;
sensitivity, cathode luminous;
sensitivity, cathode radiant;
sensitivity, dynamic;
sensitivity, illumination;
sensitivity, luminous;
sensitivity, radiant;
signal output current;
spectral characteristic;
spectral quantum yield;
spectral sensitivity characteristic;
transit time;
transit-time spread;
wavelength shifter.

See also: **field-enhanced photoelectric emission; tube definitions.** 42A70/E160/E175-15E6

phototube housing (industrial control). An enclosure containing a phototube and an optical system. *See also:* **photoelectric control.** 42A25-34E10

phototube, gas. *See:* **gas phototube.**

phototube, multiplier. *See:* **multiplier phototube.**

phototube, vacuum. *See:* **vacuum phototube.**

photovaristor. A varistor in which the current-voltage relation may be modified by illumination, for example, cadmium sulphide or lead telluride. *See also:* **semiconductor device.** E102/42A70-0

photovoltaic effect. For certain combinations of transparent conducting films separated by thin layers of semiconducting materials, electromagnetic radiation incident on one of the films can create no-load potential differences. *See:* **photoconductive effect (photoconductivity); photoemissive effect; photoelectric effect.** E270-0

photovoltaic power system.

See:
albedo;
dose;
dose rate;
emissivity;
emittance;
flux;
gamma ray;
heat sink;
ionization;
ionizing radiation;
n-on-*p* solar cells;
photon;
p-on-*n* solar cells;
rad;
reflectance;
reflectivity;
solar array;
solar cells;
solar panel;
substrate;
transmittance. 0-10E1

physical concept. Anything that has existence or being in the ideas of man pertaining to the physical world. Examples are magnetic fields, electric currents, electrons. E270-0

physical entity. *See:* **physical quantity.**

physical photometer. An instrument containing a physical receptor (photoemissive cell, barrier-layer cell, thermopile, etcetera) and associated filters, that is calibrated so as to read photometric quantities directly. Z7A1-0

physical property. Any one of the generally recognized characteristics of a physical system by which it can be described. E270-0

physical quantity (physical entity) (concrete quantity). A particular example of a measurable physical property of a physical system. It is characterized by both a qualitative and a quantitative attribute (that is, kind and magnitude). It is independent of the system of units and equations by which it and its relation to other physical quantities are described quantitatively. E270-0

physical system. A part of the real physical world that is directly or indirectly observed or employed by mankind. E270-0

physical unit. *See:* **unit.**

pickle (corrosion) (electroplating). A solution or process used to loosen or remove corrosion products such as oxides, scale, and tarnish from a metal. *See also:* **electroplating.** 42A60/CM-34E2

pickling (electroplating) (1) (chemical). The removal of oxides or other compounds from a metal surface by means of a solution that acts chemically upon the compounds.

(2) (electrolytic). Pickling during which a current is passed through the pickling solution to the metal (cathodic pickling) or from the metal (anodic pickling). *See also:* **electroplating.** 42A60-0

pickup (1) (electronics). A device that converts a sound, scene, or other form of intelligence into corresponding electric signals (for example, a microphone, a television camera, or a phonograph pickup). *See also:* **microphone; phonograph pickup; television.** 42A65/E188-2E2

(2) (interference terminology)*. Interference arising from sources external to the signal path. *See also:* **interference.**

*Deprecated 0-13E6

(3) (relay). (A) The action of a relay as it makes designated response to increase of input. (B) As a qualifying term, the state of a relay when all response to increase of input has been completed. (C) Also used to identify the minimum current, voltage, power, or other value of an input quantity reached by progressive increases that will cause the relay to reach the pickup state from reset. *Note:* In describing the performance of relays having multiple inputs, pickup has been used to denote contact closing, in which case pickup value of any input is meaningful only when related to all other inputs. *See also:* **microphone; phonograph pickup; railway signal and interlocking; relay; television.** 0-31E6

pickup and seal voltage (magnetically operated device) (industrial control). The minimum voltage at which the device moves from its deenergized into its fully energized position. *See also:* **initial contact pressure.** E74/IC1-34E10

pickup current. *See:* **pickup value.**

pickup factor, direction-finder antenna system. An index of merit expressed as the voltage across the receiver input impedance divided by the signal field strength to which the antenna system is exposed, the direction of arrival and polarization of the wave being such as to give maximum response. *See also:* **navigation.** E173-10E6

pickup spectral characteristic (color television). The set of spectral responses of the device, including the optical parts, that converts radiation to electric signals, as measured at the output terminals of the pickup tubes. *Note:* Because of nonlinearity, the spectral characteristics of some kinds of pickup tubes depend upon the magnitude of radiance used in the measurement. *See also:* **color terms.** E201-2E2

pickup tube*. *See:* **camera tube.**

*Deprecated

pickup value. The minimum input that will cause a device to complete contact operation or similar designated action. *Note:* In describing the performance of devices having multiple inputs, the pickup value of an input is meaningful only when related to all other inputs. 37A100-31E11

pickup voltage (or current) (magnetically operated device). The voltage (or current) at which the device starts to operate when its operating coil is energized under conditions of normal operating temperature. *See also:* **contactor.** 42A25/E16-34E10

picture element. The smallest area of a television picture capable of being delineated by an electric signal passed through the system or part thereof. *Note:* It has three important properties, namely P_v, the vertical height of the picture element; P_h, the horizontal length of the picture element, and P_a, the aspect ratio of the picture element. In addition, N_p, the total number of picture elements in a complete picture, is of interest since this number provides a convenient way of comparing systems. For convenience P_v and P_h are normalized for V, the vertical height of the picture; that is, P_v or P_h must be multiplied by V to obtain the actual dimension in a particular picture. P_v is defined as $P_v = 1/N$, where N is the number of active scanning lines in the raster. P_h is defined as $P_h = t_r A/t_e$, where t_r is the average value of the rise and delay times (10 percent to 90 percent) of the most rapid transition that can pass through the system or part thereof, t_e is the duration of the part of a scanning line that carries picture information, and A is the aspect ratio of the picture. (At present all broadcast television systems have a horizontal to vertical aspect ratio of 4/3.) P_a is defined as $P_a = P_h/P_v = t_r AN/t_e$ and N_p is defined as $N_p = (1/P_v) \times (A/P_h) = Nt_e/T_r$. *See:* **television.** 0-2E1

picture frequencies (facsimile). The frequencies that result solely from scanning subject copy. *Note:* This does not include frequencies that are part of a modulated carrier signal. *See also:* **scanning (facsimile).** E168-0

picture inversion (facsimile). A process that causes reversal of the black and white shades of the recorded copy. *See also:* **facsimile transmission.** E168-0

picture signal (television or facsimile). The signal resulting from the scanning process. *See also:* **television.** 42A65/E203-2E2

picture transmission (telephotography). The electric transmission of a picture having a gradation of shade values. *See also:* **communication.** 42A65-0

picture tube (kinescope) (television). A cathode-ray tube used to produce an image by variation of the beam intensity as the beam scans a raster. *See also:* **television; tube definitions.** 42A70-2E2

Pierce gun (microwave tubes). A gun that delivers an initially convergent electron beam. If a magnetic focusing scheme is used, the beam is made to enter the field at the minimum beam diameter or else, if the magnetic field threads through the cathode, the magnetic field must have a shape that is consistent with the desired beam that imparts certain flow characteristics to the electron beam. In Brillouin flow, angular electron velocity about the axis is imparted to the beam on entry into the magnetic field and the resulting inwardly directed force balances both the space charge and centrifugal forces. In practice, values of field up to twice the theoretical equilibrium value may be found necessary. In confined flow there is no overall angular velocity of the beam about the beam axis. Individual electron trajectories are tight helices (of radius small compared to beam radius) whose axis is along a magnetic-field line. The required magnetic field is several times greater than the Brillouin value and the flux must intersect the cathode surface. *See also:* **microwave tube or valve.** 0-15E6

Pierce oscillator. An oscillator that includes a piezoelectric crystal connected between the input and the output of a three-terminal amplifying element, the feedback being determined by the internal capacitances of the amplifying elements. *Note:* This is basically a Colpitts oscillator. *See also:* **Colpitts oscillator; oscillatory circuit.** E145/E182A/42A65-0

piezoelectric crystal cut, type. *See:* **type of piezoelectric crystal cut.**

piezoelectric-crystal element. A piece of piezoelectric material cut and finished to a specified geometrical shape and orientation with respect to the crystallographic axes of the material. *See also:* **crystal.** 42A65-21E0

piezoelectric-crystal plate. A piece of piezoelectric material cut and finished to specified dimensions and orientation with respect to the crystallographic axes of the material, and having two major surfaces that are essentially parallel. *See also:* **crystal.** 42A65-21E0

piezoelectric-crystal unit. A complete assembly, comprising a piezoelectric-crystal element mounted, housed, and adjusted to the desired frequency, with means provided for connecting it in an electric circuit. Such a device is commonly employed for purposes of frequency control, frequency measurement, electric wave filtering, or interconversion of electric waves and elastic waves. *Note:* Sometimes a piezoelectric-crystal unit may be an assembly having in it more than one piezoelectric-crystal plate. Such an assembly is called a **multiple-crystal unit.** *See also:* **crystal.** 42A65-21E0

piezoelectric effect. Some materials become electrically polarized when they are mechanically strained. The direction and magnitude of the polarization depend upon the nature and amount of the strain, and upon the direction of the strain. In such materials the converse effect is observed, namely, that a strain results from the application of an electric field. E270-0

piezoelectric loudspeaker. *See:* **crystal loudspeaker.**

piezoelectric microphone. *See:* **crystal microphone.**

piezoelectric pickup. *See:* **crystal pickup.**

piezoelectric transducer. A transducer that depends for its operation on the interaction between electric charge and the deformation of certain materials having piezoelectric properties. *Note:* Some crystals and specially processed ceramics have piezoelectric properties. 0-1E1

pigtail. A flexible metallic conductor, frequently stranded, attached to a terminal of a circuit component and used for connection into the circuit. *See also:* **circuits and devices.** 42A65-0

pileup. *See:* **relay pileup.**

pill-box antenna. A cylindrical reflector enclosed by two parallel conducting plates perpendicular to the cylinder, so spaced and excited as to permit the propagation of a single mode in the narrow dimension of the parallel-plate region. It is fed on the focal line. *See also:* **antenna.** 0-3E1

pilot (transmission system). A signal wave, usually a single frequency, transmitted over the system to indicate or control its characteristics. 42A65-0

pilotage. The process of directing a vehicle by reference to recognizable landmarks or soundings, or to electronic or other aids to navigation. Observations may be by any means including optical, aural, mechanical, or electronic. *See also:* **navigation.** 0-10E6

pilot cell (storage battery). A selected cell whose condition is assumed to indicate the condition of the entire battery. *See also:* **battery (primary or secondary).** 42A60-0

pilot channel. A channel over which a pilot is transmitted. 42A65-0

pilot circuit (industrial control). The portion of a control apparatus or system that carries the controlling signal from the master switch to the controller. *See:* **control.** 42A25-34E10

pilot director indicator. A device that indicates to the pilot information as to whether or not the aircraft has departed from the target track during a bombing run. *See also:* **air-transportation instruments.** 42A41-0

pilot exciter (rotating machinery). The source of all or part of the field current for the excitation of another exciter. *See:* **direct-current commutating machine; synchronous machine.** 42A10-31E8

pilot fit (spigot fit) (rotating machinery). A clearance hole and mating projection used to guide parts during assembly. *See:* **cradle base (rotating machinery).** 0-31E8

pilot-house control (illuminating engineering). A mechanical means for controlling the elevation and train of a searchlight from a position on the other side of the bulkhead or deck on which it is mounted. *See also:* **searchlight.** Z7A1-0

pilot lamp. A lamp that indicates the condition of an associated circuit. In telephone switching, a pilot lamp is a switchboard lamp that indicates a group of line lamps, one of which is or should be lit. *See also:* **circuits and devices.** 42A65-0

pilot light. A light, associated with a control, that by means of position or color indicates the functioning of the control. *See also:* **circuits and devices.** 42A65-0

pilot protection. A method of line protection in which an internal fault is identified by using a communication channel for relays to compare electric conditions at the terminals of a circuit. 37A100-31E11/31E6

pilot streamer (lightning). The initial low-current discharge that begins when the voltage gradient exceeds the breakdown voltage of air. *See also:* **direct-stroke protection (lightning).** 0-31E13

pilot wire. An auxiliary conductor used in connection with remote measuring devices or for operating apparatus at a distant point. *See also:* **center of distribution.** 42A35-31E13

pilot-wire-controlled network. A network whose switching devices are controlled by means of pilot wires. *See also:* **alternating-current distribution.** 42A35-31E13

pilot-wire protection. *See:* **wire-pilot protection.**

pilot-wire regulator. An automatic device for controlling adjustable gains or losses associated with transmission circuits to compensate for transmission changes caused by temperature variations, the control usually depending upon the resistance of a conductor or pilot wire having substantially the same temperature conditions as the conductors of the circuits being regulated. *See also:* **transmission regulator.** 42A65-0

pi (π) mode (magnetrons). The mode of operation for which the phases of the fields of successive anode openings facing the interaction space differ by π radians. *See also:* **magnetrons.** E160/42A70-15E6

pinboard. A perforated board that accepts manually inserted pins to control the operation of equipment. *See also:* **electronic digital computer.** X3A12-16E9

pinch (electron tubes). The part of the envelope of an electron tube or valve carrying the electrodes and through which pass the connections to the electrodes. *See also:* **electron tube.** 50I07-15E6

pinch effect (1) (rheostriction). The phenomenon of transverse contraction and sometimes momentary rupture of a fluid conductor due to the mutual attraction of the different parts carrying currents. *See also:* **electrothermics; induction heating.** 42A60/E54/E169/E270-0

(2) (disk recording). A pinching of the reproducing stylus tip twice each cycle in the reproduction of lateral recordings due to a decrease of the groove angle cut by the recording stylus when it is moving across the record as it swings from a negative to a positive peak. *See also:* **electroacoustics.** E157-1E1

pi (π) network. A network composed of three branches connected in series with each other to form a mesh, the three junction points forming an input terminal, an output terminal, and a common input and output terminal, respectively. See accompanying figure. *See also:* **network analysis.** E153/E270-0

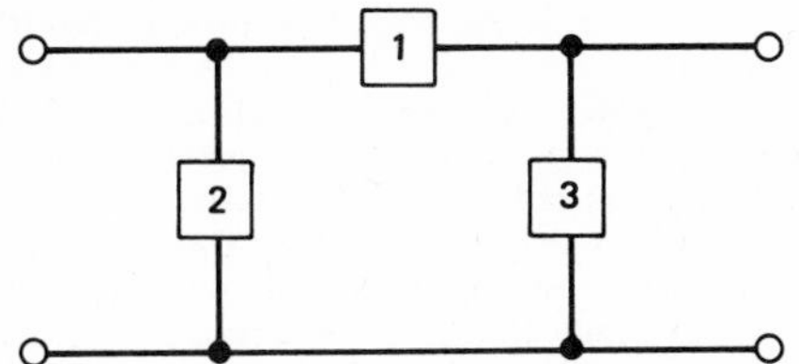

Pi network. The junction point between branches 1 and 2 forms an input terminal, that between branches 1 and 3 forms an output terminal, and that between branches 2 and 3 forms a common input and output terminal.

pin insulator. A complete insulator, consisting of one insulating member or an assembly of such members without tie wires, clamps, thimbles, or other accessories, the whole being of such construction that when mounted on an insulator pin it will afford insulation and mechanical support to a conductor that has been properly attached with suitable accessories. *See also:* **insulator; tower.** 42A35-31E13

pin jack. A single-conductor jack having an opening for the insertion of a plug of very small diameter. *See also:* **circuits and devices.** 42A65-21E0

pins (electron tube or valve). Metal pins connected to the electrodes that plug into the holder. They ensure the electric connection between the electrodes and the external circuit and also mechanically fix the tube in its holder. *See also:* **electron tube.** 50I07-15E6

pip. A popular term for a sharp deflection in a visible trace. *See also:* **radar.** 42A65-0

pipe cable. A pressure cable in which the container for the pressure medium is a loose-fitting rigid metal pipe. *See:* **pressure cable; oil-filled pipe cable; gas-filled pipe cable.** *See also:* **power distribution, underground construction.** 42A35-31E13

pipe-ventilated (rotating machinery). *See:* **duct-ventilated.**

pip-matching display (electronic navigation). A display in which the received signal appears as a pair of blips, the comparison of the characteristics of which provides a measure of the desired quantity. *See:* ***K, L,* or *N* display.** *See also:* **navigation.** E172-10E6

pi (π) point. A frequency at which the insertion phase shift of an electric structure is 180 degrees or an integral multiple thereof. *See also:* **transmission characteristics.** 42A65-0

Pirani gauge. A bolometric vacuum gauge that depends for its operation on the thermal conduction of the gas present; pressure being measured as a function of the resistance of a heated filament ordinarily over a pressure range of 10^{-1} to 10^{-4} conventional millimeter of mercury. *See also:* **instrument.** 42A30-0

piston (high-frequency communication practice) (plunger). A conducting plate movable along the inside of an enclosed transmission path and acting as a short-circuit for high-frequency currents. *See also:* **waveguide.** 42A65-0

piston attentuator (waveguide). A variable cutoff attentuator in which one of the coupling devices is carried on a sliding member like a piston. *See:* **waveguide.** 50I62-3E1

pistonphone. A small chamber equipped with a reciprocating piston of measurable displacement that permits the establishment of a known sound pressure in the chamber. *See also:* **loudspeaker.** E157-1E1

pit (rotating machinery). A depressed area in a foundation under a machine. *See:* **cradle base (rotating machinery).** 0-31E8

pitch (acoustics) (audio and electroacoustics). The attribute of auditory sensation in terms of which sounds may be ordered on a scale extending from low to high, such as a musical scale. *Notes:* (1) Pitch depends primarily upon the frequency of the sound stimulus, but it also depends upon the sound pressure and wave form of the stimulus. (2) The pitch of a sound may be described by the frequency of that simple tone, having a specified sound pressure or loudness level, that seems to the average normal ear to produce the same pitch. (3) The unit of pitch is the mel. *See also:* **electroacoustics.** 0-1E1

pitch angle (electronic navigation). *See:* **pitch attitude.**

pitch attitude (electronic navigation). The angle between the longitudinal axis of the vehicle and the horizontal. *See also:* **navigation.** 0-10E6

pitch factor (rotating machinery). The ratio of the resultant voltage induced in a coil to the arithmetic sum of the magnitudes of the voltages induced in the two coil sides. *See:* **armature.** 0-31E8

pits. Depressions produced in metal surfaces by nonuniform electrodeposition or from electrodissolution; for example, corrosion. *See also:* **electrodeposition.** 42A60-0

pitting (corrosion). Localized corrosion taking the form of cavities at the surface. *See:* **corrosion terms.** CM-34E2

pitting factor (corrosion). The depth of the deepest pit resulting from corrosion divided by the average penetration as calculated from weight loss. *See:* **corrosion terms.** CM-34E2

PIV. *See:* **peak inverse voltage.** *See also:* **peak reverse voltage (semiconductor rectifier).**

pivot-friction error. Error caused by friction between the pivots and the jewels; it is greatest when the instrument is mounted with the pivot axis horizontal. *Note:* This error is included with other errors into a combined error defined in **repeatability.** *See also:* **moving element (instrument).** 39A1-0

place. In positional notation, a position corresponding to a given power of the base, a given cumulated product, or a digit cycle of a given length. It can usually be specified as the *n*th character from one end of the numerical expression. *See also:* **electronic computation; electronic digital computer.** E162-0

plain conductor. A conductor consisting of one metal only. *See also:* **conductors.** 42A35-31E13

plain flange (plane flange) (plain connector) (waveguide). A coupling flange with a flat face. *See:* **waveguide.** 0-3E1

planar array (antenna). An array antenna having the centers of the radiating elements lying in a plane. *See also:* **antenna.** 0-3E1

planar network. A network that can be drawn on a plane without crossing of branches. *See also:* **network analysis.** E153/E270

Planckian locus. The locus of chromaticities of Planckian (blackbody) radiators having various temperatures. *See:* **chromaticity diagram.** *See also:*

color terms. E201-2E2

Planck radiation law. An expression representing the spectral radiance of a blackbody as a function of the wavelength and temperature. This law commonly is expressed by the formula

$$L_\lambda = I_\lambda / A' = c_{1L} \lambda^{-5} [e^{(c_2/\lambda T)} - 1]^{-1}$$

in which L_λ is the spectral radiance, I_λ is the spectral radiant intensity, A' is the projected area ($A \cos \theta$) of the aperture of the blackbody, e is the base of natural logarithms (2.718+), T is absolute temperature, c_{1L} and c_2 are constants designated as the first and second radiation constants. *Note:* The designation c_{1L} is used to indicate that the equation in the form given here refers to the radiance L, or to the intensity I per unit projected area A', of the source. Numerical values are commonly given not for c_{1L} but for c_1 which applies to the total flux radiated from a blackbody aperture, that is, in a hemisphere (2π steradians), so that, with the Lambert cosine law taken into account, $c_1 = \pi c_{1L}$. The currently recommended value of c_1 is 3.7415×10^{-16} $\mathrm{W{\cdot}m^2}$ or 3.7415×10^{-12} $\mathrm{W{\cdot}cm^2}$. Then c_{1L} is 1.1910×10^{-16} $\mathrm{W{\cdot}m^2{\cdot}sr^{-1}}$ or 1.1910×10^{-12} $\mathrm{W{\cdot}cm^2{\cdot}sr^{-1}}$. If, as is more convenient, wavelengths are expressed in micrometers and area in square centimeters, $c_{1L} = 1.1910 \times 10^1 \mathrm{W{\cdot}\mu m^4{\cdot}cm^{-2}{\cdot}sr^{-1}}$, L_λ being given in $\mathrm{W{\cdot}cm^{-2}{\cdot}sr^{-1}{\cdot}\mu m^{-1}}$ The presently recommended value of c_2 is 1.43879 cm kelvin. The Planck law in the following form gives the energy radiated from the blackbody in a given wavelength interval ($\lambda_1 - \lambda_2$):

$$Q = \int_{\lambda_1}^{\lambda_2} Q_\lambda d\lambda = Atc_1 \int_{\lambda_1}^{\lambda_2} \lambda^{-5} (e^{(c_2/\lambda T)} - 1)^{-1} d\lambda$$

If A is the area of the radiation aperture or surface in square centimeters, t is time in seconds, λ is wavelength in micrometers, and $c_1 = 3.7415 \times 10^4 \mathrm{W{\cdot}\mu m^4{\cdot}cm^{-2}}$, then Q is the total energy in watt-seconds emitted from this area (that is, in the solid angle 2π), in time t, within the wavelength interval ($\lambda_1 - \lambda_2$). *Note:* It often is convenient, as is done here, to use different units of length in specifying wavelengths and areas, respectively. If both quantities are expressed in centimeters and the corresponding value for c_1 (3.7415×10^{-5} $\mathrm{erg{\cdot}cm^2{\cdot}sec^{-1}}$) is used, this equation gives the total emission of energy in ergs from area A (that is in the solid angle 2π), for time t, and for the interval $\lambda_1 - \lambda_2$ in centimeters. *See also:* **radiant energy (in illuminating engineering).** Z7A1-0

plane-earth factor (radio wave propagation). The ratio of the electric field strength that would result from propagation over an imperfectly conducting plane earth to that which would result from propagation over a perfectly conducting plane. *See also:* **radio wave propagation.** E211-3E2

plane flange. *See:* **plain flange.**

plane of polarization (plane-polarized wave). The plane containing the electric intensity and the direction of propagation. *See also:* **radiation; radio wave propagation.** 42A65-0

plane of propagation (electromagnetic wave). The plane containing the attenuation vector and the wave normal; in the common degenerate case where these vectors have the same direction, the plane containing the electric vector and the phase vector and the wave normal. *See also:* **radio wave propagation.** 0-3E2

plane-polarized wave (homogeneous isotropic medium). A wave whose electric intensity at all times lies in a fixed plane that contains the direction of propagation. *See also:* **linearly polarized wave; radiation.** 42A65-0

plane wave (radio wave propagation). A wave whose equiphase surfaces form a family of parallel planes. *See also:* **radiation; radio wave propagation.** E211/42A65-3E2

planned stop. *See:* **optional stop.**

plan-position indicator (PPI). A type of presentation on a radar indicator in which the signal appears as a bright spot with range indicated by distance from the center of the screen and bearing by its radial angle. *Note:* Hence blips produced by signals from reflecting objects and transponders are shown in plan position, thus forming a maplike display. *See also:* **radar.** 42A65/E172-10E6

plan-position-indicator scope. A cathode-ray oscilloscope arranged to present a plan-position-indicator display. *See also:* **radar.** 42A65/E172-10E6

plant (control system). The part of the system to be controlled whose parameters are usually unalterable by the designer. *See also:* **control system. 0-23E0**

plant-capacity factor. *See:* **plant factor.**

plant dynamics (control system). Equations which describe the behavior of the plant. *See also:* **control system.** 0-23E0

Planté plate (storage cell). A formed lead plate of large area, the active material of which is formed in thin layers at the expense of the lead itself. *See also:* **battery (primary or secondary).** 42A60-0

plant factor (plant-capacity factor). The ratio of the average load on the plant for the period of time considered to the aggregate rating of all the generating equipment installed in the plant. *See also:* **generating station.** 42A35-31E13

plasma. A gas made up of charged particles. *Note:* Usually plasmas are neutral, but not necessarily so, as, for example, the space charge in an electron tube. E270-0

plasma frequency (radio wave propagation). A natural frequency of oscillation of charged particles in a plasma given by

$$f_N = \frac{1}{2\pi} \left(\frac{Nq^2}{\epsilon_0 m} \right)^{1/2}$$

where q is the charge per particle, m is the particle mass, N is the charge density, and ϵ_0 is the permittivity of free space, all quantities being expressed in meter-kilogram-second (MKS) units. *Note:* For electrons, $f_N = 8.979\ N^{1/2}$ in the International System of Units (SI). *See also:* **radio wave propagation.** 0-3E2

plasma sheath. A layer of charged particles of substantially one sign that accumulates around a body in a plasma. *See also:* **radio wave propagation.** 0-3E2

plastic (rotating machinery). A material that contains as an essential ingredient an organic substance of large molecular weight, is solid in its finished state, and, at some stage in its manufacture or in its processing into finished articles, can be shaped by flow. AD883-31E8

plate (electron tubes). A common name for an anode in an electron tube. *See also:* **electrode (electron tube).** 42A70-0

plateau (radiation-counter tubes). The portion of the counting-rate-versus-voltage characteristic in which the counting rate is substantially independent of the applied voltage. *See also:* **gas-filled radiation counter tubes.** 42A70-15E6

plateau length (radiation-counter tubes). The range of applied voltage over which the plateau of a radiation-counter tube extends. *See also:* **gas-filled radiation-counter tubes.** 42A70-15E6

plateau slope, normalized (radiation counter tubes). The slope of the substantially straight portion of the counting rate versus voltage characteristic divided by the quotient of the counting rate by the voltage at the Geiger-Mueller threshold. *See also:* **gas-filled radiation counter tubes.** 42A70-15E6

plateau slope, relative (radiation-counter tubes). The average percentage change in the counting rate near the midpoint of the plateau per increment of applied voltage. *Note:* Relative plateau slope is usually expressed as the percentage change in counting rate per 100-volt change in applied voltage (see accompanying figure). *See also:* **gas-filled radiation-counter tubes.** 42A70-15E6

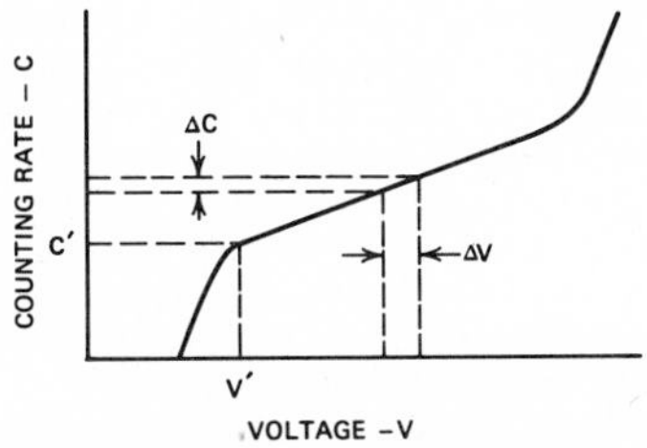

Counting rate–voltage characteristic in which

$$\text{Relative plateau slope} = 100\,\frac{\Delta C/C}{\Delta V}$$

$$\text{Normalized plateau slope} = \frac{\Delta C/\Delta V}{C'/V'} = \frac{\Delta C/C'}{\Delta V/V'}$$

plate-circuit detector. A detector functioning by virtue of a nonlinearity in its plate-circuit characteristic. *See also:* **modulating systems.** 42A65-0

plated-through hole (electronic and electrical applications). Deposition of metal on the side of a hole and on both sides of a base to provide electric connection, and an enlarged portion of conductor material surrounding the hole on both sides of the base. *See also:* **soldered connections (electronic and electrical applications).** 99A1-0

plate (anode) efficiency. The ratio of load circuit power (alternating current) to the plate power input (direct current). *See also:* **network analysis.** E145/42A65-0

plate keying. Keying effected by interrupting the plate-supply circuit. *See also:* **modulating systems; telegraphy.** E145/42A65-0

plate (anode) load impedance. The total impedance between anode and cathode exclusive of the electron stream. *See also:* **circuits and devices; network analysis.** E145/42A65-31E3

plate (anode) modulation. Modulation produced by introducing the modulating signal into the plate circuit of any tube in which the carrier is present. *See also:* **modulating systems; modulator.** 42A65/E145/E182A-0

plate (anode) neutralization. A method of neutralizing an amplifier in which a portion of the plate-to-cathode alternating voltage is shifted 180 degrees and applied to the grid-cathode circuit through a neutralizing capacitor. *See also:* **amplifier, feedback.** 42A65/E145/E182A-0

plate (anode) power input. The power delivered to the plate (anode) of an electron tube by the source of supply. *Note:* The direct-current power delivered to the plate of an electron tube is the product of the mean plate voltage and the mean plate current. E145/42A65-31E3

plate (anode) pulse modulation. Modulation produced in an amplifier or oscillator by application of externally generated pulses to the plate circuit. *See also:* **modulating systems.** E145-0

platform erection (alignment of inertial systems). The process of bringing the vertical axis of a stable platform system into agreement with the local vertical. *See also:* **navigation.** E174-10E6

plating rack (electroplating). Any frame used for suspending one or more electrodes and conducting current to them during electrodeposition. *See also:* **electroplating.** 42A60-0

playback. A term used to denote reproduction of a recording. 42A65-0

playback loss. *See:* **translation loss.**

plenary capacitance (between two conductors). The capacitance between two conductors when the changes in the charges on the two are equal in magnitude but opposite in sign and the other $n-2$ conductors are isolated conductors. *See:* **direct capacitances of a system of conductors.** E270-0

pliotron. A hot-cathode vacuum tube having one or more grids. *Note:* This term is used primarily in the industrial field. *See also:* **tube definitions.** 42A70-15E6

plotting board. *See:* **recorder, X-Y (plotting board).**

plug. A device, usually associated with a cord, that by insertion in a jack or receptacle establishes connection between a conductor or conductors associated with the plug and a conductor or conductors connected to the jack or receptacle. *See also:* **circuits and devices.** 42A65-21E0

plug adapter (plug body*). A device that by insertion in a lampholder serves as a receptacle. *See also:* **interior wiring.**
*Deprecated 42A95-0

plug adapter lampholder (current tap*). A device that by insertion in a lampholder serves as one or more receptacles and a lampholder. *See also:* **interior wiring.**
*Deprecated 42A95-0

plugboard. A perforated board that accepts manually inserted plugs to control the operation of equipment. *See:* **control panel.** *See also:* **electronic digital computer.** X3A12-16E9

plug body. *See:* **plug adapter.**

plug braking (rotating machinery). A form of electric braking of an induction motor obtained by reversing the phase sequence of its supply. *See:* **asynchronous machine.** 0-31E8

plugging (industrial control). A control function that provides braking by reversing the motor line voltage polarity or phase sequence so that the motor develops a counter-torque that exerts a retarding force. *See also:* **electric drive.** 42A65/IC1-34E10

plug-in. A communication device when it is so designed that connections to the device may be completed through pins, plugs, jacks, sockets, receptacles, or other forms of ready connectors. 42A65-0

plug-in-type bearing (rotating machinery). A complete journal bearing assembly, consisting of a bearing liner and bearing housing and any supporting structure that is intended to be inserted into a machine end-shield. *See also:* **bearing.** 0-31E8

plumb-bob vertical (electronic navigation). The direction indicated by a simple, ideal, frictionless pendulum that is motionless with respect to the earth; it indicates the direction of the vector sum of the gravitational and centrifugal accelerations of the earth at the location of the observer. *See also:* **navigation.** E174-10E6

plumbing (in communication practice). A colloquial expression term employed to designate coaxial lines or waveguides and accessory equipment for radio-frequency transmission. *See also:* **waveguide.** 42A65/E172-10E6/3E1

plunger relay. A relay operated by a movable core or plunger through solenoid action. *See also:* **relay.** 83A16-0

plunger, waveguide (waveguide). A longitudinally movable obstacle that reflects essentially all the incident energy. *See also:* **waveguide.** E147-3E1

plural service. *See:* **dual service.**

plural tap (cube tap). *See:* **multiple plug.**

PM. *See:* **permanent magnet.**

PM. *See:* **phase modulation.**

PM (microwave tubes). *See:* **permanent-magnet focusing.**

pneumatically release-free (pneumatically operated switching device) (pneumatically trip-free). A term indicating that by pneumatic control the switching device is free to open at any position in the closing stroke if the release is energized. *Note:* This release-free feature is operative even though the closing control switch is held closed. 37A100-31E11

pneumatic bellows, relay. *See:* **relay pneumatic bellows.**

pneumatic controller. A pneumatically supervised device or group of devices operating electric contacts in a predetermined sequence. *See also:* **multiple-unit control.** E16-0

pneumatic loudspeaker. A loudspeaker in which the acoustic output results from controlled variation of an air stream. *See also:* **loudspeaker.** 42A65-0

pneumatic operation. Power operation by means of compressed gas. 37A100-31E11

pneumatic switch. A pneumatically supervised device opening or closing electric contacts, and differs from a pneumatic controller in being purely an ON and OFF type device. *See also:* **control switch.** E16-0

pneumatic transducer. A unilateral transducer in which the sound output results from a controlled variation of an air stream. 0-1E1

pneumatic tubing system (protective signaling). An automatic fire-alarm system in which the rise in pressure of air in a continuous closed tube, upon the application of heat, effects signal transmission. *Note:* Most pneumatic tubing systems contain means for venting slow pressure changes resulting from temperature fluctuations and therefore operate on the so-called rate-of-rise principle. *See also:* **protective signaling.** 42A65-0

pocket-type plate (of a storage cell). A plate of an alkaline storage battery consisting of an assembly of perforated oblong metal pockets containing active material. *See also:* **battery (primary or secondary).** 42A60-0

poid. The curve traced by the center of a sphere when it rolls or slides over a surface having a sinusoidal profile. *See also:* **electroacoustic.** E157-1E1

point (1) (positional notation). (A) The character, or the location of an implied symbol, that separates the integral part of a numerical expression from its fractional part. For example, it is called the binary point in binary notation and the decimal point in decimal notation. If the location of the point is assumed to remain fixed with respect to one end of the numerical expressions, a fixed-point system is being used. If the location of the point does not remain fixed with respect to one end of the numerical expression, but is regularly recalculated, then a floating-point system is being used. *Note:* A fixed-point system usually locates the point by some convention, while a floating-point system usually locates the point by expressing a power of the base. *See:* **branchpoint; breakpoint; checkpoint; entry point; fixed point; floating point; rerun point; variable point.** E270/X3A12-16E9
(B) The character, or implied location of such a character, that separates the integral part of a numerical expression from the fractional part. Since the place to the left of the point has unit weight in the most commonly used systems, the point is sometimes called the units point, although it is frequently called the binary point in binary notation and the decimal point in decimal notation. *See:* **breakpoint; fixed point; floating point.** *See also:* **electronic digital computer.** E162-0
(2) (lightning protection). The pointed piece of metal used at the upper end of the elevation rod to receive a lightning discharge. *See also:* **lightning protection and equipment.** 42A95-0
(3) (supervisory control or indication or telemeter selection). All of the supervisory control or indication devices, in a system, exclusive of the common devices, in the master station and in the remote station that are necessary for (A) energizing the closing, opening, or other positions of a unit, or set of units of switchgear or other equipment being controlled; (B) automatic indication of the closed or open or other positions of a unit, or set of units of switchgear or other equipment for which indications are being obtained; or (C) connecting a telemeter transmitting equipment into the circuit to be measured and to transmit the telemeter reading over a channel to a telemeter receiving equipment. *Note:* A point may serve for any two or all three of the purposes described above; for example, when a supervisory system is used for the combined control and indication of remotely operated equipment, point (for supervisory control) and point (for supervisory indication) are combined into a single control and indication point. 37A100-31E11

point-by-point method (lighting calculation). A lighting design procedure for predetermining the illumination at various locations in lighting installations, by use of luminaire photometric data. *Note:* Since interreflections are not taken into account, the point-by-point method may indicate lower levels of illumination than are actually realized. *See also:* **inverse-square law (illuminating engineering).** Z7A1-0

point contact (semiconductors). A pressure contact between a semiconductor body and a metallic point. *See also:* **semiconductor; semiconductor device.** 42A70/E102/E270/E216-34E17

point detector. A device that is a part of a switch-operating mechanism and is operated by a rod connected to a switch, derail, or movable-point frog to indicate that the point is within a specified distance of the stock rail. *See also:* **railway signal and interlocking.** 42A42-0

pointer pusher (demand meter). The element that advances the maximum demand pointer in accordance with the demand and in integrated-demand meters is reset automatically at the end of each demand interval. *See also:* **demand meter.** 42A30-0

pointer shift due to tapping. The displacement in the position of a moving element of an instrument that occurs when the instrument is tapped lightly. The displacement is observed by a change in the indication of the instrument. *See also:* **moving element (instrument).** 39A1-0

point of fixation. A point or object in the visual field at which the eyes look and upon which they are focused. *See also:* **visual field.** Z7A1-0

point of observation. The midpoint of the base line connecting the centers of rotation of the two eyes. For practical purposes, the center of the pupil of the eye often is taken as the point of observation. *See also:* **visual field.** Z7A1-0

point-to-point control system. *See:* **positioning control system.**

point-to-point radio communication. Radio communication between two fixed stations. *See also:* **radio transmission.** 42A65-0

point transposition. A transposition, usually in an open wire line, that is executed within a distance comparable to the wire separation, without material distortion of the normal wire configuration outside this distance. *See also:* **open wire.** 42A65-0

Poisson's equation. In rationalized form:

$$\nabla^2 V = -\frac{\rho}{\epsilon_0 \epsilon}$$

where $\epsilon_0\epsilon$ is the absolute capacitivity of the medium, V the potential, and ρ the charge density at any point. E270-0

polar contact. A part of a relay against which the current-carrying portion of the movable polar member is held so as to form a continuous path for current. *See also:* **railway signal and interlocking.** 42A42-0

polar direct-current telegraph system. A system that employs positive and negative currents for transmission of signals over the line. *See also:* **telegraphy.** 42A65-0

polarential telegraph system. A direct-current telegraph system employing polar transmission in one direction and a form of differential duplex transmission in the other direction. *Note:* Two kinds of polarential systems, known as types *A* and *B*, are in use. In half-duplex operation of a type-*A* polarential system the direct-current balance is independent of line resistance. In half-duplex operation of a type-*B* polarential system the direct-current balance is substantially independent of the line leakage. *See also:* **telegraphy.** 42A65-19E4

polarity (1) (battery). An electrical condition determining the direction in which current tends to flow on discharge. By common usage the discharge current is said to flow from the positive electrode through the external circuit. *See also:* **battery (primary or secondary).** 42A60-0

(2) (television) (picture signal). The sense of the potential of a portion of the signal representing a dark area of a scene relative to the potential of a portion of the signal representing a light area. Polarity is stated as black negative or black positive. *See also:* **television.** E203/42A65-2E2

polarity and angular displacement (regulator). Relative lead polarity of a regulator or a transformer is a designation of the relative instantaneous direction of current in its leads. In addition to its main transformer windings, a regulator commonly has auxiliary transformers or auxiliary windings as an integral part of the regulator. The same principles apply to the polarity of all transformer windings. *Notes:* (1) Primary and secondary leads are said to have the same polarity when at a given instant the current enters an identified secondary lead in the same direction as though the two leads formed a continuous circuit. (2) The relative lead polarity of a single-phase transformer may be either additive or subtractive. If one pair of adjacent leads from the two windings is connected together and voltage applied to one of the windings, then: (A) The relative lead polarity is additive if the voltage across the other two leads of the windings is greater than that of the higher-voltage winding alone. (B) The relative lead polarity is subtractive if the voltage across the other two leads of the winding is less than that of the higher-voltage winding alone. (3) The polarity of a polyphase transformer is fixed by the internal connections between phases as well as by the relative locations of leads; it is usually designated by means of a vector line diagram showing the angular displacement of windings and a sketch showing the marking of leads. The vector lines of the diagram represent induced voltages and the recognized counterclockwise direction of rotation is used. The vector line representing any phase voltage of a given winding is drawn parallel to that representing the corresponding phase voltage of any other winding under consideration. *See also:* **voltage regulator.** 57A15-0

polarity marks (instrument transformer). The identifications used to indicate the relative instantaneous polarities of the primary and secondary current and voltages. *Notes:* (1) On voltage transformers during most of each half cycle in which the identified primary terminal is positive with respect to the unidentified primary terminal, the identified secondary terminal is also positive with respect to the unidentified secondary terminal. (2) The polarity marks are so placed on current transformers that during most of each half-cycle, when the direction of the instantaneous current is into the identified primary terminal, the direction of the instantaneous secondary current is out of the correspondingly identified secondary terminal. (3) This convention is in accord with that by which standard terminal markings H_1, X_1, etcetera, are correlated. *See:* **instrument transformer.** 42A15-31E12;42A30-0

polarity test (rotating machinery). A test taken on a machine to demonstrate that the relative polarities of the windings are correct. *See also:* **asynchronous machine; direct-current commutating machine; synchronous machine.** 0-31E8

polarizability. The average electric dipole moment produced per molecule per unit of electric field strength. E270-0

polarization (1) (ferroelectric material). The electric moment per unit volume. *Note:* The polarization *P* may be expressed as the bound charge per unit area, which in macroscopic electric theory is equal to the component of the polarization normal to the surface. The polarization is related to the electric displacement through the expression

$$D = P + \epsilon_0 E$$

where the electric constant ϵ_0 (sometimes called permittivity of free space) equals 8.854×10^{-12} coulomb per volt-meter. In practice the electric field is supplied along a ferroelectric axis of the crystal. This expression then may be regarded as a scalar equation, since D, E, and P all point along the same direction. Then D, E, and P can be regarded simply as the scalar magnitudes of the corresponding vectors. In ferroelectric materials D is a nonlinear function of E. When the term $\epsilon_0 E$ in the above expression is negligible (as is the case for most ferroelectric materials), D is nearly equal to P; therefore, the D-versus-E and P-versus-E plots of the hysteresis loop become practically equivalent. *See also:* **ferroelectric domain.** E180-0

(2) (radiated wave). That property of a radiated electromagnetic wave describing the time-varying direction and amplitude of the electric field vector; specifically, the figure traced as a function of time by the extremity of the vector at a fixed location in space, as observed along the direction of propagation. *Note:* In general the figure is elliptical and it is traced in a clockwise or counterclockwise sense. The commonly referenced circular and linear polarizations are obtained when the ellipse becomes a circle or a straight line, respectively. Clockwise sense rotation of the electric vector is designated **right-hand polarization** and counterclockwise sense rotation is designated **left-hand polarization.** *See also:* **radiation.** 0-3E1

(3) (desired) (electronic navigation). The polarization of the radio wave, ordinarily either vertical or horizontal, for which an antenna system is designed. *See also:* **navigation.** E173-10E6

(4) (antenna). In a given direction, the polarization of the radiated wave, when the antenna is excited. Alternatively, the polarization of an incident wave from the given direction that results in maximum available power at the antenna terminals. *Note:* When the direction is not stated the polarization is taken to be the polarization in the direction of maximum gain. *See also:* **radiation.** 0-3E1

(5) (electrolytic). A change in the potential of an electrode produced during electrolysis, such that the potential of an anode always becomes more positive (more noble), or that of a cathode becomes more negative (less noble), than their respective static electrode potentials. The polarization is equal to the difference between the static electrode potential for the specified electrode reaction and the dynamic potential (that is, the potential when current is flowing) at a specified current density. *See also:* **electrochemistry.** 42A60-0

(6) (battery). The change in voltage at the terminals of the cell or battery when a specified current is flowing, and is equal to the difference between the actual and the equilibrium (constant open-circuit condition) potentials of the plates, exclusive of the IR drop. *See:* **polarization (electrolytic).** *See also:* **battery (primary or secondary).** 42A60-0

(7) (relay). A term identifying the input that provides a reference for establishing the direction of system phenomena such as direction of power or reactive flow, or direction to a fault or other disturbance on a power system. 37A100-31E11/3E6

polarization capacitance (biological). The reciprocal of the product of electrode capacitive reactance and 2π times the frequency.

$$C_p = \frac{1}{2\pi f X_p}$$

See also: **electrode impedance (biological).** 42A80-18E1

polarization diversity reception. That form of diversity reception that utilizes separate vertically and horizontally polarized receiving antennas. *See also:* **radio receiver.** 42A65-0

polarization ellipse (field vector). The locus of positions for variable time of the terminus of an instantaneous field vector of one frequency at a point in space. *See also:* **waveguide.** E146-3E1

polarization error (electronic navigation). The error arising from the transmission or reception of an electromagnetic wave having a polarization other than that intended for the system. *See also:* **navigation.** E172-10E6

polarization index. The insulation resistance of a machine winding measured at one minute after voltage has been applied; divided into the measurement at ten minutes. *See also:* **insulation testing (large alternating-current rotating machinery).** E95-0

polarization index test (rotating machinery). A test for measuring the ohmic resistance of insulation at specified time intervals for the purpose of determining the polarization index. *See also:* **asynchronous machine; direct-current commutating machine; synchronous machine.** 0-31E8

polarization potential (biological). The boundary potential over an interface. *See also:* **electrobiology.** 42A80-18E1

polarization reactance (biological). The impedance multiplied by the sine of the angle between the potential vector and the current vector.

$$X_p = Z_p \sin \theta$$

See also: **electrode impedance (biological).** 42A80-18E1

polarization receiving factor. The ratio of the power received by an antenna from a given plane wave of arbitrary polarization to the power received by the same antenna from a plane wave of the same power density and direction of propagation, whose state of polarization has been adjusted for the maximum received power. *Note:* It is equal to the square of the absolute value of the scalar product of the polarization unit vector of the given plane wave with that of the radiation field of the antenna along the direction opposite to the direction of propagation of the plane wave. *See also:* **waveguide.** E146-0

polarization resistance (biological). The impedance multiplied by the cosine of the phase angle between the potential vector and the current vector.

$$R_p = Z_p \cos \theta$$

See also: **electrode impedance (biological).** 42A80-18E1

polarization unit vector (field vector) (at a point). A complex field vector divided by its magnitude. *Notes:* (1) For a field vector of one frequency at a point, the polarization unit vector completely describes the state of polarization, that is, the axial ratio and orientation of the polarization ellipse and the sense of rotation on the ellipse. (2) A complex vector is one each of whose components is a complex number. The magnitude is the positive square root of the scalar product of the vector and its complex conjugate. *See also:* **waveguide.** E146-0

polarized electrolytic capacitor. An electrolytic capacitor in which the dielectric film is formed adjacent to only one metal electrode and in which the impedance to the flow of current in one direction is greater than in the other direction. *See also:* **electrolytic capacitor.** 42A60-0

polarized relay. A relay that consists of two elements, one of which operates as a neutral relay and the other of which operates as a polar relay. *See:* **neutral relay; polar relay.** 42A42-0

polarizer. A substance that when added to an electrolyte increases the polarization. *See also:* **electrochemistry.** 42A60-0

polar mode. The mode of operation that produces a transformation from rectangular to polar coordinates. *See:* **resolver.** *See also:* **electronic analog computer.** E165-0

polar relay. A relay in which the direction of movement of the armature depends upon the direction of the current in the circuit controlling the armature. *See also:* **electromagnetic relay; neutral relay; polarized relay; railway signal and interlocking.** 42A65-0

pole (pole unit) (1) (switching device or fuse). That portion of the device associated exclusively with one electrically separated conducting path of the main circuit of the device. *Notes:* (A) Those portions that provide a means for mounting and operating all poles together are excluded from the definition of a pole. (B) A switching device or fuse is called single-pole if it has only one pole. If it has more than one pole, it may be called multipole (two-pole, three-pole, etcetera) and provided, in the case of a switching device, that the poles are or can be coupled in such a manner as to operate together. 37A100-31E11

(2) (electric power or communication). A column of wood or steel, or some other material, supporting overhead conductors, usually by means of arms or brackets. *See:* **field pole; pole shoe; tower.** 42A35-31E13

(3) (network function). (A) Any value p_j of p, real or complex, for which the network function is infinite, provided that there exists some positive integer m such that, if the network function is multiplied by $(p-p_j)^m$, the resulting function is not infinite or zero when $p=p_j$. (B) The corresponding point in the P plane. *See also:* **control system, feedback; Laplace transform; network analysis; zero.** E270/85A1-23E0

pole body (rotating machinery). The part of a field pole around which the field winding is fitted. *See also:* **asynchronous machine; direct-current commutating machine; synchronous machine.** 0-31E8

pole-body insulation (rotating machinery). Insulation between the pole body and the field coil. *See also:* **asynchronous machine; direct-current commutating machine; synchronous machine.** 0-31E8

pole bolt (rotating machinery). A bolt used to fasten a pole to the spider. *See:* **cradle base (rotating machinery).** 0-31E8

pole-cell insulation (salient pole) (rotating machinery). Insulation that constitutes the liner between the field pole coil and the salient pole body. *See:* **rotor (rotating machinery).** 0-31E8

pole-changing winding (rotating machinery). A winding so designed that the number of poles can be changed by simple changes in the coil connections at the winding terminals. *See also:* **rotor (rotating machinery); stator.** 0-31E8

pole end plate (rotating machinery). A plate or structure at each end of a laminated pole to maintain axial pressure on the laminations. *See also:* **asynchronous machine; direct-current commutating machine; synchronous machine.** 0-31E8

pole face (rotating machinery). The surface of the pole shoe or nonsalient pole forming one boundary of the air gap. *See also:* **asynchronous machine; direct-current commutating machine; synchronous machine.** 0-31E8

pole-face bevel (rotating machinery). The portion of the pole shoe that is beveled so as to increase the length of the radial air gap. *See also:* **asynchronous machine; direct-current commutating machine; synchronous machine.** 0-31E8

pole face, relay. *See:* **relay pole face.**

pole-face shaping (rotating machinery). The contour of the pole shoe that is shaped other than by being beveled, so as to produce nonuniform radial length of the air gap. *See:* **rotor (rotating machinery); stator.** 0-31E8

pole fixture. A structure installed in lieu of a single pole to increase the strength of a pole line or to provide better support for attachments than would be provided by a single pole. Examples are *A* fixtures, *H* fixtures, etcetera. *See also:* **open wire.** 42A65-0

pole guy. A tension member having one end securely anchored and the other end attached to a pole or other structure that it supports against overturning. *See also:* **tower.** 42A35-31E13

pole line. A series of poles arranged to support conductors above the surface of the ground; and the structures and conductors supported thereon. *See also:* **open wire.** 42A65-0

pole piece. A piece or an assembly of pieces of ferromagnetic material forming one end of a magnet and so shaped as to appreciably control the distribution of the magnetic flux in the adjacent medium. E270-0

pole pitch (rotating machinery). The peripheral distance between corresponding points on two consecutive poles; also expressed as a number of slot positions. *See:* **armature; rotor (rotating machinery); stator.** 0-31E8

pole shoe. The portion of a field pole facing the armature that serves to shape the air gap and control its reluctance. *Note:* For round-rotor fields, the effective pole shoe includes the teeth that hold the field coils and wedges in place. *See:* **field pole; rotor (rotating machinery); stator.** 42A10-31E8

pole slipping (rotating machinery). The process of the secondary member of a synchronous machine slipping one pole pitch with respect to the primary magnetic flux. *See also:* **synchronous machine.** E230-31E8

pole steps. Devices attached to the side of a pole, conveniently spaced to provide a means for climbing the pole. *See also:* **tower.** 42A35-31E13

pole tip (rotating machinery). The leading or trailing extremity of the pole shoe. *See:* **rotor (rotating machinery); stator.** 0-31E8

pole-type (transformers, regulators, etcetera). Suitable for mounting on a pole or similar structure. *See also:* **constant-current transformer.** 57A14-0

pole-unit mechanism (switching device). That part of the mechanism that actuates the moving contacts of one pole. 37A100-31E11

policy (control system). *See:* **control law.**

poling. The adjustment of polarity. Specifically, in wire line practice, it signifies the use of transpositions between transposition sections of open wire or between lengths of cable to cause the residual crosstalk couplings in individual sections or lengths to oppose one another. *See also:* **open wire.** 42A65-0

polishing (electroplating). The smoothing of a metal surface by means of abrasive particles attached by adhesive to the surface of wheels or belts. *See also:* **electroplating.** 42A60-0

polyphase (relay). A descriptive term indicating that the relay is responsive to polyphase alternating electric input quantities. *Note:* A multiple-unit relay with individual units responsive to single-phase electric inputs is not a polyphase relay even though the several single-phase units constitute a polyphase set. 37A100-31E11/31E6

polyphase circuit. An alternating-current circuit consisting of more than two intentionally interrelated conductors that enter (or leave) a delimited region at more than two terminals of entry and that are intended to be so energized that in the steady state the alternating voltages between successive pairs of terminals of entry of the phase conductors, selected in a systematic chosen sequence, have: (1) the same period, (2) definitely related and usually equal amplitudes, and (3) definite and usually equal phase differences. If a neutral conductor exists, it is intended also that the voltages from the successive phase conductors to the neutral conductor be equal in amplitude and equally displaced in phase. *Note:* For all polyphase circuits in common use except the two-phase three-wire circuit, it is intended that the voltage amplitudes and the phase differences of the systematically chosen voltages between phase conductors be equal. For a two-phase three-wire circuit it is intended that voltages between two successive pairs of terminals be equal and have a phase difference of $\pi/2$ radians, but that the voltage between the third pair of terminals have an amplitude $(2)^{1/2}$ times as great as the other two, and a phase difference from each of the other two of $3\pi/4$ radians. *See:* **mesh connection of polyphase circuit; star connection of polyphase circuit; zig-zag connection of polyphase circuits.** *See also:* **center of distribution; network analysis.** E270-0

polyphase machine (rotating machinery). A machine that generates or utilizes polyphase alternating-current power. These are usually three-phase machines with three voltages displaced 120 electrical degrees with respect to each other. *See also:* **asynchronous machine; synchronous machine.** 0-31E8

polyphase symmetrical set (1) (polyphase voltages). A symmetrical set of polyphase voltages in which the angular phase difference between successive members of the set is not zero, π radians, or a multiple thereof. The equations of **symmetrical set of polyphase voltages** represent a polyphase symmetrical set of polyphase voltages if k/m is not zero, ½, or a multiple thereof. (The symmetrical set of voltages represented by the equations of **symmetrical set of polyphase voltages** may be said to have polyphase symmetry if k/m is not zero, ½, or a multiple of ½.) *Note:* This definition may be applied to a two-phase four-wire or five-wire circuit if m is considered to be 4 instead of 2. It is not applicable to a two-phase three-wire circuit. *See:* **symmetrical set of polyphase voltages.** *See also:* **network analysis.** E270-0

(2) (polyphase currents). This definition is obtained from the corresponding definitions for voltage by substituting the word current for voltage, the symbol I for E, and β for α wherever they appear. The subscripts are unaltered. *See also:* **network analysis.** E270-0

polyphase synchronous generator. A generator whose alternating-current circuits are so arranged that two or more symmetrical alternating electromotive forces with definite phase relationships are produced at its terminals. Polyphase synchronous generators are usually two-phase, producing two electromotive forces displaced 90 electrical degrees with respect to one another, or three-phase, with three electromotive forces displaced 120 electrical degrees with respect to each other. *See also:* **synchronous machine.** 42A10-0

polyplexer (radar). Equipment combining the functions of duplexing and lobe switching. *See also:* **navigation.** E172-10E6

pondage station. A hydroelectric generating station with storage sufficient only for daily or weekend regulation of flow. *See also:* **generating station.** 42A35-31E13

***p*-on-*n* solar cells (photovoltaic power system).** Photovoltaic energy-conversion cells in which a base of *n*-type silicon (having fixed positive holes in a silicon lattice and electrons that are free to move) is overlaid with a surface layer of *p*-type silicon (having fixed electrons in a silicon lattice and positive holes that are free to move). *See also:* **photovoltaic power system; solar cells (photovoltaic power system).** 0-10E1

pool-cathode mercury-arc converter. A frequency converter using a mercury-arc pool-type discharge device. *See also:* **industrial electronics.** E54/E169-0

pool rectifier. A gas-filled rectifier with a pool cathode, usually mercury. *See also:* **gas-filled rectifier.** 50I07-15E6

pool tube. A gas tube with a pool cathode. *See also:* **electronic controller.** 42A70-15E6

pores (electroplating). Micro discontinuities in a metal coating that extend through to the base metal or underlying coating. *See also:* **electroplating.** 42A60-0

port (1) (electronic devices or networks). A place of access to a device or network where energy may be supplied or withdrawn or where the device or network variables may be observed or measured. *Notes:* (A) In any particular case, the ports are determined by the way the device is used and not by its structure alone. (B) The terminal pair is a special case of a port. (C) In the case of a **waveguide** or **transmission line**, a port is characterized by a specified mode of propagation and a specified reference plane. (D) At each place of access, a separate port is assigned to each significant independent mode of propagation. (E) In frequency

changing systems, a separate port is also assigned to each significant independent frequency response. *See also:* **network analysis; optoelectronic device; waveguide.** E222-15E7/15E6/9E4
(2) (rotating machinery). An opening for the intake or discharge of ventilating air. *See:* **cradle base (rotating machinery).** 0-31E8

portable battery. A storage battery designed for convenient transportation. *See also:* **battery (primary or secondary).** 42A60-0

portable concentric mine cable. A double-conductor cable with one conductor located at the center and with the other conductor strands located concentric to the center conductor with rubber or synthetic insulation between conductors and over the outer conductor. *See also:* **mine feeder circuit.** 42A85-0

portable lighting. Lighting involving lighting equipment designed for manual portability. *See also:* **general lighting.** Z7A1-0

portable luminaire. A lighting unit that is not permanently fixed in place. *See also:* **luminaire.** Z7A1-0

portable mine blower. A motor-driven blower to provide secondary ventilation into spaces inadequately ventilated by the main ventilating system and with the air directed to such spaces through a duct. *See also:* **mining.** 42A85-0

portable mine cable. An extra-flexible cable, used for connecting mobile or stationary equipment in mines to a source of electric energy when permanent wiring is prohibited or impracticable. *See also:* **mine feeder circuit.** 42A85-0

portable mining-type rectifier transformer. A rectifier transformer that is suitable for transporting on skids or wheels in the restrictive areas of mines. *See also:* **rectifier transformer.** 57A18-0

portable parallel duplex mine cable. A double- or triple-conductor cable with conductors laid side by side without twisting, with rubber or synthetic insulation between conductors and around the whole. The third conductor, when present, is a safety ground wire. *See also:* **mine feeder circuit.** 42A85-0

portable shunts (electric power systems). Instrument shunts with insulating bases that may be laid on or fastened to any flat surface. *Note:* They may be used also for switchboard applications where the current is relatively low and connection bars are not used. *See also:* **power system, low-frequency and surge testing.** 0-31E5

portable standard meter. A portable form of meter principally used as a standard for testing other meters. It is usually provided with several current and voltage ranges and with dials indicating revolutions and fractions of a revolution of the rotor. *See also:* **electricity meter (meter).** 12A0/42A30-0

portable station (mobile communication). A mobile station designed to be carried by or on a person. **Personal** or **pocket** stations are special classes of portable stations. *See also:* **mobile communication system.** 0-6E1

portable traffic-control light. A signaling light producing a controllable distinctive signal for purposes of directing aircraft operations in the vicinity of an aerodrome. *See also:* **signal lighting.** Z7A1-0

portable transmitter. A transmitter that can be carried on a person and may or may not be operated while in motion. *Notes:* (1) This has been called a transportable transmitter, but the designation portable is preferred. (2) This includes the class of so-called **walkie-talkies, handy-talkies,** and **personal** transmitters. *See also:* **radio transmission; radio transmitter; transportable transmitter.** E182A/E145/42A65-0

port difference (hybrid). A port that yields an output proportional to the difference of the electric field quantities existing at two other ports of the hybrid. *See also:* **waveguide.** 0-9E4

port sum (hybrid). A port that yields an output proportional to the sum of the electric-field quantities existing at two other ports of the hybrid. *See also:* **waveguide.** 0-9E4

position. *See:* **punch position, sign position.**

position (navigation). The location of a point with respect to a specific or implied coordinate system. *See also:* **navigation.** 0-10E6

positional crosstalk (multibeam cathode-ray tubes). The variation in the path followed by any one electron beam as the result of a change impressed on any other beam in the tube. *See:* **beam tubes.** E160-15E6

positional notation (1) (general). A number representation that makes use of an ordered set of digits, such that the value contributed by each digit depends on its position as well as on the digit value. *Note:* The Roman numeral system, for example, does not use positional notation. *See:* **binary system; binary-coded-decimal system; biquinary system; decimal system; Gray code.**
(2) One of the schemes for representing numbers, characterized by the arrangement of digits in sequence, with the understanding that successive digits are to be interpreted as coefficients of successive powers of an integer called the base (or radix) of the number system. *Notes:* (A) In the binary number system the successive digits are interpreted as coefficients of the successive powers of the base two, just as in the decimal number system they relate to successive powers of the base ten. (B) In the ordinary number systems each digit is a character that stands for zero or for a positive integer smaller than the base. (C) The names of the number systems with bases from 2 to 20 are: binary, ternary, quaternary, quinary, senary, septenary, octonary (also octal), novenary, decimal, unidecimal, duodecimal, terdenary, quaterdenary, quindenary, sexadecimal (also hexadecimal), septendecimal, octodenary, novendenary, and vicenary. The sexagenary number system has the base 60. The commonly used alternative of saying **base-3, base-4,** etcetera, in place of ternary, quaternary, etcetera, has the advantage of uniformity and clarity. (D) In the most common form of positional notation the expression $\pm a_n a_{n-1} \ldots a_2 a_1 a_0 \cdot a_{-1} a_{-2} \ldots a_{-m}$ is an abbreviation for the sum

$$\pm \sum_{i=-m}^{n} a_i r^i$$

where the point separates the positive powers from the negative powers, the a_i are integers $(0 \leq a_i < r)$ called digits, and r is an integer, greater than one, called the base (or radix). *See also:* **electronic computation; base; radix.**
(3) A number-representation system having the property that each number is represented by a sequence of characters such that successive characters of the sequence represent integral coefficients of accumulated products of a sequence of integers (or reciprocals of

integers) and such that the sum of these products, each multiplied by its coefficient, equals the number. Each occurrence of a given character represents the same coefficient value. *Note:* The biquinary system is an example of (3).
(4) A number-representation system such that if the representations are arranged vertically in order of magnitude with digits of like significance in the same column, then each column of digits consists of recurring identical cycles (for numbers sufficiently large in absolute value) whose length is an integral multiple of the cycle length in the column containing the next-less-significant digits. *Note:* (2), (3), and (4) are not mutually exclusive. The biquinary system is an example of (3) and (4); whereas the Gray code system is an example of (4) only. The binary and decimal systems are examples of (2), (3), and (4). *See also:* **electronic digital computer.** E162/E270-16E9

positional response (close-talking pressure microphone). The response-frequency measurements conducted with the principal axis of a microphone collinear with the axis of the artificial voice and the combination of microphone and artificial voice placed at various angles to the horizontal plane. *Note:* Variations in positional response of carbon microphones may be due to gravitational forces. *See also:* **close-talking pressure-type microphone.** E258-0

position-control system (industrial control). A control system that attempts to establish and/or maintain an exact correspondence between the reference input and the directly controlled variable, namely physical position. *See also:* **control system, feedback.** AS1-34E10

position index. The position index is a factor that represents the relative average luminance (photometric brightness), for a sensation at the borderline between comfort and discomfort for a source located anywhere within the visual field. *See also:* **inverse-square law (illuminating engineering).** Z7A1-0

position indicator (elevators). A device that indicates the position of the elevator car in the hoistway. It is called a hall position indicator when placed at a landing or a car position indicator when placed in the car. *See:* **control.** 42A45-0

position influence (electric instrument). The change in the indication of an instrument that is caused solely by a position departure from the normal operating position. *Note:* Unless otherwise specified, the maximum change in the recorded value caused solely by an inclination in the most unfavorable direction from the normal operating position. *See also:* **accuracy rating (instrument).** 39A1/39A2/39A4-0

positioning-control system (numerically controlled machines). A system in which the controlled motion is required only to reach a given end point, with no path control during the transition from one end point to the next. *See also:* **numerically controlled machines.** EIA3B-34E12

position lights. Aircraft aeronautical lights that form the basic or internationally recognized navigation light system. *Note:* The system is composed of a red light showing from dead ahead to 110 degrees to the left, a green light showing from dead ahead to 110 degrees to the right, and a white light showing to the rear through 140 degrees. Position lights are also called **navigation lights.** *See also:* **signal lighting.** Z7A1-0

position light signal. A fixed signal in which the indications are given by the position of two or more lights. *See also:* **railway signal and interlocking.** 42A42-0

position of the effective short (microwave switching tube). The distance between a specified reference plane and the apparent short-circuit of the fired tube in its mount. *See:* **gas tube.** E160-15E6

position sensor or position transducer (numerically controlled machines). A device for measuring a position and converting this measurement into a form convenient for transmission. *See also:* **numerically controlled machines.** EIA3B-34E12

position stopping (industrial control). A control function that provides for stopping the driven equipment at a preselected position. *See:* **electric controller.** IC1-34E10

position-type telemeter. *See:* **ratio-type telemeter.**

positive after-potential (electrobiology). Relatively prolonged positivity that follows the negative after-potential. *See also:* **contact potential.** 42A80-18E1

positive column (gas tube). The luminous glow, often striated, in a glow-discharge cold-cathode tube between the Faraday dark space and the anode. *See:* **gas tube.** 42A70-15E6

positive conductor. A conductor connected to the positive terminal of a source of supply. *See also:* **center of distribution.** 42A35-31E13

positive creep effect (semiconductor rectifier). The gradual increase in reverse current with time, that may occur when a direct-current reverse voltage is applied to a semiconductor rectifier cell. *See also:* **rectification.** E59-34E17

positive electrode (1) (primary cell). The cathode when the cell is discharging. The positive terminal is connected to the positive electrode. *See also:* **electrolytic cell.** 42A60-0
(2) (metallic rectifier). The electrode to which the forward current flows within the metallic rectifying cell. *See also:* **rectification.** 42A15-0

positive feedback. The process by which a part of the power in the output circuit of an amplifying device reacts upon the input circuit in such a manner as to reinforce the initial power, thereby increasing the amplification. *See also:* **control system, feedback; feedback.** 42A65-0

positive grid. *See:* **retarding field (positive-grid) oscillator.**

positive-grid oscillator tube (Barkhausen tube). A triode operating under oscillating conditions such that the quiescent voltage of the grid is more positive than that of either of the other electrodes. *See also:* **tube definitions.** 42A70-15E6

positive matrix (positive). A matrix with a surface like that which is to be ultimately produced by electroforming. *See also:* **electroforming.** 42A60-0

positive modulation (in an amplitude-modulation television system). That form of modulation in which an increase in brightness corresponds to an increase in transmitted power. *See also:* **television.** 42A65-0

positive nonconducting period (rectifier element). The nonconducting part of an alternating-voltage cycle during which the anode has a positive potential with respect to the cathode. *See also:* **power rectifier; rectification.** 42A15-34E24

positive noninterfering and successive fire-alarm system. A manual fire-alarm system employing stations and circuits such that, in the event of simultaneous operation of several stations, one of the operated stations will take control of the circuit, transmit its full signal, and then release the circuit for successive transmission by other stations that are held inoperative until they gain circuit control. *See also:* **protective signaling.** 42A65-0

positive-phase-sequence reactance (rotating machinery). The quotient of the reactive fundamental component of the positive-sequence primary voltage due to the sinusoidal positive-sequence primary current of rated frequency, and the value of this current, the machine running at rated speed. *See also:* **asynchronous machine; synchronous machine.** 0-31E8

positive-phase-sequence relay. A relay that responds to the positive-phase-sequence component of a polyphase input quantity. *See also:* **relay.** 0-31E6

positive-phase-sequence resistance (rotating machinery). The quotient of the in-phase component of positive-sequence primary voltage corresponding to direct load losses in the primary winding and stray load losses due to sinusoidal positive-sequence primary current, and the value of this current, the machine running at rated speed. 0-31E8

positive-phase-sequence symmetrical components (of an unsymmetrical set of polyphase voltages or currents of m phases). The set of symmetrical components that have the first phase sequence. That is, the angular phase lag from the first member of the set to the second, from every other member of the set to the succeeding one, and from the last member to the first, is equal to the characteristic angular phase difference, or $2\pi/m$ radians. The members of this set will reach their positive maxima uniformly in their designated order. The positive-phase-sequence symmetrical components for a three-phase set of unbalanced sinusoidal voltages ($m=3$), having the primitive period, are represented by the equations

$$e_{a1} = (2)^{1/2}\ E_{a1} \cos(\omega t + \alpha_{a1})$$

$$e_{b1} = (2)^{1/2}\ E_{a1} \cos\left(\omega t + \alpha_{a1} - \frac{2\pi}{3}\right)$$

$$e_{c1} = (2)^{1/2}\ E_{a1} \cos\left(\omega t + \alpha_{a1} - \frac{4\pi}{3}\right)$$

derived from the equation of **symmetrical components of a set of polyphase (alternating) voltages.** Since in this case $r=1$ for every component (of 1st harmonic) the third subscript is omitted. Then k is 1 for 1st sequence and s takes on the algebraic values 1, 2, and 3 corresponding to phases a, b, and c. The sequence of maxima occurs in the order a, b, c. *See also:* **network analysis.** E270-0

positive plate (storage cell). The grid and active material from which current flows to the external circuit when the battery is discharging. *See also:* **battery (primary or secondary).** 42A60-0

positive-polarity lightning stroke. A stroke resulting from a positively charged cloud that lowers positive charge to the earth. *See also:* **direct-stroke protection (lightning).** 0-31E13

positive-sequence impedance. The quotient of that component of positive-sequence rated-frequency voltage, assumed to be sinusoidal, that is due to the positive-sequence component of current, divided by the positive-sequence component of current. *See also:* **asynchronous machine; synchronous machine.** 0-31E8

positive-sequence resistance. That value of resistance that, when multiplied by the square of the fundamental positive-sequence rated-frequency component of armature current and by the number of phases, is equal to the sum of the copper loss in the armature and the load loss resulting from that current, when the machine is operating at rated speed. Positive-sequence resistance is normally that corresponding to rated armature current. *Note:* Inasmuch as the load loss may not vary as the square of the current, the positive-sequence resistance applies accurately only near the current for which it was determined. *See:* **synchronous machine.** 42A10-0

positive terminal (battery). The terminal from which the positive electric charge flows through the external circuit to the negative terminal when the cell discharges. *Note:* The flow of electrons in the external circuit is to the positive terminal and from the negative terminal. *See also:* **battery (primary or secondary).** 42A60-0

post (waveguide). A cylindrical rod placed in a transverse plane of the waveguide and behaving substantially as a shunt susceptance. *See also:* **waveguide.** E147-3E1

post-accelerating (deflection) electrode (intensifier electrode). An electrode to which a potential is applied to produce post-acceleration. *See also:* **electrode (electron tube).** 42A70-15E6

post acceleration (electron-beam tube). Acceleration of the beam electrons after deflection. *See also:* **beam tubes.** 42A70-15E6

post-deflection acceleration. *See:* **post-acceleration.**

post emphasis. *See:* **deemphasis.**

post equalization. *See:* **deemphasis.**

post insulator. An insulator of generally columnar shape, having means for direct and rigid mounting. *See also:* **insulator.** 29A1-0

post-mortem (computing systems). Pertaining to the analysis of an operation after its completion. X3A12-16E9

post-mortem dump (computing systems). A static dump used for debugging purposes that is performed at the end of a machine run. X3A12-16E9

post processor (numerically controlled machines). A set of computer instructions that transform tool centerline data into machine motion commands using the proper tape code and format required by a specific machine/control system. Instructions such as feedrate calculations, spindle-speed calculations, and auxiliary-function commands may be included. *See also:* **numerically controlled machines.** EIA3B-34E12

post puller. An electric vehicle having a powered drum handling wire rope used to pull mine props, after coal has been removed, for the recovery of the timber. *See also:* **mining.** 42A85-0

potential diagram (electron-optical system). A diagram showing the equipotential curves in a plane of symmetry of an electron-optical system. *See also:* **electron optics.** 50I07-15E6

potential difference, electrostatic. *See:* **electrostatic potential difference.**

potential, electrostatic. *See:* **electrostatic potential.**

potential energy (of a body or of a system of bodies, in a given configuration with respect to an arbitrarily chosen reference configuration). The work required to bring the system from an arbitrarily chosen reference configuration to the given configuration without change in other energy of the system. E270-0

potential false-proceed operation. The existence of a condition of vehicle or roadway apparatus in an automatic train control or cab-signal installation under which a false-proceed operation would have occurred had a vehicle approached or entered a section where normally a restrictive operation would occur. *See also:* **railway signal and interlocking.** 42A42-0

potential gradient. A vector of which the direction is normal to the equipotential surface, in the direction of decreasing potential, and of which the magnitude gives the rate of variation of the potential. 50I05-31E3

potential profile. A plot of potential as a function of distance along a specified path. *See also:* **ground.** E81-0

potential transformer (voltage transformer). An instrument transformer that is intended to have its primary winding connected in shunt with a power-supply circuit, the voltage of which is to be measured or controlled. *See also:* **instrument transformer.** 12A0/42A30-0

potential transformer, cascade-type. A single high-voltage line-terminal potential transformer with the primary winding distributed on several cores with the cores electromagnetically coupled by coupling windings and the secondary winding on the core at the neutral end of the high-voltage winding. Each core is insulated from the other cores and is maintained at a fixed potential with respect to ground and the line-to-ground voltage. *See also:* **instrument transformer.** 57A13-31E12

potential transformer, double-secondary. One that has two secondary windings on the same magnetic circuit insulated from each other and the primary. Either or both of the secondary windings may be used for measurement or control. *See also:* **instrument transformer.** 57A13-31E12

potential transformer, fused-type. One that is provided with the means for mounting a fuse, or fuses, as an integral part of the transformer in series with the primary winding. *See also:* **instrument transformer.** 57A13-31E12

potential transformer, grounded-neutral terminal type. One that has the neutral end of the high-voltage winding connected to the case or mounting base. *See also:* **instrument transformer.** 57A13-31E12

potential transformer, insulated-neutral terminal type. One that has the neutral end of the high-voltage winding insulated from the case or base and connected to a terminal that provides insulation for a lower-voltage insulation class than required for the rated insulation class of the transformer. *See also:* **instrument transformer.** 57A13-31E12

potential transformer, single-high-voltage line terminal. One that has the line end of the primary winding connected to a terminal insulated from ground for the rated insulation class. The neutral end of the winding may be (1) insulated from ground but for a lower insulation class than the line end (insulated neutral) or (2) connected to the case or base (grounded neutral). *See also:* **instrument transformer.** 42A15/57A13-31E12

potential transformer, two-high-voltage line terminal. One that has both ends of the high-voltage winding connected to separate terminals that are insulated from each other, and from other parts of the transformer, for the rated insulation class of the transformer. *See also:* **instrument transformer.** 57A13-31E12

potentiometer (1) (general). A three-terminal rheostat, or a resistor with one or more adjustable sliding contacts, that functions as an adjustable voltage divider. *See:* **potentiometer, function; normal linearity; potentiometer, multiplier; potentiometer, parameter.** *See also:* **attenuator; electronic analog computer.** E165/42A25-34E10-16E9

(2) (measurement techniques). An instrument for measuring an unknown electromotive force or potential difference by balancing it, wholly or in part, by a known potential difference produced by the flow of known currents in a network of circuits of known electrical constants. *See:* **instrument.** 42A30-0

potentiometer, follow-up. A servo potentiometer that generates the signal for comparison with the input signal. *See also:* **electronic analog computer.** E165-0

potentiometer, function. A multiplier potentiometer in which the voltage at the movable contact follows a prescribed functional relationship to the displacement of the contact. *See:* **linearity.** *See also:* **electronic analog computer.** E165-16E9

potentiometer granularity. The physical inability of a potentiometer to produce an output voltage that varies in other than discrete steps, due either to contacting individual turns of wire in a wire-wound potentiometer or to discrete irregularities of the resistance element of composition or film potentiometers. *See also:* **electronic analog computer.** E165-0

potentiometer, grounded. A potentiometer with one end terminal attached directly to ground. *See also:* **electronic analog computer.** E165-0

potentiometer, linear. A potentiometer in which the voltage at a movable contact is a linear function of the displacement of the contact. *See:* **linearity.** *See also:* **electronic analog computer.** E165-0

potentiometer, multiplier. Any of the ganged potentiometers of a servo multiplier that permit the multiplication of one variable by a second variable. *See also:* **electronic analog computer.** E165-16E9

potentiometer, parameter (scale-factor potentiometer) (coefficient potentiometer). A potentiometer used in an analog computer to represent a problem parameter such as a coefficient or a scale factor. *See also:* **electronic analog computer.** E165-16E9

potentiometer, servo. A potentiometer driven by a positional servomechanism. *See also:* **electronic analog computer.** E165-0

potentiometer set. A computer control state that supplies the same operating potentiometer loading as under computing conditions and thus allows correct potentiometer adjustment. *See also:* **electronic analog computer; problem check.** E165-0

potentiometer, sine-cosine. A function potentiometer with movable contacts attached to a rotating shaft so that the voltages appearing at the contacts are proportional to the sine and cosine of the angle of rotation of the shaft, the angle being measured from a fixed referenced position. *See also:* **electronic analog computer.** E165-0

potentiometer, tapered. A function potentiometer that achieves a prescribed functional relationship by means of a nonuniform winding. *See also:* **electronic analog computer.** E165-0

potentiometer, tapped. A potentiometer, usually a servo potentiometer, that has a number of fixed contacts (or taps) to the resistance element in addition to the end and movable contacts. *See also:* **electronic analog computer.** E165-0

potentiometer, ungrounded. A potentiometer with neither end terminal attached directly to ground. *See also:* **electronic analog computer.** E165-0

pothead. A device that seals the end of a cable and provides insulated egress for the conductor or conductors.
See also:
aerial lug;
cable entrance fitting;
cable sheath insulator;
design test;
disconnect-type pothead;
external connector;
field test;
flashover;
indoor pothead;
internal connector;
outdoor pothead;
pothead body;
pothead bracket or mounting plate;
pothead bracket or mounting plate insulator;
pothead entrance fitting;
pothead insulator;
pothead insulator lid;
pothead sheath insulator;
pressure-type pothead;
routine tests;
switchgear pothead;
transformer;
withstand test voltage.
57A12.75/57A12.76/E48-0

pothead body. The part of a pothead that joins the entrance fitting to the insulator or to the insulator lid. *See also:* **pothead; transformer.** 57A12.75/57A12.76/E48-0

pothead bracket or mounting plate. The part of the pothead used to attach the pothead to the supporting structure. *See also:* **pothead; transformer.** E48-0

pothead bracket or mounting-plate insulator. An insulator used to insulate the pothead from the supporting structure for the purpose of controlling cable sheath currents. *See also:* **pothead; transformer.** E48-0

pothead entrance fitting. A fitting used to seal or attach the cable sheath, armor, or other coverings to the pothead. *See also:* **pothead; transformer.** E48-0

pothead insulator. An insulator used to insulate and protect each conductor passing through the pothead. *See also:* **pothead; transformer.** 57A12.75/57A12.76/E48-0

pothead insulator lid. The part of a multi-conductor pothead used to join two or more insulators to the body. *See also:* **pothead; transformer.** 57A12.75/57A12.76/E48-0

pothead mounting plate. The part of the pothead used to attach the pothead to the supporting structure. *See also:* **transformer.** 57A12.75/57A12.76-0

pothead mounting-plate insulator. An insulator used to insulate the pothead from the supporting structure for the purpose of controlling cable sheath currents. *See also:* **transformer.** 57A12.75-0/57A12.76-0

pothead sheath insulator. An insulator used to insulate an electrically conductive cable sheath or armor from the metallic parts of the pothead in contact with the supporting structure for the purpose of controlling cable sheath currents. *See also:* **pothead.** E48-0

Potier reactance (rotating machinery). An equivalent reactance used in place of the primary leakage reactance to calculate the excitation on load by means of the Potier method. *Note:* It takes into account the additional leakage of the excitation winding on load and in the overexcited region; it is greater than the real value of the primary leakage reactance. It is useful for the calculation of excitation of the machine at other loads and power factors.† The height of a Potier reactance triangle determines the reactance drop, and the reactance X_p is equal to the reactance drop divided by the current. The value of Potier reactance is that obtained from the no-load normal-frequency saturation curve; and normally with the excitation for rated voltage and current at zero power factor (overexcited), and at rated frequency. Approximate values of Potier reactance may be obtained from test load excitations at loads differing from rated load, and at power factors other than zero. *See also:* **synchronous machine.**

†The excitation results in the range from zero power factor overexcited to unity power factor are close enough to the test values for most practical applications. 42A10-31E8

power (1) (noun). The time rate of transferring or transforming energy. E270-0
(2) (adjective). A general term used, by reason of specific physical or electrical characteristics, to denote application or restriction or both, to generating stations, switching stations, or substations. The term may also denote use or application to energy purposes as contrasted with use for control purposes. 37A100-31E11

power, active (polyphase circuit) (power†). At the terminals of entry of a polyphase circuit into a delimited region, the algebraic sum of the active powers for the individual terminals of entry when the voltages are all determined with respect to the same arbitrarily selected common reference point in the boundary surface (which may be the neutral terminal of entry). *Notes:* (1) The active power for each terminal of entry is determined by considering each conductor and the common reference point as a single-phase two-wire circuit and finding the active power for each in accordance with the definition of **power, active (single-phase two-wire circuit).** If the voltages and currents are sinusoidal and of the same period, the active power P for a three-phase circuit is given by

$$P=E_aI_a\cos(\alpha_a-\beta_a)+E_bI_b\cos(\alpha_b-\beta_b)+E_cI_c\cos(\alpha_c-\beta_c)$$

where the symbols have the same meaning as in **power, instantaneous (polyphase circuit).** (2) If there is no neutral conductor and the common point for voltage measurement is selected as one of the phase terminals of entry, the expression will be changed in the same way as that for **power, instantaneous (polyphase circuit).** (3) If both the voltages and the currents in the preceding equations constitute symmetrical sets of the same phase sequences

$$P=3\,E_aI_a\cos(\alpha_a-\beta_a).$$

(4) In general the active power P at the $(m+1)$ terminals of entry of a polyphase circuit of m phases to a delimited region, when one of the terminals is the neutral terminal of entry, is expressed by the equation

$$P=\sum_{s=1}^{s=m}\sum_{r=1}^{r=\infty}E_{sr}I_{sr}\cos(\alpha_{sr}-\beta_{sr})$$

where E_{sr} is the root-mean-square amplitude of the rth harmonic of the voltage e_s, from phase conductor to neutral. I_{sr} is the root-mean-square amplitude of the rth harmonic of the current i_s through terminal s. α_{sr} is the phase angle of the rth harmonic of e_s with respect to a common reference. β_{sr} is the phase angle of the rth harmonic of i_s with respect to the same reference as the voltages. The indexes s and r have the same meaning as in **power, instantaneous (polyphase circuit).** (5) The active power can also be stated in terms of the root-mean-square amplitudes of the symmetrical components of the voltages and currents as

$$P=m\sum_{k=0}^{k=m-1}\sum_{r=1}^{r=\infty}E_{kr}I_{kr}\cos(\alpha_{kr}-\beta_{kr})$$

where m is the number of phase conductors, k denotes the number of the symmetrical component, and r denotes the number of the harmonic component. (6) When the voltages and currents are quasi-periodic and the amplitudes of the voltages and currents are slowly varying, the active power for the circuit of each conductor may be determined for this condition as in **power, active (single-phase two-wire circuit).** The active power for the polyphase circuit is the sum of the active power values for the individual conductors. The active power is also the time average of the instantaneous power for the polyphase circuit. (7) Mathematically the active power at any time t_0 is

$$P=\frac{1}{T}\int_{t_0-T/2}^{t_0+T/2}p\,dt$$

where p is the instantaneous power and T is the period. This formulation may be used when the voltage and current are periodic or quasi-periodic so that the period is defined. The active power is expressed in watts when the voltages are in volts and the currents in amperes.

†When it is clear that average power and not instantaneous power is meant, **power** is often used for **active power.**

E270-0

power, active (single-phase two-wire circuit) (power)†. At the terminals of entry of a single-phase, two-wire circuit into a delimited region, when the voltage and current are periodic or quasi-periodic, the time average of the values of the instantaneous power, the average being taken over one period. *Notes:* (1) Mathematically, the active power P at a time t_0 is given by the equation

$$P=\frac{1}{T}\int_{t_0-T/2}^{t_0+T/2}p\,dt$$

where T is the period, and p is the instantaneous power. (2) If both the voltage and current are sinusoidal and of the same period the active power P is given by

$$P=EI\cos(\alpha-\beta)$$

in which the symbols have the same meaning as in **power, instantaneous (two-wire circuit).** (3) If both the voltage and current are sinusoidal, the active power P is also equal to the real part of the product of the phasor voltage and the conjugate of the phasor current, or to the real part of the product of the conjugate of the phasor voltage and the phasor current. Thus

$$\begin{aligned}P&=\mathrm{Re}\,\mathbf{EI}^*\\&=\mathrm{Re}\,\mathbf{E}^*\mathbf{I}\\&=\frac{1}{2}[\mathbf{EI}^*+\mathbf{E}^*\mathbf{I}]\end{aligned}$$

in which **E** and **I** are the root-mean-square phasor voltage and root-mean-square phasor current, respectively (see **phasor quantity**), and the * denotes the conjugate of the phasor to which it is applied. (4) If the voltage is an alternating voltage and the current is an alternating current (see **alternating voltage and alternating current**), the active power is given by the equations

$$\begin{aligned}P&=E_1I_1\cos(\alpha_1-\beta_1)+E_2I_2\cos(\alpha_2-\beta_2)+\cdots\\&=\sum_{r=1}^{r=\infty}\mathbf{E}_r\mathbf{I}_r\cos(\alpha_r-\beta_r)\\&=\mathrm{Re}\sum_{r=1}^{r=\infty}E_rI_r\\&=\frac{1}{2}\sum_{r=1}^{r=\infty}[E_rI_r+E_rI_r]\end{aligned}$$

in which r is the order of the harmonic component of the voltage (see **harmonic components (harmonics)**) and r is also the order of the harmonic component of the current. $\mathbf{E}_r$ and $\mathbf{I}_r$ are the phasors corresponding to the r th harmonic of the voltage and current, respectively. (5) If the voltage and current are quasi-periodic functions of the form given in **power, instantaneous (two-wire circuit),** the integral over the period T will not result in the simple expressions that are obtained when E_r and I_r are constant. However, if the relative rates of change of the quantities are so small that each may be considered to be constant during any one period, but to have slightly different values in successive periods, the active power at any time t is very closely approximated by

$$P=\sum_{r=1}^{r=\infty}E_r(t)I_r(t)\cos(\alpha_r-\beta_r)$$

which is analogous to the preceding expression. When the amplitudes of voltage and current are slowly changing, the active power may be represented by this expression. (6) Active power is expressed in watts when the voltage is in volts and the current in amperes.

†When it is clear that average power and not instantaneous power is meant, **power** is often used for **active power.**

E270-0

power amplification (1) (general). The ratio of the power level at the output terminals of an amplifier to that at the input terminals. Also called **power gain.** *See also:* **amplifier; power gain.** E145-0

(2) (magnetic amplifier). The product of the voltage amplification and the current amplification. *See also:* **rating and testing magnetic amplifiers.** E107-0

power amplifier. *See:* **amplifier, power.**

power, apparent (rotating machinery). The product of the root-mean-square current and the root-mean-square voltage. *Note:* It is a scalar quantity equal to the magnitude of the phasor power. *See also:* **asynchronous machine; synchronous machine.** 0-31E8

power, apparent (polyphase circuit). At the terminals of entry of a polyphase circuit, a scalar quantity equal to the magnitude of the vector power. *Notes:* (1) In determining the apparent power, the reference terminal for voltage measurement shall be taken as the neutral terminal of entry, if one exists, otherwise as the true neutral point. (2) If the ratios of the components of the vector power, for each of the terminals of entry, to the corresponding apparent power are the same for every terminal of entry, the total apparent power is equal to the arithmetic apparent power for the polyphase circuit; otherwise the apparent power is less than the arithmetic apparent power. (3) If the voltages have the same wave form as the corresponding currents, the apparent power is equal to the amplitude of the phasor power. (4) Apparent power is expressed in volt-amperes when the voltages are in volts and the currents in amperes. E270-0

power, apparent (single-phase two-wire circuit). At the two terminals of entry of a single-phase two-wire circuit into a delimited region, a scalar equal to the product of the root-mean-square voltage between one terminal of entry and the second terminal of entry, considered as the reference terminal, and the root-mean-square value of the current through the first terminal. *Notes:* (1) Mathematically the apparent power U is given by the equation

$$U=EI$$
$$=(\pm)(E_1^2+E_2^2+\ldots+E_r^2+\ldots)^{1/2} \times (I_1^2+I_r^2+\ldots+I_q^2+\ldots)^{1/2}$$

in which E and I are the root-mean-square amplitudes of the voltage and current, respectively. E_r and I_q are the root-mean-square amplitudes of the rth harmonic of voltage and the qth harmonic of current, respectively. (2) If both the voltage and current are sinusoidal and of the same period, so that the distortion power is zero, the apparent power becomes

$$U=EI=E_1I_1$$

in which E_1 and I_1 are the root-mean-square amplitudes of voltage and current of the primitive period. The apparent power is equal to the amplitúde of the phasor power. (3) If the voltage and current are quasi-periodic and the amplitude of the voltage and current components are slowly varying, the apparent power at any instant may be taken as the value derived from the amplitudes of the components at that instant. (4) Apparent power is expressed in volt-amperes when the voltage is in volts and the current in amperes. Because apparent power has the property of magnitude only and its sign is ambiguous, it does not have a definite direction of flow. For convenience it is usually treated as positive. E270-0

power, available (1) (audio and electroacoustics). The maximum power obtainable from a given source by suitable adjustment of the load. *Note:* For a source that is equivalent to a constant sinusoidal electromotive force in series with an impedance independent of amplitude, the available power is the mean-square value of the electromotive force divided by four times the resistive part of the impedance of the source. *See also:* **transmission characteristics.** E151/E196/E270-0

(2) (at a port) The maximum power that can be transferred from the port to a load. *Note:* At a specified frequency, maximum power transfer will take place when the impedance of the load is the conjugate of that of the source. The source impedance must have a positive real part. *See also:* **network analysis.** E160-15E6

(3) (of a sound field with a given object placed in it). The power that would be extracted from the acoustic medium by an ideal transducer having the same dimensions and the same orientation as the given object. The dimensions and the orientation with respect to the sound field must be specified. *Note:* The acoustic power available to an electroacoustic transducer, in a plane-wave sound field of given frequency, is the product of the free-field intensity and the effective area of the transducer. For this purpose the effective area of an electroacoustic transducer, for which the surface velocity distribution is independent of the manner of excitation of the transducer, is $\frac{1}{4}\pi$ times the product of the receiving directivity factor and the square of the wavelength of a free progressive wave in the medium. If the physical dimensions of the transducer are small in comparison with the wavelength, the directivity factor is near unity, and the effective area varies inversely as the square of the frequency. If the physical dimensions are large in comparison with the wavelength, the directivity factor is nearly proportional to the square of the frequency, and the effective area approaches the actual area of the active face of the transducer. *See also:* **transmission characteristics; electroacoustics.** 0-1E1

(4) (signal generators). The power at the output port supplied by the generator into a specified load impedance. *See also:* **signal generator.** 0-9E4

power, average phasor (single-phase two-wire, or polyphase circuit). A phasor of which the real component is the average active power and the imaginary component is the average reactive power. The amplitude of the phasor power is

$$S_{av}=[(P_{av})^2+(Q_{av})^2]^{1/2}$$

in which P_{av} and Q_{av} are the active power and the reactive power, respectively. E270-0

power, carrier-frequency, peak pulse. *See:* **peak pulse power, carrier-frequency.**

power-circuit limit switch (industrial control). A limit switch the contacts of which are connected into the power circuit. *See also:* **switch.** 42A25-34E10

power-closed car door or gate (elevators). A door or gate that is closed by a car-door or gate power closer or by a door or gate power operator. *See also:* **elevators.** 42A45-0

power control center. The location where the area-control error of a control area is computed for the purpose of controlling area generation. *See also:* **power system, low-frequency and surge testing.** E94-0

power density. Value of the Poynting vector at a point in space. *See also:* **electromagnetic compatibility.** 0-27E1

power-density spectrum. A plot of power density per unit frequency as a function of frequency. *See also:* **electromagnetic compatibility.** 0-27E1

power detection. That form of detection in which the power output of the detecting device is used to supply a substantial amount of power directly to a device such as a loudspeaker or recorder. *See also:* **detection.** 42A65-0

power distribution, overhead construction. *Note:* For an extensive list of cross references, see *Appendix A.*

power distribution underground cables. *See:* **cable bedding; cable separator; base ambient temperature; aluminum-covered steel wire.**

power distribution, underground construction. *Note:* For an extensive list of cross references, see *Appendix A.*

power divider (waveguide). A device for producing a desired distribution of power at a branch point. *See also:* **waveguide.** 50I62-3E1

power, effective radiated (mobile communication). The product in a given direction of the effective gain of the antenna in that direction over a half-wave dipole antenna, and the antenna power input. *See also:* **mobile communication system.** 0-6E1

power elevator. An elevator utilizing energy other than gravitational or manual to move the car. *See also:* **elevators.** 42A45-0

power factor (1) (general). The ratio of total watts to the total root-mean-square (RMS) volt-amperes.

$$F_P = \frac{\Sigma \text{ Watts per Phase}}{\Sigma \text{ RMS Volt-Amperes per Phase}} = \frac{\text{Active Power}}{\text{Apparent Power}}$$

Note: If the voltages have the same waveform as the corresponding currents, power factor becomes the same as phasor power factor. If the voltages and currents are sinusoidal and, for polyphase circuits, form symmetrical sets, $F_P = \cos(\alpha - \beta)$. *See also:* **asynchronous machine; synchronous machine.** E270/50I05-10E1/31E8

(2) (rectifier or rectifier unit). The ratio of the total watts input to the total volt-ampere input to the rectifier, or rectifier unit. *Notes:* (A) This definition includes the effect of harmonic components of current and voltage, the effect of phase displacement between the current and voltage, and the exciting current of the transformer. Volt-ampere is the product of root-mean-square volts and root-mean-square amperes. (B) It is determined at the alternating-current line terminals of the rectifier unit. *See also:* **power rectifier; rectification.** 42A15-34E24

power-factor angle. The angle whose cosine is the power factor. *See also:* **asynchronous machine; synchronous machine.** 0-31E8

power factor, arithmetic. The ratio of the active power to the arithmetic apparent power. The arithmetic power factor is expressed by the equation

$$F_{pa} = \frac{P}{U_a}$$

where F_{pa} = arithmetic power factor
P = active power
U_a = arithmetic apparent power.

Note: Normally power factor, rather than arithmetic power factor, will be specified, but in particular cases, especially when the determination of the apparent power for a polyphase circuit is impracticable with the available instruments, arithmetic power factor may be used. When arithmetic power factor and power factor differ, arithmetic power factor is the smaller. E270-0

power-factor-corrected mercury-lamp ballast. A ballast of the multiple-supply type that has a power-factor-correcting device, such as a capacitor, so that the input current is at a power factor in excess of that of an otherwise comparable low-power-factor ballast design, but less than 90 percent, when the ballast is operated with center rated voltage impressed upon its input terminals and with a connected load, consisting of the appropriate reference lamp(s), operated in the position for which the ballast is designed. The minimum input power factor of such a ballast should be specifically stated. 82A9-0

power-factor influence (electric instruments). The change in the recorded value that is caused solely by a power-factor departure from a specified reference power factor maintaining constant power (or vars) at rated voltage, and not exceeding 120 percent of rated current. It is to be expressed as a percentage of the full-scale value. *See also:* **accuracy rating (instrument).** 39A1/39A2-0

power-factor meter. A direct-reading instrument for measuring power factor. It is provided with a scale graduated in power factor. *See also:* **instrument.** 42A30-0

power-factor tip-up (rotating-machinery stator-coil insulation). (1) The difference between the power-factors measured at two different designated voltages applied to an insulation system, other conditions being constant. *Note:* Used mainly as a measure of discharges, and hence of voids, within the system at the higher voltage. (2) The incremental change in power factor divided by incremental change in voltage applied to an insulation system. *See also:* **synchronous machine, asynchronous machine.** (3) Tip-up tests may be made using dissipation factor (tan δ) instead of power factor. In this case the tip-up is often identified as Δ tan δ or delta tan delta. E286-31E8

power-factor tip-up test (rotating machinery). A test applied to insulation to determine the power-factor tip-up. *See also:* **asynchronous machine; synchronous machine.** 0-31E8

power-factor–voltage characteristic (rotating machinery stator-coil insulation). The relation between the magnitude of the applied test voltage and the measured power factor of the insulation. The characteristic is usually shown as a curve of power factor plotted against test voltage. *See also:* **asynchronous machine; synchronous machine.** E286-31E8

power feeder. A feeder supplying principally a power or heating load. *See also:* **feeder.** 42A95-0

power flux density (electromagnetic). *See:* **power density.**

power-frequency current-interrupting rating (expulsion-type arrester). The range of minimum to maximum prospective currents within which the arrester is designed to operate at its rated voltage under prescribed conditions of rate-of-rise of restriking voltage, amplitude factor, and power-factor (or X/R ratio). *Notes:* (1) The follow current passed by the arrester will be lower than the prospective current when the arrester has current-limiting characteristics. (2) An expulsion arrester is given a maximum current-interrupting rating and may also have a minimum current-interrupting rating. *See:* **lightning arrester (surge diverter).** *See also:* **current rating, 60-hertz (arrester).** E28/99I2-31E7

power-frequency sparkover voltage (lightning arrester). The root-mean-square value of the lowest power-frequency sinusoidal voltage, applied between the terminals of an arrester, that causes sparkover of all the series gaps. *See:* **lightning arrester (surge diverter).** *See also:* **current rating, 60-hertz (arrester).** E28-31E7

power-frequency withstand voltage (lightning arresters). The root-mean-square value of a power-frequency sinusoidal voltage that the equipment is capable of withstanding without breakdown under specified conditions of test. *See:* **lightning arrester (surge diverter).** *See also:* **lightning protection and equipment.** E28/50I25-31E7

power fuse. A fuse consisting of an assembly of a fuse support and a fuse unit or fuseholder which may or may not include the refill unit or fuse link. *Note:* The power fuse is identified by the following characteristics: (1) Dielectric withstand (basic impulse insulation level) strengths at power levels; (2) Application primarily in stations and substations; (3) Mechanical construction basically adapted to station and substation mountings. 37A100-31E11

power gain (1) (general). The ratio of the signal power that a transducer delivers to its load to the signal power absorbed by its input circuit. *Notes:* (A) Power gain is usually expressed in decibels. (B) If more than one component is involved in the input or output, the particular components used must be specified. (C) If the output signal power is at a frequency other than the input signal power, the gain is a conversion gain. *See also:* **transmission characteristics.** E151/42A65-31E3

(2) (antenna) (in a given direction). 4π times the ratio of the radiation intensity in that direction to the net power accepted by the antenna from the connected transmitter. *Notes:* (A) When the direction is not stated, the power gain is usually taken to be the power gain in the direction of its maximum value. (B) Power gain does not include reflection losses arising from mismatch of impedance. (C) Power gain is fully realized on reception only when the incident polarization is the same as the polarization of the antenna on transmission. *See also:* **antenna; radiation; transmission characteristics.** 0-3E1

power, instantaneous (1) (circuit). At the terminals of entry into a delimited region the rate at which electric energy is being transmitted by the circuit into or out of the region. *Note:* Whether power into the region or out of the region is positive is a matter of convention and depends upon the selected reference direction of energy flow. *See:* **sign of power or energy.** E270-0

(2) (polyphase circuit). At the terminals of entry into a delimited region, the algebraic sum of the products obtained by multiplying the voltage between each terminal of entry and some arbitrarily selected common point in the boundary surface (which may be the neutral terminal of entry) by the current through the corresponding terminal of entry. *Notes:* (A) The reference direction for each current must be the same, either into or out of the delimited region. The reference polarity for each voltage must be consistently chosen, either with all the positive terminals at the terminals of entry and all negative terminals at the common reference point, or vice-versa. If the reference direction for currents is into the delimited region and the positive reference terminals for voltage are at the phase terminals of entry, the power will be positive when the energy flow is into the delimited region and negative when the flow is out of the delimited region. Reversal of either the reference direction or the reference polarity will reverse the relation between the sign of the power and the direction of energy flow. (B) When the circuit has a neutral terminal of entry, it is usual to select the neutral terminal as the common point for voltage measurement, because one of the voltages is then always zero, and, when both the currents and voltages form symmetrical polyphase sets of the same phase sequence, the average power for each single-phase circuit consisting of one phase conductor and the neutral conductor, will be the same. When the voltages and currents are sinusoidal and the voltages are measured to the neutral terminal of entry as the common point, the instantaneous power at the four points of entry of a three-phase circuit with neutral is given by

$$\begin{aligned} p = E_a I_a\ &[\cos(\alpha_a - \beta_a) + \cos(2\omega t + \alpha_a + \beta_a)] \\ + E_b I_b\ &[\cos(\alpha_b - \beta_b) + \cos(2\omega t + \alpha_b + \beta_b)] \\ + E_c I_c\ &[\cos(\alpha_c - \beta_c) + \cos(2\omega t + \alpha_c + \beta_c)] \end{aligned}$$

where E_a, E_b, E_c are the root-mean-square amplitudes of the voltages from the phase conductors *a*, *b*, and *c*, respectively, to the neutral conductor at the terminals of entry. I_a, I_b, I_c are the root-mean-square amplitudes of the currents in the phase conductors *a*, *b*, and *c*. α_a, α_b, α_c are the phase angles of the voltages E_a, E_b, E_c with respect to a common reference. β_a, β_b, β_c are the phase angles of the currents I_a, I_b, I_c with respect to the same reference as the voltages. (C) If there is no neutral conductor, so that there are only three terminals of entry, the point of entry of one of the phase conductors may be chosen as the common voltage point, and the voltage from that conductor to the common point becomes zero. If, in the preceding, the terminal of entry of phase conductor *b* is chosen as the common point, E_a is replaced by E_{ab} in the first line, E_c is replaced by E_{cb} in the third line, and the second line, being zero, is omitted. (D) If both the voltages and currents in the preceding equations constitute symmetrical polyphase sets of the same phase sequence, then $p = 3E_a I_a \cos(\alpha_a - \beta_a)$. Because this expression and similar expressions for m phases are independent of time, it follows that the instantaneous power is constant when the voltages and currents constitute polyphase symmetrical sets of the same phase sequence. (E) However, if the polyphase sets have single-phase symmetry or zero-phase symmetry rather than polyphase symmetry, the higher frequency terms do not cancel, and the instantaneous power is not a constant. (F) In general, the instantaneous power p at the $(m+1)$ terminals of entry of a polyphase circuit of m phases to a delimited area, when one of the terminals is that of the neutral conductor, is expressed by the equation

$$p = \sum_{s=1}^{s=m} e_s i_s$$

$$= \sum_{s=1}^{s=m} \sum_{r=1}^{r=\infty} \sum_{q=1}^{q=\infty} E_{sr} I_{sq} \Big[\cos[(r-q)\omega t + \alpha_{sr} - \beta_{sq}] + \cos[(r+q)\omega t + \alpha_{sr} + \beta_{sq}]\Big]$$

where e_s is the instantaneous alternating voltage between the sth terminal of entry and the terminal of voltage reference, which may be the true neutral point, the neutral conductor, or another point in the boundary surface. i_s is the instantaneous alternating current through the sth terminal of entry. E_{sr} is the root-mean-square amplitude of the rth harmonic of voltage e_s. I_{sq} is the root-mean-square amplitude of the qth harmonic of current i_s. α_{sr} is the phase angle of the rth harmonic of e_s with respect to a common reference. β_{sq} is the phase angle of the qth harmonic of i_s with respect to the same reference as the voltages. The index s runs through the phase letters identifying the m-phase conductor of an m-phase system, a, b, c, etcetera, and then concludes with the neutral conductor n if one exists. The indexes r and q identify the order of the harmonic term in each e_s and i_s, respectively, and run through all the harmonics present in the Fourier series representation of each alternating voltage and current. If the terminal of voltage reference is that of the neutral conductor, the terms for $s=n$ will vanish. If the voltages and current are quasi-periodic, of the form given in **power, instantaneous (two-wire circuit),** this expression is still valid but E_{sr} and I_{sq} become aperiodic functions of time. (G) Instantaneous power is expressed in watts when the voltages are in volts and the currents in amperes. *See:* **single-phase symmetrical set of polyphase voltages; zero-phase symmetrical set of polyphase voltage; single-phase symmetrical set of polyphase currents; zero-phase symmetrical set of polyphase currents.** E270-0

(3) (two-wire circuit). At the two terminals of entry into a delimited region, the product of the instantaneous voltage between one terminal of entry and the second terminal of entry, considered as the reference terminal, and the current through the first terminal. *Notes:* (A) The entire path selected for the determination of each voltage must lie in the boundary surface of the delimited region or be so selected that the voltage is the same as that along such a path. (B) Mathematically the instantaneous power p is given by $p=ei$ in which e is the voltage between the first terminal of entry and the second (reference) terminal of entry and i is the current through the first terminal of entry in the reference direction. (C) If both the voltage and current are sinusoidal and of the same period, the instantaneous power at any instant t is given by the equation

$$\begin{aligned} p = ei &= [(2)^{1/2} E \cos(\omega t+\alpha)]\,[(2)^{1/2} I \cos(\omega t+\beta)] \\ &= 2EI \cos(\omega t+\alpha)\cos(\omega t+\beta) \\ &= EI\,[\cos(\alpha-\beta) + \cos(2\omega t+\alpha+\beta)] \end{aligned}$$

in which E and I are the root-mean-square amplitudes of voltage and current, respectively, and α and β are the phase angles of the voltage and current, respectively, from the same reference. (D) If the voltage is an alternating voltage and the current is an alternating current of the same primitive period (see **alternating voltage, alternating current,** and **period (primitive period) of a function**), the instantaneous power is given by the equation

$$\begin{aligned} p &= ei \\ &= E_1 I_1 [\cos(\alpha_1-\beta_1) + \cos(2\omega t+\alpha_1+\beta_1)] \\ &+ E_2 I_2 [\cos(\omega t+\alpha_2-\beta_1) + \cos(3\omega t+\alpha_2+\beta_1)] \\ &+ E_1 I_2 [\cos(\omega t-\alpha_1+\beta_2) + \cos(3\omega t+\alpha_1+\beta_2)] \\ &+ E_2 I_2 [\cos(\alpha_2-\beta_2) + \cos(4\omega t+\alpha_2+\beta_2)] \\ &+ \ldots \end{aligned}$$

This equation can be written conveniently as a double summation

$$p = \sum_{r=1}^{r=\infty} \sum_{q=1}^{q=\infty} E_r I_q \{\cos[(r-q)\omega t + \alpha_r - \beta_q] + \cos[(r+q)\omega t + \alpha_r - \beta_q]\}$$

in which r is the order of the harmonic component of the voltage and q is the order of the harmonic component of the current (see **harmonic components (harmonics)**), and E, I, α, and β apply to the harmonic denoted by the subscript. (E) If the voltage and current are quasi-periodic functions of the form

$$e = (2)^{1/2} \sum_{r=1}^{r=\infty} E_r(t) \cos(r\omega t + \alpha_r)$$

$$i = (2)^{1/2} \sum_{q=1}^{q=\infty} I_q(t) \cos(q\omega t + \beta_r)$$

where $E_r(t)$, $I_q(t)$ are aperiodic functions of t, the instantaneous power is given by the equation

$$\begin{aligned} p &= ei \\ &= E_1(t) I_1(t)\,[\cos(\alpha_1-\beta_1) + \cos(2\omega t+\alpha_1+\beta_1)] \\ &+ E_2(t) I_1(t)\,[\cos(\omega t+\alpha_2-\beta_1) + \cos(3\omega t+\alpha_2+\beta_1)] \\ &+ E_1(t) I_2(t)\,[\cos(\omega t-\alpha_1+\beta_2) + \cos(3\omega t+\alpha_1+\beta_2)] \\ &+ E_2(t) I_2(t)\,[\cos(\alpha_2-\beta_2) + \cos(4\omega t+\alpha_2+\beta_2)] \\ &+ \ldots \end{aligned}$$

(F) Instantaneous power is expressed in watts when the voltage is in volts and the current in amperes. *Note:*

See note 3 of **reference direction of energy.** The sign of the energy will be positive if the flow of power is in the reference direction and negative if the flow is in the opposite direction. E270-0

power inverter. A rectifier unit in which the direction of average energy flow is from the direct-current circuit to the alternating-current circuit. *See also:* **power rectifier; rectification.** 42A15-34E24

power klystron (microwave tubes). A klystron, usually an amplifier, with two or more cavities uncoupled except by the beam, designed primarily for power amplification or generation. *See also:* **microwave tube or valve.** 0-15E6

power level. The magnitude of power averaged over a specified interval of time. *Note:* Power level may be expressed in units in which the power itself is measured or in decibels indicating the ratio to a reference power. This ratio is usually expressed either in decibels referred to one milliwatt, abbreviated dBm, or in decibels referred to one watt, abbreviated dBW. *See also:* **level.** E151-42A65

power-line carrier. The use of radio-frequency energy, generally below 600 kilohertz, to transmit information over transmission lines whose primary purpose is the transmission of power.
See:
carrier bypass;
carrier-current coupling capacitor;
carrier-current drain coil;
carrier-current lead;
carrier-current line trap;
carrier-current line trap, single-frequency;
carrier-current line trap, two-frequency;
carrier-frequency choke coil, power line;
carrier reinsertion;
carrier start time;
carrier stop time;
carrier-terminal grounding switch;
carrier-terminal protective gap;
carrier test switch;
channel-failure alarm;
coupling, broadband, aperiodic;
coupling capacitor;
coupling, common;
coupling (phase-to-ground);
coupling (phase-to-phase);
drain coil;
frequency lock;
hybrid coil;
power-line carrier receiver;
power-line carrier relaying;
power-line carrier transmitter;
resonance curve, carrier current line trap. 0-31E3

power-line carrier receiver. A receiver for power-line carrier signals. *See also:* **power-line carrier.** 0-31E3

power-line carrier relaying. The use of power-line carrier for protective relaying. *See also:* **power-line carrier.** 0-31E3

power-line carrier transmitter. A device for producing radio-frequency power for purposes of transmission on power lines. *See also:* **power-line carrier.** 0-31E3

power loss (1) (from a circuit, in the sense that is is converted to another form of power not useful for the purpose at hand (for example, RI^2 loss). A physical quantity measured in watts in the meter-kilogram-second (MKS) system and having the dimensions of power. For a given R, it will vary with the current in R. E196-0

(2) (defined as the ratio of two powers). If P_o is the output power and P_i the input power of a transducer or network under specified conditions, P_i/P_o is a dimensionless quantity that would be unity if $P_o = P_i$. 42A65-31E1/31E3

(3) (logarithmic). Loss may also be defined as the logarithm, or a quantity directly proportional to the logarithm of a power ratio, such as P_o/P_i. Thus if loss $= 10 \log_{10}(P_o/P_i)$ the loss is zero when $P_o = P_i$. This is the standard for measuring loss in decibels. *Notes:* (A) It should be noted that in cases (2) and (3) the loss for a given linear system is the same whatever may be the power levels. Thus (2) and (3) give characteristics of the system, and do not depend, as (1) does, on the value of the current or other dependent quantity. (B) If more than one component is involved in the input or output, the particular components used must be specified. This ratio is usually expressed in decibels. (C) If the output signal power is at a frequency other than the input signal power, the loss is a conversion loss. *See also:* **transducer; transmission; transmission loss.** 0-42A65

(4) (electric instrument) (watt loss). In the circuit of a current- or voltage-measuring instrument, the active power at its terminals for end-scale indication. *Note:* For other than current- or voltage-measuring instruments, for example, wattmeters, the power loss of any circuit is expressed at a stated value of current or of voltage. *See also:* **accuracy rating (instrument).** 39A1/42A30-0

power, nonreactive (polyphase circuit). At the terminals of entry of a polyphase circuit, a vector equal to the (vector) sum of the nonreactive powers for the individual terminals of entry. *Note:* The nonreactive power for each terminal of entry is determined by considering each phase conductor and the common reference point as a single-phase circuit, as described for distortion power. The sign given to the distortion power in determining the nonreactive power for each single-phase circuit shall be the same as that of the total active power. Nonreactive power for a polyphase circuit has as its two rectangular components the active power and the distortion power. If the voltages have the same waveform as the corresponding currents, the magnitude of the nonreactive power becomes the same as the active power. Nonreactive power is expressed in volt-amperes when the voltages are in volts and the currents in amperes. E270-0

power, nonreactive (single-phase two-wire circuit). At the two terminals of entry of a single-phase two-wire circuit into a delimited region, a vector quantity having as its rectangular components the active power and the distortion power. Its magnitude is equal to the square root of the difference of the squares of the apparent power and the amplitude of the reactive power. Its magnitude is also equal to the square root of the sum of the squares of the amplitudes of the active power and the distortion power. If voltage and current have the same waveform, the magnitude of the nonreactive power is equal to the active power. The amplitude of the nonreactive power is given by the equation

$$N = (U^2 - Q^2)^{1/2} = (P^2 + D^2)^{1/2}$$

$$= \left\{ \sum_{r=1}^{r=\infty} \sum_{q=1}^{q=\infty} [E_r^2 I_q^2 - E_r E_q I_r I_q \sin(\alpha_r - \beta_r) \sin(\alpha_q - \beta_q)] \right\}^{1/2}$$

where the symbols are those in **power, apparent (single-phase two-wire circuit).** In determining the vector position of the nonreactive power, the sign of the distortion power component must be assigned arbitrarily. Nonreactive power is expressed in volt-amperes when the voltage is in volts and the current in amperes. *See:* **distortion power (single-phase two-wire circuit).** E270-0

power-operated door or gate (elevators). A hoistway door and/or a car door or gate that is opened and closed by a door or gate power operator. *See also:* **hoistway (elevator or dumbwaiter).** 42A45-0

power operation. Operation by other than hand power. 37A100-31E11

power output, instantaneous. The rate at which energy is delivered to a load at a particular instant. *See also:* **radio transmitter.** E145-0

power pack. A unit for converting power from an alternating-current or direct-current supply into alternating-current or direct-current power at voltages suitable for supplying an electronic device.
See:
anode power supply;
circuits and devices;
filament power supply;
grid voltage supply;
plate efficiency;
plate power input;
power supply, direct-current;
ripple;
tuned transformer. 42A65-0

power, peak pulse. *See:* **peak pulse power.**

power, phasor (polyphase circuit). At the terminals of entry of a polyphase circuit into a delimited region, a phasor (or plane vector) that is equal to the (phasor) sum of the phasor powers for the individual terminals of entry when the voltages are all determined with respect to the same arbitrarily selected common reference point in the boundary surface (which may be the neutral terminal of entry). The reference direction for the currents and the reference polarity for the voltages must be the same as for instantaneous power, active power, and reactive power. The phasor power for each terminal of entry is determined by considering each conductor and the common reference point as a single-phase, two-wire circuit and finding the phasor power for each in accordance with the definition of **power, phasor (single-phase two-wire circuit).** The phasor power **S** is given by $\mathbf{S} = P + jQ$ where P is the active power for the polyphase circuit and Q is the reactive power for the same terminals of entry. If the voltages and currents are sinusoidal and of the same period, the phasor power **S** for a three-phase circuit is given by

$$\mathbf{S} = \mathbf{E}_a \mathbf{I}_a^* + \mathbf{E}_b \mathbf{I}_b^* + \mathbf{E}_c \mathbf{I}_c^*$$

where $\mathbf{E}_a$, $\mathbf{E}_b$, and $\mathbf{E}_c$ are the phasor voltages from the phase conductors *a, b,* and *c*, respectively, to the neutral conductor at the terminals of entry. $\mathbf{I}_a$, $\mathbf{I}_b$, and $\mathbf{I}_c$ are the conjugate of the phasor currents in the phase conductors at the terminals of entry. If there is no neutral conductor, so that there are only three terminals of entry, the point of entry of one of the phase conductors may be chosen as the common voltage point, and the phasor from that conductor to the common voltage point becomes zero. If the terminal of entry of phase conductor b is chosen as the common point, the phasor power of a three-phase, three-wire circuit becomes

$$\mathbf{S} = \mathbf{E}_{ab} \mathbf{I}_a^* + \mathbf{E}_{cb} \mathbf{I}_c^*$$

where $\mathbf{E}_{ab}$, $\mathbf{E}_{cb}$ are the phasor voltages from phase conductor a to b and from c to b, respectively. If both the voltages and currents in the preceding equations constitute symmetrical sets of the same phase sequence $\mathbf{S} = 3\mathbf{E}_a\mathbf{I}_a^*$. In general the phasor power at the $(m+1)$ terminals of entry of a polyphase circuit of m phases to a delimited region, when one of the terminals is the neutral terminal of entry, is expressed by the equation

$$\mathbf{S} = \sum_{s=1}^{s=m} \sum_{r=1}^{r=\infty} \mathbf{E}_{sr} \mathbf{I}_{sr}^*$$

where $\mathbf{E}_{sr}$ is the phasor representing the rth harmonic of the voltage from phase conductor s to neutral at the terminals of entry. $\mathbf{I}_{sr}^*$ is the conjugate of the phasor representing the rth harmonic of the current through the sth terminal of entry. The phasor power can also be stated in terms of the symmetrical components of the voltages and currents as

$$\mathbf{S} = m \sum_{k=0}^{k=m-1} \sum_{r=1}^{r=\infty} \mathbf{E}_{kr} \mathbf{I}_{kr}^*$$

where $\mathbf{E}_{kr}$ is the phasor representing the symmetrical component of kth sequence of the rth harmonic of the line-to-neutral set of polyphase voltages at the terminals of entry. $\mathbf{I}_{kr}^*$ is the conjugate of the phasor representing the symmetrical component of the kth sequence of the rth harmonic of the polyphase set of currents through the terminals of entry. Phasor power is expressed in volt-amperes when the voltages are in volts and the currents in amperes. *Note:* This term was once defined as **vector power.** With the introduction of the term **phasor quantity,** the name of this term has been altered to correspond. The definition has also been altered to agree with the change in the sign of reactive power. *See:* **power, reactive (magner) (single-phase two-wire circuit); power, reactive (magner) (polyphase circuit).** E270-0

power, phasor (single-phase two-wire circuit). At the two terminals of entry of a single-phase two-wire circuit into a delimited region, a phasor (or plane vector) of which the real component is the active power and the imaginary component is the reactive power at the same two terminals of entry. When either component of phasor power is positive, the direction of that component is in the reference direction. The phasor power **S** is given by $\mathbf{S}=P+jQ$ where P and Q are the active and reactive power, respectively. If both the voltage and current are sinusoidal, the phasor power is equal to the product of the phasor voltage and the conjugate of the phasor current.

$$\mathbf{E} = Ee^{j\alpha}; \qquad \mathbf{I} = Ie^{j\beta};$$

the phasor power is

$$\begin{aligned}\mathbf{S} &= P + jQ = \mathbf{EI}^* = EIe^{j(\alpha-\beta)}\\ &= EI[\cos(\alpha-\beta) + j\sin(\alpha-\beta)]\,.\end{aligned}$$

If the voltage is an alternating voltage and the current is an alternating current, the phasor power for each harmonic component is defined in the same way as for the sinusoidal voltage and sinusoidal current. Mathematically the phasor power of the rth harmonic component $\mathbf{S}_r$ is given by

$$\begin{aligned}\mathbf{S}_r &= P_r+jQ_r = \mathbf{E}_r\mathbf{I}_r^* = E_rI_re^{j(\alpha r-\beta r)}\\ &= E_rI_r[\cos(\alpha_r-\beta_r) + j\sin(\alpha_r-\beta_r)]\,.\end{aligned}$$

The phasor power at the two terminals of entry of a single-phase two-wire circuit into a delimited region, for an alternating voltage and current, is equal to the (phasor) sum of the values of the phasor power for every harmonic. Mathematically, this relation may be expressed

$$\begin{aligned}\mathbf{S} &= \mathbf{S}_1+\mathbf{S}_2+\mathbf{S}_3 + \cdots = \Sigma\,\mathbf{S}_r\\ &= \mathbf{E}_1\mathbf{I}_1^*+\mathbf{E}_2\mathbf{I}_2^*+\mathbf{E}_3\mathbf{I}_3^* \cdots = \Sigma\,\mathbf{E}_r\mathbf{I}_r^*\\ &= (P_1+P_2+P_3+\cdots)\\ &\quad + j(Q_1+Q_2+Q_3+\cdots) = \Sigma\,(P_r+jQ_r).\end{aligned}$$

The amplitude of the phasor power is equal to the square root of the sum of the squares of the active power and the reactive power. Mathematically, if S is the amplitude of the phasor power and θ is the angle between the phasor power and the real-power axis,

$$\begin{aligned}\mathbf{S} &= Se^{j\theta}\\ S &= (P^2+Q^2)^{1/2} = [(P_1+P_2+P_3+\cdots)^2\\ &\qquad + (Q_1+Q_2+Q_3+\cdots)^2]^{1/2}\end{aligned}$$

$$\theta = \tan^{-1}\frac{Q}{P} = \tan^{-1}\frac{Q_1+Q_2+Q_3+\cdots}{P_1+P_2+P_3+\cdots}\,.$$

If the voltage and current are quasi-periodic and the amplitude of the voltage and current components are slowly varying, the phasor power may still be taken as the phasor having P and Q as its components, the values of P and Q being determined for these conditions, as specified in **power, active (single-phase two-wire circuit) (average power) (power)** and **power, reactive (magner) (single-phase two-wire circuit),** respectively. For this condition the phasor power will be a function of time. If the voltage and current have the same waveform, the amplitude of the phasor power is equal to the apparent power, but they are not the same for all other cases. The phasor power is expressed in volt-amperes when the voltage is in volts and the current in amperes. *Note:* This term was once defined as **vector power.** With the introduction of the term **phasor quantity,** the name of this term has been altered to correspond. The definition has also been altered to agree with the change in the sign of reactive power. *See:* **power, reactive (magner) (single-phase two-wire circuit).** *See also:* **alternating current.** E270-0

power primary detector (electric power systems). A power-measuring device for producing an output proportional to power input. *See also:* **speed-governing system.** E94-0

power quantities (single-phase three-wire circuit) and (two-phase circuit). The definitions of the power quantities for a single-phase circuit of more than two wires and of a two-phase circuit are essentially the same as those given for a polyphase circuit. *Note:* Where mathematical expressions involve m, the number of phases or phase conductors, the numeral 2 should be used for single-phase, three-wire systems, and the numeral 4 for two-phase, four-wire and five-wire systems. *See:* **polyphase symmetrical sets (polyphase voltages).** E270-0

power rating (waveguide attenuator). The maximum power that, if applied under specified conditions of environment and duration, will not produce a permanent change that causes any performance characteristics to be outside of specifications. This includes characteristic insertion loss and standing-wave ratio. *See also:* **waveguide.** 0-9E4

power rating or voltage rating (line and connectors) (coaxial transmission line). That value of transmitted power or voltage that permits satisfactory operation of the line assembly and provides an adequate safety factor below the point where injury or appreciably shortened life will occur. *See also:* **transmission line.** 83A14-0

power, reactive. The product of voltage and the out-of-phase component of alternating current. In a passive network, reactive power represents the alternating exchange of stored energy (inductive or capacitive) between two areas. *See also:* **power, reactive (magner) (polyphase circuit); power, reactive (magner) (single-phase two-wire circuit).** 0-31E8

power, reactive (magner) (polyphase circuit). At the terminals of entry into a delimited region, the algebraic sum of the reactive powers for the individual terminals of entry when the voltages are all determined with respect to the same arbitrarily selected common reference point in the boundary surface (which may be the neutral terminal of entry). The reference direction for the currents and the reference polarity for the voltages must be the same as for the instantaneous power and the active power. The reactive power for each terminal of entry is determined by considering each conductor and the common reference point as a single-phase two-wire circuit and finding the reactive power for each in accordance with the definition of **power, reactive (magner) (single-phase two-wire circuit).** If the voltages and currents are sinusoidal and of the same period, the reactive power Q for a three-phase circuit is given by

$$\begin{aligned}Q &= E_aI_a\sin(\alpha_a-\beta_a) + E_bI_b\sin(\alpha_b-\beta_b)\\ &\quad + E_cI_c\sin(\alpha_c-\beta_c)\end{aligned}$$

where the symbols have the same meaning as in **power, instantaneous (polyphase circuit).** If there is no neutral conductor and the common point for voltage measurement is selected as one of the phase terminals of entry, the expression will be changed in the same way as that for **power, instantaneous (polyphase circuit).** If both the voltages and currents in the preceding equations constitute symmetrical polyphase set of the same phase sequence

$$Q = 3E_aI_a \sin(\alpha_a - \beta_a).$$

In general the reactive power Q at the $(m+1)$ terminals of entry of a polyphase circuit of m phases to a delimited region, when one of the terminals is the neutral terminal of entry, is expressed by the equation

$$Q = \sum_{s=1}^{s=m} \sum_{r=1}^{r=\infty} E_{sr}I_{sr} \sin(\alpha_{sr} - \beta_{sr})$$

where the symbols have the same meaning as in **power, active (polyphase circuit).** The reactive power can also be stated in terms of the root-mean-square amplitudes of the symmetrical components of the voltages and currents as

$$Q = m \sum_{k=0}^{k=m-1} \sum_{r=1}^{r=\infty} E_{kr}I_{kr} \sin(\alpha_{kr} - \beta_{kr})$$

where the symbols have the same meaning as in **power, active (polyphase circuit).** When the voltages and currents are quasi-periodic and the amplitudes of the voltages and currents are slowly varying, the reactive power for the circuit of each conductor may be determined for this condition as in **power, reactive (magner) (single-phase two-wire circuit).** The reactive power for the polyphase circuit is the sum of the reactive power values for the individual conductors. Reactive power is expressed in vars when the voltages are in volts and the currents in amperes. *Note:* The sign of reactive power resulting from the above definition is the opposite of that given by the definition in the 1941 edition of the American Standard Definitions of Electrical Terms. The change has been made in accordance with a recommendation approved by the Standards Committee of the Institute of Electrical and Electronics Engineers, by the American National Standards Institute, and by the International Electrotechnical Commission. E270-0

power, reactive (magner) (single-phase two-wire circuit). At the two terminals of entry of a single-phase two-wire circuit into a delimited region, for the special case of a sinusoidal voltage and a sinusoidal current of the same period, is equal to the product obtained by multiplying the root-mean-square value of the voltage between one terminal of entry and the second terminal of entry, considered as the reference terminal, by the root-mean-square value of the current through the first terminal and by the sine of the angular phase difference by which the voltage leads the current. The reference direction for the current and the reference polarity for the voltage must be the same as for active power at the same two terminals. Mathematically the reactive power Q, for the case of sinusoidal voltage and current, is given by

$$Q = EI \sin(\alpha - \beta)$$

in which the symbols have the same meaning as in **power, instantaneous (two-wire circuit).** For the same conditions, the reactive power Q is also equal to the imaginary part of the product of the phasor voltage and the conjugate of the phasor current, or to the negative of the imaginary part of the product of the conjugate of the phasor voltage and the phasor current. Thus

$$\begin{aligned} Q &= \text{Im } \mathbf{EI}^* \\ &= -\text{Im } \mathbf{E}^*\mathbf{I} \\ &= \frac{1}{2j}[\mathbf{EI}^* - \mathbf{E}^*\mathbf{I}] \end{aligned}$$

in which $\mathbf{E}$ and $\mathbf{I}$ are the phasor voltage and phasor current, respectively, and * denotes the conjugate of the phasor to which it is applied. If the voltage is an alternating voltage and the current is an alternating current, the reactive power for each harmonic component is equal to the product obtained by multiplying the root-mean-square amplitude of that harmonic component of the voltage by the root-mean-square amplitude of the same harmonic component of the current and by the sine of the angular phase difference by which that harmonic component of the voltage leads the same harmonic component of the current. Mathematically the reactive power of the rth harmonic component of Q_r is given by

$$\begin{aligned} Q_r &= E_rI_r \sin(\alpha_r - \beta_r) \\ &= \text{Im } \mathbf{E}_r\mathbf{I}_r^* \\ &= \frac{1}{2j}[\mathbf{E}_r\mathbf{I}_r^* - \mathbf{E}_r^*\mathbf{I}_r] \\ &= -\text{Im } \mathbf{E}_r^*\mathbf{I}_r \end{aligned}$$

in which the symbols have the same meaning as in **power, instantaneous (two-wire circuit)** and **power, active (single-phase two-wire circuit) (average power) (power).** The reactive power at the two terminals of entry of a single-phase two-wire circuit into a delimited region, for an alternating voltage and current, is equal to the sum of the values of reactive power for every harmonic component. Mathematically the reactive power Q for an alternating voltage and current, is given by

$$\begin{aligned} Q &= Q_1 + Q_2 + Q_3 + Q_4 + \cdots + Q_r + \cdots \\ &= E_1I_1 \sin(\alpha_1 - \beta_1) + E_2I_2 \sin(\alpha_2 - \beta_2) + \cdots \\ &= \sum_{r=1}^{r=\infty} Q_r \\ &= \sum_{r=1}^{r=\infty} E_rI_r \sin(\alpha_r - \beta_r) \end{aligned}$$

in which the symbols have the same meaning as in **power, instantaneous (two-wire circuit).** If the voltage and current are quasi-periodic functions of the form given in **power, instantaneous (two-wire circuit),** and the amplitudes are slowly varying, so that each may be considered to be constant during any one period, but to have slightly different values in successive periods, the reactive power at any time t may be taken as

$$Q = \sum_{r=1}^{r=\infty} E_r(t)I_r(t) \sin(\alpha_r - \beta_r)$$

by analogy with the expression for active power. When the reactive power is positive, the direction of flow of quadergy is in the reference direction of energy flow. Because the reactive power for each harmonic may have either sign, the direction of the reactive power for a harmonic component may be the same as or opposite to the direction of the total reactive power. The value of reactive power is expressed in vars when the voltage is in volts and the current in amperes. *Notes:* (1) The sign of reactive power resulting from the above definition is the opposite of that given by the definition in the 1941 edition of the American Standard Definitions of Electrical Terms. The change has been made in accordance with a recommendation approved by the Standards Committee of the Institute of Electrical and Electronics Engineers, by the American National Standards Institute, and by the Electrotechnical Commission. (2) Any designation of positive reactive power as inductive reactive power and negative reactive power as capacitive reactive power is deprecated. If the reference direction is from the generator toward the load, reactive power is positive if the load is predominantly inductive and negative if the load is predominantly capacitive. Thus a capacitor is a source of quadergy and an inductor is a consumer of quadergy. Designations of two kinds of reactive power are unnecessary and undesirable. E270-0

power rectifier. A rectifier unit in which the direction of average energy flow is from the alternating-current circuit to the direct-current circuit.
See:
cascade rectifier;
ceiling direct voltage;
displacement power factor;
double-way rectifier;
form factor;
frequently repeated overload rating;
harmonic content;
inherent voltage regulation;
nonreversible power converter;
parallel rectifier;
positive nonconducting period;
power factor;
power inverter;
rectification;
reverse voltage dividers;
reversible power converter;
ripple amplitude;
semiconductor rectifier stack;
single-way rectifier;
total voltage regulation;
water treatment equipment. 42A15-34E24

power relay. (1) A relay that responds to power flow in an electric circuit. 37A100-31E11
(2) A relay that responds to a suitable product of voltage and current in an electric circuit. *See also:* **relay.** 0-31E6

power response (close-talking pressure-type microphone). The ratio of the power delivered by a microphone to its load, to the applied sound pressure as measured by a Laboratory Standard Microphone placed at a stated distance from the plane of the opening of the artificial voice. *Note:* The power response is usually measured as a function of frequency in decibels (dB) above 1 milliwatt per newton per square centimeter [mW/(N/m^2)] or 1 milliwatt per 10 microbars [mW/10μbar]. *See also:* **close-talking pressure-type microphone.** E258-0

power selsyn (synchros or selsyns). An inductive type of positioning system having two or more similar mechanically independent slip-ring machines with corresponding slip rings of all machines connected together and the stators fed from a common power source. *See also:* **synchro system.** 0-31E8

power storage. That portion of the water stored in a reservoir available for generating electric power. *See also:* **generating station.** 42A35-31E13

power supply, direct-current (power-system communication) (1) (direct-current general). A device for converting available electric service energy into direct-current energy at a voltage suitable for electronic components. *See:* **power pack; power supply, direct-current regulated (power-system communication).** 0-31E3
(2) (alternating-current to direct-current). Generally, a device consisting of a transformer, rectifier, and filter for converting alternating current to a prescribed direct voltage or current. *Note:* For an extensive list of cross references, see *Appendix A.* KPSH-10E1

power supply, direct-current regulated. A direct-current power supply whose output voltage is automatically controlled to remain within specified limits for specified variations in supply voltage and load current. *See also:* **power supply, direct-current (power-system communication).** 0-31E3

power-supply voltage range (transmitter performance). The range of voltages over which there is not significant degradation in the transmitter or receiver performance. *See also:* **audio-frequency distortion.** 0-6E1

power switchboard. A type of switchboard including primary power-circuit switching and interrupting devices together with their interconnections. *Note:* Knife switches, fuses, and air circuit breakers are the commonly used switching and interrupting devices. 37A100-31E11

power system (electric). A group of one or more generating sources and/or connecting transmission lines operated under common management or supervision to supply load.
See:
computer control;
control metering point;
direct digital control;
frequency bias;
frequency bias setting;
hybrid control;
interconnected system;
time bias.
See also: **power system, low-frequency and surge testing.** E94-31E4

power system, low-frequency and surge testing. *Note:* For an extensive list of cross references, see *Appendix A.*

power transfer relay. A relay so connected to the normal power supply that the failure of such power supply causes the load to be transferred to another power supply. 42A42-0

power type relay. A term for a relay designed to have heavy-duty contacts usually rated 15 amperes or higher. Sometimes called a **contactor.** 0-21E0

power, vector (polyphase circuit). At the terminals of entry of a polyphase circuit, a vector of which the three rectangular components are, respectively, the active power, the reactive power, and the distortion power at the same terminals of entry. In determining the components, the reference terminals for voltage measurement shall be taken as the neutral terminal of entry, if one exists, otherwise as the true neutral point. The vector power is also the (vector) sum of the vector powers for the individual terminals of entry. The vector power for each terminal of entry is determined by considering each phase conductor and the common reference point as a single-phase circuit, as described for distortion power. The sign given to the distortion power in determining the vector power for each single-phase circuit shall be the same as that of the total active power. The magnitude of the vector power is the apparent power. If the voltages have the same waveform as the corresponding currents, the magnitude of the vector power is equal to the amplitude of the phasor power. Vector power is expressed in volt-amperes when the voltages are in volts and the currents in amperes. *See also:* **network analysis.** E270-0

power, vector (single-phase two-wire circuit). At the two terminals of entry of a single-phase two-wire circuit into a delimited region, a vector whose magnitude is equal to the apparent power, and the three rectangular components of which are, respectively, the active power, the reactive power, and the distortion power at the same two terminals of entry. Mathematically the vector power **U** is given by

$$\mathbf{U} = \mathbf{i}P + \mathbf{j}Q + \mathbf{k}D$$

where **i**, **j**, and **k** are unit vectors along the three perpendicular axes, respectively. P, Q, and D are the active power, reactive power, and distortion power, respectively. The direction cosines of the angles between the vector power **U** and the three rectangular axes are

$$\cos\phi = \frac{P}{U}$$

$$\cos\Psi = \frac{Q}{U}$$

$$\cos\theta = \frac{D}{U}$$

The magnitude of the vector power is the apparent power, or

$$U = (P^2 + Q^2 + D^2)^{1/2}$$

$$= \left(\sum_{r=1}^{r=\infty} \sum_{q=1}^{q=\infty} E_r^2 I_q^2\right)^{1/2}$$

where the symbols are those of the preceding definitions. The geometric power diagram shows the relationships among the different types of power. Active power, reactive power, and distortion power are represented in the directions of the three rectangular axes. The accompanying diagram corresponds to a case in which all three are positive.

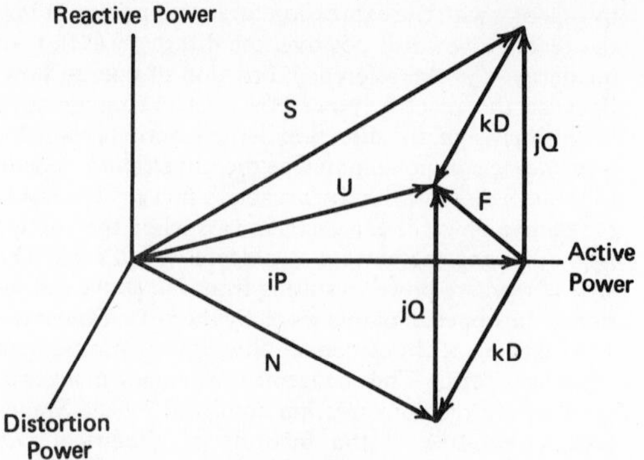

Vector power.

Since the sign of D is not definitely determined, **k**D may be drawn in either direction along the **k** axis. The position of **U** is thus also ambiguous, as it may occupy either of two positions, for D positive or negative. When the sign of D has been assumed, the vector positions of the fictitious power **F** and the nonreactive power **N** are determined. They have been shown in the figure, with the assumption that D has the same sign as P. Vector power is expressed in volt-amperes when the voltage is in volts and the current in amperes. *Notes:* (1) The vector power becomes a plane vector having the same magnitude as the phasor power if the voltage and the current have the same wave form. This condition is fulfilled as a special case when the voltage and current are sinusoidal and of the same period. (2) The term **vector power** as defined in the 1941 edition of the American Standard Definitions of Electrical Terms has now been called **phasor power** (*see* **power, phasor (single-phase two-wire circuit)**) and the present definition of vector power is new. *See also:* **network analysis.** E270-0

power winding (saturable reactor). A winding to which is supplied the power to be controlled. Commonly the functions of the output and power windings are accomplished by the same winding, which is then termed the output winding. *See also:* **magnetic amplifier.** 42A65-0

Poynting's vector. If there is a flow of electromagnetic energy into or out of a closed region, the rate of flow of this energy is, at any instant, proportional to the surface integral of the vector product of the electric field strength and the magnetizing force. This vector product is called Poynting's vector. If the electric field strength is **E** and the magnetizing force is **H**, then Poynting's vector is given by

$$\mathbf{U} = \mathbf{E} \times \mathbf{H} \quad \text{and} \quad \mathbf{U} = \mathbf{E} \times \mathbf{H}/4\pi$$

in rationalized and unrationalized systems, respectively. Poynting's vector is often assumed to be the local surface density of energy flow per unit time. E270-0

PPI (electronic navigation). *See:* **plan-position indicator.**

PPM. *See:* **periodic permanent-magnet focusing.**

PPM. *See:* **pulse-position modulation.**

preamplifier. An amplifier connected to a low-level signal source to present suitable input and output impedances and provide gain so that the signal may be further processed without appreciable degradation in the signal-to-noise ratio. *Notes:* (1) A preamplifier may include provision for equalizing and/or mixing. (2) Further processing frequently includes further amplification in a main amplifier. *See also:* **amplifier.** E151/42A65-31E3

precision (1) (general). The quality of being exactly or sharply defined or stated. A measure of the precision of a representation is the number of distinguishable alternatives from which it was selected, which is sometimes indicated by the number of significant digits it contains. *See also:* **accuracy; double precision; electronic analog computer; electronic digital computer.** E162-0;E162;13E6;E165;0/E270-34E10

(2) (measurement process). The quality of coherence or repeatability of measurement data, customarily expressed in terms of the standard deviation of the extended set of measurement results from a well-defined (adequately specified) measurement process in a state of statistical control. The standard deviation of the conceptual population is approximated by the standard deviation of an extended set of actual measurements. *See also:* **accuracy; reproducibility.** E172-10E6/9E3

(3) (measurement) (transmission lines and waveguides). The degree of combined repeatability and resolution. *See also:* **measurement system.** 0-9E4

precision approach radar (PAR). A radar system located on an airfield for observation of the position of an aircraft with respect to an approach path and specifically intended to provide guidance to the aircraft in the approach. *See also:* **navigation.** E172-10E6

precision wound (rotating machinery). A coil wound so that maximum nesting of the conductors occurs, usually with all crossovers at one end, with conductor aligned and positioned with respect to each adjacent conductor. *See also:* **rotor (rotating machinery); stator.** 0-31E8

preconditioning (industrial control). A control-function that provides for manually or automatically establishing a desired condition prior to normal operation of the system. *See also:* **control system, feedback.** IC1-34E10

precursor. *See:* **undershoot.**

predicted reliability. *See:* **reliability, predicted.**

predissociation. A process by which a molecule that has absorbed energy dissociates before it has had an opportunity to lose energy by radiation. *See also:* **gas-filled radiation-counter tubes.** 42A70-15E6

predistortion (pre-emphasis) (system) (transmitter performance). A process that is designed to emphasize or de-emphasize the magnitude of some frequency components with respect to the magnitude of others. *See:* **pre-emphasis.** 0-6E1

pre-emphasis (pre-equalization) (general). A process in a system designed to emphasize the magnitude of some frequency components with respect to the magnitude of others, to reduce adverse effects, such as noise, in subsequent parts of the system. *Note:* After transmitting the pre-emphasized signal through the noisy part of the system, de-emphasis may be applied to restore the original signal with a minimum loss of signal-to-noise ratio. *See also:* **circuits and devices; modulating systems.** 0-42A65

(2) (recording). An arbitrary change in the frequency response of a recording system from its basic response (such as constant velocity or amplitude) for the purpose of improvement in signal-to-noise ratio, or the reduction of distortion. E157-1E1

pre-emphasis network. A network inserted in a system in order to emphasize one range of frequencies with respect to another. *See also:* **network analysis.** E145-2E2

pre-envelope. *See:* **analytic signal.**

preference (power-system communication) (1) (channel supervisory control). An assembly of devices arranged to prevent the transmission of any signals over a channel other than supervisory control signals when supervisory control signals are being transmitted. *See also:* **supervisory control system.** 0-31E3

(2) (protective relaying). An assembly of devices arranged to prevent the transmission of any signals other than protective relaying signals over a channel when protective relaying signals are being transmitted. *See also:* **relay.** 0-31E3

preferred basic impulse insulation level (insulation strength). A basic impulse insulation level that has been adopted as a preferred American National Standard voltage value. *See:* **basic impulse insulation level (BIL) (insulation strength).** 92A1-0

prefix multipliers. The prefixes listed in the following table, when applied to the name of a unit, serve to form the designation of a unit greater or smaller than the original by the factor indicated.

Prefix	Abbreviation	Factor
tera- (megamega-*)	T (MM*)	10^{12}
giga- (kilomega-*)	G (kM*)	10^{9}
mega-	M	10^{6}
myria-		10^{4}
kilo-	k	10^{3}
hecto-	h	10^{2}
deka-		10
deci-	d	10^{-1}
centi-	c	10^{-2}
milli-	m	10^{-3}
decimilli-	dm	10^{-4}
micro-	μ	10^{-6}
nano- (millimicro-*)	n (mμ*)	10^{-9}
pico- (micromicro-*)	p ($\mu\mu$*)	10^{-12}

*Deprecated E270-0

preform (biscuit*) (disk recording) (electroacoustics). A small slab of record stock material as it is prepared for use in the record presses. *See also:* **phonograph pickup.**

*Deprecated E157-1E1

preformed coil or coil side (rotating machinery). An element of a preformed winding, composed of conductor strands, usually insulated and sometimes transposed, cooling ducts in some designs, turn insulation where number of turns exceeds one, and coil insulation. *See also:* **rotor (rotating machinery); stator.** 0-31E8

preformed winding (rotating machinery). A winding consisting of complete form-wound coils that are given their final shape before being assembled in the machine. *See also:* **asynchronous machine; direct-current commutating machine; synchronous machine.** 0-31E8

preheating time (mercury-arc valve). The time required for all parts of the valve to attain operating temperature. *See also:* **gas-filled rectifier.** 50I07-15E6

preheat-starting (fluorescent lamps) (switch-starting systems). The designation given to those systems in which hot-cathode electric discharge lamps are started from preheated cathodes through the use of a starting switch, either manual or automatic in its operation. *Note:* The starting switch, when closed, connects the two cathodes in series in the ballast circuit so that current flows to heat the cathodes to emission temperature. When the switch is opened, a voltage surge is produced that initiates the discharge. Only the arc current flows through the cathodes after the lamp is in operation. *See also:* **fluorescent lamp.** 82A1/Z7A1-0

preparatory function (numerically controlled machines). A command changing the mode of operation of the control such as from positioning to contouring or calling for a fixed cycle of the machine. *See also:* **numerically controlled machines.** EIA3B-34E12

prepatch panel. *See:* **problem board.**

preregister operation (elevators). Operation in which signals to stop are registered in advance by buttons in the car and at the landings. At the proper point in the car travel, the operator in the car is notified by a signal, visual, audible, or otherwise, to initiate the stop, after which the landing stop is automatic. *See also:* **control.** 42A45-0

preselector. (1) A device placed ahead of a frequency converter or other device, that passes signals of desired frequencies and reduces others. (2) In automatic switching, a device that performs its selecting operation before seizing an idle trunk. *See also:* **telephone switching system; circuits and devices.** 42A65-0

preset. To establish an initial condition, such as the control values of a loop. *See also:* **electronic digital computer.** X3A12-16E9

preset guidance. That form of missile guidance wherein the control mechanism is set, prior to launching, for a predetermined path, with no provision for subsequent adjustment. *See also:* **guided missile.** 42A65-0

preset speed (industrial control). A control function that establishes the desired operating speed of a drive before initiating the speed change. *See also:* **electric drive.** IC1-34E10

preshoot (pulse techniques). *See:* **distortion pulse.**

pressing (disk recording). A pressing is a record produced in a record-molding press from a master or stamper. *See also:* **phonograph pickup.** E157-1E1

pressure cable. An oil-impregnated paper-insulated cable in which positive gauge pressure is maintained on the insulation under all operating conditions.
See:
gas-filled cable;
gas-filled pipe cable;
oil-filled cable;
oil-filled pipe cable;
pipe cable;
self-contained pressure cable.
See also: **power distribution, underground construction.** 42A35-31E13

pressure-containing terminal box (rotating machinery). A terminal box so designed that the products of an electric breakdown within the box are completely contained inside the box. *See also:* **cradle base (rotating machinery).** 0-31E8

pressure-lubricated bearing (rotating machinery). A bearing in which a continuous flow of lubricant is forced into the space between the journal and the bearing. *See also:* **bearing.** 0-31E8

pressure microphone. A microphone in which the electric output substantially corresponds to the instantaneous sound pressure of the impressed sound waves. *Note:* A pressure microphone is a gradient microphone of zero order and is nondirectional when its dimensions are small compared to a wavelength. *See also:* **microphone.** 42A65-0

pressure relay. A relay that responds to liquid or gas pressure. 37A100-31E11/31E6

pressure-relief device (arrester). A means for relieving internal pressure in an arrester and preventing explosive shattering of the housing, following prolonged passage of follow current or internal flashover of the arrester. *See also:* **lightning arrester (surge diverter).** 0-31E7

pressure-relief terminal box (rotating machinery). A terminal box so designed that the products of an electric breakdown within the box are relieved through a pressure-relief diaphragm. *See also:* **cradle base (rotating machineery).** 0-31E8

pressure-relief test (lightning arresters). A test made to ascertain that an arrester failure will not cause explosive shattering of the housing. *See also:* **lightning arrester (surge diverter).** 0-31E7

pressure switch (industrial control). A switch in which actuation of the contacts is effected at a predetermined liquid or gas pressure. 50I16-34E10

pressure system (protective signaling). A system for protecting a vault by maintaining a predetermined differential in air pressure between the inside and outside of the vault. Equalization of pressure resulting from opening the vault or cutting through the structure initiates an alarm condition in the protection circuit. *See also:* **protective signaling.** 42A65-0

pressure-type pothead. A pressure-type pothead is a pothead intended for use on positive-pressure cable systems. *See:* **multipressure zone pothead; single pressure zone potheads.** E48-0

pressure wire connector. A device that establishes the connection between two or more conductors or between one or more conductors and a terminal by means of mechanical pressure and without the use of solder. *See also:* **interior wiring.** 42A95-0

prestrike current (lightning). The current that flows in a lightning stroke prior to the return stroke current. *See also:* **direct-stroke protection (lightning).** 0-31E13

pretersonic. Ultrasonic and with frequency higher than 500 megahertz. 0-20E0

pretransmit-receive tube. A gas-filled radio-frequency switching tube used to protect the transmit-receive tube from excessively high power and the receiver from frequencies other than the fundamental. *See also:* **gas tubes.** E160-15E6

PRF. *See:* **pulse-repetition frequency.**

primaries (color). The colors of constant chromaticity and variable amount that, when mixed in proper proportions, are used to produce or specify other colors. *Note:* Primaries need not be physically realizable. *See also:* **color terms.** E201-2E2

primary (1) (electric machines and devices). The part of a machine having windings that are connected to the power supply line (for a motor or transformer) or to the load (for a generator). 0-31E8

(2) (color technology). Any one of three lights in terms of which a color is specified by giving the amounts required to duplicate it by additive combination. *See also:* **color.** Z7A1-0

(3) (adjective). (A) First to operate; for example, primary arcing contacts, primary detector; (B) First in preference; for example, primary protection; (C) Referring to the main circuit as contrasted to auxiliary or control circuits, for example, primary disconnecting devices; (D) Referring to the energy input side of transformers, or the conditions (voltages) usually encountered at this location; for example, primary unit substation. 37A100-31E11

primary arcing contacts (switching device). The contacts on which the initial arc is drawn and the final current, except for the arc-shunting-resistor current, is interrupted after the main contacts have parted. 37A100-31E11

primary battery. *See also:* **battery (primary or secondary); electrochemistry; primary cell.**

primary cell. A cell that produces electric current by electrochemical reactions without regard to the reversibility of those reactions. Some primary cells are reversible to a limited extent. *See also:* **electrochemistry.** 42A60-0

primary center (telephony). A toll switching point to which toll centers and toll points may be connected. Primary centers are classified as Class-3 offices. *See also:* **telephone switching system.** 0-19E1

primary circuit (voltage regulator). The circuit on the input side at the regulator. *See also:* **voltage regulator.** 57A15-0

primary-color unit (television). The area within a color cell occupied by one primary color. *See also:* **beam tubes; television.** E160-2E1/15E6

primary current ratio (electroplating). The ratio of the current densities produced on two specified parts of an electrode in the absence of polarization. It is equal to the reciprocal of the ratio of the effective resistances from the anode to the two specified parts of the cathode. *See also:* **electroplating.** 42A60-0

primary detector (sensing element) (initial element). The first system element or group of elements that responds quantitatively to the measurand and performs the initial measurement operation. A primary detector performs the initial conversion or control of measurement energy and does not include transformers, amplifiers, shunts, resistors, etcetera, when these are used as auxiliary means. *See also:* **measurement system.** 37A100/42A30-31E11

primary disconnecting devices (switchgear assembly). Self-coupling separable contacts provided to connect and disconnect the main circuits between the removable element and the housing. 37A100-31E11

primary distribution feeder. A feeder operating at primary voltage supplying a distribution circuit. *Note:* A primary feeder is usually considered as that portion of the primary conductors between the substation or point of supply and the center of distribution. 42A35-31E13

primary distribution mains. The conductors that feed from the center of distribution to direct primary loads or to transformers that feed secondary circuits. *See also:* **center of distribution.** 42A35-31E13

primary distribution network. A network consisting of primary distribution mains. *See also:* **center of distribution.** 42A35-31E13

primary distribution system. A system of alternating-current distribution for supplying the primaries of distribution transformers from the generating station or substation distribution buses. *See also:* **alternating-current distribution; center of distribution.** 42A35-31E13

primary distribution trunk line. A line acting as a main source of supply to a distribution system. *See also:* **center of distribution.** 42A35-31E13

primary electron (thermionics). An electron in a primary emission. *See also:* **electron emission.** 50I07-15E6

primary emission. Electron emission due directly to the temperature of a surface, irradiation of a surface, or the application of an electric field to a surface. 50I07-15E6

primary fault. The initial breakdown of the insulation of a conductor, usually followed by a flow of power current. *See also:* **center of distribution.** 42A35-31E13

primary flow (carriers). A current flow that is responsible for the major properties of the device. *See also:* **semiconductor device.** E102/42A70-0

primary line of sight. The line connecting the point of observation and the point of fixation. *See also:* **visual field.** Z7A1-0

primary line-to-ground voltage (coupling capacitors and capacitance potential devices). Refers to the high-tension root-mean-square line-to-ground voltage of the phase to which the coupling capacitors or potential device, in combination with its coupling capacitor or bushing, is connected. *See:* **rated primary line-to-ground voltage.** *See also:* **outdoor coupling capacitor.** E31-0

primary network. A network supplying the primaries of transformers whose secondaries may be independent or connected to a secondary network. *See:* **primary distribution network.** *See also:* **center of distribution.** 42A35-31E13

primary protection (relay system). First-choice relay protection in contrast with backup relay protection. 37A100-31E11/31E6

primary radar. A radar in which the reply signal is a portion of the transmitted energy reflected by the target. *See also:* **navigation.** 0-10E6

primary radiator. That portion of an antenna system that is energized by a transmitter either directly or through a feeder. *See also:* **antenna.** 50I62-3E1

primary reactor starter (industrial control). A starter that includes a reactor connected in series with the primary winding of an induction motor to furnish reduced voltage for starting. It includes the necessary switching mechanism for cutting out the reactor and connecting the motor to the line. *See also:* **starter.** IC1-34E10

primary resistor starter (industrial control). A starter that includes a resistor connected in series with the primary winding of an induction motor to furnish reduced voltage for starting. It includes the necessary switching mechanism for cutting out the resistor and connecting the motor to the line. *See also:* **starter.** 42A25-34E10

primary service area (radio broadcast transmitter). The area within which reception is not normally subject to objectionable interference or fading. *See also:* **radio transmitter.** 42A65-0

primary standards (illuminating engineering). A light source by which the unit of light is established and from which the values of other standards are derived. This order of standard also is designated as the national standard. *Note:* A satisfactory primary (national) standard must be reproducible from specifications. Primary (national) standards usually are found in national physical laboratories.
See:
comparison lamp;
lamp;
photometry;
radiometry;
reference ballast;
secondary standards;
spectroradiometer;
working standard. Z7A1-0

primary supply voltage (mobile communication). The voltage range over which a radio transmitter, a radio receiver, or selective signaling equipment is designed to operate without degradation in performance. *See also:* **mobile communication system.** 0-6E1

primary switchgear connections. *See:* **main switchgear connections.**

primary transmission feeder. A feeder connected to a primary transmission circuit. *See also:* **center of distribution.** 42A35-31E13

primary unit substation. *See:* **unit substation.**

primary voltage rating (transformer). The input-circuit voltage for which the primary winding is designed. *See also:* **duty.** 89A1-0

primary winding (1) (rotating machinery) (motor or generator). The winding carrying the current and voltage of incoming power (for a motor) or power output (for a generator). The choice of what constitutes a primary circuit is arbitrary for certain machines having bilateral power flow. In a synchronous or direct-current machine, this is more commonly called the armature winding. *See also:* **armature.** 0-31E8
(2) (voltage regulator). The shunt winding. *See also:* **voltage regulator.** 57A15-0
(3) (general). The winding on the energy input side. 42A15-31E12

prime (charge-storage tubes). To charge storage elements to a potential suitable for writing. *Note:* This is a form of erasing. *See also:* **charge-storage tube; television.** E158-15E6

prime power. The maximum potential power (chemical, mechanical, or hydraulic) constantly available for transformation into electric power. *See also:* **generating station.** 42A35-31E13

priming rate (charge-storage tubes). The time rate of priming a storage element, line, or area from one specified level to another. Note the distinction between this and **priming speed.** *See also:* **charge-storage tube.** E158-15E6

priming speed (charge-storage tubes). The lineal scanning rate of the beam across the storage surface in priming. Note the distinction between this and **priming rate.** *See also:* **charge-storage tube.** E158-15E6

primitive period (function). *See:* **period (function).**

principal axis (1) (close-talking pressure-type microphone). The axis of a microphone normal to the plane of the principal acoustic entrance of a microphone, and that passes through the center of the entrance. *See also:* **close-talking pressure-type microphone.** E258-0
(2) (transducer used for sound emission or reception). A reference direction used in describing the directional characteristics of the transducer. It is usually an axis of structural symmetry, or the direction of maximum response, but if one of these does not coincide with the reference direction, it must be described explicitly. *See also:* **electroacoustics.** 0-1E1

principal current (thyristor). A generic term for the current through the collector junction. *Note:* It is the current through the main terminals.
See:
breakover current;
critical rate of rise of ON-state current;
forward current;
forward gate current;
gate nontrigger current;
gate trigger current;
gate turn-off current;
holding current;
OFF-state current;
ON-state current;
overload ON-state current;
peak value of the ON-state current;
reverse blocking current;
reverse breakdown current;
reverse current;
reverse gate current;
semiconductor device;
surge ON-state current. E223-34E17/34E24/15E7

principal voltage (thyristor). The voltage between the main terminals. *Note:* In the case of reverse-blocking and reverse-conducting thyristors, the principal voltage is called positive when the anode potential is higher than the cathode potential and called negative when the anode potential is lower than the cathode potential. *See also:* **principal voltage-current characteristic.** E223-34E17/34E24/15E7

principal voltage-current characteristic (principal characteristic) (thyristor). The function, usually represented graphically, relating the principal voltage to the principal current with gate current, where applicable, as a parameter.
See:
anode-to-cathode voltage;
anode-to-cathode voltage-current characteristic;
breakover point;
breakover voltage;
circuit-commutated turn-OFF time;
critical rate-of-rise of OFF-state voltage;
forward gate voltage;
gate-controlled delay time;
gate-controlled rise time;
gate-controlled turn-OFF time;
gate-controlled turn-ON time;
gate nontrigger voltage;
gate trigger voltage;
gate turn-OFF voltage;
gate voltage;

minimum ON-state voltage;
negative-differential-resistance region;
nonrepetitive peak OFF-state voltage;
nonrepetitive peak reverse voltage;
OFF impedance;
OFF state;
OFF-state voltage;
ON impedance;
ON state;
ON-state voltage;
principal voltage;
repetitive peak OFF-state voltage;
repetitive peak reverse voltage;
reverse blocking impedance;
reverse blocking state;
reverse breakdown voltage;
reverse gate voltage;
reverse recovery time;
reverse voltage;
semiconductor device;
thermal resistance;
transient thermal impedance;
working peak OFF-state voltage;
working peak reverse voltage. E223-34E17/15E7

printed circuit (soldered connections). A pattern comprising printed wiring formed in a predetermined design in, or attached to, the surface or surfaces of a common base. *See also:* **soldered connections (electronic and electrical applications).** 99A1-0

printed-circuit assembly. A printed-circuit board on which separately manufactured component parts have been added. 99A1-0

printed-circuit board (1) (general). A board for mounting of components on which most connections are made by printed circuitry. *See also:* **circuits and devices.** 0-31E3

(2) (double-sided). A board having printed circuits on both sides. *See also:* **circuits and devices.** 0-31E3

(3) (single-sided). A board having printed circuits on one side only. *See also:* **circuits and devices.** 0-31E3

printed wiring (soldered connections). A portion of a printed circuit comprising a conductor pattern for the purpose of providing point-to-point electric connections only. *See also:* **soldered connections (electronic and electrical applications).** 99A1-0

printer (teleprinter) (teletypewriter). A printing telegraph instrument having a signal-actuated mechanism for automatically printing received messages. It may have a keyboard similar to that of a typewriter for sending messages. The term receiving-only is applied to a printer having no keyboard. *See also:* **telegraphy.** 42A65-19E4

printing. *See:* **line printing.**

printing demand meter. An integrated demand meter that prints on a paper tape the demand for each demand interval and indicates the time during which the demand occurred. *See also:* **electricity meter (meter).** 42A30-0

printing recorder (protective signaling). An electromechanical recording device that accepts electric signal impulses from transmitting circuits and converts them to a printed record of the signal received. *See also:* **protective signaling.** 42A65-0

printing telegraphy. That method of telegraph operation in which the received signals are automatically recorded in printed characters. *See also:* **telegraphy.** 42A65-19E4

print-through. The undesirable transfer of a recorded signal from a section of a magnetic recording medium to another section of the same medium when these sections are brought into proximity. *Note:* The resulting copy usually is distorted. 0-1E1

priority string (power-system communication). A series connection of logic circuits such that inputs are accommodated in accordance with their position in the string, one end of the string corresponding to the highest priority. *See also:* **digital.** 0-31E3

privacy system (radio transmission). A system designed to make unauthorized reception difficult. *See also:* **radio transmission.** 42A65-0

private automatic branch exchange (PABX). A private branch exchange (PBX) in which connections are made by remotely controlled switches. *See also:* **telephone system.** 42A65-0

private automatic exchange (PAX) (telephony). A private telephone exchange in which connections are made by apparatus controlled from remote calling devices. *See also:* **telephone system.** 0-19E1

private branch exchange (PBX). A telephone exchange serving a single organization and having connections to a public telephone exchange. *See also:* **telephone system.** 42A65-19E1

private-branch-exchange trunk (PBX trunk). A subscriber line used as a trunk between a private branch exchange and the central office that serves it. *See also:* **telephone system.** 42A65-19E1

private exchange. A telephone exchange serving a single organization and having no means for connection with a public telephone exchange. *See also:* **telephone system.** 42A65-19E1

private residence. A separate dwelling or a separate apartment in a multiple dwelling that is occupied only by the members of a single family unit. 42A45-0

private residence elevator. A power passenger electric elevator, installed in a private residence, and that has a rated load not in excess of 700 pounds, a rated speed not in excess of 50 feet per minute, a net inside platform area not in excess of 12 square feet, and a rise not in excess of 50 feet. *See also:* **elevators.** 42A45-0

private-residence inclined lift. A power passenger lift, installed on a stairway in a private residence, for raising and lowering persons from one floor to another. *See also:* **elevator.** 42A45-0

probe (potential) (gas). An auxiliary electrode of small dimensions compared with the gas volume, that is placed in a gas tube to determine the space potential. *See also:* **discharge (gas).** 50I07-15E6

probe loading. The effect of a probe on a network, for example, on a slotted line, the loading represented by a shunt admittance or a discontinuity described by a reflection coefficient. *See also:* **measurement system.** 0-9E4

probe pickup, residual. *See:* **residual probe pickup.**

problem. *See:* **benchmark problem.**

problem board (patch board, patch panel, prepatch panel) (analog computer). A removable frame of receptacles, for patch cords and plugs that, through a patch bay, allows convenient preparation, use, and storage of problem interconnections. *See also:* **electronic analog computer.** E165-16E9

problem check. One or more tests used to assist in obtaining the correct machine solution to a problem.

Static check consists of one or more tests of computing elements, their interconnections, or both, performed under static conditions. **Dynamic check** consists of one or more tests of computing elements, their interconnections, or both, performed under dynamic conditions. **Rate test** is a test that verifies that the time constants of the integrators are correct. **Dynamic problem check** is any dynamic check used to ascertain that the computer solution satisfies the given system of equations. **Dynamic computer check** is any dynamic check used to ascertain the correct performance of some or all of the computer components. *See also:* **computer control state.** *See also:* **electronic analog computer.** E165-16E9

problem oriented language (computing systems). A programming language designed for the convenient expression of a given class of problems. X3A12-16E9

problem variable. A variable appearing in the mathematical model of the problem. *See also:* **electronic analog computer; scale factor.** E165-0

procedure (computing systems). The course of action taken for the solution of a problem. X3A12-16E9

procedure-oriented language (computing systems). A programming language designed for the convenient expression of procedures used in the solution of a wide class of problems. X3A12-16E9

process (control system). Synonymous with plant. 0-23E0

process control. Control imposed upon physical or chemical changes in a material. *See also:* **control system, feedback.** 85A1-23E0

processing. *See:* **data processing; information processing; multiprocessing; parallel processing.** *See also:* **electronic digital computer.**

processor (computing systems) (1) (hardware). A data processor.

(2) (software). A computer program that includes the compiling, assembling, translating, and related functions for a specific programming language, for example, **Cobol** processor, **Fortran** processor. *See:* **data processor; multiprocessor.** X3A12-16E9

production tests (switchgear). Those tests made to check the quality and uniformity of the workmanship and materials used in the manufacture of switchgear or its components. 37A100-31E11

product modulator. A modulator whose modulated output is substantially equal to the product of the carrier and the modulating wave. *Note:* The term implies a device in which intermodulation between components of the modulating wave does not occur. *See also:* **modulation.** 42A65-0

product relay. A relay that operates in response to a suitable product of two alternating electrical input quantities. *See also:* **relay.** 37A100-31E11

program (1) (general). A sequence of signals transmitted for entertainment or information. *See also:* **communication.** E151-42A65

(2) (electronic computation). (A) A plan for solving a problem. (B) Loosely, a routine. (C) To devise a plan for solving a problem. (D) Loosely, to write a routine. *See also:* **acceleration, programmed; communication; computer program; electronic digital computer; object program; source program; target program.** E162/E270-0;X3A12-16E9

program amplifier. *See:* **amplifier, line.**

program level. The magnitude of program in an audio system expressed in volume units. E151-0

program library (computing systems). A collection of available computer programs and routines. *See also:* **electronic digital computer.** X3A12-16E9

programmed check. A check procedure designed by the programmer and implemented specifically as a part of his program. *See:* **check, automatic; check problem; mathematical check.** *See also:* **electronic computation; electronic digital computer.** X3A12-16E9

programmed control (industrial control). A control system in which the operations are determined by a predetermined input program from cards, tape, plug boards, cams, etcetera. *See also* : **control system, feedback.** AS1-34E10

programmer. An arrangement of operating elements or devices that initiates, and often controls, one or a series of operations in a given sequence. 37A100-31E11

programming (1) (electronic computation). The ordered listing of a sequence of events designed to accomplish a given task. *See:* **linear programming; multiprogramming; automatic programming.** *See also:*

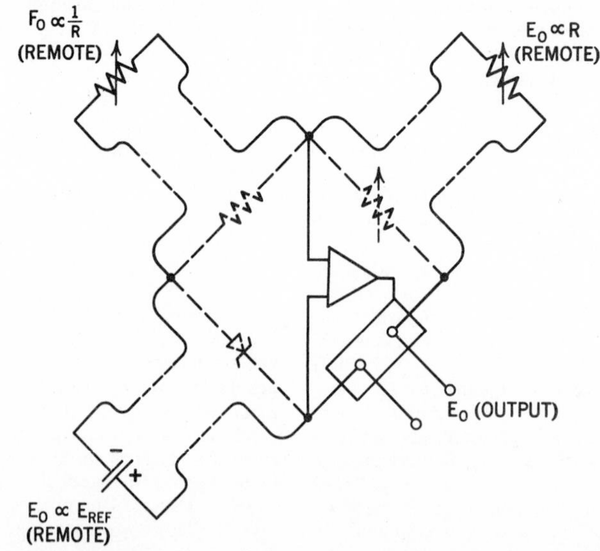

Remote programming connection showing programming of power supplies.

electronic digital computer. EIA3B-34E12
(2) (power supplies). The control of any power-supply functions, such as output voltage or current, by means of an external or remotely located variable control element. Control elements may be variable resistances, conductances, or variable voltage or current sources. *See also:* **power supply.** KPSH-10E1

programming language (computing systems). A language used to prepare computer programs. *See also:* **electronic digital computer.** X3A12-16E9

programming, linear (1) general. Optimization problem characterization in which a set of parameter values are to be determined, subject to given linear constraints, optimizing a cost function that is linear in the parameter. *See also:* **system.** 0-35E2
(2) (computing systems). The analysis or solution of problems in which linear function of a number of variables is to be maximized or minimized when those variables are subject to a number of constraints in the form of linear inequalities. *See also:* **electronic digital computer.** X3A12-16E9

programming, nonlinear. Optimization problem in which any or all of the following are nonlinear in the variables: (1) The objective functions. (2) The defining interrelationships among the variables, the plant description. (3) The constraints. *See also:* **system.** 0-35E2

programming, quadratic. Optimization problem in which: (1) The objective function is a quadratic function of the variable. (2) The plant description is linear. *See also:* **system.** 0-35E2

programming relay. A relay whose function is to establish or detect electrical sequences. 37A100-31E11/31E6

programming speed (power supplies). Describes the time required to change the output voltage of a power supply from one value to another. The output voltage must change across the load and because the supply's filter capacitor forms a resistance-capacitance network with the load and internal source resistance, programming speed can only be described as a function of load. Programming speed is the same as the recovery-time specification for current-regulated operation; it is not related to the recovery-time specification for voltage-regulated operation. *See also:* **power supply.** KPSH-10E1

program-sensitive fault (computing systems). A fault that appears in response to some particular sequence of program steps. *See also:* **electronic digital computer.** X3A12-16E9

program stop (numerically controlled machines). A miscellaneous function command to stop the spindle, coolant, and feed after completion of other commands in the block. It is necessary for the operator to push a button in order to continue with the remainder of the program. *See also:* **numerically controlled machines.** EIA3B-34E12

progressive scanning (television). A rectilinear process in which adjacent lines are scanned in succession. *See also:* **television.** E204-0

projected peak point (tunnel-diode characteristic). The point on the forward current-voltage characteristic where the current is equal to the peak-point current and where the voltage is greater than the valley-point voltage. *See also:* **peak point (tunnel-diode characteristic).** E253-15E7

projected peak-point voltage (tunnel-diode characteristic). The voltage at which the projected peak point occurs. *See also:* **peak point (tunnel-diode characteristic).** E253-15E7

projection tube (electron device). A cathode-ray tube specifically designed for use with an optical system to produce a projected image. *See also:* **tube definitions.** 0-15E6

projector. A lighting unit that, by means of mirrors and lenses, concentrates the light to a limited solid angle so as to obtain a high value of luminous intensity. *See also:* **luminaire.** Z7A1-0

proof (suffix). Apparatus is designated as splashproof, dustproof, etcetera, when so constructed, protected, or treated that its successful operation is not interfered with when subjected to the specified material or condition. 42A95/IC1-34E10;37A100-31E11/31E6

proof test (1) (general). A test made to demonstrate that the item of equipment is in satisfactory condition in one or more respects. E270-0
(2) (insulated winding) (withstand test). A fail or no-fail test of the insulation system of a rotating machine made to demonstrate whether the electrical strength of the insulation is above a predetermined minimum value. *See also:* **asynchronous machine; direct-current commutating machine; insulation testing (large alternating-current rotating machinery); synchronous machine.** E95-31E8

propagated potential (biological). A change of potential involving depolarization progressing along excitable tissue. *See also:* **medical electronics.** 0-18E1

propagation (electrical practice). The travel of waves through or along a medium. *See also:* **signal wave.** 42A65-0

propagation constant (1) (traveling electromagnetic wave in a homogeneous medium). The negative of the natural logarithmic partial derivative, with respect to distance in the direction of the wave normal, of the phasor quantity describing the wave. *Notes:* (A) In the case of cylindrical or spherical traveling waves, the amplitude factors $1/(r)^{1/2}$ and $1/r$, respectively, are not to be included in the phasor quantity. (B) This is a complex quantity whose real part is the attenuation constant in nepers per unit length and whose imaginary part is the phase constant in radians per unit length. *See also:* **radio transmission; radio wave propagation; transmission characteristics.** 0-3E2
(2) (transmission lines and transducers). (A) (per unit length of a uniform line). The natural logarithm of the ratio of the phasor current at a point of the line, to the phasor current at a second point, at unit distance from the first point along the line in the direction of transmission, when the line is infinite in length or is terminated in its characteristic impedance. **(B) (per section of a periodic line).** The natural logarithm of the ratio of the phasor current entering a section, to the phasor current leaving the same section, when the periodic line is infinite in length or is terminated in its iterative impedance. **(C) (of an electric transducer).** The natural logarithm of the ratio of the phasor current entering the transducer, to the phasor current leaving the transducer, when the transducer is terminated in its iterative impedance. *See also:* **transmission characteristics.** 42A65/E270-0

propagation factor (electromagnetic wave propagating from one point to another). The ratio of the complex electric field strength at the second point to that value that would exist at the second point if propagation took place in a vacuum. 0-3E2

propagation loss. The total reduction in radiant power surface density. The propagation loss for any path traversed by a point on a wave front is the sum of the spreading loss and the attenuation loss for that path. *See also:* **radio transmission.** E270-0

propagation model. An empirical or mathematical expression used to compute propagation path loss. *See also:* **electromagnetic compatibility.** 0-27E1

propagation vector (traveling electromagnetic wave at a given frequency). The complex vector whose real part is the attenuation vector and whose imaginary part is the phase vector. *See also:* **radio wave propagation.** 0-3E2

propeller-type blower (rotating machinery). An axial-flow fan with air-foil-shaped blades. *See also:* **fan (rotating machinery).** 0-31E8

proper operation. The functioning of the train control or cab signaling system to create or continue a condition of the vehicle apparatus that corresponds with the condition of the track of the controlling section when the vehicle apparatus is in operative relation with the track elements of the system. 42A42-0

proportional amplifier (industrial control). An amplifier in which the output is a single value and an approximately linear function of the input over its operating range. *See also:* **control system, feedback.** AS1-34E10

proportional control action (electric power systems). *See:* **control action, proportional.**

proportional counter tube. A radiation-counter tube designed to operate in the proportional region. *See also:* **gas-filled radiation-counter tubes.** 42A70-0

proportionality. *See:* **linearity.**

proportional region (radiation-counter tubes). The range of operating voltage for a counter tube in which the gas amplification is greater than unity and is independent of the amount of primary ionization. *Notes:* (1) In this region the pulse size from a counter tube is proportional to the number of ions produced as a result of the initial ionizing event. (2) The proportional region depends on the type and energy of the radiation. *See:* **gas-filled radiation-counter tubes.** 42A70-15E6

proprietary system (protective signaling). A local system sounding and/or recording alarm and supervisory signals at a control center located within the protected premises, the control center being under the supervision of employees of the proprietor of the protected premises. *Note:* According to the United States Underwriters' rules, a proprietary system must be a recording system. *See also:* **protective signaling.** 42A65-0

propulsion-control transfer switch. Apparatus in the engine room for transfer of control from engine room to bridge and vice versa. *Note:* Engine-room control is provided on all ships. Bridge control with a transfer switch is optional and is used principally on small vessels such as tugs or ferries, usually with a direct-current propulsion system. 42A43-0

propulsion set-up switch. Apparatus providing ready means to set up for operation under varying conditions where practicable; for example, cutout of one or more generators when multiple units are provided. *See also:* **electric propulsion system.** 42A43-0

prorated section (lightning arrester). A complete, suitably housed section of an arrester including series gaps and nonlinear series resistors or valve-element components in the same proportion as in the complete arrester. *See:* **arresters; lightning arrester (surge diverter).** E28-31E7

prorated unit (arrester). A completely housed prorated section of an arrester that may be connected in series with other prorated units to construct an arrester of higher voltage rating. *See also:* **lightning arrester (surge diverter).** 0-31E7

prospective current (available current) (1) (lightning arresters). The root-mean-square symmetrical short-circuit current that would flow at a given point in a circuit if the arrester(s) at that point were replaced by links of zero impedance. *See also:* **lightning arrester (surge diverter).** 99I2-31E7

(2) (circuit) (switching device situated therein). *See:* **available current (circuit) (switching device situated therein) (prospective current).**

prospective peak (crest) value (of a chopped impulse) (lightning arresters). The peak (crest) value of the full-wave impulse voltage from which a chopped impulse voltage is derived. *See also:* **lightning arrester (surge diverter).** 0-31E7

prospective short-circuit current (at a given point in a circuit). *See:* **available short-circuit current (at a given point in a circuit) (prospective short-circuit current).**

prospective short-circuit test current (at the point of test). *See:* **available short-circuit test current (at the point of test) (prospective short-circuit test current).**

protected enclosure. An enclosure in which all openings are protected with wire screen, expanded metal, or perforated covers. *Note:* A common form of specification for protected enclosure is: "The openings should not exceed ½ square inch (323 square millimeters) in area and should be of such shape as not to permit the passage of a rod larger than ½ inch (12.7 millimeters) in diameter, except where the distance of exposed live parts from the guard is more than 4 inches (101.7 millimeters) the openings may be ¾ square inch (484 square millimeters) in area and must be of such shape as not to permit the passage of a rod larger than ¾ inch (19 millimeters) in diameter." E45-0

protected location (computing systems). A storage location reserved for special purposes in which data cannot be stored without undergoing a screening procedure to establish suitability for storage therein. *See also:* **electronic digital computer.** X3A12-16E9

protected machine. *See:* **guarded machine.**

protected zone. *See:* **cone of protection.**

protection (computing systems). *See:* **storage protection.**

protective gap. A gap placed between live parts and ground to limit the maximum overvoltage that may occur. 37A100-31E11

protective lighting. A system intended to facilitate the nighttime policing of industrial and other properties. *See also:* **floodlighting.** Z7A1-0

protective margin (lightning arresters). The value of the protective ratio minus one expressed in percent ((PR − 1) × 100). *See also:* **lightning arrester (surge diverter).** 0-31E7

protective ratio (lightning arresters). The ratio of the insulation withstand characteristics of the protected equipment to the arrester protective level, expressed as a multiple of the latter figure. *See also:* **lightning ar-**

rester (surge diverter). 0-31E7

protective relay. A relay whose function is to detect defective lines or apparatus or other power-system conditions of an abnormal or dangerous nature and to initiate appropriate control circuit action. *Note:* A protective relay may be classified according to its input quantities, operating principle, or performance characteristics. 37A100-31E11/31E6

protective screen (burglar-alarm system). A lightweight barrier of either solid strip or lattice construction, carrying electric protection circuits, and barring access through a normal opening to protected premises. *See also:* **protective signaling.** 42A65-0

protective signaling. Protective signaling comprises the initiation, transmission, and reception of signals involved in the detection and prevention of property loss or damage due to fire, burglary, robbery, and other destructive conditions, and in the supervision of persons and of equipment concerned with such detection and prevention. *Note:* For an extensive list of cross references, see *Appendix A.* 42A65-0

protector tube (1) (lightning arresters). An expulsion arrester used primarily for the protection of line and switch insulation. *See:* **arresters; lightning arrester (surge diverter).** E28/62A1-31E7

(2) (electron-tube type). A glow-discharge cold-cathode tube that employs a low-voltage breakdown between two or more electrodes to protect circuits against over voltage. *See also:* **tube definitions.** 42A70-15E6

proton microscope. A device similar to the electron microscope but in which the charged particles are protons. *See also:* **electron optics.** 50I07-15E6

prototype standard. A concrete embodiment of a physical quantity having arbitrarily assigned magnitude, or a replica of such embodiment. *Note:* As an illustration of the distinction between prototype standard and unit, the length of the United States Prototype Meter Bar is not exactly one meter. E270-0

proximity effect (electric circuits and lines). The phenomenon of change of current distribution over the cross section of a conductor caused by the time variation of the current in a neighboring conductor. *See also:* **induction heating.** E54/E270-0

proximity-effect error (electronic navigation systems). An error in determination of system performance caused by improper use of measurements made in the near field of the antenna system. *See also:* **navigation.** 0-10E6

proximity-effect ratio (power distribution, underground cables). The quotient obtained by dividing the alternating-current resistance of a cable conductor subject to proximity effect, by the alternating-current resistance of an identical conductor free of proximity effect. *See also:* **power distribution, underground construction.** 0-31E1

proximity influence. The percentage change in indication caused solely by the fields produced from two edgewise instruments mounted in the closest possible proximity, one on each side (or above and below for horizontal-scale instruments). *Note:* Proximity influence of alternating-current instruments on either alternating-current or direct-current types is determined by energizing two instruments, one on each side of the test instrument (or above and below) at 90 percent of end-scale value (in phase with the current in the instrument under test, if the latter is alternating current). The current in the two outside instruments only shall be reversed. For rating purposes, the proximity influence shall be taken as one-half the difference in the readings in percentage of full scale. In direct-current permanent-magnet moving-coil instruments the field produced by the current in the instrument is small compared with the field from the permanent magnet. The proximity influence on either an alternating-current or direct-current test instrument will be the difference in reading, expressed as a percentage of full-scale value, of the instrument under test mounted alone on the panel, compared with the reading when two direct-current instruments are mounted in closest possible proximity, each with current applied to give 90-percent end-scale deflection. All three instruments shall be of the same manufacture and size. *See also:* **accuracy rating (instrument).** 39A1-0

proximity switch (industrial control). A device that reacts to the proximity of an actuating means without physical contact or connection therewith. *See also:* **switch.** IC1-34E10

***P* scan (electronic navigation).** *See:* **plan-position indicator.**

***P* scope (electronic navigation).** *See:* **plan-position indicator.**

pseudolatitude (navigation). A latitude in a coordinate system that has been arbitrarily displaced from the earth's conventional latitude system so as to move the meridian convergence zone (polar region) away from the place of intended operation. *See also:* **navigation.** E174-10E6

pseudolongitude (navigation). A longitude in a coordinate system that has been arbitrarily displaced from the earth's conventional longitude system so as to move the meridian convergence zone (polar region) away from the place of intended operation. *See also:* **navigation.** E174-10E6

pseudorandom number sequence. A sequence of numbers, determined by some defined arithmetic process, that is satisfactorily random for a given purpose, such as by satisfying one or more of the standard statistical tests for randomness. Such a sequence may approximate any one of several statistical distributions, such as uniform distribution or normal Gaussian distribution. *See also:* **electronic digital computer.** X3A12-16E9

PSK. *See:* **phase-shift keying.**

PTM. *See:* **pulse-time modulation.**

***p*-type crystal rectifier.** A crystal rectifier in which forward current flows when the semiconductor is positive with respect to the metal. *See also:* **rectifier.** 42A65-0

***p*-type semiconductor.** *See:* **semiconductor, *p*-type.**

public-address system. A system designed to pick up and amplify sounds for an assembly of people. 42A65-0

public telephone station (pay station). A station available for use by the public, generally on the payment of a fee that is deposited in a coin collector or is paid to an attendant. *See also:* **telephone station.** 42A65-0

pull box. A box with a blank cover that is inserted in one or more runs of raceway to facilitate pulling in the conductors, and may also serve the purpose of distributing the conductors. *See also:* **cabinet.** 42A95-0

pulley (sheave) (rotating machinery). A shaft-mounted wheel used to transmit power by means of a belt, chain, band, etcetera. *See also:* **rotor (rotating machinery).** 0-31E8

pulling eye. A device that may be fastened to the conductor or conductors of a cable or fomred by or fastened to the wire armor and to which a hook or rope may be directly attached in order to pull the cable into or from a duct. *Note:* Pulling eyes are sometimes equipped, like test caps, with facilities for oil feed or vacuum treatment. *See also:* **power distribution, underground construction.** 42A35-31E13

pulling figure (oscillator). The difference between the maximum and minimum values of the oscillator frequency when the phase angle of the load-impedance reflection coefficient varies through 360 degrees, while the absolute value of this coefficient is constant and equal to a specified value, usually 0.20. (Voltage standing-wave ratio 1.5.) *See also:* **oscillatory circuit; waveguide.** 42A65-0;E160-15E6

pulling into synchronism (rotating machinery). The process of synchronizing by changing from asynchronous speed to synchronous. *See also:* **synchronous machine.** 0-31E8

pulling out of synchronism (rotating machines). The process of losing synchronism by changing from synchronous speed to a lower asynchronous speed (for a motor) or higher asynchronous speed (for a generator). *See also:* **synchronous machines.** 0-31E8

pull-in test (synchronous machine). A test taken on a machine that is pulling into synchronism from a specified slip. *See also:* **synchronous machine.** 0-31E8

pull-in torque (synchronous motor). The maximum constant torque under which the motor will pull its connected inertia load into synchronism, at rated voltage and frequency, when its field excitation is applied. *Note:* The speed to which a motor will bring its load depends on the power required to drive it and whether the motor can pull the load into step from this speed depends on the inertia of the revolving parts, so that the pull-in torque cannot be determined without having the Wk^2 as well as the torque of the load. *See also:* **synchronous machine.** 42A10-31E8

pull or transfer box. A box without a distribution panel, within which one or more corresponding electric circuits are connected or branched. 43A35-31E13

pull-out test (rotating machinery). A test to determine the conditions under which an alternating-current machine develops maximum torque while running at specified voltage and frequency. *See also:* **asynchronous machine; synchronous machine.** 0-31E8

pull-out torque (synchronous machine). The maximum sustained torque which the machine will develop at synchronous speed with rated voltage applied at rated frequency and with normal excitation. *See also:* **synchronous machine.** 42A10-31E8

pull-up torque (alternating-current motor). The minimum torque developed by the motor during the period of acceleration from rest to the speed at which breakdown torque occurs with rated voltage applied at rated frequency. 0-31E8

pulsating function. A periodic function whose average value over a period is not zero. For example, $f(t)=A+B\sin\omega t$ is a pulsating function where neither A nor B is zero. E270-0

pulse (1) (general). A brief excursion of a quantity from normal. 42A65-31E3

(2)(modulation). A single disturbance characterized by the rise and decay in time or space or both of a quantity whose value is normally constant. *Notes:* (A) In these modulation definitions, a radio-frequency carrier, amplitude-modulated by a pulse, is not considered to be a pulse. (B) This definition is broad so that it covers almost any transient phenomenon. The only features common to all pulses are rise, finite duration, and decay. It is necessary that the rise, duration, and decay be of a quantity that is constant (not necessarily zero) for some time before the pulse and has the same constant value for some time afterwards. The quantity has a normally constant value and is perturbed during the pulse. No relative time scale can be assigned. E145/E203-0

(3) (industrial control). A signal of relatively short duration. AS1-34E10

(4) (pulse) (impulse*). A sudden change of brief duration produced in the current or voltage of a circuit in order to actuate or control an instrument, meter, or relay. *Note:* For an extensive list of cross references, see *Appendix A.*

*Deprecated 12A0-0

pulse accumulator (register) (telemeter system). A device that accepts and stores pulses and makes them available for readout on demand. 37A100-31E11

pulse amplifier (1) (pulse techniques). An amplifier designed specifically for the amplification of electric pulses. *See also:* **pulse.** E175-0

(2) (watthour meter). A device used to change the amplitude or waveform of a pulse for retransmission to another pulse device. *See also:* **auxiliary device to an instrument.** 12A0-0

pulse amplitude. A general term indicating the magnitude of a pulse. *Notes:* (1) For specific designation, adjectives such as average, instantaneous, peak, root-mean-square (effective), etcetera, should be used to indicate the particular meaning intended. (2) Pulse amplitude is measured with respect to the normally constant value unless otherwise stated. *See also:* **average absolute pulse-amplitude; average pulse amplitude; pulse terms; signal.** E194-2E2/13E6

pulse-amplitude modulation (PAM). Modulation in which the modulating wave is caused to amplitude modulate a pulse carrier. *See also:* **modulating systems; pulse terms.** E145/E194-19E4;42A65-0

pulse amplitude, peak. *See:* **peak pulse amplitude.**

pulse amplitude, root-mean-square (effective). *See:* **root-mean-square (effective) pulse amplitude.**

pulse bandwidth. The smallest continuous frequency interval outside of which the amplitude of the spectrum does not exceed a prescribed fraction of the amplitude at a specified frequency. Caution: This definition permits the spectrum amplitude to be less than the prescribed amplitude within the interval. *Notes:* (1) Unless otherwise stated, the specified frequency is that at which the spectrum has its maximum amplitude. (2) This term should really be pulse spectrum bandwidth because it is the spectrum and not the pulse itself that has a bandwidth. However, usage has caused the contraction and for that reason the term has been accepted. *See also:* **pulse terms; signal.** E194-2E2/13E6

pulse base (pulse waveform) (pulse techniques). That major segment having the lesser displacement in amplitude from the baseline, excluding major transitions. *See also:* **pulse.** 0-9E4

pulse, bidirectional. *See:* **bidirectional pulse.**

pulse carrier. A carrier consisting of a series of pulses. *Note:* Usually, pulse carriers are employed as subcarriers. *See also:* **carrier; pulse terms.** E145/E194/42A65-0

pulse, carrier-frequency. *See:* **carrier-frequency pulse.**

pulse code. (1) A pulse train modulated so as to represent information. (2) Loosely, a code consisting of pulses, such as Morse code, Baudot code, binary code. *See also:* **pulse; pulse terms.** E194-9E4

pulse-code modulation. A modulation process involving the conversion of a waveform from analog to digital form by means of coding. *Notes:* (1) This is a generic term, and additional specification is required for a specific purpose. (2) The term is commonly used to signify that form of pulse modulation in which a code is used to represent quantized values of instantaneous samples of the signal wave. *See also:* **modulating systems; pulse terms.** E170/E194/42A65-0

pulse coder (electronic navigation). A device for varying one or more of the characteristics of a pulse or of a pulse train so as to transmit information. *See also:* **navigation.** 0-10E6

pulse counter (pulse techniques). A device that indicates or records the total number of pulses that it has received during a time interval. *See also:* **pulse.** E175-0

pulsed Doppler radar. A pulsed radar system that utilizes the Doppler effect for obtaining velocity information relative to the target. *See also:* **navigation.** 0-10E6

pulse decay time (general). The interval between the instants at which the instantaneous amplitude last reaches specified upper and lower limits, namely, 90 percent and 10 percent of the peak pulse amplitude unless otherwise stated. *See also:* **pulse; pulse terms; television.** E175/E203-2E2;42A65-31E3

pulse decoder (electronic navigation). A device for extracting information from a pulse-coded signal. *See also:* **navigation.** 0-10E6

pulse delay, transducer. *See:* **delay, pulse.**

pulse delay, transmitter. *See:* **delay, pulse.**

pulse devices (watthour meter). The functional units for initiating, transmitting, retransmitting, and receiving electric pulses, representing finite quantities of energy normally transmitted from some form of watthour meter to a receiver unit for billing purposes. *See also:* **auxiliary device to an instrument.** 12A0-0

pulsed oscillator. An oscillator that is made to operate during recurrent intervals by self-generated or externally applied pulses. *See also:* **oscillatory circuit; pulse terms.** E145/42A65-0

pulsed-oscillator starting time. The interval between the leading-edge pulse time of the pulse at the oscillator control terminals and the leading-edge pulse time of the related output pulse. *See also:* **pulse terms.** E194-0

pulse droop (television). A distortion of an otherwise essentially flat-topped rectangular pulse characterized by a decline of the pulse top. *See also:* **pulse terms; television.** E194/E203-2E2

pulse duration (1) (loosely). The duration of a rectangular pulse whose energy and peak power equal those of the pulse in question. *Note:* When determining the peak power, any transients of relatively short duration are frequently ignored. *See also:* **phase modulation; pulse.** E145-0

(2) (television) (radiation counters) (telecommunications). The time interval between the first and last instants at which the instantaneous amplitude reaches a stated fraction of the peak pulse amplitude. *See also:* **pulse; pulse terms; signal.** E175/E194-2E2/13E6;E203/42A65-31E3/19E4

pulse-duration discriminator (electronic navigation). A circuit in which the output is a function of the deviation of the input pulse duration from a reference. *See also:* **navigation.** E172-10E6

pulse-duration modulation (PDM). Pulse-time modulation in which the value of each instantaneous sample of the modulating wave is caused to modulate the duration of a pulse. *Notes:* (1) The deprecated terms **pulse-width modulation** and **pulse-length modulation** also have been used to designate this system of modulation. (2) In pulse-duration modulation, the modulating wave may vary the time of occurrence of the leading edge, the trailing edge, or both edges of the pulse. *See also:* **pulse terms; modulating systems.** E145/E170/E194/42A65-0

pulse-duration telemetering (pulse-width modulation) (electric power systems). A type of telemetering in which the duration of each transmitted pulse is varied as a function of the magnitude of the measured quantity. *See also:* **telemetering.** E94-0

pulse duty factor. The ratio of the average pulse duration to the average pulse spacing. *Note:* This is equivalent to the product of the average pulse duration and the pulse-repetition rate. *See also:* **pulse terms.** E194-0

pulse-forming line (radar modulators). An electric network whose parameters are selected to give a specified shape to the modulation pulse. *See also:* **navigation.** 0-10E6

pulse-frequency modulation (PFM). A form of pulse-time modulation in which the pulse repetition rate is the characteristic varied. *Note:* A more precise term for pulse-frequency modulation would be pulse-repetition-rate modulation. *See also:* **modulation systems; pulse terms.** E170/E194-19E4

pulse, Gaussian. A pulse shape tending to follow the Gaussian curve corresponding to $A(t)=e^{-a(b-t)^2}$. See also: **pulse.** 0-9E4

pulse-height analyzer (radiation counters). An instrument capable of indicating the number or rate of occurrence of pulses falling within each of one or more specified amplitude ranges. *See also:* **pulse.** E175-0

pulse-height resolution constant, electron (photomultipliers). The product of the square of the electron (photomultiplier) pulse-height resolution expressed as the fractional full width at half maximum (FWHM/A_l), and the mean number of electrons per pulse from the photocathode. *See also:* **phototube.** E175-0

pulse-height resolution, electron (photomultiplier). A measure of the smallest change in the number of electrons in a pulse from the photocathode that can be discerned as a change in height of the output pulse. Quantitatively, it is the fractional standard deviation (σ/A_1) of the pulse-height distribution curve for output pulses resulting from a specified number of electrons per pulse from the photocathode. *Note:* The fractional full width at half maximum of the pulse-height distribution curve (FWHM/A_1) is frequently used as a measure of this resolution, where A_1 is the pulse height corresponding to the maximum of the distribution curve. *See also:* **pulse.** E175-0

pulse-height selector (pulse techniques). A circuit that produces a specified output pulse when and only when it receives an input pulse whose amplitude lies between two assigned values. *See:* **pulse.** E175-0

pulse initator. Any device, mechanical or electrical, used with a meter to initiate pulses, the number of which are proportional to the quantity being measured. *Note:* The complete pulse initiator may include an external amplifier or auxiliary relay assembly or both. *See also:* **auxiliary device to an instrument.** 12A0-0

pulse initiator gear ratio (watthour meter). The ratio of disk shaft revolutions to revolutions of pulse initiating shaft. *Note:* It is commonly denoted by the symbol P_g. *See also:* **auxiliary device to an instrument.** 12A0-0

pulse-initiator ratio (watthour meter). The ratio of revolutions of first gear of pulse device to revolutions of pulse initiating shaft. *Note:* It is commonly denoted by the symbol P_r. *See also:* **auxiliary device to an instrument.** 12A0-0

pulse-initiator shaft reduction (watthour meter). The ratio of meter disk shaft revolutions to the revolutions of the first gear of pulse initiator. *Note:* It is commonly denoted by the symbol P_s. *See also:* **auxiliary device to an instrument.** A12A0-0

pulse interleaving. A process in which pulses from two or more sources are combined in time-division multiplex for transmission over a common path. *See also:* **modulating systems; pulse terms.** E194/42A65-0

pulse interrogation. The triggering of a transponder by a pulse or pulse mode. *Note:* Interrogations by means of pulse modes may be employed to trigger a particular transponder or group of transponders. *See also:* **pulse terms.** E194-0

pulse interval.* *See:* **pulse spacing.**
*Deprecated

pulse-interval modulation. A form of pulse-time modulation in which the pulse spacing is varied. *See also:* **modulating systems; pulse terms.** E170/E194-0

pulse jitter. A relatively small variation of the pulse spacing in a pulse train. *Note:* The jitter may be random or systematic, depending on its origin, and is generally not coherent with any pulse modulation imposed. *See also:* **pulse terms.** E194-0

pulse-length modulation. *See:* **pulse-duration modulation.**

pulse mode. (1) A finite sequence of pulses in a prearranged pattern used for selecting and isolating a communication channel. (2) The prearranged pattern. *See also:* **pulse terms.** E194-0

pulse-mode multiplex. A process or device for selecting channels by means of pulse modes. *Note:* This process permits two or more channels to use the same carrier frequency. *See also:* **pulse terms.** E194-0

pulse mode, spurious. *See:* **spurious pulse mode.**

pulse-modulated radar. A form of radar in which the radiation consists of a series of discrete pulses. *See also:* **radar.** 42A65-0

pulse modulation (continuous-wave). (1) Modulation of a carrier by a pulse train. *Note:* In this sense, the term is used to describe the process of generating carrier-frequency pulses. (2) Modulation of one or more characteristics of a pulse carrier. *Note:* In this sense, the term is used to describe methods of transmitting information on a pulse carrier. *See also:* **modulating systems.** E170-0

pulse modulator. A device that applies pulses to the element in which modulation takes place. *See also:* **modulation.** E145-0

pulse multiplex*. *See:* **pulse-mode multiplex.**
*Deprecated

pulse operation. The method of operation in which the energy is delivered in pulses. *Note:* Pulse operation is usually described in terms of the pulse shape, the pulse duration, and the pulse-recurrence frequency. *See also:* **pulse.** 42A65-31E3

pulse packet (pulse-radar system). The volume of space occupied by a single pulse. The dimensions of this volume are determined by the angular width of the beam, the duration of the pulse, and the distance from the antenna. *See also:* **radio navigation.** E172-10E6

pulse-phase modulation (PPM). *See:* **pulse-position modulation.**

pulse-position modulation (PPM). Pulse-time modulation in which the value of each instantaneous sample of a modulating wave is caused to modulate the position in time of a pulse. *See also:* **modulating systems; pulse terms.** E145/42A65-0

pulse power, carrier-frequency, peak. *See:* **peak pulse power, carrier-frequency.**

pulse power, peak. *See:* **peak pulse power.**

pulse, radio-frequency. *See:* **radio-frequency pulse.**

pulse rate (1) (electronic navigation). *See:* **pulse-repetition frequency.**
(2) (watthour meter). The number of pulses per demand interval at which a pulse device is nominally rated. *See also:* **auxiliary device to an instrument.** 12A0-0

pulse rate telemetering (electric power systems). A type of telemetering in which the number of unidirectional pulses per unit time is varied as a function of the magnitude of the measured quantity. *See also:* **telemetering.** E94-0

pulse receiver (watthour meter). The unit that receives and registers the pulses. It may include a periodic resetting mechanism, so that a reading proportional to demand may be obtained. *See also:* **auxiliary device of an instrument.** 12A0-0

pulse regeneration. The process of restoring a series of pulses to their original timing, form, and relative magnitude. *See also:* **pulse; pulse terms.** 42A65-19E4

pulse repeater (transponder). A device used for receiving pulses from one circuit and transmitting corresponding pulses into another circuit. It may also change the frequency and waveforms of the pulses and perform other functions. *See also:* **pulse; repeater.** E145/42A65-0

pulse-repetition frequency (PRF) (system using recurrent pulses). The number of pulses per unit of time. 0-10E6/19E4

pulse-repetition rate (electronic navigation). See: **pulse-repetition frequency.**

pulse reply. The transmission of a pulse or pulse mode by a transponder as the result of an interrogation. *See also:* **pulse terms.** E194-0

pulse response characteristics (pulse response curve) (electromagnetic compatibility). The relationship between the indication of a quasi-peak voltmeter and the repetition rate of regularly repeated pulses of constant amplitude. *See also:* **electromagnetic compatibility.** CISPR-27E1

pulse rise time (general). The interval between the instants at which the instantaneous amplitude first reaches specified lower and upper limits, namely 10 percent and 90 percent of the peak pulse amplitude unless otherwise stated. *See also:* **pulse.** E175/E203/42A65-9E4/19E4/31E3

pulse separation. The interval between the trailing-edge pulse time of one pulse and the leading-edge pulse time of the succeeding pulse. *See also:* **leading-edge pulse time; pulse; pulse terms; trailing-edge pulse time.** E194-2E2/9E4

pulses, equalizing. *See:* **equalizing pulses.**

pulse shape. The identifying geometrical and/or mathematical characteristics of a pulse waveform used for descriptive purposes. *Note:* For heuristic or tutorial purposes (1) the practical shape of a pulse may be imprecisely described with the following adjectives: exponential, Gaussian, irregular, rectangular, rounded, trapezoidal, triangular, half sine, etcetera. These or similar adjectives describe the general shape of a pulse and no precise distinctions are defined. (2) A hypothetical pulse may be precisely described by the addition of the adjective ideal; for example, ideal rectangular pulse. *See also:* **pulse.** 0-9E4

pulse shaper (pulse techniques). Any transducer used for changing one or more characteristics of a pulse. *Note:* This term includes pulse regenerators. *See also:* **pulse.** E175/E194-2E2

pulse shaping. Intentionally changing the shape of a pulse. *See also:* **pulse terms.** E194-2E2

pulse, single-polarity. *See:* **unidirectional pulse.**

pulse spacing (pulse interval). The interval between the corresponding pulse times of two consecutive pulses. *Note:* The term pulse interval is deprecated because it may be taken to mean the duraction of the pulse instead of the space or interval from one pulse to the next. Neither term means the space between pulses. *See also:* **pulse terms.** E194-2E2

pulse spectrum (signal-transmission system). The frequency distribution of the sinusoidal components of the pulse in relative amplitude and in relative phase. *See also:* **pulse terms; signal.** E194-13E6

pulse stretcher (pulse techniques). A pulse shaper that produces an output pulse whose duration is greater than that of the input pulse and whose amplitude is proportional to that of the peak amplitude of the input pulse. *See also:* **pulse.** E175-0

pulse terms. *Note:* For an extensive list of cross references, see *Appendix A.*

pulse techniques. *See:* **burst.**

pulse tilt. A distortion in an otherwise essentially flat-topped rectangular pulse characterized by either a decline or a rise of the pulse top. *See also:* **pulse terms; television.** E203/E194-2E2

pulse time, leading-edge. *See:* **leading-edge pulse time.**

pulse time, mean. *See:* **mean pulse time.**

pulse-time modulation (PTM). Modulation in which the values of instantaneous samples of the modulating wave are caused to modulate the time of occurrence of some characteristic of a pulse carrier. *Note:* Pulse-duration modulation, pulse-position modulation, and pulse-interval modulation are particular forms of pulse-time modulation. *See also:* **modulating systems; pulse terms.** E145/E194/42A65-0

pulse time, trailing-edge. *See:* **trailing-edge pulse time.**

pulse timing of video pulses (television). The determination of an occurrence of a pulse or a specified portion thereof at a particular time. *See:* **pulse width of video pulses (television); television; time of rise (decay) of video pulses (television).** E207-0

pulse top. That major segment of a pulse waveform having the greater displacement in amplitude from the baseline. *See also:* **pulse.** 0-9E4

pulse train (signal-transmission system) (industrial control). A sequence of pulses. *See also:* **pulse; pulse terms; signal.** E194/E270/85A1-2E2/13E6/34E10

pulse train, bidirectional. *See:* **bidirectional pulse train.**

pulse train, periodic. *See:* **periodic pulse train.**

pulse-train spectrum (pulse-train frequency-spectrum). The frequency distribution of the sinusoidal components of the pulse train in amplitude and in phase angle. *See also:* **pulse terms.** E194-0

pulse train, unidirectional. *See:* **unidirectional pulse train.**

pulse transmitter. A pulse-modulated transmitter whose peak power-output capabilities are usually large with respect to average power-output rating. *See also:* **pulse; radio transmitter.** E145-0

pulse-type telemeter. A telemeter that employs characteristics of intermittent electric signals other than their frequency, as the translating means. *Note:* These pulses may be utilized in any desired manner to obtain the final indication, such as periodically counting the total number of pulses; or measuring their ON time, their OFF time, or both. 37A100-31E11

pulse, unidirectional. *See:* **unidirectional pulse.**

pulse width (1) (television)*. *See:* **pulse duration.**

*Deprecated

(2) (data transmission). The time interval between the points on the leading and trailing edges at which the instantaneous value bears a specified relation to the maximum instantaneous value of the pulse. *See also:* **data transmission.** 0-19E4

pulse-width modulation. *See:* **pulse-duration modulation.**

pulse width of video pulses (television). The duration of a pulse measured at a specified level. *See also:* **pulse timing of video pulses (television).** E207-0

pulsing circuit (peaking circuit) (industrial control). A circuit designed to provide abrupt changes in voltage or current of some characteristic pattern. *See also:* **electronic controller.** 42A25-34E10

pulsing transformer (industrial control). Supplies pulses of voltage or current. *See also:* **electronic controller.** 42A25-34E10

pump (parametric device). The source of alternating-current power that causes the nonlinear reactor to behave as a time-varying reactance. *See also:* **parametric device.** E254-15E7

pump-back test (electrical back-to-back test) (rotating machinery). A test in which two identical machines are mechanically coupled together, and they are both connected electrically to a power system. The total losses of both machines are taken as the power input drawn from the system. *See also:* **asynchronous machine; direct-current commutating machine; synchronous machine.** 0-31E8

pumped-storage hydro capability (electric power supply). The capability supplied by hydroelectric sources using a pond that is alternately filled by pumping and depleted by generating. *See also:* **generating station.** 0-31E4

pumped tube. An electron tube that is continuously connected to evacuating equipment during operation. *Note:* This term is used chiefly for pool-cathode tubes. *See also:* **tube definitions.** 42A70-15E6

pump-free control. *See:* **antipump (pump-free device) device.**

pump-free device. *See:* **antipump device.**

punch (computing systems). *See:* **keypunch, zone punch.**

punched card. A card on which a pattern of holes or cuts is used to represent data. 0-16E9

punched tape (computing systems). A tape on which a pattern of holes or cuts is used to represent data. *See also:* **electronic digital computer.** X3A12-16E9

punching (rotating machinery). A lamination made from sheet material using a punch and die. *See also:* **rotor (rotating machinery); stator.** 0-31E8

punch position (computing systems). A site on a punched tape or card where holes are to be punched. *See also:* **electronic digital computer.** X3A12-16E9

puncture (voltage testing). A disruptive discharge through the body of a solid dielectric. *See also:* **lightning arrester (surge diverter); test voltage and current.** 68A1-31E5

puncture voltage (lightning arresters). The voltage at which the test specimen is electrically punctured. *See also:* **lightning arrester (surge diverter); power systems, low-frequency and surge testing.** 42A35-31E13/31E7

pure tone. *See:* **simple tone.**

purification of electrolyte. The treatment of a suitable volume of the electrolyte by which the dissolved impurities are removed in order to keep their content in the electrolyte within desired limits. *See also:* **electrorefining.** 42A60-0

purity (excitation purity) (color). *See:* **excitation purity (purity) (color).**

Purkinje phenomenon. The reduction in subjective brightness of a red light relative to that of a blue light when the luminances (photometric brightnesses) are reduced in the same proportion without changing the respective spectral distributions. In passing from photopic to scotopic vision, the curve of spectral luminous efficiency changes, the wavelength of maximum efficiency being displaced toward the shorter wavelengths. *See also:* **visual field.** Z7A1-0

purple boundary. The straight line drawn between the ends of the spectrum locus on a chromaticity diagram. *See also:* **color; color terms.** Z7A1/E201-2E2

push brace. A supporting member, usually of timber, placed between a pole or other structural part of a line and the ground or a fixed object. *See also:* **tower.** 42A35-31E13

pushbutton (industrial control). Part of an electric device, consisting of a button that must be pressed to effect an operation. *See also:* **switch.** 50I16-34E10

pushbutton station (industrial control). A unit assembly of one or more externally operable pushbutton switches, sometimes including other pilot devices such as indicating lights or selector switches, in a suitable enclosure. *See also:* **switch.** 42A25-34E10

pushbutton switch (pushbutton). A master switch, usually mounted behind an opening in a cover or panel, and having an operating plunger or button extending forward in the opening. Operation of the switch is normally obtained by pressure of the finger against the end of the button. *See also:* **switch.** 42A25-0

pushbutton switching. A reperforator switching system in which selection of the outgoing channel is initiated by an operator. 0-19E4

pushdown list (computing systems). A list that is constructed and maintained so that the next item to be retrieved is the most recently stored item in the list, that is, last in, first out. *See also:* **electronic digital computer; pushup list.** X3A12-16E9

pushing figure (oscillator). The change of oscillator frequency with a specified change in current, excluding thermal effects. *See:* **tuning sensitivity, electronic.** *See also:* **oscillatory circuit; television.** E160-15E6

push-pull amplifier circuit. *See:* **amplifier, balanced.**

push-pull circuit. A circuit containing two like elements that operate in 180-degree phase relationship to produce additive output components of the desired wave, with cancellation of certain unwanted products. *Note:* Push-pull amplifiers and push-pull oscillators are examples. *See also:* **amplifier; circuits and devices.** 42A65-31E3

push-pull currents. Balanced currents. *See also:* **waveguide.** E146-0

push-pull microphone. A microphone that makes use of two like microphone elements actuated by the same sound waves and operating 180 degrees out of phase. *See also:* **microphone.** 42A65-0

push-pull operation (electron device). The operation of two similar electron devices or of an equivalent double-unit device, in a circuit such that equal quantities in phase opposition are applied to the input electrodes, and the two outputs are combined in phase. *See also:* **circuits and devices.** 0-15E6

push-pull oscillator. A balanced oscillator employing two similar tubes in phase opposition. *See also:* **balanced oscillator; oscillatory circuit.** E145-0

push-pull voltages. Balanced voltages. *See also:* **waveguide.** E146-0

push-push circuit. A circuit employing two similar tubes with grids connected in phase opposition and plates in parallel to a common load, and usually used as a frequency multiplier to emphasize even-order harmonics. *See also:* **circuits and devices.** E145-0

push-push currents. Currents flowing in the two conductors of a balanced line that, at every point along the line, are equal in magnitude and in the same direction. *See also:* **waveguide.** E146-0

push-push voltages. Voltages (relative to ground) on the two conductors of a balanced line that, at every point along the line, are equal in magnitude and have the same polarity. *See also:* **waveguide.** E146-0

push-to-talk circuit. A method of communication over a speech channel in which transmission occurs in only one direction at a time, the talker being required to keep a switch operated while he is talking. *See also:* **telephone system.** 42A65-31E3

push-to-type operation. That form of telegraph operation, employing a one-way reversible circuit, in which the operator must keep a switch operated in order to send from his station. It is generally used in radio transmission where the same frequency is employed for transmission and reception. *See also:* **telegraphy.** 42A65-0

pushup list (computing systems). A list that is constructed and maintained so that the next item to be retrieved and removed is the oldest item still in the list, that is, first in, first out. *See also:* **electronic digital computer; pushdown list.** X3A12-16E9

pyramidal horn antenna. A horn antenna the sides of which form a pyramid. *See also:* **antenna.** 0-3E1

pyroconductivity. Electric conductivity that develops with rising temperature, and notably upon fusion, in solids that are practically nonconductive at atmospheric temperatures. 42A60-0

pyrometer. A thermometer of any kind usable at relatively high temperatures (above 500 degrees Celsius). *See also:* **electric thermometer (temperature meter).** 42A30-0

Q

***Q*-meter (quality-factor meter).** An instrument for measuring the quality factor Q of a circuit or circuit element. *See also:* **instrument.** 42A30-0

quad. A structural unit employed in cable, consisting of four separately insulated conductors twisted together. *See also:* **cable.** 42A65-0

quadded cable. A cable in which at least some of the conductors are arranged in the form of quads. *See also:* **cables.** 42A65-0

quadergy. Delivered by an electric circuit during a time interval when the voltages and currents are periodic, the product of the reactive power and the time interval, provided the time interval is one or more complete periods or is quite long in comparison with the time of one period. If the reference direction for energy flow is selected as into the region the net delivery of quadergy will be into the region when the sign of the quadergy is positive and out of the region when the sign is negative. If the reference direction is selected as out of the region, the reverse will apply. The quadergy is expressed by

$$K = Qt$$

where Q is the reactive power and t is the time interval. If the voltages and currents form polyphase symmetrical sets, there is no restriction regarding the relation of the time interval to the period. If the voltages and currents are quasi-periodic and the amplitudes of the voltages and currents are slowly varying, the quadergy is the integral with respect to time of the reactive power, provided the integration is for a time that is one or more complete periods or that is quite long in comparison with the time of one period. Mathematically

$$K = \int_{t_0}^{t_0+t} Q\,\mathrm{d}t$$

where Q is the reactive power determined for the condition of voltages and currents having slowly varying amplitudes. Quadergy is expressed in var-seconds or var-hours when the voltages are in volts and the currents in amperes, and the time is in seconds or hours, respectively. *See also:* **network analysis.** E270-0

quadrantal error (electronic navigation). An angular error in measured bearing caused by characteristics of the vehicle or station that adversely affect the direction of signal propagation; the error varies in a sinusoidal manner throughout the 360 degrees and has two positive and two negative maximums. *See also:* **navigation.** 0-10E6

quadrature. The relation between two periodic functions when the phase difference between them is one-fourth of a period. *See also:* **network analysis.** E270-0

quadrature axis (synchronous machine). The axis that represents the direction of the radial plane along which the main field winding produces no magnetization, normally coinciding with the radial plane midway between adjacent poles. *Notes:* (1) The positive direction of the quadrature axis is 90 degrees ahead of the positive direction of the direct axis, in the direction of rotation of the field relative to the armature. (2) The definitions of currents and voltages given in the terms listed below are applicable to balanced load conditions and for sinusoidal currents and voltages. They may also be applied under other conditions to the positive-sequence fundamental-frequency components of currents and voltages. More generalized definitions, applicable under all conditions, have not been agreed upon. *See:*
direct-axis component of armature current;
direct-axis component of armature voltage;
direct-axis component of magnetic flux;
phasor diagram;
quadrature-axis component;
subtransient internal voltage;
synchronous internal voltage;
transient internal voltage. 42A10-31E8

quadrature-axis component (1) (armature voltage). That component of the armature voltage of any phase that is in time phase with the quadrature-axis component of current in the same phase. *Note:* A quadrature-axis component of voltage may be reproduced by: (A) rotation of the direct-axis component magnetic flux; (B) variation (if any) of the quadrature-axis component of magnetic flux; (C) resistance drop caused by flow of the quadrature-axis component of armature current. The quadrature-axis component of terminal voltage is related to the synchronous internal voltage by

$$\mathbf{E}_{aq} = \mathbf{E}_{i} - R\mathbf{I}_{aq} - jX_{d}\mathbf{I}_{ad}.$$

See also: **phasor diagram (synchronous machine).** 42A10-0

(2) (armature current). That component of the armature current that produces a magnetomotive-force distribution that is symmetrical about the quadrature axis. *See also:* **quadrature-axis component of armature voltage; quadrature-axis component of magnetomotive force; synchronous machine.** 42A10-0

(3) (magnetomotive force) (rotating machinery). The component of a magnetomotive force that is directed along an axis in quadrature with the axis of the poles. *See also:* **asynchronous machine; direct-axis synchronous impedance; synchronous machine.** 0-31E8

quadrature-axis current (rotating machinery). The current that produces quadrature-axis magnetomotive force. *See also:* **direct-axis synchronous impedance.** 0-31E8

quadrature-axis magnetic flux (rotating machinery). The magnetic-flux component directed along the quadrature axis. *See also:* **direct-axis synchronous impedance; synchronous machine.** 0-31E8

quadrature-axis subtransient impedance (rotating machinery). The operator expressing the relation between the initial change in armature voltage and a sudden change in quadrature-axis armature current, with only the fundamental-frequency components considered for both voltage and current, with no change in the voltage applied to the field winding, and with the rotor running at steady speed. In terms of network theory it corresponds to the quadrature-axis impedance the machine displays against disturbances (modulations) with infinite frequency. *Note:* If no rotor winding is along the quadrature axis and/or the rotor is not made out of solid steel, this impedance equals the quadrature-axis synchronous impedance. *See also:* **asynchronous machine; direct-axis synchronous impedance; synchronous machine.** 0-31E8

quadrature-axis subtransient open-circuit time constant. The time in seconds required for the rapidly decreasing component (negative) present during the first few cycles in the quadrature-axis component of symmetrical armature voltage under suddenly removed symmetrical short-circuit conditions with the machine running at rated speed, to decrease to $1/e \approx 0.368$ of its initial value. *See also:* **synchronous machine.** 42A10-0

quadrature-axis subtransient reactance. The ratio of the fundamental component of reactive armature voltage due to the initial value of the fundamental quadrature-axis component of the alternating-current component of the armature current, to this component of current under suddenly applied balanced load conditions and at rated frequency. Unless otherwise specified, the quadrature-axis subtransient reactance will be that corresponding to rated armature current. *See also:* **synchronous machine.** 42A10-0

quadrature-axis subtransient short-circuit time constant. The time in seconds required for the rapidly decreasing component present during the first few cycles in the quadrature-axis component of the alternating-current component of the armature current under suddenly applied symmetrical short-circuit conditions, with the machine running at rated speed to decrease to $1/e \approx 0.368$ of its initial value. *See also:* **synchronous machine.** 42A10-0

quadrature-axis subtransient voltage (rotating machinery). The quadrature-axis component of the armature voltage that appears immediately after the sudden opening of the external circuit when running at a specified load. *See also:* **direct-axis synchronous impedance; synchronous machine.** 0-31E8

quadrature-axis synchronous impedance (synchronous machine) (rotating machinery). The impedance of the armature winding under steady-state conditions where the axis of the armature current and magnetomotive force coincides with the quadrature axis. In large machines where the armature resistance is negligibly small, the quadrature-axis synchronous impedance is equal to the quadrature-axis synchronous reactance. *See also:* **synchronous machine.** 0-31E8

quadrature-axis synchronous reactance. The ratio of the fundamental component of reactive armature voltage, due to the fundamental quadrature-axis component of armature current, to this component of current under steady-state conditions and at rated frequency. Unless otherwise specified, the value of quadrature-axis synchronous reactance will be that corresponding to rated armature current. *See also:* **synchronous machine.** 42A10-0

quadrature-axis transient impedance (rotating machinery). The operator expressing the relation between the initial change in armature voltage and a sudden change in quadrature-axis armature current component, with only the fundamental frequency components considered for both voltage and current, with no change in the voltage applied to the field winding, with the rotor running at steady speed, and by considering only the slowest decaying component and the steady-state component of the voltage drop. In terms of network theory it corresponds to the quadrature-axis impedance the machine displays against disturbances (modulation) with infinite frequency by considering only two pole pairs* (or poles*), namely those with smallest (including zero) real parts, of the impedance function. *Note:* If no rotor winding is along the quadrature axis and/or the rotor is not made out of solid steel, this impedance equals the quadrature-axis synchronous impedance. *See also:* **asynchronous machine; direct-axis synchronous impedance; synchronous machine.**

*The term pole refers here to the roots of the denominator of the impedance function. 0-31E8

quadrature-axis transient open-circuit time constant. The time in seconds required for the root-mean-square alternating-current value of the slowly decreasing component present in the quadrature-axis component of symmetrical armature voltage on open-circuit to decrease to $1/e \approx 0.368$ of its initial value when the quadrature field winding (if any) is suddenly short-circuited with the machine running at rated speed. *Note:* This time constant is important only in turbine generators. *See also:* **synchronous machine.** 42A10-0

quadrature-axis transient reactance . The ratio of the fundamental component of reactive armature voltage, due to the fundamental quadrature-axis component of the alternating-current component of the armature current, to this component of current under suddenly applied load conditions and at rated frequency, the value of current to be determined by the extrapolation of the envelope of the alternating-current component of the current wave to the instant of the sudden application of load, neglecting the high-decrement currents during the first few cycles. *Note:* The quadrature-axis transient reactance usually equals the quadrature-axis synchronous reactance except in solid-rotor machines, since in general there is no really effective field current in the quadrature axis. *See also:* **synchronous machine.** 42A10-0

quadrature-axis transient short-circuit time constant. The time in seconds required for the root-mean-square alternating-current value of the slowly decreasing component present in the quadrature-axis component of the alternating-current component of the armature current under suddenly applied short-circuit conditions with the machine running at rated speed to decrease to $1/e \approx 0.368$ of its initial value. *See also:* **synchronous machine.** 42A10-0

quadrature-axis transient voltage (rotating machinery). The quadrature-axis component of the terminal voltage that appears immediately after the sudden opening of the external circuit when running at a specified load, neglecting the components with very rapid decay that may exist during the first few cycles. *See also:* **asynchronous machine; direct-axis synchronous impedance; synchronous machine.** 0-31E8

quadrature-axis voltage (rotating machinery). The component of voltage that would produce quadrature-axis current when resistance limited. *See also:* **synchronous machine.** 0-31E8

quadrature modulation. Modulation of two carrier components 90 degrees apart in phase by separate modulating functions. *See also:* **modulating systems.** E170-0

quadri pole. *See:* **two-terminal pair network.**

quadrupole parametric amplifier. A beam parametric amplifier having transverse input and output couplers for the signal, separated by a quadrupole structure that is excited by a pump to obtain parametric amplification of a cyclotron wave. *See also:* **parametric device.** E254-15E7

quality area. The area of the cathode-ray-tube phosphor screen that is limited by the cathode-ray tube and instrument specification. *Note:* If the quality area and the graticule area are not equal, this must be specified. *See:* **graticule area; viewing area.** *See also:* **oscillograph.** 0-9E4

quality factor (Q) (network, structure, or material). 2π times the ratio of the maximum stored energy to the energy dissipated per cycle at a given frequency. *Notes:* (1) The Q of an inductor at any frequency is the magnitude of the ratio of its reactance to its effective series resistance at that frequency. (2) The Q of a capacitor at any frequency is the magnitude of the ratio of its susceptance to its effective shunt conductance at that frequency. (3) The Q of a simple resonant circuit comprising an inductor and a capacitor is given by

$$Q = \frac{Q_L Q_C}{Q_L + Q_C}$$

where Q_L and Q_C are the Q's of the inductor and capacitor, respectively, at the resonance frequency. If the resonant circuit comprises an inductance L and a capacitance C in series with an effective resistance R, the value of Q is

$$Q = \frac{1}{R}\left(\frac{L}{C}\right)^{1/2}.$$

(4) An approximate equivalent definition, which can be applied to other types of resonant structures, is that the Q is the ratio of the resonance frequency to the bandwidth between those frequencies on opposite sides of the resonance frequency (known as half-power points) where the response of the resonant structure differs by 3 decibels from that at resonance. (5) The Q of a magnetic or dielectric material at any frequency is equal to 2π times the ratio of the maximum stored energy to the energy dissipated in the material per cycle. (6) For networks that contain several elements, and distributed parameter systems, the Q is generally evaluated at a frequency of resonance. (7) The nonloaded Q of a system is the value of Q obtained when only the incidental dissipation of the system elements is present. The loaded Q of a system is the value of Q obtained when the system is coupled to a device that dissipates energy. (8) The **period** in the expression for Q is that of the driving force, not that of energy storage, which is usually half that of the driving force. *See also:* **network analysis; transmission characteristics.** E270-42A65

quality of lighting. The distribution of brightness in a visual environment. The term is used in a positive sense and implies that all brightnesses contribute favorably to visual performance, visual comfort, ease of seeing, safety, and esthetics for the specific visual tasks involved. *See also:* **visual field.** Z7A1-0

quantity equations. Equations in which the quantity symbols represent mathematico-physical quantities possessing both numerical values and dimensions. E270-0

quantity of electricity. *See:* **electric charge.**

quantity of light (luminous energy). The product of the luminous flux by the time it is maintained. It is the time integral of luminous flux. *See:* **light and luminous flux.** *See also:* **light.** Z7A1-0

quantization (telecommunication). A process in which the continuous range of values of an input signal is divided into nonoverlapping subranges, and to each subrange a discrete value of the output is uniquely assigned. Whenever the signal value falls within a given subrange, the output has the corresponding discrete value. *Note:* Quantized may be used as an adjective modifying various forms of modulation, for example, quantized pulse- amplitude modulation. *See:* **communication; quantization distortion (quantization noise); quantization level).** *See also:* **electronic digital computer; modulating systems.** E170-0;X3A12-16E9;42A65-19E4

quantization distortion (quantization noise). The inherent distortion introduced in the process of quantization. *See also:* **modulation systems; quantization.** 42A65-0

quantization level (telecommunication). The discrete value of the output designating a particular subrange of the input. *See also:* **modulating systems; quantization.** E170-0

quantization noise. *See:* **quantization distortion.**

quantize. To subdivide the range of values of a variable into a finite number of nonoverlapping subranges or intervals, each of which is represented by an assigned value within the subrange, for example, to represent a person's age as a number of whole years. X3A12-16E9

quantized pulse modulation. Pulse modulation that involves quantization. *See also:* **modulating systems.** 42A65-0

quantizing operation (control system). An operation that converts a signal into one that has only discrete values of amplitude. *See also:* **control system.** 0-23E0

quantum efficiency (photocathodes). The average number of electrons photoelectrically emitted from the photocathode per incident photon of a given wavelength. *Note:* the quantum efficiency varies with the wavelength, angle of incidence, and polarization of the incident radiation. *See also:* **photocathodes; photoelectric converter; phototubes; semiconductors.** E160-15E6;42A70-10E1

quarter-phase or two-phase circuit. A combination of circuits energized by alternating electromotive forces that differ in phase by a quarter of a cycle, that is, 90 degrees. *Note:* In practice the phases may vary several degrees from the specified angle. *See also:* **center of distribution.** 42A35-31E13

quarter-thermal-burden ambient-temperature rating. The maximum ambient temperature at which the transformer can be safely operated when the transformer is energized at rated voltage and frequency and is carrying 25 percent of its thermal-burden rating without exceeding the specified temperature limitations. 42A30-0

quasi-impulsive noise (electromagnetic compatibility). A superposition of impulsive and continuous noise. *See also:* **electromagnetic compatibility.** CISPR-27E1

quasi-peak detector (electromagnetic compatibility). A detector having specified electrical time constants that, when regularly repeated pulses of constant amplitude are applied to it, delivers an output voltage that is a fraction of the peak value of the pulses, the fraction increasing towards unity as the pulse repetition rate is increased. *See also:* **electromagnetic compatibility.** CISPR-27E1

quasi-peak voltmeter (electromagnetic compatibility). A quasi-peak detector coupled to an indicating instrument having a specific mechanical time-constant. *See also:* **electromagnetic compatibility.** CISPR-27E1

quenched spark gap converter (dielectric heater usage). A generator or power source that utilizes the oscillatory discharge of a capacitor through an inductor and a spark gap as a source of radio-frequency power. The spark gap comprises one or more closely spaced gaps operating in series. *See also:* **induction heating; industrial electronics.** E54/E169-0

quenching (gas-filled radiation-counter tube). The process of terminating a discharge in a Geiger-Mueller radiation-counter tube by inhibiting reignition. *Note:* This may be effected internally (internal quenching or self-quenching) by use of an appropriate gas or vapor filling, or externally (external quenching) by momentary reduction of the applied potential difference. *See:* **gas-filled radiation-counter tubes.** 42A70-15E6

quenching circuit (radiation counters). A circuit that reduces the voltage applied to a Geiger-Mueller tube after an ionizing event, thus preventing the occurrence of subsequent multiple discharges. Usually the original voltage level is restored after a period that is longer than the natural recovery time of the Geiger-Mueller tube. *See:* **anticoincidence (radiation counters).** 0-15E6

quick-break. A term used to describe a device that has a high contact-opening speed independent of the operator. 37A100-31E11

quick charge (storage battery). *See:* **boost charge.**

quick-flashing light. A rhythmic light exhibiting very rapid regular alternations of light and darkness. There is no restriction on the ratio of the durations of the light to the dark periods. *See also:* **signal lighting.** Z7A1-0

quick-make. A term used to describe a device that has a high contact-closing speed independent of the operator. 37A100-31E11

quick release (control brakes). The provision for effecting more rapid release than would inherently be obtained. *See also:* **electric drive.** IC1-34E10

quick set (control brakes). The provision for effecting more rapid setting than would inherently be obtained. *See also:* **electric drive.** IC1-4E10

quiescent-carrier telephony. That form of carrier telephony in which the carrier is suppressed whenever there are no modulating signals to be transmitted. *See also:* **modulating systems.** 42A65-0

quiescent current (electron tubes). The electrode current corresponding to the electrode bias voltage. *See also:* **electronic tube.** 50I07-15E6

quiescent operating point (magnetic amplifier). The output obtained under any specified external conditions when the signal is non-time-varying and zero. *See also:* **rating and testing magnetic amplifiers.** E107-0

quiescent point (amplifier). That point on its characteristic that represents the conditions existing when the signal input is zero. *Note:* The quiescent values of the parameters are not in general equal to the average values existing in the presence of the signal unless the characteristic is linear and the signal has no direct-current component. *See:* **operating point.** 42A65-0

quiet automatic volume control. Automatic volume control that is arranged to be operative only for signal strengths exceeding a certain value, so that noise or other weak signals encountered when tuning between strong signals are suppressed. *See also:* **radio receiver.** 42A65-0

quieting sensitivity (frequency-modulation receivers). The minimum unmodulated signal input for which the output signal-noise ratio does not exceed a specified limit, under specified conditions. *See also:* **radio receiver; receiver performance.** E188-0;42A65-0;0-6E1

quiet tuning. A circuit arrangement for silencing the output of a radio receiver except when the receiver is accurately tuned to an incoming carrier wave. *See also:* **radio receiver.** 42A65-0

quill drive. A form of drive in which a motor or generator is geared to a hollow cylindrical sleeve, or quill, or the armature is directly mounted on a quill, in either case, the quill being mounted substantially concentrically with the driving axle and flexibly connected to the driving wheels. *See also:* **traction motor.** 42A42-0

quinary. *See:* **biquinary.**

quinhydrone electrode. *See:* **quinhydrone half cell.**

quinhydrone half cell (quinhydrone electrode). A half cell with an electrode of an inert metal (such as platinum or gold) in contact with a solution saturated with quinhydrone. *See also:* **electrochemistry.** 42A60-0

quotation board. A manually or automatically operated panel equipped to display visually the price quotations received by a ticker circuit. Such boards may provide displays of large size or may consist of small automatic units that ordinarily display one item at a time under control of the user. *See also:* **telegraphy.** 42A65-0

quotient. *See:* **phasor product.**

quotient relay. A relay that operates in response to a suitable quotient of two alternating electrical input quantities. *See also:* **relay.** 37A100-31E11/31E6

R

rabbet, mounting (rotating machinery). A male or female pilot on a face or flange type of end shield of a machine, used for mounting the machine with a mating rabbet. The rabbet may be circular, or of other configuration, and need not be continuous. 0-31E8

raceway (electric system). Any channel for enclosing, and for loosely holding wires, cables, or busbars in interior work that is designed expressly for, and used solely for, this purpose. *Note:* Raceways may be of metal or insulating material and the term includes rigid metal conduit, rigid nonmetallic conduit, flexible metal conduit, electrical metallic tubing, underfloor raceways, cellular concrete-floor raceways, cellular metal-floor raceways, surface metal raceways, structural raceways, wireways and busways, and auxiliary gutters or moldings.
See:
access fitting;
cellular metal-floor raceway;
cleat;
conduit fitting;
coupling;
elbow;
electrical metallic tubing;
fitting;
flexible metal conduit;
flexible nonmetallic tubing;
interior wiring;
knob;
multioutlet assembly;
nipple;
rigid metal conduit;
split fitting;
surface metal raceway;
tube;
underfloor raceway. 1A0/2A2/42A95-0

rack mounting (1) (general). A method of mounting electric equipment in which metal panels supporting the equipment are attached to predrilled tapped vertical steel channel rails or racks. The dimensions of panels, the spacing and height of rails, and the spacing and size of mounting screws are standardized. 0-31E3
(2) (cabinet). A method of rack mounting in which the vertical rail or rack is enclosed in a metal cabinet that may have hinged doors on the front and back. 0-31E3
(3) (enclosed). A method of rack mounting in which boxlike covers are attached to the panels to protect and shield the electric equipment. 0-31E3
(4) (fixed). A method of cabinet rack mounting in which the rack or rails are fixed to the inside of the cabinet. Equipment is accessible through front and rear doors of the cabinet. 0-31E3
(5) (swinging). A method of cabinet rack mounting in which a frame containing the rails or rack is hinged on one side to allow the rack and the equipment on it to be swung out of the front door of the cabinet. This makes both sides of the rack accessible when the back of the cabinet is not accessible. 0-31E3

racon. *See:* **radar beacon.**

rad (photovoltaic power system). An absorbed radiation unit equivalent to 100 ergs/gram of absorber. *See also:* **photovoltaic power system; solar cells (photovoltaic power system).** 0-10E1

radar. A radio detecting and ranging system that detects the presence of objects and their distance (range) by the transmission and return of electromagnetic energy; direction is usually obtained also, through the use of a movable or rotating directive antenna pattern. *Notes:* (1) The name is derived from the initial letters of the expression **radio direction and ranging**. (2) The motion of the directive antenna pattern may be produced by electronic scanning; without physically apparent motion. (3) For an extensive list of cross references, see *Appendix A.* 0-10E6

radar altimeter (electronic navigation). *See:* **radio altimeter.**

radar beacon (racon). A device that receives signals from a radar transmitter and transmits signals for enabling the radar receiver to determine its position. *Note:* Such a device is also termed a **transponder** and is said to be replying to interrogations from primary radars. *See also:* **navigation; radar.** 42A65-10E6

radar camouflage. The art, means, or result of concealing the presence or the nature of an object from radar detection by the use of coverings or surfaces that considerably reduce the radio energy reflected toward a radar. *See also:* **radar.** 42A65 -0

radar cross section. The portion of the back-scattering cross section of a target associated with a specified polarization component of the scattered wave. *See also:* **antenna; circular scanning; effective echo area.** E149-3E1

radar equation. A mathematical expression for primary radar that, in its basic form, relates radar transmitter power, antenna gain, wavelength, effective echo area of the target, distance to the target, and receiver input power; the equation may be modified to take into account other factors, such as attenuation of energy caused by a radome, attenuation due to unusual atmospheric losses or precipitation, and scanning losses that might affect performance of the radar. *See also:* **navigation.** 0-10E6

radar performance figure. The ratio of the pulse power of the radar transmitter to the power of the minimum signal detectable by the receiver. *See also:* **navigation.** E172-10E6

radar range equation. *See:* **radar equation.**

radar relay (radar). An equipment for relaying the radar video and appropriate synchronizing signals to a remote location. *See also:* **navigation.** E172-10E6

radar shadow. A region shielded from radar illumination by an intervening reflecting or absorbing medium, this region appearing as an area void of targets on a radar display. *See also:* **navigation; radar.** 42A65-10E6

radar transmitter. The transmitter portion of a radio detecting and ranging system. E145-6E1

radial (navigation). One of a number of lines of position defined by an azimuthal navigational facility; the radial is identified by its bearing (usually the magnetic bearing) from the facility. *See also:* **navigation.** 0-10E6

radial air gap (rotating machinery). *See:* **air gap (gap) (rotating machinery).**

radial-blade blower (rotating machinery). A fan made with flat blades mounted so that the plane of the blades passes through the axis of rotation of the rotor. *See also:* **fan (blower) (rotating machinery).** 0-31E8

radial distribution feeder. *See:* **radial feeder.**

radial feeder. A feeder supplying electric energy to a substation or a feeding point that receives energy by no other means. *Note:* The normal flow of energy in such a feeder is in one direction only. *See also:* **center of distribution.** 42A35-31E13

radially outer coil side. *See:* **bottom coil slot.**

radial probable error (RPE) (electronic navigation). *See:* **circular probable error.**

radial system. A system in which independent feeders branch out radially from a common source of supply. *See also:* **alternating-current distribution; direct-current distribution.** 42A35-31E13

radial-time-base display (electronic navigation). *See:* **plan-position indicator.**

radial transmission feeder. *See:* **radial feeder.**

radial transmission line. Basically a pair of parallel conducting planes used for propagating circularly cylindrical waves having their axes normal to the planes, sometimes applied to tapered versions, such as biconical lines. *See also:* **waveguide.** E146-3E1

radiance (in a direction, at point of the surface of a source, of a receiver, or of any other real or virtual surface). The quotient of (1) the radiant flux leaving, passing through, or arriving at an element of the surface surrounding the point, and propagated in directions defined by an elementary cone containing the given direction, by (2) the product of the solid angle of the cone and the area of the orthogonal projection of the element of the surface on a plane perpendicular to the given direction. *Note:* The usual unit is the watt per steradian per square meter. This is the radiant analog of luminance. *See also:* **color terms; radiant energy (illuminating engineering).** E201-2E2;Z7A1-0

radiant density. Radiant energy per unit volume; for example, joules/cubic meter. *See also:* **radiant energy (illuminating engineering).** Z7A1-0

radiant emittance. *See:* **radiant flux density at a surface.**

radiant energy (illuminating engineering). Energy traveling in the form of electromagnetic waves. It is measured in units of energy such as joules, ergs, or kilowatt-hours.

See:
blackbody;
graybody;
irradiance;
light;
Planck radiation law;
radiance;
radiant density;
radiant emittance;
radiant exitance;
radiant flux;
radiant-flux density at a surface;
radiant intensity;
radiator;
regions of electromagnetic spectrum;
spectral emissivity;
spectral radiance;
spectral-radiant energy;
spectral-radiant flux;
spectral-radiant intensity;
Stefan-Boltzmann law;
temperature radiator;
units of wavelength;
Wien radiation law. Z7A1-0

radiant exitance. *See:* **radiant flux density (at an element of surface).**

radiant flux. The time rate of flow of radiant energy. *Note:* It is expressed preferably in watts, or in ergs per second. *See also:* **color terms; radiant energy (illuminating engineering).** E201-2E2;Z7A1-0

radiant flux density (at an element of a surface). The quotient of radiant flux at that element of surface to the area of that element; for example, watts per centimeter squared. When referring to radiant flux "emitted" from a surface, this has been called **radiant emittance***. The preferred term for radiant flux "leaving" a surface is **radiant exitance.** When referring to radiant flux incident on a surface, it is called **irradiance.** *See also:* **radiant energy (illuminating engineering).**
*Deprecated Z7A1-0

radiant gain (optoelectronic device). The ratio of the emitted radiant flux to the incident radiant flux. *Note:* The emitted and incident radiant flux are both determined at specified ports. *See also:* **optoelectronic device.** E222-15E7

radiant heater. A heater that dissipates an appreciable part of its heat by radiation rather than by conduction or convection. *See also:* **appliances (including portable).** 42A95-0

radiant intensity (light source). The radiant flux proceeding from the source per unit solid angle in the direction considered; for example, watts/steradian. *See also:* **color terms; radiant energy (illuminating engineering).** Z7A1-0

radiant sensitivity (phototube). *See:* **sensitivity, radiant.**

radiated interference (electromagnetic compatibility). Radio interference resulting from radiated noise or unwanted signals. *See also:* **electromagnetic compatibility.** 0-27E1

radiated noise (electromagnetic compatibility). Radio noise energy in the form of an electromagnetic field, including both the radiation and induction components of the field. *See also:* **electromagnetic compatibility.** 0-27E1

radiated power output (transmitter performance). The average power output available at the antenna terminals, less the losses of the antenna, for any combination of signals transmitted when averaged over the longest repetitive modulation cycle. *See also:* **audio-frequency distortion.** 0-6E1

radiating element. A basic subdivision of an antenna that in itself is capable of effectively radiating or receiving radio waves. *Note:* Typical examples of a radiating element are a slot, horn, or dipole antenna. *See also:* **antenna.** 0-3E1

radiation (1) (radio communication). The emission of energy in the form of electromagnetic waves. The term is also used to describe the radiated energy. *Note:* For an extensive list of cross references, see *Appendix A.* 42A65-0

(2) (nuclear) (nuclear work). The usual meaning of radiation is extended to include moving nuclear particles, charged or uncharged. E160-15E6

radiation counter. An instrument used for detecting or measuring radiation by counting action. *See also:* **gas-filled radiation-counter tubes.** 42A70-15E6

radiation coupling (interference terminology). *See:* **coupling, radiation.**

radiation detector. Any device whereby radiation produces some physical effect suitable for observation and/or mesurement. *See:* **anticoincidence (radiation counters).** 0-15E6

radiation efficiency (antenna). The ratio of (1) the total power radiated by an antenna to (2) the net power accepted by the antenna from the connected transmitter. *See also:* **antenna.** 42A65-3E1

radiation intensity (in a given direction). The power radiated from an antenna per unit solid angle in that direction. *See also:* **antenna.** E149/42A65-3E1

radiation lobe (antenna pattern). A portion of the radiation pattern bounded by regions of relatively weak radiation intensity. *See also:* **antenna.** E149-3E1

radiation loss (transmission system). That part of the transmission loss due to radiation of radio-frequency power. *See also:* **waveguide.** E146-3E1

radiation pattern (antenna pattern). A graphical representation of the radiation properties of the antenna as a function of space coordinates. *Notes:* (1) In the usual case the radiation pattern is determined in the far field and is represented as a function of directional coordinates. (2) Radiation properties include radiation intensity, field strength, phase, and polarization. *See also:* **antenna; radiation.** 0-3E1

radiation protection guide (electrobiology). Radiation level that should not be exceeded without careful consideration of the reasons for doing so. *See also:* **electrobiology.** 95A1-0

radiation pyrometer (radiation thermometer). A pyrometer in which the radiant power from the object or source to be measured is utilized in the measurement of its temperature. The radiant power within wide or narrow wavelength bands filling a definite solid angle impinges upon a suitable detector. The detector is usually a thermocouple or thermopile or a bolometer responsive to the heating effect of the radiant power, or a photosensitive device connected to a sensitive electric instrument. *See also:* **electric thermometer (temperature meter).** 42A30-0

radiation resistance (antenna). The ratio of the power radiated by an antenna to the square of the toor-mean-square antenna current referred to a specified point. *See also:* **antenna.** E149-3E1

radiation thermometer. *See:* **radiation pyrometer.**

radiator (1) (telecommunication). Any antenna or radiating element, usually a discrete physical and functional identity. *See also:* **antenna.** 0-3E1
(2) (illumination). An emitter of radiant energy. *See also:* **radiant energy (illuminating engineering).** Z7A1-0

radio altimeter. An altimeter using radar principles for height measurement; height is determined by measurement of propagation time of a radio signal transmitted from the vehicle and reflected back to the vehicle from the terrain below. *See also:* **navigation.** 0-10E6

radio-autopilot coupler (electronic navigation). Equipment providing means by which electric signals from navigation receivers control the vehicle autopilot. *See also:* **navigation.** 0-10E6

radio beacon (electronic navigation). A radio transmitter, usually nondirectional, that emits identifiable signals intended for radio direction finding. *Note:* In air operations a radio beacon is called an aerophare. *See also:* **radio direction-finder; radio navigation.** E172-10E6;42A43/42A65-0

radio broadcasting. Radio transmission intended for general reception. *See also:* **radio transmission.** E145/E182/42A65-0

radio channel. A band of frequencies of a width sufficient to permit its use for radio communication. *Note:* The width of a channel depends upon the type of transmission and the tolerance for the frequency of emission. Normally allocated for radio transmission in a specified type of service or by a specified transmitter. *See also:* **radio transmission.** E145-0

radio circuit. A means for carrying out one radio communication at a time in either or both directions between two points. *See also:* **radio channel; radio transmission.** 42A65-0

radio compass. A direction-finder used for navigational purposes. *See also:* **radio navigation.** 42A65-0

radio compass indicator. A device that, by means of a radio receiver and rotatable loop antenna, provides a remote indication of the relationship between a radio bearing and the heading of the aircraft. *See also:* **air-transportation instruments.** 42A41-0

radio compass magnetic indicator. A device that provides a remote indication of the relationship between a magnetic bearing, radio bearing, and the aircraft's heading. *See also:* **air-transportation instruments.** 42A41-0

radio control. The control of mechanism or other apparatus by radio waves. *See also:* **radio transmission.** 42A65-0

radio detection (radio warning). The detection of the presence of an object by radiolocation without precise determination of its position. *See also:* **radio transmission.** 42A65-0

radio direction-finder. A device used to determine the direction of arrival of radio signals. *Note:* In practice, the term is usually applied to a device for ascertaining the apparent direction of an object by means of its own independent emissions (that is, excluding reflected or retransmitted waves). *See also:* **loran; navigational radar (surface search radar); radio beacon; radio receiver.** 42A65-10E6

radio direction finding. A procedure for determining the bearing, at a receiving point, of the source of a radio signal by observing the direction of arrival and other properties of the signal. *See also:* **navigation.** 0-10E6

radio distress signal (SOS). Radiotelegraph distress signal consists of the group ········· in Morse code, transmitted on prescribed frequencies. The radiotelephone distress signal consists of the spoken words **May Day.** *Note:* By international agreement, the effect of the distress signal is to silence all radio traffic that may interfere with distress calls. 42A43-0

radio disturbance (electromagnetic compatibility). An electromagnetic disturbance in the radio-frequency range. *See:* **radio interference; radio noise.** *See also:* **electromagnetic compatibility.** 27E1

radio Doppler. The direct determination of the radial component of the relative velocity of an object by an observed frequency change due to such velocity. *See also:* **radio transmission.** 42A65-0

radio fadeout (Dellinger effect). A phenomenon in radio propagation during which substantially all radio waves that are normally reflected by ionospheric layers in or above the *E* region suffer partial or complete absorption. *SEe also:* **radiation.** 42A65-0

radio field strength (1). The electric or magnetic field strength at a given location associated with the passage of radio waves. It is commonly expressed in terms of the electric field strength in microvolts, millivolts, or volts per meter. In the case of a sinusoidal wave, the root-mean-square value is commonly stated. Unless otherwise stated, it is taken in the direction of maximum field strength. *See also:* **radiation.** 42A65-0
(2) (radio wave propagation). The electric or magnetic field strength at radio frequency. *See also:* **radio wave propagation.** 0-3E2

radio frequency (1) (loosely). A frequency in the portion of the electromagnetic spectrum that is between the audio-frequency portion and the infrared portion. 0-13E6
(2) A frequency useful for radio transmission. *Note:* The present practicable limits of radio frequency are roughly 10 kilohertz to 100 000 megahertz. Within this frequency range, electromagnetic radiation may be detected and amplified as an electric current at the wave frequency. *See also:* **radio transmission; radio wave propagation; signal.** 42A65-3E2

radio-frequency alternator. A rotating-type generator for producing radio-frequency power. E145-0

radio-frequency attenuator (signal-transmission system). A low-pass filter that substantially reduces the radio-frequency power at its output relative to that at its input, but transmits lower-frequency signals with little or no power loss. *See also:* **signal.** 0-13E6

radio-frequency converter (industrial electronics). A power source for producing electric power at a frequency of 10 kilohertz and above. *See also:* **industrial electronics.** E169-0

radio-frequency generator (signal-transmission system). A source of radio-frequency energy.
See also:
Colpitts oscillator;
dielectric heating;
Hartley oscillator;
induction heating;
magnetron;
radio frequency;
signal;
tuned-grid–tuned-plate oscillator. 0-13E6

radio-frequency pulse. A radio-frequency carrier amplitude modulated by a pulse. The amplitude of the modulated carrier is zero before and after the pulse. *Note:* Coherence of the carrier (with itself) is not implied. *See also:* **pulse terms.** E194-0

radio-frequency radiation. The propagation of electromagnetic energy in space. E54-0

radio-frequency switching relay. A relay designed to switch frequencies that are higher than commercial power frequencies with low loss. 0-21E0

radio-frequency system loss (mobile communication). The ratio expressed in decibels of (1) the power delivered by the transmitter to its transmission line to (2) the power required at the receiver-input terminals that is just sufficient to provide a specified signal-to-noise ratio at the audio output of the receiver. *See also:* **mobile communication system.** 0-6E1

radio-frequency transformer. A transformer for use with radio-frequency currents. *Note:* Radio-frequency transformers used in broadcast receivers are generally shunt-tuned devices that are tunable over a relatively broad range of frequencies. *See also:* **radio transmission.** 42A65-21E1

radio gain (radio system). The reciprocal of the system loss. *See also:* **radio wave propagation.** 0-3E2

radio horizon (antenna). The locus of the farthest points at which direct rays from the antenna become tangential to a planetary surface. *Note:* On a spherical surface the horizon is a circle. The distance to the horizon is affected by atmospheric refraction. *See also:* **radiation; radio wave propagation.** 0-3E2

radio-influence field (RIF) (electromagnetic compatibility). Radio-influence field is the radio noise field emanating from an equipment or circuit, as measured using a radio noise meter in accordance with specified methods. *See also:* **electromagnetic compatibility.** 0-27E1

radio-influence tests. Tests that consist of the application of voltage and the measurement of the corresponding radio-influence voltage produced by the device being tested. 37A100-31E11

radio influence voltage (RIV) (1) (general) (electromagnetic compatibility). The radio noise voltage appearing on conductors of electric equipment or circuits, as measured using a radio noise meter as a two-terminal voltmeter in accordance with specified methods. *See also:* **electromagnetic compatibility.** 0-27E1
(2) (insulator). The radio-frequency voltage produced, under specified conditions, by the application of an alternating voltage of 60 hertz ± 5 percent. *Note:* A radio influence voltage is a high-frequency voltage generated as a result of ionization and may be propagated by conduction, induction, radiation, or a combined effect of all three. *See also:* **insulator; lightning arrester; power distribution, overhead construction.** 29A1/76A1-0;LA1-31E7

radio interference (electromagnetic compatibility). Impairment of the reception of a wanted radio signal caused by an unwanted radio signal or a radio disturbance. *Note:* For example, the words interference and disturbance are also commonly applied to a radio disturbance. *See:* **radio disturbance; radio noise.** *See also:* **electromagnetic compatibility.** 0-27E1

radiolocation. Radio determination used for purposes other than those of radio navigation. *Note:* This may include the determination of the relative direction, position, or motion of an object or its detection, by means of the constant-velocity or rectilinear-propagation characteristics of radio waves. *See also:* **navigation; radio transmission.** E172-10E6;42A65-0

radio magnetic indicator (RMI). A combined indicating instrument that converts omnibearing indications to a display resembling an automatic-direction-finder display, one in which the indicator points toward the omnirange station; it combines omnibearing, vehicle heading, and relative bearing. *See also:* **radio navigation.** 42A65-10E6

radiometric sextant (electronic navigation). An instrument that measures the direction to a celestial body by detecting and tracking the nonvisible natural radiation of the body; such radiation includes radio, infrared, and ultraviolet. *See also:* **navigation.** 0-10E6

radiometry. The measurement of quantities associated with radiant energy. *See also:* **primary standards (illuminating engineering).** Z7A1-0

radio navigation. The use, in navigation, of radiolocation for the determination of position or direction or for warning of obstruction.
See:
aided tracking;
air-position indicator;
***A-N* radio range;**
approach path;
aural radio range;
boundary marker;
directional homing;
distance-measuring equipment;
equiphase zone;
equisignal localizer;
equisignal zone;
fan marker;
fix;
glide-slope facility;
ground-controlled approach;
ground-position indicator;
homing;
instrument landing system;
leader cable;
lobe switching;
localizer;
localizer on-course line;
loran;
magnetic direction indicator;
marker;
master station;
navigation;
omnibearing;
omnidirectional radio range;
phase localizer;
phosphor;
pulse packet;
radio beacon;
radio compass;
radio magnetic indicator;
radio range;
shoran;
slave;
slave station;
star chain;
terrain clearance;
to-from indicator;
track homing;
visual-aural radio range;
visual radio range;
***Z* marker.** 42A65-0

radio noise. An electromagnetic noise in the radio-frequency range. *See:* **radio disturbance; radio interference.** *See also:* **electromagnetic compatibility.** 0-27E1

radio-noise field strength. A measure of the field strength, at a point (as a radio receiving station), of electromagnetic waves of an interfering character. *Notes:* (1) In practice the quantity measured is not the field strength of the interfering waves, but some quantity that is proportional to, or bears a known relation to the field strength. (2) It is commonly measured in average microvolts, quasi-peak microvolts, peak microvolts, or peak microvolts in a unit band per meter, according to which detector function of a radio-noise meter is used. *See also:* **broad-band radio noise; narrow-band radio noise; radiation.** 42A65/63A4-0

radiophare (electronic navigation). Term often used in international terminology, meaning radio beacon. *See also:* **navigation.** 0-10E6

radio propagation path (mobile communication). For a radio wave propagating from one point to another, the great-circle distance between the transmitter and receiver antenna sites. *See also:* **mobile communication system.** 0-6E1

radio proximity fuse. A radio device contained in a missile to detonate it within predetermined limits of distance from a target by means of electromagnetic interaction with the target. *See also:* **radio transmission.** E145/42A65-0

radio range (electronic navigation). A radio facility that provides radial lines of position by having characteristics in its emission that are convertible to bearing information and useful in lateral guidance of aircraft. *See also:* **navigation; radio navigation.** 42A65-10E6

radio range-finding. The determination of the range of an object by means of radio waves. *See also:* **radio transmission.** 42A65-0

radio receiver. A device for converting radio-frequency power into perceptible signals.
See:
balanced mixer;
balancer;
band spreading;
bandwidth;
beacon receiver;
beat note;
coaxial transmission line;
compensated-loop direction-finder;
compensator;
cross talk;
differential-gain control;
diversity reception;
double-superheterodyne reception;
dummy antenna;
electric field strength;
frequency-diversity reception;
image frequency;
image ratio;
image-response reradiation;
intermediate-frequency response ratio;
limiter;
limiting;
maximum output;
maximum undistorted output;
monitoring radio receiver;
noise factor;
noise temperature;
polarization-diversity reception;
quiet automatic volume control;
quieting sensitivity;
quiet tuning;
radio direction-finder;
radio transmission;
reconditioned-carrier reception;
root-sum-square;
second-channel interference;
selectivity;
sense finder;
sensitivity;
sidebands;
space-diversity reception;
spurious-response ratio;
superheterodyne reception;

superregeneration;
tone control. 0-6E1

radio relay system (radio relay). A point-to-point radio transmission system in which the signals are received and retransmitted by one or more intermediate radio stations. *See also:* **radio transmission.** 42A65-0

radio shielding. A metallic covering in the form of conduit and electrically continuous housings for airplane electric accessories, components, and wiring, to eliminate radio interference from aircraft electronic equipment. *See also:* **air-transportation electric equipment.** 42A41-0

radiosonde. An automatic radio transmitter in the meteorological-aids service, usually carried on an aircraft, free balloon, kite, or parachute, that transmits meteorological data. *See also:* **radio transmitter.** E145/42A65-0

radio station. A complete assemblage of equipment for radio transmission or reception, or both. *See also:* **radio transmission.** 42A65-0

radio transmission. The transmission of signals by means of radiated electromagnetic waves other than light or heat waves. *Note:* For an extensive list of cross references, see *Appendix A.* 42A65-0

radio transmitter. A device for producing radio-frequency power, for purposes of radio transmission. *See:*
alternator transmitter;
amplitude-modulated transmitter;
automatic frequency control;
automatic grid bias;
automatic load control;
average power output;
carrier-frequency range of a transmitter;
carrier-frequency stability of a transmitter;
crystal-controlled transmitter;
crystal-stabilized transmitter;
double-sideband transmitter;
fixed-frequency transmitter;
fixed transmitter;
frequency-modulated transmitter;
frequency tolerance of a radio transmitter;
instantaneous power output;
jamming;
local control;
magnetic storm;
marking wave;
mobile transmitter;
modulated amplifier;
multichannel transmitter;
multifrequency transmitter;
peak power output;
phase-modulated transmitter;
portable transmitter;
primary service area;
pulse transmitter;
radiosonde;
radio transmission;
remote control;
secondary service area;
semiremote control;
spacing wave;
spark transmitter;
spurious output;
spurious transmitter output, conducted;
spurious transmitter output, extraband;
spurious transmitter output, inband;
spurious transmitter output, radiated;
squelch circuit;
static;
strays;
transportable transmitter;
tuner;
vacuum-tube transmitter;
vestigial-sideband transmitter. E145/42A65-0

radio warning. *See:* **radio detection.**

radio wave. An electromagnetic wave of radio frequency. *See also:* **radio frequency; radio wave propagation.** 0-3E2

radio wave propagation. The transfer of energy by electromagnetic radiation at radio frequencies. *Note:* For an extensive list of cross references, see *Appendix A.* 0-3E2

radix. A quantity whose successive integral powers are the implicit multipliers of the sequence of digits that represent a number in some positional-notation systems. For example, if the radix is 5, then 143.2 means 1 times 5 to the second power, plus 4 times 5 to the first power, plus 3 times 5 to the zero power, plus 2 times 5 to the minus one power. Synonymous with base. *See:* **mixed radix.** *See also:* **electronic digital computer.** 0-16E9

radix complement. A numeral in radix notation that can be derived from another by subtracting each digit from one less than the radix and then adding one to the least-significant digit of the difference, executing all carries required, for example, tens complement in decimal notation, twos complement in binary notation. *See:* **true complement.** *SEe also:* **electronic digital computer.** X3A12-16E9

radix-minus-one complement. A numeral in radix notation that can be derived from another by subtracting each digit from one less than the radix, for example, nines complement in decimal notation, ones complement in binary notation. X3A12-16E9

radome (antenna). An enclosure for protecting an antenna from the harmful effects of its physical environment, generally intended to leave the electric performance of the antenna unaffected. *See also:* **antenna, circular scanning.** E149-3E1

rail (soleplate) (rotating machinery). A support fastened to a foundation on which a frame foot can be mounted. *See also:* **cradle base (rotating machinery).** 0-31E8

rail clamp (mining). A device for connecting a conductor or a portable cable to the track rails that serve as the return power circuit in mines. *See also:* **mine feeder circuit.** 42A85-0

railway signal and interlocking. *Note:* For an extensive list of cross references, see *Appendix A.*

rain clutter (radar). Return from rain that impairs or obscures returns from targets. *See also:* **navigation.** 0-10E6

raintight (weathertight). So constructed or protected that exposure to a beating rain will not result in the entrance of water. 42A95-0

raise-lower control point. *See:* **supervisory control point, raise-lower.**

ramp (1) (electronic circuits). A voltage or current that varies at a constant rate; for example, that portion of the output waveform of a time–linear-sweep generator used as a time base for an oscilloscope display. *See:* **signal, unit-ramp.** *See also:* **oscillograph.** 0-9E4

(2) (railway control). A roadway element consisting of a metal bar of limited length, with sloping ends, fixed on the roadway, designed to make contact with and raise vertically a member supported on the vehicle. 42A42-0

ramp response (null-balancing electric instrument). A criterion of the dynamic response of an instrument when subjected to a measured signal that varies at a constant rate. *See also:* **accuracy rating (instrument).** 39A4-0

ramp response time (null-balancing electric instrument). The time lag, expressed in seconds, between the measured signal and the equivalent positioning of the end device when the measured signal is varying at constant rate. *See also:* **accuracy rating (instrument).** 39A4-0

ramp response-time rating (null-balancing electric instrument). The maximum ramp response time for all rates of change of measured signal not exceeding the average velocity corresponding to the span step-response-time rating of the instrument when the instrument is used under rated operating conditions. *Example:* If the span step-response-time rating is 4 seconds, the ramp response-time rating shall apply to any rate of change of measured signal not exceeding 25 percent of span per second. *See also:* **accuracy rating (instrument).** 39A4-0

ramp shoe. *See:* **shoe.**

random access (computing systems). (1) Pertaining to the process of obtaining data from, or placing data into storage where the time required for such access is independent of the location of the data most recently obtained or placed in storage. (2) Pertaining to a storage device in which the access time is effectively independent of the location of the data. *See also:* **electronic digital computer.** X3A12-16E9

random errors (electronic navigation). The errors that cannot be predicted except on a statistical basis. *See also:* **navigation.** 0-10E6

random failure (reliability). *See:* **failure, random.**

random noise (fluctuation noise). Noise that comprises transient disturbances occurring at random. *Note:* The part of the noise that is unpredictable except in a statistical sense. The term is most frequently applied to the limiting case where the number of transient disturbances per unit time is large, so that the spectral characteristics are the same as those of thermal noise. Thermal noise and shot noise are special cases of random noise. *See also:* **drift; signal-to-noise ratio.** E165 /42A65-0

random number. *See:* **pseudorandom number sequence.**

random paralleling (rotating machinery). Paralleling of an alternating-current machine by adjusting its voltage to be equal to that of the system, but without adjusting the frequency and phase angle of the incoming machine to be sensibly equal to those of the system. *See also:* **asynchronous machine; synchronous machine.** 0-31E8

random winding (rotating machinery). A winding in which the individual conductors of a coil side occupy random position in a slot, and the ends are shaped during winding. *Note:* Synonymous with **mush-wound; pancake-wound.** *See also:* **rotor (rotating machinery); stator.** 0-31E8

random-wound motorette. A motorette for random-wound coils. *See also:* **direct-current commutating machine.** 0-31E8

range (1) (instrument). The region between the limits within which a quantity is measured, expressed by stating the lower- and upper-range values. *Notes:* (1) For example: (A) 0 to 150 degrees Fahrenheit, (B) 20 to 200 degrees Fahrenheit, and (C) −20 to 150 degrees Fahrenheit. (2) The following compound terms are used with suitable modifications in the units: measured-variable range, measured-signal range, indicating-scale range, chart-scale range, etcetera. (3) For multirange devices, this definition applies to the particular range that the device is set to measure. *See also:* **instrument.** 0-31E8

(2) (orientation range) (printing telegraphy). That fraction of a perfect signal element through which the time of selection may be varied so as to occur earlier or later than the normal time of selection, without causing errors while signals are being received. The range of a printing-telegraph receiving device is commonly measured in percent of a perfect signal element by adjusting the indicator. *See also:* **instrument; telegraphy.** 42A65-19E4

(3) (computing systems). (A) The set of values that a quantity or function may assume. (B) The difference between the highest and lowest value that a quantity or function may assume. *See:* **error range.** *See also:* **dynamic range; electronic digital computer; instrument.** X3A12-16E9

(4) (electronic navigation). An ambiguous term meaning either: (A) a distance, as in artillery techniques and radar measurements or (B) a line of position, located with respect to ground references, such as a very-high-frequency omnidirectional radio range (VOR) station, or a pair of lighthouses, or an aural radio range (A-N) radio beacon. *Note:* In electronic navigation, the reader must be particularly wary, since the two meanings of the word **range** often occur in close proximity. *See:* **radio range.** 0-E

(5) (radar techniques and artillery-fire control). Deprecated, because the term invites confusion with **radio range**. Use the common synonym, **distance**.

range and elevation guidance for approach and landing (regal). A ground-based navigation system used in conjunction with a localizer to compute vertical guidance for proper glide-slope and flare-out during an instrument approach and landing; it uses a digitally coded vertically scanning fan bean that provides data for both elevation angle and distance. *See also:* **navigation.** 0-10E6

range lights. Groups of color-coded boundary lights provided to indicate the direction and limits of a preferred landing path normally on an aerodrome without runways but exceptionally on an aerodrome with runways. *See also:* **signal lighting.** Z7A1-0

range mark (electronic navigation). *See:* **distance mark.**

range noise (radar). The noiselike variation in the apparent distance of a target because of change of target aspect; a component of scintillation. *See also:* **navigation.** 0-10E6

range resolution (electronic navigation). *See:* **distance resolution.**

rank (network) (degrees of freedom on a node basis). The number of independent cut-sets that can be selected in a network. The rank R is equal to the number of nodes V minus the number of separate parts P. $R=V-P$. *See also:* **network analysis.** E153/E270-0

rapid-start fluorescent lamp. A fluorescent lamp designed for operation with a ballast that provides a low-voltage winding for preheating the electrodes and initiating the arc without a starting switch or the application of high voltage. *See also:* **fluorescent lamp.** Z7A1-0

rapid-starting systems (fluorescent lamps). The designation given to those systems in which hot-cathode electric discharge lamps are operated with cathodes continuously heated through low-voltage heater windings built as part of the ballast, or through separate low-voltage secondary transformers. Sufficient voltage is applied across the lamp and between the lamp and fixture to initiate the discharge when the cathodes reach a temperature high enough for adequate emission. The cathode-heating current is maintained even after the lamp is in full operation. *Note:* In Europe this system is sometimes referred to as an instant-start system. 82A1-0

raster (cathode-ray tubes). A predetermined pattern of scanning lines that provides substantially uniform coverage of an area. *See also:* **beam tubes; oscillograph; television.** E160-2E2/15E6;E204-2E2/9E4;42A65-2E2

raster burn (camera tubes). A change in the characteristics of that area of the target that has been scanned, resulting in a spurious signal corresponding to that area when a larger or tilted raster is scanned. *See also:* **beam tubes.** E160-2E2/15E6

ratchet demand (electric power utilization). The maximum past or present demands that are taken into account to establish billings for previous or subsequent periods. *See also:* **alternating-current distribution.** 0-31E4

ratchet demand clause (electric power utilization). A clause in a rate schedule that provides that maximum past or present demands be taken into account to establish billings for previous or subsequent periods. *See also:* **alternating-current distribution.** 0-31E4

ratchet relay. A stepping relay actuated by an armature-driven ratchet. *See also:* **relay.** 83A16-0

rate control action (electric power systems). Action in which the output of the controller is proportional to the input signal and the first derivative of the input signal. Rate time is the time interval by which the rate action advances the effect of the proportional control action. *Note:* Applies only to a controller with proportional control action plus derivative control action. *See also:* **speed-governing system.** E94-0

rated. A qualifying term that, applied to an operating characteristic, indicates the designated limit or limits of the characteristic for application under specified conditions. *Note:* The specific limit or limits applicable to a given device is specified in the standard for that device and included in the title of the rated characteristic, that is, rated maximum voltage, rated frequency range, etcetera. 37A100-31E11

rated accuracy (automatic null-balancing electric instrument). The limit that errors will not exceed when the instrument is used under any combination of rated operating conditions. *Notes:* (1) It is usually expressed as a percent of the span. It is preferred that a + sign or − sign or both precede the number or quantity. The absence of a sign infers a ± sign. (2) Rated accuracy does not include accuracy of sensing elements or intermediate means external to the instrument. *See also:* **accuracy rating (instrument).** 39A4-0

rated accuracy of instrument shunts (electric power systems). The limit of error, expressed as a percentage of rated voltage drop, with two-thirds rated current applied for one-half hour to allow for self-heating. *Note:* Practically, it represents the expected accuracy of the shunt obtainable over normal operating current ranges. *See also:* **accuracy rating (instrument); power system, low-frequency and surge testing.** 0-31E5

rated alternating-current winding voltages (rectifier). The root-mean-square voltages between the alternating-current line terminals that are specified as the basis for rating. *Note:* When the alternating-current winding of the rectifier transformer is provided with taps, the rated voltage shall refer to a specified tap that is designated as the rated-voltage tap. *See also:* **rectifier transformer.** 57A18-0

rated alternating voltage (rectifier unit). The root-mean-square voltage between the alternating-current line terminals that are specified as the basis for rating. *Note:* When the alternating-current winding of the rectifier transformer is provided with taps, the rated voltage shall refer to a specified tap that is designated as the rated-voltage tap. *See also:* **rectification.** 34A1-34E24

rated average tube current. The current capacity of a tube, in average amperes, as assigned to it by the manufacturer for specified circuit conditions. *See also:* **rectification.** 34A1-0

rated burden (capacitance potential device). The maximum unity-power-factor burden, specified in watts at rated secondary voltage, that can be carried for an unlimited period when energized at rated primary line-to-ground voltage, without causing the established limitations to be exceeded. *See also:* **outdoor coupling capacitor.** E31-0

rated circuit voltage. Used to designate the rated, root-mean-square, line-to-line, voltage of the circuit on which coupling capacitors or the capacitance potential device in combination with its coupling capacitor or bushing is designed to operate. *See also:* **outdoor coupling capacitor.** E31-0

rated continuous current (bushing for use in oil circuit breakers and transformers). The current expressed in amperes, root-mean-square, that it can carry continuously under specified service conditions without exceeding the allowable temperature rise. *See also:* **power distribution, overhead construction.** E45/76A1-0

rated current (1) (machine or apparatus). The value of the current that is used in the specification and from which the conditions of temperature rise, as well as the operation of the equipment, are calculated.

(2) For equipment where the rated output is mechanical power or some related quantity such as air flow per minute, it is determined from the current at rated output under standard test conditions. E270-0

(3) (instrument) (wattmeter or power-factor meter). The normal value of current used for design purposes. 39A1/42A30-0

(4) (equipment or of a winding). The current for which the equipment, or winding, is designed, and to which certain operating and performance characteristics are referred. 42A15-31E12

(5) (neutral grounding device) (electric power). The thermal current rating. The rated current of resistors whose rating is based on constant voltage is the initial root-mean-square symmetrical value of the current that will flow when rated voltage is applied. *See also:* **grounding device.** E32/42A35-31E13

rated direct-current winding voltage (rectifier). The root-mean-square voltage of the direct-current winding obtained by turns ratio from the rated alternating-current winding voltage of the rectifier transformer. *See also:* **rectifier transformer.** 57A18-0

rated direct voltage (power inverter). The nominal direct input voltage. *See also:* **self-commutated inverters.** 0-34E24

rated duty. That duty that the particular machine or apparatus has been designed to comply with. E270-0

rated excitation-system voltage (rotating machinery). The main exciter rated voltage. *See also:* **synchronous machine.** 0-31E8

rated frequency (1) (general). The frequency used in the specification of the apparatus and from which the test conditions and the frequency limits for the use of the equipment are calculated. E270-0

(2) (power system or interconnected system). The normal frequency in hertz for which alternating-current generating equipment operating on such system is designed. *See also:* **power system, low-frequency and surge testing.** E94-0

(3) (arrester). The frequency, or range of frequencies, of the power systems on which the arrester is designed to be used. *See also:* **lightning arrester (surge diverter).** 99I2-31E7

(4) (grounding device). The frequency of the alternating current for which it is designed. *Note:* Some devices, such as neutral wave traps, may have two or more rated frequencies: the rated frequency of the circuit and the frequencies of the harmonic or harmonics the devices are designed to control. *See also:* **grounding devices.** E32/42A35-31E13

rated impulse protective level (arrester). The impulse protective level with the residual voltage referred to the nominal discharge current. *See also:* **lightning arrester (surge diverter).** 0-31E7

rated impulse withstand voltage (apparatus). An assigned crest value of a specified impulse voltage wave that the apparatus must withstand without flashover, disruptive discharge, or other electric failure. *See also:* **lightning arrester (surge diverter); power systems, low-frequency and surge testing.** E270/42A35-31E13/31E7

rated insulation class (neutral grounding device) (electric power). An insulation class expressed in root-mean-square kilovolts, that determines the dielectric tests that the device shall be capable of withstanding. *See also:* **grounding device; outdoor coupling capacitor.** E31/E32-0

rated kilovolt-ampere (kVA) (transformer or reactor) (1) (transformer). The output that can be delivered for the time specified at rated secondary voltage and rated frequency without exceeding the specified temperature-rise limitations under prescribed conditions of test, and within the limitations of established standards. *See also:* **duty.** 42A15-31E12

(2) (current-limiting reactor). The kilovolt-amperes that can be carried for the time specified at rated frequency without exceeding the specified temperature limitations, and within the limitations of established standards. *See also:* **reactor.** 57A16-0

(3) (grounding transformer). The short-time kilovolt-ampere rating is the product of the rated line-to-neutral voltage at rated frequency, and the maximum constant current that can flow in the neutral for the specified time without causing specified temperature-rise limitations to be exceeded, and within the limitations of established standards for such equipment. *See also:* **duty.** 42A15-31E12

rated kilovolt-ampere (kVA) tap (transformer winding). A tap through which the transformer can deliver its rated kilovolt-ampere output without exceeding the specified temperature rise. *Note:* The term **rated kilowatt output** is deprecated, unless power factor is specified. *See:* **windings, high-voltage and low-voltage.** 42A15-31E12

rated kilowatts (constant-current transformer). The kilowatt output at the secondary terminals with rated primary voltage and frequency, with rated secondary current and power factor, and within the limitations of established standards. *See also:* **duty.** 42A15-31E12;57A14-0

rated life (glow lamp). The length of operating time, expressed in hours, that produces specified changes in characteristics. *Note:* In lamps for indicator use the characteristic usually is light output; the end of useful life is considered to be when light output reaches 50 percent of initial, or when the lamp becomes inoperative at line voltage. In lamps used as circuit components, the characteristic is usually voltage; life is determined as the length of time for a specified change from initial. 78A385-0

rated line kilovolt-ampere (kVA) rating (rectifier transformer). The kilovolt-ampere rating assigned to it by the manufacturer corresponding to the kilovolt-ampere drawn from the alternating-current system at rated voltage and kilowatt load on the rectifier under the normal mode of operation. *See also:* **rectifier transformer.** 57A18-0

rated load (1) (elevator, dumbwaiter, escalator, or private residence inclined lift). The load which the device is designed and installed to lift at the rated speed. *See also:* **elevator.** 42A45-0

(2)(rectifier unit). The kilowatt power output that can be delivered continuously at the rated output voltage. It may also be designated as the one-hundred-percent-load or full-load rating of the unit. *Note:* Where the rating of a rectifier unit does not designate a continuous load it is considered special. *See:* **continuous rating.** *See also:* **rectification.** 42A15-34E24

rated-load field voltage (nominal collector ring voltage) (rotating machinery). The voltage required across the terminals of the field winding of an electric machine under rated continuous-load conditions with the field winding at: (1) 75 degrees Celsius for field windings designed to operate at rating with a temperature rise of 60 degrees Celsius or less. (2) 100 degrees Celsius for field windings designed to operate at rating with a temperature rise greater than 60 degrees Celsius. 42A10-0;0-31E8

rated-load impedance (magnetic amplifier). The load impedance specified by the manufacturer that is determined by the specifications to which the amplifier is designed. *See also:* **rating and testing magnetic amplifiers.** E107-0

rated-load torque (rated torque) (rotating machinery). The shaft torque necessary to produce rated power output at rated-load speed. *See also:* **asynchronous machine; direct-current commutating machine; synchronous machine.** 42A10-31E8

rated OFF voltage (magnetic amplifier). The output voltage existing with trip OFF control signal applied. *See also:* **rating and testing magnetic amplifiers.** E107-0

rated ON voltage (magnetic amplifier). The output voltage existing with trip ON control signal applied. Rated ON voltage shall be specified either as root-mean-square or average. *Note:* While specification may be either root-mean-square or average it remains fixed for a given amplifier. *See also:* **rating and testing magnetic amplifiers.** E107-0

rated operating conditions (automatic null-balancing electric instrument). The limits of specified variables or conditions within which the performance ratings apply. *See also:* **measurement system.** 39A4-0

rated output capacity (inverters). The kilovolt-ampere output at specified load power-factor conditions. *See also:* **self-commutated inverters.** 0-34E24

rated output current (1) (general). The output current at the continuous rating for a specified circuit. E59-34E17

(2) (rectifier unit). The current derived from the rated load and the rated output voltage. *See also:* **average forward-current rating (rectifier circuit); rectification.** 42A15-34E24

(3) (magnetic amplifier). Rated output current that the amplifier is capable of supplying to the rated load impedance, either continuously or for designated operating intervals, under nominal conditions of supply voltage, supply frequency, and ambient temperature such that the intended life of the amplifier is not reduced or a specified temperature rise is not exceeded. Rated output current shall be specified either as root-mean-square or average. *Notes:* (1) When other than rated load impedance is used, the root-mean-square value of the rated output current should not be exceeded. (2) While specification may be either root-mean-square or average, it remains fixed for a given amplifier. *See also:* **rating and testing magnetic amplifiers.** E107-0

(4) (power inverter). The nominal effective (total root-mean-square) current that can be obtained at rated output voltage. *See also:* **self-commutated inverters.** 0-34E24

rated output frequency (inverters). The fundamental frequency or the frequency range over which the output fundamental frequency may be adjusted. *See also:* **self-commutated inverters.** 0-34E24

rated output voltage (1) (general). The output voltage at the continuous rating for a specified circuit. E59-34E17

(2) (rectifier unit). The voltage specified as the basis of rating. *See also:* **average forward-current rating (rectifier circuit); rectification.** 42A15-34E24

(3) (magnetic amplifier). The voltage across the rated load impedance when rated output current flows. Rated output voltage shall be specified by the same measure as rated output current (that is, both shall be stated as root-mean-square or average). *Note:* While specification may be either root-mean-square or average, it remains fixed for a given amplifier. *See also:* **rating and testing magnetic amplifiers.** E107-0

(4) (power inverter). The nominal effective (total root-mean-square) voltage that is used to establish the power-inverter rating. *See also:* **self-commutated inverters.** 0-34E24

rated output volt-amperes (magnetic amplifier). The product of the rated output voltage and the rated output current. *See also:* **rating and testing magnetic amplifiers.** E107-0

rated performance (automatic null-balancing electric instrument). The limits of the values of certain operating characteristics of the instrument that will not be exceeded under any combination of rated operating conditions. *See also:* **to test an instrument or meter.** 39A4-0

rated power output (gas turbines (1) (normal rated power). The rated or guaranteed power output of the gas turbine when it is operated with an inlet temperature of 80 degrees Fahrenheit (26.67 degrees Celsius) and with inlet and exhaust absolute pressures of 14.17 pounds-force per square inch (9.770 newtons per square centimeter). *See also:* **direct-current commutating machine; synchronous machine; asynchronous machine.** E282-31E2

(2) (site rated power). The rated or guaranteed power output of the gas turbine when it is operated under specified conditions of compressor inlet temperature, compressor inlet pressure, and gas-turbine exhaust pressure. It is measured at, or is referred to, the output of the generator terminals, the gas-turbine generator unit, or the specified generator if separately purchased. E282-31E2

rated primary current (current transformer). Current selected for the basis of performance specifications. *See also:* **instrument transformer.** 57A13-31E12;42A15/42A30-0

rated primary line-to-ground voltage. The root-mean-square line-to-ground voltage for which the potential device, in combination with its coupling capacitor or bushing, is designed to deliver rated burden at rated secondary voltage. The rated primary line-to-ground voltage is equal to the rated circuit voltage (line-to-line) divided by $(3)^{1/2}$. *See:* **primary line-to-ground voltage.** E31-0

rated primary voltage (1) (constant-current transformer). The primary voltage for which the transformer is designed, and to which operation and performance characteristics are referred. *See also:* **constant-current transformer.** 42A15-0

(2) (constant-voltage transformer). The voltage obtained from the rated secondary voltage by turn ratio. *Notes:* (A) See **turn ratio of a transformer** and its note, for the definition of the turn ratio to be used. (B) In the case of a multiwinding transformer, the rated voltage of any other winding is obtained in a similar manner. *See also:* **duty.** 42A15-31E12

(3) (instrument transformer). (A) The rated primary voltage (of a potential (voltage) transformer) is the voltage selected for the basis of performance guarantees. (B) The rated primary voltage (of a current transformer) designates the insulation class of the primary winding. *Note:* A current transformer can be applied on a circuit having a nominal system voltage corresponding to or less than the rated primary voltage of the current transformer. *See also:* **instrument transformer.** 42A15/42A30-0;57A13-31E12

rated range of regulation (voltage regulator). The amount that the regulator will raise or lower its rated voltage. *Note:* The rated range may be expressed in per unit, or in percent, of rated voltage; or it may be expressed in kilovolts. *See also:* **voltage regulator.** 57A15-0

rated secondary current (1) (constant-current transformer). The secondary current for which the transformer is designed and to which operation and performance characteristics are referred. *See also:* **constant-current transformer.** 42A15-31E12
(2) (constant-voltage transformer). The secondary current obtained by dividing the rated kilovolt-ampere, by the rated secondary voltage. *Note:* This applies only to single-phase transformers. For polyphase transformers, as well as for single-phase transformers, the rated secondary current is the secondary current existing when the transformer is delivering rated kilovolt-amperes, at rated secondary voltage. *See also:* **duty.** 42A15-0

rated secondary voltage (1) (constant-voltage transformer). The voltage at which the transformer is designed to deliver rated kilovolt-ampere and to which operating and performance characteristics are referred. *See also:* **duty.** 42A15-31E12
(2) (capacitance potential device). This is the root-mean-square secondary voltage for which the potential device, in combination with its coupling capacitor or bushing, is designed to deliver its rated burden when energized at rated primary line-to-ground voltage. *See:* **secondary voltage.** *See also:* **outdoor coupling capacitor.** E31-0

rated speed (1) (gas turbines). The speed of the power output shaft corresponding to rated generator speed at which rated power output is developed. E282-31E2
(2) (elevators). The speed at which the elevator, dumbwaiter, escalator, or inclined lift is designed to operate under the following conditions: (A) **Elevator or dumbwaiter:** The speed in the up direction with rated load in the car. (B) **Escalator or private-residence inclined lift:** The rate of travel of the steps or carriage, measured along the angle of inclination, with rated load on the steps or carriage. In the case of a reversible escalator the rated speed shall be the rate of travel of the steps in the up direction, measured along the angle of inclination, with rated load on the steps. *See also:* **elevator.** 42A45-0

rated supply current (magnetic amplifier). The root-mean-square current drawn from the supply when the amplifier delivers rated output current. *See also:* **rating and testing magnetic amplifiers.** E107-0

rated system deviation (mobile communication). The greatest frequency deviation specified in the operation of the radio system. *See also:* **mobile communication system.** 0-6E1

rated system voltage (current-limiting reactor). The voltage to which operations and performance characteristics are referred. It corresponds to the nominal system voltage of the circuit on which the reactor is intended to be used. *See also:* **reactor.** 57A16-0

rated time (grounding device) (electric power). The time during which it will carry its rated current, or withstand its rated voltage, or both, under standard conditions without exceeding standard limitations, unless otherwise specified. *See also:* **grounding device.** E32/42A35-31E13

rated torque (rotating machinery). *See:* **rated-load torque.**

rated voltage (1) (general). The voltage to which operating and performance characteristics of apparatus and equipment are referred. *Note:* Deviation from rated voltage may not impair operation of equipment, but specified performance characteristics are based on operation under rated conditions. However, in many cases apparatus standards specify a range of voltage within which successful performance may be expected. *See also:* **power systems, low-frequency and surge testing.** 24A1/67A1/57A15/E270-0;42A15-31E12
(2) (step-voltage or induction-voltage regulator). The voltage between terminals of the series winding, with rated voltage applied to the regulator, when the regulator is in the maximum raise position and is delivering rated kilovolt-ampere output at 80-percent power factor. *See also:* **voltage regulator.** 57A15-0
(3) (electric instrument). Of an instrument such as a wattmeter, power-factor meter, or frequency meter, the rated voltage is the normal continuous operating voltage of the voltage circuit and the normal value of applied voltage for test purposes. *Note:* This quantity should not be confused with the voltage rating of any auxiliary motors, amplifiers, or other devices. *See also:* **instrument.** 39A1/42A30-0
(4) (arrester). The designated maximum permissible root-mean-square value of power-frequency voltage between its line and earth terminals at which it is designed to operate correctly. *See also:* **lightning arrester (surge diverter).** 0-31E7
(5) (grounding device) (electric power). The root-mean-square voltage, at rated frequency, that may be impressed between its terminals under standard conditions for its rated time without exceeding standard limitations, unless otherwise specified. *See also:* **grounding device.** E32/42A35-31E13

rated watts input (household electric ranges). The power input in watts (or kilowatts) that is marked on the range nameplate, heating units, etcetera. *See also:* **appliance outlet.** 71A1-0

rated withstand current (surge current) (lightning arresters). The crest value of a surge, of given wave shape and polarity, to be applied under specified conditions without causing disruptive discharge on the test specimen. *See also:* **power systems, low-frequency and surge testing; lightning arrester (surge diverter).** 42A35-31E13/31E7

rated withstand voltage (insulation strength). The voltage that electric equipment is required to withstand without failure or disruptive discharge when tested under specified conditions and within the limitations of established standards. *See also:* **basic impulse insulation level (insulation strength).** 92A1-0

rate-of-change protection. A form of protection in which an abnormal condition causes disconnection or inhibits connection of the protected equipment in accordance with the rate of change of current, voltage, power, frequency, pressure, etcetera. 37A100-31E11/31E6

rate-of-change relay. A relay that responds to the rate of change of current, voltage, power, frequency, pressure, etcetera. 37A100-31E11/31E6

rate of decay (audio and electroacoustics). The time rate at which the sound pressure level (or other stated characteristic) decreases at a given point and at a given time. *Note:* Rate of decay is frequently expressed in decibels per second. *See also:* **electroacoustics.** 0-1E1

rate-of-rise current tripping. *See:* **rate-of-rise release (rate-of-rise trip).**

rate-of-rise fire-alarm thermostat. A fire-alarm thermostat designed to operate when the rate of temperature increase exceeds a predetermined value. *See also:* **protective signaling.** 42A65-0

rate of rise of restriking voltage (transient recovery voltage rate) (usually abbreviated to rrrv) (lightning arresters). The rate, expressed in volts per microsecond, that is representative of the increase of the restriking voltage. *See also:* **lightning arrester (surge diverter).** 99I2-31E7

rate-of-rise release (rate-of-rise trip). A release that operates when the rate of rise of the actuating quantity in the main circuit exceeds the release setting. 37A100-31E11

rate-of-rise suppressors (semiconductor rectifiers). Devices used to control the rate of rise of current and/or voltage to the semiconductor devices in a semiconductor power converter. *See also:* **semiconductor rectifier stack.** 0-34E24

rate-of-rise trip. *See:* **rate-of-rise release.**

rate signal (industrial control). A signal that is the time derivative of a specified variable. *See also:* **control system, feedback.** AS1-34E10

rate test. *See:* **problem check.**

rating (rating of electric equipment) (1) (general). The whole of the electrical and mechanical quantities assigned to the machine, apparatus, etcetera, by the designer, to define its working in specified conditions indicated on the rating plate. *Note:* The rating of electric apparatus in general is expressed in volt-amperes, horsepower, kilowatts, or other appropriate units. Resistors are generally rated in ohms, amperes, and class of service. E96/E270/42A25-SCC4

(2) (rotating machinery). The numerical values of electrical quantities (frequency, voltage, current, apparent and active power, power factor) and mechanical quantities (power, torque), with their duration and sequences, that express the capability and limitations of a machine. The rated values are usually associated with a limiting temperature rise of insulation and metallic parts. *See also:* **asynchronous machine; direct-current commutating machine; synchronous machine.** 0-31E8

(3) (device). The designated limit(s) of the rated operating characteristic(s). *Note:* Such operating characteristics as current, voltage, frequency, etcetera, may be given in the rating. 37A100-31E11/31E6

(4) (controller). An arbitrary designation of an operating limit. It is based on power governed, the duty and service required. *Note:* A rating is arbitrary in the sense that it must necessarily be established by definite field standards and cannot, therefore, indicate the safe operating limit under all conditions that may occur. *See also:* **electric controller.** 42A25-34E10

(5) (current-limiting reactor). The volt-amperes that it can carry, together with any other characteristics, such as system voltage, current, and frequency assigned to it by the manufacturer. *Note:* It is regarded as a test rating that defines an output that can be carried under prescribed conditions of test, and within the limitations of established standards. *See also:* **reactor.** 57A16-0

(6) (arc-welding apparatus). A designated limit of output capacity in applicable terms, such as load, voltage, current, and duty cycle, together with any other characteristics necessary to define its performance. It shall be regarded as a test rating obtained within the limitations of established test standards. *See also:* **electric arc-welding apparatus.** 87A1-0

(7) (mechanical or thermal short-circuit) (instrument transformer). Short-time emergency rating for fault conditions. Short-circuit rating of a current transformer defines the ability of the transformer to withstand an overload current resulting from a short-circuit or other fault condition in the line in which primary winding is connected. Short-circuit rating of a potential transformer defines the ability of the transformer to withstand a short-circuit directly across the secondary terminals. 57A13-31E12

(8) (interphase transformer). The root-mean-square current, root-mean-square voltage, and frequency at the terminals of each winding, when the rectifier unit is operating at rated load and with a designated amount of phase control. *See also:* **duty; rectifier transformer.** 57A18-0

(9) (rectifier transformer). The kilovolt-ampere output, voltage, current, frequency, and number of phases at the terminals of the alternating-current winding; the voltage (based on turn ratio of the transformer), root-mean-square current, and number of phases at the terminals of the direct-current winding, to correspond to the rated load of the rectifier unit. *Notes:* (1) Because of the current wave shapes in the alternating- and direct-current windings of the rectifier transformer, these windings may have individual ratings different from each other and from those of power transformers in other types of service. The ratings are regarded as test ratings that define the output that can be taken from the transformer under prescribed conditions of test without exceeding any of the limitations of the standards. (2) For rectifier transformers covered by established standards, the root-mean-square current ratings and kilovolt-ampere ratings of the windings are based on values derived from rectangular rectifier circuit element currents without overlap. *See also:* **rectifier transformer.** 42A15/57A18-31E12

(10) (lightning arrester). The designation of an operating limit. *See also:* **current rating, 60-hertz (arrester).** E28-0

(11) (power-inverter unit). The kilovolt-amperes, power output, voltages, currents, number of phases, frequency, etcetera, assigned to it by the manufacturer. *See also:* **self-commutated inverters.** 0-34E24

(12) (rating of storage batteries) (storage cell). The number of ampere-hours that the batteries are capable of delivering when fully charged and under specified conditions as to temperature, rate of discharge, and final voltage. For particular classes of service different time rates are frequently specified. *See also:* **charge; storage battery; time rate.** 42A60-0

(13) (relay). *See:* **relay rating.**

rating and testing magnetic amplifiers. *Note:* For an extensive list of cross references, see *Appendix A.*

rating plate (rotating machinery). *See:* **nameplate.**

ratio (magnetic storage). *See:* **squareness ratio.** *See also:* **static magnetic storage.**

ratio correction factor (instrument transformer). The factor by which the marked ratio of a current or a potential transformer must be multiplied to obtain the true ratio. *Note:* This factor is expressed as the ratio of true ratio to marked ratio. If both a current transformer and a potential transformer are used in conjunction with a wattmeter or watthour meter, the resultant ratio correction factor, for use with a wattmeter or a watthour meter, is the product of the individual ratio correction factors. *See also:* **instrument transformer.** 12A0/42A30-31E12

ratio meter. An instrument that measures electrically the quotient of two quantities. A ratio meter generally has no mechanical control means, such as springs, but operates by the balancing of electromagnetic forces that are a function of the position of the moving element. *See also:* **instrument.** 42A30-0

rationalized system of equations. A rationalized system of electrical equations is one in which the proportionality factors in the equations that relate (1) the surface integral of electric flux density to the enclosed charge, and (2) the line integral of magnetizing force to the linked current, are each unity. *Notes:* (A) By these choices, some formulas applicable to configurations having spherical or circular symmetry contain an explicit factor of 4π or 2π; for example, Coulomb's law is $f=q_1q_2/(4\pi\ \epsilon_0 r^2)$. (B) The differences between the equations of a rationalized system and those of an unrationalized system may be considered to result from either (a) the use of a different set of units to measure the same quantities or (b) the use of the same set of units to measure quantities that are quantitatively different (though of the same physical nature) in the two systems. The latter consideration, which represents a changed relation between certain mathematico-physical quantities and the associated physical quantities, is sometimes called **total rationalization.** E270-0

ratio-type (position-type) telemeter. A telemeter that employs the relative phase position between, or the magnitude relation between, two or more electrical quantities as the translating means. *Note:* Examples of ratio-type telemeters include alternating-current or direct-current position-matching systems. *See also:* **telemetering.** 37A100-31E11;42A30-0

ratproof electric installation. Apparatus and wiring designed and arranged to eliminate harborage and runways for rats. *See also:* **marine electric apparatus.** 42A43-0

rat race*. *See:* **hybrid ring.**

*Deprecated

Rayleigh disk. A special form of acoustic radiometer that is used for the fundamental measurement of particle velocity. *See also:* **electroacoustics.** E157-1E1

RDF (electronic navigation). *See:* **radio direction finding.** *Note:* At one time this term was used by the British to mean radio distance finding, that is, radar. *See also:* **navigation.** 0-10E6

***R* display (radar).** An *A* display with a segment of the time base expanded near the blip for greater accuracy in distance measurement. *See also:* **navigation.** 0-10E6

reach (relay). The extent of the protection afforded by a relay in terms of the impedance or circuit length as measured from the relay location. *Note:* The measurement is usually to a point of fault, but excessive loading or system swings may also come within reach or operating range of the relay. 37A100-31E11/31E6

reactance (1) (general). The imaginary part of impedance. *See also:* **reactor.** E270-9E4

(2) (portion of a circuit for a sinusoidal current and potential difference of the same frequency). The product of the sine of the angular phase difference between the current and potential difference times the ratio of the effective potential difference to the effective current, there being no source of power in the portion of the circuit under consideration. The reactance of a circuit is different for each component of an alternating current. If

$$e = E_{1m}\sin(\omega t+\alpha_1) + E_{2m}\sin(2\omega t+\alpha_2) + \cdots$$

and

$$i = I_{1m}\sin(\omega t+\beta_1) + I_{2m}\sin(2\omega t+\beta_2) + \cdots$$

then the reactances, X_1, X_2, etcetera, for the different components are

$$X_1 = \frac{E_{1m}\sin(\alpha_1-\beta_1)}{I_{1m}} = \frac{E_1\sin(\alpha_1-\beta_1)}{I_1}$$

$$X_2 = \frac{E_{2m}\sin(\alpha_2-\beta_2)}{I_{2m}} = \frac{E_2\sin(\alpha_2-\beta_2)}{I_2}$$

etcetera. *Note:* The reactance for the entire periodic current is not the sum of the reactance of the components. A definition of reactance for a nonsinusoidal periodic current has not been agreed upon. E45-0

reactance amplifier. *See:* **parametric amplifier.**

reactance drop (general). The voltage drop in quadrature with the current. 42A15-31E12;57A15-0

reactance, effective synchronous. An assumed value of synchronous reactance used to represent a machine in a system study calculation for a particular operating condition. *See also:* **synchronous machine.** 0-31E8

reactance frequency divider. A frequency divider whose essential element is a nonlinear reactor. *Note:* The nonlinearity of the reactor is utilized to generate subharmonics of a sinusoidal source. *See also:* **parametric device.** E254-15E7

reactance frequency multiplier. A frequency multiplier whose essential element is a nonlinear reactor. *Note:* The nonlinearity of the reactor is utilized to generate harmonics of a sinusoidal source. *See also:* **parametric device.** E254-15E7

reactance function (linear passive networks). The driving-point impedance of a lossless network. *Note:* This is an odd function of the complex frequency. *See also:* **linear passive networks.** E156-0

reactance grounded (electric power). Grounded through impedance, the principal element of which is reactance. *Note:* The reactance may be inserted either directly, in the connection to ground, or indirectly by increasing the reactance of the ground return circuit. The latter may be done by intentionally increasing the zero-sequence reactance of apparatus connected to ground or by omitting some of the possible connections from apparatus neutrals to ground. *See also:* **ground.** E32-0;42A15-31E12;42A35-31E13

reactance modulator. A device, used for the purpose of modulation, whose reactance may be varied in accordance with the instantaneous amplitude of the modulating electromotive force applied thereto. *Note:* Such a device is normally an electron-tube circuit and is commonly used to effect phase or frequency modulation. *See also:* **frequency modulation (telecommunication); modulation; phase modulation.** E145/E182A-42A65

reactance relay. A linear-impedance form of distance relay for which the operating characteristic of the distance unit on an *R-X* diagram is a straight line of constant reactance. *Note:* The operating characteristic may be described by either equation $X=K$, or $Z\sin\theta=K$, where K is a constant and θ is the angle by which the input voltage leads the input current. 37A100-31E11/31E6

reaction frequency meter. *See:* **absorption frequency meter.**

reaction time. The interval between the beginning of a stimulus and the beginning of the response of an observer. *See also:* **visual field.** Z7A1-0

reactive attenuator (waveguide). An attenuator that absorbs no energy. *See also:* **waveguide.** 50I62-3E1

reactive current (rotating machinery). The component of a current in quadrature with the voltage. *See also:* **asynchronous machine; synchronous machine.** 50I05-31E8

reactive-current compensator (rotating machinery). A compensator that acts to modify the functioning of a voltage regulator in accordance with reactive current. *See also:* **synchronous machine.** 0-31E8

reactive factor. The ratio of the reactive power to the apparent power. The reactive factor is expressed by the equation

$$F_q = \frac{Q}{U}$$

where F_q = reactive factor
Q = reactive power
U = apparent power.

If the voltages have the same waveform as the corresponding currents, reactive factor becomes the same as phasor reactive factor. If the voltages and currents are sinusoidal and for polyphase circuits form symmetrical sets

$$F_q = \sin(\alpha-\beta).$$

See also: **network analysis.** E270-0

reactive-factor meter. An instrument for measuring reactive factor. It is provided with a scale graduated in reactive factor. *See also:* **instrument.** 42A30-0

reactive field (antenna). Electric and magnetic fields surrounding an antenna and resulting in storage of electromagnetic energy. *See also:* **radiation.** 0-3E1

reactive ignition cable (electromagnetic compatibility). High-tension ignition cable, the core of which is so constructed to give a high reactive impedance at radio frequencies. *See also:* **electromagnetic compatibility.** CISPR-27E1

reactive near-field region (antenna). That region of the field of an antenna immediately surrounding the antenna wherein the reactive field predominates. *Note:* For most antennas the outer boundary of the region is commonly taken to exist at $\lambda/2\pi$ where λ is the wavelength. *See also:* **antenna.** E149-0

reactive-power relay. A power relay that responds to reactive power. *See also:* **relay.** 0-31E6

reactive volt-ampere-hour meter. *See:* **varhour meter.**

reactive volt-ampere meter. *See:* **varmeter.**

reactor. A device, the primary purpose of which is to introduce reactance into a circuit. *Notes:* (1) A reactor is a device used for introducing reactance into a circuit for purposes such as motor starting, paralleling transformers, and control of current. (2) **Inductive reactance** is frequently abbreviated **inductor.**
See:
anode paralleling reactor;
circuits and devices;
commutating reactor;
compound-filled current-limiting reactor;
dry-type;
impedance voltage;
losses;
neutral grounding reactor;
oil-immersed;
rated kilovolt-amperes;
rated system voltage;
rating;
reactor, alternating-current saturable;
reactor, amplistat;
reactor, bus;
reactor, current-balancing;
reactor, diode-current-balancing;
reactor, current-limiting;
reactor, feeder;
reactor, filter;
reactor, paralleling;
reactor, shunt;
reactor, starting;
reactor, synchronizing;
short-time rating;
tap;
transformer. E45-0;42A65-21E1

reactor, amplistat. A reactor conductively connected between the direct-current winding of a rectifier transformer and rectifier circuit elements that when operating in conjunction with other similar reactors, provides a relatively small controlled direct-current voltage range at the rectifier output terminals. *See also:* **reactor.** 0-31E12

reactor, bus. A current-limiting reactor for connection between two different buses or two sections of the same bus for the purpose of limiting and localizing the disturbance due to a fault in either bus. *See also:* **reactor.** 42A15-31E12

reactor, current-balancing. A reactor used in semiconductor rectifiers to achieve satisfactory division of current among parallel-connected semiconductor diodes. *See also:* **reactor.** 0-31E12

reactor, current-limiting. A reactor intended for limiting the current that can flow in a circuit under short-circuit conditions, or under other operating conditions such as starting, synchronizing, etcetera. *See also:* **reactor.** 42A15-31E12

reactor, diode-current-balancing. A reactor with a set of mutually coupled windings that, operating in conjunction with other similar reactors, forces substantially equal division of current among the parallel paths of a rectifier circuit element. *See also:* **reactor.** 0-31E12

reactor, feeder. A current-limiting reactor for connection in series with an alternating-current feeder circuit for the purpose of limiting and localizing the disturbance due to faults on the feeder. *See also:* **reactor.** 42A15-31E12

reactor, filter. A reactor used to reduce harmonic voltage in alternating-current or direct-current circuits. *See also:* **reactor.** 42A15-31E12

reactor, neutral grounding (lightning arresters). A current-limiting inductive reactor for connection in the neutral for the purpose of limiting and neutralizing disturbances due to ground faults. *See also:* **lightning arrester (surge diverter).** 42A15-31E7

reactor, paralleling. A current-limiting reactor for correcting the division of load between parallel-connected transformers that have unequal impedance voltages. *See also:* **reactor.** 42A15-31E12

reactor, shunt. A reactor intended for connection in shunt to an electric system for the purpose of drawing inductive current. *Note:* The normal use for shunt reactors is to compensate for capacitive currents from transmission lines, cable, or shunt capacitors. The need for shunt reactors is most apparent at light load. *See also:* **reactor.** 42A15-31E12

reactor, starting. A current-limiting reactor for decreasing the starting current of a machine or device. *See also:* **reactor.** 42A15-31E12

reactor-start motor. A single-phase induction motor of the split-phase type with a main winding connected in series with a reactor for starting operation and an auxiliary winding with no added impedance external to it. For running operation, the reactor is short-circuited or otherwise made ineffective, and the auxiliary winding circuit is opened. *See also:* **asynchronous machine.** 42A10-31E8

reactor, synchronizing. A current-limiting reactor for connecting momentarily across the open contacts of a circuit-interrupting device for synchronizing purposes. *See also:* **reactor.** 42A15-31E12

read (electronic computation). To acquire information, usually from some form of storage. *See also:* **destructive read; electronic computation; nondestructive read; write.** E162/E270-0;X3A12-16E9

read-around number (storage tubes). The number of times reading operations are performed on storage elements adjacent to any given storage element without more than a specified loss of information from that element. *Note:* The sequence of operations (including priming, writing, or erasing), and the storage elements on which the operations are performed, should be specified. *See also:* **storage tube.** E158-15E6

read-around ratio* (storage tubes). *See:* **read-around number.**

*Deprecated

readily accessible. Capable of being reached quickly, for operation, renewal, or inspection, without requiring those to whom ready access is requisite to climb over or remove obstacles or to resort to portable ladders, chairs, etcetera. 42A95-0

reading (recording instrument). The value indicated by the position of the index that moves over the indicating scale. *See also:* **accuracy rating (instrument).** 42A30-0

reading rate (storage tubes). The rate of reading successive storage elements. *See also:* **storage tube.** E158-15E6

reading speed (storage tubes). The lineal scanning rate of the beam across the storage surface in reading. Note the distinction between this and **reading rate.** *See also:* **data processing; storage tube.** E158-15E6

reading speed, minimum usable (storage tubes). The slowest scanning rate under stated operating conditions before a specified degree of decay occurs. *Note:* The qualifying adjectives **minimum usable** are frequently omitted in general usage when it is clear that the minimum usable reading speed is implied. *See also:* **storage tube.** E158-15E6

reading time (storage tubes). The time during which stored information is being read. *See also:* **storage tube.** E158-15E6

reading time, maximum usable (storage tubes). The length of time a storage element, line, or area can be read before a specified degree of decay occurs. *Notes:* (1) This time may be limited by static decay, dynamic decay, or a combination of the two. (2) It is assumed that rewriting is not done. (3) The qualifying adjectives **maximum usable** are frequently omitted in general usage when it is clear that the maximum usable reading time is implied. *See also:* **storage tube.** E158-15E6

read number (storage tubes). The number of times a storage element, line, or area is read without rewriting. *See also:* **storage tube.** E158-15E6

read number, maximum usable (storage tubes). The number of times a storage element, line, or area can be read without rewriting before a specified degree of decay results. *Note:* The qualifying adjectives **maximum usable** are frequently omitted in general usage when it is clear that the maximum usable read number is implied. *See also:* **storage tube.** E158-15E6

read-only storage (computing systems). *See:* **fixed storage.**

readout, command (numerically controlled machines). Display of absolute position as derived from position command. *Note:* In many systems the readout information may be taken directly from the dimension command storage. In others it may result from the summation of command departures. *See also:* **numerically controlled machines.** EIA3B-34E12

readout, position (numerically controlled machines). Display of absolute position as derived from a position transducer. *See also:* **numerically controlled machines.** EIA3B-34E12

read pulse. A pulse that causes information to be acquired from a magnetic cell or cells. *See:* **ONE state.** *See also:* **static magnetic storage.** E163-0

ready-to-receive signal (facsimile). A signal sent back to the facsimile transmitter indicating that a facsimile receiver is ready to accept the transmission. *See also:* **facsimile signal (picture signal).** E168-0

realizable function (linear passive networks). A response function that can be realized by a network containing only positive resistance, inductance, capacitance, and ideal transformers. *Note:* This is the sense of realizability in the theory of linear, passive, reciprocal, time-invariant networks. *See also:* **linear passive networks.** E156-0

real time. (1) Pertaining to the actual time during which a physical process transpires. (2) Pertaining to the performance of a computation during the actual time that the related physical process transpires in order that results of the computation can be used in guiding the physical process. *See also:* **electronic digital computer.** X3A12-16E9

recalescent point (metal). The temperature at which there is a sudden liberation of heat as the metal is lowered in temperature. *See also:* **dielectric heating;**

inducting heating. E54/E169-0

received power (mobile communication). The root-mean-square value of radio-frequency power that is delivered to a load that correctly terminates an isotropic reference antenna. The reference antenna most commonly used is the half-wave dipole. *See also:* **mobile communication system.** 0-6E1

receive-only equipment. Data communication equipment capable of receiving signals, but not arranged to transmit signals. *See also:* **data transmission.** 0-19E4

receiver (1). A device on a vehicle so placed that it is in position to be influenced inductively or actuated by the train-stop, train-control, or cab-signal roadway element. 42A42-0
(2) (facsimile). The apparatus employed to translate the signal from the communications channel into a facsimile record of the subject copy. *See also:* **facsimile (in electrical communication).** E168-0

receiver gating (electronic navigation). The application of selector pulses to one or more stages of a receiver only during that part of a cycle of operation when reception is desired. *See also:* **navigation.** 0-10E6

receiver ground (signal-transmission system). The potential reference at the physical location of the signal receiver. *See also:* **signal.** 0-13E6

receiver performance.
See:
adjacent-channel selectivity and desensitization;
audio-frequency response;
audio output power;
hum and noise;
impulse-noise selectivity;
intermodulation and spurious-response attentuation;
modulation acceptance bandwidth;
noise quieting;
radio receiver;
sinad ratio;
sinad sensitivity;
squelch sensitivity;
usable sensitivity.

receiver, power-line carrier. *See:* **power-line carrier receiver.**

receiver primaries (color television). *See:* **display primaries.**

receiver pulse delay. *See:* **transducer pulse delay.**

receiver relay. An auxiliary relay whose function is to respond to the output of a communications set such as power-line carrier, wire-line audio or carrier, radio, or microwave receiver. 37A100-31E11/31E6

receiver, telephone. An earphone for use in a telephone system. 42A65-1E1

receiving converter, facsimile (frequency-shift to amplitude-modulation converter). A device that changes the type of modulation from frequency shift to amplitude. *See also:* **facsimile transmission.** E168-0

receiving-end crossfire (telegraph channel). The crossfire from one or more adjacent telegraph channels at the end remote from the transmitting end. *See also:* **telegraphy.** 42A65-0

receiving loop loss. That part of the repetition equivalent assignable to the station set, subscriber line, and battery supply circuit that are on the receiving end. *See also:* **transmission loss.** 42A65-0

receiving voltage sensitivity. *See:* **free-field voltage response.**

receptacle (electric distribution) (convenience outlet). A contact device installed at an outlet for the connection of an attachment plug and flexible cord to supply portable equipment. 1A0/42A95-0

receptacle circuit. A branch circuit to which only receptacle outlets are connected. *See also:* **branch circuit.** 42A95-0

receptacle outlet. An outlet where one or more receptacles are installed. 42A95-0

receptive field (medical electronics). The region in which activity is observed by means of the pickup electrode. *See also:* **medical electronics.** 0-18E1

reciprocal bearing (navigation). The opposite direction to a bearing. *See also:* **navigation.** 0-10E6

reciprocal transducer. A transducer in which the principle of reciprocity is satisfied. *Note:* The use of the term **reversible transducer** as a synonym for reciprocal transducer is deprecated. *See also:* **transducer.** E196/E270/42A65-0

reciprocity (multiport network). The property described by symmetry of the impedance matrix. In the case of a network with identical ports, it is also described by symmetry of the scattering matrix. *See also:* **transmission characteristics.** 0-9E4

reciprocity theorem. States that if an electromotive force E at one point in a network produces a current I at a second point in the network, then the same voltage E acting at the second point will produce the same current I at the first point. *See also:* **communication.** 42A65-0

reclosing fuse. A combination of two or more fuseholders, fuse units, or fuse links mounted on a fuse support or supports, mechanically or electrically interlocked, so that one fuse can be connected into the circuit at a time and the functioning of that fuse automatically connects the next fuse into the circuit, with or without intentionally added time delay, thereby permitting one or more service restorations without replacement of fuse links, refill units, or fuse units. 37A100-31E11

reclosing interval (automatic circuit recloser). The open-circuit time between an automatic opening and the succeeding automatic reclosure. 37A100-31E11

reclosing relay. A programming relay whose function is to initiate the automatic reclosing of a circuit breaker. 37A100-31E11/31E6

reclosing time (circuit breaker). The interval between the time when the actuating quantity of the release (trip) circuit reaches the operating value (the breaker being in the closed position) and the reestablishment of the circuit on the primary arcing contacts on the reclosing stroke. 37A100-31E11

recognition. *See:* **character recognition; magnetic-ink character recognition; optical character recognition; pattern recognition.** *See also:* **electronic digital computer.**

recombination rate (semiconductor) (1) (surface). The time rate at which free electrons and holes recombine at the surface of a semiconductor. *See also:* **semiconductor device.** E102-0
(2) volume). The time rate at which free electrons and holes recombine within the volume of a semiconductor. *See also:* **semiconductor device.** E102-10E1

recombination velocity (semiconductor surface). The quotient of the normal component of the electron (hole) current density at the surface by the excess electron (hole) charge density at the surface. *See also:* **semiconductor device.** E102/E216-34E17;E270-10E1

reconditioned carrier reception (exalted-carrier reception). The method of reception in which the carrier is separated from the sidebands for the purpose of eliminating amplitude variations and noise, and then added at increased level to the sideband for the purpose of obtaining a relatively undistorted output. This method is frequently employed, for example, when a reduced-carrier single-sideband transmitter is used. *See also:* **radio receiver.** 42A65-0

reconstituted mica (integrated mica) (rotating machinery). *See:* **mica paper.**

reconstruction. The replacement or rearrangement of any portion of an existing installation by new equipment or construction. *Note:* It does not include ordinary maintenance replacements. 2A2/42A95-0

record. A collection of related items of data, treated as a unit. *See:* **file.** *See also:* **electronic digital computer.** X3A12-16E9

recorded spot (facsimile). The image of the recording spot on the record sheet. *See also:* **recording (facsimile).** E168-0

recorded spot, *X* dimension (facsimile). The effective recorded-spot dimension measured in the direction of the recorded line. *Notes:* (1) By effective dimension is meant the largest center-to-center spacing between recorded spots that gives minimum peak-to-peak variation of density of the recorded line. (2) This term applies to that type of equipment that responds to a constant density in the subject copy by a succession of discrete recorded spots. *See also:* **recording (facsimile).** E168-0

recorded spot, *Y* dimension (facsimile). The effective recorded-spot dimension measured perpendicularly to the recorded line. *Note:* By effective dimension is meant the largest center-to-center distance between recorded lines that gives minimum peak-to-peak variation of density across the recorded lines. *See also:* **recording (facsimile).** E168-0

recorded value. The value recorded by the marking device on the chart, with reference to the division lines marked on the chart. *See also:* **accuracy rating (instrument).** 42A30-0

recorder (1). A device that makes a permanent record, usually visual, of varying signals, usually voltages. *See also:* **electronic analog computer.** E165-16E9

(2) (facsimile). That part of the facsimile receiver that performs the final conversion of electric picture signal to an image of the subject copy on the record medium. *See also:* **facsimile (electrical communication); recording (facsimile).** E168-0

recorder, strip-chart. A recorder in which one or more records are made simultaneously as a function of time. *See also:* **electronic analog computer.** E165-0

recorder, *X-Y* (plotting board). A recorder that makes a record of any one voltage with respect to another. *See also:* **electronic analog computer.** E165-0

record gap (computing systems) (storage medium). An area used to indicate the end of a record. *See also:* **electronic digital computer.** X3A12-16E9

recording (facsimile). The process of converting the electric signal to an image on the record medium. *See:*
black recording;
carbon pressure recording;
definition;
direct recording;
drive pattern;
drum speed;
echo;
electrochemical recording;
electrolytic recording;
electromechanical recording;
electrostatic recording;
electrothermal recording;
facsimile;
framing;
grouping;
halftone characteristic;
index of cooperation, scanning or recording line;
ink-vapor recording;
jitter;
Kendall effect;
magnetic recording;
nominal linewidth;
overlap *X*;
overlap *Y*;
photosensitive recording;
record medium;
record sheet;
recorded spot;
recorded spot, *X* dimension;
recorded spot, *Y* dimension;
recorder;
recording spot;
reproduction speed;
skew;
spot projection;
spot speed;
stagger;
stroke speed;
underlap *X*;
underlap *Y*;
white recording. E168-0

recording channel (electroacoustics). The term refers to one of a number of independent recorders in a recording system or to independent recording tracks on a recording medium. *Note:* One or more channels may be used at the same time for covering different ranges of the transmitted frequency band, for multichannel recording, or for control purposes. *See also:* **phonograph pickup.** E157-1E1

recording-completing trunk (combined line and recording trunk). A trunk for extending a connection from a local line to a toll operator, used for recording the call and for completing the toll connection. *See:* **telephone system.** 42A65-0

recording demand meter. A demand meter that records on a chart the demand for each demand interval. *See also:* **electricity meter (meter).** 42A30-0

recording, instantaneous. A phonograph recording that is intended for direct reproduction without further processing. *See also:* **phonograph pickup.** 0-1E1

recording instrument (recorder) (graphic instrument). An instrument that makes a graphic record of the value of one or more quantities as a function of another variable, usually time. *See also:* **instrument.** 42A30-0

recording loss (mechanical recording). The loss in recorded level whereby the amplitude of the wave in the recording medium differs from the amplitude executed by the recording stylus. *See also:* **phonograph pickup.** E157-1E1

recording spot (facsimile). The image area formed at the record medium by the facsimile recorder. *See also:* **recording (facsimile).** E168-0

recording stylus. A tool that inscribes the groove into the recording medium. *See also:* **phonograph pickup.** 0-1E1

recording trunk. A trunk extending from a local central office or private branch exchange to a toll office, that is used only for communication with toll operators and not for completing toll connections. *See also:* **telephone system.** 42A65-0

record medium (facsimile). A physical medium on which the facsimile recorder forms an image of the subject copy. *See also:* **recording (facsimile).** E168-0

record sheet (facsimile). The medium that is used to produce a visible image of the subject copy in record form. The record medium and the record sheet may be identical. *See also:* **recording (facsimile).** E168-0

recovery current (semiconductor rectifier). The transient component of reverse current associated with a change from forward conduction to reverse voltage. *See also:* **rectification.** E59-34E17

recovery cycle (electrobiology). The sequence of states of varying excitability following a conditioning stimulus. The sequence may include periods such as absolute refractoriness, relative refractoriness, supernormality, and subnormality. *See also:* **excitability (electrobiology).** 42A80-18E1

recovery time (1) (automatic gain control). The time interval required, after a sudden decrease in input signal amplitude to a system or transducer, to attain a stated percentage (usually 63 percent) of the ultimate change in amplification or attenuation due to this decrease. *See also:* **radar.** E151-42A65

(2) (power supplies). Specifies the time needed for the output voltage or current to return to a value within the regulation specification after a step load or line change. *Notes:* (A) Recovery time, rather than response time, is the more meaningful and therefore preferred way of specifying power-supply performance, since it relates to the regulation specification. (B) For load change, current will recover at a rate governed by the rate-of-change of the compliance voltage across the load. This is governed by the resistance-capacitance time constant of the output filter capacitance, internal source resistance, and load resistance. *See:* **programming speed.** *See also:* **radar.** KPSH-10E1

(3) (radar or component thereof). The time required, after the end of the transmitted pulse, for recovery to a specified relation between receiving sensitivity or received signal and the normal value. *Note:* The time for recovery to 6 decibels below normal sensitivity is frequently specified. *See also:* **radar.** 42A65-0

(4) (Geiger-Mueller counters). The minimum time from the start of a counted pulse to the instant a succeeding pulse can attain a specific percentage of the maximum value of the counted pulse. *See also:* **gas-filled radiation-counter tubes; radar.** 42A70-15E6

(5) (antitransmit-receive tubes). The time required for a fired tube to deionize to such a level that the normalized conductance and susceptance of the tube in its mount are within specified ranges. *Note:* Normalization is with respect to the characteristic admittance of the transmission line at its junction with the tube mount. *See also:* **gas tubes; radar.** E160-15E6

(6) (microwave gas tubes). The time required for a fired tube to deionize to such a level that the attenuation of a low-level radio-frequency signal transmitted through the tube is decreased to a specified value. *See also:* **gas tubes; radar.** E160-15E6

(7) (gas tubes). The time required for the control

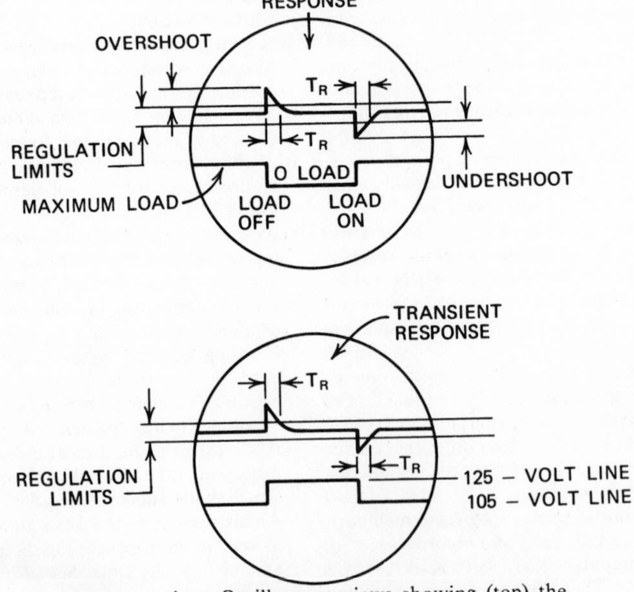

Recovery time. Oscilloscope views showing (top) the effects of a step load change, and (bottom) the effects of a step line change. T_R = recovery time.

electrode to regain control after anode current interruption. *Note:* To be exact, the deionization and recovery time of a gas tube should be presented as families of curves relating such factors as condensed-mercury temperature, anode current, anode and control electrode voltages, and control-circuit impedance. *See also:* **radar.** E160-15E6

(8) (regulators). The elapsed time from the initiation of a transient until the output returns to within the regulation limits. See the figures attached to the definition of **response time.** *See also:* **electrical conversion; radar; response time.** 0-10E1

(9) (gas turbines). The interval between two conditions of speed occurring with a specified sudden change in the steady-state electric load on the gas-turbine–generator unit. It is the time in seconds from the instant of change from the initial load condition to the instant when the decreasing oscillation of speed finally enters a specified speed band. *Note:* The specified speed band is taken with respect to the midspeed of the steady-state speed band occurring at the subsequent steady-state load condition. The recovery time for a specified load increase and the same specified load decrease may not be identical and will vary with the magnitude of the load change. E282-31E2

(10) (relay). *See:* **relay recovery time.**

recovery voltage (lightning arresters). The voltage that occurs across the terminals of a pole of a circuit-interrupting device upon the interruption of the current. *See also:* **lightning arrester (surge diverter).** 37A100-31E11;62A1-31E7

rectangular impulse (lightning arresters). An impulse that rises rapidly to a maximum value, remains substantially constant for a specified period, and then falls rapidly to zero. The parameters that define a rectangular impulse wave are polarity, peak value, duration of the peak, total duration. *See also:* **lightning arrester (surge diverter).** 0-31E7

rectangular mode. The mode of operation that produces a transformation from polar to rectangular coordinates or a rotation of rectangular coordinates. *See:* **resolver.** *See also:* **electronic analog computer.** E165-0

rectangular wave. A periodic wave that alternately assumes one of two fixed values, the time of transition being negligible in comparison with the duration of each fixed value. *See also:* **network analysis; wavefront.** E270/42A65-0

rectification. The term used to designate the process by which electric energy is transferred from an alternating-current circuit to a direct-current circuit. *Note:* For an extensive list of cross references, see *Appendix A.* E59/42A15-0

rectification factor. The quotient of the change in average current of an electrode by the change in amplitude of the alternating sinusoidal voltage applied to the same electrode, the direct voltages of this and other electrodes being maintained constant. *See also:* **circuit characteristics of electrodes; conductance for rectification; transrectification factor.** 42A70-15E6

rectification of an alternating current. Process of converting an alternating current to a unidirectional current. *See also:* **electronic rectifier; inverse voltage (rectifier); semiconductor device.** 50I07-15E6

rectified value (alternating quantity). The average of all the positive values of the quantity during an integral number of periods. Since the positive values of a quantity y are represented by the expression

$$\frac{1}{2}[y+|y|],$$

$$y_r = \frac{1}{T}\int_0^T \frac{1}{2}[y+|y|]\,dt.$$

Note: The word *positive* and the sign + may be replaced by the word *negative* and the sign −. *See also:* **network analysis.** E270-0

rectifier. (1) A rectifier circuit element. (2) An integral assembly of two or more rectifier circuit elements, with their essential auxiliaries.
See:
bridge rectifier;
contact rectifier;
crystal diode;
full-wave rectifier;
grid-controlled mercury-arc rectifier;
half-wave rectifier;
linear rectifier;
***n*-type crystal rectifier;**
power rectifier;
***p*-type crystal rectifier;**
ripple voltage;
self-commutated inverters;
transrectifier;
voltage doubler;
voltage multiplier. *See also:* **circuits and devices; electric controller; rectification; rectifier circuit element; static converter.** 42A15/42A25-0

rectifier anode. An electrode of the rectifier from which the current flows into the arc. *Note:* The direction of current flow is considered in the conventional sense from positive to negative. The cathode is the positive direct-current terminal of the apparatus and is usually a pool of mercury. The neutral of the transformer secondary system is the negative direct-current terminal of the rectifier unit. *See also:* **rectification.** 42A15-0

rectifier assembly. A complete unit containing rectifying components, wiring, and mounting structure capable of converting alternating-current power to direct-current power. *See also:* **converter.** 0-31E8

rectifier cathode. The electrode of the rectifier into which the current flows from the arc. *Note:* The direction of current flow is considered in the conventional sense from positive to negative. The cathode is the positive direct-current terminal of the rectifier unit and is usually a pool of mercury. The neutral of the transformer secondary system is the negative direct-current terminal of the rectifier unit. *See also:* **rectification.** 42A15-0

rectifier circuit element. A circuit element bounded by two circuit terminals that has the characteristic of conducting current substantially in one direction only. *Note:* The rectifier circuit element may consist of more than one semiconductor rectifier cell, rectifier diode, or rectifier stack connected in series or parallel or both, to operate as a unit.
See:
blocking period;
cascade rectifier circuit;
commutating group;

commutating reactance;
commutating reactance factor;
commutating voltage;
commutation;
conducting period;
double-way rectifier circuit;
forward period;
full-wave rectifier circuit;
half-wave rectifier circuit;
light transition load;
multiple rectifier circuit;
number of rectifier circuit phases;
parallel rectifier circuit;
phase control;
reverse period;
series rectifier circuit;
set of commutating groups;
simple rectifier circuit;
single-way rectifier circuit;
transition load.
See also: **rectification; semiconductor rectifier stack.** E59-34E17

rectifier electric locomotive. An electric locomotive that collects propulsion power from an alternating-current distribution system and converts this to direct current for application to direct-current traction motors by means of rectifying equipment carried by the locomotive. *Note:* A rectifier electric locomotive may be defined by the type of rectifier used on the locomotive, such as **ignitron electric locomotive.** *See also:* **electric locomotive.** 42A42-0

rectifier electric motor car. An electric motor car that collects propulsion power from an alternating-current distribution system and converts this to direct current for application to direct-current traction motors by means of rectifying equipment carried by the motor car. *Note:* A rectifier electric motor car may be defined by the type of rectifier used on the motor car, such as **ignitron electric motor car.** *See also:* **electric motor car.** 42A42-0

rectifier instrument. The combination of an instrument sensitive to direct current and a rectifying device whereby alternating currents or voltages may be measured. *See also:* **instrument.** 42A30-0

rectifier junction (semiconductor rectifier cell). The junction in a semiconductor rectifier cell that exhibits asymmetrical conductivity. *See also:* **semiconductor; semiconductor rectifier stack.** E59-34E17;0-34E24

rectifier transformer. A transformer that operates at the fundamental frequency of an alternating-current system and designated to have one or more output windings conductively connected to the main electrodes of a rectifier.
See:
alternating-current winding;
angular displacement;
commutating impedance;
commutating reactance;
commutating resistance;
current rating;
direct-current winding;
excitation losses;
interphase transformer;
interphase-transformer loss;
percent impedance;
portable mining-type rectifier transformer;
rated alternating-current winding voltages;
rated direct-current winding voltages;
rated line kilovolt-ampere rating;
rating;
regulating autotransformer;
service rating;
transformer;
turn ratio. 42A15-31E12

rectifier tube (valve). An electronic tube or valve designed to rectify alternating current. 0-15E6

rectifier unit. An operative assembly consisting of the rectifier, or rectifiers, together with the rectifier auxiliaries, the rectifier transformer equipment, and the essential switchgear. *See also:* **rectification; rectifier transformer.** 42A15-34E24

rectifying device. An elementary device, consisting of one anode and its cathode, that has the characteristic of conducting current effectively in only one direction. *See also:* **rectification.** 42A15-0

rectifying element. A circuit element that has the property of conducting current effectively in only one direction. *Note:* When a group of rectifying devices is connected, either in parallel or series arrangement, to operate as one circuit element, the group of rectifying devices should be considered as a rectifying element. *See also:* **rectification; rectifying device; rectifier circuit element; rectifying junction; metallic rectifying cell.** 42A15-0

rectifying junction (barrier layer) (blocking layer). The region in a metallic rectifier cell that exhibits the asymmetrical conductivity. *See also:* **rectification.** 42A15-0

rectilinear scanning (television). The process of scanning an area in a predetermined sequence of straight parallel scanning lines. *See also:* **television.** E204-2E2/42A65

recurrence rate (pulse techniques). *See:* **pulse repetition frequency.**

recurrent sweep. A sweep that repeats or recurs regularly. It may be free-running or synchronized. *See also:* **oscillograph.** 0-9E4

redirecting surfaces and media (illuminating engineering). Those that change the direction of the flux without scattering the redirected flux. *See also:* **lamp.** Z7A1-0

redistribution (storage or camera tubes). The alteration of the charge pattern on an area of a storage surface by secondary electrons from any other part of the storage surface. *See also:* **charge-storage tube.** E158-15E6

reduced generator efficiency (thermoelectric device). The ratio of (1) a specified generator efficienty to (2) the corresponding Carnot efficiency. *See also:* **thermoelectric device.** E221-15E7

reduced kilovolt-ampere tap (transformer). A tap through which the transformer can only deliver an output less than rated kilovolt-ampere without exceeding the specified temperature rise. *See also:* **windings, high-voltage and low-voltage.** 42A15-0

reduced-voltage starter (industrial control). A starter, the operation of which is based on the application of a reduced voltage to the motor. *See also:* **starter.** 42A25-0;50I16-34E10

reducing joint. A joint between two lengths of cable the conductors of which are not the same size. *See:* **branch joint; cable joint; straight joint.** *See also:* **pow-**

er distribution, underground construction. 42A35-31E13

reduction (1) (electrochemical cells) (corrosion). The gain of electrons by a constituent of a chemical reaction. *See also:* **corrosion terms; electrochemical cell.** CV1-10E1;CM-34E2

(2) (data processing). *See:* **data reduction.**

redundancy (1) (transmission of information). The fraction of the gross information content of a message that can be eliminated without loss of essential information. *Note:* Numerically, it is one minus the ratio of the net information content to the gross information content, expressed in percent. 42A65-0

(2) (source) (information theory). The amount by which the logarithm of the number of symbols available at the source exceeds the average information content per symbol of the source. *Note:* The term redundancy has been used loosely in other senses. For example, a source whose output is normally transmitted over a given channel has been called redundant, if the channel utilization index is less than unity. *See also:* **information theory.** E171-0

(3) (reliability). The introduction of auxiliary elements and components to a system to perform the same functions as other elements in the system for the purpose of improving reliability and safety. *Note:* The term **active redundancy** denotes that redundancy wherein all redundant items are operating simultaneously, rather than being switched on when needed. *See also:* **reliability; system.** 0-7E1/35E2

redundancy check (1) (data transmission). A check based on the insertion of extra bits for the purpose of error control. *See also:* **data transmission.** 0-19E4

(2) (electronic computation). *See:* **check, forbidden-combination.**

redundancy, standby (reliability). The redundancy wherein the alternative means of performing the function is inoperative until needed and is switched on upon failure of the primary means of performing the function. *See also:* **reliability.** 0-7E1

redundant. Pertaining to characters that do not contribute to the information content. Redundant characters are often used for checking purposes or to improve reliability. *See also:* **parity; self-checking code; error-detecting code; check, forbidden-combination; check digit.** *See also:* **electronic digital computer.** E162-0

redundant code (data transmission). A code using more signal elements than necessary to represent the intrinsic information. *See also:* **data transmission.** 0-19E4

reed relay. A relay using glass-enclosed, magnetically closed reeds as the contact members. Some forms are mercury wetted. 0-21E0

reentrant beam (microwave tubes). An unterminated recirculating electron beam. *See also:* **microwave tube (or valve).** 0-15E6

reentrant-beam crossed-field amplifier (amplitron) (microwave tubes). A crossed-field amplifier in which the beam is reentrant and interacts with either a forward or a backward wave. *See also:* **microwave tube (or valve).** 0-15E6

reentrant circuit (microwave tubes). A slow-wave structure that closes upon itself. *See also:* **microwave tube (or valve).** 0-15E6

referee test. A test made by or in the presence of a representative of a regulatory body or other disinterested agency. *Note:* Such tests usually are made under specific provisions relating thereto, promulgated by the regulator body. *See also:* **demand meter.** 12A0/42A30-0

reference (computing systems). *See:* **linearity.**

reference accuracy (automatic null-balancing electric instrument). A number or quantity that defines the limit of error under reference operating conditions. *Notes:* (1) It is usually expressed as a percent of the span. It is preferred that a + sign or − sign or both precede the number or quantity. The absence of a sign infers a ± sign. (2) Reference accuracy does not include accuracy of sensing elements or intermediate means external to the instrument. *See:* **error and correction.** *See also:* **accuracy rating (instrument).** 39A4-0

reference air line. A uniform section of air-dielectric transmission line of accurately calculable characteristic impedance used as a standard of immittance. *See also:* **transmission line.** 0-9E4

reference ballasts. Specially constructed series ballasts having certain prescribed characteristics. *Note:* They serve as comparison standards for use in testing ballasts or lamps and are used also in selecting the reference lamps that are necessary for the testing of ballasts. Reference ballasts are characterized by a constant impedance over a wide range of operating current. They also have constant characteristics that are relatively uninfluenced by time and temperature. *See also:* **fixed-impedance type (reference ballast); primary standards (illuminating engineering); variable-impedance type (reference ballast).** Z7A1/82A3-0

reference black level (television). The picture-signal level corresponding to a specified maximum limit for black peaks. *See also:* **television.** E203-2E2/42A65

reference block (numerically controlled machines). A block within the program identified by an *o* (letter *o*) in place of the word address *n* and containing sufficient data to enable resumption of the program following an interruption. This block should be located at a convenient point in the program that enables the operator to reset and resume operation. *See also:* **numerically controlled machines.** EIA3B-34E12

reference boresight (antenna). A direction defined by an optical, mechanical, or electrical axis of an antenna, established as a reference for the purposes of beam-direction or tracking-axis alignment. *See also:* **antenna.** E149-3E1

reference conditions. The values assigned for the different influence quantities at which or within which the instrument complies with the requirements concerning errors in indication. *See also:* **accuracy rating (instrument).** 39A1-0

reference current (fluorescent lamp). The value of current specified in a specific lamp standard. *Note:* It is normally the same as the value of current for which the corresponding lamp is rated. Since the reference ballast is a standard that is representative of the impedance of lamp power sources installed, it is not necessary to change this current value unless major changes in lamp standards require modification of the ballast impedance. For this reason, reference ballast characteristics are specified in terms of, and with reference to, reference current. 82A3-0

reference designation (abbreviation). Numbers, or letters and numbers, used to identify and locate units, portions thereof, and basic parts of a specific set. *Compare with:* **functional designation** and **symbol for a quantity.** *See also:* **abbreviation.** E267-0

reference direction (1) (navigation). A direction from which other directions are reckoned; for example, true north, grid north, and so on. *See also:* **navigation.** 0-10E6

(2) (voltage). The assigned direction of the positive line element in the integral for voltage. *See also:* **network analysis; radio transmission.** E270-0

(3) (current). In a conductor, the assigned direction of the positive normal to the cross section of the conductor. *Note:* The reference direction for a current is usually denoted on a circuit diagram by an arrow placed near the line that represents the conductor. *See also:* **network analysis.** E270-0

(4) (energy) (specified circuit). With reference to the boundary of a delimited region, the arbitrarily selected direction in which electric energy is assumed to be transmitted past the boundary, into or out of the region. *Notes:* (1) When the actual direction of energy flow is the same as the reference direction, the sign of power is positive, and when the actual flow is in the opposite direction, the sign is negative. (2) Unless specifically stated to the contrary, it shall be assumed that the reference direction for all power, energy, and quadergy quantities associated with the circuit is the same as the reference direction of the energy flow. (3) In these definitions it will be assumed that the reference direction of the current in each conductor of the circuit is the same as the reference direction of energy flow. E270-0

reference excursion (analog computer). The range from zero voltage to nominal full-scale operating voltage. *See also:* **electronic analog computer.** E165-16E9

reference frequency. A selected frequency from which frequency departure is measured. *Note:* Reference frequency is not necessarily synonymous with carrier frequency. Carrier frequency is the frequency of a periodic wave upon which modulation is imposed. *See also:* **network analysis.** 0-12E1

reference input signal (industrial control). *See:* **signal, reference input.**

reference lamp (1) (mercury). A seasoned lamp that under stable burning conditions, in the specified operating position (usually vertical, base up), and in conjunction with the reference ballast specified for the lamp size and rating, and at reference ballast rated input voltage, operates at values of lamp volts, watts, and amperes, each within ± 2 percent of the nominal values. 82A7/82A9-0

(2) (fluorescent). Seasoned lamps that under stable burning conditions, in conjunction with the reference ballast specified for the lamp size and rating, and at the rated reference ballast supply voltage, operate at values of lamp volts, watts, and amperes each within $\pm 2\frac{1}{2}$ percent of the values, and under conditions established by present standards. *See:* **reference ballasts.** 82A1-0

reference line (1) (navigation). A line from which angular or linear measurements are reckoned. *See also:* **navigation.**

(2) (illuminating engineering). Either of two radial lines where the surface of the cone of maximum candlepower is intersected by a vertical plane parallel to the curb line and passing through the lightcenter of the luminaire. *See also:* **street lighting luminaire.** 0-10E6

reference modulation (very-high-frequency omnidirectional radio range). The modulation of the ground station radiation that produces a signal in the airborne receiver whose phase is independent of the bearing of the receiver; the reference signal derived from this modulation is used for comparison with the variable signal. *See also:* **navigation.** 0-10E6

reference noise. The magnitude of circuit noise that will produce a circuit noise meter reading equal to that produced by 10^{-12} watt of electric power at 1000 hertz. *See also:* **signal-to-noise ratio.** 42A65-0

reference operating conditions (automatic null-balancing electric instrument). The conditions under which reference performance is stated and the base from which the values of operating influences are determined. *See also:* **measurement system.** 39A4-0

reference performance (1) (automatic null-balancing electric instrument). The limits of the values of certain operating characteristics of the instrument that will not be exceeded under any combination of reference operating conditions. *See also:* **electricity meter (meter); test (instrument or meter).** 39A4-0

(2) (meter). Its percentage registration under standard conditions with which its percentage registration under a particular condition is to be compared to determine the influence on the meter of the particular condition. *See also:* **electricity meter (meter); service test; test (instrument or meter).** 12A0/42A30-0

reference plane. A plane perpendicular to the direction of propagation in a waveguide or transmission line, to which measurement of immittance, electrical length, reflection coefficients, scattering coefficients, and other parameters may be referred. *See also:* **waveguide.** 0-9E4

reference power supply. A regulated, electronic power supply furnishing the reference voltage. *See also:* **electronic analog computer.** E165-0

reference test field (direction-finder testing). The field strength, in microvolts per meter, numerically equal to the direction-finder sensitivity. *See also:* **navigation.** E173-10E6

reference time (magnetic storage). An instant near the beginning of switching chosen as an origin for time measurements. It is variously taken as the first instant at which the instantaneous value of the drive pulse, the voltage response of the magnetic cell, or the integrated voltage response reaches a specified fraction of its peak pulse amplitude. *See also:* **static magnetic storage.** E163-16E9

reference voltage (analog computer). A voltage used as a standard of reference, usually the nominal full scale of the computer; also a constant unit of computation used in normalizing and scaling for machine solution. *See also:* **electronic analog computer.** E165-16E9

reference volume (acoustics). The volume that gives a reading of 0 (volume units) on a standard volume indicator. *See also:* **transmission characteristics.** E151-42A65

reference waveguide. A uniform section of waveguide with accurately fabricated internal cross-sectional dimensions used as a standard of immittance. *See also:* **waveguide.** 0-9E4

reference white. The light from a nonselective diffuse reflector that is lighted by the normal illumination of the scene. *Notes:* (1) Normal illumination is not intended to include lighting for special effects. (2) In the reproduction of recorded material, the word **scene** refers to the original scene. E201-2E2

reference white level (television). The picture-signal level corresponding to a specified maximum limit for white peaks. *See also:* **color terms; television.** E203-2E2/42A65

refill unit (high-voltage fuse unit). An assembly comprised of a conducting element, the complete arc-extinguishing medium, and parts normally required to be replaced after each circuit interruption to restore the fuse unit to its original operating condition. 37A100-31E11

reflectance (1) (surface or medium). The ratio of the reflected flux to the incident flux. *Note:* Measured values of reflectance depend upon the angles of incidence and view and on the spectral character of the incident flux. Because of this dependence, the angles of incidence and view and the spectral characteristics of the source should be specified. *See also:* **lamp.** Z7A1-0

(2) (photovoltaic power system). The fraction of radiation incident on an object that is reflected. *See also:* **photovoltaic power system; solar cells (photovoltaic power system).** 0-10E1

reflected binary code. *See:* **Gray code.**

reflected glare. Glare resulting from specular reflections of high brightnesses in polished or glossy surfaces in the field of view. It usually is associated with reflections from within a visual task or areas in close proximity to the region being viewed. *See:* **veiling reflection; visual field.** Z7A1-0

reflected harmonics (electrical conversion). Harmonics produced in the prime source by operation of the conversion equipment. These harmonics are produced by current-impedance (*IZ*) drop due to nonsinusoidal load currents, and by switching or commutating voltages produced in the conversion equipment. *See also:* **electrical conversion.** 0-10E1

reflected-light scanning (industrial control). The scanning of changes in the magnitude of reflected light from the surface of an illuminated web. 42A25-34E10

reflected wave (1) (wave propagation). When a wave in one medium is incident upon a discontinuity or a different medium, the reflected wave is the wave component that results in the first medium in addition to the incident wave. *Note:* The reflected wave includes both the reflected rays of geometrical optics and the diffracted wave. *See also:* **radiation; waveguide.** E146-3E1;42A65-0

(2) (lightning arresters). A wave, produced by an incident wave, that returns in the opposite direction to the incident wave after reflection at the point of transition. *See also:* **lightning arrester (surge diverter); radiation.** 50I25-31E7

(3) (reverse wave). A wave traveling along a waveguide or transmission line from a reflective termination, discontinuity, or reference plane in the direction usually opposite to the incident wave. *See:* **incident wave.** *See also:* **waveguide.** 0-9E4

reflection. A general term for the process by which the incident flux leaves a surface or medium from the incident side. *Note:* **Reflection** is usually a combination of regular and diffuse reflection. *See also:* **lamp.** Z7A1-0

reflection coefficient (at any given point on a transmission line or in a medium). The ratio of the phasor magnitude of the reflected wave to the phasor magnitude of the incident wave under conditions that are specified. *Notes:* (1) Specification of conditions includes defining incident wave and reflected wave; definition applies to one type of physical quantity such as current, voltage, pressure, electric field strength, etcetera. (2) The voltage reflection coefficient is most commonly used and is defined as the ratio of the complex electric field strength (or voltage) of the reflected wave to that of the incident wave. (3) At any specified plane in a uniform transmission line between a source of power and an absorber of power, the voltage reflection coefficient is the vector ratio of the electric field associated with the reflected wave, to that associated with the incident wave. It is given by the formula $(Z_2-Z_1)/(Z_2+Z_1)$, where Z_1 is the impedance of the source and Z_2 is the impedance of the load. *See also:* **transmission characteristics; transmission line; waveguide.** E146/E270/42A65-0

reflection color tube. A color-picture tube that produces an image by means of electron reflection techniques in the screen region. *See also:* **beam tubes; television.** E160-15E6

reflection error (electronic navigation). The error due to the fact that some of the total received signal arrives from a reflection rather than all via the direct path. *See also:* **navigation.** 0-10E6

reflection factor (1) (electric circuits) (mismatch factor). The reflection factor between two impedances Z_1 and Z_2 is

$$\frac{(4Z_1Z_2)^{1/2}}{Z_1+Z_2}$$

Notes: (A) Physically the reflection factor is the ratio of the current delivered to a load, whose impedance is not matched to the source, to the current that would be delivered to a load of matched impedance. (B) In this context, the synonym **mismatch factor** may also be used. *See also:* **network analysis; transmission characteristics.** 42A65-0

(2) (reflex klystron). The ratio of the number of electrons of the reflected beam to the total number of electrons that enter the reflector space in a given time. *See also:* **transmission characteristics; velocity-modulated tube.** 50I07-15E6

reflectionless termination. *See:* **termination, reflectionless.**

reflectionless transmission line. A transmission line having no reflected wave at any transverse section. *See also:* **transmission line.** 0-9E4

reflectionless waveguide. A waveguide having no reflected wave at any transverse section. *See also:* **waveguide.** 0-9E4

reflection loss (given frequency). The reflection loss at the junction of a source of power and a load is given by the formula

$$20 \log_{10} \frac{Z_1+Z_2}{(4Z_1Z_2)^{1/2}} \text{ decibels}$$

where Z_1 is the impedance of the source of power and Z_2 is the impedance of the load. Physically, the reflection loss is the ratio, expressed in decibels, of the scalar values of the volt-amperes delivered to the load to the volt-amperes that would be delivered to a load of the same impedance as the source. The reflection loss is equal to the number of decibels that corresponds to the scalar value of the reciprocal of the reflection factor. *Note:* When the two impedances have opposite phases and appropriate magnitudes, a reflection gain may be obtained. *See also:* **network analysis; transmission loss.** 42A65-0

reflection modulation (storage tubes). A change in character of the reflected reading beam as a result of the electrostatic fields associated with the stored signal. A suitable system for collecting electrons is used to extract the information from the reflected beam. *Note:* Typically the beam approaches the target closely at low velocity and is then selectively reflected toward the collection system. *See also:* **charge-storage tube.** E158-15E6

reflectivity (photovoltaic power system). The reflectance of an opaque, optically smooth, clean portion of material. *See also:* **photovoltaic power system; solar cells (photovoltaic power system).** 0-10E1

reflectometer (1). A photometer for measuring reflectance. *Note:* Reflectometers may be visual or physical instruments. *See also:* **photometry.** Z7A1-0

(2) An instrument for the measurement of the ratio of reflected-wave to incident-wave amplitudes in a transmission system. *Note:* Many instruments yield only the magnitude of this ratio. *See also:* **instrument.** 0-9E4

reflectometer, time-domain. An instrument designed to indicate and to measure reflection characteristics of a transmission system connected to the instrument by monitoring the step-formed signals entering the test object and the superimposed reflected transient signals on an oscilloscope equipped with a suitable time-base sweep. The measuring system, basically, consists of a fast-rise step-function generator, a tee coupler, and an oscilloscope connected to the probing branch of the coupler. *See also:* **instrument.** 0-9E4

reflector (1) (wave propagation). A reflector comprises one or more conductors or conducting surfaces for reflecting radiant energy. *See also:* **antenna; reflector element.** 0-42A65

(2) (illuminating engineering). A device used to redirect the luminous flux from a source by the process of reflection. *See also:* **bare (exposed) lamp.** 27A1-0

reflector antenna. An antenna consisting of a reflecting surface and a feed. *See also:* **antenna.** 0-3E1

reflector element (antenna). A parasitic element located in a direction other than forward of the driven element of an antenna intended to increase the directive gain of the antenna in the forward direction. *See also:* **antenna.** 0-3E1

reflector space (reflex klystron). The part of the tube following the buncher space, and terminated by the reflector. *See also:* **velocity-modulated tube.** 50I07-15E6

reflex baffle (audio and electroacoustics). A loudspeaker enclosure in which a portion of the radiation from the rear of the diaphragm is propagated outward after controlled shift of phase or other modification, the purpose being to increase the useful radiation in some portion of the frequency spectrum. *See also:* **loudspeaker.** 42A65-1E1

reflex bunching. The bunching that occurs in an electron stream that has been made to reverse its direction in the drift space. *See also:* **beam tubes; electron devices, miscellaneous.** E160-15E6

reflex circuit. A circuit through which the signal passes for amplification both before and after a change in its frequency. *See also:* **circuits and devices.** 42A65-0

reflex klystron (microwave tubes). A single-resonator oscillator klystron in which the electron beam is reversed by a negative electrode so that it passes twice through the resonator, thus providing feedback. *See also:* **microwave tube.** 0-15E6

reforming (semiconductor rectifier). The operation of restoring by an electrical or thermal treatment, or both, the effectiveness of the rectifier junction after loss of forming. *See also:* **rectification.** E59-34E17

refracted wave. That part of an incident wave that travels from one medium into a second medium. *Note:* This is also called the **transmitted wave.** *See also:* **radiation; radio wave propagation; waveguide.** E146/E211-3E2;42A65-0

refraction error (electronic navigation). Error due to the bending of one or more wave paths by undesired refraction. *See also:* **navigation.** E 172-10E6

refraction loss. That part of the transmission loss due to refraction resulting from nonuniformity of the medium. *See also:* **electroacoustics.** E157-1E1

refractive index (1) (wave transmission medium). The ratio of the phase velocity in free space to that in the medium. *See also:* **radiation; radio wave propagation.** E211-3E2;42A65-0

(2) (dielectric for an electromagnetic wave) (refractivity). The ratio of the sine of the angle of incidence to the sine of the angle of refraction as the wave passes from a vacuum into the dielectric. The angle of incidence θ_i is the angle between the direction of travel of the wave in vacuum and the normal to the surface of the dielectric. The angle of refraction θ_r is the angle between the direction of travel of the wave after it has entered the dielectric and the normal to the surface. Refractive index is related to the dielectric constant through the following relation

$$n = \frac{\sin \theta_i}{\sin \theta_r} = (\epsilon')^{1/2}$$

where ϵ' is the real dielectric constant. Since ϵ' and n vary with frequency, the above relation is strictly correct only if all quantities are measured at the same frequency. The refractive index is also equal to the ratio of the velocity of the wave in the vacuum to the velocity in the dielectric medium. *See also:* **radiation.** E270-0

refractive modulus (radio wave propagation) (troposphere). The excess over unity of the modified index of refraction, expressed in millionths. It is represented by M and is given by the equation

$$M = (n + h/a - 1)\,10^6$$

where a is the mean geometrical radius of the surface and n is the index of refraction at a height h above the local surface. *See also:* **radiation; radio wave propagation.** 42A65-3E2

refractivity. *See:* **refractive index.**

refractor (illuminating engineering). A device used to redirect the luminous flux from a source, primarily by the process of refraction. *See also :* **bare (exposed) lamp.** Z7A1-0

refractory. A nonmetallic material highly resistant to fusion and suitable for furnace roofs and linings. 42A60-0

regal. *See:* **range and elevation guidance for approach and landing.**

regenerate (electronic storage devices) (1). To bring something into existence again after decay of its own accord or after intentional destruction.
(2) (storage tubes). To replace charge to overcome decay effects, including loss of charge by reading.
(3) (storage devices in which physical states used to represent data deteriorate). To restore the device to its latest undeteriorated state. *See also:* **electronic digital computer; rewrite.** E162-0

regenerated leach liquor (electrometallurgy). The solution that has regained its ability to dissolve desired constituents from the ore by the removal of those constituents in the process of electrowinning. *See also:* **electrowinning.** 42A60-0

regeneration (1) (electronic computers). In a storage device whose information storing state may deteriorate, the process of restoring the device to its latest undeteriorated state. *See also:* **electronic computation; rewrite.** E270-0
(2) (storage tubes). The replacing of stored information lost through static decay and dynamic decay. *See also:* **storage tube.** E158-15E6
(3) *See:* **positive feedback.**

regeneration of electrolyte. The treatment of a depleted electrolyte to make it again fit for use in an electrolytic cell. *See also:* **electrorefining.** 42A60-0

regenerative braking (industrial control). A form of dynamic braking in which the kinetic energy of the motor and driven machinery is returned to the power supply system. *See also:* **asynchronous machine; dynamic breaking; electric drive; synchronous machine.** 42A25-34E10;42A42-31E8

regenerative divider (regenerative modulator). A frequency divider that employs modulation, amplification, and selective feedback to produce the output wave. *See also:* **circuits and devices.** 42A65-0

regenerative fuel-cell system. A system in which the reactance may be regenerated using an external energy source. *See also:* **fuel cell.** CV1-10E1

regenerative repeater. A repeater that performs pulse regeneration. *Note:* the retransmitted signals are practically free from distortion. *See also:* **data transmission; repeater.** 42A65-19E4

regional center (telephony). A toll switching point to which a number of sectional centers are connected. Regional centers are classified as class 1 offices. *See also:* **telephone switching system.** 0-19E1

region, Geiger-Mueller (radiation-counter tubes). *See:* **Geiger-Mueller region.**

region of limited proportionality (radiation-counter tubes). The range of applied voltage below the Geiger-Mueller threshold, in which the gas amplification depends upon the charge liberated by the initial ionizing event. *See also:* **gas-filled radiation-counter tubes.** 42A70-0

region, proportional (radiation-counter tubes). *See:* **proportional region.**

regions of electromagnetic spectrum. For convenience of reference the electromagnetic spectrum near the visible spectrum is divided as follows.

Spectrum	Wavelength in nanometers
far ultraviolet	10–280
middle ultraviolet	280–315
near ultraviolet	315–380
visible	380–780
infrared	780–10^5

Note: The spectral limits indicated above should not be construed to represent sharp delineations between the various regions. There is a gradual transition from region to region. The above ranges have been established for practical purposes. *See also:* **radiant energy (illuminating engineering).** Z7A1-0

register (1) (electronic computation). A device capable of retaining information, often that contained in a small subset (for example, one word), of the aggregate information in a digital computer. *See:* **address register; circulating register; index register; instruction register; shift register.** *See also:* **electronic digital computer; storage.** E162/E270/X3A12-16E9
(2) (telephone switching system). A part of a telephone switching system that receives and stores signals from a calling device or other source for interpretation and action. 0-19E1
(3) (protective signaling) (signal recorder). An electromechanical recording device that marks a paper tape in response to electric signal impulses received from transmitting circuits. A register may be driven by a prewound spring mechanism, by an electric motor, or by a combination thereof. *Note:* An inking register marks the tape with ink; a punch register marks the tape by cutting holes therein; a slashing register marks the tape by cutting V-shaped slashes therein. *See also:* **protective signaling.** 42A65-0
(4) (meter). That part of the meter that registers the revolutions of the rotor or the number of impulses received from or transmitted to a meter in terms of units of electric energy or other quantity measured. *See also:* **moving element (instrument); watthour meter.** 12A0/42A30-0

register constant. The factor by which the register reading must be multiplied in order to provide proper consideration of the register, or gear, ratio and of the instrument transformer ratios to obtain the registration in the desired units. *Note:* It is commonly denoted by the symbol K_r. *See also:* **electricity meter (meter); moving element (instrument).** 12A0/42A30-0

register length (electronic computation). The number of characters that a register can store. E270-0

register marks or lines (industrial control). Any mark or line printed or otherwise impressed on a web of material, and which is used as a reference to maintain register. *See also:* **photoelectric control.** 42A25-34E10

register, mechanical (pulse techniques). An electromechanical indicating pulse counter. *See also:* **pulse.** E175-0

register ratio (watthour meters). The number of revolutions of the first gear of the register, for one revolution of the first dial pointer. *Note:* This is commonly denoted by the symbol Rr. *See also:* **electricity meter (meter); moving element (instrument).** 12A0-0

register reading. The numerical value indicated by the register. Neither the register constant nor the test dial (or dials), if any exist, is considered. *See also:* **electricity meter (meter).** 42A30-0

registration (1) (general). Accurate positioning relative to a reference. *See also:* **electronic digital computer.** X3A12-16E9

(2) (display device). The condition in which corresponding elements of the primary-color images are in geometric coincidence. *See:* **registration (camera device).** 0-42A65

(3) (camera device). The condition in which corresponding elements of the primary-color images are scanned in time sequence. *See:* **registration (display device).** 0-42A65

(4) (meter). The apparent amount of electric energy (or other quantity being measured) that has passed through the meter, as shown by the register reading. It is equal to the product of the register reading and the register constant. The registration during a given period is equal to the product of the register constant and the difference between the register readings at the beginning and the end of the period. 42A30-0

regressed luminaire. A luminaire that is mounted above the ceiling with the opening of the luminaire above the ceiling line. *See also:* **suspended (pendant) luminaire.** Z7A1-0

regular (specular) reflectance. The ratio of the flux leaving a surface or medium by regular (specular) reflection to the incident flux. *See also:* **lamp.** Z7A1-0

regular (specular) reflection. Regular (specular) reflection is that process by which incident flux is redirected at the specular angle. *See:* **specular angle.** *See also:* **lamp.** Z7A1-0

regular transmittance (illuminating engineering). The ratio of the regularly transmitted flux leaving a surface or medium to the incident flux. *See also:* **transmission (illuminating engineering).** Z7A1-0

regular transmission (illuminating engineering). That process by which incident flux passes through a surface or medium without scattering. *See also:* **transmission (illuminating engineering).** Z7A1-0

regulated circuit (voltage regulator). The circuit on the output side of the regulator, and in which it is desired to control the voltage, or the phase relation, or both. *Note:* The voltage may be held constant at any selected point on the regulated circuit. *See also:* **voltage regulator.** 57A15-0

regulated frequency. Frequency so adjusted that the average value does not differ from a predetermined value by an appreciable amount. *See also:* **generating station.** 42A35-31E13

regulated power supply. A power supply that maintains a constant output voltage (or current) for changes in the line voltage, output load, ambient temperature, or time.
See:
output-capacitor discharge time;
output impedance;
output ripple voltage;
output voltage regulation;
output voltage stabilization;
regulated-power-supply efficiency;
regulation pull-out;
unit warmup time;
unregulated voltage.
See also: **power supply; voltage regulator.** KPSH-10E1

regulated-power-supply efficiency. The ratio of the regulated output power to the input power. *See also:* **regulated power supply.** E209-0

regulating autotransformer (rectifier). A transformer used to vary the voltage applied to the alternating-current winding of the rectifier transformer by means of de-energized autotransformer taps, and with load-tap-changing equipment to vary the voltage over a specified range on any of the autotransformer taps. *See also:* **rectifier transformer.** 57A18-0

regulating limit setter (speed-governing system). A device in the load-frequency-control system for limiting the regulating range on a station or unit. *See also:* **speed-governing system.** E94-0

regulating range (load-frequency control). A range of power output within which a generating unit is permitted to operate. *See also:* **speed-governing system.** E94-0

regulating relay. A relay whose function is to detect a departure from specified system operating conditions and to restore normal conditions by acting through supplementary equipment. 37A100-31E11/31E6

regulating system, synchronous-machine. An electric-machine regulating system consisting of one or more principal synchronous electric machines and the associated excitation system. *See also:* **synchronous machine.** 0-31E8

regulating transformer (rectifier). A transformer used to vary under load within specified limits the voltage applied to the alternating-current winding of the rectifier transformer for the purpose of varying the direct-current output voltage of the rectifier or to maintain constant direct-current output voltage with varying alternating-current supply voltage. *See also:* **rectifier transformer.** 57A18-0

regulation (1) (rotating machinery). The amount of change in voltage or speed resulting from a load change. *See also:* **asynchronous machine; direct-current commutating machine; synchronous machine.** 0-31E8

(2) (overall) (power supplies). The maximum amount that the output will change as a result of the specified change in line voltage, output load, temperature, or time. *Note:* Line regulation, load regulation, stability, and temperature coefficient are defined and usually specified separately. *See also:* **line regulation; load regulation; power supply; stability; temperature coefficient.** KPSH-10E1

(3) (gas tubes). The difference between the maximum and minimum anode voltage drop over a range of anode current. *See also:* **gas tubes.** 42A70-15E6

(4) (electrical conversion). The change of one of the controlled or regulated output parameters resulting from a change of one or more of the unit's variables within specification limits. *See also:* **electrical conversion.** 0-10E1

regulation curve (generator). A characteristic curve between voltage and load. The speed is either held constant, or varied according to the speed characteristics of the prime mover. The excitation is held constant for separately excited fields, and the rheostat setting is held constant for self-excited machines. *See also:* **direct-current commutating machine; synchronous machine.** 42A10-0

regulation pull-out (regulation drop-out) (power supply). The load currents at which the power supply fails to regulate when the load current is gradually increased or decreased. *See also:* **regulated power supply.** E209-0

regulator, synchronous-machine. An electric-machine regulator that controls the excitation of a synchronous machine. *See also:* **synchronous machine.** 0-31E8

reguline. A word descriptive of electrodeposits that are firm and coherent. *See also:* **electrodeposition.** 42A60-0

reignition (1) (relay). A resumption of current between the contacts of a switching device during an opening operation after an interval of zero current of less than ¼ cycle at normal frequency. 37A100-31E11

(2) (radiation-counter tubes). A process by which multiple counts are generated within a counter tube by atoms or molecules excited or ionized in the discharge accompanying a tube count. *See also:* **gas-filled radiation-counter tubes.** 42A70-15E6

reinforced plastic (rotating machinery). A plastic with some strength properties greatly superior to those of the base resin, resulting from the presence of high-strength fillers imbedded in the composition. *Note:* The reinforcing fillers are usually fibers, fabrics, or mats made of fibers. The plastic laminates are the most common and strongest. AD883-31E8

rejection band (1) (rejection filter, in a signal-transmission system). A band of frequencies between which the transmission is less than a specified fraction of that at a reference frequency. The reference frequency is usually that of maximum transmission. *See also:* **signal.** 0-13E6

(2) (uniconductor waveguide). The frequency range below the cutoff frequency. *See also:* **waveguide.** E146-3E1

rejection filter (signal-transmission system). A filter that attenuates alternating currents between given upper and lower cutoff frequencies and transmits substantially all others. Also, a filter placed in the signal transmission path to attenuate interference. *See also:* **signal.** 0-13E6

related transmission terms (loss and gain). The term **loss** used with different modifiers has different meanings, even when applied to one physical quantity such as power. In view of definitions containing the word **loss** (as well as others containing the word **gain**), the following brief explanation is presented. (1) Power loss from a circuit, in the sense that it is converted to another form of power not useful for the purpose at hand (for example, i^2R loss) is a physical quantity measured in watts in the International System of Units (SI) and having the dimensions of power. For a given R, it will vary with the square of the current in R. (2) Loss may be defined as the ratio of two powers, for example: if P_o is the output power and P_i the input power of a network under specified conditions, P_i/P_o is a dimensionless quantity that would be unity if $P_o = P_i$. Thus, no power loss in the sense of (1) means a loss, defined as the ratio P_i/P_o, of unity. The concept is closely allied to that of efficiency. (3) Loss may also be defined as the logarithm, or as directly proportional to the logarithm, of a power ratio such as P_i/P_o. Thus if loss $= 10 \log_{10} P_i/P_o$ the loss is zero when $P_o = P_i$. This is the standard for measuring loss in decibels. It should be noted that in cases (2) and (3) the **loss** (for a given linear system) is the same whatever may be the power levels. Thus (2) and (3) give characteristics of the system and do not depend (as (1) does) on the value of the current or other dependent quantity. **Power** refers to average power, not instantaneous power. *See also:* **network analysis.** E270-0

relative address (computing systems). The number that specifies the difference between the absolute address and the base address. *See also:* **electronic digital computer.** X3A12-16E9

relative bearing (navigation). Bearing relative to heading. *See also:* **navigation.** 0-10E6

relative capacitivity. *See:* **relative dielectric constant.**

relative coding (computing systems). Coding that uses machine instructions with relative addresses. *See also:* **electronic digital computer.** X3A12-16E9

relative complex dielectric constant (homogeneous isotropic material) (complex capacitivity) (complex permittivity). The ratio of the admittance between two electrodes of a given configuration of electrodes with the material as a dielectric to the admittance between the same two electrodes of the configuration with vacuum as dielectric or

$$\epsilon^* \equiv \epsilon' - j\epsilon'' = Y/(j\omega C_V)$$

where Y is the admittance with the material and $j\omega C_v$ is the admittance with vacuum. Experimentally, vacuum must be replaced by the material at all points where it makes a significant change in the admittance. *Note:* The word **relative** is frequently dropped. *See:* **dielectric loss index; relative dielectric constant.** E270-0

relative damping (specific damping) (instrument). Under given conditions, the ratio of the damping torque at a given angular velocity of the moving element to the damping torque that, if present at this angular velocity, would produce the condition of critical damping. *See also:* **accuracy rating (instrument).** 42A30-0

relative dielectric constant (homogeneous isotropic material) (relative capacitivity) (relative permittivity). The ratio of the capacitance of a given configuration of electrodes with the material as a dielectric to the capacitance of the same electrode configuration with a vacuum (or air for most practical purposes) as the dielectric or

$$\epsilon' = C_X/C_V$$

where C_x is the capacitance with the material and C_v is the capacitance with vacuum. Experimentally, vacuum must be replaced by the material at all points where it makes a significant change in the capacitance. *See:* **electric flux density; relative capacitivity.** E270-0

relative gain (antenna). The ratio of the power gain in a given direction to the power gain of a reference antenna in its reference direction. *Note:* Common reference antennas are half-wave dipoles, electric dipoles, monopoles, and calibrated horn radiators. *See also:* **antenna.** 0-3E1

relative interfering effect (single-frequency electric wave in an electroacoustic system). The ratio, usually expressed in decibels, of the amplitude of a wave of specified reference frequency to that of the wave in question when the two waves are equal in interfering effect. The frequency of maximum interfering effect is usually taken as the reference frequency. Equal interfering effects are usually determined by judgment tests or intelligibility tests. *Note:* When applied to complex waves, the relative interfering effect is the ratio, usually expressed in decibels, of the power of the reference wave to the power of the wave in question when the two waves are equal in interfering effect. *See also:* **transmission characteristics.** 42A65-0

relative lead polarity (transformer). A designation of the relative instantaneous directions of current in its leads. *Notes:* (1) Primary and secondary leads are said to have the same polarity when at a given instant during most of each half cycle, the current enters an identified, or marked, primary lead and leaves the similarly identified, or marked, secondary lead in the same direction as though the two leads formed a continuous circuit. (2) The relative lead polarity of a single-phase transformer may be either additive or subtractive. If one pair of adjacent leads from the two windings is connected together and voltage applied to one of the windings, then: (A) The relative lead polarity is additive if the voltage across the other two leads of the windings is greater than that of the higher-voltage winding alone. (B) The relative lead polarity is subtractive if the voltage across the other two leads of the windings is less than that of the higher-voltage winding alone. (3) The relative lead polarity is indicated by identification marks on primary and secondary leads of like polarity, or by other appropriate identification. *See:* **routine test.** *See also:* **constant-current transformer.** 42A15-31E12

relative luminosity. The ratio of the value of the luminosity at a particular wavelength to the value at the wavelength of maximum luminosity. *See also:* **color terms.** E201-2E2

relatively refractory state (electrobiology). The portion of the electric recovery cycle during which the excitability is less than normal. *See also:* **excitability (electrobiology).** 42A80-18E1

relative permittivity. *See:* **relative dielectric constant.**

relative permeability. The ratio of normal permeability to the magnetic constant. *Note:* In anisotropic media, relative permeability becomes a matrix. E270-0

relative redundancy (of a source) (information theory). The ratio of the redundancy of the source to the logarithm of the number of symbols available at the source. *See also:* **information theory.** E171-0

relative refractive index (two media). The ratio of their refractive indexes. *See also:* **radiation; radio wave propagation.** 42A65/E211-3E2

relative response (audio and electroacoustics). The ratio, usually expressed in decibels, of the response under some particular conditions to the response under reference conditions. Both conditions should be stated explicitly. *See also:* **electroacoustics.** 0-1E1

relaxation oscillator. Any oscillator whose fundamental frequency is determined by the time of charging or discharging of a capacitor or inductor through a resistor, producing waveforms that may be rectangular or sawtooth. *Note:* The frequency of a relaxation oscillator may be self-determined or determined by a synchronizing voltage derived from an external source. *See also:* **electronic controller; oscillatory circuit.** E145-0

relay. An electric device that is designed to interpret input conditions in a prescribed manner and after specified conditions are met to respond to cause contact operation or similar abrupt change in associated electric control circuits. *Notes:* (1) Inputs are usually electric, but may be mechanical, thermal, or other quantities. Limit switches and similar simple devices are not relays. (2) A relay may consist of several units, each responsive to specified inputs, the combination providing the desired performance characteristic. (3) For an extensive list of cross references, see *Appendix A.* 37A100/E270-31E6/31E11/34E10

relay actuation time. The time at which a specified contact functions. 83A16-0

relay actuation time, effective. The sum of the initial actuation time and the contact chatter intervals following such actuation. 83A16-0

relay actuation time, initial. The time of the first closing of a previously open contact or the first opening of a previously closed contact. 83A16-0

relay actuation time, final. The time of termination of chatter following contact actuation. 83A16-0

relay actuator. The part of the relay that converts electric energy into mechanical work. *See also:* **railway signaling and interlocking.** 83A16-0

relay adjustment. The modification of the shape or position of relay parts to affect one or more of the operating characteristics, that is, armature gap, restoring spring, contact gap. 83A16-0

relay air gap. Air space between the armature and the pole piece. This is used in some relays instead of a nonmagnetic separator to provide a break in the magnetic circuit. 83A16-0

relay, alternating-current. A relay designed for operation from an alternating-current source. *See also:* **relay.** 83A16-0

relay amplifier. An amplifier that drives an electromechanical relay. *See also:* **electronic analog computer.** E165-16E9

relay antifreeze pin. Sometimes used for **relay armature stop, nonmagnetic.** 83A16-0

relay armature (electromechanical relay). The moving element that contributes to the designed response of the relay and that usually has associated with it a part of the relay contact assembly. 37A100-31E11

relay armature, balanced. An armature that is approximately in equilibrium with respect to both static and dynamic forces. 83A16-0

relay armature bounce. *See:* **relay armature rebound.** 83A16-0

relay armature card. An insulating member used to link the movable springs to the armature. 83A16-0

relay armature contact. (1) A contact mounted directly on the armature. (2) Sometimes used for **relay contact, movable.** 83A16-0

relay armature, end-on. An armature whose motion is in the direction of the core axis, with the pole face at the end of the core and perpendicular to this axis. 83A16-0

relay armature, flat-type. An armature that rotates about an axis perpendicular to that of the core, with the pole face on a side surface of the core. 83A16-0

relay armature gap. The distance between armature and pole face. 83A16-0

relay armature hesitation. Delay or momentary reversal of armature motion in either the operate or release stroke. 83A16-0

relay armature lifter. *See:* **relay armature stud.**

relay armature, long-lever. An armature with an armature ratio greater than 1:1. 83A16-0

relay armature overtravel. The portion of the available stroke occurring after the contacts have touched. 83A16-0

relay armature ratio. The ratio of the distance through which the armature stud or card moves to the armature travel. 83A16-0

relay armature rebound. Return motion of the armature following impact on the backstop. 83A16-0

relay armature, short-lever. An armature with an armature ratio of 1:1 or less. 83A16-0

relay armature, side. An armature that rotates about an axis parallel to that of the core, with the pole face on a side surface of the core. 83A16-0

relay armature stop. Sometimes used for **relay backstop.** 83A16-0

relay armature stop, nonmagnetic. A nonmagnetic member separating the pole faces of core and armature in the operated position, used to reduce and stabilize the pull from residual magnetism in release. 83A16-0

relay armature stud. An insulating member that transmits the motion of the armature to an adjacent contact member. 83A16-0

relay armature travel. The distance traveled during operation by a specified point on the armature. 83A16-0

relay, automatic reset. (1) A stepping relay that returns to its home position either when it reaches a predetermined contact position, or when a pulsing circuit fails to energize the driving coil within a given time. May either pulse forward or be spring reset to the home position. (2) An overload relay that restores the circuit as soon as an overcurrent situation is corrected. 83A16-0

relay back contacts. Sometimes used for **relay contacts, normally closed.** *See also:* **railway signal and interlocking.** 83A16-0

relay backstop. The part of the relay that limits the movement of the armature away from the pole face or core. In some relays a normally closed contact may serve as backstop. 83A16-0

relay backup. That part of the backup protection that operates in the event of failure of the primary relays. 37A100-31E11

relay bank. *See:* **relay level.**

relay bias winding. An auxiliary winding used to produce an electric bias. 83A16-0

relay blades. Sometimes used for **relay contact springs.** 83A16-0

relay bracer spring. A supporting member used in conjunction with a contact spring. 83A16-0

relay bridging. (1) A result of contact erosion, wherein a metallic protrusion or bridge is built up between opposite contact faces to cause an electric path between them. (2) A form of contact erosion occurring on the break of a low-voltage, low-inductance circuit, at the instant of separation, that results in melting and resolidifying of contact metal in the form of a metallic protrusion or bridge. (3) Make-before-break contact action, as when a wiper touches two successive contacts simultaneously while moving from one to the other. 83A16-0

relay brush. *See:* **relay wiper.**

relay bunching time. The time during which all three contacts of a bridging contact combination are electrically connected during the armature stroke. 83A16-0

relay bushing. Sometimes used for **relay spring stud.** 83A16-0

relay chatter time. The time interval from initial actuation of a contact to the end of chatter. 83A16-0

relay coil. One or more windings on a common form. 83A16-0

relay coil, concentric-wound. A coil with two or more insulated windings, wound one over the other. 83A16-0

relay-coil dissipation. The amount of electric power consumed by a winding. For the most practical purposes, this equals the I^2R loss. 83A16-0

relay-coil resistance. The total terminal-to-terminal resistance of a coil at a specified temperature. 83A16-0

relay-coil serving. A covering, such as thread or tape, that protects the winding from mechanical damage. 83A16-0

relay-coil temperature rise. The increase in temperature of a winding above the ambient temperature when energized under specified conditions for a given period of time, usually the time required to reach a stable temperature. 83A16-0

relay-coil terminal. A device, such as a solder lug, binding post, or similar fitting, to which the coil power supply is connected. 83A16-0

relay-coil tube. An insulated tube upon which a coil is wound. 83A16-0

relay comb. An insulating member used to position a group of contact springs. 83A16-0

relay contact actuation time. The time required for any specified contact on the relay to function according to the following subdivisions. When not otherwise specified, contact actuation time is **relay initial actuation time.** For some purposes, it is preferable to state the actuation time in terms of final actuation time or effective actuation time. 83A16-0

relay contact arrangement. The combination of contact forms that make up the entire relay switching structure. 83A16-0

relay contact bounce. Sometimes used for **relay contact chatter**, when internally caused. 83A16-0

relay contact chatter. The undesired intermittent closure of open contacts or opening of closed contacts. It may occur either when the relay is operated or released or when the relay is subjected to external shock or vibration. 83A16-0

relay contact chatter, armature hesitation. Chatter ascribed to delay or momentary reversal in direction of the armature motion during either the operate or the release stroke. 83A16-0

relay contact chatter, armature impact. Chatter ascribed to vibration of the relay structure caused by impact of the armature on the pole piece in operation, or on the backstop in release. 83A16-0

relay contact chatter, armature rebound. Chatter ascribed to the partial return of the armature to its operated position as a result of rebound from the backstop in release. 83A16-0

relay contact chatter, externally caused. Chatter resulting from shock or vibration imposed on the relay by external action. 83A16-0

relay contact chatter, external shock. Chatter ascribed to impact experienced by the relay or by the apparatus of which it forms a part. 83A16-0

relay contact chatter, initial. Chatter ascribed to vibration produced by opening or closing the contacts themselves, as by contact impact in closure. 83A16-0

relay contact chatter, internally caused. Chatter resulting from the operation or release of the relay. 83A16-0

relay contact chatter, transmitted vibration. Chatter ascribed to vibration originating outside the relay and transmitted to it through its mounting. 83A16-0

relay contact combination. (1) The total assembly of contacts on a relay. (2) Sometimes used for contact form. 83A16-0

relay contact, fixed. *See:* **relay contact, stationary.**

relay contact follow. The displacement of a stated point on the contact-actuating member following initial closure of a contact. 83A16-0

relay contact follow, stiffness. The rate of change of contact force per unit contact follow. 83A16-0

relay contact form. A single-pole contact assembly. 83A16-0

relay contact functioning. The establishment of the specified electrical state of the contacts as a continuous condition. 83A16-0

relay contact gap. *See:* **relay contact separation.**

relay contact, movable. The member of a contact pair that is moved directly by the actuating system. 83A16-0

relay contact pole. Sometimes used for **relay contact, movable.** 83A16-0

relay contact rating. A statement of the conditions under which a contact will perform satisfactorily. 83A16-0

relay contacts. The current-carrying parts of a relay that engage or disengage to open or close electric circuits. 83A16-0

relay contacts, auxiliary. Contacts of lower current capacity than the main contacts; used to keep the coil energized when the original operating circuit is open, to operate an audible or visual signal indicating the position of the main contacts, or to establish interlocking circuits, etcetera. 83A16-0

relay contacts, back. Sometimes used for **relay contacts, normally closed.** 83A16-0

relay contacts, break. *See:* **relay contacts, normally closed.**

relay contacts, break-make. A contact form in which one contact opens its connection to another contact and then closes its connection to a third contact. 83A16-0

relay contacts, bridging. A contact form in which the moving contact touches two stationary contacts simultaneously during transfer. 83A16-0

relay contacts, continuity transfer. Sometimes used for **relay contacts, make-break.** 83A16-0

relay contacts, double break. A contact form in which one contact is normally closed in simultaneous connection with two other contacts. 83A16-0

relay contacts, double make. A contact form in which one contact, which is normally open, makes simultaneous connection when closed with two other independent contacts. 83A16-0

relay contacts, dry. (1) Contacts which neither break nor make current. (2) Erroneously used for **relay contacts, low level.** 83A16-0

relay contacts, early. Sometimes used for **relay contacts, preliminary.** 83A16-0

relay contacts, front. Sometimes used for **relay contacts, normally open.** 83A16-0

relay contacts, interrupter. An additional set of contacts on a stepping relay, operated directly by the armature. 83A16-0

relay contacts, late. Contacts that open or close after other contacts when the relay is operated. 83A16-0

relay contacts, low-capacitance. A type of contact construction providing low intercontact capacitance. 83A16-0

relay contacts, low-level. Contacts that control only the flow of relatively small currents in relatively low-voltage circuits; for example, alternating currents and voltages encountered in voice or tone circuits, direct currents and voltages of the order of microamperes and microvolts, etcetera. 83A16-0

relay contacts, make. *See:* **relay contacts, normally open.**

relay contacts, make-break. A contact form in which one contact closes connection to another contact and then opens its prior connection to a third contact. 83A16-0

relay contacts, multiple-break. Contacts that open a circuit in two or more places. 83A16-0

relay contacts, nonbridging. A contact arrangement in which the opening contact opens before the closing contact closes. 83A16-0

relay contacts, normally closed. A contact pair that is closed when the coil is not energized. 83A16-0

relay contacts, normally open. A contact pair that is open when the coil is not energized. 83A16-0

relay contacts, off-normal. Contacts on a multiple switch that are in one condition when the relay is in its normal position and in the reverse condition for any other position of the relay. 83A16-0

relay contacts, preliminary. Contacts that open or close in advance of other contacts when the relay is operating. 83A16-0

relay contacts, sealed. A contact assembly that is sealed in a compartment separate from the rest of the relay. 83A16-0

relay contact separation. The distance between mating contacts when the contacts are open. 83A16-0

relay contacts, snap-action. A contact assembly having two or more equilibrium positions, in one of which the contacts remain with substantially constant contact pressure during the initial motion of the actuating member, until a condition is reached at which stored energy snaps the contacts to a new position of equilibrium. 83A16-0

relay contact spring. (1) A current-carrying spring to which the contacts are fastened. (2) A non-current-carrying spring that positions and tensions a contact-carrying member. *See also:* **railway signal and interlocking.** 83A16-0

relay contact, stationary. The member of a contact pair that is not moved directly by the actuating system. 83A16-0

relay contact wipe. The sliding or tangential motion between two contact surfaces when they are touching. 83A16-0

relay core. The magnetic member about which the coil is wound. 83A16-0

relay critical voltage (current). That voltage (current) that will just maintain thermal relay contacts operated. 83A16-0

relay cycle timer. A controlling mechanism that opens or closes contacts according to a preset cycle. 83A16-0

relay damping ring, mechanical. A loose member mounted on a contact spring to reduce contact chatter. 83A16-0

relay, direct-current. A relay designed for operation from a direct-current source. *See also:* **relay.** 83A16-0

relay, double-pole. A term applied to a contact arrangement to denote that it includes two separate contact forms; that is, two single-pole contact assemblies. 83A16-0

relay, double-throw. A term applied to a contact arrangement to denote that each contact form included is a break-make. 83A16-0

relay driving spring. The spring that drives the wipers of a stepping relay. 83A16-0

relay drop-out. *See:* **relay release.**

relay, dry circuit. Erroneously used for a relay with either dry or low-level contacts. *See:* **relay contacts, low-level.** 83A16-0

relay duty cycle. A statement of energized and de-energized time in repetitious operation, as: 2 seconds on, 6 seconds off. 83A16-0

relay electric bias. An electrically produced force tending to move the armature towards a given position. 83A16-0

relay, electric reset. A relay that may be reset electrically after an operation. 83A16-0

relay, electromagnetic. A relay, controlled by electromagnetic means, that opens and closes electric contacts. *See also:* **relay.** 83A16-0

relay, electrostatic. A relay in which the actuator element consists of nonconducting media separating two or more conductors that change their relative positions because of the mutual attraction or repulsion by electric charges applied to the conductors. *See also:* **relay.** 83A16-0

relay, electrostrictive. A relay in which an electrostrictive dielectric serves as the actuator. *See also:* **relay.** 83A16-0

relay electrothermal expansion element. An actuating element in the form of a wire strip or other shape having a high coefficient of thermal expansion. 83A16-0

relay element. A subassembly of parts. *Note:* The combination of several relay elements constitutes a relay unit. 37A100-31E11/31E6

relay finish lead. The outer termination of the coil. 83A16-0

relay, flat-type. *See:* **relay armature, flat-type.**

relay frame. The main supporting portion of a relay. This may include parts of the magnetic structure. 83A16-0

relay freezing, magnetic. Sticking of the relay armature to the core as a result of residual magnetism. 83A16-0

relay fritting. Contact erosion in which the electrical discharge makes a hole through the film and produces molten matter that is drawn into the hole by electrostatic forces and solidifies there to form a conducting bridge. 83A16-0

relay front contacts. Sometimes used for **relay contacts, normally open.** 83A16-0

relay functioning time. The time between energization and operation or between de-energization and release. 83A16-0

relay functioning value. The value of applied voltage, current, or power at which the relay operates or releases. 83A16-0

relay header. The subassembly that provides support and insulation to the leads passing through the walls of a sealed relay. 83A16-0

relay heater. A resistor that converts electric energy into heat for operating a thermal relay. 83A16-0

relay heel piece. The portion of a magnetic circuit of a relay that is attached to the end of the core remote from the armature. 83A16-0

relay, high, common, low (HCL). A type of relay control used in such devices as thermostats and in relays operated by them, in which a momentary contact between the common lead and another lead operates the relay, that then remains operated until a momentary contact between the common lead and a third lead causes the relay to return to its original position. 83A16-0

relay hinge. The joint that permits movement of the armature relative to the stationary parts of the relay structure. 83A16-0

relay hold. A specified functioning value at which no relay meeting the specification may release. 83A16-0

relay housing. An enclosure for one or more relays, with or without accessories, usually providing access to the terminals. 83A16-0

relay hum. The sound emitted by relays when their coils are energized by alternating current or in some cases by unfiltered rectified current. 83A16-0

relay inside lead. *See:* **relay start lead.**

relay inverse time. A qualifying term applied to a relay indicating that its time of operation decreases as the magnitude of the operating quantity increases. 83A16-0

relay just-operate value. The measured functioning value at which a particular relay operates. 83A16-0

relay just-release value. The measured functioning value for the release of a particular relay. 83A16-0

relay leakage flux. The portion of the magnetic flux that does not cross the armature-to-pole-face gap. 83A16-0

relay level. A series of contacts served by one wiper in a stepping relay. 83A16-0

relay load curves. The static force displacement characteristic of the total load of the relay. 83A16-0

relay magnetic bias. A steady magnetic field applied to the magnetic circuit of a relay. 83A16-0

relay magnetic gap. Nonmagnetic portion of a magnetic circuit. 83A16-0

relay, manual reset. A relay that may be reset manually after an operation. 83A16-0

relay mechanical bias. A mechanical force tending to move the armature towards a given position. 83A16-0

relay mounting plane. The plane to which the relay mounting surface is fastened. 83A16-0

relay must-operate value. A specified functioning value at which all relays meeting the specification must operate. 83A16-0

relay must-release value. A specified functioning value, at which all relays meeting the specification must release. 83A16-0

relay nonfreeze pin. Sometimes used for **relay armature stop, nonmagnetic.** 83A16-0

relay nonoperate value. A specified functioning value at which no relay meeting the specification may operate. 83A16-0

relay normal condition. The de-energized condition of the relay. 83A16-0

relay operate. The condition attained by a relay when all contacts have functioned. *See also:* **relay contact actuation time.** 83A16-0

relay operate time. The time interval from coil energization to the functioning time of the last contact to function. Where not otherwise stated the functioning time of the contact in question is taken as its initial functioning time. 83A16-0

relay operate time characteristic. The relation between the operate time of an electromagnetic relay and the operate power. 83A16-0

relay operating frequency. The rated alternating-current frequency of the supply voltage at which the relay is designed to operate. 83A16-0

relay outside lead. *See:* **relay finish lead.**

relay overtravel. Amount of contact wipe. *See:* **relay armature overtravel; relay contact wipe.** 83A16-0

relay pickup value. Sometimes used for **relay must-operate value.** 83A16-0

relay pileup. A set of contact arms, assemblies, or springs, fastened one on top of the other with insulation between them. 83A16-0

relay pneumatic bellows. Gas-filled bellows, sometimes used with plunger-type relays to obtain time delay. 83A16-0

relay pole face. The part of the magnetic structure at the end of the core nearest the armature. 83A16-0

relay pole piece. The end of an electromagnet, sometimes separable from the main section, and usually shaped so as to distribute the magnetic field in a pattern best suited to the application. 83A16-0

relay pull curves. The force-displacement characteristics of the actuating system of the relay. 83A16-0

relay pull-in value. Sometimes used for **relay must-operate value.** 83A16-0

relay pusher. Sometimes used for **relay armature stud.** *See also:* **relay.** 83A16-0

relay rating. A statement of the conditions under which a relay will perform satisfactorily. 83A16-0

relay recovery time. A cooling time required from heater de-energization of a thermal time-delay relay to subsequent re-energization that will result in a new operate time equal to 85 percent of that exhibited from a cold start. 83A16-0

relay recovery time, instantaneous. Recovery time of a thermal relay measured when the heater is de-energized at the instant of contact operation. 83A16-0

relay recovery time, saturated. Recovery time of a thermal relay measured after temperature saturation has been reached. 83A16-0

relay release. The condition attained by a relay when all contacts have functioned and the armature (where applicable) has reached a fully opened position. 83A16-0

relay release time. The time interval from coil de-energization to the functioning time of the last contact to function. Where not otherwise stated the functioning time of the contact in question is taken as its initial functioning time. *See also:* **railway signal and interlocking.** 83A16-0

relay reoperate time. Release time of a thermal relay. 83A16-0

relay reoperate time, instantaneous. Reoperate time of a thermal relay measured when the heater is de-energized at the instant of contact operation. 83A16-0

relay reoperate time, saturated. Reoperate time of a thermal relay measured when the relay is de-energized after temperature saturation (equilibrium) has been reached. 83A16-0

relay residual gap. Sometimes used for **relay armature stop, nonmagnetic.** 83A16-0

relay restoring spring. A spring that moves the armature to the normal position and holds it there when the relay is de-energized. 83A16-0

relay retractile spring. Sometimes used for **relay restoring spring.** 83A16-0

relay return spring. Sometimes used for **relay restoring spring.** 83A16-0

relay saturation. The condition attained in a magnetic material when an increase in field intensity produces no further increase in flux density. 83A16-0

relay sealing. Sometimes used for **relay seating.** 83A16-0

relay seating. The magnetic positioning of an armature in its final desired location. 83A16-0

relay seating time. The elapsed time after the coil has been energized to the time required to seat the armature of the relay. 83A16-0

relay shading coil. Sometimes used for **relay shading ring.** 83A16-0

relay shading ring. A shorted turn surrounding a portion of the pole of an alternating-current magnet, producing a delay of the change of the magnetic field in that part, thereby tending to prevent chatter and reduce hum. 83A16-0

relay shields, electrostatic spring. Grounded conducting members located between two relay springs to minimize electrostatic coupling. 83A16-0

relay shim, nonmagnetic. Sometimes used for **relay armature stop, nonmagnetic.** 83A16-0

relay, single-pole. A relay in which all contacts connect, in one position or another, to a common contact. 83A16-0

relay, single-throw. A relay in which each contact form included is a single contact pair. 83A16-0

relay sleeve. A conducting tube placed around the full length of the core as a short-circuited winding to retard the establishment or decay of flux within the magnetic path. 83A16-0

relay slow-release time characteristic. The relation between the release time of an electromagnetic relay and the conductance of the winding circuit or of the conductor (sleeve or slug) used to delay release. The conductance in this definition is the quantity N^2/R, where N is the number of turns and R is the resistance of the closed winding circuit. (For a sleeve or slug $N=1$). 83A16-0

relay slug. A conducting tube placed around a portion of the core to retard the establishment or decay of flux within the magnetic path. 83A16-0

relay soak. The condition of an electromagnetic relay when its core is approximately saturated. 83A16-0

relay soak value. The voltage, current, or power applied to the relay coil to insure a condition approximating magnetic saturation. 83A16-0

relay spool. A flanged form upon which a coil is wound. 83A16-0

relay spring buffer. Sometimes used for **relay spring stud.** 83A16-0

relay spring curve. A plot of spring force on the armature versus armature travel. 83A16-0

relay spring stop. A member that controls the position of a pretensioned spring. 83A16-0

relay spring stud. An insulating member that transmits the motion of the armature from one movable contact to another in the same pileup. 83A16-0

relay stack. Sometimes used for **relay pileup.** 83A16-0

relay stagger time. The time interval between the actuation of any two contact sets. 83A16-0

relay starting switch (rotating machinery). A relay, actuated by current, voltage, or the combined effect of current and voltage, used to perform a circuit-changing function in the primary winding of a single-phase induction motor within a predetermined range of speed as the rotor accelerates; and to perform the reverse circuit-changing operation when the motor is disconnected from the supply line. One of the circuit changes that is usually performed is to open or disconnect the auxiliary-winding circuit. *See also:* **starting-switch assembly.** 0-31E8

relay start lead. The inner termination of the coil. 83A16-0

relay static characteristic. The static force–displacement characteristic of the spring system or of the actuating system. 83A16-0

relay station (mobile communication). A radio station used for the reception and retransmission of the signals from another radio station. *See also:* **mobile communication system.** 0-6E1

relay, thermal. A relay that is actuated by the heating effect of an electric current. *See also:* **relay.** 83A16-0

relay, three-position. Sometimes used for a center-stable polar relay. *See also:* **relay.** 83A16-0

relay transfer contacts. Sometimes used for **relay contacts, break-make.** 83A16-0

relay transfer time. The time interval between opening the closed contact and closing the open contact of a break-make contact form. 83A16-0

relay unit. An assembly of relay elements that in itself can perform a relay function. *Note:* One or more relay units constitutes a relay. 37A100-31E11/31E6

relay winding. Sometimes used for **relay coil.** 83A16-0

relay wiper. The moving contact on a rotary stepping switch or relay. 83A16-0

release (trip) coil (mechanical switching device). A coil used in the electromagnet that initiates the action of a release (trip). 37A100-31E11

release (tripping) delay (mechanical switching device). Intentional time delay introduced into contact-parting time in addition to opening time. *Note:* In devices employing a shunt release, release delay includes the operating time of protective and auxiliary relays external to the device. In devices employing direct or indirect release, release delay consists of intentional delay introduced into the function of the release. 37A100-31E11

release-delay (trip-delay) setting. A calibrated setting of the time interval between the time when the actuating value reaches the release setting and the time when the release operates. 37A100-31E11

release-free (trip-free) (1) (mechanical switching device). A descriptive term that indicates that the opening operation can prevail over the closing operation during specified parts of the closing operation. 37A100-31E11

(2) (trip-free in any position). A descriptive term indicating that a switching device is release-free at any part of the closing operation. *Note:* If the release circuit is completed through an auxiliary switch, electric release will not take place until such auxiliary switch is closed. 37A100-31E11

release-free (trip-free) relay. An auxiliary relay whose function is to open the closing circuit of an electrically operated switching device so that the opening operation can prevail over the closing operation. 37A100-31E11/31E6

release (tripping) mechanism (mechanical switching device). A device, mechanically connected to the mechanical switching device, that releases the holding means and permits the opening or closing of the switching device. 37A100-31E11

release (trip) setting. A calibrated point at which the release is set to operate. 37A100-31E11

release time, relay. *See:* **relay release time.**

reliability (1) (general). (A) The ability of an item to perform a required function under stated conditions for a stated period of time. *Note:* For an extensive list of cross references, see *Appendix A.* 0-7E1
(B) The probability that a device will function without failure over a specified time period or amount of usage. *Notes:* (1) Definition (B) is most commonly used in engineering applications. In any case where confusion may arise, specify the definition being used. (2) The probability that the system will perform its function over the specified time should be equal to or greater than the reliability. X3A12-16E9/35E2

(2) (relay or relay system). A measure of the degree of certainty that the relay, or relay system, will perform correctly. *Note:* Reliability denotes certainty of correct operation together with assurance against incorrect operation from all extraneous causes. *See also:* **dependability; security.** 37A100-31E11/31E6

reliability, assessed. The reliability of an item determined within stated confidence limits from tests or failure data on nominally identical items. The source of the data shall be stated. Results can only be accumulated (combined) when all the conditions are similar. *Note:* Alternatively, point estimates may be used, the basis of which shall be defined. *See also:* **reliability.** 0-7E1

reliability, extrapolated. Extension by a defined extrapolation or interpolation of the assessed reliability for durations or stress conditions different from those applying to the conditions of that assessed reliability. *See also:* **reliability.** 0-7E1

reliability, inherent. The potential reliability of an item present in its design. *See also:* **reliability.** 0-7E1

reliability, operational. The assessed reliability of an item based on field data. *See also:* **reliability.** 0-7E1

reliability, predicted. The reliability of an equipment computed from its design considerations and from the reliability of its parts in the intended conditions of use. *See also:* **reliability.** 0-7E1

reliability, test. The assessed reliability of an item based on a particular test with stated stress and stated failure criteria. *See also:* **reliability.** 0-7E1

relief door (rotating machinery). A pressure-operated door to prevent excessive gas pressure within a housing. *See also:* **cradle base (rotating machinery).** 0-31E8

relieving (electroplating). The removal of compounds from portions of colored metal surfaces by mechanical means. *See also:* **electroplating.** 42A60-0

relieving anode (pool-cathode tube). An auxiliary anode that provides an alternative conducting path for reducing the current to another electrode. *See also:* **electrode (electron tube).** 42A70-0

relocate (computing systems) (programming). To move a routine from one portion of storage to another and to adjust the necessary address references so that the routine, in its new location, can be executed. *See also:* **electronic digital computer.** X3A12-16E9

reluctance (magnetic circuit). The ratio of the magnetomotive force to the magnetic flux through any cross section of the magnetic circuit. E270-0

reluctance motor. A synchronous motor similar in construction to an induction motor, in which the member carrying the secondary circuit has salient poles, without permanent magnets or direct-current excitation. It starts as an induction motor, is normally provided with a squirrel-cage winding, but operates normally at synchronous speed. *See also:* **synchronous machine.** 42A10-31E8

reluctance torque (synchronous motor). The torque developed by the motor due to the flux produced in the field poles by action of the armature-reaction magnetomotive force. *See also:* **synchronous machine.** 0-31E8

reluctivity. The reciprocal of permeability. *Note:* In anisotropic media, reluctivity becomes a matrix. E270-0

remanence. The magnetic flux density that remains in a magnetic circuit after the removal of an applied magnetomotive force. *Note:* This should not be confused with **residual flux density.** If the magnetic circuit has an air gap, the remanence will be less than the residual flux density. *See also:* **static magnetic storage.** E163-0

remanent charge (ferroelectric device). The charge remaining when the applied voltage is removed. *Note:* The remanent charge is essentially independent of the previously applied voltage, provided this voltage was sufficient to cause saturation. If the device was not or cannot be saturated, the value of the previously applied voltage should be stated when measurements of remanent charge are reported. See the figure under **total charge.** *See also:* **ferroelectric domain.** E180-0

remanent induction (magnetic material). The induction when the magnetomotive force around the complete magnetic circuit is zero. *Note:* If there are no air gaps or other inhomogeneities in the magnetic circuit, the remanent induction will equal the residual induction; if there are air gaps or other inhomogeneities, the remanent induction will be less than the residual induction. E270-0

remanent polarization (ferroelectric material). The value of the electric displacement upon removal of the field after saturation. *Note:* This is illustrated in the accompanying figure as point P_r.

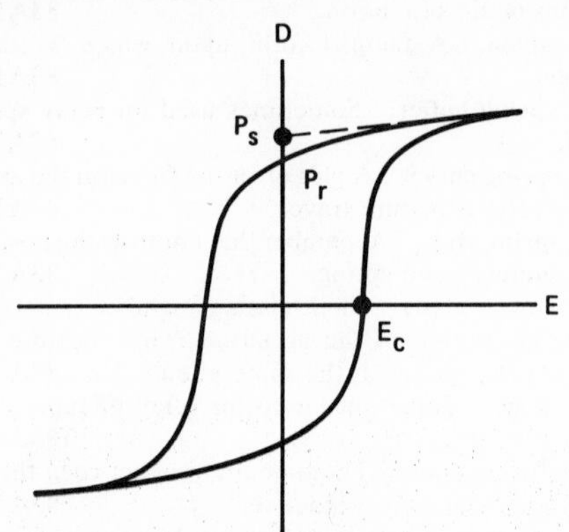

Ferroelectric hysteresis loop.

remote backup. A form of backup protection in which the protection is at a station or stations other than that which has the primary protection. 37A100-31E11/31E6

remote control (general). Control of an operation from a distance: this involves a link, usually electrical, between the control device and the apparatus to be operated. *Note:* Remote control may be over (1) direct wire, (2) other types of interconnecting channels such as carrier-current or microwave, (3) supervisory control, or (4) mechanical means. *See also:* **control; railway signal and interlocking.** 37A100/50I16-31E11/34E10

remote-control circuits. Any electric circuit that controls any other circuit through a relay or an equivalent device. 1A0-0

remote-cutoff tube. *See:* **variable μ tube.**

remote data-logging equipment. Equipment for the numerical recording of selected telemetered quantities on log sheets or paper or magnetic tape or the like by means of an electric typewriter, teletype, or other suitable devices. 37A100-31E11

remote error-sensing (power supplies). A means by which the regulator circuit senses the voltage directly at the load. This connection is used to compensate for voltage drops in the connecting wires. *See also:* **power supply.** KPSH-10E1

remote indication. Indication of the position or condition of remotely located devices. *Note:* Remote indication may be over (1) direct wire, (2) other types of interconnecting channels such as carrier-current or microwave, (3) supervisory indication, or (4) mechanical means. 37A100-31E11

remote line (electroacoustics). A program transmission line between a remote pickup point and the studio or transmitter site. *See also:* **transmission line.** E151-0

remote manual operation. *See:* **indirect manual operation.**

remote metering. *See:* **telemetering.**

remote operation. *See:* **remotely controlled operation.**

remote release (remote trip) (relay installation). A general term applied to indicate that the switching device is located physically at a point remote from the initiating protective relay, device, or source of release power or all these. *Note:* This installation is commonly called **transfer trip** when a communication channel is used to transmit the signal for remote release. 37A100-31E11

remote-station supervisory equipment (single supervisory system). That part that includes all supervisory control relays and associated devices located at the remote station for selection, control, indication, telemetering, and other functions to be performed. 37A100-31E11

remote trip. *See:* **remote release.**

remotely controlled operation. Operation of a device by remote control. 37A100-31E11

removable element (switchgear assembly). That portion that normally carries the circuit-switching and circuit-interrupting devices and the removable part of the primary and secondary disconnecting devices. 37A100-31E11

renewable (field-renewable) fuse. A fuse that, after circuit interruption, may be restored readily for service by the replacement of the renewal element, fuse link, or refill unit. 37A100-31E11

renewal element (low-voltage fuse). That part of a renewable fuse that is replaced after each interruption to restore the fuse to operating condition. 37A100-31E11

renewal parts. Those parts that it is necessary to replace during maintenance as a result of wear. 37A100-31E11

reoperate time, relay. *See:* **relay reoperate time.**

repairable item (reliability). *See:* **item, repaired.**

repeatability (measurement) (control equipment) (1) (general). The closeness of agreement among repeated measurements of the same variable under the same conditions. *See also:* **measurement system; power systems, low-frequency and surge testing.** E94-0
(2) (zero-center instruments). The ability of an instrument to repeat its readings taken when the pointer is deflected upscale, compared to the readings taken when the pointer is deflected downscale, expressed as a percentage of the full-scale value. *See also:* **measurement system; moving element (instrument).** 39A1-0

repeatability (voltage regulator, or voltage-reference tubes). The ability of a tube to attain the same voltage drop at a stated time after the beginning of any conducting period. *Note:* The lack of repeatability is measured by the change in this voltage from one conducting period to any other, the operating conditions remaining unchanged. *See also:* **circuit characteristics of electrodes; transmission regulator.** E160-15E6

repeater. A combination of apparatus for receiving either one-way or two-way communication signals and delivering corresponding signals that are either amplified or reshaped or both. A repeater for one-way communication signals is termed a **one-way repeater** and one for two-way communication signals a **two-way repeater.**
See:
carrier repeater;
cord-circuit repeater;
data transmission;
four-wire repeater;
intermediate repeater;
negative-resistance repeater;
pulse repeater;
repeater, 21-type;
repeater, 22-type;
telephone repeater;
television repeater;
terminal repeater;
two-wire repeater. 42A65-31E3

repeater, 21-type. A two-wire telephone repeater in which there is one amplifier serving to amplify the telephone currents in both directions, the circuit being arranged so that the input and output terminals of the amplifier are in one pair of conjugate branches, while the lines in the two directions are in another pair of conjugate branches. *See also:* **repeater.** 42A65-0

repeater, 22-type. A two-wire telephone repeater in which there are two amplifiers, one serving to amplify the telephone currents being transmitted in one direction and the other serving to amplify the telephone currents being transmitted in the other direction. *See also:* **repeater.** 42A65-0

repeating coil* (repeat coil*). A term sometimes used in telephone practice to designate a transformer. *See also:* **telephone system.**
*Obsolescent 42A65-31E3

repeller (reflector) (electron tubes). An electrode whose primary function is to reverse the direction of an electron stream. *See also:* **beam tube; electrode (electron tube).** E160-15E6

reperforator. *See:* **perforator.**

reperforator switching center. A message-relaying center at which incoming messages are received on a reperforator that perforates a storage tape from which the message is retransmitted into the proper outgoing circuit. The reperforator may be of t he type that also prints the message on the same tape, and the selection of the outgoing circuit may be manual or under control of selection characters at the head of the message. *See also:* **telegraphy.** 42A65-19E4

repertory. *See:* **instruction repertory.**

repetition equivalent (of a complete telephone connection, including the terminating telephone set). A measure of the grade of transmission experienced by the subscribers using the connection. It includes the combined effects of volume, distortion, noise, and all other subscriber reactions and usages. The repetition equivalent of a complete telephone connection is expressed numerically in terms of the trunk loss of a working reference system when the latter is adjusted to given an equal repetition rate. *See also:* **transmission characteristics.** 42A65-0

repetition instruction. An instruction that causes one or more instructions to be executed an indicated number of times. X3A12-16E9

repetition rate. Repetition rate signifies broadly the number of repetitions per unit time. Specifically, in appraising the effectiveness of transmission over a telephone connection, repetition rate is the number of repetitions per unit time requested by users of the connection. *See also:* **volume equivalent.** 42A65-0

repetitive operation (analog computer). A condition in which the computer operates as a repetitive device; the solution time may be a small fraction of a second after which the problem is automatically and repetitively cycled through reset, hold, and operate. *See also:*

electronic analog computer. E165-16E9

repetitive peak OFF-state voltage. The maximum instantaneous value of the OFF-state voltage that occurs across a thristor, including all repetitive transient voltages, but excluding all nonrepetitive transient voltages. E223-34E17/34E24/15E7

repetitive peak reverse voltage (semiconductor rectifier). The maximum instantaneous value of the reverse voltage, including all repetitive transient voltages but excluding all nonrepetitive transient voltages, that occurs across a semiconductor rectifier cell, rectifier diode, or rectifier stack. *See also:* **principal voltage-current characteristic; rectification; semiconductor rectifier stack.** E59/E223-34E17/34E24/15E7

repetitive peak reverse-voltage rating (rectifier circuit). The maximum value of repetitive peak reverse voltage permitted by the manufacturer under stated conditions. *See also:* **average forward-current rating (rectifier circuit).** E59-34E17/34E24

replica temperature relay. A thermal relay whose internal operating temperature rise is proportional to a specified external temperature rise. 37A100-31E11/31E6

reply (transponder operation). A radio-frequency signal or combination of signals transmitted as a result of interrogation. *See also:* **navigation.** E172-10E6

reproducibility (1) (general). The ability of a system or element to maintain its output/input precision over a relatively long period of time. *See also:* **accuracy; numerically controlled machines; precision.** 85A1-34E12

(2) (transmission lines and waveguides). The degree to which a given set of conditions or observations, using different components or instruments each time, can be reproduced. *See also:* **measurement system.** 0-9E4

(3) (automatic null-balancing electric instrument). The closeness of agreement among repeated measurements by the instrument for the same value of input made under the same operating conditions, over a long period of time, approaching from either direction. *Notes:* (1) It is expressed as a maximum nonreproducibility in percent of span for a specified time. (2) Reproducibility includes drift, repeatability, and dead band. *See also:* **measurement system.** 39A4-0

reradiation. (1) The scattering of incident radiation, or (2) the radiation of signals amplified in a radio receiver. *See also:* **radio receiver.** 42A65-0

rerecording (electroacoustics). The process of making a recording by reproducing a recorded sound source and recording this reproduction. *See also:* **dubbing.** E157-1E1

rerecording system (electroacoustics). An association of reproducers, mixers, amplifiers, and recorders capable of being used for combining or modifying various sound recordings to provide a final sound record. *Note:* Recording of speech, music, and sound effects may be so combined. *See also:* **dubbing; phonograph pickup.** E157-1E1

rerun point (computing systems). The location in the sequence of instructions in a computer program at which all information pertinent to the rerunning of the program is available. *See also:* **electronic digital computer.** X3A12-16E9

reproduction speed (facsimile). The area of copy recorded per unit time. *See also:* **recording (facsimile).** E168-0

reproducing stylus. A mechanical element adapted to following the modulations of a record groove and transmitting the mechanical motion thus derived to the pickup mechanism. *See also:* **phonograph pickup.** 0-1E1

repulsion-induction motor. A motor with repulsion-motor windings and short-circuited brushes, without an additional device for short-circuiting the commutator segments, and with a squirrel-cage winding in the rotor in addition to the repulsion motor winding. 0-31E8

repulsion motor. A single-phase motor that has a stator winding arranged for connection to a source of power and a rotor winding connected to a commutator. Brushes on the commutator are short-circuited and are so placed that the magnetic axis of the rotor winding is inclined to the magnetic axis of the stator winding. This type of motor has a varying-speed characteristic. *See also:* **asynchronous machine.** 42A10-31E8

repulsion-start induction motor. A single-phase motor with repulsion-motor windings and brushes, having a commutator-short-circuiting device that operates at a predetermined speed of rotation to convert the motor into the equivalent of a squirrel-cage motor for running operation. For starting operation, this motor performs as a repulsion motor. *See also:* **asynchronous machine.** 0-31E8

request test. A test made at the request of a customer. *See also:* **service test (field test).** 12A0/42A30-0

reserve cell. A cell that is activated by shock or other means immediately prior to use. *See also:* **electrochemistry.** 42A60-0

reserve equipment. The installed equipment in excess of that required to carry peak load. *Note:* Reserve equipment not in operation is sometimes referred to as **standby equipment.** *See also:* **generating station.** 42A35-31E13

reset (1) (electronic digital computation) (industrial control). (A) To restore a storage device to a prescribed state, not necessarily that denoting zero. (B) To place a binary cell in the initial or zero state. *See also:* **electronic computation; electronic digital computer; set.** E162/E270/X3A12-16E9/34E10

(2) (electronic analog computation). The computer control state in which integrators are held inoperative and the proper initial-condition voltages or charges are applied or reapplied. *See also:* **electronic analog computer; initial condition.** E165-16E9

(3) (relay). The action of a relay as it makes designated response to decreases in input. Reset as a qualifying term denotes the state of a relay when all response to decrease of input has been completed. Reset is also used to identify the maximum value of an input quantity reached by progressive decreases that will permit the relay to reach the state of complete reset from pickup. *Note:* In defining the designated performance of relays having multiple inputs, reset describes the state when all inputs are zero and also when some input circuits are energized, if the resulting state is not altered from the zero-input condition. 37A100-31E11/31E6

reset control action (electric power systems). Action in which the controller output is proportional to the input signal and the time integral of the input signal. The number of times per minute that the integral control action repeats the proportional control action is called the reset rate. *Note:* Applies only to a controller with proportional control action plus integral control action. *See also:* **speed-governing system.** E94-0

reset device. A device whereby the brakes may be released after an automatic train-control brake application. *See also:* **railway signal and interlocking.** 42A42-0

reset dwell time. The time spent in reset. In cycling the computer from reset, to operate, to hold, and back to reset, this time must be long enough to permit the computer to recover from any overload and to charge or discharge all integrating capacitors to appropriate initial voltages. *See also:* **electronic analog computer.** E165-0

reset pulse. A drive pulse that tends to reset a magnetic cell. *See also:* **static magnetic storage.** E163-0

reset switch (industrial control). A machine-operated device that restores normal operation to the control system after a corrective action. *See also:* **photoelectric control.** 42A25-34E10

resettability (oscillator). The ability of the tuning element to retune the oscillator to the same operating frequency for the same set of input conditions. *See also:* **tunable microwave oscillators.** E158-15E6

reset time (1) (relay). The time interval from occurrence of specified conditions to reset. *Note:* When the conditions are not specified it is intended to apply to a picked-up relay and to be a sudden change from pickup value of input to zero input. 37A100-31E11/31E6

(2) (sectionalizer). The time required, after one counting operation, for the counting mechanism to return to the starting position of that counting operation. 37A100-31E11

residual-component telephone-influence factor (three-phase synchronous machine). The ratio of the square root of the sum of the squares of the weighted residual harmonic voltages to three times the root-mean-square no-load phase-to-neutral voltage. *See also:* **synchronous machine.** 42A10-31E8

residual current (1) (electric supply circuit). The vector sum of the currents in the several wires. *Note:* When the vector sum is of the currents in the phase conductors only it is sometimes termed the **circuit residual current,** to distinguish it from the vector sum of the currents in all conductors on a line termed the **line residual current** or more commonly, **ground return current.** *See also:* **inductive coordination.** 42A65-0

(2) (thermionics). The value of the current in the residual-current state. *See also:* **electron emission; inductive coordination.** 50I07-15E6

residual-current state (thermionics). The state of working of an electronic valve or tube in the absence of an accelerating field from the anode of a diode or equivalent diode, in which the cathode current is due to the nonzero velocity of emission of electrons. *See also:* **electron emission; inductive coordination.** 50I07-15E6

residual error (electronic navigation). The sum of the random errors and the uncorrected systematic errors. *See also:* **navigation.** E172-10E6

residual-error rate. *See:* **undetected-error rate.**

residual flux density. The magnetic flux density at which the magnetizing force is zero when the material is in a symmetrically cyclically magnetized condition. *See also:* **remanence; static magnetic storage.** E163-0

residual induction (1) (magnetic material). The magnetic induction corresponding to zero magnetizing force in a material that is in a symmetrically cyclically magnetized condition. E270-0

(2) (residual flux density) (toroidal magnetic amplifier cores). The magnetic induction at which the magnetizing force is zero when the magnetic core is cyclically magnetized with a half-wave sinusoidal magnetizing force of a specified peak magnitude. *Note:* This use of the term **residual induction** differs from the standard definition that requires symmetrically cyclically magnetized conditions. E106-0

residual magnetism (ferromagnetic bodies). A property by which they retain a certain magnetization (induction) after the magnetizing force has been removed. *See:* **remanent induction; residual induction.** 50I05/E270-0

residual modulation. *See:* **carrier noise; carrier noise level.**

residual probe pickup (constancy of probe coupling) (slotted line). The noncyclical variation of the amplitude of the probe output over its complete range of travel when reflected waves are eliminated on the slotted section by proper matching at the output and the input, discounting attenuation along the slotted section. It is defined by the ratio of one-half of the total variation to the average value of the probe output, assuming linear amplitude response of the probe, at a specified frequency(ies) within the range of usage. *Note:* This quantity consists of two parts of which one is reproducible and the other is not. The repeatable part can be eliminated by subtraction in repeated measurements, while the nonrepeatable part must cause an error. The residual probe pickup depends to some extent on the insertion depth of the probe. *See also:* **measurement system; residual standing-wave ratio.** 0-9E4

residual reflected coefficient (reflectometer). The erroneous reflection coefficient indicated when the reflectometer is terminated in reflectionless terminations. *See also:* **measurement system.** 0-9E4

residual relay. A relay that is so applied that its input, derived from external connections of instrument transformers, is proportional to the zero phase-sequence component of a polyphase quantity. *See also:* **relay.** 37A100-31E11/31E6

residual standing-wave ratio (SWR) (slotted line). The standing-wave ratio measured when the slotted line is terminated by a reflectionless termination and fed by a signal source that provides a nonreflecting termination for waves reflected toward the generator. *Note:* Residual standing-wave ratio does not include the residual noncyclical probe pickup or the attenuation encountered as the probe is moved along the line. *See:* **residual probe pickup.** *See also:* **waveguide.** 0-9E4

residual voltage (1) (electric supply circuit). The vector sum of the voltages to ground of the several phase wires of the circuit. *See also:* **inductive coordination.** 42A65-0

(2) (arrester) (discharge voltage). The voltage that appears between the line and ground terminals of an arrester during the passage of discharge current. *See also:* **inductive coordination; lightning arrester (surge diverter).** 99I2-31E7

residue check (computing systems). A check in which each operand is accompanied by the remainder obtained by dividing this number by *n*, the remainder then being used as a check digit or digits. *See also:* **electronic digital computer; modulo *n* check.** X3A12-16E9

resin-bonded paper-insulated bushing (outdoor electric apparatus). A bushing in which the major insulation is provided by paper bonded with resin. *See also:* **power distribution, overhead construction.** 76A1-0

resist (electroplating). Any material applied to part of a cathode or plating rack to render the surface nonconducting. *See also:* **electroplating.** 42A60-0

resistance (1) (network analysis). (1) That physical property of an element, device, branch, network, or system that is the factor by which the mean-square conduction current must be multiplied to give the corresponding power lost by dissipation as heat or as other permanent radiation or loss of electromagnetic energy from the circuit. (2) The real part of impedance. *Note:* Definitions (1) and (2) are not equivalent but are supplementary. In any case where confusion may arise, specify definition being used. *See:* **resistor.** E270-9E4/34E10

(2) (shunt). The quotient of the voltage developed across the instrument terminals to the current passing between the current terminals. In determining the value, account should be taken of the resistance of the instrument and the measuring cable. The resistance value is generally derived from a direct-current measurement such as by means of a double Kelvin bridge. *See also:* **test voltage and current.** 68A1-31E5

resistance box. A rheostat consisting of an assembly of resistors of definite values so arranged that the resistance of the circuit in which it is connected may be changed by known amounts. E270-0

resistance braking. A system of dynamic braking in which electric energy generated by the traction motors is dissipated by means of a resistor. *See also:* **dynamic braking.** 42A42-0

resistance-capacitance (RC) coupling. Coupling between two or more circuits, usually amplifier stages, by means of a combination of resistance and capacitance elements. *See also:* **coupling.** 42A65-0

resistance-capacitance oscillator (RC oscillator). Any oscillator in which the frequency is determined principally by resistance-capacitance elements. *See also:* **oscillatory circuit.** 0-42A65

resistance drop. The voltage drop in phase with the current. *See also:* **efficiency.** 57A15/42A15-31E12

resistance furnace. An electrothermic apparatus, the heat energy for which is generated by the flow of electric current against ohmic resistance internal to the furnace. 42A60-0

resistance grading (rotating machinery). A form of corona shielding embodying high -resistance material on the surface of the coil immediately outside the ends of the slot, to reduce gradients of electric potential to harmless values. *See also:* **asynchronous machine; direct-current commutating machine; synchronous machine.** 0-31E8

resistance grounded (electric power). Grounded through impedance, the principal element of which is resistance. *Note:* The resistance may be inserted either directly in the connection to the ground or indirectly, as for example, in the secondary of a transformer, the primary of which is connected between neutral and ground, or in series with the delta-connected secondary of a wye-delta grounding transformer. *See also:* **ground; grounded.** 42A15/42A35/E32-31E12/31E13

resistance lamp. An electric lamp used to prevent the current in a circuit from exceeding a desired limit. *See also:* **circuits and devices.** 42A65-0

resistance magnetometer. A magnetometer that depends for its operation upon the variation of electrical resistance of a material immersed in the field to be measured. *See also:* **magnetometer.** 42A30-0

resistance method of temperature determination (electric power). The determination of temperature by comparison of the resistance of a winding at the temperature to be determined, with the resistance at a known temperature. *See also:* **power sytems, low-frequency and surge testing; routine test.** 42A15/57A15/E32-31E12

resistance relay. A linear-impedance form of distance relay for which the operating characteristic on an *R-X* diagram is a straight line of constant resistance. *Note:* The operating characteristic may be described by the equation $R=K$, or $Z\cos\theta=K$, where K is a constant, and θ is the angle by which the input voltage leads the input current. 37A100-31E11/31E6

resistance-start motor. A form of split-phase motor having a resistance connected in series with the auxiliary winding. The auxiliary circuit is opened when the motor has attained a predetermined speed. *See also:* **asynchronous machine.** 42A10-0

resistance starting (industrial control). A form of reduced-voltage starting employing resistances that are short-circuited in one or more steps to complete the starting cycle. *See:* **resistance starting, generator-field; resistance starting, motor-armature.** IC1-34E10

resistance starting, generator-field. Field resistance starting provided by one or more resistance steps in series with the shunt field of a generator, the output of which is connected to a motor armature. *See:* **resistance starting; resistance starting, motor-armature.** IC1-34E10

resistance starting, motor-armature. Motor resistance starting provided by one or more resistance steps connected in series with the motor armature. *See:* **resistance starting; resistance starting, generator-field.** IC1-34E10

resistance temperature detector (resistance thermometer detector) (resistance thermometer resistor). A resistor made of some material for which the electrical resistivity is a known function of the temperature and that is intended for use with a resistance thermometer. It is usually in such a form that it can be placed in the region where the temperature is to be determined. *Note:* A resistance temperature detector with its support and enclosing envelope, is often called a resistance thermometer bulb. *See also:* **electric thermometer (temperature meter); embedded temperature detector.** 42A30-0

resistance thermometer (resistance temperature meter). An electric thermometer that operates by measuring the electric resistance of a resistor, the resistance of which is a known function of its temperature. The temperature-responsive element is usually called a **resistance temperature detector.** *See also:* **electric**

thermometer (temperature meter); instrument. 42A30-0

resistance to ground (lightning arresters). The ratio, at a point in a grounding system, of the component of the voltage to ground that is in phase with the ground current, to the ground current that produces it. *See:* **lightning arrester (surge diverter).** 50I25-31E7

resistant (suffix) (rotating machinery). Material or apparatus so constructed, protected or treated, that it will not be injured readily when subjected to the specified material or condition, for example, fire-resistant, moisture-resistant. *See also:* **asynchronous machine; direct-current commutating machine; synchronous machine.** 42A10/42A95/37A100/IC1-31E6/31E8/31E11/34E10

resistive attenuator (waveguide). A length of waveguide designed to introduce a transmission loss by the use of some dissipative material. *See also:* **absorptive attenuator (waveguide); waveguide.** 50I62-3E1

resistive conductor. A conductor used primarily because it possesses the property of high electric resistance. *See also:* **power distribution, underground construction.** 42A35/IC1-34E10/31E13

resistive coupling. The association of two or more circuits with one another by means of resistance mutual to the circuits. *See also:* **coupling.** 42A65-0

resistive distributor brush (electromagnetic compatibility). Resistive pickup brush in an ignition distributor cap. *See also:* **electromagnetic compatibility.** CISPR-27E1

resistive ignition cable (electromagnetic compatibility). High-tension ignition cable, the core of which is made of resistive material. *See also:* **electromagnetic compatibility.** CISPR-27E1

resistivity (material). A factor such that the conduction-current density is equal to the electric field in the material divided by the resistivity. E270-0

resistor. A device the primary purpose of which is to introduce resistance into an electric ci rcuit. *Note:* A resistor as used in electric circuits for purposes of operation, protection, or control, commonly consists of an aggregation of units. Resistors as commonly supplied consist of wire, metal ribbon, cast metal, or carbon compounds supported by or imbedded in an insulating medium. The insulating medium may enclose and support the resistance material as in the case of the porcelain-tube type, or the insulation may be provided only at the points of support as in the case of heavy-duty ribbon or cast iron grids mounted in metal frames. E45/E270-34E10/21E0

resistor furnace. A resistance furnace in which the heat is developed in a resistor that is not a part of the charge. 42A60-0

resistor-start motor. A single-phase induction motor with a main winding and an auxiliary winding connected in series with a resistor, with the auxiliary winding circuit opened for running operation. 0-31E8

resolution (1) (general communication). (A) The act of deriving from a sound, scene, or other form of intelligence, a series of discrete elements wherefrom the original may subsequently be synthesized. (B) The degree to which nearly equal values of a quantity can be discriminated. (C) The fineness of detail in a reproduced spatial pattern. (D) The degree to which a system or a device distinguishes fineness of detail in a spatial pattern. *See also:* **communication.** 42A65-0

(2) (storage tube). A measure of the quantity of information that may be written into and read out of a storage tube. *Notes:* (A) Resolution can be specified in terms of number of bits, spots, lines, or cycles. (B) Since the relative amplitude of the output may vary with the quantity of information, the true representation of the resolution of a tube is a curve of relative amplitude versus quantity. *See also:* **storage tube.** E158-15E6

(3) (television). A measure of ability to delineate picture detail. *Note:* Resolution is usually expressed in terms of a number of lines discriminated on a test chart. For a number of lines N (normally alternate black and white lines) the width of each line is $1/N$ times the picture height. In television practice, resolution is measured vertically in a raster having a four-to-three aspect ratio. *See also:* **beam tubes; limiting resolution; line number; resolution response; television.** E158/E160/E204-2E2/15E6

(4) (oscilloscopes). A measure of the total number of trace lines discernible along the coordinate axes, bounded by the extremities of the graticule or other specific limits. *See also:* **oscillograph.** 0-9E4

(5) (power supplies). The minimum voltage (or current) increment within which the power supply's output can be set using the panel controls. For continuous controls, the minimum increment is taken to be the voltage (or current) change caused by one degree of shaft rotation. *See also:* **power supply.** KPSH-10E1

(6) (numerically controlled machines). The least interval between two adjacent discrete details that can be distinguished one from the other. *See also:* **numerically controlled machines.** 85A1-34E12

(7) (transmission lines and waveguides). The degree to which nearly equal values of a quantity can be discriminated. 0-9E4

(8) (industrial control). The smallest distinguishable increment into which a quantity is divided in a device or system. *See also:* **communication; control system, feedback.** AS1-34E10

re-solution (electrodeposition). The passing back into solution of metal already deposited on the cathode. *See also:* **electrodeposition.** 42A60-0

resolution element (radar). A spatial and velocity region contributing echo energy that can be separated from that of adjacent regions by action of the antenna or receiving system; in conventional radar its dimensions are given by the beamwidths of the antenna, the transmitter pulse width, and the receiver bandwidth; its dimensions may be increased by the presence of spurious response regions (sidelobes), or decreased by use of specially coded transmissions and appropriate processing techniques (such as moving-target indicator). *See also:* **navigation.** 0-10E6

resolution error. The error due to the inability of a transducer to manifest changes of a variable smaller than a given increment. *See also:* **electronic analog computer.** E165-16E9

resolution of output adjustment (of any output parameter, voltage, frequency, etcetera) (inverters). The minimum increment of change in setting. *See also:* **self-commutated inverters.** 0-34E24

resolution response (square-wave) (television) (1) (camera device). The ratio of (A) the peak-to-peak signal amplitude given by a test pattern consisting of alternate black and white vertical* bars of equal widths corresponding to a specified line number to (B) the peak-to-peak signal amplitude given by large-area blacks and large-area whites having the same luminances as the black and white bars in the test pattern. *Note: Horizontal scanning lines are assumed. *See also:* **television.**

(2) (display device). The ratio of (A) the peak-to-peak luminance given by a square-wave test signal whose half-period corresponds to a specified line number to (B) the peak-to-peak luminances given by a test signal producing large-area blacks and large-area whites having the same amplitudes as the signal of (A). *Note:* In a display device, **resolution response,** at relatively high line numbers, is sometimes called **detail contrast.** 42A65/E208-2E2

resolution, structural (color-picture tubes). The resolution as limited by the size and shape of the screen elements. *See also:* **beam tubes; television.** E160-2E2/15E6

resolution time (counter tube or counting system) (radiation counters). The minimum time interval between two distinct events that will permit both to be counted. *See also:* **anticoincidence (radiation counters).** 0-15E6

resolution time correction (radiation counters). Correction to the observed counting rate to allow for the probability of the occurrence of events within the resolution time. *See also:* **anticoincidence (radiation counters).** 0-15E6

resolution wedge. A narrow-angle wedge-shaped pattern calibrated for the measurement of resolution and composed of alternate contrasting strips that gradually converge and taper individually to preserve equal widths along any given line at right angles to the axis of the wedge. *Note:* Alternate strips may be black and white of maximum contrast or strips of different colors. *See also:* **television.** 42A65/E204-2E2

resolver. A device whose input is a vector quantity and whose outputs are components of the vector. *Note:* The process being reversible, the term **resolver** may also be used to denote a device or computing element used for vector composition, as well as for vector resolution. *See also:* **electronic analog computer.** E165/X3A12-16E9

resolving time (1) (electronic navigation). The minimum time interval by which two events must be separated to be distinguishable in a navigation system by the time measurement alone. *See also:* **navigation.** E172-10E6

(2) (radiation counters). The time from the start of a counted pulse to the instant a succeeding pulse can assume the minimum strength to be detected by the counting circuit. *Note:* This quantity pertains to the combination of tube and recording circuit. *See also:* **gas-filled radiation-counter tubes; ionizing radiation.** 42A70-15E6

resonance. The result of resonating. *See also:* **network analysis.** E270-0

resonance bridge. A 4-arm alternating-current bridge in which both an inductor and a capacitor are present in one arm, the other three arms being (usually) nonreactive resistors, and the adjustment for balance includes the establishment of resonance for the applied frequency. *Note:* Normally used for the measurement of inductance, capacitance, or frequency. Two general types can be distinguished according as the inductor and capacitor are effectively in series or in parallel. *See also:* **bridge.** 42A30-0

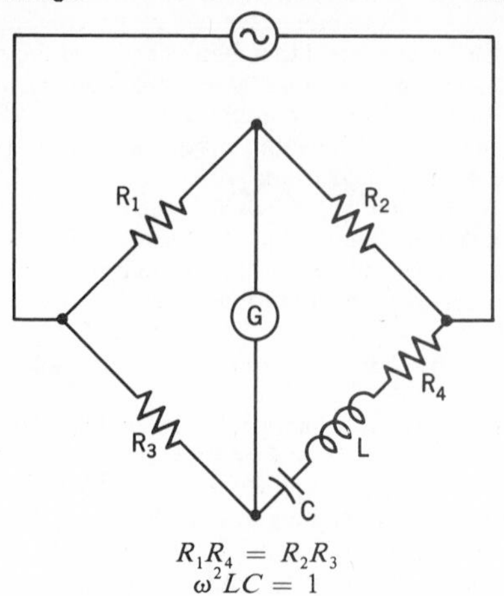

$$R_1R_4 = R_2R_3$$
$$\omega^2LC = 1$$

Series resonance bridge.

resonance curve, carrier-current line trap (power-system communication). A graphical plot of the ohmic impedance of a carrier current line trap with respect to frequency at frequencies near resonance. *See also:* **power-line carrier.** 0-31E3

resonance frequency (resonant frequency*) (1) network. Any frequency at which resonance occurs. *Note:* For a given network, resonance frequencies may differ for different quantities, and almost always differ from natural frequencies. For example, in a simple series resistance-inductance-capacitance circuit there is a resonance frequency for current, a different resonance frequency for capacitor voltage, and a natural frequency differing from each of these. *See also:* **network analysis.**

*Deprecated E270-31E3

(2) (crystal unit). The frequency for a particular mode of vibration to which, discounting dissipation, the effective impedance of the crystal unit is zero. *See also:* **crystal.** 42A65-0

resonant gap (microwave gas tubes). The small region in a resonant structure interior to the tube, where the electric field is concentrated. *See also:* **gas tubes.** E160-15E6

resonant grounded. *See:* **ground-fault neutralizer grounded.**

resonant grounded system (arc-suppression coil) (lightning arresters). A system grounded through a reactor, the reactance being of such value that during a single line-to-ground fault, the power-frequency inductive current passed by this reactor essentially neutralizes the power-frequency capacitive component of the ground-fault current. *Note:* With resonant grounding of a system, the net current in the fault is limited to such an extent that an arc fault in air would be self-extinguishing. *See also:* **ground; lightning arrester (surge diverter).** E32/50I25-31E7

resonant line oscillator. An oscillator in which the principal frequency-determining elements are one or more resonant transmission lines. *See also:* **oscillatory circuit.** 42A65/E145/E182A-0

resonant mode (1) (general). A component of the response of a linear device that is characterized by a certain field pattern, and that when not coupled to other modes is representable as a single-tuned circuit. *Note:* When modes are coupled together, the combined behavior is similar to that of the corresponding single-tuned circuits correspondingly coupled. *See also:* **waveguide.** 42A65-0
(2) (cylindrical cavities). When a metal cylinder is closed by two metal surfaces perpendicular to its axis a cylindrical cavity is formed. The resonant modes in this cavity are designated by adding a third subscript to indicate the number of half-waves along the axis of the cavity. When the cavity is a rectangular parallelepiped the axis of the cylinder from which the cavity is assumed to be made should be designated since there are three possible cylinders out of which the parallelepiped may be made. *See also:* **guided wave.** E210-0

resonant wavelengths (cylindrical cavities). Those given by $\lambda_r = 1[(1/\lambda_c)^2+(l/2c)^2]^{1/2}$ where λ_c is the cutoff wavelength for the transmission mode along the axis, l is the number of half-period variations of the field along the axis, and c is the axial length of the cavity. *See also:* **guided wave.** E210-0

resonating (steady-state quantity or phasor). The maximizing or minimizing of the amplitude or other characteristic provided the maximum or minimum is of interest. *Notes:* (1) Unless otherwise specified, the quantity varied to obtain the maximum or minimum is to be assumed to be frequency. (2) Phase angle is an example of a quantity in which there is usually no interest in a maximum or a minimum. (3) In the case of amplification, transfer ratios, etcetera, the amplitude of the phasor is maximized or minimized; in the case of currents, voltages, charges, etcetera, it is customary to think of the amplitude of the steady-state simple sine-wave quantity as being maximized or minimized. *See also:* **network analysis.** E270-0

resonator. A device, the primary purpose of which is to introduce resonance into a system. *See also:* **network analysis.** E270-0

resonator grid (electron tubes). An electrode, connected to a resonator, that is traversed by an electron beam and that provides the coupling between the beam and the resonator. *See also:* **velocity-modulated tube.** 50I07-15E6

resonator mode (oscillator). A condition of operation corresponding to a particular field configuration for which the electron stream introduces a negative conductance into the coupled circuit. *See also:* **oscillatory circuit.** E160-15E6

resonator, waveguide (resonant element). A waveguide or transmission-line device primarily intended for storing oscillating electromagnetic energy. *See also:* **waveguide.** E147-3E1

responder beacon (electronic navigation). *See:* **transponder.**

response (device or system). A quantitative expression of the output as a function of the input under conditions that must be explicitly stated. *Note:* The response characteristic, often presented graphically, gives the response as a function of some independent variable such as frequency or direction. *See also:* **transmission characteristics.** 42A65-21E1

response function (linear passive networks). The ratio of response to excitation, both expressed as functions of the complex frequency, $s=\sigma+j\omega$. *Note:* The response function is the Laplace transform of the response due to unit impulse excitation. *See also:* **linear passive networks.** E156-0

response, Gaussian. *See:* **Gaussian response.**

response, instrument (1) (dynamic). The behavior of the instrument output as a function of the measured signal, both with respect to time. *See:* **damping characteristic; frequency response; ramp response; step response.** *See also:* **accuracy rating (instrument).** 39A4-0
(2) (forced). The total steady-state plus transient time response resulting from an external input. 0-23E0

response, sinusoidal (sine-force). The forced response due to a sinusoidal stimulus. *Note:* A set of steady-state sinusoidal responses for sinusoidal inputs at different frequencies is called the frequency-response characteristic. *See also:* **control system, feedback.** 0-23E0

response, steady-state (system or element). The part of the time response remaining after transients have expired. *Note:* The term **steady-state** may also be applied to any of the forced-response terms; for example **steady-state sinusoidal response.** *See also:* **control system, feedback; response, sinusoidal.** 0-23E0

response time (1) (magnetic amplifier). The time (preferably in seconds; may also be in cycles of supply frequency) required for the output quantity to change by some agreed-upon percentage of the differential output quantity in response to a step change in control signal equal to the differential control signal. *Note:* The initial and final output quantities shall correspond to the test output quantities. The response time shall be the maximum obtained including differences arising from increasing or decreasing output quantity or time phase of signal application. *See also:* **rating and testing magnetic amplifiers.** E107-0
(2) (turn ON response time) (control devices). The time required for the output voltage to change from rated OFF voltage to rated ON voltage in response to a step change in control signal equal to 120 percent of the differential trip signal. *Note:* The absolute magnitude of the initial signal condition shall be the absolute magnitude of the trip OFF control signal plus 10 percent of the differential trip signal. E107-0
(3) (turn OFF response time) (control devices). The time required for the output voltage to change from rated ON voltage to rated OFF voltage in response to a step change in control signal equal to 120 percent of the differential trip signal. *Note:* The absolute magnitude of the initial signal condition shall be the absolute magnitude of the trip ON control signal minus 10 percent of the differential trip signal. E107-0
(4) (electrically tuned oscillator). The time following a change in the input to the tuning element required for a characteristic to reach a predetermined range of values within which it remains. *See also:* **tunable microwave oscillators.** E158-15E6
(5) (instrument). The time required after an abrupt change has occurred in the measured quantity to a new constant value until the pointer, or indicating means, has first come to apparent rest in its new position. *Note:* (1) Since, in some instruments, the response time depends on the magnitude of the deflection, a

value corresponding to an initial deflection from zero to end scale is used in determining response time for rating purposes. (2) The pointer is at apparent rest when it remains within a range on either side of its final position equal to one-half the accuracy rating, when determined as specified above. *See also:* **accuracy rating (instrument); moving element (instrument).** 39A2/42A30/39A1-0

(6) (bolometric power meter). The time required for the bolometric power indication to reach 90 percent of its final value after a fixed amount of radio-frequency power is applied to the bolometer unit. 0-9E4

(7) (thermal converter). The time required for the output electromotive force to come to its new value after an abrupt change has occurred in the input quantity (current, voltage, or power) to a new constant value. *Notes:* (A) Since, in some thermal converters, the response time depends upon the magnitude and direction of the change, the value obtained for an abrupt change from zero to rated input quantity is used for rating purposes. (B) The output electromotive force shall be considered to have come to its new value when all but 1 percent of the change in electromotive force has been indicated. *See also:* **thermal converter.** 39A1-0

(8) (industrial control). The time required, following the initiation of a specified stimulus to a system, for an output going in the direction of necessary corrective action to first reach a specified value. *Note:* The response time is expressed in seconds. *See also:* **control system, feedback.** AS1-34E10

(9) (electrical conversion). The elapsed time from the initiation of a transient until the output has recovered to 63 percent of its maximum excursion. See accompanying figures. *See also:* **electrical conversion equipment; recovery time.** 0-10E1

Response and recovery time for a critically damped circuit.

(10) (arc-welding apparatus). The time required to attain conditions within a specified amount of their final value in an automatically regulated welding circuit after a definitely specified disturbance has been initiated. *See also:* **photoelectric control; electric arc-welding apparatus.** 87A1-0

(11) (photoelectric lighting control) (industrial control). The time required for operation following an abrupt change in illumination from 50 percent above to 50 percent below the minimum illumination sensitivity. *See also:* **photoelectric control.** 42A25/AS1-34E10

(12) (control system or element) (time of response) (control system, feedback). The time required for an output to make the change from an initial value to a large specified percentage of the steady state, either before overshoot or in the absence of overshoot. *Note:* If the term is unqualified, time of response of a first-order system to a unit-step stimulus is generally understood; otherwise the pattern and magnitude of the stimulus should be specified. Usual percentages are 90, 95, or 99. 85A1-23E0

responsor (electronic navigation). The receiving component of an interrogator. *Note:* It is, therefore, a receiver intended to receive and interpret the signals from a transponder. *See also:* **radio transmission.** 42A65/E172-10E6

rest and de-energized (rotating machinery). The complete absence of all movement and of all electric or mechanical supply. *See also:* **asynchronous machine; direct-current commutating machine; synchronous machine.** 0-31E8

restart (computing systems). To reestablish the execution of a routine, using the data recorded at a checkpoint. *See also:* **electronic digital computer.** X3A12-16E9

resting potential (biological). The voltage existing between the two sides of a living membrane or interface in the absence of stimulation. *See also:* **medical electronics.** 0-18E1

restoring force gradient (direct-acting recording instrument). The rate of change, with respect to the displacement, of the resultant of the electric, or of the electric and mechanical, forces tending to restore the

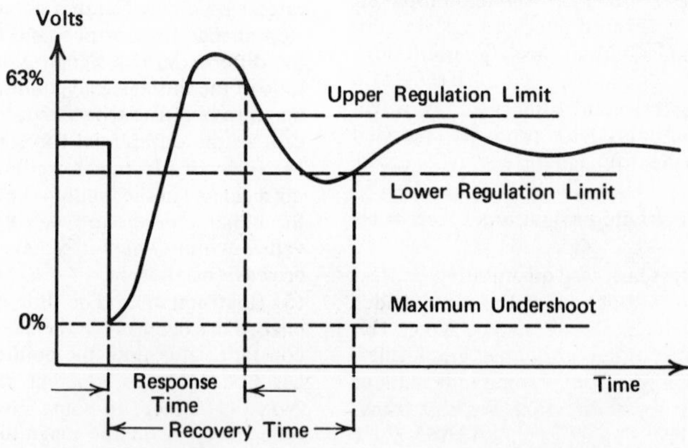

Response and recovery time for an underdamped circuit.

marking device to any position of equilibrium when displaced from that position. *Note:* The force gradient may be constant throughout the entire travel of the marking device or it may vary greatly over this travel, depending upon the operating principles and the details of construction. *See also:* **accuracy rating (instrument).** 42A30-0

restoring torque gradient (instrument). The rate of change, with respect to the deflection, of the resultant of the electric, or electric and mechanical, torques tending to restore the moving element to any position of equilibrium when displaced from that position. *See also:* **accuracy rating (instrument).** 42A30-0

restraint relay. A relay that is so constructed that its operation is dependent on the proportion of one input quantity with respect to a second input quantity that restrains operation. *See also:* **relay.** 37A100-31E11/31E6

restricted radiation frequencies for industrial, scientific, and medical equipment. Center of a band of frequencies assigned to industrial, scientific, and medical equipment either nationally or internationally and for which a power limit is specified. *See also:* **electromagnetic compatibility.** CISPR-27E1

restrike (switching device). A resumption of current between the contacts during an opening operation after an interval of zero current of ¼ cycle at normal frequency or longer. 37A100-31E11

restriking voltage (1) (gas tube). The anode voltage at which the discharge recommences when the supply voltage is increasing before substantial deionization has occurred. *See also:* **gas-filled rectifier.** 50I07-15E6

(2) (industrial control). The voltage that appears across the terminals of a switching device immediately after the breaking of the circuit. *Note:* This voltage may be considered as composed of two components. One, which subsists in steady-state conditions, is direct current or alternating current at service frequency, according to the system. The other is a transient component that may be oscillatory (single or multifrequency) or nonoscillatory (for example, exponential) or a combination of these depending on the characteristics of the circuit and the switching device. *See also:* **switch.** 50I16-34E10

retained image (image burn). A change produced in or on the target that remains for a large number of frames after the removal of a previously stationary light image and that yields a spurious electric signal corresponding to that light image. *See also:* **beam tubes; camera tube.** E160-2E2/15E6

retainer. *See:* **separator (storage cell).**

retaining ring (rotating machinery) (1) (steel). A mechanical structure surrounding parts of a rotor to restrain radial movement due to centrifugal action. *See also:* **rotor (rotating machinery).** 0-31E8

(2) (insulation). The insulation forming a dielectric and mechanical barrier between the rotor end windings and the high-strength steel retaining ring. *See also:* **rotor (rotating machinery).** 0-31E8

retard coil*. *See:* **inductor.**

*Deprecated

retard transmitter. A transmitter in which a delay period is introduced between the time of actuation and the time of transmission. *See also:* **protective signaling.** 42A65-0

retardation (deceleration) (industrial control). The operation of reducing the motor speed from a high level to a lower level or zero. *See also:* **electric drive.** 42A25-34E10

retardation coil. *See:* **inductor.**

retardation test (rotating machinery). A test in which the losses in a machine are deduced from the rate of deceleration of the machine when only these losses are present. *See also:* **asynchronous machine; direct-current commutating machine; synchronous machine.** 0-31E8

retarding-field (positive-grid) oscillator. An oscillator employing an electron tube in which the electrons oscillate back and forth through a grid maintained positive with respect to the cathode and the plate. The frequency depends on the electron-transit time and may also be a function of the associated circuit parameters. The field in the region of the grid exerts a retarding effect that draws electrons back after passing through it in either direction. Barkhausen-Kurz and Gill-Morell oscillators are examples. *See also:* **oscillatory circuit.** E145-0

retarding magnet (braking magnet) (drag magnet). A magnet used for the purpose of limiting the speed of the rotor of a motor-type meter to a value proportional to the quantity being integrated. *See also:* **watthour meter.** 42A30-0

retention time, maximum (storage tubes). The maximum time between writing into a storage tube and obtaining an acceptable output by reading. *See also:* **storage tube.** E158-15E6

retentivity (magnetic material). That property that is measured by its maximum residual induction. *Note:* The maximum residual induction is usually associated with a hysteresis loop that reaches saturation, but in special cases this is not so. *See also:* **electronic digital computer; static magnetic storage.** E270-0

retrace (oscillography). Return of the spot on the cathode-ray tube to its starting point after a sweep; also that portion of the sweep waveform that returns the spot to its starting point. *See also:* **oscillograph.** 0-9E4

retrace blanking. *See:* **blanking.**

retrace interval* (television). *See:* **return interval.**

*Deprecated

retrace line. The line traced by the electron beam in a cathode-ray tube in going from the end of one line or field to the start of the next line or field. E188-0

retrieval. *See:* **information retrieval.** *See also:* **electronic digital computer.**

retro-reflector (reflex reflector). A device designed to reflect light in a direction close to that at which it is incident, whatever the angle of incidence. *See also:* **headlamp.** Z7A1-0

return. *See:* **carriage return.** *See also:* **electronic digital computer.**

return-beam mode (camera tube). A mode of operation in which the output current is derived, usually through an electron multiplier, from that portion of the scanning beam not accepted by the target. *See also:* **camera tubes.** 0-15E6

return difference (network analysis). One minus the loop transmittance. *See also:* **linear signal flow graphs.** E155-0

return interval (television) (retrace interval*). The interval corresponding to the direction of sweep not used for delineation. *See also:* **television.**

*Deprecated 42A65/E204-2E2

return loss. (1) At a discontinuity in a transmission system, the difference between the power incident upon the discontinuity and the power reflected from the discontinuity. (2) The ratio in decibels of the power incident upon the discontinuity to the power reflected from the discontinuity. *Note:* This ratio is also the square of the reciprocal of the magnitude of the reflection coefficient. Return loss $= 20 \log_{10}(1/\Gamma)$. *See also:* **transmission loss; waveguide.** (3) More broadly, the return loss is a measure of the dissimilarity between two impedances, being equal to the number of decibels that corresponds to the scalar value of the reciprocal of the reflection coefficient, and hence being expressed by the formula:

$$20 \log_{10} \left| \frac{Z_1+Z_2}{Z_1-Z_2} \right| \text{ decibel}$$

where Z_1 and Z_2 are the two impedances. *See also:* **transmission loss.** 42A65/E146-3E1/9E4/31E3

return stroke (lightning). The luminescent, high-current discharge that is initiated after the stepped leader and pilot streamer have established a highly ionized path between charge centers. *See also:* **direct-stroke protection (lightning).** 0-31E13

return-to-zero (RZ) code (power-system communication). A code form having two information states termed ZERO and ONE and having a third state or an at-rest condition to which the signal returns during each period. *See also:* **digital.** 0-31E3

return trace (television). The path of the scanning spot during the return interval. *See also:* **oscillograph; television.** E204/42A65-2E2/9E4

reverberation. The presence of reverberant sound. *See also:* **electroacoustics.** 0-1E1

reverberation chamber. An enclosure especially designed to have a long reverberation time and to produce a sound field as diffuse as possible. *See also:* **anechoic chamber; electroacoustics.** 0-1E1

reverberation room. *See:* **reverberation chamber.**

reverberant sound. Sound that has arrived at a given location by a multiplicity of indirect paths as opposed to a single direct path. *Notes:* (1) Reverberation results from multiple reflections of sound energy contained within an enclosed space. (2) Reverberation results from scattering from a large number of inhomogeneities in the medium or reflection from bounding surfaces. (3) Reverberant sound can be produced by a device that introduces time delays that approximate a multiplicity of reflections. *See:* **echo; electroacoustics; flutter echo.** 0-1E1

reverberation time. The time required for the mean-square sound pressure level, or electric equivalent, originally in a steady state, to decrease 60 decibels after the source output is stopped. *See also:* **electroacoustics.** 0-1E1

reverberation-time meter. An instrument for measuring the reverberation time of an enclosure. *See also:* **instrument.** 42A30-0

reversal (storage battery) (storage cell). A change in normal polarity of the cell or battery. *See also:* **charge.** 42A60-0

reverse battery supervision. A form of supervision in which supervisory signals are furnished from the terminating end to the originating end by reversing the direction of current flow over the trunk. *See also:* **telephone switching system.** 42A65-19E1

reverse-blocking current (reverse-blocking thyristor). The reverse current when the thyristor is in the reverse-blocking state. *See also:* **principal current.** E223-34E17/34E24/15E7

reverse-blocking diode-thyristor. A two-terminal thyristor that switches only for positive anode-to-cathode voltages and exhibits a reverse-blocking state for negative anode-to-cathode voltages. *See also:* **thyristor.** E223-34E17/34E24/15E7

reverse-blocking impedance (reverse-blocking thyristor). The differential impedance between the two terminals through which the principal current flows, when the thyristor is in the reverse-blocking state at a stated operating point. *See also:* **principal voltage-current characteristic (principal characteristic).** E223-34E17/15E7

reverse-blocking state (reverse-blocking thyristor). The condition of a reverse-blocking thyristor corresponding to the portion of the anode-to-cathode voltage-current characteristic for reverse currents of lower magnitude than the reverse-breakdown current. *See also:* **principal voltage-current characteristic (principal characteristic).** E223-34E17/34E24/15E7

reverse-blocking triode-thyristor. A three-terminal thyristor that switches only for positive anode-to-cathode voltages and exhibits a reverse-blocking state for negative anode-to-cathode voltages. *See also:* **thyristor.** E223-34E17/34E24/15E7

reverse-breakdown current (reverse-blocking thyristor). The principal current at the reverse-breakdown voltage. *See also:* **principal current.** E223-34E17/34E24/15E7

reverse-breakdown voltage (reverse-blocking thyristor). The value of negative anode-to-cathode voltage at which the differential resistance between the anode and cathode terminals changes from a high value to a substantially lower value. *See also:* **principal voltage-current characteristic (principal characteristic).** E223-34E17/34E24/15E7

reverse-conducting diode-thyristor. A two-terminal thyristor that switches only for positive anode-to-cathode voltages and conducts large currents at negative anode-to-cathode voltages comparable in magnitude to the ON-state voltages. *See also:* **thyristor.** E223-34E17/34E24/15E7

reverse-conducting triode-thyristor. A three-terminal thyristor that switches only for positive anode-to-cathode voltages and conducts large currents at negative anode-to-cathode voltages comparable in magnitude to the ON-state voltages. *See also:* **thyristor.** E223-34E17/34E24/15E7

reverse contact. A contact that is closed when the operating unit is in the reverse position. *See also:* **railway signal and interlocking.** 42A42-0

reverse current (1) (general). Current that flows upon application of reverse voltage. *See also:* **circuits and devices.** 42A65-0
(2) (reverse-blocking or reverse-conducting thyristor). The principal current for negative anode-to-cathode voltage. *See also:* **circuits and devices; principal current.** E223-34E17/34E24/15E7
(3) (metallic rectifier). The current that flows through a metallic rectifier cell in the reverse direction. *See also:* **circuits and devices; rectification.** 42A15-0
(4) (semiconductor rectifier). The total current that flows through a semiconductor rectifier cell in the reverse direction. *See also:* **circuits and devices; rectification.** E59/E216/E270-34E17

(5) (capacitive). The current that flows through a semiconductor rectifier cell during the blocking period due to the capacitance of the cell. E59-34E17
(6) (resistive) (semiconductor rectifier). The in-phase current that flows through a semiconductor rectifier cell during the blocking period exclusive of the reverse recovery current. *See also:* **rectification.** E59-34E17

reverse-current cleaning. *See:* **anode cleaning.**

reverse-current cutout. A magnetically operated direct-current device that operates to close an electric circuit when a predetermined voltage condition exists and operates to open an electric circuit when more than a predetermined current flows through it in the reverse direction. (1) **Fixed-voltage type:** A reverse-current cutout that closes an electric circuit whenever the voltage at the cutout terminal exceeds a predetermined value and is of the correct polarity. It opens the circuit when more than a predetermined current flows through it in the reverse direction. (2) **Differential-voltage type:** A reverse-current cutout that closes an electric circuit when a predetermined differential voltage appears at the cutout terminal, provided this voltage is of the correct polarity and exceeds a predetermined value. It opens the circuit when more than a predetermined current flows through it in the reverse direction. *See also:* **air-transportation electric equipment.** 42A41-0

reverse-current relay (direct-current circuit). A relay that operates on a current flow in a direction opposite to a predetermined reference direction described as normal. *See also:* **relay.** 50I16/37A100-31E6/31E11/34E10

reverse-current release (reverse-current trip) (direct current). A release that operates upon reversal in the main circuit from a predetermined direction. 37A100-31E11

reverse-current tripping; reverse-power tripping. *See:* **reverse-current release (reverse-current trip).**

reverse direction (semiconductor rectifier diode). The direction of greater resistance to steady direct-current flow through the diode, for example, from cathode to anode. *See also:* **rectification; semiconductor; semiconductor device; semiconductor rectifier; semiconductor rectifier stack.** E59/E216/E270-34E17/34E26

reverse emission (back emission) (vacuum tubes). The inverse electrode current from an anode during that part of a cycle in which the anode is negative with respect to the cathode. *See also:* **circuit characteristics of electrodes; electron emission.** E160-15E6

reverse gate current (thyristor). The gate current when the junction between the gate region and the adjacent anode or cathode region is reverse biased. *See also:* **principal current.** E223-34E17/34E24/15E7

reverse gate voltage (thyristor). The voltage between the gate terminal and the terminal of an adjacent region resulting from reverse gate current. *See also:* **principal voltage-current characteristic (principal characteristic).** E223-34E17/34E24/15E7

reverse period (rectifier circuit) (rectifier circuit element). The part of an alternating-voltage cycle during which the current flows in the reverse direction. See note under **blocking period.** *See also:* **rectifier circuit element.** E59-34E17/34E24

reverse position (device). The opposite of the normal position. 42A42-0

reverse power loss (semiconductor rectifier). The power loss resulting from the flow of reverse current. *See also:* **rectification; semiconductor rectifier stack.** E59-34E17/34E24

reverser. A switching device for interchanging electric circuits to reverse the direction of motor rotation. *See also:* **multiple-unit control.** 42A42-0

reverse recovery current (semiconductor rectifier). The transient component of reverse current associated with a change from forward conduction to reverse voltage. *See also:* **rectification.** E59-34E17

reverse recovery time (reverse-blocking thyristor or semiconductor diode). The time required for the principal current or voltage to recover to a specified value after instantaneous switching from an ON state to a reverse-voltage or current. *See also:* **principal voltage-current characteristic (principal characteristic); rectification.** E59/E223-34E17/34E24/15E7

reverse resistance (metallic rectifier). The resistance measured at a specified reverse voltage or a specified reverse current. *See also:* **rectification.** 42A15-0

reverse voltage (rectifier). Voltage of that polarity that produces the smaller current. *See also:* **circuits and devices; principal voltage-current characteristic; rectification.** 42A15/42A65-0

reverse voltage dividers (rectifier). Devices employed to assure satisfactory division of reverse voltage among series-connected semiconductor rectifier diodes. Transformers, bleeder resistors, capacitors, or combinations of these may be employed. *See also:* **power rectifier.** 0-34E24

reverse wave. *See:* **reflected wave.**

reversible capacitance (nonlinear capacitor). The limit, as the amplitude of an applied sinusoidal capacitor voltage approaches zero, of the ratio of the amplitude of the resulting in-phase fundamental-frequency component of transferred charge to the amplitude of the applied voltage, for a given constant bias voltage superimposed on the sinusoidal voltage. *See also:* **nonlinear capacitor.** E226-15E7

reversible-capacitance characteristic (nonlinear capacitor). The function relating the reversible capacitance to the bias voltage. *See also:* **nonlinear capacitor.** E226-15E7

reversible motor. A motor whose direction of rotation can be selected by change in electric connections or by mechanical means but the motor will run in the selected direction only if it is at a standstill or rotating below a particular speed when the change is initiated. *See also:* **asynchronous machine; direct-current commutating machine; synchronous machine.** 42A10-0

reversible permeability. The limit of the incremental permeability as the incremental change in magnetizing force approaches zero. *Note:* In anisotropic media, reversible permeability becomes a matrix. E270-0

reversible permittivity (ferroelectric material). The change in displacement per unit field when a very small relatively high-frequency alternating signal is applied to a ferroelectric at any point of a hysteresis loop. *See also:* **ferroelectric domain.** E180-0

reversible potential. *See:* **equilibrium potential.**

reversible power converter. An equipment containing thyristor converter assemblies connected in such a way that energy transfer is possible from the alternating-current side to the direct-current side and from the direct-current side to the alternating-current side with or without reversing the current in the direct-current circuit. *See also:* **power rectifier.** 0-34E24

reversible process. An electrochemical reaction that takes place reversibly at the equilibrium electrode potential. *See also:* **electrochemistry.** 42A60-0

reversing (industrial control). The control function of changing motor rotation from one direction to the opposite direction. *See also:* **electric drive.** 42A25-34E10

reversing motor. One the torque and hence direction of rotation of which can be reversed by change in electric connections or by other means. These means may be initiated while the motor is running at full speed, upon which the motor will come to a stop, reverse, and attain full speed in the opposite direction. *See also:* **asynchronous machine; direct-current commutating machine; synchronous machine.** 42A10-0

reversing starter (industrial control). An electric controller for accelerating a motor from rest to normal speed in either direction of rotation. *See also:* **electric controller; starter.** 42A25-34E10

reversing switch. A switch intended to reverse the connections of one part of a circuit. *See also:* **switch.** 42A25-34E10

reverting call. A telephone call between two subscriber stations on the same party line. *See also:* **telephone system.** 42A65-19E1

rewrite. (1) To write again. (2) In a destructive-read storage device, to return the data to the state it had prior to reading. *See also:* **electronic computation; electronic digital computer; regenerate.** E162/E270-0

RF. *See:* **radio frequency.**

rheobase (medical electronics). The intensity of the steady cathodal current of adequate duration that when suddenly applied just suffices to excite a tissue. *See also:* **medical electronics.** 0-18E1

rheostat. An adjustable resistor so constructed that its resistance may be changed without opening the circuit in which it may be connected. E270-34E10

rheostat loss (synchronous machine). The I^2R loss in the rheostat controlling the field current. *See also:* **synchronous machine.** 50A10-31E8

rheostatic braking (industrial control). A form of dynamic braking in which electric energy generated by the traction motors is controlled and dissipated by means of a resistor whose value of resistance may be varied. *See also:* **dynamic braking; electric drive.** 42A42/42A25-34E10

rheostatic control (elevators). A system of control that is accomplished by varying resistance and/or reactance in the armature and/or field circuit of the driving-machine motor. *See also:* **control (elevators).** 42A45-0

rheostatic-type voltage regulator (rotating machinery). A regulator that accomplishes the regulating function by mechanically varying a resistance. *See also:* **synchronous machine.** 0-31E8

rheostriction. *See:* **pinch effect.**

rho rho (electronic navigation). A generic term referring to navigation systems based on the measurement of only two distances for determination of position. *See also:* **navigation.** 0-10E6

rho theta (electronic navigation). A generic term referring to polar-coordinate navigation systems for determination of position of a vehicle through measurement of distance and direction. *See also:* **navigation.** 0-10E6

rhombic antenna. An antenna composed of long-wire radiators comprising the sides of a rhombus. The antenna usually is terminated in a resistance. The side of the rhombus, the angle between the sides, the elevation and the termination are proportioned to give the desired radiation properties. *See also:* **antenna.** 0-3E1

rhumbatron (electron tube) (microwave tube). A resonator, usually in the form of a torus. *See also:* **velocity-modulated tube.** 50I07-15E6

rhythmic light. A light that when observed from a fixed point has a luminous intensity that changes periodically. *See also:* **signal lighting.** Z7A1-0

ribbon microphone. A moving-conductor microphone in which the moving conductor is in the form of a ribbon that is directly driven by the sound waves. *See also:* **microphone.** 42A65-0

ribbon transducer. A moving-conductor transducer in which the movable conductor is in the form of a thin ribbon. 0-1E1

Richardson-Dushmann equation (thermionics). An equation representing the saturation current of a metallic thermionic cathode in the saturation-current state:

$$J = A_0(1-r)T^2 \exp\left(-\frac{b}{T}\right)$$

where

J = density of the saturation current

T = absolute temperature

A_0 = universal constant equal to 120 amperes per centimeter^2 kelvin^2

b = absolute temperature equivalent to the work function

r = reflection coefficient, which allows for the irregularities of the surface.

See: **work function.** *See also:* **electron emission.** 50I07-15E6

Richardson effect. *See:* **thermionic emission.**

ridge waveguide. A waveguide with interior projections extending along the length and in contact with the boundary wall. *See also:* **waveguide.** 0-3E1

Rieke diagram (oscillator performance). A chart showing contours of constant power output and constant frequency drawn on a polar diagram whose coordinates represent the components of the complex reflection coefficient at the oscillator load. *See:* **load impedance diagram.** *See also:* **oscillatory circuit.** E160-15E6

RIF (electromagnetic compatibility). *See:* **radio-influence field.**

right-handed (clockwise) polarized wave (radio wave propagation). An elliptically polarized electromagnetic wave in which the rotation of the electric field vector with time is clockwise for a stationary observer looking in the direction of the normal wave. *Note:* For an observer looking from a receiver toward the apparent source of the wave, the direction of rotation is reversed. *See also:* **radiation; radio wave propagation.** 42A65-3E2

rigid metal conduit. A raceway specially constructed for the purpose of the pulling in or the withdrawing of wires or cables after the conduit is in place and made of metal pipe of standard weight and thickness permitting the cutting of standard threads. *See also:* **raceway.** 42A95-0

rigid tower. A tower that depends only upon its own structural members to withstand the load that may be placed upon it. *See:* **angle tower; dead-end-tower; flexible tower.** *See also:* **tower.** 42A35-31E13

rim (spider rim) (rotating machinery). The outermost part of a spider. A rotating yoke. *See also:* **rotor (rotating machinery).** 0-31E8

ring (plug). A ring-shaped contacting part, usually placed in back of the tip but insulated therefrom. *See also:* **telephone switching system.** 42A65-0

ring around (secondary radar). (1) The undesired triggering of a transponder by its own transmitter. (2) The triggering of a transponder at all bearings causing a ring-type presentation on a plan-position indicator. *See also:* **navigation.** E172-10E6

ring circuit (waveguide practice). A hybrid *T* having the physical configuration of a ring with radial branches. *See also:* **waveguide.** 42A65-0

ring counter. A re-entrant multistable circuit consisting of any number of stages arranged in a circle so that a unique condition is present in one stage, and each input pulse causes this condition to transfer one unit around the circle. *See also:* **electronic digital computer; trigger circuit.** 42A65-0

ringdown (telephony). The method of signaling an operator in which telephone ringing current is sent over the line to operate a device or circuit to produce a steady signal (normally a visual signal). *See also:* **telephone switching system.** 0-19E1

ringer (station ringer). *See:* **telephone ringer.**

ringer box. *See:* **bell box.**

ring feeder. *See:* **loop-service feeder.**

ring head (electroacoustics). A magnetic head in which the magnetic material forms an enclosure with one or more air gaps. The magnetic recording medium bridges one of these gaps and is in contact with or in close proximity to the pole pieces on one side only. E157-1E1

ringing. (1) The production of an audible or visible signal at a station or switchboard by means of an alternating or pulsating current. (2) A damped oscillation occurring in the output signal of a system as a result of a sudden change in input signal. *See also:* **telephone switching system; signal.** 42A65-31E3

ringing key. A key whose operation sends ringing current over the circuit to which the key is connected. *See also:* **telephone switching system.** 42A65-0

ring oscillator. An arrangement of two or more pairs of tubes operating as push-pull oscillators around a ring, usually with alternate successive pairs of grids and plates connected to tank circuits. Adjacent tubes around the ring operate in phase opposition. The load is supplied by coupling to the plate circuits. *See also:* **oscillatory circuit.** E145/E182/42A65-0

ring time (radar). The time during which the indicated output of an echo box remains above a specified signal-to-noise level. The ring time is used in measuring the performance of radar equipment. *See also:* **navigation.** E172-10E6

ripple (general). The alternating-current component from a direct-current power supply arising from sources within the power supply. *Notes:* (1) Unless specified separately, ripple includes unclassified noise. (2) In electrical-conversion technology, ripple is expressed in peak, peak-to-peak, root-mean-square volts, or as percent root-mean-square. (3) Unless otherwise specified, **percent ripple** is the ratio of the root-mean-square value of the ripple voltage to the absolute value of the total voltage, expressed in percent. *See:* **percent ripple.** E188/E270/KPSH-10E1

ripple amplitude. The maximum value of the instantaneous difference between the average and instantaneous values of a pulsating unidirectional wave. *Note:* The amplitude is a useful measure of ripple magnitude when a single harmonic is dominant. Ripple amplitude is expressed in percent or per unit referred to the average value of the wave. *See also:* **power rectifier; rectification.** 34A1-34E24

ripple current. *See:* **ripple voltage or current.**

ripple factor. The ratio of the ripple magnitude to the arithmetic mean value of the voltage. *See also:* **electrical conversion; interference; power pack; power supply; radio receiver.** 68AI-31E5

ripple filter. A low-pass filter designed to reduce the ripple current, while freely passing the direct current, from a rectifier or generator. *See also:* **filter.** 42A65-31E3

ripple voltage (rectifier or generator). The alternating-voltage component of the unidirectional voltage from a direct-current power supply arising from sources within the power supply. *See:* **interference.** *See also:* **rectifier.** E145/E182A-13E6

ripple voltage or current. The alternating component whose instantaneous values are the difference between the average and instantaneous values of a pulsating unidirectional voltage or current. *See also:* **rectification.** 42A15/34A1/E59-34E17/34E24

rise. *See:* **travel.**

rise time (1) (industrial control). The time required for the output of a system (other than first-order) to make the change from a small specified percentage (often 5 or 10) of the steady-state increment to a large specified percentage (often 90 or 95), either before overshoot or in the absence of overshoot. *Note:* If the term is unqualified, response to a step change is understood; otherwise the pattern and magnitude of the stimulus should be specified. *See also:* **control system, feedback.** AS1-34E10

(2) (instrument). The time, in seconds, for the pointer to reach 0.9 plus or minus a specified tolerance of the end scale when constant electric power is suddenly applied from a source of sufficiently high impedance so as not to influence damping (100 times the impedance of the instrument). *See also:* **moving element of an instrument.** 39A1-0

rise time, pulse. *See:* **pulse rise time.**

rise-and-fall pendant. A pendant the height of which can be regulated by means of a cord adjuster. *See also:* **interior wiring.** 42A95-0

riser cable (communication practice). The vertical portion of a house cable extending from one floor to another. In addition, the term is sometimes applied to other vertical sections of cable. *See also:* **cable.** 42A65-0

rising-sun magnetron. A multicavity magnetron in which resonators of two different resonance frequencies are arranged alternately for the purpose of mode separation. *See also:* **magnetrons.** 42A70-15E6

RLC circuit. *See:* **simple series circuit.**

RMI (electronic navigation). *See:* **radio magnetic indicator.**

RMS. *See:* **root-mean-square value (of a periodic quantity).**

roadway element (track element). That portion of the roadway apparatus associated with automatic train stop, train control, or cab signal systems, such as a ramp, trip arm, magnet, inductor, or electric circuit, to which the locomotive apparatus is directly responsive. *See also:* **automatic train control.** 42A42-0

rock-dust distributor. *See:* **rock duster.**

rock duster (rock-dust distributor). A machine that distributes rock dust over the interior surfaces of a coal mine by means of air from a blower or pipe line or by means of a mechanical contrivance, to prevent coal dust explosions. *See also:* **mining.** 42A85-0

rodding a duct. *See:* **duct rodding.**

Roebel transposition (rotating machinery). An arrangement of strands occupying two heightwise tiers in a bar (half coil), wherein at regular intervals through the core length, one top strand and one bottom strand cross over to the other tier in such a way that each strand occupies every vertical position in each tier so as to equalize the voltage induced in each of the strands, thereby eliminating current that would otherwise circulate among the strands. Looking from one end of the slot, the strands are seen to progress in a clockwise direction through the core length through what may be interpreted as an angle of 360 degrees so that the strands occupy the same position at both ends of the core. There are several variations of the Roebel transposition in use. In a bar having four tiers of copper, the two pairs of tiers would each have a Roebel transposition. The uninsulated bar, then, would be assembled as two Roebel-transposed bars, side-by-side. In order to transpose against voltages induced by end-winding flux, various modifications of the transposition in the slot, and extension of the Roebel transposition into the end winding have been used. *See also:* **rotor (rotating machinery; stator.** 0-31E8

roll-in-jewel error. Error caused by the pivot rolling up the side of the jewel and then falling to a lower position when tapped. This effect is not present when instruments are mounted with the axis of the moving element in a vertical position. (Roll-in-jewel error includes pivot-friction error that is small compared to the roll-in-jewel error.) *See also:* **moving element of an instrument.** 39A1-0

roll-off (electroacoustics). A gradually increasing loss or attenuation with increase or decrease of frequency beyond the substantially flat portion of the amplitude-frequency response characteristic of a system or transducer. *See also:* **transmission characteristics.** 0-42A65

roller bearing (rotating machinery). A bearing incorporating a peripheral assembly of rollers. *See also:* **bearing.** 0-31E8

rolling contacts (industrial control). A contact arrangement in which one cooperating member rolls on the other. *See also:* **contactor.** 50I16-34E10

rolling transposition. A transposition in which the conductors of an open wire circuit are physically rotated in a substantially helical manner. With two wires a complete transposition is usually executed in two consecutive spans. *See also:* **open wire.** 42A65-0

roof bushing. A bushing intended primarily to carry a circuit through the roof, or other grounded barriers of a building, in a substantially vertical position. Both ends must be suitable for operating in air. At least one end must be suitable for outdoor operation. *See also:* **bushing.** E49-0

roof conductor. The portion of the conductor above the eaves running along the ridge, parapet, or other portion of the roof. *See also:* **lightning protection and equipment.** 42A95-0

room coefficient. A number computed from wall and floor areas. It is used to indicate room proportions in tables of luminance factors of room surfaces. *Note:* The room coefficient is computed from

$$K_r = \frac{\text{height} \times (\text{length} + \text{width})}{2 \times \text{length} \times \text{width}}$$

See also: **inverse-square law (illuminating engineering).** Z7A1-0

room index. The room index is a letter designation for a range of room ratios. *See also:* **inverse-square law (illuminating engineering).** Z7A1-0

room ratio. A number indicating room proportions, calculated from the length, width, and ceiling height (or luminaire mounting height) above the work plane. It is used to simplify lighting design tables by expressing the equivalence of room shapes with respect to the utilization of direct or interreflected light. *Note:* The room ratio depends upon the light distribution characteristics of luminaires. The following formulas are used (1) For direct, semidirect, and general diffuse distributions,

$$\text{Numerical room ratio} = wl/[h_m(w+l)]$$

(2) For semiindirect and indirect distributions,

$$\text{Numerical room ratio} = 3wl/[2h_c(w+l)]$$

where w is the width, l is the length, h_m is the mounting height above the work plane, and h_c is the ceiling height above the work plane. *See also:* **inverse-square law (illuminating engineering).** Z7A1-0

room utilization factor (utilance). The ratio of the luminous flux (lumens) received on the work plane to that emitted by the luminaire. *Note:* This ratio sometimes is called **interflectance.** Room utilization factor is based on the flux emitted by a complete luminaire, whereas coefficient of utilization is based on the rated flux generated by the lamps in a luminaire. *See also:* **inverse-square law (illuminating engineering).** Z7A1-0

root locus (control system, feedback) (for a closed loop whose characteristic equation is $KG(s)H(s) + 1 = 0$). A plot in the s plane of all those values of s that make $G(s)H(s)$ a negative real number; those points that make the loop transfer function $KG(s)H(s) = -1$ are roots. *Note:* The locus is conveniently sketched from the factored form of $KG(s)H(s)$; each branch starts at a pole of that function with $K = 0$. With increasing K, the locus proceeds along its several branches toward a zero of that function and, often asymptotic to one of several equiangular radial lines, toward infinity. Roots lie at points on the locus for which (1) the sum of the phase angles of component $G(s)H(s)$ vectors totals 180 degrees, and for which (2) $1/K = |G(s)H(s)|$. Critical damping of the closed loop occurs when the locus breaks away from the real axis; instability when it crosses the imaginary axis. *See also:* **control system, feedback.** 85A1-23E0

root-mean-square detector. A detector, the output voltage of which approximates the root-mean-square values of an applied signal. *See also:* **electromagnetic compatibility.** CISPR-27E1

root-mean-square (effective) burst magnitude (audio and electroacoustics). The square root of the average square of the instantaneous magnitude of the voltage or current taken over the burst duration. See the figure attached to the definition of **burst duration.** *See also:* **burst (audio and electroacoustics).** E257-1E1

root-mean-square (effective) pulse amplitude. The square root of the average of the square of the instantaneous amplitude taken over the pulse duration. *See also:* **pulse terms.** E194-0

root-mean-square reverse-voltage rating (rectifier circuit). The maximum sinusoidal root-mean-square reverse voltage permitted by the manufacturer under stated conditions. *See also:* **average forward-current rating (rectifier circuit).** E59-34E17

root-mean-square ripple. The effective value of the instantaneous difference between the average and instantaneous values of a pulsating unidirectional wave integrated over a complete cycle. *Note:* The root-mean-square ripple is expressed in percent or per unit referred to the average value of the wave. *See also:* **rectification.** 34A1-34E24

root-mean-square sound pressure. *See:* **effective sound pressure.**

root-mean-square value (1) (periodic function) (effective value*). The square root of the average of the square of the value of the function taken throughout one period. Thus, if y is a periodic function of t

$$Y_{\text{rms}} = \left[\frac{1}{T} \int_a^{a+T} y^2 \, dt \right]^{1/2}$$

where Y_{rms} is the root-mean-square value of y, a is any value of time, and T is the period. If a periodic function is represented by a Fourier series, then:

$$Y_{\text{rms}} = \frac{1}{(2)^{1/2}} \left(\frac{1}{2}A_0^2 + A_1^2 + A_2^2 + \cdots + B_1^2 + B_2^2 + \cdots \right)^{1/2}$$

$$= \frac{1}{(2)^{1/2}} \left(\frac{1}{2}A_0^2 + C_1^2 + C_2^2 + \cdots + C_n^2 \right)^{1/2}$$

*Deprecated E270-0

(2) (alternating voltage or current). The square root of the mean of the square of the voltage, or current, during a complete cycle. 60I0-31E7

root-sum-square. The square root of the sum of the squares. *Note:* Commonly used to express the total harmonic distortion. *See also:* **radio receiver.** E188-0

rope-lay conductor or cable. A cable composed of a central core surrounded by one or more layers of helically laid groups of wires. *Note:* This kind of cable differs from a concentric-lay conductor in that the main strands are themselves stranded. In the most common type of rope-lay conductor or cable, all wires are of the same size and the central core is a concentric-lay conductor. *See also:* **conductor.** E30/42A35-31E13

roped-hydraulic driving machine (elevators). A machine in which the energy is applied by a piston, connected to the car with wire ropes, that operates in a cylinder under hydraulic pressure. It includes the cylinder, the piston, and the multiplying sheaves if any and their guides. *See:* **roped-hydraulic elevator.** *See also:* **driving machine (elevators).** 42A45-0

roped-hydraulic elevator. A hydraulic elevator having its piston connected to the car with wire ropes. *See:* **roped-hydraulic driving machine (elevators).** *See also:* **elevators.** 42A45-0

rosette. An enclosure of porcelain or other insulating material, fitted with terminals and intended for connecting the flexible cord carrying apendant to the permanent wiring. *See also:* **cabinet.** 42A95-0

rotary attenuator (waveguide). A variable attenuator in circular waveguide having absorbing vanes fixed diametrically across one section; the attenuation is varied by rotation of this section about the common axis. *See also:* **waveguide.** 0-3E1

rotary converter. A machine that combines both motor and generator action in one armature winding connected to both a commutator and slip rings, and is excited by one magnetic field. It is normally used to change alternating-current power to direct-current power. *See also:* **synchronous machine.** 0-31E8

rotary generator (induction heating). An alternating-current generator adapted to be rotated by a motor or prime mover. *See also:* **dielectric heating; industrial electronics.** E54/E169-0

rotary inverter. A machine that combines both motor and generator action in one armature winding. It is excited by one magnetic field and changes direct-current power to alternating-current power. (Usually it has no amortisseur winding.) *See also:* **synchronous machine.** 0-31E8

rotary joint (rotating joint) (waveguide). A coupling for transmission of electromagnetic energy between two waveguide or transmission-line structures designed to permit mechanical rotation of one structure. *See also:* **waveguide.** 0-3E1

rotary phase changer (rotary phase shifter) (waveguide). A phase changer that alters the phase of a transmitted wave in proportion to the rotation of one of its waveguide sections. 50I62-3E1

rotary relay. (1) A relay whose armature moves in rotation to close the gap between two or more pole faces (usually with a balanced armature). (2) Sometimes used for stepping relay. *See also:* **relay.** 83A16-0

rotary solenoid relay. A relay in which the linear motion of the plunger is converted mechanically into rotary motion. *See also:* **relay.** 83A16-0

rotary switch. A bank-and-wiper switch whose wipers or brushes move only on the arc of a circle. *See also:* **telephone switching system; switch.** 42A65-0

rotary system. An automatic telephone switching system that is generally characterized by the following features: (1) The selecting mechanisms are rotary switches. (2) The switching pulses are received and stored by controlling mechanisms that govern the subsequent operations necessary in establishing a telephone connection. *See also:* **telephone switching system.** 42A65-0

rotary voltmeter. *See:* **generating voltmeter.**

rotatable frame (rotating machinery). A stator frame that can be rotated by a limited amount about the axis of the machine shaft. *See also:* **stator.** 0-31E8

rotatable phase-adjusting transformer (phase-shifting transformer). A transformer in which the secondary voltage may be adjusted to have any desired phase relation with the primary voltage by mechanically orienting the secondary winding with respect to the primary. The primary winding of such a transformer usually consists of a distributed symmetrical polyphase winding and is energized from a polyphase circuit. *See also:* **auxiliary device to an instrument.** 42A30-0

rotating amplifier. An electric machine in which a small energy change in the field is amplified to a large energy change at the armature terminals. *See also:* **asynchronous machine; synchronous machine.** 0-31E8

rotating-anode tube (X-ray). An X-ray tube in which the anode rotates. *Note:* The rotation continually brings a fresh area of its surface into the beam of electrons, allowing greater output without melting the target. *See also:* **electron devices, miscellaneous.** 42A70-15E6

rotating control assembly (rotating machinery). The complete control circuits for a brushless exciter mounted to permit rotation. *See also:* **rotor (rotating machinery).** 0-31E8

rotating field. A variable vector field that appears to rotate with time. E270-0

rotating-insulator switch. A switch in which the opening and closing travel of the blade is accomplished by the rotation of one or more of the insulators supporting the conducting parts of the switch. 37A100-31E11

rotating joint (waveguides). A coupling for transmission of electromagnetic energy between two waveguide structures designed to permit mechanical rotation of one structure. *See also:* **waveguide.** E147-0

rotating machinery. *See:* **machine, electric.**

rotation plate (rotating machinery). A plaque showing the proper direction of rotor rotation. *See also:* **rotor (rotating machinery).** 0-31E8

rotation test (rotating machinery). A test to determine that the rotor rotates in the specified direction when the voltage applied agrees with the terminal markings. *See:* **asynchronous machine; direct-current commutating machine; synchronous machine.** 0-31E8

rotor (1) (rotating machinery). The rotating member of a machine, with shaft. *Note:* In a direct-current machine with stationary field poles, universal, alternating-current series, and repulsion-type motors, it is commonly called the armature. *Note:* For an extensive list of cross references, see *Appendix A.* 42A10-31E8
(2) (meter) (rotating element). That part of the meter that is directly driven by electromagnetic action. *See also:* **watthour meter.** 12A0/42A30-0

rotor bar (rotating machinery). A solid conductor that constitutes an element of the slot section of a squirrel-cage winding. *See also:* **rotor (rotating machinery).** 0-31E8

rotor bushing (rotating machinery). A ventilated or nonventilated piece or assembly used for mounting onto a shaft, an assembled rotor core whose inside opening is larger than the shaft. *See also:* **rotor (rotating machinery).** 0-31E8

rotor coil (rotating machinery). A unit of a rotor winding of a machine. *See also:* **rotor (rotating machinery).** 0-31E8

rotor core (rotating machinery). That part of the magnetic circuit that is integral with, or mounted on, the rotor shaft. It frequently consists of an assembly of laminations. 0-31E8

rotor-core assembly (rotating machinery). The rotor core with a squirrel-cage or insulated-conductor winding, put together as an assembly. *See also:* **rotor (rotating machinery).** 0-31E8

rotor-core lamination (rotating machinery). A sheet of material containing teeth, slots, and other perforations required by design, which forms the rotor core when assembled with other identical or similar laminations. *See also:* **rotor (rotating machinery).** 0-31E8

rotor displacement angle (load angle) (rotating machinery). The displacement caused by load between the terminal voltage and the armature voltage generated by that component of flux produced by the field current. *See also:* **rotor (rotating machinery).** 0-31E8

rotor end ring (rotating machinery). The conducting structure of a squirrel-cage winding that short-circuits all of the rotor bars at one end. *See also:* **rotor (rotating machinery).** 0-31E8

rotor-resistance starting (rotating machinery). The process of starting a wound-rotor induction motor by connecting the rotor initially in series with starting resistors that are short-circuited for the running operation. *See also:* **asynchronous machine.** 0-31E8

rotor slot armor (cylindrical-rotor synchronous machine) (rotating machinery). Main ground insulation surrounding the slot or core portions of a field coil assembled on a slotted rotor. *See also:* **rotor (rotating machinery).** 0-31E8

rotor spider. *See:* **spider.**

rotor winding (rotating machinery). A winding on the rotor of a machine. *See also:* **rotor (rotating machinery).** 0-31E8

roughness (navigational-system display). Irregularities resembling scalloping, but distinguished by their random, noncyclic nature; sometimes called course roughness. *See also:* **navigation.** 0-10E6

round conductor. Either a solid or stranded conductor of which the cross section is substantially circular. *See also:* **conductor.** E30/42A35-31E13

round rotor (cylindrical rotor) (rotating machinery). A rotor of cylindrical shape in which the coil sides of the winding are contained in axial slots. *See also:* **rotor (rotating machinery).** 42A10-31E8

rounding (pulse techniques). *See:* **distortion, pulse.**

round off. To delete the least-significant digit or digits of a numeral and to adjust the part retained in accordance with some rule. *See also:* **electronic digital computer.** X3A12-16E9

route locking. Locking effective when a train passes a signal and adapted to prevent manipulation of levers that would endanger the train while it is within the limits of the route entered. It may be so arranged that a train in clearing each section of the route releases the locking affecting that section. *See also:* **interlocking (interlocking plant).** 42A42-0

routine (electronic computation). A set of instructions arranged in proper sequences to cause a computer to perform a desired operation, such as the solution of a mathematical problem.
See:
check routine;
executive routine;
library routine;
service routine;

subroutine;
supervisory routine;
tracing routine;
utility routine.
See also: **electronic computation; electronic digital computer.** E162/E270;X3A12-16E9

routine tests (1) (general). Tests made for quality control by the manufacturer on every device or representative samples, or on parts or materials as required to verify during production that the product meets the design specifications.
See:
acceptance test;
ambient temperature;
angular displacement;
arresters;
design test;
dielectric tests;
excitation current;
excitation losses;
impulse tests;
load losses;
pothead;
relative lead polarity of a transformer;
resistance methods of temperature determination;
thermometer methods of temperature determination;
total loss;
transformer. E48/E49/E270/42A15-31E12
(2) (rotating machinery). The tests applied to a machine to show that it has been constructed and assembled correctly, is able to withstand the appropriate high-voltage tests, is in sound working order both electrically and mechanically, and has the proper electrical characteristics. *See also:* **asynchronous machine; direct-current commutating machine; limiting insulation temperature (limiting hottest-spot temperature); synchronous machine.** 0-31E8
(3) (switchgear). *See:* **production tests (switchgear).**

roving (rotating machinery). A loose assemblage of fibers drawn or rubbed into a single strand with very little twist. In spun yarn systems, the product of the stage or stages just prior to spinning. AD123-31E8

row. A path perpendicular to the edge of a tape along which information may be stored by presence or absence of holes or magnetized areas. EIA3B-34E12

row binary. Pertaining to the binary representation of data on punched cards in which adjacent positions in a row correspond to adjacent bits of data, for example, each row in an 80-column card may be used to represent 80 consecutive bits of two 40-bit words. *See also:* **electronic digital computer.** X3A12-16E9

RPE. (radial probable error). *See:* **circular probable error.**

RRRV. *See:* **rate of rise of restriking voltage.**

***R* scan (electronic navigation).** *See:* ***R* display.**

***R* scope (electronic navigation).** *See:* ***R* display.**

RT box* (electronic navigation). *See:* **antitransmit-receive switch.**

*Deprecated

rubber tape. A tape composed of rubber or rubberlike compounds that provides insulation for joints. 42A95-0

rudder-angle-indicator system. A system consisting of an indicator (usually in the wheel house) so controlled by a transmitter connected to the rudder stock as to show continually the angle of the rudder relative to the center line of the ship. 42A43-0

rumble (electroacoustics). Low-frequency vibration of the recording or reproducing drive mechanism superimposed on the reproduced signal. *See also:* **electroacoustics; phonograph pickup.** 0-1E1

rumble, turntable. *See:* **turntable rumble.**

run (computing systems). A single, continuous performance of a computer routine. *See also:* **electronic digital computer.** X3A12-16E9

run-of-river station. A hydroelectric generating station that utilizes all or a part of the stream flow without storage. *See also:* **generating station.** 42A35-31E13

running-light-indicator panel (telltale). A panel in the wheelhouse providing audible and visible indication of the failure of any running light connected thereto. 42A43-0

running lights (navigation lights). Lanterns constructed and located as required by navigation laws, to permit the heading and approximate course of a vessel to be determined by an observer on a nearby vessel. *Note:* Usual running lights are port side, starboard side, mast-head, range, and stern lights. 42A43-0

running open-phase protection (industrial control). The effect of a device operative on the loss of current in one phase of a polyphase circuit to cause and maintain the interruption of power in the circuit. 42A25-34E10

running operation (single-phase motor). (1) For a motor employing a starting switch or relay: operation at speeds above that corresponding to the switching operation. (2) For a motor not employing a starting switch or relay: operation in the range of speed that includes breakdown-torque speed and above. *See also:* **asynchronous machine; synchronous machine.** 0-31E8

running tension control (industrial control). A control function that maintains tension in the material at operating speeds. *See also:* **control system, feedback.** IC1-34E10

runout rate (industrial control). The velocity at which the error in register accumulates. 42A25-34E1

runway alignment indicator. An alignment indicator consisting of a group of aeronautical ground lights arranged and located to provide early direction and roll guidance on the approach to a runway. *See also:* **signal lighting.** Z7A1-0

runway centerline lights. Runway lights installed in the surface of the runway along the centerline indicating the location and direction of the runway centerline and are of particular value in conditions of very poor visibility. *See also:* **signal lighting.** Z7A1-0

runway edge lights. Lights installed along the edges of a runway marking its lateral limits and indicating its direction. *See also:* **signal lighting.** Z7A1-0

runway end identification light. A pair of flashing aeronautical ground lights symmetrically disposed on each side of the runway at the threshold to provide additional threshold conspicuity. *See also:* **signal lighting.** Z7A1-0

runway exit lights. Lights placed on the surface of a runway to indicate a path to the taxiway centerline. *See also:* **signal lighting.** Z7A1-0

runway lights. Aeronautical ground lights arranged along or on a runway. *See also:* **signal lighting.** Z7A1-0

runway visibility. The meteorological visibility along an identified runway. Where a transmissometer is used for measurement, the instrument is calibrated in terms of a human observer; that is, the sighting of dark objects against the horizon sky during daylight and the sighting of moderately intense unfocused lights of the order of 25 candelas at night. *See also:* **signal lighting.** Z7A1-0

runway visual range (RVR) (navigation). (1) The forward distance a human pilot can see along the runway during an approach to landing; this distance is derived from electro-optical instruments operated on the ground and it is improved (increased) by the use of lights (such as high-intensity runway lights). *See also:* **navigation.** 0-10E6
(2) (in the United States). An instrumentally derived value, based on standard calibrations, that represents the horizontal distance a pilot will see down the runway from the approach end; it is based on the sighting of either high-intensity runway lights or on the visual contrast of other targets, whichever yields the greater visual range. *See also:* **signal lighting.** Z7A1-0

rural line. A line serving one or more subscribers in a rural area. *See also:* **telephone system.** 42A65-0

rust (corrosion). A corrosion product consisting primarily of hydrated iron oxide. (*Note:* This term is properly applied only to iron and ferrous alloys). *See also:* **corrosion terms.** CM-34E2

RVR. *See:* **runway visual range.**

***R-X* diagram (relay unit).** A graphical presentation of the characteristics in terms of the ratio of voltage to current and the angle between them. *Note:* For example, if a relay just operates with ten volts and ten amperes in phase, one point on the operating curve of the relay would be plotted as one ohm on the R axis, that is, $R = 1, X = 0$, where R is the abscissa and X is the ordinate. 0-31E6

S

sabin (audio and electroacoustics). A unit of absorption having the dimensions of area. *Notes:* (1) The metric sabin has dimensions of square meters. (2) When used without a modifier, the sabin is the equivalent of one square foot of a perfectly absorptive surface. *See also:* **electroacoustics.** 0-1E1

sacrificial protection (corrosion). Reduction or prevention of corrosion of a metal in an environment acting as an electrolyte by coupling it to another metal that is electrochemically more active in that particular electrolyte. *See also:* **stray-current corrosion.** CM-34E2

safe working voltage to ground (electric recording instrument). The highest safe voltage in terms of maximum peak value that should exist between any circuit of the instrument and its case. *See also:* **test (instrument or meter).** 39A2-0

safety control feature (deadman's feature). That feature of a control system that acts to reduce or cut off the current to the traction motors or to apply the brakes, or both, if the operator relinquishes personal control of the vehicle. *See also:* **multiple-unit control.** 42A42-0

safety control handle (deadman's handle). A safety attachment to the handle of a controller, or to a brake valve, causing the current to the traction motors to be reduced or cut off, or the brakes to be applied, or both, if the pressure of the operator's hand on the handle is released. *Note:* This function may be applied alternatively to a foot-operated pedal or in combination with attachments to the controller or the brake valve handles, or both. *See also:* **multiple-unit control.** 42A42-0

safety outlet.* *See:* **grounding outlet.**
*Deprecated

SAFI. *See:* **semiautomatic flight inspection.**

sag (1) (of a span). The difference in elevation between the highest point of support of the conductor and the lowest point of the conductor in the span. *See also:* **tower.** 42A35-31E13
(2) (apparent sag at any point). The departure of the wire at the particular point in the span from the straight line between the two points of support of the span, at 60 degrees Fahrenheit, with no wind loading.
See:
apparent sag of a span;
final unloaded sag;
initial unloaded sag;
maximum total sag;
total sag;
unloaded sag. 2A2-0

sal ammoniac cell. A cell in which the electrolyte consists primarily of a solution of ammonium chloride. *See also:* **electrochemistry.** 42A60-0

salient pole (rotating machinery). A field pole that projects from the yoke or hub towards the primary winding core. *See also:* **rotor (rotating machinery).** 0-31E8

salient-pole machine. An alternating-current machine in which the field poles project from the yoke toward the armature and/or the armature winding self-inductance undergoes a significant single cyclic variation for a rotor displacement through one pole pitch. *See also:* **asynchronous machine; synchronous machine.** 0-31E8

salinity indicator system. A system, based on measurement of varying electric resistance of the solution, to indicate the amount of salt in boiler feed water, the output of an evaporator plant, or other fresh water. *Note:* Indication is usually in grains per gallon. 42A43-0

sampled data. Data in which the information content can be, or is, ascertained only at discrete intervals of time. *Note:* Sampled data can be analog or digital. *See also:* **control system feedback.** EIA3B-34E12

sampled-data control system (industrial control). A system that operates with sampled data. *See also:* **control system, feedback.** AS1-34E10

sampled signal. The sequence of values of a signal taken at discrete instants. *See also:* **control system.** 0-23E0

sampling (modulation systems). The process of obtaining a sequence of instantaneous values of a wave; at regular or intermittent intervals. *See also:* **modulating systems.** E170/X3A12-16E9

sampling circuit (sampler). A circuit whose output is a series of discrete values representative of the values of the input at a series of points in time. *See also:* **circuits and devices.** 42A65-0

sampling control. *See:* **control system, sampling.**

sampling gate (electronic navigation). A device that extracts information from the input wave only when activated by a selector pulse. *See also:* **navigation.** E172-10E6

sampling, instantaneous. The process for obtaining a sequence of instantaneous values of a wave. *Note:* These values are called instantaneous samples. *See also:* **modulating systems.** E145/42A65

sampling interval (automatic control). The time between samples in a sampling control system. *See also:* **control system, feedback.** 0-23E0

sampling period (automatic control). The time interval between samples in a periodic sampling control system. *See also:* **control system, feedback.** 0-23E0

sampling tests. Tests carried out on a few samples taken at random out of one consignment. *See also:* **asynchronous machine; direct-current commutating machine; synchronous machine.** 0-31E8

saturable-core magnetometer. A magnetometer that depends for its operation on the changes in permeability of a ferromagnetic core as a function of the field to be measured. *See also:* **magnetometer.** 42A30-0

saturable-core reactor. *See:* **saturable reactor.**

saturable reactor (saturable-core reactor). (1) A magnetic-core reactor whose reactance is controlled by changing the saturation of the core through variation of a superimposed unidirectional flux. (2) A magnetic-core reactor operating in the region of saturation without independent control means. *Note:* Thus, a reactor whose impedance varies cyclically with the alternating current (or voltage). *See also:* **magnetic amplifier.** 42A15/42A25/42A65/E270-34E10/31E12

saturated signal*. *See:* **saturating signal.**

*Deprecated

saturated sleeving. A flexible tubular product made from braided cotton, rayon, nylon, glass, or other fibers, and coated or impregnated with varnish, lacquer, a combination of varnish and lacquer, or other electrical insulating materials. The impregnant or coating need not form a continuous film. 42A95-0

saturating reactor. A magnetic-core reactor operating in the region of saturation without independent control means. *See also:* **magnetic amplifier.** 42A65-0

saturating signal (electronic navigation). A signal of an amplitude greater than can be accommodated by the dynamic range of a circuit. *See also:* **navigation.** 0-10E6

saturation (1) (signal-transmission system). A natural phenomenon or condition in which any further change of input no longer results in appreciable change of output. AS1-34E10
(2) (automatic control) A condition caused by the presence of a signal or interference large enough to produce the maximum limit of response, resulting in loss of incremental response. *See also:* **control system, feedback; signal.** 85A1-13E6
(3) (perceived object color). The attribute of any color perception possessing a hue that determines the degree of its difference from the achromatic color perception most resembling it. *Notes:* (A) This is a subjective term corresponding, or nearly so, to the psychophysical term purity. (B) The description of saturation is not commonly undertaken beyond the use of rather vague terms, such as vivid, strong, and weak. The terms brilliant, pastel, pale, and deep, which are sometimes used as descriptive of saturation, have connotations descriptive also of brightness. *See also:* **color terms; light.** Z7A1/50I45/AS1/E201-2E2/34E10
(4) (perceived light-source color). The attribute used to describe its departure from a light-source color of the same brightness perceived to have no hue. *See also:* **color.** Z7A1-0

saturation current (1) (thermionics). The value of the current in the saturation state. *See also:* **electron emission.** 50I07-15E6
(2) (semiconductor diode). That portion of the steady-state reverse current that flows as a result of the transport across the junction of minority carriers thermally generated within the regions adjacent of the junction. *See also:* **semiconductor device.** E270/E216-34E17

saturation curve (machine or other apparatus). A characteristic curve that expresses the degree of magnetic saturation as a function of some property of the magnetic excitation. *Note:* For a direct-current or synchronous machine the curve usually expresses the relation between armature voltage and field current for no load or some specified load current, and for specified speed. *See also:* **direct-current commutating machine; synchronous machine.** 42A10-31E8

saturation factor (1) (direct-current or synchronous machine). The ratio of a small percentage increase in field excitation to the corresponding percentage increase in voltage thereby produced. *Note:* Unless otherwise specified, the saturation factor of a machine refers to the no-load excitation required at rated speed and voltage. *See also:* **asynchronous machine; direct-current commutating machine; synchronous machine.** 42A10-31E8
(2) (rotating machinery). The ratio of the unsaturated value of a quantity to its saturated value. The reciprocal of this definition is also used. 0-31E8

saturation flux density. *See:* **saturation induction.**

saturation induction. The maximum intrinsic induction possible in a material. *Note:* Saturation induction is sometimes loosely referred to as saturation flux density. *See:* **intrinsic induction.** *See also:* **static magnetic storage.** E163/E270-0

saturation level (storage tubes). The output level beyond which no further increase in output is produced by further writing (then called write saturation) or reading (then called read saturation). *Note:* The word saturation is frequently used alone to denote saturation level. *See also:* **storage tube.** E158-15E6

saturation state (thermionics). The state of working of an electron tube or valve in which the current is limited by the emission from the cathode. *See also:* **electron emission.** 50I07-15E6

sawtooth. *See:* **sawtooth waveform.**

sawtooth sweep. A sweep generated by the ramp portion of a sawtooth waveform. *See also:* **oscillograph.** 0-9E4

sawtooth wave. A periodic wave whose instantaneous value varies substantially linearly with time between two values, the interval required for one direction of progress being longer than that for the other. *See also:* **television.** E204/42A65-2E2

sawtooth waveform. A waveform containing a ramp and a return to initial value, the two portions usually of unequal duration. *See also:* **oscillograph.** 0-9E4

scalar. A quantity that is completely specified by a single number. E270-0

scalar field. The totality of scalars in a given region represented by a scalar function $S(x,y,z)$ of the space coordinates x,y,z. E270-0

scalar function. A functional relationship that results in a scalar. E270-0

scalar product (of two vectors) (dot product). The scalar obtained by multiplying the product of the magnitudes of the two vectors by the cosine of the angle between them. The scalar product of the two vectors **A** and **B** may be indicated by means of a dot $\mathbf{A}\cdot\mathbf{B}$. If the two vectors are given in terms of their rectangular components, then

$$\mathbf{A}\cdot\mathbf{B} = A_xB_x + A_yB_y + A_zB_z.$$

Example: Work is the scalar product of force and displacement. E270-0

scale (1) (acoustics). A musical scale is a series of notes (symbols, sensations, or stimuli) arranged from low to high by a specified scheme of intervals, suitable for musical purposes. E157-1E1

(2) (computing systems). To change a quantity by a factor in order to bring its range within prescribed limits. *See also:* **electronic digital computer.** X3A12-16E9

(3) (instrument scale). *See:* **full scale.**

scale factor (1) (instrument or device). The factor by which the number of scale divisions indicated or recorded by an instrument should be multiplied to compute the value of the measurand. *Note:* Deflection factor is a more general term than scale factor in that the instrument response may be expressed alternatively in units other than scale divisions. *See also:* **accuracy rating (instrument) test voltage and current.** 42A30/68A1-31E5

(2) (computing systems). A number used as a multiplier, so chosen that it will cause a set of quantities to fall within a given range of values. *Note:* To scale the values 856, 432, −95, and −182 between −1 and +1, a scale factor of 1/1000 would be suitable. X3A12-16E9

(3) (electronic computation). The multiplication factor necessary to transform problem variables to computer variables. *See also:* **electronic analog computer.** E165-0

scale-factor potentiometer. *See:* **parameter potentiometer.**

scale length (electric instrument). The length of the path described by the indicating means or the tip of the pointer in moving from one end of the scale to the other. *Notes:* (1) In the case of knife-edge pointers and others extending beyond the scale division marks, the pointer shall be considered as ending at the outer end of the shortest scale division marks. In multiscale instruments the longest scale shall be used to determine the scale length. (2) In the case of antiparallax instruments of the step-scale type with graduations on a raised step in the plane of and adjacent to the pointer tip, the scale length shall be determined by the end of the scale divisions adjacent to the pointer tip. *See also:* **accuracy rating (instrument); instrument.** 39A1/42A30-0

scale-of-two counter. A flip-flop circuit in which successive similar pulses, applied at a common point, cause the circuit to alternate between its two conditions of permanent stability. *See also:* **trigger circuit.** 42A65-0

scaler (radiation counters). An instrument incorporating one or more scaling circuits and used for registering the number of counts received. *See also:* **anticoincidence (radiation counters).** 0-15E6

scaler, pulse (pulse techniques). A device that produces an output signal whenever a prescribed number of input pulses has been received. It frequently includes indicating devices for interpolation. *See also:* **pulse.** E175-0

scale span (instrument). The algebraic difference between the values of the actuating electrical quantity corresponding to the two ends of the scale. *See also:* **instrument.** 42A30-0

scaling (corrosion). (1) The formation at high temperatures of thick corrosion product layer(s) on a metal surface. (2) The deposition of water-insoluble constituents on a metal surface (as on the interior of water boilers). *See also:* **corrosion terms.** CM-34E2

scaling circuit (radiation counters). A device that produces an output pulse whenever a prescribed number of input pulses has been received. *See also:* **anticoincidence (radiation counters).** 0-15E6

scalloping (in a navigation system such as localizer, glide-slope, very-high-frequency omnidirectional radio range, Tacan). The irregularities in the field pattern of the ground facility due to unwanted reflections from obstructions or terrain features, exhibited in flight as cyclical variations in bearing error. Also called **course scalloping.** *See also:* **navigation.** 0-10E6

scan (1) (general). To examine sequentially part by part. X3A12-16E9

(2) (oscillography). The process of deflecting the electron beam. *See:* **graticule area; uniform luminance area; phosphor screen.** *See also:* **oscillograph.** 0-9E4

scanner (1) (facsimile). That part of the facsimile transmitter that systematically translates the densities of the subject copy into signal waveform. *See also:* **scanning (facsimile).** E168-0

(2) (industrial control). (A) A multiplexing arrangement that sequentially connects one channel to a number of channels. (B) An arrangement that progressively examines a surface for information. *See also:* **control system, feedback.** AS1-34E10

(3) *See:* **flying-spot scanner; optical scanner; visual scanner.** *See also:* **electronic digital computer.**

scanning (1) (radar). The process of directing a beam of radio-frequency energy successively over the elements of a given region, or the corresponding process in reception. *See also:* **radar.** 42A65-0

(2) (television). The process of analyzing or synthesizing successively, according to a predetermined method, the light values or equivalent characteristics of elements constituting a picture area. *See also:* **television.** 42A65-2E2

(3) (navigation aids). A periodic motion given to the major lobe of an antenna. *See also:* **antenna; navigation.** 0-3E1/10E6

(4) (facsimile). The process of analyzing successively the densities of the subject copy according to the elements of a predetermined pattern. *Note:* The normal scanning is from left to right and top to bottom of the subject copy as when reading a page of print. Reverse direction is from right to left and top to bottom of the subject copy.
See:
density;
drum speed;
electronic line scanning;
electronic raster scanning;
elemental area;
facsimile;
flood projection;
index of cooperation, scanning or recording line;
light carrier injection;
maximum keying frequency;
multiple-spot scanning;
nominal line width;
phasing;
picture frequencies;
scanner;
scanning line length;
scanning spot;
scanning spot, *X* dimension;
scanning spot, *Y* dimension;
simple scanning;
spot projection;
spot speed;
stroke speed. E168-0

scanning, high-velocity (electron tube). The scanning of a target with electrons of such velocity that the secondary-emission rate is greater than unity. *See also:* **beam tube; television.** E160-2E2/15E6

scanning line (television). A single continuous narrow strip that is determined by the process of scanning. *Note:* In most television systems, the scanning lines that occur during the return intervals are blanked. The total number of scanning lines is numerically equal to the ratio of line frequency to frame frequency. *See also:* **television.** E204/42A65-2E2

scanning linearity (television). A measure of the uniformity of scanning speed during the unblanked trace interval. *See also:* **television.** E204/42A65-2E2

scanning line frequency (facsimile). *See:* **stroke speed (scanning or recording line frequency).**

scanning line length (facsimile). The total length of scanning line is equal to the spot speed divided by the scanning line frequency. *Note:* This is generally greater than the length of the available line. *See also:* **scanning (facsimile).** E168-0

scanning loss (radar system employing a scanning antenna). The reduction in sensitivity, usually expressed in decibels, due to scanning across a target, compared with that obtained when the beam is directed constantly at the target. *See also:* **antenna.** E145/42A65-0

scanning, low-velocity (electron tube). The scanning of a target with electrons of velocity less than the minimum velocity to give a secondary-emission ratio of unity. *See also:* **beam tube; television.** E160-2E2/15E6

scanning speed (television). The time rate of linear displacement of the scanning spot. *See also:* **television.** E204/42A65-2E2

scanning spot (1) (television). The area with which the scanned area is being explored at any instant in the scanning process. *See also:* **television.** 42A65-0
(2) (facsimile). The area on the subject copy viewed instantaneously by the pickup system of the scanner. *See also:* **scanning (facsimile).** E168-0

scanning spot, *X* dimension (facsimile). The effective scanning-spot dimension measured in the direction of the scanning line on the subject copy. *Note:* The numerical value of this will depend upon the type of system used. *See also:* **scanning (facsimile).** E168-0

scanning spot, *Y* dimension (facsimile). The effective scanning-spot dimension measured perpendicularly to the scanning line on the subject copy. *Note:* The numerical value of this will depend upon the type of system used. *See also:* **scanning (facsimile).** E168-0

scatterband (interrogation systems). The total bandwidth occupied by the various received signals from interrogators operating with carriers on the same nominal radio frequency; the scatter results from the individual deviations from the nominal frequency. *See also:* **navigation.** 0-10E6

scattering. The production of waves of changed direction, frequency, or polarization when radio waves encounter matter. *Note:* The term is frequently used in a narrower sense, implying a disordered change in the incident energy. *See also:* **radiation.** 42A65-0

scattering coefficient. Element of the scattering matrix. *See:* **scattering matrix.** *See also:* **transmission characteristics.** 0-9E4

scattering cross section (of an object in a given orientation). 4π times the ratio of the radiation intensity of the scattered wave in a specified direction to the power per unit area in an incident plane wave of a specified polarization. *Note:* The term **bistatic cross section** denotes the scattering cross section in any specified direction other than back towards the source. *See also:* **antenna; radiation.** 0-3E1

scattering loss (acoustics). That part of the transmission loss that is due to scattering within the medium or due to roughness of the reflecting surface. *See also:* **electroacoustics.** E157-1E1

scattering matrix. A square array of complex numbers consisting of the transmission and reflection coefficients of a waveguide component. As most commonly used, each of these coefficients relates the complex electric field strength (or voltage) of a reflected or transmitted wave to that of an incident wave. The subscripts of a typical coefficient S_{ij} refer to the output and input ports related by the coefficient. These coefficients, which may vary with frequency, apply at a specified set of input and output reference planes. E148-3E1

scheduled frequency (electric power systems). The frequency that a power system or an interconnected system attempts to maintain. *See also:* **power systems, low frequency and surge testing.** E94-0

scheduled frequency offset (electric power systems). The amount, usually expressed in hundredths of a hertz, by which the frequency schedule is changed from rated frequency in order to correct a previously accumulated time deviation. *See also:* **power systems, low-frequency and surge testing.** E94-0

scheduled interruption (electric power systems). An interruption caused by a scheduled outage. *See also:* **outage.** 0-31E4

scheduled net interchange (electric power systems) (control area). The mutually prearranged intended net power and/or energy on the area tie lines. *See also:* **power systems, low-frequency and surge testing.** E94-0

scheduled outage (electric power systems). An outage that results when a component is deliberately taken out of service at a selected time, usually for purposes of construction, preventive maintenance, or repair. *See also:* **outage.** 0-31E4

scheduled outage duration (electric power systems). The period from the initiation of the outage until construction, preventive maintenance, or repair work is completed and the affected component is made available to perform its intended function. *See also:* **outage.** 0-31E4

schedule setter or set-point device (speed-governing system). A device for establishing or setting the desired value of a controlled variable. *See also:* **speed-governing system.** E94-0

schematic diagram (elementary diagram) (industrial control). A diagram that shows, by means of graphic symbols, the electric connections and functions of a specific circuit arrangement. The schematic diagram facilitates tracing the circuit and its functions without regard to the actual physical size, shape, or location of the component devices or parts. E270-34E10

Scherbius machine (rotating machinery). A polyphase alternating-current commutator machine capable of generator or motor action, intended for connection in the secondary circuit of a wound-rotor induction motor supplied from a fixed-frequency polyphase power system, and used for speed and/or power-factor control. The magnetic circuit components are laminated and may be of the salient-pole type or of the cylindrical-rotor uniformly slotted type; either type having a series-connected armature reaction compensating winding as part of the field system. The control field winding may be separately or shunt-excited with or without an additional series-excited field winding. *See also:* **asynchronous machine.** 0-31E8

Schering bridge. A 4-arm alternating-current bridge in which the unknown capacitor and a standard loss-free capacitor form two adjacent arms, while the arm adjacent to the standard capacitor consists of a resistor and a capacitor in parallel, and the fourth arm is a nonreactive resistor. *Note:* Normally used for the measurement of capacitance and dissipation factor. Usually, one terminal of the source is connected to the junction of the unknown capacitor with the standard capacitor. With this connection, if the impedances of the capacitance arms are large compared to those of the resistance arms, most of the applied voltage appears across the former, the maximum test voltage being limited by the rating of the standard capacitor. If the detector and the source of electromotive force are interchanged the resulting circuit is called a **conjugate Schering bridge**. The balance is independent of frequency. *See also:* **bridge.** 42A30-0

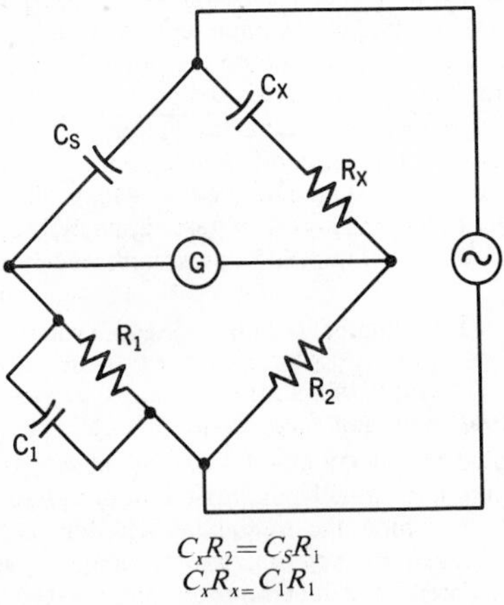

$C_xR_2 = C_SR_1$
$C_xR_x = C_1R_1$

Schering bridge.

Schlieren method (acoustics). The technique by which light refracted by the density variations resulting from acoustic waves is used to produce a visible image of a sound field. *See also:* **electroacoustics.** 0-1E1

Schottky effect. *See:* **Schottky emission.**

Schottky emission (electron tubes). The increased thermionic emission resulting from an electric field at the surface of the cathode. *See also:* **electron emission.** E160-15E6

Schottky noise (electron tubes). The variation of the output current resulting from the random emission from the cathode. *See also:* **electronic tube.** 50I07-15E6

Schuler tuning (design of inertial navigation equipment). The application of parameter values such that accelerations do not deflect the platform system from any vertical to which it has been set; a Schuler-tuned system, if fixed to the mean surface of a nonrotating earth, exhibits a natural period of 84.4 minutes. *See also:* **navigation.** E174-10E6

scintillation (1) (radio propagation). A random fluctuation of the received field about its mean value, the deviations usually being relatively small. *Note:* This use of the term scintillation is an extension of the astronomical term for the twinkling of stars, and the underlying explanation may be similar. *See also:* **radiation.** 42A65-0

(2) (radar). Variations in the signal reflected from a complex target due to changes in the aspect of the target. *Note:* Because of its noiselike properties, scintillation is sometimes called target noise and, in turn, is sometimes divided into amplitude noise, angle noise, and range noise. Other terms such as glint and wander are sometimes used synonymously with scintillation or its various components but usage varies. *See also:* **navigation; radiation.** 0-10E6

(3) (scintillators). The optical photons emitted as a result of the incidence of a particle or photon of ionizing radiation on a scintillator. *Note:* Optical photons unless otherwise specified are photons with energies corresponding to wavelengths between 2000 and 15 000 angstroms. *See also:* **ionizing radiation; radiation.** E175-0

scintillation counter. The combination of scintillation-counter heads and associated circuitry for detection and measurement of ionizing radiation.
See:
count;
counting efficiency;
counting-rate meter;

full width at half maximum;
ionizing radiation;
scintillation-counter cesium resolution;
scintillation-counter energy resolution;
scintillation-counter energy resolution constant;
scintillation counter head;
scintillation-counter time discrimination;
scintillation decay time;
scintillation duration;
scintillation rise time;
scintillator;
scintillator conversion efficiency;
scintillator material;
scintillator-material total conversion efficiency;
scintillator photon distribution;
spurious count;
spurious pulse;
time-interval selector. E175-0

scintillation-counter cesium resolution. The scintillation-counter energy resolution for the gamma ray or conversion electron from cesium-137. *See also:* **scintillation counter.** E175-0

scintillation-counter energy resolution. A measure of the smallest difference in energy between two particles or photons of ionizing radiation that can be discerned by the scintillation counter. Quantitatively it is the fractional standard deviation (σ / E_1) of the energy distribution curve. *Note:* The fractional full width at half maximum of the energy distribution curve (FWHM/E_1) is frequently used as a measure of the scintillation-counter energy resolution where E_1 is the mode of the distribution curve. See the accompanying figure. *See also:* **scintillation counter.** E175-0

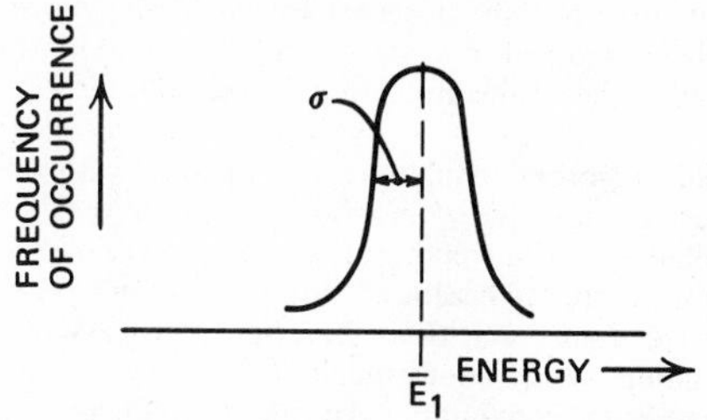

Scintillation-counter energy resolution.

scintillation-counter energy-resolution constant. The product of the square of the scintillation-counter energy resolution, expressed as the fractional full width at half maximum (FWHM/E_1), and the specified energy. *See also:* **scintillation counter.** E175-0

scintillation counter head. The combination of scintillators and phototubes or photocells that produces electric pulses or other electric signals in response to ionizing radiation. *See also:* **phototube; scintillation counter.** E175-0

scintillation-counter time discrimination. A measure of the smallest interval of time between two individually discernible events. Quantitatively it is the standard deviation σ of the time-interval curve. See the accompanying figure. *Note:* The full width at half maximum of the time-interval curve is frequently used as a measure of the time discrimination. *See also:* **scintillation counter.** E175-0

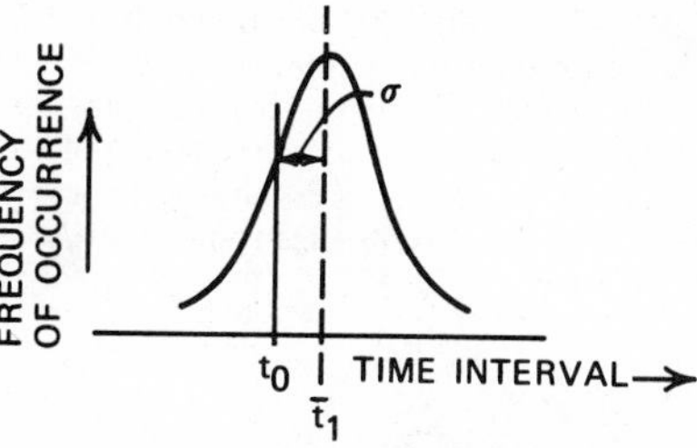

Scintillation-counter time discrimination.

scintillation decay time. The time required for the rate of emission of optical photons of a scintillation to decrease from 90 percent to 10 percent of its maximum value. *Note:* Optical photons, for the purpose of this Standard, are photons with energies corresponding to wavelengths between 2000 and 15 000 angstroms. *See also:* **scintillation counter.** E175-0

scintillation duration. The time interval from the emission of the first optical photon of a scintillation until 90 percent of the optical photons of the scintillation have been emitted. *Note:* Optical photons are photons with energies corresponding to wavelengths between 2000 and 15 000 angstroms. *See also:* **scintillation counter.** E175-0

scintillation rise time. The time required for the rate of emission of optical photons of a scintillation to increase from 10 percent to 90 percent of its maximum value. *Note:* Optical photons are photons with energies corresponding to wavelengths between 2000 and 15 000 angstroms. *See also:* **scintillation counter.** E175-0

scintillator. The body of scintillator material together with its container. *See also:* **scintillation counter.** E175-0

scintillator conversion efficiency. The ratio of the optical photon energy emitted by a scintillator to the incident energy of a particle or photon of ionizing radiation. *Note:* The efficiency is generally a function of the type and energy of ionizing radiation. Optical photons are photons with energies corresponding to wavelengths between 2000 and 15 000 angstroms. *See also:* **scintillation counter.** E175-0

scintillator material. A material that emits optical photons in response to ionizing radiation. *Notes:* (1) There are five major classes of scintillator materials, namely: (A) inorganic crystals such as NaI(Tl) single crystals, ZnS(Ag) screens, (B) organic crystals (such as, anthracene, *trans*-stilbene), (C) solution scintillators: (1) liquid, (2) plastic, (3) glass, (D) gaseous scintillators, (E) Cerenkov scintillators. (2) Optical photons are photons with energies corresponding to wavelengths between 2000 and 15 000 angstroms. *See also:* **scintillation counter.** E175-0

scintillator-material total conversion efficiency. The ratio of the optical photon energy produced to the energy of a particle or photon of ionizing radiation that is totally absorbed in the scintillator material. *Note:* The efficiency is generally a function of the type and energy of the ionizing radiation. Optical photons are photons with energies corresponding to wavelengths between 2000 and 15 000 angstroms. *See also:* **scintillation counter.** E175-0

scintillator photon distribution (in number). The statistical distribution of the number of optical photons produced in the scintillator by total absorption of monoenergetic particles. *Note:* Optical photons are photons with energies corresponding to wavelengths between 2000 and 15 000 angstroms. *See also:* **scintillation counter.** E175-0

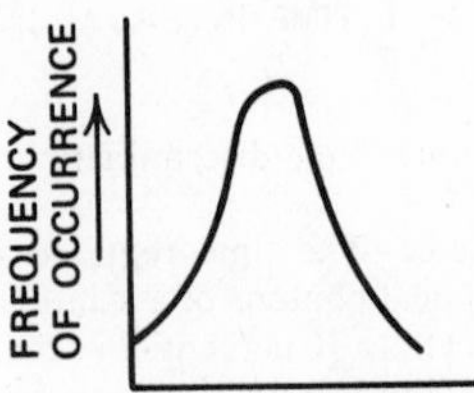

Scintillator photon distribution.

scope. The face of a cathode-ray tube or a colloquial expression for a cathode-ray oscilloscope. 0-10E6

scoring system (electroacoustics) (motion-picture production). A recording system used for recording music to be reproduced in timed relationship with a motion picture. *See also:* **electroacoustics.** E157-1E1

scotopic vision. Vision mediated essentially or exclusively by the rods. It is generally associated with adaptation to a luminance below about 0.01 footlambert (0.03 nit). *See also:* **visual field.** Z7A1-0

Scott-connected transformer assembly. An assembly for transforming from three phase to two phase or from two phase to three phase. *Note:* It consists of a main transformer with a tap at its midpoint connected directly between two of the phase wires of a three-phase circuit, and of a teaser transformer connected between the midtap of the main transformer and the third-phase wire of the three-phase circuit. The other windings of the transformers are connected in a two-phase circuit. *See also:* **autotransformer; transformer.** 42A15-31E12;57A18-0

Scott-connected transformer, interlacing impedance voltage. The single-phase voltage applied from the midtap of the main transformer winding to both ends, connected together, that is sufficient to circulate in the supply lines a current equal to the three-phase line current. The current in each half of the winding is 50 percent of this value. *See:* **efficiency.** 42A15-31E12

Scott-connected transformer per-unit resistance. The measured watts expressed in per-unit on the base of the rated kilovolt-ampere of the teaser winding. 42A15-31E12

SCR. *See:* **semiconductor controlled rectifier.**

scraper hoist. A power-driven hoist operating a scraper to move material (generally ore or coal) to a loading point. *See also:* **mining.** 42A85-0

screen (1) (rotating machinery). A port cover with multiple openings used to limit the entry of foreign objects. *See also:* **cradle base (rotating machinery).** 0-31E8

(2) (cathode-ray tubes). The surface of the tube upon which the visible pattern is produced. *See also:* **electrode (electron tube).** 42A70-15E6

screen factor (electron-tube grid). The ratio of the actual area of the grid structure to the total area of the surface containing the grid. *See also:* **electron tube.** 50I07-15E6

screen grid. A grid placed between a control grid and an anode, and usually maintained at a fixed positive potential, for the purpose of reducing the electrostatic influence of the anode in the space between the screen grid and the cathode. *See also:* **electrode (of an electron tube); grid.** 42A70-15E6

screen-grid modulation. Modulation produced by application of a modulating voltage between the screen grid and the cathode of any multigrid tube in which the carrier is present. *See also:* **modulating systems.** E182A/42A65-0

screening test (reliability). A test or combination of tests intended to remove unsatisfactory items or those likely to exhibit early failures. *See also:* **reliability.** 0-7E1

screen protected. *See:* **guarded.**

screen, viewing. *See:* **viewing area.**

screw machine (elevators). An electric driving machine, the motor of which raises and lowers a vertical screw through a nut with or without suitable gearing, and in which the upper end of the screw is connected directly to the car frame or platform. The machine may be of direct or indirect drive type. 42A45-0

seal (window) (in a waveguide). A gastight or watertight membrane or cover designed to present no obstruction to radio-frequency energy. *See also:* **waveguide.** 0-3E1

sea return (radar). The aggregate of received echoes due to reflection from the surface of the sea. *See also:* **clutter; radar.** E171/42A65-10E6

sealable equipment (electric system). Equipment so arranged or enclosed that it may be sealed or locked to prevent operation or access to live parts. *Note:* Enclosed equipment may or may not be operable without opening the enclosure. *See also:* **cabinet.** 1A0/42A95-0

sealed (rotating machinery). Provided with special seals to minimize either the leakage of the internal coolant out of the enclosure or the leakage of medium surrounding the enclosure into the machine. *See also:* **asynchronous machine; direct-current commutating machine; synchronous machine.** 0-31E8

sealed-beam headlamp. An integral optical assembly designed for headlighting purposes, identified by the name **Sealed Beam** branded on the lens. *See also:* **headlamp.** Z7A1-0

sealed end (cable) (shipping seal). The end fitted with a cap for protection against the loss of compound or the entrance of moisture. 42A35-31E13

sealed refrigeration compressor (hermetic type). A mechanical compressor consisting of a compressor and a motor, both of which are enclosed in the same sealed housing, with no external shaft or shaft seals, the motor operating in the refrigerant atmosphere. *See also:* **appliances.** 1A0-0

sealed-tank system (regulator) (transformer). A method of oil preservation in which the interior of the tank is sealed from the atmosphere and in which the gas plus the oil volume remains constant over the temperature range. *See:* **oil-immersed self-cooled transformer (class OA).** 42A15/57A15-31E12

sealed tube. An electron tube that is hermetically sealed. *Note:* This term is used chiefly for pool-cathode tubes. *See also:* **tube definitions.** 42A70-15E6

sealing gap (industrial control). The distance between the armature and the center of the core of a magnetic circuit-closing device when the contacts first touch each other. *See also:* **electric controller; initial contact pressure.** E74/IC1-34E10

sealing voltage (or current) (contactors). The voltage (or current) necessary to complete the movement of the armature of a magnetic circuit-closing device from the position at which the contacts first touch each other. *See also:* **contactor; control switch.** E16/42A25-34E10

seal-in relay. An auxiliary relay that remains picked up through one of its own contacts that bypasses the initiating circuit until de-energized by some other device. 37A100-31E11/31E6

search. To examine a set of items for those that have a desired property. *See:* **binary search; dichotomizing search.** *See also:* **electronic digital computer.** X3A12-16E9

searchlight. A projector designed to produce an approximately parallel beam of light and having an optical system with an aperture of eight inches or more. *See:*
elevation;
horizontal plane of a searchlight;
light;
pilot-house control;
signal shutter;
train;
vertical plane of a searchlight. Z7A1-0

searchlighting (radar). Projecting a beam of radio-frequency energy continuously at an object, as contrasted to scanning. *See also:* **radar.** E172-10E6;42A65-0

search radar. A radar used primarily for the detection of targets entering a particular area of interest. *See also:* **radar.** 0-10E6

season cracking (corrosion). Cracking resulting from the combined effect of corrosion and internal stress. A term usually applied to stress-corrosion cracking of brass. *See also:* **corrosion terms.** CM-34E2

SEC. *See:* **secondary-electron conduction.**

second. The duration of 9 192 631 770 periods of the radiation corresponding to the transition between the two hyperfine levels of the ground state of the cesium-133 atom. CGPM-SCC14

secondary (adjective). (1) Operates after the primary device; for example, secondary arcing contacts. (2) Second in preference. (3) Referring to auxiliary or control circuits as contrasted with the main circuit; for example, secondary disconnecting devices, secondary and control wiring. (4) Referring to the energy output side of transformers or the conditions (voltages) usually encountered at this location; for example, secondary fuse, secondary unit substation. 37A100-31E11

secondary and control wiring (switchgear assemblies) (small wiring). Wire used for control circuits and for connections between instrument transformer secondaries, instruments, meters, relays, or other equipment. 37A100-31E11

secondary arcing contacts (switching device). The contacts on which the arc of the arc-shunting resistor current is drawn and interrupted. 37A100-31E11

secondary current rating (transformer). The secondary current existing when the transformer is delivering rated kilovolt-amperes at rated secondary voltage. *See also:* **transformer.** 0-31E12

secondary disconnecting devices (switchgear assembly). Self-coupling separable contacts provided to connect and disconnect the auxiliary and control circuits between the removable element and the housing. 37A100-31E11

secondary distribution feeder. A feeder operating at secondary voltage supplying a distribution circuit. 42A35-31E13

secondary distribution mains. The conductors connected to the secondaries of distribution transformers from which consumers' services are supplied. *See also:* **center of distribution.** 42A35-31E13

secondary distribution network. A network consisting of secondary distribution mains. *See also:* **center of distribution.** 42A35-31E13

secondary distribution system. A low-voltage alternating-current system that connects the secondaries of distribution transformers to the consumers' services. *See:* **alternating-current distribution.** *See also:* **center of distribution.** 42A35-31E13

secondary distribution trunk line. A line acting as a main source of supply to a secondary distribution system. *See also:* **center of distribution.** 42A35-31E13

secondary electron (thermionics). An electron detached from a surface during secondary emission by an incident electron. *See also:* **electron emission.** 50I07-15E6

secondary-electron conduction (SEC). The transport of charge under the influence of an externally applied field in low-density structured materials by free secondary electrons traveling in the interparticle spaces (as opposed to solid-state conduction). *See also:* **camera tube.** 0-15E6

secondary-electron conduction (SEC) camera tube. A camera tube in which an electron image is generated by a photocathode and focused on a target composed of (1) a backplate and (2) a secondary-electron-conduction layer that provides charge amplification and storage. *See also:* **camera tube.** 0-15E6

secondary emission. Electron emission from solids or liquids due directly to bombardment of their surfaces by electrons or ions. *See also:* **electron emission.** 42A70/E160-15E6

secondary-emission characteristic (surface) (thermionics). The relation, generally shown by a graph, between the secondary-emission rate of a surface and the voltage between the source of the primary emission and the surface. *See also:* **electron emission.** 50I07-15E6

secondary-emission ratio (electrons). The average number of electrons emitted from a surface per incident primary electron. *Note:* The result of a sufficiently large number of events should be averaged to ensure that statistical fluctuations are negligible. E160-15E6

secondary failure (reliability). *See:* **failure, secondary.**

secondary fault. An insulation breakdown occurring as a result of a primary fault. *See also:* **center of distribution.** 42A35-31E13

secondary fuse (transformers). A fuse used on the secondary-side circuits. *Note:* In high-voltage fuse parlance such a fuse is restricted for use on a low-voltage secondary distribution system that connects the secondaries of distribution transformers to consumers' services. 37A100-31E11

secondary neutral grid. A network of neutral conductors, usually grounded, formed by connecting together within a given area all the neutral conductors of individual transformer secondaries of the supply system. *See also:* **center of distribution.** 42A35-31E13

secondary power. The excess above firm power to be furnished when, as, and if available. *See also:* **generating station.** 42A35-31E13

secondary radar. A radar in which the reply is a signal from a transponder triggered by the incident radar signal. *See also:* **navigation.** 0-10E6

secondary radiator. A portion of an antenna system that is not connected to the transmitter by a feeder and that, when transmitting, is excited only by fields of other radiators. *See also:* **antenna.** 50I62-3E1

secondary service area (radio broadcast station). The area within which satisfactory reception can be obtained only under favorable conditions. *See also:* **radio transmitter.** 42A65-0

secondary, single-phase induction motor. The rotor or stator member that does not have windings that are connected to the supply line. *See also:* **asynchronous machine; induction motor.** 0-31E8

secondary standard (illuminating engineering). A constant and reproducible light source calibrated directly or indirectly by comparison with a primary standard. This order of standard also is designated as a reference standard. *Note:* National secondary (reference) standards are maintained at national physical laboratories; laboratory secondary (reference) standards are maintained at other photometric laboratories. *See also:* **primary standard (illuminating engineering).** Z7A1-0

secondary unit substation. See note under **unit substation.**

secondary voltage (1) (capacitance potential device). The root-mean-square voltage obtained from the main secondary winding, and when provided, from the auxiliary secondary winding. *See:* **rated secondary voltage.** *See also:* **outdoor coupling capacitor.** E31-0
(2) (wound-rotor motor). The open-circuit voltage at standstill, measured across the collector rings with rated voltage applied to the primary winding. *See:* **asynchronous machine; synchronous machine.** 42A10-0

secondary voltage rating (transformer). The load-circuit voltage for which the secondary winding is designed. *See also:* **duty; transformer, specialty; transformer.** 42A15/89A1-0

secondary winding (1) (general). The winding on the energy output side. 42A15-31E12
(2) (voltage regulator). The series winding. *See also:* **voltage regulator.** 57A15-0

secondary winding (rotating machinery). Any winding that is not a primary winding. *See:* **asynchronous machine; synchronous machine; voltage regulator.** 0-31E8

second-channel attenuation. *See:* **selectance.**

second-channel interference. Interference in which the extraneous power originates from a signal of assigned (authorized) type in a channel two channels removed from the desired channel. *See:* **interference.** *See also:* **radio receiver.** E188-0

second-time-around echo (radar). An echo received after an interval exceeding the pulse repetition interval. *See also:* **navigation.** 0-10E6

second Townsend discharge (gas). A semi-self-maintained discharge in which the additional ionization is due to the secondary electrons emitted by the cathode under the action of the bombardment by the positive ions present in the gas. *See also:* **discharge (gas).** 50I07-15E6

second voltage range (railway signal). *See:* **voltage range.**

section (rectifier unit). A part of a rectifier unit with its auxiliaries that may be operated independently. *See also:* **rectification.** 42A15-34E24

sectional center (telephony). A toll switching point to which may be connected a number of primary centers, toll centers, or toll points. Sectional centers are classified as Class 2 offices. *See also:* **telephone switching system.** 0-19E1

section locking. Locking effective while a train occupies a given section of a route and adapted to prevent manipulation of levers that would endanger the train while it is within that section. *See also:* **interlocking (interlocking plant).** 42A42-0

sectoral horn. A horn two opposite sides of which are parallel and the two remaining sides of which diverge. *See also:* **antenna.** 42A65-3E1

sector cable. A multiple-conductor cable in which the cross section of each conductor is substantially a sector of a circle, an ellipse, or a figure intermediate between them. *Note:* Sector cables are used in order to obtain decreased overall diameter and thus permit the use of larger conductors in a cable of given diameter. *See also:* **power distribution; underground construction.** E30/42A35-31E13

sector display (1) (radar). A limited display in which only a sector of the total service area of the radar system is shown; usually the sector to be displayed is selectable.
(2) (continuously rotating radar-antenna system). A range-amplitude display used with a radar set, the antenna system of which is continuously rotating. The screen, which is of the long-persistence type, is excited only while the beam of the antenna is within a narrow sector centered on the object. *See also:* **radar.** 42A65-10E6

sector impedance relay. A form of distance relay that by application and design has its operating characteristic limited to a sector of its operating circle on the *R-X* diagram. 37A100-31E11/31E6

sector scanning. A modification of circular scanning in which only a portion of the plane or flat cone is generated. *See also:* **antenna; circular scanning.** 42A65-3E1

security (relay or relay system). A facet of reliability pertaining to the ability of the relay or relay system not to operate due to false information supplied to the relay or relay system. *Note:* The distinction between the two aspects of reliability, dependability and security, is usually made when a communication channel is involved in the relay system and noise or extraneous signals are a potential hazard to the correct performance of the system. This distinction may be extended to other systems. 37A100-31E11/31E6

sedimentation potential (electrobiology). The electrokinetic potential gradient resulting from unity velocity of a colloidal or suspended material forced to move by gravitational or centrifugal forces through a liquid electrolyte. *See also:* **electrobiology.** 42A80-18E1

sediment separator (rotating machinery). Any device, used to collect foreign material in the lubricating oil. *See:* **oil cup (rotating machinery).** 0-31E8

Seebeck coefficient (of a couple) (for homogeneous conductors). The limit of the quotient of (1) the Seebeck electromotive force by (2) the temperature difference between the junctions as the temperature difference approaches zero; by convention, the Seebeck coefficient of a couple is positive if the first-named conductor has a positive potential with respect to the second conductor at the cold junction. *Note:* The Seebeck coefficient of a couple is the algebraic difference of either the relative or absolute Seebeck coefficients of the two conductors. *See also:* **thermoelectric device.** E221-15E7

Seebeck coefficient, absolute. The integral, from absolute zero to the given temperature, of the quotient of (1) the Thomson coefficient of the material by (2) the absolute temperature. *See also:* **thermoelectric device.** E221-15E7

Seebeck coefficient, relative. The Seebeck coefficient of a couple composed of the given material as the first-named conductor and a specified standard conductor. *Note:* Common standards are platinum, lead, and copper. *See also:* **thermoelectric device.** E221-15E7

Seebeck effect. The generation of an electromotive force by a temperature difference between the junctions in a circuit composed of two homogeneous electric conductors of dissimilar compostion; or, in a nonhomogeneous conductor, the electromotive force produced by a temperature gradient in a nonhomogeneous region. *See:* **thermoelectric effect.** *See also:* **thermoelectric device.** E221-15E7

Seebeck electromotive force. The electromotive force resulting from the Seebeck effect. *See also:* **thermoelectric device.** E221-15E7

segmental conductor. A stranded conductor consisting of three or more stranded conducting elements, each element having approximately the shape of the sector of a circle, assembled to give a substantially circular cross section. The sectors are usually lightly insulated from each other and, in service, are connected in parallel. *Note:* This type of conductor is known as type-*M* conductor in Canada. *See also:* **conductor.** 42A35-31E13

segmental-rim rotor (rotating machinery). A rotor in which the rim is composed of interleaved segmental plates bolted together. *See also:* **rotor (rotating machinery).** 0-31E8

segment shoe (bearing shoe) (rotating machinery). A pad that is part of the bearing surface of a pad-type bearing. *See also:* **bearing.** 0-31E8

segregated-phase bus. A bus in which all phase conductors are in a common metal enclosure, but are segregated by metal barriers between phases. 37A100-31E11

selectance (1) (general). A measure of the falling off in the response of a resonant device with departure from resonance. It is expressed as the ratio of the amplitude of response at the resonance frequency, to the response at some frequency differing from it by a specified amount.

(2) (radio receivers). The reciprocal of the ratio of the sensitivity of a receiver tuned to a specified channel to its sensitivity at another channel separated by a specified number of channels from the one to which the receiver is tuned. *Notes:* (1) Unless otherwise specified, selectance should be expressed as a voltage or field-strength ratio. (2) Selectance is often expressed as adjacent-channel attenuation (ACA) or second-channel attenuation (2 ACA). *See also:* **cutoff frequency.** E188/42A65-0

select before operate, supervisory control. *See:* **supervisory control system, select before operate.**

selection (computing systems). *See:* **amplitude selection; coincident-current selection.** *See also:* **static magnetic storage.**

selection check (electronic computation). A check (usually an automatic check) to verify that the correct register, or other device, is selected in the interpretation of an instruction. *See also:* **electronic digital computer.** E162/X3A12-16E9

selection ratio. The least ratio or a magnetomotive force used to select a cell to the maximum magnetomotive force used that is not intended to select a cell. *See:* **coincident-current selection.** *See also:* **static magnetic storage.** E163-0

selective collective automatic operation (elevators). Automatic operation by means of one button in the car for each landing level served and by UP and DOWN buttons at the landings, wherein all stops registered by the momentary actuation of the car buttons are made as defined under **nonselective collective automatic operation,** but wherein the stops registered by the momentary actuation of the landing buttons are made in the order in which the landings are reached in each direction of travel after the buttons have been actuated. With this type of operation, all UP landing calls are answered when the car is traveling in the up direction and all DOWN landing calls are answered when the car is traveling in the down direction, except in the case of the uppermost or lowermost calls, which are answered as soon as they are reached irrespective of the direction of travel of the car. *See also:* **control (elevators).** 42A45-0

selective dump (computing systems). A dump of a selected area of storage. *See also:* **electronic digital computer.** X3A12-16E9

selective fading (radio wave propagation). Fading that is different at different frequencies in a frequency band occupied by a modulated wave. *See also:* **radiation; radio wave propagation.** E211/42A65-3E2

selective opening (selective tripping) (devices carrying fault current). The application of switching devices in series such that only the device nearest the fault will open and the devices closer to the source will remain closed and carry the remaining load. 37A100-31E11

selective overcurrent trip. *See:* **selective release (selective trip); overcurrent release (overcurrent trip).**

selective overcurrent tripping. *See:* **overcurrent release (overcurrent trip); selective opening (selective tripping).**

selective release (selective trip). A delayed release with selective settings that will automatically reset if the actuating quantity falls and remains below the release setting for a specified time. 37A100-31E11

selective ringing. Party-line ringing that rings only the bell of the desired station. *See also:* **telephone switching system.** 42A65-19E1

selective signaling equipment (mobile communication). Arrangements for signaling, selective from a base station, of any one of a plurality of mobile stations associated with the base station for communication purposes. *See also:* **mobile communication system.** 0-6E1

selectivity (1) (receiver performance). A measure of the extent to which a receiver is capable of differentiating between the desired signal and disturbances at other frequencies. *See also:* **radio receiver.** E188-0

(2) (signal-transmission system). The characteristic of a filter that determines the extent to which the filter is capable of differentiating between desired signals and undesired interference. *See also:* **receiver performance; signal.** 0-6E1/13E6

(3) (protective system). A general term describing the interrelated performance of relays and breakers, and other protective devices; complete selectivity being obtained when a minimum amount of equipment is removed from service for isolation of a fault or other abnormality. 37A100-31E11/31E6

selector, amplitude (pulse techniques). *See:* **selector, pulse-height.**

selector pulse (electronic navigation). A pulse that is used to identify, for selection, one event in a series of events. *See also:* **navigation.** E172-10E6

selector, pulse-height (pulse techniques). *See:* **pulse-height selector.**

selector switch (1) (general). A switch arranged to permit connecting a conductor to any one of a number of other conductors. 37A100-31E11

(2) (automatic telephone switching). A remotely-controlled switch for selecting a group of trunk lines fixed by part of the call number and connecting to an idle trunk in that group. *See also:* **telephone switching system; switch.** 42A65-0

(3) (industrial control). A manually operated multiposition switch for selecting alternative control circuits. 42A25-34E10

(4) (control systems). A machine-operated device that establishes definite zones in the operating cycle within which corrective functions can be initiated. 42A25-0

self-adapting. Pertaining to the ability of a system to change its performance characteristics in response to its environment. X3A12-16E9

self-aligning bearing (rotating machinery). A sleeve bearing designed so that it can move in the end shield to align itself with the journal of the shaft. *See also:* **bearing.** 0-31E8

self-capacitance (conductor) (grounded capacitance) (total capacitance). In a multiple-conductor system, the capacitance between this conductor and the other $(n-1)$ conductors connected together. *Note:* The self-capacitance of a conductor equals the sum of its $(n-1)$ direct capacitances to the other $(n-1)$ conductors. E270-0

self-checking code (electronic computation). A code that uses expressions such that one (or more) error(s) in a code expression produces a forbidden combination. Also called an error-detecting code. *See also:* **check, forbidden combination; electronic digital computer; error-detecting code; parity.** E162-0

self-closing door or gate (elevators). A manually opened hoistway door and/or a car door or gate that closes when released. *See also:* **hoistway (elevator or dumbwaiter).** 42A45-0

self-commutated inverters. An inverter in which the commutation elements are included within the power inverter.

See:

circuit switching element;
commutating period;
continuous-load rating of a power inverter unit;
externally commutated inverters;
input transient energy;
inversion efficiency;
off period;
output frequency stability;
periodic frequency modulation;
period output voltage modulation;
rated direct voltage;
rated output capacity;
rated output current of a power inverter;
rated output frequency;
rated output voltage of a power inverter;
rating of a power inverter;
rectifier;
resolution of output adjustment;
short time rating of a power inverter;
supply impedance;
supply transient voltage;
tracking error. 0-34E24

self-contained instrument. An instrument that has all the necessary equipment built into the case or made a corporate part thereof. *See also:* **instrument.** 39A1/42A30-0

self-contained navigation aid. An aid that consists only of facilities carried by the vehicle. *See also:* **navigation.** 0-10E6

self-contained pressure cable. A pressure cable in which the container for the pressure medium is an impervious flexible metal sheath, reinforced if necessary, that is factory assembled with the cable core. *See:* **gas-filled cable; oil-filled cable; pressure cable.** *See also:* **power distribution, underground construction.** 42A35-31E13

self-coupling separable contacts (switchgear assembly disconnecting device). Contacts, mounted on the stationary and removable elements of a switchgear assembly, that align and engage or disengage automatically when the two elements are brought into engagement or disengagement. 0-31E11

self-excited. A qualifying term applied to a machine to denote that the excitation is supplied by the machine itself. *See also:* **direct-current commutating machine; synchronous machine.** 42A10-31E8

self-impedance (1) (mesh). The impedance of a passive mesh or loop with all other meshes of the network open-circuited. *See also:* **transmission characteristics.** E270-0

(2) (radiating element). The input impedance of a radiating element of an array antenna with all other elements in the array open-circuited. *Note:* In general, the self-impedance of a radiating element in an array is not the same as the input impedance of the same element with the other elements absent. *See also:* **antenna.** 0-3E1

(3) (network). At any pair of terminals, the ratio of an applied potential difference to the resultant current at these terminals, all other terminals being open.

See:

blocked impedance;
characteristic impedance;
conjugate impedance;
driving-point impedance;

free impedance;
free motional impedance.
image impedance;
impedance matching;
input impedance;
iterative impedance;
loaded impedance;
load impedance;
motional impedance;
mutual impedance;
open-circuit impedance;
output impedance;
sending-end impedance;
source impedance;
transfer impedance;
wave impedance.
See also: **transmission characteristics.** 42A65-0

self-inductance. The property of an electric circuit whereby an electromotive force is induced in that circuit by a change of current in the circuit. *Notes:* (1) The coefficient of self-inductance L of a winding is given by the following expression:

$$L = \frac{\partial \lambda}{\partial i}$$

where λ is the total flux-linkage of the winding and i is the current in the winding. (2) The voltage e induced in the winding is given by the following equation:

$$e = -\left[L\frac{di}{dt} + i\frac{dL}{dt}\right]$$

If L is constant

$$e = -L\frac{di}{dt}.$$

(3) The definition of self-inductance L is restricted to relatively slow changes in i, that is, to low frequencies, but by analogy with the definitions, equivalent inductances may often be evolved in high-frequency applications such as resonators, waveguide equivalent circuits, etcetera. Such inductances, when used, must be specified. The definition of self-inductance L is also restricted to cases in which the branches are small in physical size compared with a wavelength, whatever the frequency. Thus in the case of a uniform 2-wire transmission line it may be necessary even at low frequencies to consider the parameters as distributed rather than to have one inductance for the entire line. E270-2E1

self-information. *See:* **information content.**

self-lubricating bearing (rotating machinery). A bearing lined with a material containing its own lubricant such that little or no additional lubricating fluid need be added subsequently to ensure satisfactory lubrication of the bearing. *See also:* **bearing.** 0-31E8

self-maintained discharge (gas). A discharge characterized by the fact that it maintains itself after the external ionizing agent is removed. *See also:* **discharge (gas).** 50I07-15E6

self-organizing. Pertaining to the ability of a system to arrange its internal structure. X3A12-16E9

self-propelled electric car. An electric car requiring no external source of electric power for its operation. *Note:* Diesel-electric, gas-electric, and storage-battery-electric cars are examples of self-propelled cars. The prefix self-propelled is also applied to buses. *See also:* **electric motor car.** 42A42-0

self-propelled electric locomotive. An electric locomotive requiring no external source of electric power for its operation. *Note:* Storage-battery, diesel-electric, gas-electric and turbine-electric locomotives are examples of self-propelled electric locomotives. *See also:* **electric locomotive.** 42A42-0

self-pulse modulation. Modulation effected by means of an internally generated pulse. *See:* **blocking oscillator.** *See also:* **oscillatory circuit.** E145-0

self-quenched counter tube. A radiation counter tube in which reignition of the discharge is inhibited by internal processes. *See also:* **gas-filled radiation-counter tubes.** 42A70-0

self-rectifying X-ray tube. An X-ray tube operating on alternating anode potential. *See also:* **electron devices, miscellaneous.** 42A70-15E6

self-reset manual release (control) (industrial control). A manual release that is operative only while it is held manually in the release position. *See:* **electric controller.** IC1-34E10

self-reset relay (automatically reset relay). A relay that is so constructed that it returns to its reset position following an operation after the input quantity is removed. 37A100-31E11/31E6

self-saturation (magnetic amplifier). The saturation obtained by rectifying the output current of a saturable reactor. 42A65-0

self-supporting aerial cable. A cable consisting of one or more insulated conductors factory assembled with a messenger that supports the assemblage, and that may or may not form a part of the electric circuit. *See also:* **conductor.** 42A35-31E13

self-surge impedance. *See:* **surge impedance.**

self-ventilated (rotating machinery). Applied to a machine which has its ventilating air circulated by means integral with the machine. *See also:* **asynchronous machine; direct-current commutating machine; synchronous machine.** E45/E67-31E8

semantics. The relationships between symbols and their meanings. X3A12-16E9

semaphore signal. A signal in which the day indications are given by the vertical angular position of a blade known as the semaphore arm. *See also:* **railway signal and interlocking.** 42A42-0

semianalytic inertial navigation equipment. The same as geometric inertial navigation equipment except that the horizontal measuring axes are not maintained in alignment with a geographic direction. *Note:* The azimuthal orientations are automatically computed. *See also:* **navigation.** E174-10E6

semiautomatic. Combining manual and automatic features so that a manual operation is required to supply to the automatic feature the actuating influence that causes the automatic feature to function. 42A95-0

semiautomatic controller. An electric controller in which the influence directing the performance of some of its basic functions is automatic. *See also:* **electric controller.** 42A25-34E10

semiautomatic flight inspection (SAFI) (electronic navigation). A specialized and largely automatic system for evaluating the quality of information in signals from ground-based navigational aids; data from navigational aids along and adjacent to any selected air route are simultaneously received by a specially equipped semiautomatic-flight-inspection aircraft as it proceeds under automatic control along the route, evaluated at once for gross errors and recorded for subsequent processing and detailed analysis at a computer-equipped central ground facility. *Note:* Flight inspection means the evaluation of performance of navigational aids by means of in-flight measurements. *See also:* **navigation.** 0-10E6

semiautomatic gate (elevators). A gate that is opened manually and that closes automatically as the car leaves the landing. *See also:* **hoistway (elevator or dumbwaiter).** 42A45-0

semiautomatic holdup-alarm system. An alarm system in which the signal transmission is initiated by the indirect and secret action of the person attacked or of an observer of the attack. *See also:* **protective signaling.** 42A65-0

semiautomatic plating. Mechanical plating in which the cathodes are conveyed automatically through only one plating tank. *See also:* **electroplating.** 42A60-0

semiautomatic signal. A signal that automatically assumes a stop position in accordance with traffic conditions, and that can be cleared only by cooperation between automatic and manual controls. *See also:* **railway signal and interlocking.** 42A42-0

semiautomatic telephone system. A telephone system in which operators receive orders orally from the calling parties and establish connections by means of automatic apparatus. *See also:* **telephone switching system.** 42A65-0

semiconducting jacket. A jacket of such resistance that its outer surface can be maintained at substantially ground potential by contact at frequent intervals with a grounded metallic conductor, or when buried directly in the earth. *See also:* **power distribution, underground construction.** 42A35-31E13/31E1

semiconducting material. A conducting medium in which the conduction is by electrons, and holes, and whose temperature coefficient of resistivity is negative over some temperature range below the melting point. *See:* **semiconductor.** *See also:* **semiconductor device.** E270-0

semiconducting paint (rotating machinery). A paint in which the pigment or portion of pigment is a conductor of electricity and the composition is such that when converted into a solid film, the electrical conductivity of the film is in the range between metallic substances and electrical insulators. *See also:* **cradle base (rotating machinery).** 0-31E8

semiconducting tape (power distribution, underground cables). A tape of such resistance that when applied between two elements of a cable the adjacent surfaces of the two elements will maintain substantially the same potential. Such tapes are commonly used for conductor shielding and in conjunction with metallic shielding over the insulation. *See also:* **power distribution, underground construction.** 0-31E1

semiconductive ignition cable (electromagnetic compatibility). High-tension ignition cable, the core of which is made of semiconductive material. *Note:* Semiconductive is understood here as referring to conductivity and no other physical properties. *See also:* **electromagnetic compatibility.** CISPR-27E1

semiconductor. An electronic conductor, with resistivity in the range between metals and insulators, in which the electric-charge-carrier concentration increases with increasing temperature over some temperature range. *Note:* Certain semiconductors possess two types of carriers, namely, negative electrons and positive holes. *Note:* For an extensive list of cross references, see *Appendix A.* E59/E102/E216/E270-10E1/34E17

semiconductor, compensated. A semiconductor in which one type of impurity or imperfection (for example, donor) partially cancels the electrical effects of the other type of impurity or imperfection (for example, acceptor). *See also:* **semiconductor.** E102/E216/E270-34E17

semiconductor controlled rectifier (SCR). An alternative name used for the reverse-blocking triode-thyristor. *Note:* The name of the actual semiconductor material (selenium, silicon, etcetera) may be substituted in place of the word **semiconductor** in the name of the components. *See also:* **thyristor.** E102/E216/E270-34E17

semiconductor device. An electron device in which the characteristic distinguishing electronic conduction takes place within a semiconductor. *See also:* **semiconductor.** 47A70/E59/E102/E270-34E17

semiconductor device, multiple unit. A semiconductor device having two or more sets of electrodes associated with independent carrier streams. *Note:* It is implied that the device has two or more output functions that are independently derived from separate inputs, for example, a duo-triode transistor. *See also:* **semiconductor.** E102/42A70-0

semiconductor device, single unit. A semiconductor device having one set of electrodes associated with a single carrier stream. *Note:* It is implied that the device has a single output function related to a single input. *See also:* **semiconductor.** E102-42A70

semiconductor diode. A semiconductor device having two terminals and exhibiting a nonlinear voltage-current characteristic. *See also:* **semiconductor; semiconductor rectifier cell.** E59-34E17

semiconductor-diode parametric amplifier. A parametric amplifer using one or more varactors. *See also:* **parametric device.** E254-15E7

semiconductor, extrinsic. A semiconductor with charge-carrier concentration dependent upon impurities. *See also:* **semiconductor.** E102/E216/E270-34E17

semiconductor frequency changer. A complete equipment employing semiconductor devices for changing from one alternating-current frequency to another. *See also:* **semiconductor rectifier stack.** 0-34E24

semiconductor, intrinsic. A semiconductor whose charge-carrier concentration is substantially the same as that of the ideal crystal. *See also:* **semiconductor.** E102/E216/E270-34E17

semiconductor, *n*-type. An extrinsic semiconductor in which the conduction electron concentration exceeds the mobile hole concentration. *Note:* It is implied that the net ionized impurity concentration is donor type. *See also:* **semiconductor.** E102/E216/E270-34E17

semiconductor, n^+-type. An *n*-type semiconductor in which the excess conduction electron concentration is very large. *See also:* **semiconductor.** E216/E270-34E17

semiconductor, *p*-type. An extrinsic semiconductor in which the mobile hole concentration exceeds the conduction electron concentration. *Note:* It is implied that the net ionized impurity concentration is acceptor type. *See also:* **semiconductor.** E102/E216/E270-34E17

semiconductor, p^+-type. A *p*-type semiconductor in which the excess mobile hole concentration is very large. *See also:* **semiconductor.** E216/E270-34E17

semiconductor power converter. A complete equipment employing semiconductor devices for the transformation of electric power. *See also:* **semiconductor rectifier stack.** 0-34E24

semiconductor rectifier. An integral assembly of semiconductor rectifier diodes or stacks including all necessary auxiliaries such as cooling equipment, current balancing, voltage divider, surge suppression equipment, etcetera, and housing, if any. *See also:* **semiconductor rectifier stack.** 0-34E24

semiconductor rectifier cell. A semiconductor device consisting of one cathode, one anode, and one rectifier junction. *See also:* **semiconductor; semiconductor rectifier stack.** E59-34E17

semiconductor rectifier cell combination. The arrangement of semiconductor rectifier cells in one rectifier circuit, rectifier diode, or rectifier stack. The semiconductor rectifier cell combination is described by a sequence of four symbols written in the order 1-2-3-4 with the following significances: (1) Number of rectifier circuit elements. (2) Number of semiconductor rectifier cells in series in each rectifier circuit element. (3) Number of semiconductor rectifier cells in parallel in each rectifier circuit element. (4) Symbol designating circuit. If a semiconductor rectifier stack consists of sections of semiconductor rectifier cells insulated from each other, the total semiconductor rectifier cell combination becomes the sum of the semiconductor rectifier cell combinations of the individual insulated sections. If the insulated sections have the same semiconductor rectifier cell combination, the total semiconductor rectifier cell combination may be indicated by the semiconductor rectifier cell combination of one section preceded by a figure showing the number of insulated sections. Example: 4(4-1-1-B) indicates 4 single-phase full-wave bridges insulated from each other assembled as one semiconductor rectifier stack. *Notes:* (1) The total number of semiconductor rectifier cells in each semiconductor rectifier cell combination is the product of the numbers in the combination. (2) This arrangement can also be applied by analogy to give a semiconductor rectifier diode combination.

Symbol	Circuit	Example
H	half wave	1-1-1-H
C	center tap	2-1-1-C
B	bridge	4-1-1-B
		6-1-1-B
Y	wye	3-1-1-Y
S	star	6-1-1-S
D	voltage doubler	2-1-1-D

See also: **semiconductor rectifier cell.** E59-34E17

semiconductor rectifier diode. A semiconductor diode having an asymmetrical voltage-current characteristic, used for the purpose of rectification, and including its associated housing, mounting, and cooling attachments if integral with it. *See also:* **semiconductor; semiconductor rectifier cell.** E59-34E17/34E24

semiconductor rectifier stack. An integral assembly, with terminal connections, of one or more semiconductor rectifier diodes, and includes its associated mounting and cooling attachments if integral with it. *Note:* It is a subassembly of, but not a complete semiconductor rectifier.
See:
alternating-current root-mean-square voltage rating;
angle of advance;
angle of retard;
commutation elements;
crest working voltage;
current-balancing reactors;
differential resistance;
diode fuses;
direct liquid-cooling system;
direct liquid-cooling system with recirculation;
forward direction;
forward power loss;
forward voltage drop;
heat sink;
nonrepetitive peak reverse voltage;
power rectifier;
rate-of-rise suppressors;
rectifier circuit element;
rectifier junction;
repetitive peak reverse voltage;
reverse direction;
reverse power loss;
semiconductor frequency changer;
semiconductor power converter;
semiconductor rectifier;
semiconductor rectifier cell;
temperature derating;
thermal resistance, effective;
threshold voltage;
transient thermal impedance;
voltage regulation;
voltage surge suppressors. E59-34E17/34E24

semidirect lighting. Lighting that involves luminaires that distribute 60 to 90 percent of the emitted light downward and the balance upward. *See also:* **general lighting.** Z7A1-0

semienclosed. (1) Having the ventilating openings in the case protected with wire screen, expanded metal, or perforated covers or (2) having a solid enclosure except for a slot for an operating handle or small openings for ventilation, or both. 42A95-0

semienclosed brake (industrial control). A brake that is provided with an enclosure that covers the brake shoes and the brake wheel but not the brake actuator. *See also:* **control.** IC1-34E10

semiflush-mounted device. One in which the body of the device projects in front of the mounting surface a specified distance between the distances specified for flush-mounted and surface-mounted devices. 37A100-31E11

semiguarded machine (rotating machinery). One in which part of the ventilating openings, usually in the top half, are guarded as in the case of a guarded machine but the others are left open. *See also:* **synchronous machine; direct-current commutating machine; synchronous machine.** 42A10-31E8

semi-indirect lighting. Lighting involving luminaires that distribute 60 to 90 percent of the emitted light upward and the balance downward. *See also:* **general lighting.** Z7A1-0

semimagnetic controller. An electric controller having only part of its basic functions performed by devices that are operated by electromagnets. *See also:* **electric controller.** 42A25-34E10

semioutdoor reactor. A reactor suitable for outdoor use provided that certain precautions in installation (specified by the manufacturer) are observed. For example, protection against rain. 57A16-0

semiprotected enclosure. An enclosure in which all other openings, usually in the top half, are protected as in the case of a protected enclosure, but the others are left open. *See:* **protected enclosure.** *See also:* **asynchronous machine; direct-current commutating machine; synchronous machine.** E45-0

semiremote control. A system or method of radio-transmitter control whereby the control functions are performed near the transmitter by means of devices connected to but not an integral part of the transmitter. *See also:* **radio transmitter.** E145-0

semiselective ringing. Party-line ringing wherein the bells of two stations are rung simultaneously, differentiation being by the number of rings, either one or two. *See also:* **telephone switching system.** 42A65-19E1

semistrain insulator (semitension assembly). Two insulator strings at right angles, each making an angle of about 45 degrees with the line conductor. *Note:* These assemblies are used at intermediate points where it may be desirable to partially anchor the conductor to prevent too great movement in case of a broken wire. *See also:* **tower.** 42A35-31E13

semit (half-step). *See:* **semitone.**

semitone (semit) (half-step). The interval between two sounds having a basic frequency ratio approximately the twelfth root of two. *Note:* In equally tempered semitones, the interval between any two frequencies is 12 times the logarithm to the base 2 (or 39.86 times the logarithm to the base 10) of the frequency ratio. *See also:* **electroacoustics.** 0-1E1

semitransparent photocathode (camera tube or phototube). A photocathode in which radiant flux incident on one side produces photoelectric emission from the opposite side. *See also:* **electrode (electron tube); phototubes.** E160/E175/42A70-15E6

sender (telephony). A part of an automatic switching system that generates signals in response to information received from another part of the system. *See also:* **telephone switching system.** 0-19E1

sending-end crossfire. The crossfire in a telegraph channel from one or more adjacent telegraph channels transmitting from the end at which the crossfire is measured. *See also:* **telegraphy.** 42A65-0

sending-end impedance (line). The ratio of an applied potential difference to the resultant current at the point where the potential difference is applied. The sending-end impedance of a line is synonymous with the driving-point impedance of the line. *Note:* For an infinite uniform line the sending-end impedance and the characteristic impedance are the same; and for an infinite periodic line the sending-end impedance and the iterative impedance are the same. *See also:* **self-impedance; waveguide.** 42A65-0

send-only equipment. Data communication channel equipment capable of transmitting signals, but not arranged to receive signals. *See also:* **data transmission.** 0-19E4

sensation level (sound) (acoustics). *See:* **level above threshold.**

sense (navigation). The pointing direction of a vector representing some navigation parameter. *See also:* **navigation.** 0-10E6

sense finder. That portion of a direction-finder that permits determination of direction without 180-degree ambiguity. *See also:* **radio receiver.** 42A65-0

sensing (electronic navigation). The process of finding the sense, as, for example, in direction finding, the resolution of the 180-degree ambiguity in bearing indication; and, as in phase- or amplitude-comparison systems like instrument landing systems and omnidirectional radio range, the establishment of a relation between course displacement signal and the proper response in the control of the vehicle. *See also:* **navigation.** 0-10E6

sensing element (initial element). *See:* **primary detector.**

sensitive volume (radiation-counter tubes). That portion of the tube responding to specific radiation. *See also:* **gas-filled radiation-counter tubes.** 42A70-15E6

sensitive relay. A relay that operates on comparatively low input power, commonly defined as 100 milliwatts or less. *See also:* **relay.** 83A16-21E0

sensitivity (1) (general comment). Definitions of sensitivity fall into two contrasting categories. In some fields, sensitivity is the ratio of response to cause. Hence increasing sensitivity is denoted by a progressively larger number. In other fields, sensitivity is the ratio of cause to response. Hence increasing sensitivity is denoted by a progressively smaller number.

(2) (measuring device). The ratio of the magnitude of its response to the magnitude of the quantity measured. *Notes:* (A) It may be expressed directly in divisions per volt, millimeters per volt, milliradians per microampere, etcetera, or indirectly by stating a property from which sensitivity can be computed (for example, ohm per volt for a stated deflection. (B) In the case of mirror galvanometers it is customary to express sensitivity on the basis of a scale distance of 1 meter. *See also:* **accuracy rating (instrument).** E175/42A30-0

(3) (radio receiver or similar device). Taken as the minimum input signal required to produce a specified output signal having a specified signal-to-noise ratio. *Note:* This signal input may be expressed as power or as voltage, with input network impedance stipulated. 42A65-31E3

(4) (transmission lines, waveguides, and nuclear techniques). The least signal input capable of causing an output signal having desired characteristics. *See:* **ionizing radiation; transmission characteristics.** 0-9E4

(5) (bolometric detection). The sensitivity is given by the change of resistance in ohms per milliwatt of energy dissipated in the bolometer. E270-9E4

(6) (close-talking pressure-type microphone). The voltage or power response of a microphone measured at a stated frequency. *See also:* **close-talking pressure-type microphone.** E258-0

(7) (camera tube or phototube). The quotient of output current by incident luminous flux at constant electrode voltages. *Notes:* (1) The term output current as here used does not include the dark current. (2) Since luminous sensitivity is not an absolute characteristic but depends on the special distribution of the incident flux, the term is commonly used to designate the sensitivity to light from a tungsten-filament lamp operating at a color temperature of 2870 kelvins. *See:* **sensitivity, cathode luminous.** *See also:* **phototubes.** 42A70-0

sensitivity, cathode luminous (photocathodes). The quotient of photoelectric emission current from the photocathode by the incident luminous flux under specified conditions of illumination. *Notes:* (1) Since cathode luminous sensitivity is not an absolute characteristic but depends on the spectral distribution of the incident flux, the term is commonly used to designate the sensitivity to radiation from a tungsten filament lamp operating at a color temperature of 2870 kelvins. (2) Cathode luminous sensitivity is usually measured with a collimated beam at normal incidence. *See also:* **phototube.** E160/E175-15E6

sensitivity, cathode radiant (photocathodes). The quotient of the photoelectric emission current from the photocathode by the incident radiant flux at a given wavelength under specified conditions of irradiation. *Note:* Cathode radiant sensitivity is usually measured with a collimated beam at normal incidence. E160/E175-15E6

sensitivity coefficient (control system). The partial derivative of a system signal with respect to a system parameter. *See also:* **control system.** 0-23E0

sensitivity, dynamic (phototubes). The quotient of the modulated component of the output current by the modulated component of the incident radiation at a stated frequency of modulation. *Note:* Unless otherwise stated the modulation wave shape is sinusoidal. *See also:* **phototubes.** E158-15E6

sensitivity, incremental (instrument) (nuclear techniques). A measure of the smallest change in stimulus that produces a statistically significant change in response. Quantitatively it is usually expressed as the change in the stimulus that produces a change in response equal to the standard deviation of the response. *See also:* **ionizing radiation.** E175-0

sensitivity level (response level) (sensitivity) (response) (in electroacoustics) (of a transducer) (in decibels). 20 times the logarithm to the base 10 of the ratio of the amplitude sensitivity S_A to the reference sensitivity S_0, where the amplitude is a quantity proportional to the square root of power. The kind of sensitivity and the reference sensitivity must be indicated. *Note:* For a microphone, the free-field voltage/pressure sensitivity is the kind often used and a common reference sensitivity is $S_0 = 1$ volt per newton per square meter. The square of the sensitivity is proportional to a power ratio. The free-field voltage sensitivity-squared level, in decibels, is therefore $S_A = 10 \log (S_A^2/S_0^2) = 20 \log (S_A/S_0)$. Often, **sensitivity-squared level** in decibels can be shortened, without ambiguity, to **sensitivity level** in decibels, or simply **sensitivity** in decibels. 0-1E1

sensitivity, radiant (camera tube or phototube). The quotient of signal output current by incident radiant flux at a given wavelength, under specified conditions of irradiation. *Note:* Radiant sensitivity is usually measured with a collimated beam at normal incidence. *See also:* **beam tubes; luminous flux; phototubes; radiant flux.** E160-15E6

sensitivity, threshold (instrument) (nuclear techniques). A measure of the smallest stimulus that produces a significant response. *See also:* **ionizing radiation.** E175-0

sensitivity time control. The portion of a system that varies the amplification of a radio receiver in a predetermined manner as a function of time. *See also:* **navigation.** E172-10E6

sensitizing (electrostatography). The act of establishing an electrostatic surface charge of uniform density on an insulating medium. *See also:* **electrostatography.** E224-15E7

sensitometry. The measurement of the light response characteristics of photographic film under specified conditions of exposure and development. E157-1E1

sentinel (computing systems). *See:* **flag.** *See also:* **electronic digital computer.**

separate parts of a network. The parts that are not connected. *See also:* **network analysis.** E153/E270-0

separate terminal enclosure (rotating machinery). A form of termination in which the ends of the machine winding are connected to the incoming supply leads inside a chamber that need not be fully enclosed and may be formed by the foundations beneath the machine. *See also:* **cradle base (rotating machinery).** 0-31E8

separately excited (rotating machinery). A qualifying term applied to a machine to denote that the excitation is obtained from a source other than the machine itself. *See also:* **direct-current commutating machine; synchronous machine.** 42A10-0;0-31E8

separately ventilated machine (rotating machinery). A machine that has its ventilating air supplied by an independent fan or blower external to the machine. *See also:* **asynchronous machine; direct-current commutating machine; externally ventilated machine; open-pipe ventilated machine; synchronous machine.** E45/E67-31E8

separation criteria (electromagnetic compatibility). Curves that relate the frequency displacement to the minimum distance between a receiver and an undesired transmitter to insure that the signal-to-interference ratio does not fall below a specified value. *See also:* **electromagnetic compatibility.** 0-27E1

separator (1) (storage cell). A spacer employed to prevent metallic contact between plates of opposite polarity within the cell. (Perforated sheets are usually called retainers.) *See also:* **battery (primary or secondary).** 42A60-0
(2) (computing systems). *See:* **delimiter.** *See also:* **electronic digital computer.**

separator, insulation slot (rotating machinery). Insulation member placed in a slot between individual coils, such as between main and auxiliary windings. *See also:* **rotor (rotating machinery); stator.** 0-31E8

sequence. *See:* **calling sequence; collating sequence; pseudorandom number sequence.** *See also:* **electronic digital computer.**

sequence number. A number identifying the relative location of blocks or groups of blocks on a tape. EIA3B-34E12

sequence-number readout. Display of the sequence number punched on the tape. *See:* **block-count readout.** EIA3B-34E12

sequence switch. A remotely controlled power-operated switching device used as a secondary master controller. *See also:* **multiple-unit control.** 42A42-0

sequence table (electric controller). A table indicating the sequence of operation of contactors, switches, or other control apparatus for each step of the periodic duty. *See also:* **multiple-unit control.** E16-0

sequential (formatted system) (telecommunication). If the signal elements are transmitted successively in time over a channel, the transmission is said to be **sequential.** If the signal elements are transmitted at the same time over a multiwire circuit, the transmission is said to be **coincident.** *See also:* **bit.** 0-19E4

sequential control (computing systems). A mode of computer operation in which instructions are executed consecutively unless specified otherwise by a jump. *See also:* **electronic digital computer.** X3A12-16E9

sequential lobing (electronic navigation). A direction-determining technique utilizing the signals of partially overlapped lobes occurring in sequence. *See also:* **antenna.** E149-3E1

sequential logic element. A device having at least one output channel and one or more input channels, all characterized by discrete states, such that the state of each output channel is determined by the previous states of the input channels. *See also:* **electronic digital computer.** X3A12-16E9

sequential operation. Pertaining to the performance of operations one after the other. *See also:* **electronic digital computer.** X3A12-16E9

sequential relay. A relay that controls two or more sets of contacts in a predetermined sequence. *See also:* **relay.** 83A16-0

sequential scanning (television). A rectilinear scanning process in which the distance from center to center of successively scanned lines is equal to the nominal line width. *See also:* **television.** 42A65-0

serial. (1) Pertaining to the time sequencing of two or more processes. (2) Pertaining to the time sequencing of two or more similar or identical processes, using the same facilities for the successive processes. (3) Pertaining to the time-sequential processing of the individual parts of a whole, such as the bits of a character, the characters of a word, etcetera, using the same facilities for successive parts. *See also:* **electronic computation; electronic digital computer; serial-parallel.** E162/X3A12-16E9

serial access (computing systems). Pertaining to the process of obtaining data from, or placing data into, storage when there is a sequential relation governing the access time to successive storage locations. *See also:* **electronic digital computer.** X3A12-16E9

serial by bit. *See:* **serial transmission (data transmission) (telecommunications).**

serial digital computer. A digital computer in which the digits are handled serially. Mixed serial and parallel machines are frequently called serial or parallel according to the way arithmetic processes are performed. An example of a serial digital computer is one that handles decimal digits serially although it might handle the bits that comprise a digit either serially or in parallel. *See also:* **electronic computation; parallel digital computer.** E270-0

serial operation (telecommunication). The flow of information in time sequence, using only one digit, word, line, or channel at a time. *See also:* **bit; electronic digital computer.** 0-19E4

serial-parallel. Pertaining to processing that includes both serial and parallel processing, such as one that handles decimal digits serially but handles the bits that comprise a digit in parallel. *See also:* **electronic digital computer.** E162-0

serial transmission (data transmission) (telecommunication). Used to identify a system wherein the bits of a character occur serially in time. Implies only a single transmission channel. Also called **serial by bit.** *See also:* **bit; data transmission.** 0-19E4

series circuit. A circuit supplying energy to a number of devices connected in series, that is, the same current passes through each device in completing its path to the source of supply. *See also:* **center of distribution.** 42A35-31E13

series connection. The arrangement of cells in a battery made by connecting the positive terminal of each successive cell to the negative terminal of the next adjacent cell so that their voltages are additive. *See also:* **battery (primary or secondary).** 42A60-0

series distribution system. A distribution system for supplying energy to units of equipment connected in series. *See also:* **alternating-current distribution; direct-current distribution.** 42A35-31E13

series elements (network). (1) Two-terminal elements are connected in series when they form a path between two nodes of a network such that only elements of this path, and no other elements, terminate at intermediate nodes along the path. (2) Two-terminal elements are connected in series when any mesh including one must include the others. *See also:* **network analysis.** E153/E270-0

series-fed vertical antenna. A vertical antenna that is insulated from ground and energized at the base. *See also:* **antenna.** E145/42A65-3E1

series gap (lightning arrester). An intentional gap(s) between spaced electrodes; it is in series with the valve nonlinear series resistor of expulsion element of the arrester, substantially isolating the element from line or ground or both under normal line voltage conditions. *See also:* **arresters; lightning arrester (surge diverter).** E28/42A20-31E7

series loading. Loading in which reactances are inserted in series with the conductors of a transmission circuit. *See also:* **loading.** 42A65-0

series-mode interference (signal-transmission system). *See:* **interference, differential-mode.**

series modulation. Modulation in which the plate circuits of a modulating tube and a modulated amplifier tube are in series with the same plate voltage supply. *See also:* **modulating systems.** 42A65-0

series operation (power supplies). The output of two or more power supplies connected together to obtain a total output voltage equal to the sum of their individual voltages. Load current is equal and common through each supply. The extent of series connection is limited by the maximum specified potential rating between any output terminal and ground. For series connection of current regulators, master/slave (compliance extension) or automatic crossover is used. *See:* **isolation voltage.** *See also:* **power supply.** KPSH-10E1

series overcurrent tripping. *See:* **direct release (series trip); overcurrent release (overcurrent trip).**

series-parallel connection. The arrangement of cells in a battery made by connecting two or more series-connected groups, each having the same number of cells so that the positive terminals of each group are connected together and the negative terminals are connected together in a corresponding manner. *See also:* **battery (primary or secondary).** 42A60-0

series-parallel control. A method of controlling motors wherein the motors, or groups of them, may be connected successively in series and in parallel. *See also:* **multiple-unit control.** 42A42-0

series-parallel network. Any network, containing only two-terminal elements, that can be constructed by successively connecting branches in series and/or in parallel. *Note:* An elementary example is the parallel combination of two branches, one containing resistance and inductance in series, the other containing capacitance. This network is sometimes called a **simple parallel circuit.** *See also:* **network analysis.**
*Deprecated E270-0

series-parallel starting (rotating machinery). The process of starting a motor by connecting it to the supply with the primary winding phase circuits initially in series, and changing them over to a parallel connection for running operation. *See also:* **asynchronous machine; synchronous machine.** 0-31E8

series rectifier circuit. A rectifier circuit in which two or more simple rectifier circuits are connected in such a way that their direct voltages add and their commutations coincide. *See also:* **rectification; rectifier circuit element.** E59/42A15-34E17

series regulator (power supplies). A device placed in series with a source of power that is capable of controlling the voltage or current output by automatically varying its series resistance. *See:* **passive element.** *See also:* **power supply.** KPSH-10E1

series relay. *See:* **current relay.** *See also:* **relay.** 83A16-0

series resistor (electric instrument). A resistor that forms an essential part of the voltage circuit of an instrument and generally is used to adapt the instrument to operate on some designated voltage or voltages. The series resistor may be internal or external to the instrument. *Note:* Inductors, capacitors, or combinations thereof are also used for this purpose. *See also:* **auxiliary device to an instrument.** 39A1/42A30-0

series resonance. The steady-state condition that exists in a circuit comprising inductance and capacitance connected in series, when the current in the circuit is in phase with the voltage across the circuit. 42A65-31E3

series system. The arrangement in a multielectrode electrolytic cell whereby in each cell an anode connected to the positive bus bar is placed at one end and a cathode connected to the negative bus bar is placed at the other end, with the intervening unconnected electrodes acting as bipolar electrodes. *See also:* **electrorefining.** 42A60-0

series tee junction (waveguides). A tee junction having an equivalent circuit in which the impedance of the branch guide is predominantly in series with the impedance of the main guide at the junction. *See also:* **waveguide.** E147-3E1

series-trip recloser. A recloser in which main-circuit current above a specified value, flowing through a solenoid or operating coil, provides the energy necessary to open the main contacts. 37A100-31E11

series two-terminal pair networks. Two-terminal pair networks are connected in series at the input or at the output terminals when their respective input or output terminals are in series. *See also:* **network analysis.** E153-0

series undercurrent tripping. *See:* **direct release (series trip); undercurrent release (undercurrent trip).**

series winding (autotransformer). The portion of the autotransformer winding that is not common to both the primary and the secondary circuits, but is connected in series between the input and output circuits. *See also:* **autotransformer.** 57A15-0

series-wound (rotating machinery). A qualifying term applied to a machine to denote that the excitation is supplied by a winding or windings connected in series with or carrying a current proportional to that in the armature winding. *See:* **asynchronous machine; direct-current commutating machine.** 42A10-31E8

series-wound motor. A commutator motor in which the field circuit and armature circuit are connected in series. *See:* **asychonous machine; direct-current commutating machine.** 42A10-31E8

service (1) (electric systems). The conductors and equipment for delivering electric energy from the secondary distribution or street main, or other distribution feeder, or from the transformer, to the wiring system of the premises served. *Note:* For overhead circuits, it includes the conductors from the last line pole to the service switch or fuse. The portion of an overhead service between the pole and building is designated as service drop.
See:
circuit;
dual service;
duplicate service;
emergency service;
loop service;
service cable;
service conductors;
service drop;
service entrance conductors;
service equipment;
service lateral;
service pipe;
service raceway;
single service. 1A0/2A2-0

(2) (controller) (industrial control). The specific application in which the controller is to be used, for example: (A) general purpose, (B) definite purpose, for example, crane and hoist, elevator, machine tool, etcetera. *See:* **electric controller.** IC1-34E10

service area (navigation). The area within which a navigational aid provides either generally satisfactory service or a specific quality of service. *See also:* **navigation.** 0-10E6

service band. A band of frequencies allocated to a given class of radio service. *See also:* **radio transmission.** E145/E182/42A65-0

service bits (telecommunication). Those bits that are neither check nor information bits. *See also:* **bit.** 0-19E4

service cable. Service conductors made up in the form of a cable. *See:* **service conductors.** *See also:* **service.** 42A35-31E13

service capacity (cell or battery). The electric output (expressed in ampere-hours, watthours, or similar units) on a service test before its working voltage falls to a specified cutoff voltage. *See also:* **battery (primary**

or secondary). 42A60-0

service conductors. That portion of the supply conductors that extends from the street mains or feeder or transformer to the service equipment of the premises served. For an overhead system it includes the conductors from the last line pole to the service equipment. *See also:* **service.** 42A35-31E13

service corrosion (dry cell). The consumption of the negative electrode as a result of useful current delivered by the cell. *See also:* **electrolytic cell.** 42A60-0

service drop. That portion of overhead service conductors between the last pole and the premises served, extending from the pole to the junction with the service entrance conductors. *See also:* **service.** 42A35-31E13

service entrance conductors (electric system) (1) (overhead system). The service conductors between the terminals of the service equipment and a point usually outside the building, clear of building walls, where joined by tap or splice to the service drop.

(2) (underground system). The service conductors between the terminals of the service equipment and the point of connection to the service lateral. *Note:* Where service equipment is located outside the building walls, there may be no service-entrance conductors, or they may be entirely outside the building. 42A35/1A0-31E13

service equipment (electric system). The necessary equipment, usually consisting of circuit breaker or switch and fuses, and their accessories, located near point of entrance of supply conductors to a building and intended to constitute the main control and means of cutoff for the supply to that building. *See also:* **distribution center; service.** 1A0/42A35-31E13

service factor (general-purpose alternating-current motor). A multiplier that, when applied to the rated power, indicates a permissible power loading that may be carried under the conditions specified for the service factor. *See also:* **asynchronous machine; direct-current commutating machine; synchronous machine.** MG1-31E8

service ground. A ground connection to a service equipment or a service conductor or both. *See:* **ground.** 42A35-31E13

service lateral (electric system). The underground service conductors between the street main, including any risers at a pole or other structure or from transformers, and the first point of connection to the service entrance conductors in a terminal box inside or outside the building wall. Where there is no terminal box, the point of connection shall be considered to be the point of entrance of the service conductors into the building. *See also:* **service.** 1A0-0

service life (1) (primary cell or battery). The period of useful service before its working voltage falls to a specified cutoff voltage.

(2) (storage cell or battery). The period of useful service under specified conditions, usually expressed as the period elapsed before the ampere-hour capacity has fallen to a specified percentage of the rated capacity. *See also:* **battery (primary or secondary); charge.** 42A60-0

service period (illuminating engineering). The number of hours per day for which the day lighting provides a specified illumination level. It often is stated as a monthly average. *See also:* **sunlight.** Z7A1-0

service pipe. The pipe or conduit that contains underground service conductors and extends from the junction with outside supply wires into the customer's premises. *See:* **distributor duct.** *See also:* **service.** 42A35-31E13

service raceway (electric system). The rigid metal conduit, electrical metallic tubing, or other raceway, that encloses the service entrance conductors. *See also:* **service.** 1A0-0

service rating (rectifier transformer). The maximum constant load that, after a transformer has carried its continuous rated load until there is no further measurable increase in temperature rise, may be applied for a specified time without injury. *See also:* **rectifier transformer.** 57A18-0

service routine (computing systems). A routine in general support of the operation of a computer, for example, an input-output, diagnostic, tracing, or monitoring routine. *See:* **utility routine.** *See also:* **electronic digital computer.** X3A12-16E9

service test (1) (primary battery). A test designed to measure the capacity of a cell or battery under specified conditions comparable with some particular service for which such cells are used. 42A60-0

(2) (field test) (meter). A test made during the period that the meter is in service. *Note:* A service test may be made on the consumer's premises without removing the meter from its support, or by removing the meter for test, either on the premises or in a laboratory or meter shop.
See also:
approval test;
battery;
dielectric tests;
inspection;
installation test;
office test;
periodic test;
referee test;
reference performance;
request test;
shop test;
watthour meter. 12A0/42A30-0

servicing time (electric drive). The portion of down time that is necessary for servicing due to breakdowns or for preventive servicing measures. *See:* **electric drive.** AS1-34E10

serving (cable). A wrapping applied over the core of a cable before the cable is leaded, or over the lead if the cable is armored. *Note:* Materials commonly used for serving are jute or cotton. The serving is for mechanical protection and not for insulating purposes. *See also:* **power distribution, underground construction.** 42A35-31E13

servo. *See:* **servomechanism.**

servo amplifier. An amplifier, used as part of a servomechanism, that supplies power to the electric input terminals of a mechanical actuator. *See also:* **electronic analog computer.** E165-16E9

servomechanism. (1) A feedback control system in which at least one of the system signals represents mechanical motion. (2) Any feedback control system. (3) An automatic feedback control system in which the controlled variable is mechanical position or any of its time derivatives. *See also:* **control system, feedback.** 85A1/X3A12-16E9/23E0/34E12

servomechanism, positional. A servomechanism in which a mechanical shaft is positioned, usually in the angle of rotation, in accordance with one or more input signals. *Note:* Frequently, the shaft is positioned (excluding transient motion) in a manner linearly related to the value of the input signal. However, the term also applies to any servomechanism in which a loop input signal generated by a transmitting transducer can be compared to a loop feedback signal generated by a compatible or identical receiving transducer to produce a loop error signal that, when reduced to zero by movement of the receiving transducer, results in a shaft position related in a prescribed and repeatable manner to the position of the transmitting transducer. *See also:* **electronic analog computer; servomechanism, repeater.** E165-0

servomechanism, rate. A servomechanism in which a mechanical shaft is translated or rotated at a rate proportional to an input signal amplitude. *See also:* **electronic analog computer.** E165-0

servomechanism, repeater. A positional servomechanism in which loop input signals from a transmitting transducer are compared with loop feedback signals from a compatible or identical receiving transducer mechanically coupled to the servomechanism to produce a mechanical shaft motion or position linearly related to motion or position of the transmitting transducer. *See also:* **electronic analog computer.** E165-0

servomechanism type number. In control systems in which the loop transfer function is

$$\frac{K(1+a_1s+a_2s^2+\cdots+a_is^i)}{s^n(1+b_1s+b_2s^2+\cdots+b_ks^k)}$$

where K, a_1, b_1, b_2, etcetera, are constant coefficients, the value of the integer n. *Note:* The value of n determines the low-frequency characteristic of the transfer function. The log-gain–log-frequency curve (Bode diagram) has a zero-frequency slope of zero for $n = 0$, slope -1 for $n = 1$, etcetera. *See also:* **control system, feedback.** 85A1-23E0

set (1) (electronic computation). (A) To place a storage device into a specified state, usually other than that denoting ZERO or BLANK. (B) To place a binary cell into the state denoting ONE. *See:* **preset; reset.** *See also:* **electronic computation; electronic digital computer.** E162/E270/X3A12-16E9

(2) (instruments). To position the various adjusting devices so as to secure the desired operating characteristic. *Note:* Typical adjustment devices are taps, dials, levers, and scales suitably marked, rheostats that may be adjusted during tests, and switches with numbered positions that refer to recorded operating characteristics. 37A100-31E11/31E6

(3) (polyphase currents) (of *m* phases). A group of m interrelated alternating currents, each in a separate phase conductor, that have the same primitive period but normally differ in phase. They may or may not differ in amplitude and waveform. The equations for a set of m-phase currents, when each is sinusoidal, and has the primitive period, are

$$i_a = (2)^{1/2} I_a \cos(\omega t+\beta_{a1})$$
$$i_b = (2)^{1/2} I_b \cos(\omega t+\beta_{b1})$$
$$i_c = (2)^{1/2} I_c \cos(\omega t+\beta_{c1})$$
$$\vdots$$
$$i_m = (2)^{1/2} I_m \cos(\omega t+\beta_{m1})$$

where the symbols have the same meaning as for the general case given later. The general equations for a set of m-phase alternating currents are

$$i_a = (2)^{1/2}[I_{a1}\cos(\omega t+\beta_{a1}) + I_{a2}\cos(2\omega t+\beta_{a2}) + \cdots + I_{aq}\cos(q\omega t+\beta_{aq}) + \cdots]$$
$$i_b = (2)^{1/2}[I_{b1}\cos(\omega t+\beta_{b1}) = I_{b2}\cos(2\omega t+\beta_{b2}) + \cdots + I_{bq}\cos(q\omega t+\beta_{bq}) + \cdots]$$
$$\vdots$$
$$i_m = (2)^{1/2}[I_{m1}\cos(\omega t+\beta_{m1}) + I_{m2}\cos(2\omega t+\beta_{m2}) + \cdots + I_{mq}\cos(q\omega t+\beta_{mq}) + \cdots]$$

where i_a, i_b, .. , i_m are the instantaneous values of the currents, and I_{a1}, I_{a2}, . . , I_{aq} are the root-mean-square amplitudes of the harmonic components of the individual currents. The first subscript designates the individual current and the second subscript denotes the number of the harmonic component. If there is no second subscript, the quantity is assumed to be sinusoidal. $\beta_{a\ 1}$ β_{a2}, .. , β_q are the phase angles of the components of the same subscript determined with relation to a common reference. *Notes:* (1) If the circuit has a neutral conductor, the current in the neutral conductor is generally not considered as a separate current of the set, but as the negative of the sum of all the other currents (with respect to the same reference direction). (2) See Note (3) of **voltage sets (polyphase circuit).** *See also:* **network analysis.** E270-0

(4) (polyphase voltages) (*m* phases). A group of m interrelated alternating voltages that have the same primitive period but normally differ in phase. They may or may not differ in amplitude and wave form. The equations for a set of m-phase voltages, when each is sinusoidal and has the primitive period, are

$$e_a = (2)^{1/2} E_a \cos(\omega t+\alpha_{a1})$$
$$e_b = (2)^{1/2} E_b \cos(\omega t+\alpha_{b1})$$
$$e_c = (2)^{1/2} E_c \cos(\omega t+\alpha_{c1})$$
$$\vdots$$
$$e_m = (2)^{1/2} E_m \cos(\omega t+\alpha_{m1})$$

where the symbols have the same meaning as for the general case given below. The general equations for a set of m-phase alternating voltages are

$$e_a = (2)^{1/2}[E_{a1}\cos(\omega t+\alpha_{a1}) + E_{a2}\cos(2\omega t+\alpha_{a2}) + \cdots + E_{ar}\cos(r\omega t+\alpha_{ar}) + \cdots]$$
$$e_b = (2)^{1/2}[E_{b1}\cos(\omega t+\alpha_{b1}) + E_{b2}\cos(2\omega t+\alpha_{b2}) + \cdots + E_{br}\cos(r\omega t+\alpha_{br}) + \cdots]$$
$$\vdots$$
$$e_m = (2)^{1/2}[E_{m1}\cos(\omega t+\alpha_{m1}) + E_{m2}\cos(2\omega t+\alpha_{m2}) + \cdots + E_{mr}\cos(r\omega t+\alpha_{mr}) + \cdots]$$

where e_a , e_b, . . ., e_m are the instantaneous values of the voltages, and E_{a1}, $E_{a\ 2}$, . . ., E_{ar} the root-mean-square amplitudes of the harmonic components of the individual voltages. The first subscript designates the individual voltage and the second subscript denotes the number of the harmonic component. If there is no second subscript, the quantity is assumed to be sinusoidal. α_{a1}, $\alpha_{a79\ 2}$, . . ., a_{ar79} are the phase angles of the components with the same subscript determined with relation to a common reference. *Note:* This definition may be applied to a two-phase four-wire or five-wire circuit if *m* is considered to be 4 instead of 2. A two-phase three-wire circuit should be treated as a special case. *See also:* **network analysis.** E270-0

set light (television). The separate illumination of the background or set, other than that provided for principal subjects or areas. *See also:* **television lighting.** Z7A1-0

set of commutating groups (rectifier). Two or more commutating groups that have simultaneous commutations. *See also:* **rectification; rectifier circuit element.** 42A15/E59-34E17/34E24

set point (process control systems). A fixed or constant (for relatively long time periods) command. *See also:* **control system, feedback.** 85A1-23E0

set pulse. A drive pulse that tends to set a magnetic cell. *See also:* **static magnetic storage.** E163-0

setting (noun) (1) (general). The desired characteristic, obtained as a result of having set a device, stated in terms of calibration markings or of actual performance bench marks such as pickup current and operating time at a given value of input. *Note:* When the setting is made by adjusting the device to operate as desired in terms of a measured input quantity, the procedure may be the same as in calibration. However, since it is for the purpose of finding one particular position of an adjusting device, which in the general case may have several marked positions that are not being calibrated, the word setting is to be preferred over the word calibration. 37A100-31E11/31E6

(2) (circuit breaker). The value of the current at which it is set to trip. *See also:* **contactor.** 1A0-0

setting error. The departure of the actual performance from the desired performance resulting from errors in adjustment or from limitations in testing or measuring techniques. 37A100-31E11/31E6

setting limitation. The departure of the actual performance from the desired performance resulting from limitations of adjusting devices. 37A100-31E11/31E6

settling time (control) (industrial control). The time required, following the initiation of a specified stimulus to a system, for a specified variable to enter and remain within a specified narrow band centered on its final value. Settling time is expressed in seconds. *See also:* **control system, feedback.** AS1-34E10

setup (television). The ratio between reference black level and reference white level, both measured from blanking level. It is usually expressed in percent. *See also:* **television.** E203/42A65-2E2

sexadecimal. (1) Pertaining to a characteristic or property involving a selection, choice, or condition in which there are sixteen possibilities. (2) Pertaining to the numeration system with a radix of sixteen. *Note:* More commonly called **hexadecimal.** *See:* **positional notation.** *See also:* **electronic digital computer.** 0-16E9

shade. A screen made of opaque or diffusing material that is designed to prevent a light source from being directly visible at normal angles of view. *See also:* **bare (exposed) lamp.** Z7A1-0

shaded-pole motor. A single-phase induction motor with a main winding and one or more short-circuited windings displaced in magnetic position from the main winding. *See:* **asynchronous machine.** 42A10-31E8

shading (1) (storage tubes). The type of spurious signal, generated within a tube, that appears as a gradual variation or a small number of gradual variations in the amplitude of the output signal. These variations are spatially fixed with reference to the target area. Note the distinction between this and **disturbance.** *See also:* **beam tubes; storage tube; television.** E158-15E6

(2) (television). The process of compensating for the spurious signal generated in a camera tube during the trace intervals. *See also:* **television.** 42A65-0

(3) (audio and electroacoustics). A method of controlling the directional response pattern of a transducer through control of the distribution of phase and amplitude of the transducer action over the active face. *See also:* **electroacoustics; television.** 0-1E1

shading coil (rotating machinery). (1) The short-circuited winding used in a shaded-pole motor, for the purpose of producing a rotating component of magnetic flux.

(2) (direct-current motors and generators). A short-circuited winding used on a main (excitation) pole to delay the shift in flux caused by transient armature current. Transient commutation is aided by the use of this coil. *See also:* **rotor (rotating machinery); stator.** 0-31E8

shading wedge (rotating machinery). A strip of magnetic material placed between adjacent pole tips of a shaded-pole motor to reduce the effective separation between the pole tips. The shading wedge usually has a slot running most of its length to provide some separation effect. *See also:* **rotor (rotating machinery); stator.** 0-31E8

shadow factor (radio wave propagation). The ratio of the electric field strength that would result from propagation over a sphere to that which would result from propagation over a plane, other factors being the same. *See also:* **radio wave propagation.** E211-3E2

shadowing (shielding). The interference of any part of an anode, cathode, rack, or tank with uniform current distribution upon a cathode. 42A60-0

shadow loss (mobile communication). The attenuation to a signal caused by obstructions in the radio propagation path. *See also:* **mobile communication system.** 0-6E1

shadow mask (color-picture tubes). A color-selecting-electrode system in the form of an electrically conductive sheet containing a plurality of holes that uses masking to effect color selection. *See also:* **beam tubes; television.** E160-15E6

shaft (rotating machinery). That part of a rotor that carries other rotating members and that is supported by bearings in which it can rotate. *See:* **rotor (rotating machinery).** 42A10-31E8

shaft current (rotating machinery). Electric current that flows from one end of the shaft of a machine through bearings, bearing supports, and machine framework to the other end of the shaft, driven by a voltage between the shaft ends that results from flux linking the shaft caused by irregularities in the magnetic circuit. *See:* **rotor (rotating machinery).** 0-31E8

shaft encoder. A transducer whose input is the turning of a shaft and whose output is a measure of the position of the shaft. *See also:* **transducer.** 0-31E3

shaft extension (rotating machinery). The portion of a shaft that projects beyond the bearing housing and away from the core. *See:* **armature.** 0-31E8

shaft revolution indicator. A system consisting of a transmitter driven by a propeller shaft and one or more remote indicators to show the speed of the shaft in revolutions per minute, the direction of rotation and (usually) the total number of revolutions made by the shaft. *See also:* **electric propulsion system.** 42A43-0

shaft voltage test (rotating machinery). A test taken on an energized machine to detect the induced voltage that is capable of producing shaft currents. *See also:* **rotor (rotating machinery).** 0-31E8

shaped-beam antenna. An antenna whose directional pattern over a certain angular range is designed to a special shape for some particular use. *See also:* **antenna.** 42A65-3E1

shaping pulse. The intentional processing of a pulse waveform to cause deviation from a reference waveform. *See also:* **pulse.** 0-9E4

sharing. *See:* **time sharing.** *See also:* **electronic digital computer.**

shear pin (rotating machinery). A dowel designed to shear at a predetermined load and thereby prevent damage to other parts. *See also:* **rotor (rotating machinery).** 0-31E8

shear wave (acoustics) (rotational wave). A wave in an elastic medium that causes an element of the medium to change its shape without a change of volume. *Notes:* (1) Mathematically, a shear wave is one whose velocity field has zero divergence. (2) A shear plane wave in an isotropic medium is a transverse wave. (3) When shear waves combine to produce standing waves, linear displacements may result. *See also:* **electroacoustics.** 0-1E1

shearing machine. An electrically driven machine for making vertical cuts in coal. *See also:* **mining.** 42A85-0

sheave (rotating machinery). *See:* **pulley.**

shelf corrosion (dry cell). The consumption of the negative electrode as a result of local action. *See also:* **electrolytic cell.** 42A60-0

shelf depreciation. The depreciation in service capacity of a primary cell as measured by a shelf test. *See also:* **battery (primary or secondary).** 42A60-0

shelf test. A storage test designed to measure retention of service ability under specified conditions of temperature and cutoff voltage. *See also:* **battery (primary or secondary).** 42A60-0

shell (1) (insulators). A single insulating member, having a skirt or skirts without cement or other connecting devices intended to form a part of an insulator or an insulator assembly. *See also:* **insulator.** 29A1/42A35-31E13

(2) (electrolysis). The external container in which the electrolysis of fused electrolyte is conducted. *See also:* **fused electrolyte.** 42A60-0

(3) (electrotyping). A layer of metal (usually copper or nickel) deposited upon, and separated from, a mold. *See also:* **electroforming.** 42A60-0

shell, stator (rotating machinery). A cylinder in tight assembly around the wound stator core, all or a portion of which is machined or otherwise made to a specific outer dimension so that the stator may be mounted into an appliance, machine, or other end product. *See also:* **stator.** 0-31E8

shell-type motor. A stator and rotor without shaft, end shields, bearings or conventional frame. *Note:* A shell-type motor is normally supplied by a motor manufacturer to an equipment manufacturer for incorporation as a built-in part of the end product. Separate fans or fans larger than the rotor are not included. *See:* **asynchronous machine; direct-current commutating machine; synchronous machine.** 42A10-31E8

shield (1) (electromagnetic). A housing, screen, or other object, usually conducting, that substantially reduces the effect of electric or magnetic fields on one side thereof, upon devices or ciruits on the other side. *See:* **dielectric heating; induction heating; industrial electronics; signal.** 42A65-21E0

(2) (mechanical protection) (rotating machinery). An internal part used to protect rotating parts or parts of the electric circuit. In general, the word **shield** will be preceded by the name of the part that is being protected. *See also:* **cradle base (rotating machinery).** 0-31E8

(3) (magnetrons). *See:* **end shield.**

shielded conductor cable. A cable in which the insulated conductor or conductors is/are enclosed in a conducting envelope or envelopes. *See also:* **power distribution, underground construction.** E30-0

shielded ignition harness. A metallic covering for the ignition system of an aircraft engine, that acts as a shield to eliminate radio interference with aircraft electronic equipment. The term includes such items as ignition wiring and distributors when they are manufactured integral with an ignition shielding assembly. *See:* **air transportation electric equipment.** 42A41-0

shielded joint. A cable joint having its insulation so enveloped by a conducting shield that substantially every point on the surface of the insulation is at ground potential or at some predetermined potential with respect to ground. *See also:* **power distribution, underground construction.** 42A35-31E13

shielded pair (signal-transmission system). A two-wire transmission line surrounded by a sheath of conducting material to protect it from the effects of external fields, or to confine fields produced by the transmission line. *See:* **signal.** *See also:* **waveguide.** E146-13E6

shielded strip transmission line. A strip conductor between two ground planes. Some common designations are: Stripline (trade mark); Tri-plate (trade mark); slab line (round conductor); balanced strip line.* *See:* **strip (-type) transmission line; unshielded strip transmission line.**

*Deprecated 0-3E1

shielded transmission line (signal-transmission system). A transmission line surrounded by a sheath of conducting material to protect it from the effects of external fields, or to confine fields produced by the transmission line. *See:* **signal.** *See also:* **waveguide.** 0-13E6

shielded-type cable. A cable in which each insulated conductor is enclosed in a conducting envelope so constructed that substantially every point on the surface of the insulation is at ground potential or at some predetermined potential with respect to ground under normal operating conditions. *See also:* **power distribution, underground construction.** 42A35-31E13

shield factor (telephone circuit). The ratio of noise, induced current, or voltage when a source of shielding is present, to the corresponding quantity when the shielding is absent. *See also:* **induction coordination.** 42A65-0

shield grid (gas tubes). A grid that shields the control electrode in a gas tube from the anode or the cathode, or both, with respect to the radiation of heat and the deposition of thermionic activating material and also reduces the electrostatic influence of the anode. It may be used as a control electrode. *See also:* **electrode (electron tube); grid.** 42A70-15E6

shielding angle (1) (lightning protection). The angle between the vertical line through the overhead ground wire and a line connecting the overhead ground wire with the shielded conductor. *See also:* **direct stroke protection (lightning).** 0-31E13

(2) (luminaire). The angle between a horizontal line through the light center and the line of sight at which the bare source first becomes visible. *See also:* **bare (exposed) lamp.** Z7AI-0

shielding effectiveness (electromagnetic compatibility). For a given external source, the ratio of electric or magnetic field strength at a point before and after the placement of the shield in question. *See also:* **electromagnetic compatibility.** 0-27E1

shielding failure (lightning protection). The occurrence of a lightning stroke that bypasses the overhead ground wire and terminates on the phase conductor. *See also:* **direct-stroke protection (lightning).** 0-31E13

shield wire (electromagnetic fields). A wire employed for the purpose of reducing the effects on electric supply or communication circuits from extraneous sources. *See also:* **inductive coordination.** 42A65-0

shift (electronic computation). A displacement of an ordered set of characters one or more places to the left or right. If the characters are the digits of a numerical expression, a shift may be equivalent to multiplying by a power of the base. *See also:* **arithmetic shift; cyclic shift; electronic computation; electronic digital computer; logic shift.** E162/E270-0

shift pulse. A drive pulse that initiates shifting of characters in a register. *See also:* **static magnetic storage.** E163-0

shift register (computing systems). (1) A logic network consisting of a series of memory cells such that a binary code can be caused to shift into the register by serial input to only the first cell. *See also:* **digital.** 0-31E3

(2) A register in which the stored data can be moved to the right or left. *See also:* **electronic digital computer.** X3A12-16E9

shim (rotating machinery). A lamination usually machined to a close-tolerance thickness, for assembly between two parts to control spacing. *See also:* **rotor (rotating machinery); stator.** 0-31E8

shingle (photoelectric converter). Combination of photoelectric converters in series in a shingle-type structure. *See also:* **semiconductor.** 0-10E1

ship control telephone system. A system of sound-powered telephones (requiring no external power supply for talking) with call bells, exclusively for communication among officers responsible for control and operation of a ship. *Note:* Call bells are usually energized by hand-cranked magneto generators. 42A43-0

shipping brace (rotating machinery). Any structure provided to reduce motion or stress during shipment, that must be removed before operation. *See:* **cradle base (rotating machinery).** 0-31E8

shipping seal (cable). *See:* **sealed end.**

ship's service electric system. On any vessel, all electric apparatus and circuits for power and lighting, except apparatus provided primarily either for ship propulsion or for the emergency system. *Note:* Emergency and interior communication circuits are normally supplied with power from the ship's service system, upon failure of which they are switched to an independent emergency generator or other sources of supply. *See also:* **marine electric apparatus.** 42A43-0

shock excitation (1) (oscillatory systems). The excitation of natural oscillations in an oscillatory system due to a sudden acquisition of energy from an external source or a sudden release of energy stored within the oscillatory system. *See also:* **oscillatory circuit.** 42A65-0

(2) (signal-transmission system). The type of excitation supplied by a voltage, current, temperature, etcetera, variation of relatively short duration. *See:* **signal.** 0-13E6

shock motion (mechanical system). Transient motion that is characterized by suddenness, by significant relative displacements, and by the development of substantial internal forces in the system. *See:* **mechanical shock.** 0-1E1

shockproof electric apparatus. Electric apparatus designed to withstand, to a specified degree, shock of specified severity. *Note:* The severity is stated in foot-pounds impact on a special test stand equivalent to shock of gunfire, explosion of mine or torpedo, etcetera. *See also:* **marine electric apparatus.** 42A43-0

shoe (ramp shoe). Part of a vehicle-carried apparatus that makes contact with a ramp. 42A42-0

shop instruments. Instruments and meters that are used in regular routine shop or field operations. 12A0-0

shop test (laboratory test). A test made upon the receipt of a meter from a manufacturer, or prior to reinstallation. Such tests are made in a shop or a laboratory of a meter department. *See also:* **service test (field test).** 42A30-0

shoran. A precision position-fixing system using a pulse transmitter and receiver in connection with two transponder beacons at fixed points. *See also:* **radio navigation.** 42A65-0

shore feeder. Permanently installed conductors from a distribution switchboard to a connection box (or boxes) conveniently located for the attachment of portable leads for supply of power to a ship from a source on shore. *See also:* **marine electric apparatus.** 42A43-0

short circuit. An abnormal connection of relatively low resistance, whether made accidentally or intentionally, between two points of different potential in a circuit. *Note:* The term is often applied to the group of phenomena that accompany a short circuit between points at different potentials. *See also:* **center of distribution; network analysis.** E270/42A35-31E13;37A100-31E11

short-circuit driving-point admittance. *See:* **admittance, short-circuit driving-point.**

short-circuit duration rating (magnetic amplifier). The length of time that a short circuit may be applied to the load terminals nonrecurrently without reducing the intended life of the amplifier or exceeding the specified temperature rise. *See also:* **rating and testing magnetic amplifiers.** E107-0

short-circuiter. A device designed to short circuit the commutator bars when the motor has attained a predetermined speed in some forms of single-phase commutator-type motors. *See:* **asynchronous machine.** 42A10-0

short-circuit feedback admittance. *See:* **admittance, short-circuit feedback.**

short-circuit forward admittance. *See:* **admittance, short-circuit forward.**

short-circuit impedance (1) (general). A qualifying adjective indicating that the impedance under consideration is for the network with a specified pair or group of terminals short-circuited. *See also:* **network analysis; self-impedance.** E270-0
(2) (line or four-terminal network). The driving-point impedance when the far-end is short-circuited. *See also:* **self-impedance.** 42A65-0

short-circuit inductance. The apparent inductance of a winding of a transformer with one or more specified windings short circuited. 0-21E1

short-circuit input admittance. *See:* **admittance, short-circuit input.**

short-circuit loss (rotating machinery). The difference in power required to drive a machine at normal speed, when excited to produce a specified balanced short-circuit armature current, and the power required to drive the unexcited machine at the same speed. *See also:* **asynchronous machine; direct-current commutating machine; synchronous machine.** 0-31E8

short-circuit output admittance. *See:* **admittance, short-circuit output.**

short-circuit output capacitance. *See:* **capacitance, short-circuit output.**

short-circuit protection (power supplies) (automatic). Any automatic current-limiting system that enables a power supply to continue operating at a limited current, and without damage, into any output overload including short circuits. The output voltage must be restored to normal when the overload is removed, as distinguished from a fuse or circuit-breaker system that opens at overload and must be closed to restore power. *See:* **current limiting.** *See also:* **power supply.** KPSH-10E1

short-circuit ratio (synchronous machine). The ratio of the field current for rated open-circuit armature voltage and rated frequency to the field current for rated armature current on sustained symmetrical short-circuit at rated frequency. *See:* **synchronous machine.** 42A10-31E8

short-circuit saturation curve (synchronous machine). The relationship between the current in the short-circuited armature winding and the field current. *See:* **synchronous machine.** 0-31E8

short-circuit time constant (1) (armature winding). The time in seconds for the asymmetrical (direct-current) component of armature current under suddenly applied short-circuit conditions, with the machine running at rated speed, to decrease to $1/e \approx 0.368$ of its initial value. The rated current value of the short-circuit time constant of the armature winding will be that obtained from the test specified for the rated current value of direct-axis transient reactance. The rated voltage value of the short-circuit time constant of the armature winding will be that obtained from a short-circuit test at the terminals of the machine at no load and rated speed and at rated armature voltage. *See:* **synchronous machine.** 42A10-0
(2) (primary winding) (rotating machinery). The time required for the direct-current component present in the short-circuit primary-winding current following a sudden change in operating conditions to decrease to $1/e \approx 0.368$ of its initial value, the machine running at rated speed. *See also:* **asynchronous machine; synchronous machine.** 0-31E8

short-circuit transfer admittance. *See:* **admittance, short-circuit transfer.**

short-circuit transfer capacitance. *See:* **capacitance, short-circuit transfer.**

short dimension. Incremental dimensions whose number of digits is the same as normal dimensions except the first digit shall be zero, that is 0.XXXX for the example under normal dimension. *See:* **dimension; incremental dimension; long dimension; normal dimension.** EIA3B-34E12

short-distance navigation. Navigation utilizing aids usable only at comparatively short distances; this term covers navigation between approach navigation and long-distance navigation, there being no distinct, universally accepted demarcation between them. *See also:* **navigation.** 0-10E6

short field (tapped field*). Where two field strengths are required for a series machine, short field is the minimum-strength field connection. *See:* **asynchronous machine; direct-current commutating machine.**
*Deprecated 42A10-0

short-pitch winding (rotating machinery). A winding in which the coil pitch is less than the pole pitch. *See also:* **rotor (rotating machinery); stator.** 0-31E8

short-time current. The current carried by a device, an assembly, or a bus for a specified short time interval. 37A100-31E11

short-time duty (rating of electric equipment). A duty that demands operation at a substantially constant load for a short and definitely specified time. *Note:* The specified conditions normally require duty at constant load during a given time less than that required to obtain constant temperature in continuous duty at the same load, followed by a rest of sufficient duration to re-establish equality of temperature with the cooling medium. *See also:* **asynchronous machine; direct-current commutating machine; industrial control; synchronous machine; voltage regulator.** E270/51A15/87A1;E96-SCC4;42A15-31E12; 50I16-34E10

short-time operation influence (electric instrument). The operation influence arising from continuous operation over a period of 15 minutes. *See also:* **accuracy rating (of an instrument).** 39A1-0

short-time rating (1) (general). Defines the load that can be carried for a short and definitely specified time, the machine, apparatus, or device being at approximately room temperature at the time the load is applied. *See also:* **duty.** E96/E270-SCC4;IC1-34E10
(2) (rotating machinery). The statement of the load, duration, and conditions assigned to a machine by the manufacturer at which the machine may be operated for a limited period, starting at ambient temperature. *See also:* **asynchronous machine; direct-current com-**

mutating machine; synchronous machine. 0-31E8

(3) (transformer). Defines the maximum constant load that can be carried for a specified short time without exceeding established temperature-rise limitations, under prescribed conditions of test and when the transformer is approximately at room temperature at the time the load is applied. 42A15-31E12;57A14-0

(4) (voltage regulator). Defines the maximum constant load that can be carried for a specified short time, the regulator being approximately at room temperature at the time the load is applied without exceeding the specified temperature limitation, and within the limitations of established standards. 57A15-0

(5) (reactor). Defines the maximum constant load that can be carried for a specified short time, the reactor being approximately at room temperature at the time the load is applied, without exceeding the specified temperature limitation, and within the limitations of established standards. 57A16-0

(6) (arc-welding apparatus). Defines the load that can be carried for a short and definitely specified time, the apparatus being approximately at room temperature at the time the temperature test is started. *See also:* **electric arc-welding apparatus.** 87A1-0

(7) (power inverter unit). Defines the maximum load that can be carried for a specified short time, without exceeding the specified limitations under prescribed conditions of test, and within the limitation of established standards. *See also:* **self-commutated inverters.** 0-34E24

(8) (rectifier unit). The maximum load that can be carried for a specified short time, without exceeding the specified temperature-rise limitations under prescribed conditions of test, and within the limitations of established standards. *Note:* The short-time rating includes loads of two hours duration. *See also:* **rectification.** 42A15-34E24

(9) (neutral grounding device) (electric power). A rating in which the rated time is a short and definitely specified time. *See also:* **grounding device.** E32-0

shot effect (electron tubes). The variations in the output of an electron valve or tube due to: (1) Random variation in the emission of electrons from the cathode. (2) Instantaneous variations in the distribution of the electrons among the electrodes. *See also:* **electron tube.** 50I07-15E6

shot-firing (blasting) cord. A two-conductor cable used for completing the circuit between the electric blasting cap (or caps) and the blasting unit or other source of electric energy. *See also:* **blasting unit.** 42A85-0

shot-firing unit. *See:* **blasting unit.**

shot noise, full (electron tubes) (interference terminology). The fluctuation in the current of charge carriers passing through a surface at statistically independent times. *Notes:* (1) **Shot noise** has a uniform spectral density W_i given by

$$W_i = \frac{eI_0}{2\pi}$$

where e is the charge of the carrier and I_0 is the average current. (2) The mean-square noise current $\overline{i^2}$ of full shot noise within a frequency increment Δf is

$$\overline{i^2} = 2eI_0\Delta f.$$

(3) The mean-square noise current $\overline{i^2}$ within a frequency increment Δf associated with an average current I_0 is often expressed in terms of full shot noise through a shot noise reduction factor Γ^2, in general a function of frequency, by the formula:

$$\overline{i^2} = \Gamma^2 2eI_0\Delta f.$$

When $\Gamma^2 < 1$, $\overline{i^2}$ is called **reduced shot noise.** *See also:* **interference; signal-to-noise ratio.** E160-13E6/15E6

shot noise, reduced. *See:* **shot noise, full.**

shrink link (rotating machinery). A bar with an enlarged head on each end for use like a rivet but slipped into place after expansion by heat. It tightens on cooling by shrinkage only. *See:* **cradle base (rotating machinery).** 0-31E8

shunt. A device having appreciable resistance or impedance connected in parallel across other devices or apparatus, and diverting some (but not all) of the current from it. Appreciable voltage exists across the shunted device or apparatus and an appreciable current may exist in it. E270-0

shunt control. A method of controlling motors employing the shunt method of transition from series to parallel connections of the motors. *See also:* **multiple-unit control.** 42A42-0

shunt-fed vertical antenna. A vertical antenna connected to ground at the base and energized at a point suitably positioned above the grounding point. *See also:* **antenna.** E145-3E1

shunting transition. *See:* **shunt transition.**

shunt leads (instrument). Those leads that connect a circuit of an instrument to an external shunt. The resistance of these leads is taken into account in the adjustment of the instrument. *See also:* **auxiliary device to an instrument; instrument.** 39A1/42A30-0

shunt loading. Loading in which reactances are applied in shunt across the conductors of a transmission circuit. *See also:* **loading.** 42A65-0

shunt noninterfering fire-alarm system. A manual fire-alarm system employing stations and circuits such that, in case two or more stations in the same premises are operated simultaneously, the signal from the operated box electrically closest to the control equipment is transmitted and other signals are shunted out. *See also:* **protective signaling.** 42A65-0

shunt regulator (power supplies). A device placed across the output that controls the current through a series dropping resistance to maintain a constant voltage of current output. *See also:* **power supply.** KPSH-10E1

shunt release (shunt trip). A release energized by a source of voltage. *Note:* The voltage may be derived either from the main circuit or from an independent source. 37A100-31E11

shunt tee junction (waveguides). A tee junction having an equivalent circuit in which the impedance of the branch guide is predominantly in parallel with the impedance of the main guide at the junction. *See also:* **waveguide.** E147-3E1

shunt transition (shunting transition). A method of changing the connection of motors from series to parallel in which one motor, or group of motors, is first shunted or short circuited, then open circuited, and finally connected in parallel with the other motor or motors. *See also:* **multiple-unit control.** 42A42-0

shunt trip. *See:* **shunt release.**

shunt-trip recloser. A recloser in which the tripping mechanism, by releasing the holding means, permits the main contacts to open, with both the tripping mechanism and the contact-opening mechanism deriving operating energy from other than the main circuit. 37A100-31E11

shunt-wound. A qualifying term applied to a direct-current machine to denote that the excitation is supplied by a winding connected in parallel with the armature in the case of a motor, with the load in the case of a generator, or is connected to a separate source of voltage. 41A10-0

shunt-wound generator. A direct-current generator in which ordinarily the entire field excitation is derived from one winding consisting of many turns with a relatively high resistance. This one winding is connected in parallel with the armature circuit for a self-excited generator and to the load side of another generator or other source of direct current for a separately excited generator. *See also:* **direct-current commutating machine.** E45-0

shunt-wound motor. A direct-current motor in which the field circuit and armature circuit are connected in parallel. *See also:* **direct-current commutating machine.** 42A10-0

shutter (1) (electric machine). A protective covering used to close, or to close partially, an opening in a stator frame or end shield. In general, the word **shutter** will be preceded by the name of the part to which it is attached. As used for an electric machine, a shutter is rigid and hence not adjustable. *See also:* **stator.** 0-31E8

(2) (switchgear assembly). A device that is automatically operated to completely cover the stationary portion of the primary disconnecting devices when the removable element is either in the disconnected position, test position, or has been removed. 37A100-31E11

shuttle car. A vehicle on rubber tires or caterpillar treads and usually propelled by electric motors, electric energy for which is supplied by a diesel-driven generator, by storage batteries, or by a power distribution system through a portable cable. Its chief function is the transfer of raw materials, such as coal and ore, from loading machines in trackless areas of a mine to the main transportation system. *See also:* **mining.** 42A85-0

shuttle car, explosion-tested. A shuttle car equipped with explosion-tested equipment. *See also:* **mining.** 42A85-0

side back light (television). Illumination from behind the subject in a direction not parallel to a vertical plane through the optical axis of the camera. *See also:* **television lighting.** Z7A1-0

sideband attenuation. That form of attenuation in which the transmitted relative amplitude of some component(s) of a modulated signal (excluding the carrier) is smaller than that produced by the modulation process. *See also:* **wave front.** E145-0

sideband null (rectilinear navigation system). The surface of position along which the resultant energy from a particular pair of sideband antennas is zero. *See also:* **navigation.** 0-10E6

sideband-reference glide slope (instrument landing systems). A modified null-reference glide-slope antenna system in which the upper (sideband) antenna is replaced with two antennas, both at lower heights, and fed out of phase, so that a null is produced at the desired glide-slope angle. *Note:* This system is used to reduce unwanted reflections of energy into the glide-slope sector at locations where rough terrain exists in front of the approach end of the runway, by producing partial cancellation of energy at low elevation angles. *See also:* **navigation.** 0-10E6

sidebands. (1) The frequency bands on both sides of the carrier frequency within which fall the frequencies of the wave produced by the process of modulation. (2) The wave components lying within such bands. *Note:* In the process of amplitude modulation with a sine-wave carrier, the upper sideband includes the sum (carrier plus modulating) frequencies; the lower sideband includes the difference (carrier minus modulating) frequencies. *See also:* **amplitude modulation; radio receiver.** E145/42A65-0;E188-31E3/13E6

sideband suppression (power-system communication). A process that removes the energy of one of the sidebands from the modulated carrier spectrum. *See also:* **modulating systems.** 0-31E3

side-break switch. A switch in which the travel of the blade is in a plane parallel to the base of the switch. 37A100-31E11

side circuit. A circuit arranged for deriving a phantom circuit. *Note:* In the case of two-wire side circuits, the conductors of each side circuit are placed in parallel to form a side of the phantom circuit. In the case of four-wire side circuits, the lines of the two side circuits that are arranged for transmission in the same direction provide a one-way phantom channel for transmission in that same direction, the two conductors of each line being placed in parallel to provide a side for that phantom channel. Similarly the conductors of the other two lines provide a phantom channel for transmission in the opposite direction. *See also:* **transmission line.** 42A65-0

side-circuit loading coil. A loading coil for introducing a desired amount of inductance in a side circuit and a minimum amount of inductance in the associated phantom circuit. *See also:* **loading.** 42A65-0

side-circuit repeating coil (side-circuit repeat coil). A repeating coil that functions simultaneously as a transformer at a terminal of a side circuit and as a device for superposing one side of a phantom circuit on that side circuit. *See also:* **telephone system.** 42A65-0

sideflash (lightning). A spark occurring between nearby metallic objects or from such objects to the lightning protection system or to ground. *See also:* **arresters; direct-stroke protection (lightning).** 5A1-31E13

side flashover (lightning). A flashover of insulation resulting from a direct lightning stroke that bypasses the overhead ground wire and terminates on a phase conductor of a transmission line. *See also:* **direct-stroke protection (lightning).** 0-31E13

side frequency. One of the frequencies of a sideband. *See also:* **amplitude modulation.** E145/E170/42A65-0

side lobe (antenna). A radiation lobe in any direction other than that of the intended lobe. *See also:* **radia-**

tion. 0-3E1

side-lobe level, maximum relative (antenna). The relative level of the highest side lobe. *See also:* **radiation.** 0-3E1

side lobe, relative level of (antenna). The ratio of the radiation intensity of a side lobe in the direction of its maximum value to that of the intended lobe, usually expressed in decibels. *See also:* **radiation.** 0-3E1

side-lock. Spurious synchronization in an automatic frequency synchronizing system by a frequency component of the applied signal other than the intended component. *See:* **television.** 0-2E2/42A65

side marker lamp. A light indicating the presence of a vehicle when seen from the front and sometimes serving to indicate its width. *See also:* **headlamp.** Z7A1-0

side panel (rotating machinery). A structure enclosing or partly enclosing one side of a machine. *See also:* **cradle base (rotating machinery).** 0-31E8

side thrust (skating force) (disk recording). The radial component of force on a pickup arm caused by the stylus drag. *See also:* **phonograph pickup.** 0-1E1

sidetone. The transmission and reproduction of sounds through a local path from the transmitter to the receiver of the same telephone station. *See also:* **telephone switching system.** 42A65-0

sidetone telephone set. A telephone set that does not include a balancing network for the purpose of reducing sidetone. *See also:* **telephone station.** 42A65-0

sidewalk elevator. A freight elevator that operates between a sidewalk or other area exterior to the building and floor levels inside the building below such area, that has no landing opening into the building at its upper limit of travel, and that is not used to carry automobiles. *See also:* **elevators.** 42A45-0

siemens. *See:* **mho.**

sigma (σ). The term **sigma** designates a group of telephone wires, usually the majority or all wires of a line, that is treated as a unit in the computation of noise or in arranging connections to ground for the measurement of noise or current balance ratio. *See also:* **induction coordination.** 42A65-0

sign (power or energy). Positive, if the actual direction of energy flow agrees with the stated or implied reference direction; negative, if the actual direction is opposite to the reference direction. *See also:* **network analysis.** E270-0

signal (1). A visual, audible, or other indication used to convey information.

(2). The intelligence, message, or effect to be conveyed over a communication system.

(3). A signal wave; the physical embodiment of a message. *See also:* **communication; information theory; modulating systems.** E151-27E1;E170/E151/92A65-0

(4) (computing systems). The event or phenomenon that conveys data from one point to another. *See also:* **communication; electronic digital computer.** X3A12-16E9

(5) (control) (industrial control). Information about a variable that can be transmitted in a system. *Note:* For an extensive list of cross references, see *Appendix A.* AS1-34E10

signal, actuating (control system, feedback). *See:* **actuating signal.**

signal aspect. The appearance of a fixed signal conveying an indication as viewed from the direction of an approaching train; the appearance of a cab signal conveying an indication as viewed by an observer in the cab. *See also:* **railway signal and interlocking.** 42A42-0

signal back light. A light showing through a small opening in the back of an electrically lighted signal, used for checking the operation of the signal lamp. *See also:* **railway signal and interlocking.** 42A42-0

signal charge (ferroelectric device). The charge that flows when the condition of the device is changed from that of zero applied voltage (after having previously been saturated with either a positive or negative voltage) to at least that voltage necessary to saturate in the reverse sense. *Note:* The signal charge Q_s equals the sum of Q_r and Q_t, as illustrated in the accompanying figure. It is dependent on the magnitude of the applied voltage, which should be specified in describing this characteristic of ferroelectric devices. *See also:* **ferroelectric domain.** E180-0

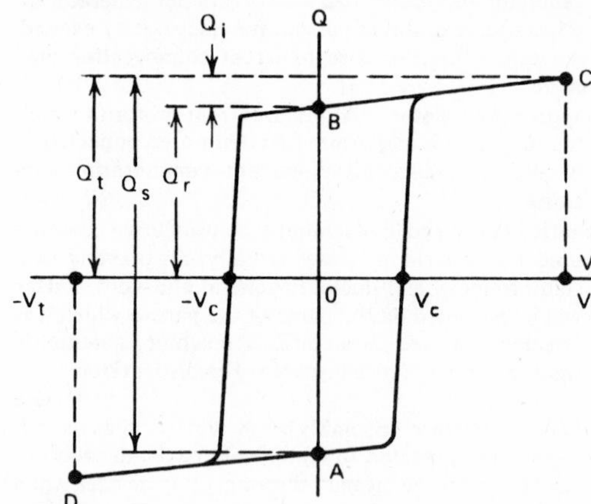

Hysteresis loop for a ferroelectric device.

signal circuit. Any electric circuit that supplies energy to an appliance that gives a recognizable signal. Such circuits include circuits for door bells, buzzers, code-calling systems, signal lights, and the like. *See also:* **appliances.** 1A0-0

signal contrast (facsimile). The ratio expressed in decibels between white signal and black signal. *See also:* **facsimile signal (picture signal).** E168-0

signal delay. The transmission time of a signal through a network. The time is always finite, may be undesired, or may be purposely introduced. *See:* **delay line.** *See also:* **oscillograph.** 0-9E4

signal distance (computing systems). The number of digit positions in which the corresponding digits of two binary words of the same length are different. *See:* **Hamming distance.** *See also:* **electronic digital computer.** X3A12-16E9

signal electrode (camera tube). An electrode from which the signal output is obtained. *See also:* **electrode (electron tube).** 42A70-2E2

signal element (unit interval). The part of a signal that occupies the shortest interval of the signaling code. It is considered to be of unit duration in building up signal combinations. *See also:* **telegraphy; data transmission.** 42A65-31E3/19E4

signal, error (1) (automatic control device) (general). A signal whose magnitude and sign are used to correct the alignment between the controlling and the controlled elements. 42A65-0

(2) (power supplies). The difference between the output voltage and a fixed reference voltage compared in ratio by the two resistors at the null junction of the comparison bridge. The error signal is amplified to drive the pass elements and correct the output. *See also:* **power supply.** KPSH-10E1

(3) (closed loop) (control system, feedback). The signal resulting from subtracting a particular return signal from its corresponding input signal. See the accompanying figure. *See also:* **control system, feedback.** 85A1-23E0

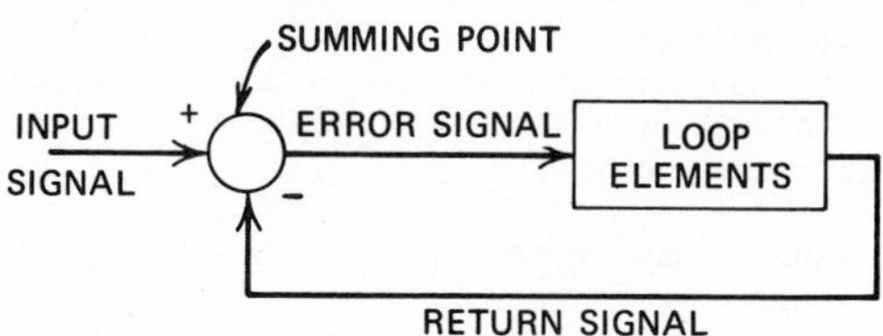

Block diagram of a closed loop.

signal, feedback (1) (general). A function of the directly controlled variable in such form as to be used at the summing point. *See also:* **control system, feedback.** AS1-34E10

(2) (control system, feedback). The return signal that results from the reference input signal. See the accompanying figure. *See also:* **control system, feedback.** 85A1-23E0

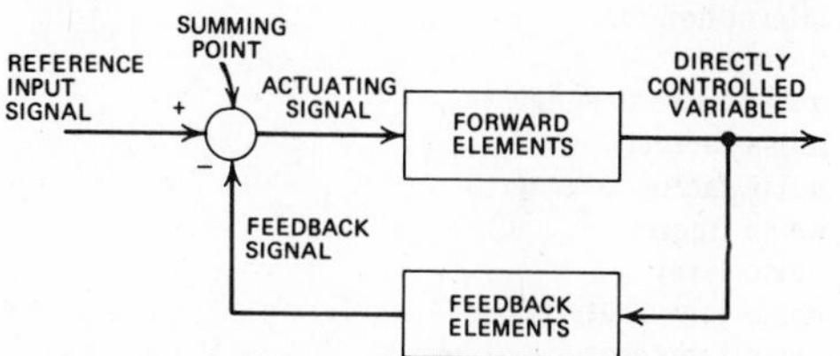

Simplified block diagram indicating essential elements of an automatic control system.

signal flow graph (network analysis). A network of directed branches in which each dependent node signal is the algebraic sum of the incoming branch signals at that node. *Note:* Thus,

$$x_1 t_{1k} + x_2 t_{2k} + \cdots + x_n t_{nk} = x_k$$

at each dependent node k, where t_{jk} is the branch transmittance of branch jk. *See also:* **linear signal flow graphs.** E155-0

signal frequency shift (frequency-shift facsimile system). The numerical difference between the frequencies corresponding to white signal and black signal at any point in the system. *See also:* **facsimile signal (picture signal).** E168-0

signal generator. A shielded source of voltage or power, the output level and frequency of which are calibrated, and usually variable over a range. *Note:* The output of known waveform is normally subject to one or more forms of calibrated modulation.
See:
absolute error;
automatic frequency control;
auxiliary device to an instrument;
available power;
backlash;
calibration level;
frequency deviation;
incidental amplitude-modulation factor;
incidental frequency modulation;
incidental phase modulation;
output attenuation;
spurious output. 42A30-9E4

signal indication. The information conveyed by the aspect of a signal. *See also:* **railway signal and interlocking.** 42A42-0

signaling. *See:* **ringing.**

signaling light. A projector used for directing light signals toward a designated target zone. *See also:* **signal lighting.** Z7A1-0

signal, input (control system, feedback). A signal applied to a system or element. See the figure attached to the definition of **signal, error.** *See also:* **control system, feedback.** 85A1-23E0

signal integration. The summation of a succession of signals by writing them at the same location on the storage surface. *See also:* **storage tube.** E158-15E6

signal level (electroacoustics). The magnitude of a signal, especially when considered in relation to an arbitrary reference magnitude. *Note:* Signal level may be expressed in the units in which the quantity itself is measured (for example, volts or watts) or in units expressing a logarithmic function of the ratio of the two magnitudes. *See:* **level.** E151-42A65

signal lighting. *Note:* For an extensive list of cross references, see *Appendix A.* E151-42A65

signal lines. The passive transmission lines through which the signal passes from one to another of the elements of the signal transmission system. *See:* **signal.** 0-13E6

signal operation (elevators). Operation by means of single buttons or switches (or both) in the car, and up-or-down direction buttons (or both) at the landings, by which predetermined landing stops may be set up or registered for an elevator or for a group of elevators. The stops set up by the momentary actuation of the car buttons are made automatically in succession as the car reaches those landings, irrespective of its direction of travel or the sequence in which the buttons are actuated. The stops set up by the momentary actuation of the up-and-down buttons at the landing are made automatically by the first available car in the group approaching the landing in the corresponding direction, irrespective of the sequence in which the buttons are actuated. With this type of operation, the car can be started only by means of a starting switch or button in the car. *See also:* **control (elevators).** 42A45-0

signal, output (control system, feedback). A signal delivered by a system or element. *See also:* **control system, feedback.** 85A1-23E0

signal output current (camera tubes or phototubes). The absolute value of the difference between output current and dark current. *See:* **phototubes.** *See also:* **beam tubes.** E160-15E6/2E2

signal-processing antenna system. An antenna system having circuit elements associated with its radiating elements that perform functions such as multiplication, storage, correlation, and time modulation of the input signals. *See also:* **antenna.** 0-3E1

signal, reference input (1) (general). The command expressed in a form directly usable by the system. The reference input signal is in the terms appropriate to the form in which the signal is used, that is, voltage, current, ampere-turns, etcetera. *See also:* **control system feedback.** ASI-34E10

(2) (control system, feedback). A signal external to a control loop that serves as the standard of comparison for the directly controlled variable. See the figure attached to the definition of **signal, feedback.** *See also:* **control system, feedback.** 85A1-23E0

signal relay. *See:* **alarm relay.**

signal repeater lights. A group of lights indicating the signal displayed for humping and trimming. 42A42-0

signal, return (control system, feedback) (closed loop). The signal resulting from a particular input signal, and transmitted by the loop and to be subtracted from that input signal. See the figure attached to the definition of **signal, error.** *See also:* **control system, feedback.** 85A1-23E0

signal-shaping amplifier (telegraph practice). An amplifier and associated electric networks inserted in the circuit, usually at the receiving end of an ocean cable, for amplifying and improving the waveshape of the signals. *See also:* **telegraphy.** 42A65-0

signal-shaping network (wave-shaping set). An electric network inserted (in a telegraph circuit) for improving the waveshape of the received signals. *See also:* **telegraphy.** 42A65-0

signal shutter (illuminating engineering). A device that modulates a beam of light by mechanical means for the purpose of transmitting intelligence. *See also:* **searchlight.** Z7A1-0

signal-to-clutter ratio (radar moving-target-indicator). The ratio of mean target echo power to the mean power received from clutter sources lying within the same resolution element. *See also:* **navigation.** 0-10E6

signal-to-interference ratio. The ratio of the magnitude of the signal to that of the interference or noise. *Note:* The ratio may be in terms of peak values or root-mean-square values and is often expressed in decibels. The ratio may be a function of the bandwidth of the system. *See:* **signal.** 0-13E6

signal-to-noise ratio (1) (general). The ratio of the value of the signal to that of the noise. *Notes:* (A) This ratio is usually in terms of peak values in the case of impulse noise and in terms of the root-mean-square values in the case of the random noise. (B) Where there is a possibility of ambiguity, suitable definitions of the signal and noise should be associated with the term; as, for example: peak-signal to peak-noise ratio; root-mean-square signal to root-mean-square noise ratio; peak-to-peak signal to peak-to-peak noise ratio, etcetera. (C) This ratio may be often expressed in decibels. (D) This ratio may be a function of the bandwidth of the transmission system. *See also:* **transmission characteristics; signal; signal-to-interference ratio; television.** E145-2E2

(2) (camera tubes). The ratio of peak-to-peak signal output current to root-mean-square noise in the output current. *Note:* Magnitude is usually not measured in tubes where the signal output is taken from target. *See also:* **beam tubes; camera tube; television; transmission characteristics.**

(3) (television transmission). The signal-to-noise ratio at any point is the ratio in decibels of the maximum peak-to-peak voltage of the video television signal, including synchronizing pulse, to the root-mean-square voltage of the noise. *Note:* The signal-to-noise ratio is defined in this way because of the difficulty of defining the root-mean-square value of the video signal or the peak-to-peak value of random noise. *See also:* **transmission characteristics; television.** 42A65-0

(4) (mobile communication). The ratio of a specified speech-energy spectrum to the energy of the noise in the same spectrum. *See also:* **transmission characteristics; television.** 0-6E1

(5) (sound recording and reproducing system). The ratio of the signal power output to the noise power in the entire pass band.

See:

available signal-to-noise ratio;
average noise figure;
babble;
circuit noise;
circuit noise level;
effective input-noise temperature;
equivalent noise conductance;
equivalent noise current;
equivalent noise resistance;
exchangeable power;
exchangeable power gain;
ideal noise diode;
impulse noise;
interference;
microphonism;
noise;
noise current generator;
noise factor;
noise factor, average;
noise figure;
noise level;
noise temperature;
noise temperature, standard;
noise unit;
noise voltage generator;
random noise;
reference noise;
shot noise, full;
spot noise figure;
television;
thermal noise;
thump.

See also: **transmission characteristics.** E191-0

signal transmission system. *See:* **carrier.**

signal, unit-impulse (automatic control). A signal that is an impulse having unity area. *See also:* **control system, feedback.** E270/85A1-23E0

signal, unit-ramp (automatic control). A signal that is zero for all values of time prior to a certain instant and equal to the time measured from that instant. *Note:* The unit-ramp signal is the integral of the unit-step signal. *See also:* **control system-feedback.** E270/85A1-23E0

signal, unit-step (automatic control). A signal that is zero for all values of time prior to a certain instant and unity for all values of time following. *Note:* The unit-step signal is the integral of the unit-impulse signal. *See also:* **control system, feedback.** E270/85A1-23E0

signal wave. A wave whose shape conveys some intelligence, message, or effect.
See:
audio;
beating;
beats;
communication;
electrical distance;
electrical length;
frequency band;
frequency range;
fundamental component;
fundamental frequency;
infrasonic frequency;
instantaneous frequency;
light microsecond;
propagation;
sideband attenuation;
signal wave;
spectrum;
subharmonic;
ultrasonic frequency;
video;
voice frequency;
wave front. 42A65-27E1

signal winding (input winding) (saturable reactor). A control winding to which the independent variable (signal wave) is applied. 42A65-0

sign digit (1) (electronic computation). A character used to designate the algebraic sign of a number. *See also:* **electronic digital computer.** E162/E270-0
(2). The digit in the sign position. *See also:* **electronic digital computer.** X3A12-16E9

significant digit. A digit that contributes to the accuracy or precision of a numeral. The number of significant digits is counted beginning with the digit contributing the most value, called the most-significant digit, and ending with the one contributing the least value, called the least-significant digit. X3A12-16E9

sign position. The position at which the sign of a number is located. *See also:* **electronic digital computer.** X3A12-16E9

silvering (electrotyping). The application of a thin conducting film of silver by chemical reduction upon a plastic or wax matrix. *See also:* **electroforming.** 42A60-0

silver oxide cell. A cell in which depolarization is accomplished by oxide of silver. *See also:* **electrochemistry.** 42A60-0

silver storage battery. An alkaline storage battery in which the positive active material is silver oxide and the negative contains zinc. *See also:* **battery (primary or secondary).** 42A60-0

silver-surfaced or equivalent. The term indicates metallic materials having satisfactory long-term performance and that operate within the temperature rise limits established for silver-surfaced electric contact parts and conducting mechanical joints. 37A100-31E11

simple *GCL* circuit. *See:* **simple parallel circuit.**

simple parallel circuit (simple *GCL* circuit). A linear, constant-parameter circuit consisting of resistance, inductance, and capacitance in parallel. *See also:* **network analysis.** E270-0

simple rectifier. A rectifier consisting of one commutating group if single-way or two commutating groups if double-way. *See also:* **rectification.** 34A1-34E24

simple rectifier circuit. A rectifier circuit consisting of one commutating group if single-way, or two commutating groups if double-way. *See also:* **rectification; rectifier circuit element.** E59/42A15-34E17

simple *RLC* circuit. *See:* **simple series circuit.**

simple series circuit (simple *RLC* circuit). A resistance, inductance, and capacitance in series. *See also:* **network analysis.** E270-0

simple sine-wave quantity. A physical quantity that is varying with time t as either $A \sin(\omega t + \theta_A)$ or $A \cos(\omega + \theta_B)$ where A, ω, θ_A, θ_B are constants. (**Simple** denotes that A, ω, θ_A, θ_B are constants.) *Notes:* (1) It is immaterial whether the sin or cos form is used, so long as no ambiguity or inconsistency is introduced. (2) A is the amplitude or maximum value, $\omega t + \theta_A$ (or $\omega t + \theta_B$) the phase, θ_A (or θB) the phase angle, However, when no ambiguity may arise, **phase angle** may be abbreviated **phase**. (3) In certain special applications, for example, modulation, $\omega t + \theta$ is called the angle (of a sine wave), (not phase angle) in order to clarify particular uses of the word "phase." Another permissible term for $(\omega t + \theta)$ is argument (sine wave). E270-0

simple sound source. A source that radiates sound uniformly in all directions under free-field conditions. E157-1E1

simple target (radar). A target having a reflecting surface such that the amplitude of the reflected signal does not vary with the aspect of the target; for example, a metal sphere. *See also:* **navigation.** E172-10E6

simple tone (pure tone). (1) A sound wave, the instantaneous sound pressure of which is a simple sinusoidal function of the time. (2) A sound sensation characterized by its singleness of pitch. *Note:* Whether or not a listener hears a tone as simple or complex is dependent upon the ability, experience, and listening attitude. *See:* **complex tone.** *See also:* **electroacoustics.** 0-1E1

simplex circuit. A circuit derived from a pair of wires by using the wires in parallel with ground return. *See also:* **transmission line.** 42A65-0

simplexed circuit. A two-wire metallic circuit from which a simplex circuit is derived, the metallic and simplex circuits being capable of simultaneous use. *See also:* **transmission line.** 42A65-0

simplex operation. A method of operation in which communication between two stations takes place in one direction at a time. *Note:* This includes ordinary transmit-receive operation, press-to-talk operation, voice-operated carrier and other forms of manual or automatic switching from transmit to receive. *See also:* **radio transmission; telegraphy.** 42A65-31E3;E145/E182

simplex supervision. The use of a simplex signaling channel for transmitting supervisory signals between two points in a connection. *See also:* **telephone switching system.** 42A65-19E1

simply connected region (two-dimensional space). A region, such that any closed curve in the region encloses points all of which belong to the region. E270-0

simply mesh-connected circuit. A circuit in which two, and only two, current paths extend from the terminal of entry of each phase conductor, one to the terminal of entry that precedes and the other to the terminal of entry that follows the first terminal in the normal sequence, and from which the amplitude of the voltages to the first terminal is normally the smallest (when the number of phases is greater than three). *See also:* **network analysis.** E270-0

simulate (computing systems). To represent the functioning of one system by another, for example, to represent one computer by another, to represent a physical system by the execution of a computer program, to represent a biological system by a mathematical model. *See also:* **electronic digital computer; electronic analog computer.** X3A12-16E9

simulation (1) (general). The representation of an actual system by the analogous characteristics of some device easier to construct, modify, or understand. *See also:* **electronic analog computer; system.**
(2) (physical). The use of a model of a physical system in which computing elements are used to represent some but not all of the subsystems. *See also:* **electronic analog computer.**
(3) (mathematical). The use of a model of mathematical equations in which computing elements are used to represent all of the subsystems. E165-0

simultaneous lobing (radar) (electronic navigation). A direction-determining technique utilizing the received energy of two concurrent and partially overlapped signal lobes; the relative phase, or the relative power, of the two signals received from a target is a measure of the angular displacement of the target from the equiphase or equisignal direction. Compare with lobe switching. *See also:* **antenna; navigation.** E172-10E6;E149-3E1

sinad ratio (mobile communication). A measure expressed in decibels of the ratio of (1) the signal plus noise plus distortion to (2) noise plus distortion produced at the output of a receiver that is the result of a modulated-signal input. *See also:* **mobile communication system; receiver performance.** 0-6E1

sinad sensitivity (receiver performance). The minimum standard modulated carrier-signal input required to produce a specified sinad ratio at the receiver output. *See also:* **receiver performance.** 0-6E1

sine-current coercive force (toroidal magnetic amplifier cores). The instantaneous value of sine-current magnetizing force at which the dynamic hysteresis loop passes through zero induction. E106-0

sine-current differential permeability (toroidal magnetic amplifier cores). The slope of the sides of the dynamic hysteresis loop obtained with a sine-current magnetizing force. E106-0

sine-current magnetizing force (toroidal magnetic amplifier cores). The applied magnetomotive force per unit length for a core symmetrically cyclicly magnetized with sinusoidal current. E106-0

sine wave. A wave that can be expressed as the sine of a linear function of time, or space, or both. E145-0

sine-wave generator. An alternating-current generator whose output voltage waveform contains a single main frequency with low harmonic content of prescribed maximum level. *See also:* **asynchronous machine; synchronous machine.** 0-31E8

sine-wave response (camera tubes). *See:* **amplitude response.**

sine-wave sweep. A sweep generated by a sine function. *See:* **oscillograph.** 0-9E4

singing. An undesired self-sustained oscillation existing in a transmission system or transducer. *Note:* Very-low-frequency oscillation is sometimes called motor-boating. *See also:* **transmission characteristics; oscillatory circuit.** E145/E151-2E2;42A65-31E3;0-42A65

singing margin (gain margin). The excess of loss over gain around a possible singing path at any frequency, or the minimum value of such excess over a range of frequencies. *Note:* Singing margin is usually expressed in decibels. *See also:* **transmission characteristics.** E151-0;42A65-31E3;0-42A65

singing point (circuit coupled back to itself). The point at which the gain is just sufficient to make the circuit break into oscillation. *See also:* **oscillatory circuit; transmission characteristics.** 42A65-31E3

single-address. Pertaining to an instruction that has one address part. In a typical single-address instruction the address may specify either the location of an operand to be taken from storage, the destination of a previously prepared result, the location of the next instruction to be interpreted, or an immediate address operand. Synonymous with one-address. *See also:* **electronic digital computer.** 0-16E9

single-address code (electronic computation). *See:* **instruction code.**

single-anode tank (single-anode tube). An electron tube having a single main anode. *Note:* This term is used chiefly for pool-cathode tubes. *See also:* **tube definitions.** 42A70-15E6

single automatic operation (elevators). Automatic operation by means of one button in the car for each landing level served and one button at each landing so arranged that if any car or landing button has been actuated, the actuation of any other car or landing operating button will have no effect on the operation of the car until the response to the first button has been completed. *See also:* **control (elevators).** 42A45-0

single-break switch (circuit). A switch that opens each conductor at one point only. 37A100-31E11

single-circuit system (protective signaling). A system of protective wiring that employs only the nongrounded side of the battery circuit, and consequently depends primarily on an open circuit in the wiring to initiate an alarm. *See also:* **protective signaling.** 42A65-0

single-degree-freedom gyro. A gyro in which the rotor is free to precess (relative to the case) about only the axis orthogonal to the rotor spin axis. *See also:* **navigation.** E174-10E6

single-element fuse. A fuse having a current-responsive element comprising one or more parts with single fusing characteristic. 37A100-31E11

single-element relay. An alternating-current relay having a set of coils energized by a single circuit. 42A42-0

single-end control (single-station control). A control system in which provision is made for operating a vehicle from one end or one location only. *See also:* **multiple-unit control.** 42A42-0

single-ended amplifier. An amplifier in which each stage normally employs only one active element (tube, transistor, etcetera) or, if more than one active element is used, in which they are connected in parallel so that operation is asymmetric with respect to ground. *See also:* **amplifier.** E145/E151/42A65-0

single-ended push-pull amplifier circuit (electroacoustics). An amplifier circuit having two transmission paths designed to operate in a complementary manner and connected so as to provide a single unbalanced output without the use of an output transformer. *See:* **amplifier.** E151-0

single-faced tape. Fabric tape finished on one side with rubber or synthetic compound. 42A35-31E13

single feeder. A feeder that forms the only connection between two points along the route considered. 42A35-31E13

single-frequency signal-to-noise ratio (sound recording and reproducing system). The ratio of the single-frequency signal power output to the noise power in the entire pass band. *See also:* **noise (sound recording and reproducing system).** E191-0

single-frequency simplex operation (radio communication). The operation of a two-way radio-communication circuit on the same assigned radio-frequency channel, which necessitates that intelligence can be transmitted in only one direction at a time. *See also:* **channel spacing.** 0-6E1

single hoistway (elevators). A hoistway for a single elevator or dumbwaiter. *See also:* **hoistway (elevator or dumbwaiter).** 42A45-0

single-layer winding (rotating machinery). A winding in which there is only one actual coil side in the depth of the slot. (Also known as one-coil-side-per-slot winding). *See also:* **asynchronous machine; direct-current commutating machine; synchronous machine.** 0-31E8

single-line diagram. *See:* **one-line diagram.**

single-office exchange. A telephone exchange served by a single central office. *See also:* **telephone system.** 42A65-19E1

single-operator arc welder. An arc-welding power supply designed to deliver current to only one welding arc. *See also:* **electric arc-welding apparatus.** 87A1-0

single-phase circuit. An alternating-current circuit consisting of two or three intentionally interrelated conductors that enter (or leave) a delimited region at two or three terminals of entry. If the circuit consists of two conductors, it is intended to be so energized that, in the steady state, the voltage between the two terminals of entry is an alternating voltage. If the circuit consists of three conductors, it is intended to be so energized that, in the steady state, the alternating voltages between any two terminals of entry have the same period and are in phase or in phase opposition. *See also:* **center of distribution; network analysis.** E270-0

single-phase electric locomotive. An electric locomotive that collects propulsion power from a single phase of an alternating-current distribution system. *See also:* **electric locomotive.** 42A42-0

single-phase machine. A machine that generates or utilizes single-phase alternating-current power. *See also:* **asynchronous machine; synchronous machine.** 0-31E8

single-phase motor (rotating machinery). A machine that converts single-phase alternating-current electric power into mechanical power, or that provides mechanical force or torque. *See also:* **synchronous machine.** 0-31E8

single-phase symmetrical set (1) (polyphase voltages). A symmetrical set of polyphase voltages in which the angular phase difference between successive members of the set is π radians or odd multiples thereof. The equations of **symmetrical set (polyphase voltages)** represent a single-phase symmetrical set of polyphase voltages if k/m is ½ or an odd multiple thereof. (The symmetrical set of voltages represented by the equations of **symmetrical set (polyphase voltages)** may be said to have single-phase symmetry if k/m is an odd (positive or negative) multiple of ½.) *Notes:* (1) A set of polyphase voltages may have single-phase symmetry only if m, the number of members of the set, is an even number. (2) This definition may be applied to a two-phase four-wire or five-wire circuit if m is considered to be 4 instead of 2. It is not applicable to a two-phase three-wire circuit. *See also:* **network analysis.** E270-0

(2) (polyphase currents). This definition is obtained from the corresponding definitions for voltage by substituting the word **current** for **voltage**, and the symbol I for E and β for α wherever they appear in the equations of **symmetrical set (polyphase voltages).** The subscripts are unaltered. *See also:* **network analysis.** E270-0

single-phase synchronous generator. A generator that produces a single alternating electromotive force at its terminals. It delivers electric power that pulsates at double frequency. *See:* **synchronous machine.** 42A10-0

single-phase three-wire circuit. A single-phase circuit consisting of three conductors, one of which is identified as the neutral conductor. *See also:* **network analysis.** E270-0

single-phase two-wire circuit. A single-phase circuit consisting of only two conductors. *See also:* **network analysis.** E270-0

single-phasing (rotating machinery). An abnormal operation of a polyphase machine when its supply is effectively single-phase. *See also:* **asynchronous machine; synchronous machine.** 0-31E8

single-polarity pulse. A pulse in which the sense of the departure from normal is in one direction only. *See:* **unidirectional pulse.** *See also:* **pulse.** 42A65-0

single-pole relay. *See:* **relay, single-pole.**

single-pressure-zone potheads. A pressure-type pothead intended to operate with one pressure zone. *See:* **multipressure-zone pothead; pressure-type pothead.** E48-0

single service. One service only supplying a consumer. *Note:* Either or both lighting and power load may be connected to the service. *See also:* **service.** 42A35-31E13

single-shot blasting unit. A unit designed for firing only one explosive charge at a time. *See also:* **blasting unit.** 42A85-0

single-shot blocking oscillator. A blocking oscillator modified to operate as a single-shot trigger circuit. *See also:* **trigger circuit.** 42A65-0

single-shot multivibrator (single-trip multivibrator). A multivibrator modified to operate as a single-shot trigger circuit. *See also:* **trigger circuit.** 42A65-0

single-shot trigger circuit (single-trip trigger circuit). A trigger circuit in which a triggering pulse intiates one complete cycle of conditions ending with a stable condition. *See also:* **trigger circuit.** 42A65-0

single-sideband modulation (SSB). Modulation whereby the spectrum of the modulating function is translated in frequency by a specified amount either with or without inversion. *See also:* **modulating systems; modulation.** E145/E170/42A65-31E3

single-sideband transmission. The method of operation in which one sideband is transmitted and the other sideband is suppressed. The carrier wave may be either transmitted or suppressed. E145-31E3

single-sideband transmitter. A transmitter in which one sideband is transmitted and the other is effectively eliminated. *See also:* **modulating systems.** E145-0

single-sided printed-circuit board. *See:* **printed-circuit board.**

single-station control. *See:* **single-end control.**

single step (computing systems). Pertaining to a method of operating a computer in which each step is performed in response to a single manual operation. *See also:* **electronic digital computer.** X3A12-16E9

single-stroke bell. An electric bell that produces a single stroke on its gong each time its mechanism is actuated. *See also:* **protective signaling.** 42A65-0

single-sweep mode (oscilloscopes). Operating mode for a triggered-sweep oscilloscope in which the sweep must be reset for each operation, thus preventing unwanted multiple displays. Particularly useful for trace photography. In the interval after the sweep is reset and before it is triggered, it is said to be armed. *See:* **oscillograph.** 0-9E4

single-throw (switching device). A qualifying term indicating that the circuit can be opened or closed by the operation of only one set of contacts. 37A100-31E11

single-tone keying (modulation systems). That form of keying in which the modulating function causes the carrier to be modulated by a single tone for one condition, which may be either a mark or a space, the carrier being unmodulated for the other condition. *See also:* **modulating systems; telegraphy.** E145/E170-0;42A65-19E4

single-track (standard track) (electroacoustics). A variable-density or variable-area sound track in which both positive and negative halves of the signal are linearly recorded. *See also:* **phonograph pickup.** E157-1E1

single-trip multivibrator. *See:* **single-shot multivibrator.**

single-trip trigger circuit. *See:* **single-shot trigger circuit.**

single-tuned amplifier. An amplifier characterized by resonance at a single frequency. *See also:* **amplifier.** 42A65-0

single-tuned circuit. A circuit that may be represented by a single inductance and a single capacitance, together with associated resistances. *See also:* **circuits and devices.** 42A65-0

single-valued function. A function u is single valued when to every value of x (or set of values of x_1, x_2, $\cdots$, x_n) there corresponds one and only one value of u. Thus $u = ax$ is single valued if a is in arbitrary constant. E270-0

single-way rectifier. A rectifier in which the current between each terminal of the alternating-voltage circuit and the rectifier circuit element or elements conductively connected to it flows only in one direction. *See also:* **double-way rectifier circuit; power rectifier; rectification; rectifier circuit element.** E59-34E17/34E24;34A1/42A15-0

single-winding multispeed motor. A type of multispeed motor having a single winding capable of reconnection in two or more pole groupings. *See:* **asynchronous machine; direct-current commutating machine; synchronous machine.** 42A10-0

single-wire line (waveguides). A surface-wave transmission line consisting of a single conductor so treated as to confine the propagated energy to the neighborhood of the wire. The treatment may consist of a coating of dielectric. *See:* **waveguides.** 0-3E1

singular point (control system). Synonymous with **equilibrium point.** *See also:* **control system.** 0-23E0

sink (1) (oscillator). The region of a Rieke diagram where the rate of change of frequency with respect to phase of the reflection coefficient is maximum. Operation in this region may lead to unsatisfactory performance by reason of cessation or instability of oscillations. *See:* **oscillatory circuit.** E160-15E6

(2) (communication practice). (A) A device that drains off energy from a system. (B) A place where energy from several sources is collected or drained away. 42A65-0

sink node (network analysis). A node having only incoming branches. *See also:* **linear signal flow graphs.** E155-0

sinusoidal electromagnetic wave (homogeneous medium). A wave whose electric field vector is proportional to the sine (or cosine) or an angle that is a linear function of time, or a distance, or of both. *See also:* **radio propagation.** 0-3E2

sinusoidal field. A field in which the field quantities vary as a sinusoidal function of an independent variable, such as space or time. E270-0

sinusoidal function. A function of the form $A \sin(x+a)$. A is the amplitude, x is the independent variable, and a the phase angle. Note that $\cos(x)$ may be expressed as $\sin[x + (\pi/2)]$. *See also:* **simple sine-wave quantity.** E270-0

siphon recorder. A telegraph recorder comprising a sensitive moving-coil galvanometer with a siphon pen that is directed by the moving coil across a traveling strip of paper. *See also:* **telegraphy.** 42A65-0

site error (electronic navigation). Error due to the distortion in the electromagnetic field by objects in the vicinity of the navigational equipment. *See also:* **navigation.** 0-10E6

six-phase circuit. A combination of circuits energized by alternating electromotive forces that differ in phase by one-sixth of a cycle, that is, 60 degrees. *Note:* In practice the phases may vary several degrees from the specified angle. *See also:* **center of distribution.** 42A35-31E13

size threshold. The minimum perceptible size of an object. It also is defined as the size that can be detected some specific fraction of the times it is presented to an observer, usually 50 percent. It usually is measured in minutes of arc. *See also:* **visual field.** Z7A1-0

skate machine. A mechanism, electrically controlled, for placing on, or removing from, the rails a skate that, if allowed to engage with the wheels of a car, provides continuous braking until the car is stopped and that may be electrically or pneumatically operated. 42A42-0

skating force. *See:* **side thrust.**

skeleton frame (rotating machinery). A stator frame consisting of a simple structure that clamps the core but does not enclose it. *See also:* **cradle base (rotating machinery).** 0-31E8

skew (1) (facsimile). The deviation of the received frame from rectangularity due to asynchronism between scanner and recorder. *Note:* Skew is expressed numerically as the tangent of the angle of this deviation. *See also:* **recording (facsimile); transmission characteristics.** E168/42A65-0
(2) (magnetic storage). The angular displacement of an individual printed character, group of characters, or other data, from the intended or ideal placement. *See also:* **static magnetic storage; transmission characteristics.** X3A12-16E9

skewed slot (rotating machinery). A slot of a rotor or stator of an electric machine, placed at an angle to the shaft so that the angular location of the slot at one end of the core is displaced from that at the other end. Slots are commonly skewed in many types of machines to provide more uniform torque, less noise, and better voltage waveform. *See also:* **rotor (rotating machinery); stator.** 0-31E0

skiatron (electronic navigation). (1) A dark-trace storage-type cathode-ray tube. (2) A display employing an optical system with a dark-trace tube. *See:* **dark-trace tube.** *See also:* **tube definition.** E172-10E6

skid wire (pipe-type cable) (power distribution, underground cables). Wire or wires, usually D shaped, applied open spiral with curved side outward with a suitable spacing between turns over the outside surface of the cable. Its purpose is to facilitate cable pulling and to provide mechanical protection during installation. *See also:* **power distribution, underground construction.** 0-31E1

skim tape. Filled tape coated on one or both sides with a thin film of uncured rubber or synthetic compound to produce a coating suitable for vulcanization. *See also:* **power distribution, underground construction.** 42A35-31E13

skin depth. For a conductor carrying currents at a given frequency as a result of the electromagnetic waves acting upon its surface, the depth below the surface at which the current density has decreased one neper below the current density at the surface. *Note:* Usually the skin depth is sufficiently small so that for ordinary configurations of good conductors, the value obtained for a plane wave falling on a plane surface is a good approximation. E146-3E1

skin effect (conductor). The phenomenon of nonuniform current distribution over the cross section caused by the time variation of the current in the conductor itself. *See also:* **transmission line.** E270-0

skip (computing systems). To ignore one or more instructions in a sequence of instructions. *See also:* **electronic digital computer.** X3A12-16E9

skip distance. The minimum separation for which radio waves of a specified frequency can be transmitted at a specified time between two points on the earth by reflection from the regular ionized layers of the ionosphere. *See also:* **radiation.** 42A65-0

sky factor. The ratio of the illumination on a horizontal plane at a given point inside a building due to the light received directly from the sky, to the illumination due to an unobstructed hemisphere of sky of uniform luminance (photometric brightness) equal to that of the visible sky. *See also:* **sunlight.** Z7A1-0

sky light. Visible radiation from the sun redirected by the atmosphere. *See also:* **sunlight.** Z7A1-0

sky wave. Ionospheric wave. 0-3E2

sky-wave contamination (electronic navigation). Degradation of the received ground-wave signal, or of the desired sky-wave signal, by the presence of delayed ionospheric-wave components of the same transmitted signal. *See also:* **navigation.** 0-10E6

sky-wave correction (electronic navigation). A correction for sky-wave propagation errors applied to measured position data; the amount of the correction is established on the basis of an assumed position and an assumed ionosphere height. *See also:* **navigation.** E172-10E6

sky-wave station-error (sky-wave synchronized loran). The error of station synchronization due to the effect of variations of the ionosphere on the time of transmission of the synchronizing signal from one station to the other. *See also:* **navigation.** E172-10E6

slabbing or arcwall machine. A power-driven mobile-cutting machine that is a single-purpose cutter in that it cuts only a horizontal kerf at variable heights. *See also:* **mining.** 42A85-0

slack-rope switch (elevators). A device that automatically causes the electric power to be removed from the elevator driving-machine motor and brake when the hoisting ropes of a winding-drum machine become slack. *See:* **control.** 42A45-0

slant distance (electronic navigation). The distance between two points not at the same elevation. Used in contrast to ground distance. *See also:* **navigation.** E172-10E6

slant range (radar). A term often used to mean slant distance. *See:* **slant distance.** *See also:* **navigation.** E172-10E6

slave drive. *See:* **electric drive; follower drive.**

slaved tracking (power supplies). A system of interconnection of two or more regulated supplies in which one (the master) operates to control the others (the slaves). The output voltage of the slave units may be equal or proportional to the output voltage of the master unit. (The slave output voltages track the master output voltage in a constant ratio.) *See:* **complementary tracking, master/slave.** *See also:* **power supply.** KPSH-10E1

slave relay. *See:* **auxiliary relay.** *See also:* **relay.**

slave station (electronic navigation). A station of a synchronized group whose emissions are controlled by a master station. *See also:* **radio navigation.** 42A65-0;E172-10E6

sleet hood (switch). A cover for the contacts to prevent sleet from interfering with successful operation of the switch. 37A100-31E11

sleetproof. So constructed or protected that the accumulation of sleet will not interfere with successful operation. 42A95-0;IC1-34E10;37A100-31E11

sleeve (1) (plug) (three-wire telephone-switchboard plug). A cylindrically shaped contacting part, usually placed in back of the tip or ring but insulated therefrom. *See also:* **telephone switching system.** 42A65-0
(2) (rotating machinery). A tubular part designed to fit around another part. *Note:* In a sleeve bearing, the sleeve is that component that includes the cylindrical

inner surface within which the shaft journal rotates. 0-31E8

sleeve bearing (rotating machinery). A bearing with a cylindrical inner surface in which the journal of a rotor (or armature) shaft rotates. *See:* **rotor (rotating machinery).** 0-31E8

sleeve conductor. *See:* **sleeve wire.**

sleeve-dipole antenna. A dipole antenna surrounded in its central portion by a coaxial conducting sleeve. *See also:* **antenna.** 0-3E1

sleeve-stub antenna. An antenna consisting of half of a sleeve-dipole antenna projecting from an extended conducting surface. *See also:* **antenna.** 42A65-3E1

sleeve supervision. The use of the sleeve circuit for transmitting supervisory signals. *See also:* **telephone switching system.** 42A65-0

sleeve-type suppressor (electromagnetic compatibility). A suppressor designed for insertion in a high-tension ignition cable. *See also:* **electromagnetic compatibility.** CISPR-27E1

sleeve wire (telephony) (sleeve conductor). The conductor, usually accompanying the tip and ring leads of a switched connection, that provides for miscellaneous functions necessary to the control and supervision of the connection. In cord-type switchboards, the sleeve wire is that conductor that is associated with the sleeve contacts of the jacks and plugs. *See also:* **telephone switching systems.** 0-9E1

slewing rate (power supplies). A measure of the programming speed or current-regulator-response timing. The slewing rate measures the maximum rate-of-change of voltage across the output terminals of a power supply. Slewing rate is normally expressed in volts per second ($\Delta E/\Delta T$) and can be converted to a sinusoidal frequency-amplitude product by the equation $f(E_{pp})$ = slewing rate$/\pi$, where E_{pp} is the peak-to-peak sinusoidal volts. Slewing rate $= \pi f(E_{pp})$. *See:* **high-speed regulator.** *See also:* **power supply.** KPSH-10E1

slicer (amplitude gate) (clipper-limiter*). A transducer that transmits only portions of an input wave lying between two amplitude boundaries. *Note:* The term is used especially when the two amplitude boundaries are close to each other as compared with the amplitude range of the input. *See also:* **circuits and devices.**

*Deprecated 42A65-0

slide-screw tuner (transmission lines and waveguides). An impedance or matching transformer that consists of a slotted waveguide or coaxial-line section and an adjustable screw or post that penetrates into the guide or line and can be moved axially along the slot. *See also:* **waveguide.** 0-9E4

sliding contact. An electric contact in which one conducting member is maintained in sliding motion over the other conducting member. *See also:* **contactor; railway signal and interlocking.** 42A42-0

sliding load. *See:* **load, sliding.**

sliding short circuit. A short-circuit termination that consists of a section of waveguide or transmission line fitted with a sliding short-circuiting piston (contacting or noncontacting) that ideally reflects all the energy back toward the source. *See also:* **waveguide.** 0-9E4

slime. Finely divided insoluble metal or compound forming on the surface of an anode or in the solution during electrolysis. *See also:* **electrodeposition.** 42A60-0

slinging wire. A wire used to suspend and carry current to one or more cathodes in a plating tank. *See also:* **electroplating.** 42A60-0

slip (rotating machinery) (1). The quotient of (A) the difference between the synchronous speed and the actual speed of a rotor, to (B) the synchronous speed, expressed as a ratio, or as a percentage.

(2). The difference between the speed of a rotating magnetic field and that of a rotor, expressed in revolutions per minute.

(3) (electric couplings). The difference between the speeds of the two rotating members. *See:* **asynchronous machine; synchronous machine.** 42A10-31E8

slip relay. A relay arranged to act when one or more pairs of driving wheels increase or decrease in rotational speed with respect to other driving wheels of the same motive power unit. *See also:* **multiple-unit control.** 42A42-0

slip ring. *See:* **collector ring.**

slip-ring induction motor. *See:* **wound-rotor induction motor.**

slope angle (electronic navigation). *See:* **glide-slope angle.** *See also:* **navigation.**

slot (rotating machinery). A channel or tunnel opening onto or near the air gap and passing essentially in an axial direction through the rotor or stator core. A slot usually contains the conductors of a winding, but may be used exclusively for ventilation. *See also:* **rotor (rotating machinery); stator.** 0-31E8

slot antenna. A radiating element formed by a slot in a conducting surface. *See also:* **antenna.** E145/E149/42A65-3E1

slot array. An antenna array formed of slot radiators. *See also:* **antenna.** 50I62-3E1

slot cell (rotating machinery). A sheet of insulation material used to line a slot before the winding is placed in it. *See also:* **rotor (rotating machinery); stator.** 0-31E8

slot coupling factor (slot-antenna array). The ratio of the desired slot current to the available slot current, controlled by changing the depth of penetration of the slot probe into the waveguide. *See also:* **navigation.** 0-10E6

slot current ratio (slot-antenna array). The relative slot currents in the slots of the waveguide reading from its center to its end, with the maximum taken as 1; this ratio is dependent upon the slot spacing factor and the slot coupling factor. *See also:* **navigation.** 0-10E6

slot discharge (rotating machine). Sparking between the outer surface of coil insulation and the grounded slot surface, caused by capacitive current between conductors and iron. The resulting current pulses have a fundamental frequency of a few kilohertz. *See also:* **asynchronous machine; direct-current commutating machine; synchronous machine.** 0-31E8

slot-discharge analyzer (rotating machinery). An instrument designed for connection to an energized winding of a rotating machine, to detect pulses caused by slot discharge, and to discriminate between them and pulses otherwise caused. *See also:* **asynchronous machine; direct-current commutating machine; synchronous machine.** 0-31E8

slot insulation (rotating machinery). A sheet or deposit of insulation material used to line a slot before the winding is placed in it. *See also:* **asynchronous machine; direct-current commutating machine; synchronous machine.** 0-31E8

slot liner (rotating machinery). Separate insulation between an embedded insulated coil side and the slot, which can provide mechanical protection and additional dielectric strength. *See also:* **rotor (rotating machinery); stator.** 0-31E8

slot packing (filler) (rotating machinery). Additional insulation used to pack embedded coil sides to ensure a tight fit in the slots. *See also:* **rotor (rotating machinery); stator.** 0-31E8

slot separator insulation. *See:* **separator insulation, slot.**

slot space factor (rotating machinery). The ratio of the cross-sectional area of the conductor metal in a slot to the total cross-sectional area of the slot. *See also:* **asynchronous machine; direct-current commutating machine; synchronous machine.** 42A10-31E8

slot spacing factor (slot-antenna array). A value proportional to the size of the angle between the slot location and the null of the internal standing wave; this factor is dependent upon frequency. 0-10E6

slotted armature (rotating machine). An armature with the winding placed in slots. *See also:* **armature.** 50I10-31E8

slotted line. *See:* **slotted section.**

slotted section (slotted line) (slotted waveguide). A section of a waveguide or shielded transmission line the shield of which is slotted to permit the use of a carriage and travelling probe for examination of standing waves. *See also:* **auxiliary device to an instrument.** 42A30/50I62-3E1/9E4

slotted waveguide. *See:* **slotted section.**

slot-type antenna (aircraft). A slot in the normal streamlined metallic surface of an aircraft, excited electromagnetically by a structure within the aircraft. Radiation is thus obtained without projections that would disturb the aerodynamic characteristics of the aircraft. Radiation from a slot is essentially directive. 42A41-0

slot wedge (rotating machinery). The element placed above the turns or coil sides in a stator or rotor slot, and held in place by engagement of wedge (slots) grooves along the sides of the coil slot, or by projections from the sides of the slot tending to close the top of the slot. *Note:* A wedge may be a thin strip of material provided solely as insulation or to provide temporary retention of the coils during the manufacturing process. It may be a piece of structural insulating material or high-strength metal to hold the coils in the slot. Slots in laminated cores are normally wedged with insulating material. *See also:* **rotor (rotating machinery); stator.** 0-31E8

slow-operate relay. A slugged relay that has been specifically designed for long operate time but not for long release time. *Caution:* The usual slow-operate relay has a copper slug close to the armature, making it also at least partially slow to release. 0-21E0

slow-operating relay. A relay that has an intentional delay between energizing and operation. *See also:* **electromagnetic relay.** 42A65-0

slow-release relay. A relay that has an intentional delay between de-energizing and release. *Note:* The reverse motion need not have any intentional delay. *See also:* **electromagnetic relay.** 42A65-21E0

slow release time characteristic, relay. *See:* **relay slow release time characteristic.**

slow-speed starting (industrial control). A control function that provides for starting an electric drive only at the minimum-speed setting. *See:* **starter.** ICI-34E10

slow-wave circuit (microwave tubes). A circuit whose phase velocity is much slower than the velocity of light. For example, for suitably chosen helixes the wave can be considered to travel on the wire at the velocity of light but the phase velocity is less than the velocity of light by the factor that the pitch is less than the circumference. *See also:* **microwave tube or valve.** 0-15E6

slug, relay. *See:* **relay slug.**

slug tuner (waveguide). A waveguide or transmission-line tuner containing one or more longitudinally adjustable pieces of metal or dielectric. *See:* **waveguide.** E147-3E1

slug tuning. A means for varying the frequency of a resonant circuit by introducing a slug of material into either the electric or magnetic fields or both. *See also:* **network analysis; radio transmission.** E145/42A65-0

slush compound (corrosion). A non-drying oil, grease, or similar organic compound that, when coated over a metal, affords at least temporary protection against corrosion. *See:* **corrosion terms.** CM-34E2

small-signal forward transadmittance. The value of the forward transadmittance obtained when the input voltage is small compared to the beam voltage. *See:* **electron-tube admittances.** E160-15E6

small-signal resistance (semiconductor rectifier). The resistive part of the quotient of incremental voltage by incremental current under stated operating conditions. *See also:* **rectification.** E59-34E17

small wiring. *See:* **secondary and control wiring.**

smashboard signal. A signal so designed that the arm will be broken when passed in the stop position. *See also:* **railway signal and interlocking.** 42A42-0

smooth. To apply procedures that decrease or eliminate rapid fluctuations in data. *See also:* **electronic digital computer.** X3A12-16E9

smothered-arc furnace. A furnace in which the arc or arcs is covered by a portion of the charge. 42A60-0

snake. *See:* **fish tape.**

snapover. When used in connection with alternating-current testing, a quasi-flashover or quasi-sparkover, characterized by failure of the alternating- current power source to maintain the discharge, thus permitting the dielectric strength of the specimen to recover with the test voltage still applied. *See also:* **test voltage and current.** 68AI-31E5

snapshot dump (computing systems). A selective dynamic dump performed at various points in a machine run. *See also:* **electronic digital computer.** X3A12-16E9

snow (intensity-modulated display). A varying speckled background caused by noise. *See also:* **radar; television.** 42A65-2E2

soak, relay. *See:* **relay soak.**

socket*. *See:* **lampholder.**

*Deprecated

soft limiting. *See:* **limiter circuit.** *See also:* **electronic analog computer.**

software (electronic computers). (1) Computer programs, routines, programming languages and systems. (2) The collection of related utility, assembly, and other programs that are desirable for properly presenting a given machine to a user. (3) Detailed procedures to be followed, whether expressed as programs for a computer or as procedures for an operator or other person.

(4) Documents, including hardware manuals and drawings, computer-program listings and diagrams, etcetera. (5) Items such as those in (1), (2), (3), and (4) as contrasted with **hardware.** *See also:* **electronic digital computer.** E162-0

solar array (photovoltaic power system). A group of electrically interconnected solar cells assembled in a configuration suitable for oriented exposure to solar flux. *See also:* **photovoltaic power system; solar cells (photovoltaic power system).** 0-10E1

solar cells (photovoltaic power system).
See:
albedo;
dose;
dose rate;
emissivity;
emittance;
flux;
gamma ray;
heat sink;
ionization;
ionizing radiation;
***n*-on-*p* solar cells;**
photon;
***p*-on-*n* solar cells;**
rad;
reflectance;
reflectivity;
solar array;
substrate;
transmittance.
See also: **photovoltaic power system.**

solar panel (photovoltaic power system). *See:* **solar array.**

solderability. That property of a metal surface to be readily wetted by molten solder. *See also:* **soldered connections (electronic and electric applications).** 99A1-0

soldered connections (electronic and electric applications.
See:
bridging;
dip soldering;
eyelet;
flux;
interfacial connection;
jumper;
mechanical wrap or connection;
plated-through hole;
printed circuit;
printed-circuit assembly;
printed wiring;
solder projections;
solder splatter;
solderability;
soldered joints;
stress relief;
visual inspection;
wetting;
wicking.

soldered joints. The connection of similar or dissimilar metals by applying molten solder, with no fusion of the base metals. *See also:* **soldered connections (electronic and electric applications).** 99A1-0

solder projections. Icicles, nubs, and spikes are undesirable protrusions from a solder joint. *See also:* **soldered connections (electronic and electric applications).** 99A1-0

solder splatter. Unwanted fragments of solder. *See also:* **soldered connections (electronic and electric applications).** 99A1-0

solenoid. An electric conductor wound as a helix with a small pitch, or as two or more coaxial helixes. *See:* **solenoid magnet.** E270-0

solenoid magnet (solenoid) (industrial control). An electromagnet having an energizing coil approximately cylincrical in form, and an armature whose motion is reciprocating within and along the axis of the coil. E270/42A25-34E10

solenoid relay. *See:* **plunger relay.** *See also:* **relay.**

soleplate (rotating machinery). *See:* **rail.**

solid angle. A ratio of the area on the surface of a sphere to the square of the radius of the sphere. It is expressed in steradians. *Note:* Solid angle is a convenient way of expressing the area of light sources and luminaires for computation of discomfort glare factors. It combines into a single number the projected area A_p of the luminaire and the distance D between the luminaire and the eye. It usually is computed by means of the approximate formula

$$\omega = \frac{A_p}{D^2}$$

in which A_p and D^2 are expressed in the same units. This formula is satisfactory when the distance D is greater than about three times the maximum linear dimension of the projected area of the source. Larger projected areas should be subdivided into several elements. *See also:* **inverse-square law (illuminating engineering).** Z7A1-0

solid bushing (outdoor electric apparatus). A bushing in which the major insulation is provided by a ceramic or analogous material. *See also:* **power distribution, overhead construction.** 76A1-0

solid conductor. A conductor consisting of a single wire. *See also:* **conductor.** 42A35-31E13

solid contact. A contact having relatively little inherent flexibility and whose contact pressure is supplied by another member. 37A100-31E11

solid coupling (rotating machinery). A coupling that makes a rigid connection between two shafts. *See also:* **rotor (rotating machinery).** 0-31E8

solid electrolytic capacitor. A capacitor in which the dielectric is primarily an anodized coating on one electrode, with the remaining space between the electrodes filled with a solid semiconductor. 0-21E0

solid enclosure. An enclosure that will neither admit accumulations of flyings or dust nor transmit sparks or flying particles to the accumulations outside. 42A95-0

solid-iron cylindrical-rotor generator. *See:* **cylindrical-rotor generator.**

solid-material fuse unit. A fuse unit in which the arc is drawn through a hole in solid material. 37A100-31E11

solid-pole synchronous motor. A salient-pole synchronous motor having solid steel pole shoes, and either laminated or solid pole bodies. *See also:* **synchronous machine.** 0-31E8

solid rotor (rotating machinery). (1) A rotor, usually constructed of a high-strength forging, in which slots may be machined to accommodate the rotor winding. (2) A spider-type rotor in which spider hub is not split. *See also:* **rotor (rotating machinery).** 0-31E8

solid-state. An adjective used to describe a device, circuit, or system whose operation is dependent upon any combination of optical, electrical, or magnetic phenomena within a solid. Specifically excluded are devices, circuits, or systems dependent upon the macroscopic physical movement, rotation, contact, or noncontact of any combination of solids, liquids, gases, or plasmas. *See also:* **circuits and devices.** 0-31E3

solid-state component. A component whose operation depends on the control of electric or magnetic phenomena in solids, for example, a transistor, crystal diode, ferrite core. X3A12-16E9

solid-state device (control equipment). A device that may contain electronic components that do not depend on electronic conduction in a vacuum or gas. The electrical function is performed by semiconductors or the use of otherwise completely static components such as resistors, capacitors, etcetera. *See also:* **power systems, low-frequency and surge testing.** E94-0

solid-state relay. A static relay constructed exclusively of solid-state components. *See also:* **relay.** 37A100-31E11/31E6

solid-state scanning (facsimile). A method in which all or part of the scanning process is due to electronic commutation of a solid-state array of thin-film photosensitive elements. *See also:* **facsimile (electrical communication).** 0-19E4

solid-type paper-insulated cable. Oil-impregnated, paper-insulated cable, usually lead covered, in which no provision is made for control of internal pressure variations. *See also:* **power distribution, underground construction.** 42A35-31E13

solution. *See:* **check solution.**

solvent cleaning (electroplating). Cleaning by means of organic solvents. *See also:* **electroplating.** 42A60-0

solventless (rotating machinery). A term applied to liquid or semiliquid varnishes, paints, impregnants, resins, and similar compounds that have essentially no change in weight or volume when converted into a solid or semisolid. 0-31E8

sonar. Apparatus or techniques whereby underwater acoustic energy is employed to obtain information regarding objects or events below the surface of the water. *Note:* The name is derived from the initial letters of the phrase **sound navigation and ranging.** A device that radiates underwater acoustic energy and utilizes reflection of this energy is termed an **active sonar.** A device that merely receives underwater acoustic energy generated at a distant source is termed a **passive sonar.** *See also:* **loudspeaker.** 42A65-0

sonic delay line. *See:* **acoustic delay line.** *See also:* **electronic digital computer.**

sonne. A radio navigation aid that provides a number of characteristic signal zones that rotate in a time sequence; a bearing may be determined by observation (by interpolation) of the instant at which transition occurs from one zone to the following zone. *See:* **consol.** *See also:* **navigation.** E172-10E6

sort. To arrange data or items in an ordered sequence by applying specific rules. *See also:* **electronic digital computer.** X3A12-16E9

sorter. A person, device, or computer routine that sorts. *See also:* **electronic digital computer.** X3A12-16E9

SOS. *See:* **radio distress signal.**

sound. (1) An oscillation in pressure, stress, particle displacement, particle velocity, etcetera, in a medium with internal forces (for example, elastic, viscous), or the superposition of such propagated oscillations. (2) An auditory sensation evoked by the oscillation described above. *Notes:* (A) In case of possible confusion, the term sound wave or elastic wave may be used for concept (1) and the term sound sensation for concept (2). Not all sound waves can evoke an auditory sensation, for example, an ultrasonic wave. (B) The medium in which the sound exists is often indicated by an appropriate adjective, for example, air-borne, water-borne, structure-borne. *See also:* **electroacoustics.** 0-1E1

sound absorption. (1) The change of sound energy into some other form, usually heat, in passing through a medium or on striking a surface. (2) The property possessed by material and objects, including air, of absorbing sound energy. *See also:* **electroacoustics.** 0-1E1

sound-absorption coefficient (surface). The ratio of sound energy absorbed or otherwise not reflected by the surface, to the sound energy incident upon the surface. Unless otherwise specified, a diffuse sound field is assumed. *See also:* **electroacoustics.** 0-1E1

sound analyzer. A device for measuring the band pressure level, or pressure spectrum level, of a so und at various frequencies. *Notes:* (1) A sound analyzer usually consists of a microphone, an amplifier and wave analyzer, and is used to measure amplitude and frequency of the components of a complex sound. (2) The band pressure level of a sound for a specified frequency band is the effective root-mean-square sound pressure level of the sound energy contained within the bands. *See also:* **electroacoustics; instrument.** E157-42A30-0

sound articulation (percent sound articulation). The percent articulation obtained when the speech units considered are fundamental sounds (usually combined into meaningless syllables). *See also:* **volume equivalent.** 42A65-0

sound-detection system (protective signaling). A system for the protection of vaults by the use of sound-detecting devices and relay equipment to pick up and convert noise, caused by burglarious attack on the structure, to electric impulses in a protection circuit. *See also:* **protective signaling.** 42A65-0

sound-effects filter. *See:* **filter, sound-effects.**

sound energy. Of a given part of a medium, the total energy in this part of the medium minus the energy that would exist in the same part of the medium with no sound waves present. E157-1E1

sound field. A region containing sound waves. E157-1E1

sound intensity (sound-energy flux density) (sound power density) (in a specified direction at a point). The average rate of sound energy transmitted in the specified direction through a unit area normal to this direction at the point considered. *Notes:* (1) The sound intensity in any specified direction a of a sound field is the sound-energy flux through a unit area normal to that direction. This is given by the expression

$$I_a = \frac{1}{T}\int_0^T p v_a \, dt$$

where

T = an integral number of periods or a time long compared to a period
p = the instantaneous sound pressure
v_a = the component of the instantaneous particle velocity in the direction a
t = time.
(2) In the case of a free plane or spherical wave having an effective sound pressure p, the velocity of propagation c, in a medium of density ρ, the intensity in the direction of propagation is given by

$$I = \frac{p^2}{\rho c}.$$

See also: **electroacoustics.** 0-1E1

sound level. A weighted sound pressure level obtained by the use of metering characteristics and the weightings *A, B,* or *C* specified in American National Standard Specification for General-Purpose Sound-Level Meters, S1.4-1961 (or latest revision thereof). The weighting employed must always be stated. The reference pressure is 2×10^{-5} newton per meter ^{2}. *Notes:* (1) The meter reading (in decibels) corresponds to a value of the sound pressure integrated over the audible frequency range with a specified frequency weighting and integration time. (2) A suitable method of stating the weighting is, for example, "The sound level (*A*) was 43 decibels." *See also:* **electroacoustics; level.** 42A65-1E1

sound-level meter. An instrument including a microphone, an amplifier, an output meter, and frequency-weighting networks for the measurement of noise and sound levels in a specified manner. *Notes:* (1) The measurements are intended to approximate the loudness level of pure tones that would be obtained by the more-elaborate ear balance method. (2) Loudness level in phons of a sound is numerically equal to the sound pressure level in decibels relative to 0.0002 microbar of a simple tone of frequency 1000 hertz that is judged by the listeners to be equivalent in loudness. (3) Specifications for sound-level meters are given in American National Standard Specification for General-Purpose Sound-Level Meters, S1.4-1961 (or latest revision thereof). *See also:* **electroacoustics; instrument.** 42A30-1E1

sound power (source). The total sound energy radiated by the source per unit of time. *See also:* **electroacoustics.** 0-1E1

sound-powered telephone set. A telephone set in which the transmitter and receiver are passive transducers. *See also:* **telephone station.** 42A65-0

sound pressure (at a point). The total instantaneous pressure at that point, in the presence of a sound wave, minus the static pressure at that point. E269-19E8

sound pressure, effective (root-mean-square sound pressure). At a point over a time interval, the root-mean-square value of the instantaneous sound pressure at the point under consideration. In the case of periodic sound pressures, the interval must be an integral number of periods or an interval long compared to a period. In the case of nonperiodic sound pressures, the interval should be long enough to make the value obtained essentially independent of small changes in the length of the interval. *Note:* The term **effective sound pressure** is frequently shortened to **sound pressure.** 0-1E1

sound pressure, instantaneous (at a point). The total instantaneous pressure at that point minus the static pressure at that point. *Note:* The commonly used unit is the newton per square meter. *See also:* **electroacoustics.** 0-1E1

sound pressure level. The sound pressure level, in decibels, is 20 times the logarithm to the base 10 of the ratio of the pressure of this sound to the reference pressure. The reference pressure must be explicitly stated. *Notes:* (1) Unless otherwise explicitly stated, it is to be understood that the sound pressure is the effective root-mean-square sound pressure. (2) The following reference pressures are in common use: (A) 2×10^{-4} dyne per square centimeter. (B) 1 dyne per square centimeter. Reference pressure (A) has been in general use for measurements dealing with hearing and sound-level measurements in air and liquids, while (B) has gained widespread use for calibrations and many types of sound-level measurements in liquids. It is to be noted that in many sound fields the sound pressure ratios are not proportional to the square root of corresponding power ratios and hence cannot be expressed in decibels in the strict sense; however, it is common practice to extend the use of the decibel to these cases. *See also:* **level; electroacoustics.** 42A65-1E1;E269-19E8

sound probe. A device that responds to some characteristic of an acoustic wave (for example, sound pressure, particle velocity) and that can be used to explore and determine this characteristic in a sound field without appreciably altering the field. *Note:* A sound probe may take the form of a small microphone or a small tubular attachment added to a conventional microphone. *See also:* **electroacoustics; instrument.** 42A30-1E1

sound recording and reproducing.
See:
drift;
electroacoustics
flutter rate;
percent flutter;
percent total flutter;
phonograph pickup.

sound recording system. A combination of transducing devices and associated equipment suitable for storing sound in a form capable of subsequent reproduction. *See also:* **phonograph pickup.** E157/42A65-1E1

sound reflection coefficient (surface). The ratio of the sound reflected by the surface to the sound incident upon the surface. Unless otherwise specified, reflection of sound energy in a diffuse sound field is assumed. *See also:* **electroacoustics.** 0-1E1

sound reproducing system. A combination of transducing devices and associated equipment for reproducing recorded sound. *See also:* **loudspeaker.** E157/42A65-1E1

sound spectrum analyzer (sound analyzer). A device or system for measuring the band pressure level of a sound as a function of frequency. *See also:* **electroacoustics.** 0-1E1

sound track (electroacoustics). A band that carries the sound record. In some cases, a plurality of such bands may be used. In sound film recording, the band is usually along the margin of the film. *See also:* **phonograph pickup.** 0-1E1

sound transmission coefficient (interface or partition). The ratio of the transmitted to incident sound energy. Unless otherwise specified, transmission of sound energy between two diffuse sound fields is assumed. *See also:* **electroacoustics.** 0-1E1

source (electroacoustics). That which supplies signal power to a transducer or system. *Note:* Modifiers are usually used to differentiate between sources of different kinds as signal source, interference source, common-mode source, etcetera. *See:* **circuits and devices; signal.** E151-42A65/13E6

source ground (signal-transmission system). Potential reference at the physical location of a source, usually the signal source. *See:* **signal.** 0-13E6

source impedance. *See:* **impedance, source.** *See also:* **self-impedance.**

source language. A language that is an input to a given translation process. X3A12-16E9

source node (network analysis). A node having only outgoing branches. *See also:* **linear signal flow graphs.** E155-0

source program (computing systems). A program written in a source language. *See also:* **electronic digital computer.** X3A12-16E9

source resistance. The resistance presented to the input of a device by the source. *See also:* **measurement system.** 39A4-0

source resistance rating. The value of source resistance that, when injected in an external circuit having essentially zero resistance, will either (1) double the dead band, or (2) shift the dead band by one-half its width. *See also:* **measurement system.** 39A4-0

space (computing devices). (1) A site intended for the storage of data, for example, a site on a printed page or a location in a storage medium. (2) A basic unit of area, usually the size of a single character. (3) One or more blank characters. (4) To advance the reading or display position according to a prescribed format, for example, to advance the printing or display position horizontally to the right or vertically down. 0-16E9

space charge (1) (general). A net excess of charge of one sign distributed throughout a specified volume.
(2) (thermionics). Electric charge in a region of space due to the presence of electrons and/or ions. *See also:* **electron emission.** E270/42A70/50I07-15E6

space-charge-control tube. *See:* **density-modulated tube.**

space-charge debunching. Any process in which the mutual interactions between electrons in the stream disperse the electrons of a bunch. *See:* **beam tubes.** *See also:* **electron devices, miscellaneous.** E160-15E6

space-charge density (thermionics). The space charge per unit volume. *See also:* **electron emission.** 50I07-15E6

space-charge grid. A grid, usually positive, that controls the position, area, and magnitude of a potential minimum or of a virtual cathode in region adjacent to the grid. *See also:* **electrode (electron tube); grid.** E160-15E6;42A70-0

space-charge-limited current (electron vacuum tubes). The current passing through an interelectrode space when a virtual cathode exists therein. *See also:* **circuit characteristics of electrodes; electrode current (electron tube).** E160-15E6

space-charge region (semiconductor device). A region in which the net charge density is significantly different from zero. *See also:* **depletion layer; semiconductor device.** E102/E216/E270-34E17

space correction (industrial control). A method of register control that takes the form of a sudden change in the relative position of the web. 42A25-34E10

space current (electron tubes). Synonym in a diode or equivalent diode of cathode current. *See:* **electrode current (electron tube); leakage current (electron tubes); load current (electron tubes); quiescent current (electron tubes).** 50I07-15E6

space diversity. *See:* **space diversity reception.**

space-diversity reception (space diversity). That form of diversity reception that utilizes receiving antennas placed in different locations. *See also:* **radio receiver.** 42A65-0

space factor (rotating machinery). The ratio of (1) the sum of the cross-sectional areas of the active or specified material to (2) the cross-sectional area within the confining limits specified. *See:* **asynchronous machine; direct-current commutating machine; slot space factor; synchronous machine.** 42A10-31E8

space heater (1) (general). A heater that warms occupied spaces.
(2) (rotating machinery). A device that warms the ventilating air within a machine and prevents condensation of moisture during shut-down periods. *See also:* **appliances (including portable); cradle base (rotating machinery).** 42A95-31E8

space pattern (television). A geometrical pattern appearing on a test chart designed for the measurement of geometric distortion. *See also:* **television.** E203-0

space-referenced navigation data. Data in terms of a coordinate system referenced to inertial space. *See also:* **navigation.** 0-10E6

spacer shaft (rotating machinery). A separate shaft connecting the shaft ends of two machines. *See:* **armature.** 0-31E8

spacing pulse (data transmission). A spacing pulse or **space** is the signal pulse that, in direct-current neutral operation, corresponds to a **circuit open** or **no current** condition. *See also:* **data transmission; pulse.** 0-19E4

spacing wave (back wave) (telegraph communication). The emission that takes place between the active portions of the code characters or while no code characters are being transmitted. *See also:* **radio transmitter.** E145-0

spalling (corrosion). Spontaneous separation of a surface layer from a metal. *See:* **corrosion terms.** CM-34E2

span (1) (measuring devices). The algebraic difference between the upper and lower values of a range. *Notes:* (A) For example: (a) Range 0 to 150, span 150; (b) Range −20 to 200, span 220; (c) Range 20 to 150, span 130; (d) Range −100 to −20, span 80. (B) The following compound terms are used with suitable modifications in the units: measured variable span, measured signal span, etcetera. (C) For multirange devices, this definition applies to the particular range that the device is set to measure. *See also:* **instrument.** 39A4-0
(2) (overhead conductors). (A) The horizontal distance between two adjacent supporting points of a conductor. (B) That part of any conductor, cable, suspension strand, or pole line between two consecutive points of support. *See also:* **cable; open wire.** 42A35-31E13;42A65-0

span frequency-response rating. The maximum frequency in cycles per minute of sinusoidal variation of measured signal for which the difference in amplitude between output and input represents an error no greater than five times the accuracy rating when the instrument is used under rated operating conditions. The peak-to-peak amplitude of the sinusoidal variation of measured signal shall be equivalent to full span of the instrument. It must be recognized that the span frequency-response rating is a measure of dynamic behavior under the most adverse conditions of measured signal (that is, the maximum sinusoidal excursion of the measured signal). The frequency response for an amplitude of measured signal less than full span is not proportional to the frequency response for full span. The relationship between the frequency response of different instruments at any particular amplitude of measured signal is not indicative of the relationship that will exist at any other amplitude. *See also:* **accuracy rating of an instrument.** 39A4-0

span length. The horizontal distance between two adjacent supporting points of a conductor. *See also:* **power distribution, overhead construction.** 2A2-0

span step-response-time rating. The time that the step-response time will not exceed for a change in measured signal essentially equivalent to full span when the instrument is used under rated operating conditions. The actual span step-response time shall not be less than 2/3 of the span step-response-time rating. (For example, for an instrument of 3-second span step-response-time rating, the span step-response time, under rated operating conditions, will be between 3 and 2 seconds.) It must be recognized that the step-response time for smaller steps is not proportional to the step-response time for full span. *Note:* The end device shall be considered to be at rest when it remains within a band of plus and minus the accuracy rating from its final position. *See also:* **accuracy rating (instrument).** 39A4-0

spare equipment. Equipment complete or in parts, on hand for repair or replacement. *See:* **reserve equipment.** 42A35-31E13

spare point (for supervisory control or indication or telemeter selection). A point that is not being utilized but is fully equipped with all of the necessary devices for a point. 37A100-31E11

spark. A brilliantly luminous phenomenon of short duration that characterizes a disruptive discharge. *Note:* A disruptive discharge is the sudden and large increase in current through an insulating medium due to the complete failure of the medium under electric stress. *See:* **disruptive discharge.** *See also:* **discharge (gas).** E270-0

spark capacitor (spark condenser*). A capacitor connected across a pair of contace points, or across the inductance that causes the spark, for the purpose of diminishing sparking at these points. *See also:* **circuits and devices.** 42A65-21E0

*Deprecated

spark condenser*. *See:* **spark capacitor.**

*Deprecated

spark gap. Any short air space between two conductors electrically insulated from or remotely electrically connected to each other. 5A1-0

spark-gap modulation. A modulation process that produces one or more pulses or energy by means of a controlled spark-gap breakdown for application to the element in which modulation takes place. *See also:* **modulating systems; oscillatory circuit.** E145/42A65-0

spark killer. An electric network, usually consisting of a capacitor and resistor in series, connected across a pair of contact points, or across the inductance that causes the spark, for the purpose of diminishing sparking at these points. *See also:* **network analysis.** 42A65-0

sparkover (1) (general). A disruptive discharge between preset electrodes in either a gaseous or a liquid dielectric. *See also:* **spark gap; test voltage and current.** 68A1-31E5

(2) (lightning arrester). A disruptive discharge between electrodes of a measuring gap, voltage control gap, or protective device. *See also:* **lightning arrester (surge diverter); lightning protection and equipment.** E28-0

spark-plug suppressor (electromagnetic compatibility). A suppressor designed for direct connection to a spark plug. *See also:* **electromagnetic compatibility.** 27E1

spark transmitter. A radio transmitter that utilizes the oscillatory discharge of a capacitor through an inductor and a spark gap as the source of its radio-frequency power. *See also:* **radio transmitter.** E145/E182/42A65-0

speaker. *See:* **loudspeaker.**

special character (character set). A character that is neither a numeral, a letter, nor a blank, for example, virgule, asterisk, dollar sign, equals sign, comma, period. *See also:* **electronic digital computer.** X3A12-16E9

special-purpose computer. A computer that is designed to solve a restricted class of problems. *See also:* **electronic digital computer.** X3A12-16E9

special-purpose motor. A motor with special operating characteristics or special mechanical construction, or both designed for a particular application and not falling within the definition of a general-purpose or definite-purpose motor. *See:* **asynchronous machine; direct-current commutating machine; synchronous machine.** 42A10/MG1-31E8

specific acoustic impedance (unit area acoustic impedance) (at a point in the medium). The complex ratio of sound pressure to particle velocity. *See:* **Note 2 under acoustic impedance.** E157-1E1

specific acoustic reactance. The imaginary component of the specific acoustic impedance. E157-1E1

specific acoustic resistance. The real component of the specific acoustic impedance. E157-1E1

specific coordinated methods. Those additional methods applicable to specific situations where general coordinated methods are inadequate. *See also:* **inductive coordination.** 42A65-0

specific emission. The rate of emission per unit area. 50I07-15E6

specific inductive capacitance. *See:* **relative capacitivity.**

specific repetition frequency (loran). One of a set of closely spaced pulse repetition frequencies derived from the basic repetition frequency and associated with a specific set of synchronized stations. *See also:* **navigation.** 0-10E6

specific repetition rate (electronic navigation). *See:* **specific repetition frequency.** *See also:* **navigation.**

specific unit capacity purchases (electric power supply). That capacity that is purchased or sold in transactions with other utilities and that is from a designated unit on the system of the seller. It is understood that the seller does not provide reserve capacity for this type of capacity transaction. *See also:* **generating station.** 0-31E4

specified achromatic lights. (1) Light of the same chromaticity as that having an equi-energy spectrum. (2) The standard illuminants of colorimetry *A, B,* and *C,* the spectral energy distributions of which were specified by the International Commission on Illumination (CIE) in 1931, with various scientific applications in view. Standard *A:* incandescent electric lamp of color temperature 2854 kelvins. Standard *B:* Standard *A* combined with a specified liquid filter to give a light of color temperature approximately 4800 kelvins. Standard *C:* Standard *A* combined with a specified liquid filter to give a light of color temperature approximately 6500 kelvins. (3) Any other specified white light. *See also:* **color.** 50I45-2E2

spectral characteristic (1) (color television). The set of spectral responses of the color separation channels with respect to wavelength. *Notes:* (A) The channel terminals at which the characteristics apply must be specified and an appropriate modifier may be added to the term, such as pickup spectral characteristic or studio spectral characteristic. (B) Because of nonlinearity, some spectral characteristics depend upon the magnitude of radiance used in the measurement. (C) Nonlinearizing and matrixing operations may be performed within the channels. *See also:* **color terms.** E201-2E2

(2) (camera tube). A relation, usually shown by a graph, between wavelength and sensitivity per unit wavelength interval *See:* **spectral sensitivity characteristic.** *See also:* **beam tubes.** 42A70-0

(3) (luminescent screen). The relation, usually shown by a graph, between wavelength and emitted radiant power per unit wavelength interval. *Note:* The radiant power is commonly expressed in arbitrary units. *See also:* **beam tubes.** E160-15E6/2E2;42A70-0

(4) (phototube). A relation, usually shown by a graph, between the radiant sensitivity and the wavelength of the incident radiant flux. *See:* **spectral sensitivity characteristic.** *See also:* **phototubes.** 42A70-0

spectral-conversion luminous gain (optoelectronic device). The luminous gain for specified wavelength-intervals of both incident and emitted luminous flux. *See also:* **optoelectronic device.** E222-15E7

spectral-conversion radiant gain (optoelectronic device). The radiant gain for specified wavelength intervals of both incident and emitted radiant flux. *See also:* **optoelectronic device.** E222-15E7

spectral emissivity (element of surface of a temperature radiator at any wavelength). The ratio of its radiant flux density per unit wavelength interval (spectral radiant exitance) at that wavelength to that of a blackbody at the same temperature. *See also:* **radiant energy (illuminating engineering).** Z7A1-0

spectral luminous efficacy (radiant flux). The quotient of the luminous flux at a given wavelength by the radiant flux at that wavelength. It is expressed in lumens per watt. *Note:* This term formerly was called **luminosity factor.** The reciprocal of the maximum luminous efficacy of radiant flux is sometimes called **mechanical equivalent of light**; that is, the watts per lumen at the wavelength of maximum luminous efficacy. The most probable value is 0.00147 watt per lumen, corresponding to 680 lumens per watt as the maximum possible luminous efficacy. These values are based on the candela, on the 1948 International Temperature Scale by using Planck's equation with $c_2 = 1.438$ centimeter kelvin, $c_1 = 3.741 \times 10^{-12}$ watt centimeter^2 and 2042 kelvins as the temperature of freezing platinum, and on the standard CIE photopic spectral luminous efficiency values given in the accompanying table. If the standard CIE scotopic spectral luminous efficiency values are used, the maximum luminous efficacy is 1746 "scotopic" lumens per watt. *See:* **radiant flux; spectral radiant flux.** *See also:* **light.** Z7A1-0

spectral luminous efficiency (radiant flux). The ratio of the luminous efficacy for a given wavelength to the value of the wavelength of maximum luminous efficacy. It is dimensionless. *Notes:* (1) The term **spectral luminous efficiency** replaces the previously used terms **relative luminosity** and **relative luminosity factor.** (2) Values of spectral luminous efficiency for photopic vision at 10-nanometer intervals were provisionally adopted by the International Commission on Illumination in 1924 and were adopted in 1933 by the International Committee on Weights and Measures as a basis for the establishment of photometric standards of types of sources differing from the primary standard in spectral distribution of radiant flux. These standard values of spectral luminous efficiency were determined by observations with a two-degree photometric field having a moderately high luminance (photometric brightness), and photometric evaluations based upon them consequently do not apply exactly to other conditions of observation. Watts weighted in accord with these standard values are often referred to as **lightwatts.** See the accompanying table.
(3) Values of spectral luminous efficiency for scotopic vision at 10-nanometer intervals were provisionally adopted by the International Commission on Illumination in 1951. These values of sp ctral luminous efficiency were determined by observation by young dark-adapted observers using extra-foveal vision at near-threshold luminance. *See also:* **light.** Z7A1-0

spectral luminous gain (optoelectronic device). Luminous gain for a specified wavelength interval of either the incident or the emitted flux. *See also:* **optoelectronic device.** E222-15E7

spectral-noise density (sound recording and reproducing system). The limit of the ratio of the noise output within a specified frequency interval to the frequency interval, as that interval approaches zero. *Note:* This is approximately the total noise within a narrow frequency band divided by that bandwidth in hertz. *See also:* **noise (sound recording and reproducing system).** E191-0

spectral power density (radio wave propagation). The power density per unit bandwidth. *See also:* **radio wave propagation.** 0-3E2

spectral power flux density (radio wave propagation). *See:* **spectral power density.** *See also:* **radio wave propagation.** 0-3E2

spectral quantum yield (photocathode). The average number of electrons photoelectrically emitted from the photocathode per incident photon of a given wavelength. *Note:* The spectral quantum yield may be a function of the angle of incidence and of the direction of polarization of the incident radiation. *See also:* **phototube.** E175-0

Photopic spectral luminous efficiency $V(\lambda)$.
(Unity at wavelength of maximum luminous efficacy.)

λ in nanometers	Standard values	Standard values interpolated at intervals of 1 nanometer								
		1	2	3	4	5	6	7	8	9
380	0.00004	0.000045	0.000049	0.000054	0.000059	0.000064	0.000071	0.000080	0.000090	0.000104
390	0.00012	0.000138	0.000155	0.000173	0.000193	0.000215	0.000241	0.000272	0.000308	0.000350
400	0.0004	0.00045	0.00049	0.00054	0.00059	0.00064	0.00071	0.00080	0.00090	0.00104
410	0.0012	0.00138	0.00156	0.00174	0.00195	0.00218	0.00244	0.00274	0.00310	0.00352
420	0.0040	0.00455	0.00515	0.00581	0.00651	0.00726	0.00806	0.00889	0.00976	0.01066
430	0.0116	0.01257	0.01358	0.01463	0.01571	0.01684	0.01800	0.01920	0.02043	0.02170
440	0.023	0.0243	0.0257	0.0270	0.0284	0.0298	0.0313	0.0329	0.0345	0.0362
450	0.038	0.0399	0.0418	0.0438	0.0459	0.0480	0.0502	0.0525	0.0549	0.0574
460	0.060	0.0627	0.0654	0.0681	0.0709	0.0739	0.0769	0.0802	0.0836	0.0872
470	0.091	0.0950	0.0992	0.1035	0.1080	0.1126	0.1175	0.1225	0.1278	0.1333
480	0.139	0.1448	0.1507	0.1567	0.1629	0.1693	0.1761	0.1833	0.1909	0.1991
490	0.208	0.2173	0.2270	0.2371	0.2476	0.2586	0.2701	0.2823	0.2951	0.3087
500	0.323	0.3382	0.3544	0.3714	0.3890	0.4073	0.4259	0.4450	0.4642	0.4836
510	0.503	0.5229	0.5436	0.5648	0.5865	0.6082	0.6299	0.6511	0.6717	0.6914
520	0.710	0.7277	0.7449	0.7615	0.7776	0.7932	0.8082	0.8225	0.8363	0.8495
530	0.862	0.8739	0.8851	0.8956	0.9056	0.9149	0.9238	0.9320	0.9398	0.9471
540	0.954	0.9604	0.9661	0.9713	0.9760	0.9803	0.9840	0.9873	0.9902	0.9928
550	0.995	0.9969	0.9983	0.9994	1.0000	1.0002	1.0001	0.9995	0.9984	0.9969
560	0.995	0.9926	0.9898	0.9865	0.9828	0.9786	0.9741	0.9691	0.9638	0.9581
570	0.952	0.9455	0.9386	0.9312	0.9235	0.9154	0.9069	0.8981	0.8890	0.8796
580	0.870	0.8600	0.8496	0.8388	0.8277	0.8163	0.8046	0.7928	0.7809	0.7690
590	0.757	0.7449	0.7327	0.7202	0.7076	0.6949	0.6822	0.6694	0.6565	0.6437
600	0.631	0.6182	0.6054	0.5926	0.5797	0.5668	0.5539	0.5410	0.5282	0.5156
610	0.503	0.4905	0.4781	0.4658	0.4535	0.4412	0.4291	0.4170	0.4049	0.3929
620	0.381	0.3690	0.3570	0.3449	0.3329	0.3210	0.3092	0.2977	0.2864	0.2755
630	0.265	0.2548	0.2450	0.2354	0.2261	0.2170	0.2082	0.1996	0.1912	0.1830
640	0.175	0.1672	0.1596	0.1523	0.1452	0.1382	0.1316	0.1251	0.1188	0.1128
650	0.107	0.1014	0.0961	0.0910	0.0862	0.0816	0.0771	0.0729	0.0688	0.0648
660	0.061	0.0574	0.0539	0.0506	0.0475	0.0446	0.0418	0.0391	0.0366	0.0343
670	0.032	0.0299	0.0280	0.0263	0.0247	0.0232	0.0219	0.0206	0.0194	0.0182
680	0.017	0.01585	0.01477	0.01376	0.01281	0.01192	0.01108	0.01030	0.00956	0.00886
690	0.0082	0.00759	0.00705	0.00656	0.00612	0.00572	0.00536	0.00503	0.00471	0.00440
700	0.0041	0.00381	0.00355	0.00332	0.00310	0.00291	0.00273	0.00256	0.00241	0.00225
710	0.0021	0.001954	0.001821	0.001699	0.001587	0.001483	0.001387	0.001297	0.001212	0.001130
720	0.00105	0.000975	0.000907	0.000845	0.000788	0.000736	0.000688	0.000644	0.000601	0.000560
730	0.00052	0.000482	0.000447	0.000415	0.000387	0.000360	0.000335	0.000313	0.000291	0.000270
740	0.00025	0.000231	0.000214	0.000198	0.000185	0.000172	0.000160	0.000149	0.000139	0.000130
750	0.00012	0.000111	0.000103	0.000096	0.000090	0.000084	0.000078	0.000074	0.000069	0.000064

spectral radiance. *See:* **Planck radiation law; radiant energy (illuminating engineering); spectral radiant intensity.** Z7A1-0

spectral radiant energy. Radiant energy per unit wavelength interval at wavelength λ; for example, joules/nanometer. *See also:* **radiant energy (illuminating engineering).** Z7A1-0

spectral radiant flux. Radiant flux per unit wavelength interval at wavelength λ; for example, watts/nanometer. *See also:* **radiant energy (illuminating engineering).** Z7A1-0

spectral radiant gain (optoelectronic device). Radiant gain for a specified wavelength interval of either the incident or the emitted radiant flux. *See also:* **optoelectronic device.** E222-15E7

spectral radiant intensity. Radiant intensity per unit wavelength interval; for example, watts/steradian nanometer. *See also:* **radiant energy (illuminating engineering).** Z7A1-0

spectral reflectance (surface or medium). The ratio of the reflected flux to the incident flux at a particular wavelength λ or within a small band of wavelengths Δλ about λ. *Note:* The terms **hemispherical, regular,** or **diffuse reflectance** may each be considered restricted to a specific region of the spectrum and may be so designated by the addition of the adjective **spectral.** *See also:* **lamp.** Z7A1-0

spectral response characteristic (photoelectric devices). *See:* **spectral sensitivity characteristic.**

spectral selectivity (photoelectric device). The change of photoelectric current with the wavelength of the irradiation. *See also:* **photoelectric effect.** 50I07-15E6

spectral sensitivity characteristic (camera tubes or phototubes). The relation between the radiant sensitivity and the wavelength of the incident radiation, under specified conditions of irradiation. *Note:* Spectral sensitivity characteristic is usually measured with a collimated beam at normal incidence. *See:* **phototubes.** *See also:* **beam tubes.** E160/E175-15E6/2E2

spectral transmittance (medium). The ratio of the transmitted flux to the incident flux at a particular wavelength γ or within a small band of wavelengths Δγ about γ. *Note:* The terms **hemispherical, regular,** or **diffuse transmittance** may each be considered restricted to a specific region of the spectrum and may be so designated by the addition of the adjective **spectral.** *See also:* **transmission (illuminating engineering).** Z7A1-0

spectral tristimulus values. Values per unit wavelength interval and unit spectral radiant flux. *Note:* Spectral tristimulus values have been adopted by the International Commission on Illumination (CIE). They are tabulated as functions of wavelength throughout the spectrum and are the basis for the evaluation of radiant energy as light. Z7A1-0

spectrophotometer. An instrument for measuring the transmittance and reflectance of surfaces and media as a function of wavelength. *See also:* **photometry.** Z7A1-0

spectroradiometer. An instrument for measuring radiant flux as a function of wavelength. *See also:* **primary standards (illuminating engineering).** Z7A1-0

spectrum. (1) The distribution of the amplitude (and sometimes phase) of the components of the wave as a function of frequency. (2) A continuous range of components, usually wide in extent, within which waves have some specified common characteristic; for example, audio-frequency spectrum. *Note:* The term **spectrum** can also be applied to functions of variables other than time, such as distance. *See also:* **signal wave.** 42A65-31E3/1E1

spectrum amplitude. The voltage spectrum of a pulse can be expressed as

$$V(\omega) = R(\omega)+jX(\omega) = \int_{-\infty}^{+\infty} v(t)e^{-j\omega t}\,dt$$

where

$$R(\omega) = \int_{-\infty}^{+\infty} v(t)\cos \omega t\,dt$$

$$X(\omega) = -\int_{-\infty}^{+\infty} v(t)\sin \omega t\,dt$$

and

$$\omega = 2\pi f.$$

The spectrum then has the amplitude

$$A(\omega) = [R^2(\omega) + X^2(\omega)]^{1/2}$$

and a phase characteristic

$$\phi(\omega) = \tan^{-1}[X(\omega)/R(\omega)]$$

Note: The inverse transform can be written

$$v(t) = \frac{1}{\pi}\int_0^{\infty} A(\omega)\cos[\omega t+\phi(\omega)]\,d\omega \text{ for real } v(t).$$

An impulse is a function of short time duration compared with the reciprocals of all frequencies of interest. Its spectrum has an amplitude that is substantially uniform (in this frequency range), and its spectrum amplitude $A(\omega)$ is the area under the impulse time function and has dimensions of volt-seconds. The spectrum amplitude is also expressible in volts per hertz as follows:

$$S(f) = \frac{1}{\pi}A(\omega) \text{ volts/hertz.}$$

It is this form that is used as the basis for calibration of commercially available impulse generators. E263-27E1

spectrum level (spectrum density level) (acoustics) (specified signal at a particular frequency). The level of that part of the signal contained within a band 1 hertz wide, centered at the particular frequency. Ordinarily this has significance only for a signal having a continuous distribution of components within the frequency range under consideration. The words **spectrum level** cannot be used alone but must appear in combination with a prefatory modifier; for example, pressure, velocity, voltage. *Note:* For illustration, if L_{ps} be a desired pressure spectrum level, p the effective pressure measured through the filter system, p_0 reference sound pressure, Δf the effective bandwidth of the filter system, and $\Delta_0 f$ the reference bandwidth (1 hertz), then

$$L_{ps} = \log_{10} \frac{p^2/\Delta f}{p_0{}^2/\Delta_0 f}$$

For computational purposes, if L_{ps} is the band pressure level observed through a filter of bandwidth Δf, the above relation reduces to

$$L_{ps} = L_p - 10 \log_{10} \frac{\Delta f}{\Delta_0 f}$$

See: **electroacoustics.** 0-1E1

spectrum locus (color). The locus of points representing the chromaticities of spectrally pure stimuli in a chromaticity diagram. *See also:* **color.** E201-2E2

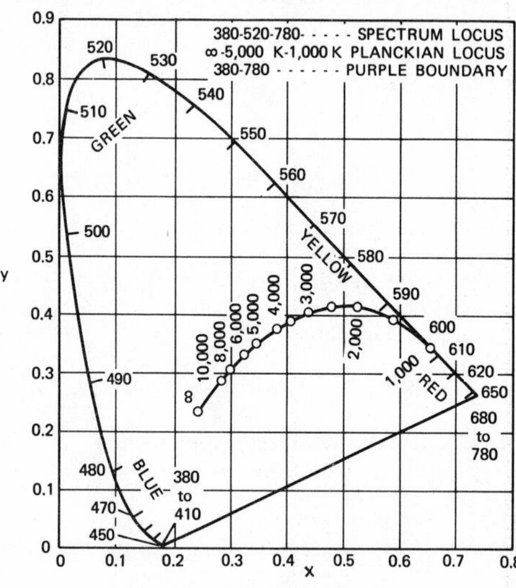

Chromaticity diagram.

specular angle. The angle between the perpendicular to the surface and the reflected ray that is numerically equal to the angle of incidence and that lies in the same plane as the incident ray and the perpendicular but on the opposite side of the perpendicular to the surface. *See also:* **lamp.** Z7A1-0

specular surface. One from which the reflection is predominantly regular. *See:* **regular (specular) reflection.** *See also:* **bare (exposed) lamp.** Z7A1-0

speech interpolation. The method of obtaining more than one voice channel per voice circuit by giving each subscriber a speech path in the proper direction only at times when his speech requires it. *See also:* **telephone system.** 42A65-0

speed adjustment (control). A speed change of a motor accomplished intentionally through action of a control element in the apparatus or system governing the performance of the motor. *Note:* For an adjustable-speed direct-current motor, the speed adjustment is expressed in percent (or per unit) of base speed. Speed adjustment of all other motors is expressed in percent (or per unit) of rated full-load speed. *See also:* **adjustable-speed motor; base speed of an adjustable-speed motor; electric drive.** 42A25-34E10

speed changer (gas turbines). A device by means of which the speed-governing system is adjusted to change the speed or power output of the turbine during operation. *See also:* **asynchronous machine; direct-current commutating machine; synchronous machine.** E282-31E2

speed-changer high-speed stop (gas turbines). A device that prevents the speed changer from moving in the direction to increase speed or power output beyond the position for which the device is set. *See also:* **asynchronous machine; direct-current commutating machine; synchronous machine.** E282-31E2

speed-control mechanism (electric power systems). Includes all equipment such as relays, servomotors, pressure or power-amplifying devices, levers, and linkages between the speed governor and the governor-controlled valves. *See also:* **speed-governing system.** E94-0

speed-governing system. Control elements and devices for the control of the speed or power output of a gas turbine. This includes a speed governor, speed changer, fuel-control mechanism, and other devices and control elements.

See:

area assist action;
base load control;
cam-shaft position;
command control;
constant-frequency control;
constant net-interchange control;
continuous-type control;
controlling means;
derivative control action;
final controlling element;
frequency standard;
function generator;
generating station;
governor-controlled gates,
governor-controlled valves;
governor dead band;
governor speed changer;
governor speed-changer position;
integral control action;
load-limit changer;
master controller;
net interchange schedule programmer;
permissible response rate;
permissive control;
power primary detector;
proportional control action;
rate control action;
regulating range;
reset control action;
schedule setter or set-point device;
speed-control mechanism;
speed governor;
steady-state incremental speed regulation;
steady-state speed regulation;
tie-line bias control;
unit rate-limiting controller;
valve point loading control. E94/E282-31E1

speed governor (electric power system). Includes only those elements that are directly responsive to speed and that position or influence the action of other elements of the speed-governing system. *See also:* **asynchronous machine; direct-current commutating machine; gas turbines; speed-governing system; synchronous machine.** E94/E282-31E2

speed limit (industrial control). A control function that prevents a speed from exceeding prescribed limits. Speed-limit values are expressed as percent of maximum rated speed. If the speed-limit circuit permits the limit value to change somewhat instead of being a single value, it is desirable to provide either a curve of the limit value of speed as a function of some variable, such as load, or to give limit values at two or more conditions of operation. *See also:* **control system, feedback.** AS1-34E10

speed-limit indicator. A series of lights controlled by a relay to indicate the speeds permitted corresponding to the track conditions. 42A42-0

speed of transmission (telecommunication). The instantaneous rate at which information is transferred over a transmission facility. This quantity is usually expressed in characters per unit time or bits per unit time. **Rate of transmission** is more common usage. *See also:* **communication.** 42A65-19E4

speed of transmission, effective. Speed, less than rated, of information transfer that can be averaged over a significant period of time and that reflects effects of control codes, timing codes, error detection, retransmission, tabbing, hand keying, etcetera. 0-19E4

speed of vision. The reciprocal of the duration of the exposure time required for something to be seen. *See also:* **visual field.** Z7A1-0

speed range (industrial control). All the speeds that can be obtained in a stable manner by action of part (or parts) of the control equipment governing the performance of the motor. The speed range is generally expressed as the ratio of the maximum to the minimum operating speed. *See also:* **electric drive.** 42A25-34E10

speed ratio (fuse). *See:* **melting-speed ratio (fuse).**

speed ratio control (industrial control). A control function that provides for operation of two drives at a preset ratio of speed. *See also:* **control system, feedback.** IC1-34E10

speed-regulating rheostat (industrial control). A rheostat for the regulation of the speed of a motor. *See:* **control.** 50I16-34E10

speed regulation characteristic (rotating machinery). The relationship between speed and the load of a motor under specified conditions. *See also:* **asynchronous machine; direct-current commutating machine; synchronous machine.** 0-31E8

speed regulation of a constant-speed direct-current motor. The change in speed when the load is reduced gradually from the rated value to zero with constant applied voltage and field-rheostat setting, expressed as a percent of speed at rated load. *See:* **direct-current commutating machine.** 42A10-0

speed variation (industrial control). Any change in speed of a motor resulting from causes independent of the control-system adjustment, such as line-voltage changes, temperature changes, or load changes. *See also:* **electric drive.** 42A25-34E10

spherical hyperbola (electronic navigation). The locus of the points on the surface of a sphere having a specified constant difference in great-circle distances from two fixed points on the sphere. E172-10E6

spherical reflector. A reflector that is a portion of a spherical surface. *See also:* **antenna.** 0-3E1

spherical-seated bearing (self-aligning bearing) (rotating machinery). A journal bearing in which the bearing liner is supported in such a manner as to permit the axis of the journal to be moved through an appreciable angle in any direction. *See also:* **bearing.** 0-31E8

spherical support seat (rotating machinery). A support for a journal bearing in which the inner surface that mates with the bearing shell is spherical in shape, the center of the sphere coinciding approximately with the shaft centerline, permitting the axis of the bearing to be aligned with that of the shaft. *See also:* **bearing.** 0-31E8

spherical wave. A wave whose equiphase surfaces form a family of concentric spheres. *See also:* **radiation; radio wave propagation.** E211/42A65-3E2

spider (rotor spider)(rotating machinery). A structure supporting the core or poles of a rotor from the shaft, and typically consisting of a hub, spokes, and rim, or some modified arrangement of these. 0-31E8

spider rim (rotating machinery). *See:* **rim.**

spider web (rotating machinery). The component of a rotor that provides radial separation between the hub or shaft and the rim or core. *See also:* **rotor (rotating machinery).** 0-32E8

spike (pulse techniques). A transient of short duration, comprising part of a pulse, during which the amplitude considerably exceeds the average amplitude of the pulse. *See:* **distortion, pulse.** *See also:* **pulse.** 42A65-0

spike leakage energy (microwave gas tubes). The radio-frequency energy per pulse transmitted through the tube before and during the establishment of the steady-state radio-frequency discharge. *See:* **gas tubes.** E160-15E6

spike train (electrotherapy) (courant iteratif). A regular succession of pulses of unspecified shape, frequency, duration, and polarity. *See also:* **electrotherapy.** 42A80-18E1

spindle speed (numerically controlled machines). The rate of rotation of the machine spindle usually expressed in terms of revolutions per minute. *See also:* **numerically controlled machines.** EIA3B-34E12

spindle wave (electrobiology). A sharp, rather large wave considered of diagnostic importance in the electroencephalogram. *See also:* **electrocardiogram.** 42A80-0

spinner (radar). Rotating part of a radar antenna, together with directly associated equipment, used to impart any subsidiary motion, such as conical scanning, in addition to the primary slewing of the beam. *See also:* **radar.** E172-10E6;42A65-0

spinning reserve. That reserve generating capacity connected to the bus and ready to take load. *See also:* **generating station.** 42A35-31E13

spiral four (star quad). A quad in which the four conductors are twisted about a common axis, the two sets of opposite conductors being used as pairs. *See also:* **cable.** 42A65-0

spiral scanning (electronic navigation). Scanning in which the direction of maximum response describes a portion of a spiral. *See also:* **antenna.** 42A65-3E1

SPL. *See:* **sound pressure level.**

***s* plane (control system, feedback).** Plane of the complex variable s; often in control usage, $s = \sigma + j\omega$. *See also:* **control system, feedback.** 85A1-23E0

splashproof (industrial control). So constructed and protected that external splashing will not interfere with successful operation. *See also:* **traction motor.** E16/37A100/42A95-31E11;IC1-34E10

splashproof enclosure. An enclosure in which the openings are so constructed that drops of liquid or solid particles falling on the enclosure or coming towards it in a straight line at any angle not greater than 100 degrees from the vertical cannot enter the enclosure either directly or by striking and running along a surface. E45-0

splashproof machine. An open machine in which the ventilating openings are so constructed that drops of liquid or solid particles falling on the machine or coming towards it in a straight line at any angle not greater than 100 degrees downward from the vertical cannot enter the machine either directly or by striking and running along a surface. *See:* **asynchronous machine; direct-current commutating machine; synchronous machine.** 42A10-31E8

splice (straight-through joint). A joint used for connecting in series two lengths of conductor or cable. 42A95-0

splice box (mine type). An enclosed connector permitting short sections of cable to be connected together to obtain a portable cable of the required length. *See also:* **mine feeder circuit.** 42A85-0

splicing chamber. *See:* **cable vault; manhole.**

split-anode magnetron. A magnetron with an anode divided into two segments; usually by slots parallel to its axis. *See also:* **magnetrons.** 42A70-15E6

split-beam cathode-ray tube (double-beam cathode-ray tube). A cathode-ray tube containing one electron gun producing a beam that is split to produce two traces on the screen. *See also:* **tube definitions.** 0-15E6

split brush (electric machines). Either an industrial or fractional-horsepower brush consisting of two pieces that are used in place of one brush. The adjacent sides of the split brush are parallel to the commutator bars. *Note:* A split brush is normally mounted so that the plane formed by the adjacent contacting brush sides is parallel to or passes through the rotating axis of the rotor. *See:* **asynchronous machine.** *See also:* **brush; brush (rotating machinery);direct-current commutating machine.** 64A1-0

split collector ring (rotating machinery). A collector ring that can be separated into parts for mounting or removal without access to a shaft end. *See also:* **rotor (rotating machinery).** 0-31E8

split-conductor cable. A cable in which each conductor is composed of two or more insulated conductors normally connected in parallel. *See:* **segmental conductor.** *See also:* **power distribution, underground construction.** E30/42A35-31E13

split-core-type current transformer. *See:* **current transformer.**

split fitting. A conduit fitting split longitudinally so that it can be placed in position after the wires have been drawn into the conduit, the two parts being held together by screws or other means. *See also:* **raceway.** 42A95-0

split hub (rotating machinery). A hub that can be separated into parts for ease of mounting on removal from a shaft. *See:* **rotor (rotating machinery).** 0-31E8

split hydrophone (audio and electroacoustics). *See:* **split transducer.**

split node (network analysis). A node that has been separated into a source node and a sink node. *Notes:* (1) Splitting a node interrupts all signal transmission through that node. (2) In splitting a node, all incoming branches are associated with the resulting sink node, and all outgoing branches with the resulting source node. *See also:* **linear signal flow graphs.** E155-0

split-phase electric locomotive. A single-phase electric locomotive equipped with electric devices to change the single-phase power to polyphase power without complete conversion of the power supply. *See also:* **electric locomotive.** 42A42-0

split-phase motor. A single-phase induction motor having a main winding and an auxiliary winding, designed to operate with no external impedance in either winding. The auxiliary winding is energized only during the starting operation of the auxiliary-winding circuits and is open-circuited during running operation. *See:* **asynchronous machine.** 0-31E8

split projector (audio and electroacoustics). *See:* **split transducer.**

split rotor (rotating machinery). A rotor that can be separated into parts for mounting or removal without access to a shaft end. *See also:* **rotor (rotating machinery).** 0-31E8

split-sleeve bearing (rotating machinery). A journal bearing having a bearing sleeve that is split for assembly. *See also:* **bearing.** 0-31E8

split-throw winding (rotating machinery). A winding wherein the conductors that constitute one complete coil side in one slot do not all appear together in another slot. *See:* **asynchronous machine; direct-current commutating machine; synchronous machine.** 42A10-31E8

split transducer (audio and electroacoustics). A directional transducer in which electroacoustic transducing elements are so divided and arranged that each division is electrically separate. *See also:* **electroacoustics.** 0-1E1

split-winding protection (relays). A form of differential protection in which the current in the total winding is compared to the normally proportional current in part of the winding. 0-31E6

spoiler resistors (power supplies). Resistors used to spoil the load regulation of regulated power supplies to permit parallel operation when not otherwise provided for. *See also:* **power supply.** KPSH-10E1

***s* pole.** *See:* **junction pole.**

sponge (electrodeposition). A loose cathode deposit that is fluffy and of the nature of a sponge, contrasted with a reguline metal. *See also:* **electrodeposition.** 42A60-0

spontaneous polarization (ferroelectric material). The displacement due to the polarization vector alone with no induced moments. *Note:* The value of spontaneous polarization (P_s) may be obtained by extrapolating the linear saturated portion of the hysteresis loop back to the axis $E = 0$. In the meter-kilogram-second-ampere (MKSA) system, spontaneous polarization is expressed in coulombs per square meter (C/m^2). It is also frequently expressed in microcoulombs per square centimeter ($\mu C/cm^2$), which unit is equal to $10^{-4} C/m^2$. *See also:* **ferroelectric domain.** E180-0

spool insulator. An insulating element of generally cylindrical form having an axial mounting hole and a circumferential groove or grooves for the attachment of a conductor. *See also:* **insulator; tower.** 42A35-31E13;29A1-0

spot (oscilloscopes) (cathode-ray tube). The illuminated area that appears where the primary electron beam strikes the phosphor screen of a cathode-ray tube. *Note:* The effect of the impact on this small area of the screen is practically instantaneous. *See:* **beam tubes; cathode-ray tubes; oscillograph.** 42A70/50I07-15E69E4

spot noise figure (transducer at a selected frequency) (spot noise factor). The ratio of the output noise power per unit bandwidth to the portion thereof attributable to the thermal noise in the input termination per unit-bandwidth, the noise temperature of the input termination being standard (290 kelvins). The spot noise figure is a point function of input frequency. *See:* **noise figure.** *See also:* **signal-to-noise ratio.** 42A65-0

spot projection (facsimile). The optical method of scanning or recording in which the scanning or recording spot is defined in the path of the reflected or transmitted light. *See also:* **scanning (facsimile); recording (facsimile).** E168-0

spot speed (facsimile). The speed of the scanning or recording spot within the available line. *Note:* This is generally measured on the subject copy or on the record sheet. *See also:* **recording (facsimile); scanning (facsimile).** E168-0

spotting (electroplating). The appearance of spots on plated or finished metals. 42A60-0

spot wobble (television). A process wherein a scanning spot is given a small periodic motion transverse to the scanning lines at a frequency above the picture signal spectrum. *See also:* **television.** E204/42A65-2E2

spreading loss (wave propagation). The reduction in radiant-power surface density due to spreading. E270-0

spring (relay). *See:* **relay spring.**

spring attachment (burglar-alarm system) (spring contact) (trap). A device designed for attachment to a movable section of the protected premises, such as a door, window, or transom, so as to carry the electric protective circuit in or out of such section, and to indicate an open- or short-circuit alarm signal upon opening of the movable section. *See also:* **protective signaling.** 42A65-0

spring barrel. The part that retains and locates the short-circuiter. *See:* **rotor (rotating machinery).** 42A10-0

spring buffer. A buffer that stores in a spring the kinetic energy of the descending car or counterweight. *See also:* **elevators.** 42A45-0

spring-buffer load rating (spring buffer) (elevators). The load required to compress the spring an amount equal to its stroke. *See also:* **elevators.** 42A45-0

spring-buffer stroke (elevators). The distance the contact end of the spring can move under a compressive load until all coils are essentially in contact. *See also:* **elevators.** 42A45-0

spring contact. An electric contact that is actuated by a spring. *See also:* **railway signal and interlocking.** 42A42-0

spring-loaded bearing (rotating machinery). A ball bearing provided with a spring to ensure complete angular contact between the balls and inner and outer races, thereby removing the effect of diametral clearance in both bearings of a machine provided with ball bearing at each end. *See also:* **bearing.** 0-31E8

spring operation. Stored-energy operation by means of spring-stored energy. 37A100-31E11

sprinkler supervisory system. A supervisory system attached to an automatic sprinkler system that initiates signal transmission automatically upon the occurrence of abnormal conditions in valve positions, air or water pressure, water temperature or level, the operability of power sources necessary to the proper functioning of the automatic sprinkler, etcetera. *See also:* **protective signaling.** 42A65-0

spurious count (nuclear techniques). A count from a scintillation counter other than (1) one purposely generated or (2) one due directly to ionizing radiation. *See also:* **scintillation counter.** E175-0

spurious emissions (transmitter performance). Any part of the radio-frequency output that is not a component of the theoretical output, as determined by the type of modulation and specified bandwidth limitations. *See also:* **audio-frequency distortion.** 0-6E1

spurious output (nonharmonic) (signal generator). Those signals in the output of a source that have a defined amplitude and frequency and are not harmonically related to the fundamental frequency. This definition excludes sidebands due to residual and intentional modulation. *See also:* **signal generator.** 0-9E4

spurious pulse (nuclear techniques). A pulse in a scintillation counter other than (1) one purposely generated or (2) one due directly to ionizing radiation. *See also:* **scintillation counter.** E175-0

spurious pulse mode. An unwanted pulse mode, formed by the chance combination of two or more pulse modes, that is indistinguishable from a pulse interrogation or pulse reply. *See also:* **pulse terms.** E194-0

spurious radiation. Any emission from a radio transmitter at frequencies outside its communication band. *See also:* **radio transmission.** E145/42A65-31E13

spurious response (1) (general). Any response, other than the desired response, of an electric transducer or device. *See also:* **transmission characteristics.** 42A65-31E3

(2) (mobile communication or electromagnetic compatibility). Output, from a receiver, due to a signal or signals having frequencies other than that to which the receiver is tuned. *See also:* **electromagnetic compatibility.** 0-6E1

spurious-response ratio (radio receiver). The ratio of (1) the field strength at the frequency that produces the spurious response to (2) the field strength at the desired frequency, each field being applied in turn, under specified conditions, to produce equal outputs. *Note:* Image ratio and intermediate-frequency-response ratio are special forms of spurious response ratio. *See also:* **radio receiver.** E188-0

spurious transmitter output. Any part of the radio-frequency output that is not implied by the type of modulation (amplitude modulation, frequency modulation, etcetera) and specified bandwidth. *See also:* **radio transmitter.** E182-0

spurious transmitter output, conducted. Any spurious output of a radio transmitter conducted over a tangible transmission path. *Note:* Power lines, control leads, radio-frequency transmission lines and waveguides are all considered as tangible paths in the foregoing definition. Radiation is not considered a tangible path in this definition. *See also:* **radio transmitter.** E182-0

spurious transmitter output, extraband. Spurious output of a transmitter outside of its specified band of transmission. *See also:* **radio transmitter.** E182-0

spurious transmitter output, inband. Spurious output of a transmitter within its specified band of transmission. *See also:* **radio transmitter.** E182-0

spurious transmitter output, radiated. Any spurious output radiated from a radio transmitter. *Note:* The radio transmitter does not include the associated antenna and transmission lines. *See also:* **radio transmitter.** E182-0

spurious tube counts (radiation-counter tubes). Counts in radiation-counter tubes, other than background counts and those caused by the source measured. *Note:* Spurious counts are caused by failure of the quenching process, electric leakage, and the like. Spurious counts may seriously affect measurement of background counts. *See also:* **gas-filled radiation-counter tubes.** 42A70-15E6

sputtering (electroacoustics) (cathode sputtering). A process sometimes used in the production of the metal master wherein the original is coated with an electric conducting layer by means of an electric discharge in a vacuum. *Note:* This is done prior to electroplating a heavier deposit. *See also:* **phonograph pickup.** E157-1E1

square-law detection. That form of detection in which the output voltage is substantially proportional, over the useful range of the detecting device, to the square of the voltage of the input wave. *See also:* **detection.** 42A65-0

squareness ratio (material in a symmetrically cyclically magnetized condition) (magnetic storage). The ratio of (1) the flux density at zero magnetizing force to the maximum flux density. (2) The ratio of the flux density when the magnetizing force has changed halfway from zero toward its negative limiting value, to the maximum flux density. *Note:* Both these ratios are functions of the maximum magnetizing force. *See:* **static magnetic storage.** E163-16E9

square wave (1) (regarded as a wave). A periodic wave that alternately for equal lengths of time assumes one of two fixed values, the time of transition being negligible in comparison. *See also:* **wavefront.** E188/E270/42A65-0

(2) (regarded as a pulse train) (pulse techniques). A pulse train that has a normally rectangular shape and a nominal duty factor of 0.5. *See also:* **pulse.** 0-9E4

square-wave response (camera tubes). The ratio of (1) the peak-to-peak signal amplitude given by a test pattern consisting of alternate black and white bars of equal widths to (2) the difference in signal between large-area blacks and large-area whites having the same illuminations as the black and white bars in the test pattern. *Note:* Horizontal square-wave response is measured if the bars run perpendicular to the direction of horizontal scan. Vertical square-wave response is measured if the bars run parallel to the direction of horizontal scan. *See:* **amplitude response; beam tubes.** E160-15E6

square-wave response characteristic (camera tubes). The relation between square-wave response and the ratio of (1) a raster dimension to (2) the bar width in the square-wave response test pattern. *Note:* Unless otherwise specified, the raster dimension is the vertical height. *See:* **amplitude response characteristic; television.** E160-15E6

squeezable waveguide (radar). A variable-width waveguide for shifting the phase of the radio-frequency wave traveling through it. *See also:* **navigation.** E172-10E6

squeeze section (transmission lines and waveguides). A length of rectangular waveguide so constructed as to permit alteration of the broad dimension with a corresponding alteration in electrical length. *See also:* **waveguide.** 0-9E4

squeeze trace (electroacoustics). A variable-density sound track wherein, by means of adjustable masking of the recording light beam and simultaneous increase of the electric signal applied to the light modulator, a track having variable width with greater signal-to-noise ratio is obtained. *See also:* **phonograph pickup.** E157-1E1

squelch (radio receivers). To automatically quiet a receiver by reducing its gain in response to a specified characteristic of the input. E188-0

squelch circuit. A circuit for preventing a radio receiver from producing audio-frequency output in the absence of a signal having predetermined characteristics. A squelch circuit may be operated by signal energy in the receiver pass band, by noise quieting, or by a combination of the two (ratio squelch). It may also be operated by a signal having special modulation characteristics (selective squelch). *See also:* **radio receiver.** 42A65-31E3

squelch sensitivity (receiver performance). A measure of the ability of the squelch circuit to distinguish between a signal and a no-signal condition. *See also:* **receiver performance.** 0-6E1

squint (radar). An ambiguous term, meaning either: (1) the angle between the two major lobe axes in a lobe-switching antenna, or (2) the angular difference between the axis of antenna radiation and a selected geometric axis, such as the axis of the reflector. *See also:* **navigation.** E172-10E6

squint angle. A small difference in pointing angle between a reference beam direction and the direction of maximum radiation. *See also:* **antenna.** 0-3E1

squirrel-cage induction motor. An induction motor in which a primary winding on one member (usually the stator) is connected to an alternating-current power source and a secondary cage winding on the other member (usually the rotor) carries alternating current produced by electromagnetic induction. *Note:* The cage winding is suitably disposed in slots in the secondary core. *See also:* **asynchronous machine.** 42A10-31E8

squirrel-cage rotor (rotating machinery). A rotor core assembly having a squirrel-cage winding. *See also:* **rotor (rotating machinery).** 0-31E8

squitter (electronic navigation). Random output pulses from a transponder caused by ambient noise, or by an intentional random triggering system, but not by the interrogation pulses. *See also:* **navigation.** 0-10E6

SSB. *See:* **single-sideband modulation.**

stability (1) (general). The property of a system or element by virtue of which its output will ultimately attain a steady state. *See also:* **control system, feedback.** AS1-34E10

(2) (systems). An aspect of system behavior associated with systems having the general property that bounded input perturbations result in bounded output perturbations. *Notes:* (A) The multiplicity of existing stability definitions results from the variety of input variables initial states, system inputs and system parameters, and from the intricate behavior of nonlinear systems. The output perturbations considered in the various stability definitions generally consist of the system state. *See also:* **control system.** (B) A measure of the size of a vector state $\mathbf{x}(t)$ or vector output $\mathbf{y}(t)$ is taken as the norm of the vector in question, denoted $\|\mathbf{x}\|$, having the usual properties associated with a norm. A vector $\mathbf{x}$ is said to be bounded if its norm is bounded, that is, $\|\mathbf{x}\| \leq C < \infty$. The particular norm chosen for any given example is a function of the properties desired of the vector and its components. (C) The system behavior is characterized by a state vector solution that results from an initial state $\mathbf{x}(t_0)$, a set of system parameters $\mathbf{a}$, and a system input $\mathbf{u}(t)$ (defined over some interval $t_0 \leq t \leq T$). The vector solution is denoted by

$$\mathbf{x}(t) = \varphi(\mathbf{x}(t_0),\mathbf{a},\mathbf{u};t).$$

0-23E0

(3) (perturbations). For convenience in defining various stability concepts, only those parameters or signals that are perturbed are explicitly exhibited, or mentioned, that is, for perturbations in initial states, a perturbed solution is denoted

$$\varphi(\mathbf{x}(t_0)+\Delta\mathbf{x}(t_0);t),$$

where $\Delta\mathbf{x}(t_0)$ represents the perturbation in initial state. Finally, the perturbed-state solution is denoted

$$\Delta\varphi = \varphi(\mathbf{x}(t_0)+\Delta\mathbf{x}(t_0);t)-\varphi(\mathbf{x}(t_0);t).$$

0-23E0

(4) (power system stability). In a system of two or more synchronous machines connected through an electric network, the condition in which the difference of the angular positions of the rotors of the machines either remains constant while not subjected to a disturbance, or becomes constant following an aperiodic disturbance. *Note:* If automatic devices are used to aid stability, their use will modify the steady-state and transient stability terms to: **steady-state stability with automatic devices; transient stability with automatic devices.** Automatic devices as defined for this purpose are those devices that are operating to increase stability during the period preceding and following a disturbance as well as during the disturbance. Thus relays and circuit breakers are excluded from this classification and all forms of voltage regulators included. Devices for inserting and removing shunt or series impedance may or may not come within this classification depending upon whether or not they are operating during the periods preceding and following the disturbance. *See:* **steady-state stability; transient stability.** *See also:* **alternating-current distribution.**

42A35-31E13

(5) (electric machines). The change in output voltage or current as a function of time, at constant line voltage, load, and ambient temperature (sometimes referred to as drift). 0-10E1

(6) (oscilloscopes). Property of retaining defined electrical characteristics for a prescribed period. *Note:* Deviations from a stable state may be called drift or jitter. In triggered sweep systems, triggering stability may refer to the ability of the trigger and sweep systems to maintain jitter-free display of high-frequency waveforms for long (seconds to hours) periods of time. Also, the name of the control used on some oscilloscopes to adjust the sweep for triggered, free-running, or synchronized operation. *See:* **oscillograph; sweep mode.** 0-9E4

stability, absolute (control system). Global asymptotic stability maintained for all nonlinearities within a given class. *Note:* A typical problem to which the concept of absolute stability has been applied consists of a system with dynamics described by the vector differential equation

$$\dot{\mathbf{x}} = A\mathbf{x}+\mathbf{b}f(\sigma),$$
$$\sigma = \mathbf{c}^T\mathbf{x},$$

with a nonlinearity class defined by the conditions

$$f(0) = 0,$$
$$k_1 \leqslant f(\sigma)/\sigma \leqslant k_2.$$

The solution $\mathbf{x}(t) = \mathbf{0}$ is said to be absolutely stable if it is globally asymptotically stable for all nonlinear functions $f(\sigma)$ in the above class. *See also:* **control system.** 0-23E0

stability, asymptotic (control system) (of a solution $\phi(\mathbf{x}(t_0);t)$). The solution is (1) Lyapunov stable, (2) such that

$$\lim_{t\to\infty} \|\Delta\phi\| = 0,$$

where $\Delta\phi$ is a change in the solution due to an initial state perturbation. See **stability** for explanation of symbols. *Notes:* (A) The solution $\mathbf{x} = 0$ of the system $\dot{\mathbf{x}} = \mathbf{ax}$ is asymptotically stable for $\mathbf{a} < 0$, but not for $\mathbf{a} = 0$. In this case

$$\varphi(\mathbf{x}(t_0);t) = \mathbf{x}(t_0)\exp(-\mathbf{a}(t-t_0)).$$

(B) In some cases the rate of convergence to zero depends on both the initial state $\mathbf{x}(t_0)$ and the initial time t_0. See stability, equiasymptotic for stability concepts where this rate of convergence is independent of either $\mathbf{x}(t_0)$ or t_0. *See also:* **control system.** 0-23E0

stability, bounded-input–bounded-output (control system). Driven stability when the solution of interest is the output solution. *See also:* **control system.**

0-23E0

stability, conditional (linear feedback control system). A property such that the system is stable for prescribed operating values of the frequency-invariant factor of the loop gain and becomes unstable not only for higher values, but also for some lower value. *See also:* **control system, feedback.** 85A1-23E0

stability, driven (control system) (solution $\phi(\mathbf{u};t)$). For each bounded system input perturbation $\Delta\mathbf{u}(t)$ the output perturbation $\Delta\phi$ is also bounded for $t \geq t_0$. *Note:* A necessary and sufficient condition for a solution of a linear system to be driven-stable is that the solution be uniformly asymptotically stable. See **stability** for explanation of symbols. *See also:* **control system.**

0-23E0

stability, equiasymptotic (control system). Asymptotic stability where the rate of convergence to zero of the perturbed-state solution is independent of all initial states in some region $\|\Delta\mathbf{x}(t_0)\| \leq \nu$. *See also:* **control system.** 0-23E0

stability factor. The ratio of a stability limit (power limit) to the nominal power flow at the point of the system to which the stability limit is referred. *Note:* In determining stability factors it is essential that the nominal power flow be specified in accordance with the one of several bases of computation, such as rating or capacity of, or average or maximum load carried by, the equipment or the circuits. *See also:* **alternating-current distribution.** 42A35-31E13

stability, finite-time (control system) (solutions). For all initial states that originate in a specified region R at time t_0, the resulting solutions remain in another specified region R_ϵ over the given time interval $t_0 \leq t \leq T$. *Notes:* (1) In the definition of finite-time stability the quantities R_δ, R_ϵ, and T are prespecified. Obviously, R must be included in R_ϵ. (2) A system may be Lyapunov unstable and still be finite-time stable. For example, a system with dynamics $\mathbf{x} = \mathbf{ax}$, $\mathbf{a} > 0$, is Lyapunov unstable, but if

$$R_\delta : |\mathbf{x}| \leqslant \delta,$$
$$R_\epsilon : |\mathbf{x}| \leqslant \epsilon,$$

and $T < \mathbf{a}^{-1}\ln(\epsilon/\delta)$, the system is finite-time stable (relative to the given values of δ, ϵ, and T). *See also:* **control system.** 0-23E0

stability, global (control system) (solution $\phi(\mathrm{x}(t_0);t)$). Stable for all initial perturbations, no matter how large they may be. See **stabil**for explanation of symbols. *See also:* **control system.** 0-23E0

stability in-the-whole (control system). Synonymous with **global stability.** *See:* **stability, global.** *See also:* **control system.** 0-23E0

stability, Lagrange (system) (control system). Every solution that is generated by a finite initial state is bounded. *Note:* An example of a system that is Lagrange stable is a second-order system with a single stable limit cycle. Although this system must contain a point inside the limit cycle that is Lyapunov unstable, the system is still Lagrange stable because every solution remains bounded. *See also:* **control system.** 0-23E0

stability limit (1) (general). A condition of a linear system or one of its parameters that places the system on the verge of instability. 85A1-23E0

(2) (power limit) (electric systems). The maximum power flow possible through some particular point in the system when the entire system or the part of the system to which the stability limit refers is operating with stability. *See also:* **alternating-current distribution.** 42A35-31E13

stability, long-term (LTS) (power supplies). The change in output voltage or current as a function of time, at constant line voltage, load, and ambient temperature (sometimes referred to as drift). *See also:* **power supply.** KPSH-10E1

stability, Lyapunov (control system) (of a solution $\phi(\mathbf{x}(t_0);t)$). For every given $\epsilon > 0$ there exists a $\delta > 0$ (which, in general, may depend on ϵ and on t_0) such that $||\Delta\mathbf{x}(t_0)|| \leq \delta$ implies $||\Delta\phi|| \leq \epsilon$ for $t \geq t_0$. *Notes:* (1) The solution $\mathbf{x} = 0$ of the system $\dot{\mathbf{x}} = \mathbf{ax}$ is Lyapunov stable if $\mathbf{a} < 0$ and is Lyapunov unstable if $\mathbf{a} > 0$. (2) For a linear system with an irreducible transfer function $T(s)$, Lyapunov stability implies that all the poles of $T(s)$ are in the left-half s plane and that those on the $j\omega$ axis are simple. See **stability** for explanation of symbols. *See also:* **control system.** 0-23E0

stability of a limit cycle (control system). Synonymous with orbital stability. *See also:* **control system.** 0-23E0

stability of the speed-governing system (gas turbines). A characteristic of the system that indicates that the speed-governing system is capable of actuating the turbine fuel-control valve so that sustained oscillations in turbine speed, or rate of energy input to the turbine, are limited to acceptable values by the speed-governing system. E282-31E2

stability of the temperature-control system (gas turbines). A characteristic of the system that indicates that the temperature-control system is capable of actuating the turbine fuel-control valve so that sustained oscillations in rate of energy input to the turbine are limited to acceptable values by the temperature-control system during operation under constant system frequency. E282-31E2

stability, orbital (control system) (closed solution curve denoted Γ). Implies that for every given $\epsilon > 0$ there exists a $\delta > 0$ (which, in general, may depend on ϵ and on t_0) such that $\rho(\Gamma, \mathbf{x}(t_0)) \leq \delta$ implies $\rho(\Gamma, \phi(\mathbf{x}(t_0);t)) \leq \epsilon$ for $t \geq t_0$, where $\rho(\Gamma, \mathbf{a})$ denotes the minimum distance between the curve Γ and the point $\mathbf{a}$. Here the point $\mathbf{x}(t_0)$ is assumed to be off the curve Γ. *Notes:* (1) Orbital stability does not imply Lyapunov stability of a closed solution curve, since a point on the closed curve may not travel at the same speed as a neighboring point off the curve. (2) Only nonlinear systems can produce the type of solutions for which the concept of orbital stability is applicable. See **stability** for explanation of symbols. *See also:* **control system.** 0-23E0

stability, practical (control system). Synonymous with **finite-time stability.** *See:* **stability, finite-time.** *See also:* **control system.** 0-23E0

stability, quasi-asymptotic (control system) (solution $\phi\mathrm{x}(\mathrm{x}(t_0);t)$). Implies

$$\lim_{t \to \infty} ||\Delta\phi|| = 0.$$

Notes: (1) Quasi-asymptotic stability is condition (2) in the definition of asymptotic stability and, hence, need not imply Lyapunov stability. (2) An example of a solution that is quasi-asymptotically stable but not asymptotically stable is the solution $\mathbf{x}(t) = 0$ of the system $\dot{\mathbf{x}} = \mathbf{x}^2$. The solution of the above system for a perturbation $\Delta x(t_0)$ from 0 is

$$\varphi(\Delta\mathbf{x}(t_0);t) = \Delta\mathbf{x}(t_0)/[1 - (t - t_0)\Delta\mathbf{x}(t_0)].$$

Obviously, $\phi(\Delta\mathbf{x}(t_0);t)$ approaches zero as t approaches ∞, yet is not Lyapunov stable since it is unbounded at $t = t_0 + (1/\Delta\mathbf{x}(t_0))$. See **stability** for explanation of symbols. *See also:* **control system.** 0-23E0

stability, relative (automatic control) (stable underdamped system). The property measured by the relative settling times when parameters are changed. *See also:* **control system, feedback.** 0-23E0

stability, short-time (control system). Synonymous with **finite-time stability.** *See:* **stability, finite-time.** *See also:* **control system.** 0-23E0

stability, synchronous-machine regulating-system. The property of a synchronous-machine-regulating system in which a change in the controlled variable, resulting from a stimulus, decays with time if the stimulus is removed. *See:* **synchronous machine.** 0-31E8

stability, total (control system) (solution $\phi = \phi(\mathbf{x}(t_0);t)$ of the system $\dot{\mathbf{x}} = \mathbf{f}(\mathbf{x},t)$). Implies that for every given $\epsilon > 0$ there exist a $\delta_1 > 0$ and a $\delta_2 > 0$ (both of which, in general, may depend on ϵ and t_0) such that $\|\Delta\mathbf{x}(t_0)\| \leq \delta_1$ and $\|g(\mathbf{x},t)\| \leq \delta_2$ imply $\|\phi - \Psi\| \leq \epsilon$ for $t \geq t_0$, where $\Psi = \Psi(\mathbf{x}(t_0) + \Delta\mathbf{x}(t_0);t)$ is a solution of the system $\dot{\mathbf{x}} = \mathbf{f}(\mathbf{x},t) + \mathbf{g}(\mathbf{x},t)$. See **stability** for explanation of symbols. *See also:* **control system.** 0-23E0

stability, trajectory (control system). Orbital stability where the solution curve is not closed. *See also:* **control system.** 0-23E0

stability, uniform-asymptotic (control system). Asymptotic stability where the rate of convergence to zero of the perturbed-state solution is independent of the initial time t_0. *Note:* An example of a solution that is asymptotically stable but not uniformly asymptotically stable is the solution

$$\varphi(\mathbf{x}(t_0);t) = \mathbf{x}(t_0)t_0/t$$

of the system $\dot{\mathbf{x}} = -\mathbf{x}/t,\ t_0 > 0$. Note that the initial rate of decay,

$$\dot{\mathbf{x}}(t_0)/\mathbf{x}(t_0) = -1/t_0,$$

is clearly a function of t_0. Compare with the time-invariant system $\dot{\mathbf{x}} = \mathbf{a}\mathbf{x}$ where $\dot{\mathbf{x}}(t_0)/\mathbf{x}(t_0) = \mathbf{a}$ is independent of t_0. The concept of uniformity with respect to the initial time t_0 applies only to time-varying systems. All stable time-invariant systems are uniformly stable. *See also:* **control system.** 0-23E0

stabilization (1) (control system, feedback). Act of attaining stability or of improving relative stability. *See also:* **control system, feedback.** 85A1-23E0
(2) (navigation). Maintenance of a desired orientation of a vehicle or device with respect to one or more reference directions. *See also:* **navigation.** 0-10E6
(3) (direct-current amplifier). *See:* **drift stabilization.**

stabilization network. As applied to operational amplifiers and servomechanisms, a network used to shape the transfer characteristics to eliminate or minimize oscillations when feedback is provided. *See also:* **electronic analog computer.** E165-16E9

stabilized feedback. Feedback employed in such a manner as to stabilize the gain of a transmission system or section thereof with respect to time or frequency or to reduce noise or distortion arising therein. *Note:* The section of the transmission system may include amplifiers only, or it may include modulators. *See also:* **feedback.** E145-0

stabilized flight. The type of flight that obtains control information from devices that sense orientation with respect to external references. *See also:* **navigation.** 0-10E6

stabilized shunt-wound generator. A type of compound-wound generator with a series field winding of such proportion and polarity as to minimize voltage regulation or provide sufficient droop so that machines may be operated in parallel without equalizers. *Note:* The voltage regulation of such a generator is always drooping; that is, the voltage at rated load is less than at no load. This definition is not intended to apply to stability of machines operating through a voltage range on a straight-line portion of the saturation curve. *See also:* **direct-current commutating machine.** 42A10-0

stabilized shunt-wound motor. A shunt-wound motor having a light series winding added to prevent a rise in speed or to obtain a slight reduction in speed, with increase of load. *See also:* **direct-current commutating machine.** 42A10-0

stabilizing winding (transformer). An auxiliary winding used particularly in Y-connected transformers for such purposes as the following: (1) to stabilize the neutral point of the fundamental frequency voltages; (2) to minimize third-harmonic voltage and the resultant effects on the system; (3) to minimize telephone influence due to third harmonic currents and voltages; and (4) to minimize the residual direct-current magnetomotive force on the core. *See also:* **transformer; windings, high-voltage and low-voltage.** 42A15/57A18-0

stable (control system, feedback). Possessing stability. *See also:* **control system, feedback.** 85A1-23E0

stable element (electronic navigation). An instrument or device that maintains a desired orientation independently of the motion of the vehicle. *See also:* **navigation.** E172-10E6

stable oscillation. A response that does not increase indefinitely with increasing time; an unstable oscillation is the converse. *Note:* The response must be specified or understood; a steady current in a pure resistance network would be stable, although the total charge passing any cross section of a network conductor would be increasing continuously. E270-0

stable platform (electronic navigation). A gimbal-mounted platform, usually containing gyros and accelerometers, whose purpose is to maintain a desired orientation in inertial space independent of the motion of the vehicle. *See also:* **navigation.** E174-10E6

stack. A rigid assembly of two or more switch and bus insulating units. *See also:* **tower.** 29A1/42A35-31E13

stacker (computing systems). *See:* **card stacker.**

stage (1) (communication practice). One step, especially if part of a multistep process, or the apparatus employed in such a step. The term is usually applied to an amplifier. *See also:* **amplifier.** 42A65-0
(2) (thermoelectric device). One thermoelectric couple or two or more similar thermoelectric couples arranged thermally in parallel and electrically connected. *See also:* **thermoelectric device.** E221-15E7

stage efficiency. The ratio of useful power delivered to the load (alternating current) and the plate power input (direct current). *See also:* **network analysis.** E145/42A65-0

stagger (facsimile). Periodic error in the position of the recorded spot along the recorded line. *See also:* **recording (facsimile).** E168-0

staggering. The offsetting of two channels of different carrier systems from exact sideband frequency coincidence in order to avoid mutual interference. *See also:* **transmission characteristics.** 42A65-0

staggering advantage. The effective reduction, in decibels, of interference between carrier channels, due to staggering. *See also:* **transmission characteristics.** 42A65-0

stagger time, relay. *See:* **relay stagger time.**

stagger-tuned amplifier. An amplifier consisting of two or more single-tuned stages that are tuned to different frequencies. *See also:* **amplifier.** 42A65-0

stain spots (electroplating). Spots produced by exudation, from pores in the metal, of compounds absorbed from cleaning, pickling plating solutions. The appearance of stain spots is called spotting out. 42A60-0

staircase signal (television). A waveform consisting of a series of discrete steps resembling a staircase. *Note:* The staircase signal is the electrical equivalent of a gray scale. The height of the steps may vary, monotonically increasing or decreasing. *See also:* **television.** 42A65/E204-2E2

stairstep sweep (oscilloscopes). An incremental sweep in which each step is equal. The electric deflection waveform producing a stairstep sweep is usually called a staircase or stairstep waveform. *See:* **incremental sweep.** *See also:* **oscillograph.** 0-9E4

stalled tension control (industrial control). A control function that maintains tension in the material at zero speed. *See also:* **electric drive.** IC1-34E10

stalled torque control (industrial control). A control function that provides for the control of the drive torque at zero speed. *See also:* **electric drive.** IC1-34E10

stalo (moving-target-indicator radar). A highly stable local radio-frequency oscillator used for heterodyning signals to produce an intermediate frequency. *See also:* **navigation.** E172-10E6

stamper (electroacoustics). A negative (generally made of metal by electroforming) from which finished pressings are molded. *See also:* **phonograph pickup.** E157-1E1

standard (transmission lines and waveguides). A device having stable, precisely defined characteristics that may be used as a reference. 0-9E4

standard cable*. The standard cable formerly used for specifying transmission losses had, in American practice, a linear series resistance and linear shunt capacitance of 88 ohms and 0.054 microfarad, respectively, per loop mile, with no inductance or shunt conductance. *See also:* **transmission characteristics.**
*Obsolete 42A65-0

standard cell. A cell that serves as a standard of electromotive force. *See:* **Weston normal cell; unsaturated standard cell.** *See also:* **auxiliary device to an instrument; electrochemistry.** 42A60-0

standard code. The operating, block signal, and interlocking rules of the Association of American Railroads. *See also:* **railway signal and interlocking.** 42A42-0

standard compass. A magnetic compass so located that the effect of the magnetic mass of the vessel and other factors that may influence compass indication is the least practicable. 42A43-0

standard-dimensioned motor. A general or definite-application motor so dimensioned that it is mechanically interchangeable as a whole with any other motor of the same frame size and complying with the same standard specification. *See also:* **asynchronous machine; direct-current commutating machine; synchronous machine.** 0-31E8

standard electrode potential. An equilibrium potential for an electrode in contact with an electrolyte, in which all of the components of a specified electrochemical reaction are in their standard states. The standard state for a gas is the pressure of one atmosphere, for an ionic constituent it is unit ion activity, and it is a constant for a solid. *See also:* **electrochemistry.** 42A60-0

standard frequency (electric power systems). A precise frequency intended to be used for a frequency reference. *See also:* **power systems, low-frequency and surge testing.** E94-0

standard full impulse voltage wave (1) (insulation strength). An impulse that rises to crest value of voltage in 1.2 microseconds (virtual time) and drops to 0.5 crest value of voltage in 50 microseconds (virtual time), both times being measured from the same origin and in accordance with established standards of impulse testing techniques. *Note:* The virtual value for the duration of the wavefront is 1.67 times the time taken by the voltage to increase from 30 percent to 90 percent of its crest value. The origin from which time is measured is the intersection with the zero axis of a straight line drawn through points on the front of the voltage wave at 30-percent and 90-percent crest value. 92A1-0

(2) (mercury lamp transformers). An impulse that rises to crest value of voltage in 1.5 microseconds (nominal time) and drops to 0.5 crest value of voltage in 40 microseconds (nominal time), both times being measured from the same time origin and in accordance with established standards of impulse testing techniques. *See also:* **basic impulse insulation level (BIL) (insulation strength).** 82A7-0

standard insulation class (instrument transformer). Denotes the maximum voltage in kilovolts that the insulation of the primary winding is designed to withstand continuously. *See also:* **instrument transformer.** 12A0-0

standardization. *See:* **echelon; interlaboratory working standards; laboratory reference standards.**

standardize. *See:* **check (instrument or meter).**

standard microphone. A microphone the response of which is accurately known for the condition under which it is to be used. *See also:* **instrument.** 42A30-0

standard noise temperature (interference terminology). The temperature used in evaluating signal transmission systems for noise factor 290 kelvins (27 degrees Celsius). *See also:* **interference.** 0-13E6

standard observer (color). A hypothetical observer who requires standard amounts of primaries in a color mixture to match every color. *Note:* Standard amounts of a particular set of primaries used by this observer can be computed, by established methods, from standard amounts of the standard (and usually nonphysical) primaries. The present standard primaries and standard amounts of them required to match various wavelengths of the spectrum were established in 1931 by the International Commission on Illumination (CIE). *See also:* **color terms.** E201-0

standard operating duty. *See:* **operating duty.**

standard pitch. *See:* **standard tuning frequency.**

standard potential (standard electrode potential). The reversible potential for an electrode process when all products and reactants are at unit activity on a scale in which the potential for the standard hydrogen half-cell is zero. CM-34E2

standard propagation (radio waves). The propagation of radio waves over a smooth spherical earth of specific uniform dielectric constant and conductivity, under conditions of standard refraction in the atmosphere.

See also: **radiation; radio wave propagation.** E211-3E2;42A65-0

standard reference position (contact). The nonoperated or de-energized position of the associated main device to which the contact position is referred. *Note:* Standard reference positions of typical devices are listed in American National Standard C37.2-1962, for example:

Device	Standard Reference Position
Circuit breaker	Main contacts open
Disconnecting switch	Main contacts open
Relay	De-energized position
Contactor	De-energized position
Valve	Closed position

37A100-31E11

standard refraction (radio waves). The refraction that would occur in an idealized atmosphere in which the refractive index decreases uniformly with height above the earth at the rate of 39×10^{-6} per meter *Note:* Standard refraction may be included in ground wave calculations by use of an effective earth radius of 8.5×10^6 meters, or 4/3 the geometrical radius of the earth. Refraction exceeding standard refraction is called superrefraction, and refraction less than standard refraction is called subrefraction. *See also:* **radiation.** 42A65-3E2

standard register (motor meter) (dial register). A four- or five-dial register, each dial of which is divided into ten equal parts, the division marks being numbered from zero to nine, and the gearing between the dial pointers being such that the relative movements of the adjacent dial pointers are in opposite directions and in a 10-to-1 ratio. *See also:* **watthour meter.** 42A30-0

standard rod gap. A gap between the ends of two one-half-inch square rods cut off squarely and mounted on supports so that a length of rod equal to or greater than one-half the gap spacing overhangs the inner edge of each support. It is intended to be used for the approximate measurement of crest voltages. *See also:* **instrument.** 42A30-0

standard resistor (resistance standard). A resistor that is adjusted with high accuracy to a specified value, is but slightly affected by variations in temperature, and is substantially constant over long periods of time. *See also:* **auxiliary device to an instrument.** 42A30-0

standard source (illuminating engineering). One that has a specified spectral distribution and is used as a standard for colorimetry. *Note:* In 1931 the International Commission on Illumination (CIE) specified the spectral energy distributions for three standard sources, *A*, *B* and *C*. *See:* **color; standard source *A*; standard source *B*; standard source *C*.** Z7A1-0

standard source *A*. A tungsten filament lamp operated at a color temperature of 2854 degrees Celsius, and approximates a blackbody operating at that temperature. *See also:* **standard source (illuminating engineering).** Z7A1-0

standard source *B*. An approximation of noon sunlight having a correlated color temperature of approximately 4870 degrees Celsius. It is obtained by a combination of Source *A* and a special filter. *See also:* **standard source (illuminating engineering).** Z7A1-0

standard source *C*. An approximation of daylight provided by a combination of direct sunlight and clear sky having a correlated color temperature of approximately 6770 degrees Celsius. It is obtained by a combination of Source *A* and a special filter. *See also:* **standard source (illuminating engineering).** Z7A1-0

standard sphere gap. A gap between two metal spheres of standard dimensions, mounted and used in a specified manner. It is intended to be used for the measurement of the crest value of a voltage by observing the maximum gap spacing at which sparkover occurs when the voltage is applied under known atmospheric conditions. *See also:* **instrument.** 42A30-0

standard television signal. A signal that conforms to certain accepted specifications. *See also:* **television.** 42A65-0

standard track (electroacoustics). *See:* **single track (electroacoustics).**

standard tuning frequency (standard musical pitch). The frequency for the note A_4, namely, 440 hertz. See the accompanying table. *See also:* **electroacoustics.** 0-1E1

standard volume indicator (electroacoustics). *See:* **volume indicator, standard.** *See also:* **circuits and devices; instruments.**

standard-wave error (direction-finder measurements). The bearing error produced by a wave whose vertically and horizontally polarized electric fields are equal and phased so as to give maximum error in the direction finder, and whose incidence direction is arranged to be 45 degrees. *See also:* **navigation.** 0-10E6

standby. *See:* **reserve equipment.**

standby redundancy (reliability). *See:* **redundancy, standby.**

standing-on-nines carry (parallel addition of decimal numbers). A high-speed carry in which a carry input to a given digit place is bypassed to the next digit place if the current sum in the given place is nine. *See also:* **carry; electronic digital computer.** X3A12-16E9

standing wave. A wave in which, for any component of the field, the ratio of its instantaneous value at one point to that at any other point does not vary with time. *Notes:* (1) Commonly a periodic wave in which the amplitude of the displacement in the medium is a periodic function of the distance in the direction of any line of propagation of the wave. (2) A standing wave results when the electric variation in a circuit takes the form of periodic exchange of energy between current and potential forms without translation of energy. (3) Also termed stationary wave. *See also:* **power systems, low-frequency and surge testing; radio; radio wave propagation; transmission line; waveguide.**
E146-3E1/9E4;E211-3E2; E270/42A35-31E13

standing-wave detector. *See:* **standing-wave meter.**

standing-wave indicator. *See:* **standing-wave meter.**

standing-wave loss factor. The ratio of the transmission loss in an unmatched waveguide to that in the same waveguide when matched. *See also:* **waveguide.** E146-0

standing-wave machine. *See:* **standing-wave meter.**

standing-wave meter (standing-wave indicator) (standing-wave detector) (standing-wave machine). An instrument for measuring the standing-wave ratio in a transmission line. In addition a standing-wave meter

Frequencies and Frequency Levels of the Usual Equally Tempered Scale; Based on A_4 = 440 Hz

Note Name	Frequency Level (semits)	Frequency (hertz)
C_0	0	16.352
	1	17.324
D_0	2	18.354
	3	19.445
E_0	4	20.602
F_0	5	21.827
	6	23.125
G_0	7	24.500
	8	25.957
A_0	9	27.500
	10	29.135
B_0	11	30.868
C_1	12	32.703
	13	34.648
D_1	14	36.708
	15	38.891
E_1	16	41.203
F_1	17	43.654
	18	46.249
G_1	19	48.999
	20	51.913
A_1	21	55.000
	22	58.270
B_1	23	61.735
C_2	24	65.406
	25	69.296
D_2	26	73.416
	27	77.782
E_2	28	82.407
F_2	29	87.307
	30	92.499
G_2	31	97.999
	32	103.83
A_2	33	110.00
	34	116.54
B_2	35	123.47
C_3	36	130.81
	37	138.59
D_3	38	146.83
	39	155.56

Note Name	Frequency Level (semits)	Frequency (hertz)
E_3	40	164.81
F_3	41	174.61
	42	185.00
G_3	43	196.00
	44	207.65
A_3	45	220.00
	46	233.08
B_3	47	246.94
C_4	48	261.63
	49	277.18
D_4	50	293.66
	51	311.13
E_4	52	329.63
F_4	53	349.23
	54	369.99
G_4	55	392.00
	56	415.30
A_4	57	440.00
	58	466.16
B_4	59	493.88
C_5	60	523.25
	61	554.37
D_5	62	587.33
	63	622.25
E_5	64	659.26
F_5	65	698.46
	66	739.99
G_5	67	783.99
	68	830.61
A_5	69	880.00
	70	932.33
B_5	71	987.77
C_6	72	1046.5
	73	1108.7
D_6	74	1174.7
	75	1244.5
E_6	76	1318.5
F_6	77	1396.9
	78	1480.0
G_6	79	1568.0
	80	1661.2

Note Name	Frequency Level (semits)	Frequency (hertz)
A_6	81	1760.0
	82	1864.7
B_6	83	1975.5
C_7	84	2093.0
	85	2217.5
D_7	86	2349.3
	87	2489.0
E_7	88	2637.0
F_7	89	2793.8
	90	2960.0
G_7	91	3136.0
	92	3322.4
A_7	93	3520.0
	94	3729.3
B_7	95	3951.1
C_8	96	4186.0
	97	4434.9
D_8	98	4698.6
	99	4978.0
E_8	100	5274.0
F_8	101	5587.7
	102	5919.9
G_8	103	6271.9
	104	6644.9
A_8	105	7040.0
	106	7458.6
B_8	107	7902.1
C_9	108	8372.0
	109	8869.8
D_9	110	9397.3
	111	9956.1
E_9	112	10548.
F_9	113	11175.
	114	11840.
G_9	115	12544.
	116	13290.
A_9	117	14080.
	118	14917.
B_9	119	15084.
C_{10}	120	16744.

may include means for finding the location of maximum and minimum amplitudes. *See also:* **instrument.** 42A30-0

standing-wave ratio (1) (general). The ratio of the amplitude of a standing wave at an antinode to the amplitude at a node. *Note:* The standing-wave ratio in a uniform transmission line is

$$\frac{1+p}{1-p}$$

where p is the reflection coefficient. *See also:* **transmission characteristics; wave front.** 42A65/83A14-0

(2) (voltage-standing-wave ratio*) (at a given frequency in a uniform waveguide or transmission line). The ratio of the maximum to the minimum amplitudes of corresponding components of the field (or the voltage or current) appearing along the guide or line in the direction of propagation. *Notation:* SWR (VSWR*). *Note:* Alternatively, the standing-wave ratio may be expressed as the reciprocal of the ratio defined above.
*Deprecated E146/E148-3E1/9E4

(3) (at a port of a multiport network). The standing-wave ratio (SWR) when reflectionless terminations are connected to all other ports.

standing-wave-ratio indicator (standing-wave-ratio meter). A device or part thereof used to indicate the standing-wave ratio. *Note:* In common terminology, it is the combination of amplifier and meter as a supplement to the slotted line or bridge, etcetera, when performing impedance or reflection measurements. 0-9E4

standing-wave-ratio meter. *See:* **standing-wave-ratio indicator.**

standstill locking (rotating machinery). The occurrence of zero or unusably small torque in an energized polyphase induction motor, at standstill, for certain rotor positions. *See also:* **asynchronous machine.** 0-31E8

star chain (electronic navigation). A group of navigational radio transmitting stations comprising a master station about which three or more slave stations are symmetrically located. *See also:* **radio navigation.** E172-10E6;42A65-0

star-connected circuit. A polyphase circuit in which all the current paths of the circuit extend from a terminal of entry to a common terminal or conductor (which may be the neutral conductor). *Note:* In a three-phase system this is sometimes called a Y (or wye) connection. *See:* **star network.** *See also:* **network analysis.** E270-0

star-delta starter (industrial control). A switch for starting a three-phase motor by connecting its windings first in star and then in delta. *See also:* **starter.** 50I16-34E10

star-delta starting. The process of starting a three-phase motor by connecting it to the supply with the primary winding initially connected in star, then reconnected in delta for running operation. 0-31E8

star network. A set of three or more branches with one terminal of each connected at a common node. *See also:* **network analysis.** E153/E270-0

star quad. *See:* **spiral four.**

star rectifier circuit. A circuit that employs six or more rectifying elements with a conducting period of 60 electrical degrees plus the commutating angle. *See also:* **rectification.** 42A15-0

start-dialing signal (semiautomatic or automatic working) (telecommunication). A signal transmitted from the incoming end of a circuit, following the receipt of a seizing signal, to indicate that the necessary circuit conditions have been established for receiving the numerical routing information. *See also:* **communication.** 0-19E4

starter (1) (industrial control). An electric controller for accelerating a motor from rest to normal speed and to stop the motor.
See:
across-the-line starter;
automatic starter;
autotransformer starter;
electric controller;
full-voltage starter;
increment starter;
part-winding starter;
primary reactor starter;
primary resistor starter;
reduced-voltage starter;
resistance starting;
reversing starter;
slow-speed starting;
star-delta starter. 42A25-34E10

(2) (illuminating engineering). A device used in conjunction with a ballast for the purpose of starting an electric-discharge lamp. *See also:* **fluorescent lamp.** Z7A1-0

(3) (glow-discharge cold-cathode tube). An auxiliary electrode used to initiate conduction. *See also:* **electrode (of an electron tube).** 42A70-0

starter gap (gas tube). The conduction path between a starter and the other electrode to which starting voltage is applied. *Note:* Commonly used in the glow-discharge cold-cathode tube. *See also:* **gas-filled rectifier.** 50I07-15E6;42A70-0

starters (fluorescent lamps). Devices that first connect a fluorescent or similar discharge lamp in a circuit to provide for cathode preheating and then open the circuit so that the starting voltage is applied across the lamp to establish an arc. Starters also include a capacitor for the purpose of assisting the starting operation and for the suppression of radio interference during lamp starting and lamp operation. They may also include a circuit-opening device arranged to disconnect the preheat circuit if the lamp fails to light normally. 78A180-0

starter voltage drop (glow-discharge cold-cathode tube). The starter-gap voltage drop after conduction is established in the starter gap. *See also:* **electrode voltage (electron tube); gas tubes.** E160-15E6;42A70-0

starting (rotating machinery). The process of bringing a motor up to speed from rest. *Note:* This includes breaking away, accelerating and if necessary, synchronizing with the supply. *See also:* **asynchronous machine; direct-current commutating machine; synchronous machine.** 0-31E8

starting amortisseur. An amortisseur the primary function of which is the starting of the synchronous machine and its connected load. *See also:* **synchronous machine.** 42A10-0

starting anode. An electrode that is used in establishing the initial arc. *See also:* **rectification.** 42A15-0

starting capacitance (capacitor motor). The total effective capacitance in series with the auxiliary winding for starting operation. *See also:* **asynchronous machine.** 0-31E8

starting current (1) (rotating machinery). The current drawn by the motor during the starting period. (A function of speed or slip). *See also:* **asynchronous machine; direct-current commutating machine; synchronous machine.** 0-31E8

(2) (oscillator). The value of electron-stream current through an oscillator at which selfsustaining oscillations will start under specified conditions of loading. *See:* **magnetrons.** *See also:* **circuit characteristics of electrodes.** E160-15E6

starting electrode (1) (gas tube). An auxiliary electrode used to initiate conduction. *See also:* **gas-filled rectifier.** 50I07-15E6

(2) (pool-cathode tube). An electrode used to establish a cathode spot. *See also:* **electrode (electron tube).** 42A70-15E6

starting motor. An auxiliary motor used to facilitate the starting and accelerating of a main machine to which it is mechanically connected. *See also:* **asynchronous machine; direct-current commutating machine; synchronous machine.** 0-31E8

starting open-phase protection. The effect of a device operative to prevent connecting the load to the supply unless all conductors of a polyphase system are energized. 42A25-34E10

starting operation (single-phase motor). (1) The range of operation between locked rotor and switching for a motor employing a starting-switch or relay. (2) The range of operation between locked rotor and a point just below but not including breakdown-torque speed for a motor not employing a starting switch or relay. *See also:* **asynchronous machine; synchronous machine.** 0-31E8

starting resistor (rotating machinery). A resistor connected in a secondary or field circuit to modify starting performance of an electric machine. *See also:* **rotor (rotating machinery); stator.** 0-31E8

starting rheostat (industrial control). A rheostat that controls the current taken by a motor during the period of starting and acceleration, but does not control the speed when the motor is running normally. 42A25-34E10

starting sheet (electrorefining). A thin sheet of refined metal introduced into an electrolytic cell to serve as a cathode surface for the deposition of the same refined metal. *See also:* **electrorefining.** 42A60-0

starting-sheet blank (electrorefining). A rigid sheet of conducting material designed for introduction into an electrolytic cell as a cathode for the deposition of a thin temporarily adherent deposit to be stripped off as a starting sheet. *See also:* **electrorefining.** 42A60-0

starting-switch assembly. The make-and-break contacts, mechanical linkage, and mounting parts necessary for starting or running, or both starting and running, split-phase and capacitor motors. *Note:* The starting-switch assembly may consist of a stationary-contact assembly and a contact that moves with the rotor.
See:
asynchronous machine;
automatic-reset thermal protector;
control module;
manual-reset thermal protector;
relay starting switch;
stationary-contact assembly;
switch sleeve. 42A10-0

starting switch, centrifugal (rotating machinery). *See:* **centrifugal starting switch.**

starting switch, relay. *See:* **relay starting switch.**

starting test (rotating machinery). A test taken on a machine while it is accelerating from standstill under specified conditions. *See also:* **asynchronous machine; direct-current commutating machine; synchronous machine.** 0-31E8

starting torque (synchronous motor). The torque exerted by the motor during the starting period. (A function of speed or slip). *See also:* **synchronous machine.** 0-31E8

starting voltage (radiation counters). The voltage applied to a Geiger-Mueller tube at which pulses of 1 volt amplitude appear across the tube when irradiated. *See:* **anticoincidence (radiation counters).** 0-15E6

starting winding (rotating machinery). A winding, the sole or main purpose of which is to set up or aid in setting up a magnetic field for producing the torque to start and accelerate a rotating electric machine. *See also:* **asynchronous machine; direct-current commutating machine; synchronous machine.** 0-31E8

start-record signal. A signal used for starting the process of converting the electric signal to an image on the record sheet. *See also:* **facsimile signal (picture signal).** E168-0

start signal (1) (start-stop system). Signal serving to prepare the receiving mechanism for the reception and registration of a character, or for the control of a function. 0-19E4

(2) (facsimile). A signal that initiates the transfer of a facsimile equipment condition from standby to active. *See also:* **facsimile signal (picture signal).** E168-0

start-stop printing telegraphy. That form of printing telegraphy in which the signal-receiving mechanisms are started in operation at the beginning and stopped at the end of each character transmitted over the channel. *See also:* **telegraphy.** 42A65-0

start-stop system (data transmission). A system in which each group of code elements corresponding to character is preceded by a start signal that serves to prepare the receiving mechanism for the reception and registration of a character and is followed by a stop signal that serves to bring the receiving mechanism to rest in preparation for the reception of the next character. *See also:* **data transmission.** 0-19E4

startup (relay). The action of a relay as it just departs from complete reset. Startup as a qualifying term is also used to identify the minimum value of the input quantity that will permit this condition. 37A100-31E11/31E6

statcoulomb. The unit of charge in the centimeter-gram-second electrostatic system. It is that amount of charge that repels an equal charge with a force of one dyne when they are in a vacuum, stationary, and one centimeter apart. One statcoulomb is approximately 3.335×10^{-10} coulomb. E270-0

state (control system). The values of a minimal set of functions that contain information about the past history of a system sufficient to determine the future behavior, given knowledge of future inputs. *See:* **computer control state.** *See also:* **control system.** 0-23E0

statement (computer programming). A meaningful expression or generalized instruction in a source language. X3A12-16E9

state space (control system). A space that contains the state vectors of a system. *Note:* The number of state variables in the system determines the dimension of the state space. *See also:* **control system.** 0-23E0

state variable (control system). One of the functions whose values determine the state. *See also:* **control system.** 0-23E0

state vector (control system). One whose components are the state variables in some arbitrary order. *See also:* **control system.** 0-23E0

static (atmospherics). Interference caused by natural electric disturbances in the atmosphere, or the electromagnetic phenomena capable of causing such interference. *See also:* **radio transmitter.** 42A65-0

static accuracy. Accuracy determined with a constant output. Contrast with **dynamic accuracy.** *See also:* **electronic analog computer.** E165-16E9

static breeze. *See:* **convective discharge.**

static characteristic (electron tubes). A relation, usually represented by a graph, between a pair of variables such as electrode voltage and electrode current, with all other voltages maintained constant. *See also:* **circuit characteristics of electrodes.** 42A70-15E6

static characteristic, relay. *See:* **relay static characteristic.**

static check. *See:* **problem check.** *See also:* **electronic analog computer.**

static converter. A unit that employs static switching devices such as metallic controlled rectifiers, transistors, electron tubes, or magnetic amplifiers. *See also:* **rectifier.** E45-0

static dump (computing systems). A dump that is performed at a particular point in time with respect to a machine run, frequently at the end of a run. *See also:* **electronic digital computer.** X3A12-16E9

static electrode potential. The electrode potential that exists when no current is flowing between the electrode and the electrolyte. *See also:* **electrolytic cell.** 42A60-0

static error. An error independent of the time-varying nature of a variable. *See:* **dynamic error.** *See also:* **electronic analog computer.** E165-16E9

static friction. *See:* **stiction.**

static induced current*. The charging and discharging current of a pair of Leyden jars or other capacitors, which current is passed through a patient. *See also:* **electrotherapy.**
*Deprecated 42A80-18E1

staticize (electronic digital computation). (1) To convert serial or time-dependent parallel data into static form. (2) Occasionally, to retrieve an instruction and its operands from storage prior to its execution. E162/X3A12-16E9

staticizer (electronic computation). A storage device for converting time-sequential information into static parallel information. *See also:* **electronic computation.** E270-0

static Kramer system (rotating machinery). A system of speed control below synchronous speed for wound-rotor induction motors. Slip power is recovered through the medium of a static converter equipment electrically connected between the secondary winding of the induction motor and a power system. *See also:* **asynchronous machine.** 0-31E8

static load line (electron device). The locus of all simultaneous average values of output electrode current and voltage, for a fixed value of direct-current load resistance. *See also:* **circuit characteristics of electron tube.** 0-15E6

static magnetic storage. *Note:* For an extensive list of cross references, see *Appendix A.*

static overvoltage (lightning arresters). An overvoltage due to an electric charge on an isolated conductor or installation. *See also:* **lightning arrester (surge diverter).** 50A25-31E7

static pressure (acoustics) (audio and electroacoustics) (at a point in a medium). The pressure that would exist at that point in the absence of sound waves. *See also:* **electroacoustics.** 0-1E1

static regulation. Expresses the change from one steady-state condition to another as a percentage of the final steady-state condition.

$$\text{Static Regulation} = \frac{E_{\text{initial}} - E_{\text{final}}}{E_{\text{final}}} (100\%).$$

0-10E1

static regulator. A transmission regulator in which the adjusting mechanism is in self-equilibrium at any setting and requires control power to change the setting. *See also:* **transmission regulator.** 42A65-0

static relay. A relay or relay unit in which there is no armature or other moving element, the designed response being developed by electronic, solid-state, magnetic, or other components without mechanical motion. *Note:* A relay that is composed of both static and electromechanical units in which the designed response is accomplished by static units may be referred to as a static relay. 37A100-31E11/31E6

static resistance (forward or reverse) (semiconductor rectifier cell). The quotient of the voltage by the current at a stated point on the static characteristic curve. *See also:* **rectification.** E59-34E17

static short-circuit ratio (arc-welding apparatus). The ratio of the steady-state output short-circuit current of a welding power supply at any setting to the output current at rated load voltage for the same setting. *See also:* **electric arc-welding apparatus.** 87A1-0

static volt-ampere characteristic (arc-welding apparatus). The curve or family of curves that gives the terminal voltage of a welding power supply as ordinate, plotted against output load current as abscissa, is the static volt-ampere characteristic of the power supply. *See also:* **electric arc-welding apparatus.** 87A1-0

static wave current* (electrotherapy). The current resulting from the sudden periodic discharging of a patient who has been raised to a high potential by means of an electrostatic generator. *See also:* **electrotherapy.**
*Deprecated 42A80-18E1

station. *See:* **tape unit.** *See also:* **electronic digital computer.**

stationary battery. A storage battery designed for service in a permanent location. *See also:* **battery (primary or secondary).** 42A60-0

stationary-contact assembly. The fixed part of the starting-switch assembly. *See also:* **starting-switch assembly.** 42A10-0

stationary-mounted device. A device that cannot be removed except by the unbolting of connections and mounting supports. *Note:* Compare with **drawout-mounted device.** 37A100-31E11

stationary wave. *See:* **standing wave.**

station-control error (electric power systems). The station generation minus the assigned station generation. *Note:* Refer to note on polarity under **area control error.** *See also:* **power systems, low-frequency and surge testing.** E94-0

station ringer (ringer). *See:* **telephone ringer.**

station-type (transformers and regulators). Designed for installation in a station. *See also:* **constant-current transformer; transformer; voltage regulator.** 42A15/57A14/57A15-31E12

station-type cubicle switchgear. Metal-enclosed power switchgear composed of the following equipment: (1) Primary power equipment for each phase segregated and enclosed by metal; (2) Stationary-mounted power circuit breakers; (3) Group-operated switches, interlocked with the circuit breakers, for isolating the circuit breakers; (4) Bare bus and connections; (5) Instrument transformers; (6) Control wiring and accessory devices. 37A100-31E11

station-type regulator. *See:* **station type.**

statistical delay (gas tubes) (electron device). The time lag from (1) the application of the specified voltage to initiate the discharge to (2) the beginning of breakdown. *See:* **gas tubes.** 0-15E6

stator (1) (rotating machinery). The portion that includes and supports the stationary active parts. The stator includes the stationary portions of the magnetic circuit and the associated winding and leads. It may, depending on the design, include a frame or shell, winding supports, ventilation circuits, coolers, and temperature detectors. A base, if provided, is not ordinarily considered to be part of the stator. *Note:* For an extensive list of cross references, see *Appendix A.* 0-31E0

(2) (induction watthour meter). A voltage circuit, one or more current circuits, and a combined magnetic circuit so arranged that their joint effect when energized is to exert a driving torque on the rotor by the reaction with currents induced in an individual, or a common, conducting disk. *Note:* In a 2-wire stator, only one current circuit is present. In a 3-wire stator, two current circuits of the same number of turns are wound on a common core. The term **stator** may be similarly applied to the corresponding parts of certain motor-type meters designed to measure other quantities. *See also:* **watthour meter.** 12A0/42A30-0

stator coil (rotating machinery). A unit of a winding on the stator of a machine. *See also:* **stator.** 0-31E8

stator coil pin (rotating machinery). A rod through an opening in the stator core, extending beyond the faces of the core, for the purpose of holding coils of the stator winding to a desired position. *See also:* **stator.** 0-31E8

stator core (rotating machinery). The stationary magnetic-circuit of an electric machine. It is commonly an assembly of laminations of magnetic steel, ready for winding. *See also:* **stator.** 42A10-31E8

stator-core lamination (rotating machinery). A sheet of material usually of magnetic steel, containing teeth and winding slots, or containing pole structures, that forms the stator core when assembled with other identical or similar laminations. *See also:* **stator.** 0-31E8

stator frame (rotating machinery). The supporting structure holding the stator core or core assembly. *Note:* In certain types of machines, the stator frame may be made integral with one end shield. *See also:* **stator.** 42A10-31E8

stator iron (rotating machinery). A term commonly used for the magnetic steel material or core of the stator of a machine. *See also:* **stator.** 0-31E8

stator-resistance starting. The process of starting a squirrel-cage induction motor by connecting the stator initially in series with starting resistors, which are short-circuited for running operation. 0-31E8

stator shell. *See:* **shell, stator.**

stator winding (rotating machinery). A winding on the stator of a machine. *See also:* **stator.** 0-31E8

stator winding copper (rotating machinery). A term commonly used for the material or conductors of a stator winding. *See also:* **stator.** 0-31E8

status point, supervisory control (power-system communication). *See:* **supervisory control point, status.** *See also:* **supervisory control system.**

STC (electronic navigation). *See:* **sensitivity time control.**

steady current. A current that does not change with time. E270-0

steady state (1) (signal-transmission system). That in which some specified characteristic of a condition, such as value, rate, periodicity, or amplitude, exhibits only negligible change over an arbitrarily long period of time. *See also:* **signal.**

(2) (industrial control). The condition of a specified variable at a time when no transients are present. *Note:* For the purpose of this definition, drift is not considered to be a transient. *See also:* **control system, feedback.** AS1-34E10

steady-state condition (gas turbines). A condition or value with limited deviations having a constant mean value over an arbitrarily long interval of time. E282-31E2

steady-state deviation (control). *See:* **deviation, steady-state.**

steady-state governing speedband (gas turbines). The total magnitude of the variation in turbine speed when the turbine-generator unit is operating isolated and under steady-state conditions with sustained load, or following a change to a new sustained load. It is expressed in percent of rated speed. E282-31E2

steady-state incremental speed regulation (excluding the effects of deadband) (gas turbines). At a given steady-state speed and power is the rate of change of the steady-state speed with respect to the power output. It is the slope of the tangent to the steady-state speed versus power curve at the point of power output under consideration. It is the difference in steady-state speed, expressed in percent of rated speed, for any two points on the tangent, divided by the corresponding difference in power output, expressed as a fraction of the rated power output. For the basis of comparison, the several points of power output at which the values of steady-state incremental speed regulation are derived are based upon rated speed being obtained at each point of power output. E94/E282-31E2

steady-state oscillation. A condition in which some aspect of the oscillation is a continuing periodic function. *See also:* **electroacoustics.** 0-1E1

steady-state short-circuit current (synchronous machine). The steady-state current in the armature winding when short-circuited. *See also:* **synchronous machine.** 0-31E8

steady-state speed regulation (straight condensing and noncondensing steam turbines, nonautomatic extraction turbines, hydro-turbines, and gas turbines). The change in steady-state speed, expressed in percent of rated speed, when the power output of the turbine operating isolated is gradually reduced from rated power output to zero power output with unchanged settings of all adjustments of the speed-governing system. *Note:* Speed regulation is considered positive when the speed increases with a decrease in power output. *See also:* **asynchronous machine; direct-current commutating machine; speed-governing system; synchronous machine.** E94;E282-31E2

steady-state stability. A condition that exists in a power system if it operates with stability when not subjected to an aperiodic disturbance. *Note:* In practical systems, a variety of relatively small aperiodic disturbances may be present without any appreciable effect upon the stability, as long as the resultant rate of change in load is relatively slow in comparison with the natural frequency of oscillation of the major parts of the system or with the rate of change in field flux of the rotating machines. *See also:* **alternating-current distribution.** 42A35-31E13

steady-state stability factor (system or part of a system). The ratio of the steady-state stability limit to the nominal power flow at the point of the system to which the stability limit is referred. *See:* **stability factor.** *See also:* **alternating-current distribution.** 42A35-31E13

steady-state stability limit (steady-state power limit). The maximum power flow possible through some particular point in the system when the entire system or the part of the system to which the stability limit refers is operating with steady-state stability. *See also:* **alternating-current distribution.** 42A35-31E13

steady-state value. The value of a current or voltage after all transients have decayed to a negligible value. For an alternating quantity, the root-mean-square value in the steady state does not vary with time. *See also:* **asynchronous machine; direct-current commutating machine; synchronous machine.** 0-31E8

steady voltage. *See:* **steady current.**

steam capability (electric power supply). The maximum net rating that can be obtained under normal operating practices for at least four continuous hours as demonstrated by total plant tests. Units can be rated at overpressure and for top heaters but providing one is willing to operate under these conditions and has developed normal procedures for such operation. The effects of any seasonal changes in cooling-water temperature are to be taken into account. *See also:* **generating station.** 0-31E4

steam turbine-electric drive. A self-contained system of power generation and application in which the power generated by a steam turbine is transmitted electrically by means of a generator and a motor (or multiples of these) for propulsion purposes. *Note:* The prefix steam turbine-electric is applied to ships, locomotives, cars, buses, etcetera, that are equipped with this drive. *See also:* **electric locomotive; electric propulsion system.** 42A40-0

steel container (storage cell). The container for the element and electrolyte of a nickel-alkaline storage cell. This steel container is sometimes called a can. *See also:* **battery (primary or secondary).** 42A60-0

steerable antenna. A directional antenna whose major lobe can be readily shifted in direction. *See also:* **antenna.** E/145-3E1

steering compass. A compass located within view of a steering stand, by reference to which the helmsman holds a ship on the set course. 42A43-0

Stefan-Boltzmann law. The statement that the radiant exitance of a blackbody is proportional to the fourth power of its absolute temperature; that is,

$$M = \sigma T^4.$$

See: National Bureau of Standards Technical News Bulletin, October 1963 (corrected) for constants. *See also:* **radiant energy (illuminating engineering).** Z7A1-0

step (1) (pulse techniques). A waveform that, from the observer's frame of reference, approximates a Heaviside (unit step) function. *See:* **signal, unit step.** *See also:* **pulse.** 0-9E4

(2) (computing systems). (A) One operation in a computer routine. (B) To cause a computer to execute one operation. *See also:* **electronic digital computer; single step.** X3A12-16E9

step-back relay (motors and industrial control). A relay that functions to limit the current peaks of a motor when the armature or line current increases. A stepback relay may, in addition, remove this limitation when the cause of the high current has been removed. *Note:* When used with a motor having a flywheel, the relay causes the momentary transfer of stored energy from the flywheel to the load but does not limit the current peaks if the load is sustained for a considerable interval after the relay operates. *See also:* **relay.** E45/IC1-34E10

step-by-step switch. A bank-and-wiper switch in which the wipers are moved by electromagnet ratchet mechanisms individual to each switch. *Note:* This type of switch may have either one or two types of motion. *See also:* **telephone switching system.** 42A65-0

step-by-step system. An automatic telephone switching system that is generally characterized by the following features. (1) The selecting mechanisms are step-by-step switches. (2) The switching pulses may either actuate the successive selecting mechanisms directly or may be received and stored by controlling mechanisms that, in turn, actuate the selecting mechanisms. *See also:* **telephone switching system.** 42A65-0

step change (control) (industrial control) (step function). An essentially instantaneous change of an input variable from one value to another. *See also:* **control system, feedback.** AS1-34E10

step compensation (correction) (industrial control). The effect of a control function or a device that will cause a step change in an other function when a predetermined operating condition is reached. IC1-34E10

step line-voltage change (power supplies). An instantaneous change in line voltage (for example, 105-125 volts alternating current); for measuring line regulation and recovery time. *See also:* **power supply.** KPSH-10E1

step load change (power supplies). An instantaneous change in load current (for example, zero to full load); for measuring the load regulation and recovery time. *See also:* **power supply.** KPSH-10E1

stepped leader (lightning). A series of discharges emanating from a region of charge concentration at short time intervals. Each discharge proceeds with a luminescent tip over a greater distance than the previous one. *See also:* **direct-stroke protection (lightning).** 0-31E13

stepping relay (1) (general). A multiposition relay in which moving wiper contacts mate with successive sets of fixed contacts in a series of steps, moving from one step to the next in successive operations of the relay. *See also:* **relay.** 83A16-0

(2) (rotary type). A relay having many rotary positions, ratchet actuated, moving from one step to the next in successive operations, and usually operating its contacts by means of cams. There are two forms: (A) **directly driven,** where the forward motion occurs on energization, and (B) **indirectly (spring) driven,** where a spring produces the forward motion on pulse cessation. *Note:* The term is also incorrectly used for **stepping switch.** 0-21E0

stepping relay, spring-actuated. A stepping relay that is cocked electrically and operated by spring action. *See also:* **relay.** 83A16-0

step response. A criterion of the dynamic response of an instrument when subjected to an instantaneous change in measured quantity from one value to another. *See also:* **accuracy rating of an instrument.** 39A4-0

step-response time. The time required for the end device to come to rest in its new position after an abrupt change to a new constant value has occurred in the measured signal. *See also:* **accuracy rating of an instrument.** 39A4-0

step speed adjustment (industrial control). The speed drive can be adjusted in rather large and definite steps between minimum and maximum speed. *See also:* **electric drive.** 42A25-34E10

step-stress test (reliability). A test consisting of several stress levels applied sequentially, for periods of equal duration, to a sample. During each period a stated stress-level is applied and the stress level is increased from one step to the next. *See also:* **reliability.** 0-7E1

step voltage (safety). The potential difference between two points on the earth's surface, separated by a distance of one pace, that will be assumed to be one meter, in the direction of maximum potential gradient. *See also:* **ground.** E81-0

step-voltage regulator (regulating transformer). A voltage regulator in which the voltage and phase angle of the regulated circuit are controlled in steps by means of taps, and without interrupting the load. *Note:* The regulator has one or more windings excited from the system circuit or a separate source and one or more windings connected in series with the system circuit and has 2500 kilovoltamperes (kvA) for three-phase units or not exceeding 250 kilovoltamperes (kvA) for single-phase units. 42A15/57A15-0

step-voltage test (rotating machinery). A controlled overvoltage test in which designated voltage increments are applied at designated times. Time increments may be constant or graded. *See:* **graded-time step-voltage test.** *See also:* **asynchronous machine; direct-current commutating machine; synchronous machine.** 0-31E8

step wedge* (television). *See:* **gray scale.**

*Deprecated

stereophonic system. A sound reproducing system in which a plurality of microphones, transmission channels, and loudspeakers are arranged so as to afford the listener a sense of the spatial distribution of the sound sources. *See also:* **loudspeaker; phonograph pickup.** 42A65/E157-1E1

stick circuit. A circuit used to maintain a relay or similar unit energized through its own contact. *See also:* **railway signal and interlocking.** 42A42-0

stickiness. The condition caused by physical interference with the rotation of the moving element. *See also:* **moving element of an instrument.** 39A1-0

sticking voltage (luminescent screen). The voltage applied to the electron beam below which the rate of secondary emission from the screen is less than unity. The screen then has a negative charge that repels the primary electrons. *See also:* **cathode-ray tubes.** 50I07-15E6

stick operation (switching device) (hook operation). Manual operation by means of a switch stick. 37A100-31E11

stiction (1). The force in excess of the coulomb friction required to start relative motion between two surfaces in contact. 0-23E0
(2) (static friction) (industrial control). The total friction that opposes the start of relative motion between elements in contact. *See also:* **control system, feedback.** AS1-34E10

stiffness (industrial control). The ability of a system or element to resist deviations resulting from loading at the output. *See also:* **control system, feedback.** 85A1-23E0;AS1-34E10

stiffness coefficient. The factor K (also called spring constant) in the differential equation for oscillatory motion $M\ddot{x} + B\dot{x} + Kx = 0$. *See also:* **control system, feedback.** 85A1-23E0

stilb. The unit of luminance (photometric brightness) equal to one candela per square centimeter. *Note:* The name **stilb** has been adopted by the International Commission on Illumination (CIE) and is commonly used in European publications. In the United States and Canada the preferred practice is to use self-explanatory terms such as candela per square inch and candela per square centimeter. *See also:* **light.** Z7A1-0

Stiles-Crawford effect. The reduced luminous efficiency of rays entering the peripheral portion of the pupil of the eye. *See also:* **visual field.** Z7A1-0

stimulus (industrial control). Any change in signal that affects the controlled variable; for example, a disturbance or a change in reference input. *See also:* **control system, feedback.** AS1-34E10

stirring effect (induction heater usage). The circulation in a molten charge due to the combined forces of motor and pinch effects. *See also:* **induction heating.** E54/E169-0

stop (limit stop). A mechanical or electric device used to limit the excursion of electromechanical equipment. *See also:* **limiter circuit.** *See also:* **electronic analog computer.** E165-0

stop dowel (rotating machinery). A pin fitted into a hole to limit motion of a second part. *See also:* **cradle base (rotating machinery).** 0-31E8

stop lamp. A lamp giving a steady warning light to the rear of a vehicle or train of vehicles, to indicate the intention of the operator to diminish speed or to stop. *See also:* **headlamp.** Z7A1-0

stop-motion switch (elevators). *See:* **machine final-terminal stopping device.**

stopping off. The application of a resist to any part of a cathode or plating rack. *See also:* **electroplating.** 42A60-0

stop-record signal (facsimile). A signal used for stopping the process of converting the electric signal to an image on the record sheet. *See also:* **facsimile signal (picture signal).** E168-0

stop signal (data transmission) (1) (start-stop system). Signal serving to bring the receiving mechanism to rest in preparation for the reception of the next telegraph signal. *See also:* **data transmission.** 0-19E4
(2) (facsimile). A signal that initiates the transfer of a facsimile equipment condition from active to standby. *See also:* **facsimile signal (picture signal).** E168-0

storage (electronic computation). (1) The act of storing information. (2) Any device in which information can be stored, sometimes called a memory device. (3) In a computer, a section used primarily for storing information. Such a section is sometimes called a memory or store (British). *Notes:* (A) The physical means of storing information may be electrostatic, ferroelec-

tric, magnetic, acoustic, optical, chemical, electronic, electric, mechanical, etcetera, in nature. (B) Pertaining to a device in which data can be entered, in which it can be held, and from which it can be retrieved at a later time.
See:
associative storage;
auxiliary storage;
content-addressed storage;
electrostatic storage;
fixed storage;
magnetic storage;
nonerasable storage;
parallel search storage;
parallel storage;
permanent storage;
read-only storage;
temporary storage;
volatile storage;
working storage.
See also: **electronic computation; electronic digital computer; store.** E270/X3A12-16E9

storage allocation (computing systems). The assignment of blocks of data to specified blocks of storage. *See also:* **electronic digital computer.** X3A12-16E9

storage assembly (storage tubes). An assembly of electrodes (including meshes) that contains the target together with electrodes used for control of the storage process, those that receive an output signal, and other members used for structural support. *See also:* **storage tube.** E158-15E6

storage battery (secondary battery or accumulator). A battery consisting of two or more storage cells electrically connected for producing electric energy. Common usage permits this designation to be applied also to a single storage cell used independently. *See also:* **battery (primary or secondary).** 42A60-0

storage capacity. The amount of data that can be contained in a storage device. *Notes:* (1) The units of capacity are bits, characters, words, etcetera. For example, capacity might be "32 bits," "10 000 decimal digits," "16 384 words with 10 alphanumeric characters each." (2) When comparisons are made among devices using different character sets and word lengths, it may be convenient to express the capacity in equivalent bits, which is the number obtained by taking the logarithm to the base 2 of the number of usable distinguishable states in which the storage can exist. (3) The storage (or memory) capacity of a computer usually refers only to the internal storage section. *See also:* **electronic digital computer.** E162/E270/X3A12-16E9

storage cell (secondary cell or accumulator) (1) (electric energy). A galvanic cell for the generation of electric energy in which the cell, after being discharged, may be restored to a fully charged condition by an electric current flowing in a direction opposite to the flow of current when the cell discharges. *See also:* **electrochemistry.** 42A60-0
(2) (computing systems) (information). An elementary unit of storage, for example, a binary cell, a decimal cell. *See also:* **electrochemistry; electronic digital computer.** X3A12-16E9

storage device. A device in which data can be stored and from which it can be copied at a later time. The means of storing data may be chemical, electrical, mechanical, etcetera. *See also:* **electronic digital computer; storage.** E162-0

storage element (storage tubes). An area of a storage surface that retains information distinguishable from that of adjacent areas. *Note:* The storage element may be a portion of a continuous storage surface or a discrete area such as a dielectric island. *See also:* **storage tube.** E158-15E6

storage-element equilibrium voltage (storage tubes). A limiting voltage toward which a storage element charges under the action of primary electron bombardment and secondary emission. At equilibrium voltage the escape ratio is unity. *Note:* **Cathode equilibrium voltage, second-crossover equilibrium voltage,** and **gradient-established equilibrium voltage** are typical examples. *See also:* **charge-storage tube.** E158-15E6

storage-element equilibrium voltage, cathode (storage tubes). The storage-element equilibrium voltage near cathode voltage and below first-crossover voltage. *See also:* **charge-storage tube.** E158-15E6

storage-element equilibrium voltage, collector (storage tubes). *See:* **charge storage tube.**

storage-element equilibrium voltage, gradient established (storage tubes). The storage-element equilibrium voltage, between first- and second-crossover voltages, at which the escape ratio is unity. *See also:* **charge-storage tube.** E158-15E6

storage-element equilibrium voltage, second-crossover (storage tubes). The storage-element equilibrium voltage at the second-crossover voltage. *See also:* **charge-storage tube.** E158-15E6

storage integrator (analog computer). An integrator used to store a voltage in the hold condition for future use while the rest of the computer assumes another computer control state. *See also:* **electronic analog computer.** E165-0

storage light-amplifier (optoelectronic device). *See:* **image-storage panel.**

storage medium. Any device or recording medium into which data can be stored and held until some later time, and from which the entire original data can be obtained. EIA3B-34E12

storage protection (computing systems). An arrangement for preventing access to storage for either reading or writing, or both. *See also:* **electronic digital computer.** 0-16E9

storage station. A hydroelectric generating station having storage sufficient for seasonal or hold-over operation. *See also:* **generating station.** 42A35-31E13

storage surface (storage tubes). The surface upon which information is stored. *See also:* **storage tube.** E158-15E6

storage temperature (power supply). The range of environmental temperatures in which a power supply can be safely stored (for example, -40 to $+85$ degrees Celsius). *See also:* **power supply.** 0-10E1

storage time* (storage tubes). See: **retention time, maximum; decay time.** *See also:* **storage tube.**
*Deprecated

storage tube. An electron tube into which information can be introduced and read at a later time. *Note:* The output may be an electric signal and/or a visible image corresponding to the stored information. *Note:* For an extensive list of cross references, see *Appendix A.* E158-15E6

store. (1) To retain data in a device from which it can be copied at a later time. (2) To put data into a storage device. (3) British synonym for storage. *See also:* **elec-**

tronic computation; electronic digital computer; storage. E162/E270/X3A12-16E9

stored-energy operation. Operation by means of energy stored in the mechanism itself prior to the completion of the operation and sufficient to complete it under predetermined conditions. *Note:* This kind of operation may be subdivided according to: (1) How the energy is stored (spring, weight, etcetera); (2) How the energy originates (manual, electric, etcetera); (3) How the energy is released (manual, electric, etcetera). 37A100-31E11

stored-program computer. A digital computer that, under control of internally stored instructions, can synthesize, alter, and store instructions as though they were data and can subsequently execute these new instructions. *See also:* **electronic digital computer.** X3A12-16E9

storm guys. Anchor guys, usually placed at right angles to direction of line, to provide strength to withstand transverse loading due to wind. *See also:* **tower.** 42A35-31E13

storm loading. The mechanical loading imposed upon the components of a pole line by the elements, that is, wind and/or ice, combined with the weight of the components of the line. *Note:* The United States has been divided into three loading districts, light, medium, and heavy, for which the amounts of wind and/or ice have been arbitrarily defined. *See also:* **cable; open wire.** 42A65-0

straight-cut control system (numerically controlled machines). A system in which the controlled cutting action occurs only along a path parallel to linear, circular, or other machine ways. *See also:* **numerically controlled machines.** EIA3B-34E12

straightforward trunking (manual telephone switchboard system). That method of operation in which the *A* operator gives the order to the *B* operator over the trunk on which talking later takes place. *See also:* **telephone switching system.** 42A65-0

straight joint. A joint used for connecting two lengths of cable in approximately the same straight line in series. *Note:* A straight joint is made between two like cables, for example, between two single-conductor cables, between two concentric cables, or between two triplex cables. *See:* **branch joint; cable joint; reducing joint.** *See also:* **power distribution, underground construction.** 42A35-31E13

straight-line coding (computing systems). Coding in which loops are avoided by the repetition of parts of the coding when required. *See also:* **electronic digital computer.** X3A12-16E9

straight-seated bearing (rotating machinery) (cylindrical bearing). A journal bearing in which the bearing liner is constrained about a fixed axis determined by the supporting structure. *See also:* **bearing.** 0-31E8

straight storage system (electric power supply). A system in which the electrical requirements of a car are supplied solely from a storage battery carried on the car. *See also:* **axle generator system.** 42A42-0

strain element (fuse) (strain wire). That part of the current-responsive element, connected in parallel with the fusible element in order to relieve it of tensile strain. *Note:* The fusible element melts and severs first and then the strain element melts during circuit interruption. 37A100-31E11

strain insulator (1). An insulator generally of elongated shape, with two transverse holes or slots. 29A1-0

(2). A single insulator, an insulator string, or two or more strings in parallel, designed to transmit to the tower or other support the entire pull of the conductor and to insulate it therefrom. *See also:* **insulator.** 42A35-31E13

strain wire. *See:* **strain element (fuse).**

strand. (1) One of the wires, or groups of wires, of any stranded conductor. (2) One of a number of paralleled uninsulated conducting elements of a conductor which is stranded to provide flexibility in assembly or in operation. (3) One of a number of paralleled insulated conducting elements which constitute one turn of a coil in rotating machinery. The strands are usually separated electrically through all the turns of a multi-turn coil. Various types of transposition are commonly employed to reduce the circulation of current among the strands. A strand has a solid cross section, or it may be hollow to permit the flow of cooling fluid in intimate contact with the conductor (one form of "conductor cooling"). *See also:* **conductor; rotor (rotating machinery; stator.** E30-0;42A35-31E13;0-31E8

stranded conductor. A conductor composed of a group of wires or of any combination of groups of wires. *Note:* The wires in a stranded conductor are usually twisted or braided together. *See also:* **conductor; power distribution, underground construction.** E30/42A35-31E13

stranded wire. *See:* **stranded conductor.**

strand insulation (conductor insulation). The insulation surrounding a strand or single metallic conductor to prevent electric contact between paralleled strands in a coil. *See also:* **stator.** 0-31E8

strap, anode (magnetron). *See:* **anode strap.**

strap key. A pushbutton circuit controller that is biased by a spring metal strip and is used for opening or closing a circuit momentarily. *See also:* **railway signal and interlocking.** 42A42-0

strapped-down (gimbal-less) inertial navigation equipment. Inertial navigation equipment wherein the inertial devices (gyros and accelerometers) are attached directly to the carrier, eliminating the stable platform and gimbal system. *Note:* In this equipment a computer utilizes gyro information to resolve the accelerations that are sensed along the carrier axes and to refer these accelerations to an inertial frame of reference. Navigation is then accomplished in the same manner as in systems using a stable platform. *See also:* **navigation.** E174-10E6

strapping. *See:* **jumper.**

strapping (multiple-cavity magnetrons). *See:* **anode strap.**

stray-current corrosion. Corrosion caused by current through paths other than the intended circuit or by an extraneous current in the earth. *See:* **cathodic corrosion; long-line current (corrosion); noble potential; sacrificial protection.** *See also:* **corrosion terms.** CM-34E2

stray load loss (synchronous machine). The losses due to eddy currents in copper and additional core losses in the iron, produced by distortion of the magnetic flux by the load current, not including that portion of the core loss associated with the resistance drop. *See also:* **synchronous machine.** 50A10-31E8

strays*. Electromagnetic disturbances in radio reception other than those produced by radio transmitting systems. *See also:* **radio transmitter.**
*Obsolete 42A65-0

streamer (voltage testing). When used in connection with high-voltage testing, an incomplete disruptive discharge in a gaseous or liquid dielectric that does not completely bridge the test piece or gap. *See also:* **test voltage and current.** 68A1-31E1

stream flow. The quantity rate of water passing a given point. *See also:* **generating station.** 42A35-31E13

streaming (audio and electroacoustics). Unidirectional flow currents in a fluid that are due to the presence of acoustic waves. *See also:* **electroacoustics.** 0-1E1

streaming potential (electrobiology). The electrokinetic potential gradient resulting from unit velocity of liquid forced to flow through a porous structure or past an interface. *See also:* **electrobiology.** 42A80-18E1

street-lighting luminaire. A complete lighting device consisting of a light source together with its direct appurtenances such as globe, reflector, refractor, housing, and such support as is integral with the housing. The pole, post, or bracket is not considered part of the luminaire.
See:
bracket;
lamp post;
lateral width of a light distribution;
pole;
reference line;
street-lighting unit;
width line.
See also: **luminaire.** Z7A1-0

street-lighting unit. The assembly of a pole or lamp post with a bracket and a luminaire. *See also:* **street lighting luminaire.** Z7A1-0

strength-duration (time-intensity) curve (medical electronics). A graph of the intensity curve of applied electrical stimuli as a function of the duration just needed to elicit response in an excitable tissue. *See also:* **medical electronics.** 0-18E1

strength of a sound source (strength of a simple source). The maximum instantaneous rate of volume displacement produced by the source when emitting a wave with sinusoidal time variation. *Note:* The term is properly applicable only to sources of dimension small with respect to the wavelength. *See also:* **electroacoustics.** 0-1E1

stress-accelerated corrosion. Corrosion that is accelerated by stress. *See also:* **corrosion terms.** CM-34E2

stress corrosion cracking. Spontaneous cracking produced by the combined action of corrosion and static stress (residual or applied). *See also:* **corrosion terms.** CM-34E2

stress relief. A predetermined amount of slack to relieve tension in component or lead wires. *See also:* **soldered connections (electronic and electric applications).** 99A1-0

strike deposit (1) (electroplating). A thin film of deposited metal to be followed by other coatings. *See also:* **electroplating.** 42A60-0
(2) (bath) (electroplating). An electrolyte used to deposit a thin initial film of metal. *See also:* **electroplating.** 42A60-0

striking (1) (arc) (spark) (gas). The process of establishing an arc or a spark. *See also:* **discharge (gas).** 50I07-15E6
(2) (electroplating). The electrodeposition of a thin initial film of metal, usually at a high current density. *See also:* **electroplating.** 42A60-0

striking current (gas tube). The starter-gap current required to initiate conduction across the main gap for a specified anode voltage. *See also:* **gas-filled rectifier.** 50I07-15E6

striking distance. The shortest distance, measured through air, between parts of different polarities. 37A100-31E11

string. A connected sequence of entities such as characters or physical elements. *See also:* **electronic digital computer.** X3A12-16E9

string-shadow instrument. An instrument in which the indicating means is the shadow (projected or viewed through an optical system) of a filamentary conductor, the position of which in a magnetic or an electric field depends upon the measured quantity. *See also:* **instrument.** 42A30-0

strip (electroplating). A solution used for the removal of a metal coating from the base metal. *See also:* **electroplating.** 42A60-0

stripper tank (electrorefining). An electrolytic cell in which the cathode deposit, for the production of starting sheets, is plated on starting-sheet blanks. *See also:* **electrorefining.** 42A60-0

stripping (1) (electroplating) (mechanical.) The removal of a metal coating by mechanical means. 42A60-0
(2) (chemical). The removal of a metal coating by dissolving it. 42A60-0
(3) (electrolytic). The removal of a metal coating by dissolving it or an underlying coating anodically with the aid of a current. *See also:* **electroplating.** 42A60-0

stripping compound (electrometallurgy). Any suitable material for coating a cathode surface so that the metal electrodeposited on the surface can be conveniently stripped off in sheets. *See also:* **electrowinning.** 42A60-0

strip terminals (rotating machinery). A form of terminal in which the ends of the machine winding are brought out to terminal strips mounted integral with the machine frame or assembly. *See also:* **cradle base (rotating machinery).** 0-31E8

strip-type transmission line (waveguides). A transmission line consisting of a conductor above or between extended conducting surfaces. *See:* **shielded strip transmission line; unshielded strip transmission line.** *See also:* **waveguides.** 0-3E1

stroboscopic tube. A gas tube designed for the periodic production of short light flashes. *See also:* **gas tubes.** 50I07-15E6

stroke speed (scanning or recording line frequency) (facsimile). The number of times per minute, unless otherwise stated, that fixed line perpendicular to the direction of scanning is crossed in one direction by a scanning or recording spot. *Note:* In most conventional mechanical systems this is equivalent to drum speed. In systems in which the picture signal is used while scanning in both directions, the stroke speed is twice the above figure. *See also:* **recording (facsimile); scanning (facsimile).** E168-0

structurally dual networks. A pair of networks such that their branches can be marked in one-to-one correspondence so that any mesh of one corresponds to a cut-set of the other. Each network of such a pair is said to be the dual of the other. *See also:* **network analysis.** E153/E270-0

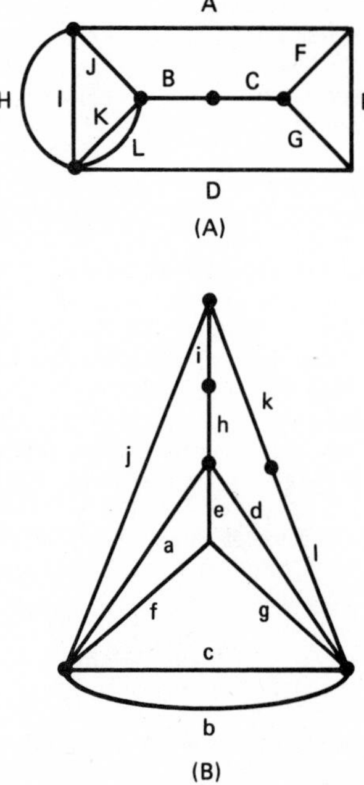

Structurally dual networks. For example, the mesh EFG in (A) corresponds to the cut-set efg in (B), the mesh bc in (B) to the cut-set BC in (A), and the mesh JAEGCB in (A) to the cut-set jaegcb in (B).

structurally symmetrical network. A network that can be arranged so that a cut through the network produces two parts that are mirror images of each other. *See also:* **network analysis.** E153/E270-0

structure conflict (pole line). A line so situated with respect to a second line, that the overturning (at the ground line) of the first line will result in contact between its poles or conductors and the conductors of the second line, assuming that no conductors are broken in either line. *Exceptions:* Lines are not considered as conflicting under the following conditions: (1) Where one line crosses another. (2) Where two lines are on opposite sides of a highway, street, or alley and are separated by a distance not less than 60 percent of the height of the taller pole line and not less than 20 feet. *See also:* **tower.** *See also:* **conflict (wiring system).** 2A2/42A35-31E13

stub (communication practice). A short length of transmission line or cable that is joined as a branch to another transmission line or cable. *See also:* **cable.** 42A65-0

stub feeder (radial feeder). A feeder that connects a load to its only source of power. 37A100-31E11

stub-multiple feeder. A feeder that operates as either a stub or a multiple feeder. 37A100-31E11

stub shaft (rotating machinery). A separate shaft not carried in its own bearings and connected to the shaft of a machine. *See also:* **rotor (rotating machinery).** 0-31E8

stub-supported coaxial. A coaxial whose inner conductor is supported by means of short-circuited coaxial stubs. *See also:* **waveguide.** 42A65-0

stub tuner. A stub that is terminated by movable short-circuiting means and used for matching impedance in the line to which it is joined as a branch. *See also:* **waveguide.** 42A65-3E1

stub, waveguide. An auxiliary section of a waveguide or transmission line with an essentially nondissipative termination and joined at some angle with the main section of the waveguide or transmission line. *See also:* **transmission line; waveguide.** E147-3E1

stud (circuit breaker). A rigid conductor between a terminal and a contact. 37A100-31E11

stuffing box (watertight gland). A device for use where a cable passes into a junction box or other piece of apparatus and is so designed as to render the joint watertight. *See also:* **power distribution, underground construction.** 42A35-31E13

stylus (electroacoustics). A mechanical element that provides the coupling between the recording or the reproducing transducer and the groove of a recording medium. *See:* **phonograph pickup.** 0-1E1

stylus drag (needle drag) (electroacoustics). An expression used to denote the force resulting from friction between the surface of the recording medium and the reproducing stylus. *See also:* **phonograph pickup.** E157-1E1

stylus force (electroacoustics). The vertical force exerted on a stationary recording medium by the stylus when in its operating position. *See also:* **phonograph pickup.** E157-1E1

subcarrier. A carrier used to generate a modulated wave that is applied, in turn, as a modulating wave to modulate another carrier. *See also:* **carrier; facsimile transmission.** E170-0

subdivided capacitor (condenser box*). A capacitor in which several capacitors known as sections are so mounted that they may be used individually or in combination.

*Deprecated E270-0

subfeeder. A feeder originating at a distribution center other than the main distribution center and supplying one or more branch-circuit distribution centers. *See:* **feeder.** 42A95-0

subharmonic. A sinusoidal quantity having a frequency that is an integral submultiple of the frequency of some other sinusoidal quantity to which it is referred. For example, a wave, the frequency of which is half the fundamental frequency of another wave, is called the second subharmonic of that wave. *See also:* **signal wave.** E151/E145-13E6;42A65-31E3

subject copy (facsimile). The material in graphic form that is to be transmitted for facsimile reproduction. *See also:* **facsimile (electrical communication).** E168-0

subjective brightness. The subjective attribute of any light sensation giving rise to the percept of luminous intensity, including the whole scale of qualities of being bright, light, brilliant, dim, or dark. *Note:* The term brightness often is used when referring to the measurable **photometric brightness.** While the context usually makes it clear as to which meaning is intended, the preferable term for the photometric quantity is **luminance,** thus reserving **brightness** for the subjective sensation. *See:* **luminance.** *See also:* **visual field.** Z7A1-0

submarine cable. A cable designed for service under water. *Note:* Submarine cable is usually a lead-covered cable with a steel armor applied between layers of jute. *See also:* **power distribution, underground cable.** 42A35-31E13

submerged-resistor induction furnace. A device for melting metal comprising a melting hearth, a depending melting channel closed through the hearth, a primary induction winding, and a magnetic core that links the melting channel and the primary winding. *See also:* **induction heating.** E54/E169-0

submersible (rotating machinery) (industrial control) (transformers, regulators, enclosures). So constructed as to be successfully operable when submerged in water under specified conditions of pressure and time. *See also:* **cradle base (rotating machinery); constant current transformer; direct-current commutating machine; synchronous machine; transformer.** 37A100-31E11;42A95-31E8;IC1-34E10

submersible entrance terminals (distribution oil cutouts) (cableheads). A hermetically sealable entrance terminal for the connection of cable having a submersible sheathing or jacket. 37A100-31E11

submersible fuse (subway oil cutout). *See:* **submersible; fuse.**

subnormality (electrical depression) (electrobiology). The state of reduced electrical sensitivity after a response or succession of responses. *See also:* **excitability (electrobiology).** 42A80-18E1

subpanel (photoelectric converter). Combination of photoelectric converters in parallel mounted on a flat supporting structure. *See also:* **semiconductor.** 0-10E1

subpost car frame (elevators). A car frame all of whose members are located below the car platform. *See also:* **hoistway (elevator or dumbwaiter).** 42A45-0

subroutine (electronic computation). (1) In a routine, a portion that causes a computer to carry out a well-defined mathematical or logic operation. (2) A routine that is arranged so that control may be transferred to it from a master routine and so that at the conclusion of the subroutine, control reverts to the master routine. *Note:* Such a subroutine is usually called a closed subroutine. A single routine may simultaneously be both a subroutine with respect to another routine and a master routine with respect to a third. Usually control is transferred to a single subroutine from more than one place in the master routine and the reason for using the subroutine is to avoid having to repeat the same sequence of instructions in different places in the master routine. *See also:* **electronic computation; electronic digital computer.** E270-0

subscriber equipment (protective signaling). That portion of a system installed in the protected premises or otherwise supervised. *See also:* **protective signaling.** 42A65-0

subscriber loop (telephony). A telephone line between a central office and a telephone station, private branch exchange, or other end equipment. *See also:* **telephone system.** 0-19E1

subscriber multiple. A bank of jacks in a manual switchboard providing outgoing access to subscriber lines, and usually having more than one appearance across the face of the switchboard. *See also:* **telephone switching system.** 42A65-0

subscriber set (customer set). An assembly of apparatus for use in originating or receiving calls on the premises of a subscriber to a communication or signaling service. *See also:* **voice-frequency telephony.** 42A65-0

subsidiary conduit (lateral). A terminating branch of an underground conduit run, extending from a manhole or handhole to a nearby building, handhole, or pole. *See also:* **cable.** 42A65-0

subsonic frequency*. *See:* **infrasonic frequency.**

*Deprecated

substantial (electric systems). So constructed and arranged as to be of adequate strength and durability for the service to be performed under the prevailing conditions. *See also:* **power distribution overhead construction.** 2A2/42A95-0

substitution error, direct-current–radio-frequency (bolometers). The error arising in the bolometric measurement technique when a quantity of direct-current or audio-frequency power is replaced by a quantity of radio-frequency power with the result that the different current distributions generate different temperature fields that give the bolometer element different values of resistance for the same amounts of power. This error is expressed as

$$\epsilon_s = \frac{\eta_e - \eta}{\eta}$$

where η_e is the effective efficiency of the bolometer unit and η is the efficiency of the bolometer unit. *See also:* **bolometric power meter.** 0-9E4

substitution error, dual-element. A substitution error peculiar to dual-element bolometer units that results from a different division of direct-current (or audio-frequency) and radio-frequency powers between the two elements. 0-9E4

substitution power (bolometers). The difference in bias power required to maintain the resistance of a bolometer at the same value before and after radio-frequency power is applied. Commonly, a bolometer is placed in one arm of a Wheatstone bridge that is balanced when the bias current (direct current and/or audio frequency) holds the bolometer at its nominal operating resistance. Following the application of the radio-frequency signal, the reduction in bias power is taken as a measure of the radio-frequency power. This reduction in the bias power is the substitution power and is given by

$$P = I_1^2R - I_2^2R$$

where I_1 and I_2 are the bias currents before and after radio-frequency power is applied and R is the nominal operating resistance of the bolometer. *See also:* **bolometric power meter.** 0-9E4

substrate (1) (integrated circuit). The supporting material upon or within which an integrated circuit is fabricated or to which an integrated circuit is attached. E274-15E7/21E1

(2) (photovoltaic power system). Supporting material or structure for solar cells in a panel assembly. Solar cells are attached to the substrate. *See also:* **photovoltaic power system; solar cells (in a photovoltaic power system).** 0-10E1

subsurface corrosion. Formation of isolated particles of corrosion product(s) beneath the metal surface. This results from the preferential reaction of certain alloy constituents by inward diffusion of oxygen, nitrogen, sulfur, etcetera (internal oxidation). *See:* **corrosion terms.** CM-34E2

subsynchronous reluctance motor. A form of reluctance motor that has the number of salient poles greater than the number of electrical poles of the primary winding, thus causing the motor to operate at a constant average speed that is a submultiple of its apparent synchronous speed. *See:* **asynchronous machine; synchronous machine.** 42A10-31E8

subsystem. A division of a system that in itself has the characteristics of a system. *See also:* **circuits and devices.** 0-31E3

subtransient current (rotating machinery). The initial alternating component of armature current following a sudden short circuit. *See also:* **armature.** 0-31E8

subtransient internal voltage (synchronous machine) (specified operating condition). The fundamental-frequency component of the voltage of each armature phase that would appear at the terminals immediately following the sudden removal of the load. *Note:* The subtransient internal voltage, as shown in the phasor diagram, is related to the terminal-voltage and phase-current phasors by the equation:

$$E''_1 = E_a + RI_a + jX''_d I_{ad} + jX''_q I_{aq}$$

For a machine subject to saturation, the reactances should be determined for the degree of saturation applicable to the specified operating conditions. *See:* **synchronous machine.** 42A10-31E8

subway-type transformer. A submersible constant current transformer suitable for installation in an underground vault. *See also:* **constant-current transformer; transformers.** 42A15/57A14-31E12

sudden-pressure relay. A relay that operates by the rate of rise of fluid pressure. *See also:* **relay.** 37A100-31E11/31E6

sudden short-circuit test (synchronous machine). A test in which a short-circuit is suddenly applied to the armature winding of the machine under specified operating conditions. *See also:* **synchronous machine.** 0-31E8

Suez Canal searchlight. A searchlight constructed to the specifications of the Canal Administration that by regulation of the Administration, must be carried by every ship traversing the canal, so located as to illuminate the banks. 42A43-0

suicide control (adjustable-speed drive). A control function that reduces and automatically maintains the generator voltage at approximately zero by negative feedback. *See also:* **control system, feedback.** AS1-34E10

sum frequency (parametric device). The sum of a harmonic (nf_p) of the pump frequency (f_p) and the signal frequency (f_s), where n is a positive integer. *Note:* Usually n is equal to one. *See also:* **parametric device.** E254-15E7

sum-frequency parametric amplifier*. *See:* **noninverting parametric device.** *See also:* **parametric device.**

*Deprecated

summation check (computing systems). A check based on the formation of the sum of the digits of a numeral. The sum of the individual digits is usually compared with a previously computed value. *See also:* **electronic digital computer.** X3A12-16E9

summer (computing systems). *See:* **summing amplifier.**

summing amplifier. An operational amplifier that produces an output signal equal to a weighted sum of the input signals. *Note:* In an analog computer, the term **summer** is synonymous with **summing amplifier.** *See also:* **electronic analog computer.** E165-16E9

summing junction. The junction common to the input and feedback impedances used with an operational amplifier. *See also:* **electronic analog computer.** E165-16E9

summing point (1). Any point at which signals are added algebraically. *Note:* For example the null junction of a power supply is a summing point because, as the input to a high-gain direct-current amplifier, operational summing can be performed at this point. As a virtual ground, the summing point decouples all inputs so that they add linearly in the output, without other interaction. *See:* **operational programming.** *See also:* **null junction; power supply.** 85A1-23E0;KPSH-10E1

(2) (industrial control). The point in a feedback control system at which the algebraic sum of two or more signals is obtained. *See also:* **control system, feedback.** AS1-34E10

sun bearing. The angle measured in the plane of the horizon between a vertical plane at a right angle to the window wall and the position of this plane after it has been rotated to contain the sun. *See also:* **sunlight.** Z7A1-0

sunlight. Direct visible radiation from the sun.
See:
altitude;
azimuth;
clear sky;
clerestory;
daylight factor;
fenestration;
ground light;
night;
orientation;
service period;
sky factor;
sky light;
sun bearing.

superconducting. The state of a superconductor in which it exhibits superconductivity. *Example:* Lead is superconducting below a critical temperature and at sufficiently low operating frequencies. *See:* **normal (state of a superconductor).** *See also:* **superconductivity.** E217-15E7

superconductive. Pertaining to a material or device that is capable of exhibiting superconductivity. *Example:* Lead is a superconductive metal regardless of temperature. The cryotron is a superconductive computer component. *See also:* **superconductivity.** E217-15E7

superconductivity. A property of a material that is characterized by zero electric resistivity and, ideally, zero permeability.
See:
control;
critical controlling current;
critical current;
critical magnetic field;
critical temperature;

cryotron;
electron devices, miscellaneous;
gate;
normal;
persistent current;
superconductive;
superconductor;
trapped flux. E217-15E7

superconductor. Any material that is capable of exhibiting superconductivity. *Example:* Lead is a superconductor. *See also:* **superconductivity.** E217-15E7

superdirectivity (antenna). The directivity of an antenna when its value exceeds the value that could be expected from the antenna on the basis of its dimensions and the excitation that would have yielded inphase addition in the direction of maximum radiation intensity. *Note:* Superdirectivity is obtained only at the cost of a sharp increase in the ratio of average stored energy to power radiated per hertz. *See also:* **radiation.** 0-3E1

supergroup. *See:* **channel supergroup.**

superheterodyne reception. A method of receiving radio waves in which the process of heterodyne reception is used to convert the voltage of the received wave into a voltage of an intermediate, but usually superaudible, frequency, that is then detected. *See also:* **radio receiver.** 42A65-0

superimposed ringing. Party-line ringing which utilizes a combination of alternating and direct currents, both positive and negative direct current being provided to obtain selectivity. *See also:* **telephone switching system.** 42A65-19E1

superposed circuit. An additional channel obtained from one or more circuits, normally provided for other channels, in such a manner that all the channels can be used simultaneously without mutual interference. *See also:* **transmission line.** 42A65-0

superposition theorem. States that the current that flows in a linear network, or the potential difference that exists between any two points in such a network, resulting from the simultaneous application of a number of voltages distributed in any manner whatsoever throughout the network is the sum of the component currents at the first point, or the component potential differences between the two points, that would be caused by the individual voltages acting separately. *See also:* **communication.** 42A65-0

superregeneration. A form of regenerative amplification, frequently used in radio receiver detecting circuits, in which oscillations are alternately allowed to build up and are quenched at a superaudible rate. *See also:* **radio receiver.** 42A65-0

supersonic frequency*. *See:* **ultrasonic frequency.**
*Obsolescent

supervised circuit (protective signaling). A closed circuit having a current-responsive device to indicate a break in the circuit, and, in some cases, to indicate an accidental ground. *See also:* **protective signaling.** 42A65-0

supervisory control. A form of remote control comprising an arrangement for the selective control of remotely located units by electrical means over one or more common interconnecting channels. 37A100-31E11

supervisory control (power-system communication). A discrete portion of a supervisory control system associated with the control and/or indication of the status of a remote device. *See also:* **supervisory control system.** 0-31E3

supervisory control point, control and indication (C & I) (power-system communication). A supervisory control point permitting control of an end device with report back and display of its status. *See also:* **supervisory control system.** 0-31E3

supervisory control point, indication (power-system communication). A point on a supervisory control system providing indication only of the status of a remote device. *See also:* **supervisory control system.** 0-31E3

supervisory control point, jog control (power-system communication). A point on a supervisory control system that permits changing a controlled device one increment each time a raise or a lower command is executed. *See also:* **supervisory control system.** 0-31E3

supervisory control point, raise-lower (power-system communication). A supervisory control point capable of performing a raise or lower control for as long as a manual raise or lower control switch is operated. *See also:* **supervisory control system.** 0-31E3

supervisory control point, status (power-system communication). A supervisory control point displaying the one of two or three possible conditions of a remote device. If the device has four or more possible conditions, the remote indication of the condition will be termed telemetering. *See also:* **supervisory control system.** 0-31E3

supervisory control receiver (power-system communication). A supervisory control subsystem used to receive codes. *See also:* **supervisory control system.** 0-31E3

supervisory control system (1). A remote-control system exchanging coded signals over communications channels so as to effect control and display the status of remote equipment with far fewer than one channel per control point.
See:
alarm point;
communication;
preference, channel supervisory control;
select before operate, supervisory control;
status point, supervisory control;
supervisory control point;
supervisory control point, control and indication;
supervisory control point, indication;
supervisory control point, jog control;
supervisory control point, raise-lower;
supervisory control point, status;
supervisory control receiver;
supervisory control system, continuous scan;
supervisory control system, quiescent;
supervisory control system, select before operate;
supervisory control transmitter. 0-31E3

(2) (power-system communication) (continuous scan). A supervisory control system that continually reports the position of controlled devices except when interrupted for control, telemeter, or other discrete operations. *See also:* **supervisory control system.** 0-31E3

supervisory control system, quiescent (power system communication). A supervisory control system that does not transmit codes in an at-rest condition. *See also:* **supervisory control system.** 0-31E3

supervisory control system, select before operate (power-system communication). A supervisory control system in which a point is first selected and a return displayed proving the selection after which one of several operations can be performed on that point. *See also:* **supervisory control system.** 0-31E3

supervisory control transmitter (power-system communication). A supervisory control subsystem used to transmit codes. *See also:* **supervisory control system.** 0-31E3

supervisory indication. A form of remote indication comprising an arrangement for the automatic indication of the position or condition of remotely located units by electrical means over one or more common interconnecting channels. 37A100-31E11

supervisory relay. A relay that, during a call, is generally controlled by the transmitter current supplied to a subscriber line in order to receive, from the associated station, directing signals that control the actions of operators or switching mechanisms with regard to the connection. *See also:* **telephone switching system.** 42A65-0

supervisory routine (computing systems). *See:* **executive routine.** *See also:* **electronic digital computer.**

supervisory signals. Signals used to indicate the various operating states of circuits or circuit combinations. *See also:* **telephone switching system.** 42A65-19E1

supervisory station check. The automatic selection in a definite order, by means of a single initiation at the master station, of all of the supervisory points associated with one remote station of a system; and the transmission to the master station of indications of positions or conditions of the individual equipment or device associated with each point. 37A100-31E11

supervisory system. All supervisory control, indicating, and telemeter selection devices in the master station and all of the complementary devices in the remote station, or stations, that utilize a single common interconnecting channel for the transmission of the control or indication signals between these stations. *Note:* The supervisory system may be designed to detect and signal the deviation of supervised persons or devices from an established norm. *See also:* **protective signaling.** 37A100-31E11;42A65-0

supervisory system check. The automatic selection in a definite order, by means of a single initiation at the master station, of all supervisory points associated with all of the remote stations in a system; and the transmission to the master station of indications of positions or conditions of the individual equipment or device associated with each point. *See also:* **protective signaling.** 37A100-31E11

supervisory telemeter selection. A form of remote telemeter selection comprising an arrangement for the selective connection of telemeter transmitting equipment to an appropriate telemeter receiving equipment over one or more common interconnecting channels. 37A100-31E11

supervisory tones (telephony). The audible signals that indicate to the caller or other relevant party that a particular state in the call has been reached, and may indicate the need for action to the party concerned. The terms used for the various supervisory tones are usually self-explanatory. *See also:* **telephone switching system.** 0-19E1

supplementary lighting. Lighting used to provide an additional quantity and quality of illumination that cannot readily be obtained by a general lighting system and that supplements the general lighting level, usually for specific work requirements. *See also:* **general lighting.** Z7A1-0

supply circuit (household electric ranges). The circuit that is the immediate source of the electric energy used by the range. *See also:* **appliance outlet.** 71A1-0

supply impedance (inverters). The impedance appearing across the input lines to the power inverter with the power inverter disconnected. *See:* **self-commutated inverters.** 0-34E24

supply line, motor (rotating machinery). The source of electric power to which the windings of a motor are connected. *See also:* **asynchronous machine; direct-current commutating machine; synchronous machine.** 0-31E8

supply transient voltage (inverters). The peak instantaneous voltage appearing across the input lines to the power inverter with the inverter disconnected. *See:* **self-commutated inverters.** 0-34E24

supply voltage (electrode) (electron tubes). The voltage, usually direct, applied by an external source to the circuit of an electrode. *See:* **electrode voltage (electron tube).** 50I07-15E6

support ring (rotating machinery). A structure for the support of a winding overhang; either constructed of insulating material, carrying support-ring insulation, or separately insulated before assembly. *See also:* **stator.** 0-31E8

support-ring insulation (rotating machinery). Insulation between the winding overhang or end winding and the winding support rings. *See also:* **rotor (rotating machinery); stator.** 0-31E8

suppressed-carrier modulation. Modulation in which the carrier is suppressed. *Note:* By **carrier** is meant that part of the modulated wave that corresponds in a specified manner to the unmodulated wave. *See also:* **modulating systems.** E170-0

suppressed-carrier operation. That form of amplitude-modulation carrier transmission in which the carrier wave is suppressed. *See also:* **amplitude modulation.** 42A65-0

suppressed time delay (electronic navigation). A deliberate displacement of the zero of the time scale with respect to time of emission of a pulse. *See also:* **navigation.** E172-10E6

suppressed-zero instrument. An indicating or recording instrument in which the zero position is below the end of the scale markings. *See also:* **instrument.** 42A30-0

suppressed-zero range. A range where the zero value of the measured variable, measured signal, etcetera, is less than the lower range value. Zero does not appear on the scale. *Note:* For example: 20 to 100. 39A4-0

suppression (computing systems). *See:* **zero suppression.** *See also:* **electronic digital computer.**

suppression distributor rotor. Rotor of an ignition distributor with a built-in suppressor. *See also:* **electromagnetic compatibility.** CISPR-27E1

suppression ratio (suppressed-zero range). The ratio of the lower range-value to the span. *Note:* For example: Range 20 to 100

$$\text{Suppression Ratio} = \frac{20}{80} = 0.25$$

39A4-0

suppressor grid. A grid that is interposed between two positive electrodes (usually the screen grid and the plate), primarily to reduce the flow of secondary electrons from one electrode to the other. *See also:* **electrode (electron tube); grid.** 42A70-15E6

suppressor spark plug. A spark plug with a built-in interference suppressor. *See also:* **electromagnetic compatibility.** 27E1

surface active agent. *See:* **wetting agent (electroplating).**

surface duct (radio wave propagation). An atmospheric duct for which the lower boundary is the surface of the earth. *See also:* **radiation; radio wave propagation.** 42A65-3E2

surface leakage. The passage of current over the surface of a material rather than through its volume. E270-0

surface metal raceway (metal molding). A raceway consisting of an assembly of backing and capping. *See also:* **raceways.** 42A95-0

surface-mounted device. A device, the entire body of which projects in front of the mounting surface. 37A100-31E11

surface-mounted luminaire. A luminaire that is mounted directly on the ceiling. *See also:* **suspended (pendant) luminaire.** Z7A1-0

surface noise (mechanical recording). The noise component in the electric output of a pickup due to irregularities in the contact surface of the groove. *See also:* **phonograph pickup.** 0-1E1

surface of position (electronic navigation). Any surface defined by a constant value of some navigation quantity. *See also:* **navigation.** E172-10E6

surface-potential gradient. The slope of a potential profile, the path of which intersects equipotential lines at right angles. *See also:* **ground.** E81-0

surface search radar. *See:* **navigational radar.**

surface-wave antenna. An antenna that radiates power from discontinuities in the structure that interrupt a bound wave on the antenna surface. *See also:* **antenna.** 0-3E1

surface-wave transmission line (waveguides). A transmission line in which propagation in other than a *TEM* mode is constrained to follow the external face of a guiding structure. *See:* **waveguide.** 0-3E1

surge (electric power). A transient wave of current, potential, or power in the electric circuit. *Note:* A transient has a high rate of change of voltage (current) in the system. It will be propagated along the length of the circuit. *See also:* **lightning; power systems, low-frequency and surge testing; pulse.** E28-0;42A35-31E13;50I25-31E7

surge-crest ammeter. A special form of magnetometer intended to be used with magnetizable links to measure the crest value of transient electric currents. *See also:* **instrument.** 42A30-0

surge diverter. *See:* **lightning arrester.**

surge electrode current. *See:* **fault electrode current.**

surge generator (impulse generator) (lightning generator*). An electric apparatus suitable for the production of surges. *Note:* Surge generator types common in the art are: transformer-capacitor; transformer-rectifier; transformer-rectifier-capacitor, parallel charging, series discharging. *See also:* **lightning arrester (surge diverter); power systems, low-frequency and surge testing.** 42A35-31E7/31E13

*Deprecated

surge impedance (self-surge impedance). The ratio between voltage and current of a wave that travels on a line of infinite length and of the same characteristics as the relevant line. *See:* **characteristic impedance.** 50I25-31E7

surge (nonrepetitive) ON-state current (thyristor). An ON-state current of short-time duration and specified wave shape. *See also:* **principal current.** E223-34E1/34E24/15E7

surge protection. *See:* **rate-of-change protection.**

surge suppressor (industrial control). A device operative in conformance with the rate of change of current, voltage, power, etcetera, to prevent the rise of such quantity above a predetermined value. 42A25-34E10

surge voltage recorder (klydonograph). *See:* **Lichtenberg figure camera.**

surveillance radar. A search radar used to maintain cognizance of selected traffic within a selected area, such as an airport terminal area or air route. 0-10E6

susceptance. The imaginary part of admittance. E270-9E4

susceptance function (linear passive networks). The driving-point admittance of a lossless network. *Note:* This is an odd function of the complex frequency. *See also:* **linear passive networks.** E156-0

susceptance relay. A Siemen(s) type distance relay for which the center of the operating characteristic on the R-X diagram is on the X axis. *Note:* The equation that describes such a characteristic is $Z = K \sin \theta$ where K is a constant and θ is the phase angle by which the energizing voltage leads the energizing current. 37A100-31E11/31E6

susceptibility meter. A device for measuring low values of magnetic susceptibility. One type is the Curie balance. *See also:* **magnetometer.** 42A30-0

suspended (pendant) luminaire. A luminaire that is hung from a ceiling by supports. *See:* **flush mounted or recessed; regressed; surface mounted.** Z7A1-0

suspended-type handset telephone (bracket-type handset telephone). *See:* **hang-up hand telephone.**

suspension insulator. One or a string of suspension-type insulators assembled with the necessary attaching members and designed to support in a generally vertical direction the weight of the conductor and to afford adequate insulation from tower of other structure. *See also:* **insulator; tower.** 42A35-31E13

suspension-insulator unit. An assembly of a shell and hardware, having means for nonrigid coupling to other units or terminal hardware. 29A1-0

suspension-insulator weights. Devices, usually cast iron, hung below the conductor on a special spindle supported by the conductor clamp. *Note:* Suspension insulator weights will limit the swing of the insulator string, thus maintaining adequate clearances. In practice, weights of several hundreds of pounds are sometimes used. *See also:* **tower** 42A35-31E13

suspension strand (messenger). A stranded group of wires supported above the ground at intervals by poles or other structures and employed to furnish within these intervals frequent points of support for conductors or cables. *See also:* **cable; open wire.** 42A65-0

sustained interruption (electric power systems). Any interruption not classified as a momentary interruption. *See also:* **outage.** 0-31E4

sustained-operation influence. The change in the recorded value, including zero shift, caused solely by energizing the instrument over extended periods of time, as compared to the indication obtained at the end of the first 15 minutes of the application of energy. It is to be expressed as a percentage of the full-scale value. *Note:* The coil used in the standard method shall be approximately 80 inches in diameter, not over 5 inches long, and shall carry sufficient current to produce the required field. The current to produce a field to an accuracy of ± 1 percent in air shall be calculated without the instrument in terms of the specific dimensions and turns of the coil. In this coil, 800 ampere-turns will produce a field of approximately 5 oersteds. The instrument under test shall be placed in the center of the coil. *See also:* **accuracy rating (instrument).** 39A2-0

sustained oscillation (sustained vibration) (1) (system). The oscillation when forces controlled by the system maintain a periodic oscillation of the system. Example: Pendulum actuated by a clock mechanism. E270-0

(2) (gas turbines). Those oscillations in which the amplitude does not decrease to zero, or to a negligibly small, final value. E282-31E2

sustained short-circuit test (synchronous machine). A test in which the machine is run as a generator with its terminals short-circuited. *See also:* **synchronous machine.** 0-31E8

sweep. A traversing of a range of values of a quantity for the purpose of delineating, sampling, or controlling another quantity. *Notes:* (1) Examples of swept quantities are (A) the displacement of a scanning spot on the screen of a cathode-ray tube, and (B) the frequency of a wave. (2) Unless otherwise specified, a linear time function is implied; but the sweep may also vary in some other controlled and desirable manner. 42A65-2E2/9E4

sweep accuracy (oscilloscopes). Accuracy of the horizontal (vertical) displacement of the trace compared with the reference independent variable, usually expressed in terms of average rate error as a percent of full scale. *See:* **sweep linearity.** *See also:* **oscillograph.** 0-9E4

sweep, delayed. *See:* **delayed sweep.**

sweep duration (sawtooth sweep). The time required for the sweep ramp. *See also:* **oscillograph.** 0-9E4

sweep duty factor. For repetitive sweeps, the ratio of the sweep duration to the interval between the start of one sweep and the start of the next. *See also:* **oscillograph.** 0-9E4

sweep, expanded. *See:* **magnified sweep.**

sweep, free-running. *See:* **free-running sweep.**

sweep frequency (oscilloscopes). The sweep repetition rate. *See also:* **oscillograph.** 0-9E4

sweep gate (oscilloscopes). Rectangular waveform used to control the duration of the sweep; usually also used to unblank the cathode-ray tube for the duration of the sweep. *See also:* **oscillograph.** 0-9E4

sweep, gated. *See:* **gated sweep.**

sweep holdoff interval. The interval between sweeps during which the sweep and/or trigger circuits are inhibited. 0-9E4

sweep, incremental. *See:* **incremental sweep.**

sweep lockout (oscilloscopes). Means for preventing multiple sweeps when operating in a single-sweep mode. *See also:* **oscillograph.** 0-9E4

sweep magnifier (oscilloscopes). Circuit or control for expanding part of the sweep display. Sometimes known as **sweep expander.** *See also:* **oscillograph.** 0-9E4

sweep oscillator. An oscillator in which the output frequency varies continuously and periodically between two frequency limits. *See also:* **telephone station.** E269-19E8

sweep range (oscilloscopes). The set of sweep-time/division settings provided. *See also:* **oscillograph.** 0-9E4

sweep recovery time (oscilloscopes). The minimum possible time between the completion of one sweep and the initiation of the next, usually the sweep holdoff interval. *See also:* **oscillograph.** 0-9E4

sweep, recurrent. *See:* **recurrent sweep.**

sweep reset (oscilloscopes). In oscilloscopes with single-sweep operation, the arming of the sweep generator to allow it to cycle once. *See also:* **oscillograph.** 0-9E4

switch (1) (electrical systems). A device for opening and closing or for changing the connection of a circuit. *Note:* A switch is understood to be manually operated, unless otherwise stated.

See:

alternating-current general-use snap switch;
alternating-current–direct-current general-use snap switch;
cam-operated switch;
circuit breaker;
control-circuit limit switch;
control cutout switch;
disconnecting means;
drum switch;
electrically interlocked manual release;
enclosed switch;
float switch;
foot switch;
fusible enclosed (safety) switch;
general-use switch;
hand operation;
ignition switch;
isolating switch;
limit switch;
master switch;
molded-case circuit breaker;
motor-circuit switch;
multiposition switches;
oil-tight pilot devices;
power-circuit limit switch;
proximity switch;
push-button;
pushbutton station;
push-button switch (pushbutton);
rated continuous current;
restriking-voltage
reversing switch;
rotary switch;
selector switch;
service;
thermostatic switch;
transfer switch. 2A2/42A25-34E10

(2) (computing systems). A device or programming technique for making a selection, for example, a toggle, a conditional jump. *See also:* **electronic digital computer.** X3A12-16E9

switch-and-lock movement. A device, the complete operation of which performs the three functions of unlocking, moving, and locking a switch, movable-point frog, or derail. 42A42-0

switchboard (electric power systems). A large single panel, frame, or assembly of panels, on which are mounted, on the face or back or both, switches, overcurrent and other protective devices, buses, and usually instruments. *Note:* Switchboards are generally accessible from the rear as well as from the front and are not intended to be installed in cabinets. *See also:* **center of distribution; distribution center; panelboard.** 1A0/2A2

switchboard cord. A cord that is used in conjunction with switchboard apparatus to complete or build up a telephone connection. *See also:* **telephone switching system.** 42A65-0

switchboard lamp (switchboard). A small electric lamp associated with the wiring in such a way as to give a visual indication of the status of a call or to give information concerning the condition of trunks, subscriber lines, and apparatus. *See also:* **telephone switching system.** 42A65-0

switchboard position (telephony). Usually the portion of a manual switchboard normally provided for the use of one operator. *See also:* **telephone switching system.** 0-19E1

switchboards and panels (marine transportation). A **generator and distribution switchboard** receives energy from the generating plant and distributes directly or indirectly to all equipment supplied by the generating plant. A **subdistribution switchboard** is essentially a section of the generator and distribution switchboard (connected thereto by a bus feeder and remotely located for reasons of convenience or economy) that distributes energy for lighting, heating, and power circuits in a certain section of the vessel. A **distribution panel** receives energy from a distribution or subdistribution switchboard and distributes energy to energy-consuming devices or other distribution panels or panelboards. A **panelboard** is a distribution panel enclosed in a metal cabinet. *See also:* **generating station.** E45-0

switchboard section. A structural unit, providing for one or more operator positions. A complete switchboard may consist of one or more sections. *See also:* **telephone switching system.** 42A65-19E1

switchboard supervisory lamp (cord circuit or trunk circuit). A lamp that is controlled by one or other of the users to attract the attention of the operator. *See also:* **telephone switching system.** 42A65-0

switchboard supervisory relay. A relay that controls a switchboard supervisory lamp. *See also:* **telephone switching system.** 42A65-0

switchgear. A general term covering switching and interrupting devices and their combination with associated control, metering, protective, and regulating devices, also assemblies of these devices with associated interconnections, accessories, enclosures, and supporting structures, used primarily in connection with the generation, transmission, distribution, and conversion of electric power. 37A100-31E11

switchgear assembly. An assembled equipment (indoor or outdoor) including, but not limited to, one or more of the following: switching; interrupting; control; metering; protective and regulating devices, together with their supporting structures, enclosures, conductors, electric interconnections, and accessories. 37A100-31E11

switchgear pothead. A pothead intended for use in a switchgear where the inside ambient air temperature may exceed 40 degrees Celsius. It may be an indoor or outdoor pothead that has been suitably modified by silver surfacing (or the equivalent) the current-carrying parts and incorporates sealing materials suitable for the higher operating temperatures. *See also:* **pothead.** E48-0

switchhook (hookswitch). A switch on a telephone set, associated with the structure supporting the receiver or handset. It is operated by the removal or replacement of the receiver or handset on the support. *See also:* **telephone station.** 42A65-0

switchhook. *See:* **switch stick.**

switch indicator. A device used at a noninterlocked switch to indicate the presence of a train in a block. *See also:* **railway signal and interlocking.** 42A42-0

switching (single-phase motor). The point in the starting operation at which the stator-winding circuits are switched from one connection arrangement to another. *See also:* **asynchronous machine; synchronous machine.** 0-31E8

switching amplifier (industrial control). An amplifier which is designed to be applied so that its output is sustained at one of two specified states dependent upon the presence of specified inputs. *See also:* **control system, feedback.** AS1-34E10

switching coefficient. The derivative of applied magnetizing force with respect to the reciprocal of the resultant switching time. It is usually determined as the reciprocal of the slope of a curve of reciprocals of switching times versus values of applied magnetizing forces. The magnetizing forces are applied as step functions. *See also:* **static magnetic storage.** E163-16E9

switching device (switch). A device designed to close or open, or both, one or more electric circuits. *Note:* The term **switch** in International Electrotechnical Commission (IEC) practice refers to a mechanical switching device capable of opening and closing rated continuous load current. *See:* **mechanical switching device; nonmechanical switching device.** 37A100-31E11

switching-impulse sparkover voltage (arrester). The impulse sparkover voltage with an impulse having a virtual duration of wavefront greater than 30 microseconds. *See also:* **lightning arrester (surge diverter).** 0-31E7

switching structure. An open framework supporting the main switching and associated equipment, such as instrument transformers, buses, fuses, and connections. It may be designed for indoor or outdoor use and may be assembled with or without switchboard panels carrying the control equipment. 37A100-31E11

switching-surge protective level (arrester). The highest value of switching-surge voltage that may appear across the terminals under the prescribed conditions. *Note:* The switching-surge protective levels are given numerically by the maximums of the following quantities: (1) discharge voltage at a given discharge current, and (2) switching-impulse sparkover voltage. *See also:* **lightning arrester (surge diverter).** 0-31E7

switching time (1) (electric power circuits). The period from the time a switching operation is required due to a forced outage until that switching operation is performed. Switching operations include reclosing a circuit breaker after a trip-out, opening or closing a sectionalizing switch or circuit breaker, or replacing a fuse link. *See also:* **outage.** 0-31E4

(2) (magnetic storage cells). (A) T_s, the time interval between the reference time and the last instant at which the instantaneous voltage response of a magnetic cell reaches a stated fraction of its peak value. (B) T_x, the time interval between the reference time and the first instant at which the instantaneous integrated voltage response reaches a stated fraction of its peak value. *See also:* **static magnetic storage.** E163-16E9

switching torque (motor having an automatic connection change during the starting period). The minimum external torque developed by the motor as it accelerates through switch operating speed. *Note:* It should be noted that if the torque on the starting connection is never less than the switching torque, the pull-up torque is identical with the switching torque; however, if the torque on the starting connection falls below the switching torque at some speed below switch operating speed, the pull-up and switching torques are not identical. *See also:* **asynchronous machine; synchronous machine.** 42A10-0

switching torque, single-phase motor. The minimum torque which a motor will provide at switching at normal operating temperature, with rated voltage applied at rated frequency. *See also:* **asynchronous machine; synchronous machine.** 0-31E8

switch machine. A quick-acting mechanism, electrically controlled, for positioning track switch points, and so arranged that the accidental trailing of the switch points does not cause damage. A switch machine may be electrically or pneumatically operated. *See also:* **car retarder.** 42A42-0

switch machine lever lights. A group of lights indicating the position of the switch machine. 42A42-0

switch-machine point detector. *See:* **point detector.**

switch room. The part of a central office building that houses an assemblage of switching mechanisms and associated apparatus. *See also:* **telephone system.** 42A65-19E1

switch signal. A low two-indication horizontal color light signal with electric lamps for indicating position of switch or derail . *See also:* **railway signal and interlocking.** 42A42-0

switch sleeve. A component of the linkage between the centrifugal mechanism and the starting-switch assembly. *See also:* **starting-switch assembly.** 42A10-0

switch starting. *See:* **preheat starting.**

switch stick (switchhook). A device with an insulated handle and a hook or other means for performing stick operation of a switching device. 37A100-31E11

switch train. A series of switches in tandem. *See also:* **telephone switching system.** 42A65-0

SWR. *See:* **standing-wave-ratio indicator.**

syllabic companding (modulation systems). Companding in which the gain variations occur at a rate comparable to the syllabic rate of speech; but do not respond to individual cycles of the audio-frequency signal wave. *See also:* **modulating systems; transmission characteristics.** E170/42A65-0

syllable articulation (percent syllable articulation). The percent articulation obtained when the speech units considered are syllables (usually meaningless and usually of the consonant-vowel-consonant type). *See:* **articulation (percent articulation).** *See also:* **volume equivalent.** 42A65-0

symbol. A representation of something by reason of relationship, association, or convention. *See:* **logic symbol.** *See also:* **electronic digital computer.** X3A12-16E9

symbol for a quantity (quantity symbol) (abbreviation). A letter (which may have letters or numbers, or both, as subscripts or superscripts, or both), used to represent a physical quantity or a relationship between quantities. *Compare with:* **abbreviation, functional designation, mathematical symbol, reference designation, and symbol for a unit.** *See also:* **abbreviation.** E267-0

symbol for a unit (unit symbol) (abbreviation). A letter, a character, or combinations thereof, that may be used in place of the name of the unit. With few exceptions, the letter is taken from the name of the unit. *Compare with:* **abbreviation, mathematical symbol, symbol for a quantity.** *See also:* **abbreviation.** E267-0

symbolic address (computing systems). An address expressed in symbols convenient to the programmer. *See also:* **electronic digital computer.** X3A12-16E9

symbolic coding (computing systems). Coding that uses machine instructions with symbolic addresses. *See also:* **electronic digital computer.** X3A12-16E9

symbolic logic. The discipline that treats formal logic by means of a formalized artificial language or symbolic calculus whose purpose is to avoid the ambiguities and logical inadequacies of natural languages. X3A12-16E9

symbolic quantity. *See:* **mathematico-physical quantity.**

symmetrical alternating current. A periodic alternating current in which points one-half a period apart are equal and have opposite signs. *See also:* **alternating function; network analysis.** E270-0

symmetrical component (total current) (alternating-current component). That portion of the total current that constitutes the symmetry. 37A100-31E11

symmetrical components (1) (set of polyphase alternating voltages). The symmetrical components of an unsymmetrical set of sinusoidal polyphase alternating voltages of m phases are the m symmetrical sets of polyphase voltages into which the unsymmetrical set can be uniquely resolved, each component set having an angular phase lag between successive members of the set that is a different integral multiple of the characteristic angular phase difference for the number of phases. The successive component sets will have phase differences that increase from zero for the first set to $(m-1)$ times the characteristic angular phase difference for the last set. The phase sequence of each component set is identified by the integer that denotes the number of times the angle of lag between successive members of the component set contains the characteristic angular phase difference. If the members of an unsymmetrical set of alternating polyphase voltages are not sinusoidal, each voltage is first resolved into its harmonic components, then the harmonic components of the same period are grouped to form unsymmetrical sets of sinusoidal voltages, and finally each harmonic set of sinusoidal voltages is uniquely resolved into its symmetrical components. Because the resolution of a set of polyphase voltages into its harmonic components is also unique, it follows that the resolution of an unsymmetrical set of polyphase voltages into its symmetrical components is unique. There may be a symmetrical-component set of voltages for each of the possible phase sequences from zero to $(m-1)$ and for each of the harmonics present from 1 to r, where r may approach infinity in particular cases. Each member of a set of symmetrical component voltages of kth phase sequence and rth harmonic may be denoted by

$$e_{ski} = (2)^{1/2} E_{akr} \cos\left(r\omega t + \alpha_{akr} - (s-1) K \frac{2\pi}{m}\right)$$

where e_{skr} is the instantaneous voltage component of phase sequence k and harmonic r in phase s. E_{akr} is the root-mean-square amplitude of the voltage component of phase sequence k and harmonic r, using phase a as reference. α_{akr} is the phase angle of the first member of the set, selected as phase a, with respect to a common reference. The letter s as the first subscript denotes the phase identification of the individual member, a, b, c, etcetera for successive members, and a denotes that the first phase, a, has been used as a reference from which other members are specified. The second subscript k denotes the phase sequence of the component, and may run from 0 to $m-1$. The third subscript denotes the order of the harmonic, and may run from 1 to ∞. The letter s as an algebraic quantity denotes the member of the set and runs from 1 for phase a to m for the last phase. Of the m symmetrical component sets for each harmonic, one will be of zero phase sequence, one of positive phase sequence, and one of negative phase sequence. If the number of phases m $(m>2)$ is even, one of the symmetrical component sets for $k=m/2$ will be a single-phase symmetrical set (polyphase voltages). The zero-phase-sequence component set will constitute a zero-phase symmetrical set (polyphase voltages), and the remaining sequence components will constitute polyphase symmetrical sets (polyphase voltages). *See also:* **network analysis.** E270-0

(2) (set of polyphase alternating currents). Obtained from the corresponding definition for **symmetrical components (set of polyphase alternating voltages)** by substituting the word **current** for **voltage** wherever it appears. *See also:* **network analysis.** E270-0

symmetrical fractional-slot winding (rotating machinery). A distributed winding in which the average number of slots per pole per phase is not integral, but in which the winding pattern repeats after every pair of poles, for example, 3½ slots per pole per phase. *See also:* **rotor (rotating machinery); stator.** 0-31E8

symmetrically cyclically magnetized condition. A condition of a magnetic material when it is in a cyclically magnetized condition and the limits of the applied magnetizing forces are equal and of opposite sign, so that the limits of flux density are equal and of opposite sign. *See also:* **static magnetic storage.** E163/E270-0

symmetrical network. *See:* **structurally symmetrical network.**

symmetrical periodic function. A function having the period 2π is symmetrical if it satisfies one or more of the following identities.

(1) $f(x) = -f(-x)$	(4) $f(x) = f(-x)$
(2) $f(x) = -f(\pi + x)$	(5) $f(x) = f(\pi + x)$
(3) $f(x) = -f(\pi - x)$	(6) $f(x) = f(\pi - x)$

See also: **network analysis.** E270-0

symmetrical set (1) (polyphase voltages). A symmetrical set of polyphase volt ages of m phases is a set of polyphase voltages in which each voltage is sinusoidal and has the same amplitude, and the set is arranged in such a sequence that the angular phase difference between each member of the set and the one following it, and between the last member and the first, can be expressed as the same multiple of the characteristic angular phase difference $2\pi/m$ radians. A symmetrical set of polyphase voltages may be expressed by the equations

$$e_a = (2)^{1/2} E_{ar} \cos(r\omega t + \alpha_{ar})$$

$$e_b = (2)^{1/2} E_{ar} \cos\left(r\omega t + \alpha_{ar} - k\frac{2\pi}{m}\right)$$

$$e_c = (2)^{1/2} E_{ar} \cos\left(r\omega t + \alpha_{ar} - 2k\frac{2\pi}{m}\right)$$

$$e_m = (2)^{1/2} E_{ar} \cos\left(r\omega t + \alpha_{ar} - (m-1)k\frac{2\pi}{m}\right)$$

where E_{ar} is the root-mean-square amplitude of each member of the set, r is the order of the harmonic of each member, with respect to a specified period. α_{ar} is the phase angle of the first member of the set with respect to a selected reference. k is an integer that denotes the phase sequence. *Notes:* (1) Although sets of polyphase voltages that have the same amplitude and waveform but that are not sinusoidal possess some of the characteristics of a symmetrical set, only in special cases do the several harmonics have the same phase sequence. Since phase sequence is an important feature in the use of symmetrical sets, the definition is limited to sinusoidal quantities. This represents a change from the corresponding definition in the 1941 edition of the American Standard Definitions of Electrical Terms. (2) This definition may be applied to a two-phase four-wire or five-wire circuit if m is considered to be 4 instead of 2. The concept of symmetrical sets is not directly applicable to a two-phase three-wire circuit. E270-0

(2) (polyphase currents). This definition is obtained from the corresponding definitions for voltage by substituting the word **current** for **voltage**, and the symbol I for E and β for α wherever they appear. The subscripts are unaltered. *See also:* **network analysis.** E270-0

symmetrical terminal voltage (electromagnetic compatibility). Terminal voltage measured in a delta network across the mains lead. *See also:* **electromagnetic compatibility.** 0-27E1

symmetrical transducer (1) (specified pair of terminations). A transducer in which the interchange of that pair of terminations will not affect the transmission. *See also:* **transducer.** 42A65-0

(2) (specified terminations in general). A transducer in which all possible pairs of specified terminations may be interchanged without affecting transmission. *See also:* **transducer.** E196/E270-0

synapse. The junction between two neural elements, which has the property of one-way propagation. *See also:* **biological.** 42A80-18E1

synchro control transformer (synchro or selsyn devices). A transformer with relatively rotatable primary and secondary windings. The primary input is a set of two or more voltages from a synchro transmitter that define an angular position relative to that of the transmitter. The secondary output voltage varies with the relative angular alignment of primary and secondary windings, of the control transformer and the position of the transmitter. The output voltage is substantially zero in value at a position known as correspondence. *See also:* **synchro system.** 0-31E8

synchro differential receiver (motor) (synchro or selsyn devices). A transformer identical in construction to a synchro differential transmitter but used to develop a torque increasing with the difference in the relative angular displacement (up to about 90 electrical degrees) between the two sets of voltage input signals to its primary and secondary windings, the torque being in a direction to reduce this difference to zero. *See also:* **synchro system.** 0-31E8

synchro differential transmitter (generator) (rotating machinery). A transformer with relatively rotatable primary and secondary windings. The primary input is a set of two or more voltages that define an angular position. The secondary output is a set of two or more voltages that represent the sum or difference, depending upon connections, of the position defined by the primary input and the relative angular displacement between primary and secondary windings. *See also:* **synchro system.** 42A10-31E8

synchronism (rotating machinery). The state where connected alternating-current systems, machines, or a combination operate at the same frequency and where the phase-angle displacements between voltages in them are constant, or vary about a steady and stable average value. *See:* **asynchronous machine; synchronous machine.** 0-31E8

synchronism-check relay. A verification relay whose function is to operate when two input voltages are within predetermined phasor limits. 37A100-31E11/31E6

synchronization error (electronic navigation). The error due to imperfect timing of two operations; this may or may not include signal transmission time. *See also:* **navigation.** E172-10E6

synchronized sweep (oscilloscopes). A sweep that would free run in the absence of an applied signal but in the presence of the signal is synchronized by it. *See also:* **oscillograph.** 0-9E4

synchronizing (1) (rotating machinery). The process whereby a synchronous machine, with its voltage and phase suitably adjusted, is paralleled with another synchronous machine or system. *See also:* **asynchronous machine; synchronous machine.** 0-31E8

(2) (facsimile). The maintenance of predetermined speed relations between the scanning spot and the recording spot within each scanning line. *See also:* **facsimile (electrical communication).** E168-0

(3) (television). Maintaining two or more scanning processes in phase. E203-2E2

synchronizing coefficient (rotating machinery). The quotient of the shaft power and the angular displacement of the rotor. *Note:* It is expressed in kilowatts per electrical radian. Unless otherwise stated, the value will be for rated voltage, load, power-factor, and frequency. *See also:* **asynchronous machine; synchronous machine.** 0-31E8

synchronizing relay. A programming relay whose function is to initiate the closing of a circuit breaker between two alternating-current sources when the voltages of these two sources have a predetermined relationship of magnitude, phase angle, and frequency. 37A100-31E11/31E6

synchronizing signal (1) (television). The signal employed for the synchronizing of scanning. *Note:* In television, this signal is composed of pulses at rates related to the line and field frequencies. The signal usually originates in a central synchronizing generator and is added to the combination of picture signal and blanking signal, comprising the output signal from the pickup equipment, to form the composite picture signal. In a television receiver, this signal is normally separated from the picture signal and is used to synchronize the deflection generators. E203/42A65-2E2

(2) (facsimile). A signal used for maintenance of predetermined speed relations between the scanning spot and recording spot within each scanning line. *See also:* **facsimile signal (picture signal).** E168-0

(3) (oscillograph). A signal used to synchronize repetitive functions. *See:* **oscillograph.** 0-9E4

(4) (telecommunication). A special signal which may be sent to establish or maintain a fixed relationship in synchronous systems. 0-19E4

synchronizing signal compression (television). The reduction in gain applied to the synchronizing signal over any part of its amplitude range with respect to the gain at a specified reference level. *Notes:* (1) The gain referred to in the definition is for a signal amplitude small in comparison with the total peak-to-peak composite picture signal involed. A quantitative evaluation of this effect can be obtained by a measurement of differential gain. (2) Frequently the gain at the level of the peaks of synchronizing pulses is reduced with respect to the gain at the levels near the bases of the synchronizing pulses. Under some conditions, the gain over the entire synchronizing signal region of the composite picture signal may be reduced with respect to the gain in the region of the picture signal. *See also:* **television.** E203-2E2

synchronizing signal level (television). The level of the peaks of the synchronizing signal. *See also:* **television.** E203-2E2

synchronizing torque (synchronous machine). The torque produced, primarily through interaction between the armature currents and the flux produced by the field winding, tending to pull the machine into synchronism with a connected power system or with another synchronous machine. *See also:* **synchronous machine.** 0-31E8

synchronous booster converter. A synchronous converter having a mechanically connected alternating-current reversible booster connected in series with the alternating-current supply circuit for the purpose of adjusting the output voltage. *See also:* **converter.** 42A10-0

synchronous booster inverter. An inverter having a mechanically connected reversible synchronous booster connected in series for the purpose of adjusting the output voltage. *See also:* **converter.** 42A10-0

synchronous capacitor (synchronous condenser)* (rotating machinery). A synchronous machine running without mechanical load and supplying or absorbing reactive power to or from a power system. *See also:* **converter.**

*Deprecated 0-31E8

synchronous computer. A computer in which each event, or the performance of each operation, starts as a result of a signal generated by a clock. X3A12-16E9

synchronous converter. A converter that combines both motor and generator action in one armature winding and is excited by one magnetic field. It is normally used to change alternating-current power to direct-current power. *See:* **converter.** 42A10-0

synchronous gate. A time gate wherein the output intervals are synchronized with an incoming signal. *See also:* **circuits and devices; modulating systems.** E145/42A65

synchronous generator. A synchronous alternating-current machine that transforms mechanical power into electric power. *Notes:* (1) A synchronous machine is one in which the average speed of normal operation is exactly proportional to the frequency of the system to which it is connected. (2) Unless otherwise stated, it is generally understood that a synchronous generator (or motor) has field poles excited with direct current. (3) As a synonym, use of the term **alternator** is now deprecated. *See also:* **synchronous machine.** E95/42A10-31E8

synchronous impedance (per unit direct-axis). The ratio of the field current at rated armature current on sustained symmetrical short-circuit to the field current at normal open-circuit voltage on the air-gap line. *Note:* This definition of synchronous impedance is used to a great extent in electrical literature and corresponds to the definition of direct-axis synchronous reactance as determined from open-circuit and sustained short-circuit tests. *See also:* **positive phase-sequence reactance (rotating machinery); synchronous machine.** 42A10-0

synchronous internal voltage (synchronous machine for any specified operating conditions). The fundamental-frequency component of the voltage of each armature phase that would be produced by the steady (or very slowly varying) component of the current in the main field winding (or field windings) acting alone provided the permeance of all parts of the magnetic circuit remained the same as for the specified operating condition. *Note:* The synchronous internal voltage, as shown in the phasor diagram, is related to the terminal-voltage and phase-current phasors by the equation

$$\mathbf{E}_i=\mathbf{E}_a+R\,\mathbf{I}_a+jX_d\mathbf{I}_{ad}+jX_q\mathbf{I}_{aq}$$

For a machine subject to saturation, the reactances should be determined for the degree of saturation applicable to the specified operating condition. *See also:* **synchronous machine.** 42A10-31E8

synchronous inverter. An inverter that combines both motor and generator action in one armature winding. It is excited by one magnetic field and changes direct-current power to alternating-current power. *Note:* Usually it has no amortisseur winding. *See also:* **converter.** 42A10-0

synchronous machine. Note: For an extensive list of cross references, see *Appendix A.*

synchronous machine, ideal. A hypothetical synchronous machine that has certain idealized characteristics that facilitate analysis. *Note:* The results of the analysis of ideal machines may be applied to similar actual machines by making, when necessary, approximate corrections for the deviations of the actual machine from the ideal machine. The ideal machine has, in general, the following properties: (1) the resistance of each winding is constant throughout the analysis, independent of current magnitude or its rate of change; (2) the permeance of each portion of the magnetic circuit is constant throughout the analysis, regardless of the flux density; (3) the armature circuits are symmetrical with respect to each other; (4) the electric and magnetic circuits of the field structure are symmetrical about the direct axis or the quadrature axis; (5) the self-inductance of the field, and every circuit on the field structure, is constant; (6) the self-inductance of each armature circuit is a constant or a constant plus a second-harmonic sinusoidal function of the angular position of the rotor relative to the stator; (7) the mutual inductance between any circuit on the field structure and any armature circuit is a fundamental sinusoidal function of the angular position of the rotor relative to the stator; (8) the mutual inductance between any two armature circuits is a constant or a constant plus a second-harmonic sinusoidal function of the angular position of the rotor relative to the stator; (9) the amplitude of the second-harmonic component of variation of the self-inductance of the armature circuits and of the mutual inductances between any two armature circuits is the same; (10) effects of hysteresis are negligible; (11) effects of eddy currents are negligible or, in the case of solid-rotor machines, may be represented by hypothetical circuits on the field structure symmetrical about the direct axis and the quadrature axis. 42A10-31E8

synchronous motor. A synchronous machine that transforms electric power into mechanical power. Unless otherwise stated, it is generally understood that it has field poles excited by direct current. *See:* **asynchronous machine; synchronous machine.** 0-31E8

synchronous operation (of a machine). Operation where the speed of the rotor is equal to that of the rotating magnetic flux and where there is a stable phase relationship between the voltage generated in the primary winding and the voltage of a connected power system or synchronous machine. *See also:* **synchronous machine.** 0-31E8

synchronous speed (rotating machinery). The speed of rotation of the magnetic flux, produced by or linking the primary winding. *See also:* **synchronous machine.** 0-31E8

synchronous system (telecommunication). A system in which the sending and receiving instruments are operating continuously at substantially the same rate and are maintained by means of correction if necessary, in a fixed relationship. *See also:* **communication.** 0-19E4

synchronous voltage (traveling-wave tubes). The voltage required to accelerate electrons from rest to a velocity equal to the phase velocity of a wave in the absence of electron flow. *See:* **magnetron.** *See also:* **electron devices, miscellaneous.** E160-15E6

synchro receiver (or motor) (rotating machinery). A transformer electrically similar to a synchro transmitter and that, when the secondary windings of the two devices are interconnected, develops a torque increasing with the difference in angular alignment of the transmitter and receiver rotors and in a direction to reduce the difference toward zero. *See also:* **synchro**

system. 42A10-31E8

synchroscope. An instrument for indicating whether two periodic quantities are synchronous. It usually embodies a continuously rotatable element the position of which at any time is a measure of the instantaneous phase difference between the quantities; while its speed of rotation indicates the frequency difference between the quantities; and its direction of rotation indicates which of the quantities is of higher frequency. *Note:* This term is also used to designate a cathode-ray oscilloscope providing either (1) a rotating pattern giving indications similar to that of the conventional synchroscope, or (2) a triggered sweep, giving an indication of synchronism. *See also:* **instrument.** 42A30-0

synchro system (alternating current). An electric system for transmitting angular position or motion. It consists of one or more sychro transmitters, one or more synchro receivers or synchro control transformers and may include differential synchro machines.
See:
direct-current, self-synchronous system;
permanent-magnet, second-harmonic, self-synchronous system;
power selsyn;
synchro control transformer;
synchro differential receiver;
synchro differential transmitter;
synchro receiver;
synchro transmitter. 42A10-31E8

synchro transmitter (or generator) (rotating machinery). A transformer with relatively rotatable primary and secondary windings, the output of the secondary winding being two or more voltages that vary with and completely define the relative angular position of the primary and secondary windings. *See also:* **synchro system.** 42A10-31E8

synchrotron. A device for accelerating charged particles (for example, electrons) to high energies in a vacuum. The particles are guided by a changing magnetic field while they are accelerated many times in a closed path by a radio-frequency electric field. *See also:* **electron devices, miscellaneous.** 42A70-15E6

sync signal. *See:* **synchronizing signal.**

syntax. (1) The structure of expressions in a language. (2) The rules governing the structure of a language. X3A12-16E9

system (1) (general). An integrated whole even though composed of diverse, interacting, specialized structures and subjunctions. *Notes:* (1) Any system has a number of objectives and the weights placed on them may differ widely from system to system. (2) A system performs a function not possible with any of the individual parts. Complexity of the combination is implied.
See:
availability;
circuits and devices;
constraints;
environment;
functional unit;
mission;
model;
modeling;
optimization;
performance index;
programming, linear;
programming, nonlinear;
programming, quadratic;
redundancy;
reliability;
sensitivity;
simulation;
system science. 0-31E3/35E2
(2) An organized collection of men, machines, and methods required to accomplish a specific objective. *See:* **number system; numeral system; numeration system; operation system.** X3A12-16E9
(3) (control). A controller and a plant combined to perform specific functions. *See also:* **control system.** 0-23E0
(4) (continuous control system). A system that is not discrete and that does not contain sampling or quantization. 0-23E0
(5) (electric power). Designates a combination of lines, and associated apparatus connected therewith, all connected together without intervening transforming apparatus. *See also:* **power systems, low-frequency and surge testing.**
(6) *See:* **adjoint.**
(7) (controlling). *See:* **controlling system.** E32-0

systematic error (1) (general). The inherent bias (offset) of a measurement process or of one of its components. 0-9E3
(2) (electronic navigation). Error capable of identification due to its orderly character. *See also:* **navigation.** E172-10E6

system delay time (mobile communication). The time required for the transmitter associated with the system to provide rated radio-frequency output after activation of the local control (push to talk) plus the time required for the system receiver to provide useful output. *See also:* **mobile communication system.** 0-6E1

system demand factor. *See:* **demand factor.**

system deviation (control). *See:* **deviation, system.**

system, directly controlled. That portion of the controlled system that is directly guided or restrained by the final controlling element to achieve a prescribed value of the directly controlled variable. *See also:* **control system, feedback.** AS1-23E0/34E10

system, discrete (control system). A system whose signals are inherently discrete. *See also:* **control system.** 0-23E0

system, discrete-state (control system) (system, finite-state). A system whose state is defined only for discrete values of time and amplitude. *See also:* **control system.** 0-23E0

system diversity factor. *See:* **diversity factor.**

system element. One or more basic elements with other components and necessary parts to form all or a significant part of one of the general functional groups into which a measurement system can be classified. While a system element must be functionally distinct from other such elements it is not necessarily a separate measurement device. Typical examples of system elements are: a thermocouple, a measurement amplifier, a millivoltmeter. *See also:* **measurement system.** 42A30-0

system, finite-state (control system). *See:* **system, discrete state.**

system frequency (electric power system). Frequency in hertz of the power system alternating voltage. *See* *also:* **power system, low-frequency and surge testing.** E94-0

system frequency stability (radio system) (mobile communication). The measure of the ability of all stations, including all transmitters and receivers, to remain on an assigned frequency-channel as determined on both a short-term and long-term basis. *See also:* **mobile communication system.** 0-6E1

system ground (lightning arresters). The connection between a grounding system and a point of an electric circuit (for example, a neutral point). *See also:* **lightning arrester (surge diverter).** 50I25-31E7

system grounding conductor. An auxiliary solidly grounded conductor that connects together the individual grounding conductors in a given area. *Note:* This conductor is not normally a part of any current-carrying circuit including the system neutral. *See also:* **ground.** 42A35-31E13

system, idealized (automatic control). An imaginary system whose ultimately controlled variable has a stipulated relationship to specified commands. *Note:* It is a basis for performance standards. *See also:* **control system, feedback.** 85AI-23E0

system, indirectly controlled (industrial control). The portion of the controlled system in which the indirectly controlled variable is changed in response to changes in the directly controlled variable. *See also:* **control system, feedback.** AS1/85A1-23E0/34E10

system interconnection. The connecting together of two or more power systems. *See also:* **alternating-current distribution; direct-current distribution.** 42A35-31E13

system load (electric power utilization). The summation of load served by a given system. *See also:* **generating station.** 0-31E4

system loss (radio system). The transmission loss plus the losses in the transmitting and receiving antennas. *See also:* **radio wave propagation.** 0-3E2

system matrix (control system). A matrix of transfer functions that relate the Laplace transforms of the system outputs and of the system inputs. *See also:* **control system.** 0-23E0

system maximum hourly load. The maximum hourly integrated load served by a system. *Note:* An energy quantity usually expressed in kilowatthours per hour. *See also:* **generating station.** 0-31E4

system noise (sound recording and reproducing system). The noise output that arises within or is generated by the system or any of its components, including the medium. *See also:* **noise (sound recording and reproducing system).** E191-0

system of units. A set of interrelated units for expressing the magnitudes of a number of different quantities. E270-0

system overshoot (control) (industrial control). The largest value of system deviation following the first dynamic crossing of the ideal value in the direction of correction, after the application of a specified stimulus. *See also:* **control system, feedback.** AS1-34E10

system, quantized (control system). A system in which at least one quantizing operation is present. *See also:* **control system.** 0-23E0

system recovery time (mobile communication). The elapsed time from deactivation of the local transmitter control until the local receiver is capable of producing useful output. *See also:* **mobile communication system.** 0-6E1

system reserve. The capacity, in equipment and conductors, installed on the system in excess of that required to carry the peak load. *See also:* **generating station.** 42A35-31E13

system, sampled-data (control system). A system in which at least one sampled signal is present. *See also:* **control system.** 0-23E0

system science. The branch of organized knowledge dealing with systems and their properties, the systematized knowledge of systems. *See also:* **adaptive system; cybernetics; learning system; system; systems engineering; tradeoff.** 0-35E2

systems engineering. The application of the mathematical and physical sciences to develop systems that utilize economically the materials and forces of nature for the benefit of mankind. *See:* **system science.** 0-35E2

system utilization factor. *See:* **utilization factor.**

system (circuit) voltage (lightning arrester). The root-mean-square power-frequency voltage from line to line as distinguished from the voltage from line to neutral. *See also:* **lightning protection and equipment.** E28-0

T

table. A collection of data, each item being uniquely identified either by some label or by its relative position. *See:* **decision table; truth table.** X3A12-16E9

table lamp. A portable luminaire with a short stand suitable for standing on furniture. *See also:* **luminaire.** Z7A1-0

table look-up. A procedure for obtaining the function value corresponding to an argument from a table of function values. X3A12-16E9

tab sequential format (numerically controlled machines). A means of identifying a word by the number of tab characters preceding the word in the block. The first character in each word is a tab character. Words must be presented in a specific order but all characters in a word, except the tab character, may be omitted when the command represented by that word is not desired. *See also:* **numerically controlled machines.** EIA3B-34E12

tabulate. (1) To form data into a table. (2) To print totals. X3A12-16E9

tacan (tactical air navigation). A complete ultra-high-frequency polar coordinate (rho-theta) navigation system using pulse techniques; the distance (rho) function operates as distance-measuring equipment (DME) and the bearing function is derived by rotating the ground transponder antenna so as to obtain a rotating multilobe pattern for coarse and fine bearing information. *See also:* **navigation.** 0-10E6

tachometer. A device to measure speed or rotation. *See also:* **rotor (rotating machinery).** 0-31E8

tachometer electric indicator. A device that provides an indication of the speed of an aircraft engine, of a helicopter rotor, of a jet engine, and of similar rotating apparatus used in aircraft. Such tachometer indicators may be calibrated directly in revolutions per minute or in percent of some particular speed in revolutions per minute. *See also:* **air-transportation instruments.** 42A41-0

tachometer generator (rotating machinery). A generator, mechanically coupled to an engine, whose main function is to generate a voltage, the magnitude or frequency of which is used either to determine the speed of rotation of the common shaft or to supply a signal to a control circuit to provide speed regulation. *See also:* **air-transportation electric equipment; instrument.** 42A41-31E8

tachometric relay (industrial control). A relay in which actuation of the contacts is effected at a predetermined speed of a moving part. *See also:* **relay.** 50I16-34E10

taffrail log. A device that indicates distance traveled based on the rotation of a screw-type rotor towed behind a ship which drives, through the towing line, a counter mounted on the taffrail. *Note:* An electric contact made (usually) each tenth of a mile causes an audible signal to permit ready calculation of speed. 42A43-0

tag. (1) Same as flag. (2) Same as label. 0-16E9

tags. Men-at-work tags of distinctive appearance, indicating that the equipment or lines so marked are being worked on. 2A2-0

tailing (hangover) (1) (facsimile). The excessive prolongation of the decay of the signal. *See also:* **facsimile signal (picture signal).** E168-0
(2) (electrometallurgy) (hydrometallurgy and ore concentration). The discarded residue after treatment of an ore to remove desirable minerals. *See also:* **electrowinning.** 42A60-0

tail lamp. A lamp used to designate the rear of a vehicle by a warning light. *See also:* **headlamp.** Z7A1-0

tail-of-wave (chopped wave) impulse test voltage (insulation strength). The crest voltage of a standard impulse wave that is chopped by flashover at or after crest. *See also:* **power distribution, overhead construction.** 76A1-0

talker echo. Echo that reaches the ear of the talker. *See also:* **transmission characteristics.** 42A65-0

talking key. A key whose operation permits conversation over the circuit to which the key is connected. *See also:* **telephone switching system.** 42A65-0

talk-ringing key (listening and ringing key). A combined talking key and ringing key operated by one handle. *See also:* **telephone switching system.** 42A65-0

tandem (cascade) (network). Two terminal pair networks are in tandem when the output terminals of one network are directly connected to the input terminals of the other network. *See also:* **network analysis.** E153/E270-0

tandem central office (tandem office). A central office used primarily as a switching point for traffic between other central offices. *See also:* **telephone system.** 42A65-19E1

tandem-completing trunk. A trunk, extending from a tandem office to a central office, used as part of a telephone connection between stations. *See also:* **telephone system.** 42A65-0

tandem control (electric power systems). A means of control whereby the area control error of an area or areas *A,* connected to the interconnected system *B* only through the facilities of another area *C,* is included in control of area *C* generation. *See also:* **power system, low-frequency and surge testing.** E94-0

tandem drive (industrial control). Two or more drives that are mechanically coupled together. *See also:* **electric drive.** AS1-34E10

tandem office. *See:* **tandem central office.**

tandem trunk. A trunk extending from a central office or a tandem office to a tandem office and used as part of a telephone connection between stations. *See also:* **telephone system.** 42A65-19E1

tangential wave path (radio wave propagation over the earth). A path of propagation of a direct wave that is tangential to the surface of the earth. The tangential wave path is curved by atmospheric refraction. *See also:* **radiation.** 42A65-0

tank (storage cell). A lead container, supported by wood, for the element and electrolyte of a storage cell. *Note:* This is restricted to some relatively large types of lead-acid cells. *See also:* **battery (primary or secondary).** 42A60-0

tank circuit (signal-transmission system). A circuit consisting of inductance and capacitance, capable of storing electric energy over a band of frequencies continuously distributed about a single frequency at which the circuit is said to be resonant, or tuned. *Note:* The selectivity of the circuit is proportional to the ratio of the energy stored in the circuit to the energy dissipated. The ratio is often called the *Q* of the circuit. *See also:* **circuits and devices; dielectric heating; oscillatory circuit; *Q* quality factor; signal.** E145-13E6

tank, single-anode. *See:* **single-anode tank.**

tank voltage. The total potential drop between the anode and cathode bus bars during electrodeposition. *See also:* **electroplating.** 42A60-0

tap (1) (general). An available connection that permits changing the active portion of the device in the circuit. *See also:* **grounding device.** E32-0
(2) (transformer). A connection brought out of a winding at some point between its extremities to permit changing the voltage, or current, ratio. *See:* **windings, high-voltage and low-voltage.** 42A15/57A15-31E12
(3) (reactor). A connection brought out of a winding at some point between its extremities, to permit changing the impedance. *See also:* **reactor.** 57A16-0
(4) (rotating machinery). A connection made at some intermediate point in a winding. *See also:* **rotor (rotating machinery); stator; voltage regulator.** 0-31E8

tape (1) (rotating machinery). A relatively narrow, long, thin, flexible fabric, mat, or film, or a combination of them with or without binder, not over 20 centimeters in width. *See also:* **rotor (rotating machinery); stator.**
(2) (electronic computation). *See:* **magnetic tape.** 0-31E8

taped insulation. Insulation of helically wound tapes applied over a conductor or over an assembled group of insulated conductors. (1) When successive convolutions of a tape overlie each other for a fraction of the tape width, the taped insulation is lap wound. This is also called positive lap wound. (2) When a tape is applied so that there is an open space between successive convolutions, this construction is known as open butt or negative lap wound. (3) When a tape is applied so that the space between successive convolutions is too small to measure with the unaided eye, it is a closed butt taping. *See also:* **power distribution, underground construction.** 42A35-31E13

tape drive. A device that moves tape past a head. *See:* **tape transport.** X3A12-16E9

tape preparation. The act of translating command information into punched or magnetic tape. EIA3B-34E12

taper (communication practice). A continuous or gradual change in electrical properties with length, as obtained, for example, by a continuous change of cross-section of a waveguide. *See also:* **transmission line.** E270/42A65-0

tape recorder. *See:* **magnetic recorder.**

tapered key (rotating machinery). A wedge-shaped key to be driven into place, in a matching hole or recess. *See also:* **cradle base (rotating machinery).** 0-31E8

tapered transmission line. *See:* **tapered waveguide.**

tapered waveguide. A waveguide in which a physical or electrical characteristic increases continuously with distance along the axis of the guide. *See also:* **waveguide.** E146/E147-3E1

taper, waveguide. A section of tapered waveguide. *See also:* **waveguide.** E147-0

tape station. *See:* **tape unit.** *See also:* **electronic digital computer.**

tape thickness. The lesser of the cross-sectional dimensions of a length of ferromagnetic tape. *See:* **tape-wound core.** *See also:* **static magnetic storage.** E163-0

tape to card. Pertaining to equipment or methods that transmit data from either magnetic tape or punched tape to punched cards. *See also:* **electronic digital computer.** X3A12-16E9

tape transmitter (telegraphy). A machine for keying telegraph code signals previously recorded on tape. *See also:* **telegraphy.** 42A65-19E4

tape transport. *See:* **tape drive.** *See also:* **electronic digital computer.**

tape unit. A device containing a tape drive, together with reading and writing heads and associated controls. *See also:* **electronic digital computer.** X3A12-16E9

tape width. The greater of the cross-sectional dimensions of a length of ferromagnetic tape. *See:* **tape-wound core.** E163-0

tape-wound core. A length of ferromagnetic tape coiled about an axis in such a way that one convolution falls directly upon the preceding convolution. *See:* **wrap thickness.** *See also:* **static magnetic storage.** E163-0

tapped field. *See:* **short field.**

tapped field control. A system of regulating the tractive force of an electrically driven vehicle by changing the number of effective turns of the traction motor series-field windings by means of an intermediate tap or taps in those windings. *See also:* **multiple-unit control.** 42A42-0

tapper bell. A single-stroke bell having a gong designed to produce a sound of low intensity and relatively high pitch. *See also:* **protective signaling.** 42A65-0

target (1) (radar). (A) Specifically, an object of radar search or surveillance. (B) Broadly, any discrete object that reflects energy back to the radar equipment. *See also:* **navigation; radar.** 42A65/E172-10E6
(2) (camera tube). A structure employing a storage surface that is scanned by an electron beam to generate a signal output current corresponding to a charge-density pattern stored thereon. *Note:* The structure may include the storage surface that is scanned by an electron beam, the backplate, and the intervening dielectric. *See also:* **beam tube; radar; television.** E160-15E6/2E2
(3) (relay) (operation indicator). A supplementary device, operated either mechanically or electrically, to visibly indicate that the relay has operated or completed its function. *Notes:* (1) A mechanically operated target indicates the physical operation of the relay. (2) An electrically operated target, when not further described, is actuated by the current in the control circuit associated with the relay and hence indicates that the relay has not only operated, but also that it has completed its function by causing current to flow in the associated control circuit. (3) A shunt-energized target only indicates operation of the relay contacts and does not necessarily show that current has actually flowed in the associated control circuit. 37A100-31E11/31E6
(4) (storage tubes). The storage surface and its immediate supporting electrodes. *See also:* **radar; storage tube.** E158-15E6

target (anode) (anticathode*) (X-ray tube). An electrode, or part of an electrode, on which a beam of electrons is focused and from which X-rays are emitted. *See also:* **electrode (electron tube); radar.**
*Deprecated 42A70-15E6

target capacitance (camera tubes). The capacitance between the scanned area of the target and the backplate. *See also:* **beam tubes; television.** E160-2E2/15E6

target cutoff voltage (camera tubes). The lowest target voltage at which any detectable electric signal corresponding to a light image on the sensitive surface of the tube can be obtained. *See also:* **beam tubes; television.** E160-15E6

target glint (electronic navigation). *See:* **scintillation.**

target language. A language that is an output from a given translation process. *See:* **object language.** *See also:* **electronic digital computer.** X3A12-16E9

target program. A program written in a target language. *See:* **object program.** *See also:* **electronic digital computer.** X3A12/16E9

target transmitter (electronic navigation). A source of radio-frequency energy suitable for providing test signals at a test site. *See also:* **navigation.** E173-10E6

target voltage (camera tube with low-velocity scanning). The potential difference between the thermionic cathode and the backplate. *See also:* **beam tube; television.** E160-2E215E6

tarnish (corrosion). Surface discoloration of a metal caused by formation of a thin film of corrosion product. *See also:* **corrosion terms.** CM-34E2

taut-band suspension (electric instrument). A mechanical arrangement whereby the moving element of an instrument is suspended by means of ligaments, usually in the form of a thin flat conducting ribbon, at each of its ends. The ligaments normally are in tension sufficient to restrict the lateral motion of the moving element to within limits that permit freedom of useful motion when the instrument is mounted in any position. A restoring torque is produced within the ligaments with rotation of the moving element. *See also:* **moving element (instrument).** 39A1-0

taxi light. An aircraft aeronautical light designed to provide necessary illumination for taxiing. *See also:* **signal lighting.** Z7A1-0

taxi-channel light. Aeronautical ground lights arranged along a taxi-channel of a water aerodrome to indicate the route to be followed by taxiing aircraft. *See also:* **signal lighting.** Z7A1-0

taxiway centerline lights. Taxiway lights placed along the centerline of a taxiway except that on curves or corners having fillets, these lights are placed a distance equal to half the normal width of the taxiway from the outside edge of the curve or corner. *See also:* **signal lighting.** Z7A1-0

taxiway-edge lights. Taxiway lights placed along or near the edges of a taxiway. *See also:* **signal lighting.** Z7A1-0

taxiway holding-post light. A light or group of lights installed at the edge of a taxiway near an entrance to a runway, or to another taxiway, to indicate the position at which the aircraft should stop and obtain clearance to proceed. *See also:* **signal lighting.** Z7A1-0

taxiway lights. Aeronautical ground lights provided to indicate the route to be followed by taxiing aircraft. *See also:* **signal lighting.** Z7A1-0

TB cell. *See:* **transmitter-blocker cell.**

TE. *See:* **transverse electric.**

tearing (television). An erratic lateral displacement of some scanning lines of a raster caused by disturbance of synchronization. *See also:* **television.** 0-2E2

teaser transformer (polyphase power) two single-phase Scott-connected units for three-phase to two-phase, or two-phase to three-phase operation. The transformer that is connected between the midpoint of the main transformer and the third phase wire of the three-phase system. *See also:* **autotransformer.** 42A15-31E12

tee junction (waveguide techniques). *See:* **T junction.**

teed feeder. A feeder that supplies two or more feeding points. *See also:* **center of distribution.** 42A35-31E13

telautograph. A system in which writing movement at the transmitting end causes corresponding movement of a writing instrument at the receiving end. *See also:* **communication; telegraphy.** 42A65-0

telecommunication (electrical practice). The transmission of information from one point to another. 0-19E4

telegraph (marine transportation). A mechanized or electric device for the transmission of stereotyped orders or information from one fixed point to another. *Note:* The usual form of telegraph is a transmitter and a receiver, each having a circular dial in sectors upon which are printed standard orders. When the index of the transmitter is placed at any order, the pointer of the receiver designates that order. Dual mechanism is generally provided to permit repeat back or acknowledgment of orders. 42A43-0

telegraph channel. A channel suitable for the transmission of telegraph signals. *Note:* Three basically different kinds of telegraph channels used in multichannel telegraph transmissions are (1) one of a number of paths for simultaneous transmission in the same frequency range as in bridge duplex, differential duplex and quadruplex telegraphy; (2) one of a number of paths for simultaneous transmission in different frequency ranges as in carrier telegraphy; (3) one of a number of paths for successive transmission as in multiplex printing telegraphy. Combinations of these three types may be used on the same circuit. *See also:* **channel.** 42A65-0

telegraph concentrator. A switching arrangement by means of which a number of branch or subscriber lines or station sets may be connected to a lesser number of trunk lines or operating positions or instruments through the medium of manual or automatic switching devices in order to obtain more efficient use of facilities. *See also:* **telegraphy.** 42A65-19E4

telegraph distortion (telecommunication). The condition in which the significant intervals have not all exactly their theoretical durations. *See also:* **telegraphy.** 0-19E4

telegraph distributor. A device that effectively associates one direct-current or carrier telegraph channel in rapid succession with the elements of one or more signal sending or receiving devices. *See also:* **telegraphy.** 42A65-0

telegraph key. A hand-operated telegraph transmitter used primarily in Morse telegraphy. *See also:* **telegraphy.** 42A65-0

telegraph repeater. An arrangement of apparatus and circuits for receiving telegraph signals from one line and retransmitting corresponding signals into another line. *See also:* **telegraphy.** 42A65-0

telegraph selector. A device that performs a switching operation in response to a definite signal or group of successive signals received over a controlling circuit. *See also:* **telegraphy.** 42A65-0

telegraph sender. A transmitting device for forming telegraph signals. Examples are a manually operated telegraph key and a printer keyboard. *See also:* **telegraphy.** 42A65-0

telegraph signal (telecommunication). The set of conventional elements established by the code to enable the transmission of a written character (letter, figure, punctuation sign, arithmetic sign, etcetera) or the control of a particular function (spacing, shift, line-feed, carriage return, phase correction, etcetera); this set of elements being characterized by the variety, the duration and the relative position of the component elements or by some of these features. 0-19E4

telegraph signal distortion. Time displacement of transitions between conditions, such as marking and spacing, with respect to their proper relative positions in perfectly timed signals. *Note:* The total distortion is the algebraic sum of the bias and the characteristic and fortuitous distortions. *See also:* **telegraphy.** 42A65-0

telegraph sounder. A telegraph receiving instrument by means of which Morse signals are interpreted aurally (or read) by noting the intervals of time between two diverse sounds. *See also:* **telegraphy.** 42A65-0

telegraph speed (telecommunication). *See:* **modulation rate.**

telegraph transmission speed. The rate at which signals are transmitted, and may be measured by the equivalent number of dot cycles per second or by the average number of letters or words transmitted, and received per minute. *Note:* A given speed in dot cycles per second (often abbreviated to dots per second) may be converted to **bauds** by multiplying by 2. The baud is the unit of signaling transmission speed recommended by the International Consultative Committee on Telegraph Communication. Where words per minute are used as a measure of transmission speed, five letters and a space per word are assumed. *See also:* **telegraphy.** 42A65-0

telegraph transmitter. A device for controlling a source of electric power so as to form telegraph signals. *See also:* **telegraphy.** 42A65-0

telegraph word (conventional). A word comprising five letters together with one letter-space, used in computing telegraph speed in words per minute or traffic capacity. *See also:* **telegraphy.** 0-19E4

telegraphy. A system of telecommunication for the transmission of graphic symbols, usually letters or numerals, by the use of a signal code. It is used primarily for record communication. The term may be extended to include any system of telecommunication for the transmission of graphic symbols or images for reception in record form, usually without gradation of shade values. *Note:* For an extensive list of cross references, see *Appendix A.* 42A65-0

telemeter service. Metered telegraph transmission between paired telegraph instruments over an intervening circuit adapted to serve a number of such pairs on a shared-time basis. *See:* **electric metering.** *See also:* **telegraphy.** 42A65-0

telemetering (remote metering*). Measurement with the aid of intermediate means that permit the measurement to be interpreted at a distance from the primary detector. *Note:* The distinctive feature of telemetering is the nature of the translating means, which includes provision for converting the measurand into a representative quantity of another kind that can be transmitted conveniently for measurement at a distance. The actual distance is irrelevant.
See:
current-type telemeter;
electric telemeter;
electric telemetering;
frequency-type telemeter;
measurement system;
mobile telemetering;
phase-modulation telemetering;
pulse-duration telemetering;
pulse-rate telemetering;
ratio-type telemeter;
variable-frequency telemetering.
*Deprecated 37A100/42A30-31E11

telephone air-to-air input-output characteristic. The acoustical output level of a telephone set as a function of the acoustical input level of another telephone set to which it is connected. The output is measured in an artificial ear, and the input is measured free-field at a specified location relative to the reference point of an artificial mouth. *See also:* **telephone station.** E269-19E8

telephone booth. A booth, closet, or stall for housing a telephone station. *See also:* **telephone station.** 42A65-0

telephone central office. A telephone switching unit, installed in a telephone system providing service to the general public, having the necessary equipment and operating arrangements for terminating and interconnecting lines and trunks. *Note:* There may be more than one central office in the same building. *See also:* **telephone system.** 42A65-19E1

telephone channel. A channel suitable for the transmission of telephone signals. *See also:* **channel.** 42A65-0

telephone connection. A two-way telephone channel completed between two points by means of suitable switching apparatus and arranged for the transmission of telephone currents, together with the associated arrangements for its functioning with the other parts of a telephone system in switching and signaling operations. *Note:* The term is also sometimes used to mean a two-way telephone channel permanently established between two telephone stations. *See also:* **telephone system.** 42A65-0

telephone electrical impedance. The complex ratio of the voltage to the current at the line terminals at any given single frequency. *See also:* **telephone station.** E269-19E8

telephone equalization. A property of a telephone circuit that ideally causes both transmit and receive responses to be inverse functions of current, thus tending to equalize variations in loop loss. *See also:* **telephone system.** E269-19E8

telephone exchange. A unit of a telephone communication system for the provision of communication service in a specified area that usually embraces a city, town, or village, and its environs. *See also:* **telephone system.** 42A65-19E1

telephone feed circuit. An arrangement for supplying direct-current power to a telephone set and an alternating-current path between the telephone set and a terminating circuit. *See also:* **telephone system.** E269-19E8

telephone frequency characteristics. Electrical and acoustical properties as functions of frequency. *See also:* **telephone system.** E269-19E8

telephone handset. A telephone transmitter and receiver combined in a unit with a handle. *See also:* **telephone station.** E269-19E8

telephone influence factor (TIF). Of a voltage or current wave in an electric supply circuit, the ratio of the square root of the sum of the squares of the weighted root-mean-square values of all the sine-wave components (including in alternating-current waves both fundamental and harmonics) to the root-mean-square value (unweighted) of the entire wave. *Note:* This factor was formerly known as **telephone interference factor,** which term is still used occasionally when referring to values based on the original (1919) weighting curve. *See also:* **inductive coordination.** 42A65-0

telephone line. (1) The conductors and circuit apparatus associated with a particular communication channel. 0-19E4
(2) A multiplicity of conductors between telephone stations and a central office or between central offices. *See also:* **telephone system.** 0-19E1

telephone modal distance. The distance between the center of the grid of a telephone handset transmitter and the center of the lips of a human talker (or the reference point of an artificial mouth) when the handset is in the modal position. *See also:* **telephone station.** E269-19E8

telephone modal position. The position a telephone handset assumes when the receiver of the handset is held in close contact with the ear of a person with head dimensions that are modal for a population. *See also:* **telephone station.** E269-19E8

telephone operator. A person who handles switching and signaling operations needed to establish telephone connections between stations or who performs various auxiliary functions associated therewith. *See also:* **telephone system.** 42A65-19E1

telephone receive input-output characteristic. The acoustical output level of a telephone set as a function of the electric input level. The output is measured in an artificial ear, and the input is measured across a specified termination connected to the telephone feed circuit. *See also:* **telephone station.** E269-19E8

telephone receiver. *See:* **receiver, telephone.**

telephone repeater. A repeater for use in a telephone circuit. *See also:* **telephone switching system; repeater.** 42A65-0

telephone ringer (station ringer)(ringer). An electric bell designed to operate on low-frequency alternating or pulsating current and associated with a telephone station for indicating a telephone call to the station. *See also:* **telephone station.** 42A65-0

telephone set (general)(telephone). An assemblage of apparatus including a telephone transmitter, a telephone receiver, and usually a switch, and the immediately associated wiring and signaling arrangements. *See also:* **telephone station.** 42A65-0

telephone set, common battery. A telephone set for which both transmitter current and the current for signaling by the telephone station are supplied from a centralized direct-current power source. 42A65-31E3

telephone set, local battery. A telephone set for which the the transmitter current is supplied from a battery or other direct-current supply circuit, individual to the telephone set. Current for signaling by the telephone station may be supplied from a local hand-operated generator or from a centralized direct-current power source. *See also:* **telephone station.** 42A65-31E3

telephone sidetone. The ratio of the acoustical output of the receiver of a given telephone set to the acoustical input of the transmitter of the same telephone set. *See:* **telephone air-to-air input-output characteristic.** *See also:* **telephone station.** E269-19E8

telephone speech network. An electric circuit that connects the transmitter and the receiver to a telephone line or telephone test loop and to each other. *See also:* **telephone station.** E269-19E8

telephone station. An installed telephone set and associated wiring and apparatus, in service for telephone communication. *Note:* As generally applied, this term does not include the telephone sets employed by central-office operators and by certain other personnel in the operation and maintenance of a telephone system. *Note:* For an extensive list of cross references, see *Appendix A.* 42A65-0

telephone subscriber. A customer of a telephone system who is served by the system under a specific agreement or contract. *See also:* **telephone system.** 42A65-19E1

telephone switchboard. A switchboard for interconnecting telephone lines and associated circuits. *See also:* **telephone switching system.** 42A65-19E1

telephone switching system. *Note:* For an extensive list of cross references, see *Appendix A.*

telephone system. An assemblage of telephone stations, lines, channels, and switching arrangements for their interconnection, together with all the accessories for providing telephone communication. *Note:* For an extensive list of cross references, see *Appendix A.* 42A65-19E1

telephone test connection. Two telephone sets connected together by means of telephone test loops and a telephone feed circuit. *See also:* **telephone system.** E269-19E8

telephone test loop. A circuit that is interposed between a telephone set and a telephone feed circuit to simulate a real telephone line. *See also:* **telephone system.** E269-19E8

telephone transmit input-output characteristic. The electric output level of a telephone set as a function of the acoustical input level. The output is measured across a specified impedance connected to the telephone feed circuit, and the input is measured free-field at a specified location relative to the reference point of an artificial mouth. *See also:* **telephone station.** E269-19E8

telephone transmitter. A microphone for use in a telephone system. *See also:* **telephone station.** 42A65-1E1

telephone-type relay. A type of electromechanical relay in which the significant structural feature is a hinged armature mechanically separate from the contact assembly. This assembly usually consists of a multiplicity of stacked leaf-spring contacts. 37A100-31E11/31E6

telephony. *See:* **telephone switching system; sleeve conductor; sleeve wire.**

telephotography. *See:* **picture transmission.**

teleprinter (teletypewriter). *See:* **printer.**

teletypewriter (teleprinter). *See:* **printer.**

television (TV). The electric transmission and reception of transient visual images. *Note:* For an extensive list of cross references, See *Appendix A.* 42A65-0

television broadcast station. A radio station for transmitting visual signals, and usually simultaneous aural signals, for general reception. *See also:* **television.** 42A65-0

television camera. A pickup unit used in a television system to convert into electric signals the optical image formed by a lens. *See also:* **television.** 42A65-0

television channel. A channel suitable for the transmission of television signals. The channel for associated sound signals may or may not be considered a part of the television channel. *See also:* **channel.** 42A65-0

television lighting.
See:
back light;
base light;
cross light;
eye light;
fill light;
high-key lighting;
key light;
low-key lighting;
set light;
side back light. Z7A1-0

television line number. The ratio of the raster height to the half period of a periodic test pattern. *Example:* In a test pattern composed of alternate equal-width black and white bars, the television line number is the ratio of the raster height to the width of each bar. *Note:* Both quantities are measured at the camera-tube sensitive surface. *See also:* **television.** E158-15E6

television picture tube. *See:* **picture tube.**

television receiver. A radio receiver for converting incoming electric signals into television pictures and customarily associated sound. *See also:* **television.** 42A65-0

television repeater. A repeater for use in a television circuit. *See also:* **repeater; television.** 42A65-0

television transmitter. The aggregate of such radio-frequency and modulating equipment as is necessary to supply to an antenna system modulated radio-frequency power by means of which all the component parts of a complete television signal (including audio, video, and synchronizing signals) are concurrently transmitted. *See also:* **television.** E145/42A65-0

telltale. *See:* **running light indicator panel.**

TEM. *See:* **transverse electromagnetic.**

temperature coefficient (1) (rotating machinery). The variation of the quantity considered, divided by the difference in temperature producing it. Temperature coefficient may be defined as an average over a temperature range or an incremental value applying to a specified temperature. *See:* **asynchronous machine; direct-current commutating machine; synchronous machine.** 0-31E8

(2) (power supplies). The percent change in the output voltage or current as a result of a 1 degree-Celsius change in the ambient operating temperature (percent per degree Celsius). *See also:* **power supply.** KPSH-10E1

temperature coefficient of capacity (storage cell or battery). The change in delivered capacity (ampere-hour or watthour capacity) per degree Celsius relative to the capacity of the cell or battery at a specified temperature. *See also:* **initial test temperature.** 42A60-0

temperature coefficient of electromotive force (storage cell or battery). The change in open-circuit voltage per degree Celsius relative to the electromotive force of the cell or battery at a specified temperature. *See also:* **initial test temperature.** 42A60-0

temperature coefficient of resistance (rotating machinery). The temperature coefficient relating a change in electric resistance to the difference in temperature producing it. *See also:* **asynchronous machine; direct-current commutating machine; synchronous machine.** 0-31E8

temperature coefficient of voltage drop (glow-discharge tubes). The quotient of the change of tube voltage drop (excluding any voltage jumps) by the change of ambient (or envelope) temperature. *Note:* It must be indicated whether the quotient is taken with respect to ambient or envelope temperature. *See also:* **gas tube.** E160-15E6

temperature-compensated overload relay. A device that functions at any current in excess of a predetermined value essentially independent of the ambient temperature. *See also:* **relay.** E45-0

temperature control system (gas turbines). The devices and elements, including the necessary temperature detectors, relays, or other signal-amplifying devices and control elements, required to actuate directly or indirectly the fuel-control valve, speed of the air compressor, or stator blades of the compressor so as to limit or control the rate of fuel input or air flow inlet to the gas turbine. By this means the temperature in the combustion system or the temperatures in the turbine stages or turbine exhaust may be limited or controlled. E282-31E2

temperature derating (rectifier circuit). The reduction in reverse-voltage or forward-current rating, or both, assigned by the manufacturer under stated conditions of higher ambient temperatures. *See also:* **average forward-current rating (rectifier circuit); semiconductor rectifier stack.** E59-34E17/34E24

temperature detectors (gas turbines and rotating electric machinery). The primary temperature-sensing elements that are directly responsive to temperature. *See also:* **asynchronous machine; direct-current commutating machines; electric thermometer; synchronous machine.** E282-31E2;0-31E8

temperature meter. *See:* **electric thermometer.**

temperature, operating (power supply). *See:* **operating temperature.**

temperature radiator. A radiator whose radiant flux density (radiant exitance) is determined by its temperature and the material and character of its surface, and is independent of its previous history. Z7A1-0

temperature-regulating equipment (rectifier). Any equipment used for heating and cooling a rectifier, together with the devices for controlling and indicating its temperature. *See also:* **rectification.** 42A15-34E24

temperature relays (gas turbines). Devices by means of which the output signals of the temperature detectors are enabled to control directly or indirectly the rate of fuel energy input, the air flow input, or both, to the combustion system. *Note:* Operation of a temperature relay is caused by a specified external temperature; whereas operation of a thermal relay is caused by the heating of a part of the relay. *See:* **thermal relay.** E282-31E2;37A100-31E11-31E6;E282/37A100-31E2/31E6/31E11

temperature-rise rate, locked rotor, winding. *See:* **locked-rotor temperature-rise rate, winding.**

temperature-rise test (rotating machinery). A test undertaken to determine the temperature rise above ambient of one or more parts of a machine under specified operating conditions. *Note:* The specified conditions may refer to current, load, etcetera. *See also:* **asynchronous machine; direct-current commutating machine; synchronous machine.** 37A100-31E11/31E8

temperature stability (electrical conversion). Static regulation caused by a shift or change in output that was caused by temperature variation. This effect may be produced by a change in the ambient or by self-heating. *See also:* **electrical conversion.** 0-10E1

temporary emergency circuits (marine transportation). Circuits arranged for instantaneous automatic transfer to a storage-battery supply upon failure of a ship's service supply. *See also:* **emergency electric system.** 42A43-0

temporary emergency lighting (marine transportation). The lighting of exits and passages to permit passengers and crew, upon failure of a ship's service lighting, readily to find their way to the lifeboat embarkation deck. *See also:* **emergency electric system.** 42A43-0

temporary ground. A connection between a grounding system and parts of an installation that are normally alive, applied temporarily so that work may be safely carried out in them. *See also:* **lightning arrester (surge diverter).** 50I25-31E7

temporary storage (programming). Storage locations reserved for intermediate results. *See:* **working storage.** *See also:* **electronic digital computer.** X3A12-16E9

tens complement. The radix complement of a numeral whose radix is ten. *Note:* Using synonyms, the definition may also be expressed as "a true colement with a base of ten." *See also:* **base; electronic digital computer; radix.** E162/X3A12-16E9

tension. *See:* **final unloaded conductor tension; initial conductor tension.** *See also:* **conductor.**

tenth-power width (in a plane containing the direction of the maximum of a lobe). The full angle between the two directions in that plane about the maximum in which the radiation intensity is one-tenth the maximum value of the lobe. *See also:* **antenna.** E145-3E1

terminal (1) (general). (A) A conducting element of an equipment or a circuit intended for connection to an external conductor. (B) A device attached to a conductor to facilitate connection with another conductor. E270-0

(2) (network). A point at which any element may be directly connected to one or more other elements. *See also:* **network analysis.** E153-0

(3) (semiconductor device) (industrial control). The externally available point of connection to one or more electrodes. *See also:* **anode; semiconductor; semiconductor rectifier cell.**
E59/E216/E223/E270-34E10/34E17/34E24/15E7

(4) (communication channels) (A) (general). A point in a system or communication network at which data can either enter or leave. *See also:* **electronic digital computer.** X3A12-16E9

(B) (telegraph circuits). A general term referring to the equipment at the end of a telegraph circuit, modems, input-output and associated equipment. *See also:* **telegraph.** 0-19E4

(5) (rotating machinery). A conducting element of a winding intended for connection to an external electrical conductor. *See also:* **stator.** 0-31E8

terminal block. *See:* **terminal board.**

terminal board (rotating machinery). A plate of insulating material that is used to support terminations of winding leads. *Notes:* (1) The terminations, which may be mounted studs or blade connectors, are used for making connections to the supply line, the lead, other external circuits, or among the windings of the machine. (2) Small terminal boards may also be termed **terminal blocks,** or **terminal strips.** 0-31E8;IC1-34E10

terminal-board cover (rotating machinery). A closure that permits access to the terminal board and prevents accidental contact with the terminals. 42A10-31E8

terminal box (conduit box) (rotating machinery). A form of termination in which the ends of the machine winding are connected to the incoming supply leads inside a box that virtually encloses the connections, and is of minimum size consistent with adequate access and with clearance and creepage-distance requirements. The box is provided with a removable cover plate for access. *See also:* **stator.** 0-31E8

terminal connector (industrial control). A connector for attaching a conductor to a lead, terminal block, or stud of electric apparatus. *See:* **terminal.** IC1-34E10

terminal interference voltage (electromagnetic compatibility). *See* **terminal voltage.**

terminal of entry (for a conductor entering a delimited region). That cross section of the conductor that coincides with the boundary surface of the region and that is perpendicular to the direction of the electric field intensity at its every point within the conductor. In a conventional circuit, in which the conductors have a cross section that is uniform and small by comparison of the largest dimension with the length, the terminal of entry is a cross section perpendicular to the axis of the conductor. If the cross section of the conductor is infinitesimal, the terminal of entry becomes the point at which the conductor cuts the surface. *Notes:* (1) It follows from this definition and **delimited region** that the algebraic sum of the currents directed into a delimited region through all the terminals of entry is zero at every instant. (2) The term **terminal of entry** has been introduced because of the need in precise definitions of indicating definitely the terminations of the paths along which voltages are determined. The terms **phase conductor** and **neutral conductor** refer to a portion of a conductor rather than to a particular cross section although they may be considered by a practical engineer as representing a portion along which the integral of the electric intensity is negligibly small. Hence he may treat these terms as synonymous with **terminal of entry** in particular cases. *See also:* **network analysis.** E270-0

terminal pad. A usually flat conducting part of a device to which a terminal connector is fastened. 37A100-31E11

terminal pair (network). An associated pair of accessible terminals, such as input pair, output pair, and the like. *See also:* **network analysis.** E153/E270-0

terminal repeater. A repeater for use at the end of a trunk or line. *See also:* **repeater.** 42A65-0

terminal room. A room, associated with a central office, private branch exchange, or private exchange, that contains distributing frames, relays, and similar apparatus except switchboard equipment. *See also:* **telephone system.** 42A65-18E1

terminals (1) (lightning arrester). The conducting parts provided for connecting the arrester across the insulation to be protected. *See also:* **arresters.** E28-0

(2) (storage battery) (storage cell). The parts to which the external circuit is connected. *See also:* **battery (primary or secondary).** 42A60-0

terminal screw. *See:* **binding screw.**

terminal, stator winding (rotating machinery). The end of a lead cable or a stud or blade of a terminal board to which connections are normally made during installation. *See also:* **stator.** 0-31E8

terminal strip. *See:* **terminal board.**

terminal voltage (terminal interference voltage) (electromagnetic compatibility). Interference voltage measured between two terminals of an artificial mains network. *See also:* **electromagnetic compatibility.** CISPR-27E1

terminating (line or transducer). The closing of the circuit at either end by the connection of some device thereto. Terminating does not imply any special condition, such as the elimination of reflection. *See also:* **transmission characteristics.** 42A65-0

termination (1) (general). A one-port load that terminates a section of a transmission system in a specified manner. *See also:* **transmission line.** 0-9E4

(2) (rotating machinery). The arrangement for making the connections between the machine terminals and the external conductors. *See also:* **stator.** 0-31E8

termination, conjugate. A termination whose input impedance is the complex conjugate of the output impedance of the source or network to which it is connected. *See also:* **transmission line.** 0-9E4

termination, matched. A termination matched with regard to the impedance in a prescribed way; for example, (A) a reflectionless termination, or (B) a conjugate termination. *See:* **termination, reflectionless.** *See also:* **transmission line.** 0-9E4

termination, reflectionless. A termination that terminates a waveguide or transmission line without causing a reflected wave at any transverse section. *See also:* **transmission line; waveguide.** E146-3E1/9E4

ternary. (1) Pertaining to a characteristic or property involving a selection, choice, or condition in which there are three possibilities. (2) Pertaining to the numeration system with a radix of three. *See also:* **base; electronic digital computer; positional notation; radix.** E162/X3A12-16E9

ternary code (information theory). A code employing three distinguishable types of code elements. *See also:* **information theory; positional notation.** E171/42A65-0

terrain-clearance indicator. An absolute altimeter using the measurement of height above the sea or ground to alert the pilot of danger. *See also:* **radio navigation.** 42A65-10E6

terrain echoes. *See:* **ground clutter.**

terrain error (in navigation). The error resulting from the use of a wave that has become distorted by the terrain over which it has propagated. 0-10E6

terrestrial-reference flight. The type of stabilized flight that obtains control information from terrestrial phenomena, such as earth's magnetic field, atmospheric pressure, etcetera. E172-10E6

tesla. The unit of magnetic induction in the International System of Units (SI). The tesla is a unit of magnetic induction equal to 1 weber per square meter. E270-0

Tesla current* (electrotherapy) (coagulating current*). A spark discharge having a drop of 5 to 10 kilovolts in air, from monopolar or bipolar electrodes, generated by a special arrangement of transformers, spark gaps, and capacitors, delivered to a tissue surface, and dense enough to precipitate and oxidize (char) tissue proteins. *Note:* The term **Tesla current** is appropriate if the emphasis is on the method of generation; a **coagulating current,** if the emphasis is on the physiological effects. *See also:* **electrotherapy.**

*Deprecated 42A80-18E1

test (1) (item of equipment). To determine its performance characteristics while functioning under controlled conditions. E270-0

(2) (electronic digital computation). (A) To ascertain the state or condition of an element, device, program, etcetera. (B) Sometimes used as a general term to include both check and diagnostic procedures. (C) Loosely, same as check. *See also:* **check; electronic digital computer; check problem.** E162-0

(3) (instrument or meter). To ascertain its performance characteristics while functioning under controlled conditions.

See:

acceptance test;
adjust;
calibrate;
calibration voltage;
check;
rated performance;
reference performance;
safe working voltage to ground. 42A30-0

test block. *See:* **test switch.**

test board. A switchboard equipped with testing apparatus so arranged that connections can be made from it to telephone lines or central-office equipment for testing purposes. *See also:* **telephone switching system.** 42A65-19E1

test cabinet (switchgear assembly). An assembly of a cabinet containing permanent electric connections, with cable connections to a contact box arranged to make connection to the secondary contacts on an electrically operated removable element, permitting operation and testing of the removable element when removed from the housing. It includes the necessary control switch and closing relay, if required. 37A100-31E11

test cap. A protective structure that is placed over the exposed end of the cable to seal the sheath or other covering completely against the entrance of dirt, moisture, air, or other foreign substances. *Note:* Test caps are often provided with facilities for vacuum treatment, oil filling, or other special field operations. *See:* **live cable test cape.** *See also:* **power distribution, underground construction.** 42A35-31E13

test current (watthour meter). The current marked on the nameplate by the manufacturer (identified as *TA* on meters manufactured since 1960) and is the current in amperes that is used as the base for adjusting and determining the percent registration of a watthour meter at heavy and light loads. *See also:* **watthour meter.** 12A0-0

test handset (hand test telephone). A handset used for test purposes in a central office or in the outside plant. It may contain in the handle other components in addition to the transducer, as for example a dial, keys, capacitors, and resistors. *See also:* **telephone station.** 42A65-0

test position (switchgear assembly removable element). That position in which the primary disconnecting devices of the removable element are separated by a safe distance from those in the housing and some or all of the secondary disconnecting devices are in operating contact. *Notes:* (1) A set of test jumpers or mechanical movement of secondary disconnecting devices may be used to complete all secondary connections for test in the test position. This may correspond with the disconnected position. (2) Safe distance, as used here, is a distance at which the equipment will meet its withstand-voltage ratings, both low-frequency and impulse, between line and load terminals with the switching device in the closed position. 37A100-31E11

test routine. (1) Usually a synonym for check routine. (2) Sometimes used as a general term to include both check routine and diagnostic routine. E270-0

test signal, telephone channel (power-system communication). A signal of specified power and frequency that is applied to the input terminals of a telephone channel for measurement purposes. At any other point in the system, the test signal may be specified in terms applicable to the specific system. *See also:* **telephone system.** 0-31E3

test site (electromagnetic compatibility). A site meeting specified requirements suitable for measuring radio interference fields radiated by an appliance under test. *See also:* **electromagnetic compatibility.** CISPR-27E1

test specimen (insulator). An insulator that is representative of the product being tested; it is a specimen that is undamaged in any way that would influence the result of the test. 29A1-0

test stimulus (electrical). A single shock or succession of shocks, used to characterize or determine the state of excitability or the threshold of a tissue. 42A80-18E1

test switch (test block). A combination of connection studs, jacks, plugs, or switch parts arranged conveniently to connect the necessary devices for testing instruments, meters, relays, etcetera. 37A100-31E11

test voltage and current.
See:
chopped impulse voltage;
disruptive discharge;
disruptive-discharge voltage;
disruptive-discharge voltage; 50 percent
disruptive-discharge voltage, 100 percent
flashover;
full impulse voltage;
impulse current (in current testing);
peak value of alternating voltage;
peak value of impulse current;
peak value of impulse voltage;
puncture;
resistance of a shunt;
ripple of direct voltage;
root-mean-square value of alternating voltage;
scale factor;
shunt; snapover;
sparkover;
streamer;
value;
virtual duration;
virtual front time;
virtual instant of chopping;
virtual origin;
virtual rate of rise of the front;
virtual steepness of voltage during chopping;
virtual time of voltage collapse during chopping;
virtual time to chopping;
virtual time to half value;
voltage divider;
withstand voltage.

tetanizing current (electrotherapy). The current that, when applied to a muscle or to a motor nerve connected with a muscle stimulates the muscle with sufficient intensity and frequency to produce a smoothly sustained contraction as distinguished from a succession of twitches. *See also:* **electrotherapy.** 42A80-18E1

tetrode. A four-electrode electron tube containing an anode, a cathode, a control electrode, and one additional electrode that is ordinarily a grid. *See also:* **tube definitions.** 42A70-15E6

theoretical cutoff. *See:* **theoretical cutoff frequency.**

theoretical cutoff frequency (theoretical cutoff) (electric structure). A frequency at which, disregarding the effects of dissipation, the attenuation constant changes from zero to a positive value or vice versa. *See also:* **cutoff frequency.** 42A65-0

theory. *See:* **information theory.**

thermal burden rating (potential transformer). The volt-amperes that the potential transformer will carry continuously at rated voltage and frequency without causing the specified temperature limitations to be exceeded. *See:* **instrument transformer.** 12A0/42A15/42A30-31E12

thermal cell. A reserve cell that is activated by the application of heat. *See also:* **electrochemistry.** 42A60-0

thermal conduction. The transport of thermal energy by processes having rates proportional to the temperature gradient and excluding those processes involving a net mass flow. *See also:* **thermoelectric device.** E221-15E7

thermal conductivity. The quotient of (1) the conducted heat through unit area per unit time by (2) the component of the temperature gradient normal to that area. *See also:* **thermoelectric device.** E221-15E7

thermal conductivity, electronic. The part of the thermal conductivity resulting from the transport of thermal energy by electrons and holes. *See also:* **thermoelectric device.** E221-15E7

thermal converter (electric instrument) (thermocouple converter) (thermoelement*). A device that consists of one or more thermojunctions in thermal contact with an electric heater or integral therewith, so that the electromotive force developed at its output terminals by thermoelectric action gives a measure of the input current in its heater. *Note:* The combination of two or more thermal converters when connected with appropriate auxiliary equipment so that its combined direct-current output gives a measure of the active power in the circuit is called a thermal watt converter.
See:
apparent time constant;
electric thermometer;
ninety-percent response time of a thermal converter;
response time;
thermal current converter;
thermal power converter;
thermal voltage converter;
thermoelement.
*Deprecated 42A30-0

thermal current converter (electric instrument). A type of thermal converter in which the electromotive force developed at the output terminals gives a measure of the current through the input terminals. *See also:* **thermal converter.** 39A1-0

thermal current rating (neutral grounding device) (electric power). The root-mean-square neutral current in amperes that it will carry under standard conditions for its rated time without exceeding standard temperature limitations, unless otherwise specified. *See also:* **grounding device.** E32/42A35-31E13

thermal cutout (industrial control). An overcurrent protective device that contains a heater element in addition to and affecting a renewable fusible member that opens the circuit. *Note:* It is not designed to interrupt short-circuit currents. *See also:* **relay.** 1A0/42A25-34E10

thermal electromotive force. Alternative term for **Seebeck electromotive force.** *See also:* **thermoelectric device.** E221-15E7

thermal endurance (electric insulation) (rotating machinery). The relationship, between temperature and time spent at that temperature, required to produce such degradation of an electrical insulation that it fails under specified conditions of stress, electric or mechanical, in service or under test. For most of the chemical reactions encountered, this relationship is a straight line when plotted with ordinates of logarithm of time against abscissae of reciprocal of absolute temperature (Arrhenius plot). *See also:* **asynchronous machine; direct-current commutating machine; synchronous machine.** 0-31E8

thermal equilibrium (rotating machinery). The state reached when the observed temperature rise of the several parts of the machine does not vary by more than 2 degrees Celsius over a period of one hour. *See:* **asynchronous machine; direct-current commutating machine; synchronous machine.** 0-31E8

thermal flow switch. *See:* **flow relay.**

thermal-mechanical cycling (rotating machinery). The experience undergone by rotating-machine windings, and particularly their insulation, as a result of differential movement between copper and iron on heating and cooling. Also denotes a test in which such actions are simulated for study of the resulting behavior of an insulation system, particularly for machines having a long core length. *See also:* **asynchronous machine; direct-current commutating machine; synchronous machine.** 0-31E8

thermal noise (circuit) (resistance noise). Random noise associated with the thermodynamic interchange of energy necessary to maintain thermal equilibrium between the circuit and its surroundings. *Note:* The average square of the open-circuit voltage across the terminals of a passive two-terminal network of uniform temperature, due to thermal agitation, is given by

$$V_{\mathrm{T}}^2 = 4kT\int R(f)\,\mathrm{d}f$$

where T is the absolute temperature in degrees Celsius, R is the resistance component in ohms of the network impedance at the frequency f measured in hertz, and k is the Boltzmann constant, 1.38×10^{-23}. *See also:* **Johnson noise; signal-to-noise ratio.** 42A65-0

thermal power converter (thermal watt converter*) (electric instrument). A complex type of thermal converter having both potential and current input terminals. It usually contains both current and potential transformers or other isolating elements, resistors, and a multiplicity of thermoelements. The electromotive force developed at the output terminals gives a measure of the power at the input terminals. *See also:* **thermal converter.**

*Deprecated 39A1-0

thermal protection (motor). The words **thermal protection** appearing on the nameplate of a motor indicate that the motor is provided with a thermal protector. *See also:* **contactor.** 1A0-0

thermal protector (rotating machinery). A protective device for assembly as an integral part of a machine that protects the machine against dangerous overheating due to overload and, in a motor, failure to start. *Notes:* (1) It may consist of one or more temperature-sensing elements integral with the machine and a control device external to the machine. (2) When a thermal protector is designed to perform its function by opening the circuit to the machine and then automatically closing the circuit after the machine cools to a satisfactory operating temperature, it is an automatic-reset thermal protector. (3) When a thermal protector is designed to perform its function by opening the circuit to the machine but must be reset manually to close the circuit, it is a manual-reset thermal protector. *See also:* **contactor.** 1A0/42A10-31E8

thermal relay (industrial control). A relay in which the displacement of the moving contact member is produced by the heating of a part of the relay under the action of electric currents. *Note:* Compare with **temperature relay.** *See also:* **relay.** 50I16-34E10;37A100-31E11-31E6

thermal resistance (1) (cable). The resistance offered by the insulation and other coverings to the flow of heat from the conductor or conductors to the outer surface. *Note:* The thermal resistance of the cable is equal to the difference of temperature between the conductor or conductors and the outside surface of the cable divided by rate of flow of heat produced thereby. It is preferably expressed by the number of degrees Celsius per watt per foot of cable. *See also:* **power distribution, underground construction.** 42A35-31E13

(2) (semiconductor device). The temperature difference between two specified points or regions divided by the power dissipation under conditions of thermal equilibrium. *See also:* **principal voltage-current characteristic (principal characteristic).** E223-34E17/34E24/15E7

thermal resistance, effective (semiconductor rectifier) (semiconductor device). The effective temperature rise per unit power dissipation of a designated junction, above the temperature of a stated external reference point under conditions of thermal equilibrium. *Note:* Thermal impedance is the temperature rise of the junction above a designated point on the case, in degrees Celsius per watt of heat dissipation. *See also:* **rectification; semiconductor; semiconductor rectifier stack.** E59/E216/E270-34E17/34E24

thermal short-time current rating (current transformer). The root-mean-square symmetrical primary current that may be carried for a stated period (five seconds or less) with the secondary winding short-circuited, without exceeding a specified maximum temperature in any winding. *See also:* **instrument transformer.** 42A15/42A30/57A13-31E12

thermal telephone receiver (thermophone). A telephone receiver in which the temperature of a conductor is caused to vary in response to the current input, thereby producing sound waves as a result of the expansion and contraction of the adjacent air. *See also:* **loudspeaker.** 42A65-0

thermal tuning. *See:* **tuning, thermal.**

thermal voltage converter (electric instrument). A thermoelement of low-current input rating with an associated series impedance or transformer, such that the electromotive force developed at the output terminals gives a measure of the voltage applied to the input terminals. *See also:* **thermal converter.** 39A1-0

thermal watt converter. *See:* **thermal power converter.** *See also:* **thermal converter.**

thermally delayed overcurrent trip. *See:* **thermally delayed release.**

thermally delayed release (trip). A release delayed by a thermal device. 37A100-31E11

thermionic arc (gas). An electric arc characterized by the fact that the thermionic cathode is heated by the arc current itself. *See also:* **discharge (gas).** 50I07-15E6

thermionic emission (Edison effect) (Richardson effect). The liberation of electrons or ions from a solid or liquid as a result of its thermal energy. *See also:* **electron emission.** E160/E270-15E6

thermionic generator. A thermoelectric generator in which a part of the circuit, across which a temperature difference is maintained, is a vacuum or a gas. *See also:* **thermoelectric device.** E221-15E7

thermionic grid emission. Current produced by electrons thermionically emitted from a grid. *See:* **electron emission.** E160-15E6

thermionic tube. *See:* **hot-cathode tube.**

thermistor. An electron device that makes use of the change of resistivity of a semiconductor with change in temperature. *See also:* **bolometric detector; circuits and devices; electron devices, miscellaneous; semiconductor.** E102/E216/E270/42A70-9E4/34E17

thermistor mount (bolometer mount) (waveguide). A waveguide termination in which a thermistor (bolometer) can be incorporated for the purpose of measuring electromagnetic power. *See also:* **waveguide.** 50I62-3E1

thermocouple. A pair of dissimilar conductors so joined at two points that an electromotive force is developed by the thermoelectric effects when the junctions are at different temperatures. *See also:* **electric thermometer (temperature meter); thermoelectric effect.** 42A30-13E6

thermocouple converter (thermoelement). *See:* **thermal converter.**

thermocouple instrument. An electrothermic instrument in which one or more thermojunctions are heated directly or indirectly by an electric current or currents and supply a direct current that flows through the coil of a suitable direct-current mechanism, such as one of the permanent-magnet moving-coil type. *See also:* **instrument.** 42A30-0

thermocouple vacuum gauge. A vacuum gauge that depends for its operation on the thermal conduction of the gas present, pressure being measured as a function of the electromotive force of a thermocouple the measuring junction of which is in thermal contact with a heater that carries a constant current. It is ordinarily used over a pressure range of 10^{-1} to 10^{-3} conventional millimeter of mercury. *See also:* **instrument.** 42A30-0

thermoelectric cooling device. A thermoelectric heat pump that is used to remove thermal energy from a body. *See also:* **thermoelectric device.** E221-15E7

thermoelectric device. A generic term for thermoelectric heat pumps and thermoelectric generators.
See:
electric thermometer;
junction resistance;
leg, thermoelectric;
Lorenz number;
over-all generator efficiency;
Peltier coefficient;
Peltier coefficient, absolute;
Peltier coefficient, relative;
Peltier effect;
Peltier heat;
reduced generator efficiency;
Seebeck coefficient;
Seebeck coefficient, absolute;
Seebeck coefficient, relative;
Seebeck effect;
Seebeck electromotive force;
stage;
thermal conduction;
thermal conductivity;
thermal conductivity, electronic;
thermal electromotive force;
thermal generator;
thermoelectric cooling device;
thermoelectric generator;
thermoelectric heating device;
thermoelectric heat pump;
Thomson coefficient;
Thomson effect;
Thomson heat;
weight coefficient;
Wiedemann-Franz ratio. E221-15E7

thermoelectric effect. *See:* **Seebeck effect.**

thermoelectric effect error (bolometric power meter). An error arising in bolometric power meters that employ thermistor elements in which the majority of the bias power is alternating current and the remainder direct current. The error is caused by thermocouples at the contacts of the thermistor leads to the metal oxides of the thermistors. *See also:* **bolometric power meter.** 0-9E4

thermoelectric generator. A device that converts thermal energy into electric energy by direct interaction of a heat flow and the charge carriers in an electric circuit, and that requires for this process the existence of a temperature difference in the electric circuit. *See also:* **thermoelectric device.** E221-15E7

thermoelectric heating device. A thermoelectric heat pump that is used to add thermal energy to a body. *See also:* **thermoelectric device.** E221-15E7

thermoelectric heat pump. A device that transfers thermal energy from one body to another by the direct interaction of an electric current and the heat flow. *See also:* **thermoelectric device.** E221-15E7

thermoelectric power*. *See:* **Seebeck coefficient.** *See also:* **thermoelectric device.**

*Deprecated

thermoelement (electric instrument). The simplest type of thermal converter. It consists of a thermocouple, the measuring junction of which is in thermal contact with an electric heater or integral therewith. *See also:* **thermal converter; thermoelectric arm; thermoelectric couple.** 39A1-0

thermoelectric thermometer (thermocouple thermometer). An electric thermometer that employs one or more thermocouples of which the set of measuring junctions is in thermal contact with the body, the temperature of which is to be measured, while the temperature of the reference junctions is either known or otherwise taken into account. *See also:* **electric thermometer (temperature meter).** 42A30-0

thermogalvanic corrosion. Corrosion resulting from a galvanic cell caused primarily by a thermal gradient. *See also:* **electrolytic cell.** CM-34E2

thermojunction. One of the surfaces of contact between the two conductors of a thermocouple. The thermojunction that is in thermal contact with the body under measurement is called the measuring junction, and the other thermojunction is called the reference junction. *See also:* **electric thermometer (temperature meter).** 42A30-0

thermometer method of temperature determination (electric power). The determination of the temperature by mercury, alcohol, resistance, or thermocouple thermometer, any of these instruments being applied to the hottest accessible part of the device. *See also:* **power systems, low-frequency and surge testing; routine test.** E32/42A15/57A15-31E12

thermophone. An electroacoustic transducer in which sound waves of calculable magnitude result from the expansion and contraction of the air adjacent to a conductor whose temperature varies in response to a current input. *Note:* When used for the calibration of pressure microphones, a thermophone is generally used in a cavity the dimensions of which are small compared to a wavelength. *See also:* **microphone.** E157-1E1

thermopile. A group of thermocouples connected in series aiding. This term is usually applied to a device used either to measure radiant power or energy or as a source of electric energy. *See also:* **electric thermometer (temperature meter).** 42A30-0

thermoplastic (1) (noun). A plastic that is thermoplastic in behavior.

(2) (adjective). Having the quality of softening when heated above a certain temperature range and of returning to its original state when cooling below that range. 0-31E8

thermoplastic insulating tape. A tape composed of a thermoplastic compound that provides insulation for joints. 42A95-0

thermoplastic insulations and jackets (power distribution, underground cables). Insulations and jackets made of materials that are softened by heat for application to the cable and then become firm, tough and resilient upon cooling. Subsequent heating and cooling will reproduce similar changes in the physical properties of the material. *See also:* **power distribution, underground construction.** 0-31E1

thermoplastic protective tape. A tape composed of a thermoplastic compound that provides a protective covering for insulation. 42A95-0

thermoset. A plastic that, when cured by the application of heat or chemical means, changes into a substantially infusible and insoluble product. D883-31E8

thermosetting insulations and jackets (power distribution, underground cables). Insulations and jackets made of materials that may be applied to the cable in a relatively soft state and then vulcanized under heat and pressure to develop firm, tough, and resilient properties. *See also:* **power distribution, underground construction.** 0-31E1

thermostat. A temperature-sensitive device that automatically opens and closes an electric circuit to regulate the temperature of the space with which it is associated. *See also:* **appliance outlet.** 71A1-0

thermostatic switch (thermostat). A form of temperature-operated switch that receives its operating energy by thermal conduction or convection from the device being controlled or operated. *See also:* **switch.** 42A25-34E10

theta (θ) polarization. The state of the wave in which the E vector is tangential to the meridian lines of a given spherical frame of reference. *Note:* The usual frame of reference has the polar axis vertical and the origin at or near the antenna. Under these conditions, a vertical dipole will radiate only theta (θ) polarization and the horizontal loop will radiate only phi (ϕ) polarization. *See also:* **antenna.** E149/42A65/50I62-3E1

thévenin's theorem. States that the current that will flow through an impedance Z', when connected to any two terminals of a linear network between which there previously existed a voltage E and an impedance Z, is equal to the voltage E divided by the sum of Z and Z'. *See also:* **communication.** 42A65-0

thickener (hydrometallurgy) (electrometallurgy). A tank in which suspension of solid material can settle so that the solid material emerges from a suitable opening with only a portion of the liquid while the remainder of the liquid overflows in clear condition at another part of the thickener. *See also:* **electrowinning.** 42A60-0

thin film. Loosely, magnetic thin film. *See also:* **electronic digital computer.** X3A12-16E9

thin-wall counter (radiation counters). A counter tube in which part of the envelope is made thin enough to permit the entry of radiation of low penetrating power. *See:* **anticoincidence (radiation counters).** 0-15E6

third-rail clearance line (railroad). The contour that embraces all cross sections of third rail and its insulators, supports, and guards located at an elevation higher than the top of the running rail. *See also:* **electric locomotive.** 42A42-0

third-rail electric car. An electric car that collects propulsion power through a third-rail system. *See also:* **electric motor car.** 42A42-0

third-rail electric locomotive. An electric locomotive that collects propulsion power from a third-rail system. *See also:* **electric locomotive.** 42A42-0

third voltage range. *See:* **voltage range.**

Thomson bridge (double bridge). *See:* **Kelvin bridge.**

Thomson coefficient (Thomson heat coefficient). The quotient of (1) the rate of Thomson heat absorption per unit volume of conductor by (2) the scalar product of the electric current density and the temperature gradient. The Thomson coefficient is positive if Thomson heat is absorbed by the conductor when the component of the electric current density in the direction of the temperature gradient is positive. *See also:* **thermoelectric device.** E221-15E7

Thomson effect. The absorption or evolution of thermal energy produced by the interaction of an electric current and a temperature gradient in a homogeneous electric conductor. *Notes:* (1) An electromotive force exists between two points in a single conductor that are at different temperatures. The magnitude and direction of the electromotive force depend on the material of the conductor. A consequence of this effect is that if a current exists in a conductor between two points at different temperatures, heat will be absorbed or liberated depending on the material and on the sense of the current. (2) In a nonhomogeneous conductor, the Peltier effect and the Thomson effect cannot be separated. *See also:* **thermoelectric device.** E221/E270-15E7

Thomson heat. The thermal energy absorbed or evolved as a result of the Thomson effect. *See also:* **thermoelectric device.** E221-15E7

thread (control) (industrial control). A control function that provides for maintained operation of a drive at a preset reduced speed such as for setup purposes. *See:* **electric drive.** IC1-34E10

threaded coupling (rigid steel conduit). An internally threaded steel cylinder for connecting two sections of rigid steel conduit. 80A1-0

three-address. Pertaining to an instruction code in which each instruction has three address parts. Also called triple-address. In a typical three-address instruction the addresses specify the location of two operands and the destination of the result, and the instructions are taken from storage in a preassigned order. *See also:* **two-plus-one-address.** *See also:* **electronic digital computer.** E162-0

three-address code. *See:* **instruction code.**

three-dimensional (3D) radar (volumetric radar). A radar capable of producing three-dimensional position data on a multiplicity of targets. *See also:* **radar.** 42A65-10E6

three-phase circuit. A combination of circuits energized by alternating electromotive forces that differ in phase by one-third of a cycle, that is, 120 degrees. *Note:* In practice the phases may vary several degrees from the specified angle. *See also:* **center of distribution.** E45-0/42A35-31E13

three-phase electric locomotive. An electric locomotive that collects propulsion power from three phases of an alternating-current distribution system. *See also:* **electric locomotive.** 42A42-0

three-phase four-wire system. A system of alternating-current supply comprising four conductors, three of which are connected as in a three-phase three-wire system, the fourth being connected to the neutral point of the supply, which may be grounded. *See also:* **alternating-current distribution.** 42A35-31E13

three-phase seven-wire system. A system of alternating-current supply from groups of three single-phase transformers connected in Y so as to obtain a three-phase four-wire grounded-neutral system for lighting and a three-phase three-wire grounded-neutral system of a higher voltage for power, the neutral wire being common to both systems. *See also:* **alternating-current distribution.** 42A35-31E13

three-phase three-wire system. A system of alternating-current supply comprising three conductors between successive pairs of which are maintained alternating differences of potential successively displaced in phase by one-third of a period. *See also:* **alternating-current distribution.** 42A35-31E13

three-plus-one-address. Pertaining to a four-address code in which one address part always specifies the location of the next instruction to be interpreted. *See also:* **electronic digital computer.** E162-0

three-position relay. A relay that may be operated to three distinct positions. *See also:* **railway signal and interlocking.** 42A42-0

three-terminal capacitor. Two conductors (the active electrodes) insulated from each other and from a surrounding third conductor that constitutes the shield. When the capacitor is provided with properly designed terminals and used with shielded leads, the direct capacitance between the active electrodes is independent of the presence of other conductors. (Specialized usage.) E270-0

three-terminal network. A four-terminal network having one input and one output terminal in common. *See also:* **network analysis.** 0-31E3

three-wire control (industrial control). A control function that utilizes a momentary-contact pilot device and a holding-circuit contact to provide undervoltage protection. *See:* **undervoltage protection.** *See also:* **relay.** IC1-34E10

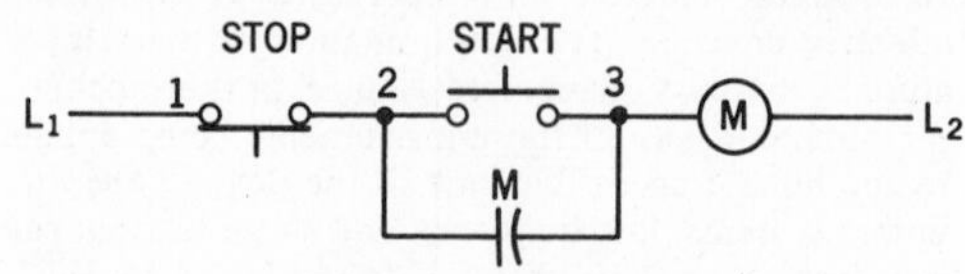

Three-wire control.

three-wire system (direct current or single-phase alternating current). A system of electric supply comprising three conductors, one of which (known as the **neutral wire**) is maintained at a potential midway between the potential of the other two (referred to as the outer conductors). *Note:* Part of the load may be connected directly between the outer conductors, the remainder being divided as evenly as possible into two parts each of which is connected between the neutral and one outer conductor. There are thus two distinct voltages of supply, the one being twice the other. *See also:* **alternating-current distribution; direct-current distribution.** 42A35-31E13

threshold (1) (general). The value of a physical stimulus (such as size, luminance, contrast, or time) that permits an object to be seen a specific percentage of the time or at a specific accuracy level. *Note:* In many psychophysical experiments, thresholds are presented in terms of 50-percent accuracy or accurately 50 percent of the time. However, the threshold also is expressed as the value of the physical variable that permits the object to be just barely seen. The threshold may be determined by merely detecting the presence of an object or it may be determined by discriminating certain details of the object. Z7A1-0

(2) (ultimate sensitivity*). The least change in the measurand that can be detected with a specified degree of certainty by use of the instrument under specified conditions in a specified time. The limiting factor that fixes the threshold may be any of a number of effects such as the uncertainty of reading, friction, spurious external disturbances or internal thermal agitation, depending upon the type of instrument and the conditions of use. *See also:* **accuracy rating (instrument).**
*Deprecated 42A30-0

(3) (audiology) (medical electronics). The lowest amplitude of sound that may be heard by an individual. *See also:* **medical electronics.** 0-18E1

(4) (modulation systems). The smallest value of carrier-to-noise ratio at the input to the demodulator for all values above which a small percentage change in the input carrier-to-noise ratio produces a substantially equal or a smaller percentage change in the output signal-to-noise ratio. *Note:* Where precision is required, the method of determining the value of the threshold must be specified. *See also:* **modulating systems.** E170-0

(5) (electrobiology) (limen). (A) The least current or voltage needed to elicit a minimal response. (B) That needed to elicit a minimal response in 50 percent of the trials. *See also:* **biological; excitability (electrobiology).** 42A80-18E1

(6) (logic operations). (A) A logic operator having the property that if P is a statement, Q is a statement, R is a statement, ..., then the threshold of P, Q, R, ..., is true if at least N statements are true, false if less than N statements are true, where N is a specified nonnegative integer called the threshold condition. (B) The threshold condition as in (A). X3A12-16E9

threshold audiogram. *See:* **audiogram.**

threshold current (current-limiting fuse). A current magnitude of specified wave shape at which the melting of the current-responsive element occurs at the first instantaneous peak current for that wave shape. *Note:* The current magnitude is usually expressed in root-mean-square amperes. 37A100-31E11

threshold element. (1) A combinational logic element such that the output channel is in its ONE state if and only if at least n input channels are in their ONE states, where n is a specified fixed nonnegative integer, called the threshold of the element. (2) By extension, a similar element whose output channel is in its ONE state if and only if at least n input channels are in states specified for them, not necessarily the ONE state but a fixed state for each input channel. *See also:* **electronic digital computer.** E162-0

(3) A device that performs the logic threshold operation but in which the truth of each input statement contributes to the output determination a weight associated with that statement. X3A12-16E9

threshold field. The least magnetizing force in a direction that tends to decrease the remanence, that, when applied either as a steady field of long duration or as a pulsed field appearing many times, will cause a stated fractional change of remanence. *See also:* **static magnetic storage.** E163-16E9

threshold frequency (photoelectric device) (photoelectric tubes). The frequency of incident radiant energy below which there is no photoemissive effect.

$$\nu_0 = p/h.$$

See also: **photoelectric effect.** 50I07-15E6

threshold lights. Runway lights so placed as to indicate the longitudinal limits of that portion of a runway, channel or landing path usable for landing. *See also:* **signal lighting.** Z7A1-0

threshold of audibility (threshold of detectability) (specified signal). The minimum effective sound pressure level of the signal that is capable of evoking an auditory sensation in a specified fraction of the trials. The characteristics of the signal, the manner in which it is presented to the listener, and the point at which the sound pressure is measured must be specified. *Notes:* (1) Unless otherwise specified, the ambient noise reaching the ears is assumed to be negligible. (2) The threshold is usually given as a sound pressure level in decibels relative to 20 micronewtons per square meter. (3) Instead of the method of constant stimuli, which is implied by the phrase in a specified fraction of the trials, another psychophysical method (which should be specified) may be employed. *See also:* **electroacoustics.** 0-1E1

threshold of discomfort (for a specified signal) (audio and electroacoustics). The minimum effective sound pressure level at the entrance to the external auditory canal that, in a specified fraction of the trials, will stimulate the ear to a point at which the sensation of feeling becomes uncomfortable. *See also:* **electroacoustics.** 0-1E1

threshold of feeling (tickle) (for a specified signal) (audio and electroacoustics). The minimum effective sound pressure level at the entrance to the external auditory canal that, in a specified fraction of the trials, will stimulate the ear to a point at which there is a sensation of feeling that is different from the sensation of hearing. *See also:* **electroacoustics.** 0-1E1

threshold of pain (for a specified signal) (audio and electroacoustics). The minimum effective sound pressure level at the entrance to the external auditory canal that, in a specified fraction of the trials, will stimulate the ear to a point at which the discomfort gives way to definite pain that is distinct from the mere nonnoxious feeling of discomfort. *See also:* **electroacoustics.** 0-1E1

threshold ratio (current-limiting fuse). The ratio of the threshold current to the fuse current rating. 37A100-31E11

threshold signal (1) (signal-transmission system). The minimum increment in signal magnitude that can be distinguished from the interference variations. *See:* **signal.** 0-13E6

(2) (electronic navigation). The smallest signal capable of effecting a recognizable change in navigational information. *See also:* **navigation.** E172-10E6

threshold signal-to-interference ratio (TSI) (electromagnetic compatibility). The minimum signal to interference power, described in a prescribed way, required to provide a specified performance level. *See also:* **electromagnetic compatibility.** 0-27E1

threshold voltage (semiconductor rectifiers). The zero-current–voltage intercept of a straight-line approximation of the forward current-voltage characteristic over the normal operating range. *See:* **semiconductor rectifier stack.** 0-34E24

threshold wavelength (photoelectric tubes) (photoelectric device). The wavelength of the incident radiant energy above which there is no photoemission effect. *See also:* **photoelectric effect.** 50I07-15E6

throat microphone. A microphone normally actuated by mechanical contact with the throat. *See also:* **microphone.** 42A65-0

through bolt (rotating machinery). A bolt passing axially through a laminated core, that is used to apply pressure to the end plates. *See also:* **cradle base (rotating machinery).** 0-31E8

through supervision. The automatic transfer of supervisory signals through one or more trunks in a manual telephone switchboard. *See also:* **telephone switching system.** 42A65-19E1

throwing power (of a solution) (electroplating). A measure of its adaptability to deposit metal uniformly upon a cathode of irregular shape. In a given solution under specified conditions it is equal to the improvement (in percent) of the metal distribution ratio above the primary-current distribution ratio. *See also:* **electroplating.** 42A60-0

thrust bearing (rotating machinery). A bearing designed to carry an axial load so as to prevent or to limit axial movement of a shaft, or to carry the weight of a vertical rotor system. *See also:* **bearing.** 0-31E8

thrust block (rotating machinery). A support for a thrust-bearing runner. *See also:* **bearing.** 0-31E8

thrust collar (rotating machinery). The part of a shaft or rotor that contacts the thrust bearing and transmits the axial load. *See:* **rotor (rotating machinery).** 0-31E8

thump. A low-frequency transient disturbance in a system or transducer characterized audibly by the onomatopoeic connotation of the word. *Note:* In telephony, thump is the noise in a receiver connected to a telephone circuit on which a direct-current telegraph channel is superposed caused by the telegraph currents. *See also:* **signal-to-noise ratio.** E151/42A65-0

Thury transmission system. A system of direct-current transmission with constant current and a variable high voltage. *Note:* High voltage used on this system is obtained by connecting series direct-current generators in series at the generating station; and is utilized by connecting series direct-current motors in series at the substations. *See also:* **direct-current distribution.** 42A35-31E13

thyratron. A hot-cathode gas tube in which one or more control electrodes initiate but do not limit the anode current except under certain operating conditions. *See also:* **tube definitions.** 42A70-15E6

thyristor. A bistable semiconductor device comprising three or more junctions that can be switched from the OFF state to the ON state or vice versa, such switching occurring within at least one quadrant of the principal voltage-current characteristic.
See:
bidirectional diode-thyristor;
***n*-gate thyristor;**
***p*-gate thyristor;**
reverse-blocking diode-thyristor;
reverse-blocking triode-thyristor;
reverse-conducting diode-thyristor;
semiconductor;
semiconductor controlled rectifier;
turn-off thyristor. E223-15E7/34E17/34E24

ticker. A form of receiving-only printer used in the dissemination of information such as stock quotations and news. *See also:* **telegraphy.** 42A65-19E4

tie (rotating machinery). A binding of the end turns used to hold a winding in place or to hold leads to windings for purpose of anchoring. *See also:* **rotor (rotating machinery); stator.** 0-31E8

tie feeder. A feeder that connects together two or more independent sources of power and has no tapped load between the terminals. *Notes:* (1) If a feeder has any tapped load between the two sources, it is designated as a multiple feeder. (2) The source of power may be a generating system, substation, or feeding point. The normal flow of energy in such a feeder may be in either direction. *See also:* **center of distribution.** 37A100/42A35-31E11/31E13

tie line (electric power systems). A transmission line connecting two or more power systems. *See also:* **transmission line.** E94-0

tie-line bias control (control area). A mode of operation under load-frequency control in which the area control error is determined by the net interchange minus the biased scheduled net interchange. *See also:* **speed-governing system.** E94-0

tie point (tie line) (electric power systems). The location of the switching facilities that, when closed, permit power to flow between the two power systems. *See also:* **center of distribution.** E94-0

tie trunk. A telephone line or channel directly connecting two private branch exchanges. *See also:* **telephone system.** 42A65-19E1

tie wire. A short piece of wire used to bind an overhead conductor to an insulator or other support. *See also:* **conductor; tower.** 42A35-31E13

TIF. *See:* **telephone influence factor.**

tight (suffix). Apparatus is designated as watertight, dusttight, etcetera, when so constructed that the enclosing case will exclude the specified material under specified conditions. 37A100/42A95/IC1-31E6/31E11/34E10

tight coupling. *See:* **close coupling.**

tilt (1) (in a directive antenna). The angle that the antenna axis forms with the horizontal. *See also:* **navigation.** 0-10E6
(2) (pulse techniques). *See:* **distortion, pulse.**

tilt angle (in radar). The vertical angle between the axis of radiation and a reference axis; the reference is normally horizontal. *See also:* **navigation.** E172-10E6

tilt error (electronic navigation). *See:* **ionospheric tilt error.**

tilting-insulator switch. A switch in which the opening and closing travel of the blade is accomplished by a tilting movement of one or more of the insulators supporting the conducting parts of the switch. 37A100-31E11

tilting-pad bearing (Kingsbury bearing) (rotating machinery). A pad-type bearing in which the pads are capable of moving in such a manner as to improve the flow of lubricating fluid between the bearing and the shaft journal or collar (runner). *See:* **bearing.** 0-31E8

timbering machine. An electrically driven machine to raise and hold timbers in place while supporting posts are being set after being cut to length by the machine's power-driven saw. *See also:* **mining.** 42A85-0

timbre. The attribute of auditory sensation in terms of which a listener can judge that two sounds similarly presented and having the same loudness and pitch are dissimilar. *Note:* Timbre depends primarily upon the spectrum of the stimulus, but it also depends upon the waveform, the sound pressure, the frequency location of the spectrum, and the temporal characteristics of the stimulus. *See also:* **electroacoustics; loudspeaker.** 0-1E1

time (1) (reliability). Any duration of observations of the considered items either in actual operation or in storage, readiness, etcetera, but excluding down time due to a failure. *Note:* In definitions where time is used, this parameter may be replaced by distance, cycles, or other measures of life as may be appropriate. This refers to terms such as acceleration factor, wear-out failure, failure rate, mean life, mean time between failures, mean time to failure, reliability, and useful life. *See also:* **reliability.** 0-7E1
(2) (electronic computation). *See:* **access time; downtime; real time; reference time; switching time; word time.** *See also:* **electronic digital computer.**

time base (oscilloscopes). The sweep generator in an oscilloscope. *See:* **oscillograph.** 0-9E4

time bias (electric power systems). An offset in the scheduled net interchange power of a control area that varies in proportion to the time deviation. This offset is in a direction to assist in restoring the time deviation to zero. *See also:* **power system.** 0-31E4

time bias setting (electric power systems). For a control area, a factor with negative sign that is multiplied by the time deviation to yield the time bias. 0-31E4

time constant (automatic control). The value T in an exponential response term $A \exp(-1/T)$, or in one of the transform factors. *Notes:* (1) For the output of a first-order system forced by a step or an impulse, T is the time required to complete 63.2 percent of the total rise or decay. In higher-order systems, there is a time constant for each of the first-order factors of the process. In a Bode diagram, breakpoints occur at $\omega = 1/T$. *See also:* **control system, feedback.** (2) In terms of Laplace transforms it is the absolute value of the reciprocal of the real part of pole's position. The term **pole** refers here to the roots of the denominator of the Laplace transform. 0-23E0/31E8

time constant of an exponential function (ae^{-bt}). $1/b$, if t represents time and b is real. E270-0

time constant of fall (pulse). The time required for the pulse to fall from 70.7 percent to 26.0 percent of its maximum amplitude excluding spike. *See also:* **pulse.** 42A65-0

time constant of integrator (for each input). The ratio of the input to the corresponding time rate of change of the output. *See also:* **electronic analog computer.** E165-16E9

time constant of rise (pulse). The time required for the pulse to rise from 26.0 percent to 70.7 percent of its maximum amplitude excluding spike. *See also:* **pulse.** 42A65-0

time-current characteristic (fuse). *See:* **fuse time-current characteristic.**

time-current tests (fuse). *See:* **fuse time-current tests.**

time delay (1) (general). The time interval between the manifestation of a signal at one point and the manifestation or detection of the same signal at another point. *Notes:* (1) Generally, the term **time delay** is used to describe a process whereby an output signal has the same form as an input signal causing it but is delayed in time; that is, the amplification of all frequency components of the output are related by a single constant to those of corresponding input frequency components but each output component lags behind the corresponding input component by a phase angle proportional to the frequency of the component. (2) Transport delay is synonymous with time delay but usually is reserved for applications that involve the flow of material. *See also:* **electronic analog computer.** E165-0

(2) (industrial control). A time interval is purposely introduced in the performance of a function. *See also:* **control system, feedback.** IC1-34E10

time-delay relay. *See:* **delay relay.** *See also:* **relay.**

time-domain reflectometer. *See:* **reflectometer, time domain.**

time deviation (power system). The integrated or accumulated difference in cycles between system frequency and rated frequency. This is usually expressed in seconds by dividing the deviation in cycles by the rated frequency. *See also:* **power system, low-frequency and surge testing.** E94-0

time dial (relay) (time lever). An adjustable, graduated element by which, under fixed input conditions, the prescribed relay operating time can be varied. 37A100-31E11/31E6

time discriminator (electronic navigation). A circuit in which the sense and magnitude of the output is a function of the time difference of the occurrence, and relative time sequence, of two pulses. *See also:* **navigation.** E172-10E6

time distribution analyzer (nuclear techniques). An instrument capable of indicating the number or rate of occurrence of time intervals falling within one or more specified time interval ranges. The time interval is delineated by the separation between pulses of a pulse pair. *See also:* **ionizing radiation.** E175-0

time-division multiplex. The process or device in which each modulating wave modulates a separate pulse subcarrier, the pulse subcarriers being spaced in time so that no two pulses occupy the same time interval. *Note:* Time division permits the transmission of two or more signals over a common path by using different time intervals for the transmission of the intelligence of each message signal. *See also:* **modulating systems.** E145/E170/42A65-0

time gain control (electronic navigation). *See:* **differential gain-control circuit.**

time gate. A transducer that gives output only during chosen time intervals. *See also:* **circuits and devices; modulating systems.** E145/42A65-0

time-interval selector (nuclear techniques). A circuit that produces a specified output pulse when and only when the time interval between two pulses lies between specified limits. *See also:* **scintillation counter.** E175-0

time lag. *See:* **lag.**

time lag of impulse flashover (lightning arresters). The time between the instant when the voltage of the impulse wave first exceeds the power-frequency flashover crest voltage and the instant when the impulse flashover causes the abrupt drop in the testing wave. *See also:* **power systems, low-frequency and surge testing; lightning arrester (surge diverter).** 42A35-31E7/31E13

time-load withstand strength (of an insulator). The mechanical load that, under specified conditions, can be continuously applied without mechanical or electrical failure. *See also:* **insulator.** 29A1-0

time locking. A method of locking, either mechanical or electric, that, after a signal has been caused to display an aspect to proceed, prevents, until after the expiration of a predetermined time interval after such signal has been caused to display its most restrictive aspect, the operation of any interlocked or electrically locked switch, movable-point frog, or derail in the route governed by that signal, and that prevents an aspect to proceed from being displayed for any conflicting route. *See also:* **interlocking (interlocking plant).** 42A42-0

time of response. *See:* **response time.**

time-overcurrent relay. An overcurrent relay in which the input current and operating time are inversely related throughout a substantial portion of the performance range. 37A100-31E11/31E6

time of rise (decay) of video pulses (television). The duration of the rising (decaying) portion of a pulse measured between specified levels. *See also:* **pulse tim-**

ing of video pulses (television). E207-0

time pattern (television). A picture tube presentation of horizontal and vertical lines or dot rows generated by two stable frequency sources operating at multiples of the line and field frequencies. *See also:* **television.** E203-0

time per point (multiple-point recorders). The time interval between successive points on printed records. *Note:* For some instruments this interval is variable and depends on the magnitude of change in measured signal. For such instruments, time per point is specified as the minimum and maximum time intervals. 39A4-0

time rate (storage cell). The current in amperes at which a storage battery will be discharged in a specified time, under specified conditions of temperature and final voltage. *See also:* **battery (primary or secondary).** 42A60-0

time release. A device used to prevent the operation of an operative unit until after the expiration of a predetermined time interval after the device has been actuated. *See also:* **railway signal and interlocking.** 42A42-0

time, response. *See:* **response time.**

time response (1) (control system, feedback). An output, expressed as a function of time, resulting from the application of a specified input under specified operating conditions. *Note:* It consists of a transient component that depends on the initial conditions of the system, and a steady-state component that depends on the time pattern of the input. AS1/85A1-23E0/34E10

(2) (synchronous-machine regulator). The output of the synchronous-machine regulator (that is, voltage, current, impedance, or position) expressed as a function of time following the application of prescribed inputs under specified conditions. *See:* **synchronous machine.** 0-31E8

time share. To use a device for two or more interleaved purposes. *See also:* **electronic digital computer.** X3A12-16E9

time sharing. Pertaining to the interleaved use of the time of a device. *See also:* **electronic digital computer.** X3A12-16E9

time sorter (nuclear techniques). *See:* **time distribution analyzer.**

time to half-value on the wavetail (virtual duration of an impulse) (lightning arresters). *See:* **virtual time to half-value (on the wavetail).**

time to impulse flashover. The time between the initial point of the voltage impulse causing flashover and the point at which the abrupt drop in the voltage impulse takes place. *See also:* **power systems, low-frequency and surge testing.** 42A35-31E13

time to impulse-sparkover (lightning arresters). The time between virtual zero of the voltage impulse causing sparkover and the point on the voltage wave at which sparkover occurs. *Note:* The time is expressed in microseconds. *See:* **lightning arrester (surge diverter).** *See also:* **lightning protection and equipment.** E28/62A1-31E7

time-undervoltage protection. A form of undervoltage protection that disconnects the protected equipment upon a deficiency of voltage after a predetermined time interval. 37A100-31E11/31E6

timing deviation (demand meter). The difference between the elapsed time indicated by the timing element and the true elapsed time, expressed as a percentage of the true elapsed time. *See also:* **demand meter.** 12A0-0

timing mechanism (1) (demand meter). That mechanism through which the time factor is introduced into the result. The principal function of the timing mechanism of a demand meter is to measure the demand interval, but it has a subsidiary function, in the case of certain types of demand meters, to provide also a record of the time of day at which any demand has occurred. A timing mechanism consists either of a clock or its equivalent, or of a lagging device that delays the indications of the electric mechanism. In thermally lagged meters the time factor is introduced by the thermal time lag of the temperature responsive elements. In the case of curve-drawing meters, the timing element merely provides a continuous record of time on a chart or graph. *See also:* **demand meter.** 42A30-0

(2) (recording instrument). The time-regulating device usually includes the motive power unit necessary to propel the chart at a controlled rate (linear or angular). *See also:* **moving element (instrument).** 39A2-0

timing relay (or relay unit). An auxiliary relay or relay unit whose function is to introduce one or more definite time delays in the completion of an associated function. *See also:* **relay.** 37A100-31E6/31E11

timing table. That portion of central-station equipment at which means are provided for operators' supervision of signal reception. *See also:* **protective signaling.** 42A65-0

tinning (electrotyping). The melting of lead-tin foil or tin plating upon the back of shells. *See also:* **electroforming.** 42A60-0

tinsel cord. A flexible cord in which the conducting elements are thin metal ribbons wound helically around a thread core. *See also:* **transmission line.**

(1) tip (plug). The contacting part at the end of the plug. *See also:* **telephone switching system.** 42A65-0

(2) (electron tubes) (pip). A small protuberance on the envelope resulting from the sealing of the envelope after evacuation. *See also:* **electronic tube.** 50I07-15E6

tip and ring wires (telephony) (tip and ring conductors). A pair of conductors associated with the transmission portions of circuits and apparatus. Tip or ring designation of the individual conductors is arbitrary except when applied to cord-type switchboard wiring in which case the conductors are designated according to their association with tip or ring contacts of the jacks and plugs. *See also:* **telephone switching system.** 0-19E1

T junction (waveguide). A junction of waveguides in which the longitudinal guide axes form a T. *Note:* The guide that continues through the junction is the main guide; the guide that terminates at a junction is the branch guide. *See also:* **waveguide.** 42A65/E147-3E1

TM. *See:* **transverse magnetic.**

T network. A network composed of three branches with one end of each branch connected to a common junction point, and with the three remaining ends connected to an input terminal, an output terminal, and a common input and output terminal, respectively. *See also:* **network analysis.** E153/E270/42A65-0

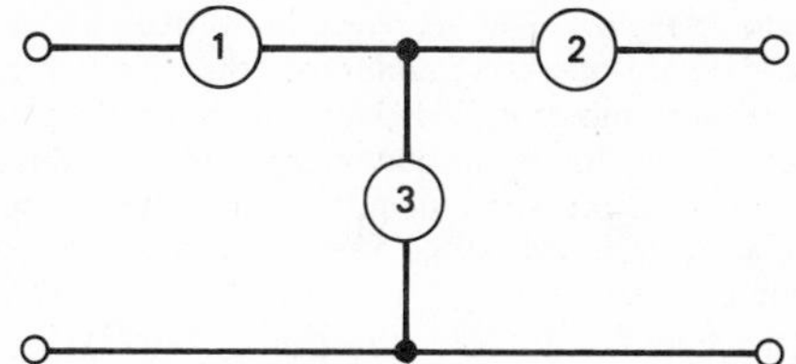

T network. One end of each of the branches 1, 2, and 3 is connected to a common point. The other ends of branches 1 and 2 form, respectively, an input and an output terminal, and the other end of branch 3 forms a common input and output terminal.

to-from indicator (omnirange receiver). An instrument, forming part of the omnirange facilities, that resolves the 180-degree ambiguity. *See also:* **radio navigation.** 42A65-10E6

toe and shoulder (of a Hurter and Driffield (H and D) curve) (photographic techniques). The terms applied to the nonlinear portions of the H and D curve that lie, respectively, below and above the straight portion of this curve. 0-1E1

toggle. Pertaining to any device having two stable states. *See:* **flip-flop.** *See also:* **electronic digital computer.** X3A12-16E9

toll board. A switchboard used primarily for establishing connections over toll lines. *See also:* **telephone switching system.** 42A65-0

toll call (telephony). Any telephone call for a destination outside the local service area of the calling station that is subject to a separate charge. *See also:* **telephone system.** 0-19E1

toll center (telephony). A switching point where trunks from end offices are connected to the distance-dialing network and where operators are present and assistance in completing incoming calls is provided in addition to other traffic operating functions. Toll centers are classified as Class 4C offices. *See also:* **telephone switching system.** 0-19E1

toll line. A telephone line or channel between two central offices in different telephone exchanges. *See also:* **telephone system.** 42A65-0

toll office. A central office primarily arranged for terminating toll lines, toll switching trunks, recording trunks, and recording-completing trunks and for their interconnection with each other as necessary for the purpose of establishing connections over toll lines. *See also:* **telephone system.** 42A65-0

toll point (telephone networks). A switching point where trunks from end offices are connected to the distance-dialing network and where operators handle only outward calls or where there are no operators present. Examples are decentralized outward switchboards, outward and terminating tandem offices, and offices where centralized machine ticketing only is provided for outward calls. Toll points are classified as Class 4P offices. *See also:* **telephone system.** 0-19E1

toll station. A public telephone station connected directly to a toll telephone switchboard. *See also:* **telephone station.** 42A65-0

toll switching trunk. A trunk extending from a toll office to a local central office for connecting toll lines to subscriber lines. *See also:* **telephone system.** 42A65-19E1

toll switch train (toll train). A switch train that carries a connection from a toll board to a subscriber line. *See also:* **switching system.** 42A65-0

toll terminal loss (toll connection). That part of the over-all transmission loss that is attributable to the facilities from the toll center through the tributary office to and including the subscriber's equipment. *Note:* The toll terminal loss at each end of the circuit is ordinarily taken as the average of the transmitting loss and the receiving loss between the subscriber and the toll center. *See also:* **transmission loss.** 42A65-0

toll train. *See:* **toll switch train.**

toll transmission selector. A selector in a toll switch train that furnishes toll-grade transmission to the subscriber and controls the ringing. *See also:* **telephone switching system.** 42A65-0

tone. (1) A sound wave capable of exciting an auditory sensation having pitch. (2) A sound sensation having pitch. E157-1E1

tone control. A means for altering the frequency response at the audio-frequency output of a circuit, particularly of a radio receiver or hearing aid, for the purpose of obtaining a quality more pleasing to the listener. *See also:* **amplifier; radio receiver.** 42A65-0

tone-modulated waves. Waves obtained from continuous waves by amplitude modulating them at audio frequency in a substantially periodic manner. *See also:* **telegraphy.** 42A65-0

tone-operated net-loss adjuster. *See:* **tonlar.**

toner (electrostatography). The image-forming material in a developer that, deposited by the field of an electrostatic-charge pattern, becomes the visible record. *See also:* **electrostatography.** E224-15E7

tone receiver (power-system communication). A device for receiving a specific voice-frequency carrier telegraph signal and converting it into direct current. *See also:* **telegraphy.** 0-31E3

tone transmitter (power-system communication). A device for transmitting a voice-frequency telegraph signal. *See also:* **telegraphy.** 0-31E3

tonlar. A system for stabilizing the net loss of a telephone circuit by means of a tone transmitted between conversations. The name is derived from the initial letters of the expression **tone-operated net-loss adjuster.** *See also:* **tone-frequency telephony.** 42A65-0

tool function (numerically controlled machines). A command identifying a tool and calling for its selection either automatically or manually. The actual changing of the tool may be initiated by a separate tool-change command. *See also:* **numerically controlled machines.** EIA3B-34E12

tool offset (numerically controlled machines). A correction for tool position parallel to a controlled axis. *See also:* **numerically controlled machines.** EIA3B-34E12

tooth (rotating machinery). A projection from a core, separating two adjacent slots, the tip of which forms part of one surface of the air gap. *See also:* **rotor (rotating machinery); stator.** 0-31E8

tooth tip (rotating machinery). That portion of a tooth that forms part of the inner or outer periphery of the air gap. It is frequently considered to be the section of a tooth between the radial location of the wedge and the air gap. *See also:* **rotor (rotating machinery); sta-**

tor. 0-31E8

top cap (side contact) (electron tubes). A small metal shell on the envelope of an electron tube or valve used to connect one electrode to an external circuit. *See also:* **electron tube.** 50I07-15E6

top car clearance (elevators). The shortest vertical distance between the top of the car crosshead, or between the top of the car where no car crosshead is provided, and the nearest part of the overhead structure or any other obstruction when the car floor is level with the top terminal landing. *See also:* **hoistway (elevator or dumbwaiter).** 42A45-0

top coil side (radially inner coil side) (rotating machinery). The coil side of a stator slot nearest the bore of the stator or nearest the slot wedge. *See also:* **stator.** 0-31E8

top counterweight clearance (elevator counterweight) (elevator). The shortest vertical distance between any part of the counterweight structure and the nearest part of the overhead structure or any other obstruction when the car floor is level with the bottom terminal landing. *See also:* **hoistway (elevator or dumbwaiter).** 42A45-0

top half bearing (rotating machinery). The upper half of a split sleeve bearing. *See also:* **bearing.** 0-31E8

top-loaded vertical antenna. A vertical antenna so constructed that, because of its greater size at the top, there results a modified current distribution giving a more desirable radiation pattern in the vertical plane. A series reactor may be connected between the enlarged portion of the antenna and the remaining structure. *See also:* **antenna.** E145-3E1

top terminal landing (elevators). The highest landing served by the elevator that is equipped with a hoistway door and hoisting-door locking device that permits egress from the hoistway side. *See also:* **elevator landing.** 42A45-0

torchere. An indirect floor lamp that sends all or nearly all of its light upward. *See also:* **luminaire.** Z7A1-0

toroid (doughnut) (electron device). A toroidal-shaped vacuum envelope in which electrons are accelerated. *See also:* **electron device.** 50I07-15E6

toroidal coil. A coil wound in the form of a toroidal helix. *See also:* **circuits and devices.** 42A65-21E0

torque (instrument). The turning moment on the moving element produced by the quantity to be measured or some quantity dependent thereon acting through the mechanism. This is also termed the deflecting torque and in many instruments is opposed by the controlling torque, which is the turning moment produced by the mechanism of the instrument tending to return it to a fixed position. *Note:* Full-scale torque is the particular value of the torque for the condition of full-scale deflection and as an index of performance should be accompanied by a statement of the angle corresponding to this deflection. *See also:* **accuracy rating (instrument).** 42A30-0

torque-coil magnetometer. A magnetometer that depends for its operation on the torque developed by a known current in a coil that can turn in the field to be measured. *See also:* **magnetometer.** 42A30-0

torque control (relay). A method of constraining the pickup of a relay by preventing the torque-producing element from developing operating torque until another associated relay unit operates. 37A100-31E11/31E6

torque margin. The increase in torque, under any steady-state operating condition, that a synchronous propulsion motor will deliver without pulling out of step. *Note:* Rated torque margin is the additional torque available when the propulsion system is operating at its designed rating. *See also:* **converter; electric propulsion system.** 42A43-0

torque motor. A motor designed primarily to exert torque through a limited travel or in a stalled position. *Note:* Such a motor may be capable of being stalled continuously or only for a limited time. *See:* **asynchronous machine; direct-current commutating machine; synchronous machine.** 42A1031E8

torquing rate (inertial navigation equipment). The angular rate at which the orientation of a gyro, with respect to inertial space, is changed in response to a command. *See also:* **navigation.** E174-10E6

torsional critical speed (rotating machinery). The speed at which the amplitudes of the angular vibrations of a machine rotor due to shaft torsional vibration reach a maximum. *See:* **rotor (rotating machinery).** 0-31E8

torsionmeter. A device to indicate the torque transmitted by a propeller shaft based on measurement of the twist of a calibrated length of the shaft. *See also:* **electric propulsion system.** 42A43-0

total break time (mechanical switching device). *See:* **interrupting time.**

total capability for load (electric power supply). The capability available to a system from all sources including purchases. *See also:* **generating station.** 0-31E4

total capacitance. *See:* **self-capacitance (conductor).**

total charge (ferroelectric device). One-half of the charge that flows as the condition of the device is changed from that of full applied positive voltage to that of full negative voltage (or vice versa). *Note:* Total charge is dependent on the amplitude of the applied voltage which should be stated when measurements of total charge are reported. *See also:* **ferroelectric domain.** E180-0

total clearing time (fuse). *See:* **clearing time (fuse).**

total current (asymmetrical current). The combination of the symmetrical component and the direct-current component of the current. 37A100-31E11

total-current regulation (axle generator). That type of automatic regulation in which the generator regulator controls the total current output of the generator. *See also:* **axle generator system.**

total cyanide (in a solution for metal deposition)(electroplating). The total content of the cyanide radical (CN), whether present as the simple or complex cyanide of an alkali or other metal *See also:* **electroplating.** 42A60-0

total electric current density. At any point, the vector sum of the conduction-current density vector, the convection-current density vector, and the displacement-current density vector at that point. E270-0

total electrode capacitance (electron tubes). The capacitance of one electrode to all other electrodes connected together. 50I07/15E6

total emissivity (element of surface of a temperature radiator). The ratio of its radiant-flux density (radiant exitance) to that of a blackbody at the same temperature. Z7A1-0

totalizing relay. A device used to receive and totalize pulses from two or more sources for proportional transmission to another totalizing relay or to a receiver. *See* *also:* **auxiliary device to an instrument.** 12A0-0

total losses (transformer or voltage regulator). The sum of the excitation losses and the load losses. *See also:* **efficiency; voltage regulator.** 42A15/57A14/57A15-31E12

totally enclosed (rotating machinery). A term applied to apparatus with an integral enclosure that is constructed so that while it is not necessarily airtight, the enclosed air has no deliberate connection with the external air except for the provision for draining and breathing. 42A95-31E8

totally enclosed fan-cooled (totally enclosed fan-ventilated). A term applied to a totally enclosed apparatus equipped for exterior cooling by means of a fan or fans, integral with the apparatus but external to the enclosing parts. *See:* **asynchronous machine; direct-current commutating machine; synchronous machine.** 42A10-31E8

totally enclosed machine. A machine so enclosed as to prevent the free exchange of air between the inside and the outside of the case but not sufficiently enclosed to be termed airtight. *See:* **asynchronous machine; direct-current commutating machine; synchronous machine.** 42A10-0

totally enclosed nonventilated (rotating machinery). A term applied to a totally enclosed apparatus that is not equipped for cooling by means external to the enclosing parts. *See:* **asynchronous machine; direct-current commutating machine; synchronous machine.** 42A10-31E8

totally enclosed pipe-ventilated machine. A totally enclosed machine except for openings so arranged that inlet and outlet ducts or pipes may be connected to them for the admission and discharge of the ventilating air. This air may be circulated by means integral with the machine or by means external to and not a part of the machine. In the latter case, these machines shall be known as separately ventilated or forced ventilated machines. *See:* **asynchronous machine; closed air-circuit; direct-current commutating machine; synchronous machine.** 42A10-0

totally enclosed ventilated apparatus. Apparatus totally enclosed in which the cooling air is carried through the case and apparatus by means of ventilating tubes and the air does not come in direct contact with the windings of the apparatus. 42A95-0

totally unbalanced currents (balanced line). Push-push currents. *See also:* **waveguide.** E146-0

total power loss (semiconductor rectifier). The sum of the forward and reverse power losses. *See also:* **rectification.** E59-34E17/34E24

total range (instrument). The region between the limits within which the quantity measured is to be indicated or recorded and is expressed by stating the two end-scale values. *Notes:* (1) If the span passes through zero, the range is stated by inserting zero or 0 between the end-scale values. (2) In specifying the range of multiple-range instruments, it is preferable to list the ranges in descending order, for example, 750/300/150. *See also:* **instrument.** 39A2/42A30-0

total sag. The distance measured vertically from any point of a conductor to the straight line joining its two points of support, under conditions of ice loading equivalent to the total resultant loading for the district in which it is located. *See:* **power distribution, overhead construction.** 2A2-0

total start-stop telegraph distortion. Refers to the time displacement of selecting-pulse transitions from the beginning of the start pulse expressed in percent of unit pulse. E145-0

total switching time (ferroelectric device). The time required to reverse the signal charge. *Note:* Total switching time is measured from the time of application of the voltage pulse, which must have a rise time much less than and a duration greater than the total switching time. The magnitude of the applied voltage pulse should be specified as part of the description of this characteristic. *See also:* **ferroelectric domain.** E180-0

total telegraph distortion. Telegraph transmission impairment, expressed in terms of time displacement of mark-space and space-mark transitions from their proper positions relative to one another, in percent of the shortest perfect pulses called the unit pulse. (Time lag affecting all transitions alike does not cause distortion). Telegraph distortion is specified in terms of its effect on code and terminal equipment. Total Morse telegraph distortion for a particular mark or space pulse is expressed as the algebraic sum of time displacements of space-mark and mark-space transitions determining the beginning and end of the pulses, measured in percent of unit pulse. Lengthening of mark is positive, and shortening, negative. *See also:* **distortion.** E145-0

total voltage regulation (rectifier). The change in output voltage, expressed in volts, that occurs when the load current is reduced from its rated value to zero or light transition load with rated sinusoidal alternating voltage applied to the alternating-current line terminals, but including the effect of the specified alternating-current system impedance as if it were inserted between the line terminals and the transformer, with the rectifier transformer on the rated tap. *Note:* The measurement shall be made with zero phase control and shall exclude the corrective action of any automatic voltage-regulating means, but not impedance. *See also:* **power rectifier; rectification.** 34A1-34E24

touch voltage (safety). The potential difference between a grounded metallic structure and a point on the earth's surface separated by a distance equal to the normal maximum horizontal reach, approximately one meter. *See also:* **ground.** E81-0

touchdown tone lights. Barettes of runway lights installed in the surface of the runway between the runway edge lights and the runway centerline lights to provide additional guidance during the touchdown phase of a landing in conditions of very poor visibility. *See also:* **signal lighting.** Z7A1-0

tower. A broad-base latticed steel support for line conductors. *Note:* For an extensive list of cross references, see *Appendix A.* 42A35-31E13

tower footing resistance (lightning protection). The resistance between the tower grounding system and true ground. *See also:* **direct-stroke protection (lightning).** 0-31E13

tower loading. The load placed on a tower by its own weight, the weight of the wires with or without ice covering, the insulators, the wind pressure normal to the line acting both on the tower and the wires and the pull from the wires in the direction of the line. *See also:* **tower.** 42A35-31E13

towing light. A lantern or lanterns fixed to the mast or hung in the rigging to indicate that a ship is towing another vessel or other objects. 42A43-0

Townsend coefficient (gas). The number of ionizing collisions per centimeter of path in the direction of the applied electric field. *See also:* **discharge (gas).** 50I07-15E6

TR. *See:* **transmit-receive.**

trace. The cathode-ray-tube display produced by a moving spot. *See:* **spot.** *See also:* **oscillograph.** 0-9E4

trace interval (television). The interval corresponding to the direction of sweep used for delineation. *See also:* **television.** E204/42A65-2E2

trace width (oscilloscope). The distance between two points on opposite sides of a trace perpendicular to the direction of motion of the spot, at which luminance is 50 percent of maximum. With one setting of the beam controls, the width of both horizontally and vertically going traces within the quality area should be stated. *See:* **oscillograph.** 0-9E4

tracing distortion. The nonlinear distortion introduced in the reproduction of mechanical recording because the curve traced by the motion of the reproducing stylus is not an exact replica of the modulated groove. For example, in the case of a sine-wave modulation in vertical recording the curve traced by the center of the tip of a stylus is a poid. *See also:* **phonograph pickup.** E157-1E1

tracing routine (computing systems). A routine that provides a historical record of specified events in the execution of a program. *See also:* **electronic digital computer.** X3A12-16E9

track (1) (in navigation). (A) The resultant direction of actual travel projected in the horizontal plane and expressed as a bearing. (B) The component of motion that is in the horizontal plane and represents the history of accomplished travel. *See also:* **navigation.** 0-10E6

(2) (in electronic computers). The portion of a moving-type storage medium that is accessible to a given reading station; for example, as on film, drum, tapes, or discs. *See also:* **band; electronic computation; electronic digital computer.** E62/E270/X3A12/EIA3B-16E9/34E12

track circuit. An electric circuit that includes the rails of a track relay as essential parts. *See also:* **railway signal and interlocking.** 42A42-0

track element. *See:* **roadway element.**

track homing (navigation). The process of following a line of position known to pass through an objective. *See also:* **radio navigation.** E172/42A65-10E6

track indicator chart. A maplike reproduction of railway tracks controlled by track circuits so arranged as to indicate automatically for defined sections of track whether such sections are or are not occupied. *See also:* **railway signal and interlocking.** 42A42-0

tracking (1) (radar). The process of keeping a radio beam, or the cross hairs of an optical system, set on a target, and usually determining the range of the target simultaneously.

(2) (electric). The maintenance of proper frequency relations in circuits designed to be simultaneously varied by gang operation.

(3) (phonographic technique). The accuracy with which the stylus of a phonograph pickup follows a prescribed path. E145/E188/42A65-10E6

(4) (instrument). The ability of an instrument to indicate at the division line being checked, when energized by corresponding proportional value of actual end-scale excitation, expressed as a percentage of actual end-scale value. 39A1-0

tracking error (1) (general). The deviation of a dependent variable with respect to a reference function. *Note:* As applied to power inverters, tracking error may be the deviation of the output volts per hertz from a prescribed profile or the deviation of the output frequency from a given input synchronizing signal or others. *See:* **self-commutated inverters.** 0-34E24

(2) (phonographic techniques) (lateral mechanical recording). The angle between the vibration axis of the mechanical system of the pickup and a plane containing the tangent to the unmodulated record groove that is perpendicular to the surface of the recording medium at the point of needle contact. *See also:* **phonograph pickup.** 0-1E1

track instrument. A device in which the vertical movement of the rail or the blow of a passing wheel operates a contact to open or close an electric circuit. *See also:* **railway signal and interlocking.** 42A42-0

trackless trolley coach. *See:* **trolley coach (trolley bus).**

track relay. A relay receiving all or part of its operating energy through conductors of which the track rails are an essential part and that responds to the presence of a train on the track. *See also:* **railway signal and interlocking.** 42A42-0

traction machine (elevators). A direct-drive machine in which the motion of a car is obtained through friction between the suspension ropes and a traction sheave. *See also:* **driving machine (elevators).** 42A45-0

traction motor. An electric propulsion motor used for exerting tractive force through the wheels of a vehicle. *See:*
auxiliary generator;
axle bearing;
axle-hung motor;
control generator;
fixed motor connections;
gearless motor;
nose suspension;
overload relay;
quill drive;
splashproof. 42A42-0

tractive force (tractive effort) (electrically propelled vehicle). The total propelling force measured at the rims of the driving wheels, or at the pitch line of the gear rack in the case of a rack vehicle. *Note:* Tractive force of an electrically propelled vehicle is commonly qualified by such terms as: maximum starting tractive force; short-time-rating tractive force; continuous-rating tractive force. *See also:* **electric locomotive.** 42A42-0

tradeoff. Parametric analysis of concepts or components for the purpose of optimizing the system or some trait of the system. *See:* **system science.** 0-35E2

traffic-control system. A block signal system under which train movements are authorized by block signals whose indications supersede the superiority of trains for both opposing and following movements on the same track. *See also:* **centralized traffic-control system.** 42A42-0

traffic locking. Electric locking adapted to prevent the manipulation of levers or other devices for changing the direction of traffic on a section of track while that section is occupied or while a signal is displayed for a train to proceed into that section. *See also:* **interlocking (interlocking plant).** 42A42-0

trailing edge (television). The major portion of the decay of a pulse. *See also:* **television.** E203-2E2

trailing edge, pulse. The major transition towards the pulse baseline occurring before a reference time. *See also:* **pulse.** 0-9E4

trailing-edge pulse time. The time at which the instantaneous amplitude last reaches a stated fraction of the peak pulse amplitude. *See also:* **pulse terms.** E194-2E2

trailing-type antenna (aircraft). A flexible conductor usually wound on a reel within the aircraft passing through a fairlead to the outside of the aircraft, terminated in a streamlined weight or wind sock and fed out to the proper length for the desired radio frequency of operation. It has taken other forms such as a capsule that when exploded releases the antenna. 42A41-0

train (illuminating engineering). The angle between the vertical plane through the axis of the searchlight drum and the plane in which this plane lies when the searchlight is in a position designated as having zero train. *See also:* **searchlight.** Z7A1-0

train-control territory. That portion of a division or district equipped with an automatic train-control system. *See also:* **automatic train control.** 42A42-0

train describer. An instrument used to give information regarding the origin, destination, class, or character of trains, engines, or cars moving or to be moved between given points. *See also:* **railway signal and equipment.** 42A42-0

train-line coupler. A group of devices that connects the electric train-line circuits of two adjacent vehicles in the train and consists of the following: (1) A **jumper,** which is the removable member of a circuit connecting two adjacent cars or locomotives together, although sometimes one end of the connecting cable is permanently fixed. *Note:* This term is usually preceded by a designating name, such as control jumper, bus line jumper, train line jumper, etcetera. (2) A **coupler** plug, which is that portion of a jumper that serves to directly connect the jumper wiring to the coupler socket or receptacle. (3) A **coupler socket** or **receptacle,** which is that portion of the jumper attached to the locomotive or car and serves to directly connect the car or locomotive wiring to the jumper coupler plug. *See also:* **car wiring apparatus.** E16-0

transadmittance. For harmonically varying quantities at a given frequency, the ratio of the complex amplitude of the current at one pair of terminals of a network to the complex amplitude of the voltage across a different pair of terminals. *See:* **interelectrode transadmittance (*j*—*l* interelectrode transadmittance of an *n*-electrode electron tube).** *See also:* **transmission characteristics.** 0-9E4

transceiver. The combination of radio transmitting and receiving equipment in a common housing, usually for portable or mobile use, and employing common circuit components for both transmitting and receiving. *See also:* **radio transmission.** 42A65-0

transconductance. The real part of the transadmittance. *Note:* Transconductance is, as most commonly used, the interelectrode transconductance between the control grid and the plate. At low frequencies, transconductance is the slope of the control-grid-to-plate transfer characteristic. *See also:* **electron-tube admittances; interelectrode transconductance; transmission characteristics.** 42A70-9E4/15E6

transconductance meter (mutual-conductance meter). An instrument for indicating the transconductance of a grid-controlled electron tube. *See also:* **instrument.** 42A30-0

transcribe (electronic computation). To convert data recorded in a given medium to the medium used by a digital computing machine or vice versa. *See also:* **electronic digital computer.** E162-0

transcriber (electronic computation). Equipment associated with a computing machine for the purpose of transferring input (or output) data from a record of information in a given language to the medium and the language used by a digital computing machine (or from a computing machine to a record of information). *See also:* **electronic computation.** E270-0

transducer (communication and power transmission). A device by means of which energy can flow from one or more transmission systems or media to one or more other transmission systems or media. *Note:* The energy transmitted by these systems or media may be of any form (for example, it may be electric, mechanical, or acoustical), and it may be of the same form or different forms in the various input and output systems or media. *Note:* For an extensive list of cross references, see *Appendix A.* E145/E151/E196/E270/42A65/85A1/X3A12-13E6/16E9/23E0/31E3/34E10

transducer, active. A transducer whose output waves are dependent upon sources of power, apart from that supplied by any of the actuating waves, which power is controlled by one or more of the waves. *Note:* The definition of active transducer is a restriction of the more general **active network;** that is, one in which there is an impressed driving force. *See also:* **transducer.** E151/E196/E270/42A65-0

transducer gain. The ratio of the power that the transducer delivers to the specified load under specified operating conditions to the available power of the specified source. *Notes:* (1) If the input and/or output power consist of more than one component, such as multifrequency signals or noise, then the particular components used and their weighting must be specified. (2) This gain is usually expressed in decibels. *See also:* **transducer.** E151/E196/E270/42A65-0

transducer ideal (for connecting a specified source to a specified load). A hypothetical passive transducer that transfers the maximum available power from the source to the load. *Note:* In linear transducers having only one input and one output, and for which the impedance concept applies, this is equivalent to a transducer that (1) dissipates no energy and (2) when connected to the specified source and load presents to each its conjugate impedance. *See also:* **transducer.** E151/E196/E270/42A65-0

transducer, line. *See:* **line transducer.**

transducer loss. The ratio of the available power of the specified source to the power that the transducer delivers to the specified load under specified operating conditions. *Notes:* (1) If the input and/or output power consist of more than one component, such as multifrequency signals or noise, then the particular components used and their weighting must be specified. (2) This loss is usually expressed in decibels. *See also:* **transducer.** E151/E196/E270/42A65-21E1

transducer, passive. A transducer that has no source of power other than the input signal(s), and whose output signal-power cannot exceed that of the input. *Note:* The definition of a passive transducer is a restriction of the more general **passive network,** that is, one containing no impressed driving forces. *See also:* **transducer.** E151/E196/E270-0

transfer (1) (electronic computation). (A) To transmit, or copy, information from one device to another. (B) To jump. (C) The act of transferring. *See also:* **electronic computation; electronic digital computer; jump; transmit.** E162/E270-0
(2) (electrostatography). The act of moving a developed image, or a portion thereof, from one surface to another, as by electrostatic or adhesive forces, without altering the geometric configuration of the image. *See also:* **electrostatography.** E224-15E7

transfer admittance (1) (linear passive networks, general). A transmittance for which the excitation is a voltage and the response is a current. *See also:* **linear passive networks.** E156-0
(2) (from the *j*th terminal to the *l*th terminal of an *n*-terminal network). The quotient of (A) the complex alternating component of the current flowing to the *l*th terminal from the *l*th external termination by (B) the complex alternating component of the voltage applied to the *j*th terminal with respect to the reference point when all other terminals have arbitrary external terminations. *See also:* **electron-tube admittances; network analysis.** E160/42A70-15E6

transfer characteristic (1) (electron tubes). A relation, usually shown by a graph, between the voltage of one electrode and the current to another electrode, all other electrode voltages being maintained constant. *See also:* **circuit characteristics of electrodes; electrode (electron tube).** E188/42A70-0
(2) (camera tubes). A relation between the illumination on the tube and the corresponding signal output current, under specified conditions of illumination. *Note:* The relation is usually shown by a graph of the logarithm of the signal output current as a function of the logarithm of the illumination. *See also:* **beam tubes; illumination; sensitivity; television.** E160-2E2/15E6

transfer check (electronic computation). A check (usually an automatic check) on the accuracy of a data transfer. *Note:* In particular, a check on the accuracy of the transfer of a word. *See:* **electronic digital computer.** E162/E270/X3A12-16E9

transfer constant (electric transducer). *See:* **image transfer constant.**

transfer control (electronic computation). *See:* **jump.**

transfer current (glow-discharge cold-cathode tube). The starter-gap current required to cause conduction across the main gap. *Note:* The transfer current is a function of the anode voltage. *See:* **gas tubes.** 42A70-15E6

transfer-current ratio (linear passive network). A transmittance for which the variables are currents. *Note:* The word **transfer** is frequently dropped in present usage. *See also:* **linear passive networks.** E156-0

transfer function. *See:* **function, transfer.**

transfer immittance. *See:* **transmittance.**

transfer impedance (linear passive networks). A transmittance for which the excitation is a current and the response is a voltage. *Note:* It is therefore the impedance obtained when the response is determined at a point other than that at which the driving force is applied, all terminals being terminated in any specified manner. In the case of an electric circuit, the response would be determined in any branch except that in which the driving force is. *See also:* **self-impedance; linear passive networks; network analysis.** E156/E270/42A65-0

transfer locus (linear system or element). A plot of the transfer function as a function of frequency in any convenient coordinate system. *Note:* A plot of the reciprocal of the transfer function is called the inverse transfer locus. *See:* **amplitude frequency locus, locus; phase locus.** *See also:* **control system, feedback.** 0-23E0

transfer of control. Same as jump. *See also:* **electronic digital computer.**

transfer ratio. A dimensionless transfer function. E270-0

transfer ratio correction (correction to setting). The deviation of the output phasor from nominal, in proportional parts of the input phasor.

$$\frac{\text{Output}}{\text{Input}} = A + \alpha + j\beta$$

INPUT A OUTPUT

A = setting
α = in-phase transfer ratio correction
β = quadrature transfer ratio correction. 0-9E3

transfer standards, alternating-current–direct current. Devices used to establish the equality of a root-mean-square current or voltage (or the average value of alternating power) with the corresponding steady-state direct-current quantity that can be referred to the basic standards through potentiometric techniques. *See also:* **auxiliary device to an instrument.** 12A0-0

transfer switch (high-voltage switch). A switch arranged to permit transferring a conductor connection from one circuit to another without interrupting the current. (1) A **tandem transfer switch** is a switch with two blades, each of which can be moved into or out of only one contact. (2) A **double-blade double-throw transfer switch** is a switch with two blades, each of which can be moved into or out of either of two contacts. *Note:* In contrast to high-voltage switches, many low-voltage control and instrument transfer switches interrupt current during transfer. Compare with **selector switch.** Also compare with **automatic transfer (or throw-over) equipment** that connects a load to an alternative source after failure of an original source. *See also:* **switch.** 37A100/42A25-31E11/34E10

transfer time, relay. *See:* **relay transfer time.**

transfer trip. A form of remote release in which a communication channel is used to transmit the signal for release from the relay location to a remote location. 37A100-31E6/31E11

transfer voltage ratio (linear passive networks). A transmittance for which the variables are voltages. *Note:* The word **transfer** is frequently dropped in present usage. *See also:* **linear passive networks.** E156-0

transfer winding (rotating machinery). A winding for which coils are form-wound to a suitable shape and then inserted into slots or around poles by a mechanical means. *See also:* **rotor (rotating machinery); stator.** 0-31E8

transferred charge (capacitor). The net electric charge transferred from one terminal of a capacitor to another via an external circuit. *See also:* **nonlinear capacitor.** E226-15E7

transferred-charge characteristic (nonlinear capacitor). The function relating transferred charge to capacitor voltage. *See also:* **nonlinear capacitor.** E226-15E7

transferred information. *See:* **transinformation.**

transform (computing systems). To change the form of data according to specific rules. *See also:* **electronic digital computer.** X3A12-16E9

transformation (impedance or admittance) (rotating machinery). The result of one of many mathematical processes that transforms the original impedance or admittance of a machine or system into a more manageable form. During the process all currents and voltages undergo consistent transformations. *Note:* Following are some of the better-known transformations: (1) Symmetrical component transformation, also known as Fortescue transformation, sequence component transformation, and phase-sequence transformation. (2) Synchronously rotating reference frame transformation, also known as Park transformation and *d-q* transformation. (3) Clark transformation, also known as *a-b* component transformation, and equivalent two-phase transformations. (4) Complex rotating transformation, also known as *Ku* transformation, *f-b* transformation, and forward-backward component transformation. *See also:* **asynchronous machine; direct-current commutating machine; synchronous machine.** 0-31E8

transformer. A device consisting of a winding with tap or taps, or two or more coupled windings, with or without a magnetic core, for introducing mutual coupling between electric circuits.
See:
autotransformer;
auxiliary power transformer;
compound-filled transformer;
connection diagram;
constant-current transformer;
constant-voltage transformer;
dry-type;
grounding transformer;
indoor;
instrument transformer;
moisture-resistant;
network analysis;
oil-immersed;
phase-shifting transformer;
pole-type;
protected outdoor transformer;
reactor;
rectifier transformer;
routine test;
Scott-connected transformer assembly;
secondary voltage rating;
stabilizing winding;
station-type;
submersible;
subway-type;
transformer, network;
transformer, outdoor;
transformer removable cable-terminating box;
transformer secondary current rating;
transformer, series;
transformer, shunt;
transformer, specialty;
transformer, step-down;
transformer, step-up;
vault-type transformer;
weatherproof. 0-21E0

transformer, alternating-current arc welder. A transformer with isolated primary and secondary windings and suitable stabilizing, regulating, and indicating devices required for transforming alternating current from normal supply voltages to an alternating-current output suitable for arc welding. *See also:* **electric arc-welding apparatus.** 87A1-0

transformer class designations. *See:* **transformer, oil-immersed.**

transformer, constant-voltage (constant-potential transformer). *See:* **constant-voltage transformer.**

transformer correction factor. *See:* **instrument transformer correction factor.**

transformer, dry-type. *See:* **dry-type.** *See also:* **transformer, oil-immersed.**

transformer, energy-limiting. A transformer that is intended for use on an approximately constant-voltage supply circuit and that has sufficient inherent impedance to limit the output current to a thermally safe maximum value. *See:* **transformer, specialty.** 42A15-31E12

transformer equipment rating. A volt-ampere output together with any other characteristics, such as voltage, current, frequency, and power factor, assigned to it by the manufacturer. *Note:* It is regarded as a test rating that defines an output that can be taken from the item of transformer equipment without exceeding established temperature-rise limitations, under prescribed conditions of test and within the limitations of established standards. *See:* **duty.** 42A15-31E12

transformer, grounding. *See:* **grounding transformer.**

transformer grounding switch and gap (capacitance potential device). Consists of a protective gap connected across the capacitance potential device and transformer unit to limit the voltage impressed on the transformer and the auxiliary or shunt capacitor, when used; and a switch that when closed removes voltage from the potential device to permit adjustment of the potential device without interrupting high-voltage line operation and carrier-current operation when used. *See also:* **outdoor coupling capacitor.** E31-0

transformer, group-series loop insulating. An insulating transformer whose secondary is arranged to operate a group of series lamps and/or a series group of individual-lamp transformers. *See:* **transformer, specialty.** 42A15-31E12

transformer, high-power-factor. A high-reactance transformer that has a power-factor-correcting device such as a capacitor, so that the input current is at a power factor of not less than 90 percent when the transformer delivers rated current to its intended load device. *See:* **transformer, specialty.** 89A2/89A1-0

transformer, high-reactance (1) (output limiting). An energy-limiting transformer that has sufficient inherent reactance to limit the output current to a maximum value . *See:* **transformer, specialty.** 42A15-31E12

(2) secondary short-circuit current rating. The current in the secondary winding when the primary winding is connected to a circuit of rated primary voltage and frequency and when the secondary terminals are short-circuited. *See:* **transformer, specialty.** 42A15-31E12

(3) (kilovolt-ampere or voltampere short-circuit input rating). The input kilovolt-amperes or volt-amperes at rated primary voltage with the secondary terminals short-circuited. *See:* **transformer, specialty.** 42A15-31E12

transformer, ideal. A hypothetical transformer that neither stores nor dissipates energy. *Note:* An ideal transformer has the following properties: (1) Its self- and mutual impedances are equal and are pure inductances of infinitely great value. (2) Its self-inductances have a finite ratio. (3) Its coefficient of coupling is unity. (4) Its leakage inductance is zero. (5) The ratio of the primary to secondary voltage is equal to the ratio of secondary to primary current. *See also:* **circuits and devices.** E151-21E1

transformer, individual-lamp insulating. An insulating transformer used to protect the secondary circuit, casing, lamp, and associated luminaire of an individual street light from the high-voltage hazard of the primary circuit. *See:* **transformer, specialty.** 42A15-31E12

transformer, insulating. A transformer used to insulate one circuit from another. *See:* **transformer, specialty.** 42A15-31E12

transformer integrally mounted cable terminating box. A weatherproof air-filled compartment suitable for enclosing the sidewall bushings of a transformer and equipped with any one of the following entrance devices: (1) Single or multiple-conductor potheads with couplings or wiping sleeves. (2) Wiping sleeves. (3) Couplings with or without stuffing boxes for conduit-enclosed cable, metallic-sheathed cable, or rubber-covered cable. 57A12.76-0

transformer, isolating (1) (signal-transmission system). A transformer inserted in a system to separate one section of the system from undesired influences of the other sections. *Example:* A transformer having electrical insulation and electrostatic shielding between its windings such that it can provide isolation between parts of the system in which it is used. It may be suitable for use in a system that requires a guard for protection against common-mode interference. *See:* **signal.** 0-13E6

(2) (electroacoustics). A transformer inserted in a system to separate one section of the system from undesired influences of other sections. *Note:* Isolating transformers are commonly used to isolate system grounds and prevent the transmission of undesired currents. *See:* **circuits and devices.** E151-0

transformer, line. *See:* **line transformer.**

transformer loss (communication). The ratio of the signal power that an ideal transformer would deliver to a load, to the power delivered to the same load by the actual transformer, both transformers having the same impedance ratio. *Note:* Transformer loss is usually expressed in decibels. *See also:* **transmission loss.** E151/42A65-0

transformer-loss compensator. A passive electric network that is connected in series-parallel with a meter to add to or to subtract from the meter registration active or reactive components of registration proportional to predetermined iron and copper losses of transformers and transmission lines. *See also:* **auxiliary device to an instrument.** 12A0-0

transformer, low-power-factor. A high-reactance transformer that does not have means for power-factor correction. *See:* **transformer, specialty.** 89A1-0

transformer, network. A transformer designed for use in a vault to feed a variable-capacity system of interconnected secondaries. *See:* **transformer.** 42A15-31E12

transformer, nonenergy-limiting. A constant-potential transformer that does not have sufficient inherent impedance to limit the output current to a thermally safe maximum value. *See:* **transformer, specialty.** 42A15-31E12

transformer, oil-immersed (1) (general). A transformer in which the core and coils are immersed in an insulating oil.

See:
askarel;
conservator system;
gas-oil sealed system;
inert-gas pressure system;
inhibited oil;
oil;
oil-immersed;
sealed-tank system;
uninhibited oil.
See also: **transformer.** 42A15-31E12

(2) class designations (oil-immersed transformers, reactors, regulators, etcetera):

FA (forced-air-cooled). Cooled by the forced circulation of air over the cooling surface.

FO (oil-immersed forced-oil-cooled). Cooled by forced circulation of the cooling oil through some external cooling means.

FOA (oil-immersed forced-oil-cooled, with forced-air cooler). Cooled by the forced circulation of oil through external oil-to-air heat-exchanger equipment utilizing forced circulation of air over its cooling surface.

FOW (oil-immersed forced-oil-cooled with forced-water cooler). Cooled by the forced circulation of the oil through external oil-to-water heat exchanger equipment utilizing forced circulation of water over its cooling surface.

OA (oil-immersed self-cooled). Cooled by natural circulation of the cooling air over the cooling surface.

OW (oil-immersed water-cooled). Cooled by the natural circulation of the cooling oil over the water-cooled surface.

OFA (oil-immersed forced-air-cooled). Cooled by forced circulation of the cooling air over the cooling surface. *Note:* As installed, oil-immersed transformers, reactors, regulators, etcetera, are frequently provided with a choice of several methods of cooling the oil. In such cases, multiple class codes such as OA/FA/FOA or OA/FOA/FOA designate the several available options, each option having a corresponding power rating. For example, OA/FOA/FOA means having (1) a self-cooled rating with cooling obtained by natural circulation of the cooling air over the air-cooled surface, and (2) two forced-oil-cooled ratings with auxiliary cooling controls arranged to start a portion of the oil pumps and fans for the first auxiliary rating and the remainder of the pumps and fans for the second auxiliary rating. In addition, transformers, regulators, reactors, etcetera, not immersed in oil (dry-type) may have the following class designations:

AA (dry-type self-cooled). Cooled by the natural circulation of the cooling air.

AFA (dry-type forced-air-cooled). Cooled by the forced circulation of the cooling air through the core and coils.

AA/FA (dry-type self-cooled/forced-air-cooled). Having two ratings corresponding to the cooling methods used, as described for oil-immersed.

transformer, outdoor. A transformer of weatherproof construction. *See:* **transformer.** 42A15-31E12

transformer overcurrent tripping. *See:* **indirect release and overcurrent release.**

transformer, protected outdoor. A transformer that is not of weatherproof construction but that is suitable for outdoor use if it is so installed as to be protected from rain or immersion in water. *See also:* **transformer.** 42A15-31E12

transformer, phase-shifting. *See:* **phase-shifting transformer.**

transformer, pole-type. *See:* **pole-type.**

transformer primary voltage rating. *See:* **primary voltage rating.**

transformer-rectifier, alternating-current–direct-current arc welder. A combination of static rectifier and the associated isolating transformer, reactors, regulators, control, and indicating devices required to produce either direct or alternating current suitable for arc-welding purposes. *See also:* **electric arc-welding apparatus.** 87A1-0

transformer-rectifier, direct-current arc welder. A combination of static rectifiers and the associated isolating transformer, reactors, regulators, control, and indicating devices required to produce direct current suitable for arc welding. *See also:* **electric arc-welding apparatus.** 87A1-0

transformer relay. A relay in which the coils act as a transformer. *See also:* **railway signal and interlocking.** 42A42-0

transformer removable cable-terminating box. A weatherproof air-filled compartment suitable for enclosing the sidewall bushings of a transformer and equipped with mounting flange(s) (one or two) to accommodate either single-conductor or multiconductor potheads or entrance fittings, depending upon the type of cable termination to be used and the number of three-phase cable circuits (one or two) to be terminated.
See:
cable entrance fitting;
cable sheath insulator;
compartment-cover mounting plate;
couplings;
external connector;
internal connector;
pothead;
pothead insulator;
pothead insulator lid;
pothead mounting plate;
pothead mounting-plate insulator;
wiping gland. *See also:* **transformer.** 57A12.75-0

transformer, assembly, Scott-connected. *See:* **Scott-connected transformer assembly.**

transformer secondary current rating. *See:* **secondary current rating (transformer).**

transformer, series. A transformer in which the primary winding is connected in series with a power-supply circuit, and that transfers energy to another circuit at the same or different current from that in the primary circuit. *See:* **transformer.** 42A15-31E12

transformer, series street-lighting. A series transformer that receives energy from a current-regulating series circuit and that transforms the energy to another winding at the same or different current from that in the primary. *See:* **transformer, specialty.** 42A15-31E12

transformer, series street-lighting, rating. The lumen rating of the series lamp, or the wattage rating of the multiple lamps, that the transformer is designed to operate. *See:* **specialty transformer.** 42A15-31E12

transformer, shunt. A transformer in which the primary winding is connected in shunt with a power-supply circuit, and that transfers energy to another circuit at the same or different voltage from that of the primary circuit. *See:* **transformer.** 42A15-31E12

transformer, specialty. A transformer generally intended to supply electric power for control, machine tool, Class 2, signaling, ignition, luminous-tube, cold-cathode lighting, series street-lighting, low-voltage general purpose, and similar applications.
See:
autotransformer, individual-lamp;
primary voltage rating;
secondary voltage rating;
transformer, energy-limiting;
transformer, group-series loop insulating;
transformer, high-power-factor;
transformer, high-reactance;
transformer, individual-lamp insulating;
transformer, insulating;
transformer, low-power-factor;
transformer, nonenergy-limiting;
transformer, series street lighting;
transformer, series street-lighting, rating.
See also: **transformer.** 42A15-31E12

transformer, station-type. *See:* **station-type.**

transformer, step-down. A transformer in which the energy transfer is from a high-voltage circuit to a low-voltage circuit. *See:* **transformer.** 42A15-31E12

transformer, step-up. A transformer in which the energy transfer is from a low-voltage circuit to a high-voltage circuit. *See:* **transformer.** 42A15/31E12

transformer, submersible. *See:* **submersible.**

transformer, subway-type. *See:* **subway-type transformer.**

transformer undercurrent tripping. *See:* **indirect release and undercurrent release.**

transformer, unit-substation. *See:* **unit substation.**

transformer vault. An isolated enclosure either above or below ground, with fire-resistant walls, ceiling, and floor, for unattended transformers and their auxiliaries. *See also:* **power distribution, underground construction.** 2A2/42A35-31E13

transformer, vault-type. *See:* **vault-type transformer.**

transformer voltage (network protector). The voltage between phases or between phase and neutral on the transformer side of a network protector. 37A100-31E11

transformer, variable-voltage. A voltage regulator in which the output voltage can be changed (essentially from turn to turn) by means of a movable contact device. *See:* **voltage regulator.** 42A15-31E12

transformer, waveguide. A device, usually fixed, added to a waveguide or transmission line for the purpose of impedance transformation. *See also:* **waveguide.** E147-3E1

transient (industrial control). That part of the variation in a variable that ultimately disappears during transition from one steady-state operating condition to another. *Note:* Using the term to mean the total variation during the transition between two steady states is deprecated. *See also:* **control system, feedback; signal.** AS1;85A1-13E6/23E0/34E10

transient analyzer. An electronic device for repeatedly producing in a test circuit a succession of equal electric surges of small amplitude and of adjustable waveform, and for presenting this waveform on the screen of an oscilloscope. *See also:* **oscillograph.** 42A30-0

transient-cause forced outage (electric power systems). A component outage whose cause is immediately self-clearing so that the affected component can be restored to service either automatically or as soon as a switch or circuit breaker can be reclosed or a fuse replaced. *Note:* An example of a transient-cause forced outage is a lightning flashover that does not permanently disable the flashed component. *See also:* **outage.** 0-31E4

transient-cause forced outage duration (electric power systems). The period from the initiation of the outage until the affected component is restored to service by switching or fuse replacement. *See also:* **outage .** 0-31E4

transient current (rotating machinery). (1) The current under nonsteady conditions. (2) The alternating component of armature current immediately following a sudden short-circuit, neglecting the rapidly decaying component present during the first few cycles. *See:* **synchronous machine.** 0-31E8

transient-decay current (photoelectric device). The decreasing current flowing in the device after the irradiation has been abruptly cut off. *See also:* **phototubes.** 0-15E6

transient deviation (control). *See:* **deviation, transient.**

transient fault (lightning arresters). A fault that disappears of its own accord. *See also:* **lightning arrester (surge diverter).** 50I25-31E7

transient internal voltage (synchronous machine) (for any specified operating condition). The fundamental-frequency component of the voltage of each armature phase that would be determined by suddenly removing the load, without changing the excitation voltage applied to the field, and extrapolating the envelope of the voltage back to the instant of load removal, neglecting the voltage components of rapid decrement that may be present during the first few cycles after removal of the load. *Note:* The transient internal voltage, as shown in the phasor diagram, is related to the terminal-voltage and phase-current phasors by the equation

$$\mathbf{E}_i'=\mathbf{E}_a+R\mathbf{I}_a+jX_d'\,\mathbf{I}_{ad}+jX_q'\,\mathbf{I}_{aq}\,.$$

For a machine subject to saturation, the reactances should be determined for the degree of saturation applicable to the specified operating condition. *See:* **direct-axis synchronous reactance; phasor diagram.** *See also:* **synchronous machine.** 42A10-31E8

transient motion (audio and electroacoustics). Any motion that has not reached or that has ceased to be a steady state. *See also:* **electroacoustics.** 0-1E1

transient overshoot. An excursion beyond the final steady-state value of output as the result of a step-input change. *Note:* It is usually referred to as the first such excursion; expressed as a percent of the steady-state output step. *See also:* **accuracy rating (instrument); control system, feedback.** 39A4-0

transient performance (synchronous-machine regulating system). The performance under a specified stimulus, before the transient expires. *See also:* **synchronous machine.** 0-31E8

transient phenomena (rotating machinery). Phenomena appearing during the transition from one operating condition to another. 50I05-31E8

transient recovery voltage (circuit-switching device). The voltage transient that occurs across the terminals of a pole upon interruption of the current. *Notes:* (1) It is the difference between the transient voltages to ground occurring on the terminals. It may be a circuit transient recovery voltage, a modified circuit transient recovery voltage, or an actual transient recovery voltage. (2) In a multipole circuit breaker, the term is usually applied to the voltage across the first pole to interrupt. For circuit breakers having several interrupting units in series, the term may be applied to the voltage across units or groups of units. 37A100-31E11

transient recovery voltage rate (circuit switching device). The rate at which the voltage rises across the terminals of a pole upon interruption of the current. *Note:* It is usually determined by dividing the voltage at one of the crests of the transient recovery voltage by the time from current zero to that crest. In case no definite crest exists, the rate may be taken to some stated value usually arbitrarily selected as a certain percentage of the crest value of the normal-frequency recovery voltage. In case the transient is an exponential function the rate may also be taken at the point of zero voltage. It is the rate of rise of the algebraic difference between the transient voltages occurring on the terminals of the switching device upon interruption of the current. The transient recovery voltage rate may be a circuit transient recovery voltage rate or a modified circuit transient recovery voltage rate, or an actual transient recovery voltage rate according to the type of transient from which it is obtained. When giving actual transient recovery voltage rates, the points between which the rate is measured should be definitely stated. 37A100-31E11

transient response (pulse techniques). The time response of a system or device under test to a stated input stimulus. *Note:* Step functions, impulse functions, and ramps are the commonly used stimuli. *See also:* **pulse.** 0-9E4

transient speed deviation (1) (load decrease) (gas turbines). The maximum instantaneous speed above the steady-state speed occurring after the sudden decrease from one specified steady-state electric load to another specified steady-state electric load having values within limits of the rated output of the gas-turbine–generator unit. It is expressed in percent of rated speed. E282-31E2

(2) (load increase) (gas turbines). The minimum instantaneous speed below the steady-state speed occurring after the sudden increase from one specified steady-state electric load to another specified steady-state electric load having values within the limits of rated output of the gas-turbine–generator unit. It is expressed in percent of rated speed. E282-31E2

transient stability. A condition that exists in a power system if, after an aperiodic disturbance, the system regains steady-state stability. *See also:* **alternating-current distribution.** 42A35-31E13

transient stability factor (system or part of a system). The ratio of the transient stability limit to the nominal power flow at the point of the system to which the stability limit is referred. *See:* **stability factor.** *See also:* **alternating-current distribution.** 42A35-31E13

transient stability limit (transient power limit). The maximum power flow possible through some particular point in the system when the entire system or the part of the system to which the stability limit refers is operating with transient stability. *See also:* **alternating-current distribution.** 42A35-31E13

transient thermal impedance (semiconductor device). The change of temperature-difference between two specified points or regions at the end of a time interval, divided by the step-function change in power dissipation at the beginning of the same time interval causing the change of temperature-difference. *Notes:* (1) Such thermal measurements are commonly made at a junction of dissimilar materials. For example, in semiconductor rectifiers. (2) It is the thermal impedance of the junction under conditions of change and is generally given in the form of a curve as a function of the duration of an applied pulse. *See also:* **principal voltage-current characteristic (principal characteristic); semiconductor rectifier stack.** E223-15E7/34E17/34E24

transimpedance (1) (general). For harmonically varying quantities at a given frequency, the ratio of the complex amplitude of the voltage at one pair of terminals of a network to the complex amplitude of the current across a different pair of terminals. *See also:* **transmission characteristics.** 0-9E4

(2) (of a magnetic amplifier). The ratio of differential output voltage to differential control current. *See also:* **rating and testing magnetic amplifiers.** E107-0

transinformation (of an output symbol about an input symbol) (information theory). The difference between the information content of the input symbol and the conditional information content of the input symbol given the output symbol. *Notes:* (1) If x_i is an input symbol and y_j is an output symbol, the transinformation is equal to

$$[-\log p(x_i)] - [-\log p(x_i | y_j)] = \log \frac{p(x_i | y_j)}{p(x_i)} = \log \frac{p(x_i, y_j)}{p(x_i)\, p(y_j)}$$

where $p(x_i|y_j)$ is the conditional probability that x_i was transmitted when y_j is received, and $p(x_i, y_j)$ is the joint probability of x_i and y_j. (2) This quantity has been called **transferred information, transmitted information,** and **mutual information.** *See also:* **information theory.** E171-0

transistor. An active semiconductor device with three or more terminals.

See:
base;
base electrode;
base region;
boundary, $p-n$;
collector;
emitter;
emitter, majority;
emitter, minority;
junction, collector (of a transistor);
junction, emitter (of a transistor);
transistor, conductivity modulation;
transistor, filamentary;
transistor, junction;
transistor, point-contact;
transistor, point-junction;
transistor, unipolar;
transition region. *See also:* **semiconductor.**
E102/E216/E270/42A70-34E17

transistor, conductivity-modulation. A transistor in which the active properties are derived from minority-carrier modulation of the bulk resistivity of a semiconductor. *See also:* **semiconductor; transistor.** E102/E216/E270/42A70-34E17

transistor, filamentary. A conductivity-modulation transistor with a length much greater than its transverse dimensions. *See also:* **semiconductor; transistor.** E102/E216/E270/42A70-34E17

transistor, junction. A transistor having a base electrode and two or more junction electrodes. *See also:* **transistor.** E102/E216/E270/42A70-34E17

transistor, point-contact. A transistor having a base electrode and two or more point-contact electrodes. *See also:* **semiconductors; transistor.** E102/E216/E270/42A70-34E17

transistor, point-junction. A transistor having a base electrode and both point-contact and junction electrodes. *See also:* **transistor.** E102/42A70-0

transistor, unipolar. A transistor that utilizes charge carriers of only one polarity. *See also:* **semiconductor; transistor.** E102/E216/E270/42A70-34E17

transit (electronic navigation). A radio navigation system using low-orbit satellites to provide world-wide coverage, with transmissions from the satellites at very- and ultra-high frequencies, in which fixes are determined from measurements of the Doppler shift of the continuous-wave signal received from the moving satellite. *See also:* **navigation.** 0-10E6

transit angle. The product of angular frequency and the time taken for an electron to traverse a given path. *See also:* **electron emission.** 42A70-15E6

transition (1) (motor control). The procedure of changing from one scheme of motor connections to another scheme of connections, such as from series to parallel. *See also:* **multiple-unit control.**

(2) (signal transmission). The change from one circuit condition to the other, that is, the change from mark to space or from space to mark. *See also:* **data transmission; multiple-unit control.**

(3) (waveform) (pulse techniques). A change of the instantaneous amplitude from one amplitude level to another amplitude level. *See also:* **pulse.**

(4) *See:* **adapter.** 42A42-9E4/19E4

transition frequency (disk recording system) (crossover frequency) (turnover frequency). The frequency corresponding to the point of intersection of the asymptotes to the constant-amplitude and the constant-velocity portions of its frequency response curve. This curve is plotted with output voltage ratio in decibels as the ordinate and the logarithm of the frequency as the abscissa. *See also:* **phonograph pickup.** 0-1E1

transition load (rectifier circuit). The load at which a rectifier unit changes from one mode of operation to another. *Note:* The load current corresponding to a transition load is determined by the intersection of extensions of successive portions of the direct-current voltage-regulation curve where the curve changes shape or slope. *See also:* **rectification; rectifier circuit element.** E59-34E17;42A15-0

transition loss (1) (wave propagation). (A) At a transition or discontinuity between two transmission media, the difference between the power incident upon the discontinuity and the power transmitted beyond the discontinuity that would be observed if the medium beyond the discontinuity were match-terminated. (B) The ratio in decibels of the power incident upon the discontinuity to the power transmitted beyond the discontinuity that would be observed if the medium beyond the discontinuity were match terminated. *See:* **waveguide.** E146-0

(2) (junction between a source and a load). The ratio of the available power to the power delivered to the load. Transition loss is usually expressed in decibels. *See:* **waveguide.** *See also:* **transmission loss.** E146/E151/42A65-21E1

transition point (circuit). A point in a transmission system at which there is change in the surge impedance. 50I25-31E7

transition pulse (pulse waveform). That segment comprising a change from one amplitude level to another amplitude level. *See also:* **pulse.** 0-9E4

transition region (semiconductor). The region, between two homogeneous semiconductor regions, in which the impurity concentration changes. *See also:* **semiconductor; transistor.** E102-10E1

transitron oscillator. A negative-transconductance oscillator employing a screen-grid tube with negative transconductance produced by a retarding field between the negative screen grid and the control grid that serves as the anode. *See also:* **oscillatory circuit.** E145/42A65-0

transit time (1) (electron tube). The time taken for a charge carrier to traverse a given path. *See also:* **electron emission.** 42A70-15E6

(2) (multiplier-phototube). The time interval between the arrival of a delta-function light pulse at the entrance window of the tube and the time at which the output pulse at the anode terminal reaches peak amplitude. *See also:* **electron emission; phototube.** E158-15E6

transit-time spread. The time interval between the half-amplitude points of the output pulse at the anode terminal, arising from a delta function of light incident on the entrance window of the tube. *See also:* **phototube.** E158-15E6

translate. (1) To convert expressions in one language to synonymous expressions in another language. (2) To encode or decode. *See also:* **electronic digital computer; matrix; translator.** E162-0;X3A12-16E9

translation (telecommunication). The process of converting information from one system of representation into equivalent information in another system of representation. *See also:* **communication.** 0-19E4

translation loss (playback loss) (reproduction of a mechanical recording). The loss whereby the amplitude of motion of the reproducing stylus differs from the recorded amplitude in the medium. *See also:* **phonograph pickup.** E157-1E1

translator (1) (general). Equipment capable of interpreting and converting information from one form to another form. *See also:* **telephone switching system.** 0-19E1

(2) (electronic computation). A network or system having a number of inputs and outputs and so connected that signals representing information expressed in a certain code, when applied to the inputs, cause output signals to appear that are a representation of the input information in a different code. Sometimes called **matrix.** *See also:* **electronic computation; matrix.** E270-0

transliterate. To convert the characters of one alphabet to the corresponding characters of another alphabet. X3A12-16E9

transmission (illuminating engineering). A general term for the process by which incident flux leaves a surface or medium on a side other than the incident side. *Note:* Transmission through a medium is often a combination of regular and diffuse transmission.
See:
diffuse transmission;
diffuse transmittance;
hemispherical transmittance;
lamp;
regular transmission;
regular transmittance;
spectral transmittance;
transmittance. Z7A1-0

transmission band (uniconductor waveguide). The frequency range above the cutoff frequency. *See also:* **waveguide.** E146-3E1

transmission characteristics. *Note:* For an extensive list of cross references, see *Appendix A.*

transmission coefficient (1) (transition or discontinuity between two transmission media, at a given frequency). The ratio of some quantity associated with the transmitted wave at a specified point in the second medium to the same quantity associated with the incident wave at a specified point in the first medium, the second medium being match terminated. *See also:* **waveguide.** E146-0

(2) (transmission medium) (given frequency, at a given point, and for a given mode of transmission). The ratio of some quantity associated with the resultant field, which is the sum of the incident and reflected waves, to the corresponding quantity in the incident wave. *Note:* The transmission coefficient may be different for different associated quantities, and the chosen quantity must be specified. The **voltage transmission coefficient** is commonly used and is defined as the complex ratio of the resultant electric field strength (or voltage) to that of the incident wave. *See also:* **waveguide.** E146-0

(3) (multiport). Ratio of the complex amplitude of the wave emerging from a port of a multiport terminated by reflectionless terminations to the complex amplitude of the wave incident upon another port. *See also:* **reflection coefficient; scattering coefficient; transmission characteristics.** 0-9E4

transmission feeder. A feeder forming part of a transmission circuit. *See also:* **center of distribution.** 42A35-31E13

transmission frequency meter (waveguide). A cavity frequency meter that, when tuned, couples energy from a waveguide into a detector. *See also:* **waveguide.** 50I62-3E1

transmission gain stability, environmental (power-system communication). The variation in the gain or loss of a transmission medium with change in ambient temperature, ambient humidity, supply-voltage variations, and frequency variation in supply voltage. *See also:* **transmission characteristics.** 0-31E3

transmission level (signal power at any point in a transmission system). The ratio of the power at that point to the power at some point in the system chosen as a reference point. *Note:* This ratio is usually expressed in decibels. The transmission level at the transmitting switchboard is frequently taken as the zero level reference point. *See also:* **level.** E145/42A65-0

transmission line (1) (signal-transmission system). (A) The conductive connections between system elements which carry signal power. (B) A waveguide consisting of two or more conductors. *See also:* **communication; transmission characteristics; waveguide.** E146/E270-13E6

(2) (electric power). A line used for electric power transmission. *Note:* For an extensive list of cross references, see *Appendix A.* 42A35-31E13

(3) (electromagnetic wave guidance). A system of material boundaries or structures for guiding electromagnetic waves. Frequently, such a system for guiding electro-magnetic waves, in the TEM mode. Commonly a two-wire or coaxial system of conductors. *See also:* **waveguide.** 0-3E1/9E4

transmission-line capacity (electric power supply). ,The maximum continuous rating of a transmission line. The rating may be limited by thermal considerations, capacity of associated equipment, voltage regulation, system stability, or other factors. *See also:* **generating station.** 0-31E4

transmission line, coaxial. *See:* **coaxial transmission line.**

transmission loss (1) (electric power system). (A) The power lost in transmission between one point and another. It is measured as the difference between the net power passing the first point and the net power passing the second. (B) The ratio in decibels of the net power passing the first point to the net power passing the second. *See also:* **transmission characteristics; related transmission terms.** E146-0

(2) (communication). A general term used to denote a decrease in power in transmission from one point to another. This loss is usually expressed in decibels. *See also:* **waveguide.**

See:
bridging loss;
discrimination;
insertion loss;
junction loss;
net loss;
power loss;
receiving loop loss;
reflection loss;
return loss;
toll terminal loss;
transformer loss;
transition loss;
transmitting loop loss;
trunk loss;
tuner. 42A65-21E1/27E1/31E3

(3) (radio system). In a system consisting of a transmitting antenna, receiving antenna, and the intervening propagation medium, the ratio of the power radiated from the transmitting antenna to the resultant power that would be available from an equivalent loss-free receiving antenna. *See also:* **radio wave propagation; transmission characteristics.** 0-3E2

transmission-loss coefficients (electric power systems). Mathematically derived constants to be combined with source powers to provide incremental transmission losses from each source to the composite system load. These coefficients may also be used to calculate total system transmission losses. *See also:* **power system, low-frequency and surge testing.** E94-0

transmission measuring set. A measuring instrument comprising a signal source and a signal receiver having known impedances, that is designed to measure the insertion loss or gain of a network or transmission path connected between those impedances. *Note:* This name also applies to the signal receiver as a separate unit when it is used at a location remote from the signal source. *See also:* **instrument.** 42A30-0

transmission mode. A form of propagation along a transmission line characterized by the presence of any one of the elemental types of TE, or TM, or TEM waves. *Note:* Waveguide transmission modes are designated by integers (modal numbers) associated with the orthogonal functions used to describe the waveform. These integers are known as waveguide mode subscripts. They may be assigned from observations of the transverse field components of the wave and without reference to mathematics. A waveguide transmission mode is commonly described as a $TE_{m,n}$ or $TM_{m,n}$ mode, $_{m,n}$ being numerics according to the following system. **(1) (waves in rectangular waveguides).** If a single wave is transmitted in a rectangular waveguide, the field that is everywhere transverse may be resolved into two components, parallel to the wide and narrow walls respectively. In any transverse section, these components vary periodically with distance along a path parallel to one of the walls. $m =$ the total number of half-period variations of either component of field along a path parallel to the wide walls. $n =$ the total number of half-period variations of either component of field along a path parallel to the narrow walls. **(2) (waves in circular waveguides).** If a single wave is transmitted in a circular waveguide, the transverse field may be resolved into two components, radial and angular, respectively. These components vary periodically along a circular path concentric with the wall and vary in a manner related to the Bessel function of order m along a radius, where $m =$ the total number of full-period variations of either component of field along a circular path concentric with the wall. $n =$ one more than the total number of reversals of sign of either component of field along a radial path. This system can be used only if the observed waveform is known to correspond to a single mode. *See also:* **waveguide.** 42A65-0

transmission modulation (storage tubes). Amplitude modulation of the reading-beam current as it passes through apertures in the storage surface, the degree of modulation being controlled by the charge pattern stored on that surface. *See also:* **storage tube.** E158-15E6

transmission network. A group of interconnected transmission lines or feeders. *See also:* **transmission line.** 42A35-31E13

transmission primaries (color television). The set of three primaries, either physical or nonphysical, so chosen that each corresponds in amount to one of the three independent signals contained in the color-picture signal. *Note:* The chromaticities of two possible sets of transmission primaries are: (1) those of the display primaries (receiver primaries) and (2) those of a specified luminance primary and two chrominance primaries. *See also:* **color terms.** E201-0

transmission quality (mobile communication). The measure of the minimum usable speech-to-noise ratio, with reference to the number of correctly received words in a specified speech sequence. *See also:* **mobile communication system.** 0-6E1

transmission regulator (electric communication). A device that functions to maintain substantially constant transmission over a transmission system.
See:
backward-acting regulator;
dynamic regulator;
forward-acting regulator;
pilot-wire regulator;
repeatability;
static regulator;
transmission characteristics. 42A65-0

transmission route. The route followed by a transmission circuit. *See also:* **transmission line.** 42A35-31E13

transmission system (communication practice). An assembly of elements capable of functioning together to transmit signal waves. *See also:* **communication.** 42A65-0

transmission throughput. *See:* **speed of transmission, effective.**

transmission time (signal). The absolute time interval from transmission to reception. 42A65-0

transmissometer. A photometer for measuring transmittance. *Note:* Transmissometers may be visual or physical instruments. *See also:* **photometry.** Z7A1-0

transmit (computing machines). To move data from one location to another location. *See:* **transfer (2).** *See also:* **electronic digital computer.** X3A12-16E9

transmit-receive cavity (radar). The resonant portion of a transmit-receive switch. *See also:* **navigation.** E172-10E6

transmit-receive cell (tube) (waveguide). A gas-filled waveguide cavity that acts as a short circuit when ionized but is transparent to low-power energy when un-ionized. It is used in a transmit-receive switch for protecting the receiver from the high power of the transmitter but is transparent to low-power signals received from the antenna. *See also:* **waveguide.** 0-3E1

transmit-receive switch, duplexer. A switch, frequently of the gas discharge type, employed when a common transmitting and receiving antenna is used, that automatically decouples the receiver from the antenna during the transmitting period. *See also:* **navigation.** E172-10E6

transmit-receive switch (TR switch) (TR box). An automatic device employed in a radar for substantially preventing the transmitted energy from reaching the receiver but allowing the received energy to reach the receiver without appreciable loss. *See also:* **radar.** 42A65-0

transmit-receive (TR) tube. A gas-filled radio-frequency switching tube used to protect the receiver in pulsed radio-frequency systems. *See also:* **gas tube.** E160-15E6

transmittance (medium) (1) (illuminating engineering). The ratio of the transmitted flux to the incident flux. *Note:* Measured values of transmittance depend upon the angle of incidence, the method of measurement of the transmitted flux, and the spectral character of the incident flux. Because of this dependence, complete information on the technique and conditions of measurement should be specified. It should be noted that transmittance refers to the ratio of flux emerging to flux incident; therefore, reflections at the surface as well as absorption within the material operate to reduce the transmittance. *See also:* **transmission (illuminating engineering).** Z7A1-0

(2) (photovoltaic power system). The fraction of radiation incident on an object that is transmitted through the object. *See also:* **photovoltaic power system; solar cells (photovoltaic power system).** 0-10E1

(3) (transfer function) (linear passive networks). A response function for which the variables are measured at different ports (terminal pairs). *See also:* **linear passive networks.** E156-0

transmitted-carrier operation. That form of amplitude-modulation carrier transmission in which the carrier wave is transmitted. *See also:* **amplitude modulation.** 42A65-0

transmitted information. *See:* **transinformation.**

transmitted light scanning (industrial control). The scanning of changes in the magnitude of light transmitted through a web. *See also:* **photelectric control.** 42A25-34E10

transmitted harmonics (induced harmonics) (electrical conversion). Harmonics that are transformed or pass through the conversion device from the input to the output. *See also:* **electrical conversion.** 0-10E1

transmitted wave (1) (circuit). A wave (or waves) produced by an incident wave that continue(s) beyond the transition point. 50I25-31E75

(2) (at a discontinuity). When a wave in a medium of certain propagation characteristics is incident upon a discontinuity or a second medium, the forward traveling wave that results in the second medium. *Note:* In a single medium the transmitted wave is that wave which is traveling in the forward direction. *See also:* **waveguide.** E146-3E1

(3) (radio wave propagation). Refracted wave. *See:* **reflected wave; refracted wave.** 0-3E2

transmitter (protective signaling). A device for transmitting a coded signal when operated by any one of a group of actuating devices. *See also:* **protective signaling.** 42A65-0

transmitter-blocker cell (TB cell) (antitransmit-receive tube) (with reference to a waveguide). A gas-filled waveguide cavity that acts as a short circuit when ionized but as an open circuit when un-ionized. It is used in a transmit-receive switch for directing the energy received from the aerial to the receiver, no matter what the transmitter impedance may be. *See also:* **waveguide.** 0-3E1

transmitter, facsimile. The apparatus employed to translate the subject copy into signals suitable for delivery to the communication system. *See also:* **facsimile (in electrical communication).** E168-0

transmitter, pulse delay. *See:* **pulse delay transducer.**

transmitter performance. *See:* **audio input power; audio input signal.**

transmitter, telephone. *See:* **telephone transmitter.**

transmitting converter (facsimile) (amplitude-modulation to frequency-shift-modulation converter). A device that changes the type of modulation from amplitude to frequency shift. *See also:* **facsimile transmission.** E168-0

transmitting current response (electroacoustic transducer used for sound emission). The ratio of the sound pressure apparent at a distance of 1 meter in a specified direction from the effective acoustic center of the transducer to the current flowing at the electric input terminals. *Note:* The sound pressure apparent at a distance of 1 meter can be found by multiplying the sound pressure observed at a remote point (where the sound field is spherically divergent) by the number of meters from the effective acoustic center of the transducer to that point. *See also:* **loudspeaker.** 0-1E1

transmitting efficiency (electroacoustic transducer) (projector efficiency). The ratio of the total acoustic power output to the electric power input. *Note:* In computing the electric power input, it is customary to omit any electric power supplied for polarization or bias. *See also:* **electroacoustics.** 0-1E1

transmitting loop loss. That part of the repetition equivalent assignable to the station set, subscriber line, and battery supply circuit that are on the transmitting end. *See also:* **transmission loss.** 42A65-0

transmitting power response (projector power response) (electroacoustic transducer used for sound emission). The ratio of the mean-square sound pressure apparent at a distance of 1 meter in a specified direction from the effective acoustic center of the transducer to the electric power input. *Note:* The sound pressure apparent at a distance of 1 meter can be found by multiplying the sound pressure observed at a remote point (where the sound field is spherically divergent) by the number of meters from the effective acoustic center of the transducer to that point. *See also:* **loudspeaker.** 0-1E1

transmitting voltage response (electroacoustic transducer used for sound emission). The ratio of the sound pressure apparent at a distance of 1 meter in a specified direction from the effective acoustic center of the transducer to the signal voltage applied at the electric input terminals. *Note:* The sound pressure apparent at a distance of 1 meter can be found by multiplying the sound pressure observed at a remote point (where the sound field is spherically divergent) by the number of meters from the effective acoustic center of the transducer to that point. *See also:* **loudspeaker.** 0-1E1

trans-μ-factor (multibeam electron tubes). The ratio of (1) the magnitude of an infinitesimal change in the voltage at the control grid of any one beam to (2) the magnitude of an infinitesimal change in the voltage at the control grid of a second beam. The current in the second beam and the voltage of all other electrodes are maintained constant. *See also:* **beam tubes.** E160-15E6

transponder (electronic navigation). A transmitter-receiver facility the function of which is to transmit signals automatically when the proper interrogation is received. *See also:* **navigation; radio transmission.** E172/42A65-10E6

transponder beacon (electronic navigation). *See:* **transponder.**

transponder, crossband (electronic navigation). A transponder that replies in a different frequency band from that of the received interrogation. *See also:* **navigation.** E172-10E6

transponder reply efficiency (electronic navigation). The ratio of the number of replies emitted by a transponder to the number of interrogations that the transponder recognizes as valid; the interrogations recognized as valid include those accidentally combined to form recognizable codes, a statistical computation of them normally being made. *See also:* **navigation.** E172-10E6

transport (computing machines). *See:* **tape transport.** *See also:* **electronic digital computer.**

transportable transmitter. A transmitter designed to be readily carried or transported from place to place, but which is not normally operated while in motion. *Note:* This has been commonly called a portable transmitter, but the term transportable transmitter is preferred. *See also:* **radio transmitter; radio transmission.** E145/E182A/42A65-0

transportation and storage conditions. The conditions to which a device may be subjected between the time of construction and the time of installation. Also included are the conditions that may exist during shutdown. *Note:* No permanent physical damage or impairment of operating characteristics shall take place under these conditions, but minor adjustments may be needed to restore performance to normal. 39A4-0

transportation lag. *See:* **lag, distance/velocity.**

transport delay. *See:* **time delay.**

transport lag. *See:* **lag, distance/velocity.**

transport standards. Standards of the same nominal value as the basic reference standards of a laboratory (and preferably of equal quality), that are regularly intercompared with the basic group but that are reserved for periodic interlaboratory comparison tests that act as checks on the stability of the basic reference group. *See also:* **measurement system.** 12A0-0

transport time (industrial control) (feedback system). The time required to move an object, element or information from one predetermined position to another. *See also:* **control system, feedback.** AS1/34E10

transposition (1) (transmission lines). An interchange of positions of the several conductors of a circuit between successive lengths. *Notes:* (1) It is normally used to reduce inductive interference on communication or signal circuits by cancellation. (2) The term is most frequently applied to open wire circuits. *See also:* **open wire; signal; tower.** 42A35/42A65-13E6/31E13

(2) (rotating machinery). An arrangement of the conductors comprising a turn or coil whereby they take different relative positions for the purpose of reducing eddy-current losses. *See:* **asynchronous machine; direct-current commutating machine; synchronous machine.** 0-31E8

transposition section. A length of open wire line to which a fundamental transposition design or pattern is applied as a unit. *See also:* **open wire.** 42A65-0

transreactance. The imaginary part of the transimpedance. *See also:* **transmission characteristics.** 0-9E4

transrectification factor. The quotient of the change in average current of an electrode by the change in the amplitude of the alternating sinusoidal coltage applied to another electrode, the direct voltages of this and other electrodes being maintained constant. *Note:* Unless otherwise stated, the term refers to cases in which the alternating sinusoidal voltage is of infinitesimal magnitude. *See also:* **circuit characteristics of electrodes; rectification factor.** E160/42A70-15E6

transrectifier. A device, ordinarily a vacuum tube in which rectification occurs in one electrode circuit when an alternating voltage is applied to another electrode. *See also:* **rectifier.** 42A65-0

transresistance. The real part of the transimpedance. *See also:* **transmission characteristics.** 0-9E4

transsusceptance. The imaginary part of the transadmittance. *See also:* **transmission characteristics.** 0-9E4

transverse-beam traveling-wave tube. A traveling-wave tube in which the direction of motion of the electron beam is transverse to the average direction in which the signal wave moves. *See also:* **beam tube; miscellaneous electron devices.** E160-15E6

transverse crosstalk coupling (between a disturbing and a disturbed circuit in any given section). The vector summation of the direct couplings between adjacent short lengths of the two circuits, without dependence on intermediate flow in other nearby circuits. *See also:* **coupling.** 42A65-0

transverse-electric hybrid wave. An electromagnetic wave in which the electric field vector is linearly polarized normal to the plane of propagation and the magnetic field vector is elliptically polarized in this plane. *See also:* **radio wave propagation.** 0-3E2

transverse electric ($TE_{m,n,p}$) resonant mode (cylindrical cavity). In a hollow metal cylinder closed by two plane metal surfaces perpendicular to its axis, the resonant mode whose transverse field pattern is similar to the $TE_{m,n}$ wave in the corresponding cylindrical waveguide and for which p is the number of half-period field variations along the axis. *Note:* When the cavity is a rectangular parallelepiped, the axis of the cylinder from which the cavity is assumed to be made should be designated since there are three such axes possible. *See also:* **waveguide.** E146-3E1

transverse electric wave (TE wave) (1) (general). In a homogeneous isotropic medium, an electromagnetic wave in which the electric field vector is everywhere perpendicular to the direction of propagation. *See also:* **waveguide.** E146/E210/42A65-3E1

(2) ($TE_{m,n}$ wave) (rectangular waveguide) (hollow rectangular metal cylinder). The transverse electric wave for which m is the number of half-period variations of the field along the x coordinate, which is assumed to coincide with the larger transverse dimension, and n is the number of half-period variations of the field along the y coordinate, which is assumed to coincide with the smaller transverse dimension. *Note:* The dominant wave in a rectangular waveguide is $TE_{1,0}$; its electric lines are parallel to the shorter side. *See also:* **guided waves; waveguide.** E146/E210-3E1

(3) ($TE_{m,n}$ wave) (circular waveguide) (hollow circular metal cylinder). The transverse electric wave for which m is the number of axial planes along which the normal component of the electric vector vanishes, and n is the number of coaxial cylinders (including the boundary of the waveguide) along which the tangential component of electric vector vanishes. *Notes:* (1) $TE_{0,n}$ waves are circular electric waves of order n. The $TE_{0,1}$ wave is the circular electric wave with the lowest cutoff frequency. (2) The $TE_{1,1}$ wave is the dominant wave. Its lines of electric force are approximately parallel to a diameter. *See also:* **waveguide.** E146-3E1

transverse electromagnetic (TEM) mode (waveguide). A mode in which the longitudinal components of the electric and magnetic fields are everywhere zero. *See also:* **waveguide.** 50I62-3E1

transverse electromagnetic wave (TEM wave). In a homogeneous isotropic medium, an electromagnetic wave in which both the electric and magnetic field vectors are everywhere perpendicular to the direction of propagation. *See also:* **radio-wave propagation; waveguide.** E146/42A65-3E1/3E2

transverse-field traveling-wave tube. A traveling-wave tube in which the traveling electric fields that interact with electrons are essentially transverse to the average motion of the electrons. *See also:* **beam tube; electron devices, miscellaneous.** E160-15E6

transverse interference (signal-transmission system). *See:* **interference, differential-mode; interference, normal-mode.** *See also:* **accuracy rating (instrument); signal.**

transverse magnetic hybrid wave. An electromagnetic wave in which the magnetic field vector is linearly polarized normal to the plane of propagation and the electric field vector is elliptically polarized in this plane. *See also:* **radio wave propagation.** 0-3E2

transverse magnetic ($TM_{m,n,p}$) resonant mode (cylindrical cavity). In a hollow metal cylinder closed by two plane metal surfaces perpendicular to its axis, the resonant mode whose transverse field pattern is similar to the $TM_{m,n}$ wave in the corresponding cylindrical waveguide and for which p is the number of half-period field variations along the axis. *Note:* When the cavity is a rectangular parallelepiped, the axis of the cylinder from which the cavity is assumed to be made should be designated since there are three such axes possible. *See also:* **waveguide.** E146-3E1

transverse magnetic wave (TM wave) (1) (general). In a homogeneous isotropic medium, an electromagnetic wave in which the magnetic field vector is everywhere perpendicular to the direction of propagation. *See also:* **waveguide.** E146/42A65-3E1/3E2

(2) ($TM_{m,n}$ wave) (circular waveguide) (hollow circular metal cylinder). The transverse magnetic wave for which m is the number of axial planes along which the normal component of the magnetic vector vanishes, and n is the number of coaxial cylinders to which the electric vector is normal. *Note:* $TM_{0,n}$ waves are circular magnetic waves of order n. The $TM_{0,1}$ wave is the circular magnetic wave with the lowest cutoff frequency. *See also:* **guided wave; circular magnetic wave; waveguide.** E146-3E1

(3) ($TM_{m,n}$ wave) (rectangular waveguide) (hollow rectangular metal cylinder). The transverse magnetic wave for which m is the number of half-period variations of the magnetic field along the longer transverse dimension, and n is the number of half-period variations of magnetic field along the shorter transverse dimension. *See also:* **guided wave; circular magnetic wave; waveguide.** E146-3E1

transverse magnetization (magnetic recording). Magnetization of the recording medium in a direction perpendicular to the line of travel and parallel to the greatest cross-sectional dimension. *See also:* **phonograph pickup.** E157-1E1

transverse-mode interference (signal-transmission system). *See:* **interference, differential-mode.** *See also:* **signal.**

transverse wave. A wave in which the direction of displacement at each point of the medium is perpendicular to the direction of propagation. *Note:* In those cases where the displacement makes an acute angle with the direction of propagation, the wave is considered to have longitudinal and transverse components. E270-0

trap (1) (computing machines). An unprogrammed conditional jump to a known location, automatically activated by hardware, with the location from which the jump occurred recorded. *See also:* **electronic digital computer.** X3A12-16E9

(2) (burglar-alarm system). An automatic device applied to a door or window frame for the purpose of producing an alarm condition in the protective circuit whenever a door or window is opened. *See also:* **protective signaling.** 42A65-0

trap circuit. A circuit used at locations where it is desirable to protect a section of track on which it is impracticable to maintain a track circuit. It usually consists of an arrangement of one or more stick circuits so connected that when a train enters the trap circuit the stick relay drops and cannot be picked up again until the train has passed through the other end of the trap circuit. *See also:* **railway signal and interlocking.** 42A42-0

trapezium distortion (cathode-ray tube). A fault characterized by a variation of the sensitivity of the deflection parallel to one axis (vertical or horizontal) as a function of the deflection parallel to the other axis and having the effect of transforming an image that is a rectangle into one which is a trapezium. *See also:* **cathode-ray tubes.** 50I07-15E6

trapped flux (superconducting material). Magnetic flux that links with a closed superconducting loop. *See also:* **superconductivity.** E217-15E7

travel (1) (relay). The amount of movement in either direction (towards pickup or reset) of the principal responsive element or a contact part of it. *Note:* Travel may be specified in linear, angular, or other measure. 37A100-31E6/31E11

(2) (rise) (elevators). Of an elevator, dumbwaiter, escalator, or of a private-residence inclined lift, the vertical distance between the bottom terminal landing and the top terminal landing. *See also:* **elevators.** 42A45-0

traveling cable (elevators). A cable made up of electric conductors that provides electric connection between an elevator or dumbwaiter car and fixed outlet in the hoistway. *See:* **control.** 42A45-0

traveling overvoltage (lightning arresters). A surge propagated along a conductor. *See also:* **lightning arrester (surge diverter).** 50I25-31E7

traveling plane wave. A plane wave each of whose frequency components has an exponential variation of amplitude and a linear variation of phase in the direction of propagation. *See also:* **radio wave propagation; waveguide.** E146-3E2

traveling wave. The resulting wave when the electric variation in a circuit takes the form of translation of energy along a conductor, such energy being always equally divided between current and potential forms. *See also:* **direction of propagation; lightning arrester (surge diverter); power systems, low-frequency and surge testing; traveling plane; waveguide.** 42A35-31E13/31E7

traveling-wave magnetron. *See:* **slow-wave circuit (microwave tube).**

traveling-wave magnetron oscillations. Oscillations sustained by the interaction between the space-charge cloud of a magnetron and a traveling electromagnetic field whose phase velocity is approximately the same as the mean velocity of the cloud. *See:* **magnetron, slow-wave circuit.** 42A70-15E6

traveling-wave parametric amplifier. A parametric amplifier that has a continuous or iterated structure incorporating nonlinear reactors and in which the signal, pump, and difference-frequency waves are propagated along the structure. *See also:* **parametric device.** E254-15E7

traveling-wave tube (microwave tubes). An *O*-type tube characterized by continuous interaction of the beam with one or more forward-wave circuits that are in sequence along the beam and may be separated by attenuators or severs (circuit breaks). *See also:* **microwave tube (or valve).** 0-15E6

traveling-wave-tube interaction circuit. *See:* **slow-wave circuit; periodic slow-wave circuit.**

tray (storage cell) (storage battery). A support or container for one or more storage cells. *See also:* **battery (primary or secondary).** 42A60-0

treated fabric (treated mat) (rotating machinery). A fabric or mat in which the elements have been essentially coated but not filled with an impregnant such as a compound or varnish. *See also:* **rotor (rotating machinery); stator.** 0-31E8

treated mat (rotating machinery). *See:* **treated fabric.**

treble boost. An adjustment of the amplitude-frequency response of a system or transducer to accentuate the higher audio frequencies. E151/42A65-0

tree. A set of connected branches including no meshes. *See also:* **network analysis.** E153/E270-0

trees and nodules. Projections formed on a cathode during electrodeposition. Trees are branched whereas nodules are rounded. *See also:* **electrodeposition.** 42A60-0

tree wire. A conductor with an abrasion-resistant outer covering, usually nonmetallic, and intended for use on overhead lines passing through trees. *See also:* **armored cable; conductor.** 42A35-31E13

triboelectrification (electrification by friction). The mechanical separation of electric charges of opposite sign by processes such as (1) the separation (as by sliding) of dissimilar solid objects; (2) interaction at a solid-liquid interface; (3) breaking of a liquid-gas interface. E270-0

tributary office. A telephone central office that passes toll traffic to, and receives toll traffic from, a toll center. *See also:* **telephone system.** 42A65-0

trickle charge (storage battery) (storage cell). A continuous charge at a low rate approximately equal to the internal losses and suitable to maintain the battery in a fully charged condition. *Note:* This term is also applied to very low rates of charge suitable not only for compensating for internal losses but to restore intermittent discharges of small amount delivered from time to time to the load circuit. *See:* **floating.** *See also:* **charge.** 42A60-0

trigatron. (1) A triggered spark-gap switch on which control is obtained by a voltage applied to a trigger electrode. *Note:* This voltage distorts the field between the two main electrodes converting the sphere-to-sphere gap to a point to sphere gap. *See also:* **electron device.** 50I07-15E6
(2) An electronic switch in which conduction is initiated by the breakdown of an auxiliary gap. *See also:* **tube definitions.** E172-10E6

trigger (1) (verb). To start action in another circuit which then functions for a period of time under its own control. *See also:* **circuits and devices.** 42A65-0
(2) (noun). A pulse used to initiate some function, for example, a triggered sweep or delay ramp. *Note:* Trigger may loosely refer to a waveform of any shape used as a signal from which a trigger pulse is derived as in **trigger source, trigger input,** etcetera. *See:* **triggering signal.** *See also:* **circuits and devices; oscillograph.** 0-9E4

trigger circuit. A circuit that has two conditions of stability, with means for passing from one to the other when certain conditions are satisfied, either spontaneously or through application of an external stimulus. *See:*
counter;
Eccles-Jordan circuit;
flip-flop circuit;
ring counter;
scale-of-two counter;
single-shot blocking oscillator;
single-shot multivibrator;
single-shot trigger circuit. *See also:* **circuits and devices.** 42A65-0

trigger countdown. A process that reduces the repetition rate of a triggering signal. *See also:* **oscillograph.** 0-9E4

triggered sweep. A sweep that can be initiated only by a trigger signal, not free running. *See also:* **oscillograph.** 0-9E4

triggering level. The instantaneous level of a triggering signal at which a trigger is to be generated. Also, the name of the control that selects the level. *See also:* **oscillograph.** 0-9E4

triggering signal. The signal from which a trigger is derived. *See also:* **oscillograph.** 0-9E4

triggering slope. The positive-going (+slope) or negative-going (−slope) portion of a triggering signal from which a trigger is to be derived. Also, the control that selects the slope to be employed. *Note:* + and − slopes apply to the slope of the waveform only and not to the absolute polarity. *See also:* **oscillograph.** 0-9E4

triggering stability. *See:* **stability.**

trigger level (transponder). The minimum input to the receiver that is capable of causing the transmitter to emit a reply. *See also:* **navigation.** E172-10E6

trigger lockout. *See:* **sweep lockout.**

trigger pickoff. A process or a circuit for extracting a triggering signal. *See also:* **oscillograph.** 0-9E4

trigger-starting systems (fluorescent lamps). Applied to systems in which hot-cathode electric discharge lamps are started with cathodes heated through low-voltage heater windings built into the ballast. Sufficient voltage is applied across the lamp and between the lamp and fixture to initiate the discharge when the cathodes reach a temperature high enough for adequate emission. The ballast is so designed that the cathode-heating current is greatly reduced as soon as the arc is struck. 82A1-0

trigger tube (electron device). A cold-cathode gas-filled tube in which one or more electrodes initiate, but do not control, the anode current. *See also:* **tube definitions.** 0-15E6

trimmer capacitor (trimming capacitor). A small adjustable capacitor associated with another capacitor and used for fine adjustment of the total capacitance of an element or part of a circuit. *See also:* **circuits and devices.** 42A65-21E0

trimmer signal. A signal that gives indication to the engineman concerning movements to be made from the classification tracks into the switch and retarder area. 42A42-0

triode. A three-electrode electron tube containing an anode, a cathode, and a control electrode. *See also:* **tube definitions.** 42A70-15E6

trip (1) (verb). (A) To release in order to initiate either an opening or a closing operation or other specified action. (B) To release in order to initiate an opening operation only. (C) To initiate and complete an opening operation. *Note:* All terms employing **trip, tripping,** or their derivatives are referred to the term that expresses the intent of the usage.
(2) (noun). (A) A release that initiates either an opening or a closing operation or other specified action. (B) A release that initiates an opening operation only. (C) A complete opening operation. 37A100-31E11/31E6
(3) (tripping) (adjective). (A) Pertaining to a release that initiates either an opening or a closing operation or other specified action. (B) Pertaining to a release that initiates an opening operation only. (C) Pertaining to a complete opening operation. *Note:* All terms employing **trip, tripping,** or their derivatives are referred to the term that expresses the intent of the usage. 37A100-31E11

trip arm. *See:* **mechanical trip.**

trip coil. *See:* **release coil.**

trip delay setting. *See:* **release delay setting.**

trip-free. *See:* **release-free.**

trip-free in any position. *See:* **release-free in any position.**

trip-free relay. *See:* **release-free relay.**

trip lamp. A removable self-contained mine lamp, designed for marking the rear end of a train (trip) of mine cars. 42A85-0

triple-address. Same as three-address. *See also:* **electronic digital computer.**

triple detection. *See:* **double superheterodyne reception.**

triplen (rotating machinery). An order of harmonic that is a mulitple of three. *See also:* **asynchronous machine; direct-current commutating machine; synchronous machine.** 0-31E8

triplet (electronic navigation systems). Three radio stations operated as a group for the determination of positions. *See also:* **navigation.** E172-10E6

triplex cable. A cable composed of three insulated single-conductor cables twisted together. *Note:* The assembled conductors may or may not have a common covering of binding or protecting material. *See also:* **power distribution, underground construction.** E30-0;42A35-31E13

trip OFF control signal (magnetic amplifier). The final value of signal measured when the amplifier has changed from the ON to the OFF state as the signal is varied so slowly that an incremental increase in the speed with which it is varied does not affect the measurement of the trip OFF control signal. That is, the change in trip OFF control signal is below the sensitivity of the measuring instrument. *See also:* **rating and testing magnetic amplifiers.** E107-0

trip ON control signal (magnetic amplifier). The final value of signal measured when the amplifier has changed from the OFF to the ON state as the signal is varied so slowly that an incremental increase in the speed with which it is varied does not affect the measurement of the trip ON control signal. That is, the change in trip ON control signal is below the sensitivity of the measuring instrument. *See also:* **rating and testing magnetic amplifiers.** E107-0

tripping delay. *See:* **release delay.**

tripping mechanism. *See:* **release (tripping) mechanism).**

trip-point repeatability (magnetic amplifier). The change in trip point (either trip OFF or trip ON, as specified) control signal due to uncontrollable causes over a specified period of time when all controllable quantities are held constant. *See also:* **rating and testing magnetic amplifiers.** E107-0

trip point repeatability coefficient (magnetic amplifier). The ratio of (1) the maximum change in trip point control signal due to uncontrollable causes to (2) the specified time period during which all controllable quantities have been held constant. *Note:* The units of this coefficient are the control signal units per the time period over which the coefficient was determined. *See also:* **rating and testing magnetic amplifiers.** E107-0

trip setting. *See:* **release setting (trip setting).**

tristimulus values (light). The amounts of the three reference or matching stimuli required to give a match with the light considered, in a given trichromatic system. *Note:* In the standard colorimetric system of the Commission Internationale de l'Eclairage (CIE) (1931) the symbols x, y, z are recommended for the tristimulus values. These values may be obtained by multiplying the spectral concentration of the radiation at each wavelength by the distribution coefficients and integrating these products over the whole spectrum. *See:* **color; color terms.** 50I45-2E2

troffer (illuminating engineering). A long recessed lighting unit usually installed with the opening flush with the ceiling. The term is derived from **trough** and **coffer.** *See also:* **luminaire.** Z7A1-0

troland. A unit used for expressing the magnitude of the external light stimulus applied to the eye. When the eye is viewing a surface of uniform luminance, the number of trolands is equal to the product of the area in square millimeters of the limiting pupil, natural or artificial, and the luminance of the surface in candelas per square meter. *See also:* **visual field.** Z7A1-0

trolley. A current collector, the function of which is to make contact with a trolley wire. *See also:* **contact conductor.** E16-0

trolley bus. *See:* **trolley coach.**

trolley car. An electric motor car that collects propulsion power from a trolley system. *See also:* **electric motor car.** 42A42-0

trolley coach (trolley bus) (trackless trolley coach). An electric bus that collects propulsion power from a trolley system. *See also:* **electric bus.** 42A42-0

trolley locomotive. An electric locomotive that collects propulsion power from a trolley system. *See also:* **electric locomotive.** 42A42-0

trombone line (transmission lines and waveguides). A U-shaped length of waveguide or transmission line of adjustable length. *See also:* **waveguide.** 0-9E4

troposphere. That part of the earth's atmosphere in which temperature generally decreases with altitude, clouds form and convection is active. *Note:* Experiments indicate that the troposphere occupies the space above the earth's surface up to a height ranging from about 6 kilometers at the poles to about 18 kilometers at the equator. *See also:* **radiation; radio wave propagation.** 42A65-3E2

tropospheric wave. A radio wave that is propagated by reflection from a place of abrupt change in the dielectric constant or its gradient in the troposphere. *Note:* In some cases the ground wave may be so altered that new components appear to arise from reflections in regions of rapidly changing dielectric constant. When these components are distinguishable from the other components, they are called tropospheric waves. *See also:* **radiation.** 42A65-0

troubleshoot. *See:* **debug.**

troughing. An open channel of earthenware, wood, or other material in which a cable or cables may be laid and protected by a cover. *See also:* **power distribution, underground construction.** 42A35-31E13

truck generator suspension. A design of support for an axle generator in which the generator is supported by the vehicle truck. 42A42-0

true bearing (navigation). Bearing relative to true north. *See also:* **navigation.** 0-10E6

true complement. *See:* **radix complement.**

true complement. A number representation that can be derived from another by subtracting each digit from one less than the base and then adding one to the least significant digit and executing all carries required. Tens complements and twos complements are true complements. *See also:* **electronic digital computer.** E162-0

true course (navigation). Course relative to true north. *See also:* **navigation.** 0-10E6

true heading (navigation. Heading relative to true north. *See also:* **navigation.** 0-10E6

true neutral point (at terminals of entry). Any point in the boundary surface that has the same voltage as the point of junction of a group of equal nonreactive resistors placed in the boundary surface of the region and connected at their free ends to the appropriate terminals of entry of the phase conductors of the circuit, provided that the resistance of the resistors is so great that the voltages are not appreciably altered by the introduction of the resistors. *Notes:* (1) The number of resistances required is two for direct-current or single-phase alternating-current circuits, four for two-phase four-wire or five-wire circuits, and is equal to the number of phases when the number of phases is three or more. Under normal symmetrical conditions the number of resistors may be reduced to three for six- or twelve-phase systems when the terminals are properly selected, but the true neutral point may not be obtained by this process under all abnormal conditions. The concept of a true neutral point is not considered applicable to a two-phase, three-wire circuit. (2) Under abnormal conditions the voltage of the true neutral point may not be the same as that of the neutral conductor. *See also:* **network analysis.** E270-0

true north (navigation). The direction of the north geographical pole. *See also:* **navigation.** 0-10E6

true ratio (instrument transformer). The ratio of root-mean-square primary voltage, or current, as the case may be, to the root-mean-square secondary voltage, or current, under specified conditions. *Note:* The true ratio may be determined by test or by calculation. *See also:* **instrument transformer.** 12A0/42A15/42A30-0;57A13-31E12

truncate. To terminate a computational process in accordance with some rule, for example, to end the evaluation of a power series at a specified term. *See also:* **electronic digital computer.** X3A12-16E9

trunk (telephony). One- or two-way channel provided as a common traffic artery between central offices or a central office and other switching equipment. *See also:* **telephone system.** 0-19E1

trunk circuit, combined line and recording (CLR). (1) Name given to a class of trunk circuits that provide access to operator positions generally referred to by abbreviation only. (2) Recording-completing trunk circuit for operator recording and completing of toll calls originated by subscribers of central offices. 0-19E1

trunk feeder. A feeder connecting two generating stations or a generating station and an important substation. *See also:* **center of distribution.** 42A35-31E13

trunk group. All of the trunks of a given type or characteristic that extend between two switching points. *See also:* **telephone system.** 42A65-19E1

trunk hunting. The operation of a selector or other similar device, to establish connection with an idle circuit of a chosen group. This is usually accomplished by successively testing terminals associated with this group until a terminal is found that has an electrical condition indicating it to be idle. *See also:* **telephone switching system.** 42A65-0

trunk-line conduit. A duct bank provided for main or trunk-line cables. *See also:* **power distribution, underground construction.** 42A35-31E13

trunk loss. That part of the repetition equivalent assignable to the trunk used in the telephone connection. *See also:* **transmission loss.** 42A65-0

trunk transmission line. A transmission line acting as a source of main supply to a number of other transmission circuits. *See also:* **transmission line.** 42A35-31E13

trussed blade (switching device). A blade that is reinforced by truss construction to provide stiffness. 37A100-31E11

truth table. A table that describes a logic function by listing all possible combinations of input values and indicating, for each combination, the true output values. X3A12-16E9

TSI. *See:* **threshold signal-to-interference ratio.**

tube (1) (interior wiring). A hollow cylindrical piece of insulating material having a head or shoulder at one end, through which an electric conductor is threaded where passing through a wall, floor, ceiling, joist, stud, etcetera. *See also:* **raceways.** 42A95-0

(2) (primary cell). A cylindrical covering of insulating material, without closure at the bottom. *See also:* **electrolytic cell.** 42A60-0

tube count (radiation-counter tubes). A terminated discharge produced by an ionizing event in a radiation-counter tube. *See also:* **gas-filled radiation-counter tubes.** 42A70-15E6

tube current averaging time. The time interval over which the current is averaged in defining the operating capability of the tube. *See also:* **rectification.** 34A1-0

tube definitions. *Note:* For an extensive list of cross references, see *Appendix A.*

tube, display. *See:* **display tube.**

tube, electron. *See:* **electron tube.**

tube fault current. The current that flows through a tube under fault conditions, such as arc-back or short circuit. *See also:* **rectification.** 34A1-0

tube, fuse. *See:* **fuse tube.**

tube heating time (mercury-vapor tube). The time required for the coolest portion of the tube to attain operating temperature. *See:* **preheating time.** *See also:* **electronic controller; gas tubes.** 42A70-0

tubelet (soldered connections). *See:* **eyelet.**

tuberculation (corrosion). The formation of localized corrosion products scattered over the surface in the form of knoblike mounds. *See:* **corrosion terms.** CM-34E2

tube-type plate (storage cell). A plate of an alkaline storage battery consisting of an assembly of metal tubes filled with active material. *See also:* **battery (primary or secondary).** 42A60-0

tube, vacuum. *See:* **vacuum tube.**

tube voltage drop (electron tube). The anode voltage during the conducting period. *See also:* **electrode voltage (electron tube); electronic controller.** 42A70-15E6

tubing (rotating machinery). A tubular flexible insulation, extruded or made of layers of film plastic, into which a conductor is inserted to provide additional insulation. Tubing is frequently used to insulate connections and crossovers. *See also:* **asynchronous machine; direct-current commutating machine; synchronous machine.** 0-31E8

Tudor plate (storage cell). A lead storage battery plate obtained by molding and having a large area. *See also:* **battery (primary and secondary).** 42A60-0

tunable microwave oscillator.
See:
backlash;
electrically tuned oscillators;
oscillatory circuit;
resettability;
response time;
tuning creep;
tuning hysteresis;
tuning range;
tuning sensitivity. E158-15E6

tuned-grid oscillator. An oscillator whose frequency is determined by a parallel-resonance circuit in the grid circuit coupled to the plate to provide the required feedback. *See also:* **oscillatory circuit.** E145/42A65-0

tuned-grid–tuned-plate oscillator. An oscillator having parallel-resonance circuits in both plate and grid circuits, the necessary feedback being obtained by the plate-to-grid interelectrode capacitance. *See also:* **os-**

cillatory circuit. E54/E145/42A65-0

tuned-plate oscillator. An oscillator whose frequency is determined by a parallel-resonance circuit in the plate circuit coupled to the grid to provide the required feedback. *See also:* **oscillatory circuit.** E145/42A65-0

tuned transformer. A transformer, the associated circuit elements of which are adjusted as a whole to be resonant at the frequency of the alternating current supplied to the primary, thereby causing the secondary voltage to build up to higher values than would otherwise be obtained. *See also:* **power pack.** 42A65-21E0

tuner (1) (radio receiver). In the broad sense, a device for tuning. Specifically, in radio receiver practice, it is (A) a packaged unit capable of producing only the first portion of the functions of a receiver and delivering either radio-frequency, intermediate-frequency, or demodulated information to some other equipment, or (B) that portion of a receiver that contains the circuits that are tuned to resonance at the received-signal frequency and those that are tuned to local oscillator frequency. *See also:* **radio receiver.** 42A65-0
(2) (transmission line) (waveguide). An ideally lossless, fixed or adjustable, network capable of transforming a given impedance into a different impedance. *See also:* **transmission loss; waveguide.** E147-3E1/9E4

tuning (of circuits). The adjustment in relation to frequency of a circuit or system to secure optimum performance; commonly the adjustment of a circuit or circuits to resonance. *See also:* **radio transmission.** E270/42A65-31E3

tuning creep (oscillator). The change of an essential characteristic as a consequence of repeated cycling of the tuning element. *See also:* **tunable microwave oscillators.** E158-15E6

tuning, electronic. The process of changing the operating frequency of a system by changing the characteristics of a coupled electron stream. Characteristics involved are, for example: velocity, density, or geometry. *See also:* **circuit characteristics of electrodes; oscillatory circuit.** E160-15E6

tuning hysteresis (microwave oscillator). The difference in a characteristic when a tuner position, or input to the tuning element, is approached from opposite directions. *See also:* **tunable microwave oscillator.** E158-15E6

tuning indicator (electron device). An electron-beam tube in which the signal supplied to the control electrode varies the area of luminescence of the screen. *See also:* **tube definitions.** 0-15E6

tuning probe (waveguides). An essentially lossless probe of adjustable penetration extending through the wall of the waveguide or cavity resonator. *See also:* **waveguide.** E147-0

tuning range (1) (switching tubes). The frequency range over which the resonance frequency of the tube may be adjusted by the mechanical means provided on the tube or associated cavity. *See:* **gas tubes.** E160-15E6
(2) (oscillator). The frequency range of continuous tuning within which the essential characteristics fall within prescribed limits. *See also:* **tunable microwave oscillators.** E158-15E6

tuning range, electronic. The frequency range of continuous tuning between two operating points of specified minimum power output for an electronically tuned oscillator. *Note:* The reference points are frequently the half-power points, but should always be specified. *See also:* **circuit characteristics of electrodes; oscillatory circuit.** E160-15E6

tuning rate, thermal. The initial time rate of change in frequency that occurs when the input power to the tuner is instantaneously changed by a specified amount. *Note:* This rate is a function of the power input to the tuner as well as the sign and magnitude of the power change. *See also:* **circuit characteristics of electrodes; oscillatory circuit.** E160-15E6

tuning screw (waveguide technique). An impedance-adjusting element in the form of a rod whose depth of penetration through the wall into a waveguide or cavity is adjustable by rotating the screw. *See also:* **waveguide.** 42A65-3E1

tuning sensitivity (oscillator). The rate of change of frequency with the control parameter (for example, the position of mechanical tuner, electric tuning voltage, etcetera) at a given operating point. *See also:* **tunable microwave oscillators.** E158-15E6

tuning sensitivity, electronic. At a given operating point, the rate of change of oscillator frequency with the change of the controlling electron stream. For example, this change may be expressed in terms of an electrode voltage or current. *See:* **pushing figure (oscillator.)** *See also:* **circuit characteristics of electrodes; oscillatory circuit.** E160-15E6

tuning sensitivity, thermal. The rate of change of resonator equilibrium frequency with respect to applied thermal tuner power. *See also:* **circuit characteristics of electrodes.** E160-15E6

tuning, thermal. The process of changing the operating frequency of a system by using a controlled thermal expansion to alter the geometry of the system. *See:* **oscillatory circuit.** E160-15E6

tuning time constant, thermal. The time required for the frequency to change by a fraction $(1-1/e)$ of the change in equilibrium frequency after an incremental change of the applied thermal tuner power. *Notes:* (1) If the behavior is not exponential, the initial conditions must be stated. (2) Here e is the base of natural logarithms. *See:* **oscillatory circuit.** E160-15E6

tuning time thermal (1) (cooling). The time required to tune through a specified frequency range when the tuner power is instantaneously changed from the specified maximum to zero. *Note:* The initial condition must be one of equilibrium. *See also:* **circuit characteristics of electrodes; electron emission.** E160-15E6
(2) (heating). The time required to tune through a specified frequency range when the tuner power is instantaneously changed from zero to the specified maximum. *Note:* The initial condition must be one of equilibrium. *See:* **electron emission.** *See also:* **circuit characteristics of electrodes.** E160-15E6

turbine-driven generator. An electric generator driven by a turbine. *See also:* **direct-current commutating machine; synchronous machine.** 42A10-0

turbine end (rotating machinery). The driven or power-input end of a turbine-driven generator. *See also:* **cradle base (rotating machinery).** 0-31E8

turbine-generator. *See:* **cylindrical-rotor generator.**

turbine-generator unit. An electric generator with its driving turbine. *See also:* **direct-current commutating machine; synchronous machine.** 42A10-0

turbine-nozzle control system (gas turbines). A means by which the turbine diaphragm nozzles are adjusted to vary the nozzle angle or area, thus varying the rate of energy input to the turbine(s). E282-31E2

turbine-type (rotating machinery). Applied to alternating-current machines designed for high-speed operation and having an excitation winding embedded in slots in a cylindrical steel rotor made from forgings or thick discs. *See:* **asynchronous machine; synchronous machine.** 0-31E8

Turing machine. A mathematical model of a device that changes its internal state and reads from, writes on, and moves a potentially infinite tape, all in accordance with its present state, thereby constituting a model for computerlike behavior. *See:* **universal Turing machine.** *See also:* **electronic digital computer.** X3A12-16E9

turn (rotating machinery). The basic coil element that forms a single conducting loop comprising one insulated conductor. The conductor may consist of a number of parallel-connected insulated strands or laminations, each strand or lamination being in the form of wire, rod, strip, or bar depending on its cross section. *See also:* **rotor (rotating machinery); stator.** 0-31E8

turnbuckle. A threaded device inserted in a tension member to provide minor adjustment of tension or sag. *See also:* **tower.** 42A35-31E13

turning gear (rotating machinery). A separate drive to rotate a machine at very low speed for the purpose of thermal equalization at a time when it would otherwise be at rest. *See also:* **rotor (rotating machinery).** 0-31E8

turn insulation (salient pole) (rotating machinery). The insulation placed between adjacent turns of field coils. *Note:* The insulation does not necessarily encircle the strap of edge-wound coils but may leave outer edges exposed to facilitate cooling. *See also:* **armature; rotor (rotating machinery); stator.** 0-31E8

turn-off thyristor. A thyristor that can be switched from the ON state to the OFF state and vice versa by applying control signals of appropriate polarities to the gate terminal, with the ratio of triggering power to triggered power appreciably less than one. *See also:* **thyristor.** E223-34E17/15E7

turnover frequency. *See:* **transition frequency.**

turn ratio (1) (transformer). The ratio of the number of turns in the high-voltage winding to that in the low-voltage winding. *Note:* In the case of a constant-voltage transformer having taps for changing its voltage ratio, the turn ratio is based on the number of turns corresponding to the normal rated voltage of the respective windings to which operating and performance characteristics are referred. *See:* **current ratio (transformer); voltage ratio (transformer); voltage regulation (constant-voltage transformer).** 42A15-31E12

(2) (constant-current transformer). The ratio of the number of turns in the primary winding to that in the secondary winding. *Note:* In case of a constant-current transformer having taps for changing its voltage ratio, the turn ratio is based on the number of turns corresponding to the normal rated voltage of the respective windings, to which operation and performance characteristics are referred. 57A14-0

(3) (potential transformer). The ratio of the primary winding turns to the secondary winding turns. 57A13-31E12

(4) (rectifier transformer). The ratio of the number of turns in the alternating-current winding to that in the direct-current winding. *Note:* The turn ratio is based on the number of turns corresponding to the normal rated voltage of the respective windings to which operating and performance characteristics are referred. *See also:* **rectifier transformer.** 57A18-0

turn separator (rotating machinery). An insulation strip between turns; a form of turn insulation. *See also:* **rotor (rotating machinery); stator.** 0-31E8

turn-signal operating unit. That part of a signal system by which the operator of a vehicle indicates the direction a turn will be made, usually by a flashing light. *See also:* **headlamp.** Z7A1-0

turns per phase, effective (rotating machinery). The number of fictitious full-pitch coils in a concentrated winding having the same level of flux linkages with fundamental air-gap flux as the actual phase winding. It is calculated as the actual number of series turns per circuit multiplied by the winding factor. *See also:* **asynchronous machine; synchronous machine.** 0-31E8

turnstile antenna. An antenna composed of two dipole antennas, normal to each other, with their axes intersecting at their midpoints. Usually, the currents are equal and in phase quadrature. *See also:* **antenna.** E145-3E1

turn-to-turn test (interturn test) (rotating machinery). A test that is designed to apply or develop a voltage of specified amplitude and waveform between turns of a winding for the purpose of determining the integrity of the turn insulation. *See also:* **rotor (rotating machinery); stator.** 0-31E8

turn-to-turn voltage (rotating machinery). The voltage existing between adjacent turns of a coil. *See also:* **rotor (rotating machinery); stator.** 0-31E8

turntable rumble (audio and electroacoustics). Low-frequency vibration mechanically transmitted to the recording or reproducing turntable and superimposed on the reproduction. *See also:* **phonograph pickup; rumble.** E188-0

TV. *See:* **television.**

twin cable. A cable composed of two insulated conductors laid parallel and either attached to each other by the insulation or bound together with a common covering. *See also:* **power distribution, underground construction.** 42A35-31E13

twin-T network. *See:* **parallel-T network.**

twin wire. A cable composed of two small insulated conductors laid parallel, having a common covering. *See also:* **conductor.** E30-0

twisted-lead transposition (rotating machinery). A form of transposition used on a distributed armature winding wherein the strands comprising each turn are kept insulated from each other throughout all the coils in a phase belt, and the last half turn of each coil is given a 180-degree twist prior to connecting it to the first half turn of the next coil in the series. *See also:* **rotor (rotating machinery); stator.** 0-31E8

twisted pair. A cable composed of two small insulated conductors, twisted together without a common covering. *Note:* The two conductors of a twisted pair are usually substantially insulated, so that the combination is a special case of a cord. *See also:* **conductors.** E30/42A35-31E13

twist, waveguide. A waveguide section in which there is a progressive rotation of the cross section about the longitudinal axis. *See also:* **waveguide.** E147-3E1

two-address. Pertaining to an instruction code in which each instruction has two address parts. Some two-address instructions use the addresses to specify the location of one operand and the destination of the result, but more often they are one-plus-one-address instructions. *See also:* **electronic digital computer.** E162-0

two-degree-freedom gyro. A gyro in which the rotor axis is free to move in any direction. *See also:* **navigation.** E174-10E6

two-element relay. An alternating-current relay that is controlled by current from two circuits through two cooperating sets of coils. *See also:* **railway signal and interlocking.** 42A42-0

two-fluid cell. A cell having different electrolytes at the two electrodes. *See also:* **electrochemistry.** 42A60-0

two-frequency simplex operation (radio communication). The operation of a two-way radio-communication circuit utilizing two radio-frequency channels, one for each direction of transmission, in such manner that intelligence can be transmitted in only one direction at a time. *See also:* **channel spacing.** 0-6E1

two-layer winding (two-coil-side-per-slot winding). A winding in which there are two coil sides in the depth of a slot. *See also:* **rotor (rotating machinery); stator.** 0-31E8

two-out-of-five code. A code in which each decimal digit is represented by five binary digits of which two are one kind (for example, ones) and three are the other kind (for example, zeros). X3A12-16E9

two-phase circuit. A polyphase circuit of three, four, or five distinct conductors intended to be so energized that in the steady state the alternating voltages between two selected pairs of terminals of entry, other than the neutral terminal when one exists, have the same periods, are equal in amplitude, and have a phase difference of $\pi/2$ radians. When the circuit consists of five conductors, but not otherwise, one of them is a neutral conductor. *Note:* A **two-phase circuit** as defined here does not conform to the general pattern of polyphase circuits. Actually a two-phase, four-wire or five-wire circuit could more properly be called a four-phase circuit, but the term **two-phase** is in common usage. A two-phase, three-wire circuit is essentially a special case, as it does not conform to the general pattern of other polyphase circuits. *See also:* **network analysis.** E270-0

two-phase five-wire system. A system of alternating-current supply comprising five conductors, four of which are connected as in a four-wire two-phase system, the fifth being connected to the neutral points of each phase. *Note:* The neutral is usually grounded. Although this type of system is usually known as the two-phase five-wire system, it is strictly a four-phase five-wire system. *See also:* **alternating-current distribution; network analysis.** 42A35-31E13

two-phase four-wire system. A system of alternating-current supply comprising two pairs of conductors between one pair of which is maintained an alternating difference of potential displaced in phase by one-quarter of a period from an alternating difference of potential of the same frequency maintained between the other pair. *See also:* **alternating-current distribution; network analysis.** 42A35-31E13

two-phase three-wire system. A system of alternating-current supply comprising three conductors between one of which (known as the common return) and each of the other two are maintained alternating differences of potential displaced in phase by one-quarter of a period with relation to each other. *See also:* **alternating-current distribution; network analysis.** 42A35-31E13

two-plus-one address (electronic computation). Pertaining to an instruction that contains two operand addresses and a control address. *See:* **control address; electronic digital computer; instruction; operand; three-address code.** E162/X3A12-16E9

two-rate meter. A meter having two sets of register dials with a changeover arrangement such that integration of the quantity will be registered on one set of dials for a specified number of hours each day and on the other set of dials for the remaining hours. *Note:* This is a special case of a multirate meter. *See also:* **electricity meter.** 12A0-0

twos complement. The radix complement of a numeral whose radix is two. *Note:* Using approved synonyms, the same definition may be expressed in the alternative form: "A true complement with a base of two." *See also:* **electronic digital computer.** E162/X3A12-16E9

two-source frequency keying. That form of keying in which the modulating wave abruptly shifts the output frequency between predetermined values, where the values of output frequency are derived from independent sources. *Note:* Therefore, the output wave is not coherent and, in general, will have a phase discontinuity. *See also:* **modulating systems; telegraphy.** E145/42A65-0

two-speed alternating-current control. A control for two-speed driving-machine induction motor that is arranged to run near two different synchronous speeds by connecting the motor windings so as to obtain different numbers of poles. *See also:* **control (elevators).** 42A45-0

two-terminal capacitor. Two conductors separated by a dielectric. The construction is usually such that one conductor essentially surrounds the other and therefore the effect of the presence of other conductors, except in the immediate vicinity of the terminals, is elminiated. (Specialized usage.) E270-0

two-terminal pair network (quadripole) (four-pole). A network with four accessible terminals grouped in pairs. In such a network one terminal of each pair may coincide with a network node. *See also:* **network analysis.** E153/E270-0

two-tone keying. That form of keying in which the modulating wave causes the carrier to be modulated with a single tone for the marking condition and modulated with a different single tone for the spacing condition. *See also:* **modulating system; telegraphy.** E145/E170/42A65-19E4

two-value capacitor motor. A capacitor motor using different values of effective capacitance for the starting and running conditions. *See also:* **asynchronous machine.** 42A10-31E8

two-way automatic maintaining leveling device. A device that corrects the car level on both underrun and overrun, and maintains the level during loading and unloading. *See also:* **elevator-car leveling device.** 42A45-0

two-way automatic nonmaintaining leveling device. A device that corrects the car level on both underrun and overrun, but will not maintain the level during loading and unloading. *See also:* **elevator-car leveling device.** 42A45-0

two-way correction (industrial control). A method of register control that effects a correction in register in either direction. 42A25-34E1

two-wire circuit. A metallic circuit formed by two adjacent conductors insulated from each other. *Note:* Also used in contrast with **four-wire circuit** to indicate a circuit using one line or channel for transmission of electric waves in both directions. *See also:* **center of distribution; transmission line.** 42A35/42A65/31E13

two-wire control (industrial control). A control function which utilizes a maintained-contact type of pilot device to provide undervoltage release. *See:* **undervoltage release.** *See also:* **control.** IC1-34E10

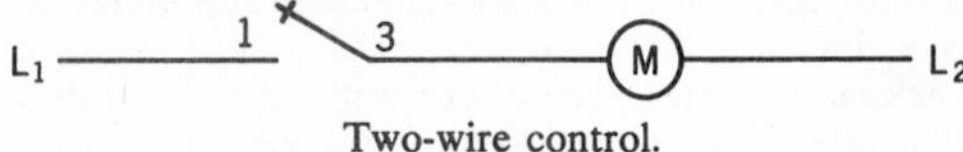

Two-wire control.

two-wire repeater. A telephone repeater which provides for transmission in both directions over a two-wire telephone circuit. *Note:* In practice this may be either a 21-type repeater or a 22-type repeater. *See also:* **repeater.** 42A65-31E3

two-wire system. *See:* **two-wire circuit.**

type-*A* display; type-*B* display; etcetera (radar). *See:* ***A* display; *B* display; etcetera.**

type font. A type face of a given size, for example, 10-point Bodoni Book Medium; 9-point Gothic. X3A12-16E9

type of emission (mobile communication). A system of designating emission, modulation, and transmission characteristics of radio-frequency transmissions, as defined by the Federal Communications Commission. *See also:* **mobile communication system.** 0-6E1

type of piezoelectric crystal cut. The orientation of a piezoelectric crystal plate with respect to the axes of the crystal. It is usually designated by symbols. For example, *GT, AT, BT, CT,* and *DT* identify certain quartz crystal cuts having very low temperature coefficients. *See also:* **crystal.** 42A65-0

type of service (industrial control). The specific type of application in which the controller is to be used, for example: (1) general purpose; (2) special purpose, namely, crane and hoist, elevator, steel mill, machine tool, printing press, etcetera. *See also:* **electric controller.** 42A25-34E10

type tests (rotating machinery). The performance tests taken on the first machine of each type of design. *See:* **asynchronous machine; direct-current commutating machine; synchronous machine.** 0-31E8

U

ultimately controlled variable (control) (industrial control). The variable the control of which is the end purpose of the automatic control system. *See also:* **control system, feedback.** AS1-34E10/23E0

ultimate mechanical strength (insulator). The load at which any part of the insulator fails to perform its function of providing a mechanical support without regard to electrical failure. *See also:* **insulator.** 29A1-0

ultra-audible frequency (supersonic frequency)*. *See:* **ultrasonic frequency.**

*Obsolescent

ultra-audion oscillator. *See:* **Colpitts oscillator.**

ultrasonic cross grating (grating). A space grating resulting from the crossing of beams of ultrasonic waves having different directions of propagation. *Note:* The grating may be two- or three-dimensional. *See also:* **electroacoustics.** 0-1E1

ultrasonic delay line. A transmission device, in which use is made of the propagation time of sound to obtain a time delay of a signal. *See also:* **electroacoustics.** 0-1E1

ultrasonic frequency (ultra-audible frequency) (supersonic frequency)*. A frequency lying above the audio-frequency range. The term is commonly applied to elastic waves propagated in gases, liquids, or solids. *Note:* The word **ultrasonic** may be used as a modifier to indicate a device or system employing or pertaining to ultrasonic frequencies. The term **supersonic,** while formerly applied to frequency, is now generally considered to pertain to velocities above those of sound waves. Its use as a synonym of ultrasonic is now deprecated. *See also:* **electroacoustics; signal wave.**

*Obsolescent 42A65-0;0-1E1

ultrasonic generator. A device for the production of sound waves of ultrasonic frequency. *See also:* **electroacoustics.** 42A65-0

ultrasonic grating constant. The distance between diffracting centers of the sound wave that is producing particular light diffraction spectra. *See also:* **electroacoustics.** E157-1E1

ultrasonic light diffraction. Optical diffraction spectra or the process that forms them when a beam of light is passed through the field of a longitudinal wave. *See also:* **electroacoustics.** 0-1E1

ultrasonic space grating (grating). A periodic spatial variation of the index of refraction caused by the presence of acoustic waves within the medium. *See also:* **electroacoustics.** E157-1E1

ultrasonic stroboscope. A light interrupter whose action is based on the modulation of a light beam by an ultrasonic field. *See also:* **electroacoustics.** E157-1E1

ultraviolet radiation. For practical purposes, any radiant energy within the wavelength range 10 to 380 nanometers. *Note:* On the basis of practical applications and the effect obtained, the ultraviolet region often is divided into the following bands: ozone-producing, 180-220 nanometers; bactericidal (germicidal), 220-300 nanometers; erythemal, 280-320 nanometers; black light, 320-400 nanometers. There are no sharp demarcations between these bands, the indicated effects usually being produced to a lesser extent by longer and shorter wave-lengths. For engineering purposes, the black-light region extends slightly into the visible portion of the spectrum.
See:
bactericidal efficiency of radiant flux;

bactericidal exposure;
bactericidal flux;
bactericidal flux density;
black light;
black-light flux;
black-light-flux density;
erythema;
erythemal effectiveness;
erythemal efficiency of radiant flux;
erythemal exposure;
erythemal flux;
erythemal-flux density;
filter factor;
glow factor;
light;
minimal perceptible erythema;
ozone-producing radiation;
photochemical radiation;
units of wavelength. Z7A1-0

unary operation (computing machines). *See:* **monadic operation.** *See also:* **electronic digital computer.**

unattended automatic exchange (CDO or CAX). A normally unattended telephone exchange, wherein the subscribers, by means of calling devices, set up in the central office the connections to other subscribers or to a distant central office. *See also:* **telephone system.** 42A65-0

unbalance. A differential mutual impedance or mutual admittance between two circuits that ideally would have no coupling. *See also:* **coupling.** 42A65-0

unbalanced circuit. A circuit, the two sides of which are inherently electrically unlike with respect to a common reference point, usually ground. *Note:* Frequently, unbalanced signifies a circuit, one side of which is grounded. E151/42A65-13E6

unbalanced modulator (signal-transmission system). *See:* **modulator, asymmetrical.** *See also:* **signal.**

unbalanced strip line. *See:* **strip (strip-type) transmission line.**

unbalanced wire circuit. A wire circuit whose two sides are inherently electrically unlike. *See also:* **transmission line.** 42A65-31E13

unbiased telephone ringer. A telephone ringer whose clapper-driving element is not normally held toward one side or the other, so that the ringer will operate on alternating current. Such a ringer does not operate reliably on pulsating current. *Note:* A ringer that is weakly biased so as to avoid tingling when dial pulses pass over the lines may be referred to as an unbiased ringer. *See also:* **telephone station.** 42A65-0

unblanking. Turning on of the cathode-ray-tube beam. *See:* **oscillograph.** 0-9E4

uncertainty. (1) General term for the estimated amount by which the observed or calculated value of a quantity may depart from the true value. *Note:* The uncertainty is often expressed as the average deviation, the probable error, or the standard deviation. *See also:* **measurement system.** 0-9E4
(2) The assigned allowance for the systematic error, together with the random error attributed to the imprecision of the measurement process. 0-9E3

unconditional jump (unconditional transfer of control) (electronic computation). An instruction that interrupts the normal process of obtaining instructions in an ordered sequence and specifies the address from which the next instruction must be taken. *See also:* **electronic computation; jump.** E162/E270-0

unconditional transfer of control. *See:* **unconditional jump.** *See also:* **electronic digital computer.**

underbunching. A condition representing less than optimum bunching. *See also:* **electron devices, miscellaneous.** E160-15E6

undercounter dumbwaiter. A dumbwaiter that has its top terminal landing located underneath a counter and that serves only this landing and the bottom terminal landing. 42A45-0

undercurrent relay. A relay that operates when the current through the relay is equal to or less than its setting. *See also:* **relay.** 37A100/83A16-31E6/31E11

undercurrent release (undercurrent trip). A release that operates when the current in the main circuit is equal to or less than the release setting. 37A100-31E11

undercurrent trip. *See:* **undercurrent release.**

underdamped. Damped insufficiently to prevent oscillation of the output following an abrupt input stimulus. *Note:* In an underdamped linear second-order system, the roots of the characteristic equation have complex values. *See:* **damped harmonic system.** AS1-34E10;0-23E0

underdamped period (instrument) (periodic time). The time between two consecutive transits of the pointer or indicating means in the same direction through the rest position, following an abrupt change in the measurand. 42A30-0

underdamping (periodic damping*). The special case of damping in which the free oscillation changes sign at least once. A damped harmonic system is underdamped if $F^2 < MS$. See **damped harmonic system** for equation, definitions of letter symbols, and referenced terms.
*Deprecated E270-0

underdome bell. A bell whose mechanism is mostly concealed within its gong. *See also:* **protective signaling.** 42A65-0

underfilm corrosion. Corrosion that occurs under films in the form of randomly distributed hairlines (filiform corrosion). *See:* **corrosion terms.** CM-34E2

underfloor raceway. A raceway suitable for use in the floor. *See also:* **raceway.** 42A95-0

underflow (computing machines). Pertaining to the condition that arises when a machine computation yields a nonzero result that is smaller than the smallest nonzero quantity that the intended unit of storage is capable of storing. *See also:* **electronic digital computer.** X3A12-16E9

underground cable. A cable installed below the surface of the ground. *Note:* This term is usually applied to cables installed in ducts or conduits or under other conditions such that they can readily be removed without disturbing the surrounding ground. *See also:* **cable; tower.** 42A35/42A65-31E13

underground collector or plow. A current collector, the function of which is to make contact with an underground contact rail. *See also:* **contact conductor.** E16-0

underlap, *X* (facsimile). The amount by which the center-to-center spacing of the recorded spots exceeds the recorded spot *X* dimension. *Note:* This effect arises in that type of equipment that responds to a constant density in the subject copy by a succession of discrete recorded spots. *See also:* **recording (facsimile).** E168-0

underlap, Y (facsimile). The amount by which the nominal line width exceeds the recorded spot Y dimension. *See also:* **recording (facsimile).** E168-0

underreaching protection. A form of protection in which the relays at a given terminal do not operate for faults at remote locations on the protected equipment, the given terminal being cleared either by other relays with different performance characteristics or by a transferred trip signal from a remote terminal similarly equipped with underreaching relays. 37A100-31E6/31E11

undershoot (television). The initial transient response to a unidirectional change in input, that precedes the main transition and is opposite in sense. *See also:* **distortion, pulse; television.** E204/42A65-2E2

underslung car frame. A car frame to which the hoisting-rope fastenings or hoisting rope sheaves are attached at or below the car platform. *See also:* **hoistway (elevator or dumbwaiter).** 42A45-0

undervoltage protection (industrial control) (low-voltage protection) (industrial control). The effect of a device, operative on the reduction or failure of voltage, to cause and maintain the interruption of power to the main circuit. *Note:* The principal objective of this device is to prevent automatic restarting of the equipment. Standard undervoltage or low-voltage protection devices are not designed to become effective at any specific degree of voltage reduction. 37A100/42A25-31E6/31E11/34E10

undervoltage relay. A relay that operates when the voltage applied to the relay is equal to or less than its setting. *See:* **relay.** 37A100/42A25/83A16-31E6/31E11/34E10

undervoltage release (low-voltage release) (industrial control). The effect of a device, operative on the reduction or failure of voltage, to cause the interruption of power to the main circuit but not to prevent the re-establishment of the main circuit on return of voltage. *Note:* Standard undervoltage or low-voltage release protection devices are not designed to become effective at any specific degree of voltage reduction. 42A25-34E10

underwater log. A device that indicates a ship's speed based on the pressure differential, resulting from the motion of the ship relative to the water, as developed in a Pitot tube system carried by a retractable support extending through the ship's hull. Continuous integration provides indication of total distance travelled. The ship's draft is indicated, based on static pressure. 42A43-0

underwater sound projector. A transducer used to produce sound in water. *Notes:* (1) There are many types of underwater sound projectors whose definitions are analogous to those of corresponding loudspeakers, for example, crystal projector, magnetic projector, etcetera. (2) Where no confusion will result, the term underwater sound projector may be shortened to projector. *See also:* **microphone.** 42A65/E157-1E1

undetected error rate (data transmission). The ratio of the number of bits, unit elements, characters, blocks incorrectly received but undetected or uncorrected by the error-control equipment, to the total number of bits, unit elements, characters, blocks sent. *See also:* **data transmission.** 0-19E4

undisturbed-ONE output (magnetic cell). A ONE output to which no partial-read pulses have been applied since that cell was last selected for writing. *See also:* **coincident-current selection; static magnetic storage.** E163-0

undisturbed-ZERO output (magnetic cell). A ZERO output to which no partial-write pulses have been applied since that cell was last selected for reading. *See also:* **coincident-current selection; static magnetic storage.** E163-0

unfired tube (microwave gas tubes). The condition of the tube during which there is no radio-frequency glow discharge at either the resonant gap or resonant window. *See:* **gas tubes.** E160-15E6

ungrounded (electric power). A system, circuit, or apparatus without an intentional connection to ground except through potential-indicating or measuring devices or other very-high-impedance devices. *See:* **ground; grounded.** *See also:* **alternating-current distribution; direct-current distribution.** E22/42A15/42A35/31E12/31E13

uniconductor waveguide. A waveguide consisting of a cylindrical metallic surface surrounding a uniform dielectric medium. *Note:* Common cross-sectional shapes are rectangular and circular. *See also:* **waveguide.** E146/E148-3E1

unidirectional. A connection between telegraph sets, one of which is a transmitter and the other a receiver. 0-19E4

unidirectional antenna. An antenna that has a single well-defined direction of maximum gain. *See also:* **antenna.** E145-3E1

unidirectional current. A current that has either all positive or all negative values. E270-0

unidirectional microphone. A microphone that is responsive predominantly to sound incident from a single solid angle of one hemisphere or less. *See also:* **microphone.** 0-1E1

unidirectional pulse (signal-transmission system). A pulse in which pertinent departures from the normally constant value occur in one direction only. *See also:* **modulating systems; pulse; pulse terms; signal.** E194-13E6

unidirectional transducer (unilateral transducer). A transducer that cannot be actuated at its output by waves in such a manner as to supply related waves at its input. *See also:* **transducer.** E196/E270/42A65-0

uniform current density. A current density that does not change (either in magnitude or direction) with position within a specified region. (A uniform current density may be a function of time.) E270-0

uniform field. A field (scalar or vector or other) in which the field quantities do not vary with position. E270-0

uniform line. A line that has substantially identical electrical properties throughout its length. *See also:* **transmission line.** 42A65-0

uniform linear array (antenna). A linear array of identically oriented and equally spaced radiating elements having equal current amplitudes and equal phase increments between excitation currents. *See also:* **antenna.** 0-3E1

uniform luminance area. The area in which a display on a cathode-ray tube retains 70 percent or more of its luminance at the center of the viewing area. *Note:* The corners of the rectangle formed by the vertical and horizontal boundaries of this area may be below the

70-percent luminance level. *See:* **oscillograph.** 0-9E4

uniform plane wave (radio wave propagation). A plane wave in which the electric and magnetic intensities have constant amplitude over the equiphase surfaces. *Note:* Such a wave can only be found in free space at an infinite distance from the source. *See also:* **radiation; radio wave propagation.** E211/42A65-3E2

uniform waveguide. A waveguide or transmission line in which the physical and electrical characteristics do not change with distance along the axis of the guide. *See:* **waveguide.** E146-3E1

unilateral area track. A sound track in which one edge only of the opaque area is modulated in accordance with the recorded signal. There may, however, be a second edge modulated by a noise-reduction device. *See also:* **phonograph pickup.** E157-1E1

unilateral connection (control system, feedback). A connection through which information is transmitted in one direction only. *See also:* **control system, feedback.** 0-23E0

unilateral network. A network in which any driving force applied at one pair of terminals produces a nonzero response at a second pair but yields zero response at the first pair when the same driving force is applied at the second pair. *See also:* **network analysis.** E270-0

unilateral transducer. *See:* **unidirectional transducer.**

unipolar (power supplies). Having but one pole, polarity, or direction. Applied to amplifiers or power supplies, it means that the output can vary in only one polarity from zero and, therefore, must always contain a direct-current component. *See:* **bipolar**. *See also:* **power supply.** KPSH-10E1

unipolar electrode system (electrobiology) (monopolar electrode system). Either a pickup or a stimulating system, consisting of one active and one dispersive electrode. *See also:* **electrobiology.** 42A80-18E1

unipole*. *See:* **antenna; isotropic antenna.**

*Deprecated

unipotential cathode. *See:* **cathode, indirectly heated.**

unit (1) (physical unit). An amount of a physical quantity arbitrarily assigned magnitude unity and determined by a specified relationship to one or more prototype standards. E270-0

(2) (general). A device having a special function. X3A12-16E9

(3) (electronic computation). A portion of a computer that constitutes the means of accomplishing some inclusive operation or function, as **arithmetic unit.** *See also:* **arithmetic unit; processing unit; control unit; electronic computation; electronic digital computer; tape unit.** E162/E270-0

(4) (relay). *See:* **relay unit.**

unit-area capacitance (electrolytic capacitor). The capacitance of a unit area of the anode surface at a specified frequency after formation at a specified voltage. *See also:* **electrolytic capacitor.** 42A60-0

unit cable construction. That method of cable manufacture in which the pairs of the cable are stranded into groups (units) containing a certain number of pairs and these groups are then stranded together to form the core of the cable. *See also:* **cable.** 42A65-0

unit-control error (electric power systems). The unit generation minus the assigned unit generation. *Note:* Refer to note on polarity under **area control error.** *See also:* **power system, low-frequency and surge testing.** E94-0

unit-impulse function. *See:* **signal, unit-impulse.**

unit interval. *See:* **signal element.**

unit operation (1) (arrester). Consists of initiating follow current by discharging a surge through the arrester while the arrester is energized. *See also:* **arrester; lightning arrester (surge diverter); current rating, 60-hertz (arrester).** E28/42A20/62A1-31E7

(2) (CO) (circuit breaker). *See:* **close-open operation (switching device).**

(3) (recloser). An interrupting operation followed by a closing operation. The final interruption is also considered one unit operation. 37A100-31E11

unit-ramp function. *See:* **signal, unit-ramp.**

unit rate-limiting controller (electric power systems). A controller that limits rate of change of generation of a generating unit to an assigned value or values. *Note:* The limiting action is normally based on a measured megawatt-per-minute rate. *See also:* **speed-governing system.** E94-0

units of luminace (photometric brightness). *See:* **luminance.**

units of wavelength. The distance between two successive points of a periodic wave in the direction of propagation, in which the oscillation has the same phase. The three commonly used units are listed in the following table:

Name	Symbol	Value
micrometer	μm	1 μm = 10^{-3} millimeters
nanometer	nm	1 nm = 10^{-6} millimeters
angstrom	Å	1 Å = 10^{-7} millimeters

See also: **radiant energy.** Z7A1-0

unit-step function. *See:* **signal, unit-step.**

unit substation. A substation consisting primarily of one or more transformers mechanically and electrically connected and coordinated in design with one or more switchgear or motor control assemblies or combinations thereof. *Note:* A unit substation may be described as primary or secondary depending on the voltage rating of the low-voltage section: primary, of 1000 volts and above; secondary, of less than 1000 volts. 37A100-31E11

unit symbol (abbreviation). *See:* **symbol for a unit.**

unit vector. A vector whose magnitude is unity. E270-0

unit warmup time (power supply). The interval between the time of application of input power to the unit and the time at which the regulated power supply is supplying regulated power at rated output voltage. *See also:* **regulated power supply.** E209-0

unity-gain bandwidth (power supplies). A measure of the gain-frequency product of an amplifier. Unity-gain

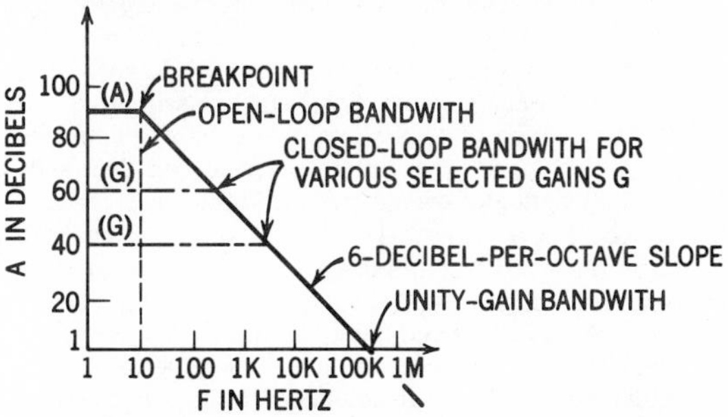

Typical gain–frequency (Bode) plot, showing unity-gain bandwidth.

bandwidth is the frequency at which the open-loop gain becomes unity, based on a 6-decibel-per-octave crossing. See the accompanying figure. *See also:* **power supply.** KPSH-10E1

unity power-factor test (synchronous machine). A test in which the machine is operated as a motor under specified operating conditions with its excitation adjusted to give unity power factor. *See:* **synchronous machine.** 0-31E8

univalent function. If to every value of u there corresponds one and only one value of x (or one and only one set of values of $x_1, x_2, \cdots, x_n$) then u is a univalent function. Thus $u^2 = ax + b$ is univalent, within the interval of definition. E270-0

universal fuse links. Fuse links that, for each rating, provide mechanical and electrical interchangeability within prescribed limits over the specified time-current range. 37A100-31E11

universal motor. A series-wound or compensated series-wound motor designed to operate at approximately the same speed and output on either direct current or single-phase alternating current within a specified frequency range and at the same root-mean-square voltage. *See:* **asynchronous machine; direct-current commutating machine; synchronous machine.** 0-31E8

universal-motor parts (rotating machinery). A term applied to a set of parts of a universal motor. Rotor shaft, conventional stator frame (or shell), end shields, or bearings may not be included, depending on the requirements of the end product into which the universal-motor parts are to be assembled. *See also:* **asynchronous machine; direct-current commutating machine.** 0-31E8

universal or arcshear machine. A power-driven cutter that will not only cut horizontal kerfs, but will also cut vertical kerfs or at any angle, and is designed for operation either on track, caterpillar treads, or rubber tires. *See also:* **mining.** 42A85-0

universal Turing machine. A Turing machine that can simulate any other Turing machine. *See also:* **electronic digital computer.** X3A12-16E9

unloaded applicator impedance (dielectric heating usage). The complex impedance measured at the point of application, without the load material in position, at a specified frequency. *See also:* **dielectric heating.** E54/E169-0

unloaded (intrinsic) *Q* (switching tubes). The Q of a tube unloaded by either the generator or the termination. *Note:* As here used, Q is equal to 2π times the energy stored at the resonance frequency divided by the energy dissipated per cycle in the tube or, for cell-type tubes, in the tube and its external resonant circuit. *See also:* **circuit characteristics of electrodes; gas tubes.** E160-15E6

unloaded sag (conductor at any point in a span). The distance measured vertically from the particular point in the conductor to a straight line between its two points of support, without any external load. *See:* **power distribution overhead construction; sag.** 2A2-0

unloading amplifier. An amplifier that is capable of reproducing or amplifying a given voltage signal while drawing negligible current from the voltage source. *Note:* In an analog computer, the term **buffer amplifier** is sometimes synonymous with unloading amplifier. *See also:* **electronic analog computer.** E165-16E9

unloading circuit (analog computer). A computing element or combination of computing elements capable of reproducing or amplifying a given voltage signal while drawing negligible current from the voltage source, thus decreasing the loading errors. *See also:* **unloading amplifier.** *See also:* **electronic analog computer.** E165-0

unloading point (electric transmission system used on self-propelled electric locomotives or cars). The speed above or below which the design characteristics of the generators and traction motors or the external control system, or both, limit the loading of the prime mover to less than its full capacity. *Note:* The unloading point is not always a sharply defined point, in which case the unloading point may be taken as the useful point at which essentially full load is provided. *See also:* **traction motor.** 42A42-0

unmodulated groove (blank groove) (mechanical recording). A groove made in the medium with no signal applied to the cutter. *See also:* **phonograph pickup.** E157-1E1

unpack. To separate various sections of packed data. *See also:* **electronic digital computer.** X3A12-16E9

unpropagated potential (electrobiology). An evoked transient localized potential not necessarily associated with changed excitability. *See also:* **excitability (electrobiology).** 42A80-0

unregulated voltage (electronically regulated power supply). The voltage at the output of the rectifier filter. *See also:* **regulated power supply.** E209-0

unsaturated standard cell. A cell in which the electrolyte is a solution of cadmium sulphate at less than saturation at ordinary temperatures. (This is the commercial type of cadmium standard cell commonly used in the United States). *See also:* **electrochemistry.** 42A60-0

unshielded strip transmission line. A strip conductor above a single ground plane. Some common designations are: **microstrip (flat-strip conductor); unbalanced strip line.*** *See:* **strip (type) transmission line; shielded strip transmission line; waveguides.**
*Deprecated 0-3E1

unstable (control system, feedback). Not possessing stability. *See also:* **control system, feedback.** 85A1-23E0

unusual service conditions. Environmental conditions that may affect the constructional or operational requirements of a machine. This includes the presence of moisture and abrasive, corrosive, or explosive atmosphere. It also includes external structures that limit ventilation, unusual conditions relating to the electrical supply, the mechanical loading, and the position of the machine. 0-31E8

upper (driving) beams. One or more beams intended for distant illumination and for use on the open highway when not meeting other vehicles. Formally **country beam.** *See also:* **headlamp.** Z7A1-0

upper bracket (rotating machinery). A bearing bracket mounted above the core of a vertical machine. *See also:* **cradle base (rotating machinery).** 0-31E8

upper burst reference (audio and electroacoustics). A selected multiple of the long-time average magnitude of the quantity mentioned in the definition of **burst.** See the figure attached to the definition of **burst duration.** *See also:* **burst (audio and electroacoustics).** E257-1E1

upper coil support (rotating machinery). A coil support to restrain field-coil motion in the direction toward the air gap. *See also:* **rotor (rotating machinery); stator.** 0-31E8

upper frequency limit (coaxial transmission line). The limit determined by the cutoff frequency of higher-order waveguide modes of propagation, and the effect that they have on the impedance and transmission characteristics of the normal TEM coaxial-transmission-line mode. The lowest cutoff frequency occurs with the $TE_{1,1}$ mode, and this cutoff frequency in air dielectric line is the upper frequency limit of a practical transmission line. How closely the $TE_{1,1}$ mode cutoff frequency can be approached depends on the application. *See also:* **waveguide.** 83A14-0

upper guide bearing (rotating machinery). A guide bearing mounted above the core of a vertical machine. *See also:* **bearing.** 0-31E8

upper half bearing bracket (rotating machinery). The top half of a bracket that can be separated into halves for mounting or removal without access to a shaft end. *See also:* **bearing.** 0-31E8

upper range-value. The highest quantity that a device is adjusted to measure. *Note:* The following compound terms are used with suitable modifications in the units: **measured variable upper range-value, measured signal upper range-value,** etcetera. *See also:* **instrument.** 39A4-0

upper-sideband parametric down-converter. A noninverting parametric device used as a parametric down-converter. *See also:* **parametric device.** E254-15E7

upper-sideband parametric up-converter. A noninverting parametric device used as a parametric up-converter. *See also:* **parametric device.** E254-15E7

upset duplex system. A direct-current telegraph system in which a station between any two duplex equipments may transmit signals by opening and closing the line circuit, thereby causing the signals to be received by upsetting the duplex balance. *See also:* **telegraphy.** 42A65-0

upward component (illuminating engineering). That portion of the luminous flux from a luminaire that is emitted at angles above the horizontal. *See also:* **inverse-square law (illuminating engineering).** Z7A1-0

usable sensitivity. The minimum standard modulated carrier-signal power required to produce usable receiver output. *See also:* **receiver performance.** 0-6E1

useful life (reliability). The length of time an item operates with an acceptable failure rate. *See also:* **reliability.** 0-7E1

useful line. *See:* **available line.**

useful output power (electron device). That part of the output power that flows into the load proper. *See:* **circuits and devices.** 0-15E6

usual service conditions. Environmental conditions in which standard machines are designed to operate. The temperature of the cooling medium does not exceed 40 degrees Celsius and the altitude does not exceed 3300 feet. 0-31E8

utilance. *See:* **room utilization factor.**

utility routine. *See:* **service routine.** *See also:* **electronic digital computer.**

utilization equipment. Equipment, devices, and connected wiring that utilize electric energy for mechanical, chemical, heating, lighting, testing, or similar useful purposes and are not a part of supply equipment, supply lines, or communication lines. *See also:* **distribution center.** E270/42A95/2A2-0

utilization factor (system utilization factor). The ratio of the maximum demand of a system to the rated capacity of the system. *Note:* The utilization factor of a part of the system may be similarly defined as the ratio of the maximum demand of the part of the system to the rated capacity of the part of the system under consideration. *See also:* **alternating-current distribution; direct-current distribution.** 42A35-31E13

utilization time (hauptnutzzeit) (medical electronics). (1) The minimum duration that a stimulus of rheobasic strength must have to be just effective. (2) The shortest latent period between stimulus and response obtainable by very strong stimuli.* (3) The latent period following application of a shock of rheobasic intensity.* *See also:* **biological; medical electronics.**

*Deprecated 0-18E1

V

vacuum envelope (electron tube). The airtight envelope that contains the electrodes. *See also:* **electrode (of an electron tube).** 42A70-0

vacuum gauge. A device that indicates the gas pressure in an evacuated system. *Note:* Pressures in vacuum systems are usually expressed in microns absolute, one micron being the pressure that will support a column of mercury 1/1000 of a millimeter high. There are two types of vacuum gauges in common use in rectifier practice; the McLeod type, that operates on the principle that pressure multiplied by volume is constant, and measures the pressure of a gas by compressing a sample of known volume into a calibrated measuring tube; the hot-wire type, that operates on the principle that the heat conductivity of a gas is a function of its pressure. *See also:* **rectification.** 42A15-0

vacuum phototube. A phototube that is evacuated to such a degree that its electrical characteristics are essentially unaffected by gaseous ionization. *See also:* **tube definitions.** 42A70-15E6

vacuum-pressure impregnation (VPI) (rotating machinery). The filling of voids in a coil or insulation system by withdrawal of air and solvent, if any, from the contained voids by vacuum, admission of a resin or resin solution followed by pressurization, and finally cure, usually with the application of heat. *See also:* **rotor (rotating machinery); stator.** 0-31E8

vacuum seal. The airtight junction between component parts of the evacuated system. *See also:* **rectification.** 42A15-0

vacuum switch. A switch whose contacts are enclosed in an evacuated bulb, usually to minimize sparking. *See also:* **circuits and devices.** 42A65-0

vacuum tank. The airtight metal chamber that contains the electrodes and in which the rectifying action takes place. *See also:* **rectification.** 42A15-0

vacuum tube. An electron tube evacuated to such a degree that its electrical characteristics are essentially unaffected by the presence of residual gas or vapor. *See also:* **tube definitions.** E188/42A70-15E6

vacuum-tube amplifier. An amplfiier employing electron tubes to effect the control of power from the local source. *See also:* **amplifier.** E145-0

vacuum-tube transmitter. A radio transmitter in which electron tubes are utilized to convert the applied electric power into radio-frequency power. *See also:* **radio transmitter.** E145-0

vacuum-tube voltmeter. *See:* **electronic voltmeter.**

vacuum valve. A device for sealing and unsealing the passage between two parts of an evacuated system. *See also:* **rectification.** 42A15-0

valance. A longitudinal shielding member mounted across the top of a window or along a wall, usually parallel to the wall, to conceal light sources giving both upward and downward distributions. *See also:* **luminaire.** Z7A1-0

valance lighting. Lighting comprising light sources shielded by a panel parallel to the wall at the top of a window. *See also:* **general lighting.** Z7A1-0

valence band. The range of energy states in the spectrum of a solid crystal in which lie the energies of the valence electrons that bind the crystal together. *See also:* **electron devices, miscellaneous; semiconductor.** E102/E216/E270-10E1/34E17

valley point (tunnel-diode characteristic). The point on the forward current-voltage characteristic corresponding to the second-lowest positive (forward) voltage at which $di/dV = 0$. *See also:* **peak point (tunnel-diode characteristic).** E253-15E7

valley-point current (tunnel-diode characteristic). The current at the valley point. *See also:* **peak point (tunnel-diode characteristic).** E253-15E7

valley-point voltage (tunnel-diode characteristic). The voltage at which the valley point occurs. *See also:* **peak point (tunnel-diode characteristic).** E253-15E7

value (1) (several) (automatic control). The quantitative measure of a signal or variable. *See also:* **control system, feedback.** 0-23E0

(2) (direct-current through test object). The arithmetic mean value. *See also:* **test voltage and current.** 68A1-31E6

(3) (test direct voltages). The arithmetic mean value; that is, the integral of the voltage over a full period of the ripple divided by the period. *Note:* The maximum value of the test voltage may be taken approximately as the sum of the arithmetic mean value plus the ripple magnitude. *See also:* **test voltage and current.** 68A1-31E5

(4) (alternating test voltage). The peak value divided by $(2)^{1/2}$. *See also:* **test voltage and current.** 68A1-31E5

value, ideal. *See:* **ideal value.**

value, Munsell. *See:* **Munsell value.**

valve. *See:* **electron tube.**

valve action (electrochemical). The process involved in the operation of an electrochemical valve. *See also:* **electrochemical valve.** 42A60-0

valve arrester. A lightning arrester that includes a valve element. *See:* **nonlinear-resistor type arrester; valve-type arrester.** *See also:* **arrester.** E28-0;62A1-31E7

valve element (arrester). A resistor that, because of its nonlinear volt-ampere characteristic, limits the voltage across the arrester terminals during the flow of discharge current and contributes to the limitation of follow current at normal power-frequency voltage. *See:* **valve-type arrester.** *See also:* **arrestor; lightning arrester (surge diverter).** 62A1-31E7

valve-point loading control (electric power system). A control means for making a unit operate in the more efficient portions of the range of the governor-controlled valves. *See also:* **speed-governing system.** E94-0

valve ratio (electrochemical valve). The ratio of the impedance to current flowing from the valve metal to the compound or solution, to the impedance in the opposite direction. *See also:* **electrochemical valve.** 42A60-0

valve tube. *See:* **kenotron.**

valve-type arrester. *See:* **arrester, valve-type.**

vane-type relay. A type of alternating-current relay in which a light metal disc or vane moves in response to a change of the current in the controlling circuit or circuits. *See also:* **railway signal and interlocking.** 42A42-0

V antenna. A V-shaped arrangement of conductors, balanced-fed at the apex and with included angle, length, and elevation proportioned to give the desired directive properties. *See also:* **antenna.** 0-3E1

vapor openings. Openings through a tank shell or roof above the surface of the stored liquid. Such openings may be provided for tank breathing, tank gauging, fire fighting, or other operating purposes. *See also:* **lightning protection and equipment.** 5A1-0

vapor-safe electric equipment. A unit so constructed that it may be operated without hazard to its surroundings in an atmosphere containing fuel, oil, alcohol, or other vapors that may occur in aircraft; that is, the unit is capable of so confining any sparks, flashes, or explosions of the combustible vapors within itself that ignition of the surrounding atmosphere is prevented. *Note:* This definition closely parallels that given for **explosionproof**; however, it is believed that the new term is needed in order to avoid the connotation of compliance with Underwriter's standards that are now associated with **explosionproof** in the minds of most engineers who are familiar with the use of that term applied to industrial motors and control equipment. *See:* **air transportation electric equipment.** 42A41-0

vaportight (1) (general). So enclosed that vapor will not enter the enclosure. 37A100/42A95-31E11

(2) (luminaire). A luminaire designed and approved for installation in damp or wet locations. *Note:* It also is described as **enclosed and gasketed.** *See also:* **luminaire.** Z7A1-0

var (electric power circuits). The unit of reactive power in the International System of Units (SI). The var is the reactive power at the two points of entry of a single-phase, two-wire circuit when the product of the root-mean-square value in amperes of the sinusoidal current by the root-mean-square value in volts of the sinusoidal voltage and by the sine of the angular phase difference by which the voltage leads the current is equal to one. E270-0

VAR. *See:* **visual-aural range.**

varactor. A two-terminal semiconductor device in which the electrical characteristic of primary interest is a voltage-dependent capacitance. E254-15E7

varhour. The unit of a quadrature-energy quadergy in the International System of Units (SI). The varhour is the quadrature energy that is considered to have flowed past the points of entry of a reactive circuit when a reactive power of one var has been maintained at the terminals of entry for one hour. E270-0

varhour meter (reactive volt-ampere-hour meter). An electricity meter that measures and registers the integral, with respect to time, of the reactive power of the circuit in which it is connected. The unit in which this integral is measured is usually the kilovarhour. *See also:* **electricity meter (meter).** 12A0/42A30-0

variable. (1) A quantity or condition that is subject to change. *See also:* **control system, feedback.** 85A1-23E0
(2) A quantity that can assume any of a given set of values. X3A12-16E9

variable-area track (electroacoustics). A sound track divided laterally into opaque and transparent areas, a sharp line of demarcation between these areas forming an oscillographic trace of the wave shape of the recorded signal. *See also:* **phonograph pickup.** E157-1E1

variable-block format. A format that allows the number of words in successive blocks to vary. EIA3B-34E12

variable carrier. *See:* **controlled carrier.**

variable-density track (electroacoustics). A sound track of constant width, usually but not necessarily of uniform light transmission on any instantaneous transverse axis, on which the average light transmission varies along the longitudinal axis in proportion to some characteristic of the applied signal. *See also:* **phonograph pickup.** 0 -1E1

variable, directly controlled. *See:* **directly controlled variable.**

variable field. One that varies with time. E270-0

variable-frequency telemetering (electric power systems). A type of telemetering in which the frequency of the alternating-voltage signal is varied as a function of the magnitude of the measured quantity. *See also:* **telemetering.** E94-0

variable, indirectly controlled. *See:* **indirectly controlled variable.**

variable inductor.* *See:* **continuously adjustable inductor.**

*Deprecated

variable, input. A variable applied to a system or element. *See also:* **control system, feedback.** 85A1-23E0

variable, manipulated. *See:* **manipulated variable.**

variable modulation (in very-high-frequency omnidirectional radio ranges). The modulation of the ground station radiation that produces a signal in the airborne receiver whose phase with respect to a radiated reference modulation corresponds to the bearing of the receiver. *See also:* **navigation.** 0-10E6

variable-mu tube (variable-μ tube) (remote-cutoff tube). An electron tube in which the amplification factor varies in a predetermined way with control-grid voltage. *See also:* **tube definitions.** E160/42A70-15E6

variable, output. A variable delivered by a system or element. *See also:* **control system, feedback.** 85A1-23E0

variable point. Pertaining to a numeration system in which the position of the point is indicated by a special character at that position. *See:* **fixed point; floating point.** X3A12-16E9

variable-reluctance microphone (magnetic microphone). A microphone that depends for its operation on variations in the reluctance of a magnetic circuit. *See also:* **microphone.** 42A65-0

variable-reluctance pickup (magnetic pickup). A phonograph pickup that depends for its operation on the variation in the reluctance of a magnetic circuit. *See also:* **phonograph pickup.** 42A65-0

variable-reluctance transducer. An electroacoustic transducer that depends for its operation on the variation in the reluctance of a magnetic circuit. 0-1E1

variable-speed axle generator. An axle gnerator in which the speed of the generator varies directly with the speed of the car. *See also:* **axle generator system.** 42A42-0

variable-speed drive (industrial control). An electric drive so designed that the speed varies through a considerable range as a function of load. *See also:* **electric drive.** 42A25-34E10

variable-torque motor. (1) A multispeed motor whose rated load torque at each speed is proportional to the speed. Thus the rated power of the motor is proportional to the square of the speed. (2) An adjustable-speed motor in which the specified torque increases with speed. It is common to provide a variable-torque adjustable-speed motor in which the torque varies as the square of the speed and hence the power output varies as the cube of the speed. *See also:* **asynchronous machine.** 0-31E8

variable, ultimately controlled. *See:* **ultimately controlled variable.**

varindor. An inductor whose inductance varies markedly with the current in the winding. *See also:* **circuits and devices.** 42A65-0

variocoupler (radio practice). A transformer, the self-impedance of whose windings remains essentially constant while the mutual impedance between the windings is adjustable. *See also:* **circuits and devices.** 42A65-21E0

variolosser. A device whose loss can be controlled by a voltage or current. *See also:* **circuits and devices.** 42A65-0

variometer. A variable inductor in which the change of inductance is effected by changing the relative position of two or more coils. *See also:* **circuits and devices.** 42A65-21E0

varioplex. A telegraph switching system that establishes connections on a circuit-sharing basis between a multiplicity of telegraph transmitters in one locality and respective corresponding telegraph receivers in another locality over one or more intervening telegraph channels. Maximum usage of channel capacity is secured by momentarily storing the signals and allocating circuit time in rotation among those transmitters having intelligence in storage. *See also:* **telegraphy.** 42A65-0

varistor. (1) A two-terminal resistive element, composed of an electronic semiconductor and suitable contacts, that has a markedly nonlinear volt-ampere characteristic. (2) A two-terminal semiconductor device having a voltage-dependent nonlinear resistance. *Note:* **Varistors** may be divided into two groups, symmetrical and nonsymmetrical, based on the symmetry or lack of symmetry of the volt-ampere curve. *See also:* **circuits and devices; semiconductor.** E102/E216/E270/42A65/42A70-34E17

varmeter (reactive volt-ampere meter). An instrument for measuring reactive power. It is provided with a scale usually graduated in either vars, kilovars, or megavars. If the scale is graduated in kilovars or megavars, the instrument is usually designated as a kilovarmeter or megavarmeter. *See also:* **instrument.** 42A30-0

varnish (rotating machinery). A liquid composition that is converted to a transparent or translucent solid film after application as a thin layer. AD16-31E8

varnished fabric (varnished mat) (rotating machinery). A fabric or mat in which the elements and interstices have been essentially coated and filled with an impregnant such as a compound or varnish and that is relatively homogeneous in structure. *See also:* **rotor (rotating machinery); stator.** 0-31E8

varnished tubing. A flexible tubular product made from braided cotton, rayon, nylon, glass, or other fibers, and coated, or impregnated and coated, with a continuous film or varnish, lacquer, a combination of varnish and lacquer, or other electrical insulating materials. 42A95-0

varying duty (rating of electric equipment). A requirement of service that demands operation at loads, and for periods of time, both of which may be subject to wide variation. *See also:* **asynchronous machine; direct-current commutating machine; duty; synchronous machine; voltage regulator.** E96/E270/42A15/57A15-31E12/34E10

varying parameter. *See:* **linear varying parameter.** E270-0

varying-speed motor. A motor the speed of which varies appreciably with the load, ordinarily decreasing when the load increases, for example, a series or repulsion motor. *See:* **asynchronous machine; direct-current commutating machine; synchronous machine.** 42A10-31E8

varying-voltage control. A form of armature-voltage control obtained by impressing on the armature of the motor a voltage that varies considerably with change in load, with a consequent change in speed, such as may be obtained from a differentially compound-wound generator or by means of resistance in the armature circuit. *See also:* **control.** 42A25-34E10

VASIS. *See:* **visual approach slope indicator system.**

vault-type transformer. A submersible transformer that is so constructed as to be suitable for occasional submerged operation in water under specified conditions of time and external pressure. *See also:* **constant-current transformer; transformer.** 42A15/57A14-31E12

***V*-beam radar.** A ground-based three-dimensional radar system for the determination of distance, bearing and, uniquely, the height of the target. It uses two fan-shaped beams, one vertical and the other inclined (intersecting at ground level), that rotate together in azimuth so as to give two responses from the target, the time difference between these responses, together with distance, being factors used in determining the height of the target. *See also:* **navigation; radar.** 42A65-10E6

***V* curve (synchronous machine).** The load characteristic giving the relationship between the armature current and the field current for constant values of load, power, and armature voltage. *See also:* **asynchronous machine.** 0-31E8

vector. A mathematico-physical quantity that represents a vector quantity. *See:* **mathematico-physical quantity (mathematical quantity) (abstract quantity).** E270-0

vectorcardiogram. *See:* **vector electrocardiogram.**

vector electrocardiogram (electrobiology) (vectorcardiogram). The 2-dimensional or 3-dimensional presentation of cardiac electric activity that results from displaying lead pairs against each other rather than against time. More strictly, it is a loop pattern taken from leads placed orthogonally. *See also:* **electrocardiogram.** 42A80-18E1

vector field. The totality of vectors in a given region represented by a vector function $\mathbf{V}(x,y,z)$ of the space coordinates x,y,z. E270-0

vector function. A functional relationship that results in a vector. E270-0

vector operator del ∇. A differential operator defined as follows in terms of Cartesian coordinates:

$$\nabla = \mathbf{i}\frac{\partial}{\partial x} + \mathbf{j}\frac{\partial}{\partial y} + \mathbf{k}\frac{\partial}{\partial z}.$$

E270-0

vector power. *See:* **power, vector.**

vector product (cross product). The vector product of vector **A** and a vector **B** is a vector **C** that has a magnitude obtained by multiplying the product of the magnitudes of **A** and **B** by the sine of the angle between them; the direction of **C** is that traveled by a right-hand screw turning about an axis perpendicular to the plane of **A** and **B**, in the sense in which **A** would move into **B** by a rotation of less than 180 degrees; it is assumed that **A** and **B** are drawn from the same point. The vector product of two vectors **A** and **B** may be indicated by using a small cross: **A** × **B.** The direction of the vector product depends on the order in which the vectors are multiplied, so that **A** × **B** = −**B** × **A.** If the two vectors are given in terms of their rectangular components, then

$$\mathbf{A}\times\mathbf{B} = \begin{vmatrix} \mathbf{i} & \mathbf{j} & \mathbf{k} \\ A_x & A_y & A_z \\ B_x & B_y & B_z \end{vmatrix} = \mathbf{i}(A_yB_z - A_zB_y) + \mathbf{j}(A_zB_x - A_xB_z) + \mathbf{k}(A_xB_y - A_yB_x).$$

Example: The linear velocity **V** of a particle in a rotating body is the vector product of the angular velocity ω and the radius vector **r** from any point on the axis to the point in question, or

$$\mathbf{V} = \omega \times \mathbf{r} = -\mathbf{r} \times \omega$$

E270-0

vector quantity. Any physical quantity whose specification involves both magnitude and direction and that obeys the parallelogram law of addition. E270-0

vehicle. That in or on which a person or thing is being or may be carried. *See also:* **navigation.** E172-10E6

vehicle-derived navigation data. Data obtained from measurements made at a vehicle. *See also:* **navigation.** 0-10E6

veiling brightness. A brightness superimposed on the retinal image that reduces its contrast. It is this veiling effect produced by bright sources or areas in the visual field that results in decreased visual performance and visibility. *See also:* **visual field.** Z7A1-0

veiling reflection. Regular reflections that are superimposed upon diffuse reflections from an object that partially or totally obscure the details to be seen by reducing the contrast. This sometimes is called **reflect-**

ed glare. *See also:* **visual field.** Z7A1-0

velocity correction (industrial control). A method of register control that takes the form of a gradual change in the relative velocity of the web. 42A25-34E1

velocity level in decibels of a sound (acoustics). Twenty times the logarithm to the base 10 of the ratio of the particle velocity of the sound to the reference particle velocity. The reference particle velocity shall be stated explicitly. *Note:* In many sound fields the particle velocity ratios are not proportional to the square root of corresponding power ratios and hence cannot be expressed in decibels in the strict sense; however, it is common practice to extend the use of the decibel to these cases. *See also:* **electroacoustics.** E157-1E1

velocity microphone. A microphone in which the electric output substantially corresponds to the instantaneous particle velocity in the impressed sound wave. *Note:* A velocity microphone is a gradient microphone of order one, and it is inherently bidirectional. *See:* **gradient microphone.** *See also:* **microphone.** E157/42A65-1E1

velocity-modulated amplifier (velocity-variation amplifier). An amplifier that employs velocity modulation to amplify radio frequencies. *See also:* **amplifier.** 42A65-0

velocity-modulated oscillator. An electron-tube structure in which the velocity of an electron stream is varied (velocity-modulated) in passing through a resonant cavity called a buncher. Energy is extracted from the bunched electron stream at a higher energy level in passing through a second cavity resonator called the catcher. Oscillations are sustained by coupling energy from the catcher cavity back to the buncher cavity. *See also:* **oscillatory circuit.** E145-0

velocity-modulated tube. An electron-beam tube in which the velocity of the electron stream is alternately increased and decreased with a period comparable with the total transit time.
See:
acceleration space;
buncher space;
catcher space;
conversion efficiency;
density modulation;
drift tunnel;
electron collector;
input resonator buncher;
output resonator catcher;
reflection factor;
reflector space;
resonator grid;
rhumbatron;
velocity modulation. 50I07-15E6

velocity modulation (velocity variation) (of an electron beam). The modification of the velocity of an electron stream by the alternate acceleration and deceleration of the electrons with a period comparable with the transit time in the space concerned. *See also:* **circuits and devices; velocity-modulated oscillator; velocity modulated tube.** 42A65-0

velocity response factor (radar moving-target indicator). The ratio of voltage gain at a specific target velocity (or Doppler frequency) to the root-mean-square voltage gain evaluated over the entire velocity spectrum; this ratio for the conventional single-delay canceller varies sinusoidally from zero at the blind speeds to 1.414 at the optimum speeds. *See also:* **navigation.** 0-10E6

velocity shock. A mechanical shock resulting from a nonoscillatory change in velocity of an entire system. 0-1E1

velocity sorting (electronic). Any process of selecting electrons according to their velocities. *See also:* **electron devices, miscellaneous.** E160-15E6

velocity variation. *See:* **velocity modulation (electron beam).**

velocity-variation amplifier. *See:* **velocity-modulated amplifier.**

Venn diagram. A diagram in which sets are represented by closed regions. X3A12-16E9

vent (1) (fuse). The means provided for the escape of the gases developed during circuit interruption. *Note:* In distribution oil cutouts, the vent may be an opening in the housing or an accessory attachable to a vent opening in the housing with suitable means to prevent loss of oil. *See also:* **arrester; lightning arrester (surge diverter).** 37A100/42A20/62A1-31E7/31E11
(2) (rotating machinery). An opening that will permit the flow of air. *See also:* **cradle base (rotating machinery).** 0-31E8

vented fuse (or fuse unit). A fuse with provision for the escape of arc gases, liquids, or solid particles to the surrounding atmosphere during circuit interruption. 37A100-31E11

vent finger. *See:* **duct spacer.**

ventilated. Provided with a means to permit circulation of the air sufficiently to remove an excess of heat, fumes, or vapors. 42A95-0

ventilated enclosure. An enclosure provided with means to permit circulation of sufficient air to remove an excess of heat, fumes, or vapors. *Note:* For outdoor applications ventilating openings or louvers are usually filtered, screened, or restricted to limit the entrance of dust, dirt, or other foreign objects. 37A100-31E11

ventilating and cooling loss (synchronous machine). Any power required to circulate the cooling medium through the machine and cooler (if used) by fans or pumps that are driven by external means (such as a separate motor) so that their power requirements are not included in the friction and windage loss. It does not include power required to force ventilating gas through any circuit external to the machine and cooler. *See also:* **synchronous machine.** 50A10-31E8

ventilating duct (cooling duct) (rotating machinery). A passage provided in the interior of a magnetic core in order to facilitate circulation of air or other cooling agent. *See also:* **cradle base (rotating machinery).** 50I10-31E8

ventilating passage (rotating machinery). A passage provided for the flow of cooling medium. *See also:* **cradle base (rotating machinery).** 0-31E8

ventilating slot (rotating machinery). A slot provided for the passage of cooling medium. *See also:* **cradle base (rotating machinery).** 0-31E8

verification (electronic computation). The process of checking the results of one data transcription against the results of another data transcription. Both transcriptions usually involve manual operations. *See also:* **check; electronic computation.** E270-0

verification relay. A monitoring relay restricted to functions pertaining to power-system conditions and not involving opening circuit breakers during fault condition. *Note:* Such a relay is sometimes referred to as a check or checking relay. 37A100-31E6/31E11

verify. (1) To check, usually automatically, one typing or recording of data against another in order to minimize human and machine errors in the punching of tape or cards. E162/EIA3B-34E12
(2) To check the results of keypunching. *See also:* **electronic digital computer.** X3A12-16E9

vernier control (industrial control). A method for improving resolution. The amount of vernier control is expressed as either the percent of the total operating range or of the actual operating value, whichever is appropriate to the circuit in use. *See also:* **control system, feedback.** AS1-34E10

vertex. *See:* **node.**

vertex plate (reflector). A plate placed near the vertex of a reflector to prevent undesired reflection back to the primary radiator. 50I62-3E1

vertical amplifier (oscilloscope). An amplifier for signals intended to produce vertical deflection. *See also:* **oscillograph.** 0-9E4

vertical-break switch. A switch in which the travel of the blade is in a plane perpendicular to the plane of the mounting base. 37A100-31E11

vertical conductor (pole line work). A wire or cable extending in an approximately vertical direction on the supporting pole or structure. *See also:* **conductor; tower.** 2A2/42A35-31E13

vertical, gravity. *See:* **mass-attraction vertical.**

vertical-hold control (television). A synchronization control that varies the free-running period of the vertical-deflection oscillator. *See also:* **television.** E204-0

vertically polarized wave. A linearly polarized wave whose electric field vector is vertical. *Note:* The term **vertical polarization** is commonly employed to characterize ground-wave propagation in the medium-frequency broadcast band; these waves, however, have a small component of electric field in the direction of propagation due to finite ground conductivity. *See also:* **radio wave propagation.** 0-3E2

vertical machine (rotating machinery). A machine whose axis of rotation is approximately vertical. *See also:* **cradle base (rotating machinery).** 0-31E8

vertical plane (searchlight). The plane that is perpendicular to the train axis and in which the elevation axis lies. *See also:* **searchlight.** Z7A1-0

vertical recording. A mechanical recording in which the groove modulation is in a direction perpendicular to the surface of the recording medium. *See also:* **electroacoustics.** E157/42A65-0

vertical riser cable. Cable designed for use in long vertical runs, as in tall buildings. 0-31E1

vertical switchboard. A switchboard composed only of vertical panels. *Note:* This type of switchboard has an open rear. 37A100-31E11

very-high-frequency omnidirectional radio range (VOR). A specific type of range operating at very-high-frequency and providing radial lines of position in any direction as determined by bearing selection within the receiving equipment; it emits a (varying) modulation whose phase relative to a reference modulation is different for each bearing of the receiving point from the station. *See also:* **navigation.** 0-10E6

very-low-frequency high-potential test. An alternating-voltage high-potential test performed at a frequency equal to or less than 1 hertz. *See also:* **asynchronous machine; synchronous machine.** 0-31E8

vestigial sideband (amplitude-modulated transmission). The transmitted portion of the sideband that has been largely suppressed by a transducer having a gradual cutoff in the neighborhood of the carrier frequency, the other sideband being transmitted without much suppression. *See also:* **amplitude modulation; facsimile transmission.** E168/42A65-0

vestigial-sideband modulation. A modulation process involving a prescribed partial suppression of one of the two sidebands. E170-0

vestigial-sideband transmission (facsimile). That method of signal transmission in which one normal sideband and the corresponding vestigial sideband are utilized. *See also:* **amplitude modulation; facsimile transmission.** E168/42A65-0

vestigial-sideband transmitter. A transmitter in which one sideband and a portion of the other are intentionally transmitted. *See also:* **radio transmitter.** E145-0

VF. *See:* **voice frequency.**

vibrating bell. A bell having a mechanism designed to strike repeatedly when and as long as actuated. *See also:* **protective signaling.** 42A65-0

vibrating circuit (telegraph circuit). An auxiliary local timing circuit associated with the main line receiving relay for the purpose of assisting the operation of the relay when the definition of the incoming signals is indistinct. *See also:* **telegraphy.** 42A65-0

vibrating-contact machine regulator (electric machine). A regulator that varies the excitation by changing the average time of engagement of vibrating contacts in the field circuit. 37A100-31E11

vibrating-reed relay. A relay in which the application of an alternating or a self-interrupted voltage to the driving coil produces an alternating or pulsating magnetic field that causes a reed to vibrate and operate contacts. *See also:* **relay.** 83A16-21E0

vibration. An oscillation wherein the quantity is a parameter that defines the motion of a mechanical system. *See also:* **electroacoustics; oscillation.** 0-1E1

vibration detection system (protective signaling). A system for the protection of vaults by the use of one or more detector buttons firmly fastened to the inner surface in order to pick up and convert vibration, caused by burglarious attack on the structure, to electric impulses in a protection circuit. *See also:* **protective signaling.** 42A65-0

vibration meter. An apparatus including a vibration pickup, calibrated amplifier, and output meter for the measurement of displacement, velocity, and acceleration of vibrations. *See also:* **electroacoustics; instrument.** 42A30-0

vibration relay. A relay that responds to the magnitude and frequency of a mechanical vibration. 37A100-31E6/31E11

vibration test (rotating machinery). A test taken on a machine to measure the vibration of any part of the machine under specified conditions. *See also:* **cradle base (rotating machinery).** 0-31E8

vibrato. A family of tonal effects in music that depend upon periodic variations in one or more charactericts of the sound wave. *Note:* When the particular characteristics are known, the term **vibrato** should be modified accordingly, for example, **frequency vibrato; amplitude vibrato; phase vibrato** and so forth. 0-1E1

video (television). A term pertaining to the bandwidth and spectrum position of the signal resulting from television scanning. *Note:* In present usage, video means a bandwidth of the order of several megahertz, and a spectrum position that goes with a direct-current carrier. *See also:* **signal wave.** E203/E188/42A65-0

video-frequency amplifier. A device capable of amplifying such signals as comprise periodic visual presentation. *See also:* **television.** E145-0

video integration (electronic navigation). A method of utilizing the redundancy of repetitive video signals to improve the output signal-to-noise ratio, by summing successive signals. *See also:* **navigation.** E172-10E6

video mapping (electronic navigation). The electronic superposition of geographic or other data on a radar display. *See also:* **navigation.** 0-10E6

video stretching (electronic navigation). The increasing of the duration of a video pulse. *See also:* **navigation.** 0-10E6

vidicon. A camera tube in which a charge-density pattern is formed by photoconduction and stored on that surface of the photoconductor that is scanned by an electron beam, usually of low-velocity electrons. *See also:* **beam tubes; television.** E160-2E2/15E6

viewing area (oscilloscope). The area of the phosphor screen of a cathode-ray tube that can be excited to emit light by the electron beam. *See:* **oscillograph; screen, viewing.** 0-9E4

viewing time (storage tubes). The time during which the storage tube is presenting a visible output corresponding to the stored information. *See also:* **storage tube.** E158-15E6

viewing time, maximum usable (storage tubes). The length of time during which the visible output of a storage tube can be viewed, without rewriting, before a specified decay occurs. *Note:* The qualifying adjectives **maximum usable** are frequently omitted in general usage when it is clear that maximum usable viewing time is implied. *See also:* **storage tube.** E158-15E6

virtual cathode (potential-minimum surface) (electron tubes). A region in the space charge where there is a potential minimum that, by reason of the space charge density, behaves as a source of electrons. *See also:* **circuit characteristics of electrodes; electronic tube.** 50I07-15E6

virtual duration (of the peak of a rectangular-wave current or voltage impulse) (lightning arresters). The time during which the amplitude of the wave is greater than 90 percent of its peak value. *See also:* **lightning arrester (surge diverter).** 99I1-31E7

virtual duration of wavefront (virtual front time) (impulse) (lightning arresters). The time, in microseconds, equal to (1) for voltage waves with front durations equal to or less than 30 microseconds, 1.67 times the time taken by the voltage to increase from 30 percent to 90 percent of its peak value; (2) for voltage waves with front durations greater than 30 microseconds, 1.05 times the time taken by the voltage to increase from zero to 95 percent of its peak value; (3) for current waves, 1.25 times the time taken by the current to increase from 10 percent to 90 percent of its peak value. *Note:* If oscillations are present on the front, the reference points at 10, 30, 90, and 95 percent should be taken on the mean curve drawn through the oscillations. *See also:* **lightning arrester (surge diverter).** 0-31E7

virtual front time. *See:* **virtual duration of wavefront.**

virtual height (radio wave propagation). The apparent height of an ionized layer determined from the time interval between the transmitted signal and the ionospheric echo at vertical incidence, assuming that the velocity of propagation is the velocity of light in a vacuum over the entire path. *See also:* **radiation; radio wave propagation.** E211/42A65-3E2

virtual instant of chopping (voltage testing). The instant preceding point C on the accompanying figures by 0.3 times the (estimated) virtual time of voltage collapse during chopping. *See also:* **test voltage and current.** 68A1-31E5

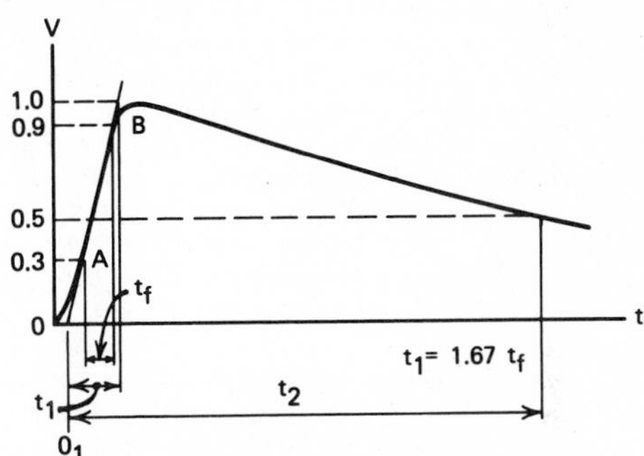

Full impulse voltage.

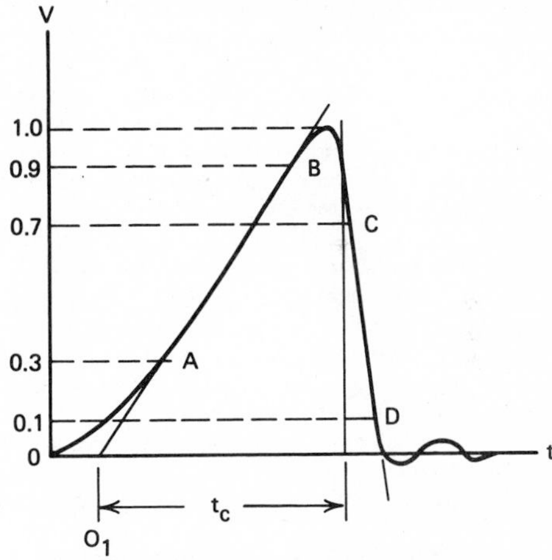

Impulse voltage chopped on the front.

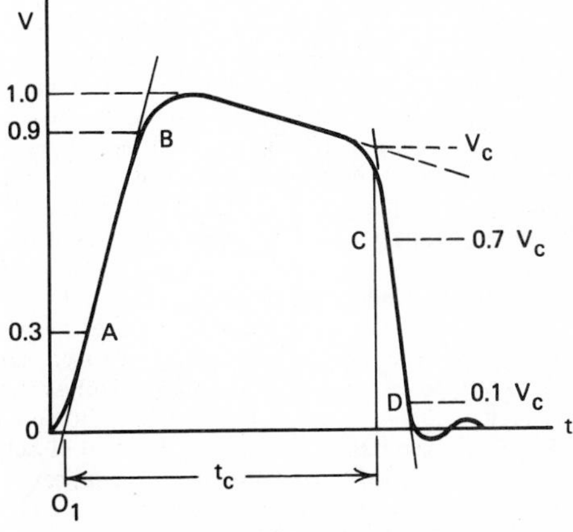

Impulse voltage chopped on the tail.

virtual origin (impulse current or voltage). *See:* **virtual zero time.**

virtual peak value. *See:* **peak value.**

virtual rate of rise of the front (impulse voltage). The quotient of the peak value and the virtual front time. *Note:* The term **peak value** is to be understood at including the term **virtual peak value** unless otherwise stated. *See also:* **test voltage and current.** 68A1-31E5

virtual steepness of voltage during chopping (lightning arresters). The quotient of the estimated voltage at the instant of chopping and the virtual time of voltage collapse. *See also:* **lightning arrester (surge diverter); test voltage and current.** 60I0/68A1-31E5/31E7

virtual steepness of wavefront of an impulse (lightning arresters). The slope of the line that determines the virtual-zero time. It is expressed in kilovolts per microsecond or kiloamperes per microsecond. *See also:* **lightning arrester (surge diverter).** 99I1-31E7

virtual time of voltage collapse during chopping. 1.67 times the time interval between points C and D on the figures attached to the definition of **virtual instant of chopping.** *See also:* **test voltage and current.** 68A1-31E5

virtual time to chopping (impulse voltage). The time interval between the virtual origin and the virtual instant of chopping. *See also:* **test voltage and current.** 68A1-31E5

virtual time to half-value (on the wavetail) (current or voltage impulse) (lightning arresters). The time interval between virtual zero and the instant when the voltage or current has decreased to half its peak value. *Note:* This time is expressed in microseconds. The term **peak value** is to be understood as including the term **virtual peak value** unless otherwise stated. *See also:* **lightning arrester (surge diverter).** 99I1-31E7

virtual total duration (rectangular-wave current or voltage impulse) (lightning arresters). The time during which the amplitude of the wave is greater than 10 percent of its peak value. *Note:* If small oscillations are present on the wavefront, a mean curve should be drawn in order to determine the time at which the 10 percent value is reached. *See also:* **lightning arrester (surge diverter).** 0-31E7

virtual zero point (impulse in a conductor). *See:* **virtual zero time.**

virtual zero time (impulse voltage or current in a conductor) (lightning arresters) (conventional origin) (virtual origin). The point on a graph of voltage-time or current-time determined by the intersection with the zero voltage or current axis, of a straight line drawn through two points on the front of the wave: (1) for full voltage waves and voltage waves chopped on the front, peak, or tail, the reference points shall be 30 percent and 90 percent of the peak value, and (2) for current waves the reference points shall be 10 percent and 90 percent of the peak value. *See also:* **lightning arrester (surge diverter).** 99I2-31E7

viscous friction (industrial control). The component of friction that is due to the viscosity of a fluid medium, usually idealized as a force proportional to velocity, and that opposes motion. *See also:* **control system, feedback.** AS1-23E0/34E10

visibility (1) (general). The quality or state of being perceivable by the eye. In many outdoor applications, visibility is defined in terms of the distance at which an object can be just perceived by the eye. In indoor applications it usually is defined in terms of the contrast or size of a standard test object, observed under standardized viewing conditions, having the same threshold as the given object. *See also:* **visual field.** Z7A1-0

(2) (meteorological). A term that denotes the greatest distance, expressed in miles, that selected objects (visibility markers) or lights of moderate intensity (25 candelas) can be seen and identified under specified conditions of observation. *See also:* **signal lighting.** Z7A1-0

visibility factor (radar) (1) (pulsed radar). The ratio of single-pulse signal energy to noise power per unit bandwidth that provides stated probabilities of detection and false alarm on display, measured in the intermediate-frequency portion of the receiver under conditions of optimum bandwidth and viewing environment. *See also:* **navigation.** 0-10E6

(2) (continuous-wave radar). The ratio of single-look energy to noise power per unit bandwidth using a filter matched to the time on target. *Note:* The equivalent term for radar using automatic detection is **detectability factor;** for operation in a clutter environment a **clutter visibility factor** is defined. 0-10E6

visual acuity. A measure of the ability to distinguish fine details. Quantitatively, it is the reciprocal of the angular size in minutes of the critical detail that is just large enough to be seen. *See also:* **visual field.** Z7A1-0

visual angle. The angle that an object or detail subtends at the point of observation. It usually is measured in minutes of arc. *See also:* **visual field.** Z7A1-0

visual approach slope indicator system (VASIS). The system of angle-of-approach lights, accepted as a standard by the International Civil Aviation Organization, comprising two bars of lights located at each side of the runway near the threshold and showing red or white or a combination of both (pink) to the approaching pilot depending upon his position with respect to the glide path. *See also:* **signal lighting.** Z7A1-0

visual-aural range (VAR) (electronic navigation). A special type of very-high-frequency radio range that provides (A) two reciprocal radial lines of position displayed to the pilot visually on a course-deviation indicator, and (B) two reciprocal radial lines of position presented to the pilot as interlocked and alternate *A* and *N* aural code signals; the aural lines of position are displaced 90 degrees from the visual and either may be used to resolve the ambiguity of the other. *See also:* **navigation; electronic navigation.** 42A65-10E6

visual field. The locus of objects or points in space that can be perceived when the head and eyes are kept fixed. The field may be monocular or binocular. *Note:* For an extensive list of cross references, see *Appendix A.* Z7A1-0

visual inspection. Qualitative observation of physical characteristics utilizing the unaided eye or with stipulated levels of magnification. 99A1-0

visual perception. The interpretation of impressions transmitted from the retina to the brain in terms of information about a physical world displayed before the eye. *Note:* Visual perception involves any one or more of the following: recognition of the presence of something (object, aperture, or medium); identifying it; locating it in space; noting its relation to other things; identifying its movement, color, brightness, or form. *See also:* **visual field.** Z7A1-0

visual performance. The quantitative assessment of the performance of a task taking into consideration speed and accuracy. *See also:* **visual field.** Z7A1-0

visual photometer. A visual photometer is one in which the equality of brightness of two surfaces is established visually. *Note:* The two surfaces usually are viewed simultaneously side by side. This method is used in portable visual luminance (photometric brightness) meters. This is satisfactory when the color difference between the test source and comparison source is small. However, when there is a color difference, a flicker photometer provides more precise measurements. In this type of photometer the two surfaces are viewed alternately at such a rate that the color sensations either nearly or completely blend and the flicker due to brightness difference is balanced by adjusting the comparison source. Z7A1-0

visual radio range. Any radio range (such as very-high-frequency omnidirectional radio range) whose primary function is to provide lines of position to be flown by visual reference to a course-deviation indicator. *See also:* **radio navigation.** 0-10E6

visual range (of a light or object). The maximum distance at which that particular light (or object) can be seen and identified. *See also:* **signal lighting.** Z7A1-0

visual scanner (character recognition). *See:* **optical scanner.** *See also:* **electronic digital computer.**

visual signal device (protective signaling). A general term for pilot lights, annunciators, and other devices providing a visual indication of the condition supervised. *See also:* **protective signaling.** 42A65-0

visual surround. The visual surround includes all portions of the visual field except the visual task. *See also:* **visual field.** Z7A1-0

visual task. Those details and objects that must be seen for the performance of a given activity including the immediate background of the details or objects. *See also:* **visual field.** Z7A1-0

visual transmitter. All parts of a television transmitter that handle picture signals, whether exclusively or not. *See also:* **television.** E145-0

visual transmitter power. The peak power output during transmission of a standard television signal. *See also:* **television.** 42A65-0

vital circuit. Any circuit the function of which affects the safety of train operation. *See also:* **railway signal and interlocking.** 42A42-0

V-network (electromagnetic compatibility). An artificial mains network of specified disymmetric impedance used for two-wire mains operation and comprising resistors in V formation connected between each conductor and earth. *See also:* **electromagnetic compatibility.** 0-27E1

vodas. A system for preventing the over-all voice-frequency singing of a two-way telephone circuit by disabling one direction of transmission at all times. The name is derived from the initial letters of the expression **voice-operated device anti-sing.** *See also:* **voice-frequency telephony.** 42A65-0

vogad. A voice-operated device used to give a substantially constant volume output for a wide range of inputs. The name is derived from the initial letters of the expression **voice-operated gain-adjusting device.** *See also:* **voice-frequency telephony.** 42A65-0

voice channel (mobile communication). A transmission facility defined by the constraints of the human voice. For mobile-communication systems, a voice channel may be considered to have a range of approximately 250 to 3000 hertz; since the Rules and Regulations of the Federal Communications Commission do not authorize the use of modulating frequencies higher than 3000 hertz for radiotelephony or tone signaling on radio frequencies below 500 megahertz. *See also:* **channel spacing.** 0-6E1

voice frequency (VF). A frequency lying within that part of the audio range that is employed for the transmission of speech. *Note:* Voice frequencies used for commercial transmission of speech usually lie within the range 200 to 3500 hertz. *See also:* **signal wave.** 42A65-19E4/31E3

voice-frequency carrier telegraphy. That form of carrier telegraphy in which the carrier currents have frequencies such that the modulated currents may be transmitted over a voice-frequency telephone channel. *See also:* **telegraphy.** 42A65-19E4/31E3

voice-frequency telephony. That form of telephony in which the frequencies of the components of the transmitted electric waves are substantially the same as the frequencies of corresponding components of the actuating acoustical waves.
See:
echo suppressor;
subscriber set;
tonlar;
vodas;
vogad;
volcas. *See also:* **communication.** 42A65-0

volatile (electronic data processing). Pertaining to a storage device in which data cannot be retained without continuous power dissipation, for example, an acoustic delay line. *Note:* Storage devices or systems employing nonvolatile media may or may not retain data in the event of planned or accidental power removal. *See also:* **electronic computation; electronic digital computer.** E162/E270-0

volatile flammable liquid. A flammable liquid having a flash point below 100 degrees Fahrenheit or whose temperature is above its flash point. 1A0-0

volcas. A voice-operated device that switches loss out of the transmitting branch and inserts loss in the receiving branch under control of the subscriber's speech. The name is derived from the initial letters of the expression **voice-operated loss control and suppressor.** *See also:* **voice-frequency telephony.** 42A65-0

volt. The unit of voltage or potential difference in SI units. The volt is the voltage between two points of a conducting wire carrying a constant current of one ampere, when the power dissipated between these points is one watt. E270-0

volta effect. *See:* **contact potential.**

voltage (electromotive force*) (1) (general) (along a specified path in an electric field). The dot product line integral of the electric field strength along this path. *Notes:* (A) Voltage is a scalar and therefore has no spatial direction. (B) As here defined, voltage is synonymous with potential difference only in an electrostatic field. (C) In cases in which the choice of the specified path may make a significant difference, the path is taken in an equiphase surface unless otherwise noted. (D) If is often convenient to use an adjective with voltage, for example, phase voltage, electrode voltage, line voltage, etcetera. The basic definition of

voltage applies and the meaning of adjectives should be understood or defined in each particular case. *See also:* **reference voltage.**
*Deprecated E270-0
(2) (effectively grounded circuit). The highest effective voltage between any conductor and ground unless otherwise indicated. *See also:* **power systems, low-frequency and surge testing; ground; grounding.** 2A2-0
(3) (circuit not effectively grounded). The highest effective voltage between any two conductors unless otherwise indicated. *Notes:* (A) If one circuit is directly connected to another circuit of higher voltage (as in the case of an autotransformer), both are considered as of the higher voltage, unless the circuit of lower voltage is effectively grounded, in which case its voltage is not determined by the circuit of higher voltage. Direct connection implies electric connection as distinguished from connection merely through electromagnetic or electrostatic induction. (B) Where safety considerations are involved, the voltage to ground that may occur in an ungrounded circuit is usually the highest voltage normally existing between the conductors of the circuit, but in special circumstances higher voltages may occur. *See also:* **ground; grounded; power system, low-frequency and surge testing.** 2A2-0
(4) (lightning arresters). The voltage between a part of an electric installation connected to a grounding system and points on the ground at an adequate distance (theoretically at an infinite distance) from any earth electrodes. *See also:* **lightning arrester (surge diverter).** 50I25-31E7

voltage amplification (1) (general). An increase in signal voltage magnitude in transmission from one point to another or the process thereof. *See also:* **amplifier.** E270-0
(2) (transducer). The scalar ratio of the signal output voltage to the signal input voltage. *Warning:* By incorrect extension of the term decibel, this ratio is sometimes expressed in decibels by multiplying its common logarithm by 20. It may be correctly expressed in decilogs. *Note:* If the input and/or output power consist of more than one component, such as multifrequency signal or noise, then the particular components used and their weighting must be specified. *See also:* **transducer.** E151/E270/42A65-0
(3) (magnetic amplifier). The ratio of differential output voltage to differential control voltage. *See also:* **rating and testing magnetic amplifiers; transmission characteristics.** E107-0

voltage attenuation. *See:* **attenuation, voltage.**

voltage buildup (rotating machinery). The inherent establishment of the excitation current and induced voltage of a generator. *See also:* **direct-current commutating machine.** 0-31E8

voltage circuit (instrument). That combination of conductors and windings of the instrument to which is applied the voltage of the circuit in which a given electrical quantity is to be measured, or a definite fraction of that voltage, or a voltage or current dependent upon it. *See also:* **instrument; moving element (instrument); watthour meter.** 39A1/42A30-0

voltage coefficient of capacitance (nonlinear capacitor). The derivative with respect to voltage of a capacitance characteristic, such as a differential capacitance characteristic or a reversible capacitance characteristic, at a point, divided by the capacitance at that point. *See also:* **nonlinear capacitor.** E226-15E7

voltage corrector (power supplies). An active source of regulated power placed in series with an unregulated supply to sense changes in the output voltage (or current); also to correct for the changes by automatically varying its own output in the opposite direction, thereby maintaining the total output voltage (or current) constant. See the accompanying figure.

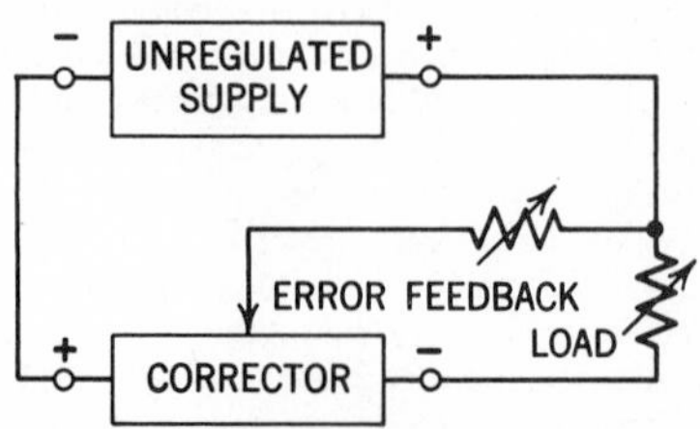

Circuit used to sense output voltage changes.

See also: **power supply.** KPSH-10E1

voltage divider. A network consisting of impedance elements connected in series, to which a voltage is applied, and from which one or more voltages can be obtained across any portion of the network. *Notes:* (1) Dividers may have parasitic impedances affecting the response. These impedances are, in general, the series inductance and the capacitance to ground and to neighboring structures at ground or at other potentials. (2) An adjustable voltage divider of the resistance type is frequently referred to as a potentiometer. *See also:* **circuits and devices; power system, low-frequency and surge testing.** 42A65/68A1-21E0/31E5

voltage doubler. A voltage multiplier that separately rectifies each half cycle of the applied alternating voltage and adds the two rectified voltages to produce a direct voltage whose amplitude is approximately twice the peak amplitude of the applied alternating voltage. *See also:* **rectifier.** 42A65-0

voltage drop (1) (general). The difference of voltages at the two terminals of a passive impedance. 0-31E8
(2) (supply system). The difference between the voltages at the transmitting and receiving ends of a feeder, main, or service. *Note:* With alternating current the voltages are not necessarily in phase and hence the voltage drop is not necessarily equal to the algebraic sum of the voltage drops along the several conductors. *See also:* **alternating-current distribution.** 42A35-31E13

voltage efficiency (specified electrochemical process). The ratio of the equilibrium reaction potential to the bath voltage. 42A60-0

voltage endurance (rotating machinery). A characteristic of an insulation system, obtained by plotting voltage against time to failure, for a number of samples tested to destruction at each of several sustained voltages. Constant conditions of frequency, waveform, temperature, mechanical restraint, and ambient atmosphere are required. Ordinate scales of arithmetical or logarithmic voltage, and abscissa scales of multicycle logarithmic time, normally give approximately linear characteristics. *See also:* **asynchronous machine; direct-current commutating machine; synchronous machine.** 0-31E8

voltage endurance test (rotating machinery). A test designed to determine the effect of voltage on the useful life of electric equipment. When this test voltage exceeds the normal design voltage for the equipment, the test is voltage accelerated. When the test voltage is alternating and the frequency of alternation exceeds the normal voltage frequency for the equipment, the test is frequency accelerated. *See:* **asynchronous machine; direct-current commutating machine; synchronous machine.** 0-31E8

voltage factor (electron tubes). The magnitude of the ratio of the change in one electrode voltage to the change in another electrode voltage, under the conditions that a specified current remains unchanged and that all other electrode voltages are maintained constant. *See also:* **ON period.** 50I07-15E6

voltage generator (network analysis and signal-transmission system). A two-terminal circuit element with a terminal voltage substantially independent of the current through the element. *Note:* An ideal voltage generator has zero internal impedance. *See also:* **circuit characteristics of electrodes; network analysis; signal.** E160-13E6/15E6

voltage impulse. A voltage pulse of sufficiently short duration to exhibit a frequency spectrum of substantially uniform amplitude in the frequency range of interest. As used in electromagnetic compatibility standard measurements, the voltage impulse has a uniform frequency spectrum over the frequency range 25 to 1000 megahertz. E263-27E1

voltage influence (electric instrument). In instruments, other than indicating voltmeters, wattmeters, and varmeters, having voltage circuits, the percentage change (of full-scale value) in the indication of an instrument that is caused solely by a voltage departure from a specified reference voltage. *See also:* **accuracy rating (instrument).** 39A1/39A2-0

voltage jump (glow-discharge tube). An abrupt change or discontinuity in tube voltage drop during operation. *Note:* This may occur either during life under constant operating conditions or as the current or temperature is varied over the operating range. *See also:* **gas-filled radiation-counter tube; gas tube.** E160-15E6

voltage level (transmission system). At any point, the ratio of the voltage existing at that point to an arbitrary value of voltage used as a reference. Specifically, in systems such as television systems, where wave shapes are not sinusoidal or symmetrical about a zero axis and where the arithmetical sum of the maximum positive and negative excursions of the wave is important in system performance, the voltage level is the ratio of the peak-to-peak voltage existing at any point in the transmission system to an arbitrary peak-to-peak voltage used as a reference. This ratio is usually expressed in decibels referred to one volt peak-to-peak (dBv). *See also:* **level.** 42A65-0

voltage limit (industrial control). A control function that prevents a voltage from exceeding prescribed limits. Voltage limit values are usually expressed as percent of rated voltage. If the voltage-limit circuit permits the limit value to increase somewhat instead of being a single value, it is desirable to provide either a curve of the limit value of voltage as a function of some variable such as current or to give limit values at two or more conditions of operation. *See also:* **control system, feedback.** AS1-34E10

voltage loss (electric instrument)(current circuits). In a current-measuring instrument, the value of the voltage between the terminals when the applied current corresponds to nominal end-scale deflection. In other instruments the voltage loss is the value of the voltage between the terminals at rated current. *Note:* By convention, when an external shunt is used, the voltage loss is taken at the potential terminals of the shunt. The overall voltage drop resulting may be somewhat higher owing to additional drop in shunt lugs and connections. *See also:* **accuracy rating (instrument).** 39A1/42A30-0

voltage multiplier. A rectifying circuit that produces a direct voltage whose amplitude is approximately equal to an integral multiple of the peak amplitude of the applied alternating voltage. *See also:* **rectifier.** 42A65-0

voltage, nominal (system or circuit). *See:* **nominal system voltage.**

voltage overshoot (arc-welding apparatus). The ratio of transient peak voltage substantially instantaneously following the removal of the short circuit to the normal steady-state voltage value. *See also:* **voltage recovery time.** 87A1-0

voltage overshoot, effective (arc-welding apparatus). The area under the transient voltage curve during the time that the transient voltage exceeds the steady-state value. *See also:* **voltage recovery time.** 87A1-0

voltage phase-angle method (economic dispatch) (electric power systems). Considers the actual measured phase-angle difference between the station bus and a reference bus in the determination of incremental transmission losses. *See also:* **power systems, low-frequency and surge testing.** E94-0

voltage-phase-balance protection. A form of protection that disconnects or prevents the connection of the protected equipment when the voltage unbalance of the phases of a normally balanced polyphase system exceeds a predetermined amount. 37A100-31E6/31E11

voltage range (electrically propelled vehicle). Divided into five voltage ranges, as follows. **first voltage range:** 30 volts or less. **second voltage range:** over 30 volts to and including 175 volts. **third voltage range:** over 175 volts to and including 250 volts. **fourth voltage range:** over 250 volts to and including 660 volts. **fifth voltage range:** over 660 volts. *See also:* **railway signal and interlocking.** 42A42-0

voltage range multiplier (instrument multiplier). A particular type of series resistor or impedor that is used to extend the voltage range beyond some particular value for which the measurement device is already complete. It is a separate component installed external to the measurement device. *See also:* **auxiliary device (instrument).** 39A2/42A30-0

voltage rating (1) (arrester). The designated maximum permissible operating voltage (60 hertz root-mean-square) between the line and ground terminals at which it is designated to perform its duty cycle. It is the voltage rating specified on the nameplate. *See:* **lightning arrester (surge diverter); current rating, 60-hertz (arrester).** 42A20-0;62A1-31E7

(2) (household electric ranges). The voltage limits within which the range is intended to be used. *See:* **appliance outlet.** 7IA1-0

voltage ratio (1) (transformer). The ratio of the root-mean-square primary terminal voltage to the root-mean-square secondary terminal voltage under specified conditions of load. *See:* **turn ratio (transformer).** 42A15-31E12

(2) (capacitance potential device, in combination with its coupling capacitor or bushing). The overall ratio between the root-mean-square primary line-to-ground voltage and the root-mean-square secondary voltage. *Note:* It is not the turn ratio of the transformer used in the network. *See also:* **outdoor coupling capacitor.** E31-0

voltage recovery time (arc-welding apparatus). With a welding power supply delivering current through a short-circuiting resistor whose resistance is equivalent to the normal load at that setting on the power supply, and measurement being made when the short circuit is suddenly removed, the time measured in seconds between the instant the short circuit is removed and the instant when voltage has reached 95 percent of its steady-state value. *See:* **voltage overshoot; voltage overshoot, effective.** *See also:* **electric arc welding apparatus.** 87A1-0

voltage reference (power supplies). A separate, highly regulated voltage source used as a standard to which the output of the power supply is continuously referred. *See also:* **power supply.** 0-10E1

voltage-reference tube. A gas tube in which the tube voltage drop is approximately constant over the operating range of current and relatively stable with time at fixed values of current and temperature. *See also:* **tube definitions.** E160-15E6

voltage reflection coefficient. The ratio of the complex number (phasor) representing the phase and magnitude of the electric field of the backward-traveling wave to that representing the forward-traveling wave at a cross section of a waveguide. The term is also used to denote the magnitude of this complex ratio. *See also:* **waveguide.** 0-3E1

voltage-regulating relay. A voltage-sensitive device that is used on an automatically operated voltage regulator to control the voltage of the regulated circuit. *See also:* **voltage regulator.** 42A15/57A15-31E12

voltage-regulating transformer (step-voltage regulator). A voltage regulator in which the voltage and phase angle of the regulated circuit are controlled in steps by means of taps and without interrupting the load. *See also:* **voltage regulator.** 42A15-31E12

voltage-regulating transformer, two-core. A voltage-regulating transformer consisting of two separate core and coil units in a single tank. *See also:* **voltage regulator.** 42A15-31E12

voltage-regulating transformer, two-core, excitation-regulating winding. In some designs, the main unit will have one winding operating as an autotransformer that performs both functions listed under regulating winding and excitation winding. Such a winding is called the excitation-regulating winding. *See also:* **voltage regulator.** 42A15-31E12

voltage-regulating transformer, two-core, excitation winding. The winding of the main unit that draws power from the system to operate the two-core transformer. *See also:* **voltage regulator.** 42A15-31E12

voltage-regulating transformer, two-core, excited winding. The winding of the series unit that is excited from the regulating winding of the main unit. *See also:* **voltage regulator.** 42A15-31E12

voltage-regulating transformer, two-core, regulating winding. The winding of the main unit in which taps are changed to control the voltage or phase angle of the regulated circuit through the series unit. *See also:* **voltage regulator.** 42A15-31E12

voltage-regulating transformer, two-core, series unit. The core and coil unit that has one winding connected in series in the line circuit. *See also:* **voltage regulator.** 42A15-31E12

voltage-regulating transformer, two-core, series winding. The winding of the series unit that is connected in series in the line circuit. *Note:* If the main unit of a two-core transformer is an autotransformer, both units will have a series winding. In such cases, one is referred to as the series winding of the autotransformer and the other, the series winding of the series unit. *See also:* **voltage regulator.** 42A15-31E12

voltage regulation (1) (constant-voltage transformer). The change in output (secondary) voltage that occurs when the load is reduced from rated value to zero, with the applied (primary) voltage maintained constant. *Note:* In case of multiwinding transformers, the loads on all windings, at specified power factors, are to be reduced from rated kilovolt-amperes to zero simultaneously. The regulation may be expressed in per unit, or percent, on the base of rated output (secondary voltage at full load). *See also:* **turn ratio (transformer).** 42A15-31E12

(2) (rectifier). The change in output voltage that occurs when the load is reduced from rated value to zero, or light transition load, with the values of all other quantities remaining unchanged. *Note:* The regulation may be expressed in volts or percentage of rated output voltage at full load. *See also:* **rectification; semiconductor rectifier stack.** 42A15-34E24

(3) (outdoor coupling capacitor). The variation in voltage ratio and phase angle of the secondary voltage of the capacitance potential device as a function of primary line-to-ground voltage variation over a specified range, when energizing a constant, linear impedance burden. *See also:* **outdoor coupling capacitor.** E31-0

(4) (direct-current generator). The final change in voltage with constant field-rheostat setting when the specified load is reduced gradually to zero, expressed as a percent of rated-load voltage, the speed being kept constant. *Note:* In practice it is often desirable to specify the over-all regulation of the generator and its driving machine thus taking into account the speed regulation of the driving machine. *See also:* **direct-current commutating machine.** 42A10-0

(5) (induction frequency converter). The rise in secondary voltage when the rated load at rated power factor is reduced to zero, expressed in percent of rated secondary voltage, the primary voltage, primary frequency, and the speed being held constant. *See also:* **asynchronous machine.** 42A10-0

(6) (synchronous generator). The rise in voltage with constant field current, when, with the synchronous generator operated at rated voltage and rated speed, the specified load at the specified power factor is reduced to zero, expressed as a percent of rated voltage. *See also:* **synchronous machine.** 42A10-31E8

(7) (line regulator circuits). *See:* **zener diode; pulse-width modulation.**

voltage regulation curve (voltage regulation characteristic) (synchronous generator). The relationship between the armature winding voltage and the load on the generator under specified conditions and constant field current. *See also:* **synchronous machine.** 0-31E8

voltage regulator (transformer type). An induction device having one or more windings in shunt with, and excited from, the primary circuit, and having one or more windings in series between the primary circuit and the regulated circuit, all suitably adapted and arranged for the control of the voltage, or the phase-angle, or of both, of the regulated circuit. *Notes:* (1) A voltage regulator is basically an autotransformer. (2) For an extensive list of cross references, see *Appendix A. See:* **autotransformer.** 42A15/57A15-31E12/10E1

voltage regulator, continuously acting type (rotating machinery). A regulator that initiates a corrective action for a sustained infinitesimal change in the controlled variable. *See also:* **synchronous machine.** 0-31E8

voltage regulator, direct-acting type (rotating machinery). A rheostatic-type regulator that directly controls the excitation of an exciter by varying the input to the exciter field circuits. *See also:* **direct-current commutating machine; synchronous machine.** 0-31E8

voltage regulator, dynamic type (rotating machinery). A continuously acting regulator that does not require mechanical acceleration of parts to perform the regulating function. *Note:* Dynamic-type voltage regulators utilize magnetic amplifiers, rotating amplifiers, electron tubes, semiconductor elements, and/or other static components. *See also:* **direct-current commutating machine; synchronous machine.** 0-31E8

voltage regulator, indirect-acting type (rotating machinery). A rheostatic-type regulator that controls the excitation of the exciter by acting on an intermediate device not considered part of the voltage regulator or exciter. *See also:* **direct-current commutating machine; synchronous machine.** 0-31E8

voltage regulator, noncontinuously acting type (rotating machinery). A regulator that requires a sustained finite change in the controlled variable to initiate corrective action. *See also:* **synchronous machine.** 0-31E8

voltage regulator, synchronous-machine (rotating machinery). A synchronous-machine regulator that functions to maintain the voltage of a synchronous machine at a predetermined value, or to vary it according to a predetermined plan. *See also:* **synchronous machine.** 0-31E8

voltage-regulator tube. A glow-discharge cold-cathode tube in which the voltage drop is approximately constant over the operating range of current, and that is designed to provide a regulated direct-voltage output. *See also:* **tube definitions.** E160/42A70-15E6

voltage relay (industrial control). A relay that functions at a predetermined value of voltage. *Note:* It may be an overvoltage relay, an undervoltage relay, or a combination of both. *See also:* **relay.** 42A20-34E10

voltage response (close-talking pressure-type microphone). The ratio of the open-circuit output voltage to the applied sound pressure, measured by a laboratory standard microphone placed at a stated distance from the plane of the opening of the artificial voice. *Note:* The voltage response is usually measured as a function of frequency. *See also:* **close-talking pressure-type microphone.** E258-0

voltage response, exciter. The rate of increase or decrease of the exciter voltage when a change in this voltage is demanded. It is the rate determined from the exciter voltage response curve that if maintained constant would develop the same exciter voltage-time area as is obtained from the curve for a specified period. The starting point for determining the rate of voltage change shall be the initial value of the exciter voltage-time response curve. *See also:* **asynchronous machine; direct-current commutating machine; synchronous machine.** 0-31E8

voltage response ratio, excitation-system (rotating machinery). The numerical value that is obtained when the excitation-system voltage response in volts per second, measured over the first 1/2-second interval unless otherwise specified, is divided by the rated-load field voltage of the synchronous machine. *Note:* This response, if maintained constant, would develop, in 1/2 second, the same excitation voltage-time area as attained by the actual response. *See also:* **synchronous machine.** 0-31E8

voltage response, synchronous-machine excitation-system. The rate of increase or decrease of the excitation-system output voltage, determined from the synchronous machine excitation-system voltage-time response curve, that if maintained constant would develop the same excitation-system voltage-time areas as are obtained from the curve for a specified period. The starting point for determining the rate of voltage change shall be the initial value of the synchronous-machine excitation-system voltage-time response curve. *See also:* **synchronous machine.** 0-31E8

voltage restraint (relay). A method of restraining the operation by means of a voltage input that opposes the typical response of the relay to other inputs. 37A100-31E6/31E11

voltage sensing relay. (1) A term correctly used to designate a special-purpose voltage-rated relay that is adjusted by means of a voltmeter across its terminals in order to secure pickup at a specified critical voltage without regard to coil or heater resistance and resulting energizing current at that voltage. (2) A term erroneously used to describe a general-purpose relay for which operational requirements are expressed in voltage. 0-21E0

voltage sensitivity (nonlinear capacitor). *See:* **voltage coefficient of capacitance.** *See also:* **nonlinear capacitor.**

voltage sets (polyphase circuit). The voltages at the terminals of entry to a polyphase circuit into a delimited region are usually considered to consist of two sets of voltages: the line-to-line voltages, and the line-to-neutral voltages. If the phase conductors are identified in a properly chosen sequence, the voltages between the terminals of entry of successive pairs of phase conductors form the set of line-to-line voltages, equal in number to the number of phase conductors. The voltage from the successive terminals of entry of the phase conductors to the terminal of entry of the neutral conductor, if one exists, or to the true neutral point, form the set of line-to-neutral voltages, also equal in number to the number of phase conductors. In case of doubt, the set intended must be identified. In the absence of other information, stated or implied, the line-to-neutral-conductor set is understood. *Notes:* (1) Under abnormal conditions the voltage of the neutral conductor and of the true neutral point may not be the same. Therefore it may become necessary to designate which is intended when the line-to-neutral voltages are being specified. (2) The set of line-to-line voltages may be determined by taking the differences in pairs of the successive line-to-neutral voltages. The line-to-neutral voltages can be determined from the line-to-line voltages by an inverse process only when the voltage between the neutral conductor and the true neutral point is completely specified, or equivalent additional information is available. If instantaneous voltages are used, algebraic differences are taken, but if root-mean-square voltages are used, information regarding relative phase angles must be available, so that the voltages may be expressed in phasor form and the phasor differences taken. (3) This definition may be applied to a two-phase, four-wire or five-wire circuit. A two-phase, three-wire circuit should be treated as a special case. *See also:* **network analysis.** E270-0

voltage spread. The difference between maximum and minimum voltages. *See also:* **power systems, low-frequency and surge testing.** 84A1-0

voltage-stabilizing tube. *See:* **voltage regulator tube.**

voltage standing-wave ratio (VSWR) (mode in a waveguide). The ratio of the magnitude of the transverse electric field in a plane of maximum strength to the magnitude at the equivalent point in an adjacent plane of minimum field strength. *See also:* **waveguide.** 0-3E1

voltage surge suppressor (semiconductor rectifier). A device used in the semiconductor rectifier to attenuate surge voltages of internal or external origin. Capacitors, resistors, nonlinear resistors, or combinations of these may be employed. Nonlinear resistors include electronic and semiconductor devices. *See also:* **semiconductor rectifier stack.** 0-34E24

voltage test. *See:* **controlled overvoltage test.**

voltage-time response, synchronous-machine excitation-system. The output voltage of the excitation system, expressed as a function of time, following the application of prescribed inputs under specified conditions. *See also:* **synchronous machine.** 0-31E8

voltage-time response, synchronous-machine voltage-regulator. The voltage output of the synchronous-machine voltage regulator expressed as a function of time following the application of prescribed inputs under specified conditions. *See also:* **synchronous machine.** 0-31E8

voltage to ground. *See:* **voltage.**

voltage-tunable magnetron (microwave tubes). A magnetron in which the resonant circuit is heavily loaded ($Q_L = 1$ to 10) and in which the supply of electrons to the interaction space is restricted whereby the frequency of oscillation becomes proportional to the plate voltage. *See also:* **magnetrons.** 0-15E6

voltage-type telemeter. A telemeter that employs the magnitude of a single voltage as the translating means. *See also:* **telemetering.** 37A100/42A30-31E11

voltage-withstand test (1) (insulation). The application of a voltage higher than the rated voltage for a specified time for the purpose of determining the adequacy against breakdown of insulation materials and spacing under normal conditions. *See also:* **dielectric tests (voltage-withstand tests).** 12A0-0

(2) (rotating machinery). *See:* **overvoltage test.** *See also:* **asynchronous machine; direct-current commutating machine; synchronous machine.**

voltaisation (galvanization) (electrotherapy). *See:* **galvanism.**

voltameter (coulometer). *See:* **coulometer; instrument.**

volt-ammeter. An instrument having circuits so designed that the magnitude either of voltage or of current can be measured on a scale calibrated in terms of each of these quantities. *See also:* **instrument.** 42A30-0

volt-ampere. The unit of apparent power in the International System of Units (SI). The volt-ampere is the apparent power at the points of entry of a single-phase, two-wire system when the product of the root-mean-square value in amperes of the current by the root-mean-square value in volts of the voltage is equal to one. E270-0

volt-ampere-hour meter. An electricity meter that measures and registers the integral, with respect to time, of the apparent power in the circuit in which it is connected. The unit in which the integral is measured is usually the kilovolt-ampere-hour. *See also:* **electricity meter (meter).** 42A30-0

volt-ampere loss (electric instrument). *See:* **apparent power loss (electric instrument).**

volt-ampere meter. An instrument for measuring the apparent power in an alternating-current circuit. It is provided with a scale graduated in volt-amperes or in kilovolt-amperes. *See also:* **instrument.** 42A30-0

volt efficiency (storage battery) (storage cell). The ratio of the average voltage during the discharge to the average voltage during the recharge. *See also:* **electrochemistry; charge.** 42A60-0

voltmeter. An instrument for measuring the magnitude of electric potential difference. It is provided with a scale, usually graduated in either volts, millivolts, or kilovolts. If the scale is graduated in millivolts or kilovolts the instrument is usually designated as a millivoltmeter or a kilovoltmeter. *See also:* **instrument.** 42A30-0

voltmeter-ammeter. The combination in a single case, but with separate circuits, of a voltmeter and an ammeter. *See also:* **instrument.** 42A30-0

volt-time curve (lightning arresters) (1) (impulses with fronts rising linearly) (lightning arresters). The curve relating the disruptive-discharge voltage of a test object to the virtual time to chopping. The curve is obtained by applying voltages that increase at different rates in approximately linear manner. *See also:* **lightning arrester (surge diverter).** 60I0-31E7

(2) (standard impulses). A curve relating the peak value of the impulse causing disruptive discharge of a test object to the virtual time to chopping. The curve is obtained by applying standard impulse voltages of different peak values. *See also:* **lightning arrester (surge diverter).** 60I0-31E7

volume (electric circuit). The magnitude of a complex audio-frequency wave as measured on a standard volume indicator. *Notes:* (1) Volume is expressed in volume units (vu). (2) The term volume is used loosely to signify either the intensity of a sound or the magnitude of an audiofrequency wave. *See also:* **transmission characteristics.** E151/42A65-0

volume control. *See:* **gain control.**

volume density of magnetic pole strength. At any point of the medium in a magnetic field, the negative of the divergence of the magnetic polarization vector there. E270-0

volume equivalent (complete telephone connection, including the terminating telephone sets). A measure of the loudness of speech reproduced over it. The volume equivalent of a complete telephone connection is expressed numerically in terms of the trunk loss of a working reference system when the latter is adjusted to give equal loudness. *Note:* For engineering purposes, the volume equivalent is divided into volume losses assignable to (1) the station set, subscriber line, and battery-supply circuit that are on the transmitting end; (2) the station set, subscriber line, and battery supply that are on the receiving end; (3) the trunk; and (4) interaction effects arising at the trunk terminals.
See:
articulation;
articulation equivalent;
consonant articulation;
discrete-sentence intelligibility;
discrete-word intelligibility;
percentage immediate appreciation;
repetition rate;
sound articulation;
syllable articulation;
vowel articulation.
See also: **transmission characteristics.** 42A65-0

volume indicator (standard volume indicator). A standardized instrument having specified electric and dynamic characteristics and read in a prescribed manner, for indicating the volume of a complex electric wave such as that corresponding to speech or music. *Notes:* (1) The reading in volume units is equal to the number of decibels above a reference volume. The sensitivity is adjusted so that the reference volume or zero volume unit is indicated when the instrument is connected across a 600-ohm resistor in which there is dissipated a power of 1 milliwatt at 1000 hertz. (2) Specifications for a volume indicator are given in American National Standard Volume Measurements of Electrical Speech and Program Waves, C16.5. *See also:* **instrument; volume unit.** E269/42A30-1E1/19E8

volume limiter. A device that automatically limits the output volume of speech or music to a predetermined maximum value. *See also:* **circuits and devices; peak limiter.** 42A65-0

volume-limiting amplifier. An amplifier containing an automatic device that functions when the input volume exceeds a predetermined level and so reduces the gain that the output volume is thereafter maintained substantially constant notwithstanding further increase in the input volume. *Note:* The normal gain of the amplifier is restored when the input volume returns below the predetermined limiting level. *See also:* **amplifier.** E145/42A65-0

volume range (1) (transmission system). The difference, expressed in decibels between the maximum and minimum volumes that can be satisfactorily handled by the system.

(2) (complex audio-frequency signal). The difference, expressed in decibels, between the maximum and minimum volumes occurring over a specified period of time. *See also:* **transmission characteristics.** 42A65-0

volume unit (vu). The unit in which the standard volume indicator is calibrated. *Note:* One volume unit equals one decibel for a sine wave but volume units should not be used to express results of measurements of complex waves made with devices having characteristics differing from those of the standard volume indicator. *See:* **volume indicator.** *See also:* **transmission characteristics.** E151/42A65-0

volumetric radar. *See:* **three-dimensional radar.**

VOR. *See:* **very-high-frequency onmidirectional radio range.**

vortac (electronic navigation). A designation applied to certain navigation stations (primarily in the United States) in which both VOR and tacan are used; the distance function in tacan is used with VOR to provide bearing and distance (VOR/DME) (rho theta) navigation. *See also:* **navigation.** 0-10E6

vowel articulation (percent vowel articulation). The percent articulation obtained when the speech units considered are vowels (usually combined with consonants into meaningless syllables). *See also:* **articulation (percent articulation); volume equivalent.** 42A65-1E1

VPI. *See:* **vacuum-pressure impregnation.**

VSWR. *See:* **voltage standing-wave ratio.**

***V*-terminal voltage (electromagnetic compatibility).** Terminal voltage measured with a *V* network between each mains conductor and earth. *See also:* **electromagnetic compatibility.** CISPR-27E1

vu. *See:* **volume unit.**

W

waiting-passenger indicator (elevators). An indicator that shows at which landings and for which direction elevator-hall stop or signal calls have been registered and are unanswered. *See also:* **control.** 42A45-0

walkie-talkie. A two-way radio communication set designed to be carried by one person, usually strapped over the back, and capable of operation while in motion. *See also:* **radio transmission.** 42A65-0

wall bushing. A bushing intended primarily to carry a circuit through a wall or other grounded barrier in a substantially horizontal position. Both ends must be suitable for operating in air. *See also:* **bushing.** E49-0

wall telephone set. A telephone set arranged for wall mounting. *See also:* **telephone station.** 42A65-0

wander (electronic navigation). *See:* **scintillation.**

warble-tone generator. A voice-frequency oscillator, the frequency of which is varied cyclically at a subaudio rate over a fixed range. It is usually used with an integrating detector to obtain an averaged transmission or crosstalk measurement. *See also:* **auxiliary device to an instrument.** 42A30-0

warm-up time (power supplies). The time (after power turn on) required for the output voltage or current to reach an equilibrium value within the stability specification. *See also:* **power supply.** 0-10E1

warning whistle. *See:* **audible cab indicator.**

washer (rotating machinery). *See:* **collar.**

watchman's reporting system. A supervisory system arranged for the transmission of a patrolling watchman's regularly recurrent report signals to a central supervisory agency from stations along his patrol route. *See also:* **protective signaling.** 42A65-0

water. *See:* **acoustic properties of water.**

water-air-cooled machine. A machine that is cooled by circulating air that in turn is cooled by circulating water. *Note:* The machine is so enclosed as to prevent the free exchange of air between the inside and outside of the enclosure, but not sufficiently to be termed airtight. It is provided with a water-cooled heat exchanger for cooling the ventilating air and a fan or fans, integral with the rotor shaft or separate, for circulating the ventilating air. *See:* **asynchronous machine; cradle base (rotating machinery); direct-current commutating machine; synchronous machine.** 42A10-31E8

water-cooled (rotating machinery). (1) A term applied to apparatus cooled by circulating water, the water or water ducts coming in direct contact with major parts of the apparatus. (2) In certain types of machine, it is customary to apply this term to the cooling of the major parts by enclosed air or gas ventilation, where water removes the heat through an air-to-water or gas-to-water heat exchanger. *See:* **asynchronous machine; direct-current commutating machine; synchronous machine.** 42A10-31E8

water cooler (rotating machinery). A cooler using water as one of the fluids. *See also:* **cradle base (rotating machinery).** 0-31E8

waterflow-alarm system (protective signaling). An alarm system in which signal transmission is initiated automatically by devices attached to an automatic sprinkler system and actuated by the flow through the sprinkler system pipes of water in excess of a predetermined maximum. *See also:* **protective signaling.** 42A65-0

water load (high-frequency circuits). A matched termination in which the electromagnetic energy is absorbed in a stream of water for the purpose of measuring power by continuous-flow calorimetric methods. *See also:* **waveguide.** 50I62-3E1

water-motor bell. A vibrating bell operated by a flow of water through its water-motor striking mechanism. *See also:* **protective signaling.** 42A65-0

waterproof electric blasting cap. A cap specially insulated to secure reliability of firing when used in wet work. *See also:* **blasting unit.** 42A85-0

waterproof enclosure. An enclosure so constructed that any moisture or water leakage that may occur into the enclosure will not interfere with its successful operation. In the case of motor or generator enclosures, leakage that may occur around the shaft may be considered permissible provided it is prevented from entering the oil reservoir and provision is made for automatically draining the motor or generator enclosure. E45-0

waterproof machine (rotating machinery). A machine so constructed that water directed on it under prescribed conditions cannot cause interference with satisfactory operation. *See also:* **asynchronous machine; direct-current commutating machine; synchronous machine.** 0-31E8

watertight. So constructed that water will not enter the enclosing case under specified conditions. *Note:* A common form of specification for watertight is: "So constructed that there shall be no leakage of water into the enclosure when subjected to a stream from a hose with a 1-inch nozzle and delivering at least 65 gallons per minute, with the water directed at the enclosure from a distance of not less than 10 feet for a period of 5 minutes, during which period the water may be directed in one or more directions as desired." *See also:* **distribution center.**
1A0/37A100/42A25/42A95/IC1-31E11/34E10

watertight door-control system. A system of control for power-operated watertight doors providing individual local control of each door and, at a remote station in or adjoining the wheelhouse, individual control of any door, collective control of all doors, and individual indication of open or closed condition. *See also:* **marine electric apparatus.** 42A43-0

watertight enclosure. An enclosure so constructed that a stream of water from a hose not less than 1 inch in diameter under a head of 35 feet from a distance of 10 feet can be played on the enclosure from any direction for a period of 15 minutes without leakage. The hose nozzle shall have a uniform inside diameter of 1 inch. E45-0

water treatment equipment. Any apparatus such as deionizers, electrolytic targets, filters, or other devices employed to control electrolysis, corrosion, scaling, or clogging in water systems. *See also:* **power rectifier.** 0-34E24

watt. The unit of power in the International System of Units (SI). The watt is the power required to do work at the rate of 1 joule per second. E270-0

watthour. 3600 joules. E270-0

watthour capacity (storage battery) (storage cell). The number of watthours that can be delivered under specified conditions as to temperature, rate of discharge, and final voltage. *See also:* **battery (primary or secondary).** 42A60-0

watthour constant (meter). The registration expressed in watthours corresponding to one revolution of the rotor. *Note:* It is commonly denoted by the symbol K_h. When a meter is used with instrument transformers, the watthour constant is expressed in terms of primary watthours. For a secondary test of such a meter, the constant is the primary watthour constant divided by the product of the nominal ratios of transformation. *See also:* **electricity meter (meter); watthour meter.** 12A0/42A30-0

watthour-demand meter. A watthour meter and a demand meter combined as a single unit. *See also:* **electricity meter.** 42A30-0

watthour efficiency (storage battery) (storage cell). The energy efficiency expressed as the ratio of the watthours output to the watthours of the recharge. *See also:* **charge.** 42A60-0

watthour meter. An electricity meter that measures and registers the integral, with respect to time, of the active power of the circuit in which it is connected. This power integral is the energy delivered to the circuit during the interval over which the integration extends, and the unit in which it is measured is usually the kilowatthour.
See:
class designation of a watthour meter;
current circuit;
cyclometer register;
dial train of a register;

electricity meter;
first dial;
load range;
phase-angle correction factor;
register;
retarding magnet;
rotor of a meter;
service test;
standard register;
stator of an induction watthour meter;
test current of a watthour meter;
voltage circuit;
watthour constant. 12A0/42A30-0

watt loss (electric instrument). *See:* **power loss (electric instrument).**

wattmeter. An instrument for measuring the magnitude of the active power in an electric circuit. It is provided with a scale usually graduated in either watts, kilowatts, or megawatts. If the scale is graduated in kilowatts or megawatts, the instrument is usually designated as a kilowattmeter or megawattmeter. *See also:* **instrument.** 42A30-0

watt-second constant (meter). The registration in wattseconds corresponding to one revolution of the rotor. *Note:* The wattsecond constant is 3600 times the watthour constant and is commonly denoted by the symbol K_s. *See also:* **electricity meter (meter).** 42A30-0

wave (1). A disturbance that is a function of time or space or both. *See also:* **modulating systems.** E145-0

(2). A disturbance propagated in a medium or through space. *Notes:* (A) Any physical quantity that has the same relationship to some independent variable (usually time) that a propagated disturbance has, at a particular instant, with respect to space, may be called a wave. (B) Disturbance, in this definition, is used as a generic term indicating not only mechanical displacement but also voltage, current, electric field strength, temperature, etcetera. E270-0

(3) (electric circuit). The variation of current, potential, or power at any point in the electric circuit. *See also:* **lightning arrester (surge diverter); power systems, low-frequency and surge testing.** E28/42A35/62A1-31E7/31E13

wave analyzer. An electric instrument for measuring the amplitude and frequency of the various components of a complex current or voltage wave. *See also:* **instrument.** 42A30-0

wave antenna. *See:* **Beverage antenna.**

wave clutter. Clutter caused by echoes from waves of the sea. *See also:* **radar.** 42A65-0

wave filter. *See:* **filter.**

waveform (wave) (pulse techniques). The geometrical shape as obtained by displaying a characteristic of the wave as a function of some variable, usually time, when it is plotted over one primitive period. *Note:* It may be expressed mathematically by dividing the function by the maximum absolute value of the function that occurs during the period. *See also:* **pulse.** E270-9E4

waveform-amplitude distortion. Nonlinear distortion in the special case where the desired relationship is direct proportionality between input and output. *Note:* Also sometimes called **amplitude distortion.** *See:* **nonlinear distortion.** E154-0

waveform distortion (oscilloscopes). A displayed deviation from the representation of the input reference signal. *See:* **oscillograph.** 0-9E4

waveform influence of root-mean-square (RMS) responding instruments. The change in indication produced in an RMS responding instrument by the presence of harmonics in the alternating electrical quantity under measurement. In magnitude it is the deviation between an indicated RMS value of an alternating electrical quantity and the indication produced by the measurement of a pure sine-wave form of equal RMS value. *See also:* **instrument.** 39A1-0

waveform pulse. A waveform or a portion of a waveform containing one or more pulses or some portion of a pulse. *See also:* **pulse.** 0-9E4

waveform reference. A specified waveform, not necessarily ideal, relative to which waveform measurements, derivations, and definitions may be referred. 0-9E4

waveform test (rotating machinery). A test in which the waveform of any quantity associated with a machine is recorded. *See also:* **asynchronous machine; synchronous machine.** 0-31E8

wave front (1) (progressive wave in space). A continuous surface that is a locus of points having the same phase at a given instant. 42A65-0

(2) (impulse in a conductor). That part (in time or distance) between the virtual-zero point and the point at which the impulse reaches its crest value. *See also:* **power systems, low-frequency and surge testing.** 42A35-31E13

(3) (lightning arresters). The rising part of an impulse wave. 50I25-31E7

(4) (signal wave envelope). That part (in time or distance) between the initial point of the envelope and the point at which the envelope reaches its crest. 42A65-0

See:
angle or phase;
attenuation per unit length;
coherent interrupted wave;
continuous wave;
damped waves;
gating;
interference pattern;
lightning arrester;
modulated continuous wave;
rectangular wave;
signal wave;
square wave;
standing wave ratio;
wave interference.

waveguide. (1) Broadly, a system of material boundaries capable of guiding electromagnetic waves. (2) More specifically, a transmission line comprising a hollow conducting tube within which electromagnetic waves may be propagated or a solid dielectric or dielectricfilled conductor for the same purpose. (3) A system of material boundries or structures for guiding transverse-electromagnetic mode, often and originally a hollow metal pipe for guiding electromagnetic waves. *Note:* For an extensive list of cross reference, see *Appendix A. See also:* **transmission line.** 0-9E4

waveguide component. A device designed to be connected at specified ports in a waveguide system. E148-3E1

waveguide cutoff frequency (critical frequency). *See:* **cutoff frequency.**

waveguide joint. A connection between two sections of waveguide. *See also:* **waveguide.** 0-9E4

waveguide plunger. *See:* **sliding short-circuit.**

waveguide switch (waveguide system). A device for stopping or diverting the flow of high-frequency energy as desired. *See also:* **waveguide.** 50I62-3E1

waveguide-to-coaxial transition. A mode changer for converting coaxial line transmission to rectangular waveguide transmission. *See also:* **waveguide.** 50I62-3E1

waveguide wavelength (traveling wave at a given frequency). The distance along a uniform guide between points at which a field component (or the voltage or current) differs in phase by 2π radians. *Note:* It is equal to the quotient of phase velocity divided by frequency. For a waveguide with air dielectric, the waveguide wavelength is given by the formula:

$$\lambda_g = \frac{\lambda}{(1 - (\lambda^2/\lambda_c))^{1/2}}$$

where λ is the free-space wavelength and λ_c is the cutoff wavelength of the guide. *See also:* **waveguide.** 42A65/E146/E148-3E1/9E4

wave heater (dielectric heating). A heater in which heating is produced by energy absorption from a traveling electromagnetic wave. *See also:* **dielectric heating.** E54-0

wave heating. The heating of a material by energy absorption from a traveling electromagnetic wave. *See also:* **dielectric heating; induction heating.** E54/E169-0

wave impedance (waveguides including transmission lines). The complex factor relating the transverse component of the magnetic field to the transverse component of the electric field at every point in any specified plane, for a given mode. *Note:* Both incident and reflected waves may be present. *See also:* **self-impedance; transmission line; waveguide.** 42A65-3E1

wave impedance, characteristic (traveling wave) (waveguides). (1) The wave impedance with the sign so chosen that the real part is positive. *Note:* In a given mode, in a homogeneously filled waveguide, this is constant for all points and all cross sections. *See also:* **self-impedance; waveguide.** 0-3E1
(2) Wave impedance for purely progressing waves. *See also:* **transmission characteristics.** 0-9E4

wave interference. The variation of wave amplitude with distance or time, caused by the superposition of two or more waves. *Notes:* (1) As most commonly used, the term refers to the interference of waves of the same or nearly the same frequency. (2) Wave interference is characterized by a spatial or temporal distribution of amplitude of some specified characteristic differing from that of the individual superposed waves. *See also:* **wave front.** E211/42A65-3E2

wavelength (1) (sinusoidal wave). The distance between points of corresponding phase of two consecutive cycles. *Note:* The wavelength λ is related to the phase velocity v and the frequency f by $\lambda = v/f$. *See also:* **radio wave propagation.**
(2) (periodic wave). 2π times the limit of the ratio of the distance between two equiphase surfaces to the phase difference as these quantities go to zero. E270-3E2

wavelength constant. *See:* **phase constant.**

wavelength shifter (scintillator). A photofluorescent compound used with a scintillator material to absorb photons and emit related photons of a longer wavelength. *Note:* The purpose is to cause more efficient use of the photons by the phototube or photocell. *See also:* **phototube.** E175-0

wavemeter. An instrument for measuring the wavelength of a radio-frequency wave. The following are representative types: resonant-cavity, resonant-circuit, and standing-wave. *Note:* The standing-wave type is exemplified by a Lecher wire. Wavemeters may also be classified as (1) transmission type in which the resonant element and the detector are so arranged as to give a maximum response at resonance; and (2) suppression (absorption or reaction) type, which provide minimum response at resonance. *See also:* **frequency meter; instrument.** 42A30-0

wave normal. A unit vector normal to an equiphase surface with its positive direction taken on the same side of the surface as the direction of propagation. *Note:* In isotropic media, the wave normal is in the direction of propagation. *See also:* **radio wave propagation; waveguide.** E146-0

waves, electrocardiographic, *P, Q, R, S,* and *T* (medical electronics) (in electrocardiograms obtained from electrodes placed on the right arm and left leg). The characteristic tracing consists of five consecutive waves; *P,* a prolonged, low, positive wave; *Q,* brief, low, negative; *R,* brief, high, positive; *S,* brief, low, negative, and *T,* prolonged, low, positive. *See also:* **medical electronics.** 0-18E1

wave shape (impulse test wave) (1) (graphical representation). The graph of the current or voltage as a function of time or distance. *See:* **waveform.** *See also:* **lightning; lightning arrester (surge diverter); power systems, low-frequency and surge testing.** E28/42A35/62A1-31E7/31E13
(2) (numerical representation). (A) The designation of current or voltage, other than rectangular impulse, by a combination of two numbers. The first, an index of the wave front, is the virtual duration of the wave front in microseconds. The second, an index of the wave tail, is the time in microseconds from virtual zero to the instant at which one-half of the crest value is reached on the wave tail. Examples are 1.2 × 50 and 8 × 20 waves. (B) The wave shape of a rectangular impulse of current or voltage is designated by two numbers. The first designates the minimum value of current or voltage which is sustained for the time in microseconds designated by the second number. An example is the 75A × 1000 wave. *Note:* The signs × have *no* mathematical meaning. *See:* **virtual duration of wavefront.** *See also:* **lightning; lightning arrester (surge diverter).** E28-0

wave, square (pulse techniques). *See:* **square wave.**

wave tail (impulse in a conductor). That part (in time or distance) between the point of crest value and the end of the impulse. *See also:* **lightning; lightning arrester (surge diverter); power systems, low-frequency and surge testing.** E28/42A35/50I25-31E7/31E13

wave tilt. The forward inclination of a radio wave due to its proximity to ground. *See also:* **radiation.** 42A65-0

wave train. A limited series of wave cycles. E270-0

wave winding. A winding that progresses around the armature by passing successively under each main pole of the machine before again approaching the starting point. In commutator machines, the ends of individual coils are not connected to adjacent commutator bars. *See also:* **asynchronous machine; direct-current commutating machine; synchronous machine.** 42A10-0

way point (navigation). A selected point on or near a course line and having significance with respect to navigation or traffic control. *See also:* **navigation.** 0-10E6

wear-out failure (reliability). *See:* **failure, wear-out.**

wear-out-failure period (reliability). The period during which the failure rate of some items is rapidly increasing due to deterioration processes. *Note:* The accompanying figure shows the failure pattern when this definition applies to an item. *See also:* **reliability.** 0-7E1

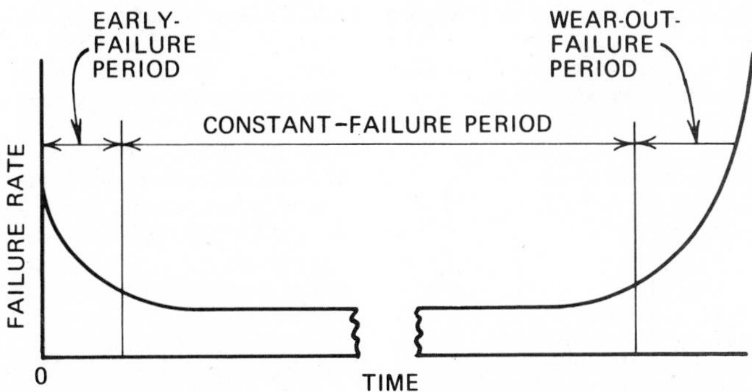

Wear-out failure period.

weatherproof (1) (outside exposure). So constructed or protected that exposure to the weather will not interfere with successful operation. *See:* **outdoor.** 37A100/42A95/IC1-31E11/34E10

(2) (conductor covering). Made up of braids of fibrous material that are thoroughly impregnated with a dense moistureproof compound after they have been placed on the conductor, or an equivalent protective covering designed to withstand weather conditions. 42A95-0

weatherproof enclosure. An enclosure for outdoor application designed to protect against weather hazards such as rain, snow, or sleet. *Note:* Condensation is minimized bv use of space heaters. 37A100-31E11

weather-protected machine. A guarded machine whose ventilating passages are so designed as to minimize the entrance of rain, snow, and airborne particles to the electric parts. *See also:* **asynchronous machine; direct-current commutating machine; synchronous machine.** 42A10-31E8

weathertight. *See:* **raintight.**

weber. The unit of magnetic flux in the International System of Units (SI). The weber is the magnetic flux whose decrease to zero when linked with a single turn induces in the turn a voltage whose time integral is one volt-second. E270-0

wedge (rotating machinery). A tapered shim or key. *See:* **slot wedge.** *See also:* **cradle base (rotating machinery); rotor (rotating machinery); stator.** 0-31E8

wedge groove (wedge slot) (rotating machinery). A groove, usually in the side of a coil slot, to permit the insertion of and to retain a slot wedge. *See also:* **cradle base (rotating machinery).** 0-31E8

wedge, slot. *See:* **wedge groove.**

wedge washer (salient pole) (rotating machinery). Insulation triangular in cross section placed underneath the inner ends of field coils and spanning between field coils. *See also:* **cradle base (rotating machinery).** 0-31E8

weight coefficient (thermoelectric generator) (thermoelectric generator couple). The quotient of the electric power output by the device weight. *See also:* **thermoelectric device.** E221-15E7

weighting. The artificial adjustment of measurements in order to account for factors that in the normal use of the device, would otherwise be different from the conaitions during measurement. For example, background noise measurements may be weighted by applying factors or by introducing networks to reduce measured values in inverse ratio to their interfering effects. E145-0

weight transfer compensation. A system of control wherein the tractive forces of individual traction motors may be adjusted to compensate for the transfer of weight from one axle to another when exerting tractive force. *See also:* **multiple-unit control.** 42A42-0

weld decay (corrosion). Localized corrosion at or adjacent to a weld. *See also:* **corrosion terms.** CM-34E2

welding arc voltage. The voltage across the welding arc. *See also:* **electric arc-welding apparatus.** 87A1-0

Weston normal cell. A standard cell of the cadmium type containing a saturated solution of cadmium sulphate as the electrolyte. *Note:* Strictly speaking this cell contains a neutral solution, but acid cells are now in more common use. *See also:* **electrochemistry.** 42A60-0

wet cell. A cell whose electrolyte is in liquid form. *See also:* **electrochemistry.** 42A60-0

wet contact. A contact through wnich direct current flows. *Note:* The term has significance because of the healing action of direct current flowing through unsoldered contacts. *See also:* **telephone switching system.** 42A65-19E1

wet electrolytic capacitor. A capacitor in which the dielectric is primarily an anodized coating on one electrode, with the remaining space between the electrodes filled with a liquid electrolytic solution. 0-21E0

wet location. A location subject to saturation with water or other liquids, such as locations exposed to weather, washrooms in garages, and like locations. Installations underground or in concrete slabs or masonry in direct contact with the earth shall be considered as wet locations. *See also:* **distribution center.** 1A0-0

wetting. The free flow of solder alloy, with proper application of heat and flux, on a metallic surface to produce an adherent bond. *See also:* **soldered connections (electronic and electrical applications).** 99A1-0

wetting agent (electroplating) (surface active agent). A substance added to a cleaning, pickling or plating solution to decrease its surface tension. *See also:* **electroplating.** 42A60-0

wet-wound (rotating machinery). A coil in which the conductors are coated with wet resin in passage to the winding form, or on to which a bonding or insulating resin is applied on each successive winding layer to produce an impregnated coil. *See also:* **rotor (rotating machinery); stator.** 0-31E8

Wheatstone bridge. A 4-arm bridge, all arms of which are predominantly resistive. *See also:* **bridge.** 42A30-0

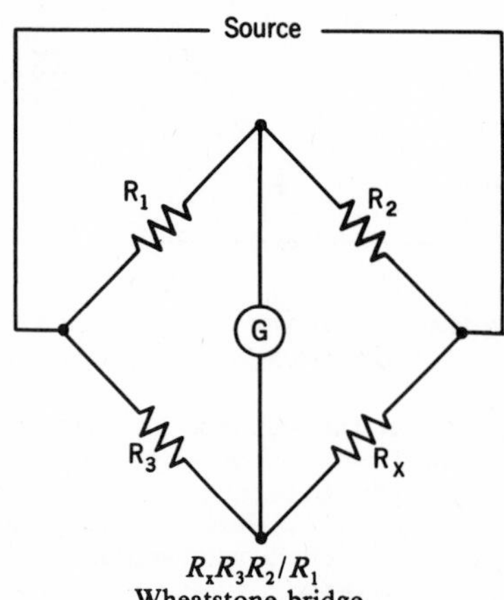

$R_x R_3 R_2 / R_1$
Wheatstone bridge.

whip antenna. A simple vertical antenna consisting of a slender whiplike conductor supported on a base insulator. *See also:* **antenna.** 42A65-3E1

whistle operator. A device to provide automatically the timed signals required by navigation laws when underway in fog, and also manual control of electrical operation of a whistle or siren, or both, for at-will signals. 42A43-0

white (color television). Used most commonly in the nontechnical sense. More specific usage is covered by the term **achromatic locus,** and this usage is explained in the note under the term **achromatic locus.** *See also:* **color terms.** E201-0

white compression (white saturation) (television). The reduction in gain applied to a picture signal at those levels corresponding to light areas in a picture with respect to the gain at that level corresponding to the midrange light value in the picture. *Notes:* (1) The gain referred to in the definition is for a signal amplitude small in comparison with the total peak-to-peak picture signal involved. A quantitative evaluation of this effect can be obtained by a measurement of differential gain. (2) The overall effect of white compression is to reduce contrast in the highlights of the picture as seen on a monitor. *See also:* **television.** E203-2E2

white noise (interference terminology). Noise, either random or impulsive type, that has a flat frequency spectrum at the frequency range of interest. *See:* **Johnson noise; thermal noise.** *See also:* **electromagnetic compatibility.** 0-27E1

white object (color). An object that reflects all wavelengths of light with substantially equal high efficiencies and with considerable diffusion. *See also:* **color terms.** E201-2E2

white peak (television). A peak excursion of the picture signal in the white direction. *See also:* **television.** E203/42A65-2E2

white recording (1) (amplitude-modulation facsimile system). That form of recording in which the maximum received power corresponds to the minimum density of the record medium.

(2) (frequency-modulation facsimile system). That form of recording in which the lowest received frequency corresponds to the minimum density of the record medium. *See also:* **recording (facsimile).** E168-0

white saturation. *See:* **white compression.**

white signal (at any point in a facsimile system). The signal produced by the scanning of a minimum-density area of the subject copy. *See also:* **facsimile signal (picture signal).** E168-0

white transmission (1) (amplitude-modulation facsimile system). The form of transmission in which the maximum transmitted power corresponds to the minimum density of the subject copy.

(2) (frequency-modulation facsimile system). That form of transmission in which the lowest transmitted frequency corresponds to the minimum density of the subject copy. *See also:* **facsimile transmission.** E168-0

whole body irradiation (electrobiology). Pertains to the case in which the entire body is exposed to the incident electromagnetic energy or in which the cross section of the body is smaller than the cross section of the incident radiation beam. *See also:* **electrobiology.** 95A1-0

wicking. The flow of solder along the strands and under the insulation of stranded lead wires. *See also:* **soldered connections (electronic and electrical applications).** 99A1-0

wick-lubricated bearing (rotating machinery). (1) A sleeve bearing in which a supply of lubricant is provided by the capillary action of a wick that extends into a reservoir of free oil or of oil-saturated packing material. (2) A sleeve bearing in which the reservoir and other cavities in the bearing region are packed with a material that holds the lubricant supply and also serves as a wicking. *See also:* **bearing.** 0-31E8

wide-angle diffusion. Diffusion in which flux is scattered at angles far from the direction that the flux would take by regular reflection or transmission. *See also:* **lamp.** Z7A1-0

wide-angle luminaire. A luminaire that distributes the light through a comparatively wide solid angle. *See also:* **luminaire.** Z7A1-0

wide-band axis (color television) (phasor representation of the chrominance signal). The direction of the phasor representing the fine chrominance primary. *See* *also:* **color terms.** E201-2E2

wide-band improvement. The ratio of the signal-to-noise ratio of the system in question to the signal-to-noise ratio of a reference system. *Note:* In comparing frequency-modulation and amplitude-modulation systems, the reference system usually is a double-sideband amplitude-modulation system with a carrier power, in the absence of modulation, that is equal to the carrier power of the frequency-modulation system. *See also:* **modulating systems.** E145-0

wide-band ratio. The ratio of the occupied frequency bandwidth to the intelligence bandwidth. *See also:* **modulating systems.** E145-0

width line (illuminating engineering). The radial line (the one that makes the larger angle with the reference line) that passes through the point of one-half maximum candlepower on the lateral candlepower distribution curve plotted on the surface of the cone of maximum candlepower. *See also:* **street-lighting luminaire.** Z7A1-0

Wiedemann-Franz ratio. The quotient of the thermal conductivity by the electric conductivity. *See also:* **thermoelectric device.** E221-15E7

Wien bridge oscillator. An oscillator whose frequency of oscillation is controlled by a Wien bridge. *See also:* **oscillatory circuit.** 42A65-0

Wien capacitance bridge. A 4-arm alternating-current bridge characterized by having in two adjacent arms capacitors respectively in series and in parallel with resistors, while the other two arms are normally non-reactive resistors. *Note:* Normally used for the measurement of capacitance in terms of resistance and frequency. The balance depends upon frequency, but from the balance conditions the capacitance of either or both capacitors can be computed from the resistances of all four arms and the frequency. *See also:* **bridge.** 42A30-0

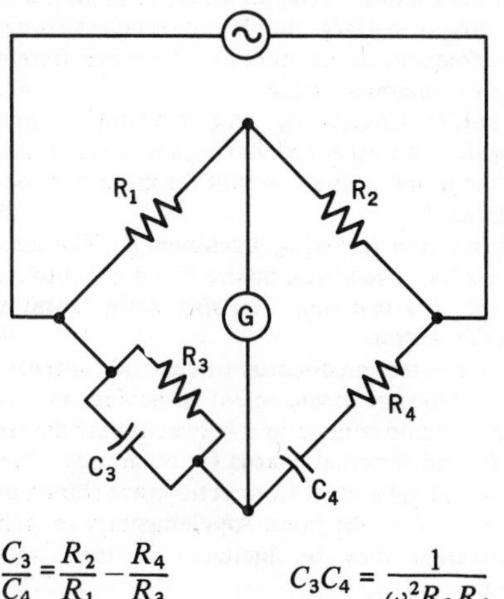

$$\frac{C_3}{C_4} = \frac{R_2}{R_1} - \frac{R_4}{R_3} \qquad C_3C_4 = \frac{1}{\omega^2 R_3 R_4}$$

Wien capacitance bridge.

Wien displacement law. An expression representing, in a functional form, the spectral radiance L_λ of a blackbody as a function of the wavelength λ and the temperature T:

$$L_\lambda = I_\lambda / A'$$
$$= c_1 \lambda^{-5} f(\lambda T),$$

where the symbols are those used in the definition of **Planck radiation law.** The two principal corollaries of this law are:

$$\lambda_m T = b$$
$$L_m / T^5 = b'$$

which show how the maximum spectral radiance L_m and the wavelength λ_m at which it occurs are related to the absolute temperature T. *Note:* The currently recommended value of b is 2.8978×10^{-3} m • K or 2.8978×10^{-1} cm • K. From the definition of the **Planck radiation law,** and with the use of the value of b, as given above, b' is found to be 4.10×10^{-12} W • cm^{-3} • K^{-5} • sr^{-1}. Z7A1-0

Wien inductance bridge. A 4-arm alternating-current bridge characterized by having in two adjacent arms inductors respectively in series and in parallel with resistors, while the other two arms are normally non-reactive resistors. *Note:* Normally used for the measurement of inductance in terms of resistance and frequency. The balance depends upon frequency, but from the balance conditions the inductances of either or both inductors can be computed from the resistances of the four arms and the frequency. *See also:* **bridge.** 42A30-0

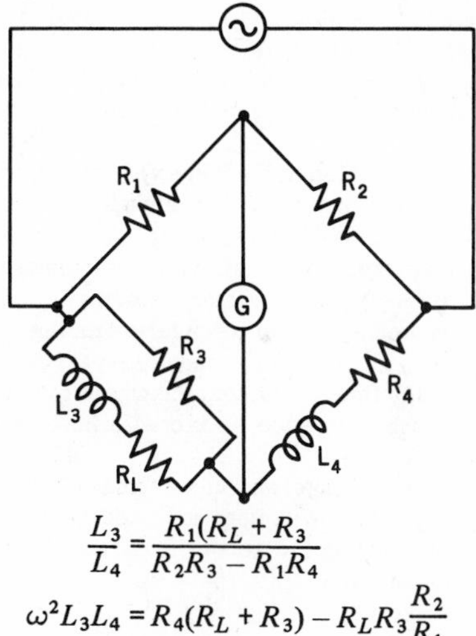

$$\frac{L_3}{L_4} = \frac{R_1(R_L + R_3)}{R_2R_3 - R_1R_4}$$

$$\omega^2 L_3 L_4 = R_4(R_L + R_3) - R_L R_3 \frac{R_2}{R_1}$$

Wien inductance bridge.

Wien radiation law. An expression representing approximately the spectral radiance of a blackbody as a function of its wavelength and temperature. It commonly is expressed by the formula

$$L_\lambda = I_\lambda / A'$$
$$= c_{1L} \lambda^{-5} e^{-(c_2/\lambda T)}$$

where the symbols are those used in the definition of **Planck radiation law.** This formula is accurate to one percent or better for values of λT less than 3000 micrometer kelvins. *See also:* **radiant energy.** Z7A1-0

wigwag signal. A railroad-highway crossing signal, the indication of which is given by a horizontally swinging disc with or without a red light attached. *See also:* **railway signal and interlocking.** 42A42-0

Williams-tube storage (electronic computation). A type of electrostatic storage. E270-0

Wilson center (limb center) (*V* potential) (medical electronics) (electrocardiography). An electric reference contact; the junction of three equal resistors to the limb leads. *See also:* **medical electronics.** 0-18E1

wind-driven generator for aircraft. A generator used on aircraft that derives its power from the air stream applied on its own air screw or impeller during flight. *See also:* **air-transportation electric equipment.** 42A41-0

winding (1) (rotating machinery). An assembly of coils designed to act in consort to produce a magnetic flux field or to link a flux field. 0-31E8
(2) (data processing). A conductive path, usually of wire, inductively coupled to a magnetic core or cell. *Note:* When several windings are employed, they may be designated by the functions performed. Examples are: sense, bias, and drive windings. Drive windings include read, write, inhibit, set, reset, input, shift, and advance windings. *See also:* **static magnetic storage.** E163-0

winding, autotransformer series. *See:* **series winding.**

winding, control power. The winding (or transformer) that supplies power to motors, relays, and other devices used for control purposes. *See:* **windings, high-voltage and low-voltage.** 42A15-31E12

winding-drum machine (elevators). A geared-drive machine in which the hoisting ropes are fastened to and wind on a drum. *See also:* **driving machine (elevators).** 42A45-0

winding factor (rotating machinery). The product of the distribution factor and the pitch factor. *See also:* **rotor (rotating machinery); stator.** 0-31E8

winding inductance. *See:* **air core inductance.**

winding, primary. *See:* **primary winding.**

winding, secondary. *See:* **secondary winding.**

winding shield (rotating machinery). A shield secured to the frame to protect the windings but not to support the bearing. *See also:* **cradle base (rotating machinery).** 42A10-31E8

windings, high-voltage and low-voltage. The terms high-voltage and low-voltage are used to distinguish the winding having the greater from that having the lesser voltage rating.
See:
impedance voltage;
primary winding;
rated kilovolt-ampere tap;
reduced kilovolt-ampere tap;
secondary winding;
stablizing winding;
tap;
winding, control power. 42A15-31E12

winding, stabilizing. *See:* **stabilizing winding.**

winding voltage rating. The voltage for which the winding is designed. *See also:* **duty.** 89A1-0

window (counter tube) (radiation-counter tubes). That portion of the wall that is made thin enough for radiation of low penetrating power to enter. *See:* **anticoincidence (radiation counters).** 0-15E6

windshield wiper for aircraft. A motor-driven device for removing rain, sleet, or snow from a section of an aircraft windshield, window, navigation dome, or turret. *See:* **air-transportation electric equipment.** 42A41-0

windup. Lost motion in a mechanical system that is proportional to the force or torque applied. E1A3B-34E12

wing-clearance lights. A pair of aircraft lights provided at the wing tips to indicate the extent of the wing span when the navigation lights are located an appreciable distance inboard of the wing tips. *See also:* **signal lighting.** Z7A1-0

wiper (brush). That portion of the moving member of a selector or other similar device, that makes contact with the terminals of a bank. *See also:* **telephone switching system.** 42A65-0

wiper relay. *See:* **relay wiper.**

wiping gland (wiping sleeve). A projecting sleeve on a junction box, pothead, or other piece of apparatus serving to make a connection to the lead sheath of a cable by means of a plumber's wiped joint. *See also:* **tower; transformer removable cable-terminating box.** 42A35-31E13

wire. A slender rod or filament of drawn metal. *Note:* The definition restricts the term to what would be ordinarily understood by the term solid wire. In the definition, the word slender is used in the sense that the length is great in comparison with the diameter. If a wire is covered with insulation, it is properly called an insulated wire; while primarily the term wire refers to the metal, nevertheless when the context shows that the wire is insulated, the term wire will be understood to include the insulation. *See also:* **car wiring apparatus; conductors.** E16/E30/42A35-31E13

wire-band serving (power distribution underground cables). A short closed helical serving of wire applied tightly over the armor of wire-armored cables spaced at regular intervals, such as on vertical riser cables, to bind the wire armor tightly over the core to prevent slippage. *See also:* **power distribution, underground construction.** 0-31E1

wire broadcasting. The distribution of programs over wire circuits to a large number of receivers, using either voice frequencies or modulated carrier frequencies. *See also:* **communication.** 42A65-0

wire holder (insulator). An insulator of generally cylindrical or pear shape, having a hole for securing the conductor and a screw or bolt for mounting. *See also:* **insulator.** 29A1-0

wire insulation (rotating machinery). The insulation that is applied to a wire before it is made into a coil or inserted in a machine. *See also:* **rotor (rotating machinery); stator.** 0-31E8

wireless connection diagram (industrial control). The general physical arrangement of devices in a control equipment and connections between these devices, terminals, and terminal boards for outgoing connections to external apparatus. Connections are shown in tabular form and not by lines. An elementary (or schematic) diagram may be included in the connection diagram. E270-34E10

wire-pilot protection. Pilot protection in which an auxiliary metallic circuit is used for the communicating means between relays at the circuit terminals. 37A100-31E6/31E11

wire spring relay. A relay design in which the contacts are attached to round wire springs instead of the conventional flat or leaf spring. 0-21E0

withstand current (surge). The crest value attained by a surge of a given wave shape and polarity that does not cause disruptive discharge on the test specimen. *See also:* **power systems, low-frequency and surge testing.** 42A35-31E13

withstand test voltage. The voltage that the device must withstand without flashover, disruptive discharge, puncture, or other electric failure when voltage is applied under specified conditions. *Note:* For low-frequency voltage the values are expressed as root-mean-square and for a specified time. For impulse voltages the values are expressed in crest value of a specified wave. *See also:* **lightning arrester (surge diverter); pothead; test voltage and current.** E48/E49/E270-0

withstand voltage (1) (general). The voltage that electric equipment is capable of withstanding without failure or disruptive discharge when tested under specified conditions. 92A1/37A100-31E11
(2) (impulse) (electric power). The crest value attained by an impulse of any given wave shape, polarity, and amplitude, that does not cause disruptive discharge on the test specimen. *See also:* **power systems, low-frequency and surge testing.** E32-0
(3) (lightning arresters). A specified voltage that is to be applied to a test object in a withstand test under specified conditions. During the test, in general no disruptive discharge should occur. *See also:* **basic impulse insulation level (insulation strength); lightning arrester (surge diverter); test voltage and current.** 60I0/68AI-31E5/31E7

word. An ordered set of characters that is the normal unit in which information may be stored, transmitted, or operated upon within a given computer. *Example:* The set of characters 10692 is a word that may give a command for a machine element to move to a point 10.692 inches from a specified zero. *See also:* **computer word; electronic computation; electronic digital computer; machine word.** E162/E270/EIA3B-34E12

word address format. Addressing each word of a block by one or more characters that identify the meaning of the word. EIA3B-34E12

word length. The number of bits or other characters in a word. *See also:* **electronic digital computer.** X3A12-16E9

word time (electronic computation). In a storage device that provides serial access to storage locations, the time interval between the appearance of corresponding parts of successive words. *See also:* **electronic digital computer; minor cycle.** X3A12-16E9

work. The work done by a force is the dot-product line integral of the force. *See:* **line integral.** E270-0

work coil. *See:* **load coil (induction heating usage).**

work function. The minimum energy required to remove an electron from the Fermi level of a material into field-free space. *Note:* Work function is commonly expressed in electron volts. *See also:* **circuit characteristics of electrodes.** E160-15E6

working (electrolysis). The process of stirring additional solid electrolyte or constituents of the electrolyte into the fused electrolyte in order to produce a uniform solution thereof. *See also:* **fused electrolyte.** 42A60-0

working peak OFF-state voltage (thyristor). The maximum instantaneous value of the OFF-state voltage that occurs across a thyristor, excluding all repetitive and nonrepetitive transient voltages. *See also:* **principal voltage-current characteristic (principal characteristic).** E223-15E7/34E17/34E24

working peak reverse voltage (semiconductor rectifier). The maximum instantaneous value of the reverse voltage, excluding all repetitive and nonrepetitive transient voltages, that occurs across a semiconductor rectifier cell, rectifier diode, rectifier stack, or reverse-blocking thyristor. *See also:* **principal voltage-current characteristic (principal characteristic); rectification.** E59/E223-15E7/34E17/34E24

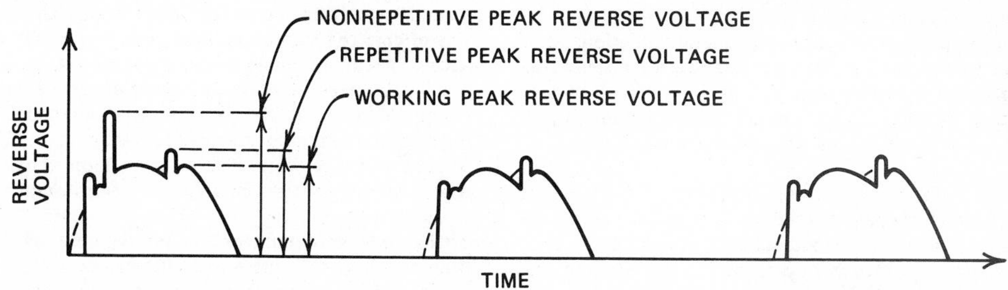

Working peak reverse voltage.

working peak reverse voltage rating (rectifier circuit). The maximum value of working peak reverse voltage permitted by the manufacturer under stated conditions. *See also:* **average forward current rating (rectifier circuit).** E59-34E17/34E24

working point. *See:* **operating point.**

working pressure. The pressure, measured at the cylinder of a hydraulic elevator, when lifting the car and its rated load at rated speed. *See also:* **elevators.** 42A45-0

working reference system. A secondary reference telephone system consisting of a specified combination of telephone sets, subscriber lines, and battery supply circuits connected through a variable distortionless trunk and used under specified conditions for determining, by comparison, the transmission performance of other telephone systems and components. *See also:* **transmission characteristics.** 42A65-0

working standard (illuminating engineering). A standardized light source for regular use in photometry. *See also:* **primary standards (illuminating engineering).** Z7A1-0

working storage (computing machines). *See:* **temporary storage.** *See also:* **electronic digital computer.**

working value. The electrical value that when applied to an electromagnetic instrument causes the movable member to move to its fully energized position. This value is frequently greater than pick-up. *See:* **pick-up.** *See also:* **railway signal and interlocking.** 42A42-0

working voltage to ground (electric instrument). The highest voltage, in terms of maximum peak value, that should exist between any terminal of the instrument proper on the panel, or other mounting surface, and ground. *See also:* **instrument.** 39A1-0

work plane. The plane at which work is usually done, and at which the illumination is specified and measured. Unless otherwise indicated, this is assumed to be a horizontal plane 30 inches above the floor. *See also:* **inverse-square law (illuminating engineering).** Z7A1-0

worm-geared machine (elevators). A direct-drive machine in which the energy from the motor is transmitted to the driving sheave or drum through worm gearing. *See also:* **driving machine (elevators).** 42A45-0

wound rotor (rotating machinery). A rotor core assembly having a winding made up of individually insulated wires. *See also:* **asynchronous machine; direct-current commutating machine; synchronous machine.** 0-31E8

wound-rotor induction motor. An induction motor in which a primary winding on one member (usually the stator) is connected to the alternating-current power source and a secondary polyphase coil winding on the other member (usually the rotor) carries alternating current produced by electromagnetic induction. *Note:* The terminations of the rotor winding are usually connected to collector rings. The brush terminals may be either short-circuited or closed through suitable adjustable circuits. *See also:* **asynchronous machine.** 42A10-31E8

wound stator core (rotating machinery). A stator core into which the stator winding, with all insulating elements and lacing has been placed, including any components imbedded in or attached to the winding, and including the lead cable when this is used. *See also:* **stator.** 0-31E8

wow (sound reproduction). A slow periodic change in pitch, or low-frequency flutter, observable in recorded sound reproduction, due to nonuniform rate of reproduction of the original sound. *See also:* **phonograph pickup.** E188/42A65-0

wrap. One convolution of a length of ferromagnetic tape about the axis. *See:* **tape-wound core.** E163-0

wrapper (rotating machinery). (1) A relatively thin flexible sheet material capable of being formed around the slot section of a coil to provide complete enclosure. (2) The outer cylindrical frame component used to contain the ventilating gas. *See also:* **rotor (rotating machinery); stator.** 0-31E8

wrap thickness. The distance between corresponding points on two consecutive wraps, measured parallel to the ferromagnetic tape thickness. *See:* **tape-wound core.** E163-0

wrap width (tape width). *See:* **tape-wound core.** *See also:* **static magnetic storage.**

write. To introduce data, usually into some form of storage. *See:* **read.** *See also:* **electronic digital computer.** E158-15E6;E162/E270-0;X3A12-16E9

write pulse. *See:* **ONE state.** *See also:* **static magnetic storage.**

writing rate (storage tubes). The time rate of writing on a storage element, line, or area to change it from one specified level to another. Note the distinction between this and **writing speed.** *See also:* **storage tube.** E158-15E6

writing speed (storage tubes). Lineal scanning rate of the beam across the storage surface in writing. Note the distinction between this and **writing rate.** *See:* **information writing speed.** *See also:* **circuit characteristics of electrodes; storage tube.** E158-15E6

writing speed, maximum usable (storage tubes). The maximum speed at which information can be written under stated conditions of operation. Note the qualifying adjectives **maximum usable** are frequently omitted in general usage when it is clear that the maximum usable writing speed is implied. *See also:* **storage tube.** E158-15E6

writing time, minimum usable (storage tubes). The time required to write stored information from one specified level to another under stated conditions of operation. *Note:* The qualifying adjectives **minimum usable** are frequently omitted in general usage when it is clear that the minimum usable writing time is implied. *See also:* **storage tube.** E158-15E6

wye. *See:* **Y.**

wye junction (waveguides). A junction of waveguides such that the longitudinal guide axes form a Y. *See also:* **waveguide.** E147-3E1

wye rectifier circuit. A circuit that employs three or more rectifying elements with a conducting period of 120 electrical degrees plus the commutating angle. *See also:* **rectification.** 42A15-0

X

***X*-axis amplifier.** *See:* **horizontal amplifier.**

xerography. The branch of electrostatic electrophotography that employs a photoconductive insulating medium to form, with the aid of infrared, visible, or ultraviolet radiation, latent electrostatic-charge patterns for producing a viewable record. *See also:* **electrostatography.** E224-15E7

xeroprinting. The branch of electrostatic electrography that employs a pattern of insulating material on a conductive medium to form electrostatic-charge patterns for duplicating purposes. *See also:* **electrostatography.** E224-15E7

xeroradiography. The branch of electrostatic electrophotography that employs a photoconductive insulating medium to form, with the aid of X rays or gamma rays, latent electrostatic-charge patterns for producing a viewable record. *See also:* **electrostatography.** E224-15E7

X-ray tube. A vacuum tube designed for producing X-rays by accelerating electrons to a high velocity by means of an electrostatic field and then suddenly stopping them by collision with a target. *See also:* **tube definitions.** 42A70-15E6

***X* wave (radio wave propagation).** Extraordinary-wave component. 0-3E2

***X-Y* display.** A rectilinear coordinate plot of two variables. *See also:* **oscillograph.** 0-9E4

***XY* switch.** A remotely controlled bank-and-wiper switch arranged in a flat manner, in which the wipers are moved in a horizontal plane, first in one direction and then in another. *See also:* **telephone switching system.** 42A65-0

Y

***Y* amplifier.** *See:* **vertical amplifier.**

***Y*-axis amplifier.** *See:* **vertical amplifier.**

Y-connected circuit. A three-phase circuit that is star connected. *See also:* **network analysis.** E270-0

Y network. A star network of three branches. *See also:* **network analysis.** E153/E270-0

yoke (magnetic) (rotating machinery). The element of ferromagnetic material, not surrounded by windings, used to connect the cores of an electromagnet, or of a transformer, or the poles of a machine, or used to support the teeth of stator or rotor. *Note:* A yoke may be of solid material or it may be an assembly of laminations. *See also:* **rotor (rotating machinery); stator.** E270-31E8

***Y-T* display.** An oscilloscope display in which a time-dependent variable is displayed against time. *See also:* **oscillograph.** 0-9E4

Z

***Z*-axis amplifier (oscilloscopes).** An amplifier for signals controlling a display perpendicular to the *X-Y* plane, commonly intensity of the spot. *See:* **intensity amplifier.** *See also:* **oscillograph.** 0-9E4

Zeeman effect. If an electric discharge tube, or other light source emitting a bright-line spectrum, is placed between the poles of a magnet, each spectrum line is split by the action of the magnetic field into three or more close-spaced but separate lines. The amount of splitting or the separation of the lines, is directly proportional to the strength of the magnetic field. E270-0

Zener breakdown (semiconductor device). A breakdown that is caused by the field emission of charge carriers in the depletion layer. *See also:* **semiconductor; semiconductor device.** E216/E270-34E17

Zener diode (semiconductor). A class of silicon diodes that exhibit in the avalanche-breakdown region a large change in reverse current over a very narrow range of reverse voltage. *Note:* This characteristic permits a highly stable reference voltage to be maintained across the diode despite a relatively wide range of current through the diode. *See:* **Zener breakdown; avalanche breakdown.** 0-10E1

Zener diode regulator. A voltage regulator that makes use of the constant-voltage characteristic of the Zener diode to produce a reference voltage that is compared with the voltage to be regulated to initiate correction when the voltage to be regulated varies through changes in either load or input voltage. See the accompanying figure. *See:* **Zener diode.** *See also:* **electrical conversion.** 0-10E1

Current and voltage characteristics for a typical Zener diode regulator $|V_A| \gg |V_B|$.

Zener impedance* (semiconductor diode). *See:* **breakdown impedance.** *See also:* **semiconductor.**

*Deprecated

Zener voltage*. *See:* **breakdown voltage.** *See also:* **semiconductor.**

*Deprecated

zero (1) (function) (root of an equation). A zero of a function $f(x)$ is any value of the argument X for which $f(x)=0$. *Note:* Thus the zeros of sin x are $x_1 = 0$, $x_2 = \pi$, $x_3 = 2\pi$, $x_4 = 3\pi$, ..., $x_n = (n-1)\pi$, ... The roots of the equation $f(x) = 0$ are the zeros of $f(x)$. E270-0

(2) (transfer function in the complex variable *s*). (A) A value of s that makes the function zero. (B) The corresponding point in the s plane. *See also:* **control system, feedback; pole (network function).** 85A1-23E0

(3) (network function). Any value of p, real or complex, for which the network function is zero. *See also:* **network analysis.** E270-0

zero adjuster. A device for bringing the indicator of an electric instrument to a zero or fiducial mark when the electrical quantity is zero. *See also:* **moving element (instrument).** 42A30-0

zero-based linearity. *See:* **linearity.** *See also:* **electronic analog computer.**

zero-beat reception. *See:* **homodyne reception.**

zero-error (device operating under the specified conditions of use). The indicated output when the value of the input presented to it is zero. *See also:* **control system, feedback.** 0-23E0

zero-error reference. *See:* **linearity.** *See also:* **electronic analog computer.**

zero guy. A line guy installed in a horizontal position between poles to provide clearance and transfer strain to an adjacent pole. *See also:* **tower.** 42A35-31E13

zero lead (medical electronics). *See:* **biolectric null.**

zero-level address (computing machines). *See:* **immediate address.** *See also:* **electronic digital computer.**

zero-modulation medium noise (sound recording and reproducing system). The noise that is developed in the scanning or reproducing device during the reproducing process when a medium is scanned in the zero-modulation state. *Note:* For example, zero-modulation medium noise is produced in magnetic recording by undesired variations of the magnetomotive force in the medium, that are applied across the scanning gap of a demagnetized head, when the medium moves with the desired motion relative to the scanning device. Medium noise can be ascribed to nonuniformities of the magnetic properties and to other physical and dimensional properties of the medium. *See also:* **noise (sound recording and reproducing system).** E191-0

zero-modulation state (sound recording medium). The state of complete preparation for playback in a particular system except for omission of the recording signal. *Notes:* (1) Magnetic recording media are considered to be in the zero-modulation state when they have been subjected to the normal erase, bias, and duplication printing fields characteristic of the particular system with no recording signal applied. (2) Mechanical recording media are considered to be in the zero-modulation state when they have been recorded upon and processed in the customary specified manner to form the groove with no recording signal applied. (3) Optical recording media are considered to be in the zero-modulation state when all normal processes of recording and processing, including duplication, have been performed in the customary specified manner, but with no modulation input to the light modulator. *See also:* **noise (sound recording and reproducing system).** E191-0

zero offset (1) (industrial control). A control function for shifting the reference point in a control system. *See also:* **control system, feedback.** AS1-34E10
(2) (numerically controlled machines). A characteristic of a numerical machine control permitting the zero point on an axis to be shifted readily over a specified range. The control retains information on the location of the permanent zero. *See:* **floating zero.** *See also:* **numerically controlled machines.** EIA3B-34E12

zero output. (1) The voltage response obtained from a magnetic cell in a ZERO state by a reading or resetting process. (2) The integrated voltage response obtained from a magnetic cell in a ZERO state by a reading or resetting process. *See:* **ONE state; ZERO state.** *See also:* **static magnetic storage.** E163-0

zero-phase-sequence relay. A relay that responds to the zero-phase-sequence component of a three-phase input quantity. 37A100-31E6/31E11

zero-phase-sequence symmetrical components (unsymmetrical set of polyphase voltages or currents of *m* phases). That set of symmetrical components that have zero phase sequence. That is, the angular phase lag from each member to every other member is 0 radians. The members of this set will all reach their positive maxima simultaneously. The zero-phase-sequence symmetrical components for a three-phase set of unbalanced sinusoidal voltages ($m=3$) having the primitive period are represented by the equations

$$e_{a0} = e_{b0} = e_{c0} = (2)^{1/2} E_{a0}\cos(\omega t + \alpha_{a0})$$

derived from the equation of **symmetrical components (set of polyphase alternating voltages).** Since in this case $r=1$ for every component (of first harmonic), the third subscript is omitted. Then k is 0 for the zero sequence, and s takes on the values 1, 2, and 3 corresponding to phases *a, b,* and *c.* These voltages have no phase sequence since they all reach their positive maxima simultaneously. E270-0

zero-phase symmetrical set (1) (polyphase voltage). A symmetrical set of polyphase voltages in which the angular phase difference between successive members of the set is zero or a multiple of 2π radians. The equations of **symmetrical set (polyphase voltages)** represent a zero-phase symmetrical set of polyphase voltages if k/m is zero or an integer. (The symmetrical set of voltages represented by the equations of **symmetrical set of polyphase voltages** may be said to have zero-phase symmetry if k/m is zero or an integer (positive or negative).) *Note:* This definition may be applied to a two-phase four-wire or five-wire system if m is considered to be 4 instead of 2. E270-0
(2) (polyphase currents). This definition is obtained from the corresponding definitions for voltage by substituting the word current for voltage, and the symbol I for E and β for α wherever they appear. The subscripts are unaltered. E270-0

zero-power-factor saturation curve (zero-power-factor characteristic) (synchronous machine). The saturation curve of a machine supplying constant current with a power-factor of approximately zero, overexcited. *See also:* **synchronous machine.** 0-31E8

zero-power-factor test (synchronous machine). A no-load test in which the machine is overexcited and operates at a power-factor very close to zero. *See also:* **synchronous machine.** 0-31E8

zero-sequence impedance (rotating machinery). The quotient of the zero-sequence component of the voltage, assumed to be sinusoidal, supplied to a synchronous machine, and the zero-sequence component of the current at the same frequency. *See also:* **direct-axis synchronous reactance; synchronous machine.** 0-31E8

zero-sequence reactance (rotating machinery). The ratio of the fundamental component of reactive armature voltage, due to the fundamental zero-sequence component of armature current, to this component at rated frequency, the machine running at rated speed. *Note:* Unless otherwise specified, the value of zero-sequence reactance will be that corresponding to a zero-sequence current equal to rated armature current. *See also:* **direct-axis synchronous reactance; synchronous machine.** 42A10-31E8

zero-sequence resistance. The ratio of the fundamental in-phase component of armature voltage, resulting from fundamental zero-sequence current, to this component of current at rated frequency. *See:* **synchronous machine.** 42A10-0

ZERO shift error. Error measured by the difference in deflectiion as between an initial position of the pointer, such as at zero, and the deflection after the instrument has remained deflected upscale for an extended length of time, expressed as a percentage of the end-scale deflection. *See also:* **moving element (instrument).** 39A1-0

ZERO state. A state wherein the magnetic flux has a negative value, when similarly determined. *See:* **ONE state.** *See also:* **static magnetic storage.** E163-0

zero-subcarrier chromaticity (color television). The chromaticity that is intended to be displayed when the subcarrier amplitude is zero. *See also:* **color terms.** E201-2E2

zero suppression. The elimination of nonsignificant zeros in a numeral. *See also:* **electronic digital computer.** X3A12-16E9

zero synchronization (numerically controlled machines). A technique that permits automatic recovery of a precise position after the machine axis has been approximately positioned by manual control. *See also:* **numerically controlled machines.** EIA3B-34E12

zero vector. A vector whose magnitude is zero. E270-0

zeta potential. *See:* **electrokinetic potential.**

zig-zag connection of polyphase circuits (zig-zag or interconnected star). The connection in star of polyphase windings, each branch of which is made up of windings that generate phase-displaced voltage. *See also:* **connections of polyphase circuits; polyphase circuits; polyphase systems.** 50I05-31E8

zig-zag leakage flux. The high-order harmonic air-gap flux attributable to the location of the coil sides in discrete slots. *See also:* **rotor (rotating machinery); stator.** 0-31E8

***Z* marker (zone marker) (electronic navigation).** A marker designed to radiate vertically and used to define a zone above a radio range station. *See also:* **navigation; radio navigation.** 42A65-10E6

zonal constant. A factor by which the mean candlepower emitted by a source of light in a given angular zone is multiplied to obtain the lumens in the zone. Z7A1-0

zonal factor interflection method (lighting calculation). A procedure for calculating coefficients of utilization that takes into consideration the ultimate disposition of luminous flux from every 10-degree zone from luminaires. *See also:* **inverse-square law (illuminating engineering).** Z7A1-0

zonal factor method (lighting calculation). A procedure for predetermining, from typical luminaire photometric data in discrete angular zones, the proportion of luminaire output that would be incident initially (without interreflections) on the work plane, ceiling, walls, and floor of a room. *See also:* **inverse-square law (illuminating engineering).** Z7A1-0

zoning (stepping) (lens or reflector). The displacement of various portions (called zones or steps) of the lens or surface of the reflector so that the resulting phase front in the near field remains unchanged. *See also:* **antenna (aerial).** 42A65-3E1

***z* transform, advanced (data processing).** The advanced z transform of $f(t)$ is the z transform of $f(t+\Delta T)$; that is,

$$\sum_{n=0}^{\infty} f(nT+\Delta T)z^{-n} \qquad 0<\Delta<1.$$

See also: **electronic digital computer.** 0-23E3

***z* transform, delayed (data processing).** The delayed z transform of $f(t)$, denoted $F(z,\Delta)$, is the z transform of $f(t-\Delta T)u(T-\Delta T)$, where $u(t)$ is the unit step function; that is,

$$F(z,\Delta)=\sum_{n=0}^{\infty} f(nT-\Delta T)u(nT-\Delta T)z^{-n} \qquad 0<\Delta<1.$$

See also: **electronic digital computer.** 0-23E3

***z* transform, modified (data processing).** The modified z transform of $f(t)$, denoted $F(z,m)$, is the delayed z transform of $f(t)$ with the substitution $\Delta=1-m$; that is,

$$F(z,m)=\sum_{n=0}^{\infty} f[nT-(1-m)T]\,u[nT-(1-m)T]\,z^{-n} \qquad 0<m<1.$$

See also: **electronic digital computer.** 0-23E3

zone (relay). *See:* **reach (relay).**

zone comparison protection. A form of pilot protection in which the response of fault-detector relays, adjusted to have a zone of response commensurate with the protected line section, is compared at each line terminal to determine whether a fault exists within the protected line section. 37A100-31E6/31E11

zone leveling (semiconductor processing). The passage of one or more molten zones along a semiconductor body for the purpose of uniformly distributing impuritites throughout the material. *See also:* **semiconductor device.** E102-0

zone of protection (1) (general). The part of an installation guarded by a certain protection. 37A100-31E11

(2) (relays). That segment of the power system in which the occurrence of assigned abnormal conditions should cause the protective relay system to operate. *See also:* **relay.** 0-31E6

zone punch. A punch in the 0, 11, or 12 row on a Hollerith punched card. 0-16E9

zone purification (semiconductor processing). The passage of one or more molten zones along a semiconductor for the purpose of reducing the impurity concentration of part of the ingot. *See also:* **semiconductor device.** E102-0

***z* transform, one-sided (data processing).** Let T be a fixed positive number, and let $f(t)$ be defined for $t\geq 0$. The z transform of $f(t)$ is the function

$$[f(t)]=F(z)=\sum_{n=0}^{\infty} f(nT)z^{-n}, \qquad \text{for } |z|>R=1/\rho$$

where ρ is the radius of convergence of the series and z is a complex variable. If $f(t)$ is discontinuous at some instant $t=kT$, k an integer, the value used for $f(kT)$ in the z transform is $f(kT^{+})$. The z transform for the sequence $\{f_n\}$ is:

$$[\{f_n\}]=F(z)=\sum_{n=0}^{\infty} f_n z^{-n}.$$

See also: **electronic digital computer.** 0-23E3

***z* transform, two-sided (data processing).** The two-sided z transform of $f(t)$ is

$$F(z)=\sum_{n=\infty}^{-1} f(nT)z^{-n}+\sum_{n=0}^{\infty} f(nT)z^{-n}$$

where the first summation is for $f(t)$ over all negative time and the second summation is for $f(t)$ over all positive time. *See also:* **electronic digital computer.** 0-23E3

Appendix A

If included in the Dictionary proper, the large number of alphabetically arranged references for each of the following terms could be confused with the main alphabetizing of all terms. To avoid this, the longer lists of cross references have been transferred to this Appendix. For convenience, the major terms appearing in the Appendix are given alone in the immediately following short list.

accuracy rating (instrument);
air-transportation instruments;
alternating-current instruments;
amplifier;
antenna;
arrester;
asynchronous machine;
auxiliary device to an instrument;
battery;
beam tube;
bearing;
brush;
burst;
cable;
center of distribution;
charge-storage tube;
circuit characteristics of electrodes;
circuits and devices;
color terms;
communication;
control;
control system;
control system, feedback;
corrosion terms;
cradle base;
data transmission;
dielectric heating;
direct-axis synchronous impedance;
direct-current commutating machine;
discharge;
electric arc-welding apparatus;
electric controller;
electric drive;
electricity meter;
electroacoustics;
electrochemistry;
electrode;
electrolytic cell;
electromagnetic compatibility;
electron devices, miscellaneous;
electron emission;
electronic analog computer;
electronic computation;
electronic controller;
electronic digital computer;
electron tube;
electron-tube admittances;
electroplating;
elevator;
gas-filled radiation-counter tube;
gas tube;
generating station;
ground;
grounding device;
induction heater;
instrument;
instrument transformer;
interference;
interior wiring;
inverse-square law;
lamp;
lightning arrester;
loudspeaker;
measurement system;
medical electronics;
microwave tube;
mining;
mobile communication system;
modulating systems;
navigation;
network analysis;
numerically controlled machine;
oscillatory circuit;
oscillograph;
phonograph pickup;
power distribution, overhead construction;
power distribution, underground construction;
power supply;
power system, low-frequency and surge testing;
protective signaling;
pulse;
pulse terms;
radar;
radiation;
radio transmission;
radio wave propagation;
railway signal and interlocking;
rating and testing magnetic amplifiers;
rectification;
relay;
reliability;
rotor;
semiconductor;
signal;
signal lighting;
static magnetic storage;
stator;
storage tube;
synchronous machine;
telegraphy;
telephone station;
telephone switching system;
telephone system;
television;
tower;
transducer;
transmission characteristics;
transmission line;
tube definitions;
visual field;
voltage regulator;
waveguide.

accuracy rating.
See:
accuracy;
apparent-power loss;
common-mode interference;
correction;
creep;
current loss;
damping;
damping factor;
dead band;
deflecting force;
deflection factor;
end-scale value;
error;
error and correction;
external field influence;
external temperature;
external-temperature influence;
frequency influence;
full-scale value;
influence;
inkwell influence;
interelement influence;
interference;
interference, differential-mode;
interference, normal-mode;
longitudinal interference;
magnetic field interference;
magnetic platform influence;
mean of reversed direct-current values;
operating influence;

spherical reflector;
spiral scanning;
squint angle;
steerable antenna;
surface-wave antenna;
tenth-power width;
theta (θ) polarization;
top-loaded vertical antenna;
tracking;
turnstile antenna;
unidirectional antenna;
uniform linear array;
unipole;
V antenna;
whip antenna;
zoning.

arrester.
See:
air terminal;
arrester discharge capacity;
arrester ground;
arrester, valve-type;
cable;
characteristic element;
conductor;
conformance tests;
copper-clad steel;
current rating, 60-hertz;
deflector;
design tests;
discharge counter;
discharge indicator;
expulsion element;
expulsion-type arrester;
fastener;
grading or control ring;
ground rod;
ground terminal;
lightning;
lightning arrester;
line terminal;
operating duty cycle;
prorated section;
protector tube;
routine tests;
series gap;
sideflash;
unit operation;
vent of an arrester.

asynchronous machine.
See:
accelerating;
acceptance test;
active current;
active power;
adjustable constant-speed motor;
adjustable-speed motor;
adjustable varying-speed motor;
admittance;
air;
air-cooled;
aircraft induction motor;
alternating-current commutator motor;
ampere-turn;
aperiodic component of short-circuit current;
aperiodic time constant;
apparent power;
armature coil;
armature winding;
asynchronous impedance;
asynchronous operation;
automatically regulated;
autotransformer starting;
auxiliary winding, single-phase induction motor;
B stage;
balanced condition;
base speed;
base value;
Blondel diagram;
blow-off valve;
breakaway;
breakaway torque;
breakdown;
breakdown torque speed;
breaking test;
brushless exciter;
calibrated-driving-machine test;
calorimetric test;
can loss;
capacitor braking;
capacitor motor;
capacitor start-and-run motor;
capacitor-start motor;
characteristic curves;
coil space factor;
collector;
collector ring;
commissioning test;
commutating-field winding;
commutator;
compensated repulsion motor;
compensated series-wound motor;
compensating-field winding;
compensator;
complex power;
compressor-stator blade-control system;
concentrated winding;
concentric winding;
connections of polyphase circuits;
constant-horsepower motor;
constant-speed motor;
constant-torque motor;
continuous duty;
continuous rating;
continuous voltage-rise test;
control exciter;
control winding;
controlled overvoltage test;
corona shielding;
crawling;
creepage surface;
current pulsation;
cyclic duration factor;
cyclic irregularity;
deadband;
definite-purpose motor;
direct-axis component of magnetomotive force;
direct-axis subtransient voltage;
direct-axis synchronous impedance;
direct-connected exciter;
direct on-line starting;
direct-voltage high-potential test;
discharge energy test;
discharge extinction voltage;
discharge inception test;
discharge inception voltage;
double-fed asynchronous machine;
double squirrel cage;
dripproof machine;
dust-ignitionproof machine;
dust seal;
duty;
duty cycle;
duty-cycle rating;
dynamic braking;
dynamometer test;
eddy-current braking;
effective number of turns per phase;
efficiency;
elastomer;
electric braking;
electric generator;
electric motor;
electromagnetic braking;
encapsulated;
enclosed ventilated;
end winding;
equalizer;
excitation response;
excite;
exciter, alternating-current;
explosionproof machine;
facing;
fan duty resistor;
field coil;
field-lead insulation;
field pole;
field-spool;
field system;
field-turn insulation;
film;
fluctuating power;
fluid loss;
form-wound motorette;

fractional-horsepower motor;
fractional-slot winding;
frequency converter, commutator-type;
front of a motor or generator;
full-pitch winding;
gasproof or vaporproof;
general-purpose encloser;
general-purpose induction motor;
general-purpose motor;
generated voltage;
generator;
generator, alternating-current;
generator motor;
generator set;
generette;
graded-time-step-voltage test;
ground insulation;
guarded machine;
harmonic test;
heteropolar machine;
high-potential test;
hydro-generator;
ideal paralleling;
impedance;
impulse test;
inching;
induction generator;
induction machine;
induction motor;
inherent regulation;
initial excitation response;
input;
insulating material;
insulation power factor;
insulation resistance test;
insulation resistance versus voltage test;
integral-horsepower motor;
intermittent duty;
intermittent-duty rating;
internal impedance;
internal impedance drop;
interspersing;
interturn insulation;
inverted;
isochronous speed governing;
Kramer system;
lap winding;
line terminal;
linkage voltage test, direct-current;
load;
load limit changer;
locked-rotor current;
locked-rotor test;
locked-rotor torque;
loss tangent;
loss tangent test;
loss, total;
main load;
main terminal;
main winding;
maximum continuous rating;
maximum instantaneous fuel change;
maximum rate of fuel change;
mechanical fatigue test;
minimum fuel limiter;
modified Kramer system;
motor;
motor reduction unit;
motor synchronizing;
motorette;
multispeed motor;
negative phase-sequence impedance;
negative phase-sequence reactance;
negative phase-sequence resistance;
neutral;
neutral lead;
neutral terminal;
noise-level test;
no-load;
nonsalient pole;
ohmic resistance test;
oilproof enclosure;
open-ended coil;
open externally ventilated machine;
open machine;
open pipe-ventilated machine;
operating conditions;
oscillation;
output;
overcurrent;
overspeed and overtemperature protection system;
overspeed governor;
overtemperature detector;
overvoltage;
overvoltage test;
paralleling;
part-winding starting;
peak load;
peak switching-current;
performance tests;
periodic rating;
permanent-split capacitor motor;
per-unit quantity;
per-unit system;
phase coil insulation;
phase sequence;
phase-sequence test;
phase voltage of a winding;
phasor power;
plug braking;
polarity test;
polarization-index test;
pole body;
pole-body insulation;
pole end-plate;
pole face;
pole-face bevel;
polyphase machine;
positive phase-sequence reactance;
positive sequence impedance;
power factor;
power-factor angle;
power-factor tip-up;
power-factor tip-up test;
power-factor voltage characteristic;
preformed winding;
proof test;
pull-out test;
pull-up torque;
pump-back test;
quadrature-axis component of magnetomotive force;
quadrature axis transient impedance;
quadrature-axis transient voltage;
random paralleling;
rated-load torque;
rated power output;
rated speed;
rating;
rotating test;
reactive current;
reactor-start motor;
reactor starting;
recovery time;
regenerative braking;
regulation;
repulsion-start induction motor;
resistance grading;
resistance-start motor;
resistant;
rest and de-energized;
retardation test;
reversible motor;
reversing motor;
rotating amplifier;
rotor-resistance starting;
routine test;
running operation;
salient-pole machine;
sampling tests;
saturation factor;
Scherbius machine;
sealed;
secondary;
secondary voltage;
secondary winding;
self-ventilated:
semiguarded machine;
semiprotected enclosure;
separately ventilated;

series-parallel starting;
series-wound;
series-wound motor;
service factor;
shaded-pole motor;
shell-type motor;
short-circuiter;
short-circuit loss;
short-circuit time constant of primary winding;
short field;
short-time rating;
single-layer winding;
single-phase machine;
single-phase motor;
single-phase rotary machine;
single phasing;
single-winding multispeed motor;
slip;
slot discharge;
slot-discharge analyzer;
slot insulating;
space factor;
special-purpose motor;
speed changer;
speed-changer high-speed stop;
speed-governing system;
speed governor;
speed-regulation characteristic;
splashproof machine;
split brush;
split-phase motor;
split-throw winding;
squirrel-cage induction motor;
squirrel-cage winding;
stability;
standard dimensioned motor;
standstill locking;
star-delta starting;
starting;
starting capacitance;
starting current;
starting motor;
starting operating;
starting-switch assembly;
starting test;
starting winding;
static Kramer system;
stator coil;
stator-resistance starting;
steady-state condition;
steady-state governing speed-band;
steady-state incremental speed regulation;
steady-state speed regulation;
steady-state value;
step-voltage;
submersible;
subsynchronous reluctance motor;
supply line, motor;
sustained oscillation;
switching;
switching torque;
synchronism;
synchronizing;
synchronizing coefficient;
synchronous motor;
temperature coefficient;
temperature coefficient of resistance;
temperature control system;
temperature detectors;
temperature relays;
temperature-rise test;
thermal endurance;
thermal equilibrium;
thermal-mechanical cycling;
torque motor;
totally enclosed fan-cooled machine;
totally enclosed machine;
totally enclosed nonventilated machine;
totally enclosed pipe-ventilated machine;
transformation impedance or admittance;
transient speed deviation;
transposition;
triplen;
turbine;
turbine nozzle control system;
turbine type;
two-value capacitor motor;
type tests;
universal motor;
universal motor parts;
V-curve;
variable-torque motor;
varying duty;
varying-speed motor;
very-low-frequency high-potential test;
voltage drop;
voltage endurance;
voltage endurance test;
voltage regulation;
voltage response, exciter;
voltage withstand test;
water-air-cooled machine;
water-cooled machine;
waterproof machine;
waveform test;
wave winding;
weather-protected machine;
wound rotor;
wound-rotor induction motor.

auxiliary device to an instrument.
See:
compensatory leads;
connector;
connector precision;
directivity of a directional coupler;
detector;
immittance comparator;
instrument multiplier;
instrument shunt;
line-drop voltmeter compensator;
measurement voltage divider;
measuring modulator;
phase shifter;
phase-shifting transformer;
pulse amplifier or relay;
pulse devices;
pulse initiator;
pulse initiator gear ratio;
pulse initiator ratio;
pulse initiator shaft reduction;
pulse rate;
pulse receiver;
rotatable phase-adjusting transformer;
series resistor;
shunt leads;
signal generator;
slotted section;
standard cell;
standard resistor;
totalizing relay;
transfer standards;
transformer-loss compensator;
voltage range multiplier;
warble-tone generator.

battery.
See:
A battery;
active materials;
alkaline storage battery;
ampere-hour capacity;
average voltage;
battery, electric;
battery eliminator;
battery, power station, communications;
battery, power station, control;
B battery;
case;
C battery;
cell connector;
charge;
closed-circuit voltage;
continuous test;
counter electromotive force cells;
couple;
covered plate;
cutoff voltage;
delayed test;
drain;
Edison storage battery;
element;

emergency cells;
end cells;
Faure plate;
flashlight battery;
framed plate;
grid;
group;
helical plate;
initial output test;
initial voltage;
intermittent test;
internal resistance;
ironclad plate;
jar;
lateral insulator;
lead storage battery;
local action;
negative plate;
negative terminal;
nickel-cadmium storage battery;
open-circuit voltage;
parallel connection;
pasted sintered plate;
pilot cell;
Plante plate;
pocket-type plate;
polarity;
polarization;
portable battery;
positive plate;
positive terminal;
separator;
series connection;
series-parallel connection;
service capacity;
service life;
service test;
shelf depreciation;
shelf test;
silver storage battery;
stationary battery;
steel container;
storage battery;
tank;
terminals;
time rate;
tray;
tube-type plate;
Tudor plate;
watthour capacity.

beam tube.
See:
angle, maximum-deflection;
astigmatism;
backplate;
beam alignment;
beam bending;
beam-deflection tube;
beam-indexing color tube;
beam modulation, percentage;
beam power tube;
cathode-ray tube;
charge-storage tube;
cloud pulse;
color cell;
color center;
color-field corrector;
color-picture tube;
color plane;
color-purity magnet;
color-selecting-electrode system transmission;
color triad;
convergence;
convergence, dynamic;
convergence, electrode;
convergence magnet;
convergence plane;
convergence surface;
deflection center;
deflection factor;
deflection plane;
deflection sensitivity;
deflection yoke;
deflection-yoke pull-back;
density;
distortion, barrel;
distortion, keystone;
distortion, pincushion;
distortion, spiral;
drift space;
dynode spots;
electrode dark current;
electron beam;
electron gun;
electron-gun density multiplication;
electron-wave tube;
electrostatic focusing;
focusing;
focusing, dynamic;
focusing magnet;
forward wave;
gamma;
gas focusing;
insertion gain;
interaction gap;
interaction impedance;
ion spot;
knee of transfer characteristic;
lag;
magnetic focusing;
monoscope;
moire;
optimum bunching;
oscilloscope tube;
output gap;
overbunching;
peak-alternation gap-voltage;
persistence characteristic;
porthold;
positional crosstalk;
post-acceleration;
primary-color tube unit;
raster;
raster burn;
reflection color-tube;
reflex bunching;
repeller;
resolution;
resolution, structural;
retained image;
scanning, high-velocity;
scanning, low-velocity;
sensitivity;
sensitivity, radiant;
shading;
shadow mask;
signal;
signal output current;
signal-to-noise ratio;
space-charge debunching;
spectral characteristic;
spectral-sensitivity characteristic;
spot;
square-wave response;
square-wave response characteristic;
target;
target capacitance;
target cutoff voltage;
target voltage;
transfer characteristic;
trans-μ factor;
transverse-beam traveling-wave tube;
velocity-modulated tube.

bearing.
See:
antifriction bearing;
ball bearing;
bearing bracket;
bearing cap;
bearing cartridge;
bearing clearance;
bearing dust cap;
bearing housing;
bearing indicating thermometer;
bearing insulation;
bearing liner;
bearing lining;
bearing locknut;
bearing lock washer;
bearing pedestal;
bearing pedestal cap;
bearing seat;
bearing shell;
bearing temperature detector;
bearing temperature relay;
bearing thermometer;
bottom half bearing;
cartridge-type bearing;
disc- and wiper-lubricated bearing;
end rail;

feed groove;
flood-lubricated bearing;
guide bearing;
insulated bearing;
insulated bearing-housing;
insulated bearing-pedestal;
Jordan bearing;
journal;
journal bearing;
labyrinth seal ring;
locating bearing;
oil-lift bearing;
oil-ring lubricated bearing;
pad-type bearing;
plug-in type bearing;
pressure-lubricated bearing;
roller bearing;
segment shoe;
self-aligning bearing;
self-lubrication bearing;
sleeve;
spherical seated bearing;
spherical support seat;
split-sleeve bearing;
spring-loaded bearing;
straight-seated bearing;
thrust bearing;
thrust block;
tilting-pad bearing;
top half bearing;
upper guide bearing;
upper half bearing bracket;
wick lubricated bearing.

brush.
See:
beveled brush corners;
beveled brush edges;
beveled brush ends and toes;
bronze leaf brush;
brush box;
brush chamfers;
brush contact loss;
brush corners;
brush diameter;
brush edges;
brush ends;
brush friction loss;
brush hammer, lifting or guide clips;
brush holder;
brush-holder bolt insulation;
brush-holder insulating barriers;
brush-holder spindle insulation;
brush-holder spring;
brush-holder stud;
brush-holder stud insulation;
brush-holder support;
brush-holder yoke;
brush length;
brushless;
brush rigging;
brush rocker;
brush-rocker gear;
brush shoulders;
brush shunt;
brush-shunt length;
brush sides;
brush slots, grooves, and notches;
brush thickness;
brush width;
carbon brush;
carbon-graphite brush;
convex and concave brush ends;
copper brush;
electrographitic brush;
fractional-horsepower brush;
graphite brush;
headed brushes;
industrial brush;
metal-graphite brush;
metallized brush;
miniature brush;
paired brushes;
split brush.

burst.
See:
average absolute burst magnitude;
bandwidth;
burst build-up interval;
burst decay interval;
burst duration;
burst duty factor;
burst fall-off interval;
burst leading edge time;
burst quiet interval;
burst repetition rate;
burst rise interval;
burst safeguard interval;
burst spacing;
burst trailing edge time;
burst train;
center frequency;
convolution function;
electroacoustics ;
energy density spectrum;
instantaneous burst magnitude;
interpolation function;
lower burst reference;
modified autocorrelation function;
peak burst magnitude;
pulse;
root-mean-square burst magnitude;
upper burst reference.

cable.
See:
aerial cable;
arresters;
block cable;
buried cable;
cable complement;
cable fill;
cable splice;
cable terminal;
coaxial;
coaxial cable;
composite cable;
conduit;
cord;
distribution cable;
drop wave;
duct;
feeder cable;
house cable;
line fill;
multiple twin cable;
pair;
paired cable;
quad;
quadded cable;
riser cable;
span;
spiral four;
storm loading;
stranded wire;
stub;
subsidiary conduit;
suspension strand;
underground cable;
unit cable construction.

center of distribution.
See:
alternating-current distribution;
balanced circuit;
circuit;
common return;
control-metering point;
current-carrying part;
direct-current distribution;
direct-current neutral grid;
direct feeder;
distribution center;
distribution feeder;
distribution main;
distribution trunk line;
duplicate lines;
electric supply lines;
energy metering point;
fault;
fault current;
feeder;
feeding point;
floating neutral;
guarded;
interconnection tie;
leakage current;
leg of a circuit;
loop feeder;
mains;
multiple circuit;
multiple feeder;
negative conductor;
network feeder;
neutral conductor;
neutral point;
parallel feeder;

pilot wire;
polyphase circuit;
positive conductor;
primary distribution mains;
primary distribution trunk line;
primary fault;
primary transmission feeder;
quarter-phase or two-phase circuit;
radial distribution feeder;
radial feeder;
radial transmission feeder;
secondary distribution feeder;
secondary distribution mains;
secondary distribution trunk line;
secondary fault;
secondary neutral grid;
series circuit;
short-circuit;
single feeder;
single-phase circuit;
six-phase circuit;
switchboard;
teed feeder;
three-phase circuit;
tie feeder;
tie point (electric power systems);
transmission feeder;
trunk feeder;
two-wire circuit.

charge-storage tube.
See:
abnormal decay;
barrier grid;
beam tubes;
bistable operation;
cathode-ray;
cloud pulse;
crossover voltage, secondary emission;
crosstalk, electron beam;
decay, static;
destructive reading;
escape ratio;
flood;
hold;
ion charging;
ion repeller;
nondestructive reading;
overwriting;
prime;
priming rate;
priming speed;
re-distribution;
reflection modulation;
storage element;
storage-element equilibrium voltage;
storage-element equilibrium voltage, cathode;
storage-element equilibrium voltage, collector;
storage-element equilibrium voltage gradient;
storage-element equilibrium voltage second-cross-over.

circuit characteristics of electrodes.
See:
amplification factor;
angle, maximum-deflection;
capacitance;
capacitance, output;
cathode coating impedance;
cathode interface capacitance;
cathode interface impedance;
cathode interface resistance;
cathode preheating time;
circuit efficiency;
composite controlling voltage;
conductance for rectification;
constant-current characteristic;
control electrode;
control-electrode discharge recovery time;
convection current;
convection current modulation;
coupling coefficient, small-signal;
decay;
decay time;
diode characteristic;
direct-current electron-stream resistance;
drift rate;
drift space;
dynamic characteristic;
electrode capacitance;
electrode characteristic;
electrode-current averaging time;
electron efficiency;
electron stream potential;
electron stream transmission efficiency;
electronic efficiency;
equivalent diode;
equivalent noise current;
equivalent noise resistance;
fiber-optic plate;
flection-point emission current;
frequency pulling;
grid-drive characteristic;
grid-driving power;
heterodyne conversion transducer;
hiss;
inflection-point emission current;
input capacitance;
interelectrode capacitance;
load characteristic;
microchannel plate;
microphonism;
noise temperature;
ON period;
operation time;
perveance;
radiation;
rectification factor;
repeatability;
reverse emission;
space-charge-limited current;
starting current;
static characteristic;
thermal tuning time (cooling);
thermal tuning time (heating);
transfer characteristic;
transrectification factor;
tuning, electronic;
tuning range, electronic;
tuning rate, thermal;
tuning sensitivity, electronic;
tuning sensitivity, thermal;
unloaded Q:
virtual cathode;
voltage generator;
work function;
writing speed.

circuits and devices.
See:
amplifier;
annunciator;
artificial load;
audio frequency transformer;
automatic frequency control;
automatic gain control;
automatic grid bias;
automatic volume control;
banana plug;
band switch;
bank winding;
bass boost;
bleeder;
blocking capacitor;
bootstrap circuit;
bridging connection;
building-out capacitor;
buzzer;
bypass capacitor;
carrier isolating choke coil;
cathode follower;
catwhisker;
chopper;
circuit;
circuit element;
clamping circuit;
clipper;
closed-circuit signaling;
coder;
component;
composite set;
converter;
cord;
crystal;
crystal filter;
cue circuit;

de-emphasis;
delay line;
delay pickoff;
detector;
device;
differentiator;
direct-current restorer;
discriminator;
double-tuned circuit;
driver;
drop;
electric bell;
electromagnetic relay;
equalizer;
excitation (drive);
filter;
forward current;
forward voltage;
four-wire terminating set;
free-running frequency;
frequency divider;
frequency doubler;
frequency monitor;
frequency multiplier;
frequency tripler;
gate;
grounded-cathode amplifier;
honeycomb coil;
hybrid coil;
hybrid set;
inductor;
integrator;
jack;
light valve;
limiter;
line transformer;
litz wire;
load;
load circuit;
locking-in;
magnetic amplifier;
mixer;
network analysis;
oscillatory circuit;
pancake coil;
peak limiter;
peaking circuit;
phase shifter;
phase splitter;
pilot lamp;
pilot light;
pin jack;
plate load impedance;
plug;
power pack;
pre-emphasis;
preselector;
printed circuit board;
printed circuit board, double-sided;
printed circuit board, single-sided;
push-pull circuit;
push-push circuit;
reactor;
rectifier;
reflex circuit;
regenerative divider;
resistance lamp;
reverse current;
reverse voltage;
sampling circuit;
single-tuned circuit;
slicer;
solid-state;
source;
spark capacitor;
standard volume indicator;
synchronous gate;
system;
tank circuit;
thermistor;
time gate;
transducer;
transformer, ideal;
transformer, isolating;
trigger;
trigger circuit;
trimmer capacitor;
vacuum switch;
varindor;
variolosser;
variometer;
varistor;
velocity modulation;
voltage divider;
volume limiter.

color terms.
See:
achromatic locus (achromatic region);
black and white;
blackbody locus;
brightness;
brightness of a perceived light-source;
CIE;
CIE standard chromaticity diagram;
CIE standard colorimetric observer;
candle;
candlepower;
chromaticity;
chromaticity coordinate;
chromaticity diagram;
chromaticity flicker;
chrominance demodulator;
chrominance modulator;
chrominance primary;
chrominance signal;
chrominance subcarrier;
coarse chrominance primary;
color;
color breakup;
color burst;
color coder;
color contamination;
color coordinate transformation;
color decoder;
color-difference signal;
color flicker;
color fringing;
color match;
color mixture;
color picture signal;
color signal;
color temperature;
color transmission;
color triangle;
colorimetric purity;
colorimetry;
compatibility;
complementary wavelength;
composite color signal;
composite color synchronization;
constant luminance transmission;
correlated color temperature;
display primaries;
distribution coefficients;
distribution temperature of a light-source;
dominant wavelength;
dot-sequential;
equal-energy source;
excitation purity;
field-sequential;
fine chrominance primary;
flicker;
footcandle;
footlambert;
frequency interlace;
gamma;
gamma correction;
hue;
illuminance;
lambert;
light;
lightness;
light-source color;
lumen;
luminance;
luminance flicker;
luminance primary;
luminance signal;
luminosity;
luminosity coefficients;
luminous efficiency;
luminous flux;
luminous intensity;
matrix;
matrix unit;
metamer;
mixed highs;
moiré;
monochromatic;
monochrome;

monochrome signal;
Munsell chroma;
Munsell color system;
Munsell hue;
Munsell value;
narrow-band axis;
nonphysical primary;
object color;
perceived light-source color;
perceived object color;
photometry;
pickup spectral characteristic;
Planckian locus;
primaries;
primary;
purity;
purple boundary;
radiance;
radiant flux;
radiant intensity;
reference white;
relative luminosity;
saturation;
specified achromatic lights;
spectral characteristic;
spectral tristimulus values;
spectrum locus;
standard observer;
transmission primaries;
tristimulus values;
white;
white object;
wide-band axis;
zero-subcarrier chromaticity.

communication.
See:
building out;
carrier telephony;
communication;
communication lines;
communication theory;
compensation theorem;
facility;
facsimile transmission;
Foster's reactance theorem;
full duplex;
full-duplex operation;
half duplex;
inside plant;
Norton's theorem;
outside plant;
picture transmission;
program;
quantization;
reciprocity theorem;
resolution;
signal;
signal wave;
speed of transmission;
start-dialing signal;
synchronous system;
telautograph;
Thevenin's theorem;
tracking;
translation;
transmission line;
transmission system;
voice-frequency telephony;
wire broadcasting.

control.
See:
adjustable varying-voltage control;
adjustable-voltage control;
armature-voltage control;
automatic control;
automatic elevators;
braking capacity;
car-switch automatic floor-stop operation;
car-switch operation;
closed-loop control system;
computer control state;
continuous-pressure operation;
control;
control apparatus;
control circuit;
control-circuit limit switch;
control-circuit transformer;
control cutout switch;
control desk;
control device;
control system;
controller;
critically damped;
current limiter;
current-limiting reactor;
deadman's release;
delay line;
device;
electronic digital computer;
elevator;
elevator automatic dispatching device;
elevator-car-flash signal device;
elevator-landing stopping device;
elevator parking device;
elevator separate-signal system;
elevator signal transfer switch;
elevator starter's control panel;
elevator truck zone;
elevator truck-zoning device;
emergency stop switch;
emergency-terminal stopping device;
excitation-system stability;
field control;
field-discharge protection;
field forcing;
field-limiting adjusting means;
field protection;
final-terminal stopping device;
final value;
fuse;
gate;
generator-field control;
hoistway access switch;
illustrative diagram;
indicating circuit;
industrial control;
load circuit;
load rheostat;
machine final-terminal stopping device;
manual control;
motor-field control;
multivoltage control;
nonselective collective automatic operation;
nonstop switch;
normal-terminal stopping device;
numerical control;
open-loop control system;
operating device;
operation;
overdamped;
pilot circuit;
position indicator;
pre-register operation;
pulse;
remote control;
rheostatic control;
selective collective automatic operation;
semienclosed break;
sequential control;
short-time duty;
signal operation;
single automatic operation;
slack-rope switch;
speed-regulating rheostat;
transducer;
transfer function;
traveling cable;
two-speed alternating-current control;
two-wire control;
underdamped;
varying-voltage control;
waiting-passenger indicator.

control system.
See:
adjoint system;
admissible control input set;
constraint;
controllable;
controllable, completely;
control law;
control law, closed-loop;
control law, open-loop;
controller;
control problem;
control system, feedback;
co-state;
equilibrium point;
loss function;
Lyapunov function;

resolution;
response, sinusoidal;
response, steady-state;
response time;
rise time;
root locus;
running tension control;
S-plane;
sampled data;
sampled data control system;
sampling interval;
sampling period;
saturation;
scanner;
servomechanism;
servomechanism type number;
set point;
settling time;
signal;
signal, error;
signal, input;
signal, output;
signal, reference-input;
signal, return;
signal, unit-impulse;
signal, unit-ramp;
signal, unit-step;
speed limit;
speed ratio control;
stability;
stability, conditional;
stability, relative;
stabilization;
stable;
steady state;
step change;
stiction;
stiffness;
stiffness coefficient;
stimulus;
suicide control;
summing point;
switching amplifier;
system, directly controlled;
system, idealized;
system, indirectly controlled;
system overshoot;
time constant;
time delay;
time response;
transducer;
transfer locus;
transient;
transient overshoot;
transport time;
ultimately controlled variable;
underdamped;
unilateral connection;
unstable;
value;
value, ideal;
variable;
variable, input;
variable, output;
vernier control;
viscous friction;
voltage limit;
zero;
zero error;
zero offset.

corrosion terms.
See:
active;
aggressive carbon dioxide;
anaerobic;
anode corrosion efficiency;
anti-fouling;
austenitic;
cathode protection;
caustic embrittlement;
cavitation damage;
chalking;
chemical conversion coating;
corrosion;
corrosion fatigue;
corrosion fatigue limit;
corrosion rate;
crevice corrosion;
critical humidity;
deactivation;
demineralization;
deposit attack;
drainage;
electrolytic cleaning;
embrittlement;
endurance limit;
erosion;
exfoliation;
fatigue;
ferritic;
fouling;
fretting;
graphitic corrosion;
graphitization;
impingement attack;
interdendritic corrosion;
intergranular corrosion;
ion;
mill scale;
oxidation;
parting;
parting limit;
passivation;
passivator;
patina;
pitting;
pitting factor;
reduction;
rust;
scaling;
season cracking;
slush compound;
spalling;
stray-current corrosion;
stress-accelerated corrosion;
subsurface corrosion;
tarnish;
tuberculation;
underfilm corrosion;
weld decay.

cradle base.
See:
air duct;
air guide;
air-insulated terminal box;
air opening;
air pipe;
air shield;
air trunking;
auxiliary lead;
auxiliary terminal;
backing;
barrier;
base;
belt;
binder;
bonded motor;
brushing, insulating;
building bolt;
building pin;
bushing, electrical;
cable coupler;
canned;
carbon-dioxide system;
closed air circuit;
conducting paint;
conductive coating;
conductor-cooled;
conventionally cooled;
cooler;
cooling coil;
cooling medium;
cooling water system;
cover;
cover plate;
cushion;
cushion clamp;
D dimension of a motor;
discharge opening;
dowel;
drive strip;
duct spacer;
duct ventilated;
end bracket;
end-winding cover;
end-winding support;
exciter dome;
exciter platform;
eye bolt;
fan shroud;
flameproof terminal box;
flash barrier;
flexible mounting;
forced-lubricated bearing;
foundation bolt;
foundation bolt cone;
frame ring;
frame split;
frame work;

frame yoke;
gas;
gas dryer;
gas seal;
gas system;
gland seal;
grounding cable;
grounding cable connector;
grout;
heat sink;
holding-down bolt;
horizontal machine;
hose;
housing;
housing fan;
hydrogen-cooled machine;
imbedded temperature-detector insulation;
impregnant;
impregnate;
impregnation, winding;
inner frame;
inspection opening;
insulation sleeving;
intake opening;
jack bolt;
keyway;
lead box;
leveling plate;
lifting bar;
lifting hole;
liner;
locator joint;
locking ring;
lockplate;
lower bracket;
lower guide bearing;
lower-half bearing bracket;
mechanical part;
motor parts;
mounting ring;
nameplate;
open terminal box;
orifice plate;
outer frame;
package, core;
pancake motor;
pedestal bearing insulation;
phase-insulated terminal box;
phase-segregated terminal box;
phase-separated terminal box;
pilot fit;
pit;
pole bolt;
port;
pressure-containing thermal box;
pressure-relief terminal box;
rail;
relief door;
screen;
semiconducting plate;
separate terminal enclosure;
shield;
shipping brace;
side panel;
skeleton frame;
space heater;
stop dowel;
strip terminals;
submersible;
tapered key;
terminal block;
through bolt;
turbine end;
upper bracket;
vent;
vent plate;
ventilating duct;
ventilating passage;
ventilating slot;
vertical machine;
vibrating test;
water-air-cooled machine;
wedge;
wedge groove;
wedge washer;
winding shield.

data transmission.
See:
accuracy;
address;
baud;
bit;
block;
error detecting and feedback system;
error detecting code;
error detecting system;
error rate;
format;
input-output;
interface;
internal bias;
local side;
modulo N check;
nonsynchronous transmission;
parallel transmission;
pulse rise time;
pulse spacing;
pulse width;
receive-only equipment;
redundancy;
redundancy check;
redundant code;
regenerative repeater;
repeater;
send-only equipment;
serial transmission;
signal;
signal element;
start signal;
start-stop system;
stop signal;
transition;
undetected error rate.

dielectric heating.
See:
applicators or electrodes;
autoregulation induction heater;
coil Q;
coil shape factor;
Colpitts oscillator;
coreless type induction heater or furnace;
core-type induction heater or furnace;
coupling;
Curie point;
decalescent point;
depth of current penetration;
depth of heating;
dielectric constant;
dielectric dissipation factor;
dielectric heaters;
dielectric phase angle;
dielectric power factor;
dielectric strength;
glue-line heating;
industrial electronics;
interference measurement;
load circuit;
load circuit efficiency;
loaded applicator impedance;
load leads or transmission line;
load matching;
load matching network;
load-matching switch;
load or charge;
load transfer switch;
loss factor;
low-frequency induction heater or furnace;
normal dielectric heating applications;
over-all electrical efficiency;
pad electrode;
radio-frequency generator;
recalescent point;
shield;
stray field dielectric applications;
unloaded applicator impedance;
wave heaters;
wave heating.

direct-axis synchronous impedance.
See:
asynchronous machine;
deviation factor of a wave;
direct-axis component of magnetomotive force;
direct-axis subtransient open-circuit constant;
direct-axis synchronous reactance;

high-potential test;
hydro-generator;
impulse test;
inching;
inertia consant;
inherent regulation;
initial excitation response;
insulating material;
insulation power factor;
insulation resistance;
insulation-resistance test;
insulation-resistance versus voltage test;
integral-horsepower motor;
intermittent duty;
interturn insulation;
isochronous speed governor;
lap winding;
linkage voltage test;
load;
load limit changer;
locked-rotor current of a motor;
locked-rotor test;
locked-rotor torque;
loss tangent;
loss-tangent test;
loss, total;
magnetic loading;
magnetoelectric generator;
main exciter;
main exciter response ratio;
maximum continuous rating;
maximum instantaneous fuel change;
mechanical fatigue test;
minimum fuel limiter;
motor;
motor reduction unit;
motorette;
multiple-current generator;
multispeed motor;
noise level test;
no-load;
no-load field voltage;
nominal band of regulated voltage;
oilproof enclosure;
open-ended coil;
open externally ventilated machine;
open machine;
open pipe-ventilated machine;
operating conditions;
oscillation;
output;
over-compounded;
overcurrent;
overspeed and overtemperature protection system;
overspeed governor;
overtemperature detector;
overvoltage;
overvoltage test;
paralleling;
peak load;
peak switching-current;
per-unit quantity;
per-unit system;
performance tests;
periodic rating;
pilot exciter;
polarity test;
polarization index test;
pole body;
pole-body insulation;
pole end-plate;
pole face;
pole-face bevel;
preformed winding;
proof test;
pump-back test;
random-wound motorette;
rated-load field voltage;
rated-load torque;
rated power output;
rated speed;
rating;
recovery time;
regenerative braking;
regulation;
regulation curve;
resistance grading;
resistant;
rest and deenergized;
retardation test;
reversible motor;
reversing motor;
rotating test;
routine test;
sampling tests;
saturation curve;
saturation factor;
sealed;
self-excited;
self-ventilated;
semiguarded machine;
semiprotected enclosure;
separately excited;
separately ventilated;
series wound;
series-wound motor;
service factor;
shell-type motor;
short-circuit loss;
short field;
short-time duty;
short-time rating;
shunt-wound;
shunt-wound generator;
shunt-wound motor;
single-layer winding;
single winding multispeed motor;
slot discharge;
slot-discharge analyzer;
slot insulation;
slot space factor;
space factor;
special-purpose motor;
speed changer;
speed changer high-speed stop;
speed governing system;
speed governor;
speed regulation characteristic;
speed regulation;
splashproof machine;
split brush;
split-throw winding;
stability;
stabilized shunt-wound generator;
stabilized shunt-wound motor;
standard dimensioned motor;
starting;
starting current;
starting motor;
starting test;
starting winding;
stator coil;
steady-state condition;
steady-state governing speed-band;
steady-state incremental speed regulation;
steady-state speed regulation;
steady-state valve;
step-voltage test;
submersible;
supply line, motor;
sustained oscillations;
temperature coefficient;
temperature-coefficient of resistance;
temperature control system;
temperature detectors;
temperature relays;
temperature-rise test;
thermal endurance;
thermal equilibrium;
thermal-mechanical cycling;
torque motor;
totally enclosed fan-cooled machine;
totally enclosed nonventilated machine;
totally enclosed pipe-ventilated machine;
transformation impedence or admittance;
transient speed deviation;
transposition;
triplen;
tubing;
turbine-driven generator;
turbine-generator unit;
turbine nozzle control system;

stalled torque control;
step speed adjustment;
tandem drive;
thread;
variable-speed drive.

electricity meter.
See:
ampere-hour meter;
commutator motor meter;
coulometer;
cumulative demand meter;
demand interval;
demand meter;
excess meter;
gear ratio;
indicating demand meter;
induction-motor meter;
instrument;
integrated-demand meter;
lagged demand meter;
mercury motor meter;
motor meter;
multirate meter;
percentage error;
percentage registration;
portable standard meter;
printing demand meter;
recording demand meter;
reference performance;
register constant;
register ratio;
register reading;
registration;
two-rate meter;
varhour meter;
volt-ampere-hour meter;
watthour constant;
watthour-demand meter;
watthour meter;
wattsecond constant.

electroacoustics.
See:
absorption loss;
acceleration;
acetate disks;
acoustic, acoustical;
acoustical ohm;
acoustical reciprocity theorem;
acoustical units;
acoustic compliance;
acoustic dispersion;
acoustic impedance;
acoustic interferometer;
acoustic mass;
acoustic radiating element;
acoustic radiation pressure;
acoustic radiometer;
acoustic reactance;
acoustic refraction;
acoustic resistance;
acoustics;
acoustic scattering;
acoustic stiffness;
acoustic transmission system;
advance ball;
air conduction;
alternating-current erasing head;
alternating-current magnetic bias;
amplitude range;
anechoic chamber;
angular deviation loss;
applied shock;
articulation;
artificial ear;
attack time;
audio frequency;
audio-frequency spectrum;
audiogram;
audiometer;
auditory sensation area;
aural harmonic;
available power;
basic frequency;
blocked impedance;
bone conduction;
cavitation;
cent;
complex tone;
compressional wave;
dead room;
de-emphasis;
difference limen;
diffracted wave;
diffraction;
directivity index;
discrete sentence intelligibility;
discrete word intelligibility;
Doppler effect;
Doppler shift;
drive-pin hole;
echo;
electroacoustic transducer;
electromechanical recorder;
electrophonic effect;
energy density;
energy flux;
equally tempered scale;
flutter echo;
force factor;
free field;
free motional impedance;
free progressive wave;
frequency-response equalization;
fundamental frequency;
fundamental tone;
ground noise;
grouping;
H and D curve;
harmonic;
harmonic series of sounds;
hearing loss;
horn:
infrasonic frequency;
instantaneous sound pressure;
intensity level;
interval;
just scale;
lateral recording;
level above threshold;
light modulator;
loaded impedance;
loudspeaker;
masking;
maximum sound pressure;
mean free path;
mel;
microbar;
microphone;
mixer;
modulation noise;
motional impedance;
musical echo;
note;
octave;
octave-band pressure level;
opacity;
oscillation;
partial;
particle velocity;
peak sound pressure;
percent impairment of hearing;
phonograph pickup;
photographic sound recorder;
photographic sound reproducer;
pitch;
poid;
pressure spectrum level;
pretersonic;
principal axis;
program level;
Pythagorean scale;
rate of decay;
Rayleigh disk;
refraction loss;
relative response;
relative velocity;
reverberant sound;
reverberation;
reverberation chamber;
reverberation time;
reverberation time meter;
rumble;
sabin;
scattering loss;
Schlieren method;
scoring system;
semitone;
shading;
shear wave;
simple tone;
sound;
sound absorption;
sound-absorption coefficient;
sound analyzer;
sound articulation;

sound intensity;
sound level;
sound-level meter;
sound power;
sound-pressure level;
sound probe;
sound-reflection coefficient;
sound-spectrum-analyzer;
sound-transmission coefficient;
spectrum level;
split projector;
split transducer;
standard sea-water conditions;
standard tuning frequency;
static pressure;
steady-state oscillation;
streaming;
strength of a sound source;
syllable articulation;
threshold of audibility;
threshold of discomfort;
threshold of feeling;
threshold of pain;
timbre;
transducer equivalent noise pressure;
transient motion;
transmitting efficiency;
ultrasonic coagulation;
ultrasonic cross grating;
ultrasonic delay line;
ultrasonic detector;
ultrasonic frequency;
ultrasonic generator;
ultrasonic grating constant;
ultrasonic light diffraction;
ultrasonic material dispersion;
ultrasonic space grating;
ultrasonic stroboscope;
ultrasonics;
velocity level;
vertical recording;
vibration;
vibration meter;
volume indication;
volume velocity.

electrochemistry.
See:
activation polarization;
air cell;
anode efficiency;
anodic polarization;
battery;
calomel half-cell;
carbon-consuming cell;
cathode efficiency;
cathodic polarization;
caustic soda cell;
concentration cell;
concentration polarization;
cuprous chloride cell;
current efficiency;
decomposition potential;
depolarization;
depolarizer;
dichromate cell;
dry cell;
dynamic electrode potential;
electrochemical equivalent;
electrolytic cell;
electrolytic dissociation;
electrolytic oxidation;
electrolytic reduction;
electromotive series;
energy efficiency;
equilibrium electrode potential;
equilibrium reaction potential;
equivalent conductance;
Faraday;
fused-electrolyte cell;
galvanic cell;
gas cell;
irreversible process;
magnesium cell;
mercury oxide cell;
molar conductance;
one-fluid cell;
overvoltage;
passivity;
polarization;
polarizer;
primary battery;
primary cell;
quinhydrone half-cell;
reserve cell;
reversible process;
sal ammoniac cell;
silver oxide cell;
standard cell;
standard electrode potential;
storage cell;
thermal cell;
two-fluid cell;
unsaturated standard cell;
voltage efficiency;
Weston normal cell;
wet cell.

electrode.
See:
accelerating electrode;
anode;
backplate;
cathode;
cold cathode;
collector;
color-selecting-electrode system;
control electrode;
control grid;
convergence electrode;
decelerating electrode;
deflecting electrode;
dynode;
electron gun;
excitation anode;
filament;
focusing electrode;
grid;
hot cathode;
ignitor electrode;
indirectly heated cathode;
intensifier electrode;
ionic-heated cathode;
main anode;
modulating electrode;
photocathode;
plate;
pool cathode;
post-accelerating electrode;
relieving anode;
screen grid;
semitransparent photocathode;
shield grid;
signal electrode;
space-charge grid;
starter;
starters;
starting electrode;
suppressor grid;
target.

electrolytic cell.
See:
anode;
bag-type construction;
bath voltage;
battery;
bipolar electrode;
bobbin;
can;
cathode;
cell constant;
counter electromotive force;
creepage;
current density;
depolarizer;
depolarizing mix;
differential aeration cell;
electrochemistry;
electrode;
electrode potential;
electrolysis;
electrolyte;
flash current;
gassing;
half cell;
ir drop;
jacket;
liner;
local cell;
negative electrode;
nonlined construction;
oxygen concentration cell;
passive-active cell;
positive electrode;
surface corrosion;
shelf corrosion;
static electrode potential;
thermogalvanic corrosion;
tube.

secondary emission ratio;
secondary grid emission;
shot noise, full;
space charge;
space charge density;
thermal tuning time (cooling);
thermal tuning time (heating);
thermionic cathode;
thermionic emission;
thermionic grid emission;
transit angle;
transit time.

electronic analog computer.
See:
absolute-value device;
accuracy;
accuracy, dynamic;
accuracy, static;
adder;
adjoint;
amplifier;
amplifier, buffer;
amplifier, high-gain direct-current;
amplifier, integrating;
amplifier, inverting;
amplifier, operational;
amplifier, optional;
amplifier, relay;
amplifier, servo;
amplifier, summing;
amplifier, unloading;
amplitude selection;
analog;
analog computer;
analog computer, alternating-current;
analog computer, direct-current;
attentuator;
automatic hold;
automatic programming;
balance check;
balancing;
bandwidth;
bay;
board;
boldface;
boost;
bridge limiter;
check solution;
chopper;
clamping circuit;
coefficient potentiometer;
comparator;
component;
computer;
computer component;
computer control state;
computer diagram;
computer equation;
computer time;
computer variable;
computing element;
conformity;
control panel;
curve follower;
differential analyzer;
differentiator;
digital differential analyzer;
divider;
drift;
drift, zero;
dynamic check;
dynamic computer check;
dynamic problem check;
dynamic range;
error;
error, dynamic;
error, linearity;
error loading;
error matching;
error resolution;
error static;
feedback limiter;
full scale;
function generator;
function generator bivariant;
function generator curve follower;
function generator diode;
function generator map reader;
function generator servo;
function generator switch type;
function relay;
function switch;
grid current;
ground loop;
hard limiting;
hold;
impedance, feedback;
impedance input;
implicit computation;
initial condition;
input limiter;
input-output table;
integrator;
inverter;
leakage;
limit circuit;
linearity;
machine time;
malfunction;
mathematical mode;
mistake;
multiple;
multiplier;
multiplier, constant;
multiplier, electronic;
multiplier, four-quadrant;
multiplier, one-quadrant;
multiplier, servo;
multiplier, two-quadrant;
network analyzer;
noise;
noise generator;
normal linearity;
offset;
operate;
operational relay;
optical scanner;
oven;
overload;
panel;
partial-system test;
patch bay;
patch board;
patch panel;
photoformer;
plotting board;
polar mode;
potentiometer;
potentiometer, followup;
potentiometer, function;
potentiometer, granularity;
potentiometer, grounded;
potentiometer, linear;
potentiometer, multiplier;
potentiometer, parameter;
potentiometer, servo;
potentiometer, set;
potentiometer, sine cosine;
potentiometer, tapered;
potentiometer, tapped;
potentiometer, ungrounded;
precision;
prepatch panel;
problem board;
problem check;
problem variable;
proportionality;
rate test;
recorder;
recorder, strip chart;
recorder, XY;
rectangular mode;
reference excursion;
reference power supply;
reference voltage;
relay;
repeatability;
repetitive operation;
reset;
reset dwell time;
resolver;
scale factor;
servomechanism, positional;
servomechanism, rate;
servomechanism, repeater;
simulate;
simulation;
simulation, mathematical;
simulation, physical;
simulator;
stabilization, drift;
stabilization network;
static check;
stop;
storage integrator;
sub limiting;

summing junction;
time constant of integrator;
time delay;
time, real;
time scale;
transport delay;
unloading circuit;
zero-based linearity;
zero-error reference

electronic computation.
See:
access time;
accumulator;
adder;
address;
address part;
analog;
analog computer;
AND circuit;
AND gate;
arithmetic unit;
band;
binary cell;
binary-coded-decimal system;
bit;
block;
break point;
buffer;
bus;
carry;
cell;
channel;
character;
check;
check, automatic;
check, forbidden-combination;
check problem;
circulation register;
clear;
code;
command;
complement;
computer;
conditional jump;
counter;
counter ring;
decoder;
delay-line storage;
digital computer;
digital data;
double-length number;
error;
error-detecting code;
excess-three code;
extract;
flow diagram;
gate;
half adder;
inhibiting input;
instruction code;
integrator;
jump;
language;
logic design;
logic diagram;
logic element;
logic operation;
logic symbol;
machine language;
major cycle;
marginal checking;
matrix;
minor cycle;
multiple-address code;
multiplier;
number;
operation code;
operation part;
order;
OR gate;
overflow;
parallel;
parallel digital computer;
place;
point;
positional notation;
program;
programmed check;
pulse regeneration;
read;
regeneration;
reset;
rewrite;
routine;
selection check;
serial;
serial digital computer;
set;
shift;
sign digit;
staticizer;
storage;
storage capacity;
store;
subroutine;
track;
transcriber;
transfer;
transfer check;
translator;
unconditional jump;
unit;
verification;
word;
word time;
write.

electronic controller.
See:
acceptance angle;
angle of extinction;
angle of ignition;
anode circuit;
anode current;
anode firing;
anode supply voltage;
anode voltage;
auxiliary anode;
auxiliary power supply;
cathode current;
controlled rectifier;
electronic direct-current motor controller;
electronic direct-current motor drive;
electronic rectifier;
filament current;
filament voltage;
filter capacitor;
filter inductor;
grid circuit;
grid control;
grid current;
grid resistor;
grid transformer;
grid voltage;
heater current;
heater transformer;
heater voltage;
ignition control;
ignitor;
ignitron;
independent firing;
inverse parallel connection;
lens and aperture;
peak anode current;
peak forward anode voltage;
peak inverse anode voltage;
phase-shift circuit;
pool tube;
pulsing;
pulsing transformer;
rectifier;
relaxation oscillator;
tube heating time;
tube voltage drop.

electronic digital computer.
See:
absolute address;
access;
access time;
accumulator;
accuracy;
acoustic delay line;
adder;
address;
address format;
address part;
address register;
allocation;
analog-to-digital converter;
AND gate;
arithmetic shift;
arithmetic unit;
assemble;
associative storage;
asynchronous computer;
automatic carriage;
automatic check;
band;

base;
base-on-ones complement;
baud;
benchmark problem;
binary;
binary cell;
binary code;
binary-coded decimal;
binary digit;
binary number;
binary numeral;
binary search;
bit;
block;
bootstrap;
borrow;
branch;
branchpoint;
break point;
buffer;
built-in check;
bus;
call;
calling sequence;
capacity;
card;
card hopper;
card image;
card stacker;
carriage return;
carry;
cascade carry;
cell;
central processing unit;
chad;
chadded;
chadless;
chain code;
channel;
character;
character recognition;
check;
check bit;
check digit;
checkpoint;
check problem;
circulating register;
clear;
clock;
closed subroutine;
code;
coded;
coding;
collate;
collating sequence;
collator;
column;
column binary;
combination;
combinational logic element;
command;
compile;
complement;
complete carry;
component;
computer;
computer code;
computer instruction;
computer network;
computer program;
computing element;
conditional jump;
conditional transfer of control;
content address storage;
control;
control panel;
control unit;
convert;
copy;
core;
correction;
counter;
cycle;
cyclic shift;
data-hold;
data reconstruction;
decimal;
decision instruction;
decision table;
deck;
decode;
decoder;
delay;
delay line;
delimiter;
design;
destructive read;
device;
diagnostic;
diagram;
dichotomizing research;
digit;
digital;
digital computer;
digital controller;
digital-to-analog converter;
disc;
double length;
double precision;
drum;
dump;
duodecimal;
dynamic dump;
echo check;
edit;
electromagnetic delay line;
element;
encode;
end-around carry;
entry point;
equivalent binary digits;
error;
error correction code;
error-detecting code;
escape character;
excess-three code;
expression;
extract instruction;
feed;
field;
file;
file maintenance;
film;
fixed point;
flag;
flip-flop;
floating point;
flying spot scanner;
font;
forbidden combination;
formal;
four-address;
four-plus-one address;
full duplex;
gate;
generate;
generator;
gray code;
half-adder;
Hamming distance;
hardware;
head;
high-speed carry;
hopper;
hybrid system;
identifier;
image;
image dissector;
immediate address;
incremental computer;
index;
indirect address;
information processing;
information retrieval;
inhibit;
input;
instruction;
instruction code;
instruction counter;
instruction register;
instruction repertory;
integrator;
jump;
keyhole;
language;
length;
library routine;
line;
linear programming;
linkage;
list;
load-and-go;
location;
logic;
logic design;
logic diagram;
logic element;
logic operation;
logic operator;

logic symbol;
look-up;
loop;
machine;
machine address;
machine check;
machine code;
machine instruction;
machine language;
macro instruction;
magazine;
magnetic core;
magnetic delay line;
magnetic disc;
magnetic drum;
magnetic storage;
magnetic tape;
magnetic thin film;
major cycle;
malfunction;
manual input;
marginal check;
mark;
mask;
mathematical check;
matrix;
memory;
merge;
minimum distance code;
minor cycle;
mistake;
modulo N check;
monadic operation;
monitor;
multilevel address;
multiple-address;
multiprocessing;
multiprocessor;
multiprogramming;
N-level address;
negate;
network;
nines complement;
no op;
noise;
nondestructive read;
nonerasable storage;
normalize;
notation;
number;
numerical control;
octonary;
octal;
odd-even check;
offline;
one-address;
one-level address;
one-plus-one address;
ones complement;
open ended;
open subroutine;
operand;
operating system;
operation;
operation code;
operation part;
operator;
optical character recognition;
optical scanner;
order;
OR gate;
output;
overflow;
overlay;
pack;
packing density;
panel;
parallel;
parallel processing;
parallel search storage;
parallel storage;
parity;
parity bit;
parity check;
part;
partial carry;
patch;
pattern recognition;
pattern sensitive fault;
permanent storage;
pinboard;
place;
plugboard;
point;
positional notation;
precision;
preset;
printing;
problem;
processing;
program;
program library;
programmed check;
programming;
programming language;
program-sensitive fault;
protected location;
protection;
pseudorandom number sequence;
punch;
punch position;
punch tape;
pushdown list;
pushup list;
quantization;
radix;
radix complement;
random access;
random number;
range;
read;
read-only storage;
real time;
recognition;
record;
record gap;
redundant;
reflected binary code;
regenerate;
register;
registration;
relative address;
relative coding;
relocate;
reperforator;
repertory;
rerun point;
reset;
residue check;
restart;
retrieval;
return;
rewrite;
ring counter;
roundoff;
routine;
row binary;
run;
scale;
scanner;
search;
selection check;
selective dump;
self-checking code;
sentinel;
separator;
sequence;
sequential control;
sequential logic element;
sequential operation;
serial;
serial access;
serial operation;
serial-parallel;
service routine;
set;
sexadecimal;
sharing;
shift;
shift register;
signal;
signal distance;
sign digit;
sign position;
simulate;
single-address;
single step;
skip;
smooth;
snapshot dump;
software;
sonic delay line;
sort;
sorter;
source program;
special character;
special-purpose computer;
standing-on-nines carry;

static dump;
staticize;
station;
step;
storage;
storage allocation;
storage capacity;
storage cell;
storage device;
storage protection;
store;
stored program computer;
straight-line coding;
string;
subroutine;
summation check;
supervisory routine;
suppression;
switch;
symbol;
symbolic address;
symbolic coding;
tape;
tape station;
tape to card;
tape transport;
tape unit;
target language;
target program;
temporary storage;
tens complement;
terminal;
ternary;
test;
thin film;
three-address;
three-plus-one address;
threshold element;
time;
time sharing;
toggle;
tracing routine;
track;
transcribe;
transfer;
transfer check;
transfer of control;
transform;
translate;
transmit;
transport;
trap;
triple-address;
true complement;
truncate;
Turing machine;
two address;
two-plus-one address;
twos complement;
unary operation;
unconditional jump;
unconditional transfer of control;
underflow;
unit;
universal Turing machine;
unpack;
utility routine;
verify;
visual scanner;
volatile storage;
word;
word length;
word time;
working storage;
write;
z-transform, advanced;
z-transform, delayed;
z-transform, modified;
z-transform, one-sided;
z-transform, two-sided;
zero-level address;
zero suppression.

electron tube.
See:
aligned-grid tube or valve;
base;
bulb envelope vacuum;
degassing;
dispenser cathode;
dynatron effect;
dynode;
electrode radiator;
electron-beam tube or valve;
flicker effect;
floating grid;
getter;
grid;
grid number;
grid pitch;
heat shield;
input electrode;
island effect;
lead-in wire;
microphonics;
output electrode;
pinch;
pins;
plate;
Schottky noise;
screen factor;
shot-effect;
tip;
top-cap;
tube definitions;
virtual cathode.

electron-tube admittances.
See:
admittance, short-circuit driving-point;
admittance, short-circuit feedback;
admittance, short-circuit forward;
admittance, short-circuit input;
admittance, short-circuit output;
admittance, short-circuit transfer;
amplification factor;
capacitance;
capacitance, input;
capacitance, output;
capacitance, short-circuit input;
capacitance, short-circuit output;
capacitance, short-circuit transfer;
driving-point admittance;
electrode admittance;
electrode capacitance;
electrode impedance;
equivalent conductance;
equivalent noise conductance;
external termination;
factor;
gap admittance;
gap admittance circuit;
gap capacitance, effective;
gap loading, multipactor;
gap loading, primary transit-angle;
gap loading, secondary electron;
input admittance, effective;
interelectrode capacitance;
interelectrode transadmittance;
interelectrode transconductance;
output admittance, effective;
small-signal forward transadmittance;
transadmittance compression ratio;
transadmittance, forward;
transconductance;
transfer admittance.

electroplating.
See:
addition agent;
alkaline cleaning;
alloy plate;
anode cleaning;
ball burnishing;
barrel plating;
base;
blue dip;
bright dip;
brightener;
brush or sponge plating;
buffing;
building up;
burnishing;
burnt deposit;
cathode cleaning;
cleaner;
cleaning;
coating;
coloring;
composite plating;
conducting salts;

geissler tube;
glow discharge;
hexode;
high-level firing time;
high-level radio frequency signal;
high-level voltage standing-wave ratio;
hysteresis;
ignitor-current temperature drift;
ignitor discharge;
ignitor electrode;
ignitor firing time;
ignitor interaction;
ignitor leakage resistance;
ignitor oscillations;
ignitor voltage drop;
input gap;
insertion loss;
ionization time;
leakage power;
loaded *Q;*
low-level radio-frequency signal;
minimum firing power;
misfire;
mode purity;
mount;
negative glow;
neon indicator;
overvoltage;
phase recovery time;
phase-tuned tube;
position of the effective short;
positive column;
pre-transmit-receive tube;
recovery time;
regulation;
resonant gap;
spike leakage energy;
starter voltage drop;
stroboscopic tube;
temperature coefficient of voltage drop;
transfer current;
transmit-receive tube;
tube heating time;
unfired time;
unloaded *Q;*
voltage jump.

generating station.
See:
apparatus;
automatic station;
availability factor;
balanced polyphase load;
base load;
capability;
capability margin;
capacity;
capacity factor;
coincidence factor;
cold reserve;
connected curve;
cylindrical rotor generator;
dump power;
economy power;
efficiency, station system;
electrical supply station;
emergency transfer capability;
excitation voltage;
extended capability;
firm capacity purchase or sales;
firm power;
firm transfer capability:
frequency control;
fuel economy;
generating auxiliary power;
generating station auxiliaries;
gross demonstrated capacity;
gross rated capacity;
high-speed excitation system;
holding frequency;
holding load;
hot reserve;
house turbine;
hydro capability;
integrated energy curve;
interruptible loads;
interruptible power;
load;
load center;
load curve;
load curves, daily;
load diversity;
load diversity power;
load duration curve;
load factor;
load-frequency control;
load variation within the hour;
loss factor;
low-voltage protection;
low-voltage release;
monthly peak duration curves;
net assured capability;
net dependable capability;
net load capability;
net rated capacity;
off-peak power;
on-peak power;
operation factor;
output factor;
peak load;
peak load station;
peak responsibility;
plant factor;
pondage station;
power storage;
prime power;
pumped-storage hydro capability;
regulated frequency;
reserve equipment;
run-of-river station;
secondary power;
spare equipment;
specific unit capacity purchases;
spinning reserve;
steam capability;
storage station;
stream flow;
switchboards and panels;
system load;
system maximum hourly load;
system reserve;
total capability for load;
transmission line capacity.

ground.
See:
counterpoise;
effectively grounded;
equipment ground;
equipotential line or contour;
ground bus;
ground cable bond;
ground clamp;
ground conduit;
ground current;
ground detector;
grounded;
grounded capacities;
grounded circuit;
grounded concentric wiring system;
grounded conductor;
grounded parts;
grounded system;
ground-fault neutralizer grounded;
ground indication;
grounding conductor;
grounding connection;
grounding device;
grounding electrode;
grounding outlet;
ground lug;
ground plate;
ground relay;
ground resistance;
ground-return circuit;
ground switch;
ground system of an antenna;
ground transformer;
ground wire of an overhead line;
guard wire;
impedance grounded;
interior wiring system ground;
neutral ground;
potential profile;
rating of a ground transformer;
reactance grounded;
resistance grounded;
resonant grounded;
service ground;
solidly grounded;
step voltage;
surface-potential gradient;
system grounding conductor;
touch voltage;

ungrounded;
voltage to ground.

grounding device.
See:
extended-time rating;
ground;
ground end;
ground-fault neutralizer;
ground grid;
grounding system;
grounding transformer;
ground mat;
line end;
line end and ground end;
losses;
mechanical current rating;
mutual resistance;
neutral grounding capacitor;
neutral grounding device;
neutral grounding impedor;
neutral grounding reactor;
neutral grounding resistor;
neutral wave trap;
rated current;
rated frequency;
rated insulation class;
rated time;
rated voltage;
short-time rating;
tap;
thermal current rating.

induction heater.
See:
autoregulation induction heater;
coreless-type induction heater or furnace;
core-type induction heater or furnace;
coupling;
Curie point;
decalescent point;
depth of penetration;
domestic induction heater;
dual-frequency induction heater or furnace;
field strength meter;
flux guide;
gaseous tube generator;
high-frequency induction heater or furnace;
horizontal ring induction furnace;
hysteresis heater;
induction-conduction heater;
induction current;
induction ring heater;
industrial electronics;
interference;
interference measurement;
load;
load circuit;
load circuit efficiency;
load coil;
load leads or transmission lines;
load matching;
load matching network;
load matching switch;
load switch or contactor;
load transfer switch;
load, work or heater coil;
low-frequency induction heater or furnace;
melting channel;
mercury-hydrogen spark-gap converter;
motor effect;
over-all electrical efficiency;
pinch effect;
proximity effect;
quenched spark-gap converter;
radio-frequency generator;
radio-frequency generator, electron tube type;
recalescent point;
rotary generator;
shield;
stirring effect;
submerged resistor induction furnace;
tank circuit;
wave heating.

instrument.
See:
acoustic interferometer;
acoustic radiometer;
ammeter;
audiometer;
bolometric detector;
bolometric instrument;
bridge;
capacitance meter;
chart scale length;
circuit noise meter;
corona voltmeter;
crest voltmeter;
current circuit;
deadband;
decibel meter;
detecting means;
differential analyzer;
direct-acting recording instrument;
discharge detector;
discharge probe;
echo box;
electrical range;
electric hydrometer;
electricity meter;
electric stroboscope;
electric tachometer;
electric thermometer;
electrodynamic instrument;
electrometer;
electron beam instrument;
electronic instrument;
electronic voltmeter;
electroscope;
electrostatic instrument;
electrostatic voltmeter;
electrothermic instrument;
elevated-zero range;
end-scale value;
ferrodynamic instrument;
frequency meter;
frequency monitor;
full-scale value;
galvanometer;
generating electric field meter;
generating voltmeter;
harmonic analyzer;
hinged-iron ammeter;
hot-wire instrument;
indicating instrument;
indirect-acting recording instrument;
induction instrument;
instrument shunt;
instrument transformer;
intermediate means;
ionization vacuum gauge;
Lichtenberg figure camera;
light-beam instrument;
lower range value;
magnetically shielded type instrument;
magnetometer;
maximum current;
maximum voltage;
measurement range of an instrument;
measurement system;
mechanism;
modulation meter;
moving element;
moving-iron instrument;
moving-magnet instrument;
ohmmeter;
optical ammeter;
oscillograph;
permanent-magnet moving-coil instrument;
permanent-magnet moving-iron instrument;
phase meter;
phase sequence indicator;
Philips gauge;
Pirani gauge;
potentiometer;
power-factor meter;
quality-factor meter;
rated current;
rated voltage;
ratio meter;
reactive factor meter;
recording instrument;
rectifier instrument;
reflectometer;
reflectometer, time-domain;
resistance thermometer;

interflected component;
interflection;
lamp;
lumen method;
luminaire efficiency;
luminance factor for room surfaces;
maintenance factor;
point-by-point method;
position index;
room coefficient;
room index;
room ratio;
room utilization factor;
solid angle;
upward component;
work plane;
zonal factor interflection method;
zonal factor method.

lamp.
See:
absorptance;
absorption;
arc discharge;
beam spread;
candlepower distribution curve;
carbon-arc lamp;
characteristic curve;
coefficient of attenuation;
cold-cathode lamp;
colorimetry;
complete diffusion;
diffuse reflectance;
diffuse reflection;
diffusing surfaces and media;
electric-discharge lamp;
electroluminescence;
fluorescence;
fluorescent lamp;
fluorescent-mercury lamp;
gaseous discharge;
glow discharge;
glow lamp;
hemispherical reflectance;
hot-cathode lamp;
incandescence;
incandescence filament lamp;
incomplete diffusion;
interior wiring;
inverse-square law;
isocandela line;
isolux line;
Lambert's cosine law;
life performance curve;
life tests of lamps;
light;
light center;
light center length;
luminance factor;
luminescence;
narrow-angle diffusion;
perfect diffusion;
phosphorescence;
photoflash lamp;
primary standards;
redirecting surfaces and media;
reflectance;
reflection;
regular reflectance;
regular reflection;
spectral reflectance;
specular angle;
transmission;
wide-angle diffusion;
zonal constant.

lightning arrester.
See:
amplitude factor;
angle of protection;
arching chamber of expulsion-type arrester;
arrester alternating sparkover-voltage;
arrester discharge capacity;
arrester discharge voltage-current characteristic;
arrester discharge voltage-time curve;
arrester disconnector;
arrester, expulsion-type;
arrester ground;
arrester, valve-type;
attenuation of a traveling wave;
background ionization voltage;
basic-impulse-insulation level;
breakdown;
characteristic element;
charging circuit of a surge generator;
chopped impulse wave;
coefficient of grounding;
conformance tests;
contact voltage;
coordination of insulation;
crest of a wave, surge, or impulse;
critical-impulse flashover voltage;
critical withstand current;
critical withstand voltage;
deflector;
design tests;
direct stroke;
discharge circuit;
discharge current;
discharge-inception voltage;
discharge-indicator;
discharge voltage;
discharge-withstand current rating;
disruptive discharge;
disruptive-discharge voltage, 50 percent;
disruptive-discharge voltage, 100 percent;
equipment ground;
exposed installation;
expulsion element;
external insulation of apparatus;
external series gap;
fault;
fault current;
fault resistance;
flashover;
follow current;
front-of-wave pulse sparkover voltage;
full impulse wave;
full-wave voltage impulse;
grading ring;
ground bar;
ground conductor;
ground current;
grounded-neutral system;
ground electrode;
ground fault;
grounding switch;
grounding system;
ground-return system;
ground terminal;
ground-wire;
impulse;
impulse flashover voltage;
impulse inertia;
impulse protection level;
impulse ratio;
impulse sparkover voltage;
impulse sparkover voltage-time curve;
impulse test;
impulse voltage;
impulse wave;
impulse withstand voltage;
incident wave;
independent ground electrode;
indirect stroke;
inherent restriking voltage;
insulation fault;
insulation level;
intermittent fault;
internal insulation of apparatus;
ionization;
ionization voltage;
isolated neutral system;
lightning;
lightning surge;
maximum system voltage;
minimum flashover voltage;
mutual surge impedance;
natural frequency;
nominal discharge current;
nominal rate of rise;
nominal system voltage;
nonexposed installation;
nonlinear-resistor-type arrester;
nonlinear series resistor;
operating-duty test;
oscillatory surge;

double modulation;
effective percentage modulation;
electronic keying;
expansion;
frequency-division multiplex;
frequency modulation;
frequency-shift keying;
gating;
grid leak detector;
grid modulation;
grid pulse modulation;
heterodyne reception;
high-level modulation;
homodyne reception;
impulse noise;
instantaneous companding;
instantaneous frequency;
instantaneous sampling;
integrating network;
intelligence bandwidth;
interchannel interference;
intermodulation;
interrupted continuous wave;
intersymbol interference;
keying;
keying interval;
keying rate;
low-level modulation;
modulated amplifier;
modulating function;
modulation;
modulation capability;
multiple modulation;
multiplexing;
noise;
on-off keying;
percentage modulation;
phase modulation;
phase-shift keying;
plate-circuit detector;
plate keying;
plate modulation;
plate pulse modulation;
pre-emphasis;
pulse;
pulse-amplitude modulation;
pulse-code modulation;
pulse decay time;
pulse-duration modulation;
pulse-frequency modulation;
pulse interleaving;
pulse-interval modulation;
pulse modulation;
pulse-position modulation;
pulse-time modulation;
quadrature modulation;
quantization;
quantization distortion;
quantization level;
quantized pulse modulation;
quiescent carrier telephony;
random noise;
sampling;
screen-grid modulation;
series modulation;
sideband suppression;
signal;
single-sideband modulation;
single-sideband transmitter;
single-tone keying;
spark-gap modulation;
suppressed-carrier modulation;
syllabic companding;
synchronous gate;
threshold;
time-division multiplex;
time-gate;
two-source frequency keying;
two-tone keying;
undirectional pulses;
wave;
wide-band improvements;
wide-band ratio.

navigation.

See:
A and *R* display;
absolute delay;
accelerometer;
accuracy;
A display;
aerophare;
aided tracking;
airborne intercept radar;
air-derived navigation data;
airport surface detection equipment;
airport surveillance radar;
air-position indicator;
air speed;
Alford loop;
alignment;
ambiguity;
amplitude balance control;
amplitude discriminator;
amplitude noise;
analytic inertial-navigation equipment;
angle noise;
angle cut;
angular deviation sensitivity;
angular resolution;
angular width;
A-N radio range;
antenna effect;
anticlutter circuits;
apparent bearing;
apparent vertical;
approach navigation;
approach path;
area moving-target indicator;
A-trace;
attitude-effect error;
automatic chart-line follower;
automatic direction-finder;
automatic pilot;
autonavigator;
autopilot coupler;
autoradar plot;
azimuth;
back course;
back scatter;
baseline;
baseline delay;
basic repetition frequency;
B display;
beam error;
beam noise;
bearing;
bearing accuracy;
bearing accuracy, instrumental;
bearing-error curve;
bearing offset, indicated;
bearing sensitivity;
bend;
bend amplitude;
bend frequency;
bend reduction factor;
blinking;
blink speed;
blip;
blooming;
blur;
boresighting;
B trace;
calibration marks;
cancellation ratio;
canceller;
carrier-controlled-approach system;
C display;
celestial inertial-navigation equipment;
chain;
chart comparison unit;
circular probability error;
capture effect;
canceled video;
clearance;
clearance antenna array;
clearance sector;
clutter attenuation;
clutter reflectivity;
clutter residue;
clutter visibility factor;
coding delay;
cohered video;
coherent oscillator;
commutated-antenna direction finder;
compass bearing;
compass course;
compass heading;
complex target;
composite pulse;
compound target;
cone of ambiguity;
cone of silence;
consol;

consolan;
Coriolis correction;
cosecant-squared pattern;
countdown;
coupler;
course;
course line;
course linearity;
course-line computer;
course-line deviation;
course-line-deviation indicator;
course made good;
course push;
course roughness;
course sector;
course-sector width;
course sensitivity;
course softening;
course width;
crossover characteristic curve;
crossover region;
dark-trace tube;
data stabilization;
D display;
dead reckoning;
dead time;
decca;
decision gate;
dectra;
delayed plan-position indicator;
derived envelope;
derived pulse;
detectability factor;
deviation sensitivity;
difference detector;
difference in depth of modulation;
directed reference flight;
direction;
directional localizer;
direction-finder;
direction-finder antenna;
direction-finder antenna system;
direction-finder deviation;
direction-finder noise level;
direction-finder sensitivity;
distance resolution (radar);
Doppler-inertial navigation equipment;
Doppler navigator;
Doppler radar;
Doppler very-high-frequency omnidirectional radio range;
drift angle;
drift correction angle;
dummy-antenna system;
duty cycle;
duty ratio;
echo;
echo box;
echo suppressor;
E display;
effective echo area;
electrical distance;
element of a fix;
enabling pulse;
fast-time-constant circuit;
F display;
flag alarm;
flare-out;
flarescan;
flight path;
flight-path computer;
flight-path deviation;
flight-path-deviation indicator;
flight track;
gain time control;
gain turn down;
gap coding;
gate;
G display;
G drift;
G^2 drift;
geocentric latitude;
geodesic;
geodesic latitude;
geographic vertical;
geoid;
geometric;
geometric factor;
ghost signals;
glide path;
glide slope;
glide-slope angle;
glide-slope deviation;
glide-slope facility;
glide-slope sector;
grid bearing;
grid course;
grid heading;
grid north;
ground-based navigation aid;
ground-controlled approach;
ground-derived navigation data;
ground position indicator;
ground-referenced navigation data;
ground return;
ground speed;
ground surveillance radar;
gyro;
gyrocompass alignment;
H-beacon;
H display;
heading;
heading-effect error;
hybrid navigation equipment;
identification;
I display;
indicated bearing;
inertial-navigation equipment;
inertial navigator;
inertial space;
initial conditions;
instrumental error;
instrument approach;
instrument-approach system;
instrument landing system marker beacon;
instrument landing system reference point;
intensity modulation;
interrogation;
ionospheric error;
ionospheric-height error;
ionospheric-tilt error;
J display;
jitter;
K display;
keep-alive circuit;
lane;
leader-cable system;
limited signal;
linearity region;
line of position;
lin-log receiver;
local level;
local vertical;
localizer sector;
long-distance navigation;
lorhumb line;
low-clearance area;
low-clearance points;
magnetic bearing;
magnetic course;
magnetic deviation;
magnetic heading;
magnetic north;
M-array glide-slope;
mass-attraction vertical;
M display;
mechanical modulator;
microwave early warning;
middle marker;
minimum en-route altitude;
minimum reception altitude;
misalignment drift;
modulation eliminator;
most-probable position;
moving-base-derived navigation data;
moving-base navigation aid;
moving-base-referenced navigation data;
moving-target indicator;
moving-target-indicator improvement factor;
multipath error;
multiple rho;
navigation coordinate;
navigation parameter;
navigation quantity;
N display;
near field;
night effect;
nondirectional beacon;
north-stabilized plan-position indicator;
null;

null-reference glide slope;
octantal error;
O display;
omega;
omnibearing;
omnibearing converter;
omnibearing distance facility;
omnibearing-distance navigation;
omnibearing indicator;
omnibearing selector;
omnidirectional range;
on-course curvature;
outer marker;
overload point, signal;
path;
pencil beam;
phantom target;
pick-up factor;
pilotage;
pip-matching display;
pitch altitude;
plan-position indicator;
platform erection;
plumb-bob vertical;
polarization, desired;
polarization error;
polyplexer;
position;
precision approach radar;
primary radar;
proximity-effect error;
pseudo-latitude;
pseudo-longitude;
pulse coder;
pulsed Doppler radar;
pulse decoder;
pulse-duration discriminator;
pulse-forming line;
pulse rate;
pulse repetition rate;
quadrantal error;
radar;
radar altimeter;
radar beacon;
radar equation;
radar performance figure;
radar relay;
radar shadow;
radial;
radial-time-base display;
radio altimeter;
radio-autopilot coupler;
radio direction-finder;
radiolocation;
radiometric sextant;
radio navigation;
radiophare;
radio range;
rain clutter;
random errors;
range and elevation guidance for approach and landing;
range noise;
R display;
receiver gating;
reciprocal bearing;
reference direction;
reference line;
reference modulation;
reference test field;
reflection error;
refraction error;
relative bearing;
reply;
residual error;
resolution element;
resolving time;
rho-rho;
rho-theta;
ring around;
ring time;
roughness;
runway visual range;
safe semiautomatic flight inspection;
sampling gate;
saturated signal;
saturating signal;
scalloping;
scanning;
scatter band;
Schuler tuning;
scintillation;
secondary radar;
second-time-around echo;
selector pulse;
self-contained navigation aid;
semianalytic inertial-navigation equipment;
sense;
sensing;
sensitivity time control;
service area;
short-distance navigation;
sideband null;
sideband reference glide slope;
signal-to-clutter ratio;
simple target;
simultaneous lobing;
single-degree-freedom gyro;
site error;
sky-wave contamination;
sky-wave correction;
sky-wave station error;
slant distance;
slant range;
slope angle;
slot coupling factor;
slot current ratio;
slot spacing factor;
sonne;
space-referenced navigation data;
specific repetition rate;
specified repetition frequency;
squeezable waveguide;
squint;
squitter;
stabilization;
stabilized flight;
stable element;
stable platform;
stalo;
standard wave error;
strapped down;
suppressed time delay;
surface of position;
synchronization error;
systematic errors;
tacan;
target;
target transmitter;
threshold signal;
tilt;
tilt angle;
time discriminator;
to-from indicator;
torquing rate;
track;
tracking;
transit;
transmit-receive cavity;
transmit-receive switch;
transponder;
transponder, crossband;
transponder reply efficiency;
trigger level;
triplet;
true bearing;
true heading;
true north;
two-degree-freedom gyro;
variable modulation;
V-beam radar;
vehicle-derived navigation data;
velocity response factor;
very-high-frequency omnidirectional range;
video integration;
video mapping;
visibility factor;
visual-aural range;
vortac;
way point;
Z marker.

network analysis.
See:
accessible terminal;
active electric network;
all-pass network;
analytic signal;
angular frequency;
arm;
artificial line;
attenuation equalizer;
available power;
available power gain;
balanced termination;

balancing network;
basic network;
bilateral network;
branch;
branch current;
branch impedance;
branch point;
branch voltage;
bridge network;
bridged-*T* network;
building-out network;
circuit;
circuit efficiency;
circuit parameters;
circuits and devices;
closed-loop control system;
C network;
coefficients of capacitance;
coefficients of elastance of a system of conductors;
completely mesh-connected circuit;
conjugate branches of a network;
conjugate impedance;
connected;
constant-*K* network;
control system;
corrective network;
coupled circuit;
coupling;
coupling coefficient;
critical damping;
current generator;
cutoff frequency;
cut-set;
delay equalizer;
delimited region;
delta-connected circuit;
delta network;
differentiating circuit;
differentiator;
driving-point admittance;
dynamic equilibrium;
effective bandwidth;
effective cutoff frequency;
element;
energy density spectrum;
equivalent circuit;
equivalent network;
filter impedance compensator;
forced oscillation;
gain, available power;
gain, available power, maximum;
gain, insertion;
gain, insertion voltage;
gain, power;
gain transducer;
Hilbert transform;
H network;
impedance;
impedance compensator;
impedance function;
impedor;
indicial admittance;
inductive coupling;
inductively coupled circuit;
input, driving-point, or sending-end impedance;
inverse network;
Kirchhoff's laws;
ladder network;
lattice network;
linear passive network;
linear signal flow graphs;
L network;
load circuit;
load-circuit efficiency;
load-circuit power input;
low-frequency impedance corrector;
matched impedance;
mesh (or loop);
mesh-connected circuits;
mesh-current;
mesh or loop equations;
minimum-phase network;
M-phase circuit;
mutual impedance;
network;
network function;
network synthesis;
neutral conductor;
node;
node equations;
noise-current generator;
noise-voltage generator;
nonlinear network;
nonplanar network;
N-terminal network;
N-terminal pair network;
nullity;
O network;
open-circuit impedance;
open-loop control system;
overdamping;
parallel elements;
parallel resonance;
parallel-*T* network;
parallel two-terminal pair networks;
passive electric network;
peaking network;
phase corrector;
pi network;
plate efficiency;
plate load impedance;
plate power input;
planar network;
pole;
polyphase circuit;
polyphase symmetrical set;
port;
positive-phase-sequence symmetrical components;
power gain;
power, vector;
pre-emphasis network;
Q;
quadergy;
quadrature;
radio transmission;
rank;
reactive factor;
reactor;
rectangular wave;
rectified value;
rectifier;
reference direction;
reference frequency;
reflection factor;
reflection loss;
related transmission terms;
relay;
resonance;
resonance frequency;
resonating;
resonator;
self-impedance;
separate parts;
series element;
series-parallel network;
series two-terminal pair networks;
set;
short-circuit;
short-circuit impedance;
sign;
simple mesh-connected circuit;
simple parallel circuit;
simple series circuit;
single-phase circuit;
single-phase symmetrical set;
single-phase three-wire circuit;
single-phase two-wire circuit;
slug tuning;
source impedance;
spark killer;
stage efficiency;
star-connected circuit;
star network;
structurally dual network;
structurally symmetrical network;
symmetrical alternating current;
symmetrical components;
symmetrical periodic function;
symmetrical set;
tandem;
terminal;
terminal of entry;
terminal pair;
three-terminal network;
T network;
transfer admittance;
transfer function;
transfer impedance;
transformer;

tree;
true neutral point;
two-phase circuit;
two-phase, five-wire system;
two-phase, four-wire system;
two-phase, three-wire system;
two-terminal pair network;
unilateral network;
voltage generator;
voltage sets;
Y-connected circuit;
Y network;
zero.

numerically controlled machine.
See:
accuracy;
analog and digital data;
arc clockwise;
arc counterclockwise;
automatic programmed tools;
auxiliary function;
backlash;
cancel;
canned cycle;
circular interpolation;
closed-loop system;
constant cutting speed;
contouring control system;
control position accuracy, precision, or reproducibility;
coordinate dimension word;
cutter compensation;
dead band;
deceleration, programmed;
dwell;
end of clock signal;
end of program;
end of tape;
feed function;
feed rate bypass;
feed rate override;
feedback loop;
fixed block format;
fixed cycle;
floating zero;
format classification;
format detail;
incremental dimension;
incremental feed;
linear interpolation;
machine positioning accuracy, precision, or reproducibility;
manual data input;
manuscript;
measuring accuracy, precision, or reproducibility;
miscellaneous function;
numerical control system;
optional stop;
part programming, computer;
part programming, manual;
position sensor or position transducer;
positioning control system;
post processor;
preparatory function;
program stop;
readout, command;
readout, position;
reference block;
reproducibility;
resolution;
spindle speed;
straight-cut control system;
tab sequential format;
tool function;
tool offset;
zero offset;
zero synchronization.

oscillatory circuit.
See:
aperiodic circuit;
audio-frequency oscillator;
automatic phase control;
balanced oscillator;
Barkhausen-Kurz oscillator;
blocking oscillator;
circuits and devices;
Colpitts oscillator;
crystal oscillator;
dynatron oscillation;
dynatron oscillator;
electron-coupled oscillator;
feedback oscillator;
free oscillations;
frequency pulling;
gas-tube generator;
gas-tube relaxation oscillator;
Gill-Morrell oscillator;
Hartley oscillator;
hysteresis;
impulse excitation;
load impedance diagram;
local oscillator;
magnetostriction oscillator;
magnetron;
master oscillator;
Meissner oscillator;
mercury arc converter, pool-cathode;
mode;
modes, degenerate;
mode separation;
modulation;
multivibrator;
negative-resistance oscillator;
negative-transconductance oscillator;
parasitic oscillations;
phase locking;
phase-shift oscillator;
Pierce oscillator;
pulling figure;
pulsed oscillator;
pushing figure;
push-pull oscillator;
relaxation oscillator;
resistance-capacitance oscillator;
resonant-line oscillator;
resonator mode;
retarding-field oscillator;
Rieke diagram;
ring oscillator;
self-pulse modulation;
shock excitation;
singing;
singing point;
sink;
spark-gap modulation;
tank circuit;
transitron oscillator;
tunable microwave oscillators;
tuned-grid, oscillator;
tuned-grid–tuned-plate oscillator;
tuned-plate oscillator;
tuning, electronic;
tuning range, electronic;
tuning rate, thermal;
tuning sensitivity, electronic;
tuning sensitivity, thermal;
tuning, thermal;
tuning time constant, thermal;
velocity-modulated oscillator;
Wien bridge.

oscillograph.
See:
accelerating voltage;
alternate display;
astigmatism;
automatic triggering;
cathode-ray oscilloscope;
chopped display;
chopping rate;
common-mode rejection ratio;
common-mode signal;
compression;
deflection blanking;
deflection factor;
deflection polarity;
deflection sensibility;
deflection sensitivity;
delaying sweep;
differential signal;
dual-beam oscilloscope;
dual trace;
electromagnetic oscillograph;
fluorescence;
focus;
free-running sweep;
gated sweep;
geometry;
graticule;
graticule area;
horizontal amplifier;
incremental sweep;
information writing speed;
intensity;

intensity amplifier;
intensity modulation;
internal triggering;
jitter;
line triggering;
Lissajous figure;
luminance;
magnified sweep;
mixed sweep;
multibeam oscilloscope;
multitrace;
network analyzer;
orthogonality;
oscillogram;
oscillography;
oscilloscope;
phosphor decay;
phosphor screen;
phosphorescence;
quality area;
ramp;
raster;
recurrent sweep;
resolution;
retrace;
return trace;
sawtooth sweep;
sawtooth waveform;
scan;
signal delay;
sine-wave sweep;
single-sweep mode;
spot;
stability;
stairstep sweep;
sweep;
sweep accuracy;
sweep duration;
sweep duty factor;
sweep frequency;
sweep gate;
sweep lockout;
sweep magnifier;
sweep range;
sweep recovery time;
sweep reset;
synchronized sweep;
synchronizing signal;
time base;
trace;
trace width;
transient analyzer;
trigger;
trigger countdown;
trigger pickoff;
triggered sweep;
triggering level;
triggering signal;
triggering slope;
unblanking;
uniform-luminance area;
vertical amplifier;
viewing area;
waveform distortion;
X-Y display;
Y-T display;
Z-axis amplifier.

phonograph pickup.
See:
acoustic pickup;
backed stamper;
background noise;
bilateral area track;
binder;
burnishing surface;
capacitor pickup;
carbon-contact pickup;
chip;
class-A push-pull sound track;
class-B push-pull sound track;
constant-amplitude recording;
constant-velocity recording;
control track;
crystal pickup;
cutter;
cutting stylus;
difference limen;
direct-current erasing head;
direct-current magnetic biasing;
disk recorder;
double-pole-piece magnetic head;
drive pinhole;
eccentric groove;
eccentricity;
electroacoustics;
embossing stylus;
erasing head;
fast groove;
filler;
frequency record;
gap length;
grain;
groove;
groove angle;
groove shape;
groove speed;
grouping;
hum;
instantaneous recording;
lacquer disks;
lacquer original;
lacquer recording;
laminated record;
land;
lead-over groove;
light modulator;
light valve;
locked groove;
longitudinal magnetization;
magnetic biasing;
magnetic head;
magnetic powder-impregnated tape;
magnetic recording;
magnetic recording head;
magnetic recording medium;
magnetic recording reproducer;
magnetic reproducing head;
mechanical transmission system;
mold;
multiple sound track;
multitrack recording system;
noise reduction;
number 1 mold;
number 2, number 3, etcetera master;
number 2, number 3, etcetera mold;
offset angle;
original master;
permanent-magnet erasing head;
perpendicular magnetization;
photographic transmission density;
pickup;
pinch effect;
preform;
pressing;
recording channel;
recording loss;
recording stylus;
re-recording;
re-recording system;
ring head;
rumble;
side thrust;
single track;
sound recording system;
sound track;
sputtering;
squeeze track;
stamper;
stereophonic system;
stylus;
stylus drag;
stylus force;
surface noise;
tracing distortion;
tracking error;
transition frequency;
translation loss;
transmission;
transverse magnetization;
turntable rumble;
unilateral-area track;
unmodulated groove;
variable-area track;
variable-density track;
variable-reluctance pickup;
wow.

power distribution overhead construction.
See:
apparent sag at any point;
apparent sag of a span;
bushing;

filters;
frequency response;
full-wave rectification;
half-wave rectification;
high-speed regulator;
hybrid;
inverting amplifier;
isolation voltage;
lag networks;
lead networks;
linearity programming;
line regulation;
load regulation;
loop current;
loop gain;
master–slave operation;
mean time between failures;
modular;
null junction;
offset voltage;
open-loop gain;
operating temperature;
operational power supply;
operational programming;
output impedance;
overshoot;
overtemperature protection;
parallel operation;
parallel padding;
parallel programming;
pass element;
programming;
programming speed;
recovery time;
regulated power supply;
regulation, overall;
remote error sensing;
resolution;
response time;
ripple;
series operation;
series regulator;
short-circuit protection;
shunt regulator;
signal, error;
slaved tracking;
slewing rate;
spoiler resistors;
stability, long-term;
step line-voltage change;
step load change;
storage temperature;
summing point;
temperature coefficient;
unipolar;
unity-gain bandwidth;
voltage corrector;
voltage reference;
warmup time.

power system, low-frequency and surge testing.
See:
accuracy;
adjustment accuracy;
analog device;
applied-potential tests;
area control error;
area frequency-response characteristic;
area load-frequency characteristic;
area supplementary control;
automatic dispatching system;
biased scheduled-net-interchange;
breakdown;
bus-type shunts;
charging circuit;
chopped wave;
circuit voltage class;
control area;
cost;
critical impulse flashover voltage;
critical withstand current;
critical withstand voltage;
current terminals of instrument shunts;
dead band;
demand factor;
dielectric tests;
digital device;
discharge circuit;
distributive discharge;
economic dispatch;
electromechanical device;
electron device;
error;
flashover;
frequency bias;
frequency deviation;
gross generation;
guard electrode;
hysteresis;
impulse;
impulse flashover voltage;
impulse flashover volt-time characteristic;
impulse inertia;
impulse ratio;
impulse test;
impulse withstand voltage;
inadvertent interchange;
incremental cost;
incremental delivered power;
incremental fuel cost;
incremental generating cost;
incremental heat rate;
incremental loading;
incremental maintenance cost;
incremental transmission loss;
incremental worth;
induced-potential tests;
instrument shunts;
instrument terminals of shunts;
interconnected system;
line;
low-frequency dry-flashover voltage;
low-frequency wet-flashover voltage;
minimum flashover voltage;
mode voltage;
net generation;
net interchange;
net-interchange deviation;
nominal voltage;
nonconforming load;
oscillatory surge;
penalty factor;
portable shunts;
power control center;
power system;
puncture voltage;
rated accuracy;
rated frequency;
rated impulse withstand voltage;
rated voltage;
rated withstand current;
repeatability;
resistance method of temperature determination;
scheduled frequency;
scheduled-frequency offset;
scheduled net interchange;
shield;
solid-state device;
sparkover;
standard frequency;
standing wave;
station-control error;
surge;
surge generator;
system;
system frequency;
tandem control;
thermometer method of temperature determination;
time deviation;
time lag of impulse flashover;
time to impulse flashover;
transmission-loss coefficients;
traveling wave;
unit-control error;
virtual zero point;
voltage;
voltage divider;
voltage–phase-angle method;
voltage spread;
voltage to ground;
wave;
wavefront;
wave shape;
wave tail;
withstand current;
withstand voltage.

protective signaling.
See:
actuating device;

air horn;
alarm system;
audible signal device;
automatic fire-alarm system;
automatic holdup-alarm system;
automatic smoke alarm;
body-capacitance alarm system;
burglar-alarm system;
cabinet-for-safe;
central station;
central-station equipment;
central-station switchboard;
central-station system;
coded fire-alarm system;
coding siren;
combination watch-report and fire-alarm system;
contactless vibrating bell;
direct-wire circuit;
door contact;
double-circuit system;
electric horn;
electromechanical bell;
fire-alarm system;
fire-alarm thermostat;
fixed-point fire-alarm thermostat;
floor tape;
foil (foil tape);
headquarters system;
heat detector;
holdup-alarm attachment;
holdup-alarm system;
industrial-process supervisory system;
local manual fire-alarm system general-alarm type;
local manual fire-alarm system presignal type;
local system;
manual fire-alarm system;
manual holdup-alarm system;
McCullough circuit;
modulated photoelectric system;
municipal fire-alarm system;
municipal police-report system;
noncode fire-alarm system;
photoelectric system;
pneumatic-tubing system;
positive noninterfering and successive fire-alarm system;
pressure system;
printing recorder;
proprietary system;
protective screen;
rate-of-rise fire-alarm thermostat;
register;
retard transmitter;
semiautomatic holdup-alarm system;
shunt noninterfering fire-alarm system;
single-circuit system;
single-stroke bell;
sound detection system;
spring attachment;
sprinkler supervisory system;
subscriber's equipment;
supervised circuit;
supervisory system;
tapper bell;
timing table;
transmitter;
trap;
underdome bell;
vibrating bell;
vibration-detection system;
visual signal device;
watchman's reporting system;
water-motor bell;
waterflow system.

pulse.
See:
alternating-current pulse;
amplitude pulse;
amplitude reference level;
anticoincidence circuit;
baseline;
baseline offset;
bidirectional pulses;
binary code;
burst;
code element;
coherent pulse operation;
coincidence circuit;
control;
delay circuit;
delay coincidence circuit;
delay line;
delay pulse;
discriminator, pulse-height;
distortion pulse;
duration pulse;
duty factor;
enabling pulse;
fall time, pulse;
frequency, pulse-repetition;
impulse;
jitter;
kick-sorter;
leading edge pulse;
linear pulse amplifier;
modulation pulse;
peak pulse amplitude;
pedestal;
pulse amplifier;
pulse base;
pulse code;
pulse counter;
pulse decay time;
pulse duration;
pulse, Gaussian;
pulse-height analyzer;
pulse-height resolution, electron;
pulse operation;
pulse regeneration;
pulse repeater;
pulse rise time;
pulse separation;
pulse shape;
pulse shaper;
pulse spacing;
pulse stretcher;
pulse top;
pulse train;
pulse transmitter;
register, mechanical;
rise time, pulse;
scaler, pulse;
selector, pulse-height;
shaping pulse;
single-polarity pulse;
spike;
step;
time constant of fall;
time constant of rise;
trailing edge;
transient response;
transition;
transition pulse;
unidirectional pulses;
wave, square;
waveform;
waveform pulse.

pulse terms.
See:
average absolute pulse amplitude;
average pulse amplitude;
carrier-frequency pulse;
crest factor of a pulse;
fruit pulse;
leading-edge pulse time;
main bang;
mean pulse time;
modulating system;
peak pulse amplitude;
peak pulse power;
peak pulse power, carrier-frequency;
pulse amplitude;
pulse-amplitude modulation;
pulse bandwidth;
pulse carrier;
pulse code;
pulse-code modulation;
pulse decay time;
pulse droop;
pulse duration;
pulse-duration modulation;
pulse duty factor;
pulse-frequency modulation;
pulse interleaving;
pulse interrogation;
pulse-interval modulation;
pulse jitter;
pulse mode;

back-shunt keying;
back wave;
break-in keying;
carrier noise;
circularly polarized wave;
co-channel interference;
continuous waves;
diplex radio transmission;
dummy load;
elliptically polarized wave;
frequency band of emission;
frequency departure;
guard band;
identify friend or foe;
intermediate frequency;
interrogator;
interrogator-responder;
interrupted continuous wave;
keyer;
leakage radiation;
linearly polarized wave;
maximum system deviation;
microwaves;
mobile radio service;
multiplex radio transmission;
node;
percentage modulation;
point-to-point radio communication;
portable transmitter;
privacy system;
propagation constant;
propagation loss;
radio broadcasting;
radio channel;
radio circuit;
radio control;
radio detection;
radio Doppler;
radio frequency;
radio-frequency transformer;
radio location;
radio proximity fuse;
radio range-finding;
radio receiver;
radio relay system;
radio station;
radio transmitter;
responder;
service band;
simplex operation of a radio system;
slug tuning;
spurious radiation;
transceiver;
transponder;
transportable transmitter;
tuning;
walkie-talkie.

radio wave propagation.
See:
absorption;
atmospheric duct;
attenuation;
attenuation constant;
attenuation ratio;
attenuation vector;
circularly polarized wave;
collision frequency;
conical wave;
critical frequency;
cylindrical wave;
Debye length;
direct wave;
direction of polarization;
direction of propagation;
D layer;
D region;
effective radius of the earth;
E layer;
electric field strength;
electric field vector;
electric flux density;
electric vector;
electrical length;
electromagnetic wave;
envelope delay;
equiphase surface;
E region;
fading;
Faraday rotation;
F layer;
F 1 layer;
F 2 layer;
F region;
ground wave;
group velocity;
guided wave;
gyro frequency;
horizontally polarized wave;
hybrid wave;
incident wave;
incoherent scattering;
ionosphere;
ionospheric wave;
left-handed polarized wave;
limiting polarization;
linearly polarized wave;
lowest useful high frequency;
magnetic field vector;
magnetic vector;
magneto-ionic medium;
magneto-ionic mode;
magneto-ionic wave component;
maximum usable frequency;
modified index of refraction;
optimum working frequency;
ordinary-wave component;
O wave;
penetration frequency;
periodic electromagnetic wave;
phase constant;
phase velocity;
plane-earth factor;
plane of polarization;
plane of propagation;
plane wave;
plasma frequency;
plasma sheath;
polarization;
power flux density;
propagation constant;
propagation vector;
radio field strength;
radio frequency;
radio gain;
radio horizon;
radio transmission;
radio wave;
refracted wave;
refractive index;
refractive modulus;
relative refractive index;
right-handed polarized wave;
selective fading;
shadow factor;
sinusoidal electromagnetic wave;
spectral power-flux-density;
spherical wave;
standard propagation;
standing wave;
surface duct;
system loss;
transmission loss;
transverse electric hybrid wave;
transverse electric wave;
transverse electromagnetic wave;
transverse magnetic hybrid wave;
transverse magnetic wave;
traveling plane wave;
troposphere;
uniform plane wave;
vertically polarized wave;
virtual height;
wave;
wavelength;
wave normal.

railway signal and interlocking.
See:
absolute block;
absolute permissive block;
alternating-current floating-storage-battery system;
approach circuit;
approach indicator;
approach lighting;
approach-lighting relay;
approach signal;
automatic signal;
back contact;
ballast leakage;
ballast resistance;
battery chute;
block;
block indicator;
block signal;

block station;
bootleg;
centrifugal relay;
circuit controller;
clearance point;
clearing circuit;
closed-circuit principal;
coded track circuit;
coder;
color light signal;
color position light signal;
continuous lighting;
cross protection;
current of traffic;
cutout;
decoder;
dependent contact;
drop-away;
dual control;
electropneumatic valve;
fixed signal;
flasher relay;
flashing light signal;
floating battery;
focusing device;
fouling point;
front contact;
highway-crossing back light;
highway-crossing bell;
highway-crossing signal;
home signal;
impedance bond;
independent contact;
insulated rail joint;
interlocking relay;
interlocking signals;
interlocking station;
lever blocking device;
lever indication;
line circuit;
line relay;
marker light;
movable-bridge coupler;
normal clear;
normal contact;
normally closed contact;
normally open contact;
operated unit;
overlap;
permissive block;
pick-up;
point detector;
polar contact;
polarized relay;
polar relay;
position light signal;
railroad grade crossing;
remote control;
reverse contact;
semaphore signal;
semiautomatic signal;
signal aspect;
signal back light;
signal indication;
sliding contact;
smashboard signal;
spring contact;
standard code;
stick circuit;
strap key;
switch indicator;
switch signal;
three-position relay;
time release;
track circuit;
track indicator chart;
track instrument;
track relay;
train describer;
transformer relay;
trap circuit;
two-element relay;
vane-type relay;
vital circuit;
voltage range;
wigwag signal;
working value.

rating and testing magnetic amplifiers.

See:
bistable amplifier;
current amplification;
differential control current;
differential control voltage;
differential output current;
differential output voltage;
differential trip signal;
drift band of amplification;
drift offset;
environmental change of amplification;
environmental coefficient of amplification;
environmental coefficient of offset;
environmental coefficient of trip-point stability;
figure of merit;
linear magnetic amplifier;
maximum control current;
maximum OFF voltage;
maximum output voltage;
maximum test output voltage;
minimum ON voltage;
minimum output voltage;
minimum test output voltage;
normalized transimpedance;
output;
power amplification;
quiescent operating point;
rated load impedance;
rated OFF voltage;
rated output current;
rated output voltage;
rated output volt-amperes;
rated supply current;
response time;
short-circuit-duration rating;
time constant;
transimpedance;
trip OFF control signal;
trip ON control signal;
trip-point repeatability;
trip-point repeatability coefficient;
voltage amplification.

rectification.

See:
aging;
angle of advance;
angle of retard;
arc-back;
arc suppression of a rectifier;
arc through;
aster rectifier circuit;
average current;
average voltage;
basic alternating voltage;
basic metallic rectifier;
blocking;
breakdown;
breakdown region;
capacitance;
ceiling direct potential;
ceiling direct voltage;
commutating angle;
commutating group;
commutating reactance;
commutating-reactance factor;
commutating-reactance transformation constant;
commutation;
commutation factor;
conducting period;
continuous rating;
controlled rectifier;
conversion efficiency;
cooling system;
cross rectifier circuit;
current density;
degassing;
diametric rectifier circuit;
direct-raw-water cooling system;
direct-raw-water cooling system with recirculation;
displacement power factor;
double-way rectifier;
electrolytic rectifier;
electronic frequency changer;
electronic power converter;
evacuating equipment;
excitation anode;
excitation equipment;
form factor;
forming;
forward current;
forward direction;
forward power loss;

forward recovery time;
forward resistance;
forward voltage drop;
full-wave rectifier;
full-wave rectifier circuit;
grids;
half-wave rectification;
harmonic content;
heat-exchanger cooling system;
high-pressure vacuum pump;
inherent voltage regulation of a power rectifier;
initial inverse voltage;
initial reverse voltage;
inverse period of a rectifier element;
light transition load;
loss of forming;
low-pressure vacuum pump;
mechanical rectifier;
mercury-arc rectifier;
metallic rectifier;
metallic rectifier cell;
metallic rectifier stack assembly;
metallic rectifier unit;
misfire;
mode of operation;
multiple metallic rectifying cell;
multiple rectifier circuit;
negative electrode;
nonrepetitive peak reverse voltage;
number of rectifier phases;
parallel rectifier circuit;
peak forward voltage;
peak inverse voltage;
peak reverse voltage;
percent ripple voltage;
percent ripple voltage or current;
positive creep effect;
positive electrode;
positive nonconducting period;
power factor;
power inverter;
rectifier;
rated alternating voltage;
rated average tube current;
rated load;
rated output current;
rated output voltage;
rating of a rectifier unit;
rectifier;
rectifier anode;
rectifier cathode;
rectifier circuit element;
rectifier unit;
rectifying device;
rectifying element;
rectifying junction;
reforming;
repetitive peak reverse voltage;
reverse current;
reverse direction;
reverse power loss;
reverse recovery current;
reverse recovery time;
reverse resistance;
reverse voltage;
ripple amplitude;
ripple voltage or current;
section of rectifier unit;
semiconductor device;
series rectifier circuit;
set of commutating groups;
short-time rating;
simple rectifier circuit;
single-way rectifier;
small-signal resistance;
star rectifier circuit;
starting anode;
static resistance;
temperature-regulating equipment;
thermal resistance, effective;
total power loss;
total voltage regulation;
transition load;
tube current averaging time;
tube fault current;
vacuum gauge;
vacuum seal;
vacuum tank;
vacuum valve;
voltage regulation;
working peak reverse voltage;
wye rectifier circuit.

relay.
See:
active-power relay;
add-and-subtract relay;
alarm relay;
alternating-current relay;
annunciator relay;
auxiliary relay;
back contact;
blocking;
burden;
calibration error;
coaxial relay;
contact current-carrying rating;
contact current-closing rating;
current-balance relay;
current relay;
decelerating relay;
definite time;
definite-time relay;
delay relay;
differential relay;
direct-current relay;
dropout;
enclosed relay;
end-on armature relay;
failure to trip;
false tripping;
field decelerating relay;
field failure relay;
field forcing relay;
field protective relay;
flasher relay;
frequency relay;
frequency-sensitive relay;
full-field relay;
function relay;
gasket-sealed relay;
general-purpose relay;
generator-field accelerating relay;
generator-field decelerating relay;
harmonic-restraint relay;
hermetically sealed relay;
high-speed relay;
homing relay;
hot-wire relay;
impulse relay;
impulse time margin;
impulse transmitting relay;
inertia relay;
instantaneous;
instrument relay;
integrating relay;
interlock relay;
inverse time;
latching relay;
local backup;
lock-up relay;
magnetic control relay;
magnetic overload relay;
magnetostrictive relay;
marginal relay;
mechanically timed relay;
memory relay;
mercury relay;
meter relay;
motor-driven relay;
motor-field accelerating relay;
motor-field failure relay;
motor-field protection relay;
multiposition relay;
negative-phase-sequence relay;
network analysis;
neutral relay;
notching relay;
open-phase relay;
open relay;
operational relay;
overcurrent relay;
overload relay;
overspeed protection;
overtravel;
overvoltage relay;
phase-sequence relay;
phase-sequence reversal;
phase-undervoltage relay;
pick-up;
plunger relay;
polarized relay;
positive phase-sequence relay;

power relay;
preference, protective relaying;
product relay;
quotient relay;
ratchet relay;
reactive power relay;
relay, alternating-current;
relay armature;
relay, direct-current;
relay, dry circuit;
relay, electromagnetic;
relay, electrostatic;
relay, electrostrictive;
relay, pusher;
relay, thermal;
relay, three-position;
residual relay;
restraint relay;
reverse-current relay;
rotary relay;
rotary solenoid relay;
security;
sensitive relay;
sequential relay;
series relay;
slave relay;
solenoid relay;
solid-state relay;
static relay;
step-back relay;
stepping relay;
stepping relay, spring-actuated;
sudden-pressure relay;
synchronism-check relay;
tachometric relay;
temperature-compensated overload relay;
temperature relay;
thermal cutout;
thermal relay;
three-wire control;
time-delay relay;
timing relay;
transfer trip;
travel;
undercurrent relay;
undervoltage relay;
vibrating-reed relay;
voltage relay;
zero-phase-sequence relay;
zone of protection.

reliability.

See:
accelerated test;
acceleration factor;
burn-in;
constant-failure period;
debugging;
derating;
early failure period;
failure;
failure cause;
failure criteria;
failure mechanism (reliability);
failure mode (reliability);
failure rate;
failure-rate acceleration factor;
failure rate, assessed;
failure rate, extrapolated;
failure rate, observed;
failure rate, predicted;
failure, wear-out;
instantaneous failure rate;
item;
item, nonrepaired;
item, repaired;
mean life;
mean life, assessed;
mean life, extrapolated;
mean life, observed;
mean life, predicted;
mean time between failures;
mean time between failures, assessed;
mean time between failures, extrapolated;
mean time between failures, observed;
mean time between failures, predicted;
mean time to failure;
mean time to failure, assessed;
mean time to failure, extrapolated;
mean time to failure, observed;
mean time to failure, predicted;
redundancy;
redundancy, active;
redundancy, standby;
reliability, assessed;
reliability, extrapolated;
reliability, inherent;
reliability, operational;
reliability, predicted;
reliability, test;
screening test;
step-stress test;
time;
useful life;
wear-out-failure period.

rotor.

See:
accelerating torque;
air gap factor;
amortisseur bar;
ampere-conductor;
balance test;
banding insulation;
barring hole;
belt insulation;
binding band;
bobbin;
bolt leakage;
bore-hole lead insulation;
box frame;
brake assembly;
brake ring;
coated fabric;
cogging;
coil;
coil-end bracing;
coil insulation;
coil lashing;
coil side;
coil-side separator;
coil span;
coil-support bracket;
collar cheek, field-coil flange;
collector-ring lead insulation;
collector-ring shaft insulation;
compound;
conductor insulation;
core ducts;
core length;
core test;
coupling;
coupling flange;
critical speed;
critical torsional speed;
cylindrical-rotor machine;
damper segment;
deep-bar rotor;
distributed winding;
distribution factor;
double squirrel cage;
dummy coil;
electric loading;
enamel, wire;
end finger;
end ring rotor;
end winding;
end wire insulation;
end wire winding;
field-coil flange;
field terminal;
filler;
flexible coupling;
flux belt leakage;
flywheel ring;
form wound;
ground detection rings;
hand winding;
high-impedance rotor;
high-reactance rotor;
hub;
idle bar;
induced voltage;
insulated coupling;
insulated flange;
insulating cell;
insulating spacer;
integral coupling;
integral-slot winding;
inverted-turn transposition;
jack shaft;
jack system;
key;
ladder-winding insulation;
laminated core;

lateral critical speeds;
lead;
lead clamp;
lead collar;
light-load test;
line terminal;
locked rotor;
locked-rotor temperature-rise rate, rotor end ring;
locked-rotor temperature-rise rate, winding;
lower coil support;
machine winding;
magnet wire;
magnetic axis;
mat;
mica flake;
mica folium;
mica paper;
mica sheet;
mica splitting;
mica tape;
overhang packing;
overspeed test;
preformed coil or coil side;
peripheral air-gap leakage flux;
phase belt;
phase connectors;
phase separator;
pole cell insulation;
pole-changing winding;
pole-face shaping;
pole pitch;
pole shoe;
pole tip;
precision wound;
pulley;
punching;
random winding;
random wound;
retaining ring;
retaining-ring insulation;
retaining ring, rotor end winding;
rim;
Roebel transposition;
rotating control assembly;
rotating plate;
rotating test;
rotor bar;
rotor bushing;
rotor coil;
rotor-core;
rotor-core assembly;
rotor-core lamination;
rotor displacement angle;
rotor end ring;
rotor slot armor;
rotor winding;
round rotor;
salient pole;
segmental rim rotor;
separator insulation, slot;
shading coil;
shading wedge;
shaft;
shaft current;
shaft voltage test;
shear pin;
shim;
short-circuiter;
short-pitch winding;
skewed slot;
sleeve bearing;
slot;
slot cell;
slot liner;
slot packing;
slot wedge;
solid coupling;
solid rotor;
spider rotor;
spiderweb;
split collector ring;
split hub;
split rotor;
spring barrel;
squirrel-cage rotor;
standstill locking;
starting resistor;
strand;
stub shaft;
support-ring insulation;
symmetrical fractional slot winding;
tachometer;
tap;
tape;
thrust collar;
tie;
tooth;
tooth tip;
torsional critical speed;
transfer winding;
treated fabric, treated mat;
turn;
turning;
turning gear;
turn insulation;
turn separator;
turn-to-turn test;
turn-to-turn voltage;
twisted-load transposition;
two-layer winding;
upper coil support;
vacuum-pressure impregnation;
varnished fabric;
wedge;
wet-wound;
winding factor;
wire insulation;
wrapper;
yoke;
zig-zig leakage flux.

semiconductor.
See:
acceptor;
active area;
active area (sensitive area):
anode;
anode terminal;
array;
avalanche breakdown;
avalanche impedance;
average forward-current rating;
base;
boundary;
boundary, *p-n*;
breakdown;
breakdown impedance;
breakdown region;
breakdown voltage;
capacitance;
carrier;
cathode;
cathode terminal;
charge carrier;
collector grid;
conduction band;
conduction electron;
conductivity modulation;
conductivity, *n*-type;
conductivity, *p*-type;
contact area;
contact, high recombination rate;
conversion efficiency, overall;
depletion layer;
diffused junction;
diffusion capacitance;
diffusion constant, charge-carrier;
diffusion length, charge carrier;
diode, semiconductor;
donor;
doping;
doping compensation;
drift mobility;
electrode;
electrons, conduction;
element;
emitter;
energy gap;
extrinsic properties;
extrinsic semiconductor;
forming, electric;
forward direction;
generation rate;
grown junction;
Hall coefficient;
Hall mobility;
hole;
imperfection;
impurity;
impurity, acceptor;
impurity, donor;
impurity, stoichiometric;
intrinsic properties;
intrinsic semiconductor;

atmospheric transmissivity;
bar;
barette;
beacon;
boundary lights;
circling guidance lights;
conspicuity;
equal-interval light;
fixed light;
flashing light;
formation light;
fuselage lights;
group flashing light;
hazard or obstruction beacon;
ice detection light;
identification beacon;
interrupted quick-flashing light;
landing direction indicator;
landing light;
landmark beacon;
linear light;
navigation light system;
obstruction lights;
occulting light;
perimeter lights;
portable traffic-control light;
position lights;
quick-flashing light;
range light;
rhythmic light;
runway alignment indicator;
runway centerline lights;
runway edge lights;
runway end identification light;
runway exit lights;
runway lights;
runway visibility;
signaling light;
taxi channel lights;
taxi light;
taxiway centerline lights;
taxiway-edge lights;
taxiway holding-post light;
threshold lights;
touchdown zone lights;
visibility;
visual approach slope-indicator system;
visual range;
wing clearance lights.

static magnetic storage.
See:
bobbin core;
coefficient;
coercive force;
coercivity;
coincident-current selection;
cyclically magnetized condition;
delta;
density;
disturbed-ONE output;
distrubed-ZERO output;
drive;
drive pulse;
fixed storage;
groove diameter;
groove width;
hysteresis loop;
inhibit pulse;
intrinsic induction;
leader;
magnetic card;
magnetic ink;
magnetic ink character;
magnetic-ink character recognition;
nondestructive read;
ONE output;
ONE state;
ONE-to-partial-select ratio,
ONE-to-ZERO ratio;
partial read pulse;
partial select output;
partial write pulse;
path length;
peak flux density;
peak magnetizing force;
ratio;
read pulse;
reference time;
remanence;
reset pulse;
residual flux density;
retentivity;
saturation flux density;
saturation induction;
selection;
selection ratio;
set pulse;
shift pulse;
squareness ratio;
switching coefficient;
switching time;
symmetrically cyclically magnetized condition;
tape thickness;
tape-wound core;
threshold field;
undisturbed-ONE output;
undisturbed-ZERO output;
winding;
wrap;
wrap thickness;
wrap width;
write pulse;
zero output;
zero state.

stator.
See:
air-gap factor;
amortisseur bar;
ampere-conductors;
banding insulation;
belt insulation;
belt leakage flux;
blocking;
bobbin;
bolt leakage;
bore;
bottom coil slot;
coated fabric;
cogging;
coil;
coil brace;
coil end bracing;
coil insulation;
coil lashing;
coil side;
coil-side separator;
coil span;
coil support bracket;
compound;
conductor insulation;
connection insulation;
core length;
core-loss test;
core test;
distributed winding;
distribution factor;
dovetail projection;
dovetail slot;
drive strip;
dummy coil;
electric loading;
enamel wire;
end finger;
end shift frame;
end winding;
end-winding support;
end-wire insulation;
end-wire winding;
field-coil flange;
field terminal;
filler;
foot;
form wound;
grounding pad;
grounding terminal;
hand winding;
induced voltage;
insulating cell;
insulation;
insulation spacer;
integral slot winding;
inverted-turn transposition;
lacing, stator-winding end wire;
ladder-winding insulation;
laminated core;
laminated frame;
lead;
lead cable;
line terminal;
lower coil support;
lug, stator mounting;
machine winding;
magnetic axis;
magnet wire;
manifold insulation;
mat;

brushless synchronous machine;
B stage;
calibrated-driving-machine test;
calorimetric test;
can loss;
characteristic curves;
circle diagram;
circuit;
closed armortisseur;
coil space factor;
collector;
collector ring;
commissioning test;
compensator;
complex power;
compressor stator-blade control system;
concentrated winding;
concentric winding;
connections of polyphase circuits;
constant-speed motor;
constant torque;
continuous duty;
continuous rating;
continuous-voltage-rise test;
control exciter;
controlled overvoltage test;
control winding;
core loss;
core loss, open-circuit;
corona shielding;
crawling;
creepage surface;
current pulsation;
cyclic duration factor;
cyclic irregularity;
damping amortisseur;
damping torque;
damping torque coefficient;
deadband;
definite-purpose machine;
deviation factor of a wave;
direct axis;
direct-axis component of armature current;
direct-axis component of armature voltage;
direct-axis component of magnetic flux;
direct-axis component of magnetomotive force;
direct-axis subtransient open-circuit time constant;
direct-axis subtransient reactance;
direct-axis subtransient short-circuit time constant;
direct-axis subtransient voltage;
direct-axis synchronous impedance;
direct-axis synchronous reactance;
direct-axis transient open-circuit time constant;
direct-axis transient reactance;
direct-axis transient short-circuit time constant;
direct-axis voltage;
direct-connected exciter;
direct on-line starting;
direct-voltage high-potential test;
discharge energy test;
discharge extinction voltage;
discharge inception test;
discharge inception voltage;
distortion factor of a wave;
double-winding synchronous generator;
drift;
dripproof;
dripproof machine;
dust-ignitionproof;
dust-ignitionproof machine;
dust seal;
duty;
duty cycle;
duty-cycle rating;
dynamic braking;
dynamometer test;
eddy-current braking;
effective number turns per phase;
effective synchronous reactance;
efficiency;
elastomer;
electrical degree;
electric braking;
electric dynamometer;
electric generator;
electric machine regulating system;
electric machine regulator;
electric motor;
electromagnetic braking;
embedded temperature detector;
encapsulated;
end winding;
equalizer;
excitation system;
excitation-system stability;
excite;
exciter;
exciter, ceiling voltage;
exciter ceiling voltage, nominal;
exciter losses;
explosionproof machine;
externally ventilated machine;
facing;
fan-duty resistor;
field coil;
field *FR* loss;
field-lead insulation;
field pole;
field spool;
field-spool insulation;
field system;
field-turn insulation;
field winding;
film;
fluctuating power;
fluid loss;
form-wound motorette;
fractional-horsepower motor;
fractional-slot winding;
frequency converter, commutator type;
friction and windage loss;
front;
full-pitch winding;
gasproof or vaporproof;
general-purpose enclosure;
general-purpose motor;
generated voltage;
generator;
generator, alternating-current;
generator motor;
generator set;
generette;
graded-time step-voltage test;
ground insulation;
guarded machine;
harmonic test;
heteropolar machine;
high-potential test;
hunting;
hydro-generator;
hysteresis motor;
ideal paralleling;
ideal synchronous machine;
ideal value;
impedance of an electric machine;
impulse test;
inching;
inductor alternator;
inductor machine;
inductor synchronous motor;
inductor-type synchronous generator;
inductor-type synchronous motor;
inertia constant;
inherent regulation;
initial excitation response;
input;
insulating material insulant;
insulating-resistance-versus-voltage test;
insulation power factor;
insulation resistance;
integral-horsepower motor;
intermittent duty;
internal impedance;
internal impedance drop;
interspersing;

interturn insulation;
isochronous speed governing;
lap winding;
limit;
limiter;
line-charging capacity;
line terminal;
linkage voltage, direct-current test;
load;
load angle;
load-angle curve;
load-band of regulated voltage;
load limit changer;
load saturation curve;
locked-rotor current;
locked-rotor test;
locked-rotor torque;
loss tangent;
loss-tangent test;
loss total;
machine, electric;
magnetic loading;
magnetoelectric generator;
main exciter;
main-exciter response ratio;
main load;
main terminal;
maximum asymmetric short-circuit current;
maximum continuous rating;
maximum instantaneous fuel change;
maximum rate of fuel change;
mechanical fatigue test;
motor;
motor reduction unit;
motor synchronizing;
motorette;
multiple-current generator;
multispeed motor;
negative-sequence reactance;
neutral;
neutral lead;
neutral terminal;
noise-level test;
no load;
no-load field voltage;
no-load saturation curve;
no-load test;
nominal band of regulated voltage;
nominal exciter ceiling voltage;
nominal pull-in torque;
nominal synchronous-machine excitation-system ceiling voltage;
nonsalient pole;
ohmic resistance test;
oilproof enclosure;
open amortisseur;
open-circuit saturation curve;
open-circuit test;
open-ended coil;
open machine;
open pipe-ventilated machine;
operating conditions;
oscillation;
output;
overcurrent;
overspeed governor;
overtemperature detector;
overvoltage;
overvoltage test;
paralleling;
part winding starting;
peak load;
peak switching-current;
performance tests;
periodic rating;
permanent-field synchronous motor;
permanent-magnet synchronous motor;
per-unit quantity;
per-unit system;
phase characteristic of a synchronous machine;
phase coil insulation;
phase sequence;
phase-sequence test;
phase swinging;
phase voltage of a winding;
phasor diagram of a synchronous machine;
phasor power;
pilot exciter;
polarity test;
polarization index test;
pole body;
pole-body insulation;
pole end-plate;
pole face;
pole-face bevel;
pole slipping;
polyphase machine;
polyphase synchronous generator;
positive-phase-sequence reactance;
positive-sequence impedance;
positive-sequence resistance;
power factor;
power-factor angle;
power-factor tip-up;
power-factor tip-up test;
power-factory–voltage characteristic;
preformed winding;
proof test;
pull-in test;
pull-in torque;
pulling into synchronism;
pulling out of synchronism;
pull-out test;
pull-out torque;
pull-up torque;
pump-back test;
quadrature axis;
quadrature-axis component;
quadrature-axis component of armature current;
quadrature-axis component of magnetomotive force;
quadrature-axis magnetic flux;
quadrature-axis subtransient impedance;
quadrature-axis subtransient open-circuit time constant;
quadrature-axis subtransient reactance;
quadrature-axis subtransient short-circuit time constant;
quadrature-axis subtransient voltage;
quadrature-axis synchronous impedance;
quadrature-axis synchronous reactance;
quadrature-axis transient impedance;
quadrature-axis transient open-circuit time constant;
quadrature-axis transient reactance;
quadrature-axis transient short-circuit time constant;
quadrature-axis transient voltage;
quadrature-axis voltage;
random paralleling;
rated excitation system voltage;
rated-load field voltage;
rated-load torque;
rated power output;
rated speed;
rating;
reactive current;
reactive-current compensator;
reactor starting;
recovery time;
regenerative braking;
regulating system, synchronous-machine;
regulation;
regulation curve of a generator;
regulator, synchronous-machine;
reluctance motor;
reluctance torque;
residual-component telephone-influence factor;
resistance grading;
resistant;
rest and deenergized;
retardation test;
reversible motor;
reversing motor;
rheostat loss;

rheostatic-type voltage regulator;
rotary converter;
rotary inverter;
rotating amplifier;
rotating test;
routine test;
running operation ;
salient-pole machine;
sampling tests;
saturation curve;
saturation factor;
sealed;
secondary voltage;
secondary winding;
self-excited;
self-ventilated;
semiguarded machine;
semiprotected enclosure;
separately excited;
separately ventilated;
series-parallel starting;
series-wound;
service factor;
shell-type motor;
short-circuit loss;
short-circuit ratio;
short-circuit saturation curve;
short-circuit time constant;
short-circuit time constant of primary winding;
short-time duty;
short-time rating;
sine-wave generator;
single-layer winding;
single-phase machine;
single-phase rotary machine;
single-phase synchronous generator;
single phasing;
single-winding multispeed motor;
slot discharge;
slot discharge analyzer;
slot insulation;
slot space factor;
solid-pole synchronous motor;
space factor;
special-purpose motor;
speed changer;
speed-changer high-speed stop;
speed governor;
speed-governing system;
speed regulation characteristic;
splashproof machine;
split-throw winding;
stability of the speed-governing system;
stability of the temperature control system;
stability, synchronous-machine regulating-system;
standard-dimensioned motor;
star-delta starting;
starting;
starting amortisseur;
starting current;
starting motor;
starting operation, single-phase motor;
starting test;
starting torque;
starting winding;
stator coil;
steady-state condition;
steady-state governing speed-band;
steady-state incremental speed regulation;
steady-state short-circuit current;
steady-state speed regulation;
steady-state value;
step voltage test;
stray load loss;
submersible;
subsynchronous reluctance motor;
subtransient internal voltage;
sudden short-circuit test;
supply line, motor;
sustained oscillations;
sustained short-circuit test;
switching, single-phase motor;
switching torque;
switching torque, single-phase motor;
synchronism;
synchronizing;
synchronizing coefficient;
synchronizing torque;
synchronous double-fed machine;
synchronous generator;
synchronous impedance;
synchronous internal voltage;
synchronous motor;
synchronous operation;
synchronous speed;
temperature coefficient;
temperature coefficient of resistance;
temperature control system;
temperature relays;
temperature-rise test;
thermal endurance;
thermal equilibrium;
thermal-mechanical cycling;
time response;
torque motor;
totally enclosed fan-cooled;
totally enclosed machine;
totally enclosed nonventilated machine;
totally enclosed pipe-ventilated machine;
transformation impedance or admittance;
transient current;
transient internal voltage;
transient performer;
transient speed deviation;
transposition;
triplen;
tubing;
turbine-driven generator;
turbine-generator unit;
turbine nozzle control system;
turbine type;
type tests;
unity power-factor test;
universal motor;
varying duty;
varying-speed motor;
ventilating and cooling loss;
very-low-frequency high-potential test;
voltage endurance;
voltage-endurance test;
voltage regulation;
voltage-regulation curve;
voltage regulator;
voltage regulator, continuously acting type:
voltage regulator, direct-acting type;
voltage regulator, dynamic type;
voltage regulator, indirect-acting type;
voltage regulator, noncontinuously acting type;
voltage response exciter;
voltage response ratio, excitation system;
voltage-time response;
voltage withstand test;
water–air-cooled machine;
water-cooled machine;
waterproof machine;
waveform test;
wave winding;
weather-protected machine;
wound rotor;
zero-power-factor saturation-curve;
zero-power-factor test;
zero-sequence impedance;
zero-sequence reactance.

telegraphy.

See:
automatic telegraphy;
baud;
bias;
break;
bridge duplex system;
bug;
cable Morse code;
carrier telegraphy;
case shift;

characteristic distortion;
crossfire;
current margin;
differential duplex system;
diplex operation;
direct-current quadruplex system;
direct-current telegraphy;
direct-point repeater;
dot cycle;
duplex artificial line;
duplex operation;
duplex service;
electronic keying;
end distortion;
facsimile transmission;
fortuitous distortion;
frequency-shift keying;
full-duplex operation;
half-duplex;
half-duplex operation;
half-duplex repeater;
high-frequency carrier telegraphy;
impulse transmission;
international Morse code;
interpolation;
interrupted continuous waves;
inverse neutral telegraphy transmission;
keying;
lag;
level compensator;
local correction;
margin;
marking and spacing intervals;
Morse code;
Morse telegraphy;
multiplex operation;
multiplex printing telegraphy;
neutral direct-current telegraphy system;
noise killer;
on-off keying;
peak distortion;
perforator;
phase correction;
plate keying;
polar direct-current telegraphy system;
polarential telegraphy system;
printer;
printer telegraphy;
push-to-type operation;
quotation board;
range;
receiving-end crossfire;
reperforator switching center;
sending-end crossfire;
signal element;
signal-shaping amplifier;
signal-shaping network;
simplex operation;
single-tone keying;
siphon recorder;
start-stop printing telegraphy;
tape transmitter;
telautograph;
telegraph concentrator;
telegraph distortion;
telegraph distributor;
telegraph key;
telegraph repeater;
telegraph selector;
telegraph sender;
telegraph signal;
telegraph-signal distortion;
telegraph sounder;
telegraph transmission speed;
telegraph transmitter;
telegraph word;
telemeter service;
terminal;
ticker;
tone-modulated waves;
tone receiver;
tone transmitter;
two-source frequency keying;
two-tone keying;
upset duplex operation;
varioplex;
vibrating circuit;
voice-frequency carrier telegraphy.

telephone station.
See:
acoustical output;
ambient noise;
antisidetone induction coil;
antisidetone telephone set;
artificial ear;
artificial mouth;
bell box;
biased telephone ringer;
carbon telephone transmitter;
coin box;
combined telephone set;
constant available power source;
deskstand;
dial telephone set;
extension station;
handset;
hand telephone set;
hang-up hand telephone set;
harmonic telephone ringer;
induction coil;
local-battery talking, common-battery signaling telephone set;
magneto telephone set;
main station;
operator telephone set;
public telephone station;
sidetone telephone set;
sound-powered telephone set;
sweep oscillator;
switchhook;
telephone air-to-air input-output characteristic;
telephone booth;
telephone electric impedance;
telephone handset;
telephone modal distance;
telephone modal position;
telephone receive input-output characteristic;
telephone receiver;
telephone ringer;
telephone set;
telephone set, common-battery;
telephone set, local-battery;
telephone sidetone;
telephone speech network;
telephone transmit input-output characteristic;
telephone transmitter;
test handset;
toll station;
unbiased telephone ringer;
volume indicator;
wall telephone set.

telephone switching system.
See:
alternating-current–direct-current ringing system;
alarm signal;
all-relay system;
announcement system;
answering plug and cord;
A switchboard;
audible busy signal;
audible ringing signal;
automatic message accounting system;
automatic switchboard;
automatic switching system;
bank-and-wiper switch;
basic number plan;
B switchboard;
call announcer;
call indicator;
calling device;
calling plug and cord;
cord circuit;
clearing-out drop;
clipping;
code ringing;
common battery switchboard;
composite supervision;
conference connection;
connector switch;
console;
cordless;
cordless switchboard;
crossbar switch;
crossbar system;
cutoff relay;
decoder;
dial;

blanking level;
blanking signal;
bounce;
breezeway;
brightness channel;
brightness control;
burst flag;
burst gate;
camera tube;
chrominance channel;
chrominance-channel bandwidth;
chrominance signal component;
color synchronization signal;
color tracking;
color transmission;
composite color-picture signal;
composite picture signal;
compression;
contrast;
contrast control;
contrast ratio;
deflection yoke;
differential gain;
differential phase;
direct-current component;
direct-current restorer;
direct-current restoration;
direct-current transmission;
driving signals;
envelope delay;
equalizing pulses;
field;
field frequency;
flicker;
flyback;
frame;
frame frequency;
front porch;
gamma;
geometric distortion;
ghost;
gray scale;
high peaking;
iconoscope;
image dissector tube;
image iconoscope;
intensity modulation;
intercarrier sound system;
interference;
interlace factor;
interlaced scanning;
ion spot;
keying signal;
keystone distortion;
knee of transfer characteristic;
lag;
leading edge;
leading-edge pulse time;
level;
limiting resolution;
linearity control;
line frequency;
line number;
luminance channel;
luminance-channel bandwidth;
masking;
maximum average power output;
moiré;
monochrome channel;
monochrome-channel bandwidth;
monochrome transmission;
monoscope;
negative modulation;
nominal line pitch;
nominal line width;
noncomposite color-picture signal;
number of scanning lines;
orthicon;
overshoot;
pairing;
peak power output;
persistence characteristic;
phosphor;
pickup;
picture signal;
picture tube;
polarity;
positive modulation;
prime;
progressive scanning;
pulse-decay time;
pulse droop;
pulse tilt;
raster;
raster burn;
rectilinear scanning;
reference black level;
reference white level;
reflection color tube;
resolution;
resolution response;
resolution, structural;
resolution wedge;
return interval;
return trace;
ringing;
sawtooth wave;
scanning;
scanning, high-velocity;
scanning line;
scanning linearity;
scanning, low-velocity;
scanning, picture;
scanning speed;
scanning spot;
sensitivity;
sequential scanning;
setup;
shading;
shadow mask;
side look;
signal-to-noise ratio;
snow;
space pattern;
spot wobble;
square-wave response characteristic;
staircase signal;
standard television signal;
synchronizing;
synchronizing signal;
synchronizing-signal compression;
synchronizing-signal level;
target;
target capacitance;
target cutoff voltage;
target voltage;
tearing;
television broadcast station;
television camera;
television line number;
television receiver;
television repeater;
television transmitter;
time pattern;
trace interval;
trailing edge;
transfer characteristic;
undershoot;
vertical-hold control;
video-frequency amplifier;
vidicon;
visual transmitter;
visual-transmitter power;
white compression;
white peak.

tower.
See:
aerial cable;
anchor log;
anchor rod;
angle tower;
apparatus insulator;
apparent sag;
back arm;
bull ring;
buzz stick;
clearance;
climbing space;
combined mechanical and electrical strength;
conductor;
conductor loading;
corona;
crossarm;
crossarm guy;
dead-end guy;
dead-end tower;
flexible tower;
footings;
guy;
guy anchor;
guy insulator;
guy wire;

insulated turnbuckle;
insulator;
insulator arcing horn;
insulator arcing ring;
insulator arcing shield;
insulator arcover;
insulator grading shield;
insulator string;
lateral conductor;
lateral working space;
line conductor;
line guy;
line tap;
mast;
open wire;
pin insulator;
pole;
pole guy;
pole steps;
push brace;
rigid tower;
sag;
semistrain insulator;
shell;
span;
spool insulator;
stack;
storm guys;
strain insulator;
structure conflict;
suspension insulating weights;
suspension insulator;
tie wire;
tower loading;
transposition;
turnbuckle;
vertical conductor;
wiping gland;
zero guy.

transducer.
See:
all-pass network;
attenuation, voltage;
attenuator;
available conversion power gain;
available power;
available power gain;
bidirectional transducer;
circuits and devices;
circulator;
clipper;
clipper limiter;
control;
control system, feedback;
conversion transconductance;
conversion transducer;
dissymetrical transducer;
electric transducer;
electroacoustic transducer;
electromechanical transducer;
frequency selectivity;
gain, available conversion;
gain, insertion voltage;
gain, power;
harmonic-conversion transducer;
heterodyne-conversion transducer;
image attenuation constant;
image impedances;
image phase constant;
image transfer constant;
input noise temperature, effective;
insertion gain;
insertion loss;
integration circuit;
interaction loss;
iterative impedance;
limiter;
linear transducer;
mode transducer;
noise factor;
noise factor, spot;
noise temperature, standard;
operating-noise temperature;
pad;
phase-versus-frequency response characteristic;
power loss;
pulse-delay transducer;
reciprocal transducer;
shaft encoder;
signal;
symmetrical transducer;
transducer, active;
transducer gain;
transducer, ideal;
transducer loss;
transducer, passive;
transformer, ideal;
unidirectional transducer;
voltage amplification.

transmission characteristics.
See:
absorption;
admittance;
amplification;
amplification, current;
amplification, voltage;
amplitude-frequency response characteristic;
attenuation;
attenuation constant;
available-power gain;
background noise;
body capacitance;
bridge gain;
capacitance effective;
circuit noise;
companding;
compression;
cross modulation;
cutoff frequency;
decibel;
decilog;
distortion;
disturbance;
dynamic range;
echo;
electromagnetic compatibility;
envelope delay;
expansion;
fidelity;
filter transmission band;
flutter;
force factor;
gain;
group velocity;
hybrid balance;
immitance;
impedance;
impedance characteristic;
impedance, image;
impedance, input;
impedance, iterative;
impedance load;
impedance matrix;
impedance, normalized;
impedance, output;
impedance, source;
impedance, wave;
inductance, effective;
insertion gain;
insertion phase shift;
instantaneous companding;
interaction factor;
interference;
intermodulation;
isolation;
jitter;
leakage;
level;
line noise;
listener echo;
load, dummy;
loaded Q;
loss tangent;
master reference system;
matched condition;
mile of standard cable;
mismatch;
motorboating;
neper;
noise transmission impairment;
nonloaded Q;
nonuniformity;
Nyquist interval;
parasitic oscillations;
phase constant;
phase delay;
phase shift;
phase velocity;
pi point;
power gain;
power level;
power loss;
Q;
reciprocity;

visual field.
See:
absolute threshold;
accommodation;
adaptation;
after image;
artificial pupil;
blinding glare;
central vision;
central visual field;
color discrimination;
contrast sensitivity;
contrast threshold;
dark adaptation;
direct glare;
disability glare;
discomfort glare;
flicker fusion frequency;
glare;
light;
light adaptation;
luminance contrast;
luminance difference;
luminance ratio;
luminance threshold;
monocular visual field;
peripheral-visual field;
point of fixation;
point of observation;
primary line of sight;
pupil (pupillary aperture);
photopic vision;
Purkinje phenomenon;
quality of lighting;
reaction time;
reflected glare;
scotopic vision;
size threshold;
speed of vision;
Stiles-Crawford effect;
stray light;
subjective brightness;
threshold;
troland;
veiling brightness;
veiling reflection;
visibility;
visual acuity;
visual angle;
visual perception;
visual performance;
visual surround;
visual task.

voltage regulator.
See:
angular displacement;
continuous duty;
continuous rating;
dry-type;
duty;
electrical conversion;
excitation losses;
impedance drop;
impedance kilovolt-amperes;
impedance voltage;
indoor regulator;
induction voltage regulator;
intermittent duty;
kilovolt-amperes rating;
line-drop compensator;
liquid-immersed regulator;
load losses;
main unit of a two-core voltage-regulating transformer;
nominal voltage;
oil immersed;
outdoor regulator;
periodic duty;
phase-shifting transformer;
polarity and angular displacement;
pole-type regulator;
rated voltage;
reactance drop;
regulated power supply;
resistance drop;
short-time duty;
short-time rating;
station-type;
step-voltage regulator;
tap;
total losses;
transformer, oil-immersed;
transformer, variable-voltage;
varying duty;
voltage, nominal;
voltage-regulating relay;
voltage-regulating transformer;
voltage-regulating transformer, series unit of a two-core;
voltage-regulating transformer, two-core, excitation-regulating winding;
voltage-regulating transformer, two-core excitation winding;
voltage-regulating transformer, two-core, excited winding;
voltage-regulating transformer, two-core, regulating winding;
voltage-regulating transformer, two-core, series winding;
voltage regulator, primary circuit;
voltage regulator, primary winding;
voltage regulator, rated range of regulation;
voltage regulator, rated voltage of the series winding;
voltage regulator, rating of a single-phase;
voltage regulator, rating of a three-phase;
voltage regulator, regulated circuit.
voltage regulator, secondary winding;

waveguide.
See:
absorption frequency meter;
adapter, waveguide;
artificial dielectric;
attenuation;
attenuation band;
attenuator, waveguide;
balanced currents;
balanced mixer;
balanced voltages;
beam waveguide;
bend, waveguide;
butt joint;
capacitance discontinuity;
cavity resonator;
cavity-resonator frequency meter;
characteristic impedance;
characteristic insertion loss;
characteristic insertion loss, incremental;
characteristic insertion loss, residual;
characteristic wave impedance;
choke;
choke flange;
choke joint;
choke piston;
circular electric wave;
circular magnetic wave;
circulator;
coaxial stub;
connector;
contact piston;
corner;
coupling, aperture;
coupling loop;
coupling probe;
critical dimension;
crystal mixer;
crystal receiver;
curvature;
cutoff attenuator;
cutoff frequency;
cutoff waveguide;
cutoff wavelength;
degeneracy;
dielectric guide;
dielectric lens;
dielectric waveguide;
directional coupler;
direction of propagation;
discontinuity;
distributed constant;
dominant mode;
dominant wave;
E-bend;
E-H tee;
E-H tuner;
electrical length;
electric field vector;
electromagnetic wave;

elliptically polarized wave;
ellipticity;
E-plane *T*-junction;
evanescent mode;
exchangeable power;
exchangeable power gain;
exponential transmission line;
flexible waveguide;
frequency pulling;
guided wave;
guide wavelength;
gyrator;
heat loss;
H bend;
high-order mode;
H-plane *T*-junction;
hybrid;
hybrid coupler;
hybrid electromagnetic wave;
hybrid junction;
hybrid ring;
hybrid tee;
impedance;
incident wave;
input impedance;
input noise temperature, effective;
insertion loss;
iris;
isolator;
line stretcher;
load;
load sliding;
matched transmission line;
matched waveguide;
matching section;
mode coupler;
mode filter;
mode of propagation;
mode of resonance;
mode transducer;
mode transformer;
multimode waveguide;
noise temperature;
normalized admittance;
normalized impedance;
orifice;
overmoded waveguide;
oversized waveguide;
periodic waveguide;
phase constant;
phase-corrected horn;
phase shifter, waveguides;
phase velocity;
piston;
piston attenuator;
plain flange;
plumbing;
plunger, waveguides;
polarization ellipse;
polarization receiving factor;
polarization unit vector;
port;
port difference of a hybrid;
port sum;
post, waveguides;
power divider;
power rating;
propagation constant;
pulling figure;
push-pull currents;
push-pull voltages;
radial transmission line;
radiation loss;
reactive attenuator;
reference plane;
reference waveguide;
reflected wave;
reflection coefficient;
reflectionless waveguide;
reflection loss;
refracted wave;
rejection band;
residual standing-wave ratio;
resistive attenuator;
resonant mode;
resonator, waveguide;
return loss;
ridge waveguide;
ring circuit;
rotary attenuator;
rotary joint;
rotary phase changer;
scattering matrix;
seal;
sending end impedance;
series T junction;
shielded pair;
shielded transmission line;
single-wire line;
skin depth;
slide-screw tuner;
sliding short-circuit;
slotted section;
slug tuner;
standing wave;
standing-wave loss factor;
standing-wave ratio;
strip transmission line;
stub-supported coaxial;
stub tuner;
stub, waveguide;
squeeze section;
surface-wave transmission line;
tapered transmission line;
tapered waveguide;
taper, waveguide;
TB cell;
TEM mode;
$TE_{m,n,p}$ resonant mode;
$TE_{m,n}$ wave;
termination, reflectionless;
thermistor mount;
T junction;
totally unbalanced currents;
transformer, waveguide;
transition loss;
transmission band;
transmission coefficient;
transmission frequency meter;
transmission line;
transmission loss;
transmission mode;
transmission wave;
transmit-receive cell;
transverse electric wave;
transverse electromagnetic mode;
transverse electromagnetic wave;
transverse magnetic wave;
traveling plane wave;
traveling wave;
trombone line;
tuner, waveguide;
tuning probe;
tuning screw;
twist, waveguide;
uniconductor waveguide;
uniform waveguide;
upper frequency limit;
voltage reflection coefficient;
voltage standing-wave ratio;
water load;
waveguide cutoff frequency;
waveguide joint;
waveguide switch;
waveguide-to-coaxial transition;
waveguide wavelength;
wave impedance, characteristic;
wave normal;
Y junction.